Handbook of Homotopy Theory

CRC Press/Chapman and Hall Handbooks in Mathematics Series

Series Editor: Steven G. Krantz

AIMS AND SCOPE STATEMENT FOR HANDBOOKS IN MATHEMATICS SERIES

The purpose of this series is to provide an entree to active areas of mathematics for graduate students, beginning professionals, and even for seasoned researchers. Each volume should contain lively pieces that introduce the reader to current areas of interest. The writing will be semi-expository, with some proofs included for texture and substance. But it will be lively and inviting. Our aim is to develop future workers in each field of study.

These handbooks are a unique device for keeping mathematics up-to-date and vital. And for involving new people in developing fields. We anticipate that they will have a distinct impact on the development of mathematical research.

Handbook of Analytic Operator Theory
Kehe Zhu

Handbook of Homotopy Theory
Haynes Miller

https://www.crcpress.com/CRC-PressChapman-and-Hall-Handbooks-in-Mathematics-Series/book-series/CRCCHPHBKMTH

Handbook of Homotopy Theory

Edited by

Haynes Miller

CRC Press is an imprint of the
Taylor & Francis Group, an **informa** business
A CHAPMAN & HALL BOOK

CRC Press
Taylor & Francis Group
6000 Broken Sound Parkway NW, Suite 300
Boca Raton, FL 33487-2742

Printed on acid-free paper

International Standard Book Number-13: 978-0-815-36970-7 (Hardback)

Visit the Taylor & Francis Web site at
http://www.taylorandfrancis.com

and the CRC Press Web site at
http://www.crcpress.com

KD 12.16.2019 1041

Contents

Preface vii

1 **Goodwillie calculus** 1
Gregory Arone and Michael Ching

2 **A factorization homology primer** 39
David Ayala and John Francis

3 **Polyhedral products and features of their homotopy theory** 103
Anthony Bahri, Martin Bendersky, and Frederick R. Cohen

4 **A guide to tensor-triangular classification** 145
Paul Balmer

5 **Chromatic structures in stable homotopy theory** 163
Tobias Barthel and Agnès Beaudry

6 **Topological modular and automorphic forms** 221
Mark Behrens

7 **A survey of models for (∞, n)-categories** 263
Julia E. Bergner

8 **Persistent homology and applied homotopy theory** 297
Gunnar Carlsson

9 **Algebraic models in the homotopy theory of classifying spaces** 331
Natàlia Castellana

10 **Floer homotopy theory, revisited** 369
Ralph L. Cohen

11 **Little discs operads, graph complexes and Grothendieck–Teichmüller groups** 405
Benoit Fresse

12 **Moduli spaces of manifolds: a user's guide** 443
Søren Galatius and Oscar Randal-Williams

13 An introduction to higher categorical algebra 487
David Gepner

14 A short course on ∞-categories 549
Moritz Groth

15 Topological cyclic homology 619
Lars Hesselholt and Thomas Nikolaus

16 Lie algebra models for unstable homotopy theory 657
Gijs Heuts

17 Equivariant stable homotopy theory 699
Michael A. Hill

18 Motivic stable homotopy groups 757
Daniel C. Isaksen and Paul Arne Østvær

19 E_n-spectra and Dyer-Lashof operations 793
Tyler Lawson

20 Assembly maps 851
Wolfgang Lück

21 Lubin-Tate theory, character theory, and power operations 891
Nathaniel Stapleton

22 Unstable motivic homotopy theory 931
Kirsten Wickelgren and Ben Williams

Index 973

Preface

This Handbook collects twenty-two essays by some thirty-one authors on a wide range of topics of current interest in homotopy theory. It is intended to present a snapshot of some key elements of contemporary work in this area, along with indications of potential avenues for future research. I hope it is useful to those interested in pursuing research in this field, as well as to the growing class of mathematicians in neighboring areas who find themselves employing the methods of homotopy theory in their own research.

This volume may be regarded as a successor to the "Handbook of Algebraic Topology," edited by Ioan James and published a quarter of a century ago. In calling it the "Handbook of Homotopy Theory," I am recognizing that the discipline has expanded and deepened, and traditional questions of topology, as classically understood, are now only one of many distinct mathematical disciplines in which it has had a profound impact and which serve as sources of motivation for research directions within homotopy theory proper.

The topics represented here range from axiomatic to applied, from applications in symplectic and algebraic geometry to the modern chromatic development of stable homotopy theory, from local methods in group theory to the theory and use of ∞-categories. It would nevertheless have been easy to come up with as many more topics in the subject that are active and promising today, and as many more authors engaged in trail-blazing research and ready to write about it. The chapters vary widely in style and in the demands placed on the reader, but taken together form a surprisingly highly connected narrative.

The strong response to my call for participation in this project is a reflection of the excitement and vitality of research today in homotopy theory. In fact, every author took the topic I proposed in unanticipated directions, often adding authors to the team, and in the aggregate providing a much broader and more vibrant coverage than I imagined possible at the outset. I am very grateful to them all. I also thank the many others who gave advice and assistance of all sorts during the assembly of this book. In particular I thank Ashley Fernandes for his unflagging assistance with the LaTeX, CRC editor Bob Ross for his assistance in bringing this together, Niles Johnson for allowing us to use his beautiful image of the Hopf fibration (a seminal example in our subject), and Steve Krantz, who proposed this project in the first place.

1

Goodwillie calculus

Gregory Arone and Michael Ching

Goodwillie calculus is a method for analyzing functors that arise in topology. One may think of this theory as a categorification of the classical differential calculus of Newton and Leibnitz, and it was introduced by Tom Goodwillie in a series of foundational papers [44, 45, 46].

The starting point for the theory is the concept of an *n-excisive* functor, which is a categorification of the notion of a polynomial function of degree n. One of Goodwillie's key results says that every homotopy functor F has a universal approximation by an n-excisive functor P_nF, which plays the role of the n-th Taylor approximation of F. Together, the functors P_nF fit into a tower of approximations of F: the *Taylor tower*

$$F \longrightarrow \cdots \longrightarrow P_nF \longrightarrow \cdots \longrightarrow P_1F \longrightarrow P_0F.$$

It turns out that 1-excisive functors are the ones that represent generalized homology theories (roughly speaking). For example, if $F = I$ is the identity functor on the category of based spaces, then P_1I is the functor $P_1I(X) \simeq \Omega^\infty\Sigma^\infty X$. This functor represents stable homotopy theory in the sense that $\pi_*(P_1I(X)) \cong \pi_*^s(X)$. Informally, this means that the best approximation to the homotopy groups by a generalized homology theory is given by the stable homotopy groups. The Taylor tower of the identity functor then provides a sequence of theories, satisfying higher versions of the excision axiom, that interpolate between stable and unstable homotopy.

The analogy between Goodwillie calculus and ordinary calculus reaches a surprising depth. To illustrate this, let D_nF be the homotopy fiber of the map $P_nF \to P_{n-1}F$. The functors D_nF are the homogeneous pieces of the Taylor tower. They are controlled by Taylor "coefficients" or derivatives of F. This means that for each n there is a spectrum with an action of Σ_n that we denote ∂_nF, and there is an equivalence of functors

$$D_nF(X) \simeq \Omega^\infty \left(\partial_nF \wedge X^{\wedge n}\right)_{h\Sigma_n}.$$

Here for concreteness F is a homotopy functor from the category of pointed spaces to itself; similar formulas apply for functors to and from other categories. The spectrum ∂_nF plays the role of the n-th derivative of F, and the spectra ∂_nF are relatively easy to calculate. There is an obvious similarity between the formula for D_nF and the classical formula for the n-th term of the Taylor series of a function.

Mathematics Subject Classification. 55P65, 55P48, 55P42, 19D10.

Key words and phrases. calculus of functors, identity functor, operads, algebraic K-theory.

Of course there are differences between the classical differential calculus and Goodwillie calculus, and they are just as interesting as the similarities. One place where the analogy breaks down is in the complex ways that the homogeneous pieces can be "added up" to create the full Taylor tower. Homogeneous layers only determine the Taylor tower up to extensions. Theorems of Randy McCarthy, Nick Kuhn and the present authors reveal situations in which these extension problems can be understood via interesting connections to Tate spectra. Considerable simplification occurs by passing to chromatic homotopy theory, a fact that forms the basis for recent work of Gijs Heuts on the classification of unstable v_n-periodic homotopy theory via spectral Lie algebras.[1]

Also unlike ordinary calculus, a crucial example is provided by the identity functor (for based spaces or, more generally, for any ∞-category of interest).[2] As hinted at above, the identity functor typically has an interesting and non-trivial Taylor tower controlled by its own sequence of Taylor coefficients. For based spaces, these derivatives were first calculated by Brenda Johnson. Mark Mahowald and the first author used a detailed description of these objects to get further information about the Taylor tower of the identity in unstable v_n-periodic homotopy. The derivatives of the identity also play an important theoretical role in the calculus; in particular, by a result of the second author they form an operad that encodes structure possessed by the derivatives of any functor to or from a given ∞-category. For topological spaces, this operad is a topological analogue of the Lie operad, explaining the role of Lie algebras in Quillen's work on rational homotopy theory, and in Heuts's work mentioned above. Structures related to the Lie operad also form the basis of a classification of functors up to Taylor tower equivalence given by the present authors.

The nature of Goodwillie calculus lends itself to both computational and conceptual applications. Goodwillie originally developed the subject in order to understand more systematically certain calculations in algebraic K-theory, and this area remains a compelling source of specific examples. However, the deeply universal nature of these constructions gives functor calculus a crucial role in the foundations of homotopy theory, especially given the expanding role therein of higher category theory.

Indeed it seems that the calculus has not yet found its most general form. The similarities to Goodwillie calculus borne by the manifold and orthogonal "calculi" of Michael Weiss suggest some deeper structure that is still to be properly worked out. There are also important but not fully understood connections to manifolds and factorization homology of E_n-algebras,[3] and a properly equivariant version of Goodwillie calculus has been hinted at by work of Dotto, but much remains to be explored.

Notwithstanding such future developments, the fundamental role of Goodwillie calculus in homotopy theory is as clear as that of ordinary calculus in other areas of mathematics: it provides a systematic interpolation between the linear (homotopy-theoretically, this usually means the stable) and nonlinear (or unstable) worlds, and thus brings our deep intuition of the nature of change to bear on our understanding of homotopy theory.

---

[1] See Gijs Heuts's chapter in this Handbook for an account of this development.

[2] See Moritz Groth's chapter in this Handbook for an introduction to the theory of ∞-categories.

[3] See the chapter in this Handbook by David Ayala and John Francis for a survey of factorization homology.

1.1 Polynomial Approximation and the Taylor Tower

Goodwillie's calculus of functors is modelled after ordinary differential calculus with the role of smooth maps between manifolds played by *homotopy functors* $F : \mathcal{C} \to \mathcal{D}$, i.e. those that preserve some notion of weak equivalence. In Goodwillie's original formulation [44], the categories $\mathcal{C}$ and $\mathcal{D}$ were each taken to be some category of topological spaces or spectra, but modern higher-category-theoretic technology allows the theory to be developed for functors F between any $(\infty, 1)$-categories that are suitably well-behaved. The basic tenets of the theory are independent of any particular model for $(\infty, 1)$-categories. In this paper, we will mostly use the language of ∞-categories (i.e. quasicategories) from [63], and the details of Goodwillie calculus have been developed in the greatest generality by Lurie in that context, see [64, Sec. 6]. Thus our typical assumption will be that $F : \mathcal{C} \to \mathcal{D}$ is a functor between ∞-categories. The reader can equally, however, view F as a functor between model categories that preserves weak equivalences. In that setting, many of the basic constructions are described by Kuhn in [58].

We will make considerable use of the notions of (homotopy) limit and colimit inside an ∞-category. We will refer to these simply as limits and colimits, though everywhere in this paper the appropriately homotopy-invariant concepts are intended. When working with a functor $F : \mathcal{C} \to \mathcal{D}$, we will usually require that $\mathcal{C}$ and $\mathcal{D}$ admit limits and colimits of particular shapes and that (especially in $\mathcal{D}$) certain limits and colimits commute. The relevant conditions will be made explicit when necessary.

Polynomial functors in the homotopy calculus

To some extent, the theory of the calculus of functors is completely determined by making a choice as to which functors $F : \mathcal{C} \to \mathcal{D}$ are to be considered the analogues of degree n polynomials. In Goodwillie's version, this choice is described in terms of cubical diagrams.

Definition 1.1.1. An *n-cube* in an ∞-category $\mathcal{C}$ is a functor $\mathcal{X} : \mathcal{P}(I) \to \mathcal{C}$, where $\mathcal{P}(I)$ is the poset of subsets of some finite set I of cardinality n. An n-cube $\mathcal{X}$ is *cartesian* if the canonical map

$$\mathcal{X}(\emptyset) \to \operatorname{holim}_{\emptyset \neq S \subseteq I} \mathcal{X}(S)$$

is an equivalence, and *cocartesian* if

$$\operatorname{hocolim}_{S \subsetneq I} \mathcal{X}(S) \to \mathcal{X}(I)$$

is an equivalence. When $n = 2$, these conditions reduce to the familiar notions of pullback and pushout, respectively. We also say that an n-cube $\mathcal{X}$ is *strongly cocartesian* if every 2-dimensional face is a pushout. Note that a strongly cocartesian n-cube is also cocartesian if $n \geq 2$.

We can now give Goodwillie's condition on a functor that plays the role of "polynomial of degree $\leq n$".

Definition 1.1.2. Let $\mathcal{C}$ be an ∞-category that admits pushouts. A functor $F : \mathcal{C} \to \mathcal{D}$ is *n-excisive* if it takes every strongly cocartesian $(n+1)$-cube in $\mathcal{C}$ to a cartesian $(n+1)$-cube in $\mathcal{D}$. We will say that F is *polynomial* if it is n-excisive for some integer n.

Let $\mathrm{Fun}(\mathcal{C}, \mathcal{D})$ be the ∞-category of functors from $\mathcal{C}$ to $\mathcal{D}$, and let $\mathrm{Exc}_n(\mathcal{C}, \mathcal{D})$ denote the full subcategory whose objects are the n-excisive functors.

Example 1.1.3. In the somewhat degenerate case $n = 0$, Definition 1.1.2 reduces to the statement that F is 0-excisive if and only if it is homotopically constant, i.e. F takes every morphism in $\mathcal{C}$ to an equivalence in $\mathcal{D}$. (In an even more degenerate case, F is (-1)-excisive if and only if $F(X)$ is a terminal object of $\mathcal{D}$ for all X in $\mathcal{C}$.)

Example 1.1.4. A functor $F : \mathcal{C} \to \mathcal{D}$ is 1-excisive if and only if it takes pushout squares in $\mathcal{C}$ to pullback squares in $\mathcal{D}$. The prototypical example of such a functor (when $\mathcal{C}$ and $\mathcal{D}$ are both the category $\mathcal{T}op_*$ of based topological spaces) is

$$X \mapsto \Omega^\infty(E \wedge X)$$

where E is some spectrum. In fact, these examples constitute a classification of those functors that are 1-excisive, *reduced* (i.e. preserve the null object) and *finitary* (i.e. preserve filtered colimits). This fact illustrates the key role played by stable homotopy theory in Goodwillie calculus.

Remark 1.1.5. It is notable that the identity functor $I : \mathcal{C} \to \mathcal{C}$ is typically *not* 1-excisive (or n-excisive for any n) unless $\mathcal{C}$ is a stable ∞-category. One might naturally think it would make more sense if a 1-excisive functor were defined to preserve either pushouts or pullbacks, rather than to mix the two notions. Indeed, one can make such a definition and explore its properties. However, the notion defined by Goodwillie has turned out to be much more useful. This is partly because of the close connection to stable homotopy theory hinted at in Example 1.1.4, but also because the fact that the identity functor is not 1-excisive makes the theory *more* useful rather than less, since, as we shall see, the Taylor tower of the identity functor provides an interesting decomposition of a space that we would not have if the identity were automatically linear.

Remark 1.1.6. The property of a functor F being n-excisive can also be described as a condition on sequences of $n+1$ morphisms in $\mathcal{C}$ with a common source, say $f_i : A \to X_i$ for $i = 0, \dots, n$. The condition relates the value of F on the pushouts, over A, of all possible subsets of the sequence $(f_0, \dots, f_n)$. This can be viewed as the analogue of a way to specify when a function $f : \mathbb{R} \to \mathbb{R}$ is a degree $\leq n$ polynomial by considering the values of f on sums of subsets of a set $(x_0, \dots, x_n)$ of real numbers.

Remark 1.1.7. Johnson and McCarthy [53] have given a different, slightly broader definition for a functor $F : \mathcal{C} \to \mathcal{D}$ to be *degree* $\leq n$: that the $(n+1)$-th cross-effect of F vanishes. This choice leads to a different version of the Taylor tower described in the next section, although the difference seems not to be important in most cases of interest. In particular, for "analytic" functors, the two towers agree within the 'radius of convergence'.

As one might expect, the conditions of being n-excisive for varying n are nested.

Lemma 1.1.8. *If $F : \mathcal{C} \to \mathcal{D}$ is n-excisive, then F is also $n+1$-excisive. We therefore have a sequence of inclusions of subcategories*

$$\mathrm{Exc}_0(\mathcal{C}, \mathcal{D}) \subseteq \mathrm{Exc}_1(\mathcal{C}, \mathcal{D}) \subseteq \cdots \subseteq \mathrm{Exc}_n(\mathcal{C}, \mathcal{D}) \subseteq \mathrm{Exc}_{n+1}(\mathcal{C}, \mathcal{D}) \subseteq \cdots .$$

Polynomial approximation in the homotopy calculus

The fundamental construction in ordinary differential calculus is that of polynomial approximation: given a smooth function $f : \mathbb{R} \to \mathbb{R}$ and a real number x, there is a unique "best" degree $\leq n$ polynomial that approximates f in a "neighbourhood" of x. To transfer this idea to the calculus of functors, we need to be able to compare the values of functors on objects in $\mathcal{C}$ that are related in some sense. In particular, we require a map between the objects in order to make this comparison. Thus the n-excisive approximation to a functor $F : \mathcal{C} \to \mathcal{D}$ in a neighbourhood of an object $X \in \mathcal{C}$ is only defined on objects Y that come equipped with a map $Y \to X$ in $\mathcal{C}$, that is, on Y in the slice ∞-category $\mathcal{C}_{/X}$.

Definition 1.1.9. We say that functors $\mathcal{C} \to \mathcal{D}$ *admit n-excisive approximations* at X in $\mathcal{C}$ if the composite

$$\mathrm{Exc}_n(\mathcal{C}_{/X}, \mathcal{D}) \hookrightarrow \mathrm{Fun}(\mathcal{C}_{/X}, \mathcal{D}) \to \mathrm{Fun}(\mathcal{C}, \mathcal{D})$$

has a left adjoint, which, when it exists, we write P_n^X. Here (as everywhere in this article) we mean an adjunction in the $(\infty, 1)$-categorical sense: see, for example Lurie [63, 5.2.2.1] or Riehl-Verity [74, 1.1].

Remark 1.1.10. Biedermann, Chorny and Röndigs showed in [27] that the n-excisive approximation is a left Bousfield localization of a suitable category of functors, and it provides a best approximation *on the right* to a given functor $F : \mathcal{C} \to \mathcal{D}$ by one that is n-excisive. Explicitly, the n-excisive approximation to F at X, if it exists, consists of a natural transformation

$$F \to P_n^X F$$

of functors $\mathcal{C}_{/X} \to \mathcal{D}$ that is initial (up to homotopy) among natural transformations from F (restricted to $\mathcal{C}_{/X}$) to an n-excisive functor.

The first main theorem of Goodwillie calculus is that such n-excisive approximations exist under mild conditions on $\mathcal{C}$ and $\mathcal{D}$. The following result is stated by Lurie, but the proof is no different than that given originally by Goodwillie in the context of functors of topological spaces and spectra.

Theorem 1.1.11 (Goodwillie [46, 1.13], Lurie [64, 6.1.1.10])**.** *Let $\mathcal{C}$ and $\mathcal{D}$ be ∞-categories, and suppose that $\mathcal{C}$ has pushouts, and that $\mathcal{D}$ has sequential colimits, and finite limits, which commute. Then functors $\mathcal{C} \to \mathcal{D}$ admit n-excisive approximations at any object $X \in \mathcal{C}$.*

Example 1.1.12. The 0-excisive approximation to F at X is, as you would expect, equivalent to the constant functor with value $F(X)$.

There is no loss of generality (and the notation is simpler) if we focus on the case where X is a terminal object of $\mathcal{C}$. In this setting, the n-excisive approximation to $F : \mathcal{C} \to \mathcal{D}$ is another functor from $\mathcal{C}$ to $\mathcal{D}$, which we simply denote by $P_n F$.

Example 1.1.13. Goodwillie gives an explicit construction of P_nF which is easiest to describe when $n = 1$ and where $\mathcal{C}$ and $\mathcal{D}$ are both pointed with $F : \mathcal{C} \to \mathcal{D}$ *reduced*, i.e. preserving the null object. In this case, $P_1F(Y)$ can be written as the colimit of the following sequence of maps

$$F(Y) \to \Omega F(\Sigma Y) \to \Omega^2 F(\Sigma^2 Y) \to \cdots$$

where Σ and Ω are the suspension and loop-space functors for $\mathcal{C}$ and $\mathcal{D}$ respectively. For $F : \mathcal{T}op_* \to \mathcal{T}op_*$, we have

$$P_1F(Y) \simeq \Omega^\infty \bar{F}(\Sigma^\infty Y)$$

where $\bar{F}$ is a reduced 1-excisive functor from spectra to spectra. For Y a finite CW-complex, such a functor can be written in the form of Example 1.1.4:

$$P_1F(Y) \simeq \Omega^\infty(\partial_1 F \wedge Y)$$

where $\partial_1 F$ is a spectrum which we refer to as the *(first) derivative of* F (at the one-point space $*$).

Example 1.1.14. The 1-excisive approximation to the identity functor on *based* spaces is simply the stable homotopy functor

$$P_1I(Y) \simeq \Omega^\infty \Sigma^\infty Y =: Q(Y)$$

or equivalently, $\partial_1 I \simeq S^0$, the sphere spectrum.

The unbased case is slightly more subtle: the 1-excisive approximation to the identity functor in that context can be written as

$$P_1I(Y) \simeq \mathrm{hofib}(Q(Y_+) \to Q(S^0)) =: \tilde{Q}(Y)$$

where the homotopy fibre is calculated over the point in $Q(S^0)$ corresponding to the *identity* map on the sphere spectrum (as opposed to the null map).

Taylor tower and convergence

The explicit description of P_nF is hard to make use of for $n > 1$, even in the case of the identity functor. The real power of the calculus of functors derives from the tower formed by the n-excisive approximations for varying n.

Definition 1.1.15. The *Taylor tower* (or *Goodwillie tower*) of $F : \mathcal{C} \to \mathcal{D}$ at $X \in \mathcal{C}$ is the sequence of natural transformations (of functors $\mathcal{C}_{/X} \to \mathcal{D}$):

$$F \to \cdots \to P_{n+1}^X F \to P_n^X F \to \cdots \to P_1^X F \to P_0^X F \simeq F(X)$$

where it follows from the universal property of $P_{n+1}^X F$, and Lemma 1.1.8, that each $F \to P_n^X F$ factors as shown.

For a given map $f : Y \to X$ in $\mathcal{C}$, the Taylor tower provides a sequence of factorizations of $F(f) : F(Y) \to F(X)$. Ideally, we would be able to recover the value $F(Y)$ from this sequence of approximations $P_n^X F(Y)$ in the following way.

Definition 1.1.16. The Taylor tower of $F : \mathcal{C} \to \mathcal{D}$ *converges* at $Y \in \mathcal{C}_{/X}$ if the induced map

$$F(Y) \to \operatorname{holim}_n P_n^X F(Y)$$

is an equivalence in $\mathcal{D}$.

Arguably the question of convergence of the Taylor tower is the most important step in actually applying the calculus of functors to a particular functor F. Very general approaches to proving convergence seem rare, but Goodwillie has developed an extensive set of tools, based on connectivity estimates, in the contexts of topological spaces and spectra.

These tools are based on measuring the failure of a functor F to be n-excisive via connectivity. This can be done by applying F to a strongly cocartesian $(n+1)$-cube, and examining the failure of the resulting cube to be cartesian, in terms of the connectivity of the map from the initial vertex to the homotopy limit of the rest of the diagram.

Roughly speaking, a functor F is *stably n-excisive* if this connectivity is controlled relative to the connectivities of the maps in the original cocartesian cube. Goodwillie then says that F is *ρ-analytic*, for some real number ρ, if it is stably n-excisive for all n where these connectivity estimates depend linearly on n with slope ρ. See [45] for complete details.

The upshot of these definitions is the following theorem.

Theorem 1.1.17 (Goodwillie [46, 1.13]). *Let $F : \mathcal{C} \to \mathcal{D}$ be a ρ-analytic functor where $\mathcal{C}$ and $\mathcal{D}$ are each either spaces or spectra. Then the Taylor tower of F at $X \in \mathcal{C}$ converges on those objects Y in $\mathcal{C}_{/X}$ whose underlying map $Y \to X$ is ρ-connected.*

Examples 1.1.18. The identity functor $I : \mathcal{T}op \to \mathcal{T}op$ is 1-analytic [45, 4.3]. This depends on higher dimensional versions of the Blakers-Massey Theorem relating pushouts and pullbacks in $\mathcal{T}op$. Indeed, the usual Blakers-Massey Theorem implies that I is stably 1-excisive. Waldhausen's algebraic K-theory of spaces functor $A : \mathcal{T}op \to \mathcal{S}p$ is also 1-analytic [45, 4.6]. In particular, the Taylor towers at $*$ of both of these functors converge on simply connected spaces.

Definition 1.1.19. Let $F : \mathcal{C} \to \mathcal{T}op_*$ be a homotopy functor with values in pointed spaces. The *Goodwillie spectral sequence* associated to F at $Y \in \mathcal{C}_{/X}$ is the homotopy spectral sequence of the tower of pointed spaces $(P_n^X F(Y))_{n \geq 1}$ [32, section IX.4]. This spectral sequence converges to

$$\pi_* P_\infty^X F(Y)$$

where $P_\infty^X F := \operatorname{holim}_n P_n^X F$, and has E^1-term given by the homotopy groups of the *layers* of the Taylor tower, i.e.

$$E^1_{s,t} \cong \pi_{t-s} D_s^X F(Y)$$

where $D_s^X F := \operatorname{hofib}(P_s^X F \to P_{s-1}^X F)$. If F is analytic (and $Y \to X$ is suitably connected), then this spectral sequence converges strongly [32, section IX.5].

Example 1.1.20. For the identity functor on based spaces, the Goodwillie spectral sequence at $Y \in \mathcal{T}op_*$ takes the form

$$E^1_{s,t} = \pi_{t-s} D_s I(Y)$$

and converges to the homotopy groups of Y when Y is simply-connected (or, more generally, when the Taylor tower of the identity converges for Y. This spectral sequence has been studied extensively by Behrens [25]. The spectral sequence also motivates the study of the layers $D_n^X F$ of the Taylor tower in general, and we turn to these now.

1.2 The Classification of Homogeneous Functors

Let $F : \mathcal{C} \to \mathcal{D}$ be a homotopy functor, where $\mathcal{C}$ and $\mathcal{D}$ are as in Theorem 1.1.11, and suppose further that $\mathcal{D}$ is a *pointed* ∞-category. Then we make the following definition, generalizing that given at the end of the previous section.

Definition 1.2.1. The n-th *layer* of the Taylor tower of F at X is the functor $D_n^X F$: $\mathcal{C}_{/X} \to \mathcal{D}$ given by

$$D_n^X F(Y) := \mathrm{hofib}(P_n^X F(Y) \to P_{n-1}^X F(Y)).$$

These layers play the role of *homogeneous* polynomials in the theory of calculus, and satisfy the following definition.

Definition 1.2.2. Let $F : \mathcal{C} \to \mathcal{D}$ be a homotopy functor that admits n-excisive approximations, and where $\mathcal{D}$ has a terminal object $*$. We say that F is *n-homogeneous* if F is n-excisive and $P_{n-1}F \simeq *$.

Another of Goodwillie's main theorems from [46] is the existence of a natural delooping, and consequently a classification, of homogeneous functors. To state this in the generality of ∞-categories, we first recall that any suitable pointed ∞-category $\mathcal{C}$ admits a *stabilization*, that is a stable ∞-category $\mathcal{S}p(\mathcal{C})$ together with an adjunction

$$\Sigma^\infty_{\mathcal{C}} : \mathcal{C} \rightleftarrows \mathcal{S}p(\mathcal{C}) : \Omega^\infty_{\mathcal{C}}$$

that generalizes the suspension spectrum / infinite-loop space adjunction, which we write simply as $(\Sigma^\infty, \Omega^\infty)$, for $\mathcal{C} = \mathcal{T}op_*$.

Theorem 1.2.3. *Let $F : \mathcal{C} \to \mathcal{D}$ be an n-homogeneous functor between pointed ∞-categories. Then there is a* symmetric multilinear *functor $H : \mathcal{S}p(\mathcal{C})^n \to \mathcal{S}p(\mathcal{D})$, and a natural equivalence*

$$F(X) \simeq \Omega^\infty_{\mathcal{D}} [H(\Sigma^\infty_{\mathcal{C}} X, \dots, \Sigma^\infty_{\mathcal{C}} X)_{h\Sigma_n}]$$

where we are taking the homotopy orbit construction with respect to the action of the symmetric group Σ_n that permutes the entries of H.

Example 1.2.4. For functors $\mathcal{T}op_* \to \mathcal{T}op_*$, a symmetric multilinear functor is uniquely determined (on finite CW-complexes at least) by a single spectrum with a symmetric group action. Applying this classification to the layers of the Taylor tower of $F : \mathcal{T}op_* \to \mathcal{T}op_*$, we get an equivalence

$$D_n F(X) \simeq \Omega^\infty(\partial_n F \wedge (\Sigma^\infty X)^{\wedge n})_{h\Sigma_n}$$

where $\partial_n F$ is a spectrum with a (naive) action of the symmetric group Σ_n, which we refer to as the *n-th derivative* of F (at $*$). Similar formulas apply when $\mathcal{C}$ and/or $\mathcal{D}$ is $\mathcal{S}p$ instead of $\mathcal{T}op_*$, and sense can be made of the object $\partial_n F$ for more general $\mathcal{C}$ and $\mathcal{D}$, though in such cases $\partial_n F$ is a diagram of spectra (indexed by generators for the stable ∞-categories $\mathcal{S}p(\mathcal{C})$ and $\mathcal{S}p(\mathcal{D})$) rather than a single spectrum, e.g. see [37, 1.1].

Example 1.2.5. The n-th derivative of the identity functor $I : \mathcal{T}op_* \to \mathcal{T}op_*$ was calculated by Brenda Johnson in [52]. The spectrum $\partial_n I$ is equivalent to a wedge of $(n-1)!$ copies of the $(1-n)$-sphere spectrum. In particular, $\partial_1 I \simeq S^0$ (as mentioned above) and $\partial_2 I \simeq S^{-1}$ (with trivial Σ_2-action), so

$$D_2 I(X) \simeq \Omega^\infty \Sigma^{-1} (\Sigma^\infty X)^{\wedge 2}_{h\Sigma_2}.$$

We can now attempt to calculate $P_2 I$ using the fibre sequence $D_2 I \to P_2 I \to P_1 I$. This takes the form

$$\Omega^\infty \Sigma^{-1} (\Sigma^\infty X)^{\wedge 2}_{h\Sigma_2} \to P_2 I(X) \to \Omega^\infty \Sigma^\infty X.$$

Goodwillie's results imply that this sequence deloops, so that $P_2 I(X)$ can be written as the fibre of a certain natural transformation.

Example 1.2.6. The n-th derivative of the functor $\Sigma^\infty \Omega^\infty : \mathcal{S}p \to \mathcal{S}p$ is equivalent to S^0 (with trivial Σ_n-action) [1, Corollary 1.3]. Therefore

$$D_n(\Sigma^\infty \Omega^\infty)(X) \simeq X^{\wedge n}_{h\Sigma_n}.$$

This tells us that, for a spectrum X, the spectrum $\Sigma^\infty \Omega^\infty X$ is given by piecing together the extended powers $X^{\wedge n}_{h\Sigma_n}$. When X is a 0-connected suspension spectrum, Snaith splitting provides an equivalence

$$\Sigma^\infty \Omega^\infty X \simeq \bigvee_{n \geq 1} X^{\wedge n}_{h\Sigma_n}$$

which can be interpreted as the splitting of the Taylor tower. For arbitrary X, the layers of the tower are pieced together in a less trivial way.

Splitting Results for the Taylor Tower

As illustrated in Example 1.2.6, the simplest situation holds when the Taylor tower is a product of its layers:

$$P_n F(X) \simeq \prod_{k=0}^{n} D_k F(X).$$

Example 1.2.7. Let K be a d-dimensional based finite CW-complex and consider the functor $\Sigma^\infty \operatorname{Hom}_*(K, -) : \mathcal{T}op_* \to \mathcal{S}p$ where $\operatorname{Hom}_*(-,-)$ denotes the space of basepoint-preserving maps between two based spaces. The first author, in [6], proved (1) that

$$\partial_n(\Sigma^\infty \operatorname{Hom}_*(K, -)) \simeq \operatorname{Map}(\Sigma^\infty K^{\wedge n}/\Delta^n K, S^0)$$

where $\operatorname{Map}(-,-)$ is the mapping spectrum construction, and $\Delta^n K$ denotes the "fat diagonal", i.e. the subspace of $K^{\wedge n}$ consisting of those points $(k_1, \dots, k_n)$ where $k_i = k_j$ for some $i \neq j$, (2) that the Taylor tower converges on d-connected X, and (3) when K is a parallelizable d-dimensional manifold and X is a d-fold suspension, that the Taylor tower splits. Thus in the latter case we have:

$$\Sigma^\infty \operatorname{Hom}_*(K, X) \simeq \prod_{n=1}^{\infty} \operatorname{Map}(\Sigma^\infty K^{\wedge n}/\Delta^n K, \Sigma^\infty X^{\wedge n})_{h\Sigma_n}.$$

Kuhn has proved, in [56], the following splitting result which reveals some of the interesting interaction between Goodwillie calculus, Tate cohomology and chromatic homotopy theory.

Theorem 1.2.8 (Kuhn). *Let $F : \mathcal{S}p \to \mathcal{S}p$ be a homotopy functor. Then the Taylor tower of F splits after $T(k)$-localization. (Here $T(k)$ denotes the spectrum given by the telescope of a v_k-self-map of a finite type-k complex.) In other words*

$$L_{T(k)} P_n F(X) \simeq \prod_{j=1}^{n} L_{T(k)} D_j F(X).$$

The proof of Theorem 1.2.8 relies on a number of interesting ingredients, in particular the vanishing of the $T(k)$-localization of the Tate construction associated to a finite group action on a spectrum. The specific part coming from functor calculus, however, is the following result of McCarthy [66].

Theorem 1.2.9 (McCarthy). *For a functor $F : \mathcal{S}p \to \mathcal{S}p$ that preserves filtered colimits, there is a natural homotopy pullback square of the form*

$$\begin{array}{ccc} P_n F(X) & \longrightarrow & (\partial_n F \wedge X^{\wedge n})^{h\Sigma_n} \\ \downarrow & & \downarrow \\ P_{n-1} F(X) & \longrightarrow & (\partial_n F \wedge X^{\wedge n})^{t\Sigma_n}. \end{array}$$

Here Y^{tG} denotes the Tate construction of the action of a finite group G on a spectrum Y, that is the cofibre of the norm map $N : Y_{hG} \to Y^{hG}$, and the right-hand vertical map above is the canonical map from the homotopy fixed points to the Tate construction.

Note that the induced map between the homotopy fibres of the vertical maps in this diagram is the equivalence $D_n F(X) \xrightarrow{\sim} (\partial_n F \wedge X^{\wedge n})_{h\Sigma_n}$ that appears in the classification of n-homogeneous functors from spectra to spectra.

Kuhn's theorem, roughly speaking, follows from McCarthy's by taking $T(k)$-localization of the homotopy pullback square, and using the vanishing of the Tate construction.

The results quoted here begin to address the question of how the layers of the Taylor tower of a homotopy functor are pieced together to form the tower itself, and they start to illustrate the role of the Tate spectrum construction in that picture. We will return to this topic in Section 1.4.

1.3 The Taylor tower of the identity functor for based spaces

Let I be the identity functor from the category of based spaces to itself. From the perspective of functor calculus, this is a highly non-trivial object. As noted above, I is not an n-excisive functor for any n, and its derivatives have a lot of structure. Applying homotopy groups to the Taylor tower, we get a sequence

$$\pi_* X \to \cdots \to \pi_* P_n I(X) \to \pi_* P_{n-1} I(X) \to \cdots \to \pi_* P_1(X) = \pi_*^s X$$

interpolating between the unstable and stable homotopy groups of a space X.

The first step to understanding the Taylor tower of a functor is to calculate the derivatives (and hence the layers). For the identity functor, the derivatives were, as mentioned above, calculated by Johnson in [52]. Here we give the reformulation of her result produced in [14].

Definition 1.3.1. Let $\mathcal{P}_n$ be the poset of partitions of the set $\{1, \ldots, n\}$, ordered by refinement. Let $|\mathcal{P}_n|$ be the geometric realization of $\mathcal{P}_n$. Note that $\mathcal{P}_n$ has both an initial and a final object. It follows in particular that $|\mathcal{P}_n|$ is contractible. Let $\partial|\mathcal{P}_n|$ be the subcomplex of $|\mathcal{P}_n|$ spanned by simplices that do not contain both the initial and final element as vertices. Let $T_n = |\mathcal{P}_n|/\partial|\mathcal{P}_n|$.

It is easy to see that T_n is an $n-1$-dimensional complex with an action of the symmetric group Σ_n. It is well known that non-equivariantly T_n is equivalent to $\bigvee_{(n-1)!} S^{n-1}$. In fact, this equivalence holds already after restricting the action from Σ_n to Σ_{n-1}, where Σ_{n-1} is considered a subgroup of Σ_n in the standard way. This means that there is a Σ_{n-1}-equivariant equivalence $T_n \simeq \Sigma_{n-1+} \wedge S^{n-1}$ [8].

Theorem 1.3.2. [52, 14] *There is a Σ_n-equivariant equivalence of spectra*

$$\partial_n I \simeq \mathbb{D}(T_n)$$

between the n-th derivative of the identity functor and the Spanier-Whitehead dual of the complex T_n.

It turns out that the Taylor tower of the identity functor has some rather special properties when evaluated at a sphere. The results are cleanest to state for an odd-dimensional sphere, so we will mostly focus on this case.

We begin by noting that rationally the tower is constant.

Theorem 1.3.3. [14, Proposition 3.1] *Let X be an odd-dimensional sphere. The spectrum*

$$(\partial_n I \wedge X^{\wedge n})_{h\Sigma_n}$$

is rationally contractible for $n > 1$.

Proof sketch. Since $\partial_n I \simeq \mathbb{D}(T_n)$, it is enough to prove that $\Sigma^\infty X^{\wedge n}_{hG}$ has trivial rational homology when G is any isotropy group of T_n. This follows by an easy spectral sequence argument. □

This strengthens the following classical computation of Serre:

Corollary 1.3.4 (Serre, 1953). *When X is an odd-dimensional sphere, the map $X \to \Omega^\infty\Sigma^\infty X$ is a rational homotopy equivalence.*

It follows from the theorem that for X an odd sphere, the homology of $(\partial_n I \wedge X^{\wedge n})_{h\Sigma_n}$ is torsion for all $n > 1$. The following theorem gives considerably more information about how the torsion is distributed among the layers.

Theorem 1.3.5. [14, 3] *Let X be an odd-dimensional sphere, and let p be a prime. The homology with mod p coefficients of the spectrum $(\partial_n I \wedge X^{\wedge n})_{h\Sigma_n}$ is non-trivial only if n is a power of p.*

Note that it follows that if n is not a prime power then the spectrum $(\partial_n I \wedge X^{\wedge n})_{h\Sigma_n}$ is contractible, since it is a connective spectrum of finite type whose homology is trivial with rational and mod p coefficients for all primes p. Theorem 1.3.5 implies that if one is willing to pick a prime p and localize all spaces at p, then the only non-trivial layers in the tower are the ones numbered by powers of p. Thus the tower converges exponentially faster than usual in this case.

Theorem 1.3.5 was first proved in [14], by a brute force calculation of the homology with mod p coefficients. A more conceptual proof was given in [3]. Furthermore, when $n = p^k$, it turns out that the spectrum $(\partial_n I \wedge X^{\wedge n})_{h\Sigma_n}$ is closely related to well-studied spectra in the literature (remember that X is an odd sphere throughout the discussion). In particular, this spectrum is equivalent to (the k-fold desuspension of) the direct summand of the spectrum $\Sigma^\infty X^{\wedge p^k}_{h(\mathbb{Z}/p)^k}$ split off by the Steinberg idempotent. Such wedge summands were studied by Mitchell, Kuhn, Priddy and others. In the case $X = S^1$, this spectrum is equivalent (up to a suspension) to the n-th subquotient in the filtration of $H\mathbb{Z}$ by symmetric powers of the sphere spectrum.

It turns out that Serre's theorem on rational homotopy groups of spheres (corollary 1.3.4) admits a rather dramatic generalization to v_k-periodic homotopy for all k (rational homotopy fits in as the case $k = 0$). We will now state the result, and then spend the rest of the section explaining what it says and outlining its proof.

Theorem 1.3.6. [14] *Let X be an odd-dimensional sphere, and work p-locally for a prime p. For $k \geq 0$, the map $X \mapsto P_{p^k} I(X)$ is a v_k-periodic equivalence.*

We will now take a detour to review some background on v_k-periodic homotopy. For an omnibus reference on this material we suggest Ravenel's orange book [73] (a revised version of which is available online).[4] Fix a prime p and let all spaces be implicitly localized at p. Homology and cohomology groups will be taken with mod p coefficients. Recall that for each integer $n \geq 0$ there is a generalized homology theory $K(n)$ called the n-th Morava K-theory. For $n = 0$, $K(0) = H\mathbb{Q}$, and for $n > 0$ the coefficient ring of $K(n)$ is $K(n)_* = \mathbb{F}_p[v_n, v_n^{-1}]$, where $|v_n| = 2p^n - 2$.

Definition 1.3.7. A finite complex V is said to be of *type k* if $K(n)_*V$ is trivial for $n < k$ and non-trivial for $n = k$.

[4]See also the chapters in this Handbook by Gijs Heuts and by Tobias Barthel and Agnès Beaudry.

By the Periodicity Theorem [51] [73, Theorem 1.5.4], for each $k \geq 0$ there exists a finite complex of type k. Furthermore, suppose $k \geq 1$, and V_k is a complex of type k. Then there exists a self-map $f\colon \Sigma^{d|v_k|+i}V_k \to \Sigma^i V_k$ for some $i \geq 0, d \geq 1$, whose effect on $K(n)_*$ is an isomorphism for $n = k$ and zero for $n > k$. A map with these properties is called a v_k-periodic map. By the uniqueness part of the Periodicity Theorem, any two v_k-periodic self-maps of V_k are equivalent after taking some suspensions and iterations.

Suppose X is either a pointed space or a spectrum, and V is a finite complex with a basepoint. If X is a space, let X^V denote the space of basepoint-preserving maps from V to X. If X is a spectrum, then let X^V denote the mapping spectrum from $\Sigma^\infty V$ to X. Clearly, if X is a spectrum then $\Omega^\infty(X^V) \cong (\Omega^\infty X)^V$. Let V_k be a complex of type k. By replacing V_k with some suspension thereof if necessary, we may assume that V_k has a self-map of the form $\Sigma^{d|v_k|}V_k \to V_k$. This map induces a map $X^{V_k} \to \Omega^{d|v_k|}X^{V_k}$, where X is still either a pointed space or a spectrum. We can form a mapping telescope as follows

$$v_k^{-1}X^{V_k} := \operatorname{hocolim}\left(X^{V_k} \to \Omega^{d|v_k|}X^{V_k} \to \Omega^{2d|v_k|}X^{V_k} \to \cdots\right).$$

The mapping telescope serves as the definition of $v_k^{-1}X^{V_k}$. If X is a space, then $v_k^{-1}X^{V_k}$ is, by definition, a space, that is easily seen to be an infinite loop space. If X is a spectrum then $v_k^{-1}X^{V_k}$ is a spectrum. Clearly, if X is a spectrum then

$$\Omega^\infty(v_k^{-1}X^{V_k}) \simeq v_k^{-1}(\Omega^\infty X)^{V_k}.$$

We define the v_k-periodic homotopy groups of X with coefficients in V_k as follows.

$$v_k^{-1}\pi_*(X; V_k) := \pi_*\left(v_k^{-1}X^{V_k}\right) \cong \operatorname{colim}(\pi_*(X^{V_k}) \to \pi_{*+d|v_k|}(X^{V_k}) \to \cdots).$$

As before, the groups $v_k^{-1}\pi_*(X; V_k)$ are defined if X is either a space or a spectrum. It is easy to see that if X is a spectrum then there is a canonical isomorphism

$$v_k^{-1}\pi_*(X; V_k) \cong v_k^{-1}\pi_*(\Omega^\infty X; V_k). \tag{1.3.1}$$

The groups $v_k^{-1}\pi_*(X; V_k)$ are periodic with period that divides $d|v_k|$. They depend on the choice of V_k, but by the uniqueness statement above they do not depend on the self-map of V_k (up to a dimension shift).

While the groups $v_k^{-1}\pi_*(X; V_k)$ depend on a choice of V_k, the following is a well-known consequence of the Thick Subcategory Theorem [51].

Proposition 1.3.8. *Let $f\colon X \to Y$ be a map of spaces. If there exists one complex V_k of type k for which f induces an isomorphism $v_k^{-1}\pi_*(X; V_k) \to v_k^{-1}\pi_*(Y; V_k)$ then f induces an isomorphism for every such V_k.*

We say that a map $X \to Y$ is a v_k-equivalence if it induces an isomorphism on v_k-periodic groups for some, and therefore every, choice of a complex V_k of type k. This explains the use of the term in Theorem 1.3.6.

A convenient subclass of complexes of type k are ones that are *strongly of type k*. They are defined by certain freeness properties of the Steenrod algebra action, and the requirement that the Atiyah-Hirzebruch spectral sequence for Morava K-theory collapses. For a precise definition see [73, Definition 6.2.3]. Proposition 1.3.9 summarizes the relevant facts about complexes that are strongly of type k. Recall that the Adams spectral sequence has the following form, where E and F are spectra

$$\mathrm{Ext}^{s,t}_{\mathcal{A}}(H^*(F), H^*(E)) \Rightarrow \pi_{t-s}(\mathrm{Map}(E, F)^{\wedge}_p).$$

Proposition 1.3.9. [73, Sections 6.2–6.4] *For every k there exists a complex strongly of type k. Let V_k be such a complex. After some suspension, V_k has a v_k-periodic self-map $\Sigma^{d|v_k|}V_k \to V_k$ whose stabilization is represented in the second page of the Adams spectral sequence by an element of $\mathrm{Ext}^{d,d|v_k|+d}_{\mathcal{A}}(H^*(V_k), H^*(V_k))$. In particular, the self-map has Adams filtration d.*

It is often convenient to bigrade the Adams spectral sequence by $(t-s, s)$, so that the horizontal axis corresponds to the topological degree and the vertical axis is the Adams filtration. Vanishing lines in the spectral sequence are calculated in these coordinates. Thus a vanishing line gives an upper bound on the possible Adams filtration of a non-zero element of the homotopy groups of a spectrum in terms of its topological dimension. If one uses the grading $(t-s, s)$ then Proposition 1.3.9 says the self-map of V_k is represented by multiplication by an element on a line of slope $\frac{1}{|v_k|}$ and intercept zero.

Let $\mathrm{End}(\Sigma^\infty V_k) \simeq \Sigma^\infty V_k \wedge D(\Sigma^\infty V_k)$ be the endomorphism spectrum of $\Sigma^\infty V_k$. The self-map $\Sigma^{d|v_k|}V_k \to V_k$ gives rise to an element of $\pi_{d|v_k|}(\mathrm{End}(\Sigma^\infty V_k))$. Proposition 1.3.9 says that this element has Adams filtration d. It follows that if E is any spectrum, then the induced map $E^{V_k} \to \Omega^{d|v_k|}E^{V_k}$ also has Adams filtration at least d. In fact, a slightly stronger statement is true: the self-map of E^{V_k} is represented in Adams spectral sequence by an operation that raises topological degree by $d|v_k|$ and raises Adams filtration by d. In other words, it moves elements in the Adams spectral sequence for E^{V_k} along a line of slope $\frac{1}{|v_k|}$. It follows that if the Adams spectral sequence for E^{V_k} has a vanishing line of slope smaller than $\frac{1}{|v_k|}$ then the action of the self-map of V_k on the homotopy groups of E^{V_k} is nilpotent.

It turns out that layers in the Goodwillie tower of the identity have good vanishing lines. So now it is time to get back to the Goodwillie tower.

We remind the reader that X is an odd sphere and everything is localised at a prime p. By Theorem 1.3.5, the only non-trivial layers of the Goowillie tower of the identity at X are the ones indexed by powers of p. Their underlying spectra have the form $(\partial_{p^k}I \wedge X^{\wedge p^k})_{h\Sigma_{p^k}}$. It turns out that these spectra have interesting properties in v_k-periodic homotopy. Our way to see it is via their cohomology with the Steenrod action. For an integer $k \geq 0$ let A_k be the subalgebra of the Steenrod algebra generated by $\{Sq^1, Sq^2, Sq^4, \ldots, Sq^{2^k}\}$ for $p=2$ and by $\{\beta, P^1, P^p, \ldots, P^{p^{k-1}}\}$ for $p>2$.

Proposition 1.3.10. *The cohomology of $(\partial_{p^k}I \wedge X^{\wedge p^k})_{h\Sigma_{p^k}}$ is free over A_{k-1}.*

Again, this proposition was first proved in [14] by a brute force computation. The cohomology of the spectrum $(\partial_{p^k}I \wedge X^{\wedge p^k})_{h\Sigma_{p^k}}$ was calculated explicitly, and it was observed that in the case $X = S^1$ it is isomorphic (up to degree shift) to the cohomology of the quotient of symmetric product spectra

$$\mathrm{Sp}^{p^k}(S^0)/\mathrm{Sp}^{p^k-1}(S^0).$$

The cohomology of the latter is A_{k-1}-free by a theorem of Welcher [83], and Welcher's argument can be adapted to the case of a more general X.

It was pointed out to us by Nick Kuhn that there is a more direct way to deduce proposition 1.3.10 from Welcher's result. We know from [3] that the spectrum $(\partial_{p^k} I \wedge S^{p^k})_{h\Sigma_{p^k}}$ is homotopy equivalent (up to a suspension) to

$$\mathrm{Sp}^{p^k}(S^0)/\mathrm{Sp}^{p^k-1}(S^0),$$

and for a general odd sphere X, the spectrum $(\partial_{p^k} I \wedge X^{\wedge p^k})_{h\Sigma_{p^k}}$ is (roughly speaking) a Thom spectrum over the case $X = S^1$. Thus there is a Thom isomorphism between the cohomologies of the two spectra:

$$H^*\left(\left(\partial_{p^k} I \wedge S^{p^k}\right)_{h\Sigma_{p^k}}\right) \cong H^{*+2lp^k}\left(\left(\partial_{p^k} I \wedge S^{(2l+1)p^k}\right)_{h\Sigma_{p^k}}\right).$$

Furthermore, one can show that A_{k-1} acts trivially on the Thom class. This can be done by identifying the Thom class with a power of the top Dickson invariant, and using the formulas in [84]. From here it follows that in our case the Thom isomorphism respects the A_{k-1}-module structure.

Remark 1.3.11. It seems likely that Proposition 1.3.10 can also be proved by adapting the methods of Steve Mitchell [69].

Proposition 1.3.10 has consequences regarding the v_k-periodic homotopy thanks to the following theorem, due to Anderson-Davis [2] for $p = 2$ and Miller-Wilkerson [68] for $p > 2$.

Theorem 1.3.12. *Let M be a connected A-module that is free over A_k. Then $\mathrm{Ext}_A(M, \mathbb{F}_p)$ has a vanishing line of slope that is strictly smaller than $\frac{1}{|v_k|}$ and an intercept that is bounded above by a number that depends only on k. This bound is a slowly growing function of k.*

Corollary 1.3.13. *Suppose E is a spectrum for which $H^*(E)$ is free over A_{l-1}. Then E and $\Omega^\infty E$ have trivial v_k-periodic homotopy for $k < l$.*

Proof. It is enough to show that for a complex V_k that is strongly of type k, $v_k^{-1}\pi_*(E; V_k) = 0$. Since $H^*(E)$ is free over A_{l-1}, it follows that $H^*(E^{V_k})$ is free over A_{l-1}. By theorem 1.3.12, the Adams spectral sequence for $\pi_*(E^{V_k})$ has a vanishing line of slope smaller than $\frac{1}{|v_{l-1}|}$, and therefore smaller than $\frac{1}{|v_k|}$. By proposition 1.3.9 and subsequent comments, multiplication by a self-map of V_k is represented, on the level of second page of the Adams spectral sequence, by multiplication by an element on a line of slope $\frac{1}{|v_k|}$, which is larger than the slope of the vanishing line. Therefore, every element of $\pi_*(E^{V_k})$ is annihilated by some power of the self-map of V_k. It follows that inverting the self-map kills everything in the homotopy groups. □

Corollary 1.3.14. *If X is an odd sphere, then the only layers of Goodwillie tower of the identity evaluated at X that are non-trivial in v_k-periodic homotopy are $D_1 I(X), D_p I(X), D_{p^2} I(X), \ldots, D_{p^k} I(X)$.*

Proof. Everything is localized at p, and by theorem 1.3.5, $D_n I(X)$ is non-trivial at p only if n is a power of p. We need to show that if $l > k$ then $D_{p^l} I(X)$ is trivial in v_k-periodic homotopy. Recall that

$$D_{p^l} I(X) \simeq \Omega^\infty \left(\left(\partial_{p^l} I \wedge S^{p^l} \right)_{h\Sigma_{p^l}} \right).$$

By Proposition 1.3.10, $H^* \left(\left(\partial_{p^l} I \wedge S^{p^l} \right)_{h\Sigma_{p^l}} \right)$ is free over $\mathcal{A}_{l-1}$. The result follows by Corollary 1.3.13. □

Corollary 1.3.14 suggests that the map $X \to P_{p^k} I(X)$ induces an isomorphism on v_k-periodic homotopy groups. This conclusion is not automatic, because inverting v_k does not always commute with inverse homotopy limits, but it turns out to be true in this case. So we can now sketch the proof of Theorem 1.3.6.

Proof of Theorem 1.3.6. Let k be fixed. Let $F_k(X)$ be the homotopy fibre of the map $X \to P_{p^k} I(X)$. For $n \geq k$ let us define $\overline{P_{p^n}}(X)$ to be the homotopy fibre of the map $P_{p^n} I(X) \to P_{p^k} I(X)$. Note that the homotopy fibre of the map $\overline{P_{p^n}}(X) \to \overline{P_{p^{n-1}}}(X)$ is $D_{p^n} I(X)$. Choose a complex V_k that is strongly of type k with a self-map $\nu_k \colon \Sigma^{d|v_k|} V_k \to V_k$ of Adams filtration d, as per proposition 1.3.9. There is a tower of the following form, where $F_k(X)^{V_k}$ is the homotopy inverse limit of the top row:

$$\begin{array}{ccccccc} F_k(X)^{V_k} \cdots \to & \overline{P_{p^n}}(X)^{V_k} & \to & \overline{P_{p^{n-1}}}(X)^{V_k} & \longrightarrow \cdots \longrightarrow & \overline{P_{p^{k+1}}}(X)^{V_k} \\ & \uparrow & & \uparrow & & \uparrow \sim \\ & D_{p^n} I(X)^{V_k} & & D_{p^{n-1}} I(X)^{V_k} & & D_{p^{k+1}} I(X)^{V_k}. \end{array}$$

We need to prove that $v_k^{-1} \pi_*(F_k; V_k)$ is zero. Let $\alpha \in \pi_*(F_k^{V_k})$. We need to show that some power of the self-map ν_k annihilates α. Let us say that α has height $\geq m$ if the image of α in $\pi_* \left(\overline{P_{p^n}}^{V_k} \right)$ is zero for $n < m$. Suppose α has height $\geq m$. Then the image of α in $\pi_* \left(\overline{P_{p^m}}^{V_k} \right)$ is in the image of some element of $\pi_* \left(D_{p^m} I(X)^{V_k} \right)$. We know that ν_k acts along a line of higher slope than the vanishing line of the Adams spectral sequence for the underlying spectrum of $D_{p^m} I(X)^{V_k}$. Because of this, some power ν_k^l of ν_k annihilates this element of $\pi_* \left(D_{p^m} I(X)^{V_k} \right)$. This means that $v_k^l(\alpha)$ has height $\geq m + 1$. Repeating the process, one finds that for every $n > k$, there exists some l_n for which $v_k^{l_n}$ has height $\geq n$. Let l_n be the lowest such l. Using Theorem 1.3.12, it is not very difficult to write down an effective upper bound on the topological dimension of $\nu_k^{l_n}(\alpha)$ as a function of n. Crucially, it is not very difficult to show that this function grows slower than the connectivity of $D_{p^n} I(X)^{V_k}$ (the calculation is done in [14]). It follows that there exists an n for which $\nu_k^{l_n}(\alpha)$ has lower topological dimension than the connectivity of $D_{p^n} I(X)^{V_k}$. It follows that the image of $\nu_k^{l_n}(\alpha)$ in $\pi_* \left(\overline{P_{p^m}}^{V_k} \right)$ is zero for all m. It follows that $\nu_k^{l_n}(\alpha) = 0$. □

Theorem 1.3.6 says that the unstable v_k-periodic homotopy type of an odd sphere can be resolved into $(k + 1)$ stable homotopy types. The case $k = 0$ is essentially 1.3.4.

A Bousfield-Kuhn functor reformulation

Theorem 1.3.6 has a more modern reformulation in terms of the Bousfield-Kuhn functor. Let us recall what this functor is. Choose a v_k self-map $\Sigma^{d|v_k|}V_k \to V_k$. Use it to form a direct system of spectra

$$\Sigma^\infty V_k \to \Omega^{d|v_k|}\Sigma^\infty V_k \to \cdots.$$

Let $T(k)$ be the homotopy colimit of this system, and $L_{T(k)}$ be the Bousfield localization with respect to $T(k)$. It follows from the uniqueness part of the Periodicity Theorem that the functor $L_{T(k)}$ does not depend on the choice of V_k or the self-map.

The Bousfield-Kuhn functor Φ_k is a functor from pointed spaces to spectra, whose main property is that there is an equivalence $\Phi_k(\Omega^\infty E) \simeq L_{T(k)}(E)$. Here E is any spectrum, the equivalence is natural in E. The functor Φ_k is constructed as an inverse homotopy limit

$$\Phi_k(X) = \mathrm{holim}_\alpha\, v_k^{-1} X^{V_k^\alpha},$$

where $\{V_k^\alpha\}$ is a direct system of complexes of type k with certain properties. See [59] for more details. The following is now an immediate corollary of Theorem 1.3.6.

Theorem 1.3.15. *When X is an odd sphere, the map $X \to P_{p^k}I(X)$ becomes an equivalence after applying the Bousfield-Kuhn functor Φ_k.*

The case of even-dimensional spheres, and beyond

There is a version of Theorems 1.3.6 and 1.3.15 that holds for even-dimensional spheres:

Theorem 1.3.16. [14, Theorems 4.4 and 4.5] *Fix a prime p and localize everything at p. Let X be an even-dimensional sphere. Then $D_nI(X) \simeq *$ unless n is a power of p or twice a power of p. The map*

$$X \to P_{2p^k}I(X)$$

is a v_k-periodic equivalence.

The easiest way to prove this theorem that we know is to use the EHP sequence.

$$X \mapsto \Omega\Sigma X \to \Omega\Sigma(X \wedge X).$$

This is a fibration sequence if X is an odd-dimensional sphere, and one can show that it induces a fibration sequence of Taylor towers. Given this, it is easy to deduce Theorem 1.3.16 from Theorem 1.3.6.

Remark 1.3.17. The connection between the Goodwillie tower and the EHP sequence was investigated much more deeply by M. Behrens at [25].

Theorems 1.3.6 and 1.3.16 tell us that the behavior of the Taylor tower of the identity in v_k-periodic homotopy has two non-obvious properties when evaluated at spheres:

1. It is finite.

2. It converges.

It is therefore reasonable to ask if there exist other spaces for which the Taylor tower of the identity has these properties. Regarding the finiteness property, it seems clear that spheres (or at best homology spheres) are the only spaces for which the Taylor tower is finite in v_k-periodic homotopy. For example, it is easy to show that the only spaces for which the tower is rationally finite are rational homology spheres. On the other hand, there do exist other spaces for which the tower converges in v_k-periodic homotopy. Such results were obtained by Behrens-Rezk [26, section 8] and by Heuts [50]. For example, convergence holds for products of spheres and special unitary groups $SU(k)$. It would be interesting to characterize the spaces for which v_k-periodic convergence holds.

The case $X = S^1$. Theorems of Behrens and Kuhn

We noted above that there is a relationship between the layers of the Taylor tower of the identity evaluated at S^1, and the subquotients of the filtration of the Eilenberg-Mac Lane spectrum $H\mathbb{Z} = \mathrm{Sp}^\infty(S^0)$ by the symmetric powers of the sphere spectrum. More precisely, if n is not a power prime, then both layers are trivial, and if $n = p^k$ then there is an equivalence

$$D_{p^k} I(S^1) \simeq \Omega^{\infty+2k-1}\mathrm{Sp}^{p^k}(S^0)/\mathrm{Sp}^{p^k-1}(S^0).$$

In fact, there is a deeper connection between the two filtrations. Roughly speaking, one can recast each filtration as a "chain complex" of infinite loop spaces or spectra, and each chain complex can serve as a kind of contracting homotopy for the other one. This implies, in particular, that each filtration is trivial in the sense that its homotopy spectral sequence collapses at the second page. The triviality result for the symmetric powers filtration used to be known as the Whitehead conjecture. It was proved by Kuhn [55] at the prime 2 and by Kuhn and Priddy [61] at odd primes. The triviality result for the Goodwillie tower of the identity, and the connection between the two filtrations was proved by Behrens [24] at the prime 2 and in another way by Kuhn [60] at all primes.

1.4 Operads and Tate data: the Classification of Taylor towers

The layers of a Taylor tower (say, of a functor $F : \mathcal{T}op_* \to \mathcal{T}op_*$) are homogeneous functors classified by the spectra $\partial_* F = (\partial_n F)_{n\geq 1}$, along with the action of Σ_n on $\partial_n F$, i.e. a *symmetric sequence* of spectra. As we have seen, however, further information is needed to encode the full Taylor tower, and hence under convergence conditions, the functor F itself.

Roughly speaking, there are two approaches to understanding this extra information: (1) inductive techniques based on the delooped fibre sequences

$$P_n F \to P_{n-1} F \to \Omega^{-1} D_n F;$$

and (2) analysis of operad/module structures on the symmetric sequence $\partial_* F$ in its entirety, based on a chain rule philosophy for the calculus of functors.

We saw approach (1) applied to $P_2 I$ in Example 1.2.5 and, for functors from spectra to spectra, in McCarthy's Theorem (1.2.9). Here we focus on (2).

Chain Rules in Functor Calculus

The first version of a "Chain Rule" for the calculus of functors was proved by Klein and Rognes in [54]. The simplest version of this states the following: for reduced functors $F, G : \mathcal{T}op_* \to \mathcal{T}op_*$, we have

$$\partial_1(FG) \simeq \partial_1 F \wedge \partial_1 G.$$

More generally, Klein and Rognes provided a formula for the first derivative of FG at an arbitrary space X, in terms of the first derivatives of G (at X) and F (at $G(X)$).

For higher derivatives, the Chain Rule for functors of spectra is much simpler than that for spaces. Suppose first that $F, G : \mathcal{S}p \to \mathcal{S}p$ are given by the formulas

$$F(X) \simeq \bigvee_{k=1}^{\infty} (\partial_k F \wedge X^{\wedge k})_{h\Sigma_k}, \quad G(X) \simeq \bigvee_{l=1}^{\infty} (\partial_l G \wedge X^{\wedge l})_{h\Sigma_l}.$$

A simple calculation then shows that

$$\partial_n(FG) \simeq \bigvee_{\text{partitions of } \{1,\ldots,n\}} \partial_k F \wedge \partial_{n_1} G \wedge \cdots \wedge \partial_{n_k} G$$

where $n_1, \ldots, n_k \geq 1$ are the sizes of the terms in a given partition. This is also the formula for the composition product of symmetric sequences, and, more succinctly, we can write

$$\partial_*(FG) \simeq \partial_* F \circ \partial_* G. \tag{1.4.1}$$

The second author proved in [35] that (1.4.1) holds for any reduced functors $F, G : \mathcal{S}p \to \mathcal{S}p$ where F preserves filtered colimits.

For functors of based spaces (or more general ∞-categories), the formula (1.4.1) does not hold, but there is nonetheless a natural map

$$l : \partial_* F \circ \partial_* G \to \partial_*(FG)$$

making ∂_*, at least up to homotopy, into a lax monoidal functor from a suitable ∞-category of functors to the ∞-category of symmetric sequences. Specializing the map l to the case when one or both of F, G is the identity, we obtain a number of important consequences. The following results were proved in [9] for functors between based spaces and spectra, and (with the exception of the final statement of the chain rule, though a version of this appears in [64, 6.3]) in [37] for arbitrary ∞-categories.

Theorem 1.4.1. *The derivatives $\partial_* I_{\mathcal{C}}$ of the identity functor on an ∞-category $\mathcal{C}$ have a canonical operad structure, and for an arbitrary functor $F : \mathcal{C} \to \mathcal{D}$, the derivatives $\partial_* F$ form a $(\partial_* I_{\mathcal{D}}, \partial_* I_{\mathcal{C}})$-bimodule. Moreover, if F preserves filtered colimits and $G : \mathcal{B} \to \mathcal{C}$ is reduced, then there is an equivalence (of bimodules):*

$$\partial_*(FG) \simeq \partial_* F \circ_{\partial_* I_{\mathcal{C}}} \partial_* G$$

where the right-hand side involves a (derived) relative composition product of bimodules over the operad $\partial_* I_{\mathcal{C}}$.

Remark 1.4.2. When $\mathcal{C} = \mathcal{T}op_*$, the ∞-category of based spaces, the operad $\partial_* I_{\mathcal{T}op_*}$ from Theorem 1.4.1 can be viewed as the analogue in stable homotopy theory of the operad encoding the structure of a Lie algebra. This perspective can be justified in a number of ways. Firstly, it follows from Johnson's calculation that taking homology groups of the spectra $\partial_* I_{\mathcal{T}op_*}$ recovers precisely the ordinary Lie operad (up to a shift in degree). A deeper connection is given by viewing $\partial_* I_{\mathcal{T}op_*}$ as an example of bar-cobar (or Koszul) duality for operads of spectra. Ginzburg and Kapranov developed the theory of bar-cobar duality for differential-graded operads in [43], and identified the Lie operad as the dual of the commutative cooperad. The following result was proved in [34] (or in the given form in [9]) and justifies viewing $\partial_* I_{\mathcal{T}op_*}$ as a version of the Lie operad.

Theorem 1.4.3. *The operad $\partial_* I_{\mathcal{T}op_*}$ is equivalent to the cobar construction, or (derived) Koszul dual, of the cooperad $\partial_*(\Sigma^\infty \Omega^\infty)$, which itself can be identified with the commutative cooperad of spectra.*

Remark 1.4.4. Theorem 1.4.3 can actually be understood quite easily from the point of view of calculus. For simply-connected X, the adjunction $(\Sigma^\infty, \Omega^\infty)$ determines an equivalence

$$X \xrightarrow{\simeq} \mathrm{Tot}(\Omega^\infty (\Sigma^\infty \Omega^\infty)^\bullet \Sigma^\infty X)$$

which connects the identity functor to a cobar construction on the comonad $\Sigma^\infty \Omega^\infty$. Taking derivatives of each side, and applying the chain rule for spectra from (1.4.1) we recover 1.4.3 with a little work. Similar arguments were used in [9] to understand the bimodule structures on the derivatives of functors to/from based spaces.

Remark 1.4.5. A version of Theorem 1.4.3 seems likely to be valid for the identity functor on an arbitrary $\mathcal{C}$. Lurie constructs in [64, 6.3.0.14] a cooperad that represents the derivatives $\partial_*(\Sigma^\infty_{\mathcal{C}} \Omega^\infty_{\mathcal{C}})$. (In Lurie's language, this object is actually a "stable corepresentable ∞-operad", but the connection with cooperads is explained in [64, 6.3.0.12].) A similar proof to that of 1.4.3 should imply that the cobar construction on this cooperad recovers the operad $\partial_* I_{\mathcal{C}}$.

Heuts's theorem on spectral Lie algebras in chromatic homotopy theory

The calculations of Arone and Mahowald [14] described above already show that the Taylor tower of the identity functor on based spaces has interesting structure when viewed through the lens of v_k-periodic homotopy theory. Recent work of Heuts [50] has further developed this connection, taking the Lie operad structure on $\partial_* I_{\mathcal{T}op_*}$ into account.

First recall that one of Quillen's models for rational homotopy theory is in terms of Lie algebras. Specifically, he constructs in [72] a Quillen equivalence between simply connected rational spaces and 0-connected differential-graded rational Lie algebras. Heuts's work extends Quillen's to higher chromatic height in the following way.

Theorem 1.4.6 (Heuts)**.** *Fix a prime p and positive integer k. Let $\mathcal{M}_k^f$ denote the ∞-category obtained from that of p-local based spaces by inverting those maps that induce an equivalence on v_k-periodic homotopy groups. Then there is an equivalence of ∞-categories*

$$\mathcal{M}_k^f \simeq \mathcal{S}p_{T(k)}(\partial_* I_{\mathcal{T}op_*})$$

between $\mathcal{M}_k^f$ and the category of $T(k)$-local spectra with an algebra structure over the operad $\partial_* I_{\mathcal{T}op_*}$. Moreover, to a space X this equivalence assigns a $\partial_* I_{\mathcal{T}op_*}$-algebra whose underlying spectrum is given by applying the Bousfield-Kuhn functor Φ_k to X.

Theorem 1.4.6 should be compared to Kuhn's Theorem 1.2.8 which also described a simplification to Goodwillie calculus that appears in the presence of chromatic localization. Both results illustrate the general principle that analysis in Goodwillie calculus typically comprises two pieces: (1) an operadic part related to the derivatives of the identity functor, and (2) something related to the Tate spectrum construction, which vanishes chromatically. Notice that in Kuhn's result, the operadic part is absent because the derivatives of the identity on $\mathcal{S}p$ form the trivial operad.

Tate data and the classification of Taylor towers

We have already seen in McCarthy's Theorem 1.2.9 that the extra information for reconstructing the Taylor tower of a functor $F : \mathcal{S}p \to \mathcal{S}p$ can be described in terms of Tate spectrum constructions for the Σ_n-action on $\partial_n F$. For functors $F : \mathcal{T}op_* \to \mathcal{T}op_*$, we instead think of the Taylor tower information being given by a combination of such "Tate data" with the $\partial_* I$-bimodule structure of Theorem 1.4.1.

To understand this perspective, we use the fact the functor ∂_* (when viewed with values in $\partial_* I$-bimodules) admits a right adjoint which we denote Ψ, i.e. there is an adjunction

$$\partial_* : \mathrm{Exc}_n^*(\mathcal{C}, \mathcal{D}) \rightleftarrows \mathrm{Bimod}_{\leq n}(\partial_* I_{\mathcal{D}}, \partial_* I_{\mathcal{C}}) : \Psi \tag{1.4.2}$$

between the ∞-category of reduced n-excisive functors $\mathcal{C} \to \mathcal{D}$, and the ∞-category of n-truncated bimodules over the derivatives of the identity on $\mathcal{C}$ and $\mathcal{D}$. (The right adjoint was denoted Φ in [10] but we use Ψ here to avoid confusion with the Bousfield-Kuhn functor.) In [10], we proved the following:

Theorem 1.4.7. *The adjunction (∂_*, Ψ) of (1.4.2) is comonadic. In particular, an n-excisive functor $F : \mathcal{C} \to \mathcal{D}$ is classified by the bimodule $\partial_* F$ together with an action of the comonad $\mathbf{C} := \partial_* \Psi$.*

Understanding the full structure on the derivatives of a functor $F : \mathcal{C} \to \mathcal{D}$ thus involves a calculation of the comonad $\mathbf{C}$ of Theorem 1.4.7. This is most easily described in the case of functors $F : \mathcal{T}op_* \to \mathcal{S}p$ where we obtain the following consequence.

Theorem 1.4.8. *The Taylor tower of $F : \mathcal{T}op_* \to \mathcal{S}p$ is determined by the right $\partial_* I$-module structure on $\partial_* F$ together with (suitably compatible) lifts $\psi_{n_1,\dots,n_k}$ of the form*

$$\begin{array}{ccc} & & \mathrm{Map}(\partial_{n_1} I \wedge \cdots \wedge \partial_{n_k} I, \partial_n F)_{h\Sigma_{n_1} \times \cdots \times \Sigma_{n_k}} \\ & \overset{\psi_{n_1,\dots,n_k}}{\nearrow} & \big\downarrow N \\ \partial_k F & \xrightarrow[\phi_{n_1,\dots,n_k}]{} & \mathrm{Map}(\partial_{n_1} I \wedge \cdots \wedge \partial_{n_k} I, \partial_n F)^{h\Sigma_{n_1} \times \cdots \times \Sigma_{n_k}} \end{array}$$

where N is the norm map from homotopy orbits to homotopy fixed points, and $\phi_{n_1,\dots,n_k}$ is the map associated to the right $\partial_* I$-module structure.

Such lifts make $\partial_* F$ into what we call a *divided power right $\partial_* I$-module.*

The data of the lifts $\psi_{n_1,\dots,n_k}$ in Theorem 1.4.8 can be reframed as choices of nullhomotopies for maps

$$\partial_k F \to \mathrm{Map}(\partial_{n_1} I \wedge \dots \wedge \partial_{n_k} I, \partial_n F)^{t\Sigma_{n_1} \times \dots \times \Sigma_{n_k}}$$

into the corresponding Tate construction. We think of this as the *Tate data* corresponding to the Taylor tower of the functor F.

Problem 1.4.9. *It is still unclear how to describe the structure on the derivatives of a functor $F : \mathcal{T}op_* \to \mathcal{T}op_*$ as explicitly as that in Theorem 1.4.8. The comonad guaranteed by Theorem 1.4.7 is hard to understand in this case. In [10] we gave a concrete description of this structure for 3-excisive functors, but a more general picture was too elusive.*

Vanishing Tate data and applications

Another way to see how the Tate spectrum construction comes up is via the unit map $\eta : P_n F \to \Psi \partial_{\leq n} F$ of the adjunction (1.4.2). The right adjoint Ψ can be written in terms of mapping spaces for the ∞-category of bimodules. This leads to the existence of the following diagram for, for example, a functor $F : \mathcal{T}op_* \to \mathcal{T}op_*$:

$$\begin{array}{ccc} \Omega^\infty(\partial_n F \wedge X^{\wedge n})_{h\Sigma_n} & \xrightarrow{N} & \Omega^\infty(\partial_n F \wedge X^{\wedge n})^{h\Sigma_n} \\ \downarrow & & \downarrow \\ P_n F(X) & \xrightarrow{\eta} & \mathrm{Bimod}_{\partial_* I}(\partial_*(\mathrm{Hom}(X,-)), \partial_{\leq n} F) \\ \downarrow & & \downarrow \\ P_{n-1} F(X) & \xrightarrow{\eta} & \mathrm{Bimod}_{\partial_* I}(\partial_*(\mathrm{Hom}(X,-)), \partial_{\leq n-1} F) \end{array} \tag{1.4.3}$$

where the columns are fibration sequences, and $\mathrm{Bimod}_{\partial_* I}(-,-)$ denotes the space of maps of $\partial_* I$-bimodules between the given derivatives. Here $\partial_{\leq n} F$ denotes the *n-truncation* of a $\partial_* I$-bimodule, given by setting all terms in degree larger than n to be the trivial spectrum.

We think of (1.4.3) as a generalization of McCarthy's Theorem 1.2.9, and there is a similar picture for functors from spectra to spectra that reduces to McCarthy's result. For functors $F : \mathcal{T}op_* \to \mathcal{S}p$, we can similarly deduce the following version, which is from [10, 4.17].

Theorem 1.4.10. *Let $F : \mathcal{T}op_* \to \mathcal{S}p$ be a reduced functor that preserved filtered colimits. Then there is a pullback square of the form*

$$\begin{array}{ccc} P_n F(X) & \longrightarrow & (\partial_n F \wedge X^{\wedge n}/\Delta^n X)_{h\Sigma_n} \\ \downarrow & & \downarrow \\ P_{n-1} F(X) & \longrightarrow & (\partial_n F \wedge \Sigma \Delta^n X)_{h\Sigma_n} \end{array}$$

where $\Delta^n X$ is the fat diagonal inside $X^{\wedge n}$.

All the results mentioned here have similar consequences to McCarthy's Theorem in cases where the Tate spectra vanish. In such situations, the Taylor tower of a functor is completely determined by the relevant module or bimodule structure.

Proposition 1.4.11. *Suppose $F : \mathcal{T}op_* \to \mathcal{T}op_*$ preserves filtered colimits and X has the property that*

$$(\partial_n F \wedge X^{\wedge n})^{t\Sigma_n} \simeq *$$

for $2 \leq n \leq N$. Then

$$P_N F(X) \simeq \mathrm{Bimod}_{\partial_* I}(\partial_* \mathrm{Hom}(X, -), \partial_{\leq N} F).$$

For $F : \mathcal{T}op_* \to \mathcal{S}p$ satisfying the same assumptions

$$P_N F(X) \simeq \mathrm{RMod}_{\partial_* I}(\partial_* \Sigma^\infty \mathrm{Hom}(X, -), \partial_{\leq N} F)$$

where $\mathrm{RMod}_{\partial_* I}(-,-)$ is the mapping spectrum for the ∞-category of right $\partial_* I$-modules.

Since the Tate construction vanishes for any rational spectrum, we see that the hypothesis of Proposition 1.4.11 holds if either:

- $\partial_n F$ is a rational spectrum for $2 \leq n \leq N$; or
- X is a rational (simply-connected) topological space.

In particular, if F is a functor either to or from the category of rational based spaces, then the Taylor tower of F is given by the formula in Proposition 1.4.11.

Proposition 1.4.11 allows us to extend Kuhn's Theorem 1.2.8 to functors from $\mathcal{T}op_*$ to $\mathcal{S}p$.

Proposition 1.4.12. *Let F be a functor from $\mathcal{T}op_*$ to $T(n)$-local spectra. Then there is an equivalence*

$$P_N F(X) \simeq \mathrm{RMod}_{\partial_* I}(\mathbb{D}(X^{\wedge *}/\Delta^* X), \partial_{\leq N} F).$$

Similarly, if F is a functor from $\mathcal{T}op_*$ to $\mathcal{T}op_*$, and V_k is a finite complex of type k, then there is an equivalence

$$P_N v_k^{-1} F(X)^{V_k} \simeq \mathrm{RMod}_{\partial_* I}(\mathbb{D}(X^{\wedge *}/\Delta^* X), v_k^{-1} \partial_{\leq N} F^{V_k}).$$

The spectra $\mathbb{D}X^{\wedge *}/\Delta^* X$ form a right $\partial_* I$-module, by identifying them as the derivatives of the functor $\Sigma^\infty \mathrm{Hom}(X, -) : \mathcal{T}op_* \to \mathcal{S}p$.

It seems less straightforward to obtain an analogous result for functors from spaces to spaces than in the spectrum-valued case, yet Heuts's Theorem 1.4.6 simplifies this to the case of functors between spectral Lie algebras.

Functors from spaces to spectra and modules over the little disc operads

There is a close connection between the spectral Lie operad $\partial_* I$ and the little disc operads. Write E_n for the operad of spectra formed by taking suspension spectra of the terms in the little n-discs operad of May [65], and write KE_n for the (derived) Koszul dual of E_n in the sense of [36]. Then $\partial_* I$ can be expressed as the inverse limit of the sequence of Koszul duals:

$$\partial_* I \to \cdots \to KE_n \to KE_{n-1} \to \cdots \to KE_1.$$

It is conjectured that KE_n is equivalent as an operad to a desuspension of E_n itself. This is proved in the context of chain complexes by Fresse [42].

The connection between $\partial_* I$ and the sequence KE_n has powerful consequences. Any divided power module over KE_n determines, by pulling back along the map $\partial_* I \to KE_n$, a divided power module over $\partial_* I$. Moreover, the terms in the operad KE_n have free symmetric group actions, so the norm map for the Σ_k-action on $\mathrm{Map}(KE_n(k), X^{\wedge k})$ is an equivalence. It follows that any right KE_n-module has a unique divided power structure and hence determines a divided power $\partial_* I$-module.

This allows us to ask the following question: for a given functor $F : \mathcal{T}op_* \to \mathcal{S}p$, is the Taylor tower of F determined by a right KE_n-module structure on $\partial_* F$?

The following result of [11] gives a criterion for this to be the case.

Theorem 1.4.13. *A polynomial functor $F : \mathcal{T}op_* \to \mathcal{S}p$ is determined by a KE_n-module structure on $\partial_* F$ if and only if F is the left Kan extension of a functor $f\mathcal{M}an_n \to \mathcal{S}p$ along the inclusion $f\mathcal{M}an_n \to \mathcal{T}op_*$, where $f\mathcal{M}an_n$ is the subcategory of $\mathcal{T}op_*$ consisting of certain "pointed framed n-dimensional manifolds" (that is, one-point compactifications of framed n-manifolds) and "pointed framed embeddings".*

In the next section we will see an application of 1.4.13 to the Taylor tower of algebraic K-theory of spaces.

Remark 1.4.14. The category $f\mathcal{M}an_n$ in Theorem 1.4.13 is related to the "zero-pointed manifolds" of Ayala and Francis [18], and this result suggests deeper connections between Goodwillie calculus and factorization homology that are yet to be explored.[5]

1.5 Applications and calculations in algebraic K-theory

Much of Goodwillie's initial motivation for developing the calculus of functors came from algebraic K-theory, and aside from the identity functor, most of the calculations and applications of calculus have been in this area, largely by Randy McCarthy and coauthors. We review here what is known about the Taylor tower of algebraic K-theory both in the context of spaces and ring spectra.

Algebraic K-theory of spaces

Let $A : \mathcal{T}op \to \mathcal{S}p$ denote Waldhausen's functor calculating the algebraic K-theory of a topological space X, i.e. of the category of spaces over and under X; see [79]. For calculus to be at all relevant to the study of the functor A, we first have to know that the Taylor tower converges for some spaces X.

[5]See the chapter in this Handbook by David Ayala and John Francis for information about factorization homology.

Theorem 1.5.1 (Goodwillie [45]). *The functor $A : \mathcal{T}op \to \mathcal{S}p$ is 1-analytic. Thus the Taylor tower of A converges on simply connected spaces.*

Goodwillie's initial application of this result, however, was not to the convergence of the Taylor tower, but to the cyclotomic trace map τ from $A(X)$ to the *topological cyclic homology* $TC(X)$ of Bökstedt, Hsiang and Madsen [31].

Theorem 1.5.2 (Böksted, Carlsson, Cohen, Goodwillie, Hsiang, Madsen [30]). *For a simply connected finite CW-complex X, there is a pullback square*

$$\begin{array}{ccc} A(X) & \xrightarrow{\tau} & TC(X) \\ \downarrow & & \downarrow \\ A(*) & \xrightarrow{\tau} & TC(*). \end{array}$$

The main step in the proof of Theorem 1.5.2 is to show that the cyclotomic trace τ induces equivalences between the first derivatives of A and TC (at an arbitrary space X). The general theory of calculus then implies that the fibre of τ is "locally constant", i.e. takes a suitably connected map of spaces to an equivalence of spectra.

The first derivative of A at the space X can be thought of as a parametrized spectrum over the space X with fibre over $x \in X$ given by

$$\partial_1 A(X)_x \simeq \Sigma^\infty(\Omega_x X)_+.$$

In [46], Goodwillie generalizes this formula to higher derivatives. The n-th derivative is a spectrum parametrized over X^n and for simplicity, we will only describe the case $X = *$, where

$$\partial_n A \simeq \Sigma^\infty(\Sigma_n/C_n)_+ \tag{1.5.1}$$

and where C_n is an order n cyclic subgroup of the symmetric group Σ_n.

What can we now say about the Taylor tower of A? Firstly, a choice of basepoint for a space Y determines a splitting

$$A(Y) \simeq A(*) \times \tilde{A}(Y)$$

where $\tilde{A} : \mathcal{T}op_* \to \mathcal{S}p$ is the corresponding reduced functor. Next, Waldhausen's splitting result [79] implies that the 1-excisive approximation to A splits off too, so we have

$$A(Y) \simeq A(*) \times \Sigma^\infty Y \times \tilde{\mathrm{Wh}}(Y)$$

where $\tilde{\mathrm{Wh}}(Y)$ is the "reduced Whitehead spectrum" of Y, which contains all of the higher degree information from the Taylor tower of A.

The formula (1.5.1) implies the following simple calculation of the layers of the Taylor tower for A: for $n \geq 2$ we have (using the chosen basepoint for Y):

$$D_n A(Y) \simeq (\Sigma^\infty Y)^{\wedge n}_{hC_n}.$$

How are these layers attached to each other to form $\tilde{\mathrm{W}}\mathrm{h}(Y)$?

Recall that Theorem 1.4.13 gave us conditions for the Taylor tower of a functor F: $\mathcal{T}op_* \to \mathcal{S}p$ to be determined by a KE_n-module structure on $\partial_* F$. The following result of [11] applies this to the functor A.

Theorem 1.5.3. *The divided power module structure on $\partial_* A$, and hence the Taylor tower of A, is determined by a certain KE_3-module structure on $\partial_* A$.*

Note that the KE_3-module structure on $\partial_* A$ pulls back to a KE_n structure for any $n > 3$, but we do not know if in fact it comes from a KE_2-, or KE_1-module.

Problem 1.5.4. *Extract from the proof of Theorem 1.5.3 an explicit description of the KE_3-module structure on $\partial_* A$, and a corresponding construction of the Taylor tower of A.*

Algebraic K-theory of rings

For an A_∞-ring spectrum R, the K-theory spectrum $K(R)$ is defined as the algebraic K-theory of the subcategory of compact objects in the ∞-category of R-modules [41, VI.3.2]. Our goal in this section is to describe what is known about the Taylor towers of the functor K defined in this way.

It is easiest to describe results in the pointed case, i.e. for augmented algebras. Let us fix an A_∞-ring spectrum R, and let $\mathcal{A}lg_R^{\mathrm{aug}}$ denote the ∞-category of augmented R-algebras. We are interested in the Taylor tower (at the terminal object R of $\mathcal{A}lg_R^{\mathrm{aug}}$) of the functor

$$\tilde{K}_R : \mathcal{A}lg_R^{\mathrm{aug}} \to \mathcal{S}p$$

given by $\tilde{K}_R(A) := \mathrm{hofib}(K(A) \to K(R))$.

The first calculation of part of the Taylor tower of $\tilde{K}_R$ was made by Dundas and McCarthy in [40].

Theorem 1.5.5 (Dundas-McCarthy). *Let R be the Eilenberg-Mac Lane spectrum of a discrete ring. Then*

$$\partial_1 \tilde{K}_R \simeq \Sigma\,\mathrm{THH}(R)$$

the topological Hochschild homology of R of Bökstedt.

To recover the first layer $D_1 \tilde{K}_R$ of the Taylor tower from Theorem 1.5.5, we need to know the stabilization of the category of augmented R-algebras. By work of Basterra and Mandell [23], this is equivalent to the category of R-bimodules. The suspension spectrum construction takes an augmented R-algebra A to its *topological André-Quillen homology* $\mathrm{taq}_R(A)$, a derived version of I/I^2 where I is the augmentation ideal of A. We then have

$$D_1 \tilde{K}_R(A) \simeq \Sigma\,\mathrm{THH}(R; \mathrm{taq}_R(A))$$

where $\mathrm{THH}(R; M) := R \wedge_{R \wedge R^{op}} M$ is the topological Hochschild homology with coefficients.

Lindenstrauss and McCarthy [62] have extended Theorem 1.5.5 to higher layers in the following way. The generalization to all connective ring spectra is due to Pancia [70].

Theorem 1.5.6 (Lindenstrauss-McCarthy, Pancia). *Let R be a connective ring spectrum. Then for an R-bimodule A*

$$D_n \tilde{K}_R(A) \simeq \Sigma U^n(R; \mathrm{taq}_R(A))_{hC_n}$$

where $U^n(R; M)$ is a generalization of $\mathrm{THH}(R; M)$ *given by the cyclic tensoring of n copies of a bimodule M over R, with action of the cyclic group C_n given by permutation.*

The proof of Theorem 1.5.6 is a consequence of Lindenstrauss and McCarthy's calculation of the complete Taylor tower of the functor

$$\tilde{K}_R(T_R(-)) : \mathrm{Bimod}_R \to \mathcal{S}p$$

where $T_R : \mathrm{Bimod}_R \to \mathcal{A}lg_R^{\mathrm{aug}}$ is the free tensor R-algebra functor given by

$$T_R(M) := \bigoplus_{n \geq 0} M^{\otimes_R}.$$

It is shown in [62] that

$$P_n(\tilde{K}_R T_R)(M) \simeq \mathrm{holim}_{k \leq n} \Sigma U^k(R; M)^{C_k}$$

where the homotopy limit is formed over a diagram of restriction maps

$$U^k(R; M)^{C_k} \to U^l(R; M)^{C_l}$$

for positive integers l dividing k. It is reasonable to expect that more information about the Taylor tower of $\tilde{K}_R$ itself, beyond just the layers, can be extracted from this description, but this has not been done.

One way to approach this question is via the general theory of Section 1.4. According to Theorem 1.4.1, one would expect the derivatives of $\tilde{K}_R$ to be a right module over the derivatives of the identity functor on $\mathcal{A}lg_R^{\mathrm{aug}}$. Theorem 1.4.7 then tells us that the Taylor tower of $\tilde{K}_R$ is determined by the action of a certain comonad on that right module.

Interestingly, and unlike the case for topological spaces, the Taylor tower for the identity functor on $\mathcal{A}lg_R^{\mathrm{aug}}$ is rather easy to describe:

$$P_n I_{\mathcal{A}lg_R^{\mathrm{aug}}}(A) \simeq I/I^{n+1} \tag{1.5.2}$$

where the right-hand side is a derived version of the quotient of the augmentation ideal I by its $(n+1)$-th power, generalizing the construction of $\mathrm{taq}_R(A) \simeq I/I^2$. This formula seems to have been written down first by Kuhn [57], and a more formal version is developed by Pereira [71]. With this calculation it should now be possible to make more progress with the calculation of the Taylor tower of $\tilde{K}_R$.

Remark 1.5.7. One final remark on the connection between algebraic K-theory and Goodwillie calculus is worth making here. Barwick [22] identifies the process of forming algebraic K-theory itself as the first layer of a Taylor tower. He defines an ∞-category $\mathcal{W}ald_\infty$ of

Waldhausen ∞-categories that are ∞-category-theoretic analogues of Waldhausen's categories with cofibrations. He then identifies the algebraic K-theory functor as

$$K \simeq P_1(\iota)$$

where $\iota : \mathcal{W}ald_\infty \to \mathcal{T}op$ is the functor that sends a Waldhausen ∞-category to its underlying ∞-groupoid of objects. The 1-excisive property for K is a version of Waldhausen's Additivity Theorem, and this result identifies K-theory as the universal example of an additive theory on $\mathcal{W}ald_\infty$ together with a natural transformation from ι.

1.6 Taylor towers of infinity-categories

Recent work of Heuts has taken Goodwillie calculus in a new direction. In [49] he constructs, for each pointed compactly-generated ∞-category $\mathcal{C}$, a *Taylor tower of ∞-categories* for $\mathcal{C}$. This is a sequence of adjunctions

$$\mathcal{C} \rightleftarrows \cdots \rightleftarrows \mathcal{P}_n\mathcal{C} \rightleftarrows \mathcal{P}_{n-1}\mathcal{C} \rightleftarrows \cdots \rightleftarrows \mathcal{P}_1\mathcal{C} \tag{1.6.1}$$

where $\mathcal{P}_n\mathcal{C}$ is a universal approximation to $\mathcal{C}$ by an *n-excisive ∞-category.*

This approximation has the property that the identity functor on $\mathcal{P}_n\mathcal{C}$ is n-excisive, and both the unit and counit of the adjunction $\mathcal{C} \rightleftarrows \mathcal{P}_n\mathcal{C}$ are P_n-equivalences. In the case $n = 1$, we have $\mathcal{P}_1\mathcal{C} \simeq \mathcal{S}p(\mathcal{C})$, the stabilization of $\mathcal{C}$. Thus, the sequence above can be viewed as interpolating between $\mathcal{C}$ and its stabilization via ∞-categories that better and better approximate the potentially unstable ∞-category $\mathcal{C}$.

Heuts's main results concern the classification of n-excisive ∞-categories such as $\mathcal{P}_n\mathcal{C}$. Just as in the classification of Taylor towers of functors, this process is broken down into two parts: one *operadic* and the other related to the *Tate* construction.

The first part associates to the ∞-category $\mathcal{C}$ the *cooperad* constructed by Lurie that represents the derivatives of the functor $\Sigma^\infty_\mathcal{C}\Omega^\infty_\mathcal{C}$, as described in Remark 1.4.5. (Recall that this cooperad is actually an *∞-operad* in Lurie's terminology.) The notation for this cooperad in [49] is $\mathcal{S}p(\mathcal{C})^\otimes$, but we will denote it by $\partial_*(\Sigma^\infty_\mathcal{C}\Omega^\infty_\mathcal{C})$.

Proposition 1.6.1. *Let $\mathcal{C}$ be a pointed compactly generated ∞-category. Then the suspension spectrum construction for $\mathcal{C}$ can be made into a functor*

$$\Sigma^\infty_\mathcal{C} : \mathcal{C} \to \mathrm{Coalg}(\partial_*(\Sigma^\infty_\mathcal{C}\Omega^\infty_\mathcal{C}))$$

where the right-hand side is the ∞-category of non-unital coalgebras over the cooperad $\partial_*(\Sigma^\infty_\mathcal{C}\Omega^\infty_\mathcal{C})$.

Example 1.6.2. For $\mathcal{C} = \mathcal{T}op_*$, the cooperad $\partial_*(\Sigma^\infty_\mathcal{C}\Omega^\infty_\mathcal{C})$ is the commutative cooperad in $\mathcal{S}p$. In this case the functor in Proposition 1.6.1 associates to a based space X the commutative coalgebra structure on $\Sigma^\infty X$ given by the diagonal on X.

The category of coalgebras described in Proposition 1.6.1 can be thought of as the best approximation to the ∞-category $\mathcal{C}$ based on the information in the cooperad $\partial_*(\Sigma^\infty_\mathcal{C}\Omega^\infty_\mathcal{C})$.

The main focus of [49] is an understanding of the additional information needed to recover $\mathcal{C}$ itself (or at least its Taylor tower) from this category of coalgebras.

Heuts identifies $\mathcal{P}_n\mathcal{C}$ with an ∞-category of *n-truncated Tate coalgebras* over the cooperad $\partial_*(\Sigma^\infty_\mathcal{C}\Omega^\infty_\mathcal{C})$. We do not have space here to provide a precise definition of Tate coalgebra, but we can give the general idea. Full details are in [49].

An n-truncated Tate coalgebra over a cooperad $\mathcal{Q}$ consists first of an ordinary (truncated) $\mathcal{Q}$-coalgebra structure on an object E in $\mathcal{P}_1\mathcal{C} \simeq \mathcal{S}p(\mathcal{C})$. That is, for $k \leq n$, we have structure maps of the form

$$c_k : E \to [\mathcal{Q}(k) \wedge E^{\wedge k}]^{h\Sigma_k}.$$

Each map c_k is then required to be compatible (via the canonical map from fixed points to the Tate construction) with a certain natural transformation

$$t_k : E \to [\mathcal{Q}(k) \wedge E^{\wedge k}]^{t\Sigma_k},$$

that Heuts calls the *Tate diagonal* for $\mathcal{C}$.

Crucially, the natural transformation t_k is defined only when E is already a $(k-1)$-truncated Tate $\mathcal{Q}$-coalgebra, and t_k must be compatible, via the cooperad structure on $\mathcal{Q}$, with the maps $c_1, \dots, c_{k-1}$. Also note that the Tate diagonal t_k depends on the ∞-category $\mathcal{C}$ and not just the cooperad $\mathcal{Q}$. It is the choice of these maps that carries the extra information needed to reconstruct the Taylor tower of $\mathcal{C}$.

We can now state Heuts's main result from [49].

Theorem 1.6.3. *Let $\mathcal{C}$ be a pointed compactly generated ∞-category with $\mathcal{Q} := \partial_*(\Sigma^\infty_\mathcal{C}\Omega^\infty_\mathcal{C})$, the corresponding cooperad in the stable ∞-category $\mathcal{S}p(\mathcal{C})$. Then there is a unique sequence of Tate diagonals, i.e. natural transformations*

$$t_k : E \to [\mathcal{Q}(k) \wedge E^{\wedge k}]^{t\Sigma_k}$$

where t_k is defined for $E \in \mathcal{P}_{k-1}\mathcal{C}$, such that $\mathcal{P}_k\mathcal{C}$ is equivalent to the ind-completion of the ∞-category of k-truncated Tate $\mathcal{Q}$-coalgebras whose underlying object of $\mathcal{S}p(\mathcal{C})$ is compact.

Note that this is intrinsically an inductive result: the construction of $\mathcal{P}_{k-1}\mathcal{C}$ needs to be made in order to understand the Tate diagonal t_k that allows for the definition of $\mathcal{P}_k\mathcal{C}$. Here is how this process plays out for based spaces.

Example 1.6.4. We have $\mathcal{P}_1\mathcal{T}op_* \simeq \mathcal{S}p$ and $\partial_*(\Sigma^\infty\Omega^\infty) \simeq \mathcal{C}om$, the commutative cooperad of spectra with $\mathcal{C}om(n) \simeq S^0$ for all n. The first Tate diagonal has no compatibility requirements and is simply a natural transformation

$$t_2 : Y \to (Y \wedge Y)^{t\Sigma_2}$$

defined for all $Y \in \mathcal{S}p$.

Since both the source and target are 1-excisive functors, such a natural transformation is determined by its value on the sphere spectrum, where it takes the form of a map

$$t_2 : S^0 \to (S^0)^{t\Sigma_2} \simeq (S^0)^\wedge_2$$

whose target is the 2-complete sphere by the Segal Conjecture [33]. The specific Tate diagonal that corresponds to the 2-excisive ∞-category $\mathcal{P}_2\mathcal{T}op_*$ is then the ordinary 2-completion map $S^0 \to (S^0)^{\wedge}_2$.

According to Theorem 1.6.3, a (compact) object of $\mathcal{P}_2\mathcal{T}op_*$ consists of a finite spectrum Y with map

$$c_2 : Y \to (Y \wedge Y)^{h\Sigma_2}$$

and a homotopy between t_2 and the composite

$$Y \to (Y \wedge Y)^{h\Sigma_2} \to (Y \wedge Y)^{t\Sigma_2}.$$

Now consider a Tate diagonal for $\mathcal{P}_2\mathcal{T}op_*$. This should be a natural transformation

$$t_3 : Y \to (Y^{\wedge 3})^{t\Sigma_3}$$

that is compatible with the map

$$Y \to \left(\prod_3 Y \wedge (Y \wedge Y)\right)^{t\Sigma_3}$$

that is built from iterating c_2, where the three copies of $Y \wedge (Y \wedge Y)$ are indexed by the three binary trees with leaves labelled $\{1, 2, 3\}$, and composing with the map from fixed points to the Tate spectrum. As in the $n = 2$ case, there is a specific such natural transformation t_3 that corresponds to $\mathcal{P}_3\mathcal{T}op_*$ though it doesn't seem to be as easy to describe explicitly for general $Y \in \mathcal{P}_2\mathcal{T}op_*$. (For objects Y that are of the form $\Sigma^\infty X$ for a space X, t_3 is induced by the diagonal on X.)

An object of $\mathcal{P}_3\mathcal{T}op_*$ then consists of an object of $\mathcal{P}_2\mathcal{T}op_*$ together with a map

$$c_3 : Y \to (Y^{\wedge 3})^{h\Sigma_3}$$

that lifts t_3 and is compatible with c_2.

As in the case of Taylor towers of functors, a substantial simplification occurs when the Tate data vanishes, in which case the Taylor tower of the ∞-category $\mathcal{C}$ is completely determined by the cooperad $\partial_*(\Sigma^\infty_{\mathcal{C}}\Omega^\infty_{\mathcal{C}})$.

Corollary 1.6.5. [49, 6.14] *Let $\mathcal{C}$ be an ∞-category such that for any object $X \in \mathcal{S}p(\mathcal{C})$ with Σ_k-action, the Tate construction $X^{t\Sigma_k}$ is trivial, for all $k \geq 2$. Then $\mathcal{P}_n\mathcal{C}$ is equivalent to the ind-completion of the category of n-truncated $\partial_*(\Sigma^\infty_{\mathcal{C}}\Omega^\infty_{\mathcal{C}})$-coalgebras whose underlying object is compact in $\mathcal{S}p(\mathcal{C})$.*

Remark 1.6.6. The Tate diagonals for an ∞-category $\mathcal{C}$ bear a close relationship to the Taylor tower of the functor $\Sigma^\infty_{\mathcal{C}}\Omega^\infty_{\mathcal{C}} : \mathcal{S}p(\mathcal{C}) \to \mathcal{S}p(\mathcal{C})$. In particular t_n can be written as a composite

$$E \to P_{n-1}(\Sigma^\infty_{\mathcal{C}}\Omega^\infty_{\mathcal{C}})(E) \xrightarrow{t'_n} (\partial_n(\Sigma^\infty_{\mathcal{C}}\Omega^\infty_{\mathcal{C}}) \wedge E^{\wedge n})^{t\Sigma_n}$$

where the second map here is the bottom horizontal map in McCarthy's square (1.2.9), and the first map contains the structure of an $(n-1)$-truncated Tate coalgebra on E. The map on derivatives induced by t'_n can furthermore be identified with the coalgebra structure from Theorem 1.4.7 that classifies the Taylor tower of the functor $\Sigma^\infty_{\mathcal{C}}\Omega^\infty_{\mathcal{C}}$.

1.7 The manifold and orthogonal calculi

While this article has been focused on Goodwillie's calculus of homotopy functors, there are two other theories of "calculus" developed by Michael Weiss, that are inspired by, and related to, Goodwillie calculus to varying degrees. They are called *manifold calculus* (initially known as embedding calculus) and *orthogonal calculus*. In this section we give a review of these theories and some applications they have had.

Manifold calculus

Manifold calculus was initially developed in [81, 48], see also [29]. It concerns *contravariant* functors on manifolds. Of all the brands of calculus, this one is closest to classical sheaf theory, and therefore may look the most familiar.

Suppose M is an m-manifold. Let F be a presheaf of spaces on M. In other words, F is a contravariant functor $F\colon \mathcal{O}(M) \to \mathcal{T}op$, where $\mathcal{O}(M)$ is the poset of open subsets of M.[6] We assume that F takes isotopy equivalences to homotopy equivalences and filtered colimits to inverse limits. The motivating example is the presheaf $U \mapsto \mathrm{Emb}(U, N)$, where N is a fixed manifold, and $\mathrm{Emb}(-,-)$ denotes the space of smooth embeddings.

The notion of excisive functor in manifold calculus is analogous to that in Goodwillie's homotopy calculus. We say that a presheaf is *n-excisive* if it takes strongly cocartesian $(n+1)$-cubes in $\mathcal{O}(M)^{op}$ to cartesian cubes in $\mathcal{T}op$. For example, F is 1-excisive if for any open sets $U, V \subset M$, the following diagram is a (homotopy) pull-back square

$$\begin{array}{ccc} F(U\cup V) & \longrightarrow & F(V) \\ \downarrow & & \downarrow \\ F(U) & \longrightarrow & F(U\cap V). \end{array}$$

Thus we see that 1-excisive presheaves are, essentially, sheaves, except that the sheaf condition has to be interpreted in a homotopy invariant way. Sometimes 1-excisive functors are called homotopy sheaves.

Similarly, n-excisive functors can be interpreted as homotopy sheaves with respect to a different Grothendieck topology, where one says that $V_1, \ldots, V_k$ cover U if $U^n = V_1^n \cup \cdots \cup V_k^n$.

Just as in Goodwillie calculus, the inclusion of the category of n-excisive presheaves into the category of all presheaves has a left adjoint. The adjoint is constructed by a process of sheafification, rather than stabilization. There is another, slightly different procedure for constructing the approximation that is often useful. We will describe it now.

Definition 1.7.1. For an m-manifold U, let $\mathcal{O}_n(U) \subset \mathcal{O}(U)$ be the poset of open subsets of U that are diffeomorphic to a disjoint union of at most n copies of $\mathbb{R}^m$. Given a presheaf F on M, define a new presheaf T_nF by the formula

$$T_nF(U) = \mathrm{holim}_{V\in\mathcal{O}_n(U)}\, F(V).$$

[6] For concreteness we focus on presheaves in $\mathcal{T}op$, but it seems likely that the theory can be extended without difficulty to presheaves in a general $(\infty, 1)$-category.

Theorem 1.7.2 (Weiss [81]). *Assume that F takes isotopy equivalences to homotopy equivalences and filtered colimits to inverse homotopy limits. Then T_nF is n-excisive, and the natural transformation $F \to T_nF$ is initial among all natural transformations from F to an n-excisive functor.*

Thus T_nF is the universal n-excisive approximation of F. As in homotopy calculus, there are natural transformations $T_nF \to T_{n-1}F$. So the approximations fit into a "tower" of functors under F

$$F \to \cdots \to T_nF \to T_{n-1}F \to \cdots$$

sometimes called the *embedding tower* when F is $\mathrm{Emb}(-, N)$ for a manifold N.

Continuing the analogy with homotopy calculus, there is a classification theorem for homogeneous functors that can be used to describe the layers in the tower of approximations. Unlike in homotopy calculus, a homogeneous functor is not classified by a spectrum, but by a fibration over a space of configurations of points in M. We will give below a description of the homotopy fibre of the map $T_nF \to T_{n-1}F$ (for space-valued F).

Let $\binom{M}{n}$ be the space of unordered n-tuples of pairwise distinct points of M. Given a point $\mathbf{x} = [x_1, \ldots, x_n]$ of $\binom{M}{n}$, let $U_\mathbf{x}$ be a tubular neighborhood of $\{x_1, \ldots, x_n\}$ in M. In particular, $U_\mathbf{x} = \coprod_{i=1}^n U_i$ is diffeomorphic to a disjoint union of n copies of $\mathbb{R}^n$. We have an n-dimensional cubical diagram, indexed by subsets of $\{1, \ldots, n\}$ and opposites of inclusions, that sends a subset $S \subset \{1, \ldots, n\}$ to $F(\coprod_{i \in S} U_i)$. Let $\widehat{F}(U_\mathbf{x})$ be the total homotopy fibre of this diagram. One can naturally construct a fibred space over $\binom{M}{n}$ where the fibre at $\mathbf{x}$ is equivalent to $\widehat{F}(U_\mathbf{x})$.

Theorem 1.7.3 (Weiss [81]). *A choice of point in $T_{n-1}F(M)$ gives a germ of a section of this fibration near the fat diagonal. The homotopy fibre of $T_nF(M) \to T_{n-1}F(M)$ is equivalent to the space of sections of the fibration, that agree near the fat diagonal with the local section prescribed by the choice of point in $T_{n-1}F(M)$.*

Remark 1.7.4. Suppose $f \colon \mathbb{R} \to \mathbb{R}$ is a function. For each $n \geq 0$, let t_nf be the unique polynomial of degree n for which $f(i) = t_nf(i)$ for $i = 0, 1, \ldots, n$. There is a well known formula for t_nf. Namely $t_nf(m) = \sum_{i=0}^n \hat{f}(i) \cdot \binom{m}{i}$. Here $\hat{f}(i) = \sum_{j=0}^i (-1)^j f(i-j)$ is the i-th cross-effect of f. In particular, the n-th term of $t_nf(m)$ is $\hat{f}(n) \cdot \binom{m}{n}$. Note the formal similarity of this formula with the formula for the n-th layer in the manifold calculus tower in Theorem 1.7.3. This suggests that the construction T_nF of manifold calculus is analogous to the interpolating polynomial t_nf rather than to the Taylor polynomial of F. In fact, it is true that the map $F \to T_nF$ is characterized by the property that it is an equivalence on objects of $\mathcal{O}_n(M)$. This confirms the intuition that the approximations in manifold calculus are analogous to interpolation polynomials.

Just as in Goodwillie calculus, the question of convergence is important, and there is a general theory of analytic functors, for which the tower of approximations converges strongly. The main result on the subject is the following deep theorem of Goodwillie and Klein. Let Emb denote the space of smooth embeddings.

Theorem 1.7.5. [47] *Let M^m, N^d be smooth manifolds. Consider the presheaf on M given by the formula $U \mapsto \mathrm{Emb}(U, N)$. The map $\mathrm{Emb}(M, N) \to T_n \mathrm{Emb}(M, N)$ is $(d-m-2)n-m+1$-connected.*

The theorem says in particular that if $d - m \geq 3$, then the connectivity of the map goes to infinity linearly in n, and the tower converges strongly. Even in situations where there is no strong convergence, the tower may be useful for constructing invariants of embeddings and obstructions to existence of embeddings. For example, in the case of $m = 1$, $d = 3$, there is a close connection between the manifold tower and finite type knot invariants [78].

It is often possible to express the tower T_kF in terms of mapping spaces between modules over the little discs operad. This idea probably first appeared in [76] in the context of spaces of knots, and then was developed in [12, 15, 77], and finally and definitively in [29]. Thanks to this connection, facts about the homotopy of the little disc operads (most notably Kontsevich's formality theorem) have consequences regarding the homotopy type of spaces of embeddings [12, 5, 16].

Recently manifold calculus was used by Michael Weiss in the course of proving some striking results about Pontryagin classes of topological bundles [82].

Orthogonal calculus

Let $\mathcal{J}$ be the category of finite-dimensional Euclidean spaces and linear isometric inclusions. Orthogonal calculus [80] concerns continuous functors from $\mathcal{J}$ to the category of topological spaces. It seems likely that topological spaces can be replaced in a routine way with a more general topologically enriched category. The paper [20] develops orthogonal calculus using the perspective of Quillen model categories. Formally, orthogonal calculus is more similar to homotopy calculus than manifold calculus is, and the paper [19] explores this similarity in some depth. Orthogonal calculus has probably received the least attention of the three main brands, and we believe it is ripe for exploration.

Let V denote a generic Euclidean space. Here are some examples of functors to which one might profitably apply orthogonal calculus: $V \mapsto BO(V)$, $V \mapsto G(V)$ (the space of homotopy self equivalences of the unit sphere of V), $V \mapsto BTop(V)$, $V \mapsto \mathrm{Emb}(M, N \times V)$, etc.

Such functors tend to be easier to understand when evaluated at high dimensional vector spaces. The rough idea of orthogonal calculus is to analyze the behavior of a functor on high-dimensional vector spaces and then to extrapolate to low-dimensions. As a result one obtains a kind of Taylor expansion at infinity. Thus to a functor $F\colon \mathcal{J} \to \mathcal{T}op$ one associates a tower of functors under F

$$F \to \cdots \to P_nF \to P_{n-1}F \to \cdots \to P_0F$$

where P_nF plays the role of the Taylor tower of F at infinity. In particular, the constant term P_0F is equivalent to $\mathrm{colim}_{n\to\infty} F(\mathbb{R}^n)$. The higher Taylor approximations are usually difficult to describe explicitly, but the layers in the tower can be described in terms of certain spectra that play the role of derivatives. This is similar to homotopy calculus, but unlike in homotopy calculus, the sequence of derivatives of a functor in orthogonal calculus is an *orthogonal* sequence of spectra, rather than a symmetric one. This means that the n-th derivative is a spectrum with an action of $O(n)$. The following theorem summarizes some results from [80] about homogeneous functors and layers in orthogonal calculus. Compare with Theorem 1.2.3 and Example 1.2.4.

Theorem 1.7.6. *To a continuous functor $F\colon \mathcal{J} \to \mathcal{T}op$ one can associate an orthogonal sequence of spectra, $\partial_1 F, \ldots, \partial_n F, \ldots$. The homotopy fibre of the map $P_nF(V) \to P_{n-1}F(V)$*

is naturally equivalent to the functor that sends V to $\Omega^\infty\left(\partial_n F \wedge S^{nV}\right)_{hO(n)}$. Here S^{nV} is the one-point compactification of the vector space nV.

Examples 1.7.7. Recall that $\mathcal{J}$ is the category of Euclidean spaces and linear isometric inclusions. Suppose V_0, V are Euclidean spaces. Let $\mathrm{Hom}_{\mathcal{J}}(V_0, V)$ denote the space of linear isometric inclusions from V_0 to V. Now fix V_0 and consider the functor $\Omega^\infty\Sigma^\infty \mathrm{Hom}_{\mathcal{J}}(V_0, -)_+$. The Taylor tower of this functor was described in [7]. This is a rare example of a functor whose Taylor tower can be described rather explicitly. When $\dim(V_0) \leq \dim(V)$, the Taylor tower splits as a product of its layers, and in fact this splitting is equivalent to a classical stable splitting of Stiefel manifolds discovered by Haynes Miller [67, 4].

For another example, consider the functor $V \mapsto BO(V)$. The n-th derivative of this functor can be described in terms of the complex of direct-sum decompositions of $\mathbb{R}^n$. A (proper) direct-sum decomposition is a collection (at least two) of pairwise-orthogonal non-zero subspaces of $\mathbb{R}^n$ whose direct sum is $\mathbb{R}^n$. We can partially order such decompositions by refinement and obtain a topological poset whose realization we denote by L_n. Let $L_n^\diamond$ be the unreduced suspension of that realization. The space $L_n^\diamond$ is analogous to the complex of partitions T_n from definition 1.3.1. The n-th derivative of the functor $BO(V)$ is equivalent to $\mathbb{D}(L_n^\diamond) \wedge S^{ad_n}$, where S^{ad_n} is the one-point compactification of the adjoint representation of $O(n)$. Compare with Theorem 1.3.2 about the Goodwillie derivatives of the identity functor. The space $L_n^\diamond$, and even more so its complex analogue, has many striking properties. It appeared in several recent works in an interesting way. For example, it was used in the study of the stable rank filtration of complex K-theory [13, 17], and in the study of the Balmer spectrum of the equivariant stable homotopy category [21]. [7]

It would be especially interesting to understand the derivatives of the functor $V \mapsto \mathrm{Emb}(M, N \times V)$. We speculate that the n-th derivative of this functor can be expressed in terms of moduli space of connected graphs with points marked in M, for which the quotient space of the graph by the subset of marked points is homotopy equivalent to a wedge of n circles.

1.8 Further directions

The basic concepts of Goodwillie calculus are very general and can be applied to a wide variety of homotopy-theoretic settings. Nonetheless, not many calculations have been done outside of representable functors, the identity functor and algebraic K-theory. We have seen that the Taylor tower of the identity functor plays a key role in the theory, so a calculation of that Taylor tower, or at least its layers, would be valuable in other contexts. Obvious candidates for further exploration include:

- unpointed and parametrized homotopy theory: that is, a more detailed understanding of Taylor towers at base objects other than the one-point space;
- unstable equivariant homotopy theory: presumably the derivatives of the identity here are some equivariant version of the spectral Lie operad;

[7] See Paul Balmer's chapter in this Handbook for more on the spectrum of a tensor-triangulated category.

- unstable motivic homotopy theory: as far as we know, no work has been done in this direction despite the wealth of structure the Taylor tower of the identity should possess in this case.

In each of these cases calculations of the derivatives of the identity, their operad structure and Heuts's Tate diagonals, could reveal something deep about how unstable information is built from stable.

The equivariant setting is potentially very interesting. Dotto [39] has generalized Goodwillie calculus to a G-equivariant context, for a finite group G, in which the Taylor tower of a functor is replaced by a diagram of approximations indexed by the finite G-sets. This is based on an idea of Blumberg [28] for a G-equivariant version of excision.

Dotto's approach seems to be better able to make use of the power of modern equivariant stable homotopy theory, something rather neglected by Goodwillie's original version. In particular, the derivatives of a functor in this setting are genuine G-spectra. Dotto calculates the derivatives of the identity functor on pointed G-spaces as equivariant Spanier-Whitehead duals of the partition poset complexes of [14]. It seems reasonable to expect that much of the theory described in this article could be extended the equivariant setting, but little has been done yet.

Manifold and orthogonal calculus have been applied to problems in geometric topology, but there surely is much that remains to be done. In particular orthogonal calculus should have a lot of potential that has barely began to be tapped. We mentioned above that it would be interesting to apply orthogonal calculus to the study of embedding spaces by considering the functor $V \mapsto \mathrm{Emb}(M, N \times V)$. An even more interesting and challenging example is given by the functor $V \mapsto \mathrm{BTop}(V)$ and the closely related functor $V \mapsto B^{V+1}\mathrm{Diff}_c(V)$. Here B^{V+1} stands for $V+1$-fold delooping, and $\mathrm{Diff}_c(V)$ stands for the space of diffeomorphisms of V that equal the identity outside a compact set. We speculate that the n-th derivative of these functors is related to the moduli space of graphs homotopy equivalent to a wedge of n circles (without marked points).

The close analogies that connect Goodwillie calculus with the manifold and orthogonal versions suggests the existence of a more encompassing framework that could also provide new instances of "calculus" for other kinds of functors. One such framework is the theory of tangent categories of Rosický [75] and Cockett and Cruttwell [38], developed to axiomatize the structure of the category of manifolds and smooth maps. Lurie's notion of tangent bundle for an ∞-category [64, 7.3.1] fits Goodwillie calculus into this story, and opens up a way to make precise ideas of Goodwillie on applying differential-geometric ideas such as connections and curvature directly to homotopy theory. There is a helpful intuition that stable ∞-categories such as spectra are "flat", whereas unstable worlds are "curved". However, manifold and orthogonal calculus do not appear to fit into this same picture, so perhaps an even more general theory is still waiting to be discovered.

Acknowledgements

The first author was partially supported by the Swedish Research Council (grant number 2016-05440), and the second author was partially supported by the National Science Foundation (grant DMS-1709032) and the Isaac Newton Institute (EPSRC grant number

EP/R014604/1 and the HHH programme), during the writing of this article. Useful feedback and corrections to an earlier draft were provided by Gijs Heuts and Haynes Miller.

Bibliography

[1] Stephen Ahearn and Nicholas Kuhn, *Product and other fine structure in polynomial resolutions of mapping spaces*, Algebr. Geom. Topol. **2** (2002), 591–647.

[2] Donald W. Anderson and Donald M. Davis, *A vanishing theorem in homological algebra*, Comment. Math. Helv. **48** (1973), 318–327.

[3] G. Z. Arone and W. G. Dwyer, *Partition complexes, Tits buildings and symmetric products*, Proc. London Math. Soc. (3) **82** (2001), no. 1, 229–256.

[4] Greg Arone, *The Mitchell-Richter filtration of loops on Stiefel manifolds stably splits*, Proc. Amer. Math. Soc. **129** (2001), no. 4, 1207–1211.

[5] Greg Arone, Pascal Lambrechts, Victor Turchin, and Ismar Volić, *Coformality and rational homotopy groups of spaces of long knots*, Math. Res. Lett. **15** (2008), no. 1, 1–14.

[6] Gregory Arone, *A generalization of Snaith-type filtration*, Trans. Amer. Math. Soc. **351** (1999), no. 3, 1123–1150.

[7] Gregory Arone, *The Weiss derivatives of $BO(-)$ and $BU(-)$*, Topology **41** (2002), no. 3, 451–481.

[8] Gregory Arone and Lukas Brantner, *The action of Young subgroups on the partition complex*, arXiv:1801.01491, 2018.

[9] Gregory Arone and Michael Ching, *Operads and chain rules for the calculus of functors*, Astérisque (2011), no. 338.

[10] Gregory Arone and Michael Ching, *A classification of Taylor towers of functors of spaces and spectra*, Adv. Math. **272** (2015), 471–552.

[11] Gregory Arone and Michael Ching, *Manifolds, K-theory and the calculus of functors*, Manifolds and K-theory, Contemp. Math. **82**, Amer. Math. Soc., 2017, pp. 1–37.

[12] Gregory Arone, Pascal Lambrechts, and Ismar Volić, *Calculus of functors, operad formality, and rational homology of embedding spaces*, Acta Math. **199** (2007), no. 2, 153–198.

[13] Gregory Arone and Kathryn Lesh, *Filtered spectra arising from permutative categories*, J. Reine Angew. Math. **604** (2007), 73–136.

[14] Gregory Arone and Mark Mahowald, *The Goodwillie tower of the identity functor and the unstable periodic homotopy of spheres*, Invent. Math. **135** (1999), no. 3, 743–788.

[15] Gregory Arone and Victor Turchin, *On the rational homology of high-dimensional analogues of spaces of long knots*, Geom. Topol. **18** (2014), no. 3, 1261–1322.

[16] Gregory Arone and Victor Turchin, *Graph-complexes computing the rational homotopy of high dimensional analogues of spaces of long knots*, Ann. Inst. Fourier (Grenoble) **65** (2015), no. 1, 1–62.

[17] Gregory Z. Arone and Kathryn Lesh, *Augmented Γ-spaces, the stable rank filtration, and a bu analogue of the Whitehead conjecture*, Fund. Math. **207** (2010), no. 1, 29–70.

[18] D. Ayala and J. Francis, *Zero-pointed manifolds*, arXiv:1409.2857, 2014.

[19] David Barnes and Rosona Eldred, *Comparing the orthogonal and homotopy functor calculi*, J. Pure Appl. Algebra **220** (2016), no. 11, 3650–3675.

[20] David Barnes and Peter Oman, *Model categories for orthogonal calculus*, Algebr. Geom. Topol. **13** (2013), no. 2, 959–999.

[21] Tobias Barthel, Markus Hausmann, Niko Naumann, Thomas Nikolaus, Justin Noel, and Nathaniel Stapleton, *The Balmer spectrum of the equivariant homotopy category of a finite abelian group*, Invent. Math. **216** (2019), no. 1, 215–240.

[22] Clark Barwick, *On the algebraic K-theory of higher categories*, J. Topol. **9** (2016), no. 1, 245–347.

[23] Maria Basterra and Michael A. Mandell, *Homology and cohomology of E_∞ ring spectra*, Math. Z. **249** (2005), no. 4, 903–944.

[24] Mark Behrens, *The Goodwillie tower for S^1 and Kuhn's theorem*, Algebr. Geom. Topol. **11** (2011), no. 4, 2453–2475.

[25] Mark Behrens, *The Goodwillie tower and the EHP sequence*, Mem. Amer. Math. Soc. **218** (2012), no. 1026.

[26] Mark Behrens and Charles Rezk, *Spectral algebra models of unstable v_n-periodic homotopy theory*, arXiv:1703.02186, 2017.

[27] Georg Biedermann, Boris Chorny, and Oliver Röndigs, *Calculus of functors and model categories*, Adv. Math. **214** (2007), no. 1, 92–115.

[28] Andrew J. Blumberg, *Continuous functors as a model for the equivariant stable homotopy category*, Algebr. Geom. Topol. **6** (2006), 2257–2295.

[29] Pedro Boavida de Brito and Michael Weiss, *Manifold calculus and homotopy sheaves*, Homology Homotopy Appl. **15** (2013), no. 2, 361–383.

[30] M. Bökstedt, G. Carlsson, R. Cohen, T. Goodwillie, W. C. Hsiang, and I. Madsen, *On the algebraic K-theory of simply connected spaces*, Duke Math. J. **84** (1996), no. 3, 541–563.

[31] M. Bökstedt, W. C. Hsiang, and I. Madsen, *The cyclotomic trace and algebraic K-theory of spaces*, Invent. Math. **111** (1993), no. 3, 465–539.

[32] A. K. Bousfield and D. M. Kan, *Homotopy limits, completions and localizations*, Lecture Notes in Math. **304**.

[33] Gunnar Carlsson, *Equivariant stable homotopy and Segal's Burnside ring conjecture*, Ann. of Math. (2) **120** (1984), no. 2, 189–224.

[34] Michael Ching, *Bar constructions for topological operads and the Goodwillie derivatives of the identity*, Geom. Topol. **9** (2005), 833–933.

[35] Michael Ching, *A chain rule for Goodwillie derivatives of functors from spectra to spectra*, Trans. Amer. Math. Soc. **362** (2010), no. 1, 399–426.

[36] Michael Ching, *Bar-cobar duality for operads in stable homotopy theory*, J. Topol. **5** (2012), no. 1, 39–80.

[37] Michael Ching, *Infinity-operads and Day convolution in Goodwillie calculus*, arXiv:1801.03467 (2018).

[38] J. R. B. Cockett and G. S. H. Cruttwell, *Differential structure, tangent structure, and SDG*, Appl. Categ. Structures **22** (2014), no. 2, 331–417.

[39] Emanuele Dotto, *Higher equivariant excision*, Adv. Math. **309** (2017), 1–96.

[40] Bjørn Ian Dundas and Randy McCarthy, *Stable K-theory and topological Hochschild homology*, Ann. of Math. (2) **140** (1994), no. 3, 685–701.

[41] A. D. Elmendorf, I. Kriz, M. A. Mandell, and J. P. May, *Rings, modules, and algebras in stable homotopy theory*, Math. Surveys Monogr. **47**, Amer. Math. Soc., 1997, With an appendix by M. Cole.

[42] Benoit Fresse, *Koszul duality of E_n-operads*, Selecta Math. (N.S.) **17** (2011), no. 2, 363–434.

[43] Victor Ginzburg and Mikhail Kapranov, *Koszul duality for operads*, Duke Math. J. **76** (1994), no. 1, 203–272.

[44] Thomas G. Goodwillie, *Calculus. I. The first derivative of pseudoisotopy theory*, K-Theory **4** (1990), no. 1, 1–27.

[45] Thomas G. Goodwillie, *Calculus. II. Analytic functors*, K-Theory **5** (1991/92), no. 4, 295–332.

[46] Thomas G. Goodwillie, *Calculus. III. Taylor series*, Geom. Topol. **7** (2003), 645–711 .

[47] Thomas G. Goodwillie and John R. Klein, *Multiple disjunction for spaces of smooth embeddings*, J. Topol. **8** (2015), no. 3, 651–674.

[48] Thomas G. Goodwillie and Michael Weiss, *Embeddings from the point of view of immersion theory. II*, Geom. Topol. **3** (1999), 103–118.

[49] Gijs Heuts, *Goodwillie approximations to higher categories*, arXiv:1510.03304, 2015.

[50] Gijs Heuts, *Lie algebras and v_n-periodic spaces*, arXiv:1803.06325, 2018.

[51] Michael J. Hopkins and Jeffrey H. Smith. *Nilpotence and stable homotopy theory. II.*, Ann. of Math. (2) 148 (1998), no. 1, 1–49.

[52] Brenda Johnson, *The derivatives of homotopy theory*, Trans. Amer. Math. Soc. **347** (1995), no. 4, 1295–1321.

[53] B. Johnson and R. McCarthy, *Deriving calculus with cotriples*, Trans. Amer. Math. Soc. **356** (2004), no. 2, 757–803.

[54] John R. Klein and John Rognes, *A chain rule in the calculus of homotopy functors*, Geom. Topol. **6** (2002), 853–887.

[55] Nicholas J. Kuhn, *A Kahn-Priddy sequence and a conjecture of G. W. Whitehead*, Math. Proc. Cambridge Philos. Soc. **92** (1982), no. 3, 467–483.

[56] Nicholas J. Kuhn, *Tate cohomology and periodic localization of polynomial functors*, Invent. Math. **157** (2004), no. 2, 345–370.

[57] Nicholas J. Kuhn, *Localization of André-Quillen-Goodwillie towers, and the periodic homology of infinite loopspaces*, Adv. Math. **201** (2006), no. 2, 318–378.

[58] Nicholas J. Kuhn, *Goodwillie towers and chromatic homotopy: an overview*, Proceedings of Nishida Fest (Kinosaki 2003), Geom. Topol. Monogr. **10** (2007), pp. 245–279.

[59] Nicholas J. Kuhn, *A guide to telescopic functors*, Homology, Homotopy Appl. **10** (2008), no. 3, 291–319.

[60] Nicholas J. Kuhn, *The Whitehead conjecture, the tower of S^1 conjecture, and Hecke algebras of type A*, J. Topol. **8** (2015), no. 1, 118–146.

[61] Nicholas J. Kuhn and Stewart B. Priddy, *The transfer and Whitehead's conjecture*, Math. Proc. Cambridge Philos. Soc. **98** (1985), no. 3, 459–480.

[62] Ayelet Lindenstrauss and Randy McCarthy, *On the Taylor tower of relative K-theory*, Geom. Topol. **16** (2012), no. 2, 685–750.

[63] Jacob Lurie, *Higher topos theory*, Ann. of Math. Stud. **170**, Princeton University Press, 2009.

[64] Jacob Lurie, *Higher algebra*, available at `http://www.math.harvard.edu/~lurie/`.

[65] J. P. May, *The geometry of iterated loop spaces*, Lecture Notes in Math. **271**, Springer-Verlag, 1972.

[66] Randy McCarthy, *Dual calculus for functors to spectra*, in *Homotopy methods in algebraic topology (Boulder, CO, 1999)*, Contemp. Math. **271**, Amer. Math. Soc., 2001, pp. 183–215.

[67] Haynes Miller, *Stable splittings of Stiefel manifolds*, Topology **24** (1985), no. 4, 411–419.

[68] Haynes Miller and Clarence Wilkerson, *Vanishing lines for modules over the Steenrod algebra*, J. Pure Appl. Algebra **22** (1981), no. 3, 293–307.

[69] Stephen A. Mitchell, *Finite complexes with $A(n)$-free cohomology*, Topology **24** (1985), no. 2, 227–246.

[70] Matthew Pancia, *The Goodwillie tower of free augmented algebras over connective ring spectra*, Ph.D. thesis, University of Texas at Austin, 2014, pp. 1–103.

[71] Luís Pereira, *Goodwillie calculus and algebras over a spectral operad*, Ph.D. thesis, MIT, 2013.

[72] Daniel Quillen, *Rational homotopy theory*, Ann. of Math. (2) **90** (1969), 205–295.

[73] Douglas C. Ravenel, *Nilpotence and periodicity in stable homotopy theory*, Ann. of Math. Stud. **128**, Princeton University Press, 1992, Appendix C by Jeff Smith.

[74] Emily Riehl and Dominic Verity, *Homotopy coherent adjunctions and the formal theory of monads*, Adv. Math. **286** (2016), 802–888.

[75] J. Rosický, *Abstract tangent functors*, Diagrammes **12** (1984), JR1–JR11.

[76] Dev P. Sinha, *Operads and knot spaces*, J. Amer. Math. Soc. **19** (2006), no. 2, 461–486.

[77] Victor Turchin, *Context-free manifold calculus and the Fulton-MacPherson operad*, Algebr. Geom. Topol. **13** (2013), no. 3, 1243–1271.

[78] Ismar Volić, *Finite type knot invariants and the calculus of functors*, Compos. Math. **142** (2006), no. 1, 222–250.

[79] Friedhelm Waldhausen, *Algebraic K-theory of spaces*, in Algebraic and Geometric Topology (New Brunswick, N.J., 1983), Lecture Notes in Math. **1126**, Springer, 1985, pp. 318–419.

[80] Michael Weiss, *Orthogonal calculus*, Trans. Amer. Math. Soc. **347** (1995), no. 10, 3743–3796.

[81] Michael Weiss, *Embeddings from the point of view of immersion theory. I*, Geom. Topol. **3** (1999), 67–101.

[82] Michael S. Weiss, *Dalian notes on rational Pontryagin classes*, arXiv:1507.00153, 2015.

[83] Peter J. Welcher, *Symmetric fibre spectra and $K(n)$-homology acyclicity*, Indiana Univ. Math. J. **30** (1981), no. 6, 801–812.

[84] Clarence Wilkerson, *A primer on the Dickson invariants*, Proceedings of the Northwestern Homotopy Theory Conference (Evanston, Ill., 1982), Contemp. Math. **19**, Amer. Math. Soc., 1983, pp. 421–434.

Department of Mathematics, Stockholm University, Stockholm 106 91, Sweden

E-mail address: `gregory.arone@math.su.se`

Department of Mathematics and Statistics, Amherst College, Amherst, MA 01002, U.S.A.

E-mail address: `mching@amherst.edu`

2

A factorization homology primer

David Ayala and John Francis

2.1 Introduction

This article is an introduction to factorization homology—or factorization algebras—in the topological setting; see also Lurie [43]. For introductions in the closely related algebro-geometric or Riemannian settings, see Beilinson–Drinfeld [9], Francis–Gaitsgory [22], and Costello–Gwilliam [17].

Factorization homology takes:

- a geometric input: an n-manifold M;
- an algebraic input: an n-disk algebra A, or a stack X over n-disk algebras, in a symmetric monoidal ∞-category $\mathcal{V}$;[1]

the resulting factorization homology

$$\int_M A$$

is an object of $\mathcal{V}$. It can be thought of as the integral of the algebra A over the manifold M, in the same sense that ordinary homology $\mathsf{H}_*(M, A)$ is given by integrating an abelian group A over a space M.[2] The functor $\int A$ is covariant with respect to open embeddings in the manifold variable: for each fixed M, the functor on the poset of opens $\int A : \mathsf{Opens}(M) \to \mathcal{V}$, sending $U \subset M$ to $\int_U A$, defines a *factorization algebra* on M, both in the definitions of Beilinson–Drinfeld [9] and Costello–Gwilliam [17].

Factorization homology has three essential features making it technically advantageous:

1. **Local-to-global principle**: $\otimes$-excision, generalizing the Eilenberg–Steenrod axioms;
2. **Filtration**: a generalization of the Goodwillie–Weiss embedding calculus;
3. **Duality**: Poincaré/Koszul duality.

2010 *Mathematics Subject Classification.* primary: 55N40, secondary: 57R56, 57N35.

Key words and phrases. factorization homology, topological quantum field theory, derived algebraic geometry, Koszul duality, manifold calculus, Hochschild homology, stratified spaces, factorization algebras.

[1] $\mathcal{V}$ is usually $\mathsf{Mod}_{\Bbbk}$, $\mathsf{Spectra}$, or Spaces—respectively, chain complexes over a commutative ring $\Bbbk$, spectra, or spaces. See §2.1.2 of [43] for a thorough development of *symmetric monoidal ∞-categories.*

[2] This is the alpha version of the theory; there is a beta version, developed in [6], which takes more general geometric and algebraic inputs.

The algebraic input A can arise from a variety of sources: as an n-fold loop space in classical algebraic topology; or as a formal deformation of an n-Poisson algebra. This latter class of examples arises in higher quantization in mathematical physics, a major source of contemporary interest in factorization methods.

The following is a table of values for factorization homology, as the algebra input varies, which we will explain through this article.

Manifold input	**Algebra input**	**Factorization homology output**
$M = S^1$ the circle	A an associative algebra	$\int_{S^1} A \simeq \mathsf{HH}_\bullet(A)$ Hochschild homology
M an n-manifold	A an abelian group	$\int_M A \simeq \mathsf{H}_*(M, A)$ ordinary homology
M an n-manifold	A a spectrum	$\int_M A \simeq \Sigma^\infty_* M \wedge A$ generalized homology
M an n-manifold	A a commutative algebra (e.g., $\mathsf{Sym}(V)$)	$\int_M A \simeq M \boxtimes A$ categorical tensoring (e.g., $\int_M \mathsf{Sym}(V) \simeq \mathsf{Sym}(\mathsf{C}_*(M) \otimes V)$)
M an n-manifold	$A = \mathbb{F}_n V$ a free n-disk algebra	$\int_M \mathbb{F}_n V \simeq \bigoplus_{k \geq 0} \mathsf{C}_*(\mathsf{Conf}^{\mathsf{fr}}_k(M)) \underset{\Sigma_k \wr \mathsf{O}(n)}{\otimes} V^{\otimes k}$
M an n-manifold	$A = \Omega^n K$ the n-fold loop space of an n-connective space	$\int_M \Omega^n K \simeq \mathsf{Map}_{\mathsf{c}}(M, K)$ space of compactly-supported maps
M an n-manifold	$A = \mathsf{U}_n(\mathfrak{g})$ an n-disk enveloping algebra of a Lie algebra	$\int_M \mathsf{U}_n(\mathfrak{g}) \simeq \mathsf{C}^{\mathsf{Lie}}_*(\mathsf{C}^*_{\mathsf{c}}(M, \mathfrak{g}))$ Lie algebra homology
M an n-manifold	$A = \mathsf{Obs}(\mathbb{R}^n)$ the local observables in a TQFT	$\int_M \mathsf{Obs}(\mathbb{R}^n) \longrightarrow \mathsf{Obs}(M)$ $\otimes$-excisive left-approximation to global observables

The last row is not a genuine equivalence in general: the result of alpha factorization homology is not always the genuine global observables, but only those observables that are determined by the point-local ones. Additionally, the alpha version of factorization homology does not compute the state space of a field theory. This motivates a beta version of factorization homology, developed in [6], where the algebraic input is a more general object, an (∞, n)-category, and which satisfies a more general, and nuanced, form of excision.[3] As this beta version of factorization homology is more involved, it will not be discussed in this article.

[3]See Julia Bergner's chapter in this Handbook for a discussion of (∞, n)-categories.

While the subject of factorization homology is relatively new in name, it has important roots and antecedents. It derives foremost from the algebro-geometric theory of chiral and factorization algebras developed by Beilinson and Drinfeld in conformal field theory in [9]. Secondly, it has an antecedent in the labeled configuration space models of mapping spaces dating to the 1970s; it is closest to the models of Salvatore [51] and Segal [56], but see also [31], [14], [46], [45], and [53]. Aspects appear implicitly in other works, particularly Bott–Segal [15], in realizing the Gelfand–Fuks cohomology of vector fields as the homology of a section space, and in foliation theory [47]. Factorization homology thus arises from the broad nexus of Segal's ideas on conformal field theory [55], on mapping spaces [53] and [56], on foliations [54], and on Lie algebra homology [15].[4]

Outline of contents

We briefly outline the contents of this article. Section 2 concerns n-manifolds and certain topological structures thereon. In §2.2-2.5 we discuss a classification of sheaves on an ∞-category $\mathcal{M}\mathsf{fld}_n$ of n-manifolds and embeddings among them: *sheaves on* $\mathcal{M}\mathsf{fld}_n$ *are n-dimensional tangential structures.* In §2.5-2.7 we undergo a similar examination concerning *Weiss sheaves*, thereby motivating and introducing an ∞-category $\mathcal{D}\mathsf{isk}_n$, which plays a key role in factorization homology. In §2.8 we extend these developments to manifolds with boundary; in §2.9 we record a technical result concerning localizing with respect to isotopies, which plays a key role in essentially all of the technical results concerning factorization homology.

Section 3 concerns homology theories for manifolds, and introduces factorization homology. In §3.2 we relate factorization homology to factorization algebras. The remainder of this section is devoted to establishing a pushforward formula for factorization homology, and characterizing homology theories in terms of factorization homology. Section 4 applies this characterization to prove nonabelian Poincaré duality. Section 5 is devoted to a few formal calculations of factorization homology.

Section 6 concerns filtrations and cofiltrations of factorization homology, whose layers are explicit in terms of configuration spaces. These (co)filtrations offer access to identifying and controlling factorization homology. Section 7 demonstrates this with a statement for how factorization homology intertwines Poincaré duality and Koszul duality. This *Poincaré/Koszul duality* (Theorem 2.7.8) is the deepest result discussed in this article and supplies a physical interpretation to factorization homology as observables of perturbative sigma-models.

Section 8 is a synopsis of an adaptation of factorization homology for singular manifolds with coefficients in *disk*-algebras and module data among such.

Implementation of ∞-categories

In this work, we use Joyal's *quasi-category* model of ∞-category theory [30]. Boardman and Vogt first introduced these simplicial sets in [13], as weak Kan complexes, and their and Joyal's theory has been developed in great depth by Lurie in [42] and [43], our primary references; see the first article of [42] for an introduction. We use this model, rather than

[4]More recent works include [1], [2], [10], [11], [18], [25], [26], [27], [29], [34], [36], [35], among many others.

model categories or simplicial categories, because of the great technical advantages for constructions involving categories of functors, which are ubiquitous in this work.[5]

More specifically, we work inside of the quasi-category associated to this model category of Joyal's. In particular, each map between quasi-categories is understood to be an iso- and inner-fibration; (co)limits among quasi-categories are equivalent to homotopy (co)limits with respect to Joyal's model structure. As we work in this way, we refer the reader to these sources for ∞-categorical versions of numerous familiar results and constructions among ordinary categories. In particular, we will make repeated use of the ∞-categorical adjoint functor theorem (Corollary 5.5.2.9 of [42]); the straightening-unstraightening equivalence between Cartesian fibrations over an ∞-category $\mathcal{C}$ and Cat_∞-valued contravariant functors from $\mathcal{C}$ (Theorem 3.2.0.1 of [42]), and likewise between right fibrations over $\mathcal{C}$ and space-valued presheaves on $\mathcal{C}$ (Theorem 2.2.1.2 of [42]); the ∞-categorical version of the Yoneda functor $\mathcal{C} \to \mathsf{PShv}(\mathcal{C}) \simeq \mathsf{RFib}_{\mathcal{C}}$ as it evaluates on objects as $c \mapsto \mathcal{C}_{/c}$ (see §5.1 of [42]).

We will also make use of topological categories, such as $\mathcal{M}\mathsf{fld}_n$ of n-manifolds and embeddings among them. By a functor $\mathcal{S} \to \mathcal{C}$ from a topological category to an ∞-category $\mathcal{C}$ we will always mean a functor $\mathsf{N}\,\mathsf{Sing}\,\mathcal{S} \to \mathcal{C}$ from the simplicial nerve of the Kan-enriched category obtained by applying the product preserving functor Sing to the morphism topological spaces.

The reader uncomfortable with this language can substitute the words "topological category" for "∞-category" wherever they occur in this paper to obtain the correct sense of the results, but they should then bear in mind the proviso that technical difficulties may then abound in making the statements literally true. The reader only concerned with algebras in chain complexes, rather than spectra, can likewise substitute "pre-triangulated differential graded category" for "stable ∞-category" wherever those words appear, with the same proviso.

2.2 Manifolds with tangential structure

In this section, we first introduce a topological category of n-manifolds and embeddings among them. We then define the *tangent classifier* functor. Using this tangent classifier, we then introduce tangential structures, and define an ∞-category of such structured n-manifolds.

2.2.1 Manifolds and embeddings

Definition 2.2.1. A smooth manifold M is *finitary* if it admits a *good* cover, which is to say it admits a finite open cover $\mathcal{U} := \{U \subset M\}$ with the property that, for each finite subset $S \subset \mathcal{U}$, the intersection $\bigcap_{U \in S} U$ is either empty or diffeomorphic to Euclidean space.

Observation 2.2.2. A smooth manifold M is finitary if and only if it is the interior of a compact smooth manifold with (possibly empty) boundary.

Terminology 2.2.3. In this article, by "manifold" we will mean "finitary smooth manifold", unless otherwise stated.

[5] See Moritz Groth's chapter in this Handbook for a development of the theory of quasi-categories.

Remark 2.2.4. The size restriction of Terminology 2.2.3 is not an essential requirement. Indeed, each non-compact manifold is built as sequential colimit of finitary manifolds. Correspondingly, this smallness condition could be removed and one could instead add to Definition 2.3.28 the requirement that a homology theory preserves sequential colimits. With this modification to that definition, the main results are still valid.

Definition 2.2.5. The symmetric monoidal topological category $\mathcal{M}\mathsf{fld}_n$ has as objects smooth n-manifolds. The space of morphisms in $\mathcal{M}\mathsf{fld}_n$ from M to N is

$$\mathsf{Map}_{\mathcal{M}\mathsf{fld}_n}(M,N) := \mathsf{Emb}(M,N)$$

the set of embeddings of M into N equipped with the compact-open C^∞ topology. The symmetric monoidal structure is disjoint union.

Note that disjoint union is not the coproduct in $\mathcal{M}\mathsf{fld}_n$; in fact, there are almost no nontrivial colimits in $\mathcal{M}\mathsf{fld}_n$.

We will make ongoing use of the following result.

Proposition 2.2.6. *The continuous homomorphism of topological monoids* $\mathsf{O}(n) \to \mathsf{Emb}(\mathbb{R}^n,\mathbb{R}^n)$ *is a homotopy equivalence.*

Proof. This continuous homomorphism factors as a composite of continuous homomorphisms:

$$\mathsf{O}(n) \longrightarrow \mathsf{GL}(n) \longrightarrow \mathsf{Emb}_0(\mathbb{R}^n,\mathbb{R}^n) \longrightarrow \mathsf{Emb}(\mathbb{R}^n,\mathbb{R}^n);$$

where $\mathsf{GL}(n)$ is the topological group of linear automorphisms of $\mathbb{R}^n$, and $\mathsf{Emb}_0(\mathbb{R}^n,\mathbb{R}^n)$ is the submonoid of $\mathsf{Emb}(\mathbb{R}^n,\mathbb{R}^n)$ consisting of those smooth embeddings that preserve the origin. The result is established upon showing each of these continuous homomoprhisms is a section of a deformation retraction. Well, the Gram–Schmidt process demonstrates a deformation retraction to the inclusion $\mathsf{O}(n) \to \mathsf{GL}(n)$. Translation $(\mathbb{R}^n \xrightarrow{x\mapsto f(x)} \mathbb{R}^n) \mapsto (\mathbb{R}^n \xrightarrow{x\mapsto f(x)-tf(0)} \mathbb{R}^n)$ demonstrates a deformation retraction to the inclusion $\mathsf{Emb}_0(\mathbb{R}^n,\mathbb{R}^n) \to \mathsf{Emb}(\mathbb{R}^n,\mathbb{R}^n)$. The expression

$$f \mapsto \begin{cases} \frac{f(tx)}{t} & \text{for } t > 0 \\ D_0 f(x) & \text{for } t = 0 \end{cases}$$

demonstrates a deformation retraction to the inclusion $\mathsf{GL}(n) \to \mathsf{Emb}_0(\mathbb{R}^n,\mathbb{R}^n)$. □

Now, temporarily consider the full ∞-subcategory

$$\mathcal{E}\mathsf{uc}_n \subset \mathcal{M}\mathsf{fld}_n$$

consisting of the object $\mathbb{R}^n$; this ∞-category is a delooping of the topological monoid $\mathsf{Emb}(\mathbb{R}^n,\mathbb{R}^n)$ We draw an immediate consequence of Proposition 2.2.6. First, note that there exists a natural functor

$$\mathsf{BO}(n) \longrightarrow \mathcal{E}\mathsf{uc}_n, \qquad * \mapsto \mathbb{R}^n,$$

from the classifying space of the orthogonal group, defined by the inclusion $\mathsf{O}(n) \hookrightarrow \mathsf{Emb}(\mathbb{R}^n, \mathbb{R}^n)$.

Corollary 2.2.7. *The functor*

$$\mathsf{BO}(n) \longrightarrow \mathcal{E}\mathsf{uc}_n,$$

is an equivalence between ∞-categories. In particular, there is a canonical equivalence between the ∞-category of presheaves

$$\mathsf{PShv}(\mathcal{E}\mathsf{uc}_n) \simeq \mathsf{Spaces}_{/\mathsf{BO}(n)}$$

and spaces over $\mathsf{BO}(n)$.

2.2.2 Sheaves on n-manifolds

Proposition 2.2.6 offers a classification of sheaves on the ∞-category of B-framed n-manifolds, with respect to a standard Grothendieck topology.

Definition 2.2.8. The symmetric monoidal category Mfld_n has objects n-manifolds; its morphisms are smooth embeddings; the symmetric monoidal structure is disjoint union. The full subcategory $\mathsf{Euc}_n \subset \mathsf{Mfld}_n$ consists of the object $\mathbb{R}^n$. In the *standard* Grothendieck topology on Mfld_n, a sieve $\mathcal{U} \subset \mathsf{Mfld}_{n/M}$ is a covering sieve exactly if, for each element $x \in M$, there is an object $(U \xrightarrow{e} M) \in \mathcal{U}$ for which $\{x\} \subset e(U)$. The ∞-category of *sheaves* on Mfld_n is the full ∞-subcategory

$$\mathsf{Shv}(\mathsf{Mfld}_n) \subset \mathsf{PShv}(\mathsf{Mfld}_n)$$

consisting of those presheaves $\mathcal{F}\colon \mathsf{Mfld}_n^{\mathsf{op}} \to \mathsf{Spaces}$ for which, for each standard covering sieve $\mathcal{U} \subset \mathsf{Mfld}_{n/M}$, the canonical functor

$$(\mathcal{U}^{\mathsf{op}})^{\triangleleft} \simeq (\mathcal{U}^{\triangleright})^{\mathsf{op}} \longrightarrow (\mathsf{Mfld}_{n/M})^{\mathsf{op}} \longrightarrow \mathsf{Mfld}_n^{\mathsf{op}} \xrightarrow{\mathcal{F}} \mathsf{Spaces}$$

is a limit diagram, where

$$\mathcal{U}^{\triangleright} := \mathcal{U} \times [1] \coprod_{\mathcal{U}\times\{1\}} *$$

is the right-cone of $\mathcal{U}$ and likewise $\mathcal{U}^{\triangleleft}$ is the left-cone.

Note the evident functor between ∞-categories

$$\mathsf{Mfld}_n \longrightarrow \mathcal{M}\mathsf{fld}_n$$

defined by the natural identity mapping from sets of embeddings, with the discrete topology, to the same sets of embeddings endowed the compact-open C^∞ topology.

Definition 2.2.9. The ∞-category of *sheaves* on $\mathcal{M}\mathsf{fld}_n$ is the pullback among ∞-categories:

$$\begin{array}{ccc} \mathsf{Shv}(\mathcal{M}\mathsf{fld}_n) & \longrightarrow & \mathsf{PShv}(\mathcal{M}\mathsf{fld}_n) \\ \downarrow & & \downarrow{\scriptstyle \text{restriction}} \\ \mathsf{Shv}(\mathsf{Mfld}_n) & \longrightarrow & \mathsf{PShv}(\mathsf{Mfld}_n). \end{array}$$

The next result serves to contextualize the recurring role of spaces over $\mathsf{BO}(n)$. This result is not essential for the overall logic of this paper, so we do not supply a proof. The result and its proof is, however, analogous to (and simpler than) Proposition 2.2.22 and its proof.

Proposition 2.2.10. [7] *Restriction along the functor* $\mathsf{BO}(n) \to \mathcal{M}\mathsf{fld}_n$ *defines an equivalence between ∞-categories*

$$\mathsf{Shv}(\mathcal{M}\mathsf{fld}_n) \xrightarrow{\simeq} \mathsf{PShv}\big(\mathsf{BO}(n)\big) \simeq \mathsf{Spaces}_{/\,\mathsf{BO}(n)}\,.$$

Remark 2.2.11. Proposition 2.2.10 is notable in that the∞-topos $\mathsf{Shv}(\mathcal{M}\mathsf{fld}_n)$ is *free* on its completely compact objects , which is a connected ∞-groupoid: $\mathsf{BO}(n)$. Consequently, a sheaf on the ∞-category $\mathcal{M}\mathsf{fld}_n$ is uniquely determined by its value on $\mathbb{R}^n$ as an $\mathsf{O}(n)$-space, or equivalently a space over $\mathsf{BO}(n)$, without reference to a sheaf condition.

Remark 2.2.12. Proposition 2.2.10 asserts that a sheaf on $\mathcal{M}\mathsf{fld}_n$ is precisely the datum of a space B equipped with a map $B \to \mathsf{BO}(n)$, which is equivalent to a rank-n vector bundle $E \to B$ over B. In this equivalence, the value of the sheaf corresponding to $B \to \mathsf{BO}(n)$ on an object $M \in \mathcal{M}\mathsf{fld}_n$ is the space $\mathsf{Map}_{/\,\mathsf{BO}(n)}\big(M, B\big)$ of lifts:

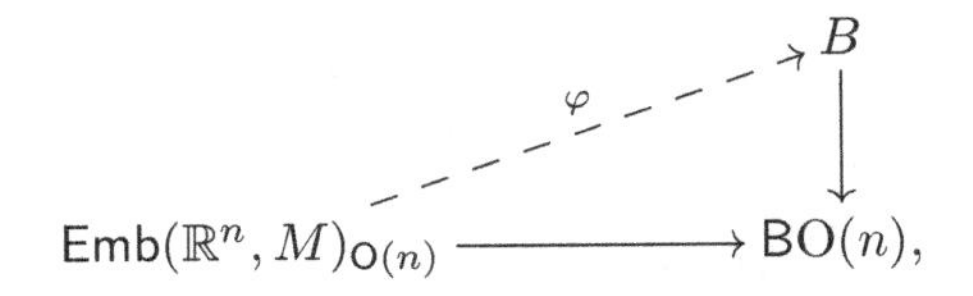

where the bottom left space is the $\mathsf{O}(n)$-coinvariants of the space of morphisms in $\mathcal{M}\mathsf{fld}_n$ from $\mathbb{R}^n$ to M. Corollary 2.2.31 will identify the bottom horizontal arrow as the tangent classifier, $M \xrightarrow{TM} \mathsf{BO}(n)$. Through that identification, such a lift is equivalent data to a map $\varphi\colon M \to B$ together with an isomorphism $TM \cong \varphi^* E$ between vector bundles over M.

Remark 2.2.13. The canonical functor $\mathsf{Mfld}_n \to \mathcal{M}\mathsf{fld}_n$ carries morphisms $U \hookrightarrow V$ that are isotopy equivalences to equivalences in the ∞-category $\mathcal{M}\mathsf{fld}_n$. It follows that the restriction functor factors

$$\mathsf{Shv}(\mathcal{M}\mathsf{fld}_n) \longrightarrow \mathsf{Shv}^{\mathsf{l.c.}}(\mathsf{Mfld}_n) \subset \mathsf{Shv}(\mathsf{Mfld}_n)$$

through the ∞-subcategory of locally constant sheaves. We warn the reader that this first functor is *not* an equivalence. Indeed, it follows from a result of Segal ([54]) that the shape of the∞-topos $\mathsf{Shv}(\mathsf{Mfld}_n)$ is the classifying space of the groupoid completion $\mathsf{BEmb}^\delta(\mathbb{R}^n, \mathbb{R}^n)$

of the discrete monoid of smooth self-embeddings of $\mathbb{R}^n$. As such, through Proposition 2.2.10, this functor can be identified as

$$\mathsf{PShv}(\mathsf{BO}(n)) \longrightarrow \mathsf{PShv}(\mathsf{BEmb}^\delta(\mathbb{R}^n,\mathbb{R}^n)),$$

which is restriction along the canonical functor $\mathsf{BEmb}^\delta(\mathbb{R}^n,\mathbb{R}^n) \to \mathsf{BEmb}(\mathbb{R}^n,\mathbb{R}^n) \simeq \mathsf{BO}(n)$. The work [28] establishes that this map is not an equivalence.

The following result, proved in [7], contrasts with Remark 2.2.13. Indeed, the shape of the∞-topos $\mathsf{Shv}(\mathsf{Mfld}_{n/M})$ is the underlying space of M, while $\mathsf{Shv}(\mathcal{M}\mathsf{fld}_{n/M})$ is identified, as a consequence of Corollary 2.2.31, as the ∞-category $\mathsf{Spaces}_{/M}$ of local systems on M.

Proposition 2.2.14. *For each n-manifold M, the restriction functor* $\mathsf{Shv}(\mathcal{M}\mathsf{fld}_{n/M}) \to \mathsf{Shv}(\mathsf{Mfld}_{n/M})$ *factors through an equivalence between ∞-categories*

$$\mathsf{Shv}(\mathcal{M}\mathsf{fld}_{n/M}) \xrightarrow{\simeq} \mathsf{Shv}^{\mathsf{l.c.}}(\mathsf{Mfld}_{n/M}).$$

2.2.3 Tangent classifier

We will be particularly interested in n-manifolds that are equipped with the additional structure of a section of a given sheaf on $\mathcal{M}\mathsf{fld}_n$. Through Proposition 2.2.10, such a section is a continuous system of linear structures on each tangent space, such as an orientation, spin structure, or a framing. Tangential structure of this sort can be swiftly accommodated by way of the *tangent classifier*: each n-manifold M has a tangent bundle, and it is classified by a map $\tau_M \colon M \to \mathsf{BO}(n)$ to the classifying space of the topological group $\mathsf{O}(n)$ of linear isometries of $\mathbb{R}^n$. For $B \to \mathsf{BO}(n)$ a map between spaces, a B-framing on M is a homotopy commutative diagram among spaces

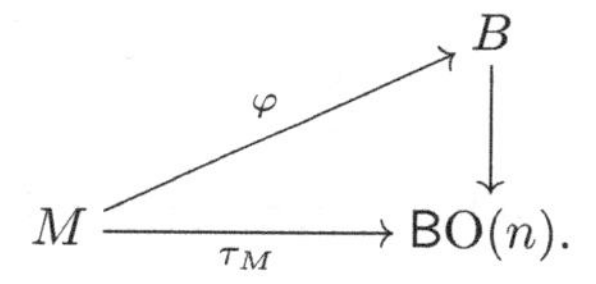

Example 2.2.15. Consider the surjective homomorphism

$$\mathsf{O}(n) \xrightarrow{\text{determinant}} \mathsf{O}(1).$$

The kernel of this homomorphism is $\mathsf{SO}(n) \subset \mathsf{O}(n)$, and a $\mathsf{BSO}(n)$-framing on a topological n-manifold is precisely an orientation.

Toward defining an ∞-category $\mathcal{M}\mathsf{fld}_n^B$ of B-framed n-manifolds, we next explain how to make the tangent classifier continuously functorial among open embeddings. Namely, through Corollary 2.2.7, we define the tangent classifier as the restricted Yoneda functor:

$$\tau \colon \mathcal{M}\mathsf{fld}_n \xrightarrow{\text{Yoneda}} \mathsf{PShv}(\mathcal{M}\mathsf{fld}_n) \xrightarrow{\text{restriction}} \mathsf{PShv}(\mathcal{E}\mathsf{uc}_n) \underset{\text{Cor } 2.2.7}{\simeq} \mathsf{Spaces}_{/\,\mathsf{BO}(n)}\,. \quad (2.2.1)$$

We postpone to Corollary 2.2.31 justification that the value τ_M is indeed the familiar tangent classifier, namely that the functor τ sends a manifold M to the map of spaces $M \to \mathsf{BO}(n)$, homotopy-coherently among embeddings in the manifold variable.

Observation 2.2.16. Because $\mathbb{R}^n$ is connected, this tangent classifier $\tau\colon \mathcal{M}\mathsf{fld}_n \to \mathsf{Spaces}_{/\,\mathsf{BO}(n)}$ is symmetric monoidal with respect to coproducts in the codomain. In other words, τ carries finite disjoint unions to finite coproducts over $\mathsf{BO}(n)$.

2.2.4 *B*-framed manifolds

In this section, we fix a space B as well as a map $B \to \mathsf{BO}(n) \simeq \mathsf{BGL}(n)$, which is equivalent to a rank-n vector bundle $E \to B$ over B. Through Proposition 2.2.10, such data defines a sheaf on $\mathcal{M}\mathsf{fld}_n$. We now consider an ∞-category $\mathcal{M}\mathsf{fld}_n^B$, of n-manifolds equipped with sections of this sheaf.

Definition 2.2.17. The symmetric monoidal ∞-category $\mathcal{M}\mathsf{fld}_n^B$ of B-framed n-manifolds is the limit in the following diagram:

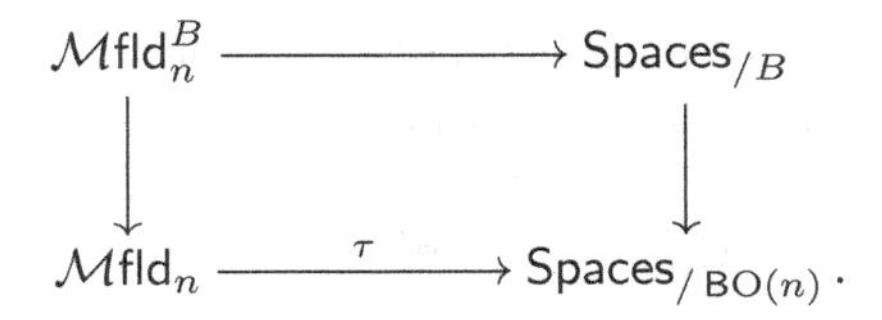

With coproduct as the symmetric monoidal structure on $\mathsf{Spaces}_{/B}$ and $\mathsf{Spaces}_{/\,\mathsf{BO}(n)}$, the right vertical functor is canonically symmetric monoidal. Observation 2.2.16 grants that the bottom horizontal functor τ in the diagram in Definition 2.2.17 is canonically symmetric monoidal. Formally, the forgetful functor $\mathsf{Cat}_\infty^\otimes \to \mathsf{Cat}_\infty$ from symmetric monoidal ∞-categories to ∞-categories creates and preserves limits. We conclude that Definition 2.2.17 indeed defines an ∞-category equipped with a symmetric monoidal structure, as claimed.

Remark 2.2.18. We describe the objects and spaces of morphism spaces in the ∞-category $\mathcal{M}\mathsf{fld}_n^B$.

- An object is a *B-framed n-manifold*, which is the data of an n-manifold M and a lift

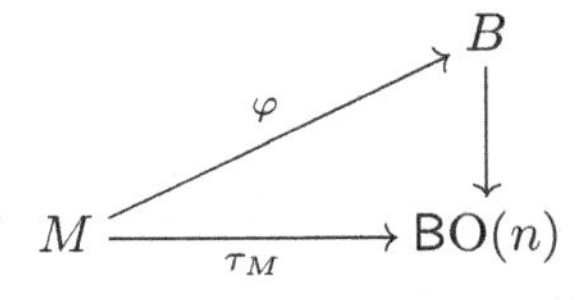

  of its tangent classifier. Here, we understand that this is a commutative diagram in the ∞-category Spaces, i.e. a homotopy coherently commutative diagram of spaces. Equivalently, for $E \to B$ the rank-n vector bundle over B classified by the given map $B \to \mathsf{BO}(n)$, such a commutative diagram is the data of a map $M \xrightarrow{\varphi} B$ between spaces together with an isomorphism $TM \cong \varphi^* E$ between vector bundles over M.

- Let (M, φ) and (N, ψ) be B-framed n-manifolds, as above. The space of morphisms in $\mathcal{M}\mathsf{fld}_n^B$ from (M, φ) to (N, ψ) is the space of *B-framed embeddings*, which is the limit space:

$$\begin{array}{ccc} \mathsf{Emb}^B(M,N) & \longrightarrow & \mathsf{Map}_{/B}(M,N) \\ \downarrow & & \downarrow \\ \mathsf{Emb}(M,N) & \longrightarrow & \mathsf{Map}_{/\,\mathsf{BO}(n)}(M,N). \end{array} \tag{2.2.2}$$

Here, for $M \xrightarrow{\varphi} X \xleftarrow{\psi} N$ a diagram in Spaces, then $\mathsf{Map}_{/X}(M,N)$ is the space of maps from M to N over X: it is the limit space:

$$\begin{array}{ccc} \mathsf{Map}_{/X}(M,N) & \longrightarrow & \mathsf{Map}(M,N) \\ \downarrow & & \downarrow{\scriptstyle -\circ\psi} \\ * & \xrightarrow{\{\varphi\}} & \mathsf{Map}(M,X). \end{array}$$

That is, a point in $\mathsf{Map}_{/X}(M,N)$ is represented by a map $M \to N$ and a homotopy between the two resulting maps from M to X

Proposition 2.2.6 yields the following.

Observation 2.2.19. The space of B-framings on $\mathbb{R}^n$ is the fiber of the given map $B \to \mathsf{BO}(n)$ over the point $* \xrightarrow{\{\mathbb{R}^n\}} \mathsf{BO}(n)$ selecting the vector space $\mathbb{R}^n$. For φ a B-framing on $\mathbb{R}^n$, there is a canonical equivalence between monoid-objects in Spaces:

$$\Omega_\varphi B \;\simeq\; \mathsf{Emb}^B(\mathbb{R}^n,\mathbb{R}^n);$$

here, $\Omega_\varphi B$ is the monoid in Spaces, via concatenation, of loops in B based at the point $* \xrightarrow{\varphi} \mathsf{fiber}\big(B \to \mathsf{BO}(n)\big) \to B$.

2.2.5 Examples and discussion of B-framings

Framings. In the case that $\big(B \to \mathsf{BO}(n)\big) := \big(* \xrightarrow{\{\mathbb{R}^n\}} \mathsf{BO}(n)\big)$, then a B-framing on an n-manifold M is a *framing* of M, i.e., a trivialization of its tangent bundle. Such a trivialization is a sequence of n linearly independent vector fields $(X_1,\dots,X_n)$ on M.

Now, let (M,φ) and (N,ψ) be framed n-manifolds, with corresponding vector fields $(X_i)_{1\leq i\leq n}$ and $(Y_i)_{1\leq i\leq n}$. A naive definition of the space of *framed embeddings* might be the subspace,

$$\left\{M \overset{e}{\hookrightarrow} N \mid De(X_i) = Y_i \text{ for each } 1 \leq i \leq n\right\} \;\subset\; \mathsf{Emb}(M,N),$$

of those smooth embeddings that strictly respect the framings. However, for the purposes of defining manifold invariants from algebraic input (e.g., factorization homology), this naive definition of framed embeddings is deficient:

1. for (N,ψ) a framed n-manifold with N compact, it receives no framed embeddings from $\mathbb{R}^n$ (with its standard framing) since it receives no isometric embeddings from $\mathbb{R}^n$;

2. for each $i \geq 0$, this space of framed embeddings from $(\mathbb{R}^n)^{\sqcup i}$ to $\mathbb{R}^n$ (each with its standard framing) is *not* equivalent to the space $\mathcal{E}_n(i)$ of i-ary operations of the little n-disks operad.

One might try to fix the above naive definition of framed embeddings by weakening the condition $De(X_i) = Y_i$ to the *data* of a sequence of smooth functions $M \xrightarrow{\lambda_i} \mathbb{R}_{>0}$ subject to the condition of an equality $De(X_i) = \lambda_i Y_i$. But this, too, has similar shortcomings.

Practically, the essential feature of the 'correct' space of framed embeddings is so that there is a homotopy equivalence

$$\mathsf{Emb}^{\mathsf{fr}}\big((\mathbb{R}^n)^{\sqcup i}, (N,\psi)\big) \;\simeq\; \mathsf{Conf}_i(N)$$

with the *configuration space* of injections from $\{1,\dots,i\}$ into N. Lemma 2.2.30, to come, grants that this is the case for the definition (2.2.2) of framed embeddings (in this case that $B = *$) as a homotopy pullback.

Linear structures. More generally, for G a Lie group and $\rho\colon G \to \mathsf{GL}(n)$ a smooth homomorphism, consider the composite map $\mathsf{B}G \to \mathsf{BGL}(n) \simeq \mathsf{BO}(n)$. A $\mathsf{B}G$-framing on an n-manifold M is a compatible system of lifts $D(\alpha^{-1}\beta)\colon U \cap V \to G$ along ρ of the derivatives of the transition maps of a smooth atlas for M. So a $\mathsf{BSO}(n)$-framing on M is a smooth structure on M together with an orientation on M; a $\mathsf{B}*$-framing on M is a framing, as discussed above; a $\mathsf{BSpin}(n)$-framing on M is a spin structure on M.

General structures. In general, for B a space, the space of maps $B \to \mathsf{BO}(n)$ is a moduli space of rank-n vector bundles on B. So, given such a vector bundle $E \to B$, a B-framing on an n-manifold M is a continuous map $f\colon M \to B$ together with an isomorphism $TM \cong f^*E$ between vector bundles over M. In the case that $\big(B \to \mathsf{BO}(n)\big) \simeq \big(X \times \mathsf{BO}(n) \xrightarrow{\mathsf{pr}} \mathsf{BO}(n)\big)$, then a B-framing on an n-manifold M is simply a map $M \to X$ to X from the underlying space of M.

2.2.6 Weiss sheaves on n-manifolds

Definition 2.2.20. [59] In the *Weiss* Grothendieck topology on the category Mfld_n, a sieve $\mathcal{U} \subset \mathsf{Mfld}_{n/M}$ is a covering sieve if, for each finite subset $S \subset M$, there is an object $(U \hookrightarrow M) \in \mathcal{U}$ for which $S \subset e(U)$. The ∞-category of *Weiss* sheaves on Mfld_n is the full ∞-subcategory

$$\mathsf{Shv}^{\mathsf{Weiss}}(\mathsf{Mfld}_n) \;\subset\; \mathsf{PShv}(\mathsf{Mfld}_n)$$

consisting of those presheaves $\mathcal{F}\colon \mathsf{Mfld}_n^{\mathsf{op}} \to \mathsf{Spaces}$ for which, for each Weiss covering sieve $\mathcal{U} \subset \mathsf{Mfld}_{n/M}$, the canonical functor

$$(\mathcal{U}^{\mathsf{op}})^{\triangleleft} \;\simeq\; (\mathcal{U}^{\triangleright})^{\mathsf{op}} \longrightarrow (\mathsf{Mfld}_{n/M})^{\mathsf{op}} \longrightarrow \mathsf{Mfld}_n^{\mathsf{op}} \xrightarrow{\mathcal{F}} \mathsf{Spaces}$$

is a limit diagram. The ∞-category of *Weiss* sheaves on $\mathcal{M}\mathsf{fld}_n$ is the pullback among ∞-categories:

$$\begin{array}{ccc}
\mathsf{Shv}^{\mathsf{Weiss}}(\mathcal{M}\mathsf{fld}_n) & \longrightarrow & \mathsf{PShv}(\mathcal{M}\mathsf{fld}_n) \\
\downarrow & & \downarrow{\scriptstyle \text{restriction}} \\
\mathsf{Shv}^{\mathsf{Weiss}}(\mathsf{Mfld}_n) & \longrightarrow & \mathsf{PShv}(\mathsf{Mfld}_n).
\end{array}$$

We next exhibit generators the for the ∞-category of Weiss sheaves.

Definition 2.2.21. The symmetric monoidal ∞-categories

$$\mathsf{Disk}_n \subset \mathsf{Mfld}_n \qquad \text{and} \qquad \mathcal{D}\mathsf{isk}_n \subset \mathcal{M}\mathsf{fld}_n$$

are the full ∞-subcategories consisting of disjoint unions of n-dimensional Euclidean spaces.

Notice the evident symmetric monoidal functors

$$\mathsf{Disk}_n \longrightarrow \mathcal{D}\mathsf{isk}_n \qquad \text{and} \qquad \mathsf{Mfld}_n \longrightarrow \mathcal{M}\mathsf{fld}_n\,.$$

We insert the next result to contextualize the fundamental role of disks, though this result is not essential for the overall logic of this paper. Its proof makes use of results established later on, which of course do not logically depend on it.

Proposition 2.2.22. *Restriction along $\mathcal{D}\mathsf{isk}_n \hookrightarrow \mathcal{M}\mathsf{fld}_n$ defines an equivalence between ∞-categories:*

$$\mathsf{Shv}^{\mathsf{Weiss}}(\mathcal{M}\mathsf{fld}_n) \xrightarrow{\simeq} \mathsf{PShv}(\mathcal{D}\mathsf{isk}_n).$$

Proof. Denote the fully-faithful inclusion $\iota\colon \mathcal{D}\mathsf{isk}_n \hookrightarrow \mathcal{M}\mathsf{fld}_n$. This functor determines an adjunction

$$\iota^*\colon \mathsf{PShv}(\mathcal{M}\mathsf{fld}_n) \rightleftarrows \mathsf{PShv}(\mathcal{D}\mathsf{isk}_n)\colon \iota_*$$

in which the left adjoint is restriction along ι and the right adjoint is right Kan extension along ι. It is enough to show that this adjunction restricts as an equivalence:

$$\iota^*\colon \mathsf{Shv}^{\mathsf{Weiss}}(\mathcal{M}\mathsf{fld}_n) \rightleftarrows \mathsf{PShv}(\mathcal{D}\mathsf{isk}_n)\colon \iota_*.$$

This amounts to verifying two assertions.

1. For each $\mathcal{F} \in \mathsf{Shv}^{\mathsf{Weiss}}(\mathcal{M}\mathsf{fld}_n)$, and each $M \in \mathcal{M}\mathsf{fld}_n$, the unit map

$$\mathcal{F}(M) \xrightarrow{\text{unit}} \lim\Big((\mathcal{D}\mathsf{isk}_{n/M})^{\mathsf{op}} \to \mathcal{D}\mathsf{isk}_n^{\mathsf{op}} \xrightarrow{\iota} \mathcal{M}\mathsf{fld}_n^{\mathsf{op}} \xrightarrow{\mathcal{F}} \mathsf{Spaces}\Big)$$

is an equivalence between spaces.

2. For each $\mathcal{G} \in \mathsf{PShv}(\mathcal{D}\mathsf{isk}_n)$, and for each Weiss covering sieve $\mathcal{U} \subset \mathsf{Mfld}_{n/M}$, the canonical map

$$\iota_*\mathcal{G}(M) \longrightarrow \lim\Big(\mathcal{U}^{\mathsf{op}} \to \mathsf{Mfld}_n^{\mathsf{op}} \to \mathcal{M}\mathsf{fld}_n^{\mathsf{op}} \xrightarrow{\iota_*\mathcal{G}} \mathsf{Spaces}\Big)$$

is an equivalence between spaces.

We first establish (1). The canonical functor $\mathsf{Disk}_{n/M} \to \mathcal{D}\mathsf{isk}_{n/M}$ determines the sequence of maps among spaces

$$\begin{array}{ccc} \mathcal{F}(M) \xrightarrow{\text{unit}} & \lim\big((\mathcal{D}\mathsf{isk}_{n/M})^{\mathsf{op}} \to \mathcal{D}\mathsf{isk}_n^{\mathsf{op}} \xrightarrow{\iota} \mathcal{M}\mathsf{fld}_n^{\mathsf{op}} \xrightarrow{\mathcal{F}} \mathsf{Spaces}\big) \\ \searrow & \downarrow \\ & \lim\big((\mathsf{Disk}_{n/M})^{\mathsf{op}} \to \mathsf{Disk}_n^{\mathsf{op}} \to \mathcal{D}\mathsf{isk}_n^{\mathsf{op}} \xrightarrow{\iota} \mathcal{M}\mathsf{fld}_n^{\mathsf{op}} \xrightarrow{\mathcal{F}} \mathsf{Spaces}\big). \end{array}$$

Because smooth open embeddings from disjoint unions of n-dimensional Euclidean spaces to each smooth n-manifold is a basis for the Weiss Grothendieck topology thereon, the full subcategory $\mathsf{Disk}_n \subset \mathsf{Mfld}_n$ is a basis for the Weiss Grothendieck topology. It follows that the diagonal map in the above diagram is an equivalence between spaces. Furthermore,

Proposition 2.2.38 implies the functor $(\mathsf{Disk}_{n/M})^{\mathsf{op}} \to (\mathcal{D}\mathsf{isk}_{n/M})^{\mathsf{op}}$ is initial. It follows that the downward map in the above diagram is an equivalence. We conclude that the horizontal map in the above diagram is an equivalence, as desired.

We now establish (2). The Weiss covering sieve $\mathcal{U} \subset \mathsf{Mfld}_{n/M}$ determines a functor between ∞-categories

$$\mathsf{colim}\Big(\mathcal{U} \to \mathsf{Mfld}_n \xrightarrow{\mathcal{D}\mathsf{isk}_{n/-}} \mathsf{Cat}_\infty\Big) \longrightarrow \mathcal{D}\mathsf{isk}_{n/M} \tag{2.2.3}$$

from the colimit indexed by $\mathcal{U}$. Through the standard formula computing values of right Kan extension as limits indexed by ∞-undercategories, the map in (2) is canonically identified as the map between spaces

$$\mathsf{lim}\Big((\mathcal{D}\mathsf{isk}_{n/M})^{\mathsf{op}} \to \mathcal{D}\mathsf{isk}_n^{\mathsf{op}} \xrightarrow{\mathcal{G}} \mathsf{Spaces}\Big) \longrightarrow$$

$$\mathsf{lim}\Big(\mathsf{colim}\big(\mathcal{U} \to \mathsf{Mfld}_n \xrightarrow{\mathcal{D}\mathsf{isk}_{n/-}} \mathsf{Cat}_\infty\big)^{\mathsf{op}} \to (\mathcal{D}\mathsf{isk}_{n/M})^{\mathsf{op}} \to \mathcal{D}\mathsf{isk}_n^{\mathsf{op}} \xrightarrow{\mathcal{G}} \mathsf{Spaces}\Big).$$

It is therefore enough to show that the functor (2.2.3) is final. Observe that, for each n-manifold W, the ∞-category $\mathcal{D}\mathsf{isk}_{n/W}$ is canonically a right fibration over $\mathcal{D}\mathsf{isk}_n$. To assess finality of (2.2.3), it is enough to compute the relevant colimit in the ∞-category $\mathsf{RFib}_{\mathcal{D}\mathsf{isk}_n}$ of right fibrations over $\mathcal{D}\mathsf{isk}_n$, and show that the morphism between right fibrations over $\mathcal{D}\mathsf{isk}_n$

$$\mathsf{colim}\Big(\mathcal{U} \to \mathsf{Mfld}_n \xrightarrow{\mathcal{D}\mathsf{isk}_{n/-}} \mathsf{RFib}_{\mathcal{D}\mathsf{isk}_n}\Big) \longrightarrow \mathcal{D}\mathsf{isk}_{n/M}$$

is an equivalence. A morphism between right fibrations over an ∞-category is an equivalence if and only if it is so when base changed over the maximal ∞-subgroupoid of the base. Through Lemma 2.2.30, the maximal ∞-subgroupoid of $\mathcal{D}\mathsf{isk}_n$ is the coproduct $\coprod_{i\geq 0} \mathsf{B}\big(\Sigma_i \wr \mathsf{O}(n)\big)$. For each n-manifold W, Lemma 2.2.30 identifies the base change of $\mathcal{D}\mathsf{isk}_{n/M}$ over the i-cofactor as $\mathsf{Conf}_i(W)_{\Sigma_i}$, the unordered configuration space. So let $i \geq 0$. So we are to show that the canonical map between spaces

$$\mathsf{colim}\Big(\mathcal{U} \to \mathsf{Mfld}_n \xrightarrow{\mathsf{Conf}_i(-)_{\Sigma_i}} \mathsf{Spaces}_{/\mathsf{B}(\Sigma_i \wr \mathsf{O}(n))}\Big) \longrightarrow \mathsf{Conf}_i(M)_{\Sigma_i} \tag{2.2.4}$$

is an equivalence between spaces over $\mathsf{B}\big(\Sigma_i \wr \mathsf{O}(n)\big)$. This is simply to show that this map is an equivalence between spaces, ignorant to the structure maps to $\mathsf{B}\big(\Sigma_i \wr \mathsf{O}(n)\big)$. Consider the smallest sieve $\mathcal{U}_i \subset \mathsf{Mfld}_{ni/\,\mathsf{Conf}_i(M)_{\Sigma_i}}$ containing, for each $U \in \mathcal{U}$, the object $\big(\mathsf{Conf}_i(U)_{\Sigma_i} \hookrightarrow \mathsf{Conf}_i(M)_{\Sigma_i}\big) \in \mathsf{Mfld}_{ni/\,\mathsf{Conf}_i(M)_{\Sigma_i}}$. Precisely because $\mathcal{U}$ is a Weiss covering sieve, the sieve $\mathcal{U}_i$ is a standard covering sieve. Using that the underlying topological space of $\mathsf{Conf}_i(M)_{\Sigma_i}$ is paracompact and Hausdorff, Theorem A.3.1 [43] can be applied to $\mathcal{U}_i$ with the result that the above map (2.2.4) between ∞-groupoids is an equivalence. This completes the proof. □

Definition 2.2.23. Let $\mathcal{C}$ be a presentable ∞-category. The ∞-category of *$\mathcal{C}$-valued Weiss cosheaves* on Mfld_n is the full ∞-subcategory

$$\mathsf{cShv}^{\mathsf{Weiss}}_{\mathcal{C}}(\mathsf{Mfld}_n) \subset \mathsf{Fun}(\mathsf{Mfld}_n, \mathcal{C})$$

consisting of those functors $\mathcal{A}\colon \mathsf{Mfld}_n \to \mathcal{C}$ for which, for each Weiss covering sieve $\mathcal{U} \subset \mathsf{Mfld}_{n/M}$, the canonical functor

$$\mathcal{U}^{\triangleright} \longrightarrow \mathsf{Mfld}_{n/M} \longrightarrow \mathsf{Mfld}_n \xrightarrow{\mathcal{A}} \mathcal{C}$$

is a colimit diagram. The ∞-category of *$\mathcal{C}$-valued Weiss cosheaves* on $\mathcal{M}\mathsf{fld}_n$ is the pullback among ∞-categories:

$$\begin{array}{ccc} \mathsf{cShv}^{\mathsf{Weiss}}_{\mathcal{C}}(\mathcal{M}\mathsf{fld}_n) & \longrightarrow & \mathsf{Fun}(\mathcal{M}\mathsf{fld}_n, \mathcal{C}) \\ \downarrow & & \downarrow{\scriptstyle \text{restriction}} \\ \mathsf{cShv}^{\mathsf{Weiss}}_{\mathcal{C}}(\mathsf{Mfld}_n) & \longrightarrow & \mathsf{Fun}(\mathsf{Mfld}_n, \mathcal{C}). \end{array}$$

The next result is routine consequence of Proposition 2.2.22.

Corollary 2.2.24. *Let $\mathcal{C}$ be a presentable ∞-category. Restriction along $\mathcal{D}\mathsf{isk}_n \to \mathcal{M}\mathsf{fld}_n$ defines an equivalence between ∞-categories:*

$$\mathsf{cShv}^{\mathsf{Weiss}}_{\mathcal{C}}(\mathcal{M}\mathsf{fld}_n) \xrightarrow{\simeq} \mathsf{Fun}(\mathcal{D}\mathsf{isk}_n, \mathcal{C}).$$

Remark 2.2.25. Corollary 2.2.24 says that a $\mathcal{C}$-valued Weiss cosheaf on $\mathcal{M}\mathsf{fld}_n$ is simply a functor $\mathcal{D}\mathsf{isk}_n \to \mathcal{C}$, without regard to a cosheaf condition.

Remark 2.2.26. We follow up on Remark 2.2.13 and Remark 2.2.14. Namely, restriction along $\mathsf{Mfld}_n \to \mathcal{M}\mathsf{fld}_n$ factors

$$\mathsf{cShv}^{\mathsf{Weiss}}_{\mathcal{C}}(\mathcal{M}\mathsf{fld}_n) \longrightarrow \mathsf{cShv}^{\mathsf{Weiss,l.c.}}_{\mathcal{C}}(\mathsf{Mfld}_n) \subset \mathsf{cShv}^{\mathsf{Weiss}}_{\mathcal{C}}(\mathsf{Mfld}_n)$$

through the ∞-subcategory of locally constant Weiss cosheaves. This first functor is not an equivalence. However, for each n-manifold M, the restriction functor factors as an equivalence between ∞-categories,

$$\mathsf{cShv}^{\mathsf{Weiss}}_{\mathcal{C}}(\mathcal{M}\mathsf{fld}_{n/M}) \xrightarrow{\simeq} \mathsf{cShv}^{\mathsf{Weiss,l.c}}_{\mathcal{C}}(\mathsf{Mfld}_{n/M}) \subset \mathsf{cShv}^{\mathsf{Weiss}}_{\mathcal{C}}(\mathsf{Mfld}_{n/M}),$$

involving the locally constant Weiss cosheaves on M.

2.2.7 Disks

Prompted by the results in §2.2.6, we now turn to consider B-framed n-disks, as they organize as a symmetric monoidal ∞-category. We identify the maximal ∞-subgroupoid of the ∞-category $\mathcal{D}\mathsf{isk}^B_{n/M}$ in terms of configuration spaces in a B-framed n-manifold M.

Definition 2.2.27. [3] The symmetric monoidal ∞-category $\mathcal{D}\mathsf{isk}^B_n \subset \mathcal{M}\mathsf{fld}^B_n$ is the full ∞-subcategory of $\mathcal{M}\mathsf{fld}^B_n$ whose objects are disjoint unions of B-framed n-dimensional Euclidean spaces.

Remark 2.2.28. Consider the framing structure, which is the map $* \to \mathsf{BO}(n)$ selecting the basepoint of $\mathsf{BO}(n)$. As discussed in §2.2.5, a $*$-framing on an n-manifold M is a trivialization of the tangent bundle of M. We denote the associated ∞-category of framed n-disks as $\mathcal{D}\mathsf{isk}_n^{\mathsf{fr}}$. This symmetric monoidal ∞-category $\mathcal{D}\mathsf{isk}_n^{\mathsf{fr}}$ is equivalent to the PROP associated to the operad $\mathcal{E}_n$, of Boardman-Vogt [13]. This is the case because the inclusion of rectilinear embeddings as framed embeddings determines an equivalence $\mathcal{E}_n(i) \xrightarrow{\simeq} \mathsf{Emb}^{\mathsf{fr}}\big((\mathbb{R}^n)^{\sqcup i}, \mathbb{R}^n\big)$ from the space of i-ary operations of the operad $\mathcal{E}_n$; see, e.g., [3] for a presentation of this equivalence.

Example 2.2.29. Consider the structure $\big(B \to \mathsf{BO}(n)\big) := \big(\mathsf{BO}(n) \xrightarrow{\mathsf{id}} \mathsf{BO}(n)\big)$. The symmetric monoidal ∞-category $\mathcal{D}\mathsf{isk}_n = \mathcal{D}\mathsf{isk}_n^{\mathsf{BO}(n)}$ is equivalent to the PROP associated to the unoriented version of the ribbon, or "framed," $\mathcal{E}_n$-operad; see [52] for a treatment of this operad.[6] This equivalence follows from Proposition 2.2.6.

Given a topological space X and a finite cardinality i, the *configuration space* of i (ordered) points in X is the subspace

$$\mathsf{Conf}_i(X) := \Big\{\{1,\dots,i\} \xrightarrow{c} X \mid c \text{ is injective}\Big\} \subset X^{\times i}.$$

This configuration space has an evident free action of the symmetric group Σ_i, as given by precomposition. The *unordered* configuration space is the Σ_i-coinvariants: $\mathsf{Conf}_i(X)_{\Sigma_i}$.

In the next result, for M a B-framed n-manifold, we consider the ∞-overcategory

$$\mathcal{D}\mathsf{isk}_{n/M}^B := \mathcal{D}\mathsf{isk}_n^B \underset{\mathcal{M}\mathsf{fld}_n^B}{\times} \mathcal{M}\mathsf{fld}_{n/M}^B .$$

An object in this ∞-category is given by a B-framed embedding $(\mathbb{R}^n)^{\sqcup i} \hookrightarrow M$ for some i.

Lemma 2.2.30. [3] *The maximal ∞-subgroupoid of $\mathcal{D}\mathsf{isk}_n^B$ is canonically identified as the space*

$$\coprod_{i\geq 0} {B^{\times i}}_{\Sigma_i} \simeq \big(\mathcal{D}\mathsf{isk}_n^B\big)^{\sim}$$

where the coproduct is indexed by finite cardinalities and each cofactor is the Σ_i-homotopy coinvariants of the i-fold product of the space B. In particular, the symmetric monoidal functor $[-]$: $\mathcal{D}\mathsf{isk}_n^B \to \mathsf{Fin}$, given by taking sets of connected components of underlying manifolds, is conservative.

For M a B-framed n-manifold, the maximal ∞-subgroupoid of $\mathcal{D}\mathsf{isk}_{n/M}^B$ is canonically identified as the space

$$\coprod_{i\geq 0} \mathsf{Conf}_i(M)_{\Sigma_i} \simeq \big(\mathcal{D}\mathsf{isk}_{n/M}^B\big)^{\sim}$$

where the coproduct is indexed by finite cardinalities, and each cofactor is an unordered configuration space.

We conclude this section by justifying the term *tangent classifier* for the functor $\mathcal{M}\mathsf{fld}_n \xrightarrow{\tau} \mathsf{Spaces}_{/\,\mathsf{BO}(n)}$ from (2.2.1).

Corollary 2.2.31. *The value of the tangent classifier (2.2.1) on an n-manifold M is the map between spaces $M \xrightarrow{\tau M} \mathsf{BO}(n)$ classifying its tangent bundle.*

[6] The historical use of "framed" here is potentially misleading, since in the "framed" $\mathcal{E}_n$ operad the embeddings do not preserve the framing, while in the usual $\mathcal{E}_n$ operad the embeddings do preserve the framing (up to scale). It might lead to less confusion to replace the term "framed $\mathcal{E}_n$ operad" with "unoriented $\mathcal{E}_n$ operad."

Proof. First, recognize $\mathcal{E}\mathsf{uc}_n \subset \mathcal{D}\mathsf{isk}_n$ as the full ∞-subcategory consisting of the connected n-manifolds. Next, specialize the second statement of Lemma 2.2.30 at $i = 1$ to obtain an identification

$$M \underset{\text{Lem } 2.2.30}{\simeq} \mathcal{E}\mathsf{uc}_{n/M} \underset{\text{Prop } 2.2.6}{\simeq} \mathsf{Emb}(\mathbb{R}^n, M)_{\mathsf{O}(n)} \longrightarrow \mathsf{BO}(n)$$

involving the homotopy $\mathsf{O}(n)$-coinvariants. Unwinding the equivalences above recognizes this map in terms of the frame bundle for M, as

$$M \simeq \mathsf{Fr}(M)_{\mathsf{O}(n)} \longrightarrow \mathsf{BO}(n).$$

Evidently, this map classifies the tangent bundle of M. □

2.2.8 Manifolds with boundary

We will also employ the category of n-manifolds with boundary.

Definition 2.2.32. A smooth manifold M with boundary is *finitary* if it admits a *good* cover, which is to say it admits a finite open cover $\mathcal{U} := \{U \subset M\}$ with the property that, for each finite subset $S \subset \mathcal{U}$, the intersection $\bigcap_{U \in S} U$ is either empty, diffeomorphic to Euclidean space, or diffeomorphic to Euclidean half-space, $\mathbb{R}_{\geq 0} \times \mathbb{R}^{n-1}$.

Terminology 2.2.33. In this article, by "manifold with boundary" we mean "finitary smooth manifold with boundary," unless otherwise stated.

Definition 2.2.34. $\mathcal{M}\mathsf{fld}_n^\partial$ is the symmetric monoidal topological category of n-manifolds, possibly with boundary. The topological space of morphisms between two is the set of smooth open embeddings equipped with the compact-open C^∞ topology. The symmetric monoidal structure is disjoint union. The full symmetric monoidal topological category $\mathcal{D}\mathsf{isk}_n^\partial \subset \mathcal{M}\mathsf{fld}_n^\partial$ is that consisting of finite disjoint unions of $\mathbb{R}^n$ and $\mathbb{R}_{\geq 0} \times \mathbb{R}^{n-1}$.

Remark 2.2.35. The category $\mathcal{D}\mathsf{isk}_n^\partial$ is minimal with respect to the condition that any finite subset of an n-manifold with boundary has an open neighborhood diffeomorphic to an object of $\mathcal{D}\mathsf{isk}_n^\partial$. In particular, the closed n-disk $\mathbb{D}^n$ is not an object of $\mathcal{D}\mathsf{isk}_n^\partial$.

The following result is an application of the Alexander trick.

Proposition 2.2.36. [3] *The symmetric monoidal functor*

$$\mathbb{R}_{\geq 0} \times - : \mathcal{M}\mathsf{fld}_{n-1} \longrightarrow \mathcal{M}\mathsf{fld}_n^\partial$$

is fully faithful. Namely, for each pair of $(n-1)$-manifolds M and N, the map

$$\mathsf{Emb}(M, N) \longrightarrow \mathsf{Emb}(\mathbb{R}_{\geq 0} \times M, \mathbb{R}_{\geq 0} \times N)$$

is an equivalence between spaces.

Remark 2.2.37. Together with Proposition 2.2.6, the previous proposition implies that the continuous homomorphism between topological monoids

$$\mathsf{O}(n-1) \hookrightarrow \mathsf{Emb}(\mathbb{R}_{\geq 0} \times \mathbb{R}^{n-1}, \mathbb{R}_{\geq 0} \times \mathbb{R}^{n-1})$$

is a homotopy equivalence. Consequently, $\mathcal{D}\mathsf{isk}_n^{\partial}$ is an unoriented variant of the Swiss cheese operad of Voronov [58]. Specifically, the framed variant $\mathcal{D}\mathsf{isk}_n^{\partial,\mathsf{fr}}$ is equivalent to the PROP associated to the Swiss cheese operad.

2.2.9 Localizing with respect to isotopy equivalences

Here we show that the ∞-category $\mathcal{D}\mathsf{isk}_{n/M}^{B}$ is a localization of its more discrete version $\mathsf{Disk}_{n/M}^{B}$ on the collection of those embeddings that are isotopic to isomorphisms. This comparison plays a fundamental role in recognizing certain colimit expressions in this theory, for instance those that support the pushforward formula of §2.3.5.

Recall Definition 2.2.21, and the symmetric monoidal functors:

$$\mathsf{Disk}_n \longrightarrow \mathcal{D}\mathsf{isk}_n \qquad \text{and} \qquad \mathsf{Mfld}_n \longrightarrow \mathcal{M}\mathsf{fld}_n.$$

We denote the pullback symmetric monoidal ∞-categories:

$$\begin{array}{ccc} \mathsf{Disk}_n^B & \longrightarrow & \mathcal{D}\mathsf{isk}_n^B \\ \downarrow & & \downarrow \\ \mathsf{Disk}_n & \longrightarrow & \mathcal{D}\mathsf{isk}_n \end{array} \qquad \text{and} \qquad \begin{array}{ccc} \mathsf{Mfld}_n^B & \longrightarrow & \mathcal{M}\mathsf{fld}_n^B \\ \downarrow & & \downarrow \\ \mathsf{Mfld}_n & \longrightarrow & \mathcal{M}\mathsf{fld}_n. \end{array}$$

For each n-manifold M, we denote the ∞-subcategory

$$\mathcal{I}_M \subset \mathsf{Disk}_{n/M}^{B} := \mathsf{Disk}_n^B \underset{\mathsf{Mfld}_n^B}{\times} \mathsf{Mfld}_{n/M}^{B} \tag{2.2.5}$$

that consists of the same objects but only those morphisms $(U \hookrightarrow M) \hookrightarrow (V \hookrightarrow M)$ whose image in $\mathcal{D}\mathsf{isk}_{n/M}^{B}$ is an equivalence.

Proposition 2.2.38. [3] *The functor* $\mathsf{Disk}_{n/M}^{B} \longrightarrow \mathcal{D}\mathsf{isk}_{n/M}^{B}$ *witnesses a localization between ∞-categories:*

$$\left(\mathsf{Disk}_{n/M}^{B}\right)[\mathcal{I}_M^{-1}] \simeq \mathcal{D}\mathsf{isk}_{n/M}^{B}.$$

Let M be a B-framed n-manifold. Each of the ∞-categories $\mathsf{Disk}_{n/M}^{B}$ and $\mathcal{D}\mathsf{isk}_{n/M}^{B}$ is naturally the active ∞-subcategory of an ∞-operad,[7] each of which we again denote as $\mathsf{Disk}_{n/M}^{B}$ and $\mathcal{D}\mathsf{isk}_{n/M}^{B}$, respectively.

- The ∞-operad structure on $\mathsf{Disk}_{n/M}^{B}$ is such that the ∞-category of colors is the poset in which an object is an open subset $U \subset M$ that is abstractly diffeomorphic to Euclidean space. There is a unique i-ary morphism from an i-fold collection $(U_k)_{1\leq k\leq i}$ of such to

[7] Recall, §2 of [43], that a morphism of finite based sets $I_* \xrightarrow{f} J_*$ is *active* if $f^{-1}\{*\} = \{*\}$ and is *inert* if the restriction $f_| : f^{-1}J \to J$ is injective. A morphism in an ∞-operad $\mathcal{O} \to \mathsf{Fin}_*$ is active or inert if its image in Fin_* is.

another V precisely if the U_k are pairwise disjoint and their union $\bigcup_{1\leq k\leq i} U_k \subset V$ is contained in V.

- The ∞-operad structure on $\mathcal{D}\mathsf{isk}^B_{n/M}$ is such that the ∞-category of colors is the ∞-groupoid $\mathsf{Emb}(\mathbb{R}^n, M)_{\mathsf{O}(n)} \simeq M$ of an open disk in M. The space of i-ary morphisms from an i-fold collection $(U_j)_{1\leq j\leq i}$ of such to another V is the fiber of the composite map
$$\mathsf{Emb}\Big(\bigsqcup_{j=1}^{i} U_j, V\Big) \longrightarrow \prod_{j=1}^{i} \mathsf{Emb}(U_j, V) \longrightarrow \prod_{j=1}^{i} \mathsf{Emb}(U_j, M)$$
over the point selecting the given sequence of embeddings $(U_j)_{1\leq j\leq i}$.

The next result also appears in [43] as Theorem 5.4.5.9.

Corollary 2.2.39. *Let $\mathcal{V}$ be a symmetric monoidal ∞-category. For each B-framed n-manifold M, restriction along the morphism $\mathsf{Disk}^B_{n/M} \to \mathcal{D}\mathsf{isk}^B_{n/M}$ defines a fully faithful functor*
$$\mathsf{Alg}_{\mathcal{D}\mathsf{isk}^B_{n/M}}(\mathcal{V}) \xrightarrow{\text{f.f.}} \mathsf{Alg}_{\mathsf{Disk}^B_{n/M}}(\mathcal{V}).$$

In the case $M = \mathbb{R}^n$, Corollary 2.2.39 can be used to prove that locally constant factorization algebras on $\mathbb{R}^n$ are equivalent to $\mathcal{E}_n$-algebras—see §2.3.2 for a discussion of factorization algebras.

The next construction is made possible using Proposition 2.2.38.

Construction 2.2.40. Let M be a B-framed n-manifold, and let N be a B'-framed k-manifold possibly with boundary. Let $f\colon M \to N$ be a continuous map that satisfies the following regularity condition:

Each of the restrictions
$$f_|\colon f^{-1}(N \smallsetminus \partial N) \to N \smallsetminus \partial N \qquad \text{and} \qquad f_|\colon f^{-1}(\partial N) \to \partial N$$
is a smooth fiber bundle.

We produce a composite morphism between ∞-operads
$$f^{-1}\colon \mathsf{Disk}^{\partial, B'}_{k/N} \longrightarrow \mathsf{Mfld}^B_{n/M} \longrightarrow \mathcal{M}\mathsf{fld}^B_{n/M},$$
in which the second morphism is the standard one. We now describe the first functor. For formal reasons, we can assume the maps $B \to \mathsf{BO}(n)$ and $B' \to \mathsf{BO}(k)$ are equivalences. For this case, the first functor is given by $(U \hookrightarrow N) \mapsto (U \times_N M \hookrightarrow M)$, which is evidently a multi-functor. By inspection, this functor f^{-1} carries isotopy equivalences to equivalences. Through Proposition 2.2.38, there results a multi-functor
$$f^{-1}\colon \mathcal{D}\mathsf{isk}^{B'}_{k/N} \longrightarrow \mathcal{M}\mathsf{fld}^B_{n/M}, \tag{2.2.6}$$
as desired.

2.3 Homology theories for manifolds

The value of factorization homology over an n-manifold M of an n-disk algebra A is a sort of average, indexed by n-disks embedded in M, of the value of A on such n-disks. We make this definition precise, as well as observe a property of this definition for which it is universal, by defining factorization homology as left Kan extension of an n-disk algebra $A\colon \mathcal{D}\mathsf{isk}_n \to \mathcal{V}$ along the inclusion $\mathcal{D}\mathsf{isk}_n \hookrightarrow \mathcal{M}\mathsf{fld}_n$.

Throughout this entire section, we fix a space $B \to \mathsf{BO}(n)$ over $\mathsf{BO}(n)$ as well as a symmetric monoidal ∞-category $\mathcal{V}$ that is $\otimes$-*presentable* in the following sense.

Definition 2.3.1. A symmetric monoidal ∞-category $\mathcal{V}$ is $\otimes$-*presentable* if it satisfies both of the following conditions.

- The underlying ∞-category of $\mathcal{V}$ is presentable: $\mathcal{V}$ admits colimits and every object is a filtered colimit of compact objects.[8]
- The symmetric monoidal structure of $\mathcal{V}$ distributes over colimits: for each object $V \in \mathcal{V}$, the functor $V \otimes -\colon \mathcal{V} \to \mathcal{V}$ carries colimit diagrams to colimit diagrams.

Example 2.3.2. Let $\mathcal{S}$ be a presentable ∞-category. Consider the Cartesian symmetric monoidal ∞-category $(\mathcal{S}, \times)$, in which the symmetric monoidal structure is categorical product. Provided $\mathcal{S}$ is Cartesian closed, then this symmetric monoidal ∞-category is $\otimes$-presentable. In particular, $(\mathsf{Spaces}, \times)$ is $\otimes$-presentable; for $\mathcal{X}$ any ∞-topos, then $(\mathcal{X}, \times)$ is $\otimes$-presentable; also, $(\mathsf{Cat}_\infty, \times)$ is $\otimes$-presentable.

Example 2.3.3. Let R be a commutative ring. The symmetric monoidal ∞-category $\big(\mathsf{Mod}_R, \otimes_R\big)$ of R-modules with tensor product over R is $\otimes$-presentable. Note, however, that its opposite $\big(\mathsf{Mod}_R^{\mathsf{op}}, \otimes_R\big)$ is not $\otimes$-presentable.

Remark 2.3.4. The results in §2.3 (in particular, the Eilenberg–Steenrod axioms for factorization homology) only require that the symmetric monoidal structure of $\mathcal{V}$ distributes over sifted colimits. (This generality is established as a special case of §2 of [8].) However, the calculations of §2.5 onwards require the symmetric monoidal structure to distribute over all colimits, so for simplicity of exposition we enforce this stronger hypothesis throughout.

2.3.1 Disk algebras

Definition 2.3.5. The ∞-category of $\mathcal{D}\mathsf{isk}_n^B$-*algebras* in $\mathcal{V}$ is that of symmetric monoidal functors from $\mathcal{D}\mathsf{isk}_n^B$ to $\mathcal{V}$:

$$\mathsf{Alg}_{\mathcal{D}\mathsf{isk}_n^B}(\mathcal{V}) \ := \ \mathsf{Fun}^{\otimes}(\mathcal{D}\mathsf{isk}_n^B, \mathcal{V}).$$

Remark 2.3.6. Let $G \to \mathsf{O}(n)$ be a morphism between group-objects in the ∞-category Spaces (equivalently, a map of loop spaces). Through this representation, change-of-framing

[8] This is with respect to an understood fixed uncountable cardinal κ, i.e., $\mathcal{V}$ admits κ-small colimits and every object is a κ-filtered colimit of κ-compact objects.

defines an action of G on the symmetric monoidal ∞-category $\mathcal{D}\mathsf{isk}_n^{\mathsf{fr}}$. As such, the symmetric monoidal forgetful functor

$$\mathcal{D}\mathsf{isk}_n^{\mathsf{fr}} \longrightarrow \mathcal{D}\mathsf{isk}_n^{BG}$$

witnesses the G-coinvariants: $(\mathcal{D}\mathsf{isk}_n^{\mathsf{fr}})_G \xrightarrow{\simeq} \mathcal{D}\mathsf{isk}_n^{BG}$. Consequently, for each symmetric monoidal ∞-category $\mathcal{V}$, the restriction functor canonically factors through the G-invariants,

$$\mathsf{Alg}_{\mathcal{D}\mathsf{isk}_n^{\mathsf{B}G}}(\mathcal{V}) \xrightarrow{\simeq} (\mathsf{Alg}_{\mathcal{E}_n}(\mathcal{V}))^G \longrightarrow \mathsf{Alg}_{\mathcal{E}_n}(\mathcal{V})$$

as an equivalence. In particular, there is a canonical equivalence from the ∞-category of $\mathcal{D}\mathsf{isk}_n$-algebras

$$\mathsf{Alg}_{\mathcal{D}\mathsf{isk}_n}(\mathcal{V}) \xrightarrow{\simeq} (\mathsf{Alg}_{\mathcal{E}_n}(\mathcal{V}))^{\mathsf{O}(n)}$$

and that of $\mathsf{O}(n)$-invariant cE_n-algebras.

We denote the restricted Yoneda functor

$$\mathbb{E}\colon \mathcal{M}\mathsf{fld}_n^B \xrightarrow{\text{Yoneda}} \mathsf{PShv}(\mathcal{M}\mathsf{fld}_n^B) \xrightarrow{\text{restriction}} \mathsf{PShv}(\mathcal{D}\mathsf{isk}_n^B).$$

Definition 2.3.7. Let M be a B-framed n-manifold. Let A be a $\mathcal{D}\mathsf{isk}_n^B$-algebra in $\mathcal{V}$. The *factorization homology (of M with coefficients in A)* is the object in $\mathcal{V}$ given either as the colimit (provided it exists), or the coend (provided it exists):

$$\begin{aligned}\int_M A \quad &:= \quad \mathsf{colim}\big(\mathcal{D}\mathsf{isk}_{n/M}^B \to \mathcal{D}\mathsf{isk}_n^B \xrightarrow{A} \mathcal{V}\big)\\ &\simeq \quad \mathbb{E}_M \bigotimes_{\mathcal{D}\mathsf{isk}_M^B} A.\end{aligned}$$

Remark 2.3.8. We comment on the two equivalent expressions defining factorization homology: as a colimit indexed by an overcategory and as a coend. This is analogous to the familiar fact that, for $X_\bullet\colon \Delta^{\mathsf{op}} \to \mathsf{Spaces}$ a simplicial space, its geometric realization can be expressed

$$|X_\bullet| \;:=\; \mathsf{colim}(\Delta_{/X_\bullet} \to \Delta \xrightarrow{\Delta^\bullet} \mathsf{Spaces}) \;\simeq\; X_\bullet \bigotimes_{\Delta} \Delta^\bullet$$

as a colimit of topological simplices indexed by the category of simplices in $X_\bullet$, or as the standard coend expression which is a quotient of $\coprod_{p\geq 0} X_p \times \Delta^p$.

The fully faithful symmetric monoidal functor $\iota\colon \mathcal{D}\mathsf{isk}_n^B \hookrightarrow \mathcal{M}\mathsf{fld}_n^B$ gives the restriction functor

$$\mathsf{Alg}_{\mathcal{D}\mathsf{isk}_n^B}(\mathcal{V}) := \mathsf{Fun}^{\otimes}(\mathcal{D}\mathsf{isk}_n^B, \mathcal{V}) \;\longleftarrow\; \mathsf{Fun}^{\otimes}(\mathcal{M}\mathsf{fld}_n^B, \mathcal{V}) : \iota^*.$$

The next result, proved in [3], identifies factorization homology as the values of a left adjoint to this restriction functor, provided $\mathcal{V}$ is $\otimes$-presentable.

Proposition 2.3.9. [3] *Let $\mathcal{V}$ be $\otimes$-presentable ∞-category. The restriction functor ι^* admits a left adjoint,*

$$\iota_!\colon \mathsf{Alg}_{\mathcal{D}\mathsf{isk}_n^B}(\mathcal{V}) \rightleftarrows \mathsf{Fun}^{\otimes}(\mathcal{M}\mathsf{fld}_n^B, \mathcal{V}) : \iota^*,$$

over a left adjoint $\iota_!\colon \mathsf{Fun}(\mathcal{D}\mathsf{isk}_n^B, \mathcal{V}) \rightleftarrows \mathsf{Fun}(\mathcal{M}\mathsf{fld}_n^B, \mathcal{V})\colon \iota^*$*. Furthermore, this left adjoint evaluates on a* $\mathcal{D}\mathsf{isk}_n^B$*-algebra* A *as factorization homology:*

$$\iota_!(A)\colon M \mapsto \int_M A.$$

Remark 2.3.10. Proposition 2.3.9 implies factorization homology can be expressed as a symmetric monoidal left Kan extension, at least when $\mathcal{V}$ is $\otimes$-presentable. Consequently, the definition of factorization homology given above is equivalent to an operadic left Kan extension (after parsing Definitions 3.1.1.2 and 3.1.2.2 of [43]), which is the definition of factorization homology, or topological chiral homology, given by Lurie [43] (Definition 5.5.2.6).

The next result explains how factorization homology transforms under change of tangential structure.

Proposition 2.3.11. *Let* $B \xrightarrow{\alpha} B'$ *be a map between spaces over* $\mathsf{BO}(n)$*. Let* $M = (M, \varphi)$ *be a* B*-framed* n*-manifold. Consider the* B'*-framed* n*-manifold* $\alpha M := (M, \alpha\varphi)$*. Let* A *be a* $\mathcal{D}\mathsf{isk}_n^{B'}$*-algebra* $A\colon \mathcal{D}\mathsf{isk}_n^{B'} \to \mathcal{V}$*. Consider the* $\mathcal{D}\mathsf{isk}_n^B$*-algebra* $\alpha A\colon \mathcal{D}\mathsf{isk}_n^B \to \mathcal{D}\mathsf{isk}_n^{B'} \xrightarrow{A} \mathcal{V}$*. The canonical morphism in* $\mathcal{V}$

$$\int_M \alpha A \xrightarrow{\simeq} \int_{\alpha M} A$$

is an equivalence.

Proof. Note the canonical commutative diagram among ∞-categories:

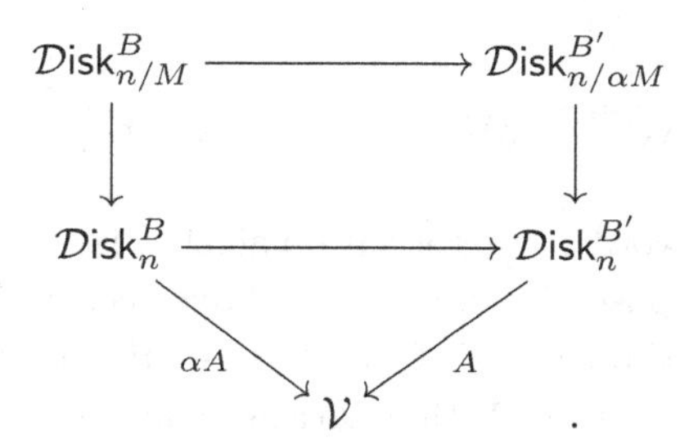

For formal reasons, the top horizontal functor is an equivalence between ∞-categories. It follows that the canonical morphism in $\mathcal{V}$

$$\int_M \alpha A = \mathsf{colim}\Big(\mathcal{D}\mathsf{isk}_{n/M}^B \to \mathcal{D}\mathsf{isk}_n^B \xrightarrow{\alpha A} \mathcal{V}\Big) \longrightarrow \mathsf{colim}\Big(\mathcal{D}\mathsf{isk}_{n/\alpha M}^{B'} \to \mathcal{D}\mathsf{isk}_n^{B'} \xrightarrow{A} \mathcal{V}\Big) = \int_{\alpha M} A$$

is an equivalence, as desired. □

2.3.2 Factorization algebras

We now use factorization homology to construct *factorization algebras* over each B-framed n-manifold, as in the sense used by Costello–Gwilliam (see Definition 2.3.12). Namely, Proposition 2.3.14 shows that $\mathcal{D}\mathsf{isk}_n^B$-algebra A in $\mathcal{V}$ determines a factorization algebra $\mathcal{F}_A$ on each B-framed n-manifold M whose value $\mathcal{F}_A(U)$ on an open subset $U \subset M$

is factorization homology of A,

$$\mathcal{F}_A(U) \simeq \int_U A,$$

over U (with its B-framing inherited from that of M).

For the next definition, for each n-manifold M, we observe the following multi-category structure on the poset $\mathsf{Opens}(M) := \mathsf{Mfld}_{n/M}$ of open subsets in M ordered by inclusion. Namely, an object is an open subset of M, and there is a multi-morphism from $(U_i)_{i\in I}$ to V, which is unique, provided the collection $\{U_i\}_{i\in I}$ is pairwise disjoint and provided $\bigcup_{i\in I} U_i \subset V$. Recall from §2.2.6 the definition of (locally constant) Weiss cosheaves.

Definition 2.3.12. [17] Let $\mathcal{V}$ be a symmetric monoidal ∞-category. Let M be an n-manifold. The ∞-category of ($\mathcal{V}$-valued) *factorization algebras* (on M) is the full ∞-subcategory of algebras in $\mathcal{V}$ over the multi-category $\mathsf{Opens}(M)$ as in the pullback among ∞-categories:

$$\begin{array}{ccc} \mathsf{Alg}_M(\mathcal{V}) & \longrightarrow & \mathsf{Alg}_{\mathsf{Opens}(M)}(\mathcal{V}) \\ \downarrow & & \downarrow \\ \mathsf{cShv}^{\mathsf{Weiss}}_{\mathcal{V}}(M) & \longrightarrow & \mathsf{Fun}(\mathsf{Opens}(M), \mathcal{V}). \end{array}$$

The ∞-category of *locally constant* factorization algebras is the full ∞-subcategory of factorization algebras as in the pullback diagram among ∞-categories:

$$\begin{array}{ccc} \mathsf{Alg}^{\mathsf{l.c.}}_M(\mathcal{V}) & \longrightarrow & \mathsf{Alg}_M(\mathcal{V}) \\ \downarrow & & \downarrow \\ \mathsf{cShv}^{\mathsf{Weiss,l.c.}}_{\mathcal{V}}(M) & \longrightarrow & \mathsf{cShv}^{\mathsf{Weiss}}_{\mathcal{V}}(M). \end{array}$$

Remark 2.3.13. In other words, a factorization algebra is a multi-functor $\mathcal{F}\colon \mathsf{Opens}(M) \to \mathcal{V}$ whose restriction to the poset $\mathsf{Opens}(M)$ is a Weiss cosheaf on M. Informally, a factorization algebra is likewise a functor $\mathcal{F}\colon \mathsf{Opens}(M) \to \mathcal{V}$ from the poset of open subsets of M to the underlying ∞-category of $\mathcal{V}$ that satisfies codescent with respect to Weiss covers, together with a system of compatible equivalences in $\mathcal{V}$: for each finite sequence $(U_i)_{i\in I}$ of pairwise disjoint open subsets of M, an equivalence $\mathcal{F}(\bigcup_{i\in I} U_i) \simeq \bigotimes_{i\in I} \mathcal{F}(U_i)$.

Proposition 2.3.14. *Let $\mathcal{V}$ be a $\otimes$-presentable ∞-category, and let B be a space over* $\mathsf{BO}(n)$. *For each B-framed n-manifold M, factorization homology defines a functor*

$$\mathsf{Alg}_{\mathcal{D}\mathsf{isk}^B_n}(\mathcal{V}) \longrightarrow \mathsf{Alg}^{\mathsf{l.c.}}_M(\mathcal{V}), \qquad A \mapsto \left(U \mapsto \int_U A \right)$$

from $\mathcal{D}\mathsf{isk}^B_n$-algebras to locally constant factorization algebras on M.

Proof. Notice the evident diagram among ∞-operads

$$\mathcal{D}\mathsf{isk}^B_n \xrightarrow{\iota} \mathcal{M}\mathsf{fld}^B_n \longleftarrow \mathsf{Mfld}^B_n \longleftarrow \mathsf{Opens}(M).$$

This diagram determines the diagram among ∞-categories:

$$\begin{array}{ccccccc}
\mathsf{Alg}_{\mathcal{D}\mathsf{isk}_n^B}(\mathcal{V}) & \xleftarrow{\iota^*} & \mathsf{Fun}^{\otimes}(\mathcal{M}\mathsf{fld}_n^B, \mathcal{V}) & \longrightarrow & \mathsf{Fun}^{\otimes}(\mathsf{Mfld}_n^B, \mathcal{V}) & \longrightarrow & \mathsf{Alg}_{\mathsf{Opens}(M)}(\mathcal{V}) \\
\downarrow{\scriptstyle \text{forget}} & & \downarrow{\scriptstyle \text{forget}} & & \downarrow{\scriptstyle \text{forget}} & & \downarrow{\scriptstyle \text{forget}} \\
\mathsf{Fun}(\mathcal{D}\mathsf{isk}_n^B, \mathcal{V}) & \xleftarrow{\iota^*} & \mathsf{Fun}(\mathcal{M}\mathsf{fld}_n^B, \mathcal{V}) & \longrightarrow & \mathsf{Fun}(\mathsf{Mfld}_n^B, \mathcal{V}) & \longrightarrow & \mathsf{Fun}\big(\mathsf{Opens}(M), \mathcal{V}\big).
\end{array}$$

Proposition 2.3.9 gives a commutative diagram involving left adjoints to each instance of ι^*:

$$\begin{array}{ccccccc}
\mathsf{Alg}_{\mathcal{D}\mathsf{isk}_n^B}(\mathcal{V}) & \xrightarrow{\iota_!} & \mathsf{Fun}^{\otimes}(\mathcal{M}\mathsf{fld}_n^B, \mathcal{V}) & \longrightarrow & \mathsf{Fun}^{\otimes}(\mathsf{Mfld}_n^B, \mathcal{V}) & \longrightarrow & \mathsf{Alg}_{\mathsf{Opens}(M)}(\mathcal{V}) \\
\downarrow{\scriptstyle \text{forget}} & & \downarrow{\scriptstyle \text{forget}} & & \downarrow{\scriptstyle \text{forget}} & & \downarrow{\scriptstyle \text{forget}} \\
\mathsf{Fun}(\mathcal{D}\mathsf{isk}_n^B, \mathcal{V}) & \xrightarrow{\iota_!} & \mathsf{Fun}(\mathcal{M}\mathsf{fld}_n^B, \mathcal{V}) & \longrightarrow & \mathsf{Fun}(\mathsf{Mfld}_n^B, \mathcal{V}) & \longrightarrow & \mathsf{Fun}\big(\mathsf{Opens}(M), \mathcal{V}\big).
\end{array}$$

Corollary 2.2.24 gives the factorization of the bottom horizontal sequence of functors:

$$\begin{array}{ccccccc}
\mathsf{Alg}_{\mathcal{D}\mathsf{isk}_n^B}(\mathcal{V}) & & \xrightarrow{\text{restrict} \circ \iota_!} & & & & \mathsf{Alg}_{\mathsf{Opens}(M)}(\mathcal{V}) \\
\downarrow{\scriptstyle \text{forget}} & & & & & & \downarrow{\scriptstyle \text{forget}} \\
\mathsf{Fun}(\mathcal{D}\mathsf{isk}_n^B, \mathcal{V}) & \xrightarrow{\iota_!} & \mathsf{cShv}_{\mathcal{V}}^{\mathsf{Weiss}}(\mathcal{M}\mathsf{fld}_n^B) & \longrightarrow & \mathsf{cShv}_{\mathcal{V}}^{\mathsf{Weiss,l.c.}}(\mathsf{Mfld}_n^B) & \longrightarrow & \mathsf{cShv}_{\mathcal{V}}^{\mathsf{Weiss,l.c.}}(M).
\end{array}$$

The result follows. □

Remark 2.3.15. Proposition 2.3.14 grants the commutative diagram among ∞-categories:

$$\begin{array}{ccc}
\mathsf{Alg}_{\mathcal{D}\mathsf{isk}_n}(\mathcal{V}) & \xrightarrow[\text{Prop 2.3.14}]{A \mapsto \mathcal{F}_A} & \mathsf{Alg}_M(\mathcal{V}) \\
& {\scriptstyle \int_M} \searrow \quad \swarrow {\scriptstyle \text{global cosections}} & \\
& \mathcal{V} & .
\end{array}$$

In other words, for $\mathcal{F}_A$ the factorization algebra determined by the $\mathcal{D}\mathsf{isk}_n$-algebra A, its global cosections is factorization homology:

$$\int_M A \;\simeq\; \mathcal{F}_A(M).$$

In the case that M is equipped with a framing, A need only be a $\mathcal{E}_n$-algebra (see Remark 2.3.6).

Terminology 2.3.16. For X a stratified space, a factorization algebra $\mathcal{F}$ on X is *constructible* if, for each stratum $X_p \subset X$ of X, the restricted factorization algebra $\mathcal{F}_{|X_p}$ on X_p is locally constant.

Remark 2.3.17. Let X be a stratified space. Consider the ∞-category $\mathcal{B}_X$ of singularity-types in X, and stratified open embeddings among them. Consider the symmetric monoidal ∞-category $\mathcal{D}\mathsf{isk}(\mathcal{B}_X)$ in which an object is a finite-fold disjoint union of objects in $\mathcal{B}_X$, and a morphism between them is a stratified open embedding. Established in [8] is a similar construction to that of Proposition 2.3.14 as it concerns symmetric monoidal functors from $\mathcal{D}\mathsf{isk}(\mathcal{B}_X)$ to constructible factorization algebras on X. See that reference for a thorough development, and §2.8 of this article for a synopsis.

2.3.3 Factorization homology over oriented 1-manifolds with boundary

We show that factorization homology over a closed interval is a two-sided bar construction.

Recall from §2.2.8 the symmetric monoidal ∞-category $\mathcal{M}\mathsf{fld}_1^\partial$ and its symmetric monoidal full ∞-subcategory $\mathcal{D}\mathsf{isk}_1^\partial$ consisting of finite disjoint unions of Euclidean and half-Euclidean spaces. In this subsection we consider, similarly, the symmetric monoidal ∞-category and its symmetric monoidal full ∞-subcategory,

$$\mathcal{D}\mathsf{isk}_1^{\partial,\mathsf{or}} \subset \mathcal{M}\mathsf{fld}_1^{\partial,\mathsf{or}}.$$

An object in the latter is a 1-manifold with boundary equipped with an orientation, while such an object belongs to the smaller if each connected component of its underlying 1-manifold with boundary is diffeomorphic to Euclidean space or half-Euclidean space. The space of morphisms between two such objects therein is the space of smooth open embeddings that preserve orientations, equipped with the compact-open C^∞ topology. The symmetric monoidal structure is disjoint union.

Remark 2.3.18. The 1-manifold with boundary $[-1,1]$, equipped with its standard orientation determined by the non-vanishing vector field ∂_t, is an object in $\mathcal{M}\mathsf{fld}_1^{\partial,\mathsf{or}}$ that does *not* belong to $\mathcal{D}\mathsf{isk}_1^{\partial,\mathsf{or}}$.

We next articulate a sense in which $\mathcal{D}\mathsf{isk}_1^{\partial,\mathsf{or}}$ is entirely combinatorial.

Definition 2.3.19. [8] $\mathsf{Assoc}^{\mathsf{RL}}$ is the ∞-operad corepresenting triples $(A;P,Q)$ consisting of an associative algebra together with a unital right and a unital left module. Specifically, it is a unital multi-category whose space of colors is the three-element set $\{M,R,L\}$, and with spaces of multi-morphisms given as follows. Let $I \xrightarrow{\sigma} \{M,R,L\}$ be a map from a finite set.

- $\mathsf{Assoc}^{\mathsf{RL}}(\sigma,M)$ is the set of linear orders on I for which no element is related to an element of $\sigma^{-1}(\{R,L\})$. In other words, should $\sigma^{-1}(\{R,L\})$ be empty, then there is one multi-morphism from σ to M for each linear order on $\sigma^{-1}(M)$; should $\sigma^{-1}(\{R,L\})$ not be empty, then there are no multi-morphisms from σ to M.

- $\mathsf{Assoc}^{\mathsf{RL}}(\sigma,L)$ is the set of linear orders on I for which each element of $\sigma^{-1}(L)$ is a minimum, and no element is related to an element in $\sigma^{-1}(R)$. In other words, should $\sigma^{-1}(\{R\})$ be empty and $\sigma^{-1}(\{L\})$ have cardinality at most 1, then there is one multi-morphism from σ to M for each linear order on $\sigma^{-1}(M)$; should $\sigma^{-1}(\{R\})$ not be empty or $\sigma^{-1}(\{L\})$ have cardinality greater than 1, then there are no multi-morphisms from σ to M.

- $\mathsf{Assoc}^{\mathsf{RL}}(\sigma, R)$ is the set of linear orders on I for which each element of $\sigma^{-1}(R)$ is a maximum, and no element is related to an element in $\sigma^{-1}(L)$. In other words, should $\sigma^{-1}(\{L\})$ be empty and $\sigma^{-1}(\{R\})$ have cardinality at most 1, then there is one multi-morphism from σ to M for each linear order on $\sigma^{-1}(M)$; should $\sigma^{-1}(\{L\})$ not be empty or $\sigma^{-1}(\{R\})$ have cardinality greater than 1, then there are no multi-morphisms from σ to M.

Composition of multi-morphisms is given by concatenating linearly ordered sets.

The next result references the *symmetric monoidal envelope* of the colored operad $\mathsf{Assoc}^{\mathsf{RL}}$. It is initial among symmetric monodal ∞-categories equipped with an $\mathsf{Assoc}^{\mathsf{RL}}$-algebra. In other words, it is a symmetric monoidal ∞-category $\mathsf{Env}(\mathsf{Assoc}^{\mathsf{RL}})$ corepresenting the copresheaf on the ∞-category of symmetric monoidal ∞-categories

$$\mathsf{Map}^{\otimes}\Big(\mathsf{Env}(\mathsf{Assoc}^{\mathsf{RL}}), -\Big)\colon \mathsf{Cat}_{\infty}{}^{\otimes} \longrightarrow \mathsf{Spaces}, \qquad \mathcal{V} \mapsto \big(\mathsf{Alg}_{\mathsf{Assoc}^{\mathsf{RL}}}(\mathcal{V})\big)^{\sim},$$

whose value on a symmetric monoidal ∞-category is the moduli space of $\mathsf{Assoc}^{\mathsf{RL}}$-algebras in it.

Lemma 2.3.20. *Taking connected components defines an equivalence between symmetric monoidal ∞-categories,*

$$[-]\colon \mathcal{D}\mathsf{isk}_1^{\partial,\mathsf{or}} \xrightarrow{\simeq} \mathsf{Env}(\mathsf{Assoc}^{\mathsf{RL}}) \tag{2.3.1}$$

to the symmetric monoidal envelope, where the values on the symmetric monoidal generators are: $[\mathbb{R}] = M$, $[\mathbb{R}_{\geq 0}] = R$, and $[\mathbb{R}_{\leq 0}] = L$.

Proof. Evidently, this defines a symmetric monoidal functor. We now show that it is an equivalence. Because it is so on symmetric monoidal generators, (2.3.1) is essentially surjective on spaces of objects. It remains to show that (2.3.1) is fully faithful. So let U and V be objects in $\mathcal{D}\mathsf{isk}_1^{\partial,\mathsf{or}}$. We must show that the map between spaces

$$\mathsf{Map}_{\mathcal{D}\mathsf{isk}_1^{\partial,\mathsf{or}}}(U,V) \longrightarrow \mathsf{Map}_{\mathsf{Env}(\mathsf{Assoc}^{\mathsf{RL}})}([U],[V]) \tag{2.3.2}$$

is an equivalence. For $V = \bigsqcup_{\alpha\in[V]} V_\alpha$ the partition as connected components, direct inspection of the definition of these two symmetric monoidal ∞-categories yields an identification of this map is as the $[V]$-indexed product of such maps

$$\begin{aligned}
\mathsf{Map}_{\mathcal{D}\mathsf{isk}_1^{\partial,\mathsf{or}}}(U,V) \quad &\xrightarrow{\simeq} \quad \coprod_{[U]\xrightarrow{f}[V]} \prod_{\alpha\in[V]} \mathsf{Map}_{\mathcal{D}\mathsf{isk}_1^{\partial,\mathsf{or}}}(U_{|f^{-1}\alpha}, V_\alpha) \\
&\longrightarrow \quad \coprod_{[U]\xrightarrow{f}[V]} \prod_{\alpha\in[V]} \mathsf{Map}_{\mathsf{Env}(\mathsf{Assoc}^{\mathsf{RL}})}(f^{-1}\alpha, [V_\alpha]) \\
&\xleftarrow{\simeq} \quad \mathsf{Map}_{\mathsf{Env}(\mathsf{Assoc}^{\mathsf{RL}})}([U],[V]).
\end{aligned}$$

We are therefore reduced to the case that V is non-empty and connected. By direct inspection, the space of morphisms in $\mathcal{D}\mathsf{isk}_1^{\partial,\mathsf{or}}$ from U to V is a 0-type. So we are left to show that the map (2.3.2) is a bijection (on connected components). There are three cases to consider.

- In the case that $V \cong (-1,1)$ is oriented-diffeomorphic to an open interval, this 0-type is empty if U has non-empty boundary, and otherwise it is the set of linear orders on the set $[U]$ of connected components.

- In the case that $V \cong [-1,1)$ is oriented-diffeomorphic to a left-closed/right-open interval, this 0-type is empty if U has non-empty outward-pointing boundary, and otherwise it is the set of linear orders on $[U]$ for which each connected component with inward-pointing boundary is a minimum.

- In the case that $V \cong (-1,1]$ is oriented-diffeomorphic to a left-open/right-closed interval, this 0-type is empty if U has non-empty inward-pointing boundary, and otherwise it is the set of linear orders on $[U]$ for which each connected component with outward-pointing boundary is a maximum.

Inspecting Definition 2.3.19, and the description of the symmetric monoidal functor $[-]$ under examination, reveals that the map (2.3.2) between 0-types is an equivalence, as desired. □

Example 2.3.21. Note the functor $\mathsf{Alg}^{\mathsf{aug}}_{\mathsf{Assoc}}(\mathcal{V}) \to \mathsf{Alg}_{\mathsf{Assoc}^{\mathsf{RL}}}(\mathcal{V})$ from augmented associative algebras, given by $(A \to \mathbb{1}) \mapsto (A; \mathbb{1}, \mathbb{1})$. Concatenating with the equivalence of Lemma 2.3.20 results in a functor $\mathsf{Alg}^{\mathsf{aug}}_{\mathsf{Assoc}}(\mathcal{V}) \to \mathsf{Fun}^{\otimes}(\mathcal{D}\mathsf{isk}_1^{\partial,\mathsf{or}}, \mathcal{V})$.

Recall from §2.2.8 the ∞-operad structure on $\mathcal{D}\mathsf{isk}_{n/M}$.

Observation 2.3.22. Consider the ordinary category O^{RL} in which an object is a finite linearly ordered set $(I, \leq)$ together with a pair of disjoint subsets $R \subset I \supset L$ for which each element in R is a minimum and each element in L is a maximum; and in which a morphism $(I, \leq, R, L) \to (I', \leq', R', L')$ is an order preserving map $I \xrightarrow{f} I'$ for which $f(R \sqcup L) \subset R' \sqcup L'$. Concatenating linear orders endows O^{RL} with the structure of a multi-category. Note the evident forgetful morphism between multi-categories $\mathsf{O}^{\mathsf{RL}} \to \mathsf{Env}(\mathsf{Assoc}^{\mathsf{RL}})$. By direct inspection, the equivalence (2.3.1) lifts to an equivalence between ∞-operads:

$$\begin{array}{ccc} \mathcal{D}\mathsf{isk}^{\partial,\mathsf{or}}_{1/[-1,1]} & \xrightarrow[{[-]}]{\simeq} & \mathsf{O}^{\mathsf{RL}} \\ \Big\downarrow{\scriptstyle \mathrm{forget}} & & \Big\downarrow \\ \mathcal{D}\mathsf{isk}^{\partial,\mathsf{or}}_1 & \xrightarrow[{[-]}]{\simeq} & \mathsf{Env}(\mathsf{Assoc}^{\mathsf{RL}}). \end{array} \tag{2.3.3}$$

Corollary 2.3.23. *Let $(A; P, Q)$ be an $\mathsf{Assoc}^{\mathsf{RL}}$-algebra in $\mathcal{V}$; which is to say an associative algebra A together with a unital left and a unital right A-module. Through Lemma 2.3.20, regard $(A; P, Q)$ as a symmetric monoidal functor $(A; P, Q)\colon \mathcal{D}\mathsf{isk}_1^{\partial,\mathsf{or}} \to \mathcal{V}$. There is a canonical equivalence in $\mathcal{V}$ from the balanced tensor product to factorization homology over the closed interval:*

$$Q \underset{A}{\otimes} P \xrightarrow{\simeq} \int_{[-1,1]} (A; P, Q).$$

Proof. Recognize the opposite of the simplex category as the full subcategory $\Delta^{\mathsf{op}} \subset \mathsf{O}^{\mathsf{RL}}$ consisting of those objects $(I, \leq, R, L)$ for which $R \neq \emptyset \neq L$. Adjoining minima and maxima gives a left adjoint $\mathsf{O}^{\mathsf{RL}} \to \Delta^{\mathsf{op}}$ to the inclusion. Therefore, the inclusion $\Delta^{\mathsf{op}} \to \mathsf{O}^{\mathsf{RL}}$ is final. Concatenating this final functor with the equivalence of Observation 2.3.22 results in a final functor $\mathbf{\Delta}^{\mathsf{op}} \to \mathcal{D}\mathsf{isk}^{\partial,\mathsf{or}}_{1/[-1,1]}$. By inspection, the resulting simplicial object

$$\mathsf{Bar}_\bullet(Q,A,P)\colon \mathbf{\Delta}^{\mathsf{op}} \to \mathcal{D}\mathsf{isk}^{\partial,\mathsf{or}}_{1/[-1,1]} \to \mathcal{D}\mathsf{isk}^{\partial,\mathsf{or}}_1 \xrightarrow{[-]} \mathsf{Env}(\mathsf{Assoc}^{\mathsf{RL}}) \xrightarrow{(A;P,Q)} \mathcal{V}$$

is identified as the two-sided bar construction, as indicated. We conclude the equivalence in $\mathcal{V}$:

$$Q \underset{A}{\otimes} P \simeq \left|\mathsf{Bar}_\bullet(Q,A,P)\right| \xrightarrow{\simeq} \mathsf{colim}\left(\mathbf{\Delta}^{\mathsf{op}} \xrightarrow{\mathsf{Bar}_\bullet(Q,A,P)} \mathcal{V}\right) \xrightarrow{\simeq} \int_{[-1,1]} (A;P,Q).$$

□

Note that the given multi-functor $\mathsf{O}^{\mathsf{RL}} \to \mathsf{Env}(\mathsf{Assoc}^{\mathsf{RL}})$ witness an equivalence $\mathsf{Env}(\mathsf{O}^{\mathsf{RL}}) \xrightarrow{\simeq} \mathsf{Env}(\mathsf{Assoc}^{\mathsf{RL}})$ between symmetric monoidal ∞-categories. After Observation 2.3.22, this offers the following consequence, which refers to Terminology 2.3.16.

Corollary 2.3.24. *Factorization homology defines an equivalence of ∞-categories,*

$$\int\colon \mathsf{Alg}_{\mathsf{Assoc}^{\mathsf{RL}}}(\mathcal{V}) \xrightarrow{\simeq} \mathsf{Alg}_{\mathcal{D}\mathsf{isk}^{\partial,\mathsf{or}}_{1/[-1,1]}}(\mathcal{V}),$$

between associative algebra equipped with a unital left and right module and constructible factorization algebras over the closed interval.

2.3.4 Homology theories: definition

We now define homology theories for B-framed n-manifolds.

Definition 2.3.25 (Collar-gluing)**.** Let M be a B-framed n-manifold. A *collar-gluing* of M is a continuous map

$$f\colon M \to [-1,1]$$

to the closed interval for which the restriction $f_|\colon M_{|(-1,1)} \to (-1,1)$ is a smooth fiber bundle. We will often denote a collar-gluing $M \xrightarrow{f} [-1,1]$ simply as the open cover

$$M_- \bigcup_{M_0\times\mathbb{R}} M_+ \cong M,$$

where $M_- = f^{-1}[-1,1)$ and $M_+ = f^{-1}(-1,1]$ and $M_0 = f^{-1}\{0\}$.

Remark 2.3.26. We think of a collar-gluing of M as a codimension-1 properly embedded submanifold $M_0 \subset M$ whose complement is partitioned by connected components: $M \smallsetminus M_0 = M_- \sqcup M_+$. Such data is afforded by gluing two manifolds with boundary along a common boundary. The actual data of a collar-gluing specifies that named just above, in addition to a bi-collaring of the common boundary.

Construction 2.2.40 and the results of §2.3.3 give the following.

Corollary 2.3.27. *Let $\mathcal{F}\colon \mathcal{M}\mathsf{fld}_n^B \longrightarrow \mathcal{V}$ be a symmetric monoidal functor. Let M be a B-framed n-manifold. A collar-gluing $M_- \bigcup_{M_0\times\mathbb{R}} M_+ \cong M$ determines an associative algebra $\mathcal{F}(M_0)$ together with a unital left module structure on the object $\mathcal{F}(M_+)$ and a unital right module structure on $\mathcal{F}(M_-)$, as well as a morphism in $\mathcal{V}$:*

$$\mathcal{F}(M_-) \bigotimes_{\mathcal{F}(M_0)} \mathcal{F}(M_+) \longrightarrow \mathcal{F}(M). \tag{2.3.4}$$

Proof. The collar-gluing is a continuous map $M \xrightarrow{f} [-1,1]$. Construction 2.2.40 gives the the first morphism in the composible sequence of morphisms among of ∞-operads:

$$f_*\mathcal{F}\colon \mathcal{D}\mathsf{isk}^{\partial,\mathsf{or}}_{1/[-1,1]} \xrightarrow{f^{-1}} \mathcal{M}\mathsf{fld}^B_{n/M} \to \mathcal{M}\mathsf{fld}^B_n \xrightarrow{\mathcal{F}} \mathcal{V}.$$

Through Corollary 2.3.24, this morphism between ∞-operads is equivalent to an algebra in $\mathcal{V}$ over the operad $\mathsf{Assoc}^{\mathsf{RL}}$. Unwinding that equivalence reveals that its underlying associative algebra is $\mathcal{F}(M_0)$, whose underlying object is the value $\mathcal{F}(M_0\times\mathbb{R})$ and whose associative algebra structure is given by oriented embeddings among the $\mathbb{R}$-coordinate, and its underlying unital modules are the values $\mathcal{F}(M_\pm)$. Furthermore, the definition of factorization homology as a colimit supplies the canonical morphism in $\mathcal{V}$:

$$\mathcal{F}(M_-) \bigotimes_{\mathcal{F}(M_0)} \mathcal{F}(M_+) \underset{\text{Cor } 2.3.23}{\simeq} \int_{[-1,1]} f_*\mathcal{F} \longrightarrow \mathcal{F}(f^{-1}([-1,1])) = \mathcal{F}(M).$$

□

Definition 2.3.28. A symmetric monoidal functor $\mathcal{F} : \mathcal{M}\mathsf{fld}_n^B \to \mathcal{V}$ satisfies $\otimes$-*excision* if, for each collar-gluing $M_- \bigcup_{M_0\times\mathbb{R}} M_+ \cong M$ of B-framed n-manifolds, the canonical morphism in $\mathcal{V}$,

$$\mathcal{F}(M_-) \bigotimes_{\mathcal{F}(M_0\times\mathbb{R})} \mathcal{F}(M_+) \underset{(2.3.4)}{\xrightarrow{\simeq}} \mathcal{F}(M),$$

is an equivalence. The ∞-category of $\mathcal{V}$-valued *homology theories* for B-framed n-manifolds is the full ∞-subcategory

$$\mathbf{H}(\mathcal{M}\mathsf{fld}_n^B, \mathcal{V}) \subset \mathsf{Fun}^{\otimes}(\mathcal{M}\mathsf{fld}_n^B, \mathcal{V})$$

consisting of those symmetric monoidal functors that satisfy $\otimes$-excision.

Remark 2.3.29. We emphasize that a $\mathcal{V}$-valued homology theory depends on the symmetric monoidal structure $\otimes$ of $\mathcal{V}$. For instance, let $\Bbbk$ a field and consider the ∞-category $\mathsf{Mod}_\Bbbk$ of $\Bbbk$-modules. There are two natural symmetric monoidal structures on $\mathsf{Mod}_\Bbbk$: direct sum $\oplus$; tensor product $\otimes_\Bbbk$. As established in §2.5.1, a homology theory valued in $(\mathsf{Mod}_\Bbbk, \oplus)$ evaluates on an n-manifold M as the $\Bbbk$-chains, $\mathsf{C}_*(M;\Bbbk)$. On the other hand, in §2.5.2, specifically Remark 2.5.17, shows that a homology theory valued in $(\mathsf{Mod}_\Bbbk, \otimes_\Bbbk)$ typically does *not* factor through the forgetful functor $\mathcal{M}\mathsf{fld}_n \to \mathsf{Spaces}$ to the underlying homotopy type of manifolds.

2.3.5 Pushforward

We prove that factorization homology satisfies $\otimes$-excision in the sense of Definition 2.3.28. We realize this as an instance of a general construction of a pushforward.

The following result is the technical crux of the later results of this article; through this result one can access the values of factorization homology. We state the the result now, and prove it at the end of this section.

Lemma 2.3.30. ([21], [3]) *Let M be a B-framed n-manifold, and let A be a $\mathcal{D}\mathsf{isk}_n^B$-algebra in $\mathcal{V}$, for $\mathcal{V}$ a $\otimes$-presentable ∞-category. For each collar-gluing $M_- \bigcup_{M_0\times\mathbb{R}} M_+ \cong M$, the canonical morphism in $\mathcal{V}$,*

$$\int_{M_-} A \bigotimes_{\int_{M_0\times\mathbb{R}} A} \int_{M_+} A \xrightarrow[(2.3.4)]{\simeq} \int_M A$$

is an equivalence.

Lemma 2.3.30 specializes to to the following special case of $n = 1$, $B \simeq *$, and $M = S^1$, which demonstrates the utility of $\otimes$-excision.

Corollary 2.3.31. ([3], [43]) *Let A be an associative algebra in a $\otimes$-presentable ∞-category $\mathcal{V}$. There is a canonical equivalence between objects in $\mathcal{V}$:*

$$\mathsf{HH}_\bullet(A) \simeq \int_{S^1} A$$

between the Hochschild homology of A and factorization homology of A over the circle.

Proof. Consider the collar-gluing $\mathbb{R}^{\mathsf{op}} \bigcup_{\mathbb{R}^{\mathsf{op}}\sqcup\mathbb{R}} \mathbb{R} \cong S^1$, where the 'op' superscripts denote the opposite orientation from the standard orientation on Euclidean space, which is the map $\mathsf{prom}\colon S^1 \to [-1,1]$ which projects unit vectors in $\mathbb{R}^2$ to the first coordinate. Lemma 2.3.30, applied to this collar-gluing, states the last of the equivalences in the sequence of equivalences in $\mathcal{V}$:

$$\mathsf{HH}_\bullet(A) \simeq A \bigotimes_{A\otimes A^{\mathsf{op}}} A \simeq \int_{\mathbb{R}} A \bigotimes_{\int_{S^0\times\mathbb{R}} A} \int_{\mathbb{R}} A \xrightarrow{\simeq} \int_{S^1} A.$$

The middle equivalence is by inspecting values, while the first is definitional. □

The next definition makes use of the multi-functor $f^{-1}\colon \mathcal{D}\mathsf{isk}_{k/N}^{\partial} \to \mathcal{M}\mathsf{fld}_{n/M}^B$ of Construction 2.2.40 associated to each continuous map $M \to N$ for which each of the restrictions, $M_{|N\smallsetminus\partial N} \to N \smallsetminus \partial N$ and $M_{|\partial N} \to \partial N$, are smooth manifold bundles.

Definition 2.3.32. [3] Let M be an B-framed n-manifold, and let N be an oriented k-manifold, possibly with boundary. For $f : M \to N$ a map such that the restrictions of f over both the interior of N and the boundary of N are fiber bundles, then the ∞-category

$\mathcal{D}\mathsf{isk}_f$ is the limit of the diagram among ∞-categories

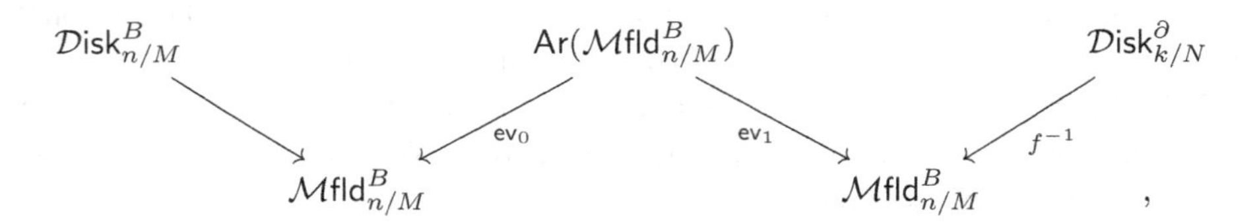

where $\mathsf{Ar}(\mathcal{M}\mathsf{fld}^B_{n/M})$ is the ∞-category of functors $[1] \to \mathcal{M}\mathsf{fld}^B_{n/M}$.

Remark 2.3.33. Informally, an object in the ∞-category $\mathcal{D}\mathsf{isk}_f$ is a triple (V, U, β) for which: U is an open subset of N that is diffeomorphic to a disjoint union of Euclidean spaces; V is an open submanifold of M that is diffeomorphic to a disjoint union of Euclidean spaces; and β is an isotopy among embeddings of V into M from the given inclusion to one that factors through $f^{-1}U \hookrightarrow M$.

The following result is stated and proved in §3 of [3]. Its proof amounts to a partition-of-unity argument (specifically Theorem 1.1 [19] or Theorem A.3.1 [43]), applied to unordered configuration spaces.

Lemma 2.3.34. [3] *In the situation of Definition 2.3.32, the functor*

$$\mathsf{ev}_0 : \mathcal{D}\mathsf{isk}_f \to \mathcal{D}\mathsf{isk}^B_{n/M}, \qquad (U, V, \beta) \mapsto V,$$

is final.

Lemma 2.3.34 applied to the fold map $M \sqcup M \to M$ gives the following important technical property to the ∞-category of disks in M.

Corollary 2.3.35 (Corollary 3.22 in [3], Proposition 5.5.2.16 in [43]). *For M a B-framed n-manifold, the ∞-overcategory $\mathcal{D}\mathsf{isk}^B_{n/M}$ is sifted.*

We now apply Lemma 3.32 to factorization homology in the next result. In the statement of this result we use the notation

$$f_*A : \mathcal{D}\mathsf{isk}^\partial_{k/N} \xrightarrow{f^{-1}} \mathcal{M}\mathsf{fld}^B_{n/M} \xrightarrow{\int A} \mathcal{V}$$

for the composite functor, where f^{-1} is as in Construction 2.2.40.

Proposition 2.3.36 (Proposition 3.23 in [3]). *Fix a map of spaces $B \to \mathsf{BO}(n)$. Let M be a B-framed manifold, and let A be a $\mathcal{D}\mathsf{isk}^B_n$-algebra in a $\otimes$-presentable ∞-category $\mathcal{V}$. Let $f\colon M \to N$ be a continuous map to a smooth k-manifold with boundary whose restriction over the boundary and interior of N is a smooth fiber bundle. There is a canonical map in $\mathcal{V}$,*

$$\int_N f_*A := \mathsf{colim}\Big(\mathcal{D}\mathsf{isk}^\partial_{k/N} \xrightarrow{f_*A} \mathcal{V}\Big) \xrightarrow{\simeq} \int_M A,$$

which is an equivalence.

Proof of Lemma 2.3.30. Let $f : M \to [-1,1]$ be a collar-gluing. There are canonical morphisms in $\mathcal{V}$

$$\int_{M_-} A \bigotimes_{\int_{M_0\times\mathbb{R}} A} \int_{M_+} A \xrightarrow[\text{Cor } 2.3.23]{\simeq} \int_{[-1,1]} f_* A \xrightarrow[\text{Prop } 2.3.36]{\simeq} \int_M A.$$

Corollary 2.3.23 states that the first arrow is an equivalence. Proposition 2.3.36 states that the second morphism is an equivalence. □

2.3.6 Homology theories: characterization

We characterize factorization homology, analogously to Eilenberg–Steenrod's characterization of homology theories for spaces.

Theorem 2.3.37. [3] *Let $\mathcal{V}$ be $\otimes$-presentable ∞-category. Let $B \to \mathsf{BO}(n)$ be a map from a space. There is a canonical equivalence between ∞-categories,*

$$\int : \mathsf{Alg}_{\mathcal{D}\mathsf{isk}_n^B}(\mathcal{V}) \overset{\simeq}{\rightleftarrows} \mathbf{H}(\mathcal{M}\mathsf{fld}_n^B, \mathcal{V}) : \mathsf{ev}_{\mathbb{R}^n},$$

between $\mathcal{D}\mathsf{isk}_n^B$-algebras in $\mathcal{V}$ and homology theories for B-framed n-manifolds with coefficients in $\mathcal{V}$. This equivalence is implemented by the factorization homology functor $\int$ and the functor of evaluation on $\mathbb{R}^n$.

Proof. Proposition 2.3.9 identifies factorization homology as the values of symmetric monoidal left Kan extension, thereby implementing the left adjoint in an adjunction

$$i_! : \mathsf{Alg}_{\mathcal{D}\mathsf{isk}_n^B}(\mathcal{V}) \rightleftarrows \mathsf{Fun}^{\otimes}(\mathcal{M}\mathsf{fld}_n^B, \mathcal{V}) : i^*.$$

The unit of this adjunction is an equivalence because $\mathcal{D}\mathsf{isk}_n^B \to \mathcal{M}\mathsf{fld}_n^B$ is fully faithful, and Kan extension along a fully faithful functor restricts as the original functor.

Now let $\mathcal{F}\colon \mathcal{M}\mathsf{fld}_n^B \to \mathcal{V}$ be a symmetric monoidal functor. The counit of this adjunction evaluates on $\mathcal{F}$ as a morphism $\int A \to \mathcal{F}$, where $A = \mathcal{F}_{|\mathbb{R}^n}$ is the $\mathcal{D}\mathsf{isk}_n^B$-algebra defined by the values of $\mathcal{F}$ on disjoint unions of B-framed Euclidean n-spaces. It remains to verify that this counit is an equivalence. To that end, consider the full ∞-subcategory $\mathcal{M} \subset \mathcal{M}\mathsf{fld}_n^B$ consisting of those objects M for which this counit $\int_M A \xrightarrow{\simeq} \mathcal{F}(M)$ is an equivalence. We wish to show the inclusion $\mathcal{M} \hookrightarrow \mathcal{M}\mathsf{fld}_n^B$ is an equivalence.

Both $\int A$ and $\mathcal{F}$ are symmetric monoidal functors. So the full ∞-subcategory $\mathcal{M} \subset \mathcal{M}\mathsf{fld}_n^B$ is symmetric monoidal. By definition, $\mathcal{M}$ contains B-framed Euclidean spaces. We conclude the containment $\mathcal{D}\mathsf{isk}_n^B \subset \mathcal{M}$.

Now, for each $0 \leq k \leq n$, consider the base change B_k of $B \to \mathsf{BO}(n)$ along the map $\mathsf{BO}(k) \times \mathsf{BO}(n-k) \xrightarrow{\oplus} \mathsf{BO}(n)$ then projected $B_k \to \mathsf{BO}(k) \times \mathsf{BO}(n-k) \xrightarrow{\mathsf{proj}} \mathsf{BO}(k)$. We will show, by induction on k, that $\mathcal{M}$ contains the image of the evident functor $\mathcal{M}\mathsf{fld}_k^{B_k} \to \mathcal{M}\mathsf{fld}_n^B$. The previous paragraph gives that this is the case for $k = 0$.

By assumption, $\mathcal{F}$ satisfies $\otimes$-excision, while Lemma 2.3.30 states that $\int A$ satisfies $\otimes$-excision as well. We conclude that $\mathcal{M}$ is closed under collar-gluings. Therefore $\mathcal{M}$

contains the image of any B_k-framed k-manifold that can be witnessed by a finite iteration of collar-gluings from disjoint unions of B_{k-1}-framed $(k-1)$-manifolds. By induction on k, we are therefore reduced to showing each B_k-framed k-manifold can be witnessed as a finite iteration of collar-gluings involving B_k-framed k-manifolds in the image of $\mathcal{M}\mathsf{fld}_{k-1}^{B_{k-1}} \to \mathcal{M}\mathsf{fld}_k^{B_k} \to \mathcal{M}\mathsf{fld}_n^{B}$.

Well, the reigning assumption that manifolds be finitary guarantees, for each B_k-framed k-manifold M, the existence of a compactification $\overline{M}$ of the underlying manifold of M as a smooth k-manifold with boundary. Choose a proper Morse function $f\colon \overline{M} \to [-R, R]$ for which $f^{-1}(\{\pm R\}) = \partial\overline{M}$. Such a Morse function witnesses M as a finite collar-gluing involving B_k-framed k-manifolds of whose underlying manifold is diffeomorphic to $M_0 \times \mathbb{R}$ for some $(k-1)$-manifold M_0. Such a $(k-1)$-manifold is therefore isomorphic to one that is equipped with a B_{k-1}-framing. □

Remark 2.3.38. In [8] a version of Theorem 2.3.37 is established for structured singular manifolds equipped with a stratified tangential structure – this will be surveyed in §2.8. The result, as it appears there, is slightly more general than Theorem 2.3.37 above: the assumption that $\mathcal{V}$ be $\otimes$-presentable can be weakened to the assumption that the underlying ∞-category of $\mathcal{V}$ is presentable yet the symmetric monoidal structure distributes only over geometric realizations and filtered colimits (i.e., over sifted colimits).

Remark 2.3.39. In [3] a version of Theorem 2.3.37 is established for topological manifolds in place of smooth manifolds. In essence, the only feature of the smooth category used in the proof of Theorem 2.3.37 was the existence of open handlebody decompositions, granted by Morse functions. But such open handlebody decompositions are guaranteed in the topological setting, through the results of [48] for $n = 3$, of [50] as it concerns the complement of a point for $n = 4$ together with smoothing theory [32], of [50] for $n = 5$, and of [32] for $n > 5$.

For much of this article, we will primarily concern ourselves with homology theories in the sense of Definition 2.3.28. We point out, however, that there are very interesting functors symmetric monoidal functors $\mathcal{M}\mathsf{fld}_n \to \mathcal{V}$ that are not $\otimes$-excisive. For instance, as detailed in [12], for X a derived (commutative) stack over a fixed field $\Bbbk$, cotensoring to X followed by taking ring of functions defines a symmetric monoidal functor:

$$\mathcal{M}\mathsf{fld}_n \xrightarrow{M \mapsto X^M} \mathsf{Stacks}(\Bbbk) \xrightarrow{\mathcal{O}} \mathsf{Mod}_{\Bbbk}\,.$$

As the case $n = 1$, the value of this functor on the circle is the Hochschild homology of X: $\mathcal{O}(X^{S^1}) \simeq \mathsf{HH}_\bullet(X)$ (compare with Corollary 2.3.31). If X is not affine, the above functor will generically fail to be $\otimes$-excisive. Now, the cotensor X^M only depends on the underlying homotopy type of M, as we shall see in Proposition 2.5.7. This is a feature of X being a derived *commutative* stack. Should X be a derived stack over $\mathcal{D}\mathsf{isk}_n$-algebras as in [20], there is a refined version of the cotensor X^M that is sensitive to the manifold structure of M. Derived stacks over $\mathcal{D}\mathsf{isk}_n$-algebras arise as the outcome of deformation quantization of a shifted symplectic derived (commutative) stacks, such as in Rozansky–Witten theory.

2.4 Nonabelian Poincaré duality

Applying Theorem 2.3.37, we offer a slightly different perspective, and proof, of the nonabelian Poincaré duality of Salvatore [51], Segal [56], and Lurie [43], which calculates factorization homology of iterated loop spaces as compactly-supported mapping spaces.

Definition 2.4.1. For a space B, $\mathsf{Spaces}_B := (\mathsf{Spaces}_{/B})^{B/}$ is the ∞-category of *retractive* spaces over B, i.e., spaces over B equipped with a section. The ∞-category $\mathsf{Spaces}_B^{\geq n}$ is the full ∞-subcategory of Spaces_B consisting of those $X \rightleftarrows B$ for which the retraction is n-connective, that is, the homomorphism $\pi_q X \to \pi_q B$ is an isomorphism for $q < n$ with any choice of base-point of B.

Remark 2.4.2. Equivalently, an object $X \in \mathsf{Spaces}_B^{\geq n}$ may be thought of as a fibration $X \to B$ equipped with a section for which, for each $b \in B$, the fiber X_b is n-connective.

Definition 2.4.3. Fix $X \in \mathsf{Spaces}_B$. Let M be a manifold, equipped with a map $M \to B$. The space of *compactly supported sections* of X over M is the colimit in the ∞-category Spaces:

$$\Gamma_{\mathsf{c}}(M, X) \;:=\; \underset{\substack{K \subset M \\ \text{compact}}}{\mathsf{colim}} \mathsf{Map}_{/B}\big((M \smallsetminus K \to M), (B \to X)\big),$$

where the colimit is indexed by the filtered poset of compact codimension-0 submanifold with boundary in M, and where the space indexed by a compact subset $K \subset M$ is the pullback:

$$\begin{array}{ccc} \mathsf{Map}_{/B}\big((M \smallsetminus K \to M), (B \to X)\big) & \longrightarrow & \mathsf{Map}_{/B}(M, X) \\ \downarrow & & \downarrow \\ * \simeq \mathsf{Map}_{/B}(M \smallsetminus K, B) & \longrightarrow & \mathsf{Map}_{/B}(M \smallsetminus K, X). \end{array}$$

For $B \to \mathsf{BO}(n)$ a map between spaces, note that $\Gamma_{\mathsf{c}}(-, X)$ defines a covariant functor $\mathcal{M}\mathsf{fld}_n^B \to \mathsf{Spaces}$. By inspection, this functor carries finite disjoint unions to finite products of spaces, which is to say that $\Gamma_{\mathsf{c}}(-, X)$ is symmetric monoidal with respect to the Cartesian monoidal structure on the ∞-category Spaces of spaces. In particular, the restriction

$$\Omega_B^n X \colon \mathcal{D}\mathsf{isk}_n^B \hookrightarrow \mathcal{M}\mathsf{fld}_n^B \xrightarrow{\Gamma_{\mathsf{c}}(-,X)} \mathsf{Spaces} \tag{2.4.1}$$

is a symmetric monoidal functor, i.e., a $\mathcal{D}\mathsf{isk}_n^B$-algebra in Spaces. Each point $* \xrightarrow{\{b\}} B$ determines a B-framed n-dimensional vector space (V, b), which can be regarded as an object $(V, b) \in \mathcal{D}\mathsf{isk}_n^B$. The value of the above functor $\Omega_B^n X$ on this object is equivalent to the n-fold loop space $\Omega^n X_b$ of the fiber X_b of the map $X \to B$ over $b \in B$.

We introduce the following terminology in order to state Theorem 2.4.5. Consider the base point $* \xrightarrow{\{\mathbb{R}^n\}} \mathsf{BO}(n)$. Each lift of this point $* \xrightarrow{\{b\}} B$, which is a point in the fiber $\mathsf{fiber}(B \to \mathsf{BO}(n))$, canonically determines a symmetric monoidal functor $\mathcal{D}\mathsf{isk}_n^{\mathsf{fr}} \to \mathcal{D}\mathsf{isk}_n^B$. Thereafter, each symmetric monoidal functor $A \colon \mathcal{D}\mathsf{isk}_n^B \to \mathsf{Spaces}$ restricts as an associative monoid

$$\pi_0(A,b)\colon \mathcal{D}\mathsf{isk}_1^{\mathsf{fr}} \xrightarrow{\mathbb{R}^{n-1}\times -} \mathcal{D}\mathsf{isk}_n^{\mathsf{fr}} \longrightarrow \mathcal{D}\mathsf{isk}_n^B \xrightarrow{A} \mathsf{Spaces} \xrightarrow{\pi_0} \mathsf{Set}.$$

We say A is *group-like* if, for each point $b \in \mathsf{fiber}\big(B \to \mathsf{BO}(n)\big)$, this monoid $\pi_0(A,b)$ is a group.

Example 2.4.4. In the case $B \simeq *$, a B-framing is a framing in the standard sense and $\mathcal{D}\mathsf{isk}_n^B = \mathcal{D}\mathsf{isk}_n^{\mathsf{fr}}$. A symmetric monoidal functor $A\colon \mathcal{D}\mathsf{isk}_n^{\mathsf{fr}} \to \mathsf{Spaces}$ is then the data of an $\mathcal{E}_n$-algebra, and this $\mathcal{E}_n$-algebra is group-like in the sense of [45] if and only if A is group-like in the the above sense.

Theorem 2.4.5. [3] *Restricting the functor* $\Gamma_{\mathsf{c}} : \mathsf{Spaces}_B \to \mathsf{Fun}^{\otimes}(\mathcal{M}\mathsf{fld}_n^B, \mathsf{Spaces})$ *defines a fully-faithful inclusion*

$$\mathsf{Spaces}_B^{\geq n} \hookrightarrow \mathbf{H}(\mathcal{M}\mathsf{fld}_n^B, \mathsf{Spaces})$$

of n-connective retractive spaces over B into homology theories. The essential image consists of those $\mathcal{F}$ for which the restriction $\mathcal{F}_{|\mathcal{D}\mathsf{isk}_n^B}$ is group-like.

Remark 2.4.6. The fully faithfulness of the above functor is given by Theorem 5.1.3.6 of [43]. This is a parametrized form of May's theorem from [45], identifying n-connective spaces as $\mathcal{D}\mathsf{isk}_n^{\mathsf{fr}}$-algebras.

The following result is the technical crux in the proof of Theorem 2.4.5; it asserts that, for each n-connective retractive space $X \rightleftarrows B$, the assignment of compactly supported sections $\Gamma_{\mathsf{c}}(-,X)$ is $\otimes$-excisive.

Lemma 2.4.7. [3] *Let $X \in \mathsf{Spaces}_B^{\geq n}$ be an n-connective retractive space over B. Let M be B-framed n-manifold. For $M' \bigcup_{M_0\times\mathbb{R}} M'' \cong M$ a collar-gluing, the canonical map between spaces,*

$$\Gamma_{\mathsf{c}}(M',X) \times_{\Gamma_{\mathsf{c}}(M_0\times\mathbb{R},X)} \Gamma_{\mathsf{c}}(M'',X) \xrightarrow{\simeq} \Gamma_{\mathsf{c}}(M,X),$$

from the quotient of the product $\Gamma_{\mathsf{c}}(M',X) \times \Gamma_{\mathsf{c}}(M'',X)$ by the diagonal action of $\Gamma_{\mathsf{c}}(M_0 \times \mathbb{R},X)$, is an equivalence.

Proof given in [3]. Since $M_0 \hookrightarrow M$ is a proper embedding, a compactly supported section over M can be restricted to obtain a compactly-supported section over M_0, as well as over $M \smallsetminus M'$ and over $M \smallsetminus M''$. Namely, there is a diagram among spaces of compactly-supported sections

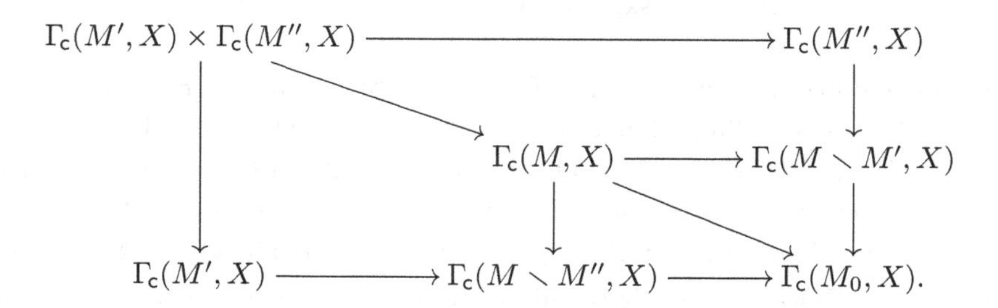

By inspection, the bottom horizontal sequence is a fiber sequence, as is the right vertical sequence, as is the diagonal sequence. Also, the inner square is pullback because

$M' \bigcup_{M_0 \times \mathbb{R}} M'' \cong M$ is a pushout. Because $M_0 \subset M$ is equipped with a regular neighborhood, these fiber sequences are in fact Serre fibration sequences, and so the inner square is a weak homotopy pullback square. In particular, there is a right homotopy coherent action of $\Omega\,\Gamma_{\mathsf{c}}(M_0, X)$ on $\Gamma_{\mathsf{c}}(M', X)$, a left homotopy coherent action of $\Omega\,\Gamma_{\mathsf{c}}(M_0, X)$ on $\Gamma_{\mathsf{c}}(M'', X)$, and a continuous map of topological spaces

$$\Gamma_{\mathsf{c}}(M', X) \underset{\Omega\,\Gamma_{\mathsf{c}}(M_0,X)}{\times} \Gamma_{\mathsf{c}}(M'', X) \longrightarrow \Gamma_{\mathsf{c}}(M, X) \tag{2.4.2}$$

from the balanced homotopy coinvariants. Because $X \to B$ is n-connective and M_0 is $(n-1)$-dimensional, the base $\Gamma_{\mathsf{c}}(M_0, X)$ is connected. It follows that the map (2.4.2) is in fact a weak homotopy equivalence. The assertion follows after the canonical identification $\Omega\,\Gamma_{\mathsf{c}}(M_0, X) \cong \Gamma_{\mathsf{c}}(M_0 \times \mathbb{R}, X)$ as group-like $\mathcal{E}_1$-spaces. □

Theorem 2.3.37 implies the nonabelian Poincaré duality theorem of Salvatore [51], Segal [56], and Lurie [43]. First, recall from (2.4.1) the $\mathcal{D}\mathsf{isk}_n^B$-algebra $\Omega^n_B X$ determined by a retractive space $X \rightleftarrows B$.

Corollary 2.4.8 (Nonabelian Poincaré duality [3]). *Let $B \to \mathsf{BO}(n)$ be a map from a space. Let $X \rightleftarrows B$ be an n-connective retractive space over B. For each B-framed n-manifold M, there is a canonical equivalence*

$$\int_M \Omega^n_B X \xrightarrow{\simeq} \Gamma_{\mathsf{c}}(M, X)$$

from the factorization homology over M of $\Omega^n_B X$ to the space of compactly-supported sections of X over M.

We now explain how this result specializes to familiar Poincaré duality between homology and compactly-supported cohomology. Let A be an abelian group. Consider the product space $X \simeq \mathsf{BO}(n) \times K(A, n)$. Observe that the values of the n-disk algebra

$$\Omega^n_{\mathsf{BO}(n)} X \colon \mathcal{D}\mathsf{isk}_n \longrightarrow \mathsf{Spaces}, \quad \bigsqcup_{0 \leq i \leq k} V_i \mapsto \prod_{1 \leq i \leq k} \mathsf{Map}_*(V_i^+, K(A, n)) \underset{\text{noncanonical}}{\simeq} A^{\times k},$$

are non-canonically identified as products of A. Furthermore, the restriction of this n-disk algebra along the standard symmetric monoidal functor $\mathcal{D}\mathsf{isk}_n^{\mathsf{fr}} \to \mathcal{D}\mathsf{isk}_n$, which selects Euclidean spaces, is the $\mathcal{E}_n$-algebra forgotten from the commutative algebra structure on A. The $\mathsf{O}(n)$-action on A is that inherited from the infinite-loop structure of the Eilenberg–Mac Lane spectrum $\mathsf{H}A$ through the J-homomorphism. Let M be an n-manifold. An A-orientation of M defines a natural transformation making the diagram of ∞-categories commute:

$$\begin{array}{ccc} \mathcal{D}\mathsf{isk}_{n/M} & \longrightarrow & \mathcal{D}\mathsf{isk}_n \\ {\scriptstyle \pi_0}\downarrow & & \downarrow{\scriptstyle \Omega^n_{\mathsf{BO}(n)}X} \\ \mathsf{Fin} & \xrightarrow{I \mapsto A^I} & \mathsf{Spaces}; \end{array} \tag{2.4.3}$$

here, the bottom horizontal functor is the symmetric monoidal functor from finite sets with disjoint union, which is the free symmetric monoidal ∞-category generated by a point, determined by the commutative algebra A in the Cartesian symmetric monoidal ∞-category Spaces. So choose an A-orientation of M (supposing one exists). Corollary 2.4.8 then yields an equivalence between spaces

$$\int_M A \underset{(2.4.3)}{\simeq} \int_M \Omega^n_{\mathsf{BO}(n)} X \underset{\text{Cor } 2.4.8}{\simeq} \mathsf{Map_c}(M, K(A,r)). \tag{2.4.4}$$

Note that each space in this display is equipped with a base point, and the maps in this display are canonically based maps. As the $\mathcal{E}_n$-algebra A is forgotten from a commutative algebra, we recognize $\int_M A$ as the free A-module in the ∞-category Spaces generated by the underlying space of M. Therefore, applying homotopy groups to (2.4.4) gives an isomorphism between graded abelian groups,

$$\overline{\mathsf{H}}_*(M;A) \cong \pi_* \int_M A \cong \pi_* \mathsf{Map_c}(M, K(A,n)) \cong \overline{\mathsf{H}}^{n-*}_{\mathsf{c}}(M;A),$$

from reduced homology to reduced cohomology.

Remark 2.4.9. The factorization homology $\int_M \Omega^n_B X$ is built from configuration spaces of disks in M with labels defined by $X \to B$, and the preceding result thereby has roots in the configuration space models of mapping spaces dating to the work of Segal, May, McDuff and others in the 1970s; see [53], [45], [46], and [14]. However, the classical configuration-space-with-labels, as described in [14], models a mapping space with target the n-fold suspension of X, rather than into X itself. Factorization homology thus more closely generalizes the configuration spaces with *summable* or *amalgamated* labels of Salvatore [51] and Segal [56].

Proof of Theorem 2.4.5. Corollary 2.4.8 identifies the functor

$$\Gamma_{\mathsf{c}} \colon \mathsf{Spaces}^{\geq n}_B \to \mathbf{H}(\mathcal{M}\mathsf{fld}^B_n, \mathsf{Spaces})$$

as the composition $\Gamma_{\mathsf{c}} \colon \mathsf{Spaces}^{\geq n}_B \xrightarrow{\Omega^n_B} \mathcal{D}\mathsf{isk}^B_n \xrightarrow{\int} \mathbf{H}(\mathcal{M}\mathsf{fld}^B_n, \mathsf{Spaces})$. Theorem 2.3.37 gives that $\int$ is fully faithful, so it remains to argue that Ω^n_B is fully faithful with essential image the group-like $\mathcal{D}\mathsf{isk}^B_n$-algebras in spaces. This is immediate because, for instance, $\mathsf{Spaces}_{/B}$ is an ∞-topos (Theorem 5.1.3.6 of [43]). □

2.5 Calculations

In §2.4, we identified values of factorization homology of $\mathcal{D}\mathsf{isk}_n$-algebras in spaces, with its Cartesian monoidal structure, as twisted compactly supported mapping spaces. Here we identify more values of factorization homology of $\mathcal{D}\mathsf{isk}_n$-algebras in chain complexes, and spectra, notably with monoidal structure given by tensor product – these cases are closest to the physical motivation given in the introduction.

In this section, we fix a map $B \to \mathsf{BO}(n)$ from a space. Two cases of notable interest are $* \xrightarrow{\{\mathbb{R}^n\}} \mathsf{BO}(n)$ and $\mathsf{BO}(n) \xrightarrow{\mathsf{id}} \mathsf{BO}(n)$.

2.5.1 Factorization homology for direct sum

We now show that in the simple case of $\mathcal{D}\mathsf{isk}_n$-algebras in chain complexes with direct sum, factorization homology recovers ordinary homology.

Fix a presentable ∞-category $\mathcal{V}$, such as the ∞-category $\mathsf{Spectra}$ of spectra, or the ∞-category $\mathsf{Mod}_{\Bbbk}$ of chain complexes over a fixed commutative ring $\Bbbk$. Coproduct, which in the case that $\mathcal{V}$ is stable is direct sum, endows $\mathcal{V}$ with the coCartesian symmetric monoidal structure: $(\mathcal{V}, \amalg)$. Consider the commutative diagram among ∞-categories, in which each functor is implemented by the evident restriction:

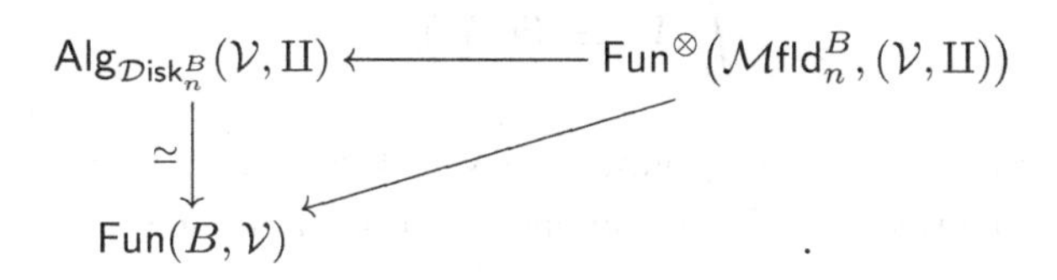

$$\begin{array}{ccc} \mathsf{Alg}_{\mathcal{D}\mathsf{isk}_n^B}(\mathcal{V}, \amalg) & \longleftarrow & \mathsf{Fun}^{\otimes}(\mathcal{M}\mathsf{fld}_n^B, (\mathcal{V}, \amalg)) \\ \simeq \downarrow & \swarrow & \\ \mathsf{Fun}(B, \mathcal{V}) & & \end{array}$$

In this case of a coCartesian symmetric monoidal structure, the downward functor is an equivalence, as indicated. In other words, a $\mathcal{D}\mathsf{isk}_n^B$-algebra in $(\mathcal{V}, \amalg)$ is precisely a functor $B \to \mathcal{V}$. [9]

Example 2.5.1. Let G be a topological group equipped with a continuous representation $G \xrightarrow{\rho} \mathsf{GL}(n) \simeq \mathsf{O}(n)$. Then a $\mathcal{D}\mathsf{isk}_n^{\mathsf{B}G}$-algebra in $(\mathcal{V}, \amalg)$ is a G-module $\mathsf{B}G \to \mathcal{V}$. In the case that $G = e$ is the trivial group, such a G-module is simply an object in $\mathcal{V}$.

Through Proposition 2.3.9, which identifies factorization homology as a left adjoint to the top horizontal functor in the preceding diagram, we conclude that factorization is identified as a left adjoint to the downleftward functor, which is restriction along the fully faithful inclusion $B \to \mathcal{M}\mathsf{fld}_n^B$ of B-framed Euclidean spaces. Left adjoints to restriction functors evaluate as left Kan extensions. The standard formula for left Kan extensions therefore identifies, for each B-framed n-manifold $M = (M, \varphi)$ and each $\mathcal{D}\mathsf{isk}_n^B$-algebra $B \xrightarrow{V} \mathcal{V}$, the value of factorization homology as the colimit in $\mathcal{V}$:

$$\int_M V \simeq \mathsf{colim}\big(B_{/M} \to B \xrightarrow{V} \mathcal{V}\big).$$

Corollary 2.2.31 identifies the functor $B_{/M} \to B$ appearing in this expression as the given functor $M \xrightarrow{\varphi} B$ from the underlying ∞-groupoid, or homotopy type, of M:

$$\int_M V \simeq \mathsf{colim}\big(M \xrightarrow{\varphi} B \xrightarrow{V} \mathcal{V}\big).$$

Remark 2.5.2. Note that the value of factorization homology in $(\mathcal{V}, \amalg)$ over a B-framed manifold $M \xrightarrow{\varphi} B$ only depends on the underlying space of M equipped with its given map to B.

Example 2.5.3. Let G be a topological group equipped with a continuous representation $G \xrightarrow{\rho} \mathsf{GL}(n) \simeq \mathsf{O}(n)$. Let V be a G-module in $\mathcal{V}$. Let M be an n-manifold. A $\mathsf{B}G$-framing

[9]Here and always, the space B is taken as an ∞-groupoid and thereafter as an ∞-category. In this way, we make sense of a functor $B \to \mathcal{V}$, which is simply a B-indexed local system of objects in $\mathcal{V}$.

on M is a principal G-bundle $P \to M$ equipped with a ρ-equivariant map $P \to \mathsf{Fr}(M)$ over M to the frame-bundle of M. Given such a $\mathsf{B}G$-framing on M, we identify the factorization homology

$$\int_M V \simeq P \underset{G}{\otimes} V$$

as the diagonal G-quotient of the tensor of $V \in \mathcal{V}$ with the space P. In particular, in the case that $G \xrightarrow{=} \mathsf{O}(n)$, the factorization homology is the diagonal $\mathsf{O}(n)$-quotient with the frame-bundle of M tensored with the $\mathsf{O}(n)$-module V in $\mathcal{V}$:

$$\int_M V \simeq \mathsf{Fr}(M) \underset{\mathsf{O}(n)}{\otimes} V.$$

In the case that $G = e$ is a trivial group, then $V \in \mathcal{V}$ is simply an object and the principal bundle $P \xrightarrow{=} M$ is identical to M, so that factorization homology

$$\int_M V \simeq M \otimes V$$

is simply the tensor of the object $V \in \mathcal{V}$ with the underlying space of M.

Example 2.5.4. Suppose $B = *$, so that a B-framing on an n-manifold is exactly a framing. Suppose $\mathcal{V} = \mathsf{Mod}_{\Bbbk}$ is the ∞-category of chain complexes over a fixed commutative ring $\Bbbk$. As discussed above, a $\mathcal{E}_n$-algebra in $(\mathsf{Mod}_{\Bbbk}, \oplus)$ is simply a chain complex, $V \in \mathsf{Mod}_{\Bbbk}$. Furthermore, the factorization homology

$$\int_M V \simeq M \otimes V \simeq \mathsf{C}_*(M; V) \in \mathsf{Mod}_{\Bbbk}$$

is the singular chain complex of M with coefficients in V.

Example 2.5.5. Again suppose $B = *$, so that a B-framing on an n-manifold is exactly a framing. Suppose $\mathcal{V} = \mathsf{Spectra}$ is the ∞-category of spectra. As discussed above, an $\mathcal{E}_n$-algebra in $(\mathsf{Spectra}, \oplus)$ is simply a spectrum $V \in \mathsf{Spectra}$. Furthermore, the factorization homology

$$\int_M V \simeq M \otimes V \simeq \Sigma^\infty_+ M \wedge V \in \mathsf{Spectra}$$

is the smash product of V with the suspension spectrum of M.

Remark 2.5.6. Taking products with Euclidean spaces defines a sequence of symmetric monoidal ∞-categories:

$$\mathcal{M}\mathsf{fld}_0^{\mathsf{fr}} \longrightarrow \mathcal{M}\mathsf{fld}_1^{\mathsf{fr}} \longrightarrow \cdots \longrightarrow \mathcal{M}\mathsf{fld}_n^{\mathsf{fr}} \longrightarrow \cdots.$$

Forgetting to underlying spaces defines, for each k, a symmetric monoidal functor $\mathcal{M}\mathsf{fld}_k^{\mathsf{fr}} \to \mathsf{Spaces}^{\mathsf{fin}}$ to the ∞-category of finite spaces (i.e., finite CW complexes) and coproduct among them. Via the canonical equivalence between underlying spaces $M \simeq M \times \mathbb{R}$ for each framed manifold M, these forgetful functors assemble as a symmetric monoidal functor from the colimit

$$\underset{k \geq 0}{\mathsf{colim}}\, \mathcal{M}\mathsf{fld}_k^{\mathsf{fr}} \longrightarrow \mathsf{Spaces}^{\mathsf{fin}}. \tag{2.5.1}$$

We argue that this symmetric monoidal functor is an equivalence.

- **Essentially surjective.** Each finite space is equivalent to the underlying space of the geometric realization of a finite simplicial complex. Such a geometric realization can be piecewise-linearly embedded into Euclidean space. A regular open neighborhood of such an embedding is a finitary smooth manifold that is homotopy equivalent to this geometric realization. In this way we conclude that the functor (2.5.1) is essentially surjective.

- **Fully faithful.** Let M and N be n-manifolds. It follows from Whitney's embedding theorem that the map between spaces

$$\underset{k\geq 0}{\mathsf{colim}}\, \mathsf{Emb}^{\mathsf{fr}}(M \times \mathbb{R}^{k-n}, N \times \mathbb{R}^{k-n}) \longrightarrow \mathsf{Map}_{\mathsf{Spaces}}(M, N)$$

  is a weak homotopy equivalence.

Through the equivalence (2.5.1), Theorem 2.3.37 applied to $\mathcal{V} = (\mathsf{Mod}_{\Bbbk}, \oplus)$ recovers to the formulation of the Eilenberg-Steenrod axioms given in the introduction. If one sets $\mathcal{V}$ to be the opposite $(\mathsf{Mod}_{\Bbbk}^{\mathsf{op}}, \oplus)$, then one likewise recovers the Eilenberg–Steenrod axioms for cohomology.

2.5.2 Factorization homology with coefficients in commutative algebras

Here, we examine factorization homology of commutative algebras, otherwise known as $\mathcal{E}_\infty$-algebras.

Fix a symmetric monoidal ∞-category $\mathcal{V}$ that is $\otimes$-presentable.

There is a canonical identification of the symmetric monoidal envelope of the commutative ∞-operad

$$\mathsf{Env}(\mathsf{Comm}) \;\simeq\; \mathsf{Fin} = (\mathsf{Fin}, \amalg)$$

with the coCartesian symmetric monoidal category of finite sets. [10] This is to say that restriction along the commutative algebra in Fin whose underlying object is $* \in \mathsf{Fin}$ defines an equivalence between ∞-categories:

$$\mathsf{Fun}^{\otimes}(\mathsf{Fin}, \mathcal{V}) \xrightarrow{\simeq} \mathsf{Alg}_{\mathsf{Com}}(\mathcal{V}).$$

So restriction along the symmetric monoidal functor given by taking connected components

$$[-] : \mathcal{D}\mathsf{isk}_n^B \to \mathcal{D}\mathsf{isk}_n \xrightarrow{[-]} \mathsf{Fin}$$

defines a forgetful functor

$$\mathsf{fgt}\colon\ \mathsf{Alg}_{\mathsf{Com}}(\mathcal{V}) \longrightarrow \mathsf{Alg}_{\mathcal{D}\mathsf{isk}_n^B}(\mathcal{V}).$$

[10]See the discussion just above Lemma 2.3.20 for a quick description of a *symmetric monoidal envelope.*

Via Corollary 3.2.3.3 of [43], the assumed $\otimes$-presentability of $\mathcal{V}$ implies the ∞-category $\mathsf{Alg}_{\mathsf{Com}}(\mathcal{V})$ is presentable. In particular, it is tensored over the ∞-category of spaces:

$$\mathsf{Spaces} \times \mathsf{Alg}_{\mathsf{Com}}(\mathcal{V}) \xrightarrow{\otimes} \mathsf{Alg}_{\mathsf{Com}}(\mathcal{V}), \qquad (X, A) \mapsto \mathsf{colim}\big(X \xrightarrow{!} * \xrightarrow{\{A\}} \mathsf{Alg}_{\mathsf{Com}}(\mathcal{V})\big).$$

Proposition 2.5.7. *The following diagram among ∞-categories canonically commutes:*

$$\begin{array}{ccccc}
\mathcal{M}\mathsf{fld}_n^B \times \mathsf{Alg}_{\mathsf{Com}}(\mathcal{V}) & \xrightarrow{\text{forget}\times\mathsf{id}} & \mathsf{Spaces} \times \mathsf{Alg}_{\mathsf{Com}}(\mathcal{V}) & \xrightarrow{\otimes} & \mathsf{Alg}_{\mathsf{Com}}(\mathcal{V}) \\
\downarrow{\scriptstyle \mathsf{id}\times\mathsf{fgt}} & & \int & & \downarrow{\scriptstyle \text{forget}} \\
\mathcal{M}\mathsf{fld}_n^B \times \mathsf{Alg}_{\mathcal{D}\mathsf{isk}_n^B}(\mathcal{V}) & & \longrightarrow & & \mathcal{V}.
\end{array}$$

In particular, for each B-framed n-manifold M, and each commutative algebra A, there is a canonical equivalence

$$\int_M A \simeq M \otimes A$$

between the factorization homology of A over M and the tensor of the commutative algebra A with the underlying space of M.

Proof. Observe the commutative diagram among symmetric monoidal ∞-categories:

$$\begin{array}{ccc}
\mathcal{D}\mathsf{isk}_n^B & \longrightarrow & \mathcal{M}\mathsf{fld}_n^B \\
\downarrow{\scriptstyle [-]} & & \downarrow{\scriptstyle \text{forget}} \\
\mathsf{Fin} & \longrightarrow & \mathsf{Spaces}
\end{array}$$

in which the bottom two are endowed with their coCartesian symmetric monoidal structures. Restriction defines the commutative diagram among ∞-categories of symmetric monoidal functors

$$\begin{array}{ccc}
\mathsf{Alg}_{\mathcal{D}\mathsf{isk}_n^B}(\mathcal{V}) & \longleftarrow & \mathsf{Fun}^{\otimes}(\mathcal{M}\mathsf{fld}_n^B, \mathcal{V}) \\
\uparrow{\scriptstyle \mathsf{fgt}} & & \uparrow \\
\mathsf{Alg}_{\mathsf{Com}}(\mathcal{V}) & \longleftarrow & \mathsf{Fun}^{\otimes}(\mathsf{Spaces}, \mathcal{V}).
\end{array} \tag{2.5.2}$$

It follows from Proposition 3.2.4.7 of [43] that the forgetful functor $\mathsf{Alg}_{\mathsf{Com}}(\mathcal{V}) \to \mathcal{V}$ canonically lifts as a symmetric monoidal functor $\big(\mathsf{Alg}_{\mathsf{Com}}(\mathcal{V}), \amalg\big) \to \big(\mathcal{V}, \otimes\big)$ from the coCartesian symmetric monoidal structure on $\mathsf{Alg}_{\mathsf{Com}}(\mathcal{V})$. Postcomposing by this symmetric monoidal functor determines the commutative diagram among ∞-categories:

$$\begin{array}{ccc}
\mathsf{Alg}_{\mathsf{Com}}(\mathsf{Alg}_{\mathsf{Com}}(\mathcal{V}), \amalg) & \longleftarrow & \mathsf{Fun}^{\otimes}(\mathsf{Spaces}, (\mathsf{Alg}_{\mathsf{Com}}(\mathcal{V}), \amalg)) \\
\downarrow & & \downarrow \\
\mathsf{Alg}_{\mathsf{Com}}(\mathcal{V}) & \longleftarrow & \mathsf{Fun}^{\otimes}(\mathsf{Spaces}, \mathcal{V}).
\end{array} \tag{2.5.3}$$

It follows from §2.5.1 that the right downward functor in this diagram is an equivalence. From the same proposition cited in [43] above, the right downward functor preserves colimits.

From §2.5.1, we conclude that the left adjoint to the bottom horizontal functor in (2.5.2) is the composite functor

$$\mathsf{Alg}_{\mathsf{Com}}(\mathcal{V}) \xrightarrow{A\mapsto(X\mapsto X\otimes A)} \mathsf{Fun}^{\otimes}(\mathsf{Spaces}, \mathsf{Alg}_{\mathsf{Com}}(\mathcal{V})) \xrightarrow{\text{forget}} \mathsf{Fun}^{\otimes}(\mathsf{Spaces}, \mathcal{V})$$

of first tensoring followed by the forgetful functor. Now, taking left adjoints of the horizontal functors results in a lax-commutative diagram among ∞-categories:

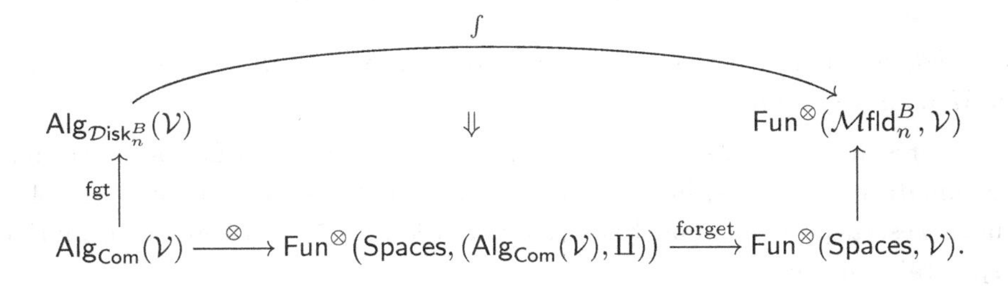

The proof is complete once we show the 2-cell in this diagram is invertible. This 2-cell is invertible if and only if, for each commutative algebra $A \in \mathsf{Alg}_{\mathsf{Com}}(\mathcal{V})$ and each B-framed n-manifold M, the resulting morphism in $\mathcal{V}$,

$$\int_M A \longrightarrow M \otimes A, \tag{2.5.4}$$

is an equivalence. (Here, and in what follows, we are not giving notation to the functor fgt.) Because the underlying space of each B-framed Euclidean n-dimensional space is contractible, this morphism (2.5.4) is an equivalence in such cases. By construction, the tensoring functor $\mathsf{Spaces} \xrightarrow{X\mapsto X\otimes A} \mathsf{Alg}_{\mathsf{Com}}(\mathcal{V})$ preserves colimits. Because collar-gluings among B-framed n-manifolds forget as pushouts among their underlying spaces, it follows that the functor $\mathcal{M}\mathsf{fld}_n^B \xrightarrow{M\mapsto M\otimes A} \mathsf{Alg}_{\mathsf{Com}}(\mathcal{V}) \xrightarrow{\text{forget}} \mathcal{V}$ is $\otimes$-excisive, and in particular it is symmetric monoidal. It follows from Theorem 2.3.37 that the morphism (2.5.4) is an equivalence, as desired. □

Proposition 2.5.7, together with the universal property of tensoring over spaces, yields the following.

Corollary 2.5.8. *Let $A \in \mathsf{Alg}_{\mathsf{Com}}(\mathcal{V})$ be a commutative algebra in a symmetric monoidal ∞-category that is $\otimes$-presentable. Let M be a B-framed n-manifold. There is a canonical commutative algebra structure on the factorization homology $\int_M A$. Furthermore, as a commutative algebra, it corepresents the copresheaf*

$$\mathsf{Map}_{\mathsf{Com}}\Big(\int_M A, -\Big)\colon \mathsf{Alg}_{\mathsf{Com}}(\mathcal{V}) \longrightarrow \mathsf{Spaces}, \qquad C \mapsto \mathsf{Map}_{\mathsf{Com}}(A, C)^M,$$

whose value on a commutative algebra is the the cotensor, or mapping space, of M to the space of morphisms from A to C.

Proposition 2.5.7 and Corollary 2.5.8, together with the fact that left adjoints compose, yields the following.

Corollary 2.5.9. *In the situation of Corollary 2.5.8, for $V \in \mathcal{V}$ an object with $\mathsf{Sym}(V) \in \mathsf{Alg}_{\mathsf{Com}}(\mathcal{V})$ the free commutative algebra on V, there is a canonical equivalence between commutative algebras in $\mathcal{V}$:*

$$\int_M \mathsf{Sym}(V) \ \simeq\ \mathsf{Sym}(M \otimes V).$$

Example 2.5.10. In the case that $\mathcal{V} = (\mathsf{Mod}_{\Bbbk}, \otimes_{\Bbbk})$, Corollary 2.5.9 specializes as an equivalence

$$\int_M \mathsf{Sym}(V) \ \simeq\ \mathsf{Sym}\big(\mathsf{C}_*(M; V)\big),$$

the graded-commutative algebra on the chain complex of singular chains on the underlying space of M with coefficients in V.

The next result, proved as Proposition 5.3 in [3], further identifies factorization homology of commutative algebras, in a suitably infinitesimal case: the cohomology of suitably nilpotent spaces. Its proof amounts to using Proposition 2.5.7, then arguing via an Eilenberg–Moore spectral sequence.

Proposition 2.5.11. [3] *Let M be an n-manifold. Let X be a nilpotent space whose first n homotopy groups are finite, $|\pi_i X| < \infty$ for $i \leq n$, and let $\Bbbk$ be a commutative ring. There is a natural equivalence of $\Bbbk$-modules*

$$\int_M \mathsf{C}^*(X; \Bbbk) \ \simeq\ \mathsf{C}^*(X^M; \Bbbk)$$

between the factorization homology of M with coefficient in the $\Bbbk$-cohomology of X and the $\Bbbk$-cohomology of the space of maps to X from the underlying space of M.

Remark 2.5.12. In the work [25], the authors lay out a similar approach, by way of Hochschild homology-type invariants, to the study of the cohomology of mapping spaces.

2.5.3 Factorization homology from Lie algebras

We now identify factorization homology of $\mathcal{D}\mathsf{isk}_n$-algebras coming from Lie algebras. The results that follow are closely analogous to those above concerning the factorization homology of $\mathcal{D}\mathsf{isk}_n$-algebras coming from based topological spaces. For simplicity, we assume our Lie algebras are defined over a fixed field $\Bbbk$ of characteristic zero.

As we proceed, we use that the ∞-category $\mathsf{Alg}_{\mathsf{Lie}}(\mathsf{Mod}_{\Bbbk})$ of Lie algebras over $\Bbbk$ are cotensored over spaces:

$$\mathsf{Spaces}^{\mathsf{op}} \times \mathsf{Alg}_{\mathsf{Lie}}(\mathsf{Mod}_{\Bbbk}) \xrightarrow{(X,\mathfrak{g}) \mapsto \mathfrak{g}^X} \mathsf{Alg}_{\mathsf{Lie}}(\mathsf{Mod}_{\Bbbk}),\ (X, \mathfrak{g}) \mapsto \lim\Big(X \xrightarrow{!} * \xrightarrow{\{\mathfrak{g}\}} \mathsf{Alg}_{\mathsf{Lie}}(\mathsf{Mod}_{\Bbbk})\Big).$$

This limit indeed exists because, for instance, the ∞-category $\mathsf{Alg}_{\mathsf{Lie}}(\mathsf{Mod}_{\Bbbk})$ admits products, cofiltered limits, and totalizations. More explicitly, $\mathfrak{g}^X \simeq \mathsf{C}^*(X; \mathfrak{g}) = \mathsf{Hom}_{\Bbbk}\big(\mathsf{C}_*(X; \Bbbk), \mathfrak{g}\big)$, with a canonically inherited Lie algebra structure. For M an n-manifold, we define the kernel Lie algebra,

$$\begin{array}{ccc} \mathsf{Map}_{\mathsf{c}}(M, \mathfrak{g}) & \longrightarrow & \mathfrak{g}^{M^+} \\ \downarrow & & \downarrow{\scriptstyle \mathsf{ev}_+} \\ 0 & \longrightarrow & \mathfrak{g}, \end{array}$$

where M^+ is the 1-point compactification of M. More explicitly, $\mathsf{Map_c}(M, \mathfrak{g}) \simeq \mathsf{C}^*_\mathsf{c}(M; \mathfrak{g})$, the compactly-supported cochains on M with coefficients in $\mathfrak{g}$. In the case that $M = \mathbb{R}^n$, we notate $\Omega^n\mathfrak{g} := \mathsf{Map_c}(\mathbb{R}^n, \mathfrak{g})$, and observe that this Lie algebra assembles as a $\mathcal{D}\mathsf{isk}_n$-algebra,

$$\Omega^n\mathfrak{g} \colon \mathcal{D}\mathsf{isk}_n \longrightarrow (\mathsf{Alg_{Lie}}(\mathsf{Mod}_\Bbbk), \times),$$

to the Cartesian symmetric monoidal structure on Lie algebras. Postcomposing this symmetric monoidal functor by that of Lie algebra chains, which indeed carries finite products of Lie algebras to finite tensor products (over $\Bbbk$) of $\Bbbk$-modules, results in a composite symmetric monoidal functor

$$\mathsf{C}^{\mathsf{Lie}}_*(\Omega^n\mathfrak{g}) \colon \mathcal{D}\mathsf{isk}_n \xrightarrow{\Omega^n\mathfrak{g}} (\mathsf{Alg_{Lie}}(\mathsf{Mod}_\Bbbk), \times) \xrightarrow{\mathsf{C}^{\mathsf{Lie}}_*} (\mathsf{Mod}_\Bbbk, \underset{\Bbbk}{\otimes}).$$

The next identification is also discussed in [27] and [17], and appears in the present form in [3]. Its proof amounts to applying Lie algebra chains, which is a geometric realization-preserving functor, to the Poincaré/Koszul duality equivalence $\int_M \Omega^n\mathfrak{g} \simeq \mathsf{Map_c}(M, \mathfrak{g})$ of [4].

Proposition 2.5.13. *Let M be an n-manifold. Let $\mathfrak{g}$ be a Lie algebra over $\Bbbk$. There is a canonical identification between $\Bbbk$-modules,*

$$\int_M \mathsf{C}^{\mathsf{Lie}}_*(\Omega^n\mathfrak{g}) \simeq \mathsf{C}^{\mathsf{Lie}}_*(\mathsf{Map_c}(M, \mathfrak{g})).$$

Remark 2.5.14. Let $\mathfrak{g}$ be a Lie algebra over $\Bbbk$. The $\mathcal{D}\mathsf{isk}_n$-algebra $\mathsf{C}^{\mathsf{Lie}}_*(\Omega^n\mathfrak{g})$ appearing above has an interesting interpretation, established by Knudsen [35]. There is a forgetful functor from $\mathcal{E}_n$-algebras to Lie algebras,

$$\mathsf{Alg_{Lie}}(\mathsf{Mod}_\Bbbk) \longleftarrow \mathsf{Alg}_{\mathcal{E}_n}(\mathsf{Mod}_\Bbbk) : \mathsf{fgt}.$$

For $n = 1$, there is the standard forgetful functor; for $n > 1$, Kontsevich's formality result (see [37] for a sketched outline, and [39] for a more detailed account) identifies the operad $\mathsf{C}_*(\mathcal{E}_n; \mathbb{R})$ in $\mathsf{Mod}_\Bbbk$ is equivalent to the Poisson operad P_n, to which the Lie operad in $\mathsf{Mod}_\mathbb{R}$ maps.[11] (See [16] for an account at the level of homology.) The adjoint functor theorem (Corollary 5.5.2.9 of [42]) applies to this forgetful functor, thereby offering a left adjoint:

$$\mathsf{U}_n \colon \mathsf{Alg_{Lie}}(\mathsf{Mod}_\Bbbk) \rightleftarrows \mathsf{Alg}_{\mathcal{E}_n}(\mathsf{Mod}_\Bbbk) : \mathsf{fgt}.$$

In the case $n = 1$, this left adjoint U_1 agrees with the familiar universal enveloping algebra functor. In general, there is an identification between $\mathcal{E}_n$-algebras,

$$\mathsf{U}_n\mathfrak{g} \simeq \mathsf{C}^{\mathsf{Lie}}_*(\Omega^n\mathfrak{g})$$

[11]The operad P_n in $\mathsf{Mod}_\Bbbk$ represents: a $\mathsf{Mod}_\Bbbk$-module with a commutative and associative multiplication $\cdot$; an $(n-1)$-shifted binary operation $[-,-]$ that satisfies the Jacobi identity; the structure of this Lie bracket $[-,-]$ being a derivation over $\cdot$ in each variable.

proved by Knudsen [35]. Through this identification, Proposition 2.5.13 can be reformulated as an equivalence of chain complexes over $\Bbbk$,

$$\int_M \mathsf{U}_n\mathfrak{g} \;\simeq\; \mathsf{C}^{\mathsf{Lie}}_*(\mathsf{Map}_{\mathsf{c}}(M,\mathfrak{g})),$$

in the case that M is equipped with a framing.

2.5.4 Factorization homology of free $\mathcal{D}\mathsf{isk}_n^B$-algebras

We now identify the factorization homology of free $\mathcal{D}\mathsf{isk}_n^B$-algebras. The results here are extracted from §5.2 of [3], §4.3 of [8], and §2.4 of [5].

For this section, we fix a symmetric monoidal ∞-category $\mathcal{V}$ that is $\otimes$-presentable. Also, for simplicity restrict our attention to the case the space B is connected, and equipped with a base point. This condition, and structure, on B identify the tangential structure

$$(B \to \mathsf{BO}(n)) \;=\; (\mathsf{B}G \xrightarrow{\mathsf{B}\rho} \mathsf{BO}(n))$$

as the classifying space of a morphism $\rho\colon G \to \mathsf{O}(n)$ between group-objects in Spaces.

For each B-framed n-manifold M, and for each finite cardinality $i \geq 0$, consider the configuration space

$$\mathsf{Conf}_i(M) \;:=\; \Big\{\{1,\dots,i\} \xrightarrow{c} M \mid c \text{ is injective}\Big\} \;\subset\; M^{\times i},$$

which is an open subspace of the i-fold product of M. As an open subspace, it inherits the structure of a smooth ni-manifold. The $\mathsf{B}G$-framing on M then determines a $\mathsf{B}(\Sigma_i \wr G)$-framing on $\mathsf{Conf}_i(M)$:

$$\begin{array}{ccc} & & \mathsf{B}(\Sigma_i \wr G) \\ & \nearrow & \downarrow \\ \mathsf{Conf}_i(M) & \xrightarrow{\tau_{\mathsf{Conf}_i(M)}} & \mathsf{BO}(ni). \end{array}$$

This lift of the tangent classifier of $\mathsf{Conf}_i(M)$ selects a principal $\Sigma_i \wr G$-bundle

$$\mathsf{Conf}_i^G(M) \longrightarrow \mathsf{Conf}_i(M),$$

which we refer to as a *G-framed* configuration space. In the extremal case of $G \xrightarrow{=} \mathsf{O}(n)$, then we have $\mathsf{Conf}_i^G(M) = \mathsf{Conf}_i(M)$. In the extremal case that $G = e$ is the trivial group then $\mathsf{Conf}_i^G(M) = \mathsf{Conf}_i^{\mathsf{fr}}(M)$ is the *framed* configuration space, a point in which is an injection $c\colon \{1,\dots,i\} \hookrightarrow M$ together with a choice of the basis for each tangent space $T_{c(j)}M$.

The *free $\mathcal{D}\mathsf{isk}_n^B$-algebra* functor is the left adjoint

$$\mathsf{Free}_n^B\colon \mathsf{Mod}_G(\mathcal{V}) \;:=\; \mathsf{Fun}(B,\mathcal{V}) \;\rightleftarrows\; \mathsf{Alg}_{\mathcal{D}\mathsf{isk}_n^B}(\mathcal{V})$$

to the restriction along the fully faithful inclusion $B \to \mathcal{D}\mathsf{isk}_n^B$ of the B-framed Euclidean spaces.

The next result identifies the factorization homology of free $\mathcal{D}\mathsf{isk}_n^{\mathsf{B}G}$-algebras in terms of G-framed configuration spaces.

Proposition 2.5.15. *Let $\rho\colon G \to \mathsf{O}(n)$ be a morphism between group-objects in* Spaces. *Let M be a $\mathsf{B}G$-framed n-manifold. Let $V \in \mathsf{Mod}_G(\mathcal{V})$ be a G-module in $\mathcal{V}$. There is a canonical equivalence in $\mathcal{V}$,*

$$\int_M \mathsf{Free}_n^{\mathsf{B}G}(V) \;\simeq\; \coprod_{i\geq 0} \mathsf{Conf}_i^G(M) \underset{\Sigma_i \wr G}{\otimes} V^{\otimes i}$$

between the factorization homology of the free $\mathcal{D}\mathsf{isk}_n^{\mathsf{B}G}$-algebra and the coproduct of G-framed configuration spaces of M labeled by V.

The proof of Proposition 2.5.15, which appears in the above-mentioned citations, amalgamates the following key observations. First, the result is true for M a Euclidean space. Second, a hypercover-type argument (as in Theorem A.3.1 of [43]), applied to V-labeled configuration spaces, gives that the right-hand term satisfies $\otimes$-excision.

Using that left adjoints compose, Proposition 2.5.15 applied to a free G-module on an object $W \in \mathcal{V}$ gives the following result.

Corollary 2.5.16. *In the case that $V \simeq G \otimes W$ is the free G-module on an object $W \in \mathcal{V}$, there is a canonical equivalence in $\mathcal{V}$:*

$$\int_M \mathsf{Free}_n^{\mathsf{B}G}(G\otimes W) \;\simeq\; \coprod_{i\geq 0} \mathsf{Conf}_i(M) \underset{\Sigma_i}{\otimes} W^{\otimes i}.$$

Remark 2.5.17 (On homotopy invariance)**.** The work of Longoni–Salvatore [41] shows that configuration spaces of a manifold are sensitive to simple-homotopy type of the manifold. Based on Proposition 2.5.15, this has the following consequence, that factorization homology is not a homotopy invariant of manifolds, even in the weakest sense. The strongest form of homotopy invariance for the functor $\int A$, for a fixed $\mathcal{D}\mathsf{isk}_n$-algebra A, would be the structure of a factorization

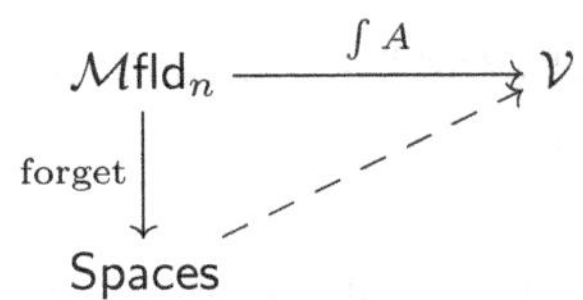

of $\int A$ through the forgetful functor $\mathcal{M}\mathsf{fld}_n \to \mathsf{Spaces}$. This structure of homotopy invariance is equivalent to the $\mathcal{D}\mathsf{isk}_n$-algebra A being commutative. Likewise, a strong form of proper-homotopy invariance would be a factorization

$$\begin{array}{ccc} \mathcal{M}\mathsf{fld}_n & \xrightarrow{\int A} & \mathcal{V} \\ {\scriptstyle (-)^+}\big\downarrow & \nearrow & \\ \mathsf{Spaces}_*^{\mathsf{op}} & & \end{array}$$

where $(-)^+\colon \mathcal{M}\mathsf{fld}_n \to \mathsf{Spaces}_*^{\mathsf{op}}$ is 1-point compactification: sending a manifold M to the 1-point compactification M^+, and an open embedding $M \hookrightarrow N$ to the collapse map $N^+ \to M^+$. The existence of this second factorization is roughly equivalent to A being the n-disk enveloping algebra of a Lie algebra.

2.6 Filtrations

In this section, we establish two filtrations of factorization homology: in the manifold variable, one can filter by bounding the cardinality of the number of embedded disks; in the algebra variable, one can filter by bounding the number of multiplications allowed by elements of the algebra. Filtrations of the first kind generalize the Goodwillie–Weiss embedding calculus [59]; filtrations of the second kind are an instance of the Goodwillie functor calculus [24]. These filtrations are exchanged under Koszul duality, as is described in the following section. For simplicity, we have written these sections for the case of framed n-manifolds—see [5] and [4] for the case of unoriented n-manifolds, and the case of B-framed n-manifolds is analogous.

For the rest of this section, we fix a symmetric monoidal ∞-category $\mathcal{V}$ which is stable and $\otimes$-presentable.

2.6.1 Cardinality filtrations

Definition 2.6.1. The ∞-category $\mathcal{M}\mathsf{fld}_n^{\mathsf{fr},\mathsf{surj}}$ is the ∞-subcategory of $\mathcal{M}\mathsf{fld}_n^{\mathsf{fr}}$ whose objects are nonempty framed n-manifolds and whose morphisms consist of those embeddings which induce surjections on the set of connected components. $\mathcal{M}\mathsf{fld}_n^{\mathsf{fr},\leq k} \subset \mathcal{M}\mathsf{fld}_n^{\mathsf{fr},\mathsf{surj}}$ is the full ∞-subcategory whose objects have at most k components. The ∞-categories $\mathcal{D}\mathsf{isk}_n^{\mathsf{fr},\mathsf{surj}}$ and $\mathcal{D}\mathsf{isk}_n^{\mathsf{fr},\leq k}$ are the further ∞-subcategories of $\mathcal{M}\mathsf{fld}_n^{\mathsf{fr},\mathsf{surj}}$ whose objects are nonempty finite disjoint unions of Euclidean spaces.

That is, an embedding $f\colon M \hookrightarrow N$ belongs to $\mathcal{M}\mathsf{fld}_n^{\mathsf{fr},\mathsf{surj}}$ if and only if $\pi_0 f\colon \pi_0 M \to \pi_0 N$ is surjective.

Definition 2.6.2. For a functor $A : \mathcal{D}\mathsf{isk}_n^{\mathsf{fr}} \to \mathcal{V}$, the kth truncation of factorization homology is the colimit of the composite functor

$$\tau_{\leq k} \int_M A := \mathsf{colim}\Big(\mathcal{D}\mathsf{isk}_{n/M}^{\mathsf{fr},\leq k} \xrightarrow{A} \mathcal{V}\Big).$$

The difference between the subsequent truncations has a concrete description in terms of configuration spaces:

Lemma 2.6.3. [5] *Let A be an $\mathcal{E}_n$-algebra in $\mathcal{V}$. For each closed framed n-manifold M, there is a canonical cofiber sequence in $\mathcal{V}$:*

$$\tau_{\leq k-1} \int_M A \longrightarrow \tau_{\leq k} \int_M A \longrightarrow \mathsf{Conf}_k(M)^+ \bigotimes_{\Sigma_k} A^{\otimes k}.$$

Here, $\mathsf{Conf}_k(M)^+ \in \mathsf{Spaces}_*$ is the 1-point compactification of $\mathsf{Conf}_k(M)$, regarded as a based Σ_k-space; the cofiber is the reduced tensor with this based Σ_k-space.

Example 2.6.4. Consider the case that $\mathcal{V} = \mathsf{Mod}_{\Bbbk}$ and $n = 1$, so that $\int_{S^1} A \simeq \mathsf{HC}_*(A)$ is the Hochschild chain complex of the associative algebra A. After Lemma 2.6.3, Poincaré duality applied to the the framed k-manifold $\mathsf{Conf}_k(S^1)$ with coefficients in the local system $A^{\otimes k}$, identifies, for $k > 0$, the cofibers of the cardinality filtration as

$$\mathsf{Conf}_k(S^1)^+\bigotimes_{\Sigma_k} A^{\otimes k} \simeq \mathsf{cInd}^{\mathbb{T}}_{\mathsf{C}_k}(A[1]^{\otimes k}).$$

Here, $A[1]$ is chain complex over $\Bbbk$ which is the suspension of A; C_k is the cyclic group of order k, regarded as the closed subgroup of kth roots of unity in the Lie group $\mathbb{T}$ of unit complex numbers; $\mathsf{cInd}^{\mathbb{T}}_{\mathsf{C}_k}$ is the coinduction from the standard action of C_k on the k-fold tensor product $A[1]^{\otimes k}$.

The next series of results lead to the proof of Proposition 2.6.9, that factorization homology can be computed as a sequential colimit of its truncated values. To do this, we will use a comparison of our disk categories with the Ran space $\mathsf{Ran}(M)$. This is the space of finite subsets of M, topologized so that points can collide.

Definition 2.6.5. $\mathsf{Fin}^{\mathsf{surj}}$ is the category of finite nonempty sets and surjections among them. The Ran space of a topological space M is the topological space

$$\mathsf{Ran}(M) := \mathsf{colim}\Big(\mathsf{Fin}^{\mathsf{surj,op}} \xrightarrow{M} \mathsf{Top}\Big)$$

which is the colimit of the diagram sending a set J to the space M^J, and a surjection $J \to I$ to the diagonal map $M^I \hookrightarrow M^J$. For finite subset $S \subset M$, the topological space

$$\mathsf{Ran}(M)_S \subset \mathsf{Ran}(M)$$

is the subspace consisting of those finite subsets of M which contain S. The cardinality stratification of $\mathsf{Ran}(M)_S \to \mathbb{N}$ sends a finite subset T to its cardinality $|T|$.

We will make use of the exit-path ∞-category of a stratified space: see [43] or [7].

Lemma 2.6.6. [5] *There is contravariant equivalence*

$$\mathcal{D}\mathsf{isk}^{\mathsf{fr,surj}}_{n/M} \simeq \mathsf{Exit}\big(\mathsf{Ran}(M)\big)^{\mathsf{op}}$$

with the opposite of the exit-path ∞-category of the Ran space of M. More generally, fix an embedding $S \subset \coprod_S \mathbb{R}^n \hookrightarrow M$ for $S \subset M$. Then there is an equivalence

$$\mathcal{D}\mathsf{isk}^{\mathsf{fr,surj}}_{n/M} \underset{\mathcal{D}\mathsf{isk}^{\mathsf{fr}}_{n/M}}{\times} \mathcal{D}\mathsf{isk}^{\mathsf{fr},\coprod_S \mathbb{R}^n/\mathsf{inj}}_{n/M} \simeq \mathsf{Exit}\big(\mathsf{Ran}(M)_S\big)^{\mathsf{op}}$$

where $\mathcal{D}\mathsf{isk}^{\mathsf{fr},\coprod_S \mathbb{R}^n/\mathsf{inj}}_{n/M}$ is the ∞-subcategory of $(\mathcal{D}\mathsf{isk}_{n/M})^{\coprod_S \mathbb{R}^n/}$ whose objects are those $\coprod_S \mathbb{R}^n \hookrightarrow U \hookrightarrow M$ such that $S \to \pi_0 U$ is injective.

Lemma 2.6.7 (Lemma 2.3.2 [5]). *The functor*

$$\mathcal{D}\mathsf{isk}^{\mathsf{fr,surj}}_{n/M} \longrightarrow \mathcal{D}\mathsf{isk}^{\mathsf{fr}}_{n/M}$$

is final.

Proof. By Quillen's Theorem A, it suffices to show that for each $g : \coprod_k \mathbb{R}^n \hookrightarrow M$, an object of $\mathcal{D}\mathsf{isk}_{n/M}$, the ∞-undercategory

$$\mathcal{D}\mathsf{isk}_{n/M}^{\mathsf{fr,surj}} \underset{\mathcal{D}\mathsf{isk}_{n/M}^{\mathsf{fr}}}{\times} \left(\mathcal{D}\mathsf{isk}_{n/M}^{\mathsf{fr}}\right)^{\coprod_k \mathbb{R}^n/}$$

has a contractible classifying space. There exists an adjunction

$$\mathcal{D}\mathsf{isk}_{n/M}^{\mathsf{fr,surj}} \underset{\mathcal{D}\mathsf{isk}_{n/M}^{\mathsf{fr}}}{\times} \mathcal{D}\mathsf{isk}_{n/M}^{\mathsf{fr},\coprod_S \mathbb{R}^n/^{\mathsf{inj}}} \leftrightarrows \mathcal{D}\mathsf{isk}_{n/M}^{\mathsf{fr,surj}} \underset{\mathcal{D}\mathsf{isk}_{n/M}^{\mathsf{fr}}}{\times} \left(\mathcal{D}\mathsf{isk}_{n/M}^{\mathsf{fr}}\right)^{\coprod_k \mathbb{R}^n/}$$

hence an equivalence between their classifying spaces. By Lemma 2.6.7, we obtain

$$\mathsf{B}\left(\mathcal{D}\mathsf{isk}_{n/M}^{\mathsf{fr,surj}} \underset{\mathcal{D}\mathsf{isk}_{n/M}^{\mathsf{fr}}}{\times} \mathcal{D}\mathsf{isk}_{n/M}^{\mathsf{fr},\coprod_S \mathbb{R}^n/^{\mathsf{inj}}}\right) \simeq \mathsf{B}\,\mathsf{Exit}(\mathsf{Ran}(M)_S) \simeq \mathsf{Ran}(M)_S.$$

The result now follows from the contractibility of this form of the Ran space—to see this, we use a now standard argument from [9]: the Ran space carries a natural H-space structure, given by taking unions of subsets, for which the composition of the diagonal and the H-space multiplication is the identity. Consequently, its homotopy groups must be zero. □

Remark 2.6.8. Combining the equivalence

$$\mathcal{D}\mathsf{isk}_{n/M}^{\mathsf{fr,surj}} \simeq \mathsf{Exit}(\mathsf{Ran}(M))^{\mathsf{op}}$$

together with the finality of Lemma 2.6.7, we obtain that factorization homology (or any colimit indexed by $\mathcal{D}\mathsf{isk}_{n/M}^{\mathsf{fr}}$) is equivalent to the global sections of an associated cosheaf on the Ran space of M.

Together with Lemma 2.6.3, the following is the main conclusion of this section, describing a complete filtration of factorization homology with layers given by twisted homologies of configuration spaces.

Proposition 2.6.9. *The natural map from the sequential colimit of truncations*

$$\varinjlim_k \tau_{\leq k} \int_M A \longrightarrow \int_M A$$

is an equivalence.

Proof. The finality of Lemma 2.6.7 implies the equivalence

$$\int_M A := \mathsf{colim}\left(\mathcal{D}\mathsf{isk}_{n/M}^{\mathsf{fr}} \xrightarrow{\int A} \mathcal{V}\right) \simeq \mathsf{colim}\left(\mathcal{D}\mathsf{isk}_{n/M}^{\mathsf{fr,surj}} \xrightarrow{A} \mathcal{V}\right).$$

There is then a sequence of equivalences

$$\begin{aligned}\mathsf{colim}\left(\mathcal{D}\mathsf{isk}_{n/M}^{\mathsf{fr,surj}} \xrightarrow{A} \mathcal{V}\right) &\simeq \mathsf{colim}\left(\left(\varinjlim_k \mathcal{D}\mathsf{isk}_{n/M}^{\mathsf{fr},\leq k}\right) \xrightarrow{A} \mathcal{V}\right) \simeq \varinjlim_k \mathsf{colim}\left(\mathcal{D}\mathsf{isk}_{n/M}^{\mathsf{fr},\leq k} \xrightarrow{A} \mathcal{V}\right) \\ &= \varinjlim_k \tau_{\leq k} \textstyle\int_M A\end{aligned}$$

from which the result follows. □

A dual result holds for factorization cohomology with coefficients in an $\mathcal{E}_n$-coalgebra.

Definition 2.6.10. For C an $\mathcal{E}_n$-coalgebra in $\mathcal{V}$ (i.e., an $\mathcal{E}_n$-algebra in $\mathcal{V}^{\mathsf{op}}$), the factorization cohomology of M with coefficients in C is the limit

$$\int^M C := \lim\left(\mathcal{D}\mathsf{isk}_{n/M}^{\mathsf{op}} \xrightarrow{C} \mathcal{V}\right).$$

Proposition 2.6.11. *For M a closed framed n-manifold and C an $\mathcal{E}_n$-coalgebra in $\mathcal{V}$, there exists a cofiltration of $\int^M C$,*

$$\int^M C \longrightarrow \tau^{\leq \bullet} \int^M C$$

whose associated graded is

$$\coprod_k \left((C^{\otimes k})^{\mathsf{Conf}_k(M)^+}\right)^{\Sigma_k}.$$

2.6.2 Goodwillie filtrations

Fixing a manifold M, one can apply the Goodwillie functor calculus, as developed in [24], [38], and [43], to the functor

$$\int_M : \mathsf{Alg}_{\mathcal{E}_n}^{\mathsf{aug}}(\mathcal{V}) \longrightarrow \mathcal{V}.$$

The output is a cofiltration of factorization homology whose kth term is the polynomial approximation $P_k \int_M$.

Proposition 2.6.12. [5] *Let A an augmented $\mathcal{E}_n$-algebra in $\mathcal{V}$ with cotangent space LA. For each closed framed n-manifold M, the following is a fiber sequence*

$$\begin{array}{ccc} \mathsf{Conf}_k(M) \underset{\Sigma_k}{\otimes} LA^{\otimes k} & \longrightarrow & P_k \int_M A \\ \downarrow & & \downarrow \\ \mathbb{1} & \longrightarrow & P_{k-1} \int_M A \end{array}$$

where P_k is Goodwillie's kth polynomial approximation [24] applied to the functor $\int_M$, and LA is the cotangent space of A at the augmentation (see [20]). In particular, the Goodwillie derivative $\partial_k \int_M$ is canonically equivalent with $\mathsf{Conf}_k(M) \otimes \mathbb{1}$, the tensor of the space $\mathsf{Conf}_k(M)$ with the unit object in the symmetric monoidal ∞-category $\mathcal{V}$.

2.7 Poincaré/Koszul duality

Koszul duality, in its most basic form, interchanges two forms of algebra, determined by the form of algebraic structure that the cotangent space at an augmentation naturally inherits. There are two classical forms of Koszul duality: commutative algebra is Koszul dual to Lie algebra; associative algebra is dual to associative algebra. In particular, associative algebra is Koszul self-dual. This second assertion generalizes to the situation of $\mathcal{E}_n$-algebra, where the Koszul self-duality dates to Getzler–Jones [23]. This Koszul duality for $\mathcal{E}_n$-algebra can be expressed in terms of factorization homology for manifolds with boundary. See [44] for another treatment of Koszul self-duality of $\mathcal{E}_n$-algebra, in terms of twisted arrow categories.

Theorem 2.7.1. ([5], [4]) *Let $\mathcal{V}$ be a stable $\otimes$-presentable ∞-category. There is a functor*

$$\int_{\mathbb{D}^n/\partial\mathbb{D}^n} : \mathsf{Alg}^{\mathsf{aug}}_{\mathcal{E}_n}(\mathcal{V}) \longrightarrow \mathsf{cAlg}^{\mathsf{aug}}_{\mathcal{E}_n}(\mathcal{V})$$

sending an augmented $\mathcal{E}_n$-algebra A to an $\mathcal{E}_n$-coalgebra structure on

$$\int_{\mathbb{D}^n/\partial\mathbb{D}^n} A \simeq A \bigotimes_{\int_{S^{n-1}\times\mathbb{R}} A} \mathbb{1}\,.$$

There is a natural transformation, the Poincaré/Koszul duality map, in $\mathsf{Fun}(\mathbb{N}^{\triangleright}, \mathcal{V})$:

$$\begin{array}{ccc} P_\bullet \displaystyle\int_M A & \longrightarrow & \tau^{\leq\bullet} \displaystyle\int^M \Big(\int_{\mathbb{D}^n/\partial\mathbb{D}^n} A\Big) \\ \downarrow & & \downarrow \\ \displaystyle\int_M A & \longrightarrow & \displaystyle\int^M \Big(\int_{\mathbb{D}^n/\partial\mathbb{D}^n} A\Big). \end{array}$$

This map is an equivalence when restricted to $\mathsf{Fun}(\mathbb{N}, \mathcal{V})$.

Remark 2.7.2. The coalgebra structure of $\int_{\mathbb{D}^n/\partial\mathbb{D}^n} A$ is given in [4] by showing that factorization homology with coefficients in an augmented $\mathcal{E}_n$-algebra is functorial with respect to the fold map

$$\mathbb{D}^n/\partial\mathbb{D}^n \longrightarrow \mathbb{D}^n/\partial\mathbb{D}^n \vee \mathbb{D}^n/\partial\mathbb{D}^n.$$

More generally, the theory of zero-pointed manifolds of [4] exhibits an additional functoriality for factorization homology of augmented algebras, of extension-by-zero maps.

In particular, the Goodwillie completion $\varprojlim P_\bullet \int_M A$ is equivalent to factorization cohomology with the coefficients in the coalgebra which is Koszul-dual to A. The natural map to the completion

$$\int_M A \longrightarrow \varprojlim P_\bullet \int_M A$$

need not be an equivalence, however.[12] A generalization of factorization homology, taking coefficients in formal moduli problems, gives a way to correct the failure of Poincaré/Koszul duality to be an equivalence in general.

[12]One should only expect this map to be an equivalence if $\mathcal{V}$ carries a t-structure, suitably compatible with the monoidal structure, for which the augmentation ideal of A is either connected or n-coconnected.

Definition 2.7.3. Let $\mathcal{V}$ be a $\otimes$-presentable ∞-category. Let $B \to \mathsf{BO}(n)$ be a map from a space. Let $X\colon \mathsf{Alg}_{\mathcal{D}\mathsf{isk}_n^B}(\mathcal{V}) \to \mathsf{Spaces}$ be a functor. Let M be a B-framed n-manifold M. The factorization homology of M with coefficients in X is the object in $\mathcal{V}$

$$\int_M X := \lim_{A \in \mathsf{Aff}^{\mathsf{op}}_{/X}} \int_M A,$$

where $\mathsf{Aff} := \mathsf{Alg}_{\mathcal{D}\mathsf{isk}_n^B}(\mathcal{V})^{\mathsf{op}} \subset \mathsf{Fun}\big(\mathsf{Alg}_{\mathcal{D}\mathsf{isk}_n^B}(\mathcal{V}), \mathsf{Spaces}\big)$ is the image of the Yoneda embedding.

Remark 2.7.4. Intuitively, the object $\int_M X$ is $\Gamma(X, \int_M \mathcal{O})$, the global sections of the presheaf on X obtained by applying factorization homology of M to the structure sheaf of X. We interpret this form of factorization homology as the observables in a topological quantum field theory: a sigma-model with algebraic target X.

There is a modification of the above definition for a *formal* moduli problem, i.e., a space-valued functor on an ∞-category of Artin $\mathcal{E}_n$-algebras.

Definition 2.7.5 ($\mathsf{Artin}_{\mathcal{E}_n}$)**.** The full ∞-subcategory $\mathsf{Triv}_{\mathcal{E}_n} \subset \mathsf{Alg}^{\mathsf{aug},\geq 0}_{\mathcal{E}_n}(\mathsf{Mod}_{\Bbbk})$ is the essential image of connective perfect $\Bbbk$-modules $\mathsf{Perf}^{\geq 0}_{\Bbbk}$ under the functor that assigns a complex V the associated square-zero extension $\Bbbk \oplus V$. The ∞-category $\mathsf{Artin}_{\mathcal{E}_n}$ of local Artin $\mathcal{E}_n$-algebras is the smallest full ∞-subcategory of $\mathsf{Alg}^{\mathsf{aug},\geq 0}_{\mathcal{E}_n}(\mathsf{Mod}_{\Bbbk})$ that contains $\mathsf{Triv}_{\mathcal{E}_n}$ and is closed under small extensions. That is:

If B is in $\mathsf{Artin}_{\mathcal{E}_n}$, $\Bbbk \oplus V$ is in $\mathsf{Triv}_{\mathcal{E}_n}$, and the following diagram

$$\begin{array}{ccc} A & \longrightarrow & B \\ \downarrow & & \downarrow \\ \Bbbk & \longrightarrow & \Bbbk \oplus V \end{array}$$

forms a pullback square in $\mathsf{Alg}^{\mathsf{aug},\geq 0}_{\mathcal{E}_n}(\mathsf{Mod}_{\Bbbk})$, then A is in $\mathsf{Artin}_{\mathcal{E}_n}$.

Definition 2.7.6. Fix a field $\Bbbk$, and let $X\colon \mathsf{Artin}_{\mathcal{E}_n} \to \mathsf{Spaces}$ be a functor of local Artin $\mathcal{E}_n$-algebras over $\Bbbk$. The factorization homology of X over M is the $\Bbbk$-module

$$\int_M X := \lim_{R \in \mathsf{Artin}^{\mathsf{op}}_{\mathcal{E}_n/X}} \int_M R.$$

Such formal moduli problems X are defined by the generalization of the Maurer–Cartan functor from classical Lie theory. In the following definition, we use the Koszul duality functor $\mathbb{D}^n$ given as the composite of linear duality with factorization homology of $\mathbb{D}^n/\partial\mathbb{D}^n$.

$$\mathbb{D}^n : \mathsf{Alg}^{\mathsf{aug}}_{\mathcal{E}_n}(\mathsf{Mod}_{\Bbbk}) \longrightarrow \mathsf{cAlg}^{\mathsf{aug}}_{\mathcal{E}_n}(\mathsf{Mod}_{\Bbbk}) \longrightarrow \mathsf{Alg}^{\mathsf{aug}}_{\mathcal{E}_n}(\mathsf{Mod}_{\Bbbk})^{\mathsf{op}} .$$

That is, the Koszul-dual $\mathcal{E}_n$-algebra is

$$\mathbb{D}^n A = \Big(\int_{\mathbb{D}^n/\partial\mathbb{D}^n} A\Big)^{\vee}.$$

Definition 2.7.7. For an Artin $\mathcal{E}_n$-algebra R and an augmented $\mathcal{E}_n$-algebra A over a field $\Bbbk$, the Maurer–Cartan space

$$\mathsf{MC}_A(R) := \mathsf{Alg}^{\mathsf{aug}}_{\mathcal{E}_n}(\mathbb{D}^n R, A)$$

is the space of maps from the Koszul dual of R to A. The Maurer–Cartan functor $\mathsf{MC}\colon \mathsf{Alg}^{\mathsf{aug}}_{\mathcal{E}_n} \to \mathsf{Fun}(\mathsf{Artin}_{\mathcal{E}_n}, \mathsf{Spaces})$ is the adjoint of the pairing

$$\mathsf{Artin}_{\mathcal{E}_n} \times \mathsf{Alg}^{\mathsf{aug}}_{\mathcal{E}_n}(\mathsf{Mod}_\Bbbk) \xrightarrow{\mathbb{D}^n \times \mathsf{id}} \mathsf{Alg}^{\mathsf{aug}}_{\mathcal{E}_n}(\mathsf{Mod}_\Bbbk)^{\mathsf{op}} \times \mathsf{Alg}^{\mathsf{aug}}_{\mathcal{E}_n}(\mathsf{Mod}_\Bbbk) \xrightarrow{\mathsf{Map}(-,-)} \mathsf{Spaces}\,.$$

MC sends A to the functor $\mathsf{MC}_A : \mathsf{Artin}_{\mathcal{E}_n} \to \mathsf{Spaces}$.

The following is the framed version of the main theorem of [5].

Theorem 2.7.8 (Poincaré/Koszul duality [5])**.** *Let M be a closed framed n-manifold; let A be an augmented $\mathcal{E}_n$-algebra over a field $\Bbbk$ with MC_A the associated formal moduli functor of $\mathcal{E}_n$-algebras. There is a natural equivalence*

$$\Big(\int_M A\Big)^\vee \simeq \int_M \mathsf{MC}_A$$

between the $\Bbbk$-linear dual of the factorization homology of M with coefficients in A and the factorization homology of M with coefficients in the moduli functor MC_A.

Remark 2.7.9. One can regard the failure of the duality map in Theorem 2.7.1 to be an equivalence to be related to the non-affineness of the formal moduli problem MC_A. Were MC_A to be affine, it would be equivalent to the formal spectrum of the Koszul dual, $\mathsf{Spf}(\mathbb{D}^n A)$, in which the case the duality map of Theorem 2.7.1 would be an equivalence. In terms of topological quantum field theory, we regard MC_A as determining a sigma-model whose target $X = \mathsf{MC}_A$ is formal, but not necessarily affine. Theorem 2.7.8 then describes the observables $\int_M X$ in this sigma-model in terms of the factorization homology with coefficients in a shift of the cotangent space of the point $\mathsf{Spec}(\Bbbk) \to X$.

2.8 Factorization homology for singular manifolds

We briefly outline an extension of factorization homology to singular manifolds, such as manifolds equipped with *defects*. The notion of singular manifolds, and structured versions thereof, is developed in [7]. The development of factorization homology of such can be found in [8].

2.8.1 Singular manifolds

Heuristically, a singular manifold is a stratified space in which each stratum is endowed with the structure of a smooth manifold, and in which each stratum has a canonically associated link which is itself a singular manifold. Informally, the ∞-category $\mathcal{S}\mathsf{nglr}$ of singular manifolds and open (stratified) embeddings among such is minimal with respect to the following features.

1. The empty set $\emptyset$ is a singular manifold.

2. For L a compact singular manifold, the open cone

$$\mathsf{C}(L) := * \coprod_{L\times\{0\}} L\times[0,1)$$

is a singular manifold, with the cone-point $*$ a stratum.

3. For X and Y singular manifolds, their product $X\times Y$ is a singular manifold.

4. For $U\subset X$ an open subset of a singular manifold, then U is canonically a singular manifold.

5. For $\mathcal{U}$ an open cover by singular manifolds of a paracompact Hausdorff space X, then X is a singular manifold.

6. For L a compact singular manifold, the continuous homomorphism

$$\mathsf{Aut}(L)\xrightarrow{\simeq}\mathsf{Aut}(\mathsf{C}(L)),\qquad f\mapsto\mathsf{C}(f),$$

is equipped with a deformation retraction whose retraction $D_*\colon \mathsf{Aut}(\mathsf{C}(L))\to\mathsf{Aut}(L)$ is a continuous homomorphism.

A rigorous definition of $\mathcal{S}\mathsf{nglr}$ is given in §1 of [7]. That definition is distractingly inductive, and so we elect to not spell it out here.

Terminology 2.8.1. In this article, by "singular manifold" we will mean "finitary singular manifold with boundary," in the sense of [7].

Remark 2.8.2. The first two axioms give that $*$ is a singular manifold, and thereafter that $\mathsf{C}(*)=[0,1)$ is a singular manifold with strata $\{0\}$ and $(0,1)$. The third and fifth axiom gives that $(0,1)^{\times n}\cong\mathbb{R}^n$ is a singular manifold. The fifth axiom thereafter gives that each manifold is a singular manifold. The second and third axiom then give that $\mathbb{R}^k\times\mathsf{C}(L)$ is a singular manifold for each compact manifold L. In general, an arbitrary singular manifold is a paracompact Hausdorff topological space that admits a basis for its topology consisting of open embeddings from singular manifolds of the form $\mathbb{R}^k\times\mathsf{C}(L)$, where k is an integer and L is a compact singular manifold.

A *basic* is a singular manifold of the form $\mathbb{R}^k\times\mathsf{C}(L)$ for k an integer and L a compact singular manifold.

Remark 2.8.3. We follow up on the previous remark. Namely, the definition of a singular manifold detailed in [7] is in analogy with that of a smooth manifold: it is a paracompact Hausdorff topological space equipped with a maximal atlas by *basics*.

Remark 2.8.4. The sixth axiom above reflects the regularity of singular manifolds. The retraction map can be interpreted as taking the 'derivative at the cone-point'. A consequence of this structure is that one can 'resolve singularities' of a singular manifold. Specifically, for X a singular manifold, and for $X_d\subset X$ a stratum that is closed as a subspace, there is a well-defined *blow-up* of X along X_d, as well as *link* of X_d in X, which assemble as the pullback and pushout diagram consisting of singular manifolds and continuous stratified maps among them:

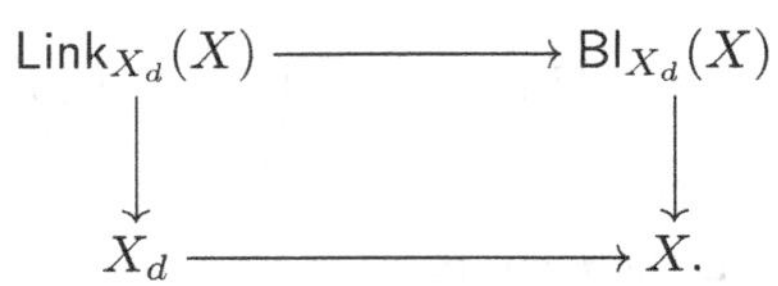

$$\begin{array}{ccc} \mathsf{Link}_{X_d}(X) & \longrightarrow & \mathsf{Bl}_{X_d}(X) \\ \downarrow & & \downarrow \\ X_d & \longrightarrow & X. \end{array}$$

This diagram has the feature that the left vertical map is a bundle of singular manifolds, while the right vertical map restricts as a bundle of singular manifolds over each stratum of X, and over the complement $X \smallsetminus X_d$ the right vertical map is an isomorphism.

Example 2.8.5. Let $\overline{M}$ be an n-manifold with boundary. Regard $\overline{M}$ as a stratified space, whose strata are the connected components of the boundary $\partial\overline{M}$ and the connected components of of the interior $M := \overline{M} \smallsetminus \partial\overline{M}$. As a stratified space, $\overline{M}$ is a singular manifold. Indeed, it admits an atlas by basics of the form $\mathbb{R}^n$ and $\mathbb{R}^{n-1} \times \mathsf{C}(*)$.

Example 2.8.6. Let M be an n-manifold. Let $W \subset M$ be a properly embedded d-submanifold. Consider the stratified space $(W \subset M)$ whose strata are the connected components of W and the connected components of $M \smallsetminus W$. This stratified space $(W \subset M)$ is a singular manifold. Indeed, it admits an atlas by basics of the form $\mathbb{R}^n$ and $\mathbb{R}^d \times \mathsf{C}(S^{n-d-1})$. A singular manifold of this form is a *defect*, or *refinement*, of the manifold M. In particular, a knot in a 3-manifold determines a stratified space that refines onto the given 3-manifold.

Example 2.8.7. Following up on Example 2.8.6, consider a sequence $W_0 \subset W_1 \subset M$ of properly embedded d_0-submanifold in a properly embedded d_1-submanifold in an n-manifold. Consider the stratified space $(W_0 \subset W_1 \subset M)$ whose strata are the connected components of W_0, the connected components of $W_1 \smallsetminus W_0$, and the connected components of $M \smallsetminus W_1$. This stratified space $(W_0 \subset W_1 \subset M)$ is a singular manifold. Indeed, it admits an atlas by basics of the form $\mathbb{R}^n$, $\mathbb{R}^{d_1} \times \mathsf{C}(S^{n-d_1-1})$, and $\mathbb{R}^{d_0} \times \mathsf{C}(\mathbf{S}^{n-d_0-1})$ where $\mathbf{S}^{n-d_0-1} = (S^{d_1-d_0-1} \subset S^{n-d_0-1})$ is the singular manifold of Example 2.8.6 applied to an equatorially embedded sphere.

Example 2.8.8. Let $p \geq 0$, and consider the topological simplex Δ^p. Consider the stratification of Δ^p in which the i-dimensional strata comprise the connected components of the complement of skeleta $\mathsf{sk}_i(\Delta^n) \smallsetminus \mathsf{sk}_{i-1}(\Delta^n)$. This stratified space is a singular manifold. Indeed, it admits an atlas by basics of the form $\mathbb{R}^{n-p} \times \mathsf{C}(\Delta^{p-1})$ for $0 \leq p \leq n$. As a consequence, the geometric realization $|Z|$ of a finite simplicial complex is canonically endowed with the structure of a singular manifold.

Example 2.8.9. The Grassmannian $\mathsf{Gr}_k(n)$ of k-planes in $\mathbb{R}^n$ admits the *Schubert* stratification, for which, for each cardinality k-subset $S \subset \{1, \ldots, n\}$, the S-stratum consists of those V for which S is minimal (in the Schubert poset) for which the linear map $V \hookrightarrow \mathbb{R}^n \xrightarrow{\mathsf{proj}} \mathbb{R}^S$ is an isomorphism. This Schubert stratification of $\mathsf{Gr}_k(n)$ is a singular manifold.

Example 2.8.10. The orthogonal group $\mathsf{O}(n)$ admits the *Bruhat* stratification, for which, for each element $\varphi \in \Sigma_n \wr (\mathbb{Z}/2\mathbb{Z}) = \mathsf{Aut}(n\mathsf{Cube}) \subset \mathsf{O}(n)$ of the group of symmetries of the n-cube, the φ-stratum consists of those orthogonal matrices that, via row elimination and scaling by positive scalars, have φ as its matrix of pivots. This Bruhat stratification of $\mathsf{O}(n)$ is a singular manifold.

Recall from §2.2.2 that we classified sheaves on the ∞-category $\mathcal{M}\mathsf{fld}_n$ as space over $\mathsf{BO}(n)$; for $B \to \mathsf{BO}(n)$ such a thing, the space of sections of the associated sheaf on an n-manifold M is the space of lifts of the tangent classifier:

$$\begin{array}{ccc} & & B \\ & \overset{\varphi}{\nearrow} & \downarrow \\ M & \xrightarrow{\tau_M} & \mathsf{BO}(n). \end{array}$$

This was the context supporting the constant consideration of spaces over $\mathsf{BO}(n)$ in the developments above.

In [7] there appears a similar classification. Namely, consider the full ∞-subcategory $\mathcal{B}\mathsf{sc} \subset \mathcal{S}\mathsf{nglr}$ consisting of the basic singular manifolds.

Theorem 2.8.11. [7] *Restriction along the inclusion $\mathcal{B}\mathsf{sc} \hookrightarrow \mathcal{S}\mathsf{nglr}$ defines an equivalence between ∞-categories*

$$\mathsf{Shv}(\mathcal{S}\mathsf{nglr}) \xrightarrow{\simeq} \mathsf{PShv}(\mathcal{B}\mathsf{sc}).$$

Through the straightening-unstraightening equivalence of §2 of [42], Theorem 2.8.11 gives that a sheaf on $\mathcal{S}\mathsf{nglr}$ is equivalent data to a right fibration among ∞-categories:

$$\mathcal{B} \longrightarrow \mathcal{B}\mathsf{sc};$$

a section of the space of sections of the associated sheaf on a singular manifold X is the space of lifts

$$\begin{array}{ccc} & & \mathcal{B} \\ & \overset{\varphi}{\nearrow} & \downarrow \\ \mathsf{Exit}(X)^{\mathsf{op}} & \xrightarrow{\tau_X} & \mathcal{B}\mathsf{sc}; \end{array}$$

we call such a lift a *$\mathcal{B}$-structure.* Here, $\mathsf{Exit}(X)$ is the exit-path ∞-category of X, and the functor τ_X carries a point in X to a basic neighborhood of it. In [7] it is shown that the functor $\mathsf{Exit}(X)^{\mathsf{op}} \xrightarrow{\tau_X} \mathcal{B}\mathsf{sc}$ is the unstraightening of the presheaf $\mathcal{B}\mathsf{sc}^{\mathsf{op}} \xrightarrow{U \mapsto \mathsf{Emb}(U,X)} \mathsf{Spaces}$. In particular, $\mathsf{Exit}(X)^{\mathsf{op}} \to \mathcal{B}\mathsf{sc}$ is a right fibration.

Example 2.8.12. Let $B \to \mathsf{BO}(n)$ be a map from a space. The composite map $B \to \mathsf{BO}(n) \to \mathcal{B}\mathsf{sc}$ is a right fibration. A singular manifold X admits a B-structure if and only if X is an n-manifold. Furthermore, should X be an n-manifold, the space B-structures is the space of B-framings on that n-manifold.

Example 2.8.13. Let $\mathcal{D}_{d\subset n} \subset \mathcal{B}\mathsf{sc}$ be the full ∞-subcategory consisting of the two objects $\mathbb{R}^d \times \mathsf{C}(S^{n-d-1})$ and $\mathbb{R}^n$. The inclusion $\mathcal{D}_{d\subset n} \hookrightarrow \mathcal{B}\mathsf{sc}$ is a right fibration. A singular manifold X admits a $\mathcal{D}_{d\subset n}$-structure if and only if the singular manifold is of the form $(W^d \subset M^n)$, a properly embedded d-submanifold in an n-manifold (see Example 2.8.6). Furthermore, if X admits a $\mathcal{D}_{d\subset n}$-structure, then it is unique.

Example 2.8.14. Choose, once and for all, an orientation-preserving diffeomorphism $\mathbb{R} \cong (0,1)$. Consider the right fibration $\mathcal{D}^{\mathsf{fr}}_{n-1,n} := \{n-1 < n\} \to \mathcal{B}\mathsf{sc}$ that selects out the morphism $\mathbb{R}^n = \mathbb{R}^{n-1} \times \mathbb{R} \cong \mathbb{R}^{n-1} \times (0,1) \hookrightarrow \mathbb{R}^{n-1} \times \mathsf{C}(*)$, which is the standard open embedding of the complement of the cone-locus. A $\mathcal{D}^{\mathsf{fr}}_{n-1,n}$-manifold is an n-manifold with boundary, equipped with a framing of its interior and a splitting of its framing along the boundary for which the last vector field points inward.

Example 2.8.15. Let $\mathcal{D}^{\mathsf{fr}}_{d\subset n} \subset \mathcal{B}\mathsf{sc}$ be the right fibration $\mathsf{Exit}\big(\mathbb{R}^d \times \mathsf{C}(S^{n-d-1})\big)^{\mathsf{op}} \xrightarrow{\tau} \mathcal{B}\mathsf{sc}$. A singular manifold X admits a $\mathcal{D}^{\mathsf{fr}}_{d\subset n}$-structure if and only if the singular manifold is of the form $(W^d \subset M^n)$, a properly embedded d-submanifold in an n-manifold (see Example 2.8.6). Furthermore, for $X = (W \subset M)$ such a singular manifold, a $\mathcal{D}^{\mathsf{fr}}_{d\subset n}$-structure on it is a framing of M and a splitting of the framing along W via the first d-coordinates.

Example 2.8.16. The representable right fibrations over $\mathcal{B}\mathsf{sc}$ are particularly tractable. Let $U = \mathbb{R}^d \times \mathsf{C}(L)$ be a basic. Notate the right fibration $\mathsf{Exit}(U)^{\mathsf{op}} \xrightarrow{\tau_U} \mathcal{B}\mathsf{sc}$ as $\mathcal{D}_U \to \mathcal{B}\mathsf{sc}$. A singular manifold X admits a $\mathcal{D}_U$-structure if and only if X admits an atlas by open subsets of U. Furthermore, for X such a singular manifold, a $\mathcal{D}_U$-structure on X determines the structure of a framed d-manifold on the lowest-dimensional strata X_d of X, as well as a trivialization of the link,

$$\mathsf{Link}_{X_d}(X) \underset{\text{over } X_d}{\cong} X_d \times L,$$

and thereafter, upon the choice of a collar-neighborhood of $\mathsf{Link}_{X_d}(X)$ in the blow-up $\mathsf{Bl}_{X_d}(X)$, an open stratified embedding

$$X_d \times L \times \mathbb{R} \hookrightarrow X \smallsetminus X_d.$$

Observation 2.8.17. Let $U = \mathbb{R}^d \times \mathsf{C}(L)$ be a basic. Consider the right fibration $\mathcal{D}_U \to \mathcal{B}\mathsf{sc}$ of Example 2.8.16. Consider the full ∞-subcategory $\mathsf{Exit}(U \smallsetminus \mathbb{R}^d)^{\mathsf{op}} := \mathcal{D}_{>U} \to \mathcal{D}_U$ consisting of those objects, which are points in U, that do not belong to the cone-locus $\mathbb{R}^d$. The composite functor $\mathcal{D}_{>U} \to \mathcal{D}_U \to \mathcal{B}\mathsf{sc}$ is again a right fibration. For X a singular manifold equipped with a $\mathcal{D}_U$-structure, the complement $X \smallsetminus X_d$ of its lowest-dimensional strata is a singular manifold, and it canonically inherits a $\mathcal{D}_{>U}$-structure from the given $\mathcal{D}_U$-structure on X. Through the conclusion of Example 2.8.16, the product $\mathbb{R}^d \times L \times \mathbb{R} \cong \mathbb{R}^{d+1} \times L$ is canonically endowed with a $\mathcal{D}_{>U}$-structure. Thereafter, taking products with Euclidean-$(d+1)$-spaces defines a symmetric monoidal functor

$$\mathcal{D}\mathsf{isk}^{\mathsf{fr}}_{d+1} \xrightarrow{-\times L} \mathcal{M}\mathsf{fld}(\mathcal{D}_{>U}), \qquad (\mathbb{R}^{d+1})^{\sqcup I} \mapsto (\mathbb{R}^{d+1})^{\sqcup I} \times L \cong (\mathbb{R}^d \times L \times \mathbb{R})^{\sqcup I}. \quad (2.8.1)$$

Definition 2.8.18. Let $\mathcal{B} \to \mathcal{B}\mathsf{sc}$ be a right fibration. The ∞-category $\mathcal{M}\mathsf{fld}(\mathcal{B})$ of *$\mathcal{B}$-manifolds* is the pullback among ∞-categories:

$$\begin{array}{ccc} \mathcal{M}\mathsf{fld}(\mathcal{B}) & \longrightarrow & \mathsf{PShv}(\mathcal{B}\mathsf{sc})_{/\mathcal{B}} \\ \downarrow & & \downarrow \\ \mathcal{S}\mathsf{nglr} & \xrightarrow{\tau\colon X \mapsto \left(\mathsf{Exit}(X)^{\mathsf{op}} \xrightarrow{\tau_X} \mathcal{B}\mathsf{sc}\right)} & \mathsf{PShv}(\mathcal{B}\mathsf{sc}). \end{array}$$

The ∞-category of *$\mathcal{B}$-disks* is the full ∞-subcategory

$$\mathcal{D}\mathsf{isk}(\mathcal{B}) \subset \mathcal{M}\mathsf{fld}(\mathcal{B})$$

consisting of those $\mathcal{B}$-structured singular manifolds X for which X is a finite disjoint union of basics. Disjoint union of underlying singular manifolds makes both $\mathcal{D}\mathsf{isk}(\mathcal{B})$ and $\mathcal{M}\mathsf{fld}(\mathcal{B})$ into symmetric monoidal ∞-categories.

So a $\mathcal{B}$-manifold is a pair (X, φ) consisting of a singular manifold together with a $\mathcal{B}$-structure on it. The space of morphisms between two such are open (stratified) embeddings that respect $\mathcal{B}$-structures.

Example 2.8.19. Consider the basic $U = \mathsf{C}(\Delta^{n-1})$, where Δ^{n-1} is the compact singular manifold of Example 2.8.8. A $\mathcal{D}_U$-manifold is an n-manifold $\overline{M}$ with corners equipped with a framing as well as compatible splittings of the framing along each face. We call an n-manifold with corners equipped with such framing data a *framed $\langle n\rangle$-manifold.* (This notion of $\langle n\rangle$-manifolds, which are n-manifolds with corners with certain corner structure, is thoroughly developed in [40].) For this case of U, we use the simplified notation:

$$\mathcal{D}^{\mathsf{fr}}_{\langle n\rangle} := \mathcal{D}_U \qquad \text{and} \qquad \mathcal{D}\mathsf{isk}^{\mathsf{fr}}_{\langle n\rangle} := \mathcal{D}\mathsf{isk}(\mathcal{D}^{\mathsf{fr}}_{\langle n\rangle}).$$

2.8.2 Homology theories for structured singular manifolds

In analogy with §2.3.6, one can define factorization homology of a $\mathcal{D}\mathsf{isk}(\mathcal{B})$-algebra over a $\mathcal{B}$-manifold, and factorization homology can be characterized as homology theories for $\mathcal{B}$-manifolds.

In this subsection, we fix a right fibration $\mathcal{B} \to \mathcal{B}\mathsf{sc}$, as well as a symmetric monoidal ∞-category $\mathcal{V}$ that is $\otimes$-presentable.

Definition 2.8.20. The ∞-category of *$\mathcal{D}\mathsf{isk}(\mathcal{B})$-algebras* in $\mathcal{V}$ is

$$\mathsf{Alg}_{\mathcal{D}\mathsf{isk}(\mathcal{B})}(\mathcal{V}) := \mathsf{Fun}^{\otimes}\big(\mathcal{D}\mathsf{isk}(\mathcal{B}), \mathcal{V}\big).$$

Remark 2.8.21. Recall the identification with the symmetric monoidal envelope $\mathsf{Env}(\mathcal{E}_n) \simeq \mathcal{D}\mathsf{isk}^{\mathsf{fr}}_n$. This identification can be read as $\mathcal{D}\mathsf{isk}^{\mathsf{fr}}_n$ being initial among symmetric monoidal ∞-categories equipped with an $\mathcal{E}_n$-algebra therein. There is a likewise identification $\mathsf{Env}\big(\mathcal{E}(\mathcal{B})\big) \simeq \mathcal{D}\mathsf{isk}(\mathcal{B})$ where $\mathcal{E}(\mathcal{B})$ is an ∞-operad whose underlying ∞-category of 1-ary operations is $\mathcal{B}$, and whose space I-ary morphisms from $(U_i)_{i\in I} \in \mathcal{B}^I$ to $V \in \mathcal{B}$ is the space of $\mathcal{B}$-structured embeddings from the I-fold disjoint union: $\mathsf{Map}_{\mathcal{D}\mathsf{isk}(\mathcal{B})}\big(\bigsqcup_{i\in I} U_i, V\big)$.

The Definition 2.3.7 of factorization homology for n-manifolds is adequately formal, defined through a universal property, to imitate for singular manifolds.

Definition 2.8.22. For $A \in \mathsf{Alg}_{\mathcal{D}\mathsf{isk}(\mathcal{B})}(\mathcal{V})$, and X a $\mathcal{B}$-manifold, the *factorization homology* of A over X is the colimit in $\mathcal{V}$,

$$\int_X A := \mathsf{colim}\big(\mathcal{D}\mathsf{isk}(\mathcal{B})_{/X} \to \mathcal{D}\mathsf{isk}(\mathcal{B}) \xrightarrow{A} \mathcal{V}\big),$$

should it exist.

Restriction along the fully faithful symmetric monoidal functor $\iota\colon \mathcal{D}\mathsf{isk}(\mathcal{B}) \hookrightarrow \mathcal{M}\mathsf{fld}(\mathcal{B})$ defines a functor

$$\mathsf{Alg}_{\mathcal{D}\mathsf{isk}(\mathcal{B})}(\mathcal{V}) \longleftarrow \mathsf{Fun}^{\otimes}\big(\mathcal{M}\mathsf{fld}(\mathcal{B}),\mathcal{V}\big) : \iota^*. \tag{2.8.2}$$

Proposition 2.8.23. *Factorization homology exists, and defines a left adjoint to (2.8.2):*

$$\int : \mathsf{Alg}_{\mathcal{D}\mathsf{isk}(\mathcal{B})}(\mathcal{V}) \rightleftarrows \mathsf{Fun}^{\otimes}\big(\mathcal{M}\mathsf{fld}(\mathcal{B}),\mathcal{V}\big) : \iota^*.$$

Furthermore, the a priori lax-commutative diagram among ∞-categories involving a left adjoint $\iota_!$ of the restriction functor ι^*,

$$\begin{array}{ccc} \mathsf{Alg}_{\mathcal{D}\mathsf{isk}(\mathcal{B})}(\mathcal{V}) & \xrightarrow{\quad\int\quad} & \mathsf{Fun}\big(\mathcal{D}\mathsf{isk}(\mathcal{B}),\mathcal{V}\big) \\ {\scriptstyle\text{forget}}\big\downarrow & & \big\downarrow{\scriptstyle\text{forget}} \\ \mathsf{Fun}\big(\mathcal{D}\mathsf{isk}(\mathcal{B}),\mathcal{V}\big) & \xrightarrow{\quad\iota_!\quad} & \mathsf{Fun}\big(\mathcal{M}\mathsf{fld}(\mathcal{B}),\mathcal{V}\big), \end{array}$$

is in fact a commutative diagram.

The Definition 2.3.25 of a collar-gluing for n-manifolds can be imitated for singular manifolds: for X a singular manifold, a *collar-gluing* of X is a continuous map $f\colon X \to [-1,1]$ for which the restriction $f_|\colon f^{-1}(-1,1) \to (-1,1)$ is a fiber bundle among singular manifolds. We denote such a collar-gluing as $X_- \bigcup_{X_0\times\mathbb{R}} X_+ \cong X$ where we understand that $X_- := f^{-1}[-1,1)$, $X_+ := f^{-1}(-1,1]$, and $X_0 := f^{-1}(0)$, while $X_0 \times \mathbb{R} \cong X_0 \times (-1,1) \cong f^{-1}(-1,1)$. Such a collar-gluing of a $\mathcal{B}$-manifold X determines an algebra in the symmetric monoidal ∞-category $\mathcal{M}\mathsf{fld}(\mathcal{B})$ over the multi-category $\mathsf{Assoc}^{\mathsf{RL}}$:

$$\big(X_0; X_+, X_-\big) \in \mathsf{Alg}_{\mathsf{Assoc}^{\mathsf{RL}}}\big(\mathcal{M}\mathsf{fld}(\mathcal{B})\big).$$

Thereafter, each symmetric monoidal functor $\mathcal{F}\colon \mathcal{M}\mathsf{fld}(\mathcal{B}) \to \mathcal{V}$ determines, for each such collar-gluing, a canonical morphism in $\mathcal{V}$ from the 2-sided bar construction:

$$\mathcal{F}(X_-) \underset{\mathcal{F}(X_0)}{\otimes} \mathcal{F}(X_+) \longrightarrow \mathcal{F}(X). \tag{2.8.3}$$

Definition 2.8.24. The ∞-category of *homology theories* for $\mathcal{B}$-manifolds is the full ∞-subcategory

$$\mathbf{H}\big(\mathcal{M}\mathsf{fld}(\mathcal{B}),\mathcal{V}\big) \subset \mathsf{Fun}^{\otimes}\big(\mathcal{M}\mathsf{fld}(\mathcal{B}),\mathcal{V}\big)$$

consisting of those symmetric monoidal functors $\mathcal{F}$ that are $\otimes$-*excisive*:

For which, for each collar-gluing $X_- \bigcup_{X_0\times\mathbb{R}} X_+ \cong X$ of a $\mathcal{B}$-manifold, the canonical morphism in $\mathcal{V}$,

$$\mathcal{F}(X_-) \underset{\mathcal{F}(X_0)}{\otimes} \mathcal{F}(X_+) \xrightarrow{(2.8.3)} \mathcal{F}(X)$$

is an equivalence.

The following foundational characterizing factorization homology is result is proved in [8].

Theorem 2.8.25. *Factorization homology defines an equivalence between ∞-categories,*

$$\int : \mathsf{Alg}_{\mathcal{D}\mathsf{isk}(\mathcal{B})}(\mathcal{V}) \simeq \mathbf{H}(\mathcal{M}\mathsf{fld}(\mathcal{B}), \mathcal{V}),$$

with inverse given by restriction.

2.8.3 Characterizing some $\mathcal{D}\mathsf{isk}(\mathcal{D}_U)$-algebras

Recall from §2.2.5 that, for $B = *$ so that a B-framing on an n-manifold is a framing thereof, then a $\mathcal{D}\mathsf{isk}_n^{\mathsf{fr}}$-algebra was precisely an $\mathcal{E}_n$-algebra, which is a rather algebraic entity. In the same spirit, we characterize some $\mathcal{D}\mathsf{isk}(\mathcal{B})$-algebras in algebraic terms. Through Theorem 2.8.25, this gives algebraic input for invariants of $\mathcal{B}$-manifolds.

In this subsection, we fix a commutative ring spectrum $\Bbbk$, and consider its symmetric monoidal ∞-category $(\mathsf{Mod}_\Bbbk, \otimes_\Bbbk)$ of $\Bbbk$-modules and tensor product over $\Bbbk$ among them.

The next result is an immediate consequence of the main result of [57] after the observation that $\mathcal{D}\mathsf{isk}(\mathcal{D}^{\mathsf{fr}}_{n-1,n})$ is the symmetric monoidal envelope of the Swiss cheese operad. To state the result, recall that, for B a $\mathcal{E}_k$-algebra in $(\mathsf{Mod}_\Bbbk, \otimes_\Bbbk)$, its (derived) center is

$$\mathsf{Z}(B) := \mathsf{Hom}_{\int_{S^{k-1}} B}(B, B),$$

the endomorphisms of B as a module over the $\mathcal{E}_1$-algebra $\int_{S^{k-1}} B$. Deligne's conjecture, as stated and proved in §5 of [43], endows $\mathsf{Z}(B)$ with a canonical structure of a $\mathcal{E}_{k+1}$-algebra structure in $(\mathsf{Mod}_\Bbbk, \otimes_\Bbbk)$.

Proposition 2.8.26. *Recall from Example 2.8.5 the right fibration $\mathcal{D}^{\mathsf{fr}}_{n-1,n} \to \mathcal{B}\mathsf{sc}$. A $\mathcal{D}\mathsf{isk}(\mathcal{D}^{\mathsf{fr}}_{n-1,n})$-algebra in $(\mathsf{Mod}_\Bbbk, \otimes_\Bbbk)$ is equivalent to the following data.*

1. *An $\mathcal{E}_n$-algebra A in $(\mathsf{Mod}_\Bbbk, \otimes_\Bbbk)$.*

2. *An $\mathcal{E}_{n-1}$-algebra B in $(\mathsf{Mod}_\Bbbk, \otimes_\Bbbk)$.*

3. *An action of A on B, instantiated as a morphism between $\mathcal{E}_n$-algebras*

$$A \longrightarrow \mathsf{Z}(B)$$

to the (derived) center of B.

The above result is typical of its kind. The next result appears in [8].

Proposition 2.8.27. *Consider the right fibration $\mathcal{D}^{\mathsf{fr}}_{d\subset n} \to \mathcal{B}\mathsf{sc}$ of Example 2.8.15. A $\mathcal{D}\mathsf{isk}(\mathcal{D}^{\mathsf{fr}}_{d\subset n})$-algebra in $(\mathsf{Mod}_\Bbbk, \otimes_\Bbbk)$ is equivalent to the following data.*

1. *A $\mathcal{E}_n$-algebra A in $(\mathsf{Mod}_\Bbbk, \otimes_\Bbbk)$.*

2. *A $\mathcal{E}_d$-algebra B in $(\mathsf{Mod}_\Bbbk, \otimes_\Bbbk)$.*

3. *An action of* $\int_{S^{n-d-1}} A$ *on* B, *instantiated as a morphism between* $\mathcal{E}_{d+1}$*-algebras*

$$\int_{S^{n-d-1}} A \longrightarrow \mathsf{Z}(B).$$

Example 2.8.28. Consider the case $(d,n) = (1,3)$. A compact $\mathcal{D}^{\mathsf{fr}}_{1\subset 3}$-manifold is a link in a 3-manifold, equipped with a framing of the 3-manifold and a splitting of this framing along the link via the first coordinate. A $\mathcal{D}\mathsf{isk}(\mathcal{D}^{\mathsf{fr}}_{1\subset 3})$-algebra is an $\mathcal{E}_3$-algebra A, a $\mathcal{E}_1$-algebra B, and a morphism between $\mathcal{E}_2$-algebras

$$\alpha\colon \mathsf{HH}_{\bullet}(A) \longrightarrow \mathsf{HH}^{\bullet}(B)$$

from the Hochschild chains to the Hochschild cochains. Through Theorem 2.8.25, such algebraic data determines, via factorization homology, a $\Bbbk$-module associated to each framed link in a framed 3-manifold, $L \subset M$:

$$\int_{(L\subset M)} (A, B, \alpha) \;\in\; \mathsf{Mod}_{\Bbbk}\,.$$

The above two results are specific instances of a more general paradigm.

Proposition 2.8.29. *Let* $U = \mathbb{R}^d \times \mathsf{C}(L)$ *be a basic. Consider the right fibration* $\mathcal{D}_U \to \mathcal{B}\mathsf{sc}$ *from Example 2.8.16. Consider the right fibration* $\mathcal{D}_{>U} \to \mathcal{B}\mathsf{sc}$ *from Observation 2.8.17. A* $\mathcal{D}\mathsf{isk}(\mathcal{D}_U)$*-algebra in* $(\mathsf{Mod}_{\Bbbk}, \otimes_{\Bbbk})$ *is equivalent to the following data.*

1. *A* $\mathcal{D}\mathsf{isk}(\mathcal{D}_{>U})$*-algebra* A *in* $(\mathsf{Mod}_{\Bbbk}, \otimes_{\Bbbk})$.

   Note how restriction along (2.8.1) in Observation 2.8.17, followed by factorization homology, defines an $\mathcal{E}_{d+1}$*-algebra in* $(\mathsf{Mod}_{\Bbbk}, \otimes_{\Bbbk})$:

$$\int_L A\colon\; \mathcal{D}\mathsf{isk}^{\mathsf{fr}}_{d+1} \xrightarrow[(2.8.1)]{\times L} \mathcal{M}\mathsf{fld}(\mathcal{D}_{>U}) \xrightarrow{\int A} \mathsf{Mod}_{\Bbbk}\,.$$

2. *A* $\mathcal{E}_d$*-algebra* B *in* $(\mathsf{Mod}_{\Bbbk}, \otimes_{\Bbbk})$.

3. *An action of* $\int_L A$ *on* B, *instantiated as a morphism between* $\mathcal{E}_{d+1}$*-algebras*

$$\int_L A \longrightarrow \mathsf{Z}(B).$$

The next result references Example 2.8.19, notably the right fibration $\mathcal{D}^{\mathsf{fr}}_{\langle n\rangle} \to \mathcal{B}\mathsf{sc}$ for which a $\mathcal{D}^{\mathsf{fr}}_{\langle n\rangle}$-manifold is a framed $\langle n\rangle$-manifold.

Corollary 2.8.30. *The following data canonically determines a* $\mathcal{D}\mathsf{isk}^{\mathsf{fr}}_{\langle n\rangle}$*-algebra in* $(\mathsf{Mod}_{\Bbbk}, \otimes_{\Bbbk})$.

1. *For each* $0 \leq i \leq n$, *a* $\mathcal{E}_i$*-algebra* A_i *in* $(\mathsf{Mod}_{\Bbbk}, \otimes_{\Bbbk})$.

2. *For each* $0 < i \leq n$, *a morphism between* $\mathcal{E}_i$*-algebras*

$$\alpha_i\colon A_i \longrightarrow \mathsf{Z}(A_{i-1}).$$

Remark 2.8.31. Suppose $\Bbbk$ is an algebraically closed field whose characteristic is 0. Recall from [49] the notions of a shifted symplectic algebraic stack over $\Bbbk$, and of a Lagrangian map to one. We also use the result therein which states that the derived intersection of two Lagrangian maps to an n-shifted symplectic stack canonically inherits the structure of a $(n-1)$-shifted symplectic stack. We expect that the data of Corollary 2.8.30 can be supplied through a parametrized version of deformation quantization from the following input.

- A sequence

$$L_0 \longrightarrow L_1 \longrightarrow \cdots \longrightarrow L_n$$

  of morphisms between algebraic stacks over $\Bbbk$.

  - Now, denote $X_n := L_n$, and for each $0 < i < n$, inductively denote the derived intersection $X_i := L_i \times_{X_{i+1}} L_i$.

- L_n is equipped with an n-shifted structure.

- For each $0 \leq i < n$, the diagonal map

$$L_{i-1} \longrightarrow L_i \underset{X_{i+1}}{\times} L_i$$

  is equipped with a Lagrangian structure.

Acknowledgements

This project was partially supported by NSF grants, for which the authors are grateful: DA, under award numbers 1507704 and 1812055; JF under award number 1508040 and 1812057.

Bibliography

[1] Yinghua Ai, Liang Kong, and Hao Zheng. Topological orders and factorization homology. Adv. Theor. Math. Phys. 21 (2017), no. 8, 1845–1894.

[2] Ricardo Andrade. From manifolds to invariants of E_n-algebras. Thesis (PhD) – Massachusetts Institute of Technology. 2010.

[3] David Ayala and John Francis. Factorization homology of topological manifolds. J. Topol. 8 (2015), no. 4, 1045–1084.

[4] David Ayala and John Francis. Zero-pointed manifolds. arXiv:1409.2857, 2014.

[5] David Ayala and John Francis. Poincaré/Koszul duality. Comm. Math. Phys. 365 (2019), no. 3, 847–933.

[6] David Ayala, John Francis, and Nick Rozenblyum. Factorization homology I: Higher categories. Adv. Math. 333 (2018), 1042–1177.

[7] David Ayala, John Francis, and Hiro Lee Tanaka. Local structures on stratified spaces. Adv. Math. 307 (2017), 903–1028.

[8] David Ayala, John Francis, and Hiro Lee Tanaka. Factorization homology of stratified spaces. Selecta Math. (N.S.) 23 (2017), no. 1, 293–362.

[9] Alexander Beilinson and Vladimir Drinfeld. Chiral algebras. Amer. Math. Soc. Colloq. Publ., 51, 2004.

[10] David Ben-Zvi, Adrien Brochier, and David Jordan. Integrating quantum groups over surfaces. J. Topol. 11 (2018), no. 4, 873–916.

[11] David Ben-Zvi, Adrien Brochier, and David Jordan. Quantum character varieties and braided module categories. Selecta Math. (N.S.) 24 (2018), no. 5, 4711–4748.

[12] David Ben-Zvi, John Francis, and David Nadler. Integral transforms and Drinfeld centers in derived algebraic geometry. J. Amer. Math. Soc. 23 (2010), no. 4, 909–966.

[13] J. Michael Boardman and Rainer Vogt. Homotopy invariant algebraic structures on topological spaces. Lecture Notes in Math., Vol. 347. Springer-Verlag, 1973.

[14] C.-F. Bödigheimer. Stable splittings of mapping spaces. Algebraic topology (Seattle, Wash., 1985), pp. 174–187, Lecture Notes in Math., Vol. 1286. Springer, 1987.

[15] Raoul Bott and Graeme Segal. The cohomology of the vector fields on a manifold. Topology 16 (1977), no. 4, 285–298.

[16] Frederick Cohen. The homology of $\mathcal{C}_{n+1}$-spaces, $n \geq 0$. The homology of iterated loop spaces, pp. 207–351, Lecture Notes in Math., Vol. 533. Springer-Verlag, 1976.

[17] Kevin Costello and Owen Gwilliam. Factorization algebras in quantum field theory. Vol. 1. New Mathematical Monographs, 31. Cambridge University Press, 2017.

[18] Gabriel Drummond-Cole and Ben Knudsen. Betti numbers of configuration spaces of surfaces. J. Lond. Math. Soc. (2) 96 (2017), no. 2, 367–393.

[19] Daniel Dugger and Daniel Isaksen. Topological hypercovers and $\mathbf{A}^1$-realizations. Math. Z. 246 (2004), no. 4, 667–689.

[20] John Francis. Derived algebraic geometry over $\mathcal{E}_n$-rings. Thesis (PhD) – Massachusetts Institute of Technology. 2008.

[21] John Francis. The tangent complex and Hochschild cohomology of $\mathcal{E}_n$-rings. Compos. Math. 149 (2013), no. 3, 430–480.

[22] John Francis and Dennis Gaitsgory. Chiral Koszul duality. Selecta Math. (N.S.) 18 (2012), no. 1, 27–87.

[23] Ezra Getzler and John Jones. Operads, homotopy algebra and iterated integrals for double loop spaces. arXiv:hep-th/9403055, 1994.

[24] Thomas Goodwillie. Calculus. III. Taylor series. Geom. Topol. 7 (2003), 645–711.

[25] Grégory Ginot, Thomas Tradler, and Mahmoud Zeinalian. A Chen model for mapping spaces and the surface product. Ann. Sci. Éc. Norm. Supér. (4) 43 (2010), no. 5, 811–881.

[26] Grégory Ginot, Thomas Tradler, and Mahmoud Zeinalian. Higher Hochschild homology, topological chiral homology and factorization algebras. Comm. Math. Phys. 326 (2014), no. 3, 635–686.

[27] Owen Gwilliam. Factorization algebras and free field theories. Thesis (PhD), Northwestern University. 2012.

[28] André Haefliger. Homotopy and integrability. Manifolds–Amsterdam 1970 (Proc. Nuffic Summer School), pp. 133–163, Lecture Notes in Math., Vol. 197, Springer, 1971.

[29] Quoc Ho. Free factorization algebras and homology of configuration spaces in algebraic geometry. Selecta Math. (N.S.) 23 (2017), no. 4, 2437–2489.

[30] André Joyal. Quasi-categories and Kan complexes. Special volume celebrating the 70th birthday of Professor Max Kelly. J. Pure Appl. Algebra 175 (2002), no. 1-3, 207–222.

[31] Sadok Kallel. Spaces of particles on manifolds and generalized Poincaré dualities. Q. J. Math. 52 (2001), no. 1, 45–70.

[32] Robion Kirby and Laurence Siebenmann. Foundational essays on topological manifolds, smoothings, and triangulations. With notes by John Milnor and Michael Atiyah. Ann. of Math. Stud., No. 88. Princeton University Press; University of Tokyo Press, 1977.

[33] James Kister. Microbundles are fibre bundles. Ann. of Math. (2) 80 (1964), 190–199.

[34] Inbar Klang. The factorization theory of Thom spectra and twisted nonabelian Poincaré duality. Algebr. Geom. Topol. 18 (2018), no. 5, 2541–2592.

[35] Ben Knudsen. Higher enveloping algebras and configuration spaces of manifolds. Thesis (Ph.D.), Northwestern University. 2016.

[36] Ben Knudsen. Betti numbers and stability for configuration spaces via factorization homology. Algebr. Geom. Topol. 17 (2017), no. 5, 3137–3187.

[37] Maxim Kontsevich. Operads and motives in deformation quantization. Moshé Flato (1937–1998). Lett. Math. Phys. 48 (1999), no. 1, 35–72.

[38] Nicholas Kuhn. Goodwillie towers and chromatic homotopy: an overview. Proceedings of the Nishida Fest (Kinosaki 2003), pp. 245–279, Geom. Topol. Monogr., 10, Geom. Topol. Publ., 2007.

[39] Pascal Lambrechts and Ismar Volić. Formality of the little N-disks operad. Mem. Amer. Math. Soc. 230 (2014), no. 1079..

[40] Gerd Laures. On cobordism of manifolds with corners. Trans. Amer. Math. Soc. 352 (2000), no. 12, 5667–5688.

[41] Riccardo Longoni and Paolo Salvatore. Configuration spaces are not homotopy invariant. Topology 44 (2005), no. 2, 375–380.

[42] Jacob Lurie. Higher topos theory. Ann. of Math. Stud., 170. Princeton University Press, 2009.

[43] Jacob Lurie. Higher algebra. Available at `http://www.math.harvard.edu/~lurie/`

[44] Jacob Lurie. Derived algebraic geometry X: Formal moduli problems. Available at `http://www.math.harvard.edu/~lurie/`

[45] J. Peter May. The geometry of iterated loop spaces. Lecture Notes in Math., Vol. 271. Springer-Verlag, 1972.

[46] Dusa McDuff. Configuration spaces of positive and negative particles. Topology 14 (1975), 91–107.

[47] Dusa McDuff. Foliations and monoids of embeddings. Geometric topology (Proc. Georgia Topology Conf., Athens, Ga., 1977), pp. 429–444, Academic Press, 1979.

[48] Edwin Moise. Affine structures in 3-manifolds. V. The triangulation theorem and Hauptvermutung. Ann. of Math. (2) 56, (1952). 96–114.

[49] Tony Pantev, Bertrand Toen, Michel Vaquie, and Gabrielle Vezzosi. Shifted symplectic structures. Publ. Math. Inst. Hautes Études Sci. 117 (2013) 271–328.

[50] Frank Quinn. Ends of maps. III. Dimensions 4 and 5. J. Differential Geom. 17 (1982), no. 3, 503–521.

[51] Paolo Salvatore. Configuration spaces with summable labels. Cohomological methods in homotopy theory (Bellaterra, 1998), pp. 375–395, Progr. Math., 196, Birkhäuser, 2001.

[52] Paolo Salvatore and Nathalie Wahl. Framed discs operads and Batalin–Vilkovisky algebras. Q. J. Math. 54 (2003), no. 2, 213–231.

[53] Graeme Segal. Configuration-spaces and iterated loop-spaces. Invent. Math. 21 (1973), 213–221.

[54] Graeme Segal. Classifying spaces related to foliations. Topology 17 (1978), no. 4, 367–382.

[55] Graeme Segal. The definition of conformal field theory. Topology, geometry and quantum field theory, 421–577, London Math. Soc. Lecture Note Ser., 308, Cambridge Univ. Press, 2004.

[56] Graeme Segal. Locality of holomorphic bundles, and locality in quantum field theory. The many facets of geometry, 164–176, Oxford Univ. Press, 2010.

[57] Justin Thomas. Kontsevich's Swiss cheese conjecture. Geom. Topol. 20 (2016), no. 1, 1-48.

[58] Alexander Voronov. The Swiss-cheese operad. Homotopy invariant algebraic structures (Baltimore, MD, 1998), pp. 365–373, Contemp. Math., 239, Amer. Math. Soc., 1999.

[59] Michael Weiss. Embeddings from the point of view of immersion theory. I. Geom. Topol. 3 (1999), 67–101.

Department of Mathematics, Montana State University, Bozeman, MT 59717, U.S.A.

E-mail address: `david.ayala@montana.edu`

Department of Mathematics, Northwestern University, Evanston, IL 60208, U.S.A.

E-mail address: `jnkf@northwestern.edu`

3

Polyhedral products and features of their homotopy theory

Anthony Bahri, Martin Bendersky, and Frederick R. Cohen

3.1 Introduction

A polyhedral product is a natural topological subspace of a Cartesian product, determined by a simplicial complex K and a family of pointed pairs of spaces (X_i, A_i), one for each vertex of K. As noted in [126], in the special case that A_i is a basepoint for X_i, the polyhedral product mediates between the product $X_1 \times X_2 \times \cdots \times X_m$ when K is the full $(m-1)$-simplex, and the wedge $X_1 \vee X_2 \vee \cdots \vee X_m$ when K consists of m discrete points.

We begin with a brief historical note. The development of the theory of polyhedral products was guided by their inextricable link to spaces known as moment–angle manifolds which arose within the subject of toric topology. This subject, which incorporates ideas from geometry, symplectic geometry, combinatorics and commutative algebra, has now swelled well beyond its original confines as a topological generalization of toric geometry.

Toric topology has precipitated investigations into new areas of manifold and orbifold theory, centered around toric actions, [53, 38, 40, 61, 10, 69, 75, 48, 112, 93, 21, 20], and the unpublished notes of N. Strickland, [123]. An excellent overview by V. Buchstaber and N. Ray is to be found in [41].

The study of polyhedral products in particular has matured to the point that it extends now into a wide variety of fields of mathematics distant from its origin, as detailed in the table below. Consequently, the development of their homotopy theory has become useful in this context.

In order to be more specific, it is now necessary to give a definition. Begin by setting $[m] = \{1, 2, \ldots, m\}$ and define a category $\mathcal{C}([m])$ whose objects are pairs $(\underline{X}, \underline{A})$ where

$$(\underline{X}, \underline{A}) \;=\; \{(X_1, A_1), (X_2, A_2), \ldots, (X_m, A_m)\}$$

is a family of based CW-pairs. A morphism

$$\underline{f}\colon (\underline{X}, \underline{A}) \longrightarrow (\underline{Y}, \underline{B})$$

2010 *Mathematics Subject Classification.* primary: 55P42, 55Q15, 52C35, 52B11, 35S22, 13F55. secondary: 14M25, 14F45, 55U10, 55R20, 55N10.

Key words and phrases. polyhedral product, moment–angle complex, cohomology, arrangements, stable splitting, simplicial wedge, Davis–Januszkiewicz space, Golodness, monomial ideal ring.

consists of m continuous maps $f_i\colon X_i \longrightarrow Y_i$ satisfying $f_i(A_i) \subset B_i$. Next, let K be a simplicial complex on the vertex set $[m]$. We consider K to be a category where the objects are the simplices of K and the morphisms $d_{\sigma,\tau}$ are the inclusions $\sigma \subset \tau$.

Definition 3.1.1. Let K be a simplicial complex on the vertex set $[m]$ as above. A polyhedral product is a functor $Z(K;-)\colon \mathcal{C}([\boldsymbol{m}]) \longrightarrow \mathbf{Top}$ satisfying

$$Z(K;(\underline{X},\underline{A})) \subseteq \prod_{i=1}^{m} X_i$$

that is given as the colimit of a diagram $D: K \to CW_*$, where at each $\sigma \in K$, we set

$$D(\sigma) = \prod_{i=1}^{m} W_i, \quad \text{where} \quad W_i = \begin{cases} X_i & \text{if} \quad i \in \sigma \\ A_i & \text{if} \quad i \in [m] - \sigma. \end{cases} \tag{3.1.1}$$

Though here, the colimit is a union given by

$$Z(K;(\underline{X},\underline{A})) = \bigcup_{\sigma\in K} D(\sigma),$$

the full colimit structure is used heavily in the development of the elementary theory of polyhedral products. Notice that when $\sigma \subset \tau$ then $D(\sigma) \subseteq D(\tau)$. In the case that K itself is a simplex,

$$Z(K;(\underline{X},\underline{A})) = \prod_{i=1}^{m} X_i.$$

Remark 3.1.2. When all the pairs (X_i, A_i) are the same pair (X, A), the family $(\underline{X},\underline{A})$ is written simply as (X, A). Notice that in this case $Aut(K)$ acts naturally on $Z(K;(X,A))$.

In a way entirely similar to that above, a related space $\widehat{Z}(K;(\underline{X},\underline{A}))$, called the *smash polyhedral product*, is defined by replacing the Cartesian product everywhere in Definition 3.1.1 by the smash product. That is,

$$\widehat{D}(\sigma) = \bigwedge_{i=1}^{m} W_i \quad \text{and} \quad \widehat{Z}(K;(\underline{X},\underline{A})) = \bigcup_{\sigma\in K} \widehat{D}(\sigma)$$

with

$$\widehat{Z}(K;(\underline{X},\underline{A})) \subseteq \bigwedge_{i=1}^{m} X_i.$$

Polyhedral products generalize the spaces called *moment–angle complexes* which were developed by V. Buchstaber and T. Panov [37] and correspond to the case

$$(X_i, A_i) = (D^2, S^1), \quad i = 1, 2, \ldots, m.$$

As mentioned above, the construction of moment–angle complexes originated within the subject of toric topology. Specifically, M. Davis and T. Januszkiewicz [53] constructed spaces $\mathcal{Z}$ as quotients of a product of an m-dimensional torus and an n-dimensional polytope that is *simple*, meaning that it satisfies a certain homogeneity condition. (See Section 3.2.) They were able to realize a class of toric manifolds as quotients of $\mathcal{Z}$ by free torus actions. The space $\mathcal{Z}$ was then reformulated as a moment–angle manifold by V. Buchstaber and T. Panov [37].

The moment–angle complex construction (3.3.2) was extended to pairs (X, A) replacing (D^2, S^1) by V. Buchstaber and T. Panov [37], N. Strickland (unpublished), V. Baskakov [23], G. Denham and A. Suciu [56] and T. Panov [109]. The construction in the generality of Definition 3.1.1 appeared in [11].

The table below lists a selection of current applications of polyhedral products.

$(\underline{X}, \underline{A})$	$Z(K; (\underline{X}, \underline{A}))$
(D^2, S^1)	toric geometry and topology, arachnid mechanisms
(D^1, S^0)	surfaces, number theory, representation theory, linear programming
$(S^1, *)$	right–angled Artin groups, robotics
$(\mathbb{R}\mathrm{P}^\infty, *)$	right–angled Coxeter groups
$(\mathbb{C}, \mathbb{C}^*)$	complements of coordinate arrangements
$(\underline{\mathbb{R}}^n, (\underline{\mathbb{R}}^n)^*)$	complements of certain non-coordinate arrangements
$(\underline{\mathbb{C}\mathrm{P}}^\infty, \underline{\mathbb{C}\mathrm{P}}^k)$	monomial ideal rings
$(\underline{EG}, \underline{G})$	free groups and monodromy
$(\underline{BG}, \underline{*})$	combinatorics, aspherical spaces
$(\underline{PX}, \underline{\Omega X})$	homotopy theory, Whitehead products
$(S^{2k+1}, *)$	graph products, quadratic algebras

By way of motivation, we illustrate the Hopf map and the Whitehead product from the point of view of polyhedral products. The 3–sphere S^3 is realized as a polyhedral product by taking K to be the simplicial complex consisting of two discrete points,

$$\partial(D^2 \times D^2) \cong S^1 \times D^2 \cup_{S^1 \times S^1} D^2 \times S^1 = Z(K; (D^2, S^1)) \subset D^2 \times D^2 \subset \mathbb{C} \times \mathbb{C}.$$

The action of the compact two-torus T^2 on the product $D^2 \times D^2$ leaves invariant the component pieces: $S^1 \times D^2$, $S^1 \times S^1$ and $D^2 \times S^1$, which comprise the basic building blocks of the polyhedral product, moment–angle complex, structure in this case. The Hopf map

$$S^3 \simeq S^1 \times D^2 \cup_{S^1 \times S^1} D^2 \times S^1 \longrightarrow D^2 \cup_{S^1} D^2 \simeq S^2$$

is the quotient by the diagonal subgroup S^1 in T^2 acting in a natural way.

Consider next the Whitehead product $[f, g]$ defined by

$$S^3 \xrightarrow{\omega} S^2_\alpha \vee S^2_\beta \xrightarrow{f \vee g} X$$

where ω is a map attaching the top cell of $S^2 \times S^2$ and $f\colon S^2_\alpha \longrightarrow X$ and $g\colon S^2_\beta \longrightarrow X$ are two basepoint preserving maps, [74, page 381]. The particular case $X = S^2_\alpha \vee S^2_\beta$ and $f = \iota_\alpha$, $g = \iota_\beta$, being the respective inclusions of S^2_α and S^2_β into $S^2_\alpha \vee S^2_\beta$, gives the Whitehead product $[\iota_\alpha, \iota_\beta]$. (Note here that $\iota_\alpha \vee \iota_\beta$ is the identity map.) From the point of view of polyhedral products[1], $[\iota_\alpha, \iota_\beta]$ is induced by the map of pairs $(D^2, S^1) \longrightarrow (S^2, *)$ that collapses $S^1 \subset D^2$ to a point,

$$S^3 \simeq D^2_\alpha \times S^1_\beta \cup_{S^1_\alpha \times S^1_\beta} S^1_\alpha \times D^2_\beta \longrightarrow (S^2_\alpha \times *) \cup_{* \times *} (* \times S^2_\beta) \simeq S^2_\alpha \vee S^2_\beta. \tag{3.1.2}$$

Here, the space on the right-hand side is the polyhedral product $Z(K; (\underline{S}^2, *))$ where again, K consists of two discrete points. Within this subject, wedge products appear often, thus Whitehead products appear in profusion. These topics are discussed in more detail in Section 3.14.

Organization

This paper is organized as follows. A brief technical discussion of the origin of polyhedral products as moment–angle complexes in Sections 3.2 and 3.3, includes the original topological construction of toric manifolds by M. Davis and T. Januszkiewicz [53] and the reformulation of one of their main spaces as a moment–angle complex by V. Buchstaber and T. Panov [37, 38].

A discussion follows in Section 3.4 about the independent discovery of moment–angle complexes as intersections of certain quadrics. We describe briefly the work of S. López de Medrano [100], S. López de Medrano and A. Verjovsky [101], F. Bosio and L. Meersseman [31], followed by that of S. López de Medrano and S. Gitler [65]. This approach has proved effective in identifying classes of moment–angle manifolds that are connected sums of products of spheres. This is followed in Section 3.5 by a sketch of the computation of the cohomology ring of moment–angle complexes.

We begin our focus on more general polyhedral products in Section 3.6, by describing their behaviour with respect to the exponentiation of CW-pairs. This has led to an application to the study of toric manifolds [58, 128, 129, 124, 71, 68, 47], and orbifolds [19]. An application to topological joins is included as an example of the utility of this property with respect to exponentiation.

The behaviour of polyhedral products with respect to fibrations, due to G. Denham and A. Suciu [56], is surveyed briefly in Section 3.7. We take advantage of the formalism to describe briefly the early work in this area by G. Porter [117, 118], T. Ganea [64] and A. Kurosh [92]. and G. W. Whitehead. Instances of polyhedral products appear also in the work of D. Anick [6].

In Section 3.8, we review the various fundamental unstable and stable splitting theorems for the polyhedral product. These theorems give access to the homotopy type of the polyhedral product in many of the most important cases and drive a number of applications. The identification of the various wedge summands which appear in the stable decompositions, occupies the second part of this section.

A fine theorem of A. Al-Raisi [3] shows that, in certain cases, the stable splitting of the polyhedral product can be chosen to be equivariant with respect to the action of $Aut(K)$. This and other related results, described in Section 3.9, represent the first foray of the

[1] Lukas Katthän alerted us to this fact.

subject into representation theory. An application of Al-Raisi's theorem to surfaces yields a cyclotomic identity.

The important case $Z(K;(\underline{X},\underline{*}))$ is discussed in Section 3.10. It arises in algebraic combinatorics, the study of free groups and monodromy representations, geometric group theory, right–angled Coxeter and Artin groups and asphericity,

In this section also, we use a recent result of T. Panov and S. Theriault [113] to give a short proof that if K is a flag simplicial complex then, for a discrete group G, $Z(K;(BG,*))$ is an Eilenberg–Mac Lane space, a known result, [45, 54, 122].

Results on the cohomology of polyhedral products are presented in Section 3.11. The discussion begins with a description of the cup-product structure as a consequence of the stable splitting Theorem 3.8.6. A spectral sequence corresponding to a natural filtration of $Z(K;(\underline{X},\underline{A}))$ by the simplices of K is described next and followed by several examples.

Section 3.12 describes an entirely geometric approach to the calculation of the cohomology. The notion of *wedge decomposable pairs* is introduced and the corresponding polyhedral products are determined explicitly as wedges of identifiable spaces. Moreover, it is shown in Theorem 3.12.7, that over a field, every CW-pair $(\underline{X},\underline{A})$ of finite type is cohomologically wedge decomposable. This suffices to give the additive structure of the homology of any $Z(K;(X,A))$ for CW-pairs (X,A) of finite type, and any field coefficients.

The application of polyhedral products to questions concerning the Golod properties of certain rings is described in Section 3.13. Higher Whitehead products, constructed using polyhedral products, are discussed in Section 3.14. In cases when a moment–angle complex splits into a wedge of spheres, higher Whitehead products describe certain canonical maps related to the inclusions of the summands, [111, 72, 83, 1]. They arise also in the analysis of $\Omega Z(K;(D^2,S^1))$.

Though the KO–theory of certain polyhedral products and toric manifolds is discussed in the literature [10, 7], it is omitted from this article. Applications to robotics ([76, 88], for example) are also omitted.

3.2 The origin of polyhedral products in toric topology

A *polytope* is the convex hull of a finite set of points in some $\mathbb{R}^n$. The *dimension* of the polytope is the dimension of its affine hull, [38, Chapter 1]. We shall assume that an n-dimensional polytope is a subset of $\mathbb{R}^n$. A codimension-one face of a polytope is called a *facet*. A polytope P^n is called *simple* if at each vertex, exactly n facets intersect. The convention is to let m denote the number of facets of P^n.

In order to introduce the notion of a *toric manifold*, we recall that the real torus T^n acts on $\mathbb{C}^n$ in a standard way, and the quotient is

$$\mathbb{C}^n/T^n \cong \mathbb{R}^n_+ = \{(x_1,\ldots,x_n) \in \mathbb{R}^n : x_i \geq 0 \text{ for } i=1,\ldots,n\}.$$

A toric manifold M^{2n}, (sometimes called a *quasitoric manifold* in the literature), is a compact manifold covered by local charts $\mathbb{C}^n$, with a global action of a real n-dimensional torus T^n, that restricts to the standard action on each $\mathbb{C}^n$ component. Under this action, each copy of $\mathbb{C}^n$ must have a quotient $\mathbb{C}^n/T^n \cong \mathbb{R}^n_+$ that is a neighborhood of a vertex of a simple polytope $P^n \cong M^{2n}/T^n$.

Remark 3.2.1. Smooth projective toric varieties are examples of toric manifolds in the sense above.

A topological approach to the construction of toric manifolds was developed by M. Davis and T. Januszkiewicz in [53]. They begin with a simple polytope P^n and a function

$$\lambda\colon \mathcal{F} = \{F_1, F_2, \ldots, F_m\} \longrightarrow \mathbb{Z}^n$$

from the set of facets of P^n into an n-dimensional integer lattice. The function λ must satisfy a *regularity* condition, namely, whenever $F = F_{i_1} \cap F_{i_2} \cap \cdots \cap F_{i_l}$ is a codimension-l face of P^n, then $\{\lambda(F_{i_1}), \lambda(F_{i_2}), \ldots, \lambda(F_{i_l})\}$ spans an l-dimensional submodule of $\mathbb{Z}^n$ that is a direct summand. (The fact that every such face can be written uniquely as such an intersection is a consequence of the polytope being simple.) Such a map is called a *characteristic function* associated to P^n.

Next, regarding $\mathbb{R}^n$ as the Lie algebra of T^n, the map λ is used to associate to each codimension-l face F of P^n a rank-l subgroup $G^\lambda_F \subset T^n$, namely the subgroup of T^n determined by the span of the $\lambda(F_{i_j})$. (Specifically, writing $\lambda(F_{i_j}) = (\lambda_{1i_j}, \lambda_{2i_j}, \ldots, \lambda_{ni_j})$ describes the subgroup G^λ_F as

$$\big\{\big(e^{2\pi i(\lambda_{1i_1}t_1+\lambda_{1i_2}t_2+\cdots+\lambda_{1i_l}t_l)}, \ldots, e^{2\pi i(\lambda_{ni_1}t_1+\lambda_{ni_2}t_2+\cdots+\lambda_{ni_l}t_l)}\big) \in T^n\big\} \tag{3.2.1}$$

where $t_i \in \mathbb{R}$, $i = 1, 2, \ldots, l$.)

Finally, let $p \in P^n$ and $F(p)$ be the unique face with p in its relative interior. Define an equivalence relation $\sim_\lambda$ on $T^n \times P^n$ by $(g,p) \sim_\lambda (h,q)$ if and only if $p = q$ and $g^{-1}h \in G^\lambda_{F(p)} \cong T^l$. Then

$$M^{2n} \cong M^{2n}(\lambda) = T^n \times P^n/\sim_\lambda \tag{3.2.2}$$

is a smooth, closed, connected, $2n$-dimensional toric manifold with T^n action induced by left translation [53, page 423]. A projection

$$\pi\colon M^{2n} \to P^n \cong M^{2n}/T^n$$

onto the polytope is induced from the projection $T^n \times P^n \to P^n$.

Example 3.2.2. Consider the case for which $P^n = [0,1]$, the one-simplex having dimension $n = 1$ and two facets, so $m = 2$. Here, the set of facets $\mathcal{F} = \{F_1, F_2\}$ consists of the two vertices of the one-simplex. Define $\lambda : \mathcal{F} \longrightarrow \mathbb{Z}^1$ by $\lambda(F_1) = -1$ and $\lambda(F_2) = 1$. Topologically, $T^1 \times P^1$ is the cylinder $T^1 \times [0,1]$. In this case, the equivalence relation $\sim_\lambda$ collapses to points each circle $T^1 \times \{0\}$ and $T^1 \times \{1\}$, yielding

$$T^1 \times P^1/\sim_\lambda \;\cong\; S^2.$$

In fact, this S^2 has the structure of the toric variety $\mathbb{C}\mathrm{P}^1$.

Remark 3.2.3. Every projective non-singular toric variety has the description (3.2.2); P^n and λ encode topologically the information in the defining fan, [39, Chapter 5].

As part of their approach to the construction of toric manifolds, the authors of [53] introduced a *second* manifold

$$\mathcal{Z} = T^m \times P^n/\!\sim \tag{3.2.3}$$

where T^m denotes the real m–torus, its coordinate circles indexed by the m facets of the simple polytope P^n. Notice here a distinction between (3.2.3) and (3.2.2) given by the fact that T^m has replaced T^n; the equivalence relation will be different too (see below). This space was introduced into the theory of toric manifolds in [53] in order to facilitate the calculation of the cohomology of the toric manifold $M^{2n}(\lambda)$ in (3.2.2).) The equivalence relation $\sim$ is given by defining $\theta\colon \mathcal{F} \longrightarrow \mathbb{Z}^m$ by $\theta(F_i) = \underline{e}_i \in \mathbb{Z}^m$, the standard basis vector.

Remark 3.2.4. Unlike a characteristic map λ (above) the map θ indexes the coordinate circles in T^m by the facets of P^n, *and so depends on the combinatorics of the polytope* P^n *only*.

Next, by exact analogy with the construction of the subgroups $G_F^\lambda \subset T^n$ above for each codimension-l face F of P^n, the map θ determines a rank-l subgroup $G_F^\theta \subset T^m$. The explicit description of this group is similar to (3.2.1). Finally, the equivalence relation $\sim$ is defined in a manner entirely analogous to $\sim_\lambda$ above. That is, for $p \in P^n$, and $F(p)$, the unique face with p in its relative interior, define $\sim$ on $T^n \times P^n$ by $(g,p) \sim (h,q)$ if and only if $p = q$ and $g^{-1}h \in G^\theta_{F(p)} \cong T^l$.

It follows that $\mathcal{Z}$ is a smooth, closed, connected, $(m+n)$–dimensional manifold endowed with a T^m action induced by left translation [53, page 423]. A projection $\mathcal{Z} \to \mathcal{Z}/T^m \cong P^n$ onto the polytope is induced from the projection $T^m \times P^n \to P^n$. (See also [8] and [87].)

Example 3.2.5. Consider the case for which P^n is the one-simplex again so that $n = 1$ and $m = 2$. Topologically, $T^m \times P^n$ is the cylinder $T^2 \times [0,1]$. The equivalence relation collapses one of the coordinate circles in $T^2 \times \{0\}$ to a point as a group, as well as the *other* coordinate circle in $T^2 \times \{1\}$. This gives

$$\mathcal{Z} \cong S^1 * S^1 \cong S^3.$$

A reformulation of the ideas above indicates that every toric manifold M^{2n} can be described as a quotient

$$M^{2n} \cong M^{2n}(\lambda) \cong \mathcal{Z}/T^{m-n} \tag{3.2.4}$$

where here, $T^{m-n} \subset T^m$ is the torus determined by the kernel of the map λ, which the regularity condition of λ ensures acts freely. In the next section, the space $\mathcal{Z}$ is reformulated as a *moment–angle manifold*.

3.3 The introduction of moment–angle complexes

Every simple polytope P^n with m facets, embeds naturally into the union of the facets of the m-cube I^m by virtue of its *cubical decomposition*, [38, Construction 4.5]. V. Buchstaber and T. Panov constructed a T^m–invariant natural subspace $Z_{K_{P^n}}$ of $\mathbb{C}^m$, that completes a commutative diagram:

$$\begin{array}{ccc} Z_{K_{P^n}} & \xrightarrow{\alpha} & (D^2)^m \subset \mathbb{C}^m \\ \big\downarrow /T^m & & \big\downarrow /T^m \\ P^n & \xrightarrow{\iota} & I^m. \end{array} \tag{3.3.1}$$

Here, $D^2 \subset \mathbb{C}$ is the unit disc with the standard circle action, the vertical maps are quotients by the standard action of T^m, and K_{P^n} denotes the boundary complex of the dual of P, (the vertices of K_{P^n} are the facets of P^n and simplices correspond to intersections of facets), [39, Example 2.2.4]. They recognized the space $Z_{K_{P^n}}$ as the natural subspace of $(D^2)^m$ defined as a colimit by a diagram $D : K_{P^n} \to CW_*$, which at each $\sigma \in K_{P^n}$, is given by

$$D(\sigma) = \prod_{i=1}^{m} W_i, \quad \text{where} \quad W_i = \begin{cases} D^2 & \text{if} \quad i \in \sigma \\ S^1 & \text{if} \quad i \in [m] - \sigma. \end{cases} \tag{3.3.2}$$

Here, the colimit is $Z_{K_{P^n}} = \bigcup_{\sigma \in K_{P^n}} D(\sigma)$, and is now written $Z(K_{P^n}; (D^2, S^1))$; it was called a *moment–angle manifold* by V. Buchstaber and T. Panov.

In order to verify that the moment–angle complex $Z(K_{P^n}; (D^2, S^1))$ is in fact the space $\mathcal{Z}$ constructed by M. Davis and T. Januszkiewicz, we work locally. Construction (3.2.3) can be analyzed in the neighborhood of a vertex, where the simple polytope P^n is PL-homeomorphic to $\mathbb{R}^n_+$, as illustrated in Figure 3.3.1.

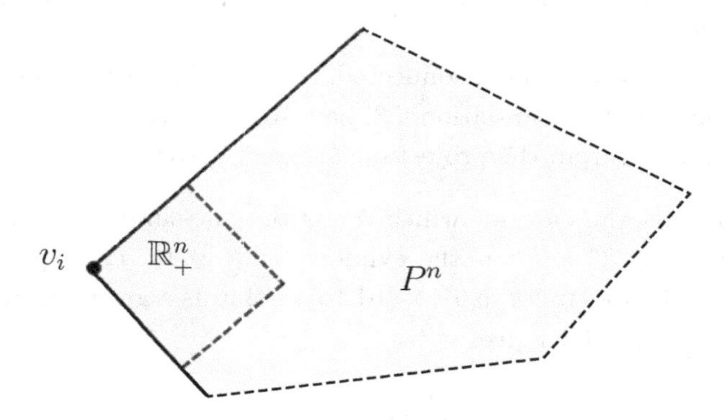

FIGURE 3.3.1. The local structure of a simple polytope P^n

The polytope can be given a *cubical* structure as in [38, Construction 5.8 and Lemma 6.6]. A cube I^n, *anchored* by the vertex v_i, sits inside the copy of $\mathbb{R}^n_+$ that arises by deleting all faces of P^n that do not contain v_i. Locally, $T^m \times P^n$ is

$$T^m \times I^n \cong (S^1 \times I)^n \times (S^1)^{m-n}. \tag{3.3.3}$$

Recall that the map θ, defined in Section 3.2, indexes all the circles in T^m by the facets of the polytope, so the order of factors here has been shuffled naturally. The factors S^1 that are paired with a copy of I are those corresponding to the facets of P^n that meet at v_i. The effect of the equivalence relation $\sim$ in (3.2.3) on $T^m \times I^n$ is to convert every $S^1 \times I$ on the right-hand side of (3.3.3) into a disc D^2, by collapsing $S^1 \times \{0\}$ to a point. So

$$T^m \times I^n/\sim \;\;\cong\;\; (D^2)^n \times (S^1)^{m-n}. \tag{3.3.4}$$

The vertices of P^n correspond to the maximal simplices of the simplicial complex K_{P^n} so, assembling the blocks (3.3.4), gives the moment-angle manifold

$$\mathcal{Z} \;=\; T^m \times P^n/\sim \;\;\cong\;\; Z(K_{P^n}; (D^2, S^1)). \tag{3.3.5}$$

Remark 3.3.1. In a natural way, the Buchstaber–Panov formulation generalizes from K_{P^n} to any simplicial complex K, but the result is not in general a manifold. For general K the space $Z(K; (D^2, S^1))$ is called a moment–angle *complex*. The condition that K be a polytopal sphere, (a triangulated sphere that is isomorphic to the boundary complex of a polytope that has all its faces simplices, a simplicial polytope, [39, Definition 2.5.7]), ensures that $Z(K; (D^2, S^1))$ is a manifold. This was weakened in [42, Corollary 2.10] to: the moment–angle complex is a topological $(n+m)$–manifold if and only if the realization of K is a *generalized homology* $(n-1)$*–sphere*. The latter is defined to be a triangulated polyhedron that has the homology of an $(n-1)$–sphere for which the link of each p-simplex has the homology of an $(n-2-p)$–sphere [42].

Currently, the best result implying that the moment–angle complex is a smooth manifold, requires that K be the simplicial complex underlying a complete simplicial fan; more details are to be found in [114, Theorem 2.2] and [110, Theorem 9.2].

The construction of moment–angle complexes was extended to simplicial posets by Z. Lü and T. Panov in [102] where they are used to study the face rings of posets topologically.

Figure 3.3.2 exhibits the 2-torus as the polyhedral product $Z(K; (D^1, S^0)) \subset (D^1)^4$, where K is the boundary complex of a square on vertices $\{1, 2, 3, 4\}$. According to Definition 3.1.1, the empty simplex of K contributes $(S^0)^4$ which is identified with the 16 vertices of $(D^1)^4$, and the four vertices of K contribute the one-skeleton consisting of 32 edges. The maximal simplices of K are: $\sigma_1 = \{12\}, \sigma_2 = \{14\}, \sigma_3 = \{23\}, \sigma_4 = \{34\}$.

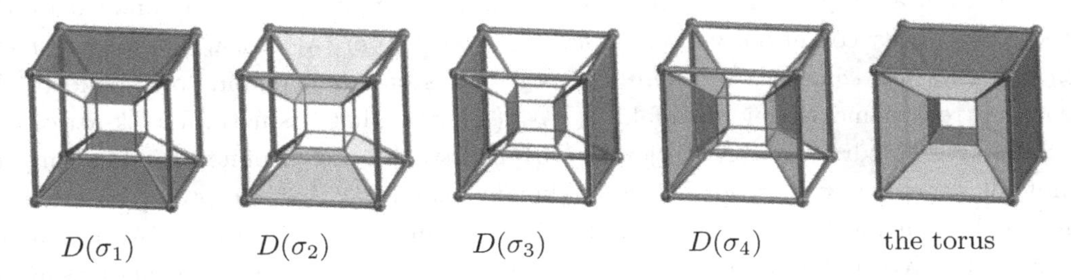

FIGURE 3.3.2. The 2–torus as the polyhedral product $Z(K; (D^1, S^0))$. This figure is reproduced by kind permission of Alvise Trevisan, cf. [127].

Remark 3.3.2. For clarity, we have included the whole one-skeleton in each $D(\sigma_i)$ in Figure 3.3.2, but, strictly speaking, the *whole* one-skeleton should appear in the final picture only.

3.4 Moment-angle complexes as intersections of quadrics

The purpose of this section is to provide an overview of the independent topological development of moment–angle complexes arising from beautiful work of F. Bosio and L. Meersseman [31], Y. Barreto, S. López de Medrano and A. Verjovsky [22], S. López de Medrano [100], as well as his subsequent joint work with S. Gitler [65], and S. López de Medrano and A. Verjovsky [101]. Closely related work was developed earlier, and independently, by C. T. C. Wall [130]. All of these authors considered quadrics specified by certain natural choices of homogeneous quadratic polynomials intersected with the unit sphere. One consequence is the López de Medrano–Gitler proof of a conjecture of Bosio–Meersseman in the case that the polytope P^n is even dimensional and "dual neighborly". For such polytopes P^n, they show that $Z(K_{P^n};(D^1,S^0))$ is a connected sum of products of spheres. In addition to the proof of this conjecture, the authors of [65] also give further computations of various cup-products in moment-angle manifolds.

Let m and n be positive integers satisfying $m > n$. Set $k = 2m-n-1$. A certain manifold, $M \subset \mathbb{C}^m$ associated to an n-dimensional polytope, is defined next as the intersection of k quadrics. Specifically with $\lambda_i \in \mathbb{R}^k, i = 1,\dots,m$ the manifold M is the intersection of the solution of a homogeneous system of equations with the unit sphere in $\mathbb{C}^m$,

$$\Big\{(z_1,z_2,\dots,z_m) \in \mathbb{C}^m : \sum_{i=1}^{m} \lambda_i |z_i|^2 = 0\Big\} \cap \Big\{(z_1,z_2,\dots,z_m) \in \mathbb{C}^m : \sum_{i=1}^{m} |z_i|^2 = 1\Big\}.$$

The manifold M admits an action of the m–torus T^m by multiplication of unit complex numbers on each of the coordinates. The quotient by this action can be identified with the n-dimensional convex polytope P^n given by

$$\sum_{i=1}^{2m} \lambda_i x_i = 0, \quad \sum_{i=1}^{2m} x_i = 1, \quad x_i \geq 0.$$

This construction is essentially equivalent to the construction (3.2.3) of a moment-angle manifold from a simple polytope in Section 3.2. The simplicity of P^n is implied by the weak hyperbolicity condition: *the convex hull of any subset of* $\{\lambda_1,\dots,\lambda_m\}$, *with* k *or fewer elements does not contain the origin.* The real points of this manifold correspond to the associated real moment-angle manifold $Z(K_{P^n};(D^1,S^0))$ that has natural a $\mathbb{Z}_2^n$-action.

S. López de Medrano and A. Verjovsky [101] constructed a class of non-Kähler, compact, manifolds generalizing Hopf and Calabi-Eckmann manifolds that are projectivizations of some of these moment–angle manifolds. L. Meersseman [103] generalized this construction and also observed that all even-dimensional moment–angle manifolds are themselves in this class.

S. López de Medrano [100] showed that when $k = 2$, the moment–angle manifolds constructed above and their real parts are either three-fold products of spheres or connected sums of two-fold products of spheres.

These works culminated in the article by S. Gitler and S. López de Medrano [65], showing that for all values of k, these manifolds are frequently connected sums of products of spheres.

One result is that if K is a polytopal sphere (see Remark 3.3.1) then any manifold $Z(K;(D^1,S^0))$ that is $2c$-dimensional, and $(c-1)$-connected for $c \geq 3$, must be diffeomorphic to a connected sum of products of spheres. This is a proof of a conjecture of Bosio and Meersseman [31, page 115], in the case of even dimensional dual-neighborly polytopes mentioned earlier in this section. The authors also introduce operations including *corner cutting* or *edge cutting* of K and they analyze the effect of these operations on the manifolds $Z(K;(D^1,S^0))$. One result in this direction is that they obtain new, infinite families of manifolds that are diffeomorphic to connected sums of products of spheres.

The methods in [65] influence the argument used by S. Theriault in [125] to prove Panov's Conjecture. The number of wedge summands in a homotopy decomposition of the moment-angle complex corresponding to the simplicial complex consisting of l discrete points, is described in (3.8.1). Panov's conjecture relates this to the connected sum factors in a decomposition of the moment–angle manifold, up to diffeomorphism, corresponding to the simple polytope obtained by making l corner cuts on a standard simplex.

The varieties discussed above, appearing in work of S. López de Medrano et alia, split after suspending once. Other varieties, given by classical Stiefel manifolds, also stably decompose by work of I. M. James [86], and H. Miller [105]. It is natural to consider the variety given by intersections of these two types of quadratic varieties, as well as examine whether, and how they stably decompose. A result of Cohen and Peterson [49] shows that the stable decompositions for Stiefel varieties generically require many suspensions in order to split.

3.5 The cohomology of moment–angle complexes

The integral cohomology ring of a moment–angle complex was computed by M. Franz [61] and independently by V. Baskakov, V. Buchstaber and T. Panov [24]. In the latter, the two-disc $D^2 \subset \mathbb{C}$ is decomposed into three cells: the zero-cell is the point $1 \in D^2$, the one-cell, denoted by T, is the complement $\partial D^2 \smallsetminus \{1\}$ and the two-cell, denoted by D is the interior of $D^2 \subset \mathbb{C}$. Taking products yields a cellular decomposition of $D^m \subset \mathbb{C}^m$.

Below we summarize the computation, following the development in [39]. The cells of D^m are parametrized by pairs of subset $I, J \subset [m] = \{1,2,3,\ldots\}$ satisfying $I \cap J = \varnothing$. The set J parametrizes the T-cells and the set I parametrizes the D-cells. The remaining positions $[m] \smallsetminus (I \cup J)$ are occupied by 0-cells. The cell corresponding to the pair J, I is denoted $\langle J, I\rangle$. Next, we introduce K, a simplicial complex on the vertices $[m]$. In this decomposition, the moment–angle complex $Z(K;(D^2,S^1))$ embeds in D^m as a cellular subcomplex; the cell $\langle I, J\rangle$ is in $Z(K;(D^2,S^1))$ whenever I corresponds to a simplex in K.

The cellular cochains $C^*\big(Z(K;(D^2,S^1))\big)$ have a basis $\langle I, J\rangle^*$ dual to the cells. These are bigraded by $\deg\langle I, J\rangle^* = (-|J|,\ 2|I| + 2|J|)$. The cellular differential preserves the *second* grading and so the complex splits as follows:

$$C^*\big(Z(K;(D^2,S^1))\big) = \bigoplus_{q=0}^{m} C^{*,2q}\big(Z(K;(D^2,S^1))\big).$$

The *bigraded Betti numbers* are defined by

$$b^{-p,2q}(K) \; = \; \operatorname{rank} H^{-p,2q}\big(Z(K;(D^2,S^1))\big)$$

and so the ordinary Betti numbers of the moment-angle complex satisfy

$$b^k(K) = \sum_{-p+2q=k} b^{-p,2q}(K).$$

Recall that the integral Stanley–Reisner ring of K, $\mathbb{Z}(K)$, is a polynomial ring on two-dimensional generators v_i, modulo the ideal I generated by monomials corresponding to the minimal non-faces of K,

$$\mathbb{Z}(K) \cong \mathbb{Z}[v_1,\ldots,v_m]/I. \tag{3.5.1}$$

A *minimal non-face* of K is a sequence of vertices of K that is not a simplex of K, but such that any proper subset is a simplex of K.
There is an isomorphism of cochain complexes, [39, Lemma 4.5.1],

$$g\colon R^*(K) \longrightarrow C^*\big(Z(K;(D^2,S^1))\big)$$

where $R^*(K)$ is the quotient of a Koszul algebra

$$\Lambda[u_1,\ldots,u_m] \otimes \mathbb{Z}(K)/(v_i^2 = u_iv_i = 0, 1 \leq i \leq m)$$

with bigrading and differential given by

$$\operatorname{deg} u_i = (-1,2), \;\; \operatorname{deg} v_i = (0,2), \;\; du_i = v_i, \;\; dv_i = 0. \tag{3.5.2}$$

The isomorphism g is given by $u_Jv_I \mapsto \langle I,J\rangle^*$. It follows that there is an additive isomorphism

$$H\big(R^*(K)\big) \cong H^*\big(Z(K;(D^2,S^1))\big).$$

The properties of polyhedral products are used then to show that the algebra $R^*(K)$ is weakly equivalent to the Koszul algebra as follows;

$$\Lambda[u_1,\ldots,u_m] \otimes \mathbb{Z}(K) = C^*\big(Z(K;(S^\infty,S^1))\big) \cong C^*\big(Z(K;(D^2,S^1))\big) = R^*(K).$$

Furthermore, a cellular diagonal approximation is constructed in [39], which allows for an extension to the multiplicative structure of the cohomology.

Theorem 3.5.1. [39, Theorem 4.5.4] *There are isomorphisms of bigraded algebras*

$$H^{*,*}\big(Z(K;(D^2,S^1))\big) \cong \operatorname{Tor}_{\mathbb{Z}[v_1,\ldots,v_m]}\big(\mathbb{Z}(K),\mathbb{Z}\big) \cong H\big(\Lambda[u_1,\ldots,u_m] \otimes \mathbb{Z}(K), d\big)$$

where the bigrading and differential are as in (3.5.2)*. The isomorphisms are natural for simplicial morphisms of K, which are assumed to be monomorphisms.*

Notice now that the bigraded Betti numbers satisfy

$$b^{-p,2q}(K) = \dim \operatorname{Tor}^{-p,2q}_{\mathbb{Z}[v_1,\ldots,v_m]}(\mathbb{Z}(K),\mathbb{Z})$$

and their calculation is simplified by the following theorem due to M. Hochster.

Theorem 3.5.2. [78] *Let K_J denote the full subcomplex of K corresponding to $J \subset [m]$. Then*

$$\operatorname{Tor}^{-p,2q}_{\mathbb{Z}[v_1,\ldots,v_m]}(\mathbb{Z}(K),\mathbb{Z}) \;=\; \bigoplus_{J\subset[m],|J|=q} \widetilde{H}^{q-p-1}(|K_J|;\mathbb{Z})$$

with the convention $\widetilde{H}^{-1}(K_\varnothing;\mathbb{Z}) = \mathbb{Z}$.

This leads to a formula for the bigraded Betti numbers in terms of the full subcomplexes

$$b^{-p,2q}(K) = \sum_{J\subset[m],|J|=q} \dim \widetilde{H}^{q-p-1}(|K_J|;\mathbb{Z}). \tag{3.5.3}$$

From Theorem 3.5.1, (cf. Theorem 3.8.1), we conclude that the bigraded Betti numbers of the Tor module are the ranks of the contributions of $H^*(|K_J|)$ to the cohomology of $Z(K;(D^2,S^1))$. Further computations and properties of the bigraded and *multigraded* Betti numbers may be found in [9] [97] [133] and [44]. The calculations in [44] are used to verify the Halperin–Carlsson conjecture, ([44, Remark 3]), in the case of free torus actions on moment–angle complexes. This conjecture bounds the ranks of these tori by the sum of the ranks of the cohomology groups.

An analogue of Theorem 3.5.1 is obtained in [102] for a generalization of moment–angle complexes arising from simplicial posets.

The question whether the cohomology of a moment–angle manifold determines its diffeomorphism type is currently an active area of research, [36].

3.6 The exponentiation property of polyhedral products

A product is defined on CW-pairs by

$$(X,A)\times(U,V) \;=\; \big(X\times U,(A\times U)\cup_{A\times V}(X\times V)\big).$$

Let K be a simplicial complex on vertices $[m]$ and $(\underline{X},\underline{A})$ a family of CW-pairs. For a sequence of positive integers $J=(j_1,j_2,\ldots,j_m)$, define a new family of pairs $(\underline{Y},\underline{B})$ by

$$(Y_i,B_i) \;=\; (X_i,A_i)^{j_i}. \tag{3.6.1}$$

It turns out that the combinatorics of a polyhedral product can detect this change. Let K be a simplicial complex of dimension $n-1$ on vertices $\{v_1,v_2,\ldots,v_m\}$. From K and a sequence J as above, a new simplicial complex $K(J)$ exists on $j_1+j_2+\cdots+j_m$ vertices, labelled

$$v_{11},v_{12},\ldots,v_{1j_1},\; v_{21},v_{22},\ldots,v_{2j_2},\;\;\cdots\;\;,v_{m1},v_{m2},\ldots,v_{mj_m}.$$

It is characterized by the property that

$$\{v_{i_1 1}, v_{i_1 2}, \ldots, v_{i_1 j_{i_1}}, v_{i_2 1}, v_{i_2 2}, \ldots, v_{i_2 j_{i_2}}, \ldots, v_{i_k 1}, v_{i_k 2}, \ldots, v_{i_k j_{i_k}}\}$$

is a minimal non-face of $K(J)$ if and only if $\{v_{i_1}, v_{i_2}, \ldots, v_{i_k}\}$ is a minimal non-face of K. Moreover, all minimal non-faces of $K(J)$ have this form.

An alternative explicit construction of the simplicial complex $K(J)$ may be found in either [16, Construction 2.2] or [119]. The construction of $K(J)$ is known variously in the literature as the *simplicial wedge construction* or the *J-construction*.

Next, we adopt the convention of denoting by $Z\big(K(J); (\underline{X}, \underline{A})\big)$ the polyhedral product determined by the simplicial complex $K(J)$ and the family of pairs obtained from $(\underline{X}, \underline{A})$ by repeating each (X_i, A_i), j_i times in sequence. The main structure theorem is the following.

Theorem 3.6.1. [16] *Let K be a simplicial complex with m vertices and let $(\underline{X}, \underline{A})$ denote a family of CW-pairs and the family $(\underline{Y}, \underline{B})$ be defined as in* (3.6.1). *Then*

$$Z\big(K(J); (\underline{X}, \underline{A})\big) = Z\big(K; (\underline{Y}, \underline{B})\big)$$

as subspaces of $X_1^{j_1} \times X_2^{j_2} \times \cdots \times X_m^{j_m}$.

This theorem has applications in toric topology and geometry. In particular, since we have $(D^1, S^0)^2 \cong (D^2, S^1)$, it implies that each moment-angle complex $Z(K; (D^2, S^1))$ is homeomorphic to a real moment-angle complex $Z(K(J); (D^1, S^0))$ where $J = (2, 2, \ldots, 2)$. Three other applications are discussed below.

The cohomology of toric manifolds given by (3.2.4) and (3.3.5) has a presentation determined by the combinatorics of K and relations arising from the function λ. In cases when $\mathcal{Z}$ in (3.2.4) has the form $Z\big(K(J); (D^2, S^1)\big)$, for some K and J, Theorem 3.6.1 can be used to give a simpler presentation of the cohomology of associated toric manifolds which uses the combinatorics of K only. More details can be found in [16].

Remark 3.6.2. A generalization of the Davis–Januszkiewicz construction (3.2.2) which accommodates the construction here can be found in [15].

An application to topological joins begins by defining a family of CW-pairs

$$(\underline{C(\ast_J X)}, \underline{\ast_J X}) := \{(C(\underbrace{X_i * X_i * \cdots * X_i}_{j_i}), \underbrace{X_i * X_i * \cdots * X_i}_{j_i})\}_{i=1}^{m}$$

for each sequence $J = (j_1, j_2, \ldots, j_m)$ and family of CW-complexes $\{X_i\}_{i=1}^m$. Here, $C(-)$ denotes the unreduced cone. It is shown in [16] that the equivalence $X * X \xrightarrow{\simeq} (CX \times X) \cup (X \times CX)$ iterates appropriately so that the next theorem follows from Theorem 3.6.1.

Corollary 3.6.3. [16] *There is a homeomorphism of polyhedral products,*

$$Z\Big(K; (\underline{C(\ast_J X)}, \underline{\ast_J X})\Big) \longrightarrow Z\big(K(J); (\underline{CX}, \underline{X})\big).$$

Remark 3.6.4. Notice that in the case $X_i = S^1$ for all i, this corollary yields a homeomorphism

$$Z\Big(K; (\underline{D}^{2J},\ \underline{S}^{2J-1})\Big) \longrightarrow Z\big(K(J); (D^2, S^1)\big), \tag{3.6.2}$$

where $\left(\underline{D}^{2J},\ \underline{S}^{2J-1}\right) = \left\{(D^{2j_i}, S^{2j_i-1})\right\}_{i=1}^{m}$.

Theorem 3.6.1 also implies an observation about the action of the Steenrod algebra.

Corollary 3.6.5. ([16, 13]) *There is an isomorphism of ungraded rings*

$$H^*(Z(K; (D^2, S^1)); \mathbb{Z}/2) \longrightarrow H^*\big(Z\big(K(J); (D^2, S^1)\big); \mathbb{Z}/2\big)$$

which commutes with the action of the Steenrod algebra.

Remark 3.6.6. The construction of Theorem 3.6.1 has been generalized by Ayzenberg in [9]; see also [15, Section 6].

3.7 Fibrations

A direct extension of a result of Denham and Suciu [56, Lemma 2.9] yields the next theorem detailing the behaviour of polyhedral products with respect to fibrations.

Theorem 3.7.1. *Let $p_i : (E_i, E_i') \to (B_i, B_i')$ be a map of pairs, such that $p : E_i \to B_i$ and $p|E_i' : E_i' \to B_i'$ are fibrations over path connected CW complexes with fibres F_i and F_i' respectively. Let K be a simplicial complex with m vertices. If either $\underline{B} = \underline{B}'$ or $\underline{F} = \underline{F}'$, then the following is a fibration:*

$$Z(K; (\underline{F}, \underline{F}')) \to Z(K; (\underline{E}, \underline{E}')) \to Z(K; (\underline{B}, \underline{B}')).$$

Special cases of this fibration were developed earlier by G. Porter, [117]. He proved that these fibrations exist in the special cases where $(X, A) = (X, *)$ with $A = *$ the base-point, and K is the q-skeleton of the full m-simplex for $0 \le q \le m-1$.

Natural consequences of Theorem 3.7.1 also arose earlier in the work of T. Ganea [64] and A. Kurosh [92]. G. W. Whitehead had defined a filtration of the product $X_1 \times \cdots \times X_m$, where the jth-filtration is given by the space

$$W_j(X_1, X_2, \ldots, X_m) \ = \ \big\{(y_1, \ldots, y_m) \mid y_i = *_i \in X_i, \text{ for at least } m-j \text{ values of } i\big\}.$$

This space is the polyhedral product

$$W_j(X_1, X_2, \ldots, X_m) \ = \ Z\big(\Delta[m-1]_{j-1}; (\underline{X}, \underline{*})\big),$$

where $\Delta[m-1]_q$ denotes the q-skeleton of the $(m-1)$-dimensional simplex. In this case, Porter's result gives an identification of the homotopy theoretic fibre of the inclusion

$$W_j(X_1, X_2, \ldots, X_m) \subset \prod_{1 \le i \le m} X_i,$$

which follows directly from the classical path-loop fibration over X and Theorem 3.7.1. To see this, we shall extend to the right the fibration of pairs

$$(\Omega X, \Omega X) \longrightarrow (PX, \Omega X) \xrightarrow{e_1} (X, *),$$

where e_1 is the map evaluating the endpoint of a path. Consider again the cone CY over a space Y. There is a map

$$\begin{aligned} \kappa\colon \quad X &\longrightarrow X \times C(PX) \\ x &\mapsto (x, [0, f_x]) \end{aligned}$$

where $f_x\colon [0,1] \to X$ is the constant path satisfying $f_x(t) = x$. The map κ is evidently a homotopy equivalence. Consider now PX to be the subspace of $X \times C(PX)$ given by pairs $\big(f(1), [0, f]\big)$ for $f \in PX$. This gives a pair $\big(X \times C(PX), PX\big)$. Consequently, we get a fibration of pairs

$$\big(C(PX), \Omega X\big) \longrightarrow \big(X \times C(PX), PX\big) \xrightarrow{\pi_X \times e_1} (X, X)$$

for which $\pi_X : X \times C(PX) \to X$ is the natural projection. Applying Theorem 3.7.1 yields the fibration

$$Z(K; (\underline{PX}, \underline{\Omega X})) \longrightarrow Z(K; (\underline{X \times PX}, \underline{PX})) \xrightarrow{\pi_X \times e_1} Z(K; (\underline{X}, \underline{X})).$$

Moreover, the equivalence of pairs $(X, *) \to \big(X \times C(PX), PX\big)$, gives the associated homotopy fibration

$$Z(K; (\underline{PX}, \underline{\Omega X})) \longrightarrow Z(K; (\underline{X}, \underline{*})) \xrightarrow{\pi_X \times e_1} Z(K; (\underline{X}, \underline{X})), \qquad (3.7.1)$$

which we record as the next corollary, which generalizes Porter's identification.

Corollary 3.7.2. [12, Corollary 2.32] *If all of the X_i are path-connected, the homotopy theoretic fibre of the inclusion $Z(K; (\underline{X}, \underline{*})) \subset Z(K; (\underline{X}, \underline{X}))$ is $Z(K; (\underline{PX}, \underline{\Omega X}))$. In particular, the homotopy theoretic fibre of the inclusion $W_j(X_1, X_2, \ldots, X_m) \subset \prod_{i=1}^{m} X_i$ is*

$$Z\big(\Delta[m-1]_{j-1}; (\underline{PX}, \underline{\Omega X})\big).$$

For a more general version of this result, see [112, Proposition 5.1]. See also Section 3.10, and for related information about generalized Whitehead products, [113, Section 7], [72] and [83], where the techniques used involve the *fat wedge filtration*.

The structure of the fibre $Z\big(\Delta[m-1]_{j-1}; (\underline{PX}, \underline{\Omega X})\big)$ is given by a second theorem of G. Porter [117]. After one suspension, this result also follows from the next theorem which gives a decomposition of the suspension of $Z\big(\Delta[m-1]_q; (\underline{CY}, \underline{Y})\big)$, for a family of connected, pointed CW-complexes $\{Y_i, *_i\}_{i=1}^{m}$. Recall that $\widehat{Y}^I$ is $Y_{i_1} \wedge \cdots \wedge Y_{i_k}$ if $I = (i_1, \ldots, i_k)$.

Theorem 3.7.3. *Let $\{Y_i, *_i\}_{i=1}^{m}$ be a family of connected, pointed CW-complexes. Then there is a homotopy equivalence*

$$\Sigma\big(Z(\Delta[m-1]_q; (\underline{CY}, \underline{Y}))\big) \to \Sigma\Big(\bigvee_{q+1<|I|\leq m} \bigvee_{t_I} (\Sigma^{q+1}\widehat{Y}^I)\Big),$$

where t_I is the binomial coefficient $\binom{|I|-1}{q+1}$.

This result is deduced from [12, Theorem 2.19], by a counting argument which uses the fact that here, the full sub-complex K_I is an $(|I|-1)$-simplex, and then enumerates the spheres in the q-skeleton of the $(|I|-1)$-simplex.

When $q = m-2$, $Z\big(\Delta[m-1]_{m-2}; (\underline{X}; \underline{*})\big)$ is the fat wedge $W_{m-1}(X_1, X_2, \ldots, X_m)$ and the homotopy fibre is $Z\big(\Delta[m-1]_{m-2}; (\underline{PX}, \underline{\Omega X})\big)$. In this case, it follows that there is a homotopy equivalence

$$\Sigma\big(Z\big(\Delta[m-1]_{m-2}; (\underline{PX}, \underline{\Omega X})\big)\big) \to \Sigma(\Omega X_1 * \ldots * \Omega X_m).$$

T. Ganea [64] identified the homotopy theoretic fibre of the natural inclusion $X \vee Y \to X \times Y$ as the join $\Omega(X) * \Omega(Y)$ in the case X and Y are path-connected, pointed spaces having the homotopy type of a CW-complex. This example is a special case of the homotopy theoretic fibre for the inclusion of a polyhedral product into a product corresponding to the simplicial complex K given by two discrete points.

3.8 Unstable and stable decompositions of the polyhedral product

A Cartesian product $(X_1 \times \cdots \times X_m)$ of pointed, connected CW-complexes splits stably by a homotopy equivalence

$$H : \Sigma(X_1 \times \cdots \times X_m) \to \Sigma\Big(\bigvee_{I \subseteq [m]} \widehat{X}^I\Big)$$

where I runs over all the non-empty sub-sequences of $(1, 2, \ldots, m)$. It is natural then to ask if a similar splitting holds for the polyhedral product

$$Z(K; (\underline{X}, \underline{A})) \subset X_1 \times \cdots \times X_m.$$

In other words, does a stable splitting of the polyhedral product exist for simplicial complexes $K \neq \Delta^m$? Indeed, in the setting of moment–angle complexes, such a splitting was suggested by the cohomology result of V. Baskakov [23] following.

Theorem 3.8.1. *There is an isomorphism of rings*

$$H^k\big(Z_K; (D^2, S^1)\big) \cong \bigoplus_{I \subset [m]} \widetilde{H}^{k-|I|-1}(|K_I|),$$

where the product on the right-hand side is that described in Section 3.11.

This result is deduced from those in Section 3.5. This decomposition is related to a splitting of the cohomology of the complements of subspace arrangements as follows. A simplicial complex K determines a complex coordinate arrangement of subspaces whose complement is homeomorphic to $Z(K; (\mathbb{C}, \mathbb{C}^*))$. The moment–angle complex $Z(K; (D^2, S^1))$ is a strong deformation retract of $Z(K; (\mathbb{C}, \mathbb{C}^*))$, by a result of N. Strickland [123, Proposition 20], (cf. V. Buchstaber and T. Panov [37, Theorem 5.2.5]). In [67], M. Goresky and R. MacPherson derive a decomposition for the cohomology of the complements of subspace arrangements which gives a cohomology splitting analogous to Theorem 3.8.1. A more direct proof was given by G. Ziegler and R. Živaljević in [135].

The first geometric result in this direction is the *unstable* splitting due to J. Grbić and S. Theriault, [69, Theorem 1]. They show that when K consists of m discrete points, then there is a homotopy equivalence

$$Z(K;(D^2,S^1)) \simeq Z(K;(\mathbb{C},\mathbb{C}^*)) \longrightarrow \bigvee_{k=2}^{m} (k-1)\binom{m}{k} S^{k+1}. \tag{3.8.1}$$

Their proof uses an inductive argument based on a version of Corollary 3.7.2. As noted above, the space $Z(K;(\mathbb{C},\mathbb{C}^*))$ corresponds to the complement of a codimension-two coordinate subspace arrangement. Later in [70], they extended this result to a class of simplicial complexes called *shifted*, and to simplicial complexes that can be obtained from shifted complexes by certain elementary topological operations.

Definition 3.8.2. A simplicial complex K is called *shifted* if there is an ordering of its vertices such that whenever v is a vertex of $\sigma \in K$ and $v' < v$, $(\sigma \smallsetminus v) \cup v' \in K$.

Theorem 3.8.3. [70, Theorem 1.1] *If K is a shifted complex, then $Z(K;(D^2,S^1))$ is homotopy equivalent to a wedge of spheres.*

Following a conjecture made by the authors in [12], this result has been extended in [71] and [79] to the case $Z(K;(\underline{CX},\underline{X}))$, as described in Theorem 3.8.4 below. Indeed, under a condition on K, known as *dual sequentially Cohen-Macaulay over* $\mathbb{Z}$, there is the following improvement over previous results, including those of [73]. (Recognizing that shifted implies dual shifted, arguments from [29] and [30] can be amalgamated to deduce that K shifted implies that it is dual sequentially Cohen-Macaulay over $\mathbb{Z}$.)

Theorem 3.8.4. [82, Theorems 1.3 and 1.4] *If K is dual sequentially Cohen-Macaulay over $\mathbb{Z}$, then*

$$Z(K;(\underline{CX},\underline{X})) \simeq \bigvee_{\varnothing \neq I \subset [m]} \Sigma|K_I| \wedge \widehat{X}^I. \tag{3.8.2}$$

Notice that if $X = S^1$, the right-hand side is a wedge of spheres.

Example 3.8.5. A simple example of the splitting in Theorem 3.8.4 occurs for the simplicial complex consisting of two discrete points, $K = \{\varnothing, \{v_1\}, \{v_2\}\}$. Here $m = 2$ and

$$(\underline{CX},\underline{X}) = \{(CX_1, X_1), (CX_2, X_2)\}.$$

In this case, (3.8.2) sees the well known homotopy equivalence $X_1 * X_2 \simeq \Sigma(X_1 \wedge X_2)$ in the form

$$CX_1 \times X_2 \cup_{X_1 \times X_2} X_1 \times CX_2 \simeq CX_1 \vee CX_2 \vee \Sigma(X_1 \wedge X_2).$$

For general $(\underline{X},\underline{A})$ there is the stable splitting of [11] and [12].

Theorem 3.8.6. [12, Theorem 2.10] *Let K be an abstract simplicial complex on vertices $[m]$. Given $(\underline{X},\underline{A}) = \{(X_i,A_i)\}_{i=1}^m$ where (X_i,A_i,x_i) are pointed pairs of CW-complexes.*

then there is a natural pointed homotopy equivalence

$$H\colon \Sigma(Z(K;(\underline{X},\underline{A}))) \longrightarrow \Sigma\Big(\bigvee_{I\subseteq[m]} \widehat{Z}(K_I;(\underline{X},\underline{A})_I)\Big) \tag{3.8.3}$$

where $(\underline{X},\underline{A})_I$ *denotes the restricted family of CW-pairs* $\{(X_i,A_i)\}_{i\in I}$.

Remark 3.8.7. The spaces $Z\big(K;(\underline{X},\underline{A})\big)$ generally do not decompose as a wedge before suspending. One example is the simplicial complex K determined by a square, with four vertices and four edges, for which $Z\big(K;(D^2,S^1)\big)$ is $S^3\times S^3$.

The homotopy equivalence (3.8.3) is induced from the following decomposition which is well known, [106, 85].

Theorem 3.8.8. *Let* (Y_i,y_i) *be pointed CW-complexes. There is a pointed, natural homotopy equivalence*

$$H:\Sigma(Y_1\times\cdots\times Y_m)\to\Sigma\Big(\bigvee_{I\subseteq[m]}\widehat{Y}^I\Big)$$

where I *runs over all the nonempty subsequences of* $(1,2,\ldots,m)$. *Furthermore, the map* H *commutes with colimits.*

It is shown in [12] that the map H of Theorem 3.8.8 induces an equivalence between the diagrams that define the colimits on either side of (3.8.3), yielding Theorem 3.8.6.

In many important cases, the spaces that appear on the right-hand side of (3.8.3) can be identified explicitly. Below are a few of the most common of these. We begin with a definition.

Definition 3.8.9. For σ a simplex in K, $\mathrm{lk}_\sigma(K)$ *the link of* σ *in* K, is defined to be the simplicial complex for which

$$\tau\in\mathrm{lk}_\sigma(K)\quad\text{if and only if}\quad\tau\cup\sigma\in K.$$

Theorem 3.8.10. [12] *Let* K *be an abstract simplicial complex on vertices* $[m]$ *and* $I\subseteq[m]$. *If the family* $(\underline{X},\underline{A})$ *has the additional property that the inclusion* $A_i\subset X_i$ *is null-homotopic for all* i, *then there is a homotopy equivalence*

$$\widehat{Z}(K_I;(\underline{X},\underline{A})_I)\longrightarrow\bigvee_{\sigma\in K_I}|\mathrm{lk}_\sigma(K_I)|*\widehat{D}(\sigma) \tag{3.8.4}$$

which, when combined with (3.8.3), *gives a homotopy equivalence*

$$\Sigma(Z(K;(\underline{X},\underline{A})))\longrightarrow\Sigma\Big(\bigvee_{I\subseteq[m]}\Big(\bigvee_{\sigma\in K_I}|\mathrm{lk}_\sigma(K_I)|*\widehat{D}(\sigma)\Big)\Big).$$

(Here, $\widehat{D}(\sigma)$ is defined in Section 3.1.) The homotopy equivalence (3.8.4) is a generalization of the wedge lemma of Welker-Ziegler-Živaljević in [131], applied to the smash polyhedral product.

Next, consider all homology to be taken with coefficients in a field k. Define the Poincaré series of a pointed space X of finite type by

$$P(X,t) = \sum d_n(X)t^n$$

where $d_n(X) = \text{dimension}_{\mathbb{F}} H_n(X;\mathbb{F})$, and the *reduced* Poincaré series by

$$\overline{P}(X,t) = -1 + P(X,t).$$

This series behaves well with respect to wedges and smash products. Hence, (for example), for pairs satisfying the hypotheses of Theorem 3.8.10 we have the next result.

Lemma 3.8.11. *Assume that homology is taken with field coefficients $\mathbb{F}$ and that path connected pairs of pointed CW-complexes $(\underline{X},\underline{A})$, of finite type, have the property that the inclusion $A_i \subset X_i$ is null-homotopic for all i. Then*

$$\overline{P}(\widehat{Z}(K;(\underline{X},\underline{A})),t) = \sum_{\sigma\in K} t\overline{P}(|\mathrm{lk}_\sigma(K_I)|,t)\overline{P}(\widehat{D}(\sigma),t).$$

Notice that if in Theorem 3.8.10 all the spaces A_i are contractible, then the spaces $\widehat{D}(\sigma)$ are contractible unless I represents a simplex σ in K, that is K_I is an $(|I|-1)$–simplex; in this case, $\widehat{D}(\sigma) = \widehat{X}^I$ and the next theorem follows.

Theorem 3.8.12. [12] *Let K be an abstract simplicial complex with m vertices and $(\underline{X},\underline{A})$ has the property that all the A_i are contractible. Then there is a homotopy equivalence*

$$\widehat{Z}(K_I;(\underline{X},\underline{A})_I) \longrightarrow \widehat{X}^I$$

yielding a homotopy equivalence

$$\Sigma(Z(K;(\underline{X},\underline{A}))) \longrightarrow \Sigma(\bigvee_{I\in K}\widehat{X}^I). \tag{3.8.5}$$

The next example is the one for which all the spaces X_i are contractible. Now, for each $I \subseteq [m]$, the spaces $\widehat{D}(\sigma)$ in (3.8.4) are contractible with the possible exception of the case $\sigma = \varnothing$, the empty simplex, which gives $\widehat{D}(\sigma) = \widehat{A}^I$. In this case $|\mathrm{lk}_\varnothing(K_I)| \simeq |K_I|$ and (3.8.4) simplifies to yield the next theorem.

Theorem 3.8.13. [12] *Let K be an abstract simplicial complex with m vertices and suppose that the collection $(\underline{X},\underline{A})$ has the property that all the X_i are contractible. Then there are homotopy equivalences*

$$\widehat{Z}(K_I;(\underline{X},\underline{A})_I) \longrightarrow |K_I| * \widehat{A}^I \longrightarrow \Sigma(|K_I| \wedge \widehat{A}^I)$$

yielding a homotopy equivalence

$$\Sigma(Z(K;(\underline{X},\underline{A}))) \longrightarrow \Sigma(\bigvee_{I\notin K}|K_I| * \widehat{A}^I).$$

The last statement of this theorem follows on observing that if $I \in K$ then $|K_I|$ is contractible. The case $(X_i, A_i, x_i) = (D^{n+1}, S^n, *)$ for all i, $n \geq 1$, yields the next corollary.

Corollary 3.8.14. *There are homotopy equivalences*

$$\Sigma(Z(K;(D^{n+1},S^n))) \to \Sigma(\bigvee_{I\notin K} |K_I| * S^{n|I|}) \to \bigvee_{I\notin K} \Sigma^{2+n|I|}|K_I|.$$

In the case $n = 1$, which corresponds to the case of complements of complex coordinate subspace arrangements, this corollary implies the Goresky–MacPherson cohomological splitting mentioned earlier. To observe this, the reduced cohomology of full sub-complexes of K needs to be related to the reduced homology of certain links in the Alexander dual of K. This is done in, for example, [39, Corollary 2.4.6].

3.9 Equivariance of the stable splitting and an application to number theory

In the case that all the pairs (X_i, A_i) are the same, the automorphism group of K, $Aut(K) \subseteq \Sigma_m$, acts in a natural way on both sides of the splitting (3.8.3). Using James–Hopf invariants, A. Al-Raisi [3] shows that there is a stable splitting, as in (3.8.3), which is equivariant with respect to the action of $Aut(K)$.

Theorem 3.9.1. [3] *The adjoint of the stable decomposition* (3.8.3),

$$Z(K;(X,A)) \longrightarrow J\Big(\bigvee_{I\subseteq[m]} \widehat{Z}(K_I;(X,A)_I)\Big) \longrightarrow \Omega\Sigma\Big(\bigvee_{I\subseteq[m]} \widehat{Z}(K_I;(X,A)_I)\Big)$$

is $Aut(K)$-equivariant.

When $K = K_n$ is the boundary dual of an n-gon, (see Section 3.3), a theorem of Coxeter identifies the real moment–angle manifold $Z(K_n;(D^1,S^0))$.

Theorem 3.9.2. [50] *The real moment-angle manifold $Z(K_n;(D^1,S^0))$ corresponding to an n-gon, $n \geq 4$, is an oriented surface of genus $g = 1 + (n-4)2^{n-3}$.*

Remark 3.9.3. The simplicial complex K_n has also been used by S. Das [51], to show directly that the genus of the *hypercube graph*, the one-skeleton of the n-cube, is $1 + (n - 4)2^{n-3}$. This is a result due originally to G. Ringel [120] and L. Beineke and F. Harary [26]. The computation is done by embedding equivariantly the hypercube graph into the real moment-angle complex $Z(K_n;(D^1,S^0))$.

In this simple case, where $Aut(K_n) = D_{2n}$, the dihedral group with cyclic subgroup C_n, Theorem 3.9.1 has interesting consequences. The real moment–angle manifold is the conduit through which A. Al-Raisi uses Theorem 3.9.1 to link combinatorics of K_n and the action of C_n on a choice of basis for the cohomology of a surface of genus $g = 1 + (n-4)2^{n-3}$.

Briefly, one first counts the number of orbits of the action of C_n by considering, up to rotation, the boundary of an n-gon as a 2-colored beaded necklace with n beads. Certain colorings are shown to correspond to generators of $H^*(Z(K_n;(D^1,S^0)))$. A theorem of Burnside gives the number of orbits of the action of C_n arising from this choice of basis as

$$\frac{1}{n}\Big(\sum_{d|n}\phi(d)2^{n/d} - (n+1)\Big)$$

where $\phi(d)$ is the Euler phi function.

Next, we think of the sequence of beads as a word made from two letters. A word of length n is called an *aperiodic word* if it has n distinct *rotations*. Here a rotation of a word $W = a_1a_2\ldots a_n$ is given by $r(W) = a_na_1a_2\ldots a_{n-1}$. An equivalence class of an aperiodic word under rotation is called a *primitive necklace*. A. Al-Raisi shows that the rank of $H_1(Z(K_n;(D^1,S^0)))$ is given in terms of the number of primitive necklaces. The number of primitive n-necklaces on an alphabet of size k, denoted by $M(k,n)$, was computed by Moreau in the nineteenth century, [107]

$$M(k,n) = \frac{1}{n}\sum_{d|n}\mu(d)k^{\frac{n}{d}}$$

where μ is the Möbius inversion function. The number of orbits of size d, where $d|n$ and $1 < d \leq n$, can be counted with the aid of this theorem. The number of orbits with exactly n elements is given by

$$\frac{1}{n}\Big(\sum_{d|n}\mu(d)2^{n/d} - (n-1)\Big).$$

The number of orbits of size $d < n$, where $d|n$, is given by

$$\frac{1}{d}\sum_{d_1|d}\mu(d_1)2^{d/d_1}.$$

Finally, equating the two different orbit counts, we arrive at Al-Raisi's proof of the cyclotomic identity below.

Theorem 3.9.4. [3] *For $n \geq 4$ there is an identity*

$$\Big(\sum_{d|n}\mu(d)2^{n/d} - (n-1)\Big) + \sum_{\substack{d|n\\1<d<n}}\frac{n}{d}\Big(\sum_{d_1|d}\mu(d_1)2^{d/d_1}\Big) = \Big(\sum_{d|n}\phi(d)2^{n/d} - (n+1)\Big)$$

relating the Möbius function $\mu(d)$ and the Euler phi function $\phi(d)$.

The rank of H_1 above is incidentally given by the rank of certain Lie tensors in a free Lie algebra as first computed by E. Witt [132]. Moreau's formula and Witt's formula agree with one another. This connection between polyhedral products and free Lie algebras is as yet not clearly understood.

These structures together with connections to Hochschild homology as well as new filtrations of classifying spaces will appear in [4]. Additional interesting work subsequent to [3], but from a different point of view, was developed by X. Fu and J. Grbić. They construct a sequence of simplicial complexes $\{L_m\}$ with each L_m obtained from the m-cube by a vertex cut, [63, Construction 4.4] and verify a form of Σ_m-representation stabilty in characteristic zero, for the homology of $Z(L_m;(X,A))$.

In a recent preprint, S. Cho, S. Choi and S. Kaji [46], also address the representations afforded by this action. They compute some actions and the associated representations on the homology of $Z(K_n;(D^1,S^0))$ in characteristic zero.

3.10 The case that $A_i = *$ for all i

The stable structure of the polyhedral product $Z(K;(\underline{X},\underline{*}))$ is given by Theorem 3.8.12. The importance of this particular case of the polyhedral product first came to light in the development of toric topology with the following observation.

Theorem 3.10.1. [39, Theorem 4.3.2] *The inclusion*

$$i\colon Z\big(K;(\mathbb{C}P^{\infty},*)\big) \longrightarrow (\mathbb{C}P^{\infty})^m$$

factors into a composition of a homotopy equivalence

$$Z\big(K;(\mathbb{C}P^{\infty},*)\big) \;\longrightarrow\; ET^m \times_{T^m} Z(K;(D^2,S^1)) \tag{3.10.1}$$

and the fibration $ET^m \times_{T^m} Z(K;(D^2,S^1)) \longrightarrow BT^m = (\mathbb{C}P^{\infty})^m$. The action of the torus T^m here is that of (3.3.1).

This gives one of the most important fibrations in this subject;

$$Z(K;(D^2,S^1)) \xrightarrow{\widetilde{\omega}} Z(K;(\mathbb{C}P^{\infty},*)) \longrightarrow (\mathbb{C}P^{\infty})^m. \tag{3.10.2}$$

The cohomology of the right-hand side of (3.10.1) was determined by geometers in the context of toric geometry, [32, 28]. In the setting of toric topology, it appears in the paper of M. Davis and T. Januszkiewicz, [53, Theorem 4.8]. This calculation, together with the theorem, implies the following result in algebraic combinatorics, for which independent proofs exist.

Corollary 3.10.2. [37, 12] *Let K be a simplicial complex. Then*

$$H^*\big(Z(K;(\mathbb{C}P^{\infty},*));\mathbb{Z}\big) \;\cong\; \mathbb{Z}(K)$$

the Stanley–Reisner ring of K.

Recall that Theorem 3.5.1 tells us that $\mathbb{Z}(K)$, determines the cohomology of the moment–angle complex $Z(K;(D^2,S^1))$.

An analogue of Corollary 3.10.2 holds over any ring R for the algebra $R(K)$. Somewhat surprisingly, for any nonzero ring R, $R(K)$ determines K by a theorem of W. Bruns and J. Gubeladze, [34], [35, Theorem 5.27 and Example 5.28].

Theorem 3.10.3. [34] *Let K_1 and K_2 be simplicial complexes satisfying $R(K_1) \cong R(K_2)$ as R-algebras, where R is any nonzero ring. Then K_1 and K_2 are isomorphic as simplicial complexes.*

Over a field k there is also the *exterior* Stanley–Reisner ring over k, which we denote by $\Lambda_k(K)$. It is the quotient of a graded exterior algebra on m one-dimensional generators, by the same two-sided ideal as in (3.5.1). In a manner entirely analogous to the proof of Corollary 3.10.2 from Theorem 3.8.12, we have

$$\Lambda_k(K) \cong H^*\big(Z(K;(S^1,*));k\big). \tag{3.10.3}$$

A proof that (3.10.3) also determines K up to isomorphism, follows by an argument similar to that for Theorem 3.10.3, see [35, Example 5.28(b), Exercise 5.12]. Independently, using the theory of algebraic groups, C. Stretch, (unpublished), asserts the same result. In order to compare the Stanley–Reisner ring to (3.10.3) from a geometric perspective, it is convenient to work over $k = \mathbb{Z}_2$ and to use the ungraded isomorphism of rings, (Theorem 3.8.12, again),

$$\Lambda_{\mathbb{Z}_2}(K) \cong H^*\big(Z(K;(S^2,*));\mathbb{Z}_2\big), \quad \text{(as ungraded rings).} \tag{3.10.4}$$

The fibrations associated to (3.7.1) or (3.14.4), give a commutative diagram

$$\begin{array}{ccccc} Z(K;(D^2,S^1)) & \longrightarrow & Z(K;(\mathbb{C}\mathrm{P}^\infty,*)) & \longrightarrow & (\mathbb{C}\mathrm{P}^\infty)^m \\ \uparrow{\scriptstyle \delta} & & {\scriptstyle i}\uparrow & & \uparrow{\scriptstyle i} \\ Z(K;(C(\Omega S^2),\Omega S^2)) & \longrightarrow & Z(K;(S^2,*) & \longrightarrow & (S^2)^m \end{array} \tag{3.10.5}$$

where the right hand vertical maps are determined by the inclusion $S^2 \xrightarrow{i} \mathbb{C}\mathrm{P}^\infty$. In cohomology with $\mathbb{Z}_2$ coefficients, the middle map determines the natural map from the ring $\mathbb{Z}_2(K)$ to the algebra $\Lambda_{\mathbb{Z}_2}(K)$ of (3.10.4). The map

$$\Omega S^2 \xrightarrow{\Omega i} \Omega\mathbb{C}\mathrm{P}^\infty \simeq \Omega BS^1 \simeq S^1$$

induces the vertical map δ, and is given by the projection

$$\Omega S^2 \simeq S^1 \times \Omega S^3 \longrightarrow S^1$$

since this detects the generator of $\pi_1(\Omega S^2)$. It follows from (3.10.1) that the Borel construction is a CW-subcomplex of $(\mathbb{C}\mathrm{P}^\infty)^m$. In particular it has cells in even degree only. In the literature the space $Z(K;(\mathbb{C}\mathrm{P}^\infty,*))$ is often denoted by $DJ(K)$ or DJ_K and is called the *Davis–Januszkiewicz space*.

The integral and rational formality of the Davis–Januszkiewicz spaces is studied by M. Franz in [62], and further by D. Notbohm and N. Ray in [108], using model category theory. They show that the rationalization of $DJ(K)$ is unique for a class of simplicial complexes K.

Remark 3.10.4. Under suitable freeness conditions, $H^*\big(Z(K;(X,*));\mathbb{Z}\big)$ is an analogue of the Stanley–Reisner ring for any space X and simplicial complex K, that specializes to the classical Stanley–Reisner ring in case X is $\mathbb{C}\mathrm{P}^\infty$. The result follows from Theorem 3.8.12, and for more details, [11, Theorem 2.35]. Example 3.11.5 in Section 3.11 gives a different perspective and a more concise description of the ring under discussion here. It is in fact the case that *all* monomial ideal rings can be realized as cohomology rings of spaces related to certain polyhedral products, [14].

Unstable phenomena in the homotopy of the left-hand side of (3.10.1) has been investigated by D. Allen and J. La Luz [94, 5] using the unstable Novikov Spectral Sequence. In particular, their study of the derived functors of the indecomposables of the Stanley–Reisner ring, leads to an interesting and mysterious invariant of simplicial complexes.

Next, we consider the relationship between the space $Z(K;(S^1,*))$ and the right–angled Artin group determined by K.

Definition 3.10.5. Given a simplicial graph Γ with vertex set S and a family of groups $\{G_s\}_{s\in S}$, their graph product $\prod_\Gamma G_s$ is the quotient of the free product of the groups G_s by the relations that elements of G_s and G_t commute whenever $\{s,t\}$ is an edge of Γ. In the case that $G_i = \mathbb{Z}$ and $\Gamma = SK$, the one-skeleton of K, the graph product is called the *right–angled Artin group*, RA_K, corresponding to K. When $G_i = \mathbb{Z}_2$, the graph product is called the *right–angled Coxeter group*, RC_K. A simplicial complex for which every collection of pairwise adjacent vertices spans a simplex, is called a *flag complex*.

Theorem 3.10.6. [33, 90, 54, 45, 115] *The space $Z(K;(S^1,*))$ is aspherical if and only if K is a flag complex, in which case, $\pi_1\big(Z(K;(S^1,*))\big) = RA_K$.*

Related results, [52, 115], identify $\pi_1\big(Z(K;(\mathbb{R}\mathrm{P}^\infty,*))\big)$ as a right–angled Coxeter groups RC_K. From this follows the fact that $\pi_1\big(Z(K;(D^1,S^0))\big)$ is the commutator subgroup $[RC_K, RC_K]$.

A general result in this direction is due to M. Stafa. A proof is sketched below using a recent result about the homotopy type of $\Omega Z(K;(\underline{X},\underline{*}))$ due to T. Panov and S. Theriault, [113].

Theorem 3.10.7. [122] *Let $G_1, G_2, \ldots, G_m$ be non–trivial discrete groups and K be a simplicial complex with m vertices. Then $Z(K;(\underline{BG},\underline{*}))$ is an Eilenberg–Mac Lane space if and only if K is a flag complex. Equivalently, $Z(K;(\underline{EG},\underline{G}))$ is an Eilenberg–Mac Lane space if and only if K is a flag complex.*

Proof. (SKETCH.) For the simplicial complex V_K, consisting of the vertices of K, there is a natural map

$$X_1 \vee X_2 \vee \cdots \vee X_m \xrightarrow{\simeq} Z(V_K;(\underline{X},\underline{*})) \xrightarrow{i} Z(K;(\underline{X},\underline{*}))$$

where the first equivalence is that of (3.8.5) and the second map is induced by the inclusion $V_K \hookrightarrow K$. For a flag complex K, T. Panov and S. Theriault show in [113] that Ωi has a right homotopy inverse

$$\Omega(X_1 \vee X_2 \vee \cdots \vee X_m) \longleftarrow \Omega Z\big(K;(\underline{X},\underline{*})\big). \tag{3.10.6}$$

so that $\Omega Z\big(K;(\underline{X},\underline{*})\big)$ is a retract of $\Omega(X_1 \vee X_2 \vee \cdots \vee X_m)$.

When $X_i = BG_i$ it is known that

$$BG_1 \vee BG_2 \vee \cdots \vee BG_m \simeq B(G_1 * G_2 * \cdots * G_m)$$

and so for K a flag complex, it follows immediately that $Z(K;(\underline{BG},\underline{*}))$ is an Eilenberg–Mac Lane space.

If K is not a flag complex, then $\partial\Delta^n$, the boundary of an n–simplex, is a full subcomplex of K for some $n > 1$. In this case, [56, Proposition 3.3.1] implies that $Z\big(\partial\Delta^n;(\underline{BG},\underline{*})\big)$ splits off from $Z\big(K;(\underline{BG},\underline{*})\big)$. Hence it suffices to show that $Z\big(\partial\Delta^n;(\underline{BG},\underline{*})\big)$ has non-trivial higher homotopy groups. For the simplicial complex $\partial\Delta^n$, the fibration of Theorem 3.7.1

takes the form

$$Z(\partial\Delta^n;(\underline{EG},\underline{G})) \longrightarrow Z(\partial\Delta^n;(\underline{BG}),\underline{*})) \longrightarrow BG_1 \times BG_2 \times \cdots \times BG_{n+1}. \tag{3.10.7}$$

Next, since $\partial\Delta^n$ is a shifted simplicial complex, (Definition 3.8.2), and EG is contractible, Theorem 3.8.4 gives

$$Z(\partial\Delta^n;(\underline{EG},\underline{G})) \simeq S^n \wedge G_1 \wedge G_2 \wedge \cdots \wedge G_{n+1}.$$

Since the groups G_i are discrete for all i, $Z(\partial\Delta^n;(\underline{EG},\underline{G}))$ becomes a wedge of n-spheres. The assumption $n > 1$, the Hilton-Milnor theorem, and the homotopy sequence of (3.10.7) imply that $Z(\partial\Delta^n;(\underline{BG},\underline{*}))$ has non-trivial higher homotopy groups which completes the proof.

M. Stafa [121] also uses polyhedral products to construct monodromy representations of a finite product of discrete groups into outer automorphisms of free groups. We close this section by noting that the classifying spaces of certain groups derived from right-angled Coxeter groups through quandles, are described in terms of polyhedral products, with $X_i \neq *$ and $A_i \neq *$, by D. Kishimoto in [91].

3.11 The cohomology of polyhedral products and a spectral sequence

The right adjoint of the stable splitting Theorem 3.8.6 induces the product structure in cohomology of a polyhedral product by observing the consequences of suspending the diagonal map. We begin by defining

$$\mathcal{H}^q(K;(\underline{X},\underline{A})) = \bigoplus_{I\subseteq[m]} H^q(\widehat{Z}(K_I;(\underline{X},\underline{A})_I)).$$

Next, we impose an algebra structure on $\mathcal{H}^q(K;(\underline{X},\underline{A}))$. In [13, Section 3], it is shown that whenever $I = J \cup L$, there are well defined maps, *partial diagonals*,

$$\widehat{\Delta}_I^{J,L}\colon \widehat{Z}(K_I;(\underline{X},\underline{A})_I) \longrightarrow \widehat{Z}(K_J;(\underline{X},\underline{A})_J) \wedge \widehat{Z}(K_L;(\underline{X},\underline{A})_L)$$

making the diagram below commute

$$\begin{array}{ccc} Z(K;(\underline{X},\underline{A})) & \xrightarrow{\Delta_K} & Z(K;(\underline{X},\underline{A})) \wedge Z(K;(\underline{X},\underline{A})) \\ \big\downarrow \widehat{\Pi}_I & & \widehat{\Pi}_J\wedge\widehat{\Pi}_L \big\downarrow \\ \widehat{Z}(K_I;(\underline{X},\underline{A})_I) & \xrightarrow{\widehat{\Delta}_I^{J,L}} & \widehat{Z}(K_J;(\underline{X},\underline{A})_J) \wedge \widehat{Z}(K_L;(\underline{X},\underline{A})_L), \end{array} \tag{3.11.1}$$

where Δ_K is the reduced diagonal map. The maps Π_I, Π_J and Π_L are induced by the appropriate projection maps

$$\Pi_S\colon Y^{[m]} = Y_1 \times Y_2 \times \cdots \times Y_m \longrightarrow Y_{s_1} \wedge Y_{s_2} \wedge \cdots \wedge Y_{s_k}$$

where $S = \{s_1, s_2, \ldots, s_k\}$ corresponds in turn to I, J and L. This allows us to define a product on $\mathcal{H}^q(K; (\underline{X}, \underline{A}))$ as follows. Given cohomology classes $u \in H^p\big(\widehat{Z}\big(K_J; (\underline{X}, \underline{A})_J\big)\big)$ and $v \in H^q\big(\widehat{Z}\big(K_L; (\underline{X}, \underline{A})_L\big)\big)$, define

$$u * v = (\widehat{\Delta}_I^{J,L})^*(u \otimes v) \;\in\; H^{p+q}(\widehat{Z}(K_I)).$$

The element $u * v \in H^{p+q}(\widehat{Z}(K_I))$ is called the $*$-product of u and v. Moreover, the commutativity of diagram (3.11.1) gives

$$\widehat{\Pi}_I^*(u * v) = \widehat{\Pi}_J^*(u) \cup \widehat{\Pi}_L^*(v) \tag{3.11.2}$$

where $\cup$ is the cup product for the CW-complex $Z(K; (\underline{X}, \underline{A}))$. Consequently the $*$-product gives $\mathcal{H}^q(K; (\underline{X}, \underline{A}))$ a ring structure. Next, consider the map

$$\eta = \bigoplus_{I \subseteq [m]} \Pi_I^* \colon\; \mathcal{H}^*(K; (\underline{X}, \underline{A})) \longrightarrow H^*\big(Z(K; (\underline{X}, \underline{A}))\big),$$

which the stable splitting (3.8.3) ensures is an additive isomorphism. We can now state the main theorem of this section.

Theorem 3.11.1. [13] *Let K be an abstract simplicial complex with m vertices and assume that $(\underline{X}, \underline{A}) = \{(X_i, A_i, x_i)\}_{i=1}^m$ is a family of based CW-pairs. Then*

$$\eta : \mathcal{H}^*(K; (\underline{X}, \underline{A})) \to H^*\big(Z(K; (\underline{X}, \underline{A}))\big)$$

is a ring isomorphism.

It follows that the $*$-product gives $\mathcal{H}^q(K; (\underline{X}, \underline{A}))$ an algebra structure. More details can be found in [13, Section 3]. This theorem generalizes the result of V. Baskakov [23] who proved it for the case of moment–angle complexes, beginning with the cohomological splitting of Theorem 3.8.1.

Remark 3.11.2. As discussed in [13], various corollaries follow from Theorem 3.11.1. Among these are to be found:

1. If $(\underline{X}, \underline{A})$ consists of suspension pairs and $J \cap L \neq \varnothing$, then $u * v = 0$ for all $u \in H^p\big(\widehat{Z}\big(K_J; (\underline{X}, \underline{A})_J\big)\big)$ and $v \in H^q\big(\widehat{Z}\big(K_L; (\underline{X}, \underline{A})_L\big)\big)$.

2. Whenever $(\underline{CX}, \underline{X}) = \{(CX_i, X_i, x_i)\}_{i=1}^m$ is such that any finite product of the X_i with the spaces $Z(K_I; (D^1, S^0))$ satisfies the strong form of the Künneth theorem, the cup product structure for the cohomology algebra $H^*(Z(K; (\underline{CX}, \underline{X})))$ is a functor of the cohomology algebras of X, and $Z(K_I; (D^1, S^0)))$ for all I. Real moment–angle complexes $Z(L; (D^1, S^0)))$ have been studied extensively in the work of A. Suciu and A. Trevisan [124], A. Trevisan [127], L. Cai [42] and L. Cai and S. Choi [43].

We turn our attention now to the computation of the cohomology groups of a polyhedral product $H^*\big(Z(K; (\underline{X}, \underline{A}))\big)$. For a suitable collection of pairs of spaces $(\underline{X}, \underline{A})$, the cohomology of the polyhedral product $Z\big(K; (\underline{X}, \underline{A})\big)$ is computed using a spectral sequence by the authors in [17]. A computation using different methods by Q. Zheng can be found in [134]. The family of pairs $(\underline{X}, \underline{A})$ amenable to a cohomology calculation of $Z\big(K; (\underline{X}, \underline{A})\big)$ satisfy a *strong freeness condition*, [17, Definition 2.2].

Definition 3.11.3. The homology of $(\underline{X}, \underline{A})$ with coefficients in a ring k, is said to be *strongly free* if the long exact sequence

$$\xrightarrow{\delta} \widetilde{H}^*(X_i/A_i) \xrightarrow{\ell} \widetilde{H}^*(X_i) \xrightarrow{\iota} \widetilde{H}^*(A_i) \xrightarrow{\delta} \widetilde{H}^{*+1}(X_i/A_i) \rightarrow \tag{3.11.3}$$

satisfies the condition that there exist isomorphisms

1. $\widetilde{H}^*(A_i) \cong E_i \oplus B_i$.
2. $\widetilde{H}^*(X_i) \cong B_i \oplus C_i$, where $B_i \underset{\simeq}{\xrightarrow{\iota}} B_i$, $\iota|_{C_i} = 0$
3. $\widetilde{H}^*(X_i/A_i) \cong C_i \oplus W_i$, where $C_i \underset{\simeq}{\xrightarrow{\ell}} C_i$, $\ell|_{W_i} = 0$, $E_i \underset{\simeq}{\xrightarrow{\delta}} W_i$

for graded modules E_i, B_i, C_i and W_i of finite type and free over k.

In particular, for finite dimensional complexes, this happens when coefficients are taken in a field. In order to describe the cohomology, more notation is required. For a simplicial complex K with vertices in $[m] = \{1, 2, 3, \ldots, m\}$, and a subset $J = \{j_1, \ldots, j_r\} \subset [m]$, set $E^J = E_{j_1} \otimes \cdots \otimes E_{j_r}$. Let K_J be the full subcomplex with vertices in J. For $I = \{i_1, \cdots, i_t\} \subset [m]$, and for $\sigma \subset I$, set $Y^{I,\sigma} = Y_1 \otimes \cdots \otimes Y_t$ where

$$Y_j = \begin{cases} C_{i_j} & \text{if } i_j \in \sigma \\ B_{i_j} & \text{if } i_j \notin \sigma. \end{cases}$$

and for $I = \varnothing$, set $Y^{\varnothing,\varnothing} = k$.

Theorem 3.11.4. *If $(\underline{X}, \underline{A})$ satisfies the strong freeness condition there is a direct sum decomposition of the cohomology group*

$$\widetilde{H}^*\big(\widehat{Z}(K; (\underline{X}, \underline{A}))\big) = \bigoplus_{\substack{\sigma \subset I,\ \sigma \in K \\ I \cup J = [m],\ I \cap J = \varnothing}} E^J \otimes \widetilde{H}^*\big(\Sigma|\mathrm{lk}_\sigma(K_J)|\big) \otimes Y^{I,\sigma}$$

with the convention that $\widetilde{H}^*(\Sigma\varnothing) = k$. (Recall that $\mathrm{lk}_\sigma(K_J)$ is defined in Definition 3.8.9.)

Combined with the splitting Theorem 3.8.6 in Section 3.8, this theorem gives a complete description of the cohomology of the topological spaces $Z\big(K; (\underline{X}, \underline{A})\big)$ with appropriate coefficients. Moreover, the theorem generalizes to any multiplicative cohomology theory h^* for which $(\underline{X}, \underline{A})$ satisfies the natural formulation of the strong freeness condition for h^*. More information about the cohomology ring may be found in [17] and [134]. The use of the methods developed in [17] is illustrated next with a few examples, each having a different flavor.

Example 3.11.5. In the particular case that H^* is cohomology with coefficients in a field k, (or over $\mathbb{Z}$, under suitable freeness conditions), and the map $H^*(X_i) \longrightarrow H^*(A_i)$ is surjective for all $i = 1, 2, \ldots, m$, the spectral sequence constructed in [17] collapses by [17,

Proposition 3.8], and allows for a more concise statement of Theorem 3.11.4. The assumption allows us to consider $H^*(X_i/A_i)$ as a subring of $H^*(X_i)$ and then [17, Corollary 3.9] gives an isomorphism of rings

$$H^*\big(Z(K;(\underline{X},\underline{A}))\big) \cong H^*(X_1)\otimes H^*(X_2)\otimes\cdots\otimes H^*(X_m)/I$$

where I is the ideal generated by $H^*(X_{j_1}/A_{j_1})\otimes H^*(X_{j_2}/A_{j_2})\otimes\cdots\otimes H^*(X_{j_t}/A_{j_t})$ with $(j_1,j_2,\ldots,j_t)$ not spanning a simplex in K.

Remark 3.11.6. Notice that this generalizes Corollary 3.10.2:

$$H^*\big(Z(K;(\mathbb{C}P^\infty,*);k)\big) \;\cong\; k(K)$$

where $k(K)$ denotes the Stanley–Reisner ring of the simplicial complex K.

Example 3.11.7. Another tractable example is the important case

$$(\underline{X},\underline{A}) = (\underline{CA},\underline{A})$$

which includes the example of moment–angle complexes. In this case, with $H^*(A_i)$ free, the modules in the strong freeness condition of Theorem 3.11.4 are

$$E_i = \widetilde{H}^*(A_i),\quad B_i=0,\quad C_i=0, \text{ and } W_i=\widetilde{H}^*(\Sigma A_i).$$

The modules $Y^{I,\sigma}$ of Theorem 3.11.4, take a particularly simple form: $Y^{I,\sigma}=0$ if $I\neq\varnothing$ and is equal to k if $I=\varnothing$, (see (3.11.3), item (1)). As a consequence, the only links that appear in Theorem 3.11.4 are $\mathrm{lk}_\varnothing(K)=K$ and we have

$$\widetilde{H}^*\big(\widehat{Z}(K;(\underline{CA},\underline{A})\big) = \widetilde{H}^*(\Sigma\big|K\big|)\otimes\widetilde{H}^*(A_1)\otimes\cdots\otimes\widetilde{H}^*(A_m). \tag{3.11.4}$$

This result agrees with the one given by the wedge lemma [131] as described in [12]. Finally, the splitting theorem [12, Theorem 2.10] gives the cohomology of the polyhedral product as

$$\begin{aligned} H^*\big(Z(K;(\underline{CA},\underline{A}))\big) &= \bigoplus_{I\subset[m]} H^*(\Sigma|K_I|)\otimes\widetilde{H}^*(A_{i_1})\otimes\cdots\otimes\widetilde{H}^*(A_{i_t}) \\ &\subset\; H^*(Z(K;(D^1,S^0)))\otimes H^*(A_1)\otimes\cdots\otimes H^*(A_m). \end{aligned}$$

There is now an evident product on $H^*\big(Z(K;(\underline{CA},\underline{A}))\big)$ induced by coordinate-wise multiplication and by the product in $H^*\big(Z(K;(D^1,S^0))\big)$, which is the case corresponding to $A_i=S^0$. The latter is described in [42]. In [17] it is shown, using the results of [13], that this is indeed the ring structure in $H^*\big(Z(K;(\underline{CA},\underline{A}))\big)$.

Remark 3.11.8. The ring structure for general $(\underline{X},\underline{A})$ satisfying the strong freeness condition may be found in [13], [17] and [134].

The next example is of a different nature.

Example 3.11.9. Consider a CW-pair (X, A) with cohomology satisfying

$$H^*(X) = \mathbb{Z}\{b_4, c_6\} \quad \text{and} \quad H^*(A) = \mathbb{Z}\{e_2, b_4\} \tag{3.11.5}$$

where the dimensions of the classes are given by the subscripts. A trivial example of $(\underline{X}, \underline{A})$ is given by wedges of spheres in the appropriate dimensions. A more interesting example is obtained from the mapping cylinder of the composite map

$$f \colon \mathbb{C}P^2 \hookrightarrow \mathbb{C}P^3 \to \mathbb{C}P^3/\mathbb{C}P^1. \tag{3.11.6}$$

We denote the mapping cylinder of (3.11.6) by M_f and consider the pair $(M_f, \mathbb{C}P^2)$, which satisfies the cohomology condition above.

Remark 3.11.10. Notice here that Theorem 3.11.4 will give the *same* cohomology for $\widehat{Z}(K; (\underline{X}, \underline{A}))$ whether we realize condition (3.11.5) with a pair made from appropriate wedges of spheres, or from the projective spaces above. Given K, the cohomology of $\widehat{Z}(K; (\underline{X}, \underline{A}))$ depends on the modules E_i, B_i and C_i only.

For the example at hand, let K be the simplicial complex with three vertices and edges $\{1,3\}, \{1,2\}$. The cases in this example are indexed by the I in Theorem 3.11.4 starting with $I = \varnothing$ and building up to $I = \{1,2,3\}$. For each I there are the sub-cases indexed by the simplices $\sigma \subset I$. Theorem 3.11.4 reduces the calculation to bookkeeping, as it does with every example, but the bookkeeping can become quite complicated.

1. $I = \varnothing$, $\sigma = \varnothing$.
   Here $J = \{1,2,3\}$ and $|\mathrm{lk}_\varnothing(K_J)| = |K|$ is contractible. So, there is no contribution to the Poincaré series for $H^*\big(\widehat{Z}(K; (\underline{X}, \underline{A}))\big)$ in this case.

2. $I = \{1\}$, $\sigma = \varnothing$.
   Now, $J = \{2,3\}$, so E^J contributes $(t^2)^2$, $Y^{I,\varnothing} = b_4$ which gives a t^4, and $|\mathrm{lk}_\varnothing(K_J)| = |\{\{2\},\{3\}\}| = S^0$. So for this case, Theorem 3.11.4 specifies a total contribution of t^9 to the Poincaré series for $H^*\big(\widehat{Z}(K; (\underline{X}, \underline{A}))\big)$.

3. $I = \{1\}$, $\sigma = \{1\}$.
   Again, $J = \{2,3\}$, so E^J contributes $(t^2)^2$, $Y^{I,\{1\}} = c_6$ which gives a t^6, and $|\mathrm{lk}_{\{1\}}(K_J)| = |\{\{2\},\{3\}\}| = S^0$. So for this case, Theorem 3.11.4 specifies a total contribution of t^{11} to the Poincaré series for $H^*\big(\widehat{Z}(K; (\underline{X}, \underline{A}))\big)$.

Continuing in this way, we arrive at the (reduced) Poincaré series

$$\overline{P}(H^*(\widehat{Z}(K; (\underline{X}, \underline{A})))) = t^9 + t^{11} + 3t^{12} + 5t^{14} + 2t^{16}.$$

Remark 3.11.11. This illustrative example lends itself to direct calculation. For K as above, it follows directly from the definition that

$$\widehat{Z}(K; (X, A)) = X \wedge \big((X \wedge A) \cup (A \wedge X)\big).$$

The Künneth theorem now reduces the calculation to $H^*\big((X \wedge A) \cup (A \wedge X)\big)$, which is direct for the pair (X, A) in Example 3.11.9 by the Mayer-Vietoris sequence.

3.12 A geometric approach to the cohomology of polyhedral products

In the forthcoming paper [18], the authors show that for certain pairs $(\underline{X}, \underline{A})$, called *wedge decomposable*, the algebraic decomposition given by Theorem 3.11.4 is a consequence of an underlying geometric splitting. *Moreover, the consequences of this observation extend to more general based CW-pairs.* (The results of this section are from the authors' unpublished preprint from 2014, which in turn originated from an earlier preprint from 2010, and is currently being revised.)

Definition 3.12.1. Based CW-pairs of the form

$$(\underline{X}, \underline{A}) = (\underline{B \vee C}, \underline{B \vee E})$$

so that, for all i, $(X_i, A_i) = (B_i \vee C_i, B_i \vee E_i)$, where $E_i \hookrightarrow C_i$ is a null homotopic inclusion, are called wedge decomposable.

For such pairs, there is a decomposition of the smash polyhedral product. Let

$$J = \{j_1, j_2, \ldots, j_k\} \subset [m]$$

and set $\widehat{B}^J = B_{j_1} \wedge B_{j_2} \wedge \cdots \wedge B_{j_k}$. Similarly, define $\widehat{C}^J$.

Theorem 3.12.2. [18] *Let $(\underline{X}, \underline{A}) = (\underline{B \vee C}, \underline{B \vee E})$ be a wedge decomposable pair. Then there is a homotopy equivalence*

$$\widehat{Z}(K; (\underline{X}, \underline{A})) \longrightarrow \bigvee_{I \leq [m]} \widehat{Z}(K_I; (\underline{C}, \underline{E})_I) \wedge \widehat{B}^{([m]-I)}$$

which is natural with respect to maps of wedge decomposable pairs.

Since the inclusion $E_i \hookrightarrow C_i$ is null homotopic, the terms $\widehat{Z}(K_I; (\underline{C}, \underline{E})_I)$ are completely determined by the the *wedge lemma*, [11], as follows.

Proposition 3.12.3. *For pairs $(\underline{C}, \underline{E})$ as above there is a homotopy equivalence*

$$\widehat{Z}(K_I; (\underline{C}, \underline{E})_I) \to \bigvee_{\sigma \in K_I} |\mathrm{lk}_\sigma(K_I)| * \widehat{D}^I_{\underline{C},\underline{E}}(\sigma)$$

where $|\mathrm{lk}_\sigma(K_I)|$ is the realization of the link of σ in the full subcomplex K_I and

$$\widehat{D}^I_{\underline{C},\underline{E}}(\sigma) = \bigwedge_{j=1}^{|I|} W_{i_j}, \quad \text{with} \quad W_{i_j} = \begin{cases} C_{i_j} & \text{if} \quad i_j \in \sigma \\ E_{i_j} & \text{if} \quad i_j \in I - \sigma. \end{cases} \tag{3.12.1}$$

Combined with the splitting Theorem 3.8.6, these results give a complete description of the topological spaces $Z(K; (\underline{X}, \underline{A}))$ for wedge decomposable pairs $(\underline{X}, \underline{A})$.

The case $E_i \simeq *$ simplifies further by Theorem 3.8.12 to give the next corollary.

Corollary 3.12.4. *For wedge decomposable pairs of the form $(\underline{B \vee C}, \underline{B})$, corresponding to $E_i \simeq *$ for all $i = 1, 2, \ldots, m$, there are homotopy equivalences*

$$\widehat{Z}(K_I; (\underline{C}, \underline{E})_I) \simeq \widehat{Z}(K_I; (\underline{C}, *)_I) \simeq \widehat{C}^I,$$

and so Theorem 3.12.2 gives $\widehat{Z}(K; (\underline{B \vee C}, \underline{B})) \simeq \bigvee_{I \subseteq [m]} (\widehat{C}^I \wedge \widehat{B}^{([m]-I)})$.

Notice that the Poincaré series for the space $\widehat{Z}(K; (\underline{B \vee C}, \underline{B}))$ follows easily from Corollary 3.12.4.

Remark 3.12.5. In comparing these observations with Theorem 3.11.4, notice that the links appear in the terms $\widehat{Z}(K_I; (\underline{C}, \underline{E})_I)$. Also, while Theorem 3.12.2 and Proposition 3.12.3 give a geometric underpinning for the cohomology calculation in Theorem 3.11.4 for wedge decomposable pairs only, the geometric splitting does not require that E, B or C have torsion-free cohomology

Theorem 3.12.2 applies particularly well in cases where spaces have unstable attaching maps.

Example 3.12.6. The homotopy equivalence $S^1 \wedge Y \simeq \Sigma(Y)$ implies homotopy equivalences

$$\Sigma^{mq}(\widehat{Z}(K; (\underline{X}, \underline{A}))) \longrightarrow \widehat{Z}(K; (\underline{\Sigma^q(X)}, \underline{\Sigma^q(A)})) \tag{3.12.2}$$

where as usual, m is the number of vertices of K. Next, recalling that $SO(3) \cong \mathbb{RP}^3$, consider the pair

$$(X, A) = (SO(3), \mathbb{RP}^2)$$

for which there is a well known homotopy equivalence of pairs

$$(\Sigma^2(SO(3)),\ \Sigma^2(\mathbb{RP}^2)) \longrightarrow (\Sigma^2(\mathbb{RP}^2) \vee \Sigma^2(\mathrm{S}^3),\ \Sigma^2(\mathbb{RP}^2)), \tag{3.12.3}$$

making the pair $(SO(3), \mathbb{RP}^2)$ *stably wedge decomposable.* Now, combining (3.12.2) and (3.12.3), we get a homotopy equivalence

$$\Sigma^{2m}(\widehat{Z}(K; (SO(3), \mathbb{RP}^2))) \longrightarrow \widehat{\mathrm{Z}}(\mathrm{K}; (\Sigma^2(\mathbb{RP}^2) \vee \Sigma^2(\mathrm{S}^3),\ \Sigma^2(\mathbb{RP}^2)).$$

Finally, Theorem 3.12.2 allows us to conclude that $\widehat{Z}(K; (SO(3), \mathbb{RP}^2)))$, and hence the polyhedral product $Z(K; (SO(3), \mathbb{RP}^2))$, is stably a wedge of smash products of S^3 and $\mathbb{RP}^2$. Similar splitting results follow for the polyhedral product whenever the spaces X and A split after finitely many suspensions. In particular, the fact that $\Omega^2 S^3$ splits stably into a wedge of Brown–Gitler spectra implies that the polyhedral product $Z(K; (\Omega^2 S^3, *))$ splits stably into a wedge of smash products of Brown–Gitler spectra.

The result of the previous section can be exploited to give information about the groups $H^*(\widehat{Z}(K; (\underline{X}, \underline{A})))$ over a field k for pointed, finite, path connected finite pairs of CW-complexes $(\underline{X}, \underline{A})$, which are ***not*** wedge decomposable. In so doing, we explain further the remark in Example 3.11.9.

Given $(\underline{X}, \underline{A})$, let B_i, C_i and E_i be the k–modules specified in items (1) and (2) of Definition 3.11.3. Now, wedges of spheres and Moore spaces B'_i, C'_i and E'_i exist realizing the modules B_i, C_i and E_i so that

$$(\underline{B' \vee C'}, \underline{B' \vee E'})$$

satisfies the criterion for a wedge decomposable pair as in Definition 3.12.1. Moreover, the diagram below commutes:

$$\begin{array}{ccc} H^*(B'_j \vee C'_j) & \xrightarrow[\cong]{} & H^*(X_j) \\ \downarrow & & \downarrow \\ H^*(B'_j \vee E'_j) & \xrightarrow{\cong} & H^*(A_j). \end{array}$$

This leads to the following result.

Theorem 3.12.7. [18] *Under the conditions stated above, the following isomorphism of groups holds for cohomology with coefficients in a field k*

$$H^*\big(\widehat{Z}(K; (\underline{X}, \underline{A}))\big) \cong H^*\big(\widehat{Z}(K; (\underline{B' \vee C'}, \underline{B' \vee E'}))\big)$$

where the right-hand side is determined by Theorem 3.12.2 and Proposition 3.12.3. (This is not necessarily an isomorphism of modules over the Steenrod algebra.)

As a consequence, the additive structure of $H^*\big(Z(K; (X, A)); k\big)$ over a field k is determined by $H^*(X)$, $H^*(A)$, and the ranks of the restriction maps appearing in the extension

$$0 \to V_i \to H^i(X) \to H^i(A) \to W_i \to 0,$$

(cf. [134]). Information about the ring structure requires additional assumptions.

3.13 Polyhedral products and the Golodness of monomial ideal rings

We begin with the definition of a Golod ring, [60, 89].

Definition 3.13.1. Let $S = k[x_1, x_2, \ldots, x_m]$ be a polynomial ring in m variables over a field k, and let $I = (m_1, m_2, \ldots, m_r)$ be an ideal generated by monomials. The monomial ideal ring $R = S/I$ is called *Golod* if

$$\sum_{j=0}^{\infty} \dim \mathrm{Tor}_j^R(k, k) t^j = \frac{(1+t)^m}{1 - t(\sum_{j=0}^{\infty} \dim \mathrm{Tor}_j^S(R, k) t^j - 1)}. \tag{3.13.1}$$

Remark 3.13.2. According to [89] and [66], Serre had observed that the coefficients on the left-hand side in (3.13.1) are always less than or equal to the corresponding coefficients on the right.

In [66], Golod showed that (3.13.1) holds if and only if all products and all higher Massey products vanish in $\mathrm{Tor}^S(R,K)$, (see also [89]). Higher Massey products in $H^*(X;\mathbb{Q})$ obstruct the rational formality of X. As is noted in [96], this has a bearing on whether a complex manifold can admit a Kähler structure, an observation about the formality of Kähler manifolds which can be found in the paper by P. Deligne, Ph. A. Griffiths, J. W. Morgan, and D. Sullivan [55]. Golodness is relevant also in symplectic geometry and the theory of subspace arrangements; an overview of these connections can be found in [56] and[96].

The study of Golodness is particularly important in the case that R is the Stanley–Reisner ring $k(K)$ of a simplicial complex K, (see (3.5.1)). The property has been much investigated by algebraic combinatorial theorists. The arrival of moment–angle complexes has invigorated this line of research. Moment–angle complexes become relevant via the split fibration arising from Theorem 3.10.1, namely

$$\Omega Z(K;(D^2,S^1)) \longrightarrow \Omega Z(K;(\mathbb{C}\mathrm{P}^\infty,*)) \longrightarrow T^m. \tag{3.13.2}$$

As noted in Section 3.5, there is an isomorphism of rings

$$H^*\big(Z(K;(D^2,S^1))\big) \cong \mathrm{Tor}^*_S(k(K),k). \tag{3.13.3}$$

Using these ideas, and the isomorphism

$$H^*\big(\Omega Z(K;(\mathbb{C}\mathrm{P}^\infty,*)\big) \cong \mathrm{Tor}^*_{k(K)}(k,k),$$

due to Buchstaber and Panov [38], in conjunction with an Eilenberg–Moore spectral sequence argument, J. Grbić and S. Theriault, [70, Theorem 11.1], were able to recapture for Stanley–Reisner rings the inequality involving the terms in (3.13.1), mentioned in the remark above and attributed to Serre in the general case. Moreover, extending Theorem 3.8.3, they showed that for a certain class $\mathcal{F}_0$, of simplicial complexes, obtained from shifted simplicial complexes by elementary topological operations, $Z(K;(D^2,S^1))$ has the homotopy type of a wedge of spheres, implying Golodness. G. Denham and A. Suciu also study obstructions to Golodness, and in particular, triple Massey products in lowest degree in [56].

Theorem 3.13.3. [70, Theorem 1.2] *If $K\in\mathcal{F}_0$, then $k(K)$ is a Golod ring over k.*

K. Iriye and D. Kishimoto [82], begin with the stable decomposition of Theorem 3.8.13 applied to the case $(\underline{CX},\underline{X})$ to get

$$\Sigma Z(K;(\underline{CX},\underline{X})) \simeq \Sigma \bigvee_{\varnothing\neq I\subset[m]} \Sigma K_I \wedge \widehat{X}^I. \tag{3.13.4}$$

They notice that if (3.13.4) desuspends for $X = S^1$, the space $Z(K;(D^2,S^1))$ remains a suspension and hence (3.13.3) implies that $k(K)$ is Golod over k, [118, Corollary 3.11].

They introduce a structure on $Z(K;(D^1,S^0))$ which they call the *fat wedge filtration*. This then allows them to give sufficient conditions on K ensuring the desuspension of (3.13.4) for any X and concluding that $k(K)$ is a Golod ring. Among these results is the next theorem.

Theorem 3.13.4. [82] *If the Alexander dual of Kis shellable, then $Z(K;(D^n,S^{n-1}))$ has the homotopy type of a wedge of spheres, and hence $k(K)$ is Golod over k.*

They are able also to recover a result of Herzog, Reiner and Welker.

Theorem 3.13.5. [77, 82] *If the Alexander dual of the simplicial complex K is sequentially Cohen–Macaulay over k, then $k(K)$ is Golod over k.*

Fat wedge filtration techniques are also employed by K. Iriye and D. Kishimoto to prove that if K is a triangulated surface orientable over k, then $k(K)$ is Golod over k if and only if it is *2–neighborly*, that is, any two vertices are connected by an edge, [80, Theorem 1.3]. Using similar ideas, K. Iriye and D. Kishimoto [81] find a two-dimensional simplicial complex K with $k(K)$ Golod over $\mathbb{Q}$ but not over $\mathbb{Z}/p$.

Remark 3.13.6. The fat wedge filtration methods of K. Iriye and D. Kishimoto [82] allow them to recover a variety of stable and unstable homotopy decompositions of moment–angle complexes and other polyhedral products.

K. Iriye and T. Yano [84], also take the approach of desuspending (3.13.4). They use a simplicial complex constructed from a triangulated Hopf map $\eta\colon S^3 \to S^2$, and Alexander duality to prove the existence of a simplicial complex K for which $k(K)$ is a Golod ring but $Z(K;(D^2,S^1))$ is not a suspension.

J. Grbić, T. Panov, S. Theriault and J. Wu give an example, [68, Example 3.3], of a simplicial complex K, the standard 6-vertex triangulation of $\mathbb{R}\mathrm{P}^2$, which is Golod over any field but $Z(K;(D^2,S^1))$ is not a wedge of spheres because it has torsion. I. Limonchenko gives a similar example arising from a 9-vertex triangulation of $\mathbb{C}\mathrm{P}^2$, [95, Theorem 2.5]. He does this by explicitly computing the Betti numbers of $Z(K;(D^2,S^1))$ using (3.5.3), and then checking that the attaching maps in the stable splitting Corollary 3.8.14 are all stable maps. In [96], I. Limonchenko determines conditions on the bigraded Betti numbers (3.5.3) that imply the non-Golodness of K over k. He uses the simplicial wedge construction (3.6.2), to find a family of generalized moment–angle complexes, such that for any $l, r \geq 2$, the family contains an l-connected manifold M with a non-trivial r-fold Massey product in $H^*(M,\mathbb{Q})$.

Another approach to the Golodness problem for K via the homotopy theory of the moment–angle complex $Z(K;(D^2,S^1))$, can be found in the work of P. Beben and J. Grbić, [25]. A. Berglund gives a combinatorial condition in [27] which ensures the Golodness of a monomial ideal ring.

3.14 Higher Whitehead products and loop spaces

Higher Whitehead products were introduced into the homotopy theory of moment–angle complexes in the work of T. Panov and N. Ray [111]. They were concerned with the problem of realizing sphere wedge summands in $Z(K;(D^2,S^1))$ by *higher* Whitehead products. This work is summarized in [39, Sections 8.4 and 8.5]. Cases of this problem are discussed also in [72] and [83]. Recently, S. Abramyan [1] showed that not all sphere summands can be realized in this way using higher Whitehead products. These ideas are developed further in [2].

In the language of moment–angle complexes, the maps of (3.1.2) can be reformulated as follows. Let K be the simplicial complex consisting of two discrete vertices, the boundary of a one-simplex $\partial\Delta^1$, and consider the corresponding fibration (3.10.1),

$$Z(\partial\Delta^1;(D^2,S^1)) \xrightarrow{\widetilde{\omega}} Z(K;(\mathbb{C}\mathrm{P}^\infty,*)) \longrightarrow \mathbb{C}\mathrm{P}^\infty \times \mathbb{C}\mathrm{P}^\infty. \tag{3.14.1}$$

We have

$$\begin{aligned} Z(\partial\Delta^1;(D^2,S^1)) &= S^1 \times D^2 \cup_{S^1\times S^1} D^2 \times S^1 \;\simeq\; S^3 \\ Z(\partial\Delta^1;(S^2,*)) &= (S^2 \times *) \cup_{*\times *} (* \times S^2) \;=\; S^2 \vee S^2 \\ Z(\partial\Delta^1;(\mathbb{C}\mathrm{P}^\infty,*)) &= \mathbb{C}\mathrm{P}^\infty \vee \mathbb{C}\mathrm{P}^\infty \\ Z(\Delta^1;(\mathbb{C}\mathrm{P}^\infty,*)) &= \mathbb{C}\mathrm{P}^\infty \times \mathbb{C}\mathrm{P}^\infty. \end{aligned}$$

Then (3.14.1) factors as

$$\begin{aligned} S^3 \;\simeq\; Z(\partial\Delta^1;(D^2,S^1)) &\longrightarrow Z(\partial\Delta^1;(S^2,*)) \\ &\longrightarrow Z(\partial\Delta^1;(\mathbb{C}\mathrm{P}^\infty,*)) \longrightarrow \mathbb{C}\mathrm{P}^\infty \times \mathbb{C}\mathrm{P}^\infty. \end{aligned} \tag{3.14.2}$$

Here, the first map is induced by the map of pairs $(D^2,S^1) \to (S^2,*)$, the second by the inclusion $(S^2,*) \to (\mathbb{C}\mathrm{P}^\infty,*)$, and the third by the inclusion of simplicial complexes $\partial\Delta^1 \hookrightarrow \Delta^1$.

Now suppose that $\partial\Delta^{k-1}$ is a minimal missing face of a simplicial complex K, then we get the generalization

$$\begin{aligned} S^{2k-1} \;\simeq\; Z(\partial\Delta^{k-1};(D^2,S^1)) &\longrightarrow Z(\partial\Delta^{k-1};(S^2,*)) \\ &\longrightarrow Z(\partial\Delta^{k-1};(\mathbb{C}\mathrm{P}^\infty,*)) \longrightarrow Z(K;(\mathbb{C}\mathrm{P}^\infty,*)). \end{aligned} \tag{3.14.3}$$

Here, the last map is induced by the inclusion $\partial\Delta^{k-1} \hookrightarrow K$ and $Z(\partial\Delta^{k-1};(\mathbb{C}\mathrm{P}^\infty,*))$ retracts off $Z(K;(\mathbb{C}\mathrm{P}^\infty,*))$ by [56, Proposition 3.3.1]. Following [39, page 339], the group $\pi_2\big(Z(K;(\mathbb{C}\mathrm{P}^\infty,*))\big) \cong \mathbb{Z}^m$ has m generators

$$\widehat{\mu}_i\colon S^2 \to \mathbb{C}\mathrm{P}^\infty \xrightarrow{i} \mathbb{C}\mathrm{P}^\infty \vee \mathbb{C}\mathrm{P}^\infty \vee \cdots \vee \mathbb{C}\mathrm{P}^\infty \to Z(K;(\mathbb{C}\mathrm{P}^\infty,*)),$$

where the second map is the inclusion of the ith wedge summand and the last map is induced by the inclusion of the zero-skeleton into K. Next, labelling the vertices of the missing face $\partial\Delta^{k-1}$ by $\{i_1, i_2, \ldots, i_k\}$, we call the composite map (3.14.3) the *k-fold higher Whitehead product* and denote it by the symbol $[\widehat{\mu}_{i_1}, \widehat{\mu}_{i_2}, \ldots, \widehat{\mu}_{i_k}]_w$.

These maps, and associated Samelson products, play an important role in the study of the homotopy theory of $\Omega Z(K;(D^2,S^1))$ and $\Omega Z(K;(\mathbb{C}\mathrm{P}^\infty,*))$, particularly cases for which $Z(K;(D^2,S^1))$ is homotopy equivalent to a wedge of spheres, see [39, Section 8.4]. In the case of a flag complex K, T. Panov and N. Ray [111] compute the rational Pontrjagin ring of $\Omega Z(K;(\mathbb{C}\mathrm{P}^\infty,*))$ by introducing various algebraic and geometric models. Motivated in part by this, and the work of S. Papadima and A. Suciu [116], N. Dobrinskaya [57] also addresses $\Omega Z(K;(X,*))$ from a different point of view, and relates the computation to diagonal arrangements. (Recall from Section 3.8 that $Z(K;(D^2,S^1))$ is a deformation retract of $Z(K;(\mathbb{C},\mathbb{C}^*))$, the complement of a complex coordinate arrangement determined by K.)

K. Iriye and D. Kishimoto, [83], call a simplicial complex K *totally fillable* if each of its full subcomplexes has the property that it becomes contractible if some of its minimal missing faces are added. They employ their theory of *fat wedge filtrations* [82], to show that if K is totally fillable, the moment–angle complex $Z(K;(D^2,S^1))$ decomposes as a wedge of spheres. For each such sphere S^α, they recognize the map

$$S^\alpha \hookrightarrow Z(K;(D^2,S^1)) \xrightarrow{\widetilde{\omega}} Z(K;(\mathbb{C}\mathrm{P}^\infty,*))$$

in terms of higher and iterated Whitehead product, [83, Theorem 1.3].

K. Iriye and D. Kishimoto then extend the discussion, as do J. Grbić and S. Theriault in [72], to the more general fibration

$$Z\big(K;(C(\Omega\underline{X})),\Omega\underline{X})\big) \xrightarrow{\widetilde{\omega}} Z(K;(\underline{X},*)) \longrightarrow \prod_{i=1}^{m} X_i, \tag{3.14.4}$$

(which was studied also by G. Porter, [118]). In both papers, the map $\widetilde{\omega}$ is described fully in particular cases. More comprehensive information about Whitehead products in the context of (3.14.4) is to be found in the paper by S. Theriault [126]. The rational type of the fibre in (3.14.4) is studied also in the paper of Y. Félix and D. Tanré [59].

Acknowledgements

The authors would like to thank Haynes Miller, Peter Landweber, Taras Panov and Jelena Grbić for their careful reading of the manuscript and for their many valuable suggestions. Our thanks also to Santiago López de Medrano, Daisuke Kishimoto, Matthias Franz, Alex Suciu, Alvise Trevisan, Mentor Stafa, Stephen Theriault, Kouyemon Iriye, and Graham Denham, for comments and recommendations which have led to a discernible improvement in the presentation.

This work was supported in part by grant 426160 from the Simons Foundation.

Bibliography

[1] S. Abramyan, *Iterated higher Whitehead products in topology of moment–angle complexes*, arXiv:1708.01694, 2017.

[2] S. Abramyan and T. Panov, *Higher Whitehead products in moment–angle complexes and substitution of simplicial complexes*, arXiv:1901.07918, 2019.

[3] A. Al–Raisi, *Branched covers, Strickland maps and cohomology related to the polyhedral product functor*, Thesis, University of Rochester, (2014).

[4] A. Al-Raisi, F. R. Cohen and E. Vidaurre, *Automorphisms of polyhedral products and their applications*, in preparation.

[5] D. Allen and J. La Luz, *Local face rings and diffeomorphisms of quasitoric manifolds*, Homology Homotopy Appl., **21**(1), 2019, 1–20.

[6] D. Anick, *Connections between Yoneda and Pontrjagin algebras*, Algebraic Topology, Aarhus 1982, Springer-Verlag Lecture Notes in Math., **1051** (1984), 331–350.

[7] L. Astey, A. Bahri, M. Bendersky, F. R. Cohen, D. Davis, S. Gitler, M. Mahowald, N. Ray and R. Wood, *The KO^*-rings of BT^m, the Davis-Januszkiewicz spaces and certain toric manifolds.* J. Pure Apl. Algebra, **218**, (2014), 303–320.

[8] M. F. Atiyah, *Convexity and commuting Hamiltonians*, Bull. Lond. Math. Soc., **14**, (1982), 1–15.

[9] A. Ayzenberg, *Substitutions of polytopes and of simplicial complexes, and multigraded Betti numbers*, Trans. Moscow Math. Soc., (2013), 175–202.

[10] A. Bahri and M. Bendersky, *The KO-theory of toric manifolds.* Trans. Amer. Math. Soc., **352**, (2000), 1191–1202.

[11] A. Bahri, M. Bendersky, F. R. Cohen, and S. Gitler, *Decompositions of the polyhedral product functor with applications to moment-angle complexes and related spaces*, Proc. Natl. Acad. Sci. USA, **106**, July 2009, 12241–12244.

[12] A. Bahri, M. Bendersky, F. R. Cohen and S. Gitler, *The Polyhedral Product Functor: a method of computation for moment-angle complexes, arrangements and related spaces.* Adv. Math., **225**, (2010), 1634–1668.

[13] A. Bahri, M. Bendersky, F. R. Cohen and S. Gitler, *Cup products in generalized moment-angle complexes.* Math. Proc. Cambridge Phil. Soc., **153**, (2012), 457–469.

[14] A. Bahri, M. Bendersky, F. R. Cohen and S. Gitler, *The geometric realization of monomial ideal rings and a theorem of Trevisan*, Homology Homotopy Appl., **14**(1), (2012), 1–8.

[15] A. Bahri, M. Bendersky, F. R. Cohen and S. Gitler, *A generalization of the Davis-Januszkiewicz construction and applications to toric manifolds and iterated polyhedral products*, "Perspectives in Lie Theory", F. Callegaro, G. Carnovale, F. Caselli, C. De Concini, A. De Sole (Eds.), Springer INdAM series, (2017), 369–388.

[16] A. Bahri, M. Bendersky, F. R. Cohen and S. Gitler, *Operations on polyhedral products and a new topological construction of infinite families of toric manifolds.* Homology, Homotopy and Appl., **17**, (2015), 137–160.

[17] A. Bahri, M. Bendersky, F. R. Cohen and S. Gitler, *A spectral sequence for polyhedral products.* Adv. Math., **308**, (2017), 767–814

[18] A. Bahri, M. Bendersky, F. R. Cohen and S. Gitler, *A Cartan formula for polyhedral products*, in preparation.

[19] A. Bahri, S. Sarkar and J. Song, *Infinite families of equivariantly formal toric orbifolds*, Forum Math., **31**(2), 283–301.

[20] A. Bahri, D. Notbohm, S. Sarkar and J. Song, *On the integral cohomology of certain orbifolds*, Int. Math. Res. Not., `https://dx.doi.org/10.1093/imrn/rny283`.

[21] A. Bahri, S. Sarkar and J. Song, *On the integral cohomology ring of toric orbifolds and singular toric varieties*, Algebr. Geom. Topol., **17**(6), 3779–3810

[22] Y. Barreto, S. López de Medrano and A. Verjovsky, *Some open book and contact structures on moment-angle manifolds*,. Bol. Soc. Mat. Mex., **23**, (2017), 423–437.

[23] V. Baskakov, *Cohomology of K-powers of spaces and the combinatorics of simplicial divisions*, Russian Math. Surveys, **57**, (2002), no. 5, 989–990.

[24] I. Baskakov, V. Buchstaber and T. Panov, *Cellular cochain complexes and torus actions*, Uspeckhi. Mat. Nauk, **59**, (2004), no. 3, 159–160 (Russian); Russian Math. Surveys, **89**, (2004), no. 3, 562–563 (English translation).

[25] P. Beben and J. Grbić, *Configuration spaces and polyhedral products.*, Adv. Math., (2017), 378-425.

[26] L. W. Beineke and F. Harary, *The genus of the n-cube,* Can. J. Math., **17**, (1965), 494–496.

[27] A. Berglund, *Poincaré series and homotopy Lie algebras of monomial rings*, Research Report, no. 6, Stockholm University, (2005).

[28] E. Bifet, C. De Concini and C. Procesi, *Cohomology of regular embeddings*, Adv. Math., **82**, (1990), 1–34.

[29] A. Björner and M. L. Wachs, *Shellable nonpure complexes and posets II,* Trans. Amer. Math. Soc., **349**, (1997), 3945–3975.

[30] A. Björner, M. L. Wachs, and V. Welker, *On sequentially Cohen-Macaulay complexes and posets*, Israel J. Math. **169**, (2009), 295–316.

[31] F. Bosio and L. Meersseman, *Real quadrics in $\mathbb{C}^n$, complex manifolds and convex polytopes*, Acta Math., 197 (2006), 53–127,

[32] M. Brion, *Piecewise polynomial functions, convex polytopes and enumerative geometry*, P. Pragacz (ed.), Parameter Spaces. Banach Cent. Publ., **36** (Warszawa 1996), 25–44.

[33] K. Brown, Cohomology of Groups, Grad. Texts in Math., **87**, Springer, (1982).

[34] W. Bruns and J. Gubeladze, *Combinatorial invariance of Stanley–Reisner rings*, Georgian Math. J., **3**(4), (1996), 315–318.

[35] W. Bruns and J. Gubeladze, *Polytopes Rings and K–Theory*, Springer, (2009).

[36] V. Buchstaber, N. Erokhovets, M. Masuda, T. Panov and S. Park, *Cohomological rigidity of manifolds defined by right–angled 3–dimensional polytopes*, Uspekhi Mat. Nauk **72**(2), (2017), 3–66 (Russian); Russian Math. Surveys, **72**(2), (2017), 199–256 (English translation).

[37] V. Buchstaber, and T. Panov, *Actions of tori, combinatorial topology and homological algebra*, Russian Math. Surveys, **55**, (2000), no. 5, 825–921.

[38] V. Buchstaber and T. Panov, *Torus actions and their applications in topology and combinatorics*, Univ. Lecture Ser., **24**, (2002), Amer. Math. Soc.

[39] V. Buchstaber and T. Panov, Toric Topology, Math. Surveys and Monogr., **204**, (2015), Amer. Math. Soc.

[40] V. Buchstaber, T. Panov and N. Ray, *Spaces of polytopes and cobordism of quasitoric manifolds*, Mosc. Math. J., **7**, (2007), no. 2, 219–242.

[41] V. Buchstaber and N. Ray, *An invitation to toric topology: vertex four of a remarkable tetrahedron.* In: Toric Topology (M. Harada et al., eds.). Contemp. Math., 460, Amer. Math. Soc., (2008), 1–27.

[42] L. Cai, *On products in a real moment-angle manifold*, J. Math. Soc. Japan, **69**(2), (2017), 503–528.

[43] L. Cai and S. Choi, *Integral cohomology groups of real toric manifolds and small covers*, arXiv:1604.06988, 2016.

[44] X. Cao and Z. Lü, *Möbius transform, moment–angle complexes and Halperin–Carlsson conjecture*, J. of Algebraic Combin., **35**, (2012), 121–140.

[45] R. Charney and M. Davis, *The $K(\pi,1)$-problem for hyperplane complements associated to infinite reflection groups*, J. Amer. Math. Soc., **8**, (1995), 597–627.

[46] S. Cho, S. Choi and S. Kaji, *Geometric representations of finite groups on real toric spaces*, arXiv:1704.08591, 2017.

[47] S. Choi and H. Park, *Wedge operations and torus symmetries*, Tohoku Math. J. (2), **68**(1) (2016), 91–138.

[48] Y. Civan and N. Ray, *Homotopy decompositions and real K-Theory of Bott towers*, K-Theory, **34**, (2005), 1–33.

[49] F. R. Cohen and F. P. Peterson, *Suspensions of Stiefel manifolds*, Quart. J. Math. Oxford Ser. (2), **35**(138), (1984), 115–119.

[50] H. S. M. Coxeter, *Regular skew polyhedra in three and four dimension, and their topological analogues*, Proc. London Math. Soc. , (2) **43** (1), (1938), 33–62.

[51] S. Das, *Genus of the hypercube graph and real moment–angle complex*, Topology Appl., **258**, (2019), 415–424.

[52] M. W. Davis, *Groups generated by reflections and aspherical manifolds not covered by Euclidean space*, Ann. of Math., **117**, (1983), no. 2, 293–324.

[53] M. Davis and T. Januszkiewicz, *Convex polytopes, Coxeter orbifolds and torus actions*, Duke Math. J., **62**, (1991), no. 2, 417–451.

[54] M. Davis and B. Okun, *Cohomology computations for Artin groups, Bestvina-Brady groups, and graph products*, Groups Geom. Dyn. **6**, (2012), 485-531 Bull. Soc. Math. France, **116**, (1988), no. 3, 315–339.

[55] P. Deligne, Ph. A. Griffiths, J. W. Morgan, and D. Sullivan, *Real homotopy theory of Kähler manifolds*, Invent. Math., **29**, (1975), 245–274.

[56] G. Denham and A. Suciu, *Moment-angle complexes, monomial ideals and Massey products*, Pure Appl. Math. Q., **3** (2007), no. 1, 25–60.

[57] N. Dobrinskaya, *Loops on polyhedral products and diagonal arrangements*, arXiv:0901.2871, 2009.

[58] N. Erokhovets, *Buchstaber invariant of simple polytopes*, Uspekhi Mat. Nauk, **63**(5), (2008), 187–188 (Russian); Russian Math. Surveys, **63**(5), (2008), 962–964, (English translation).

[59] Y. Félix and D. Tanré, *Rational homotopy of the polyhedral product functor*, Proc. Amer. Math. Soc., **137**, (2009), 891–898.

[60] R. Frankhuizen, *A^{∞}–resolutions and the Golod property for monomial rings*, Algebr. Geom. Topol., 18 (2018), no. 6, 3403–3424.

[61] M. Franz, *On the integral cohomology of smooth toric varieties*, arXiv:0308253, 2003.

[62] M. Franz, *The integral cohomology of toric manifolds*, Proc. Steklov Inst. Math., **252** (2006), 53–62, Proceedings of the Keldysh Conference, Moscow 2004.

[63] X. Fu and J. Grbić, *Simplicial G-complexes and representational stability of polyhedral products*, arXiv:1803.11047, 2018.

[64] T. Ganea, *A generalization of the homology and homotopy suspension*, Comment. Math. Helv., **39** (1965), 295–322.

[65] S. Gitler and S. López de Medrano, *Intersections of quadrics, moment-angle manifolds and connected sums*, Geom. Topol., **17** (3), 1497–1534.

[66] E. S. Golod, *On the cohomology of some local rings*, (Russian); Soviet Math. Dokl. **3**, (1962), 745–749.

[67] M. Goresky and R. MacPherson, *Stratified Morse theory*, Ergeb. Math. Grenzgeb. 3rd series, **14**, Springer-Verlag, 1988.

[68] J. Grbić, T. Panov, S. Theriault and J. Wu, *The homotopy types of moment-angle complexes for flag manifolds*, Trans. Amer. Math. Soc., **368**(9), (2016), 6663–6682.

[69] J. Grbić and S. Theriault, *Homotopy type of the complement of a coordinate subspace arrangement of codimension two*, Russian Math. Surveys, **59**, (2004), no. 3, 1207–1209.

[70] J. Grbić and S. Theriault, *The homotopy type of the complement of a coordinate subspace arrangement*, Topology, **46**, (2007), 357–396.

[71] J. Grbić and S. Theriault, *The homotopy type of the polyhedral product for shifted complexes*, Adv. Math., **245**, (2013), 690-715.

[72] J. Grbić and S. Theriault, *Homotopy theory in toric topology*, Russian Math. Surveys **71**, (2016), 185–251.

[73] V. Grujić and V. Welker, *Moment-angle complexes of pairs (D^n, S^{n-1}) and simplicial complexes with vertex-decomposable duals*, Monatsh. Math., **176**(2), (2015), 255–273.

[74] A. Hatcher, *Algebraic topology*, Cambridge Univ. Press, 2001.

[75] A. Hattori and M. Masuda, *Theory of multi-fans*, Osaka J. Math., **40**, (2003) 1–68.

[76] G. C. Haynes, F. R. Cohen, and D. Koditschek, *Gait transitions for quasi-static hexapedal locomotion on level ground*, 14-th International Symposium on Robotics Research (ISRR 2009).

[77] J. Herzog, V. Reiner, V. Welker, *Componentwise linear ideals and Golod rings*, Michigan Math. J., **46**,(1999), 211–223.

[78] M. Hochster, *Cohen–Macaulay rings, combinatorics, and simplicial complexes*, in: Ring Theory II (Proc. Second Oklahoma Conference), B. R. McDonald and R. Morris, eds. Dekker, (1977), 171–223.

[79] K. Iriye and D. Kishimoto, *Decompositions of polyhedral products for shifted complexes*, Adv. Math., **245** (2013), 716–736.

[80] K. Iriye and D. Kishimoto, *Golodness and polyhedral products for two dimensional simplicial complexes.* Forum Math., **30**(2), (2018), 527–532.

[81] K. Iriye and D. Kishimoto, *Golodness and polyhedral products of simplicial complexes with minimal Taylor resolutions*, Homology Homotopy Appl., **20**(1), (2018), 69–78.

[82] K. Iriye and D. Kishimoto, *Fat wedge filtrations and decomposition of polyhedral products*, Kyoto J. Math., 59 (2019), no. 1, 1–51.

[83] K. Iriye and D. Kishimoto, *Whitehead products in moment–angle complexes*, arXiv:1807.00087, 2018.

[84] K. Iriye and T. Yano, *A Golod complex with non-suspension moment-angle complex*, Topology Appl. **225**, (2017), 145–163.

[85] I. M. James, *Reduced product spaces*, Ann. of Math., **62**, No. 1 (1955), 170-197.

[86] I. M. James, *The topology of Stiefel manifolds*, London Math. Soc. Lecture Note Ser., No. 24. Cambridge University Press, (1977).

[87] J. Jurkiewicz, *On the complex projective torus embedding, the associated variety with corners and Morse functions*, Bull. Acad. Pol. Sci., Ser. Sci. Math., **29**, (1981), 21–27.

[88] Y. Kamiyama and S. Tsukuda, *The configuration space of the n-arms machine in the Euclidean space*, Topol. Appl., **154** (2007), 1447–1464.

[89] L. Katthän, *A non-Golod ring with a trivial product on its Koszul homology*, J. Algebra, **479**, (2017), 244–262

[90] K. H. Kim and F. W. Roush, *Homology of certain algebras defined by graphs*, J. Pure Appl. Algebra, **17**, (1980), 179–186.

[91] D. Kishimoto, *Right-angled Coxeter quandles and polyhedral products*, arXiv:1706.06209v2, 2017.

[92] A. Kurosh, *Lectures on general algebra*, published in Russian in 1960 and an English translation published in 1963, the Chelsea Publishing Company.

[93] H. Kuwata, M. Masuda, H. Zeng, *Torsion in the cohomology of torus orbifolds*, Chin. Ann. Math., **38**(6), (2016), 1247–1268.

[94] J. La Luz and D. Allen, *Certain generalized higher derived functors associated to quasitoric manifolds*, J. Homotopy Relat. Struct., **13**, (2018), no. 2, 395–421.

[95] I. Limonchenko, *Families of minimally non-Golod complexes and their polyhedral products*, Far Eastern Mathematical Journal, **15**(2), (2015), 222–237.

[96] I. Limonchenko, *On higher Massey products and rational formality for moment–angle manifolds over multiwedges*, arXiv:1711.00461v2, 2017.

[97] I. Limonchenko, *Bigraded Betti numbers of certain simple polytopes*, Math. Notes, **3**(3–4), (2013), 373–388 .

[98] J. L. Loday, *Cyclic Homology*, Springer, (1998).

[99] J. L. Loday and D. Quillen, *Cyclic homology and the Lie algebra homology of matrices*, Comment. Math. Helv., **59**, (1984), 565–591.

[100] S. López de Medrano, *Topology of the intersection of quadrics in* $\mathbb{R}^n$, Algebraic Topology, Arcata California, (1986), Lecture Notes in Math., **1370** Springer, (1989), 280–292.

[101] S. López de Medrano and A. Verjovsky, *A new family of complex, compact, non-symplectic manifolds* Bol. Soc. Bras. Mat., **28**(2), (1997), 253–269

[102] Z. Lü and T. Panov, *Moment–angle complexes from simplicial posets* Central European Journal of Mathematics, **9**(4), (2011), 715–730.

[103] L. Meersseman, *A new geometric construction of compact complex manifolds in any dimension*, Math. Ann., **317**, (2000), 79–115.

[104] L. Meersseman and A. Verjovsky, *Holomorphic principal bundles over projective topic varieties*, J. Reine Angew. Math. **572**, (2004), 57–96.

[105] H. Miller, *Stable splittings of Stiefel manifolds*, Topology, **24**(4), (1985), 411–419.

[106] J. Milnor, *On the construction F[K]*, A Student's Guide to Algebraic Topology, J. F. Adams, editor, London Math. Soc. Lecture Note Ser., **4**(1972), 119-136. Also in: Collected Papers of John Milnor IV, Homotopy, Homology and Manifolds, John McCleary editor, Amer. Math. Soc., 2009, pp. 45–53.

[107] C. Moreau, *Sur les permutations circulaires distinctes*, Nouvelles Annales de Mathématiques, journal des candidats aux écoles polytechnique et normale, Sér. 2, **11**, (1872) 309–331.

[108] D. Notbohm and N. Ray, *On Davis-Januszkiewicz homotopy types I; formality and rationalisation*, Algebr. Geom. Topol., **5**, (2005), 31–51.

[109] T. Panov, *Cohomology of face rings, and torus actions*, Surveys in Contemporary Mathematics, London Math. Soc. Lecture Note Ser., **347**, (2008), 165–201.

[110] T. Panov, *Geometric structures on moment–angle manifolds*, Uspekhi Mat. Nauk **68**(3), (2013), 111–186 (Russian); Russian Math. Surveys, **68**(3), (2013), 503–568 (English translation).

[111] T. Panov and N. Ray, *Categorical aspects of toric topology*, Toric Topology, Contemp. Math. **460**, Amer. Math. Soc. (2008), 293–322.

[112] T. Panov, N. Ray, and R. Vogt *Colimits, Stanley-Reisner algebras, and loop spaces*, in Categorical decomposition techniques in algebraic topology, Prog. Math., **215** (2004), 261–291, Birkhäuser.

[113] T. Panov and S. Theriault, *The homotopy theory of polyhedral products associated with flag complexes.* Compos. Math., **155**, (2019), no. 1, 206–228.

[114] T. Panov and Y. Ustinovsky. *Complex-analytic structures on moment-angle manifolds.* Moscow Math. J., **12** (2012), no. 1, 149–172.

[115] T. Panov and Y. Veryovkin, *Polyhedral products and commutator subgroups of right–angled Artin and Coxeter groups*, Mat. Sbornik, **207**(11), (2016), 105–126 (Russian); Sbornik Math., **207**(11), (2016), 1582–1600 (English).

[116] S. Papadima and A. Suciu, *Algebraic invariants for right-angled Artin groups,* Math. Ann., **334**(3), (2006), 533–555.

[117] G. Porter, *The homotopy groups of wedges of suspensions*, Amer. J. Math., **88**, (1966), 655–663.

[118] G. Porter, *Higher products*, Trans. Amer. Math. Soc., **148**, (1970), 315–345.

[119] J. S. Provan and L. J. Billera, *Decompositions of simplicial complexes related to diameters of convex polyhedra*, Math. Oper. Res., **5**, (1980), 576–594.

[120] G. Ringel, *Über drei kombinatorische Probleme am n-dimensionalen Würfel und Würfelgitter*, Abh. Math. Semin. Univ. Hambg., **20**(1), (1955), 10–19.

[121] M. Stafa, *Polyhedral products, flag complexes and monodromy representations*, Topology Appl., **244**, (2018), 12–30.

[122] M. Stafa, *On the fundamental group of certain polyhedral products*, J. Pure Appl. Algebra **219**, (2015), no. 6, 2279–2299.

[123] N. Strickland, *Notes on toric spaces*, preprint 1999, (unpublished).

[124] A. Suciu, *The rational homology of real toric manifolds*, Oberwolfach Reports, **9**(4), (2012), 2972–2976.

[125] S. Theriault, *Moment-angle manifolds and Panov's problem*, Int. Math. Res. Not., **2015**(20), 10154–10175.

[126] S. Theriault, *The dual polyhedral product, cocategory and nilpotence*, Adv. Math., **340**, (2018), 138–192.

[127] A. Trevisan, *Generalized Davis-Januszkiewicz spaces and their applications to algebra and topology*, Thesis, Vrije Universiteit Amsterdam, (2012).

[128] Y. Ustinovsky, *Doubling operation for polytopes and torus actions*. Uspekhi Mat. Nauk, **64**(5), (2009), 181–182 (Russian); Russian Math. Surveys, **64**(5), (2009), 952–954 (English).

[129] Y. Ustinovsky, *Toral rank conjecture for moment-angle complexes*, Mat. Zametki **90**(2), (2011), 300—305 (Russian); Math. Notes **90**(1–2), (2011), 279–283 (English). **5** (1971), 1083–1119.

[130] C. T. C. Wall, *Stability, pencils and polytopes*, Bull. London Math. Soc., **12**, (1980), 401-421.

[131] V. Welker, G. Ziegler and R. Živaljević, *Homotopy colimits–comparison lemmas for combinatorial applications*, J. Reine Angew. Math., **509** (1999), 117–149.

[132] E. Witt, *Treue Darstellungen Lieschen Ringe*, J. Reine Angew. Math., **177**, (1937), 152–160.

[133] L. Yu, *On lower bounds of the sum of multigraded Betti numbers of simplicial complexes*, arXiv:1811.04398, 2018.

[134] Q. Zheng, *The homology coalgebra and cohomology algebra of generalized moment-angle complexes*, arXiv:1201.4917, 2012.

[135] G. Ziegler and R. Živaljević, *Homotopy types of sub-space arrangements via diagrams of spaces*, Math. Ann., **295**, (1993), 527–548.

Department of Mathematics, Rider University, Lawrenceville, NJ 08648, U.S.A.

E-mail address: bahri@rider.edu

Department of Mathematics, Hunter College, CUNY, New York, NY 10065, U.S.A.

E-mail address: mbenders@hunter.cuny.edu

Department of Mathematics, University of Rochester, Rochester, NY 14625, U.S.A.

E-mail address: fred.cohen@rochester.edu

4

A guide to tensor-triangular classification

Paul Balmer

4.1 Introduction

Stable homotopy theory shines across pure mathematics, from topology to analysis, from algebra to geometry. While its liturgy invokes Quillen model structures and ∞-categories, profane users around the world often speak the vernacular of *triangulated categories*, as we shall do in this chapter.

Perhaps the first salient fact about stable homotopy categories is that in almost all cases they turn out to be *wild* categories – beyond the trivial examples of course. Dade famously began his paper [28] with the admonition "There are just too many modules over p-groups!" and this truth resonates in all other fields as well: no hope to classify topological spaces up to stable homotopy equivalence; no more hope with complexes of sheaves, nor with equivariant C^*-algebras, nor with motives, etc., etc. One might dream that things improve with 'small' objects (compact, rigid, or else) but the problem persists even there: Stable homotopy theory is just too complicated!

Faced with the complexity of stable homotopy categories, we are led to the following paradigm shift. A classification *up to isomorphism* makes sense in any category, *i.e.* as soon as we can speak of *isomorphism*. But stable homotopy categories are more than mere categories: They carry additional structures, starting with the *triangulation*. In the case of a *tensor*-triangulated category (*tt-category* for short), as we consider in this chapter, we have two basic tools at hand: triangles and tensor. Instead of ignoring these additional structures, we should include them in the concept of

tt-classification

which is our nickname for *classification up to the tensor-triangular structure*.

More precisely, we want to decide when two objects X and Y can be obtained from one another by using tensor with anything, direct sums, summands, cones, suspension, etc. In mathematical terms, we ask when X and Y generate the same thick triangulated tensor-ideals. Heuristically, if you can build Y out of X by using the tt-structure then X contains at least as much information as Y. If you can go back and forth between X and Y, then they contain the same amount of information.

The remarkable gain is that the tt-classification of an essentially small (rigid) tensor-triangulated category can *always* be achieved by means of a geometric object, more precisely

Mathematics Subject Classification. 18F99, 55P42, 55U35.
Key words and phrases. tensor-triangulated category, spectrum, classification.

a spectral topological space, called its *tensor-triangular spectrum.*[1] Let us highlight this starting point:

Fundamental fact: *Although almost every symmetric monoidal stable homotopy category $\mathcal{K}$ is 'wild' as a category, we always have a tt-classification of its objects, via a topological space,* $\mathrm{Spc}(\mathcal{K})$, *called the spectrum of* $\mathcal{K}$.[2]

This chapter is dedicated to a survey of tt-classifications across different examples, as far as they are known to the author at this point in time.

The original idea of classifying objects up to the ambient structure was born in topology, around Ravenel's conjectures [65] and the 'chromatic' theorems of Devinatz-Hopkins-Smith [33, 46]; this relied on Morava's work, among many other contributions. The ground-breaking insight of transposing from topology to other fields began with Hopkins [45]. It is arguably Thomason [73] who first understood how essential the tensor was in the global story. We recall in Remark 4.4.6 why such a *geometric* classification does not exist for mere triangulated categories, *i.e.* without the tensor.

The tt-spectrum was introduced in [4] and is reviewed in Section 4.2. The survey begins in Section 4.3, with the initial example of topological stable homotopy theory. Section 4.4 touches commutative algebra and algebraic geometry. Section 4.5 is dedicated to stable module categories in modular representation theory and beyond. Section 4.6 discusses equivariant stable homotopy theory and Kasparov's equivariant KK-theory. Section 4.7 pertains to motives and $\mathbb{A}^1$-homotopy theory.

Everywhere, we have tried to give some idea of the actual tt-categories that come into play. When the amount of specialized definitions appears too high for this chapter, we simply point to the bibliographical references.

Finally, let us say a word about the bigger picture. In commutative algebra, the Zariski spectrum is not meant to be explicitly computed for every single commutative ring in the universe; instead, it serves as a stepping stone towards the geometric reasonings of algebraic geometry. In the same spirit, the tt-spectrum opens up a world of mathematical investigation, called *tensor-triangular geometry*, which reaches far beyond classical algebraic geometry into the broad kingdom of stable homotopy theory. The short final Section 4.8 points to further reading in that direction.

4.2 The tt-spectrum and the classification of tt-ideals

Definition 4.2.1. A *tt-category*, short for *tensor-triangulated category*, is a triangulated category $\mathcal{K}$ together with a symmetric monoidal structure

$$\otimes\colon \mathcal{K} \times \mathcal{K} \longrightarrow \mathcal{K}$$

which is exact in each variable. See details in [47, App. A] or [62, 49]. The $\otimes$-unit is denoted $\mathbb{1}$.

[1] Our use of the word 'spectrum' comes from commutative algebra, as in the 'Zariski spectrum', and should not be confused with the suspension-inverting 'spectra' of topology.

[2] $\mathrm{Spc}(\mathcal{K})$ is a space in the universe containing the 'set' of isomorphism classes of $\mathcal{K}$.

Assumption 4.2.2. Unless otherwise stated, we always assume that $\mathcal{K}$ is *essentially small*, *i.e.* has a set of isomorphism classes of objects. Subcategories $\mathcal{J} \subseteq \mathcal{K}$ are always assumed full and replete (*i.e.* closed under isomorphisms).

Definition 4.2.3. A *triangulated* subcategory $\mathcal{J} \subseteq \mathcal{K}$ is a non-empty subcategory such that whenever $X \to Y \to Z \to \Sigma X$ is an exact triangle in $\mathcal{K}$ and two out of X, Y and Z belong to $\mathcal{J}$ then so does the third. A *thick* subcategory $\mathcal{J} \subseteq \mathcal{K}$ is a triangulated subcategory closed under direct summands: if $X \oplus Y \in \mathcal{J}$ then $X, Y \in \mathcal{J}$. A *tt-ideal* $\mathcal{J} \subseteq \mathcal{K}$, short for *thick tensor-ideal*, is a thick subcategory closed under tensoring with any object: $\mathcal{K} \otimes \mathcal{J} \subseteq \mathcal{J}$. A tt-ideal $\mathcal{J} \subseteq \mathcal{K}$ is called *radical* if $X^{\otimes n} \in \mathcal{J}$ for $n \geq 2$ forces $X \in \mathcal{J}$.

Remark 4.2.4. When every object of $\mathcal{K}$ is rigid (*i.e.* admits a dual, a. k. a. *strongly dualizable* [47, § 2.1]), then we say that $\mathcal{K}$ is *rigid* and we can show that every tt-ideal $\mathcal{J}$ is automatically radical. See [5, Prop. 2.4]. So, for simplicity, we assume that every tt-category $\mathcal{K}$ that we discuss below is either rigid or that the phrase 'tt-ideal' means 'radical tt-ideal'.

Notation 4.2.5. For a class $\mathcal{E} \subseteq \mathcal{K}$ of objects, the tt-ideal generated by $\mathcal{E}$ is $\langle \mathcal{E} \rangle = \bigcap_{\mathcal{J} \supseteq \mathcal{E}} \mathcal{J}$, where $\mathcal{J}$ runs through the tt-ideals containing $\mathcal{E}$.

Definition 4.2.6. A *prime* $\mathcal{P} \subset \mathcal{K}$ is a proper tt-ideal such that $X \otimes Y \in \mathcal{P}$ forces $X \in \mathcal{P}$ or $Y \in \mathcal{P}$. We denote the set of prime tt-ideals by

$$\operatorname{Spc}(\mathcal{K}) = \{\, \mathcal{P} \subset \mathcal{K} \mid \mathcal{P} \text{ is prime} \,\}$$

and call it the *spectrum* of $\mathcal{K}$. The *support* of an object $X \in \mathcal{K}$ is the subset

$$\operatorname{supp}(X) = \{\, \mathcal{P} \in \operatorname{Spc}(\mathcal{K}) \mid X \notin \mathcal{P} \,\}.$$

The topology of $\operatorname{Spc}(\mathcal{K})$ is defined to have $\{\operatorname{supp}(X)\}_{X \in \mathcal{K}}$ as basis of closed subsets. Explicitly, for each set of objects $\mathcal{E} \subseteq \mathcal{K}$ the subset $U(\mathcal{E}) = \{\, \mathcal{P} \in \operatorname{Spc}(\mathcal{K}) \mid \mathcal{E} \cap \mathcal{P} \neq \varnothing \,\}$ is an open of $\operatorname{Spc}(\mathcal{K})$, and all open subsets are of this form, for some $\mathcal{E}$.

Remark 4.2.7. The above construction is introduced in [4], where the pair $(\operatorname{Spc}(\mathcal{K}), \operatorname{supp})$ is characterized by a universal property: It is the *final support data.* See [4, Thm. 3.2]. We shall not make this explicit but intuitively it means that the space $\operatorname{Spc}(\mathcal{K})$ is the best possible one carrying *closed* supports for objects of $\mathcal{K}$ with the following rules for all X, Y, Z in $\mathcal{K}$:

(1) $\operatorname{supp}(0)$ is empty and $\operatorname{supp}(\mathbb{1})$ is the whole space;

(2) $\operatorname{supp}(X \oplus Y) = \operatorname{supp}(X) \cup \operatorname{supp}(Y)$;

(3) $\operatorname{supp}(\Sigma X) = \operatorname{supp}(X)$;

(4) $\operatorname{supp}(Z) \subseteq \operatorname{supp}(X) \cup \operatorname{supp}(Y)$ for each exact triangle $X \to Y \to Z \to \Sigma X$;

(5) $\operatorname{supp}(X \otimes Y) = \operatorname{supp}(X) \cap \operatorname{supp}(Y)$.

Example 4.2.8. Dually to the Zariski topology, in the tt-spectrum $\operatorname{Spc}(\mathcal{K})$ the closure of a point $\mathcal{P} \in \operatorname{Spc}(\mathcal{K})$ consists of all the primes *contained* in it: $\overline{\{\mathcal{P}\}} = \{\, \mathcal{Q} \in \operatorname{Spc}(\mathcal{K}) \mid \mathcal{Q} \subseteq \mathcal{P} \,\}$. See [4, Prop. 2.9].

Remark 4.2.9. The tt-spectrum $\operatorname{Spc}(\mathcal{K})$ is always a *spectral space* in the sense of Hochster [44]: it is quasi-compact, it admits a basis of quasi-compact open subsets, and each of its non-empty irreducible closed subsets has a unique generic point. See [4, § 2].

Remark 4.2.10. The construction $\mathcal{K} \mapsto \operatorname{Spc}(\mathcal{K})$ is a contravariant functor. Every exact $\otimes$-functor $F\colon \mathcal{K} \to \mathcal{K}'$ between tt-categories induces a continuous (spectral) map $\varphi = \operatorname{Spc}(F)\colon \operatorname{Spc}(\mathcal{K}') \to \operatorname{Spc}(\mathcal{K})$ defined by $\mathcal{Q} \mapsto F^{-1}(\mathcal{Q})$. It satisfies $\varphi^{-1}(\operatorname{supp}(X)) = \operatorname{supp}(F(X))$ for all $X \in \mathcal{K}$. See [4, § 3].

To express tt-classification via the spectrum, we need some preparation.

Definition 4.2.11. To every subset $V \subseteq \operatorname{Spc}(\mathcal{K})$ we can associate a tt-ideal

$$\mathcal{K}_V = \{\, X \in \mathcal{K} \,|\, \operatorname{supp}(X) \subseteq V \,\}$$

of $\mathcal{K}$. (In fact, this tt-ideal is always radical. See Remark 4.2.4.)

Definition 4.2.12. A subset $V \subseteq \operatorname{Spc}(\mathcal{K})$ is called a *Thomason subset* if it is the union of the complements of a collection of quasi-compact open subsets: $V = \cup_\alpha V_\alpha$ where each V_α is closed with quasi-compact complement. In the terminology of Hochster [44], these are the *dual-open* subsets.

Example 4.2.13. If the space $\operatorname{Spc}(\mathcal{K})$ is topologically noetherian (*i.e.* all open subsets are quasi-compact), then V being Thomason is just being closed under specialization ($x \in V \Rightarrow \overline{\{x\}} \subseteq V$), *i.e.* being a union of closed subsets.

Theorem 4.2.14 (Classification of tt-ideals, [4, Thm. 4.10])**.** *The assignment $V \mapsto \mathcal{K}_V$ of Definition 4.2.11 defines an order-preserving bijection between the Thomason subsets $V \subseteq \operatorname{Spc}(\mathcal{K})$ and the (radical) tt-ideals $\mathcal{J} \subseteq \mathcal{K}$ of $\mathcal{K}$, whose inverse is given by $\mathcal{J} \mapsto \operatorname{supp}(\mathcal{J}) := \cup_{X\in\mathcal{J}} \operatorname{supp}(X) = \{\, \mathcal{P} \,|\, \mathcal{J} \not\subseteq \mathcal{P} \,\}$.*

Specifically for the tt-classification of objects $X, Y \in \mathcal{K}$ (see Remark 4.2.4):

Corollary 4.2.15. *Two objects $X, Y \in \mathcal{K}$ generate the same tt-ideals $\langle X \rangle = \langle Y \rangle$ if and only if they have the same support $\operatorname{supp}(X) = \operatorname{supp}(Y)$. More precisely, Y belongs to $\langle X \rangle$ if and only if $\operatorname{supp}(Y) \subseteq \operatorname{supp}(X)$.*

The following converse to Theorem 4.2.14 holds. See [4] for details.

Theorem 4.2.16 (Balmer/Buan-Krause-Solberg)**.** *Suppose that a spectral space $\mathcal{S}$ carries a support data $\sigma(X) \subseteq \mathcal{S}$ for $X \in \mathcal{K}$ in the sense of [4] and suppose that the assignment $\mathcal{S} \supseteq V \mapsto \{\, X \in \mathcal{K} \,|\, \sigma(X) \subseteq V \,\}$ induces a bijection between Thomason subsets V of $\mathcal{S}$ and (radical) tt-ideals of $\mathcal{K}$. Then the canonical map $\mathcal{S} \to \operatorname{Spc}(\mathcal{K})$ of Remark 4.2.7 is a homeomorphism.*

This result was established in [4] under the additional assumption that $\mathcal{S}$ be noetherian. It was proved in the above maximal generality in [25]. (See Remark 4.2.9.)

Remark 4.2.17. Theorems 4.2.14 and 4.2.16 allow for a compact reformulation of tt-classifications, including the ones anterior to [4]. Thus most classifications for the tt-categories $\mathcal{K}$ discussed in Sections 4.3-4.7 are phrased in the simple form of a description of $\mathrm{Spc}(\mathcal{K})$. The tt-classification is then always the same, in terms of subsets of $\mathrm{Spc}(\mathcal{K})$, as in Theorems 4.2.14 and Corollary 4.2.15, *and we shall not repeat these corollaries.*

On the other hand, approaching tt-classification via $\mathrm{Spc}(\mathcal{K})$ buys us some flexibility, for partial results about $\mathrm{Spc}(\mathcal{K})$ can be interesting while a 'partial classification' is an odd concept. For instance, one can know $\mathrm{Spc}(\mathcal{K})$ *as a set* in some examples, with partial information on the topology. Or one can describe $\mathrm{Spc}(\mathcal{K}) = U \cup Z$ with a complete description of the closed subset Z and its open complement U without knowing exactly how they attach. And so on.

In recent years, the geometric study of the tt-spectrum *per se* has led to new computations of $\mathrm{Spc}(\mathcal{K})$, from which the tt-classification can be deduced a posteriori. This will be illustrated in the later sections.

Remark 4.2.18. Some of the above results connect to lattice theory; see [25, 52]. It is a non-trivial property of a lattice, like that of tt-ideals in $\mathcal{K}$, to be *spatial*, *i.e.* in bijection with the open subsets of a topological space. In fact, without the tensor this fails in general (Remark 4.4.6).

The tt-classification of Theorem 4.2.14 tacitly assumes that $\mathcal{K}$ consists of 'small enough' objects. Assumption 4.2.2 and Remark 4.2.4 belong to this logic too. Another indication of the smallness of $\mathcal{K}$ is that we do not mention infinite coproducts in $\mathcal{K}$, and we only discuss thick subcategories, not localizing ones (*i.e.* those closed under arbitrary coproducts). When dealing with a 'big' tt-category $\mathcal{T}$, the natural candidate for a 'small' $\mathcal{K}$ is the subcategory of rigid objects in $\mathcal{T}$, which may or may not coincide with compact ones.

There are also 'big' subcategories of 'big' tt-categories worth investigating, most famously *smashing* subcategories. It is an open problem whether the lattice of smashing $\otimes$-ideals is spatial or not. We prove in [12] that it is a *frame*, thus it at least 'spatial' in the quirky sense of pointless topology.

The connection between thick subcategories of compact objects and smashing subcategories is a topic in its own right, often dubbed the *Telescope Conjecture*. We shall not attempt to discuss it systematically here but will mention it in a few examples. See Krause [53] for a beautiful abstract answer via ideals of morphisms.

4.3 Topology

As already said, tt-classification (or at least 't-classification') was born in topology, more precisely in *chromatic homotopy theory*.[3] The tt-category we consider here is the topological stable homotopy category SH, *i.e.* the homotopy category of topological spectra, and more specifically its subcategory SH^c of compact objects. See for instance [66]. In other words, SH^c is the Spanier-Whitehead stable homotopy category of finite pointed CW-complexes.

The first operation one can do on SH is to p-localize it at a prime p, *i.e.* invert multiplication by every prime different from p. On compacts, this gives us $\mathrm{SH}^c_{(p)}$. Both SH^c and $\mathrm{SH}^c_{(p)}$ are essentially small rigid tt-categories.

Remark 4.3.1. Something special happens in SH^c and therefore in $\mathrm{SH}^c_{(p)}$ as well: The unit $\mathbb{1} = S^0$, a. k. a. the sphere spectrum, generates the category as a thick triangulated

[3]See the chapter in this Handbook by Tobias Barthel and Agnès Beaudry for much more on this topic.

subcategory. Consequently, every thick subcategory is automatically a tt-ideal. In such situations, the tensor is not essential in the tt-classification and we are equivalently classifying thick subcategories.

Remark 4.3.2. A critical ingredient in chromatic theory is the countable family of so-called *Morava K-theories*, which are homology theories $K_{p,n}$, for $n \geq 1$, defined on $\mathrm{SH}^c_{(p)}$ and taking values in graded modules over the 'graded field' $\mathbb{F}_p[v_n, v_n^{-1}]$, with v_n in degree $2(p^n - 1)$. See [66, § 1.5].

Theorem 4.3.3 (Hopkins-Smith [46]). *The spectrum of the classical stable homotopy category* SH^c *is the following topological space:*

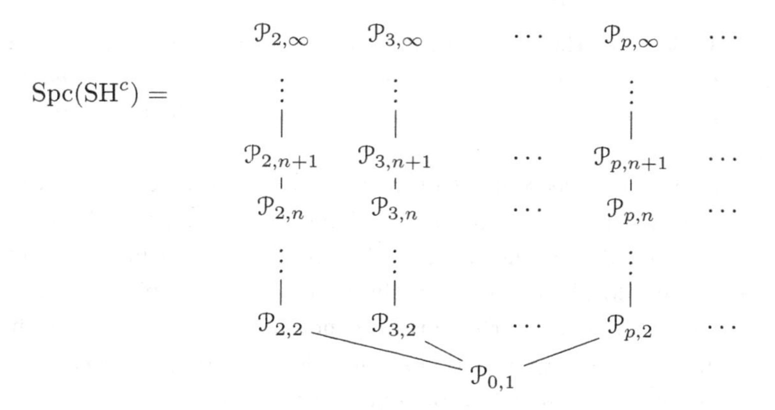

in which every line indicates that the higher point belongs to the closure of the lower one (Example 4.2.8). More precisely:

(a) *The tt-prime* $\mathcal{P}_{0,1}$ *is the kernel of rationalization* $\mathrm{SH}^c \to \mathrm{SH}^c_{\mathbb{Q}} \cong \mathrm{D^b}(\mathbb{Q})$*, that is, the subcategory of torsion spectra. It is the dense point of* $\mathrm{Spc}(\mathrm{SH}^c)$.

(b) *For each prime number* p*, the tt-prime* $\mathcal{P}_{p,\infty}$ *is the kernel of localization* $\mathrm{SH}^c \to \mathrm{SH}^c_{(p)}$. *These* $\mathcal{P}_{p,\infty}$ *are exactly the closed points of* $\mathrm{Spc}(\mathrm{SH}^c)$.

(c) *For each prime number* p *and each integer* $2 \leq n < \infty$*, the tt-prime* $\mathcal{P}_{p,n}$ *is the kernel of the composite* $\mathrm{SH}^c \to \mathrm{SH}^c_{(p)} \to \mathbb{F}_p[v_{n-1}^{\pm 1}]\text{-grmod}$ *of localization at* p *and* $(n-1)^{st}$ *Morava* K*-theory* $K_{p,n-1}$ *(Remark 4.3.2).*

(d) *The support of an object* X *is either empty when* $X = 0$*, or the whole of* $\mathrm{Spc}(\mathrm{SH}^c)$ *when* X *is non-torsion, or a finite union of 'columns'*

$$\overline{\{\mathcal{P}_{p,m_p}\}} = \{\, \mathcal{P}_{p,n} \mid m_p \leq n \leq \infty \,\}$$

for integers $2 \leq m_p < \infty$ *varying with the primes* p.

(e) *A closed subset is either empty, or the whole* $\mathrm{Spc}(\mathrm{SH}^c)$*, or a finite union of closed points* $\{\mathcal{P}_{p,\infty}\}$ *and of columns* $\overline{\{\mathcal{P}_{p,m_p}\}}$ *with* $m_p \geq 2$ *as in (d).*

(f) *A Thomason subset of* $\mathrm{Spc}(\mathrm{SH}^c)$ *is either empty, or the whole* $\mathrm{Spc}(\mathrm{SH}^c)$*, or an arbitrary union of columns* $\overline{\{\mathcal{P}_{p,m_p}\}}$ *with* $m_p \geq 2$ *as in (d).*

The above Theorem 4.3.3 is not the way the chromatic filtration is expressed in the original literature; see the translation in [6, § 9].

Example 4.3.4. An object $X \in \mathrm{SH}^c$ has support contained in the p-th column, $\mathrm{supp}(X) \subseteq \overline{\{\mathcal{P}_{p,2}\}}$, if and only if it is '$p$-primary torsion', *i.e.* it satisfies $p^\ell \cdot X = 0$ for some $\ell \geq 1$.

Example 4.3.5. The support of the tt-ideal $\mathcal{J} = \mathcal{P}_{0,1}$ of torsion spectra is exactly the Thomason subset $\mathrm{Spc}(\mathrm{SH}^c) \setminus \{\mathcal{P}_{0,1}\}$ and is therefore the disjoint union of all columns $\sqcup_p \overline{\{\mathcal{P}_{p,2}\}}$. This reflects the fact that a torsion object in SH^c is the direct sum of p-primary torsion objects as in Example 4.3.4.

Remark 4.3.6. The fact that the closed point $\{\mathcal{P}_{p,\infty}\}$ cannot be the support of an object reflects the fact that an object in $\mathrm{SH}^c_{(p)}$ that is killed by all Morava K-theories $K_{p,n}$ for $n \geq 1$ must be zero. It also shows that $\mathrm{Spc}(\mathcal{K})$ is not noetherian, already in this initial case of $\mathcal{K} = \mathrm{SH}^c$ (see Example 4.2.13).

Remark 4.3.7. In this setting, the Telescope Conjecture is open (again). See [53] and further references therein.

4.4 Commutative algebra and algebraic geometry

As already indicated, Hopkins [45] initiated the transposition of the chromatic classification from topology to algebra. The correct statement for noetherian rings was proved by Neeman [61] and the perfect version for general schemes, not necessarily noetherian, is due to Thomason in his last published paper [73]. In terms of tt-spectra it becomes the following very beautiful result.

Theorem 4.4.1 (Thomason [73]). *Let $\mathcal{X}$ be a scheme that is quasi-compact and quasi-separated. Then the spectrum of the derived category $\mathrm{D}^{\mathrm{perf}}(\mathcal{X})$ of perfect complexes (with $\otimes = {}^{\mathrm{L}}\!\otimes_{\mathcal{O}_\mathcal{X}}$) is isomorphic to the underlying space $|\mathcal{X}|$ itself, via the homeomorphism*

$$\begin{aligned} |\mathcal{X}| &\xrightarrow{\ \cong\ } \mathrm{Spc}(\mathrm{D}^{\mathrm{perf}}(\mathcal{X})) \\ x &\longmapsto \mathcal{P}(x) \end{aligned}$$

where, for each point x of $\mathcal{X}$, the tt-prime $\mathcal{P}(x) = \{ Y \in \mathrm{D}^{\mathrm{perf}}(\mathcal{X}) \mid Y_x \cong 0 \}$ is the kernel of localization $\mathrm{D}^{\mathrm{perf}}(\mathcal{X}) \to \mathrm{D}^{\mathrm{perf}}(\mathcal{O}_{\mathcal{X},x})$ at x.

Remark 4.4.2. Equivalently, $\mathcal{P}(x)$ can be described as the kernel of the residue functor $\mathrm{D}^{\mathrm{perf}}(\mathcal{X}) \to \mathrm{D}^{\mathrm{b}}(\kappa(x))$ to the residue field $\kappa(x)$ of $\mathcal{X}$ at x.

Remark 4.4.3. Recall that a scheme $\mathcal{X}$ is quasi-compact and quasi-separated if the underlying space $|\mathcal{X}|$ admits a basis of quasi-compact open subsets (including $|\mathcal{X}|$ itself). This purely topological condition is equivalent to $|\mathcal{X}|$ being spectral. Hence this condition is the maximal generality in which the above result can hold in view of Remark 4.2.9. Noetherian schemes and affine schemes are quasi-compact and quasi-separated.

The affine case vindicates our use of the word 'spectrum':[4]

[4]If this creates confusion with Example 4.2.8, note that the map of Corollary 4.4.4, $\mathfrak{p} \mapsto \mathcal{P}(\mathfrak{p}) = \mathrm{Ker}(\mathrm{D}^{\mathrm{perf}}(A) \to \mathrm{D}^{\mathrm{perf}}(A_\mathfrak{p}))$, *reverses* inclusions: $\mathfrak{p} \subseteq \mathfrak{q} \Rightarrow \mathcal{P}(\mathfrak{p}) \supseteq \mathcal{P}(\mathfrak{q})$.

Corollary 4.4.4. *Let A be a commutative ring. Then the tt-spectrum of the homotopy category $\mathrm{K^b}(A\text{-proj}) \cong \mathrm{D^{perf}}(A)$ of bounded complexes of finitely generated projective A-modules is homeomorphic to the Zariski spectrum of A*

$$\mathrm{Spec}(A) \xrightarrow{\sim} \mathrm{Spc}(\mathrm{K^b}(A\text{-proj})).$$

Remark 4.4.5. An error in [45], corrected in [61], was not to assume A noetherian. However we see that Thomason's Corollary 4.4.4 does not assume A noetherian. The point is that the tt-classification (Theorem 4.2.14) which is equivalent to Corollary 4.4.4 involves actual Thomason subsets not mere specialization-closed subsets, whereas [45] and [61] are phrased in terms of specialization-closed subsets. The assumption that A is noetherian is only useful to replace 'Thomason' by 'specialization-closed' (Example 4.2.13).

Remark 4.4.6. As we saw in the topological example of Section 4.3, when the unit $\mathbb{1}$ generates the tt-category $\mathcal{K}$ as a thick subcategory we do not really need the tensor. This is also the case for $\mathcal{K} = \mathrm{K^b}(A\text{-proj})$ for instance.

But in general the tensor is essential for classification by means of subsets of $\mathrm{Spc}(\mathcal{K})$. Indeed, the lattice of thick subcategories of a triangulated category $\mathcal{K}$ cannot be classified in terms of the lattice of subsets of pretty much anything because it may not satisfy *distributivity*: $\mathcal{J}_1 \wedge (\mathcal{J}_2 \vee \mathcal{J}_3) = (\mathcal{J}_1 \wedge \mathcal{J}_2) \vee (\mathcal{J}_1 \wedge \mathcal{J}_3)$. Already for $\mathcal{K} = \mathrm{D^{perf}}(\mathcal{X})$ over the projective line $\mathcal{X} = \mathbb{P}^1_k$ distributivity fails with $\mathcal{J}_i$ the thick subcategory generated by $\mathcal{O}(i)$. See [12, Rem. 5.10].

Remark 4.4.7. An application of Theorem 4.4.1 is the reconstruction of every quasi-compact and quasi-separated scheme $\mathcal{X}$ from the *tensor*-triangulated category $\mathrm{D^{perf}}(\mathcal{X})$. Indeed, one can equip the tt-spectrum $\mathrm{Spc}(\mathcal{K})$ with a sheaf of commutative rings, which in the case of $\mathcal{K} = \mathrm{D^{perf}}(\mathcal{X})$ recovers the structure sheaf $\mathcal{O}_{\mathcal{X}}$. See details in [4, § 6]. By contrast, Mukai [60] proved earlier that such a reconstruction is impossible from the triangulated structure alone.

Remark 4.4.8. The above $\mathcal{K} = \mathrm{D^{perf}}(R)$ are the compact and rigid objects in the big derived category $\mathcal{T} = \mathrm{D}(R)$. The Telescope Conjecture holds when R is noetherian by Neeman [61] but fails in general by Keller [48].

Remark 4.4.9. Other tt-categories can be associated to schemes, or commutative rings, for instance right-bounded derived categories. For first results in this direction, see work of Matsui and Takahashi [57, 56].

One can generalize Theorem 4.4.1 almost verbatim to reasonable stacks:

Theorem 4.4.10 (Hall [40, Thm. 1.2]). *Let $\mathcal{X}$ be a quasi-compact algebraic stack with quasi-finite and separated diagonal and whose stabilizer groups at geometric points are finite linearly reductive group schemes ($\mathcal{X}$ is 'tame'). Then* $\mathrm{Spc}(\mathrm{D^{perf}}(\mathcal{X})) \cong |\mathcal{X}|$.

We refer to [40] for terminology. Note earlier work of Krishna [54] in characteristic 0, and of Dubey-Mallick [34] for finite groups acting on smooth schemes in characteristic prime to the order of the groups.

One can also consider the graded version of Corollary 4.4.4:

Theorem 4.4.11 (Dell'Ambrogio-Stevenson [31, Thm. 4.7]). *Let A be a graded-commutative ring (graded over any abelian group), then there is a canonical isomorphism* $\mathrm{Spc}(\mathrm{D}^{\mathrm{perf}}(A)) \cong \mathrm{Spec}^{\mathrm{h}}(A)$, *between the tt-spectrum of* $\mathrm{D}^{\mathrm{perf}}(A)$ *and the spectrum of homogeneous prime ideals of A.*

Let us mention a variation relating to singularities.

Theorem 4.4.12 (Stevenson [70, Thm. 7.7]). *Let $\mathcal{X}$ be a noetherian separated scheme with only hypersurface singularities. Then there is an order-preserving bijection between the specialization-closed subsets of the singular locus of $\mathcal{X}$ and the thick* $\mathrm{D}^{\mathrm{perf}}(\mathcal{X})$*-submodules of the singularity category* $\mathrm{D}^{\mathrm{b}}(\mathrm{coh}\,\mathcal{X})/\,\mathrm{D}^{\mathrm{perf}}(\mathcal{X})$.

Here the singularity category is not itself a tt-category but a triangulated category with an action by the tt-category $\mathrm{D}^{\mathrm{perf}}(\mathcal{X})$. As such, this result is an application of Stevenson's *relative tt-geometry* [69]. Another application of Stevenson's theory is the tt-classification for derived categories of matrix factorizations in Hirano [43], which extends earlier result of Takahashi [72].

4.5 Modular representation theory and related topics

4.5.1. Let G be a finite group and let k be a field. Maschke's Theorem says that the order of G is invertible in k if and only if kG is semisimple. In that case, all kG-modules are projective. *Modular* representation theory refers to the non-semisimple situation. Then the *stable module category* is the additive quotient [41]

$$kG\text{-}\mathrm{stmod} = \frac{kG\text{-}\mathrm{mod}}{kG\text{-}\mathrm{proj}}$$

which precisely measures how far kG is from being semisimple. It is a tt-category whose objects are all finitely generated kG-modules and its Hom groups $\mathrm{Hom}_{kG\text{-}\mathrm{stmod}}(X,Y)$ are given by the quotient of the abelian group of kG-linear maps $\mathrm{Hom}_{kG}(X,Y)$ modulo the subgroup of those maps factoring via a projective module. Tensor is over k with diagonal G-action: $g\cdot(x\otimes y) = (gx)\otimes(gy)$ in $X\otimes_k Y$. The $\otimes$-unit is $\mathbb{1} = k$ with trivial G-action.

4.5.2. We can also consider the derived category $\mathrm{D}^{\mathrm{b}}(kG\text{-}\mathrm{mod})$, with the 'same' tensor. Every non-zero tt-ideal $\mathcal{J} \subseteq \mathrm{D}^{\mathrm{b}}(kG\text{-}\mathrm{mod})$ contains $\mathrm{D}^{\mathrm{perf}}(kG)$ because $kG\otimes - \cong \mathrm{Ind}_1^G \mathrm{Res}_1^G$ and $\mathrm{D}^{\mathrm{b}}(k\text{-}\mathrm{mod}) = \mathrm{D}^{\mathrm{perf}}(k)$ is semisimple. Hence the tt-classification of $\mathrm{D}^{\mathrm{b}}(kG\text{-}\mathrm{mod})$ and of its Verdier quotient by $\mathrm{D}^{\mathrm{perf}}(kG)$ are very close. (The former has just one more tt-ideal: zero.) By Rickard [67], that quotient is equivalent to the stable module category:

$$\frac{\mathrm{D}^{\mathrm{b}}(kG\text{-}\mathrm{mod})}{\mathrm{D}^{\mathrm{perf}}(kG)} \cong kG\text{-}\mathrm{stmod}\,.$$

Theorem 4.5.3 (Benson-Carlson-Rickard [21]). *There is a homeomorphism between the spectrum of the stable module category and the so-called projective support variety*

$$\mathrm{Spc}(kG\text{-}\mathrm{stmod}) \cong \mathrm{Proj}(\mathrm{H}^{\bullet}(G,k))$$

which can be extended (by adding one closed point) to a homeomorphism

$$\operatorname{Spc}(\mathrm{D^b}(kG\text{-mod})) \cong \operatorname{Spec^h}(\mathrm{H}^\bullet(G,k)).$$

Explicitly, to every homogeneous prime $\mathfrak{p}^\bullet \subset \mathrm{H}^\bullet(G,k)$ *corresponds the tt-prime*

$$\mathcal{P}(\mathfrak{p}^\bullet) = \big\{\, X \,\big|\, \text{ there is a homogeneous } \zeta \notin \mathfrak{p}^\bullet \text{ such that } \zeta \cdot X = 0 \,\big\}.$$

Again, the above does not appear verbatim in the source. See details in [4] or [6, Prop. 8.5]. A more recent proof can be found in [26].

Remark 4.5.4. The reader interested in the related derived category of cochains on the classifying space BG is referred to [19] for finite groups and to the comprehensive recent work [15] for p-local compact groups; see comments and references therein about earlier work by Benson-Greenlees.

4.5.5. For finite group *schemes* G, the following generalization of Theorem 4.5.3 would follow from claims made in [35] but a flaw was found in [35, Thm. 5.3], which was eventually fixed in the recent [20]; see in particular [20, Rem. 5.4 and Thm. 10.3].

Theorem 4.5.6 (Benson-Friedlander-Iyengar-Krause-Pevtsova). *For a finite group scheme G over k, there is a canonical homeomorphism* $\operatorname{Spc}(kG\text{-stmod}) \cong \operatorname{Proj}(\mathrm{H}^\bullet(G,k))$.

Remark 4.5.7. Stable module categories of finite group schemes over a field are very 'noetherian' and several other results are known about the 'big' stable module category as well, like the Telescope Conjecture. See details in [20]. The technique of *stratification* has led to the tt-classification (of small and large subcategories) in several 'noetherian enough' derived settings. See the survey in [22] and further references in [20].

4.5.8. Extending beyond field coefficients to other rings R, we can consider the *relative stable module category* $RG\text{-strel}$, obtained from the Frobenius exact structure on the exact category of finitely generated RG-modules with R-split exact sequences. Already in small Krull dimension, interesting phenomena can be observed, as in the following result.

Theorem 4.5.9 (Baland-Chirvasitu-Stevenson [3, Thm. 1.1]). *Let S be a discrete valuation ring having residue field k and uniformizing parameter t and let $R_n = S/t^n$. Let G be a finite group. Then the tt-spectrum of the relative stable module category*

$$\operatorname{Spc}(R_nG\text{-strel}) \cong \sqcup_{i=1}^{n} \operatorname{Spc}(kG\text{-stmod}),$$

is a coproduct of n copies of the projective support variety of Theorem 4.5.3.

On the topic of singularity categories, let us mention [77] and its recent generalization (recall that a category is EI if any endomorphism is invertible):

Theorem 4.5.10 (Wang [76, Thm. 5.2]). *Let $\mathcal{C}$ be a finite EI category, projective over a field k and* $\mathrm{D}_{sing}(k\mathcal{C}) = \mathrm{D^b}(k\mathcal{C}\text{-mod})/\,\mathrm{D^b}(k\mathcal{C}\text{-proj})$ *its singularity category. Then there is a homeomorphism*

$$\operatorname{Spc}(\mathrm{D}_{sing}(k\mathcal{C})) \cong \sqcup_{x\in\mathcal{C}} \operatorname{Spc}(kG_x\text{-stmod}),$$

where $G_x = \operatorname{Aut}_{\mathcal{C}}(x)$.

4.5.11. Antieau-Stevenson [1] consider further derived categories of representations of small categories over commutative noetherian rings. They obtain several interesting classifications, including for localizing subcategories, and in particular for simply laced Dynkin quivers. See also the earlier [55].

4.5.12. Let us now turn our attention to stable module categories related to Lie algebras. Boe-Kujawa-Nakano [24] prove several results about classical Lie superalgebras. In particular for the general linear Lie superalgebra $\mathfrak{g} = \mathfrak{gl}(m|n) = \mathfrak{g}_{\bar{0}} \oplus \mathfrak{g}_{\bar{1}}$ and $\mathcal{K} = \underline{\mathcal{F}}$ the stable category of the category $\mathcal{F}$ of finite dimensional $\mathfrak{g}$-modules that admit a compatible action by $G_{\bar{0}}$ and are completely reducible as $G_{\bar{0}}$-modules (where $\operatorname{Lie} G_{\bar{0}} = \mathfrak{g}_{\bar{0}}$). They prove in [24, Thm. 5.2.2] that the spectrum $\operatorname{Spc}(\underline{\mathcal{F}})$ is homeomorphic to the N-homogenous spectrum $N - \operatorname{Proj}(S^\bullet(\mathfrak{f}_{\bar{1}}))$ where $\mathfrak{f}$ is the detecting subalgebra of $\mathfrak{g}$ and $N = \operatorname{Norm}_{G_{\bar{0}}}(\mathfrak{f}_{\bar{1}})$.

The same authors more recently considered quantum groups:

Theorem 4.5.13 (Boe-Kujawa-Nakano [23, Thm. 7.6.1])**.** *Let G be a complex simple algebraic group over $\mathbb{C}$ with $\mathfrak{g} = \operatorname{Lie} G$. Assume that ζ is a primitive ℓth root of unity where ℓ is greater than the Coxeter number for $\mathfrak{g}$. Then the tt-spectrum of the stable module category for the quantum group $U_\zeta(\mathfrak{g})$ is*

$$\operatorname{Spc}(U_\zeta(\mathfrak{g})\,\text{-stmod}) \cong G - \operatorname{Proj}(\mathbb{C}[\mathcal{N}])$$

where $\mathcal{N}$ is the nullcone, i.e. the set of nilpotent elements of $\mathfrak{g}$.

Example 4.5.14. Another example where $\operatorname{Spc}(A\text{-stmod})$ is isomorphic to $\operatorname{Proj}(H^\bullet(A,k))$ is the algebra $A = k[X_1,\ldots,X_n]/(X_1^\ell,\ldots,X_n^\ell) \rtimes (\mathbb{Z}/\ell\mathbb{Z})^{\times n}$ which appears in Pevtsova-Witherspoon [64, Thm. 1.2].

4.6 Equivariant stable homotopy and KK-theory

4.6.1. Let G be a compact Lie group, e.g. a finite group, and let $\mathrm{SH}(G)$ be the equivariant stable homotopy category of genuine G-spectra. The tensor-triangulated category of compact (rigid) objects in $\mathrm{SH}(G)$ is denoted $\mathrm{SH}(G)^c$. In general, the spectrum of $\mathrm{SH}(G)^c$ is not quite known but significant progress occurred in recent years. It relies in an essential way on the non-equivariant case $G = 1$ of Section 4.3.

4.6.2. For every chromatic tt-prime $\mathcal{P}_{p,n} \in \operatorname{Spc}(\mathrm{SH}^c)$ in the stable homotopy category SH^c (Theorem 4.3.3) and every closed subgroup $H \leq G$, let

$$\mathcal{P}(H,p,n) = (\Phi^H)^{-1}(\mathcal{P}_{p,n})$$

be its preimage under geometric H-fixed points $\Phi^H\colon \mathrm{SH}(G)^c \to \mathrm{SH}^c$, which is a tt-functor. This 'equivariant' tt-prime $\mathcal{P}(H,p,n)$ is the image of the chromatic $\mathcal{P}_{p,n}$ under $\operatorname{Spc}(\Phi^H)\colon \operatorname{Spc}(\mathrm{SH}^c) \to \operatorname{Spc}(\mathrm{SH}(G)^c)$ as in Remark 4.2.10.

It is convenient to use the convention that $\mathcal{P}_{p,1}$ means $\mathcal{P}_{0,1}$ for all p. And similarly, to read $\mathcal{P}(H,p,1)$ as $\mathcal{P}(H,0,1)$.

Let us first discuss the case where G is a finite group. Varying the subgroup $H \leq G$, the maps $\mathrm{Spc}(\Phi^H)$ cover $\mathrm{Spc}(\mathrm{SH}(G)^c)$ – a fact that is also true for general compact Lie groups, see Theorem 4.6.9.

Theorem 4.6.3 (Balmer-Sanders [14]). *Let G be a finite group. Then every tt-prime in $\mathrm{SH}(G)^c$ is of the form $\mathcal{P}(H,p,n)$ for a unique subgroup $H \leq G$ up to conjugation and a unique chromatic tt-prime $\mathcal{P}_{p,n} \in \mathrm{Spc}(\mathrm{SH}^c)$. Understanding inclusions between tt-primes completely describes the topology on $\mathrm{Spc}(\mathrm{SH}(G)^c)$.*

If $K \lhd H$ is a normal subgroup of index $p > 0$, then $\mathcal{P}(K,p,n+1) \subset \mathcal{P}(H,p,n)$ for every $n \geq 1$. There is no inclusion $\mathcal{P}(K,q,n) \subseteq \mathcal{P}(H,p,m)$ unless the corresponding chromatic tt-primes are included $\mathcal{P}_{q,n} \subseteq \mathcal{P}_{p,m}$ (which forces $n \geq m$, and $p = q$ if $m > 1$) and *K is conjugate to a q-subnormal subgroup of H (see 4.6.5).*

4.6.4. For finite groups of square-free order, like $G = C_p$ for instance, the above result completely describes $\mathrm{Spc}(\mathrm{SH}(G)^c)$, with its topology, and thus gives the tt-classification. This result was a first major example where $\mathrm{Spc}(\mathcal{K})$ was determined first and the tt-classification deduced as a corollary.

4.6.5. For other groups, the question is to decide when $\mathcal{P}(K,p,n) \subset \mathcal{P}(H,p,m)$, in terms of $n - m$, for $K \leq H$ a p-subnormal subgroup of H (*i.e.* one such that there exists a tower of normal subgroups of index p from K to H). Theorem 4.6.3 implies that this inclusion holds when $n - m \geq \log_p([H:K])$.

The case of abelian groups (and a little more) was recently tackled in [17], showing that the above $\log_p([H:K])$ is not the sharpest bound.

Theorem 4.6.6 (Barthel-Hausmann-Naumann-Nikolaus-Noel-Stapleton). *Let G be a finite abelian group, let $K \leq H \leq G$ be subgroups, let p be a prime and let $1 \leq n < \infty$ be an integer. Then the minimal i such that $\mathcal{P}(K,p,n) \subseteq \mathcal{P}(H,p,n-i)$ is $i = \mathrm{rk}_p(H/K)$ the p-rank of the quotient.*

See [17, Cor. 1.3]. The precise topology of $\mathrm{Spc}(\mathrm{SH}(G)^c)$ for general finite groups remains an open problem.

4.6.7. Let us now consider the case of an arbitrary compact Lie group G but after rationalization $\mathrm{SH}(G)^c_{\mathbb{Q}}$. Of course, tensoring with $\mathbb{Q}$ hides the 'chromatic direction' but this is an essential step in understanding the 'equivariant direction', by which we mean the role played by the subgroups of G.

A closed subgroup $K \leq H$ is *cotoral* if K is normal and H/K is a torus. Every closed subgroup $H \leq G$ defines a tt-prime $\mathcal{P}_H$ in the spectrum $\mathrm{Spc}(\mathrm{SH}(G)^c_{\mathbb{Q}})$, namely $\mathcal{P}_H := \mathrm{Ker}(\Phi^H)$ the kernel of geometric H-fixed points, *i.e.* the image of the unique prime (0) under the map $\mathrm{Spc}(\mathrm{D^b}(\mathbb{Q})) \to \mathrm{Spc}(\mathrm{SH}(G)^c)$ associated to the tt-functor $\Phi^H\colon \mathrm{SH}(G)^c_{\mathbb{Q}} \to \mathrm{SH}^c_{\mathbb{Q}} \cong \mathrm{D^b}(\mathbb{Q})$.

Theorem 4.6.8 (Greenlees [39, Thm. 1.3]). *Let G be a compact Lie group. Every tt-prime of the rational equivariant stable homotopy category $\mathrm{SH}(G)^c_{\mathbb{Q}}$ is equal to $\mathcal{P}_H = \mathrm{Ker}(\Phi^H)$ for a closed subgroup $H \leq G$, unique up to conjugation. Furthermore, specialization of tt-primes corresponds to cotoral inclusions: We have $\mathcal{P}_K \subseteq \mathcal{P}_H$ if and only if K is conjugate to a cotoral subgroup of H. The topology on $\mathrm{Spc}(\mathrm{SH}(G)^c_{\mathbb{Q}})$ corresponds to the f-topology of [38].*

Recently there has been further progress for arbitrary compact Lie group:

Theorem 4.6.9 (Barthel-Greenlees-Hausmann [16]). *Let G be a compact Lie group. Then every tt-prime of $\mathrm{SH}(G)^c$ is of the form $\mathcal{P}(H,p,n)$ as in 4.6.2. Moreover, the topology is completely understood in terms of inclusions of tt-primes.*

Barthel-Greenlees-Hausmann more precisely track the inclusion of primes, in terms of functions on the compact and totally-disconnected Hausdorff orbit space $\mathrm{Sub}(G)/G$ of G acting by conjugation on its closed subgroups. Furthermore, they give a complete description of the topology in the case of an *abelian* compact Lie group, extending Theorem 4.6.6; see [16, Thm. 1.4].

4.6.10. The closest to analysis that tt-geometry has gone so far is in the theory of C^*-algebras, via Kasparov's KK-theory. Although this is not strictly speaking equivariant homotopy theory, we include it in this section as KK-theory belongs to the broad topic of noncommutative topology.

One begins with the 'cellular' subcategory, a. k. a. the 'bootstrap' category.

Theorem 4.6.11 (Dell'Ambrogio [29, § 6]). *Let G be a finite group and $\mathcal{K}^G$ be the thick subcategory of the G-equivariant Kasparov category KK^G generated by the unit. Then the comparison map $\rho\colon \mathrm{Spc}(\mathcal{K}^G) \to \mathrm{Spec}(R(G))$ to the spectrum of the complex representation ring $R(G)$, as in Definition 4.8.2, is surjective and admits a continuous section. In the non-equivariant case, the above map induces a homeomorphism between the tt-spectrum of the so-called 'bootstrap category' $\mathcal{K}^1 = \mathrm{Boot}$ and $\mathrm{Spec}(\mathbb{Z})$.*

4.6.12. Dell'Ambrogio conjectures that $\rho\colon \mathrm{Spc}(\mathcal{K}^G) \to \mathrm{Spec}(R(G))$ is a homeomorphism for all finite groups. The tt-classification for larger KK-categories, or for infinite groups, is another interesting open problem.

4.7 Motives and $\mathbb{A}^1$-homotopy

4.7.1. We consider here two classes of 'motivic' tensor-triangulated categories: First, we have the derived category of motives $\mathrm{DM}(F;R)$, over a base field F and with coefficients in a ring R, as first introduced by Voevodsky [75]; see also [2, 27]. Secondly, we have $\mathrm{SH}^{\mathbb{A}^1}(F)$ the stable $\mathbb{A}^1$-homotopy category over the base F introduced by Morel and Voevodsky [59, 98, 58]. (Other base schemes can be considered.)

4.7.2. Let us begin with $\mathrm{DM}(F;R)$. Like in KK-theory (see 4.6.10), one can first consider the 'cellular' tt-subcategory of *(mixed) Tate motives* $\mathrm{DTM}(F;R)$ generated as a localizing subcategory by the invertible Tate objects $R(i)$ for $i \in \mathbb{Z}$. Its subcategory of compact objects is the rigid tt-category $\mathrm{DTM}(F;R)^c$ we shall discuss now.

4.7.3. Peter [63] established the tt-bridgehead into motivic territory, when he proved that the spectrum $\mathrm{Spc}(\mathrm{DTM}(F;\mathbb{Q})^c) = *$ reduces to a point, for F a field satisfying the Beilinson-Soulé vanishing conjecture and a less standard restriction on rational motivic cohomology, namely $\mathrm{H}^i_{\mathrm{mot}}(F;\mathbb{Q}(j)) = 0$ for $j \geq i \geq 2$. For instance, this applies to $F = \bar{\mathbb{Q}}$. In fact, $\mathrm{Spc}(\mathrm{DTM}(F;\mathbb{Q})^c) = *$ would follow from $\mathrm{Spc}(\mathrm{DM}(F;\mathbb{Q})^c) = *$, a conjecture which is supported by:

Theorem 4.7.4 (Kelly [50, Thm. 36]). *Let F be a finite field such that every connected smooth projective variety X over F satisfies the Beilinson-Parshin conjecture and agreement of rational and numerical equivalence. Then* $\mathrm{Spc}(\mathrm{DM}(F;\mathbb{Q})^c) = *$ *is a point.*

The above are rational results. Our understanding of the integral picture recently evolved thanks to the following breakthrough.

Theorem 4.7.5 (Gallauer [36, Thm. 8.6]). *Let F be an algebraically closed field of characteristic zero*[5] *whose rational motivic cohomology $\mathrm{H}^i_{\mathrm{mot}}(F;\mathbb{Q}(j))$ vanishes for $i \leq 0 < j$ (Beilinson-Soulé) and for $j \geq i \geq 2$. Then the spectrum of* $\mathrm{DTM}(F;\mathbb{Z})^c$ *is the following:*

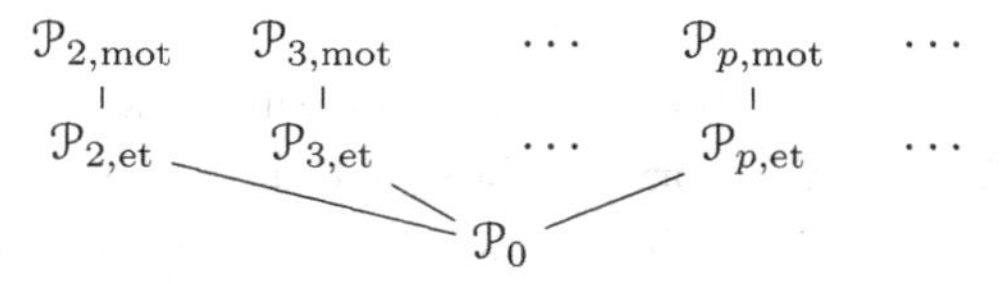

where $\mathcal{P}_0 = \mathrm{Ker}(\mathrm{DTM}(F;\mathbb{Z})^c \to \mathrm{DTM}(F;\mathbb{Q})^c)$ *consists of the torsion objects and, for every prime number p, the tt-primes $\mathcal{P}_{p,\mathrm{mot}}$ and $\mathcal{P}_{p,\mathrm{et}}$ are the kernels of motivic and étale cohomology with $\mathbb{Z}/p$ coefficients, respectively.*

Example 4.7.6. Peter's conditions 4.7.3 on rational motivic cohomology are satisfied for $F = \bar{\mathbb{Q}}$ for instance. This also provides an example for Theorem 4.7.5. Indeed, Gallauer's result uses Peter's theorem rationally.

Remark 4.7.7. Gallauer [36] proves more general results about the derived category $\mathrm{DTM}(F;\mathbb{Z}/p)^c$ of Tate motives with $\mathbb{Z}/p$ coefficients without any condition about rational motivic cohomology. The latter theorem follows from the study of the derived categories of filtered modules in [37].

4.7.8. In the case of Theorem 4.7.4, because -1 is a sum of squares in F, the rational derived category of motives coincides with the rational stable $\mathbb{A}^1$-homotopy category $\mathrm{DM}(F;\mathbb{Q}) \cong \mathrm{SH}^{\mathbb{A}^1}(F;\mathbb{Q})$. Let us now mention some integral information about $\mathrm{SH}^{\mathbb{A}^1}(F)$. The first partial results about its spectrum were obtained in [6, § 10]. The most advanced information is currently:

Theorem 4.7.9 (Heller-Ormsby [42, Thm. 1.1]). *Let F be a field of characteristic different from 2. Then the comparison map (Definition 4.8.2) to the homogeneous spectrum of Milnor-Witt K-theory is surjective:*

$$\mathrm{Spc}(\mathrm{SH}^{\mathbb{A}^1}(F)^c) \twoheadrightarrow \mathrm{Spec}^{\mathrm{h}}(K_*^{MW}(F)).$$

The exact computation of the tt-spectrum of $\mathrm{SH}(F)^c$ is a major open challenge, which involves understanding the fibers of the above map.

Remark 4.7.10. Partial results have also been obtained by Dell'Ambrogio-Tabuada [32] for non-commutative (dg-)motives.

[5] In positive characteristic ℓ, replace the coefficients $\mathbb{Z}$ by $\mathbb{Z}[1/\ell]$ and only allow $p \neq \ell$.

4.8 Pointers to tt-geometry

4.8.1. A snapshot of tensor-triangular geometry as of the year 2010 can be found in [7]. For a more recent survey, see [71]. Beyond those references, let us simply highlight some aspects close to the author's own research.

A very useful basic tool introduced in [6] is the following *comparison map* between tt-spectra and Zariski spectra of suitable graded rings:

Definition 4.8.2. Let $u \in \mathcal{K}$ be a $\otimes$-invertible object and $R^\bullet_{\mathcal{K},u} = \oplus_{n\in\mathbb{Z}} \mathrm{Hom}_{\mathcal{K}}(\mathbb{1}, u^{\otimes n})$ the associated graded-commutative graded ring. Then

$$\mathcal{P} \mapsto \{\, f \in R^\bullet_{\mathcal{K},u} \,|\, \mathrm{cone}(f) \notin \mathcal{P} \,\}$$

defines a continuous map $\rho^\bullet\colon \mathrm{Spc}(\mathcal{K}) \to \mathrm{Spec}^{\mathrm{h}}(R^\bullet_{\mathcal{K},u})$. Without grading, one can similarly define $\rho\colon \mathrm{Spc}(\mathcal{K}) \to \mathrm{Spec}(\mathrm{End}_{\mathcal{K}}(\mathbb{1}))$.

4.8.3. Dell'Ambrogio-Stanley [30] give a class of cellular tt-categories for which $\rho^\bullet$ is a homeomorphism, namely when the ring $R^\bullet_{\mathcal{K},\mathbb{1}}$ is concentrated in even degrees and is 'regular' in a weak sense, which includes noetherian.

4.8.4. The comparison map was generalized in two directions. First by Dell'Ambrogio-Stevenson [31], by allowing grading by a collection of invertible objects instead of a single one. Secondly, higher comparison maps were defined by Sanders [68] in order to refine the analysis of the fibers of 'lower' comparison maps, through an inductive process.

4.8.5. The above comparison map is still very much concerned with the computation of $\mathrm{Spc}(\mathcal{K})$. Moving away from this preoccupation, some first 'geometric' results were established in [5], like the decomposition of an object associated to a decomposition of its support, and applications to filtrations of $\mathcal{K}$ by (co)dimension of support. These ideas naturally led to *tensor-triangular Chow groups* in [8] and further improvements by Klein [51] and Belmans-Klein [18], using the already mentioned relative tt-geometry of [69].

4.8.6. In recent years, a great deal of progress followed from the development of the idea of *separable extensions* of tt-categories. The ubiquity of this notion through stable homotopy theory, in connection with equivariant ideas, can be seen in [11]. As a slogan, this theory extends tt-geometry from the Zariski setting to the étale setting. Implications for the spectrum are discussed in [9].

4.8.7. Another area of tt-geometry which seems promising is the theory of *homological residue field*, which aims at abstractly understanding the various 'fields' which appear in examples: Morava K-theories, ordinary residue fields, π-points, etc. The reader can enter this ongoing project via [13, 10].

Acknowledgements

I would like to thank Tobias Barthel, Ivo Dell'Ambrogio, Martin Gallauer, Beren Sanders and Greg Stevenson for their help in assembling this survey. I apologize to anyone whose tt-geometric results are not mentioned here: For the sake of pithiness, I chose to restrict myself to the topic of tt-classification.

Bibliography

[1] Benjamin Antieau and Greg Stevenson. Derived categories of representations of small categories over commutative Noetherian rings. *Pacific J. Math.*, 283(1):21–42, 2016.

[2] Joseph Ayoub. A guide to (étale) motivic sheaves. In *Proc. of the International Congress of Mathematicians, Seoul (2014). Vol. II*, pages 1101–1124, 2014.

[3] Shawn Baland, Alexandru Chirvasitu, and Greg Stevenson. The prime spectra of relative stable module categories. Trans. Amer. Math. Soc. 371 (2019), no. 1, 489–503.

[4] Paul Balmer. The spectrum of prime ideals in tensor triangulated categories. *J. Reine Angew. Math.*, 588:149–168, 2005.

[5] Paul Balmer. Supports and filtrations in algebraic geometry and modular representation theory. *Amer. J. Math.*, 129(5):1227–1250, 2007.

[6] Paul Balmer. Spectra, spectra, spectra – tensor triangular spectra versus Zariski spectra of endomorphism rings. *Algebr. Geom. Topol.*, 10(3):1521–1563, 2010.

[7] Paul Balmer. Tensor triangular geometry. In *International Congress of Mathematicians, Hyderabad (2010), Vol. II*, pages 85–112. Hindustan Book Agency, 2010.

[8] Paul Balmer. Tensor triangular Chow groups. *J. Geom. Phys.*, 72:3–6, 2013.

[9] Paul Balmer. Separable extensions in tensor-triangular geometry and generalized Quillen stratification. *Ann. Sci. Éc. Norm. Supér. (4)*, 49(4):907–925, 2016.

[10] Paul Balmer. Nilpotence theorems via homological residue fields, arXiv:1710.04799, 2017.

[11] Paul Balmer, Ivo Dell'Ambrogio, and Beren Sanders. Restriction to finite-index subgroups as étale extensions in topology, KK-theory and geometry. *Algebr. Geom. Topol.*, 15(5):3025–3047, 2015.

[12] Paul Balmer, Henning Krause, and Greg Stevenson. The frame of smashing tensor-ideals. arXiv:1701.05937, 2017.

[13] Paul Balmer, Henning Krause, and Greg Stevenson. Tensor-triangular fields: Ruminations. Selecta Math. (N.S.) 25 (2019), no. 1, Art. 13.

[14] Paul Balmer and Beren Sanders. The spectrum of the equivariant stable homotopy category of a finite group. *Invent. Math.*, 208(1):283–326, 2017.

[15] Tobias Barthel, Natalia Castellana, Drew Heard, and Gabriel Valenzuela. Stratification and duality for homotopical groups, arXiv:1711.03491, 2017.

[16] Tobias Barthel, John Greenlees, and Markus Hausmann. On the Balmer spectrum of compact Lie groups. arXiv:1810.04698, 2018.

[17] Tobias Barthel, Markus Hausmann, Niko Naumann, Thomas Nikolaus, Justin Noel, and Nathaniel Stapleton. The Balmer spectrum of the equivariant homotopy category of a finite abelian group. *Invent. Math.* 216(1):215–240, 2019.

[18] Pieter Belmans and Sebastian Klein. Relative tensor triangular Chow groups for coherent algebras. *J. Algebra*, 487:386–428, 2017.

[19] Dave Benson, Srikanth B. Iyengar, and Henning Krause. Localising subcategories for cochains on the classifying space of a finite group. *C. R. Math. Acad. Sci. Paris*, 349(17-18):953–956, 2011.

[20] Dave Benson, Srikanth B. Iyengar, Henning Krause, and Julia Pevtsova. Stratification for module categories of finite group schemes. *J. Amer. Math. Soc.*, 31(1):265–302, 2018.

[21] David J. Benson, Jon F. Carlson, and Jeremy Rickard. Thick subcategories of the stable module category. *Fund. Math.*, 153(1):59–80, 1997.

[22] David J. Benson, Srikanth Iyengar, and Henning Krause. *Representations of finite groups: Local cohomology and support*, volume 43 of *Oberwolfach Seminars*. Birkhäuser/Springer, 2012.

[23] Brian Boe, Jonathan Kujawa, and Daniel Nakano. Tensor triangular geometry for quantum groups. arXiv:1702.01289, 2017.

[24] Brian D. Boe, Jonathan R. Kujawa, and Daniel K. Nakano. Tensor triangular geometry for classical Lie superalgebras. *Adv. Math.*, 314:228–277, 2017.

[25] Aslak Bakke Buan, Henning Krause, and Øyvind Solberg. Support varieties: an ideal approach. *Homology, Homotopy Appl.*, 9(1):45–74, 2007.

[26] Jon F. Carlson and Srikanth B. Iyengar. Thick subcategories of the bounded derived category of a finite group. *Trans. Amer. Math. Soc.*, 367(4):2703–2717, 2015.

[27] Denis-Charles Cisinski and Frédéric Déglise. Triangulated categories of mixed motives. arXiv:0912.2110, 2009.

[28] Everett C. Dade. Endo-permutation modules over p-groups. I. *Ann. of Math. (2)*, 107(3):459–494, 1978.

[29] Ivo Dell'Ambrogio. Tensor triangular geometry and KK-theory. *J. Homotopy Relat. Struct.*, 5(1):319–358, 2010.

[30] Ivo Dell'Ambrogio and Donald Stanley. Affine weakly regular tensor triangulated categories. *Pacific J. Math.*, 285(1):93–109, 2016.

[31] Ivo Dell'Ambrogio and Greg Stevenson. Even more spectra: tensor triangular comparison maps via graded commutative 2-rings. *Appl. Categ. Structures*, 22(1):169–210, 2014.

[32] Ivo Dell'Ambrogio and Gonçalo Tabuada. Tensor triangular geometry of non-commutative motives. *Adv. Math.*, 229(2):1329–1357, 2012.

[33] Ethan S. Devinatz, Michael J. Hopkins, and Jeffrey H. Smith. Nilpotence and stable homotopy theory. I. *Ann. of Math. (2)*, 128(2):207–241, 1988.

[34] Umesh V. Dubey and Vivek M. Mallick. Spectrum of some triangulated categories. *J. Algebra*, 364:90–118, 2012.

[35] Eric M. Friedlander and Julia Pevtsova. Π-supports for modules for finite group schemes. *Duke Math. J.*, 139(2):317–368, 2007.

[36] Martin Gallauer Alves de Souza. tt-geometry of Tate motives over algebraically closed fields. arXiv:1708.00834, 2017.

[37] Martin Gallauer Alves de Souza. Tensor triangular geometry of filtered modules. *Algebra Number Theory*, 12(8):1975–2003, 2018.

[38] J. P. C. Greenlees. Rational Mackey functors for compact Lie groups. I. *Proc. London Math. Soc. (3)*, 76(3):549–578, 1998.

[39] J. P. C. Greenlees. The Balmer spectrum of rational equivariant cohomology theories. *J. Pure Appl. Algebra*, 223:2845–2871, 2019.

[40] Jack Hall. The Balmer spectrum of a tame stack. *Ann. K-Theory*, 1(3):259–274, 2016.

[41] Dieter Happel. *Triangulated categories in the representation theory of finite-dimensional algebras*, volume 119 of *London Math. Soc. Lecture Note Ser.* Cambridge University Press, 1988.

[42] Jeremiah Heller and Kyle Ormsby. Primes and fields in stable motivic homotopy theory. *Geom. Topol.*, 22(4):2187–2218, 2018.

[43] Yuki Hirano. Relative singular locus and Balmer spectrum of matrix factorizations. *Trans. Amer. Math. Soc.*, 371(7):4993–5021, 2019.

[44] M. Hochster. Prime ideal structure in commutative rings. *Trans. Amer. Math. Soc.*, 142:43–60, 1969.

[45] Michael J. Hopkins. Global methods in homotopy theory. In *Homotopy theory (Durham, 1985)*, volume 117 of *London Math. Soc. Lecture Note Ser.*, pages 73–96. Cambridge University Press, 1987.

[46] Michael J. Hopkins and Jeffrey H. Smith. Nilpotence and stable homotopy theory. II. *Ann. of Math. (2)*, 148(1):1–49, 1998.

[47] Mark Hovey, John H. Palmieri, and Neil P. Strickland. Axiomatic stable homotopy theory. *Mem. Amer. Math. Soc.*, 128(610), 1997.

[48] Bernhard Keller. A remark on the generalized smashing conjecture. *Manuscripta Math.*, 84(2):193–198, 1994.

[49] B. Keller and A. Neeman. The connection between May's axioms for a triangulated tensor product and Happel's description of the derived category of the quiver D_4. *Doc. Math.*, 7:535–560, 2002.

[50] Shane Kelly. Some observations about motivic tensor triangulated geometry over a finite field. arXiv:1608.02913, 2016.

[51] Sebastian Klein. Chow groups of tensor triangulated categories. *J. Pure Appl. Algebra*, 220(4):1343–1381, 2016.

[52] Joachim Kock and Wolfgang Pitsch. Hochster duality in derived categories and point-free reconstruction of schemes. *Trans. Amer. Math. Soc.*, 369(1):223–261, 2017.

[53] Henning Krause. Smashing subcategories and the telescope conjecture – an algebraic approach. *Invent. Math.*, 139(1):99–133, 2000.

[54] Amalendu Krishna. Perfect complexes on Deligne-Mumford stacks and applications. *J. K-Theory*, 4(3):559–603, 2009.

[55] Yu-Han Liu and Susan J. Sierra. Recovering quivers from derived quiver representations. *Comm. Algebra*, 41(8):3013–3031, 2013.

[56] Hiroki Matsui. Connectedness of the Balmer spectrum of the right bounded derived category. *J. Pure Appl. Algebra*, 222(11):3733–3744, 2018.

[57] Hiroki Matsui and Ryo Takahashi. Thick tensor ideals of right bounded derived categories. *Algebra Number Theory*, 11(7):1677–1738, 2017.

[58] Fabien Morel. $\mathbb{A}^1$-algebraic topology. In *International Congress of Mathematicians. Vol. II*, pages 1035–1059. Eur. Math. Soc., 2006.

[59] Fabien Morel and Vladimir Voevodsky. $\mathbb{A}^1$-homotopy theory of schemes. *Inst. Hautes Études Sci. Publ. Math.*, (90):45–143 (2001), 1999.

[60] Shigeru Mukai. Duality between $D(X)$ and $D(\hat{X})$ with its application to Picard sheaves. *Nagoya Math. J.*, 81:153–175, 1981.

[61] Amnon Neeman. The chromatic tower for $D(R)$. *Topology*, 31(3):519–532, 1992.

[62] Amnon Neeman. *Triangulated categories*, volume 148 of *Ann. of Math. Stud.*. Princeton University Press, 2001.

[63] Tobias J. Peter. Prime ideals of mixed Artin-Tate motives. *Journal of K-theory*, 11(2):331–349, 2013.

[64] Julia Pevtsova and Sarah Witherspoon. Tensor ideals and varieties for modules of quantum elementary abelian groups. *Proc. Amer. Math. Soc.*, 143(9):3727–3741, 2015.

[65] Douglas C. Ravenel. Localization with respect to certain periodic homology theories. *Amer. J. Math.*, 106(2):351–414, 1984.

[66] Douglas C. Ravenel. *Nilpotence and periodicity in stable homotopy theory*, volume 128 of *Ann. of Math. Stud.*. Princeton University Press, 1992.

[67] Jeremy Rickard. Derived categories and stable equivalence. *J. Pure Appl. Algebra*, 61(3):303–317, 1989.

[68] Beren Sanders. Higher comparison maps for the spectrum of a tensor triangulated category. *Adv. Math.*, 247:71–102, 2013.

[69] Greg Stevenson. Support theory via actions of tensor triangulated categories. *J. Reine Angew. Math.*, 681:219–254, 2013.

[70] Greg Stevenson. Subcategories of singularity categories via tensor actions. *Compos. Math.*, 150(2):229–272, 2014.

[71] Greg Stevenson. A tour of support theory for triangulated categories through tensor triangular geometry. In *Building Bridges between Algebra and Topology*, Adv. Courses Math. CRM Barcelona, pages 63–101. Birkhäuser/Springer, 2018.

[72] Ryo Takahashi. Classifying thick subcategories of the stable category of Cohen-Macaulay modules. *Adv. Math.*, 225(4):2076–2116, 2010.

[73] R. W. Thomason. The classification of triangulated subcategories. *Compos. Math.*, 105(1):1–27, 1997.

[74] Vladimir Voevodsky. $\mathbb{A}^1$-homotopy theory. In *Proceedings of the International Congress of Mathematicians, Vol. I (Berlin, 1998)*, pages 579–604, 1998.

[75] Vladimir Voevodsky. Triangulated categories of motives over a field. In *Cycles, Transfers, and Motivic Homology Theories*, volume 143 of *Ann. of Math. Stud.*, pages 188–238. Princeton University Press, 2000.

[76] Ren Wang. The spectrum of the singularity category of a category algebra. arXiv:1803.09439, 2018.

[77] Fei Xu. Spectra of tensor triangulated categories over category algebras. *Arch. Math. (Basel)*, 103(3):235–253, 2014.

MATHEMATICS DEPARTMENT, UCLA, LOS ANGELES, CA 90095, U.S.A.
E-mail address: balmer@math.ucla.edu

5

Chromatic structures in stable homotopy theory

Tobias Barthel and Agnès Beaudry

5.1 Introduction

At its core, chromatic homotopy theory provides a natural approach to the computation of the stable homotopy groups of spheres π_*S^0. Historically, the first few of these groups were computed geometrically through the classification of stably framed manifolds, using the Pontryagin–Thom isomorphism $\pi_*S^0 \cong \Omega_*^{\mathrm{fr}}$. However, beginning with the work of Serre, it soon turned out that algebraic tools were more effective, both for the computation of specific low-degree values as well as for establishing structural results. In particular, Serre proved that π_*S^0 is a degreewise finitely generated abelian group with $\pi_0S^0 \cong \mathbb{Z}$ and that all higher groups are torsion.

Serre's method was essentially inductive: starting with the knowledge of the first n groups $\pi_0S^0, \ldots, \pi_{n-1}S^0$, one can in principle compute π_nS^0. Said differently, Serre worked with the Postnikov filtration of π_*S^0, in which the $(n+1)$st filtration quotient is given by π_nS^0. The key insight of chromatic homotopy theory is that π_*S^0 comes naturally equipped with a completely different filtration—*the chromatic filtration*—which systematically exhibits the large scale symmetries hidden in the stable homotopy category.

Chromatic homotopy theory is the study of the chromatic filtration and the structures that arise from it, both on π_*S^0 but also on the category of spectra itself. As with many young and active fields, points of views are evolving rapidly and there are few surveys that keep up with the developments. Our goal for this chapter is to present our perspective on the subject and, in the process, to draw one of the possible maps of the field in its current state.

We would like to emphasize that our exposition is in many ways revisionistic and certainly far from comprehensive, but rather reflects our own understanding of and point of view on the subject. We apologize to those who would have preferred us to present the material from a different point of view, or for us to include important topics we have left untouched. Hopefully, they will take this as a cue to write an expository piece of their own as we feel there is a great need for more background literature in this vibrant field.

Mathematics Subject Classification. 55N20, 55N22, 55P42, 55P60, 55Q45, 55Q51.

Key words and phrases. chromatic filtration, Morava E-theory, Morava stabilizer group, finite resolutions, $K(1)$-local and $K(2)$-local homotopy theory, chromatic splitting conjecture, Picard groups, algebraicity.

Overview

In the rest of this short introduction, we give a brief overview of the content of the chapter.

The goal of Section 5.2 is to introduce and study the chromatic filtration and its consequences from an abstract point of view. More precisely, we will:

(1) Explain that the chromatic filtration arises *canonically* from the global structure of the stable homotopy category. See Section 5.2.1.

(2) Describe the geometric origins of the chromatic filtration through the relation with the stack of formal groups. See Section 5.2.2.

(3) Demonstrate that many geometric structures have homotopical manifestations in the chromatic picture that motivate and guide the past and recent developments in the subject. See Section 5.2.3 and Section 5.2.4.

While Section 5.2 focuses mostly on the global picture, in Section 5.3 we zoom in on $K(n)$-local homotopy theory. In Section 5.3, we introduce Morava E-theory E_n and the Morava stabilizer group $\mathbb{G}_n$, which play a central role in this story because of their relationship to the $K(n)$-local sphere via the equivalence $L_{K(n)}S^0 \simeq E_n^{h\mathbb{G}_n}$. The resulting descent spectral sequence, whose E_2-term is expressed in terms of group cohomology, is one of the most important computational tools in the subject. For this reason, Section 5.3.2 is devoted to the study of $\mathbb{G}_n$ and its homological algebraic properties.

At this point, we go on a hiatus and give an overview of the chromatic story at height $n = 1$. This is the content of Section 5.4, whose goal is to provide the reader with a concrete example to keep in mind for the rest of the chapter.

The most technical part of this overview of chromatic homotopy theory is Section 5.5, which presents the theory of finite resolutions. These are finite sequences of spectra that approximate the $K(n)$-local sphere by spectra of the form E_n^{hF} for F finite subgroups of $\mathbb{G}_n$. The advantage of this approach is that the spectra E_n^{hF} are computationally tractable. Finite resolutions have been one of the most important tools in computations at height $n = 2$ and we give detailed examples in this case in Section 5.5.5.

In the last part, Section 5.6, we provide an overview of three topics in chromatic homotopy theory that have seen recent breakthroughs:

(1) In Section 5.6.1, we discuss chromatic reassembly, which describes the passage from the $K(n)$-local to the p-local picture. The main open problem is the *chromatic splitting conjecture* and we give an overview of the current state of affairs on this question.

(2) In Section 5.6.2, we turn to the problem of computing the group of invertible objects in the symmetric monoidal category of $K(n)$-local spectra. We also touch upon the closely related topic of $K(n)$-local dualities.

(3) In Section 5.6.3 we talk about the asymptotic behavior of local chromatic homotopy theory when $p \to \infty$.

These developments demonstrate how chromatic homotopy theory uncovers structures in the stable homotopy category that reveal the many interactions between homotopy theory and other areas of mathematics.

Conventions and prerequisites

We will assume that the reader is familiar with basic stable homotopy theory and category theory, as for example contained in the appendices to Ravenel [107]. Throughout this chapter, Sp will denote a good symmetric monoidal model for the category of spectra, as for example S-modules [43], symmetric spectra [75], orthogonal spectra [90], or the ∞-category of spectra [86].[1] Note that all of these categories model the stable homotopy category, i.e., their associated homotopy categories are equivalent to the stable homotopy category, so the homotopical constructions in this chapter will be model-independent. In fact, Schwede's rigidity theorem [111] justifies that we may work in a model-independent fashion.

In particular, we freely use the theory of ring spectra in Sp and module spectra over them, formal groups, and spectral sequences. A full triangulated subcategory of a triangulated category $\mathcal{T}$ is called thick if it is closed under suspensions and desuspensions, fiber sequences, and retracts. If $\mathcal{T}$ is cocomplete, then a thick subcategory is called localizing if it closed under all set-indexed direct sums, and we write $\mathrm{Loc}(S)$ for the smallest localizing subcategory of $\mathcal{T}$ containing a given set S of objects in $\mathcal{T}$. Further, recall that an object $C \in \mathcal{T}$ is said to be compact (or small) if $\mathrm{Hom}_{\mathcal{T}}(C, -)$ commutes with arbitrary direct sums in $\mathcal{T}$; we will write $\mathcal{T}^{\omega}$ for the full subcategory spanned by the compact objects in $\mathcal{T}$. If $\mathcal{C}$ denotes a model (i.e., a stable model category or stable ∞-category) for $\mathcal{T}$, then the corresponding notions for $\mathcal{C}$ are defined analogously.

5.2 A panoramic view of the chromatic landscape

The goal of this section is to give an overview of the global structure of the stable homotopy category from the chromatic perspective. Motivated by the analogy with abelian groups and the geometry of the moduli stack of formal groups, we will explain how the solution of the Ravenel Conjectures by Devinatz, Hopkins, Ravenel, and Smith leads to a canonical filtration in stable homotopy theory. The construction as well as the coarse properties of the resulting chromatic filtration are then summarized in the remainder of this section, which prepares for the in-depth study of the local filtration quotients in Section 5.3.

Remark 5.2.1. The global point of view taken in this section goes back to Hopkins' original account [67] of his work with Devinatz and Smith on the nilpotence conjectures. It has subsequently led to the study of the global structure of more general tensor-triangulated categories and the systematic development of tt-geometry by Balmer and his coauthors. We refer to Balmer's chapter in this handbook for background and a plethora of further examples.

5.2.1 From abelian groups to spectra

As expressed in Waldhausen's vision of *brave new algebra*, the category Sp of spectra should be thought of as a homotopical enrichment of the derived category $\mathcal{D}_{\mathbb{Z}}$ of abelian

[1] See David Gepner's chapter in this Handbook for a précis of Lurie's "spectral algebra."

groups. Consequently, before beginning with our analysis of the global structure of the stable homotopy category, we may consider the case of abelian groups as a toy example. The starting point is the *Hasse square* for the integers, displayed as the pullback square on the left:

$$\begin{array}{ccc} \mathbb{Z} & \longrightarrow & \prod_p \mathbb{Z}_p \\ \downarrow & & \downarrow \\ \mathbb{Q} & \longrightarrow & \mathbb{Q} \otimes \prod_p \mathbb{Z}_p \end{array} \qquad\qquad \begin{array}{ccc} M & \longrightarrow & \prod_p M_p^{\wedge} \\ \downarrow & & \downarrow \\ \mathbb{Q} \otimes M & \longrightarrow & \mathbb{Q} \otimes \prod_p M_p^{\wedge}. \end{array} \tag{5.2.1}$$

This is a special case of a local-to-global principle for any chain complex $M \in \mathcal{D}_{\mathbb{Z}}$, expressed by the homotopy pullback square on the right, in which $M_p^{\wedge}$ denotes the derived p-completion of M. While the remaining terms in this square seem to be more complicated than M itself, they are often easier from a structural point of view. This is the reason that problems in arithmetic geometry—for example finding integer valued solutions to a set of polynomial equations—can often be divided into two steps: First solve the usually simpler question at individual primes p, and then attempt to globalize the solutions.

This approach is tied closely to the global structure of the category $\mathcal{D}_{\mathbb{Z}}$. Let $\mathcal{D}_{\mathbb{Q}}$ be the derived category of $\mathbb{Q}$-vector spaces and write $(\mathcal{D}_{\mathbb{Z}})_p^{\wedge}$ for the category of derived p-complete complexes of abelian groups. (Recall that a complex C of abelian groups is derived p-complete if it is p-local and $\mathrm{Ext}^i(\mathbb{Q}, C) = 0$ for $i = 0, 1$ or, equivalently, if C is in the image of the zeroth left derived functor of p-completion on $\mathcal{D}_{\mathbb{Z}}$.) We highlight three fundamental properties of these subcategories of $\mathcal{D}_{\mathbb{Z}}$:

(1) The category $(\mathcal{D}_{\mathbb{Z}})_p^{\wedge}$ is compactly generated by $\mathbb{Z}/p$. In particular, an object $X \in (\mathcal{D}_{\mathbb{Z}})_p^{\wedge}$ is trivial if and only if $X \otimes \mathbb{Z}/p$ is trivial.

(2) The only proper localizing subcategory of $(\mathcal{D}_{\mathbb{Z}})_p^{\wedge}$ is (0), i.e., if X is any non-trivial object in $(\mathcal{D}_{\mathbb{Z}})_p^{\wedge}$, then $\mathrm{Loc}(X) = (\mathcal{D}_{\mathbb{Z}})_p^{\wedge}$, i.e., the smallest full triangulated subcategory of $(\mathcal{D}_{\mathbb{Z}})_p^{\wedge}$ closed under shifts and colimits that contains X is $(\mathcal{D}_{\mathbb{Z}})_p^{\wedge}$ itself.

(3) Any object $M \in \mathcal{D}_{\mathbb{Z}}$ can be reassembled from its derived p-completions $M_p^{\wedge} \in (\mathcal{D}_{\mathbb{Z}})_p^{\wedge}$, its rationalization $\mathbb{Q} \otimes M \in \mathcal{D}_{\mathbb{Q}}$, together with the gluing information specified in the pullback square displayed on the right of (5.2.1).

Therefore, we may think of $(\mathcal{D}_{\mathbb{Z}})_p^{\wedge}$ as an irreducible building block of $\mathcal{D}_{\mathbb{Z}}$. In fact, we can promote these observations to a natural bijection between the residue fields of $\mathbb{Z}$, which are parametrized by the points of $\mathrm{Spec}(\mathbb{Z})$, and the irreducible subcategories of $\mathcal{D}_{\mathbb{Z}}$ they detect:

$$\left\{ \begin{matrix} \text{Prime fields} \\ \mathbb{Q} \text{ and } \mathbb{F}_p \text{ for } p \text{ prime} \end{matrix} \right\} \xleftrightarrow{\sim} \left\{ \begin{matrix} \text{Minimal localizing subcategories} \\ \mathcal{D}_{\mathbb{Q}} \text{ and } (\mathcal{D}_{\mathbb{Z}})_p^{\wedge} \text{ for } p \text{ prime} \end{matrix} \right\}. \tag{5.2.2}$$

A convenient language and framework for describing the global structure of categories like $\mathcal{D}_{\mathbb{Z}}$ and Sp is provided by Balmer's *tensor triangular geometry.* Roughly speaking, the Balmer spectrum $\mathrm{Spc}(\mathcal{T})$ of a tensor triangulated category $\mathcal{T}$ has as points the thick $\otimes$-ideal of $\mathcal{T}^{\omega}$ (where $\mathcal{T}^{\omega}$ denotes the subcategory of compact objects), equipped with a topology that encodes the inclusions among these subcategories. Whenever $\mathcal{T}$ is compactly generated by its $\otimes$-unit, as is the case for example for Sp, thick $\otimes$-ideals coincide with thick subcategories of $\mathcal{T}^{\otimes}$. We refer to Balmer's chapter in this Handbook for precise definitions and many examples.

With this terminology at hand, we are now ready to make the slogan at the beginning of this section more precise. First note that we can truncate the homotopy groups S^0 above degree 0 to obtain a ring map $\phi\colon S^0 \to \tau_{\leq 0}S^0 \simeq H\mathbb{Z}$, which is the Hurewicz map for integral homology. Base-change along ϕ then provides a functor

$$\mathrm{Sp} \simeq \mathrm{Mod}_{S^0}(\mathrm{Sp}) \xrightarrow{\phi^*} \mathrm{Mod}_{H\mathbb{Z}}(\mathrm{Sp}) \simeq \mathcal{D}_{\mathbb{Z}}$$

which represents the passage from higher algebra to classical algebra; here, the second equivalence was established by Shipley in [115]. Moreover, identifying $\mathbb{Z} \cong [S^0, S^0]$, Balmer constructs a canonical comparison map ρ from the Balmer spectrum of Sp to the Zariski spectrum of $\mathbb{Z}$. The bijection (5.2.2) implies that the composite

$$\mathrm{Spc}(\mathcal{D}_{\mathbb{Z}}) \xrightarrow{\mathrm{Spc}(\phi^*)} \mathrm{Spc}(\mathrm{Sp}) \xrightarrow{\quad\rho\quad} \mathrm{Spec}(\mathbb{Z})$$

is an isomorphism, so Spc(Sp) contains Spec($\mathbb{Z}$) as a retract. This leads to the following natural question: For $p \in \mathrm{Spec}(\mathbb{Z})$, what is the fiber $\rho^{-1}(p)$ in Spc(Sp)? We will see in Theorem 5.2.7 below that, for each prime ideal $(p) \in \mathrm{Spec}(\mathbb{Z})$, there is an infinite family of points in Spc(Sp) that interpolates between (p) and $(0) \in \mathrm{Spec}(\mathbb{Z})$, the so-called *chromatic primes*. In other words, the global structure of the stable homotopy category refines the global structure of $\mathcal{D}_{\mathbb{Z}}$; see Theorem 4.3.3 in Paul Balmer's chapter of this Handbook for a picture.

Let $\mathrm{Sp}_{(p)}$ be the category of p-local spectra, i.e., those spectra whose homotopy groups are p-local abelian groups. It turns out that $\rho^{-1}(p)$ is determined by $\mathrm{Spc}(\mathrm{Sp}_{(p)})$. We will address the following two problems:

(1) Classify the thick subcategories of $\mathrm{Sp}_{(p)}^{\omega}$.

(2) Find the analogues of prime fields of $\mathrm{Sp}_{(p)}$.

As we will see, the classification of thick subcategories is a consequence of the answer to Problem 2, but before we can get there, we will exhibit a geometric model that serves as a good approximation to stable homotopy theory.

Convention 5.2.2. From here onwards, we fix a prime p and only consider the category of p-local spectra. We write Sp for $\mathrm{Sp}_{(p)}$ and assume without further mention that our spectra have been localized at p.

5.2.2 A geometric model for stable homotopy theory

In order to prepare for the resolution of the questions above, we first exhibit a geometric model for the stable homotopy category whose main structural features will turn out to reflect that of Sp rather closely.

Recall that the mod p singular cohomology $H^*(X, \mathbb{F}_p)$ of any space or spectrum X is endowed with an action of cohomology operations $\pi_*\mathrm{Hom}(H\mathbb{F}_p, H\mathbb{F}_p)$, which form the mod

p Steenrod algebra $\mathcal{A}_p$. In other words, singular cohomology naturally factors through the functor that forgets the $\mathcal{A}_p$-module structure and only remembers the underlying $\mathbb{F}_p$-vector space of $H^*(X, \mathbb{F}_p)$:

$$\begin{array}{ccc} & & \mathrm{Mod}_{\mathcal{A}_p}^{\mathrm{graded}} \\ & \overset{H^*(-,\mathbb{F}_p)}{\nearrow} & \downarrow \text{forget} \\ \mathrm{Sp}^{\mathrm{op}} & \xrightarrow[H^*(-,\mathbb{F}_p)]{} & \mathrm{Mod}_{\mathbb{F}_p}^{\mathrm{graded}}. \end{array}$$

The Adams spectral sequence, first introduced in [1], can then be interpreted as a device that attempts to go back, or at least recover partial information about X: There is a spectral sequence

$$E_2^{s,t} \cong \mathrm{Ext}^s_{\mathcal{A}_p}(H^*(Y), H^*(X))_t \Longrightarrow [X, Y_p^\wedge]_{t-s},$$

that converges whenever X and Y are spectra of finite type with X finite. Here, finite type means that the mod p cohomology is finitely generated in each degree, and $Y_p^\wedge$ denotes the p-completion of Y. The subscript t on the Ext-indicates the internal grading, arising from the grading of cohomology groups involved. Informally speaking, this spectral sequence measures to what extent $\mathrm{Mod}_{\mathcal{A}_p}$ deviates from being a perfect model for Sp. See [30] for a general study of the convergence properties of (generalized) Adams spectral sequences.

Remark 5.2.3. Paraphrasing, the *Mahowald uncertainty principle* asserts that any spectral sequence that computes the stable homotopy groups of a finite spectrum with a machine computable E_2-term will be infinitely far from the actual answer. In practical terms this means that the Adams spectral sequence for $X = S^0$ and Y a finite spectrum contains many differentials that require additional input to be determined.

Building on the work of Novikov [99] and Quillen [104], Morava [97] realized that replacing $H\mathbb{F}_p$ by the Brown–Peterson spectrum BP gives rise to a geometric model for Sp that resembles its global structure more closely. To describe it, recall that BP is an irreducible additive summand in the p-localized complex cobordism spectrum MU with coefficients $BP_* = \mathbb{Z}_{(p)}[v_1, v_2, \ldots]$ and $\deg(v_n) = 2p^n - 2$. The generator v_{i+1} is uniquely determined only modulo the ideal $(p, v_1, \ldots, v_i)$ and there are different choices available, for example the Araki or Hazewinkel generators. See, for example, [106, A2.2]. The corresponding Hopf algebroid (BP_*, BP_*BP) is a presentation of the moduli stack of (p-typical) formal groups $\mathcal{M}_{fg}$ and the category of evenly graded comodules over (BP_*, BP_*BP) is equivalent to the category of quasi-coherent sheaves over $\mathcal{M}_{fg}$, see for example [98] for a general treatment. Miller [93] explains how this equivalence can be extended to all graded comodules by replacing $\mathcal{M}_{fg}$ by a moduli stack of *spin formal groups*, see also [45]. Taking BP-homology induces a functor

$$\mathrm{Sp} \longrightarrow \mathrm{Comod}^{\mathrm{even}}_{BP_*BP} \simeq \mathrm{QCoh}(\mathcal{M}_{fg}), \qquad X \mapsto BP_*(X),$$

where $\mathrm{Comod}^{\mathrm{even}}_{BP_*BP}$ denotes the abelian category of evenly graded BP_*BP-comodules. The associated Adams–Novikov spectral sequence has signature

$$E_2^{s,t} \cong H^s(\mathcal{M}_{fg}; BP_*(X))_t \cong \mathrm{Ext}_{BP_*BP}(BP_*, BP_*(X)) \Longrightarrow \pi_{t-s}X.$$

The structure of this spectral sequence, whose computational exploitation was a major impetus in the development of chromatic homotopy theory (see [95]), is governed by the particularly simple geometric structure of $\mathcal{M}_{fg}$, which we describe next.

As explained in great detail in [45], the height filtration of formal groups manifests itself in a descending filtration by closed substacks

$$\mathcal{M}_{fg} \supset \mathcal{M}(1) \supset \mathcal{M}(2) \supset \cdots \tag{5.2.3}$$

where $\mathcal{M}(n)$ is cut out locally by the ideal defined by the regular sequence $(p, v_1, v_2, \ldots, v_{n-1})$. Note that this filtration is not separated, as the additive formal group has height ∞. Write

- $\mathcal{M}_{fg}^{\leq n}$ for the open complement of $\mathcal{M}(n+1)$ representing formal groups of height at most n with $i_n \colon \mathcal{M}_{fg}^{\leq n} \to \mathcal{M}_{fg}$ the inclusion,
- $\mathcal{H}(n) = \mathcal{M}(n) \cap \mathcal{M}_{fg}^{\leq n}$ for the locally closed substack of formal groups of height exactly n, and
- $\widehat{\mathcal{H}}(n)$ for its formal completion.

If Γ is any formal group of height n over $\mathbb{F}_p$, then $\mathcal{H}(n)$ is equivalent as a stack to $B\mathrm{Aut}_{\overline{\mathbb{F}}_p}(\Gamma)$, so the filtration quotients of the height filtration (5.2.3) contain a single geometric point. Furthermore, there is a (pro-)Galois extension

$$\mathrm{Def}(\overline{\mathbb{F}}_p, \Gamma) \longrightarrow \widehat{\mathcal{H}}(n) \tag{5.2.4}$$

with Galois group $\mathrm{Gal}(\overline{\mathbb{F}}_p/\mathbb{F}_p) \ltimes \mathrm{Aut}_{\overline{\mathbb{F}}_p}(\Gamma)$, with $\mathrm{Def}(\overline{\mathbb{F}}_p, \Gamma)$ being the Lubin–Tate deformation space. See Remark 5.3.4 below.

In light of (5.2.3), any quasi-coherent sheaf $\mathcal{F} \in \mathrm{QCoh}(\mathcal{M}_{fg})$ can be approximated by its restrictions to the open substacks $\mathcal{M}_{fg}^{\leq n}$, so the geometric filtration on $\mathcal{M}_{fg}$ gives rise to a filtration of $\mathrm{QCoh}(\mathcal{M}_{fg})$. It follows that the computation of the cohomology of a quasi-coherent sheaf $\mathcal{F}$ on $\mathcal{M}_{fg}$ can be restricted to the computation of the cohomology of $\mathcal{F}$ reduced to the strata $\mathcal{H}(n)$ together with the gluing data between different strata. The insight of Bousfield, Morava, and Ravenel was that the resulting structure on the E_2-term of the Adams–Novikov spectral sequence is in fact manifest in $\pi_* S^0$ and Sp as well, as we shall see in the next sections.

Remark 5.2.4. An early hint that there is such a close relation between Sp and $\mathrm{QCoh}(\mathcal{M}_{fg})$ is the Landweber exact functor theorem, which shows that any flat map $f \colon \mathrm{Spec}(R) \to \mathcal{M}_{fg}$ can be lifted to a complex orientable ring spectrum with formal group classified by f. We refer to Behrens's chapter in this Handbook for more details.

5.2.3 The chromatic filtration: construction

The goal of this section is to answer the questions raised at the end of Section 5.2.1 and to construct the chromatic filtration. We continue to work in the category Sp of p-local spectra for a fixed prime p as in Convention 5.2.2.

In loose analogy with algebra, a ring spectrum $K \in \mathrm{Sp}$ is said to be a field if every K-module splits into a wedge of shifted copies of K. In particular, for any spectra X and Y, there is a Künneth isomorphism

$$K_*(X \wedge Y) \cong K_*(X) \otimes_{K_*} K_*(Y).$$

There exists a family of distinct field spectra $K(n)$ for $0 \leq n \leq \infty$ called the *Morava K-theories*, whose construction will be reviewed in Section 5.3.1.

As a result of the seminal nilpotence theorem proven by Devinatz, Hopkins, and Smith [40, 71], we obtain a classification of fields in Sp.

Theorem 5.2.5 (Hopkins–Smith). *Any field object in* Sp *splits (additively) into a wedge of shifted copies of Morava K-theories. Moreover, if R is a ring spectrum such that $K(n)_*(R) = 0$ for all $0 \leq n \leq \infty$, then $R \simeq 0$.*

For example, $K(0) = H\mathbb{Q}$, $K(\infty) = H\mathbb{F}_p$, and $K(1)$ is an Adams summand of mod p K-theory. Informally speaking, the spectra $K(n)$ may be thought of as the homotopical residue fields of the sphere spectrum.

Remark 5.2.6. As remarked in [71], this theorem can be interpreted as providing a classification of prime fields of Sp. The existence and uniqueness of $\mathbb{A}_\infty$-structures on $K(n)$ was established in Angeltveit's paper [4], while Hopkins and Mahowald have proved that none of these multiplicative structures on $K(n)$ refine to an $\mathbb{E}_2$-ring structure (e.g., [6, Corollary 5.4]).

In light of this theorem, there is a natural notion of support for a spectrum $X \in \mathrm{Sp}$, namely

$$\mathrm{supp}(X) = \{n \mid K(n)_*(X) \neq 0\} \subseteq \mathbb{Z}_{\geq 0} \cup \{\infty\}.$$

This notion of support turns out to be particularly well-behaved for the category of finite spectra Sp^ω. Since $K(\infty)_*F \cong H_*(F, \mathbb{F}_p) = 0$ implies $F \simeq 0$ for finite F, for any non-trivial F there exists an $n \in \mathbb{N}$ such that $n \in \mathrm{supp}(F)$. Ravenel [105] further proved that $n \in \mathrm{supp}(F)$ implies $(n+1) \in \mathrm{supp}(F)$, so the only subsets of $\mathbb{Z}_{\geq 0} \cup \{\infty\}$ that can be realized as the support of a finite spectra are the sets $\{n, n+1, n+2, \ldots, \infty\}$ with $n \in \mathbb{N}$. A result of Mitchell's [96] implies that all of these subsets can be realized by a finite spectrum.

Write $\mathcal{C}_0 = \mathrm{Sp}^\omega$ and, for $n \geq 1$, let $\mathcal{C}_n \subseteq \mathrm{Sp}^\omega$ be the thick subcategory of finite spectra F with $\mathrm{supp}(F) \subseteq \{n, n+1, n+2, \ldots, \infty\}$ for $n \in \mathbb{N}$. The following consequence of Theorem 5.2.5 is often called the *thick subcategory theorem*, proven in [71]. It says in particular that the support function defined above detects the thick subcategories of Sp^ω:

Theorem 5.2.7 (Hopkins–Smith). *If $\mathcal{C} \subseteq \mathrm{Sp}^\omega$ is a nonzero thick subcategory, then there exists an $n \geq 0$ such that $\mathcal{C} = \mathcal{C}_n$. Moreover, there is a sequence of proper inclusions*

$$\mathrm{Sp}^\omega = \mathcal{C}_0 \supset \mathcal{C}_1 \supset \mathcal{C}_2 \supset \cdots \supset (0),$$

which completely describes Spc(Sp).

This categorical filtration gives rise to a sequence of functorial approximations of any finite spectrum F by spectra that are supported on $\{0, 1, \ldots, n\}$ for varying n, where the zeroth approximation is given by the rationalization $F \wedge H\mathbb{Q}$. This filtration should be understood as a homotopical incarnation of the geometric filtration of $\mathcal{M}_{fg}$, so that the approximations of F correspond to the restriction of the associated sheaf $BP_*(F) \in \mathrm{QCoh}(\mathcal{M}_{fg})$ to $\mathcal{M}_{fg}^{\leq n}$.

The tool required to formulate this notion of approximation rigorously is provided by Bousfield localization, which we briefly review here for the convenience of the reader. Let E be a spectrum and consider the full subcategory $\langle E \rangle \subseteq \mathrm{Sp}$ of E-acyclic spectra, i.e., those spectra A with $E \wedge A \simeq 0$. Bousfield [30] proved that there exists a fiber sequence

$$C_E \longrightarrow \mathrm{id} \longrightarrow L_E$$

of functors on Sp satisfying the following properties:

(1) For any $X \in \mathrm{Sp}$, $C_E X$ is in $\langle E \rangle$.

(2) For any $X \in \mathrm{Sp}$, $L_E X$ is E-local, i.e., it does not admit any nonzero maps from an E-acyclic spectrum.

It follows formally that $L_E X$ is the initial E-local spectrum equipped with a map from X, and it is called the E-localization of X. The full subcategory of Sp whose objects are the E-local spectra will be denoted by Sp_E; by construction, it is the quotient of Sp by $\langle E \rangle$.

In order to extract the part of a spectrum X that is supported on $\{0, 1, \ldots, n\}$, i.e., the information of X that is seen by the residue fields $K(0), K(1), \ldots, K(n)$, it is natural to consider the following Bousfield localization

$$X \longrightarrow L_n X := L_{K(0) \vee K(1) \vee \cdots \vee K(n)} X.$$

In fact, for every n there exists a spectrum $E(n)$ with coefficients $E(n)_* = \mathbb{Z}_{(p)}[v_1, \ldots, v_n][v_n^{-1}]$ called *Johnson–Wilson spectrum* (of height n) that has the property that $\langle E(n) \rangle = \langle K(0) \vee K(1) \vee \ldots \vee K(n) \rangle$, hence $L_n = L_{E(n)}$. We let $\mathrm{Sp}_n = \mathrm{Sp}_{E(n)}$ denote the category of $E(n)$-local spectra.

By construction, these localization functors fit into a *chromtic tower* under X as follows

$$\begin{array}{ccccccccc} & M_n X & & & M_2 X & & M_1 X & & M_0 X \simeq H\mathbb{Q} \wedge X \\ & \downarrow & & & \downarrow & & \downarrow & & \| \\ \cdots \longrightarrow & L_n X & \longrightarrow \cdots \longrightarrow & & L_2 X & \longrightarrow & L_1 X & \longrightarrow & L_0 S^0 \simeq H\mathbb{Q} \wedge X, \end{array} \tag{5.2.5}$$

where the *monochromatic layers* $M_n X$ are defined by the fiber sequence

$$M_n X \longrightarrow L_n X \longrightarrow L_{n-1} X.$$

Specializing to the sphere spectrum and applying homotopy groups, we arrive at the definition of the chromatic filtration.

Definition 5.2.8. The chromatic filtration on π_*S^0 is given by the descending filtration

$$\pi_*S^0 \supseteq \mathbb{C}_0\pi_*S^0 \supseteq \mathbb{C}_1\pi_*S^0 \supseteq \cdots \tag{5.2.6}$$

defined as $\mathbb{C}_n\pi_*S^0 = \mathrm{Ker}(\pi_*S^0 \to \pi_*L_nS^0)$.

There is an important subtlety in the definition of the chromatic filtration, as there is an a priori different way of constructing a filtration of Sp from the thick subcategory theorem (Theorem 5.2.7). Indeed, without relying on the Morava K-theories $K(n)$, one may instead take the quotient of Sp by the localizing subcategories $\mathrm{Loc}(\mathcal{C}_n) \subseteq \mathrm{Sp}$ for each n. The resulting localization functors L_n^f can then be used as above to construct a descending filtration

$$\pi_*S^0 \supseteq \mathbb{C}_0^f\pi_*S^0 \supseteq \mathbb{C}_1^f\pi_*S^0 \supseteq \cdots$$

with $\mathbb{C}_n^f\pi_*S^0 = \mathrm{Ker}(\pi_*S^0 \to \pi_*L_n^fS^0)$, known as the *geometric filtration*, see Ravenel [107, Section 7.5]. If X is a spectrum such that $L_n^fX \simeq 0$, then also $L_nX \simeq 0$, so there are natural comparison transformations $L_n^f \to L_n$, leading to the following optimistic conjecture about the comparison between the two filtrations:

Conjecture 5.2.9 (Telescope conjecture). *The natural transformation $L_n^f \to L_n$ is an equivalence.*

A number of equivalent formulations of this conjecture and the current state of knowledge about it can be found in Mahowald–Ravenel–Schick [88] and [10]. The smash product theorem of Hopkins and Ravenel [107, Section 8] states that L_n is smashing, i.e., L_n as an endofunctor on Sp commutes with colimits, while the analogous fact for L_n^f was proven by Miller [92]. It therefore suffices to show the telescope conjecture for S^0. This has been verified by explicit computations for $n = 0$ and $n = 1$ by work of Mahowald [87] for $p = 2$ and Miller [94, 3] for odd p, but the telescope conjecture is open in all other cases. It is known however that $L_n^fM \to L_nM$ is an equivalence for many spectra M, including BP-modules [72, Corollary 1.10] and $E(m)$-local spectra [76, Corollary 6.10] for any $m \geq 0$.

5.2.4 The chromatic filtration: disassembly and reassembly

The goal of this subsection is to first demonstrate how the chromatic filtration decomposes the stable homotopy groups of spheres into periodic families and then to explain how these irreducible pieces reassemble into π_*S^0. The starting point is the chromatic convergence theorem due to Hopkins and Ravenel, proven in [107], whose content is that the chromatic tower (5.2.5) does not lose any information about S^0. In particular, the chromatic filtration (5.2.6) on π_*S^0 is exhaustive. We continue to follow Convention 5.2.2.

Theorem 5.2.10 (Hopkins–Ravenel). *The canonical map $X \to \lim_n L_nX$ is an equivalence for all finite spectra X.*

Remark 5.2.11. For general X, this map can be far from being an equivalence. For example, the chromatic tower of $H\mathbb{F}_p$ or the Brown–Comenetz dual IS^0 of the sphere is identically zero. However, chromatic convergence is known to hold for a class of spectra larger than just finite ones, including $\mathbb{CP}^\infty$. See [8].

We now turn to the filtration quotients of the chromatic filtration, which correspond homotopically to the monochromatic layers M_nX. Much of the material in this section can be found in [76].

The layers M_nX decompose into spectra that are periodic of periods a multiple of $2(p^n - 1)$, thereby resembling the decomposition of light into waves of different frequencies. (This is the origin of the term *chromatic* homotopy theory, coined by Ravenel.) More precisely, if X is any spectrum, then its nth monochromatic layer is equivalent to a filtered colimit of spectra F_α,

$$\operatorname{colim}_\alpha F_\alpha \xrightarrow{\sim} M_nX,$$

such that each F_α is periodic. That is, for each α there exists a natural number $\lambda(\alpha)$ and a homotopy equivalence $F_\alpha \simeq \Sigma^{2(p^n-1)p^{\lambda(\alpha)}} F_\alpha$. This follows from the fact that M_n is equivalent to the colocalization of the $E(n)$-local category with respect to the $E(n)$-localization of a finite type n spectrum, see for example [76, Proposition 7.10], together with the periodicity theorem of Hopkins and Smith [71, Theorem 9]

Having resolved S^0 into its irreducible chromatic pieces M_nS^0, it is now time to consider the question of how to reassemble the pieces. For this, it is more convenient to consider the $K(n)$-localizations instead of the monochromatic layers, as we shall explain next.

Write $\mathfrak{M}_n \subset \mathrm{Sp}$ for the essential image of the functor M_n and let $\mathrm{Sp}_{K(n)}$ be the category of $K(n)$-local spectra. For any n, the functors $L_{K(n)}$ and M_n restrict to an adjunction on the category Sp_n (with M_n as the left adjoint) and then further to a symmetric monoidal equivalence [76, Theorem 6.19]

$$M_n \colon \mathrm{Sp}_{K(n)} \rightleftarrows \mathfrak{M}_n : L_{K(n)}.$$

So we may equivalently work with $L_{K(n)}S^0$ in place of M_nS^0.

Remark 5.2.12. The more categorically minded reader may think of the situation as follows: The descending filtration of Sp^ω of Theorem 5.2.7 extends to two descending filtrations of Sp:

$$\mathrm{Sp} = \mathrm{Ker}(0) \supset \mathrm{Ker}(L_0) \supset \mathrm{Ker}(L_1) \supset \cdots \supset \mathrm{Ker}(\mathrm{id}) = (0)$$

and

$$\mathrm{Sp} = \mathrm{Loc}(\mathcal{C}_0) \supset \mathrm{Loc}(\mathcal{C}_1) \supset \mathrm{Loc}(\mathcal{C}_2) \supset \cdots \supset (0),$$

which are equivalent if the telescope conjecture holds for all n. Focusing on the first filtration for concreteness and writing Sp_n for the essential image of L_n as before, we could equivalently pass to the associated ascending filtration

$$(0) = \mathrm{im}(0) \subset \mathrm{Sp}_0 \subset \mathrm{Sp}_1 \subset \mathrm{Sp}_2 \subset \cdots \subset \mathrm{im}(\mathrm{id}) = \mathrm{Sp}.$$

The consecutive subquotients $\mathrm{Sp}_n/\mathrm{Sp}_{n-1}$ can then be realized in two different ways as subcategories of Sp, namely either as a localizing subcategory $\mathfrak{M}_n$ or as a colocalizing subcategory $\mathrm{Sp}_{K(n)}$. The resulting equivalence between $\mathfrak{M}_n$ and $\mathrm{Sp}_{K(n)}$ is an instance of a phenomenon called *local duality*, see [15].

Suppose X is a spectrum for which we have determined $L_{n-1}X$ and $L_{K(n)}X$, and we are interested in reassembling them to obtain L_nX. Motivated by the geometric model of Section 5.2.2, we expect this process to be analogous to the way a sheaf on the open subset $\mathcal{M}_{fg}^{\leq n-1}$ and another sheaf on the stratum $\mathcal{H}_n$ are glued together along the formal neighborhood $\widehat{\mathcal{H}}_n$ of $\mathcal{H}_n$ inside $\mathcal{M}_{fg}^{\leq n-1}$ to produce a sheaf on $\mathcal{M}_{fg}^{\leq n}$. This picture turns out to be faithfully reflected in stable homotopy theory: The chromatic reassembly process for $X \in \mathrm{Sp}$ is governed by the homotopy pullback square displayed on the left, usually called the *chromatic fracture square* (see for example [51]):

$$\begin{array}{ccc} L_nX & \longrightarrow & L_{K(n)}X \\ \downarrow & & \downarrow \\ L_{n-1}X & \xrightarrow[\iota_n(X)]{} & L_{n-1}L_{K(n)}X \end{array} \qquad \begin{array}{ccc} \mathrm{Sp}_n & \xrightarrow{L_{K(n)}} & \mathrm{Sp}_{K(n)} \\ {\scriptstyle X\mapsto\iota_n(X)}\downarrow & & \downarrow{\scriptstyle L_{n-1}} \\ \mathrm{Fun}(\Delta^1, \mathrm{Sp}_{n-1}) & \xrightarrow[\text{target}]{} & \mathrm{Sp}_{n-1}. \end{array} \tag{5.2.7}$$

In fact, by [5] the category Sp_n itself admits a decomposition into chromatically simpler pieces, see the pullback square on the right of (5.2.7). Here, $\mathrm{Fun}(\Delta^1, \mathrm{Sp}_{n-1})$ is the arrow category of Sp_{n-1} and the pullback is taken in a suitably derived sense. The labels of the arrows in this diagram indicate how to translate from the chromatic fracture square of a spectrum X to the categorical decomposition on the right of (5.2.7).

Based on computations of Shimomura–Yabe [114], Hopkins [72] conjectured that the chromatic reassembly process which recovers L_nX from $L_{K(n)}X$ and $L_{n-1}X$ takes a particularly simple form:

Conjecture 5.2.13 (Weak Chromatic Splitting). *The map*

$$\iota_n(S_p^0)\colon L_{n-1}S_p^0 \longrightarrow L_{n-1}L_{K(n)}S_p^0$$

in (5.2.7) *is split, i.e., it admits a section. Here, S_p^0 is the p-complete sphere spectrum.*

This conjecture, its variations, and its consequences are discussed in more detail in Section 5.6.1. For now we note that Conjecture 5.2.13 is known to hold for $n \leq 2$ and all primes p, and is wide open otherwise.

We can now summarize the chromatic approach as follows:

Chromatic Approach. *The chromatic approach to $\pi_*S_{(p)}^0$ consists of three steps:*

(1) Compute $\pi_*L_{K(n)}S^0$ for each n.

(2) Understand the gluing in the chromatic fracture square (5.2.7).

(3) Use chromatic convergence (Theorem 5.2.10) to recover $S_{(p)}^0$.

Finally, the p-local sphere spectrum $S_{(p)}^0$ determines S_p^0 by p-completion. We can thus reassemble the sphere spectrum S^0 itself via the following homotopical analogue of the

Hasse square (5.2.1):

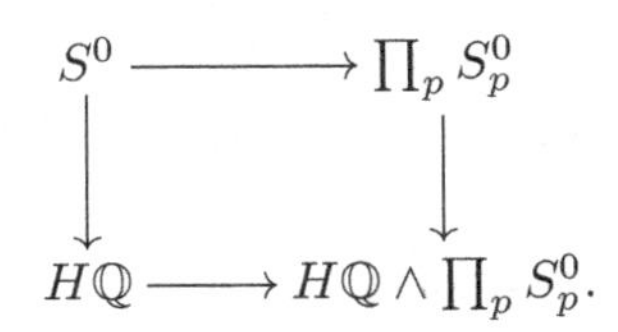

In the next section, we discuss the first two steps of the chromatic approach.

Remark 5.2.14. As mentioned earlier, the deconstructive analysis of the stable homotopy category based on its spectrum Spc(Sp) can be carried out in any tensor triangulated category; many examples can be found in Paul Balmer's chapter in this Handbook. This is the subject of *prismatic algebra*. An especially interesting example is the stable module category StMod_{kG} of a finite p-group G and field k of characteristic p, whose spectrum $\mathrm{Spc}(\mathrm{StMod}_{kG})$ is homeomorphic to $\mathrm{Proj}(H^*(G;k))$, the Proj construction of the graded ring $H^*(G;k)$. This category is a good test case for chromatic questions: for instance, the analogues of both the telescope conjecture and the weak chromatic splitting conjecture are known to hold in StMod_{kG}, see [28] and [14].

5.3 Local chromatic homotopy theory

We begin this section by introducing the main players of local chromatic homotopy theory: Morava E-theory E_n, the Morava stabilizer group $\mathbb{G}_n$ and its action on E_n, and the resulting descent spectral sequence computing $\pi_* L_{K(n)} S^0$. We then summarize the key algebraic features of the Morava stabilizer group, its continuous cohomology, and its action on the coefficients of Morava E-theory. In order to have a toy case in mind for the general constructions to follow, we study in detail the case of height 1.

5.3.1 Morava E-theory and the descent spectral sequence

The chromatic program has led us naturally and inevitably to the study of the $K(n)$-local categories, which should be thought of as an analog of $(\mathcal{D}_{\mathbb{Z}})_p^{\wedge}$ for abelian groups. Formally, we note that $\mathrm{Sp}_{K(n)}$ is a closed symmetric monoidal stable category. Moreover, in close analogy with Section 5.2.1, the $K(n)$-local categories have the following properties:

(1) The category $\mathrm{Sp}_{K(n)}$ is compactly generated by $L_n F(n)$ for any $F(n) \in \mathcal{C}_n \setminus \mathcal{C}_{n+1}$ for $\mathcal{C}_n$ as in Theorem 5.2.7, and an object $X \in \mathrm{Sp}_{K(n)}$ is trivial if and only if $X \wedge K(n)$ is trivial.

(2) The only proper localizing subcategory of $\mathrm{Sp}_{K(n)}$ is (0).

(3) A spectrum $X \in \mathrm{Sp}_n$ can be reassembled from $L_{K(n)}X$, $L_{n-1}X$, together with the gluing information specified in the pullback square displayed on the right of (5.2.7).

This confirms the idea that the $K(n)$-local categories play the role of the irreducible pieces of Sp. With the techniques developed so far, both the finer structural properties of $\mathrm{Sp}_{K(n)}$

as well as any concrete calculations would be essentially inapproachable: Incipit *Morava E-theory.*

We let Γ_n denote the Honda formal group law of height n. It is the formal group law classified by the map

$$BP_* \cong \mathbb{Z}_{(p)}[v_1, v_2, v_3, \ldots] \longrightarrow \mathbb{F}_p$$

that sends v_n to 1 and $(p, v_1, \ldots, v_{n-1}, v_{n+1}, v_{n+2}, \ldots)$ to zero. In fact, it is the unique p-typical formal group law over $\mathbb{F}_{p^n}$ whose p-series satisfies $[p]_{\Gamma_n}(x) = x^{p^n}$. A good reference on formal group laws for homotopy theorists is [106, Appendix A2].

Let $\mathbb{W}_n = W(\mathbb{F}_{p^n})$ be the ring of Witt vectors of $\mathbb{F}_{p^n}$, which is isomorphic to the ring of integers in an unramified extension of $\mathbb{Q}_p$ of degree n. Lubin and Tate [84] showed that there exists a p-typical universal deformation F_n of Γ_n to complete local rings with residue field $\mathbb{F}_{p^n}$, whose formal group law $F_n(x, y)$ is defined over the ring

$$(E_n)_0 = \mathbb{W}_n[\![u_1, \ldots, u_{n-1}]\!] \quad u_i \in (E_n)_0. \tag{5.3.1}$$

Introducing a formal variable u in degree -2 then allows to extend F_n to a graded formal group law $F_{n,\mathrm{gr}}(x, y) = uF_n(u^{-1}x, u^{-1}y)$ defined over $(E_n)_* = (E_n)_0[u^{\pm 1}]$, classified by the ring homomorphism

$$BP_* \cong \mathbb{Z}_{(p)}[v_1, v_2, v_3, \ldots] \longrightarrow (E_n)_*$$

that sends $(v_{n+1}, v_{n+2}, \ldots)$ to zero, v_n to u^{1-p^n}, and v_k to $u_k u^{1-p^k}$ for $k < n$. Here, we are using the Araki generators for BP_*. See [106, A2.2] for more details.

In order to lift this construction to stable homotopy theory, one first shows that the functor

$$X \mapsto (E_n)_* \otimes_{BP_*} BP_*(X)$$

is a homology theory, represented by a complex orientable ring spectrum $E_n = E(\mathbb{F}_{p^n}, \Gamma_n)$, called Morava E-theory or the *Lubin–Tate spectrum* because of its connection to Lubin–Tate deformation theory, see Rezk [108]. This is an instance of the Landweber exact functor theorem mentioned in Remark 5.2.4. The spectrum E_n is a completed and 2-periodized version of the Johnson–Wilson spectrum $E(n)$ from Section 5.2.3, and it turns out that $L_{E(n)} = L_{E_n}$ for all n; in particular, the terms $E(n)$-local and E_n-local are synonymous.

Since $(E_n)_*$ is a regular graded commutative ring that is concentrated in even degrees, reduction modulo the maximal ideal $\mathfrak{m} = (p, u_1, \ldots, u_{n-1})$ can be realized by a (homotopy) ring map

$$E_n \longrightarrow E_n/\mathfrak{m} =: K_n$$

with $\pi_* K_n \cong \mathbb{F}_{p^n}[u^{\pm 1}]$, see for example Chapter V of [43]. The spectrum K_n splits as a wedge of equivalent spectra, which are shifts of the Morava K-theory $K(n)$ of Theorem 5.2.5, with homotopy groups $K(n)_* \cong \mathbb{F}_p[v_n^{\pm 1}]$ for $v_n = u^{1-p^n}$.

Definition 5.3.1. The *small Morava stabilizer group* $\mathbb{S}_n := \mathrm{Aut}_{\mathbb{F}_{p^n}}(\Gamma_n)$ is the group of automorphisms of Γ_n with coefficients in $\mathbb{F}_{p^n}$

$$\mathbb{S}_n = \{f(x) \in \mathbb{F}_{p^n}[\![x]\!] : f(\Gamma_n(x, y)) = \Gamma_n(f(x), f(y)),\ f'(0) \neq 0\}.$$

Since Γ_n is defined over $\mathbb{F}_p$, the Galois group $\mathrm{Gal} = \mathrm{Gal}(\mathbb{F}_{p^n}/\mathbb{F}_p)$ acts on $\mathbb{S}_n$ by acting on the coefficients of an automorphism. The *big Morava Stabilizer group* $\mathbb{G}_n$ is the extension $\mathbb{S}_n \rtimes \mathrm{Gal}$. Equivalently, $\mathbb{G}_n$ is the group of automorphisms of the pair $(\mathbb{F}_{p^n}, \Gamma_n)$.

The construction $E(\mathbb{F}_{p^n}, \Gamma_n)$ is natural in the pair $(\mathbb{F}_{p^n}, \Gamma_n)$, so there is an up to homotopy action of $\mathrm{Aut}_{\mathbb{F}_{p^n}}(\Gamma_n)$ on $E(\mathbb{F}_{p^n}, \Gamma_n)$. This action can be promoted to an action through $\mathbb{E}_\infty$-ring maps in a unique way: By Goerss–Hopkins–Miller obstruction theory [70, 50], E_n admits an essentially unique structure of an $\mathbb{E}_\infty$-ring spectrum and $\mathbb{G}_n$ acts on it through $\mathbb{E}_\infty$-ring maps. In fact, $\mathbb{G}_n$ gives essentially all such automorphisms of E_n. A new proof of these results from the perspective of derived algebraic geometry has recently appeared in Lurie [85]. The connection between $K(n)$-local homotopy theory and Morava E-theory is then illustrated in the diagram

$$L_{K(n)}S^0 \longrightarrow E_n \longrightarrow K_n. \tag{5.3.2}$$

The first map is a pro-Galois extension of ring spectra with Galois group $\mathbb{G}_n$ in the sense of Rognes. In particular, $L_{K(n)}S^0 \simeq E_n^{h\mathbb{G}_n}$ and the extension $L_{K(n)}S^0 \to E_n$ behaves like an unramified field extension. The second map in (5.3.2) corresponds to the passage to the residue field. See [109, 7] for precise definitions on pro-Galois extensions and [39] for a definition of homotopy fixed points for profinite groups. Further results and alternative approaches to the construction of (continuous) homotopy fixed points in the generality needed for chromatic homotopy theory can be found in [34, 103, 35] and the references therein.

Remark 5.3.2. Note also that the extension $L_{K(n)}S^0 \to E_n$ can be broken into two pro-Galois extensions

$$L_{K(n)}S^0 \xrightarrow{\quad \mathrm{Gal} \quad} E_n^{h\mathbb{S}_n} \simeq L_{K(n)}S^0(\omega) \xrightarrow{\quad \mathbb{S}_n \quad} E_n,$$

where the arrows are labelled by the structure group of the extension. Here, $L_{K(n)}S^0(\omega)$ is an $\mathbb{E}_\infty$-ring obtained by adjoining a primitive (p^n-1)th root of unity ω to the $K(n)$-local sphere. See [109, 5.4.6] and [29, Section 1.6] for details on this.

From the fact that the first map of (5.3.2) is a Galois extension, it follows that

$$\pi_* L_{K(n)}(E_n \wedge E_n) \cong \mathrm{Map}^c(\mathbb{G}_n, (E_n)_*), \tag{5.3.3}$$

where Map^c denotes the continuous functions as profinite sets. See for example [73, Theorem 4.11]. In fact, the functor $(E_n)^\vee_*(-) := \pi_* L_{K(n)}(E_n \wedge -)$ takes values in a category of *Morava modules*, which are $(E_n)_*$-modules equipped with a continuous action by $\mathbb{G}_n$ (see Definition 5.3.20). Furthermore, a map f is a $K(n)$-local equivalence (i.e., $K(n)_*(f)$ is an isomorphism) if and only if $(E_n)^\vee_*(f)$ is an isomorphism. The resulting relationship between the topological category $\mathrm{Sp}_{K(n)}$ and the algebraic category of Morava modules provides very powerful tools for computations in the $K(n)$-local category. In particular, it gives rise to a *homotopy fixed point spectral sequence*, also called the *descent spectral sequence*.

Theorem 5.3.3 (Hopkins–Ravenel [107], Devinatz–Hopkins [39], Rognes [109])**.** *The unit map $L_{K(n)}S^0 \to E_n$ is a pro-Galois extension with Galois group $\mathbb{G}_n$. There is a convergent* *descent spectral sequence*

$$E_2^{s,t} \cong H_c^s(\mathbb{G}_n, (E_n)_t) \Longrightarrow \pi_{t-s} L_{K(n)} S^0, \tag{5.3.4}$$

which collapses with a horizontal vanishing on a finite page.

The spectral sequence (5.3.4) is the *$K(n)$-local E_n-based Adams–Novikov spectral sequence*, which for a general X has the form

$$E_1^{s,t} = \pi_t L_{K(n)}(E_n \wedge E_n^s \wedge X) \Longrightarrow \pi_{t-s} L_{K(n)} X. \tag{5.3.5}$$

It is constructed in [39, Appendix A]. The description of the E_2-page in terms of continuous group cohomology H_c^* for $X = S^0$ uses (5.3.3) to identify the E_1-term with the cobar complex. More generally, if the $(E_n)_*$-module $(E_n)_*^\vee(X)$ is flat, or finitely generated, or if there exists $k \geq 1$ such that $\mathfrak{m}^k[(E_n)_*^\vee(X)] = 0$, then there is an isomorphism [13]

$$E_2^{s,t} \cong H_c^s(\mathbb{G}_n, (E_n)_t^\vee(X)).$$

Section 5.3.2.3 below further discusses homological algebra over profinite groups and properties of this spectral sequence.

In fact, as discussed in [91], the $\mathbb{G}_n$-action on E_n lifts to an action on the ∞-category Mod_{E_n}, which yields a categorical reformulation of the theorem as a canonical equivalence

$$\mathrm{Sp}_{K(n)} \xrightarrow{\sim} \mathrm{Mod}_{E_n}^{h\mathbb{G}_n},$$

where the right-hand side denotes the homotopy fixed points taken in the ∞-category of ∞-categories. These observations demonstrate the fundamental role of E_n-theory and the Morava stabilizer group $\mathbb{G}_n$ together with its cohomology in chromatic homotopy theory.

Remark 5.3.4. Other choices for Morava E-theory are possible. For any perfect field k of characteristic p and formal group law Γ of height n defined over $\mathbb{F}_p$, there is an associated spectrum $E(k, \Gamma)$ whose formal group law is a universal deformation of Γ to complete local rings with residue field isomorphic to k. There is an associated Morava K-theory $K(k, \Gamma)$, stabilizer group $\mathbb{G}(k, \Gamma)$, and $\mathbb{G}(k, \Gamma)$-Galois extension $L_{K(k,\Gamma)} S^0 \to E(k, \Gamma)$. The localization functor $L_{K(k,\Gamma)}$ is independent of the choice of formal group law Γ and extension k of $\mathbb{F}_p$, so one can make any convenient choice to study $L_{K(k,\Gamma)} S^0$.

Recall the Galois extension $\mathrm{Def}(\overline{\mathbb{F}}_p, \Gamma) \longrightarrow \widehat{\mathcal{H}}(n)$ from (5.2.4). By definition, $\mathbb{G}(\overline{\mathbb{F}}_p, \Gamma)$ is the group $\mathrm{Aut}_{\overline{\mathbb{F}}_p}(\Gamma) \rtimes \mathrm{Gal}(\overline{\mathbb{F}}_p/\mathbb{F}_p)$, and the Galois extension $E(\overline{\mathbb{F}}_p, \Gamma) \leftarrow L_{K(\overline{\mathbb{F}}_p,\Gamma)} S^0$ is a homotopical lift of the pro-Galois extension (5.2.4). The coefficients of the Lubin–Tate spectrum $E(\overline{\mathbb{F}}_p, \Gamma)$ correspond to the global sections of $\mathrm{Def}(\overline{\mathbb{F}}_p, \Gamma)$ (hence the reversal of the arrow direction). A more thorough discussion of Morava E-theory is given in Nat Stapleton's chapter in this Handbook. See also Remark 5.3.6 for more on this point.

The first step in the Chromatic Approach described in Section 5.2.4 is to compute the homotopy groups of $L_{K(n)} S^0 \simeq E_n^{h\mathbb{G}_n}$. As for any Galois extension, it makes sense to first study intermediate extensions. In general, if H and K are closed subgroups of $\mathbb{G}_n$ and H is normal in K, then $E_n^{hK} \to E_n^{hH}$ is a K/H-Galois extension and there is a descent spectral sequence

$$E_2^{s,t} \cong H_c^s(K/H, (E_n^{hH})_t) \Longrightarrow \pi_{t-s} E_n^{hK}. \tag{5.3.6}$$

See for example Devinatz [37]. It seems natural to consider the following kinds of intermediate extensions:

(a) The $\mathbb{G}_n/K$ Galois extensions $L_{K(n)}S^0 \to E_n^{hK}$ for $K \subseteq \mathbb{G}_n$ normal closed subgroups.

(b) The F-Galois extensions $E_n^{hF} \to E_n$ for finite subgroups F of $\mathbb{G}_n$.

An important example of an intermediate extension of the form (a) is given in Remark 5.3.8 below. These kinds of extensions are conceptually important, but the homotopy groups of spectra of the form E_n^{hK} are generally out of reach at heights $n \geq 3$. An exception is when $K = F \subseteq \mathbb{G}_n$ is finite, which brings us to extensions of the form (b), in which case the intermediate extensions $E_n^{hF} \to E_n$ and computations of the homotopy groups of E_n^{hF} are more accessible.

In fact, there are many computations of the homotopy groups of E_n^{hF} at various heights and the recent developments in equivariant homotopy theory by Hill, Hopkins and Ravenel [63], followed by the work on real orientations for E-theory of Hahn and Shi [52] make these computations even more accessible. A non-exhaustive list of reference related to these types of computations is given by [18, 21, 27, 29, 52, 54, 64, 65, 89], and Mike Hill's chapter in this Handbook.

In view of this, an approach to studying the $K(n)$-local sphere is to approximate it by spectra of the form E_n^{hF} for finite subgroups $F \subseteq \mathbb{G}_n$. These approximations fit together to form so-called *finite resolutions.* This is the philosophy established by Goerss, Henn, Mahowald, and Rezk (GHMR) in [49]. It has proven to be very effective for organizing computations and clarifying the structure of the $K(2)$-local category. In the next sections, we will describe the study of chromatic homotopy theory using the finite resolution perspective, starting with explicit examples at height $n = 1$. In particular, finite resolutions will be discussed at length in Section 5.5.

5.3.2 The Morava stabilizer group

In this section, we give more details on the structure of the Morava stabilizer groups $\mathbb{G}_n$ and $\mathbb{S}_n$, which were introduced in Definition 5.3.1. We also discuss homological algebra in this context. More detail on this material can be found, for example, in [58].

5.3.2.1 The structure of $\mathbb{G}_n$

Recall that Γ_n denotes the Honda formal group law of height n. We write

$$x +_{\Gamma_n} y := \Gamma_n(x, y).$$

By definition, $\mathbb{S}_n$ is the group of the units in $\mathrm{End}_{\mathbb{F}_{p^n}}(\Gamma_n)$. In fact,

$$\mathrm{End}_{\overline{\mathbb{F}}_p}(\Gamma_n) \cong \mathrm{End}_{\mathbb{F}_{p^n}}(\Gamma_n).$$

That is, all endomorphisms of Γ_n have coefficients in $\mathbb{F}_{p^n}$. We give a brief description of the endomorphism ring here, originally due to Dieudonné [41] and Lubin [83]. A good reference for this material from the perspective of homotopy theory is [106, Appendix A2.2].

Recall that $\mathbb{W}_n$ denotes the ring of Witt vectors on $\mathbb{F}_{p^n}$. It is isomorphic to the ring of integers $\mathbb{Z}_p(\omega)$ of the unramified extension of $\mathbb{Q}_p(\omega)$ obtained from $\mathbb{Q}_p$ by adjoining a primitive (p^n-1)th root of unity ω. The residue field is $\mathbb{F}_{p^n}$ and we also let ω denote its reduction in $\mathbb{F}_{p^n}$.

The series $\omega(x) = \omega x$ and $\xi(x) = x^p$ are elements of $\mathrm{End}_{\mathbb{F}_{p^n}}(\Gamma_n)$ and, in fact, the endomorphism ring is a $\mathbb{Z}_p$-module generated by these elements:

$$\mathrm{End}_{\mathbb{F}_{p^n}}(\Gamma_n) \cong \mathbb{W}_n\langle\xi\rangle/(\xi\omega - \omega^p\xi, \xi^n - p) \tag{5.3.7}$$

The identification of (5.3.7) is given explicitly as follows. An element of the right hand side can be written uniquely as

$$f = \sum_{j=0}^{n-1} f_j \xi^j \tag{5.3.8}$$

for $f_j \in \mathbb{W}_n$. Further, $f_j = \sum_{i=0}^{\infty} a_{j+in}p^i$ for unique elements $a_i \in \mathbb{W}_n$ such that $a_i^{p^n} - a_i = 0$. Using the fact that $\xi^n = p$, the element f can also be written uniquely as

$$f = \sum_{i=0}^{\infty} a_i \xi^i.$$

The series

$$[f](x) = {\sum_{i\geq 0}}^{\Gamma_n} a_i x^{p^i} = a_0 x +_{\Gamma_n} a_1 x^p +_{\Gamma_n} \cdots$$

is the endomorphism of Γ_n corresponding to f.

Let $\mathbb{D}_n = \mathbb{Q} \otimes \mathrm{End}_{\mathbb{F}_{p^n}}(\Gamma_n)$. There is a valuation $v\colon \mathbb{D}_n^\times \to \frac{1}{n}\mathbb{Z}$ normalized so that $v(\xi) = 1/n$. The ring $\mathbb{D}_n$ is a central division algebra algebra over $\mathbb{Q}_p$ of Hasse invariant $1/n$. The ring of integers of $\mathcal{O}_{\mathbb{D}_n}$ is defined to be those $x \in \mathbb{D}_n$ such that $v(x) \geq 0$, so that $\mathcal{O}_{\mathbb{D}_n} \cong \mathrm{End}_{\mathbb{F}_{p^n}}(\Gamma_n)$.

The element ξ is invertible in $\mathbb{D}_n$ and conjugation by ξ preserves $\mathcal{O}_{\mathbb{D}_n}$. In fact, conjugation by ξ corresponds to the action of a generator of $\mathrm{Gal} = \mathrm{Gal}(\mathbb{F}_{p^n}/\mathbb{F}_p)$ on $\mathbb{S}_n \cong \mathcal{O}_{\mathbb{D}_n}^\times$. From this, we get a presentation

$$\mathbb{D}_n^\times/(\xi^n) \cong \mathbb{G}_n.$$

The problem of determining the isomorphism types and conjugacy classes of maximal finite subgroups of $\mathbb{S}_n$ was studied by Hewett [61, 62] and was revisited by Bujard [32]. We have listed the conjugacy classes of maximal finite subgroups of $\mathbb{S}_n$ in Table 5.3.1. Note that the list is rather restricted and that the groups that appear all have periodic cohomology in characteristic p.

The kind of finite subgroups of $\mathbb{G}_n$ that have appeared in the construction of finite resolutions so far are extensions of finite subgroups of $\mathbb{S}_n$ in the following sense.

Prime p	Height n: $k \geq 1$, $p \nmid m$	Isomorphism Types of Maximal Finite Subgroups in $\mathbb{S}_n$
$p \neq 2$	n not divisible by $p-1$	C_{p^n-1}
$p \neq 2$	$n = (p-1)p^{k-1}m$	C_{p^n-1}, and $C_{p^\ell} \rtimes C_{(p^{p^{k-\ell}m}-1)(p-1)}$, $1 \leq \ell \leq k$
$p = 2$	n odd	$C_{2(2^n-1)}$
$p = 2$	$n = 2^{k-1}m$ and $k \neq 2$	$C_{2^\ell(2^{2^{k-\ell}m}-1)}$, $1 \leq \ell \leq k$
$p = 2$	$n = 2m$ and $m \neq 1$	$C_{2(2^m-1)}$, and $T_{24} \times C_{2^m-1}$
$p = 2$	$n = 2$	T_{24}

Table 5.3.1: The table above lists isomorphism types of maximal finite subgroups of $\mathbb{S}_n$ at various heights and primes. Each isomorphism type listed below belongs to a unique conjugacy class. Here, C_q denotes a cyclic group of order q and $T_{24} \cong Q_8 \rtimes C_3$ is the binary tetrahedral group (the action of C_3 on Q_8 permutes a choice of generators i, j and k). See Hewett [61, 62] and Bujard [32] for more details. In particular, see [61] for the isomorphism type of the semi-direct product in the table.

Definition 5.3.5. For F_0 a finite subgroup of $\mathbb{S}_n$, an *extension of* F_0 *to* $\mathbb{G}_n$ is a subgroup F of $\mathbb{G}_n$ that contains F_0 as a normal subgroup and such that the following diagram commutes:

$$\begin{array}{ccccccccc} 0 & \longrightarrow & F_0 & \longrightarrow & F & \longrightarrow & \mathrm{Gal} & \longrightarrow & 0 \\ & & \downarrow & & \downarrow & & \downarrow \cong & & \\ 0 & \longrightarrow & \mathbb{S}_n & \longrightarrow & \mathbb{G}_n & \longrightarrow & \mathrm{Gal} & \longrightarrow & 0. \end{array}$$

Here, the rows are exact, the left and middle vertical arrows are the inclusions, and the induced right vertical map is an isomorphism.

The question of when a finite subgroup F_0 of $\mathbb{S}_n$ extends to a finite subgroup of $\mathbb{G}_n$ is subtle and largely addressed by Bujard in [32]. We do not give it much attention here.

Remark 5.3.6. For any formal group law Γ of height n defined over a perfect field extension k of $\mathbb{F}_p$, one can define the group

$$\mathbb{G}(k, \Gamma) = \{(f, i) : \sigma \in \mathrm{Gal}(k/\mathbb{F}_p),\ f \in k[\![x]\!] : \sigma^*\Gamma \xrightarrow{\cong} \Gamma\}.$$

With this definition, $\mathbb{G}_n = \mathbb{G}(\mathbb{F}_{p^n}, \Gamma_n)$. This group was mentioned in Remark 5.3.4. The group $\mathbb{S}(k, \Gamma) = \mathrm{End}_k(\Gamma)^\times$ is the subgroup of $\mathbb{G}(k, \Gamma)$ consisting of pairs for which $\sigma = \mathrm{id}$.

In general, both $\mathbb{S}(k, \Gamma)$ and $\mathbb{G}(k, \Gamma)$ depend on the pair (k, Γ). However, since any two formal group laws of height n are isomorphic over $\overline{\mathbb{F}}_p$, $\mathrm{End}_{\overline{\mathbb{F}}_p}(\Gamma)$ is independent of Γ, and hence so are $\mathbb{G}(\overline{\mathbb{F}}_p, \Gamma)$ and $\mathbb{S}(\overline{\mathbb{F}}_p, \Gamma)$. Since

$$\mathbb{S}_n = \mathbb{S}(\mathbb{F}_{p^n}, \Gamma_n) \cong \mathbb{S}(\overline{\mathbb{F}}_p, \Gamma_n),$$

it follows that for any formal group law Γ as above, there is an isomorphism $\mathbb{S}_n \cong \mathbb{S}(\overline{\mathbb{F}}_p, \Gamma)$. So, Table 5.3.1 is canonical in the sense that it classifies conjugacy classes of finite subgroups of $\mathbb{S}(\overline{\mathbb{F}}_p, \Gamma)$ for any formal group law Γ of height n defined over $\overline{\mathbb{F}}_p$.

However, even if all of the automorphisms of Γ are defined over $\mathbb{F}_{p^n}$, so that

$$\mathbb{S}(\mathbb{F}_{p^n}, \Gamma) \cong \mathbb{S}(\overline{\mathbb{F}}_p, \Gamma) \cong \mathbb{S}_n,$$

it can still be the case that $\mathbb{G}(\mathbb{F}_{p^n}, \Gamma)$ and $\mathbb{G}_n$ are not isomorphic. If this is the case, extensions of a finite subgroup of $\mathbb{S}_n \cong \mathbb{S}(\mathbb{F}_{p^n}, \Gamma)$ to $\mathbb{G}(\mathbb{F}_{p^n}, \Gamma)$ and $\mathbb{G}_n$ can have different isomorphism types.

We now turn to the definition of a few group homomorphisms that play a role in the rest of this paper.

Definition 5.3.7. The *determinants*

$$\det\colon \mathbb{G}_n \longrightarrow \mathbb{Z}_p^\times \qquad\qquad \det\colon \mathbb{S}_n \longrightarrow \mathbb{Z}_p^\times$$

are the homomorphisms defined as follows. The group $\mathbb{S}_n$ acts on $\mathcal{O}_{\mathbb{D}_n}$ by right multiplication. This action gives a representation $\rho\colon \mathbb{S}_n \to GL_n(\mathbb{W}_n)$. The composite $\det \circ \rho$ has image in the Galois invariants of $\mathbb{W}_n^\times$ (see [58, Section 5.4]), so it induces a homomorphism $\mathbb{S}_n \to \mathbb{Z}_p^\times$, which we also denote by det. We extend this homomorphism to $\mathbb{G}_n$ via the composite

$$\det\colon \mathbb{G}_n \cong \mathbb{S}_n \rtimes \mathrm{Gal} \xrightarrow{\det \times \mathrm{id}} \mathbb{Z}_p^\times \times \mathrm{Gal} \to \mathbb{Z}_p^\times,$$

where the second map is the projection.

Composing $\det\colon \mathbb{G}_n \to \mathbb{Z}_p^\times$ with the quotient map to $\mathbb{Z}_p^\times/\mu \cong \mathbb{Z}_p$ gives a homomorphism

$$\zeta_n\colon \mathbb{G}_n \longrightarrow \mathbb{Z}_p \tag{5.3.9}$$

where $\mu = C_2$ if $p = 2$ and $\mu = C_{p-1}$ if p is odd. This corresponds to a class

$$\zeta_n \in H_c^1(\mathbb{G}_n, \mathbb{Z}_p) \cong \mathrm{Hom}^c(\mathbb{G}_n, \mathbb{Z}_p),$$

where Hom^c denotes continuous group homomorphisms and H_c^1 the continuous cohomology (see Section 5.3.2.3). If $p = 2$, the determinant also induces a map

$$\chi_n\colon \mathbb{G}_n \to (\mathbb{Z}_2/4)^\times \cong \mathbb{Z}/2. \tag{5.3.10}$$

which then represents a class $\chi_n \in H_c^1(\mathbb{G}_n, \mathbb{Z}/2)$. Let $\widetilde{\chi}_n \in H_c^2(\mathbb{G}_n, \mathbb{Z}_2)$ be the Bockstein of χ_n, and note that $2\widetilde{\chi}_n = 0$.

Denote by $\mathbb{G}_n^1$ the kernel of ζ_n and let $\mathbb{S}_n^1 = \mathbb{S}_n \cap \mathbb{G}_n^1$. The homomorphism ζ_n is surjective, and necessarily split since $\mathbb{Z}_p$ is topologically free. Therefore,

$$\mathbb{G}_n \cong \mathbb{G}_n^1 \rtimes \mathbb{Z}_p, \qquad\qquad \mathbb{S}_n \cong \mathbb{S}_n^1 \rtimes \mathbb{Z}_p. \tag{5.3.11}$$

If n is coprime to p, then the splitting is trivial and this is a product.

Remark 5.3.8. As a consequence of the fact that $\mathbb{G}_n/\mathbb{G}_n^1 \cong \mathbb{Z}_p$, there is an equivalence $L_{K(n)}S^0 \simeq (E_n^{h\mathbb{G}_n^1})^{h\mathbb{Z}_p}$. If $\psi \in \mathbb{G}_n$ is such that $\zeta_n(\psi)$ is a topological generator of $\mathbb{Z}_p$, then we get an exact triangle

$$L_{K(n)}S^0 \longrightarrow E_n^{h\mathbb{G}_n^1} \xrightarrow{\psi-1} E_n^{h\mathbb{G}_n^1} \xrightarrow{\delta} \Sigma L_{K(n)}S^0.$$

We also denote by ζ_n its image $H_c^*(\mathbb{G}_n, (E_n)_0)$. It is known that ζ_n is a permanent cycle in the homotopy fixed point spectral sequence, see [39, Section 8]. It detects the composite $S^0 \to E_n^{h\mathbb{G}_n^1} \xrightarrow{\delta} \Sigma L_{K(n)}S^0$ (where the first map is the unit), which is also denoted by $\zeta_n \in \pi_{-1}L_{K(n)}S^0$.

5.3.2.2 The action of the Morava stabilizer group

We now discuss the action of $\mathbb{G}_n$ on $(E_n)_*$. Most notably, this problem was first attacked in depth by Devinatz and Hopkins in [38] using the Gross–Hopkins period map (Remark 5.3.11). A very nice summary of this approach is given by Kohlhaase [79] and we discuss some of the consequences here.

Let F_n be the formal group law over $(E_n)_0$ that is a universal deformation of Γ_n and was defined in Section 5.3.1. For $\alpha \in \mathbb{G}_n$ given by a pair (f, σ) where $\sigma \in \mathrm{Gal}(\mathbb{F}_{p^n}/\mathbb{F}_p)$ and $f \in \mathbb{S}_n$, the universal property of the deformation F_n implies that there exists a unique pair (f_α, α_*) consisting of a continuous ring isomorphism $\alpha_* \colon (E_n)_0 \to (E_n)_0$ and an isomorphism of formal group laws $f_\alpha \colon \alpha_* F_n \to F_n$ such that

$$(f_\alpha, \alpha_*) \equiv (f, \sigma) \mod (p, u_1, \dots, u_{n-1}). \tag{5.3.12}$$

The isomorphism α_* is extended to $(E_n)_*$ by defining $\alpha_*(u) = f'(0)u$. The assignment $\alpha \mapsto \alpha_*$ gives a left action of $\mathbb{G}_n$ on $(E_n)_*$. The action of an element (id, σ) corresponds to the natural action of the Galois group on the coefficients $\mathbb{W}_n$ in $(E_n)_* \cong \mathbb{W}_n[\![u_1, \dots, u_{n-1}]\!][u^{\pm 1}]$, and we denote it by σ_*. Similarly, if $\alpha = (f, \mathrm{id})$, we let f_* denote the isomorphism α_*.

Computing the action explicitly is difficult and there exists no general formula. However, three cases are fairly simple to deduce from the general description above:

(a) If $\alpha = (\mathrm{id}, \sigma)$ for $\sigma \in \mathrm{Gal}(\mathbb{F}_{p^n}/\mathbb{F}_p)$, then σ_* is the action of the Galois group on the coefficients $\mathbb{W}_n$. For $x \in \mathbb{W}_n$, we write $x^\sigma = \sigma_*(x)$.

(b) If $\omega \in \mathbb{S}_n$ is a primitive $(p^n - 1)$th root of unity, then $\omega_*(u_i) = \omega^{p^i-1}u_i$ and $\omega_*(u) = \omega u$.

(c) If $\psi \in \mathbb{Z}_p^\times \subseteq \mathbb{S}_n$ is in the center, then $\psi_*(u_i) = u_i$ and $\psi_*(u) = \psi u$.

Understanding the action more generally is difficult, but we say a few words on this here.

For $f \in \mathbb{S}_n$, write $f = \sum_{j=0}^{n-1} f_j \xi^j$ for $f_j \in \mathbb{W}_n$ with $f_0 \in \mathbb{W}_n^\times$ as in (5.3.8). The following results due to Devinatz and Hopkins [38] are also given in Theorem 1.3 and Theorem 1.19 of [79].

Theorem 5.3.9 (Devinatz–Hopkins). *Let $1 \le i \le n-1$ and f_j be as above. Then, modulo $(p, u_1, \dots, u_{n-1})^2$,*

$$f_*(u) \equiv f_0 u + \sum_{j=1}^{n-1} f_{n-j}^{\sigma^j} u u_j, \qquad f_*(uu_i) \equiv \sum_{j=1}^{i} f_{i-j}^{\sigma^j} u u_j + \sum_{j=i+1}^{n} p f_{n+i-j}^{\sigma^j} u u_j.$$

Further, if $f \in \mathbb{W}_n^\times \subseteq \mathbb{S}_n$, so that $f = f_0$ then $f_*(u_i) \equiv f_0^{\sigma^i} f_0^{-1} u_i$ modulo $(u_1, \dots, u_{n-1})^2$.

An example of an immediate consequence of Theorem 5.3.9 is the following result. See [29, Lemma 1.33] for a surprisingly simple proof.

Corollary 5.3.10. *For all primes p and all heights n, the unit $\mathbb{Z}_p \to (E_n)_*$ induces an isomorphism on $\mathbb{G}_n$ fixed points $\mathbb{Z}_p \cong (E_n)_*^{\mathbb{G}_n}$.*

Remark 5.3.11 (Gross–Hopkins period map)**.** The proof of Theorem 5.3.9 relies on one of the deepest results in chromatic homotopy theory, due to Gross and Hopkins [68] , which points towards the mysterious interplay between this subject and arithmetic geometry. Let K be the quotient field of $\mathbb{W}_n$ and $\mathrm{Spf}((E_n)_0)_K$ be the generic fiber of the formal scheme associated to $(E_n)_0$. Since the division algebra $\mathbb{D}_n$ splits over K, i.e., $\mathbb{D}_n \otimes_{\mathbb{Q}_p} K$ is isomorphic to a matrix algebra $M_n(K)$, there is a natural n-dimensional $\mathbb{G}_n$-representation V_K. It follows that $\mathbb{G}_n$ acts on the corresponding projective space $\mathbf{P}(V_K)$ through projective linear transformations. In [68, 66], Gross and Hopkins construct a *period mapping* that linearizes the action of $\mathbb{G}_n$ on $\mathrm{Spf}((E_n)_0)_K$: They prove that there is an étale and $\mathbb{G}_n$-equivariant map of rigid analytic varieties

$$\Phi\colon \mathrm{Spf}((E_n)_0)_K \longrightarrow \mathbf{P}(V_K). \tag{5.3.13}$$

Devinatz and Hopkins use this map to prove Theorem 5.3.9 and it also features in the computations of Kohlhaase [79].

One often needs more precision than that provided by Theorem 5.3.9. Since f_α is a morphism of formal group laws, it follows that

$$f_\alpha([p]_{\alpha_* F_n}(x)) = [p]_{F_n}(f_\alpha(x)).$$

This relation contains a lot of information. In practice, it gives a recursive formula to compute the morphism α_* as a function of the α_js. This method is applied explicitly in Section 4 of the paper [60] by Henn–Karamanov–Mahowald.

However, even with these methods, it is difficult to get good approximations for the action of $\mathbb{G}_n$. If one restricts attention to finite subgroups $F \subseteq \mathbb{G}_n$, it is sometimes possible to do much better than these kinds of approximations. Recent developments suggest that working with a formal group law other than the Honda formal group law Γ_n may be better suited to this task. For example, when $n = 2$, one can choose to work with the formal group law of a super-singular elliptic curve. The automorphisms of the curve embed in the associated Morava stabilizer group and one can use geometric information to write explicit formulas for their action on the associated E-theory. See Strickland [116] and [20, Section 2]. In fact, the spectra E_2^{hF} at height 2 are the $K(2)$-localizations of various spectra of topological modular forms with level structures. See, for example, [24]. Elliptic curves are not available at higher heights, but there is a hope that the theory of automorphic forms will provide a replacement. This is the subject of [26]; see also Behrens's chapter in this Handbook. Finally, equivariant homotopy theory also seems to provide better choices of formal group laws for studying the action of the finite subgroups. See, for example, [63, 64] together with [52], [65], and [21].

5.3.2.3 Morava stabilizer group: homological algebra

Recall that the E_2-term of the descent spectral sequence in Theorem 5.3.3 is given by the continuous cohomology of the Morava stabilizer group with coefficients in $(E_n)_*$. The goal of this section is to summarize the homological algebra required to construct these cohomology groups and to then discuss some features specific to $\mathbb{G}_n$. An important subtlety arising from the homotopical applications we have in mind is that we have to study the continuous cohomology of $\mathbb{G}_n$ with profinite coefficients, and not merely discrete ones. This theory has been systematically developed by Lazard [81]; our exposition follows the more modern treatment of Symonds and Weigel [118].

Let $G = \lim_i G_i$ be a profinite group, given as an inverse limit of a system of finite groups (G_i) and write $\mathfrak{C}_p(G)$ for the category of profinite modules over

$$\mathbb{Z}_p[\![G]\!] = \lim_{i,j} \mathbb{Z}/p^j[G_i]$$

and continuous homomorphisms. The category $\mathfrak{C}_p(G)$ is abelian and has enough projective objects. Moreover, the completed tensor product equips $\mathfrak{C}_p(G)$ with the structure of a symmetric monoidal category with unit $\mathbb{Z}_p$. In order to define a well-behaved notion of continuous cohomology for G, assume that G is a compact p-analytic Lie group in the sense of [81]. A good reference for properties of p-adic analytic groups is [42]. Lazard then shows that:

- G is of type p-FP_∞, i.e., $\mathbb{Z}_p$ admits a resolution by finitely generated projective $\mathbb{Z}_p[\![G]\!]$-modules. It follows that the continuous cohomology of G with coefficients in $M \in \mathfrak{C}_p(G)$, defined as
$$H_c^*(G, M) = \mathrm{Ext}^*_{\mathbb{Z}_p[\![G]\!]}(\mathbb{Z}_p, M),$$
is a well-behaved cohomological functor, where the (continuous) Ext-group is computed in $\mathfrak{C}_p(G)$. In particular, there is a Lyndon–Hochschild–Serre spectral sequence and an Eckmann–Shapiro type lemma for open normal subgroups [118, Theorem 4.2.6 and Lemma 4.2.8]. Similarly, continuous homology is defined as
$$H_*^c(G, M) = \mathrm{Tor}_*^{\mathbb{Z}_p[\![G]\!]}(\mathbb{Z}_p, M)$$
where the (continuous) Tor-group is computed in $\mathfrak{C}_p(G)$.

- G is a virtual Poincaré duality group in dimension $d = \dim(G)$ [118, Theorem 5.1.9], i.e., there exists an open subgroup H in G such that
$$H_c^*(H, \mathbb{Z}_p[\![H]\!]) \cong \begin{cases} \mathbb{Z}_p & \text{if } * = d, \\ 0 & \text{otherwise,} \end{cases}$$
and the length of a projective resolution of $\mathbb{Z}_p \in \mathfrak{C}_p(H)$ can be taken to be d. The second property is referred to by saying that the cohomological dimension of H is d and that the virtual cohomological dimension of G is d; in symbols, $\mathrm{cd}_p(H) = d$ and $\mathrm{vcd}_p(G) = d$. The Poincaré duality property gives rise to a non-degenerate pairing
$$H_c^*(H, \mathbb{F}_p) \otimes H_c^{d-*}(H, \mathbb{F}_p) \longrightarrow H_c^d(H, \mathbb{F}_p) \cong \mathbb{F}_p,$$
thereby justifying the terminology.

The key theorem, proved by Morava [97, §2.2] and relying on work by Lazard [81], allows us to apply this theory to the Morava stabilizer group:

Theorem 5.3.12 (Lazard, Morava). *The Morava stabilizer group $\mathbb{S}_n$ is a compact p-analytic virtual Poincaré duality group of dimension n^2. Further, the group $\mathbb{S}_n$ is p-torsion-free if and only if $p-1$ does not divide n, and in this case $\mathrm{vcd}_p(\mathbb{S}_n) = \mathrm{cd}_p(\mathbb{S}_n) = n^2$.*

We note an important immediate consequence of this theorem, which is the underlying reason for the small prime vs. large prime dichotomy in chromatic homotopy theory. See also Figure 5.3.2:

Corollary 5.3.13. *If $p > 2$ is such that $2(p-1) > n^2$, then the descent spectral sequence (5.3.4) for S^0 collapses at the E_2-page with a horizontal vanishing line of intercept $s = n^2$ (meaning that $E_2^{s,t} = 0$ for $s > n^2$) and there are no non-trivial extensions.*

Remark 5.3.14. The condition $2(p-1) > n^2$ can be improved to $2(p-1) \geq n^2$ using Corollary 5.3.10.

Remark 5.3.15. An extremely powerful result of Devinatz–Hopkins is that, for any prime p and any height n, there exists an integer N such that, for all spectra X, the $K(n)$-local E_n-based Adams–Novikov spectral sequence for X (see (5.3.5)) has a horizontal vanishing line on the E_∞-term at $s = N$, although the minimal such N may be greater than n^2. For example, when $n = 1$ and $p = 2$, the homotopy fixed point spectral sequence (5.3.4) has non-trivial elements on the $s = 2 > 1^2$ line at E_∞. See [23, Section 2.3] for a proof of the existence of the vanishing line.

Note further that it follows from Corollary 5.3.10 and the existence of the vanishing line that the natural map $\mathbb{Z}_p \to \pi_0 L_{K(n)} S^0$ is a nilpotent extension of rings.

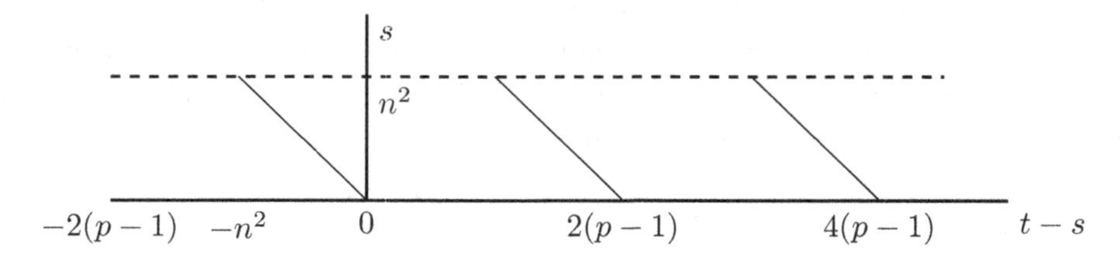

FIGURE 5.3.2. The E_2-term of $E_2^{s,t} \cong H_c^s(\mathbb{G}_n, (E_n)_t) \Rightarrow \pi_{t-s} L_{K(n)} S^0$ for $p > 2$ and $2(p-1) > n^2$. The dashed line indicates the horizontal vanishing line at E_2, that is, $E_2^{s,t} = 0$ when $s > n^2$. The nonzero contributions are concentrated on the lines of slope -1 that intercept the $(t-s)$-axis at multiples of $2(p-1)$.

In order to run the descent spectral sequence computing $\pi_* L_{K(n)} S^0$, we have to come to grips with $H_c^*(\mathbb{G}_n, (E_n)_*)$, an extremely difficult problem. However, if one restricts attention to $H_c^*(\mathbb{G}_n, (E_n)_0)$, the computation appears to radically simplify in a completely unexpected way. Let $\iota\colon \mathbb{W}_n \to (E_n)_0$ be the natural inclusion. The following has been shown to be true at all primes when $n \leq 2$, see [114, 25, 79, 48, 23, 22]:

Conjecture 5.3.16 (Vanishing conjecture). *Let p be any prime and n be any height. The map ι induces an isomorphism*

$$\iota_*\colon H_c^*(\mathbb{G}_n, \mathbb{W}_n) \xrightarrow{\cong} H_c^*(\mathbb{G}_n, (E_n)_0).$$

Remark 5.3.17. The conjecture is so named because it implies that the cohomology of the $\mathbb{G}_n$-module $(E_n)_0/\mathbb{W}_n$ vanishes in all degrees. Note further that if one proves that $\mathbb{W}_n/p \to (E_n)_0/p$ induces an isomorphism on cohomology, then Conjecture 5.3.16 follows formally.

As we will see in Section 5.6.1, this conjecture and the accompanying computations inform our understanding of $L_{K(n)}S^0$, the essence of which is distilled in the formulation of the chromatic splitting conjecture. In fact, what makes Conjecture 5.3.16 particularly appealing is the fact that $H_c^*(\mathbb{G}_n, \mathbb{W}_n)$ appears to be rather simple when p is large with respect to n.

Rationally, we have some partial understanding due to work of Lazard [81, Chapter V, (2.4)] and [97, Remark 2.2.5], who established an isomorphism for all heights and primes

$$H_c^*(\mathbb{G}_n, \mathbb{W}_n) \otimes \mathbb{Q} \cong \Lambda_{\mathbb{Z}_p}(x_1, \ldots, x_n) \otimes \mathbb{Q}, \tag{5.3.14}$$

where $\Lambda_{\mathbb{Z}_p}(x_1, \ldots, x_n)$ is the exterior algebra over $\mathbb{Z}_p$ on n generators in degrees $\deg(x_i) = 2i-1$. Here, the class x_1 is ζ_n as defined in (5.3.9). Furthermore, when p is large with respect to n, it is believed that there is an isomorphism (5.3.14) before rationalization.

Conjecture 5.3.18. *If $p \gg 0$, then $H_c^*(\mathbb{G}_n, \mathbb{W}_n) \cong \Lambda_{\mathbb{Z}_p}(x_1, \ldots, x_n)$.*

Remark 5.3.19. For our chromatic applications, we need a mild extension of the setup presented above. Here and below, $\mathbb{W}_n = W(\mathbb{F}_{p^n})$ denotes the $\mathbb{G}_n$-module whose action is the restriction along $\mathbb{G}_n \to \mathrm{Gal}$ of the natural action of Gal on $W(\mathbb{F}_{p^n})$. We write $w^g = g(w)$ for this action. For $G \subseteq \mathbb{G}_n$, define the *twisted group ring* to be

$$\mathbb{W}_n\langle\langle G\rangle\rangle := \lim_{i,j} \mathbb{W}_n/\mathfrak{m}^j\langle G_i\rangle$$

with G-twisted multiplication determined by the relations $g \cdot r = g(r)g$ for $r \in \mathbb{W}_n$ and $g \in G$. We let $\mathrm{Mod}_{\mathbb{W}_n\langle\langle G\rangle\rangle}$ be the category of profinite left $\mathbb{W}_n\langle\langle G\rangle\rangle$-modules. These are profinite abelian groups $M = \lim_k M_k$ with a continuous action $\mathbb{W}_n\langle\langle G\rangle\rangle \times M \to M$. If $H \subseteq G$ is a closed subgroup and M is a left $\mathbb{W}_n\langle\langle H\rangle\rangle$-module, then

$$M\uparrow_H^G := \mathbb{W}_n\langle\langle G\rangle\rangle \otimes_{\mathbb{W}_n\langle\langle H\rangle\rangle} M = \lim_{i,j,k} \left(\mathbb{W}_n/\mathfrak{m}^j[G_i] \otimes_{\mathbb{W}_n\langle\langle H\rangle\rangle} M_k\right) \tag{5.3.15}$$

is a left $\mathbb{W}_n\langle\langle G\rangle\rangle$-module.

One can show that the homological algebra summarized above also works in the context of profinite modules over twisted group rings. Note that there is a functor from $\mathrm{Mod}_{\mathbb{Z}_p[\![G]\!]}$ to $\mathrm{Mod}_{\mathbb{W}_n\langle\langle G\rangle\rangle}$ that sends a $\mathbb{Z}_p[\![G]\!]$-module M to the $\mathbb{W}_n\langle\langle G\rangle\rangle$-module $\mathbb{W}_n \otimes_{\mathbb{Z}_p} M$ with action given by $g(w \otimes m) = w^g \otimes g(m)$. This allows us to transport constructions in $\mathrm{Mod}_{\mathbb{Z}_p\langle\langle G\rangle\rangle}$ to constructions in $\mathrm{Mod}_{\mathbb{W}_n\langle\langle G\rangle\rangle}$.

We now come to another important construction in chromatic homotopy theory, namely the $(E_n)_*$-module

$$(E_n)_*^\vee X = \pi_* L_{K(n)}(E_n \wedge X)$$

associated to a spectrum X. The action of $\mathbb{G}_n$ on E_n induces an action on $(E_n)_*^\vee X$ compatible with the $(E_n)_*$-action. Moreover, let $\mathfrak{m} = (p, u_1, \ldots, u_{n-1})$ be the maximal ideal of $(E_n)_0$ and, for $s \geq 0$, let $\mathbb{L}_s$ be the sth left derived functor of $\mathfrak{m}$-adic completion on $\mathrm{Mod}_{(E_n)_*}$. There is a strongly convergent spectral sequence

$$\mathbb{L}_s(\pi_*(E_n \wedge X))_t \Longrightarrow (E_n)^\vee_{s+t}X$$

which in particular implies that the canonical map $(E_n)^\vee_* X \to \mathbb{L}_0((E_n)^\vee_* X)$ is an isomorphism. Such $(E_n)_*$-modules are called $\mathbb{L}$-complete and we refer the interested reader to [76, Appendix A] for a more thorough treatment. Taken together, this structure is called the *Morava module of X*:

Definition 5.3.20 (Morava modules). A *Morava module M* is an $\mathbb{L}$-complete $(E_n)_*$-module equipped with an action by $\mathbb{G}_n$ in $\mathbb{L}$-complete modules that is compatible with the action on $(E_n)_*$. That is, for every $g \in \mathbb{G}_n$, $e \in (E_n)_*$ and $m \in M$, $g(em) = g(e)g(m)$. A morphism of Morava modules is a continuous map of $(E_n)_*$-modules that preserves the action. We denote the category of Morava modules by $\mathrm{Mod}^{\mathbb{G}_n}_{(E_n)_*}$.

By the discussion above, $(E_n)^\vee_* X$ is a Morava module for any spectrum X and we obtain a functor

$$(E_n)^\vee_*(-) := \pi_* L_{K(n)}(E_n \wedge -)\colon \mathrm{Sp} \longrightarrow \mathrm{Mod}^{\mathbb{G}_n}_{(E_n)_*}. \tag{5.3.16}$$

This functor detects and reflects $K(n)$-local equivalences, but has the advantage that $(E_n)^\vee_*(-)$ comes equipped with an action of $\mathbb{G}_n$. This extra structure proves to be extremely powerful for computations, and is one of the reasons why Morava modules play a central role in the field.

For more information on Morava modules, we refer the reader to [29, Section 1.3] and [49, Section 2], noting that authors often simply write $(E_n)_*X = \pi_* L_{K(n)}(E_n \wedge X)$ as opposed to the non-completed homology $(E_n)_*X = \pi_*(E_n \wedge X)$, but we will not do so here. Note also that, if X is finite, then $(E_n)_*X \cong (E_n)^\vee_* X$.

Remark 5.3.21. For F a finite subgroup of $\mathbb{G}_n$, the action of $\mathbb{G}_n$ on $(E_n)_*$ restricts to an action of F. We can also consider the category $\mathrm{Mod}^F_{(E_n)_*}$ of $\mathbb{L}$-complete $(E_n)_*$-modules equipped with an action of F. Then $(E_n)_*$ is periodic as an object in $\mathrm{Mod}^F_{(E_n)_*}$ since the element $N = \prod_{g \in F} g(u)$ for $u \in (E_n)_{-2}$ as in (5.3.1) is an invariant unit. Let d_F^{alg} be the smallest integer such that $(E_n)_* \cong (E_n)_{*+d_F^{\mathrm{alg}}}$ in $\mathrm{Mod}^F_{(E_n)_*}$. This leads to an isomorphism of Morava modules

$$(E_n)^\vee_* E_n^{hF} \cong \mathrm{Map}^c(\mathbb{G}_n/F, (E_n)_*) \cong \mathrm{Hom}^c_{\mathbb{W}_n}(\mathbb{W}_n\!\uparrow_F^{\mathbb{G}_n}, (E_n)_*)$$

closely related to (5.3.3) and it implies that

$$(E_n)^\vee_* E_n^{hF} \cong (E_n)^\vee_* \Sigma^{d_F^{\mathrm{alg}}} E_n^{hF}.$$

However, E_n^{hF} need not be equivalent to $\Sigma^{d_F^{\mathrm{alg}}} E_n^{hF}$. Nonetheless, because of the strong vanishing line discussed in Remark 5.3.15, some power of N is a permanent cycle and gives rise to a periodicity generator for E_n^{hF}, so for some multiple d_F^{top} of d_F^{alg}, there is an equivalence $E_n^{hF} \simeq \Sigma^{d_F^{\mathrm{top}}} E_n^{hF}$.

For example, at $p = 2$, E_1 is 2-complete complex K-theory and $E_1^{hC_2}$ is the 2-complete real K-theory spectrum KO. We have:

$$K^\vee_* KO \cong K^\vee_* \Sigma^4 KO, \qquad KO \not\simeq \Sigma^4 KO, \qquad KO \simeq \Sigma^8 KO.$$

5.4 $K(1)$-local homotopy theory

In this section, we tell a part of the chromatic story at height $n = 1$ as a warm up for the more complicated ideas needed to study higher heights. The contents of this section are classical and can be found in various forms throughout the literature, for example, Adams and Baird [2], Bousfield [30, 31], Ravenel [105, Theorem 8.10, 8.15]. See [58, Section 6] for a more recent treatment, and [23, Section 4] for more details on the case $p = 2$.

5.4.1 Morava E-theory and the stabilizer group at $n = 1$

At height $n = 1$, Morava E-theory is the p-completed complex K-theory spectrum, which we simply denote by K. There is an isomorphism $K_* \cong \mathbb{Z}_p[u^{\pm 1}]$ for a unit $u \in K_{-2}$ which can be chosen so that $u^{-1} \in K_2$ is the Bott element.

In this case, $\mathbb{G}_1 = \mathbb{S}_1 \cong \mathbb{Z}_p^\times$ corresponds to the p-completed Adams operations. The action of $\mathbb{S}_1$ on K_* is the $\mathbb{Z}_p$-algebra isomorphism determined by

$$\alpha_*(u) = \alpha u \tag{5.4.1}$$

for $\alpha \in \mathbb{Z}_p^\times$. The keen reader will notice that this is the inverse of the action of the Adams operations, which is given by $\alpha_*(u) = \alpha^{-1} u$. This comes from switching a right action to a left action.

The map $L_{K(1)}S^0 \to K^{h\mathbb{Z}_p^\times}$ of (5.3.2) is a $\mathbb{Z}_p^\times$ pro-Galois extension. We use this extension to compute the homotopy groups of $\pi_* L_{K(1)}S^0$. One can take the direct approach of computing the spectral sequence of (5.3.4)

$$E_2^{s,t} = H_c^s(\mathbb{Z}_p^\times, K_t) \Longrightarrow \pi_{t-s} L_{K(1)}S^0. \tag{5.4.2}$$

In fact, this spectral sequence collapses at the E_2-page at odd primes and at the E_4-page at the prime 2. This is not a hard computation, but we take a different path in order to illustrate the finite resolution philosophy.

5.4.2 Finite resolution at height $n = 1$

Here, we describe our first example of a finite resolution. Let C_m denote a cyclic group of order m, $\mu = C_2$ if $p = 2$, and $\mu = C_{p-1}$ if p is odd. Then, $\mathbb{Z}_p^\times \cong \mu \times \mathbb{Z}_p$, where the $\mathbb{Z}_p$ corresponds to the subgroup of units congruent to 1 modulo p if p is odd, and to those congruent to 1 modulo 4 is $p = 2$. We let ψ be a topological generator for this factor of $\mathbb{Z}_p$. The notation is meant to be reminiscent of the Adams operations. We will make a choice for ψ below in (5.4.8).

Remark 5.4.1. The spectrum $K^{hC_{p-1}}$ is the unit component in the splitting of the p-completed complex K-theory spectrum K into Adams summands if p is odd, and K^{hC_2} is the 2-completed real K-theory spectrum if $p = 2$.

The $K(1)$-local sphere can be obtained by an iterated fixed points construction:

$$L_{K(1)}S^0 \simeq K^{h\mathbb{Z}_p^\times} \simeq (K^{h\mu})^{h\mathbb{Z}_p}.$$

Since $\psi \in \mathbb{Z}_p$ is a topological generator, taking homotopy fixed point with respect to $\mathbb{Z}_p$ is equivalent to taking the homotopy fiber of the map $\psi - 1$. Therefore, there is a fiber sequence

$$L_{K(1)}S^0 \longrightarrow K^{h\mu} \xrightarrow{\psi-1} K^{h\mu} \longrightarrow \Sigma L_{K(1)}S^0 . \tag{5.4.3}$$

This is a *finite resolution* of $L_{K(1)}S^0$ as will be defined in Definition 5.5.1 below.

To construct finite resolutions at higher heights where the structure of the Morava stabilizer group is more intricate, we start by attacking the problem in algebra and then we transfer algebraic constructions to topology. We give a quick overview of how this would happen at height 1 to give the reader something to think of while reading Section 5.5.

Step 1: Algebraic resolution. The group $\mathbb{Z}_p$ is topologically free of rank one and there is an exact sequence of left $\mathbb{Z}_p^\times$-modules

$$0 \longrightarrow \mathbb{Z}_p\uparrow_\mu^{\mathbb{Z}_p^\times} \xrightarrow{\psi-1} \mathbb{Z}_p\uparrow_\mu^{\mathbb{Z}_p^\times} \longrightarrow \mathbb{Z}_p \cong \mathbb{Z}_p\uparrow_{\mathbb{Z}_p^\times}^{\mathbb{Z}_p^\times} \longrightarrow 0. \tag{5.4.4}$$

Here, $\mathbb{Z}_p[\![\mathbb{Z}_p^\times]\!] = \lim_{i,j} \mathbb{Z}/p^i[(\mathbb{Z}/p^j)^\times]$ is the completed group ring, which was discussed in Section 5.3.2.3, and $\mathbb{Z}_p\uparrow_\mu^{\mathbb{Z}_p^\times} \cong \mathbb{Z}_p[\![\mathbb{Z}_p^\times]\!] \otimes_{\mathbb{Z}_p[\mu]} \mathbb{Z}_p$. This is a projective resolution of $\mathbb{Z}_p$ as a $\mathbb{Z}_p^\times$-module if and only if $p > 2$. See Remark 5.4.5 below on this point. Applying $\mathrm{Hom}^c_{\mathbb{Z}_p}(-, K_*)$ to (5.4.4) gives a short exact sequence of Morava modules

$$K_* \longrightarrow \mathrm{Hom}^c_{\mathbb{Z}_p}(\mathbb{Z}_p\uparrow_\mu^{\mathbb{Z}_p^\times}, K_*) \xrightarrow{\mathrm{Hom}^c_{\mathbb{Z}_p}(\psi-1,K_*)} \mathrm{Hom}^c_{\mathbb{Z}_p}(\mathbb{Z}_p\uparrow_\mu^{\mathbb{Z}_p^\times}, K_*) . \tag{5.4.5}$$

Step 2: Topological Resolution. The second step is to prove that the algebraic resolution has a topological realization. More precisely, (5.4.5) is an exact sequence in the category of Morava modules $\mathrm{Mod}^{\mathbb{G}_1}_{(E_1)_*}$. As was described in (5.3.16), there is a functor

$$K_*^\vee(-) = (E_1)_*^\vee(-)\colon \mathrm{Sp} \longrightarrow \mathrm{Mod}^{\mathbb{G}_1}_{(E_1)_*}.$$

When we have an algebraic resolution of length 1, a *topological realization of* (5.4.4) is a choice of fiber sequence in the category of $K(1)$-local spectra

$$\mathcal{E}_{-1} \xrightarrow{\delta_{-1}} \mathcal{E}_0 \xrightarrow{\delta_0} \mathcal{E}_1 \tag{5.4.6}$$

where $\mathcal{E}_0$ and $\mathcal{E}_1$ are finite wedges of suspensions of spectra of the form K^{hF} for $F \subseteq \mathbb{G}_1$ a finite subgroup, such that, up to isomorphism of complexes, (5.4.5) is the complex of

Morava modules obtained from (5.4.6) by applying $K_*^\vee(-)$. If $\mathcal{E}_{-1} = L_{K(1)}S^0$, then this is an example of a finite resolution of the $K(1)$-local sphere.

Remark 5.4.2. The case when the algebraic resolution has length $d \geq 1$ is discussed in the next section. We will see in Definition 5.5.1 that the definition of a finite topological resolution of length greater than 1 is more subtle. See also Example 5.5.2.

There is no algorithm for finding a topological realization. A priori, one may not exist, and if it does, it may not be unique. Without a priori knowledge of the existence of (5.4.3), the key observations for finding a topological realization of (5.4.4) are

- the isomorphism of Morava modules

$$K_*^\vee K^{h\mu} \cong \mathrm{Map}^c(\mathbb{Z}_p^\times/\mu, K_*) \cong \mathrm{Hom}^c_{\mathbb{Z}_p}(\mathbb{Z}_p\uparrow_\mu^{\mathbb{Z}_p^\times}, K_*),$$

  and

- the fact that $\mathrm{Hom}^c_{\mathbb{Z}_p}(\psi - 1, K_*) = K_*^\vee(\psi - 1)$.

Knowing these facts, (5.4.5) can be identified with the short exact sequence of Morava modules

$$K_* \longrightarrow K_*^\vee K^{h\mu} \xrightarrow{\quad K_*^\vee(\psi-1) \quad} K_*^\vee K^{h\mu}\ . \tag{5.4.7}$$

Given this, we let $\mathcal{E}_{-1} = L_{K(1)}S^0$, $\mathcal{E}_0 = \mathcal{E}_1 = K^{h\mu}$. We let δ_{-1} be the unit and δ_0 be $\psi - 1$. It follows that the fiber sequence

$$L_{K(1)}S^0 \longrightarrow K^{h\mu} \xrightarrow{\quad \psi-1 \quad} K^{h\mu}$$

is a topological realization as it gives rise to (5.4.5) under the functor $K_*^\vee(-)$. This is our first example of a finite resolution of $L_{K(1)}S^0$.

Remark 5.4.3. We *did* make choices here and different choices could have given a different topological realization. For example, for $p = 2$, $K^{hC_2} \simeq KO$ and $K_*^\vee KO \cong K_*^\vee \Sigma^4 KO$, yet $KO \not\simeq \Sigma^4 KO$. In fact, we could have constructed a topological realization using $\Sigma^4 KO$ instead of KO. Such a resolution is described below in (5.6.2). The resolution described there is a topological realization of the algebraic resolution (5.4.4), but it is not a finite resolution of the sphere as $\mathcal{E}_{-1} = P_1 \not\simeq L_{K(1)}S^0$.

5.4.3 Homotopy groups and chromatic reassembly

The long exact sequence on homotopy groups associated to (5.4.3) allows one to compute $\pi_*L_{K(1)}S^0$ from $\pi_*K^{h\mu}$ and knowledge of the action of ψ. The homotopy groups of $K^{h\mu}$ are computed using the homotopy fixed point spectral sequence

$$E_2^{s,t} \cong H^s(\mu, \pi_t K) \Longrightarrow \pi_{t-s}K^{h\mu}.$$

Recall that $\mu = C_{p-1}$ if p is odd and C_2 if $p = 2$. So computing group cohomology with coefficients in $K_* = \mathbb{Z}_p[u^{\pm 1}]$ is not so bad given the explicit formula (5.4.1). We get

$$H^*(\mu, \pi_* K) \cong \begin{cases} \mathbb{Z}_p[u^{\pm(p-1)}] & p \neq 2 \\ \mathbb{Z}_2[\eta, u^{\pm 2}]/(2\eta) & p = 2, \end{cases}$$

where the (s,t) bidegree of η is $(1,2)$. The element η detects the Hopf map in $\pi_1 S^0$. For p odd, the spectral sequence collapses for degree reasons. For $p = 2$, the fact that $\eta^4 = 0$ in $\pi_* S^0$ implies a differential $d_3(u^{-2}) = \eta^3$, and the spectral sequence collapses at E_4 for degree reasons. So, we have

$$\pi_* K^{h\mu} \cong \begin{cases} \mathbb{Z}_p[\beta^{\pm 1}] & p \neq 2,\ |\beta| = 2(p-1) \\ \mathbb{Z}_p[\eta, \alpha, \beta^{\pm 1}]/(2\eta, \eta^3, \alpha^2 - 4\beta) & p = 2,\ |\eta| = 1,\ |\alpha| = 4,\ |\beta| = 8. \end{cases}$$

If p is odd, $\beta \in \pi_{2(p-1)}$ is detected by u^{1-p}. If $p = 2$, $\eta \in \pi_1$ is detected by the same-named class on the E_2-page, $\alpha \in \pi_4$ is detected by $2u^{-2}$ and $\beta \in \pi_8$ is detected by u^{-4}.

Remark 5.4.4. The differential $d_3(u^{-2}) = \eta^3$ can be obtained as a consequence of the *slice differentials theorem* [63, Theorem 9.9]. This is overkill for this particular example which follows from classical considerations. However, we mention this here since the slice differentials theorem also implies differentials at higher heights in spectral sequences computing $\pi_* E_n^{hF}$ for finite subgroups $F \subseteq \mathbb{G}_n$.

Now, we turn to computing the long exact sequence on homotopy groups associated to (5.4.3). Choose an element ψ of $\mathbb{Z}_p^\times$ that satisfies

$$\psi^{-1} = \begin{cases} (1+p) & p \neq 2 \\ 5 & p = 2. \end{cases} \tag{5.4.8}$$

There are other possible choices: One could choose any element in $\mathbb{Z}_p^\times$ such that the image of ψ^{-1} in $\mathbb{Z}_p^\times/\mu$ is a topological generator. The outcome of these calculations is independent of the choice.

From (5.4.1), we deduce that the action of ψ is then given by

$$\psi_*(\beta) = \begin{cases} (1+p)^{p-1}\beta & p \neq 2 \\ 5^4\beta & p = 2 \end{cases}, \qquad \psi_*(\alpha) = 5^2\alpha, \qquad \psi_*(\eta) = \eta.$$

Let $v_p(k)$ denote the p-adic valuation of $k \in \mathbb{Z}$. For p odd, the long exact sequence on homotopy groups gives

$$\pi_* L_{K(1)} S^0 = \begin{cases} \mathbb{Z}_p & * = 0, -1 \\ \mathbb{Z}/p^{v_p(k)+1} & * = 2k(p-1) - 1 \\ 0 & \text{otherwise.} \end{cases}$$

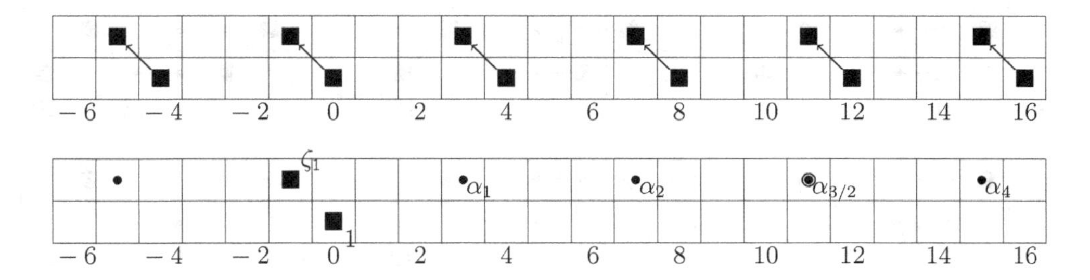

FIGURE 5.4.1. The long exact sequence on homotopy groups associated to (5.4.3) and computing $\pi_* L_{K(1)} S^0$ at $p = 3$. It is drawn as a spectral sequence in Adams grading $(t-s, s)$ with $E_1^{s,t} \cong \pi_t K^{h\mu}$ for $s = 0, 1$. The arrows denote the d_1-differentials, $d_1 \colon E_1^{0,t} \to E_1^{1,t}$, which is just the connecting homomorphism. The top chart is the E_1-term and the bottom chart the E_∞-term. A ■ is a $\mathbb{Z}_3$, a • a $\mathbb{Z}/3$ and a ◉ a $\mathbb{Z}/9$.

This is depicted in Figure 5.4.1 for $p = 3$. For $p = 2$, we have

$$\pi_* L_{K(1)} S^0 = \begin{cases} \mathbb{Z}_2 & * = -1 \\ \mathbb{Z}_2 \oplus \mathbb{Z}/2 & * = 0 \\ \mathbb{Z}/2 & * = 0, 2 \mod 8, * \neq 0 \\ \mathbb{Z}/2 \oplus \mathbb{Z}/2 & * = 1 \mod 8 \\ \mathbb{Z}/8 & * = 3 \mod 8 \\ \mathbb{Z}/2^{v_2(k)+4} & * = -1 + 8k, k \neq 0 \\ 0 & * = 4, 5, 6 \mod 8. \end{cases}$$

This is depicted in Figure 5.4.2. One has to argue that there is no additive extension in degrees 1 mod 8 but we do not do this here.

Remark 5.4.5. The dichotomy between $p = 2$ and odd primes in the computations is an instance of the general phenomena which was discussed in Section 5.3.2.3 and is revisited in Section 5.6.3 below. That is, when p is large with respect to n, chromatic homotopy theory becomes algebraic (see for example Corollary 5.3.13). On the other hand, when p is small the stabilizer group $\mathbb{S}_n$ might contain p-torsion and this appears to reflect interesting topological phenomena. Here, $\mathbb{S}_1 \cong \mathbb{Z}_2^\times$ contains 2-torsion at $p = 2$ and there are differentials in the spectral sequence computing the homotopy groups of $KO \simeq K^{hC_2}$, a much more intricate spectrum than the Adams summand $K^{hC_{p-1}}$ at odd primes whose homotopy fixed point spectral sequence collapses at the E_2-page.

Remark 5.4.6. The $\mathbb{Z}_p$ summand in $\pi_{-1} L_{K(1)} S^0$ is generated by the image of the composite $S^0 \to K^{h\mu} \to \Sigma^{-1} L_{K(1)} S^0$ where the first map is the unit and the second is the connecting homomorphism of (5.4.3). We call this map and the homotopy class it represents $\zeta_1 \in \pi_{-1} L_{K(1)} S^0$. It is detected in (5.4.2) by the same-named class

$$\zeta_1 \in H_c^1(\mathbb{Z}_p^\times, K_0) \cong \mathrm{Hom}^c(\mathbb{Z}_p^\times, \mathbb{Z}_p)$$

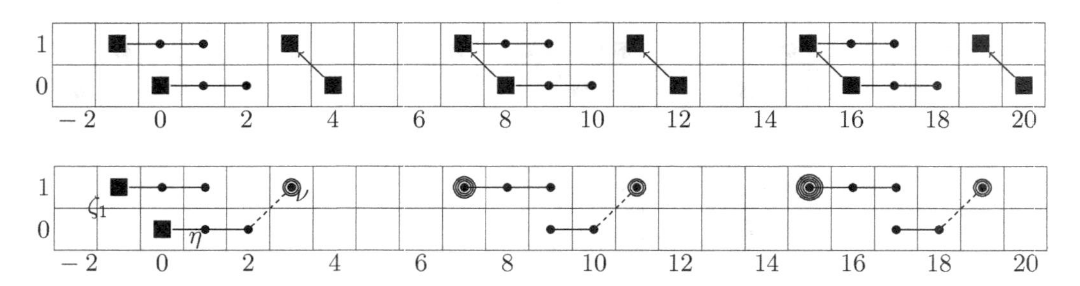

FIGURE 5.4.2. The long exact sequence on homotopy groups associated to (5.4.3) and computing $\pi_* L_{K(1)} S^0$ at $p = 2$. It is drawn using the same convention as in Figure 5.4.1, except that a ■ is a $\mathbb{Z}_2$, a • a $\mathbb{Z}/2$, a ◉ a $\mathbb{Z}/4$, etc. Dashed arrows denote exotic multiplications by η.

corresponding to the projection $\mathbb{Z}_p^\times \to \mathbb{Z}_p^\times/\mu \cong \mathbb{Z}_p$. See (5.3.9) and Remark 5.3.8 for analogues at higher heights.

Remark 5.4.7. An easy computation that will be relevant later is that of $\pi_*(L_{K(1)} S^0/p)$ for p odd. The descent spectral sequence

$$H_c^s(\mathbb{Z}_p^\times, K_t/p) \Longrightarrow \pi_{t-s}(L_{K(1)} S^0/p).$$

collapses with no extensions and

$$\pi_*(L_{K(1)} S^0/p) \cong \mathbb{F}_p[v_1^{\pm 1}] \otimes \Lambda_{\mathbb{F}_p}(\zeta_1),$$

where $v_1 = u^{1-p}$. We abuse notation by denoting the composite $\mathbb{Z}_p^\times \xrightarrow{\zeta_1} \mathbb{Z}_p^\times/\mu \cong \mathbb{Z}_p \to \mathbb{Z}/p$ also by $\zeta_1 \in H_c^1(\mathbb{Z}_p^\times, K_0/p)$.

Finally, we turn to the problem of chromatic reassembly at height $n = 1$. The chromatic fracture square (5.2.7) in this case gives

$$\begin{array}{ccccc} F_1 & \longrightarrow & L_1 S^0 & \longrightarrow & L_{K(1)} S^0 \\ \simeq\downarrow & & \downarrow & & \downarrow \\ F_1 & \longrightarrow & L_0 S_p^0 & \longrightarrow & L_0 L_{K(1)} S^0 \end{array}$$

where F_1 is the fiber of the horizontal maps. In particular, it is the fiber of the map $L_0 S_p^0 \to L_0 L_{K(1)} S^0$ induced by the unit. Since L_0 is rationalization, there is an isomorphism $\pi_* L_0 L_{K(1)} S^0 \cong p^{-1} \pi_* L_{K(1)} S^0$. From the above calculations, we see that the map $1 \vee \zeta_1 \colon S^0 \vee S^{-1} \to L_{K(1)} S^0$ induces an equivalence

$$L_0 L_{K(1)} S^0 \simeq L_0 S_p^0 \vee L_0 S_p^{-1}. \tag{5.4.9}$$

In particular, $L_0 S_p^0 \to L_0 L_{K(1)} S^0$ is split and $\Sigma F_1 \simeq L_0 S_p^{-1}$. This proves the *strong chromatic splitting conjecture* for $n = 1$, which will be stated in general in Section 5.6.1.

We get the following diagram from the long exact sequence on homotopy groups associated to the fiber sequence $L_1 S_p^0 \to L_{K(1)} S^0 \to \Sigma F_1 \simeq L_0 S_p^{-1}$:

$$\begin{array}{ccccccccc}
\pi_{-1} L_1 S_p^0 & \longrightarrow & \pi_{-1} L_{K(1)} S^0 & \longrightarrow & \pi_{-2} F_1 & \longrightarrow & \pi_{-2} L_1 S_p^0 & \longrightarrow & \pi_{-2} L_{K(1)} S^0 \\
\| & & \| & & \| & & \| & & \| \\
0 & \longrightarrow & \mathbb{Z}_p & \longrightarrow & \mathbb{Q}_p & \longrightarrow & \mathbb{Q}_p/\mathbb{Z}_p & \longrightarrow & 0.
\end{array}$$

Piecing the rest of the long exact sequence on homotopy groups together gives

$$\pi_* L_1 S_p^0 \cong \mathbb{Z}_p \oplus \Sigma^{-2} \mathbb{Q}_p/\mathbb{Z}_p \oplus \operatorname{Tor}(\pi_* L_{K(1)} S^0),$$

where the $\mathbb{Z}_p$ is in degree 0 and comes from the summand $\mathbb{Z}_p \subseteq \pi_0 L_{K(1)} S^0$, this inclusion being an isomorphism when p is odd. The summand $\mathbb{Q}_p/\mathbb{Z}_p$ is in degree -2 and Tor denotes the torsion subgroup.

In the next sections, we review these topics at higher heights. While we are not able to do such an explicit analysis for $n \geq 2$, the tools and ideas described above do generalize and we give an overview of some of the techniques available to study the $K(n)$-local category and the $K(n)$-local sphere.

5.5 Finite resolutions and their spectral sequences

We now describe a *recipe* for the construction of finite resolutions of the $K(n)$-local sphere. We note that every step of this procedure requires hard work specific to the height and the prime. We then illustrate the general formalism with many examples at height $n = 2$ in Section 5.5.5. Some applications of these finite resolutions will then be discussed in the next section on the chromatic splitting conjecture and local Picard groups. References for this material are [49, 57, 58].

5.5.1 What is a finite resolution?

Definition 5.5.1. A *finite resolution of* $L_{K(n)} S^0$ of length d is a diagram

$$\begin{array}{ccccccccc}
L_{K(n)} S^0 = X_d & \xrightarrow{i_d} & X_{d-1} & \xrightarrow{i_{d-1}} & \cdots & \xrightarrow{i_2} & X_1 & \xrightarrow{i_1} & X_0 \\
\uparrow{\scriptstyle j_d} & & \uparrow{\scriptstyle j_{d-1}} & & & & \uparrow{\scriptstyle j_1} & & \uparrow{\scriptstyle j_0}\,\simeq \\
\Sigma^{-d} \mathcal{E}_d & & \Sigma^{-(d-1)} \mathcal{E}_{d-1} & & & & \Sigma^{-1} \mathcal{E}_1 & & \mathcal{E}_0
\end{array} \tag{5.5.1}$$

in the $K(n)$-local category such that

(a) the sequences

$$\Sigma^{-k} \mathcal{E}_k \xrightarrow{j_k} X_k \xrightarrow{i_k} X_{k-1} \xrightarrow{\ell_k} \Sigma^{-k+1} \mathcal{E}_k \tag{5.5.2}$$

are exact triangles, and

(b) the $\mathcal{E}_k$s are finite wedges of suspensions of spectra of the form E_n^{hF} for finite subgroups F of $\mathbb{G}_n$.

In other words, a finite resolution is a tower of fibrations resolving $L_{K(n)}S^0$ by spectra of the form E_n^{hF} in a finite number of steps using a finite number of pieces. Typically, $d = n^2$. Note that the tower (5.5.1) gives a diagram

$$\begin{array}{ccccccccc} L_{K(n)}S^0 & \xrightarrow{\delta_0} & \mathcal{E}_0 & \xrightarrow{\delta_1} & \mathcal{E}_1 & \xrightarrow{\delta_2} & \mathcal{E}_2 \longrightarrow \cdots & \xrightarrow{\delta_d} & \mathcal{E}_d \\ & & \simeq \downarrow j_0 & \nearrow \ell_1 & \downarrow \Sigma j_1 & \nearrow \Sigma\ell_2 & & & \downarrow \Sigma^d j_d \\ & & X_0 & & \Sigma X_1 & & & & \Sigma^d L_{K(n)}S^0 \end{array} \tag{5.5.3}$$

where δ_0 is defined so that $j_0\delta_0 = i_1 \dots i_d$. We often denote the finite resolution by the top line of this diagram.

We can also smash (5.5.1) (in the $K(n)$-local category) with a spectrum Y to obtain a tower of fibrations resolving $L_{K(n)}Y$.

For any $X \in \mathrm{Sp}$, a resolution of the form (5.5.1) gives rise to a strongly convergent spectral sequence

$$E_1^{s,t} = [X, \mathcal{E}_s \wedge Y]_t \Longrightarrow [X, L_{K(n)}Y]_{t-s},$$

with differentials $d_r \colon E_r^{s,t} \to E_r^{s+r,t+r-1}$ that collapses at the E_{d+1}-page. There is also a similar spectral sequence computing $[L_{K(n)}Y, X]$.

Example 5.5.2. The proto-example of such a resolution is the resolution (5.4.3). Recall that E_1 is p-completed K-theory and let μ be as in Section 5.4. The fiber sequence (5.4.3) can be rearranged into a (very short) tower of fibrations

$$\begin{array}{ccc} L_{K(1)}S^0 & \xrightarrow{i_1} & E_1^{h\mu} = X_0 \\ j_1 \uparrow & & j_0 \uparrow \simeq \\ \Sigma^{-1}E_1^{h\mu} = \Sigma^{-1}\mathcal{E}_1 & & E_1^{h\mu} = \mathcal{E}_0. \end{array}$$

In this case, the associated Bousfield–Kan spectral sequence degenerates to the long exact sequence on homotopy groups.

For the rest of this section, we give an overview of how such resolutions are constructed. Note that the art of building finite resolutions has evolved in the last fifteen years. For a long time, the role of the Galois group was not as clear as it has become recently in the work of Henn in [59], so we give a revised recipe here.

5.5.2 Algebraic resolutions

In practice, the first step to constructing a finite topological resolution is to construct its algebraic "reflection". These are the finite algebraic resolution. In practice, experts do not work from a definition, but rather know a finite algebraic resolution when they see one. Because of this, we give the following loose *description* as opposed to *definition.*

Description 5.5.3. *A* finite algebraic resolution *of length d is an exact sequence*

$$0 \longrightarrow \mathcal{C}_d \xrightarrow{\partial_d} \mathcal{C}_{d-1} \xrightarrow{\partial_{d-1}} \cdots \xrightarrow{\partial_1} \mathcal{C}_0 \xrightarrow{\partial_0} \mathcal{C}_{-1} = \mathbb{W}_n \longrightarrow 0, \tag{5.5.4}$$

where the $\mathcal{C}_k$s are $\mathbb{W}_n\langle\!\langle \mathbb{G}_n \rangle\!\rangle$-modules that have the property that, for some $\mathcal{E}_k$ as in Definition 5.5.1 (b), there is an isomorphism

$$(E_n)^\vee_* \mathcal{E}_k \cong \mathrm{Hom}^c_{\mathbb{W}_n}(\mathcal{C}_k, (E_n)_*). \tag{5.5.5}$$

Roughly, a topological resolution *realizes* an algebraic topological resolution if there is an isomorphism of exact sequences

$$\mathrm{Hom}^c_{\mathbb{W}_n}(\mathcal{C}_\bullet, (E_n)_*) \cong (E_n)^\vee_*(\mathcal{E}_\bullet).$$

Here $\mathcal{C}_\bullet$ is as in (5.5.4) and $\mathcal{E}_\bullet$ is the top row of (5.5.3). In this sense, the algebraic resolution is a "reflection" of the topological resolution.

Remark 5.5.4. Recall from (5.3.15) that $M\uparrow_F^{\mathbb{G}_n} = \mathbb{W}_n\langle\!\langle \mathbb{G}_n \rangle\!\rangle \otimes_{\mathbb{W}_n\langle F\rangle} M$. Typical examples for the modules $\mathcal{C}_k$ are among the following:

- If $\mathcal{C}_k$ is a direct sum of modules of the form $\mathbb{W}_n\uparrow_F^{\mathbb{G}_n}$ for F a finite subgroup of $\mathbb{G}_n$, then $\mathcal{C}_k$ satisfies (5.5.5). Indeed, it was mentioned in (5.3.21) that for any $m \in \mathbb{Z}$ and F a finite subgroup of $\mathbb{G}_n$, there are isomorphisms

$$\mathrm{Hom}^c_{\mathbb{W}_n}(\mathbb{W}_n\uparrow_F^{\mathbb{G}_n}, (E_n)_*) \cong \mathrm{Map}^c(\mathbb{G}_n/F, (E_n)_*) \cong (E_n)^\vee_* \Sigma^{md_F^{\mathrm{alg}}} E_n^{hF}$$

- By a *character* χ of $\mathbb{W}_n\langle F\rangle$, we will mean a $\mathbb{W}_n\langle F\rangle$-module that has rank one over $\mathbb{W}_n$. Suppose that χ is a summand (as a $\mathbb{W}_n\langle F\rangle$-module) in $\mathbb{W}_n\langle F\rangle$ and that e_χ is an idempotent of $\mathbb{W}_n\langle F\rangle$ that picks up χ. Let E_n^χ be wedge summand of E_n associated to this idempotent, obtained as the telescope on $e_\chi\colon E_n \to E_n$. Then,

$$(E_n)^\vee_* E_n^\chi \cong \mathrm{Hom}^c_{\mathbb{W}_n}(\chi\uparrow_F^{\mathbb{G}_n}, (E_n)_*).$$

  In existing examples, some of the summands of the terms $\mathcal{C}_k$s are built out of projective characters χ of $\mathbb{W}\langle F\rangle$, such that E_n^χ is a suspension of E_n^{hF}. See, for example, [49, Section 5] and Section 5.5.5 below.

Remark 5.5.5. One reason for using $\mathbb{W}_n$-coefficients (which don't seem to play a role in the topological story) rather than $\mathbb{Z}_p$-coefficients in these constructions is that, if p divides n, $\mathbb{G}_n$ is "cohomologically larger" than $\mathbb{S}_n$ over $\mathbb{Z}_p$, but not over $\mathbb{W}_n$ since the later is free over Gal. So, if one wants to construct a resolution of length n^2 for $L_{K(n)}S^0 \simeq E_n^{h\mathbb{G}_n}$ in cases when p divides n, the right approach appears to be to work over $\mathbb{W}_n$, and not over $\mathbb{Z}_p$. See also Remark 5.5.13 below.

We now give an outline of the steps one follows to construct a finite algebraic resolution. In practice, to construct such a resolution, it is essential to have some control over the homology $H_*^c(U, \mathbb{W}_n)$ for an open subgroup U of $\mathbb{G}_n$ of finite cohomological dimension. In fact, all the examples of finite algebraic resolutions which we describe below restrict to a projective resolution of $\mathbb{W}_n$ as a $\mathbb{W}_n\langle\langle U\rangle\rangle$-module for some choice of U. This motivates the name of *resolutions* for these exact sequences. In practice, if p is large with respect to n so that $\mathbb{S}_n$ has finite cohomological dimension, the finite algebraic resolutions are projective resolutions of $\mathbb{W}_n$ as a $\mathbb{W}_n\langle\langle \mathbb{G}_n\rangle\rangle$-modules.

The process is inductive and goes as follows. Suppose that the $\mathbb{W}_n\langle\langle \mathbb{G}_n\rangle\rangle$-modules $\mathcal{C}_i$ for $i \leq k-1$ together with maps $\partial_{k-1}\colon \mathcal{C}_{k-1} \to \mathcal{C}_{k-2}$ of $\mathbb{W}_n\langle\langle \mathbb{G}_n\rangle\rangle$-modules have been defined so that

$$\mathcal{C}_{k-1} \xrightarrow{\partial_{k-1}} \mathcal{C}_{k-2} \xrightarrow{\partial_{k-2}} \cdots \xrightarrow{\partial_0} \mathcal{C}_{-1} = \mathbb{W}_n \longrightarrow 0$$

is an exact sequence. Suppose further that each term restricts to a projective $\mathbb{W}_n\langle\langle U\rangle\rangle$-module. Let $N_{k-1} = \mathrm{Ker}(\partial_{k-1})$. The projectivity assumption implies that

$$\mathrm{Tor}_0^{\mathbb{W}_n\langle\langle U\rangle\rangle}(\mathbb{W}_n, N_{k-1}) = H_0^c(U, N_{k-1}) \cong H_k^c(U, \mathbb{W}_n).$$

This isomorphism, the knowledge of $H_k^c(U, \mathbb{W}_n)$ and a generalized form of Nakayama's Lemma [49, Lemma 4.3] allow us to identify a set of $\mathbb{W}_n\langle\langle \mathbb{G}_n\rangle\rangle$-generators for $N_{k-1} \subseteq \mathcal{C}_{k-1}$. The trick then is to choose a $\mathbb{W}_n\langle\langle \mathbb{G}_n\rangle\rangle$-module $\mathcal{C}_k$ of the desired form (preferably as "small" as possible) and to construct a map $f\colon \mathcal{C}_k \to N_{k-1}$ that surjects onto this set of generators. The map f is surjective by construction since it is chosen to make $\mathrm{Tor}^0_{\mathbb{W}_n\langle\langle U\rangle\rangle}(f, \mathbb{F}_{p^n})$ surjective. The map $\partial_k\colon \mathcal{C}_k \to \mathcal{C}_{k-1}$ is then defined to be the composite $\mathcal{C}_k \to N_{k-1} \to \mathcal{C}_{k-1}$, completing the inductive step.

The process stops once ∂_{d-1} has been defined. At this point, we define $\mathcal{C}_d = N_{d-1} = \mathrm{Ker}(\partial_{d-1})$ and prove that $\mathcal{C}_d$ is a $\mathbb{W}_n\langle\langle \mathbb{G}_n\rangle\rangle$-module of the required type. Of course, this need not be the case and proving that this happens for some series of choices of modules $\mathcal{C}_k$ and maps ∂_k is usually difficult.

Remark 5.5.6 (Algebraic resolution spectral sequence)**.** If one resolves (5.5.4) into a double complex $P_{\bullet,\bullet}$ where $P_{\bullet,k} \to \mathcal{C}_k$ for $0 \leq k \leq d$ is a projective resolution as $\mathbb{W}_n\langle\langle \mathbb{G}_n\rangle\rangle$-modules, then the totalization of the double complex $P_{\bullet,\bullet}$ is a projective resolution of $\mathbb{W}_n$. For a (graded) profinite $\mathbb{W}_n\langle\langle \mathbb{G}_n\rangle\rangle$-module $M = \{M_t\}_{t\in\mathbb{Z}}$, let $E_0^{s,k,t} = \mathrm{Hom}^c_{\mathbb{W}_n\langle\langle \mathbb{G}_n\rangle\rangle}(P_{k,s}, M_t)$ and take the vertical cohomology (i.e., with k fixed). The result is the E_1-term of a spectral sequence

$$E_1^{s,k,t} = \mathrm{Ext}^s_{\mathbb{W}_n\langle\langle \mathbb{G}_n\rangle\rangle}(\mathcal{C}_k, M_t) \Longrightarrow H_c^{s+k}(\mathbb{G}_n, M_t)$$

with differentials $d_r\colon E_1^{s,k,t} \to E_1^{s+r,k+r-1,t}$. If the $\mathcal{C}_k$s are direct sums of modules of the form $\chi\!\uparrow_F^{\mathbb{G}_n}$ for characters χ, then the E_1-term is easy to compute since by a version of Shapiro's lemma, we have

$$\mathrm{Ext}^s_{\mathbb{W}_n\langle\langle \mathbb{G}_n\rangle\rangle}(\chi\!\uparrow_F^{\mathbb{G}_n}, M_t) \cong \mathrm{Ext}^s_{\mathbb{W}_n\langle F\rangle}(\chi, M_t).$$

We call this an *algebraic resolution spectral sequence.*

Finally, applying the functors $\mathrm{Hom}^{c}_{\mathbb{W}_n}(-,(E_n)_*)$ to (5.5.4) gives an exact sequence in the category of Morava modules $\mathrm{Mod}^{\mathbb{G}_n}_{(E_n)_*}$:

$$0 \to (E_n)_* \xrightarrow{\partial^0} \mathrm{Hom}^{c}_{\mathbb{W}_n}(\mathcal{C}_0,(E_n)_*) \xrightarrow{\partial^1} \cdots \xrightarrow{\partial^d} \mathrm{Hom}^{c}_{\mathbb{W}_n}(\mathcal{C}_d,(E_n)_*) \to 0, \tag{5.5.6}$$

where the maps ∂^k are induced by ∂_k.

5.5.3 Topological resolutions

With an algebraic resolution (5.5.4) in hand, the next step is to prove that it has a topological realization which is a finite resolution of $L_{K(n)}S^0$. That is, one wants to construct a finite resolution

$$\mathcal{E}_{-1} = L_{K(n)}S^0 \xrightarrow{\delta_0} \mathcal{E}_0 \xrightarrow{\delta_1} \cdots \xrightarrow{\delta_{d-1}} \mathcal{E}_d \tag{5.5.7}$$

in the sense of Definition 5.5.1 with the property that applying the functor

$$(E_n)^\vee_*(-)\colon \mathrm{Sp} \longrightarrow \mathrm{Mod}^{\mathbb{G}_n}_{(E_n)_*}$$

to this sequence gives rise to a complex of Morava modules isomorphic to (5.5.6).

By our choice of $\mathcal{C}_k$s (see Description 5.5.3), there are isomorphisms of Morava modules $(E_n)^\vee_*\mathcal{E}_k \cong \mathrm{Hom}^{c}_{\mathbb{W}_n}(\mathcal{C}_k,(E_n)_*)$ for non-uniquely determined spectra $\mathcal{E}_k$ of the form specified in part (b) of Definition 5.5.1. The non-uniqueness of the $\mathcal{E}_k$s comes from the freedom in choosing the values of m above. (Note that the spectrum E_n^{hF} itself is periodic with periodicity some multiple d_F^{top} of d_F^{alg} so there is a limited number of choices.) Fixing some choice of $\mathcal{E}_k$s, we can identify (5.5.6) with

$$0 \longrightarrow (E_n)^\vee_* \xrightarrow{\partial^0} (E_n)^\vee_*\mathcal{E}_0 \xrightarrow{\partial^1} (E_n)^\vee_*\mathcal{E}_1 \longrightarrow \cdots \xrightarrow{\partial^d} (E_n)^\vee_*\mathcal{E}_d \longrightarrow 0.$$

To obtain a topological realization, one must also show that the maps ∂^k are of the form $(E_n)^\vee_*(\delta_k)$ for maps of spectra $\delta_k\colon \mathcal{E}_{k-1} \to \mathcal{E}_k$. Note that this being the case can depend on the choices of $\mathcal{E}_k$s. The existence of δ_k is established using a Hurewicz homomorphism

$$[\mathcal{E}_{k-1},\mathcal{E}_k] \longrightarrow \mathrm{Hom}_{(E_n)^\vee_*E_n}((E_n)^\vee_*\mathcal{E}_{k-1},(E_n)^\vee_*\mathcal{E}_k).$$

See Proposition 2.7 [49] for more details.

Even if the δ_ks exist, it still does not imply that any choice of $\mathcal{E}_k$s and δ_ks gives a finite resolution in the sense of Definition 5.5.1. For this to be the case, one must have first that the compositions $\delta_k \circ \delta_{k-1}$ are null-homotopic. If such choices exists, then we inductively define spectra X_k and maps ℓ_k so that

$$\Sigma^{k-1}X_{k-1} \xrightarrow{\Sigma^{k-1}\ell_k} \mathcal{E}_k \xrightarrow{\Sigma^k j_k} \Sigma^k X_k \xrightarrow{\Sigma^k i_k} \Sigma^k X_{k-1}$$

are exact triangles (see (5.5.2)). That is, if the map ℓ_k can be chosen so that $\delta_{k+1} \circ \Sigma^{k-1}\ell_k$ is null-homotopic, then X_{k+1} is defined as the cofiber of $\Sigma^{-1}\ell_k\colon \Sigma^{-1}X_{k-1} \to \Sigma^{-k}\mathcal{E}_k$ and there exists a map $\Sigma^k X_k \xrightarrow{\Sigma^k\ell_{k+1}} \mathcal{E}_{k+1}$ that factorizes δ_{k+1}.

To prove that $X_d \simeq L_{K(n)}S^0$, one needs to check that the map δ_0 lifts along the tower

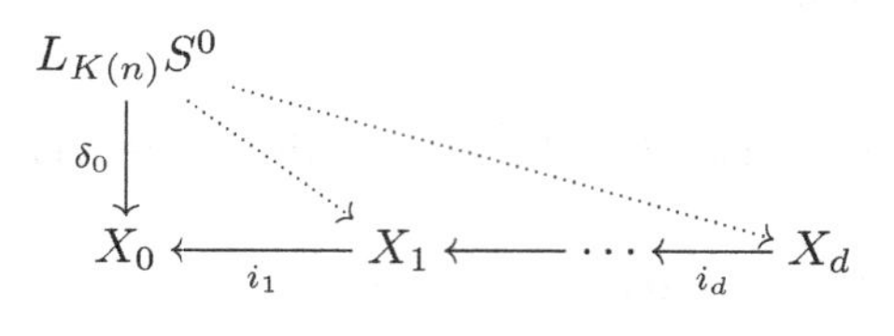

to a map $L_{K(n)}S^0 \to X_d$. If the lift exists, it will induce an isomorphism $(E_n)^\vee_*(L_{K(n)}S^0) \xrightarrow{\cong} (E_n)^\vee_*(X_d)$ so will be a $K(n)$-local equivalence.

Remark 5.5.7 (Doubling up)**.** In (5.3.11) above, we defined a normal subgroup $\mathbb{G}_n^1 \subseteq \mathbb{G}_n$ with the property that $\mathbb{G}_n \cong \mathbb{G}_n^1 \rtimes \mathbb{Z}_p$ where the extension splits whenever n is coprime to p. In practice, one first constructs a finite resolution of $\mathbb{W}_n$ as a $\mathbb{W}_n\langle\langle\mathbb{G}_n^1\rangle\rangle$-module and then upgrades it to a resolution of $\mathbb{W}_n$ as a $\mathbb{W}_n\langle\langle\mathbb{G}_n\rangle\rangle$-module. See Corollary 4.2 of [49] for an example.

5.5.4 Diagram of resolution spectral sequences

The resolutions whose construction is described above give rise to spectral sequences which fit in a diagram:

$$\begin{array}{ccc} E_1^{s,k,t} = \mathrm{Ext}^s_{\mathbb{W}_n\langle\langle\mathbb{G}_n\rangle\rangle}(\mathcal{C}_k, (E_n)_t) & \xRightarrow{\text{ARSS}} & H_c^{s+k}(\mathbb{G}_n, (E_n)_t) \cong E_2^{s+k,t} \\ \Downarrow{\scriptstyle \text{LHFPSS}} & & \Downarrow{\scriptstyle \text{HFPSS}} \\ E_1^{k,t-s} = \pi_{t-s}\mathcal{E}_k & \xRightarrow[\text{TRSS}]{} & \pi_{t-(s+k)}L_{K(n)}S^0. \end{array}$$

Here ARSS stands for *algebraic resolution spectral sequence*, TRSS for *topological resolution spectral sequence*, HFPSS for *homotopy fixed point spectral sequence* and in LHFPSS, the L is for *level-wise.* The horizontal spectral sequence have the advantage of being first quadrant spectral sequences which are zero in degrees $k > d$ and so collapse at the E_{d+1}-page. By Remark 5.3.15, the vertical spectral sequences also collapse at some finite stage with a horizontal vanishing line.

5.5.5 Finite resolutions at height $n = 2$

Now we give examples of some of the finite resolutions at height $n = 2$ that exist in the literature. In the references cited, the algebraic resolutions are usually constructed in the category $\mathbb{Z}_p[\![G]\!]$ for $G = \mathbb{G}_2$ or $G = \mathbb{G}_2^1$. As is explained Remark 5.3.19, we can transport the constructions to the category of $\mathrm{Mod}_{\mathbb{W}_2\langle\langle G\rangle\rangle}$ and this is what we do here. The reason for this change is explained in Remark 5.5.13.

Notation 5.5.8 (Finite subgroups and their modules)**.** The maximal finite subgroups of $\mathbb{S}_n$ were given in Table 5.3.1. Here, we discuss them more specifically in the case $n = 2$. Note that in the cases $p = 2, 3$, $\mathbb{S}_2$ contains p-torsion and so has more interesting finite subgroups (see (2) and (3) below).

(1) Let p be odd. Let $\sigma \in \mathrm{Gal} = \mathrm{Gal}(\mathbb{F}_{p^2}/\mathbb{F}_p)$ be the Frobenius and $\omega \in \mathbb{F}_{p^2}^\times$ be a primitive $(q = p^2 - 1)$th root of unity. The group Gal acts on $\mathbb{F}_{p^2}^\times$ by $\sigma(\omega) = \omega^\sigma = \omega^p$. We define

$$F = F_{2q} := \mathbb{F}_{p^2}^\times \rtimes \mathrm{Gal}.$$

For example, if $p = 3$, then $F \cong SD_{16}$, the semi-dihedral group of order 16. We let $\omega \in \mathbb{W}_2 \cong \mathbb{Z}_p(\omega)$ denote the Teichmüller lift of the same named class in $\mathbb{F}_{p^2}$. The Teichmüller lifts then specify an embedding of $\mathbb{F}_{p^2}^\times$ in $\mathbb{S}_n$, and so of F in $\mathbb{G}_n$.

We define $\mathbb{W}_2\langle F\rangle$-modules χ_i^+ and χ_i^- for $0 \le i \le q-1$ as follows. The underlying $\mathbb{W}_2$-module of $\chi_i^\pm$ is $\mathbb{W}_2$. For $x \in \chi_i^\pm$, define $\omega_*(x) = \omega^i x$ and

$$\sigma_*(x) = \begin{cases} x^\sigma & x \in \chi_i^+ \\ -x^\sigma & x \in \chi_i^-. \end{cases}$$

Let $\chi_i = \chi_i^+ \oplus \chi_i^-$. The twisted group ring completely decomposes as a $\mathbb{W}_2\langle F\rangle$-modules as

$$\mathbb{W}_2\langle F\rangle \cong \bigoplus_{i=0}^{q-1} \chi_i = \bigoplus_{i=0}^{q-1} \chi_i^+ \oplus \chi_i^-.$$

To see this isomorphism, let $x_i \in \mathbb{W}_2\langle F\rangle$ be given by

$$x_i = [e] + \omega^{-i}[\omega] + \omega^{-2i}[\omega^2] + \cdots + \omega^{-(q-2)i}[\omega^{q-2}]$$

for $0 \le i \le q-2$. The elements x_i together with the elements $x_i[\sigma]$ generated $\mathbb{W}_2\langle F\rangle$ as a $\mathbb{W}_2$-module. Furthermore, $\omega_*(x_i) = \omega^i x_i$ and $\sigma_*(x_i) = x_i[\sigma]$. So, there are isomorphisms $\chi_i^+ \cong \mathbb{W}_2\{x_i + x_i[\sigma]\}$ and $\chi_i^- \cong \mathbb{W}_2\{x_i - x_i[\sigma]\}$.

(Note that the $\mathbb{Z}_p$-module λ_i of [57] has the property that

$$\mathbb{W}_2 \otimes_{\mathbb{Z}_p} \lambda_i \cong \chi_{-i}^+ \oplus \chi_{-pi}^+$$

when viewed as a $\mathbb{W}_2\langle F\rangle$-module as described in Remark 5.3.19.)

(2) For $p = 3$, let G_{24} be an extension of $C_3 \rtimes C_4 \subseteq \mathbb{S}_2$ to $\mathbb{G}_2$ in the sense of Definition 5.3.5. We can give an explicit choice as follows. The subgroup of $\mathbb{S}_2$ generated by $s = \frac{1}{2}(1+\omega\xi)$ and $t = \omega^2$ is isomorphic to $C_3 \rtimes C_4$. Here, s is of order 3, t is of order 4, and $tst^{-1} = s^2$. We let G_{24} be the group generated by s, t, $\psi = \omega\xi$ in $\mathbb{D}_2^\times/\xi^2 \cong \mathbb{G}_2$. Note that $\psi s = s\psi$ and $t\psi = \psi t^3$. The group G_{24} is an extension of $C_3 \rtimes C_4$ in $\mathbb{G}_2$. See Section 1.1 of [49]. Note that C_3 is normal in G_{24}. Therefore, the $\chi_i^{\pm 1}$ inherit a G_{24}-module structure via the map

$$G_{24} \to G_{24}/C_3 \cong (\mathbb{F}_9^\times)^2 \times \mathrm{Gal} \xrightarrow{\subseteq} SD_{16},$$

where $(\mathbb{F}_9^\times)^2 \cong C_4$ denotes the subgroup of squares in $\mathbb{F}_9^\times$.

(3) For $p = 2$, the group $\mathbb{S}_2$ contains a unique conjugacy class of maximal finite subgroups isomorphic to the binary tetrahedral group $T_{24} \cong Q_8 \rtimes \mathbb{F}_4^\times$. There is a choice of T_{24} generated by $\omega \in \mathbb{F}_4^\times$ and an element of order four which we denote by $i \in Q_8$ with the property that $i^2 = -1 \in \mathbb{S}_2$. For $j = \omega i \omega^{-1}$, the elements i and j satisfy the

usual quaternion relations. We let G_{48} be the extension of $T_{24} \subseteq \mathbb{S}_2$ to $\mathbb{G}_2$ given by $G_{48} = \langle \omega, 1+i \rangle \subseteq \mathbb{D}_2^\times/\xi^2 \cong \mathbb{G}_2$. The group G_{48} is isomorphic to the binary octahedral group. See [59, Lemma 2.1, 2.2].

We also let $C_2 = (\pm 1) \subseteq \mathbb{S}_2$. Define $V_4 = C_2 \times \mathrm{Gal}$ and $G_{12} = (C_2 \times \mathbb{F}_4^\times) \rtimes \mathrm{Gal} \cong C_2 \times \Sigma_3$. The group $C_4 = \langle i \rangle \subseteq \mathbb{S}_2$ also extends to a finite subgroup of $\mathbb{G}_2$ that is a cyclic group of order eight given by $C_8 = \langle 1+i \rangle \subseteq \mathbb{D}_2^\times/\xi^2$.

Finally, we let $\pi = 1 + 2\omega$, which has the property that $\det(\pi) = \pi\pi^\sigma = 3 \in \mathbb{Z}_2^\times/(\pm 1)$ is a topological generator. We let $G_{48}' = \pi G_{48} \pi^{-1}$. This group is conjugate to G_{48} in $\mathbb{G}_2$, but not in $\mathbb{G}_2^1$.

Recall from Remark 5.5.7 that in practice, one begins by constructing a finite resolution of the group $\mathbb{G}_2^1$. (The groups $\mathbb{G}_2^1$ and $\mathbb{S}_2^1$ were defined right before (5.3.11).)

Example 5.5.9 (The Duality Resolutions)**.** The following examples of resolutions of $\mathbb{W}_2$ as a $\mathbb{W}_2\langle\!\langle \mathbb{G}_2^1 \rangle\!\rangle$-module have been coined the *duality resolutions*. They are self-dual in a suitable sense (see [60, Section 3.4] or [19, Section 3.3]). In fact, this duality is related to the virtual Poincaré duality of the group $\mathbb{G}_2^1$. They are given by exact sequences

$$0 \longrightarrow \mathcal{D}_3 \longrightarrow \mathcal{D}_2 \longrightarrow \mathcal{D}_1 \longrightarrow \mathcal{D}_0 \longrightarrow \mathbb{W}_2 \longrightarrow 0 \tag{5.5.8}$$

such that each $\mathcal{D}_i$ is isomorphic to a direct sum of modules of the form

$$\chi\!\uparrow_H^{\mathbb{G}_2^1} := \mathbb{W}_2\langle\!\langle \mathbb{G}_2^1 \rangle\!\rangle \otimes_{\mathbb{W}_2\langle H \rangle} \chi \tag{5.5.9}$$

for H an extension to $\mathbb{G}_2^1$ of a finite subgroup of $\mathbb{S}_2^1$ and χ is $\mathbb{W}_2\langle H \rangle$-module that restricts to a free module of rank one over $\mathbb{W}_2$.

They are minimal in the sense that their associated algebraic resolution spectral sequence

$$E_1^{r,q} = \mathrm{Ext}^q_{\mathbb{W}_2\langle\!\langle \mathbb{G}_2^1 \rangle\!\rangle}(\mathcal{D}_r, \mathbb{F}_{p^2}) \Longrightarrow H_c^{p+q}(\mathbb{G}_2^1, \mathbb{F}_{p^2}) \tag{5.5.10}$$

collapses at the E_1-term. They can be realized as finite resolutions of $E_2^{h\mathbb{G}_2^1}$

$$E_2^{h\mathbb{G}_2^1} \longrightarrow \mathcal{E}_0^{\mathcal{D}} \longrightarrow \mathcal{E}_1^{\mathcal{D}} \longrightarrow \mathcal{E}_2^{\mathcal{D}} \longrightarrow \mathcal{E}_3^{\mathcal{D}}. \tag{5.5.11}$$

(a) Let $p \geq 5$. There is an exact sequence (5.5.8) with

$$\mathcal{D}_0 \cong \mathcal{D}_3 \cong \mathbb{W}_2\!\uparrow_F^{\mathbb{G}_2^1}, \qquad \mathcal{D}_1 \cong \mathcal{D}_2 \cong (\chi_{p-1}^+ \oplus \chi_{1-p}^+)\!\uparrow_F^{\mathbb{G}_2^1}.$$

This can be realized as a finite resolution (5.5.11) with

$$\mathcal{E}_0^{\mathcal{D}} \simeq \mathcal{E}_3^{\mathcal{D}} \simeq E_2^{hF}, \qquad \mathcal{E}_1^{\mathcal{D}} \simeq \mathcal{E}_2^{\mathcal{D}} \simeq \Sigma^{2(p-1)} E_2^{hF} \vee \Sigma^{2(1-p)} E_2^{hF}.$$

These were constructed by Henn [57, Theorem 5]. See also Lader [80].

(b) Let $p = 3$. There is an exact sequence (5.5.8) with

$$\mathcal{D}_0 \cong \mathcal{D}_3 \cong \mathbb{W}_2\!\uparrow_{G_{24}}^{\mathbb{G}_2^1}, \qquad \mathcal{D}_1 \cong \mathcal{D}_2 \cong \chi_4^-\!\uparrow_{SD_{16}}^{\mathbb{G}_2^1}.$$

This can be realized as a finite resolution (5.5.11) with

$$\mathcal{E}_0^{\mathcal{D}} \simeq E_2^{hG_{24}}, \qquad \mathcal{E}_1^{\mathcal{D}} \simeq \Sigma^8 E_2^{hSD_{16}}, \qquad \mathcal{E}_2^{\mathcal{D}} \simeq \Sigma^{40} E_2^{hSD_{16}}, \qquad \mathcal{E}_3^{\mathcal{D}} \simeq \Sigma^{48} E_2^{hG_{24}}.$$

These were constructed by Goerss, Henn, Mahowald and Rezk in [49].

(c) Let $p = 2$. There is an exact sequence (5.5.8) with

$$\mathcal{D}_0 \cong \mathbb{W}_2\uparrow_{G_{48}}^{\mathbb{G}_2^1}, \qquad \mathcal{D}_1 \cong \mathcal{D}_2 \cong \mathbb{W}_2\uparrow_{G_{12}}^{\mathbb{G}_2^1}, \qquad \mathcal{D}_3 \cong \mathbb{W}_2\uparrow_{G_{48}'}^{\mathbb{G}_2^1}.$$

This can be realized as a finite resolution (5.5.11) with

$$\mathcal{E}_0^{\mathcal{D}} \simeq E_2^{hG_{48}}, \qquad \mathcal{E}_1^{\mathcal{D}} \simeq E_2^{hG_{12}}, \qquad \mathcal{E}_2^{\mathcal{D}} \simeq \Sigma^{48} E_2^{hG_{12}}, \qquad \mathcal{E}_3^{\mathcal{D}} \simeq \Sigma^{48} E_2^{hG_{48}}.$$

These were constructed by Beaudry, Bobkova, Goerss, Henn, Mahowald, and Rezk in [57, 19, 29]. In these references, the resolution is constructed for $E_2^{h\mathbb{S}_2}$. However, using the ideas of [59] it is now straightforward to construct it for $E_2^{h\mathbb{G}_2}$.

Remark 5.5.10. If $p \neq 2$, the algebraic resolution can be doubled up in the sense of Remark 5.5.7 and the result can be realized topologically. For $p \geq 5$, this gives a resolution

$$L_{K(2)}S^0 \longrightarrow E_2^{hF} \xrightarrow{\delta_0} X \vee E_2^{hF} \xrightarrow{\delta_1} X \vee X \xrightarrow{\delta_2} E_2^{hF} \vee X \xrightarrow{\delta_3} E_2^{hF}$$

where $X = \Sigma^{2(p-1)} E_2^{hF} \vee \Sigma^{2(1-p)} E_2^{hF}$, and for $p = 3$, we get

$$L_{K(2)}S^0 \to E_2^{hG_{24}} \xrightarrow{\delta_0} E_2^{hSD_{16}} \vee E_2^{hG_{24}} \xrightarrow{\delta_1} \Sigma^{48} E_2^{hSD_{16}} \vee E_2^{hSD_{16}}$$
$$\xrightarrow{\delta_2} \Sigma^{48}(E_2^{hG_{24}} \vee E_2^{hSD_{16}}) \xrightarrow{\delta_3} \Sigma^{48} E_2^{hG_{24}}.$$

However, the duality resolution at $p = 2$ cannot be doubled up.

Example 5.5.11 (The Centralizer Resolutions)**.** The following two resolutions of the trivial $\mathbb{W}_2\langle\langle\mathbb{G}_2^1\rangle\rangle$-modules are called *centralizer resolutions* because their construction has as a key input Henn's Centralizer Approximation Theorem [56, Theorem 1.4]. They are given by exact sequences

$$0 \longrightarrow \mathcal{C}_3 \longrightarrow \mathcal{C}_2 \longrightarrow \mathcal{C}_1 \longrightarrow \mathcal{C}_0 \longrightarrow \mathbb{W}_2 \longrightarrow 0, \tag{5.5.12}$$

where the $\mathcal{C}_i$s are of the form described in (5.5.9). They can be realized as finite resolutions

$$E_2^{h\mathbb{G}_2^1} \longrightarrow \mathcal{E}_0^{\mathcal{C}} \longrightarrow \mathcal{E}_1^{\mathcal{C}} \longrightarrow \mathcal{E}_2^{\mathcal{C}} \longrightarrow \mathcal{E}_3^{\mathcal{C}}. \tag{5.5.13}$$

The algebraic centralizer exact sequences (5.5.12) described below are *$\mathcal{F}$-resolutions* in the sense of [57, §3.5.1] and [59, §1.2]. We will not explain what this means, but a consequence of this fact is that the algebraic centralizer resolutions can be doubled up in the sense of Remark 5.5.7. The downside of the centralizer resolutions is that they are "larger" than the duality resolutions. For example, the analogues of (5.5.10) for the centralizer resolutions do not collapse. As a consequence, the associated algebraic and topological spectral sequences are less efficient for computations. Nonetheless, having different resolutions offers different perspectives and the centralizer resolutions have been crucial in recent computations. See for example [29, 47].

(a) Let $p = 3$. There is an exact sequence (5.5.12) with

$$\mathcal{C}_0 \cong \mathbb{W}_2\uparrow_{G_{24}}^{\mathbb{G}_2^1}, \qquad \mathcal{C}_1 \cong \chi_4^-\uparrow_{SD_{16}}^{\mathbb{G}_2^1} \oplus \chi_2^+\uparrow_{G_{24}}^{\mathbb{G}_2^1},$$
$$\mathcal{C}_2 \cong (\chi_2^+ \oplus \chi_{-2}^+)\uparrow_{SD_{16}}^{\mathbb{G}_2^1}, \qquad \mathcal{C}_3 \cong \mathbb{W}_2\uparrow_{SD_{16}}^{\mathbb{G}_2^1}.$$

This can be realized as a finite resolution (5.5.13) with

$$\mathcal{E}_0^{\mathcal{C}} \simeq E_2^{hG_{24}}, \qquad \mathcal{E}_1^{\mathcal{C}} \simeq \Sigma^8 E_2^{hSD_{16}} \vee \Sigma^{36} E_2^{hG_{24}},$$
$$\mathcal{E}_2^{\mathcal{C}} \simeq \Sigma^{36} E_2^{hSD_{16}} \vee \Sigma^{44} E_2^{hSD_{16}}, \qquad \mathcal{E}_3^{\mathcal{C}} \simeq \Sigma^{48} E_2^{hSD_{16}}.$$

This is constructed by Henn in [57]. See also [47, Section 4]. Since $\mathbb{G}_2 \cong \mathbb{G}_2^1 \times \mathbb{Z}_3$, Remark 5.5.7 applies and we get a resolution of $L_{K(2)}S^0$.

(b) Let $p = 2$. There is an exact sequence (5.5.12) with

$$\mathcal{C}_0 \cong \mathbb{W}_2\uparrow_{G_{48}}^{\mathbb{G}_2^1} \oplus \mathbb{W}_2\uparrow_{G'_{48}}^{\mathbb{G}_2^1}, \qquad \mathcal{C}_1 \cong \mathbb{W}_2\uparrow_{C_8}^{\mathbb{G}_2^1} \oplus \mathbb{W}_2\uparrow_{G_{12}}^{\mathbb{G}_2^1},$$
$$\mathcal{C}_2 \cong \mathbb{W}_2\uparrow_{V_4}^{\mathbb{G}_2^1}, \qquad \mathcal{C}_3 \cong \mathbb{W}_2\uparrow_{G_{12}}^{\mathbb{G}_2^1}.$$

This can be realized as a finite resolution (5.5.13) with

$$\mathcal{E}_0^{\mathcal{C}} \simeq E_2^{hG_{48}} \vee E_2^{hG'_{48}}, \qquad \mathcal{E}_1^{\mathcal{C}} \simeq E_2^{hC_8} \vee E_2^{hG_{12}}, \qquad \mathcal{E}_2^{\mathcal{C}} \simeq E_2^{hV_4}, \qquad \mathcal{E}_3^{\mathcal{C}} \simeq E_2^{hG_{12}}.$$

This is constructed by Henn [59]. Note again that, as opposed to the duality resolution at $p = 2$, the algebraic centralizer resolution *can* be doubled up and the resulting sequence can be realized topologically to give a finite resolution of $L_{K(2)}S^0$. See [59, Theorem 1.1, 1.5].

Remark 5.5.12. The doubled up centralizer resolution at $n = p = 2$ is very large compared to the duality resolution available at odd primes. However, there is a handicraft way to glue a duality resolution with a centralizer resolution to obtain a much smaller resolution called the *hybrid resolution.* It can be realized as a resolution of $L_{K(2)}S^0$

$$L_{K(2)}S^0 \to E_2^{hG_{48}} \xrightarrow{\delta_0} E_2^{hC_8} \vee E_2^{hG_{12}} \xrightarrow{\delta_1} E_2^{hV_4} \vee E_2^{hG_{12}}$$
$$\xrightarrow{\delta_2} E_2^{hG_{12}} \vee \Sigma^{48} E_2^{hG_{12}} \xrightarrow{\delta_3} \Sigma^{48} E_2^{hG_{48}}.$$

The construction of this resolution is not published but is due to the second author and Henn.

Remark 5.5.13. Until recently, at $n = p = 2$, topological and algebraic resolutions existed only for $\mathbb{S}_2$ and not $\mathbb{G}_2$. In some loose sense, the reason for this was our inability to average over the action of Gal. By an insight of Hans-Werner Henn [57] if one switches to $\mathbb{W}_n$-coefficients, one can form "weighted averages" in $\mathbb{W}_n$, and this has allowed us to upgrade our resolutions for $\mathbb{S}_2$ to resolutions for $\mathbb{G}_2$. Note further that the description of the duality resolution at $n = 2$ and $p \geq 5$ is much cleaner if one works over the Witt vectors.

5.6 Chromatic splitting, duality, and algebraicity

This section discusses two of the major areas of applications of the techniques developed above in local chromatic homotopy theory: the chromatic splitting conjecture and the study of duality phenomena in $K(n)$-local homotopy theory. In both cases, we start with an outline of the general picture, before specializing to a summary of the known results at height 2. We then conclude with a brief outlook to the asymptotic behavior of chromatic homotopy theory for primes large with respect to the height.

5.6.1 Chromatic splitting and reassembly

In this section, we discuss the chromatic splitting conjecture in more detail, putting an emphasis on new developments and points that were not discussed in [72].

The chromatic splitting conjecture (CSC) gives a fairly simple prediction of $L_{n-1}L_{K(n)}S^0$ in the chromatic fracture square. Although the original conjecture does not hold for the prime $p = 2$ and $n = 2$ as it was stated in [72], the fundamental idea behind the conjecture remains intact. The philosophy behind the CSC is that chromatic reassembly is governed by the structure of $H_c^*(\mathbb{G}_n, \mathbb{W})$, via the map $H_c^*(\mathbb{G}_n, \mathbb{W}) \to H_c^*(\mathbb{G}_n, (E_n)_*)$ to the E_2-term of the homotopy fixed point spectral sequence (5.3.4). As discussed in Conjecture 5.3.16, this map is expected to be an isomorphism onto $H_c^*(\mathbb{G}_n, (E_n)_0)$.

The isomorphism (5.3.14) implies at the very least that there is an inclusion

$$\Lambda_{\mathbb{Z}_p}(x_1, x_2, \ldots, x_n) \subseteq H_c^*(\mathbb{G}_n, \mathbb{W}_n)$$

for generators x_i of cohomological degree $2i - 1$ at all primes and heights. Here, we always choose $x_1 = \zeta_n$ for ζ_n as in (5.3.9) and choose the other x_is so that they do not map to zero in $H_c^*(\mathbb{G}_n, \mathbb{W}_n)/p$. Conjecture 5.3.16 then implies the existence of nonzero classes $x_i \in E_2^{2i-1,0}$ in

$$E_2^{s,t} \cong H_c^s(\mathbb{G}_n, (E_n)_t) \Longrightarrow \pi_{t-s}L_{K(n)}S^0.$$

Further, at heights $n \leq 2$ the following phenomena have been observed.

Conjecture 5.6.1. *If p is odd, or if $p = 2$ and n is odd, then*

$$E_\infty^{*,0} \cong \Lambda_{\mathbb{Z}_p}(e_1, e_2, \ldots, e_n) \subseteq \pi_* L_{K(n)}S^0$$

for some choice of classes e_i detected by a multiple of x_i. If $p = 2$ and n is even, then

$$E_\infty^{*,0} \cong \Lambda_{\mathbb{Z}_2}(f, e_1, e_2, \ldots, e_n)/(2f, e_n f) \subseteq \pi_* L_{K(n)}S^0$$

for e_i as above and some choice of class f detected by $\widetilde{\chi}_n$ (see (5.3.10)).

Remark 5.6.2. If p is large with respect to n, then there is no ambiguity about the choice of classes e_i because of the sparsity of the spectral sequence. At $n = 2$ and $p = 3$, one can choose e_2 to be detected by $3x_2$ and at $n = p = 2$, by $4x_2$.

The dichotomy between odd and even heights for the prime 2 comes from the following observations. If n is odd, then the inclusion of $C_2 \cong (\pm 1)$ in $\mathbb{G}_n$ splits the map χ_n and

$$H^*(C_2, \mathbb{Z}_2) \cong \mathbb{Z}_2[\widetilde{\chi}_n]/(2\widetilde{\chi}_n) \to H^*(C_2, \mathbb{W}_n) \xrightarrow{\chi_n^*} H_c^*(\mathbb{G}_n, \mathbb{W}_n)$$

is an inclusion. However, if the image of $\widetilde{\chi}_n$ in $H_c^2(\mathbb{G}_n, (E_n)_0)$ is non-trivial, then it must support a non-trivial d_3 differential since its image in the HFPSS computing $E_n^{hC_2}$ has this property by [52, Theorem 1.3]. (The image of $\widetilde{\chi}_n$ would be the class $u_{2\sigma}^{-1} a_\sigma^2$ that supports a non-zero d_3 differential.)

At $n = 2$, what we observe is that $(\widetilde{\chi}_2^2)$ is the kernel of χ_2^* so that the latter induces the inclusion of $\mathbb{Z}_2[\widetilde{\chi}_2]/(2\widetilde{\chi}_2, \widetilde{\chi}_2^2)$ into $H_c^*(\mathbb{G}_2, \mathbb{W}_2)$. We do not know how this generalizes at even heights $n > 2$.

Problem 5.6.3. Determine the nilpotence order of $\widetilde{\chi}_n \in H_c^*(\mathbb{G}_n, \mathbb{W}_n)$ when n is even.

As is discussed in [72], the induced maps $e_i \colon S^{1-2i} \to L_{n-1}S^0$ (for some choice of e_i) are conjectured to factor through $L_{n-i}S^0$. Since f has order 2, it induces a map $f \colon S^{-2}/2 \to L_{K(n)}S^0$, so after localizing at E_{n-1}, we get a map $f \colon L_{n-1}S^{-2}/2 \to L_{n-1}L_{K(n)}S^0$. The CSC as stated in [72] did not take into account this class f. Based on what we see in the case $n = p = 2$, Conjecture 5.6.4 below is a suggestion for a revised version of the CSC in its strongest form which includes f. Below, for spectra X_i, we let $\Lambda_{L_{n-1}S_p^0}(X_1, \ldots, X_n)$ be the wedge of $L_{n-1}S_p^0$ and of $X_{i_1} \wedge \ldots \wedge X_{i_j}$ for $1 \le i_1 < \ldots < i_j \le n$. Let

$$\iota \colon L_{n-1}S_p^0 \to \Lambda_{L_{n-1}S_p^0}(X_1, \ldots, X_n)$$

be the inclusion of the $L_{n-1}S_p^0$ summand.

Conjecture 5.6.4 (Strong CSC). *There is an equivalence in the category of E_{n-1}-local spectra*

$$L_{n-1}L_{K(n)}S^0 \simeq \Lambda_{L_{n-1}S_p^0}\left(L_{n-i}S^{1-2i} : 1 \le i \le n\right)$$

if $p \neq 2$, or $p = 2$ and n odd. The map ι corresponds to the unit $L_{n-1}S_p^0 \to L_{n-1}L_{K(n)}S_p^0$. If $p = 2$ and n is even, there is an E_{n-1}-local equivalence

$$L_{n-1}L_{K(n)}S^0 \simeq \Lambda_{L_{n-1}S_2^0}\left(L_{n-i}S^{1-2i} : 1 \le i \le n\right) \wedge \Lambda_{L_{n-1}S_2^0}\left(L_{n-1}S^{-2}/2\right).$$

In this case, the map $\iota \wedge \iota$ corresponds to the unit $L_{n-1}S_p^0 \to L_{n-1}L_{K(n)}S_p^0$.

Remark 5.6.5. A criterion for this revision is for the conjecture at odd primes to remain as stated in [72]. We have made what we think is a minimal modification to reflect what we see at $n = p = 2$. However, other reformulations are possible and we concede that this is a somewhat arbitrary choice.

Note that Conjecture 5.6.4 implies the weak CSC (Conjecture 5.2.13), saying that $L_{n-1}S^0 \to L_{n-1}L_{K(n)}S^0$ splits. Further, both Conjecture 5.6.4 and Conjecture 5.2.13 imply similar statements with sphere replaced by any finite complex X. However, even Conjecture 5.2.13 does not hold for arbitrary spectra. In [36], Devinatz proves that it fails for the p-completion of BP.

Before giving examples we would like to point out that, among its many consequences, the strong form of the chromatic splitting conjecture would also imply

Conjecture 5.6.6. *For any $n \geq 0$ and any prime p, the homotopy groups $\pi_* L_{K(n)}S^0$ are degreewise finitely generated $\mathbb{Z}_p$-modules.*

Example 5.6.7. At height $n = 1$, the equivalence $L_0 L_{K(1)}S^0 \simeq L_0(S_p^0 \vee S_p^{-1})$ holds for all primes and was discussed in Section 5.4.3, see (5.4.9).

Theorem 5.6.8. *At height $n = 2$, if p is odd, then*

$$L_1 L_{K(2)}S^0 \simeq L_1(S_p^0 \vee S_p^{-1}) \vee L_0(S_p^{-3} \vee S_p^{-4}).$$

See [25, 48]. If $p = 2$, there is an equivalence

$$L_1 L_{K(2)}S^0 \simeq L_1(S_2^0 \vee S_2^{-1} \vee S^{-2}/2 \vee S^{-3}/2) \vee L_0(S_2^{-3} \vee S_2^{-4}).$$

See [23].

Remark 5.6.9. The fact that the CSC in its original form would most likely fail at $n = p = 2$ was first noticed by Mark Mahowald. His intuition was based on the computations of Shimomura and Wang [113], who identify v_1-torsion-free summands in the E_2-term of the Adams-Novikov Spectral Sequence for $\pi_* L_{K(2)}V(0)$ that are not predicted by the CSC. Their work, however, did not preclude the possibility of differentials that could have eliminated the summands not accounted for in the original statement of the CSC.

The CSC is one of the key inputs for chromatic reassembly, which recovers $L_n S^0$ from $L_{K(n)}S^0$ and $L_{n-1}S^0$ via the chromatic fracture square (5.2.7). We discuss this a little more here.

Let F_n be the fiber of the map $L_{n-1}S^0 \to L_{n-1}L_{K(n)}S^0$, whose homotopy type is predicted by the CSC. The chromatic fracture square implies that F_n is also the fiber of $L_n S^0 \to L_{K(n)}S^0$. The homotopy groups of $L_n S^0$ can be reassembled from the long exact sequence on homotopy groups

$$\pi_k \Sigma^{-1} L_{K(n)}S^0 \longrightarrow \pi_k F_n \longrightarrow \pi_k L_n S^0 \longrightarrow \pi_k L_{K(n)}S^0 \longrightarrow \pi_k \Sigma F_n.$$

We explained chromatic reassembly at height $n = 1$ in Section 5.4.3. At this point, we would like to at least give the reader an idea of chromatic reassembly at chromatic level $n = 2$. A description of reassembly for $L_2 S_p^0$ itself would be very technical, so instead, we describe the reassembly process for $L_2 S^0/p$ for primes $p \geq 5$, which is significantly simpler. To do this, we first give a qualitative description of the homotopy groups of $\pi_*(L_{K(2)}S^0/p)$.

For primes $p \geq 5$, the homotopy fixed point spectral sequence is too sparse for differentials or extensions, so collapses to give

$$\pi_m(L_{K(2)}S^0/p) \cong \bigoplus_{m=t-s} H_c^s(\mathbb{G}_2, (E_2)_t/p).$$

The groups on the right side of this isomorphism can be deduced from the computation by Shimomura and Yabe in [114] and were discussed in Sadofsky [110]. They are computed directly using the finite resolution (1) of Example 5.5.9 by Lader in [80]. See [80, Corollaire 4.4] and the discussion before it for an explicit description. We make a few observations about the answer:

(a) The homotopy groups $\pi_*(L_{K(2)}S^0/p)$ form a module over $\mathbb{F}_p[v_1] \otimes \Lambda_{\mathbb{F}_p}(\zeta_2)$ for a class $v_1 = u^{1-p} \in \pi_{2(p-1)}(L_{K(2)}S^0/p)$ and $\zeta_2 \in \pi_{-1}(L_{K(2)}S^0/p)$ as in Remark 5.3.8. The group $\pi_m(L_{K(2)}S^0/p)$ is zero if

$$2k(p-1) < m < 2(k+1)(p-1) - 4.$$

(b) Since $L_0S^0/p \simeq *$, the chromatic fracture square gives an equivalence $L_1S^0/p \simeq L_{K(1)}S^0/p$ and, similarly, $L_1L_{K(2)}S^0/p \simeq L_{K(1)}L_{K(2)}S^0/p$. Furthermore, on homotopy groups, the effect of E_1-localization on $L_{K(2)}S^0/p$ is to invert v_1.

(c) There is unbounded v_1-torsion in $\pi_*(L_{K(2)}S^0/p)$. However, the homotopy groups are finite in each degree m. In fact, for any class x detected in $H_c^s(\mathbb{G}_2, (E_2)_t/p)$ for $t < 0$, if $t + 2k(p-1) \geq 0$, then $v_1^k x = 0$. That is, multiplication by v_1 never "crosses" the $s = t$ line in $H_c^s(\mathbb{G}_2, (E_2)_t/p)$.

(d) The only homotopy classes in $\pi_* L_{K(2)}S^0$ that are not v_1-torsion are given by $\mathbb{F}_p[v_1] \otimes \Lambda_{\mathbb{F}_p}(\zeta_2, h_0)$ for a class $h_0 \in \pi_{2(p-1)-1}(L_{K(2)}S^0/p)$ that is the image of the homotopy class $\alpha_1 \in \pi_{2(p-1)-1}S^0_{(p)}$. Furthermore,

$$\pi_*(L_1L_{K(2)}S^0/p) \cong v_1^{-1}\pi_*(L_{K(2)}S^0/p) \cong \mathbb{F}_p[v_1^{\pm 1}] \otimes \Lambda_{\mathbb{F}_p}(\zeta_2, h_0).$$

Under the canonical map $\pi_*(L_1S^0/p) \to \pi_*(L_1L_{K(2)}S^0/p)$, the class $\zeta_1 \in \pi_{-1}(L_1S^0/p)$ maps to $v_1^{-1}h_0$ up to a unit in $\mathbb{F}_p$.

We use the long exact sequence on homotopy groups associated to the fiber sequence

$$F_2/p \to L_2S^0/p \to L_{K(2)}S^0/p$$

to deduce that

$$\pi_*(L_2S^0/p) \cong \mathrm{Tor}_{v_1}(\pi_*(L_{K(2)}S^0/p)) \oplus \mathbb{F}_p[v_1]\{1, h_0\} \oplus \Sigma^{-1}\mathbb{F}_p[v_1]/(v_1^\infty)\{\zeta_2, h_0\zeta_2\},$$

where $\mathrm{Tor}_{v_1}(\pi_*(L_{K(2)}S^0/p))$ is the v_1-power torsion subgroup of $\pi_*(L_{K(2)}S^0/p)$ and $\mathbb{F}_p[v_1]/(v_1^\infty)$ is the cokernel of the canonical map in the following short exact sequence

$$0 \longrightarrow \mathbb{F}_p[v_1] \longrightarrow \mathbb{F}_p[v_1^{\pm 1}] \longrightarrow \mathbb{F}_p[v_1]/(v_1^\infty) \longrightarrow 0.$$

All available proofs of the CSC, even in its weakest form, have been brutally computational. Short of simply computing $\pi_* L_{K(2)} S^0$ explicitly, the steps have been:

(a) Prove that there are nonzero homotopy classes e_1 and e_2, and if $p = 2$ an additional class f detecting a non-trivial class of order 2. This gives the map

$$S^0 \vee S^{-1} \vee S^{-3} \vee S^{-4} \xrightarrow{\varphi} L_{K(2)} S^0,$$

where $\varphi = 1 \vee e_1 \vee e_2 \vee e_1 e_2$. If $p = 2$, there is an additional factor of $S^{-2}/2 \vee S^{-3}/2 \xrightarrow{f \vee e_1 f} L_{K(2)} S^0$.

(b) Compute $v_1^{-1} \pi_* L_{K(2)}(\varphi \wedge X)$ for a finite type 1 complex, usually $X = S^0/p$.

(c) Compute $p^{-1} \pi_* L_{K(2)} S^0$.

(d) Reassemble the fracture square.

Remark 5.6.10. Let p be an odd prime. In general, the CSC predicts that $1 \vee \zeta_n$ induces an equivalence $L_{K(n-1)} S^0 \vee L_{K(n-1)} S^{-1} \xrightarrow{\simeq} L_{K(n-1)} L_{K(n)} S^0$. In particular, it implies that $L_{K(n-1)} S^0 \simeq L_{K(n-1)} E_n^{h\mathbb{G}_n^1}$.

There is another conjecture of Hopkins related to chromatic splitting called the *algebraic chromatic splitting conjecture* [100, Section 14]. It states:

Conjecture 5.6.11 (Algebraic CSC). *Let p be an odd prime, possibly large with respect to n. Then*

$$\lim_i{}^s_{(E_n)_* E_n} (E_n)_t/(p, v_1, \ldots, v_{n-2}, v_{n-1}^i) \cong \begin{cases} (E_n)_*/(p, v_1, \ldots, v_{n-2}) & s = 0 \\ v_{n-1}^{-1}(E_n)_*/(p, v_1, \ldots, v_{n-2}) & s = 1 \\ 0 & s > 1. \end{cases}$$

Here, the limit (and its derived functors) is taken in the category of $(E_n)_* E_n$-comodules, where $(E_n)_* E_n = \pi_*(E_n \wedge E_n)$ is the group of non-completed E_n-cooperations. Provided that $(E_n)_* \zeta_n$ generates the $\lim^1$-term, Conjecture 5.6.11 implies that $1 \vee \zeta_n$ is an E_n-local equivalence, thereby verifying the chromatic splitting conjecture at height n for finite type $n-1$ complexes.

Remark 5.6.12. As explained above, the chromatic splitting conjecture is a fundamentally transchromatic statement. In a series of papers, Torii uses generalized character theory to study the relation between adjacent strata in the chromatic filtration. In particular, he shows in [119] that under the canonical map

$$\pi_{-1} L_{K(n-1)} S^0 \longrightarrow \pi_{-1} L_{K(n-1)} L_{K(n)} S^0$$

the class ζ_{n-1} maps non-trivially.

5.6.2 Invertibility and duality

In analogy with the problem of computing the group of units of a classical ring, an important aspect of understanding a symmetric monoidal category $(\mathcal{C}, \otimes, I)$ with unit I is to classify its invertible objects. An object $X \in \mathcal{C}$ is invertible if there exists another object $Y \in \mathcal{C}$ such that $X \otimes Y \cong I$ where I is the unit of the symmetric monoidal structure. If the collection of invertible objects forms a set, then it is an abelian group under $\otimes$ and this group is called the *Picard group* of $\mathcal{C}$, denoted $\mathrm{Pic}(\mathcal{C})$. The Picard group of a symmetric monoidal ∞-category $\mathcal{C}$ is defined to be the Picard group of the homotopy category of $\mathcal{C}$, i.e., we set $\mathrm{Pic}(\mathcal{C}) = \mathrm{Pic}(\mathrm{Ho}(\mathcal{C}))$.

If $\mathcal{C}$ is a triangulated category, the Picard group always contains a cyclic subgroup generated by the shift of the unit $I[1]$. For example, the Picard group of the stable homotopy category contains a copy of $\mathbb{Z}$ generated by S^1. In fact, in this case, there is nothing else and $\mathrm{Pic}(\mathrm{Sp}) \cong \mathbb{Z}\langle S^1 \rangle$, see [69]. We can view Sp_E as a symmetric monoidal category with product $L_E(- \wedge -)$ and unit $L_E S^0$. The objects $L_E S^n$ for $n \in \mathbb{Z}$ are invertible in this category. One of the fascinating aspects of E-local homotopy theory is that, for some choices of E, there are invertible object in Sp_E that are not of the form $L_E S^n$ for some $n \in \mathbb{Z}$. When $E = K(n)$ for $0 < n < \infty$, the Picard group is in fact much larger; we remark that for $E = K(n)$ and $E = E(n)$ and arbitrary n, the collection of isomorphism classes of invertible objects in Sp_E indeed forms a set, see [69, Proposition 7.6] and [74, Proposition 1.4].

The Picard group of the $K(n)$-local category (with p fixed and suppressed from the notation) is usually denoted by Pic_n. Note that, if $X \in \mathrm{Pic}_n$, its inverse is the Spanier–Whitehead dual of X in the $K(n)$-local category: $D_n X = F(X, L_{K(n)} S^0)$. By Galois descent [91, Proposition 10.10], there is an isomorphism

$$\mathrm{Pic}_n \cong \mathrm{Pic}_{K(n)}(\mathrm{Mod}_{E_n}^{\mathbb{G}_n}),$$

where $\mathrm{Mod}_{E_n}^{\mathbb{G}_n}$ is the $K(n)$-local category of $\mathbb{G}_n$-twisted E-module spectra. The right hand side has a natural algebraic analogue given by

$$\mathrm{Pic}_n^{\mathrm{alg}} := \mathrm{Pic}(\mathrm{Mod}_{(E_n)_*}^{\mathbb{G}_n})$$

where $\mathrm{Mod}_{(E_n)_*}^{\mathbb{G}_n}$ is the category of Morava modules (see Definition 5.3.20).

A Morava module M is in $\mathrm{Pic}_n^{\mathrm{alg}}$ if and only if it is free of rank one over $(E_n)_*$. Since $(E_n)_*$ is two periodic, $\mathrm{Pic}_n^{\mathrm{alg}}$ is naturally $\mathbb{Z}/2$-graded. Let $\mathrm{Pic}_n^{\mathrm{alg},0}$ be the subgroup of elements such that $M \cong (E_n)_*$ as $(E_n)_*$-modules. The latter can then be described (but not easily computed) as

$$\mathrm{Pic}_n^{\mathrm{alg},0} \cong H_c^1(\mathbb{G}_n, (E_n)_0^{\times}).$$

The functor that sends $X \in \mathrm{Sp}_{K(n)}$ to $(E_n)_*^{\vee} X$ induces a map, $\mathrm{Pic}_n \to \mathrm{Pic}_n^{\mathrm{alg}}$, and we define κ_n to be the kernel:

$$0 \longrightarrow \kappa_n \longrightarrow \mathrm{Pic}_n \longrightarrow \mathrm{Pic}_n^{\mathrm{alg}} \tag{5.6.1}$$

It is called the *exotic* Picard group fo $\mathrm{Sp}_{K(n)}$. Elements of Pic_n that are in κ_n are called exotic.

For $2(p-1) \geq n^2$, an argument that uses the sparseness of (5.3.4) shows that $\kappa_n = 0$ so that the map (5.6.1) $\mathrm{Pic}_n \to \mathrm{Pic}_n^{\mathrm{alg}}$ is an injection [69, Proposition 7.5]. However, it has been shown in many cases that κ_n is non-trivial. The following is a conjecture of Hopkins.

n	p	Pic_n	$\mathrm{Pic}_n^{\mathrm{alg}}$	κ_n	Reference
1	≥ 3	$\mathbb{Z}_p \times \mathbb{Z}/2(p-1)$	$\mathbb{Z}_p \times \mathbb{Z}/2(p-1)$	0	H–M–S [69]
1	2	$\mathbb{Z}_2 \times \mathbb{Z}/2 \times \mathbb{Z}/4$	$\mathbb{Z}_2 \times (\mathbb{Z}/2)^2$	$\mathbb{Z}/2$	H–M–S [69]
2	≥ 5	$\mathbb{Z}_p^2 \times \mathbb{Z}/2(p^2-1)$	$\mathbb{Z}_p^2 \times \mathbb{Z}/2(p^2-1)$	0	Due to Hopkins See Lader [80]
2	3	$\mathbb{Z}_3^2 \times \mathbb{Z}/16 \times (\mathbb{Z}/3)^2$	$\mathbb{Z}_3^2 \times \mathbb{Z}/16$	$(\mathbb{Z}/3)^2$	Karamanov [78] G–H–M–R [46] K–S [77]

Table 5.6.1: The table above contains some known values of Pic_n. Here, H–M–S stands for Hopkins–Mahowald–Sadofsky, G–M–H–R stands for Goerss–Henn–Mahowald–Rezk and K–S for Kamiya–Shimomura.

Conjecture 5.6.13. *The group κ_n is a finite p-group.*

In [53, Theorem 4.4.1], Heard proves that for p odd, κ_n is a direct product of cyclic p-groups. Note also that a positive answer to Conjecture 5.6.6 would imply Conjecture 5.6.13.

In [102], Pstrągowski proves that for $2(p-1) > n^2+n$, $\mathrm{Pic}_n \cong \mathrm{Pic}_n^{\mathrm{alg}}$. The question of whether or not $\mathrm{Pic}_n \to \mathrm{Pic}_n^{\mathrm{alg}}$ is surjective in general is open. Furthermore, the algebraic Picard group $\mathrm{Pic}_n^{\mathrm{alg}}$ is not known for any prime when $n > 2$. It is believed that $\mathrm{Pic}_n^{\mathrm{alg}}$ is finitely generated over $\mathbb{Z}_p$. In fact, the (folklore) expectation is that $\mathrm{Pic}_n^{\mathrm{alg}}$ is of rank two over $\mathbb{Z}_p$, with one summand generated by $L_{K(n)}S^1$ and the other by the spectrum $S\langle\det\rangle$ discussed below in Example 5.6.15 (b). Table 5.6.1 summarizes the current state of the literature on these questions. The second author, Bobkova, Goerss and Henn have been working towards identifying Pic_2 when $p=2$.

Remark 5.6.14. Analogously, there is a map from the Picard group $\mathrm{Pic}(\mathrm{Sp}_n)$ of the E_n-local category Sp_n to the category of $(E_n)_0E_n$-comodules. In [74], Hovey and Sadofsky use a variant of Theorem 5.3.12 to determine these Picard groups completely for large primes: They show that for $2(p-1) > n^2+n$, there is an isomorphism $\mathrm{Pic}(\mathrm{Sp}_n) \cong \mathbb{Z}$, generated by L_nS^1.

Example 5.6.15. We describe a few important elements in Pic_n.

(a) The spheres $L_{K(n)}S^m$ for $m \in \mathbb{Z}$ are all invertible.

(b) The determinant sphere $S\langle\det\rangle \in \mathrm{Pic}_n$. See [11] for a construction. It has the property that $(E_n)_*^\vee S\langle\det\rangle \cong (E_n)_*$ as $(E_n)_*$-modules, but with action of $\mathbb{G}_n$ twisted by the determinant. Its image in $\mathrm{Pic}_n^{\mathrm{alg},0} \cong H_c^1(\mathbb{G}_n, (E_n)_0^\times)$ is the homomorphism $\det\colon \mathbb{G}_n \to \mathbb{Z}_p^\times \subseteq (E_n)_0^\times$ of Definition 5.3.7.

(c) Given $\lambda \in (\pi_0 E_n^{h\mathbb{G}_n^1})^\times$, one can define a element S^λ via the fiber sequence

$$S^\lambda \longrightarrow E_n^{h\mathbb{G}_n^1} \xrightarrow{\psi-\lambda} E_n^{h\mathbb{G}_n^1}.$$

Some variation of this construction is discussed in Section 3.6 of [120]. If the Adams–Novikov filtration of λ is positive, one can show that S^λ is exotic. At $p=3$, the subgroup of $(\pi_0 E_2^{h\mathbb{G}_2^1})^\times$ of positive Adams–Novikov filtration is isomorphic to $\mathbb{Z}/3$. The elements in one of the factors of $\mathbb{Z}/3$ in $\kappa_2 \cong \mathbb{Z}/3 \times \mathbb{Z}/3$ are of the form S^λ.

(d) Some exotic elements cannot be constructed using (c). One can instead use finite resolutions to construct them. The first example is at $p=2$ and $n=1$. Recall that $K=E_1$ and $KO \simeq K^{hC_2}$. Since $K_*^\vee KO \cong K_*^\vee \Sigma^4 KO$, rather than choosing $\mathcal{E}_0 = \mathcal{E}_1 = KO$ to topologically realize (5.4.7), one can let $\mathcal{E}_0 = \mathcal{E}_1 = \Sigma^4 KO$ to get a fiber sequence

$$P_1 \longrightarrow \Sigma^4 KO \xrightarrow{\Sigma^4 5^2 \psi - 1} \Sigma^4 KO \tag{5.6.2}$$

where $\psi \in \mathbb{G}_1 \cong \mathbb{Z}_2^\times$ is as in (5.4.8). The fiber P_1 is a generator of κ_1. By construction, $P_1 \wedge KO \simeq \Sigma^4 KO$. See [46, Example 5.1] for more details.

Similarly, at $p=3$, there is an element $P_2 \in \kappa_2$ with the property that $P_2 \wedge E_2^{hG_{24}} \simeq \Sigma^{48} E_2^{hG_{24}}$; in fact,$P_2$ is a non-trivial exotic element that generates the other summand of $\mathbb{Z}/3 \subseteq \kappa_2$. See [46, Theorem 5.5]. It is constructed by modifying the realization of the duality resolution of Remark 5.5.10.

We now discuss another element of Pic_n which plays an important role in $K(n)$-local homotopy theory and brings us to the topic of Gross–Hopkins duality. As an application of the period map mentioned in Remark 5.3.11, Gross and Hopkins determine the dualizing complex of $\mathrm{Sp}_{K(n)}$, defined via a lift of Pontryagin duality for abelian groups. More precisely, the functor

$$I_n^*(-) = \mathrm{Hom}(\pi_{-*} M_n(-), \mathbb{Q}_p/\mathbb{Z}_p) \colon \mathrm{Sp}_{K(n)}^{\mathrm{op}} \longrightarrow \mathrm{Mod}_{\mathbb{Z}_p}^{\mathrm{graded}}$$

is cohomological and thus representable by an object $I_n \in \mathrm{Sp}_{K(n)}$, the *Gross–Hopkins dual of the sphere.* From an abstract point of view, it endows $\mathrm{Sp}_{K(n)}$ with a Serre duality functor, see [9].

By Theorem 5.3.3 and Corollary 5.3.13, I_n is determined by its Morava module $(E_n)_*^\vee(I_n) = \pi_* L_{K(n)}(E_n \wedge I_n)$ when p is large with respect to the height n; otherwise, one might have to twist by an exotic element of the $K(n)$-local Picard group. The spectrum I_n turns out to be invertible in $\mathrm{Sp}_{K(n)}$, and Gross and Hopkins use the period map (5.3.13) to show that there is an equivalence

$$I_n \simeq L_{K(n)}(S^{n^2-n} \wedge S\langle \det \rangle \wedge P_n),$$

where $P_n \in \kappa_n$ and $S\langle \det \rangle$ is as in Example 5.6.15 (b). This identification is also known as *Gross–Hopkins duality.* It turns out that P_1 for $p=2$ and P_2 for $p=3$ are the elements discussed in Example 5.6.15 (d).

It has now been shown in many cases where $\mathbb{G}_n$ has p-torsion that $P_n \not\simeq L_{K(n)} S^0$ by showing that $P_n \wedge E_n^{hF} \not\simeq E_n^{hF}$ for a suitable choice of subgroup $F \subseteq \mathbb{G}_n$. See [12] and [55]. These arguments rely on the intimate relationship between Gross–Hopkins duality and $K(n)$-local Spanier–Whitehead duality: For $X \in \mathrm{Sp}_{K(n)}$, let $I_n X = F(X, I_n)$. The invertibility of I_n implies that $I_n X \simeq D_n X \wedge I_n$ so studying $I_n X$ amounts to understanding P_n and $D_n X$.

We end this section with a few remarks on Spanier–Whitehead duality and, more specifically, on the problem of identifying $D_n E_n$. A first answer to this question due to Gross and Hopkins (see [117, Proposition 16]) states that there is a weak equivalence $\Sigma^{-n^2} E_n \to D_n E_n$

that induces an isomorphism of $\mathbb{G}_n$-modules on homotopy groups. This does not imply that D_nE_n is equivalent to $\Sigma^{-n^2}E_n$ as $\mathbb{G}_n$-equivariant spectra. But it does suggest that there is a *dualizing module*, that is, that D_nE_n is self-dual up to a twist.

In fact, the twist can be described as the $K(n)$-localization of a p-adic sphere. A first description of this sphere is given as the Spanier–Whitehead dual of

$$S^{\mathbb{G}_n} := \left(\mathrm{colim}_{m,\mathrm{tr}}\, \Sigma^\infty_+ B\mathbb{S}_n^m\right)^\wedge_p,$$

where $\mathbb{S}_n^m \subseteq \mathbb{S}_n$ is the subgroup of elements congruent to 1 modulo ξ^{nm} (as in (5.3.7)). Here, the colimit is taken over transfers and one can show that $S^{\mathbb{G}_n}$ has the homotopy type of a p-adic sphere of dimension n^2. This description is not practical for computations as the action of $\mathbb{G}_n$ on $S^{\mathbb{G}_n}$ induced by the conjugation actions on $B\mathbb{S}_n^m$ is mysterious. However, there is a conjectural description of the twist analogous to the identification of the dualizing object in the Wirthmüller isomorphism for compact Lie groups [82, Chapter III]. Let $\mathfrak{g}$ be the abelian group underlying $\mathcal{O}_{\mathbb{D}_n}$ (as defined in Section 5.3.2.1) endowed with the conjugation (or adjoint) action of $\mathbb{G}_n$. Let

$$S^{\mathfrak{g}} = \left(\mathrm{colim}_{m,\mathrm{tr}}\, \Sigma^\infty_+ Bp^m\mathfrak{g}\right)^\wedge_p.$$

Again, $S^{\mathfrak{g}}$ has the homotopy type of a p-adic sphere of dimension n^2 and the action of $\mathbb{G}_n$ on $Bp^m\mathfrak{g}$ induces an action on $S^{\mathfrak{g}}$.

Linearization Hypothesis. *There is a $\mathbb{G}_n$-equivariant equivalence*

$$S^{\mathfrak{g}} \simeq S^{\mathbb{G}_n}.$$

Inspired by Serre's definition of the dualizing module [112, Chapter I, §3.5], such a statement was first suggested to the experts by the strong connections in the $K(n)$-local category between Spanier–Whitehead duality, Brown–Comenetz duality, and Poincaré duality for the group $\mathbb{G}_n$ (see Gross–Hopkins [68] and Devinatz–Hopkins [38, Section 5]). The hypothesis is stated in work of Clausen [33, Section 6.4], not only for $\mathbb{G}_n$, but for any p-adic analytic group. Clausen has recently announced a proof of the Linearization Hypothesis in this general form.

The linearization hypothesis leads to a $\mathbb{G}_n$-equivariant equivalence

$$D_nE_n \simeq L_{K(n)}(E_n \wedge S^{-\mathfrak{g}}),$$

where $S^{-\mathfrak{g}} = F(S^{\mathfrak{g}}, S^0_p)$, a description that lends itself well to applications.

5.6.3 Compactifications and asymptotic algebraicity

We conclude this survey with a short overview of another recent direction in chromatic homotopy theory. As explained in Section 5.3.2 and demonstrated in the examples above, chromatic homotopy theory at a fixed height n simplifies when p grows large, the essential transition occurring when $p-1>n$. This leads to the question of how to isolate those phenomena that hold generically, i.e., for all primes p that are large with respect to the given height. The goal of this section is to outline a result of [17] that describes the *compactification* of chromatic homotopy theory, which is based on the notion of ultraproducts. In particular, this provides a model of the limit of the $K(n)$-local categories when $p \to \infty$ that captures the generic behavior of these categories. A different approach using Goerss–Hopkins obstruction theory that gives an algebraic triangulated model for suitably large primes has recently appeared in the work of Pstrągowski [101].

The Stone–Čech compactification of a topological space X is the initial compact Hausdorff space βX equipped with a continuous map ι from X. If X is discrete, βX can be modeled by the set of ultrafilters on X endowed with the Stone topology; recall that an *ultrafilter* on X is a set $\mathcal{U}$ of subsets of X such that whenever X is written as a disjoint union of finitely many subsets, then exactly one of them belongs to $\mathcal{U}$. The structure map $\iota\colon X \to \beta X$ sends a point $x \in X$ to the principal ultrafilter at x, i.e., the set of subsets of X that contain x. We denote the set of non-principal ultrafilters suggestively by $\partial\beta X = \beta X \setminus \iota(X)$. Assuming the axiom of choice, X is infinite if and only if $\partial\beta X$ is nonempty. (In fact, the existence of non-principal ultrafilters is weaker than the axiom of choice, but the key point here is that there is no constructive way to find non-principal ultrafilters.)

Moreover, one can show that an open subset of βX containing $\partial\beta X$ misses only finitely many points of ιX. This may be thought of as a topological manifestation of the fundamental theorem of ultraproducts due to Łoś, which says that for a collection of models $(M_i)_{i\in I}$ of a first order theory, there is an equivalence for any formula ϕ:

$$\left\{ \begin{matrix} \phi \text{ hold in } M_i \\ \text{for almost all } i \in I \end{matrix} \right\} \overset{\sim}{\Longleftrightarrow} \left\{ \begin{matrix} \phi \text{ hold in } \prod_{\mathcal{U}} M_i \\ \text{for all } \mathcal{U} \in \partial\beta I \end{matrix} \right\}, \tag{5.6.3}$$

where $\prod_{\mathcal{U}} M_i$ denotes the *ultraproduct* of the M_i at $\mathcal{U}$. While ultraproducts at non-principal ultrafilters thus capture generic information about the collection $(M_i)_{i\in I}$, they tend to also exhibit simplifying features. For example, for $\mathcal{U} \in \partial\beta\mathbf{P}$ a non-principal ultrafilter on the set of prime numbers $\mathbf{P}$, the ultraproduct of $(\mathbb{F}_p)_{p\in\mathbf{P}}$ turns out to be a rational field.

If $\mathcal{F}$ is a presheaf on a topological space X with values in a coefficient category $\mathcal{C}$ that is closed under filtered colimits and products, then one may construct a naive completion of $\mathcal{F}$ to be the presheaf on βX given as the composite

$$\widehat{\mathcal{F}}\colon \mathrm{Open}(\beta X)^{\mathrm{op}} \xrightarrow{\iota^*} \mathrm{Open}(X)^{\mathrm{op}} \xrightarrow{\mathcal{F}} \mathcal{C},$$

where ι^* denotes the inverse image functor. Assuming that X is discrete so that $\mathcal{F}$ is a just a collection of stalks $(\mathcal{F}_x)_{x\in X}$, the stalk of $\widehat{\mathcal{F}}$ at an ultrafilter $\mathcal{U} \in \beta X$ is given by

$$\widehat{\mathcal{F}}_{\mathcal{U}} \simeq \mathrm{colim}_{A\in\mathcal{U}} \prod\nolimits_{x\in A} \mathcal{F}_x, \tag{5.6.4}$$

where the filtered colimit is taken over the projection maps induced by inclusions $A \subseteq A'$ in $\mathcal{U}$. In particular, if $\mathcal{U} = \mathcal{U}_{x_0} \in \beta X$ is principal at a point $x_0 \in X$, then $\widehat{\mathcal{F}}_{\mathcal{U}_{x_0}} \simeq \widehat{\mathcal{F}}_{x_0}$. The formula (5.6.4) exhibits the stalk $\widehat{\mathcal{F}}_{\mathcal{U}}$ as a categorical generalization of ultraproducts: indeed, if all the $\mathcal{F}_x$ are nonempty and the coefficients are $\mathcal{C} = \mathrm{Set}$, then this recovers the usual notion of ultraproducts mentioned above.

In Section 5.2, we saw that the points of the spectrum Spc(Sp) are in bijective correspondence with pairs $(p, n) \in \mathbf{P} \times (\mathbb{N} \cup \{\infty\})$ with $(p, 0) \sim (q, 0)$ for all $p, q \in \mathbf{P}$. From the point of view of tensor triangular geometry, one may think of the category of spectra as behaving like a bundle of categories over the space Spc(Sp). Restricting to the discrete subspace $\mathbf{P} \times \{n\} \subset \mathrm{Spc}(\mathrm{Sp})$, this bundle should then be a disjoint union of the local categories $\mathrm{Sp}_{K(n)}$ for varying $p \in \mathbf{P}$. Therefore, the above formalism yields a diagram

$$\begin{array}{ccccc} \coprod_{p\in\mathbf{P}} \mathrm{Sp}_{K(n)} & & \widehat{\coprod_{p\in\mathbf{P}} \mathrm{Sp}_{K(n)}} & & \prod^{\flat}_{\mathcal{U}} \mathrm{Sp}_{K(n)} \\ \downarrow & & \downarrow & & \downarrow \\ \mathbf{P} \times \{n\} & \xrightarrow{\iota} & \widehat{\mathbf{P} \times \{n\}} \cong \beta\mathbf{P} & \xleftarrow{\supset} & \{\mathcal{U}\} \end{array}$$

in which the right vertical arrow exhibits $\prod^{\flat}_{\mathcal{U}} \mathrm{Sp}_{K(n)}$ as the stalk of the compactification over $\mathcal{U} \in \beta\mathbf{P}$. (Here, the superscript $\flat$ indicates that the coefficient category is $\mathcal{C} = \mathrm{Cat}^{\flat}_{\infty}$ for a suitable decoration $\flat$. For the details, we refer the interested reader to [17, 16].)

The final ingredient in the formulation of the asymptotic algebraicity of chromatic homotopy theory is the algebraic model itself. Informally speaking, this is given by the stable ∞-category of periodized ind-coherent sheaves on the formal stack $\hat{\mathcal{H}}(n)$ from Section 5.2.2. An algebraic avatar of this category has previously been studied by Franke [44]. Based on [17], the main theorem of [16] can now be stated as follows:

Theorem 5.6.16. *For any non-principal ultrafilter $\mathcal{U}$ on $\mathbf{P}$ there is a symmetric monoidal equivalence*

$$\prod^{\flat}_{\mathcal{U}} \mathrm{Sp}_{K(n)} \xrightarrow{\sim} \prod^{\flat}_{\mathcal{U}} \mathrm{IndCoh}(\hat{\mathcal{H}}(n))^{\mathrm{per}}$$

of $\mathbb{Q}$-linear stable ∞-categories.

The algebraic categories $\mathrm{IndCoh}(\hat{\mathcal{H}}(n))^{\mathrm{per}}$ admit explicit algebraic models in terms of certain comodule categories. Therefore, the above result gives a precise formulation of the empirical observation that height n chromatic homotopy theory becomes asymptotically algebraic when $p \to \infty$.

Acknowledgements

We would like to thank Paul Goerss, Hans-Werner Henn, Mike Hopkins and Vesna Stojanoska for clarifying conversations, and are grateful to Dustin Clausen and Haynes Miller for useful comments on an earlier version of this document. This material is based upon work supported by Danish National Research Foundation Grant DNRF92, the European Unions Horizon 2020 research and innovation programme under the Marie Sklodowska-Curie grant agreement No. 751794, and the National Science Foundation under Grant No. DMS-1725563. The authors would like to thank the Isaac Newton Institute for Mathematical Sciences for support and hospitality during the program Homotopy Harnessing Higher Structures when work on this chapter was undertaken. This work was supported by EPSRC Grant Number EP/R014604/1.

Bibliography

[1] J. F. Adams. On the structure and applications of the Steenrod algebra. *Comment. Math. Helv.*, 32:180–214, 1958.

[2] J. F. Adams. Operations of the nth kind in K-theory, and what we don't know about RP^{∞}. pp 1–9. London Math. Soc. Lecture Note Ser., No. 11, 1974.

[3] Michael Andrews and Haynes Miller. Inverting the Hopf map. *J. Topol.* 10(4):1145–1168, 2017.

[4] Vigleik Angeltveit. Uniqueness of Morava K-theory. *Compos. Math.*, 147(2):633–648, 2011.

[5] Omar Antolín-Camarena and Tobias Barthel. Chromatic fracture cubes. arXiv:1410.7271, 2014.

[6] Omar Antolín-Camarena and Tobias Barthel. A simple universal property of Thom ring spectra. *J. Topol.*, 12(1):56–78, 2019.

[7] Andrew Baker and Birgit Richter. Galois extensions of Lubin-Tate spectra. *Homology Homotopy Appl.*, 10(3):27–43, 2008.

[8] Tobias Barthel. Chromatic completion. *Proc. Amer. Math. Soc.*, 144(5):2263–2274, 2016.

[9] Tobias Barthel. Auslander-Reiten sequences, Brown-Comenetz duality, and the $K(n)$-local generating hypothesis. *Algebr. Represent. Theory*, 20(3):569–581, 2017.

[10] Tobias Barthel. A short introduction to the telescope and chromatic splitting conjectures. arXiv:1902.05046, 2019.

[11] Tobias Barthel, Agnès Beaudry, Paul G. Goerss, and Vesna Stojanoska. Constructing the determinant sphere using a Tate twist. arXiv:1810.06651, 2018.

[12] Tobias Barthel, Agnès Beaudry, and Vesna Stojanoska. Gross-Hopkins duals of higher real K-theory spectra. arXiv:1705.07036, 2017.

[13] Tobias Barthel and Drew Heard. The E_2-term of the $K(n)$-local E_n-Adams spectral sequence. *Topology Appl.*, 206:190–214, 2016.

[14] Tobias Barthel, Drew Heard, and Gabriel Valenzuela. The algebraic chromatic splitting conjecture for Noetherian ring spectra. *Math. Z.*, 290(3-4):1359–1375, 2018.

[15] Tobias Barthel, Drew Heard, and Gabriel Valenzuela. Local duality in algebra and topology. *Adv. Math.*, 335:563–663, 2018.

[16] Tobias Barthel, Tomer Schlank, and Nathaniel Stapleton. Monochromatic homotopy theory is asymptotically algebraic. arXiv:1903.10003, 2019.

[17] Tobias Barthel, Tomer Schlank, and Nathaniel Stapleton. Chromatic homotopy theory is asymptotically algebraic. arXiv:1711.00844, 2017.

[18] Tilman Bauer. Computation of the homotopy of the spectrum tmf. In *Groups, Homotopy and Configuration Spaces*, volume 13 of *Geom. Topol. Monogr.*, pages 11–40. Geom. Topol. Publ., 2008.

[19] Agnès Beaudry. The algebraic duality resolution at $p = 2$. *Algebr. Geom. Topol.*, 15(6):3653–3705, 2015.

[20] Agnès Beaudry. Towards the homotopy of the $K(2)$-local Moore spectrum at $p = 2$. *Adv. Math.*, 306:722–788, 2017.

[21] Agnès Beaudry, Irina Bobkova, Michael Hill, and Vesna Stojanoska. Invertible $K(2)$-Local E-Modules in C_4-Spectra. arXiv:1901.02109, 2019.

[22] Agnès Beaudry, Naiche Downey, Connor McCranie, Luke Meszar, Andy Riddle, and Peter Rock. Computations of orbits for the Lubin–Tate ring. *J. Homotopy Relat. Struct.*, 14(3):691–718, 2019.

[23] Agnès Beaudry, Paul G. Goerss, and Hans-Werner Henn. Chromatic splitting for the $K(2)$-local sphere at $p = 2$. arXiv:1712.08182, 2017.

[24] Mark Behrens. A modular description of the $K(2)$-local sphere at the prime 3. *Topology*, 45(2):343–402, 2006.

[25] Mark Behrens. The homotopy groups of $S_{E(2)}$ at $p \geq 5$ revisited. *Adv. Math.*, 230(2):458–492, 2012.

[26] Mark Behrens and Tyler Lawson. Topological automorphic forms. *Mem. Amer. Math. Soc.*, 204(958), 2010.

[27] Mark Behrens and Kyle Ormsby. On the homotopy of $Q(3)$ and $Q(5)$ at the prime 2. *Algebr. Geom. Topol.*, 16(5):2459–2534, 2016.

[28] David J. Benson, Srikanth B. Iyengar, and Henning Krause. Stratifying modular representations of finite groups. *Ann. of Math. (2)*, 174(3):1643–1684, 2011.

[29] Irina Bobkova and Paul G. Goerss. Topological resolutions in $K(2)$-local homotopy theory at the prime 2. *J. Topol.*, 11(4):917–956, 2018.

[30] A. K. Bousfield. The localization of spectra with respect to homology. *Topology*, 18(4):257–281, 1979.

[31] A. K. Bousfield. On the homotopy theory of K-local spectra at an odd prime. *Amer. J. Math.*, 107(4):895–932, 1985.

[32] Cédric Bujard. Finite subgroups of extended Morava stabilizer groups. arXiv:1206.1951, 2012.

[33] Dustin Clausen. p-adic J-homomorphisms and a product formula. arXiv:1110.5851, 2011.

[34] Daniel G. Davis. Homotopy fixed points for $L_{K(n)}(E_n \wedge X)$ using the continuous action. *J. Pure Appl. Algebra*, 206(3):322–354, 2006.

[35] Daniel G. Davis and Gereon Quick. Profinite and discrete G-spectra and iterated homotopy fixed points. *Algebr. Geom. Topol.*, 16(4):2257–2303, 2016.

[36] Ethan S. Devinatz. A counterexample to a BP-analogue of the chromatic splitting conjecture. *Proc. Amer. Math. Soc.*, 126(3):907–911, 1998.

[37] Ethan S. Devinatz. A Lyndon-Hochschild-Serre spectral sequence for certain homotopy fixed point spectra. *Trans. Amer. Math. Soc.*, 357(1):129–150, 2005.

[38] Ethan S. Devinatz and Michael J. Hopkins. The action of the Morava stabilizer group on the Lubin-Tate moduli space of lifts. *Amer. J. Math.*, 117(3):669–710, 1995.

[39] Ethan S. Devinatz and Michael J. Hopkins. Homotopy fixed point spectra for closed subgroups of the Morava stabilizer groups. *Topology*, 43(1):1–47, 2004.

[40] Ethan S. Devinatz, Michael J. Hopkins, and Jeffrey H. Smith. Nilpotence and stable homotopy theory. I. *Ann. of Math. (2)*, 128(2):207–241, 1988.

[41] Jean Dieudonné. Groupes de Lie et hyperalgèbres de Lie sur un corps de caractéristique $p > 0$. VII. *Math. Ann.*, 134:114–133, 1957.

[42] J. D. Dixon, M. P. F. du Sautoy, A. Mann, and D. Segal. *Analytic pro-p groups*, volume 61 of *Cambridge Stud. Adv. Math.* Cambridge University Press, second edition, 1999.

[43] A. D. Elmendorf, I. Kriz, M. A. Mandell, and J. P. May. *Rings, modules, and algebras in stable homotopy theory*, volume 47 of *Math. Surv. Monogr.* Amer. Math. Soc., 1997. With an appendix by M. Cole.

[44] Jens Franke. Uniqueness theorems for certain triangulated categories with an Adams spectral sequence. Available at `http://www.math.uiuc.edu/K-theory/0139/`.

[45] Paul G. Goerss. Quasi-coherent sheaves on the moduli stack of formal groups. arXiv:0802.0996, 2008.

[46] Paul Goerss, Hans-Werner Henn, Mark Mahowald, and Charles Rezk. On Hopkins' Picard groups for the prime 3 and chromatic level 2. *J. Topol.*, 8(1):267–294, 2015.

[47] Paul G. Goerss and Hans-Werner Henn. The Brown-Comenetz dual of the $K(2)$-local sphere at the prime 3. *Adv. Math.*, 288:648–678, 2016.

[48] Paul G. Goerss, Hans-Werner Henn, and Mark E. Mahowald. The rational homotopy of the $K(2)$-local sphere and the chromatic splitting conjecture for the prime 3 and level 2. *Doc. Math.*, 19:1271–1290, 2014.

[49] Paul G. Goerss, Hans-Werner Henn, Mark E. Mahowald, and Charles Rezk. A resolution of the $K(2)$-local sphere at the prime 3. *Ann. of Math. (2)*, 162(2):777–822, 2005.

[50] P. G. Goerss and M. J. Hopkins. Moduli spaces of commutative ring spectra. In *Structured ring spectra*, volume 315 of *London Math. Soc. Lecture Note Ser.*, pages 151–200. Cambridge Univ. Press, 2004.

[51] J. P. C. Greenlees. Tate cohomology in axiomatic stable homotopy theory. In *Cohomological methods in homotopy theory (Bellaterra, 1998)*, volume 196 of *Progr. Math.*, pages 149–176. Birkhäuser, 2001.

[52] Jeremy Hahn and XiaoLin Danny Shi. Real orientations of Morava E-theories. arXiv:1707.03413, 2017.

[53] Drew Heard. *Morava modules and the $K(n)$-local Picard group.* PhD thesis, University of Melbourne, 2014.

[54] Drew Heard. The Tate spectrum of the higher real K-theories at height $n = p - 1$. arXiv:1501.07759, 2015.

[55] Drew Heard, Guchuan Li, and XiaoLin Danny Shi. Picard groups and duality for Real Morava E-theories. arXiv:1810.05439, 2018.

[56] Hans-Werner Henn. Centralizers of elementary abelian p-subgroups and mod-p cohomology of profinite groups. *Duke Math. J.*, 91(3):561–585, 1998.

[57] Hans-Werner Henn. On finite resolutions of $K(n)$-local spheres. In *Elliptic cohomology*, volume 342 of *London Math. Soc. Lecture Note Ser.*, pages 122–169. Cambridge Univ. Press, 2007.

[58] Hans-Werner Henn. A mini-course on Morava stabilizer groups and their cohomology. In *Algebraic topology*, volume 2194 of *Lecture Notes in Math.*, pages 149–178. Springer, 2017.

[59] Hans-Werner Henn. The Centralizer resolution of the $K(2)$-local sphere at the prime 2. In preparation.

[60] Hans-Werner Henn, Nasko Karamanov, and Mark E. Mahowald. The homotopy of the $K(2)$-local Moore spectrum at the prime 3 revisited. *Math. Z.*, 275(3-4):953–1004, 2013.

[61] Thomas Hewett. Finite subgroups of division algebras over local fields. *J. Algebra*, 173(3):518–548, 1995.

[62] Thomas Hewett. Normalizers of finite subgroups of division algebras over local fields. *Math. Res. Lett.*, 6(3-4):271–286, 1999.

[63] M. A. Hill, M. J. Hopkins, and D. C. Ravenel. On the nonexistence of elements of Kervaire invariant one. *Ann. of Math. (2)*, 184(1):1–262, 2016.

[64] Michael A. Hill, Michael J. Hopkins, and Douglas C. Ravenel. The slice spectral sequence for the C_4 analog of real K-theory. *Forum Math.*, 29(2):383–447, 2017.

[65] Michael A. Hill, XiaoLin Danny Shi, Guozhen Wang, and Zhouli Xu. The slice spectral sequence of a C_4-equivariant height-4 Lubin- Tate theory. arXiv:1811.07960, 2018.

[66] M. J. Hopkins and B. H. Gross. Equivariant vector bundles on the Lubin-Tate moduli space. In *Topology and representation theory (Evanston, IL, 1992)*, volume 158 of *Contemp. Math.*, pages 23–88. Amer. Math. Soc., 1994.

[67] Michael J. Hopkins. Global methods in homotopy theory. In *Homotopy theory (Durham, 1985)*, volume 117 of *London Math. Soc. Lecture Note Ser.*, pages 73–96. Cambridge Univ. Press, 1987.

[68] Michael J. Hopkins and Benedict H. Gross. The rigid analytic period mapping, Lubin-Tate space, and stable homotopy theory. *Bull. Amer. Math. Soc. (N.S.)*, 30(1):76–86, 1994.

[69] Michael J. Hopkins, Mark Mahowald, and Hal Sadofsky. Constructions of elements in Picard groups. In *Topology and representation theory (Evanston, IL, 1992)*, volume 158 of *Contemp. Math.*, pages 89–126. Amer. Math. Soc., 1994.

[70] Michael J. Hopkins and Haynes R. Miller. Elliptic curves and stable homotopy I. In *Topological modular forms*, volume 201 of *Math. Surveys Monogr.*, pages 209–260. Amer. Math. Soc., 2014.

[71] Michael J. Hopkins and Jeffrey H. Smith. Nilpotence and stable homotopy theory. II. *Ann. of Math. (2)*, 148(1):1–49, 1998.

[72] Mark Hovey. Bousfield localization functors and Hopkins' chromatic splitting conjecture. In *The Čech centennial (Boston, MA, 1993)*, volume 181 of *Contemp. Math.*, pages 225–250. Amer. Math. Soc., 1995.

[73] Mark Hovey. Operations and co-operations in Morava E-theory. *Homology Homotopy Appl.*, 6(1):201–236, 2004.

[74] Mark Hovey and Hal Sadofsky. Invertible spectra in the $E(n)$-local stable homotopy category. *J. London Math. Soc. (2)*, 60(1):284–302, 1999.

[75] Mark Hovey, Brooke Shipley, and Jeff Smith. Symmetric spectra. *J. Amer. Math. Soc.*, 13(1):149–208, 2000.

[76] Mark Hovey and Neil P. Strickland. Morava K-theories and localisation. *Mem. Amer. Math. Soc.*, 139(666), 1999.

[77] Yousuke Kamiya and Katsumi Shimomura. A relation between the Picard group of the $E(n)$-local homotopy category and $E(n)$-based Adams spectral sequence. In *Homotopy theory: relations with algebraic geometry, group cohomology, and algebraic K-theory*, volume 346 of *Contemp. Math.*, pages 321–333. Amer. Math. Soc., 2004.

[78] Nasko Karamanov. On Hopkins' Picard group Pic_2 at the prime 3. *Algebr. Geom. Topol.*, 10(1):275–292, 2010.

[79] Jan Kohlhaase. On the Iwasawa theory of the Lubin-Tate moduli space. *Compos. Math.*, 149(5):793–839, 2013.

[80] Olivier Lader. *Une résolution projective pour le second groupe de Morava pour $p \geq 5$ et applications.* PhD thesis, University of Strasbourg, 2013.

[81] Michel Lazard. Groupes analytiques p-adiques. *Inst. Hautes Études Sci. Publ. Math.*, (26):389–603, 1965.

[82] L. G. Lewis, Jr., J. P. May, M. Steinberger, and J. E. McClure. *Equivariant stable homotopy theory*, volume 1213 of *Lecture Notes in Math.* Springer-Verlag, 1986. With contributions by J. E. McClure.

[83] Jonathan Lubin. One-parameter formal Lie groups over $\mathfrak{p}$-adic integer rings. *Ann. of Math. (2)*, 80:464–484, 1964.

[84] Jonathan Lubin and John Tate. Formal complex multiplication in local fields. *Ann. of Math. (2)*, 81:380–387, 1965.

[85] Jacob Lurie. Elliptic cohomology II: Orientations. Available at `http://www.math.harvard.edu/~lurie/`.

[86] Jacob Lurie. Higher algebra. Available at `http://www.math.harvard.edu/~lurie/`.

[87] Mark Mahowald. *bo*-resolutions. *Pacific J. Math.*, 92(2):365–383, 1981.

[88] Mark Mahowald, Douglas Ravenel, and Paul Shick. The triple loop space approach to the telescope conjecture. In *Homotopy methods in algebraic topology (Boulder, CO, 1999)*, volume 271 of *Contemp. Math.*, pages 217–284. Amer. Math. Soc., 2001.

[89] Mark E. Mahowald and Charles Rezk. Topological modular forms of level 3. *Pure Appl. Math. Q.*, 5(2, Special Issue: In honor of Friedrich Hirzebruch. Part 1):853–872, 2009.

[90] M. A. Mandell, J. P. May, S. Schwede, and B. Shipley. Model categories of diagram spectra. *Proc. London Math. Soc. (3)*, 82(2):441–512, 2001.

[91] Akhil Mathew. The Galois group of a stable homotopy theory. *Adv. Math.*, 291:403–541, 2016.

[92] Haynes Miller. Finite localizations. *Bol. Soc. Mat. Mexicana (2)*, 37(1-2):383–389, 1992. Papers in honor of José Adem.

[93] Haynes R. Miller. Sheaves, gradings, and the exact functor theorem. In *Homotopy Theory: Tools and Applications*, volume 729 of *Contemp. Math.*, pages 205–220. Amer. Math. Soc., 2019.

[94] Haynes R. Miller. On relations between Adams spectral sequences, with an application to the stable homotopy of a Moore space. *J. Pure Appl. Algebra*, 20(3):287–312, 1981.

[95] Haynes R. Miller, Douglas C. Ravenel, and W. Stephen Wilson. Periodic phenomena in the Adams-Novikov spectral sequence. *Ann. of Math. (2)*, 106(3):469–516, 1977.

[96] Stephen A. Mitchell. Finite complexes with $A(n)$-free cohomology. *Topology*, 24(2):227–246, 1985.

[97] Jack Morava. Noetherian localisations of categories of cobordism comodules. *Ann. of Math. (2)*, 121(1):1–39, 1985.

[98] Niko Naumann. The stack of formal groups in stable homotopy theory. *Adv. Math.*, 215(2):569–600, 2007.

[99] S. P. Novikov. Methods of algebraic topology from the point of view of cobordism theory. *Izv. Akad. Nauk SSSR Ser. Mat.*, 31:855–951, 1967.

[100] Eric Peterson. Report on E-theory conjectures seminar.

[101] Piotr Pstrągowski. Chromatic homotopy is algebraic when $p > n^2 + n + 1$. arXiv:1810.12250, 2018.

[102] Piotr Pstrągowski. Chromatic Picard groups at large primes. arXiv:1811.05415, 2018.

[103] Gereon Quick. Continuous homotopy fixed points for Lubin-Tate spectra. *Homology Homotopy Appl.*, 15(1):191–222, 2013.

[104] Daniel Quillen. On the formal group laws of unoriented and complex cobordism theory. *Bull. Amer. Math. Soc.*, 75:1293–1298, 1969.

[105] Douglas C. Ravenel. Localization with respect to certain periodic homology theories. *Amer. J. Math.*, 106(2):351–414, 1984.

[106] Douglas C. Ravenel. *Complex cobordism and stable homotopy groups of spheres*, volume 121 of *Pure Appl. Math.* Academic Press, Inc., 1986.

[107] Douglas C. Ravenel. *Nilpotence and periodicity in stable homotopy theory*, volume 128 of *Ann. of Math. Stud.*. Princeton University Press, 1992. Appendix C by Jeff Smith.

[108] Charles Rezk. Notes on the Hopkins-Miller theorem. In *Homotopy theory via algebraic geometry and group representations (Evanston, IL, 1997)*, volume 220 of *Contemp. Math.*, pages 313–366. Amer. Math. Soc., 1998.

[109] John Rognes. Galois extensions of structured ring spectra. Stably dualizable groups. *Mem. Amer. Math. Soc.*, 192(898), 2008.

[110] Hal Sadofsky. Hopkins' and Mahowald's picture of Shimomura's v_1-Bockstein spectral sequence calculation. In *Algebraic topology (Oaxtepec, 1991)*, volume 146 of *Contemp. Math.*, pages 407–418. Amer. Math. Soc., 1993.

[111] Stefan Schwede. The stable homotopy category is rigid. *Ann. of Math. (2)*, 166(3):837–863, 2007.

[112] Jean-Pierre Serre. *Galois cohomology*. *Springer Monogr. Math.* Springer-Verlag, English edition, 2002. Translated from the French by Patrick Ion and revised by the author.

[113] Katsumi Shimomura and Xiangjun Wang. The Adams-Novikov E_2-term for $\pi_*(L_2S^0)$ at the prime 2. *Math. Z.*, 241(2):271–311, 2002.

[114] Katsumi Shimomura and Atsuko Yabe. The homotopy groups $\pi_*(L_2S^0)$. *Topology*, 34(2):261–289, 1995.

[115] Brooke Shipley. $H\mathbb{Z}$-algebra spectra are differential graded algebras. *Amer. J. Math.*, 129(2):351–379, 2007.

[116] Neil Strickland. Level three structures. arXiv:1803.09962, 2018.

[117] Neil P. Strickland. Gross-Hopkins duality. *Topology*, 39(5):1021–1033, 2000.

[118] Peter Symonds and Thomas Weigel. Cohomology of p-adic analytic groups. In *New horizons in pro-p groups*, volume 184 of *Progr. Math.*, pages 349–410. Birkhäuser Boston, 2000.

[119] Takeshi Torii. $K(n)$-localization of the $K(n+1)$-local E_{n+1}-Adams spectral sequences. *Pacific J. Math.*, 250(2):439–471, 2011.

[120] Craig Westerland. A higher chromatic analogue of the image of J. *Geom. Topol.*, 21(2):1033–1093, 2017.

MAX PLANCK INSTITUTE FOR MATHEMATICS, VIVATSGASSE 7, 53111 BONN, GERMANY
E-mail address: tbarthel@mpim-bonn.mpg.de

DEPARTMENT OF MATHEMATICS, UNIVERSITY OF COLORADO AT BOULDER, BOULDER, CO 80309, U.S.A.
E-mail address: Agnes.Beaudry@colorado.edu

6

Topological modular and automorphic forms

Mark Behrens

6.1 Introduction

The spectrum of topological modular forms (TMF) was first introduced by Hopkins and Miller [53], [48], [47], and Goerss and Hopkins constructed it as an E_∞ ring spectra (see [17]). Lurie subsequently gave a conceptual approach to TMF using his theory of spectral algebraic geometry [69]. Lurie's construction relies on a general theorem [70], [71], which was used by the author and Lawson to construct spectra of topological automorphic forms (TAF) [19].

The goal of this article is to give an accessible introduction to TMF and TAF spectra. Besides the articles mentioned above, there already exist many excellent such surveys (see [51], [86], [65], [36], [37], [31]). Our intention is to give an account which is somewhat complementary to these existing surveys. We assume the reader knows about the stable homotopy category, and knows some basic algebraic geometry, and attempt to give the reader a concrete understanding of the big picture while deemphasizing many of the technical details. Hopefully, the reader does not find the inevitable sins of omission to be too grievous.

In Section 6.2 we recall the definition of a complex orientable ring spectrum E, and its associated formal group law F_E. We then explain how an algebraic group G also gives rise to a formal group law $\widehat{G}$, and define elliptic cohomology theories to be complex orientable cohomology theories whose formal group laws arise from elliptic curves. We explain how a theorem of Landweber proves the existence of certain *Landweber exact* elliptic cohomology theories.

We proceed to define topological modular forms in Section 6.3. We first begin by recalling the definition of classical modular forms as sections of powers of a certain line bundle on the compactification $\overline{\mathcal{M}}_{ell}$ of the moduli stack $\mathcal{M}_{ell}$ of elliptic curves. Then we state a theorem of Goerss-Hopkins-Miller, which states that there exists a sheaf of E_∞ ring spectra $\mathcal{O}^{top}$ on the étale site of $\overline{\mathcal{M}}_{ell}$ whose sections over an affine

$$\mathrm{spec}(R) \xrightarrow[\text{etale}]{C} \mathcal{M}_{ell}$$

recover the Landweber exact elliptic cohomology theory associated to the elliptic curve C it classifies. The spectrum Tmf is defined to be the global sections of this sheaf:

$$\mathrm{TMF} := \mathcal{O}^{top}(\overline{\mathcal{M}}_{ell}).$$

Mathematics Subject Classification. 55N34, 55N22, 55P42.
Key words and phrases. elliptic cohomology, topological modular forms, topological automorphic forms.

We compute $\pi_* \mathrm{Tmf}[1/6]$, and use that to motivate the definition of connective topological modular forms (tmf) as the connective cover of Tmf, and periodic topological modular forms (TMF) as the sections over the non-compactified moduli stack:

$$\mathrm{TMF} := \mathcal{O}^{top}(\mathcal{M}_{ell}).$$

The homotopy groups of TMF at the primes 2 and 3 are more elaborate. While we do not recount the details of these computations, we do indicate the setup in Section 6.4, and state the results in a form that we hope is compact and understandable. The computation of $\pi_* \mathrm{Tmf}_{(p)}$ for $p = 2, 3$ is then discussed in Section 6.5. We introduce the notion of the height of a formal group law, and use it to create a computable cover of the compactified moduli stack $(\overline{\mathcal{M}}_{ell})_{\mathbb{Z}_{(p)}}$. By taking connective covers, we recover the homotopy of groups of $\mathrm{tmf}_{(p)}$.

In Section 6.6, we go big picture. We explain how complex cobordism associates to certain ring spectra E a stack

$$\mathcal{X}_E \to \mathcal{M}_{fg}$$

over the moduli stack of formal group laws. The sheaf $\mathcal{O}^{top}$ serves as a partial inverse to $\mathcal{X}_{(-)}$, in the sense that where it is defined, we have

$$\mathcal{O}^{top}(\mathcal{X}_E) \simeq E,$$
$$\mathcal{X}_{\mathcal{O}^{top}(\mathcal{U})} \simeq \mathcal{U}.$$

We describe the height filtration of the moduli stack of formal groups $\mathcal{M}_{fg}$, and explain how chromatic localizations of the sphere realize this filtration in topology. The stacks associated to chromatic localizations of a ring spectrum E are computed by pulling back the height filtration to $\mathcal{X}_E$. We then apply this machinery to Tmf to compute its chromatic localizations, and explain how chromatic fracture is closely connected to the approach to $\pi_* \mathrm{Tmf}$ discussed in Section 6.5.

We then move on to discuss Lurie's theorem, which expands $\mathcal{O}^{top}$ to the étale site of the moduli space of *p-divisible groups*. After recalling the definition, we state Lurie's theorem, and explain how his theorem simultaneously recovers the Goerss-Hopkins-Miller theorem on Morava E-theory, and the Goerss-Hopkins-Miller sheaf $\mathcal{O}^{top}$ on $\mathcal{M}_{ell}$. We then discuss a class of moduli stacks of abelian varieties (*PEL Shimura stacks of type* $U(1, n-1)$) which give rise to spectra of topological automorphic forms [19].

There are many topics which should have appeared in this survey, but regrettably do not, such as the Witten orientation, the connection to 2-dimensional field theories, spectral algebraic geometry, and equivariant elliptic cohomology, to name a few. We compensate for this deficiency in Section 6.8 with a list of such topics and references to the literature for further reading.

6.2 Elliptic cohomology theories

Complex orientable ring spectra

Let E be a (homotopy associative, homotopy commutative) ring spectrum.

Definition 6.2.1. A *complex orientation* of E is an element

$$x \in \widetilde{E}^*(\mathbb{C}P^\infty)$$

such that the restriction

$$x|_{\mathbb{C}P^1} \in \widetilde{E}^*(\mathbb{C}P^1)$$

is a generator (as an E_*-module). A ring spectrum that admits a complex orientation is called *complex orientable.*

For complex oriented ring spectra E, the Atiyah-Hirzebruch spectral sequence collapses to give

$$E^*(\mathbb{C}P^\infty) = E_*[[x]], \tag{6.2.1}$$
$$E^*(\mathbb{C}P^\infty \times \mathbb{C}P^\infty) = E_*[[x_1, x_2]] \tag{6.2.2}$$

where x_i is the pullback of x under the ith projection.

Consider the map

$$\mu : \mathbb{C}P^\infty \times \mathbb{C}P^\infty \to \mathbb{C}P^\infty$$

that classifies the universal tensor product of line bundles. Novikov [82, p.853] and Quillen [85] (see also [1]) observed that because μ gives $\mathbb{C}P^\infty$ the structure of a homotopy commutative, homotopy associative H-space, the power series

$$F_E(x_1, x_2) := \mu^* x \in E_*[[x_1, x_2]]$$

is a (commutative, 1 dimensional) *formal group law* over E_*, in the sense that it satisfies

1. $F_E(x, 0) = F_E(0, x) = x$,
2. $F_E(x_1, F_E(x_2, x_3)) = F_E(F_E(x_1, x_2), x_3)$,
3. $F_E(x_1, x_2) = F_E(x_2, x_1)$.

Example 6.2.2. Let $E = H\mathbb{Z}$, the integral Eilenberg-Mac Lane spectrum. Then the complex orientation x is a generator of $H^2(\mathbb{C}P^\infty)$, and

$$F_{H\mathbb{Z}}(x_1, x_2) = x_1 + x_2.$$

This is the *additive formal group law* F_{add}.

Example 6.2.3. Let $E = KU$, the complex K-theory spectrum. Then the class

$$x := [L_{\text{can}}] - 1 \in \widetilde{KU}^0(\mathbb{C}P^\infty)$$

(where L_{can} is the canonical line bundle) gives a complex orientation for KU, and

$$F_{KU}(x_1, x_2) = x_1 + x_2 + x_1 x_2.$$

This is the *multiplicative formal group law* F_{mult}.

Example (6.2.3) above is an example of the following.

Definition 6.2.4. An *even periodic* ring spectrum is a ring spectrum E so that

$$\pi_{\text{odd}} E = 0 \tag{6.2.3}$$

and such that E_2 contains a unit.

It is easy to see using a collapsing Atiyah-Hirzebruch spectral sequence argument that (6.2.3) is enough to guarantee the complex orientability of an even periodic ring spectrum E. The existence of the unit in E_2 implies one can take the complex orientation to be a class

$$x \in \widetilde{E}^0(\mathbb{C}P^\infty)$$

giving

$$E^0(\mathbb{C}P^\infty) = E_0[[x]]. \tag{6.2.4}$$

It follows that in the even periodic case, for such choices of complex orientation, we can regard the formal group law F_E as a formal group law over E_0, and (6.2.4) can be regarded as saying

$$E^0(\mathbb{C}P^\infty) = \mathcal{O}_{F_E}$$

where the latter is the ring of functions on the formal group law. Then it follows that we have a canonical identification

$$E_2 = \widetilde{E}^0(\mathbb{C}P^1) = (x)/(x)^2 = T_0^* F_E.$$

Here (x) is the ideal generated by x in $E^0(\mathbb{C}P^\infty)$ and $T_0^* F_E$ is the cotangent space of F_E at 0. The even periodicity of E then gives

$$E_{2i} \xleftarrow{\cong} E_2^{\otimes_{E_0} i} = (T_0^* F_E)^{\otimes i}. \tag{6.2.5}$$

Here (6.2.5) even makes sense for i negative: since E_2 is a free E_0-module of rank 1, it is invertible (in an admittedly trivial manner), and $T_0^* F_E$ is invertible since it is a line bundle over $\text{spec}(E_0)$.

Formal groups associated to algebraic groups

Formal group laws also arise in the context of algebraic geometry. Let G be a 1-dimensional commutative algebraic group over a commutative ring R. If the line bundle $T_e G$ (over $\text{spec}(R)$) is trivial, there exists a coordinate x of G at the identity $e \in G$. We shall call such group schemes *trivializable*. In this case the group structure

$$G \times G \to G$$

can be expressed locally in terms of the coordinate x as a power series

$$\widehat{G}(x_1, x_2) \in R[[x_1, x_2]].$$

The unitality, associativity, and commutativity of the group structure on G makes $\widehat{G}$ a formal group law over R. The formal groups in Examples 6.2.2 and 6.2.3 arise in this manner from the additive and multiplicative groups $\mathbb{G}_a$ and $\mathbb{G}_m$ (defined over $\mathbb{Z}$) by making appropriate choices of coordinates:

$$\widehat{\mathbb{G}}_a = F_{add},$$
$$\widehat{\mathbb{G}}_m = F_{mult}.$$

It turns out that if we choose different coordinates/complex orientations, we will still get isomorphic formal group laws. A *homomorphism* $f : F \to F'$ of formal group laws over R is a formal power series

$$f(x) \in R[[x]]$$

satisfying

$$f(F(x_1, x_2)) = F'(f(x_1), f(x_2)).$$

If the power series $f(x)$ is invertible (with respect to composition) then we say that it is an isomorphism. Clearly, choosing a different coordinate on a trivializable commutative 1-dimensional algebraic group gives an isomorphic formal group law. One similarly has the following proposition.

Proposition 6.2.5. *Suppose that x and x' are two complex orientations of a complex orientable ring spectrum E, with corresponding formal group laws F and F'. Then there is a canonical isomorphism between F and F'.*

Proof. Using (6.2.1), we deduce that $x' = f(x) \in E_*[[x]]$. It is a simple matter to use the resulting change of coordinates to verify that f is an isomorphism from F to F'. □

Remark 6.2.6. A *formal group* is a formal group law where we forget the data of the coordinate. Thus Proposition 6.2.5 is asserting that a complex orientable ring spectrum gives rise to a canonical formal group.

The only 1-dimensional connected algebraic groups over an algebraically closed field are $\mathbb{G}_a$, $\mathbb{G}_m$, and elliptic curves.[1] As we have shown that there are complex orientable ring spectra that yield the formal groups of the first two, it is reasonable to consider the case of elliptic curves.

---

[1] An elliptic curve over an algebraically closed field k is a smooth proper connected curve over k of genus 1, with a chosen k-point (which serves as the identity of the resulting algebraic group).

The elliptic case

Definition 6.2.7. [3] An *elliptic cohomology theory* consists of a triple

$$(E, C, \alpha)$$

where

$$\begin{aligned} E &= \text{an even periodic ring spectrum,} \\ C &= \text{a trivializable elliptic curve over } E_0, \\ (\alpha : \widehat{C} \xrightarrow{\cong} F_E) &= \text{an isomorphism of formal group laws.} \end{aligned}$$

Remark 6.2.8. Note that every elliptic curve C that admits a Weierstrass presentation

$$C : y^2 + a_1xy + a_3y = x^3 + a_2x^2 + a_4x + a_6$$

is trivializable, since $z = x/y$ is a coordinate at $e \in C$.

For an elliptic cohomology theory (E, C, α), the map α_* gives an isomorphism

$$T_e^*C \cong T_0^*\widehat{C} \underset{\cong}{\xleftarrow{\alpha^*}} T_0^*F_E.$$

It follows that we have a canonical isomorphism

$$E_{2i} \cong (T_e^*C)^{\otimes i}. \tag{6.2.6}$$

It is reasonable to ask when elliptic cohomology theories exist. This was first studied by Landweber, Ravenel, and Stong [62] using the Landweber Exact Functor Theorem [61]. Here we state a reformulation of this theorem which appears in [81] (this perspective originates with Franke [32] and Hopkins [49]).

Theorem 6.2.9 (Landweber Exact Functor Theorem). *Suppose that F is a formal group law over R whose classifying map*

$$\mathrm{spec}(R) \xrightarrow{F} \mathcal{M}_{fg}$$

to the moduli stack of formal groups is flat. Then there exists a unique (in the homotopy category of ring spectra) even periodic ring spectrum E with $E_0 = R$ and $F_E \cong F$.

Corollary 6.2.10. *Suppose that C is a trivializable elliptic curve over R whose associated formal group law $\widehat{C}$ satisfies the hypotheses of the Landweber exact functor theorem. Then there exists an elliptic cohomology theory E_C associated to the elliptic curve C.*

Remark 6.2.11. For us, a *stack* is a functor

$$\mathrm{Rings}^{op} \to \mathrm{Groupoids}$$

that satisfies a descent condition with respect to a given Grothendieck topology. The moduli stack of formal groups $\mathcal{M}_{fg}$ associates to a ring R the groupoid whose objects are formal group schemes over R that are Zariski locally (in spec(R)) isomorphic to the formal affine line $\widehat{\mathbb{A}}^1$, and whose morphisms are the isomorphisms of such.

Remark 6.2.12. Landweber's original formulation of his exact functor theorem did not use the language of stacks, but rather gave a simple to check explicit criterion that is equivalent to the flatness condition.

The problem with Landweber's theorem is that while it gives a functor

$$\{\text{Landweber flat formal groups}\} \to \mathrm{Ho}(\text{Spectra}),$$

this functor does not refine to a point-set level functor to spectra.

6.3 Topological modular forms

Classical modular forms

Let $\mathcal{M}_{ell}$ denote the moduli stack of elliptic curves (over spec($\mathbb{Z}$)). It is the stack whose groupoid of R-points is the groupoid of elliptic curves over R and isomorphisms. Consider the line bundle ω on $\mathcal{M}_{ell}$ whose fiber over an elliptic curve C is given by the cotangent space at the identity

$$\omega_C = T_e^* C.$$

The moduli stack of elliptic curves $\mathcal{M}_{ell}$ admits a compactification $\overline{\mathcal{M}}_{ell}$ [27] where we allow our elliptic curves to degenerate to singular curves in the form of *Néron n-gons.* The line bundle ω extends over this compactification. The space of (integral) modular forms of weight k is defined to be the global sections (see [57])

$$MF_k := H^0(\overline{\mathcal{M}}_{ell}, \omega^{\otimes k}). \tag{6.3.1}$$

The complex points $\mathcal{M}_{ell}(\mathbb{C})$ admit a classical description (see, for example, [94]). Let $\mathcal{H} \subseteq \mathbb{C}$ denote the upper half plane. Then we can associate to a point $\tau \in \mathcal{H}$ an elliptic curve C_τ over $\mathbb{C}$ by defining

$$C_\tau := \mathbb{C}/(\mathbb{Z} + \mathbb{Z}\tau).$$

Every elliptic curve over $\mathbb{C}$ arises this way. Let $SL_2(\mathbb{Z})$ act on $\mathcal{H}$ through Möbius transformations:

$$\begin{pmatrix} a & b \\ c & d \end{pmatrix} \cdot \tau = \frac{a\tau + b}{c\tau + d}.$$

Two such elliptic curves C_τ and $C_{\tau'}$ are isomorphic if and only if $\tau' = A \cdot \tau$ for some A in $SL_2(\mathbb{Z})$. It follows that

$$\mathcal{M}_{ell}(\mathbb{C}) = \mathcal{H}//SL_2(\mathbb{Z}).$$

In this language, a modular form $f \in MF_k$ can be regarded as a meromorphic function on $\mathcal{H}$ that satisfies

$$f(\tau) = (c\tau + d)^{-k} f(A \cdot \tau)$$

for every

$$A = \begin{pmatrix} a & b \\ c & d \end{pmatrix} \in SL_2(\mathbb{Z}).$$

The condition of extending over the compactification $\overline{\mathcal{M}}_{ell}$ can be expressed over $\mathbb{C}$ by requiring that the Fourier expansion (a.k.a. *q-expansion*)

$$f(\tau) = \sum_{i \in \mathbb{Z}} a_i q^i \quad (q := e^{2\pi i \tau})$$

satisfies $a_i = 0$ for $i < 0$ (a.k.a. "holomorphicity at the cusp").

The Goerss-Hopkins-Miller sheaf

The following major result of Goerss-Hopkins-Miller [53], [17] gives a topological lift of the sheaf $\bigoplus_i \omega^{\otimes i}$.

Theorem 6.3.1 (Goerss-Hopkins-Miller). *There is a homotopy sheaf of E_∞-ring spectra $\mathcal{O}^{top}$ on the étale site of $\overline{\mathcal{M}}_{ell}$ with the property that the spectrum of sections*

$$E_C := \mathcal{O}^{top}(\mathrm{spec}(R) \xrightarrow{C} \mathcal{M}_{ell})$$

associated to an étale map $\mathrm{spec}(R) \to \mathcal{M}_{ell}$ *classifying a trivializable elliptic curve C/R is an elliptic cohomology theory for the elliptic curve C.*

Remark 6.3.2. Since the map

$$\begin{aligned} \mathcal{M}_{ell} &\to \mathcal{M}_{fg} \\ C &\mapsto \widehat{C} \end{aligned}$$

is flat, it follows that every elliptic cohomology theory E_C coming from the theorem above could also have been constructed using Corollary 6.2.10. The novelty in Theorem 6.3.1 is:

1. the functor $\mathcal{O}^{top}$ lands in the point-set category of spectra, rather than in the homotopy category of spectra,

2. the spectra E_C are E_∞, not just homotopy ring spectra, and

3. the functor $\mathcal{O}^{top}$ can be evaluated on non-affine étale maps of stacks

$$\mathcal{X} \to \overline{\mathcal{M}}_{ell}.$$

Elaborating on point (3) above, the "homotopy sheaf" property of $\mathcal{O}^{top}$ implies that for any étale cover

$$\mathcal{U} = \{U_i\} \to \mathcal{X}$$

the map

$$\mathcal{X} \to \mathrm{Tot}\left(\prod_{i_0} \mathcal{O}^{top}(U_{i_0}) \Rightarrow \prod_{i_0, i_1} \mathcal{O}^{top}(U_{i_0} \times_{\mathcal{X}} U_{i_1}) \Rrightarrow \cdots \right)$$

is a equivalence. The Bousfield-Kan spectral sequence [25] of the totalization takes the form

$$E_1^{s,t} = \prod_{i_0,\dots,i_s} \pi_t \mathcal{O}^{top}(U_{i_0} \times_{\mathcal{X}} \cdots \times_{\mathcal{X}} U_{i_s}) \Rightarrow \pi_{t-s}\mathcal{O}^{top}(\mathcal{X}). \tag{6.3.2}$$

Because $\overline{\mathcal{M}}_{ell}$ is a separated Deligne-Mumford stack, there exists a cover of $\mathcal{X}$ by affines, and all of their iterated pullbacks are also affine. Since every elliptic curve is locally trivializable over its base, we can refine any such cover to be a cover that classifies trivializable elliptic curves. In this context we find (using (6.2.6)) that the E_1-term above can be identified with the Čech complex

$$E_1^{s,2k} = \check{C}^s_{\mathcal{U}}(\mathcal{X}, \omega^{\otimes k})$$

and we obtain the *descent spectral sequence*

$$E_2^{s,2k} = H^s(\mathcal{X}, \omega^{\otimes k}) \Rightarrow \pi_{2k-s}\mathcal{O}^{top}(\mathcal{X}). \tag{6.3.3}$$

Non-connective topological modular forms

Motivated by (6.3.1) and (6.3.3), we make the following definition.

Definition 6.3.3. The spectrum of *(non-connective) topological modular forms* is defined to be the spectrum of global sections

$$\mathrm{Tmf} := \mathcal{O}^{op}(\overline{\mathcal{M}}_{ell}).$$

To get a feel for Tmf, we investigate the descent spectral sequence for Tmf[1/6].

Proposition 6.3.4. *We have*[2]

$$H^*((\overline{\mathcal{M}}_{ell})_{\mathbb{Z}[1/6]}, \omega^{\otimes *}) = \mathbb{Z}[1/6][c_4, c_6] \oplus \frac{\mathbb{Z}[1/6][c_4, c_6]}{(c_4^\infty, c_6^\infty)}\{\theta\}$$

where

$$c_k \in H^0((\overline{\mathcal{M}}_{ell})_{\mathbb{Z}[1/6]}, \omega^{\otimes k}),$$
$$\theta \in H^1((\overline{\mathcal{M}}_{ell})_{\mathbb{Z}[1/6]}, \omega^{\otimes -10}).$$

Thus there are no possible differentials or extensions in the descent spectral sequence, and we have

$$\pi_*\mathrm{Tmf}[1/6] \cong \mathbb{Z}[1/6][c_4, c_6] \oplus \frac{\mathbb{Z}[1/6][c_4, c_6]}{(c_4^\infty, c_6^\infty)}\{\theta\}$$

with

$$|c_k| = 2k,$$
$$|\theta| = -21.$$

[2]Here, we use the notation $\frac{\mathbb{Z}[1/6][c_4,c_6]}{(c_4^\infty,c_6^\infty)}\{\theta\}$ to mean that θ is highest degree nonzero class in this divisible pattern.

Proof. Every trivializable elliptic curve C over a $\mathbb{Z}[1/6]$-algebra R can be embedded in $\mathbb{P}^2$, where it takes the Weierstrass form (see, for example, [95, III.1])

$$C_{c_4,c_6} : y^2 = x^3 - 27c_4x - 54c_6, \quad c_4, c_6 \in R \tag{6.3.4}$$

where the *discriminant*

$$\Delta := \frac{c_4^3 - c_6^2}{1728}$$

is invertible. The isomorphisms of elliptic curves of this form are all of the form

$$\begin{aligned} f_\lambda : C_{c_4,c_6} &\to C_{c_4',c_6'} \\ (x,y) &\mapsto (\lambda^2 x, \lambda^3 y) \end{aligned}$$

with

$$c_k' = \lambda^k c_k. \tag{6.3.5}$$

We deduce that

$$(\mathcal{M}_{ell})_{\mathbb{Z}[1/6]} = \mathrm{spec}(\mathbb{Z}[1/6][c_4, c_6, \Delta^{-1}])//\mathbb{G}_m$$

where the $\mathbb{G}_m$-action is given by (6.3.5). The $\mathbb{G}_m$-action encodes a grading on $\mathbb{Z}[1/6][c_4, c_6, \Delta^{-1}]$ where

$$\deg c_k := k.$$

Using the coordinate $z = x/y$ at ∞ for the Weierstrass curve C_{c_4,c_6} (the identity for the group structure), we compute

$$f_\lambda^* dz = \lambda^{-1} dz.$$

It follows that the cohomology of $\omega^{\otimes k}$ is the kth graded summand of the cohomology of the structure sheaf of $\mathrm{spec}(\mathbb{Z}[1/6][c_4, c_6, \Delta^{-1}])$:

$$H^s((\mathcal{M}_{ell})_{\mathbb{Z}[1/6]}, \omega^{\otimes k}) \cong H^{s,k}(\mathrm{spec}(\mathbb{Z}[1/6][c_4, c_6, \Delta^{-1}])) \tag{6.3.6}$$

$$= \begin{cases} \mathbb{Z}[1/6][c_4, c_6, \Delta^{-1}]_k, & s = 0, \\ 0, & s > 0. \end{cases} \tag{6.3.7}$$

We extend the above analysis to the compactification $\overline{\mathcal{M}}_{ell}$ by allowing for nodal singularities.[3] A curve C_{c_4,c_6} has a nodal singularity if and only if $\Delta = 0$ and c_4 is invertible. We therefore compute

$$H^s((\overline{\mathcal{M}}_{ell})_{\mathbb{Z}[1/6]}, \omega^{\otimes k}) \cong H^{s,k}(\mathrm{spec}(\mathbb{Z}[1/6][c_4, c_6, \Delta^{-1}]) \cup \mathrm{spec}(\mathbb{Z}[1/6][c_4^{\pm}, c_6]))$$

as the kernel and cokernel of the map

$$\mathbb{Z}[1/6][c_4, c_6, \Delta^{-1}] \oplus \mathbb{Z}[1/6][c_4^{\pm}, c_6] \to \mathbb{Z}[1/6][c_4^{\pm}, c_6, \Delta^{-1}].$$

□

[3] A curve of the form C_{c_4,c_6} can only have nodal or cuspidal singularities [95, III.1], and the nodal case is a Néron 1-gon.

The computation above implies that the unlocalized cohomology

$$H^s(\overline{\mathcal{M}}_{ell}, \omega^{\otimes k})$$

consists entirely of 2- and 3-torsion for $s > 1$. In fact, it turns out that these groups are non-trivial for arbitrarily large values of s, resulting in 2- and 3-torsion persisting to $\pi_* \mathrm{Tmf}$. This will be discussed in more detail in Section 6.4.

Variants of Tmf

We highlight two variants of the spectrum Tmf: the *connective* and the *periodic* versions. One feature of $\pi_* \mathrm{Tmf}[1/6]$ which is apparent in Proposition 6.3.4 is that

$$\pi_k \mathrm{Tmf}[1/6] = 0, \quad -20 \le k \le -1.$$

It turns out that this gap in homotopy groups occurs in the unlocalized Tmf spectrum (see Section 6.4), and the negative homotopy groups of Tmf are related to the positive homotopy groups of Tmf by Anderson duality (at least with 2 inverted, see [96]).

We therefore isolate the positive homotopy groups by defining the *connective* tmf-*spectrum* to be the connective cover

$$\mathrm{tmf} := \tau_{\ge 0} \mathrm{Tmf}.$$

The modular form $\Delta \in MF_{12}$ is not a permanent cycle in the descent spectral sequence for unlocalized Tmf, but Δ^{24} is. It turns out that the map

$$\pi_* \mathrm{tmf} \to \pi_* \mathrm{tmf}[\Delta^{-24}]$$

is injective. Motivated by this, we define the *periodic* TMF-*spectrum*[4] by

$$\mathrm{TMF} := \mathrm{tmf}[\Delta^{-24}].$$

This spectrum is $|\Delta^{24}| = 576$-periodic. We have

$$\mathrm{TMF} \simeq \mathrm{Tmf}[\Delta^{-24}] \simeq \mathcal{O}^{top}(\mathcal{M}_{ell})$$

where the last equivalence comes from the fact that $\mathcal{M}_{ell}$ is the complement of the zero-locus of Δ in $\overline{\mathcal{M}}_{ell}$.

Another variant comes from the consideration of level structures. Given a congruence subgroup $\Gamma \le SL_2(\mathbb{Z})$, one can consider the modular forms of level Γ to be those holomorphic functions on the upper half plane that satisfy (6.3.1) for all $A \in \Gamma$, and that satisfy a holomorphicity condition at all of the cusps of the quotient

$$\mathcal{M}_{ell}(\Gamma)(\mathbb{C}) = \mathcal{H}//\Gamma.$$

[4]We warn the reader that the connective cover of TMF is *not* tmf.

Integral versions of $\mathcal{M}_{ell}(\Gamma)$ can be defined by considering moduli spaces of elliptic curves with certain types of *level structures.* The most common Γ that are considered are:

$$\Gamma_0(N) := \left\{ A \in SL_2(\mathbb{Z}) \, : \, A \equiv \begin{pmatrix} * & * \\ 0 & * \end{pmatrix} \pmod N \right\},$$
$$\Gamma_1(N) := \left\{ A \in SL_2(\mathbb{Z}) \, : \, A \equiv \begin{pmatrix} 1 & * \\ 0 & 1 \end{pmatrix} \pmod N \right\},$$
$$\Gamma(N) := \left\{ A \in SL_2(\mathbb{Z}) \, : \, A \equiv \begin{pmatrix} 1 & 0 \\ 0 & 1 \end{pmatrix} \pmod N \right\}.$$

The corresponding moduli stacks $\mathcal{M}_{ell}(\Gamma_0(N))$ and $\mathcal{M}_{ell}(\Gamma_1(N))$ (respectively $\mathcal{M}_{ell}(\Gamma(N))$) are defined over $\mathbb{Z}[1/N]$ (respectively $\mathbb{Z}[1/N, \zeta_N]$), with R-points consisting of the groupoid of pairs

$$(C, \eta)$$

where C is an elliptic curve over R, and

$$\eta = \begin{cases} \text{a cyclic subgroup of } C \text{ of order } N, & \Gamma = \Gamma_0(N), \\ \text{a point of } C \text{ of exact order } N, & \Gamma = \Gamma_1(N), \\ \text{an isomorphism } C[N] \cong \mathbb{Z}/N \times \mathbb{Z}/N, & \Gamma = \Gamma(N). \end{cases}$$

In each of these cases, forgetting the level structure results in an étale map of stacks

$$\mathcal{M}_{ell}(\Gamma) \to \mathcal{M}_{ell} \tag{6.3.8}$$

and we define the associated (periodic) spectra of *topological modular forms, with level structure* by

$$\begin{aligned} \mathrm{TMF}_0(N) &:= \mathcal{O}^{top}(\mathcal{M}_{ell}(\Gamma_0(N))), \\ \mathrm{TMF}_1(N) &:= \mathcal{O}^{top}(\mathcal{M}_{ell}(\Gamma_1(N))), \\ \mathrm{TMF}(N) &:= \mathcal{O}^{top}(\mathcal{M}_{ell}(\Gamma(N))). \end{aligned}$$

Compactifications $\overline{\mathcal{M}}_{ell}(\Gamma)$ of the moduli stacks $\mathcal{M}_{ell}(\Gamma)$ above were constructed by Deligne and Rapoport [27]. The extensions of the maps (6.3.8) to these compactifications

$$\overline{\mathcal{M}}_{ell}(\Gamma) \to \overline{\mathcal{M}}_{ell}$$

are not étale, but they are log-étale. Hill and Lawson have shown that the sheaf $\mathcal{O}^{top}$ extends to the log-étale site of $\overline{\mathcal{M}}_{ell}$ [45], allowing us to define corresponding Tmf-spectra by

$$\begin{aligned}\mathrm{Tmf}_0(N) &:= \mathcal{O}^{top}(\overline{\mathcal{M}}_{ell}(\Gamma_0(N))),\\ \mathrm{Tmf}_1(N) &:= \mathcal{O}^{top}(\overline{\mathcal{M}}_{ell}(\Gamma_1(N))),\\ \mathrm{Tmf}(N) &:= \mathcal{O}^{top}(\overline{\mathcal{M}}_{ell}(\Gamma(N))).\end{aligned}$$

6.4 Homotopy groups of TMF at the primes 2 and 3

We now give an overview of the 2- and 3-primary homotopy groups of TMF. Detailed versions of these computations can be found in [60], [8], and some very nice charts depicting the answers were created by Henriques [42]. The basic idea is to invoke the descent spectral sequence. The E_2-term is computed by imitating the argument of Proposition 6.3.4. The descent spectral sequence does not degenerate 2 or 3-locally, and differentials must be deduced using a variety of ad hoc methods similar to those used to compute differentials in the Adams-Novikov spectral sequence.

3-primary homotopy groups of TMF

Every elliptic curve over a $\mathbb{Z}_{(3)}$-algebra R can (upon taking a faithfully flat extension of R) be put in the form [8]

$$C'_{a_2,a_4} : y^2 = 4x(x^2 + a_2 x + a_4), \quad a_i \in R$$

with

$$\Delta = a_4^2(16a_2^2 - 64a_4)$$

invertible. The isomorphisms of any such are of the form

$$\begin{aligned} f_{\lambda,r} : C'_{a_2,a_4} &\to C'_{a'_2,a'_4} \\ (x,y) &\mapsto (\lambda^2(x-r), \lambda^3 y) \end{aligned} \tag{6.4.1}$$

with

$$r^3 + a_2 r^2 + a_4 r = 0$$

and

$$\begin{aligned} a'_2 &= \lambda^2(a_2 + 3r), \\ a'_4 &= \lambda^4(a_4 + 2a_2 r + 3r^2). \end{aligned} \tag{6.4.2}$$

Following the template of the proof of Proposition 6.3.4, we observe that for the coordinate $z = x/y$ at ∞, we have

$$f^*_{\lambda,r} dz = \lambda^{-1} dz. \tag{6.4.3}$$

We may therefore use the λ factor to compute the sections of $\omega^{\otimes *}$.

Specifically, by setting $\lambda = 1$, we associate to this data a graded Hopf algebroid (A', Γ') with

$$A' := \mathbb{Z}_{(3)}[a_2, a_4, \Delta^{-1}], \quad |a_i| = i,$$
$$\Gamma' := A'[r]/(r^3 + a_2 r^2 + a_4 r), \quad |r| = 2$$

with right unit given by (6.4.2) (with $\lambda = 1$) and coproduct given by the composition of two isomorphisms of the form $f_{1,r}$.

Now consider the cover

$$U = \mathrm{Proj}(A') \to (\mathcal{M}_{ell})_{\mathbb{Z}_{(3)}}.$$

We deduce from (6.4.3) that

$$\omega^{\otimes k}(U) = A'_k$$

and more generally

$$\omega^{\otimes k}(U^{\times_{\mathcal{M}_{ell}}(s+1)}) = \left((\Gamma')^{\otimes_{A'} s}\right)_k.$$

It follows that the Čech complex for $\bigoplus_k \omega^{\otimes k}$ associated to the cover U is the cobar complex for the graded Hopf algebroid (A', Γ'). We deduce that the E_2-term of the descent spectral sequence is given by the cohomology of the Hopf algebroid (A', Γ'):

$$H^s((\mathcal{M}_{ell})_{\mathbb{Z}_{(3)}}, \omega^{\otimes k}) = H^{s,k}(A', \Gamma').$$

One computes (see [8]):

Proposition 6.4.1.

$$H^{s,k}(A', \Gamma') = \frac{\mathbb{Z}_{(3)}[c_4, c_6, \Delta^{\pm}][\beta] \otimes E[\alpha]}{3\alpha, 3\beta, \alpha c_4, \alpha c_6, \beta c_4, \beta c_6, c_6^2 = c_4^3 - 1728\Delta}$$

where β is given by the Massey product

$$\beta = \langle \alpha, \alpha, \alpha \rangle$$

and the generators are in bidegrees (s, k):

$$|c_i| = (0, i), \qquad |\Delta| = (0, 12),$$
$$|\alpha| = (1, 2), \qquad |\beta| = (2, 6).$$

Figure 6.4.1 displays the descent spectral sequence

$$H^{s,k}(A', \Gamma') \Rightarrow \pi_{2k-s}\mathrm{TMF}_{(3)}.$$

Here:

- Boxes correspond to $\mathbb{Z}_{(3)}$'s.
- Dots correspond to $\mathbb{Z}/3$'s.

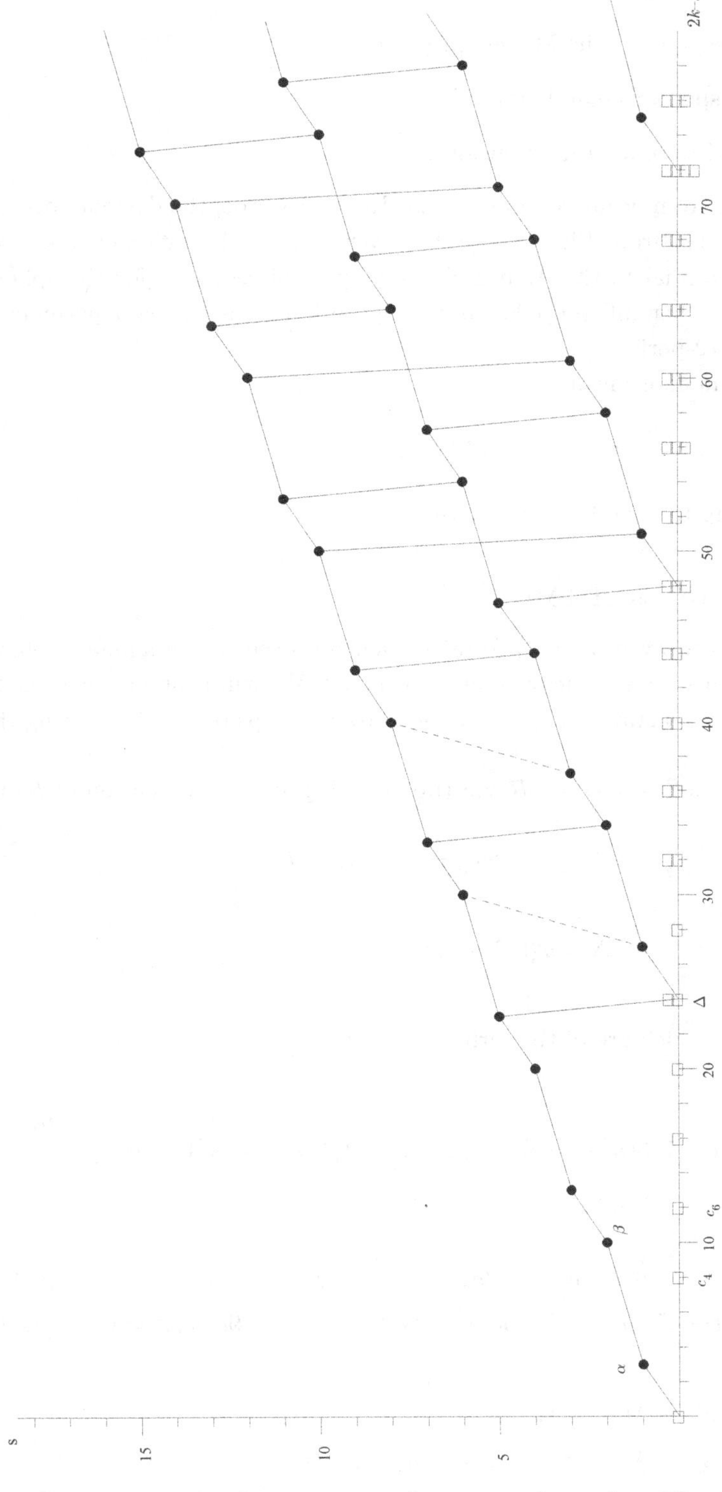

FIGURE 6.4.1. The descent spectral sequence for $\mathrm{tmf}_{(3)}$. The descent spectral sequence for $\mathrm{TMF}_{(3)}$ is obtained by inverting Δ.

- Lines of slope $1/3$ correspond to multiplication by α.
- Lines of slope $1/7$ correspond to the Massey product $\langle -, \alpha, \alpha \rangle$.
- Lines of slope $-r$ correspond to d_r-differentials.
- Dashed lines correspond to hidden α extensions.

We omit the factors coming from negative powers of Δ. In other words, the descent spectral sequence for TMF is obtained from Figure 6.4.1 by inverting Δ. The differential on $\beta\Delta$ comes from the Toda differential in the Adams-Novikov spectral sequence for the sphere, and this implies all of the other differentials. As the figure indicates, Δ^3 is a permanent cycle, and so $\pi_* \mathrm{TMF}_{(3)}$ is 72-periodic.

Under the Hurewicz homomorphism

$$\pi_* S_{(3)} \to \pi_* \mathrm{TMF}_{(3)}$$

the elements α_1 and β_1 map to α and β, respectively.

2-primary homotopy groups of TMF

The analysis of the 2-primary descent spectral sequence proceeds in a similar fashion, except that the computations are significantly more involved. We will content ourselves to summarize the set-up, and then state the resulting homotopy groups of TMF, referring the reader to [8] for the details.

Every elliptic curve over a $\mathbb{Z}_{(2)}$-algebra R can (upon taking an étale extension of R) be put in the form [8]

$$C''_{a_1,a_3} : y^2 + a_1 xy + a_3 y = x^3, \quad a_i \in R$$

with

$$\Delta = a_3^3(a_1^3 - 27a_3)$$

invertible.

The isomorphisms of any such are of the form

$$\begin{aligned} f_{\lambda,s,t} : C''_{a_1,a_3} &\to C''_{a'_1,a'_3} \\ (x,y) &\mapsto (\lambda^2(x - 1/3(s^2 + a_1 s)), \lambda^3(y - sx + 1/3(s^3 + a_1 s^2) - t)) \end{aligned} \tag{6.4.4}$$

with[5]

$$s^4 - 6st + a_1 s^3 - 3a_1 t - 3a_3 s = 0, \tag{6.4.5}$$

$$-27t^2 + 18s^3 t + 18a_1 s^2 t - 27a_3 t - 2s^6 - 3a_1 s^5 + 9a_3 s^3 + a_1^3 s^3 + 9a_1 a_3 s^2 = 0, \tag{6.4.6}$$

and

$$\begin{aligned} a'_1 &:= \lambda(a_1 + 2s), \\ a'_3 &:= \lambda^3(a_3 + 1/3(a_1 s^2 + a_1 s) + 2t). \end{aligned} \tag{6.4.7}$$

[5]We warn the reader that there may be a typo in the analog of (6.4.6) that appears in [8, Sec. 7], as even using (6.4.5), relation (6.4.6) seems to be inconsistent with what appears there.

Again, setting $\lambda = 1$, we associate to this data a graded Hopf algebroid (A'', Γ'') with

$$A'' := \mathbb{Z}_{(2)}[a_1, a_3, \Delta^{-1}], \quad |a_i| = i,$$
$$\Gamma'' := A''[s,t]/\sim, \quad |s| = 1, \ |t| = 3$$

(where $\sim$ consists of relations (6.4.5), (6.4.6)) with right unit given by (6.4.2) (with $\lambda = 1$) and coproduct given by the composition of two isomorphisms of the form $f_{1,s,t}$. The E_2-term of the descent spectral sequence takes the form

$$H^s((\mathcal{M}_{ell})_{(2)}, \omega^{\otimes k}) = H^{s,k}(A'', \Gamma'').$$

Proposition 6.4.2. ([8], [86]) *The cohomology of the Hopf algebroid (A'', Γ'') is given by*

$$H^{*,*}(A'', \Gamma'') = \mathbb{Z}_{(2)}[c_4, c_6, \Delta^{\pm}, \eta, a_1^2\eta, \nu, \varepsilon, \kappa, \bar{\kappa}]/(\sim)$$

where $\sim$ consists of the relations

$$2\eta, \ \eta\nu, \ 4\nu, \ 2\nu^2, \ \nu^3 = \eta\varepsilon,$$
$$2\varepsilon, \ \nu\varepsilon, \ \varepsilon^2, \ 2a_1^2\eta, \ \nu a_1^2\eta, \ \varepsilon a_1^2\eta, \ (a_1^2\eta)^2 = c_4\eta^2,$$
$$2\kappa, \ \eta^2\kappa, \ \nu^2\kappa = 4\bar{\kappa}, \ \varepsilon\kappa, \ \kappa^2, \ \kappa a_1^2\eta,$$
$$\nu c_4, \ \nu c_6, \ \varepsilon c_4, \ \varepsilon c_6, \ a_1^2\eta c_4 = \eta c_6, \ a_1^2\eta c_6 = \eta c_4^2,$$
$$\kappa c_4, \ \kappa c_6, \ \bar{\kappa} c_4 = \eta^4\Delta, \ \bar{\kappa} c_6 = \eta^3(a_1^2\eta)\Delta, \ c_6^2 = c_4^3 - 1728\Delta$$

and the generators are in bidegrees (s,k):

$$|c_i| = (0,i), \qquad |\Delta| = (0,12), \qquad |\eta| = (1,1),$$
$$|a_1^2\eta| = (1,3), \qquad |\nu| = (1,2), \qquad |\varepsilon| = (2,5),$$
$$|\kappa| = (2,8), \qquad |\bar{\kappa}| = (4,12).$$

There are many differentials in the descent spectral sequence

$$H^{s,k}(A'', \Gamma'') \Rightarrow \pi_{2k-s}\mathrm{TMF}_{(2)}.$$

These were first determined by Hopkins, and first appeared in the preprint "From elliptic curves to homotopy theory" by Hopkins and Mahowald [51], and we refer the reader to that paper or [8] for the details.

We content ourselves with simply stating the resulting homotopy groups of $\mathrm{TMF}_{(2)}$. These are displayed in Figure 6.4.2. Our choice of names for elements in the descent spectral sequence (and our abusive practice of giving the elements of $\pi_*\mathrm{TMF}$ they detect the same names) is motivated by the fact that the elements

$$\eta, \nu, \varepsilon, \kappa, \bar{\kappa}, q, u, w$$

in the 2-primary stable homotopy groups of spheres map to the corresponding elements in $\pi_*\mathrm{TMF}_{(2)}$. We warn the reader that there are many hidden extensions in the descent

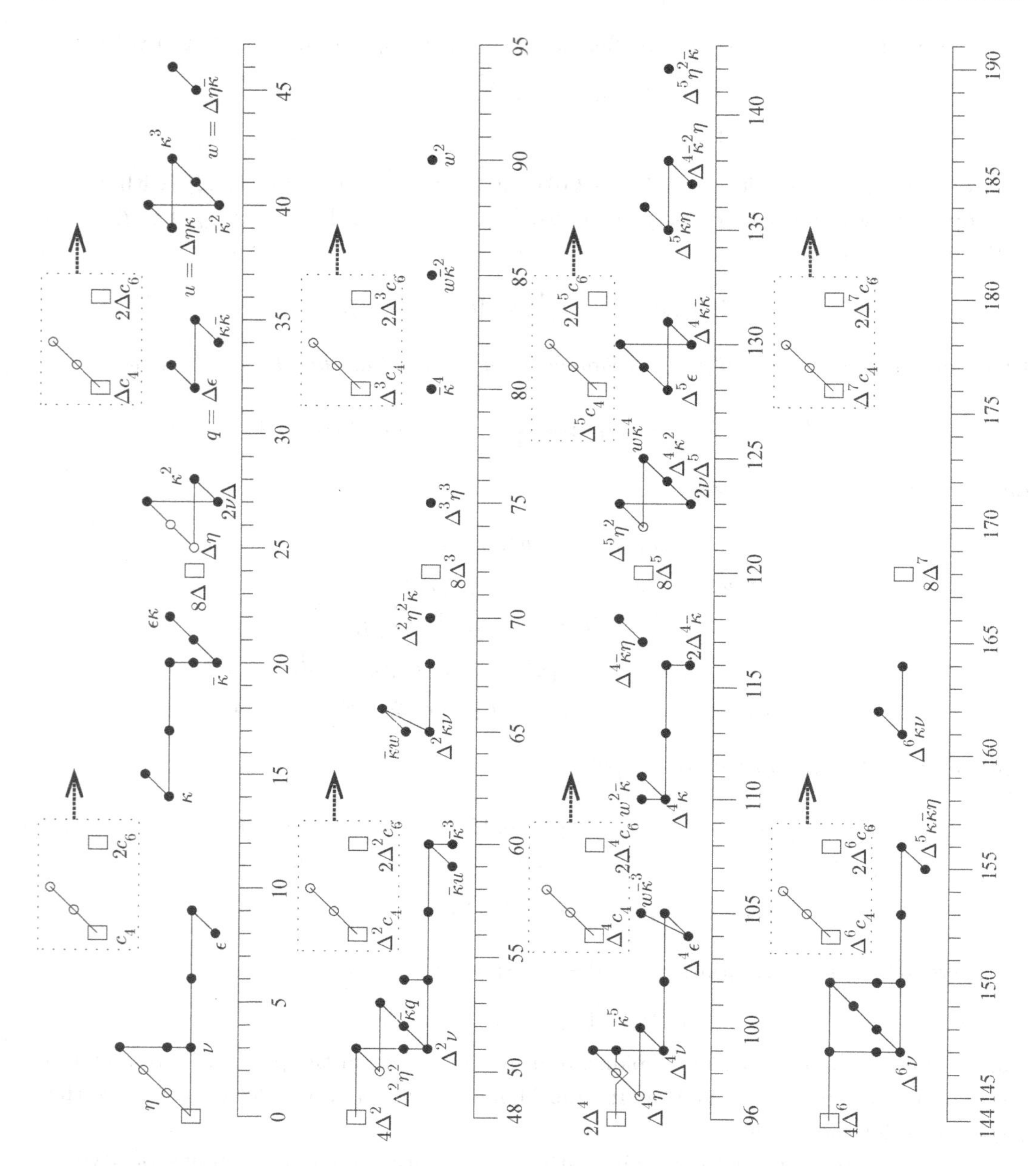

FIGURE 6.4.2. The homotopy groups of $\mathrm{tmf}_{(2)}$. $\pi_*\mathrm{TMF}_{(2)}$ is obtained by inverting Δ.

spectral sequence, so that often the names of elements in Figure 6.4.2 do not reflect the element that detects them in the descent spectral sequence because in the descent spectral sequence the product would be zero. For example, κ^2 is zero in $H^{*,*}(A'', \Gamma'')$, but nonzero in $\pi_*\mathrm{TMF}$. More complete multiplicative information can be found in [42]. In the image, the axis perpendicular to the stem degree has no theoretical significance; it is used just to separate elements.

In Figure 6.4.2:

- A series of i black dots joined by vertical lines corresponds to a factor of $\mathbb{Z}/2^i$ that is annihilated by some power of c_4.
- An open circle corresponds to a factor of $\mathbb{Z}/2$ that is not annihilated by a power of c_4.
- A box indicates a factor of $\mathbb{Z}_{(2)}$ that is not annihilated by a power of c_4.
- The non-vertical lines indicate multiplication by η and ν.
- A pattern with a dotted box around it and an arrow emanating from the right face indicates this pattern continues indefinitely to the right by c_4-multiplication (i.e. tensor the pattern with $\mathbb{Z}_{(2)}[c_4]$).

The element Δ^8 is a permanent cycle, and $\pi_*\mathrm{TMF}_{(2)}$ is 192-periodic on the pattern depicted in Figure 6.4.2. The figure does not depict powers of c_4 supported by negative powers of Δ.

6.5 The homotopy groups of Tmf and tmf

We give a brief discussion of how the analysis in Section 6.4 can be augmented to determine $\pi_*\mathrm{Tmf}$, and thus $\pi_*\mathrm{tmf}$. We refer the reader to [60] for more details. We have already described $\pi_*\mathrm{Tmf}[1/6]$ in Section 6.3, so we focus on $\pi_*\mathrm{Tmf}_{(p)}$ for $p = 2, 3$.

The ordinary locus

We first must describe a cover of $\overline{\mathcal{M}}_{ell}$. We recall that for a formal group F, the *p-series* is the formal power series

$$[p]_F(x) = \underbrace{x +_F \cdots +_F x}_{p}$$

where $x +_F y := F(x, y)$. If F is defined over a ring of characteristic p, we say it has *height* n if its p-series takes the form

$$[p]_F(x) = v_n^F x^{p^n} + \cdots$$

with v_n^F a unit.

Elliptic curves (over fields of characteristic p) have formal groups of height 1 or 2. We shall call a trivializable elliptic curve over a $\mathbb{Z}_{(p)}$-algebra R *ordinary* if the formal group $\widehat{\bar{C}}$ has height 1 (where $\bar{C}$ the base change of the curve to R/p). Let $(\mathcal{M}_{ell}^{ord})_{\mathbb{Z}_{(p)}}$ denote the moduli stack of ordinary elliptic curves, and define $(\overline{\mathcal{M}}_{ell}^{ord})_{\mathbb{Z}_{(p)}}$ to be the closure of $(\mathcal{M}_{ell}^{ord})_{\mathbb{Z}_{(p)}}$ in $(\overline{\mathcal{M}}_{ell})_{\mathbb{Z}_{(p)}}$.

We have the following lemma ([86, Sec. 21]).

Lemma 6.5.1. *Let*

$$F' = \widehat{C'}_{a_2,a_4},$$
$$F'' = \widehat{C''}_{a_1,a_3},$$

denote the formal group laws of the reductions of the elliptic curves C'_{a_2,a_4} and C''_{a_1,a_3} modulo 3 and 2, respectively. Then we have[6]

$$v_1^{F'} = -a_2,$$
$$v_1^{F''} = a_1.$$

Define

$$\mathrm{TMF}^{ord} := \mathcal{O}^{top}((\mathcal{M}_{ell}^{ord})_{\mathbb{Z}_{(p)}}).$$

Using the fact that [86, Prop.18.7]

$$c_4 \equiv \begin{cases} 16a_2^2 \pmod 3, & p = 3, \\ a_1^4 \pmod 2, & p = 2 \end{cases} \tag{6.5.1}$$

we have (for $p = 2$ or 3)

$$\mathrm{TMF}^{ord}_{(p)} = \mathrm{TMF}_{(p)}[c_4^{-1}].$$

We deduce from the computations of $\pi_*\mathrm{TMF}_{(p)}$:

Proposition 6.5.2. *We have*

$$\pi_*\mathrm{TMF}^{ord}_{(p)} = \begin{cases} \dfrac{\mathbb{Z}_{(3)}[c_4^{\pm}, c_6, \Delta^{\pm}]}{c_6^2 = c_4^3 - 1728\Delta}, & p = 3, \\ \dfrac{\mathbb{Z}_{(2)}[c_4^{\pm}, 2c_6, \Delta^{\pm}, \eta]}{2\eta, \eta^3, \eta \cdot (2c_6), (2c_6)^2 = 4(c_4^3 - 1728\Delta)}, & p = 2. \end{cases}$$

Using the covers

$$\mathrm{Proj}(\mathbb{Z}_{(3)}[a_2^{\pm}, a_4]) \to (\overline{\mathcal{M}}_{ell}^{ord})_{(3)},$$
$$\mathrm{Proj}(\mathbb{Z}_{(2)}[a_1^{\pm}, a_3]) \to (\overline{\mathcal{M}}_{ell}^{ord})_{(2)},$$

the Hopf algebroids (A', Γ') and (A'', Γ'') have variants where a_2 (respectively a_1) is inverted and Δ is not. Using these, one computes the descent spectral sequence for $\mathrm{Tmf}^{ord}_{(p)}$ at $p = 2, 3$ and finds:

Proposition 6.5.3. *We have:*

$$\pi_*\mathrm{Tmf}^{ord}_{(p)} = \begin{cases} \dfrac{\mathbb{Z}_{(3)}[c_4^{\pm}, c_6, \Delta]}{c_6^2 = c_4^3 - 1728\Delta}, & p = 3, \\ \dfrac{\mathbb{Z}_{(2)}[c_4^{\pm}, 2c_6, \Delta, \eta]}{2\eta, \eta^3, \eta \cdot (2c_6), (2c_6)^2 = 4(c_4^3 - 1728\Delta)}, & p = 2. \end{cases}$$

[6] For an elliptic curve C over a ring of characteristic p, $v_1^{\widehat{C}}$ is known as the *Hasse invariant*.

A homotopy pullback for $\mathrm{Tmf}_{(p)}$

The spectrum Tmf can be accessed at the primes 2 and 3 in a manner analogous to the case of Proposition 6.3.4: associated to the cover[7]

$$\left\{(\overline{\mathcal{M}}_{ell}^{ord})_{\mathbb{Z}_{(p)}}, (\mathcal{M}_{ell})_{\mathbb{Z}_{(p)}}\right\} \to (\overline{\mathcal{M}}_{ell})_{\mathbb{Z}_{(p)}}$$

there is a homotopy pullback (coming from the sheaf condition of $\mathcal{O}^{top}$)

$$\begin{array}{ccc} \mathrm{Tmf}_{(p)} & \longrightarrow & \mathcal{O}^{top}((\mathcal{M}_{ell})_{\mathbb{Z}_{(p)}}) \\ \downarrow & & \downarrow \\ \mathcal{O}^{top}((\overline{\mathcal{M}}_{ell}^{ord})_{\mathbb{Z}_{(p)}}) & \longrightarrow & \mathcal{O}^{top}((\mathcal{M}_{ell}^{ord})_{\mathbb{Z}_{(p)}}). \end{array} \tag{6.5.2}$$

Since we have described the homotopy groups of the spectra

$$\begin{aligned} \mathrm{TMF}_{(p)} &:= \mathcal{O}^{top}((\mathcal{M}_{ell})_{\mathbb{Z}_{(p)}}), \\ \mathrm{TMF}_{(p)}^{ord} &:= \mathcal{O}^{top}((\mathcal{M}_{ell}^{ord})_{\mathbb{Z}_{(p)}}), \\ \mathrm{Tmf}_{(p)}^{ord} &:= \mathcal{O}^{top}((\overline{\mathcal{M}}_{ell}^{ord})_{\mathbb{Z}_{(p)}}) \end{aligned}$$

at the primes 2 and 3, the homotopy groups of $\mathrm{Tmf}_{(p)}$ at these primes may be computed using the pullback square (6.5.2).

The homotopy groups of $\mathrm{tmf}_{(p)}$

Once one computes $\pi_*\mathrm{Tmf}_{(p)}$ it is a simple matter to read off the homotopy groups of the connective cover $\mathrm{tmf}_{(p)}$. We obtain:

Theorem 6.5.4. *The homotopy groups of* $\mathrm{tmf}_{(3)}$ *are given by the* E_∞*-page of the spectral sequence of Figure 6.4.1,*[8] *and the homotopy groups of* $\mathrm{tmf}_{(2)}$ *are depicted in Figure 6.4.2. These homotopy groups are* Δ^3 *(respectively* Δ^8*)-periodic.*

We end this section by stating a very useful folklore theorem which was proven rigorously in [74].

Theorem 6.5.5 (Mathew). *The mod 2 cohomology of* tmf *is given (as a module over the Steenrod algebra) by*

$$H^*(\mathrm{tmf}; \mathbb{F}_2) = A//A(2)$$

where A is the mod 2 Steenrod algebra, and $A(2)$ is the subalgebra generated by Sq^1, Sq^2, *and* Sq^4.

Corollary 6.5.6. *For a spectrum X, the Adams spectral sequence for the 2-adic* tmf*-homology of X takes the form*

$$\mathrm{Ext}^{s,t}_{A(2)}(H^*(X; \mathbb{F}_2), \mathbb{F}_2) \Rightarrow \pi_{t-s}(\mathrm{tmf} \wedge X)^\wedge_2.$$

[7]The fact that this is a cover follows from the fact that $(\mathcal{M}_{ell}^{ord})_{\mathbb{Z}_{(p)}}$ is Zariski dense in $(\mathcal{M}_{ell})_{\mathbb{Z}_{(p)}}$.
[8]There are no additive extensions in this spectral sequence.

6.6 Tmf from the chromatic perspective

We outline the essential algebro-geometric ideas behind chromatic homotopy theory, as originally envisioned by Morava [80] (see also [49], [31, Ch. 9], [36]), and apply it to understand the chromatic localizations of Tmf. We will find that the pullback (6.5.2) used to access Tmf is closely related to its chromatic fracture square.

Stacks associated to ring spectra

The material in this section is closely aligned with that of Mike Hopkins's lecture "From spectra to stacks" in [31, Ch. 9]. In this lecture, Hopkins explains how non-complex orientable ring spectra give rise to stacks over the moduli stack of formal groups. This allows certain computations to be conceptualized by taking pullbacks over the moduli stack of formal groups.

The complex cobordism spectrum MU has a canonical complex orientation. To conform better to the even periodic set-up, we utilize the even periodic variant[9]

$$\mathrm{MUP} := \bigvee_{i \in \mathbb{Z}} \Sigma^{2i}\mathrm{MU}$$

so that $\pi_0\mathrm{MUP} \cong \pi_* MU$. Quillen proved [85] (see also [1]) that the associated formal group law F_{MUP} is the universal formal group law:

$$\mathrm{spec}(\pi_0\mathrm{MUP})(R) = \{\text{formal group laws over } R\}$$

In particular

$$\mathrm{spec}(\pi_0\mathrm{MUP}) \xrightarrow{F_{MUP}} \mathcal{M}_{fg}$$

is a flat cover. In fact, we have [85], [1]

$$\begin{aligned} \mathrm{spec}(\mathrm{MUP}_0\mathrm{MUP}) &= \{\text{isomorphisms } f : F \to F' \text{ between formal group laws over } R\} \\ &= \mathrm{spec}(\pi_0\mathrm{MUP}) \times_{\mathcal{M}_{fg}} \mathrm{spec}(\pi_0\mathrm{MUP}). \end{aligned}$$

Suppose that E is a complex oriented even periodic ring spectrum whose formal group law classifying map

$$\mathrm{spec}(\pi_0 E) \xrightarrow{F_E} \mathcal{M}_{fg}.$$

is flat. We shall call such ring spectra *Landweber exact* (see Theorem 6.2.9). The formal group law F_E of such E determines E in the following sense: the classifying map

$$\pi_0\mathrm{MUP} \xrightarrow{F_E} \pi_0 E$$

[9] Just as MU is the Thom spectrum of the universal virtual bundle over BU, MUP is the Thom spectrum of the universal virtual bundle over $BU \times \mathbb{Z}$.

lifts to a map of ring spectra

$$\mathrm{MUP} \to E$$

and the associated map

$$\pi_0 E \otimes_{\pi_0 \mathrm{MUP}} \mathrm{MUP}_* X \to E_* X \tag{6.6.1}$$

is an isomorphism for all spectra X [61], [49]. Specializing to the case of $X = \mathrm{MUP}$, we have

$$\mathrm{MUP}_0 E \cong \pi_0 E \otimes_{\pi_0 \mathrm{MUP}} \mathrm{MUP}_0 \mathrm{MUP}$$

and hence the square

$$\begin{array}{ccc} \mathrm{spec}(\mathrm{MUP}_0 E) & \longrightarrow & \mathrm{spec}(\pi_0 \mathrm{MUP}) \\ \downarrow & & \downarrow{\scriptstyle F_{\mathrm{MUP}}} \\ \mathrm{spec}(\pi_0 E) & \xrightarrow[F_E]{} & \mathcal{M}_{fg} \end{array}$$

is a pullback. We deduce that $\mathrm{MUP}_0 E$ is flat as an MUP_0-module.

We shall say that a commutative ring spectrum E is *even* if

$$\mathrm{MUP}_{\mathrm{odd}} E = 0.$$

We shall say that an even ring spectrum E is *Landweber* if $\mathrm{MUP}_0 E$ is flat as an MUP_0-module. Note by the previous paragraph every Landweber exact ring spectrum is Landweber. Since $\mathrm{MUP}_0 E$ is an $\mathrm{MUP}_0\mathrm{MUP}$-comodule algebra, the morphism

$$\mathrm{spec}(\mathrm{MUP}_0 E) \to \mathrm{spec}(\mathrm{MUP}_0)$$

comes equipped with descent data to determine a stack $\mathcal{X}_E$ and a flat morphism

$$\mathcal{X}_E \to \mathcal{M}_{fg}. \tag{6.6.2}$$

We shall call $\mathcal{X}_E$ the *stack associated to* E. Let ω denote the line bundle over $\mathcal{M}_{fg}$ whose fiber over a formal group law F is the cotangent space at the identity

$$\omega_F = T_0^* F.$$

We abusively also let ω denote the pullback of this line bundle to $\mathcal{X}_E$ under (6.6.2). Then an analysis similar to that of Section 6.4 (see [28]) shows that the spectral sequence associated to the canonical Adams-Novikov resolution

$$E^{\wedge}_{MUP} := \mathrm{Tot}\,(\mathrm{MUP} \wedge E \Rightarrow \mathrm{MUP} \wedge \mathrm{MUP} \wedge E \Rrightarrow \cdots)$$

takes the form

$$H^s(\mathcal{X}_E, \omega^{\otimes k}) \Rightarrow \pi_{2k-s} E^{\wedge}_{MUP}.$$

Example 6.6.1. The spectrum TMF is Landweber [86, Sec. 20], [74, Sec. 5.1], with

$$\mathcal{X}_{\mathrm{TMF}} = \mathcal{M}_{ell}.$$

The Adams-Novikov spectral sequence is the descent spectral sequence [51]. In fact, the computations of [86, Sec. 20] also show that tmf is Landweber, with

$$\mathcal{X}_{\mathrm{tmf}} = \mathcal{M}_{\mathrm{weier}}.$$

This is the moduli stack of *Weierstrass curves*, curves that locally take the form

$$y^2 + a_1xy + a_3y = x^3 + a_2x^2 + a_4x + a_6.$$

The associated Adams-Novikov spectral sequence

$$H^s(\mathcal{M}_{weier}, \omega^{\otimes k}) \Rightarrow \pi_{2k-s}\mathrm{tmf}$$

is computed in [8]. The spectra $\mathrm{Tmf}_{(p)}^{ord}$ and $\mathrm{TMF}_{(p)}^{ord}$ are also Landweber, with

$$\mathcal{X}_{\mathrm{Tmf}_{(p)}^{ord}} = (\overline{\mathcal{M}}_{ell}^{ord})_{\mathbb{Z}_{(p)}}.$$
$$\mathcal{X}_{\mathrm{TMF}_{(p)}^{ord}} = (\mathcal{M}_{ell}^{ord})_{\mathbb{Z}_{(p)}}.$$

Unfortunately, the spectrum Tmf is not Landweber, but the pullback (6.5.2) does exhibit it as a pullback of Landweber ring spectra. The pushout of the corresponding diagram of stacks

$$\begin{array}{ccc} \mathcal{X}_{\mathrm{TMF}_{(p)}^{ord}} & \longrightarrow & \mathcal{X}_{\mathrm{TMF}_{(p)}} \\ \downarrow & & \downarrow \\ \mathcal{X}_{\mathrm{Tmf}_{(p)}^{ord}} & \longrightarrow & (\overline{\mathcal{M}}_{ell})_{(\mathbb{Z}_{(p)})} \end{array}$$

motivates us to consider $(\overline{\mathcal{M}}_{ell})_{(\mathbb{Z}_{(p)})}$ as the appropriate stack $\mathcal{X}_{\mathrm{Tmf}_{(p)}}$ over $\mathcal{M}_{fg}$ to associate to $\mathrm{Tmf}_{(p)}$. This motivates the following definition.

Definition 6.6.2. We shall call a ring spectrum E *locally Landweber* if it is given as a homotopy limit

$$E = \mathrm{holim}_{i \in \mathcal{I}} E_i$$

of Landweber ring spectra where $\mathcal{I}$ is a category whose nerve has finitely many non-degenerate simplices.[10] The colimit

$$\mathcal{X}_E := \mathrm{colim}_i\, \mathcal{X}_{E_i}$$

is the *stack associated to E*.

Remark 6.6.3. The stack $\mathcal{X}_E$ in the above definition a priori seems to to depend on the diagram $\{E_i\}$. In general, the E_2-term of the Adams spectral sequence for E is *not* isomorphic to $H^s(\mathcal{X}_E, \omega^{\otimes k})$ (as happens in the case of Landweber spectra).

[10]We specify this condition so that homotopy limits taken over $\mathcal{I}$ commute with homotopy colimits in the category of spectra — see the proof of Proposition 6.6.4.

Proposition 6.6.4. *Suppose that E and E' are locally Landweber. Then so is $E \wedge E'$, and*

$$\mathcal{X}_{E\wedge E'} \simeq \mathcal{X}_E \times_{\mathcal{M}_{fg}} \mathcal{X}_{E'}.$$

Proof. Suppose first that E and E' are Landweber. The result then follows from the fact that the Landweber hypothesis yields a Künneth isomorphism

$$\mathrm{MUP}_0(E \wedge E') \cong \mathrm{MUP}_0(E) \otimes_{\mathrm{MUP}_0} \mathrm{MUP}_0(E').$$

Now suppose that E and E' are locally Landweber, given as limits

$$E \simeq \operatorname*{holim}_{i \in \mathcal{I}} E_i,$$
$$E' \simeq \operatorname*{holim}_{j \in \mathcal{J}} E'_j.$$

Then the finiteness conditions on $\mathcal{I}$ and $\mathcal{J}$ allow us to compute

$$\begin{aligned}\operatorname*{holim}_{i,j}(E_i \wedge E'_j) &\simeq (\operatorname*{holim}_i E_i) \wedge (\operatorname*{holim}_j E'_j) \\ &\simeq E \wedge E'_j\end{aligned}$$

and we have

$$\begin{aligned}\mathcal{X}_E \times_{\mathcal{M}_{fg}} \mathcal{X}_{E'} &= (\mathrm{colim}_i\, \mathcal{X}_{E_i}) \times_{\mathcal{M}_{fg}} (\mathrm{colim}_j\, \mathcal{X}_{E_j}) \\ &= \mathrm{colim}_{i,j}\, \mathcal{X}_{E_i} \times_{\mathcal{M}_{fg}} \mathcal{X}_{E_j} \\ &\simeq \mathrm{colim}_{i,j}\, \mathcal{X}_{E_i \wedge E'_j} \\ &= \mathcal{X}_{E \wedge E'}.\end{aligned}$$

□

The stacks associated to chromatic localizations

Let

$$(\mathcal{M}_{fg})^{\leq n}_{\mathbb{Z}_{(p)}} \subset \mathcal{M}_{fg}$$

denote the substack that classifies p-local formal group laws of height $\leq n$. Let

$$(\mathcal{M}_{fg})^{[n]}_{\mathbb{Z}_{(p)}} \subset (\mathcal{M}_{fg})^{\leq n}_{\mathbb{Z}_{(p)}}$$

denote the formal neighborhood[11] of the locus of formal group laws in characteristic p of exact height n.

Over $\bar{\mathbb{F}}_p$, any two formal groups of height n are isomorphic. Lubin and Tate showed that given a height n formal group F over $\mathbb{F}_q$ (a finite extension of $\mathbb{F}_p$), its universal deformation

[11] $(\mathcal{M}_{fg})^{[n]}_{\mathbb{Z}_{(p)}}$ is technically a *formal stack*.

space is a formally affine Galois cover

$$\mathrm{Spf}(\mathbb{Z}_p[\zeta_{q-1}][[u_1, \dots, u_{n-1}]]) = \mathcal{X}_n^F \to (\mathcal{M}_{fg})^{[n]}_{\mathbb{Z}_{(p)}}.$$

If the formal group F is defined over $\mathbb{F}_p$, and q is sufficiently large, the profinite Galois group takes the form

$$\mathbb{G}_n^F = \mathbb{S}_n^F \rtimes \mathrm{Gal}(\mathbb{F}_q/\mathbb{F}_p)$$

where $\mathbb{S}_n^F = \mathrm{aut}(F)$ is the associated *Morava stabilizer group.*[12]

Fix a prime p, and let $E(n)$ denote the nth Johnson-Wilson spectrum, $K(n)$ the nth Morava K-theory spectrum, and E_n^F the nth Morava E-theory spectrum (associated to a height n formal group F over a finite extension $\mathbb{F}_q$ of $\mathbb{F}_p$),[13] [55], [56], [79] with

$$\begin{aligned}
\pi_* E(n) &= \mathbb{Z}_{(p)}[v_1, \dots, v_{n-1}, v_n^{\pm}], \\
\pi_* K(n) &= \mathbb{F}_p[v_n^{\pm}], \\
\pi_* E_n^F &= \mathbb{Z}_p[\zeta_{q-1}][[u_1, \dots, u_{n-1}][u^{\pm}].
\end{aligned}$$

Here, $|v_i| = 2(p^i - 1)$, $|u_i| = 0$, and $|u| = -2$. The spectra $E(n)$ and E_n^F are Landweber exact.

For spectra X and E, let X_E denote the E-localization of the spectrum X [24]. Then we have the following correspondence between locally Landweber spectra and associated stacks over $\mathcal{M}_{fg}$:

spectrum E	stack $\mathcal{X}_E$
S	$\mathcal{M}_{fg}$
$S_{E(n)}$	$(\mathcal{M}_{fg})^{\le n}_{\mathbb{Z}_{(p)}}$
$S_{K(n)}$	$(\mathcal{M}_{fg})^{[n]}_{\mathbb{Z}_{(p)}}$
E_n^F	$\mathcal{X}_n^F$

Remark 6.6.5. The spectrum $E = S_{K(n)}$ is really only Landweber in the $K(n)$-local category, in the sense that $(\mathrm{MUP} \wedge E)_{K(n)}$ is Landweber exact. However, similar considerations associate formal stacks $\mathcal{X}_E$ to such $K(n)$-local ring spectra, and an analog of Proposition 6.6.4 holds where

$$\mathcal{X}_{(E \wedge E')_{K(n)}} \simeq \mathcal{X}_E \widehat{\times}_{\mathcal{M}_{fg}} \mathcal{X}_{E'}.$$

The spectrum $S_{E(n)}$ is a limit of spectra that are Landweber in the above $K(i)$-local sense $i \le n$ [5].

[12] Often, the formal group F is taken to be the Honda height n formal group over $\mathbb{F}_{p^n}$, and the associated Morava stabilizer group is simply denoted $\mathbb{S}_n$.

[13] When the formal group F is the Honda height n formal group, the associated Morava E-theory spectrum is often denoted E_n.

Galois descent is encoded in the work of Goerss-Hopkins-Miller [35] and Devinatz-Hopkins [29], who showed that the group $\mathbb{G}_n^F$ acts on E_n^F with

$$S_{K(n)} \simeq (E_n^F)^{h\mathbb{G}_n^F}.$$

The following proposition follows from Proposition 6.6.4 (or its $K(n)$-local variant), and is closely related to localization formulas that appear in [39].

Proposition 6.6.6. *Suppose that E is locally Landweber. Then so is $E_{E(n)}$ and $E_{K(n)}$, and the associated stacks are given as the pullbacks*

$$\begin{array}{ccc} \mathcal{X}_{E_{E(n)}} & \longrightarrow & (\mathcal{M}_{fg})^{\le n}_{\mathbb{Z}_{(p)}} \\ \downarrow & & \downarrow \\ \mathcal{X}_E & \longrightarrow & \mathcal{M}_{fg} \end{array} \qquad \begin{array}{ccc} \mathcal{X}_{E_{K(n)}} & \longrightarrow & (\mathcal{M}_{fg})^{[n]}_{\mathbb{Z}_{(p)}} \\ \downarrow & & \downarrow \\ \mathcal{X}_E & \longrightarrow & \mathcal{M}_{fg} \end{array}$$

and

$$E_{K(n)} = (E_{E(n)})^\wedge_{I_n}$$

where $I_n = (p, v_1, \dots, v_{n-1})$ is the ideal corresponding to the locus of height n formal groups in $(\mathcal{M}_{fg})^{\le n}_{\mathbb{Z}_{(p)}}$.

For a general spectrum X, the square

$$\begin{array}{ccc} X_{E(n)} & \longrightarrow & X_{K(n)} \\ \downarrow & & \downarrow \\ X_{E(n-1)} & \longrightarrow & (X_{K(n)})_{E(n-1)}. \end{array}$$

is a homotopy pullback (the *chromatic fracture square*). If X is a locally Landweber ring spectrum, the chromatic fracture square can be regarded as the being associated to the "cover"

$$\left\{(\mathcal{M}_{fg})^{\le n-1}_{\mathbb{Z}_{(p)}}, (\mathcal{M}_{fg})^{[n]}_{\mathbb{Z}_{(p)}}\right\} \to (\mathcal{M}_{fg})^{\le n}_{\mathbb{Z}_{(p)}}.$$

$K(1)$-local Tmf

Applying Proposition 6.6.6, we find

$$\mathrm{Tmf}_{K(1)} \simeq \mathrm{tmf}_{K(1)} \simeq (\mathrm{Tmf}^{ord}_{(p)})^\wedge_p.$$

We explain the connection of $K(1)$-local Tmf to the Katz-Serre theory of p-adic modular forms.

The ring of *divided congruences* is defined to be the ring of inhomogeneous sums of rational modular forms where the sum of the q-expansions is integral:

$$D := \left\{ \sum f_k \in \bigoplus_k MF_k \otimes \mathbb{Q} \, : \, \sum f_k(\tau) \in \mathbb{Z}[[q]] \right\}.$$

This ring was studied extensively by Katz [58], who showed in [57] that there is an isomorphism

$$D \cong \Gamma\mathcal{O}_{\overline{\mathcal{M}}_{ell}^{triv}}$$

where $\overline{\mathcal{M}}_{ell}^{triv}$ is the pullback

$$\begin{array}{ccc} \overline{\mathcal{M}}_{ell}^{triv} & \longrightarrow & \overline{\mathcal{M}}_{ell} \\ \downarrow & & \downarrow \\ \mathrm{spec}(\mathbb{Z}) & \xrightarrow[\widehat{\mathbb{G}}_m]{} & \mathcal{M}_{fg}. \end{array}$$

Since complex K-theory is the Landweber exact ring spectrum associated to $\widehat{\mathbb{G}}_m$, Proposition 6.6.4 recovers the following theorem of Laures [63].

Proposition 6.6.7 (Laures). *The complex K-theory of* Tmf *is given by*

$$K_0\mathrm{Tmf} \cong D.$$

The ring of *generalized p-adic modular functions* [57], [58] is the p-completion of the ring D:

$$V_p := D_p^{\wedge}$$

and the proposition above implies that there is an isomorphism

$$\pi_0(K \wedge \mathrm{Tmf})_p^{\wedge} \cong V_p.$$

For $p \nmid \ell$, the action of the ℓth Adams operation ψ^ℓ on this space coincides with the action of $\ell \in \mathbb{Z}_p^\times$ on V described in [58], and

$$V_p\langle k\rangle = \{f \in V \,:\, \psi^\ell f = \ell^k f\}$$

is isomorphic to Serre's space of p-adic modular forms of weight k [92].

Letting p be odd, and choosing ℓ to be a topological generator of $\mathbb{Z}_p^\times$, we deduce from the fiber sequence [52, Lem. 2.5]

$$S_{K(1)}^{2k} \to K_p^{\wedge} \xrightarrow{\psi^\ell - \ell^k} K_p^{\wedge}$$

the following theorem of Baker [6].

Proposition 6.6.8. *For $p \geq 3$, the homotopy groups of* $\mathrm{Tmf}_{K(1)}$ *are given by the spaces of p-adic modular forms*

$$\pi_{2k}\mathrm{Tmf}_{K(1)} \cong V_p\langle k\rangle.$$

Remark 6.6.9. At the prime $p = 2$, Hopkins [50] and Laures [64] studied the spectrum $\mathrm{Tmf}_{K(1)}$, and showed that it has a simple construction as a finite cell object in the category of $K(1)$-local E_∞-ring spectra.

$K(2)$-local Tmf

An elliptic curve C over a field of characteristic p is called *supersingular* if its formal group $\widehat{C}$ has height 2. Over $\bar{\mathbb{F}}_p$ there are only finitely many isomorphism classes of supersingular elliptic curves. We shall let $(\mathcal{M}_{ell}^{ss})_{\mathbb{Z}_{(p)}}$ denote the formal neighborhood of the supersingular locus in $(\mathcal{M}_{ell})_{\mathbb{Z}_{(p)}}$.

Serre-Tate theory [68] implies that deformations of the supersingular elliptic curve C are in bijective correspondence with the deformations of its formal group $\widehat{C}$. Therefore the Lubin-Tate space $\widetilde{X}_2^{\widehat{C}}$ carries the universal deformation $\widetilde{C}$ of the elliptic curve C, giving us a lift:

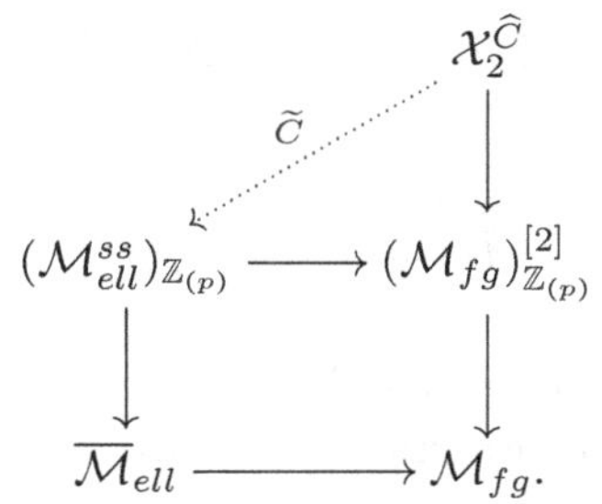

One may deduce from this that the square in the diagram above is a pullback. By Proposition 6.6.6, we have

$$\mathcal{X}_{\mathrm{Tmf}_{K(2)}} \simeq \mathcal{X}_{\mathrm{tmf}_{K(2)}} \simeq (\mathcal{M}_{ell}^{ss})_{\mathbb{Z}_{(p)}}.$$

For $p = 2, 3$ there is only one supersingular curve (defined over $\mathbb{F}_p$), and $\widetilde{C}$ is a Galois cover, with Galois group equal to

$$\mathrm{aut}(C) \rtimes \mathrm{Gal}(\mathbb{F}_{p^2}/\mathbb{F}_p) \leq \mathbb{G}_2^{\widehat{C}}$$

where

$$|\mathrm{aut}(C)| = \begin{cases} 24, & p = 2, \\ 12, & p = 3. \end{cases}$$

Since the Morava E-theory $E_2^{\widehat{C}}$ is Landweber exact with $\mathcal{X}_{E_2^{\widehat{C}}} = \mathcal{X}_2^{\widehat{C}}$, we have the following.

Proposition 6.6.10. *For $p = 2, 3$ there are equivalences*

$$\mathrm{Tmf}_{K(2)} \simeq \mathrm{TMF}_{K(2)} \simeq \mathrm{tmf}_{K(2)} \simeq (E_2^{\widehat{C}})^{h\,\mathrm{aut}(C) \rtimes Gal(\mathbb{F}_{p^2}/\mathbb{F}_p)}.$$

Remark 6.6.11. For a maximal finite subgroup $G \leq \mathbb{G}_n^F$, the homotopy fixed point spectrum spectrum $(E_2^F)^{hG}$ is sometimes denoted EO_n. Therefore the proposition above is stating that at the primes 2 and 3 there is an equivalence

$$\mathrm{Tmf}_{K(2)} \simeq EO_2.$$

Finally, we note that combining Proposition 6.6.6 with Lemma 6.5.1 and (6.5.1) yields the following.

Proposition 6.6.12. *There is an isomorphism*

$$\pi_*(\mathrm{TMF}_{K(2)}) \cong \pi_*(\mathrm{TMF})^\wedge_{p,c_4}$$

for $p = 2, 3$.

Chromatic fracture of Tmf

Since every elliptic curve in characteristic p has height 1 or 2, we deduce that the square

$$\begin{array}{ccc} (\overline{\mathcal{M}}_{ell})_{\mathbb{Z}_{(p)}} & \longrightarrow & (\mathcal{M}_{fg})^{\leq 2}_{\mathbb{Z}_{(p)}} \\ \downarrow & & \downarrow \\ \overline{\mathcal{M}}_{ell} & \longrightarrow & \mathcal{M}_{fg} \end{array}$$

is a pullback. We deduce from Proposition 6.6.6 that $\mathrm{Tmf}_{(p)}$ and $\mathrm{TMF}_{(p)}$ are $E(2)$-local.[14]

The p-completion of the chromatic fracture square for Tmf

$$\begin{array}{ccc} \mathrm{Tmf}^\wedge_p & \longrightarrow & \mathrm{Tmf}_{K(2)} \\ \downarrow & & \downarrow \\ \mathrm{Tmf}_{K(1)} & \longrightarrow & (\mathrm{Tmf}_{K(2)})_{K(1)} \end{array}$$

therefore takes the form

$$\begin{array}{ccc} \mathrm{Tmf}^\wedge_p & \longrightarrow & \mathrm{TMF}^\wedge_{(p,c_4)} \\ \downarrow & & \downarrow \\ (\mathrm{Tmf}^{ord})^\wedge_p & \longrightarrow & (\mathrm{TMF}^\wedge_{c_4}[c_4^{-1}])^\wedge_p \end{array} \tag{6.6.3}$$

and corresponds to the cover

$$\{(\mathcal{M}^{ss}_{ell})^\wedge_p, (\overline{\mathcal{M}}^{ord}_{ell})^\wedge_p\} \to (\overline{\mathcal{M}}_{ell})^\wedge_p.$$

The p-completed chromatic fracture square for Tmf is therefore a completed version of the homotopy pullback (6.5.2).

Remark 6.6.13. For an elliptic curve C in characteristic p for $p \geq 5$, we have

$$v_1^{\widehat{C}} = E_{p-1}(C)$$

where E_{p-1} is the normalized Eisenstein series of weight $p-1$. We therefore have analogs of Proposition 6.6.12 and (6.6.3) for $p \geq 5$ where we replace c_4 with the modular form E_{p-1}.

---

[14]The spectrum $\mathrm{tmf}_{(p)}$ is not $E(2)$-local, as cuspidal Weierstrass curves in characteristic p have formal groups of infinite height.

6.7 Topological automorphic forms

p-divisible groups

Fix a prime p.

Definition 6.7.1. A *p-divisible group of height n* over a ring R is a sequence of commutative group schemes

$$0 = G_0 \leq G_1 \leq G_2 \leq \cdots$$

so that each G_i is locally free of rank p^{in} over R, and such that for each i we have

$$G_i = \mathrm{Ker}([p^i] : G_{i+1} \to G_{i+1}).$$

Example 6.7.2. Suppose that A is an abelian variety over R of dimension n. Then the sequence of group schemes $A[p^\infty] := \{A[p^i]\}$ given by the p^i-torsion points of A

$$A[p^i] := \mathrm{Ker}([p^i] : A \to A)$$

is a p-divisible group of height $2n$.

Example 6.7.3. Suppose that F is a formal group law over a p-complete ring R of height n. Then the sequence of group schemes $F[p^\infty] = \{F[p^i]\}$ where

$$F[p^i] := \mathrm{spec}(R[[x]]/([p^i]_F(x)))$$

is a p-divisible group of height n.

Given a p-divisible group $G = \{G_i\}$ of height n over a p-complete ring R, the formal neighborhood of the identity

$$\widehat{G} \hookrightarrow \mathrm{colim}_i\, G_i$$

is a formal group of height $\leq n$ [77]. We define the *dimension* of G to be the dimension of the formal group $\widehat{G}$. We shall say G is *trivializable* if the line bundle $T_0^*\widehat{G}$ is trivial.

If A is an abelian variety of dimension n over R, then we have

$$\widehat{A[p^\infty]} = \widehat{A}$$

and the dimension of $A[p^\infty]$ is n.

Lurie's theorem

Let $\mathcal{M}^n_{pd}$ denote the moduli stack of 1-dimensional p-divisible groups of height n, and let $(\mathcal{M}^n_{pd})^{DM}_{et}$ denote the site of formally étale maps

$$\mathcal{X} \xrightarrow{G} \mathcal{M}^n_{pd} \tag{6.7.1}$$

where $\mathcal{X}$ is a locally Noetherian separated Deligne-Mumford stack over a complete local ring with perfect residue field of characteristic p.

Remark 6.7.4. One typically checks that a map (6.7.1) is formally étale by checking that for each closed point $x \in \mathcal{X}$, the formal neighborhood of x is isomorphic to the universal deformation space of the fiber G_x.

Lurie proved the following seminal theorem [70].

Theorem 6.7.5 (Lurie). *There is a sheaf $\mathcal{O}^{top}$ of E_∞ ring spectra on $(\mathcal{M}^n_{pd})^{DM}_{et}$ with the following property: the ring spectrum*

$$E := \mathcal{O}^{top}\left(\mathrm{spec}(R) \xrightarrow{G} \mathcal{M}^n_{pd}\right)$$

(associated to an affine formal étale open with G trivializable) is even periodic, with

$$F_E \cong \widehat{G}.$$

This theorem generalizes the Goerss-Hopkins-Miller theorem [35]. Consider the Lubin-Tate universal deformation space

$$\mathcal{X}^F_n \xrightarrow{\widetilde{F}} \mathcal{M}^{[n]}_{fg}.$$

The map classifying the p-divisible group $\widetilde{F}[p^\infty]$

$$\mathcal{X}^F_n \xrightarrow{\widetilde{F}[p^\infty]} \mathcal{M}^n_{pd}$$

is formally étale, simply because the data of a formal group is the same thing as the data of its associated p-divisible group over a p-complete ring, so they have the same deformations (see Remark 6.7.4). The associated ring spectrum is Morava E-theory:

$$\mathcal{O}^{top}(\mathcal{X}^F_n) \simeq E^F_n.$$

The functoriality of $\mathcal{O}^{top}$ implies that $\mathbb{G}^F_n$ acts on E^F_n.

Theorem 6.7.5 also generalizes (most of) Theorem 6.3.1. Serre-Tate theory states that deformations of abelian varieties are in bijective correspondence with deformations of their p-divisible groups. Again using Remark 6.7.4, this implies that the map

$$\begin{aligned}(\mathcal{M}_{ell})_{\mathbb{Z}_p} &\to \mathcal{M}^2_{pd},\\ C &\mapsto C[p^\infty]\end{aligned}$$

is formally étale. We deduce the existence of $\mathcal{O}^{top}$ on $(\mathcal{M}_{ell})_{\mathbb{Z}_p}$.

Cohomology theories associated to certain PEL Shimura stacks

The main issue that prevents us from associating cohomology theories to general n-dimensional abelian varieties is that their associated p-divisible groups are not 1-dimensional (unless $n = 1$, of course).

PEL Shimura stacks are moduli stacks of abelian varieties with the extra structure of Polarization, Endomorphisms, and Level structure [93], [41], [43]. We will now describe a class of PEL Shimura stacks (associated to a rational form of the unitary group $U(1, n-1)$) whose PEL data allow for the extraction of a 1-dimensional p-divisible group satisfying the hypotheses of Theorem 6.7.5.

In order to define our Shimura stack $\mathcal{X}_{V,L}$, we need to fix the following data.

$$\begin{aligned} F &= \text{quadratic imaginary extension of } \mathbb{Q}, \text{ such that } p \text{ splits as } u\bar{u}. \\ \mathcal{O}_F &= \text{ring of integers of } F. \\ V &= F\text{-vector space of dimension } n. \\ \langle -, - \rangle &= \mathbb{Q}\text{-valued non-degenerate hermitian alternating form on } V \\ & \qquad (\text{i.e. } \langle \alpha x, y\rangle = \langle x, \bar{\alpha} y\rangle \text{ for } \alpha \in F) \text{ of signature } (1, n-1). \\ L &= \mathcal{O}_F\text{-lattice in } V,\ \langle -, - \rangle \text{ restricts to give integer values on } L, \\ & \qquad \text{and makes } L_{(p)} \text{ self-dual.} \end{aligned}$$

The group of complex linear isometries (with respect to the form $\langle -, - \rangle$) of $V \otimes \mathbb{R}$ turns out to be a unitary group — the signature above is the signature of this unitary group.

Assume that S is a locally Noetherian scheme on which p is locally nilpotent. The groupoid $\mathcal{X}_{V,L}(S)$ consist of tuples of data (A, i, λ), and isomorphisms of such, defined as follows.

A	is an abelian scheme over S of dimension n.
$\lambda : A \to A^\vee$	is a polarization (principle at p), with Rosati involution $\dagger$ on $\mathrm{End}(A)_{(p)}$.
$i : \mathcal{O}_{F,(p)} \hookrightarrow \mathrm{End}(A)_{(p)}$	is an inclusion of rings, satisfying $i(\bar{z}) = i(z)^\dagger$.

We impose the following two conditions (one at p, one away from p) which basically amount to saying that the tuple (A, i, λ) is locally modeled on $(L, \langle -, - \rangle)$:

1. The splitting $p = u\bar{u}$ in $\mathcal{O}_F$ induces a splitting $\mathcal{O}_{F,p} = \mathcal{O}_{F,u} \times \mathcal{O}_{F,\bar{u}}$, and hence a splitting of p-divisible groups
$$A[p^\infty] \cong A[u^\infty] \oplus A[\bar{u}^\infty].$$
We require that $A[u^\infty]$ is 1-dimensional.

2. Choose a geometric point s in each component of S. We require that for each of these points there *exists* an $\mathcal{O}_F$-linear integral uniformization
$$\eta : \prod_{\ell \neq p} L_\ell^\wedge \xrightarrow{\cong} \prod_{\ell \neq p} T_\ell(A_s)$$

(where $T_\ell A = \lim_i A[\ell^i]$ is the ℓ-adic Tate module) that, when tensored with $\mathbb{Q}$, sends $\langle -, - \rangle$ to an $(\mathbb{A}^{p,\infty})^\times$-multiple of the λ-Weil pairing.[15]

Given a tuple $(A, i, \lambda) \in \mathcal{X}_{V,L}(S)$, the conditions on i and λ imply that the polarization induces an isomorphism

$$\lambda_* : A[u^\infty] \xrightarrow{\cong} A[\bar{u}^\infty]^\vee \tag{6.7.2}$$

(where $(-)^\vee$ denotes the Cartier dual). This implies that the p-divisible group $A[u^\infty]$ has height n. Serre-Tate theory [68] implies that deformations of an abelian variety are in bijective correspondence with the deformations of its p-divisible group. The isomorphism (6.7.2) therefore implies that deformations of a tuple (A, i, λ) are in bijective correspondence with deformations of $A[u^\infty]$. By Remark 6.7.4, the map

$$\mathcal{X}_{V,L} \to \mathcal{M}^n_{pd}$$

is therefore formally etale. Applying Lurie's theorem, we obtain

Theorem 6.7.6. [19] *There exists a sheaf $\mathcal{O}^{top}$ of E_∞ ring spectra on the site $(\mathcal{X}_{V,L})_{et}$, such that for each affine étale open*

$$\operatorname{spec}(R) \xrightarrow{(A,i,\lambda)} \mathcal{X}_{V,L}$$

with $A[u^\infty]$ trivializable, the associated ring spectrum

$$E := \mathcal{O}^{top}(\operatorname{spec}(R) \xrightarrow{(A,i,\lambda)} \mathcal{X}_{V,L})$$

is even periodic with

$$F_E \cong \widehat{A[u^\infty]}.$$

The spectrum of *topological automorphic forms* (TAF) for the Shimura stack $\mathcal{X}_{V,L}$ is defined to be the spectrum of global sections

$$\mathrm{TAF}_{V,L} := \mathcal{O}^{top}(\mathcal{X}_{V,L}).$$

Let ω be the line bundle over $\mathcal{X}_{V,L}$ with fibers given by

$$\omega_{A,i,\lambda} = T_0^* \widehat{A[u^\infty]}.$$

Then the construction of the descent spectral sequence (6.3.3) goes through verbatim to give a descent spectral sequence

$$E_2^{s,2k} = H^s(\mathcal{X}_{V,L}, \omega^{\otimes k}) \Rightarrow \pi_{2k-s}\mathrm{TAF}_{V,L}.$$

The motivation behind the terminology "topological automorphic forms" is that the space of sections

$$AF_k(U(V), L)_{\mathbb{Z}_p} := H^0(\mathcal{X}_{V,L}, \omega^{\otimes k})$$

---

[15] Here, $\mathbb{A}^{p,\infty} := \left(\prod_{\ell \neq p} \mathbb{Z}_\ell\right) \otimes \mathbb{Q}$ are the *adeles away from p and ∞*.

is the space of scalar valued weakly holomorphic automorphic forms for the unitary group $U(V)$ (associated to the lattice L) of weight k over $\mathbb{Z}_p$.

Remark 6.7.7. As in the modular case, the space of holomorphic automorphic forms has an additional growth condition which is analogous to the requirement that a modular form be holomorphic at the cusp. The term "weakly holomorphic" means that we drop this requirement. However, for $n \geq 3$, it turns out that every weakly holomorphic automorphic form is holomorphic [93, Sec. 5.2].

The spectra $\mathrm{TAF}_{V,L}$ are locally Landweber, with

$$\mathcal{X}_{\mathrm{TAF}_{V,L}} \simeq \mathcal{X}_{V,L}.$$

The height of the formal groups $\widehat{A[u^\infty]}$ associated to mod p points (A, i, λ) of the Shimura stack $\mathcal{X}_{V,L}$ range from 1 to n. We deduce from Proposition 6.6.6 that $\mathrm{TAF}_{V,L}$ is $E(n)$-local, and an analysis similar to that in the Tmf case (see Section 6.6) yields the following.

Proposition 6.7.8. [19] *The $K(n)$-localization of $\mathrm{TAF}_{V,L}$ is given by*

$$(\mathrm{TAF}_{V,L})_{K(n)} \simeq \left(\prod_{(A,i,\lambda)} \left(E_n^{\widehat{A[u^\infty]}}\right)^{h\,\mathrm{aut}(A,i,\lambda)} \right)^{h\mathrm{Gal}}$$

where the product ranges over the (finite, non-empty) set of mod p points (A, i, λ) of $\mathcal{X}_{V,L}$ with $\widehat{A[u^\infty]}$ of height n.

The groups $\mathrm{aut}(A, i, \lambda)$ are finite subgroups of the Morava stabilizer group. The structure of these subgroups, and the conditions under which they are maximal finite subgroups, is studied in [11].

6.8 Further reading

Elliptic genera: One of the original motivations behind tmf was Ochanine's definition of a genus of spin manifolds that takes values in the ring of modular forms for $\Gamma_0(2)$, which interpolates between the $\widehat{A}$-genus and the signature. Witten gave an interpretation of this genus in terms of 2-dimensional field theory, and produced a new genus (the *Witten genus*) of *string manifolds* valued in modular forms for $SL_2(\mathbb{Z})$ [103], [104]. These genera were refined to an orientation of elliptic spectra by Ando-Hopkins-Strickland [3], and were shown to give E_∞ orientations

$$MString \to \mathrm{tmf},$$
$$MSpin \to \mathrm{tmf}_0(2)$$

by Ando-Hopkins-Rezk [2] and Wilson [102], respectively.

Geometric models: The most significant outstanding problem in the theory of topological modular forms is to give a geometric interpretation of this cohomology theory (analogous to the fact that K-theory classes are represented by vector bundles). Motivated by the work of Witten described above, Segal proposed that 2-dimensional field theories should represent tmf-cocycles [91]. This idea has been fleshed out in detail by Stolz and Teichner, and concrete conjectures are proposed in [97]. While these conjectures have proved difficult to verify, partial results have been made by many researchers (see, for example, [22]).

Computations of the homotopy groups of $\mathrm{TMF}_0(N)$**:** Mahowald and Rezk computed the descent spectral sequence for $\pi_*\mathrm{TMF}_0(3)$ in [73], and a similar computation of $\pi_*\mathrm{TMF}_0(5)$ was performed by Ormsby and the author in [21] (see also [46]). Meier gave a general additive description of $\pi_*\mathrm{TMF}_0(N)^\wedge_2$ for all N with $4 \not| \varphi(N)$ in [76].

Self-duality: Stojanoska showed that Serre duality for the stack $\overline{\mathcal{M}}_{ell}$ lifts to a self-duality result for Tmf[1/2]. This result was extended to $\mathrm{Tmf}_1(N)$ by Meier [76]. Other self-duality results can be found in [72], [14], [23], and [26]. Closely related to duality is the study of the Picard group of modules over various tmf spectra - see [75], [13].

Detection of the divided β-family: Adams used K-theory to define his e-invariant, and deduced that the order of the image of the J-homomorphism in degree $2k-1$ is given by the denominator of the Bernoulli number $\frac{B_k}{2k}$. The divided β-family, a higher chromatic generalization of the image of J, was constructed on the 2-line of the Adams-Novikov spectral sequence by Miller-Ravenel-Wilson [78]. Laures used tmf to construct a generalization of the e-invariant, called the f-invariant [63]. This invariant relates the divided beta family to certain congruences between modular forms [16], [18], [99].

Quasi-isogeny spectra: The author showed that the Goerss-Henn-Mahowald-Rezk resolution of the 3-primary $K(2)$-local sphere [34] can be given a modular interpretation in terms of isogenies of elliptic curves [14], and conjectured that something similar happens at all primes [15].

The tmf resolution: Generalizing his seminal work on "bo-resolutions," Mahowald initiated the study of the tmf-based Adams spectral sequence. This was used in [9] to lift the 192-periodicity in $\mathrm{tmf}_{(2)}$ to a periodicity in the 2-primary stable homotopy groups of spheres, and in [10] to show coker J is non-trivial in "most" dimensions less than 140. The study of the tmf-based Adams spectral sequence begins with an analysis of the ring of cooperations $\mathrm{tmf}_*\mathrm{tmf}$. With 6 inverted, this was studied by Baker and Laures [7], [63]. The 2-primary structure of $\mathrm{tmf}_*\mathrm{tmf}$ was studied in [12].

Dyer-Lashof operations: Ando observed that power operations for elliptic cohomology are closely related to isogenies of elliptic curves [4]. Following this thread, Rezk used the geometry of elliptic curves to compute the Dyer-Lashof algebra for the Morava E-theory E_2 at the prime 2 [87]. This was generalized by Zhu to all primes [106]. Using Rezk's "modular isogeny complex" [88], Zhu was able to derive information about unstable homotopy groups of spheres [105].

Spectral algebraic geometry: As mentioned in the introduction, Lurie introduced the notion of *spectral algebraic geometry*, and used it to give a revolutionary new construction of tmf [69] (see also [70] and [71]).

Equivariant TMF: Grojnowski introduced the idea of complex analytic *equivariant elliptic cohomology* [40] (see also [89]). A variant based on K-theory was introduced by Devoto [30] (see also [33], [59], [54]). This idea was refined in the rational setting by Greenlees [38]. Lurie used his spectral algebro-geometric construction of TMF to construct equivariant TMF (this is outlined in [69], see [70] and [71] for more details). Equivariant elliptic cohomology was used by Rosu in his proof of the rigidity of the elliptic genus [90].

K3 cohomology: Morava and Hopkins suggested that cohomology theories should be also be associated to K3 surfaces. Szymik showed this can be done in [98] (see also [84]).

Computations of π_*TAF: Very little is known about the homotopy groups of spectra of topological automorphic forms, for the simple reason that, unlike the modular case, very few computations of rings of classical integral automorphic forms exist in the literature. Nevertheless, special instances have been computed in [19], [44], [20], [67], [66], [100], [101].

Acknowledgements

The author would like to thank Haynes Miller for giving numerous suggestions, as well as Sanath Devalapurkar, Lennart Meier, John Rognes, Taylor Sutton, and Markus Szymik for valuable suggestions and corrections. The author was partially supported from a grant from the National Science Foundation.

Bibliography

[1] J. F. Adams. *Stable homotopy and generalised homology.* University of Chicago Press, 1974. Chicago Lectures in Mathematics.

[2] M. Ando, M. J. Hopkins, and C. Rezk. Multiplicative orientations of KO-theory and of the spectrum of topological modular forms. Available at `https://faculty.math.illinois.edu/~mando/papers/koandtmf.pdf`, 2010.

[3] M. Ando, M. J. Hopkins, and N. P. Strickland. Elliptic spectra, the Witten genus and the theorem of the cube. *Invent. Math.*, 146(3):595–687, 2001.

[4] Matthew Ando. Power operations in elliptic cohomology and representations of loop groups. *Trans. Amer. Math. Soc.*, 352(12):5619–5666, 2000.

[5] Omar Antolín-Camarena and Tobias Barthel. Chromatic fracture cubes. arXiv:1410.7271, 2014.

[6] Andrew Baker. Elliptic cohomology, p-adic modular forms and Atkin's operator U_p. In *Algebraic topology (Evanston, IL, 1988)*, volume 96 of *Contemp. Math.*, pages 33–38. Amer. Math. Soc., 1989.

[7] Andrew Baker. Operations and cooperations in elliptic cohomology. I. Generalized modular forms and the cooperation algebra. *New York J. Math.*, 1:39–74, 1994/95.

[8] Tilman Bauer. Computation of the homotopy of the spectrum tmf. In *Groups, homotopy and configuration spaces*, volume 13 of *Geom. Topol. Monogr.*, pages 11–40. Geom. Topol. Publ., 2008.

[9] M. Behrens, M. Hill, M. J. Hopkins, and M. Mahowald. On the existence of a v_2^{32}-self map on $M(1,4)$ at the prime 2. *Homology Homotopy Appl.*, 10(3):45–84, 2008.

[10] M. Behrens, M. Hill, M. J. Hopkins, and M. Mahowald. Detecting exotic spheres in low dimensions using coker J. arXiv:1708.06854, 2017.

[11] M. Behrens and M. J. Hopkins. Higher real K-theories and topological automorphic forms. *J. Topol.*, 4(1):39–72, 2011.

[12] M. Behrens, K. Ormsby, N. Stapleton, and V. Stojanoska. On the ring of cooperations for 2-primary connective topological modular forms. arXiv:1708.06854, 2018.

[13] Agnes Beaudry, Irina Bobkova, Michael Hill, and Vesna Stojanoska. *Invertible $K(2)$-local E-modules in $C4$-spectra.* arXiv:1901.02109, 2019.

[14] Mark Behrens. A modular description of the $K(2)$-local sphere at the prime 3. *Topology*, 45(2):343–402, 2006.

[15] Mark Behrens. Buildings, elliptic curves, and the $K(2)$-local sphere. *Amer. J. Math.*, 129(6):1513–1563, 2007.

[16] Mark Behrens. Congruences between modular forms given by the divided β family in homotopy theory. *Geom. Topol.*, 13(1):319–357, 2009.

[17] Mark Behrens. The construction of *tmf*. In *Topological modular forms*, volume 201 of *Math. Surveys Monogr.*, pages 261–285. Amer. Math. Soc., 2014.

[18] Mark Behrens and Gerd Laures. β-family congruences and the f-invariant. In *New topological contexts for Galois theory and algebraic geometry (BIRS 2008)*, volume 16 of *Geom. Topol. Monogr.*, pages 9–29. Geom. Topol. Publ., 2009.

[19] Mark Behrens and Tyler Lawson. Topological automorphic forms. *Mem. Amer. Math. Soc.*, 204(958), 2010.

[20] Mark Behrens and Tyler Lawson. Topological automorphic forms on $\mathrm{U}(1,1)$. *Math. Z.*, 267(3-4):497–522, 2011.

[21] Mark Behrens and Kyle Ormsby. On the homotopy of $Q(3)$ and $Q(5)$ at the prime 2. *Algebr. Geom. Topol.*, 16(5):2459–2534, 2016.

[22] Daniel Berwick-Evans and Arnav Tripathy. *A geometric model for complex analytic equivariant elliptic cohomology.* arXiv:1805.04146, 2018.

[23] Irina Bobkova. *Spanier-Whitehead duality in $K(2)$-local category at $p = 2$.* arXiv:1809.08226, 2018.

[24] A. K. Bousfield. The localization of spectra with respect to homology. *Topology*, 18(4):257–281, 1979.

[25] A. K. Bousfield and D. M. Kan. A second quadrant homotopy spectral sequence. *Trans. Amer. Math. Soc.*, 177:305–318, 1973.

[26] Robert Bruner and John Rognes. The Adams spectral sequence for topological modular forms. Preprint.

[27] P. Deligne and M. Rapoport. Les schémas de modules de courbes elliptiques. In *Modular functions of one variable, II (Proc. Internat. Summer School, Univ. Antwerp, Antwerp, 1972)*, volume 349 of *Lecture Notes in Math.*, pages 143–316. Springer, 1973.

[28] Sanath K. Devalapurkar. Equivariant versions of Wood's theorem. Available at `http://www.mit.edu/~sanathd/wood.pdf`, 2018.

[29] Ethan S. Devinatz and Michael J. Hopkins. Homotopy fixed point spectra for closed subgroups of the Morava stabilizer groups. *Topology*, 43(1):1–47, 2004.

[30] Jorge A. Devoto. Equivariant elliptic homology and finite groups. *Michigan Math. J.*, 43(1):3–32, 1996.

[31] Christopher L. Douglas, John Francis, André G. Henriques, and Michael A. Hill, editors. *Topological modular forms*, volume 201 of *Math. Surv. Monogr.* Amer. Math. Soc., 2014.

[32] Jens Franke. Uniqueness theorems for certain triangulated categories possessing an Adams spectral sequence. Available at `http://faculty.math.illinois.edu/K-theory/0139/`, 1996.

[33] Nora Ganter. Power operations in orbifold Tate K-theory. *Homology Homotopy Appl.*, 15(1):313–342, 2013.

[34] P. Goerss, H.-W. Henn, M. Mahowald, and C. Rezk. A resolution of the $K(2)$-local sphere at the prime 3. *Ann. of Math. (2)*, 162(2):777–822, 2005.

[35] P. G. Goerss and M. J. Hopkins. Moduli spaces of commutative ring spectra. In *Structured ring spectra*, volume 315 of *London Math. Soc. Lecture Note Ser.*, pages 151–200. Cambridge Univ. Press, 2004.

[36] Paul G. Goerss. Realizing families of Landweber exact homology theories. In *New topological contexts for Galois theory and algebraic geometry (BIRS 2008)*, volume 16 of *Geom. Topol. Monogr.*, pages 49–78. Geom. Topol. Publ., 2009.

[37] Paul G. Goerss. Topological modular forms [after Hopkins, Miller and Lurie]. *Astérisque*, (332):Exp. No. 1005, viii, 221–255, 2010. Séminaire Bourbaki. Volume 2008/2009. Exposés 997–1011.

[38] J. P. C. Greenlees. Rational S^1-equivariant elliptic cohomology. *Topology*, 44(6):1213–1279, 2005.

[39] J. P. C. Greenlees and J. P. May. Completions in algebra and topology. In *Handbook of Algebraic Topology*, pages 255–276. North-Holland, 1995.

[40] I. Grojnowski. Delocalised equivariant elliptic cohomology. In *Elliptic cohomology*, volume 342 of *London Math. Soc. Lecture Note Ser.*, pages 114–121. Cambridge Univ. Press, 2007.

[41] Michael Harris and Richard Taylor. *The geometry and cohomology of some simple Shimura varieties*, volume 151 of *Ann. of Math. Stud.* Princeton University Press, 2001. With an appendix by Vladimir G. Berkovich.

[42] André Henriques. The homotopy groups of *tmf* and its localizations. In *Topological Modular Forms*, volume 201 of *Math. Surveys Monogr.*, pages 261–285. Amer. Math. Soc., 2014.

[43] Haruzo Hida. *p-adic automorphic forms on Shimura varieties.* Springer Monogr. Math. Springer-Verlag, 2004.

[44] Michael Hill and Tyler Lawson. Automorphic forms and cohomology theories on Shimura curves of small discriminant. *Adv. Math.*, 225(2):1013–1045, 2010.

[45] Michael Hill and Tyler Lawson. Topological modular forms with level structure. *Invent. Math.*, 203(2):359–416, 2016.

[46] Michael A. Hill, Michael J. Hopkins, and Douglas C. Ravenel. The slice spectral sequence for the C_4 analog of real K-theory. *Forum Math.*, 29(2):383–447, 2017.

[47] M. J. Hopkins. Algebraic topology and modular forms. In *Proceedings of the International Congress of Mathematicians, Vol. I (Beijing, 2002)*, pages 291–317. Higher Ed. Press, 2002.

[48] Michael J. Hopkins. Topological modular forms, the Witten genus, and the theorem of the cube. In *Proceedings of the International Congress of Mathematicians, Vol. 1, 2 (Zürich, 1994)*, pages 554–565. Birkhäuser, 1995.

[49] Michael J. Hopkins. Complex oriented cohomology theories and the language of stacks (COCTALOS). Available at `https://web.math.rochester.edu/people/faculty/doug/otherpapers/coctalos.pdf`.

[50] Michael J. Hopkins. $K(1)$-local E_∞-ring spectra. In *Topological modular forms*, volume 201 of *Math. Surveys Monogr.*, pages 287–302. Amer. Math. Soc., 2014.

[51] Michael J. Hopkins and Mark Mahowald. From elliptic curves to homotopy theory. In *Topological modular forms*, volume 201 of *Math. Surveys Monogr.*, pages 261–285. Amer. Math. Soc., 2014.

[52] Michael J. Hopkins, Mark Mahowald, and Hal Sadofsky. Constructions of elements in Picard groups. In *Topology and representation theory (Evanston, IL, 1992)*, volume 158 of *Contemp. Math.*, pages 89–126. Amer. Math. Soc., 1994.

[53] Michael J. Hopkins and Haynes R. Miller. Elliptic curves and stable homotopy I. In *Topological modular forms*, volume 201 of *Math. Surveys Monogr.*, pages 209–260. Amer. Math. Soc., 2014.

[54] Zhen Huan. Quasi-elliptic cohomology I. *Adv. Math.*, 337:107–138, 2018.

[55] David Copeland Johnson and W. Stephen Wilson. Projective dimension and Brown-Peterson homology. *Topology*, 12:327–353, 1973.

[56] David Copeland Johnson and W. Stephen Wilson. BP operations and Morava's extraordinary K-theories. *Math. Z.*, 144(1):55–75, 1975.

[57] Nicholas M. Katz. p-adic properties of modular schemes and modular forms. *Modular functions of one variable, III (Proc. Internat. Summer School, Univ. Antwerp, Antwerp, 1972)*, volume 350 of *Lecture Notes in Math.*, pages 69–190. Springer, 1973.

[58] Nicholas M. Katz. Higher congruences between modular forms. *Ann. of Math. (2)*, 101:332–367, 1975.

[59] Nitu Kitchloo. Quantization of the modular functor and equivariant elliptic cohomology. arXiv:1407.6698, 2014.

[60] Johan Konter. The homotopy groups of the spectrum Tmf. arXiv:1212.3656, 2012.

[61] Peter S. Landweber. Homological properties of comodules over $MU_*(MU)$ and $\mathrm{BP}_*(\mathrm{BP})$. *Amer. J. Math.*, 98(3):591–610, 1976.

[62] Peter S. Landweber, Douglas C. Ravenel, and Robert E. Stong. Periodic cohomology theories defined by elliptic curves. In *The Čech Centennial (Boston, MA, 1993)*, volume 181 of *Contemp. Math.*, pages 317–337. Amer. Math. Soc., 1995.

[63] Gerd Laures. The topological q-expansion principle. *Topology*, 38(2):387–425, 1999.

[64] Gerd Laures. $K(1)$-local topological modular forms. *Invent. Math.*, 157(2):371–403, 2004.

[65] Tyler Lawson. An overview of abelian varieties in homotopy theory. In *New topological contexts for Galois theory and algebraic geometry (BIRS 2008)*, volume 16 of *Geom. Topol. Monogr.*, pages 179–214. Geom. Topol. Publ., 2009.

[66] Tyler Lawson. The Shimura curve of discriminant 15 and topological automorphic forms. *Forum Math. Sigma*, 3:e3, 32, 2015.

[67] Tyler Lawson and Niko Naumann. Commutativity conditions for truncated Brown-Peterson spectra of height 2. *J. Topol.*, 5(1):137–168, 2012.

[68] J. Lubin and J.-P. J. Tate Serre. Lecture notes prepared in connection with the seminars held at the Summer Institute on Algebraic Geometry, Whitney Estate, Woods Hole, Massachusetts, July 6–July 31. 1964.

[69] J. Lurie. A survey of elliptic cohomology. In *Algebraic Topology*, volume 4 of *Abel Symp.*, pages 219–277. Springer, 2009.

[70] Jacob Lurie. Elliptic cohomology I.
Available for download at `www.math.harvard.edu/~lurie/papers/Elliptic-I.pdf`, 2018.

[71] Jacob Lurie. Elliptic cohomology II: orientations.
Available for download at `www.math.harvard.edu/~lurie/papers/Elliptic-II.pdf`, 2018.

[72] Mark Mahowald and Charles Rezk. Brown-Comenetz duality and the Adams spectral sequence. *Amer. J. Math.*, 121(6):1153–1177, 1999.

[73] Mark Mahowald and Charles Rezk. Topological modular forms of level 3. *Pure Appl. Math. Q.*, 5(2, Special Issue: In honor of Friedrich Hirzebruch. Part 1):853–872, 2009.

[74] Akhil Mathew. The homology of tmf. *Homology Homotopy Appl.*, 18(2):1–29, 2016.

[75] Akhil Mathew and Vesna Stojanoska. *The Picard group of topological modular forms via descent theory,* Geom. Topol. 20 (2016), no. 6, 3133-3217.

[76] Lennart Meier. Topological modular forms with level structure: decompositions and duality. arXiv:1806.06709, 2018.

[77] William Messing. *The crystals associated to Barsotti-Tate groups: with applications to abelian schemes.* Volume 264 of *Lecture Notes in Math.* Springer-Verlag, 1972.

[78] Haynes R. Miller, Douglas C. Ravenel, and W. Stephen Wilson. Periodic phenomena in the Adams-Novikov spectral sequence. *Ann. of Math. (2)*, 106(3):469–516, 1977.

[79] Jack Morava. Completions of complex cobordism. In *Geometric applications of homotopy theory (Proc. Conf., Evanston, Ill., 1977), II*, volume 658 of *Lecture Notes in Math.*, pages 349–361. Springer, 1978.

[80] Jack Morava. Noetherian localisations of categories of cobordism comodules. *Ann. of Math. (2)*, 121(1):1–39, 1985.

[81] Niko Naumann. The stack of formal groups in stable homotopy theory. *Adv. Math.*, 215(2):569–600, 2007.

[82] S. P. Novikov. Methods of algebraic topology from the point of view of cobordism theory. *Izv. Akad. Nauk SSSR Ser. Mat.*, 31:855–951, 1967.

[83] Serge Ochanine. Sur les genres multiplicatifs définis par des intégrales elliptiques. *Topology*, 26(2):143–151, 1987.

[84] Oron Propp. Constructing explicit K3 spectra. arXiv:1810.08953, 2018.

[85] Daniel Quillen. On the formal group laws of unoriented and complex cobordism theory. *Bull. Amer. Math. Soc.*, 75:1293–1298, 1969.

[86] Charles Rezk. Supplementary notes for Math 512.
Available at `https://faculty.math.illinois.edu/~rezk/512-spr2001-notes.pdf`, 2007.

[87] Charles Rezk. Power operations for Morava E-theory of height 2 at the prime 2. arXiv:0812.1320, 2008.

[88] Charles Rezk. Modular isogeny complexes. *Algebr. Geom. Topol.*, 12(3):1373–1403, 2012.

[89] Charles Rezk. Looijenga line bundles in complex analytic elliptic cohomology. arXiv:1608.03548, 2016.

[90] Ioanid Rosu. Equivariant elliptic cohomology and rigidity. *Amer. J. Math.*, 123(4):647–677, 2001.

[91] Graeme Segal. What is an elliptic object? In *Elliptic cohomology*, volume 342 of *London Math. Soc. Lecture Note Ser.*, pages 306–317. Cambridge Univ. Press, 2007.

[92] Jean-Pierre Serre. Formes modulaires et fonctions zêta p-adiques. *Modular functions of one variable, III (Proc. Internat. Summer School, Univ. Antwerp, 1972)*, volume 350 of *Lecture Notes in Math.*, pages 191–268. Springer, 1973.

[93] Goro Shimura. *Arithmeticity in the theory of automorphic forms*, volume 82 of *Math. Surveys Monogr.* Amer. Math. Soc., 2000.

[94] Joseph H. Silverman. *Advanced topics in the arithmetic of elliptic curves*, volume 151 of *Grad. Texts in Math.* Springer-Verlag, 1994.

[95] Joseph H. Silverman. *The arithmetic of elliptic curves*, volume 106 of *Grad. Texts in Math.* Springer, second edition, 2009.

[96] Vesna Stojanoska. Duality for topological modular forms. *Doc. Math.*, 17:271–311, 2012.

[97] Stephan Stolz and Peter Teichner. Supersymmetric field theories and generalized cohomology. In *Mathematical foundations of quantum field theory and perturbative string theory*, volume 83 of *Proc. Sympos. Pure Math.*, pages 279–340. Amer. Math. Soc., 2011.

[98] Markus Szymik. $K3$ spectra. *Bull. Lond. Math. Soc.*, 42(1):137–148, 2010.

[99] Hanno von Bodecker. The beta family at the prime two and modular forms of level three. *Algebr. Geom. Topol.*, 16(5):2851–2864, 2016.

[100] Hanno von Bodecker and Sebastian Thyssen. On p-local topological automorphic forms for $U(1,1;\mathbb{Z}[i])$. arXiv:1609.08869, 2016.

[101] Hanno von Bodecker and Sebastian Thyssen. Topological automorphic forms via curves. arXiv:1705.02134, 2017.

[102] Dylan Wilson. Orientations and topological modular forms with level structure. arXiv:1507.05116, 2015.

[103] Edward Witten. The index of the Dirac operator in loop space. In *Elliptic curves and modular forms in algebraic topology (Princeton, NJ, 1986)*, volume 1326 of *Lecture Notes in Math.*, pages 161–181. Springer, 1988.

[104] Edward Witten. Index of Dirac operators. In *Quantum fields and strings: a course for mathematicians, Vol. 1, 2 (Princeton, NJ, 1996/1997)*, pages 475–511. Amer. Math. Soc., 1999.

[105] Yifei Zhu. Morava E-homology of Bousfield-Kuhn functors on odd-dimensional spheres. *Proc. Amer. Math. Soc.*, 146(1):449–458, 2018.

[106] Yifei Zhu. Semistable models for modular curves and power operations for Morava E-theories of height 2. arXiv:1508.03358, 2018.

Deptartment of Mathematics, University of Notre Dame, Notre Dame, IN 46556, U.S.A.

E-mail address: mbehrens1@nd.edu

7

A survey of models for (∞, n)-categories

Julia E. Bergner

7.1 Introduction: What should an (∞, n)-category be?

When we say that we want to find "models" for (∞, n)-categories, we are looking for concrete mathematical objects that encode this desired structure. But first, we need to answer a more basic question: What is an (∞, n)-category anyway? A short answer is that it is should be a higher category that is given up to homotopy, in some sense. Alternatively, it is a higher category in which sufficiently high-level morphisms are invertible.

To give a better description of what (∞, n)-categories should be, let us first consider what is meant by a higher category more generally. Recall that a category in the usual sense consists of objects, morphisms between objects, and a composition law for morphisms, such that each object has an identity morphism, and such that composition is associative. As we move to higher categories, for specificity let us refer to morphisms as *1-morphisms*.

The essential idea behind a 2-category is that 1-morphisms that share source and target objects can have *2-morphisms* between them:

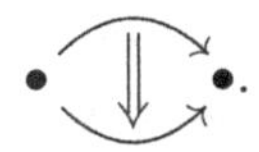

These 2-morphisms should themselves have a composition law that is associative; each 1-morphism should also have an associated identity 2-morphism. One could then imagine extending these ideas for successively higher morphisms. If we stop at some dimension n, we call the resulting structure an *n-category*; if we continue to arbitrary n we obtain an *∞-category*.

When the unitality of identity morphisms and associativity hold strictly, it is not hard to define these structures concretely. A *2-category* can be defined to be a category *enriched in categories*. In other words, it consists of objects, together with, for any pair of objects (x, y), a category of morphisms from x to y. The objects of this category define 1-morphisms $x \to y$, and the morphisms between them are 2-morphisms. More generally, we can define an *n-category* to be a category enriched in $(n-1)$-categories. Defining ∞-categories is more delicate.

Mathematics Subject Classification. primary: 55U10, 55U35, 55U40, secondary: 18D05, 18D15, 18D20, 18G30, 18G55.

Key words and phrases. (∞, n)-categories, model categories, enriched categories, simplicial objects, Θ_n-spaces.

Unfortunately, such strict conditions do not hold for many interesting examples. Often associativity, for instance, does not hold strictly, but rather only up to isomorphism. For such a structure, we need to ask that various coherence laws hold, leading to the notion of a *weak n-category*. In the case of a weak 2-category, or *bicategory*, the required coherence laws are manageable to write down. The situation gets successively more complicated for higher weak n-categories. While there are many different proposed definitions of such structures, they are, in general, not known to be equivalent to one another. A good introduction to many of these approaches can be found in [38].

Much more progress has been made when we look at higher categories from a homotopical perspective, leading to the notion of an (∞, n)-category. Heuristically, an (∞, n)-category should be a weak ∞-category such that all i-morphisms are weakly invertible for $i > n$. Since we already suggested that strict ∞-categories are not so straightforward, never mind weak ones, it is not immediately clear that such a notion should be tractable in any way. Yet, starting with $(\infty, 0)$-categories, or ∞-groupoids, which can be defined much more easily, we can again use a process of enrichment to make an inductive definition: an (∞, n)-category is a category enriched in $(\infty, n-1)$-categories. As long as we have a good model for $(\infty, 0)$-categories to begin, we have a very precise way to define higher (∞, n)-categories.

However, just as we often want weak, rather than strict, n-categories, we really want (∞, n)-categories to be defined via some kind of weak enrichment, so that composition might only be defined up to homotopy and the associativity and unit conditions need not hold in a strict way but rather in some more homotopical sense. While it might seem that we are back to the same kinds of issues as before, these structures are remarkably more manageable than weak n-categories. Such weaker models for $(\infty, 1)$-categories can be defined using homotopy-theoretic tools such as simplicial objects. The generalization of these structures to higher (∞, n)-categories is difficult work and can be done in myriad ways, some of which are still work in progress, but we have a number of models which are well-understood and known to be equivalent to one another.

In this paper, we consider some of these models from a homotopy-theoretic point of view. We look at the associated model category structures for each, and the comparisons between them in the form of Quillen equivalences.

After introducing $(\infty, 0)$-categories (and in doing so, reviewing some basic ideas in simplicial homotopy theory) and various models for $(\infty, 1)$-categories, we focus primarily on $(\infty, 2)$-categories. By emphasizing lower dimensions, we hope the reader can grasp the basic ideas without immediately drowning in the abstraction of the general case. At the end of the paper, we discuss fully general (∞, n)-categories, at least for some of the known definitions.

It is important to remark here that we by no means cover all models or approaches to the subject. To keep this paper to a reasonable length, we have chosen to focus on models given by diagrams of sets or simplicial sets, and from the point of view of model category structures. In particular, we do not look at the axiomatic treatments of Toën [53] and of Barwick and Schommer-Pries [7], nor at the model-independent approach of Riehl and Verity [49], [50]. Other models we omit include the n-relative categories of Barwick and Kan [5], [6], the n-complicial sets of Verity [54] and their recent comparison with other kinds of diagrams by Ozornova and Rovelli [42], the more geometric approach of Ayala, Francis, and Rozenblyum [3], and the variants building on marked simplicial sets of Lurie in [40]. These papers comprise important work in the subject and we hope the introduction we have given here will inspire the reader to look into them more closely.

7.2 Some model category background

Our approach to (∞, n)-categories is in the framework of model categories. The definitions we make here do not need this extra structure, but from the point of view of homotopy theory it is valuable to have it nonetheless. The reader only interested in the ways to define (∞, n)-categories themselves can safely ignore this language and regard it as means for making the formal comparisons. In this section, we summarize some of the model category language that is used.

A *model category* $\mathcal{M}$ is a category possessing all small limits and colimits, together with three distinguished classes of morphisms, called *weak equivalences*, *fibrations*, and *cofibrations*, satisfying several axioms. An *acyclic (co)fibration* is a (co)fibration that is also a weak equivalence. The original definition was given by Quillen [44]; other references for the include the surveys [26] and [27] and the books [29] and [31].

The existence of limits and colimits guarantees the existence of an initial object $\varnothing$ and a terminal object $*$.

Definition 7.2.1. An object X of a model category $\mathcal{M}$ is *cofibrant* if the unique morphism $\varnothing \to X$ is a cofibration. Dually, X is *fibrant* if the unique morphism $X \to *$ is a fibration.

For example, the category of topological spaces has a model structure $\mathcal{T}op$ in which the weak equivalences are the weak homotopy equivalences, all objects are fibrant, and the cofibrant objects are retracts of CW-complexes.

The structure of a model category enables one to have a well-defined homotopy category without running into set-theoretic obstacles. On the one hand, the critical information of a model category is the collection of weak equivalences; any two model categories with the same weak equivalences have equivalent homotopy categories. On the other hand, the fibrant and cofibrant objects are important in the construction of the specific homotopy category associated to a given model category, and are typically the main objects of interest.

The following definition is our means of comparison between model categories.

Definition 7.2.2. Let $\mathcal{M}$ and $\mathcal{N}$ be model categories. A *Quillen pair* is an adjoint pair of functors

$$F \colon \mathcal{M} \rightleftarrows \mathcal{N} \colon G$$

such that the left adjoint F preserves cofibrations and the right adjoint G preserves fibrations. It is a *Quillen equivalence* if, additionally, a morphism $FX \to Y$ is a weak equivalence in $\mathcal{N}$ if and only if the corresponding morphism $X \to GY$ is a weak equivalence in $\mathcal{M}$.

Throughout this paper, our goal is to describe model structures on certain categories in such a way that the fibrant and cofibrant objects model (∞, n)-categories. We then give Quillen equivalences between these model categories, which demonstrate that their corresponding models for (∞, n)-categories really do capture the same structure.

Many of these models use the framework of simplicial sets and more general simplicial objects, so let us briefly review these ideas. Let Δ be the category of finite ordered sets $[n] = \{0 \leq 1 \leq \cdots \leq n\}$ and order-preserving functions, and let Δ^{op} be its opposite category. Let $\mathcal{S}ets$ denote the category of sets.

Definition 7.2.3. A *simplicial set* is a functor $\Delta^{op} \to \mathcal{S}ets$.

Three critical examples of simplicial sets are the n-simplex $\Delta[n]$, which is the representable functor $\mathrm{Hom}_\Delta(-, [n])$, its boundary $\partial\Delta[n]$, in which the non-degenerate simplex in degree n has been removed, and, for any $0 \leq k \leq n$, the k-horn $V[n,k]$, for which the simplex in degree $n-1$ opposite the vertex k has also been removed.

Given any simplicial set K, one can obtain from it a topological space $|K|$ via geometric realization. This functor has a right adjoint, taking a topological space X to the singular set $\mathrm{Sing}(X)$.

Theorem 7.2.4. [44, II.3] *There is a model structure on the category of simplicial sets, which we denote by* $\mathcal{SS}ets$*, with weak equivalences the maps whose geometric realizations are weak homotopy equivalences. The adjoint pair* $|-|\colon \mathcal{SS}ets \leftrightarrows \mathcal{T}op\colon \mathrm{Sing}$ *is a Quillen equivalence of model categories.*

In light of this result, we sometimes abuse terminology and refer to simplicial sets as "spaces".

We also want to consider further structure on some of our model categories; we do not give full details of the definitions here. The first two structures are concerned with categories that are enriched either in simplicial sets or in the underlying category itself.

Definition 7.2.5. [29, 9.1.6] A model category $\mathcal{M}$ is *simplicial* if, for any two objects X and Y of $\mathcal{M}$, there is a simplicial set $\mathrm{Map}(X,Y)$ of morphisms from X to Y, and such that this simplicial structure is compatible with the model structure.

Definition 7.2.6. [46, 2.2] A model category $\mathcal{M}$ is *cartesian* if its underlying category is closed symmetric monoidal via the cartesian product and this structure is compatible with the model structure.

Many of the model structures that we consider here are given by localization of a known model structure. The idea is that we start with a model structure, then choose some set of morphisms that we would like to become weak equivalences. Doing so typically forces many more morphisms to become weak equivalences than only those in the given set. The theory of localizations of model categories began with the work of Bousfield in the setting of topological spaces [19], and conditions under which they exist more generally can be found in [4] or [29].

The definition of the localization of a model category makes use of *homotopy mapping spaces*, which can be defined for any model category $\mathcal{M}$ [24, §4], [29, §17]. The idea is that, given two objects X and Y of a model category, one can define a simplicial set $\mathrm{Map}^h(X,Y)$ that behaves like the mapping space $\mathrm{Map}(X,Y)$ in a simplicial model category yet is homotopy invariant and defined even if $\mathcal{M}$ is not actually enriched in simplicial sets, using simplicial and cosimplicial resolutions of the objects involved.

Definition 7.2.7. 1. Let $\mathcal{M}$ be a model category and S a set of morphisms in $\mathcal{M}$. A fibrant object X of $\mathcal{M}$ is *S-local* if

$$\mathrm{Map}^h(B, X) \to \mathrm{Map}^h(A, X)$$

is a weak equivalence of simplicial sets for every map $A \to B$ in S.

2. A map $C \to D$ in $\mathcal{M}$ is an *S-local equivalence* if

$$\mathrm{Map}^h(D, X) \to \mathrm{Map}^h(C, X)$$

is a weak equivalence for every S-local object X.

All of the model structures that we localize satisfy the hypotheses of the following theorem, which we do not state in full detail.

Theorem 7.2.8. [29, 4.1.1] *If $\mathcal{M}$ is a sufficiently nice model category and S is a set of maps in $\mathcal{M}$, then there exists a model structure on the same underlying category of $\mathcal{M}$ in which the weak equivalences are the S-local equivalences, the cofibrations are those of $\mathcal{M}$, and the fibrant objects are the S-local objects.*

Finally, we give a brief discussion of model structures on categories of diagrams of simplicial sets.

Theorem 7.2.9. [29, 11.6.1] *Let $\mathcal{C}$ be a small category and $\mathcal{SS}ets^{\mathcal{C}}$ the category of functors $\mathcal{C} \to \mathcal{SS}ets$. There is a model structure, called the* projective model structure, *on this category, in which the weak equivalences and fibrations $X \to Y$ are given by weak equivalences and fibrations of simplicial sets $X(c) \to Y(c)$ for every object c of $\mathcal{C}$.*

There is likewise an *injective* model structure, in which the weak equivalences and cofibrations are defined levelwise. We are primarily interested in *simplicial spaces*, or functors $\Delta^{\mathrm{op}} \to \mathcal{SS}ets$, as well as the variants $(\Delta^{\mathrm{op}})^n \to \mathcal{SS}ets$ and $\Theta_n^{\mathrm{op}} \to \mathcal{SS}ets$ which we introduce later in the paper. The categories Δ^{op}, $(\Delta^{\mathrm{op}})^n$, and Θ_n^{op} are all *Reedy categories*, so the category of functors from each to the category of simplicial sets admits a *Reedy model structure.*

In fact, these categories are furthermore *elegant* in the sense of [16], so that this Reedy structure coincides with the injective one. Using the injective structure gives us that all objects are cofibrant, while the Reedy structure has its own helpful features. While we do not describe the latter model structure explicitly here, throughout this paper we comment on some of the advantages it gives us. We refer the reader to [45] or [29, 15.3] for more details.

In what follows, we denote by $\mathcal{SS}ets^{\Delta^{\mathrm{op}}}$ the Reedy model structure on the category of simplicial spaces, and similarly when we replace Δ^{op} by $(\Delta^{\mathrm{op}})^n$ or Θ_n^{op}.

7.3 $(\infty, 0)$-categories

Using our strategy given in the introduction, our first task is to give a concrete model for $(\infty, 0)$-categories. By definition, an $(\infty, 0)$-category should be a weak ∞-groupoid: a weak ∞-category in which every i-morphism, for every $i \geq 1$, is weakly invertible.

As a first attempt at a concrete model, we propose that topological spaces are naturally $(\infty, 0)$-categories. Given a topological space X, we can think of the points of X as objects, and the paths between points as 1-morphisms. Then a homotopy between two paths with the same endpoints can be thought of as a 2-morphism, and we can continue to take homotopies between homotopies to make sense of i-morphisms for all $i \geq 1$. Since paths and homotopies are invertible up to homotopy, we get a weak ∞-groupoid.

Remark 7.3.1. We have chosen one particular approach in this regard, following work such as [2] or [41]. A general principle in higher category theory, originating with Grothendieck and called the *Homotopy Hypothesis*, is that weak n-groupoids model n-types, or topological spaces with nontrivial homotopy groups only in degrees n or lower. It stands to reason, then, that a weak ∞-groupoid should be an "∞-type", or simply a general topological space. We have chosen to take this principle so seriously that we take it as our definition. One can just as well take a more categorical definition of what an ∞-groupoid should be and then try to prove the Homotopy Hypothesis for that particular definition; for example see [17] or [20].

As in many situations in homotopy theory, it is preferable to work in the setting of simplicial sets rather than that of topological spaces. The Quillen equivalence of Theorem 7.2.4 tells us that topological spaces and simplicial sets have the same homotopy theory. Thus, we can also consider simplicial sets as models for ∞-groupoids. However, it is preferable to restrict to the simplicial sets that are both fibrant and cofibrant; all objects are cofibrant, but it is really only the fibrant objects, the Kan complexes, which best model ∞-groupoids.

Definition 7.3.2. A simplicial set K is a *Kan complex* if a lift exists in any diagram of the form

$$\begin{array}{ccc} V[n,k] & \longrightarrow & K \\ \downarrow & \nearrow & \\ \Delta[n] & & \end{array}$$

where $n \geq 1$ and $0 \leq k \leq n$.

The inclusions $V[n,k] \to \Delta[n]$ are called *horn inclusions*. To get an idea of what this lifting property means, let us look at the case when $n = 2$. The horn $V[2,0]$ can be depicted as

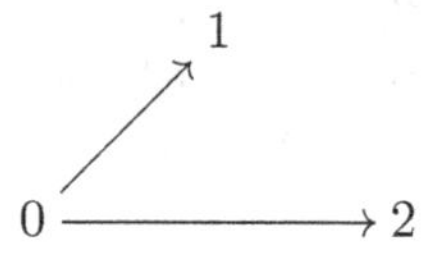

whereas $V[2,1]$ looks like

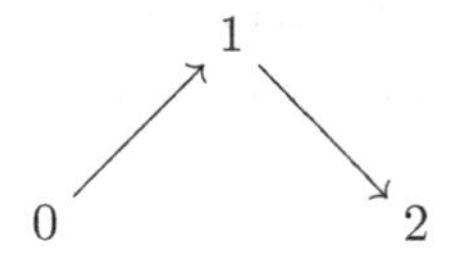

and finally $V[2,2]$ looks like

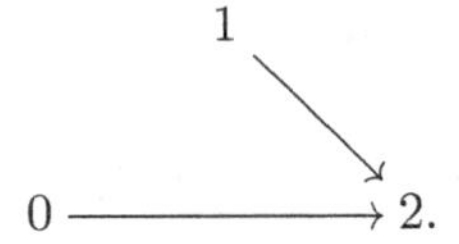

Having a lift with respect to $V[2,1]$ tells us that K has composition, in the sense that any two 1-simplices, of which the source of one is the target of the other, can be filled to a 2-simplex; we think of the additional face as a composite of the original two 1-simplices. However, this composition need not be unique. In higher dimensions, having a lift with respect to the analogous *inner horns*, for which $0 < k < n$, gives information about composites for longer strings of 1-simplices, and about associativity of composition, at least up to homotopy.

The *outer horns*, however, play a different role. For $n = 2$, the existence of lifts when $k = 0$ and $k = 2$ demonstrate the existence of left and right inverses to a given 1-simplex. Thus, these lifts show that a Kan complex not only behaves like a category up to homotopy, but moreover like a groupoid up to homotopy. This property agrees with our claim above that a Kan complex should model an ∞-groupoid. Indeed, one can make sense of paths and homotopies and homotopies between homotopies, just as we do in a topological space, to think of a Kan complex as an ∞-groupoid.

Now that we have good ways to think about $(\infty, 0)$-categories, we are ready to move up to $(\infty, 1)$-categories.

7.4 $(\infty, 1)$-categories

Following our principle that any (∞, n)-category should be a category enriched in $(\infty, n-1)$-categories, we can take categories enriched in topological spaces or categories enriched in simplicial sets as a model for $(\infty, 1)$-categories.

Definition 7.4.1. A *simplicial category* is a category enriched in simplicial sets. In other words, it has a collection of objects, together with, for any pair (x, y) of objects, a simplical set $\mathrm{Map}(x, y)$, together with a compatible composition law.

In light of the discussion in the previous section, one might protest that we want categories enriched in Kan complexes rather than arbitrary simplicial sets; we return to this subtlety momentarily. One can define topological categories analogously; we refer the reader to [32] for the corresponding homotopy theory.

Since we want to look at each of our models homotopy-theoretically, we want to show that we have a good model structure for simplicial categories. Let us first define the appropriate notion of weak equivalence, which can be thought of as a simplicial generalization of the definition of equivalence of categories. Given a simplicial category, we denote by $\pi_0(\mathcal{C})$ the category of components of $\mathcal{C}$, which has the same objects as $\mathcal{C}$ and in which

$$\mathrm{Hom}_{\pi_0(\mathcal{C})}(x, y) = \pi_0 \, \mathrm{Map}_{\mathcal{C}}(x, y).$$

Definition 7.4.2. A simplicial functor $F\colon \mathcal{C} \to \mathcal{D}$ is a *Dwyer-Kan equivalence* if:

1. for any $x, y \in \mathrm{ob}(\mathcal{C})$, the induced map

$$\mathrm{Map}_{\mathcal{C}}(x, y) \to \mathrm{Map}_{\mathcal{D}}(Fx, Fy)$$

   is a weak equivalence of simplicial sets; and

2. the induced functor $\pi_0(\mathcal{C}) \to \pi_0(\mathcal{D})$ is essentially surjective.

A functor satisfying (1) is called *homotopically fully faithful*, and one satisfying (2) is called *essentially surjective*, in analogy with terms used for functors between ordinary categories.

Theorem 7.4.3. [9, 1.1] *There is a model structure $\mathcal{SS}ets$-$\mathcal{C}at$ on the category of small simplicial categories in which the weak equivalences are the Dwyer-Kan equivalences.*

As for simplicial sets, we focus on the fibrant objects, which are precisely the simplicial categories whose mapping spaces are all Kan complexes. The cofibrant objects also have a concrete description [48, 16.2.2].

However, there are good reasons to look for alternative models for $(\infty, 1)$-categories.

- This model category does not satisfy good properties if we want to continue the process of enrichment to obtain models for $(\infty, 2)$-categories. The category of small categories enriched in small simplicial categories can be defined, but we cannot expect it to have a suitable model structure, since $\mathcal{SS}ets$-$\mathcal{C}at$ is not a cartesian model category.

- This model is too rigid to accommodate many examples. The composition law in an enriched category is required to satisfy strict associativity and unitality, and we would like models for which these properties only hold up to homotopy.

Our discussion of Kan complexes earlier lends itself to one possible way to think of certain simplicial sets as $(\infty, 1)$-categories. We can retain the conditions that encode category-like behavior but exclude the ones that impose the existence of inverses. The following definition is due to Boardman and Vogt, who instead used the term *inner Kan complex* [18].

Definition 7.4.4. A simplicial set K is a *quasi-category* if a lift exists in any diagram of the form

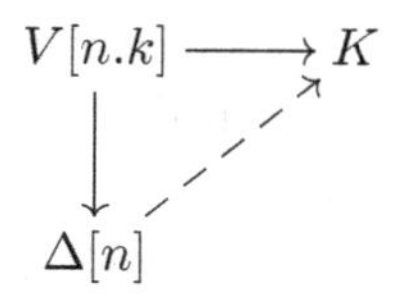

where $n \geq 1$ and $0 < k < n$.

The following theorem was proved in different ways by Joyal [36], Lurie [39, 2.2.5.1], and Dugger and Spivak [23, 2.13].

Theorem 7.4.5. *There is a cartesian model structure $\mathcal{QC}at$ on the category of simplicial sets in which the fibrant objects are the quasi-categories.*

To show that quasi-categories provide a good model for $(\infty,1)$-categories, it suffices to have a Quillen equivalence between $\mathcal{QC}at$ and $\mathcal{SS}ets$-$\mathcal{C}at$. We first need to define an adjoint pair of functors between the underlying categories. The following definition was first given by Cordier and Porter [21].

Definition 7.4.6. The *coherent nerve functor* $\widetilde{N} \colon \mathcal{SC} \to \mathcal{SS}ets$ is defined by

$$\widetilde{N}(\mathcal{C})_n = \operatorname{Hom}_{\mathcal{SC}}(F_*[n], \mathcal{C}),$$

where $F_*[n]$ denotes a simplicial resolution of the category $[n]$.

Remark 7.4.7. One way to think about the simplicial resolution $F_*[n]$ is as a simplicial object in categories, or functor $\Delta^{\mathrm{op}} \to \mathcal{C}at$, defined as follows. Forgetting the category structure on $[n]$ and regarding it as a directed graph, one can apply a free functor F to obtain a category $F[n]$. Iterating this process, we can define $F_k[n] = F^{k+1}[n]$.

Observe that the objects of $[n]$ are unchanged by this construction. One can check that a simplicial object in categories in which all face and degeneracy maps are the identity on the objects is a simplicial category in the sense in which we use the term.

Different approaches to the following result can be found in [35] and [39, §1.1.5].

Proposition 7.4.8. *The coherent nerve functor $\widetilde{N}$ admits a left adjoint $\mathfrak{C}$.*

This adjoint pair gives us our desired means of comparison.

Theorem 7.4.9. ([23, 8.2], [35], [39, 2.2.5.1]) *The adjoint pair*

$$\mathfrak{C} \colon \mathcal{QC}at \leftrightarrows \mathcal{SS}ets - \mathcal{C}at \colon \widetilde{N}$$

is a Quillen equivalence.

However, there are other approaches to defining models for $(\infty,1)$-categories with weak composition whose starting point is instead the simplicial nerve functor, which takes a simplicial category to a simplicial space, or bisimplicial set, $\Delta^{\mathrm{op}} \to \mathcal{SS}ets$. To define this functor, it is again convenient to think of a simplicial category as a simplicial object in the category of small categories for which the face and degeneracy maps are all the identity on objects.

Definition 7.4.10. Let $\mathcal{C}$ be a simplicial category, thought of as a simplicial object in categories. Its *simplicial nerve* is the simplicial space $\mathrm{nerve}^{\mathrm{s}}(\mathcal{C})$ defined by

$$\mathrm{nerve}^{\mathrm{s}}(\mathcal{C})_{*,m} = \mathrm{nerve}(\mathcal{C}_m).$$

We want to look at simplicial spaces that arise as simplicial nerves of simplicial categories and identify what properties they must have. The first thing to observe is that, since simplicial categories do not have a simplicial structure on their objects, the simplicial set $\mathrm{nerve}^{\mathrm{s}}(\mathcal{C})_0$ must be discrete. We thus make the following definition.

Definition 7.4.11. A *Segal precategory* is a simplicial space X such that X_0 is discrete.

More interesting, however, is the structure that we get from the composition of mapping spaces in a simplicial category. As many of the naming suggestions suggest, this structure originates in Segal's work on infinite loop spaces [51]. To describe it, we need to set up some notation.

For any $k \geq 0$, consider the simplicial set

$$G(k) = \underbrace{\Delta[1] \amalg_{\Delta[0]} \cdots \amalg_{\Delta[0]} \Delta[1]}_{k}$$

where the right-hand side is induced by the diagram

$$[1] \overset{d^0}{\leftarrow} [0] \overset{d^1}{\rightarrow} \cdots \overset{d^0}{\leftarrow} [0] \overset{d^1}{\rightarrow} [1]$$

in the category Δ. There is a natural inclusion $G(n) \rightarrow \Delta[n]$ for each $n \geq 0$ that is the identity when $n = 0, 1$; this map is sometimes referred to as a spine inclusion.

Since we are working with simplicial spaces, rather than simplicial sets, we want to think of $G(k)$ and $\Delta[k]$ in that context. There are two ways to think of a simplicial set K as a simplicial space: as a constant simplicial diagram given by K, or as a diagram of discrete simplicial sets given by the simplices of K. Since the former is typically still denoted by K, we denote the latter by K^t; we think of it as the "transpose" of the constant diagram, which is constant in the other simplicial direction. Thus, we have $K^t_k = K_k$, where the right-hand side denotes the constant simplicial set on the set K_k.

With this notation in place, let us consider the inclusion of simplicial spaces $G(k)^t \rightarrow \Delta[k]^t$.

Definition 7.4.12. Given any simplicial space W and any $k \geq 2$, the *Segal map* is the induced map

$$\operatorname{Map}(\Delta[k]^t, W) \rightarrow \operatorname{Map}(G(k)^t, W)$$

which can be rewritten simply as

$$W_k \rightarrow \underbrace{W_1 \times_{W_0} \cdots \times_{W_0} W_1}_{k}.$$

Now it is not hard to check the following characterization of simplicial nerves.

Proposition 7.4.13. *Let X be a simplicial space that can be obtained as the nerve of a simplicial category. Then X is a Segal precategory and, for every $k \geq 2$, the Segal maps*

$$X_k \rightarrow \underbrace{X_1 \times_{X_0} \cdots \times_{X_0} X_1}_{k}$$

are isomorphisms of simplicial sets.

Since we want a model for $(\infty, 1)$-categories that is less rigid than that of simplicial categories, we can relax the condition that the Segal maps be isomorphisms. We thus make the following definition.

Definition 7.4.14. A *Segal space* is a Reedy fibrant simplicial space W such that the Segal maps are weak equivalences of simplicial sets for all $k \geq 2$.

This requirement that the Segal maps be weak equivalences is often referred to as the *Segal condition.* Indeed, this condition often includes $k = 0, 1$ since these maps are automatically isomorphisms.

The role of the Reedy fibrancy condition is that, for such a simplicial space, the codomain of each Segal map is not just a limit but actually a homotopy limit. One could define the Segal maps in terms of a homotopy limit instead, and then omit this extra assumption.

This model structure is obtained as a localization of the Reedy model structure on simplicial spaces, where the localization is taken with respect to the maps $G(n)^t \to \Delta[n]^t$. Observe that our description of Segal spaces is essentially that they are the local objects with respect to these maps.

Observe that there is nothing particularly special about the category of simplicial sets here; one can define Segal objects in any model category as simplicial objects for which the Segal maps are weak equivalences. We return to this idea later in the paper.

Theorem 7.4.15. [47, 7.1] *There is a model structure, which we denote by $\mathcal{S}e\mathcal{S}p$, on the category of simplicial spaces such that all objects are cofibrant and the fibrant objects are precisely the Segal spaces.*

Segal spaces, with no further assumptions, do not quite model $(\infty, 1)$-categories. While the Segal condition allows us to define an up-to-homotopy composition, we have a space, rather than a set, of objects. In other words, Segal spaces model categories internal to spaces, rather than enriched in spaces. There are two approaches to remedying this difficulty.

For our first model, taking the output of the simplicial nerve as a guide, we retain the discreteness of the space in degree 0. The following definition first appeared in [25].

Definition 7.4.16. A *Segal category* is a Segal precategory for which the Segal maps are weak equivalences for all $k \geq 2$.

To show that Segal categories do indeed give a model for $(\infty, 1)$-categories, we need to define a model structure for them and show that it is Quillen equivalent to $\mathcal{SS}ets\text{-}\mathcal{C}at$. We first need a sensible notion of weak equivalence, and again we use simplicial categories as a guide.

We can apply much of the language of simplicial categories in the context of Segal categories. Given a Segal category X, we refer to the discrete space X_0 as its *set of objects.* We define *mapping spaces* between objects x and y via homotopy pullback:

$$\begin{array}{ccc} \operatorname{map}_X(x,y) & \longrightarrow & X_1 \\ \downarrow & & \downarrow \\ \{(x,y)\} & \longrightarrow & X_0 \times X_0. \end{array}$$

Using the fact that the Segal maps are weak equivalences, there is a notion of composition of mapping spaces, but it is only defined up to homotopy [47, 5.3]. Thus, we get a desired "weak composition" compared to the strict composition in a simplicial category. Taking the objects and the sets of path components of the mapping spaces, we obtain an ordinary category $\mathrm{Ho}(X)$, called the *homotopy category* associated to a Segal category X.

Although we do not go into detail here, there is a suitable functor L that takes any Segal precategory to a Segal category that is weakly equivalent to it in the model category $\mathcal{S}e\mathcal{S}p$ [12, §5]. Thus, for more general Segal precategories, we can first apply the functor L and then apply the definitions just described. In particular, we make the following definition.

Definition 7.4.17. A map $f\colon X \to Y$ of Segal precategories is a *Dwyer-Kan equivalence* if:

1. for any objects $x, y \in X_0$, the induced map

$$\mathrm{map}_{LX}(x,y) \to \mathrm{map}_{LY}(fx, fy)$$

is a weak equivalence of simplicial sets, and

2. the induced map on homotopy categories $\mathrm{Ho}(LX) \to \mathrm{Ho}(LY)$ is essentially surjective.

We can now describe the two model structures for Segal categories.

Theorem 7.4.18. ([12, 5.1, 7.1, 7.5], [43]) *There are two model structures on the category of Segal precategories, each of which has Dwyer-Kan equivalences as weak equivalences.*

1. *The first model structure, which we denote by $\mathcal{S}e\mathcal{C}at_c$, has all objects cofibrant and fibrant objects precisely the Segal categories that are fibrant in the Reedy model structure on all simplicial spaces, and this model structure is cartesian.*

2. *The second model structure, which we denote by $\mathcal{S}e\mathcal{C}at_f$, has fibrant objects precisely the Segal categories that are fibrant in the projective model structure on all simplicial spaces. The cofibrant objects are closely related to cofibrant objects in the projective model structure on simplicial spaces.*

3. *The two model structures are Quillen equivalent via the identity functors:*

$$\mathrm{id}\colon \mathcal{S}e\mathcal{C}at_f \rightleftarrows \mathcal{S}e\mathcal{C}at_c\colon \mathrm{id}\,.$$

Remark 7.4.19. The astute reader might have noticed the following incongruity in our definitions. We assume that a Segal space is Reedy fibrant, but we make no such assumption on a Segal category. In particular, what we would like to say is that a Segal category is simply a Segal space with 0-space discrete. That point of view works nicely if all we wanted was the model structure $\mathcal{S}e\mathcal{C}at_c$. Indeed, this model structure is preferable for many purposes and was the one originally developed by Pellissier in [43].

Unfortunately, the simplicial nerve functor does not induce a Quillen equivalence between $\mathcal{SS}ets\text{-}\mathcal{C}at$ and $\mathcal{S}e\mathcal{C}at_c$, essentially because there are too many cofibrations in $\mathcal{S}e\mathcal{C}at_c$ compared to $\mathcal{SS}ets\text{-}\mathcal{C}at$. The model structure $\mathcal{S}e\mathcal{C}at_f$ is designed to facilitate this comparison.

As mentioned above, for Segal spaces the Reedy fibrancy condition allows us to define the Segal maps in terms of a limit, rather than a homotopy limit. For Segal categories, the discreteness in degree zero allows us to consider strict limits, so serves this same purpose. One could take an analogous Segal space localization in the projective model structure as well. There are reasons why this model structure is not as well-behaved for comparisons as the one we have chosen; see [12, §7] for further discussion on this point.

We now give a comparison between simplicial categories and Segal categories.

Theorem 7.4.20. [12, 8.6] *The simplicial nerve functor has a left adjoint which we denote by F. This adjoint pair induces a Quillen equivalence*

$$F\colon \mathcal{S}e\mathcal{C}at_f \rightleftarrows \mathcal{S}\mathcal{S}ets - \mathcal{C}at\colon \text{nerve}^s.$$

This left adjoint functor F can be thought of as a "rigidification" of a Segal category to a simplicial category.

The model structure $\mathcal{S}e\mathcal{C}at_c$, on the other hand, is well-suited to comparison with our alternate approach to making Segal spaces models for $(\infty, 1)$-categories.

We might ask why we even want an alternative to Segal categories. The main difficulty with them is the fact that we need their degree 0 space to be discrete, which is an unnnatural condition from the perspective of homotopy theory. We could weaken this condition to homotopy discreteness, but there is another point of view, which we now describe.

For this next model, let us return to the way in which we talked about Segal categories in the language of simplicial categories. The definitions we made above still hold for more general Segal spaces. In particular, given a Segal space W, we define its *space of homotopy equivalences* to be the subspace of W_1 whose image in $\text{Ho}(W)$ consists of isomorphisms, and we denote it by W_{heq}. Observe that the degeneracy map $s_0\colon W_0 \to W_1$ factors through W_{heq}.

Definition 7.4.21. A Segal space W is *complete* if the map $W_0 \to W_{\text{heq}}$ is a weak equivalence of simplicial sets.

The idea behind this completeness condition is that the space of objects is no longer assumed to be discrete, but can be encoded as subspace of the space of morphisms. Sometimes this condition is reformulated as saying that the degree zero space is a moduli space for homotopy equivalences.

Like the Segal condition, completeness can be formulated in terms of a localization. Let E be the nerve of the category

$$0 \rightleftarrows 1$$

consisting of two objects and a single isomorphism between them. Then the unique map from this category to $[0]$ induces a map $E \to \Delta[0]$. Then a Segal space X is complete if and only if the induced map

$$X_0 \simeq \text{Map}(\Delta[0]^t, X) \to \text{Map}(E^t, X) \simeq X_{\text{heq}}$$

is a weak equivalence. The following model structure is obtained as a further localization of the Segal space model structure with respect to the map $E^t \to \Delta[0]^t$.

Theorem 7.4.22. [47, 7.2] *There is a model structure on the category of simplicial spaces, denoted by $\mathcal{CSS}$, in which all objects are cofibrant and the fibrant objects are precisely the complete Segal spaces. Furthermore, this model structure is cartesian.*

To compare this model structure to $\mathcal{S}e\mathcal{C}at_c$, we need a way to "discretize" the degree zero space of a complete Segal space to get a Segal category.

Theorem 7.4.23. [12, 6.3] *The inclusion of Segal precategories into the category of all simplicial spaces admits a right adjoint D. This adjoint pair induces a Quillen equivalence*

$$I\colon \mathcal{S}e\mathcal{C}at_c \leftrightarrows \mathcal{CSS}\colon D.$$

To understand why this theorem works, we first need to look at the role of Dwyer-Kan equivalences in the model structure $\mathcal{CSS}$ via the following theorem of Rezk.

Theorem 7.4.24. [47, 7.6, 7.7] *Let $f\colon W \to Z$ be a map of Segal spaces. Then f is a Dwyer-Kan equivalence if and only if it is a weak equivalence in $\mathcal{CSS}$.*

The right adjoint D to the inclusion functor replaces the simplicial set in degree zero with its discrete space of 0-simplices. If we apply this functor to a complete Segal space W, then the result is typically no longer complete, but it is a Segal category. If $W \to Z$ is a weak equivalence between complete Segal spaces in $\mathcal{CSS}$, establishing the Quillen equivalence above essentially reduces to showing that the functor D preserves Dwyer-Kan equivalences.

Remark 7.4.25. A natural question to ask is when complete Segal spaces and Segal categories coincide. The answer is not very often! A simplicial space X that is both a complete Segal space and a Segal category satisfies both $X_0 \simeq X_{\text{heq}}$ and X_0 is discrete. In other words, the space of homotopy equivalences of X must be homotopy discrete, and $\text{Ho}(X)$ is a category with no non-identity automorphisms of objects. It can have non-identity isomorphisms, but they must be unique between two given objects. An analogous structure is a simplicial category for which all homotopy automorphisms are homotopic to identity morphisms and whose homotopy equivalences between two given objects form a contractible space.

Joyal and Tierney proved that there are also Quillen equivalences (in both directions!) between $\mathcal{QC}at$ and $\mathcal{CSS}$, and between $\mathcal{QC}at$ and $\mathcal{S}e\mathcal{C}at_c$ [37]. We discuss one of these comparisons.

Theorem 7.4.26. [37, 4.11] *The evaluation map ev_0, taking a simplicial space W to the simplicial space $W_{*,0}$, is right adjoint to the inclusion functor taking a simplicial set K to the simplicial space Z with $Z_{*,n} = K$ for all n. This adjoint pair induces a Quillen equivalence*

$$i\colon \mathcal{QC}at \rightleftarrows \mathcal{CSS}\colon \text{ev}_0\,.$$

While we have by no means covered all possible models for $(\infty, 1)$-categories, the ones we have here give a sense of how they can be described. We now look toward moving up one more level to $(\infty, 2)$-categories.

7.5 $(\infty, 2)$-categories as enriched categories

The first approach to obtaining models for $(\infty, 2)$-categories is to take categories enriched in any one of our models for $(\infty, 1)$-categories. Simply defining such objects is not a problem for any of the models that we have, as each of the underlying categories has a well-behaved monoidal structure under cartesian product. However, as we have already seen for simplicial categories, we have no reason to expect that all of these categories have corresponding model structures. The key feature we need if we want such a model structure is that the enrichment is taken over a cartesian model category. As we saw in the previous section, three of the models have cartesian model structures: $\mathcal{QC}at$, $\mathcal{CSS}$, and $\mathcal{S}e\mathcal{C}at_c$.

In fact, we do get model structures if we enrich in any of these model structures. The general strategy is spelled out by Lurie in Appendix A of [39]. The main idea is that we want a model structure on the category of small categories enriched in a cartesian model category $\mathcal{V}$, denoted by $\mathcal{V}$-$\mathcal{C}at$, that is analogous to the model structure for simplicial categories, with a variant of Dwyer-Kan equivalences as weak equivalences. But how do we define these maps in a more general enriched category?

The first condition, that of being homotopically fully faithful, is not hard to generalize. We simply ask that the induced maps on mapping objects, which we denote by $\underline{\operatorname{Map}}_{\mathcal{C}}(x, y)$, be weak equivalences in $\mathcal{V}$. But what about essential surjectivity? For Dwyer-Kan equivalences of simplicial categories, we used the category of components $\pi_0\mathcal{C}$ of a simplicial category $\mathcal{C}$. Since there we had mapping simplicial sets, taking π_0 was a natural thing to do. More generally, we define $\pi_0\mathcal{C}$ to be the category whose objects are those of $\mathcal{C}$ and whose morphisms are given by

$$\operatorname{Hom}_{\pi_0\mathcal{C}}(x, y) = \operatorname{Hom}_{\operatorname{Ho}(\mathcal{V})}(*, \underline{\operatorname{Map}}_{\mathcal{C}}(x, y)),$$

where $*$ denotes the terminal object of $\mathcal{V}$.

Definition 7.5.1. A $\mathcal{V}$-enriched functor $f \colon \mathcal{C} \to \mathcal{V}$ is a *Dwyer-Kan equivalence* if:

1. for every $x, y \in \operatorname{ob}(\mathcal{C})$, the map

$$\underline{\operatorname{Map}}_{\mathcal{C}}(x, y) \to \underline{\operatorname{Map}}_{\mathcal{D}}(fx, fy)$$

is a weak equivalence in $\mathcal{V}$, and

2. the functor $\pi_0(\mathcal{C}) \to \pi_0(\mathcal{D})$ is essentially surjective.

Let us now show that we have the desired model categories. In the case in which we enrich in the complete Segal space model structure $\mathcal{CSS}$, a full proof applying this strategy is given by the author and Rezk in [14, 3.11].

Theorem 7.5.2. *There is a cofibrantly generated model structure $\mathcal{CSS}$-$\mathcal{C}at$ on the category of small categories enriched in $\mathcal{CSS}$ in which the weak equivalences $f \colon \mathcal{C} \to \mathcal{D}$ are Dwyer-Kan equivalences.*

Essentially the same proof technique can be used to obtain a similar result for enriching in $\mathcal{S}e\mathcal{C}at_c$.

Theorem 7.5.3. *There is a cofibrantly generated model structure on $\mathcal{S}e\mathcal{C}at_c$-$\mathcal{C}at$ in which the weak equivalences $f: \mathcal{C} \to \mathcal{D}$ are the Dwyer-Kan equivalences.*

The corresponding theorem for enriching in $\mathcal{QC}at$ is implicit in [39]; we give a brief sketch of the proof here.

Theorem 7.5.4. *There is a cofibrantly generated model structure on $\mathcal{QC}at$-$\mathcal{C}at$ in which the weak equivalences $f: \mathcal{C} \to \mathcal{D}$ are the Dwyer-Kan equivalences.*

Proof. We apply the criteria of [39, A.3.2.4]. The only condition that is not straightforward to check is that weak equivalences in $\mathcal{QC}at$ preserve filtered colimits. However, this result is proved in [22, 2.13]. □

Given these model structures, we would like to know that they are all equivalent to one another, since we are enriching in Quillen equivalent model categories. The following result is a consequence of [39, A.3.2.6].

Theorem 7.5.5. *There are Quillen equivalences*

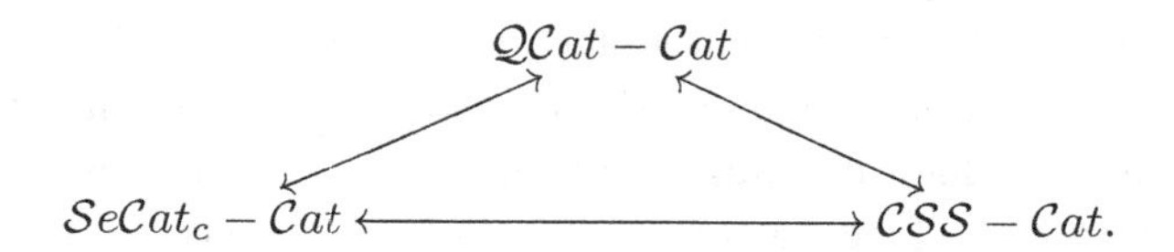

However, if we want to move to higher (∞, n)-categories by iterating this process, we have the same trouble again, as the model categories we have given here are not cartesian, for similar reasons as for the model structure on simplcial categories.

Furthermore, if we want to accommodate more flexible examples, we want to have models with less rigid composition structure. There are a number of approaches which generalize the models for $(\infty, 1)$-categories in different ways. In the next few sections, we look at some of the possibilities, starting with generalizations of complete Segal spaces and Segal categories.

7.6 Multisimplicial models for $(\infty, 2)$-categories

When trying to move to $(\infty, 2)$-categories, the first thought one might have is to add a higher categorical level by adding another simplicial level. We see this kind of intuition when we generalize from the nerves of categories (simplicial sets) to simplicial nerves of simplicial categories (bisimplicial sets).

When we consider Segal categories and complete Segal spaces, our starting point is the notion of Segal space. Thus, it is natural to start with a structure that satisfies Segal conditions in two simplicial directions. There are different ways to describe such a structure, but we begin with the following description.

Definition 7.6.1. A Reedy fibrant functor $W\colon \Delta^{op} \times \Delta^{op} \to \mathcal{SS}ets$ is a *double Segal space* if the Segal maps

$$W_{k,*} \to \underbrace{W_{1,*} \times_{W_{0,*}} \cdots \times_{W_{0,*}} W_{1,*}}_{k}$$

and

$$W_{*,k} \to \underbrace{W_{*,1} \times_{W_{*,0}} \cdots \times_{W_{*,0}} W_{*,1}}_{k}$$

are weak equivalences for any $k \geq 2$. In other words, the simplicial spaces $W_{k,*}$ and $W_{*,k}$ are Segal spaces for any fixed k.

The definition we have given treats the two simplicial directions equally, but it can be convenient to distinguish the two in the following way, using the notion of more general Segal objects in a given model category. In particular, a functor $\Delta^{op} \times \Delta^{op} \to \mathcal{SS}ets$, or *bisimplicial space*, can instead be thought of as a simplicial object in simplicial spaces, or functor $\Delta^{op} \to \mathcal{SS}ets^{\Delta^{op}}$.

Definition 7.6.2. A Reedy fibrant functor $W\colon \Delta^{op} \to \mathcal{SS}ets^{\Delta^{op}}$ is a *Segal object in Segal spaces* if the maps

$$W_k \to \underbrace{W_1 \times_{W_0} \cdots \times_{W_0} W_1}_{k}$$

are weak equivalences in $\mathcal{S}e\mathcal{S}p$ for all $k \geq 2$.

Indeed, one can check that these two definitions agree.

Proposition 7.6.3. *A bisimplicial space is a double Segal space if and only if it is a Segal object in Segal spaces.*

The following model structure appears in [13], although it was almost certainly known to experts previously.

Theorem 7.6.4. *There is a model structure on the category of bisimplicial spaces in which the fibrant objects are precisely the double Segal spaces.*

Just as Segal spaces do not quite model $(\infty, 1)$-categories, we do not expect double Segal spaces to model $(\infty, 2)$-categories. Recall that the difference for Segal spaces was that they give a model for categories internal to spaces, rather than categories enriched in spaces. In other words, we have a space of objects as well as of morphisms. The problem here is similar. A double Segal space encodes the information of a double category up to homotopy. A *double category* is a category internal to categories, and as such has objects, horizontal morphisms, vertical morphisms, and squares.

More precisely, suppose that W is a double Segal space. We can think of $X_{0,0}$ as the space of objects, $W_{0,1}$ as the space of vertical morphisms, $W_{1,0}$ as the space of horizontal morphisms, and $W_{1,1}$ as the space of squares. In other words, the 2-morphisms are encoded in squares like the following:

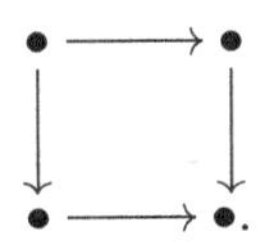

However, when we look at $(\infty, 2)$-categories, we typically want to think of 2-morphisms as being of the form

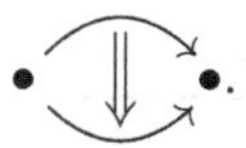

In particular, do not want to have nontrivial vertical morphisms. To model such a structure homotopically, we want to ask that the simplicial space $W_{0,*}$ be essentially constant. As in the case of $(\infty, 1)$-categories, we also want to ask either that $W_{0,0}$, the space of objects, be discrete, or to have a corresponding completeness condition. Additionally, since we want to think of $(\infty, 2)$-categories as enriched in $(\infty, 1)$-categories, we want the same kind of condition on the 1-morphisms: either that the space of such be discrete, or that we have a corresponding completeness condition. The focus of this section, then, is to describe how to impose these kinds of conditions appropriately on double Segal spaces.

Let us first consider the option of imposing two completeness conditions. As a first step, let us define Segal objects in complete Segal spaces, building on Definition 7.6.2.

Definition 7.6.5. A Reedy fibrant functor $W\colon \Delta^{op} \to \mathcal{CSS}$ is a *Segal object in complete Segal spaces* if, for any $k \geq 2$, the Segal map

$$W_k \to \underbrace{W_1 \times_{W_0} \cdots \times_{W_0} W_1}_{k}$$

is a weak equivalence in the complete Segal space model structure $\mathcal{CSS}$.

This definition, as we have given it, is quite formal, so let us investigate the structure further. We first make the following definition of mapping objects, generalizing mapping spaces in a Segal space.

Definition 7.6.6. Let W be a double Segal space. Then for any $x, y \in W_{0,0,0}$, the *mapping object* $\underline{\operatorname{map}}_W(x, y)$ is defined to be the simplicial space defined as the pullback

$$\begin{array}{ccc} \underline{\operatorname{map}}_W(x,y) & \longrightarrow & W_1 \\ \downarrow & & \downarrow{\scriptstyle (d_1,d_0)} \\ \{(x,y)\} & \longrightarrow & W_0 \times W_0. \end{array} \tag{7.6.1}$$

As for mapping spaces in Segal spaces, the fact that W is Reedy fibrant implies that this pullback is actually a homotopy pullback. We can use these mapping objects in the following alternative characterization of Segal objects in complete Segal spaces, using the bisimplicial space viewpoint.

Proposition 7.6.7. *A Reedy fibrant functor $W\colon \Delta^{op} \times \Delta^{op} \to \mathcal{SSets}$ is a Segal object in complete Segal spaces if and only if it satisfies the following conditions:*

1. *for any $m \geq 2$, the Segal map*

$$W_{m,*} \to \underbrace{W_{1,*} \times_{W_{0,*}} \cdots \times_{W_{0,*}} W_{1,*}}_{m}$$

is a weak equivalence in $\mathcal{CSS}$; and

2. *for every $x, y \in W_{0,0,0}$, $\underline{\mathrm{map}}_W(x, y)$ is a complete Segal space.*

There does not seem to be an explicit proof of this result in the literature, but we plan to include one in [10]. We use the approach of this second characterization to define complete Segal objects.

Definition 7.6.8. A Reedy fibrant functor $W \colon \Delta^{\mathrm{op}} \times \Delta^{\mathrm{op}} \to \mathcal{SS}ets$ is a *complete Segal object in complete Segal spaces*, or a *2-fold complete Segal space* if:

1. for every $k \geq 0$, the simplicial space $W_{*,k}$ is a complete Segal space;
2. for every $x, y \in W_{0,0,0}$, the simplicial space $\underline{\mathrm{map}}_W(x, y)$ is a complete Segal space; and
3. the simplicial space $W_{0,*}$ is essentially constant.

Remark 7.6.9. The definition of 2-fold complete Segal space is often stated with (1) replaced by the the seemingly weaker condition that each $W_{*,k}$ be a Segal space with only $W_{*,0}$ assumed to be complete. However, as proved by Johnson-Freyd and Scheimbauer [33, 2.8], these conditions actually imply condition (1) as we have stated it.

The following model structure, like the one for complete Segal spaces, is obtained via localization of the Reedy model structure on bisimplicial spaces.

Theorem 7.6.10. [15] *There is a model structure $\mathcal{CSS}(\mathcal{CSS})$ on the category of bisimplical spaces in which the fibrant objects are precisely the 2-fold complete Segal spaces.*

Because they play such a critical role in the theory of complete Segal spaces and their comparison with other models, and are used to define the enriched category models for $(\infty, 2)$-categories, let us look at Dwyer-Kan equivalences in this setting. We have defined mapping objects, so we can define homotopical fully faithfulness, but for essential surjectivity we need a notion of homotopy category of a 2-fold complete Segal space. Since one might think more naturally of a homotopy 2-category in this context, we need to reduce a categorical level, which we do by defining the underlying complete Segal space of a 2-fold complete Segal space.

Definition 7.6.11. Let $\tau_\Delta \colon \Delta \to \Delta \times \Delta$ be the functor defined by $[m] \mapsto ([m], [0])$. The induced functor $\tau_\Delta^* \colon \mathcal{SS}ets^{\Delta^{\mathrm{op}} \times \Delta^{\mathrm{op}}} \to \mathcal{SS}ets^{\Delta^{\mathrm{op}}}$ defines the *underlying simplicial space* of a bisimplicial space.

Proposition 7.6.12. [15] *If W is a 2-fold (complete) Segal space, then its underlying simplicial space $\tau_\Delta^* W$ is a (complete) Segal space.*

In particular, $\tau_\Delta^* W$ has an associated homotopy category $\mathrm{Ho}(\tau_\Delta^* W)$. In the following definition, let L denote a functorial fibrant replacement functor in the double Segal space model structure.

Definition 7.6.13. Let W and Z be bisimplicial spaces. A morphism $f \colon W \to Z$ is a *Dwyer-Kan equivalence* if:

- for any $x, y \in W_{0,0,0}$, the induced map $\underline{\mathrm{map}}_{LW}(x,y) \to \underline{\mathrm{map}}_{LZ}(fx, fy)$ is a weak equivalence in $\mathcal{CSS}$, and

- the induced map $\mathrm{Ho}(\tau_\Delta^* LW) \to \mathrm{Ho}(\tau_\Delta^* LZ)$ is essentially surjective.

In an analogy with Dwyer-Kan equivalences in $\mathcal{CSS}$, these morphisms play the same role in $\mathcal{CSS}(\mathcal{CSS})$.

Theorem 7.6.14. [15] *A map $f \colon W \to Z$ of Segal objects in $\mathcal{CSS}$ is a Dwyer-Kan equivalence if and only if it is a weak equivalence in $\mathcal{CSS}(\mathcal{CSS})$.*

To show that 2-fold complete Segal spaces give a good model for $(\infty, 2)$-categories, we would like to show that the model category $\mathcal{CSS}(\mathcal{CSS})$ is Quillen equivalent to $\mathcal{CSS}$-$\mathcal{C}at$. Just as in the comparison between simplicial categories and complete Segal spaces, we make use of an intermediate model structure which uses a discreteness condition. Thus, we turn to one of our variant models.

Definition 7.6.15. A *Segal category object in complete Segal spaces* is a functor $W \colon \Delta^{\mathrm{op}} \to \mathcal{CSS}$ such that W_0 is discrete and for each $k \geq 2$ the map

$$W_k \to \underbrace{W_1 \times_{W_0} \cdots \times_{W_0} W_1}_{k}$$

is a weak equivalence of complete Segal spaces.

As for Segal categories, observe that we are not simply saying that a Segal category object is a Segal object in complete Segal spaces with the appropriate discreteness, because we do not want to impose Reedy fibrancy. We once again make use of two different model structures for our desired comparison.

Theorem 7.6.16. [14] *There are two model structures on the category of functors $W \colon \Delta^{\mathrm{op}} \to \mathcal{CSS}$ with W_0 discrete such that weak equivalences are the Dwyer-Kan equivalences:*

- *the model structure $\mathcal{S}e\mathcal{C}at_c(\mathcal{CSS})$, in which the fibrant objects are the Segal category objects in complete Segal spaces that are Reedy fibrant as bisimplicial spaces, and*

- *the model structure $\mathcal{S}e\mathcal{C}at_f(\mathcal{CSS})$, in which the fibrant objects are the Segal category objects in complete Segal spaces that are projective fibrant as bisimplicial spaces.*

Furthermore, there is a Quillen equivalence

$$\mathcal{S}e\mathcal{C}at_f(\mathcal{CSS}) \rightleftarrows \mathcal{S}e\mathcal{C}at_c(\mathcal{CSS})$$

given by the identity functors.

Following the idea of the simplicial nerve functor taking a simplicial category to a Segal category, there is a bisimplicial nerve functor taking a category enriched in simplicial spaces to a bisimplicial space. The following result mirrors the Quillen equivalences between simplicial categories, Segal categories, and complete Segal spaces.

Theorem 7.6.17. [14],[15]

1. *The bisimplicial nerve functor induces a Quillen equivalence*

$$\mathcal{CSS} - \mathcal{C}at \leftrightarrows \mathcal{S}e\mathcal{C}at_f(\mathcal{CSS}).$$

2. *The inclusion functor induces a Quillen equivalence*

$$\mathcal{S}e\mathcal{C}at_c(\mathcal{CSS}) \rightleftarrows \mathcal{CSS}(\mathcal{CSS}).$$

This theorem brings together all the models we have discussed thus far, establishing that they are all equivalent to one another. While the technical points of this proof are considerably more difficult, the idea behind it is to apply the same methods as we did in the case of $(\infty, 1)$-categories, but replacing the model structure on simplicial sets by $\mathcal{CSS}$.

However, we have only considered two of the four possible combinations of completeness and discreteness conditions on double Segal spaces. In particular, one could argue that we have been giving preferential treatment to the completeness condition. Historically, however, a notion of Segal category objects in Segal categories came first, in work of Hirschowitz and Simpson [30], and are treated in Simpson's book [52]. As with Segal categories, we begin with an underlying category in which certain spaces are required to be discrete.

Definition 7.6.18. A bisimplicial space $X\colon \Delta^{op} \times \Delta^{op} \to \mathcal{SS}ets$ is a *Segal 2-precategory* if:

1. for any $m \geq 0$, the simplicial set $X_{m,0}$ is discrete; in other words, X can be thought of as a simplicial object in Segal precategories; and
2. for any $k \geq 0$, the simplicial set $X_{0,k}$ is discrete.

If, in addition, for any $k \geq 2$, the Segal map

$$X_{k,*} \to \underbrace{X_{1,*} \times_{X_{0,*}} \cdots \times_{X_{0,*}} X_{1,*}}_{k}$$

is a Dwyer-Kan equivalence in $\mathcal{S}e\mathcal{C}at_c$, then X is a *Segal 2-category.*

Alternatively, we can think of Segal 2-categories as "Segal category objects in Segal categories", in analogy with Segal space objects.

Theorem 7.6.19. [43],[10] *There is a model structure on the category of Segal 2-precategories such that the weak equivalences are Dwyer-Kan equivalences and the fibrant objects are the Reedy fibrant Segal 2-categories.*

Once again, we could take an analogous structure which is projective fibrant, rather than Reedy (or even mix and match the two in the two simplicial directions!), but we do not worry about this structure here, since their main purpose was to aid in comparison with an enriched category model, and we do not need to make such a comparison here.

Now we would like to know that all of these models are equivalent. The idea is that we can use the Quillen equivalence between $\mathcal{S}e\mathcal{C}at_c$ and $\mathcal{CSS}$ to do so. Once again, the proofs are more delicate, but the core idea comes from that original comparison.

Theorem 7.6.20. [10] *The inclusion functors and their right adjoints, which are given by appropriate discretization functors, induce Quillen equivalences*

$$\mathcal{S}e\mathcal{C}at_c(\mathcal{S}e\mathcal{C}at_c) \rightleftarrows \mathcal{S}e\mathcal{C}at_c(\mathcal{CSS}) \rightleftarrows \mathcal{CSS}(\mathcal{CSS}).$$

We can summarize the results of this section via the following diagram, in which each displayed arrow is the left adjoint of a Quillen equivalence:

$$\begin{array}{ccccccc} \mathcal{CSS}-\mathcal{C}at & \longleftarrow & \mathcal{S}e\mathcal{C}at_f(\mathcal{CSS}) & \longrightarrow & \mathcal{S}e\mathcal{C}at_c(\mathcal{CSS}) & \longrightarrow & \mathcal{CSS}(\mathcal{CSS}) \\ & & & & \uparrow & & \\ & & & & \mathcal{S}e\mathcal{C}at_c(\mathcal{S}e\mathcal{C}at_c). & & \end{array}$$

Remark 7.6.21. In theory, there is one remaining possibility, that of complete Segal objects in Segal categories. However, although such objects can be defined, they are not substantively different from Segal 2-categories. The problem goes back to the question of when Segal categories are complete Segal spaces.

To see what happens, let us modify Definition 7.6.8 to what we expect a complete Segal object in Segal categories to be. Such an object should be a functor $W\colon \Delta^{op} \times \Delta^{op} \to \mathcal{SS}ets$ with each $W_{m,0}$ discrete. It should satisfy:

1. for every $k \geq 0$, the simplicial space $W_{*,k}$ is a complete Segal space;
2. for every $x, y \in W_{0,0,0}$, the simplicial space $\underline{\operatorname{map}}_W(x,y)$ is a Segal category; and
3. the simplicial space $W_{0,*}$ is essentially constant.

We have assumed that the simplicial set $W_{0,0}$ is discrete, so by condition (3) each $W_{0,k}$ must be homotopy discrete. However, by condition (1), each of these simplicial sets must be the degree zero space of a complete Segal space. Thus, these complete Segal spaces are essentially Segal categories, with the only variation that the degree zero spaces are homotopy discrete rather than actually discrete. We give a more detailed treatment of this phenomenon in [11].

7.7 Θ_2-models for $(\infty, 2)$-categories

The models in the previous section give us a number of ways to think about $(\infty, 2)$-categories as weakly enriched categories. However, there are reasons why we might want a different approach. First, conditions such as the essential constancy of a complete Segal object suggest that somehow a bisimplicial diagram is bigger than what we need. Furthermore, such models are not cartesian, a problem which led Rezk to look for an alternative. Another approach which addresses these difficulties is given via diagrams that are modeled by the category Θ_2 rather than $\Delta \times \Delta$.

Recall from the previous section that when modeling an $(\infty, 2)$-category by a bisimplicial diagram, we have extra data, in that we do not want interesting "vertical" morphisms. So, the main idea is that we take diagrams of simplicial sets that more concisely model the data that we actually want.

Let us state the formal definition, then look at an example.

Definition 7.7.1. The category Θ_2 has objects $[m]([k_1], \ldots, [k_m])$, where $[m]$ and each $[k_i]$ are objects of Δ, and morphisms $[m]([k_1], \ldots, [k_m]) \to [p]([\ell_1], \ldots, [\ell_p])$ given by a function $\delta\colon [m] \to [p]$ in Δ, and functions $[k_i] \to [\ell_j]$ defined whenever $\delta(i-1) < j \leq \delta(i)$.

Example 7.7.2. The object $[4]([2], [3], [0], [1])$ of Θ_2 can be depicted as

$$0 \xrightarrow{[2]} 1 \xrightarrow{[3]} 2 \xrightarrow{[0]} 3 \xrightarrow{[1]} 4.$$

Since these labels, themselves objects of Δ, can also be interpreted as strings of arrows, we get a diagram such as

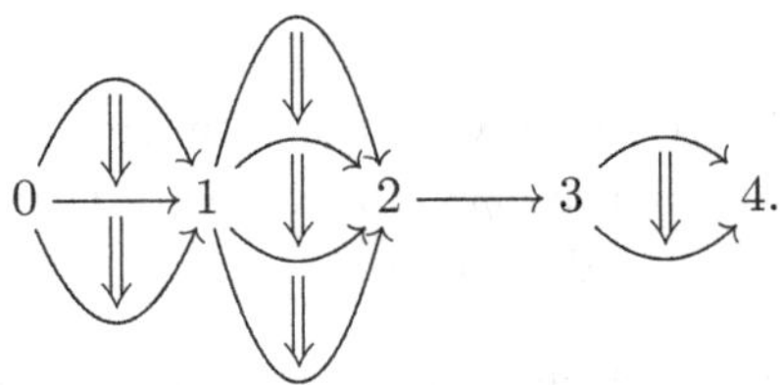

The elements of this diagram can be regarded as generating a strict 2-category by composing 1-cells and 2-cells whenever possible. In other words, the objects of Θ_2 can be seen as encoding all possible finite compositions that can take place in a 2-category, much as the objects of Δ can be thought of as listing all the finite compositions that can occur in an ordinary category.

Let us consider a morphism $[4]([2], [3], [0], [1]) \to [3]([1], [0], [2])$. The first thing we need to define such a morphism is a map $\delta\colon [4] \to [3]$ in Δ, for example the one depicted here:

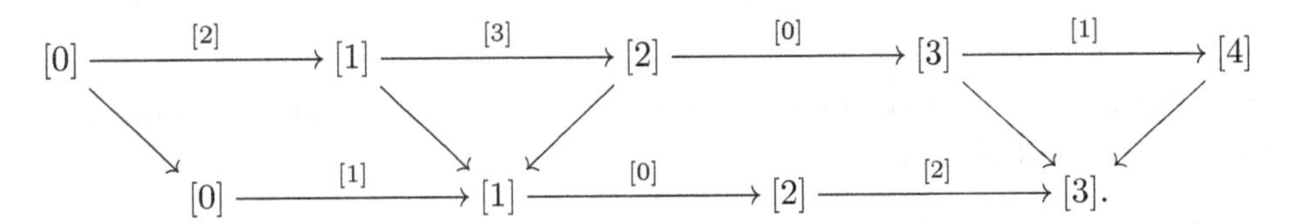

We have maps between the labeling objects when it is sensible to do so, as can be visualized in the above diagram. For example, to complete the definition here we need to specify a map $[2] \to [1]$, the unique map $[0] \to [0]$, and a map $[0] \to [2]$. We think of the unique arrow $2 \to 3$ in the top diagram as being sent to the composite of the arrow $1 \to 2$ with the arrow $2 \to 3$ specified by the chosen map $[0] \to [2]$.

More pictorially, one choice of such a morphism is given by:

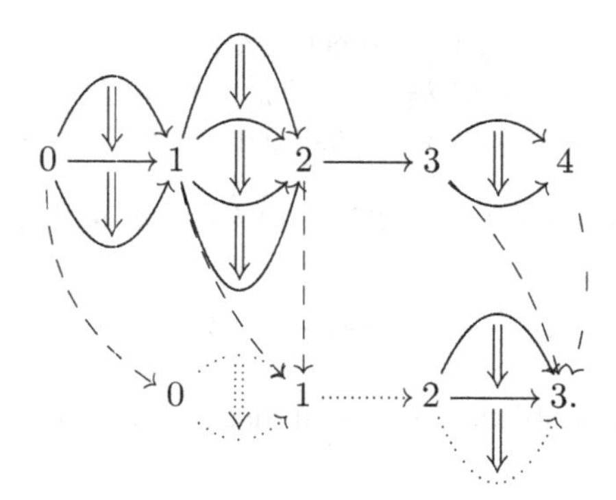

Here the map choices are indicated by giving their image as dotted arrows; we have distinguished the map δ by the dashed arrows. (Technically there are still two choices here, depending on where one sends the middle arrow $0 \to 1$.)

To model $(\infty, 2)$-categories, we want to consider functors $X \colon \Theta_2^{op} \to \mathcal{SS}ets$. Our first question is how to describe the appropriate Segal conditions, of which we expect to have two, coming from the two ways that the category Δ appears in the category Θ_2: the "outside" indexing, and the "internal" indexing, given by arrow labelings. We can conceptualize these kinds of conditions as follows. If we apply such a functor X to the object

$$0 \underset{\Downarrow}{\overset{\Downarrow}{\longrightarrow}} 1 \longrightarrow 2 \overset{\Downarrow}{\rightrightarrows} 3$$

of Θ_2^{op}, one Segal condition should give us that the resulting simplicial set should be weakly equivalent to the simplicial set

$$X\left(0 \underset{\Downarrow}{\overset{\Downarrow}{\longrightarrow}} 1\right) \times_{X(0)} X\left(1 \longrightarrow 2\right) \times_{X(1)} X\left(2 \overset{\Downarrow}{\rightrightarrows} 3\right).$$

The other Segal condition should tell us that the simplicial set

$$X\left(0 \underset{\Downarrow}{\overset{\Downarrow}{\longrightarrow}} 1\right)$$

is weakly equivalent to

$$X\left(0 \overset{\Downarrow}{\rightrightarrows} 1\right) \times_{X(0 \to 1)} X\left(0 \overset{\Downarrow}{\rightrightarrows} 1\right).$$

Once again, we have choices to make as to whether we want to require completeness conditions or discreteness of certain component spaces. Let us start by considering completeness for both, as Rezk did. We still retain the analogue of completeness for a complete Segal space, namely that $X[0]$ is weakly equivalent to a subspace of homotopy equivalences in $X[1]([0])$, which we can visualize as

$$X(0) \stackrel{\simeq}{\rightrightarrows} X\left(0 \longrightarrow 1\right)_{\text{heq}}.$$

But we additionally have a higher-dimensional analogue, which says that $X[1]([0])$ is weakly equivalent to a subspace of homotopy equivalences in $X[1]([1])$, which can be visualized as

$$X\left(0 \longrightarrow 1\right) \stackrel{\simeq}{\rightrightarrows} X\left(0 \underset{\rightarrow}{\overset{\rightarrow}{\Downarrow}} 1\right)_{\text{heq}}.$$

Let us make some of these ideas more explicit. Since Θ_2 is still a Reedy category [8], we can consider the category of functors $\Theta_2^{op} \to \mathcal{SS}ets$ with the Reedy model structure and localize so that the fibrant objects have the desired conditions.

Given an object $[m]([k_1], \ldots, [k_m])$ of Θ_2, with $m \geq 2$, let $\Theta[m]([k_1], \ldots, [k_m])$ be its associated representable functor $\Theta_2^{op} \to \mathcal{S}ets$, thought of as a discrete functor $\Theta_2^{op} \to \mathcal{SS}ets$. In analogy with the sub-simplicial space $G[n]^t \subseteq \Delta[n]^t$ used to define Segal spaces, define

$$G[m]([k_1], \ldots, [k_m]) = \Theta[1]([k_1]) \amalg_{\Theta[0]} \cdots \amalg_{\Theta[0]} \Theta[1]([k_m]),$$

and observe that there is a natural inclusion map

$$G[m]([k_1], \ldots, [k_m]) \to \Theta[m]([k_1], \ldots, [k_m]).$$

The "horizontal", or "outer", Segal condition can be made precise by asking that X be local with respect to all such maps.

The second, "internal" Segal condition is described similarly. Given $k \geq 0$, define

$$\Theta[1](G[k]) = \Theta[1]([1]) \amalg_{\Theta[1]([0])} \cdots \amalg_{\Theta[1]([0])} \Theta[1]([1]),$$

which naturally includes into $\Theta[1]([k])$. Objects that are local with respect to these maps satisfy the second Segal condition.

As we have seen in other models, we can pause here and consider such functors with no further conditions.

Definition 7.7.3. A *Segal Θ_2-space* is a Reedy fibrant functor that satisfies both the Segal conditions described above.

These objects have an associated model structure.

Theorem 7.7.4. [46] *There is a model structure on the category of functors $\Theta_2^{op} \to \mathcal{SS}ets$ in which the fibrant objects are precisely the Segal Θ_2-spaces.*

What about the completeness conditions? For the first, we use an underlying simplicial space functor, just as we did for the multisimplicial model.

Definition 7.7.5. Let $\tau_\Theta \colon \Delta \to \Theta_2$ be defined by $\tau_\Theta[m] = [m]([0], \ldots, [0])$. The *underlying simplicial space* of a functor $X \colon \Theta_2^{op} \to \mathcal{S}Sets$ is given by $\tau_\Theta^*(X)$.

This functor τ_Θ^* has a left adjoint, which we denote by $(\tau_\Theta)_\#$, To define our first completeness condition, we apply this functor to the map that we use to define complete Segal spaces. Thus, we localize with respect to the map

$$(\tau_\Theta)_\#(E^t) \to (\tau_\Theta)_\# \Delta[0]^t.$$

The second completeness condition is more subtle to define precisely. The idea is to define an object $\Theta[E]$ that models homotopy 2-equivalences, and then localize with respect to a certain map

$$\Theta[E] \to \Theta[1]([0]).$$

The object $\Theta[E]$ is defined via an intertwining functor which we do not describe here; we refer the reader to [46, 4.4] for details.

Definition 7.7.6. A *complete Segal* Θ_2*-space* is a Reedy fibrant functor $X \colon \Theta_2^{op} \to \mathcal{S}Sets$ such that these two Segal conditions and two completeness conditions hold.

Remark 7.7.7. Rezk simply refers to these objects as Θ_2*-spaces*. We have chosen to use extra adjectives here for further clarity about what conditions are assumed.

Theorem 7.7.8. [46] *There is a cartesian model structure $\Theta_2\mathcal{CSS}$ on the category of all functors $\Theta_2^{op} \to \mathcal{S}Sets$ in which the fibrant objects are precisely the complete Segal Θ_2-spaces.*

Once again, we want to have a notion of Dwyer-Kan equivalence between complete Segal Θ_2-spaces which is appropriately fully faithful and essentially surjective. To do so, we need notions of mapping objects and a homotopy category.

Definition 7.7.9. Given a functor $X \colon \Theta_2^{op} \to \mathcal{S}Sets$ and any $(x, y) \in X[0]_0 \times X[0]_0$, we define the *mapping object* $M_X^\Theta(x, y) \colon \Delta^{op} \to \mathcal{S}Sets$, evaluated at the object $[k]$ of Δ, as the pullback of the diagram

$$\{(x, y)\} \to X[0] \times X[0] \leftarrow X[1]([k]).$$

For essential surjectivity, we use the underlying simplicial space functor τ_Θ^*.

Definition 7.7.10. Let X be a Segal Θ_2-space. Its *homotopy category* $\mathrm{Ho}(X)$ has $X[0]_0$ as objects and

$$\mathrm{Hom}_{\mathrm{Ho}(X)}(x, y) = \mathrm{Hom}_{\mathrm{Ho}(\tau_\Theta^* X)}(x, y).$$

Now we can make our definitions of homotopically fully faithful and essentially surjective.

Definition 7.7.11. Let X and Y be Segal Θ_2-spaces. A morphism $f \colon X \to Y$ is *homotopically fully faithful* if for every $x, y \in X[0]$ and every $k \geq 0$ the map

$$M_X^\Theta(x, y)_k \to M_X^\Theta(fx, fy)_k$$

is a weak equivalence in $\mathcal{S}Sets$.

Definition 7.7.12. Let X and Y be Segal Θ_2-spaces. A morphism $X \to Y$ is *essentially surjective* if $\mathrm{Ho}(f)\colon \mathrm{Ho}(X) \to \mathrm{Ho}(Y)$ is an essentially surjective functor of categories.

As before, to consider these notions for more general functors $\Theta_2^{op} \to \mathcal{SS}ets$, we apply a functorial fibrant replacement functor L in the model structure for Segal Θ_2-spaces.

Definition 7.7.13. Let X and Y be functors $\Theta_2^{op} \to \mathcal{SS}ets$. A map $X \to Y$ is a *Dwyer-Kan equivalence* if the associated map $LX \to LY$ is fully faithful and essentially surjective.

As in other models, we have the following result.

Theorem 7.7.14. *Let $X, Y\colon \Theta_2^{op} \to \mathcal{SS}ets$ be Segal Θ_2-spaces. A map $X \to Y$ is a Dwyer-Kan equivalence if and only if it is a weak equivalence in $\Theta_2\mathcal{CSS}$.*

Now, we would like to compare this model structure to one of those previously developed to show that complete Segal Θ_2-spaces give good models for $(\infty, 2)$-categories. It is convenient to compare them to 2-fold complete Segal spaces, for which we need a way to relate the categories Θ_2 and $\Delta \times \Delta$.

We define the functor $d\colon \Delta \times \Delta \to \Theta_2$ by

$$([m], [k]) \mapsto [m]([k], \ldots, [k]).$$

As the name suggests, we can think of d as a kind of diagonal functor. As an example, consider the object $([2], [1])$, which maps to $[2]([1], [1])$. Visually, this assignment takes the diagram

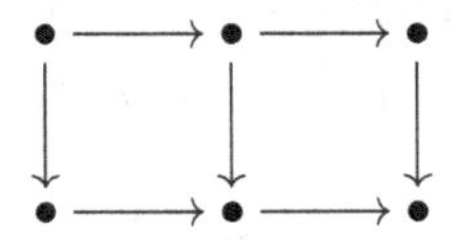

to the diagram

Observe that this functor is modeling the kind of "compression" of vertical morphisms that we wanted in the passage from a multisimplicial model to a Θ_2-model.

The induced functor

$$d^*\colon \mathcal{SS}ets^{\Theta_2^{op}} \to \mathcal{SS}ets^{\Delta^{op} \times \Delta^{op}}$$

has both a left and a right adjoint, given by left and right Kan extension. We are interested here in the right adjoint, which we denote by d_*.

Theorem 7.7.15. [15] *The adjunction (d^*, d_*) induces a Quillen equivalence*

$$d^*\colon \Theta_2\mathcal{CSS} \rightleftarrows \mathcal{CSS}(\mathcal{CSS})\colon d_*.$$

This comparison shows us, in particular, that Θ_2 does exactly what we want it to, in replacing the essential constancy condition for one simplicial direction of $\Delta \times \Delta$-diagrams by the compressed diagrams in Θ_2.

One might observe, just as in the last section, that we have been giving preferential treatment to a model that incorporates completeness conditions, rather than asking that certain component spaces be discrete. Let us remedy this situation now and consider such variants.

Definition 7.7.16. A Θ_2-*Segal precategory* is a functor $X\colon \Theta_2^{op} \to \mathcal{SS}ets$ for which the simplicial sets $X[0]$ and $X[1]([0])$ are discrete. A Θ_2-*Segal category* additionally satisfies both Segal conditions.

Theorem 7.7.17. [10] *There is a model structure $\Theta_2\mathcal{S}e\mathcal{C}at$ on the category of Θ_2-Segal precategories in which the fibrant objects are the Reedy fibrant Θ_2-Segal categories.*

But, as before, we have gone to the other extreme and required two discreteness conditions; we could instead mix and match the completeness and discreteness assumptions. As with the multisimplicial models, only some of these combinations give distinct models.

Definition 7.7.18. A Θ_2-*Segal* $[0]$-*precategory* is a functor $X\colon \Theta_2^{op} \to \mathcal{SS}ets$ such that the simplicial set $X[0]$ is discrete. It is a Θ_2-*Segal* $[0]$-*category* if additionally it satisfies both Segal conditions and the completeness condition that

$$X[1]([0]) \simeq X[1]([1])_{\text{heq}}.$$

Theorem 7.7.19. [10] *There is a model structure $\Theta_2\mathcal{S}e^{[0]}\mathcal{C}at$ on the category of Θ_2^{op}-Segal $[0]$-precategories in which the fibrant objects are the Reedy fibrant Θ_2-Segal $[0]$-categories.*

We can think of this model structure as giving an an intermediate step between Θ_2-Segal categories and complete Segal Θ_2-spaces, as given by the following theorem.

Theorem 7.7.20. [10] *The inclusion functors and their right adjoints induce Quillen equivalences*

$$\Theta_2\mathcal{S}e\mathcal{C}at \rightleftarrows \Theta_2\mathcal{S}e^{[0]}\mathcal{C}at \rightleftarrows \Theta_2\mathcal{CSS}.$$

If we try to make the opposite choice, making $X[1]([0])$ discrete and $X[0] \simeq X[1]([0])_{\text{heq}}$, we actually force $X[0]$ to be discrete, since $X[1]([0])_{\text{heq}}$ is a subspace of a discrete space and thus discrete, and $X[0]$ is a retract of it. Thus we recover the Θ_2-Segal category model rather than something new. This phenomenon mirrors the same issue that arose in the bisimplicial setting.

The model structures that we have are all compatible with their respective bisimplicial models via the adjoint pair (d^*, d_*). The Quillen equivalences below not already mentioned are to appear in [11].

Theorem 7.7.21. *The functors in the following commutative diagram are left adjoints of Quillen equivalences:*

$$\begin{array}{ccccc} \mathcal{S}e\mathcal{C}at(\mathcal{S}e\mathcal{C}at) & \longrightarrow & \mathcal{S}e\mathcal{C}at(\mathcal{CSS}) & \longrightarrow & \mathcal{CSS}(\mathcal{CSS}) \\ \downarrow & & \downarrow & & \downarrow \\ \Theta_2\mathcal{S}e\mathcal{C}at & \longrightarrow & \Theta_2\mathcal{S}e^{[0]}\mathcal{C}at & \longrightarrow & \Theta_2\mathcal{CSS}. \end{array}$$

The horizontal functors are given by inclusion, whereas the vertical ones are d_* or the appropriate restriction thereof.

7.8 Generalizations of quasi-categories

A natural question a reader might have at this point is whether there are generalizations of quasi-categories with the same kind of flavor. This question was initially asked by Joyal, who tried to describe the analogues of horn-filling conditions for functors $\Theta_2^{op} \to \mathcal{S}ets$ in an explicit combinatorial way but was not successful. With the advent of complete Segal Θ_2-spaces, however, Ara was able to exploit the method of proof for the Quillen equivalence between quasi-categories and complete Segal spaces to describe such structures and give a corresponding model structure which is Quillen equivalent to $\Theta_2\mathcal{CSS}$ [1]. We sketch the main ideas here; the methods used are substantially different than than the ones used thus far in this paper, so we do not go into great detail.

Consider the category of functors $\Theta_2^{op} \to \mathcal{SS}ets$. We can think of the objects of this category alternatively as functors $\Theta_2^{op} \times \Delta^{op} \to \mathcal{S}ets$, or in turn as simplicial objects in the category of functors $\Theta_2^{op} \to \mathcal{S}ets$. From this last perspective, there is a natural evaluation functor ev_0 taking such a simplicial object to the functor $\Theta_2^{op} \to \mathcal{S}ets$ in degree zero.

We want to have a model structure on the category of all such functors so that the adjoint pair of functors given by inclusion and evaluation at simplicial degree zero gives a Quillen equivalence of model categories with $\Theta_2\mathcal{CSS}$, in analogy with the Quillen equivalence between complete Segal spaces and quasi-categories.

The following definition is not particularly precise, but gives an idea of the flavor of how we should think of a Θ_2-analogue of quasi-categories. Ara gives a more concrete definition in [1].

Definition 7.8.1. A functor $K\colon \Theta_2^{op} \to \mathcal{S}ets$ is a Θ_2-*quasi-category* if $K \cong \mathrm{ev}_0 X$ for some complete Segal Θ_2-space $X\colon \Theta_2^{op} \to \mathcal{SS}ets$.

In line with Rezk's terminology, Ara refers to these objects as Θ_2-*sets*.

Theorem 7.8.2. [1, 8.5] *There is a cartesian model structure $\Theta_2\mathcal{QC}at$ on the category of functors $\Theta_2^{op} \to \mathcal{S}ets$ in which the fibrant objects are the Θ_2-quasi-categories.*

Ara's proof uses Cisinski's theory of A-localizers, as does the comparison with complete Segal Θ_2-spaces. Here, we sketch an argument for the comparison that more closely models the one of Joyal and Tierney for quasi-categories and complete Segal spaces.

First, we give a model structure on the category of simplicial objects in $\Theta_2\mathcal{QC}at$ so that the fibrant objects are essentially constant in the simplicial direction. Denoting this model structure by $(\Theta_2\mathcal{S}ets)_{ec}^{\Delta^{op}}$, the following comparison is not unexpected.

Theorem 7.8.3. [1, 8.4] *The inclusion and evaluation functors induce a Quillen equivalence*

$$\Theta_2\mathcal{QC}at \rightleftarrows (\Theta_2\mathcal{S}ets)_{ec}^{\Delta^{op}}.$$

The last step is to show that this right-hand model structure is precisely the one for complete Segal Θ_2-spaces.

Theorem 7.8.4. *The model structure $(\Theta_2\mathcal{S}ets)_{ec}^{\Delta^{op}}$ exactly coincides with the model structure $\Theta_2\mathcal{CSS}$.*

A natural question, given the comparisons of the previous section, is whether one can obtain a similar model using $\Delta \times \Delta$.

Question 7.8.5. *Is there an analogous model using bisimplicial sets?*

7.9 Generalizing to (∞, n)-categories

In this section we give a very brief introduction to generalizing the models of the previous sections to higher (∞, n)-categories. Conceptually, the hard work is done in knowing how to generalize from $(\infty, 1)$-categories to $(\infty, 2)$-categories, and then we can proceed in an inductive way. Of course, there are many variants and combinations, of which we can only give a hint here.

We describe two of the multisimplicial models inductively as follows.

Definition 7.9.1. A Reedy fibrant functor $(\Delta^{op})^n \to \mathcal{SS}ets$ is an *n-fold complete Segal space* if it is a complete Segal object in $(n-1)$-fold complete Segal spaces.

Definition 7.9.2. A functor $(\Delta^{op})^n \to \mathcal{SS}ets$ is a *Segal n-category* if it is a Segal category object in $(n-1)$-Segal categories.

For another approach, let us consider the generalization of Θ_2 to Θ_n. These categories were first defined by Joyal [34], but here we use the inductive approach of Berger [8], which we have essentially used already in our definition of Θ_2 above. Given a small category $\mathcal{C}$, define the category $\Theta\mathcal{C}$ to have objects $[m](c_1, \ldots, c_m)$ where $[m]$ is an object of Δ and $c_1, \ldots, c_n$ are objects of $\mathcal{C}$. A morphism between two such objects is defined analogously to the ones in Θ_2.

Let Θ_0 be the category with one object and only an identity morphism, and inductively define $\Theta_n = \Theta\Theta_{n-1}$. Observe that $\Theta_1 = \Delta$, and Θ_2 is exactly the category as we described it previously.

Consider a functor $\Theta_n^{op} \to \mathcal{SS}ets$. As we did for Θ_2-spaces, we can define n different Segal conditions: one "outermost" or "horizontal" condition, given by the inclusions

$$G[m](c_1, \ldots, c_m) \to \Theta[m](c_1, \ldots, c_m)$$

for any object $[m](c_1, \ldots c_m)$ of Θ_n, so that each c_i is an object of Θ_{n-1}. Then the other $n-1$ Segal conditions can be imported inductively from those for Θ_{n-1}-spaces.

Completeness is similar: the completeness condition that says that the space of objects is weakly equivalent to the space of homotopy equivalences is given by the usual completeness condition on the underlying simplicial space. The higher completeness conditions are given by incorporating those for complete Segal Θ_{n-1}-spaces.

Definition 7.9.3. A *complete Segal Θ_n-space* is a Reedy fibrant functor $\Theta_n^{op} \to \mathcal{SS}ets$ that satisfies these n Segal conditions and n completeness conditions.

Theorem 7.9.4. [46] *There is a cartesian model structure $\Theta_n\mathcal{CSS}$ on the category of functors $\Theta_n^{op} \to \mathcal{SS}ets$ in which the fibrant objects are the complete Segal Θ_n-spaces.*

Because this model structure is cartesian, we can hope that there is a corresponding model structure for categories enriched in it. The proof that we have a model structure on small categories enriched in $\mathcal{CSS}$ extends to this case. Then our chain of Quillen equivalences between $\mathcal{CSS}$-$\mathcal{C}at$ and $\Theta_2\mathcal{CSS}$ generalizes to the following result.

Theorem 7.9.5. [14],[15] *There is a chain of Quillen equivalences between $\Theta_{n-1}\mathcal{CSS}$-$\mathcal{C}at$ and $\Theta_n\mathcal{CSS}$.*

In particular, we have appropriate notions of complete Segal objects and Segal category objects in complete Segal Θ_n-spaces. The definitions are quite similar to the ones we have given for $n = 2$, so we instead turn our attention to some interesting things that happen when $n \geq 3$.

Recall the diagonal functor $d\colon \Delta \times \Delta \to \Theta_2$ which we used to compare 2-fold complete Segal spaces and complete Segal Θ_2-spaces. In the case of higher n, this functor actually gives us a means of going from $\Delta^n \to \Theta_n$ incrementally.

Theorem 7.9.6. [15] *Define the functor $d\colon \Delta \times \Theta_{n-1} \to \Theta_n$ by $([m], c) \mapsto [m](c, \ldots, c)$. Then the functor $d^*\colon \mathcal{SS}ets^{\Theta_n^{op}} \to \mathcal{SS}ets^{\Delta^{op} \times \Theta_{n-1}^{op}}$ and its left adjoint induce a Quillen equivalence between $\Theta_n\mathcal{CSS}$ and the model structure for complete Segal objects in complete Segal Θ_{n-1}-spaces.*

We can continue this process, successively picking copies of Δ off of Θ_n via the functor d:

$$\underbrace{\Delta \times \cdots \times \Delta}_{n} \to \underbrace{\Delta \times \cdots \Delta}_{n-2} \times \Theta_2 \to \cdots \to \Delta \times \Theta_{n-1} \to \Theta_n.$$

Theorem 7.9.7. [15] *The above chain of functors induces Quillen equivalences between a model structure for n-fold complete Segal spaces and $\Theta_n\mathcal{CSS}$.*

Another proof of this comparison, not using model categories, is given by Haugseng [28].

The various models in between can be thought of as hybrids between complete Segal Θ_n-spaces and n-fold complete Segal spaces. In all these cases, one can also replace completeness conditions with discreteness conditions, but as we saw for $(\infty, 2)$-categories, we only get distinct models when we choose discreteness conditions from the bottom up. So, for example, we could ask that the spaces of objects and of i-morphisms, for $i \leq k$, be discrete, and for the spaces of i-morphisms, for $k < i < n$, to be weakly equivalent to the space of $i+1$-homotopy equivalences. We discuss these models, and the comparisons between them, in [10] and [11].

Ara's proof of the comparison between Θ_2-quasi-categories and complete Segal Θ_2-spaces extends to an analogous comparison between Θ_n-quasi-categories and complete Segal Θ_n-spaces. Again, a natural question is whether there is an analogous way to develop a multi-simplicial model, or models corresponding to the various interpolations between $\Delta \times \cdots \times \Delta$ and Θ_n.

Acknowledgements

The author was partially supported by NSF grant DMS-1659931, the Isaac Newton Institute for Mathematical Sciences, Cambridge, during the program "Homotopy Harnessing Higher Structures", supported by EPSRC grant no EP/K032208/1, and the AWM Michler Prize.

Bibliography

[1] Dimitri Ara, Higher quasi-categories vs higher Rezk spaces, *J. K-Theory* 14 (2014), no. 3, 701–749.

[2] Dimitri Ara, On the homotopy theory of Grothendieck ∞-groupoids, *J. Pure Appl. Algebra* 217 (2013) 1237–1278.

[3] David Ayala, John Francis, and Nick Rozenblyum, Factorization homology I: Higher categories, *Adv. Math.* 333 (2018), 1042–1177.

[4] Clark Barwick, On left and right model categories and left and right Bousfield localizations, *Homology, Homotopy Appl.* 12 (2010), no. 2, 245–320.

[5] C. Barwick and D.M. Kan, Relative categories: Another model for the homotopy theory of homotopy theories, *Indag. Math. (N.S.)* 23 (2012), no. 1–2, 42–68.

[6] C. Barwick and D.M. Kan, n-relative categories: A model for the homotopy theory of n-fold homotopy theories. *Homology Homotopy Appl.* 15 (2013), no. 2, 281–300.

[7] Clark Barwick and Christopher Schommer-Pries, On the unicity of the homotopy theory of higher categories, arXiv:1112.0040, 2011.

[8] Clemens Berger, Iterated wreath product of the simplex category and iterated loop spaces, *Adv. Math.* 213 (2007), 230–270.

[9] Julia E. Bergner, A model category structure on the category of simplicial categories, *Trans. Amer. Math. Soc.* 359 (2007), 2043–2058.

[10] Julia E. Bergner, Models for (∞, n)-categories with discreteness conditions, I, in preparation.

[11] Julia E. Bergner, Models for (∞, n)-categories with discreteness conditions, II, in preparation.

[12] Julia E. Bergner, Three models for the homotopy theory of homotopy theories, *Topology* 46 (2007), 397–436.

[13] Julia E. Bergner, Angélica M. Osorno, Viktoriya Ozornova, Martina Rovelli, and Claudia I. Scheimbauer, 2-Segal objects and the Waldhausen construction, arXiv:1809.10924, 2018.

[14] Julia E. Bergner and Charles Rezk, Comparison of models for (∞, n)-categories, I, *Geom. Topol.* 17 (2013) 2163–2202.

[15] Julia E. Bergner and Charles Rezk, Comparison of models for (∞, n)-categories, II, arXiv:1406.4182, 2014.

[16] Julia E. Bergner and Charles Rezk, Reedy categories and the Θ-construction, *Math. Z.*, 274 (1), 2013, 499–514.

[17] David Blanc and Simona Paoli, Segal-type algebraic models of n-types, *Algebr. Geom. Topol.* 14 (2014), no. 6, 3419–3491.

[18] J.M. Boardman and R.M. Vogt, *Homotopy invariant algebraic structures on topological spaces, Lecture Notes in Math.*, Vol. 347. Springer-Verlag, 1973.

[19] A.K. Bousfield, The localization of spaces with respect to homology. *Topology* 14 (1975), 133–150.

[20] Denis-Charles Cisinski, Batanin higher groupoids and homotopy types, *Categories in algebra, geometry and mathematical physics, Contemp. Math.*, Vol. 431, 171–186, Amer. Math. Soc., 2007.

[21] J.M. Cordier and T. Porter, Vogt's theorem on categories of homotopy coherent diagrams, *Math. Proc. Camb. Phil. Soc.* (1986), 100, 65–90.

[22] Daniel Dugger and David I. Spivak, Mapping spaces in quasicategories, *Algebr. Geom. Topol.* 11 (2011) 263–325.

[23] Daniel Dugger and David I. Spivak, Rigidification of quasicategories, *Algebr. Geom. Topol.* 11 (2011) 225–261.

[24] W.G. Dwyer and D.M. Kan, Function complexes in homotopical algebra, *Topology* 19 (1980), 427–440.

[25] W.G. Dwyer, D.M. Kan, and J.H. Smith, Homotopy commutative diagrams and their realizations. *J. Pure Appl. Algebra* 57 (1989), 5–24.

[26] W.G. Dwyer and J. Spalinski, Homotopy theories and model categories, in *Handbook of Algebraic Topology*, 1995, 73–126, North-Holland.

[27] Paul Goerss and Kristen Schemmerhorn, Model categories and simplicial methods. *Interactions between homotopy theory and algebra, Contemp. Math.*, Vol. 436, 3–49, Amer. Math. Soc., 2007.

[28] Rune Haugseng, On the equivalence between Θ_n-spaces and iterated Segal spaces. *Proc. Amer. Math. Soc.* 146 (2018), no. 4, 1401–1415.

[29] Philip S. Hirschhorn, *Model categories and their localizations, Math. Surv. Monogr.*, Vol. 99, Amer. Math. Soc., 2003.

[30] A. Hirschowitz and C. Simpson, Descente pour les n-champs, arXiv:math.AG/9807049, 1998.

[31] Mark Hovey, *Model Categories, Math. Surv. Monogr.*, 63, Amer. Math. Soc., 1999.

[32] Amrani Ilias, Model structure on the category of small topological categories. *J. Homotopy Relat. Struct.* 10 (2015), no. 1, 63–70.

[33] Theo Johnson-Freyd and Claudia Scheimbauer, (Op)lax natural transformations, twisted quantum field theories, and "even higher" Morita categories. *Adv. Math.* 307 (2017), 147–223.

[34] A. Joyal, Disks, duality, and Θ-categories, September 1997 preprint.

[35] A. Joyal, Simplicial categories vs quasi-categories, in preparation.

[36] A. Joyal, The theory of quasi-categories I, in preparation.

[37] André Joyal and Myles Tierney, Quasi-categories vs Segal spaces, *Contemp. Math.* , Vol. 431, 277–326, Amer. Math. Soc., 2007.

[38] Tom Leinster, A survey of definitions of n-category, *Theory Appl. Categ.* 10 (2002), 1–70.

[39] Jacob Lurie, *Higher topos theory. Ann. of Math. Stud.*, 170, Princeton University Press, 2009.

[40] Jacob Lurie, $(\infty, 2)$-categories and Goodwillie calculus, arXiv:0905.0462, 2009.

[41] Jacob Lurie, On the classification of topological field theories. *Current developments in mathematics*, 2008, 129–280, Int. Press, 2009.

[42] Viktoriya Ozornova and Martina Rovelli, Model structures for (∞, n)-categories on (pre)stratified simplicial sets and prestratified simplicial spaces, arXiv:1809.10621, 2018.

[43] Regis Pellissier, Catégories enrichies faibles, arXiv:math/0308246, 2003.

[44] Daniel Quillen, *Homotopical algebra, Lecture Notes in Math.*, Vol. 43, Springer-Verlag, 1967.

[45] C.L. Reedy, Homotopy theory of model categories, available at `http://www-math.mit.edu/~psh/reedy.pdf`.

[46] Charles Rezk, A cartesian presentation of weak n-categories, *Geom. Topol.* 14 (2010), 521–571.

[47] Charles Rezk, A model for the homotopy theory of homotopy theory, *Trans. Amer. Math. Soc.* , 353(3), 973–1007.

[48] Emily Riehl, *Categorical homotopy theory, New Mathematical Monographs*, 24. Cambridge University Press, 2014.

[49] Emily Riehl and Dominic Verity, *Elements of ∞-category theory*, available at `http://www.math.jhu.edu/~eriehl/elements.pdf`.

[50] Emily Riehl and Dominic Verity, Fibrations and Yoneda's lemma in an ∞-cosmos. *J. Pure Appl. Algebra* 221 (2017), no. 3, 499–564.

[51] Graeme Segal, Categories and cohomology theories, *Topology* 13 (1974), 293–312.

[52] Carlos Simpson, *Homotopy theory of higher categories, New Mathematical Monographs* 19. Cambridge University Press, 2012.

[53] Bertrand Toën, Vers une axiomatisation de la théorie des catégories supérieures, *K-Theory* 34 (2005), no. 3, 233–263.

[54] D.R.B. Verity, Weak complicial sets I: Basic homotopy theory. *Adv. Math.* 219 (2008), no. 4, 1081–1149.

DEPARTMENT OF MATHEMATICS, UNIVERSITY OF VIRGINIA, CHARLOTTESVILLE, VA, 22904, U.S.A.

E-mail address: `jeb2md@virginia.edu`

8

Persistent homology and applied homotopy theory

Gunnar Carlsson

8.1 Introduction

Persistent homology is a technique that has been developed over the last 20 years. Initial ideas developed in the early 1990s [36], but the idea of persistence was introduced by Vanessa Robins in [60], rapidly followed by additional development ([34], [68]), and has been developing rapidly since that time. The original motivation for the method was to extend the ideas of algebraic topology from the category of spaces X to situations where we only have a sampling of the space X. Of course, a sample is a discrete space so there is nothing to be obtained unless one retains some additional information. One assumes the presence of a metric or a more relaxed "dissimilarity measure", and uses this information restricted to the sample in constructing the algebraic invariant. Over time, persistent homology has been used in other situations, for example where one has a topological space with additional information, such as a continuous real valued function, and the sublevel sets of the function determine a filtration on the space. The output of standard persistent homology (we will discuss some generalizations) is represented in two ways, via *persistence barcodes* and *persistence diagrams.* Initially persistent homology was used, as homology is used for topological spaces, to obtain a large scale geometric understanding of complex data sets, encoded as finite metric spaces. Examples of this kind of application are [21], [39], [46], [59], and [25]. Another class of applications uses persistent homology to study data sets where the *points themselves* are metric spaces, such as databases of molecule structures or images. This second set of applications is developing very rapidly and is exemplified in [15], [16], [66], and [43]. Another direction in which persistent homology is being applied is in the study of coverage and evasion problems arising in sensor net technology [27]. Example of research in this direction are [31], [32], [1], and [38].

There are a number of different active research directions in this area.

- **Coordinatization of barcodes:** Barcodes in their native form, i.e. as a set of intervals, do not lend themselves to analysis by machine learning techniques. It is therefore important to represent them in a method more amenable to analysis. This can be achieved by appropriate coordinatizations of the space of barcodes. Several methods for this task are described in Section 8.7.

Mathematics Subject Classification. 55-04, 55N35, 55U05, 55U15, 55U99.

Key words and phrases. persistent homology, applied topology, topological data analysis, random complexes, persistence diagrams, persistence barcodes.

- **Generalized persistence:** Persistent homology has as its output a diagram of complexes, parametrized by the partially ordered set of real numbers, on which algebraic computations are performed so as to produce barcodes. There are other parameter categories that are useful in the study of data sets. We discuss two examples of this notion in Section 8.8, but we would expect there to be many different types of diagrams that will shed light on finite metric spaces.

- **Stability results:** Since noise and error are key elements in the study of data, it is important to develop methods that quantify the dependence of the barcode output on small perturbations of data. This requires the imposition of metrics on the set of barcodes, and proving theorems concerning the distances between barcodes that differ only by small perturbations. The progress that has been made in this direction is described in Section 8.5.

- **Probabilistic analysis and inference:** Because many of the applications of persistent homology occur in the study of data, it becomes important to study not only stability results under small perturbations, but also perturbations that are "probabilistically small", i.e. which may be large, but where a large perturbation is a rare event. This means that one must study the distributions on barcode space that occur from applying persistent homology to complexes generated by various random models. This is a rapidly developing area within the subject.

- **Coverage and evasion problems:** This work centers around attempts to understand complements in Euclidean space of regions defined by collections of sensors. It has been approached using different methods, and appears to be a place where techniques such as Spanier-Whitehead duality and embedding analysis, applied in suitably generalized situations, should play a role. The general problem of understanding complements of objects embedded in Euclidean space of course also plays a role in robotics.

- **Symplectic geometry:** Although persistent homology is mainly used in situations where one is examining discrete approximations to continuous objects, it can be applied in any geometric situation where there is a metric, or where one is considering filtered objects. Such situations occur in symplectic geometry, and there is recent work applying the technique in studying, for example, Floer homology spectra. Examples of this kind of work are [13], [48], [56], [57], and [65].

The goal of this paper is to discuss the different research directions and applications at a high level, so that the reader can orient him/herself in the techniques. We remark that there are a number of useful surveys on persistent homology and on topological data analysis more generally. The papers [17], [35], and [55] give different perspectives on this subject.

8.2 The motivating problem

Suppose that we are given a set of points X in the plane, and believe that it is reasonable to assume that the points are sampled (perhaps with error) from a geometric object. We could ask for information about the homology of the underlying space from which X is sampled. Consider the finite set of points X in $\mathbb{R}^2$ displayed in Figure 8.2.1 below.

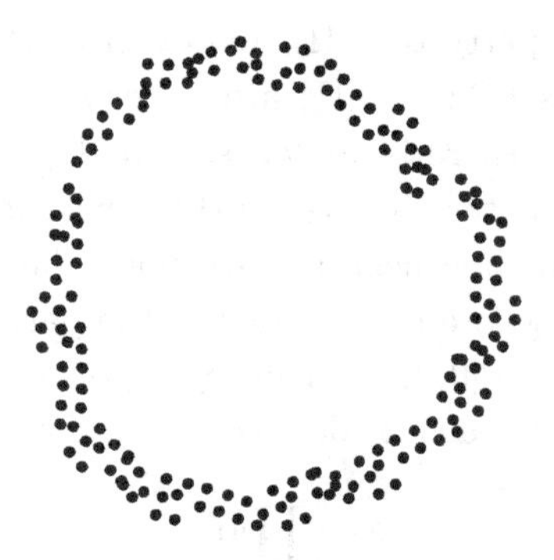

FIGURE 8.2.1. Statistical circle

When we examine X, we observe that it appears to be sampled from a loop and would like to have algebraic tools that capture the "loopy" structure of the set. Note that we are only given a discrete set of points, so direct application of homological constructions will only produce the homology of a finite set of points. However, we could attempt to construct a space from the set X, which in a sense fills in the gaps between the points. We will need to use some additional information about the points, and that will in this case be restriction of the Euclidean metric to X. One distance based construction is the *Vietoris-Rips complex.*

Definition 8.2.1. *For any finite metric space* (X, d)*, and every* $R \geq 0$*, let* $\mathcal{V}(X, R)$ *denote the simplicial complex with vertex set equal to* X*, and such that* $\{x_0, \ldots, x_k\}$ *spans a* k*-simplex if and only if* $d(x_i, x_j) \leq R$ *for all* $0 \leq i < j \leq k$*.*

Notice that if R is smaller than the smallest interpoint distance, then $\mathcal{V}(X, R)$ will be a discrete complex on the set X. On the other hand, if R is greater than the diameter of X, then $\mathcal{V}(X, R)$ is a full simplex on X. For intermediate values, one obtains other complexes. In this case, there is a range of values of R in which $\mathcal{V}(X, R)$ has the homotopy type of a circle. One could ask if there is a principled way to choose a threshold R based on only the distances between the points. After a great deal of experimentation, one finds that this is a very difficult, if not unsolvable problem. A question that one can ask is if there is a more structured object that one can study which incorporates all the thresholds in a single object, which can be analyzed in a number of different ways. Statisticians have confronted this problem in an analogous situation.

Hierarchical Clustering: The *clustering problem* in statistics is to determine ways to infer the set of connected components of a metric space X from finite samples Y. One of the approaches statisticians developed was to compute $\pi_0\mathcal{V}(Y, R)$ for a choice of threshold R, but they confronted the analogous problem to the one we discussed above, namely the selection of R. They came to the conclusion that the problem of selecting a threshold in a principled way is not a well posed problem, but managed to construct a structured output, called a *dendrogram*, which allowed them to study the behavior of all thresholds at once.

It is defined as follows. For each threshold R, we obtain a set of components $\pi_0\mathcal{V}(Y,R)$, which yields a partition Π_R of Y. If we have $R \leq R'$, then Π_R is a refinement of $\Pi_{R'}$. One definition of a dendrogram structure on a finite set Y is a parametrized family $\{\Pi_R\}_{R\geq 0}$ of partitions of Y, with the property that Π_R refines $\Pi_{R'}$ whenever $R \leq R'$, and so that for any partition Π, the intervals $\{R|\Pi_R = \Pi\}$ are either empty or closed on the left and open on the right. We assume that there is an R_∞ so that Π_{R_∞} is the partition with one block, namely Y. The reason for this terminology is that this information is equivalent to a tree $\mathfrak{D}$ with a reference map to the non-negative real line. The tree $\mathfrak{D}$ is defined as follows. The points of $\mathfrak{D}$ are pairs (c, R), where c is a block in the partition Π_R, and $0 \leq R \leq R_\infty$. We clearly have a reference map to $[0, R_\infty]$ given by $(c, R) \to R$. To define a topology on this set, we construct an auxiliary space Z, defined as

$$Z = \coprod_c [0, R_\infty]_c$$

where c is a block in the partition Π_0, and $[0, R_\infty]_c$ denotes a copy of the interval $[0, R_\infty]$ labelled by c. There is a natural map φ from Z to $\mathfrak{D}$ given by $(c, t) \to (\rho_t(c), t)$, where ρ_t denotes the projection $Y/\Pi_0 \to Y/\Pi_t$. It is clear that φ is a surjective map, and therefore that $\mathfrak{D}$ is the quotient of Z by an equivalence relation. The topology on $\mathfrak{D}$ is the quotient topology associated to the topology on Z. It is easy to check that this topology makes $\mathfrak{D}$ into a rooted tree. A tree with a reference map can be laid out in the plane, as indicated in Figure 8.2.2, and one can recover directly the clustering at any given value of R.

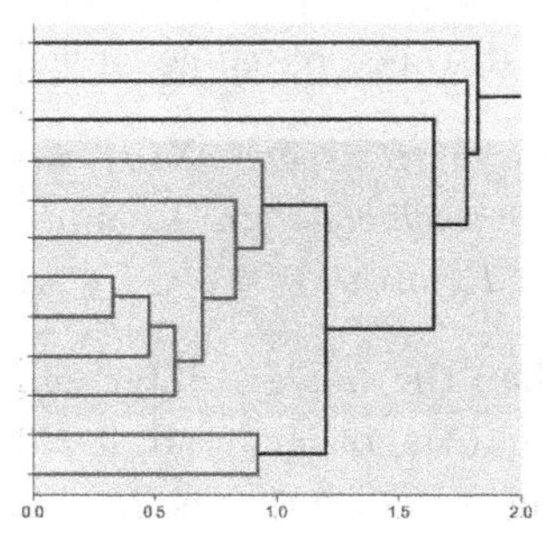

FIGURE 8.2.2. Dendrogram

The dendrogram can be regarded as the "right" version of the invariant π_0 in the statistical world of finite metric spaces. The question now becomes if there are similar invariants that can capture the notions of higher homotopy groups or homology groups. In order to define them, we need a preliminary definition.

Definition 8.2.2. *A submonoid $\mathbb{A} \subseteq \mathbb{R}_+$ is said to be* pure *if given any r_0, r_1, and r_2 in $\mathbb{R}_+$, with r_0 and r_2 in $\mathbb{A}$, so that $r_0 + r_1 = r_2$, then $r_1 \in \mathbb{A}$. $\mathbb{A}$ is a totally ordered set in its own right by restriction of the total order on $\mathbb{R}_+$.*

For example, $\mathbb{N}$ (the non-negative integers) and $\mathbb{Q}_+$ are both pure. We can now make a definition that includes the dendrogram as a special case.

Definition 8.2.3. *Let $\underline{C}$ denote any category. A persistence object in $\underline{C}$ is a functor $\mathbb{R}_+ \to \underline{C}$, where $\mathbb{R}_+$ denotes the ordered set of non-negative real numbers, regarded as a category, so that there is a unique morphism $r_0 \to r_1$ whenever $r_0 \leq r_1$. More generally, if $\mathbb{A} \subseteq \mathbb{R}_+$ is any* pure *submonoid, an $\mathbb{A}$-parametrized persistence object in $\underline{C}$ we will mean a functor from the ordered set $\mathbb{A}$ to $\underline{C}$. It is clear that the $\mathbb{A}$-parametrized persistence objects in $\underline{C}$ form a category in their own right, where the morphisms are the natural transformations of functors. We will denote this category by $\mathfrak{P}\mathbb{A}\underline{C}$, with the special case of $\mathbb{R}_+$ denoted by $\mathfrak{P}\underline{C}$.*

It is readily observed that the functor $\mathbb{R}_+ \to VR(-, R)$ is a persistence object in the category of simplicial complexes. It follows that that the functor $\mathbb{R}_+ \to \underline{\text{Sets}}$ defined by

$$R \to \pi_0(\mathcal{V}(X, R))$$

where X is a finite metric space, is a persistence object in $\underline{\text{Sets}}$, or a *persistent set.* This is the case because π_0 is a set valued functor. Similarly, if A is any abelian group, and $i \geq 0$ is an integer, then $R \to H_i(VR(-, R) : A)$ is a persistence object in the category of abelian groups, and it will be referred to as the *persistent homology with coefficients in A*. More generally, we may assign persistent abelian groups to any persistent chain complex C_*, and we will refer to them as the persistent homology of C_*. The critical question becomes whether or not the isomorphism classes of these persistence objects are in any sense understandable, and useful for distinguishing or understanding the underlying metric spaces. We will see that in the case of coefficients in a field K, persistent homology is computable and interpretable.

8.3 The structure theorem for persistence vector spaces over a field

For the entirety of this section, K will denote a field, which will be fixed throughout. We are interested in the isomorphism classification of $\mathbb{A}$-parametrized persistence K-vector spaces, for pure submonoids $\mathbb{A}$ of $\mathbb{R}_+$. This is very complicated in general, but is manageable for objects of $\mathfrak{P}\underline{\text{Vect}}(K)$ satisfying a finiteness condition, which is always satisfied for the Vietoris-Rips complexes associated with finite metric spaces. Let $\mathbb{M}$ be any commutative monoid. By an $\mathbb{M}$-graded K-vector space, we will mean a K-vector space V equipped with a decomposition

$$V \cong \bigoplus_{\mu \in \mathbb{M}} V_\mu.$$

Given two $\mathbb{M}$-graded K-vector spaces V_* and W_*, by their tensor product we will mean the tensor product $V \underset{K}{\otimes} W$ equipped with the $\mathbb{M}$-grading given by

$$(V \underset{K}{\otimes} W)_\mu = \bigoplus_{\mu_1 + \mu_2 = \mu} V_{\mu_1} \otimes W_{\mu_2}$$

and we write $V_* \underset{K}{\otimes} W_*$ for this construction. An $\mathbb{M}$-graded K-algebra is then an $\mathbb{M}$-vector space R_* together with a homomorphism $R_* \underset{K}{\otimes} R_* \to R_*$, satisfying the associativity and distributivity conditions. An important example is the monoid K-algebra $K[\mathbb{M}]_*$, for which the grading is given by

$$K[\mathbb{M}]_\mu = K \cdot \mu.$$

It will be convenient to write t^μ for elements $\mu \in K[\mathbb{M}]_*$. We define the notion of a $\mathbb{M}$-graded R_*-module M_* in a similar way.

We specialize to the situation $\mathbb{M} = \mathbb{A}$, where $\mathbb{A}$ is a pure submonoid $\mathbb{A} \subseteq \mathbb{R}_+$. We will demonstrate the classification of $\mathbb{A}$-parametrized persistence modules by using an equivalence of categories to the category of $\mathbb{A}$-graded $K[\mathbb{A}]_*$-modules.

Proposition 8.3.1. *Let $K[\mathbb{A}]_*$ denote the monoid algebra of $\mathbb{A}$ over K, for a pure submonoid $\mathbb{A} \subseteq \mathbb{R}_+$, regarded as a graded K-algebra. Let $\underline{G}(\mathbb{A}, K)$ denote the category of $\mathbb{A}$-graded $K[\mathbb{A}]_*$-modules. Then there is an equivalence of categories*

$$\mathfrak{P}\mathbb{A}\underline{\text{Vect}}(K) \cong \underline{G}(\mathbb{A}, K).$$

Proof: Given a functor $\theta : \mathbb{A} \to \underline{\text{Vect}}(K)$, we will denote by $M(\theta)$ the graded K-vector space

$$M(\theta) = \bigoplus_{\alpha \in A} \theta(\alpha)$$

where the elements of grading α are precisely the elements of the summand $\theta(\alpha)$. We now extend the vector space structure to a graded $K[\mathbb{A}]_*$-module structure by defining the action $t^\alpha \cdot : \theta(\alpha') \to \theta(\alpha + \alpha')$ to be equal to the linear transformation

$$\theta(\alpha' \to \alpha + \alpha') : \theta(\alpha') \to \theta(\alpha + \alpha')$$

where $\alpha' \to \alpha + \alpha'$ denotes the unique morphism in $\mathbb{A}$ from α to $\alpha + \alpha'$. It is clear that this defines an $\mathbb{A}$-graded $K[\mathbb{A}]_*$-module structure on $M(\theta)$, and that M is a functor. We produce an inverse functor $\eta : \underline{G}(\mathbb{A}, K) \longrightarrow \mathfrak{P}\mathbb{A}\underline{\text{Vect}}(K)$ on objects by setting

$$\eta(M_*)(\alpha) = M_\alpha$$

and on morphisms by

$$\eta(M_*)(\alpha' \to \alpha + \alpha') = t^\alpha \cdot : M_{\alpha'} \to M_{\alpha+\alpha'}.$$

The functors θ and η are clearly inverse to each other. □

Let $\mathbb{A}$ be any pure submonoid of $\mathbb{R}_+$. Then for any $\alpha \in \mathbb{A}$, we define $F(\alpha)$ to be the free $\mathbb{A}$-graded $K[\mathbb{A}]_*$-module on a single generator in grading α. For any pair $\alpha, \alpha' \in \mathbb{A}$, where $\alpha' > \alpha$, we define $F(\alpha, \alpha')$ to be the quotient

$$F(\alpha)/(t^{\alpha' - \alpha} \cdot F(\alpha)).$$

The following result describes the isomorphism classification of finitely presented graded $K[\mathbb{A}]_*$-modules.

Proposition 8.3.2. *Any finitely presented object of $\underline{G}(\mathbb{A}, K)$ is isomorphic to a module of the form*

$$\bigoplus_{s=1}^{m} F(\alpha_s) \oplus \bigoplus_{t=1}^{n} F(\alpha_t, \alpha'_t).$$

Moreover, the decomposition is unique up to reordering of summands. The kernel of any homomorphism between two finitely generated free $\mathbb{A}$-graded modules is itself a finitely generated free $\mathbb{A}$-graded module.

Remark 8.3.3. This result is formally very similar to the structure theorem for finitely generated modules over a principal ideal domain (PID). Indeed, for the case of $\mathbb{A} = \mathbb{N}$, where $K[\mathbb{A}]$ is Noetherian, the result is exactly a structure theorem for finitely generated graded modules over the graded ring $K[t]$. For other choices of $\mathbb{A}$, $K[\mathbb{A}]_*$ is not necessarily Noetherian. However, it does turn out to be *coherent*, i.e. having the property that the kernel of any homomorphism between finitely generated free modules is finitely generated.

Proof: A proof is given in [18]. A proof in a different context was given earlier in [7]. □

Remark 8.3.4. *Notice that the proof also gives an algorithm for producing a matrix in the diagonal form given above.*

The above theorem is summarized using the following definition.

Definition 8.3.5. *By an* $\mathbb{A}$-valued barcode, *we will mean a finite set of elements*

$$(\alpha, \alpha') \in \mathbb{A} \times (\mathbb{A} \cup \{+\infty\})$$

satisfying the condition $\alpha < \alpha'$. An $\mathbb{A}$-valued barcode is said to be finite *if the right-hand endpoint $+\infty$ does not occur. If $\mathbb{A} = \mathbb{R}_+$, we will simply refer to it as a barcode, without labeling by the monoid. We have shown that isomorphism classes of finitely presented $K[\mathbb{A}]_*$-modules are in bijective correspondence with $\mathbb{A}$-valued barcodes.*

Remark 8.3.6. Barcodes are often represented visually as collections of intervals. The image in Figure 8.3.1 shows barcodes in dimensions zero and one.

Notice that the zero-dimensional barcode has one infinite interval, while the one-dimensional barcode is finite. There is an equivalent visual representation called the *persistence diagram* in which each interval is encoded as a point (x, y) in the plane, where x and y are the left and right hand endpoints respectively. An example is pictured in Figure 8.3.2.

An advantage of this representation made apparent in this image is that one can represent different homology groups in the same diagram. The dark gray dots are in this case the zero-dimensional persistence diagram and the light gray ones are the one-dimensional ones. When there are infinite intervals, one often selects an upper threshold τ for the persistence parameter, and represents the infinite interval by one with left hand endpoint τ.

We also have the following.

Proposition 8.3.7. *The kernel of a homomorphism between finitely generated free $K[\mathbb{A}]_*$-modules is finitely generated free.*

Proof: This follows immediately from the matrix analysis in the proof of Proposition 8.3.2. □

Corollary 8.3.8. *Given any chain complex of finitely generated free graded $K[\mathbb{A}]_*$-modules, the homology modules are finitely presented $K[\mathbb{A}]_*$-modules.*

We also interpret this result in terms of persistence vector spaces. The category $\mathfrak{P}\mathbb{A}\underline{\mathrm{Vect}}(K)$ is clearly an abelian category. For any $\alpha_0 \in \mathbb{A}$, we define $V(\alpha)$ to be the persistence vector space $\{V_\alpha\}_{\alpha \in \mathbb{A}}$, where $V_\alpha = \{0\}$ for $\alpha < \alpha_0$, $V_\alpha = K$ for $\alpha \geq \alpha_0$, and where for any $\alpha_0 \leq \alpha \leq \alpha'$ the linear transformation $V_\alpha \to V_{\alpha'}$ is the identify on K. For any $\alpha < \alpha'$, we also define the persistence vector space $V(\alpha, \alpha')$ to be the quotient of $V(\alpha)$ by the image of the natural inclusion $V(\alpha') \hookrightarrow V(\alpha)$. Corollary 8.3.8 now has the following consequence.

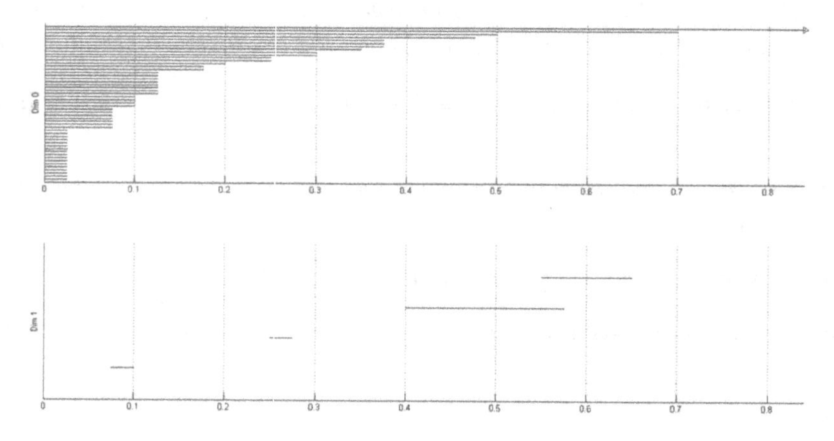

FIGURE 8.3.1. Persistence barcodes

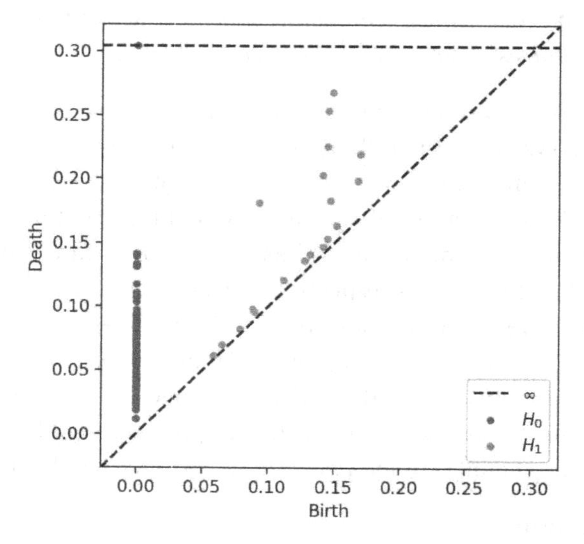

FIGURE 8.3.2. Persistence diagram

Corollary 8.3.9. *Let C_* be a chain complex of $\mathbb{A}$-persistence vector spaces, so that for every s, C_s is a finite direct sum of persistence vector spaces, each of which is of the form $V(\alpha)$ for some $\alpha \in \mathbb{A}$. Then for each s, $H_s(C_*)$ is isomorphic to a direct sum of finitely many persistence vector spaces, each of which is of the form $V(\alpha)$ or $V(\alpha, \alpha')$ for $\alpha, \alpha' \in \mathbb{A}$, and $\alpha < \alpha'$. In particular, the homology in each dimension s is uniquely described by an $\mathbb{A}$-valued barcode.*

8.4 Complex constructors

8.4.1 Introduction

All data that we consider consists of finite sets of points X. The space X itself is uninteresting topologically, since it is a discrete set of points. This means that we have to build a space using auxiliary information attached to the set of points. The auxiliary information we choose is a metric on the set X, so X is a finite metric space. In the context of data sets, metrics are often referred to as *dissimilarity measures*, since small distances between data points often reflect notions of similarity between data points. Often the metric chosen is the restriction of a well known and analyzed metric on an ambient space containing X, such as n-dimensional Euclidean space. Other choices that are often appropriate are Hamming distance, correlation distances, and normalized variants (mean centering of coordinate functions, normalizing variance to 1) of Euclidean distance. One method for constructing spaces based on metrics is the Vietoris-Rips complex that we have seen above. It is actually a persistence object in the category of simplicial complexes. All the constructions we will look at except the Mapper construction are naturally persistence objects in the category of simplicial complexes. The Mapper construction can also be equipped with many such structures, but they are not canonical. Because of the presence of a persistence structure, the Čech, Vietoris-Rips, alpha, and witness complexes all have persistence barcodes associated to them in all non-negative dimensions. It is obvious from the constructions that the zero-dimensional barcodes have a single infinite interval and that all higher dimensional barcodes are finite.

8.4.2 Čech construction

Let (X, d) denote a finite metric space. Given a threshold parameter value R, let $\mathcal{U}_R$ denote the covering of X by all balls $B_R(x) = \{x' | d(x, x') \leq R\}$. The *Čech complex at scale R* is the nerve of the covering $\mathcal{U}_R$, and we denote it by $\check{C}(X; R)$. It is clear that for $R \leq R'$, there is an inclusion $\check{C}(X; R) \hookrightarrow \check{C}(X, R')$, and that therefore $\{\check{C}(X; R)\}_R$ is a persistence object in the category of simplicial complexes. From the theoretical point of view, it has the advantage that given a covering of a topological space (with suitable point set hypotheses) by open sets, the nerve lemma (see [47], Thm. 15.21) gives a criterion that guarantees that the nerve of the covering is homotopy equivalent to the original space.

8.4.3 Vietoris-Rips complex

The Čech construction from the previous section is computationally expensive, because it involves computing simplices individually at every level. We can create another construction that has a strong relationship with the Čech construction. Recall that a simplicial complex Z is said to be a *flag complex* if for any collection σ of vertices $\{z_0, \ldots, z_k\}$ for which each pair (z_i, z_j) is an edge, σ is a k-simplex of X. From the computational point of view, flag complexes are attractive because one needs only enumerate all the edges in the complex, rather than all the higher order simplices. The Vietoris-Rips complex which was defined in Definition 8.2.1 is by definition a flag complex for every parameter value R. There is a relationship between the persistent Čech and Vietoris-Rips complexes.

Proposition 8.4.1. *There are inclusions*

$$\check{C}(X, R) \hookrightarrow \mathcal{V}(X, 2R)$$

and

$$\mathcal{V}(X, R) \hookrightarrow \check{C}(X, R).$$

For specific situations, this bound can be improved. For example, it is shown in [31] that if $X \subseteq \mathbb{R}^d$ is equipped with the restricted metric, then

$$\mathcal{V}(X, R') \subseteq \check{C}(X, R/2) \subseteq \mathcal{V}(X, \varepsilon) \text{ if } \frac{R}{R'} \geq \sqrt{\frac{2d}{d+1}}.$$

This result allows us to compare homology computed using the Čech and Vietoris-Rips methods.

8.4.4 Alpha complex

There is another kind of complex whose dimension is low and which generally has a moderate number of simplices. It is called the *alpha complex*, or the *alpha shapes complex*, and was introduced in [33], with a thorough description in [5]. It applies to data X that is embedded Euclidean space $\mathbb{R}^n$, and so that the metric on X is the restriction of the Euclidean metric to X. Typically the number n is relatively small, say ≤ 5, because the construction becomes quite expensive in higher dimensions. Also, the complex is generically of dimension $\leq n$. The notion of *generic* is the following. Given any set of points $X \subseteq \mathbb{R}^n$, it is possible for the alpha complex to have dimension higher than n, but it is possible to perturb all the points by an arbitrarily small amount and obtain a complex that is n-dimensional.

For any point $x \in X$, we define the *Voronoi cell* of x, denoted by $V(x)$, by

$$V(x) = \{y \in Y | d(x, y) \leq d(x', y) \text{ for all } x' \in X\}.$$

The collection of all Voronoi cells for a finite subset of Euclidean space is called its *Voronoi diagram.*

For each $x \in X$, we also denote by $B_\varepsilon(x)$ the set $\{y \in Y | d(x, y) \leq \varepsilon\}$. By the *$\alpha$-cell* of $x \in V(x)$ with scale parameter ε, we will mean the set $A_\varepsilon(x) = B_\varepsilon(x) \cap V(x)$. The *$\alpha$-complex with scale parameter* ε of a subset $x \in X$, denoted by $\alpha_\varepsilon(X)$ will be the abstract simplicial complex with vertex set X, and where the set $\{x_0, \ldots, x_k\}$ spans a k-simplex if

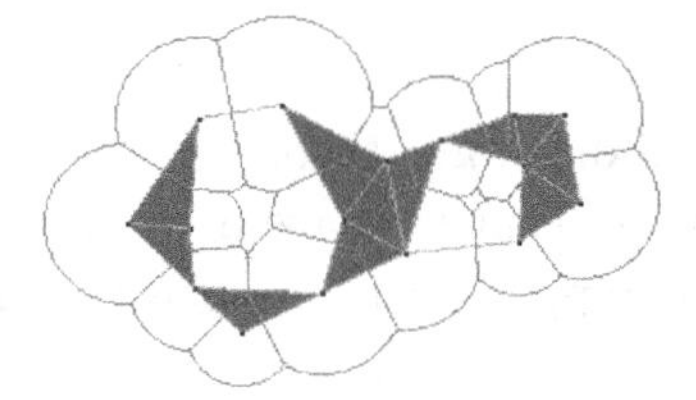

FIGURE 8.4.1. Alpha complex in $\mathbb{R}^2$

and only if

$$\bigcap_{i=0}^{k} A_\varepsilon(x_i) \neq \emptyset.$$

It is of course the nerve of the covering of $\mathbb{R}^n$ by the sets $A_\varepsilon(x_i)$. See Figure 8.4.1 for an example.

8.4.5 Witness complex

The *witness complex* was introduced in [30]. It can be thought of as an analogue of the alpha complex for non-Euclidean data. The construction of the Voronoi cells is not dependent on the fact that the embedding space is Euclidean. It can be constructed for any embedding of a data set in a larger metric space. Given a data set X, we therefore select a subset of *landmarks* $\mathcal{L} \subseteq X$. We can now build the analogues of the Voronoi cells for each of the landmark points within X, and construct the nerve of the covering. In this case, both the ambient space and the landmark set are usually taken to be finite. We will also need to introduce persistence into this picture. The construction is as follows.

Definition 8.4.2. *Given a metric space (X, d), a finite subset $\mathcal{L} \subseteq X$, called the* landmark set, *and a persistence parameter R, and for every $x \in X$ we denote by m_x the distance from x to the set $\mathcal{L}$. We define a simplicial complex $W(X, \mathcal{L}, R)$ as follows. The vertex set of $W(X, \mathcal{L}, R)$ is the set $\mathcal{L}$, and $\{l_0, l_1, \ldots, l_k\}$ spans a k-simplex if and only if there is a point $x \in X$ (the witness) so that $d(x, l_i) \leq m_x + R$ for all i. The family of complexes $\{W(X, \mathcal{L}, R)\}_R$ form a persistence simplicial complex.*

There are several variants on this construction. For example, there is the "lazy" version in which the 1-simplices are identical to the 1-simplices of the witness construction, but in which we declare that any higher dimensional simplex is an element of the complex if and only if all its one dimensional faces are. Each of the lazy complexes is a flag complex. The lazy witness complex bears the same relationship to the witness complex as the Vietoris-Rips complex bears to the Čech complex. There is also the *weak witness complex*, $\{W^{weak}(X, \mathcal{L}, R)\}_R$, which is defined as follows. For each point $x \in X$, we let δ_x denote the distance to the *second* closest element of Λ to x. We then declare that a pair (λ_1, λ_2) spans an edge of $W^{weak}(X, \mathcal{L}, R)$ if and only if there is an $x \in X$ so that

$$\max(d(\lambda_1, x), d(\lambda_2, x)) \leq \delta_x + R.$$

A higher dimensional simplex $\{\lambda_0, \ldots, \lambda_n\}$ is contained in $W^{weak}(X, \mathcal{L}, R)$ if and only if all of its edges are contained in it. This is a very useful construction because the persistence "starts faster" than the standard complex. It is often the case that one obtains the ultimate result even at $R = 0$, or for very small values of R.

8.4.6 Mapper

Another construction is based on Morse theoretic ideas. We motivate it by considering a space level construction. Suppose that we have a continuous map $r : X \to B$ of spaces, and suppose further that B is equipped with an open covering $\mathcal{U} = \{U_\alpha\}_{\alpha \in A}$. We obtain the open covering $r^{-1}\mathcal{U} = \{r^{-1}U_\alpha\}_{\alpha \in A}$ which can be refined into a new covering $r^{-1}\mathcal{U}^*$ by decomposing each set $r^{-1}U_\alpha$ into its connected components. We note that the dimension of the nerve of $r^{-1}\mathcal{U}^*$ is less than or equal to the dimension of the nerve of $\mathcal{U}$, so this construction, like the alpha complex, produces complexes of bounded dimension. The analogue of this construction for finite metric spaces is obtained by assuming that the finite metric space is equipped with a map ρ to a reference space B, and replacing the connected component construction by the output of a clustering algorithm. A simple algorithm to use is single linkage hierarchical clustering, where one makes a choice of threshold based on a simple heuristic, such as the one found in [63]. Once this is done, one obtains a covering of the finite set X by the collection of all clusters constructed in each of the sets $\rho^{-1}(U_\alpha)$, and constructs the nerve complex. This construction is referred to as *Mapper*. Usually, B is chosen to be $\mathbb{R}^n$, for a small positive integer n, and therefore the reference map is determined by an n-tuple of real valued functions on the metric space X. The reference maps can be chosen in many ways, giving different views of the data. Some standard choices are density estimators, measures of centrality, coordinates in linear algebraic algorithms such as principal component analysis and multidimensional scaling ([42]), or individual coordinates when a metric space is obtained as a subspace of $\mathbb{R}^N$ for some N. The method has been used extensively in work on life sciences data sets, see for example [51], [53], [54], and [61].

8.5 Metrics on barcode space, and stability theorems

Since persistent homology is used to analyze data sets, and data sets are often noisy in the sense that one does not want to assign meaning to small changes, it is important to analyze the stability of persistent homology outputs to small changes in the underlying data. In order to do this, it is very useful to construct metrics on the output barcodes so that one can assert the continuity of the assignment of a barcode to a finite metric space or to the graph of a function. Informally, one wants to prove that small changes in the data give rise to small changes in the barcodes. Since small changes will often result in a change in the number of bars, it will be important to construct a set of all barcodes, on which we can impose a metric. The following is a natural construction. Let n be a positive integer, and let B_n denote the set of unordered n-tuples of closed intervals $[x, y]$, where we permit $x = y$, and require $x, y \geq 0$. It is understood that B_0 consists of a single point, namely the empty set of intervals. The set B_n can be described as the orbit space of the Σ_n-action on the set $\mathcal{I}^n$ that permutes coordinates, where $\mathcal{I}$ denotes the set of closed intervals. To consider all barcodes, we form

$$\mathfrak{B}_+ = \coprod_{n \geq 0} B_n$$

and define the full barcode space $\mathfrak{B}$ to be the quotient $\mathfrak{B}_+/\sim$, where $\sim$ is the equivalence relation generated by the relations

$$\{I_1, \dots, I_{k-1}, [x_i, x_i], I_{k+1}, \dots, I_n\} \sim \{I_1, \dots, I_{k-1}, I_{k+1}, \dots, I_n\}.$$

The idea is that intervals of length zero, which do not represent nonzero vector spaces, should be ignored. We will need to construct metrics on $\mathfrak{B}$ and to prove continuity results for these metrics.

8.5.1 Metrics on barcode space

The general idea for the construction of metrics on barcode space is to consider the set of all partial matchings between the intervals in the barcodes, assign a penalty to each such matching, and finally to minimize this penalty over the set of all matchings. Partial matchings are a little awkward, so for a pair of barcodes B_1 and B_2 one instead considers actual bijections $B_1^{'} = B_1 \cup Z \to B_2 \cup Z = B_2^{'}$, where Z is the set consisting of a countable number of copies of the interval of length zero $[x, x]$ for each $x \geq 0$. To start, one assigns a penalty $\pi([x_1, x_2], [y_1, y_2]) = ||(y_1 - x_1, y_2 - x_2)||_\infty$ for every pair of intervals including those of length zero. Next, given two barcodes B_1 and B_2, let $\mathcal{D}(B_1, B_2)$ denote the set of all bijections $\theta : B_1^{'} \to B_2^{'}$ for which $\pi(I, \theta(I)) \neq 0$ for only finitely many $I \in B_1^{'}$. Given a positive number p, we extend the penalty function π from individual intervals to barcodes by forming

$$W_p(B_1, B_2) = \inf_{\theta \in \mathcal{D}(B_1, B_2)} (\sum_{I \in B_1^{'}} \pi(I, \theta(I))^p)^{\frac{1}{p}}.$$

As usual, $p = \infty$ is interpreted as the L_∞ norm.

Definition 8.5.1. *For any $p > 0$ and also for $p = \infty$, we refer to $W_p(B_1, B_2)$ as the* p-Wasserstein distance. *For $p = \infty$, this distance is often referred to as the* bottleneck distance.

It is readily verified that under this definition, W_p defines a metric on $\mathfrak{B}$.

8.5.2 Stability theorems

The metrics defined in the previous section have stability theorems associated to them, that assert that the assignment of barcodes is a continuous process. There are two situations of interest.

1. Gromov has defined (see [41]) a metric d_{GH} on the set of all isometry classes of compact metric spaces, called the Gromov-Hausdorff metric. For any integer $k \geq 0$, one can view the assignment to any finite metric space its barcode as a map from the set of isometry classes of finite metric spaces to $\mathfrak{B}$, and one can ask about its continuity properties.

2. Let $f : X \to \mathbb{R}_+$ denote a continuous function. One can assign to each such f and each integer $i \geq 0$ the persistence vector space $\{H_i(f^{-1}([0, r]))\}_{r \geq 0}$. Under suitable situations (e.g. where X is a finite simplicial complex and f is linear on simplices, or where M is a compact manifold and f is smooth), one can show that the associated barcode is of

finite type. It is then an interesting question to ask about the continuity properties of this assignment, where one assigns various metrics to the set of functions on X.

There are theorems in both these cases. The following theorem demonstrates the continuity of the assignment of a k-dimensional barcode with coefficients in a field to a finite metric space, when the metric on the set of isometry classes of finite metric spaces is the Gromov-Hausdorff distance.

Theorem 8.5.2. [26] *For any two finite metric spaces X and Y and integer $k \geq 0$, let $B(X)$ and $B(Y)$ denote the k-dimensional barcodes for the Vietoris-Rips complexes of X and Y in a field K. Then*

$$W_\infty(B(X), B(Y)) \leq d_{GH}(X, Y).$$

There is a direct analogue for continuous real valued functions on a topological space. In order to state it, we need a pair of definitions.

Definition 8.5.3. *Let X be a topological space and f a continuous real valued function on X. A real number a is said to be a* homological critical value *of f if for some k and all sufficiently small $\varepsilon > 0$ the inclusion*

$$H_k(f^{-1}(-\infty, a - \varepsilon]) \to H_k(f^{-1}(-\infty, a + \varepsilon])$$

is not an isomorphism.

Definition 8.5.4. *Let X be a topological space and f be a continuous real-valued function. We say f is* tame *if there are finitely many homological critical values and all the homology groups $H_k(f^{-1}(-\infty, a]))$ with coefficients in a field K are finite dimensional.*

Remark 8.5.5. Tameness holds in a number of familiar situations.

1. Morse functions on compact smooth manifolds.

2. Functions on finite simplicial complexes that are linear on each simplex.

3. Morse functions on compact Whitney-stratified spaces.

The theorem is as follows.

Theorem 8.5.6. [28] *Let X be triangulable space, and suppose $f, g : X \to \mathbb{R}$ are tame continuous functions. Let $B(f)$ and $B(g)$ denote the barcodes attached to $\{H_k(f^{-1}(-\infty, r]; K))\}_r$ and $\{H_k(g^{-1}(-\infty, r]; K))\}_r$, respectively. Then*

$$W_\infty(B(f), B(g)) \leq ||f - g||_\infty.$$

The situation for the p-Wasserstein distances where $p \neq \infty$ is more complex. We will need constraints on the metric space as well as on the functions. For the space X, it is required to be a triangulable compact metric space, so X is homeomorphic to a finite simplicial complex. In addition, though, there is a requirement that the number of simplices required to construct a triangulation where the diameter of the simplices is less than a threshold r.

Specifically, for a given $r > 0$, we define $N(r)$ to be the minimal number of simplices in a triangulation of X for which each simplex has diameter $\leq r$. We will assume that $N(r)$ grows polynomially with r^{-1}, i.e. that there are constants C and m so that $N(r) \leq \frac{C}{r^m}$. It is easy to observe that this result holds for a finite simplicial complex X equipped with the Euclidean metric obtained by restricting along a piecewise linear embedding $X \hookrightarrow \mathbb{R}^n$, as well as for a compact Riemannian manifold. In [29], it is proved that any metric space satisfying this condition also satisfies a homological condition. To state the homological condition, given a barcode $B = \{[x_1, y_1], \ldots, [x_n, y_n]\}$, we define $P_k(B)$ to be the sum

$$P_k(B) = \sum_i (y_i - x_i)^k.$$

Lemma 8.5.7. [29] *Suppose that X is as above, and that k is a nonnegative real number. Then there is a constant C_X so that $P_k(B(f)) \leq C_X$ for every tame function $f : X \to \mathbb{R}$ with Lipschitz constant $L(f) \leq 1$, where $B(f)$ is defined as in Theorem 8.5.6.*

Definition 8.5.8. *When the conclusion of Lemma 8.5.7 holds for a metric space X and real number $k \geq 1$, we say that X* implies bounded degree-k persistence.

The theorem is now as follows.

Theorem 8.5.9. [29] *Let X be a triangulable metric space that implies bounded degree-k persistence for some real number $k \geq 1$, and let $f, g : X \to \mathbb{R}$ be two tame Lipschitz functions. Let C_X be the constant in Definition 8.5.7, and let $L(f)$ and $L(g)$ denote the Lipschitz constants for f and g respectively. Then*

$$W_p(f, g) \leq C^{\frac{1}{p}} \cdot ||f - g||_\infty^{1-\frac{k}{p}}$$

for all $p \geq k$, where $C = C_X max\{L(f)^k, L(g)^k\}$.

8.6 Tree-like metric spaces

Persistent homology gives ways of assessing the shape of a finite metric space. One situation where this is very useful is in problems in evolution. The notion that there is a "tree of life" is a very old one which actually predates Darwin. Different organisms of the same type have attached to them sequences of the same length in a genetic alphabet A. Therefore any set of organisms produces a subset of the set of sequences of fixed length in A. One can assign a metric to the space of all such sequences using Hamming distances or variants thereof. The notion that there is a tree of the various organisms within a fixed type can be restated in mathematical terms as stating that the space $\mathcal{S}(A)$ corresponding to the organisms in the family is well modeled by a *tree-like metric space*, i.e. a metric space that is isometric to the set of nodes in a tree, possibly with weighted edges, equipped with the distance function that assigns to each pair of vertices of the tree the length of the shortest edge path between them. This approximability could be called the *phylogenetic hypothesis* for the particular class of organisms. Testing this hypothesis for particular genetic data sets has usually been done by attempting to fit trees to a given metric space and attempting to assess how well the approximation fits. Given the persistent homology construction, one is tempted to develop criteria attached to the barcodes that can distinguish between tree-like and non-tree-like metric spaces. The following theorem, proved in [25] gives such a criterion.

Theorem 8.6.1. *Let X be a finite tree-like metric space. Then the k-dimensional persistent homology of X vanishes for $k > 0$.*

Remark 8.6.2. This theorem was proved in the context of a study of data sets of viral sequences. In that paper it was also shown that representative cycles for generators of persistent homology in positive degrees gave important clues to the mechanism of the failure of the phylogenetic hypothesis.

8.7 Persistence and feature generation

8.7.1 Introduction

The output of persistent homology is an interesting data type, consisting as it does of finite collections of intervals. When humans are directly interpreting barcodes, they are typically able to interpret barcodes directly from this description. However, there is a whole class of problems where computers are used to "interpret" the barcodes. For example, suppose that we have a database of complex molecules. Each molecule is given as a collection of atoms and bonds, and the bonds may be equipped with lengths. The set of atoms in a molecule can be endowed with a metric by considering the edge-path distance using the lengths of the bonds as the lengths of the edges. What we have is now a data set in which each of the *data points* is a finite metric space, and therefore possesses a barcode. If there are many molecules, we cannot hope to interpret these barcodes by eye, and must therefore allow a computer to deal with them. Machine learning algorithms are generally not well equipped to deal with data points described as sets, and it is therefore important to encode them somehow as vectors, which are the natural input to such algorithms. In this section we will describe three distinct methods for "vectorizing" barcodes, i.e. for creating coordinates on the set of barcodes. Specifically, we will define coordinates on the space $\mathfrak{B}$ constructed in Section 8.5.

8.7.2 Algebraic functions

This method proceeds from the observation that the sets B_n can be viewed as subsets of a real algebraic variety. The set $\mathcal{I}$ embeds as a subset of the two-dimensional affine space $\mathbb{A}^2(\mathbb{R})$, and is defined by the inequalities $x, y \geq 0$ and $y \geq x$ for an interval coordinatized by the pair (x, y). Consequently, we have an embedding

$$\mathcal{I}^n \hookrightarrow \mathbb{A}^2(\mathbb{R})^n \cong \mathbb{A}^{2n}(\mathbb{R})$$

and it is equivariant with respect to the permutation actions on $\mathcal{I}^n$ and $\mathbb{A}^2(\mathbb{R})^n$. Under the identification $\mathbb{A}^2(\mathbb{R})^n \cong \mathbb{A}^{2n}(\mathbb{R})$, with $\mathbb{A}^{2n}(\mathbb{R})$ coordinatized using coordinates $(x_1, \ldots, x_n, y_1, \ldots, y_n)$, the corresponding action simply permutes the x_i's and y_i's among themselves. It is a standard result in algebraic geometry (see [52]) that for any action of a finite group on an affine algebraic variety (over $\mathbb{R}$ in this case), there is an orbit variety, whose affine coordinate ring is the ring of invariants of the group action. It is easy to verify that in this case, the orbit set of the action on the closed real points of $\mathbb{A}^{2n}(\mathbb{R})$ is exactly the symmetric product $Sp^n(\mathbb{R}^2)$. Since $B_n \subseteq Sp^n(\mathbb{R}^2)$, elements in the affine coordinate ring of the orbit variety can be regarded as functions on B_n, so we now have an algebra of functions $\mathcal{A}_n$ on B_n. This means that we can describe functions on the sets of barcodes with exactly n intervals. What one really wants is a ring of functions on all of $\mathfrak{B}$. In order to construct such a ring, we observe that $\mathfrak{B}$ can be described as a quotient of the direct limit of the system

$$B_0 \to B_1 \to B_2 \to \cdots \tag{8.7.1}$$

where the inclusion $B_n \to B_{n+1}$ is given by

$$\{[x_1, y_1], [x_2, y_2], \ldots, [x_n, y_n]\} \to \{[x_1, y_1], [x_2, y_2], \ldots, [x_n, y_n], [0, 0]\}.$$

There is a corresponding direct system of affine schemes

$$Spec(\mathcal{A}_0) \to Spec(\mathcal{A}_1) \to Spec(\mathcal{A}_2) \to \cdots \tag{8.7.2}$$

which is compatible with the system (8.7.1) above. The colimit of the system (8.7.2) is an affine scheme, whose affine coordinate ring is the inverse limit of the system

$$\mathcal{A}_0 \longleftarrow \mathcal{A}_1 \longleftarrow \mathcal{A}_2 \longleftarrow \cdots$$

which we will denote $\overline{\mathcal{A}}$. The ring $\overline{\mathcal{A}}$ can be analyzed, but is a bit too complicated to be used in applications. To define a smaller subring, we note that $Spec(\overline{\mathcal{A}})$ is equipped with an action by the algebraic group $\mathbb{G}_m$, and we can define a subring $\overline{\mathcal{A}}^{fin}$ to consist of all those functions f so that all the translates of f under the $\mathbb{G}_m$-action span a finite dimensional vector subspace within $\overline{\mathcal{A}}$. The ring $\overline{\mathcal{A}}^{fin}$ is actually a graded ring, since the $\mathbb{G}_m$-action determines a grading on it. Within $\overline{\mathcal{A}}^{fin}$ we define a subring $\mathcal{A}^{fin}$ that consists of all elements of $\overline{\mathcal{A}}^{fin}$ that respect the equivalence relation $\sim$. The main result of [3] is the following.

Theorem 8.7.1. *The rings in question have the following properties.*

1. *The ring $\overline{\mathcal{A}}^{fin}$ has the structure*

$$\overline{\mathcal{A}}^{fin} \cong \mathbb{R}[x_{i,j}; 0 \leq i, 0 \leq j, \ \textit{and}\ i + j > 0].$$

2. *The subring $\mathcal{A}^{fin}$ is identified with*

$$\mathbb{R}[x_{i,j}; 0 < i, 0 \leq j, \ \textit{and}\ i + j > 0].$$

3. The element $x_{i,j}$ is the function given on a barcode $\{[x_1, y_1], \ldots, [x_n, y_n]\}$ by

$$\sum_{s=1}^{n} (y_s - x_s)^i (y_s + x_s)^j.$$

4. The elements of $\mathcal{A}^{fin}$ separate points in $\mathfrak{B}$.

5. The ring $\mathcal{A}^{fin}$ injects into the ring of functions on $\mathfrak{B}$.

We remark that these functions are not continuous for the bottleneck distance on $\mathfrak{B}$. In [45], a tropical version of this work is presented, which gives functions that are continuous for the bottleneck distance. For the $p < \infty$ situation, the functions are continuous for the p-Wasserstein distance if one assigns $\mathfrak{B}$ the direct limit topology associated to the filtration of $\mathfrak{B}$ by the images of the spaces B_n, defined in Section 8.5.1.

8.7.3 Persistence landscapes

Persistence landscapes were introduced in [12] as another vectorization of barcodes. The vectorization in this case consists of an embedding of the set $\mathfrak{B}$ in a set of sequences of functions on the real line. Let (a, b) denote a pair of elements of $\mathbb{R}$ with $a \leq b$. Then we define a function $f_{a,b}(t)$ on the real line by setting

$$f_{(a,b)}(t) = \min(t - a, b - t)_+$$

where $c_+ = \max(c, 0)$. A quick analysis of $f_{(a,b)}$ shows that it is zero for $t \leq a$ and $t \geq b$, that on the interval $[a, \frac{a+b}{2}]$ it is equal to the graph of a line of slope 1 including the point $(a, 0)$, and on the interval $[\frac{a+b}{2}, b]$ it is the graph of a line of slope -1 including the point $(b, 0)$. The shape of the graph is that of a pyramid, as in Figure 8.7.1.

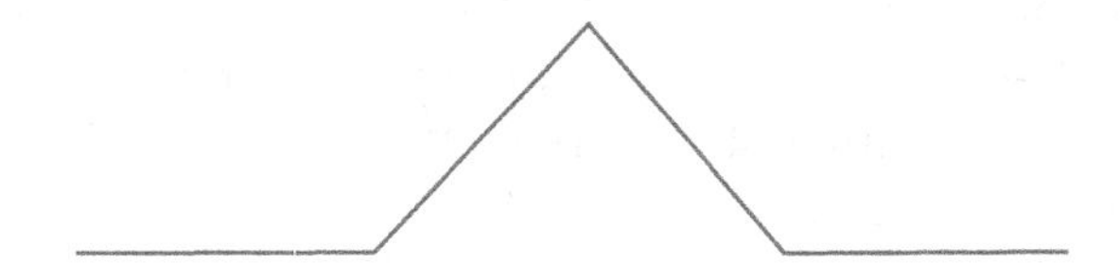

FIGURE 8.7.1. Graph of $f_{a,b}$

Remark 8.7.2. Note that for $a = b$, $f_{(a,b)}(t) \equiv 0$.

Given a persistence barcode $B = \{(a_1, b_1), \ldots, (a_n, b_n)\}$, we define a family of functions $\lambda_k(t)$ parametrized by a positive integer k. For $k = 1$, we let $\lambda_1(B)(t)$ denote the maximum of all the values $f_{(a_i,b_i)}(t)$ over all i. For $k > 1$, we set $\lambda_k(B)(t)$ equal to the k-th largest value occurring in the set $\{f_{(a_i,b_i)}(t)\}_i$. The family of functions $\{\lambda_k(B)(t)\}_{k>0}$ is defined to be the persistence landscape of the barcode B. It is clear that

$$\lambda_k(B)(t) \geq 0$$

that

$$\lambda_k(B)(t) \geq \lambda_{k+1}(B)(t)$$

and that

$$\lambda_k(B)(l) = 0 \text{ for } k > n.$$

In [12] it is also proved that each function $\lambda_k(B)$ is 1-Lipschitz, i.e. that

$$|\lambda_k(B)(t) - \lambda_k B)(t')| \leq |t - t'|.$$

To summarize, the persistence landscape lies in the vector space $\mathfrak{F}$ of real valued functions on $\mathbb{N} \times \mathbb{R}$, and it follows directly from Remark 8.7.2 that the definition gives us a function $PL : \mathfrak{B} \to \mathfrak{F}$.

One extremely useful fact about the persistence landscape is that it is compatible with the various distances assigned to barcode spaces. We let $\mathfrak{F}_p \subset \mathfrak{F}$ denote the space of functions with finite L_p-norm $|| \; ||_p$. It is clear that the function PL takes values in $\mathfrak{F}_p$ for all p. Recall the definition of the p-Wasserstein distance W_p between barcodes from Section 8.5.1. Bubenik now proves the following in [12].

Theorem 8.7.3. *The function $d_p(B, B') = ||PL(B) - PL(B')||_p$ is a metric on $\mathfrak{B}$. The two metrics W_{p+1} and d_p generate the same topology on $\mathfrak{B}$. It follows that the map PL is continuous when $\mathfrak{B}$ is equipped with the metric W_{p+1} and $\mathfrak{F}_p$ is equipped with the metric associated with the L_p norm.*

Remark 8.7.4. Bubenik also provides explicit inequalities involving the two metrics in [12]

Bubenik also proves the following continuity theorem.

Theorem 8.7.5. *The map PL is 1-Lipschitz from $\mathfrak{B}$ equipped with the bottleneck distance to the space of persistent landscapes equipped with the sup norm distance. This is equivalent to the algebraic statement*

$$|\lambda_k(B)(t) - \lambda_k(B')(t)| \leq W_\infty(B, B').$$

Remark 8.7.6. The map PL separates points.

8.7.4 Persistence images

There is another approach that proceeds by treating a barcode, recoded as a persistence diagram, as a collection of point masses and then smoothing the corresponding measure to produce and image, which is finally discretized by selecting pixels and assigning each pixel the average value of the function on a box surrounding that pixel. It is described in [2]. The detailed description is given in several steps. The input is a persistence diagram (it is more natural to use the persistence diagram view in this case), a collection of points $B = \{(x_1, y_1), \ldots, (x_n, y_n)\} \subset \mathbb{R}^2$. We assume we are given a function $\phi : \mathbb{R}^2 \times \mathbb{R}^2 \to \mathbb{R}$ so that (a) $\phi_u = \phi(u, -)$ is a probability distribution on $\mathbb{R}^2$ for each $u \in \mathbb{R}^2$ and (b) the mean of ϕ_u is u. A standard choice is that of a spherically symmetric Gaussian with mean u and a fixed variance σ. We also assume we are given a continuous and piecewise differentiable nonnegative weighting function $f : \mathbb{R}^2 \to \mathbb{R}$ that is zero along the x-axis.

- Apply the coordinate change $(x,y) \to (x, y-x) = (\xi, \eta)$ to $\mathbb{R}^2$, to obtain the new set of points $\mathcal{B} = \{(\xi_1, \eta_1) \ldots, (\xi_n, \eta_n)\}$ of the same cardinality in $\mathbb{R}^2$. The points that correspond to short intervals are now all located near the ξ-axis. The ξ-axis itself corresponds to intervals of length zero.

- Construct a new function $\rho_{\mathcal{B}}(z) : \mathbb{R}^2 \to \mathbb{R}$, called the *persistence surface of* $\mathcal{B}$, defined by

$$\rho(z) = \sum_{i=1}^{n} f(\xi_i, \eta_i)\phi((\xi_i, \eta_i), z).$$

  Notice that ρ vanishes on the x-axis.

- To construct a finite dimensional representation, we first assume that the persistence diagrams we will be dealing with will always lie in a bounded region in $\mathbb{R}^2$, and divide a box containing that region into a square grid. Construct the vector with coordinates in one to one correspondence with the squares of the grid, and assign the entry corresponding to a square to be the integral over that square of $\rho_{\mathcal{B}}$.

The following is proved in [2].

Theorem 8.7.7. *The map $PI : \mathfrak{B} \to \mathbb{R}^N$ that assigns to a persistence diagram a vector using the above procedure is continuous when the metric on $\mathfrak{B}$ is the* 1*-Wasserstein distance.*

Remark 8.7.8. In [2], estimates proving this result are given both in the case of a general choice of ϕ and the special case where ϕ is given by Gaussian distributions with fixed variance. Of course, the estimates in the latter case are stronger.

8.8 Generalized persistence

8.8.1 Introduction

Persistent homology operates on functors F from the category $\mathbb{R}_+$ to simplicial complexes, by composing them with the homology functor to obtain a persistence vector space. It is useful to consider other parameter categories for the diagrams, which can also help clarify the structure of data sets. There are at least two such constructions that have been discussed. The first is *zig-zag* persistence, introduced in [19], and the second is *multidimensional persistence*, discussed for example in [22]. The first is designed to study the relationship between homology of constructions that are not nested within each other, such as distinct samples from a given space, and the second provides invariants of situations where it is natural to study filtrations of spaces involving more than one parameter, such as filtering by both the scale parameter R and a measure of density. We describe both extensions of the standard persistent homology methods.

8.8.2 Zig-zag persistence

Consider the triangulation of $\mathbb{R}$ whose vertices are the integers. The set of vertices of the barycentric subdivision of this simplicial complex is equipped with a partial ordering, by recognizing that its elements are in one-to-one correspondence with the simplices in the original triangulation, and that that set is equipped with a partial ordering by treating it as a subset of the power set of $\mathbb{R}$. In concrete terms, we may view it as in one to one correspondence with the set all integers and half integers, with every integer n being less than or equal to the elements $n \pm \frac{1}{2}$. We'll refer to this partially ordered set as $\mathfrak{Z}$. A partially ordered set (and its corresponding category) is said to be connected if any two objects can be connected by a zig-zag sequence of morphisms. Connected partially ordered subsets of $\mathfrak{Z}$ are always determined by a pair (x, y), where x and y are both integers or half integers, via the rule that assigns to the pair (x, y) the collection of objects z for which $x \leq z \leq y$ (in the total ordering on $\mathbb{R}$).

Definition 8.8.1. *For any category $\underline{C}$, a* zig-zag persistence object *in $\underline{C}$ is a functor from a connected subcategory of $\mathfrak{Z}$ to $\underline{C}$. Suppose further that $\underline{C}$ is equipped with an object c_0 that is both initial and terminal. Then for any connected subcategory $\mathfrak{Z}_0 \subseteq \mathfrak{Z}$ and object $c \in \underline{C}$, we define the* interval object *for $\mathfrak{Z}_0$ and c to be the functor $F = F_{\mathfrak{Z}_0,c} : \mathfrak{Z} \to \underline{C}$ defined on objects by $F(x) = c$ for all $x \in \mathfrak{Z}_0$ and $F(x) = c_0$ for any $x \notin \mathfrak{Z}_0$, and on morphisms by $F(x \leq y) = id_c$ whenever $x, y \in \mathfrak{Z}_0$. The behavior on morphisms into or out of $\mathfrak{Z}_0$ is uniquely determined by the fact that c_0 is both initial and terminal. When $\underline{C}$ is the category of vector spaces over a field K, then it is understood that c will be chosen to be a one-dimensional vector space over K.*

Given a zig-zag persistence object in the category of simplicial complexes, we may apply a homology functor $H_i(-; A)$ for an abelian group A to obtain a zig-zag persistence object in the category of abelian groups. The category of abelian groups has the zero object as an object that is both initial and terminal, and so the notion of interval objects makes sense. Of course if $A = K$, where K is a field, then we obtain a zig-zag persistence vector space. It turns out that they can be classified up to isomorphism.

Theorem 8.8.2. (P. Gabriel, [37]) *Let $\mathfrak{Z}' \subset \mathfrak{Z}$ be any finite connected subcategory, and let F denote any zig-zag persistence object in the category of finite dimensional vector spaces over a field K defined on $\mathfrak{Z}'$. Then there is a finite direct sum decomposition*

$$F \cong \bigoplus_i F_i$$

where each F_i is an interval object for $\mathfrak{Z}_0$ and K, where $\mathfrak{Z}_0 \subseteq \mathfrak{Z}'$ is a connected subcategory of $\mathfrak{Z}'$. Moreover, the sum is unique up to isomorphism and reordering of the sum.

Corollary 8.8.3. *The classification of zig-zag persistence vector spaces based on a connected subcategory $\mathfrak{Z}$ is given by barcodes where the intervals have endpoints integers or half integers. We'll refer to these barcodes as the zig-zag persistence barcodes of the zig-zag persistence vector space.*

This theory with applications is discussed in [19] and [20]. Here are some particular situations in which this construction can be used.

1. **Samples:** Given a finite metric space, one can ask to what extent the persistent homology is captured on smaller samples from the data set. For example, suppose that we have taken a very large uniform sample X from the circle, and equip them with a metric by restricting the intrinsic metric (say) on the circle to these points. We will then with high probability find that the one-dimensional persistence barcode for the Vietoris-Rips construction on X will contain one long bar and many much shorter ones. Supposing that we do not actually know that the sample is coming from a circle, but simply observe that we obtain a one-dimensional barcode with one long bar and many shorter ones. A hypothesis suggested by this observation is that the data is obtained by sampling from a space with the homotopy type of the circle, but we may wonder if instead it has somehow appeared "by accident". One way to provide confirmation of our hypothesis would be to observe that we obtain the same result for various subsamples of our space, and that they are compatible in an appropriate sense. Zig-zag persistence provides a way for carrying this out. We suppose that we have chosen samples $X_1, \ldots, X_n \subseteq X$, and create a persistence object in the category of finite metric spaces and distance non-increasing maps

If we apply $\mathcal{V}(-, r)$ for a fixed choice of r, guided by the beginning and endpoints of the observed long bar in the barcode for X, we obtain a persistence object in the category of simplicial complexes. If we further apply $H_1(-, K)$ for a field K, we obtain a zig-zag persistence K-vector space. By the classification Theorem 8.8.2 above, we obtain a decomposition of the resulting zig-zag persistence K-vector space. The interpretation of the informal idea that each sample has a one-dimensional homology class and that they are consistent is the presence of an interval K-vector space for a relatively long interval within the set $\{1, \frac{3}{2}, 2, \ldots n - \frac{1}{2}, n\}$, or equivalently a relatively long bar in the zig-zag persistence barcode.

2. **Functions on spaces:** Suppose that we have a topological space X equipped with a continuous map $f : X \to \mathbb{R}_+$. Then we have the ordinary persistence K-vector spaces $\{H_i(f^{-1}([0, r], K)\}_r$, which encode information about the evolution of the homology of the sublevel sets of f as r increases. However, one might be interested in gaining information about approximations to *level sets* instead. They can provide more useful invariants in a number of cases, and are approachable through zig-zag persistent homology. We construct a zig-zag diagram of topological spaces as follows.

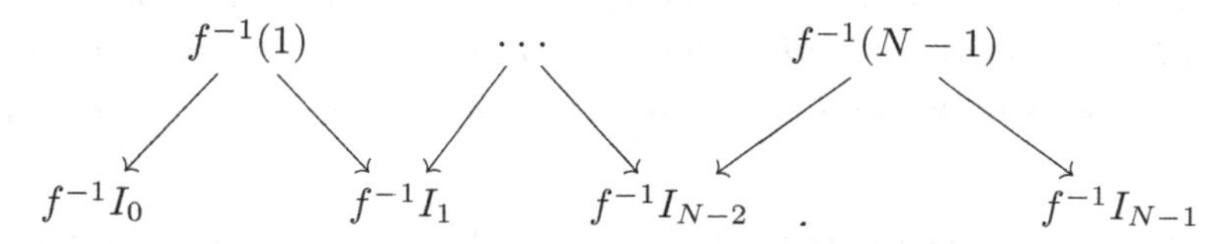

where $I_k = [k, k+1]$ for all k. Again we can apply $H_i(-, K)$ to this diagram, an obtain a zig-zag persistence barcode. It contains information about how the spaces $f^{-1}[k, k+1]$ change as k changes, and how they assemble together. This situation is studied in [20].

3. **Witness complexes:** One of the problems with the witness complex is that we have very little theory about the extent to which it reflects accurately the persistent homology of the underlying metric space. A related problem is that there is no direct relationship between the construction for two different landmark sets. Even if $\mathcal{L}_1 \subseteq \mathcal{L}_2$, there are no maps directly relating $W(X, \mathcal{L}_1, R)$ and $W(X, \mathcal{L}_2, R)$ for two different landmark sets. One approach is to attempt to assess in some manner how consistent the results of the constructions based on $\mathcal{L}_1$ and $\mathcal{L}_2$ are. It turns out that given two landmark sets $\mathcal{L}_1$ and $\mathcal{L}_2$, it is possible to construct an intermediate bivariant construction $W(X, \{\mathcal{L}_1, \mathcal{L}_2\}, R)$ for which there is an evident diagram

$$W(X, \mathcal{L}_1, R) \longleftarrow W(X, \{\mathcal{L}_1, \mathcal{L}_2\}, R) \longrightarrow W(X, \mathcal{L}_2, R).$$

This is itself a short zig-zag diagram of length three, but if we have landmark sets $\{\mathcal{L}_i\}_{i=0}^N$ we can clearly construct a longer diagram that includes the bivariant constructions $W(X, \{\mathcal{L}_i, \mathcal{L}_{i+1}\}, R)$ for $i = 0, \ldots, N-1$. The construction is quite simple. Its vertex set is $\mathcal{L}_1 \times \mathcal{L}_2$, and a subset

$$\{l_1^1 \times l_2^1, l_1^2 \times l_2^2, \ldots, l_1^k \times l_2^k\} \subseteq \mathcal{L}_1 \times \mathcal{L}_2$$

spans a simplex in $W(X, \{\mathcal{L}_i, \mathcal{L}_{i+1}\}, R)$ if and only if there is a point $x \in X$ so that x is a witness for $\{l_1^1, l_1^2, \ldots, l_1^k\}$ and $\{l_2^1, l_2^2, \ldots, l_2^k\}$ in the complexes $W(X, \mathcal{L}_1, R)$ and $W(X, \mathcal{L}_1, R)$, respectively. The projections to $W(X, \mathcal{L}_1, R)$ and $W(X, \mathcal{L}_1, R)$ are defined in the evident way.

8.8.3 Multidimensional persistence

There are many situations where it can be useful to introduce families of spaces varying with more than one real parameter. For example, in any kind of topological analysis of data sets, it usually is the case that if one considers the persistent homology of the entire data set, the presence of outliers means that we do not typically obtain the"right homology". For example, if we have data sampled from the unit circle, but a small number of points sprinkled throughout the unit disc, then persistent homology will end up reflecting the homology of the disc rather than that of the circle. This is often circumvented by selecting only the points of sufficient density, as measured by a density estimator, since outliers will typically have very low density. The question then becomes, though, how to choose the threshold for density. Also, it turns out that in general, there will be variation through different topologies as one changes the threshold. The solution to this problem is to attempt to study all thresholds at once, just as we do when considering the scale parameter in the Vietoris-Rips construction. This leads us to the following definition.

Definition 8.8.4. *Let $\underline{C}$ be a category. Then by an n-dimensional persistence object in $\underline{C}$ we mean a functor $\mathbb{R}_+^n \to \underline{C}$, where $\mathbb{R}_+^n$ denotes the n-fold product of copies of the category $\mathbb{R}_+$.*

Example 8.8.5. Let X be any metric space, and suppose that X is equipped with a function $f : X \to \mathbb{R}$. It might be a density estimator, but it might also be a measure of centrality. Then if $X[s]$ denotes the subset $\{x \in X | f(x) \leq s\}$, we obtain family of spaces $\mathcal{V}(X[s], r)$, and by applying homology with coefficients in a field K, we obtain a 2-dimensional persistence K-vector space parametrized by the pair (r, s). While density is used as described above to remove outliers or noise, the case of a centrality measure allows one to capture the presence of the analogues of ends in a finite metric space.

Example 8.8.6. Multidimensional persistent homology can also be used to capture geometric information that is not strictly topological (see [23]). For example, given a Riemannian manifold, one can compute Gaussian curvature at each point and filter by that quantity. Considering the entire manifold, one can obtain a one-dimensional persistence vector space by applying homology over a field K. For computational purposes, though, we would need to deal with a sample and use a second parameter, namely the scale parameter in a Vietoris-Rips complex. This kind of analysis can for example be used to distinguish between various ellipsoids.

The equivalence of categories in Proposition 8.3.1 has the following straightforward analogue.

Proposition 8.8.7. *The category of n-dimensional persistence vector spaces over K is equivalent to the category of $\mathbb{R}^n_+$-graded modules over the $\mathbb{R}^n_+$-graded ring $K[\mathbb{R}^n_+]$. The analogous result with $\mathbb{R}^n_+$ replaced by $\mathbb{N}^n$ also holds.*

Although useful, this result does not give us a classification of multidimensional persistence K-vector spaces analogous to the barcode classification that works in the one-dimensional case. The reason can be understood as analogous to the commutative algebraic situation, where finitely generated $K[x]$-modules can be classified because $K[x]$ is a principal ideal domain, but $K[x_1, \ldots, x_n]$-modules cannot be classified for $n \geq 2$. In a sense it is provable that there is no classification strictly analogous to the $n = 1$ case, because the classification in the case $n \geq 2$ depends on the structure of the field K, as is demonstrated in [22]. This is not the case when $n = 1$, because the classification is always by barcodes, and that is independent of the structure of K.

Although there is no complete classification of multidimensional persistence vector spaces, there exist interesting invariants. For any k-dimensional persistence K-vector space $\{V_x\}_{x \in \mathbb{R}^k}$, and pair of points $x, y \in \mathbb{R}^k$, with $x \leq y$ in the natural partial ordering on $\mathbb{R}^k$, we can define $r(x, y)$ to be the rank of the linear transformation $V_x \to V_y$. We extend the definition to all pairs $x, y \in \mathbb{R}^k$ by setting $r(x, y) = 0$ when x is not less than or equal to y. The function r is therefore defined on $\mathbb{R}^k \times \mathbb{R}^k$, and we refer to it as the *rank invariant*. In the case $k = 1$, the rank invariant is complete, in that it differentiates between distinct barcodes. There is an analogue for multidimensional persistence K-vector spaces of the results described in Section 8.7.2. To state it, we first define an analogue of barcodes. By a *cube* in $\mathbb{R}^k$, we mean a set of the form $I_1 \times I_2 \times \cdots \times I_k$, where each I_s is an interval $[a_s, b_s]$, and write $C(a_1, \ldots, a_k, b_1, \ldots, b_k)$ for this cube. We write $\mathcal{C}_k$ for the set of all k-dimensional cubes. There is a straightforward analogue to the space $\mathfrak{B}$ defined in Section 8.7.1. We first define $B_n = Sp^n(\mathcal{C}_k)$, and define $\mathfrak{B}(k)_+ = \coprod Sp^n(\mathcal{C}_k)$. Next, we say a cube $C(a_1, \ldots, a_k, b_1, \ldots, b_k)$

is *negligible* if $a_i = b_i$ for some i, and define an equivalence relation $\sim$ on $\mathfrak{B}(k)_+$ to be the equivalence relation generated by the relation

$$\{C_1, C_2, \ldots C_n\} \sim \{C_1, \ldots, C_{n-1}\} \text{ for all negligible } C_n.$$

We define $\mathfrak{B}(k)$ to be $\mathfrak{B}(k)_+/\sim$. Every equivalence class under $\sim$ has a unique minimal representative consisting entirely of non-negligible cubes. Let $\mathcal{M}(k)$ denote the set of isomorphism classes of k-dimensional persistence K-vector spaces. For each cube $C = C(a_1, \ldots, a_k, b_1, \ldots, b_k)$ we let $\mu(C)$ denote the isomorphism class of the the k-dimensional persistence K-vector space $\{V_{\vec{x}}\}_{\vec{x} \in \mathbb{R}^k_+}$ for which $V_{\vec{x}} = K$ whenever $\vec{x} \in C$, $V_{\vec{x}} = \{0\}$ whenever $\vec{x} \notin C$, and for which all induced morphisms $V_{\vec{x}} \to V_{\vec{y}}$ for $\vec{x} \leq \vec{y}$ and $\vec{x}, \vec{y} \in C$ are equal to the identity. There is an obvious map $\theta : \mathfrak{B}(k) \to \mathcal{M}(k)$ which assigns to each minimal representative $\{C_1, \ldots, C_n\}$ the direct sum $\oplus_i \mu(C_i)$.

Theorem 8.8.8. [64] *The constructions above satisfy the following properties.*

1. *The set $\mathfrak{B}(k)$ is a subset of the set of real points of an affine scheme Spec(A).*
2. *The ring A is complicated, but there is a $\mathbb{G}_m$-action on Spec(A) that allows us to define a more manageable subring $A^{fin} \subseteq A$.*
3. *A^{fin} is isomorphic to the polynomial ring $\mathbb{R}[x_{\vec{a},\vec{b}}]$ where $\vec{a}$ and $\vec{b}$ are k-vectors of integers for which $a_i \geq 1$ and $b_i \geq 0$ for all i.*
4. *The ring A^{fin} separates points in $\mathfrak{B}(k)$, and maps injectively to the ring of all real valued functions on $\mathfrak{B}(k)$.*
5. *There is a natural lift of the ring homomorphism $A^{fin} \to F(\mathfrak{B}(k), \mathbb{R})$ along θ to a ring homomorphism $j : A^{fin} \to F(\mathcal{M}(k), \mathbb{R})$. $F(X, \mathbb{R})$ denotes the ring of real-valued functions on a set X.*
6. *For any $\alpha \in A^{fin}$, the function $j(\alpha) : \mathcal{M}(k) \to \mathbb{R}$ factors through the rank invariant. For any two elements $X, Y \in \mathcal{M}(k)$, if for all $\alpha \in A^{fin}$, $j(\alpha)(X) = j(\alpha)(Y)$, then the rank invariants of X and Y are equal.*

This result gives one approach to the study of multidimensional persistence. There is a great deal of other work on this topic. The paper [24] deals with computing persistent homology using commutative algebra techniques, via the equivalence of categories from Proposition 8.8.7. They demonstrate that the multigrading yields significant simplification. In [62] and [24], an algebraic framework is constructed for dealing with the fact that there is alway noise in the applications to data analysis. In [14], multidimensional persistence is studied by examining the family of all one dimensional persistence modules obtained by considering lines with varying angles in the persistence domain. M. Lesnick in [49] has defined metric properties of the set of isomorphism classes of multidimensional persistence vector spaces, and proven uniqueness results for them. Finally, in [50], software is developed for visualization and interrogation of two-dimensional persistence vector spaces.

One generalization that has not been studied yet is to multidimensional persistence where some of the persistence directions might be "zig-zag" directions rather than ordinary persistence directions. Formally, this would mean functors from the categories of the form $\mathbb{R}_+^m \times \mathcal{3}^n$. This would be very useful in a number of situations. For example, in the zig-zag constructions discussed in Section 8.8 for samples and for witness complexes, we were forced to choose a threshold for the scale parameter. If we had a way of representing functors from $\mathbb{R}_+ \times \mathcal{3}$ to vector spaces, we would not be forced to make this selection.

8.9 Coverage and evasion problems

There is an interesting set of technologies used for sensing of various kinds called *sensor nets*. A sensor net consists of a collection of sensors distributed throughout a domain. The sensors are very primitive in the sense that they are only capable of sensing the presence of an intruder or of another sensor within a fixed detection radius R. We also assume each sensor is given an identifying label or number, and that other sensors can sense that identifier when they are within the radius R of each other. One problem of interest is whether or not the balls of radius R cover the domain, and it does not have an immediate solution due to the fact that the positions of the sensors are not available. V. de Silva, and R. Ghrist have developed a very interesting method for addressing this problem based on persistent homology (see [31] and [32]). The rough idea is as follows. Suppose that one has a domain D in the plane, with a connected and compact curve boundary ∂D, that one has a collection of points $\{v_i\}_{i \in I}$ in the region, one for each sensor, and that one knows in some way that ∂D is covered by the open balls $B_R(v_i)$. Let $\mathcal{U}$ denote the family of open subsets $\{B_R(v_i) \cap D\}_{i \in I}$ of D, and let $\mathcal{U}^\partial$ denote the covering $\{B_R(v_i) \cap \partial D\}_{i \in I}$ of ∂D. Suppose further that one knows that spaces

$$B_R(v_{i_1}) \cap \cdots \cap B_R(v_{i_s}) \cap D \quad \text{and} \quad B_R(v_{i_1}) \cap \cdots \cap B_R(v_{i_s}) \cap \partial D$$

are all either contractible or empty, for all choices of subsets $\{i_1, \ldots, i_s\} \subseteq I$. The conditions ensure that D is homotopy equivalent to the nerve of $\mathcal{U}$, and that ∂D is homotopy equivalent to the nerve of the covering $\mathcal{U}^\partial$, as a consequence of the nerve theorem. It further ensures that the pair $(N.\mathcal{U}, N.\mathcal{U}^\partial)$ is equivalent to the pair $(D, \partial D)$. For any field, the relative group $H_2(D, \partial D; K) \cong K$, since D is a connected orientable manifold with boundary ∂D, and it follows that the relative group $H_2(N.\mathcal{U}, N.\mathcal{U}^\partial; K) \cong K$. On the other hand, suppose that the sets $B_R(v_i)$ do not cover all of D, and let $D_0 \subseteq D$ denote the union

$$\bigcup_i B_r(v_i) \cap D.$$

The space D_0 is a non-compact manifold with boundary, and consequently the relative group $H_2(D_0, \partial D; K) \cong 0$. As above, it will follows that

$$H_2(N.\mathcal{U}, N.\mathcal{U}^\partial; K) \cong 0.$$

Consequently, the simplicial complex of the nerve of the covering, which can be computed using the information available from the sensors, determines whether or not we have a covering based on its simplicial homology. The conditions on the coverings given above are of course impossible to verify, but de Silva and Ghrist are able to formulate a persistent homology condition that is a reasonable substitute, and which gives a homological criterion in terms of a 2-dimensional persistent homology group which is sufficient to guarantee coverage.

In order to understand the result in [32], we first observe that the information from the sensors do not give us access to the Čech complex, since we have no way of determining the intersection of balls without precise knowledge of the distances between their centers. However, we do have access to the Vietoris-Rips complex, since for any pair of points, we can tell whether or not they are within a distance R, where R is the detection radius. We also have the comparison results for the Vietoris-Rips complex and the Čech complex given in Proposition 8.4.1. The overall idea in [31] and [32] is to leverage the relationship between the complex we have access to (Vietoris-Rips) and the complex from which we can deduce coverage (Čech). In order to formulate such a result, we assume that we are attaching a second number to each sensor, namely its *covering radius* R_c. It is understood that each sensor covers a disc of radius R_c around it, and that it can detect other sensors at the detection radius R given above. We further assume that $R_c \geq R/\sqrt{3}$. This allows us to guarantee that if $\{x_0, x_1, x_2\}$ forms a two simplex in the Vietoris-Rips complex $\mathcal{V}(X, R)$, then they span a two simplex in $\check{C}(X, R_c)$, by the second statement in Proposition 8.4.1. There are now the following assumptions made in [32].

1. The sensors lie in a compact connected domain $D \subseteq \mathbb{R}^2$, whose boundary ∂D is connected and piecewise linear with vertices called *fence nodes*.
2. Each fence node v is within R of its fence node neighbors on ∂D.

The following theorem is proved in [31] and [32].

Theorem 8.9.1. *Let $\mathcal{R}$ denote the Vietoris-Rips complex of the set of all sensors, and let $\mathcal{F}$ denote the subcomplex on the fence vertices. If the sensors satisfy the conditions above, and if there exists $[\alpha] \in H_2(\mathcal{R}, \mathcal{F})$ so that $\partial([\alpha]) \neq 0$, where $\partial : H_2(\mathcal{R}, \mathcal{F}) \to H_1(\mathcal{F})$ is the connecting homomorphism, then the balls of radius R_c around the sensors cover U.*

This theorem is in some situations not ideal, due to the strong assumptions on the boundary. In [32], it is shown that the use of the persistent homology of the pair $(\mathcal{R}, \mathcal{F})$ can be used to obtain coverage results with much weaker hypotheses.

Another interesting direction is the study of time varying situations, where the sensors move in time. In this case, there are situations where the balls around the sensors do not cover the region at any fixed time, but that no "evader" can avoid being sensed at some time. This kind of problem is referred to as an *evasion problem*, and has been studied in [1] and [38]. The two approaches are quite distinct, the approach in [1] using zig-zag persistence, and the approach in [38] develops a new kind of cohomology with semigroup coefficients. The approach in [38] yields "if and only if" results.

8.10 Probabilistic analysis

8.10.1 Random complexes

A very interesting direction of research is the study of the distributions on the space of barcodes $\mathfrak{B}$ (defined in Section 8.5) obtained by sampling points from Euclidean space using various models of randomness, i.e. sampling from various distributions on $\mathbb{R}^n$, or using the theory of random graphs [11]. Since we do not have a library of well understood distributions on $\mathfrak{B}$, one can instead study the distributions on the real line obtained by pushing forward a distribution on $\mathfrak{B}$ along a map from $\mathfrak{B}$ to $\mathbb{R}$. An interesting such map is

$$\{(x_1, y_1), (x_2, y_2), \ldots, (x_n, y_n)\} = B \to \max\{\frac{y_1}{x_1}, \ldots, \frac{y_n}{x_n}\} = \lambda(B).$$

In fact, this map is only defined for barcodes in $\mathfrak{B}$ for which the endpoints of the intervals in the barcode all lie in $\{x \in \mathbb{R} | x > 0\}$. When one is computing homology in dimensions ≥ 1 of a Čech or Vietoris-Rips complex of a set of points in Euclidean space, the barcode satisfies this property. A very interesting result in this direction is proved in [10]. They proceed by sampling from $[0, 1]^d$ using a uniform Poisson process of intensity n. This means that the sampling is done from a uniform distribution on $[0, 1]^d$, but that the number of points sampled is a governed by a Poisson distribution. The description of all these notions is beyond the scope of the present paper, but we refer the reader to [44]. The main result of [10] is the following.

Theorem 8.10.1. *We suppose that d and n are as above, that $d \geq 2$, that we are computing the k-dimensional persistence barcode B, and that $1 \leq k \leq d - 1$. Let*

$$\Delta_k(n) = \Big(\frac{\log\, n}{\log\, \log\, n}\Big)^{\frac{1}{k}}.$$

Then there exist constants A_k and B_k so that

$$\lim_{n \to \infty} \mathbb{P}\Big(A_k \leq \frac{\Pi_k(n)}{\Delta_k(n)} \leq B_k\Big) = 1$$

where $\mathbb{P}$ denotes probability, and where $\Pi_k(n)$ denotes the value of $\lambda(B)$ for a barcode generated as above.

8.10.2 Robust estimators

The stability theorem in Section 8.5.2 deals with the effect on persistence barcodes of small perturbations in the metric space, where perturbations are small in the sense of the Gromov-Hausdorff distance. In reality, though, one expects that in a perturbation of a metric space, a small number of distances may undergo relatively large perturbations. However, one believes that the number will be small, and that the points involved will be of small measure in an underlying measure. In order to deal with this problem, one incorporates a measure-theoretic component in one's definitions.

Definition 8.10.2. *By a* metric measure space *we will mean a complete separable metric space M equipped with a Borel measure μ.*

In [40], a metric d_{GPr} called the *Gromov-Prokhorov* metric is introduced on the measure preserving isometry classes of compact metric measure spaces. It is constructed by combining the Gromov-Hausdorff metric on the isometry classes of compact metric spaces with the Prokhorov metric d_{Pr} on measures on a fixed metric space (see [58]) by methods which we will not discuss here.

The paper [8] studies the distributions on the space of persistence barcodes arising from the persistence barcodes obtained by sampling from a fixed metric measure spaces. More precisely, they study distributions on the completion of the metric space $\overline{\mathfrak{B}}$ of $\mathfrak{B}$ equipped with the bottleneck distance. Let $\mu(n, k, X)$ denote the distribution on $\overline{\mathfrak{B}}$ that arises from sampling a set S of n points on a metric measure space X, and computing the k-dimensional barcode on S. We have the following.

Theorem 8.10.3. *The inequality*

$$d_{Pr}(\mu(n, k, X), \mu(n, k, X^{'})) \leq n d_{GPr}(X, X^{'})$$

holds for all compact metric measure spaces X and X'.

This result is the used in [8] to develop *robust statistics* for distinguishing the results of sampling from a fixed metric space. Robust statistics are computable quantities attached to samples from distributions that are relatively insensitive to small changes in parameter values in the distribution from which the samples are gathered, and also to the presence of outliers. An elementary example of this idea is the median, which is relatively insensitive to outliers and is considered a robust statistic, while the mean is not. For the problem at hand, [8] defines a precise notion of robustness.

Definition 8.10.4. *Let f be a function from the set of isomorphism classes of finite metric spaces to a metric space (W, d_W). We say the f is* robust with robustness coefficient $r > 0$ *if for any nonempty finite metric space (X, d_X), if for any nonempty finite metric space (X, d_X), there exists a bound δ such that for all isometric embeddings of X in a finite metric space $(X^{'}, d_{X^{'}})$ for which $\#(X^{'})/\#(X) < 1+r$, it is the case that $d(f(X), f(X^{'})) < \delta$. There is a corresponding uniform notion that states that the bound δ may be chosen universally, for all X. There is a corresponding uniform notion that states that the bound δ may be chosen universally, for all X.*

We now obtain the following result for finite metric spaces, which are being regarded as metric measure spaces by assigning to each metric space the uniform measure.

Theorem 8.10.5. [8] *For fixed n and k, $\mu(n, k, -)$ is uniformly robust with robustness coefficient r and estimate bound $\delta = nr/(1 + r)$ for any r.*

Since the space $\overline{\mathfrak{B}}$ is relatively inaccessible, it is useful to use this result to construct real valued statistics that also satisfy the robustness property. One way to do this is to consider a fixed reference distribution $\mathcal{P}$ on $\overline{\mathfrak{B}}$, and define $\Delta_{\mathcal{P}}(n, k, X)$ to be the number $d_{Pr}(\mu(n, k, X), \mathcal{P})$.

Theorem 8.10.6. [8] *For fixed n and k, and $\mathcal{P}$, $\Delta_{\mathcal{P}}(n,k,-)$ is uniformly robust with robustness coefficient r and estimate bound $nr/(1+r)$ for any r.*

It is also possible to obtain a somewhat simpler result, which does not require calculation of the full distribution $\mu(n,k,X)$. Instead of fixing a reference distribution $\mathcal{P}$, we choose a reference barcode $B \in \overline{\mathfrak{B}}$, and define $\Delta_B^{med}(n,k,X)$ to the median of the distribution of $d_B(B,-)$ applied to samples of k-dimensional barcodes attached to samples of size n taken from the metric measure space X.

Theorem 8.10.7. *For fixed n, k, and B, the function $\Delta_B^{med}(n,k,-)$ from finite metric spaces (with uniform probability measure) is robust with robustness coefficient greater than $\ln 2/n$.*

8.10.3 Random fields

Suppose that we have a function f on a manifold. We have seen in Section 8.5 that we can associate to f the persistence vector spaces $\{H_i(f^{-1}((-\infty,r],K)\}_r$ for i a nonnegative integer and K a field, and further that there are stability results that show that small changes in the function f lead to small changes, as measured by the bottleneck distance, in the corresponding sublevel set barcode. One can also ask, though, what the expected behavior of various of the features attached to barcode space is for a class of functions chosen at random. An initial question is what one means by a function chosen at random. The notion of a *random field* is defined in [4] as follows.

Definition 8.10.8. *By a* real-valued random field *on a topological space T we mean a measurable mapping*

$$F : \Omega \to \mathbb{R}^T$$

*where $\mathbb{R}^T$ denotes the set of all real-valued functions on T and $(\Omega, \mathcal{F}, \mathbb{P})$ is a complete probability space. The set $\mathbb{R}^T$ is equipped with the product σ-algebra over the set T. F creates a probability measure on $\mathbb{R}^T$, from which one can sample. Similarly, one can define an s-*dimensional vector-valued random field *as a measurable mapping*

$$F : \Omega \to (\mathbb{R}^T)^s.$$

Remark 8.10.9. Note that the definition does not use the topology of T. However, in much of the work done in this area, one studies additional hypotheses on the random fields in which the topology of T plays a role.

The idea here is that rather than being a function, a random field is an assignment to each $t \in T$ a distribution on $\mathbb{R}$, rather than a fixed value. Each of the restrictions $\mathbb{R}^T \to \mathbb{R}^{\{t\}} \cong \mathbb{R}$ produces a random variable, and therefore the corresponding distribution, which we denote F_t. In fact, for any finite set of points $t_1, \ldots, t_n \in T$, we obtain a distribution on $\mathbb{R}^n$ which we denote $F_{t_1,\ldots,t_n}$. There is a particular class of random fields called the *Gaussian random fields* that is particularly amenable to analysis.

Definition 8.10.10. *A real-valued random field F is a* Gaussian random field *if for all n, the n-dimensional distributions $F_{t_1,\ldots,t_n}$ are multivariate Gaussian distributions on $\mathbb{R}^n$.*

An s-dimensional random field is Gaussian, if all the distributions $F_{t_1,\ldots,t_n}$ are multivariate Gaussian distributions on $\mathbb{R}^{n\times s}$. Note that in Gaussian fields, the behavior of the random function is completely determined by the expectation function $m(t) = E(F_{t_1,\ldots,t_n})$, and the covariance function C.

The first example of this construction comes out of work of Wiener (see [67] and [6]), using analysis of Brownian motion. Wiener studied the case $T = \mathbb{R}_+$, and produced a Gaussian random field W, where the expected value of W_t is always $= 0$ and the variance is given by $C(s,t) = \min(s,t)$. He also showed that when one samples from the associated measure on $\mathbb{R}^{\mathbb{R}_+}$, one obtains continuous functions with probability 1, and so one calls W a *continuous Gaussian field.* Given any analytic property of functions on manifolds, such as k-th order differentiability, smoothness, or the property of being a Morse function, one can create and study Gaussian random fields whose samples have the given property with probability 1. Further, there are frequently a priori conditions on the covariance function of the random field that can be readily verified and guarantee the satisfaction of such properties.

The paper [9] proves a result concerning the persistent homology of the sublevel sets of functions sampled from Gaussian random fields. We consider the real-valued function σ on barcodes given by

$$\sigma\{[a_1,b_1],\ldots,[a_n,b_n]\} = \sum_i (b_i - a_i).$$

For any fixed $x \in \mathbb{R}$ and barcode $\beta = \{[a_1,b_1],\ldots,[a_n,b_n]\}$, we define the *$x$-truncation of* β, $\beta[x]$, to be the barcode

$$\{[a_1,\min(b_1,x)],\ldots,[a_n,\min(b_n,x)]\}$$

where it is understood that for any i such that $x \leq a_i$, the interval $[a_i,b_i]$ is simply deleted. Finally, we define

$$\chi^{pers}(M,f,x) = \sum_{i=0}^{\infty}(-1)^i\sigma(\{H_i(f^{-1}((-\infty,r],K)\}_r[x]).$$

In [9], the following result is proved concerning the distribution of $\chi^{pers}(M,f,x)$ for Gaussian random fields on Riemannian manifolds which produce Morse functions with probability one.

Theorem 8.10.11. *Let M be a closed d-dimensional Riemannian manifold, and let F be a smooth real valued Gaussian random field so that F is Morse with probability 1, and so that the mean is identically zero and the variance is identically equal to 1. Then for any $x \in \mathbb{R}$, we have*

$$\mathbb{E}\{\chi^{pers}(M,f,x)\} = \chi(M)(\varphi(x) + x\Phi(x))$$
$$+\varphi(x)\sum_{j=1}^{d}(2\pi)^{-j/2}\mathcal{L}_j(M)H_{j-2}(-x)$$

where

1. *H_n denotes the n-th Hermite polynomial.*
2. *$\varphi(x) = (2\pi)^{-1/2}e^{-x^2/2}$ is the density function for the standard Gaussian distribution.*

3. $\Phi(x) = \int_{-\infty}^{x} \varphi(u)du$.

4. $\mathcal{L}_j(M)$ *denotes the j-th Lipschitz-Killing curvature of M (defined for example in [4], Section 7.6) with respect to a metric constructed from the covariance metric attached to F.*

Remark 8.10.12. The point of this result is that it gives a theoretical estimate for the persistent Euler characteristic of sublevel sets in terms of classical invariants of the manifols. Also, the result in [9] is actually proved in a much more general context, that of regular stratified spaces and stratified Morse theory, which in particular permits the study of manifolds with boundary. It also includes the study of random fields that are of the form $G \circ (F_1, \ldots, F_k)$, where G is a deterministic function from $\mathbb{R}^k$ to $\mathbb{R}$, and $(F_1, \ldots, F_k)$ is a vector-valued Gaussian random field.

Bibliography

[1] H. Adams and G. Carlsson, *Evasion paths in mobile sensor networks*, The International Journal of Robotics Research, 34 (1), 2015, 90–104.

[2] H. Adams, T. Emerson, M. Kirby, R. Neville, C. Peterson, P. Shipman, S. Chepushtanova, E. Hanson, F. Motta, and L. Ziegelmeier, *Persistence images: a stable vector representation of persistent homology*, J. Mach. Learn. Res., 18, 2017, 1–35.

[3] A. Adcock, E. Carlsson, and G. Carlsson, *The ring of algebraic functions on persistence barcodes*, Homology, Homotopy, and Appl., 18, 2016, 381–402.

[4] R. Adler and J. Taylor, *Random fields and geometry*, Springer, 2009.

[5] N. Akkiraju, H. Edelsbrunner, M. Facello, P. Fu, E. Mucke, and C. Varela, *Alpha shapes: defintion and software*, in Proc. Internat. Comput. Geom. Software Workshop 1995.

[6] P. Baldi, *Stochastic calculus, an introduction through theory and exercises*, Springer Universitext, 2017.

[7] S. Barannikov, *The framed Morse complex and its invariants* Adv. Soviet Math., 21, 1994, 93–115.

[8] A. Blumberg, I. Gal, M. Mandell, and M. Pancia, *Robust statistics, hypothesis testing, and confidence intervals for persistent homology on metric measure spaces*, Found. Comput. Math., 14, 2014, 745–789.

[9] O. Bobrowski and M. Borman, *Euler integration of Gaussian random fields and persistent homology*, J. Topol. and Anal., 4 (1), 2012, 49–70.

[10] O. Bobrowski, M. Kahle, and P. Skraba, *Maximally persistent cycles in random geometric complexes*, Ann. Appl. Probab., 27 (4), 2017, 2032–2060.

[11] B. Bollobás, *Random graphs*, second edition, Cambridge University Press, 2011.

[12] P. Bubenik, *Statistical topological data analysis using persistence landscapes*, The Journal of Machine Learning Research 16 (1), 2015, 77–102.

[13] L.Buhovsky, V. Humilière, S. Seyfaddini. *The action spectrum and C^0 symplectic topology*, arXiv:1808.09790, 2018.

[14] F. Cagliari, B. Di Fabio, and M. Ferri, *One-dimensional reduction of multidimensional persistent homology*, Proc. Amer. Mat. Soc., 138 (8), 2010, 3003–3017.

[15] Z. X. Cang, Lin Mu and G. Wei, *Representability of algebraic topology for biomolecules in machine learning based scoring and virtual screening*, PLOS Computational Biology, 14 (1), 2018, e100592.

[16] Z. X. Cang and G. Wei, *TopologyNet: Topology based deep convolutional and multi-task neural networks for biomolecular property predictions*, PLOS Computational Biology, 13 (7), 2017, e1005690.

[17] G. Carlsson, *Topology and data*, Bull. Amer. Math. Soc., 46 (2), 2009, 255–308.

[18] G. Carlsson, *Topological pattern recognition for point cloud data*, Acta Numer., 23, 2014, 289-368

[19] G. Carlsson and V. de Silva, *Zigzag persistence*, Found. Comput. Math., 10 (4), 2010, 367–405

[20] G. Carlsson, V. de Silva, and D. Morozov, *Zigzag persistent homology and real-valued functions*, Proceedings of the twenty-fifth symposium on computational geometry, ACM, 2009, 247–256.

[21] G. Carlsson, T. Ishkhanov, V. de Silva, and A. Zomorodian, *On the local behavior of spaces of natural images*, Int. J. Comput. Vis., 76 (1) 2008, 1–12.

[22] G. Carlsson and A. Zomorodian, *The theory of multidimensional persistence*, Discrete Comput. Geometry 42 (1), 2009,71–93.

[23] G. Carlsson, A. Zomorodian, A. Collins, and L. Guibas, *Persistence barcodes for shapes*, International Journal of Shape Modeling, 11 (02), 2005, 149–187.

[24] W. Chacholski, M. Scolamiero, and F. Vaccarino, *Combinatorial presentation of multidimensional persistent homology*, J. Pure Appl. Alg., 221 (5), 2017, 1055–1075.

[25] J. Chan, G. Carlsson, and R. Rabadan, *Topology of viral evolution*, Proc. Nat. Acad. Sci. USA, 110 (46), 2013, 18566–18571.

[26] F. Chazal, D. Cohen-Steiner, L. Guibas, F. Mémoli, and S. Oudot, *Gromov-Hausdorff stable signatures for shapes using persistence*, Eurographics Symposium on Geometry Processing, 28 (5), 2009.

[27] C. Chong and S. Kumar, *Sensor networks: Evolution, opportunities, and challenges*, Proceedings of the IEEE, 91,8, 2003, 1247–1256.

[28] D. Cohen-Steiner, H. Edelsbrunner, and J. Harer, *Stability of persistence diagrams*, Discrete and Computational Geometry, 37 (1), 2007, 103–120.

[29] D. Cohen-Steiner, H. Edelsbrunner, J. Harer, and Y. Mileyko, *Lipschitz functions have L_p-stable persistence*, Found. Comput. Math., 10 (2), 2010, 127–139.

[30] V. de Silva and G. Carlsson, *Topological estimation using witness complexes*, Symposium on Point Based Graphics, ETH, Zürich, Switzerland, 2004.

[31] V. de Silva and R. Ghrist, *Coverage in sensor networks via persistent homology*, Alg. Geom. Topol. 7, 2007, 339–358.

[32] V. de Silva and R. Ghrist, *Homological sensor networks*, Notices Amer. Math. Soc., 54 (1), 2007, 10–17.

[33] H. Edelsbrunner, D. Kirkpatrick, and R. Seidel, *On the shape of a set of points in the plane*, IEEE Transactions on Information Theory, 29 (4), 1983, 551–559.

[34] H. Edelsbrunner, D. Letscher, A. Zomorodian, *Topological persistence and simplification*, Discrete and Computational Geometry, 28 (4), 2002, 511–533.

[35] H. Edelsbrunner and J. Harer, *Persistent homology - a survey*, Contemp. Math. 453, Amer. Math. Soc., 2008, 257–282.

[36] P. Frosini, *A distance for similarity classes of submanifolds of a Euclidean space*, Bull. Aust. Math. Soc. 42 (3), 1990, 407–415.

[37] P. Gabriel, *Unzerlegbare Darstellungen I*, Manuscripta Math 6, 1972, 71–103.

[38] R. Ghrist and S. Krishnan, *Positive Alexander duality for pursuit and evasion*, SIAM Journal on Applied Algebra and Geometry, 1 (1), 2017, 308–327.

[39] C. Giusti, E. Pastalkova, C. Curto, and V. Itskov, *Clique topology reveals intrinsic geometric structure in neural correlations*, Proc. Nat. Acad. Sci. USA, 112 (44), 2015, 13455–13460.

[40] A. Greven, A. Pfaffelhuber, and A. Winter, *Convergence in distribution of random metric measure spaces*, Probab. Theory Related Fields, 145, 2009, 285-322.

[41] M. Gromov. *Metric Structures for Riemannian and Non-Riemannian Spaces*, Birkhäuser, 2007.

[42] T. Hastie, R. Tibshirani, and J. Friedman, *The elements of statistical learning. Data mining, inference, and prediction*, Springer Series in Statistics, Springer, 2009.

[43] Y. Hiraoka, T. Nakamura, A. Hirata, E.G. Excolar, K. Matsue, and Y. Nishiura, *Hierarchical structures of amorphous solids characterized by persistent homology*, Proc. Nat. Acad. Sci. USA, 113 (26), 2016, 7035–7040,

[44] P. Jones and P. Smith, *Stochastic processes. An introduction*, CRC Press, 2018.

[45] S. Kalisnik, *Tropical coordinates on the space of persistence barcodes*, Found. Comput. Math., 19 (1), 2019, 101–129,

[46] L. Kanari, P. Dlotko, M. Scolamiero, R. Levi, J. C. Shillcock, K. Hess, and H. Markram, *A topological representation of branching morphologies*, Neuroinformatics, 16 (1), 2018, 3–13.

[47] D. Kozlov, *Combinatorial Algebraic Topology*, Algorithms and Computation in Mathematics, 21, Springer, 2008.

[48] F. Le Roux, S. Seyfaddini, and C. Viterbo, *Barcodes and area-preserving homeomorphisms*, arXiv:1810.03039, 2018.

[49] M. Lesnick, *The theory of the interleaving distance on multidimensional persistence modules*, Found. Comput. Math., 15 (3), 2015, 613–650.

[50] M. Lesnick and M. Wright, *Interactive visualization of 2-D persistence modules*, arXiv:1512.00180, 2015.

[51] L. Li, W. Cheng, G. Glicksberg, O. Gottesman, R. Tarnier, R. Chen, E. Bottinger, and J. Dudley, *Identification of type 2 diabetes subgroups through topological analysis of patient similarity*, Science Translational Medicine, 7(311), doi: 10.1126/scitranslmed.aaa9364, 2015.

[52] D. Mumford, J. Fogarty, and F. Kirwan, *Geometric invariant theory*, Springer Verlag, 2002.

[53] M. Nicolau, A. Levine, and G. Carlsson, *Topology based data analysis identifies a subgroup of breast cancers with a unique mutational profile and excellent survival*, Proc. Nat. Acad. Sci. USA, 108 (17), 2011, 7265–7270. doi: 10.1073/pnas.1102826108.

[54] A. Olin, E. Henckel, Y. Chen, T. Lakshmikanth, C. Pou, J. Mikes, A. Gustafsson, A. Bernhardsson, C. Zhang, K. Bohlin, and P. Brodin, *Stereotypic immune system development in newborn children*, Cell, 174 (5), 2018, 1277–1292.e14. doi: 10.1016/j.cell.2018.06.045.

[55] N. Otter, M. Porter, U. Tillmann, P. Grindrod, and H. Harrington, *A roadmap for the computation of persistent homology*, EPJ Data Science, 6 (17), 2017.

[56] L. Polterovich and E. Shelukhin, *Autonomous Hamiltonian flows, Hofer's geometry and persistence modules*, Selecta Math., 22, 2016, 227–296.

[57] L. Polterovich, E. Shelukhin, and V. Stojisavljevič, *Persistence modules with operators in Morse and Floer theory*, Moscow Math. J. 17 (4), 2017, 757–786.

[58] Y. Prokhorov, *Convergence of random processes and limit theorems in probability theory*, Theory Probab. Appl., 1, 1956, 157-214.

[59] M. W. Reimann, M. Nolte, M. Scolamiero, K. Turner, R. Perin, G. Chindemi, P. Dlotko, R. Levi, K. Hess, and H. Markram, *Cliques of neurons bound into cavities provide a missing link between structure and function*, Front. Comput. Neurosci., 12 June 2017, `https://doi.org/10.3389/fncom.2017.00048`.

[60] V. Robins, *Towards computing homology from finite approximations*, Proceedings of the 14th Summer Conference on General Topology and its Applications (Brookville, NY, 1999), Topology Proc. 24, 1999, 503–532.

[61] M. Saggar, O. Sporns, J. Gonzalez-Castillo, P. Bandettini, G. Carlsson, G. Glover, and A. Reiss, *Towards a new approach to reveal dynamical organization of the brain using topological data analysis*, Nature Communications, 11 (9), 1399, 2018, doi: 10.1038/s41467-018-03664-4.

[62] M. Scolamiero, W. Chacholski, A. Lundman, R. Ramanujam, and S. Öberg, *Multidimensional persistence and noise*, Foundations of Computational Mathematics, 17 (6), 2017, 1367–1406.

[63] G. Singh, F. Memoli, and G. Carlsson, *Topological methods for the analysis of high dimensional data sets and 3D object recognition*, Eurographics Symposium on Point-based Graphics, 2007, 91–100.

[64] J. Skryzalin and G. Carlsson, *Numeric invariants from multidimensional persistence*, J. Appl. Comput. Topol., 1, 2017, 89–119.

[65] M. Usher and J. Zhang, *Persisent homology and Floer-Novikov theory*, Geom. Topol., 6, 2016, 3333–3430.

[66] K. Xia and G. Wei, *Persistent homology analysis of protein structure, flexibility and folding*, Int. J. Numer. Methods Biomed. Eng., 30, 2014, 814–844.

[67] N. Wiener, *Nonlinear problems in random theory*, Technology Press Research Monographs, The Technology Press of the Massachusetts Institute of Technology and John Wiley & Sons, Chapman & Hall, Ltd., 1958.

[68] A. Zomorodian and G. Carlsson, *Computing persistent homology*, Discrete Comput. Geom., 33 (2), 2005, 249–274.

Department of Mathematics, Stanford University, Stanford, CA 94305, U.S.A.

E-mail address: `carlsson@stanford.edu`

9

Algebraic models in the homotopy theory of classifying spaces

Natàlia Castellana

Classically, algebraic topology refers to the study of the homotopy type of topological spaces by assigning algebraic invariants in a functorial way, but in the last decades homotopical methods have influenced other areas of mathematics. In this context, some topological spaces are more suitable for those kinds of interactions since they already encode algebraic information. This is the case of classifying spaces for groups. The classification of geometrical structures reduces to the understanding the homotopy type of maps into classifying spaces.

Homological algebra is an example of a bidirectional interaction between algebra and homotopy theory, especially group cohomology since it can be computed from the category of modules over the group ring or just as singular cohomology of the classifying space of the group. Choosing different coefficients will reflect different data and properties from the group. The other way around, to what extent this data will determine or classify the space, after a suitable localization or completion functor?

The guiding example taken in this chapter is the Stable Elements Theorem in mod p cohomology of a finite group which clearly expresses how p-subgroups and their conjugacy relations determine the mod p cohomology of a group. This information is encoded in a category, the fusion category of the group. In the 1990s Puig [107] introduced the notion of a saturated fusion system on a finite p-group abstracting the properties of the fusion category of a finite group. It was possible to describe conjugacy patterns among p-subgroups without referring to the presence of an ambient group, providing an axiomatized and uniform framework for dealing with conjugacy problems.

This algebraic nature of classifying spaces produced spectacular results in comparing algebraic and homotopical constructions. For example, for discrete groups there is a bijective correspondence between homotopy classes of pointed maps and group homomorphisms, and conjugacy relations correspond to homotopy equivalences. Starting in the 1980s, a series of developments took the homotopy theory of classifying spaces to the next level: Miller's theorem proving the Sullivan's conjecture [93], Carlsson proof of the Segal conjecture [40] and Lannes's T-functor technology to describe mapping spaces between classifying spaces [84]. It was then that the term homotopical group theory started to be used to refer to the development of the homotopy theory of classifying spaces by describing homotopical analogues of the group theoretic constructions.

The use of Bousfield-Kan p-completion functor [22] to isolate properties at a prime led to the search of homotopical analogues for classifying spaces of connected Lie groups. Compact

Mathematics Subject Classification. 55R35, 20J05, 20R35, 55P42.

Key words and phrases. classifying space, finite group, Lie group, fusion, homotopy theory.

Lie groups are finite loop spaces, but only requiring a finite loop space structure didn't give a good homotopical extension of the notion of a Lie group. But after p-completion, most of the group structure could be well determined in terms of homotopy theoretical notions. In this program, Dwyer-Wilkerson defined the notion of a p-compact group in [62]. The development of homotopical group theory in this context by many authors led to the classification of p-compact groups by work of Andersen-Grodal-Moller-Viruel [4] and Andersen-Grodal [5] by algebraic root data obtained from the loop space and similar to the classification of compact Lie groups.

A new breakthrough happened in 2003 when Broto-Levi-Oliver [26] formalized the notion of a classifying space for a saturated fusion system in a categorical way motivated by their study of self-homotopy equivalences of classifying spaces of finite groups. By work of Chermak [43] and Oliver [103], there is a unique classifying space associated to a saturated fusion system, up to homotopy. That was the essence of the Martino-Priddy conjecture proved by Oliver [101, 100]: does the fusion system associated to a finite group uniquely determine the homotopy type of the p-completion of the classifying space? From this modern point of view, results in both unstable and stable homotopy theory can be reformulated and proved by local methods only.

The goal of Sections 9.1 and 9.2 is to serve as a guiding example for the rest of the chapter. Sections 9.3 and 9.4 introduce saturated fusions systems and their classifying spaces, describing relevant results in comparison to the examples in previous sections. The notion of saturated fusion system on a p-discrete toral group is described in Section 9.5. This theory models compact Lie groups and finite loop spaces in general, among other examples. Finally, in Section 9.6 we briefly present recent instances of the interaction between homotopy theory and modular representation theory in the context of fusion systems.

This exposition in not exhaustive, but the goal is to motivate and to give a taste of some of the latest developments in the theory. An exhaustive reference which covers group theory, homotopy theory and representation theory is the book by Aschbacher, Oliver and Kessar [10]. Another complementary reference is the book by Craven [49].

Notation: We denote by $\mathrm{Syl}_p(G)$ the set of Sylow p-subgroups of a finite group G. Given $g, x \in G$, $c_g(x) = gxg^{-1}$. Given $P, Q \leq G$, the transporter set is

$$N_G(P, Q) = \{g \in G \mid gPg^{-1} = c_g(P) \leq Q\}.$$

The subgroup $O^p(G) \leq G$ is the smallest normal subgroup of G of index a power of p. The subgroup $O_p(G)$ is the largest normal p-subgroup of G. Analogously, the subgroup $O^{p'}(G) \leq G$ is the smallest normal subgroup of G of index prime to p, and $O_{p'}(G)$ is the largest normal subgroup of G of order prime to p. We will also use $\mathrm{Out}_{\mathcal{F}}(P) = \mathrm{Aut}_{\mathcal{F}}(P)/\mathrm{Inn}(P)$ and $\mathrm{Rep}(P, Q) = \mathrm{Inn}(Q)\backslash \mathrm{Hom}(P, Q)$. We will denote $H^*(-; \mathbb{F}_p)$ by $H^*(-)$.

Acknowledgements

The author was partially supported by FEDER-MEC grant MTM2016-80439-P and acknowledges financial support from the Spanish Ministry of Economy and Competitiveness through the "María de Maeztu" Programme for Units of Excellence in R&D (MDM-2014-0445). The author would furthermore like to thank the Isaac Newton Institute for Mathematical Sciences for support and hospitality during the programme *Homotopy Harnessing Higher Structures*, where work on this paper was undertaken. This work was supported by EPSRC grant no EP/K032208/1.

9.1 Key words in finite group cohomology

Let G be a finite group, R a commutative ring with unit. If M is an RG-module where RG is the group ring, the cohomology of G with coefficients in M is

$$H^*(G;M) = \mathrm{Ext}^*_{RG}(R,M).$$

The versatility of group cohomology comes from the fact that as a functor it factors through the homotopy category of topological spaces via the classifying space of G, denoted by BG. Then, $H^*(G;M) \cong H^*(BG;M)$ where the right-hand side is just singular cohomology with twisted coefficients.

If $M = R$ with the trivial action, $H^*(G;R)$ is a finitely generated graded commutative ring. Golod [70] proved this fact when G is a finite p-group and $R = \mathbb{Z}, \mathbb{Z}/p^a$, and Venkov [124] for a general G when $R = \mathbb{Z}/p$. Independently, Evens [65] showed that if R is Noetherian, $H^*(G;R)$ is a finitely generated algebra over R; even more, if M is an RG-module that is Noetherian as an R-module, then $H^*(G;M)$ is a Noetherian module over $H^*(G;R)$. Another important fact from Evens work is that if $H \leq G$, then $H^*(H;R)$ is a finitely generated $H^*(G)$-module via the induced morphism by the inclusion $\mathrm{Res}^G_H\colon H^*(H;R) \to H^*(G;R)$.

Let me sketch two different ways of proving the finite generation of $H^*(G;R)$. We fix $R = \mathbb{F}_p$ from now on unless specified, since this is the situation we will be mostly interested in the rest of the paper.

The first strategy we sketch (used by Venkov) considers a *faithful complex representation* $\rho\colon G \to U(n)$. For example, one can take the regular representation. A morphism induces a map between classifying spaces $B\rho\colon BG \to BU(n)$. The homotopy fiber of $B\rho$ is $U(n)/G$ which has the homotopy type of a finite CW-complex. The cohomology of $BU(n)$ is known to be polynomial on the Chern classes $c_1, \dots, c_n$ in degree 2, therefore finitely generated. The Serre spectral sequence for this fibration is a bounded spectral sequence of $H^*(BU(n))$-modules whose E_2-term $H^*(BU(n)) \otimes H^*(U(n)/G)$ is a finitely generated $H^*(BU(n))$-module. It follows then that the same is true for E_∞, and this fact will imply that $H^*(BG)$ is a finitely generated ring.

The second strategy (used by Evens) uses a *transfer homomorphism* argument which allows us to reduce the proof to p-groups. Let $H \leq G$, $H^*(H)$ is a $H^*(G)$-module via the homomorphism induced in cohomology $\mathrm{Res}^G_H\colon H^*(G) \to H^*(H)$. There is a transfer homomorphism in group cohomology (defined at the level of cochains)

$$\mathrm{Tr}^H_G\colon H^*(H) \longrightarrow H^*(G),$$

which has the property that is a morphism of $H^*(G)$-modules (by Frobenius reciprocity formula).

If $S \in \mathrm{Syl}_p(G)$ is a Sylow p-subgroup, the properties of the transfer homomorphism show that $\mathrm{Res}^G_S\colon H^*(G) \to H^*(S)$ admits then an $H^*(G)$-module retract Tr^S_G since $\mathrm{Tr}^S_G \circ \mathrm{Res}^G_H$ is multiplication by $[G : S]$ which is of order prime to p. Then, applying [62, Lemma 2.4] (see also [12, Lemma 2.4]), $H^*(G)$ is finitely generated if $H^*(S)$ is so.

The proof that $H^*(G)$ is a finitely generated ring when G is a p-group uses induction on the order of the group. The cohomology of the cyclic group of order p can be computed directly from the definition. To work out the induction step one uses that the center of a p-group is always non-trivial, and then applies a spectral sequence argument for a central extension of groups.

The properties of the transfer show that $\mathrm{Res}^G_S\colon H^*(G) \to H^*(S)$ is injective. Can the image of Res^G_S be identified? If c_g is the morphism induced by conjugation by $g \in G$, then $c_g^*\colon H^*(G) \to H^*(G)$ is the identity, but not when restricted to subgroups of G.

Definition 9.1.1. Let R be a commutative ring and M and RG-module. An element $x \in H^*(S, M)$ is *stable* if, for every $g \in G$, we have

$$\mathrm{Res}^S_{S\cap g^{-1}Sg}(x) = (c_g^*) \circ \mathrm{Res}^S_{gSg^{-1}\cap S}(x),$$

where c_g^* is induced by the pair $(c_g, g^{-1})\colon (G, M) \to (G, M)$.

The elements in the image of Res^G_S must satisfy the set of equations described in Definition 9.1.1. We can encode all these relations in the following category, called the transporter category.

Definition 9.1.2. The *transporter category* $\mathcal{T}_S(G)$ is the category whose objects are subgroups $P \leq S$ and a morphism from P to Q is given by those elements $g \in G$ such $gPg^{-1} = c_g(P) \leq Q$, that is, $\mathrm{Mor}_{\mathcal{T}_S(G)}(P, Q) = N_G(P, Q)$. Composition is group multiplication.

Then, the submodule of stable elements in $H^*(S)$ is precisely the inverse limit $\lim_{P\in\mathcal{T}_S(G)} H^*(P; M)$. And, $\mathrm{Im}(\mathrm{Res}^G_S) \subset \lim_{P\in\mathcal{T}_S(G)} H^*(P; M)$.

A classical result in group cohomology is the *stable elements theorem* which goes back to the work of Cartan-Eilenberg [41]. It uses mainly the properties of the transfer, especially the double coset formula which describes $\mathrm{Res}^G_S \circ \mathrm{Tr}^S_G$.

Theorem 9.1.3. *Let G be a finite group, $S \in \mathrm{Syl}_p(G)$ and M a $\mathbb{Z}_{(p)}G$- module. The morphism $\mathrm{Res}^G_S\colon H^*(G; M) \to H^*(S; M)$ is injective and*

$$\mathrm{Im}(\mathrm{Res}^G_S) = \lim_{P\in\mathcal{T}_S(G)} H^*(P; M).$$

Theorem 9.1.3 describes $H^*(G)$ as the set of solutions of a system of equations involving the cohomology of finite p-groups, and conjugacy relations among subgroups.

We go back to the situation $M = R = \mathbb{F}_p$. In that case, all the information needed in Theorem 9.1.3 to compute $H^*(G)$, or the submodule of the stable elements, is described by the group homomorphisms c_g for any $g \in G$ (it is not necessary then to keep track of the element $g \in G$ that induces c_g).

Definition 9.1.4. Let $S \in \mathrm{Syl}_p(G)$. The fusion category $\mathcal{F}_S(G)$ or *fusion system* of G on S is the subcategory of the category of groups whose objects are p-subgroups $P \leq S$ and a morphism from P to Q is given by a group homomorphism induced by an element $g \in G$ such $gPg^{-1} = c_g(P) \leq Q$, that is

$$\mathrm{Mor}_{\mathcal{F}_S(G)}(P, Q) = \mathrm{Hom}_G(P, Q) = N_G(P, Q)/C_G(P)$$

where $C_G(P)$ is the centralizer of P in G.

In that case,

$$H^*(G) \cong \lim_{P \in \mathcal{F}_S(G)} H^*(P).$$

In a course in group cohomology, the first computation obtained directly from the definition is that of a cyclic group of order p. Applying the Kunneth formula we know explicitly the cohomology of any *elementary abelian p-group* $V = (\mathbb{Z}/p)^n$. If $p = 2$, then it is polynomial $\mathbb{F}_2[x_1, \ldots, x_n]$ and if p is odd then it is a tensor product of an exterior algebra and a polynomial algebra, $E(x_1, \ldots, x_n) \otimes \mathbb{F}_p[y_1, \ldots, y_n]$, where $|x_i| = 1$ and $|y_i| = 2$.

How much information about $H^*(G)$ is captured by the collection of images of restrictions to elementary abelian subgroups?

Definition 9.1.5. Let $\mathcal{F}_S^e(G)$ be the full subcategory of $\mathcal{F}_S(G)$ whose objects are elementary abelian p-subgroups of S.

Quillen and Venkov [111, 110] studied the product of restriction homomorphisms

$$\prod \mathrm{Res}_V^G \colon H^*(G) \longrightarrow \lim_{V \in \mathcal{F}_S^e(G)} H^*(V) \subset \prod H^*(V).$$

Definition 9.1.6. Let $f \colon R \to S$ a morphism of (graded) commutative $\mathbb{F}_p$-algebras. We say that f is an *F-monomorphism* if every element in $\mathrm{Ker}(f)$ is nilpotent, *F-epimorphism* if for every $s \in S$, there exists a natural number n such that $s^{p^n} \in \mathrm{Im}(f)$. We say that f is an *F-isomorphism* if it is both an F-monomorphism and an F-epimorphism.

Theorem 9.1.7. *The morphism* $\prod \mathrm{Res}_V^G$ *has nilpotent kernel, and*

$$H^*(G) \longrightarrow \lim_{V \in \mathcal{F}_S^e(G)} H^*(V)$$

is an F-isomorphism.

Identifying $H^*(G)$ up to nilpotent elements is good enough to obtain certain type of information. For example, Quillen [110] proved that the Krull dimension of $H^*(G)$ is the maximal rank of an elementary abelian p-subgroup (using the fact that a prime ideal must contain all nilpotent elements).

Theorem 9.1.7 and some commutative algebra are the key ingredients in Quillen's strong *stratification* theorem in [110]. Denote by Spec_G^h the homogeneous prime ideal spectrum of $H^*(G)$, and $\mathrm{res}_G^H \colon \mathrm{Spec}_H^h \to \mathrm{Spec}_G^h$ induced by $H \leq G$.

For an elementary abelian p-subgroup $E \leq G$, let $\mathrm{Spec}_E^+ = \mathrm{Spec}_E^h \setminus \bigcup_{E' < E} \mathrm{res}_E^{E'}(\mathrm{Spec}_{E'}^h)$. Finally, we write $\mathrm{Spec}_{G,E}^+ = \mathrm{res}_G^E \mathrm{Spec}(E)^+$.

Theorem 9.1.8. *Let G be a finite p-group. The spectrum of homogeneous prime ideals Spec^h_G admits a decomposition as a disjoint union*

$$\mathrm{Spec}^h_G \cong \amalg_{E\in\mathcal{E}(G)} \mathrm{Spec}^+_{G,E}$$

where $\mathcal{E}(G)$ is a set if representatives of G-conjugacy classes of elementary abelian p-subgroups of S.

From this point, we go back to the challenging task of computing group cohomology using the description given by the Stable Elements Theorem 9.1.3. One way is by simplifying the set of equations by, as we do in linear algebra, eliminating redundancies: understanding how p-subgroups are conjugated among each other. This is the concept of *fusion* in group theory which has been studied since the beginning of last century. Let us state the following theorem due to Burnside as an example (see [32, Page 155]).

Theorem 9.1.9 (Burnside Fusion Theorem). *Let G be a finite group with an abelian p-Sylow subgroup S. If $x, y \in S$ are conjugate by an element in G, then they are conjugate by an element in $N_G(S)$.*

Remark 9.1.10. Theorem 9.1.9 also holds for subgroups. That is, if $P, Q \leq S$ are conjugate in G then they are also conjugate in $N_G(S)$.

We can translate this statement in terms of the corresponding fusion systems. Note that if $H \leq G$ with $T \in \mathrm{Syl}_p(H)$ and $S \in \mathrm{Syl}_p(G)$ with $T \leq S$ then $\mathcal{F}_T(H) \subset \mathcal{F}_S(G)$ as a subcategory. Then Theorem 9.1.9 says that if S is abelian then $\mathcal{F}_S(N_G(S)) = \mathcal{F}_S(G)$.

Definition 9.1.11. Let G be a finite group and $S \in \mathrm{Syl}_p(G)$. A subgroup $S \leq K \leq G$ is said to *control fusion* in G (or control G-fusion) if, whenever two subgroups $P, Q \leq G$ are conjugate in G, then they are conjugate in K.

Corollary 9.1.12. *Let $S \in \mathrm{Syl}_p(G)$. If $K \leq G$ controls G-fusion then $\mathcal{F}_S(K) = \mathcal{F}_S(G)$ and $\mathrm{Res}^G_K \colon H^*(G) \to H^*(K)$ is an isomorphism.*

A converse holds due to the following result of Mislin [94].

Theorem 9.1.13. *Let $f \colon H \to G$ be a group morphism of finite groups. Fix $S \in \mathrm{Syl}_p(G)$ and $T \in \mathrm{Syl}_p(H)$ with $f(T) \leq S$. Then $f^* \colon H^*(G) \to H^*(H)$ is an isomorphism iff the induced functor $\mathcal{F}_T(H) \to \mathcal{F}_S(G)$ is an equivalence of categories.*

An example is given by Burnside Theorem, Theorem 9.1.9: $N_G(S)$ controls fusion in G if $S \in \mathrm{Syl}_p(G)$ is abelian. It is worth noticing that Burnside fusion theorem can also be interpreted in the following way: any conjugation among p-subgroups of S is the restriction of an automorphism of S.

Normalizers of non-trivial p-subgroups are called *p-local subgroups* of G. In general one cannot find a single p-local subgroup controlling G-fusion, but a weaker statement can still hold by considering a set of p-local subgroups.

A *collection* is a set of subgroups of a group closed under conjugation. Let $\mathcal{H}$ be a collection of subgroups of S, we denote by $\mathcal{F}^{\mathcal{H}}_S(G)$ the full subcategory of $\mathcal{F}_S(G)$ with object set $\mathrm{Ob}(\mathcal{F}^{\mathcal{H}}_S(G)) = \mathcal{H}$.

Definition 9.1.14. Let G be a finite group and $S \in \mathrm{Syl}_p(G)$. Let $\mathcal{H}$ be a collection of subgroups of S. We say that $\mathcal{F}_S(G)$ is *$\mathcal{H}$-generated* if every morphism in $\mathcal{F}_S(G)$ is a composite of restrictions of automorphisms in $\mathcal{F}_S(G)$ of subgroups in $\mathcal{H}$. That is, for each isomorphism $P \to P'$ in $\mathcal{F}_S(G)$, there exist sequences of subgroups of S,

$$P = P_0, P_1, \ldots, P_k = P' \qquad \text{and} \qquad Q_1, Q_2, \ldots, Q_k$$

and morphisms $\varphi_i \in \mathrm{Aut}_{\mathcal{F}_S(G)}(Q_i)$ such that the following hold:

1. Q_i is in $\mathcal{H}$ for each i.
2. $P_{i-1}, P_i \leq Q_i$ and $\varphi_i(P_{i-1}) = P_i$ for each i.
3. $\varphi = \varphi_k \circ \varphi_{k-1} \circ \cdots \circ \varphi_1$.

Remark 9.1.15. In general, given a collection $\mathcal{H}$ there is a monomorphism

$$\lim_{P \in \mathcal{F}_S(G)} H^*(P) \hookrightarrow \lim_{P \in \mathcal{F}_S^{\mathcal{H}}(G)} H^*(P),$$

which is an isomorphism if $\mathcal{F}_S(G)$ is $\mathcal{H}$-generated.

Alperin's fusion theorem [2] is about such collections $\mathcal{H}$ of subgroups for which $\mathcal{F}_S(G)$ is $\mathcal{H}$-generated. Alperin considers certain intersections of Sylow p-sugroups (tame intersections), and Goldschmidt [69] and Puig [108] consider the collection of essential proper subgroups. We introduce one of the collections which will be used in the next section.

Definition 9.1.16. Let G be a finite group. A subgroup $P \leq S$, where $S \in \mathrm{Syl}_P(G)$, is *p-centric* if $C_G(P) \cong Z(P) \times O^p(C_G(P))$ with $O^p(C_G(P))$ of order prime to p (i.e. $Z(P) \in \mathrm{Syl}_p(C_G(P))$).

Theorem 9.1.17 (Alperin's fusion theorem). *Let G be a finite group and $S \in \mathrm{Syl}_p(G)$. If $\mathcal{H}$ is the collection of non-trivial p-centric subgroups of G, then $\mathcal{F}_S(G)$ is $\mathcal{H}$-generated.*

We finish with an example which combines the key words appearing in this section. We say that a finite group G is p-nilpotent if there is $H \trianglelefteq G$ of order prime to p and p-power index (i.e. $G \cong S \ltimes H$ where $S \in \mathrm{Syl}_p(G)$). The following two theorems characterize p-nilpotent groups in terms of fusion and mod p cohomology.

Theorem 9.1.18 (Frobenius. See 39.4 in [7]). *Let G be a finite group and $S \in \mathrm{Syl}_p(G)$. G is p-nilpotent iff S controls fusion in G.*

Theorem 9.1.19. (Quillen, [109]) *Let G be a finite group and $S \in \mathrm{Syl}_p(G)$. If $\mathrm{Res}_S^G \colon H^*(G) \to H^*(S)$ is an F-isomorphism, then S controls fusion in G if p is odd.*

Recent work of Benson-Grodal-Henke [16] have generalized Quillen's result 9.1.19 on the control of fusion by elementary abelian p-subgroups to subgroups of index prime to p.

Theorem 9.1.20. *Let G be a finite group and $H \leq G$ be an inclusion of finite groups of index prime to p, p and odd prime. If $\mathrm{Res}_H^G \colon H^*(G) \to H^*(H)$ is an F-isomorphism then H controls fusion in G.*

These results express to what extent fusion on elementary abelian subgroups controls the fusion of the group.

9.2 Decomposing classifying spaces

Let G be a finite group and $S \in \mathrm{Syl}_p(G)$. Group cohomology with coefficients in a commutative ring R can also be defined as (singular) cohomology of the classifying space for G, $H^*(BG; R)$. As in Section 9.1 we fix $R = \mathbb{F}_p$. The main idea we should keep is that mod p cohomology retains and isolates p-local information of the group G which is precisely encoded in the fusion system $\mathcal{F}_S(G)$ and the collection of p-subgroups. More precisely, if $\mathcal{H}$ is a collection of p-subgroups that generates $\mathcal{F}_S(G)$, then we only need $\mathcal{F}_S^{\mathcal{H}}(G)$ and $\{P\}_{P\in\mathcal{H}}$ for computing $H^*(G)$.

In homotopy theory, there is a functor that isolates the information of a topological space that is reflected in mod p cohomology: *Bousfield-Kan p-completion* functor (see [22] and also [10, Section III.1.4]).

It is a functor $(-)^{\wedge}_p\colon \mathrm{Top} \to \mathrm{Top}$ with a natural transformation from the identity functor $\phi_X\colon X \to X^{\wedge}_p$ with the following property: if $f\colon X \to Y$ induces an isomorphism in mod p cohomology iff $f^{\wedge}_p\colon X^{\wedge}_p \to Y^{\wedge}_p$ is a weak homotopy equivalence ([22, I.5.5]).

The natural transformation ϕ_X does not need to induce an isomorphism in mod p cohomology. If it does, ϕ_X^* is a isomorphism, we say that X is p-good. By [22], if $\pi_1(X)$ is finite then X is p-good (see [22, VII.5.1]). In particular our motivating example BG is p-good, therefore $H^*(BG) \cong H^*(BG^{\wedge}_p)$. We say that X is p-complete if $X^{\wedge}_p \simeq X$. If G is a p-group then BG is already p-complete ([10, Proposition 1.10]). In general, $BG^{\wedge}_p$ is not an Eilenberg-MacLane space anymore, but the fundamental group can still be computed in terms of G, $\pi_1(BG^{\wedge}_p) \cong G/O^p(G)$ (see [22, VII.5.1] or [10, Proposition 1.11]).

A model for BG is the nerve of the category $\mathcal{B}G$: it is a category with a single object $\bullet$ and $\mathrm{Hom}_{\mathcal{B}G}(\bullet,\bullet) = G$ where composition is given by group multiplication. We can also consider the nerve of the transporter category $\mathcal{T}_S(G)$ as suggested by group cohomology. Note that $\mathrm{Aut}_{\mathcal{T}_S(G)}(1) = G$. By [10, Page 134], the inclusion $\mathcal{B}G \subset \mathcal{T}_S(G)$ induces a homotopy equivalence on nerves $|\mathcal{B}G| \simeq |\mathcal{T}_S(G)|$, providing then a model for BG.

Definition 9.2.1. Let G be a finite group and $S \in \mathrm{Syl}_p(G)$. The *orbit category* $\mathcal{O}_S(G)$ is the category whose objects are p-subgroups $P \leq S$, and given $R, T \leq S$, $\mathrm{Mor}_{\mathcal{O}_S(G)}(R,T) = T\backslash N_G(R,T)$.

Analogously, if $\mathcal{H}$ is a collection of subgroups of S, we denote by $\mathcal{O}_S^{\mathcal{H}}(G) \subset \mathcal{O}_S(G)$ the full subcategory generated by the object set $\mathcal{H}$. Let $\mathcal{T}_S^{\mathcal{H}}(G) \subset \mathcal{T}_S(G)$ be the full subcategory of the transporter category with $\mathrm{Ob}(\mathcal{T}_S^{\mathcal{H}}(G)) = \mathcal{H}$.

Following the strategy in Section 9.1, we consider a collection $\mathcal{H}$ of subgroups of S. In [57] Dwyer studied systematically the homotopy type of $|\mathcal{T}_S^{\mathcal{H}}(G)|$ and $|\mathcal{T}_S^{\mathcal{H}}(G)|^{\wedge}_p$ for several choices of $\mathcal{H}$ in his work on *homology decompositions* of classifying spaces of finite groups. He described $|\mathcal{T}_S^{\mathcal{H}}(G)|$ in terms of the quotient categories $\mathcal{O}_S^{\mathcal{H}}(G)$ and $\mathcal{F}_S^{\mathcal{H}}(G)$.

There are functors

$$p_{\mathcal{O}}\colon \mathcal{T}_S^{\mathcal{H}}(G) \to \mathcal{O}_S^{\mathcal{H}}(G)$$

$$p_{\mathcal{F}}\colon \mathcal{T}_S^{\mathcal{H}}(G) \to \mathcal{F}_S^{\mathcal{H}}(G)$$

that are the identity on objects and quotient by the action of $P \leq N_G(P)$ (resp. $C_G(P) \leq N_G(P)$) on morphisms.

Theorem 9.2.2. [57] *Let G be a finite group, $S \in \mathrm{Syl}_p(G)$ and $\mathcal{H}$ a collection of p-subgroups of S. There are functors*

$$\alpha\colon (\mathcal{F}_S^{\mathcal{H}}(G))^{op} \to \mathrm{Top}$$

and

$$\beta\colon \mathcal{O}_S^{\mathcal{H}}(G) \to \mathrm{Top}$$

such that

$$\operatorname*{hocolim}_{\mathcal{F}_S^{\mathcal{H}}(G)^{op}} \alpha \simeq |\mathcal{T}_S^{\mathcal{H}}(G)| \simeq \operatorname*{hocolim}_{\mathcal{O}_S^{\mathcal{H}}(G)} \beta$$

and for each $P \in \mathcal{H}$, $\alpha(P) \simeq BC_G(P)$ and $\beta(P) \simeq BP$.

Remark 9.2.3. The decomposition provided by α (resp. β) is called a centralizer decomposition (resp. subgroup decomposition). In [57], Dwyer also considers a third type of decomposition for $|\mathcal{T}_S^{\mathcal{H}}(G)|$, the normalizer decomposition, with the property that the indexing category is a poset.

Remark 9.2.4. The functor β is easily described from the orbit category, $\beta(P) = EG \times_G G/P$.

For which collections $\mathcal{H}$ do we have an equivalence $|\mathcal{T}_S^{\mathcal{H}}(G)|_p^\wedge \simeq BG_p^\wedge$? Recall that there is a functor

$$\iota\colon \mathcal{T}_S^{\mathcal{H}}(G) \to \mathcal{B}G,$$

which sends every morphism set $N_G(P,Q)$ into G. A collection $\mathcal{H}$ is *ample* if $\iota_p^\wedge\colon |\mathcal{T}_S^{\mathcal{H}}(G)|_p^\wedge \simeq BG_p^\wedge$. For example, the collection of non-trivial p-subgroups (see [79]) or the collection of elementary abelian p-subgroups (see [78], [56]). We introduced the notion of p-centric subgroup in Definition 9.1.16. The collection of p-centric subgroups is also ample ([57]). More information about ample collections can be found in the work of Grodal [75].

Restricting to the collection of p-centric subgroups allows us to consider an intermediate quotient when constructing the fusion category from the transporter category by first letting $O^p(C_G(P))$ act.

Definition 9.2.5. Let G be a finite group and $S \in \mathrm{Syl}_p(G)$. The *centric linking system* $\mathcal{L}_S^c(G)$ is the category whose objects are p-centric subgroups $P \leq S$, and

$$\mathrm{Hom}_{\mathcal{L}_S^c(G)}(R,T) = N_G(R,T)/O^p(C_G(R)).$$

This category was introduced by Broto-Levi-Oliver in [25] where they prove the following equivalence.

Theorem 9.2.6. *Let G be a finite group and $S \in \mathrm{Syl}_p(G)$. There is an equivalence $|\mathcal{L}_S^c(G)|_p^\wedge \simeq BG_p^\wedge$.*

Remark 9.2.7. The key point is that the functor $\mathcal{T}_S^c(G) \to \mathcal{L}_S^c(G)$ induces a mod p equivalence on nerves, since $O^p(C_G(P))$ is of order prime to p. This inspires the following definition: $P \leq S$ is p-quasicentric if $O^p(C_G(P))$ is of order prime to p. Let $\mathcal{L}^q_S(G)$ be the category whose objects are p-quasicentric subgroups $P \leq S$, and

$$\mathrm{Hom}_{\mathcal{L}^q_S(G)}(R,T) = N_G(R,T)/O^p(C_G(R)).$$

The inclusion of categories $\mathcal{L}^c \subset \mathcal{L}^q$ also induces a homotopy equivalence on nerves ([23, Theorem 3.5]).

The next theorem shows that the centric linking category $\mathcal{L}_S(G)^c$ is an algebraic model that completely determines the homotopy type of $BG^\wedge_p$.

Theorem 9.2.8. [25] *For any prime p and any pair of finite groups G_1 and G_2, the p-completed classifying spaces $(BG_1)^\wedge_p$ and $(BG_2)^\wedge_p$ are homotopy equivalent if and only if the centric linking systems are equivalent.*

One direction is clear since an equivalence of categories induces a homotopy equivalences on nerves. The other direction depends on the description of homotopy classes of maps between classifying spaces in terms of groups morphisms which allows the authors to reconstruct the centric linking system from the homotopy type of $BG^\wedge_p$.

Theorem 9.2.9. *Let G be a finite group, and P a finite p-group. The classifying space functor induces a bijection*

$$\mathrm{Rep}(P,G) \cong [BP, BG^\wedge_p].$$

Moreover, for each $\rho \in \mathrm{Rep}(P,G)$, there is a homotopy equivalence

$$BC_G(\rho(P)) \xrightarrow{\simeq} \mathrm{Map}(BP, BG^\wedge_p)_{B\rho}.$$

The first part is a theorem of Mislin [94] and the second part is by Dwyer-Zabrodsky [64].

In order to prove Theorem 9.2.8, Broto-Levi-Oliver (see [10, Section 3.2]) define the fusion system and centric linking system of a topological space X with respect to a map $f\colon BS \to X$ where S is a finite p-group. We illustrate by describing the simpler fusion category with respect to f, $\mathcal{F}_f(X)$. Objects is given by the set of groups $P \leq S$ and $\mathrm{Hom}_{\mathcal{F}_f(X)}(P,Q)$ are group monomorphisms $\varphi\colon P \to Q$ such that $f|_{BQ} \circ B\varphi \simeq f|_{BP}$. If $f\colon BS \to BG$ is induced by the inclusion of $S \in \mathrm{Syl}_p(G)$ then they show that $\mathcal{F}_f(BG) \simeq \mathcal{F}_{f^\wedge_p}(BG^\wedge_p) \simeq \mathcal{F}_S(G)$.

But, is the fusion system enough to determine the homotopy type of $BG^\wedge_p$? This is the statement of the Martino-Priddy conjecture proved by Oliver.

Definition 9.2.10. For $i = 1, 2$, let $S_i \in \mathrm{Syl}_p(G_i)$ for finite groups G_i. A *fusion preserving isomorphism* is an isomorphism $\varphi\colon S_1 \to S_2$ that induces an isomorphism of categories from $\mathcal{F}_{S_1}(G_1)$ to $\mathcal{F}_{S_2}(G_2)$, by sending $P \leq S_1$ to $\varphi(P) \leq S_2$.

Theorem 9.2.11 (Martino-Priddy conjecture, [100], [101]). *For any prime p, and any pair of finite groups G_1 and G_2, $(BG_1)^\wedge_p \simeq (BG_2)^\wedge_p$ iff there is a fusion preserving isomorphim $\varphi\colon S_1 \to S_2$.*

There are two strong and beautiful applications of the explicit description of the homotopy type of $BG_p^\wedge$ from the centric linking system.

First, Broto-Levi-Oliver [25] describe the topological monoid of self-homotopy equivalences $\mathrm{Aut}(BG_p^\wedge)$ in terms of a subgroupoid of the groupoid of self-equivalences of the category $\mathcal{L}_S^c(G)$ and natural isomorphisms of functors, those called isotypical and which respect certain strucuture. A similar statement holds for $\mathrm{Out}(BG_p^\wedge)$, the group of homotopy classes of self-homotopy equivalences of $BG_p^\wedge$. In [25], the authors prove that $\mathrm{Out}(BG_p^\wedge)$ is isomorphic to the group of natural isomorphism classes of isotypical selfequivalences of $\mathcal{L}_S^c(G)$. Understanding the relation of these groups with the actual $\mathrm{Out}(G)$ is crucial in the study of fibrations and extensions of the fusion system and whether they can be realized by a group.

Second, Broto-Moller-Oliver [30] produce a beautiful application of techniques in homotopy theory to finite group theory. They prove equivalences between fusion systems of finite groups of Lie type by showing that their p-completed classifying spaces are equivalent. In this way they obtain equivalences that where unknown to group theorists before.

9.3 (Saturated) Fusion systems

The theory of fusion systems is a new way to solve questions in finite group theory and homotopy theory involving conjugacy relations.

In the 1990s, Lluís Puig [107] defined the notion of a saturated fusion system (Frobenius category) on a p-group S inspired by the properties of the fusion system of a finite group, but motivated by work in modular representation theory where similar categories could be constructed from the defect group of a block. His definition focuses on the properties of the morphisms and hides the role of the group G, abstracting the concept of conjugacy relations among subgroups. Next, we give the axiomatic definition introduced by Broto-Levi-Oliver in [26] which is equivalent to the original one due to Puig. We start with the basic notion of a fusion system.

Definition 9.3.1. A *fusion system* $\mathcal{F}$ on a finite p–group S is a subcategory of the category of groups whose objects are the subgroups $P \leq S$ and such that the set of morphisms $\mathrm{Hom}_{\mathcal{F}}(P, Q)$ between two subgroups P and Q satisfies the following conditions:

(a) $\mathrm{Hom}_S(P, Q) \subseteq \mathrm{Hom}_{\mathcal{F}}(P, Q) \subseteq \mathrm{Inj}(P, Q)$ for all $P, Q \leq S$.

(b) Every morphism in $\mathcal{F}$ factors as an isomorphism in $\mathcal{F}$ followed by an inclusion.

This concept is too general for the purpose of modeling the structure of p-subgroups of a given finite group G and the following axioms are the essence of the definition of a saturated fusion system.

Definition 9.3.2. Let $\mathcal{F}$ be a fusion system on a p–group S.

- We say that two subgroups $P, Q \leq S$ are *$\mathcal{F}$–conjugate* if they are isomorphic in $\mathcal{F}$. We denote by $\{P\}^{\mathcal{F}}$ the set of subgroups $P' \leq S$ that are $\mathcal{F}$-conjugate to $P \leq S$.

- A subgroup $P \leq S$ is *fully centralized* in $\mathcal{F}$ if $|C_S(P)| \geq |C_S(P')|$ for all $P' \in \{P\}^{\mathcal{F}}$.
- A subgroup $P \leq S$ is *fully normalized* in $\mathcal{F}$ if $|N_S(P)| \geq |N_S(P')|$ for all $P' \in \{P\}^{\mathcal{F}}$.
- $\mathcal{F}$ is a *saturated fusion system* if the following conditions hold:
  (I) (Sylow axiom) Each fully normalized subgroup $P \leq S$ is fully centralized and the group $\mathrm{Aut}_S(P)$ is a p–Sylow subgroup of $\mathrm{Aut}_{\mathcal{F}}(P)$.
  (II) (Extension axiom) If $P \leq S$ and $\varphi \in \mathrm{Hom}_{\mathcal{F}}(P, S)$ are such that φP is fully centralized, and if we set

$$N_\varphi = \{g \in N_S(P) \mid \varphi c_g \varphi^{-1} \in \mathrm{Aut}_S(\varphi P)\},$$

  then there is $\overline{\varphi} \in \mathrm{Hom}_{\mathcal{F}}(N_\varphi, S)$ such that $\overline{\varphi}_{|P} = \varphi$.

The first axiom is called the Sylow axiom since it is intended to model the fact that S must play the role of a Sylow p-subgroup in this structure. The second axiom is defined to model morphisms to behave like morphisms induced by conjugation.

Remark 9.3.3. The axioms of a saturated fusion system have other different equivalent formulations due to Roberts and Shpectorov [119] and to Stancu (see [90, Page 387], [83]).

The motivating example is the fusion system of a finite group G with a fixed Sylow p–subgroup $S \in \mathrm{Syl}_p(G)$ in Definition 9.1.4: $\mathcal{F}_S(G)$ is a saturated fusion system ([26], [10, Theorem 2.3]).

Definition 9.3.4. A saturated fusion system $\mathcal{F}$ on a finite p-group S is *realizable* if there exists a finite group G with $S \in \mathrm{Syl}_p(G)$ such that $\mathcal{F} \cong \mathcal{F}_S(G)$. If $\mathcal{F}$ is not realizable, we say that it is exotic.

Note that a realizable saturated fusion system can be realized by more than a finite group G. For example, if $N \trianglelefteq G$ with N of order prime to p, then G and G/N have isomorphic fusion systems.

In the original paper, Broto-Levi-Oliver [26] already described exotic examples at odd primes. The following are some examples of exotic fusion systems in the literature.

Example 9.3.5. 1. The only known examples of "simple" exotic fusions system at $p = 2$ fit in the family $\mathrm{Sol}(q)$, due to Solomon [123] but formalized by Levi-Oliver [87] (and [9]). The Sylow 2-subgroup is $S \in \mathrm{Syl}_2(\mathrm{Spin}_7(q))$ where q is an odd power of prime p. If $q \equiv \pm 3 \mod 8$ then $\mathrm{Sol}(q)$ is exotic.

2. If p is odd, Ruiz-Viruel [122] classified all possible saturated fusion systems on the extra-special p-group of order p^3 and exponent p. If $p = 7$, they describe three exotic saturated fusion systems. Later Díaz-Ruiz-Viruel [55] work out the classification when S is a p-group of rank 2, p odd. In this case, there is a family of exotic examples for $p = 3$. Ruiz [121] describes exotic examples for $p \geq 5$ of large p-rank.

3. Other exotic examples can be found in [47] by Clelland-Parker, in [27] by Broto-Levi-Oliver, and by Andersen-Oliver-Ventura [6].

Remark 9.3.6. It is worth pointing out that the way to show that the known exotic saturated fusion systems are so is by using the classification of finite simple groups, except for the Solomon example $\mathrm{Sol}(q)$. This example came out from the project of the classification of finite simple groups. This structure showed up at $p = 2$ in [123]: there is no finite group with the same Sylow 2-subgroup as $\mathrm{Spin}_7(3)$ such that $\mathcal{F}_S(\mathrm{Spin}_7(3)) \subset \mathcal{F}_S(G)$ and such that all involutions in S are G-conjugate. This example was introduced into homotopy theory by Benson (unpublished notes, [13]).

Remark 9.3.7. Leary-Stancu [85] and Robinson [120] independently showed that, given a saturated fusion system $\mathcal{F}$ over S, there is always an infinite group G with S as a maximal p-subgroup that defines $\mathcal{F}$ by conjugacy. Park [105] proved that there is always a finite group G such that $\mathcal{F}$ is isomorphic to $\mathcal{F}_S(G)$ but S is not a Sylow p-subgroup of G.

One of the main issues when constructing new saturated fusion systems is to check the saturation axioms, which have to be satisfied for subgroups and morphisms. Some results simplify this process by reducing to certain collection of subgroups.

Definition 9.3.8. Let $\mathcal{F}$ be a fusion systems on a finite p-group S, and let $\mathcal{H}$ be a collection of subgroups of S (i.e. closed under $\mathcal{F}$-conjugacy).

- $\mathcal{F}^{\mathcal{H}}$ is the full subcategory of $\mathcal{F}$ with object set $\mathcal{H}$.
- We say that $\mathcal{F}$ is $\mathcal{H}$-*generated* if any morphism in $\mathcal{F}$ is the composite of restrictions of morphisms between subgroups in $\mathcal{H}$.
- We say that $\mathcal{F}$ is $\mathcal{H}$-*saturated* if the saturation axioms are satisfied in $\mathcal{F}_{\mathcal{H}}$.

Definition 9.3.9. Let be $\mathcal{F}$ a fusion system on a finite p–group S.

- A subgroup $P \leq S$ is $\mathcal{F}$–*centric* if P and all its $\mathcal{F}$–conjugates contain their S–centralizers.
- A subgroup $P \leq S$ is $\mathcal{F}$–*radical* if $\mathrm{Out}_{\mathcal{F}}(P)$ is p–reduced, that is, if $\mathrm{Out}_{\mathcal{F}}(P) = \mathrm{Aut}_{\mathcal{F}}(P)/\mathrm{Inn}(P)$ has no proper normal p–subgroups.

We will use $\mathcal{F}^c$ to denote the full subcategory of $\mathcal{F}$ whose objects are the $\mathcal{F}$–centric subgroups and $\mathcal{F}^{cr}$ for the full subcategory of $\mathcal{F}$–centric, $\mathcal{F}$–radical subgroups.

The following theorem is a version of Alperin's fusion theorem for saturated fusion systems (Theorem A.10 in [25]). There are other versions of Alperin's fusion theorem for saturated fusion systems involving other sets of subgroups (essential subgroups), see [10, Theorem 3.5].

Theorem 9.3.10. [26] *Let $\mathcal{F}$ be a saturated fusion system on a finite p-group S. Let $\mathcal{H}$ be the set of fully normalized $\mathcal{F}$-centric $\mathcal{F}$-radical subgroups of S. Then $\mathcal{F}$ is $\mathcal{H}$-generated.*

The following theorem reduces the task of checking saturation axioms to a collection of subgroups satisfying certain hypothesis.

Theorem 9.3.11. [23] *Let $\mathcal{F}$ be a fusion system on a finite p-group S. Let $\mathcal{H}$ be a set of subgroups of S closed under $\mathcal{F}$-conjugacy and such that $\mathcal{F}$ is $\mathcal{H}$-generated and $\mathcal{H}$-saturated.*

Assume that each $\mathcal{F}$-centric subgroup not in $\mathcal{H}$ is $\mathcal{F}$-conjugate to some subgroup $P \leq S$ such that

$$\mathrm{Out}_S(P) \cap O_p(\mathrm{Out}_{\mathcal{F}}(P)) \neq 1.$$

Then $\mathcal{F}$ is saturated.

Remark 9.3.12. In Theorem 9.3.11 the hypothesis on $\mathcal{H}$ imply that $\mathcal{H}$ must contain all $\mathcal{F}$-centric $\mathcal{F}$-radical subgroups.

Both Theorem 9.3.11 and Theorem 9.3.10 allow one to construct saturated fusion systems from situations where morphisms are only explicitly described in a set of subgroups of S, by considering the fusion system generated by those morphisms.

The interest in understanding how exotic examples of saturated fusion systems arise in the theory of fusion systems is related to the classification of finite simple groups and to get a better understanding of it. In this context, one is lead to local finite group theory: developing concepts in analogy to the theory of finite groups but finding local analogues. For example, basic notions are that of the normalizer and centralizer of a subgroup.

Definition 9.3.13. Let $\mathcal{F}$ be a saturated fusion system on a finite p-group S and $P \leq S$.

- $N_{\mathcal{F}}(P)$ is the fusion system over $N_S(P)$ where for $R, S \leq N_S(P)$, $\varphi \in \mathrm{Hom}_{N_{\mathcal{F}}(P)}(R, S)$ if there exists $\varphi' \in \mathrm{Hom}_{\mathcal{F}}(RP, SP)$ such that $\varphi'|_R = \varphi$ and $\varphi|_P \in \mathrm{Aut}(P)$.
- $C_{\mathcal{F}}(P)$ is the fusion system over $C_S(P)$ where for $R, S \leq C_S(P)$, $\varphi \in \mathrm{Hom}_{C_{\mathcal{F}}(P)}(R, S)$ if there exists $\varphi' \in \mathrm{Hom}_{\mathcal{F}}(RP, SP)$ such that $\varphi'|_R = \varphi$ and $\varphi|_P = \mathrm{id}_P$.

Proposition 9.3.14. [26] *Let $\mathcal{F}$ be a saturated fusion system on a finite p-group S and $P \leq S$. If P is fully normalized (resp. centralized) then $N_{\mathcal{F}}(P)$ (resp. $C_{\mathcal{F}}(P)$) is a saturated fusion system. Moreover, if $\mathcal{F} = \mathcal{F}_S(G)$ then $N_{\mathcal{F}}(P) \cong \mathcal{F}_{N_S(P)}(N_G(P))$ and $C_{\mathcal{F}}(P) \cong \mathcal{F}_{C_S(P)}(C_G(P))$.*

With Definition 9.3.13 at hand, one can make sense of the notion of a normal subgroup of a saturated fusion system (if $N_{\mathcal{F}}(P) \cong \mathcal{F}$) or the center of a fusion system (if $C_{\mathcal{F}}(Z) \cong \mathcal{F}$).

The existence of $\mathcal{F}$-centric $\mathcal{F}$-normal subgroups in a saturated fusion system implies that $\mathcal{F}$ is realizable by a unique group G satisfying certain properties. This situation has been described in [23].

Definition 9.3.15. A saturated fusion system $\mathcal{F}$ over a finite p-group S is *constrained* if it has an $\mathcal{F}$-normal $\mathcal{F}$-centric subgroup.

Theorem 9.3.16. *Let $\mathcal{F}$ be a constrained saturated fusion system on a finite p-group S. Then there is a finite group G with $O_{p'}(G) = 1$ and $C_G(O_p(G)) \leq O_p(G)$ such that $\mathcal{F} \cong \mathcal{F}_S(G)$, and is unique satisfying this property.*

Remark 9.3.17. We can apply Theorem 9.3.16 to the case where S is abelian. Then if $\mathcal{F} = \mathcal{F}_S(G)$ for a finite group G, we recover the Frobenius Theorem 9.1.18 on the control of fusion by $N_G(S) \leq G$.

In Remark 9.2.7, we introduced the notion of a p-quasicentric subgroup in a finite group G . There is an analogue definition in the context of fusion systems.

Definition 9.3.18. Let $\mathcal{F}$ be a saturated fusion system on a finite p-group S. We say that $P \leq S$ is *$\mathcal{F}$-quasicentric* if for each $P' \in P^{\mathcal{F}}$ that is fully centralized, $C_{\mathcal{F}}(P') \cong \mathcal{F}_{C_S(P')}(C_S(P'))$.

If $\mathcal{F}^q \subset \mathcal{F}$ is the full subcategory generated by $\mathcal{F}$-quasicentric groups, then $\mathcal{F}^c \subset \mathcal{F}^q$. Recently, Ellen Henke [77] has introduced another collection of subgroups, called subcentric subgroups. When considering subsystems and quotients it is relevant to have large collections with properties that are preserved by quotients for example, making more accessible their study.

Definition 9.3.19. Let $\mathcal{F}$ a saturated fusion system on a finite p-group S. We say that $P \leq S$ is *subcentric* if, for any fully $\mathcal{F}$-normalized $P' \in \{P\}^{\mathcal{F}}$, $N_{\mathcal{F}}(P)$ is constrained.

If $\mathcal{F}^s \subset \mathcal{F}$ is the full subcategory of subcentric subgroups, we have a chain of inclusions

$$\mathcal{F}^{cr} \subset \mathcal{F}^c \subset \mathcal{F}^q \subset \mathcal{F}^s.$$

An extension theory of fusion systems is developed in several papers, for example, in [24] (where fusion subsystems of p-power index and of index prime to p where classified as well as central extensions), and [102].

There is also a theory of normal fusion subsystems developed by Aschbacher [8] and also studied by Craven in [48]. Then, a simple saturated fusion system is a saturated fusion system with no non-trivial normal fusion subsystems. One of the main goals in the theory is the understanding of simple saturated fusion systems at the prime 2. Linckelmann [90] have shown that $\mathrm{Sol}(q)$ for q an odd prime power $q \equiv \pm 3 \bmod 8$ is simple. Then, only a single family of simple exotic examples is known for $p = 2$, $\mathrm{Sol}(q)$.

To consider a category of saturated fusion systems, one needs a definition for a morphism of fusions systems, which will need to preserve some structure from the subgroups.

Definition 9.3.20. Let $\mathcal{F}$ and $\mathcal{F}'$ be two fusion systems over S and S' respectively. A *morphism of fusion systems* is a pair (α, Ψ) where $\alpha\colon S \to S'$ is a morphism and $\Psi\colon \mathcal{F} \to \mathcal{F}'$ a functor with $\alpha(P) = \Psi(P)$ for any $P \leq S$ and $\Psi(\varphi) \circ \alpha = \alpha \circ \varphi$ for any $\varphi \in \mathrm{Mor}(\mathcal{F})$.

Essentially, a morphism of fusion systems is a group homomorphism between Sylow subgroups that is compatible with the fusion/conjugacy relations.

Remark 9.3.21. In analogy to Section 9.1, Theorem 9.1.20 holds also for fusion systems ([16]). That is, if $\mathcal{F}_0 \subset \mathcal{F}$ are saturated fusion systems on the same finite p-group isomorphic on the collection of elementary abelian p-subgroups, then $\mathcal{F}_0 \cong \mathcal{F}$, if p is odd. For $p = 2$ the authors proved the result on control of fusion by enlarging the collection to include abelian subgroups of exponent at most 4.

9.4 Homotopy theory of fusion systems

The homotopy theory of fusion systems does not refer to the study of the nerve of $\mathcal{F}$, but rather the nerve of an associated "linking system" which plays the role of the classifying space of an ambient finite group G in case $\mathcal{F} = \mathcal{F}_S(G)$ for $S \in \mathrm{Syl}_p(G)$. We start by explaining what is the classifying space of a saturated fusion system in the context of unstable homotopy theory. Then, we will describe how the stable homotopy theory of the classifying space is algebraically modelled.

9.4.1 The unstable homotopy theory

Given a saturated fusion system $\mathcal{F}$, we define the stable elements of the mod p cohomology to be

$$H^*(\mathcal{F}) := \lim_{P \in \mathcal{F}} H^*(P) \subset H^*(S).$$

If $\mathcal{F} = \mathcal{F}_S(G)$ for a finite group G with $S \in \mathrm{Syl}_p(G)$, then the centric linking system (Definition 9.2.5) satisfies that $H^*(|\mathcal{L}_S(G)|_p^\wedge) \cong H^*(G) \cong H^*(\mathcal{F}_S(G))$.

Broto-Levi-Oliver [26, Definition 1.7] abstracted the main properties of $\mathcal{L}_S(G)$ in the definition of a linking system, which is the extra structure needed to obtain a classifying space that behaves like $BG_p^\wedge$ for a finite group G. The definition we present is more general allowing bigger collections of subgroups (see [23, Definition 3.3] and [102, Definition 3]).

Definition 9.4.1. Let $\mathcal{F}$ be a fusion system on a finite$p$–group S. A *linking system* associated to $\mathcal{F}$ is a finite category $\mathcal{L}$ together with functors

$$\mathcal{T}_S^{\mathrm{Ob}(\mathcal{L})}(S) \xrightarrow{\delta} \mathcal{L} \xrightarrow{\pi} \mathcal{F}^c$$

that satisfy the following conditions.

(A1) $\mathrm{Ob}(\mathcal{L}) \subseteq \mathrm{Ob}(\mathcal{F})$ is a set of subgroups $P \leq S$ closed under $\mathcal{F}$-conjugacy and over-groups that contains $\mathcal{F}$-centric and $\mathcal{F}$-radical subgroups. Each object $P \leq S$ in $\mathcal{L}$ is isomorphic to one which is fully $\mathcal{F}$-centralized.

(A2) δ is the identity on objects and π is the inclusion on objects. For each pair of objects P, Q in $\mathcal{L}$ such that P is fully $\mathcal{F}$-centralized, $C_S(P)$ acts freely on $\mathrm{Mor}_\mathcal{L}(P,Q)$ by right composition via δ and π induces a bijection

$$\mathrm{Mor}_\mathcal{L}(P,Q)/Z(P) \xrightarrow{\cong} \mathrm{Hom}_\mathcal{F}(P,Q).$$

(B) For each $\mathcal{F}$–centric subgroup $P \leq S$ and each $g \in P$, the functor π sends $\delta_P(g)$ to $c_g \in \mathrm{Aut}_\mathcal{F}(P)$.

(C) For each $f \in \mathrm{Mor}_\mathcal{L}(P,Q)$ and each $g \in \mathcal{T}_S(P,Q)$, the following square commutes in $\mathcal{L}$

$$\begin{array}{ccc} P & \xrightarrow{f} & Q \\ {\scriptstyle \delta_P(g)}\downarrow & & \downarrow{\scriptstyle \delta_Q(\pi(f)(g))} \\ P & \xrightarrow[f]{} & Q. \end{array}$$

If $\mathrm{Ob}(\mathcal{L})$ is the set of $\mathcal{F}$-centric subgroups, then $\mathcal{L}$ is called a *centric linking system*. The axioms are defined in a way that mimick the properties of morphisms in the centric linking system associated to $\mathcal{F}_S(G)$ and its relation with the fusion system, $\mathcal{L}_S^c(G) \to \mathcal{F}_S^c(G)$.

Remark 9.4.2. From Definition 9.4.1, the functor $\delta\colon \mathcal{T}_S(S) \to \mathcal{L}$ induces a map $BS \to |\mathcal{L}|_p^\wedge$ which models the inclusion of a Sylow p-subgroup and which is used to define restriction to subgroups $P \leq S$, $BP \to |\mathcal{L}|_p^\wedge$.

Example 9.4.3. [26] Let G be a finite group and $S \in \mathrm{Syl}_p(G)$. Then $\mathcal{L}_S^c(G)$ is a centric linking system associated to G.

What can we say about the collection $\mathrm{Ob}(\mathcal{L})$ for a given linking system $\mathcal{L}$? In [102, Proposition 4], Oliver shows that all objects in $\mathcal{L}$ are $\mathcal{F}$-quasicentric subgroups. Another important property is that if we have an inclusion of collections $\mathcal{H} \subset \mathcal{H}'$, then given a linking system $\mathcal{L}$ with object set $\mathcal{H}$ is contained in a linking system $\mathcal{L}'$ with object $\mathcal{H}'$ ([10, Propostion III.4.8]). But in fact, the homotopy type of the nerve of a linking system does not depend on the collection $\mathcal{H}$ ([23, Theorem 3.5]): $|\mathcal{L}| \simeq |\mathcal{L}'|$.

Next theorem shows that the structure defined provides a classifying space for a fusion system in the sense that topologically realizes $H^*(\mathcal{F})$.

Theorem 9.4.4. [26] *Let $\mathcal{F}$ be a saturated fusion system on a finite p-group S. Assume there exists a linking system $\mathcal{L}$ associated to $\mathcal{F}$. Then*

$$H^*(|\mathcal{L}^c|_p^\wedge) \cong H^*(\mathcal{F}).$$

Moreover, $H^*(\mathcal{F})$ is Noetherian.

Remark 9.4.5. A stable elements formula for cohomology with twisted coefficients has been investigated by Molinier ([95], [96]) and Levi-Ragnarsson [88].

Question 9.4.6. *Given a saturated fusion system $\mathcal{F}$, is there a linking system associated to $\mathcal{F}$?*

From the definition, there is no reason to believe that this is the case. Broto-Levi-Oliver [26] developed an obstruction theory for the existence and uniqueness of a centric linking system which is related to a question in homotopy theory. Recall from Definition 9.3.8 that $\mathcal{F}^{\mathcal{H}} \subset \mathcal{F}$ is the full subcategory generated by the objects in the collection $\mathcal{H}$.

Definition 9.4.7. Let $\mathcal{F}$ be a saturated fusion system and $\mathcal{H}$ a collection of subgroups of S. The *orbit category* of $\mathcal{F}^{\mathcal{H}}$ is the category $\mathcal{O}(\mathcal{F}^{\mathcal{H}})$ where $\mathrm{Ob}(\mathcal{O}(\mathcal{F}^{\mathcal{H}})) = \mathrm{Ob}(\mathcal{F}^{\mathcal{H}})$ and

$$\mathrm{Mor}_{\mathcal{O}(\mathcal{F}^{\mathcal{H}})}(P,Q) = \mathrm{Inn}(Q)\backslash \mathrm{Mor}_{\mathcal{F}^{\mathcal{H}}}(P,Q) = \mathrm{Rep}_{\mathcal{F}^{\mathcal{H}}}(P,Q).$$

Assuming there exists a centric linking system $\mathcal{L}^c$, one can consider the functor $\pi\colon \mathcal{L}^c \to \mathcal{F}^c \to \mathcal{O}(\mathcal{F}^c)$. The left Kan extension $L_\pi(*)$ of the constant functor to a point $*\colon \mathcal{O}(\mathcal{F}^c) \to \mathrm{Top}$ gives a homotopy equivalence

$$\operatorname*{hocolim}_{\mathcal{O}(\mathcal{F}^c)} L_\pi(*) \simeq |\mathcal{L}^c|,$$

by Segal's homotopy push-down theorem, with $L_\pi(*)(P) \simeq BP$, for any $P \in \mathcal{F}^c$ [26, Proposition 2.2]. The converse is true: given a subgroup decomposition by a functor $\beta\colon \mathcal{O}(\mathcal{F}^c) \to \mathrm{Top}$, with $\beta(P) \simeq BP$, one can recover a centric linking system by considering homotopy classes of maps from classifying spaces of finite p-groups into $(\operatorname{hocolim}_{\mathcal{O}(\mathcal{F}^c)} \beta)_p^\wedge$.

The key observation is that there is a bijection between centric linking systems associated to a saturated fusion systems and lifts of the classifying space functor $B\colon \mathcal{O}(\mathcal{F}^c) \to \mathrm{HoTop}$ to Top ([10, Proposition 5.31]).

Broto-Levi-Oliver stated the obstruction theory for Question 9.4.6 in [26, Proposition 3.1] which depends upon a functor $\mathcal{Z}\colon \mathcal{O}(\mathcal{F}^c) \to \mathrm{Ab}$ with $\mathcal{Z}(P) = Z(P)$. The obstructions to the existence lie in $\lim^3_{\mathcal{O}(\mathcal{F}^c)} \mathcal{Z}$, and $\lim^2_{\mathcal{O}(\mathcal{F}^c)} \mathcal{Z}$ acts freely and transitively on the set of isomorphic classes of centric linking systems if it is nonempty.

Remark 9.4.8. The proof of the Martino-Priddy conjecture by Oliver, Theorem 9.2.11, goes through a systematic computation of the obstruction groups using the classification of finite simple groups.

Remark 9.4.9. At this point, Broto-Levi-Oliver defined the notion of a *p–local finite group*. It is a triple $(S, \mathcal{F}, \mathcal{L})$, where $\mathcal{F}$ is a saturated fusion system on a finite p–group S and $\mathcal{L}$ is a centric linking system associated to $\mathcal{F}$. The classifying space of the p–local finite group $(S, \mathcal{F}, \mathcal{L})$ is the space $|\mathcal{L}|^\wedge_p$.

A positive answer to Question 9.4.6 was obtained by Chermak in [43] using a direct construction method.

Theorem 9.4.10. *Each saturated fusion system on a finite p-group S has an associated centric linking system, which is unique up to isomorphism.*

Chermak introduced the notion of a partial group, inspired by properties of the set $\mathrm{Mor}(\mathcal{L})$ and that of a locality in [43]. Later, Oliver [103] developed another proof, inspired by Chermak's work, by showing that the relevant obstruction groups vanish. It is worth mentioning that both proofs use the classification of finite simple groups. Finally, Glauberman-Lynd [68] provided a proof which does not depend on the classification of finite simple groups.

Remark 9.4.11. Partial groups and localities can be described in terms of simplicial sets satisfying certain properties. From this point of view, they are being studied by homotopy theorists. We mention work of Molinier [97] who proved Alperin's fusion theorem in this context. Chermak-González [44] described a unified setting to include many structures as localities, and González developed an extension theory of partial groups and localities [71].

Remark 9.4.12. Henke [77] has enlarged the collections in which a linking system and a transporter system can be defined by considering the collection of subcentric subgroups and proving existence and uniqueness of such systems.

Now, the only information needed to construct a classifying space is the saturated fusion system itself.

Definition 9.4.13. Let $\mathcal{F}$ be a saturated fusion system on a finite p-group S. The *classifying space* $B\mathcal{F}$ is $|\mathcal{L}|^\wedge_p$ where $\mathcal{L}$ is a centric linking system associated to $\mathcal{F}$.

Most of the homotopical group theory describing homotopic constructions, like mapping spaces, in terms of fusion data can be developed in this context.

Definition 9.4.14. Let $\mathcal{F}$ be a saturated fusion system, and P a finite p-group. We define $\mathrm{Rep}(P, \mathcal{F}) = \mathrm{Hom}(P, S)/\sim$ with $f \sim g$ if there is $\alpha \in \mathrm{Iso}_F(f(P), g(P))$ such that $\alpha \circ f = g$.

Theorem 9.4.15. [26] *For any finite p-group P and any saturated fusion system, the classifying space functor and composing with $\iota : BS \to B\mathcal{F}$ induces a bijection*

$$\mathrm{Rep}(P, \mathcal{F}) \xrightarrow{\cong} [BP, B\mathcal{F}].$$

Moreover, given $f \in \mathrm{Rep}(P, \mathcal{F})$ such that $f(P)$ is fully centralized then

$$\mathrm{Map}(BP, B\mathcal{F})_f \simeq BC_{\mathcal{F}}(P).$$

The proof of Theorem 9.4.15 uses the methods and T-functor technology developed by Lannes [84] and which have been applied very successfully in studying the homotopy type of p-completed classifying spaces. The description of $\mathrm{Out}(B\mathcal{F})$ in terms of equivalences of the centric linking system is also obtained in [26].

The F-isomorphism theorem also holds for classifying spaces of saturated fusion systems. We denote by $\mathcal{F}^e \subset \mathcal{F}$ the full subcategory generated by elementary abelian p-subgroups. The F-isomorphism theorem was proven by Broto-Levi-Oliver [26] and the strong stratification result by Barthel-Castellana-Heard-Valenzuela [12], and Linckelmann [91].

Theorem 9.4.16. [26] *The morphism*

$$\prod \mathrm{Res}^{\mathcal{F}}_V \colon H^*(B\mathcal{F}) \longrightarrow \lim_{\mathcal{F}^e} H^*(V)$$

is an F-isomorphism.

Denote by $\mathrm{Spec}^h_{\mathcal{F}}$ the homogeneous prime ideal spectrum of $H^*(\mathcal{F})$. If $E \subset S$ is an elementary abelian p-subgroup, let $\mathrm{Spec}^+_{\mathcal{F},E} = \mathrm{res}^E_{\mathcal{F}} \mathrm{Spec}(E)^+$ where $\mathrm{res}^E_{\mathcal{F}} \colon \mathrm{Spec}^h_E \to \mathrm{Spec}^h_{\mathcal{F}}$. Recall that $\mathrm{Spec}^+_E = \mathrm{Spec}^h_E \setminus \bigcup_{E'<E} \mathrm{res}^{E'}_E(\mathrm{Spec}^h_{E'})$.

Theorem 9.4.17. ([12], [91]) *The variety $\mathrm{Spec}^h_{\mathcal{F}}$ admits a decomposition as a disjoint union*

$$\mathrm{Spec}^h_{\mathcal{F}} \cong \amalg_{E \in \mathcal{E}(\mathcal{F})} \mathrm{Spec}^+_{\mathcal{F},E}$$

where $\mathcal{E}(\mathcal{F})$ is a set if representatives of $\mathcal{F}$-conjugacy classes of elementary abelian p-subgroups of S.

Since every saturated fusion system $\mathcal{F}$ has an associated classifying space $B\mathcal{F}$, we can ask whether this construction is functorial. Recall the Definition 9.3.20 at the end of Section 9.3 where we introduced the concept of a morphism between fusion systems. It is not clear that a morphism between fusion systems induces a map between classifying spaces.

This question has been approached by Castellana-Libman in [42]. The strategy applied is a classical one used to study maps between p-completed classifying spaces by several authors (see [82], [79], [80], [81]) Given a topological space, restriction to p-subgroups through $BS \to B\mathcal{F}$ gives a map

$$[B\mathcal{F}, X] \longrightarrow \lim_{P \in \mathcal{O}(\mathcal{F})} [BP, X].$$

In [125], Wojtkowiak described the obstructions classes for the surjectivity and injectivity for this restriction map, which lie in

$$\lim_{P\in\mathcal{O}(\mathcal{F}^{cr})}{}^{i+\varepsilon} \pi_i(\mathrm{Map}(BP, X)_{f\circ Bi_P})$$

for $\varepsilon = 0, 1$ and $i \geq 1$ where $Bi_P\colon BP \to BS$ is induced by the group inclusion.

If $X = B\mathcal{F}'$, where $\mathcal{F}'$ is a saturated fusion system on a finite p-group S', and $Bf\colon BS \to BS'$ is induced by a morphism of fusion systems, obstructions do not need to vanish. Castellana-Libman [42] showed that the obstruction groups are torsion, which is the key fact for the following partial result.

Theorem 9.4.18. *Let $(\rho, \Theta)\colon \mathcal{F} \to \mathcal{F}'$ be a morphism of saturated fusion systems where $\rho\colon S \to S'$ is a group homomorphism. Then there exists a natural number $m \geq 0$ such that the morphism $\Delta \circ \rho\colon S \to S' \to S' \wr \Sigma_m$ is fusion preserving and extends to a map $B\tilde{\rho}\colon B\mathcal{F} \to B(\mathcal{F}' \wr \Sigma_m)$.*

The next corollary can be interpreted as a proof of the existence of a regular representation for $\mathcal{F}$. It is obtained applying the previous Theorem 9.4.18 to the regular representation of the Sylow p-subgroup.

Corollary 9.4.19. *Let $\mathcal{F}$ be a saturated fusion system on a finite p-group S. There is a map $B\rho\colon B\mathcal{F} \to (B\Sigma_m)^\wedge_p$ such that its restriction to S is a sum of several copies of the regular representation.*

Remark 9.4.20. Any finite group admits a faithful unitary representation $G \hookrightarrow U(n)$ for some n. Combining the permutation matrix representation $\Sigma_n \hookrightarrow U(n)$ with the map obtained in Corollary 9.4.19, we obtain a faithful complex representation $B\rho\colon B\mathcal{F} \to BU(n)^\wedge_p$ in the homotopical sense, i.e. the homotopy fiber F of $B\rho$ has finite mod p cohomology; $H^*(F)$ is a finite $\mathbb{F}_p$-vector space. In particular, using Venkov's proof for showing that the mod p cohomology of a finite group is finitely generated, one obtains an alternative proof for the fact the mod p cohomology of a saturated fusion system is finitely generated.

Remark 9.4.21. Strategies involving the subgroup decomposition of a classifying space naturally lead to an obstruction theory which in many situations involves computing higher limits of functors over the orbit category. For finite groups, those limits have been studied in detail in many situations ([58], [80], [75]). In the context of fusion systems it is still not known whether the previous results on the vanishing of higher limits over the orbit category hold. An approach to this question has been attempted by Díaz-Park in [54] for the case of Mackey functors obtaining partial results.

We finish this part with an example of a stable elements theorem in a different context. Let $R(G)$ be the Grothendieck ring of complex representations of G. The classifying space functor $\mathrm{Rep}(G, U(n)) \to [BG, BU(n)^\wedge_p]$ induces a bijection if G is a finite p-group by work of Dwyer-Zabrodsky [64]. Then, looking at the stable virtual representations in this context, $\lim_{P\in\mathcal{F}} R(P)$ is closely related to $[B\mathcal{F}, BU(n)^\wedge_p]$ via restriction maps.

Given a topological space X one can consider a topological analogue of the representation ring at a prime p by defining $\mathbb{K}(X)$ to be the Grothendieck group completion of the monoid

$\coprod[X, BU(n)^\wedge_p]$ where the sum is induced by $U(n) \times U(m) \hookrightarrow U(n+m)$. If $X = B\mathcal{F}$, $\mathbb{K}(B\mathcal{F})$ is an homotopical analogue for the complex representation ring. We would like to compare it to the stable representations, that is, elements in $R(S)$ that satisfy the analogous relations as in Definition 9.1.1. There is a restriction morphism

$$\mathrm{Res}\colon \mathbb{K}(B\mathcal{F}) \longrightarrow \lim_{P \in \mathcal{O}(\mathcal{F})} R(P)$$

defined using the inclusion of the Sylow p-subgroup $\theta\colon BS \to B\mathcal{F}$. Jackowsky-Oliver [82] proved that Res is an isomorphism when $\mathcal{F}$ is the the fusion system of a finite group (when $B\mathcal{F} \simeq BG^\wedge_p$ for a finite group G). Cantarero-Castellana-Morales [34] showed that the same statement holds for a general saturated fusion system on a finite p-group.

Theorem 9.4.22. *If $\mathcal{F}$ is a saturated fusion system on a finite p-group S then*

$$\mathbb{K}(B\mathcal{F}) \cong \lim_{P \in \mathcal{O}(\mathcal{F}^c)} R(P).$$

9.4.2 The stable homotopy theory

Stable homotopy theory of saturated fusion systems deserves a special section since the developments in this area have been ahead of the unstable theory in solving relevant problems in the theory, like the existence of a classifying space or the functoriality of the classifying space construction.

If $S \in \mathrm{Syl}_p(G)$, the group S acts on the set G by left and right multiplication giving G the structure of an (S,S)-biset. Note that given $(s_1, s_2) \in S \times S$ and $g \in G$, the condition $s_1 g = g s_2$ is equivalent to $s_1 = c_g(s_2)$, that is, we can see conjugacy relations encoded in the (S,S)-biset G.

In an analogy with the idea of axiomatizing the properties of the fusion system $\mathcal{F}_S(G)$ in the abstract concept of a saturated fusion system, Lincklemann and Webb introduced the notion of a characteristic biset for a saturated fusion system in order to isolate the main features of G.

Let $A(G)$ be the Burnside ring of finite G-sets, i.e. the Grothendieck group completion of the monoid of isomorphism classes of finite left G-sets, where the multiplicative group structure is induced by Cartesian product. Let $A(G,H)$ be the Grothendieck group completion of the monoid of isomorphism classes of finite left free (G,H)-bisets.

Example 9.4.23. Given two finite groups G and H, we consider a pair (K, φ) where $K \leq G$ and $\varphi\colon K \to H$ is a monomorphism. Define the subgroup $\Delta(K,\varphi) = \{(\varphi(k), k) \mid k \in K\} \leq H \times G$. We denote by $H \times_{(K,\varphi)} G$ the (G,H)-biset defined by

$$(H \times G)/\Delta(K,\varphi).$$

Equivalently, $(H \times G)/\sim$ where $(x, gy) \sim (x\varphi(g), y)$ for $x \in H$, $y \in G$ and $g \in K$. Any left-free transitive (G,H)-biset is of this form. We write $[K,\varphi]$ for the isomorphism class of $H \times_{(K,\varphi)} G$ in $A(G,H)$.

Given a finite left-free (G,H)-biset X, we define a stable map $\alpha(X)\colon \Sigma^\infty_+ BG \to \Sigma^\infty_+ BH$. On transitive bisets of the form $X = H \times_{(K,\varphi)} G$, $\alpha(X) = \Sigma_+ B\varphi \circ \tau^H_G$ where $\tau^H_G\colon \Sigma^\infty_+ BG \to \Sigma^\infty_+ BK$ is the transfer map associated to the finite covering $BK \to BG$ (see [46]). Since every left-free biset is a coproduct of transitive ones, this construction extends to a homomorphism

$$\alpha\colon A(G,H) \longrightarrow \{\Sigma^\infty_+ BG, \Sigma^\infty_+ BH\}.$$

The Segal conjecture is about α being the completion with respect to the augmentation ideal $I(G)$ of the Burnside ring $A(G)$ (which acts on $A(G,H)$ by Cartesian product). When G is a finite p-group, completion can be related to p-adic completion (see [113, Section 9.1]).

Let $\varepsilon\colon A(G,H) \to \mathbb{Z}$ be the augmentation morphism defined on basis elements by $\varepsilon(X) = |X|/|H|$.

Theorem 9.4.24. ([40], [89]) *Let G, H be finite groups. Then*

$$\alpha\colon A(G,H)^\wedge_{I(G)} \xrightarrow{\cong} \{\Sigma^\infty_+ BG, \Sigma^\infty_+ BH\},$$

where $I(G)$ is the augmentation ideal $I(G) \subset A(G)$. If G and H are both finite p-groups, $A'(G,H)^\wedge_{I(G)} \cong A'(G,H)^\wedge_p \cong \mathbb{Z}^\wedge_p \otimes A'(G,H)$ where $A'(G,H)$ is the kernel of the augmentation morphism ε.

Example 9.4.25. Regard G as an (S,S)-biset. Then $\alpha([G]) \in \{\Sigma^\infty_+ BS, \Sigma^\infty_+ BS\}$ induces a morphism $H^*(\alpha([G]))\colon H^*(S) \to H^*(S)$ that corresponds to $\mathrm{Res}^G_S \circ \mathrm{Tr}^G_S$. Then, the image of $H^*(\alpha([G]))$ consists of the stable elements (Theorem 9.1.3).

Let $\varphi\colon P \hookrightarrow S$ be a morphism. If $X \in A(S,S)$, ${}_{(P,\varphi)}X \in A(S,P)$ denotes the (S,P)-biset obtained by restricting the action via φ, $X_{(P,\varphi)} \in A(P,S)$ is defined analogously.

Definition 9.4.26. Let $\mathcal{F}$ be a fusion system on a finite p-group S.

1. Let X be an (S,S)-biset. X is right (resp. left) *$\mathcal{F}$-stable* if for every $P \leq S$ and $\varphi \in \mathrm{Mor}_\mathcal{F}(P,S)$, $X_{(P,\mathrm{id})} \cong X_{(P,\varphi)}$ (resp. ${}_{(P,\mathrm{id})}X \cong {}_{(P,\varphi)}X$). We will just say that it is *$\mathcal{F}$-stable* if it is both right and left $\mathcal{F}$-stable.

2. If $P,Q \leq S$, $A_\mathcal{F}(P,Q)$ is the subgroup of $A(P,Q)$ generated by $[K,\varphi]$ where $\varphi \in \mathrm{Mor}_\mathcal{F}(P,Q)$.

Linckelmann and Webb suggested the following properties for a biset mimicking the properties of G.

Definition 9.4.27. Let $\mathcal{F}$ be a fusion system on a finite p-group S. An (S,S)-biset Ω is a *characteristic biset* for $\mathcal{F}$ if:

1. Ω is free as a left and right S-set.

2. $\Omega \in A_\mathcal{F}(S,S)$.

3. Ω is $\mathcal{F}$-stable.

4. $|\Omega|/|S|$ is prime to p.

The existence of characteristic bisets for saturated fusion systems was first proven by Broto-Levi-Oliver [26].

Theorem 9.4.28 (Proposition 5.5 in [26]). *Let $\mathcal{F}$ be a saturated fusion system on a finite p-group S. Then, there exists a characteristic biset Ω for $\mathcal{F}$. Moreover, $H^*(\alpha(\Omega))\colon H^*(S) \to H^*(S)$ is an idempotent $H^*(\mathcal{F})$-linear morphism with image $H^*(\mathcal{F})$.*

A saturated fusion system does not need to have a unique characteristic biset. For example, if $H \leq G$ controls fusion in G, then both G and H are characteristic bisets for $\mathcal{F} = \mathcal{F}_S(G) = \mathcal{F}_S(H)$.

Gelvin-Reeh [66] proved that every saturated fusion system $\mathcal{F}$ has a unique minimal characteristic biset $\Lambda_{\mathcal{F}}$ in the following sense: for any characteristic biset X we have $\Lambda_{\mathcal{F}} \subset X$. Special attention is devoted to constrained fusion systems. In that case, Theorem 9.3.16 shows that there is a group G with $O_{p'}(G) = 1$ and $C_G(O_p(G)) \leq O_p(G)$ such that $\mathcal{F} \cong \mathcal{F}_S(G)$, and is unique satisfying this property. Then the Gelvin-Reeh show that G, as a (S,S)-biset, is the minimal characteristic biset for $\mathcal{F}$.

Ragnarsson-Stancu [113] introduce an analogous notion for virtual (S,S)-bisets: $\Omega \in A(S,S)_{(p)}$ is a *characteristic element* if $\Omega \in A_{\mathcal{F}}(S,S)$ is $\mathcal{F}$-stable and $\varepsilon(\Omega)$ is prime to p, where $\varepsilon\colon A(S,S) \to \mathbb{Z}$ is the augmentation morphism defined on basis elements by $\varepsilon(X) = |X|/|S|$. By inverting $|X|/|S|$, one can always assume that $\varepsilon(X) = 1$. It is clear that a characteristic biset for $\mathcal{F}$ provides also a characteristic element, but the main result obtained by Ragnarsson in [112] is the existence of unique *idempotent characteristic element.*

Theorem 9.4.29. [112] *A saturated fusion system $\mathcal{F}$ on a finite p-group S has a unique idempotent characteristic element $\omega_{\mathcal{F}} \in A(S,S)_{(p)}$.*

By [113, Proposition 9.2], the $I(S)$-adic completion $A(S,S)^{\wedge}_{I(S)}$ injects into the p-adic completion $A(S,S)^{\wedge}_p$ with image given by the submodule of elements with augmentation in $\mathbb{Z}$. Then, since $A(S,S)_{(p)}$ is a submodule of $A(S,S)^{\wedge}_p$, a characteristic element gives an idempotent stable map $\tilde{\omega}_{\mathcal{F}}\colon \Sigma^{\infty}_{+}BS \to \Sigma^{\infty}_{+}BS$ by Theorem 9.4.24.

Remark 9.4.30. In [116] and [117], Reeh defines the Burnside ring for a saturated fusion system on a finite p-group S by considering $\mathcal{F}$-stable sets. By constructing a tranfer map from the Burnside ring of S, he gives a new construction for $\omega_{\mathcal{F}}$. The Burnside ring and the representation ring of $\mathcal{F}$ are studied in [118] and [67]. Previously, Díaz-Libman described the cohomotopy of the classifying space for saturated fusion systems (see [53], [52]).

Let $\tilde{A}(S,S)$ be the quotient module obtained from $A(S,S)$ by dividing out all the basis elements $[K,\varphi]$ where φ is the trivial homomorphisms. Then the Segal conjecture ([40], [89]) identifies stable maps between classifying spaces with no extra base point

$$\tilde{A}(S,S)^{\wedge}_p \cong \{\Sigma^{\infty}BS, \Sigma^{\infty}BS\}.$$

We consider $\tilde{\omega}_{\mathcal{F}} \in \{\Sigma^{\infty}BS, \Sigma^{\infty}BS\}$.

Definition 9.4.31. Let $\mathcal{F}$ be a saturated fusion system on a finite p-group S. The *classifying spectrum of* $\mathcal{F}$, $\mathbb{B}\mathcal{F}$, is the stable summand of $\Sigma^{\infty}BS$ obtained via the telescope construction

$$\operatorname{hocolim}(\Sigma^{\infty}BS \xrightarrow{\tilde{\omega}_{\mathcal{F}}} \Sigma^{\infty}BS \xrightarrow{\tilde{\omega}_{\mathcal{F}}} \cdots).$$

Then, there is a stable map $\mathrm{Tr}_{\mathcal{F}}\colon \mathbb{B}\mathcal{F}_+ \to \Sigma^\infty_+ BS$ (the transfer map) and $\sigma_{\mathcal{F}}\colon \Sigma^\infty_+ BS \to \mathbb{B}\mathcal{F}_+$ with $\sigma_{\mathcal{F}} \circ \mathrm{Tr}_{\mathcal{F}} \simeq \mathrm{id}_{\mathbb{B}\mathcal{F}_+}$ and $\mathrm{Tr}_{\mathcal{F}} \circ \sigma_{\mathcal{F}} \simeq \tilde{\omega}_{\mathcal{F}}$.

Remark 9.4.32. If $\mathcal{L}$ is a linking system for a saturated fusion system $\mathcal{F}$ on a finite p-group S, then $\Sigma^\infty|\mathcal{L}|^\wedge_p \simeq \mathbb{B}\mathcal{F}$ (see [112]).

Ragnarsson [112, Theorem 7.9] shows that the classifying spectrum construction is functorial with respect to morphisms of fusion systems. Let $\mathcal{F}$ and $\mathcal{F}'$ be two saturated fusion systems on finite p-groups S and S' respectively. Recall from Defintion 9.3.20 that a morphism of fusion systems is a pair (α, Ψ) where $\alpha\colon S \to S'$ is a morphism and $\Psi\colon \mathcal{F} \to \mathcal{F}'$ is a functor. Then there is a map of spectra $B\Psi\colon \mathbb{B}\mathcal{F}_+ \to \mathbb{B}\mathcal{F}'_+$ such that

$$\begin{array}{ccc} \Sigma^\infty_+ BS & \xrightarrow{B\alpha} & \Sigma^\infty_+ BS' \\ \downarrow{\scriptstyle \sigma_{\mathcal{F}}} & & \downarrow{\scriptstyle \sigma_{\mathcal{F}'}} \\ \mathbb{B}\mathcal{F}_+ & \xrightarrow[B\Psi]{} & \mathbb{B}\mathcal{F}'_+. \end{array}$$

Moreover, in [112, Theorem 7.2], the author describes a basis for stable maps between classifying spectrum of saturated fusion systems in terms of basis elements in the double Burnside ring for the corresponding Sylow p-subgroups.

The classical transfer morphism $\mathrm{Tr}_H^G\colon H^*(H) \to H^*(G)$ for given $H \leq G$ finite groups is a morphism $H^*(G)$-modules. This property is called Frobenius reciprocity.

Theorem 9.4.33. [112] *For any saturated fusion system on a finite p-group S, there is a natural transfer* $\mathrm{Tr}_{\mathcal{F}}\colon \mathbb{B}\mathcal{F}_+ \to BS_+$. *Moreover, the following diagram commutes up to homotopy*

$$\begin{array}{ccc} \mathbb{B}\mathcal{F}_+ & \xrightarrow{\Delta} & \mathbb{B}\mathcal{F}_+ \wedge \mathbb{B}\mathcal{F}_+ \\ \downarrow{\scriptstyle \mathrm{Tr}_{\mathcal{F}}} & & \downarrow{\scriptstyle 1\wedge \mathrm{Tr}_{\mathcal{F}}} \\ BS_+ & \xrightarrow[(\sigma_{\mathcal{F}}\wedge 1)\circ\Delta]{} & \mathbb{B}\mathcal{F}_+ \wedge BS_+. \end{array}$$

Remark 9.4.34. [113] The properties of the transfer map $\mathrm{Tr}_{\mathcal{F}}$ and the unique idempotent characteristic element $\tilde{\omega}_{\mathcal{F}}$ allow one to obtain a stable element theorem for any cohomology theory. That is, if E is a spectrum then $E^*(\sigma_{\mathcal{F}})$ is an split injection with image the $\mathcal{F}$-stable elements in $E^*(\Sigma^\infty_+ BS)$. If E is a ring spectrum then $E^*(\mathrm{Tr}_{\mathcal{F}})$ is map of $E^*(B\mathcal{F}_+)$-modules.

The difference with respect to previous results for finite groups, where stable elements formula for generalized cohomology theories with p-local coefficient were obtained in [82], is that the authors identified

$$E^*(BG) \longrightarrow \lim_{P\in\mathcal{O}_S(G)} E^*(BP)$$

with the edge homomorphism of the Bousfield spectral sequence for a colimit [21] and then proved that the higher limits vanish.

In Section 9.3, we pointed out that one of the main issues when constructing saturated fusion systems is to check the saturation axioms. In this context, Ragnarsson-Stancu decribed a construction of a fusion system $\mathcal{F}_\Omega$ associated to a symmetric idempotent element

$\Omega \in A(S,S)_{(p)}$. They show that saturation axioms for $\mathcal{F}_\Omega$ are encoded in the Frobenius reciprocity formula. We say that an element in $A(S,S)$ is symmetric if it is isomorphic to the one obtained by transposing the two actions.

Theorem 9.4.35. [113] *For a finite p-group S, there is a bijective correspondence between saturated fusion systems on S and symmetric idempotents in $A(S,S)_{(p)}$ of augmentation 1 that satisfy Frobenius reciprocity.*

Remark 9.4.36. In [113], the authors state a conjecture by Miller which attempts a purely homotopy theoretic description for the homotopy theory of saturated fusion systems. Let X be a connected, p-complete space with finite fundamental group and S a finite p-group. Assume there is a map $f\colon BS \to X$ such that $H^*(BS)$ is a finitely generated $H^*(X)$-module and that admits a stable transfer retract $t\colon \Sigma^\infty_+ X \to \Sigma^\infty_+ BS$ satisfying the Frobenius reciprocity relation. The conjecture says that X is then the classifying space of a saturated fusion system on S.

9.5 Algebraic models for finite loop spaces

A finite loop space is a pair of (X, BX) where $X \simeq \Omega BX$ is a pointed path-connected space with the homotopy type of finite CW-complex and BX is the classifying space. The motivating example is (G, BG) where G is a compact Lie group.

An extension to general compact Lie groups of the group cohomology of finite groups is provided by the cohomology of the classifying space. Many statements from sections 9.1 and 9.2 hold in general for compact Lie groups. For example, Quillen's results on the stratification of the mod p cohomology were originally proven for compact Lie groups. The stable elements theorem for any generalized cohomology theory with $\mathbb{Z}_{(p)}$-coefficients in the work of Jackowski-Oliver [82, Corollary 3.7] holds for compact Lie groups. Homology decompositions were described by Jackowski-McClure-Oliver in [78], [79], [80].

In order to study the homotopical properties of compact Lie groups, Rector considered their classifying spaces; but he proved that there are uncountably many distinct loop space structures on S^3 [115]. But Dwyer, Miller and Wilkerson showed that this problem goes away after p-completion, since they proved that there is a unique loop space structure on $(S^3)^\wedge_p$ [60]. This result lead to the proposal that the right category to isolate the homotopical properties of compact Lie groups is the category of p-complete loop spaces. Finally, Dwyer and Wilkerson [62] introduced the notion of a p-compact group.

Definition 9.5.1. A *p-compact group* is a loop space (X, BX) where X is an $\mathbb{F}_p$-finite space (that is, $H^*(X)$ is a finite $\mathbb{F}_p$-vector space), and BX is a p-complete space.

Examples of p-compact groups are provided by compact Lie groups such that $\pi_0(G)$ is a p-group. However, there are exotic examples of p-compact groups that are not the p-completion of any compact Lie group (see [61], [1]).

Most of the geometric structure of a connected compact Lie group can be translated into this homotopy theoretic setting, see [62]. We focus on the subgroup structure. We begin with the analogue of tori and their normalizers. A *p-compact toral group* is a p-compact group that is a finite extension of a p-compact torus (the p-completion of a torus) by a finite p-group.

Let $\mathbb{Z}/p^\infty \cong \mathbb{Z}\left[\frac{1}{p}\right]/\mathbb{Z}$ denote the union of the cyclic p-groups $\mathbb{Z}/p^n$ under the standard inclusions.

Definition 9.5.2. A *discrete p-toral group* is a group P with a normal subgroup $P_0 \lhd P$ such that P_0 is isomorphic to a finite product of copies of $\mathbb{Z}/p^\infty$, and P/P_0 is a finite p-group. The subgroup P_0 will be called the identity component of P.

The two definitions are related since, by [62, Prop. 6.9], any p-compact toral group is the $\mathbb{F}_p$-completion of a discrete p-toral group S.

Dwyer-Wilkerson proved that p-compact groups admit a maximal torus T_X, an associated Weyl group W_X, and a maximal p-compact toral subgroup N_p (see [62]) that fits into a fibration sequence

$$BT_X \to BN_p(T_X) \to B(W_X)_p$$

where $(W_X)_p \in \mathrm{Syl}_p(W_X)$. Then $N_p(T_X)$ is a p-compact toral group being an extension of a p-complete torus by a finite p-group, and it is the $\mathbb{F}_p$-completion of a discrete p-toral group S.

The main achievement in the theory of p-compact groups is the classification of connected p-compact groups by Andersen-Grodal-Møller-Viruel (p odd) and Andersen-Grodal ($p = 2$) in [5] and [4],where a bijective correspondence between connected p-compact groups and reflection data over the p-adic integers encoded in (W_X, T_X) was established.

Following the theory of saturated fusion systems on finite p-groups, Broto-Levi-Oliver introduced the notion of a saturated fusion system on a discrete p-toral group.

Definition 9.5.3. A *fusion system* $\mathcal{F}$ on a discrete p-toral group S is a subcategory of the category of groups whose objects are the subgroups of S, and whose morphism sets $\mathrm{Hom}_{\mathcal{F}}(P, Q)$ satisfy the following conditions:

(a) $\mathrm{Hom}_S(P, Q) \subseteq \mathrm{Hom}_{\mathcal{F}}(P, Q) \subseteq \mathrm{Inj}(P, Q)$ for all $P, Q \leq S$.

(b) Every morphism in $\mathcal{F}$ factors as an isomorphism in $\mathcal{F}$ followed by an inclusion.

The saturation axioms in Definition 9.3.2 include now a third axiom modeling continuity.

(III) If $P_1 \leq P_2 \leq P_3 \leq \cdots$ is an increasing sequence of subgroups of S, with $P_\infty = \bigcup_{n=1}^{\infty} P_n$, and if $\phi \in \mathrm{Hom}(P_\infty, S)$ is any homomorphism such that $\phi_{|P_n} \in \mathrm{Hom}_{\mathcal{F}}(P_n, S)$ for all n, then $\phi \in \mathrm{Hom}_{\mathcal{F}}(P_\infty, S)$.

The inspiring example for this definition is the fusion system of a compact Lie group G. If T is a maximal torus of G, let W_p be the p-Sylow subgroup of the Weyl group $W_G(T) = N_G(T)/T$. Then, the preimage of W_p in $N_G(T)$ defines a subgroup N_p that is an extension of W_p by T. The proof of Proposition 9.3 (b) in [26] shows that any compact Lie group G has a maximal discrete p-toral subgroup S which can be found as a discrete subgroup of N_p and that it is unique up to G-conjugacy. The fusion system $\mathcal{F}_S(G)$ over S defined by setting $\mathrm{Hom}_{\mathcal{F}_S(G)}(P, Q) = \mathrm{Hom}_G(P, Q)$ for all $P, Q \leq S$ is saturated.

There is also an axiomatic version of the linking system in this context in order to recover the classifying space. To do so, Broto, Levi, and Oliver in [28] axiomatized a new category $\mathcal{L}$, the centric linking system, containing the information needed to construct the classifying space. All this information together leads to the definition of a p-local compact group.

Definition 9.5.4 (Broto–Levi–Oliver). A *p-local compact group* is a triple $\mathcal{G} = (\mathcal{S}, \mathcal{F}, \mathcal{L})$ where $\mathcal{F}$ is a saturated fusion system over a discrete p-toral group S and $\mathcal{L}$ is a centric linking system associated to $\mathcal{F}$. The classifying space $B\mathcal{G}$ is defined as $|\mathcal{L}|^{\wedge}_p$, the p-completion of the nerve of the associated centric linking system.

Recently, following ideas of the proof in the finite case [103], Levi–Libman [86] proved that there is a unique p-complete classifying space associated to a saturated fusion system over a discrete p-toral group. Hence, we will often denote the classifying space of a p-local compact group simply as $B\mathcal{F}$.

Examples of p-local compact groups described in [28] are compact Lie groups, p-compact groups, and torsion linear groups. More progress on constructing new examples can be found in the work of González-Lozano-Ruiz [73].

A relevant result which unifies some of the previous examples from the point of view of homotopy theory is the following one: p-local compact groups also model p-completions of classifying spaces of finite loop spaces [29].

Theorem 9.5.5. *Let X be a connected space. For each prime p such that $H^*(\Omega X)$ is finite, the space $X^{\wedge}_p$ has the homotopy type of the classifying space of a p-local compact group. In particular, as long as $H^*(\Omega X; \mathbb{Z})$ is finite (for example if $(\Omega X, X)$ is a finite loop space), the space $X^{\wedge}_p$ has the homotopy type of the classifying space of a p-local compact group.*

There are notions of centralizer and normalizer saturated fusion systems. In particular the descriptions of mapping spaces and homotopy classes of maps from BQ where Q is a discrete p-toral group in the spirit of Theorem 9.4.15 (see also [72]), and the existence of homology decompositions (see [28]) also hold.

A version of Alperin's fusion theorem is also available in this context.

Definition 9.5.6. Let $\mathcal{F}$ be a saturated fusion system on a discrete p-toral group S. A subgroup $P \leq S$ is called *$\mathcal{F}$-centric* if P and all of its $\mathcal{F}$-conjugates contain their S-centralizers. A subgroup $Q \leq S$ is called *$\mathcal{F}$-radical* if $\mathrm{Out}_{\mathcal{F}}(Q)$ contains no nontrivial normal p-subgroup.

Remark 9.5.7. By Proposition 2.3 in [28], $\mathrm{Out}_{\mathcal{F}}(Q)$ is finite for all $Q \leq S$, so Definition 9.5.6 makes sense.

We will denote by $\mathcal{F}^c$ the full subcategory generated by $\mathcal{F}$ whose objects are the $\mathcal{F}$-centric subgroups of S and by $\mathcal{F}^{cr}$ the full subcategory whose objects are the $\mathcal{F}$-centric $\mathcal{F}$-radical subgroups of S.

Theorem 9.5.8 (Alperin's fusion theorem, [28]). *Let $\mathcal{F}$ be a saturated fusion system on a discrete p-toral group S. Then $\mathcal{F}$ is $\mathcal{F}^{cr}$-generated.*

The *orbit category* $\mathcal{O}(\mathcal{F})$ is the category whose objects are the subgroups of S and whose morphisms are given by

$$\mathrm{Hom}_{\mathcal{O}(\mathcal{F})}(P, Q) = \mathrm{Rep}_{\mathcal{F}}(P, Q) = \mathrm{Hom}_{\mathcal{F}}(P, Q)/\operatorname{Inn}(Q).$$

For any full subcategory $\mathcal{F}_0$ of $\mathcal{F}$ we also consider $\mathcal{O}(\mathcal{F}_0)$, the full subcategory of $\mathcal{O}(\mathcal{F})$ whose objects are those of $\mathcal{F}_0$. Let $B\colon \mathcal{O}(\mathcal{F}_0) \to \mathrm{HoTop}$ be the classifying space functor regarded as a functor to the homotopy category of topological spaces.

Proposition 9.5.9. [28] *Fix a saturated fusion system $\mathcal{F}$ on a discrete p-toral group S, and let $\mathcal{F}' \subseteq \mathcal{F}^c$ be any full subcategory. Given a linking system $\mathcal{L}'$ associated to $\mathcal{F}'$, the left homotopy Kan extension $\widetilde{B}\colon \mathcal{O}(\mathcal{F}') \to \mathrm{Top}$ of the constant functor $\mathcal{L}' \to \mathrm{Top}$ along the projection $\widetilde{\pi}\colon \mathcal{L}' \to \mathcal{O}(\mathcal{F}')$ is a lift of B to Top, and there is a homotopy equivalence:*

$$|\mathcal{L}'| \simeq \operatorname*{hocolim}_{\mathcal{O}(\mathcal{F}')} \widetilde{B}.$$

But there are some questions related to the key words in Section 9.1 which we haven't addressed: is $H^*(B\mathcal{F})$ finitely generated? is there a transfer map $\mathrm{Tr}\colon H^*(BS) \to H^*(B\mathcal{F})$? does the stable elements formula hold? is it true that $\mathrm{Spec}^h(H^*(B\mathcal{F}))$ is stratified by using elementary abelian p-subgroups?

The next result is a key statement. Note that the classifying space of a discrete p-toral group is a homotopy direct colimit of classifying spaces of finite p-groups. In [72, Theorem 1], González proves any p-local compact group can be approximated by a sequence of p-local finite groups in such a way that the homotopy colimit over the associated sequence of classifying spaces is equivalent to the classifying space after p-completion.

Theorem 9.5.10. *If $\mathcal{F}$ is the saturated fusion system on S of a p-local compact group, then there is a commutative diagram*

$$\begin{array}{ccccccc} \cdots & \longrightarrow & BS_i & \longrightarrow & BS_{i+1} & \longrightarrow & \cdots \\ & & \downarrow & & \downarrow & & \\ \cdots & \longrightarrow & B\mathcal{F}_i & \longrightarrow & B\mathcal{F}_{i+1} & \longrightarrow & \cdots \end{array}$$

of classifying spaces of p-local finite groups $\mathcal{F}_i$ over S_i, indexed on $i \in \mathbb{N}$ whose colimits recovers the map $BS \to B\mathcal{F}$ after p-completion.

In [72], many corollaries of the previous theorem are stated. Combining Theorem 9.5.10 with the stable elements theorem for the finite case, González proves the stable elements theorem in this context.

What about the remaining questions? They have positive answers from work of Barthel-Castellana-Heard-Valenzuela [12]. The F-isomorphim theorem with respect to elementary abelian p-subgroups and the strong Quillen's stratification hold ([12, Theorem 5.1, Theorem 5.6]), by using work of Rector on spaces with Noetherian cohomology [114]. Finally, a transfer map is constructed at the level of geometric cochains, the function spectrum $F(\Sigma^\infty X, H\mathbb{F}_p)$.

Proposition 9.5.11. *Let $\mathcal{F}$ be a saturated fusion system on a p-discrete toral group and $\theta\colon BS \to B\mathcal{F}$ the canonical map. Then, $\theta^*\colon H^*(B\mathcal{F}) \to H^*(BS)$ is split as a map of $H^*(B\mathcal{F})$-modules.*

Corollary 9.5.12. *Let $\mathcal{F}$ be a saturated fusion system on a p-discrete toral group. Then $H^*(B\mathcal{F})$ is finitely generated.*

The homotopy theory of maps between classifying spaces is less developed, as well as the stable homotopy theory. The transfer in mod p cohomology is induced from a stable map between geometric cochains but there is no analogue of a stable transfer between classifying spaces satisfying Frobenius reciprocity, for example.

Concerning homotopy classes of maps between classifying spaces, there is no analogue of Theorem 9.4.18 in this more general context but one cannot expect the obstruction groups to be always torsion. A first attempt is in the work of Cantarero-Castellana [33] where they approach the problem of constructing maps to unitary groups $U(n)$. Faithful unitary representations correspond to homotopy monomorphisms from the classifying space $B\mathcal{F}$ into $BU(n)^\wedge_p$ (the homotopy fiber is mod p finite). The first nontrivial problem is to find fusion preserving faithful complex representations.

Theorem 9.5.13. *Let $\mathcal{F}$ be a saturated fusion system on a discrete p-toral group S. Then there exists a faithful complex representation $\rho \in \mathrm{Hom}(S, U(n))$ such that $[\rho] \in \lim_{\mathcal{O}(\mathcal{F})} R(P)$.*

The second problem is to formulate the obstruction theory. In that case, only partial results were obtained which apply under certain hypothesis on the depth of the orbit category which assure vanishing of the obstruction groups.

Theorem 9.5.14. *Let (X, BX) be a finite loop space or a p-compact group. Then there exists faithful unitary representation $BX \to BU(n)^\wedge_p$ for some natural number $n \geq 1$.*

The stable homotopy theory described in 9.4.2 has yet to be developed in the context of p-local compact groups or saturated fusion systems on discrete p-toral groups.

9.6 Modular representation theory

Let k be a field and G a finite group. When attempting to classify kG-modules, there are very different situations. If the characteristic of the field $\mathrm{ch}(k)$ does not divide the order of G then Maschke's theorem says that every kG-module is a direct sum of irreducible kG-modules. Moreover, there is a finite number of such, classified by their character describing completely the picture. In contrast, when $\mathrm{ch}(k)$ does divide the order of G, it is no longer true that kG-modules split in such a way (e.g. the regular representation kG is not a direct sum of irreducible representations), and even non-isomorphic modules can have the same character. Only in special cases indecomposable can be classified, e.g. when $S \in \mathrm{Syl}_p(G)$ is cyclic or when $p = 2$ and $S = D_{2^n}, Q_{2^n}, SD_{2^n}$. When $\mathrm{ch}(k)$ divides $|G|$, the study of the category of kG-modules is called modular representation theory.

At this point one can adopt several strategies to pursue in modular representation theory: one can restrict attention to a particular class of modules that are more accessible or attempt a classification by means of a different organizational principle. We will give examples of both strategies: the classification of endotrivial modules and the stratification of the stable module category. In both cases, the fusion category will play a role to go from local to a global description.

9.6.1 Endotrivial modules

Let Mod_{kG} be the category of kG-modules where k is a field such that $\mathrm{ch}(k)$ divides $|G|$. The stable module category StMod_{kG} is the category whose objects are kG-modules but given $M, N \in \mathrm{Mod}_{kG}$, $\mathrm{Hom}_{\mathrm{StMod}_{kG}}(M, N) = \mathrm{Hom}_{kG}(M, N)/P$ where P is the linear span of morphism that factor through a projective.

Definition 9.6.1. An object $M \in \mathrm{StMod}_{kG}$ of finite dimension is called *endotrivial* if its endomorphism ring $\mathrm{End}_{kG}(M) \cong M \otimes M^*$ is equivalent in StMod_{kG} to the unit k, the one-dimensional trivial representation; equivalently if M belongs to the Picard group of StMod_{kG}.

Let $T_k(G)$ be the set of equivalence classes of endotrivial modules. The abelian group structure on $T_k(G)$ is induced by the tensor product of kG-modules. An example of an endotrivial module is given by the trivial one dimensional representation k and $\Sigma^i k$ with $i \in \mathbb{Z}$.

A general structure theorem for p-groups S due to Puig [106] proves that $T_k(S)$ is a finitely generated abelian group and Alperin [3] determined its rank. If S is abelian, Dade [51] proved that $T_k(G)$ is a cyclic infinite group generated by $\Sigma^i k$. But what it is relevant is that if S is a finite p-group, $T_k(S)$ was completely determined by work of Carlson-Thévenaz in [37], and [38].

Given a finite group G, there is a restriction map

$$\mathrm{Res}\colon T_k(G) \to T_k(S),$$

and the image of the restriction map was also determined by Carlson, Mazza, Nakano, Thévenaz ([92], [36], [35]). We define $T_k(G, S)$ to be the kernel of the morphism Res. Most of the efforts in this line of research have been then concentrated in computing $T_k(G, S)$. Carlson-Thévenaz [39] conjectured a description of it in terms of the fusion data, meaning local subgroups of G. In this direction, Balmer [11] established a connection to the equivariant topology of the Brown complex of p-subgroups of G.

Grodal [76] describes $T_k(G, S)$ in terms of the orbit category of p-subgroups in G. Let $\mathcal{O}_p^*(G)$ be the orbit category of non-trivial p-subgroups of G.

Theorem 9.6.2. [76] *Let G be a finite group and k a field of characteristic p that divides the order of G. There is an isomorphism of abelian groups*

$$\Phi\colon T_k(G, S) \to H^1(|\mathcal{O}_p^*(G)|; k^\times).$$

The morphism Φ and its inverse are explicitly described in [76]. The morphism Φ is induced by considering, for each orbit G/P, the zeroth Tate cohomology $\hat{H}^0(P; M)$ which turns out to be a 1-dimensional k-module. Consequences and different descriptions in tems of p-local data are stated as a corollary of the previous result; in particular, an affirmative solution to the Carlson-Thévenaz conjecture. To emphasize the description in terms of p-local data, Grodal describes $H_1(|\mathcal{O}_p^*(G)|; k^\times)$ as an explicit quotient of $N_G(S)$. The power of Theorem 9.6.2 is that it is suitable for explicit computations of these groups.

9.6.2 Stratification and duality

The stable module category is a tensor triangulated category. Given a tensor triangulated category, an important problem is the classification of tensor triangulated localizing subcategories, i.e. tensor triangulated subcategories that are closed under all filtered colimits.[1] Neeman's result [99] on the classification of localizing subcategories of the derived category $D(R)$ of a commutative Noetherian ring R in terms of subsets of the spectrum $\mathrm{Spec}(R)$ is such an example.

Benson-Iyengar-Krause [19] proved that if G is a finite group and k a field of characteristic p dividing $|G|$, the set of localizing subcategories of the stable module category StMod_{kG} can be parametrized in terms of subsets of the spectrum $\mathrm{Spec}^h(H^*(G))$ of homogeneous prime ideals. The idea is to describe minimal localizing subcategories in terms of certain local cohomology functors for each prime ideal. In [18], the same authors developed a general theory for triangulated categories T with an action of a graded commutative ring R, by constructing local cohomology functors and defining support functions supp_R. When R is Noetherian and T is stratified there is a one-to-one correspondence between localizing subcategories of T and subsets of $\mathrm{supp}_R(T) \subseteq \mathrm{Spec}^h(R)$.

The proof of the stratification of StMod_{kG} is based on a descent type argument which uses Quillen's strong stratification of the cohomology $H^*(G)$ and Chouinard's theorem to reduce the problem to elementary abelian p-groups.

Theorem 9.6.3 (Chouinard's Theorem, [45]). *Let R a ring, and M an RG-module. Then, M is projective iff the restriction of M is projective as an RE-module for every elementary abelian p-subgroup $E \leq G$*

How can these statements on kG-modules be translated into a homotopy theoretic setting? Let R be a commutative ring spectrum and Mod_R the category of module spectra over R. We write $C^*(X)$ for the ring spectrum of $H\mathbb{F}_p$-valued cochains on X, i.e. the function spectrum $F(\Sigma^\infty_+ X, H\mathbb{F}_p)$. Then, $C^*(X)$ is an augmented commutative $H\mathbb{F}_p$-algebra with multiplication given by the composite

$$\Sigma^\infty_+ X \xrightarrow{\Delta} \Sigma^\infty_+ X \wedge \Sigma^\infty_+ X \xrightarrow{x \wedge y} H\mathbb{F}_p \wedge H\mathbb{F}_p \xrightarrow{m} H\mathbb{F}_p$$

where the first map is induced by the diagonal of X and the last map m is the multiplication on the ring spectrum $H\mathbb{F}_p$. There is an isomorphism $\pi_* C^*(X) \cong H^{-*}(X)$.

Benson-Iyengar-Krause showed that stratification results also hold for several categories, one of which is $K(\mathrm{Inj}\, G)$, the homotopy category of complexes of injective modules. The authors define an equivalence of categories $\mathrm{StMod}_{kG} \xrightarrow{\simeq} K_{ac}(\mathrm{Inj}\, G) \subset K(\mathrm{Inj}\, G)$ where $K_{ac}(\mathrm{Inj}\, G)$ is the subcategory of acyclic complexes. And, precisely this category can be interpreted in terms of cochains on the classifying space. Benson-Krause [20] proved that there is a functor $K(\mathrm{Inj}\, G) \to \mathrm{Mod}_{C^*(BG)}$ that is an equivalence if G is a finite p-group. From the homotopy theory point of view, one is led to investigate the category $\mathrm{Mod}_{C^*(BG)}$.

The case when G compact connected Lie group was first described in the work of Benson-Greenlees [15]. Barthel-Castellana-Heard-Valenzuela [12] proved that the module category

[1] See Paul Balmer's chapter in this Handbook for a survey of classification results in the theory of tensor triangulated categories.

$\mathrm{Mod}_{C^*(B\mathcal{F})}$ is also stratified when $\mathcal{F}$ is a saturated fusion system over a finite p-group S, or $\mathcal{F} \cong \mathcal{F}_S(G)$ where G is a compact Lie group or a p-compact group.

Theorem 9.6.4. ([15], [12]) *Let $\mathcal{F}$ be a saturated fusion on a finite p-group, or the saturated fusion system of Lie group or a p-compact group. The category $\mathrm{Mod}_{C^*(B\mathcal{F})}$ is stratified by the canonical action of $H^*(B\mathcal{F})$. In particular, there is a bijection*

$$\left\{ \begin{matrix} \textit{Localizing subcategories} \\ \textit{of } \mathrm{Mod}_{C^*(B\mathcal{F})} \end{matrix} \right\} \underset{\longleftarrow}{\overset{\text{supp}}{\longrightarrow}} \left\{ \textit{Subsets of } \mathrm{Spec}^h(H^*(B\mathcal{F})) \right\}. \tag{9.6.1}$$

The proof of Theorem 9.6.4 follows a descent strategy where the role of elementary abelian p-subgroups is essential. Quillen's strong stratification Theorem 9.4.17 (see [12]) for $H^*(B\mathcal{F})$ and Chouinard's theorem are the key ingredients in the descent argument.

In general, if $f\colon R \to R'$ be a morphism of commutative ring spectra, then R' is also an R-module via f. Forgetting along f induces a restriction functor $\mathrm{Res}_f\colon \mathrm{Mod}_{R'} \to \mathrm{Mod}_R$ which admits both a left adjoint Ind_f and a right adjoint CoInd_f, given by induction (or extension of scalars) along f,

$$\mathrm{Ind}_f = R' \otimes_R (-)\colon \mathrm{Mod}_R \longrightarrow \mathrm{Mod}_{R'},$$

and coinduction, defined as

$$\mathrm{CoInd}_f = \mathrm{Hom}_R(R', -)\colon \mathrm{Mod}_R \longrightarrow \mathrm{Mod}_{R'}.$$

Restriction detects equivalences, but that is not necessarily true for the induction and coinduction functors. A functor $F\colon \mathcal{C} \to \mathcal{D}$ is conservative if it reflects equivalences, that is, $f \in \mathrm{Mor}(\mathcal{C})$ is an equivalence if and only if $F(f) \in \mathrm{Mor}(\mathcal{D})$ is so.

Definition 9.6.5. A map $f\colon R \to R'$ of commutative ring spectra is called conservative (resp. coconservative) if the associated induction functor $\mathrm{Ind}_f\colon \mathrm{Mod}_R \to \mathrm{Mod}_{R'}$ (resp. coinduction CoInd_f) is conservative. If a map is both conservative and coconservative, it will be called biconservative.

Consider

$$f := \prod \mathrm{Res}_{\mathcal{F},E}\colon C^*(B\mathcal{F}) \longrightarrow \prod_{E \in \mathcal{F}^e} C^*(BE)$$

where $C^*(B\mathcal{F}) \to C^*(BE)$ is induced by the inclusion $E \leq S$. Then Chouinard's theorem is about f being conservative. Benson-Greenlees [15] proved that f is biconservative when G is a compact Lie group. In fact, because of the existence of a transfer map $C^*(BS) \to C^*(B\mathcal{F})$ of $C^*(B\mathcal{F})$-modules (see [12], Proposition 9.5.11), in general one only needs to check that that f is biconservative when $B\mathcal{F} = BS$.

9.6.2.1 Duality

We finish with an example which comes from duality properties of the cohomology rings of finite groups. Benson–Carlson duality for cohomology rings of finite groups [14] shows that if the mod p cohomology ring of a finite group is Cohen–Macaulay, then it is Gorenstein.

One example of an explicit computation of the mod p cohomology of the classifying space of a saturated fusion system is in [74], where the mod 2–cohomology rings of the exotic 2–local finite groups $\mathrm{Sol}(q)$ constructed in Levi–Oliver [87] are described. One can check from the explicit description that these rings are Gorenstein [34, Example 6.10], suggesting Benson–Carlson duality might hold for saturated fusion systems.

Dwyer, Greenlees and Iyengar [59] expressed Benson–Carlson duality, as well as other duality properties, in the framework of ring spectra, showing that it is a particular instance of a more general situation described at the level of cochains.

From this point of view, Benson–Carlson duality is a consequence of the fact that the augmentation map $C^*(BG) \to H\mathbb{F}_p$ is Gorenstein in the sense of [59, Definition 8.1] and the existence of a local cohomology spectral sequence. Under some technical assumptions, if k is a field of characteristic p, we say that a map of ring spectra $R \to Hk$ is Gorenstein of shift a if there is an equivalence $\mathrm{Hom}_R(Hk, R) \simeq \Sigma^a Hk$ of Hk-modules. In classical commutative algebra, we recall that a commutative Noetherian local ring R is Gorenstein if and only if $\mathrm{Ext}^*_R(k, R)$ is a one dimensional k-vector space.

The proof that $C^*(BG; \mathbb{F}_p) \to H\mathbb{F}_p$ is Gorenstein follows the argument in [59, Example 10.3], where the main ingredient is the existence of a faithful unitary representation $G \hookrightarrow SU(n)$ such that $H^*(SU(n)/G)$ is an Poincaré duality $\mathbb{F}_p$-algebra.

The analogue for saturated fusion systems of the existence faithful unitary representation with Poincaré duality on the homogeneous space is the following result.

Theorem 9.6.6. [34] *Given a saturated fusion system $\mathcal{F}$ on a finite p-group, there exists a map $B\mathcal{F} \to BSU(n)^\wedge_p$ whose homotopy fiber has finite mod p cohomology satisfying Poincaré duality.*

The Gorenstein property then follows from a formal argument ([59, Example 10.3]).

Theorem 9.6.7. [34] *Let $\mathcal{F}$ be a saturated fusion system on a finite p-group. Then, the augmentation $C^*(B\mathcal{F}) \to H\mathbb{F}_p$ is Gorenstein. Moreover, if $H^*(B\mathcal{F})$ is Cohen-MacCaulay, then it is also Gorenstein.*

This is another example on how fusion systems provide a framework where cohomological and homotopical properties of classifying spaces of finite groups and Lie groups can be reinterpreted and viewed in a much broader context.

Bibliography

[1] J. Aguadé. Constructing modular classifying spaces. *Israel J. Math.*, 66(1–3):23–40, 1989.

[2] J. L. Alperin. Sylow intersections and fusion. *J. Algebra*, 6:222–241, 1967.

[3] J. L. Alperin. A construction of endo-permutation modules. *J. Group Theory*, 4(1):3–10, 2001.

[4] K. K. S. Andersen, J. Grodal, J. M. Møller, and A. Viruel. The classification of p-compact groups for p odd. *Ann. of Math. (2)*, 167(1):95–210, 2008.

[5] Kasper K. S. Andersen and Jesper Grodal. The classification of 2-compact groups. *J. Amer. Math. Soc.*, 22(2):387–436, 2009.

[6] Kasper K. S. Andersen, Bob Oliver, and Joana Ventura. Reduced, tame and exotic fusion systems. *Proc. Lond. Math. Soc. (3)*, 105(1):87–152, 2012.

[7] Michael Aschbacher. *Finite group theory*, volume 10 of *Cambridge Stud. Adv. Math.* Cambridge University Press, 1986.

[8] Michael Aschbacher. Normal subsystems of fusion systems. *Proc. Lond. Math. Soc. (3)*, 97(1):239–271, 2008.

[9] Michael Aschbacher and Andrew Chermak. A group-theoretic approach to a family of 2-local finite groups constructed by Levi and Oliver. *Ann. of Math. (2)*, 171(2):881–978, 2010.

[10] Michael Aschbacher, Radha Kessar, and Bob Oliver. *Fusion systems in algebra and topology*, volume 391 of *London Math. Soc. Lecture Note Ser.*. Cambridge University Press, 2011.

[11] Paul Balmer. Endotrivial representations of finite groups and equivariant line bundles on the Brown complex. *Geom. Topol.*, 22(7):4145–4161, 2018.

[12] Tobias Barthel, Natalia Castellana, Drew Heard, and Gabriel Valenzuela. Stratification and duality for homotopical groups. arXiv:1711.03491, 2017.

[13] David J. Benson. Cohomology of sporadic groups, finite loop spaces, and the Dickson invariants. In *Geometry and cohomology in group theory (Durham, 1994)*, volume 252 of *London Math. Soc. Lecture Note Ser.*, pages 10–23. Cambridge University Press, 1998.

[14] D. J. Benson and Jon F. Carlson. Projective resolutions and Poincaré duality complexes. *Trans. Amer. Math. Soc.*, 342(2):447–488, 1994.

[15] David Benson and John Greenlees. Stratifying the derived category of cochains on BG for G a compact Lie group. *J. Pure Appl. Algebra*, 218(4):642–650, 2014.

[16] David J. Benson, Jesper Grodal, and Ellen Henke. Group cohomology and control of p-fusion. *Invent. Math.*, 197(3):491–507, 2014.

[17] Dave Benson, Srikanth B. Iyengar, and Henning Krause. Localising subcategories for cochains on the classifying space of a finite group. *C. R. Math. Acad. Sci. Paris*, 349(17-18):953–956, 2011.

[18] Dave Benson, Srikanth B. Iyengar, and Henning Krause. Stratifying triangulated categories. *J. Topol.*, 4(3):641–666, 2011.

[19] David J. Benson, Srikanth B. Iyengar, and Henning Krause. Stratifying modular representations of finite groups. *Ann. of Math. (2)*, 174(3):1643–1684, 2011.

[20] David J. Benson and Henning Krause. Complexes of injective kG-modules. *Algebra Number Theory*, 2(1):1–30, 2008.

[21] A. K. Bousfield. On the homology spectral sequence of a cosimplicial space. *Amer. J. Math.*, 109(2):361–394, 1987.

[22] A. K. Bousfield and D. M. Kan. *Homotopy limits, completions and localizations.* volume 304 of *Lecture Notes in Math*, Springer-Verlag, 1972.

[23] Carles Broto, Natàlia Castellana, Jesper Grodal, Ran Levi, and Bob Oliver. Subgroup families controlling p-local finite groups. *Proc. London Math. Soc. (3)*, 91(2):325–354, 2005.

[24] C. Broto, N. Castellana, J. Grodal, R. Levi, and B. Oliver. Extensions of p-local finite groups. *Trans. Amer. Math. Soc.*, 359(8):3791–3858, 2007.

[25] Carles Broto, Ran Levi, and Bob Oliver. Homotopy equivalences of p-completed classifying spaces of finite groups. *Invent. Math.*, 151(3):611–664, 2003.

[26] Carles Broto, Ran Levi, and Bob Oliver. The homotopy theory of fusion systems. *J. Amer. Math. Soc.*, 16(4):779–856, 2003.

[27] Carles Broto, Ran Levi, and Bob Oliver. A geometric construction of saturated fusion systems. In *An alpine anthology of homotopy theory*, volume 399 of *Contemp. Math.*, pages 11–40. Amer. Math. Soc., 2006.

[28] Carles Broto, Ran Levi, and Bob Oliver. Discrete models for the p-local homotopy theory of compact Lie groups and p-compact groups. *Geom. Topol.*, 11:315–427, 2007.

[29] Carles Broto, Ran Levi, and Bob Oliver. An algebraic model for finite loop spaces. *Algebr. Geom. Topol.*, 14(5):2915–2981, 2014.

[30] Carles Broto, Jesper M. Møller, and Bob Oliver. Equivalences between fusion systems of finite groups of Lie type. *J. Amer. Math. Soc.*, 25(1):1–20, 2012.

[31] C. Broto and S. Zarati. Nil-localization of unstable algebras over the Steenrod algebra. *Math. Z.*, 199(4):525–537, 1988.

[32] W. Burnside. *Theory of groups of finite order.* Dover Publications, Inc., 1955. 2d ed.

[33] José Cantarero and Natàlia Castellana. Unitary embeddings of finite loop spaces. *Forum Math.*, 29(2):287–311, 2017.

[34] Jose Cantarero, Natalia Castellana, and Lola Morales. Vector bundles over classifying spaces of p-local finite groups and Benson-Carlson duality. arXiv:1701.08309, 2017.

[35] Jon F. Carlson, Nadia Mazza, and Daniel K. Nakano. Endotrivial modules for finite groups of Lie type A in nondefining characteristic. *Math. Z.*, 282(1-2):1–24, 2016.

[36] Jon F. Carlson, Nadia Mazza, and Jacques Thévenaz. Endotrivial modules over groups with quaternion or semi-dihedral Sylow 2-subgroup. *J. Eur. Math. Soc. (JEMS)*, 15(1):157–177, 2013.

[37] Jon F. Carlson and Jacques Thévenaz. The classification of endo-trivial modules. *Invent. Math.*, 158(2):389–411, 2004.

[38] Jon F. Carlson and Jacques Thévenaz. The classification of torsion endo-trivial modules. *Ann. of Math. (2)*, 162(2):823–883, 2005.

[39] Jon F. Carlson and Jacques Thévenaz. The torsion group of endotrivial modules. *Algebra Number Theory*, 9(3):749–765, 2015.

[40] Gunnar Carlsson. Equivariant stable homotopy and Segal's Burnside ring conjecture. *Ann. of Math. (2)*, 120(2):189–224, 1984.

[41] Henri Cartan and Samuel Eilenberg. *Homological algebra*. Princeton University Press, 1956.

[42] Natàlia Castellana and Assaf Libman. Wreath products and representations of p-local finite groups. *Adv. Math.*, 221(4):1302–1344, 2009.

[43] Andrew Chermak. Fusion systems and localities. *Acta Math.*, 211(1):47–139, 2013.

[44] Andy Chermak and Alex González. Discrete localities. I. arXiv:1702.02595, 2017.

[45] Leo G. Chouinard. Projectivity and relative projectivity over group rings. *J. Pure Appl. Algebra*, 7(3):287–302, 1976.

[46] Mónica Clapp. Duality and transfer for parametrized spectra. *Arch. Math. (Basel)*, 37(5):462–472, 1981.

[47] Murray Clelland and Christopher Parker. Two families of exotic fusion systems. *J. Algebra*, 323(2):287–304, 2010.

[48] David A. Craven. Normal subsystems of fusion systems. *J. Lond. Math. Soc. (2)*, 84(1):137–158, 2011.

[49] David A. Craven. *The theory of fusion systems: An algebraic approach*, volume 131 of *Cambridge Stud. Adv. Math.*. Cambridge University Press, 2011.

[50] Everett C. Dade. Endo-permutation modules over p-groups. I. *Ann. of Math. (2)*, 107(3):459–494, 1978.

[51] Everett Dade. Endo-permutation modules over p-groups. II. *Ann. of Math. (2)*, 108(2):317–346, 1978.

[52] Antonio Díaz and Assaf Libman. The Burnside ring of fusion systems. *Adv. Math.*, 222(6):1943–1963, 2009.

[53] Antonio Díaz and Assaf Libman. Segal's conjecture and the Burnside rings of fusion systems. *J. Lond. Math. Soc. (2)*, 80(3):665–679, 2009.

[54] Antonio Díaz and Sejong Park. Mackey functors and sharpness for fusion systems. *Homology Homotopy Appl.*, 17(1):147–164, 2015.

[55] Antonio Díaz, Albert Ruiz, and Antonio Viruel. All p-local finite groups of rank two for odd prime p. *Trans. Amer. Math. Soc.*, 359(4):1725–1764, 2007.

[56] W. G. Dwyer. The centralizer decomposition of BG. In *Algebraic topology: new trends in localization and periodicity (Sant Feliu de Guíxols, 1994)*, volume 136 of *Progr. Math.*, pages 167–184. Birkhäuser, 1996.

[57] W. G. Dwyer. Homology decompositions for classifying spaces of finite groups. *Topology*, 36(4):783–804, 1997.

[58] W. G. Dwyer. Sharp homology decompositions for classifying spaces of finite groups. In *Group representations: cohomology, group actions and topology (Seattle, WA, 1996)*, volume 63 of *Proc. Sympos. Pure Math.*, pages 197–220. Amer. Math. Soc., 1998.

[59] W. G. Dwyer, J. P. C. Greenlees, and S. Iyengar. Duality in algebra and topology. *Adv. Math.*, 200(2):357–402, 2006.

[60] William G. Dwyer, Haynes R. Miller, and Clarence W. Wilkerson. The homotopic uniqueness of BS^3. In *Algebraic topology, Barcelona, 1986*, volume 1298 of *Lecture Notes in Math.*, pages 90–105. Springer, 1987.

[61] W. G. Dwyer and C. W. Wilkerson. A new finite loop space at the prime two. *J. Amer. Math. Soc.* 6(1):37–64, 2993.

[62] W. G. Dwyer and C. W. Wilkerson. Homotopy fixed-point methods for Lie groups and finite loop spaces. *Ann. of Math. (2)*, 139(2):395–442, 1994.

[63] W. G. Dwyer and C. W. Wilkerson. The fundamental group of a p-compact group. *Bull. Lond. Math. Soc.*, 41(3):385–395, 2009.

[64] W. Dwyer and A. Zabrodsky. Maps between classifying spaces. In *Algebraic topology, Barcelona, 1986*, volume 1298 of *Lecture Notes in Math.*, pages 106–119. Springer, 1987.

[65] Leonard Evens. The cohomology ring of a finite group. *Trans. Amer. Math. Soc.*, 101:224–239, 1961.

[66] Matthew Gelvin and Sune Precht Reeh. Minimal characteristic bisets for fusion systems. *J. Algebra*, 427:345–374, 2015.

[67] Matthew Gelvin, Sune Precht Reeh, and Ergün Yalçin. On the basis of the Burnside ring of a fusion system. *J. Algebra*, 423:767–797, 2015.

[68] George Glauberman and Justin Lynd. Control of fixed points and existence and uniqueness of centric linking systems. *Invent. Math.*, 206(2):441–484, 2016.

[69] David M. Goldschmidt. A conjugation family for finite groups. *J. Algebra*, 16:138–142, 1970.

[70] E. Golod. The cohomology ring of a finite p-group. *Dokl. Akad. Nauk SSSR*, 125:703–706, 1959.

[71] Alex González. Extensions of partial groups and localities. arXiv:1507.04392, 2015.

[72] Alex Gonzalez. Finite approximations of p-local compact groups. *Geom. Topol.*, 20(5):2923–2995, 2016.

[73] Alex Gonzalez, Toni Lozano, and Albert Ruiz. Some new examples of simple p-local compact groups. arXiv:1512.00284, 2015.

[74] Jelena Grbić. The cohomology of exotic 2-local finite groups. *Manuscripta Math.*, 120(3):307–318, 2006.

[75] Jesper Grodal. Higher limits via subgroup complexes. *Ann. of Math. (2)*, 155(2):405–457, 2002.

[76] Jesper Grodal. Endotrivial modules for finite groups via homotopy theory. arXiv:1608.00499, 2016.

[77] Ellen Henke. Subcentric linking systems. *Trans. Amer. Math. Soc.*, 371(5):3325–3373, 2019.

[78] Stefan Jackowski and James McClure. Homotopy decomposition of classifying spaces via elementary abelian subgroups. *Topology*, 31(1):113–132, 1992.

[79] Stefan Jackowski, James McClure, and Bob Oliver. Homotopy classification of self-maps of BG via G-actions. I. *Ann. of Math. (2)*, 135(1):183–226, 1992.

[80] Stefan Jackowski, James McClure, and Bob Oliver. Homotopy classification of self-maps of BG via G-actions. II. *Ann. of Math. (2)*, 135(2):227–270, 1992.

[81] Stefan Jackowski, James McClure, and Bob Oliver. Self-homotopy equivalences of classifying spaces of compact connected Lie groups. *Fund. Math.*, 147(2):99–126, 1995.

[82] Stefan Jackowski and Bob Oliver. Vector bundles over classifying spaces of compact Lie groups. *Acta Math.*, 176(1):109–143, 1996.

[83] Radha Kessar and Radu Stancu. A reduction theorem for fusion systems of blocks. *J. Algebra*, 319(2):806–823, 2008.

[84] Jean Lannes. Sur les espaces fonctionnels dont la source est le classifiant d'un p-groupe abélien élémentaire. *Inst. Hautes Études Sci. Publ. Math.*, (75):135–244, 1992. With an appendix by Michel Zisman.

[85] Ian J. Leary and Radu Stancu. Realising fusion systems. *Algebra Number Theory*, 1(1):17–34, 2007.

[86] Ran Levi and Assaf Libman. Existence and uniqueness of classifying spaces for fusion systems over discrete p-toral groups. *J. Lond. Math. Soc. (2)*, 91(1):47–70, 2015.

[87] Ran Levi and Bob Oliver. Construction of 2-local finite groups of a type studied by Solomon and Benson. *Geom. Topol.*, 6:917–990, 2002.

[88] Ran Levi and Kári Ragnarsson. p-local finite group cohomology. *Homology Homotopy Appl.*, 13(1):223–257, 2011.

[89] L. G. Lewis, J. P. May, and J. E. McClure. Classifying G-spaces and the Segal conjecture. In *Current trends in algebraic topology, Part 2 (London, Ont., 1981)*, volume 2 of *CMS Conf. Proc.*, pages 165–179. Amer. Math. Soc., 1982.

[90] Markus Linckelmann. Simple fusion systems and the Solomon 2-local groups. *J. Algebra*, 296(2):385–401, 2006.

[91] Markus Linckelmann. Quillen's stratification for fusion systems. *Comm. Algebra*, 45(12):5227–5229, 2017.

[92] Nadia Mazza and Jacques Thévenaz. Endotrivial modules in the cyclic case. *Arch. Math. (Basel)*, 89(6):497–503, 2007.

[93] Haynes Miller. The Sullivan conjecture on maps from classifying spaces. *Ann. of Math. (2)*, 120(1):39–87, 1984.

[94] Guido Mislin. On group homomorphisms inducing mod-p cohomology isomorphisms. *Comment. Math. Helv.*, 65(3):454–461, 1990.

[95] Rémi Molinier. Cohomology with twisted coefficients of the classifying space of a fusion system. *Topology Appl.*, 212:1–18, 2016.

[96] Rémi Molinier. Cohomology of linking systems with twisted coefficients by a p-solvable action. *Homology Homotopy Appl.*, 19(2):61–82, 2017.

[97] Rémi Molinier. Alperin's fusion theorem for localities. *Comm. Algebra*, 46(6):2615–2619, 2018.

[98] Rémi Molinier. Control of fixed points over discrete p-toral groups, and existence and uniqueness of linking systems. *J. Algebra*, 499:43–73, 2018.

[99] Amnon Neeman. The chromatic tower for $D(R)$. *Topology*, 31(3):519–532, 1992. With an appendix by Marcel Bökstedt.

[100] Bob Oliver. Equivalences of classifying spaces completed at odd primes. *Math. Proc. Cambridge Philos. Soc.*, 137(2):321–347, 2004.

[101] Bob Oliver. Equivalences of classifying spaces completed at the prime two. *Mem. Amer. Math. Soc.*, 180(848),2006.

[102] Bob Oliver. Extensions of linking systems and fusion systems. *Trans. Amer. Math. Soc.*, 362(10):5483–5500, 2010.

[103] Bob Oliver. Existence and uniqueness of linking systems: Chermak's proof via obstruction theory. *Acta Math.*, 211(1):141–175, 2013.

[104] Bob Oliver and Joana Ventura. Extensions of linking systems with p-group kernel. *Math. Ann.*, 338(4):983–1043, 2007.

[105] Sejong Park. Realizing fusion systems inside finite groups. *Proc. Amer. Math. Soc.*, 144(8):3291–3294, 2016.

[106] Lluís Puig. Affirmative answer to a question of Feit. *J. Algebra*, 131(2):513–526, 1990.

[107] Lluis Puig. Frobenius categories. *J. Algebra*, 303(1):309–357, 2006.

[108] Luis Puig. Structure locale dans les groupes finis. *Bull. Soc. Math. France Suppl. Mém.*, (47):132, 1976.

[109] Daniel Quillen. A cohomological criterion for p-nilpotence. *J. Pure Appl. Algebra*, 1(4):361–372, 1971.

[110] Daniel Quillen. The spectrum of an equivariant cohomology ring. I, II. *Ann. of Math. (2)*, 94:549–572; ibid. (2) 94 (1971), 573–602, 1971.

[111] D. Quillen and B. B. Venkov. Cohomology of finite groups and elementary abelian subgroups. *Topology*, 11:317–318, 1972.

[112] Kári Ragnarsson. Classifying spectra of saturated fusion systems. *Algebr. Geom. Topol.*, 6:195–252, 2006.

[113] Kári Ragnarsson and Radu Stancu. Saturated fusion systems as idempotents in the double Burnside ring. *Geom. Topol.*, 17(2):839–904, 2013.

[114] D. L. Rector. Noetherian cohomology rings and finite loop spaces with torsion. *J. Pure Appl. Algebra*, 32(2):191–217, 1984.

[115] David L. Rector. Loop structures on the homotopy type of S^3. In *Symposium on Algebraic Topology (Battelle Seattle Res. Center, Seattle, Wash., 1971)*, volume 249 of *Lecture Notes in Math.*, pages 99–105, Springer, 1971.

[116] Sune Precht Reeh. The abelian monoid of fusion-stable finite sets is free. *Algebra Number Theory*, 9(10):2303–2324, 2015.

[117] Sune Precht Reeh. Transfer and characteristic idempotents for saturated fusion systems. *Adv. Math.*, 289:161–211, 2016.

[118] Sune Precht Reeh and Ergün Yalçın. Representation rings for fusion systems and dimension functions. *Math. Z.*, 288(1-2):509–530, 2018.

[119] K. Roberts and S. Shpectorov. On the definition of saturated fusion systems. *J. Group Theory*, 12(5):679–687, 2009.

[120] Geoffrey R. Robinson. Amalgams, blocks, weights, fusion systems and finite simple groups. *J. Algebra*, 314(2):912–923, 2007.

[121] Albert Ruiz. Exotic normal fusion subsystems of general linear groups. *J. Lond. Math. Soc. (2)*, 76(1):181–196, 2007.

[122] Albert Ruiz and Antonio Viruel. The classification of p-local finite groups over the extraspecial group of order p^3 and exponent p. *Math. Z.*, 248(1):45–65, 2004.

[123] Ronald Solomon. Finite groups with Sylow 2-subgroups of type 3. *J. Algebra*, 28:182–198, 1974.

[124] B. B. Venkov. Cohomology algebras for some classifying spaces. *Dokl. Akad. Nauk SSSR*, 127:943–944, 1959.

[125] Zdzisław Wojtkowiak. On maps from $\operatorname{ho}\varinjlim F$ to $\mathbf{Z}$. In *Algebraic topology, Barcelona, 1986*, volume 1298 of *Lecture Notes in Math.*, pages 227–236. Springer, 1987.

DEPARTAMENT DE MATEMATIQUES, UNIVERSITAT AUTÒNOMA DE BARCELONA, BELLATERRA, SPAIN

E-mail address: natalia@mat.uab.cat

10

Floer homotopy theory, revisited

Ralph L. Cohen

Introduction

In three seminal papers in 1988 and 1989 A. Floer introduced Morse theoretic homological invariants that transformed the study of low dimensional topology and symplectic geometry. In [17] Floer defined an "instanton homology" theory for 3-manifolds that, when paired with Donaldson's polynomial invariants of 4-manifolds defined a gauge theoretic 4-dimensional topological field theory that revolutionized the study of low dimensional topology and geometry. In [18], Floer defined an infinite dimensional Morse theoretic homological invariant for symplectic manifolds , now referred to as "Symplectic" or "Hamiltonian" Floer homology, that allowed him to prove a well-known conjecture of Arnold on the number of fixed points of a diffeomorphism $\phi_1 : M \xrightarrow{\cong} M$ arising from a time-dependent Hamiltonian flow $\{\phi_t\}_{0 \leq t \leq 1}$. In [16] Floer introduced "Lagrangian intersection Floer theory" for the study of interesections of Lagrangian submanifolds of a symplectic manifold.

Since that time there have been many other versions of Floer theory that have been introduced in geometric topology, including a Seiberg-Witten Floer homology [31]. This is similar in spirit to Floer's "instanton homology", but it is based on the Seiberg-Witten equations rather than the Yang-Mills equations. There were many difficult, technical analytic issues in developing Seiberg-Witten Floer theory, and Kronheimer and Mrowka's book [31] deals with them masterfully and elegantly. Another important geometric theory is Heegard Floer theory introduced by Oszvath and Szabo [44]. This is an invariant of a closed 3-manifold equipped with a $spin^c$ structure. It is computed using a Heegaard diagram of the manifold. It allowed for a related "knot Floer homology" introduced by Oszvath and Szabo [45] and by Rasmussen [47]. Khovanov's important homology theory that gave a "categorification" of the Jones polynomial [27] was eventually shown to be related to Floer theory by Seidel and Smith [55] and Abouzaid and Smith [3]. Lipshitz and Sarkar [35] showed that there is an associated "Khovanov stable homotopy". There have been many other variations of Floer theories as well.

The rough idea in all of these theories is to associate a Morse-like chain complex generated by the critical points of a functional defined typically on an infinite dimensional space. Recall that in classical Morse theory, given a Morse function $f : M \to \mathbb{R}$ on a closed

Mathematics Subject Classification. 53D12, 53D40, 55P35, 55P42, 55Q15, 57R58.

Key words and phrases. Floer homology, Floer homotopy, flow category, manifolds with corners.

Riemannian manifold, the "Morse complex" is the chain complex

$$\cdots \to C_p(f) \xrightarrow{\partial_p} C_{p-1}(f) \to \cdots$$

where $C_p(f)$ is the free abelian group generated by the critical points of f of index p, and the boundary homomorphisms can be computed by "counting" the gradient flow-lines connecting critical points of relative index one. More specfically, if $Crit_q(f)$ is the set of critical points of f of index q, then if $a \in Crit_p(f)$, then

$$\partial_p([a]) = \sum_{b \in Crit_{p-1}(f)} \#\mathcal{M}(a,b)\,[b] \tag{10.0.1}$$

where $\mathcal{M}(a,b)$ is the moduli space of gradient flow lines connecting a to b, which, since a and b have relative index one is a closed, zero dimensional oriented manifold. $\#\mathcal{M}(a,b)$ reflects the "oriented count" of this finite set. More carefully $\#\mathcal{M}(a,b)$ is the integer in the zero dimensional oriented cobordism group, $\pi_0 MSO \cong \mathbb{Z}$ represented by $\mathcal{M}(a,b)$.

In Floer's original examples, the functionals he studied were in fact $\mathbb{R}/\mathbb{Z}$-valued. In the case of Floer's instanton theory, the relevant functional is the Chern-Simons map defined on the space of connections on a principal $SU(2)$-bundle over the 3-manifold. Its critical points are flat connections and its flow lines are "instantons", i.e. anti-self-dual connections on the three-manifold Y crossed with the real line. Modeling classical Morse theory, the "Floer complex" is generated by the critical points of this functional, suitably perturbed to make them nondegenerate, and the boundary homomorphisms are computed by taking oriented counts of the gradient flow lines, i.e. anti-self-dual connections on $Y \times \mathbb{R}$ that connect critical points of relative index one.

When Floer introduced what is now called "Symplectic" or "Hamiltonian" Floer homology ("SFH") to use in his proof of Arnold's conjecture, he studied the symplectic action defined on the free loop space of the underlying symplectic manifold M,

$$\mathcal{A} : LM \to \mathbb{R}/\mathbb{Z}.$$

He perturbed the action functional by a time dependent Hamiltonian function $H : \mathbb{R}/\mathbb{Z} \times M \to \mathbb{R}$. Call the resulting functional $\mathcal{A}_H$. The critical points of $\mathcal{A}_H$ are the 1-periodic orbits of the Hamiltonian vector field. That is, they are smooth loops $\alpha : \mathbb{R}/\mathbb{Z} \to M$ satisfying the differential equaiton

$$\frac{d\alpha}{dt} = X_H(t, \alpha(t))$$

where X_H is the Hamiltonian vector field. One way to think of X_H is that the symplectic 2-form ω on M defines, since it is nondegenerate, a bundle isomorphism $\omega : TM \xrightarrow{\cong} T^*M$. This induces an identification of the periodic one-forms, that is sections of the cotangent bundle pulled back over $\mathbb{R}/\mathbb{Z} \times M$, with periodic vector fields. The Hamiltonian vector field X_H corresponds to the differential dH under this identification. Floer showed that with respect to a generic Hamiltonian, the critical points of $\mathcal{A}_H$ are nondegenerate.

Of course to understand the gradient flow lines connecting critical points, one must have a metric. This is defined using the symplectic form and a compatible choice of almost complex structure J on TM. With respect to this structure the gradient flow lines are curves

$\gamma : \mathbb{R} \to LM$, or equivalently, maps of cylinders

$$\gamma : \mathbb{R} \times S^1 \to M$$

that satisfy (perturbed) Cauchy-Riemann equations. If $\mathcal{M}(\alpha_1, \alpha_2, H, J)$ is the moduli space of such "pseudo-holomorphic" cylinders that connect the periodic orbits α_1 and α_2, then the boundary homomorphisms Morse-Floer chain complex is defined by giving an oriented count of the zero dimensional moduli spaces, in analogy with the situation in classical Morse theory described above.

The other, newer examples of Floer homology tend to be similar. There is typically a functional that can be perturbed in such a way that its critical points and zero dimensional moduli spaces of gradient flow lines define a chain complex whose homology is invariant of the choices made.

In the case of a classical Morse function $f : M \to \mathbb{R}$ closed manifold, the Morse complex can be viewed as the cellular chain complex of a CW-complex X_f of the homotopy type of M. X_f has one cell of dimension λ, for every critical point of f of index λ. The attaching maps were studied by Franks in [19], and one may view the work of the author and his collaborators in [11], [12] as a continuation of that study. This led us to ask the question:

Q1: Does the Floer chain complex arise as the cellular chain complex of a CW-complex or a $C.W$-spectrum?

More specifically we asked the the following question:

Q2: What properties of the data of a Floer functional, i.e. its critical points and moduli spaces of gradient flow lines connecting them, are needed to define a CW-spectrum realizing the Floer chain complex?

More generally one might ask the following question:

Q3: Given a finite chain complex, is there a reasonable way to classify the CW-spectra that realize this complex?

In this paper we take up these questions. We also discuss how studying Floer theory from a homotopy perspective has been done in recent years, and how it has been applied with dramatic success. We state immediately that the applications that we discuss in this paper are purely the choice of the author. There are many other fascinating applications and advances that all help to make this an active and exciting area of research. We apologize in advance to researchers whose work we will not have the time or space to discuss.

This paper is organized as follows. In Section 2 we discuss the three questions raised above. We give a new take on how these questions were originally addressed in [12], and give some of the early applications of this perspective. In Section 3 we describe how Lipshitz and Sarkar used this perspective to define the notion of "Khovanov homotopy" [35][36]. This is a stable homotopy theoretic realization of Khovanov's homology theory [27] which in turn is a categorification of the Jones polynomial invariant of knots and links. In particular we describe how the homotopy perspective Lipshitz and Sarkar used give more subtle and delicate invariants than the homology theory alone, and how these invariants have been applied. In Section 3 we describe Manolescu's work on an equivariant stable homotopy theoretic view of Seiberg-Witten Floer theory. In the case of his early work [37][38], the group acting is the circle group S^1. In his more recent work [39] [41] he studied $Pin(2)$-equivariant Floer homotopy theory and used it to give a dramatic solution (in the negative) to the longstanding question about whether all topological manifolds admit triangulations. In Section 4 we describe the Floer homotopy theoretic methods of Kragh [30] and of Abouzaid-Kragh [2] in the study of the symplectic topology of the cotangent bundle of a closed manifold, and how they were useful in studying Lagrangian immersions and embeddings inside the cotangent bundle.

Acknowledgements

The author would like to thank M. Abouzaid, A. Blumberg, R. Lipshitz, C. Manolescu, and H. Miller for helpful comments on a previous draft of this paper.

10.1 The homotopy theoretic foundations

10.1.1 Realizing chain complexes

We begin with a purely homotopy theoretic question: Given a finite chain complex, C_*,

$$C_n \xrightarrow{\partial_n} C_{n-1} \xrightarrow{\partial_{n-1}} \cdots \xrightarrow{\partial_2} C_1 \xrightarrow{\partial_1} C_0$$

how can one classify the finite CW-spectra X whose associated cellular chain complex is C_*? This is question **Q3** above.

Of course one does not need to restrict this question to finite complexes, but that is a good place to start. In particular it is motivated by Morse theory, where, given a Morse function on a closed, n-dimensional Riemannian manifold, $f : M^n \to \mathbb{R}$, one has a corresponding "Morse - Smale" chain complex C^f_*

$$C^f_n \xrightarrow{\partial_n} C^f_{n-1} \xrightarrow{\partial_{n-1}} \cdots \xrightarrow{\partial_2} C^f_1 \xrightarrow{\partial_1} C^f_0.$$

Here C^f_p is the free abelian group generated by the critical points of f of index p. The boundary homomorphisms are as described above (10.0.1). Of course in this case the Morse function together with the Riemannian metric define a CW complex homotopy equivalent to X. In Floer theory that is not the case. One does start with geometric information that allows for the definition of a chain complex, but knowing if this complex comes, in a natural way from the data of the Floer theory, is not at all clear, and was the central question of study in [12].

This homotopy theoretic question was addressed more specifically in [10]. Of central importance in this study was to understand how the attaching maps of the cells in a finite CW-spectrum can be understood geometrically, via the theory of (framed) cobordism of manifolds with corners. We will review these ideas in this section and recall some basic examples of how they can be applied.

By assumption, the chain complex C_* is finite, so each C_i is a finitely generated free abelian group. Let $\mathcal{B}_i$ be a basis for C_i. Let $\mathbb{S}$ denote the sphere spectrum. For each i, consider the free $\mathbb{S}$-module spectrum generated by $\mathcal{B}_i$,

$$\mathcal{E}_i = \bigvee_{\alpha \in \mathcal{B}_i} \mathbb{S}.$$

There is a natural isomorphism

$$H_0(\mathcal{E}_i) \simeq C_i.$$

Definition 10.1.1. We say that a finite spectrum X *realizes* the complex C_* if there exists a filtration of subspectra of X,

$$X_0 \hookrightarrow X_1 \hookrightarrow \cdots \hookrightarrow X_n = X,$$

satisfying the following properties:

1. There is an equivalence of the subquotients

$$X_i/X_{i-1} \simeq \Sigma^i \mathcal{E}_i,$$

2. The induced composition map in integral homology,

$$\begin{aligned} &\tilde{H}_i(X_i/X_{i-1}) \xrightarrow{\delta_i} \tilde{H}_i(\Sigma X_{i-1}) \xrightarrow{\rho_{i-1}} H_i(\Sigma(X_{i-1}/X_{i-2})) \\ &= H_0(\mathcal{E}_i) \to H_0(\mathcal{E}_{i-1}) \\ &= C_i \to C_{i-1} \end{aligned}$$

   is the boundary homomorphism, ∂_i.

Here the "subquotient" X_i/X_{i-1} refers to the homotopy cofiber of the map $X_{i-1} \to X_i$, the map $\rho_i : X_i \to X_i/X_{i-1}$ is the projection map, and the map $\delta_i : X_i/X_{i-1} \to \Sigma X_{i-1}$ is the Barratt-Puppe extension of the homotopy cofibration sequence $X_{i-1} \to X_i \xrightarrow{\rho_i} X_i/X_{i-1}$.

To classify finite spectra that realize a given finite complex, in [12] the authors introduced a topological category $\mathcal{J}$ whose objects are the nonnegative integers $\mathbb{Z}^+$, and whose non-identity morphisms from i to j is empty for $i \leq j$, for $i > j+1$ it is defined to be the one point compactification,

$$Mor_{\mathcal{J}}(i,j) \cong (\mathbb{R}_+)^{i-j-1} \cup \infty$$

and $Mor_{\mathcal{J}}(j+1,j)$ is defined to be the two point space, S^0. Here $\mathbb{R}_+$ is the space of nonnegative real numbers. Composition in this category can be viewed in the following way. Notice that for $i > j+1$ $Mor_{\mathcal{J}}(i,j)$ can be viewed as the one point compactification of the space $J(i,j)$ consisting of sequences of real numbers $\{\lambda_k\}_{k \in \mathbb{Z}}$ such that

$$\begin{aligned} \lambda_k &\geq 0 \quad \text{for all } k \\ \lambda_k &= 0 \quad \text{unless } i > k > j. \end{aligned}$$

For consistency of notation we write $Mor_{\mathcal{J}}(i,j) = J(i,j)^+$. Composition of morphisms $J(i,j)^+ \wedge J(j,k)^+ \to J(i,k)^+$ is then induced by addition of sequences. In this smash product the basepoint is taken to be ∞. Notice that this map is basepoint preserving. Given integers $p > q$, then there are subcategories $\mathcal{J}_q^p$ defined to be the full subcategory generated by integers $q \geq m \geq p$. The category $\mathcal{J}_q$ is the full subcategory of $\mathcal{J}$ generated by all integers $m \geq q$.

The following is a recasting of a discussion in [12].

Theorem 10.1.2. *The realizations of the chain complex C_* by finite spectra are classified by extensions of the association $j \to \mathcal{E}_j$ to basepoint preserving functors $Z : \mathcal{J}_0 \to Spectra$, with the property that for each $j \geq 0$, the map obtained by the application of morphisms,*

$$\begin{aligned} Z_{j+1,j} : J(j+1,j)^+ \wedge \mathcal{E}_{j+1} &\to \mathcal{E}_j \\ S^0 \wedge \mathcal{E}_{j+1} &\to \mathcal{E}_j \end{aligned}$$

induces the boundary homomorphism ∂_{j+1} on the level of homology groups. Here Spectra is a symmetric monoidal category of spectra (e.g. the category of symmetric spectra), and by a "basepoint preserving functor" we mean one that maps ∞ in $J(p,q)^+$ to the constant map $Z(p) \to Z(q)$. By "classified" we mean that the filtered homotopy type of any realization of C_* determines a basepoint preserving functor, $Z : \mathcal{J}_0 \to Spectra$ with these properties, and conversely every such functor determines a realization of the chain complex C_*.

We recall from [12] how a functor $Z : \mathcal{J}_0 \to Spectra$ satisfying the properties described in the theorem, defines a realization of the chain complex C_*. As described in [12], given a functor to the category of spaces, $Z : \mathcal{J}_q \to Spaces_*$, where $Spaces_*$ denotes the category of based topological spaces, one can take its geometric realization,

$$|Z| = \coprod_{q \leq j} Z(j) \wedge J(j, q-1)^+ / \sim \tag{10.1.1}$$

where one identifies the image of $Z(j) \wedge J(j,i)^+ \wedge J(i,q-1)^+$ in $Z(j) \wedge J(j,q-1)^+$ given by composition of morphisms, with its image in $Z(i) \wedge J(i,q-1)$ defined by application of morphisms $Z(j) \wedge J(j,i)^+ \to Z(i)$.

For a functor whose value is in *Spectra*, we replace the above construction by a coequalizer, in the following way:

Let $Z : \mathcal{J}_q \to Spectra$. Define two maps of spectra,

$$\iota, \mu : \bigvee_{q \leq j} Z(j) \wedge J(j,i)^+ \wedge J(i,q-1)^+ \longrightarrow \bigvee_{q \leq j} Z(j) \wedge J(j,q-1)^+. \tag{10.1.2}$$

The first map ι is induced by the composition of morphisms in $\mathcal{J}_q$, $J(j,i)^+ \wedge J(i,q-1)^+ \hookrightarrow J(j,q-1)^+$. The second map μ is the given by the wedge of maps,

$$Z(j) \wedge J(j,i)^+ \wedge J(i,q-1)^+ \xrightarrow{\mu_q \wedge 1} Z(i) \wedge J(i,q-1)^+$$

where $\mu_q : Z(j) \wedge J(j,i)^+ \to Z(i)$ is given by application of morphisms.

Definition 10.1.3. Given a functor $Z : \mathcal{J}_q \to Spectra$ we define its *geometric realization* $|Z|$ to be the coequalizer (in the category *Spectra*) of the two maps,

$$\iota, \mu : \bigvee_{q \leq j} \bigvee_{i=q}^{j-1} Z(j) \wedge J(j,i)^+ \wedge J(i, q-1)^+ \longrightarrow \bigvee_{q \leq j} Z(j) \wedge J(j, q-1)^+.$$

Technically, so that the homotopy type of the coequalizer is well-defined, the functor may have to be modified so that it takes each morphism to a cofibration.

So a functor $Z : \mathcal{J}_0 \to Spectra$ satisfying the properties specified in Theorem 10.1.2 defines a geometric realization $|Z|$ which is a finite spectrum. Consider how the data of the functor Z defines the CW-structure of $|Z|$. Clearly $|Z|$ will have one cell of dimension i for every element of $\pi_0(Z(i)) = \pi_0(\mathcal{E}_i) = \mathcal{B}_i$. The attaching maps were described in [12], [9], [10] in the following way.

In general assume that X be a finite CW-spectrum with skeletal filtration

$$X_0 \hookrightarrow X_1 \hookrightarrow \cdots \hookrightarrow X_n = X.$$

In particular each map $X_{i-1} \hookrightarrow X_i$ is a cofibration, and we call its cofiber $K_i = X_i/(X_{i-1})$. This is a wedge of (suspension spectra) of spheres of dimension i.

$$K_i \simeq \bigvee_{\mathrm{cd}_i} \Sigma^i \mathbb{S}$$

where cd_i is a finite indexing set.

As was discussed in [12] one can then "rebuild" the homotopy type of the n-fold suspension, $\Sigma^n X$, as the union of iterated cones and suspensions of the K_i's,

$$\Sigma^n X \simeq \Sigma^n K_0 \cup c(\Sigma^{n-1} K_1) \cup \cdots \cup c^i(\Sigma^{n-i} K_i) \cup \cdots \cup c^n K_n. \tag{10.1.3}$$

This decomposition can be described as follows. Define a map $\delta_i : \Sigma^{n-i} K_i \to \Sigma^{n-i+1} K_{i-1}$ to be the iterated suspension of the composition,

$$\delta_i : K_i \to \Sigma X_{i-1} \to \Sigma K_{i-1}$$

where the two maps in this composition come from the cofibration sequence, $X_{i-1} \to X_i \to K_i \to \Sigma X_{i-1} \cdots$. As was pointed out in [9], this induces a "homotopy chain complex",

$$K_n \xrightarrow{\delta_n} \Sigma K_{n-1} \xrightarrow{\delta_{n-1}} \cdots \xrightarrow{\delta_{i+1}} \Sigma^{n-i} K_i \xrightarrow{\delta_i} \Sigma^{n-i+1} K_{i-1} \xrightarrow{\delta_{i-1}} \cdots \xrightarrow{\delta_1} \Sigma^n K_0 = \Sigma^n X_0. \tag{10.1.4}$$

We refer to this as a homotopy chain complex because examination of the defining cofibrations leads to canonical null homotopies of the compositions,

$$\delta_j \circ \delta_{j+1}.$$

This canonical null homotopy defines an extension of δ_j to the mapping cone of δ_{j+1}:

$$c(\Sigma^{n-j-1}K_{j+1}) \cup_{\delta_{j+1}} \Sigma^{n-j}K_j \longrightarrow \Sigma^{n-j+1}K_{j-1}.$$

More generally, for every q, using these null homotopies, we have an extension to the iterated mapping cone,

$$\begin{aligned} c^q(\Sigma^{n-j-q}K_{j+q}) \cup c^{q-1}(\Sigma^{n-j-q+1}K_{j+q-1}) \cup \\ \cdots \cup c(\Sigma^{n-j-1}K_{j+1}) \cup_{\delta_{j+1}} \Sigma^{n-j}K_j \longrightarrow \Sigma^{n-j+1}K_{j-1}. \end{aligned} \tag{10.1.5}$$

In other words, for each $p > q$, these null homotopies define maps of spectra,

$$\phi_{p,q} : c^{p-q-1}\Sigma^{n-p}K_p \to \Sigma^{n-q}K_q. \tag{10.1.6}$$

The cell attaching data in the CW-spectrum X as in (10.1.3) is then defined via the maps $\phi_{p,q}$.

Given a functor $Z : \mathcal{J}_0 \to \mathit{Spectra}$ satisfying the hypotheses of Theorem 10.1.2, we have that

$$K_p = |Z|_p/|Z|_{p-1} \simeq \bigvee_{\mathcal{B}_p} \Sigma^p \mathbb{S}$$

and the attaching maps

$$\begin{aligned} \phi_{p,q} :& c^{p-q-1}\Sigma^{-p}K_p \to \Sigma^{-q}K_q \\ & J(p,q)^+ \wedge \Sigma^{-p}|Z|_p/|Z|_{p-1} \to \Sigma^{-q}|Z|_q/|Z|_{q-1} \\ & J(p,q)^+ \wedge Z(p) \to Z(q) \end{aligned} \tag{10.1.7}$$

are given by the application of the morphisms of the category $\mathcal{J}_0$ to the value of the functor Z.

10.1.2 Manifolds with corners and framed flow categories

In order to make use of Theorem 10.1.2 in the setting of Morse and Floer theory, one needs a more geometric way of understanding the homotopy theoretic information contained in a functor $Z : \mathcal{J}_0 \to \mathit{Spectra}$ satisfying the hypotheses of the theorem. As is common in algebraic and differential topology, the translation between homotopy theoretic information and geometric information is done via cobordism theory and the Pontrjagin-Thom construction.

In the case when the CW-structure comes from a Morse function $f : M^n \to \mathbb{R}$ on a closed n-dimensional Riemannian manifold, the attaching maps $\phi_{p,q}$ defining a functor $Z_f : \mathcal{J}_0 \to \mathit{Spectra}$, were shown in [9], [10] to come from the cobordism-type of the moduli spaces of gradient flow lines connecting critical points of index p to those of index q. The relevant cobordism theory is *framed cobordism of manifolds with corners.* Thus to extend this idea to the Floer setting, with the goal of developing a "Floer homotopy type", one needs to understand certain cobordism-theoretic properties of the corresponding moduli spaces of gradient flows of the particular Floer theory. (In [10] the author also considered how other cobordism theories give rise not to "Floer homotopy types" but rather to "Floer module spectra", or said another way, "Floer E-theory" where E is a generalized homology theory. We refer the reader to [10] for details.) In order to make this idea precise we recall some basic facts about cobordisms of manifolds with corners. The main reference for this is Laures's paper [33].

Recall that an n-dimensional manifold with corners, M, has charts which are local homeomorphisms with $\mathbb{R}^n_+$. Here $\mathbb{R}_+$ denotes the nonnegative real numbers and $\mathbb{R}^n_+$ is the n-fold cartesian product of $\mathbb{R}_+$. Let $\psi : U \to \mathbb{R}^n_+$ be a chart of a manifold with corners M. For $x \in U$, the number of zeros of this chart, $c(x)$ is independent of the chart. One defines a *face* of M to be a connected component of the space $\{m \in M \text{ such that } c(m) = 1\}$.

Given an integer k, there is a notion of a manifold with corners having "codimension k", or a $\langle k \rangle$-manifold. This notion was originally due to Janich [25], and we recall the definition from [33].

Definition 10.1.4. A $\langle k \rangle$-*manifold* is a manifold with corners, M, together with an ordered k-tuple $(\partial_1 M, ..., \partial_k M)$ of unions of faces of M satisfying the following properties.

1. Each $m \in M$ belongs to $c(m)$ faces
2. $\partial_1 M \cup \cdots \cup \partial_k M = \partial M$
3. For all $1 \leq i \neq j \leq k$, $\partial_i M \cap \partial_j M$ is a face of both $\partial_i M$ and $\partial_j M$.

The archetypical example of a $\langle k \rangle$-manifold is $\mathbb{R}^k_+$. In this case the face $F_j \subset \mathbb{R}^k_+$ consists of those k-tuples with the jth-coordinate equal to zero.

As described in [33], the data of a $\langle k \rangle$-manifold can be encoded in a categorical way as follows. Let $\underline{2}$ be the partially ordered set with two objects, $\{0, 1\}$, generated by a single nonidentity morphism $0 \to 1$. Let $\underline{2}^k$ be the product of k-copies of the category $\underline{2}$. A $\langle k \rangle$-manfold M then defines a functor from $\underline{2}^k$ to the category of topological spaces, where for an object $a = (a_1, \cdots, a_k) \in \underline{2}^k$, $M(a)$ is the intersection of the faces $\partial_i(M)$ with $a_i = 0$. Such a functor is a k-dimensional cubical diagram of spaces, which, following Laures's terminology, we refer to as a $\langle k \rangle$-diagram, or a $\langle k \rangle$-space. Notice that $\mathbb{R}^k_+(a) \subset \mathbb{R}^k_+$ consists of those k-tuples of nonnegative real numbers so that the ith-coordinate is zero for every i with $a_i = 0$. More generally, consider the $\langle k \rangle$-Euclidean space, $\mathbb{R}^k_+ \times \mathbb{R}^n$, where the value on $a \in \underline{2}^k$ is $\mathbb{R}^k_+(a) \times \mathbb{R}^n$. In general we refer to a functor $\phi : \underline{2}^k \to \mathcal{C}$ as a $\langle k \rangle$-object in the category $\mathcal{C}$.

In this section we will consider embeddings of manifolds with corners into Euclidean spaces $M \hookrightarrow \mathbb{R}^k_+ \times \mathbb{R}^n$ of the form given by the following definition.

Definition 10.1.5. A *neat embedding* of a $\langle k \rangle$-manifold M into $\mathbb{R}^k_+ \times \mathbb{R}^m$ is a natural transformation of $\langle k \rangle$-diagrams

$$e : M \hookrightarrow \mathbb{R}^k_+ \times \mathbb{R}^m$$

that satisfies the following properties:

1. For each $a \in \underline{2}^k$, $e(a)$ is an embedding.
2. For all $b < a$, the intersection $M(a) \cap \left(\mathbb{R}^k_+(b) \times \mathbb{R}^m\right) = M(b)$, and this intersection is perpendicular. That is, there is some $\varepsilon > 0$ such that

$$M(a) \cap \left(\mathbb{R}^k_+(b) \times [0, \varepsilon)^k(a - b) \times \mathbb{R}^m\right) = M(b) \times [0, \varepsilon)^k(a - b).$$

Here $a - b$ denotes the object of $\underline{2}^k$ obtained by subtracting the k-vector b from the k-vector a.

In [33] it was proved that every $\langle k \rangle$-manifold neatly embeds in $\mathbb{R}^k_+ \times \mathbb{R}^N$ for N sufficiently large. In fact it was proved there that a manifold with corners, M, admits a neat embedding into $\mathbb{R}^k_+ \times \mathbb{R}^N$ *if and only if* M has the structure of a $\langle k \rangle$-manifold. Furthermore in [24] it is shown that the connectivity of the space of neat embeddings, $Emb_{\langle k \rangle}(M; \mathbb{R}^k_+ \times \mathbb{R}^N)$ increases with the dimension N.

An embedding of manifolds with corners, $e : M \hookrightarrow \mathbb{R}^k_+ \times \mathbb{R}^m$, has a well defined normal bundle. In particular, for any pair of objects in $\underline{2}^k$, $a > b$, the normal bundle of $e(a) : M(a) \hookrightarrow \mathbb{R}^k_+(a) \times \mathbb{R}^m$, when restricted to $M(b)$, is the normal bundle of $e(b) : M(b) \hookrightarrow \mathbb{R}^k_+(b) \times \mathbb{R}^m$.

These embedding properties of $\langle k \rangle$-manifolds make it clear that these are the appropriate manifolds to study for cobordism-theoretic information. In particular, given an embedding $e : M \hookrightarrow \mathbb{R}^k_+ \times \mathbb{R}^m$ the Thom space of the normal bundle, $Th(M, e)$, has the structure of an $\langle k \rangle$-space, where for $a \in \underline{2}^k$, $Th(M, e)(a)$ is the Thom space of the normal bundle of the associated embedding, $M(a) \hookrightarrow \mathbb{R}^k_+(a) \times \mathbb{R}^N$. We can then desuspend and define the Thom spectrum, $M^\nu_e = \Sigma^{-N} Th(M, e)$, to be the associated $\langle k \rangle$-spectrum. The Pontrjagin-Thom construction defines a map of $\langle k \rangle$-spaces,

$$\tau_e : (\mathbb{R}^k_+ \times \mathbb{R}^N) \cup \infty = ((\mathbb{R}^k_+) \cup \infty) \wedge S^N \to Th(M, e).$$

Desuspending we get a map of $\langle k \rangle$-spectra, $\Sigma^\infty((\mathbb{R}^k_+) \cup \infty) \to M^\nu_e$. Notice that the homotopy type (as $\langle k \rangle$-spectra) of M^ν_e is independent of the embedding e. We denote the homotopy type of this normal Thom spectrum as M^ν, and the Pontrjagin-Thom map, $\tau : \Sigma^\infty((\mathbb{R}^k_+) \cup \infty) \to M^\nu$.

Compact manifolds with corners, and in particular $\langle k \rangle$ - manifolds naturally occur as the moduli spaces of flow lines of a Morse function, and in some cases, of a Floer function. We first recall how they appear in Morse theory.

Consider a smooth, closed n-manifold M^n, and a smooth Morse function $f : M^n \to \mathbb{R}$. Given a Riemannian metric on M, one studies the flow of the gradient vector field ∇f. In particular a flow line is a curve $\gamma : \mathbb{R} \to M$ satisfying the ordinary differential equation,

$$\frac{d}{dt}\gamma(s) + \nabla f(\gamma(s)) = 0.$$

By the existence and uniqueness theorem for solutions to ODE's, one knows that if $x \in M$ is any point then there is a unique flow line γ_x satisfying $\gamma_x(0) = x$. One then studies unstable and stable manifolds of the critical points,

$$W^u(a) = \{x \in M : \lim_{t \to -\infty} \gamma_x(t) = a\}$$
$$W^s(a) = \{x \in M : \lim_{t \to +\infty} \gamma_x(t) = a\}.$$

The unstable manifold $W^u(a)$ is diffeomorphic to a disk $D^{\mu(a)}$, where $\mu(a)$ is the index of the critical point a. Similarly the stable manifold $W^s(a)$ is diffeomorphic to a disk $D^{n-\mu(a)}$.

For a generic choice of Riemannian metric, the unstable manifolds and stable manifolds intersect transversally, and their intersections,

$$W(a,b) = W^u(a) \cap W^s(b)$$

are smooth manifolds of dimension equal to the relative index, $\mu(a) - \mu(b)$. When the choice of metric satisfies these transversality properties, the metric is said to be "Morse-Smale". The manifolds $W(a,b)$ have free $\mathbb{R}$-actions defined by "going with the flow". That is, for $t \in \mathbb{R}$, and $x \in M$,

$$t \cdot x = \gamma_x(t).$$

The "moduli space of flow lines" is the manifold

$$\mathcal{M}(a,b) = W(a,b)/\mathbb{R}$$

and has dimension $\mu(a) - \mu(b) - 1$. These moduli spaces are not generally compact, but they have canonical compactifications which we now describe.

In the case of a Morse-Smale metric, (which we assume throughout the rest of this section), there is a partial order on the finite set of critical points given by $a \geq b$ if $\mathcal{M}(a,b) \neq \emptyset$. We then define

$$\bar{\mathcal{M}}(a,b) = \bigcup_{a=a_1>a_2>\cdots>a_k=b} \mathcal{M}(a_1,a_2) \times \cdots \times \mathcal{M}(a_{k-1},a_k). \tag{10.1.8}$$

The topology of $\bar{\mathcal{M}}(a,b)$ can be described naturally, and this is done in many references including [11]. $\bar{\mathcal{M}}(a,b)$ is the space of "piecewise flow lines" emanating from a and ending at b.

The following definition of a Morse function's "flow category " was also given in [11].

Definition 10.1.6. The *flow category* $\mathcal{C}_f$ is a topological category associated to a Morse function $f : M \to \mathbb{R}$ where M is a closed Riemannian manifold. Its objects are the critical points of f. If a and b are two such critical points, then $Mor_{\mathcal{C}_f}(a,b) = \bar{\mathcal{M}}(a,b)$. Composition is determined by the maps

$$\bar{\mathcal{M}}(a,b) \times \bar{\mathcal{M}}(b,c) \hookrightarrow \bar{\mathcal{M}}(a,c),$$

which are defined to be the natural embeddings into the boundary.

The moduli spaces $\mathcal{M}(a,b)$ have natural framings on their stable normal bundles (or equivalently, their stable tangent bundles) that play an important role in this theory. These framings are defined in the following manner. Let a and b be critical points with $a > b$. Let $\varepsilon > 0$ be chosen so that there are no critical values in the half open interval $[f(a) - \varepsilon, f(a))$. Define the *unstable sphere* to be the level set of the unstable manifold,

$$S^u(a) = W^u(a) \cap f^{-1}(f(a) - \varepsilon).$$

The sphere $S^u(a)$ has dimension $\mu(a) - 1$. Notice there is a natural diffeomorphism,

$$\mathcal{M}(a,b) \cong S^u(a) \cap W^s(b).$$

This leads to the following diagram,

$$\begin{array}{ccc} W^s(b) & \xrightarrow{\hookrightarrow} & M \\ \cup\uparrow & & \uparrow\cup \\ \mathcal{M}(a,b) & \xrightarrow[\hookrightarrow]{} & S^u(a). \end{array} \tag{10.1.9}$$

From this diagram one sees that the normal bundle ν of the embedding $\mathcal{M}(a,b) \hookrightarrow S^u(a)$ is the restriction of the normal bundle of $W^s(b) \hookrightarrow M$. Since $W^s(b)$ is a disk, and therefore contractible, this bundle is trivial. Indeed an orientation of $W^s(b)$ determines a homotopy class of trivialization, or a framing. In fact this framing determines an isotopy class of diffeomorphism of the bundle to the product, $W^s(b) \times W^u(b)$. Thus these orientations give the moduli spaces $\mathcal{M}(a,b)$ canonical normal framings, $\nu \cong \mathcal{M}(a,b) \times W^u(b)$.

As was pointed out in [11], these framings extend to the boundary of the compactifications, $\bar{\mathcal{M}}(a,b)$. In order to describe what it means for these framings to be "coherent" in an appropriate sense, the following categorical approach was used in [10]. The first step is to abstract the basic properties of a flow category of a Morse function.

Definition 10.1.7. A *smooth, compact category* is a topological category $\mathcal{C}$ whose objects form a discrete set, and whose whose morphism spaces, $Mor(a,b)$ are compact, smooth $\langle k \rangle$ -manifolds, where $k = dim\, Mor(a,b)$. If $a > b > c$, the composition map, $\nu : Mor(a,b) \times Mor(b,c) \to Mor(a,c)$, is a smooth codimension one embedding (of manifolds with corners) whose image lies in the boundary. Moreover every point in the boundary of $Mor(a,c)$ is in the image under ν of a maximal sequence in $Mor(a,b_1) \times Mor(b_1,b_2) \times \cdots \times Mor(b_{k-1},b_k) \times Mor(b_k,c)$ for some objects $\{b_1,\ldots,b_k\}$.

A smooth, compact category $\mathcal{C}$ is said to be a "Morse-Smale" category if the following additional properties are satisfied.

1. The objects of $\mathcal{C}$ are partially ordered by the condition
$$a \geq b \quad \text{if} \quad Mor(a,b) \neq \emptyset.$$
2. $Mor(a,a) = \{identity\}$.
3. There is a set map, $\mu : Ob(\mathcal{C}) \to \mathbb{Z}$, which preserves the partial ordering, such that if $a > b$,
$$dim\, Mor(a,b) = \mu(a) - \mu(b) - 1.$$
The map μ is known as an "index" map. A Morse-Smale category such as this is said to have finite type, if there are only finitely many objects of any given index, and for each pair of objects $a > b$, there are only finitely many objects c with $a > c > b$. For ease of notation we write $k(a,b) = \mu(a) - \mu(b) - 1$.

The following is a folk theorem that goes back to the work of Smale and Franks [19] although a proof of this fact did not appear in the literature until much later [51].

Proposition 10.1.8. *Let $f : M \to \mathbb{R}$ be smooth Morse function on a closed Riemannian manifold with a Morse-Smale metric. Then the compactified moduli space of piecewise flow-lines, $\bar{\mathcal{M}}(a,b)$ is a smooth $\langle k(a,b)\rangle$-manifold.*

Using this result, as well as an associativity result for the gluing maps $\bar{\mathcal{M}}(a,b) \times \bar{\mathcal{M}}(b,c) \to \bar{\mathcal{M}}(a,c)$ which was eventually proved in [51], it was proven in [12] that the flow category $\mathcal{C}_f$ of such a Morse-Smale function is indeed a Morse-Smale smooth, compact category according to Definition 10.1.7.

Remark 10.1.9. The fact that [11] was never submitted for publication was due to the fact that the "folk theorem" mentioned above, as well as the associativity of gluing, both of which the authors of [11] assumed were "well known to the experts", were indeed not in the literature, and their proofs which were eventually provided in [51], were analytically more complicated than the authors imagined.

In order to define the notion of "coherent framings" of the moduli spaces $\bar{\mathcal{M}}(a,b)$, so that we may apply the Pontrjagin-Thom construction coherently, we need to study an associated category, enriched in spectra, defined using the stable normal bundles of the moduli spaces of flows.

Definition 10.1.10. Let $\mathcal{C}$ be a smooth, compact category of finite type satisfying the Morse-Smale condition. Then a *normal Thom spectrum* of the category $\mathcal{C}$ is a category, $\mathcal{C}^\nu$, enriched over spectra, that satisfies the following properties.

1. The objects of $\mathcal{C}^\nu$ are the same as the objects of $\mathcal{C}$.
2. The morphism spectra $Mor_{\mathcal{C}^\nu}(a,b)$ are $\langle k(a,b)\rangle$-spectra, having the homotopy type of the normal Thom spectra $Mor_{\mathcal{C}}(a,b)^\nu$, as $\langle k(a,b)\rangle$-spectra. The composition maps,
$$\circ : Mor_{\mathcal{C}^\nu}(a,b) \wedge Mor_{\mathcal{C}^\nu}(b,c) \to Mor_{\mathcal{C}^\nu}(a,c)$$
have the homotopy type of the maps,
$$Mor_{\mathcal{C}}(a,b)^\nu \wedge Mor_{\mathcal{C}}(b,c)^\nu \to Mor_{\mathcal{C}}(a,c)^\nu$$
of the Thom spectra of the stable normal bundles corresponding to the composition maps in $\mathcal{C}$, $Mor_{\mathcal{C}}(a,b) \times Mor_{\mathcal{C}}(b,c) \to Mor_{\mathcal{C}}(a,c)$. Recall that these are maps of $\langle k(a,c)\rangle$-spaces induced by the inclusion of a component of the boundary.
3. The morphism spectra are equipped with "Pontrjagin-Thom maps" $\tau_{a,b} : \Sigma^\infty(J(\mu(a), \mu(b))^+) = \Sigma^\infty((\mathbb{R}_+^{k(a,b)}) \cup \infty)) \to Mor_{\mathcal{C}^\nu}(a,b)$ such that the following diagram commutes:
$$\begin{array}{ccc} \Sigma^\infty(J(\mu(a),\mu(b))^+) \wedge \Sigma^\infty(J(\mu(b),\mu(c))^+) & \longrightarrow & \Sigma^\infty(J(\mu(a),\mu(c))^+) \\ {\scriptstyle \tau_{a,b}\wedge\tau_{b,c}}\downarrow & & \downarrow{\scriptstyle \tau_{a,c}} \\ Mor_{\mathcal{C}^\nu}(a,b) \wedge Mor_{\mathcal{C}^\nu}(b,c) & \longrightarrow & Mor_{\mathcal{C}^\nu}(a,c). \end{array}$$
Here the top horizontal map is defined via the composition maps in the category $\mathcal{J}$, and the bottom horizontal map is defined via the composition maps in $\mathcal{C}^\nu$.

With the notion of a "normal Thom spectrum" of a flow category $\mathcal{C}$, the notion of a coherent E^*-orientation was defined in [10]. Here E^* is a generalized cohomology theory represented by a commutative (E_∞) ring spectrum E. We recall that definition now.

First observe that a commutative ring spectrum E induces a $\langle k \rangle$-diagram in the category of spectra ("$\langle k \rangle$-spectrum"), $E\langle k \rangle$, defined in the following manner.

For $k = 1$, we let $E\langle 1 \rangle : \underline{2} \to \mathit{Spectra}$ be defined by $E\langle 1 \rangle(0) = S^0$, the sphere spectrum, and $E\langle 1 \rangle(1) = E$. The image of the morphism $0 \to 1$ is the unit of the ring spectrum $S^0 \to E$.

To define $E\langle k \rangle$ for general k, let a be an object of $\underline{2}^k$. We view a as a vector of length k, whose coordinates are either zero or one. Define $E\langle k \rangle(a)$ to be the multiple smash product of spectra, with a copy of S^0 for every every zero coordinate, and a copy of E for every string of successive ones. For example, if $k = 6$, and $a = (1, 0, 1, 1, 0, 1)$, then $E\langle k \rangle(a) = E \wedge S^0 \wedge E \wedge S^0 \wedge E$.

Given a morphism $a \to a'$ in $\underline{2}^k$, one has a map $E\langle k \rangle(a) \to E\langle k \rangle(a')$ defined by combining the unit $S^0 \to E$ with the ring multiplication $E \wedge E \to E$.

Said another way, the functor $E\langle k \rangle : \underline{2}^k \to \mathit{Spectra}$ is defined by taking the k-fold product functor $E\langle 1 \rangle : \underline{2} \to \mathit{Spectra}$ which sends $(0 \to 1)$ to $S^0 \to E$, and then using the ring multiplication in E to "collapse" successive strings of E's.

This structure allows us to define one more construction. Suppose $\mathcal{C}$ is a smooth, compact, Morse-Smale category of finite type as in Definition 10.1.7. We can then define an associated category, $E_{\mathcal{C}}$, whose objects are the same as the objects of $\mathcal{C}$ and whose morphisms are given by the spectra,

$$Mor_{E_{\mathcal{C}}}(a, b) = E\langle k(a, b) \rangle$$

where $k(a, b) = \mu(a) - \mu(b) - 1$. Here $\mu(a)$ is the index of the object a as in Definition 10.1.7. The composition law is the pairing,

$$\begin{aligned} E\langle k(a,b) \rangle \wedge E\langle k(b,c) \rangle &= E\langle k(a,b) \rangle \wedge S^0 \wedge E\langle k(b,c) \rangle \\ &\xrightarrow{1 \wedge u \wedge 1} E\langle k(a,b) \rangle \wedge E\langle 1 \rangle \wedge E\langle k(b,c) \rangle \\ &\xrightarrow{\mu} E\langle k(a,c) \rangle. \end{aligned}$$

Here $u : S^0 \to E = E\langle 1 \rangle$ is the unit. This category encodes the multiplication in the ring spectrum E.

Definition 10.1.11. An E^*-*orientation* of a smooth, compact category of finite type satisfying the Morse-Smale condition, $\mathcal{C}$, is a functor, $u : \mathcal{C}^\nu \to E_{\mathcal{C}}$, where $\mathcal{C}^\nu$ is a normal Thom spectrum of $\mathcal{C}$, such that on morphism spaces, the induced map

$$Mor_{\mathcal{C}^\nu}(a, b) \to E\langle k(a, b) \rangle$$

is a map of $\langle k(a, b) \rangle$-spectra that defines an E^* orientation of $Mor_{\mathcal{C}^\nu}(a, b) \simeq \bar{\mathcal{M}}(a, b)^\nu$.

This notion of E^*-orientation was discussed more fully in [10]. In particular it is described there how the functor $u : \mathcal{C}^\nu \to E_{\mathcal{C}}$ should be thought of as a coherent family of E^*- Thom classes for the normal bundles of the morphism spaces of $\mathcal{C}$. When $E = \mathbb{S}$, the sphere spectrum, then an E^*-orientation, as defined here, defines a coherent family of framings of the morphism spaces, and is equivalent to the notion of a *framing* of the category $\mathcal{C}$, as defined in [12].

In [12] the following was proved modulo the results of [51] which appeared much later.

Theorem 10.1.12. *Let $f : M \to \mathbb{R}$ be a Morse function on a closed Riemannian manifold satisfying the Morse-Smale condition. Then the flow category $\mathcal{C}_f$ has a canonical structure as a "$\mathbb{S}$-oriented, smooth, compact Morse-Smale category of finite type". That is, it is a "framed, smooth compact Morse-Smale category". The induced framings of the morphism manifolds $\bar{\mathcal{M}}(a, b)$ are canonical extensions of the framings of the open moduli spaces $\mathcal{M}(a, b)$ described above (10.1.9).*

The main use of the notion of compact, smooth, framed flow categories is the following result.

Theorem 10.1.13. [12, 10] *Let $\mathcal{C}$ be a compact, smooth, framed category of finite type satisfying the Morse-Smale property. Then there is an associated, naturally defined functor $Z_\mathcal{C} : \mathcal{J}_0 \to$ Spectra whose geometric realization $|Z_\mathcal{C}|$ realizes the associated "Floer complex"*

$$\to \cdots \to C_{i+1} \xrightarrow{\partial_i} C_i \xrightarrow{\partial_{i-1}} \cdots \xrightarrow{\partial_0} C_0.$$

Here C_j is the free abelian group generated by the objects $a \in Ob(\mathcal{C})$ with $\mu(a) = j$, and the boundary homomorphisms are defined by

$$\partial_{j-1}([a]) = \sum_{\mu(b)=j-1} \#Mor(a, b) \cdot [b]$$

where $\#Mor(a, b)$ is the framed cobordism type of the compact framed manifold $Mor(a, b)$ which zero dimensional (since its dimension $= \mu(a) - \mu(b) - 1 = 0$). Therefore this cobordism type is simply an integer, which can be viewed as the oriented count of the number of points in $Mor(a, b)$.

Proof. Sketch. The proof of this is sketched in [12] and is carried out in [10] in the setting of E^*-oriented compact, smooth categories, for E any ring spectrum. The idea for defining the functor $Z_\mathcal{C}$ is to use the Pontrjagin-Thom construction in the setting of framed manifolds with corners (more specifically, framed $\langle k \rangle$-manifolds). Namely, one defines

$$Z_\mathcal{C}(j) = \bigvee_{a \in Obj(\mathcal{C}) \,:\, \mu(a)=j} \mathbb{S}.$$

On the level of morphisms one needs to define, for every $i > j$, a map of spectra

$$Z_\mathcal{C}(i, j) : J(i, j)^+ \wedge Z_\mathcal{C}(i) \to Z_\mathcal{C}(j).$$

This is defined to be the wedge, taken over all $a \in Obj(\mathcal{C})$ with $\mu(a) = i$, and $b \in Obj(\mathcal{C})$ with $\mu(b) = j$, of the maps

$$Z_\mathcal{C}(a, b) : J(\mu(a), \mu(b))^+ \wedge \mathbb{S}_a \to \mathbb{S}_b$$

defined to be the composition

$$\Sigma^\infty(J(\mu(a),\mu(b)^+) \xrightarrow{\tau_{a,b}} Mor_{\mathcal{C}^\nu}(a,b) \xrightarrow{u} \mathbb{S} \tag{10.1.10}$$

where $\mathbb{S}_a$ and $\mathbb{S}_b$ are copies of the sphere spectrum indexed by a and b respectively in the definition of $Z_{\mathcal{C}}(i)$ and $Z_{\mathcal{C}}(j)$. $\tau_{a,b}$ is the Pontrjagin-Thom map, and u is $\mathbb{S}$-normal orientation class (framing). Details can be found in [10]. □

Thus to define a *"Floer homotopy type"* one is looking for a compact, smooth, framed category of finite type satisfying the Morse-Smale property. The compact framed manifolds with corners that constitute the morphism spaces, define, via the Pontrjagin-Thom construction, the attaching maps of the CW-spectrum defining this (stable) homotopy type.

We see from Theorem 10.1.12 that given a Morse-Smale function $f : M \to \mathbb{R}$, then its flow category satisfies these properties, and it was proved in [12] that, not surprisingly, its Floer homotopy type is the suspension spectrum $\Sigma^\infty(M_+)$. The CW-structure is the classical one coming from Morse theory, with one cell of dimension k for each critical point of index k. The fact that the compactified moduli spaces of flow lines, which constitute the morphism spaces in this category, together with their structure as framed manifolds with corners, define the attaching maps in the CW-structure of $\Sigma^\infty M$, can be viewed as a generalization of the well-known work of Franks in [19].

As pointed out in [12], a distinguishing feature in the flow category of a Morse-Smale function $f : M \to \mathbb{R}$ is that the framing is *canonical.* See (10.1.9) above. As also was pointed out in [12], if one chooses a *different* framing of the flow category $\mathcal{C}_f$, then the "difference" between the new framing and the canonical framing defines a functor

$$\Phi : \mathcal{C}_f \to \mathcal{G}L_1(\mathbb{S}) \tag{10.1.11}$$

where $\mathcal{G}L_1(\mathbb{S})$ is the category corresponding to the group-like monoid $GL_1(\mathbb{S})$ of "units" of the sphere spectrum (see [5]). This monoid has the homotopy time of the colimit

$$GL_1(\mathbb{S}) \simeq \mathrm{colim}_{n\to\infty} \Omega^n_{\pm 1} S^n.$$

Here the subscript denotes the path components of $\Omega^n S^n$ consisting of based self-maps of the sphere S^n of degree ± 1. By a minor abuse of notations, we let Φ denote the framing of $\mathcal{C}_f$ that defines the map (10.1.11).

Passing to the geometric realizations of these categories, one gets a map

$$\phi : M \to BGL_1(\mathbb{S}).$$

which we think of as the isomorphism type of a spherical fibration over M. The following is also a result of [12].

Proposition 10.1.14. *If $f : M \to \mathbb{R}$ is a Morse-Smale function on a closed Riemannian manifold, and its flow category $\mathcal{C}_f$ is given a framing Φ, then the Floer homotopy type of $(\mathcal{C}_f, \Phi)$ viewed as a compact, smooth, framed category, is the Thom spectrum of the corresponding stable spherical fibration, M^ϕ.*

10.1.3 The free loop space of a symplectic manifold and symplectic Floer theory

Let (N^{2n}, ω) be a symplectic manifold. Here ω is a closed, nondegenerate, skew symmetric bilinear form on the tangent bundle, TN. Let LN be its free loop space, where here and throughout this section we are taking smooth loops. One of the earliest applications of Floer theory [18] was to the (perturbed) symplectic action functional on LN.

Let $L_0N \subset LN$ be the path component of contractible loops in N. Let $\tilde{L}_0N \xrightarrow{p} L_0N$ be its universal cover. Explicitly,

$$\tilde{L}_0N = \{(\gamma, \theta) \in L_0N \times C^\infty(D^2, N) \,:\, \text{the restriction of } \theta \text{ to } S^1 = \partial D^2 \text{ is equal to } \gamma\}/\sim$$

where the equivalence relation is given by $(\gamma, \theta_1) \sim (\gamma, \theta_2)$ if, when we combine θ_1 and θ_2 to define a map of the 2-sphere,

$$\theta_{1,2} = \theta_1 \cup \theta_2 : D^2 \cup_{S^1} D^2 = S^2 \to N$$

then $\theta_{1,2}$ is null homotopic. In other words, (γ, θ_1) is equivalent to (γ, θ_2) if θ_1 and θ_2 are homotopic maps relative to the boundary.

One can then define the "symplectic action" functional,

$$\mathcal{A} : \tilde{L}_0N \to \mathbb{R}. \tag{10.1.12}$$

$$(\gamma, \theta) \to \int_{D^2} \theta^*(\omega). \tag{10.1.13}$$

There are important situations when the symplectic action descends to define an $\mathbb{R}/\mathbb{Z}$-valued function

$$\mathcal{A} : L_0N \to \mathbb{R}/\mathbb{Z}.$$

This happens, for example, when ω is an integral symplectic form. By an "integral symplectic form", we mean a symplectic form ω with the property that the real two-dimensional cohomology class it defines is in the image of integral cohomology. See [43] for details.

One needs to perturb the symplectic action functional by use of a Hamiltonian vector field in order to achieve nondegeneracy of critical points. A Morse-type complex, generated by critical points, is then studied, and the resulting symplectic Floer homology has proved to be an important invariant. We refer the reader to [43] for a thorough treatment of this theory.

We describe the situation when $N = T^*M^n$, the cotangent bundle of a closed, n-dimensional manifold, in more detail. In particular this is a situation where one has a corresponding "Floer homotopy type", defined via a compact, smooth flow category as described above. This was studied by the author in [9], making heavy use of the analysis of Abbondandolo and Schwarz in [1].

Coming from classical mechanics, the cotangent bundle of a smooth manifold has a canonical symplectic structure. It is defined as follows.

Let $p : T^*M \to M$ be the projection map. For $x \in M$ and $u \in T^*_x M$, let $z = (y, u) \in T^*M$, and consider the composition

$$\theta_z : T_z(T^*M) \xrightarrow{dp} T_y M \xrightarrow{u} \mathbb{R}.$$

This defines a 1-form θ on T^*M, called the "Liouville" 1-form, and the symplectic form ω is defined to be the exterior derivative $\omega = d\theta$. It is easy to check that ω is nondegenerate.

Let

$$H : \mathbb{R}/\mathbb{Z} \times T^*M \to \mathbb{R}$$

be a smooth function. Such a map is called a "time-dependent periodic Hamiltonian". Using the nondegeneracy of the symplectic form, this allows one to define the corresponding "Hamiltonian vector field" X_H by requiring it to satisfy the equation

$$\omega(X_H(t, z), v) = -dH_{t,z}(v)$$

for all $t \in \mathbb{R}/\mathbb{Z}$, $z \in T^*M$, and $v \in T_z(T^*M)$. We will be considering the space of 1-periodic solutions, $\mathcal{P}(H)$, of the Hamiltonian equation

$$\frac{dz}{dt} = X_H(t, z(t))$$

where $z : \mathbb{R}/\mathbb{Z} \to T^*M$ is a smooth function.

Using a periodic time-dependent Hamiltonian one can define the perturbed symplectic action functional

$$\begin{aligned} \mathcal{A}_H &: L(T^*M) \to \mathbb{R} \\ z &\to \int z^*(\theta - Hdt) = \int_0^1 (\theta(\frac{dz}{dt}) - H(t, z(t))dt. \end{aligned} \tag{10.1.14}$$

This is a smooth functional [1], and its critical points are the periodic orbits of the Hamiltonian vector field, $\mathcal{P}(H)$. Now let J be a 1-periodic, smooth almost complex structure on T^*M, so that for each $t \in \mathbb{R}/\mathbb{Z}$,

$$\langle \zeta, \xi \rangle_{J_t} = \omega(\zeta, J(t, z)\xi), \quad \zeta, \xi \in T_z T^*M, \; z \in T^*M,$$

is a loop of Riemannian metrics on T^*M. One can then consider the gradient of $\mathcal{A}_H$ with respect to the metric, $\langle \cdot, \cdot \rangle$. It is given by

$$\nabla_J \mathcal{A}_H(z) = -J(z, t)(\frac{dz}{dt} - X_H(t, z)).$$

The (negative) gradient flow equation on a smooth curve $u : \mathbb{R} \to L(T^*M)$,

$$\frac{du}{ds} + \nabla_J \mathcal{A}_H(u(s))$$

can be rewritten as a perturbed Cauchy-Riemann PDE, if we view u as a smooth map $\mathbb{R}/\mathbb{Z} \times \mathbb{R} \to T^*M$, with coordinates, $t \in \mathbb{R}/\mathbb{Z}$, $s \in \mathbb{R}$:

$$\partial_s u - J(t, u(t,s))(\partial_t u - X_H(t, u(t,s))) = 0. \tag{10.1.15}$$

Let $a, b \in \mathcal{P}(H)$. Abbondandolo and Schwarz defined the space of solutions

$$\begin{aligned} W(a,b;H,J) = \{u : \mathbb{R} \to L(T^*M) \text{ a solution to (10.1.15), such that} \\ \lim_{s \to -\infty} u(s) = a, \text{ and } \lim_{s \to +\infty} u(s) = b\}. \end{aligned} \tag{10.1.16}$$

As in the case of Morse theory, we then let $\mathcal{M}(a,b)$ the "moduli space" obtained by dividing out by the free $\mathbb{R}$-action,

$$\mathcal{M}(a,b) = W(a,b;H,J))/\mathbb{R}. \tag{10.1.17}$$

It was shown in [1] that with respect to a generic choice of Hamiltonian an almost complex structure, the spaces $W(a,b;H,J)$ and $\mathcal{M}(a,b)$ are smooth manifolds, whose dimensions are given by $\mu(a) - \mu(b)$ and $\mu(a) - \mu(b) - 1$ respectively, where μ represents the "Conley - Zehnder index" of the periodic Hamiltonian orbits a and b. Furthermore it was shown in [1] that in analogy with Morse theory, one can compactify these moduli spaces as

$$\bar{\mathcal{M}}(a,b) = \bigcup_{a=a_1>a_2>\cdots>a_k=b} \mathcal{M}(a_1,a_2) \times \cdots \times \mathcal{M}(a_{k-1},a_k).$$

The fact that that these compact moduli spaces have canonical framings was shown in [9] in part by showing that the obstruction to framing, described originally in [12], is zero in this case. This obstruction was called the *polarization class* in [12] and is defined as follows.

Let N be an almost complex manifold whose tangent bundle is classified by a map $\tau : N \to BU(n)$. Applying loop spaces, one has a composite map,

$$\rho : LN \xrightarrow{L(\tau)} L(BU(n)) \hookrightarrow LBU \simeq BU \times U \to U \to U/O. \tag{10.1.18}$$

Here the homotopy equivalence $L(BU) \simeq BU \times U$ is well defined up to homotopy, and is given by a trivialization of the fibration

$$U \simeq \Omega BU \xrightarrow{\iota} L(BU) \xrightarrow{ev} BU$$

where $ev : LX \to X$ evaluates a loop a $0 \in \mathbb{R}/\mathbb{Z}$. The trivialization is the composition

$$U \times BU \xrightarrow{\iota \times \sigma} L(BU) \times L(BU) \xrightarrow{mult} L(BU).$$

Here $\sigma : BU \to L(BU)$ is the section of the above fibration given by assigning to a point $x \in BU$ the constant loop at that point, and the "multiplication" map in this composition is induced by the infinite loop space structure of BU.

The reason we refer the map ρ as the "polarization class" of the loop space LN, is because when viewed as an infinite dimensional manifold, the tangent bundle $T(LN)$ has the structure of a "polarization" as defined in [46]. This means that this infinite dimensional tangent bundle has structure group given by the "restricted general linear group of a Hilbert space", $GL_{res}(H)$ as originally defined in [46]. As shown there, $GL_{res}(H)$ has the homotopy type (as infinite loop spaces) of $\mathbb{Z} \times BO$, and so its classifying space, $BGL_{res}(H)$ has the homotopy type of $B(\mathbb{Z} \times BO) \simeq U/O$ by Bott periodicity. The classifying map $LN \to U/O$ has the homotopy type of the "polarization class" ρ defined above. See [46] , [12], and [9] for details.

When $N = T^*M$ two things were shown in [9]. First, when one views $\bar{\mathcal{M}}(a,b)$ as a space of paths, one sees that there is a natural inclusion map into the full space of paths in $L(T^*M)$ which begin at a and end at b. This path space is equivalent to the based loop space $\Omega L(T^*M)$. The result is a map $\iota_{a,b} : \bar{\mathcal{M}}(a,b) \to \Omega L(T^*M)$, well defined up to homotopy, such that the composition

$$\tau_{a,b} : \bar{\mathcal{M}}(a,b) \xrightarrow{\iota_{a,b}} \Omega L(T^*M) \xrightarrow{\Omega\rho} \Omega U/O \simeq \mathbb{Z} \times BO \tag{10.1.19}$$

classifies the stable tangent bundle of $\bar{\mathcal{M}}(a,b)$. Second, it was shown that in this case, i.e. when $N = T^*M$, the polarization class $\rho : L(T^*M) \to U/O$ is trivial. This is essentially because the almost complex structure (i.e. $U(n)$-structure) of the tangent bundle of T^*M is the complexification of the n-dimensional real bundle (i.e. $O(n)$-structure) of the tangent bundle of M pulled back to T^*M via the projection map $T^*M \to M$. By (10.1.19) this leads to a coherent family of framings on the moduli spaces, which in turn lead to a smooth, compact, framed structure on the flow category $\mathcal{C}_H$ of the symplectic action functional, as shown in [9].

Using the methods and results of [1], which is to say, comparing the flow category of the perturbed symplectic action functional $\mathcal{C}_H$ to the Morse flow category of an energy functional on LM, the following was shown in [9].

Theorem 10.1.15. *Let M^n be a closed spin manifold of dimension n. For appropriate choices of a Hamiltonian H and a generic choice of almost complex structure J on the cotangent bundle T^*M, the Floer homotopy type determined by the smooth, compact flow category $\mathcal{C}_H$ is given by the suspension spectrum of the free loop space,*

$$Z_{\mathcal{C}_H}(T^*M) \simeq \Sigma^\infty(LM_+).$$

Remark 10.1.16. 1. This theorem generalized a result of Viterbo [58] stating that the symplectic Floer homology, $SFH_*(T^*M)$ is isomorphic to $H_*(LM)$.

2. If M is not spin, one needs to use appropriately twisted coefficients in both Viterbo's theorem and in Theorem 10.1.15 above. This was first observed by Kragh in [30], and was overlooked in all or most of the discussions of the relation between the symplectic Floer theory of the cotangent bundle and homotopy type of the free loop space before Kragh's work, including the author's work in [9].

10.2 The work of Lipshitz and Sarkar on Khovanov homotopy theory

A recent dramatic application of the ideas of Floer homotopy theory appeared in the work of Lipshitz and Sarkar on the homotopy theoretic foundations of Khovanov's homological invariants of knots and links. This work appeared in [35] and [36]. Another version of Khovanov homotopy appears in [26]. It was proved to be equivalent to the Lipshitz-Sarkar construction in [34].

The Khovanov homology of a link L is a bigraded abelian group, $Kh^{i,j}(L)$. It is computed from a chain complex denoted $KhC^{i,j}(L)$ that is defined in terms of a link diagram. However the Khovanov homology was shown to be independent of the choice of link diagram, up to isomorphism. This invariant was originally defined by Khovanov in [27] in which he viewed these homological invariants as a "categorification" of the Jones polynomial $V(L)$ in the sense that the graded Euler characteristic of this homology theory recovers $V(L)$ via the formula

$$\begin{aligned}\chi(Kh^{i,j}(L)) &= \sum_{i,j}(-1)^i q^j \, rank \, Kh^{i,j}(L) \\ &= (q+q^{-1})\, V(L).\end{aligned}$$

The goal of the work of Lipshitz and Sarkar was to associate to a link diagram L a family of spectra $X^j(L)$ whose homotopy types are invariants of the isotopy class of the link (and in particular do not depend on the particular link diagram used), and such that the Khovanov homology $Kh^{i,j}(L)$ is isomorphic to the reduced singular cohomology $\tilde{H}^i(X^j(L))$. Their basic idea is to construct a compact, smooth, framed flow category from the moduli spaces associated to a link diagram. Their construction is entirely combinatorial, and the cells of the spectrum $X(L) = \bigvee_j X^j(L)$ correspond to the standard generators of the Khovanov complex $KhC^{*,*}(L)$. That is, $X(L)$ *realizes* the Khovanov chain complex in the sense described above. $X(L)$ is referred to as the "Khovanov homotopy type" of the link L, and it has had several interesting applications.

Notice that by virtue of the existence of a Khovanov homotopy type, the Khovanov homology, when reduced modulo a prime, carries an action of the Steenrod algebra $\mathcal{A}_p$. In [36] the authors show that the Steenrod operation Sq^2 acts nontrivially on the Khovanov homology for many knots, and in particular for the torus knot, $T_{3,4}$. It is also known by work of Seed [53] that there are pairs of links with isomorphic Khovanov's homology, but distinct Khovanov homotopy types. Also, Rasmussen constructed a slice genus bound, called the s-invariant, using Khovanov homology [47]. Using the Khovanov homotopy type, Lipshitz and Sarkar produced a family of generalizations of the s-invariant, and used them to obtain even stronger slice genus bounds. Stoffregen and Zhang [54] used Khovanov homotopy theory to describe rank inequalities for Khovanov homology for prime-periodic links in S^3.

We now give a sketch of the construction of compact framed flow category of Lipshitz and Sarkar, which yields the Khovanov homotopy type.

By a "link diagram", one means the projection onto $\mathbb{R}^2$ of an embedded disjoint union of circles in $\mathbb{R}^3$ that has transverse crossings. One keeps track of the resulting "over" and "under crossings", and usually one orients the link (i.e. puts an arrow in each component). One can "resolve" a crossing in two ways. These are referred to as a 0-resolution and a 1 resolution and are shown in Figure 10.2.1.

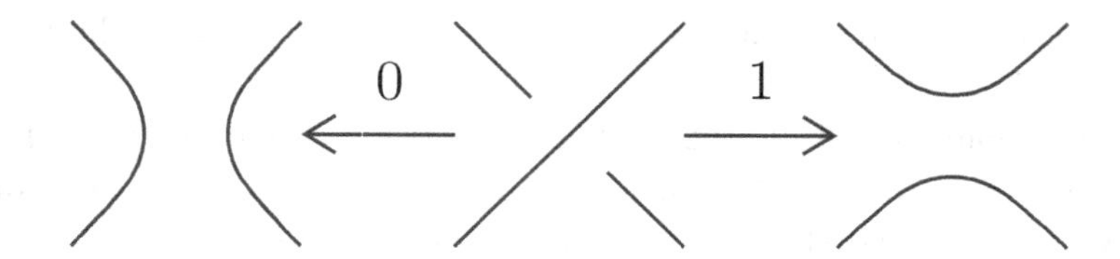

FIGURE 10.2.1. The 0-resolution and the 1-resolution

The Khovanov chain complex is generated by all possible configurations of resolutions of the crossings of a link diagram. We recall the Lipshitz-Sarkar view of its definition more carefully.

Definition 10.2.1. A *resolution configuration* D is a pair $(Z(D), A(D))$, where $Z(D)$ is a set of pairwise disjoint embedded circles in $S^2 = \mathbb{R}^2 \cup \infty$, and $A(D)$ is an ordered collection of arcs embedded in S^2 with

$$A(D) \cap Z(D) = \partial A(D).$$

The number of arcs in the resolution configuration D is called its index denoted by $ind(D)$.

Definition 10.2.2. Given a link diagram L with n crossings, an ordering of the crossings, and a vector $v \in \{0,1\}^n$, there is an associated resolution configuration $D_L(v)$ obtained by taking the resolution of L corresponding to v. That is, one takes the 0-resolution of the ith crossing if $v_i = 0$, and the 1-resolution of the ith crossing if $v_i = 1$. One then places arcs corresponding to each of the crossings labeled by zeros in v.

Figure 10.2.2 illustrates the resolution configuration corresponding to a diagram of the trefoil knot.

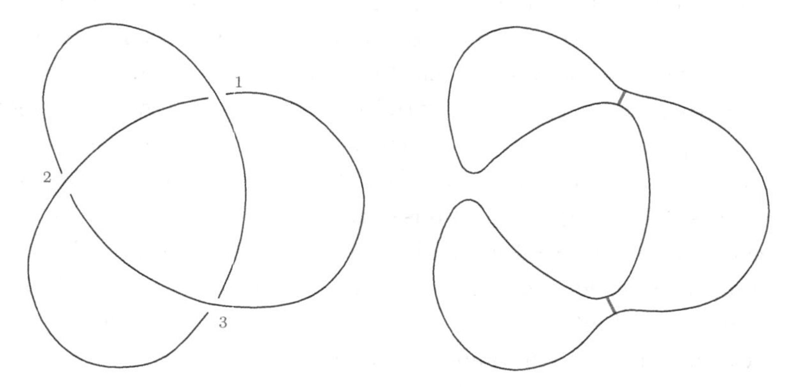

FIGURE 10.2.2. A knot diagram K for the trefoil with ordered crossings, and the resolution configuration $D_K((0,1,0))$

The following terminology is also useful.

Definition 10.2.3. 1. The *core* $c(D)$ of a resolution configuration is the resolution configuration obtained from D by deleting all the circles in $Z(D)$ that are disjoint from all arcs in $A(D)$.

2. A resolution configuration is *basic* if $D = c(D)$, i.e. every circle in $Z(D)$ intersects an arc in $A(D)$.

One can also do a *surgery* along a subset $A \subset A(D)$. The resulting resolution configuration is denoted $s_A(D)$. The surgery procedure is best illustrated in Figure 10.2.3.

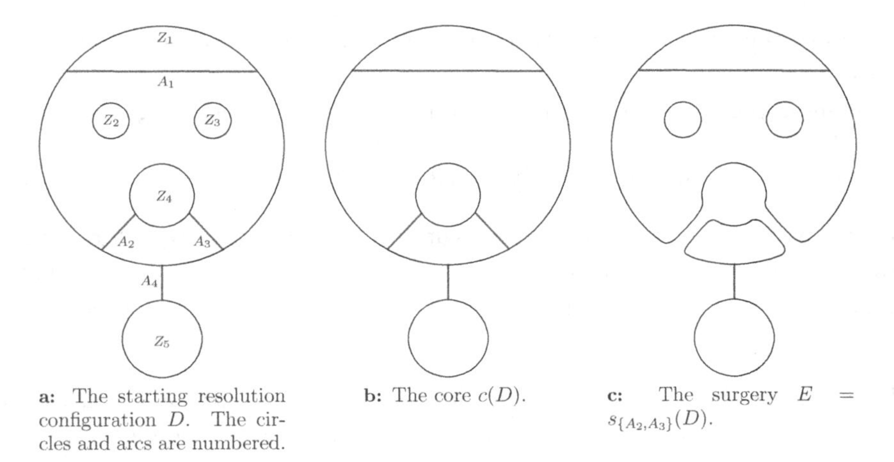

a: The starting resolution configuration D. The circles and arcs are numbered.

b: The core $c(D)$.

c: The surgery $E = s_{\{A_2,A_3\}}(D)$.

FIGURE 10.2.3. A knot diagram K for the trefoil with ordered crossings and the resolution configuration $D_K((0,1,0))$

A *labeling* $x = \{x_+, x_-\}$ of a resolution configuration D is a labeling of each circle in $Z(D)$ by either x_+ or x_-. Labeled resolutions configurations have a partial ordering defined to be the transitive closure of the following relations.

We say that $(E, y) < (D, x)$ if

1. the labelings agree on $D \cap E$.

2. D is obtained from E by surgering along a single arc of $A(E)$, In particular, either:

   (a) $Z(E \backslash D)$ contains exactly one circle, say Z_i and $Z(D \backslash E)$ contains exactly two circles, say Z_j and Z_k, or

   (b) $Z(E \backslash D)$ contains exactly two circles, say Z_i and Z_j, and $Z(D \backslash E)$ contains exactly one circle, say Z_k.

3. In case (2a), either $y(Z_i) = x(Z_j) = x(Z_k) = x_-$ or $y(Z_i) = x_+$ and $\{x(Z_j), x(Z_k)\} = \{x_+, x_-\}$.

   In case (2b), either $y(Z_i) = y(Z_j) = x(Z_k) = x_+$ or $y(Z_i) = x_+$ or $\{y(Z_i), y(Z_j)\} = \{x_+, x_-\}$ and $x(Z_k) = x_+$.

   One can now define the *Khovanov chain complex* $KhC(L)$ as follows.

Definition 10.2.4. Given an oriented link diagram L with n crossings and an ordering of the crossings in L, $KhC(L)$ is defined to be the free abelian group generated by labeled resolution configurations of the form $(D_L(u), x)$ for $u \in \{0,1\}^n$. $KhC(L)$ carries two gradings, a *homological grading* gr_h and a *quantum grading* gr_q, defined as follows:

$$\begin{aligned} gr_h((D_L(u),x)) &= -n_- + |u| \\ gr_q((D_L(u),x)) &= n_+ - 2n_- + |u| + \#\{Z \in Z(D_L(u)) : x(Z) = x_+\} \\ &\quad - \#\{Z \in Z(D_L(u)) : x(Z) = x_-\}. \end{aligned}$$

Here n_+ denotes the number of positive crossings in L, and n_- denotes the number of negative crossings.

The differential preserves the quantum grading, increases the homological grading by 1, and is defined as

$$\delta(D_L(v), y) = \sum (-1)^{s_0(\mathcal{C}_{u,v})}(D_L(u), x)$$

where the sum is taken over all labeled resolution configurations $(D_L(u), x)$ with $|u| = |v| + 1$ and $(D_L(v), y) < (D_L(u), x)$. The sign $s_0(\mathcal{C}_{u,v}) \in \mathbb{Z}/2$ is defined as follows: If $u = (\varepsilon_1, \ldots, \varepsilon_{i-1}, 1, \varepsilon_{i+1}, \ldots, \varepsilon_n)$ and $v = (\varepsilon_1, \ldots, \varepsilon_{i-1}, 0, \varepsilon_{i+1}, \ldots, \varepsilon_n)$, then $s_0(\mathcal{C}_{u,v}) = \varepsilon_1 + \cdots + \varepsilon_{i-1}$.

The homology of this chain complex is the *Khovanov homology* $Kh^{*,*}(L)$. To define the *Khovanov homotopy type* of the link L, Lipshitz and Sarkar define higher dimensional moduli spaces which have the structure of framed manifolds with corners so that they in turn define a compact, smooth, framed flow category, which by the theory described above defines the associated (stable) homotopy type.

These moduli spaces are defined as a certain covering of the moduli spaces occurring in a "framed flow category of a cube". We now sketch these constructions, following Lipshitz and Sarkar [35].

Let $f_1 : \mathbb{R} \to \mathbb{R}$ be a Morse function with one index zero critical point and one index 1 critical point. For concreteness one can use the function

$$f_1(x) = 3x^2 - 2x^3.$$

Define $f_n : \mathbb{R}^n \to \mathbb{R}$ by

$$f_n(x_1, \ldots, x_n) = f_1(x_1) + \cdots + f_1(x_n).$$

f_n is a Morse function, and we let $\mathcal{C}(n)$ denote its flow category . It is a straightforward exercise to see that the geometric realization $|\mathcal{C}(n)|$ is the n-cube $[0,1]^n$. The vertices of this cube $u \in \{0,1\}^n$ correspond to the critical points of f_n and have a grading which corresponds to the Morse index: $gr(u) = |u| = \sum_i u_i$. They also have a partial ordering coming from the ordering of $\{0,1\}$. We say that $v \leq_i u$ if $v \leq u$ and $gr(u) - gr(v) = i$. That is i is the relative index of u and v. Let $\bar{1} = (1, 1, \cdots, 1)$ and $\bar{0} = (0, \cdots, 0)$. The following is not difficult and is verified in [35].

Lemma 10.2.5. *The compactified moduli space of piecewise flows* $\bar{\mathcal{M}}_{\mathcal{C}(n)}(\bar{1}, \bar{0})$ *is*

- *a single point if* $n = 1$,
- *a closed interval if* $n = 2$,

- *a closed hexagonal disk if* $n = 3$ *(as in Figure 10.2.4), and is*
- *homeomorphic to a closed disk* D^{n-1} *for general* n.

Furthermore, given any $v < u$, $\bar{\mathcal{M}}_{\mathcal{C}(n)}(u, v) \cong \bar{\mathcal{M}}_{\mathcal{C}(gr(u)-gr(v))}(\bar{1}, \bar{0})$.

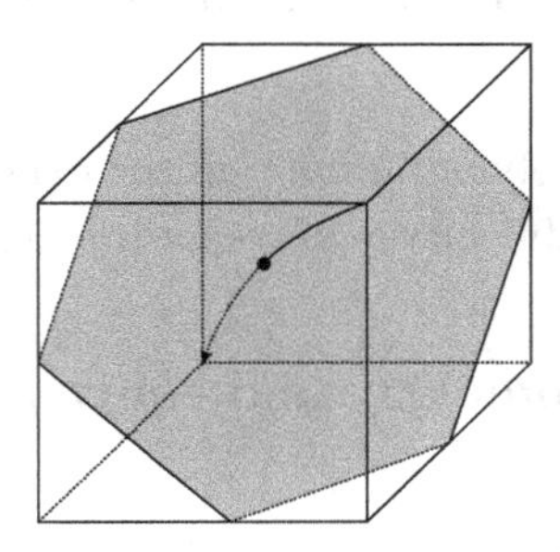

FIGURE 10.2.4. The cube moduli space $\bar{\mathcal{M}}_{\mathcal{C}(3)}(\bar{1}, \bar{0})$

Lipshitz and Sarkar then proceed to define moduli spaces of "decorated resolution configurations" that cover the moduli spaces occurring in a framed flow category of a cube.

Definition 10.2.6. A *decorated resolution configuration* is a triple (D, x, y) where D is a resolution configuration, x is a labeling of each component of $Z(s(D))$, y is a labeling of each component of $Z(D)$ such that

$$(D, y) < (s(D), x).$$

Here $s(D) = s_{A(D)}(D)$ is the maximal surgery on D.

In [35] Lipshitz and Sarkar proceed to construct moduli spaces $\mathcal{M}(D, x, y)$ for every decorated resolution configuration (D, x, y). These will be compact, framed manifolds with corners. Indeed they are $< n-1 >$-manifolds where n is the index of D. They also produce covering maps

$$\mathcal{F} : \mathcal{M}(D, x, y) \to \bar{\mathcal{M}}_{\mathcal{C}(n)}(\bar{1}, \bar{0})$$

which are maps of $< n-1 >$-spaces, trivial as a covering maps on each component of $\mathcal{M}(D, x, y)$, and are local diffeomorphisms. The framings of the moduli spaces $\mathcal{M}(D, x, y)$ are then induced from the framings of $\bar{\mathcal{M}}_{\mathcal{C}(n)}(\bar{1}, \bar{0})$. They produce composition or gluing maps for every labeled resolution configuration (E, z) with

$$(D, y) < (E, z) < (s(D), x)$$

$$\circ : \mathcal{M}(D \backslash E, z|, y|) \times \mathcal{M}(E \backslash s(D), x|, z|) \longrightarrow \mathcal{M}(D, x, y)$$

that are embeddings into the boundary of $\mathcal{M}(D, x, y)$. These moduli spaces are constructed recursively using a clever, but not very difficult argument. We refer the reader to [35] for details.

These constructions allow for the definition of a Khovanov flow category for the link L, and it is shown to be a compact, smooth, framed flow category as defined in section

1.2 above. Using the theory introduced in [12] and described in section 1.2, one obtains a spectrum realizing the Khovanov chain complex. This is the Lipshitz-Sarkar "Khovanov homotopy type" of the link L.

10.3 Manolescu's equivariant Floer homotopy and the triangulation problem

In this section the author is relying heavily on the expository article by Manolescu [40]. We refer the reader to this beautiful survey of recent topological applications of Floer theory.

10.3.1 Monopole Floer homology and equivariant Seiberg-Witten stable homotopy

One example of a dramatic application of Floer's original instanton homology theory, and in particular its topological field theory relationship to the Donaldson invariants of closed 4-manifolds, was to the study of the group of cobordism classes of homology 3-spheres. Define

$$\Theta^3_H = \{\text{oriented homology 3-spheres}\}/\sim$$

where $X_0 \sim X_1$ if and only if there exists a smooth, compact, oriented 4-manifold W with

$$\partial W = (-Y_0) \sqcup Y_1$$

and $H_1(W) = H_2(W) = 0$. The group operation is represented by connected sum, and the inverse is given by reversing orientation. The standard unit 3-sphere S^3 is the identity element. The Rokhlin homomorphism [50], [14] is the map

$$\mu : \Theta^H_3 \to \mathbb{Z}/2$$

defined by sending a homology sphere X to $\mu(X) = \sigma(X)/8\,(mod\,2)$, where W is any compact spin 4-manifold with $\partial W = X$. It is a theorem that this homomorphism is well-defined. Furthermore, using this homomorphism one knows that the group Θ^H_3 is nontrivial, since, for example, the Poincaré sphere P bounds the E_8 plumbing which has signature -8. Therefore $\mu(P) = 1$.

This result has been strengthened using Donaldson theory and Instanton Floer homology. For example Furuta and Fintushel-Stern proved that Θ^H_3 is infinitely generated [21][15]. And using the $SU(2)$ equivariance of Instanton Floer homology, in [20] Froyshov defined a surjective homomorphism

$$h : \Theta^H_3 \to \mathbb{Z}. \tag{10.3.1}$$

Monopole Floer homology is similar in nature to Floer's Instanton homology theory, but it is based on the Seiberg-Witten equations rather than the Yang-Mills equations. More precisely, let Y be a 3-manifold equipped with a $Spin^c$ structure σ. One considers the configuration space of pairs (A, ϕ), where A is a connection on the trivial $U(1)$ bundle over Y, and ϕ is a spinor. There is an action of the gauge group of the bundle on this configuration space, and one considers the orbit space of this action. One can then define the Chern-Simons-Dirac functional on this space by

$$CSD(A, \phi) = -\frac{1}{8} \int_Y (A^t - A_0^t) \wedge (F_{A^t} + F_{A_0^t}) + \frac{1}{2} \int_Y \langle D_A \phi, \phi \rangle \, dvol.$$

Here A_0 is a fixed base connection, and the superscript t denotes the induced connection on the determinant line bundle. The symbol F denotes the curvature of the connection and the symbol D denotes the covariant derivative.

Monopole Floer homology is the Floer homology associated to this functional. To make this work precisely involves much analytic, technical work, due in large part to the existence of reducible connections. Kronheimer and Mrowka dealt with this (and other issues) by considering a blow-up of this configuration space [31]. They actually defined three versions of Monopole Floer homology for every such pair (Y, σ).

Monopole Floer homology can also be used to give an alternative proof of Froyshov's theorem about the existence of a surjective homomorphism.

$$\delta : \Theta_3^H \to \mathbb{Z}.$$

This uses the S^1-equivariance of these equations. Monopole Floer homology has also been used to prove important results in knot theory [32] and in contact geometry [57].

In [37] Manolescu defined a "Monopole", or "Seiberg-Witten" Floer stable homotopy type. He did not follow the program defined by the author, Jones and Segal in [12] as outlined above, primarily because the issue of smoothness in defining a framed, compact, smooth flow category is particularly difficult in this setting. Instead, he applied Furuta's technique of "finite dimensional approximations" [22]. More specifically, the configuration space X of connections and spinors (A, ϕ) is a Hilbert space that he approximated by a nested sequence of finite dimensional subspaces X_λ. He considered the Conley index associated to the flow induced by CSD on a large ball $B_\lambda \subset X_\lambda$. Roughly, if $L_\lambda \subset \partial B_\lambda$ is that part of the boundary where the flow points in an outward direction, then Manolescu views the Conley index as the quotient space

$$I_\lambda = B_\lambda / L_\lambda.$$

The homology $H_*(I_\lambda)$ is the Morse-homology of the approximate flow on B_λ, assuming that the flow satisfies the Morse-Smale transversality condition. However Manolosecu did not need to assume the Morse-Smale condition in his work. He simply defined the Seiberg-Witten Floer homology directly as the relative homology of I_λ, with a degree shift that depends on λ. The various I_λ's fit together to give a spectrum $SWF(Y, \sigma)$ defined for every rational homology sphere Y with $Spin^c$-structure σ. Since the Seiberg-Witten equations have an S^1 symmetry, the spectrum $SWF(Y, \sigma)$ carries an S^1 action. This defines Manolescu's "S^1-equivariant Seiberg-Witten Floer stable homotopy type".

An important case is when Y is a homology sphere. In this setting there is a unique $Spin^c$ structure σ coming from a spin structure. The conjugation and S^1-action together yield an action by the group

$$Pin(2) = S^1 \oplus S^1 j \subset \mathbb{C} \oplus \mathbb{C}j = \mathbb{H}$$

where $\mathbb{H}$ is the quarternion skew-field. In [39] Manolescu defined the $Pin(2)$-equivariant Floer homology of Y to be the (Borel) equivariant homology of the spectrum,

$$SWFH_*^{Pin(2)}((Y) = \tilde{H}_*^{Pin(2)}(SWF(Y)). \tag{10.3.2}$$

This theory played a crucial role in Manolescu's resolution of the triangulation question as we will describe below. But before doing that we point out that by having the equivariant stable homotopy type, he was able to define a corresponding $Pin(2)$-equivariant Seiberg-Witten Floer K-theory, which he used in [39] to prove an analogue of Furuta's "10/8-conjecture" for 4-manifolds with boundary. That is, Furuta proved that if W is a closed, smooth spin 4-manifold then

$$b_2(W) \geq \frac{10}{8}|sign(W)| + 2$$

where b_2 is the second Betti number and $sign(W)$ is the signature. (The "11/8-conjecture" states that $b_2 \geq \frac{11}{8} sign(W)$.) Using $Pin(2)$-equivariant Floer K-theory, Manolescu proved that if Y is a homology 3-sphere, there is a number $\kappa(Y) \in \mathbb{Z}$ such that if W is a smooth, spin compact 4-manifold with boundary equal to Y, then the following analogue of Furuta's inequality holds:

$$b_2(W) \geq \frac{10}{8}|sign(W)| + 2 - 2\kappa(Y).$$

10.3.2 The triangulation problem

A famous question asked in 1926 by Kneser [29] is the following:

Question. Does every topological manifold admit a triangulation?

By "triangulation" (or "simplicial triangulation"), one means a homeomorphism to a simplicial complex. We refer to this question as the "simplicial triangulation question (or conjecture)". One can also ask the stronger question regarding whether every manifold admits a *combinatorial triangulation*, which is one in which the links of the simplices are spheres. Such a triangulation is equivalent to a piecewise linear (PL) structure on the manifold.

In the 1920s Rado proved that every surface admits a combinatorial triangulation, and in the early 1950s, Moise showed that any topological three manifold also admits a combinatorial triangulation. In the 1930s Cairns and Whitehead showed that *smooth* manifolds of any dimension admit combinatorial triangulations. And in celebrated work in the late 1960s, Kirby and Siebenmann [28] showed that there exist topological manifolds without PL-structures in every dimension greater than four. Furthermore they showed that in these dimensions, the existence of PL-structures is determined by an obstruction class $\Delta(M) \in H^4(M; \mathbb{Z}/2)$. The first counterexample to the simplicial triangulation conjecture was given by Casson [4] who showed that Freedman's four dimensional E_8-manifold, which he had proven did not have a PL-structure, did not admit a simplicial triangulation. In dimensions five or greater, a resolution of Kneser's triangulation question was not achieved until Manolescu's recent work [39] using equivariant Seiberg-Witten Floer homotopy theory.

We now give a rough sketch of Manolescu's work on this.

Let M be a closed, oriented topological n-manifold, with $n \geq 5$ that is equipped with a homeomorphism to a simplicial complex K (i.e. a simplicial triangulation). As mentioned above, the Kirby-Siebenmann obstruction to M having a combinatorial (PL) triangulation is a class $\Delta(M) \in H^4(M;\mathbb{Z}/2)$. A related cohomology class is the Sullivan-Cohen-Sato class $c(K)$ [56], [8], [52] defined by

$$c(K) = \sum_{\sigma \in K^{(n-4)}} [link_K(\sigma)] \cdot \sigma \in H_{n-4}(M;\Theta_3^H) \cong H^4(M;\Theta_3^H).$$

Here the sum is taken over all codimension 4 simplices in K. The link of each such simplex is known to be a homology 3-sphere. If this were a combinatorial triangulation the link would be an actual 3-sphere and so this class would vanish.

Consider the short exact sequence given by the Rokhlin homomorphism

$$0 \to ker(\mu) \to \Theta_3^H \xrightarrow{\mu} \mathbb{Z}/2 \to 0. \tag{10.3.3}$$

This induces a long exact sequence in cohomology

$$\cdots \to H^4(M;\Theta_3^H) \xrightarrow{\mu_*} H^4(M;\mathbb{Z}/2) \xrightarrow{\delta} H^5(M;\mathrm{Ker}(\mu_*)) \to \cdots.$$

In this sequence it is known that $\mu_*(c(K)) = \Delta(M) \in H^4(M;\mathbb{Z}/2)$. One concludes that if M is a manifold that admits a simplicial triangulation, the Kirby-Siebenmann obstruction $\Delta(M)$ is in the image of μ_* and therefore in the kernel of δ. An important result was that this necessary condition for admitting a triangulation is also a sufficient condition.

Theorem 10.3.1 (Galewski-Stern [23], Matumoto [42]). *A closed topological manifold M of dimension greater or equal to 5 is admits a triangulation if and only if $\delta(\Delta(M)) = 0$.*

Now observe that the connecting homomorphism δ in this long exact sequence would be zero if the short exact sequence (10.3.3) were split. If this were the case then by the Galewski-Stern-Matumoto theorem (10.3.1), every closed topological manifold of dimension ≥ 5 would be triangulable. The following theorem states that this is in fact a sufficient condition.

Theorem 10.3.2 (Galewski-Stern [23], Matumoto [42]). *There exist non-triangulable manifolds of every dimension greater or equal to 5 if and only if the exact sequence (10.3.3) does not split.*

Notice that this theorem reduces an important question about topological n-manifolds for $n \geq 5$ to a question in 3-manifold topology. In settling the triangulation problem Manolescu proved the following.

Theorem 10.3.3. (Manolescu [39]). *The short exact sequence (10.3.3) does not split.*

The strategy of his proof was the following. Suppose $\eta : \mathbb{Z}/2 \to \Theta_3^H$ is a splitting of the above sequence. The image under η of the nonzero element would represent a homology 3-sphere Y of order 2 in Θ_3^H with nonzero Rokhlin invariant. Notice that the fact that Y has order 2 means that $-Y$ ($= Y$ with the opposite orientation) and Y represent the same element in Θ_3^H. Thus to show that this cannot happen, Manolescu defined a lift of the Rokhlin invariant to the integers

$$\beta : \Theta_3^H \to \mathbb{Z}.$$

Given such a lift and any element $X \in \Theta_3^H$ of order 2,

$$\begin{aligned}\beta(X) &= \beta(-X) \quad \text{because } X \text{ has order two} \\ &= -\beta(X) \in \mathbb{Z}.\end{aligned}$$

Thus $\beta(X) = -\beta(X) \in \mathbb{Z}$ and therefore must be zero. Thus X has zero Rokhlin invariant.

Manolescu's strategy was therefore to construct such a lifting $\beta : \Theta_3^H \to \mathbb{Z}$ of the Rokhlin invariant $\mu : \Theta_3^H \to \mathbb{Z}/2$. His construction was modeled on Froyshov's invariant (10.3.1). But to construct β he used $Pin(2)$-equivariant Seiberg-Witten Floer homology $SWFH^{Pin(2)}$ defined using the $Pin(2)$-equivariant Seiberg-Witten Floer homotopy type (spectrum) as described above (10.3.2). We refer to [39] for the details of the construction and resulting dramatic solution of the triangulation problem.

10.4 Floer homotopy and Lagrangian immersions: the work of Abouzaid and Kragh

Another beautiful example of an application of a type of Floer homotopy theory was found by Abouzaid and Kragh [2] in their study of Lagrangian immersions of a Lagrangian manifold L into the cotangent bundle of a smooth, closed manifold, T^*N. Here, L and N must be the same dimension. The basic question is whether a Lagrangian immersion is Lagrangian isotopic to a Lagrangian embedding. This is particularly interesting when $L = N$. In this case the "Nearby Lagrangian Conjecture" of Arnol'd states that every closed exact Lagrangian submanifold of T^*N is Hamiltonian isotopic to the zero section, $\eta : N \hookrightarrow T^*N$. In [2] the authors use Floer homotopy theory and classical calculations by Adams and Quillen [49] of the J-homomorphism in homotopy theory to give families of Lagrangian immersions of S^n in T^*S^n that are regularly homotopic to the zero section embedding as smooth immersions, but are *not* Lagrangian isotopic to any Lagrangian embedding.

We now describe these constructions and results in a bit more detail. Let (M^{2n}, ω) be a symplectic manifold of dimension $2n$. Recall that a Lagrangian submanifold $L \subset M$ is a smooth submanifold of dimension n such the restriction of the symplectic form ω to the tangent space of L is trivial. That is, each tangent space of L is an isotropic subspace of the tangent space of M. A Lagrangian embedding $e : L \hookrightarrow M$ is a smooth embedding whose image is a Lagrangian submanifold. Similarly, a Lagrangian immersion $\iota : L \to M$ is a smooth immersion so that the image of each tangent space is a Lagrangian subspace of the tangent space of M.

If M is a cotangent bundle T^*N for some smooth, closed n-manifold N, then a Lagrangian immersion $\iota : L \to T^*N$ determines a map $\tau_L : L \to U/O$, which is well-defined up to homotopy. This map is defined as follows.

Let $j : N \subset \mathbb{R}^K$ be any smooth embedding with normal bundle ν. Complexifying, we get

$$\mathbb{C}^K \times N \cong (\nu \otimes \mathbb{C}) \oplus (TN \otimes \mathbb{C}) \cong (\nu \otimes \mathbb{C}) \oplus T(T^*N)_{|N}.$$

Here $T(T^*N)$ has a Hermitian structure induced by its symplectic structure and the Riemannian structure on N coming from the embedding j. So for each $x \in L$, the tangent space T_xL defines, via the immersion ι, a Lagrangian subspace of $T_{\iota(x)}(T^*N)$. By taking the direct sum with the Lagrangian $\nu = \nu \otimes \mathbb{R} \subset \nu \otimes \mathbb{C}$, one obtains a Lagrangian subspace of $\mathbb{C}^K$. The Grassmannian of Lagrangian subspaces of $\mathbb{C}^K$ is homeomorphic to $U(K)/O(K)$. One therefore has a map

$$\tau_L : L \to U(K)/O(K) \xrightarrow{\subset} U/O.$$

Now the h-principle for Lagrangian immersions states that the set of Lagrangian isotopy classes of immersions $L \to T^*N$ in the homotopy class of a fixed map $\iota_0 : L \to T^*N$ can be identified with the set of connected components of the space of injective maps of vector bundles

$$TL \to \iota_0^*(T(T^*N))$$

that have Lagrangian image. Let $Sp(T(T^*N))$ be the bundle over T^*N with fiber over $(x, u) \in T^*N$ the group of linear automorphisms of $T_{(x,u)}(T^*N))$ preserving the symplectic form. This is a principal $Sp(n, \mathbb{R})$-bundle. Then the space of all such maps of vector bundles is either empty or a principal homogeneous space over the space of sections of $\iota_0^*(Sp(T(T^*N)))$. That is, the space of sections acts freely and transitively.

Abouzaid and Kragh then consider the following special case:

$$L = N,\ TN \otimes \mathbb{C} \text{ is a trivial complex vector bundle, and} \qquad (10.4.1)$$
$$\iota_0 \text{ is the inclusion of the zero section.}$$

In this case the the bundle $Sp(T(T^*N))$ is the trivial $Sp(n; \mathbb{R})$ bundle, and so the corresponding space of sections is the (based) mapping space $Map(N, Sp(n; \mathbb{R}))$, and since $U(n) \simeq Sp(n; \mathbb{R})$ it is equivalent to $Map(N, U(n))$. Since the inclusion of $U(n) \hookrightarrow U$ is $(2n-1)$- connected, one concludes the following.

Lemma 10.4.1. *The equivalence classes of Lagrangian immersions in the homotopy class of the zero section of T^*N are classified by homotopy classes of maps from N to U.*

One therefore has the following homotopy theoretic characterization of the Lagrangian isotopy classes of Lagrangian immersions of the sphere S^n into its cotangent space.

Corollary 10.4.2. *Isotopy classes of Lagrangian immersions of the sphere S^n in T^*S^n in the homotopy class of the standard embedding of the zero section, are classified by $\pi_n(U)$.*

Of course these homotopy groups are known to be the integers $\mathbb{Z}$ when n is odd and zero when n is even. Using a type of Floer homotopy theory, Abouzaid and Kragh proved the following in [2].

Theorem 10.4.3. *Whenever n is congruent to 1, 3, or 5 modulo 8, there is a class of Lagrangian immersions of S^n in T^*S^n in the homotopy class of the zero section, that does not admit an embedded representative.*

We now sketch the Abouzaid-Kragh proof of this theorem, and in particular point out the Floer homotopy theory they used.

They first observed that given a Lagrangian immersion $j : N \to T^*N$ in the homotopy class of the zero section, satisfying condition (10.4.1), then one has a well defined (up to homotopy) classifying map $\gamma_j : N \to U$ that lifts the map $\tau_N : N \to U/O$ described above:

$$\tau_N : N \xrightarrow{\gamma_j} U \xrightarrow{project} U/O. \tag{10.4.2}$$

Now given any Lagrangian embedding $L \hookrightarrow M$ as above, Kragh [30] defined a "Maslov" (virtual) bundle η_L over the component of the free loop space consisting of contractible loops, $\mathcal{L}_0 L$. [1]

The bundle η_L is classified by the following map (which by abuse of notation we also call η_L)

$$\eta_L : \mathcal{L}_0 L \xrightarrow{\mathcal{L}\tau_L} \mathcal{L}_0 U/O \xrightarrow{\simeq} U/O \times \Omega_0 U/O \xrightarrow{project} \Omega_0 U/O \simeq BO. \tag{10.4.3}$$

Here the equivalence $\mathcal{L}_0 U/O \simeq U/O \times \Omega_0 U/O$ comes from considering the evaluation fibration

$$\Omega U/O \to LU/O \xrightarrow{ev} U/O$$

where ev evaluates a loop at the basepoint. This fibration has a canonical (up to homotopy) trivialization as infinite loop spaces because of the existence of a section as constant loops, and using the infinite loop structure of U/O. The equivalence $\Omega_0 U/O \simeq BO$ comes from Bott periodicity.

Note. Given a map of any space $f : X \to U/O$, the corresponding virtual bundle over the free loop space

$$\mathcal{L}_0 X \to BO$$

defined as this composition $\mathcal{L}_0 X \xrightarrow{\mathcal{L}f} \mathcal{L}_0 U/O \xrightarrow{\simeq} U/O \times \Omega_0 U/O \xrightarrow{project} \Omega_0 U/O \simeq BO.$ also appeared in the work of Blumberg, the author, and Schlichtkrull [6] in their work on the topological Hochschild homology of Thom spectra.

As above let $GL_1(\mathbb{S})$ be the group-like monoid of units in $\mathbb{S}$. There is a natural map $J : BO \to BGL_1(\mathbb{S})$ coming from the inclusion $O(n) \hookrightarrow \Omega^n_{\pm 1} S^n$ defined by considering the based self equivalence of $S^n = \mathbb{R}^n \cup \infty$ given by an orthogonal matrix. We call this map "J" as it induces the classical J homomorphism on the level of homotopy groups, $J : \pi_q(O) \to \pi_q(\mathbb{S})$. On the level of fibrations, J associates to a k-dimensional vector bundle $\zeta \to X$ the associated spherical fibration $S(\zeta) \to X$ defined by taking the one-point compactification of each fiber.

The following is an important result about Lagrangian *embeddings* in [2].

[1] We have changed the notation for the free loop space by using a script $\mathcal{L}$, so as not to get confused by the use of an "L" to denote a Lagrangian.

Theorem 10.4.4. *If $j : L \to T^*N$ is an exact Lagrangian embedding then the stable spherical fibration of the Maslov bundle η_L is trivial. That is, the composition*

$$\mathcal{L}_0 L \xrightarrow{\eta_L} BO \xrightarrow{J} BGL_1(\mathbb{S})$$

is null homotopic.

Before sketching how this theorem was proven in [2], we indicate how Abouzaid and Kragh used this result to detect the Lagrangian immersions of S^n in T^*S^n yielding Theorem 10.4.3.

A Lagrangian immersion $\iota : S^n \to T^*S^n$ is classified by a class $\alpha_\iota \in \pi_n(U)$ by Corollary 10.4.2. By Theorem 10.4.4 if ι is Lagrangian isotopic to a Lagrangian embedding, then the composition

$$\begin{aligned}\mathcal{L}S^n = \mathcal{L}_0 S^n \xrightarrow{\mathcal{L}_0\alpha_\iota} \mathcal{L}_0 U \to \mathcal{L}_0(U/O) \xrightarrow{\simeq} U/O \times \Omega_0 U/O \to \Omega_0 U/O \\ \simeq BO \xrightarrow{J} BGL_1(\mathbb{S})\end{aligned}$$

is null homotopic. If ι is such that $\alpha_\iota \in \pi_n(U)$ is a nonzero generator (here n must be odd), then Abouzaid and Kragh precompose this map with the composition

$$S^{n-1} \hookrightarrow \Omega S^n \to \mathcal{L}S^n$$

and show that in the dimensions given in the statement of the theorem, then classical calculations of the J-homomorphism in homotopy theory imply that this composition $S^{n-1} \to BGL_1(\mathbb{S})$ is nontrivial. Therefore by Theorem 10.4.4, the Lagrangian immersions of S^n into T^*S^n that these classes represent cannot be Lagrangian isotopic to embeddings, even though as smooth immersions, they are isotopic to the zero section embedding.

The proof of Theorem 10.4.4 is where Floer homotopy theory is used. This was based on earlier work by Kragh in [30]. The Floer homotopy theory used was a type of Hamiltonian Floer theory for the cotangent bundle. They did not directly use the constructions in [12] described above, but the spirit of the construction was similar. More technically they used finite dimensional approximations of the free loop space, not unlike those used by Manolescu [37]. We refer the reader to [30], [2] for details of this construction. In any case this construction was used to give a spectrum-level version of the Viterbo transfer map, when one has an exact Lagrangian embedding $j : L \to T^*N$. This transfer map is given on the spectrum level by a map

$$\mathcal{L}_0 j^! : \mathcal{L}_0 N^{-TN} \to (\mathcal{L}_0 L)^{-TL \oplus \eta}$$

and similarly a map

$$\mathcal{L}_0 j^! : \Sigma^\infty(\mathcal{L}_0 N_+) \to (\mathcal{L}_0 L)^{TN - TL \oplus \eta}. \tag{10.4.4}$$

An important result in [2] is that $\mathcal{L}_0 j^!$ is an equivalence. Then they make use of the result, essentially due to Atiyah, that given a finite CW-complex X with a stable spherical fibration classified by a map $\rho : X \to BGL_1(\mathbb{S})$, then the Thom spectrum X^ρ is equivalent to the suspension spectrum $\Sigma^\infty(X_+)$ if and only if ρ is null homotopic. Applying this to finite dimensional approximations to $\mathcal{L}_0 L$, they are then able to show that the equivalence (10.4.4) implies that the Maslov bundle η, when viewed as a stable spherical fibration is trivial.

We end this discussion by remarking on the recasting of the Abouzaid-Kragh results by the author and Klang in [13]. Given a Lagrangian immersion, $j : L \to T^*N$, consider the resulting class

$$\tau_L : L \to U/O$$

described above. Taking based loop spaces, one has a loop map

$$\Omega\tau_L : \Omega_0 L \to \Omega_0 U/O \simeq BO.$$

The resulting Thom spectrum $(\Omega_0 L)^{\Omega\tau_L}$ is a ring spectrum. Thus one can apply topological Hochshild homology to this ring spectrum, $THH((\Omega_0 L)^{\Omega\tau_L})$ and one obtains a homotopy theoretic invariant of the Lagrangian isotopy type of the Lagrangian immersion j. This topological Hochshild was computed by Blumberg, the author, and Schlichtkrull in [6]. It was shown that

$$THH((\Omega_0 L)^{\Omega\tau_L}) \simeq (\mathcal{L}_0 L)^{\ell(\tau_L)}$$

where $\ell(\tau_L)$ is a specific stable bundle over the free loop space $\mathcal{L}L$. In particular, as was observed in [13], in the case when τ_L factors through a map to U, as is the case when $L = N$ and it satisfies the condition (10.4.1), then there is an equivalence of stable bundles, $\ell(\tau_L) \cong \eta_L$, where η_L is the Maslov bundle as above. As a consequence of Theorem 10.4.4 of Abouzaid and Kragh, one obtains the following:

Proposition 10.4.5. *Assume $j : N \to T^*N$ is a Lagrangian immersion in the homotopy class of the zero section, and that the complexification $TN \otimes \mathbb{C}$ is stably trivial. Then if j is Lagrangian isotopic to a Lagrangian embedding, then $THH(\Omega_0 N)^{\Omega\tau_N}))$ is equivalent to $THH(\Sigma^\infty(\Omega N_+))$.*

Using the results of [6] the authors in [13] used this proposition to give a proof of Abouzaid and Kragh's Theorem 10.4.3.

Bibliography

[1] A. Abbondandolo and M. Schwarz, *On the Floer homology of cotangent bundles*, Comm. Pure Appl. Math **59,** 254–316 (2006).

[2] M. Abouzaid and T. Kragh, *On the immersion classes of nearby Lagrangians*, J. Topol. **9** (2016), no. 1, 232–244.

[3] M. Abouzaid and I. Smith, *Khovanov homology from Floer cohomology*, J. Amer. Math. Soc. **32** (2019), no. 1, 1–79.

[4] S. Akbulut and J. McCarthy, *Casson's invariant for oriented homology* 3*-spheres*, Math. Notes, vol. 36, Princeton University Press, 1990.

[5] M. Ando, A. J. Blumberg, D. J. Gepner, M. J. Hopkins, and C. Rezk, *Units of ring spectra and Thom spectra*, arXiv:0810.4535, 2008.

[6] A. Blumberg, R. L. Cohen, and C. Schlichtkrull *Topological Hochschild homology of Thom spectra and the free loop space*, Geom. Topol. **14** no.2 2010, 1165–1242.

[7] M. Chas and D. Sullivan, *String topology*, arXiv:math.GT/9911159, 1999.

[8] M. Cohen, *Homeomorphisms between homotopy manifolds and their resolutions*, Invent. Math. **10** (1970), 239–250.

[9] R. L. Cohen, *The Floer homotopy type of the cotangent bundle*, Pure Appl. Math. Q. **6** (2010), no. 2, Special Issue: In honor of Michael Atiyah and Isadore Singer, 391–438.

[10] R. L. Cohen, *Floer homotopy theory, realizing chain complexes by module spectra, and manifolds with corners,* in Proc. of Fourth Abel Symposium, Oslo, 2007, ed. N. Baas, E.M. Friedlander, B. Jahren, P. Ostvaer, Springer Verlag (2009), 39–59.

[11] R. L. Cohen, J. D. S. Jones, and G. B. Segal, *Morse theory and classifying spaces*, available at `http://math.stanford.edu/~ralph/morse.ps` (1995).

[12] R. L. Cohen, J. D. S. Jones, and G. B. Segal, *Floer's infinite-dimensional Morse theory and homotopy theory*, The Floer Memorial Volume, Progr. Math., vol. 133, Birkhäuser, 1995, pp. 297–325.

[13] R. L. Cohen and I. Klang, *Twisted Calabi-Yau ring spectra, string topology, and gauge symmetry*, Tunis. J. Math. 2 (2020), no. 1, 147–196.

[14] J. Eels, Jr. and N. Kuiper, *An invariant for certain smooth manifolds* Ann. Mat. Pura Appl. (4) **60** (1962), 93-110.

[15] R. Fintushel and R. Stern, *Instanton homology of Seifert fibred homology three spheres*, Proc. London Math. Soc. (3) **61** (1990), no. 1, 109–137.

[16] A. Floer, *Morse theory for Lagrangian intersections,* J. Differential Geom. **28** (1988), 513–547.

[17] A. Floer, *An instanton invariant for 3-manifolds,* Comm. Math. Phys., **118** (1988), 215–240.

[18] A. Floer, *Symplectic fixed points and holomorphic spheres,* Comm. Math. Phys., **120** (4) (1989), 575–611.

[19] J. M. Franks, *Morse-Smale flows and homotopy theory*, Topology **18** (1979), 119–215.

[20] K. A. Froyshov, *Equivariant aspects of Yang-Mills Floer theory*, Topology **41** (2002), no. 3, 525–552.

[21] M. Furuta, *Homology cobordism group of homology* 3*-spheres,* Invent. Math. **100** (1990), no. 2, 339–355.

[22] M. Furuta, *Monopole equation and the* $\frac{11}{8}$*-conjecture*, Math. Res. Lett. **8** (2001), no. 3, 279–291.

[23] D. Galewski and R. Stern, *Classification of simplicial triangulations of topological manifolds*, Ann. Math. (2) **111** (1980), no.1, 1–34.

[24] J. Genauer, *Cobordism categories of manifolds with corners*, Stanford University PhD thesis, (2009), arXiv.org/abs/0810.0581, 2008.

[25] K.Janich, *On the classification of* $O(n)$*-manifolds*, Math. Ann. **176** (1968), 53–76.

[26] P. Hu, D. Kriz, and I. Kriz, *Field theories, stable homotopy theory, and Khovanov homology*, Topology Proc. 48 (2016), 327–360.

[27] M. Khovanov, *A categorification of the Jones polynomial*, Duke Math. J. **101** no. 3, (2000), 359–426.

[28] R. Kirby and L. Siebenmann, *Foundational essays on topological manifolds, smoothings, and triangulations*, Ann. of Math. Stud., No. 88. Princeton University Press, 1977. With notes by J. Milnor and M. Atiyah,

[29] H. Kneser, *Die Topologie der Mannigfaltigkeiten*, Jahresbericht der Deut. Math. Verein. **34** (1926), 1–13.

[30] T. Kragh, *Parameterized ring-spectra and the nearby Lagrangian conjecture*, Geom. Topol. **17** (2013), 639–731 (Appendix by M. Abouzaid).

[31] P. Kronheimer and T. Mrowka, *Monopoles and three-manifolds,* New Mathematical Monographs, vol. 10, Cambridge University Press, 2007.

[32] P. Kronheimer, T. Mrowka, P. Oszvath, and Z. Szabo, *Monopoles and lens space surgeries*, Ann. of Math. (2) **165** (2007), no. 2, 457–546.

[33] G. Laures, *On cobordism of manifolds with corners*, Trans. Amer. Math. Soc. **352** no. 12, (2000), 5667–5688.

[34] T. Lawson, R. Lipshitz, and S. Sarkar, *Khovanov homotopy type, Burnside category, and products*, arXiv: 1505.00213v1, 2015.

[35] R. Lipshitz and S. Sarkar, *A Khovanov stable homotopy type*, J. Amer. Math. Soc. 27 (2014), no. 4, 983–1042.

[36] R. Lipshitz and S. Sarkar, *A Steenrod square on Khovanov homology*, J. Topology, 7 (2014), no. 3, 817–848.

[37] C. Manolescu, *Seiberg-Witten-Floer stable homotopy type of three-manifolds with* $b_1 = 0$, Geom. Topol. 7 (2003), 889–932.

[38] C. Manolescu, *A gluing theorem for the relative Bauer-Furuta invariants*, J. Differential Geom. 76 (2007), no. 1, 117–153.

[39] C. Manolescu, *Pin (2)-equivariant Seiberg-Witten Floer homology and the triangulation conjecture*, J. Amer. Math. Soc. 29 (2016), no. 1, 147–176.

[40] C. Manolescu, *Floer theory and its topological applications*, Jpn. J. Math. 10 (2015), no. 2, 105–133.

[41] C. Manolescu, *The Conley index, gauge theory, and triangulations*, J. Fixed Point Theory Appl. 13 (2013), no. 2, 431–457.

[42] T. Matumoto, *Triangulations of manifolds*, Algebraic and Geometric Topology (Proc. Symp. Pure Math., Stanford Univ., 1976), Part 2, Proc. Symp. Pure Math., XXXII, Amer. Math. Soc., 1978, pages 3–6.

[43] D. McDuff and D. Salamon, *Introduction to symplectic topology*, Oxford Math Monographs, Clarendon Press, 1998.

[44] P. Ozsváth and Z. Szabó, *Holomorphic disks and topological invariants for closed three-manifolds*, Ann. of Math. **159** (3) (2004) 1027–1158.

[45] P. Ozsváth and Z. Szabó, *Holomorphic disks and knot invariants*, arXiv:math/0209056v4, 2003.

[46] A. Pressley and G. Segal, *Loop groups*, Oxford Math. Monogr., Clarendon Press, 1986.

[47] J. Rasmussen, *Floer homology and knot complements*, arXiv:math/0306378, 2003.

[48] J. Rasmussen, *Khovanov homology and the slice genus*, Invent. Math. 182 (2010), no. 2, 419–447.

[49] D. Ravenel, *Complex cobordism and stable homotopy groups of spheres,* Pure Appl. Math., vol. 121, Academic Press, Inc., 1986.

[50] V. A. Rokhlin, *New results in the theory of four-dimensional manifolds*, Doklady Akad. Nauk SSSR (N.S) **84** (1952), 221–224.

[51] L. Qin, *On the associativity of gluing*, J. Topol. Anal. 10 (2018), no. 3, 585–604.

[52] H. Sato, *Constructing manifolds by homotopy equivalences, I. An obstruction to constructing PL manifolds from homology manifolds*, Ann. Inst. Fourier (Grenoble) **22** (1972), no. 1, 271–286.

[53] C. Seed, *Computations of the Lipshitz-Sarkar Steenrod square on Khovanov homology*, arXiv:1210.1882, 2012.

[54] M. Stoffregen and M. Zhang, *Localization in Khovanov homology*, arXiv:1810.04769, 2018.

[55] P. Seidel and I. Smith, *A link invariant from the symplectic geometry of nilpotent slices*, Duke Math. J. **134** (2006), 453–514.

[56] D. Sullivan, *Triangulating and smoothing homotopy equivalences and homeomorphisms*, Geometric Topology Seminar notes, The Hauptvermutung book, K-Monogr. Math., vol. 1, Kluwer Acad. Publ., 1996, pp. 69–103.

[57] C. Taubes *The Seiberg-Witten equations and the Weinstein conjecture*, Geom. Topol. **11** (2007), 2117–2202.

[58] C. Viterbo, *Functors and computations in Floer homology with applications, Part II*, arXiv:1805.01316, 2018.

Department of Mathematics, Stanford University, Stanford, CA 94305, U.S.A.

E-mail address: `rlc@stanford.edu`

11

Little discs operads, graph complexes and Grothendieck–Teichmüller groups

Benoit Fresse

Introduction

The operads of little discs (and the equivalent operads of little cubes) were introduced in topology, in the works of Boardman–Vogt [7, 8] and May [47], for the recognition of iterated loop spaces. We refer to the paper [18], in the handbook of algebraic topology, for an account of these applications. We also refer to the literature for the general definition of an operad in a category and for the precise definition of the little discs operads, which we denote by D_n throughout this chapter.

The aim of this chapter is to survey new applications of the little discs operads which were motivated by the works of Kontsevich [42, 43] and Tamarkin [53] on the deformation-quantization of Poisson manifolds and by the Goodwillie–Weiss embedding calculus in topology [34, 58]. For our purpose, we also consider the general class of E_n-operads, which consists of the operads that are weakly-equivalent to the operad of little n-discs (equivalently, to the operad of little n-cubes). Besides, we deal with E_n-operads in the category of differential graded modules, which we similarly define as the class of operads that are weakly-equivalent (quasi-isomorphic) to the operad of singular chains on the little n-discs operad (the chain little n-discs operad) $C_*(D_n)$.

In Kontsevich's approach, the proof of the existence of deformation-quantizations of Poisson manifolds reduces to the construction of a comparison map of differential graded Lie algebras between on the one hand, the Hochschild cochain complex, which governs the deformations of an associative algebra structure, and on the other hand, the algebra of polyvector fields, equipped with the Schouten-Nijenhuis bracket of polyvectors, which can be used to govern the deformations of a Poisson structure on a manifold.

Kontsevich used an explicit definition of such a comparison map in his first proof of the existence of deformation-quantizations. The theory of E_2-operads actually occurs in a second generation of proofs of this theorem. The idea is that the differential graded Lie algebra structure of both the Hochschild cochain complex and the algebra of polyvector fields can be integrated into an action of a differential graded E_2-operad and the algebra of polyvector fields is rigid (has a unique realization up to quasi-isomorphism) as an algebra over an E_2-operad.

Mathematics Subject Classification. 55P48, 55P62, 20F36.

Key words and phrases. little discs operads; formality; Grothendieck-Teichmüller groups; rational homotopy; Sullivan models.

The action of an E_2-operad on the complex of Hochschild cochains was initially conjectured by Deligne. The proof of the latter statement, now established in a wide context (which includes the topological counterpart of the Hochschild cohomology theory in the stable homotopy theory framework), can be interpreted as a measure of the degree of commutativity of the Hochschild cochain complex, regarded as a derived version of the center of associative algebras.

The algebra of polyvector fields comes actually equipped with the structure of a 2-Poisson algebra, where in general an n-Poisson algebra refers to a form of graded Poisson algebra such that the Poisson bracket is an operation of degree $n-1$ (which is actually the case of the Schouten-Nijenhuis bracket for $n=2$). The operad that governs this category of graded Poisson algebras, the n-Poisson operad Pois_n, represents the homology of the operad of little n-discs $\mathrm{H}_*(D_n)$. Therefore, the proof that the algebra of polyvector fields inherits an action of an E_2-operad, and actually, the crux of the operadic proof of the existence of deformation-quantizations, is equivalent to an operadic formality claim, which asserts that the chain operads of little 2-discs $\mathrm{C}_*(D_2)$ is quasi-isomorphic to the 2-Poisson operad Pois_2. In fact, such a statement holds for all $n \geq 2$:

$$\mathrm{C}_*(D_n) \sim \mathsf{Pois}_n,$$

and one the main objectives of this chapter will be to explain this result in details. For the moment, let us simply mention that the case $n=2$ of this formality claim was established by Tamarkin by using the theory of Drinfeld's associators.

This operadic approach gives deep insights on structures carried by the set of solutions of the deformation-quantization problem, when we consider the set of all deformation-quantization functors as a whole. Indeed, from Tamarkin's arguments, we can deduce the more precise result that a formality quasi-isomorphism for the chain operad of little 2-discs (and as a consequence, a deformation-quantization functor for Poisson manifolds) is associated to any Drinfeld associator. This observation hints that the rational version of the Grothendieck–Teichmüller group $GT(\mathbb{Q})$ acts on the moduli space of deformation-quantizations just because the set of Drinfeld's associators $Ass(\mathbb{Q})$ defines a torsor under an action of this group. We explain shortly that this connection reflects a finer identity between the Grothendieck–Teichmüller group and the group of homotopy automorphisms of E_2-operads. Recall simply for the moment that the Grothendieck–Teichmüller group models the relations that can be gained from actions of the absolute Galois group on curves. In deformation-quantization theory, we just consider a pro-algebraic version of this group.

To complete this overview, let us mention that higher dimensional generalizations of the deformation-quantization problem, which involve structures governed by any class of n-Poisson algebras, have been studied by Calaque–Pantev–Toën–Vaquié–Vezzosi in the realm of derived algebraic geometry (see [14]).

The link between the operads of little discs and the embedding calculus comes from certain descriptions of the Goodwillie–Weiss towers, which are towers of 'polynomial' approximations of the embedding spaces $\mathrm{Emb}(M, N)$, where (M, N) is any pair of smooth manifolds (see [34, 57]). We refer to Arone–Ching's paper, in this handbook volume, for a comprehensive introduction to the embedding calculus.

In what follows, we focus on the case of Euclidean spaces $M = \mathbb{R}^m$, $N = \mathbb{R}^n$, and we consider a space of embeddings with compact support $\mathtt{Emb}_c(\mathbb{R}^m, \mathbb{R}^n)$, whose elements are the embeddings $f : \mathbb{R}^m \hookrightarrow \mathbb{R}^n$ such that there exists a compact domain $K \subset \mathbb{R}^m$ with $f \mid_{\mathbb{R}^m \setminus K} = i$, where $i : \mathbb{R}^m \to \mathbb{R}^n$ denotes the standard embedding $i : (x_1, \dots, x_m) \mapsto (x_1, \dots, x_m, 0, \dots, 0)$. Then we consider an analogously defined space of immersions with compact support $\mathtt{Imm}_c(\mathbb{R}^m, \mathbb{R}^n)$ and we take the homotopy fiber of the obvious forgetful map $\mathtt{Emb}_c(\mathbb{R}^m, \mathbb{R}^n) \to \mathtt{Imm}_c(\mathbb{R}^m, \mathbb{R}^n)$. We use the notation $\overline{\mathtt{Emb}}_c(\mathbb{R}^m, \mathbb{R}^n)$ for this space.

In general, one can prove that the Goodwillie–Weiss approximations are weakly equivalent to mapping spaces of truncated (bi)modules over the little discs operads, where the notion of a truncated operadic (bi)module refers to a (bi)module that is defined up to some arity only. This result was established by Arone–Turchin in [3], after a pioneering work of Dev Sinha [50] on the particular case of the spaces of long knots $\overline{\mathtt{Emb}}_c(\mathbb{R}, \mathbb{R}^n)$. In the case of the space of embeddings with compact support modulo immersions $\overline{\mathtt{Emb}}_c(\mathbb{R}^m, \mathbb{R}^n)$, one can prove further that the Goodwillie–Weiss approximations are weakly equivalent to $m+1$-fold loop spaces of mapping spaces of truncated operads with the little m-discs operad as source object and the little n-discs operad as target object. This finer result has been established in full generality by Boavida–Weiss in [10], by an improvement of the methods used in the study of the Goodwillie–Weiss calculus of embedding spaces, while other authors have obtained general results on mapping spaces of (truncated) operadic bimodules which permit one to recover this delooping relation from the results obtained by Sinha and Arone–Turchin (see the articles of Dwyer–Hess [24] and Turchin [56] for the case $m = 1$, and the article of Ducoulombier–Turchin [23] for the case of general $m \geq 1$).

In the case $n - m \geq 3$, we can use convergence statements to deduce an equivalence of total spaces from the operadic interpretation of the Goodwillie–Weiss tower, so that we have a weak homotopy equivalence:

$$\overline{\mathtt{Emb}}_c(\mathbb{R}^m, \mathbb{R}^n) \sim \Omega^{m+1} \mathtt{Map}^h_{\mathcal{O}p}(D_m, D_n),$$

where $\mathtt{Map}^h_{\mathcal{O}p}(-,-)$ denotes a derived mapping space bifunctor on the category of operads in topological spaces.

The formality of the little discs operads over the rationals can be used to determine the rational homotopy type of the operadic derived mapping spaces $\mathtt{Map}^h_{\mathcal{O}p}(D_m, D_n)$ which occur in this description of the embedding spaces $\overline{\mathtt{Emb}}_c(\mathbb{R}^m, \mathbb{R}^n)$. For this purpose, we use the fact that we have a rational homotopy equivalence $\mathtt{Map}^h_{\mathcal{O}p}(D_m, D_n) \sim_{\mathbb{Q}} \mathtt{Map}^h_{\mathcal{O}p}(D_m, D_n^{\mathbb{Q}})$ as soon as $n - m \geq 3$, where $D_n^{\mathbb{Q}}$ denotes a rationalization of the topological operad of little n-discs D_n which is given by an operadic extension of the Sullivan rational homotopy theory of spaces (we explain this construction with more details later on). In fact, we use an improved version of the formality which implies that this rational operad $D_n^{\mathbb{Q}}$ has a model $\langle \mathtt{H}^*(D_n) \rangle$ that is determined by the rational cohomology of the operad of little n-discs $\mathtt{H}^*(D_n) = \mathtt{H}^*(D_n, \mathbb{Q})$ (equivalently, by the dual object of the n-Poisson operad Pois_n). This result gives an effective approach to compute the rational homotopy of the operadic mapping spaces $\mathtt{Map}^h_{\mathcal{O}p}(D_m, D_n)$, and hence, to compute the rational homotopy of the embedding spaces $\overline{Emb^c}(\mathbb{R}^m, \mathbb{R}^n)$ by the Goodwillie–Weiss theory of embedding calculus.

Besides the homotopy of the mapping spaces $\mathtt{Map}^h_{\mathcal{O}p}(D_m, D_n^{\mathbb{Q}})$, we can compute the rational homotopy type of the spaces of homotopy automorphisms $\mathtt{Aut}^h_{\mathcal{O}p}(D_n^{\mathbb{Q}})$ in the category of operads. We will actually see that $\mathtt{Aut}^h_{\mathcal{O}p}(D_2^{\mathbb{Q}})$ is weakly equivalent to a semi-direct product $GT(\mathbb{Q}) \ltimes \mathtt{SO}(2)^{\mathbb{Q}}$, where $GT(\mathbb{Q})$ is the rational Grothendieck–Teichmüller group, and this statement gives a theoretical explanation for the occurrence of the rational Grothendieck–Teichmüller group in deformation-quantization. We have an analogue of this result in the realm of profinite homotopy theory. We explain both statements in this chapter.

In fact, the main objective of this survey is to explain the result of these computations of mapping spaces and of homotopy automorphism spaces of operads. We mainly address this subject. We organize our account as follows.

We devote the first section of our survey to the particular case $n = 2$ of the homotopy theory of E_n-operads. We will explain that the little 2-discs operad has a model given by an operad shaped on braid groupoids. We use this model to obtain our weak equivalence between the Grothendieck–Teichmüller group and the homotopy automorphism space $\mathtt{Aut}^h_{\mathcal{O}p}(D_2^{\mathbb{Q}})$.

We explain the formality of E_n-operads in the second section, and we tackle the applications to the computation of the rational homotopy of mapping spaces of E_n-operads in the third section. We use graph complexes in the proof of the formality of E_n-operads. We therefore retrieve graph complexes, the graph complexes alluded to in the title of this paper, in our expression of the rational homotopy of the mapping spaces of E_n-operads. The ultimate goal of this survey is precisely to explain this graph complex description of the rational homotopy type of the mapping spaces of E_n-operads.

In general, in this chapter, we use the term 'differential graded module' and the language of differential graded algebra, rather than the language of chain complexes. In fact, we only use the expression '(co)chain complex' for specific constructions of differential graded modules, like the singular complex of a topological space, the Hochschild cochain complex, ... For short, we also use the prefix 'dg' for any category of structured objects that we may form within a base category of differential graded modules (like dg-algebras, dg-operads, ...).

In what follows, we generally define a differential graded module (thus, a dg-module for short) as the structure, equivalent to a (possibly unbounded) chain complex, which consists of a module M equipped with a $\mathbb{Z}$-graded decomposition $M = \bigoplus_{n\in\mathbb{Z}} M_n$ and with a differential $\delta : M \to M$ such that $\delta(M_*) \subset M_{*-1}$. If necessary, we use the phrase 'lower graded dg-modules' to refer to the objects of this category of dg-modules.

In some cases, we also deal with 'upper graded dg-modules', which are modules M equipped with a $\mathbb{Z}$-graded decomposition of the form $M = \bigoplus_{n\in\mathbb{Z}} M^n$ and with a differential $\delta : M \to M$ such that $\delta(M^{*-1}) \subset M^*$. In general, we can use the standard correspondence $M_* = M^{-*}$ to convert an upper graded dg-module structure into a lower graded dg-module structure, but we prefer to keep upper graded dg-module structures when this representation is the usual convention in the literature (for instance, in rational homotopy theory).

We equip the category of dg-modules with its standard tensor product so that this category inherits a symmetric monoidal structure, with a symmetry operator defined by using the usual sign rule of homological algebra,

In our study, we freely use the language and the results of the theory of model categories. In particular, in what follows, we rather use the generic term 'weak equivalence' for the class of quasi-isomorphisms, because the quasi-isomorphisms represent the class of weak equivalences of the usual model categories of dg-objects (dg-modules, dg-algebras, dg-operads, ...).

11.1 Braids and the homotopy theory of E_2-operads

We devote this section to the study of the homotopy of E_2-operads .

In general, we have a homotopy equivalence of spaces $D_n(r) \xrightarrow{\sim} F(\mathring{\mathbb{D}}^n, r)$, for each $r \in \mathbb{N}$, where we consider the underlying spaces of the operad of little n-discs $D_n(r)$, and $F(\mathring{\mathbb{D}}^n, r)$ denotes the configuration space of r points in $\mathring{\mathbb{D}}^n$. In the case $n = 2$, this result implies that $D_2(r)$ forms an Eilenberg–Mac Lane space $K(P_r, 1)$, where P_r is the pure braid group on r strands in $\mathring{\mathbb{D}}^2$.

The first purpose of this section is to explain that we can elaborate on this result in order to get a model of the class of E_2-operads in the category of operads in groupoids. In short, we check that we have a relation $D_2 \sim \mathtt{B}(CoB)$, where we consider the classifying spaces of a certain operad in groupoids, the operad of colored braids CoB. We will see that this operad CoB governs the category of strict braided monoidal categories as a category of algebras. We use a variant of this operad, the operad of parenthesized braids, which we associate to the category of general braided monoidal categories, in order to define the Grothendieck–Teichmüller group as a group of automorphisms of an operad in the category of groupoids. We will explain that, when we pass to topological spaces, this identity gives an equivalence between the space of homotopy automorphisms of the little 2-discs operad and a semi-direct product of the Grothendieck–Teichmüller group with the group of rotations. The statement of this result is the second and main objective of this section. To complete this survey, we also explain the definition of the notion of a Drinfeld associator from the viewpoint of the theory of operads.

In our constructions, we deal with versions of the Grothendieck–Teichmüller that are associated to various completions of operads in groupoids, and as a consequence, we actually consider various completions of the little 2-discs operad (namely, the profinite completion and the rationalization) when we examine the relationship between the Grothendieck–Teichmüller group and the homotopy of E_2-operads. In this section, we use a simple definition of these completion operations which we form at the level of the groupoid models of our operads. In the next section, we revisit the definition of the particular case of the rationalization of operads by using the Sullivan rational homotopy theory of spaces.

11.1.1 The operad of colored braids

Briefly recall that a braid on r-strands is an isotopy class of paths $\alpha : [0,1] \to F(\mathring{\mathbb{D}}^2 \times [0,1], r)$ with $\alpha(t) = (\alpha_1(t), \dots, \alpha_r(t)) \in F(\mathring{\mathbb{D}}^2 \times [0,1], r)$ such that $\alpha_i(t) = (z_i(t), t)$ for each $t \in [0,1]$, and where we assume that $\alpha(0) = (\alpha_1(0), \dots, \alpha_r(0))$ (respectively, $\alpha(1) = (\alpha_1(1), \dots, \alpha_r(1))$) is a permutation of fixed contact points $((z_1^0, 0), \dots, (z_r^0, 0))$ (respectively, $((z_1^0, 1), \dots, (z_r^0, 1))$) on the equatorial line $y = 0$ of the disc $\mathbb{D}^2 \times \{0\}$ (respectively, $\mathbb{D}^2 \times \{1\}$).

Thus, we have $z_i^0 = (x_i^0, 0)$ for $i = 1, \ldots, r$, and by convention we can also assume that the contact points are ordered so that $x_1^0 < \cdots < x_r^0$. In what follows, we use the usual representation of the isotopy class of a braid in terms of a diagram which is given by a projection onto the plane $y = 0$ in the space $\mathring{\mathbb{D}}^2 \times [0,1]$. The assumption $\alpha(t) \in F(\mathring{\mathbb{D}}^2 \times [0,1], r)$ is equivalent to the requirement that we have $z_i(t) \neq z_j(t)$ for all pairs $i \neq j$. In this definition, we assume that the strands of a braid are indexed by the set $\{1, \ldots, r\}$. This assumption is not standard in the definition of a braid, but we use this convention in our definition of colored braids. Intuitively, the indices $i \in \{1, \ldots, r\}$ are colors which we assign to the strands of our braids, as in the following picture:

$\alpha =$ 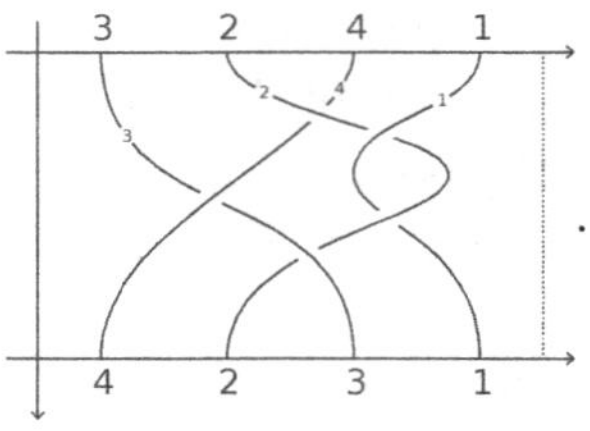 . (11.1.1)

Formally, the operad of colored braids is an operad in the category of groupoids $\mathcal{CoB} \in \mathcal{G}rd\,\mathcal{O}p$ whose components $\mathcal{CoB}(r)$ are groupoids with the permutations on r letters as objects and the isotopy classes of colored braids as morphisms. The source (respectively, the target) of a morphism is the permutation of the set $\{1, \ldots, r\}$ that corresponds to the permutation of the contact points $((z_1^0, \varepsilon), \ldots, (z_r^0, \varepsilon))$ in the sequence $\alpha(0) = (\alpha_1(0), \ldots, \alpha_r(0))$ (respectively, $\alpha(1) = (\alpha_1(1), \ldots, \alpha_r(1))$). For instance, the above example of colored braid depicts a morphism with the permutation $s = (1\ 3\ 4)$ as source object and the permutation $t = (1\ 4)$ as target object. The composition of braids is given by the standard concatenation operation on paths. Note simply that this operation preserves the indexing when we consider a pair of composable morphisms in our groupoid. Note also that our convention is to orient braids from the top to the bottom and we compose braids accordingly.

The action of the symmetric group Σ_r on $\mathcal{CoB}(r)$ is given by the obvious re-indexing operation of the strands of our braids. The operadic composition operations $\circ_i : \mathcal{CoB}(k) \times \mathcal{CoB}(l) \to \mathcal{CoB}(k+l-1)$ are functors that are defined on morphisms by a cabling operation on the strands of our braids. In brief, to define a composite $\alpha \circ_i \beta$, where $\alpha \in \mathcal{CoB}(k)$ and $\beta \in \mathcal{CoB}(l)$, we insert the braid β on the ith strand of the braid α, as in the example given in the following picture:

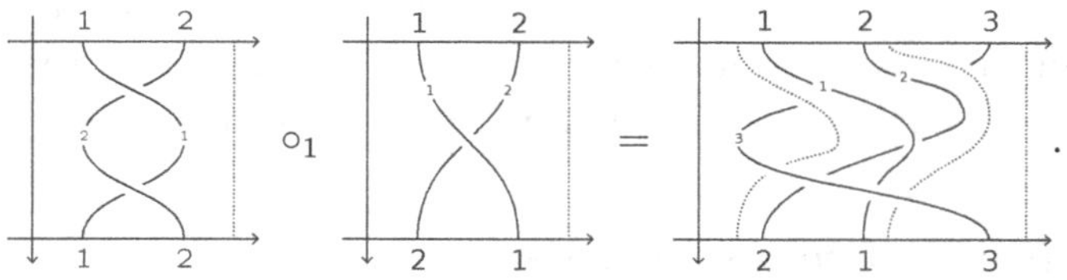 . (11.1.2)

The operadic unit $1 \in \mathcal{CoB}(1)$ is the trivial braid with one strand. By convention, we also assume that the component of arity zero of the colored braid operad is identified with the one-point set $\mathcal{CoB}(0) = *$.

In the introduction of this section, we mentioned that this operad $\mathcal{CoB}$ governs the category of strict braided monoidal categories. We give more explanations on this interpretation of the colored braid operad later on, when we explain a similar interpretation of an operad that governs the category of general braided monoidal categories (see Theorem 11.1.3).

The following theorem gives the connection between the operad of little 2-discs and the operad of colored braids:

Theorem 11.1.1 (see [26, Theorem I.5.3.4]). *We have an equivalence in the category of operads in groupoids $\pi\, D_2 \sim CoB$, where $\pi\, D_2$ is the operad in groupoids defined by the fundamental groupoids $\pi\, D_2(r)$ of the spaces of little 2-discs $D_2(r)$, $r \in \mathbb{N}$.*

In the category of operads in groupoids $\mathcal{G}rd\,\mathcal{O}p$, we say that a morphism is an equivalence $\phi : P \xrightarrow{\sim} Q$ when this morphism defines an equivalence of categories arity-wise $\phi : P(r) \xrightarrow{\sim} Q(r)$, for each $r \in \mathbb{N}$. Then we say that operads in groupoids $P, Q \in \mathcal{G}rd\,\mathcal{O}p$ are equivalent when these operads can be connected by a zigzag of equivalences $P \xleftarrow{\sim} \cdot \xrightarrow{\sim} \cdot \dots \cdot \xrightarrow{\sim} Q$ in the category of operads in groupoids.

We just use that the fundamental groupoid functor is strongly symmetric monoidal in order to equip the collection of fundamental groupoids $\pi\, D_2 = \{\pi\, D_2(r), r \in \mathbb{N}\}$ with an operad structure. We refer to the cited reference [26, Theorem I.5.3.4] for the explicit definition of a zigzag of equivalences of operads in groupoids between this object $\pi\, D_2$ and the colored braid operad CoB.

We can apply the classifying space functor $\mathtt{B}(-)$ to go back from the category of groupoids towards the category of spaces (or towards the category of simplicial sets). This functor $\mathtt{B} : \mathcal{G}rd \to \mathcal{T}op$ is also strongly symmetric monoidal, and hence, preserves operad structures. For our purpose, we consider the operad $\mathtt{B}(CoB)$ defined by the collection of the classifying spaces of the colored braid groupoids $\mathtt{B}(CoB(r))$, $r \in \mathbb{N}$. We have the following result:

Theorem 11.1.2 (Z. Fiedorowicz [25], see also [26, §I.5.2]). *We have a weak equivalence $D_2 \sim \mathtt{B}(CoB)$ in the category of topological operads.*

This theorem is established in [25] by arguments of covering theory. In [26, §I.5.3], it is explained that we can also deduce this weak equivalence relation $D_2 \sim \mathtt{B}(CoB)$ from the result of the previous theorem. In brief, the observation that each space $D_2(r)$ is an Eilenberg–Mac Lane space implies that we have a weak equivalence of spaces $D_2(r) \sim \mathtt{B}(\pi\, D_2(r))$, in each arity $r \in \mathbb{N}$. We can elaborate on the proof of this relation to establish that we actually have a weak equivalence of operads $D_2 \sim \mathtt{B}(\pi\, D_2)$ between the operad of little 2-discs D_2 and the classifying space of the fundamental groupoid operad $\pi\, D_2$. Then we just use that the equivalence of operads in groupoids of Theorem 11.1.1 induces a zigzag of weak equivalences of operads in topological spaces $\mathtt{B}(\pi\, D_2) \sim \mathtt{B}(CoB)$ when we pass to classifying spaces.

11.1.2 The operad of parenthesized braids

The operads of colored braids are not sufficient for our purpose. To define the Grothendieck–Teichmüller group, we need a variant of this operad, which we call the parenthesized braid operad PaB.

The objects of the colored braid operad form an operad in sets $\mathtt{Ob}\,\mathsf{CoB}$, which is identified with the permutation operad, the operad Π defined by the collection of the symmetric groups $\Pi = \{\Sigma_r, r \in \mathbb{N}\}$. The permutation operad also represents a set-theoretic version of the operad of unital associative algebras, or in other words, the operad in sets that governs the structure of a monoid.

To define the operad of parenthesized braids, we just take a pullback of the operad of coloured braids $\mathsf{PaB} = \omega^*\,\mathsf{CoB}$ along a morphism $\omega : \Omega \to \Pi$, where Ω is the operad in sets that governs the category of non-commutative magmas with a fixed unit element (in another terminology, the category of non-commutative non-associative monoids). This operad Ω, the magma operad, has one free generator in arity two $\mu \in \Omega(2)$, equipped with a free action of the symmetric group Σ_2, and an extra arity-zero element $* \in \Omega(0)$ such that $\mu \circ_1 * = 1 = \mu \circ_2 *$, where $1 \in \Omega(1)$ denotes the operadic unit. In positive arity, the elements of this operad $\pi \in \Omega(r)$ are formal operadic composites of the operations $\mu \in \Omega(2)$ and $(1\ 2)\mu \in \Omega(2)$. The result of these operadic composition operations can also be represented as planar binary rooted trees with r leaves indexed by the values of a permutation on r letters $\sigma = (\sigma(1), \dots, \sigma(r))$, as in the following examples:

$\sigma \cdot \mu =$ 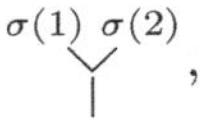

, $\sigma \cdot \mu \circ_1 \mu =$ 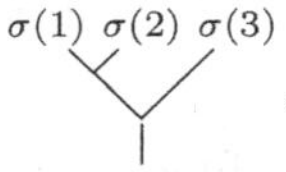

, $\sigma \cdot \mu \circ_2 \mu =$ 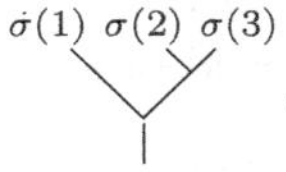

, ...

In arity zero, we just take $\Omega(0) = *$.

The morphism $\omega : \Omega \to \Pi$, which we consider in the our pullback operation $\mathsf{PaB} = \omega^*\,\mathsf{CoB}$, carries $\mu \in \Omega(2)$ to the identity permutation on two letters $id_2 \in \Sigma_2$. The groupoids $\mathsf{PaB}(r)$ underlying this operad $\mathsf{PaB} = \omega^*\,\mathsf{CoB}$ are defined by taking $\mathtt{Ob}\,\mathsf{PaB}(r) := \Omega(r)$ and $\mathtt{Mor}_{\mathsf{PaB}(r)}(p,q) := \mathtt{Mor}_{\mathsf{CoB}(r)}(\omega(p), \omega(q))$ for the morphism sets, for all $p, q \in \Omega(r)$. The operadic composition operations are defined by taking an obvious lifting of the composition operations of the operad of colored braids. In [26, §I.6.2], we represent a parenthesized braid by a braid whose contact points form the centers of a dyadic decomposition of the axis $y = 0$ in the disc $\mathring{\mathbb{D}}^2$ (a decomposition obtained by dividing intervals into equal pieces), because one can observe that such decompositions are in bijection with the elements of the magma operad. For instance, the following braid

$\beta =$ 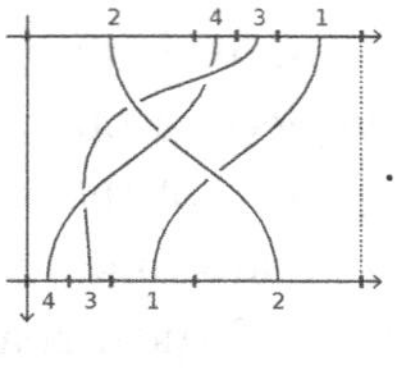

. (11.1.3)

represents a morphism of the groupoid PaB with the object $p = (1\ 2\ 4) \cdot \mu \circ_2 (\mu \circ_1 \mu)$ as source and the object $q = (1\ 4\ 2\ 3) \cdot \mu \circ_1 (\mu \circ_1 \mu)$ as target.

In the operad PaB, we consider the morphisms

$\tau =$ 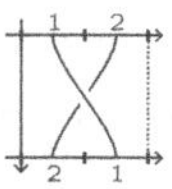

and $\alpha =$ 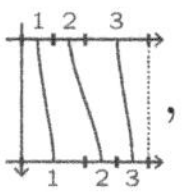

, (11.1.4)

which we call the braiding and the associator respectively.

We aim to give an interpretation of the parenthesized braid operad in classical algebraic language. The object $\mu \in \Omega(2)$ can be regarded as an abstract operation on 2 variables $\mu = \mu(x_1, x_2)$. We use the notation of a tensor product for this operation $\mu(x_1, x_2) = x_1 \otimes x_2$, because we are going to see that μ represents a universal tensor product operation within the operad PaB. The element $(1\ 2)\mu = \mu(x_2, x_1)$ represents an operation $\mu(x_2, x_1) = x_2 \otimes x_1$, where the variables (x_1, x_2) are transposed when we use this variable interpretation of our operation. We also get that $\mu \circ_1 \mu = \mu(\mu(x_1, x_2), x_3)$ represents the result of the substitution of the variable x_1 by the operation $\mu = \mu(x_1, x_2)$ in $\mu = \mu(x_1, x_2)$, while $\mu \circ_2 \mu = \mu(x_1, \mu(x_2, x_3))$ represents the result of the substitution of the second variable x_2 by the same operation $\mu = \mu(x_1, x_2)$ with an index shift of the variables. We equivalently have $\mu(\mu(x_1, x_2), x_3) = (x_1 \otimes x_2) \otimes x_3$ and $\mu(x_1, \mu(x_2, x_3)) = x_1 \otimes (x_2 \otimes x_3)$. We accordingly get that the braiding $\tau = \tau(x_1, x_2)$ represents an isomorphism such that

$$\tau(x_1, x_2) : x_1 \otimes x_2 \to x_2 \otimes x_1 \tag{11.1.5}$$

in the morphism set $\mathtt{Mor}_{\mathsf{PaB}(2)}(\mu, (1\ 2)\mu)$ of our operad in groupoids PaB, while the associator $\alpha = \alpha(x_1, x_2, x_3)$ represents an isomorphism such that

$$\alpha(x_1, x_2, x_3) : (x_1 \otimes x_2) \otimes x_3 \to x_1 \otimes (x_2 \otimes x_3) \tag{11.1.6}$$

in $\mathtt{Mor}_{\mathsf{PaB}(3)}(\mu \circ_1 \mu, \mu \circ_2 \mu)$. The operadic composition formulas $\mu \circ_1 * = 1 = \mu \circ_2 *$ are equivalent to the relations

$$x_1 \otimes * = x_1 = * \otimes x_1, \tag{11.1.7}$$

so that the arity zero object $* \in \Omega(0) = \mathtt{Ob}\,\mathsf{PaB}(0)$ can be interpreted as a unit object with respect to this tensor product operation $\mu(x_1, x_2) = x_1 \otimes x_2$.

We easily see that the braiding and the associator satisfy the following coherence relations with respect to this unit object:

$$\alpha(*, x_1, x_2) = \alpha(x_1, *, x_2) = \alpha(x_1, x_2, *) = id \quad \text{and} \quad \tau(x_1, *) = id = \tau(*, x_1). \tag{11.1.8}$$

We easily check, moreover, that the associator satisfies the pentagon relation

$$\begin{aligned} x_1 \otimes \alpha(x_2, x_3, x_4) \cdot \alpha(x_1, x_2 \otimes x_3, x_4) \cdot \alpha(x_1, x_2, x_3) \otimes x_4 \\ = \alpha(x_1, x_2, x_3 \otimes x_4) \cdot \alpha(x_1 \otimes x_2, x_3, x_4) \end{aligned} \tag{11.1.9}$$

in PaB, as well as the hexagon relations

$$\begin{aligned} x_2 \otimes \tau(x_1, x_3) \cdot \alpha(x_2, x_1, x_3) \cdot \tau(x_1, x_2) \otimes x_3 \\ = \alpha(x_2, x_3, x_1) \cdot \tau(x_1, x_2 \otimes x_3) \cdot \alpha(x_1, x_2, x_3), \end{aligned} \tag{11.1.10}$$

$$\begin{aligned} \tau(x_1, x_3) \otimes x_2 \cdot \alpha(x_1, x_3, x_2)^{-1} \cdot x_1 \otimes \tau(x_2, x_3) \\ = \alpha(x_3, x_1, x_2)^{-1} \cdot \tau(x_1 \otimes x_2, x_3) \cdot \alpha(x_1, x_2, x_3)^{-1}. \end{aligned} \tag{11.1.11}$$

Note however that the isomorphism $\tau(x_1, x_2)$ is not involutive in the sense that $\tau(x_2, x_1) \cdot \tau(x_1, x_2) \neq id$, because the braid that represents this isomorphism in the braid group is not involutive either.

From this examination, we conclude that the object $\mu(x_1, x_2) = x_1 \otimes x_2 \in \mathtt{Ob}\, \mathsf{PaB}(2)$ can be interpreted as an abstract tensor product operation that can be used to govern the structure of a braided monoidal category with a strict unit, which is given by the arity zero element of our operad $* \in \Omega(0) = \mathtt{Ob}\, \mathsf{PaB}(0)$, but where we have a general associativity isomorphism $\alpha = \alpha(x_1, x_2, x_3)$, which is depicted in Eqn. 11.1.4 together with the braiding isomorphism $\tau = \tau(x_1, x_2)$. The operad PaB, equipped with these generating elements, is actually the universal operad that governs such structures, as shown in the following statement:

Theorem 11.1.3 (see [26, Theorem I.6.2.4]). *Let $\mathsf{M} \in \mathcal{C}at\,\mathcal{O}p$ be an operad in the category of small categories $\mathcal{C} = \mathcal{C}at$. Fixing an operad morphism $\phi : \mathsf{PaB} \to \mathsf{M}$ amounts to fixing a unit object $e \in \mathtt{Ob}\, \mathsf{M}(0)$, a product object $m \in \mathtt{Ob}\, \mathsf{M}(2)$, an associativity isomorphism $a \in \mathtt{Mor}_{\mathsf{M}(3)}(m \circ_1 m, m \circ_2 m)$, and a braiding isomorphism $c \in \mathtt{Mor}_{\mathsf{M}(2)}(m, (1\ 2)m)$ that satisfy the strict unit relations $m \circ_1 e = 1 = e \circ_1 m$ together with the coherence constraints of the unit, pentagon and hexagon relations of Eqn. 11.1.8-11.1.11 in the operad M.*

This theorem is established in the cited reference. The morphism associated to the quadruple (m, e, a, c) given in the theorem is obviously determined by the formulas $\phi(\mu) = m$, $\phi(*) = e$, $\phi(\alpha) = a$ and $\phi(\tau) = c$. The claim is that this assignments determines a well-defined morphism on PaB. The proof of this result follows from a combination of an operadic interpretation of the Mac Lane coherence theorem and of the classical presentation of the braid group by generators and relations. We have an analogous statement for the operad of colored braids CoB. In this case, we just require $a = id$ in our statement, because we represent the tensor product operation by the identity permutation on two letters $id_2 \in \Sigma_2$ in the object-sets of this operad CoB and this operation satisfies a strict associativity relation.

Theorem 11.1.3 implies that the category of algebras governed by the operad PaB in the category of categories is identified with a category of braided monoidal categories with a strict unit but general associativity isomorphisms. The operad of colored braids CoB has a similar interpretation (already mentioned in the introduction), but we then consider the category of braided monoidal categories with strict associativity identities instead of associativity isomorphisms. We refer to [26, §I.6.2] for more detailed explanations on these topics.

11.1.3 The Grothendieck–Teichmüller group

The Grothendieck–Teichmüller group is defined as a group of automorphisms of the parenthesized braid operad. To be more precise, we have to consider completions of this operad in applications. These completions operations are performed at the groupoid level. In what follows, we mainly consider the case of the Malcev completion, which we denote by $G^{\wedge}_{\mathbb{Q}}$ for any groupoid $G \in \mathcal{G}rd$,and the profinite completion, which we denote by $G^{\wedge}$ (yet another natural example of completion operation is the p-profinite completion, but we do not consider this variant of the profinite completion in this survey). In all cases, the considered completion operation does not change the object sets of our groupoids and is a natural generalization, for groupoids, of the corresponding classical completion operation on groups. Recall simply that the Malcev completion of groups is an extension of the classical rationalization of abelian groups that combines a pro-nilpotent completion with a rationalization operation. In the case of a free group for instance, we can identify the elements of the Malcev completion with infinite products of iterated commutators with rational exponents. (We refer to [26, §I.8] for a detailed survey of this subject.)

To define the rationalization of our operad $\mathsf{PaB}^{\widehat{}}_{\mathbb{Q}}$, we just perform the arity-wise completion operation $\mathsf{PaB}^{\widehat{}}_{\mathbb{Q}}(r) = \mathsf{PaB}(r)^{\widehat{}}_{\mathbb{Q}}$. Then we define the rational Grothendieck–Teichmüller group $GT(\mathbb{Q})$ as the group of automorphisms of the operad $\mathsf{PaB}^{\widehat{}}_{\mathbb{Q}}$ that reduce to the identity mapping on the object sets of our operad. In principle, we regard the object $\mathsf{PaB}^{\widehat{}}_{\mathbb{Q}}$ as an operad in a category of Malcev complete groupoids, where the morphisms satisfy a continuity constraint, and we assume that our automorphisms satisfy such a condition in the definition of the Grothendieck–Teichmüller group. But all morphisms are automatically continuous in the case of the Malcev completion of the operad PaB (see [26, Proposition I.10.1.5]), and therefore, we can neglect this issue in what follows.

We use a similar construction to define the profinite completion of our operad $\mathsf{PaB}^{\widehat{}}$ and the profinite Grothendieck–Teichmüller group $GT^{\widehat{}}$. (We just need to take care of the continuity constraints in the definition of morphisms in this case.) We examine the definition of automorphisms on these completions of the parenthesized braid operad to get more insights into the definition of these Grothendieck–Teichmüller groups. We explain our constructions in full details in the case of the rational Grothendieck–Teichmüller group only, because the profinite analogues of these constructions is obvious.

Any morphism $\phi : \mathsf{PaB} \to \mathsf{PaB}^{\widehat{}}_{\mathbb{Q}}$ admits a unique extension to the completed operad $\phi^{\widehat{}} : \mathsf{PaB}^{\widehat{}}_{\mathbb{Q}} \to \mathsf{PaB}^{\widehat{}}_{\mathbb{Q}}$. By Theorem 11.1.3, such a morphism $\phi : \mathsf{PaB} \to \mathsf{PaB}^{\widehat{}}_{\mathbb{Q}}$ is fully determined by giving a triple (m, a, c) such that $m = \phi(\mu)$, $a = \phi(\alpha)$ and $c = \phi(\tau)$. Note that we automatically have $\phi(*) = *$ since $\mathsf{PaB}(0) = * \Rightarrow \mathsf{PaB}(0)^{\widehat{}}_{\mathbb{Q}} = *$. For our purpose, we also set $m = \phi(\mu) = \mu$ since we only consider morphisms that are given by the identity mapping on objects in the definition of the Grothendieck–Teichmüller group.

We necessarily have $\phi(\tau) = \tau \cdot \tau^{2\nu}$ for some parameter $\nu \in \mathbb{Q}$, where we identify $\tau^2 \in \mathtt{Mor}_{\mathsf{PaB}(2)}(\mu, \mu)$ with an element of the pure braid group on 2 strands P_2 and we use the expression $\tau^{2\nu}$ with $\nu \in \mathbb{Q}$ to represent an element in the Malcev completion of this group $(P_2)^{\widehat{}}_{\mathbb{Q}}$. We similarly have $\phi(\alpha) = \alpha \cdot f$ for some morphism $f \in \mathtt{Mor}_{\mathsf{PaB}(3)^{\widehat{}}_{\mathbb{Q}}}(\mu \circ_1 \mu, \mu \circ_1 \mu)$, which is represented by an element of the Malcev completion of the pure braid group on three strand $(P_3)^{\widehat{}}_{\mathbb{Q}}$. We have $P_3 = \langle K \rangle \times \langle x_{12}, x_{23} \rangle$, where K denotes a central element in P_3, which is defined by the expression:

$K =$ 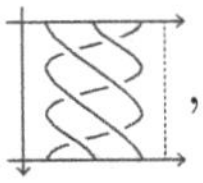 , (11.1.12)

while x_{12} and x_{23} denote the pure braids such that:

$x_{12} =$ 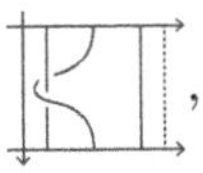 , $x_{23} =$ 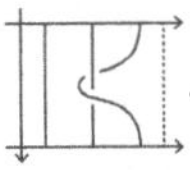 . (11.1.13)

The notation $\langle - \rangle$, which we use in this expression $P_3 = \langle K \rangle \times \langle x_{12}, x_{23} \rangle$, refers to the free group generated by a collection of elements.

We can easily deduce from the unit relation $a \circ_2 * = id$ that the associator $\phi(\alpha) = \alpha \cdot f$ has no factor in $\langle K \rangle\widehat{_{\mathbb{Q}}}$. Hence, our morphism ϕ is determined by an assignment of the form:

$$\phi(\mu) := \mu, \quad \phi(\tau) := \tau \cdot \tau^{2\nu} = \tau^{\lambda}, \quad \phi(\alpha) := \alpha \cdot f(x_{12}, x_{23}), \tag{11.1.14}$$

where we set $\lambda = 1 + 2\nu$ for $\nu \in \mathbb{Q}$ and we assume $f = f(x_{12}, x_{23}) \in \langle x_{12}, x_{23} \rangle\widehat{_{\mathbb{Q}}}$. In what follows, we use the notation of a formal series on two abstract variables $f = f(x, y)$ to represent this element f in the Malcev completion of the free group $F = \langle x_{12}, x_{23} \rangle$.

One can prove that the unit relations $a \circ_1 e = id = a \circ_3 e$ in the coherence constraints of Theorem 11.1.3 are equivalent to the identities:

$$f(x, 1) = x = f(1, x), \tag{11.1.15}$$

while the hexagon relations are equivalent to the following system of equations

$$f(x, y) \cdot f(y, x) = 1, \tag{11.1.16}$$

$$f(x, y) \cdot x^{\nu} \cdot f(z, x) \cdot z^{\nu} \cdot f(y, z) \cdot y^{\nu} = 1, \tag{11.1.17}$$

for a triple of variables (x, y, z) such that $zyx = 1$. The pentagon equation is equivalent to the following relation in the Malcev completion of the pure braid group on four strands $(P_4)\widehat{_{\mathbb{Q}}}$ (respectively, in the profinite completion $\widehat{P_4}$):

$$f(x_{23}, x_{34}) f(x_{13}x_{12}, x_{34}x_{24}) f(x_{12}, x_{23}) = f(x_{12}, x_{24}x_{23}) f(x_{23}x_{13}, x_{34}), \tag{11.1.18}$$

where in general, we use the notation x_{ij} for the pure braid group elements such that:

$$x_{ij} = \quad . \tag{11.1.19}$$

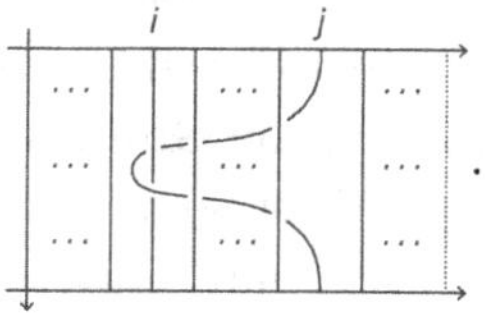

(We refer to [22] and to [26, Proof of Proposition I.11.1.4] for a more detailed analysis of these equations.)

The composition of morphisms corresponds to the following operation on this set of pairs $(\lambda, f(x, y))$:

$$(\lambda, f(x, y)) * (\mu, g(x, y)) = (\lambda\mu, f(x, y) \cdot g(x^{\lambda}, f(x, y)^{-1} \cdot y^{\lambda} \cdot f(x, y))). \tag{11.1.20}$$

Thus, an element of the Grothendieck–Teichmüller group $\gamma \in GT(\mathbb{Q})$, which corresponds to a morphism $\phi : \mathsf{PaB} \to \mathsf{PaB}\widehat{_{\mathbb{Q}}}$ that induce an isomorphism on the Malcev completion $\phi\widehat{\ }: \mathsf{PaB}\widehat{_{\mathbb{Q}}} \xrightarrow{\cong} \mathsf{PaB}\widehat{_{\mathbb{Q}}}$, can be uniquely determined by giving a pair $(\lambda, f) \in \mathbb{Q} \times \langle x_{12}, x_{23} \rangle\widehat{_{\mathbb{Q}}}$, which satisfies the constraints of Eqn. 11.1.15-11.1.18 and that is invertible with respect to this composition operation. A necessary and sufficient condition for this invertibility condition is given by $\lambda \in \mathbb{Q}^{\times}$ (see [22] and [26, Proposition I.11.1.5]).

The elements of the profinite Grothendieck–Teichmüller group $GT\widehat{\ }$ have a similar representation as pairs $(\lambda, f) \in \mathbb{Z}\widehat{\ } \times \langle x_{12}, x_{23} \rangle\widehat{\ }$, where we now consider the profinite completion of the integers $\mathbb{Z}$ for the parameter λ and the profinite completion of the free group $\langle x_{12}, x_{23}\rangle$ for the formal series $f = f(x_{12}, x_{23})$. (We just lack a simple characterization of the invertibility of morphisms in the profinite setting.)

This representation of the elements of the Grothendieck–Teichmüller group in terms of pairs (λ, f) and the equations of Eqn. 11.1.15-11.1.18 is actually Drinfeld's original definition of the Grothendieck–Teichmüller group in [22]. The correspondence between this definition and the operadic definition which we summarize in this paragraph is established with full details in the book [26, §I.11.1], but the ideas underlying this operadic interpretation were already implicitly present in Drinfeld's work [22]. We also refer to [5] for another formalization of this interpretation, which uses ideas close to the language of universal algebra. In the introduction of this chapter, we mentioned that the Grothendieck–Teichmüller group was defined by using ideas of the Grothendieck program in Galois theory. In fact, we have an embedding $Gal(\bar{\mathbb{Q}}/\mathbb{Q}) \hookrightarrow GT^{\wedge}$ that is defined by using an action of the absolute Galois group on genus zero curves with marked points (see [22]). For the rational Grothendieck–Teichmüller group, a result of F. Brown's (see [13]) implies that we have an analogous embedding $Gal_{MT(\mathbb{Z})} \hookrightarrow GT(\mathbb{Q})$, where $Gal_{MT(\mathbb{Z})}$ now denotes the motivic Galois group of a category of integral mixed Tate motives (see also [21, 55] for the definition of this group and for the definition of this mapping).

We go back to the definition of the Grothendieck–Teichmüller group $GT(\mathbb{Q})$ in terms of operad isomorphisms $\phi_{\gamma}^{\widehat{}} : \mathsf{PaB}_{\mathbb{Q}}^{\widehat{}} \xrightarrow{\simeq} \mathsf{PaB}_{\mathbb{Q}}^{\widehat{}}$. We can regard the classifying space operad $\mathsf{E}_2^{\mathbb{Q}} = \mathtt{B}(\mathsf{PaB}_{\mathbb{Q}}^{\widehat{}})$ as a model for the rationalization of the E_2-operad $\mathsf{E}_2 = \mathtt{B}(\mathsf{PaB})$. We deduce from the functoriality of the classifying space construction that any element $\gamma \in GT(\mathbb{Q})$ induces an automorphism $\phi_{\gamma}^{\widehat{}} : \mathsf{E}_2^{\mathbb{Q}} \xrightarrow{\sim} \mathsf{E}_2^{\mathbb{Q}}$ at the topological operad level, and hence, defines an element in the homotopy automorphism space $\mathtt{Aut}^h_{\mathcal{O}p}(\mathsf{E}_2^{\mathbb{Q}})$. We claim that this correspondence $\gamma \mapsto \phi_{\gamma}^{\widehat{}}$ induces a bijection when we pass to the group of homotopy classes of homotopy automorphisms. We can deduce this statement from the following more precise statement:

Theorem 11.1.4 (B. Fresse [26, Theorem III.5.2.5]). *We have* $\mathtt{Aut}^h_{\mathcal{O}p}(\mathsf{E}_2^{\mathbb{Q}}) \sim GT(\mathbb{Q}) \ltimes \mathtt{B}(\mathbb{Q})$.

The factor $\mathtt{B}(\mathbb{Q})$ in the expression of this theorem corresponds to a rationalization of the group of rotations $\mathtt{SO}(2) \sim \mathtt{B}(\mathbb{Z})$ which acts on the little 2-discs model of the class of E_2-operads $\mathsf{E}_2 = \mathsf{D}_2$ by rotating the configurations of little 2-discs in this operad D_2. We equip this factor $\mathtt{B}(\mathbb{Q})$ with the obvious additive group structure.

We have $\mathtt{Mor}_{\mathsf{PaB}(2)_{\mathbb{Q}}^{\widehat{}}}(\mu, \mu) = (P_2)_{\mathbb{Q}}^{\widehat{}}$, and as a consequence, any element of the Grothendieck–Teichmüller group $\gamma \in GT(\mathbb{Q})$ determines an automorphism of the Malcev completion of the pure braid group $(P_2)_{\mathbb{Q}}^{\widehat{}} \simeq \mathbb{Q}$ through its action on the automorphism group of the object $\mu \in \mathtt{Ob}\,\mathsf{PaB}(2)$. We use this observation to determine the action of the Grothendieck–Teichmüller group on the group $\mathtt{B}(\mathbb{Q})$ that we consider in the definition of the semi-product $GT(\mathbb{Q}) \ltimes \mathtt{B}(\mathbb{Q})$. (We refer to [26, §III.5.2] for details.)

Let us insist that we consider derived homotopy automorphism spaces in the statement of this theorem. In the model category approach, these homotopy automorphism spaces are defined by taking the actual spaces of homotopy automorphism spaces associated to a cofibrant-fibrant replacement $\mathsf{R}_2^{\mathbb{Q}}$ of our operad $\mathsf{E}_2^{\mathbb{Q}}$. To associate an element of this derived homotopy automorphism space to an element of the Grothendieck–Teichmüller group

$\gamma \in GT(\mathbb{Q})$, we use the fact that an automorphism $\phi_{\widehat{\gamma}} : E_2^{\mathbb{Q}} \to E_2^{\mathbb{Q}}$ automatically admits a lifting to this cofibrant-fibrant replacement $R_2^{\mathbb{Q}}$. The claim is that all homotopy automorphisms of this cofibrant-fibrant model $R_2^{\mathbb{Q}}$ are homotopic to such morphisms, and that this correspondence gives all the homotopy of the space $\mathtt{Aut}^h_{\mathcal{O}p}(E_2^{\mathbb{Q}})$ up to the factor $\mathtt{B}(\mathbb{Q})$.

The book [26, §§III.1-5] gives a proof of this result by using spectral sequence methods and an operadic cohomology theory which provides approximations of our homotopy automorphism spaces. This method is close to the methods that are used in the next sections, when we study the homotopy automorphism spaces of E_n-operads for any value of the dimension parameter $n \geq 2$.

We now consider the profinite version of the Grothendieck–Teichmüller group $GT^{\frown}$. We take the classifying space operad $E_2^{\frown} = \mathtt{B}(PaB^{\frown})$ as a model for the profinite completion of the E_2-operad $E_2 = \mathtt{B}(PaB)$ and we use the same construction as in the rational setting to define a mapping from the profinite Grothendieck–Teichmüller group towards the homotopy automorphism space of this object $\mathtt{Aut}^h_{\mathcal{O}p}(D_2^{\frown})$. Then we have the following analogue of the result of Theorem 11.1.5:

Theorem 11.1.5 (G. Horel, [37]). *We have* $\mathtt{Aut}^h_{\mathcal{O}p}(D_2^{\frown}) \sim GT^{\frown} \ltimes \mathtt{B}(\mathbb{Z}^{\frown})$.

The article [37] gives a proof of the result of this theorem by using the correspondence with groupoids. In short, the idea of this paper is to observe that the operad $PaB^{\frown}$ represents a cofibrant object with respect to some model structure on the category of operads in groupoids. Then we can use model category arguments (combined with higher category methods) to prove that we can transport the computation of the homotopy automorphism space $\mathtt{Aut}^h_{\mathcal{O}p}(E_2^{\frown})$ to the computation of the homotopy automorphism space associated to this object PaB in the category of operads in groupoids.

11.1.4 The Drinfeld–Kohno Lie algebra operad

Besides the colored braid and the parenthesized braid operads, which are defined by using the structures of the braid groups, we consider operads in Lie algebras that are associated to infinitesimal versions of the pure braid groups. To be explicit, in these infinitesimal versions, we consider the Drinfeld–Kohno Lie algebras (also called the Lie algebras of infinitesimal braids), which are defined by a presentation of the form:

$$\mathfrak{p}(r) = \mathbb{L}(t_{ij}, \{i,j\} \subset \{1,\dots,r\}) / < [t_{ij}, t_{kl}], [t_{ij}, t_{ik} + t_{jk}] >, \tag{11.1.21}$$

for $r \in \mathbb{N}$, where $\mathbb{L}(-)$ denotes the free Lie algebra functor, we associate a generator t_{ij} such that $t_{ij} = t_{ji}$ to each pair $\{i \neq j\} \subset \{1,\dots,r\}$, and we take the ideal generated by the commutation relations

$$[t_{ij}, t_{kl}] \equiv 0, \tag{11.1.22}$$

for all quadruples $\{i,j,k,l\} \subset \{1,\dots,r\}$ such that $\sharp\{i,j,k,l\} = 4$, together with the Yang-Baxter relations

$$[t_{ij}, t_{ik} + t_{jk}] \equiv 0, \tag{11.1.23}$$

for all triples $\{i,j,k\} \subset \{1,\dots,r\}$ such that $\sharp\{i,j,k\} = 3$. This definition makes sense over any ground ring $\Bbbk$, but from the next paragraph on, we will assume that the ground ring is a field of characteristic zero.

Note that this Lie algebra $\mathfrak{p}(r)$ inherits a weight grading from the free Lie algebra since this ideal is generated by homogeneous relations. If we use the notation $L_m = \mathbb{L}_m(-)$, for the homogeneous component of weight m of the free Lie algebra $L = \mathbb{L}(t_{ij}, \{i,j\} \subset \{1,\dots,r\})$, then we have the decomposition $\mathfrak{p}(r) = \bigoplus_{m\geq 1} \mathfrak{p}(r)_m$, where we set $\mathfrak{p}(r)_m = L_m/L_m \cap < [t_{ij}, t_{kl}], [t_{ij}, t_{ik} + t_{jk}] >$, for $m \geq 1$. In fact, we have the identity $\mathfrak{p}(r)_* = gr^{\Gamma}_* P_r$, where on the right-hand side we consider the graded Lie algebra of the sub-quotients of the central series filtration of the pure braid group $gr^{\Gamma}_* P_r$ (see [26, Theorem I.10.0.4] for a detailed proof of this statement).

The collection $\mathfrak{p} = \{\mathfrak{p}(r), r \in \mathbb{N}\}$ inherits the structure of an operad in the category of Lie algebras, where we take the direct sum of Lie algebras to define our symmetric monoidal structure. The action of the symmetric group Σ_r on the Lie algebra $\mathfrak{p}(r)$ is defined, on generators, by the obvious re-indexing operation $\sigma \cdot t_{ij} = t_{\sigma(i)\sigma(j)}$, for all $\sigma \in \Sigma_r$. The composition products are given by Lie algebra morphisms of the form

$$\circ_i : \mathfrak{p}(k) \oplus \mathfrak{p}(l) \to \mathfrak{p}(k+l-1), \tag{11.1.24}$$

defined for all $k,l \in \mathbb{N}$ and $i \in \{1,\dots,k\}$, and which satisfy the equivariance, unit and associativity relations of operads in the category of Lie algebras. For generators $t_{ab} \in \mathfrak{p}(k)$ and $t_{cd} \in \mathfrak{p}(l)$, we explicitly set:

$$t_{ab} \circ_i 0 = \begin{cases} t_{a+l-1 b+l-1}, & \text{if } i < a < b, \\ t_{a b+l-1} + \dots + t_{a+l-1 b+l-1}, & \text{if } i = a < b, \\ t_{a b+l-1}, & \text{if } a < i < b, \\ t_{ab} + \dots + t_{a j+l-1}, & \text{if } a < i = b, \\ t_{ab}, & \text{if } a < b < i, \end{cases} \tag{11.1.25}$$

and

$$0 \circ_i t_{cd} = t_{c+i-1 d+i-1} \quad \text{for all } i. \tag{11.1.26}$$

The operadic unit is just given by the zero morphism $0 : 0 \to \mathfrak{p}(1)$ with values in the zero object $\mathfrak{p}(1) = 0$.

In fact, these operations reflect the composition structure of the operad of colored braids, in the sense that we can identify the components of homogeneous weight of this operad $\mathfrak{p}(-)_m$, $m \geq 1$, with the fibers of a tower of operads $\mathsf{CoB} \,/\, \Gamma_m \,\mathsf{CoB}$, $m \geq 1$, which we deduce from the central series filtration of the pure braid group. (We refer to [26, §I.10.1] for more explanations on this correspondence.)

We call this operad $\mathfrak{p}$ the Drinfeld–Kohno Lie algebra operad. We consider generalizations of this operad when we study the Sullivan model of E_n-operads. This subsequent study is our main motivation for the recollections of this paragraph, but the Drinfeld–Kohno Lie algebra operad also occurs in the theory of Drinfeld's associators and in the definition of a graded version of the Grothendieck–Teichmüller group. We just give a brief overview of this subject to complete the account of this section.

11.1.5 The operad of chord diagrams and associators

To define the set of Drinfeld's associators, we consider an operad in groupoids, the chord diagram operad $\mathit{CD}^{\widehat{}}_{\Bbbk}$, defined over any characteristic zero field $\Bbbk$, and such that we have the relation $\mathit{CD}(r)^{\widehat{}}_{\Bbbk} = \exp \hat{\mathfrak{p}}(r)$ for each $r \in \mathbb{N}$, where we consider the exponential group associated to a completion of the Drinfeld Kohno Lie algebra $\hat{\mathfrak{p}}(r)$.

To be more precise, we explained in the previous paragraph that the Drinfeld Kohno Lie algebra admits a weight decomposition $\mathfrak{p}(r) = \bigoplus_{m \geq 1} \mathfrak{p}(r)_m$. To form the completed Lie algebra $\hat{\mathfrak{p}}(r)$, we just replace the direct sum of this decomposition by a product. Thus we have $\hat{\mathfrak{p}}(r) = \prod_{m \geq 1} \mathfrak{p}(r)_m$ so that the elements of this completed Lie algebra $\hat{\mathfrak{p}}(r)$ are represented by infinite series of Lie polynomials (modulo the ideal generated by the commutation and the Yang-Baxter relations). The exponential group $\mathit{CD}(r)^{\widehat{}}_{\Bbbk} = \exp \hat{\mathfrak{p}}(r)$ consists of formal exponential elements e^{ξ}, $\xi \in \hat{\mathfrak{p}}(r)$, together with the group operation given by the Campbell-Hausdorff formula at the level of the completed Lie algebra $\hat{\mathfrak{p}}(r)$. We identify this group $\mathit{CD}(r)^{\widehat{}}_{\Bbbk} = \exp \hat{\mathfrak{p}}(r)$ with a groupoid with a single object when we regard the chord diagram operad $\mathit{CD}^{\widehat{}}_{\Bbbk} = \{\exp \hat{\mathfrak{p}}(r), r \in \mathbb{N}\}$ as an operad in the category of groupoids. The structure operations of this operad $\mathit{CD}^{\widehat{}}_{\Bbbk}$ are induced by the structure operations of the Drinfeld–Kohno Lie algebra operad on $\hat{\mathfrak{p}}$.

We also have an identity $\mathit{CD}(r)^{\widehat{}} = \mathbb{G}(\hat{\mathbb{U}}(\hat{\mathfrak{p}}(r)))$, where we consider the set of group-like elements $\mathbb{G}(-)$ in the complete enveloping algebra of the Drinfeld–Kohno Lie algebra $\hat{\mathbb{U}}(\hat{\mathfrak{p}}(r))$. Indeed, the group-like elements are identified with actual exponential series of Lie algebra elements $\xi \in \mathfrak{p}(r)$ within the complete enveloping algebras $\hat{\mathbb{U}}(\hat{\mathfrak{p}}(r))$. The name 'chord diagram' comes from the theory of Vassiliev invariants, where a monomial $t_{i_1 j_1} \cdots t_{i_m j_m} \in \hat{\mathbb{U}}(\hat{\mathfrak{p}}(r))$ is associated to a diagram with r vertical strands numbered from 1 to r, and l chords corresponding to the factors $t_{i_k j_k}$, as in the following picture:

$$t_{12}t_{12}t_{36}t_{24} = \quad (11.1.27)$$

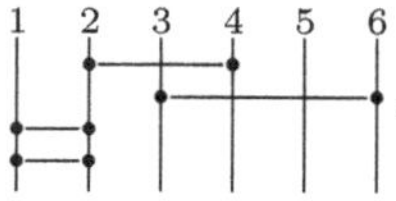

1 2 3 4 5 6

The composition products of the chord diagram operad have a simple description in terms of chord diagram insertions too.

In the previous paragraph, we explained that the components of homogeneous weight of the Drinfeld–Kohno Lie algebra operad represent the fibers of a tower decomposition of the parenthesized braid operad (and of the colored braid operad equivalently). In fact, a stronger result holds when we work over a field of characteristic zero. To be more explicit, we consider the Malcev completion of the operad PaB, and a natural extension of this construction for ground fields such that $\mathbb{Q} \subset \Bbbk$. Then we may wonder about the existence of equivalences of operads in groupoids $\phi^{\widehat{}}_{\alpha} : \mathit{PaB}^{\widehat{}}_{\Bbbk} \xrightarrow{\sim} \mathit{CD}^{\widehat{}}_{\Bbbk}$, which would be equivalent to a splitting of this tower decomposition over $\Bbbk$. By Theorem 11.1.3, the morphism of operads in groupoids $\phi_a : \mathit{PaB} \xrightarrow{\sim} \mathit{CD}^{\widehat{}}_{\Bbbk}$ which would induce such an equivalence on the completion is determined by the choice of a braiding $c \in \exp \hat{\mathfrak{p}}(2)$ and of an associativity isomorphism $a \in \exp \hat{\mathfrak{p}}(3)$. The braiding has the form $c = \exp(\kappa t_{12}/2)$, for some parameter $\kappa \in \Bbbk^{\times}$, since $\hat{\mathfrak{p}}(2) = \Bbbk t_{12}$, and one can prove that the associator is necessarily of the form $a = \exp f(t_{12}, t_{23})$, for some Lie power series $f(t_{12}, t_{23}) \in \hat{\mathbb{L}}(t_{12}, t_{23})$. Thus, the existence of an equivalence of operads in groupoids $\phi_{\alpha}^{\widehat{}} : \mathsf{PaB}_{\Bbbk}^{\widehat{}} \xrightarrow{\sim} \mathsf{CD}_{\Bbbk}^{\widehat{}}$ reduces to the existence of such a Lie power series $f(t_{12}, t_{23}) \in \hat{\mathbb{L}}(t_{12}, t_{23})$ such that $a = \exp f(t_{12}, t_{23})$ satisfies the unit, pentagon and hexagon constraints of Theorem 11.1.3, for some given parameter $\kappa \in \Bbbk^{\times}$.

The set of Drinfeld's associators precisely refers to this particular set of associators $a = \exp f(t_{12}, t_{23})$ which we associate to the chord diagram operad $\mathsf{CD}_{\Bbbk}^{\widehat{}}$. This notion was introduced by Drinfeld in the paper [22], to which we also refer for an explicit expression of the unit, pentagon and hexagon constraints (see also the survey of [26, §I.10.2]). Further reductions occur in the pentagon and hexagon constraints in the definition of Drinfeld's associators. In fact, a result of Furusho (see [30]) implies that the hexagon constraints are satisfied as soon as we have a power series that fulfills the unit and the pentagon constraints.

We have the following main result:

Theorem 11.1.6 (V.I. Drinfeld [22]). *The set of Drinfeld's associators is not empty, for any choice of field of characteristic zero as ground field $\Bbbk$ (including $\Bbbk = \mathbb{Q}$), so that we do have an operad morphism $\phi_{\alpha} : \mathsf{PaB} \to \mathsf{CD}_{\mathbb{Q}}^{\widehat{}}$ that induces an equivalence when we pass to the Malcev completion $\phi_{\alpha}^{\widehat{}} : \mathsf{PaB}_{\mathbb{Q}}^{\widehat{}} \xrightarrow{\sim} \mathsf{CD}_{\mathbb{Q}}^{\widehat{}}$.*

In [22], Drinfeld gives an explicit construction of a complex associator by using the monodromy of the Knizhnik–Zamolodchikov connection. This associator, which is usually called the Knizhnik–Zamolodchikov associator in the literature, can also be identified with a generating series of polyzeta values. Descent arguments can be used to establish the existence of a rational associator from the existence of this complex associator (see again [22] and [5] for different proofs of this descent statement), so that the result of this theorem holds over $\Bbbk = \mathbb{Q}$, and not only over $\Bbbk = \mathbb{C}$. Another explicit definition of an associator, defined over the reals, is given by Alekseev–Torossian in [1] by using constructions introduced by Kontsevich in his proof of the formality of the operads of little discs (see [43]).

11.1.6 The operad of parenthesized chord diagrams, the graded Grothendieck–Teichmüller group and other related objects

The existence of associators can be used to get insights into the structure of the rational Grothendieck–Teichmüller group $GT(\mathbb{Q})$. Indeed, the definition implies that the set of associators inherits a free and transitive action of the rational Grothendieck–Teichmüller group. To go further into the applications of associators, one introduces a parenthesized version of the chord diagram operad $\mathsf{PaCD}_{\mathbb{Q}}^{\widehat{}}$ (by using the same pullback construction as in the case of the parenthesized braid operad PaB) and a group of automorphisms, denoted by $GRT(\mathbb{Q})$, which we associate to this object $\mathsf{PaCD}_{\mathbb{Q}}^{\widehat{}}$. One can easily check that every equivalence of operads in groupoids $\phi_{a}^{\widehat{}} : \mathsf{PaB}_{\mathbb{Q}}^{\widehat{}} \xrightarrow{\sim} \mathsf{CD}_{\mathbb{Q}}^{\widehat{}}$ lifts to an isomorphism $\phi_{a}^{\widehat{}} : \mathsf{PaB}_{\mathbb{Q}}^{\widehat{}} \xrightarrow{\simeq} \mathsf{PaCD}_{\mathbb{Q}}^{\widehat{}}$ so that the existence of rational associators implies the existence of a group isomorphism $GT(\mathbb{Q}) \simeq GRT(\mathbb{Q})$ by passing to automorphism groups.

This group $GRT(\mathbb{Q})$ is usually called the graded Grothendieck–Teichmüller group in the literature, because this group is identified with the pro-algebraic group associated to a graded Lie algebra such that $\mathfrak{grt} = \bigoplus_{m \geq 0} \mathfrak{grt}_m$. We moreover have $\mathfrak{grt}_m \simeq$

$\mathsf{F}_m\, GT(\mathbb{Q})/\mathsf{F}_{m+1}\, GT(\mathbb{Q})$, for all $m \geq 1$, for some natural filtration of the Grothendieck–Teichmüller group $GT(\mathbb{Q}) = \mathsf{F}_0\, GT(\mathbb{Q}) \supset \mathsf{F}_1\, GT(\mathbb{Q}) \supset \cdots \supset \mathsf{F}_m\, GT(\mathbb{Q}) \supset \cdots$. We again refer to [22] for a proof of these results (see also the survey of [26, §I.11.4]).

The groups $GT(\mathbb{Q})$, $GRT(\mathbb{Q})$, and the Lie algebra $\mathfrak{grt}$ are also related to other objects of arithmetic geometry and group theory. In §11.1.3, we already recalled that, by a result of F. Brown (see [13]), the Grothendieck–Teichmüller group $GT(\mathbb{Q})$ contains a realization of the motivic Galois group of a category of integral mixed Tate motives $Gal_{MT(\mathbb{Z})}$. In fact, one conjectures that these groups are isomorphic (Deligne-Ihara). Furthermore, one can prove that the Galois group $Gal_{MT(\mathbb{Z})}$ reduces to the semi-direct product of the multiplicative group with the prounipotent completion of a free group on a sequence of generators $s_3, s_5, \ldots, s_{2n+1}, \ldots$ (see [21]). This result implies that the embedding $Gal_{MT(\mathbb{Z})} \hookrightarrow GT(\mathbb{Q})$ is equivalent to an embedding of the form $\Bbbk \oplus \mathbb{L}(s_3, s_5, \ldots, s_{2n+1}, \ldots) \hookrightarrow \mathfrak{grt}$ when we pass to the category of Lie algebras. In this context, one can re-express the Deligne-Ihara conjecture as the conjecture that this embedding of Lie algebras is an isomorphism.

In our comments on Theorem 11.1.6, we also explained that the Knizhnik–Zamolodchikov associator represents the generating series of polyzeta values. The polyzeta values satisfy certain equations, called the regularized double shuffle relations, which can be expressed in terms of the Knizhnik–Zamolodchikov associator, and one conjectures that all relations between polyzetas follow from the double shuffle relations and from the fact that the Knizhnik–Zamolodchikov associator defines a group-like power series. By a result of Furusho [31], the pentagon condition for associators implies the regularized double shuffle relations. This result implies that the Grothendieck–Teichmüller group embeds in a group defined by solutions of regularized double shuffle relations with a degeneration condition, and one conjectures again that this embedding is an isomorphism.

The theory of associators is also used by Alekseev–Torossian in the study of the solutions of the Kashiwara–Vergne conjecture, a problem about the Campbell-Hausdorff formula motivated by questions of harmonic analysis. These authors notably proved in [1] that the set of Drinfeld's associators embeds into the set of solutions of the Kashiwara–Vergne conjecture. In particular, one can deduce the existence of such solutions from the existence of associators. In addition, one can prove that the action of the Grothendieck–Teichmüller group on Drinfeld's associators lifts to the set of solutions of the Kashiwara–Vergne conjecture. This action is still free so that we get that the Grothendieck–Teichmüller group embeds into a group of automorphisms associated to this set of solutions. The conjecture is that this group embedding is an isomorphism, yet again.

11.2 The rational homotopy of E_n-operads and formality theorems

The goal of this section is to explain the definition of rational models of E_n-operads and a characterization of the class of E_n-operads up to rational homotopy equivalence. In what follows, we just focus on the case $n \geq 2$, because we can put the case $n = 1$ apart. Indeed, we have $D_1 \sim \Pi$, where we regard the permutation operad $\Pi = \{\Sigma_r, r \in \mathbb{N}\}$ (see §11.1.2) as a discrete operad in topological spaces. Hence, the class of E_1-operads is also identified with the class of operads that are weakly-equivalent to this discrete operad Π, and such a class of objects is fixed by the rationalization.

Recall that a map of simply connected topological spaces is a rational homotopy equivalence $f : X \xrightarrow{\sim_{\mathbb{Q}}} Y$ if this map induces a bijection on homotopy groups $f_* : \pi_*(X) \otimes_{\mathbb{Z}} \mathbb{Q} \xrightarrow{\cong} \pi_*(Y) \otimes_{\mathbb{Z}} \mathbb{Q}$. In what follows, we consider a generalization of this notion in the context of spaces that, like the underlying spaces of the little 2-disc operad, are (connected but) not necessarily simply connected. In this case, we assume that a rational homotopy equivalence also induces an isomorphism on the Malcev completion of the fundamental group $f_* : \pi_1(X,x)\hat{}_{\mathbb{Q}} \xrightarrow{\cong} \pi_1(Y, f(x))\hat{}_{\mathbb{Q}}$.

In the context of operads, we just consider operad morphisms $\phi : P \xrightarrow{\sim_{\mathbb{Q}}} Q$ that define a rational homotopy equivalence of spaces arity-wise $\phi : P(r) \xrightarrow{\sim_{\mathbb{Q}}} Q(r)$. and we write $P \sim_{\mathbb{Q}} Q$ when our objects P and Q can be connected by a zigzag of such rational homotopy equivalences. We aim to determine the class of operads such that $R \sim_{\mathbb{Q}} D_n$.

We develop a rational homotopy theory of operads to address this problem. We rely on the Sullivan rational homotopy of spaces, which we briefly review in the next paragraph. We explain the construction of an operadic extension of the Sullivan model afterwards. We eventually check that the n-Poisson cooperad, the dual structure of the n-Poisson operad, defines a Sullivan model of the little n-discs operad D_n, and as such determines a model for the class of E_n-operads up to rational homotopy. We need a cofibrant resolution of the n-Poisson cooperad to perform computations with this model. We will explain that such a cofibrant resolution is given by the Chevalley–Eilenberg cochain complex of a graded version of the Drinfeld–Kohno Lie algebra operad of the previous section. We actually consider another resolution in our construction, namely a cooperad of graphs, and we also explain the definition of this object. We use the latter model in the next section, when we explain a graph complex description of the rational homotopy type of mapping spaces of E_n-operads.

In order to apply the methods of rational homotopy theory, we take $\Bbbk = \mathbb{Q}$ as a ground ring for our categories of modules from now on, and we also consider the cohomology with coefficients in this field $\mathtt{H}^*(-) = \mathtt{H}^*(-, \mathbb{Q})$. We similarly take $\mathtt{H}_*(-) = \mathtt{H}_*(-, \mathbb{Q})$ for the homology.

11.2.1 Recollections on the Sullivan rational homotopy theory of spaces

Recall that we call 'upper graded dg-module' the structure formed by a module M equipped with a decomposition such that $M = \bigoplus_{n \in \mathbb{Z}} M^n$ and with a differential $\delta : M \to M$ such that $\delta(M^{*-1}) \subset M^*$. We say that such a dg-module M is nonnegatively graded when we have $M^n = 0$ for $n < 0$. Let $dg^* \mathcal{C}om$ be the category of commutative algebras in upper nonnegatively graded dg-modules (the category of commutative cochain dg-algebras for short). The Sullivan model for the rational homotopy of a space takes values in this category $dg^* \mathcal{C}om$ and is obtained by applying the Sullivan functor of PL differential forms, a version of the de Rham cochain complex that is defined over $\mathbb{Q}$ (instead of $\mathbb{R}$) and on the category of simplicial complexes (or simplicial sets), instead of the category of smooth manifolds (see [52]). For our purpose, we consider the simplicial set variant of this functor:

$$\Omega^* : s\mathcal{S}et^{op} \to dg^* \mathcal{C}om \,. \tag{11.2.1}$$

In the particular case of a simplex $\Delta^n = \{0 \leq x_1 \leq \cdots \leq x_n \leq 1\}$, we explicitly have:

$$\Omega^*(\Delta^n) = \mathbb{Q}[x_1, \ldots, x_n, dx_1, \ldots, dx_n], \tag{11.2.2}$$

where $dx_1, \ldots, dx_n$ represents the differential of the variables $x_1, \ldots, x_n$ in this commutative cochain dg-algebra.

The Sullivan functor $\Omega^* : sSet^{op} \to dg^*\mathcal{C}om$ has a left adjoint

$$\mathtt{G}_\bullet : dg^*\mathcal{C}om \to sSet^{op}, \tag{11.2.3}$$

which is given by the formula $\mathtt{G}_\bullet(A) = \mathtt{Mor}_{dg^*\mathcal{C}om}(A, \Omega^*(\Delta^\bullet))$, for any commutative cochain dg-algebra $A \in dg^*\mathcal{C}om$, and this pair of adjoint functors $(\mathtt{G}_\bullet, \Omega)$ defines a Quillen adjunction. (We refer to [12] for this application of the formalism of model categories to Sullivan's constructions.) Then we set:

$$\langle A \rangle := \text{derived functor of } \mathtt{G}_\bullet(A) = \mathtt{Mor}_{dg^*\mathcal{C}om}(R_A, \Omega^*(\Delta^\bullet)), \tag{11.2.4}$$

where $R_A \xrightarrow{\sim} A$ is any cofibrant resolution of A in $dg^*\mathcal{C}om$. If X satisfies reasonable finiteness and nilpotence assumptions, then the space

$$X^{\mathbb{Q}} := \langle \Omega^*(X) \rangle \tag{11.2.5}$$

defines a rationalization of the space X in the sense that we have the identities

$$\pi_*(X^{\mathbb{Q}}) := \begin{cases} \pi_*(X) \otimes_{\mathbb{Z}} \mathbb{Q}, & \text{for } * \geq 2, \\ \pi_1(X)^{\wedge}_{\mathbb{Q}}, & \text{for } * = 1, \end{cases} \tag{11.2.6}$$

where we again use the notation $(-)^{\wedge}_{\mathbb{Q}}$ for the Malcev completion functor on groups. Besides, one can prove that the unit of the derived adjunction relation between the functors $\mathtt{G}_\bullet$ and Ω^* defines a map $\eta : X \to X^{\mathbb{Q}}$ which corresponds to the usual rationalization map at the level of these homotopy groups.

11.2.2 The category of Hopf cochain dg-cooperads

To extend the Sullivan model to operads, the idea is to consider cooperads in the category of commutative cochain dg-algebras, where a cooperad is a structure that is dual to an operad in the sense of the theory of categories.

In general, a cooperad in a symmetric category $\mathcal{C}$ consists of a collection of objects $\mathsf{C} = \{\mathsf{C}(r), r \in \mathbb{N}\}$, together with an action of the symmetric group Σ_r on $\mathsf{C}(r)$, for each $r \in \mathbb{N}$, and composition coproducts

$$\circ_i^* : \mathsf{C}(k + l - 1) \to \mathsf{C}(k) \otimes \mathsf{C}(l), \tag{11.2.7}$$

defined for all $k, l \in \mathbb{N}$, $i \in \{1, \ldots, k\}$, and which satisfy equivariance, unit and coassociativity relations dual to the equivariance, unit and coassociativity axioms of operads. To handle difficulties, we consider a subcategory of cooperads such that $\mathsf{C}(0) = \mathsf{C}(1) = \mathbb{1}$ where $\mathbb{1}$

is the unit object of our base category, and we use the notation $\mathcal{O}p^c_{*1}$ for this category of cooperads. This restriction enables us to simplify some constructions, because the composition coproducts are automatically conilpotent when we put the component of arity zero apart and we assume $C(1) = \mathbb{1}$. In some cases, we consider a category of cooperads $\mathcal{O}p^c_{*N}$ such that we still have $C(0) = \mathbb{1}$, but where $C(1)$ may not reduce to the unit object. More care is necessary in this case, and we actually assume an extra conilpotence condition for the composition coproducts that involve the component of arity one. (We refer to [27] for the precise expression of this conilpotence condition.)

In [26, §II.12] the author considers a category of Λ-cooperads, whose objects have no term in arity zero, but a diagram structure over the category of finite ordinals and injective maps that extends the action of the symmetric groups on ordinary operads. This category of Λ-cooperads is isomorphic to the category of cooperads that we consider in this paragraph $\mathcal{O}p^c_{*1}$, so that the results of this reference [26] can immediately be transposed to our setting. (The structure of a Λ-cooperad is used to overcome technical difficulties that occurs in the construction of the Sullivan model of operads, but we can neglect these issues in this overview.)

We use the name 'Hopf cochain dg-cooperad' for the category of cooperads in the category of commutative cochain dg-algebras $\mathcal{C} = dg^*\mathcal{C}om$ and we also adopt the notation $dg^*\mathcal{H}opf\,\mathcal{O}p^c_{*1}$ for this category of cooperads. We also consider a category of operads in simplicial sets satisfying $P(0) = P(1) = *$ in order to deal with the restrictions imposed by the definition of our category of cooperads in our model. We use the notation $s\mathcal{S}et\,\mathcal{O}p_{*1}$ for this category of operads. We have the following statement:

Theorem 11.2.1 (B. Fresse [26, §II.10, §II.12]).

- *The letf adjoint of the Sullivan functor* $\mathtt{G}_\bullet : dg^*\mathcal{C}om \to s\mathcal{S}et^{op}$ *induces a functor* $\mathtt{G}_\bullet : dg^*\mathcal{H}opf\,\mathcal{O}p^c_{*1} \to s\mathcal{S}et\,\mathcal{O}p^{op}_{*1}$ *from the category of Hopf cochain dg-cooperads* $dg^*\mathcal{H}opf\,\mathcal{O}p^c_{*1}$ *to the category of operads in simplicial sets* $s\mathcal{S}et\,\mathcal{O}p^{op}_{*1}$. *For an object* $A \in dg^*\mathcal{H}opf\,\mathcal{O}p^c_{*1}$, *we set*
$$\mathtt{G}_\bullet(A)(r) = \mathtt{G}_\bullet(A(r))$$
and we use the fact that $\mathtt{G}_\bullet(-)$ *is strongly symmetric monoidal to equip the collection of these simplicial sets* $\mathtt{G}_\bullet(A) = \{\mathtt{G}_\bullet(A(r)), r \in \mathbb{N}\}$ *with the structure of an operad.*
- *This functor* $\mathtt{G}_\bullet : dg^*\mathcal{H}opf\,\mathcal{O}p^c_{*1} \to s\mathcal{S}et\,\mathcal{O}p^{op}_{*1}$ *admits a right adjoint*
$$\Omega^*_\sharp : s\mathcal{S}et\,\mathcal{O}p^{op}_{*1} \to dg^*\mathcal{H}opf\,\mathcal{O}p^c_{*1}$$
and the pair of functors $(\mathtt{G}_\bullet, \Omega^*_\sharp)$ *defines a Quillen adjunction.*
- *For a cofibrant operad* $P \in s\mathcal{S}et\,\mathcal{O}p_{*1}$ *such that* $\mathtt{H}^*(P(r))$ *forms a finite dimensional* $\mathbb{Q}$*-module in each arity* $r \in \mathbb{N}$ *and in each degree* $* \in \mathbb{N}$, *we have a weak equivalence*
$$\Omega^*_\sharp(P)(r) \xrightarrow{\sim} \Omega^*(P(r))$$
between the component of arity r *of the Hopf cochain dg-cooperad* $\Omega^*_\sharp(P) \in dg^*\mathcal{H}opf\,\mathcal{O}p^c_{*1}$ *and the image of the space* $P(r)$ *under the Sullivan functor* $\Omega^*(-)$, *for any* $r \in \mathbb{N}$.

The first claim of this theorem follows from the observation that the functor $\mathtt{G}_\bullet(-)$ is strongly symmetric monoidal. The functor $\Omega^*(-)$, on the other hand, is only weakly monoidal. To be more precise, in the case of this functor, we have a Künneth morphism $\nabla : \Omega^*(X) \otimes \Omega^*(Y) \to \Omega^*(X \times Y)$ which is a quasi-isomorphism but not an isomorphism. Hence, for an operad in simplicial sets P, we only get that the composition product $\circ_i$: $P(k) \times P(l) \to P(k+l-1)$ induces a morphism that fits in a zigzag of morphisms of commutative cochain dg-algebras $\Omega^*(P(k+l-1)) \xrightarrow{\circ_i^*} \Omega^*(P(k)\times P(l)) \xleftarrow{\sim} \Omega^*(P(k))\otimes\Omega^*(P(l))$. The idea is to use the adjoint lifting theorem (see for instance [11, §4.5]) to produce the functor of the second claim of the theorem $\Omega^*_\sharp : s\mathcal{S}et\,\mathcal{O}p_{*1}^{op} \to dg^*\,\mathcal{H}opf\,\mathcal{O}p_{*1}^c$ and to fix this problem. Then the crux lies in the verification of the third claim, for which we refer to the cited reference.

For an operad in simplicial sets $P \in s\mathcal{S}et\,\mathcal{O}p_{*1}$, we now set :

$$P^{\mathbb{Q}} := \langle \mathtt{R}\,\Omega^*_\sharp(P)\rangle,$$

where we use the notation $\mathtt{R}\,\Omega^*_\sharp(-)$ for the right derived functor of the functor of the previous theorem $\Omega^*_\sharp : s\mathcal{S}et\,\mathcal{O}p_{*1}^{op} \to dg^*\,\mathcal{H}opf\,\mathcal{O}p_{*1}^c$, and we again use the notation $\langle - \rangle$ for the left derived functor of the Sullivan realization on operads $\mathtt{G}_\bullet : dg^*\,\mathcal{H}opf\,\mathcal{O}p_{*1}^c \to s\mathcal{S}et\,\mathcal{O}p_{*1}^{op}$. The equivalence $\Omega^*_\sharp(P)(r) \sim \Omega^*(P(r))$ implies that we have the following result at the level of this realization:

Theorem 11.2.2 (B. Fresse [26, Theorem II.10.2.1 and Theorem II.12.2.1]). *For any operad $P \in s\mathcal{S}et\,\mathcal{O}p_{*1}$ such that $\mathtt{H}^*(P(r)) = \mathtt{H}^*(P(r), \mathbb{Q})$ forms a finite dimensional $\mathbb{Q}$-module in each arity $r \in \mathbb{N}$ and in each degree $* \in \mathbb{N}$, we have:*

$$P^{\mathbb{Q}}(r) \sim P(r)^{\mathbb{Q}},$$

where we consider the component of arity r of the operad $P^{\mathbb{Q}}$ on the left-hand side and the Sullivan rationalization of the space $P(r)$ on the right-hand side.

For the operad of little n-discs D_n, we now set $\mathtt{R}\,\Omega^*_\sharp(D_n) = \Omega^*_\sharp(E_n)$, where E_n is any cofibrant model of E_n-operad in simplicial sets such that $E_n(0) = E_n(1) = *$, and we still write $D_n^{\mathbb{Q}} = \langle \mathtt{R}\,\Omega^*_\sharp(D_n)\rangle$. To apply the rational homotopy theory to the class of E_n-operads, we aim to determine the model of these objects $\mathtt{R}\,\Omega^*_\sharp(D_n)$.

Recall that we have a homotopy equivalence $D_n(r) \sim F(\mathbb{R}^n, r)$ between the underlying spaces of the operad of little n-discs $D_n(r)$ and the configuration spaces of the Euclidean space $F(\mathbb{R}^n, r)$. (In §11.1.1, we use an equivalent homotopy equivalence $D_n(r) \sim F(\mathring{\mathbb{D}}^n, r)$, where we take the open disc $\mathring{\mathbb{D}}^n \cong \mathbb{R}^n$ rather than the Euclidean spaces $\mathbb{R}^n$.) In a first step, we recall the following result about the cohomology algebras of these spaces:

Theorem 11.2.3 (V.I. Arnold [2], F. Cohen [20]). *Let $n \geq 2$. For each $r \in \mathbb{N}$, the graded commutative algebra $\mathtt{H}^*(D_n(r)) \simeq \mathtt{H}^*(F(\mathbb{R}^n, r))$ has a presentation of the form:*

$$\mathtt{H}^*(F(\mathbb{R}^n, r)) = \frac{\bigwedge(\omega_{ij}, 1 \leq i < j \leq r)}{(\omega_{ij}^2, \omega_{ij}\omega_{ik} - \omega_{ij}\omega_{jk} + \omega_{ik}\omega_{jk})}$$

where the elements ω_{ij} correspond to cohomology classes of degree $n-1$.

In the expression of this theorem, the notation $\bigwedge(-)$ represents the free graded commutative algebra generated by the variables ω_{ij}. The result established by V.I. Arnold in [2] concerns the case $n = 2$ of this statement. The already cited work of F. Cohen [20] gives the general case $n \geq 2$. We also refer to Sinha's survey [51] for a gentle introduction to the computation of this theorem. The classes $\omega_{ij} \in \mathrm{H}^*(F(\mathbb{R}^n, r))$ represent the pullbacks of the fundamental class of the $n-1$-sphere $\omega \in \mathrm{H}^*(\mathbb{S}^{n-1})$ under the maps $\pi_{ij} : F(\mathbb{R}^n, r) \to \mathbb{S}^{n-1}$ such that $\pi_{ij}(a_1, \dots, a_r) = (a_j - a_i)/||a_j - a_i||$. We can also consider unordered pairs $\{i, j\}$ in this definition. We just have $\omega_{ij} = (-1)^n \omega_{ji}$ in this case, since ω_{ij} corresponds to the image of ω_{ji} under the action of the antipodal map on the sphere. In what follows, we refer to these cohomology classes ω_{ij} as the Arnold classes, we refer to the identity $\omega_{ij}\omega_{ik} - \omega_{ij}\omega_{jk} + \omega_{ik}\omega_{jk} = 0$ as the Arnold relation, and we refer to the presentation of the above theorem as the Arnold presentation.

The homology of the operad D_n now inherits the structure of an operad in graded modules. The cohomology $\mathrm{H}^*(D_n)$ inherits a dual cooperad structure, because the homology of the spaces $D_n(r)$ has a finite dimension as a $\mathbb{Q}$-module in each arity and in each degree, so that we have the arity-wise duality relation $\mathrm{H}^*(D_n(r)) = \mathrm{Hom}_{gr\,\mathcal{M}od}(\mathrm{H}_*(D_n(r)), \mathbb{Q})$ in the category of graded modules. Note that we have $D_n(0) = * \Rightarrow \mathrm{H}^*(D_n(0)) = \mathbb{Q}$ and $D_n(1) \sim * \Rightarrow \mathrm{H}^*(D_n(1)) = \mathbb{Q}$, so that the collection $\mathrm{H}^*(D_n) = \{\mathrm{H}^*(D_n(r), r \in \mathbb{N}\}$ satisfies our connectedness condition in the definition of a cooperad. One can easily check that this cooperad structure is compatible with the graded commutative algebra structure of the cohomology, so that the object $\mathrm{H}^*(D_n)$ actually forms a Hopf cooperad in the category of graded modules.

We aim to determine this cooperad structure. We use the following identity, already mentioned in the introduction of this chapter:

Theorem 11.2.4 (F. Cohen [20]). *For $n \geq 2$, we have an isomorphism of operads in graded modules $\mathrm{H}_*(D_n) \simeq \mathsf{Pois}_n$, where Pois_n is the operad that governs the category of n-Poisson algebras.*

Recall that the structure of an n-Poisson algebra refers to a graded version of Poisson structure where we have a commutative product $\mu(x_1, x_2) = x_1 x_2$ of degree 0 and a Poisson bracket $\lambda(x_1, x_2) = [x_1, x_2]$ of degree $n-1$. This Poisson bracket satisfies the symmetry relation $\lambda(x_1, x_2) = (-1)^n \lambda(x_2, x_1)$, a graded version of the Jacobi identity and of the Poisson distribution relation. The n-Poisson operad Pois_n is defined by the corresponding presentation by generators and relations in the category of operads. Equivalently, we can represent an element of the graded module $\mathsf{Pois}_n(r)$ as a Poisson polynomial $\pi = \pi(x_1, \dots, x_r)$ of degree one in each variable x_i.

For our purpose, we actually consider a unitary version of the n-Poisson operad, where we have an extra arity zero operation $e \in \mathsf{Pois}(0)$ such that $\mu \circ_1 e = 1 = \mu \circ_2 e$ and $\lambda \circ_1 e = 0 = \lambda \circ_2 e$. This operation corresponds to a unit in the structure of an n-Poisson algebra and reflects the identity $D_n(0) = *$ at the level of the topological operad D_n.

We get the following result when we pass to the cohomology:

Proposition 11.2.5. *The cohomology algebras $\mathrm{H}^*(D_n(r))$, $r \in \mathbb{N}$, form a Hopf cooperad in graded modules such that $\mathrm{H}^*(D_n) \cong \mathsf{Pois}_n^c$, where Pois_n^c denotes the cooperad dual to Pois_n in graded modules.*

The n-Poisson cooperad Pois^c_n is explicitly defined by taking the dual graded modules of the components of the n-Poisson operad $\mathit{Pois}^c_n(r) = \mathtt{Hom}_{gr\mathcal{M}od}(\mathit{Pois}_n(r), \mathbb{Q})$. We take the adjoint morphisms of the composition products of the n-Poisson operad to provide this collection of graded modules $\mathit{Pois}^c_n(r)$ with a cooperad structure. Therefore, the relation of this proposition $\mathtt{H}^*(D_n) \cong \mathit{Pois}^c_n$ follows from the result of the previous theorem $\mathtt{H}_*(D_n) \cong \mathit{Pois}_n$ and the duality between the homology and the cohomology $\mathtt{H}^*(D_n(r)) = \mathtt{Hom}_{gr\mathcal{M}od}(\mathtt{H}_*(D_n(r)), \mathbb{Q})$.

Let $\langle -, - \rangle : \mathit{Pois}_n(r) \otimes \mathtt{H}^*(D(r)) \to \mathbb{Q}$ denote the duality pairing that we obtain by using this relation $\mathtt{H}_*(D_n) \cong \mathit{Pois}_n$. For a Poisson monomial $\pi(x_1, \dots, x_r) \in \mathit{Pois}_n(r)$, we have the formula:

$$\langle \omega_{ij}, \pi(x_1, \dots, x_r) \rangle = \begin{cases} 1, & \text{if } \pi(x_1, \dots, x_r) = x_1 \cdots [x_i, x_j] \cdots \widehat{x_j} \cdots x_r, \\ 0, & \text{otherwise,} \end{cases}$$

where we consider the generating classes $\omega_{ij} \in \mathtt{H}^*(F(\mathbb{R}^n, r)) = \mathtt{H}^*(D(r))$ of the Arnold presentation of Theorem 11.2.3. This duality relation is immediate in arity 2, because the Poisson bracket operation $\lambda = \lambda(x_1, x_2)$ corresponds to the fundamental class of the $n-1$-sphere in the homology of $D_n(2)$, where we use the relation $D_n(2) \sim F(\mathbb{R}^n, r) \sim \mathbb{S}^{n-1}$. The general formula follows from the fact that the maps $\pi_{ij} : F(\mathbb{R}^n, r) \to F(\mathbb{R}^n, 2)$ in the definition of the classes ω_{ij} correspond to composites with the zero-ary operation $* \in D_n(0)$, which represents our algebra unit $e \in \mathit{Pois}_n(0)$ when we pass to the n-Poisson operad Pois_n. We refer to the paper [51] for a more thorough study of this duality relation between the n-Poisson polynomials and the elements of the cohomology algebras $\mathtt{H}^*(D_n(r)) = \mathtt{H}^*(F(\mathbb{R}^n, r))$ in the Arnold presentation.

We can now regard the object $\mathit{Pois}^c_n \simeq \mathtt{H}^*(D_n)$ as a Hopf cochain dg-cooperad equipped with a trivial differential. We have to make explicit a cofibrant resolution of this object for the applications of our methods of the rational homotopy theory of operads. In the next paragraphs, we explain a first definition of such a resolution by using graded analogues of the Drinfeld–Kohno Lie algebra operad of the previous section.

11.2.3 The graded Drinfeld–Kohno Lie algebra operads and the associated Chevalley–Eilenberg cochain complexes

The graded analogues of the Drinfeld–Kohno Lie algebra operad, which we define for every value of the parameter $n \geq 2$, are denoted by $\mathfrak{p}_n$. The ungraded Drinfeld–Kohno Lie algebra operad of §11.1.4 corresponds to the case $n = 2$. Thus, we have $\mathfrak{p} = \mathfrak{p}_2$ with the notation of §11.1.4.

To define the Lie algebras $\mathfrak{p}_n(r)$, we use the same presentation as in Eqn. 11.1.21:

$$\mathfrak{p}_n(r) = \mathbb{L}(t_{ij}, \{i,j\} \subset \{1, \dots, r\}) / < [t_{ij}, t_{kl}], [t_{ij}, t_{ik} + t_{jk}] >, \tag{11.2.8}$$

but we now take $\deg(t_{ij}) = n-2$ and we assume the graded symmetry relation $t_{ji} = (-1)^n t_{ij}$, for every pair $\{i,j\} \subset \{1, \dots, r\}$. Then we take the same construction as in §11.1.4 to provide these Lie algebras with an action of the symmetric groups and with additive composition products $\circ_i : \mathfrak{p}_n(k) \oplus \mathfrak{p}_n(l) \to \mathfrak{p}_n(k+l-1)$, so that the collection $\mathfrak{p}_n = \{\mathfrak{p}_n(r), r \in \mathbb{N}\}$ inherits the structure of an operad in the category of graded Lie algebras.

Note that the graded Lie algebras $\mathfrak{p}_n(r)$ still inherit a weight grading from the free Lie algebra, and hence, form weight graded objects in the category of graded modules. Besides, we can form a completed version of the operads $\hat{\mathfrak{p}}_n$, as in the case $n = 2$ in §11.1.4, but for $n \geq 3$, we trivially have $\hat{\mathfrak{p}}_n = \mathfrak{p}_n$ because the components of homogeneous weight $m \geq 1$ of the Lie algebras $\mathfrak{p}_n(r)$ are concentrated in a single degree $* = m(n-2)$ and we have $m(n-2) \to \infty$ when $n \geq 3$.

We consider the Chevalley–Eilenberg cochain complexes $\mathrm{C}^*_{CE}(\hat{\mathfrak{g}})$ associated to the complete Lie algebras $\hat{\mathfrak{g}} = \hat{\mathfrak{p}}_n(r)$. The cofibrant objects of the category of commutative cochain dg-algebras are retracts of dg-algebras of the form $R = (\mathbb{S}(V), \partial)$, where $\mathbb{S}(V)$ is the symmetric algebra on an upper graded dg-module V equipped with a filtration $\mathrm{F}_1 V \subset \mathrm{F}_2 V \subset \cdots \supset \mathrm{F}_m V \subset \cdots \subset V$ and where we have a differential ∂ such that $\partial(\mathrm{F}_m V) \subset \mathbb{S}(\mathrm{F}_{m-1} V)$. The Chevalley–Eilenberg cochain complex is precisely defined by an expression of this form $\mathrm{C}^*_{CE}(\mathfrak{g}) = (\mathbb{S}(\mathbb{Q}[-1] \otimes \hat{\mathfrak{g}}^\vee), \partial)$, where $\mathbb{Q}[-1] = \mathbb{Q}$ denotes the graded module generated by a single element in lower degree -1 (equivalently, in upper degree one) and $\hat{\mathfrak{g}}^\vee$ denotes the (continuous) dual of the completed Lie algebra $\hat{\mathfrak{g}} = \hat{\mathfrak{p}}_n(r)$. The differential ∂ is induced by the dual map of the Lie bracket $[-,-]$ on $\hat{\mathfrak{g}}$.

The commutative cochain dg-algebras $\mathrm{C}^*_{CE}(\hat{\mathfrak{p}}_n(r))$ inherit an action of the symmetric groups by functoriality of the Chevalley–Eilenberg cochain complex, as well as composition coproducts $\circ_i^* : \mathrm{C}^*_{CE}(\hat{\mathfrak{p}}_n(k+l-1)) \to \mathrm{C}^*_{CE}(\hat{\mathfrak{p}}_n(k)) \otimes \mathrm{C}^*_{CE}(\hat{\mathfrak{p}}_n(l))$, which are given by the composites of the morphisms $\circ_i^* : \mathrm{C}^*_{CE}(\hat{\mathfrak{p}}_n(k+l-1)) \to \mathrm{C}^*_{CE}(\hat{\mathfrak{p}}_n(k) \oplus \hat{\mathfrak{p}}_n(l))$ induced by the composition products of the operad $\mathfrak{p}_n$ with the Künneth isomorphisms $\mathrm{C}^*_{CE}(\hat{\mathfrak{p}}_n(k) \oplus \hat{\mathfrak{p}}_n(l)) \simeq \mathrm{C}^*_{CE}(\hat{\mathfrak{p}}_n(k)) \otimes \mathrm{C}^*_{CE}(\hat{\mathfrak{p}}_n(l))$. Hence, we get that the collection $\mathrm{C}^*_{CE}(\hat{\mathfrak{p}}_n) = \{\mathrm{C}^*_{CE}(\hat{\mathfrak{p}}_n(r)), r \in \mathbb{N}\}$ inherits the structure of a Hopf cochain dg-cooperad. In addition, one can prove that this Hopf cochain dg-cooperad is cofibrant (see [26, Theorem II.14.1.7]).

Then we have the following statement:

Theorem 11.2.6 (T. Kohno [40]). *We have a quasi-isomorphism of commutative cochain dg-algebras*

$$\mathrm{C}^*_{CE}(\hat{\mathfrak{p}}_n(r)) \xrightarrow{\sim} \mathrm{H}^*(F(\mathring{\mathbb{D}}^n, r))$$

such that $t_{ij}^\vee \mapsto \omega_{ij}$ for each pair $\{i,j\} \subset \{1,\dots,r\}$ and $p^\vee \mapsto 0$ when $p^\vee$ is the dual basis element of a homogeneous Lie polynomial $p \in \mathfrak{p}_n(r)_m$ of weight $m > 1$.

The cited reference gives the case $n = 2$ of this statement. The general result can be deduced from the observation that the cohomology algebra $\mathrm{H}^*(F(\mathring{\mathbb{D}}^n, r)$ forms a Koszul algebra with the enveloping algebra of the Lie algebra $\mathfrak{p}_n(r)$ as dual associative algebra. We refer to [26, Theorem II.14.1.14] for further explanations on this approach.

Now we can easily check that the quasi-isomorphisms of this theorem preserve cooperad structures. Hence, we get the following statement:

Proposition 11.2.7. *The quasi-isomorphisms of Theorem 11.2.6 define a weak equivalence of Hopf cochain dg-cooperads*

$$\mathrm{C}^*_{CE}(\hat{\mathfrak{p}}_n) \xrightarrow{\sim} \mathrm{H}^*(D_n) = \mathit{Pois}_n^c,$$

where we regard the cohomology of the little n-discs operad $\mathrm{H}^*(D_n)$ as a Hopf cochain dg-cooperad equipped with a trivial differential.

We deduce from this proposition that the object $\mathrm{C}^*_{CE}(\hat{\mathfrak{p}}_n)$ defines a cofibrant resolution of the object $\mathsf{Pois}^c_n = \mathrm{H}^*(D_n)$ in the category of Hopf cochain dg-cooperads. In our constructions, we actually consider a second cofibrant resolution, which is given by a Hopf cochain dg-cooperad of graphs Graphs^c_n, and we explain the definition of this object in the next paragraph.

11.2.4 The graph cooperad

The Hopf cochain dg-cooperad Graphs^c_n precisely consists of graphs $\gamma \in \mathsf{Graphs}^c_n(r)$ with unnumbered internal vertices $\bullet$ and external vertices indexed by $1, \dots, r$, as in the following picture:

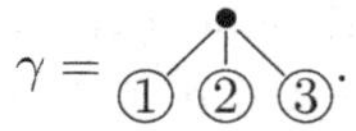

$$\gamma = \text{①} \; \text{②} \; \text{③}. \tag{11.2.9}$$

The degree of such a graph is determined by assuming that each internal vertex $\bullet$ contributes to the degree by $\deg(\bullet) = n$ and that each edge contributes to the degree by $\deg(-) = 1-n$ (in the lower grading convention). Thus, we have $\deg(\gamma) = (1-n)v + ne$ (in the lower grading convention again), where v denotes the number of internal vertices and e denotes the number of edges in the graph $\gamma \in \mathsf{Graphs}^c_n(r)$. In fact, we can regard our graphs as tensor products of symbolic elements given by the internal vertices and by the edges of our objects. In particular, we assume that graphs equipped with odd symmetries vanish in $\mathsf{Graphs}^c_n(r)$. We also assume that each edge is oriented and that a reversal of orientation is equivalent to the multiplication by a sign $(-1)^n$ in $\mathsf{Graphs}^c_n(r)$. For our purpose, we allow graphs with double edges, but not loops (edges with the same origin and endpoint) and we assume that each internal vertex is at least trivalent though the latter conditions are not essential. Besides, we assume that each connected component of our graph contains at least one external vertex.

The differential of graphs is defined by contracting edges in order to merge internal vertices together or in order to merge internal vertices with external vertices, as shown schematically in the following picture:

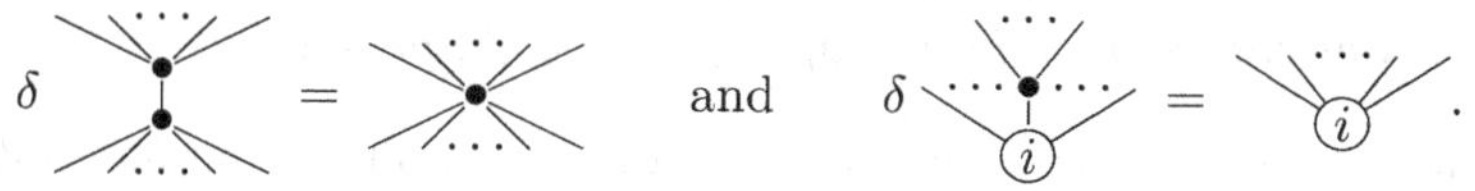

$$\delta \; \cdots \; = \; \cdots \quad \text{and} \quad \delta \; \cdots \text{ⓘ} \cdots \; = \; \cdots \text{ⓘ} \cdots. \tag{11.2.10}$$

For instance, we have the formula:

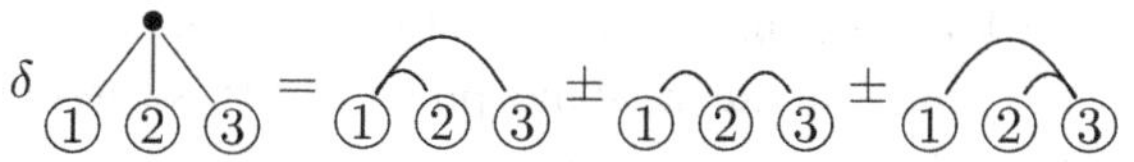

$$\delta \; \text{①} \; \text{②} \; \text{③} = \text{①} \; \text{②} \; \text{③} \pm \text{①} \; \text{②} \; \text{③} \pm \text{①} \; \text{②} \; \text{③} \tag{11.2.11}$$

in $\mathsf{Graphs}^c_n(3)$. The product is given by the amalgamated sum of graphs along external vertices. For instance, we have the formula:

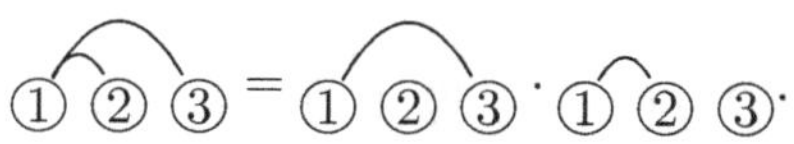

$$\text{①} \; \text{②} \; \text{③} = \text{①} \; \text{②} \; \text{③} \cdot \text{①} \; \text{②} \; \text{③}. \tag{11.2.12}$$

The cooperad coproduct $\circ_i^* : \mathsf{Graphs}_n^c(k+l-1) \to \mathsf{Graphs}_n^c(k) \otimes \mathsf{Graphs}_n^c(l)$, where we fix $k,l \in \mathbb{N}$, $i \in \{1,\dots,k\}$, has the form $\circ_i^*(\gamma) = \sum_{\alpha\subset\gamma} \gamma/\alpha \otimes \alpha$, where the sum runs over all the subgraphs $\alpha \subset \gamma$ that contain the external vertices indexed by $i,\dots,i+l-1$, and γ/α denotes the graph obtained by collapsing this subgraph to a single external vertex (which we index by i in the result of the operation, while we shift the index of the vertices such that $j > i$ by $j \mapsto j-l+1$). Note that we have $\mathsf{Graphs}_n^c(1) \neq \mathbb{Q}$ in general, so that our object Graphs_n^c belongs to the extended category of Hopf cochain dg-cooperads $dg^*\,\mathcal{H}opf\,\mathcal{O}p_{*N}^c$ but not to the category of connected Hopf cochain dg-cooperads $dg^*\,\mathcal{H}opf\,\mathcal{O}p_{*1}^c$.

We easily see that the commutative cochain dg-algebras of graphs defined in this paragraph $\mathsf{Graphs}_n^c(r)$ have a structure of the form $\mathsf{Graphs}_n^c(r) = (\mathbb{S}(\mathbb{Q}[-1] \otimes \mathsf{ICGraphs}_n^c(r)), \partial)$ (like the Chevalley–Eilenberg cochain dg-algebras of the previous paragraph), where $\mathsf{ICGraphs}_n^c(r)$ is a complex of graphs which are connected when we remove the external vertices inside $\mathsf{Graphs}_n^c(r)$. (In what follows, we refer to such graphs as internally connected graphs.) We just perform an extra degree shift in the definition of this complex of internally connected graphs in order to get a $\mathbb{Q}[-1]$ factor on the generating dg-module of our symmetric algebra (as in the definition of the Chevalley–Eilenberg cochain complex of a Lie dg-algebra). We can actually use this expression to identify the object $\mathsf{ICGraphs}_n^c(r)$ with the dual of an L_∞-algebra (a strongly homotopy Lie algebra). We can use this symmetric algebra structure $\mathsf{Graphs}_n^c(r) = (\mathbb{S}(\mathbb{Q}[-1] \otimes \mathsf{ICGraphs}_n^c(r)), \partial)$ to prove that Graphs_n^c forms a cofibrant object in the category $dg^*\,\mathcal{H}opf\,\mathcal{O}p_{*N}^c$, and we also have the following proposition:

Proposition 11.2.8 (M. Kontsevich [43]). *We have a quasi-isomorphism of Hopf dg-cooperads*

$$\mathsf{Graphs}_n^c \xrightarrow{\sim} \mathrm{H}^*(D_n) = \mathsf{Pois}_n^c$$

that carries the graph $\gamma_{ij} \in \mathsf{Graphs}_n^c(r)$ with a single edge $\overset{\frown}{(i)\,(j)}$ to the Arnold class ω_{ij} and that cancel the internally connected graphs with a nonempty set of internal vertices.

The assignment of this proposition determines the map $\mathsf{Graphs}_n^c(r) \xrightarrow{\sim} \mathrm{H}^*(D_n(r))$ as a morphism of graded commutative algebras since the internally connected graphs generate the object $\mathsf{Graphs}_n^c(r)$ as a graded commutative algebra. (Note that the graphs γ_{ij} of the proposition represent the internally connected graphs with an empty set of internal vertices.) We just check that this map preserves differentials (and hence, gives a well-defined morphism of commutative cochain dg-algebras in each arity $r \in \mathbb{N}$), as well as the cooperad structures, so that our collection of maps define a morphism of Hopf dg-cooperads. We refer to the cited reference [43] and to [46] for a proof that this morphism defines a quasi-isomorphism. Observe simply that the differential identity of Eqn. 11.2.11 is carried to the Arnold relation in $\mathrm{H}^*(D_n(3))$.

Recall that we set $\mathrm{R}\,\Omega_\sharp^*(D_n) = \Omega_\sharp^*(E_n)$ for the topological operad of little n-discs D_n, where E_n is any cofibrant model of E_n-operad in simplicial sets such that $E_n(0) = E_n(1) = *$. We have the following result:

Theorem 11.2.9 (B. Fresse and T. Willwacher [29, Theorem A']). *We have the relation*

$$\mathsf{Pois}_n^c \sim \mathrm{R}\,\Omega_\sharp^*(D_n),$$

in the category of Hopf cochain dg-cooperads.

This theorem asserts that the operad of little n-discs is formal in the sense of our operadic counterpart of the Sullivan rational homotopy theory of spaces. The cited reference [29, Theorem A'] proves an intrinsic formality theorem which implies this operadic formality result in the case $n \geq 3$. (In the next statement, we will explain that the case $n = 2$ of this theorem follows from the existence of Drinfeld's associators.)

The result of this theorem can also be deduced from Kontsevich's proof of the formality of E_n-operads when we pass to real coefficients (see [43]). Indeed, the construction of Kontsevich can be used to define a collection of quasi-isomorphisms $\mathsf{Graphs}^c_n(r) \xrightarrow{\sim} \Omega^*_{sa}(\mathsf{FM}_n(r))$, where FM_n is a model of E_n-operad given a real oriented analogue of the Fulton-MacPherson compactification of the configuration spaces (see [33]) and $\Omega^*_{sa}(-)$ denotes a cochain dg-algebra functor of semi-algebraic forms (see [36, 46]). One can observe that these morphisms can be associated to a strict morphism of Hopf cochain dg-cooperads $\mathsf{Graphs}^c_n \xrightarrow{\sim} \Omega^*_\sharp(\mathsf{FM}_n)$ (by using a general coherence statement of [26, Proposition II.12.1.3]).

The approach of the cited reference [29] does not use this construction and gives a formality quasi-isomorphism that is defined over the rationals by using obstruction theory methods. The claim of this reference is that the E_n-operads are intrinsically rationally formal for $n \geq 3$ in the sense that every Hopf cochain dg-cooperad A_n that satisfies $\mathtt{H}_*(\mathsf{A}_n) \simeq \mathsf{Pois}^c_n$ and is equipped with an extra-involution operad $J : \mathsf{A}_n \to \mathsf{A}_n$ such that $J(\lambda) = -\lambda$ in the case $4|n$ satisfies $\mathsf{A}_n \sim \mathsf{Pois}^c_n$. We apply this claim to the Hopf cochain dg-cooperad $\mathsf{A}_n = \mathtt{R}\,\Omega^*_\sharp(\mathsf{D}_n)$ to get the statement of the theorem. We have an extension of this formality result for the morphisms $\mathsf{D}_m \to \mathsf{D}_n$ that link the operads of little discs when $n - m \geq 2$ (see [29, Theorem C]).

Recall that we set

$$\mathsf{D}^{\mathbb{Q}}_n := \langle \mathtt{R}\,\Omega^*_\sharp(\mathsf{D}_n)\rangle$$

to define a model for the rationalization of the little n-discs operad in topological spaces. The result of the previous theorem has the following corollary:

Corollary 11.2.10. *We have $\mathsf{D}^{\mathbb{Q}}_n = \langle \mathsf{Pois}^c_n\rangle$, for any $n \geq 2$, where we consider the image of the dual cooperad of the n-Poisson operad Pois^c_n under the operadic upgrading of the Sullivan realization functor $\langle - \rangle$.*

We just use the implication $\mathsf{Pois}^c_n \sim \mathtt{R}\,\Omega^*_\sharp(\mathsf{D}_n) \Rightarrow \langle \mathsf{Pois}^c_n\rangle \sim \langle \mathtt{R}\,\Omega^*_\sharp(\mathsf{D}_n)\rangle$ to get the result of this corollary. This result, together with the observations of Proposition 11.2.7 and Proposition 11.2.8, implies that we can take either $\langle \mathsf{Pois}^c_n\rangle = \mathtt{G}_\bullet(\mathtt{C}^*_{CE}(\hat{\mathfrak{p}}_n))$ or $\langle \mathsf{Pois}^c_n\rangle = \mathtt{G}_\bullet(\mathsf{Graphs}^c_n)$ to get a model of the rationalization $\mathsf{D}^{\mathbb{Q}}_n$.

We have an identity $\mathtt{G}_\bullet(\mathtt{C}^*_{CE}(\hat{\mathfrak{p}}_n(r))) = \mathtt{MC}_\bullet(\hat{\mathfrak{p}}_n(r))$, for each $r \in \mathbb{N}$, where we consider a Maurer–Cartan space associated to the complete Lie algebra $\hat{\mathfrak{p}}_n(r)$ (we review the definition of this construction in the next sections). Hence, the results of this section gives a simple algebraic model of the rational homotopy type of E_n-operads. In the case $n = 2$, we have an identity $\mathtt{C}^*_{CE}(\hat{\mathfrak{p}}) \sim \mathtt{B}(\widehat{\mathsf{CD}_{\mathbb{Q}}})$, where we consider the chord diagram operad of §11.1.5 (recall also that we use the notation $\hat{\mathfrak{p}} = \hat{\mathfrak{p}}_2$ for the ungraded Drinfeld–Kohno Lie algebra operad which occurs in this case $n = 2$). Thus, since we have on the other hand $\mathsf{D}^{\mathbb{Q}}_2 = \mathtt{B}(\widehat{\mathsf{PaB}_{\mathbb{Q}}})$ (see §11.1), we can deduce the existence of a weak equivalence $\mathsf{D}^{\mathbb{Q}}_2 \sim \langle \mathsf{Pois}^c_n\rangle$ from the operadic interpretation of Drinfeld's associators given in §11.1.5 (see [26, §II.14.2]).

We now examine a counterpart of the formality result of Theorem 11.2.9 in the category of dg-modules $dg^* \mathcal{M}od$. We use the notation $\mathtt{C}_*(-)$ for both the singular complex functor from the category of topological spaces to the category of dg-modules and for the standard normalized chain complex functor on simplicial sets. These functors are lax symmetric monoidal and therefore carry operads in topological spaces (respectively, in simplicial sets) to operads in dg-modules. Furthermore, in the case of a cofibrant operad in simplicial sets R, we have the duality relation $\Omega^*_\sharp(R)^\vee \sim \mathtt{C}_*(R)$ in the category of dg-operads $dg\,\mathcal{O}p_*$ when we consider the dual in dg-modules of the Hopf cochain dg-cooperad $\Omega^*_\sharp(R)$ of Theorem 11.2.9. Therefore, the result of Theorem 11.2.9 implies the following statement, which was also obtained by the authors cited in this statement by other method:

Theorem 11.2.11 (D. Tamarkin [54], M. Kontsevich [43]). *We have the relation*

$$\mathsf{Pois}_n \sim \mathtt{C}_*(D_n),$$

in the category of dg-operads.

This result is exactly the formality theorem mentioned in the introduction of this chapter for the class of E_n-operads in dg-modules. Tamarkin's proof of this theorem, which works in the case $n = 2$, relies on the correspondence between formality equivalences and associators, whereas Kontsevich's proof, which works for every $n \geq 2$ but requires to pass to real coefficients, relies on the definition of semi-algebraic forms associated to graphs (as we explain in our survey of Theorem 11.2.9). In [9], Boavida and Horel have given a new proof of the formality result of this theorem by using a generalization of classical formality criterion of mixed Hodge theory in the context of operads (see also [48] for an application of this approach in the case $n = 2$).

11.3 The rational homotopy of mapping spaces on the operads of little discs

We now tackle the main objective of this chapter, namely the computation of the homotopy of the mapping spaces $\mathtt{Map}^h_{\mathcal{T}op\,\mathcal{O}p}(D_m, D^{\mathbb{Q}}_n)$ and of the homotopy automorphism spaces $\mathtt{Aut}^h_{\mathcal{O}p}(D^{\mathbb{Q}}_n)$, for all $n \geq 2$. Thus, we aim to generalize the computation carried out in §11.1 in the case of the automorphism space $\mathtt{Aut}^h_{\mathcal{O}p}(D^{\mathbb{Q}}_2)$. In the case of the mapping spaces $\mathtt{Map}^h_{\mathcal{T}op\,\mathcal{O}p}(D_m, D^{\mathbb{Q}}_n)$, we are also going to check that we have the relation $\mathtt{Map}^h_{\mathcal{T}op\,\mathcal{O}p}(D_m, D^{\mathbb{Q}}_n) \sim \mathtt{Map}^h_{\mathcal{T}op\,\mathcal{O}p}(D_m, D_n)^{\mathbb{Q}}$ when $n - m \geq 3$, so that the results explained in this section gives a full computation of the rational homotopy of the mapping spaces $\mathtt{Map}^h_{\mathcal{T}op\,\mathcal{O}p}(D_m, D_n)$ that occur in the operadic description of the introduction for the embedding spaces $\overline{\mathtt{Emb}}_c(\mathbb{R}^m, \mathbb{R}^n)$.

To carry out these computations, we use the graph complex model Graphs^c_n of the rational homotopy of the operad D_n. Hence, we naturally obtain, as a main outcome, a graph complex description of the homotopy type of the spaces $\mathtt{Map}^h_{\mathcal{T}op\,\mathcal{O}p}(D_m, D^{\mathbb{Q}}_n)$ and $\mathtt{Aut}^h_{\mathcal{O}p}(D^{\mathbb{Q}}_n)$. In the case of the mapping spaces $\mathtt{Map}^h_{\mathcal{T}op\,\mathcal{O}p}(D_m, D^{\mathbb{Q}}_n)$, we express the result as the Maurer–Cartan space $\mathtt{MC}_\bullet(HGC_{mn})$ associated to a Lie dg-algebra of hairy graphs HGC_{mn}. We explain the definition of this object first and we explain our computation afterwards.

In the case of the automorphism spaces $\mathtt{Aut}^h_{\mathcal{O}p}(D_n^{\mathbb{Q}})$, we get that our object is homotopy equivalent to a Cartesian product of Eilenberg–Mac Lane spaces (like any H-group in rational homotopy theory). Thus, we can focus on the computation of the homotopy groups in this case. We give a description of these groups in terms of the homology of a non-hairy graph complex GC_n of which we also explain the definition beforehand. This graph complex GC_n is a graded version of a complex introduced by Kontsevich in [41], and therefore, this complex is usually called the Kontsevich graph complex in the literature.

11.3.1 The hairy graph complex

The hairy graph complex HGC_{mn} explicitly consists of formal series of connected graphs with internal vertices $\bullet$, internal edges $\bullet\!\!-\!\!\bullet$, which link internal vertices together, and external edges $\bullet\!-$ (the hairs), which are open at one extremity, as in the following examples:

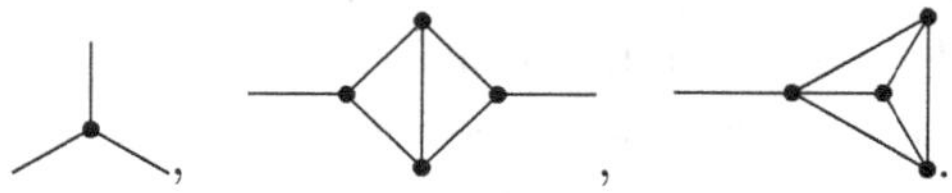

(11.3.1)

This complex HGC_{mn} is equipped with a lower grading. The degree of a graph $\gamma \in HGC_{mn}$ is determined by assuming that each vertex contributes by $\deg(\bullet) = n$, that each internal edge contributes by $\deg(\bullet\!\!-\!\!\bullet) = 1-n$, that each hair contributes by $\deg(\bullet\!-) = m-n+1$, and by adding a global degree shift by $-m$. Thus, we have $\deg(\gamma) = nv+(1-n)e+(m-n+1)h-m$, where v denotes the number of internal vertices, the letter e denotes the number of internal edges and h denotes the number of hairs of the graph $\gamma \in HGC_{mn}$. The differential of the hairy graph complex is defined by the blow-up of internal vertices:

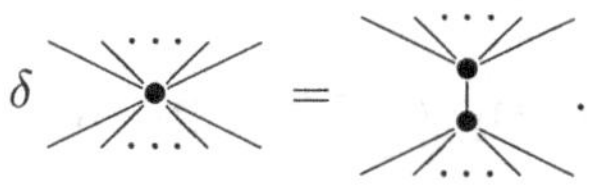

$$\delta \; (\text{vertex}) = (\text{two vertices joined by an edge}) . \tag{11.3.2}$$

We equip the hairy graph complex with the Lie bracket such that:

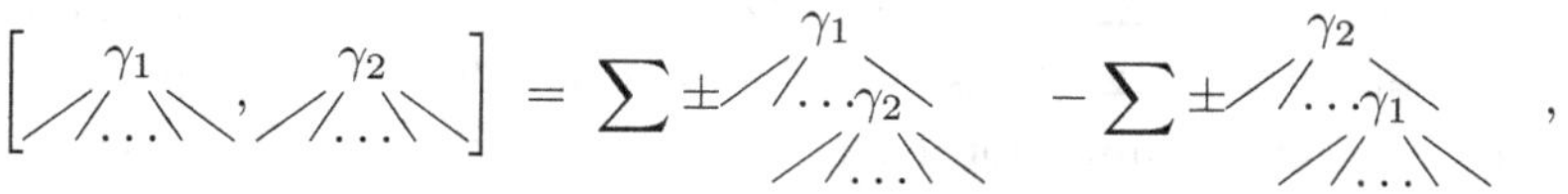

$$[\gamma_1, \gamma_2] = \sum \pm \gamma_1 \circ \gamma_2 - \sum \pm \gamma_2 \circ \gamma_1 , \tag{11.3.3}$$

where the first sum runs over the re-connections of a hair of the graph γ_1 to a vertex of the graph γ_2, and similarly in the second sum, with the role of the graphs γ_1 and γ_2 exchanged. In the case $m = 1$, we have to consider a deformation of this Lie dg-algebra structure which we call the Shoikhet L_∞-structure (a strongly homotopy Lie algebra). We just refer to [59] for the explicit definition of this structure.

In the next theorem, we consider the Maurer–Cartan space $\mathtt{MC}_\bullet(L)$ associated to the Lie dg-algebra $L = HGC_{m,n}$. This simplicial set $\mathtt{MC}_\bullet(L)$ is defined by the sets of flat L-valued PL connections on the simplices Δ^n, $n \in \mathbb{N}$. To be more precise, in the definition of this object $\mathtt{MC}_\bullet(L)$, we generally assume that L forms a complete Lie dg-algebra with respect to a filtration $L = \mathtt{F}_1 L \supset \mathtt{F}_2 L \supset \cdots \supset \mathtt{F}_k L \supset \cdots$ such that $[\mathtt{F}_k L, \mathtt{F}_l L] \subset \mathtt{F}_{k+l} L$. In the case $L = HGC_{mn}$, we assume that $\mathtt{F}_k L = \mathtt{F}_k HGC_{mn}$ consists of power series of graphs $\gamma \in \mathtt{H}_{mn}$ such that $e - v \geq k$, where e denotes the number of edges and v denotes the number of internal vertices in γ. Then we explicitly set:

$$\mathtt{MC}_n(L) = \{\omega \in (L \hat{\otimes} \Omega^*(\Delta^n))^1 \mid \delta(\omega) + \frac{1}{2}[\omega, \omega] = 0\}, \tag{11.3.4}$$

for every simplicial dimension $n \in \mathbb{N}$, where $(L \hat{\otimes} \Omega^*(\Delta^n))^1$ denotes the component of upper degree 1 in the completed tensor product of the Lie dg-algebra L with the Sullivan cochain dg-algebra of PL forms $\Omega^*(\Delta^n)$. The face and degeneracy operators of this simplicial set are inherited from the simplices. This construction has a natural extension for L_∞-algebras (see for instance [32]). We now have the following main result:

Theorem 11.3.1 (B. Fresse, V. Turchin, and T. Willwacher [28, Theorem 1]). *For any $n \geq m \geq 2$, we have the relation:*

$$\mathtt{Map}^h_{\mathcal{T}op\,\mathcal{O}p}(D_m, D_n^{\mathbb{Q}}) \sim \mathtt{MC}_\bullet(HGC_{mn}),$$

where HGC_{mn} is the hairy graph complex. This relation extends to the case $n > m = 1$ when we equip HGC_{1n} equipped with the Shoikhet L_∞-structure.

The results of the previous section imply that we have the following weak-equivalences:

$$\begin{aligned} \mathtt{Map}^h_{\mathcal{T}op\,\mathcal{O}p}(D_m, D_n^{\mathbb{Q}}) &\sim \mathtt{Map}^h_{dg^*\,\mathcal{H}opf\,\mathcal{O}p^c_{*1}}(\mathrm{R}\,\Omega^*_\sharp(D_n), \mathrm{R}\,\Omega^*_\sharp(D_m)) && (11.3.5) \\ &\sim \mathtt{Map}^h_{dg^*\,\mathcal{H}opf\,\mathcal{O}p^c_{*1}}(\mathsf{Pois}^c_n, \mathsf{Pois}^c_m), && (11.3.6) \end{aligned}$$

where $\mathtt{Map}^h_{dg^*\,\mathcal{H}opf\,\mathcal{O}p^c_{*1}}(-,-)$ denote a derived mapping space bifunctor in the category of Hopf cochain dg-cooperads, we use the Quillen adjunction between the functors $\mathtt{G}_\bullet(-)$ and $\Omega^*_\sharp(-)$ in the first equivalence (Equation 11.3.5) and the formality result of Theorem 11.2.9 in the second equivalence (Equation 11.3.6). These equivalences reduce the proof of Theorem 11.3.1 to a problem of algebra.

In order to compute the derived mapping space of Hopf cochain dg-cooperads $\mathtt{Map}^h_{dg^*\,\mathcal{H}opf\,\mathcal{O}p^c_{*1}}(\mathsf{Pois}^c_n, \mathsf{Pois}^c_m)$, we need to pick a cofibrant resolution of the object Pois^c_n on the source and a fibrant resolution of the object Pois^c_m on the target. For this purpose, we take the cofibrant Hopf cochain dg-cooperad $R_n = \mathtt{C}^*_{CE}(\hat{\mathfrak{p}}_n)$ (see §11.2.3) and we adapt the classical Boardman–Vogt W-construction of operads to define a natural fibrant resolution functor $\mathtt{W}^c(-)$ on the category of Hopf cochain dg-cooperad (see [28, §5] for a detailed definition of this functor). By analyzing the definition of maps on these Hopf cochain dg-cooperads, one sees that the mapping space $\mathtt{Map}^h_{dg^*\,\mathcal{H}opf\,\mathcal{O}p^c_{*1}}(\mathsf{Pois}^c_n, \mathsf{Pois}^c_m)$ is weakly equivalent to the Maurer–Cartan space associated to an L_∞-algebra of biderivations $\mathtt{BiDer}(\mathtt{C}^*_{CE}(\hat{\mathfrak{p}}_n), \mathtt{W}^c(\mathsf{Pois}^c_m))$ (see [28, §6]).

The object $\mathtt{C}^*_{CE}(\hat{\mathfrak{p}}_n)$ in this complex of biderivations can be replaced by the graph cooperad model of the n-Poisson cooperad $\mathsf{Graphs}^c_n \sim \mathsf{Pois}^c_n$. The connection of the derived mapping space $\mathtt{Map}^h_{dg^*\,\mathcal{H}opf\,\mathcal{O}p^c_{*1}}(\mathsf{Pois}^c_n, \mathsf{Pois}^c_m)$ with the hairy graph complex of the theorem comes from an ultimate reduction of this L_∞-algebra of biderivations, which yields a relation of the form

$$HGC_{mn} \sim \mathtt{BiDer}(\mathit{Graphs}_n^c, \mathtt{W}^c(\mathit{Pois}_m^c)) \tag{11.3.7}$$

in the category of L_∞-algebras (see [28, §8]).

The result of this theorem has the following corollary:

Corollary 11.3.2. *For any $n \geq m \geq 2$ (and for $n > m = 1$), we have the identity:*

$$\pi_*(\mathtt{Map}^h_{\mathcal{T}op\,\mathcal{O}p}(D_m, D_n^{\mathbb{Q}}), \omega) = H_{*-1}(HGC_{mn}^\omega),$$

for any $\omega \in \mathtt{MC}_0(HGC_{mn})$, where HGC_{mn}^ω is the complex HGC_{mn} equipped with the twisted differential $\delta_\omega = \delta + [\omega, -] + $ (extra terms in the L_∞-case).

The identity of this statement follows from the result of Theorem 11.3.1 and from a general result about the homotopy groups of Maurer–Cartan spaces $\mathtt{MC}_\bullet(L)$ for which we refer to [6].

A computation of the rational homotopy groups of the embedding spaces $\overline{\mathtt{Emb}}_c(\mathbb{R}^m, \mathbb{R}^n)$, analogous to the result established in this corollary, is given in [4] (see also [44] for the case $m = 1$ of these computations). These previous computations are based on the interpretation in terms of mapping spaces of operadic bimodules of the Goodwillie–Weiss tower of the embedding spaces $\overline{\mathtt{Emb}}_c(\mathbb{R}^m, \mathbb{R}^n)$ (or of the equivalent interpretation of the Goodwillie–Weiss tower in terms of Sinha's cosimplicial model in the case $m = 1$). In [45], the formality of E_n-operads in chain complexes is also used to get a description of the homology of the embedding spaces $\overline{\mathtt{Emb}}_c(\mathbb{R}^1, \mathbb{R}^n)$ in terms of a Hochschild cohomology theory for operads (we apply this Hochschild cohomology theory to the n-Poisson operad). The graph operad model of the n-Poisson can also be used to deduce a graph complex model of the homology of the embedding space $\overline{\mathtt{Emb}}_c(\mathbb{R}^1, \mathbb{R}^n)$ from this algebraic approach.

In fact, we can use the result of the above corollary and the equivalence between the embedding space $\overline{\mathtt{Emb}}_c(\mathbb{R}^m, \mathbb{R}^n)$ and the $m + 1$-fold iterated loop space of the operadic mapping space $\mathtt{Map}^h_{\mathcal{T}op\,\mathcal{O}p}(D_m, D_n)$ given in the introduction to get applications of the result of Theorem 11.3.1 in the theory embedding spaces. For this purpose, we also use the following theorem:

Theorem 11.3.3 (B. Fresse, V. Turchin, and T. Willwacher [28, Theorem 15]). *In the case $n - m \geq 3$, the space $\mathtt{Map}^h_{\mathcal{T}op\,\mathcal{O}p}(D_m, D_n)$ is $n - m - 1$ connected, and we moreover have the relation:*

$$\mathtt{Map}^h_{\mathcal{T}op\,\mathcal{O}p}(D_m, D_n)^{\mathbb{Q}} \sim \mathtt{Map}_{\mathcal{T}op\,\mathcal{O}p}(D_m, D_n^{\mathbb{Q}})$$

in the homotopy category of spaces.

We refer to the cited reference [28, §10] for the detailed proof of this statement, which relies on an analogous result for spaces, established by Haefliger in [35].

We examine the rational homotopy of the spaces of homotopy automorphisms to complete the result of this section. We first explain the definition of the Kontsevich graph complexes GC_n which occur in this computation.

11.3.2 The Kontsevich graph complex

The definition of the complex GC_n is the same as the definition of the hairy graph complex HGC_{mn}, except that we now consider graphs without hairs, as in the following examples:

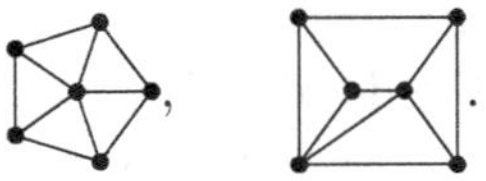

$$[\text{graph}], \quad [\text{graph}]. \tag{11.3.8}$$

We determine the degree of a graph in GC_n by assuming that each vertex contributes by $\deg(\bullet) = n$ and each edge contributes by $\deg(\bullet\!\!-\!\!\bullet) = 1-n$ as in the case of hairy graphs. We still assume that every vertex of a graph in GC_n is at least trivalent and we do not allow loops (edges with the same origin and endpoint). The differential is defined by the blow-up of vertices again.

The space of homotopy automorphisms $\mathtt{Aut}^h_{\mathcal{T}op\,\mathcal{O}p}(D_n^{\mathbb{Q}})$ is the sum of the connected components of the mapping spaces $\mathtt{Map}^h_{\mathcal{T}op\,\mathcal{O}p}(D_n, D_n^{\mathbb{Q}})$ associated to the morphisms ϕ that are invertible in the homotopy category of operads. Let $h : \mathtt{Aut}^h_{\mathcal{T}op\,\mathcal{O}p}(D_n^{\mathbb{Q}}) \to \mathtt{Aut}_{\mathcal{H}opf\,\mathcal{O}p}(\mathtt{H}_*(D_n, \mathbb{Q}))$ be the natural map that carries any such morphism to the associated homology morphism. For $n \geq 2$, we have a bijection $\mathtt{Aut}_{\mathcal{H}opf\,\mathcal{O}p}(\mathtt{H}_*(D_n, \mathbb{Q})) = \mathbb{Q}^\times$ that is determined by taking the action of an automorphism $\phi \in \mathtt{Aut}_{\mathcal{H}opf\,\mathcal{O}p}(\mathtt{H}_*(D_n, \mathbb{Q}))$ on the representative of the Poisson bracket operation $\lambda \in \mathsf{Pois}_n$ in the operad $\mathsf{Pois}_n = \mathtt{H}_*(D_n)$. We get the following result:

Theorem 11.3.4 (B. Fresse, V. Turchin, and T. Willwacher [28, Corollary 5]). *For each $\lambda \in \mathbb{Q}^\times$, we have the identity:*

$$\pi_*(h^{-1}(\lambda)) = \mathtt{H}_*(GC_n) \oplus \begin{cases} \mathbb{Q}, & \text{if } * \equiv -n-1(4), \\ 0, & \text{otherwise}, \end{cases}$$

where GC_n denotes the Kontsevich graph complex.

We deduce this statement from the result of Theorem 11.3.1, by using that the identity morphism is represented by the Maurer–Cartan element such that $\omega = |$ in the hairy graph complex HGC_{nn}. We just consider versions of the graph complexes GC_n^2 and HGC_{mn}^2 where bivalent vertices are allowed. We have $HGC_{mn}^2 \sim HGC_{mn}$ for any $n \geq m \geq 2$, whereas for the graph complex GC_n^2, we have:

$$\mathtt{H}_*(GC_n^2) = \mathtt{H}_*(GC_n) \oplus \begin{cases} \mathbb{Q}\,L_*, & \text{if } * \equiv -n-1(4), \\ 0, & \text{otherwise}, \end{cases} \tag{11.3.9}$$

where L_* denotes the homology classes of graphs of the form:

$$L_* = [\text{loop graph}] \tag{11.3.10}$$

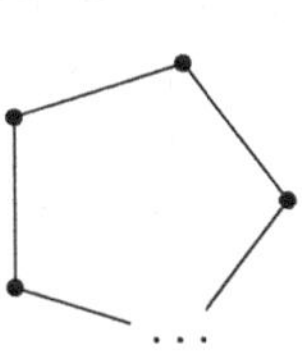

We easily see that the operation $[\omega, -]$ in the differential $\delta_\omega = \delta + [\omega, -]$ of the twisted complex $(HGC^2_{nn})^\omega$ associated to the Maurer–Cartan element $\omega = |$ is given by the addition of a hair $|$ to any graph $\gamma \in HGC^2_{nn}$. We can then use a spectral sequence to check that we have a quasi-isomorphism $\mathbb{Q} | \oplus \mathbb{Q}[-1] \otimes GC^2_n \xrightarrow{\sim} HGC^2_{nn}$ where we consider the mapping $\gamma \mapsto \gamma-$ that associates a graph with one hair $\gamma- \in HGC^2_{nn}$ to any graph $\gamma \in GC_n$. We refer to [29, Proposition 2.2.9] for the detailed line of arguments.

In the case $n = 2$, we have $\mathrm{H}_0(GC_n) = \mathfrak{grt}$ by a result of T. Willwacher, where $\mathfrak{grt}$ is the graded Grothendieck–Teichmüller Lie algebra (see [60]). Therefore, in this case, the result of Theorem 11.3.4 reflects the relation that we obtained in Theorem 11.1.4.

11.4 Outlook

Throughout this survey, we have focused on the study of the homotopy of E_n-operads themselves, but one can use variants of the definition of an E_n-operad to associate operadic right module structures to any n-manifold M.

For this purpose, we can use again the Fulton–MacPherson operad FM_n, the model of E_n-operad, given by a real oriented version of the Fulton–MacPherson compactification of the configuration spaces, which was considered by Kontsevich in his proof of the formality of E_n-operads (see §11.2). These Fulton–MacPherson compactifications have a natural generalization for the configuration spaces of manifolds $\mathsf{F}(M, r)$, and when M is a framed manifold, this construction returns a collection of spaces $\mathsf{FM}_M = \{\mathsf{FM}_M(r), r \in \mathbb{N}\}$ that inherits the structure of a right module (in the operadic sense) over the Fulton–MacPherson operad FM_n.

This object is equivalent to constructions used by Ayala–Francis in the definition of the factorization homology of manifolds (see Ayala–Francis's chapter, in this handbook volume, for a survey of this subject). In particular, one can use a relative composition product of the object FM_M over the operad FM_n to compute the factorization homology of any framed manifold M. The methods used by Kontsevich to prove the formality of E_n-operads have been used by several authors to define models of the rational homotopy type of this right module FM_M, and hence, to tackle the rational homotopy computations in factorization homology theory. To cite a few works on this subject, let us mention that a graph complex model of the object FM_M, which extends the graph cooperad of §11.2.4 when M is a simply connected compact manifold without boundary, is defined by Campos–Willwacher in [17], while an extension of Arnold's presentation is used by Idrissi in [38] to get a small model of the object FM_M. Idrissi's result provides a generalization of Knudsen's description of the factorization homology for higher enveloping algebras of Lie algebras [39]. The paper [16] provides an extension of the constructions of [17, 38] for manifolds with boundary, while the paper [15] addresses an extension of the definition of these operadic module structures by using a framed version of the operads of little discs.

In §§11.2-11.3, we entirely focus on the rational homotopy theory framework, but we may wonder which information we may still retrieve by our methods in positive characteristic. For instance, partial formality results have been obtained by Cirici–Horel in [19] when we take an arity-wise truncation of operads below the characteristic of the coefficients (see also [9] for an improvement of these partial formality results). In fact, the E_n-operads are not formal as symmetric operads in chain complexes in positive characteristic, because their components are not formal as representations of the symmetric groups. Nevertheless, we may wonder whether E_n-operads are formal as non-symmetric operads, which is enough for the study of mapping spaces over an E_1-operad. The case $n > 2$ of this question is still open, but Salvatore has proved in [49] that E_2-operads are not formal as non-symmetric operads over $\mathbb{F}_2$.

Bibliography

[1] Anton Alekseev and Charles Torossian. Kontsevich deformation quantization and flat connections. *Comm. Math. Phys.*, 300(1):47–64, 2010.

[2] V. I. Arnol'd. The cohomology ring of the group of dyed braids. *Mat. Zametki*, 5:227–231, 1969.

[3] Gregory Arone and Victor Turchin. On the rational homology of high-dimensional analogues of spaces of long knots. *Geom. Topol.*, 18(3):1261–1322, 2014.

[4] Gregory Arone and Victor Turchin. Graph-complexes computing the rational homotopy of high dimensional analogues of spaces of long knots. *Ann. Inst. Fourier (Grenoble)*, 65(1):1–62, 2015.

[5] Dror Bar-Natan. On associators and the Grothendieck-Teichmuller group. I. *Selecta Math. (N.S.)*, 4(2):183–212, 1998.

[6] Alexander Berglund. Rational homotopy theory of mapping spaces via Lie theory for L_∞-algebras. *Homology Homotopy Appl.*, 17(2):343–369, 2015.

[7] J. M. Boardman and R. M. Vogt. Homotopy-everything H-spaces. *Bull. Amer. Math. Soc.*, 74:1117–1122, 1968.

[8] J. M. Boardman and R. M. Vogt. *Homotopy invariant algebraic structures on topological spaces. Lecture Notes in Math.*, Vol. 347. Springer-Verlag, 1973.

[9] Pedro Boavida de Brito and Geoffroy Horel. On the formality of the little disks operad in positive characteristic. arXiv:1903.09191, 2019.

[10] Pedro Boavida de Brito and Michael Weiss. Spaces of smooth embeddings and configuration categories. *J. Topol.*, 11(1):65–143, 2018.

[11] Francis Borceux. *Handbook of categorical algebra. 2: Categories and structures.*, volume 51 of *Encyclopedia Math. Appl.*. Cambridge University Press, 1994.

[12] A. K. Bousfield and V. K. A. M. Gugenheim. On PL de Rham theory and rational homotopy type. *Mem. Amer. Math. Soc.*, 8(179), 1976.

[13] Francis Brown. Mixed Tate motives over $\mathbb{Z}$. *Ann. of Math. (2)*, 175(2):949–976, 2012.

[14] Damien Calaque, Tony Pantev, Bertrand Toën, Michel Vaquié, and Gabriele Vezzosi. Shifted Poisson structures and deformation quantization. *J. Topol.*, 10(2):483–584, 2017.

[15] Ricardo Campos, Julien Ducoulombier, Najib Idrissi, and Thomas Willwacher. A model for framed configuration spaces of points. arXiv:1807.08319, 2018.

[16] Ricardo Campos, Najib Idrissi, Pascal Lambrechts, and Thomas Willwacher. Configuration spaces of manifolds with boundary. arXiv:1802.00716, 2018.

[17] Ricardo Campos and Thomas Willwacher. A model for configuration spaces of points. arXiv:1604.02043, 2016.

[18] Gunnar Carlsson and R. James Milgram. Stable homotopy and iterated loop spaces. In *Handbook of Algebraic Topology*, pages 505–583. North-Holland, 1995.

[19] Joana Cirici and Geoffroy Horel. étale cohomology, purity and formality with torsion coefficients. arXiv:1806.03006, 2018.

[20] Frederick R. Cohen. The homology of $\mathcal{C}_{n+1}$-spaces, $n \geq 0$. In *The Homology of Iterated Loop Spaces, Lecture Notes in Math.*, Vol. 533, pages 207–351. Springer-Verlag,1976.

[21] Pierre Deligne and Alexander B. Goncharov. Groupes fondamentaux motiviques de Tate mixte. *Ann. Sci. École Norm. Sup. (4)*, 38(1):1–56, 2005.

[22] V. G. Drinfel'd. On quasitriangular quasi-Hopf algebras and on a group that is closely connected with Gal($\overline{\mathbf{Q}}/\mathbf{Q}$). *Algebra i Analiz*, 2(4):149–181, 1990.

[23] Julien Ducoulombier and Victor Turchin. Delooping the functor calculus tower. arXiv:1708.02203, 2017.

[24] William Dwyer and Kathryn Hess. Long knots and maps between operads. *Geom. Topol.*, 16(2):919–955, 2012.

[25] Z. Fiedorowicz. The symmetric bar construction. Preprint, 1996.

[26] Benoit Fresse. *Homotopy of operads and Grothendieck-Teichmüller groups*, volume 217 of *Math. Surveys Monogr.* Amer. Math. Soc., 2017.

[27] Benoit Fresse. The extended rational homotopy theory of operads. Georgian Math. J. 25 (2018), no. 4, 493–512.

[28] Benoit Fresse, Victor Turchin, and Thomas Willwacher. The rational homotopy of mapping spaces of E_n-operads. arXiv:1703.06123, 2017.

[29] Benoit Fresse and Thomas Willwacher. The intrinsic formality of E_n-operads. arXiv:1503.08699, 2015.

[30] Hidekazu Furusho. Pentagon and hexagon equations. *Ann. of Math. (2)*, 171(1):545–556, 2010.

[31] Hidekazu Furusho. Double shuffle relation for associators. *Ann. of Math. (2)*, 174(1):341–360, 2011.

[32] Ezra Getzler. Lie theory for nilpotent L_∞-algebras. *Ann. of Math. (2)*, 170(1):271–301, 2009.

[33] Ezra Getzler and John D.S. Jones. Operads, homotopy algebra and iterated integrals for double loop spaces. arXiv:hep-th/9403055, 1994.

[34] Thomas G. Goodwillie and Michael Weiss. Embeddings from the point of view of immersion theory. II. *Geom. Topol.*, 3:103–118, 1999.

[35] André Haefliger. Rational homotopy of the space of sections of a nilpotent bundle. *Trans. Amer. Math. Soc.*, 273(2):609–620, 1982.

[36] Robert Hardt, Pascal Lambrechts, Victor Turchin, and Ismar Volić. Real homotopy theory of semi-algebraic sets. *Algebr. Geom. Topol.*, 11(5):2477–2545, 2011.

[37] Geoffroy Horel. Profinite completion of operads and the Grothendieck-Teichmüller group. *Adv. Math.*, 321:326–390, 2017.

[38] Najib Idrissi. The Lambrechts-Stanley model of configuration spaces. *Invent. Math.* 216 (2019), no. 1, 1–68.

[39] Ben Knudsen. Higher enveloping algebras. *Geom. Topol.* 22 (2018), no. 7, 4013–4066.

[40] Toshitake Kohno. Série de Poincaré-Koszul associée aux groupes de tresses pures. *Invent. Math.*, 82(1):57–75, 1985.

[41] Maxim Kontsevich. Formal (non)commutative symplectic geometry. In *The Gelfand Mathematical Seminars, 1990–1992*, pages 173–187. Birkhäuser Boston,1993.

[42] Maxim Kontsevich. Formality conjecture. In *Deformation theory and symplectic geometry (Ascona, 1996)*, volume 20 of *Math. Phys. Stud.*, pages 139–156. Kluwer Acad. Publ., 1997.

[43] Maxim Kontsevich. Operads and motives in deformation quantization. *Lett. Math. Phys.*, 48(1):35–72, 1999.

[44] Pascal Lambrechts and Victor Turchin. Homotopy graph-complex for configuration and knot spaces. *Trans. Amer. Math. Soc.*, 361(1):207–222, 2009.

[45] Pascal Lambrechts, Victor Turchin, and Ismar Volić. The rational homology of spaces of long knots in codimension > 2. *Geom. Topol.*, 14(4):2151–2187, 2010.

[46] Pascal Lambrechts and Ismar Volić. Formality of the little N-disks operad. *Mem. Amer. Math. Soc.*, 230(1079),2014.

[47] J. P. May. *The geometry of iterated loop spaces. Lectures Notes in Math.*, Vol. 271. Springer-Verlag, 1972.

[48] Dan Petersen. Minimal models, GT-action and formality of the little disk operad. *Selecta Math. (N.S.)*, 20(3):817–822, 2014.

[49] Paolo Salvatore. Planar non-formality of the little discs operad in characteristic two. arXiv:1807.11671, 2018.

[50] Dev P. Sinha. Operads and knot spaces. *J. Amer. Math. Soc.*, 19(2):461–486, 2006.

[51] Dev P. Sinha. The (non-equivariant) homology of the little disks operad. In *OPERADS 2009*, volume 26 of *Sémin. Congr.*, pages 253–279. Soc. Math. France, 2013.

[52] Dennis Sullivan. Infinitesimal computations in topology. *Inst. Hautes Études Sci. Publ. Math.*, (47):269–331 (1978), 1977.

[53] Dmitry E. Tamarkin. Another proof of M. Kontsevich formality theorem. arXiv:math/9803025, 1998.

[54] Dmitry E. Tamarkin. Formality of chain operad of little discs. *Lett. Math. Phys.*, 66(1-2):65–72, 2003.

[55] Tomohide Terasoma. Mixed Tate motives and multiple zeta values. *Invent. Math.*, 149(2):339–369, 2002.

[56] V. Turchin. On the other side of the bialgebra of chord diagrams. *J. Knot Theory Ramifications*, 16(5):575–629, 2007.

[57] Michael Weiss. Calculus of embeddings. *Bull. Amer. Math. Soc. (N.S.)*, 33(2):177–187, 1996.

[58] Michael Weiss. Embeddings from the point of view of immersion theory. I. *Geom. Topol.*, 3:67–101, 1999.

[59] Thomas Willwacher. Deformation quantization and the Gerstenhaber structure on the homology of knot spaces. arXiv:1506.07078, 2015.

[60] Thomas Willwacher. M. Kontsevich's graph complex and the Grothendieck-Teichmüller Lie algebra. *Invent. Math.*, 200(3):671–760, 2015.

Université de Lille, Laboratoire Paul Painlevé, F-59000 Lille, France
E-mail address: Benoit.Fresse@univ-lille.fr

12

Moduli spaces of manifolds: a user's guide

Søren Galatius and Oscar Randal-Williams

12.1 Introduction

The study of manifolds and invariants of manifolds was begun more than a century ago. In this entry we shall discuss the parametrised setting: invariants of *families* of manifolds, parametrised by a base manifold X. The invariants we look for will be cohomology classes in X, characteristic classes.

This article will be structured in the following way.

(i) A discussion of the abstract classification theory. The main content here is the precise definition of the kind of families we consider, and an outline of how a classifying space may be constructed. There is one such classifying space for each pair (W, ρ_W) consisting of a closed manifold W and a tangential structure ρ_W, see §12.2. The more general case where W is compact with boundary is briefly discussed in §12.4.6.

(ii) Definition of Miller–Morita–Mumford classes, the main examples of characteristic classes. With this definition in place, we state a first version of the main result of [14, 17, 16]: a formula (see Theorem 12.3.3) for rational cohomology of the classifying spaces in a range of degrees, for many instances of (W, ρ_W).

(iii) Statement of the main result of [14, 17, 16] in their general forms. These statements require a bit more homotopy theory to formulate but apply quite generally, at least for even-dimensional manifolds, see the theorems in §12.4.

(iv) Examples of calculations and applications. The results surveyed in §§12.3–12.4 are quite well suited for explicit calculations. We believe this to be an important feature of the theory, and have included a supply of worked examples of various types to illustrate this aspect, many of which have not previously been published. In §12.5 we will focus on calculations in rational cohomology, and we include a detailed study of some complete intersections. In §12.6 we carry out some integral homology calculations, focusing on H_1.

Our theorems apply in even dimensions $2n \geq 6$, but were inspired by the breakthrough theorem of Madsen and Weiss [33] in dimension 2, building on earlier ideas of Madsen and Tillmann [32] and Tillmann [47].

2010 *Mathematics Subject Classification.* 57R90, 57R15, 57R56, 55P47

Key words and phrases. characteristic classes, diffeomorphism groups, homological stability, moduli spaces.

12.2 Smooth bundles and their classifying spaces

First, some conventions. By *smooth manifold* we shall always mean a Hausdorff, second countable topological manifold equipped with a maximal C^∞ atlas, and *smooth map* shall always mean C^∞. We shall generally use the letter $\mathcal{M}$ with various decorations for variants of classifying spaces for families of manifolds, and the letter $\mathcal{F}$ with various decorations for the functor it classifies.

12.2.1 Smooth bundles

Our notion of "family" of manifolds will be *smooth fibre bundle*, possibly equipped with extra tangent bundle structure, as follows. If V is a d-dimensional real vector bundle we shall write $\mathrm{Fr}(V)$ for the associated frame bundle, which is a principal $\mathrm{GL}_d(\mathbf{R})$-bundle.

Definition 12.2.1. Let d be a non-negative integer.

(i) A *smooth fibre bundle* of dimension d consists of smooth manifolds E and X (without boundary) and a smooth proper map $\pi : E \to X$ such that $D\pi : TE \to \pi^*TX$ is surjective and the vector bundle $T_\pi E = \mathrm{Ker}(D\pi)$ has d-dimensional fibres. The bundle $T_\pi E$ is called the *vertical tangent bundle*.

(ii) If Θ is a space with a continuous action of $\mathrm{GL}_d(\mathbf{R})$, a *smooth fibre bundle with Θ-structure* consists of a smooth fibre bundle $\pi : E \to X$, together with a continuous $\mathrm{GL}_d(\mathbf{R})$-equivariant map $\rho : \mathrm{Fr}(T_\pi E) \to \Theta$.

Typical choices of Θ include the terminal one $\Theta = \{*\}$, and $\Theta = \boldsymbol{Z}^\times = \{\pm 1\}$ on which $\mathrm{GL}_d(\mathbf{R})$ acts by multiplication by the sign of the determinant. In the former case a Θ-structure is no information, and in the latter it is the data of a continuously varying family of orientations of the d-manifolds $\pi^{-1}(x)$. (The space of equivariant maps $\mathrm{Fr}(T_\pi E) \to \Theta$ may be modelled in other ways, equivalent up to weak equivalence, see §12.4.5.)

Any smooth map $f : X' \to X$ will be transverse to any smooth fibre bundle $\pi : E \to X$ as above, and the pullback $(f^*\pi) : f^*E \to X'$ is again a smooth fibre bundle. Given a Θ-structure ρ on π, we shall write $f^*\rho$ for the induced structure on $(f^*\pi)$.

12.2.2 Classifying spaces

The natural equivalence relation between the bundles considered above is *concordance*, which we recall.

Definition 12.2.2. Let $\pi_0 : E_0 \to X$ and $\pi_1 : E_1 \to X$ be smooth bundles with Θ-structures $\rho_0 : \mathrm{Fr}(T_{\pi_0} E_0) \to \Theta$ and $\rho_1 : \mathrm{Fr}(T_{\pi_1} E_1) \to \Theta$.

(i) An *isomorphism* between (π_0, ρ_0) and (π_1, ρ_1) is a diffeomorphism $\phi : E_0 \to E_1$ over X, such that the induced map $\mathrm{Fr}(D_\pi \phi) : \mathrm{Fr}(T_{\pi_0} E_0) \to \mathrm{Fr}(T_{\pi_1} E_1)$ is over Θ.

(ii) A *concordance* between (π_0, ρ_0) and (π_1, ρ_1) is a smooth fibre bundle $\pi : E \to \mathbf{R} \times X$ with Θ-structure $\rho : \mathrm{Fr}(T_\pi E) \to \Theta$, together with isomorphisms from (π_0, ρ_0) and

(π_1, ρ_1) to the pullbacks of (π, ρ) along the two embeddings $X \cong \{0\} \times X \subset \mathbf{R} \times X$ and $X \cong \{1\} \times X \subset \mathbf{R} \times X$.

Pulling back is functorial up to isomorphism and preserves being concordant.

Definition 12.2.3. For a smooth manifold X without boundary, let $\mathcal{F}^\Theta[X]$ denote the set of concordance classes of pairs (π, ρ) of a smooth fibre bundle $\pi : E \to X$ with Θ-structure $\rho : \mathrm{Fr}(T_\pi E) \to \Theta$. (Note that X is fixed, but E is allowed to vary.)

Theorem 12.2.4. *The functor $X \mapsto \mathcal{F}^\Theta[X]$ is representable in the (weak) sense that there exists a topological space $\mathcal{M}^\Theta$ and a natural bijection*

$$\mathcal{F}^\Theta[X] \cong [X, \mathcal{M}^\Theta], \tag{12.2.1}$$

where the codomain denotes homotopy classes of continuous maps.

There are various ways to prove this representability statement. In the series [14, 17, 16] we did this by constructing explicit point-set models in terms of submanifolds of $\mathbf{R}^\infty$. Instead of repeating what we said there, let us outline an approach based on simplicial sets and the observation that the functor $\mathcal{F}^\Theta$ may be upgraded to take values in *spaces*, and the natural bijection (12.2.1) may be upgraded to a natural weak equivalence to the mapping space.

Definition 12.2.5. Let $\Delta_e^\bullet$ be the cosimplicial smooth manifold given by the extended simplices $\Delta_e^p = \{t \in \mathbf{R}^{p+1} \mid t_0 + \cdots + t_p = 1\}$.

For a smooth manifold X let $\mathcal{F}_\bullet^\Theta(X)$ denote the simplicial set whose p-simplices are all pairs (π, ρ) consisting of a smooth fibre bundle $\pi : E \to \Delta_e^p \times X$ with Θ-structure $\rho : \mathrm{Fr}(T_\pi E) \to \Theta$.

As they stand, these definitions are not quite rigorous: $\mathcal{F}^\Theta[X]$ and $\mathcal{F}_p^\Theta(X)$ are not (small) sets for size reasons, and $\mathcal{F}_p^\Theta(X)$ is not functorial in $[p] \in \Delta$ because pullback is not strictly associative on the level of underlying sets. This may be fixed in standard ways, e.g. by requiring the underlying set of E to be a subset of $X \times \Omega$ for a set Ω of sufficiently large cardinality. See e.g. [33, Section 2.1] for more detail.

Proposition 12.2.6. *Let* Man *and* sSets *be the categories of smooth manifolds and simplicial sets, and let $\mathcal{F}_\bullet^\Theta : \mathsf{Man}^{\mathrm{op}} \to \mathsf{sSets}$ be the functor defined above. Then we have a natural bijection*

$$\mathcal{F}^\Theta[X] \cong \pi_0 \mathcal{F}_\bullet^\Theta(X)$$

and a natural weak equivalence of simplicial sets

$$\mathcal{F}_\bullet^\Theta(X) \xrightarrow{\simeq} \mathrm{Maps}(\mathrm{Sing}(X), \mathcal{F}_\bullet(\{*\})). \tag{12.2.2}$$

Consequently, we may take $\mathcal{M}^\Theta := |\mathcal{F}_\bullet^\Theta(\{*\})|$ in Theorem 12.2.4.

In this statement we take $\mathrm{Sing}(X)$ to mean the smooth singular set, i.e. the simplicial set $[p] \mapsto C^\infty(\Delta_e^p, X)$. It is equivalent to the usual simplicial set made out of all continuous maps $\Delta^p \to X$ by a smooth approximation argument, and in particular the evaluation map $|\mathrm{Sing}(X)| \to X$ is a homotopy equivalence by Milnor's theorem. The codomain of (12.2.2) is the simplicial set of maps into the Kan complex $\mathcal{F}_\bullet(\{*\})$, homotopy equivalent to the space of maps $X \to |\mathcal{F}_\bullet(\{*\})|$.

Proof sketch. The first claim follows by identifying 1-simplices of $\mathcal{F}^\Theta_\bullet(X)$ with concordances between 0-simplices.

For the second claim, we define the map (12.2.2) by pulling back. To check that it is a weak equivalence one first verifies that both sides send open covers $X = U \cup V$ to homotopy pullback squares and countable increasing unions to homotopy limits. Hence it suffices to check the case $X = \mathbf{R}^n$. Then one checks that both sides send $X \times \mathbf{R} \to X$ to a weak equivalence, so it suffices to check $X = \{*\}$, which is obvious.

The third claim follows by combining the first and the second claim and the homotopy equivalence $|\mathrm{Sing}(X)| \to X$ given by evaluation. Alternatively it may be quoted directly from [33, Section 2.4]. □

The above proof of Theorem 12.2.4 gives an explicit bijection (12.2.1). Indeed, an element $(\pi, \rho) \in \mathcal{F}_0(X)$ gives a morphism of simplicial sets $\mathrm{Sing}(X) \to \mathcal{F}^\Theta_\bullet(\{*\})$ and hence a canonical zig-zag

$$X \xleftarrow{\mathrm{ev}} |\mathrm{Sing}(X)| \xrightarrow{(\pi,\rho)} |\mathcal{F}^\Theta_\bullet(\{*\})| = \mathcal{M}^\Theta. \tag{12.2.3}$$

The map $\mathrm{ev} : |\mathrm{Sing}(X)| \to X$ is a homotopy equivalence, and we may choose a homotopy inverse. For example, a smooth triangulation of X gives such an inverse, which is even a section. The resulting homotopy class of map $X \to \mathcal{M}^\Theta$ corresponds to $[(\pi, \rho)]$ under the bijection (12.2.1).

For later purposes, let us explain how a universal fibration $\pi^\Theta : \mathcal{E}^\Theta \to \mathcal{M}^\Theta$ modelling the bundle $\pi : E \to X$ may be constructed by the same simplicial method. Let $\widetilde{\mathcal{F}}^\Theta_\bullet(X)$ be the simplicial set whose p-simplices are triples (π, ρ, s) where (π, ρ) is as before and $s : \Delta^p_e \times X \to E$ is a section of π, and set $\mathcal{E}^\Theta := |\widetilde{\mathcal{F}}^\Theta_\bullet(\{*\})|$. The zig-zag (12.2.3) above associated to an element $(\pi, \rho) \in \mathcal{F}^\Theta_0(X)$ now extends to a canonical diagram

$$\begin{array}{ccccc} E & \xleftarrow{\simeq} & |\mathrm{Sing}(E)| & \longrightarrow & \mathcal{E}^\Theta \\ {\scriptstyle\pi}\downarrow & & \downarrow & & \downarrow{\scriptstyle\pi^\Theta} \\ X & \xleftarrow{\simeq} & |\mathrm{Sing}(X)| & \xrightarrow{(\pi,\rho)} & \mathcal{M}^\Theta \end{array}$$

in which both squares are homotopy cartesian, and the top right-hand map is that associated to the data $(\mathrm{pr}_1 : E \times_X E \to E, \rho \circ D\mathrm{pr}_2, \mathrm{diag} : E \to E \times_X E)$.

Finally, a p-simplex $(\pi : E \to \Delta^p_e, \rho, s)$ of $\mathcal{E}^\Theta$ gives a map $\ell = (\rho/\mathrm{GL}_d(\mathbf{R})) : E \cong (\mathrm{Fr}(T_\pi E)/\mathrm{GL}_d(\mathbf{R})) \to (\Theta/\mathrm{GL}_d(\mathbf{R}))$. Composing with the section $s|_{\Delta^p} : \Delta^p \to E$ then gives rise to a map of simplicial sets

$$\widetilde{\mathcal{F}}^\Theta_\bullet(\{*\}) \longrightarrow \mathrm{Sing}(\Theta/\mathrm{GL}_d(\mathbf{R}))$$

realising to a map $\mathcal{E}^\Theta \to |\mathrm{Sing}(\Theta/\mathrm{GL}_d(\mathbf{R}))|$. The orbit space $\Theta/\mathrm{GL}_d(\mathbf{R})$ need not be well behaved, and we would like to replace it by the homotopy orbit space $B = \Theta /\!\!/ \mathrm{GL}_d(\mathbf{R}) := (E\mathrm{GL}_d(\mathbf{R}) \times \Theta)/\mathrm{GL}_d(\mathbf{R})$. Taking homotopy orbits of the map $\Theta \to \{*\}$ yields a map $\theta : B \to B\mathrm{GL}_d(\mathbf{R})$. Repeating the construction above with $E\mathrm{GL}_d(\mathbf{R}) \times \Theta$ instead of Θ allows us to construct a zig-zag

$$\mathcal{E}^\Theta \xleftarrow{\simeq} \mathcal{E}^{E\mathrm{GL}_d(\mathbf{R})\times\Theta} \xrightarrow{\ell} |\mathrm{Sing}(B)| \xrightarrow{\mathrm{ev}} B \xrightarrow{\theta} B\mathrm{GL}_d(\mathbf{R}).$$

Slightly less precisely, we shall summarise this situation as a diagram

$$\begin{array}{ccccccc} E & \longrightarrow & \mathcal{E}^\Theta & \xrightarrow{\ell} & B & \xrightarrow{\theta} & B\mathrm{GL}_d(\mathbf{R}) \\ \pi\downarrow & & \downarrow\pi^\Theta & & & & \\ X & \longrightarrow & \mathcal{M}^\Theta & & & & \end{array} \tag{12.2.4}$$

where the square is homotopy cartesian. The vector bundle classified by the composition $\theta \circ \ell : \mathcal{E}^\Theta \to B\mathrm{GL}_d(\mathbf{R})$ is a universal instance of $T_\pi E$, and shall be denoted $T_\pi \mathcal{E}^\Theta$. The factorisation through θ gives an equivariant map $\rho^\Theta : \mathrm{Fr}(T_\pi \mathcal{E}^\Theta) \to \Theta$, a universal instance of $\rho : \mathrm{Fr}(T_\pi E) \to \Theta$.

12.2.3 Path connected classifying spaces

The bundles classified so far are inconveniently general. For example, we have not made any restrictions on the diffeomorphism type of the fibres of $\pi : E \to X$. Hence, if e.g. $\Theta = \{*\}$, the set $\pi_0(\mathcal{M}^\Theta)$ is in bijection with the set of diffeomorphism classes of compact smooth d-manifolds without boundary, a countably infinite set for any $d \geq 0$. The homotopy type of $\mathcal{M}^\Theta$ then encodes at once the classification of smooth manifolds up to diffeomorphism and the classification of smooth bundles, because it is the "moduli space" (or "∞-groupoid") of *all* d-manifolds.

We shall often study one path component of $\mathcal{M}^\Theta$ at a time, which corresponds to fixing the concordance class of the fibres of the classified bundles. We introduce the following notation.

Definition 12.2.7. Let Θ be a space with $\mathrm{GL}_d(\mathbf{R})$ action, W be a compact d-manifold without boundary, and $\rho_W : \mathrm{Fr}(TW) \to \Theta$ be an equivariant map. Considering this as a family over a point yields a $[(W, \rho_W)] \in \pi_0 \mathcal{M}^\Theta$, and we shall write $\mathcal{M}^\Theta(W, \rho_W) \subset \mathcal{M}^\Theta$ for the path component containing (W, ρ_W). This path component is a classifying space for smooth fibre bundles $\pi : E \to X$ with structure $\rho : \mathrm{Fr}(T_\pi E) \to \Theta$, whose restriction to any point $\{x\} \subset X$ is concordant to (W, ρ_W).

Remark 12.2.8. In the special case $\Theta = \{*\}$ the maps $\rho : \mathrm{Fr}(T_\pi E) \to \Theta$ are irrelevant, and in this case we shall write simply $\mathcal{M}(W)$. This space classifies smooth bundles $\pi : E \to X$ with fibres diffeomorphic to W with no further structure, and we have the weak equivalence

$$\mathcal{M}(W) \simeq B\mathrm{Diff}(W),$$

where $\mathrm{Diff}(W)$ is the diffeomorphism group of W in the C^∞ topology. Similarly, if $\Theta = \mathbf{Z}^\times$ with the orientation action and W is given an orientation ρ_W, we shall write $\mathcal{M}^{\mathrm{or}}(W)$ for $\mathcal{M}^\Theta(W, \rho_W)$ and there is a weak equivalence

$$\mathcal{M}^{\mathrm{or}}(W) \simeq B\mathrm{Diff}^{\mathrm{or}}(W),$$

where $\mathrm{Diff}^{\mathrm{or}}(W) \subset \mathrm{Diff}(W)$ is the subgroup of orientation preserving diffeomorphisms. For general Θ, the homotopy type is described as the Borel construction

$$\mathcal{M}^{\Theta}(W, \rho_W) \simeq \frac{\{\rho : \mathrm{Fr}(TW) \to \Theta, \text{ equivariantly homotopic to } \rho_W\}}{\left(\begin{smallmatrix}\text{topological group of diffeomorphisms } W \to W \\ \text{preserving the equivariant homotopy class of } \rho_W\end{smallmatrix}\right)}.$$

See [14, Definition 1.5], [14, Section 1.1], or [16, Section 1.1] for further discussion of this point of view.

12.3 Characteristic classes

The main topic of this article is the study of characteristic classes of the sort of bundles described above, i.e. the calculation of the cohomology ring of the classifying spaces $\mathcal{M}^{\Theta}(W, \rho_W)$. We first recall the conclusions in rational cohomology, which are easier to state and often gives an explicit formula for the ring of characteristic classes.

12.3.1 Characteristic classes

As before, let $B = \Theta /\!\!/ \mathrm{GL}_d(\mathbf{R})$ denote the Borel construction. For a smooth bundle $\pi : E \to X$ with Θ-structure $\rho : \mathrm{Fr}(T_\pi E) \to \Theta$, we may form the Borel construction with $\mathrm{GL}_d(\mathbf{R})$ and obtain a correspondence, i.e. a diagram

$$X \xleftarrow{\pi} E \xleftarrow{\simeq} \mathrm{Fr}(T_\pi E) /\!\!/ \mathrm{GL}_d(\mathbf{R}) \xrightarrow{\rho /\!\!/ \mathrm{GL}_d(\mathbf{R})} B. \tag{12.3.1}$$

We write $\ell : E \to B$ for the homotopy class of maps associated to $\rho /\!\!/ \mathrm{GL}_d(\mathbf{R})$. We shall let $\boldsymbol{Z}^{w_1}$ denote the coefficient system on B arising from the non-trivial action of $\pi_0(\mathrm{GL}_d(\mathbf{R})) = \boldsymbol{Z}^{\times}$ on $\boldsymbol{Z}$, and let $A^{w_1} = A \otimes \boldsymbol{Z}^{w_1}$ for an abelian group A, and we shall use the same notation for these coefficient systems pulled back to E along (12.3.1). Then we have a homomorphism

$$\ell^* : H^{k+d}(B; A^{w_1}) \longrightarrow H^{k+d}(E; A^{w_1}). \tag{12.3.2}$$

We also have a *fibre integration* homomorphism

$$\int_\pi : H^{k+d}(E; A^{w_1}) \longrightarrow H^k(X; A), \tag{12.3.3}$$

where d is the dimension of the fibres of π. It may be defined, e.g. as the composition $H^{k+d}(E; A^{w_1}) \twoheadrightarrow E_\infty^{k,d} \subset E_2^{k,d}$ in the Serre spectral sequence for the fibration π, or by a Pontryagin–Thom construction as in §12.4.2 below.

Given a class $c \in H^{d+k}(B; A^{w_1})$, we may combine (12.3.2) and (12.3.3) and define the *Miller–Morita–Mumford classes* (or MMM-classes) by the push-pull formula

$$\kappa_c(\pi) = \int_\pi \ell^* c \in H^k(X; A).$$

This class is easily seen to be natural with respect to pullback along smooth maps $X' \to X$, and in fact comes from a universal class

$$\kappa_c \in H^k(\mathcal{M}^\Theta(W, \rho_W); A)$$

for any (W, ρ_W), any A, and any $c \in H^{k+d}(B; A^{w_1})$, defined by the analogous push-pull formula in the universal instance (12.2.4).

In the special case $k = 0$ and $X = \{*\}$, the definition of $\kappa_c \in H^0(X; A) = A$ simply reproduces the usual *characteristic numbers*, c.f. [35, §16]. For example, if $d = 2n$ and e comes from the Euler class in $H^d(B\mathrm{GL}_d(\mathbf{R}); \mathbf{Z}^{w_1})$, then $\kappa_e \in H^0(X; \mathbf{Z}) = \mathbf{Z}$ is the Euler characteristic of W. Henceforth we shall mostly be interested in the case $k > 0$.

If $A = \mathbf{k}$ is a field, we may combine with cup product to get a map of graded rings

$$\mathbf{k}[\kappa_c \mid c \in \text{basis of } H^{>d}(B; \mathbf{k}^{w_1})] \longrightarrow H^*(\mathcal{M}^\Theta(W, \rho_W); \mathbf{k}), \tag{12.3.4}$$

whose domain is the free graded-commutative $\mathbf{k}$-algebra on symbols κ_c, one for each element c in a chosen basis (or more invariantly, on a degree-shifted copy of the graded $\mathbf{k}$-vector space $H^{>d}(B; \mathbf{k}^{w_1})$).

12.3.2 Genus

All of our results will hold in a range of degrees depending on *genus*, a numerical invariant that we first introduce.

Assume that $d = 2n$ and let a $\mathrm{GL}_{2n}(\mathbf{R})$-space Θ be given. We shall assume that $n > 0$ and that the homotopy orbit space $\Theta /\!\!/ \mathrm{GL}_{2n}(\mathbf{R})$ is connected, i.e. that $\pi_0(\mathrm{GL}_{2n}(\mathbf{R})) = \boldsymbol{Z}^\times$ acts transitively on $\pi_0(\Theta)$.

The genus will be defined in terms of the manifold obtained from $S^n \times S^n$ by removing a point, which plays a special role in this theory. Up to diffeomorphism this manifold may be obtained as a pushout

$$S^n \times \mathbf{R}^n \hookleftarrow \mathbf{R}^n \times \mathbf{R}^n \hookrightarrow \mathbf{R}^n \times S^n, \tag{12.3.5}$$

where the embeddings are induced by a choice of coordinate chart $\mathbf{R}^n \hookrightarrow S^n$. Following [17, Definition 1.3] a Θ-structure on $S^n \times \mathbf{R}^n$ shall be called *admissible* if it "bounds a disk", i.e. is (equivariantly) homotopic to a structure that extends over some embedding $S^n \times \mathbf{R}^n \hookrightarrow \mathbf{R}^{2n}$. Note that this is automatic if $\pi_n(\Theta) = 0$ for some basepoint. A structure on $S^n \times S^n \setminus \{*\}$ is admissible if the restriction to each piece of the gluing (12.3.5) is admissible.

Definition 12.3.1. Assume $d = 2n > 0$ and that W is connected. The *genus* $g(W, \rho_W)$ of a Θ-manifold (W, ρ_W) is the maximal number of disjoint embeddings $j : S^n \times S^n \setminus \{*\} \to W$ such that $j^*\rho_W$ is admissible.

This appeared as [17, Definition 1.3]. For example, when $n = 1$ and $\Theta = \pi_0(\mathrm{GL}_2(\mathbf{R})) = \mathbf{Z}^\times$, this is precisely the usual genus of an oriented connected 2-manifold. The admissibility condition may be illustrated by the case $\Theta = \mathrm{GL}_2(\mathbf{R})$, corresponding to framings on 2-manifolds. The Lie group framing on $\Sigma = S^1 \times S^1$ satisfies $g(\Sigma, \rho_{\mathrm{Lie}}) = 0$, but there exist other framings ρ for which $g(\Sigma, \rho) = 1$.

In §12.3.4 below we shall explain how to determine a lower bound on the number $g(W, \rho_W)$ in terms of more accessible invariants.

12.3.3 Main theorem in rational cohomology

The main results of [14, 17, 16] imply that the ring homomorphism (12.3.4) is often an isomorphism in a range of degrees when $\mathbf{k} = \mathbf{Q}$. We explain the statement.

Definition 12.3.2. Assume $\Theta /\!\!/ \mathrm{GL}_{2n}(\mathbf{R})$ is connected. We say Θ is *spherical* if S^{2n} admits a Θ-structure, i.e. if there exists an equivariant map $\mathrm{Fr}(TS^{2n}) \to \Theta$.

The condition is equivalent to existence of an $\mathrm{O}(2n)$-equivariant map $\mathrm{O}(2n+1) \to \Theta$. This obviously holds if the $\mathrm{GL}_{2n}(\mathbf{R})$-action on Θ admits an extension to an action of $\mathrm{GL}_{2n+1}(\mathbf{R})$ which holds in many cases, e.g. $\Theta = \{\pm 1\}$ with the orientation action. See [14, Section 5.1] for more information about this condition.

Theorem 12.3.3. *Let $d = 2n > 4$, W be a closed simply connected d-manifold, and $\rho_W : \mathrm{Fr}(W) \to \Theta$ be a Θ-structure which is n-connected (or, equivalently, such that the associated $\ell_W : W \to B = \Theta /\!\!/ \mathrm{GL}_d(\mathbf{R})$ is n-connected). Equip B with the local system $\mathbf{Q}^{w_1}$ as above, and assume that $H^{k+d}(B; \mathbf{Q}^{w_1})$ is finite dimensional for each $k \geq 1$. Then the ring homomorphism*

$$\mathbf{Q}[\kappa_c \mid c \in \textit{basis of } H^{>d}(B; \mathbf{Q}^{w_1})] \longrightarrow H^*(\mathcal{M}^{\Theta}(W, \rho_W); \mathbf{Q}),$$

as in (12.3.4) is an isomorphism in cohomological degrees $\leq (g(W, \rho_W) - 4)/3$. If in addition Θ is spherical, then the range may be improved to $\leq (g(W, \rho_W) - 3)/2$.

12.3.4 Estimating genus

To apply Theorem 12.3.3 we must calculate the invariant $g(W, \rho_W)$, or at least be able to give useful lower bounds for it. This section explains a general such lower bound, under the assumptions that $d = 2n > 4$, and that the homotopy orbit space $B = \Theta /\!\!/ \mathrm{GL}_d(\mathbf{R})$ is simply connected. We may then choose an equivariant map $\Theta \to \mathbf{Z}^\times$ by which any Θ-structure induces an orientation, and we shall assume given such a map.

There is an obvious upper bound for $g(W, \rho_W)$: it is certainly no larger than the number of hyperbolic summands in $H_n(W; \mathbf{Z})$ equipped with its intersection form. For odd n this is in turn no more than $b_n/2$ and for even n it is no more than $\min(b_n^+, b_n^-)$, where we write b_n for the middle Betti number of W and in the even case, $b_n = b_n^+ + b_n^-$ for its splitting into positive and negative parts. More usefully, [17, Remark 7.16] gives the following *lower* bound on genus.

Theorem 12.3.4. *Assume $d = 2n > 4$, that the homotopy orbit space $B = \Theta /\!\!/ \mathrm{GL}_d(\mathbf{R})$ is simply connected, and that $\ell_W : W \to B$ (or equivalently $\rho_W : \mathrm{Fr}(TW) \to \Theta$) is n-connected. Write $g^a(W) = \min(b_n^+, b_n^-)$ for n even and $g^a(W) = b_n/2$ for n odd. Then*

$$g^a(W) - c \leq g(W, \rho_W) \leq g^a(W), \tag{12.3.6}$$

with $c = 1 + e$, where e is the minimal number of generators of the abelian group $H_n(B; \mathbf{Z})$. If n is even or if $n \in \{3, 7\}$ then one may take $c = e$.

Remark 12.3.5. Let us briefly point out that the estimate (12.3.6) may be expressed using characteristic numbers. Indeed, writing $b_i = b_i(B) = b_i(W)$ for $i = 1, \ldots, n-1$, we have

$$g^a(W) = (-1)^n \big(\chi(W)/2 - \sum_{i=0}^{n-1} (-1)^i b_i\big) - |\sigma(W)|/2, \tag{12.3.7}$$

where $\chi(W) = \int_W e(TW)$ is the Euler characteristic and $\sigma(W) = \int_W \mathcal{L}(TW)$ is the signature (where we write $\sigma(W) = 0$ when n is odd).

12.4 General versions of main results

For some purposes, the rational cohomology statement in Theorem 12.3.3 suffices, but there are several homotopy theoretic strengthenings and variations given in [16], which we now explain.

12.4.1 Stable homotopy enhancement

To state a more robust version of Theorem 12.3.3 above, we must first introduce a space $\Omega^\infty MT\Theta$ associated to the $\mathrm{GL}_d(\mathbf{R})$-space Θ. The map $B = \Theta /\!\!/ \mathrm{GL}_d(\mathbf{R}) \to B\mathrm{GL}_d(\mathbf{R})$ classifies a d-dimensional real vector bundle over B, and we shall write $MT\Theta$ for the Thom spectrum of its virtual inverse and $\Omega^\infty MT\Theta$ for the corresponding infinite loop space. The following result is a restatement of [16, Corollary 1.7].

Theorem 12.4.1. *Let $d = 2n > 4$, W be a closed simply connected d-manifold, and $\rho_W : \mathrm{Fr}(W) \to \Theta$ be an n-connected equivariant map. Then there is a map*

$$\alpha : \mathcal{M}^\Theta(W, \rho_W) \longrightarrow \Omega^\infty MT\Theta, \tag{12.4.1}$$

inducing an isomorphism in integral homology onto the path component that it hits, in degrees $\leq (g(W, \rho_W) - 4)/3$.

If in addition Θ is spherical, then (12.4.1) induces an isomorphism in homology in degrees $\leq (g(W, \rho_W) - 3)/2$.

The statement proved in [16] is in fact stronger: the map α induces an isomorphism in homology with local coefficients in degrees up to $(g(W, \rho_W) - 4)/3$. We say that the map is *acyclic* in this range of degrees. (The induced map in π_1 is of course far from an isomorphism.)

Remark 12.4.2. Let us also briefly recall why Theorem 12.3.3 is a consequence of Theorem 12.4.1: under the Thom isomorphism $H^{k+d}(B; \mathbf{Q}^{w_1}) \cong H^k(MT\Theta; \mathbf{Q})$ each class $c \in H^{k+d}(B; \mathbf{Q}^{w_1})$ may be represented by a spectrum map $MT\Theta \to \Sigma^k H\mathbf{Q}$. If we choose a rational basis $\mathcal{B}_k \subset H^{k+d}(B; \mathbf{Q}^{w_1}) \cong H^k(MT\Theta; \mathbf{Q})$ and represent each basis element by a spectrum map $MT\Theta \to \Sigma^k H\mathbf{Q}$, we obtain

$$MT\Theta \longrightarrow \prod_{k=1}^{\infty} \prod_{c \in \mathcal{B}_k} \Sigma^k H\mathbf{Q}$$

which induces isomorphisms in rational homology and hence in rationalised homotopy in positive degrees, at least if each $\mathcal{B}_k$ is finite in which case the product in the codomain may be replaced by the wedge. It follows that the induced map of infinite loop spaces

$$\Omega^\infty MT\Theta \longrightarrow \prod_{k=1}^{\infty} \prod_{\mathcal{B}_k} K(\mathbf{Q}, k)$$

induces an equivalence in rationalised homotopy groups in positive degrees, hence in rational cohomology when restricted to any path component of its domain.

12.4.2 MMM-classes in generalised cohomology

There is a preferred map (12.4.1) in Theorem 12.4.1. Following [32], in [14, Remark 1.11] we gave an explicit point-set model. Here we shall explain the map in a conceptual but somewhat informal way. It is obtained from the following two ingredients.

(i) A smooth d-manifold W and an equivariant map $\rho_W : \mathrm{Fr}(TW) \to \Theta$ induces a continuous map

$$\ell_W = \rho_W /\!\!/ \mathrm{GL}_d(\mathbf{R}) : W \simeq \mathrm{Fr}(TW)/\!\!/\mathrm{GL}_d(\mathbf{R}) \longrightarrow B = \Theta/\!\!/\mathrm{GL}_d(\mathbf{R})$$

under which the canonical bundle γ on $B\mathrm{GL}_d(\mathbf{R}) = */\!\!/\mathrm{GL}_d(\mathbf{R})$ is pulled back to TW. By passing to Thom spectra of inverse bundles one gets a map

$$W^{-TW} \longrightarrow B^{-\gamma} = MT\Theta, \tag{12.4.2}$$

in the stable homotopy category.

(ii) If W is a closed manifold there is a canonical map

$$S^0 \longrightarrow W^{-TW} \tag{12.4.3}$$

which is Spanier–Whitehead dual to the canonical map $W \to \{*\}$ under Atiyah duality $D(W_+) \simeq W^{-TW}$. The actual map of spectra depends on certain choices, but in the right setup these choices form a contractible space. (For example, one may choose an embedding $W \hookrightarrow \mathbf{R}^\infty$ and get the map (12.4.3) by the Pontryagin–Thom collapse construction.)

If W is a smooth closed d-manifold and $\rho_W : \mathrm{Fr}(TW) \to \Theta$ is an equivariant map, we may compose (12.4.3) and (12.4.2) to get a map of spectra $S^0 \to MT\Theta$, i.e. a point in $\Omega^\infty MT\Theta$. We shall write $\alpha(W, \rho_W) \in \Omega^\infty MT\Theta$ for this point, and write

$$\Omega^\infty_{[W,\rho_W]} MT\Theta \subset \Omega^\infty MT\Theta$$

for the path component containing $\alpha(W, \rho_W)$. This is the path component that the map (12.4.1) lands in.

The map (12.4.1) is given by a parametrised version of this construction: given a family $(\pi : E \to X, \rho : \mathrm{Fr}(T_\pi E) \to \Theta)$ as in Definition 12.2.1 over some base manifold X, it associates a continuous map $\alpha : X \to \Omega^\infty MT\Theta$. Said differently, it comes from a composition of spectrum maps, the parametrised analogues of (12.4.3) and (12.4.2) respectively,

$$\begin{aligned} \Sigma^\infty_+ X &\longrightarrow E^{-T_\pi E} \\ E^{-T_\pi E} &\longrightarrow MT\Theta. \end{aligned} \tag{12.4.4}$$

In any case, (12.4.1) is a universal version of this construction.

Remark 12.4.3. Applying spectrum homology and the Thom isomorphism to the first map in (12.4.4), we get precisely the fibre integration homomorphism (12.3.3), while the second gives (12.3.2). This explains the connection to the characteristic classes in §12.3.1.

Remark 12.4.4. There is an improvement to Theorem 12.4.1, in which the domain of (12.4.1) is replaced by a disconnected space $\mathcal{M}^\Theta_n$ containing $\mathcal{M}^\Theta(W, \rho_W)$ as one of its path components, cf. §12.4.4 below and [16, Section 8]. In this improved formulation, the number $g(W, \rho_W) \in \mathbf{N}$ appearing in Theorem 12.4.1 is replaced by a function

$$\pi_0(\mathcal{M}^\Theta_n) \xrightarrow{\alpha_*} \pi_0(MT\Theta) \xrightarrow{\bar{g}^\Theta} \mathbf{Z},$$

whose value on the path component $\mathcal{M}^\Theta(W, \rho_W)$ is the *stable genus*, defined in [17, Section 5] and [16, Section 1.3]. It is at least $g(W, \rho_W)$.

If we write $\chi : \pi_0(MT\Theta) \to \boldsymbol{Z}$ and $\sigma : \pi_0(MT\Theta) \to \boldsymbol{Z}$ be the homomorphisms arising from the Euler class and (for even n) the Hirzebruch class, then (12.3.7) defines a function $g^a : \pi_0(MT\Theta) \to \mathbf{N}$. In terms of this function, the estimate (12.3.6) also holds for $\bar{g}^\Theta$.

12.4.3 General tangential structures

The requirement in Theorems 12.3.3 and 12.4.1 that $\rho_W : \mathrm{Fr}(TW) \to \Theta$ be n-connected appears quite restrictive at first sight. For example, it usually rules out the interesting special cases $\Theta = \{*\}$ and $\Theta = \{\pm 1\}$ from Remark 12.2.8, so that the cohomology of $B\mathrm{Diff}(W)$ and $B\mathrm{Diff}^{\mathrm{or}}(W)$ are not immediately calculated by Theorem 12.4.1.

A more generally useful version of Theorem 12.4.1, which holds without the connectivity assumption, may be deduced by a rather formal homotopy theoretic trick, based on the observation that the map (12.4.1) is functorial in the $\mathrm{GL}_d(\mathbf{R})$-space Θ.

In particular, any map $\Theta \to \Theta$ induces a self-map of $\Omega^\infty MT\Theta$. The following result is [16, Corollary 1.9].

Theorem 12.4.5. *For $d = 2n > 4$ and Λ a $\mathrm{GL}_d(\mathbf{R})$-space, let W be a closed simply connected smooth d-dimensional manifold, and $\lambda_W : \mathrm{Fr}(TW) \to \Lambda$ be an equivariant map. Choose an equivariant Moore–Postnikov n-stage*

$$\lambda_W : \mathrm{Fr}(TW) \xrightarrow{\rho_W} \Theta \xrightarrow{u} \Lambda, \tag{12.4.5}$$

i.e. a factorisation where u is n-co-connected equivariant fibration and ρ_W is an n-connected equivariant cofibration, and write $\mathrm{hAut}(u)$ *for the group-like topological monoid consisting of equivariant weak equivalences $\Theta \to \Theta$ over Λ.*

This topological monoid acts on $\Omega^\infty MT\Theta$, and there is a continuous map

$$\alpha : \mathcal{M}^\Lambda(W, \lambda_M) \longrightarrow (\Omega^\infty MT\Theta)/\!\!/\mathrm{hAut}(u). \tag{12.4.6}$$

which, when regarded as a map onto the path component that it hits, induces an isomorphism in homology with local coefficients in degrees $\leq (g(W, \lambda_W) - 4)/3$, and if Θ is spherical it induces an isomorphism in integral homology in degrees $\leq (g(W, \lambda_W) - 3)/2$.

We emphasise that the homotopy orbit space is formed in the category of spaces, not of infinite loop spaces (there is a comparison map $(\Omega^\infty MT\Theta)/\!\!/\mathrm{hAut}(u) \to \Omega^\infty((MT\Theta)/\!\!/\mathrm{hAut}(u))$ but it is not a weak equivalence and we shall not need its codomain). Equivariant factorisations (12.4.5) always exist, and are unique up to contractible choice.

Let us also point out that the group $\pi_0(\mathrm{hAut}(u))$ likely acts non-trivially on $\pi_0(MT\Theta)$. The construction of the map (12.4.6), outlined in §12.4.4 below, together with the orbit-stabiliser theorem, lets us re-write the relevant path component of the codomain of (12.4.6) in the following way, which is how the theorem above is typically used in practice.

Corollary 12.4.6. *Let d, Λ, W, λ_W, Θ, ρ_W, and u be as in Theorem 12.4.5, and write*

$$\mathrm{hAut}(u)_{[W,\rho_W]} \subset \mathrm{hAut}(u)$$

for the submonoid stabilising the element $[W, \rho_W] \in \pi_0(MT\Theta)$ defined in §12.4.2. The action of this submonoid on $\Omega^\infty MT\Theta$ restricts to an action on the path component $\Omega^\infty_{[W,\rho_W]}MT\Theta$ defined in §12.4.2, and (12.4.6) factors through a map of path connected spaces

$$\alpha : \mathcal{M}^\Lambda(W, \lambda_W) \longrightarrow (\Omega^\infty_{[W,\rho_W]}MT\Theta)/\!\!/(\mathrm{hAut}(u)_{[W,\rho_W]})$$

which induces an isomorphism on homology in a range, as in Theorem 12.4.5. □

Remark 12.4.7. In both Theorem 12.4.5 and Corollary 12.4.6 above, the range is expressed in terms of $g(W, \lambda_W)$ but as explained in [16, Lemma 9.4] this is equal to $g(W, \rho_W)$ when λ_W is factored equivariantly as an n-connected map $\rho_W : \mathrm{Fr}(TW) \to \Theta$ followed by an n-co-connected map $u : \Theta \to \Lambda$. Hence the estimates in §12.3.4 apply, when e is the minimal number of generators for the abelian group $H_n(\Theta/\!\!/\mathrm{GL}_{2n}(\mathbf{R}))$. The value of e likely depends on the map $\lambda_W : \mathrm{Fr}(TW) \to \Lambda$, even if W and Λ are fixed.

Remark 12.4.8. The fact that u is n-co-connected implies that $\mathrm{hAut}(u)$ is an $(n-1)$-type. Hence it is in some sense a finite problem to determine and describe $\mathrm{hAut}(u)$: finitely many homotopy groups $\pi_0, \ldots, \pi_{n-1}$ and finitely many k-invariants.

This finiteness is one of the conceptual advantages of our approach to $\mathcal{M}(W) \simeq B\mathrm{Diff}(W)$, over the more classical method which at first gives a formula for the *structure space* $\mathcal{S}(W) \simeq G(W)/\widetilde{\mathrm{Diff}}(W)$, where $G(W)$ is the monoid of homotopy equivalences from W to itself and $\widetilde{\mathrm{Diff}}(W)$ is the block diffeomorphism group. In that method, one subsequently has to study the difference between $\mathrm{Diff}(W)$ and $\widetilde{\mathrm{Diff}}(W)$, but also has to take homotopy orbits by the monoid $G(W)$ of homotopy automorphisms of W. While this last step is in some sense "purely homotopy theory", it is in practice very difficult to get a good handle on $G(W)$, even when W is relatively simple and even when one is working up to rational equivalence. See the work of Berglund and Madsen [1, 2] for a recent example.

12.4.4 Two fibre sequences

In [16, Section 9], Theorem 12.4.5 is deduced from the special case given in Theorem 12.4.1 by a rather formal argument: the homology equivalence in Theorem 12.4.1 is natural in the $\mathrm{GL}_{2n}(\mathbf{R})$-space Θ, and hence induces a homology equivalence by taking homotopy colimit over any diagram in $\mathrm{GL}_{2n}(\mathbf{R})$-spaces; in particular one may form homotopy orbits by $\mathrm{hAut}(u : \Theta \to \Lambda)$, and Theorem 12.4.5 is deduced by identifying the resulting map of homotopy orbit spaces with (12.4.6). The two fibre sequences arising from these homotopy orbit constructions, see diagram (12.4.8) below, are important for carrying out calculations in concrete examples, and hence we recall this story in slightly more detail.

Definition 12.4.9. Let $u : \Theta \to \Lambda$ be an equivariant n-co-connected fibration.

(i) Let $\mathcal{M}_n^\Theta \subset \mathcal{M}^\Theta$ be the union of those path components $\mathcal{M}^\Theta(W, \rho_W)$ for which $\rho_W : \mathrm{Fr}(TW) \to \Theta$ is n-connected.

(ii) Let $\mathcal{M}_u^\Lambda \subset \mathcal{M}^\Lambda$ be the union of those path components $\mathcal{M}^\Lambda(W, \lambda_W)$ for which λ_W *admits* a factorisation through an n-connected equivariant map $\rho_W : \mathrm{Fr}(TW) \to \Theta$.

There is an obvious forgetful map $\mathcal{M}_n^\Theta \to \mathcal{M}_u^\Lambda$ given by composing $\rho_W : \mathrm{Fr}(TW) \to \Theta$ with u. The monoid $\mathrm{hAut}(u)$ has the correct homotopy type when Θ is equivariantly cofibrant, which we shall assume. It acts on $\mathcal{M}_n^\Theta$ by postcomposing $\rho_W : \mathrm{Fr}(TW) \to \Theta$ with self-maps of Θ. This action commutes with the forgetful map, and induces a map

$$(\mathcal{M}_n^\Theta)/\!\!/\mathrm{hAut}(u) \longrightarrow \mathcal{M}_u^\Lambda. \tag{12.4.7}$$

The following lemma is proved by elementary homotopy theoretic methods, cf. [16, Section 9].

Lemma 12.4.10. *The map (12.4.7) is a weak equivalence. Hence there is an induced fibre sequence of the form*

$$\mathcal{M}_n^\Theta \longrightarrow \mathcal{M}_u^\Lambda \longrightarrow B(\mathrm{hAut}(u)).$$

The map $\mathcal{M}_n^\Theta \to \Omega^\infty MT\Theta$ explained in §12.4.2 commutes with the actions of $\mathrm{hAut}(u)$ and induces a map of fibre sequences

$$\begin{array}{ccccc} \mathcal{M}_n^\Theta & \longrightarrow & \mathcal{M}_u^\Lambda & \longrightarrow & B(\mathrm{hAut}(u)) \\ \downarrow & & \downarrow & & \| \\ \Omega^\infty MT\Theta & \longrightarrow & (\Omega^\infty MT\Theta)/\!\!/\mathrm{hAut}(u) & \longrightarrow & B(\mathrm{hAut}(u)). \end{array} \tag{12.4.8}$$

In the setup of Theorem 12.4.5, $\mathcal{M}^\Theta(W, \rho_W)$ is one path component of $\mathcal{M}_n^\Theta$, and similarly $\mathcal{M}^\Lambda(W, \lambda_W)$ is one path component of $\mathcal{M}_u^\Lambda$. A slightly stronger version of Theorem 12.4.1, which is the statement actually proved in [16, Section 8], shows that the left-most vertical map is acyclic in the range of degrees indicated in Remark 12.4.4. Theorem 12.4.5 is then deduced by a spectral sequence comparison argument.

Remark 12.4.11. This formulation has content even at the level of path components. Suppose that W is a simply connected $2n$-manifold for $2n \geq 6$, $\rho_W : \mathrm{Fr}(TW) \to \Theta$ is n-connected, and $g(W, \rho_W) \geq 3$. If $(W', \rho_{W'})$ is another such Θ-manifold and

$$\alpha(W, \rho_W) = \alpha(W', \rho_{W'}) \in \pi_0(\Omega^\infty MT\Theta) = \pi_0(MT\Theta)$$

then it follows that (W, ρ_W) and $(W', \rho_{W'})$ lie in the same path-component of $\mathcal{M}_n^\Theta$, i.e. there is a diffeomorphism from W to W' which pulls back $\rho_{W'}$ to ρ_W up to homotopy. This recovers a theorem of Kreck [28, Theorem D], though our requirement on genus is slightly stronger than Kreck's.

In practice, one usually calculates the cohomology of $\Omega^\infty_{[W,\rho_W]} MT\Theta$ first, and then uses a spectral sequence to calculate the homology or cohomology of the Borel construction in Corollary 12.4.6, or equivalently (for $g(W, \lambda_W) \geq 3$) one calculates the cohomology of $\mathcal{M}^\Theta(W, \rho_W)$ and then uses the spectral sequence for the fibre sequence

$$\mathcal{M}^\Theta(W, \rho_W) \longrightarrow \mathcal{M}^\Lambda(W, \lambda_W) \longrightarrow B(\mathrm{hAut}(u)_{[W,\rho_W]}). \tag{12.4.9}$$

We shall see examples of such calculations in §12.5.2, §12.5.3, and §12.5.4.

12.4.5 $\mathrm{GL}_d(\mathbf{R})$-spaces versus spaces over $BO(d)$

It is well known that the homotopy theory of spaces with action of $\mathrm{GL}_d(\mathbf{R})$ is equivalent to the homotopy theory of spaces over $B\mathrm{GL}_d(\mathbf{R}) \simeq BO(d)$, where the weak equivalences are the equivariant maps that are weak equivalences of underlying spaces, respectively fibrewise maps that are weak equivalences of underlying spaces. The translation goes via the space $E\mathrm{GL}_d(\mathbf{R})$ which simultaneously comes with an action of $\mathrm{GL}_d(\mathbf{R})$ and a map to $B\mathrm{GL}_d(\mathbf{R})$. Explicitly, given a $\mathrm{GL}_d(\mathbf{R})$-space Θ, the Borel construction $B = \Theta /\!/ \mathrm{GL}_d(\mathbf{R})$ comes with a map $B \to B\mathrm{GL}_d(\mathbf{R})$; conversely, given a space B and a map $\theta : B \to B\mathrm{GL}_d(\mathbf{R})$ the fibre product $\Theta = E\mathrm{GL}_d(\mathbf{R}) \times_{B\mathrm{GL}_d(\mathbf{R})} B$ comes with an action; these processes are inverse up to (equivariant/fibrewise) weak equivalence, as $E\mathrm{GL}_d(\mathbf{R})$ is contractible.

Therefore all of the theorems above that depend on a $\mathrm{GL}_d(\mathbf{R})$-space Θ may be stated in equivalent ways taking as input a space B and a map $\theta : B \to BO(d)$. In the papers [14, 17, 16] we have taken the latter point of view. In this picture, a θ-structure on a manifold W is a (fibrewise linear) vector bundle map $\hat{\ell}_W : TW \to \theta^*\gamma$, where γ denotes the universal vector bundle on $B\mathrm{GL}_d(\mathbf{R})$. As in those papers, we shall use the notation

$$MT\theta = MT\Theta,$$

when $\theta : B \to BO(d)$ is the map corresponding to the $\mathrm{GL}_d(\mathbf{R})$-space Θ; i.e., $MT\theta = B^{-\theta}$ is the Thom spectrum of the virtual inverse of the vector bundle classified by $\theta : B \to BO(d)$.

We have already seen definitions which may be stated more directly in terms of (B, θ) than of (Θ, action), e.g. the characteristic classes κ_c from §12.3.1. The constructions in

Theorem 12.4.5 form another such example, which we shall now explain. Given Λ and λ_W as in the theorem, set $B' = \Lambda /\!/ \mathrm{GL}_d(\mathbf{R})$. Up to contractible choice, λ_W induces a map $W \to B'$, which one then Moore–Postnikov factors as

$$W \longrightarrow B \longrightarrow B',$$

into an n-connected cofibration followed by an n-co-connected fibration. In this picture, $\mathrm{hAut}(u)$ is simply the group-like monoid of those self-maps of B over B' that are weak equivalences. For this to have to the correct homotopy type B should be fibrant and cofibrant in the category of spaces over B'.

In §12.5 and §12.6 we will exclusively adopt this point of view.

12.4.6 Boundary

A further generalisation, also proved in [16, Section 9], allows the compact manifolds W to have nonempty boundary. The boundary should then be a closed $(2n-1)$-manifold P, which should be equipped with an equivariant map $\rho_P : \mathrm{Fr}(\varepsilon^1 \oplus TP) \to \Theta$. The pair (P, ρ_P) should be fixed and every compact $2n$-manifold in sight should come with a specified diffeomorphism $\partial W \cong P$ compatible with a structure $\rho_W : \mathrm{Fr}(TW) \to \Theta$.

In terms of classified bundles as in §12.2.1 and §12.2.2, $\mathcal{F}^{\Theta}_{\bullet}$ should be replaced with the functor $\mathcal{F}^{\Theta,P,\rho_P}_{\bullet}$ whose value on a smooth manifold X (without boundary, possibly non-compact) has 0-simplices the smooth proper maps $\pi : E \to X$ equipped with equivariant maps $\rho : \mathrm{Fr}(T_\pi E) \to \Theta$ and a diffeomorphism $\partial E = X \times P$ such that the restriction of ρ to $\mathrm{Fr}(T_\pi E|_{\partial E}) = X \times \mathrm{Fr}(TP)$ is equal to the map arising from ρ_P.

This kind of bundle also admits a classifying space, denoted $\mathcal{N}^{\Theta}(P, \rho_P)$ and called the *moduli space of null bordisms* of (P, ρ_P). The subspace defined by the condition that $\rho_W : \mathrm{Fr}(TW) \to \Theta$ be n-connected is denoted $\mathcal{N}^{\Theta}_n(P, \rho_P)$ and is the *moduli space of highly connected null bordisms.* The path component of $\mathcal{N}^{\Theta}(P, \rho_P)$ containing (W, ρ_W) shall be denoted $\mathcal{M}^{\Theta}(W, \rho_W)$, as before. Notice that (P, ρ_P) is determined by $P = \partial W$ and ρ_W by restricting ρ_W to $TW|_P \cong \varepsilon^1 \oplus TP$. These classifying spaces are introduced in [16, Definition 1.1] using a similar notation.

Theorem 12.4.1 then has the following direct generalisation, also stated as [16, Corollary 1.8 and Section 8.4].

Theorem 12.4.12. *Let $d = 2n > 4$, let Θ be a $\mathrm{GL}_d(\mathbf{R})$-space, let P be a closed smooth $(d-1)$-manifold and $\rho_P : \mathrm{Fr}(\varepsilon^1 \oplus TP) \to \Theta$ be a $\mathrm{GL}_d(\mathbf{R})$-equivariant map. Then there is a map (canonical up to homotopy, see below)*

$$\alpha : \mathcal{N}^{\Theta}_n(P, \rho_P) \longrightarrow \Omega_{\alpha(P,\rho_P),0}(\Omega^{\infty-1} MT\Theta), \tag{12.4.10}$$

where $\Omega_{\alpha(P,\rho_P),0}$ denotes the space of paths starting at a certain point $\alpha(P, \rho_P) \in \Omega^{\infty-1} MT\Theta$ and ending at the basepoint, with the following property.

When restricted to the path component containing a particular (W, ρ_W), it is a homology equivalence onto the path component it hits, in degrees up to $(g(W, \rho_W) - 4)/3$ and possibly with twisted coefficients. If in addition Θ is spherical, then (12.4.10) induces an isomorphism in homology with constant coefficients in degrees up to $(g(W, \rho_W) - 3)/2$.

Both the point $\alpha(P, \rho_P)$ and the map (12.4.10) are constructed by the procedure in §12.4.2. If the codomain of (12.4.10) is nonempty, it is of course non-canonically homotopy equivalent to $\Omega^\infty MT\Theta$, but the map most naturally takes values in the path space. In the special case $P = \emptyset$ we have $\mathcal{N}_n^\Theta(\emptyset) = \mathcal{M}_n^\Theta$, and the map (12.4.10) is the same as the map appearing in (12.4.8).

As in §12.4.4, a version for a general tangential structure Λ may be deduced by taking homotopy orbits with respect to the monoid

$$\mathrm{hAut}(u \text{ rel } P) = \{\phi \in \mathrm{hAut}(u) \mid \phi \circ \rho_P = \rho_P\}, \tag{12.4.11}$$

provided $\rho_P : \mathrm{Fr}(\varepsilon^1 \oplus TP) \to \Theta$ is an equivariant cofibration and $u : \Theta \to \Lambda$ is an equivariant n-co-connected fibration, as can be arranged. Homotopy equivalently, factor the induced map $W \to B' = \Lambda /\!/ \mathrm{GL}_d(\mathbf{R})$ as an n-connected cofibration $W \to B$ followed by an n-co-connected fibration $B \to B'$, and define $\mathrm{hAut}(u \text{ rel } P)$ as the homotopy equivalences of B over B' and under P. We formulate the conclusion.

Theorem 12.4.13. *Let n, d, Λ and $\lambda_W : \mathrm{Fr}(TW) \to \Lambda$ be as in Theorem 12.4.5, but allow W to be a compact manifold with boundary $P = \partial W$. Let Θ, ρ_W, and u be as in Theorem 12.4.5, and ρ_P denote the restriction of ρ_W to P. Then there is a map*

$$\alpha : \mathcal{M}^\Lambda(W, \lambda_W) \longrightarrow \left(\Omega_{\alpha(P,\rho_P),0}\Omega^{\infty-1} MT\Theta\right) /\!/ \mathrm{hAut}(u \text{ rel } P) \tag{12.4.12}$$

which, when regarded as a map onto the path component that it hits, induces an isomorphism in homology in a range of degrees, exactly as in Theorem 12.4.5.

This is [16, Theorem 9.5], and the three lines following its proof. The relevant path component may again be re-written using the orbit-stabiliser theorem, as in Corollary 12.4.6.

Remark 12.4.14. As in Remark 12.2.8, in the special case $\Theta = \{*\}$ a Θ-structure contains no information and we can simply write $\mathcal{M}(W)$ for $\mathcal{M}^\Theta(W, \rho_W)$. This space classifies smooth fibre bundles with fibres diffeomorphic to W and trivialised boundary, and we have a weak equivalence

$$\mathcal{M}(W) \simeq B\mathrm{Diff}_\partial(W)$$

where $\mathrm{Diff}_\partial(W)$ denotes the group of diffeomorphisms of W which fix an open neighbourhood of the boundary, with the C^∞ topology.

There are no connectivity assumptions imposed on $\rho_P : \mathrm{Fr}(\varepsilon^1 \oplus TP) \to \Theta$, but if it happens to be $(n-1)$-connected then the monoid $\mathrm{hAut}(u \text{ rel } P)$ is contractible. More generally we have the following.

Lemma 12.4.15. *If the pair (W, P) is c-connected for some $c \leq n-1$, then the monoid $\mathrm{hAut}(u \text{ rel } P)$ is a (nonempty) $(n-c-2)$-type. In particular, it is contractible if (W, P) is $(n-1)$-connected.*

A familiar special case of this observation is the fact that if a diffeomorphism of an oriented surface with nonempty boundary is the identity on the boundary, then the diffeomorphism is automatically orientation preserving.

Proof. As before, let us write $B = \Theta /\!\!/ \mathrm{GL}_d(\mathbf{R})$ and $B' = \Lambda /\!\!/ \mathrm{GL}_d(\mathbf{R})$. We are then asking for homotopy automorphisms of B over $u : B \to B'$ and under $\ell_P : P \hookrightarrow B$. By adjunction, to give a nullhomotopy of a map $f : S^k \to \mathrm{hAut}(u \text{ rel } P)$ is to solve the relative lifting problem

$$\begin{array}{ccc} (S^k \times B) \cup_{S^k \times P} (D^{k+1} \times P) & \xrightarrow{\tilde{f} \cup (\ell_P \circ \mathrm{proj})} & B \\ \downarrow & & \downarrow u \\ D^{k+1} \times B & \xrightarrow{\mathrm{proj}} B \xrightarrow{u} & B'. \end{array}$$

The map $\ell_P : P \to B$ is c-connected because both $P \subset W$ and $\ell_W : W \to B$ are (the latter is even n-connected). Thus the pair

$$(D^{k+1} \times B, (S^k \times B) \cup_{S^k \times P} (D^{k+1} \times P))$$

is $(c + k + 1)$-connected. But the map $u : B \to B'$ is n-co-connected, so there are no obstructions to solving this lifting problem if $c + k + 1 \geq n$, i.e. $k \geq n - c - 1$. This proves that $\mathrm{hAut}(u \text{ rel } P)$ is an $(n-c-2)$-type, and it is nonempty because it contains the identity map. □

Remark 12.4.16. Formulating a statement which is valid for manifolds with nonempty boundaries is not purely for the purpose of added generality: it is essential for the strategy of proof in all three papers [14], [17], [16]. For example, the homological stability results in [17] are proved by a long *handle induction* argument, in which a compact manifold is decomposed into finitely many *handle attachments*; even if one is mainly interested in closed manifolds, this process will create boundary. Similarly, an important role in both [14] and [16] is played by *cobordism categories* as studied in [13], whose morphisms are manifolds with boundary and composition is gluing along common boundary components. For example, given a Θ-cobordism $(K, \rho_K) : (P, \rho_P) \rightsquigarrow (Q, \rho_Q)$ there is a continuous map

$$(K, \rho_K) \cup - : \mathcal{N}^\Theta(P, \rho_P) \longrightarrow \mathcal{N}^\Theta(Q, \rho_Q)$$

given by gluing on (K, ρ_K).

12.4.7 Fundamental group

The main theorem in either of the three forms given above (Theorems 12.3.3, 12.4.1, and 12.4.5) assumed the manifolds W were simply connected, but in fact it suffices that the fundamental groups $\pi = \pi_1(W, w)$ be *virtually polycyclic*, i.e. has a subnormal series with finite or cyclic quotients. In this case the *Hirsch length* $h(\pi)$ is the number of infinite cyclic quotients in such a series. The only price to pay is that the ranges of homology equivalence become offset by a constant depending on $h(\pi)$: the homology isomorphisms Theorems 12.4.1 and 12.4.5 hold in degrees $\leq (g(W, \rho_W) - (h(\pi) + 5))/2$ with integral coefficients if Θ is spherical, and in degrees $\leq (g(W, \rho_W) - (h(\pi) + 6))/3$ with local coefficients. This generalisation was established by Friedrich [11].

For arbitrary π there is a sense in which the theorems hold in "infinite genus": certain maps become acyclic after taking a colimit over forming connected sum with $S^n \times S^n$

infinitely many times. In this form the assumption $2n > 4$ is also unnecessary. See [16, Sections 1.2 and 7] for the statement and proof.

12.4.8 Outlook

We have attempted to give an overview of the methods developed in [14], [17], [16], with an emphasis on the main results from there as they may be applied in calculations in practice. This is by no means a survey of everything known; let us briefly mention some recent developments and applications that we have not covered:

(i) These results—in the form of the calculation described in Theorem 12.5.1 below—have been used by Weiss [50] to prove that $p_n \neq e^2 \in H^{4n}(B\mathrm{STop}(2n); \mathbf{Q})$ for large enough n. These methods were later used by Kupers [30] to establish the finite generation of homotopy groups of $\mathrm{Diff}_\partial(D^d)$ for $d \neq 4, 5, 7$.

(ii) These results have been used by Botvinnik, Ebert, and Randal-Williams [3], Ebert and Randal-Williams [10], and Botvinnik, Ebert, and Wraith [4] to study the topology of spaces of Riemannian metrics of positive scalar, or Ricci, curvature.

(iii) These results have been used by Krannich [24] to show that if Σ is a homotopy $2n$-sphere then $B\mathrm{Diff}(M)$ and $B\mathrm{Diff}(M\#\Sigma)$ have the same homology in the stable range with $\mathbf{Z}[\frac{1}{k}]$-coefficients, where k is the order of Σ is the group of homotopy spheres.

(iv) Progress towards a similar understanding for manifolds of *odd dimension* has been made by Perlmutter [42, 43, 44], Botvinnik and Perlmutter [6] and Hebestreit and Perlmutter [22].

(v) Progress towards versions for *topological* and *piecewise linear* manifolds has been made by Gomez-Lopez [19], Kupers [29], and Gomez-Lopez and Kupers [20].

(vi) Progress towards versions for *equivariant smooth* manifolds has been made by Galatius and Szűcs [18].

(vii) There are analogues of many of the theorems above, when the topological group $\mathrm{Diff}(W)$ is replaced by its *underlying discrete group*, by Nariman [39, 40, 38].

(viii) Progress towards understanding the homotopy equivalences of high genus manifolds have been obtained by Berglund and Madsen [2]. At present their results seem to be qualitatively quite different from the results described here.

12.5 Rational cohomology calculations

We have already advertised the feature that the general theory surveyed in §12.3 and §12.4 above is amenable to explicit calculations. In this section and the next, we back up this claim with some examples, while simultaneously illustrating how the abstract homotopy theory in §12.4 plays out in concrete examples.

In practice, given a manifold W and a structure $\lambda_W : \mathrm{Fr}(W) \to \Lambda$, one typically first estimates the genus of (W, λ_W). This step is trivial for the manifolds considered in §12.5.1 and §12.5.2 below, which are defined to have high genus, but is quite interesting for the complete intersections considered in §12.5.3. The next step would typically be to determine the associated highly connected structure $\rho_W : \mathrm{Fr}(W) \to \Theta$ and the space $B\mathrm{hAut}(u \text{ rel } P)$. This step is mostly resolved by Lemma 12.4.15 for the example given in §12.5.1, but is again interesting for the examples in §12.5.2 and especially §12.5.3. For calculations in rational cohomology, the last step would then typically be to understand the Serre spectral sequence associated to (12.4.9). The complete intersections in §12.5.3 again provide an interesting and very non-trivial illustration.

12.5.1 The manifolds $W_{g,1}$

Recall that we write $W_g = g(S^n \times S^n)$ for the g-fold connected sum, and $W_{g,1} = D^{2n} \# W_g \cong W_g \setminus \mathrm{int}(D^{2n})$. These manifolds play a distinguished role in the theory described above, as they are used to measure the genus of arbitrary $2n$-manifolds: in this sense $W_{g,1}$ is the simplest manifold of genus g. In the case $2n = 2$ the solution by Madsen and Weiss [33] of the Mumford Conjecture gave a description of $H^*(\mathcal{M}^{\mathrm{or}}(W_{g,1}); \mathbf{Q})$ in terms of Miller–Morita–Mumford classes in a stable range of degrees. In this section we wish to explain how the analogue of Madsen and Weiss' result in dimensions $2n \geq 6$ follows from the theory described above.

The Moore–Postnikov n-stage

$$\tau_{W_{g,1}} : W_{g,1} \xrightarrow{\ell_{W_{g,1}}} B \xrightarrow{\theta} BO(2n)$$

of a map classifying the tangent bundle of $W_{g,1}$ has the cofibration $\ell_{W_{g,1}}$ n-connected and the fibration θ n-co-connected. As $W_{g,1}$ is $(n-1)$-connected and the map $\tau_{W_{g,1}}$ is nullhomotopic—because $W_{g,1}$ admits a framing—we may identify θ with the n-connected cover of $BO(2n)$, which we write as

$$\theta_n : BO(2n)\langle n \rangle \xrightarrow{\theta_n^{\mathrm{or}}} B\mathrm{SO}(2n) \xrightarrow{\mathrm{or}} BO(2n).$$

As the pair $(W_{g,1}, \partial W_{g,1})$ is $(n-1)$-connected, by Lemma 12.4.15 we find that $\mathrm{hAut}(\theta_n^{\mathrm{or}}, \ell_{\partial W_{g,1}})$ is contractible. Thus by Theorem 12.4.13 there is a map

$$\alpha : \mathcal{M}^{\mathrm{or}}(W_{g,1}) \simeq \mathcal{M}^{\theta_n}(W_{g,1}, \ell_{W_{g,1}}) \longrightarrow \Omega^\infty MT\theta_n$$

which is a homology equivalence onto the path component that it hits, in degrees $* \leq \frac{g-3}{2}$, as long as $2n \geq 6$. The rational cohomology of a path component of $\Omega^\infty MT\theta_n$ is calculated as described in Remark 12.4.2, in terms of $H^*(BO(2n)\langle n \rangle; \mathbf{Q})$. To work this out we identify $BO(2n)\langle n \rangle = B\mathrm{SO}(2n)\langle n \rangle$. As

$$H^*(B\mathrm{SO}(2n); \mathbf{Q}) = \mathbf{Q}[e, p_1, p_2, \ldots, p_{n-1}]$$

is a free graded-commutative algebra the effect of taking the n-connected cover on cohomology groups is a simple as possible: it simply eliminates all free generators of degree $\leq n$.

Thus

$$H^*(B\mathrm{SO}(2n)\langle n\rangle; \mathbf{Q}) = \mathbf{Q}[e, p_{\lceil\frac{n+1}{4}\rceil}, \ldots, p_{n-1}].$$

Combining all the above, we obtain the following.

Theorem 12.5.1. *For $2n \geq 6$, let $\mathcal{B}$ denote the set of monomials in the classes e, p_{n-1}, p_{n-2}, …, $p_{\lceil\frac{n+1}{4}\rceil}$. Then the map*

$$\mathbf{Q}[\kappa_c \mid c \in \mathcal{B}, |c| > 2n] \longrightarrow H^*(\mathcal{M}^{\mathrm{or}}(W_{g,1}); \mathbf{Q})$$

is an isomorphism in degrees $* \leq \frac{g-3}{2}$.

This is [17, Corollary 1.8]. As we mentioned above, for $2n = 2$ the same statement (with a slightly different range) was earlier proved by Madsen and Weiss.

12.5.2 The manifolds W_g

For $2n = 2$ it is a theorem of Harer [21] that the map

$$\mathcal{M}^{\mathrm{or}}(W_{g,1}) \longrightarrow \mathcal{M}^{\mathrm{or}}(W_g) \tag{12.5.1}$$

given by attaching a disk induces an isomorphism on homology in a stable range of degrees. On the other hand $\mathcal{M}^{\mathrm{or}}(W_g)$ is the homotopy type of the stack $\mathbf{M}_g$ of genus g Riemann surfaces, and it is in this way that Madsen and Weiss' topological result determines $H^*(\mathbf{M}_g; \mathbf{Q})$ in a stable range.

In dimensions $2n \geq 6$ it is no longer true that the map (12.5.1) induces an isomorphism on homology in a stable range of degrees. In this section we shall explain why, by using the theory described above to calculate $H^*(\mathcal{M}^{\mathrm{or}}(W_g); \mathbf{Q})$ in a stable range.

Continuing to write

$$\theta_n : BO(2n)\langle n\rangle \xrightarrow{\theta_n^{\mathrm{or}}} B\mathrm{SO}(2n) \xrightarrow{\mathrm{or}} BO(2n)$$

as in the previous section, this tangential structure is also the Moore–Postnikov n-stage of the map classifying the tangent bundle of W_g. By Theorem 12.4.5 there is a map

$$\alpha : \mathcal{M}^{\mathrm{or}}(W_g) \longrightarrow (\Omega^\infty MT\theta_n)/\!\!/\mathrm{hAut}(\theta_n^{\mathrm{or}})$$

that induces an isomorphism on homology, onto the path component that it hits, in degrees $* \leq \frac{g-3}{2}$ as long as $2n \geq 6$.

In order to calculate the rational cohomology of $\mathcal{M}^{\mathrm{or}}(W_g)$ in a range of degrees we could try to calculate the rational cohomology of the relevant path component of $(\Omega^\infty MT\theta_n)/\!\!/\mathrm{hAut}(\theta_n^{\mathrm{or}})$, but instead we shall use the fibre sequence (12.4.9). Choosing a θ_n-structure ℓ_{W_g} on W_g, this is a fibre sequence

$$\mathcal{M}^{\theta_n}(W_g, \ell_{W_g}) \longrightarrow \mathcal{M}^{\mathrm{or}}(W_g) \xrightarrow{\xi} B(\mathrm{hAut}(\theta_n^{\mathrm{or}})_{[W_g, \ell_{W_g}]}). \tag{12.5.2}$$

As $\ell_{W_g} : W_g \to BO(2n)\langle n \rangle$ is n-connected, we may apply Theorem 12.3.3, giving that

$$\mathbf{Q}[\kappa_c \mid c \in \mathcal{B}, |c| > 2n] \longrightarrow H^*(\mathcal{M}^{\theta_n}(W_g, \ell_{W_g}); \mathbf{Q})$$

is an isomorphism in degrees $* \leq \frac{g-3}{2}$. All the classes κ_c are defined on the total space $\mathcal{M}^{\mathrm{or}}(W_g)$ of the fibration (12.5.2), which implies that this fibration satisfies the Leray–Hirsch property in the stable range. Thus the Leray–Hirsch map

$$\mathbf{Q}[\kappa_c \mid c \in \mathcal{B}, |c| > 2n] \otimes H^*(B(\mathrm{hAut}(\theta_n^{\mathrm{or}})_{[W_g, \ell_{W_g}]}); \mathbf{Q}) \longrightarrow H^*(\mathcal{M}^{\mathrm{or}}(W_g); \mathbf{Q}) \qquad (12.5.3)$$

is an isomorphism in degrees $* \leq \frac{g-3}{2}$.

To complete this calculation we must calculate the rational cohomology of the space $B(\mathrm{hAut}(\theta_n^{\mathrm{or}})_{[W_g, \ell_{W_g}]}$, and describe the map

$$\xi^* : H^*(B(\mathrm{hAut}(\theta_n^{\mathrm{or}})_{[W_g, \ell_{W_g}]}); \mathbf{Q}) \longrightarrow H^*(\mathcal{M}^{\mathrm{or}}(W_g); \mathbf{Q})$$

in terms that we understand.

12.5.2.1 Identifying $\mathrm{hAut}(\theta_n^{\mathrm{or}})$

The map $\theta_n^{\mathrm{or}} : BO(2n)\langle n \rangle \to BSO(2n)$ is a principal fibration for the group-like topological monoid $\mathrm{SO}[0, n-1]$, the truncation of SO, so is classified by the map $BSO(2n) \to BSO \to B(\mathrm{SO}[0, n-1])$, and hence there is a homomorphism

$$\iota : \mathrm{SO}[0, n-1] \longrightarrow \mathrm{hAut}(\theta_n^{\mathrm{or}})$$

given by the principal group action.

Lemma 12.5.2. *The map ι is a weak homotopy equivalence.*

Proof. If we fix a basepoint $* \in BO(2n)\langle n \rangle$, which identifies the fibre through $*$ with $\mathrm{SO}[0, n-1]$, then there is a map

$$\begin{aligned} ev : \mathrm{hAut}(\theta_n^{\mathrm{or}}) &\longrightarrow \mathrm{SO}[0, n-1] \\ \varphi &\longmapsto \varphi(*), \end{aligned}$$

and it is clear that $ev \circ \iota$ is the identity. Thus it is enough to show that ev is a weak homotopy equivalence. Suppose we are given a map $(f, g) : (D^{k+1}, S^k) \to (\mathrm{SO}[0, n-1], \mathrm{hAut}(\theta_n^{\mathrm{or}}))$. This determines a relative lifting problem

$$\begin{array}{ccc} (S^k \times BO(2n)\langle n \rangle) \cup_{S^k \times \{*\}} (D^{k+1} \times \{*\}) & \xrightarrow{g \cup f} & BO(2n)\langle n \rangle \\ \downarrow & & \downarrow \theta_n^{\mathrm{or}} \\ D^{k+1} \times BO(2n)\langle n \rangle & \xrightarrow{\theta_n^{\mathrm{or}} \circ \pi_2} & BSO(2n) \end{array}$$

and finding a nullhomotopy of (f, g) is the same as solving this relative lifting problem. The obstructions for doing so lie in the groups

$$\widetilde{H}^{i-k}(BO(2n)\langle n\rangle; \pi_{i+1}(BSO(2n), BO(2n)\langle n\rangle)),$$

but these groups are zero if $i - k \leq n$ or if $i \geq n$, so always vanish. $\square$

In particular, the submonoid $\mathrm{hAut}(\theta_n^{\mathrm{or}})_{[W_g,\ell_{W_g}]} \leq \mathrm{hAut}(\theta_n^{\mathrm{or}})$ is in fact the whole of $\mathrm{hAut}(\theta_n^{\mathrm{or}})$, so we may identify the cohomology of its classifying space with

$$H^*(B\mathrm{hAut}(\theta_n^{\mathrm{or}}); \mathbf{Q}) = H^*(B\mathrm{SO}[0,n]; \mathbf{Q}) = \mathbf{Q}[p_1, p_2, \ldots, p_{\lfloor \frac{n}{4} \rfloor}]. \tag{12.5.4}$$

12.5.2.2 Miller–Morita–Mumford class interpretation

Combining (12.5.3) and (12.5.4) gives a formula for $H^*(\mathcal{M}^{\mathrm{or}}(W_g); \mathbf{Q})$ in a range of degrees. In fact the classes obtained by pulling back $p_1, p_2, \ldots, p_{\lfloor \frac{n}{4} \rfloor}$ along the map

$$\xi : \mathcal{M}^{\mathrm{or}}(W_g) \longrightarrow B\mathrm{hAut}(\theta_n^{\mathrm{or}})$$

may be re-interpreted as Miller–Morita–Mumford classes. We shall use the following lemma to explain this.

Lemma 12.5.3. *Let $\pi^{\mathrm{or}} : \mathcal{E}^{\mathrm{or}}(W_g) \to \mathcal{M}^{\mathrm{or}}(W_g)$ denote the the path component of the fibration* (12.2.4) *modelling the universal oriented W_g-bundle, and $\tau : \mathcal{E}^{\mathrm{or}}(W_g) \to B\mathrm{SO}(2n)$ denote the map classifying the vertical tangent bundle. Then the square*

$$\begin{array}{ccccc}
\mathcal{E}^{\mathrm{or}}(W_g) & & \xrightarrow{\quad\tau\quad} & & B\mathrm{SO}(2n) \\
\downarrow{\scriptstyle \pi^{\mathrm{or}}} & & & & \downarrow \\
\mathcal{M}^{\mathrm{or}}(W_g) & \xrightarrow{\xi} & B\mathrm{SO}[0,n] & \xleftarrow{\simeq} & B\mathrm{SO}(2n)[0,n]
\end{array} \tag{12.5.5}$$

commutes up to homotopy.

Proof. Let $\pi^{\theta_n} : \mathcal{E}^{\theta_n}(W_g, \ell_{W_g}) \to \mathcal{M}^{\theta_n}(W_g, \ell_{W_g})$ denote the path component of the fibration modelling the universal W_g-bundle with θ_n-structure, which as in (12.2.4) comes with maps

$$\tau : \mathcal{E}^{\theta_n}(W_g, \ell_{W_g}) \xrightarrow{\ell} BO(2n)\langle n\rangle \xrightarrow{\theta_n^{\mathrm{or}}} B\mathrm{SO}(2n)$$

whose composition classifies the (oriented) vertical tangent bundle. This gives a commutative square

$$\begin{array}{ccc}
\mathcal{E}^{\theta_n}(W_g, \ell_{W_g}) & \xrightarrow{\ell} & BO(2n)\langle n\rangle \\
\downarrow{\scriptstyle \pi^{\theta_n}} & & \downarrow \\
\mathcal{M}^{\theta_n}(W_g, \ell_{W_g}) & \longrightarrow & *
\end{array}$$

of $\mathrm{hAut}(\theta_n^{\mathrm{or}})$-spaces and $\mathrm{hAut}(\theta_n^{\mathrm{or}})$-equivariant maps. Taking homotopy orbits, and replacing the spaces at each corner with homotopy equivalent models, we obtain the homotopy commutative square

$$\begin{array}{ccc} \mathcal{E}^{\mathrm{or}}(W_g) & \xrightarrow{\tau} & B\mathrm{SO}(2n) \\ \downarrow{\pi^{\mathrm{or}}} & & \downarrow \\ \mathcal{M}^{\mathrm{or}}(W_g) & \xrightarrow{\xi} & B\mathrm{SO}[0,n]. \end{array}$$

Here the right-hand map is the truncation $B\mathrm{SO}(2n) \to B\mathrm{SO}(2n)[0,n]$ followed by the identification $B\mathrm{SO}(2n)[0,n] \xrightarrow{\sim} B\mathrm{SO}[0,n]$, as required. □

Now we may calculate as follows: by this lemma we have

$$(\pi^{\mathrm{or}})^*\xi^*(p_i) = \tau^*(p_i),$$

and so by the projection formula (and commutativity of the cup product)

$$\kappa_{ep_i} = \int_{\pi^{\mathrm{or}}} \tau^*(e \cdot p_i) = \left(\int_{\pi^{\mathrm{or}}} \tau^* e\right) \cdot \xi^*(p_i) = \chi(W_g) \cdot \xi^*(p_i)$$

and hence, for $\chi(W_g) = 2 + (-1)^n 2g \neq 0$, we have $\xi^*(p_i) = \frac{1}{\chi(W_g)}\kappa_{ep_i}$.

Combined with the previous discussion, we obtain the following.

Theorem 12.5.4. *For $2n \geq 6$, let $\mathcal{B}$ denote the set of monomials in the classes e, p_{n-1}, $p_{n-2}, \ldots, p_{\lceil \frac{n+1}{4} \rceil}$, and $\mathcal{C}$ denote the set of the remaining Pontryagin classes $p_1, p_2, \ldots, p_{\lfloor \frac{n}{4} \rfloor}$. Then the map*

$$\mathbf{Q}[\kappa_c \mid c \in (\mathcal{B} \sqcup e \cdot \mathcal{C}), |c| > 2n] \longrightarrow H^*(\mathcal{M}^{\mathrm{or}}(W_g); \mathbf{Q})$$

is an isomorphism in degrees $* \leq (g-3)/2$.

For $2n = 2$ the same statement (with a slightly different range) holds by the theorem of Madsen and Weiss, and in this case the set $\mathcal{C}$ is empty and the result is the same as that of Theorem 12.5.1. For $2n \geq 6$ we have $\mathrm{hAut}(\theta_n^{\mathrm{or}}) \simeq \mathrm{SO}[0, n-1] \not\simeq *$ and so it follows from the discussion in this section that the map (12.5.1) is *not* an isomorphism on integral cohomology in any range of degrees (though for $2n = 6$ it is still an isomorphism on rational cohomology in a stable range, as $\mathrm{SO}[0,2] \simeq K(\mathbf{Z}/2, 1)$ is rationally acyclic; in Theorem 12.5.4 this corresponds to the fact that $\mathcal{C}$ is empty in this case).

Remark 12.5.5. In the statement of Theorem 12.5.4 we do *not* assert that $\kappa_c = 0$ for monomials $c \notin \mathcal{B} \sqcup e \cdot \mathcal{C}$. Indeed a further consequence of the homotopy commutativity of (12.5.5) is the following description of κ_c for a general monomial $c = e^i \cdot p_1^{j_1} \cdots p_{n-1}^{j_{n-1}}$ in terms of the generators of Theorem 12.5.4:

$$\kappa_c = \left(\frac{\kappa_{e \cdot p_1}}{\chi(W_g)}\right)^{j_1} \cdot \left(\frac{\kappa_{e \cdot p_2}}{\chi(W_g)}\right)^{j_2} \cdots \left(\frac{\kappa_{e \cdot p_k}}{\chi(W_g)}\right)^{j_k} \cdot \kappa_{\left(e^i \cdot p_{k+1}^{k_{k+1}} \cdots p_{n-1}^{j_{n-1}}\right)},$$

where we write $k = \lfloor \frac{n}{4} \rfloor$. This follows immediately from the observation that $p_i(T_\pi) = \pi^*\xi^*(p_i) = \pi^*(\frac{\kappa_{e \cdot p_i}}{\chi(W_g)})$ for $i \leq k$.

12.5.3 Hypersurfaces in $\mathbf{CP}^4$

If $V \subset \mathbf{CP}^{r+1}$ is a smooth hypersurface, determined by a homogeneous complex polynomial of degree d, then it is an observation of Thom that its diffeomorphism type depends only on the degree d, and not on the particular polynomial: we call the resulting $2r$-manifold V_d. As we shall explain in §12.5.3.2, these interesting manifolds tend to have large genus. More generally, for smooth complete intersections of such hypersurfaces the diffeomorphism type depends only on the degrees, and much is understood about the classification up to diffeomorphism of such manifolds in terms of these degrees, by Libgober and Wood [31], Kreck [28], and others.

We shall explain how the theory described above applies in the non-trivial example of a hypersurface $V_d \subset \mathbf{CP}^4$ of degree d, and determine a formula for the rational cohomology of $\mathcal{M}^{\mathrm{or}}(V_d)$ in a range of degrees. Let us start with outlining the steps again, and state the conclusions in this example.

(i) Determine the genus of V_d: it turns out to be $\frac{1}{2}(d^4 - 5d^3 + 10d^2 - 10d + 4)$.

(ii) Determine the Moore–Postnikov 3-stage $V_d \xrightarrow{\ell_{V_d}} B_d \xrightarrow{\theta_d} B\mathrm{SO}(6)$ of a map classifying the oriented tangent bundle of V_d.

(iii) Calculate the ring $H^*(\mathcal{M}^{\theta_d}(V_d, \ell_{V_d}); \mathbf{Q})$ in the stable range. It turns out to be the $\mathbf{Q}$-algebra

$$A = \mathbf{Q}[\kappa_{t^n c} \mid c \in \mathcal{B},\, n \geq 0,\, |c| + 2n > 6], \tag{12.5.6}$$

where $\mathcal{B}$ is the set of monomials in classes p_1, p_2, and e of degree $|p_1| = 4$, $|p_2| = 8$, and $|e| = 6$, and t is a class of degree 2.

(iv) Use the spectral sequence arising from Corollary 12.4.6 to determine the cohomology of $\mathcal{M}^{\mathrm{or}}(V_d)$ from that of $\mathcal{M}^{\theta_d}(V_d, \ell_{V_d})$ in a stable range. The result is a short exact sequence

$$0 \longrightarrow H^*(\mathcal{M}^{\mathrm{or}}(V_d); \mathbf{Q}) \longrightarrow A \xrightarrow{d_3} A \longrightarrow 0, \tag{12.5.7}$$

where $d_3 : A \to A$ is the unique derivation satisfying $d_3(\kappa_{t^n c}) = n\kappa_{t^{n-1}c}$. (The result is a scalar when $|t^{n-1}c| = 6$; this scalar is a characteristic number of V_d and therefore the derivation d_3 depends on the degree d.)

12.5.3.1 Algebraic topology of V_d

By the Lefschetz hyperplane theorem the inclusion $i : V_d \to \mathbf{CP}^4$ is 3-connected. This first implies that V_d is simply connected. Writing $H^*(\mathbf{CP}^4; \mathbf{Z}) = \mathbf{Z}[x]/(x^5)$ for $x = c_1(\mathcal{O}(1))$ and $t = i^*(x)$, we have

$$H^0(V_d; \mathbf{Z}) = \mathbf{Z} \qquad H^1(V_d; \mathbf{Z}) = 0 \qquad H^2(V_d; \mathbf{Z}) = \mathbf{Z}\{t\}$$

and hence by Poincaré duality we have

$$H^4(V_d; \mathbf{Z}) = \mathbf{Z}\{s\} \qquad H^5(V_d; \mathbf{Z}) = 0 \qquad H^6(V_d; \mathbf{Z}) = \mathbf{Z}\{u\}$$

where $\langle [V_d], u\rangle = 1$ and $s \cdot t = u$. We also have $\langle i_*[V_d], x^3\rangle = d$, obtained by intersecting V_d with a generic $\mathbf{CP}^1 \subset \mathbf{CP}^4$, giving $t^3 = d \cdot u$ and hence $t^2 = d \cdot s$. By definition, V_d is the zero locus of a transverse section of $\mathcal{O}(d) \to \mathbf{CP}^4$, so its normal bundle in $\mathbf{CP}^4$ is $i^*(\mathcal{O}(d))$ and hence as complex vector bundles we have

$$TV_d \oplus i^*(\mathcal{O}(d)) \oplus \underline{\mathbf{C}} = i^*(T\mathbf{CP}^4) \oplus \underline{\mathbf{C}} = i^*(\mathcal{O}(1))^{\oplus 5}$$

so taking total Chern classes yields $c(TV_d) = i^*\left(\frac{(1+x)^5}{(1+dx)}\right)$. We can therefore extract

$$c_3(TV_d) = \left(\binom{5}{3} - \binom{5}{2}d + \binom{5}{1}d^2 - d^3\right)t^3$$

and so compute the Euler characteristic of V_d as $\langle [V_d], c_3(TV_d)\rangle$ to be

$$\chi(V_d) = d \cdot (10 - 10d + 5d^2 - d^3).$$

We therefore find that $H^3(V_d; \mathbf{Z})$ is free of rank $4 - \chi(V_d) = d^4 - 5d^3 + 10d^2 - 10d + 4$, which finishes our calculation of the cohomology of V_d.

For later use we record two further characteristic classes of V_d, namely

$$w_2(TV_d) = (5-d)t \mod 2$$
$$p_1(TV_d) = (5-d^2)t^2,$$

obtained from the identities $w_2 \equiv c_1 \mod 2$ and $p_1 = c_1^2 - 2c_2$ among characteristic classes of complex vector bundles.

12.5.3.2 Genus of V_d

The genus of V_d may be estimated from below by the methods of §12.3.4, but in this case it turns out that an exact formula is possible. It is a theorem of Wall [49] that any simply connected smooth 6-manifold W has a decomposition $W \cong M \# g(S^3 \times S^3)$ with $H_3(M; \mathbf{Z}) = 0$, and so the genus of such a W is given by half its third Betti number. Thus we have

$$g(V_d) = \frac{1}{2}(d^4 - 5d^3 + 10d^2 - 10d + 4).$$

Similar formulae are known for higher dimensional smooth complex complete intersection varieties, see e.g. [51, 37, 7].

Remark 12.5.6. More generally, if $\mathcal{L}$ is an ample line bundle over a smooth projective complex manifold M of complex dimension $n+1$ then for all $d \gg 0$ we may consider the smooth manifolds U_d arising as the zeroes of generic holomorphic sections of $\mathcal{L}^{\otimes d}$. Writing $x := c_1(\mathcal{L})$, as $\mathcal{L}$ is ample we have $N := \int_M x^{n+1} \neq 0$. Writing $i : U_d \hookrightarrow M$ for the inclusion, and using that $i_*[U_d]$ is Poincaré dual to $e(\mathcal{L}^{\otimes d}) = dx$, it follows that $\int_{U_d} i^*(x)^n = d \cdot N \neq 0$. The analogue of the calculation above gives that $c(TU_d) = i^*(\frac{c(TM)}{(1+dx)})$ and hence we have $\chi(U_d) = (-1)^n d^{n+1} N + O(d^n)$ and so $b_n = d^{n+1}N + O(d^n)$. If n is odd then by the discussion in §12.3.4 we have

$$g(U_d) = \frac{1}{2}d^{n+1}N + O(d^n).$$

If n is even, then the analogous calculation with the total Hirzebruch $\mathcal{L}$-class gives that $\mathcal{L}(TU_d) = i^*(\frac{\mathcal{L}(TM)}{dx/\tanh(dx)})$ so $\sigma(U_d) = \frac{2^{n+2}(2^{n+2}-1)B_{n+2}}{(n+2)!} d^{n+1}N + O(d^n)$, where B_i denote the Bernoulli numbers. Hence, by the discussion in §12.3.4, we have

$$g(U_d) = \frac{1}{2}\left(1 - \frac{2^{n+2}(2^{n+2}-1)|B_{n+2}|}{(n+2)!}\right) d^{n+1}N + O(d^n).$$

The term $\frac{2^{n+2}(2^{n+2}-1)|B_{n+2}|}{(n+2)!}$ does not matter much for large n: the fact that the Taylor series for $\tanh(z)$ has convergence radius $\pi/2$ implies that that term is asymptotically smaller than $(2/\pi)^{n+\varepsilon}$ as $n \to \infty$, for any $\varepsilon > 0$; in particular it quickly becomes much smaller than 1. In the relevant cases $n \geq 4$ it is at most $2/15$.

12.5.3.3 Moore–Postnikov 3-stage of V_d

Let us write

$$\tau : V_d \xrightarrow{\ell_{V_d}} B_d \xrightarrow{\theta_d} BSO(6)$$

for the Moore–Postnikov 3-stage of a map τ classifying the oriented tangent bundle of V_d, so ℓ_{V_d} is 3-connected and θ_d is 3-co-connected. From this we easily calculate the homotopy groups of B_d, as

$$\begin{aligned} 0 = \pi_1(V_d) &\xrightarrow{\sim} \pi_1(B_d) \\ \mathbf{Z} = \pi_2(V_d) &\xrightarrow{\sim} \pi_2(B_d) \\ \pi_3(B_d) &\xrightarrow{\sim} \pi_3(BSO(6)) = 0 \\ \pi_i(B_d) &\xrightarrow{\sim} \pi_i(BSO(6)) \text{ for all } i \geq 4. \end{aligned}$$

To understand the map θ_d on homotopy groups, it remains to understand the composition

$$\tau_* : \mathbf{Z} = \pi_2(V_d) \xrightarrow{\sim} \pi_2(B_d) \longrightarrow \pi_2(BSO(6)) = \mathbf{Z}/2. \tag{12.5.8}$$

The latter group is detected by the Stiefel–Whitney class w_2, so this map is non-zero if and only if the class $w_2(TV_d) \in H^2(V_d; \mathbf{Z}/2) \cong \mathrm{Hom}(\pi_2(V_d), \mathbf{Z}/2)$ is non-zero. We have seen that $w_2(TV_d) = (5-d)t$, so (12.5.8) is surjective if and only if d is even.

Let us abuse notation by writing $t \in H^2(B_d; \mathbf{Z})$ for the unique class which pulls back to t along ℓ_{V_d}. If d is even, then t satisfies $t \equiv w_2(\theta_d^*\gamma) \mod 2$. Thus there is a Spin^c-structure on the bundle $\theta_d^*\gamma$ with $c_1 = t$, and choosing one provides a commutative diagram

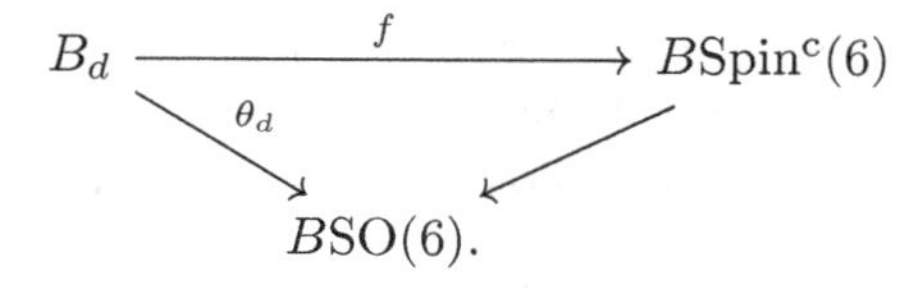

It may be directly checked using the above calculations that the map f induces an isomorphism on all homotopy groups, so is a weak equivalence (over $BSO(6)$).

If d is odd then we have $w_2(\theta_d^*\gamma) = 0$, so we may choose a Spin-structure on the bundle $\theta_d^*\gamma$, which provides a commutative diagram

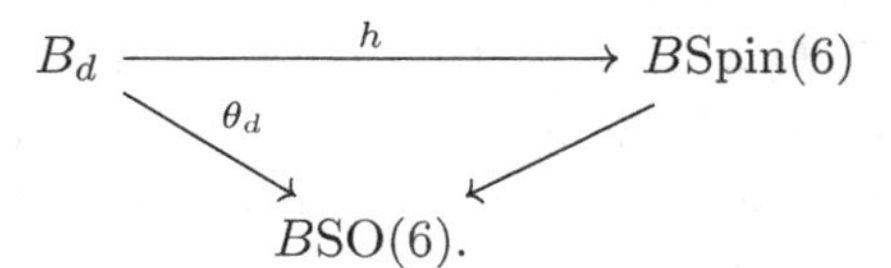

It may be directly checked using the above calculations that the map $h \times t : B_d \to B\mathrm{Spin}(6) \times K(\mathbf{Z}, 2)$ induces an isomorphism on all homotopy groups, so is a weak equivalence (over $B\mathrm{SO}(6)$).

In either case, the map

$$\theta_d \times t : B_d \longrightarrow B\mathrm{SO}(6) \times K(\mathbf{Z}, 2)$$

is a rational homotopy equivalence (over $B\mathrm{SO}(6)$), so we have

$$H^*(B_d; \mathbf{Q}) = \mathbf{Q}[t, p_1, p_2, e].$$

Writing as in the previous examples $\mathcal{B}$ for the set of monomials in p_1, p_2, and e, by Theorem 12.3.3 the map

$$\mathbf{Q}[\kappa_{t^i c} \,|\, c \in \mathcal{B}, i \geq 0, |c| + 2i > 6] \longrightarrow H^*(\mathcal{M}^{\theta_d}(V_d, \ell_{V_d}); \mathbf{Q})$$

is an isomorphism in degrees $* \leq \frac{d^4 - 5d^3 + 10d^2 - 10d + 4}{4}$, establishing (12.5.6).

12.5.3.4 Change of tangential structure

We wish to use the above to compute the rational cohomology of $\mathcal{M}^{\mathrm{or}}(V_d)$ in a range of degrees, so must analyse the forgetful map $\mathcal{M}^{\theta_d}(V_d, \ell_{V_d}) \to \mathcal{M}^{\mathrm{or}}(V_d)$. We shall do this in two stages, given by the maps of tangential structures

$$\begin{array}{ccccc} B_d & \xrightarrow{u = \theta_d \times t} & B\mathrm{SO}(6) \times K(\mathbf{Z}, 2) & \xrightarrow{\mu} & B\mathrm{SO}(6) \\ & {\scriptstyle \theta_d} \searrow & \downarrow {\scriptstyle \mu} & \swarrow {\scriptstyle \mathrm{Id}} & \\ & & B\mathrm{SO}(6). & & \end{array}$$

The space of θ_d-structures on V_d refining the μ-structure $u \circ \ell_{V_d}$ is homotopy equivalent to the space of lifts

$$\begin{array}{ccc} & & B_d \\ & \nearrow & \downarrow {\scriptstyle u} \\ V_d & \xrightarrow[u \circ \ell_{V_d}]{} & B\mathrm{SO}(6) \times K(\mathbf{Z}, 2), \end{array}$$

and $\pi_0(\mathrm{hAut}(u))$ acts on the set of homotopy classes of such lifts. If $G \leq \mathrm{hAut}(u)$ is the submonoid of those path components that preserve the θ_d-structure ℓ_{V_d} up to diffeomorphisms of V_d preserving the μ-structure $u \circ \ell_{V_d}$, then there is a fibration sequence

$$\mathcal{M}^{\theta_d}(V_d, \ell_{V_d}) \longrightarrow \mathcal{M}^{\mu}(V_d, u \circ \ell_{V_d}) \longrightarrow BG.$$

By the discussion in Remark 12.4.11, $G = \mathrm{hAut}(u)_{[V_d, \ell_{V_d}]}$ as long as $g(V_d, \ell_{V_d}) \geq 3$, and this fibration sequence is an instance of (12.4.9).

As we have seen above, the map u is a rational homotopy equivalence, and it is immediate from this that $\pi_i(\mathrm{hAut}(u)) \otimes \mathbf{Q} = 0$ for $i > 0$, so G has no higher rational homotopy groups.

We claim that $\pi_0(G)$ is also trivial, and in fact we shall show that $\pi_0(\mathrm{hAut}(u))$ is trivial (so $G = \mathrm{hAut}(u)$). To see this, let $\phi \in \mathrm{hAut}(u)$, and we must then show that the following lifting problem admits a solution:

$$\begin{array}{ccc}
\partial[0,1] \times B_d & \xrightarrow{\mathrm{Id} \sqcup \phi} & B_d \\
\downarrow & & \downarrow u \\
[0,1] \times B_d & \xrightarrow{proj} B_d \xrightarrow{u} & B\mathrm{SO}(6) \times K(\mathbf{Z}, 2).
\end{array}$$

By consideration of the cases $B_d = B\mathrm{Spin}^c(6)$ and $B_d = B\mathrm{Spin}(6) \times K(\mathbf{Z}, 2)$, we see that the homotopy fibre of u is a $K(\mathbf{Z}/2, 1)$, so there is a unique obstruction to finding the required lift, lying in

$$H^2([0,1] \times B_d, \partial[0,1] \times B_d; \mathbf{Z}/2) \cong H^1(B_d; \mathbf{Z}/2) = 0.$$

It follows that BG is simply connected and has trivial higher rational homotopy groups, so $\mathcal{M}^{\theta_d}(V_d, \ell_{V_d}) \to \mathcal{M}^{\mu}(V_d, u \circ \ell_{V_d})$ is a rational homotopy equivalence.

Analogously to the above, if $H \leq \mathrm{hAut}(\mu)$ is the submonoid of those path components that preserve the μ-structure $u \circ \ell = \tau \times t$ up to orientation-preserving diffeomorphism of V_d, then there is a fibration sequence

$$\mathcal{M}^{\mu}(V_d, u \circ \ell_{V_d}) \longrightarrow \mathcal{M}^{\mathrm{or}}(V_d) \longrightarrow BH.$$

Again, by Remark 12.4.11, $H = \mathrm{hAut}(\mu)_{[V_d, u \circ \ell_{V_d}]}$ as long as $g(V_d, u \circ \ell_{V_d}) \geq 3$, and this fibration sequence is an instance of (12.4.9). As the fibration μ is trivial, we have

$$\begin{aligned}
\mathrm{hAut}(\mu) &\simeq \mathrm{map}(B\mathrm{SO}(6), \mathrm{hAut}(K(\mathbf{Z}, 2))) \\
&\simeq \mathbf{Z}^{\times} \ltimes \mathrm{map}(B\mathrm{SO}(6), K(\mathbf{Z}, 2)) \\
&\simeq \mathbf{Z}^{\times} \ltimes K(\mathbf{Z}, 2).
\end{aligned}$$

The non-trivial path component of this monoid acts on $H^2(B\mathrm{SO}(6) \times K(\mathbf{Z}, 2); \mathbf{Z}) = \mathbf{Z}\{t\}$ as -1, but any orientation-preserving diffeomorphism of V_d fixes $t^3 \in H^6(V_d; \mathbf{Z})$ so acts as $+1$ on $H^2(V_d; \mathbf{Z}) = \mathbf{Z}\{t\}$. Thus the non-trivial path component of $\mathrm{hAut}(\mu)$ does not lie in H, so $H \simeq K(\mathbf{Z}, 2)$. Thus the fibration sequence is of the form

$$\mathcal{M}^{\mu}(V_d, u \circ \ell_{V_d}) \longrightarrow \mathcal{M}^{\mathrm{or}}(V_d) \longrightarrow K(\mathbf{Z}, 3).$$

The Serre spectral sequence for this fibration, in rational cohomology, has two columns and so a single possible non-zero differential. In the stable range, using the above, it has the form

$$E_2^{*,*} = \Lambda[\iota_3] \otimes \mathbf{Q}[\kappa_{t^i c} \mid c \in \mathcal{B}, i \geq 0, |c| + 2i > 6] \Longrightarrow H^*(\mathcal{M}^{\mathrm{or}}(V_d); \mathbf{Q}).$$

It remains to determine the d_3-differential, which by the Leibniz rule is done by the following lemma.

Lemma 12.5.7. *We have* $d_3(\kappa_{t^n c}) = \iota_3 \otimes n \cdot \kappa_{t^{n-1}c}$.

Proof. We have $d_3(\kappa_{t^n c}) = \iota_3 \otimes x$ for some x. The action map

$$a : K(\mathbf{Z}, 2) \times \mathcal{M}^{\mu}(V_d, u \circ \ell_{V_d}) \longrightarrow \mathcal{M}^{\mu}(V_d, u \circ \ell_{V_d})$$

classifies the following data: the V_d-bundle

$$\pi : K(\mathbf{Z}, 2) \times \mathcal{E}^{\mu}(V_d, u \circ \ell_{V_d}) \to K(\mathbf{Z}, 2) \times \mathcal{M}^{\mu}(V_d, u \circ \ell_{V_d})$$

pulled back by projection to the second factor, equipped with the μ-structure

$$K(\mathbf{Z}, 2) \times \mathcal{E}^{\mu}(V_d, u \circ \ell_{V_d}) \xrightarrow{\tau \times \tilde{t}} B\mathrm{SO}(6) \times K(\mathbf{Z}, 2)$$

where τ is given by projection to $\mathcal{E}^{\mu}(V_d, u \circ \ell_{V_d})$ and its vertical tangent bundle, and $\tilde{t} = \iota_2 \otimes 1 + 1 \otimes t$.

The class x is related to this action by the formula

$$a^*(\kappa_{t^n c}) = 1 \otimes \kappa_{t^n c} + \iota_2 \otimes x + \cdots.$$

Using the description above we calculate $a^*(\kappa_{t^n c})$ as

$$\pi_!((\iota_2 \otimes 1 + 1 \otimes t)^n \cdot \tau^*(c)) = \pi_! \left(\sum_{i=0}^{n} \binom{n}{i} \iota_2^i \otimes (t^{n-i} \cdot \tau^* c) \right)$$

and the Künneth factor in $H^2(K(\mathbf{Z}, 2); \mathbf{Z}) \otimes H^{|\kappa_{t^n c}|-2}(\mathcal{M}^{\mu}(V_d, u \circ \ell_{V_d}); \mathbf{Z})$ is $\iota_2 \otimes (n \cdot \kappa_{t^{n-1}c})$. It follows that $x = n \cdot \kappa_{t^{n-1}c}$, as required. □

It follows from this lemma that the differential d_3 is a surjection from the first column to the third column, so that

$$H^*(\mathcal{M}^{\mathrm{or}}(V_d); \mathbf{Q}) \cong \mathrm{Ker}(d_3 \circlearrowleft \mathbf{Q}[\kappa_{t^n c} \,|\, c \in \mathcal{B}, i \geq 0, |c| + 2n > 6])$$

in degrees $* \leq \frac{d^4 - 5d^3 + 10d^2 - 10d + 4}{4}$, establishing (12.5.7).

It may at first appear that this ring does not depend on d, but this formula is to be understood carefully. If $|\kappa_{t^n c}| = 2$ then $d_3(\kappa_{t^n c}) \in \mathbf{Q}$ is a scalar, and must be evaluated: this is a *boundary condition* for the derivation d_3, and is a characteristic number of V_d. The $\kappa_{t^n c}$ of degree 2 are given by the $t^i c$ of degree 8, so are p_2, p_1^2, te, $t^2 p_1$, and t^4, and these have

$$\begin{aligned} d_3(\kappa_{p_2}) &= 0 \\ d_3(\kappa_{p_1^2}) &= 0 \\ d_3(\kappa_{te}) &= \kappa_e = \chi(V_d) = d \cdot (10 - 10d + 5d^2 - d^3) \\ d_3(\kappa_{t^2 p_1}) &= 2\kappa_{tp_1} = 2d(5 - d^2) \\ d_3(\kappa_{t^4}) &= 4\kappa_{t^3} = 4 \cdot d. \end{aligned}$$

For the penultimate one we use the calculation of the first Pontryagin class of V_d. As an example, $H^2(\mathcal{M}^{\mathrm{or}}(V_d); \mathbf{Q})$ is 4-dimensional and is spanned by the classes

$$\kappa_{p_2}, \quad \kappa_{p_1^2}, \quad \kappa_{te} - \frac{10 - 10d + 5d^2 - d^3}{4}\kappa_{t^4}, \quad \text{and} \quad \kappa_{t^2 p_1} - \frac{5 - d^2}{2}\kappa_{t^4}.$$

Remark 12.5.8. In the Serre spectral sequence for the fibration $\pi : \mathcal{E}^{\mathrm{or}}(V_d) \to \mathcal{M}^{\mathrm{or}}(V_d)$ modelling the universal oriented V_g-bundle as in (12.2.4), the class $t \in H^2(V_d; \mathbf{Q}) = E_2^{0,2}$ must be a permanent cycle. (This may be seen as the Euler class of the vertical tangent bundle $T_\pi \mathcal{E}^{\mathrm{or}}(V_d)$ restricts to $e(TV_d) \in H^6(V_g; \mathbf{Q})$, so this must be a permanent cycle, and this is a non-zero multiple of t^3. As $d_3(t^3) = 3t^2 \cdot d_3(t)$, if $d_3(t) \neq 0$ then t^3 would not by a permanent cycle, a contradiction.) Thus there exists a class $\bar{t} \in H^2(\mathcal{M}^{\mathrm{or}}(V_d); \mathbf{Q})$ restricting to $t \in H^2(V_d; \mathbf{Q})$. We may therefore construct the class $\kappa_{\bar{t}^n c} := \pi_!(\bar{t}^n c) \in H^*(\mathcal{M}^{\mathrm{or}}(V_d); \mathbf{Q})$.

However, the class $\bar{t}$ is not uniquely determined by the above discussion: if $\delta t \in H^2(\mathcal{M}^{\mathrm{or}}(V_d); \mathbf{Q})$ is any class then $\bar{\bar{t}} = \bar{t} + \pi^*(\delta t)$ is another possible choice, and we then have

$$\kappa_{\bar{\bar{t}}^n c} = \pi_!((\bar{t} + \pi^*(\delta t))^n c) = \kappa_{\bar{t}^n c} + (\delta t)(n \cdot \kappa_{\bar{t}^{n-1} c}) + (\delta t)^2 \cdots,$$

a potentially different cohomology class.

By the formula in Lemma 12.5.7, we may think of the derivation d_3 as being $\frac{\partial}{\partial t}$. From this point of view the polynomials in the classes $\kappa_{t^n c}$ that lie in the kernel of $d_3 = \frac{\partial}{\partial t}$ are precisely those that are independent of the choice of $\bar{t}$ when evaluated in $H^*(\mathcal{M}^{\mathrm{or}}(V_d); \mathbf{Q})$ as described above.

12.5.4 Another $\mathrm{Spin}^c(6)$ example

For a simply connected manifold W of dimension $2n \geq 6$, the formula of Theorem 12.4.5 for the homology of $\mathcal{M}^{\mathrm{or}}(W)$ in a range of degrees seems at first glance as though it only depends on the equivariant homotopy type of the $\mathrm{GL}_{2n}(\mathbf{R})$-space Θ having an n-co-connected equivariant map $u : \Theta \to \mathbf{Z}^\times$ and an n-connected equivariant map $\rho_W : \mathrm{Fr}(TW) \to \Theta$. However, the codomain of the map (12.4.6) is the disconnected space $(\Omega^\infty MT\Theta)/\!\!/\mathrm{hAut}(u)$, and the different path-components of this space can have different cohomology, even rationally. In this section we give an example of this behaviour.

Construction 12.5.9. Let $V \to S^2$ be the unique non-trivial 5-dimensional real vector bundle, and $M = S(V)$ be its sphere bundle; it is an S^4-bundle over S^2 with the same homology as $S^4 \times S^2$. If we write $\pi : M \to S^2$ for the bundle projection, then there is an isomorphism $TM \cong \pi^*(V) \oplus \varepsilon^1$. In particular, the Spin^c-structure on V given by a generator of $H^2(S^2;\mathbf{Z})$ gives one on M (which is Spin^c-nullbordant), and the corresponding map $\ell_M : M \to B\,\mathrm{Spin}^c(6)$ is 3-connected. This induces a Spin^c-structure on $M_g := M\#g(S^3 \times S^3)$ such that $\ell_{M_g} : M_g \to B\,\mathrm{Spin}^c(6)$ is also 3-connected.

Let $\theta : B\,\mathrm{Spin}^c(6) \to B\mathrm{SO}(6)$. As in the last section, if $K \leq \mathrm{hAut}(\theta)$ is the submonoid of those path components that stabilise the θ-structure ℓ_{M_g} up to diffeomorphism of M_g, then there is a fibration sequence

$$\mathcal{M}^\theta(M_g, \ell_{M_g}) \longrightarrow \mathcal{M}^{\mathrm{or}}(M_g) \longrightarrow BK.$$

Lemma 12.5.10. *We have* $\mathrm{hAut}(\theta) \simeq \mathbf{Z}^\times \ltimes K(\mathbf{Z},2)$.

Proof. The fibration $\theta : B\,\mathrm{Spin}^c(6) \to B\mathrm{SO}(6)$ has fibre $K(\mathbf{Z},2)$, and is principal, so there is an action of $K(\mathbf{Z},2)$ on $B\,\mathrm{Spin}^c(6)$ fibrewise over $B\mathrm{SO}(6)$. Furthermore, writing $\mathrm{Spin}^c(6) = \mathrm{Spin}(6) \times_{\mathbf{Z}^\times} \mathrm{U}(1)$ we see that complex conjugation on the $\mathrm{U}(1)$ factor gives an involution c of $B\mathrm{Spin}^c(6)$ over $B\mathrm{SO}(6)$. Together these give a map $\mathbf{Z}^\times \ltimes K(\mathbf{Z},2) \to \mathrm{hAut}(\theta)$ which can be shown to be an equivalence by obstruction theory just as in §12.5.3.4 or the proof of Lemma 12.5.2. □

Lemma 12.5.11. *We have* $K = \mathrm{hAut}(\theta)$.

Let us give two proofs of this lemma, one in terms of the manifolds themselves, and one using the infinite loop spaces of the relevant Thom spectrum.

Proof. The proof of the previous lemma shows that if ℓ and ℓ' are two θ-structures on M_g then there is a unique obstruction to them being homotopic, namely

$$\ell^*(t) - (\ell')^*(t) \in H^2(M_g;\mathbf{Z}).$$

We therefore see that ℓ_{M_g} and $c \circ \ell_{M_g}$, where c is the involution of $B\mathrm{Spin}^c(6)$ over $B\mathrm{SO}(6)$, are not fibrewise homotopic as the obstruction is $2\ell^*_{M_g}(t) \neq 0 \in H^2(M_g;\mathbf{Z})$.

However, pulling back the vector bundle $V \to S^2$ along a diffeomorphism of S^2 of degree -1 gives an isomorphic oriented vector bundle, as $\pi_2(B\mathrm{SO}(5)) = \mathbf{Z}/2$, and so this degree -1 diffeomorphism is covered by a diffeomorphism $M \to M$ which acts as -1 on $H^2(M;\mathbf{Z})$ and as $+1$ on $H^4(M;\mathbf{Z})$, so is orientation-reversing. Composing this with the fibrewise antipodal map of $\pi : M \to S^2$ gives a diffeomorphism $\varphi : M \to M$ which acts as -1 on both $H^2(M;\mathbf{Z})$ and $H^4(M;\mathbf{Z})$, so is orientation-preserving: we may then isotope it to fix a disc, and hence extend it to a diffeomorphism $\varphi_g : M_g \to M_g$ acting as -1 on $H^2(M;\mathbf{Z})$ and on $H^4(M;\mathbf{Z})$.

Now the θ-structures $\ell_{M_g} \circ D\varphi_g$ and $c \circ \ell_{M_g}$ on M_g are homotopic, as

$$(\ell_{M_g} \circ D\varphi_g)^*(t) = \varphi_g^*(\ell^*_{M_g}(t)) = -\ell^*_{M_g}(t) = \ell^*_{M_g}(-t) = (c \circ \ell_{M_g})^*(t).$$

This shows that $c \in \mathrm{hAut}(\theta)$ lies in the submonoid K, as it preserves the θ-structure ℓ_{M_g} up to a diffeomorphism of M_g. □

Alternative proof. By the discussion in Remark 12.4.11, the submonoid $K \leq \mathrm{hAut}(\theta)$ agrees with $\mathrm{hAut}(\theta)_{[M_g,\ell_{M_g}]}$ as long as $g \geq 3$, so is the stabiliser of $\alpha(M_g,\ell_{M_g}) \in \pi_0(MT\mathrm{Spin}^c(6))$.

Thomifying the map $B\mathrm{Spin}^c(6) \to B\mathrm{Spin}^c$ gives a fibre sequence of spectra

$$F \longrightarrow MT\mathrm{Spin}^c(6) \longrightarrow \Sigma^{-6} M\mathrm{Spin}^c$$

and it is easy to check that F is connective and has $\pi_0(F) \cong \mathbf{Z}$. We therefore have an exact sequence

$$\pi_0(F) \cong \mathbf{Z} \longrightarrow \pi_0(MT\mathrm{Spin}^c(6)) \longrightarrow \pi_6(M\mathrm{Spin}^c) = \Omega_6^{\mathrm{Spin}^c},$$

and the left-hand map can be seen to send a generator to $\alpha(S^6, \ell_{S^6})$, where ℓ_{S^6} is the unique $\mathrm{Spin}^c(6)$-structure on S^6 compatible with its orientation.

As the $\mathrm{Spin}^c(6)$-manifold M is constructed as the sphere bundle of a Spin^c vector bundle, its class is trivial in $\Omega_6^{\mathrm{Spin}^c}$ as it bounds the associated disc bundle; similarly for $M_g = M \# g(S^3 \times S^3)$. Thus $\alpha(M_g, \ell_{M_g})$ is a multiple of $\alpha(S^6, \ell_{S^6})$ (by taking Euler characteristic we see that it is $2-g$ times it) and so is fixed by $\mathrm{hAut}(\theta)$, as the $\mathrm{Spin}^c(6)$-structure on S^6 is unique given its orientation. □

We may therefore develop the following diagram of fibration sequences

$$\begin{array}{ccccc}
\mathcal{M}^\theta(M_g, \ell_{M_g}) & \longrightarrow & X_g & \longrightarrow & K(\mathbf{Z},3) \\
\| & & \downarrow & & \downarrow \\
\mathcal{M}^\theta(M_g, \ell_{M_g}) & \longrightarrow & \mathcal{M}^{\mathrm{or}}(M_g) & \longrightarrow & B(\mathbf{Z}^\times \ltimes K(\mathbf{Z},2)) \\
\downarrow & & \downarrow & & \downarrow \\
* & \longrightarrow & B\mathbf{Z}^\times & = & B\mathbf{Z}^\times,
\end{array}$$

whose middle row is the fibration sequence, with lower middle arrow defined to make the bottom right-hand square commute, and X_g as its homotopy fibre, and top right-hard square homotopy cartesian.

The calculation of the previous section applies to the top row, showing that

$$H^*(X_g; \mathbf{Q}) = \mathrm{Ker}(d_3 \circlearrowleft \mathbf{Q}[\kappa_{t^n c} \,|\, c \in \mathcal{B}, i \geq 0, |c| + 2n > 6])$$

in a stable range, this time subject to the boundary conditions

$$\begin{aligned}
d_3(\kappa_{te}) &= \kappa_e = \chi(M_g) = 4 - 2g \\
d_3(\kappa_{t^2 p_1}) &= 2\kappa_{tp_1} = 0 \\
d_3(\kappa_{t^4}) &= 4\kappa_{t^3} = 0.
\end{aligned}$$

However, now the Serre spectral sequence for the middle column gives the calculation

$$H^*(\mathcal{M}^{\mathrm{or}}(M_g); \mathbf{Q}) = H^*(X_g; \mathbf{Q})^{\mathbf{Z}^\times} = \mathrm{Ker}(d_3 \circlearrowleft \mathbf{Q}[\kappa_{t^n c} \,|\, c \in \mathcal{B}, i \geq 0, |c| + 2n > 6])^{\mathbf{Z}^\times}$$

in a stable range, where the invariants are taken with respect to the involution $t \mapsto -t$.

Let us explain something of the structure of this ring in low degrees. In particular, we shall see that unlike the previous examples it not a free graded-commutative algebra, even in the stable range where our formulae apply. Before taking $\mathbf{Z}^\times$-invariants, in degree 2 the kernel is spanned by $\{\kappa_{p_2}, \kappa_{p_1^2}, \kappa_{t^2p_1}, \kappa_{t^4}\}$, and these classes are all fixed by the involution, giving

$$\dim_{\mathbf{Q}} H^2(\mathcal{M}^{\mathrm{or}}(M_g); \mathbf{Q}) = 4.$$

In degree 4 the kernel of d_3 is 16 dimensional, spanned by the 10-dimensional vector space $\mathrm{Sym}^2(\mathbf{Q}\{\kappa_{p_2}, \kappa_{p_1^2}, \kappa_{t^2p_1}, \kappa_{t^4}\})$ along with the classes

$$\begin{aligned}
&\kappa_{te}\kappa_{p_2} - (4-2g)\kappa_{tp_2}\\
&\kappa_{te}\kappa_{p_1^2} - (4-2g)\kappa_{tp_1^2}\\
&(4-2g)\kappa_{t^3p_1} - 3\kappa_{t^2p_1}\kappa_{te}\\
&(4-2g)\kappa_{t^5} - 5\kappa_{t^4}\kappa_{te}\\
&\kappa_{p_1e}\\
&\kappa_{te}^2 - (4-2g)\kappa_{t^2e}.
\end{aligned}$$

Of these, the last two classes are invariant under the involution while first four are *anti-*invariant, and hence

$$\dim_{\mathbf{Q}} H^4(\mathcal{M}^{\mathrm{or}}(M_g); \mathbf{Q}) = 12.$$

In higher degrees, we find that even though, for example, the class $\kappa_{te}\kappa_{p_2} - (4-2g)\kappa_{tp_2}$ is not invariant under the involution, its square is invariant and therefore defines a class in $H^8(\mathcal{M}^{\mathrm{or}}(M_g); \mathbf{Q})$. Similarly with products of any two classes that are anti-invariant and in the kernel of d_3. In degree 16 we find the relation

$$\begin{aligned}
&((\kappa_{te}\kappa_{p_2} - (4-2g)\kappa_{tp_2})(\kappa_{te}\kappa_{p_1^2} - (4-2g)\kappa_{tp_1^2}))^2\\
&\quad = (\kappa_{te}\kappa_{p_2} - (4-2g)\kappa_{tp_2})^2(\kappa_{te}\kappa_{p_1^2} - (4-2g)\kappa_{tp_1^2})^2
\end{aligned}$$

among squares of classes of degree 8, showing that the ring is not free.

12.6 Abelianisations of mapping class groups

The theory described above may in principle be used for calculations in integral homology and cohomology, though this is of course far more difficult. In practice such calculations are restricted to low dimensions, and have a different flavour to those described in §12.5. Here one must obtain information about the low-dimensional homology of $\Omega^\infty MT\Theta$, which is roughly the same as the low-dimensional homotopy of $\Omega^\infty MT\Theta$, which is the homotopy of the spectrum $MT\Theta$ in small positive degrees. But the spectrum $MT\Theta$ is non-connective, so computing its π_i is comparable to computing π_{i+2n} of a connective spectrum (in the alternative proof of Lemma 12.5.11 we have already engaged with this a bit, though we avoided having to actually compute).

As an example of the kinds of calculations that one is required to make, and to give some ideas of the kinds of techniques that can be used to tackle them, in this section we shall survey the calculation in [15] of $H_1(\mathcal{M}(W_{g,1}); \mathbf{Z})$, and then describe analogous calculations for certain non-simply connected 6-manifolds.

Recall that for a manifold W, possibly with boundary, its *mapping class group* is

$$\Gamma_\partial(W) := \pi_0(\mathrm{Diff}_\partial(W)).$$

Equivalently, it is the fundamental group of $B\mathrm{Diff}_\partial(W)$, so by the Hurewicz theorem we may identify its abelianisation as

$$\Gamma_\partial(W)^{ab} \cong H_1(B\mathrm{Diff}_\partial(W); \mathbf{Z}) \cong H_1(\mathcal{M}(W); \mathbf{Z}).$$

12.6.1 The manifolds $W_{g,1}$

We return to the $2n$-manifolds $W_{g,1}$ of §12.5.1. Just as in that section, there is a map

$$\mathcal{M}(W_{g,1}) \simeq \mathcal{M}^{\theta_n}(W_{g,1}, \ell_{W_{g,1}}) \longrightarrow \Omega^\infty MT\theta_n$$

that for $2n \geq 6$ is a homology isomorphism in degrees $\leq \frac{g-3}{2}$ onto the path component that it hits. In particular, as long as $g \geq 5$ we have isomorphisms

$$\Gamma_\partial(W_{g,1})^{ab} \cong H_1(\mathcal{M}(W_{g,1}); \mathbf{Z}) \cong H_1(\Omega_0^\infty MT\theta_n; \mathbf{Z}) \cong \pi_1(MT\theta_n),$$

using that all path components of $\Omega_0^\infty MT\theta_n$ are homotopy equivalent, that the Hurewicz map is an isomorphism (as this space is a loop space), and that π_1 of the space $\Omega_0^\infty MT\theta_n$ is the same as that of the spectrum $MT\theta_n$.

In [15] we attempted to calculate this group, at least in terms of other standard groups arising in geometric topology. Here we shall summarise the results and general strategy of that paper, though we refer there for more details.

To state the main result, consider the bordism theory $\Omega_*^{\langle n\rangle}$ associated to the fibration $BO\langle n\rangle \to BO$ given by the n-connected cover, and represented by the spectrum $MO\langle n\rangle$ (cf. [46, p. 51]). The natural map $BO(2n)\langle n\rangle \to BO\langle n\rangle$ covering the stabilisation map $BO(2n) \to BO$ provides a map of spectra

$$s : MT\theta_n \longrightarrow \Sigma^{-2n} MO\langle n\rangle,$$

and on π_1 this gives a homomorphism $s_* : \pi_1(MT\theta_n) \to \Omega_{2n+1}^{\langle n\rangle}$. The group $\pi_1(MT\theta_n)$ is determined in terms of this as follows.

Theorem 12.6.1. *There is an isomorphism*

$$s_* \oplus f : \pi_1(MT\theta_n) \longrightarrow \Omega_{2n+1}^{\langle n\rangle} \oplus \begin{cases} (\mathbf{Z}/2)^2 & \text{if } n \text{ is even} \\ 0 & \text{if } n \text{ is } 1,\ 3, \text{ or } 7 \\ \mathbf{Z}/4 & \text{else} \end{cases}$$

for a certain homomorphism f.

Furthermore, the groups $\Omega_{2n+1}^{\langle n\rangle}$ are related to the stable homotopy groups of spheres as follows: there is a homomorphism

$$\rho' : \mathrm{Cok}(J)_{2n+1} \longrightarrow \Omega_{2n+1}^{\langle n\rangle}$$

given by considering a stably framed manifold as a manifold with $BO\langle n\rangle$-structure, which is surjective and whose kernel is generated by the class of a certain homotopy sphere Σ_Q^{2n+1}. In several cases it follows from work of Stolz that the class of Σ_Q in $\mathrm{Cok}(J)$ is trivial—so ρ' is an isomorphism—but this is not known in general. Combining the above with known calculations of $\Omega_*^{\langle 2\rangle} = \Omega_*^{\langle 3\rangle} = \Omega_*^{\mathrm{Spin}}$ and $\Omega_*^{\langle 4\rangle} = \Omega_*^{\mathrm{String}}$ gives the following.

n	1	2	3	4	5	6	7
$\pi_1(MT\theta_n)$	0	$(\mathbf{Z}/2)^2$	0	$(\mathbf{Z}/2)^4$	$\mathbf{Z}/4$	$(\mathbf{Z}/2)^2 \oplus \mathbf{Z}/3$	$\mathbf{Z}/2.$

Let us outline the proof of Theorem 12.6.1 which was given in [15], as we shall need to refer to details of this argument in the following section. The argument combines methods from (stable) homotopy theory with Theorem 12.4.1 again. Starting with homotopy theory, we first let F denote the homotopy fibre of the map of spectra $s : MT\theta_n \to \Sigma^{-2n}MO\langle n\rangle$, and construct a map $\Sigma^{-2n}\mathrm{SO}/\mathrm{SO}(2n) \to F$ which can be shown to be n-connected, for example by computing its effect on homology. On the other hand $\mathrm{SO}/\mathrm{SO}(2n)$ is $(2n-1)$-connected, so by Freudenthal's suspension theorem the map

$$\pi_{2n+1}(\mathrm{SO}/\mathrm{SO}(2n)) \longrightarrow \pi_{2n+1}^s(\mathrm{SO}/\mathrm{SO}(2n)) \cong \pi_1^s(\Sigma^{-2n}\mathrm{SO}/\mathrm{SO}(2n))$$

is an isomorphism for $n \geq 2$, and similarly for one homotopy group lower. It follows from a calculation of Paechter [41] that $\pi_{2n+1}(\mathrm{SO}/\mathrm{SO}(2n))$ is $(\mathbf{Z}/2)^2$ if n is even and is $\mathbf{Z}/4$ if n is odd, and also that $\pi_{2n}(\mathrm{SO}/\mathrm{SO}(2n)) \cong \mathbf{Z}$. Putting the above together, we find an exact sequence

$$\Omega_{2n+2}^{\langle n\rangle} \xrightarrow{\partial} \begin{cases} (\mathbf{Z}/2)^2 & \text{if } n \text{ is even} \\ \mathbf{Z}/4 & \text{if } n \text{ is odd} \end{cases} \longrightarrow \pi_1(MT\theta_n) \longrightarrow \Omega_{2n+1}^{\langle n\rangle} \longrightarrow \mathbf{Z}. \qquad (12.6.1)$$

The rightmost map is zero (as its domain is easily seen to be a torsion group). In the cases $n \in \{1,3,7\}$ it can be shown that the images of $\mathbf{CP}^2$, $\mathbf{HP}^2$, and $\mathbf{OP}^2$ under the leftmost map are non-zero modulo 2, so the leftmost map is surjective. In the remaining cases one must show that the leftmost map is zero, and that the resulting short exact sequence is split, via a homomorphism f as in the statement of Theorem 12.6.1.

At this point is is convenient to use the isomorphism $\Gamma_\partial(W_{g,1})^{ab} \cong \pi_1(MT\theta_n)$ for some $g \gg 0$. The action of $\Gamma_\partial(W_{g,1})$ on $H_n(W_{g,1};\mathbf{Z})$ respects the $(-1)^n$-symmetric intersection form λ, and if $n \neq 1$, 3, or 7 then it also respects a certain quadratic refinement μ of this bilinear form. This yields a homomorphism

$$\Gamma_\partial(W_{g,1}) \longrightarrow \mathrm{Aut}(H_n(W_{g,1};\mathbf{Z}),\lambda,\mu).$$

These automorphism groups have been studied by other authors, and their abelianisations have been identified (for $g \gg 0$) as $(\mathbf{Z}/2)^2$ if n if even or $\mathbf{Z}/4$ if n is odd. A careful analysis of the maps involved shows that the resulting homomorphism

$$f:\pi_1(MT\theta_n) \cong \Gamma_\partial(W_{g,1})^{ab} \longrightarrow \mathrm{Aut}(H_n(W_{g,1};\mathbf{Z}),\lambda,\mu)^{ab} \cong \begin{cases} (\mathbf{Z}/2)^2 & \text{if } n \text{ is even} \\ \mathbf{Z}/4 & \text{if } n \text{ is odd} \end{cases}$$

splits the short exact sequence arising from (12.6.1), as required.

Remark 12.6.2. The Pontryagin dual of the finite abelian group $H_1(\mathcal{M}(W_{g,1});\mathbf{Z})$ calculated here is the torsion subgroup of $H^2(\mathcal{M}(W_{g,1});\mathbf{Z})$. The torsion free quotient of the latter group has been analysed in detail by Krannich and Reinhold [26]. The (unknown, at present) order of the element $[\Sigma_Q] \in \mathrm{Cok}(J)_{2n+1}$ arises there too.

12.6.2 Some non-simply connected 6-manifolds

For the example discussed in the previous section the theory described above is not the only way to calculate $\Gamma_\partial(W_{g,1})^{ab}$, because Kreck [27] has described the groups $\Gamma_\partial(W_{g,1})$ up to two extension problems, and Krannich [25] has recently resolved these extensions completely for n odd, and determined enough about them to calculate $\Gamma_\partial(W_{g,1})^{ab}$ for all $n \geq 3$ and all $g \geq 1$. However, for even slightly more complicated manifolds such an alternative approach is not available, and we suggest that the theory described above is the best way to approach the calculation of $\Gamma_\partial(W)^{ab}$. In this section we illustrate this with an example which seems inaccessible by other means.

Let G be a virtually polycyclic group, and consider a compact 6-manifold W such that

(i) a map τ_W classifying the tangent bundle of W admits a lift ℓ_W along

$$\theta : B\mathrm{Spin}(6) \times BG \xrightarrow{\mathrm{pr}_1} B\mathrm{Spin}(6) \longrightarrow BO(6)$$

such that $\ell_W : W \to B\mathrm{Spin}(6) \times BG$ is 3-connected, and

(ii) $(W, \partial W)$ is 2-connected.

Such manifolds exist for any virtually polycyclic G: these groups satisfy Wall's [48] finiteness condition (F) by [45, p.183], and so also (F_3), so W may be taken to be a regular neighbourhood of an embedding of a finite 3-skeleton of BG into $\mathbf{R}^6$. As further examples of such manifolds, one may take

$$W = (M^3 \times D^3)\#g(S^3 \times S^3) \tag{12.6.2}$$

where M^3 is a closed oriented 3-manifold that is irreducible (so that $\pi_2(M) = 0$) and has virtually polycyclic fundamental group.

Recall from §12.4.7 that the Hirsch length of a virtually polycyclic group is the number of infinite cyclic quotients in a subnormal series.

Theorem 12.6.3. *Suppose that G is virtually polycyclic of Hirsch length h, and W is a 6-manifold satisfying (i) and (ii) above, of genus $g(W) \geq 7+h$. Then there is a short exact sequence*

$$0 \longrightarrow G^{ab} \longrightarrow \Gamma_\partial(W)^{ab} \longrightarrow \mathrm{ko}_7(BG) \longrightarrow 0$$

which is (non-canonically) split.

This short exact sequence was first established by Friedrich [12], who also showed that it is split after inverting 2. We shall give a different argument to hers, which gives the splitting at the prime 2 as well.

The following three examples concern the manifolds W of construction (12.6.2) with M a 3-manifold having finite fundamental group, and $g \geq 7$ so the hypotheses of Theorem 12.6.3 are satisfied.

Example 12.6.4. Let $M^3 = L^3_{p,q}$ be the (p,q)th lens space, with fundamental group $G = \mathbf{Z}/p$ with p prime. Then we have

$$\Gamma_\partial(W)^{ab} \cong \begin{cases} \mathbf{Z}/2 \oplus \mathbf{Z}/4 & \text{if } p = 2 \\ \mathbf{Z}/3 \oplus \mathbf{Z}/9 & \text{if } p = 3 \\ (\mathbf{Z}/p)^3 & \text{if } p \geq 5. \end{cases}$$

The required calculation of $\mathrm{ko}_7(B\mathbf{Z}/p)$ may be extracted from [9, Example 7.3.1] for $p = 2$, and from the Atiyah–Hirzebruch spectral sequence, the fact that $\mathrm{ko}_*(B\mathbf{Z}/p)[\frac{1}{2}]$ is a summand of $\mathrm{ku}_*(B\mathbf{Z}/p)[\frac{1}{2}]$, and [8, Remark 3.4.6] for odd p.

Example 12.6.5. Let M^3 be the spherical 3-manifold with fundamental group $G = Q_8$. Then we have

$$\Gamma_\partial(W)^{ab} \cong (\mathbf{Z}/2)^2 \oplus (\mathbf{Z}/4)^2 \oplus \mathbf{Z}/64.$$

The required calculation of $\mathrm{ko}_7(BQ_8)$ may be extracted from [9, p. 138].

Example 12.6.6. Let $M^3 = \Sigma^3$ be the Poincaré homology 3-sphere, with fundamental group G the binary icosahedral group, isomorphic to $SL_2(\mathbf{F}_5)$. Then g we have

$$\Gamma_\partial(W)^{ab} \cong (\mathbf{Z}/5)^2 \oplus \mathbf{Z}/9 \oplus \mathbf{Z}/64.$$

As Σ^3 is a homology sphere, $H_1(BG;\mathbf{Z}) \cong H_1(\Sigma^3;\mathbf{Z}) = 0$ so we must just calculate $\mathrm{ko}_7(BG)$. The order of $G \cong SL_2(\mathbf{F}_5)$ is $120 = 2^3 \cdot 3 \cdot 5$, so we shall calculate the localisations $\mathrm{ko}_7(BG)_{(p)}$ for $p \in \{2,3,5\}$. Recall that the cohomology ring of BG is $H^*(BG;\mathbf{Z}) = \mathbf{Z}[z]/(120 \cdot z)$ with $|z| = 4$, which may be computed from the fibration sequence $S^3 \to \Sigma^3 \to BG$.

When p is odd, G has cyclic Sylow p-subgroup, so by transfer $\mathrm{ko}_7(BG)_{(p)}$ is a summand of $\mathrm{ko}_7(B\mathbf{Z}/p)_{(p)}$, which we have explained in Example 12.6.4 is $\mathbf{Z}/9$ for $p = 3$ and $(\mathbf{Z}/5)^2$ for $p = 5$. Comparing this with the Atiyah–Hirzebruch spectral sequence computing $\mathrm{ko}_*(BG)_{(p)}$ gives the claimed answer.

When $p = 2$, G has Sylow 2-subgroup Q_8, so by transfer $\mathrm{ko}_7(BG)_{(2)}$ is a summand of $\mathrm{ko}_7(BQ_8)_{(2)}$, which we have explained in Example 12.6.4 is $(\mathbf{Z}/4)^2 \oplus \mathbf{Z}/64$. By a theorem of Mitchell and Priddy [36, Theorem D] there is a stable splitting of BQ_8 as $BG_{(2)} \vee X \vee X$ for some spectrum X, from which it follows that $\mathrm{ko}_7(BG)_{(2)}$ is either $\mathbf{Z}/64$ or $(\mathbf{Z}/4)^2 \oplus \mathbf{Z}/64$; we may see that the first case occurs from the Atiyah–Hirzebruch spectral sequence computing $\mathrm{ko}_*(BG)_{(2)}$.

The rest of this section is concerned with the proof of Theorem 12.6.3.

12.6.2.1 Reduction to homotopy theory

We have assumed that $(W, \partial W)$ is 2-connected, so by Lemma 12.4.15 we have that $\mathrm{hAut}(\theta, \partial W) \simeq *$. Thus by Theorem 12.4.1 and the discussion in §12.4.7 there is a map

$$\alpha : \mathcal{M}(W) \simeq \mathcal{M}^\theta(W) \longrightarrow \Omega^\infty MT\theta = \Omega^\infty(MT\mathrm{Spin}(6) \wedge BG_+)$$

which induces an isomorphism on homology onto the path component that it hits in degrees $* \leq \frac{g(W)-h-5}{2}$, so in particular as long as $g(W) \geq 7+h$ it induces an isomorphism on first homology. As described above the first homology of $\mathcal{M}(W)$ is the abelianisation of the mapping class group $\Gamma_\partial(W)$, so to establish Theorem 12.6.3 we must establish a short exact sequence

$$0 \longrightarrow G^{ab} \longrightarrow \pi_1(MT\mathrm{Spin}(6) \wedge BG_+) \longrightarrow \mathrm{ko}_7(BG) \longrightarrow 0 \qquad (12.6.3)$$

and show that it is split.

12.6.2.2 The exact sequence

Specialising the construction in §12.6.1 to $2n = 6$, we showed that there is a cofibration sequence of spectra

$$F \longrightarrow MT\mathrm{Spin}(6) \longrightarrow \Sigma^{-6} M\mathrm{Spin}, \qquad (12.6.4)$$

that F is connective, and that $\pi_0(F) \cong \mathbf{Z}$ and $\pi_1(F) \cong \mathbf{Z}/4$. The Atiyah–Hirzebruch spectral sequence for $F \wedge BG_+$ gives isomorphisms

$$\pi_0(F \wedge BG_+) \cong \mathbf{Z} \qquad \pi_1(F \wedge BG_+) \cong \mathbf{Z}/4 \oplus H_1(BG; \mathbf{Z})$$

where the splitting in the latter is induced by the retraction $S^0 \to BG_+ \to S^0$ of pointed spaces. Smashing the cofibration sequence (12.6.4) with BG_+, the associated long exact sequence of homotopy groups has the form

$$\Omega_8^{\mathrm{Spin}}(BG) \xrightarrow{\partial} \mathbf{Z}/4 \oplus H_1(BG; \mathbf{Z}) \longrightarrow \pi_1(MT\mathrm{Spin}(6) \wedge BG_+) \longrightarrow \Omega_7^{\mathrm{Spin}}(BG) \longrightarrow \mathbf{Z}$$

and by naturality the long exact sequence for $G = \{e\}$ splits off of this one. As $\pi_1(MT\mathrm{Spin}(6)) = 0$ by Theorem 12.6.1, and $\Omega_7^{\mathrm{Spin}} = 0$ by [34], by the discussion in §12.6.1 this leaves an exact sequence

$$\widetilde{\Omega}_8^{\mathrm{Spin}}(BG) \xrightarrow{\partial} H_1(BG; \mathbf{Z}) \longrightarrow \pi_1(MT\mathrm{Spin}(6) \wedge BG_+) \longrightarrow \Omega_7^{\mathrm{Spin}}(BG) \longrightarrow 0.$$

The Atiyah–Bott–Shapiro map $M\mathrm{Spin} \to ko$ is 8-connected, so the induced map $\Omega_7^{\mathrm{Spin}}(BG) \to \mathrm{ko}_7(BG)$ is an isomorphism. To obtain the claimed short exact sequence we shall therefore prove the following.

Lemma 12.6.7. *If G is finitely generated then the connecting map $\partial : \widetilde{\Omega}_8^{\mathrm{Spin}}(BG) \to H_1(BG; \mathbf{Z})$ is trivial.*

Proof. If this connecting map were non-trivial for some finitely generated G, then because every non-trivial element of $H_1(BG; \mathbf{Z}) = G^{ab}$ remains non-trivial under some homomorphism $G^{ab} \to \mathbf{Z}/p^k$ with p prime and $k \geq 1$, by naturality this connecting map would be non-trivial for $G = \mathbf{Z}/p^k$. So it suffices to show that the map is trivial in this case.

As $M\mathrm{Spin} \to ko$ is 8-connected, the map $\widetilde{\Omega}_8^{\mathrm{Spin}}(BG) \to \widetilde{\mathrm{ko}}_8(BG)$ is an isomorphism. The result then follows as $\widetilde{\mathrm{ko}}_8(B\mathbf{Z}/p^k) = 0$, by a trivial application of the Atiyah–Hirzebruch spectral sequence for p odd and by [5, Theorem 2.4] for $p = 2$. □

12.6.2.3 Simplifying the splitting problem

Let $x : F \to H\mathbf{Z}$ be the 0-th Postnikov truncation of F. The composition

$$k : \Sigma^{-6}M\mathrm{Spin} \xrightarrow{\partial} \Sigma F \xrightarrow{\Sigma x} \Sigma H\mathbf{Z}$$

represents, under the Thom isomorphism, some element $\kappa \in H^7(B\mathrm{Spin}; \mathbf{Z})$ which we identify as follows.

Lemma 12.6.8. *We have $\kappa = \beta(w_6)$.*

Proof. We have $H^7(B\mathrm{Spin}; \mathbf{Z}/2) = \mathbf{Z}/2\{w_7\}$, but w_7 of course vanishes when restricted to $B\mathrm{Spin}(6)$. The long exact sequence on $\mathbf{Z}/2$-cohomology for (12.6.4) contains the portion

$$H^1(MT\mathrm{Spin}(6); \mathbf{Z}/2) \longleftarrow H^1(\Sigma^{-6}M\mathrm{Spin}; \mathbf{Z}/2) = \mathbf{Z}/2\{w_7 \cdot u_{-6}\} \longleftarrow H^1(\Sigma F; \mathbf{Z}/2)$$

and the left-hand map is zero, so the right-hand map is surjective. As the map x is 1-connected, the map Σx is 2-connected, so we have an isomorphism

$$(\Sigma x)^* : \mathbf{Z}/2\{\Sigma\iota\} = H^1(\Sigma H\mathbf{Z}; \mathbf{Z}/2) \xrightarrow{\sim} H^1(\Sigma F; \mathbf{Z}/2).$$

It follows that κ reduces to $w_7 \neq 0$ modulo 2. The integral cohomology of $B\mathrm{Spin}$ is known to only have torsion of order 2 (this may be deduced from [23]), so the Bockstein sequence for $B\mathrm{Spin}$ shows that $H^7(B\mathrm{Spin}; \mathbf{Z}) = \mathbf{Z}/2$, which must therefore be generated by βw_6 as this reduces modulo 2 to $\mathrm{Sq}^1(w_6) = w_7$. Thus $\kappa = \beta(w_6)$. □

As $w_6 \cdot u_{-6} = \mathrm{Sq}^6(u_{-6})$, we may write the map k as the composition

$$k : \Sigma^{-6}M\mathrm{Spin} \xrightarrow{u_{-6}} \Sigma^{-6}H\mathbf{Z}/2 \xrightarrow{\mathrm{Sq}^6} H\mathbf{Z}/2 \xrightarrow{\beta} \Sigma H\mathbf{Z}$$

and so we may form a spectrum E fitting into the diagram

$$\begin{array}{ccccc}
MT\mathrm{Spin}(6) & \longrightarrow & \Sigma^{-6}M\mathrm{Spin} & \xrightarrow{\partial} & \Sigma F \\
\downarrow & & \downarrow{\scriptstyle u_{-6}} & & \downarrow{\scriptstyle \Sigma x} \\
E & \longrightarrow & \Sigma^{-6}H\mathbf{Z}/2 & \xrightarrow{\beta\mathrm{Sq}^6} & \Sigma H\mathbf{Z}
\end{array}$$

in which the rows are homotopy cofibre sequences. Smashing with BG_+ and considering the map of long exact sequences gives

$$\begin{array}{ccc}
\Omega_8^{\mathrm{Spin}}(BG) & \longrightarrow & H_8(BG;\mathbb{Z}/2) \\
\downarrow & & \downarrow{\scriptstyle (\beta\mathrm{Sq}^6)_*=0} \\
\mathbf{Z}/4 \oplus H_1(BG;\mathbb{Z}) & \xrightarrow{\mathrm{pr}_2} & H_1(BG;\mathbb{Z}) \\
\downarrow & & \downarrow \uparrow{\scriptstyle \sigma_G} \\
\pi_1(MT\mathrm{Spin}(6) \wedge BG_+) & \longrightarrow & \pi_1(E \wedge BG_+) \\
\downarrow & & \downarrow \\
\Omega_7^{\mathrm{Spin}}(BG) & \longrightarrow & H_7(BG;\mathbb{Z}/2) \\
\downarrow & & \downarrow{\scriptstyle (\beta\mathrm{Sq}^6)_*=0} \\
0 & \longrightarrow & H_0(BG;\mathbb{Z})
\end{array}$$

where the columns are exact and the two indicated maps are zero by instability of the (co)homology operation $\beta\mathrm{Sq}^6$ in these degrees. It therefore suffices to show that the right-hand short exact sequence has a dashed splitting σ_G as indicated. (Note that the commutativity of the top square gives another proof of Lemma 12.6.7.)

12.6.2.4 Reducing the splitting problem to cyclic groups

The abelianisation homomorphism $a : G \to G^{ab}$ induces a map on the short exact sequences of the form

$$\begin{array}{ccccccccc}
0 & \longrightarrow & H_1(BG;\mathbb{Z}) & \longrightarrow & \pi_1(E \wedge BG_+) & \longrightarrow & H_7(BG;\mathbb{Z}/2) & \longrightarrow & 0 \\
\downarrow & & \downarrow{\scriptstyle =} & \overset{\sigma_{G^{ab}}}{\longleftarrow} & \downarrow{\scriptstyle a_*} & & \downarrow{\scriptstyle a_*} & & \downarrow \\
0 & \longrightarrow & H_1(BG^{ab};\mathbb{Z}) & \longrightarrow & \pi_1(E \wedge BG^{ab}_+) & \longrightarrow & H_7(BG^{ab};\mathbb{Z}/2) & \longrightarrow & 0
\end{array}$$

so if we can show that the short exact sequence is split for G^{ab}, say via the dashed homomorphism $\sigma_{G^{ab}}$, then the sequence for G is also split, via $\sigma_G := \sigma_{G^{ab}} \circ a_*$. This reduces us to studying abelian groups.

On the other hand, if such short exact sequences are split for each cyclic group of prime power order or $\mathbf{Z}$, then we may write $G^{ab} = C_1 \oplus \cdots \oplus C_n$ for cyclic groups C_i of prime power order or $\mathbf{Z}$ and combine their splittings to obtain one for G^{ab} (though of course it depends on the choice of expression for G^{ab} as a sum of cyclic subgroups, so is not canonical). We have therefore reduced the question of splitting the short exact sequences (12.6.3) to the case of such cyclic groups.

12.6.2.5 The splitting for $G = \mathbb{Z}/p^k$ or Z

If $G = \mathbf{Z}$ or $G = \mathbf{Z}/p^k$ with p odd then $H_7(BG;\mathbf{Z}/2) = 0$ and so the short exact sequence becomes

$$0 \longrightarrow G \longrightarrow \pi_1(E \wedge BG_+) \longrightarrow 0 \longrightarrow 0$$

which is certainly split.

For $G = \mathbf{Z}/2^k$ we must go to more trouble: in this case the short exact sequence becomes

$$0 \longrightarrow \mathbf{Z}/2^k \longrightarrow \pi_1(E \wedge B\mathbf{Z}/2^k_+) \longrightarrow \mathbf{Z}/2 \longrightarrow 0 \tag{12.6.5}$$

so is either split or else $\pi_1(E \wedge B\mathbf{Z}/2^k_+) \cong \mathbf{Z}/2^{k+1}$. Consider first smashing the cofibre sequence defining E with $S/2$, giving the cofibration sequence

$$S/2 \wedge E \longrightarrow S/2 \wedge \Sigma^{-6} H\mathbf{Z}/2 \simeq \Sigma^{-6} H\mathbf{Z}/2 \vee \Sigma^{-5} H\mathbf{Z}/2 \overset{\mathrm{Sq}^7 \vee \mathrm{Sq}^6}{\longrightarrow} \Sigma H\mathbf{Z}/2 \simeq S/2 \wedge H\mathbf{Z}.$$

Now further smashing this with $B\mathbf{Z}/2^k_+$ and considering the long exact sequence on homotopy groups gives an exact sequence

$$H_8(B\mathbf{Z}/2^k; \mathbf{Z}/2) \oplus H_7(B\mathbf{Z}/2^k; \mathbf{Z}/2) \xrightarrow{\mathrm{Sq}^7_* \oplus \mathrm{Sq}^6_*} H_1(B\mathbf{Z}/2^k; \mathbf{Z}/2)$$
$$\to \pi_1(S/2 \wedge E \wedge B\mathbf{Z}/2^k_+) \longrightarrow H_7(B\mathbf{Z}/2^k; \mathbf{Z}/2) \oplus H_6(B\mathbf{Z}/2^k; \mathbf{Z}/2)$$
$$\xrightarrow{\mathrm{Sq}^7_* \oplus \mathrm{Sq}^6_*} H_0(B\mathbf{Z}/2^k; \mathbf{Z}/2)$$

and so $\pi_1(S/2 \wedge E \wedge B\mathbf{Z}/2^k_+)$ has cardinality 8, because the homology operations Sq^7_* and Sq^6_* are zero in these degrees by instability. On the other hand computing with the long exact sequence as above gives an exact sequence

$$0 \longrightarrow \mathbf{Z} = H_0(B\mathbf{Z}/2^k; \mathbf{Z}) \longrightarrow \pi_0(E \wedge B\mathbf{Z}/2^k_+) \longrightarrow H_6(B\mathbf{Z}/2^k; \mathbf{Z}/2) = \mathbf{Z}/2 \longrightarrow 0$$

which is split via $B\mathbf{Z}/2^k_+ \to S^0$. Thus

$$\mathrm{Ker}(2 \cdot - : \pi_0(E \wedge B\mathbf{Z}/2^k_+) \to \pi_0(E \wedge B\mathbf{Z}/2^k_+)) = \mathbf{Z}/2$$

and so, as $\pi_1(S/2 \wedge E \wedge B\mathbf{Z}/2^k_+)$ has cardinality 8, it follows that

$$\mathrm{Coker}(2 \cdot - : \pi_1(E \wedge B\mathbf{Z}/2^k_+) \to \pi_1(E \wedge B\mathbf{Z}/2^k_+))$$

has cardinality 4. Thus $\pi_1(E \wedge B\mathbf{Z}/2^k_+)$ cannot be cyclic, so (12.6.5) is split.

Acknowledgements

The authors would like to thank M. Krannich for useful comments on a draft of this paper. S. Galatius was partially supported by the European Research Council (ERC) under the European Union's Horizon 2020 research and innovation programme (grant agreement No. 682922), by NSF grant DMS-1405001 and by the EliteForsk Prize. O. Randal-Williams was partially supported by the ERC under the European Union's Horizon 2020 research and innovation programme (grant agreement No. 756444) and by a Philip Leverhulme Prize from the Leverhulme Trust.

Bibliography

[1] A. Berglund and I. Madsen. Homological stability of diffeomorphism groups. *Pure Appl. Math. Q.*, 9(1):1–48, 2013.

[2] A. Berglund and I. Madsen. Rational homotopy theory of automorphisms of highly connected manifolds. arXiv:1401.4096, 2014.

[3] B. Botvinnik, J. Ebert, and O. Randal-Williams. Infinite loop spaces and positive scalar curvature. *Invent. Math.*, 209(3):749–835, 2017.

[4] B. Botvinnik, J. Ebert, and D. J. Wraith. On the topology of the space of Ricci-positive metrics. arXiv:1809.11050, 2018.

[5] B. Botvinnik, P. Gilkey, and S. Stolz. The Gromov-Lawson-Rosenberg conjecture for groups with periodic cohomology. *J. Differential Geom.*, 46(3):374–405, 1997.

[6] B. Botvinnik and N. Perlmutter. Stable moduli spaces of high-dimensional handlebodies. *J. Topol.*, 10(1):101–163, 2017.

[7] W. Browder. Complete intersections and the Kervaire invariant. In *Algebraic Topology Aarhus 1978*, pages 88–108. Springer, 1979.

[8] R. R. Bruner and J. P. C. Greenlees. The connective K-theory of finite groups. *Mem. Amer. Math. Soc.*, 165(785),2003.

[9] R. R. Bruner and J. P. C. Greenlees. *Connective real K-theory of finite groups*, volume 169 of *Math. Surv. Monogr.* Amer. Math. Soc., 2010.

[10] J. Ebert and O. Randal-Williams. Infinite loop spaces and positive scalar curvature in the presence of a fundamental group. *Geom. Topol.* 23(3):1549–1610, 2019.

[11] N. Friedrich. Homological stability of automorphism groups of quadratic modules and manifolds. *Doc. Math.*, 22:1729–1774, 2017.

[12] N. Friedrich. *Automorphism groups of quadratic modules and manifolds.* PhD thesis, University of Cambridge, 2018. Available at https://doi.org/10.17863/CAM.24264.

[13] S. Galatius, I. Madsen, U. Tillmann, and M. Weiss. The homotopy type of the cobordism category. *Acta Math.*, 202(2):195–239, 2009.

[14] S. Galatius and O. Randal-Williams. Stable moduli spaces of high-dimensional manifolds. *Acta Math.*, 212(2):257–377, 2014.

[15] S. Galatius and O. Randal-Williams. Abelian quotients of mapping class groups of highly connected manifolds. *Math. Ann.*, 365(1–2):857–879, 2016.

[16] S. Galatius and O. Randal-Williams. Homological stability for moduli spaces of high dimensional manifolds. II. *Ann. of Math. (2)*, 186(1):127–204, 2017.

[17] S. Galatius and O. Randal-Williams. Homological stability for moduli spaces of high dimensional manifolds. I. *J. Amer. Math. Soc.*, 31(1):215–264, 2018.

[18] S. Galatius and G. Szűcs. The equivariant cobordism category. arXiv:1805.12342, 2018.

[19] M. Gomez-Lopez. The homotopy type of the PL cobordism category. I. arXiv:1608.06236, 2016.

[20] M. Gomez-Lopez and A. Kupers. The homotopy type of the topological cobordism category. arXiv:1810.05277, 2018.

[21] J. L. Harer. Stability of the homology of the mapping class groups of orientable surfaces. *Ann. of Math. (2)*, 121(2):215–249, 1985.

[22] F. Hebestreit and N. Perlmutter. Cobordism categories and moduli spaces of odd dimensional manifolds. arXiv:1606.06168, 2016.

[23] A. Kono. On the integral cohomology of $B\mathrm{Spin}(n)$. *J. Math. Kyoto Univ.*, 26(3):333–337, 1986.

[24] M. Krannich. On characteristic classes of exotic manifold bundles. arXiv:1802.02609, 2018.

[25] M. Krannich. Mapping class groups of highly connected $(4k+2)$-manifolds. arXiv:1902.10097, 2019.

[26] M. Krannich and J. Reinhold. Characteristic numbers of manifold bundles over surfaces with highly connected fibers. arXiv:1807.11539, 2018.

[27] M. Kreck. Isotopy classes of diffeomorphisms of $(k-1)$-connected almost-parallelizable $2k$-manifolds. In *Algebraic topology, Aarhus 1978 (Proc. Sympos., Univ. Aarhus, Aarhus, 1978)*, volume 763 of *Lecture Notes in Math.*, pages 643–663. Springer, 1979.

[28] M. Kreck. Surgery and duality. *Ann. of Math. (2)*, 149(3):707–754, 1999.

[29] A. Kupers. Proving homological stability for homeomorphisms of manifolds. arXiv:1510.02456, 2015.
[30] A. Kupers. Some finiteness results for groups of automorphisms of manifolds. arXiv:1612.09475, 2016.
[31] A. S. Libgober and J. W. Wood. Differentiable structures on complete intersections. I. *Topology*, 21(4):469–482, 1982.
[32] I. Madsen and U. Tillmann. The stable mapping class group and $Q(\mathbb{C}\mathbb{P}^\infty_+)$. *Invent. Math.*, 145(3):509–544, 2001.
[33] I. Madsen and M. Weiss. The stable moduli space of Riemann surfaces: Mumford's conjecture. *Ann. of Math. (2)*, 165(3):843–941, 2007.
[34] J. Milnor. Spin structures on manifolds. *Enseignement Math. (2)*, 9:198–203, 1963.
[35] J. W. Milnor and J. D. Stasheff. *Characteristic classes*, volume 76 of *Ann. Math. Stud.* Princeton University Press, 1974.
[36] S. A. Mitchell and S. B. Priddy. Symmetric product spectra and splittings of classifying spaces. *Amer. J. Math.*, 106(1):219–232, 1984.
[37] S. Morita. The Kervaire invariant of hypersurfaces in complex projective spaces. *Comment. Math. Helv.*, 50(1):403–419, 1975.
[38] S. Nariman. On the moduli space of flat symplectic surface bundles. arXiv:1612.00043, 2016.
[39] S. Nariman. Homological stability and stable moduli of flat manifold bundles. *Adv. Math.*, 320:1227–1268, 2017.
[40] S. Nariman. Stable homology of surface diffeomorphism groups made discrete. *Geom. Topol.*, 21(5):3047–3092, 2017.
[41] G. F. Paechter. The groups $\pi_r(V_{n,m})$. I. *Quart. J. Math. Oxford Ser. (2)*, 7:249–268, 1956.
[42] N. Perlmutter. Homological stability for the moduli spaces of products of spheres. *Trans. Amer. Math. Soc.*, 368(7):5197–5228, 2016.
[43] N. Perlmutter. Linking forms and stabilization of diffeomorphism groups of manifolds of dimension $4n+1$. *J. Topol.*, 9(2):552–606, 2016.
[44] N. Perlmutter. Homological stability for diffeomorphism groups of high-dimensional handlebodies. *Algebr. Geom. Topol.*, 18(5):2769–2820, 2018.
[45] J. G. Ratcliffe. Finiteness conditions for groups. *J. Pure Appl. Algebra*, 27(2):173–185, 1983.
[46] R. E. Stong. *Notes on cobordism theory. Math. Notes*, Princeton University Press; University of Tokyo Press, 1968.
[47] U. Tillmann. On the homotopy of the stable mapping class group. *Invent. Math.*, 130(2):257–275, 1997.
[48] C. T. C. Wall. Finiteness conditions for CW-complexes. *Ann. of Math. (2)*, 81:56–69, 1965.
[49] C. T. C. Wall. Classification problems in differential topology. V. On certain 6-manifolds. *Invent. Math. 1 (1966), 355-374; corrigendum, ibid*, 2:306, 1966.
[50] M. S. Weiss. Dalian notes on rational Pontryagin classes. arXiv:1507.00153, 2015.
[51] J. W. Wood. Removing handles from non-singular algebraic hypersurfaces in CP_{n+1}. *Invent. Math.*, 31(1):1–6, 1975.

Department of Mathematics, University of Copenhagen, 1017 Copenhagen K, Denmark
E-mail address: galatius@math.ku.dk

Centre for Mathematical Sciences, University of Cambridge, Cambridge CB3 0WB, UK
E-mail address: o.randal-williams@dpmms.cam.ac.uk

13

An introduction to higher categorical algebra

David Gepner

13.1 Introduction

13.1.1 Higher structures

Higher algebra is the study of algebraic structures that arise in the setting of higher category theory. Higher algebra generalizes ordinary algebra, or algebra in the setting of ordinary category theory. Ordinary categories have sets of morphisms between objects, and elements of a set are either equal or not. Higher categories, on the other hand, have homotopy types of morphisms between objects, typically called *mapping spaces*. Sets are examples of homotopy types, namely the discrete ones, but in general it doesn't quite make sense to ask whether or not two "elements" of a homotopy type are "equal"; rather, they are equivalent if they are represented by points that can be connected by a path in some suitable model for the homotopy type. But then any two such paths might form a nontrivial loop, leading to higher automorphisms, and so on. The notion of equality only makes sense after passing to discrete invariants such as homotopy groups.

Since the higher categorical analogue of a set is a space, the higher categorical analogue of an abelian group ought to be something like a space equipped with a multiplication operation that is associative, commutative, and invertible up to coherent homotopy. While invertibility is a *property* of an associative operation, commutativity is not; rather, it is *structure*. This is because, in higher categorical contexts, it is not enough to simply permute a sequence of elements; instead, the permutation itself is recorded as a morphism. A commutative multiplication operation must also act on morphisms, so that they may also be permuted, and so on and so forth, provided we keep track of these permutations as still higher morphisms. There are a number of formalisms that make this precise, all of which are equivalent to (or obvious variations on) the notion of *spectrum* in the sense of algebraic topology.[1]

On the one hand, a spectrum is an infinite delooping of a pointed space, thereby providing an abelian group structure on all its homotopy groups (positive and negative); on the other, a spectrum represents a cohomology theory, which is to say a graded family of contravariant abelian group valued functors on pointed spaces satisfying a suspension relation and certain exactness conditions. These two notions are equivalent: the functors that

Mathematics Subject Classification. 55N20, 13D03, 13D09, 16E45, 18C10, 18C35.

Key words and phrases. (co)homology theory, structured ring spectrum, smash product, algebras, modules, localization, cotangent complex.

[1]The overuse of the term "spectrum" in mathematics is perhaps a potential cause for confusion; fortunately, it is almost always clear from context what is meant.

comprise the cohomology theory are represented by the spaces of the infinite delooping. The real starting point of higher algebra is the observation that there is a symmetric monoidal structure on the ∞-category of spectra that refines the tensor product of abelian groups, and that there are many important examples of algebras for this tensor product.

13.1.2 Overview

Ordinary algebra is set based, meaning that it is carried out in the language of ordinary categories. As mentioned, the higher categorical analogue of sets are spaces, or n-truncated spaces if one chooses to work in an $(n+1)$-categorical context, and the truncation functors allow us to switch back and forth between categorical levels. In the category of sets, limits and colimits reduce to intersections and unions in some ambient set; in higher category theory, however, these operations must be interpreted invariantly, which implies that a homotopy colimit of sets (viewed as spaces) need not be discrete.

We begin our exposition in 13.2 with some background on the behavior of colimits in higher categories, especially in ∞-categories of presheaves or certain full subcategories thereof (the presentable ∞-categories). We then turn to the Grothendieck construction, which establishes a correspondence between fibrations and functors. The fibration perspective allows for an efficient approach to the theory of (symmetric) monoidal ∞-categories and (commutative) algebras and modules therein, our main objects of interest, at least in the stable setting: a stable ∞-category is a higher categorical analogue of an abelian category, an analogy which we make precise by comparing derived categories of abelian categories with stable ∞-categories via t-structures.

Having equipped ourselves with the basic structures and language, in 13.3 we turn to a more detailed study of spectra and the smash product. An associative (respectively, commutative) ring spectrum is defined as an algebra (respectively, commutative algebra) object in the stable ∞-category of spectra, the universal stable ∞-category. The theory also allows for a notion of (left or right) module object of an ∞-category that is (left or right) tensored over a monoidal ∞-category. We conclude this section with some remarks on localizations of ring spectra, which mirror the ordinary theory save for the fact that ideals must be interpreted on the level of homotopy groups.

13.4 is devoted to module theory. In particular, monads appear as an instance of modules, allowing us to address monadicity, which plays a much more important role higher categorically due do the difficultly of ad hoc constructions. Simplicial objects and their colimits, geometric realizations, feature in the construction of the relative tensor product as well as the definition of projective module. We also study the more general class of perfect modules, which are colimits of shifted projective modules in a sense made rigorous by the theory of tor-amplitude, which acts as a substitute for projective resolutions. We also consider free algebras and isolate various finiteness properties of modules and algebras that play important roles in higher algebra.

The final 13.5 deals with deformations of commutative algebras. The formalism of the tangent bundle allows for an elegant construction of the cotangent complex, which governs derivations and square-zero extensions. The Postnikov tower of truncations of a connective commutative algebra is comprised entirely of square-zero extensions, a crucial fact which implies an obstruction theory for computing the space of maps between commutative algebra spectra. The main theorem in that all obstructions vanish in the étale case, from which it follows that the ∞-category of étale commutative algebras over of fixed commutative ring spectrum R is equivalent to an ordinary category, namely that of étale commutative algebras over its underlying ordinary ring $\pi_0 R$, a strong version of the topological invariance of étale morphisms.

We conclude this introductory section with some background on various approaches to homotopy coherence and some remarks on higher categorical "set theory": small and large spaces, categories, universes, etc.

13.1.3 Homotopy coherence

To lessen the prerequisites we avoid the use of operads or ∞-operads in this article altogether. Nevertheless, for the sake of putting the theory of higher categorical algebra into historical context, and explaining some of the standard terminology, a few remarks are in order.

A space equipped with a homotopy coherently associative, or homotopy coherently associative and commutative, multiplication operation is traditionally referred to as an $\mathbf{A}_\infty$-*monoid*, or $\mathbf{E}_\infty$-*monoid*, meaning that it admits an action by an $\mathbf{A}_\infty$ (infinitely homotopy coherently associative) or $\mathbf{E}_\infty$ (infinitely homotopy coherently "everything", i.e. associative and commutative) operad. If the multiplication operation is invertible up to homotopy, the $\mathbf{A}_\infty$-monoid is said to be *grouplike*. An $\mathbf{E}_\infty$-monoid is grouplike when viewed as an $\mathbf{A}_\infty$-monoid. These operads were originally constructed out of geometric objects like associahedra, configuration spaces, or spaces of linear isometries.

The ∞-categorical approach prefers to use small combinatorial models for associativity and commutativity, as in Segal's treatment [36], by incorporating homotopy coherence into the language itself. The result is a significantly more streamlined approach to homotopy coherent algebraic structures, as anticipated in the now extremely influential book of the same name of Boardmann–Vogt [11], which contained the original definition of ∞-category, well before the theory was systematically developed by Joyal and then Lurie. Nevertheless, there is a rich interplay between geometry, topology, and higher category theory, as evidenced by the remarkable *cobordism hypothesis*, among other things. Even the motivating example of the theory of operads, as originally developed by May [28], namely the *little n-cubes* operad, $n \in \mathbf{N}$, collectively form the most important family of ∞-operads, the so-called $\mathbf{E}_n$ operads.

As ∞-operads, $\mathbf{A}_\infty \simeq \mathbf{E}_1$ and $\mathbf{E}_1^{\otimes n} \simeq \mathbf{E}_n$; the former equivalence is easy but the latter is equivalence is hard and requires both the *tensor product* of Boardmann–Vogt [11] and the *additivity theorem* of Dunn [14]. Since we will only be concerned with $\mathbf{A}_\infty$-algebras and $\mathbf{E}_\infty$-algebras in (symmetric) monoidal ∞-categories, we choose to emphasize the analogy with ordinary algebra by referring to these objects as *associative and commutative algebras*, respectively. This drastically simplifies the terminology and also allows us to reformulate actions by the associative or commutative ∞-operad, respectively, in terms of functors from a category of ordinals or cardinals (arguably the must basic mathematical objects of all). And while the abstract theory is quite powerful, one should keep in mind that many of the most important examples come from geometry and topology via these more classical constructions.

Heuristically, a ∞-categories are generalizations of (ordinary) categories in which there is a space, instead of a set, of morphisms between any pair of objects. There are a number of ways of making this precise, but suffice it to say that *simplicially (or topologically) enriched categories* and *quasi-categories* yield equivalent models: any ∞-category is equivalent to the *homotopy coherent nerve* $\mathrm{N}(\mathcal{C})$ of a simplicially enriched category $\mathcal{C}$. We will follow the usual notational conventions and sometimes refer to simplically enriched categories as *simplicial categories*, though this should not be confused with the more general notion of simplicial object in the category of categories.

Remark 13.1.1. The theory of quasi-categories [23] has the distinct advantage that it allows for an easy construction of the ∞-category $\mathrm{Fun}(\mathcal{D},\mathcal{C})$ of functors from an ∞-category $\mathcal{D}$ to an ∞-category $\mathcal{C}$ as the exponential

$$\mathrm{Fun}(\mathcal{D},\mathcal{C}) := \mathcal{C}^{\mathcal{D}}$$

in the cartesian closed category of simplicial sets. This is a completely combinatorial object: if X and Y are simplicial sets, an m-simplex of X^Y is natural transformation $\Delta^m \times Y \to X$ of functors $\Delta^{\mathrm{op}} \to \mathrm{Set}$, so it is completely determined by a compatible family of *functions* $\Delta^m_n \times Y_n \to X_n$, $n \in \mathbf{N}$.

Remark 13.1.2. The chief issue that arises when working with simplicial categories is that the simplicial category of simplicial functors from $\mathcal{D}$ to $\mathcal{C}$ is not in general invariant under weak equivalence (simplicial functors that are fully faithful and essentially surjective up to weak homotopy equivalence). To obtain the homotopically correct simplicial category of functors we must replace $\mathcal{D}$ with a sufficiently "free" version $\mathcal{D}' \to \mathcal{D}$ of itself.

Remark 13.1.3. In practice it is usually easier to apply the *homotopy coherent nerve* functor

$$\mathrm{N} : \mathrm{Cat}_\Delta \longrightarrow \mathrm{Fun}(\Delta^{\mathrm{op}}, \mathrm{Set})$$

and work in simplicial sets. Here Cat_Δ denotes the category of simplicially enriched categories and N is the right adjoint of the colimit preserving functor $\mathfrak{C} : \mathrm{Fun}(\Delta^{\mathrm{op}}, \mathrm{Set}) \to \mathrm{Cat}_\Delta$ determined by defining $\mathrm{Map}_{\mathfrak{C}[\Delta^n]}(i,j)$ to be the simplicial set of partially ordered subsets of $\{i, i+1, \ldots, j-1, j\}$.

Theorem 13.1.4. [26, Theorem 2.2.5.1] *The homotopy coherent nerve functor* $\mathrm{N} : \mathrm{Cat}_\Delta \to \mathrm{Fun}(\Delta^{\mathrm{op}}, \mathrm{Set})$ *is a right Quillen equivalence. In particular, the homotopy coherent nerve* $\mathrm{N}(\mathcal{C})$ *of a simplically enriched category* $\mathcal{C}$ *is a quasicategory provided* $\mathcal{C}$ *is enriched in Kan complexes (every* $\mathrm{Map}_{\mathcal{C}}(A,B)$ *is a Kan complex).*

Higher categorical algebra is truly *homotopical* and not just *homological* in nature, meaning that many of its most important objects simply do not exist within the world of chain complexes or derived categories. The portion of the theory that can be formulated in these terms is *differential graded algebra*, the abstract study of which employs the language of *differential graded categories.*

Definition 13.1.5. A *differential graded category* is a category enriched over the category $\mathrm{Ch} = \mathrm{Ch}(\mathrm{Ab})$ of chain complexes of abelian groups. We write $\mathrm{Cat}_{\mathrm{dg}}$ for the category of small differential graded categories

Example 13.1.6. Let $\mathcal{A}$ be an abelian category, or an additive subcategory of an abelian category. Given complexes $A, B \in \mathrm{Ch}(\mathcal{A})$, there is a natural chain complex of abelian groups of homomorphisms $\underline{\mathrm{Hom}}(A, B)$, which in degree n is the abelian group $\underline{\mathrm{Hom}}(A, B)_n = \prod_{k \in \mathbf{Z}} \mathrm{Hom}_{\mathcal{A}}(A_k, B_{k+n})$, with differential given by the formula $(d\varphi)(a) = d(\varphi(a)) - (-1)^n \varphi(d(a))$. In this way, $\mathrm{Ch}(\mathcal{A})$ acquires a canonical structure of a differential graded category: there is a chain complex of maps between any two objects, suitably compatible with composition. Moreover, the abelian group of maps $\mathrm{Hom}(A, B) \cong \mathrm{H}_0\underline{\mathrm{Hom}}(A, B)$ falls out as the zeroth homology of this complex, so that $\mathrm{Ch}(\mathcal{A})$ is the homotopy category of this differential graded category.

Remark 13.1.7. A differential graded category determines a simplicially enriched category via the *Dold-Kan correspondence* [40], which asserts that there is a suitably monoidal equivalece of categories between connective chain complexes and simplicial abelian groups. Said differently, we may regard a chain complex, provided it is identically zero in negative degrees,[2] as a simplicial abelian group. This is done via the functor $\mathrm{Ch}_{\geq 0}(\mathrm{Ab}) \to \mathrm{Fun}(\Delta^{\mathrm{op}}, \mathrm{Ab})$ that associates to such a chain complex A and nonempty finite ordinal $[n]$ the abelian group $\bigoplus_{[n] \twoheadrightarrow [m]} A_m$.

Remark 13.1.8. The functor $\mathrm{Ch}_{\geq 0}(\mathrm{Ab}) \to \mathrm{Ab}_{\Delta}$ alluded to above admits a lax monoidal structure [40], and lax monoidal functors between enriching monoidal categories allow us to functorially change enrichment. Since the (good) truncation functor $\tau_{\geq 0} : \mathrm{Ch}(\mathrm{Ab}) \to \mathrm{Ch}_{\geq 0}(\mathrm{Ab})$ and underlying set functor $\mathrm{Ab} \to \mathrm{Set}$ also admit lax monoidal structures, we obtain a composite lax monoidal functor

$$\mathrm{Ch}(\mathrm{Ab}) \longrightarrow \mathrm{Ch}_{\geq 0}(\mathrm{Ab}) \longrightarrow \mathrm{Fun}(\Delta^{\mathrm{op}}, \mathrm{Ab}) \longrightarrow \mathrm{Fun}(\Delta^{\mathrm{op}}, \mathrm{Set}),$$

and hence a functor from differential graded to simplicially enriched categories. We write $(-)_{\Delta} : \mathrm{Cat}_{\mathrm{dg}} \to \mathrm{Cat}_{\Delta}$ for this functor and $\mathrm{N}(\mathcal{C}) := \mathrm{N}(\mathcal{C}_{\Delta})$ for the homotopy coherent nerve of the differential graded category $\mathcal{C}$. One can define a more explicit differential graded nerve functor $\mathrm{N}_{\mathrm{dg}} : \mathrm{Cat}_{\mathrm{dg}} \to \mathrm{Fun}(\Delta^{\mathrm{op}}, \mathrm{Set})$, as in [25, Construction 1.3.1.6], but it is equivalent to the homotopy coherent nerve of the associated simplicially enriched category [25, Proposition 1.3.1.17].

13.1.4 Terminology and notation

For the purposes of this article, we will always consider spaces from the point of view of ∞-category theory. This means that we will tend to think of a topological space X in terms of its associated singular simplicial set (its *fundamental ∞-groupoid*) rather than as a point-set object, and that we reserve the right to replace X with a homotopy equivalent space (or not even choose a representative of the homotopy type of X). If X has the homotopy

[2] We will use homological grading consistently throughout this article.

type of a cell complex then the realization of the singular complex of X recovers X up to homotopy equivalence; otherwise, the realization of the singular complex of X recovers X only up to weak homotopy equivalence. As we will never need actual topological spaces (up to homoemorphism) at all anywhere in this article, "space" will always mean "homotopy type" in this sense.

Definition 13.1.9. The ∞-category $\mathcal{S}$ of *spaces* is the homotopy coherent nerve of the simplicially enriched category of Kan complexes, $\mathcal{S} = \mathrm{N}(\mathrm{Kan})$.

Remark 13.1.10. As stipulated by Grothendieck's *homotopy hypothesis*, the ∞-category of spaces is equivalent to the ∞-category Gpd_∞ of ∞-groupoids. Here $\mathrm{Gpd}_\infty \subset \mathrm{Cat}_\infty$ denotes the full subcategory consisting of those ∞-categories $\mathcal{C}$ for which any morphism $\Delta^1 \to \mathcal{C}$ is an equivalence.

We will also be interested in the ∞-category $\mathcal{S}_*$ of *pointed spaces*, which can be modeled either internally as pointed spaces $\mathcal{S}_* = \mathcal{S}_{\mathrm{pt}/}$ or as the homotopy coherent nerve of the simplicial category of pointed Kan complexes.

Remark 13.1.11. The coproduct of a pair of objects X and Y of $\mathcal{S}_*$ is computed as the wedge product $X \vee Y$, the quotient of the disjoint union of X and Y obtained by identifying their basepoints. Via the colimit preserving functor $\mathcal{S} \to \mathcal{S}_*$ that freely adjoins a basepoint, the cartesian product on $\mathcal{S}$ extends to the *smash product* on $\mathcal{S}_*$. That is, the smash product sits in a cofiber sequence

$$X \vee Y \longrightarrow X \times Y \longrightarrow X \wedge Y$$

and has the property that if $X \simeq X'_+$ and $Y \simeq Y'_+$ then $X \wedge Y \simeq (X' \times Y')_+$.

In practice, it is sometimes necessary to be precise about size by bounding various classes of objects by sets of cardinality less than some infinite regular cardinal. For the purposes of this survey article, however, we will gloss over most of these distinctions and employ a very basic version of the theory of Grothendieck universes: objects that exist in our first universe will be called *small*, objects that exist in the next universe will be called *large*, and objects that exist only in a still higher universe will be called *very large*.

As one ascends the higher categorical ladder, the size of the mathematical objects under consideration tends to increase. While a space will always be assumed to be small, an ∞-category will typically be assumed to be large, unless otherwise mentioned. We write Cat_∞ for the large ∞-category of small ∞-categories and CAT_∞ for the very large ∞-category of large ∞-categories.

While some ∞-categorical statements will be formulated in the model of quasicategories, we will often employ the theory without reference to any particular model. From this perspective, a full subcategory is assumed to be closed under equivalences, a cartesian fibration is any functor that is equivalent to a cartesian fibration (see [26, Section 2.4], or 13.2.2 for an overview), and ordinary categories are ∞-categories with discrete mapping spaces.

13.1.5 Acknowledgements

The modern approach to algebra in higher categorical contexts surveyed in this article is based almost entirely on the groundbreaking work of Jacob Lurie. It is a great pleasure to thank him for his enormous contributions to the subject. Collectively, they comprise new and firm foundations for homotopy coherent mathematics.

This article is essentially a summary of the basics of Lurie's *Higher Algebra*, with results from *Higher Topos Theory* and *Spectral Algebraic Geometry* thrown in as necessary, simplified as much as possible in order to be accessible to the neophyte.[3] Nevertheless, we have tried our best to state definitions and theorems as precisely as possible and provide accurate references for these results (though in the interest of keeping the article as short as possible we have decided not to provide any proofs whatsoever, or even citations for most of the definitions). Of course, Lurie's work builds on the combined efforts of a great many mathematicians — far too many to list here, and any attempt to do so will inevitably omit many valuable contributions.

It is also a pleasure to thank Benjamin Antieau, Tobias Barthel, Jeremiah Heller, Lars Hesselholt, Achim Krause, Tyler Lawson, Haynes Miller, Thomas Nikolaus, Charles Rezk, Markus Spitzweck, and Hiro Tanaka for helpful remarks and comments on earlier versions of this draft. The author would also like to thank the Mathematical Sciences Research Institute for providing an extremely pleasant working environment while much of this article was being written, as well as the National Science Foundation for their generous support.

13.2 Category theory

13.2.1 Presheaves and colimits

Definition 13.2.1. A *presheaf* on an ∞-category $\mathcal{C}$ is a functor $\mathcal{C}^{\mathrm{op}} \to \mathcal{S}$. The ∞-category of *presheaves* on $\mathcal{C}$ is the functor ∞-category

$$\mathcal{P}(\mathcal{C}) = \mathrm{Fun}(\mathcal{C}^{\mathrm{op}}, \mathcal{S}),$$

where $\mathcal{S}$ denotes the ∞-category of (small) spaces.

Remark 13.2.2. Recall [26, Proposition 5.1.3.1] that $\mathcal{P}(\mathcal{C})$ is equipped with a fully faithful Yoneda embedding $j : \mathcal{C} \to \mathcal{P}(\mathcal{C})$, given by the formula $j(A) = \mathrm{Map}_{\mathcal{C}}(-, A) : \mathcal{C}^{\mathrm{op}} \to \mathcal{S}$. Colimits in $\mathcal{P}(\mathcal{C})$ are computed objectwise as colimits in $\mathcal{S}$. Since $\mathcal{S}$ is cocomplete, in the sense that it admits all small colimits, we see that $\mathcal{P}(\mathcal{C})$ is as well. In fact, the Yoneda embedding exhibits $\mathcal{P}(\mathcal{C})$ as the free cocompletion of $\mathcal{C}$, in the sense made precise by the following statement.

Theorem 13.2.3. [26, Theorem 5.1.5.6] *Given a small ∞-category $\mathcal{C}$ and a cocomplete ∞-category $\mathcal{D}$, precomposition with the Yonda embedding*

$$\mathrm{Fun}^{\mathrm{colim}}(\mathcal{P}(\mathcal{C}), \mathcal{D}) \subset \mathrm{Fun}(\mathcal{P}(\mathcal{C}), \mathcal{D}) \longrightarrow \mathrm{Fun}(\mathcal{C}, \mathcal{D})$$

induces an equivalence between the ∞-categories of functors $\mathcal{C} \to \mathcal{D}$ and colimit preserving functors $\mathcal{P}(\mathcal{C}) \to \mathcal{D}$.

[3]See the chapter in this Handbook by Moritz Groth for background information about ∞-categories.

Remark 13.2.4. The actual statement of [26, Theorem 5.1.5.6] is in terms of left adjoint functors $\mathcal{P}(\mathcal{C}) \to \mathcal{D}$. Left adjoint functors preserve colimits, and in this case any colimit preserving functor is a left adjoint.

We will require variants of the above construction in which we freely adjoin only certain types of colimits.

Definition 13.2.5. A map of simplicial sets $J \to I$ is *cofinal* if, for every ∞-category $\mathcal{C}$ and every I-indexed colimit diagram $I^{\triangleright} \to \mathcal{C}$, the induced cone $J^{\triangleright} \to \mathcal{C}$ is a colimit diagram. See [26, Proposition 4.1.1.8] for details.

Definition 13.2.6. Let κ be an infinite regular cardinal. A simplicial set I is *κ-small* if I has fewer than κ nondegenerate simplices. A simplicial set J is *κ-filtered* if every map $f : I \to J$ from a κ-small simplicial set I extends to a cone $f^{\triangleright} : I^{\triangleright} \to \mathcal{I}$. A simplicial set K is *sifted* if $K \to K \times K$ is cofinal.

Example 13.2.7. Filtered simplicial sets are sifted, and the simplicial indexing category Δ^{op} is sifted. A functor of ∞-categories $f : \mathcal{C} \to \mathcal{D}$ preserves sifted colimits if and only if f preserves filtered colimits and geometric realizations.

Definition 13.2.8. An object A of an ∞-category $\mathcal{C}$ that admits κ-filtered colimits is *κ-compact* if $\mathrm{Map}_{\mathcal{C}}(A, -) : \mathcal{C} \to \mathcal{S}$ commmutes with κ-filtered colimits. An object A of an ∞-category $\mathcal{C}$ that admits geometric realizations is *projective* if $\mathrm{Map}_{\mathcal{C}}(A, -) : \mathcal{C} \to \mathcal{S}$ commutes with geometric realizations.

Definition 13.2.9. An *indexing class* is a collection of simplicial sets. (This is a non-standard notion, introduced here for convenience.) Given an indexing class $\mathcal{I}$, we write $\mathcal{P}_{\mathcal{I}}(\mathcal{C}) \subset \mathcal{P}(\mathcal{C})$ for the full subcategory consisting of those presheaves $f : \mathcal{C}^{\mathrm{op}} \to \mathcal{S}$ such that, for all $I \in \mathcal{I}$, f transforms I-indexed colimits in $\mathcal{C}$ to I^{op}-indexed limits in $\mathcal{S}$.

Remark 13.2.10. The indexing classes that most commonly arise when considering algebraic structures are finite discrete, finite, filtered, and sifted. These classes often come in pairs. A functor $f : \mathcal{C} \to \mathcal{D}$ preserves all small colimits if and only if it preserves finite colimits and filtered colimits, or finite discrete colimits (that is, finite coproducts) and sifted colimits.

Proposition 13.2.11. [26, Proposition 5.3.6.2] *Let $\mathcal{C}$ and $\mathcal{D}$ be ∞-categories. If $\mathcal{C}$ admits finite coproducts and $\mathcal{D}$ admits sifted colimits, precomposition with the Yoneda embedding $\mathcal{C} \subset \mathcal{P}_{\Sigma}(\mathcal{C})$ determines an equivalence of ∞-categories*

$$\mathrm{Fun}(\mathcal{C}, \mathcal{D}) \xleftarrow{\simeq} \mathrm{Fun}^{\mathrm{sift}}(\mathcal{P}_{\Sigma}(\mathcal{C}), \mathcal{D}) \subset \mathrm{Fun}(\mathcal{P}_{\Sigma}(\mathcal{C}), \mathcal{D})$$

of functors $\mathcal{C} \to \mathcal{D}$ and sifted colimit preserving functors $\mathcal{P}_{\Sigma}(\mathcal{C}) \to \mathcal{D}$. If $\mathcal{C}$ admits κ-small colimits and $\mathcal{D}$ admits κ-filtered colimits, precomposition with the Yoneda embedding $\mathcal{C} \subset \mathcal{P}_{\kappa\text{-sm}}(\mathcal{C})$ determines an equivalence of ∞-categories

$$\mathrm{Fun}(\mathcal{C}, \mathcal{D}) \xleftarrow{\simeq} \mathrm{Fun}^{\kappa\text{-filt}}(\mathcal{P}_{\kappa\text{-sm}}(\mathcal{C}), \mathcal{D}) \subset \mathrm{Fun}(\mathcal{P}_{\kappa\text{-sm}}(\mathcal{C}), \mathcal{D})$$

of functors $\mathcal{C} \to \mathcal{D}$ and κ-filtered colimit preserving functors $\mathcal{P}_{\kappa\text{-sm}}(\mathcal{C}) \to \mathcal{D}$.

13.2.2 The Grothendieck construction

Given an ordinary category $\mathcal{C}$, the theory of categories fibered over $\mathcal{C}$ developed in [18] characterizes the essential image of the *Grothendieck construction*

$$\mathrm{Fun}(\mathcal{C}^{\mathrm{op}}, \mathrm{CAT}) \longrightarrow \mathrm{CAT}_{/\mathcal{C}}$$

as the (not full) subcategory of $\mathrm{CAT}_{/\mathcal{C}}$ consisting of the fibered categories and their morphisms. Because of the fundamental role the Grothendieck construction plays in higher algebra, we begin with a brief overview of the basic notions of fibered ∞-category theory [26]. Whenever possible we choose to phrase these notions internally inside of CAT_∞ itself, with a few exceptions: it is sometimes quite useful to represent an ∞-category by a (marked) simplicial set [26], in particular when specifying diagrams $\mathcal{D} \to \mathcal{C}$ in an ∞-category $\mathcal{C}$.

Consider the slice ∞-category $\mathrm{CAT}_{\infty/\mathcal{C}}$: objects of $\mathrm{CAT}_{\infty/\mathcal{C}}$ are functors $p : \mathcal{D} \to \mathcal{C}$ with target $\mathcal{C}$, and morphisms of $\mathrm{CAT}_{\infty/\mathcal{C}}$ from $q : \mathcal{E} \to \mathcal{C}$ to $p : \mathcal{D} \to \mathcal{C}$ are commuting triangles of CAT_∞ of the form

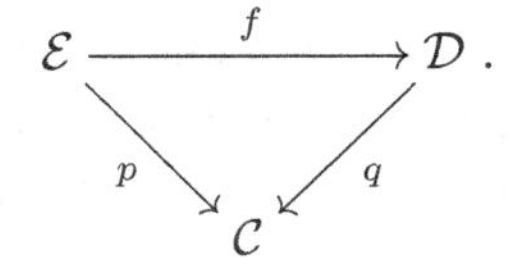

We write $\mathrm{Fun}_{\mathcal{C}}(\mathcal{E}, \mathcal{D}) \simeq \mathrm{Fun}(\mathcal{E}, \mathcal{D}) \times_{\mathrm{Fun}(\mathcal{E},\mathcal{C})} \{p\}$ for the ∞-category of functors $f : \mathcal{E} \to \mathcal{D}$ such that $q \circ f = p$.

Definition 13.2.12. Let $p : \mathcal{D} \to \mathcal{C}$ be a functor[4] of ∞-categories. An arrow $\alpha : C \to D$ of $\mathcal{D}$ is *p-cartesian* if the induced functor

$$\mathcal{D}_{/\alpha} \longrightarrow \mathcal{D}_{/D} \times_{\mathcal{C}_{/p(D)}} \mathcal{C}_{/p(\alpha)}$$

is an equivalence of ∞-categories. A *p-cartesian lift* of an arrow $f : A \to B$ in $\mathcal{C}$ is a p-cartesian arrow $\alpha : C \to D$ in $\mathcal{D}$ such that $p(\alpha) = f$. A functor $p : \mathcal{D} \to \mathcal{C}$ is a *cartesian fibration* if, for every arrow $f : A \to B$ of $\mathcal{C}$ and every object $D \in p^{-1}(B)$, there is a p-cartesian lift $\alpha : C \to D$ of f with target D.

Remark 13.2.13. When $p : \mathcal{D} \to \mathcal{C}$ is clear from context, we will refer to p-cartesian arrows simply as cartesian arrows.

A cartesian fibration $p : \mathcal{D} \to \mathcal{C}$ determines a functor $\mathrm{St}_{\mathcal{C}}(p) : \mathcal{C}^{\mathrm{op}} \to \mathrm{CAT}_\infty$, called the *straightening* of p in [26], basically by taking fibers, which happen to be contravariantly functorial in precisely this case. This operation is inverse to the *unstraightening* functor, the ∞-categorical analogue of the Grothendieck construction, which "integrates" the functor $f : \mathcal{C}^{\mathrm{op}} \to \mathrm{CAT}_\infty$ to a cartesian fibration $p : \mathrm{Un}_{\mathcal{C}}(f) \to \mathcal{C}$. There are analogous versions for cocartesian fibrations, which correspond to covariant functors $\mathcal{C} \to \mathrm{CAT}_\infty$.

[4]If we are working externally in the ordinary category of quasicategories, we would additionally require p to be a *categorical fibration* [26]; internally, however, this is a meaningless assumption, as the notion of categorical fibration is not stable under categorical equivalence.

Theorem 13.2.14. [26, Theorem 3.2.0.1] *For any small ∞-category $\mathcal{C}$, the unstraightening functor*

$$\mathrm{Un}_{\mathcal{C}}\colon \mathrm{Fun}(\mathcal{C}^{\mathrm{op}}, \mathrm{CAT}_\infty) \xrightarrow{\simeq} \mathrm{CAT}^{\mathrm{cart}}_{\infty/\mathcal{C}} \subset \mathrm{CAT}_{\infty/\mathcal{C}}$$

induces an equivalence between the ∞-category $\mathrm{Fun}(\mathcal{C}^{\mathrm{op}}, \mathrm{CAT}_\infty)$ *and the (not full) subcategory* $\mathrm{CAT}^{\mathrm{cart}}_{\infty/\mathcal{C}} \subset \mathrm{CAT}_{\infty/\mathcal{C}}$ *consisting of the cartesian fibrations over $\mathcal{C}$ and those functors over $\mathcal{C}$ that preserve cartesian arrows.*

The straightening and unstraightening equivalence will be used throughout this article, especially in our definitions of (symmetric) monoidal ∞-category and (commutative) algebra object therein. Even more fundamentally, this equivalence is used in the definition of adjunction.

Definition 13.2.15. Let $\mathcal{C}$ and $\mathcal{D}$ be ∞-categories. An *adjunction* between $\mathcal{C}$ and $\mathcal{D}$ is cartesian and cocartesian fibration $p : \mathcal{M} \to \Delta^1$ equipped with equivalences $i : \mathcal{C} \to \mathcal{M}_{\{0\}}$ and $j : \mathcal{D} \to \mathcal{M}_{\{1\}}$.

Remark 13.2.16. Given an adjunction, the left adoint $f : \mathcal{C} \to \mathcal{D}$ is determined by cocartesian lifts of $0 \to 1$, and the right adjoint $g : \mathcal{D} \to \mathcal{C}$ is determined by cartesian lifts of $0 \to 1$. It is possible to construct a unit or counit transformation [26, Proposition 5.2.2.8], and the fact that both $\mathcal{C}$ and $\mathcal{D}$ fully faithfully embed into $\mathcal{M}$ implies that we have equivalences

$$\mathrm{Map}(A, gB) \simeq \mathrm{Map}(iA, igB) \simeq \mathrm{Map}(iA, jB) \simeq \mathrm{Map}(jfA, jB) \simeq \mathrm{Map}(fA, B)$$

for any pair of objects A of $\mathcal{C}$ and B of $\mathcal{D}$. A functor $f : \Delta^1 \to \mathrm{CAT}_\infty$ is a *left adjoint* if and only if the associated cocartesian fibration $\mathcal{M} \to \Delta^1$ is also cartesian, and a functor $g : \Delta^1 \to \mathrm{CAT}_\infty^{\mathrm{op}}$ is a *right adjoint* if and only if the associated cartesian fibration $\mathcal{M} \to \Delta^1$ is also cocartesian.

13.2.3 Monoidal and symmetric monoidal ∞-categories

Informally, a symmetric monoidal ∞-category is an ∞-category $\mathcal{C}$ equipped with unit $\eta : \Delta^0 \to \mathcal{C}$ and multiplication $\mu : \mathcal{C} \times \mathcal{C} \to \mathcal{C}$ maps that are appropriately coherently associative, commutative, and unital. Making this precise requires a means to organize for us the infinite hierarchy of coherence data necessary to assert that the intermediate multiplications $\mathcal{C}^{\times m} \to \mathcal{C}^{\times n}$, which should be indexed by something like functions from the set with m elements to the set with n elements, are suitably compatible.

Remark 13.2.17. Actually, as we may first project away some of the factors in the product before we multiply, these operations are indexed by functions of *pointed* sets. For convenience we allow ourselves to specify a finite pointed set by the cardinality n of its elements, writing $\langle n \rangle = \{0, 1, \ldots, n\}$ for the pointed set with basepoint 0 and n elements in the complement $\langle n \rangle^\circ = \{1, \ldots, n\}$.

Definition 13.2.18. A morphism $\alpha : \langle m \rangle \to \langle n \rangle$ in Fin_* is *inert* if, for each element $i \in \langle n \rangle^\circ$, the preimage $\alpha^{-1}(i) \subset \langle m \rangle^\circ$ consists of a single element. A morphism $\alpha : \langle m \rangle \to \langle n \rangle$ in Fin_* is *active* if the preimage $\alpha^{-1}(\mathrm{pt}) \subset \langle m \rangle$ consists of a single element (necessarily the basepoint).

Example 13.2.19. There are exactly n inert maps $\delta_i : \langle n \rangle \to \langle 1 \rangle$, namely

$$\delta_i(j) = \begin{cases} 1 & j = i \\ 0 & j \neq i \end{cases}.$$

Definition 13.2.20. A *symmetric monoidal ∞-category* is a cocartesian fibration $p : \mathcal{C}^{\otimes} \to \mathrm{Fin}_*$ that satisfies the Segal condition: for each natural number n, the map $\mathcal{C}^{\otimes}_{\langle n \rangle} \to \prod_{i=1}^{n} \mathcal{C}^{\otimes}_{\langle 1 \rangle}$ induced by the inert morphisms $\delta_i : \langle n \rangle \to \langle 1 \rangle$, $1 \leq i \leq n$, is an equivalence. A morphism of symmetric monoidal ∞-categories, or a *symmetric monoidal functor*, from $p : \mathcal{C}^{\otimes} \to \mathrm{Fin}_*$ to $q : \mathcal{D}^{\otimes} \to \mathrm{Fin}_*$, is a functor $f : \mathcal{C}^{\otimes} \to \mathcal{D}^{\otimes}$ over Fin_* such that f preserves cocartesian edges. The ∞-category $\mathrm{CMon}(\mathrm{CAT}_{\infty})$ of symmetric monoidal ∞-categories is the full subcategory of $\mathrm{CAT}^{\mathrm{cocart}}_{\infty/\mathrm{Fin}_*}$ spanned by the symmetric monoidal ∞-categories.

Remark 13.2.21. The straightening of a symmetric monoidal ∞-category $\mathcal{C}^{\otimes} \to \mathrm{Fin}_*$ is a functor $\mathrm{Fin}_* \to \mathrm{CAT}_{\infty}$ satisfying the Segal condition. This is the data of a commutative monoid object in Cat_{∞}, hence the notation $\mathrm{CMon}(\mathrm{CAT})_{\infty}$. We could take this as the definition of a symmetric monoidal ∞-category, except that in practice these functors are often difficult to write down, an issue that already arises in ordinary category theory: for instance, the tensor product of modules is more naturally defined by a universal property than a specific choice of representative. It is usually easier to avoid making explicit choices altogether by constructing the cocartesian fibration $p : \mathcal{C}^{\otimes} \to \mathrm{Fin}_*$ instead, where one may as well take the fiber over $\langle n \rangle$ to be the ∞-category of *all* possible choices of the symmetric monoidal product of n objects of $\mathcal{C}$.

Example 13.2.22. Let $\mathcal{C}$ be an ∞-category with finite products. Then there exists a cocartesian fibration $p : \mathcal{C}^{\times} \to \mathrm{Fin}_*$ whose fiber over $\langle n \rangle$ is the ∞-category of commutative diagrams of the form $f : (\mathrm{Sub}(n), \leq)^{\mathrm{op}} \to \mathcal{C}$ such that $f(\emptyset) \simeq \mathrm{pt}$ and f carries pushout squares in $\mathrm{Sub}(n)$ to pullback squares in $\mathcal{C}$. In particular, the map $f(I) \to \prod_{i \in I} f(\{i\})$ is an equivalence in $\mathcal{C}$, from which it follows that $\mathcal{C}^{\times}_{\langle n \rangle} \to \prod_{i=1}^{n} \mathcal{C}$ is an equivalence and therefore that $p : \mathcal{C}^{\times} \to \mathrm{Fin}_*$ is a symmetric monoidal ∞-category. We refer to this as the *cartesian* symmetric monoidal structure on $\mathcal{C}$, which exists if and only if $\mathcal{C}$ has finite products.

Definition 13.2.23. Let $p : \mathcal{C}^{\otimes} \to \mathrm{Fin}_*$ be a symmetric monoidal ∞-category. A *commutative algebra object* of $\mathcal{C}^{\otimes}$ is a section $s : \mathrm{Fin}_* \to \mathcal{C}^{\otimes}$ of p that sends inert morphisms in Fin_* to cocartesian morphisms in $\mathcal{C}^{\otimes}$. The ∞-category $\mathrm{CAlg}(\mathcal{C}^{\otimes})$ of commutative algebra objects in $p : \mathcal{C}^{\otimes} \to \mathrm{Fin}_*$ is the full subcategory of the ∞-category $\mathrm{Fun}_{\mathrm{Fin}_*}(\mathrm{Fin}_*, \mathcal{C}^{\otimes})$ of sections $s : \mathrm{Fin}_* \to \mathcal{C}^{\otimes}$ of p consisting of the commutative algebra objects.

Remark 13.2.24. There is an equivalence $\mathrm{CAlg}(\mathrm{CAT}^{\times}_{\infty}) \simeq \mathrm{CMon}(\mathrm{CAT}_{\infty})$. That is, the ∞-category of commutative algebra objects in $\mathrm{CAT}^{\times}_{\infty} \to \mathrm{Fin}_*$ is equivalent to the ∞-category of commutative monoid objects in CAT_{∞}.

We will also be interested in monoidal ∞-categories. To set up the theory we need a nonsymmetric analogue on the category Fin_* of finite pointed sets.

Remark 13.2.25. To keep the prerequisites to a minimum, we purposely avoid using the language of ∞-operads. However, from that perspective, a natural choice of indexing category for monoidal structures is the "desymmetrization" $q : \mathrm{Fin}_*^{\mathrm{ord}} \to \mathrm{Fin}_*$ of Fin_*, the functor whose fiber over $\langle n \rangle$ is the set of total orderings of $\langle n \rangle^\circ$. However it will be both more convenient and elementary to simply use simplicial objects instead.

Definition 13.2.26. The functor $\mathrm{Cut} : \Delta^{\mathrm{op}} \to \mathrm{Fin}_*$ is defined by identifying the nonempty finite ordinal $[n] = \{0 \to 1 \to \cdots \to n{-}1 \to n\}$ with the set of "cuts" of the pointed cardinal $\langle n \rangle = \{0, 1, \ldots, n-1, n\}$ as follows: we can cut the string $\{0 \to 1 \to \cdots \to n-1 \to n\}$ before or after any $i \in [n]$, for a total of $n+2$ possibilities. However, cutting before 0 or after n have the same effect (nothing), so we identify the two trivial cuts with the basepoint of $\langle n \rangle$.

Definition 13.2.27. An order preserving function $f : [m] \to [n]$ is *inert* if it is it injective with convex image; that is, $f(m) = f(0) + m$.

Remark 13.2.28. The n inert morphisms $[1] \to [n]$ in Δ are the order preserving functions of the form $\sigma_i(j) = i + j$, $0 \le i < n$, $j \in [1]$.

Definition 13.2.29. A *monoidal* ∞*-category* is a cocartesian fibration $p : \mathcal{C}^\otimes \to \Delta^{\mathrm{op}}$ that satisfies the Segal condition: for each $n \in \mathbf{N}$, the map

$$\mathcal{C}^\otimes_{[n]} \longrightarrow \prod_{i=1}^{n} \mathcal{C}^\otimes_{[1]}$$

induced by the inert morphisms $\sigma_i : [1] \to [n]$, $0 \le i < n$, is an equivalence. A morphism of monoidal ∞-categories, or a *monoidal functor*, from $p : \mathcal{C}^\otimes \to \mathrm{Fin}_*$ to $q : \mathcal{D}^\otimes \to \mathrm{Fin}_*$ is a functor $f : \mathcal{C}^\otimes \to \mathcal{D}^\otimes$ over Δ^{op} such that f preserves cocartesian edges. The ∞-category $\mathrm{Mon}(\mathrm{CAT}_\infty)$ of monoidal ∞-categories is the full subcategory of $\mathrm{CAT}^{\mathrm{cocart}}_{\infty/\Delta^{\mathrm{op}}}$ spanned by the monoidal ∞-categories.

Definition 13.2.30. Let $p : \mathcal{C}^\otimes \to \Delta^{\mathrm{op}}$ be a monoidal ∞-category. An *algebra object* of $\mathcal{C}^\otimes$ is a section $s : \Delta^{\mathrm{op}} \to \mathcal{C}^\otimes$ of p that sends inert morphisms in Δ^{op} to cocartesian morphisms in $\mathcal{C}^\otimes$. The ∞-category $\mathrm{Alg}(\mathcal{C}^\otimes)$ of algebra objects in $\mathcal{C}^\otimes$ is the full subcategory of the ∞-category $\mathrm{Fun}_{\Delta^{\mathrm{op}}}(\Delta^{\mathrm{op}}, \mathcal{C}^\otimes)$ of sections $s : \Delta^{\mathrm{op}} \to \mathcal{C}^\otimes$ of p consisting of the algebra objects.

Remark 13.2.31. As in the symmetric case, there is an equivalence $\mathrm{Alg}(\mathrm{CAT}^\times_\infty) \simeq \mathrm{Mon}(\mathrm{CAT}_\infty)$: the ∞-category of associative algebra objects in $\mathrm{CAT}^\times_\infty \to \Delta^{\mathrm{op}}$ is equivalent to the ∞-category of monoid objects in CAT_∞.

Remark 13.2.32. Using [25, Construction 4.1.2.9] the theory is set up so that a symmetric monoidal ∞-category $p : \mathcal{C}^\otimes \to \mathrm{Fin}_*$ restricts to a monoidal ∞-category $q : \mathcal{C}^\otimes|_{\Delta^{\mathrm{op}}} \to \Delta^{\mathrm{op}}$. It follows that we have a forgetful functor $\mathrm{CAlg}(\mathcal{C}^\otimes) \to \mathrm{Alg}(\mathcal{C}^\otimes) := \mathrm{Alg}(\mathcal{C}^\otimes|_{\Delta^{\mathrm{op}}})$.

13.2.4 Presentable ∞-categories

Definition 13.2.33. Given a small ∞-category $\mathcal{C}$, we write $\mathrm{Ind}_\kappa(\mathcal{C}) \subset \mathcal{P}(\mathcal{C})$ for the full subcategory consisting of those functors $f : \mathcal{C}^{\mathrm{op}} \to \mathcal{S}$ such that the source $\mathcal{D}$ of the unstraightening $p : \mathcal{D} \to \mathcal{C}$ of f is κ-filtered.

Proposition 13.2.34. [26, Proposition 5.3.5.10] *Let $\mathcal{C}$ be a small ∞-category and $\mathcal{D}$ an ∞-category with κ-filtered colimits. There is an equivalence of ∞-categories*

$$\mathrm{Fun}^{\kappa\text{-filt}}(\mathrm{Ind}_\kappa(\mathcal{C}), \mathcal{D}) \simeq \mathrm{Fun}(\mathcal{C}, \mathcal{D}),$$

and hence by 13.2.11 an equivalence $\mathrm{Ind}_\kappa(\mathcal{C}) \simeq \mathcal{P}_{\kappa\text{-fin}}(\mathcal{C})$.

Definition 13.2.35. An object A of an ∞-category $\mathcal{C}$ is *κ-compact* if the corepresentable functor $\mathrm{Map}_{\mathcal{C}}(A, -) : \mathcal{C} \to \mathcal{S}$ commutes with κ-filtered colimits. We write $\mathcal{C}^\kappa \subset \mathcal{C}$ for the full subcategory of $\mathcal{C}$ consisting of the κ-compact objects.

Definition 13.2.36. An ∞-category $\mathcal{C}$ is *κ-compactly generated* if $\mathcal{C}$ admits all small colimits and, writing $\mathcal{C}^\kappa \subset \mathcal{C}$ for the full subcategory consisting of the κ-compact objects, the canonical map $\mathrm{Ind}_\kappa(\mathcal{C}^\kappa) \to \mathcal{C}$ is an equivalence. A *presentable ∞-category* is an ∞-category that is κ-compactly generated for some infinite regular cardinal κ.

There are dual notions of morphism of presentable ∞-category: namely, those functors which are left (respectively, right) adjoints. We write LPr (respectively, RPr) for these ∞-categories. A key point [26, Proposition 5.5.3.13 and Theorem 5.5.3.18] is that the inclusion of subcategories $\mathrm{LPr} \subset \mathrm{CAT}_\infty$ and $\mathrm{RPr} \subset \mathrm{CAT}_\infty$ preserves limits: in fact, given a functor $\mathcal{D} \to \mathrm{LPr}$, a cone $\mathcal{D}^\triangleleft \to \mathrm{LPr}$ is limiting in LPr (respectively, RPr) if and only if the induced cone is limiting in CAT_∞.

Remark 13.2.37. For each infinite regular cardinal κ we have subcategories (again not full) $\mathrm{LPr}_\kappa \subset \mathrm{LPr}$ (respectively, $\mathrm{RPr}_\kappa \subset \mathrm{RPr}$) consisting of the κ-compactly generated ∞-categories and those left (respectively, right) adjoint functors that preserve κ-compact objects (respectively, κ-filtered colimits).

Definition 13.2.38. A *presentable fibration* is a cartesian fibration $p : \mathcal{D} \to \mathcal{C}$ such that the straightening $\mathrm{St}_{\mathcal{C}}(p) : \mathcal{C}^{\mathrm{op}} \to \mathrm{CAT}_\infty$ factors through the subcategory $\mathrm{RPr} \subset \mathrm{CAT}_\infty$ of presentable ∞-categories and right adjoint functors.

Remark 13.2.39. The *adjoint functor theorem* [26, Corollary 5.5.2.9] states that any colimit preserving functor $L : \mathcal{A} \to \mathcal{B}$ of presentable ∞-categories admits a right adjoint $R : \mathcal{B} \to \mathcal{A}$ (this is even true more generally, for instance if $\mathcal{B}$ is only assumed to be cocomplete). Hence LPr can be equivalently described as the subcategory of presentable ∞-categories and left adjoint functors. Dually, writing $\mathrm{RPr} \subset \mathrm{CAT}_\infty$ for the subcategory consisting of the presentable ∞-categories and right adjoint functors, uniqueness of adjoints allows us to construct a canonical equivalence $\mathrm{LPr}^{\mathrm{op}} \simeq \mathrm{RPr}$.

Definition 13.2.40. A presentable symmetric monoidal ∞-category is a symmetric monoidal ∞-category $p : \mathcal{C}^\otimes \to \mathrm{Fin}_*$ such that the underlying ∞-category $\mathcal{C} \simeq \mathcal{C}^\otimes_{\langle 1 \rangle}$ is presentable and any choice of tensor product bifunctor $\otimes : \mathcal{C} \times \mathcal{C} \to \mathcal{C}$ commutes with colimits separately in each variable.

Theorem 13.2.41. [25, Proposition 4.8.1.17] *Let $\mathcal{C}$ and $\mathcal{D}$ be presentable ∞-categories. The subfunctor*

$$\mathrm{Fun}'(\mathcal{C} \times \mathcal{D}, -) \subset \mathrm{Fun}(\mathcal{C} \times \mathcal{D}, -) \colon \mathrm{LPr} \longrightarrow \mathrm{CAT}_\infty,$$

whose value at $\mathcal{E} \in \mathrm{LPr}$ consists of those functors $f : \mathcal{C} \times \mathcal{D} \to \mathcal{E}$ that preserve colimits separately in each variable, is corepresented by an object $\mathcal{C} \otimes \mathcal{D} \in \mathrm{LPr}$.

Remark 13.2.42. It is straightforward to check, with the definition of $\mathcal{C} \otimes \mathcal{D}$ as above, that for a presentable ∞-category $\mathcal{E}$, there is a canonical equivalence

$$\mathrm{LFun}(\mathcal{C} \otimes \mathcal{D}, \mathcal{E}) \simeq \mathrm{LFun}(\mathcal{C}, \mathrm{LFun}(\mathcal{D}, \mathcal{E})),$$

where $\mathrm{LFun}(\mathcal{D}, \mathcal{E})$ denotes the ∞-category of left adjoint functors $\mathcal{D} \to \mathcal{E}$, which is presentable by [25]. Dually, we write $\mathrm{RFun}(\mathcal{E}, \mathcal{D})$ for the ∞-category of right adjoint functors $\mathcal{E} \to \mathcal{D}$, and note that $\mathrm{LFun}(\mathcal{D}, \mathcal{E}) \simeq \mathrm{RFun}(\mathcal{E}, \mathcal{D})^{\mathrm{op}}$.

Proposition 13.2.43. [25, Lemma 4.8.1.16] *Let $\mathcal{C}, \mathcal{D}$ be presentable ∞-categories. Then $\mathrm{RFun}(\mathcal{D}^{\mathrm{op}}, \mathcal{C})$ is a presentable ∞-category, and*

$$\mathcal{C} \otimes \mathcal{D} \simeq \mathrm{RFun}(\mathcal{D}^{\mathrm{op}}, \mathcal{C}).$$

Theorem 13.2.44. [25, Corollary 4.8.1.12] *The functor $\mathcal{P}\colon \mathrm{Cat}_\infty \to \mathrm{LPr}$ extends to a symmetric monoidal functor $\mathcal{P}^\otimes : \mathrm{Cat}_\infty^\times \to \mathrm{LPr}^\otimes$.*

Remark 13.2.45. The symmetric monoidality of the presheaves functor is an ∞-categorical generalization of the Day convolution product. If $\mathcal{C}^\otimes$ is a small symmetric monoidal ∞-category, then $\mathcal{P}^\otimes(\mathcal{C})$ inherits a *convolution* symmetric monoidal structure [25, Remark 4.8.1.13], given by the formula

$$(X_1 \otimes \cdots \otimes X_n)(A) \simeq \mathrm{colim}_{(B_1,\ldots,B_n)\in \mathcal{C}^{\times n}_{/A}} X_1(B_1) \times \cdots \times X_n(B_n).$$

Here the colimit is taken over the ∞-category $\mathcal{C}^{\times n}_{/A} \simeq \mathcal{C}^{\times n} \times_{\mathcal{C}} \mathcal{C}_{/A}$.

Remark 13.2.46. The fact that $\mathcal{P}^\otimes$ is symmetric monoidal implies that the canonical map $\mathcal{P}(\mathcal{C}) \otimes \mathcal{P}(\mathcal{D}) \to \mathcal{P}(\mathcal{C} \times \mathcal{D})$ is an equivalence.

Definition 13.2.47. A space $X \in \mathcal{S}$ is said to be *n-truncated* if $\pi_m(X, x) \cong 0$ for all $m > n$ and $x \in X$. By convention, a space is said to be (-2)-truncated if it is contractible and (-1)-truncated if it is empty or contractible.

Definition 13.2.48. An object A of an ∞-category $\mathcal{C}$ is said to be n-truncated, $n \in \mathbf{N}$, if the associated representable functor $\mathrm{Map}_{\mathcal{C}}(-, A) : \mathcal{C}^{\mathrm{op}} \to \mathcal{S}$ factors through the full subcategory $\tau_{\leq n}\mathcal{S} \subset \mathcal{S}$ spanned by the n-truncated spaces.

Proposition 13.2.49. [26, Proposition 5.5.6.18] *Let $\mathcal{C}$ be a presentable ∞-category. The inclusion of the full subcategory of n-truncated objects $\tau_{\leq n}\mathcal{C} \subset \mathcal{C}$ is stable under limits and admits a left adjoint $\tau_{\leq n} : \mathcal{C} \to \tau_{\leq n}\mathcal{C}$.*

Proposition 13.2.50. [25, Example 4.8.1.22] *As an endofunctor of LPr, $\tau_{\leq n} : \mathrm{LPr} \to \mathrm{LPr}$ is idempotent. It therefore determines a localization of LPr with essential image the presentable n-categories. In particular, for any presentable ∞-category $\mathcal{C}$, we have a canonical equivalence $\mathcal{C} \otimes \tau_{\leq n}\mathcal{S} \simeq \tau_{\leq n}\mathcal{C}$.*

For any presentable ∞-category $\mathcal{C}$, the left adjoints of the inclusions $\tau_{\leq m}\mathcal{C} \subset \tau_{\leq n}\mathcal{C}$ for $m < n$ result in a tower $\cdots \to \tau_{\leq n}\mathcal{C} \to \cdots \to \tau_{\leq 0}\mathcal{C}$ of truncations of $\mathcal{C}$ in $\mathrm{LPr}_{\mathcal{C}/}$, and consequently a comparison map

$$\mathcal{C} \longrightarrow \lim\{\cdots \to \tau_{\leq n}\mathcal{C} \to \cdots \to \tau_{\leq 0}\mathcal{C}\}.$$

Definition 13.2.51. Let A be an object of a presentable ∞-category $\mathcal{C}$. The *Postnikov tower* of A is the tower of truncations $\cdots \to \tau_{\leq n}A \to \cdots \to \tau_{\leq 0}A$ of A, regarded as a diagram in $\mathcal{C}_{A/}$.

Definition 13.2.52. Let $\mathcal{C}$ be a presentable ∞-category. We say that *Postnikov towers converge* in $\mathcal{C}$ if the map $\mathcal{C} \to \lim\{\cdots \to \tau_{\leq n}\mathcal{C} \to \cdots \to \tau_{\leq 0}\mathcal{C}\}$ is an equivalence of ∞-categories.

More concretely, Postnikov towers converge in $\mathcal{C}$ if, for each object A, the map $A \to \lim \tau_{\leq n}A$ from A to the limit of its Postnikov tower is an equivalence.

Example 13.2.53. Postnikov towers converge in the ∞-categories $\mathcal{S}$ and $\mathcal{S}_*$.

13.2.5 Stable ∞-categories

Definition 13.2.54. A *zero object* of an ∞-category $\mathcal{C}$ is an object that is both initial and final. An ∞-category $\mathcal{C}$ is *pointed* if $\mathcal{C}$ admits a zero object.

Definition 13.2.55. Let $\mathcal{C}$ be a pointed ∞-category. A *triangle* in $\mathcal{C}$ is a commutative square $\Delta^1 \times \Delta^1 \to \mathcal{C}$ of the form

$$\begin{array}{ccc} A & \xrightarrow{f} & B \\ \downarrow & & \downarrow g \\ 0 & \longrightarrow & C \end{array}$$

where 0 is a zero object of $\mathcal{C}$. We typically write $A \xrightarrow{f} B \xrightarrow{g} C$ for a triangle in $\mathcal{C}$, though it is important to remember that a choice of composite $g \circ f$ and nullhomotopy $g \circ f \simeq 0 \in \mathrm{Map}(A, C)$ are also part of the data. We say that a triangle in $\mathcal{C}$ is *left exact* if it is cartesian (a pullback), *right exact* if it is cocartesian (a pushout), and *exact* if it is cartesian and cocartesian.

Remark 13.2.56. Let $\mathcal{C}$ be a pointed ∞-category with finite limits and colimits, and consider a triangle in $\mathcal{C}$ of the form

$$\begin{array}{ccc} A & \longrightarrow & 0 \\ \downarrow & & \downarrow \\ 0 & \longrightarrow & B. \end{array}$$

Then $A \simeq \Omega B$ if the triangle is left exact and $\Sigma A \simeq B$ is the triangle is right exact. If the triangle is exact, we have equivalences $A \simeq \Omega\Sigma A$ and $\Sigma\Omega B \simeq B$.

Definition 13.2.57. A *stable* ∞*-category* is an ∞-category $\mathcal{C}$ with a zero object, finite limits and colimits, and that satisfies the following axiom: a commutative square $\Delta^1 \times \Delta^1 \to \mathcal{C}$ in $\mathcal{C}$ is cartesian if and only if it is cocartesian (in other words, a pullback if and only if it is a pushout).

Definition 13.2.58. Let $\mathcal{C}$ and $\mathcal{D}$ be a functor of ∞-categories that admits finite limits and colimits. A functor $f : \mathcal{C} \to \mathcal{D}$ is *left exact*if f preserves finite limits, *right exact*if f preserves finite colimits, and *exact* if f preserves finite limits and finite colimits. We write $\mathrm{CAT}_\infty^{\mathrm{st}} \subset \mathrm{CAT}_\infty$ for the (not full) subcategory consisting of the stable ∞-categories and the exact functors.

Proposition 13.2.59. [25, Corollary 1.4.2.11] *Let $\mathcal{C}$ be a pointed ∞-category. The following conditions are equivalent:*

(1) *$\mathcal{C}$ is stable.*

(2) *$\mathcal{C}$ admits finite colimits and $\Sigma : \mathcal{C} \to \mathcal{C}$ is an equivalence.*

(3) *$\mathcal{C}$ admits finite limits and $\Omega : \mathcal{C} \to \mathcal{C}$ is an equivalence.*

Theorem 13.2.60. [25, Lemma 1.1.2.13] *The homotopy category of a stable ∞-category $\mathcal{C}$ admits the structure of a triangulated category such that:*

(1) *The shift functor is induced from the suspension functor $\Sigma : \mathcal{C} \to \mathcal{C}$.*

(2) *A triangle $A \xrightarrow{f} B \xrightarrow{g} C \xrightarrow{h} \Sigma A$ in the homotopy category is exact if and only if there exist exact triangles*

$$\begin{array}{ccccc} A & \xrightarrow{f'} & B & \longrightarrow & 0 \\ \downarrow & & \downarrow g' & & \downarrow \\ 0 & \longrightarrow & C & \xrightarrow{h'} & D \end{array}$$

in $\mathcal{C}$ such that f', g', and h' are representatives of f, g and h composed with the equivalence $\Sigma A \to D$, respectively.

Definition 13.2.61. Let $\mathcal{C}$ be an ∞-category with finite limits. A functor

$$F : \mathcal{S}_*^{\mathrm{fin}} \longrightarrow \mathcal{C}$$

is said to be *excisive* (respectively, *reduced*) if F sends cocartesian squares to cartesian squares (respectively, sends initial objects to final objects). We write

$$\mathrm{Exc}_*(\mathcal{S}_*^{\mathrm{fin}}, \mathcal{C}) \subset \mathrm{Fun}(\mathcal{S}_*^{\mathrm{fin}}, \mathcal{C})$$

for the full subcategory of reduced excisive functors $\mathcal{S}_*^{\mathrm{fin}} \to \mathcal{C}$.

Definition 13.2.62. Let $\mathcal{C}$ be an ∞-category with finite limits. The ∞-category $\mathrm{Sp}(\mathcal{C})$ of *spectrum objects in $\mathcal{C}$* is the ∞-category

$$\mathrm{Sp}(\mathcal{C}) = \mathrm{Exc}_*(\mathcal{S}_*^{\mathrm{fin}}, \mathcal{C})$$

of reduced excisive functors from finite spaces to $\mathcal{C}$.

The reason for the terminology "spectrum object" is that a reduced excisive functor determines, and is determined by, its value on any infinite sequence $\{S^{n_0}, S^{n_1}, S^{n_2}, \ldots\}$ of spheres of strictly increasing dimension. The most canonical choice is the sequence of all spheres $\{S^0, S^1, S^2, \ldots\}$, so that evaluation on this family of spheres induces a map

$$\mathrm{Exc}_*(\mathcal{S}_*^{\mathrm{fin}}, \mathcal{C}) \longrightarrow \lim\{\cdots \xrightarrow{\Omega} \mathcal{C}_* \xrightarrow{\Omega} \mathcal{C}_* \xrightarrow{\Omega} \mathcal{C}_*\}$$

which sends the excisive functor F to the sequence of pointed objects $\{F(S^n)\}_{n\in\mathbf{N}}$ and equivalences $F(S^n) \xrightarrow{\simeq} \Omega F(\Sigma S^n) \xrightarrow{\simeq} \Omega F(S^{n+1})$.

Proposition 13.2.63. [25, Remark 1.4.2.25] *Let $\mathcal{C}$ be an ∞-category with finite limits. The functor*

$$\mathrm{Sp}(\mathcal{C}) = \mathrm{Exc}_*(\mathcal{S}_*^{\mathrm{fin}}, \mathcal{C}) \longrightarrow \lim\{\cdots \xrightarrow{\Omega} \mathcal{C}_* \xrightarrow{\Omega} \mathcal{C}_* \xrightarrow{\Omega} \mathcal{C}_*\}$$

induced by evaluation on spheres is an equivalence of ∞-categories.

Remark 13.2.64. Since $\mathcal{C}$ has finite limits, it has a final object pt, and so it makes sense to consider the ∞-category $\mathcal{C}_* \simeq \mathcal{C}_{\mathrm{pt}/}$ of pointed objects in $\mathcal{C}$. There is a canonical equivalence $\mathrm{Sp}(\mathcal{C}_*) \simeq \mathrm{Sp}(\mathcal{C})$.

The main examples of stable ∞-categories are either presentable or embed fully faithfully inside a stable presentable ∞-category as the subcategory of κ-compact objects for some infinite regular cardinal κ. Let $\mathrm{LPr}_{\mathrm{st}} \subset \mathrm{LPr}$ denote the full subcategory spanned by the stable presentable ∞-categories.

Proposition 13.2.65. [25, Proposition 4.8.2.18] *The inclusion of the full subcategory $\mathrm{LPr}_{\mathrm{st}} \subset \mathrm{LPr}$ admits a left adjoint $\mathrm{LPr} \to \mathrm{LPr}_{\mathrm{st}}$.*

Remark 13.2.66. The left adjoint "stabilization" functor is given by tensoring with the ∞-category of spectra. By virtue of the equivalence $\mathcal{C} \otimes \mathrm{Sp} \simeq \mathrm{Sp}(\mathcal{C})$, the ∞-category of spectrum objects in $\mathcal{C}$ is a description of its stabilization.

Corollary 13.2.67. *Let $\mathcal{A}$ and $\mathcal{B}$ be presentable ∞-categories such that $\mathcal{A}$ is stable. Then $\mathcal{A} \otimes \mathcal{B}$ is stable.*

Corollary 13.2.68. *The symmetric monoidal structure $\mathrm{LPr}^{\otimes}$ on LPr induces a symmetric monoidal structure $\mathrm{LPr}_{\mathrm{st}}^{\otimes}$ on the full subcategory $\mathrm{LPr}_{\mathrm{st}} \subset \mathrm{LPr}$ consisting of the stable presentable ∞-categories.*

We write $\mathrm{LPr}_{\mathrm{st}}^{\otimes} \subset \mathrm{LPr}^{\otimes}$ for this symmetric monoidal subcategory. Note that this inclusion is lax symmetric monoidal and right adjoint to the symmetric monoidal stabilization functor $\mathrm{Sp}(-) : \mathrm{LPr}^{\otimes} \to \mathrm{LPr}_{\mathrm{st}}^{\otimes}$.

Corollary 13.2.69. *The forgetful functor $\mathrm{LPr}_{\mathrm{st}} \subset \mathrm{CAT}_\infty$ extends to a lax symmetric monoidal functor $\mathrm{LPr}_{\mathrm{st}}^{\otimes} \to \mathrm{CAT}_\infty^{\times}$. In particular, it carries (commutative) algebra objects of $\mathrm{LPr}_{\mathrm{st}}^{\otimes}$ to (symmetric) monoidal ∞-categories.*

Example 13.2.70. The ∞-category Sp of spectra is a unit of the symmetric monoidal ∞-category $\mathrm{LPr}_{\mathrm{st}}^{\otimes}$ of stable presentable ∞-categories. It therefore admits a presentable symmetric monoidal structure, called the *smash product*. We will write $\otimes^n : \mathrm{Sp}^{\times n} \to \mathrm{Sp}$ for any choice of smash product multifunctor.

Remark 13.2.71. Given spectra A and B, their smash product is usually written $A \wedge B$ is the literature. We follow the convention of [25] and write $A \otimes B$ instead, emphasizing the analogy with the tensor product of abelian groups (or more precisely the derived tensor product of chain complexes).

Remark 13.2.72. By construction, the symmetric monoidal structure on the ∞-category of spectra is compatible with the cartesian symmetric monoidal structure on the ∞-category of spaces via the suspension spectrum functor

$$\Sigma^{\infty}_{+} : \mathcal{S} \longrightarrow \mathrm{Sp}\,.$$

This means that, for spaces $X_1, \ldots, X_n$, there is a canonical equivalence

$$(\Sigma^{\infty}_{+} X_1) \otimes \cdots \otimes (\Sigma^{\infty}_{+} X_n) \simeq \Sigma^{\infty}_{+}(X_1 \times \cdots \times X_n).$$

Proposition 13.2.73. [25, Corollary 1.4.4.2] *A stable ∞-category $\mathcal{C}$ is κ-compactly generated if and only if $\mathcal{C}$ admits small coproducts, a κ-compact generator for some regular cardinal κ, and (the homotopy category of) $\mathcal{C}$ is locally small.*

Definition 13.2.74. A *Verdier sequence* is a cocartesian square

$$\begin{array}{ccc} \mathcal{A} & \longrightarrow & \mathcal{B} \\ \downarrow & & \downarrow \\ 0 & \longrightarrow & \mathcal{C} \end{array}$$

in $\mathrm{CAT}^{\mathrm{st}}_{\infty}$ such that 0 is a zero object and the top horizontal map $\mathcal{A} \to \mathcal{B}$ is fully faithful. A *semi-orthogonal decomposition* is a Verdier sequence such that the functors $\mathcal{A} \to \mathcal{B}$ and $\mathcal{B} \to \mathcal{C}$ admit right adjoints. It is common to simply write $\mathcal{A} \to \mathcal{B} \to \mathcal{C}$ for a Verdier sequence, leaving implicit the requirement that $\mathcal{A} \to \mathcal{B}$ be fully faithful with cofiber $\mathcal{C}$.

Remark 13.2.75. Given a Verdier sequence $\mathcal{A} \xrightarrow{i} \mathcal{B} \xrightarrow{j} \mathcal{C}$, the fact that i is fully faithful implies that the left adjoint functor $i_! : \mathrm{Ind}(\mathcal{A}) \to \mathrm{Ind}(\mathcal{B})$ is fully faithful with right adjoint i^*, and the right adjoint $j^* : \mathrm{Ind}(\mathcal{C}) \to \mathrm{Ind}(\mathcal{B})$ of the functor $j_! : \mathrm{Ind}(\mathcal{B}) \to \mathrm{Ind}(\mathcal{C})$ is also fully faithful. Thus the Verdier sequence $\mathrm{Ind}(\mathcal{A}) \to \mathrm{Ind}(\mathcal{B}) \to \mathrm{Ind}(\mathcal{C})$ is actually a semi-orthogonal decomposition.

13.2.6 Homotopy groups and t-structures

Unfortunately, the notion of truncation (see 13.2.48 above) considered earlier is poorly behaved in stable ∞-categories $\mathcal{C}$, even if $\mathcal{C}$ happens to be presentable. This is because it is a direct consequence of stability that an object A of $\mathcal{C}$ is n-truncated if and only if it is $(n-1)$-truncated, so that the only finitely truncated objects at all are the zero objects, all of which are canonically equivalent to one another. Thus any attempt to study $\mathcal{C}$ itself via the standard obstruction theoretic truncation type methods is doomed to fail.

The notion of a t-structure on a stable ∞-category remedies this situation, providing all kinds of other interesting and potentially effective ways of (co)filtering objects of $\mathcal{C}$ — ideally even $\mathcal{C}$ itself — as limits of "Postnikov type" towers of *truncation* functors $\tau_{\leq n} : \mathcal{C} \to \mathcal{C}$, or dually as colimits of "Whitehead type" telescopes of *connective cover* functors $\tau_{\geq n} : \mathcal{C} \to \mathcal{C}$. We emphasize that while a t-structure is *structure*, it is encoded as innocuously as possible, via a choice of "orthogonal" full subcategories of $\mathcal{C}$ in the sense made precise below.

First we need some notation. If $\mathcal{C}$ is a stable ∞-category and A is an object of $\mathcal{C}$, we also write $A[n]$ for a choice of n-fold suspension $\Sigma^n A$ of A, $n \in \mathbf{Z}$. If $\mathcal{D} \subset \mathcal{C}$ is a full subcategory of $\mathcal{C}$, we write $\mathcal{D}[n] \subset \mathcal{C}$ for the full subcategory consisting the objects of the form $B[n]$, where B is an object of the full subcategory $\mathcal{D}$. Finally, for any pair of objects A and B of $\mathcal{C}$, we write $\mathrm{Ext}^n(A, B)$ for the abelian group $\pi_0 \mathrm{Map}_{\mathcal{C}}(A, B[n])$.

Definition 13.2.76. A *t-structure* on a stable ∞-category $\mathcal{C}$ consists of a pair of full subcategories $\mathcal{C}_{\geq 0} \subset \mathcal{C}$ and $\mathcal{C}_{\leq 0} \subset \mathcal{C}$ satisfying the following conditions:

(1) $\mathcal{C}_{\geq 0}[1] \subset \mathcal{C}_{\geq 0}$ and $\mathcal{C}_{\leq 0} \subset \mathcal{C}_{\leq 0}[1]$;

(2) If $A \in \mathcal{C}_{\geq 0}$ and $B \in \mathcal{C}_{\leq 0}$, then $\mathrm{Map}_{\mathcal{C}}(A, B[-1]) = 0$;

(3) Every $A \in \mathcal{C}$ fits into an exact triangle of the form $\tau_{\geq 0} A \to A \to \tau_{\leq -1} A$ with $\tau_{\geq 0} A \in \mathcal{C}_{\geq 0}$ and $\tau_{\leq -1} A \in \mathcal{C}_{\leq 0}[-1]$.

Definition 13.2.77. An exact functor $\mathcal{C} \to \mathcal{D}$ between stable ∞-categories equipped with t-structures is *left t-exact* (respectively, *right t-exact* if it sends $\mathcal{C}_{\leq 0}$ to $\mathcal{D}_{\leq 0}$ (respectively, $\mathcal{C}_{\geq 0}$ to $\mathcal{D}_{\geq 0}$). An exact functor is *t-exact* if is both left and right t-exact. We set $\mathcal{C}_{\geq n} = \mathcal{C}_{\geq 0}[n]$ and $\mathcal{C}_{\leq n} = \mathcal{C}_{\leq 0}[n]$.

Remark 13.2.78. The data of a t-structure on a stable ∞-category $\mathcal{C}$ is equivalent to the data of a t-structure on its triangulated homotopy category $\mathrm{Ho}(\mathcal{C})$ in the sense originally defined and studied by Beilinson-Bernstein-Deligne [9].

Example 13.2.79. If $\mathcal{A}$ is a small abelian category, then the bounded derived ∞-category $\mathcal{D}^b(\mathcal{A})$ admits a canonical t-structure, where $\mathcal{D}^b(\mathcal{A})_{\geq n}$ consists of the complexes A such that $\mathrm{H}_i(A) = 0$ for $i < n$, and similarly for $\mathcal{D}^b(\mathcal{A})_{\leq n}$. If $\mathcal{A}$ is a Grothendieck abelian category, then the *unbounded* derived ∞-category $\mathcal{D}(\mathcal{A})$ admits a t-structure with the same description as the previous example. This stable ∞-category and its t-structure are studied in [25, Section 1.3.5].

Example 13.2.80. If R is a connective associative ring spectrum, then the stable presentable ∞-category LMod_R of left R-module spectra admits a t-structure with $\mathrm{LMod}_{R \geq 0} \simeq \mathrm{LMod}_R^{\mathrm{cn}}$, the ∞-category of connective left R-module spectra. We call this the *Postnikov t-structure.*

Remark 13.2.81. The mapping space $\mathrm{Map}_{\mathcal{C}}(A, B[-1])$ is actually contractible for $A \in \mathcal{C}_{\geq 0}$ and $B \in \mathcal{C}_{\leq 0}$. This is not the case for the mapping spectra, as $\pi_0 \mathrm{Map}_{\mathcal{D}(\mathcal{A})}(A[-n], B[-1]) \cong \mathrm{Ext}^{n-1}_{\mathcal{A}}(A, B)$ for any pair of objects A and B in a Grothendieck abelian category $\mathcal{A}$ (see [25, Proposition 1.3.5.6]).

Proposition 13.2.82. [25, Remark 1.2.1.12 and Warning 1.2.1.9] *The intersection $\mathcal{C}_{\geq 0} \cap \mathcal{C}_{\leq 0}$ is an abelian category equivalent to the full subcategory of $\mathcal{C}_{\geq 0}$ consisting of the discrete objects (that is, 0-truncated in the sense of 13.2.48).*

Definition 13.2.83. The abelian category $\mathcal{C}_{\geq 0} \cap \mathcal{C}_{\leq 0}$ is called the *heart* of the t-structure $(\mathcal{C}_{\geq 0}, \mathcal{C}_{\leq 0})$ on $\mathcal{C}$, and is denoted $\mathcal{C}^{\heartsuit}$.

Example 13.2.84. The hearts of the t-structures in 13.2.79 are both equivalents to the abelian category $\mathcal{A}$ itself. The heart of the t-structure in 13.2.80 is $\mathrm{LMod}^{\heartsuit}_{\pi_0 R}$, the abelian category of left $\pi_0 R$-modules.

Remark 13.2.85. It turns out that the truncations $\tau_{\geq n}$ and $\tau_{\leq n}$ are functorial in the sense that the inclusions $\mathcal{C}_{\geq n} \to \mathcal{C}$ and $\mathcal{C}_{\leq n} \to \mathcal{C}$ admit right and left adjoints, respectively, by [25, Corollary 1.2.1.6].

Definition 13.2.86. Let $\mathcal{C}$ be a stable ∞-category equipped with a t-structure $(\mathcal{C}_{\geq 0}, \mathcal{C}_{\leq 0})$ and let A be an object of $\mathcal{C}$. There is an equivalence of functors $\tau_{\geq 0}\tau_{\leq 0} \simeq \tau_{\leq 0}\tau_{\geq 0} : \mathcal{C} \to \mathcal{C}^{\heartsuit}$ [25, Proposition 1.2.1.10]. The *homotopy groups* $\pi_n A$, $n \in \mathbf{Z}$, are defined via the formula $\pi_n A = \tau_{\geq 0}\tau_{\leq 0} A[-n] \in \mathcal{C}^{\heartsuit}$.

Remark 13.2.87. In the stable setting, passing to homotopy groups is a homological functor in the sense that there are long exact sequences

$$\cdots \to \pi_{n+1} C \to \pi_n A \to \pi_n B \to \pi_n C \to \pi_{n-1} A \to \cdots$$

in $\mathcal{C}^{\heartsuit}$ whenever $A \to B \to C$ is an exact triangle in $\mathcal{C}$.

Definition 13.2.88. A t-structure $(\mathcal{C}_{\geq 0}, \mathcal{C}_{\leq 0})$ on a stable ∞-category $\mathcal{C}$ is *left complete* if the full subcategory of infinitely connected objects $\mathcal{C}_{\geq \infty} = \bigcap_{n \in \mathbf{Z}} \mathcal{C}_{\geq n} \subset \mathcal{C}$ consists only of zero objects. Right complete t-structures are defined similarly.

Definition 13.2.89. A t-structure $(\mathcal{C}_{\geq 0}, \mathcal{C}_{\leq 0})$ on a stable ∞-category $\mathcal{C}$ is *bounded* if the inclusion of the full subcategory of bounded objects $\mathcal{C}^b = \bigcup_{n \in \mathbf{N}} \mathcal{C}_{\geq -n} \cap \mathcal{C}_{\leq n} \subset \mathcal{C}$ is an equivalence. There are analogous notions of right and left bounded.

Many of the most important examples of stable ∞-categories may already be familiar from homological algebra. If $\mathcal{A}$ is an abelian category with enough projectives, the derived ∞-category $\mathcal{D}^-(\mathcal{A})$ of bounded below chain complexes in $\mathcal{A}$ admits a universal property that characterizes the stable ∞-categories that arise in this way. We will suppose that $\mathcal{A}$ has enough projective and write $\mathcal{A}^{\mathrm{proj}} \subset \mathcal{A}$ for the full subcategory consisting of the projective objects.

Definition 13.2.90. Let $\mathcal{A}$ be an abelian category with enough projectives. The bounded below *derived ∞-category* of $\mathcal{A}$ is the homotopy coherent nerve $\mathcal{D}^-(\mathcal{A}) := \mathrm{N}(\mathrm{Ch}^-(\mathcal{A}))$ of the differential graded category $\mathrm{Ch}^-(\mathcal{A})$ (viewed as a simplically enriched category as in 13.1.7) of bounded below chain complexes in $\mathcal{A}$ (those complexes $A \in \mathrm{Ch}(\mathcal{A})$ such that $\mathrm{H}_n(A) = 0$ for $n \ll 0$.

Proposition 13.2.91. [25, Corollary 1.3.2.18 and Proposition 1.3.2.19] *Let $\mathcal{A}$ be an abelian category with enough projectives. Then $\mathcal{D}^-(\mathcal{A})$ is a stable ∞-category, and the full subcategories $\mathcal{D}^-_{\geq 0}(\mathcal{A}) \subset \mathcal{D}^-(\mathcal{A})$ and $\mathcal{D}^-_{\leq 0}(\mathcal{A}) \subset \mathcal{D}^-(\mathcal{A})$ consisting of those complexes A such that $\mathrm{H}_n(A) = 0$ for $n < 0$ and $\mathrm{H}_n(A) = 0$ for $n > 0$, respectively, comprise a right bounded and left complete t-structure on $\mathcal{D}^-(\mathcal{A})$.*

Theorem 13.2.92. [25, Theorem 1.3.4.4] *Let $\mathcal{A}$ be an abelian category with enough projectives and $W \subset \mathrm{Fun}(\Delta^1, \mathrm{Ch}^-(\mathcal{A}))$ the class of quasi-isomorphisms. There is a canonical equivalence of ∞-categories $\mathcal{A}[W^{-1}] \simeq \mathcal{D}^-(\mathcal{A})$.*

Remark 13.2.93. $\mathcal{A}$ sits inside $\mathcal{D}^-(\mathcal{A})$ as the full subcategory of complexes concentrated in degree zero. Moreover, straightforward homological algebra arguments show that $\pi_0 : \mathcal{D}^-(\mathcal{A}) \to \mathcal{A}$ restricts to an equivalence $\mathcal{D}^-(\mathcal{A})^\heartsuit \simeq \mathcal{A}$.

Remark 13.2.94. Let $\mathcal{A}$ be a small abelian category with enough projectives. The category $\mathrm{Ind}(\mathcal{A})$ of inductive objects of $\mathcal{A}$ is again an abelian category with enough projective objects and is equivalent to the large abelian category $\mathrm{Fun}^\Pi((\mathcal{A}^{\mathrm{proj}})^{\mathrm{op}}, \mathrm{Set})$ of product preserving functors from $(\mathcal{A}^{\mathrm{proj}})^{\mathrm{op}}$ to Set.

Theorem 13.2.95. [25, Theorem 1.3.3.8] *Let $\mathcal{A}$ be an abelian category with enough projectives, let $\mathcal{C}$ be an ∞-category that admits geometric realizations, and let $\mathrm{Fun}^{\mathrm{geom}}(\mathcal{D}^-_{\geq 0}(\mathcal{A}), \mathcal{C}) \subset \mathrm{Fun}(\mathcal{D}^-_{\geq 0}(\mathcal{A}), \mathcal{C})$ denote the full subcategory consisting of those functors that preserve geometric realizations. Then restriction along the embedding $\mathcal{A}^{\mathrm{proj}} \subset \mathcal{D}^-_{\geq 0}(\mathcal{A})$ induces an equivalence of ∞-categories*

$$\mathrm{Fun}(\mathcal{A}^{\mathrm{proj}}, \mathcal{C}) \xleftarrow{\simeq} \mathrm{Fun}'(\mathcal{D}^-_{\geq 0}(\mathcal{A}), \mathcal{C}) \subset \mathrm{Fun}(\mathcal{D}^-_{\geq 0}(\mathcal{A}), \mathcal{C})$$

between realization preserving functors $\mathcal{D}^-_{\geq 0}(\mathcal{A}) \to \mathcal{C}$ and functors $\mathcal{A}^{\mathrm{proj}} \to \mathcal{C}$.

Theorem 13.2.96. [25, Theorem HA.1.3.3.2] *Let $\mathcal{A}$ be an abelian category with enough projectives, let $\mathcal{C}$ be a stable ∞-category with a left complete t-structure, and let $\mathrm{Fun}'(\mathcal{D}^-(\mathcal{A}), \mathcal{C}) \subset \mathrm{Fun}(\mathcal{D}^-(\mathcal{A}), \mathcal{C})$ denote the full subcategory of functors $\mathcal{D}^-(\mathcal{A}) \to \mathcal{C}$ that are right t-exact and send projective objects of $\mathcal{A}$ to $\mathcal{C}^\heartsuit$. Then restriction along $i : \mathcal{A} \to \mathcal{D}^-(\mathcal{A})$ induces an equivalence of ∞-categories*

$$\mathrm{Fun}^{\mathrm{rex}}(\mathcal{A}, \mathcal{C}^\heartsuit) \xleftarrow{\simeq} \mathrm{Fun}'(\mathcal{D}^-(\mathcal{A}), \mathcal{C}) \subset \mathrm{Fun}(\mathcal{D}^-(\mathcal{A}), \mathcal{C})$$

between right t-exact functors $\mathcal{D}^-(\mathcal{A}) \to \mathcal{C}$ that send $\mathcal{A}^{\mathrm{proj}} \subset \mathcal{A}$ to $\mathcal{C}^\heartsuit \subset \mathcal{C}$ and right exact functors $\mathcal{A} \to \mathcal{C}^\heartsuit$.

13.3 Ring theory

13.3.1 Spectra

So far we have considered the ∞-category of spectra from two seemingly different but equivalent perspectives: on the one hand, as reduced excisive functors $\mathrm{Sp} = \mathrm{Sp}(\mathcal{S}) = \mathrm{Exc}_*(\mathcal{S}^{\mathrm{fin}}_*, \mathcal{S})$, and on the other, as a unit object $\mathrm{Sp} \in \mathrm{LPr}^{\otimes}_{\mathrm{st}}$ of the symmetric monoidal ∞-category of presentable stable ∞-categories. The former is more explicit and yields lots of examples, while the latter is more abstract and suggests a universal property.

Remark 13.3.1. The reader may be wondering what any of this has to do with the classical notion of spectrum in algebraic topology. By definition, $\mathrm{Sp} = \mathrm{Exc}_*(\mathcal{S}_*^{\mathrm{fin}}, \mathcal{S})$, but evaluation on the family of spheres $\{S^n\}_{n\in\mathbf{N}}$ induces an equivalence of ∞-categories

$$\mathrm{Sp} \simeq \lim \left\{ \cdots \xrightarrow{\Omega} \mathcal{S}_* \xrightarrow{\Omega} \mathcal{S}_* \xrightarrow{\Omega} \mathcal{S}_* \right\}.$$

Remark 13.3.2. Our convention is that we work in the ∞-category $\mathcal{S}$ of spaces and its pointed variant $\mathcal{S}_*$, so that all (pointed) spaces have the homotopy type of cell complexes. If we wish to model a spectrum, however, it is often convenient to work in a category $\mathcal{C}$ equipped with a class of weak equivalences W and functor $\mathcal{C} \to \mathcal{S}_*$ that sends the maps in W to equivalences in $\mathcal{S}_*$, such as the category Top_* of pointed topological spaces. In such models we need only ask that the maps $\eta_n : A_n \to \Omega A_{n+1}$ are weak equivalences, a structure that is often referred to as an Ω-spectrum in the literature. In the case $\mathcal{C} = \mathrm{Top}_*$ we can even require the $\eta_n : A_n \to \Omega A_{n+1}$ to be homeomorphisms.

Remark 13.3.3. The forgetful functor $\mathrm{RPr} \to \mathrm{CAT}_\infty$ preserves limits, meaning we may also form this limit in RPr. Equivalently, Sp is given as the colimit

$$\mathrm{Sp} \simeq \operatorname{colim} \left\{ \mathcal{S}_* \xrightarrow{\Sigma} \mathcal{S}_* \xrightarrow{\Sigma} \mathcal{S}_* \xrightarrow{\Sigma} \cdots \right\}$$

in LPr, by virtue of the antiequivalence $\mathrm{LPr} \simeq \mathrm{RPr}^{\mathrm{op}}$ that takes the adjoint.

Remark 13.3.4. The forgetful functor $\mathrm{LPr} \to \mathrm{CAT}_\infty$ does not commute with filtered colimits; nevertheless, the ∞-category of finite — or equivalently, in this case, compact — spectra $\mathrm{Sp}^{\mathrm{fin}} \simeq \mathrm{Sp}^\omega$ is computed as the filtered colimit of the suspension functor on finite pointed spaces $\mathcal{S}_*^{\mathrm{fin}}$. This equivalence

$$\mathrm{Sp}^{\mathrm{fin}} \simeq \operatorname{colim} \left\{ \mathcal{S}_*^{\mathrm{fin}} \xrightarrow{\Sigma} \mathcal{S}_*^{\mathrm{fin}} \xrightarrow{\Sigma} \mathcal{S}_*^{\mathrm{fin}} \xrightarrow{\Sigma} \cdots \right\}$$

in Cat_∞ is the ∞-categorical analogue of the *Spanier-Whitehead* category.

Remark 13.3.5. We have taken the "coordinate free" convention that spectra are reduced excisive functors. Nevertheless, it is often the case in practice that a spectrum A is given in terms of an infinite delooping of its infinite loop space $\Omega^\infty A$, in which case the associated excisive functor is given by the formula

$$T_A(-) := \Omega^\infty(- \otimes A) \colon \mathcal{S}_*^{\mathrm{fin}} \longrightarrow \mathcal{S}.$$

Here, $\otimes$ refers to the fact that Sp, as a commutative $\mathcal{S}_*$-algebra in LPr, is canonically left tensored over $\mathcal{S}_*$, and hence over the symmetric monoidal subcategory $\mathcal{S}_*^{\mathrm{fin}} \subset \mathcal{S}_*$ as well. Alternatively, this tensor is computed by first applying $\Sigma^\infty : \mathcal{S}_* \to \mathrm{Sp}$ and then tensoring with A in $\mathrm{Sp}^\otimes$.

Remark 13.3.6. The ∞-category Sp of spectra admits a left and right complete t-structure with heart $\mathrm{Sp}^\heartsuit \simeq \mathrm{Ab}$ the category of abelian groups. The identity functor $\mathrm{Ab} \to \mathrm{Ab}$ is right exact and so determines a right t-exact functor $\mathcal{D}^-(\mathbf{Z}) \simeq \mathcal{D}^-(\mathrm{Ab}) \to \mathrm{Sp}$ with image the *Eilenberg-Mac Lane spectra.* Strictly speaking, these are the *generalized* Eilenberg-Mac Lane spectra: an Eilenberg-Mac Lane spectrum is a generalized Eilenberg-Mac Lane spectrum that has homotopy concentracted in a single degree, or equivalently is the image of a chain complex with homology concentrated in a single degree.

Definition 13.3.7. A *cohomology theory* $\{F^n\}_{n\in\mathbf{Z}}$ is a $\mathbf{Z}$-graded family of functors $F^n : \mathrm{Ho}(\mathcal{S}_*^{\mathrm{op}}) \to \mathrm{Ab}$ and natural isomorphisms $\sigma^n : F^n \to F^{n+1} \circ \Sigma$ satisfying the following exactness conditions:

(1) For any cofiber sequence $X \to Y \to Z$ integer n the sequence of abelian groups $F^n(Z) \to F^n(Y) \to F^n(X)$ is exact.

(2) For any (possibly infinite) wedge decomposition $X \simeq \bigvee_{i\in I} X_i$ and integer n, the homomorphism $F^n(X) \to \prod_{i\in I} F^n(X_i)$ is an isomorphism.

Remark 13.3.8. These are the *Eilenberg-Steenrod axioms* [15] for a (generalized) cohomology theory. The original formulation included the *dimension axiom*, which required that $F^n(S^0) = 0$ for all $n \neq 0$. One can show without much difficulty that the cohomology theories F that satisfy the dimension axiom are necessarily of the form $F^n(X) \cong H^n(X; F^0(S^0))$, which is to say cohomology with coefficients in the abelian group $F^0(S^0)$. This axiom was disregarded as interesting "generalized" cohomology theories were discovered.

Remark 13.3.9. The *Brown representability theorem* [12] asserts that any suitably left exact functor $F : \mathcal{S}^{\mathrm{op}} \to \mathrm{Set}_*$ is representable. In particular,

$$F^n \cong \pi_0 \operatorname{Map}_{\mathcal{S}_*}(-, A_n) \colon \mathcal{S}^{\mathrm{op}} \longrightarrow \mathrm{Ab}$$

for some pointed space A_n, and the resulting sequence of pointed spaces $\{A_n\}$ can be chosen so that the suspension isomorphisms $\sigma^n : F^n \to F^{n+1} \circ \Sigma$ are induced by equivalences $\eta_n : A_n \to \Omega A_{n+1}$ in $\mathcal{S}_*$ (in fact the loop space structure induces the group structure on the represented functor, and any such group structure arises in this way). This is how spectra arose in practice.

Remark 13.3.10. By choosing a representing spectrum, any cohomology theory $F = \{F^n\}$ in the classical sense as above gives rise to a *cohomological functor* in the ∞-categorical sense, by which we mean a limit preserving functor $F : \mathcal{S}_*^{\mathrm{op}} \to \mathrm{Sp}$. One can show that any such functor is necessarily a right adjoint, so that the ∞-categories of spectra and cohomological functors are canonically equivalent:

$$\mathrm{Sp} \simeq \mathcal{S}_* \otimes \mathrm{Sp} \simeq \mathrm{RFun}(\mathcal{S}_*^{\mathrm{op}}, \mathrm{Sp}).$$

Similarly, $\mathrm{Sp} \simeq \mathrm{RFun}(\mathrm{Sp}^{\mathrm{op}}, \mathrm{Sp})$, so that any cohomological functor of pointed spaces extends uniquely to a cohomological functor of spectra.

Remark 13.3.11. A cohomological functor $F : \mathrm{Sp}^{\mathrm{op}} \to \mathrm{Sp}$ induces a functor $\pi_0 F : \mathrm{Sp}^{\mathrm{op}} \to \mathrm{Ab}$ which necessarily factors through the triangulated homotopy category of spectra. There is a version of Brown representability for triangulated categories that are compactly generated in the appropriate sense, due to Neeman [30]. This is not a corollary of the corresponding formal result for compactly generated stable ∞-categories, namely that $\mathrm{Sp}(\mathcal{C}) \subset \mathrm{Fun}(\mathcal{C}^{\mathrm{op}}, \mathrm{Sp})$ is the full subcategory consisting of the cohomological functors. Indeed, triangulated categories aren't always homotopy categories of stable ∞-categories and don't necessarily even admit finite limits or colimits; rather, the requisite exactness properties are encoded by the triangulated structure.

Remark 13.3.12. Much of the classical algebraic topology literature models spectra as a full subcategory of local objects inside of a larger category. Our definition of spectra $\mathrm{Sp} \subset \mathrm{Fun}_*(\mathcal{S}_*^{\mathrm{fin}}, \mathcal{S}_*)$ is as the full subcategory of (reduced) excisive functors. The *excisive approximation* [26, Example 6.1.1.28] $\partial F : \mathcal{S}_*^{\mathrm{fin}} \to \mathcal{S}_*$ of a reduced functor $F : \mathcal{S}_*^{\mathrm{fin}} \to \mathcal{S}_*$ is given by the formula

$$(\partial F)(T) \simeq \mathrm{colim}_{n\to\infty} \Omega^n F(\Sigma^n T).$$

As any finite space admits a cell decomposition, a reduced excisive functor is determined by its values on spheres; in fact, any sequence of spheres $\{S^{n_0}, S^{n_1}, \ldots\}$ with $n_i > n_j$ whenever $i > j$. Moreover, for any reduced $F : \mathcal{S}_*^{\mathrm{fin}} \to \mathcal{S}_*$, we have maps

$$S^1 \longrightarrow \mathrm{Map}_*(S^n, S^{n+1}) \longrightarrow \mathrm{Map}_*(F(S^n), F(S^{n+1}))$$

and hence maps $S^1 \wedge F(S^n) \to F(S^{n+1})$. This motivates the next definition.

Definition 13.3.13. A *prespectrum* consists of an $\mathbf{N}$-indexed collection of pointed spaces $\{Z_n\}_{n\in\mathbf{N}}$ and structure maps $\{\varepsilon_n : \Sigma Z_{n-1} \to Z_n\}_{n\in\mathbf{N}}$.

To organize prespectra into an ∞-category PSp, it will be useful to write

$$\Sigma[1] : \mathcal{S}_*^{\mathbf{N}} \longrightarrow \mathcal{S}_*^{\mathbf{N}} \qquad \text{and} \qquad \Omega[-1] : \mathcal{S}_*^{\mathbf{N}} \longrightarrow \mathcal{S}_*^{\mathbf{N}}$$

for the "shifted" suspension and loops functors $\Sigma[1](X)_n = \Sigma(X_{n-1})$ and $\Omega[-1](X)_n = \Omega(X_{n+1})$, where we take the convention that $X_{-1} \simeq \mathrm{pt}$ is contractible. Then $\Sigma[1]$ is left adjoint to $\Omega[-1]$, and the ∞-category of prespectra is defined by forming either of following the pullbacks CAT_∞:

$$\begin{array}{ccccc}
\mathrm{Fun}(\Delta^1, \mathcal{S}_*^{\mathbf{N}}) & \longleftarrow & \mathrm{PSp} & \longrightarrow & \mathrm{Fun}(\Delta^1, \mathcal{S}_*^{\mathbf{N}}) \\
\downarrow & & \downarrow & & \downarrow \\
\mathrm{Fun}(\partial\Delta^1, \mathcal{S}_*^{\mathbf{N}}) & \xleftarrow{(\mathrm{id}, \Omega[-1])} & \mathcal{S}_*^{\mathbf{N}} & \xrightarrow{(\Sigma[1], \mathrm{id})} & \mathrm{Fun}(\partial\Delta^1, \mathcal{S}_*^{\mathbf{N}}).
\end{array}$$

Remark 13.3.14. A prespectrum $Z = \{Z_n\}$ gives rise to a sequence of representable functors $\mathrm{Map}(-, Z_n) : \mathcal{S}_*^{\mathrm{op}} \to \mathcal{S}_*$ which collectively represent a cohomology theory if and only if the map $Z \to \Omega[-1]Z$ is an equivalence. The "diagonal" map $\mathcal{S}_*^{\mathbf{N}} \to \mathrm{Fun}(\Delta^1, \mathcal{S}_*^{\mathbf{N}})$ identifies $\mathcal{S}_*^{\mathbf{N}}$ with the full subcategory of $\mathrm{Fun}(\Delta^1, \mathcal{S}_*^{\mathbf{N}})$ consisting of the equivalences, and the iterated pullback square

$$\begin{array}{ccc}
\mathrm{Sp} & \longrightarrow & \mathcal{S}_*^{\mathbf{N}} \\
\downarrow & & \downarrow \\
\mathrm{PSp} & \longrightarrow & \mathrm{Fun}(\Delta^1, \mathcal{S}_*^{\mathbf{N}}) \\
\downarrow & & \downarrow \\
\mathcal{S}_*^{\mathbf{N}} & \xrightarrow{(\mathrm{id}, \Omega[-1])} & \mathrm{Fun}(\partial\Delta^1, \mathcal{S}_*^{\mathbf{N}})
\end{array}$$

exhibits $\mathrm{Sp} \subset \mathrm{PSp}$ as the equalizer of the endofunctors $\mathrm{id}, \Omega[-1] : \mathcal{S}_*^{\mathbf{N}} \to \mathcal{S}_*^{\mathbf{N}}$.

Remark 13.3.15. This inclusion admits a left adjoint *spectrification* functor $\mathrm{PSp} \to \mathrm{Sp}$. Indeed, if $Z = \{Z_n\}$ is a prespectrum, the nth-space of the associated spectrum A is given by the formula

$$A_n \simeq \operatorname*{colim}_{m \to \infty} \Omega^m Z_{m+n}.$$

The equivalence $A_n \xrightarrow{\sim} \Omega A_{n+1}$ is induced by the maps $Z_{n+m} \to \Omega Z_{n+m+1}$, which becomes the equivalence $\operatorname{colim} \Omega^m Z_{n+m} \simeq \Omega^{m+1} Z_{n+m+1}$ after passing to the colimit. This is essentially the same formula as in 13.3.12.

Remark 13.3.16. The ∞-category of prespectra is quite useful in practice. As a source of examples, a pointed space X evidently determines a suspension presprectrum $\{\Sigma^n X\}_{n \in \mathbf{N}}$ whose associated spectrum is $\Sigma^\infty X$. Since the spectrification functor preserves colimits, the ∞-category of prespectra can be used as a tool for computing colimits and smash products of spectra.

13.3.2 The smash product

The construction of a symmetric monoidal model category of spectra was a major foundational problem in the subject for quite some time. One issue is that there isn't an obvious candidate for the smash product of prespectra; instead, given prespectra $A = \{A_m\}$ and $B = \{B_n\}$, their smash product $A \otimes B$ is most naturally indexed on the poset $\mathbf{N} \times \mathbf{N}$; that is, $(A \otimes B)_{m,n} = A_m \wedge B_n$. Adams' theory of "handicrafted smash products" [1] shows that any cofinal choice of poset $\mathbf{N} \subset \mathbf{N} \times \mathbf{N}$ results in a prespectrum representing the smash product, and verifies that this procedure descends to a symmetric monoidal structure on the homotopy category of spectra. However this is insufficient for many purposes, especially as the homotopy category doesn't admit even the most basic sorts of limits and colimits like pullbacks and pushouts.

Another more significant issue is already apparent in homological algebra: in the derived ∞-category of chain complexes of modules over a commutative ring, the derived tensor product is really only defined up to quasi-isomorphism, so there's no philosophical reason to expect this to lift to a symmetric monoidal structure on the ordinary category of chain complexes. The first resolutions of this problem in homotopy theory were the *symmetric spectra* of Hovey-Shipley-Smith [20] and the *S-modules* of Elmendorf-Kriz-Mandell-May [16].

There are morphisms of $\otimes$-idempotent objects of LPr

$$\mathcal{S} \longrightarrow \mathcal{S}_* \longrightarrow \mathrm{CMon}(\mathcal{S}) \longrightarrow \mathrm{CMon}^{\mathrm{gp}}(\mathcal{S}) \longrightarrow \mathcal{S},$$

necessarily symmetric monoidal, which enable us to calculate the tensor product in these presentable ∞-categories. Specifically, writing $\Sigma^\infty : \mathcal{S}_* \to \mathrm{Sp}$ for unique symmetric monoidal left adjoint functor, we deduce that

$$(\Sigma^\infty X_1) \otimes \cdots \otimes (\Sigma^\infty X_n) \simeq \Sigma^\infty (X_1 \wedge \cdots \wedge X_n)$$

for any finite collection of pointed spaces $X_1, \ldots, X_n$. The analogous result remains true for unpointed spaces by addition of a disjoint basepoint.

Remark 13.3.17. Using the description of spectra as the limit of tower associated to the endofunctor $\Omega : \mathcal{S}_* \to \mathcal{S}_*$, we obtain maps

$$\Omega^{\infty-n} : \mathrm{Sp} \longrightarrow \mathcal{S}_*$$

by projection to the nth factor. The reason for this notation is that we have equivalences $\Omega^\infty \simeq \Omega^n\Omega^{\infty-n}$; indeed, if A is a spectrum, $\Omega^\infty A \simeq A_0 \simeq \Omega^n A_n \simeq \Omega^n\Omega^{\infty-n}A$. The collection of functors $\{\Omega^{\infty-n}\}_{n\in\mathbf{N}}$, or any infinite subset thereof, form a conservative family of functors $\mathrm{Sp} \to \mathcal{S}_*$ in RPr. They admit left adjoints $\Sigma^{\infty-n} : \mathcal{S}_* \to \mathrm{Sp}$ which factor through $\mathrm{PSp} \to \mathrm{Sp}$.

Remark 13.3.18. By [26, Proposition 6.3.3.6], any spectrum A admits a canonical presentation by desuspended suspension spectra

$$A \simeq \mathrm{colim}_{n\to\infty} \Sigma^{\infty-n}\Omega^{\infty-n}A \simeq \mathrm{colim}_{n\to\infty} \Sigma^{-n}\Sigma^\infty A_n$$

in which the maps $\Sigma^{\infty-n}A_n \to \Sigma^{\infty-n-1}A_{n+1}$ correspond to the composites

$$A_n \to \Omega^\infty\Sigma^n\Omega^n\Sigma^\infty A_n \to \Omega^\infty\Sigma^n\Omega^{n+1}\Sigma^\infty A_{n+1} \simeq \Omega^{\infty-n}\Sigma^{\infty-n-1}A_{n+1}.$$

Using the fact that $\Sigma^{\infty-m}X \otimes \Sigma^{\infty-n}Y \simeq \Sigma^{\infty-m-n}X \wedge Y$, we can write down explicit formulas for the spaces in the smash product of any finite sequence of spectra.

Remark 13.3.19. Another characterization of the n-fold smash product functor $\otimes^n : \mathrm{Sp}^{\times n} \to \mathrm{Sp}$ is as the derivative of the n-fold cartesian product functor $\times^n : \mathcal{S}^{\times n} \to \mathcal{S}$ or its *coreduction*, the n-fold smash product functor $\wedge^n : \mathcal{S}_*^{\times n} \to \mathcal{S}_*$. See [25, Example 6.2.3.28] for further details.

Remark 13.3.20. If B is a fixed spectrum, the functor $\mathrm{Sp} \xrightarrow{(\mathrm{id},B)} \mathrm{Sp}\times\mathrm{Sp} \xrightarrow{\otimes^2} \mathrm{Sp}$ that sends A to $A\otimes B$ preserves colimits and therefore, according to the adjoint functor theorem, admits a right adjoint. This right adjoint is the *mapping spectrum* functor $\mathcal{F}(B,-) : \mathrm{Sp} \to \mathrm{Sp}$, which admits the following description: if A is a spectrum, then $\mathcal{F}(B,A)$ is the spectrum given by the formula

$$\mathcal{F}(B,A)_n \simeq \mathrm{Map}_{\mathrm{Sp}}(B, \Sigma^n A),$$

with structure maps

$$\mathrm{Map}_{\mathrm{Sp}}(B,\Sigma^n A) \simeq \mathrm{Map}_{\mathrm{Sp}}(B, \Omega\Sigma^{n+1}A) \simeq \Omega\,\mathrm{Map}_{\mathrm{Sp}}(B,\Sigma^{n+1}A),$$

where the first equivalence uses the fact that Sp is a stable ∞-category and therefore that the composite functor $\Omega\Sigma$ is equivalent to the identity.

In order to be able to work effectively with spectra, we need ordinary algebraic invariants such as homotopy groups. One way to obtain the homotopy groups of a spectrum is via the Postnikov t-structure, defined as follows.

Definition 13.3.21. Let $\mathrm{Sp}_{\leq -1} \subset \mathrm{Sp}$ denote the full subcategory of spectra consisting of those objects A such that $\Omega^\infty A$ is contractible.

Remark 13.3.22. The functor $\Omega^\infty\colon \mathrm{Sp} \to \mathcal{S}$ is corepresented by the unit object $\mathbf{S}$ of Sp, which is compact. It therefore preserves limits and filtered colimits; furthermore, limits and filtered colimits of contractible spaces are contractible. Hence the inclusion $\mathrm{Sp}_{\leq -1} \subset \mathrm{Sp}$ preserves limits and filtered colimits and therefore, by the adjoint functor theorem, it admits a left adjoint. We write $\tau_{\leq -1}\colon \mathrm{Sp} \to \mathrm{Sp}$ for the left adjoint followed by the right adjoint, so that any spectrum A admits a natural unit map $A \to \tau_{\leq -1}A$. The fiber of the unit map then determines an endofunctor $\tau_{\geq 0} : \mathrm{Sp} \to \mathrm{Sp}$, the *connective cover.* We write $\mathrm{Sp}_{\geq 0}$ for the essential image of $\tau_{\geq 0}$ and $\mathrm{Sp}_{\leq 0} = \mathrm{Sp}_{\leq -1}[1]$.

Proposition 13.3.23. [25, Proposition 1.4.3.6] *The pair of full subcategories* $\mathrm{Sp}_{\geq 0} \subset \mathrm{Sp}$ *and* $\mathrm{Sp}_{\leq 0} \subset \mathrm{Sp}$ *are the connective and coconnective parts of a t-structure on* Sp. *Moreover, this t-structure is left and right complete, and its heart* $\mathrm{Sp}^\heartsuit \simeq \mathrm{Ab}$ *is canonically equivalent to the category of abelian groups.*

Remark 13.3.24. The spectra that lie in the full subcategory $\mathrm{Sp}_{\geq 0} \subset \mathrm{Sp}$ are called the connective spectra, and we will often write $\mathrm{Sp}^{\mathrm{cn}}$ in place of $\mathrm{Sp}_{\geq 0}$.

Remark 13.3.25. The homotopy groups of a spectrum A, defined via the t-structure on Sp, collectively form a $\mathbf{Z}$-graded abelian group

$$\pi_* A = \bigoplus_{m \in \mathbf{Z}} \pi_m A.$$

Viewing A as a sequence of pointed spaces $\{A_n\}_{n\in\mathbf{N}}$ equipped with equivalences $A_n \simeq \Omega A_{n+1}$, we can also obtain the homotopy groups of A via the homotopy groups of the pointed spaces $\{A_n\}_{n\in\mathbf{N}}$. Specifically, the nonnegative homotopy groups of A are the homotopy groups of the underlying infinite loop space; that is, for $m \geq 0$,

$$\pi_m A \cong \pi_m A_0 \cong \pi_{m+1} A_1 \cong \pi_{m+2} A_2 \cong \cdots.$$

The negative homotopy groups of A, on the other hand, are the homotopy groups of a sufficiently high delooping; that is, for $m < 0$,

$$\pi_m A \cong \pi_0 A_{-m} \cong \pi_1 A_{-m+1} \cong \pi_2 A_{-m+2} \cong \cdots.$$

More generally, $\pi_m A \cong \operatorname{colim} \pi_n A_{n-m}$, where for $m > n$ we take A_{n-m} to mean the homotopy type of the space $\Omega^{m-n} A_0 \simeq \Omega^{m-n+1} A_1 \simeq \Omega^{m-n+2} A_2 \simeq \cdots$.

Definition 13.3.26. Let A and B be spectra. The A-homology of B, $A_*(B)$, is the graded abelian group $\pi_*(A \otimes B)$. Dually, the A-cohomology of B, $A^*(B)$, is the graded abelian group $\pi_*\mathcal{F}(B, A)$.

Remark 13.3.27. If X is a pointed space, we write $A_*(X) = A_*(\Sigma^\infty X)$ and $A^*(X) = A^*(\Sigma^\infty X)$. If A is a spectrum representing a generalized cohomology theory, this recovers the A-cohomology groups of the pointed space X.

13.3.3 Associative and commutative algebras

Definition 13.3.28. An *associative ring spectrum* is an algebra object of the monoidal ∞-category of spectra. A *commutative ring spectrum* is a commutative algebra object of the symmetric monoidal ∞-category of spectra.

We write $\mathrm{Alg} = \mathrm{Alg}(\mathrm{Sp}^{\otimes})$ and $\mathrm{CAlg} = \mathrm{CAlg}(\mathrm{Sp}^{\otimes})$ for the ∞-categories of associative and commutative ring spectra, respectively.

Remark 13.3.29. As the notation suggests, we often refer to associative or commutative ring spectra as associative or commutative algebra spectra, or just associative or commutative algebras when the symmetric monoidal ∞-category of spectra is clear from context. The latter terminology is especially convenient in relative contexts, such as when we wish to work over[5] a fixed base commutative ring spectrum R, where the corresponding notion is that of an associative or commutative R-algebra (spectrum here is implicit).

Remark 13.3.30. Recall that a commutative algebra object of the symmetric monoidal ∞-category of spectra is a section

$$A\langle - \rangle : \mathrm{Fin}_* \longrightarrow \mathrm{Sp}^{\otimes}$$

of the cocartesian fibration $p : \mathrm{Sp}^{\otimes} \to \mathrm{Fin}_*$ such that $A\langle m\rangle \to A\langle n\rangle \in \mathrm{Sp}^{\otimes}$ is cocartesian whenever $\langle m\rangle \to \langle n\rangle \in \mathrm{Fin}_*$ is inert. We will typically write A in place of $A\langle 1\rangle \in \mathrm{Sp}^{\otimes}_{\langle 1\rangle} \simeq \mathrm{Sp}$ and abusively refer the underlying spectrum A as the algebra object. The value of $A\langle - \rangle$ on the active map $\langle 2\rangle \to \langle 1\rangle$, pushed forward to the fiber over $\langle 1\rangle$ via this map, is the multiplication $\mu : A^{\otimes 2} \to A$.

Similarly, an associative algebra object $A \in \mathrm{Alg}(\mathrm{Sp}^{\otimes})$ is a section

$$A[-] : \Delta^{\mathrm{op}} \longrightarrow \mathrm{Sp}^{\otimes} \times_{\mathrm{Fin}_*} \Delta^{\mathrm{op}}$$

of the restricted (along the cut map $\Delta^{\mathrm{op}} \to \mathrm{Fin}_*$) fibration that carries inert arrows to cocartesian arrows. We will typically also write $A = A[1]$ for the underling spectrum of $A[-]$ and refer to A as the associative ring spectrum.

Remark 13.3.31. A section $A[-] : \Delta^{\mathrm{op}} \to \mathrm{Sp}^{\otimes} \times_{\mathrm{Fin}_*} \Delta^{\mathrm{op}}$ amounts to a diagram of the form

$$A[0] \leftleftarrows A[1] \;\substack{\longleftarrow\\\longrightarrow\\\longleftarrow\\\longrightarrow\\\longleftarrow}\; A[2] \;\substack{\longleftarrow\\\longrightarrow\\\longleftarrow\\\longrightarrow\\\longleftarrow\\\longrightarrow\\\longleftarrow}\; \cdots$$

in $\mathrm{Sp}^{\otimes}$. If it is an algebra object, the n inert maps $[1] \to [n]$ in Δ force $A[n]$ to decompose as the n-fold product $(A[1], \ldots, A[1])$ of $A[1]$ under the equivalence $\mathrm{Sp}^{\otimes}_{\langle n\rangle} \simeq \mathrm{Sp}^{\times n}$, where we identify Sp with $\mathrm{Sp}^{\otimes}_{\langle 1\rangle}$ as usual. Restricting to the active maps and pushing everything forward to the the fiber over $[1] \in \Delta$ via the inert maps, and writing $\mathbf{S}$, A, $A \otimes A$ for the images of $A[0]$, $A[1]$, $A[2]$, we obtain a diagram

$$\mathbf{S} \longrightarrow A \rightrightarrows A \otimes A \;\substack{\longrightarrow\\\longrightarrow\\\longrightarrow}\; \cdots$$

[5]Over R means over $\mathrm{Spec}\, R$, reversing the direction of the arrows. Algebraically this means working *under* R, but the terminology is influenced by geometric intuition.

in Sp, encoding exactly the maps one would expect from an algebra object. The commutative case is similar but more complex due to the permutations.

In certain situations, some of which we will encounter later, it is convenient to work with only the connective spectra. Recall that $\mathrm{Sp}^{\mathrm{cn}} \subset \mathrm{Sp}$ denotes the full subcategory consisting of the spectra A whose negative homotopy groups $\pi_n A$ vanish for all $n < 0$. In other words, $\mathrm{Sp}^{\mathrm{cn}} = \mathrm{Sp}_{\geq 0}$ as full subcategory of spectra, where $\mathrm{Sp}_{\geq 0}$ is defined via the standard Postnikov t-structure. Since the tensor product of connective spectra is again connective, we actually obtain $\mathrm{Sp}^{\otimes}_{\geq 0} \subset \mathrm{Sp}^{\otimes}$ as a symmetric monoidal subcategory. Thus we have ∞-categories $\mathrm{Alg}^{\mathrm{cn}} = \mathrm{Alg}(\mathrm{Sp}^{\mathrm{cn}})$ and $\mathrm{CAlg}^{\mathrm{cn}} = \mathrm{CAlg}(\mathrm{Sp}^{\mathrm{cn}})$ of connective associative and commutative algebra spectra.

Proposition 13.3.32. [25, Proposition 7.1.3.19] *Postnikov towers converge in the presentable ∞-categories* $\mathrm{Sp}^{\mathrm{cn}}$, $\mathrm{Alg}^{\mathrm{cn}}$, *and* $\mathrm{CAlg}^{\mathrm{cn}}$.

Definition 13.3.33. An *∞-monoid* is an algebra object of the symmetric monoidal ∞-category $\mathcal{S}^{\times}$ of spaces. An *abelian ∞-monoid* is a commutative algebra object of $\mathcal{S}^{\times}$. An ∞-monoid, or abelian ∞-monoid, G is said to an ∞-group, or abelian ∞-group, if the ordinary discrete monoid $\pi_0 G$ is a group.

Remark 13.3.34. We have ∞-categories $\mathrm{Mon}_{\infty} = \mathrm{Mon}(\mathcal{S}^{\times}) = \mathrm{Alg}(\mathcal{S}^{\times})$ and $\mathrm{Gp}_{\infty} \subset \mathrm{Mon}_{\infty}$ of ∞-monoids and ∞-groups, as well as their abelian variants $\mathrm{AbMon}_{\infty} = \mathrm{CMon}(\mathcal{S}^{\times}) = \mathrm{CAlg}(\mathcal{S}^{\times})$ and $\mathrm{AbGp}_{\infty} \subset \mathrm{AbMon}_{\infty}$. The *group completion* of an ∞-monoid M is the left adjoint of the fully faithful inclusion $\mathrm{Gp}_{\infty} \subset \mathrm{Mon}_{\infty}$, and is given by the formula $G \simeq \Omega \mathrm{B} M$.

Remark 13.3.35. The reader might be wondering why we did not define associative and commutative ring spectra as homotopy coherently associative and commutative ring objects in the ∞-category of spaces. The primary reason is that this only yields the connective ring spectra, and nonconnective spectra, even ring spectra, such as topological K-theory, are among our most important examples. Nevertheless it is true, and a good sanity check, that we have an equivalence $\mathrm{Sp}^{\mathrm{cn}} \simeq \mathrm{AbGp}_{\infty}$, which induces equivalences $\mathrm{Alg}^{\mathrm{cn}} \simeq \mathrm{Ring}_{\infty}$ and $\mathrm{CAlg}^{\mathrm{cn}} \simeq \mathrm{CRing}_{\infty}$, where the latter notions are defined as in [17] (among other algebraic theories, such as semirings and their $\mathbf{E}_n$ variants).

Definition 13.3.36. A *homotopy associative ring spectrum* is an associative algebra object in the monoidal homotopy category of spectra. A *homotopy commutative ring spectrum* is a commutative algebra object in the symmetric monoidal homotopy category of spectra.

Remark 13.3.37. Homotopy categories are ordinary categories. Hence there are no coherences to specify, and a homotopy associative ring spectrum is the data of a ring spectrum R equipped with a unit map $\eta : \mathbf{S} \to R$ and a multiplication map $\mu : R \otimes R \to R$ that are associative and unital in the sense that the diagrams

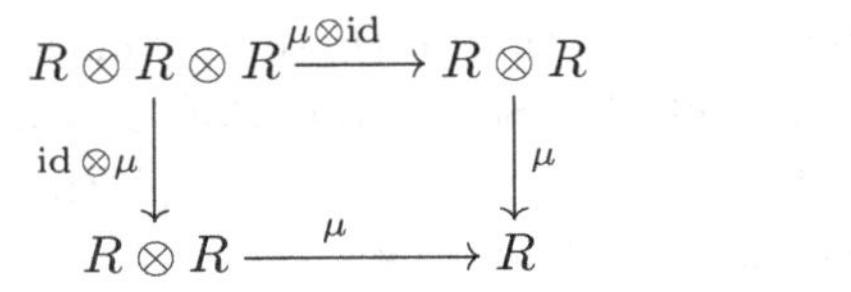

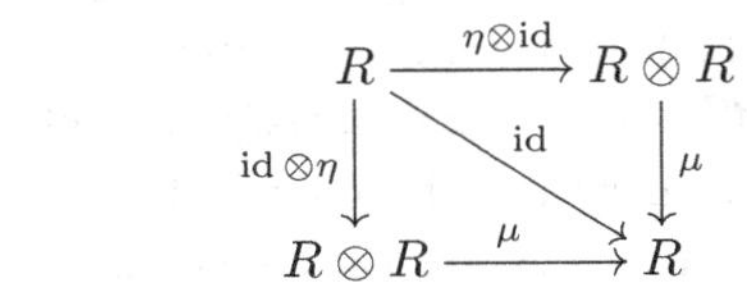

commute up to homotopy (a choice of homotopy is not part of the data). As in ordinary algebra, commutativity in this case is a *property* of a homotopy associative ring spectrum R. While the theory of homotopy associative or commutative ring spectra plays an important role in algebraic topology, the categories of left (or right) modules over homotopy associative ring spectra are poorly behaved, lacking basic structure such as finite limits and colimits.

Proposition 13.3.38. [25, Proposition 7.2.4.27] *The ∞-categories $\mathrm{Alg}^{\mathrm{cn}}$ and $\mathrm{CAlg}^{\mathrm{cn}}$ are compactly generated. Moreover, they are generated under sifted colimits by compact projective objects.*

Definition 13.3.39. A connective associative ring spectrum A is said to be *locally of finite presentation* if A is compact as an object of $\mathrm{Alg}^{\mathrm{cn}}$. A connective commutative ring spectrum A is said to be *locally of finite presentation* if A is a compact as an object of $\mathrm{CAlg}^{\mathrm{cn}}$.

13.3.4 Left and right modules

An associative (respectively, commutative) algebra A is something that exists in a monoidal (respectively, symmetric monoidal) ∞-category $\mathcal{A}^{\otimes}$. Classically it was common to take $\mathcal{A}^{\otimes}$ to be the symmetric monoidal category of abelian groups, or (left) R-modules for a commutative ring R. Recall that a monoidal ∞-category $\mathcal{A}^{\otimes}$ can be regarded as a cocartesian fibration over Δ^{op} (equivalently a functor $\Delta^{\mathrm{op}} \to \mathrm{CAT}_{\infty}$) satisfying the Segal condition, the underlying category $\mathcal{A}$ of $\mathcal{A}^{\otimes}$ is the value $\mathcal{A}^{\otimes}_{[1]}$ at the ordinal $[1]$, and the simplicial object $\mathcal{A}^{\otimes}$ can be regarded as a categorical bar construction on $\mathcal{A}$, using its monoidal structure.

The ordinary bar construction generalizes as follows: if $\mathcal{M}$ admits a left $\mathcal{A}$-action and $\mathcal{N}$ admits a right $\mathcal{A}$-action, then we may form a simplicial object that in degree n is equivalent to $\mathcal{N} \times \mathcal{A}^{\times n} \times \mathcal{M}$. The ∞-categories $\mathcal{M}$ and $\mathcal{N}$ that arise in this way are said to be left and right tensored over $\mathcal{A}$, respectively; this is the structure that, when given an algebra object A of $\mathcal{A}$, allows us to define the notions of left and right A-module object of $\mathcal{M}$ and $\mathcal{N}$. Such ∞-categories are themselves cocartesian fibrations over $\mathcal{A}^{\otimes}$, and hence over Δ^{op}, but they satisfy a slight variant of the Segal condition.

Definition 13.3.40. Let $p : \mathcal{A}^{\otimes} \to \Delta^{\mathrm{op}}$ be a monoidal ∞-category. Let $p : \mathcal{A}^{\otimes} \to \Delta^{\mathrm{op}}$ be a monoidal ∞-category. An ∞-category *left tensored over* $\mathcal{A}^{\otimes}$ is a cocartesian fibration $q : \mathcal{M}^{\otimes} \to \Delta^{\mathrm{op}}$ together with a morphism of cocartesian fibrations $f : \mathcal{M}^{\otimes} \to \mathcal{A}^{\otimes}$ over Δ^{op} that satisfies the following relative version of the Segal condition: for each natural number n, the map

$$\mathcal{M}^{\otimes}_{[n]} \longrightarrow \mathcal{A}^{\otimes}_{[n]} \times \mathcal{M}^{\otimes}_{\{n\}}$$

induced by f and the inclusion of the final vertex $\{n\} \subset [n]$ is an equivalence.

Remark 13.3.41. The morphism of cocartesian fibrations $f : \mathcal{M}^{\otimes} \to \mathcal{A}^{\otimes}$ over Δ^{op} is an example of a *left action object* of CAT_{∞}. More precisely, we write $\mathrm{LMon}(\mathrm{CAT}_{\infty}) \subset \mathrm{Fun}(\Delta^1, \mathrm{CAT}^{\mathrm{cocart}}_{\infty/\Delta^{\mathrm{op}}})$ for the full subcategory consisting of those morphisms of cocartesian fibrations $f : \mathcal{M}^{\otimes} \to \mathcal{A}^{\otimes}$ that satisfy the relative Segal condition of 13.3.40. As in [25, Notation 4.2.2.5], there is a functor $\mathrm{LMon}(\mathrm{CAT}_{\infty}) \to \mathrm{Mon}(\mathrm{CAT}_{\infty})$ that sends the left action object $f : \mathcal{M}^{\otimes} \to \mathcal{A}^{\otimes}$ to the monoidal ∞-category $\mathcal{A}^{\otimes}$.

Remark 13.3.42. The *underlying* ∞*-category* of the left tensored ∞-category $q : \mathcal{M}^{\otimes} \to \Delta^{\mathrm{op}}$ is the fiber $\mathcal{M}^{\otimes}_{[0]}$ over $[0]$, which we will typically denote $\mathcal{M}$. There are equivalences $\mathcal{M}_{\{n\}} \simeq \mathcal{M}$ for all n. The Segal condition at $n = 1$ gives an equivalence $\mathcal{M}^{\otimes}_{[1]} \simeq \mathcal{A} \times \mathcal{M}$ and the inclusion $[0] \to [1]$ of the initial vertex induces a *left action* morphism $\mathcal{A} \times \mathcal{M} \to \mathcal{M}$.

Remark 13.3.43. Dually, there is an entirely analogous notion of *right tensored* ∞-category whose definition instead involves the inclusions of the initial vertices, with the action coming from the inclusion of the final vertices.

Remark 13.3.44. A monoidal ∞-category $p : \mathcal{A}^{\otimes} \to \Delta^{\mathrm{op}}$ is canonically left (respectively, right) tensored over itself. Roughly, restricting the cocartesian fibration p along the right (respectively, left) cone functor $^{\triangleright} : \Delta \to \Delta$ that sends $[n]$ to $[n] \star [0]$ (respectively, $[n]$ to $[0] \star [n]$) yields a cocartesian fibration $q : \mathcal{A}^{\triangleright\otimes} \to \Delta^{\mathrm{op}}$ whose fiber over $[n]$ is equivalent to $\mathcal{A}^{\times n+1}$. The morphism of cocartesian fibrations $f : \mathcal{A}^{\triangleright\otimes} \to \mathcal{A}^{\otimes}$ is obtained as in [25, Example 4.2.2.4]. See [25, Variant 4.2.2.11] for details and a comparison to the operadic approach.

Definition 13.3.45. Let $p : \mathcal{A}^{\otimes} \to \Delta^{\mathrm{op}}$ be a monoidal ∞-category and let $f : \mathcal{M}^{\otimes} \to \mathcal{A}^{\otimes}$ be an ∞-category left tensored over $\mathcal{A}^{\otimes}$. A *left module object* of $\mathcal{M}^{\otimes}$ is a map $s : \Delta^{\mathrm{op}} \to \mathcal{M}^{\otimes}$ such that the composite $f \circ s$ is an algebra object of $\mathcal{A}^{\otimes}$ and, if $i : [m] \to [n]$ is an inert map in such that $i(m) = n$, $f(i)$ is a cocartesian morphism of $\mathcal{M}^{\otimes}$.

We write

$$\mathrm{LMod}(\mathcal{M}^{\otimes}) \subset \mathrm{Fun}_{\Delta^{\mathrm{op}}}(\Delta^{\mathrm{op}}, \mathcal{M}^{\otimes})$$

for the full subcategory consisting of the left module objects of $\mathcal{M}^{\otimes}$.

Remark 13.3.46. In the special case in which $\mathcal{M}^{\otimes} \simeq \mathcal{A}^{\triangleright\otimes}$ is equivalent to $\mathcal{A}$, regarded as being left tensored over itself via $f : \mathcal{A}^{\triangleright\otimes} \to \mathcal{A}^{\otimes}$, we simply write $\mathrm{LMod}(\mathcal{A}^{\otimes})$ in place of $\mathrm{LMod}(\mathcal{A}^{\triangleright\otimes})$.

In the special case in which $\mathcal{M}^{\otimes} \simeq \mathcal{A}^{\triangleright\otimes}$ is $\mathcal{A}^{\otimes}$ regarded as left tensored over itself via $f : \mathcal{A}^{\triangleright\otimes} \to \mathcal{A}^{\otimes}$, we simply write $\mathrm{LMod}(\mathcal{A}^{\otimes})$ in place of $\mathrm{LMod}(\mathcal{A}^{\triangleright\otimes})$.

Remark 13.3.47. There is an evident notion of *right module object* of an ∞-category right tensored over a monoidal ∞-category.

Definition 13.3.48. A *left module spectrum* is a left module object of the monoidal ∞-category $\mathrm{Sp}^{\otimes} \times_{\mathrm{Fin}_*} \Delta^{\mathrm{op}}$ of spectra.

We write $\mathrm{LMod} = \mathrm{LMod}(\mathrm{Sp}^{\otimes})$ for the ∞-category of left module spectra.

Remark 13.3.49. The ∞-category LMod of left module spectra comes equipped with a forgetful functor $\mathrm{LMod} \to \mathrm{Alg}$ that returns the algebra object in the definition of a left module object. Given an associative algebra spectrum A, we write LMod_A for the fiber over A of the forgetful functor. We also have a forgetful functor $\mathrm{LMod} \times_{\mathrm{Alg}} \mathrm{CAlg} \to \mathrm{CAlg}$ obtained by pulling back along the forgetful functor $\mathrm{CAlg} \to \mathrm{Alg}$. We will sometime simply write $\mathrm{LMod} \to \mathrm{CAlg}$ for this forgetful functor, when it is clear from context that we are only considering left modules over commutative algebra spectra.

13.3.5 Localization

In this section we briefly review the theory of localization of ring spectra. Recall that a multiplicatively closed subset S, containing the unit, of an associative ring R is said to satisfy the *left Ore condition* if, for or every pair of elements $r \in R$ and $s \in S$ there exist elements $r' \in R$ and $s' \in S$ such that $s'r = r's$, and if $r \in R$ is an element such that $rs = 0$ for some element $s \in S$ then there exists an element $s' \in S$ such that $s'r = 0$.

Remark 13.3.50. If A is an associative algebra spectrum, any homogeneous element $s \in \pi_n A$ can be represented by a "degree n" left A-module map $\Sigma^n A \to A$ which is unique up to homotopy.

Definition 13.3.51. Let A be an associative algebra spectrum and $S \subset \pi_* A$ a set of homogeneous elements satisfying the left Ore condition. A left A-module spectrum M is *S-local* if, for every element $s \in S$, left multiplication by s induces an isomorphism $\pi_* M \to \pi_* M$. We write $S^{-1}\,\mathrm{LMod}_A \subset \mathrm{LMod}_A$ for the full subcategory consisting of the S-local left A-module spectra.

Proposition 13.3.52. [25, Remark 7.2.3.18] *The inclusion of the full subcategory $S^{-1}\,\mathrm{LMod}_A \subset \mathrm{LMod}_A$ of S-local objects admits a left adjoint $S^{-1} : \mathrm{LMod}_A \to S^{-1}\,\mathrm{LMod}_A$, the S-localization functor.*

Remark 13.3.53. The theory of Bousfield localization generalizes that of Ore localization. In the stable setting, this is the data of a left adjoint functor $L : \mathcal{B} \to \mathcal{C}$ of stable presentable ∞-categories such that the right adjoint $R : \mathcal{C} \to \mathcal{B}$ is fully faithful. The kernel of $L : \mathcal{B} \to \mathcal{C}$ is the stable presentable subcategory $\mathcal{A} \subset \mathcal{B}$ consisting of those objects $A \in \mathcal{B}$ such that $L(A) \simeq 0$ in $\mathcal{C}$. An instance of this construction is the localization of the ∞-category $\mathcal{B} \simeq \mathrm{Sp}$ of spectra at a fixed spectrum E: in this case, a spectrum N is E-acyclic if $E \otimes N \simeq 0$ and a spectrum M is E-local if every map $N \to M$ from an E-acyclic object M is null. While it turns out that Bousfield localization preserves (commutative) algebra structures [25, Proposition 2.2.1.9], it is quite difficult to control the localization in this generality, which is why we focus on the Ore localization instead.

Remark 13.3.54. It is possible to use the left Ore condition to give a reasonable explicit description of the homotopy groups of the left Ore localization $S^{-1}M$, for $M \in \mathrm{LMod}_A$ and $S \subset \pi_* A$ a set of homogeneous elements satisfying the left Ore condition. See [25, Construction 7.2.3.19 and Proposition 7.2.3.20] for details.

Definition 13.3.55. Let A be an associative algebra spectrum and $S \subset \pi_* A$ a set of homogeneous elements. A map of associative algebra spectra $\eta : A \to A'$ exhibits A' as the *left Ore localization* of A at $S \subset \pi_* A$ if, for every associative algebra spectrum B, $\eta^* : \mathrm{Map}_{\mathrm{Alg}}(A', B) \to \mathrm{Map}_{\mathrm{Alg}}(A, B)$ is fully faithful with image those $f : A \to B$ such that $f(s)$ is invertible in $\pi_* B$ for all $s \in S$.

Theorem 13.3.56. [25, Proposition 7.2.3.27] *Let A be an associative algebra spectrum and $S \subset \pi_* A$ a set of homogeneous elements satisfying the left Ore condition. The localization $S^{-1}A$ admits the structure of an associative algebra $A[S^{-1}]$ equipped with an algebra map*

$\eta : A \to A[S^{-1}]$ such that, for any associative algebra spectrum B, precomposition with η is a fully faithful functor

$$\eta^* : \mathrm{Map}_{\mathrm{Alg}}(A[S^{-1}], B) \subset \mathrm{Map}_{\mathrm{Alg}}(A, B)$$

with image those $f : A \to B$ such that $f(s) \in \pi_* B$ is invertible for all $s \in S$.

Remark 13.3.57. If $S \subset \pi_* A$ is a set of homogeneous elements satisfying the left Ore condition, the canonical map $S^{-1}\pi_*(A) \to \pi_*(S^{-1}A)$ is an isomorphism of graded rings, where $S^{-1}\pi_* A$ denotes the localization as graded rings.

Remark 13.3.58. If A is an associative algebra spectrum such that the graded ring $\pi_* A$ is graded commutative, then the left Ore condition on a multiplicative subset $S \subset \pi_* A$ is automatically satisfied.

There is an analogous statement for commutative algebra spectra.

Definition 13.3.59. Let A be a commutative algebra spectrum and $S \subset \pi_* A$ a set of homogeneous elements. A map of commutative algebra spectra $\eta : A \to A'$ exhibits A' as the *localization* of A at $S \subset \pi_* A$ if, for every commutative algebra spectrum B, $\eta^* : \mathrm{Map}_{\mathrm{CAlg}}(A', B) \to \mathrm{Map}_{\mathrm{CAlg}}(A, B)$ is fully faithful with image those $f : A \to B$ such that $f(s)$ is invertible in $\pi_* B$ for all $s \in S$.

Theorem 13.3.60. [25, Example 7.5.0.7] *Let A be a commutative algebra spectrum and $S \subset \pi_* A$ a multiplicative set of homogeneous elements. The localization $S^{-1}A$ admits the structure of a commutative algebra $A[S^{-1}]$ equipped with a commutative algebra map $A \to A[S^{-1}]$ such that, for any commutative algebra specrum B, precomposition with η is fully faithful*

$$\eta^* : \mathrm{Map}_{\mathrm{CAlg}}(A[S^{-1}], B) \subset \mathrm{Map}_{\mathrm{CAlg}}(A, B)$$

with image those $f : A \to B$ such that $f(s) \in \pi_* B$ is invertible for all $s \in S$.

Remark 13.3.61. Given a commutative algebra spectrum A and an arbitrary subset $T \subset \pi_* A$ of homogeneous elements of A, we often write $A[T^{-1}]$ in place of $A[S^{-1}]$, where S denotes the multiplicative closure of T in $\pi_* A$. Indexing the elements of T by some ordinal I, so that $T = \{t_i\}_{i \in I}$, we have equivalences of commutative algebra spectra $\bigotimes_{i \in I} A[t_i^{-1}] \simeq A[T^{-1}]$, where the infinite tensor product is defined to be the filtered colimit of the finite tensor products.

A map of ordinary commutative rings $f : A \to B$ is a *Zariski localization* if there exists a finite collection of elements $x_i \in \pi_0 A$, defining basic Zariski open sets $A \to A[x_i^{-1}]$, such that f is isomorphic to the product map $A \to \Pi_i A[x_i^{-1}]$ as objects of CAlg_A. There is a similar notion for ring spectra.

Example 13.3.62. Let R be a commutative ring spectrum and suppose given an element $f \in \pi_0 R$, which we regard as an R-linear map $f : R \to R$ via the equivalence $R \simeq \mathrm{End}_R(R)$. Then the filtered colimit

$$R[f^{-1}] \simeq \mathrm{colim}\{R \xrightarrow{f} R \xrightarrow{f} R \xrightarrow{f} \cdots\}$$

is a $\otimes$-idempotent left R-module: that is, the relative tensor product (see 13.4.2) is an equivalence $R[f^{-1}]\otimes_R R[f^{-1}] \simeq R[f^{-1}]$. If follows that the associated Bousfield localization of LMod_R is given the formula $M \mapsto M[f^{-1}] \simeq M \otimes_R R[f^{-1}]$, and that $R[f^{-1}] \in \mathrm{CAlg}_R$ has the following universal property: $\mathrm{Map}_{\mathrm{CAlg}_R}(R[f^{-1}], A)$ is contractible if $f \in \pi_0(A)^\times$ and empty otherwise.

Example 13.3.63. Let R be a commutative ring spectrum and consider the affine scheme $X = \operatorname{Spec} \pi_0 R$. The structure sheaf $\mathcal{O}_X$ is determined by its values $\mathcal{O}_X(U_f) = \pi_0 R[f^{-1}]$ on the basic open sets $U_f = \operatorname{Spec} \pi_0 R[f^{-1}]$, $f \in \pi_0 R$. Localizing R at elements of $\pi_0 R$ allows us to enhance $\mathcal{O}_X$ to a sheaf of commutative ring spectra $\mathcal{O}_{\operatorname{Spec} R}$ on (the topological space of) X by the formula $\mathcal{O}_{\operatorname{Spec} R}(U_f) = R[f^{-1}]$. This is the *affine spectral scheme* $\operatorname{Spec} R$.

Definition 13.3.64. Let R be a commutative ring spectrum. Then R is *local* if $\pi_0 R$ is local, in the sense that there exists a unique maximal ideal $\mathfrak{m} \subset \pi_0 R$. Equivalently, R is local if, for any $f \in \pi_0 R$, either f or $1 - f$ is a unit.

13.4 Module theory

13.4.1 Monads

Given an ∞-category $\mathcal{C}$, the ∞-category $\mathrm{Fun}(\mathcal{C}, \mathcal{C})$ of endofunctors of $\mathcal{C}$ is canonically a monoidal ∞-category with respect to composition. Viewing $\mathcal{C}$ as a simplicial set and $\mathrm{Fun}(\mathcal{C}, \mathcal{C})$ as simplicial sets, where composition is already a functor on the nose, we obtain a simplicial model for this monoidal ∞-category, which moreover comes equipped with a strict left action on $\mathcal{C}$ via the evaluation pairing $\mathrm{Fun}(\mathcal{C}, \mathcal{C}) \times \mathcal{C} \to \mathcal{C}$.

Definition 13.4.1. Let $\mathcal{C}$ be an ∞-category. A *monad* T on $\mathcal{C}$ is an algebra object of the monoidal ∞-category $\mathrm{Fun}(\mathcal{C}, \mathcal{C})$ of endofunctors of $\mathcal{C}$.

Remark 13.4.2. We write $T : \mathcal{C} \to \mathcal{C}$ for the underlying endofunctor of the monad. The unit and multiplication maps are usually denoted $\eta : \mathrm{id}_{\mathcal{C}} \to T$ and $\mu : T \circ T \to T$. There are also homotopy coherent higher multiplications.

Remark 13.4.3. Let T be a monad on an ∞-category $\mathcal{C}$. As $\mathcal{C}$ is left tensored over $\mathrm{Fun}(\mathcal{C}, \mathcal{C})$ via the evaluation map $\mathrm{Fun}(\mathcal{C}, \mathcal{C}) \times \mathcal{C} \to \mathcal{C}$, it makes sense to consider left M-module objects in $\mathcal{C}$ (these are often referred to instead as M-algebras, especially in ordinary category theory). We write $\mathrm{LMod}_T(\mathcal{C})$ for the ∞-category of left T-modules.

Example 13.4.4. Let $g : \mathcal{D} \to \mathcal{C}$ be a functor of ∞-categories that admits a left adjoint $f : \mathcal{C} \to \mathcal{D}$. Then the composite functor $g \circ f : \mathcal{C} \to \mathcal{C}$ admits a canonical structure of a monad on $\mathcal{C}$. Here, the unit $\eta : \mathrm{id} \to g \circ f$ is the unit of the adjunction, and the multiplication $\mu : g \circ f \circ g \circ f \to g \circ f$ is induced by the counit $\varepsilon : f \circ g \to \mathrm{id}$. The formalism of adjunctions allows one to fill in all the coherences in a unique (up to a contractible space of choices) way.

Remark 13.4.5. Let $g : \mathcal{D} \to \mathcal{C}$ be a functor of ∞-categories that admits a left adjoint $f : \mathcal{C} \to \mathcal{D}$ and let $T = g \circ f$ denote the resulting monad on $\mathcal{C}$, as in the example above. Then $g : \mathcal{D} \to \mathcal{C}$ factors as the composite $g = p \circ g'$, where $g' : \mathcal{D} \to \mathrm{LMod}_T(\mathcal{C})$ is the functor that sends the object B to the left T-module $\varepsilon : TB \to B$, and $p : \mathrm{LMod}_M(\mathcal{C}) \to \mathcal{C}$ is the forgetful functor.

Definition 13.4.6. Let $g : \mathcal{D} \to \mathcal{C}$ be a functor of ∞-categories that admits a left adjoint $f : \mathcal{C} \to \mathcal{D}$ and let $T = g \circ f$ be the resulting monad on $\mathcal{C}$. Then g is *monadic over* $\mathcal{C}$ if the induced map $g' : \mathcal{D} \to \mathrm{LMod}_T(\mathcal{C})$ is an equivalence.

Knowing when a functor is monadic is quite important: for instance, monadic functors are conservative, and they exhibit the source as a kind of ∞-category of modules in the target. Fortunately there is a recognition principle for monadic functors. To state it we'll need the following notions.

Definition 13.4.7. The category $\Delta_{-\infty}$ has an object $[n]$ for each integer $n \geq -1$ and an arrow $\alpha : [m] \to [n]$ for each nondecreasing function $[m] \cup \{-\infty\} \to [n] \cup \{-\infty\}$ with $\alpha(-\infty) = \infty$ which compose in the obvious way. The subcategory $\Delta_+ \subset \Delta_{-\infty}$ has the same objects but only those arrows $\alpha : [m] \to [n]$ such that $\alpha^{-1}(-\infty) = \{-\infty\}$. Note that we may identify Δ with the full subcategory of Δ_+ consisting of those objects $[n]$ with $n \geq 0$; in fact, the category Δ_+ parametrizes augmented simplicial objects.

Definition 13.4.8. An augmented simplicial object $A_\bullet : \Delta_+^{\mathrm{op}} \to \mathcal{C}$ is *split* if $A_\bullet$ extends to a functor $\Delta_{-\infty}^{\mathrm{op}} \to \mathcal{C}$. A simplicial object $A_\bullet : \Delta^{\mathrm{op}} \to \mathcal{C}$ is *split* if it extends to a split augmented simplicial object. Finally, given a functor $g : \mathcal{D} \to \mathcal{C}$, an (augmented) simplicial object $A_\bullet$ of $\mathcal{D}$ is g-split if $g \circ A_\bullet$ is split as an (augmented) simplicial object of $\mathcal{C}$.

The monadicity theorem is a higher categorical analogue of the Barr-Beck Theorem. The result plays a considerably more important role higher categorically due to the difficulty of producing explicit constructions in this context.

Theorem 13.4.9. [25, Theorem 4.7.3.5] *Let $g : \mathcal{D} \to \mathcal{C}$ be a functor of ∞-categories that admits a left adjoint $f : \mathcal{C} \to \mathcal{D}$. Then g is monadic over $\mathcal{C}$ if and only if g is conservative, $\mathcal{D}$ admits colimits of g-split simplicial objects, and g preserves colimits of g-split simplicial objects.*

13.4.2 Relative tensor products

We now consider the ∞-category of left modules over a base commutative ring spectrum R (which could be the sphere itself, in the absolute case). To generalize ordinary algebra, we'd like to have a notion of (commutative) R-algebra spectrum. In order to make this notion precise, we need a (symmetric) monoidal structure on the ∞-category LMod_R of left R-modules.

Remark 13.4.10. Using the language of ∞-operads, these results can be refined to produce $\mathbf{E}_n$-monoidal ∞-categories of left R-modules when R is only an $\mathbf{E}_{n+1}$-algebra spectrum. See [25] for the details of this approach.

As in ordinary algebra, the ∞-category of (either left or right) modules for an associative ring spectrum A will not carry a symmetric monoidal structure, which has A as the unit and commutes with colimits in each variable, unless the algebra structure on A extends to commutative algebra structure. Nevertheless, given a left A-module M and a right A-module N, we may form the iterated tensor products

$$N \otimes A \otimes \cdots \otimes A \otimes M.$$

The algebra structure on A and the left and right module structures on M and N organize these into a simplicial spectrum $\mathrm{Bar}_A(M, N)$ with

$$\mathrm{Bar}_A(M, N)_n \simeq N \otimes A^{\otimes n} \otimes M.$$

Definition 13.4.11. The *relative tensor product* $N \otimes_A M$ is a spectrum equivalent to the geometric realization of the simplicial spectrum $\mathrm{Bar}_A(M, N)$:

$$N \otimes_A M \simeq |\mathrm{Bar}_A(M, N)|.$$

Remark 13.4.12. No noncanonical choices were involved in this construction, and indeed the relative tensor product can be shown to extend to a functor

$$- \otimes_A -\colon \mathrm{RMod}_A \times \mathrm{LMod}_A \longrightarrow \mathrm{Sp}$$

that preserves colimits separately in each variable. We therefore obtain a morphism $\mathrm{LMod}_{A^{\mathrm{op}} \otimes A} \simeq \mathrm{RMod}_A \otimes \mathrm{LMod}_A \to \mathrm{Sp}$ in LPr. By Morita theory, such a map is determined by a left $A \otimes A^{\mathrm{op}}$-module, which in this case is A itself.

Left adjoint functors between ∞-categories of modules determine, and are determined by, bimodules. More precisely, if A and B are associative ring spectra, an (A, B)-bimodule M determines a functor $\mathrm{LMod}_A \to \mathrm{LMod}_B$ via the relative tensor product, and conversely. We will avoid the theory of (A, B)-bimodules by simply using the equivalent ∞-category $\mathrm{LMod}_{A^{\mathrm{op}} \otimes B}$.

Theorem 13.4.13. [25, Theorem 7.1.2.4] *Let A, B be associative algebra spectra. The relative tensor product induces an equivalence of ∞-categories*

$$\mathrm{LMod}_{A^{\mathrm{op}} \otimes B} \simeq \mathrm{LFun}(\mathrm{LMod}_A, \mathrm{LMod}_B).$$

Remark 13.4.14. The above equivalence sends the left $A^{\mathrm{op}} \otimes B$-module N to the functor $M \mapsto N \otimes_A M$ and a left adjoint functor $f : \mathrm{LMod}_A \to \mathrm{LMod}_B$ to $f(A)$, viewed as a left $A^{\mathrm{op}} \otimes B$-module via its right A-action.

Remark 13.4.15. The right adjoint of the left adjoint functor $N \otimes_A - : \mathrm{LMod}_A \to \mathrm{LMod}_B$ is $\mathrm{Map}_B(N, -) : \mathrm{LMod}_B \to \mathrm{LMod}_A$.

The following version of Morita theory, originally due to Schwede-Shipley [35], is a convenient recognition principle for ∞-categories of modules.

Theorem 13.4.16. [25, Theorem 7.1.2.1] *Let $\mathcal{C}$ be a stable presentable ∞-category and let P be an object of $\mathcal{C}$. Then $\mathcal{C}$ is compactly generated by P if and only if the functor* $\mathrm{Map}_{\mathcal{C}}(P,-)\colon \mathcal{C} \to \mathrm{RMod}_{\mathrm{End}_{\mathcal{C}}(P)}$ *is an equivalence.*

Remark 13.4.17. If $f : A \to B$ is a map of associative ring spectra, then we may view B as a left $A^{\mathrm{op}} \otimes B$-module, in which case the resulting left adjoint functor $B \otimes_A - : \mathrm{LMod}_A \to \mathrm{LMod}_B$ is the basechange functor, with right adjoint $\mathrm{Map}_B(B,-) : \mathrm{LMod}_B \to \mathrm{LMod}_A$ the forgetful functor. Note that the forgetful functor preserves colimits, as they are detected on underlying spectra, so that this is the same as tensoring with the left $B^{\mathrm{op}} \otimes A$-module B, i.e. $\mathrm{Map}_B(B,-) \simeq B \otimes_B -$. In particular, there is a further right adjoint $\mathrm{Map}_A(B,-) : \mathrm{Mod}_A \to \mathrm{Mod}_B$.

Example 13.4.18. Suppose that $B \simeq A[S^{-1}]$ is a localization of A. Then the forgetful functor $\mathrm{LMod}_B \to \mathrm{LMod}_A$ is fully faithful with essential image the S-local left A-module spectra, namely those M such that $M \simeq S^{-1}M$.

13.4.3 Projective, perfect and flat modules

We now study the ∞-categories LMod_A for A an associative ring spectrum.

Remark 13.4.19. The Morita theory of the previous section can be used to identify stable ∞-categories $\mathcal{C}$ of the form LMod_A for an associative algebra spectrum A, where now $A \simeq \mathrm{End}_{\mathcal{C}}(P)^{\mathrm{op}}$ for some compact generator P of $\mathcal{C}$. It follows that LMod_A is compactly generated. Moreover, a stable ∞-category $\mathcal{C}$ is of the form LMod_A if and only if $\mathcal{C}$ admits a compact generator, and $\mathrm{LMod}_A \simeq \mathrm{Ind}(\mathrm{LMod}_A^{\omega})$ is the Ind-completion of its full subcategory LMod_A^{ω} of compact objects.

Definition 13.4.20. Let A be an associative ring spectrum. A left A-module M is *perfect* if M is a compact object of LMod_A.

Definition 13.4.21. Let A be an associative ring spectrum. A left A-module M is *free* if M is a (possibly infinite) coproduct of (unshifted) copies of A, viewed as a left module over itself.

Definition 13.4.22. A free left A-module M is *finitely generated* if it is equivalent to a finite coproduct of copies of A.

Just as in ordinary algebra, there are other notions of "freeness" corresponding to various forgetful-free adjunctions. For instance, if M is a left A-module and $f : A \to B$ to a ring map, then $B \otimes_A M$ is often referred to as the "free" left B-module associated to the left A-module M, even though $B \otimes_A M$ is rarely a free B-module (although it will be of course if M was actually a free left A-module).

Using the long exact sequence on homotopy groups, it is straightforward to check that a map $f : M \to N$ of connective left A-modules (over a connective associative ring A) is surjective if and only if $\mathrm{fib}(f)$ is connective.

Definition 13.4.23. Let A be a connective associative ring spectrum. A left A-module P is *projective* if it is connective and projective as object of the ∞-category $\mathrm{LMod}_A^{\mathrm{cn}}$ of connective left A-modules, in the sense that the corepresented functor $\mathrm{Map}(P,-)\,\mathrm{LMod}_A^{\mathrm{cn}} \to \mathcal{S}$ preserves geometric realizations (that is, colimits of simplicial diagrams). See [26, Definition 5.5.8.18] for details.

Remark 13.4.24. It is unreasonable to ask for a left A-module to be projective as an object of the ∞-category LMod_A itself, as the only projective objects of this ∞-category are the zero objects. Indeed, suppose that M is a projective object of LMod_A. For any A-module N, we can write the suspension of N as the geometric realization of the simplicial A-module

$$\Sigma N \simeq \left| 0 \leftleftarrows N \,\substack{\leftarrow\\\leftarrow\\\leftarrow}\, N \oplus N \,\substack{\leftarrow\\\leftarrow\\\leftarrow\\\leftarrow}\, \cdots \right|.$$

But

$$\mathrm{Map}_{\mathrm{LMod}_A}(\Sigma^{-1}M, N) \simeq \mathrm{Map}_{\mathrm{LMod}_A}(M, \Sigma N) \simeq B\,\mathrm{Map}_{\mathrm{LMod}_A}(M, N),$$

so $\pi_0 \mathrm{Map}_{\mathrm{LMod}_A}(\Sigma^{-1}M, N) \simeq 0$ for all left A-modules N and therefore $M \simeq 0$.

Proposition 13.4.25. [25, Proposition 7.2.2.6] *Let $\mathcal{C}$ be a stable ∞-category with a left complete t-structure $(\mathcal{C}_{\geq 0}, \mathcal{C}_{\leq 0})$ and let $P \in \mathcal{C}_{\geq 0}$ be an object. The following conditions are equivalent:*

(1) *P is projective (as an object of $\mathcal{C}_{\geq 0}$).*

(2) *For every $M \in \mathcal{C}_{\geq 0}$, $\mathrm{Ext}^1_{\mathcal{C}}(P, M) \cong 0$.*

(3) *For every $M \in \mathcal{C}_{\geq 0}$ and every integer $n > 0$, $\mathrm{Ext}^n_{\mathcal{C}}(P, M) \cong 0$.*

(4) *For every $M \in \mathcal{C}^{\heartsuit}$ and every integer $n > 0$, $\mathrm{Ext}^n_{\mathcal{C}}(P, M) \cong 0$.*

(5) *For every exact triangle $L \to M \to N$ in $\mathcal{C}_{\geq 0}$, the map $\mathrm{Ext}^0_{\mathcal{C}}(P, M) \to \mathrm{Ext}^0_{\mathcal{C}}(P, N)$ is surjective.*

Proposition 13.4.26. [25, Corollary 7.2.2.19] *Let $f : A \to B$ be a map of connective associative algebra spectra such that $\pi_0 f : \pi_0 A \to \pi_0 B$ is an isomorphism. The basechange functor $f^* = (-) \otimes_A B : \mathrm{Mod}_A \to \mathrm{Mod}_B$ restricts to an equivalence*

$$f^* : \mathrm{Ho}(\mathrm{LMod}_A^{\mathrm{proj}}) \xrightarrow{\simeq} \mathrm{Ho}(\mathrm{LMod}_B^{\mathrm{proj}})$$

on homotopy categories of projective objects. In particular, the 0-truncation map $f : A \to \pi_0 A$ induces an equivalence $\mathrm{Ho}(\mathrm{LMod}_A^{\mathrm{proj}}) \simeq \mathrm{LMod}_{\pi_0 A}^{\heartsuit\,\mathrm{proj}}$.

Proposition 13.4.27. [25, Proposition 7.2.2.7] *Let A be a connective associative algebra spectrum and P a projective left A-module. Then there exists a free left A-module M such that P is a retract of M. If additionally P is finitely generated, we may take M to be finitely generated as well.*

While the notion of projectivity really only makes sense over connective ring spectra, the notion of flatness makes sense over arbitrary ring spectra.

Definition 13.4.28. Let A be an associative ring spectrum. A left A-module spectrum M is said to be *flat* over A if $\pi_0 M$ is a flat left $\pi_0 A$-module, in the sense of ordinary algebra, and for each integer n, the canonical map

$$\pi_n A \otimes_{\pi_0 A} \pi_0 M \longrightarrow \pi_n M$$

is an isomorphism.

Nevertheless, over a connective associative ring spectrum A, the notion of flatness behaves in a manner more similar to that of ordinary algebra. For example we have the following important generalization of Lazard's theorem.

Theorem 13.4.29. [25, Theorem 7.2.2.15] *Let A be a connective associative algebra spectrum and M a connective left A-module. The following conditions are equivalent:*

(1) *M is flat.*

(2) *M is a filtered colimit of finitely generated free left A-modules.*

(3) *M is a filtered colimit of finitely generated projective left A-modules.*

(4) *The functor* $\mathrm{RMod}_A \to \mathrm{Sp}$ *given by $N \mapsto N \otimes_A M$ is left t-exact.*

(5) *If N is a discrete right A-module then $N \otimes_A M$ is discrete.*

Remark 13.4.30. A free (respectively, projective) left A-module M is a filtered colimit of finitely generated free (respectively, projective) left A-modules. Thus the statement of Lazard's theorem remains true (albeit less precise) if we disregard finite generation.

Remark 13.4.31. The *Tor spectral sequence* has E_2-page

$$E_2^{p,q} = \mathrm{Tor}_p^{\pi_* A}(\pi_* M, \pi_* N)_q$$

and converges to the homotopy groups $\pi_{p+q}(M \otimes_A N)$ of the relative tensor product. If M or N is flat over A, $E_2^{p,q}$ vanishes for $p > 0$, the spectral sequence degenerates at the E_2-page, and $\pi_*(M \otimes_A N) \cong \pi_* M \otimes_{\pi_* A} \pi_* N$ is calculated as graded tensor product of $\pi_* M$ and $\pi_* N$ over $\pi_* A$.

Remark 13.4.32. As a consequence we observe that if M and N are both flat over A, then their tensor product $M \otimes_A N$ is again flat over A. Since the unit object A of LMod_A is flat, we see that the full subcategory $\mathrm{LMod}_A^\flat \subset \mathrm{LMod}_A$ inherits the structure of a symmetric monoidal ∞-category.

In order to calculate in the ∞-category LMod_A of left A-modules, it is useful to have something analogous to a projective resolution. If A is connective, the theory of *tor-amplitude* plays this role, giving a means of construct any perfect A-module inductively, in a finite number of steps, as an iterated cofiber of maps from shifted finitely projective left A-modules.

Definition 13.4.33. Let R be a connective commutative ring spectrum. A left R-module P has *tor-amplitude contained in the interval* $[a, b]$ if for any discrete left $\pi_0 R$-module M, $\pi_i(P \otimes_R M) = 0$ for $i \notin [a, b]$. If such integers a and b exist, P is said to have *finite tor-amplitude.*

If P is an R-module, then P has tor-amplitude contained in $[a, b]$ if and only if $P \otimes_R \pi_0 R$ is a complex of left $\pi_0 R$-modules with tor-amplitude contained in $[a, b]$ in the ordinary sense. Note, however, that this definition differs from that in [10, I 5.2] as we work homologically as opposed to cohomologically.

Proposition 13.4.34. [5, Proposition 2.13] *Let A be a connective associative algebra spectrum and let M and N be left A-module spectra.*

(1) *If M is perfect then M has finite tor-amplitude.*

(2) *If $B \in \mathrm{CAlg}_A^{\mathrm{cn}}$ and M has tor-amplitude contained in $[a,b]$ then the left B-module $B \otimes_A M$ has tor-amplitude contained in $[a,b]$.*

(3) *If M and N have tor-amplitude contained in $[a,b]$ then the fiber and cofiber of a map $M \to N$ have tor-amplitude contained in $[a-1,b]$ and $[a,b+1]$.*

(4) *If M is perfect with tor-amplitude contained in $[0,b]$ then M is connective and $\pi_0 M \cong \pi_0(\pi_0 A \otimes_A M)$.*

(5) *If M is perfect with tor-amplitude contained in $[a,a]$ then M is equivalent to $\Sigma^a P$ for a finitely generated projective left A-module P.*

(6) *If M is perfect with tor-amplitude contained in $[a,b]$ then there exists an exact triangle $\Sigma^a P \to M \to Q$ with P finitely generated projective and Q perfect with tor-amplitude contained in $[a+1,b]$.*

Remark 13.4.35. If additionally A is a connective commutative algebra spectrum and M and N are left A-modules such that M has tor-amplitude contained in $[a,b]$ and N has tor-amplitude contained in $[c,d]$, then $M \otimes_A N$ has tor-amplitude contained in $[a+c, b+d]$.

13.4.4 Tensor powers, symmetric powers, and free objects

Given a map of associative algebra spectra $f : A \to B$, the basechange functor $f^* : \mathrm{LMod}_A \to \mathrm{LMod}_B$ (corresponding to the left $A^{\mathrm{op}} \otimes B$-module B) is left adjoint to the forgetful functor $f_* : \mathrm{LMod}_B \to \mathrm{LMod}_A$ (corresponding to the left $B^{\mathrm{op}} \otimes A$-module B). Indeed, the counit of the adjunction $f^* f_* \to \mathrm{id}_{\mathrm{LMod}_B}$ is induced from the left B-module action map $B \otimes_A N \to N$, and the unit of the adjunction $\mathrm{id}_{\mathrm{LMod}_A} \to f_* f^*$ is induced from the left A-module unit map $M \simeq A \otimes_A M \to B \otimes_A M$. In particular, $f^* M \simeq B \otimes_A M$ is the *free left B-module* on the left A-module M, by virtue of the equivalence

$$\mathrm{Mod}_B(f^* M, N) \simeq \mathrm{Mod}_A(M, f_* N).$$

Remark 13.4.36. Observe that $f^* M \simeq B \otimes_A M$ need not be free as a left B-module in the sense of 13.4.21 unless M is free as a left A-module.

Remark 13.4.37. The forgetful functor $f_* : \mathrm{LMod}_B \to \mathrm{LMod}_A$ is conservative and preserves all small limits and colimits. Hence it is monadic, and exhibits LMod_B as the ∞-category of left modules for the monad $T = f_* f^*$.

The forgetful functors $\mathrm{CAlg} \to \mathrm{Alg} \to \mathrm{Sp}$ preserve limits, so it is natural to ask whether or not they admit left adjoints. Using presentability and the adjoint functor theorem (see 13.2.39), this is the case if and only if they preserve κ-filtered colimits for some regular cardinal κ. As we might expect from algebraic kinds of categories, they preserve all filtered colimits (as well as geometric realizations), so the left adjoints exist, and are instances of *free algebra* functors [25]. As is ordinary algebra, it is useful to consider the relative situation, so we work over a base commutative ring spectrum R.

Proposition 13.4.38. [25, Corollaries 3.2.2.4 and 3.2.3.2] *The forgetful functors* $\mathrm{CAlg}_R \to \mathrm{Alg}_R \to \mathrm{Sp}$ *preserve small limits and sifted colimits.*

Basically by construction, these forgetful functors are also conservative, hence monadic by the Barr-Beck-Lurie 13.4.9. This means that the free-forgetful adjunctions exhibit Alg and CAlg as ∞-categories of left modules for their respective monads. The underlying endofunctors of these monads are just in ordinary algebra, the tensor and symmetric algebra constructions.

Remark 13.4.39. Straightening the symmetric monoidal ∞-category $\mathrm{LMod}_R^{\otimes}$ to a functor $\mathrm{Fin}_* \to \mathrm{CAT}_\infty$ and restricting to the active maps $\langle n\rangle \to \langle 1\rangle$ for each $n \in \mathbf{N}$, we obtain n-fold tensor power functors $\mathrm{Ten}_R^n : \mathrm{LMod}_R \to \mathrm{LMod}_R$. That is, $\mathrm{Ten}_R^n(M) \simeq M^{\otimes n}$, where the tensor product is taken over R.

Proposition 13.4.40. [25, Proposition 4.1.1.18] *Let M be a left R-module. The free associative R-algebra on M is the tensor algebra*

$$\mathrm{Ten}_R(M) \simeq \bigoplus_{k\in\mathbf{N}} \mathrm{Ten}_R^k(M) \simeq \bigoplus_{k\in\mathbf{N}} M^{\otimes k}.$$

Remark 13.4.41. This only describes the underlying left R-module of the free associative R-algebra on M. The multiplication

$$\bigoplus_{i\in\mathbf{N}} M^{\otimes i} \underset{R}{\otimes} \bigoplus_{j\in\mathbf{N}} M^{\otimes j} \simeq \bigoplus_{i,j\in\mathbf{N}} M^{\otimes i+j} \longrightarrow \bigoplus_{k\in\mathbf{N}} M^{\otimes k}$$

is given by concatenation of tensor powers. This is still only a small, but important, piece of the homotopy coherently associative algebra structure.

The free commutative algebra functor, also known as the symmetric algebra, uses the symmetric power functors $\mathrm{Sym}_R^n : \mathrm{LMod}_R \to \mathrm{LMod}_R$, given by the formula

$$\mathrm{Sym}_R^n(M) \simeq \mathrm{Ten}_R^n(M)_{h\Sigma_n} \simeq M_{h\Sigma_n}^{\otimes n},$$

where the tensor product is taken in the symmetric monoidal ∞-category $\mathrm{LMod}_R^{\otimes}$ of left R-module spectra. Here the subscript $h\Sigma_n$ refers to the *homotopy quotient* of $M^{\otimes n}$ by the permutation action of the symmetric group Σ_n, which is to say that quotient in the ∞-categorical sense. This is to distinguish from more strict versions of quotients by group actions that arise through either ordinary categorical models of the ∞-category LMod_R, where it can make sense to ask for an ordinary categorical quotient, or in the context of equivariant homotopy theory, where there are several notions of fixed points.

Proposition 13.4.42. [25, Example 3.1.3.14] *Let M be a left R-module. The free commutative R-algebra on M is the symmetric algebra*

$$\mathrm{Sym}_R(M) \simeq \bigoplus_{k\in\mathbf{N}} \mathrm{Sym}_R^n(M) \simeq \bigoplus_{k\in\mathbf{N}} M_{h\Sigma_k}^{\otimes k}.$$

Remark 13.4.43. While it isn't terribly difficult to describe the multiplication on $\mathrm{Sym}_R(M)$ explicitly, organizing all of the higher multiplications in a coherent manner seems difficult without abstract machinery. The theory of *operadic left Kan extensions* [25] is one way to make this precise.

Example 13.4.44. In the case where $M = R$, we have $\mathrm{Ten}_R(R) \simeq \bigoplus_{n\in\mathbf{N}} R$. This is often denoted $R[t]$, as we have that $\pi_*(R[t]) \cong (\pi_* R)[t]$ on homotopy groups. As the notation suggests, the free tensor algebra on one generator in degree zero, $R[t]$, happens to be a commutative R-algebra spectrum, though it is *not* the free commutative R-algebra on one generator in degree zero. Instead we have that $\mathrm{Sym}_R(R) \simeq \bigoplus_{n\in\mathbf{N}} \mathrm{Sym}^n_R(R) \simeq \bigoplus_{n\in\mathbf{N}} R^{\otimes n}_{h\Sigma_n}$, which is sometimes denoted $R\{t\}$ in order to distinguish it from $R[t]$. In this case, $\pi_*(\mathrm{Sym}_R(R)) \cong \bigoplus_{k\in\mathbf{N}} R_*(B\Sigma_n)$ is the coproduct of the R-homologies of the symmetric groups Σ_n.

Remark 13.4.45. If R is a $\mathbf{Q}$-algebra then, for all $n \in \mathbf{N}$, the map $R_*(\mathrm{pt}) \to R_*(B\Sigma_n)$ is an isomorphism. This follows from the Atiyah-Hirzebruch spectral sequence $E^2_{p,q} \cong H_p(B\Sigma_n, \pi_q R) \Rightarrow R_{p+q}(B\Sigma_n)$ and the vanishing of group cohomology in characteristic zero by Maschke's theorem. Hence the map

$$R[t] \simeq \mathrm{Ten}_R(R) \longrightarrow \mathrm{Sym}_R(R) \simeq R\{t\}$$

obtained by taking the homotopy quotient is an equivalence.

Remark 13.4.46. Away from characteristic zero the map $R_*(\mathrm{pt}) \to R_*(B\Sigma_n)$ is typically not an isomorphism. In particular, the map $\mathrm{Ten}_R(R) \to \mathrm{Sym}_R(R)$ is rarely an equivalence. Nevertheless, the fact that $\mathrm{Ten}_{\mathbf{S}}(\mathbf{S}) \simeq \mathbf{S}[t]$ admits a commutative algebra structure (although it is not free as a commutative algebra) implies, by basechange along the commutative algebra map $\mathbf{S} \to R$, that there is always a commutative R-algebra map $R\{t\} \to R[t]$, which is an equivalence when R is a $\mathbf{Q}$-algebra and not typically otherwise.

There are other sorts of free algebra functors as well. The forgetful functor $\mathrm{Sp} \to \mathcal{S}$ fails to be monadic because it isn't conservative; it is, however, monadic on the full subcategory $\mathrm{Sp}^{\mathrm{cn}} \subset \mathrm{Sp}$ of connective spectra, the obvious subcategory on which it is conservative. The resulting monad $Q \simeq \Omega^\infty\Sigma^\infty_+$ is a higher categorical analogue of the free abelian group monad.

Definition 13.4.47. Let G be an ∞-group (respectively, ∞-monoid). The *group ring* (respectively, *monoid ring*) $R[G]$ is the associative ring spectrum $R[G] \simeq R \otimes \Sigma^\infty_+ G$, with R-algebra structure induced from that of G via the symmetric monoidal functor $\Sigma^\infty_+ : \mathcal{S} \to \mathrm{Sp}$.

Remark 13.4.48. Group and monoid rings are a rich source of examples of ring spectra. For instance, toric varieties, which are locally modeled on the group and monoid rings like $R[\mathbf{Z}^{\times k}]$ and $R[\mathbf{N}^{\times k}]$, are combinatorial enough that they descend from the integers to the sphere, giving basic examples of spectral schemes such as the *projective space* $\mathbf{P}^n_R$ [27, Construction 5.4.1.3].

Definition 13.4.49. The *general linear group* $\mathrm{GL}_n(R)$ of a ring spectrum R is the ∞-group $\mathrm{Aut}_R(R^{\oplus n})$ of left R-module automorphisms of $R^{\oplus n}$.

Example 13.4.50. Let $M = \coprod_{n\in\mathbf{N}} B\Sigma_n$, the free abelian ∞-monoid on one generator. Then $\mathbf{S}[M] \simeq \bigoplus \Sigma^\infty_+ B\Sigma_n$ is the free commutative algebra spectrum on one generator in the sense that, via the equivalences

$$\mathrm{Map}_{\mathrm{CAlg}}(\mathbf{S}[M], A) \simeq \mathrm{Map}_{\mathrm{AbMon}_\infty}(M, \Omega^\infty A) \simeq \mathrm{Map}_{\mathcal{S}}(\mathrm{pt}, \Omega^\infty A) \simeq \Omega^\infty A,$$

specifying a commutative algebra map $\mathbf{S}[M] \to A$ is the same as specifying a point of $\Omega^\infty A$, the image of the generator $\mathrm{pt} \simeq B\Sigma_1 \to M$ of M.

Example 13.4.51. Let $G \simeq \Omega^\infty \mathbf{S}$ be the abelian ∞-group given by the infinite loop space of the sphere $\mathbf{S}$. Under the equivalence between the ∞-categories of connective spectra and abelian ∞-groups, it follows that G is the free abelian ∞-group on one generator, or equivalently the group completion of the free abelian ∞-monoid $M \simeq \coprod_{n\in\mathbf{N}} B\Sigma_n$ on one generator of the previous example.

By adjunction, for any commutative algebra spectrum A, we have equivalences

$$\mathrm{Map}(\mathbf{S}[G], A) \simeq \mathrm{Map}(G, \Omega^\infty A) \simeq \mathrm{GL}_1(A) \subset \Omega^\infty A,$$

where the ∞-group $\mathrm{GL}_1(A) \simeq \mathrm{Aut}_{\mathrm{Mod}_A}(A)$ is equivalently the subspace of the ∞-monoid $\Omega^\infty A \simeq \mathrm{End}_{\mathrm{Mod}_A}(A)$ consisting of the invertible components; that is, it fits into the pullback square

$$\begin{array}{ccc} \mathrm{GL}_1(A) & \longrightarrow & \Omega^\infty A \\ \downarrow & & \downarrow \\ (\pi_0 A)^\times & \longrightarrow & \pi_0 A. \end{array}$$

Since $\mathbf{S}[G]$ corepresents the functor that sends the commutative algebra A to its space of units, the "derived scheme" $\mathrm{Spec}\,\mathbf{S}[G]$ can be regarded as derived version of the multiplicative group scheme.

Example 13.4.52. Any abelian ∞-group G with $\pi_0 G \cong \mathbf{Z}$ determines a "connective spectral abelian group scheme" with underlying ordinary scheme $\mathbf{G}_m$, the multiplicative group. While the previous example represents the functor of units, it is sometimes necessary to consider more "strictly commutative" derived analogues of $\mathbf{G}_m$. At the other extreme, one can consider $\mathbf{Z}$ as an ∞-group, in which case the spherical group ring $\mathbf{S}[\mathbf{Z}]$ is a flat commutative $\mathbf{S}$-algebra such that, for any commutative ring spectrum A,

$$\mathrm{Map}_{\mathrm{CAlg}}(\mathbf{S}[\mathbf{Z}], A) \simeq \mathrm{Map}_{\mathrm{AbMon}_\infty}(\mathbf{Z}, \Omega^\infty A) \simeq \mathrm{Map}_{\mathrm{AbGp}_\infty}(\mathbf{Z}, \mathrm{GL}_1(A)).$$

This space of "strict units" of A plays a central role in elliptic cohomology [24].

Thom spectra [38, 29, 2] are spectra that occur as quotients of the sphere by the action of a group. Again, it is somewhat more useful to have the version relative to a fixed commutative ring spectrum R. By construction the ∞-group $\mathrm{GL}_1(R)$ is the universal group that acts on R by R-linear maps, so any ∞-group G over $GL_1(R)$ acts as well. The group homomorphism $f : G \to \mathrm{GL}_1(R)$ deloops to a map of pointed spaces $f : \mathrm{B}G \to \mathrm{BGL}_1(R)$.

Example 13.4.53. The equivalence $\mathrm{GL}_1(\mathbf{S}) \simeq \mathrm{colim}_{n\to\infty} \mathrm{Aut}_*(S^n)$ and compatible families of ∞-group maps $O(n) \to \mathrm{Aut}_*(S^n)$ and $U(n) \to \mathrm{Aut}_*(S^{2n})$ give ∞-group maps

$O \to \mathrm{GL}_1(\mathbf{S})$ and $U \to \mathrm{GL}_1(\mathbf{S})$. The Thom spectra MO and MU are the homotopy quotients $MO = \mathbf{S}_{hO}$ and $MU = \mathbf{S}_{hU}$. While defined in an apparently abstract and formal way, there's a surprising connection to geometry: $\pi_* MO$ is the bordism ring of unoriented manifolds, $\pi_* MU$ is the bordism ring of stably almost complex manifolds. The analogues for other tangential structures hold; for instance, $\pi_*\mathbf{S}$ is the stably framed bordism ring.

13.4.5 Smooth, proper and dualizable objects

Definition 13.4.54. A symmetric monoidal ∞-category $p : \mathcal{C}^{\otimes} \to \mathrm{Fin}_*$ is *closed* if, for any object A of $\mathcal{C} = \mathcal{C}^{\otimes}_{\langle 1\rangle}$, the right multiplication by A functor $(-) \otimes A : \mathcal{C} \to \mathcal{C}$ admits a right adjoint $\mathcal{F}_{\mathcal{C}}(A, -) : \mathcal{C} \to \mathcal{C}$.

Remark 13.4.55. This allows for the construction of *function objects* $\mathcal{F}_{\mathcal{C}}(B, C)$ of $\mathcal{C}$, naturally as a functor $\mathcal{F}_{\mathcal{C}} : \mathcal{C}^{\mathrm{op}} \times \mathcal{C} \to \mathcal{C}$. In particular, there are natural equivalences $\mathrm{Map}_{\mathcal{C}}(A \otimes B, C) \simeq \mathrm{Map}_{\mathcal{C}}(A, \mathcal{F}_{\mathcal{C}}(B, C))$.

Definition 13.4.56. Let $\mathcal{C}^{\otimes}$ be a closed symmetric monoidal ∞-category. A *dual* of an A is an object of the form $\mathcal{F}_{\mathcal{C}}(A, \mathbf{1})$, where $\mathbf{1}$ denotes a unit object.

As duals are uniquely determined, we will write $\mathrm{D}_{\mathcal{C}}A$, or $\mathrm{D}A$, for a dual of A.

Remark 13.4.57. There is a canonical *evaluation* map

$$\varepsilon : A \otimes \mathrm{D}_{\mathcal{C}}A \simeq A \otimes \mathcal{F}_{\mathcal{C}}(A, \mathbf{1}) \longrightarrow \mathbf{1},$$

any map corresponding to the identity of $\mathcal{F}_{\mathcal{C}}(A, \mathbf{1})$ under the equivalence $\mathrm{Map}(A \otimes \mathcal{F}_{\mathcal{C}}(A, \mathbf{1}), \mathbf{1}) \simeq \mathrm{Map}(\mathcal{F}_{\mathcal{C}}(A, \mathbf{1}), \mathcal{F}_{\mathcal{C}}(A, \mathbf{1}))$.

Definition 13.4.58. An object A of a closed symmetric monoidal ∞-category $\mathcal{C}^{\otimes}$ is *dualizable* if there exists a *coevaluation* map $\eta : \mathbf{1} \longrightarrow \mathrm{D}A \otimes A$ such that the compositions

$$A \xrightarrow{A \otimes \eta} A \otimes \mathrm{D}A \otimes A \xrightarrow{\varepsilon \otimes A} A$$

$$\mathrm{D}A \xrightarrow{\eta \otimes \mathrm{D}A} \mathrm{D}A \otimes A \otimes \mathrm{D}A \xrightarrow{\mathrm{D}A \otimes \varepsilon} \mathrm{D}A$$

are equivalent to the identity.

We write $\mathcal{C}^{\mathrm{dual}} \subset \mathcal{C}$ for the full subcategory consisting of the dualizable objects of $\mathcal{C}^{\otimes}$. It always contains at least one object, any unit object $\mathbf{1}$.

Example 13.4.59. Let R be a commutative ring spectrum and consider the closed symmetric monoidal ∞-category $\mathrm{LMod}_R^{\otimes}$ of left R-module spectra. Stability forces the full subcategory $\mathrm{LMod}_R^{\mathrm{dual}}$ of dualizable objects to be closed under finite limits, colimits, and retracts [19, Theorem 2.1.3]. Since LMod_R is compactly generated, it follows that the full subcategories of compact and dualizable objects coincide.

Given a commutative algebra object $\mathcal{T}$ of LPr, we may form the ∞-category $\mathrm{LMod}_{\mathcal{T}}(\mathrm{LPr})$ of left $\mathcal{T}$-module objects of LPr. For instance, $\mathrm{LPr}_{\mathrm{st}} \simeq \mathrm{LMod}_{\mathrm{Sp}}(\mathrm{LPr})$. Of course, $\mathrm{Sp} \simeq \mathrm{LMod}_{\mathbf{S}}$, so $\mathrm{LPr}_{\mathrm{st}} \simeq \mathrm{LMod}_{\mathrm{LMod}_{\mathbf{S}}}(\mathrm{LPr})$. It is useful to consider the relative version of this: given a commutative ring spectrum R, the relative tensor product equips the presentable ∞-category LMod_R with the structure of a commutative algebra object.

Definition 13.4.60. Let R be a commutative ring spectrum. An *R-linear ∞-category* $\mathcal{C}$ is a left LMod_R-module in $\mathrm{LPr}^{\otimes}$.

We write $\mathrm{Cat}_R = \mathrm{LMod}_{\mathrm{LMod}_R}(\mathrm{LPr})$ for the ∞-category of R-linear categories and R-linear functors (left LMod_R-module maps).

Remark 13.4.61. Since LMod_R is stable, we have $\mathrm{Cat}_R \simeq \mathrm{LMod}_{\mathrm{LMod}_R}(\mathrm{LPr}_{\mathrm{st}})$.

Remark 13.4.62. Cat_R is in fact *closed* symmetric monoidal: if $\mathcal{C}$ and $\mathcal{D}$ are R-linear ∞-categories, the ∞-category of R-linear functors (that is, Mod_R-module morphisms in LPr) $\mathrm{LFun}_R(\mathcal{C}, \mathcal{D})$ from $\mathcal{C}$ to $\mathcal{D}$ is again an R-linear ∞-category (a Mod_R-module in LPr). The dual

$$\mathrm{D}_R\mathcal{C} = \mathrm{LFun}_R(\mathcal{C}, \mathrm{LMod}_R)$$

of $\mathcal{C}$ is the ∞-category of R-linear functors from $\mathcal{C}$ to LMod_R. Hence an R-linear ∞-category $\mathcal{C}$ is dualizable if there exists a coevaluation map $\eta : \mathrm{LMod}_R \xrightarrow{\eta} \mathrm{D}_R\mathcal{C} \otimes_R \mathcal{C}$ such that the compositions

$$\mathcal{C} \xrightarrow{\mathcal{C}\otimes_R\eta} \mathcal{C} \otimes_R \mathrm{D}_R\mathcal{C} \otimes_R \mathcal{C} \xrightarrow{\varepsilon\otimes_R\mathcal{C}} \mathcal{C}$$

$$\mathrm{D}_R\mathcal{C} \xrightarrow{\eta\otimes\mathrm{D}_R\mathcal{C}} \mathrm{D}_R\mathcal{C} \otimes_R \mathcal{C} \otimes_R \mathrm{D}_R\mathcal{C} \xrightarrow{\mathrm{D}_R\mathcal{C}\otimes\varepsilon} \mathrm{D}_R\mathcal{C}$$

are equivalent to the identity.

Proposition 13.4.63. *Let R be a commutative ring spectrum. Then $\mathrm{Cat}_R^{\otimes}$ is a rigid symmetric monoidal ∞-category; that is, all objects are dualizable.*

Remark 13.4.64. Consider the subcategory $\mathrm{LPr}_{\mathrm{st}}^{\mathrm{cg}} \subset \mathrm{LPr}_{\mathrm{st}}$ of compactly generated stable presentable ∞-category and left adjoint functors that preserve compact objects. Then $\mathrm{LPr}_{\mathrm{st}}^{\mathrm{cg}}$ is equivalent, via the functor that restricts to the full subcategories of compact objects, to the ∞-category of idempotent complete (that is, all idempotents split) small stable ∞-categories and exact functors. The inverse equivalence is given by $\mathrm{Ind} : \mathrm{Cat}_\infty \to \mathrm{LPr}$, which when restricted to $\mathrm{Cat}_\infty^{\mathrm{st}} \subset \mathrm{Cat}_\infty$, factors through the subcategory $\mathrm{LPr}_{\mathrm{st}}^{\mathrm{cg}} \subset \mathrm{LPr}$. Since stability and idempotent completeness are properties of small ∞-categories, and a functor of small stable ∞-categories is exact if and only if it is right exact, we deduce that $\mathrm{LPr}_{\mathrm{st}}^{\mathrm{cg}}$ is equivalent to a full subcategory of $\mathrm{Cat}_\infty^{\mathrm{rex}}$ (see [26, Proposition 5.5.7.8] for details). It follows from [25, Proposition 4.8.1.4], using [25, Remark 2.2.1.2], that $\mathrm{LPr}_{\mathrm{st}}^{\mathrm{cg}}$ inherits a symmetric monoidal structure that is compatible with the symmetric monoidal structure on $\mathrm{LPr}_{\mathrm{st}}$ or LPr. The ∞-category $\mathrm{Mod}_R^{\otimes}$ is a commutative algebra object in $\mathrm{LPr}_{\mathrm{st}}^{\mathrm{cg}}$, so we have a subcategory $\mathrm{Cat}_R^{\mathrm{cg}} \subset \mathrm{Cat}_R$ of compactly generated R-linear categories and colimit and compact object preserving functors. Similar arguments show that $\mathrm{Cat}_R^{\mathrm{cg}}$ inherits the structure of a symmetric monoidal ∞-category from $\mathrm{Cat}_R^{\otimes}$.

Proposition 13.4.65. [5, Proposition 3.5] *An R-algebra A is compact in Alg_R if and only if LMod_A is compact in $\mathrm{Cat}_R^{\mathrm{cg}}$.*

Remark 13.4.66. An object $\mathcal{C}$ is dualizable in $\mathrm{Cat}_R^{\mathrm{cg}}$ if and only if it is dualizable in Cat_R and the evaluation and coevaluation morphisms lie in the (not full) subcategory $\mathrm{Cat}_R^{\mathrm{cg}} \subset \mathrm{Cat}_R$.

Definition 13.4.67. A compactly generated R-linear category $\mathcal{C}$ is *proper* if its evaluation map is in $\mathrm{Cat}_R^{\mathrm{cg}}$; it is *smooth* if it is dualizable in Cat_R and its coevaluation map is in $\mathrm{Cat}_R^{\mathrm{cg}}$. An R-algebra A is *proper* if LMod_A is proper; it is *smooth* if LMod_A is smooth.

Remark 13.4.68. The property of being compact, smooth or proper in Alg_R is invariant under Morita equivalence.

Remark 13.4.69. If A is an R-algebra, then LMod_A is proper if and only if A is a perfect R-module. Indeed, the evaluation map is the map

$$\mathrm{LMod}_{A\otimes_R A^{\mathrm{op}}} \simeq \mathrm{LMod}_A \otimes_R \mathrm{LMod}_{A^{\mathrm{op}}} \longrightarrow \mathrm{LMod}_R$$

that sends $A \otimes_R A^{\mathrm{op}}$ to A. Similarly, LMod_A is smooth if and only if the coevaluation map

$$\mathrm{LMod}_R \longrightarrow \mathrm{LMod}_{A^{\mathrm{op}}\otimes_R A},$$

which sends R to the $A^{\mathrm{op}} \otimes_R A$-module A, exists and is in $\mathrm{Cat}_R^{\mathrm{cg}}$. So we see that LMod_A is smooth if and only if A is perfect as an $A^{\mathrm{op}} \otimes_R A$-module.

Proposition 13.4.70. [5, Lemma 3.9] *If $\mathcal{C}$ is a smooth R-linear category then $\mathcal{C} \simeq \mathrm{LMod}_A$ for some R-algebra spectrum A.*

Proposition 13.4.71. [25, Proposition 7.3.5.8] *Let R be a commutative ring spectrum and A an associative R-algebra. If A is compact in Alg_R then A is smooth. If A is a smooth and proper then A is compact in Alg_R.*

Corollary 13.4.72. [25, Corollary 7.3.5.9] *Let R be a commutative ring spectrum and let A be an associative R-algebra. Then A is smooth and proper if and only if it is compact as an object of Alg_R and LMod_R.*

Remark 13.4.73. The *noncommutative cotangent complex* is the fiber

$$\Omega_{A/R} \longrightarrow A \otimes_R A^{\mathrm{op}} \longrightarrow A$$

of the multiplication map. Thus, in the noncommutative setting, A is a smooth R-algebra if and only if $\Omega_{A/R}$ is a perfect left $A \otimes_R A^{\mathrm{op}}$-module.

13.4.6 Nilpotent, local and complete objects

In this section we fix a commutative ring spectrum R and a finitely generated ideal $I = (f_1, \dots, f_n) \subset \pi_0 R$. We write $R[I^{-1}] \simeq \bigotimes_{i=1}^n R[f_i^{-1}]$ and, for any left R-module M, $M[I^{-1}] \simeq R[I^{-1}] \otimes_R M$, regarded as a left R-module via the commutative algebra map $R \to R[I^{-1}]$.

Remark 13.4.74. There does not seem to be a good notion of *ideal* in the ∞-category of commutative ring spectra. Rather, it seems that for most part, which is relevant are ideals in the discrete ring $\pi_0 R$ or graded ring $\pi_* R$. The problem is that the cofiber R/f of an R-module map $f : R \to R$ need not admit a commutative multiplication, or even any multiplication at all. A good example is the Moore spectrum $\mathbf{S}/p$ for $p \in \mathbf{Z} \cong \pi_0 \mathbf{S}$ a prime, which does not carry an associative algebra structure; in fact, $\mathbf{S}/2$ doesn't even support a unital binary multiplication map $\mathbf{S}/2 \otimes \mathbf{S}/2 \to \mathbf{S}/2$ (see [34, Proposition 4]).

Remark 13.4.75. The *nilpotence theorem* of Devinatz-Hopkins-Smith [13] states that if R is a homotopy associative ring spectrum, then the kernel of the map $\pi_*(R) \to \pi_*(MU \otimes R)$ consists entirely of nilpotent elements. This generalizes the *Nishida nilpotence theorem* [31], which states that every element of $\pi_*\mathbf{S}$ of positive degree is nilpotent. It is used to show that the only homotopy associative ring spectra R with the property that π_*R is a *graded field* are (extensions of) the Morava K-theory spectra $K(n)$ at the prime p (suppressed from the notation) at height n. Here $K(0) \simeq \mathbf{Q}$, $\pi_*K(n) \cong \mathbf{F}_p[v^{\pm}]$ for a generator v in degree $2(p^n - 1)$, and $K(\infty) \simeq \mathbf{F}_p$. By [4], there is an essentially unique associative $\mathbf{S}$-algebra structure on $K(n)$, at least at odd primes.

Remark 13.4.76. Given a left $\mathbf{S}$-module E, the *Bousfield localization* $L_E \operatorname{Sp}$ of the ∞-category of spectra at E sits in a Verdier sequence

$$K_E \operatorname{Sp} \longrightarrow \operatorname{Sp} \longrightarrow L_E \operatorname{Sp},$$

where $K_E \operatorname{Sp} \subset \operatorname{Sp}$ denotes the full subcategory of those spectra M such that $M \otimes E \simeq 0$ (the kernel of the multiplication by E map $\operatorname{Sp} \to \operatorname{Sp}$). Since the inclusion of the full subcategory $K_E(\operatorname{Sp}) \subset \operatorname{Sp}$ evidently preserves colimits, it follows that this is a semi-orthogonal decomposition. It is a recollement precisely when the right adjoint inclusion $L_E \operatorname{Sp} \to \operatorname{Sp}$ preserves colimits, which implies that the localization is smashing: $L_E M \simeq M \otimes L_E \mathbf{S}$.

Definition 13.4.77. Let R be a commutative ring spectrum, let $I \subset \pi_0 R$ be a finitely generated ideal, and let A be an associative R-algebra.

(1) A left A-module M is I-nilpotent if $M[I^{-1}] \simeq 0$.

(2) A left A-module M is I-local if $\operatorname{Map}(N, M) \simeq 0$ for every I-nilpotent N.

(3) A left A-module M is I-complete if $\operatorname{Map}(L, M) \simeq 0$ for every I-local L.

Remark 13.4.78. Taking homotopy groups, we see that N is I-nilpotent if each $\pi_k N$, $k \in \mathbf{Z}$, is an I-nilpotent left $\pi_0 R$-module. Equivalently, N is I-nilpotent if every element of each $\pi_k N$ is annihilated by some power of I.

Remark 13.4.79. It suffices to test I-locality only on those I-nilpotent objects that are also compact as left A-modules. This is because the fully faithful inclusion $i_\vee : \operatorname{LMod}_A^{I\text{-nil}} \subset \operatorname{LMod}_A$ of the I-nilpotent left A-modules is a colimit preserving functor of compactly generated stable ∞-categories [27, Proposition 7.1.1.12], which implies that it admits a right adjoint $i^\vee$. For instance, if $I = (f)$ is principle, then A/f is a compact generator of $\operatorname{LMod}_A^{I\text{-nil}} \subset \operatorname{LMod}_A$ and $A[f^{-1}]$ is generator of $\operatorname{LMod}_A^{I\text{-loc}} \subset \operatorname{LMod}_A$ which is compact as an object of $\operatorname{LMod}_A^{I\text{-loc}}$ but not typically as an object of LMod_A. In particular, $\operatorname{LMod}_A^{I\text{-loc}} \simeq \operatorname{LMod}_{A[I^{-1}]}$.

Remark 13.4.80. The fully faithful inclusion $j_* : \operatorname{LMod}_A^{I\text{-loc}} \to \operatorname{LMod}_A$ of the I-local objects admits both a left adjoint j^* and a right adjoint $j^\times$. The existence of the left adjoint follows from the fact that j_* preserves limits, essentially by definition, and filtered colimits by the previous remark. The existence of the right adjoint follows from the fact that j_* is exact and so it also preserves finite colimits, hence all colimits. It follows that the I-localization functor $j^* : \mathrm{LMod}_A \to \mathrm{LMod}_A^{I\text{-loc}}$ is given by tensoring with $j^*A \simeq A[I^{-1}]$.

Theorem 13.4.81. [27, Theorem 7.3.4.1] *A left A-module M is I-complete if each homotopy group $\pi_n M$ is a derived I-complete left $\pi_0 A$-module, in the sense that* $\mathrm{Ext}^0_{\pi_0 A}(\pi_0 A[f^{-1}], M) \cong 0 \cong \mathrm{Ext}^1_{\pi_0 A}(\pi_0 A[f^{-1}], M)$.

Remark 13.4.82. The inclusion of the I-complete objects preserves limits, again by construction, and κ-filtered colimits for $\kappa \gg 0$. Indeed, this follows from the fact that we need only test completeness on a generator $A[I^{-1}]$ of $\mathrm{LMod}_A^{I\text{-loc}}$, and this generator is κ-compact in LMod_A for some $\kappa \gg 0$.

Proposition 13.4.83. [27, Proposition 7.2.4.4 and Proposition 7.3.1.4] *The fully faithful inclusions $i_\vee : \mathrm{LMod}_A^{I\text{-nil}} \to \mathrm{LMod}_A$ and $j_* : \mathrm{LMod}_A^{I\text{-loc}} \to \mathrm{LMod}_A$ induce Verdier sequences*

$$\mathrm{LMod}_A^{I\text{-nil}} \xrightarrow{i_\vee} \mathrm{LMod}_A \xrightarrow{j^*} \mathrm{LMod}_A^{I\text{-loc}}$$
$$\mathrm{LMod}_A^{I\text{-loc}} \xrightarrow{j_*} \mathrm{LMod}_A \xrightarrow{i^\wedge} \mathrm{LMod}_A^{I\text{-cpl}}$$

which are semi-orthogonal decompositions of LMod_A.

Remark 13.4.84. In the stable setting, a *recollement* is the data of a fully faithful inclusion of a stable subcategory that admits both a left and a right adjoint [6]. An example is the inclusion $j_* : \mathrm{LMod}_A^{I\text{-loc}} \to \mathrm{LMod}_A$ of the I-local objects. The composite $i^\wedge i_\vee$ is an equivalence of categories with inverse $i^\vee i_\wedge$. A stable recollement determines a *fracture square*, a cartesian square

$$\begin{array}{ccc} \mathrm{id} & \longrightarrow & i_\wedge i^\wedge \\ \downarrow & & \downarrow \\ j_* j^* & \longrightarrow & j_* j^* i_\wedge i^\wedge \end{array}$$

of endofunctors of LMod_A. For $A \simeq \mathbf{S}$, the *arithmetic square* is the cartesian square

$$\begin{array}{ccc} M & \longrightarrow & \left(\prod_p M_p^\wedge\right) \\ \downarrow & & \downarrow \\ M \otimes \mathbf{Q} & \longrightarrow & \left(\prod_p M_p^\wedge\right) \otimes \mathbf{Q} \end{array}$$

obtained by completing a spectrum M at all primes p and rationalization.

Theorem 13.4.85. [27, Proposition 7.4.1.1] *Let R be a commutative ring spectrum and let $I \subset \pi_0 R$ be a finitely generated ideal. Suppose given a map of associative R-algebras*

$f : A \to B$, and consider the commutative square

$$\begin{array}{ccc} \mathrm{LMod}_A & \xrightarrow{f^*} & \mathrm{LMod}_B \\ \downarrow & & \downarrow \\ \mathrm{LMod}_{A[I^{-1}]} & \xrightarrow{f[I^{-1}]^*} & \mathrm{LMod}_{B[I^{-1}]} \end{array}$$

in LPr. *If $f_I^\wedge : A_I^\wedge \to B_I^\wedge$ is an equivalence, this square is cartesian.*

13.5 Deformation theory

13.5.1 The tangent bundle and the cotangent complex

The *cotangent complex formalism* is an instance of the fiberwise stabilization of a presentable fibration. Given a pair of presentable fibrations $p : \mathcal{D} \to \mathcal{C}$ and $q : \mathcal{E} \to \mathcal{C}$, we write $\mathrm{RFun}_{\mathcal{C}}(\mathcal{E}, \mathcal{D}) \subset \mathrm{Fun}_{\mathcal{C}}(\mathcal{E}, \mathcal{D})$ for the full subcategory of those functors $g : \mathcal{E} \to \mathcal{D}$ over $\mathcal{C}$ that admit a left adjoint $f : \mathcal{D} \to \mathcal{E}$ such that $p(\eta(D))$ is an equivalence in $\mathcal{C}$ for every object $D \in \mathcal{D}$, where $\eta : \mathrm{id}_{\mathcal{D}} \to gf$ denotes any choice of unit transformation exhibiting the adjunction.

Remark 13.5.1. This is the precise condition needed to ensure that $g : \mathcal{E} \to \mathcal{D}$ restricts to a right adjoint on fibers over any object of $\mathcal{C}$, or even after pullback along any morphism $\mathcal{C}' \to \mathcal{C}$.

Definition 13.5.2. A *stable envelope* of a presentable fibration $p : \mathcal{D} \to \mathcal{C}$ is a presentable fibration $q : \mathcal{E} \to \mathcal{C}$ equipped with a morphism

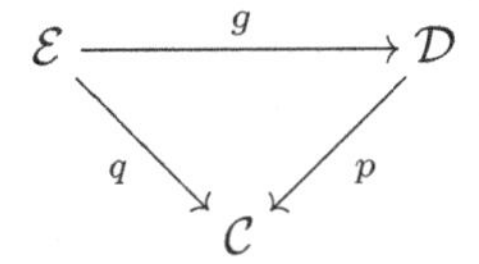

of presentable fibrations over $\mathcal{C}$ that exhibits $\mathcal{E}$ as the fiberwise stabilization of $\mathcal{D}$ over $\mathcal{C}$ in the following sense: if $q' : \mathcal{E}' \to \mathcal{C}$ is a stable presentable fibration, the induced map $g_* : \mathrm{RFun}_{\mathcal{C}}(\mathcal{E}', \mathcal{E}) \to \mathrm{RFun}_{\mathcal{C}}(\mathcal{E}', \mathcal{D})$ is an equivalence.

Definition 13.5.3. A *tangent bundle* $q : T_{\mathcal{C}} \to \mathcal{C}$ of a presentable ∞-category $\mathcal{C}$ is a stable envelope of the target fibration $p : \mathrm{Fun}(\Delta^1, \mathcal{C}) \to \mathrm{Fun}(\{1\}, \mathcal{C}) \simeq \mathcal{C}$.

Remark 13.5.4. The fiber of $q : T_{\mathcal{C}} \to \mathcal{C}$ over the object $A \in \mathcal{C}$ is the stabilization $\mathrm{Sp}(\mathcal{C}_{/A})$ of the fiber $\mathcal{C}_{/A}$ of the target fibration $p : \mathrm{Fun}(\Delta^1, \mathcal{C}) \to \mathcal{C}$.

Remark 13.5.5. The tangent bundle $T_{\mathcal{C}}$ admits an explicit construction as the ∞-category of *unreduced* excisive functors

$$\mathcal{E} \simeq \mathrm{Exc}(\mathcal{S}_*^{\mathrm{fin}}, \mathcal{C}).$$

In this case, the morphism of presentable fibrations $g : \mathcal{E} \to \mathcal{D}$ over $\mathcal{C}$ is given by evaluating at the arrow $S^0 \to \mathrm{pt}$; that is, $g(X) = \{X(S^0) \to X(\mathrm{pt})\}$.

Let $\mathcal{C}$ be a presentable ∞-category, and consider the commutative triangle

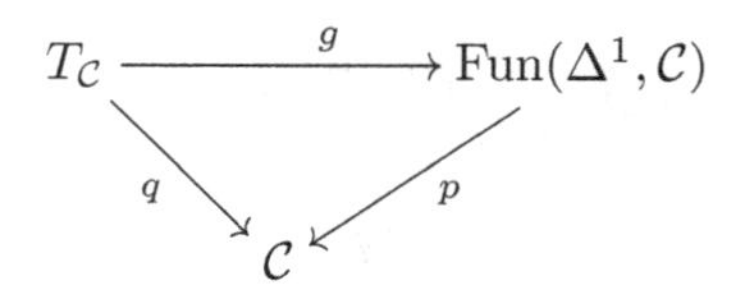

where p is the target fibration. A relative version of the adjoint functor theorem implies that f admits a left adjoint, allowing for the following construction.

Definition 13.5.6. The *absolute cotangent complex functor* $L : \mathcal{C} \to T_{\mathcal{C}}$ is the composition $\mathcal{C} \to \mathrm{Fun}(\Delta^1, \mathcal{C}) \to T_{\mathcal{C}}$, where the first map is the diagonal (constant) embedding and the second is left adjoint to $g : T_{\mathcal{C}} \to \mathrm{Fun}(\Delta^1, \mathcal{C})$. We will write L_A for the value of $L : \mathcal{C} \to T_{\mathcal{C}}$ at the object A of $\mathcal{C}$, and refer to L_A as the *(absolute) cotangent complex* of A.

Remark 13.5.7. The cotangent complex has a long history. Its first incarnation was as the sheaf of Kähler differentials, and a derived version of this was introduced by Berthelot and Illusie ([10], [21], [22]). Around the same time, Quillen ([32], [33]) and André ([3]) defined a derived version in the context of (simplicial) commutative rings, and later Basterra and Mandell ([7], [8]) developed the theory in the more general context of commutative algebra spectra, where they refer to it as *topological André-Quillen homology.*

Remark 13.5.8. Let A be an object of $\mathcal{C}$. The identification of the fiber of the tangent bundle $T_{\mathcal{C}}$ over $A \in \mathcal{C}$ with $\mathrm{Sp}(\mathcal{C}_{/A})$ is such that $L_A \in \mathrm{Sp}(\mathcal{C}_{/A})$ corresponds to the image of $\mathrm{id}_A \in \mathcal{C}_{/A}$ under $\Sigma^\infty_+ : \mathcal{C}_{/A} \to \mathrm{Sp}(\mathcal{C}_{/A})$.

Remark 13.5.9. The diagonal embedding $\mathcal{C} \to \mathrm{Fun}(\Delta^1, \mathcal{C})$ is left adjoint to the evaluation $\mathrm{Fun}(\Delta^1, \mathcal{C}) \to \mathrm{Fun}(\{0\}, \mathcal{C}) \to \mathcal{C}$ at $0 \in \Delta^1$. Hence the cotangent complex functor $L : \mathcal{C} \to T_{\mathcal{C}}$ is left adjoint to the composite functor

$$T_{\mathcal{C}} \longrightarrow \mathrm{Fun}(\Delta^1, \mathcal{C}) \longrightarrow \mathrm{Fun}(\{0\}, \mathcal{C}) \to \mathcal{C}.$$

Definition 13.5.10. The *relative cofiber over* $\mathcal{C}$ functor

$$\mathrm{cof}_{\mathcal{C}} \colon \mathrm{Fun}(\Delta^1, T_{\mathcal{C}}) \to T_{\mathcal{C}}$$

is the functor that sends the morphism $f : X \to Y$ in $T_{\mathcal{C}}$ to the pushout

$$\begin{array}{ccc} X & \longrightarrow & Y \\ \downarrow & & \downarrow \\ Z & \longrightarrow & \mathrm{cof}_{\mathcal{C}}(f) \end{array}$$

where Z is any zero object of the fiber of $T_{\mathcal{C}}$ over $p(X)$.

Definition 13.5.11. The *relative cotangent complex functor*is the composition $\mathrm{Fun}(\Delta^1, \mathcal{C}) \xrightarrow{L} \mathrm{Fun}(\Delta^1, T_{\mathcal{C}}) \xrightarrow{\mathrm{cof}_{\mathcal{C}}} T_{\mathcal{C}}$.

Remark 13.5.12. A zero object of a fiber of $q : T_{\mathcal{C}} \to \mathcal{C}$ need not be a zero object of $T_{\mathcal{C}}$ itself. Given a morphism $f : X \to Y$ in $T_{\mathcal{C}}$, the relative cofiber $\mathrm{cof}_{\mathcal{C}}(f)$ of f is an object of the fiber of $T_{\mathcal{C}}$ over $q(Y)$.

We write $L_{B/A}$ for the value of the relative cotangent complex functor on an object $f : A \to B$ of the ∞-category $\mathrm{Fun}(\Delta^1, \mathcal{C})$ of arrows in $\mathcal{C}$.

Remark 13.5.13. By construction, the relative cotangent complex of a morphism $f : A \to B$ fits into a *relative cofiber sequence*

$$\begin{array}{ccc} L_A & \longrightarrow & L_B \\ \downarrow & & \downarrow \\ 0 & \longrightarrow & L_{B/A} \end{array}$$

in $T_{\mathcal{C}}$. This induces an actual cofiber sequence $f_! L_A \to L_B \to L_{B/A}$ in the ∞-category $T_{\mathcal{C}} \times_{\mathcal{C}} \{B\} \simeq \mathrm{Sp}(\mathcal{C}_{/B})$. Here $f_! : \mathrm{Sp}(\mathcal{C}_{/A}) \to \mathrm{Sp}(\mathcal{C}_{/B})$ is a straightening of the restriction $f^* q : T_{\mathcal{C}} \times_{\mathcal{C}} \Delta^1 \to \Delta^1$ of $q : T_{\mathcal{C}} \to \mathcal{C}$ along $f : \Delta^1 \to \mathcal{C}$.

Remark 13.5.14. It follows that the commutative square in $T_{\mathcal{C}}$

$$\begin{array}{ccc} L_{B/A} & \longrightarrow & L_{C/A} \\ \downarrow & & \downarrow \\ L_{B/B} & \longrightarrow & L_{C/B} \end{array}$$

associated to a pair of composible morphisms $A \to B$ and $B \to C$ in $\mathcal{C}$ is cocartesian, hence a relative cofiber sequence since $L_{B/B} \simeq 0$ in $\mathrm{Sp}(\mathcal{C}_{/B})$.

Definition 13.5.15. The *tangent correspondence* of a presentable ∞-category $\mathcal{C}$ is the cocartesian fibration $\mathcal{M}_{\mathcal{C}} \to \Delta^1$ associated to the cotangent complex functor $L : \mathcal{C} \to T_{\mathcal{C}}$.

Remark 13.5.16. The tangent correspondence $\mathcal{M}_{\mathcal{C}} \to \Delta^1$ is also a cartesian fibration since $L : \mathcal{C} \to T_{\mathcal{C}}$ is a left adjoint.

Remark 13.5.17. The cocartesian fibration $\mathcal{M} \to \Delta^1$ associated to a functor $f : \mathcal{C} \to \mathcal{D}$ can be constructed as the pushout

$$\begin{array}{ccc} \mathcal{C} & \xrightarrow{f} & \mathcal{D} \\ \downarrow & & \downarrow \\ \Delta^1 \times \mathcal{C} & \longrightarrow & \mathcal{M} \end{array}$$

in which the left vertical map is the inclusion at $\{1\} \subset \Delta^1$. The functor $\mathcal{M} \to \Delta^1$ has fiber over 0 and 1 the full subcategories $\mathcal{C} \to \mathcal{M}$ and $\mathcal{D} \to \mathcal{M}$, and over the unique map $\varepsilon : 0 \to 1$ the functor $f : \mathcal{C} \to \mathcal{D}$; this is evidently a cocartesian fibration as we can push forward objects of $\mathcal{C}$ along ε via f. Thus a functor $h : \mathcal{M} \to \mathcal{C}$ amounts to the data of a functor $g : \mathcal{D} \to \mathcal{C}$ and a natural transformation $\eta : \Delta^1 \times \mathcal{C} \to \mathcal{C}$ from $\mathrm{id}_{\mathcal{C}}$ to gf. If $\mathcal{M}$ is also a cartesian fibration, a unit transformation $\eta : \mathrm{id}_{\mathcal{C}} \to gf$ induces a canonical functor $h : \mathcal{M} \to \mathcal{C}$.

Remark 13.5.18. A *derivation* in $\mathcal{C}$ is a morphism $\Delta^1 \to \mathcal{M}_{\mathcal{C}}$ such that the composite $\Delta^1 \to \mathcal{M}_{\mathcal{C}} \to \Delta^1$ is the identity and $\Delta^1 \to \mathcal{M}_{\mathcal{C}} \to \mathcal{C}$ is constant. More concretely, a derivation in $\mathcal{C}$ is a morphism in $\mathcal{M}_{\mathcal{C}}$ from an object A in the fiber $\mathcal{C}$ over 0 to an object M in the fiber $\mathrm{Sp}(\mathcal{C}_{/A})$ of $q : T_{\mathcal{C}} \to \mathcal{C}$ over A (see [25, Remark 7.4.1.2]). The ∞-category $\mathrm{Der}(\mathcal{C})$ of derivations in $\mathcal{C}$ is the pullback

$$\begin{array}{ccc} \mathrm{Der}(\mathcal{C}) & \longrightarrow & \mathrm{Fun}(\Delta^1, \mathcal{M}_{\mathcal{C}}) \\ \downarrow & & \downarrow \\ \mathcal{C} & \longrightarrow & \mathrm{Fun}(\Delta^1, \Delta^1 \times \mathcal{C}) \end{array}$$

in which the right vertical map is induced by the functor $\mathcal{M}_{\mathcal{C}} \to \Delta^1 \times \mathcal{C}$ and the bottom horizontal map is adjoint to the identity of $\Delta^1 \times \mathcal{C}$.

13.5.2 Derivations and square-zero extensions

We now specialize to the case in which $\mathcal{C}$ is the presentable ∞-category CAlg of commutative algebra spectra.

Remark 13.5.19. All algebras will be assumed commutative for the remainder of this article. If A is a commutative algebra object there is a canonical equivalence $A \simeq A^{\mathrm{op}}$, hence a canonical equivalence $\mathrm{LMod}_A \simeq \mathrm{LMod}_{A^{\mathrm{op}}} \simeq \mathrm{RMod}_A$, so we needn't distinguish left and right module structures in the commutative case. We thus write Mod_A in place of LMod_A and RMod_A.

Theorem 13.5.20. [25, Corollary 7.3.4.14] *The functor* p: Mod $\to$ CAlg *that sends the pair* $(A, M) \in$ Mod *to* $A \in$ CAlg *exhibits* Mod *as a tangent bundle of* CAlg. *In particular, for any commutative algebra spectrum A, we have an equivalence* $\mathrm{Mod}_A \simeq \mathrm{Sp}(\mathrm{CAlg}_{/A})$.

We will write $\mathrm{Sym}^{\leq 1} : \mathrm{Mod} \to \mathrm{Fun}(\Delta^1, \mathrm{CAlg})$ for a functor that corresponds to $\Omega^\infty : T_{\mathrm{CAlg}} \to \mathrm{Fun}(\Delta^1, \mathrm{CAlg})$. We use this notation because $\mathrm{Sym}^{\leq 1} \simeq \mathrm{Sym}^0 \oplus \mathrm{Sym}^1$ is the formula for the *split square-zero extension* in ordinary algebra, though we have obtained its augmented commutative algebra structure through abstract stabilization techniques.

Remark 13.5.21. Let A be a commutative ring spectrum and M an A-module. Then the split square-zero augmented commutative A-algebra structure on $A \oplus M$ is square-zero in the sense that the compositions

$$\mathrm{Sym}^n_A(M) \longrightarrow \mathrm{Sym}^n_A(A \oplus M) \xrightarrow{\otimes^n} A \oplus M \longrightarrow M$$

are null whenever $n > 1$.

Remark 13.5.22. On homotopy groups, the split square-zero extension is an ordinary split square-zero extension of graded commutative rings. In other words, for any pair of elements

(a_0, m_0) and (a_1, m_1) of $\pi_*(A \oplus M)$, the multiplication on $\pi_*(A \oplus M) \cong \pi_*(A) \oplus \pi_*(M)$ is given by the expected formula

$$(a_0, m_0)(a_1, m_1) = (a_0 a_1, a_0 m_1 + (-1)^{|a_1||m_0|} a_1 m_0).$$

Remark 13.5.23. Any split square-zero extension $A \oplus M$, viewed as an augmented commutative A-algebra, is canonically a spectrum object in the pointed ∞-category $\mathrm{CAlg}_{A/A}$. Indeed, in degree n, $(A \oplus M)^n \simeq A \oplus \Sigma^n M$, and the map

$$A \oplus \Sigma^n M \longrightarrow \Omega(A \oplus \Sigma^{n+1} M) \simeq A \oplus \Omega\Sigma^{n+1} M \simeq A \oplus \Sigma^n M$$

is an equivalence, as A is a zero object of $\mathrm{CAlg}_{A/A}$ and

$$\begin{array}{ccc} A \oplus \Sigma^n M & \longrightarrow & A \\ \downarrow & & \downarrow \\ A & \longrightarrow & A \oplus \Sigma^{n+1} M \end{array}$$

is a cartesian square of $\mathrm{CAlg}_{A/A}$. Moreover, as Mod_A is stable, the functor $\mathrm{Sym}_A^{\leq 1} : \mathrm{Mod}_A \to \mathrm{CAlg}_{A/A}$ factors through $\Omega^\infty : \mathrm{Sp}(\mathrm{CAlg}_{A/A}) \to \mathrm{CAlg}_{A/A}$. There is even an evident map back in the order direction given by taking the fiber, which sends the spectrum object $\{B^n\}_{n \in \mathbf{N}}$ to $\mathrm{fib}\{B^0 \to A\} \in \mathrm{Mod}_A$.

In ordinary commutative algebra, a derivation $d : A \to M$ over R is an R-module map satisfying the Leibniz rule

$$d(ab) = ad(b) + bd(a).$$

Instead of using elements, we could instead define a derivation $d : A \to M$ as a section of the projection $A \oplus M \to A$, taken in the category of commutative R-algebras. Replacing R with $\mathbf{S}$ we obtain the following notion.

Definition 13.5.24. Let A be a commutative algebra spectrum and M an A-module. A *derivation* from A to M is a section $A \to A \oplus M$ in CAlg of the projection $A \oplus M \to A$.

Remark 13.5.25. This is a special case of the notion of derivation introduced in the previous section. Here we avoid explicit mention of the tangent correspondence by pulling back to the fiber of $\mathcal{M}_{\mathrm{CAlg}} \to \Delta^1$ via the split square zero extension functor $T_{\mathrm{CAlg}} \simeq \mathrm{Mod} \to \mathrm{CAlg}$.

Remark 13.5.26. The space $\mathrm{Der}(A, M)$ of derivations $A \to A \oplus M$ is the fiber

$$\mathrm{Der}(A, M) \longrightarrow \mathrm{Map}_{\mathrm{CAlg}}(A, A \oplus M) \longrightarrow \mathrm{Map}_{\mathrm{CAlg}}(A, A)$$

over the identity $\mathrm{id}_A \in \mathrm{Map}_{\mathrm{CAlg}}(A, A)$. The composite of the commutative algebra map $A \to A \oplus M$ with the second projection $A \oplus M \to M$ is a map of spectra $d : A \to M$ which we will abusively refer to as the derivation.

Remark 13.5.27. For any connective commutative algebra A and left A-module M, there is an equivalence $\mathrm{Map}_{\mathrm{Mod}_A}(L_A, M) \to \mathrm{Der}(A, M)$. That is, the cotangent complex L_A corepresents the functor $\mathrm{Der}(A, -) : \mathrm{Mod}_A \to \mathcal{S}$.

Remark 13.5.28. Using the tangent correspondence formalism from the previous subsection, the ∞-category $\mathrm{Der} = \mathrm{Der}(\mathrm{CAlg})$ of derivations has objects derivations $d : A \to M$ and morphisms commutative squares of the form

$$\begin{array}{ccc} A & \longrightarrow & M \\ \downarrow & & \downarrow \\ B & \longrightarrow & N \end{array},$$

with $A \to B$ a commutative algebra map and $M \to N$ an A-module map.

Let A be a commutative ring spectrum, M an A-module, and $\eta : A \to \Sigma M$ a derivation with associated section $(\mathrm{id}, \eta) : A \to A \oplus \Sigma M$ of the projection $A \oplus \Sigma M \to A$. By construction, (id, η) is a morphism of commutative algebras.

Definition 13.5.29. A map of commutative algebra spectra $\varepsilon : A' \to A$ is a *square-zero extension* of A by the A-module M if there exists a derivation $\eta : A \to \Sigma M$ and a cartesian square in $\mathrm{CAlg}_{/A}$ of the form

$$\begin{array}{ccc} A' & \longrightarrow & A \\ \downarrow & & \downarrow{\scriptstyle (\mathrm{id},\eta)} \\ A & \xrightarrow{(\mathrm{id},0)} & A \oplus \Sigma M \end{array}.$$

Remark 13.5.30. There is a functor $\Phi : \mathrm{Der} \to \mathrm{Fun}(\Delta^1, \mathrm{CAlg})$ that sends the derivation $\eta : A \to \Sigma M$ to the map $A^\eta \to A$, where A^η is a pullback

$$\begin{array}{ccc} A^\eta & \longrightarrow & A \\ \downarrow & & \downarrow{\scriptstyle (\mathrm{id},\eta)} \\ A & \xrightarrow{(\mathrm{id},0)} & A \oplus \Sigma M \end{array}$$

in $\mathrm{CAlg}_{/A}$, and whose essential image consists of the square-zero extensions.

Remark 13.5.31. Note that, if $f : A^\eta \to A$ is a square-zero extension of A by M, then the fiber sequence of A-modules $M \to A \to A \oplus \Sigma M$ implies that we also have a fiber sequence $M \to A^\eta \to A$ and hence a canonical equivalence $\mathrm{fib}(f) \simeq M$. Hence derivations $\eta : A \to \Sigma M$ give rise to square-zero extensions A^η of A by M. This is why we use ΣM instead of M itself.

Remark 13.5.32. As the name suggests, square-zero extensions $f : A' \to A$ are actually square-zero. That is, the fiber $M \to A' \to A$ has the property that the n-fold symmetric power map $\mathrm{Sym}^n_A(M) \to M$ is null for any $n > 1$.

Definition 13.5.33. A morphism $\varepsilon : A' \to A$ in $\mathrm{CAlg}^{\mathrm{cn}}$ is an *n-small extension* if $\mathrm{fib}(\varepsilon)$ has homotopy concentrated in degrees $[0, 2n]$ and the multiplication map $\mathrm{fib}(\varepsilon) \otimes_{A'} \mathrm{fib}(\varepsilon) \to \mathrm{fib}(\varepsilon)$ is nullhomotopic. A derivation $\eta : A \to M$ is *n-small* if the associated square-zero extension $A^\eta \to A$ is n-small.

Theorem 13.5.34. [25, Theorem 7.4.1.26] *The composition*

$$\mathrm{Der}_n \subset \mathrm{Der} \xrightarrow{\Phi} \mathrm{Fun}(\Delta^1, \mathrm{CAlg}^{\mathrm{cn}})$$

is fully faithful with essential image the full subcategory of $\mathrm{Fun}(\Delta^1, \mathrm{CAlg}^{\mathrm{cn}})$ *consisting of the n-small extensions (here Φ is as in 13.5.30 above).*

Corollary 13.5.35. *Any n-small extension is a square-zero extension.*

One of the primary source of examples of square-zero extensions comes from the Postnikov tower of a connective commutative algebra A.

Proposition 13.5.36. [25, Corollary 7.4.1.28] *Let $A \in \mathrm{CAlg}^{\mathrm{cn}}$. For each $n > 0$, the map*

$$\tau_{\leq n} A \longrightarrow \tau_{\leq n-1} A$$

exhibits $\tau_{\leq n} A$ as a square-zero extension of $\tau_{\leq n-1} A$ by $\Sigma^n \pi_n A$.

Remark 13.5.37. The fact that the Postnikov tower is composed of square-zero extensions is one of the main reasons why the cotangent complex plays such an important role in spectral algebra and geometry. The space of maps

$$\mathrm{Map}_{\mathrm{CAlg}}(A, B) \simeq \lim \mathrm{Map}_{\mathrm{CAlg}}(A, \tau_{\leq n} B) \simeq \lim \mathrm{Map}_{\mathrm{CAlg}}(\tau_{\leq n} A, \tau_{\leq n} B)$$

between connective commutative algebra spectra A and B decomposes as the limit of the space of maps between their truncations, and for any $n > 0$ we have a pullback diagram

$$\begin{array}{ccc} \tau_{\leq n} B & \longrightarrow & \tau_{\leq n-1} B \\ \downarrow & & \downarrow \\ \tau_{\leq n-1} B & \longrightarrow & \tau_{\leq n-1} B \oplus \Sigma^{n+1} \pi_n B. \end{array}$$

This implies that the fibers of $\mathrm{Map}_{\mathrm{CAlg}}(A, B) \longrightarrow \mathrm{Map}_{\mathrm{CAlg}^{\heartsuit}}(\pi_0 A, \pi_0 B)$ are accessible via infinitesimal methods: arguing inductively up the Postnikov tower, we are reduced to understanding spaces of derivations from $A \to \Sigma^{n+1} \pi_n B$, an A-linear question concerning maps $L_A \to \Sigma^{n+1} \pi_n B$.

13.5.3 Deformations of commutative algebras

Given a connective commutative algebra spectrum A and a square-zero extension $A' \to A$ of A by a connective A-module M, it is natural to study the space of *deformations* of the square-zero extension $A' \to A$ to a connective commutative A-algebra $f : A \to B$. That is, we wish to understand the space of cocartesian squares in $\mathrm{CAlg}^{\mathrm{cn}}$ of the form

$$\begin{array}{ccc} A' & \xrightarrow{f'} & B' \\ \downarrow & & \downarrow \\ A & \xrightarrow{f} & B. \end{array}$$

(It turns out that $B' \to B$ is automatically also a square-zero extension.) As in ordinary algebra, this reduces to a module theoretic question concerning the cotangent complex: such a commutative A'-algebra B' exists if and only if the map $B \otimes_A L_A \to B \otimes_A \Sigma M$ induced by a derivation $\eta : A \to \Sigma M$ classifying $A' \to A$ factors through the absolute cotangent complex L_B of B.

Definition 13.5.38. Let A be a commutative algebra spectrum, $A' \to A$ a square-zero extension of A by an A-module M, and B a commutative A-algebra. A *deformation of B to A'* is a commutative A'-algebra B' equipped with an equivalence $B' \otimes_{A'} A \to B$ of commutative A-algebras.

Remark 13.5.39. We need not assume that B' is flat over A', as this is the case if and only if B is flat over A. Indeed, if $A' \to B'$ is flat, then the basechange $A \to B \simeq B' \otimes_{A'} A$ is flat by a spectral sequence argument. Conversely, the fact that any deformation $B' \to B$ of $A' \to A$ along a flat map $A \to B$ results in a flat map $A' \to B'$ follows from a tor-amplitude argument.

Remark 13.5.40. In the connective case, just like in ordinary commutative algebra, there are cohomological obstructions to the existence and uniqueness of deformations. Given an A-linear map $\eta : L_A \to \Sigma M$ with M a connective A-module, the associated square-zero extension $A^\eta \to A$ of A by M is again connective. Given a morphism of connective commutative algebra spectra $A \to B$, we obtain a map $\eta_B : B \otimes_A L_A \to B \otimes_A \Sigma M$, and a deformation $B' \to B$ of $A' \to A$ along $A \to B$ exists if and only if η_B factors as a composite

$$B \otimes_A L_A \xrightarrow{\varepsilon_B} L_B \xrightarrow{\eta'} B \otimes_A \Sigma M$$

of $\varepsilon_B : B \otimes_A L_A \to L_B$ and a B-module map $\eta' : L_B \to B \otimes_A \Sigma M$. In this case, the deformation is recovered as the square-zero extension $B' \simeq B^{\eta'} \to B$. Notice, though, that such a factorization exists if and only if the composite

$$\Sigma^{-1} L_{B/A} \longrightarrow B \otimes_A L_A \longrightarrow B \otimes_A \Sigma M,$$

is null, so that a deformation exists if and only if the *obstruction class* in $\mathrm{Ext}^2_B(L_{B/A}, B \otimes_A M)$ corresponding to this map vanishes. The set of equivalence classes of deformations is a torsor for the group $\mathrm{Ext}^1_B(L_{B/A}, B \otimes_A M)$; in particular, it might be empty, or only admit an element locally on $\mathrm{Spec}\, \pi_0 B$.

Definition 13.5.41. A derivation $\eta : A \to \Sigma M$ is said to be *connective* if A is a connective commutative algebra spectrum and M is an connective A-module.

Let $\mathcal{D} \subset \mathrm{Der}$ denote the subcategory consisting of the connective derivations $A \to \Sigma M$ and those morphisms of connective derivations

$$\begin{array}{ccc} A & \longrightarrow & \Sigma M \\ \downarrow & & \downarrow \\ B & \longrightarrow & \Sigma N \end{array}$$

such that the induced map $B \otimes_A M \to N$ is an equivalence. Similarly, let $\mathcal{C} \subset \mathrm{Fun}(\Delta^1, \mathrm{CAlg})$ denote the subcategory consisting of those objects $A' \to A$ such that both A and A' are connective and those morphisms the squares

$$\begin{array}{ccc} A' & \longrightarrow & A \\ \downarrow & & \downarrow \\ B' & \longrightarrow & B \end{array}$$

that are cocartesian in CAlg; in other words, $B' \otimes_{A'} A \to B$ is an equivalence. When restricted to $\mathcal{D} \subset \mathrm{Der}$, the square zero extension functor $\Phi : \mathrm{Der} \to \mathrm{Fun}(\Delta^1, \mathrm{CAlg})$ of 13.5.30 factors through $\mathcal{C} \subset \mathrm{Fun}(\Delta^1, \mathrm{CAlg})$.

Theorem 13.5.42. [25, Theorem 7.4.2.7] *The composition*

$$\mathcal{D} \subset \mathrm{Der} \xrightarrow{\Phi} \mathrm{Fun}(\Delta^1, \mathrm{CAlg})$$

factors through the subcategory $\mathcal{C} \subset \mathrm{Fun}(\Delta^1, \mathrm{CAlg})$*, and the resulting functor* $\Phi' : \mathcal{D} \to \mathcal{C}$ *is a left fibration (a cocartesian fibration with* ∞*-groupoid fibers).*

Proposition 13.5.43. [25, Proposition 7.4.2.5] *For any connective derivation* $\eta : A \to \Sigma M$*,* $\Phi : \mathrm{Der} \to \mathrm{Fun}(\Delta^1, \mathrm{CAlg})$ *induces an equivalence*

$$\Phi_{\eta/} : \mathcal{D}_{\eta/} \xrightarrow{\simeq} \mathrm{CAlg}^{\mathrm{cn}}_{A^{\eta}},$$

where $\mathcal{D} \subset \mathrm{Der}$ *denotes the subcategory defined above.*

13.5.4 Connectivity results

A *universal derivation* is any derivation $d : A \to L_A$ that is corresponds to an equivalence $L_A \to L_A$. Given a map of commutative ring spectra $f : A \to B$,

$$\begin{array}{ccc} A & \longrightarrow & L_A \\ \downarrow{\scriptstyle f} & & \downarrow \\ B & \longrightarrow & L_B \end{array}$$

is a commutative square in Der. Taking vertical cofibers, we obtain map of A-modules $\mathrm{cof}(f) \to L_{B/A}$ which is adjoint to a map of B-modules $\mathrm{cof}(f) \otimes_A B \to L_{B/A}$. We will write $\varepsilon_f : \mathrm{cof}(f) \otimes_A B \to L_{B/A}$ for this map.

Theorem 13.5.44. [25, Theorem 7.4.3.1] *Let* $f : A \to B$ *be a morphism in* $\mathrm{CAlg}^{\mathrm{cn}}$ *such that* $\mathrm{cof}(f) \in \mathrm{Mod}_A^{\geq n}$ *for some* $n \geq 0$*. Then* $\mathrm{fib}(\varepsilon_f) \in \mathrm{Mod}_B^{\geq 2n}$.

Example 13.5.45. We note the elementary fact that if $M \in \mathrm{Mod}_A^{\geq n}$ then, for any natural number m, $\mathrm{Ten}_A^m(M) \in \mathrm{Mod}_A^{\geq mn}$ and consequently $\mathrm{Sym}_A^m(M) \in \mathrm{Mod}_A^{\geq mn}$ as well. Thus if $f : \mathrm{Sym}_A M \to A$ is the projection to $A \simeq \mathrm{Sym}_A^0 M$, it is straightforward to show that $\mathrm{fib}(\varepsilon_f) \in \mathrm{Mod}_A^{\geq 2n}$. The general case is obtained from this special case by connectivity and induction arguments.

Proposition 13.5.46. [25, Corollary 7.4.3.2] *Let $f : A \to B$ be a map of connective commutative algebra spectra such that* $\mathrm{cof}(f)$ *is n-connective for some $n \geq 0$. Then the relative cotangent complex $L_{B/A}$ is n-connective, and the converse holds provided that $\pi_0 f : \pi_0 A \to \pi_0 B$ is an isomorphism.*

Remark 13.5.47. The absolute cotangent complex of a connective commutative algebra spectrum is itself connective. This follows immediately from the previous proposition since the cofiber of the unit map is connective.

Corollary 13.5.48. *Let $f : A \to B$ be a map of connective commutative algebra spectra. Then f is an equivalence if and only if $\pi_0 f : \pi_0 A \to \pi_0 B$ is an isomorphism and the relative cotangent complex $L_{B/A}$ vanishes.*

Remark 13.5.49. Let $f : A \to B$ be a map of connective commutative algebra spectra such that $\mathrm{cof}(f)$ is n-connective for some $n \geq 0$. The induced map $L_A \to L_B$ factors as the composite

$$L_A \xrightarrow{g} B \otimes_A L_A \xrightarrow{h} L_B$$

and the equivalence $\mathrm{cof}(g) \simeq \mathrm{cof}(f) \otimes_A L_A$, together with the connectivity of A and L_A, imply that $\mathrm{cof}(g)$ is n-connective. We also have an exact triangle

$$B \otimes_A \mathrm{cof}(f) \longrightarrow L_{B/A} \longrightarrow \mathrm{cof}(\varepsilon_f)$$

in which $B \otimes_A \mathrm{cof}(f)$ and $\mathrm{cof}(\varepsilon_f)$ are n-connective, so that $L_{B/A}$ is n-connective as well. It follows that the cofiber of $L_A \to L_B$ is n-connective..

Proposition 13.5.50. [25, Lemma 7.4.3.8] *Let A be a connective commutative algebra spectrum. There are canonical isomorphisms of π_0-modules*

$$\pi_0 L_A \simeq \pi_0 L_{\pi_0 A} \simeq \Omega_{\pi_0 A}.$$

Remark 13.5.51. Let $f : A \to B$ be a map of connective commutative algebra spectra such that $\mathrm{cof}(f)$ is n-connective. Tensoring the exact triangle $A \to B \to \mathrm{cof}(f)$ with the A-module $\mathrm{cof}(f)$, we obtain an exact triangle

$$\mathrm{cof}(f) \xrightarrow{\delta} B \otimes_A \mathrm{cof}(f) \longrightarrow \mathrm{cof}(f) \otimes_A \mathrm{cof}(f)$$

which exhibits δ as a $(2n-1)$-connective map. Composing with $\varepsilon_f : B \otimes_A \mathrm{cof}(f) \to L_{B/A}$, we obtain a $(2n-1)$-connective map $\mathrm{cof}(f) \to L_{B/A}$.

Proposition 13.5.52. [25, Proposition 7.4.3.9] *Let $f : A \to B$ be a morphism in* $\mathrm{CAlg}^{\mathrm{cn}}$. *Then $L_{B/A}$ is connective and $\pi_0 L_{B/A} \cong \Omega_{\pi_0 B/\pi_0 A}$.*

Theorem 13.5.53. [25, Theorem 7.4.3.18] *Let $f : A \to B$ be a morphism in* $\mathrm{CAlg}^{\mathrm{cn}}$. *If B is locally of finite presentation over A, $L_{B/A}$ is a perfect B-module. The converse holds provided $\pi_0 B$ is of finite presentation over $\pi_0 A$.*

13.5.5 Classification of étale maps

Recall that a map of discrete commutative rings $f : A \to B$ is said to be étale if B is a finitely presented flat commutative A-algebra such that the multiplication map $B \otimes_A B \to B$ is the projection onto a summand. Geometrically, this is the algebraic analogue of a (not necessarily surjective) covering space: there exists a commutative ring C and a cartesian square of schemes of the form

$$\begin{array}{ccc} \mathrm{Spec}(B) \coprod \mathrm{Spec}(C) & \longrightarrow & \mathrm{Spec}(B) \\ \downarrow & & \downarrow{\scriptstyle f} \\ \mathrm{Spec}(B) & \xrightarrow{f} & \mathrm{Spec}(A) \end{array} .$$

Étale maps of commutative rings $A \to B$ are smooth of relative dimension zero [37, Lemma 10.141.2]. In particular, there exists a presentation of B as a commutative A-algebra of the form $B \cong A[x_1, \ldots, x_n]/(f_1, \ldots, f_n)$, where the $\{f_i\}_{1 \leq i \leq n}$ are a sequence of elements of $A[x_1, \ldots, x_n]$ such that the image of the Jacobian matrix of partial derivatives $\{\partial f_i / \partial x_j\}_{1 \leq i,j \leq n}$ is invertible in B.

Definition 13.5.54. A map $f : A \to B$ of commutative algebra spectra is étale if f exhibits B as a finitely presented commutative A-algebra such that B is a flat A-module and $\pi_0 f : \pi_0 A \to \pi_0 B$ is étale.

Ideally, we would like to be able to calculate the space of étale maps between commutative algebra spectra A and B in terms of the set of étale maps between their underlying discrete commutative algebras $\pi_0 A$ and $\pi_0 B$. Since an étale map $f : A \to B$ induces an étale map $\pi_0 f : \pi_0 A \to \pi_0 B$, we can address this question directly by studying the space of étale lifts $f : A \to B$ of an étale morphism $f_0 : \pi_0 A \to \pi_0 B$. The main result of this section, one of the major results of higher algebra, is that the space of such lifts is contractible.

Proposition 13.5.55. [25, Remark 7.5.1.7] *Given a commutative triangle*

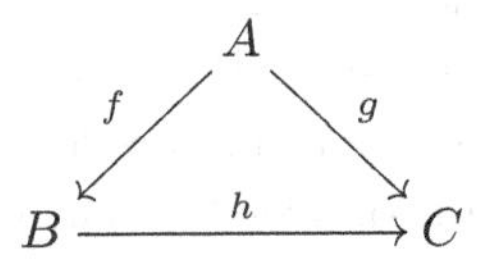

in CAlg, *if f is étale, then g is étale if and only if h is étale. In particular, any map between étale commutative A-algebras is automatically étale.*

Remark 13.5.56. Let $f : A \to B$ be a map of connective commutative ring spectra. Then f is étale if and only if each $\tau_{\leq n} f : \tau_{\leq n} A \to \tau_{\leq n} B$ is étale.

Remark 13.5.57. Since an étale map of discrete rings is smooth of relative dimension zero, its module of relative Kähler differentials vanishes, and one can show using ordinary algebraic methods that its relative cotangent complex also vanishes. This begs the question of whether or not the relative cotangent complex $L_{B/A}$ of an étale map of commutative algebra spectra $f : A \to B$ vanishes. Using flatness, one reduces to the connective case, so that $L_{B/A}$ is also connective, and if $L_{B/A} \neq 0$ there's a least $n \in \mathbf{N}$ with $\pi_n L_{B/A} \neq 0$. This is a contradiction: by connectivity considerations as in the previous section,

$$0 \cong \pi_n L_{\pi_0 B/\pi_0 A} \cong \pi_n(L_{B/A} \otimes_B \pi_0 B) \cong \pi_n L_{B/A}.$$

Hence $L_{B/A} \simeq 0$ for any étale map $f : A \to B$.

Remark 13.5.58. Said differently, this means that the absolute cotangent complex functor $L : \mathrm{CAlg} \to \mathrm{Mod}$ satisfies étale basechange: if $f : A \to B$ is an étale map of commutative algebra spectra, then the exact triangle

$$B \otimes_A L_A \longrightarrow L_B \longrightarrow L_{B/A}$$

together with the vanishing of the relative cotangent complex $L_{B/A}$ implies that $B \otimes_A L_A \simeq L_B$.

Let $\mathrm{Der}^{\text{ét}} \subset \mathrm{Der}$ denote the (not full) subcategory consisting of the connective derivations $A \to \Sigma M$ and those morphisms of connective derivations

$$\begin{array}{ccc} A & \longrightarrow & \Sigma M \\ \downarrow & & \downarrow \\ B & \longrightarrow & \Sigma N \end{array}$$

such that the induced map $B \otimes_A M \to N$ is an equivalence and the commutative algebra map $A \to B$ is étale. Let $\mathrm{CAlg}^{\text{ét}} \subset \mathrm{CAlg}^{\text{cn}}$ denote the (not full) subcategory consisting of the (connective) commutative algebra spectra and the étale maps.

Remark 13.5.59. The forgetful functor $\mathrm{Der}^{\text{ét}} \to \mathrm{CAlg}^{\text{ét}}$ is a cocartesian fibration such that, for each $A \in \mathrm{CAlg}^{\text{ét}}$, the fiber $\mathrm{Der}^{\text{ét}}_A$ is an ∞-groupoid (i.e., it is a left fibration). This is because an étale morphism of derivations is cocartesian: given connective derivations $\eta : A \to \Sigma M$ and $\eta' : B \to \Sigma N$ and a map $f : A \to B$, we have an equivalence $N \simeq B \otimes_A M$, so if $f = \mathrm{id}_A$ we obtain an equivalence $N \simeq M$. In particular, for any connective derivation $\eta : A \to \Sigma M$, the forgetful functor induces an equivalence $\mathrm{Der}^{\text{ét}}_{\eta/} \xrightarrow{\simeq} \mathrm{CAlg}^{\text{ét}}_A$.

Remark 13.5.60. Consider the functor $\Phi : \mathrm{Der} \to \mathrm{Fun}(\Delta^1, \mathrm{CAlg})$ that sends the derivation $\eta : A \to \Sigma M$ to the square-zero extension $A^\eta \to A$. Compositing with the restrictions to the set of vertices $\{0, 1\}$ of Δ^1, we obtain functors $\Phi_0 : \mathrm{Der} \to \mathrm{CAlg}$ and $\Phi_1 : \mathrm{Der} \to \mathrm{CAlg}$ such that Φ_1 restricts to the left fibration $\mathrm{Der}^{\text{ét}} \to \mathrm{CAlg}^{\text{ét}}$. For a given connective derivation $\eta : A \to \Sigma M$, Φ_0 and Φ_1 induce functors

$$\mathrm{CAlg}^{\text{ét}}_{A^\eta} \xleftarrow{\Phi_0'} \mathrm{Der}^{\text{ét}}_{\eta/} \xrightarrow{\Phi_1'} \mathrm{CAlg}^{\text{ét}}_A$$

such that Φ_0' is an equivalence by 13.5.43 and $\Phi_1' \simeq \Phi_0' \otimes_{A^\eta} A$ and Φ_0' is an equivalence by the remark above. We obtain the following corollary.

Corollary 13.5.61. *Let $f : A' \to A$ be a square-zero extension of connective commutative ring spectra. The relative tensor product*

$$(-) \otimes_{A'} A : \mathrm{CAlg}_{A'} \longrightarrow \mathrm{CAlg}_A$$

induces an equivalence from the ∞-category of étale commutative A'-algebras to the ∞-category of étale commutative A-algebras.

Theorem 13.5.62. [25, Corollary 7.5.4.3] *For any commutative algebra spectrum A, $\pi_0 : \mathrm{CAlg}_A \to \mathrm{CAlg}_{\pi_0 A}$ induces an equivalence $\mathrm{CAlg}_A^{\text{ét}} \simeq \mathrm{CAlg}_{\pi_0 A}^{\text{ét}}$.*

Remark 13.5.63. Note that any étale commutative $\pi_0 A$-algebra B is automatically in $\mathrm{CAlg}_A^{\heartsuit}$ since the flatness condition implies that B must be discrete: $\pi_n B \cong \pi_n A \otimes_{\pi_0 A} \pi_0 B \cong 0$ if $n \neq 0$. Hence $\mathrm{CAlg}_{\pi_0 A}^{\text{ét}} \simeq \mathrm{CAlg}_{\pi_0 A}^{\heartsuit\,\text{ét}}$, and the theorem asserts that the ∞-category of étale commutative A-algebras is equivalent to the *ordinary* category of étale commutative $\pi_0 A$-algebras.

Remark 13.5.64. If R is a commutative ring spectrum, the structure of the ∞-category $\mathrm{CAlg}_R^{\text{ét}}$ of étale commutative R-algebras implies that the small étale site of R is equivalent to the small étale site of the discreet ring $\pi_0 R$, and the analogous result holds for the small Zariski sites of 13.3.62. These facts form the cornerstones of *spectral algebraic geometry*, as treated in [27] and [39].

Remark 13.5.65. There are robust notions of spectral scheme and Deligne-Mumford stack. Versions of the Artin representability theorem for these higher categorical objects are formulated and proved as [27, Theorems 18.1.0.1 and 18.1.0.2], providing necessary and sufficient conditions for a functor $F : \mathrm{CAlg}^{\mathrm{cn}} \to \mathcal{S}$ to be represented by such an object. These conditions are surprisingly straightforward: the ordinary stack $F|_{\mathrm{CAlg}^{\heartsuit}} \to \mathcal{S}$ must be represented by an ordinary scheme or Deligne-Mumford stack, F must admit a cotangent complex, F must preserve limits of Postnikov towers, and F must preserve pullbacks of diagrams of the form $A \to C \leftarrow B$ in $\mathrm{CAlg}^{\mathrm{cn}}$ in which both of the maps $A \to C$ and $B \to C$ surjective on π_0 with nilpotent kernel.

Bibliography

[1] J. F. Adams, *Stable homotopy and generalised homology*, Chicago Lectures in Mathematics, University of Chicago Press, 1974 and 1995.

[2] Matthew Ando, Andrew J. Blumberg, David Gepner, Michael J. Hopkins, and Charles Rezk. An ∞-categorical approach to R-line bundles, R-module Thom spectra, and twisted R-homology. *J. Topol.*, 7(3):869–893, 2014.

[3] Michel André. *Homologie des algèbres commutatives.* Grundlehren Math. Wiss., Band 206, Springer-Verlag, 1974.

[4] Vigleik Angeltveit. Uniqueness of Morava K-theory. *Compos. Math.*, 147(2):633–648, 2011.

[5] B. Antieau and D. Gepner. *Brauer groups and étale cohomology in derived algebraic geometry.* Geom. Topol., 18 (2014), 1149-1244.

[6] Clark Barwick and Saul Glasman. A note on stable recollements. Available at `https://www.maths.ed.ac.uk/~cbarwick/papers/recoll.pdf`.

[7] M. Basterra. André-Quillen cohomology of commutative S-algebras. *J. Pure Appl. Algebra*, 144(2):111–143, 1999.

[8] Maria Basterra and Michael A. Mandell. *Homology and cohomology of E_∞ ring spectra.* Math. Z., 249(4):903–944, 2005.

[9] Aleksandr A. Beilinson, Joseph Bernstein, and Pierre Deligne. Faisceaux pervers. In *Analysis and topology on singular spaces, I (Luminy, 1981)*, volume 100 of *Astérisque*, pages 5–171. Soc. Math. France, 1982.

[10] P. Berthelot, A. Grothendieck, and L. Illusie. *Séminaire de Géométrie Algébrique du Bois Marie, SGA6, 1966-67 – Théorie des intersections et théorème de Riemann-Roch.* Lecture Notes in Math. 225. Springer-Verlag, 1971.

[11] J.M. Boardmann and R.M. Vogt. *Homotopy invariant algebraic structures on topological spaces.* Lecture Notes in Math. 347, Springer-Verlag, 1973.

[12] Edgar H. Brown, Jr. Cohomology theories. *Ann. of Math. (2)*, 75:467–484, 1962.

[13] E.S. Devinatz, M.J. Hopkins, and J.H. Smith. *Nilpotence and stable homotopy theory. I.* Ann. of Math. (2), 128(2):207-241, 1988.

[14] G. Dunn. *Tensor products of operads and iterated loop spaces.* J. Pure Appl. Algebra 50(3):237-258, 1988.

[15] Samuel Eilenberg and Norman Steenrod. *Foundations of algebraic topology.* Princeton University Press, 1952.

[16] A. D. Elmendorf, I. Kriz, M. A. Mandell, and J. P. May. *Rings, modules, and algebras in stable homotopy theory, Math. Surv. Monogr.* 47. Amer. Math. Soc., 1997. With an appendix by M. Cole.

[17] David Gepner, Moritz Groth, and Thomas Nikolaus. Universality of multiplicative infinite loop space machines. *Algebr. Geom. Topol.*, 15(6):3107–3153, 2015.

[18] A. Grothendieck, director. *Revêtements étales et groupe fondamental (SGA 1)*, volume 3 of *Doc. Math. (Paris)* Soc. Math. France, 2003. Séminaire de géométrie algébrique du Bois Marie 1960–61. With two papers by M. Raynaud, Updated and annotated reprint of the 1971 original [Lecture Notes in Math., 224, Springer].

[19] Mark Hovey, John H. Palmieri, and Neil P. Strickland. Axiomatic stable homotopy theory. *Mem. Amer. Math. Soc.*, 128(610), 1997.

[20] Mark Hovey, Brooke Shipley, and Jeff Smith. Symmetric spectra. *J. Amer. Math. Soc.*, 13(1):149–208, 2000.

[21] Luc Illusie. *Complexe cotangent et déformations. I.* Lecture Notes in Math. 239. Springer-Verlag, 1971.

[22] Luc Illusie. *Complexe cotangent et déformations. II.* Lecture Notes in Math. 283. Springer-Verlag, 1972.

[23] A. Joyal. *Quasi-categories and Kan complexes.* J. Pure Appl. Alg. 175(1–3), 207–222, 2002.

[24] J. Lurie. *Elliptic cohomology I: spectral abelian varieties.* Available at `http://www.math.harvard.edu/~lurie/papers/Elliptic-I.pdf`.

[25] J. Lurie. *Higher algebra.* Available at `http://www.math.harvard.edu/~lurie/papers/HA.pdf`.

[26] J. Lurie. *Higher topos theory*, Ann. of Math. Stud. 170, 2009. Current version available at `http://www.math.harvard.edu/~lurie/papers/HTT.pdf`.

[27] J. Lurie. *Spectral algebraic geometry.* Available at `http://www.math.harvard.edu/~lurie/papers/SAG-rootfile.pdf`.

[28] J.P. May. *The geometry of iterated loop spaces.* Lecture Notes in Math. 271. Springer-Verlag, 1972.

[29] J. Peter May. *E_∞ ring spaces and E_∞ ring spectra.* Lecture Notes in Math. 577. Springer-Verlag, 1977. With contributions by Frank Quinn, Nigel Ray, and Jørgen Tornehave.

[30] A. Neeman. *The Grothendieck duality theorem via Bousfield's techniques and Brown representability.* J. Amer. Math. Soc. 9(1):205–236, 1996.

[31] Goro Nishida. The nilpotency of elements of the stable homotopy groups of spheres. *J. Math. Soc. Japan*, 25:707–732, 1973.

[32] Daniel G. Quillen. *Homotopical Algebra.* Lecture Notes in Math. 43. Springer-Verlag, 1967.

[33] Daniel Quillen. On the (co-) homology of commutative rings. In *Applications of Categorical Algebra (Proc. Sympos. Pure Math., Vol. XVII, New York, 1968)*, pages 65–87. Amer. Math. Soc., 1970.

[34] Stefan Schwede. Algebraic versus topological triangulated categories. In *Triangulated categories*, volume 375 of *London Math. Soc. Lecture Note Ser.*, pages 389–407. Cambridge Univ. Press, 2010.

[35] S. Schwede and B. Shipley. *Stable model categories are categories of modules.* Topology 42(1):103–153, 2003.

[36] G. Segal. *Categories and cohomology theories.* Topology 13:293–312, 1974.

[37] The Stacks project authors. The Stacks project. `https://stacks.math.columbia.edu`, 2018.

[38] René Thom. Quelques propriétés globales des variétés différentiables. *Comment. Math. Helv.*, 28:17–86, 1954.

[39] B. Toën and G. Vezzosi. *Homotopical Algebraic geometry II. Geometric stacks and applications.* Mem. Amer. Math. Soc. 193 (2008), no. 902.

[40] Charles A. Weibel. *An introduction to homological algebra*, volume 38 of *Cambridge Stud. Adv. Math.*. Cambridge University Press, 1994.

THE UNIVERSITY OF MELBOURNE, PARKVILLE, VIC 3010, AUSTRALIA
E-mail address: `david.gepner@unimelb.edu.au`

14

A short course on ∞-categories

Moritz Groth

14.1 Introduction

The aim of this short course is to give a non-technical account of some ideas in the theory of ∞-categories (a.k.a. quasi-categories, inner Kan complexes, weak Kan complexes, Boardman complexes, or quategories), as originally introduced by Boardman–Vogt [23, p. 102] in their study of homotopy-invariant algebraic structures. Recently, ∞-categories have been studied intensively by Joyal [67, 68, 69], Lurie [82, 79, 80, 81], and others (this includes the Riehl–Verity program which was started in [97]). ∞-categories have applications in many areas of pure mathematics (and some of them are taken up in various chapters of this Handbook). In this chapter we are more modest. We simply try to emphasize the philosophy and some of the main ideas of ∞-category theory, and we sketch the lines along which the theory is developed. *In particular, this means that there is no claim of originality.*

Category theory is an important mathematical discipline in that it provides us with a convenient language which applies whenever we put into practice the following slogan: 'In order to study a collection of objects one should also consider suitably defined morphisms between such objects.' Many classes of mathematical objects like groups, modules over a ring, manifolds, or schemes can be organized into a category and from typical constructions one frequently abstracts the categorical character behind them. Let us recall that a category consists of objects, morphisms and a composition law which is suitably associative and unital. This allows us, in particular, to speak about *isomorphisms*, and all functorial constructions trivially preserve isomorphisms.

Category theory is a very powerful and useful language. However, it also has its limitations. To put it as a slogan, in many areas of pure mathematics we would like to identify two objects which are, while possibly not isomorphic in the purely categorical sense, 'equivalent' from a more homotopy theoretic perspective. To illustrate this, the desire of identifying different resolutions of objects in abelian categories leads us to study chain complexes up to quasi-isomorphisms. Similarly, in homotopy theory we would like to think of weak homotopy equivalences between topological spaces as actual isomorphisms. And even in category theory, often we do not have to distinguish two categories as long as they are equivalent.

These are only three examples for the fairly common situation in which we start with a pair $(\mathcal{C}, W)$ consisting of a category $\mathcal{C}$ and a class W of so-called *weak equivalences*, a class of morphisms that we would like to treat as isomorphisms. In such situations, functo-

Mathematics Subject Classification. 18Axx, 18Gxx, 55Pxx, 55U10, 55U35, 55U40.

Key words and phrases. ∞-categories, simplicial categories, coherent homotopy theory, homotopy limits and homotopy colimits, monoidal ∞-categories, stable ∞-categories, triangulated categories.

rial constructions are only 'meaningful' if they preserve weak equivalences. The search for convenient languages to study such situations has already quite some history and various different approaches have been considered. This includes triangulated categories [90], model categories [60], relative categories [9], derivators [48], simplicial categories [14], topological categories [64], and ∞-categories [82, 83].

The three last named approaches belong to a fairly large zoo of different models all of which realize 'a theory of $(\infty,1)$-categories'. While an $(\infty,1)$-category is not a well-defined mathematical notion, there is the general agreement that such a category-like concept should enjoy the following features.

(i) As part of the structure there is a class of objects.

(ii) There should be morphisms between objects, 2-morphisms between morphisms (like chain homotopies, homotopies, and natural isomorphisms), as well as 3-morphisms, 4-morphisms, and so forth, explaining the parameter '∞' in '$(\infty,1)$-categories'.

(iii) Morphisms can be composed in a suitably associative and unital way.

(iv) Higher morphisms, i.e., 2-morphisms, 3-morphisms, and so on, are supposed to be invertible in a certain sense. (All morphisms above dimension one are invertible, explaining the parameter '1' in '$(\infty,1)$-categories'.)

References for survey articles on the zoo of different axiomatizations of $(\infty,1)$-categories include [16] and [1].

The aim of these notes instead is to focus essentially on one of those different models and to describe how a good deal of classical category theory can be extended to this particular model. In these notes we follow Lurie [82, 83] in his choice of terminology and refer to these particular models for $(\infty,1)$-categories as ∞-*categories*. We refrain from giving a more detailed introduction here and instead refer the reader to the titles as well as to the short introductions of the individual sections. To conclude this introduction there are the following few remarks.

(i) In these notes we ignore essentially all set-theoretic issues (with the exception of the discussion of locally presentable categories where some care is needed).

(ii) For many of the mathematical concepts to be introduced below, there are at least two different terminologies (most frequently, one due to Joyal and one due to Lurie). Since we do not want to cause further confusion, we have to stick to one of these possible choices. As the expanded version of Lurie's thesis [82, 83] is our main reference, we stick to Lurie's terminology.

(iii) In these notes we use the environment 'Perspective' in order to include some sketches of 'the larger picture' and also to give more references to the literature.

(iv) Finally, the main prerequisites for these notes are some acquaintance with key concepts from category theory [85] as well as basics concerning simplicial sets. References for simplicial sets include the monographs [43, 46] and the more introductory account [41].

14.2 Two models for $(\infty, 1)$-categories

In this section we define ∞-*categories* as simplicial sets satisfying certain horn extension properties. These extension properties are a common generalization of extension properties enjoyed by singular complexes of topological spaces and nerves of ordinary categories. We indicate why this notion gives us a model for the theory of $(\infty, 1)$-categories. While ∞-categories have mapping spaces which can be endowed with a coherently associative and unital composition law, there is also the more rigid approach to $(\infty, 1)$-categories based on *simplicial categories* coming with strictly associative and unital composition laws. These two approaches are respectively organized by means of model structures, the *Joyal model structure* in the case of ∞-categories and the *Bergner model structure* in the case of simplicial categories. The coherent nerve construction of Cordier can be shown to be part of a Quillen equivalence between these two approaches.

14.2.1 Basics on ∞-categories

Before we give the central definition of an ∞-category, we consider two classes of examples, which one definitely wants to be covered by the definition. The actual definition of an ∞-category will then be a common generalization of these two classes of examples. The first class of examples comes from spaces.

Example 14.2.1. Given a topological space X, recall that associated to X there is the *fundamental groupoid* $\pi_1(X)$[1] of X. The objects of $\pi_1(X)$ are the points of X, and morphisms from x to y are homotopy classes of paths from x to y relative to the boundary points. Note that this is a groupoid, i.e., that all morphisms are invertible, since every path admits an inverse up to homotopy. This fundamental groupoid only depends on the 1-type of X, and hence discards a lot of information. A refined version is given by the *fundamental* ∞-*groupoid* $\pi_{<\infty}(X)$ which is roughly constructed as follows: objects are given by points of X, morphisms are paths in X, 2-morphisms are homotopies between paths, and higher morphisms are given by higher homotopies.

Before giving a precise definition of the fundamental groupoid, we include some heuristic comments. Note that $\pi_{<\infty}(X)$ seems to be an ∞-groupoid in that all morphisms are equivalences, i.e., invertible in a certain weak sense — this justifies that we refer to it as fundamental ∞-*groupoid* as opposed to as fundamental ∞-*category*. The following is a generally accepted principle of higher category theory.

'Spaces and ∞-groupoids should be the same.'

This principle is referred to as the *Grothendieck homotopy hypothesis*. Instead of working with topological spaces, one could also consider 'simplicial models for spaces', more specifically *Kan complexes* (see Definition 14.2.5). If one formalizes ∞-categories in the framework

[1] We deviate from the common notation $\Pi(X)$ since $\pi_1(X)$ matches better with the notation in Example 14.2.9.

of simplicial sets (by way of Definition 14.2.6), then the above principle tells us that ∞-groupoids should be the same as Kan complexes. Corollary 14.2.18 turns this principle into an actual mathematical statement. Using this approach to ∞-categories, a model for the fundamental ∞-groupoid of a space X is the usual singular complex $\mathrm{Sing}(X)$, which is well-known to be a Kan complex.

Let us establish some basic notation related to simplicial sets. Recall that the category $s\mathcal{S}et$ of simplicial sets is defined as the category of functors $\Delta^{\mathrm{op}} \to \mathcal{S}et$ where Δ is the category of finite ordinals

$$[n] = (0 < \cdots < n), \quad n \geq 0,$$

and order-preserving maps. Prominent maps in Δ are the coface maps d^k and codegeneracy maps s^k. The map $d^k\colon [n-1] \to [n]$, $0 \leq k \leq n$, is the unique injective map that does not have k in its image, while $s^k\colon [n+1] \to [n], 0 \leq k \leq n$, is the unique surjective map that hits k twice. We follow the usual convention and write $X_n = X([n])$ for the values of a simplicial set X, $d_k = X(d^k)$ for the face maps, and $s_k = X(s^k)$ for the degeneracy maps.

Obviously, ordinary categories should be examples of ∞-categories: we imagine all higher morphisms to be identities. To make this precise, we recall the *nerve construction* which associates a simplicial set to a category.

Example 14.2.2. Given a category $\mathcal{C}$, one can form the simplicial set $N(\mathcal{C}) \in s\mathcal{S}et$, called the *nerve* of $\mathcal{C}$. By definition, we have $N(\mathcal{C})_n = \mathrm{Fun}([n], \mathcal{C})$, where $[n]$ also denotes the ordinal number $0 < \cdots < n$ considered as a category and $\mathrm{Fun}(-,-)$ is the set of functors. Thus $N(\mathcal{C})_n$ is essentially the set of strings of n composable morphisms in $\mathcal{C}$. As a special case we have $N([n]) \cong \Delta^n, n \geq 0$, where $\Delta^n \in s\mathcal{S}et$ denotes the usual simplicial n-simplex, i.e., $\Delta^n = \hom_\Delta(-,[n])$ is the simplicial set represented by $[n] \in \Delta$.

Lemma 14.2.3. *The nerve functor $N\colon \mathcal{C}at \to s\mathcal{S}et$ is fully faithful and hence induces an equivalence onto its essential image.*

In order to describe the essential image one observes that nerves of categories enjoy certain *horn extension properties*. Recall that the k-th *n-horn* $\Lambda^n_k \subseteq \partial\Delta^n$ for $n \geq 1, 0 \leq k \leq n$, is obtained from $\partial\Delta^n$ by removing the k-th face $\partial_k\Delta^n$, i.e., the face opposite to vertex k. More formally, the horn Λ^n_k is defined as the following coequalizer

$$\coprod_{0 \leq i < j \leq n} \Delta^{n-2} \rightrightarrows \coprod_{i \neq k} \Delta^{n-1} \to \Lambda^n_k.$$

(See for example [46, p.9] which is a great reference for many more advanced aspects of simplicial homotopy theory.)

In dimension $n = 2$, horns $\lambda\colon \Lambda^2_k \to N(\mathcal{C})$ for $0 \leq k \leq 2$ hence respectively look like

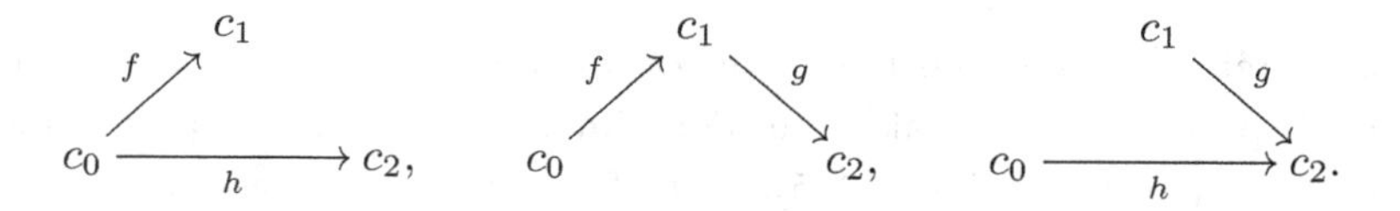

Using the composition $h = g \circ f$ we see that one can uniquely extend any horn $\lambda\colon \Lambda^2_1 \to N(\mathcal{C})$ to an entire 2-simplex $\sigma\colon \Delta^2 \to N(\mathcal{C})$, i.e., there is a unique dashed arrow making the

diagram

$$\begin{array}{ccc} \Lambda^2_1 & \longrightarrow & N(\mathcal{C}) \\ \downarrow & \nearrow_{\exists ! \sigma} & \\ \Delta^2 & & \end{array}$$

commute. The composition is given by the new face $d_1(\sigma)\colon \Delta^1 \to N(\mathcal{C})$. If instead we consider a horn $\lambda\colon \Lambda^2_0 \to N(\mathcal{C})$ in the special case that $h = \mathrm{id}$ is an identity morphism, then the existence of an extension to a 2-simplex is equivalent to the existence of a left inverse to f. Similar observations can be made for horns $\lambda\colon \Lambda^2_2 \to N(\mathcal{C})$. This different behavior extends to higher dimensions and motivates the following terminology: the horns $\Lambda^n_k, 0 < k < n$, are *inner horns* while the extremal cases Λ^n_0 and Λ^n_n are *outer horns*. It turns out that these horn extension properties are suitable to describe the essential image of the nerve functor. We denote by $\mathcal{G}rpd$ the category of groupoids.

Proposition 14.2.4. *Let X be a simplicial set.*

(i) X is isomorphic to the nerve of a category if and only if every inner horn $\Lambda^n_k \to X, 0 < k < n$, *can be* uniquely *extended to an n-simplex* $\Delta^n \to X$.

(ii) X is isomorphic to the nerve of a groupoid if and only if every horn $\Lambda^n_k \to X, 0 \leq k \leq n$, *can be* uniquely *extended to an n-simplex* $\Delta^n \to X$.

The characterization of the essential image of the nerve functor $N\colon \mathcal{G}rpd \to s\mathcal{S}et$ reminds us of the notion of a Kan complex whose definition will be recalled here for convenience.

Definition 14.2.5. A simplicial set X is a *Kan complex* if every horn $\Lambda^n_k \to X$ for $0 \leq k \leq n$ can be extended to an n-simplex $\Delta^n \to X$.

Denoting by $\mathcal{K}an \subseteq s\mathcal{S}et$ the full subcategory spanned by the Kan complexes, we thus have the following commutative diagram of fully faithful functors

$$\begin{array}{ccc} \mathcal{G}rpd & \longrightarrow & \mathcal{C}at \\ {\scriptstyle N}\downarrow & & \downarrow{\scriptstyle N} \\ \mathcal{K}an & \longrightarrow & s\mathcal{S}et. \end{array} \tag{14.2.1}$$

As a common generalization of Kan complexes and nerves of small categories there is the following definition.

Definition 14.2.6. A simplicial set $\mathcal{C}$ is an ∞-*category* if every inner horn $\Lambda^n_k \to \mathcal{C}$, $0 < k < n$, can be extended to an n-simplex $\Delta^n \to \mathcal{C}$.

Thus, by the above, spaces and ordinary categories give rise to ∞-categories. We will soon see that there are interesting ∞-categories that are not of this form and hence provide some first 'honest examples'. In particular, every simplicial model category has an underlying ∞-category but there are also other examples (see Corollary 14.2.30, Examples 14.2.31, and Perspective 14.3.7).

We begin by introducing some basic terminology. Given an ∞-category $\mathcal{C}$, the *objects* are the vertices $x \in \mathcal{C}_0$ and the *morphisms* are the 1-simplices $f \in \mathcal{C}_1$. The face map $s = d_1\colon \mathcal{C}_1 \to \mathcal{C}_0$ is the *source map*, and $t = d_0\colon \mathcal{C}_1 \to \mathcal{C}_0$ is the *target map*. As in ordinary category theory, we write $f\colon x \to y$ if $s(f) = x$ and $t(f) = y$. Slightly more formally, we define the set of morphisms $\hom_{\mathcal{C}}(x, y)$ from x to y as the pullback

$$\begin{array}{ccc} \hom_{\mathcal{C}}(x,y) & \longrightarrow & \mathcal{C}_1 \\ \downarrow & \lrcorner & \downarrow{\scriptstyle (s,t)} \\ * & \xrightarrow[(x,y)]{} & \mathcal{C}_0 \times \mathcal{C}_0. \end{array}$$

It turns out that associated to two objects in an ∞-category there is an entire *space of morphisms* (see Remark 14.2.13).

The degeneracy map $s_0\colon \mathcal{C}_0 \to \mathcal{C}_1$ sends an object x to the *identity map* $\mathrm{id}_x = s_0(x)$ of x. The simplicial identities $d_0 s_0 = d_1 s_0 = \mathrm{id}_{\mathcal{C}_0}$ imply that id_x is an endomorphism, and the terminology and notation will be justified by Proposition 14.2.12.

As for compositions, let us consider morphisms $f\colon x \to y$ and $g\colon y \to z$ in an ∞-category $\mathcal{C}$. These morphisms together define an inner horn in $\mathcal{C}$,

$$\lambda = (g, \bullet, f)\colon \Lambda^2_1 \to \mathcal{C},$$

such that $d_0\lambda = g$ and $d_2\lambda = f$. Any such horn can be *non-uniquely* extended to a 2-simplex $\sigma\colon \Delta^2 \to \mathcal{C}$. The new face $d_1(\sigma)$ opposite to vertex 1 is then a *candidate composition* of g and f. To re-emphasize, this is one of the central points in which ∞-category theory differs from ordinary category theory: one does not ask for uniquely determined compositions. Instead one demands only that there is a way to compose arrows and that any choice of such a composition is equally good: the space of all such choices is to be contractible (see the discussion of Theorem 14.2.14 for a precise statement).

We now want to describe the homotopy category of an ∞-category. This can be done in a more straightforward way, but we prefer to include a short digression in category theory as this allows us to mention a general fact which is in the background of a later construction anyhow.

Digression 14.2.7. (Yoneda extension.) Let A be a small category and let us consider the associated presheaf category $\mathrm{Fun}(A^{\mathrm{op}}, \mathcal{S}et)$, i.e., the category of contravariant set-valued functors on A. Moreover, let $\mathcal{C}$ be a cocomplete category and let us be given a functor $Q\colon A \to \mathcal{C}$. Thus, we are in the situation

$$\begin{array}{ccc} A & \xrightarrow{Q} & \mathcal{C} \\ {\scriptstyle y}\downarrow & \nearrow & \\ \mathrm{Fun}(A^{\mathrm{op}}, \mathcal{S}et) & & \end{array}$$

where y denotes the Yoneda embedding of A. Recall from [85, p.76] that every presheaf on a small category is canonically a colimit of representable ones (see also §14.4.1). The cocompleteness of $\mathcal{C}$ hence allows us to extend Q to a colimit-preserving functor

$$|-|_Q\colon \mathrm{Fun}(A^{\mathrm{op}}, \mathcal{S}et) \to \mathcal{C}.$$

(At a more conceptual level, we are thus forming the left Kan extension of Q along y.) Moreover, associated to Q, there is also a functor in the opposite direction

$$\mathrm{Sing}_Q(-)\colon \mathcal{C} \to \mathrm{Fun}(A^{\mathrm{op}}, \mathcal{S}et),$$

that is defined by $\mathrm{Sing}_Q(c)_a = \hom_{\mathcal{C}}(Qa, c)$. One observes now that this pair of functors defines an adjunction

$$(|-|_Q, \mathrm{Sing}_Q(-))\colon \mathrm{Fun}(A^{\mathrm{op}}, \mathcal{S}et) \rightleftarrows \mathcal{C}$$

with $|-|_Q$ as left adjoint and $\mathrm{Sing}_Q(-)$ as right adjoint. (Related to this see also the general discussion in §14.4.1 and in particular Theorem 14.4.5).

But even more is true: For every cocomplete category $\mathcal{C}$ the assignment that sends Q to the adjunction $(|-|_Q, \mathrm{Sing}_Q(-))$ defines an equivalence of categories

$$\mathrm{Fun}(A, \mathcal{C}) \xrightarrow{\simeq} \mathrm{Adj}\left(\mathrm{Fun}(A^{\mathrm{op}}, \mathcal{S}et), \mathcal{C}\right).$$

Here, $\mathrm{Adj}(-,-)$ is the category of adjunctions where objects are adjunctions and morphisms are, say, natural transformations of left adjoints. This construction is sometimes referred to as *Yoneda extension* since it is essentially given by left Kan extension along the Yoneda embedding. A more detailed discussion of this can for example be found in [72, pp.62-64].

In what follows we are only interested in the special case where $A = \Delta$, the category of finite ordinals. Thus, we conclude that a cosimplicial object $\Delta \to \mathcal{C}$ in a cocomplete category is equivalently specified by an adjunction $s\mathcal{S}et \rightleftarrows \mathcal{C}$. The notation employed in Digression 14.2.7 is, of course, motivated by the following example.

Example 14.2.8. Let $\mathcal{C} = \mathcal{T}op$ be the category of topological spaces and let us consider the standard cosimplicial space $|\Delta^\bullet|\colon \Delta \to \mathcal{T}op$. The associated adjunction is the usual adjunction given by the geometric realization and the singular complex functor,

$$(|-|, \mathrm{Sing})\colon s\mathcal{S}et \rightleftarrows \mathcal{T}op.$$

Example 14.2.9. Let $\mathcal{C} = \mathcal{C}at$ be the cocomplete category of small categories (see Remark 14.2.22 for the fact that $\mathcal{C}at$ is cocomplete). The inclusion $\Delta \to \mathcal{C}at$ obtained by considering a finite ordinal as a category induces by Digression 14.2.7 an adjunction

$$(\tau_1, N)\colon s\mathcal{S}et \rightleftarrows \mathcal{C}at$$

with right adjoint the nerve functor. The left adjoint τ_1 is the *fundamental category functor* or the *categorical realization functor*. The motivation for the first terminology and the notation τ_1 stems from the following. Composition with the groupoidification $\mathcal{C}at \to \mathcal{G}rpd$, i.e., the left adjoint to the forgetful functor $\mathcal{G}rpd \to \mathcal{C}at$ yields the usual *fundamental groupoid functor* $\pi_1\colon s\mathcal{S}et \to \mathcal{G}rpd$. Thus associated to the cosimplicial object $\Delta \to \mathcal{G}rpd$ which sends $[n]$ to the free groupoid on $[n]$ we obtain the adjunction

$$(\pi_1, N)\colon s\mathcal{S}et \rightleftarrows \mathcal{G}rpd.$$

If the simplicial set happens to be an ∞-category, then there is a simplification of the description of the fundamental category. It turns out that morphisms can be represented by actual 1-simplices as we shall discuss now (Proposition 14.2.12). For that purpose, we need the following definition.

Definition 14.2.10. Two morphism $f, g\colon x \to y$ in an ∞-category $\mathcal{C}$ are *homotopic* (notation: $f \simeq g$) if there is a 2-simplex $\sigma\colon \Delta^2 \to \mathcal{C}$ whose boundary $\partial\sigma = (d_0\sigma, d_1\sigma, d_2\sigma)$ is given by (g, f, id_x), i.e., the boundary looks like

$$\begin{array}{ccc} & x & \\ {}^{\mathrm{id}_x}\nearrow & & \searrow^{g} \\ x & \xrightarrow[f]{} & y. \end{array} \qquad (14.2.2)$$

Any such 2-simplex σ is a *homotopy* from f to g, denoted $\sigma\colon f \to g$.

There is a similar notion of homotopies where the identity morphism sits on the face opposite to vertex zero. However, using the inner horn extension property both notions can be shown to be the same. Moreover, there is the following result.

Proposition 14.2.11. *Let $\mathcal{C}$ be an ∞-category and let $x, y \in \mathcal{C}$. The homotopy relation is an equivalence relation on* $\hom_{\mathcal{C}}(x, y)$. *The* homotopy class *of a morphism $f\colon x \to y$ is denoted by $[f]$.*

We include a partial proof in order to give an idea of how this works. Associated to a morphism $f\colon x \to y$, let us consider $\kappa_f = s_0 f\colon \Delta^2 \to \mathcal{C}$. The simplicial identities imply that $d_0\kappa_f = d_1\kappa_f = f$ and also $d_2\kappa_f = d_2 s_0 f = s_0 d_1 f = \mathrm{id}_x$. Thus, the boundary of κ_f is $\partial\kappa_f = (f, f, \mathrm{id}_x)$ and we receive a homotopy $\kappa_f\colon f \to f$, the *constant homotopy* of f, which establishes the reflexivity of $\simeq$. For the symmetry one way to proceed is as follows. Given a homotopy $\sigma\colon f \to g$, let us form the inner horn

$$(\sigma, \kappa_g, \bullet, \kappa_{\mathrm{id}_x})\colon \Lambda^3_2 \to \mathcal{C} \qquad (14.2.3)$$

in $\mathcal{C}$. By definition of an ∞-category this horn can be extended to a 3-simplex $\tau\colon \Delta^3 \to \mathcal{C}$. The new face $\tilde{\sigma} = d_2\tau \in \mathcal{C}_2$ defines a homotopy $\tilde{\sigma}\colon g \to f$, an *inverse homotopy* of σ, showing that $\simeq$ is symmetric.

With the homotopy relation at our disposal, we would like to define the *homotopy category* $\mathrm{Ho}(\mathcal{C})$ of an ∞-category $\mathcal{C}$ by passing to homotopy classes of morphisms. The composition law in $\mathrm{Ho}(\mathcal{C})$ is obtained by representing homotopy classes by morphisms in $\mathcal{C}$, *choosing candidate compositions* of the representatives, and then passing to homotopy classes again. Of course, in order to get a well-defined category there are a lot of things to be checked, but we content ourselves by showing that all candidate compositions are homotopic. Let us consider morphisms $f\colon x \to y$ and $g\colon y \to z$ together with 2-simplices $\sigma_1, \sigma_2\colon \Delta^2 \to \mathcal{C}$ witnessing that $h_1 = d_1(\sigma_1)$ and $h_2 = d_1(\sigma_2)$ are candidate compositions of g and f. Then we can form the inner horn

$$(\sigma_1, \sigma_2, \bullet, \kappa_f)\colon \Lambda^3_2 \to \mathcal{C}$$

in $\mathcal{C}$. Again, we can find an extension to a 3-simplex $\tau\colon \Delta^3 \to \mathcal{C}$ and the new face $d_2\tau\colon \Delta^2 \to \mathcal{C}$ gives us the desired homotopy $h_2 \to h_1$. Using similar arguments, one can establish the following result which can already be found in [23].

Proposition 14.2.12. *Let $\mathcal{C}$ be an ∞-category. There is an ordinary category* $\mathrm{Ho}(\mathcal{C})$*, the* homotopy category *of $\mathcal{C}$, with the same objects as $\mathcal{C}$ and morphisms the homotopy classes of morphisms in $\mathcal{C}$. Composition and identities are given by*

$$[g] \circ [f] := [g \circ f] \qquad \textit{and} \qquad \mathrm{id}_x := [\mathrm{id}_x] = [s_0 x],$$

where $g \circ f$ is an arbitrary *candidate composition of g and f. Furthermore, there is a natural isomorphism of categories* $\mathrm{Ho}(\mathcal{C}) \cong \tau_1(\mathcal{C})$.

Remark 14.2.13. (i) One guiding principle for the theory of ∞-categories is that there should be a way to compose arrows and that the space of all such choices is contractible. Using the extension property for inner 2-horns, it is immediate that the space is non-empty. By means of the extension property for inner horns up to dimension three, we just checked that two candidate compositions are homotopic, i.e., that the space of all choices is connected. But this is only a π_0-statement of something much stronger: The extension property with respect to higher-dimensional inner horns can be thought of as guaranteeing the higher connectivity of the space of all such choices, giving finally that it is weakly contractible. See the discussion of Theorem 14.2.14 for a precise statement.

(ii) A second guiding principle for the theory of ∞-categories is that there should be morphisms of arbitrary dimensions. Let $\mathcal{C}$ be an ∞-category and let x, y be objects in $\mathcal{C}$. Then a morphism $f\colon x \to y$ is given by

$$f\colon \Delta^1 \to \mathcal{C} \qquad \text{such that} \qquad f|_{\Delta^{\{0\}}} = x \qquad \text{and} \qquad f|_{\Delta^{\{1\}}} = y.$$

Here and in the sequel, the notation is as follows: for vertices $i_0, \ldots, i_k$ in Δ^n, $\Delta^{\{i_0,\ldots,i_k\}} \subseteq \Delta^n$ denotes the k-simplex of Δ^n spanned by the given vertices. Moreover, given a vertex $x \in \mathcal{C}$ we write $x\colon \Delta^{\{i_0,\ldots,i_k\}} \to \mathcal{C}$ for the constant map with value x. A homotopy between two parallel morphisms $x \to y$ in $\mathcal{C}$ can be interpreted as a *2-morphism* from x to y. Recall that a homotopy is given by

$$\sigma\colon \Delta^2 \to \mathcal{C} \qquad \text{such that} \qquad \sigma|_{\Delta^{\{0,1\}}} = x \qquad \text{and} \qquad \sigma|_{\Delta^{\{2\}}} = y.$$

This can be generalized to higher dimensions: an *n-morphism* from x to y is a map of simplicial sets

$$\tau\colon \Delta^{n+1} \to \mathcal{C} \qquad \text{such that} \qquad \tau|_{\Delta^{\{0,\ldots,n\}}} = x \qquad \text{and} \qquad \tau|_{\Delta^{\{n+1\}}} = y.$$

For varying n, the sets of n-morphisms can be assembled in a *space of morphisms* $\mathrm{Map}^R_{\mathcal{C}}(x, y) \in s\mathcal{S}et$ which can be shown to be a Kan complex.

We already mentioned that there is a variant to our definition of a homotopy (obtained by choosing the identity morphism in (14.2.2) to sit opposite to vertex zero). More generally, there is an obvious dual way to define a space of morphisms $\mathrm{Map}^L_{\mathcal{C}}(x, y)$ which turns out to be a weakly equivalent Kan complex [82, Cor. 4.2.1.8]. Thus, the homotopy type of the mapping space is well-defined.

(iii) A third guiding principle for the theory of ∞-categories is that they should give a model for $(\infty,1)$-categories, i.e., all higher morphisms should be invertible in some weak sense. To indicate that we succeeded in establishing such a framework, let us consider a homotopy σ in an ∞-category $\mathcal{C}$,

$$\sigma\colon f \simeq g\colon x \to y.$$

In order to establish the symmetry of the homotopy relation, we considered the inner horn (14.2.3) which can be extended to a 3-simplex $\tau\colon \Delta^3 \to \mathcal{C}$. The new face $\widetilde{\sigma} = d_2\tau$ then gives us the intended homotopy $\widetilde{\sigma}\colon g \simeq f$. Note that τ satisfies $\tau|_{\Delta^{\{0,1,2\}}} = x$ and τ is hence a *3-morphism*, which can be interpreted as a 2-homotopy

$$\tau\colon \kappa_g \simeq \widetilde{\sigma} \circ \sigma.$$

Thus, every homotopy has (up to a 2-homotopy) a left inverse and a similar observation can be made for right inverses. Taking for granted that the horn extension property for higher dimensional horns allows us to deduce similar observations for higher homotopies, we are reassured that ∞-categories really provide a model for $(\infty,1)$-categories.

The following result due to Joyal [69, Prop. 2.24] makes precise that we succeeded in finding an axiomatic framework for categories with compositions determined up to contractible choices. Let $i\colon \Lambda^2_1 \to \Delta^2$ be the obvious inclusion. Moreover, let us denote by $\mathrm{Map}(-,-)\colon s\mathcal{S}et^{\mathrm{op}} \times s\mathcal{S}et \to s\mathcal{S}et$ the simplicial mapping space functor,

$$\mathrm{Map}(X,Y)_\bullet = \hom_{s\mathcal{S}et}(\Delta^\bullet \times X, Y), \tag{14.2.4}$$

so that vertices are maps, edges are homotopies, and higher dimensional simplices are 'higher homotopies' (see e.g. [46, p.20]).

Theorem 14.2.14. *A simplicial set X is an ∞-category if and only if the restriction map $i^*\colon \mathrm{Map}(\Delta^2, X) \to \mathrm{Map}(\Lambda^2_1, X)$ is an acyclic Kan fibration.*

We can think of $\mathrm{Map}(\Lambda^2_1, X)$ as the *space of composition problems* and similarly of $\mathrm{Map}(\Delta^2, X)$ as the *space of solutions to composition problems.* The theorem then tells us that the *defining feature* of an ∞-category is that these two spaces are the same from a homotopical perspective, and this motivates us to henceforth suppress the 'candidate' in 'candidate composition'.

We now turn to equivalences in an ∞-category.

Definition 14.2.15. A morphism $f\colon x \to y$ in an ∞-category $\mathcal{C}$ is an *equivalence* if $[f]\colon x \to y$ is an isomorphism in $\mathrm{Ho}(\mathcal{C})$.

It is immediate that identities are equivalences and that for two homotopic morphisms $f_1 \simeq f_2$ we have that f_1 is an equivalence if and only if f_2 is one. Moreover, it turns out that a morphism $f\colon x \to y$ in $\mathcal{C}$ is an equivalence if and only if there is a morphism $g\colon y \to x$

in $\mathcal{C}$ such that there are 2-simplices with boundaries as in

In a way one could be surprised that we can characterize equivalences in an ∞-category by these two conditions. Since ∞-category theory is some sort of 'coherent category theory' one might have expected that also higher coherence data would be necessary to characterize equivalences in an ∞-category. For a precise statement and a proof of the equivalence of these two potentially different invertibility conditions we refer to [68, Corollary 1.6] and [34, Proposition 2.2].

We mentioned already the accepted principle that all *∞-groupoids* should come from spaces. In order to make this explicit let us give the following definition.

Definition 14.2.16. An ∞-category is an *∞-groupoid* if the homotopy category is a groupoid.

Thus an ∞-category is an ∞-groupoid if and only if all morphisms are equivalences. In the motivation of the definition of an ∞-category, we saw that, in general, one should only demand the horn extension property for inner horns in order to obtain a good generalization of arbitrary categories (and not just of groupoids!). Joyal established the following result, saying that outer horns can be extended as soon as certain maps are equivalences.

Proposition 14.2.17. *Let $\mathcal{C}$ be an ∞-category. Any horn $\lambda\colon \Lambda_0^n \to \mathcal{C}$, $n \geq 2$, such that $\lambda|_{\Delta^{\{0,1\}}}$ is an equivalence can be extended to a simplex $\Delta^n \to \mathcal{C}$.*

There is of course a similar statement using the horns Λ_n^n instead. This allows us to turn the principle that all ∞-groupoids should be given by spaces into the following precise statement (see [68, Corollary 1.4] or [82, p.35]).

Corollary 14.2.18. *An ∞-category is an ∞-groupoid if and only if it is a Kan complex.*

With this result at hand, diagram (14.2.1) consisting of fully faithful functors can be refined to

$$\begin{array}{ccccc} \mathcal{G}rpd & \longrightarrow & \mathcal{C}at & \xrightarrow{N} & \\ \downarrow N & & \downarrow N & & \searrow \\ \mathcal{K}an = \mathcal{G}rpd_\infty & \longrightarrow & \mathcal{C}at_\infty & \longrightarrow & s\mathcal{S}et. \end{array} \tag{14.2.5}$$

Perspective 14.2.19. As in Corollary 14.2.18, for low values of $n \in \mathbb{N}$ there are statements using *n-types of spaces* and *n-groupoids* in a certain precise sense. The statements that higher homotopy types should be classified by higher groupoids is frequently also referred to as the *homotopy hypothesis.*

In the case of $n = 1$, a precise statement can for example be found in [58] where it is shown that such a classification is induced by the adjunction $(\pi_1, N)\colon s\mathcal{S}et \rightleftarrows \mathcal{G}rpd$. In fact, this adjunction can be seen to be a Quillen adjunction with respect to the Kan–Quillen model structure on $s\mathcal{S}et$ and the so-called *natural model stucture* on $\mathcal{G}rpd$ (related to this see

[70] or the nice short account in [93]). The slogan that 'groupoids do not carry any higher homotopical information' can be made precise as follows: the Quillen adjunction (π_1, N) induces a Quillen equivalence between the S^2-nullification of $s\mathcal{S}et$ and $\mathcal{G}rpd$. Related results in the cases of $n = 2, 3$, more precisely, in the context of *bicategories* [13] and *Gray categories* [47, 77], are made explicit in [76, §6].

14.2.2 Simplicial categories and the relation to ∞-categories

There are many alternative approaches to a theory of $(\infty, 1)$-categories including simplicial categories [14], Segal categories [57], and complete Segal spaces [94]. Besides in the original references, more details can for example be found in [15, 16, 107, 1] and in Bergner's chapter in this Handbook. Here we include a short discussion of *simplicial categories* or, more precisely, *simplicially enriched categories*. Given two objects x, y in a simplicial category $\mathcal{C}$, we write $\mathrm{Map}_{\mathcal{C}}(x, y)$ for the associated simplicial mapping space. This more rigid approach — coming with a specified strictly associative and unital composition law — gives us, by definition, a notion of a category with morphisms of arbitrary dimensions. Building on work of Joyal and Bergner, Lurie has shown that this approach and the one using ∞-categories are equivalent in a very precise sense (see Theorem 14.2.29).

We begin by describing a relation between simplicial sets and simplicial categories. First, let us recall that the nerve $N(\mathcal{C})$ of an ordinary category $\mathcal{C}$ is the simplicial set

$$N(\mathcal{C})_\bullet = \hom_{\mathcal{C}at}([\bullet], \mathcal{C}),$$

where $[\bullet]\colon \Delta \to \mathcal{C}at$ is obtained by considering the finite ordinals $[n]$ as categories. Given a *simplicial* category $\mathcal{C}$, we could simply forget the simplicial enrichment and form the nerve of the underlying ordinary category. More precisely, if we denote by $s\mathcal{C}at$ the category of (small) simplicial categories and simplicial functors, then there is the forgetful functor $s\mathcal{C}at \to \mathcal{C}at$ which we could compose with the ordinary nerve functor $\mathcal{C}at \to s\mathcal{S}et$. But this approach obviously discards too much information and instead one should proceed differently.

A better way is given by replacing $[n] \in \mathcal{C}at$ by *simplicially thickened versions* $C[\Delta^n] \in s\mathcal{C}at$ and then building the simplicial set

$$N_\Delta(\mathcal{C})_\bullet = \hom_{s\mathcal{C}at}(C[\Delta^\bullet], \mathcal{C}),$$

where $\hom_{s\mathcal{C}at}(-,-)$ denotes the set of simplicial functors. The idea behind this simplicial thickening is that $C[\Delta^n]$ encodes as objects the vertices of the standard simplex Δ^n, as morphisms all paths in increasing direction, as 2-morphisms all homotopies, and so on in higher dimensions. More conceptual comments about this construction can be found in Perspective 14.2.25. Before we give a precise definition of $C[\Delta^\bullet]$ let us describe what we expect to obtain in low dimensions.

Example 14.2.20. In dimensions 0 and 1 nothing new happens, and the simplicial categories $C[\Delta^0]$ and $C[\Delta^1]$ are just the ordinary categories [0] and [1], respectively, considered as simplicial categories with discrete mapping spaces. Thus, the pictures we have in mind are

$$C[\Delta^0]\colon\quad 0 \qquad\text{and}\qquad C[\Delta^1]\colon\quad 0 \to 1.$$

But from dimension 2 the simplicial picture is richer. In Δ^2, there are two ways to pass from 0 to 2, namely the straight path and the path passing through 1. These paths should be encoded in $C[\Delta^2]$ together with a homotopy between them. The simplicial category $C[\Delta^2]$ can hence be depicted by

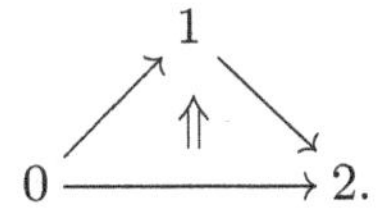

$$\begin{array}{ccc} & 1 & \\ \nearrow & \Uparrow & \searrow \\ 0 & \longrightarrow & 2. \end{array} \tag{14.2.6}$$

We now give a precise definition of $C[\Delta^n]$. The objects of $C[\Delta^n]$ are the numbers $0, 1, \ldots, n$. The strategy behind the definition of the simplicial mapping spaces is the following. Given objects $i \leq j$ we encode a path from i to j by specifying the vertices of the corresponding path. Thus, let $P_{i,j}$ be the poset

$$P_{i,j} = \{I \subseteq [i,j] \mid i,j \in I\}$$

ordered by inclusion where $[i,j]$ is short hand notation for $\{i, i+1, \ldots, j-1, j\}$. Considering these posets as categories, we can define the simplicial mapping spaces in $C[\Delta^n]$ by

$$\mathrm{Map}_{C[\Delta^n]}(i,j) = \begin{cases} NP_{i,j}, & i \leq j, \\ \emptyset, & i > j. \end{cases}$$

The composition is induced by the union of subsets, which fits fine with the strategy to encode a path by specifying the vertices one passes along. It is also immediate that identities are given by the singletons $\{i\}$. This concludes the definition of $C[\Delta^n] \in s\mathcal{C}at$.

One easily checks, that this definition specializes to the pictures we had in mind in low dimensions (14.2.20). For example in dimension $n = 2$, there is the following table of non-degenerate k-simplices in the mapping spaces $\mathrm{Map}_{C[\Delta^2]}(i,j)$:

k	$i = j = 0$	$i = 0, j = 1$	$i = 0, j = 2$
0	$\{0\}$	$\{0,1\}$	$\{0,2\}$, $\{0,1,2\}$
1			$\{0,2\} \subseteq \{0,1,2\}$

By definition $\{0,1,2\} = \{1,2\} \circ \{0,1\}$ is the composition and we see that the non-degenerate 1-simplex in $\mathrm{Map}_{C[\Delta^2]}(0,2)$ encodes the homotopy

$$\{0,2\} \to \{1,2\} \circ \{0,1\}$$

we were aiming for in (14.2.6).

It is straightforward to check that the assignment $[n] \mapsto C[\Delta^n]$ defines a cosimplicial object $C[\Delta^\bullet] \colon \Delta \to s\mathcal{C}at$. This allows us to give the following definition which appears to be due to Cordier [25].

Definition 14.2.21. The *coherent nerve* $N_\Delta(\mathcal{C})$ of a simplicial category $\mathcal{C}$ is the simplicial set

$$N_\Delta(\mathcal{C})_\bullet = \hom_{s\mathcal{C}at}(C[\Delta^\bullet], \mathcal{C}).$$

Thus we have a coherent nerve functor $N_\Delta\colon s\mathcal{C}at \to s\mathcal{S}et$. By the very definition, this coherent nerve construction takes into account the higher structure on a simplicial category given by the mapping spaces. For example a 2-simplex in such a coherent nerve is given by a homotopy as depicted in

$$\begin{array}{ccc} & y & \\ \nearrow & \Uparrow & \searrow \\ x & \longrightarrow & z. \end{array}$$

Note that such a 2-simplex is, in general, not determined by its restriction to the horn $\Lambda^2_1 \subset \Delta^2$.

Using the observation that $s\mathcal{C}at$ is cocomplete (see Remark 14.2.22) we can extend the cosimplicial object $\Delta \to s\mathcal{C}at\colon [n] \mapsto C[\Delta^n]$ to a colimit-preserving functor $C[-]\colon s\mathcal{S}et \to s\mathcal{C}at$. More explicitly, for $X \in s\mathcal{S}et$ we make the definition

$$C[X] = \mathrm{colim}_{(\Delta/X)}\, C[-] \circ p \tag{14.2.7}$$

where (Δ/X) is the category of simplices of X and $p\colon (\Delta/X) \to \Delta$ is the canonical functor. It turns out that this extension defines a left adjoint to the coherent nerve N_Δ,

$$(C[-], N_\Delta)\colon s\mathcal{S}et \rightleftarrows s\mathcal{C}at. \tag{14.2.8}$$

(In fact, this can be considered as an example of *Yoneda extensions* in the sense of Digression 14.2.7.) We observe that the notation $C[-]$ is not in conflict with the notation $C[\Delta^n]$ for the simplicial thickening of $[n]$ since the colimit-preserving extension $C[-]$ applied to Δ^n is isomorphic to what we just defined. This follows because the category of simplices (Δ/Δ^n) has $([n], \mathrm{id}\colon \Delta^n \to \Delta^n)$ as terminal object so that the defining colimit in (14.2.7) simplifies accordingly.

Remark 14.2.22. The cocompleteness of $s\mathcal{C}at$ is a special instance of a more general result. Given a symmetric monoidal category $\mathcal{M}$ let us denote by $\mathcal{C}at_\mathcal{M}$ the category of (small) $\mathcal{M}$-enriched categories and $\mathcal{M}$-enriched functors. Thus in this notation we have $s\mathcal{C}at = \mathcal{C}at_{s\mathcal{S}et}$. It turns out that if $\mathcal{M}$ is complete and cocomplete then so is $\mathcal{C}at_\mathcal{M}$. The harder part is the cocompleteness and was established by Wolff in [116]. As important examples we deduce that the categories of categories, simplicial categories, topological categories, spectral categories, differential-graded categories, and 2-categories are complete and cocomplete.

As we explain next, the adjunction (14.2.8) is in fact a Quillen equivalence with respect to the *Joyal model structure* on $s\mathcal{S}et$ and the *Bergner model structure* on $s\mathcal{C}at$. Since we will not make an intensive use of the Bergner model structure, we do not go too much into detail and instead refer to [14]. Let us recall that every simplicial category $\mathcal{C}$ has an underlying *path component category* or *homotopy category* $\pi_0\mathcal{C}$. This is an ordinary category with the same objects while the sets of morphisms are obtained by applying π_0 to the simplicial mapping spaces. For example, if we endow $s\mathcal{S}et$ with the usual simplicial enrichment given by (14.2.4), then $\pi_0 s\mathcal{S}et$ is the naive homotopy category with all simplicial sets as objects and simplicial homotopy classes as morphisms.

We can now define the weak equivalences in the Bergner model structure.

Definition 14.2.23. A simplicial functor $F\colon \mathcal{C} \to \mathcal{D}$ is a *weak equivalence* if

(i) the induced functor $\pi_0 F\colon \pi_0\mathcal{C} \to \pi_0\mathcal{D}$ is essentially surjective and

(ii) for all objects $x, y \in \mathcal{C}$ the map $\mathrm{Map}_{\mathcal{C}}(x, y) \to \mathrm{Map}_{\mathcal{D}}(Fx, Fy)$ is a weak equivalence of simplicial sets (i.e., it induces a weak equivalence on geometric realizations).

Recall that a functor between ordinary categories is an equivalence if and only if it is essentially surjective and fully faithful. The definition of a weak equivalence between simplicial categories can be read as a higher categorical generalization of equivalences since it is asking that the simplicial functor is *homotopically essentially surjective* and *homotopically fully faithful.* Such a functor is also called a *Dwyer–Kan equivalence* , attributing credit to [37]. Obviously, such a weak equivalence $\mathcal{C} \to \mathcal{D}$ induces an equivalence $\pi_0\mathcal{C} \to \pi_0\mathcal{D}$ but having a weak equivalence is, in general, a much stronger statement.

Building on work of Dwyer and Kan, Bergner [14] established the following result.

Theorem 14.2.24. *The category $s\mathcal{C}at$ carries a left proper combinatorial model structure with the Dwyer–Kan equivalences as weak equivalences. With respect to this model structure, a simplicial category is fibrant if and only if it is* locally fibrant, *i.e., if all simplicial mapping spaces are Kan complexes.*

This model structure is referred to as the *Bergner model structure* and provides us with an example of a *homotopy theory of homotopy theories.* For more details about it we refer to [14].

Perspective 14.2.25. We have already seen that the functor $C[-]\colon s\mathcal{S}et \to s\mathcal{C}at$ is essentially determined by the cosimplicial object $C[\Delta^\bullet]\colon \Delta \to s\mathcal{C}at$ given by the 'simplicial thickenings' of the finite ordinals. These simplicial thickenings arise more conceptually as follows. If we consider the categories $[n]$ as discrete simplicial categories, then we obtain a cosimplicial object $[\bullet]\colon \Delta \to s\mathcal{C}at$. The Bergner model structure of Theorem 14.2.24 induces a Reedy model structure on the category $\mathrm{Fun}(\Delta, s\mathcal{C}at)$ of cosimplicial objects in $s\mathcal{C}at$. It turns out that $C[\Delta^\bullet]\colon \Delta \to s\mathcal{C}at$ gives us a Reedy cofibrant replacement of $[\bullet]\colon \Delta \to s\mathcal{C}at$. For an additional perspective on these simplicial thickenings we refer to Perspective 14.3.5.

With a view towards the Joyal model structure on $s\mathcal{S}et$, we make the following definition.

Definition 14.2.26. A map $f\colon X \to Y$ in $s\mathcal{S}et$ is a *categorical equivalence* if the induced simplicial functor $C[f]\colon C[X] \to C[Y]$ is a Dwyer–Kan equivalence.

This terminology is not the original one of Joyal. The maps in this definition are called *weak categorical equivalences* by Joyal [67], while he has a stronger notion of categorical equivalence. However, his notions of categorical equivalence and weak categorical equivalence coincide when only maps between ∞-categories are considered.

For simplicity, a categorical equivalence between ∞-categories is called an *equivalence of ∞-categories* and we say that the ∞-categories are *equivalent.* The fact that it suffices to consider direct equivalences as opposed to more complicated zig-zags is a consequence of the following important theorem of Joyal [67].

Theorem 14.2.27. *The category sSet carries a left proper combinatorial model structure with the monomorphisms as cofibrations and the categorical equivalences as weak equivalences. Moreover, a simplicial set is fibrant with respect to this model structure if and only if it is an ∞-category.*

The model structure of this theorem is referred to as the *Joyal model structure*. We add some comments on the fibrations in the Joyal model structure. For this we first recall that in the Kan–Quillen model structure on *sSet* the fibrations are the Kan fibrations. A Kan fibration $p\colon X \to S$ gives us a family of Kan complexes, i.e., ∞-groupoids, namely the fibers X_s defined by the pullbacks

$$\begin{array}{ccc} X_s & \longrightarrow & X \\ \downarrow & \lrcorner & \downarrow p \\ \Delta^0 & \xrightarrow[s]{} & S. \end{array}$$

Similarly, there is the following class of maps giving families of ∞-categories.

Definition 14.2.28. A morphism of simplicial sets $p\colon X \to S$ is an *inner fibration* if it has the right lifting property with respect to $\Lambda^n_k \to \Delta^n$, $0 < k < n$.

Joyal uses the term *mid-fibration* instead of inner fibration. Since any class of morphisms defined by a right lifting property is closed under pullbacks this implies that the fibers of an inner fibration are ∞-categories. The dependence of the fiber on the base point is only functorial in a very weak sense, namely in the sense of *correspondences* (see [82, p. 97]), which are also known as *distributors, profunctors* or *bimodules* in the classical setting (see [18, §7.8]).

Now, the *categorical fibrations*, i.e., the fibrations in the Joyal model structure happen to be a bit difficult to describe. However, for a morphism $p\colon X \to S$ such that the target S is an ∞-category, Joyal gave the following characterization: such a map is a categorical fibration if and only if it is an inner fibration and an *isofibration*. More precisely, since p is an inner fibration and S is an ∞-category, also X is an ∞-category, and a map $X \to S$ between ∞-categories is an *isofibration* if and only if every equivalence $p(x) \to s$ in S can be lifted to an equivalence in X with domain x.

The nerve $N(F)$ of a functor $F\colon A \to B$ of ordinary categories is automatically an inner fibration; thus the notion of inner fibrations does not have a classical analogue. In particular, a functor can always be thought of as a family of categories parametrized by the objects of the target category. In the ∞-categorical world however in many definitions this condition has to be imposed. This is the case for the categorical fibrations but also for further classes of fibrations as we will see later.

As already mentioned, the original definition of categorical equivalences due to Joyal [67] is different. He gives a definition without reference to simplicial categories and his proof of the existence of the Joyal model structure is purely combinatorial. Lurie gives this alternative definition because he is heading for the following comparison result [82, Thm. 2.2.5.1].

Theorem 14.2.29. *The adjunction $(C[-], N_\Delta)\colon sSet \rightleftarrows sCat$ is a Quillen equivalence with respect to the Joyal model structure and the Bergner model structure,*

$$(C[-], N_\Delta)\colon sSet \rightleftarrows sCat.$$

A similar result can also be obtained by a combination of results due to Bergner, Joyal, Rezk, and Tierney. For more details we refer to the chapter by Bergner in this Handbook.

Theorem 14.2.29 makes precise in which sense the approaches to a theory of $(\infty, 1)$-categories given by ∞-categories and simplicial categories are equivalent. The proof of this theorem is actually hard work including a deep 'rigidification or straightening result' and can be found in [82, §2.2.5]. An alternative proof was given by Dugger and Spivak in [34, 35].

As a corollary, we have the following result which can also be obtained directly and without any mention of model structures (see for example the proof of [26, Theorem 2.1]). However, with Theorem 14.2.29 in mind the result is put into perspective. Recall that a simplicial category is called locally fibrant if all mapping spaces are Kan complexes.

Corollary 14.2.30. *The coherent nerve of a locally fibrant simplicial category is an ∞-category.*

With this corollary at our disposal we can now give some typical examples of ∞-categories that are neither nerves of categories nor singular complexes of spaces. So, these are somehow our first honest examples of ∞-categories. It turns out that the first class of examples is generic in a sense which is made precise by Theorem 14.4.15.

Examples 14.2.31. (i) Let $\mathcal{M}$ be a simplicial model category [92, §II.2] and let $\mathcal{M}_{\mathrm{cf}} \subseteq \mathcal{M}$ be the full simplicial subcategory spanned by the fibrant and cofibrant objects. It is an immediate consequence of Quillen's axiom (SM7) that $\mathcal{M}_{\mathrm{cf}}$ is a locally fibrant simplicial category. Thus, via the coherent nerve construction, we obtain the ∞-category $N_\Delta(\mathcal{M}_{\mathrm{cf}})$, the *underlying ∞-category* of the simplicial model category $\mathcal{M}$.

(ii) As a more specific example let us consider $s\mathcal{S}et$ endowed with the usual Kan–Quillen model structure. This is a simplicial model category and with respect to this model structure we have $s\mathcal{S}et_{\mathrm{cf}} = \mathcal{K}an$, where $\mathcal{K}an$ is the full simplicial subcategory spanned by the Kan complexes. The *∞-category $\mathcal{S}$ of spaces* is given by

$$\mathcal{S} = N_\Delta(\mathcal{K}an).$$

This is only one model for the ∞-category of spaces; there are others, for example the underlying ∞-category of topological spaces. For all purposes of ∞-category theory it turns out that any ∞-category equivalent to $\mathcal{S}$ is equally good and that the precise model is irrelevant. All these ∞-categories satisfy the same universal property of being the 'free cocomplete ∞-category on a single generator' (see Corollary 14.4.11).

(iii) Another class of examples is induced by additive categories. Given an additive category $\mathcal{A}$, the category $\mathrm{Ch}(\mathcal{A})$ of chain complexes in $\mathcal{A}$ can be enriched over the category $\mathrm{Ch}(\mathbb{Z})$ of chain complexes of abelian groups. The enrichment is set up in a way that if we take the 0-cycles of all the mapping complexes then we obtain the usual category of chain complexes, while cycles of positive dimensions give chain homotopies and higher chain homotopies. In what follows we restrict the enrichment to the category $\mathrm{Ch}_{\geq 0}(\mathbb{Z})$ of non-negative chain complexes.

The *Dold–Kan correspondence* gives us an equivalence of categories $\mathrm{DK}\colon \mathrm{Ch}_{\geq 0}(\mathbb{Z}) \to s\mathcal{A}b$, where $s\mathcal{A}b$ is the category of simplicial abelian groups. An inverse to DK is given by the *normalized chain complex functor* which can be shown to be lax comonoidal with respect to the levelwise tensor product on $s\mathcal{A}b$ and the usual one on $\mathrm{Ch}_{\geq 0}(\mathbb{Z})$ (in fact, this lax comonoidal structure is induced by the Alexander–Whitney maps). It follows by abstract nonsense that DK carries canonically a lax *monoidal* structure (see [74] for the abstract framework and [103] for precisely this context). Thus, we can apply DK to the morphism complexes in $\mathrm{Ch}(\mathcal{A})$ in order to obtain the category $\mathrm{DK}_*(\mathrm{Ch}(\mathcal{A}))$ enriched in simplicial abelian groups. Since simplicial abelian groups are Kan complexes this gives us a locally fibrant simplicial category, and we can define the *∞-category* $\mathrm{Ch}(\mathcal{A})$ *of chain complexes* in $\mathcal{A}$ by

$$\mathrm{Ch}(\mathcal{A}) = N_\Delta\big(\mathrm{DK}_*(\mathrm{Ch}(\mathcal{A}))\big).$$

A smaller model for these ∞-categories can be obtained by means of the *differential-graded nerve construction* (see [83, §1.3.1]).

We conclude this section by a short perspective on enriched ∞-categories.

Perspective 14.2.32. Recall that a 2-category is simply a category enriched over categories. Similarly, a strict n-category is just a category enriched over $(n-1)$-categories. This definition via iterated enrichments suggests that $(\infty,1)$-categories should be categories enriched over $(\infty,0)$-categories, i.e., Kan complexes by Corollary 14.2.18. And as we just saw, simplicial categories with the property that all mapping spaces are Kan complexes play a special role (see Theorem 14.2.24).

However, if we take the philosophy of ∞-categories seriously, then the good notion of enrichment in that context should be that of a 'weak enrichment'. We saw that for every pair of objects x, y in an ∞-category $\mathcal{C}$, there is a space of morphisms $\mathrm{Map}_{\mathcal{C}}(x,y)$. In contrast to the case of simplicial categories, for ∞-categories there is not a specified strictly associative and strictly unital composition law $\mathrm{Map}_{\mathcal{C}}(y,z)\times\mathrm{Map}_{\mathcal{C}}(x,y)\to\mathrm{Map}_{\mathcal{C}}(x,z)$. In fact, the idea of having a weak enrichment should be expressed by the existence of composition laws which are only *coherently associative.* Taking this perspective of $\mathbb{A}_\infty$-multiplications seriously, an abstract theory of enriched ∞-categories has been proposed by Gepner and Haugseng [45]. Their framework even allows one to study weak enrichments in not necessarily Cartesian contexts. For related rectification results see [53].

14.3 Categorical constructions with ∞-categories

The aim of this section is to extend some key constructions and notions from ordinary category theory to the world of ∞-categories (or more general simplicial sets) in a way that the following principles are satisfied.

(i) The concepts are extensions of the ordinary concepts in that everything is compatible with the fully faithful nerve functor $N\colon \mathcal{C}at\to s\mathcal{S}et$.

(ii) The notions are *coherent* variants of the classical notions, i.e., ∞-category theory realizes some kind of homotopy coherent category theory.

(iii) The extensions are often defined for arbitrary simplicial sets, and when applied to ∞-categories we want these extensions to again give rise to ∞-categories.

(iv) All concepts are *invariant concepts*, i.e., an application of these constructions to equivalent input ∞-categories yields equivalent output ∞-categories.

We are mostly interested in a robust theory of *limits* and *colimits* (see §14.3.5) giving us an ∞-categorical version of the more well-known homotopy (co)limits in model categories, but this needs some preparation. In §14.3.1 we discuss ∞-categories of functors. In §§14.3.2-14.3.3 we introduce join and slice constructions, which allow us to speak about ∞-categories of cones and cocones on diagrams. In §14.3.4 we introduce initial and final objects as special cases of limits and colimits, respectively. Finally, in §14.3.5 everything is put together and we introduce limits as final objects in slice categories (and dually for colimits).

The reader who is less inclined towards abstract categorical constructions is asked to consider the goal of a brief discussion of colimits as a justification for the discussion of the constructions in §14.3.2 and §14.3.3. For a more detailed discussion of the *theory* of limits we refer to [82].

14.3.1 Functors

Since ∞-categories are simply particular simplicial sets we can make the following definition.

Definition 14.3.1. Let K be a simplicial set and let $\mathcal{C}$ be an ∞-category. A *functor* $F\colon K \to \mathcal{C}$ is a map of simplicial sets. Similarly, a *natural transformation* is a map $\Delta^1 \times K \to \mathcal{C}$. More generally, the *space of functors* $\mathrm{Fun}(K, \mathcal{C})$ is

$$\mathrm{Fun}(K,\mathcal{C})_\bullet = \mathrm{Map}_{s\mathcal{S}et}(K,\mathcal{C})_\bullet = \hom_{s\mathcal{S}et}(\Delta^\bullet \times K, \mathcal{C}) \in s\mathcal{S}et.$$

This extends the classical concept of a functor because $N\colon \mathcal{C}at \to s\mathcal{S}et$ is fully faithful. More generally, there is the following refinement of this observation.

Lemma 14.3.2. *For categories A, B there is a natural isomorphism of simplicial sets*

$$N(\mathrm{Fun}(A,B)) \cong \mathrm{Fun}(NA, NB).$$

Proof. For $[n] \in \Delta$, there are the following natural bijections

$$\begin{aligned} N(\mathrm{Fun}(A,B))_n &= \hom_{\mathcal{C}at}([n], \mathrm{Fun}(A,B)) && (14.3.1)\\ &\cong \hom_{\mathcal{C}at}([n]\times A, B) && (14.3.2)\\ &\cong \hom_{s\mathcal{S}et}(N([n]\times A), NB) && (14.3.3)\\ &\cong \hom_{s\mathcal{S}et}(\Delta^n \times NA, NB) && (14.3.4)\\ &= \mathrm{Fun}(NA, NB)_n, && (14.3.5) \end{aligned}$$

given by the exponential laws, the fully faithfulness of N, the fact that N preserves products, and the isomorphism $N([n]) \cong \Delta^n$. □

This seemingly naive definition of a functor is actually the good one, as we want to indicate now (but see also Perspective 14.3.5).

Example 14.3.3. Let $A \in \mathcal{C}at$ be an ordinary category and let $\mathcal{M}$ be a locally fibrant simplicial category. By Corollary 14.2.30 we know that $N_\Delta(\mathcal{M})$ is an ∞-category and we want to unravel a bit what it means to have a functor $F\colon NA \to N_\Delta(\mathcal{M})$. The behavior on 0-simplices and 1-simplices amounts to saying that associated to each arrow $x \to y$ in A there is a morphism $Fx \to Fy$ in $\mathcal{M}$. Moreover, given a pair of composable arrows $x \xrightarrow{f} y \xrightarrow{g} z$ in A, i.e., a 2-simplex $\sigma\colon \Delta^2 \to NA$, we obtain a 2-simplex $F(\sigma)\colon \Delta^2 \to N_\Delta(\mathcal{M})$. By Definition 14.2.21 this means that we are given a simplicial functor $C[\Delta^2] \to \mathcal{M}$ which boils down to having a diagram in $\mathcal{M}$ of the form

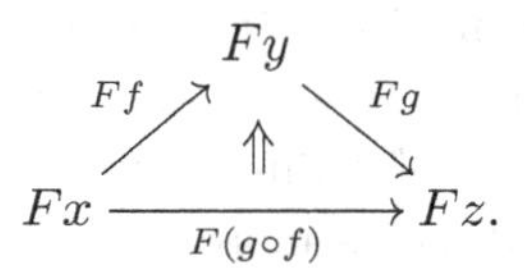

Thus, the functor F preserves compositions up to specified homotopies.

However, there is still much more information encoded by F, namely all the higher simplices obtained from longer sequences of composable arrows in A. These encode the idea that F is not only a 'functor up to homotopy' but gives us a 'functor up to *coherent* homotopy', i.e., a *homotopy coherent diagram.* For a precise statement see [25] and also Perspective 14.3.5. In order to at least give a partial justification for this claim let us consider a 3-simplex $\tau\colon \Delta^3 \to NA$, i.e., a chain consisting of three composable arrows f, g, and h. By the above we know that the four faces of the 3-simplex $F(\tau)\colon \Delta^3 \to N_\Delta(\mathcal{M})$ altogether give us two different composite homotopies from $F(h\circ g\circ f)$ to $F(h)\circ F(g)\circ F(f)$ as in the boundary of

$$\begin{array}{ccc} F(h\circ g\circ f) & \longrightarrow & F(h)\circ F(g\circ f) \\ \downarrow & & \downarrow \\ F(h\circ g)\circ F(f) & \longrightarrow & F(h)\circ F(g)\circ F(f). \end{array} \tag{14.3.6}$$

The 3-simplex $F(\tau)$ specifies one more homotopy $F(h\circ g\circ f) \to F(h)\circ F(g)\circ F(f)$ together with two 2-homotopies in $\mathcal{M}$ relating this additional homotopy to the two compositions in (14.3.6). This follows from the fact that $\mathrm{Map}_{C[\Delta^3]}(0,3)$ is isomorphic to the product $\Delta^1\times\Delta^1$. (More generally, we have $\mathrm{Map}_{C[\Delta^n]}(i,j) \cong (\Delta^1)^{\times(j-i-1)}$ for $0 \le i < j \le n$.)

Given two ordinary categories A and B, it is easy to check that the functors from A to B together with the natural transformations assemble to a category $\mathrm{Fun}(A,B)$. Moreover, if we have equivalences $A \simeq A'$ and $B \simeq B'$, then there is a canonical equivalence $\mathrm{Fun}(A,B) \simeq \mathrm{Fun}(A',B')$. Similar results also hold true in the world of ∞-categories, but there — compared to the classical context — a proof requires some work. One proof uses certain stability properties of the class of categorical equivalences and the so-called *inner anodyne maps* (see for example [69]). Using these properties, one is able to deduce the following result [82, p. 94].

Proposition 14.3.4. *Let $\mathcal{C}$ and $\mathcal{D}$ be ∞-categories and let K and L be simplicial sets.*

(i) The simplicial set $\mathrm{Fun}(K, \mathcal{C})$ *is an* ∞*-category.*

(ii) If $\mathcal{C} \to \mathcal{D}$ *is an equivalence of* ∞*-categories, then also the induced map* $\mathrm{Fun}(K, \mathcal{C}) \to \mathrm{Fun}(K, \mathcal{D})$ *is an equivalence of* ∞*-categories.*

(iii) If $K \to L$ *is a categorical equivalence of simplicial sets, then the induced map* $\mathrm{Fun}(L, \mathcal{C}) \to \mathrm{Fun}(K, \mathcal{C})$ *is an equivalence of* ∞*-categories.*

Thus this proposition tells us that the formation of functor ∞-categories is an invariant notion — as it should be the case for all categorical constructions in the world of ∞-categories. On a more conceptual side, this proposition is a consequence of the fact that with respect to the Joyal model structures the categorical product $\times\colon s\mathcal{S}et \times s\mathcal{S}et \to s\mathcal{S}et$ is a left Quillen functor of two variables.

Perspective 14.3.5. The theory of ∞-categories should really be thought of as *homotopy coherent category theory.* In particular, the basic notion of a functor is to model the idea of having a *homotopy coherent diagram.* Homotopy coherent category theory has quite some history and references include [114, 25, 26, 27]. Here we content ourselves by including a short detour only.

To begin with, whenever we have an adjunction $(L, R)\colon \mathcal{C} \rightleftarrows \mathcal{D}$ we obtain a comonad C on $\mathcal{D}$ with $C = LR\colon \mathcal{D} \to \mathcal{D}$ as underlying functor. The structure morphisms of the comonad are induced by the unit and counit of the adjunction, and the necessary relations follow directly from the triangular identities of an adjunction. These comonads can be used to form simplicial resolutions of objects in $\mathcal{D}$ (which play, for example, some role in relative homological algebra; see e.g. [115, §8]). Slightly more precisely, given an object $d \in \mathcal{D}$, the associated simplicial object $C_*(d)$ is given by $C_n(d) = C^{n+1}(d)$. (Of course, there is also a dual part of the story leading to monads and cosimplicial resolutions.)

Now, the adjunction that is in the background of our context is the adjunction between graphs and categories given by the forgetful functor U sending a category to the underlying reflexive graph (a graph in which every vertex has a specified identity edge) and the free category functor F,

$$(F, U)\colon \mathcal{G}raph \rightleftarrows \mathcal{C}at. \tag{14.3.7}$$

Note that both the free category functor and the underlying graph functor preserve the set of objects. Thus, given a category A, the resulting simplicial resolution $C_*A \in \mathrm{Fun}(\Delta^{\mathrm{op}}, \mathcal{C}at)$ also has a constant set of objects and hence defines an object $C_*A \in s\mathcal{C}at$, i.e., a simplicially enriched category as opposed to merely a simplicial object in categories.

This allows us to make the following definition which goes back to Vogt [114]. Given a small category A and a simplicial category $\mathcal{M} \in s\mathcal{C}at$, a *homotopy coherent diagram* in $\mathcal{M}$ of shape A is a simplicial functor $C_*A \to \mathcal{M}$. Thus, the set $\mathrm{coh}(A, \mathcal{M})$ of such homotopy coherent diagrams is defined by

$$\mathrm{coh}(A, \mathcal{M}) = \hom_{s\mathcal{C}at}(C_*A, \mathcal{M}). \tag{14.3.8}$$

If as a special case we consider $A = [n]$, then the resulting simplicial category $C_*[n]$ is isomorphic to the 'simplicial thickening' $C[\Delta^n]$ introduced at the beginning of §14.2.2. Thus, the coherent nerve $N_\Delta(\mathcal{M})$ of $\mathcal{M}$ is precisely obtained by considering *homotopy coherent chains of composable arrows* in the given $\mathcal{M}$.

More generally, Riehl [96, Theorem 6.7] showed that for every $A \in \mathcal{C}at$ there is a natural isomorphism of simplicial categories $C[NA] \cong C_*(A)$. In particular, for $A \in \mathcal{C}at$ and $\mathcal{M} \in s\mathcal{C}at$ we have a bijection

$$\mathrm{coh}(A, \mathcal{M}) \cong \hom_{s\mathcal{C}at}(C[NA], \mathcal{M}) \cong \hom_{s\mathcal{S}et}(NA, N_\Delta \mathcal{M}).$$

This shows that the seemingly naive Definition 14.3.1 subsumes the concept of homotopy coherent diagrams, and justifies thinking more generally of functors in the sense of that definition as homotopy coherent diagrams.

From a conceptual perspective one might argue that it is not very nice that we defined homotopy coherent diagrams in simplicial categories by means of *specific simplicial resolutions* — namely the simplicial comonad resolutions associated to (14.3.7). In fact, the actual choice of resolution should not matter, and in some modern treatments like [46, §8] homotopy coherent diagrams are defined by means of *more general* simplicial resolutions.

Remark 14.3.6. (i) Proposition 14.3.4 reveals one of the technical advantages of ∞-categories over model categories: ∞-categories are stable under the formation of functor categories without any further assumption. In this respect, model categories are less well-behaved since one has to impose certain conditions on the model categories involved to obtain this stability property: for cofibrantly-generated model categories, associated diagram categories always admit the *projective* model structure [56, p.224], whereas in the case of combinatorial model categories the *projective* and the *injective* structure both always exist on the diagram categories [82, p.824]. Note however that Proposition 14.3.4 is significantly more general since these results only tell us something about model structures on $\mathcal{M}^A$, $A \in \mathcal{C}at$. For example, we never dared to ask for the existence of a 'canonical' model category of functors between two given model categories.

(ii) A further technical advantage of ∞-categories over model categories is the following one. The 'correct' notion of equivalence for model categories is the notion of *Quillen equivalence.* Since, in general, a Quillen equivalence can not be inverted, the equivalence relation generated by this notion is quite complicated: frequently model categories are only Quillen equivalent through a zig-zag of Quillen equivalences pointing in different directions. The appropriate notion of equivalence for ∞-categories is the notion of *categorical equivalence* (Definition 14.2.26). Since ∞-categories are precisely the fibrant and cofibrant objects with respect to the Joyal model structure where categorical equivalences are the weak equivalences, it follows that a zig-zag of categorical equivalences can always be replaced by a single categorical equivalence.

(iii) As we saw in Perspective 14.3.5, the notion of homotopy coherent diagrams is quite easily established in the world of ∞-categories as a map of simplicial sets with the domain given by the nerve of an ordinary category. There will be further advantages of this flavor, i.e., where 'higher coherences' are easily encoded in the setting of ∞-categories. For example the notions of $\mathbb{A}_\infty$- and $\mathbb{E}_\infty$-algebras in monoidal and symmetric monoidal ∞-categories are conveniently introduced in this setting as specific sections of certain

∞-categorical versions of Grothendieck opfibrations, called coCartesian fibrations by Lurie. We will come back to this in §14.5.

We conclude this subsection with a perspective on how to get an ∞-category of ∞-categories and also an underlying ∞-category for arbitrary model categories.

Perspective 14.3.7. We just introduced the notion of a functor between ∞-categories and hence obtain a category of ∞-categories. But to stick more seriously to the ∞-categorical framework, we would like to have an *∞-category $\mathcal{C}at_\infty$ of ∞-categories.* Since non-invertible natural transformations play an important role in classical category theory it would be even nicer to have an $(\infty, 2)$-category of ∞-categories but let us content ourselves with such an ∞-category. Recall that the Joyal model structure is not simplicial, so that we cannot apply Corollary 14.2.30 directly in order to get such a gadget. Luckily, there is the Quillen equivalent *simplicial* model category $s\mathcal{S}et^+$ of so-called *marked simplicial sets* which hence serves the purpose (among many others), and which we want to describe very briefly.

A *marked simplicial set* is a pair $(X, \mathcal{E}_X)$ consisting of a simplicial set X together with a subset $\mathcal{E}_X \subseteq X_1$ of edges containing the degenerate ones. A morphism of marked simplicial sets $(X, \mathcal{E}_X) \to (Y, \mathcal{E}_Y)$ is a morphism $f\colon X \to Y$ of simplicial sets such that $f(\mathcal{E}_X) \subseteq \mathcal{E}_Y$. This defines the category $s\mathcal{S}et^+$ of marked simplicial sets, a category playing a key role in the relative context [82, §§3.1-3.2]. (See also Perspective 14.5.10 for further motivational remarks.)

Degenerate edges, i.e., identity morphisms in an ∞-category are equivalences and it follows that every ∞-category $\mathcal{C}$ gives us a marked simplicial set $\mathcal{C}^\natural$ by marking the equivalences. There is a simplicial model structure on $s\mathcal{S}et^+$ such that the fibrant and cofibrant objects are precisely the marked simplicial sets of the form $\mathcal{C}^\natural$ for some ∞-category $\mathcal{C}$ [82, Prop. 3.1.4.1, Cor. 3.1.4.4]. Thus, the *∞-category of ∞-categories* $\mathcal{C}at_\infty$ can be defined as the underlying ∞-category of $s\mathcal{S}et^+$,

$$\mathcal{C}at_\infty = N_\Delta(s\mathcal{S}et^+_{\mathrm{cf}}).$$

14.3.2 Join construction

The main motivation for us to study the join construction in this context is that it allows us to define the slice construction (see §14.3.3) which in turn is fundamental to the theory of (co)limits (see §14.3.5). The join construction has its origins in classical topology, but here we immediately aim for the simplicial analogue. As a preparation, we recall the classical situation in category theory.

Given categories A and B, one can form a new category $A \star B$, the *join* of A and B, as follows. The class of objects $\mathrm{obj}(A \star B)$ is given by the disjoint union of $\mathrm{obj}(A)$ and $\mathrm{obj}(B)$. For the morphisms, there are the following four different cases

$$\hom_{A\star B}(x,y) = \begin{cases} \hom_A(x,y), & x,y \in A, \\ \hom_B(x,y), & x,y \in B, \\ \quad * \quad , & x \in A, y \in B, \\ \quad \emptyset \quad , & x \in B, y \in A, \end{cases}$$

and the composition is completely determined by requiring that A and B are full subcategories of $A \star B$ in the obvious way. Note that the construction is not symmetric in A and B. To illustrate the join construction, we mention a few basic examples.

Examples 14.3.8. (i) If $A \in \mathcal{C}at$ is arbitrary and if $B = \mathbb{1}$ is the terminal category, then $A^{\triangleright} = A \star \mathbb{1}$ is the *right cone* or *cocone* on A. It is obtained by adjoining a new terminal object ∞ to A, and plays a central role in the study of colimits.

(ii) Dually, if $A = \mathbb{1}$ is terminal and $B \in \mathcal{C}at$ is arbitrary, then $B^{\triangleleft} = \mathbb{1} \star B$ is the *left cone* or *cone* on B. This category is obtained from B by adjoining a new initial object $-\infty$ and is central to the theory of limits.

(iii) As a more specific example, let A be the category occurring in the study of pushout diagrams,

$$\begin{array}{ccc} (0,0) & \longrightarrow & (1,0) \\ \downarrow & & \\ (0,1). & & \end{array}$$

Then the cocone on A is given by the commutative square $A^{\triangleright} \cong [1] \times [1]$, which can of course be depicted as

$$\begin{array}{ccc} (0,0) & \longrightarrow & (1,0) \\ \downarrow & = & \downarrow \\ (0,1) & \longrightarrow & (1,1). \end{array}$$

In particular, a diagram of this shape consists of four morphisms in the target category such that the two compositions agree. Similarly, if B is the diagram occurring in the study of pullback diagrams,

$$\begin{array}{ccc} & & (1,0) \\ & & \downarrow \\ (0,1) & \longrightarrow & (1,1), \end{array}$$

then also the cone $B^{\triangleleft}$ is the square.

The join construction can be extended to simplicial sets. There is a very conceptual approach to this construction as described by Joyal in [69] (basically as a Day convolution construction [29] applied to the ordinal addition). In [69] one can also find many 'elementary relations' about this join construction. Since we only want to rush through the theory of this notion, we instead give the following more direct 'definition'.

Definition 14.3.9. Let K and L be simplicial sets. The *join* $K \star L$ of K and L is the simplicial set defined by

$$(K \star L)_n = K_n \cup L_n \cup \bigcup_{i+1+j=n} K_i \times L_j, \quad n \geq 0.$$

We leave it to the reader to define the structure maps of $K \star L$ and to check that this defines a functor $\star\colon s\mathcal{S}et \times s\mathcal{S}et \to s\mathcal{S}et$. It follows from those details that $K \star L$ comes with canonical inclusions $K \to K \star L$ and $L \to K \star L$. The join operation for simplicial sets is in fact characterized by the following two properties.

Proposition 14.3.10. *(i) The partial join functors $K \star (-)\colon s\mathcal{S}et \to s\mathcal{S}et_{K/}$ and $(-) \star L\colon s\mathcal{S}et \to s\mathcal{S}et_{L/}$ preserve colimits.*

(ii) For the standard simplices we find $\Delta^i \star \Delta^j \cong \Delta^{i+1+j}, i, j \geq 0$, and these isomorphisms are compatible with the obvious inclusions of Δ^i and Δ^j.

The following lemma is immediate and the proof is recommended as an exercise to those readers who just saw these notions for the first time. The solution to this exercise also suggests how to define the structure maps in Definition 14.3.9.

Lemma 14.3.11. *The nerve is compatible with the join constructions in that there is a natural isomorphism $N(A) \star N(B) \to N(A \star B), A, B \in \mathcal{C}at$.*

To give some examples, we consider the (co)cone constructions and again pushout and pullback diagrams.

Examples 14.3.12. (i) If $K \in s\mathcal{S}et$ is arbitrary and $L = \Delta^0$, then $K^{\rhd} = K \star \Delta^0$ is the *right cone* or the *cocone* on K. Dually, if $L \in s\mathcal{S}et$ is arbitrary then $L^{\lhd} = \Delta^0 \star L$ is the *left cone* or *cone* on L.

(ii) Let $K = \Lambda^2_0$. Using the description of this horn as a pushout, it follows immediately from Proposition 14.3.10 that the cocone $(\Lambda^2_0)^{\rhd}$ is isomorphic to the square $\Box = \Delta^1 \times \Delta^1$, i.e., we have the picture

$$\begin{array}{ccc} (0,0) & \longrightarrow & (1,0) \\ \downarrow & \searrow \; \Uparrow \Downarrow & \downarrow \\ (0,1) & \longrightarrow & (1,1). \end{array}$$

Note that if $\mathcal{M}$ is a simplicial category then a diagram of this shape in $N_\Delta(\mathcal{M})$ consists of five morphisms and two homotopies as depicted by the diagram. In particular, in general, we do not have a commutative square but only a coherent version thereof. This example can be dualized and we obtain a similar isomorphism $(\Lambda^2_2)^{\lhd} \cong \Delta^1 \times \Delta^1$.

A careful analysis of the join construction allows one to establish the following important properties ([82, Prop. 1.2.8.3] and [82, Cor. 4.2.1.2]).

Proposition 14.3.13. *(i) If $\mathcal{C}$ and $\mathcal{D}$ are ∞-categories, then the join $\mathcal{C} \star \mathcal{D}$ is again an ∞-category.*

(ii) If $F\colon \mathcal{C} \to \mathcal{C}'$ and $G\colon \mathcal{D} \to \mathcal{D}'$ are equivalences of ∞-categories, then also the induced map $F \star G\colon \mathcal{C} \star \mathcal{D} \to \mathcal{C}' \star \mathcal{D}'$ is an equivalence.

14.3.3 Slice construction

We again begin by recalling the more classical situation of ordinary category theory. Given a category B and an object $x \in B$, one can form the *overcategory* $B_{/x}$ where objects are morphisms $x' \to x$ in B with target x. Given two such objects $x' \to x$, $x'' \to x$, a morphism $(x' \to x) \to (x'' \to x)$ in $B_{/x}$ is simply a morphism $x' \to x''$ in B making the obvious triangle

$$\begin{array}{ccc} x' & \longrightarrow & x'' \\ & \searrow \;=\; \swarrow & \\ & x & \end{array} \qquad (14.3.9)$$

commute.

One generalization of this notion is obtained by replacing the object $x\colon \mathbb{1} \to B$ by a more general diagram in B. More precisely, if we are given a functor $p\colon A \to B$, then we can form the *slice category* $B_{/p}$ of *objects over* p or *cones on* p. Thus, an object in $B_{/p}$ is simply a cone on p, i.e., an object $b \in B$ together with a natural transformation from the constant functor with value b to the given p. Morphisms in this category are simply morphisms in B that are compatible with the natural transformations. Using the join construction one can see that the slice construction satisfies a universal property: for any category C, there is a natural bijection

$$\mathrm{Fun}(C, B_{/p}) \cong \mathrm{Fun}_p(C \star A, B),$$

where the right-hand side denotes all functors $C \star A \to B$ making the following triangle commute

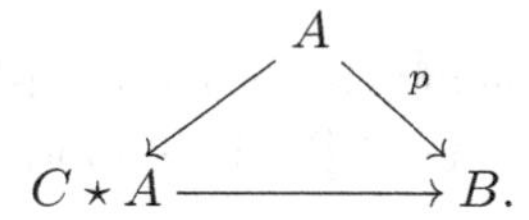

To emphasize a bit more that the right-hand side in the above universal property takes certain structure maps into account, we rewrite this as

$$\hom_{\mathcal{C}at}(C, B_{/p}) \cong \hom_{\mathcal{C}at_{A/}}(A \to C \star A, A \xrightarrow{p} B), \quad C \in \mathcal{C}at.$$

Of course this sounds like an unnecessarily complicated reformulation of something very simple. But the point of this reformulation is that it gives us an idea on how to extend these notions to ∞-categories — as it was done by Joyal in [68].

Proposition 14.3.14. *Let $p\colon L \to \mathcal{C}$ be a map of simplicial sets with $\mathcal{C}$ an ∞-category. There is an* ∞-category $\mathcal{C}_{/p}$ *characterized by the following universal property: For every simplicial set K, there is a bijection*

$$\hom_{s\mathcal{S}et}(K, \mathcal{C}_{/p}) \cong \hom_{s\mathcal{S}et_{L/}}(L \to K \star L, L \to \mathcal{C})$$

which is natural in K. The ∞-category $\mathcal{C}_{/p}$ is the ∞-category of cones on p.

To check that there is such a *simplicial set* $\mathcal{C}_{/p}$, one can use the universal property as a definition. More precisely, the special cases of the standard simplices $K = \Delta^n$ give us by the Yoneda lemma a description of the n-simplices of $\mathcal{C}_{/p}$ as

$$(\mathcal{C}_{/p})_n \cong \hom_{s\mathcal{S}et_{L/}}(L \to \Delta^n \star L, L \to \mathcal{C}). \qquad (14.3.10)$$

To show that one actually obtains an ∞-*category* requires more work and will not be done here [82, Cor. 2.1.2.2].

Remark 14.3.15. Let $p\colon L \to \mathcal{C}$ be a map of simplicial sets with $\mathcal{C}$ an ∞-category.

(i) In both the classical and the ∞-categorical situation the constructions can be dualized. For example, there is the *∞-category $\mathcal{C}_{p/}$ of cocones on p*. The ∞-categories $\mathcal{C}_{p/}$ and $\mathcal{C}_{/p}$ are also referred to as *slice ∞-categories*.

(ii) One can show that the slice construction is an *invariant* notion. First, for every equivalence of ∞-categories $q\colon \mathcal{C} \to \mathcal{D}$ the induced functor $\mathcal{C}_{p/} \to \mathcal{D}_{qp/}$ is an equivalence [82, §2.4.5]. Second, for every equivalence $v\colon L' \to L$ the induced functor $\mathcal{C}_{p/} \to \mathcal{C}_{pv/}$ is an equivalence (as a special case of [82, Prop. 4.1.1.7]).

There is the following result about the compatibility of the nerve $N\colon \mathcal{C}at \to s\mathcal{S}et$ with slice constructions.

Lemma 14.3.16. *For every functor $p\colon A \to B$ there is a natural isomorphism of simplicial sets*

$$N(B_{/p}) \cong N(B)_{/N(p)}.$$

Proof. In simplicial dimension n we have the following chain of natural bijections

$$\begin{aligned}
N(B_{/p})_n &= \hom_{\mathcal{C}at}([n], B_{/p}) && (14.3.11)\\
&\cong \hom_{\mathcal{C}at_{A/}}(A \to [n] \star A, A \to B) && (14.3.12)\\
&\cong \hom_{s\mathcal{S}et_{N(A)/}}(N(A) \to N([n] \star A), N(A) \to N(B)) && (14.3.13)\\
&\cong \hom_{s\mathcal{S}et_{N(A)/}}(N(A) \to \Delta^n \star N(A), N(A) \to N(B)) && (14.3.14)\\
&\cong \hom_{s\mathcal{S}et}(\Delta^n, N(B)_{/N(p)}), && (14.3.15)
\end{aligned}$$

where in step three we used Lemma 14.3.11 and it is obvious which canonical isomorphisms were used in the remaining steps. □

Example 14.3.17. Let $\mathcal{C}$ be an ∞-category and let $x \in \mathcal{C}$ be an object, classified by the map $\kappa_x\colon \Delta^0 \to \mathcal{C}$. Then the ∞-category $\mathcal{C}_{/\kappa_x}$ is called the *∞-category of objects over x*, and is simply denoted by $\mathcal{C}_{/x}$. Dually, the ∞-category $\mathcal{C}_{\kappa_x/}$ is called the *∞-category of objects under x*, and is denoted by $\mathcal{C}_{x/}$.

Let us spell out some of the details about $\mathcal{C}_{/x}$ in order to convince ourselves that this is the expected coherent version of the corresponding classical construction. By (14.3.10), an object in $\mathcal{C}_{/x}$ is a map $f\colon \Delta^1 \cong \Delta^0 \star \Delta^0 \to \mathcal{C}$ such that $d_0(f) = x$, i.e., we are given a map in $\mathcal{C}$ with target x. Similarly, (14.3.10) tells us that a map in $\mathcal{C}_{/x}$ is given by a 2-simplex $\sigma\colon \Delta^2 \cong \Delta^1 \star \Delta^0 \to \mathcal{C}$ such that $\sigma(2) = x$. In the case that $\mathcal{C}$ is the coherent nerve of a (locally fibrant) simplicial category, the picture of σ is

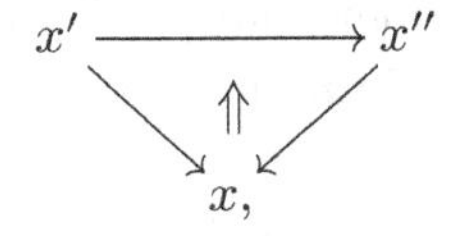

indicating that we obtained the coherent version of (14.3.9) we were aiming for.

14.3.4 Final and initial objects

We now come to the ∞-categorical variant of final and initial objects. In classical category theory, final objects are characterized by the property that for all objects there is a unique morphism to the final object. In these notes, we take a slightly different approach to the ∞-categorical generalization (but see Proposition 14.3.19). The following definition is not the original one of Joyal [68], but it is an equivalent one as shown in [68].

Definition 14.3.18. An object $x \in \mathcal{C}$ of an ∞-category $\mathcal{C}$ is a *final object* if the canonical map $\mathcal{C}_{/x} \to \mathcal{C}$ is an acyclic fibration of simplicial sets.

Thus, an object is final if 'forgetting that we lived above it does not result in a loss of information'. This notion can be reformulated in the following more usual way.

Proposition 14.3.19. *The following are equivalent for an object x of an ∞-category $\mathcal{C}$.*

(i) The object x is final.

(ii) The mapping spaces $\mathrm{Map}_{\mathcal{C}}(x', x)$ are acyclic Kan complexes for all $x' \in \mathcal{C}$.

(iii) Every simplicial sphere $\alpha\colon \partial\Delta^n \to \mathcal{C}$ such that $\alpha(n) = x$ can be filled to an entire n-simplex $\Delta^n \to \mathcal{C}$.

We expect a replacement for the fact that in classical category theory any two terminal objects are canonically isomorphic, and for this purpose we first make precise the notion of a *full* subcategory of an ∞-category. There is a more general notion of subcategories of an ∞-category but since we will not need this additional generality we stick to full subcategories. Given objects $\mathcal{D}_0 \subseteq \mathcal{C}_0$ in an ∞-category $\mathcal{C}$, let $\mathcal{D} \subseteq \mathcal{C}$ be the simplicial subset consisting precisely of those simplices $\Delta^n \to \mathcal{C}$ that have the property that all vertices belong to $\mathcal{D}_0$. It is immediate that $\mathcal{D}$ is again an ∞-category, the *full subcategory of $\mathcal{C}$ spanned by $\mathcal{D}_0$*. Obviously, $\mathcal{D}$ comes with an inclusion $\mathcal{D} \to \mathcal{C}$.

Corollary 14.3.20. *Let $\mathcal{C}$ be an ∞-category and let $\mathcal{D} \subseteq \mathcal{C}$ be the full subcategory spanned by the final objects of $\mathcal{C}$. The ∞-category $\mathcal{D}$ is either empty or a contractible Kan complex.*

Proof. This is immediate from Proposition 14.3.19. □

Remark 14.3.21. The conclusion of Corollary 14.3.20 is typical for uniqueness statements in ∞-category theory: It states that a space 'parametrizing universal objects' is empty or a contractible Kan complex. In classical category theory, if universal objects exist then they are unique up to unique isomorphism. This can be reformulated by saying that if universal objects exist then the category of such is a *contractible groupoid*: the existence of comparison maps shows that it is a connected category, while the uniqueness of these maps tells us that it is a groupoid with no nontrivial endomorphisms. Thus, the possibly non-trivial π_0 and π_1 both vanish. Now, with Corollary 14.2.18 in mind we see that the two uniqueness statements are morally very similar.

One example of such a uniqueness statement was already obtained in the discussion of Theorem 14.2.14, namely the space of compositions of two composable arrows is a contractible Kan complex. With Corollary 14.3.20 we have a further example of such a statement. In §14.3.5 we will define (co)limits as universal objects of certain slice categories and will hence again obtain such uniqueness statements. Let us mention that uniqueness results of this kind are also ubiquitous in the theory of model categories. Compare for example to [56] where many categories of choices (for example, cofibrant replacements) are shown to have contractible nerves.

We conclude by a statement about the compatibility with the nerve construction, but leave the proof to the reader.

Lemma 14.3.22. *Let A be a category. An object $a \in A$ is final if and only if $a \in N(A)$ is final.*

14.3.5 Limits and colimits

Having the basic categorical notions at our disposal, we are ready to talk about (co)limits in the framework of ∞-categories. Recall that the colimit of an ordinary functor $p\colon A \to B$ consists of an object $\operatorname{colim}_A p$ in B together with a universal cocone. Said differently, such a pair is equivalently specified by an initial object in the category $B_{p/}$ of cocones on p. This definition can now readily be extended to ∞-categories (see [68]).

Definition 14.3.23. Let K be a simplicial set and let $\mathcal{C}$ be an ∞-category. A *colimit* of a diagram $p\colon K \to \mathcal{C}$ is an initial object in $\mathcal{C}_{p/}$. An ∞-category is *cocomplete* if it admits colimits of all small diagrams. Dually, we define *limits* of diagrams and *complete* ∞-categories.

By Corollary 14.3.20 for every $p\colon K \to \mathcal{C}$ the full subcategory $\mathcal{D} \subseteq \mathcal{C}_{p/}$ spanned by the colimits of p is either empty or a contractible Kan complex. Thus, if a colimit exists, then it is unique up to a contractible choice.

Remark 14.3.24. (i) The nerve functor $N\colon \mathcal{C}at \to s\mathcal{S}et$ is compatible with the notion of (co)limits.

(ii) The notions of limits and colimits are invariant notions.

One justification for Definition 14.3.23 is that it extends the classical theory of (co)limits. Although this is certainly a convenient aspect of the notion, there is an additional justification which is more relevant for our purposes (as discussed by Lurie in [82, Theorem 4.2.4.1]). Namely, there is a precise meaning in that these notions of (co)limits coincides with the notion of *homotopy (co)limits* in simplicial categories. Joyal was fully aware of this in [68]. References to various aspects of the theory of homotopy (co)limits in other contexts include [22], [114], [56], [36], [105], and [32]. We follow Joyal and Lurie and simply speak of *(co)limits* instead of *homotopy (co)limits*, leading to a lighter terminology.

As a special case we have the following definition; see again Examples 14.3.12.

Definition 14.3.25. Let $\mathcal{C}$ be an ∞-category and let $q\colon \square \to \mathcal{C}$ be a square.

(i) The square q is a *pushout* if $q\colon (\Lambda^2_0)^{\rhd} \to \mathcal{C}$ is a colimiting cocone.

(ii) The square q is a *pullback* if $q\colon (\Lambda^2_2)^{\lhd} \to \mathcal{C}$ is a limiting cone.

Having a good definition of (co)limits available, one would now like to establish batteries of techniques to play with these notions. Many classical facts from the calculus of ordinary (co)limits can be extended to the context of ∞-categories, although the proofs of even fundamental statements are significantly harder in this framework. Nevertheless, such an extension was achieved to an impressive extent by Lurie in [82]. Instead of pursuing this any further, we conclude this section by mentioning the following important theorem [82, Cor. 4.2.4.8].

Theorem 14.3.26. *The underlying ∞-category of a simplicial, combinatorial model category is complete and cocomplete.*

14.4 Presentable ∞-categories and the relation to model categories

In this section we give a short introduction to *presentable ∞-categories*, a class of ∞-categories having very good formal properties. For example, the adjoint functor theorem of Freyd, in this context, takes the form that a functor is a left adjoint if and only if it preserves colimits, and there is also a result for right adjoint functors. Besides these good formal properties, it turns out that a good deal of mathematics is encoded by presentable ∞-categories. Many typical ∞-categories showing up in nature, e.g., in algebra, topology, or (derived) algebraic geometry are presentable.

The main reason for us to include a short discussion of presentable ∞-categories is that they play an essential role in Lurie's treatment of the *smash product* on the ∞-category of spectra (see §14.6). Moreover, they allow us to indicate a more precise relation between ∞-categories and model categories (see §14.4.2 and, in particular, Theorem 14.4.15). The reader can skip this section on a first reading since this will only affect the understanding of some of the statements in §14.6.

The theory of presentable ∞-categories has (at least) two precursors; *locally presentable categories* in the classical context as well as *combinatorial model categories* in the homotopical framework. We emphasize that, in all these three cases, there are two main ideas in the background, which help organize the material.

(i) The idea of passing from a small category to a cocomplete category in a universal way is realized by categories of presheaves in the classical case, and by model categories and ∞-categories of simplicial presheaves in the remaining two. One can think of this passage as some kind of *free generation*.

(ii) All locally presentable categories can be obtained as certain localizations of presheaf categories, and similarly in the homotopical contexts using suitable notions of (Bousfield) localizations. The passage to (Bousfield) localizations can be thought of as a way of *imposing relations*.

14.4.1 Locally presentable categories

In this subsection we give a short review of the theory of locally presentable categories. For more details we refer to [42, 87, 2] or [18, 19].

To begin with, let us again take up the idea that passage to a category of presheaves (with values in sets) can be regarded as a form of cocompletion. If A is a small category, then we denote by $y\colon A \to \mathrm{Fun}(A^{\mathrm{op}}, \mathcal{S}et)$ the Yoneda embedding which sends $a \in A$ to the represented presheaf $\hom_A(-, a)$. Associated to each presheaf $X\colon A^{\mathrm{op}} \to \mathcal{S}et$ there is the comma category (y/X). Objects of (y/X) are pairs (a, α) consisting of an object $a \in A$ and a natural transformation $\alpha\colon y(a) \to X$, and a morphism $f\colon (a, \alpha) \to (a', \alpha')$ is a morphism $f\colon a \to a'$ in A such that the diagram

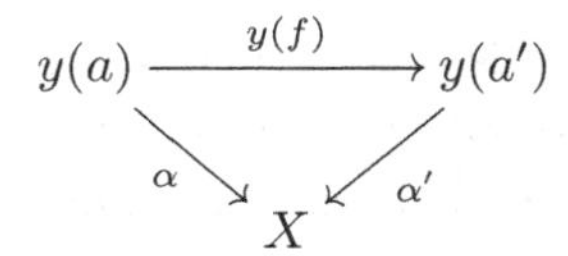

commutes. (Note that the Yoneda lemma implies that this category is isomorphic to the *category of elements* of X.) The category (y/X) comes with a projection functor $p\colon (y/X) \to A$ sending a pair (a, α) to a, and we can consider the composition

$$(y/X) \xrightarrow{p} A \xrightarrow{y} \mathrm{Fun}(A^{\mathrm{op}}, \mathcal{S}et).$$

Using these diagrams one can make precise our idea that presheaves on small categories are canonically colimits of representable ones (see for example [85, p.76]).

Proposition 14.4.1. *Let A be a small category and let $X\colon A^{\mathrm{op}} \to \mathcal{S}et$ be a set-valued presheaf. There is a canonical isomorphism*

$$\mathrm{colim}_{(y/X)}\, y \circ p \cong X.$$

Thus, we think of the passage to presheaves as a *cocompletion* or, more informally, as a *free generation*. In order to make a more precise statement, let $\mathrm{Fun}^{\mathrm{L}}(-,-)$ denote the category of colimit-preserving functors. The following result seems to go back to Ulmer [110, Rmk. 2.29].

Theorem 14.4.2. *Let A be a small category and let $\mathcal{C}$ be a cocomplete category. The restriction along the Yoneda embedding $y\colon A \to \mathrm{Fun}(A^{\mathrm{op}}, \mathcal{S}et)$ induces an equivalence of categories*

$$y^*\colon \mathrm{Fun}^{\mathrm{L}}(\mathrm{Fun}(A^{\mathrm{op}}, \mathcal{S}et), \mathcal{C}) \xrightarrow{\sim} \mathrm{Fun}(A, \mathcal{C}).$$

To motivate the notion of a locally presentable category we recall Freyd's Adjoint Functor Theorem. Let $F\colon \mathcal{C} \to \mathcal{D}$ be a functor between cocomplete categories. It is obvious that if F is a left adjoint, then F preserves all colimits, but, in general, the converse is not true. The celebrated *Adjoint Functor Theorem* of Freyd (see [40, pp. 84-86] or [85, p. 121]) gives *necessary and sufficient* conditions for the existence of a right adjoint. Slightly more precisely, the following are equivalent for a functor $F\colon \mathcal{C} \to \mathcal{D}$ between cocomplete categories.

(i) The functor F is a left adjoint.

(ii) The functor F preserves colimits and the *solution-set-condition* is satisfied.

Without going into detail, let us only mention that the solution-set-condition states that a certain *class* of arrows turns out to be small enough to actually form a *set*. Hence, one can imagine that this condition is automatically satisfied if we impose some 'smallness conditions' on the categories. Recall that a *small, cocomplete* category is necessarily a poset ([85, p. 114]). Thus, in order to not rule out interesting examples, it is essential that these 'smallness conditions' are chosen in a smart way.

Definition 14.4.3. A category is *locally presentable* if it is cocomplete and accessible.

The accessibility assumption in this definition is the smallness assumption alluded to above. The idea is that a category $\mathcal{C}$ is accessible if it admits certain filtered colimits and if it is formally determined by some small subcategory $\mathcal{D}$ consisting of small objects. To make this more precise, let us consider a regular cardinal number κ. A category $\mathcal{C}$ is *κ-accessible* if $\mathcal{C}$ admits κ-filtered colimits and if there is a small subcategory $\mathcal{D} \subseteq \mathcal{C}$ such that the following two conditions are satisfied.

(i) Every object of $\mathcal{C}$ can be canonically written as a κ-filtered colimit of objects in $\mathcal{D}$.

(ii) The set-valued functors $\hom_{\mathcal{C}}(d, -) \colon \mathcal{C} \to \mathcal{S}et$, $d \in \mathcal{D}$, preserve κ-filtered colimits — expressing the idea that objects in $\mathcal{D}$ are small.

We say that a category is *accessible* if it is κ-accessible for some κ. Similarly, a functor $F \colon \mathcal{C} \to \mathcal{D}$ is *κ-accessible* if $\mathcal{C}$ and $\mathcal{D}$ admit κ-filtered colimits and if they are preserved by F. Finally, a functor is *accessible* if it is κ-accessible for some κ.

For more details on the rich theory of accessible and locally presentable categories we again refer to [42, 87, 2] or [18, 19]. But let us emphasize that every locally presentable category is also a *complete* category. To indicate the ubiquity of locally presentable categories, we include the following list of examples.

Examples 14.4.4. (i) The category $\mathcal{S}et$ of sets is locally presentable.

(ii) If A is a small category, then the category $\mathrm{Fun}(A^{\mathrm{op}}, \mathcal{S}et)$ of presheaves on A is locally presentable. In particular, the category $s\mathcal{S}et$ of simplicial sets is locally presentable.

(iii) For a ring R the categories $\mathrm{Mod}(R)$ of R-modules and $\mathrm{Ch}(R)$ of chain complexes over R are locally presentable.

(iv) Recall that an abelian category with exact filtered colimits is *Grothendieck abelian* (see [52] and [38, §14]) if it admits a generator. It can be shown that an abelian category with exact filtered colimits is Grothendieck if and only if it is locally presentable. Important examples are given by categories of quasi-coherent $\mathcal{O}_X$-modules on any scheme X.

(v) Categories of modules of (multi-sorted) algebraic theories are locally presentable [3, §6].

(vi) If $T \colon \mathcal{C} \to \mathcal{C}$ is an accessible monad on a locally presentable category, then the category of T-algebras is locally presentable [2].

(vii) Every Grothendieck topos [6, 20, 86] is locally presentable.

(viii) The category $\mathcal{C}at$ of small categories is locally presentable. More generally, if $\mathcal{M}$ is a locally presentable, symmetric monoidal category, then the category $\mathcal{C}at_{\mathcal{M}}$ of $\mathcal{M}$-enriched categories is locally presentable (see [75]).

(ix) The category $\mathcal{T}op$ of topological spaces is *not* locally presentable, but this can be fixed by passing to the Quillen equivalent category of Δ-generated spaces. It is shown in [39] that this category is locally presentable.

Here is the simplified form of the adjoint functor theorem for locally presentable categories.

Theorem 14.4.5. *(i) A functor between locally presentable categories is a left adjoint if and only if it preserves colimits.*

(ii) A functor between locally presentable categories is a right adjoint if and only if it preserves limits and is accessible.

Other forms of representability results have appeared in various settings and these include the classical Brown representability theorem in stable homotopy theory [24], the Brown representability results for triangulated categories [90, §8], Watt's theorems in homological algebra [100, §5.3], representability theorems for Grothendieck categories [72, p.186], as well as versions of Watt's theorem in homotopical algebra [62].

It turns out that up to equivalence *all* locally presentable categories can be obtained as certain localizations of presheaf categories. To state this more precisely, let us begin by recalling that a *reflective localization* is an adjunction $(L, R)\colon \mathcal{C} \rightleftarrows \mathcal{D}$ such that the right adjoint R is fully faithful (see for example [18, §3.5 and §5.3]). In this situation it follows that $\mathcal{D}$ is equivalent to the localization $\mathcal{C}[S^{-1}]$ where S is the class of morphisms in $\mathcal{C}$ that are sent to isomorphisms by L. This is nicely illustrated by the following example.

Example 14.4.6. Let $\delta\colon \mathcal{S}et \to s\mathcal{S}et$ be the discrete simplicial set functor. The adjunction $(\pi_0, \delta)\colon s\mathcal{S}et \rightleftarrows \mathcal{S}et$ is a reflective localization. Thus, if in $s\mathcal{S}et$ we invert all maps inducing isomorphisms on π_0 then we simply get (discrete simplicial) sets.

We want to emphasize that the typical terminology from *Bousfield localization theory* [21] makes already perfectly well sense in this classical context, and that reflective localizations can be nicely described using that terminology (see Proposition 14.4.8).

Definition 14.4.7. Let $\mathcal{C}$ be a category and let S be a class of morphisms in $\mathcal{C}$.

(i) An object c in $\mathcal{C}$ is *S-local* if $f^*\colon \hom_{\mathcal{C}}(c_2, c) \to \hom_{\mathcal{C}}(c_1, c)$ is a bijection for all $f\colon c_1 \to c_2$ in S.

(ii) A morphism $f\colon c_1 \to c_2$ in $\mathcal{C}$ is an *S-local equivalence* if for all S-local objects $c \in \mathcal{C}$ the map $f^*\colon \hom_{\mathcal{C}}(c_2, c) \to \hom_{\mathcal{C}}(c_1, c)$ is a bijection.

The following is straightforward.

Proposition 14.4.8. *Let $(L, R)\colon \mathcal{C} \rightleftarrows \mathcal{D}$ be a reflective localization and let S be the morphisms in $\mathcal{C}$ that are inverted by L.*

(i) *The essential image of R consists precisely of the S-local objects.*

(ii) *The S-local equivalences are precisely the maps in S.*

With this preparation, we now state the 'classification result of locally presentable categories' (see for example [2]). Let us recall that an *accessible, reflective localization* is a reflective localization (L, R) such that the right adjoint R is accessible, i.e., preserves κ-filtered colimits for a sufficiently large regular cardinal κ.

Theorem 14.4.9. *A category is locally presentable if and only if it is equivalent to an accessible, reflective localization of* $\mathrm{Fun}(A^{\mathrm{op}}, \mathcal{S}et)$ *for some small category A.*

Thus, a category is locally presentable if and only if it can be obtained from a small category by a free generation (cocompletion) followed by imposing relations in a suitable way (accessible reflective localization). In §14.4.2 we will see that there are variants of Theorem 14.4.2 and Theorem 14.4.9 valid in the context of ∞-categories (and also in the context of model categories).

14.4.2 Presentable ∞-categories

One reason why *set-valued presheaves* play such a central role in classical category theory is that questions about the existence of universal constructions can be reformulated as representability questions for certain set-valued presheaves. In higher category theory, the representable functors take values in the category of simplicial sets. So one might expect that the central role is now taken by *simplicial presheaf categories.* In the world of model categories, these were intensively studied by Jardine (see for example [65]).

We now discuss simplicial presheaves in the framework of ∞-categories. Recall that we denote the ∞-category of spaces by $\mathcal{S} = N_\Delta(\mathcal{K}an)$ (see Examples 14.2.31). Given a simplicial set K, the ∞-category $\mathcal{P}(K)$ of *(simplicial) presheaves* on K is defined by

$$\mathcal{P}(K) = \mathrm{Fun}(K^{\mathrm{op}}, \mathcal{S}).$$

It follows from Proposition 14.3.4 that $\mathcal{P}(K)$ is an ∞-category.

In order to define the Yoneda embedding we recall that we have the adjunction $(C[-], N_\Delta)\colon s\mathcal{S}et \rightleftarrows s\mathcal{C}at$; see (14.2.8). The simplicial category $C[K]$ of an arbitrary simplicial set K is not locally fibrant. This can be fixed by choosing a product-preserving fibrant replacement functor for $s\mathcal{S}et$, like the one induced by the Quillen equivalence $s\mathcal{S}et \rightleftarrows \mathcal{T}op$ or Kan's Ex^∞-functor; see [71]. Composing the mapping space functor of $C[K]$ with such a replacement functor, we obtain a simplicial functor $C[K]^{\mathrm{op}} \times C[K] \to \mathcal{K}an$. Combining this with the canonical map $C[K^{\mathrm{op}} \times K] \to C[K]^{\mathrm{op}} \times C[K]$ and passing to adjoints, this yields a map $K^{\mathrm{op}} \times K \to N_\Delta(\mathcal{K}an) = \mathcal{S}$. The exponential law finally gives us the *Yoneda embedding*

$$y\colon K \to \mathrm{Fun}(K^{\mathrm{op}}, \mathcal{S}) = \mathcal{P}(K), \tag{14.4.1}$$

which can be shown to be fully faithful ([82, Prop. 5.1.3.1]).

The Yoneda embedding provides a model for the cocompletion. In order to make this precise let us introduce the following notation. Given ∞-categories $\mathcal{C}$ and $\mathcal{D}$, we denote by

$$\mathrm{Fun}^{\mathrm{L}}(\mathcal{C}, \mathcal{D}) \subseteq \mathrm{Fun}(\mathcal{C}, \mathcal{D})$$

the full subcategory spanned by the *colimit-preserving* functors. Lurie establishes the following result ([82, Thm. 5.1.5.6]) which is an ∞-categorical version of Theorem 14.4.2. (See also [31] and [99] for a variant in the language of model categories.)

Theorem 14.4.10. *Let K be a small simplicial set and let $\mathcal{C}$ be a cocomplete ∞-category. Restriction along the Yoneda embedding* (14.4.1) *induces an equivalence of ∞-categories*

$$\mathrm{Fun}^{\mathrm{L}}(\mathcal{P}(K), \mathcal{C}) \xrightarrow{\sim} \mathrm{Fun}(K, \mathcal{C}).$$

In particular, the *∞-category $\mathcal{S}$ of spaces is freely generated by $\Delta^0 \in \mathcal{S}$* in the sense of the following corollary.

Corollary 14.4.11. *For any cocomplete ∞-category $\mathcal{C}$ the evaluation on the 0-simplex $\Delta^0 \in \mathcal{S}$ induces an equivalence of ∞-categories*

$$\mathrm{Fun}^{\mathrm{L}}(\mathcal{S}, \mathcal{C}) \xrightarrow{\sim} \mathcal{C}.$$

In [82, §4] Lurie establishes batteries of techniques which allow us to manipulate (co)limits in the context of ∞-categories. In [82, §5] more advanced notions are introduced, including filtered colimits and small objects. Once the basic notions are in place, more advanced concepts from classical category theory can be formally extended to the ∞-categorical framework. We want to emphasize once more that although impressively many *statements* are still true in this more general framework, at least at present the *proofs*, in general, are more involved.

Definition 14.4.12. An ∞-category is *presentable* if it is cocomplete and accessible.

As in ordinary category theory, it can be shown that a presentable ∞-category is automatically also complete ([82, Cor. 5.5.2.4]). There are also ∞-categorical variants of Freyd's special adjoint functor theorem (see Theorem 14.4.5). A first step towards this of course consists of making precise the notion of an adjunction between two ∞-categories. We will not get into this here and instead refer the reader to [82, p.337] and [80, §3]. With this concept at hand, there is the following result due to Lurie ([82, Cor. 5.5.2.9]).

Theorem 14.4.13. *(i) A functor between presentable ∞-categories is a left adjoint if and only if it preserves colimits.*

(ii) A functor between presentable ∞-categories is a right adjoint if and only if it preserves limits and is accessible.

Lurie also establishes a classification result for presentable ∞-categories similar to the one given by Theorem 14.4.9. Using the concept of an adjunction of ∞-categories, one can extend additional concepts from ordinary category theory to the context of ∞-categories. A functor between two ∞-categories is a *reflective localization* if it admits a fully faithful right adjoint, and there is also the notion of an *accessible, reflective localization.* With these definitions at hand, the classification result for presentable ∞-categories takes the following form as proved by Lurie as [82, Thm. 5.5.1.1]. In that section, Lurie attributes the result to Simpson [106].

Theorem 14.4.14. *For an ∞-category $\mathcal{C}$ the following are equivalent.*

(i) The ∞-category $\mathcal{C}$ is presentable.

(ii) There is a small ∞-category $\mathcal{D}$ such that $\mathcal{C}$ is an accessible, reflective localization of $\mathcal{P}(\mathcal{D})$.

In light of Theorem 14.4.14, if one wants to understand presentable ∞-categories then it seems to be important to have good control over accessible localizations of ∞-categories of presheaves or, more generally, of presentable ∞-categories. Given a reflective localization $(L, R)\colon \mathcal{C} \rightleftarrows \mathcal{D}$ let us also write $L\colon \mathcal{C} \to \mathcal{C}$ for the *localization functor* $\mathcal{C} \to \mathcal{D} \to \mathcal{C}$, and let S_L be those maps in $\mathcal{C}$ that are sent to equivalences by L. An object $x \in \mathcal{C}$ is *S_L-local* if all maps $f^*\colon \mathrm{Map}_{\mathcal{C}}(z, x) \to \mathrm{Map}_{\mathcal{C}}(y, x)$ induced by $f \in S_L$ are weak equivalences. One can then show that the essential image of L consists precisely of the S_L-local objects (see [82, Prop. 5.5.4.2]).

In particular, an accessible localization L of a presentable ∞-category $\mathcal{C}$ is thus completely determined by the class S_L. In that case, the class S_L is closed under the formation of colimits in S_L as a subcategory of $\mathcal{C}^{[1]}$, is stable under the formation of retracts, contains the equivalences, satisfies the 2-out-of-3 property (with respect to 2-simplices), and is stable under cobase change. Lurie calls a class of morphisms with these closure properties *strongly saturated.* Intersections of strongly saturated classes are again strongly saturated as is the class of all morphisms. Thus, for each class T of morphisms in $\mathcal{C}$ there is a smallest strongly saturated class $\bar{T}$ that contains T. We call a strongly saturated class S *of small generation* if there is a *subset* $T \subseteq S$ such that $S = \bar{T}$. Lurie establishes the wonderful fact that a strongly saturated class S in a presentable ∞-category $\mathcal{C}$ is of small generation if and only if there is an accessible localization $L\colon \mathcal{C} \to \mathcal{C}$ such that $S = S_L$ (see [82, §5.5.4]).

Lurie then shows that in the case of simplicial presheaves the localization theory of ∞-categories interacts nicely with the localizaton theory of certain associated model categories (see [82, Appendix 3.7]). Having established all this theory, he is then able to build on Dugger's work [30] in order to deduce the following result (see [82, Appendix 3]). Recall that a model category is combinatorial if and only if it is cofibrantly generated and the underlying category is locally presentable (see for instance [12, 30, 26, 56, 98]).

Theorem 14.4.15. *An ∞-category $\mathcal{C}$ is presentable if and only if there is a combinatorial, simplicial model category $\mathcal{M}$ such that $\mathcal{C}$ is equivalent to the underlying ∞-category $N_\Delta(\mathcal{M}_{\mathrm{cf}})$.*

More generally, the underlying ∞-category of a combinatorial model category is presentable by [83, Prop. 1.3.4.22].

In this section we considered the 'theme of locally presentable categories' in three different frameworks, namely in ordinary category theory, in ∞-category theory, and briefly in model category theory. We conclude this section with a short perspective in which we give an outlook on a similar picture for *Grothendieck topoi* (and we refer to [82, §5.5.8] and [28, 44] for a similar picture on *algebraic categories*).

Perspective 14.4.16. Let us recall that a *Grothendieck topos* can be defined as a category equivalent to a category of set-valued sheaves on a Grothendieck site; references for this vast subject include the original [6, 7, 8] and the monographs [20, 86]. It turns out that Grothendieck topoi admit a different characterization as suitable localizations of presheaf categories. The 'theme of sheaves and topoi' was taken up again in homotopical frameworks, both using the language of model categories (see for example [65, 66, 33, 109, 95]) as well as in the ∞-categorical picture [82, §§6-7].

14.5 Monoidal and symmetric monoidal ∞-categories

In this section we give an introduction to the theory of monoidal ∞-categories. Let us recall that a monoidal structure on a category $\mathcal{M}$ consists of a monoidal pairing $\otimes\colon \mathcal{M}\times\mathcal{M} \to \mathcal{M}$ and a monoidal unit $\mathbb{S} \in \mathcal{M}$ together with three natural isomorphisms, namely the associativity and left and right unitality constraints. Moreover, to obtain the notion we have in mind we have to impose certain compatibility assumptions. In particular, we have to ask axiomatically that the two ways of comparing four-fold products $((X\otimes Y)\otimes Z)\otimes W$ and $X\otimes (Y\otimes (Z\otimes W))$ as described by the boundary in

$$\begin{array}{ccc} ((X\otimes Y)\otimes Z)\otimes W & \longrightarrow & (X\otimes Y)\otimes (Z\otimes W) \\ \downarrow & & \downarrow \\ (X\otimes (Y\otimes Z))\otimes W & & X\otimes (Y\otimes (Z\otimes W)) \\ & \searrow \quad X\otimes ((Y\otimes Z)\otimes W) \quad \nearrow & \end{array} \tag{14.5.1}$$

coincide. This leads to the *classical presentation* of monoidal categories ([84, 73]).

In the ∞-categorical setting this presentation will not model the good notion anymore. Instead one expects a monoidal structure on an ∞-category to be some kind of a monoidal pairing that is coherently associative and unital in the sense of $\mathbb{A}_\infty$-multiplications. In particular, it is therefore insufficient to consider Mac Lane's pentagon (14.5.1) and instead one expects that all Stasheff associahedra with their complicated combinatorics play a key role (see [108] or [89, §I.1.6 and §II.1.6]).

Luckily, all this structure does not have to be made explicit if one chooses a *different presentation* of ordinary monoidal categories, namely, as suitable Grothendieck opfibrations over Δ^{op} (the same observation but in the context of $\mathbb{A}_\infty$-spaces motivated Adams to refer to the category Δ as a 'storehouse of formulas' [5]). This different presentation extends more readily to ∞-categories and is obtained by combining two main ideas, namely

(i) the Segal perspective on $\mathbb{A}_\infty$-monoids as 'special simplicial objects' and

(ii) the Grothendieck construction applied to category-valued functors.

The passage to *symmetric* monoidal ∞-categories then essentially amounts to a change of combinatorics, i.e., one replaces the category Δ^{op} by a skeleton $\mathcal{F}\mathrm{in}$ of the category of finite pointed sets.

14.5.1 Monoidal categories via Grothendieck opfibrations

Let us consider the following classical situation. Let $p\colon \mathcal{C} \to \mathcal{D}$ be a functor between ordinary categories and let $d \in \mathcal{D}$ be an object. We denote by $\mathcal{C}_d$, the *fiber* of p over d, defined by the following pullback diagram

$$\begin{array}{ccc} \mathcal{C}_d & \longrightarrow & \mathcal{C} \\ \downarrow & \lrcorner & \downarrow{\scriptstyle p} \\ [0] & \underset{d}{\longrightarrow} & \mathcal{D}. \end{array}$$

Thus, $\mathcal{C}_d \subseteq \mathcal{C}$ is the (in general not full) subcategory given by the objects $c \in \mathcal{C}$ such that $p(c) = d$ and those morphisms in $\mathcal{C}$ that are sent to id_d . A functor $p\colon \mathcal{C} \to \mathcal{D}$ can always be thought of as a collection of categories $\mathcal{C}_d$ parametrized by the objects in $\mathcal{D}$.

Our first aim is to find conditions which ensure that the fiber $\mathcal{C}_d$ *depends covariantly* on the object $d \in \mathcal{D}$.

Definition 14.5.1. Let $p\colon \mathcal{C} \to \mathcal{D}$ be a functor and let $f\colon c_1 \to c_2$ be a morphism in $\mathcal{C}$ with image $p(f) = \alpha\colon d_1 \to d_2$. The morphism f is *p-coCartesian* or a *p-coCartesian lift* of α if it has the following property: For every $h\colon c_1 \to c_3$ in $\mathcal{C}$ with image $\gamma = p(h)\colon d_1 \to d_3$ and every $\beta\colon d_2 \to d_3$ such that $\gamma = \beta \circ \alpha$ there is a unique $g\colon c_2 \to c_3$ in $\mathcal{C}$ such that

$$\beta = p(g) \qquad \text{and} \qquad h = g \circ f.$$

Thus, the defining property of a p-coCartesian arrow can be described by the following diagram

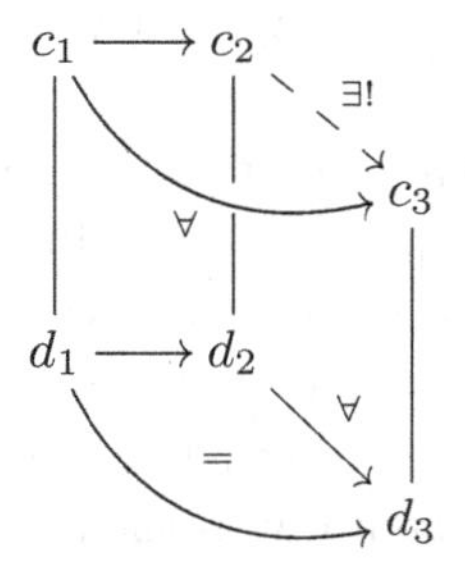

in which the vertical lines indicate the effect of an application of $p\colon \mathcal{C} \to \mathcal{D}$. Note that a morphism $f\colon c_1 \to c_2$ is p-coCartesian if and only if the diagram

$$\begin{array}{ccc} \hom_{\mathcal{C}}(c_2, c_3) & \xrightarrow{\quad f^* \quad} & \hom_{\mathcal{C}}(c_1, c_3) \\ {\scriptstyle p}\downarrow & & \downarrow{\scriptstyle p} \\ \hom_{\mathcal{D}}(p(c_2), p(c_3)) & \underset{p(f)^*}{\longrightarrow} & \hom_{\mathcal{D}}(p(c_1), p(c_3)) \end{array} \tag{14.5.2}$$

is a pullback diagram for all objects $c_3 \in \mathcal{C}$. This slightly cryptic reformulation of Definition 14.5.1 will have its uses when it comes to extending these notions to the setting of ∞-categories (see Lemma 14.5.5 and Definition 14.5.6). To develop some feeling for the notion we recommend the reader to give the easy proof of the following lemma.

Lemma 14.5.2. *Let $f'\colon c \to c'$ and $f''\colon c \to c''$ be p-coCartesian arrows with the same image $\alpha = p(f') = p(f'')$. Then there is a unique isomorphism $\phi\colon c' \to c''$ in the fiber $\mathcal{C}_{p(c')} = \mathcal{C}_{p(c'')}$ such that $\phi \circ f' = f''$.*

This lemma tells us that p-coCartesian lifts with a fixed domain are essentially unique if they exist. In particular, for $\alpha\colon d_1 \to d_2$ the targets of p-coCartesian lifts are (uniquely compatibly) isomorphic as objects in the fiber $\mathcal{C}_{d_2}$. Thus, in order to obtain a covariant dependence of the fiber it seems to be a good strategy to ask for a sufficient supply of p-coCartesian arrows.

Definition 14.5.3. A functor $p\colon \mathcal{C} \to \mathcal{D}$ is a *Grothendieck opfibration* if for all $c_1 \in \mathcal{C}$ and for all morphisms α in $\mathcal{D}$ with domain $p(c_1)$ there is a p-coCartesian lift $f\colon c_1 \to c_2$ of α.

Let $p\colon \mathcal{C} \to \mathcal{D}$ be a Grothendieck opfibration. Then we can *choose* for each $c \in \mathcal{C}$ and for each morphism $\alpha\colon p(c) \to d$ a p-coCartesian lift. We now fix a morphism $\alpha\colon d_1 \to d_2$ in $\mathcal{D}$ and define

$$\alpha_!\colon \mathcal{C}_{d_1} \to \mathcal{C}_{d_2}\colon c_1 \mapsto c_2,$$

where c_2 is the codomain of the *chosen* p-coCartesian lift $f\colon c_1 \to c_2$ of α. This defines $\alpha_!$ on objects, and we recommend the reader to check that this can be extended to define a functor $\alpha_!\colon \mathcal{C}_{d_1} \to \mathcal{C}_{d_2}$.

If we now consider an additional morphism $\beta\colon d_2 \to d_3$ in $\mathcal{D}$, then we obtain associated functors

$$\mathcal{C}_{d_1} \xrightarrow{\alpha_!} \mathcal{C}_{d_2} \xrightarrow{\beta_!} \mathcal{C}_{d_3}, \qquad \mathcal{C}_{d_1} \xrightarrow{(\beta\circ\alpha)_!} \mathcal{C}_{d_3}.$$

In general, these two functors are not equal since their definitions depend on certain choices of p-coCartesian lifts: they send an object $c_1 \in \mathcal{C}_{d_1}$ to the respective targets of two possibly different lifts of $\beta\circ\alpha$ to p-coCartesian morphisms with domain c_1 (Exercise: The composition of two p-coCartesian morphisms is again p-coCartesian.). But one can deduce from Lemma 14.5.2 that there is a unique natural isomorphism

$$\beta_! \circ \alpha_! \cong (\beta \circ \alpha)_! \tag{14.5.3}$$

of functors $\mathcal{C}_{d_1} \to \mathcal{C}_{d_3}$, i.e., all components of the natural isomorphism are sent to the identity of d_3 via p.

As an upshot, we essentially succeeded in obtaining a covariant dependence of the fiber by considering Grothendieck opfibrations. To put this in a slightly technical language, we have observed that for such functors the fibers depend *pseudo-functorially* on $d \in \mathcal{D}$, i.e., that the assignment $d \mapsto \mathcal{C}_d$ defines a pseudo-functor $\mathcal{D} \to \mathcal{C}AT$. One might be disappointed about this lack of strict functoriality, but for our purposes this is very convenient: it allows us to *encode* or, better, *hide a lot of structure* in the natural isomorphisms (14.5.3) associated to a Grothendieck opfibration. We illustrate this by the following example.

Example 14.5.4. Let $\mathcal{M}$ be a monoidal category with monoidal pairing $\otimes\colon \mathcal{M} \times \mathcal{M} \to \mathcal{M}$ and monoidal unit $\mathbb{S} \in \mathcal{M}$. We form a new category $\mathcal{M}^{\otimes}$ in the following way. The objects of $\mathcal{M}^{\otimes}$ are (possibly empty) finite sequences of objects in $\mathcal{M}$,

$$(M_1, \ldots, M_n), \quad n \geq 0,\ M_i \in \mathcal{M}.$$

Given two such sequences $(M_1, \ldots, M_n)$ and $(L_1, \ldots, L_k)$, a morphism

$$(\alpha, \{f_i\}_i) \colon (M_1, \ldots, M_n) \to (L_1, \ldots, L_k)$$

consists of a morphism $\alpha \colon [k] \to [n]$ in Δ together with morphisms

$$f_i \colon M_{\alpha(i-1)+1} \otimes \ldots \otimes M_{\alpha(i)} \to L_i, \quad i = 1, \ldots, k. \tag{14.5.4}$$

Thus given such a morphism, α encodes the domains of the f_i. In particular, if there is an $i \in [k]$ such that $\alpha(i-1) = \alpha(i)$, then by convention the corresponding map (14.5.4) is to be read as a map $f_i \colon \mathbb{S} \to L_i$. The composition of morphisms in $\mathcal{M}^{\otimes}$ is defined using the compositions in Δ and $\mathcal{M}$ together with the associativity constraints of the monoidal structure on $\mathcal{M}$. The identity of an object $(M_1, \ldots, M_n)$ is easily seen to be given by $(\mathrm{id}_{[n]}, \{\mathrm{id}_{M_i}\}_i)$.

There is an obvious projection functor $p \colon \mathcal{M}^{\otimes} \to \Delta^{\mathrm{op}}$ which sends a string $(M_1, \ldots, M_n)$ to $[n]$ and a morphism $(\alpha, \{f_i\}_i)$ to its first component α. One easily checks that $p \colon \mathcal{M}^{\otimes} \to \Delta^{\mathrm{op}}$ is a Grothendieck opfibration. Indeed, let us consider a lifting problem given by an object $(M_1, \ldots, M_n)$ in $\mathcal{M}^{\otimes}_{[n]}$ together with a morphism $\alpha^{\mathrm{op}} \colon [n] \to [k]$ in Δ^{op},

$$\begin{array}{ccc} (M_1, \ldots, M_n) & & \\ \big| & & \\ [n] & \xrightarrow[\alpha^{\mathrm{op}}]{} & [k], \end{array}$$

$$[n] \xleftarrow[\alpha]{} [k].$$

Then a p-coCartesian lift of α with domain the given string is obtained from any family of isomorphisms

$$f_i \colon M_{\alpha(i-1)+1} \otimes \cdots \otimes M_{\alpha(i)} \xrightarrow{\cong} L_i, \quad i = 1, \ldots, k.$$

More precisely, these L_i specify an object $(L_1, \ldots, L_k) \in \mathcal{M}^{\otimes}_{[k]}$, and the morphism

$$(\alpha, \{f_i\}_i) \colon (M_1, \ldots, M_n) \to (L_1, \ldots, L_k) \tag{14.5.5}$$

is the desired p-coCartesian lift.

Now, this Grothendieck opfibration $p \colon \mathcal{M}^{\otimes} \to \Delta^{\mathrm{op}}$ has the property that the fiber $\mathcal{M}_{[n]}$ is canonically equivalent to the n-fold product of $\mathcal{M}^{\otimes}_{[1]} \simeq \mathcal{M}$ in the following sense. Let $\iota_{\{i-1,i\}} \colon [1] \to [n]$ be the inclusion of the i-th length one interval, i.e., the unique monomorphism in Δ with image $\{i-1, i\}$, and let us write $\iota_i = \iota^{\mathrm{op}}_{\{i-1,i\}} \colon [n] \to [1]$ for the opposite morphism in Δ^{op}. Since $p \colon \mathcal{M}^{\otimes} \to \Delta^{\mathrm{op}}$ is a Grothendieck opfibration, we obtain induced functors

$$(\iota_i)_! \colon \mathcal{M}^{\otimes}_{[n]} \to \mathcal{M}^{\otimes}_{[1]} = \mathcal{M}, \qquad i = 1, \ldots, n, \tag{14.5.6}$$

which, taken together, induce the *Segal maps*

$$\sigma = ((\iota_1)_!, \ldots, (\iota_n)_!) \colon \mathcal{M}^{\otimes}_{[n]} \xrightarrow{\simeq} \underbrace{\mathcal{M} \times \ldots \times \mathcal{M}}_{n \text{ times}}. \tag{14.5.7}$$

Note that the explicit construction of p-coCartesian lifts in (14.5.5) implies that these Segal maps are equivalences. We refer to this observation by saying that the Grothendieck opfibration $p\colon \mathcal{M}^{\otimes} \to \Delta^{\mathrm{op}}$ satisfies the *Segal condition.* Let us emphasize that the Segal condition in simplicial degree zero amounts to saying that $\mathcal{M}^{\otimes}_{[0]}$ is equivalent to the terminal category $\mathbb{1}$.

It turns out that the monoidal product $\otimes\colon \mathcal{M} \times \mathcal{M} \to \mathcal{M}$ can be recovered up to equivalence from $p\colon \mathcal{M}^{\otimes} \to \Delta^{\mathrm{op}}$. But something seemingly more general works: Any Grothendieck opfibration $p\colon \mathcal{C} \to \Delta^{\mathrm{op}}$ satisfying the Segal condition defines a monoidal structure on the fiber $\mathcal{M} = \mathcal{C}_{[1]}$. We content ourselves by sketching a proof of this result. As usual, let $d_1\colon [2] \to [1]$ denote the face map in Δ^{op} that is opposite to the coface map $d^1\colon [1] \to [2]$. *Choosing* an inverse of the equivalence $\sigma\colon \mathcal{C}_{[2]} \to \mathcal{M} \times \mathcal{M}$ given by one of the Segal maps we can define a functor

$$\otimes\colon \mathcal{M} \times \mathcal{M} \overset{\simeq}{\leftarrow} \mathcal{C}_{[2]} \overset{(d_1)_!}{\to} \mathcal{C}_{[1]} = \mathcal{M}.$$

In order to construct an associativity constraint for $\otimes$ we will invoke the cosimplicial identity $d^2 \circ d^1 = d^1 \circ d^1\colon [1] \to [3]$. As a special instance of (14.5.3) we thus obtain a natural isomorphism

$$(d_1)_! \circ (d_2)_! \cong (d_1)_! \circ (d_1)_!\colon \mathcal{C}_{[3]} \to \mathcal{C}_{[1]} = \mathcal{M}.$$

It is an instructive exercise to use the Segal condition to translate this into a natural isomorphism

$$\alpha\colon M_1 \otimes (M_2 \otimes M_3) \cong (M_1 \otimes M_2) \otimes M_3, \qquad M_1, M_2, M_3 \in \mathcal{M}.$$

With slightly more effort, one can use the different factorizations of the map $\{0,4\}\colon [1] \to [4]$ to deduce that this associativity constraint satisfies Mac Lane's pentagon axiom; see (14.5.1). And finally, in a similar way one checks that a monoidal unit $\mathbb{S} \in \mathcal{M}$ and the remaining coherence isomorphisms are also encoded by the Grothendieck opfibration $p\colon \mathcal{C} \to \Delta^{\mathrm{op}}$ satisfying the Segal condition.

The point of this lengthy example was to show that there is an equivalent way of encoding monoidal structures. Instead of making a specific choice of a monoidal pairing, a monoidal unit, and coherence isomorphisms one can consider one 'global object' which nicely hides all this structure, namely a Grothendieck opfibration $p\colon \mathcal{C} \to \Delta^{\mathrm{op}}$ satisfying the Segal condition. In particular, one does not have to make precise the coherence axioms of a monoidal category since they will follow from the (co)simplicial identities. (In §14.5.3 we will briefly mention how monoidal functors and monoid objects fit into this opfibration picture.)

If one only cares about ordinary monoidal categories it might be arguable how large the benefit is by changing the classical perspective on monoidal categories to the Grothendieck opfibration picture. However, as mentioned in the introduction, in the ∞-categorical setting it allows us to avoid the complicated combinatorics of the Stasheff associahedra.

14.5.2 Monoidal ∞-categories via coCartesian fibrations

We now turn to ∞-categorical variants of the above concepts. The main reference for the remainder of §14.5 is Lurie's second book [83]. For the convenience of the reader we also include references to the former volumes [80, 81] which are now subsumed in [83] but which might be a bit more accessible as a first reference.

To begin with, we extend the notion of p-coCartesian morphisms to the context of ∞-categories. For this purpose it is handy to observe that the following is true (which is a refined version of the statement that (14.5.2) is a pullback square).

Lemma 14.5.5. *Let $p\colon \mathcal{C} \to \mathcal{D}$ be a functor between ordinary categories. A morphism $f\colon c_1 \to c_2$ in $\mathcal{C}$ is p-coCartesian if and only if the following functor is an isomorphism*

$$\mathcal{C}_{f/} \to \mathcal{C}_{c_1/} \underset{\mathcal{D}_{p(c_1)/}}{\times} \mathcal{D}_{p(f)/}.$$

In §14.3.3 we already introduced ∞-categorical versions of slice categories. With this lemma at hand, there is the following formal generalization of p-coCartesian arrows to the ∞-categorical setting [82, Definition 2.4.1.1].

Definition 14.5.6. Let $p\colon \mathcal{C} \to \mathcal{D}$ be a functor between ∞-categories. A morphism $f\colon c_1 \to c_2$ in $\mathcal{C}$ is *p-coCartesian* or a *p-coCartesian lift* of $\alpha = p(f)$ if the following map is an acyclic Kan fibration

$$\mathcal{C}_{f/} \to \mathcal{C}_{c_1/} \underset{\mathcal{D}_{p(c_1)/}}{\times} \mathcal{D}_{p(f)/}.$$

The ∞-categorical concept corresponding to Grothendieck opfibrations is that of a *coCartesian fibration* (Joyal [67] uses the terminology *Grothendieck opfibrations* instead). The main idea is again to axiomatically ask for a sufficient supply of coCartesian morphisms (see [82, Definition 2.4.2.1]).

Definition 14.5.7. A functor $p\colon \mathcal{C} \to \mathcal{D}$ between ∞-categories is a *coCartesian fibration* if the following two properties are satisfied.

(i) The functor p is an inner fibration (Definition 14.2.28).

(ii) For every object $c_1 \in \mathcal{C}$ and every morphism $\alpha\colon p(c_1) = d_1 \to d_2$ in $\mathcal{D}$, there is a p-coCartesian lift $f\colon c_1 \to c_2$ of α.

As in the case of categories, in this ∞-categorical context the fiber $\mathcal{C}_d$ of a functor $\mathcal{C} \to \mathcal{D}$ is defined via a pullback square,

$$\begin{array}{ccc} \mathcal{C}_d & \longrightarrow & \mathcal{C} \\ \downarrow & \lrcorner & \downarrow p \\ \Delta^0 & \xrightarrow[d]{} & \mathcal{D}. \end{array}$$

Lurie proves that a coCartesian fibration gives rise to a covariantly depending family of ∞-categories. The fact that the fibers are *∞-categories* is immediate since a coCartesian fibration is an inner fibration and inner fibrations are stable under pullbacks. The hard part is to show that they assemble to a functor; see Perspective 14.5.10. In particular, any map $\alpha\colon d_1 \to d_2$ in $\mathcal{D}$ induces an essentially unique functor $\alpha_!\colon \mathcal{C}_{d_1} \to \mathcal{C}_{d_2}$ defined by means of coCartesian lifts.

We saw in §14.5.1 that monoidal categories can be alternatively encoded by Grothendieck opfibrations satisfying the Segal condition. Having introduced the corresponding notion of coCartesian fibrations, we now turn this into a definition of monoidal ∞-categories [80, Definition 1.1.2].

Definition 14.5.8. A *monoidal ∞-category* is a coCartesian fibration $p\colon \mathcal{M}^{\otimes} \to N(\Delta^{\mathrm{op}})$ such that the Segal maps are equivalences,

$$\mathcal{M}^{\otimes}_{[n]} \xrightarrow{\sim} (\mathcal{M}^{\otimes}_{[1]})^{\times n}, \qquad n \geq 0.$$

For simplicity, we refer to the ∞-category $\mathcal{M} = \mathcal{M}^{\otimes}_{[1]}$ as a monoidal ∞-category. If one wants to be very precise, then one should call the coCartesian fibration $p\colon \mathcal{M}^{\otimes} \to N(\Delta^{\mathrm{op}})$ a *monoidal structure* on $\mathcal{M}$.

The interpretation of such a coCartesian fibration $p\colon \mathcal{M}^{\otimes} \to N(\Delta^{\mathrm{op}})$ is now similar to the one in classical category theory. To give an example we just make the following remark. One immediate consequence of the axioms is that the fiber $\mathcal{M}^{\otimes}_{[0]}$ is a contractible space. The unique map $s^0\colon [1] \to [0]$ in Δ induces a functor

$$\eta = (s_0)_!\colon \mathcal{M}^{\otimes}_{[0]} \to \mathcal{M}^{\otimes}_{[1]},$$

and we call any object in its image a *monoidal unit* of $\mathcal{M}$. We leave it to the reader to justify this terminology in the classical case.

It can be shown that given a monoidal ∞-category $\mathcal{C}$ the homotopy category $\mathrm{Ho}(\mathcal{C})$ inherits a monoidal structure (see [80, Rmk. 1.1.6]). This underlines the idea that the monoidal structure on $\mathcal{C}$ is associative and unital up to homotopy. Let us however emphasize that Definition 14.5.8 really encodes much more structure, namely that of a monoidal product that is associative and unital up to *coherent homotopy* (see Perspective 14.5.21 for a short discussion in the symmetric monoidal context).

We now turn to important examples of monoidal ∞-categories (but see also §14.6).

Examples 14.5.9. (i) Let $\mathcal{M}$ be a monoidal category and $p\colon \mathcal{M}^{\otimes} \to \Delta^{\mathrm{op}}$ be the associated Grothendieck opfibration. An application of the nerve functor yields a monoidal ∞-category

$$N(p)\colon N(\mathcal{M}^{\otimes}) \to N(\Delta^{\mathrm{op}}).$$

(ii) A special case of a monoidal category is given by the terminal category $\mathbb{1}$. The associated Grothendieck construction can be identified with the identity functor $\Delta^{\mathrm{op}} \to \Delta^{\mathrm{op}}$, and we obtain a monoidal ∞-category $N(\Delta^{\mathrm{op}}) \to N(\Delta^{\mathrm{op}})$. This examples is of some interest in the study of algebra objects (see §14.5.3).

(iii) Again, first 'honest examples' are obtained from suitable model categorical input, more precisely from suitably compatibly closed monoidal and simplicial model category structures, by passing to coherent nerves. Such a category comes with three additional structures, namely a simplicial enrichment, a monoidal structure, and a model structure, which have to be suitably compatible. The assumptions made by Lurie in [80, Prop. 1.6.5] are the following ones.

   (a) The closed monoidal structure is compatible with the enrichment in that $\otimes$ and the adjunction expressing the closedness are simplicial.

(b) The monoidal pairing $\otimes\colon \mathcal{M}\times\mathcal{M}\to\mathcal{M}$ is a *left Quillen bifunctor*.

(c) The monoidal unit $\mathbb{S}\in\mathcal{M}$ is cofibrant.

Under these assumptions, one can form a simplicial version of the category $\mathcal{M}^{\otimes}$ from §14.5.1. More precisely, given two finite strings $(M_1,\ldots,M_n)$ and $(L_1,\ldots,L_k)$ of objects in $\mathcal{M}$, the corresponding simplicial mapping space in $\mathcal{M}^{\otimes}$ is given by

$$\coprod_{\alpha\colon [k]\to[n]}\prod_{i=1}^{k}\mathrm{Map}_{\mathcal{M}}(M_{\alpha(i-1)+1}\otimes\ldots\otimes M_{\alpha(i)},L_i).$$

This simplicial category comes with an obvious simplicial functor $\mathcal{M}^{\otimes}\to\Delta^{\mathrm{op}}$ (regarding Δ^{op} as a discrete simplicial category). In order to obtain an ∞-category we consider the full simplicial subcategory

$$\mathcal{M}^{\otimes}_{\mathrm{cf}}\subseteq\mathcal{M}^{\otimes}$$

spanned by the finite strings of fibrant and cofibrant objects. It is then a consequence of the above compatibility assumptions that $\mathcal{M}^{\otimes}_{\mathrm{cf}}$ is a locally fibrant simplicial category so that $N_{\Delta}(\mathcal{M}^{\otimes}_{\mathrm{cf}})$ is an ∞-category (Corollary 14.2.30). In this situation, Lurie establishes as [80, Prop. 1.6.5] that

$$N_{\Delta}(\mathcal{M}^{\otimes}_{\mathrm{cf}})\to N_{\Delta}(\Delta^{\mathrm{op}})=N(\Delta^{\mathrm{op}})$$

endows the ∞-category $N_{\Delta}(\mathcal{M}_{\mathrm{cf}})$ with a monoidal structure.

This result and variants of it (such as [80, Thm. 1.6.16]) yield important special cases of monoidal ∞-categories. For example, the projective model structure on unbounded chain complexes over a commutative ring gives rise to the monoidal ∞-category of chain complexes [80, Ex. 1.6.17]. Similarly, the monoidal category of symmetric spectra can be endowed with a compatible model structure, and this leads to an extrinsic construction of the monoidal ∞-category of spectra [80, Ex. 1.6.18].

(iv) In ordinary category theory, important monoidal structures are given by categorical products and coproducts. These monoidal structures are referred to as the *Cartesian* and *coCartesian* monoidal structures, respectively. Given an ∞-category $\mathcal{C}$ with finite (co)products, then one can construct (co)Cartesian monoidal structures on $\mathcal{C}$, given by coCartesian fibrations

$$\mathcal{C}^{\times}\to N(\Delta^{\mathrm{op}})\qquad\text{and}\qquad\mathcal{C}^{\sqcup}\to N(\Delta^{\mathrm{op}}),$$

respectively. See [83, §2.4] for more details.

Perspective 14.5.10. CoCartesian fibrations are an ∞-categorical analogue of Grothendieck opfibrations. We saw in §14.5.1 that such an opfibration $p\colon\mathcal{C}\to\mathcal{D}$ encodes the idea of having a family of categories $\mathcal{C}_d, d\in\mathcal{D}$, that depends covariantly on the object $d\in\mathcal{D}$. More precisely, by choosing certain p-coCartesian lifts we obtain a pseudo-functor

$$F_p\colon\mathcal{D}\to\mathcal{C}AT\colon d\mapsto\mathcal{C}_d.$$

There is also a construction in the converse direction called the *Grothendieck construction* . Given a pseudo-functor $F\colon \mathcal{D} \to \mathcal{C}AT$, one can form a new category $\mathcal{E}(F)$ defined as follows.

(i) An object in $\mathcal{E}(F)$ is a pair (d, x) consisting of an object $d \in \mathcal{D}$ and an object $x \in F(d)$.

(ii) A morphism $(d, x) \to (d', x')$ in $\mathcal{E}(F)$ is a pair (α, f) consisting of a morphism $\alpha\colon d \to d'$ in $\mathcal{D}$ and a morphism $f\colon F(\alpha)(x) \to x'$ in $F(d')$.

(iii) Compositions and identities are defined in the obvious way.

The category $\mathcal{E}(F)$ comes with a forgetful functor

$$p_F\colon \mathcal{E}(F) \to \mathcal{D}$$

which projects objects and morphisms onto their respective first components, and one checks that p_F is a Grothendieck opfibration. In fact, given an object (d, x) in $\mathcal{E}(F)$ and a morphism $\alpha\colon p_F(d, x) = d \to d'$ in $\mathcal{D}$, then a p_F-coCartesian lift is given by

$$(\alpha, \mathrm{id}_{F(\alpha)(x)})\colon (d, x) \to (d', F(\alpha)(x)).$$

It turns out that if we fix a category $\mathcal{D}$, then these two constructions tell us that category-valued pseudo-functors defined on $\mathcal{D}$ and Grothendieck opfibrations with target $\mathcal{D}$ are essentially the same. References for this theory include [19, §8] and [112].

In [82, §3], Lurie has generalized these constructions to the ∞-categorical setting. Roughly speaking, he has established a result saying that giving a coCartesian fibration $p\colon \mathcal{C} \to \mathcal{D}$ is equivalent to giving a functor $\mathcal{D} \to \mathcal{C}at_\infty$, where $\mathcal{C}at_\infty$ is the ∞-category of ∞-categories introduced in Perspective 14.3.7. In fact, there is a Quillen equivalence between certain model structures making this idea precise. Let us only mention that the homotopy theory of coCartesian fibrations above $\mathcal{D}$ is encoded by the *coCartesian model structure* on $s\mathcal{S}et^+_{/\mathcal{D}}$, the category of marked simplicial sets above $\mathcal{D} = (\mathcal{D}, \mathcal{D}_1)$. An object $p\colon \mathcal{C} \to \mathcal{D}$ in this model structure is fibrant if and only if p is a coCartesian fibration and the marked edges in $\mathcal{C}$ are precisely the p-coCartesian arrows.

In ordinary category theory, there are variants to these notions for contravariant category-valued pseudo-functors. The ∞-categorical analogue is that of a *Cartesian fibration.* Moreover, in the classical context — in particular, in the theory of stacks in algebraic geometry or algebraic topology — one frequently considers *groupoid-valued* pseudo-functors (of either variance). These notions are sometimes also referred to as categories (co)fibered in groupoids. The ∞-categorical analogues of these are *left* and *right fibrations* and again there are suitable associated model categories in the background. For more details we refer to [82, §3].

14.5.3 Algebra objects and monoidal functors

Let $\mathcal{M}$ be an ordinary monoidal category with monoidal pairing $\otimes$ and unit object $\mathbb{S} \in \mathcal{M}$. An *algebra* or *monoid* in $\mathcal{M}$ is an object $A \in \mathcal{M}$ together with a multiplication map

$\mu\colon A \otimes A \to A$ and a unit map $\eta\colon \mathbb{S} \to A$ that satisfy obvious associativity and unitality conditions. We also say that (μ, η) specifies an *algebra structure* on A.

In §14.5.1 we saw how to encode monoidal structures by means of Grothendieck opfibrations $p\colon \mathcal{M}^{\otimes} \to \Delta^{\mathrm{op}}$ satisfying the Segal condition. We now want to describe algebra structures in that picture. As a first guess for a definition of algebra objects we might consider sections of $p\colon \mathcal{M}^{\otimes} \to \Delta^{\mathrm{op}}$. Given an arbitrary such section

$$A\colon \Delta^{\mathrm{op}} \to \mathcal{M}^{\otimes},$$

the equivalences given by the Segal maps (14.5.6) imply that the value $A_{[n]} \in \mathcal{M}^{\otimes}_{[n]}$ corresponds to n objects of $\mathcal{M} = \mathcal{M}^{\otimes}_{[1]}$,

$$\mathcal{M}_{[n]} \ni A_{[n]} \quad \longleftrightarrow \quad A^1_{[n]}, \ldots, A^n_{[n]} \in \mathcal{M}. \tag{14.5.8}$$

Let us again consider the face map $d_1\colon [2] \to [1]$ which we saw in §14.5.1 encodes the monoidal pairing $\otimes\colon \mathcal{M} \times \mathcal{M} \to \mathcal{M}$. The section A evaluated on d_1 gives us by means of the identification (14.5.8) a map

$$(A^1_{[2]}, A^2_{[2]}) \to A^1_{[1]} \tag{14.5.9}$$

defined on the pair of objects $(A^1_{[2]}, A^2_{[2]})$. Moreover, by the discussion in §14.5.1, we have a description of the p-coCartesian arrows for the Grothendieck opfibration $p\colon \mathcal{M}^{\otimes} \to \Delta^{\mathrm{op}}$. Applied to our situation, a p-coCartesian lift of d_1 with domain $A_{[2]}$ corresponds under (14.5.8) to a morphism

$$(A^1_{[2]}, A^2_{[2]}) \to A^1_{[2]} \otimes A^2_{[2]}.$$

The universal property of this p-coCartesian lift (as expressed by (14.5.2) being a pullback square) implies that (14.5.9) factors uniquely over this lift and we hence obtain an induced map

$$A^1_{[2]} \otimes A^2_{[2]} \to A^1_{[1]}. \tag{14.5.10}$$

If we now want to get a classical algebra object, we would like (14.5.10) to be a map of the form $M \otimes M \to M$ for some fixed $M \in \mathcal{M}$. Thus we should ensure that, among other things, the objects $A^1_{[2]}, A^2_{[2]}$, and $A^1_{[1]}$ are isomorphic.

Definition 14.5.11. A morphism $\alpha\colon [n] \to [k]$ in Δ is *convex* if it is injective and the image $\mathrm{im}(\alpha) \subseteq [k]$ is convex, i.e., the image is given by the interval $[\alpha(0), \alpha(n)]$.

It follows from the construction of the p-coCartesian arrows of $p\colon \mathcal{M}^{\otimes} \to \Delta^{\mathrm{op}}$ (compare to the discussion around (14.5.5)) that the p-coCartesian lifts of convex maps $\alpha\colon [n] \to [k]$ in Δ induce projection functors $\mathcal{M}^{\times k} \to \mathcal{M}^{\times n}$ onto some of the factors. In particular, p-coCartesian lifts defining the functors $(\iota_i)_!, i = 1, 2$, as in (14.5.6) can be identified with the maps

$$(A^1_{[2]}, A^2_{[2]}) \to A^1_{[2]} \qquad \text{and} \qquad (A^1_{[2]}, A^2_{[2]}) \to A^2_{[2]},$$

respectively. If the images of $\iota_i\colon [2] \to [1], i = 1, 2$, under A are p-coCartesian arrows, then by Lemma 14.5.2 we obtain the desired isomorphisms

$$A^1_{[2]} \cong A^1_{[1]} \qquad \text{and} \qquad A^2_{[2]} \cong A^1_{[1]}.$$

With this preparation it is straightforward to establishes the following result.

Proposition 14.5.12. *Let $p\colon \mathcal{M}^{\otimes} \to \Delta^{\mathrm{op}}$ be a monoidal structure on $\mathcal{M} = \mathcal{M}^{\otimes}_{[1]}$. Then a section $A\colon \Delta^{\mathrm{op}} \to \mathcal{M}^{\otimes}$ of p that sends convex arrows to p-coCartesian arrows encodes an algebra structure on $A_{[1]} \in \mathcal{M}$. Conversely, any algebra object in $\mathcal{M}$ determines such a section of $p\colon \mathcal{M}^{\otimes} \to \Delta^{\mathrm{op}}$.*

In the world of ∞-categories, we turn this observation into a definition ([80, Definition 1.1.14]).

Definition 14.5.13. Let $p\colon \mathcal{M}^{\otimes} \to N(\Delta^{\mathrm{op}})$ be a monoidal ∞-category. A section $A\colon N(\Delta^{\mathrm{op}}) \to \mathcal{M}^{\otimes}$ of p is an *(associative) algebra object* in $\mathcal{M}^{\otimes}$ if A sends convex morphisms to p-coCartesian arrows in $\mathcal{M}^{\otimes}$.

Note that it is already a certain abuse of language to speak of algebra objects of $\mathcal{M}^{\otimes}$ since the notion of algebra object obviously also depends on the coCartesian fibration. A further comfortable abuse of language is to simply speak of algebra objects in $\mathcal{M}$.

As in the case of monoidal structures on ∞-categories, an algebra object encodes quite a lot of structure: namely, given an algebra object A in $\mathcal{M}$, the underlying object $A_{[1]}$ is endowed with a multiplication map which is associative and unital up to *coherent homotopy* (see Perspective 14.5.21 for an explanation of this similarity in the commutative case). In particular, an algebra object in a monoidal ∞-category defines an ordinary algebra object in the underlying homotopy category, but not conversely.

Algebra objects in monoidal ∞-categories are special cases of lax monoidal functors between monoidal ∞-categories. We include the following definition and leave it to the reader to check that in the case of ordinary categories this reduces to the usual concepts.

Definition 14.5.14. Let $p\colon \mathcal{M}^{\otimes} \to N(\Delta^{\mathrm{op}})$ and $q\colon \mathcal{N}^{\otimes} \to N(\Delta^{\mathrm{op}})$ be monoidal ∞-categories. A *lax monoidal functor* $F\colon \mathcal{M}^{\otimes} \to \mathcal{N}^{\otimes}$ is a functor over $N(\Delta^{\mathrm{op}})$,

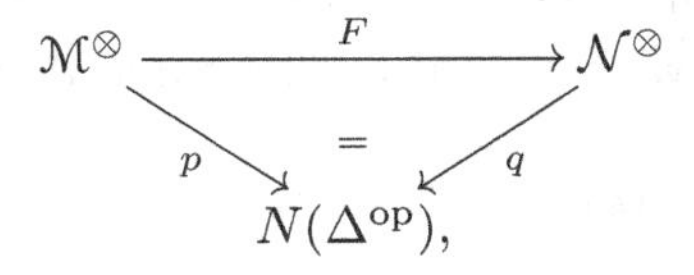

that sends p-coCartesian lifts of convex morphisms in $N(\Delta^{\mathrm{op}})$ to q-coCartesian arrows. A *monoidal functor* $F\colon \mathcal{M}^{\otimes} \to \mathcal{N}^{\otimes}$ is a functor over $N(\Delta^{\mathrm{op}})$ that sends arbitrary p-coCartesian arrows to q-coCartesian ones.

Of course we also want to consider monoidal transformations between monoidal functors. More generally, (lax) monoidal functors between monoidal ∞-categories $\mathcal{M}^{\otimes}$ and $\mathcal{N}^{\otimes}$ are organized into ∞-categories

$$\mathrm{Fun}^{\otimes,\mathrm{lax}}(\mathcal{M}^{\otimes}, \mathcal{N}^{\otimes}) \qquad \text{and} \qquad \mathrm{Fun}^{\otimes}(\mathcal{M}^{\otimes}, \mathcal{N}^{\otimes}),$$

respectively. The ∞-category $\mathrm{Fun}^{\otimes,\mathrm{lax}}(\mathcal{M}^{\otimes}, \mathcal{N}^{\otimes})$ is the full subcategory of the ∞-category $\mathrm{Map}_{N(\Delta^{\mathrm{op}})}(\mathcal{M}^{\otimes}, \mathcal{N}^{\otimes})$ of functors over $N(\Delta^{\mathrm{op}})$ spanned by the lax monoidal functors. Here,

$\mathrm{Map}_{N(\Delta^{\mathrm{op}})}(\mathcal{M}^{\otimes},\mathcal{N}^{\otimes})$ is of course defined as the pullback

$$\begin{array}{ccc} \mathrm{Map}_{N(\Delta^{\mathrm{op}})}(\mathcal{M}^{\otimes},\mathcal{N}^{\otimes}) & \longrightarrow & \mathrm{Map}(\mathcal{M}^{\otimes},\mathcal{N}^{\otimes}) \\ \downarrow & & \downarrow q_* \\ \Delta^0 & \xrightarrow[p]{} & \mathrm{Map}(\mathcal{M}^{\otimes}, N(\Delta^{\mathrm{op}})), \end{array} \tag{14.5.11}$$

which is an ∞-category because q and so q_* is an inner fibration [82, Corollary 2.3.2.5]. Similarly, $\mathrm{Fun}^{\otimes}(\mathcal{M}^{\otimes},\mathcal{N}^{\otimes}) \subseteq \mathrm{Fun}^{\otimes,\mathrm{lax}}(\mathcal{M}^{\otimes},\mathcal{N}^{\otimes})$ is the full subcategory spanned by the monoidal functors.

As a special case, we see that algebra objects in a monoidal ∞-category $\mathcal{M}^{\otimes}$ are themselves organized in an ∞-category. In fact, the *∞-category* $\mathrm{Alg}_{\mathbb{A}_\infty}(\mathcal{M}^{\otimes})$ *of algebra objects* in $\mathcal{M}^{\otimes}$ can be defined as

$$\mathrm{Alg}_{\mathbb{A}_\infty}(\mathcal{M}^{\otimes}) = \mathrm{Fun}^{\otimes,\mathrm{lax}}(N(\Delta^{\mathrm{op}}),\mathcal{M}^{\otimes}), \tag{14.5.12}$$

where $N(\Delta^{\mathrm{op}}) \to N(\Delta^{\mathrm{op}})$ is the trivial monoidal structure as in Examples 14.5.9. We will say a bit more about similar ∞-categories in the context of symmetric monoidal structures in §14.5.4.

14.5.4 Symmetric monoidal ∞-categories

In order to obtain a theory of *symmetric* monoidal ∞-categories one combines the Segal perspective on $\mathbb{E}_\infty$-monoids and the ∞-categorical Grothendieck construction, and this section is hence inspired by the theory of 'special Γ-spaces' (see [104, 102]). A good part of this section simply amounts to translating parts of §§14.5.1-14.5.3 to the context of symmetric monoidal ∞-categories, and therefore we are rather sketchy.

We again begin with the classical situation in ordinary category theory. In §14.5.1, we described monoidal categories in terms of Grothendieck opfibrations $\mathcal{M}^{\otimes} \to \Delta^{\mathrm{op}}$. In that picture, the monoidal product was encoded by the induced functor

$$(d_1)_!\colon \mathcal{M}^{\otimes}_{[2]} \to \mathcal{M}^{\otimes}_{[1]} = \mathcal{M}.$$

If we want to have a similar description of *symmetric* monoidal categories, we must be able to encode symmetry isomorphisms, thus we have to encode the flip map

$$t\colon \mathcal{M}^{\times 2} \to \mathcal{M}^{\times 2}\colon (X,Y) \mapsto (Y,X)$$

and, more generally, any permutation of n objects in $\mathcal{M}$. It is hence plausible that the role of Δ is taken by 'a category of finite sets with all maps between them'. The details are as follows.

For a natural number $n \geq 0$, let $\langle n \rangle$ be the finite pointed set

$$\langle n \rangle = \{0,1,\ldots,n\}$$

with $0 \in \langle n \rangle$ as base point. The category $\mathcal{F}\text{in}$ is the full subcategory of the category of pointed sets spanned by the objects $\langle n \rangle, n \geq 0$. Note that the natural ordering on $\langle n \rangle$ does not play a role in the definition of $\mathcal{F}\text{in}$ but it will have its uses in the formation of higher monoidal products. We denote by

$$\rho^j \colon \langle n \rangle \to \langle 1 \rangle, \quad n \geq 1, \quad j = 1, \ldots, n, \tag{14.5.13}$$

the unique pointed map $\langle n \rangle \to \langle 1 \rangle$ with $(\rho^j)^{-1}(1) = \{j\}$.

Let now $\mathcal{M}$ be a *symmetric* monoidal category with monoidal product $\otimes$ and monoidal unit $\mathbb{S} \in \mathcal{M}$. Following a pattern similar to §14.5.1, we construct a new category $\mathcal{M}^{\otimes}$ as follows. An object in $\mathcal{M}^{\otimes}$ is a finite (possibly empty) sequence of objects in $\mathcal{M}$,

$$(M_1, \ldots, M_n), \quad M_i \in \mathcal{M}, \quad n \geq 0.$$

A morphism $(M_1, \ldots, M_n) \to (L_1, \ldots, L_k)$ between two such sequences is a pair $(\alpha, \{f_i\}_i)$ consisting of a morphism $\alpha \colon \langle n \rangle \to \langle k \rangle$ in $\mathcal{F}\text{in}$ together with morphisms

$$f_i \colon \bigotimes_{j \in \alpha^{-1}(i)} M_j \to L_i, \quad i = 1, \ldots, k,$$

where the tensor product is formed according to the ordering on $\alpha^{-1}(i)$. Again, if the set $\alpha^{-1}(i)$ is empty, then this is to be read as a map $f_i \colon \mathbb{S} \to L_i$. The composition is obtained from the compositions in $\mathcal{M}$ and $\mathcal{F}\text{in}$ together with the coherence constraints of $\mathcal{M}$. There is an obvious projection functor $p \colon \mathcal{M}^{\otimes} \to \mathcal{F}\text{in}$ given by $(M_1, \ldots, M_n) \mapsto \langle n \rangle$ and $(\alpha, \{f_i\}_i) \mapsto \alpha$.

Proposition 14.5.15. *For any symmetric monoidal category $\mathcal{M}$ the functor $p \colon \mathcal{M}^{\otimes} \to \mathcal{F}\text{in}$ is a Grothendieck opfibration. Moreover, this functor satisfies the* Segal condition, *i.e., the Segal maps*

$$(\rho^1_!, \ldots, \rho^n_!) \colon \mathcal{M}^{\otimes}_{\langle n \rangle} \to \mathcal{M}^{\times n}, \quad n \geq 0,$$

are equivalences. Conversely, any Grothendieck opfibration $p \colon \mathcal{C} \to \mathcal{F}\text{in}$ satisfying the Segal condition encodes a symmetric monoidal structure on $\mathcal{M} = \mathcal{C}_{\langle 1 \rangle}$.

Here is a sketch of a proof. Given $(M_1, \ldots, M_n) \in \mathcal{M}^{\otimes}_{\langle n \rangle}$ and $\alpha \colon \langle n \rangle \to \langle k \rangle$, an associated p-coCartesian is obtained by choosing isomorphisms

$$f_i \colon \bigotimes_{j \in \alpha^{-1}(i)} M_j \to L_i, \quad i = 1, \ldots, k.$$

In the special case of ρ^j as in (14.5.13), it follows that such lifts are of the form

$$(\rho^j, \mathrm{id}_{M_j}) \colon (M_1, \ldots, M_n) \to M_j,$$

and the associated functor $\rho^j_! \colon \mathcal{M}^{\otimes}_{\langle n \rangle} \to \mathcal{M}^{\otimes}_{\langle 1 \rangle}$ can hence be identified with a projection functor, implying the Segal condition.

Conversely, let us consider a Grothendieck opfibration $p \colon \mathcal{C} \to \mathcal{F}\text{in}$ satisfying the Segal condition. We content ourselves by mentioning the following three steps towards the construction of a symmetric monoidal structure on $\mathcal{M} = \mathcal{C}_{\langle 1 \rangle}$.

(i) Let $m\colon \langle 2\rangle \to \langle 1\rangle$ be the map in $\mathcal{F}$in determined by $m(1) = m(2) = 1$. By means of the Segal condition, we can then define a functor

$$\otimes\colon \mathcal{M}\times\mathcal{M} \xleftarrow{\simeq} \mathcal{C}_{\langle 2\rangle} \xrightarrow{m_!} \mathcal{C}_{\langle 1\rangle} = \mathcal{M},$$

which will be the monoidal pairing.

(ii) As a special case of the Segal condition we obtain an equivalence $\mathcal{C}_{\langle 0\rangle} \simeq \mathbb{1}$. Thus, for the unique map $n\colon \langle 0\rangle \to \langle 1\rangle$ in $\mathcal{F}$in, the induced functor

$$n_!\colon \mathcal{C}_{\langle 0\rangle} \to \mathcal{C}_{\langle 1\rangle} = \mathcal{M}$$

essentially classifies an object in $\mathcal{M}$, which will be the monoidal unit $\mathbb{S}$.

(iii) The *twist map* $t\colon \langle 2\rangle \to \langle 2\rangle$ in $\mathcal{F}$in is the automorphism that interchanges 1 and 2. The equality $m = m\circ t\colon \langle 2\rangle \to \langle 1\rangle$ together with Lemma 14.5.2 yields a unique natural isomorphism

$$\sigma\colon m_! \cong m_!\circ t_!\colon \mathcal{C}_{\langle 2\rangle} \to \mathcal{C}_{\langle 1\rangle} = \mathcal{M}$$

over $\langle 1\rangle$, which can be shown to induce a symmetry constraint for $\otimes$.

By similar arguments, one obtains associativity and unitality constraints and establishes the coherence axioms. As in the non-symmetric case, we now turn this observation into a definition [78].

Definition 14.5.16. A *symmetric monoidal ∞-category* is a coCartesian fibration $p\colon \mathcal{M}^{\otimes} \to N(\mathcal{F}\mathrm{in})$ such that the Segal maps are equivalences,

$$(\rho^1_!,\ldots,\rho^n_!)\colon \mathcal{M}^{\otimes}_{\langle n\rangle} \xrightarrow{\simeq} (\mathcal{M}^{\otimes}_{\langle 1\rangle})^{\times n}, \qquad n\geq 0.$$

We include a few remarks which are parallel to corresponding statements in §14.5.2.

Remark 14.5.17. (i) A symmetric monoidal ∞-category $p\colon \mathcal{M}^{\otimes} \to N(\mathcal{F}\mathrm{in})$ endows the underlying ∞-category $\mathcal{M} = \mathcal{M}^{\otimes}_{\langle 1\rangle}$ with a monoidal pairing which is associative and commutative *up to coherent homotopies*; see Perspective 14.5.21. In particular, the homotopy category of a symmetric monoidal ∞-category is canonically a symmetric monoidal category.

(ii) Applying the nerve construction to Grothendieck opfibrations associated to ordinary symmetric monoidal categories we obtain symmetric monoidal ∞-categories. In particular, the identity functor $N(\mathcal{F}\mathrm{in}) \to N(\mathcal{F}\mathrm{in})$ is a symmetric monoidal ∞-category. As a variant, given a closed, symmetric monoidal, simplicial model category satisfying suitable compatibility assumptions, by means of the coherent nerve construction we obtain 'honest examples' of symmetric monoidal ∞-categories ([78, §8]).

Before we turn to algebra objects in the symmetric monoidal context, we expand a bit on the relation between monoidal and symmetric monoidal ∞-categories. For this purpose, we consider the functor

$$\phi\colon \Delta^{op} \to \mathcal{F}\mathrm{in} \tag{14.5.14}$$

that on objects is given by $[n] \mapsto \langle n \rangle$. Given a morphism $\alpha\colon [k] \to [n]$ in Δ, the induced pointed map $\phi(\alpha)\colon \langle n \rangle \to \langle k \rangle$ is defined by

$$\phi(\alpha)(j) = \begin{cases} i & \text{if there is an } i \text{ such that } j \in [\alpha(i-1)+1, \alpha(i)], \\ * & \text{otherwise.} \end{cases}$$

Since α is monotone, such an i is unique if it exists, and we leave it to the reader to check that this defines a functor. In fact, up to a restriction of the codomain, the functor ϕ is simply the simplicial circle $S^1 \in s\mathcal{S}et_*$ defined as the coequalizer

$$\Delta^0 \rightrightarrows \Delta^1 \to S^1$$

and considered as a pointed simplicial set. As a special case the image of the opposite ι_j of $\iota_{\{j-1,j\}}\colon [1] \to [n]$ under ϕ is the map ρ^j as in (14.5.13).

The role of convex maps in the theory of monoidal ∞-categories is taken by *inert* maps in the theory of symmetric monoidal ∞-categories.

Definition 14.5.18. A morphism $\alpha\colon \langle n \rangle \to \langle k \rangle$ in $\mathcal{F}\text{in}$ is *inert* if $\alpha^{-1}(i)$ is a singleton for every $1 \le i \le k$.

We note that $\alpha\colon [k] \to [n]$ in Δ is convex if and only if $\phi(\alpha)\colon \langle n \rangle \to \langle k \rangle$ in $\mathcal{F}\text{in}$ is inert, and that the maps (14.5.13) are inert. Given a symmetric monoidal ∞-category $p\colon \mathcal{M}^{\otimes} \to N(\mathcal{F}\text{in})$ we already observed that the induced functors $\rho^j_!$ are projection functors. Similarly, general inert morphisms induce projection and permutation functors. This suggests the following definition.

Definition 14.5.19. Let $p\colon \mathcal{M}^{\otimes} \to N(\mathcal{F}\text{in})$, $q\colon \mathcal{N}^{\otimes} \to N(\mathcal{F}\text{in})$ be symmetric monoidal ∞-categories and let $F\colon \mathcal{M}^{\otimes} \to \mathcal{N}^{\otimes}$ be a functor over $N(\mathcal{F}\text{in})$.

(i) The functor F is *symmetric monoidal* if it sends p-coCartesian arrows to q-coCartesian arrows.

(ii) The functor F is *lax symmetric monoidal* if it sends p-coCartesian lifts of inert morphisms to q-coCartesian arrows.

Symmetric monoidal and lax symmetric monoidal functors respectively are organized in ∞-categories, namely the corresponding full subcategories

$$\text{Fun}^{\otimes}(\mathcal{M}^{\otimes}, \mathcal{N}^{\otimes}) \subseteq \text{Fun}^{\otimes,\text{lax}}(\mathcal{M}^{\otimes}, \mathcal{N}^{\otimes}) \subseteq \text{Map}_{N(\mathcal{F}\text{in})}(\mathcal{M}^{\otimes}, \mathcal{N}^{\otimes})$$

where $\text{Map}_{N(\mathcal{F}\text{in})}(\mathcal{M}^{\otimes}, \mathcal{N}^{\otimes})$ is defined in analogy to (14.5.11). As a special case we obtain *∞-categories of commutative algebra objects* $\text{Alg}_{\mathbb{E}_\infty}(\mathcal{M}^{\otimes})$ defined by

$$\text{Alg}_{\mathbb{E}_\infty}(\mathcal{M}^{\otimes}) = \text{Fun}^{\otimes,\text{lax}}(N(\mathcal{F}\text{in}), \mathcal{M}^{\otimes}). \tag{14.5.15}$$

More explicitly, a commutative algebra object is a section $E\colon N(\mathcal{F}\text{in}) \to \mathcal{M}^{\otimes}$ of $p\colon \mathcal{M}^{\otimes} \to N(\mathcal{F}\text{in})$ sending inert morphisms to p-coCartesian ones. Such a section endows the underlying object $E_{\langle 1 \rangle} \in \mathcal{M}$ with a coherently associative and commutative pairing $E_{\langle 1 \rangle} \otimes E_{\langle 1 \rangle} \to E_{\langle 1 \rangle}$.

Remark 14.5.20. A symmetric monoidal ∞-category has an underlying monoidal ∞-category. In fact, given a symmetric monoidal ∞-category $\mathcal{M}^{\otimes} \to N(\mathcal{F}\text{in})$, then the underlying monoidal ∞-category $U(\mathcal{M}^{\otimes})$ is defined as the pullback

$$\begin{array}{ccc} U(\mathcal{M}^{\otimes}) & \longrightarrow & \mathcal{M}^{\otimes} \\ \downarrow & \lrcorner & \downarrow p \\ N(\Delta^{\mathrm{op}}) & \xrightarrow[N(\phi)]{} & N(\mathcal{F}\text{in}), \end{array}$$

where ϕ is again the simplicial circle (14.5.14). Similarly, one shows that commutative algebra objects have underlying associative algebra objects.

Perspective 14.5.21. In this section we saw that symmetric monoidal ∞-categories are endowed with a coherently associative and commutative monoidal structure and we just claimed that something similar is true for commutative algebra objects in the sense of (14.5.15). In this perspective we make the similarity between these two situations precise and refer the reader to [78, §2] for details.

Given an ∞-category $\mathcal{C}$ with finite products, there is the Cartesian monoidal structure $\mathcal{C}^{\times} \to N(\mathcal{F}\text{in})$ and there are two associated ∞-categories.

(i) As a special case of (14.5.15) we have the ∞-category $\mathrm{Alg}_{\mathbb{E}_\infty}(\mathcal{C}^{\times})$ of commutative algebra objects in $\mathcal{C}^{\times}$.

(ii) On the other hand, we can consider *commutative monoids* in $\mathcal{C}$, i.e., functors $M\colon N(\mathcal{F}\text{in}) \to \mathcal{C}$ such that the Segal maps

$$M_{\langle n\rangle} \xrightarrow{\sim} M_{\langle 1\rangle} \times \cdots \times M_{\langle 1\rangle}$$

are equivalences. The ∞-category $\mathrm{Mon}_{\mathbb{E}_\infty}(\mathcal{C})$ of commutative monoids in $\mathcal{C}$ is the full subcategory of $\mathrm{Fun}(N(\mathcal{F}\text{in}), \mathcal{C})$ spanned by the commutative monoids.

And it turns out that there is an equivalence $\mathrm{Alg}_{\mathbb{E}_\infty}(\mathcal{C}^{\times}) \simeq \mathrm{Mon}_{\mathbb{E}_\infty}(\mathcal{C})$ over $\mathcal{C}$.

We now apply this to the ∞-category $\mathcal{C}at_\infty$ of ∞-categories (Perspective 14.3.7), which is an example of an ∞-category admitting finite products. The Grothendieck construction (Perspective 14.5.10) implies that $\mathrm{Fun}(N(\mathcal{F}\text{in}), \mathcal{C}at_\infty)$ is equivalent to the ∞-category of coCartesian fibrations $p\colon \mathcal{M} \to N(\mathcal{F}\text{in})$. And under this equivalence commutative monoids in $\mathcal{C}at_\infty$ and symmetric monoidal ∞-categories correspond to each other since both are defined by similar Segal conditions. Thus, as an upshot, we obtain an equivalence of ∞-categories

$$\mathrm{Alg}_{\mathbb{E}_\infty}(\mathcal{C}at_\infty^{\times}) \simeq \mathcal{C}at_\infty^{\mathrm{sMon}},$$

where $\mathcal{C}at_\infty^{\mathrm{sMon}}$ denotes the ∞-category of symmetric monoidal ∞-categories and symmetric monoidal functors, explaining why in both cases we obtain similar coherence data. There is a similar equivalence in the case of monoidal ∞-categories and monoidal functors, namely

$$\mathrm{Alg}_{\mathbb{A}_\infty}(\mathcal{C}at_\infty^{\times}) \simeq \mathcal{C}at_\infty^{\mathrm{Mon}}.$$

Obviously, having introduced commutative algebra objects, one would now like to study modules over such algebra objects as well as the existence of limits and colimits in related ∞-categories. Such a theory exists and for more details we refer to [80, 81] and [83], as well as the chapter in this Handbook by David Gepner.

We content ourselves by concluding this section with the following result concerning *initial* objects in ∞-categories of commutative algebra objects. This result comes up again in the final section §14.6.

Proposition 14.5.22. *For every symmetric monoidal ∞-category $\mathcal{M}^{\otimes} \to N(\mathcal{F}\mathrm{in})$ the ∞-category $\mathrm{Alg}_{\mathbb{E}_\infty}(\mathcal{M})$ has an initial object. Moreover, a commutative algebra object E is initial if and only if the unit map $\mathbb{S} \to E_{\langle 1 \rangle}$ is an equivalence in $\mathcal{M}$.*

14.6 Stable ∞-categories and the universal property of spectra

In this final section we give an introduction to stable ∞-categories. By definition a finitely complete and finitely cocomplete ∞-categories is stable if it admits a zero object and if a square in it is a pullback if and only if it is a pushout. Typical examples of stable ∞-categories arise in homological algebra (∞-categories of chain complexes) and stable homotopy theory (the ∞-category of spectra). In fact, it turns out the ∞-category of spectra is the universal example of a stable ∞-category in a certain precise sense.

Stable ∞-categories are an enhancement of triangulated categories. In §14.6.1 we sketch some of ingredients involved in a proof that homotopy categories of stable ∞-categories can be turned into triangulated categories. In §14.6.2 we briefly discuss the *stabilization* of nice ∞-categories which is obtained by passing to internal spectrum objects. We conclude this subsection by a precise universal property of this stabilization process. Preparing the ground for the construction of the smash product, in §14.6.3 we discuss the tensor product of presentable ∞-categories. Following Lurie, this allows us in §14.6.4 to give a very conceptual construction of the smash product monoidal structure on spectra and hence to define associative and commutative ring spectra.

14.6.1 Stable ∞-categories

General references for the first two subsections are [79] and [83, §1]. As a first step we collect a few basics concerning *pointed* ∞-categories.

Definition 14.6.1. An ∞-category is *pointed* if it admits a zero object, i.e., an object that is initial and final.

Thus, an ∞-category $\mathcal{C}$ is pointed if there is an object $0 \in \mathcal{C}$ such that for all $x \in \mathcal{C}$ the mapping spaces $\mathrm{Map}_{\mathcal{C}}(x,0)$ and $\mathrm{Map}_{\mathcal{C}}(0,x)$ are contractible. It follows, that for any two objects $x, y \in \mathcal{C}$ there is a zero map

$$0 = 0_{x,y} \colon x \to y,$$

well-defined up to a contractible space of choices. Again by Proposition 14.3.19, if an ∞-category $\mathcal{C}$ is pointed, then the full subcategory spanned by the zero objects is a contractible Kan complex.

Examples 14.6.2. (i) Let $\mathcal{C}$ be an ∞-category with a terminal object $* \in \mathcal{C}$. The undercategory $\mathcal{C}_* = \mathcal{C}_{*/}$ (see 14.3.17) is a pointed ∞-category, called the ∞-category of *pointed objects* in $\mathcal{C}$. Adding a disjoint base point defines a functor $_+\colon \mathcal{C} \to \mathcal{C}_*$ which is left adjoint to the forgetful functor $_-\colon \mathcal{C}_* \to \mathcal{C}$,

$$(_+, _-)\colon \mathcal{C} \rightleftarrows \mathcal{C}_*.$$

(For the notion of an adjunction between ∞-categories we again refer the reader to [82, p. 337] and [80, §3].) As a special case we obtain the *∞-category $\mathcal{S}_*$ of pointed spaces* and the corresponding adjunction

$$(_+, _-)\colon \mathcal{S} \rightleftarrows \mathcal{S}_*.$$

(ii) The underlying ∞-category of a pointed simplicial model category is pointed; see Examples 14.2.31. An ordinary category is pointed if and only if the nerve is a pointed ∞-category.

The ∞-category $\mathcal{S}_*$ of pointed spaces together with the 0-sphere $S^0 \in \mathcal{S}_*$ enjoys the following universal property (which is a pointed variant of Corollary 14.4.11).

Proposition 14.6.3. *Let $\mathcal{D}$ be a pointed, presentable ∞-category. Evaluation at the 0-sphere $S^0 \in \mathcal{S}_*$ induces an equivalence of ∞-categories*

$$\mathrm{Fun}^{\mathrm{L}}(\mathcal{S}_*, \mathcal{D}) \xrightarrow{\sim} \mathcal{D}.$$

This result makes precise that $\mathcal{S}_*$ is the *free pointed, presentable ∞-category generated by S^0*.

A *triangle* τ in a pointed ∞-category $\mathcal{C}$ is a diagram $\tau\colon \square \to \mathcal{C}$,

$$\begin{array}{ccc} x & \xrightarrow{f} & y \\ \downarrow & \nearrow\!\!\!\!\swarrow \; \searrow & \downarrow g \\ 0 & \longrightarrow & z, \end{array} \tag{14.6.1}$$

that vanishes at the lower left corner. Thus, a triangle in $\mathcal{C}$ encodes a pair of composable arrows $f\colon x \to y$ and $g\colon y \to z$, a further arrow $h\colon x \to z$ together with a homotopy $h \simeq g \circ f$ and a *null-homotopy* $h \simeq 0$. Recall the definition of pullback and pushout squares in Definition 14.3.25.

Definition 14.6.4. A triangle in a pointed ∞-category is *exact* if it is a pullback square. Dually, a triangle is *coexact* if it is a pushout square.

For every finitely complete, finitely cocomplete, and pointed ∞-category $\mathcal{C}$ we denote by

$$\mathcal{C}^{\Sigma} \subseteq \mathrm{Fun}(\square, \mathcal{C})$$

the full subcategory spanned by the coexact triangles that also vanish on the upper right corner,

$$\begin{array}{ccc} x & \longrightarrow & 0' \\ \downarrow & & \ulcorner\downarrow \\ 0 & \longrightarrow & y. \end{array}$$

There is a dually defined ∞-category $\mathcal{C}^{\Omega} \subseteq \mathrm{Fun}(\square, \mathcal{C})$ of exact triangles vanishing on the upper right corner. Morally, such diagrams should be determined by the value in the upper left corner in the first and by the value in the lower right corner in the second case. This is made precise by the following result (see [83, pp. 23-24]).

Proposition 14.6.5. *Let $\mathcal{C}$ be a finitely complete, finitely cocomplete, and pointed ∞-category. The evaluation maps*

$$\mathrm{ev}_{(0,0)} \colon \mathcal{C}^{\Sigma} \to \mathcal{C} \qquad \textit{and} \qquad \mathrm{ev}_{(1,1)} \colon \mathcal{C}^{\Omega} \to \mathcal{C}$$

are acyclic Kan fibrations.

This proposition and many similar results in this section are consequences of an ∞-categorical version of the calculus of Kan extensions. One of the key facts of constant use is [82, Prop. 4.3.2.15]. In this chapter we will not pursue this calculus any further, but we only consider results from this calculus which are 'similarly plausible' as Proposition 14.6.5. For more details on the calculus of homotopy Kan extensions we also refer to [48] and references therein.

We briefly recall that given an acyclic Kan fibration $p \colon X \to Y$, then the space of sections $\Gamma(p) \in s\mathcal{S}et$ is a contractible Kan complex. In fact, for every simplicial set K the induced map $p_* \colon \mathrm{Map}(K, X) \to \mathrm{Map}(K, Y)$ between simplicial mapping spaces as defined by (14.2.4) is again an acyclic Kan fibration. Since acyclic Kan fibrations are stable under pullbacks, we conclude that $\Gamma(p)$ is a contractible Kan complex by considering the defining pullback diagram

$$\begin{array}{ccc} \Gamma(p) & \longrightarrow & \mathrm{Map}(Y, X) \\ \downarrow & \lrcorner & \downarrow p_* \\ \Delta^0 & \xrightarrow[\mathrm{id}_Y]{} & \mathrm{Map}(Y, Y). \end{array} \tag{14.6.2}$$

Thus, under the assumption of Proposition 14.6.5, we can choose sections

$$s_{\Sigma} \colon \mathcal{C} \to \mathcal{C}^{\Sigma} \qquad \text{and} \qquad s_{\Omega} \colon \mathcal{C} \to \mathcal{C}^{\Omega}$$

of the evaluation maps $\mathrm{ev}_{(0,0)}$ and $\mathrm{ev}_{(1,1)}$, respectively, and these sections are unique up to contractible spaces of choices. Consequently, the following is well-defined.

Definition 14.6.6. Let $\mathcal{C}$ be a finitely complete, finitely cocomplete, and pointed ∞-category. The *suspension functor* $\Sigma = \Sigma_{\mathcal{C}} \colon \mathcal{C} \to \mathcal{C}$ and the *loop functor* $\Omega = \Omega_{\mathcal{C}} \colon \mathcal{C} \to \mathcal{C}$ are respectively defined as

$$\Sigma \colon \mathcal{C} \xrightarrow{s_{\Sigma}} \mathcal{C}^{\Sigma} \xrightarrow{\mathrm{ev}_{(1,1)}} \mathcal{C} \qquad \text{and} \qquad \Omega \colon \mathcal{C} \xrightarrow{s_{\Omega}} \mathcal{C}^{\Omega} \xrightarrow{\mathrm{ev}_{(0,0)}} \mathcal{C}.$$

Proposition 14.6.7. *Let $\mathcal{C}$ be a finitely complete, finitely cocomplete, and pointed ∞-category. The suspension functor $\Sigma\colon \mathcal{C} \to \mathcal{C}$ is left adjoint to the loop functor $\Omega\colon \mathcal{C} \to \mathcal{C}$,*

$$(\Sigma, \Omega)\colon \mathcal{C} \rightleftarrows \mathcal{C}.$$

This is [83, Rmk. 1.1.2.8]. In a similar way one defines *cofibers* and *fibers* in pointed ∞-categories. In fact, in the case of cofibers, starting with a morphism $f\colon x \to y$, suitable combinations of Kan extensions yield a coexact triangle as in (14.6.1). Combining these Kan extensions with a restriction of such triangles to the vertical morphism on the right yields a *cofiber* functor $\mathsf{cof}\colon \mathrm{Fun}(\Delta^1, \mathcal{C}) \to \mathrm{Fun}(\Delta^1, \mathcal{C})$. Dualizing this, we obtain a *fiber* functor $\mathsf{fib}\colon \mathrm{Fun}(\Delta^1, \mathcal{C}) \to \mathrm{Fun}(\Delta^1, \mathcal{C})$.

Proposition 14.6.8. *Let $\mathcal{C}$ be a finitely complete, finitely cocomplete, and pointed ∞-category. The cofiber functor $\mathsf{cof}\colon \mathrm{Fun}(\Delta^1, \mathcal{C}) \to \mathrm{Fun}(\Delta^1, \mathcal{C})$ is left adjoint to the fiber functor $\mathsf{fib}\colon \mathrm{Fun}(\Delta^1, \mathcal{C}) \to \mathrm{Fun}(\Delta^1, \mathcal{C})$,*

$$(\mathsf{cof}, \mathsf{fib})\colon \mathrm{Fun}(\Delta^1, \mathcal{C}) \rightleftarrows \mathrm{Fun}(\Delta^1, \mathcal{C}).$$

One way of defining *stable* ∞-categories is as follows.

Definition 14.6.9. A finitely complete, finitely cocomplete, and pointed ∞-category is *stable* if a triangle in it is exact if and only if it is coexact.

This is one way of imposing the usual *linearity condition* axiomatizing stability. It turns out that this definition admits the following reformulations (see [83, Prop. 1.1.3.4] and [83, Cor. 1.4.2.27]).

Theorem 14.6.10. *The following are equivalent for a finitely complete, finitely cocomplete, and pointed ∞-category $\mathcal{C}$.*

(i) The adjunction $(\Sigma, \Omega)\colon \mathcal{C} \rightleftarrows \mathcal{C}$ is an equivalence.

(ii) The adjunction $(\mathsf{cof}, \mathsf{fib})\colon \mathrm{Fun}(\Delta^1, \mathcal{C}) \rightleftarrows \mathrm{Fun}(\Delta^1, \mathcal{C})$ is an equivalence.

(iii) The ∞-category $\mathcal{C}$ is stable.

(iv) A square in $\mathcal{C}$ is a pullback square if and only if it is a pushout square.

The last characterization can be reformulated in terms of cubical diagrams [51, 10], and we refer to [49] for additional characterizations of stability. The relation of these characterizations to abstract representation theory is discussed in [48] and references therein.

Examples 14.6.11. (i) The underlying ∞-category of a stable, combinatorial model category (Examples 14.2.31) is stable.

(ii) The ∞-category of spectra can be realized as the underlying ∞-category of the stable model category of simplicial symmetric spectra [63]. The homotopy category of this ∞-category Sp is *the* stable homotopy category $\mathcal{SHC}$ of Boardman (see [113] and [4, Part III]). We will see in §14.6.2 that there also is an intrinsic construction of this ∞-category.

(iii) In the context of homological algebra, there are stable *derived ∞-categories*; see [82, §1].

Like stable model categories also stable ∞-categories provide an enhancement of triangulated categories. In particular, the homotopy category of a stable ∞-category can be endowed with a triangulation. To this end, we define a *cofiber sequence* in a pointed ∞-category $\mathcal{C}$ to be a diagram $\Delta^2 \times \Delta^1 \to \mathcal{C}$ looking like

$$\begin{array}{ccccc} x & \xrightarrow{f} & y & \longrightarrow & 0' \\ \downarrow & & \downarrow g & & \downarrow \\ 0 & \longrightarrow & z & \xrightarrow[h]{} & w \end{array} \tag{14.6.3}$$

and such that both squares are pushout squares. (A cofiber sequence is essentially obtained by two iterations of the passage to the cofiber of a morphism.) As in the case of ordinary category theory, it follows that also the composite square is a pushout square, and, by definition of the suspension functor (Definition 14.6.6), we obtain a canonical equivalence $\phi\colon w \simeq \Sigma x$. Thus, if we pass to homotopy classes of morphisms, then associated to (14.6.3) we obtain by means of this equivalence an *incoherent* cofiber sequence or *triangle*

$$T_f\colon \qquad x \xrightarrow{f} y \xrightarrow{g} z \xrightarrow{\phi\circ h} \Sigma x,$$

which is an ordinary diagram in the homotopy category $\mathrm{Ho}(\mathcal{C})$. If $\mathcal{C}$ is a stable ∞-category, then we say that a triangle in $\mathrm{Ho}(\mathcal{C})$ is *distinguished* if it is isomorphic to T_f for some $f\colon \Delta^1 \to \mathcal{C}$.

In the following result ([83, Thm. 1.1.2.14]) we also denote by $\Sigma\colon \mathrm{Ho}(\mathcal{C}) \to \mathrm{Ho}(\mathcal{C})$ the functor induced by $\Sigma\colon \mathcal{C} \to \mathcal{C}$. We assume that the reader is familiar with the notion of a triangulated category; see the original references of Puppe [91, Satz 3.5 and §4.1] or Verdier [111] (a reprint of his 1967 thesis) as well as the monographs [90, 59].

Theorem 14.6.12. *Let $\mathcal{C}$ be a stable ∞-category. The functor $\Sigma\colon \mathrm{Ho}(\mathcal{C}) \to \mathrm{Ho}(\mathcal{C})$ and the above class of distinguished triangles endow the homotopy category $\mathrm{Ho}(\mathcal{C})$ with the structure of a triangulated category.*

For closely related discussion of these triangulations we refer to the introduction to [48] and references there. A natural class of functors between stable ∞-categories is the class of exact functors in the sense of the following definition.

Definition 14.6.13. Let $\mathcal{C}$ and $\mathcal{D}$ be finitely complete and finitely cocomplete ∞-categories.

(i) A functor $\mathcal{C} \to \mathcal{D}$ is *left exact* if it preserves pullbacks and terminal objects.

(ii) A functor $\mathcal{C} \to \mathcal{D}$ is *right exact* if it preserves pushouts and initial objects.

(iii) A functor $\mathcal{C} \to \mathcal{D}$ is *exact* if it is left exact and right exact.

Clearly, limit-preserving functors (and hence, in particular, right adjoint functors) are left exact, and dually. If $\mathcal{C}$ and $\mathcal{D}$ are stable ∞-categories, then the three classes in Definition 14.6.13 agree. It turns out that the triangulations of Theorem 14.6.12 are natural with respect to exact functors in the following sense. Given an exact functor $F\colon \mathcal{C} \to \mathcal{D}$ of stable

∞-categories, then the functor $F\colon \mathrm{Ho}(\mathcal{C}) \to \mathrm{Ho}(\mathcal{D})$ can be endowed with the structure of an exact functor.

Remark 14.6.14. Note that, by the very definition, a stable ∞-category is obtained from the general notion of an ∞-category by imposing certain (easily motivated) exactness *properties*; namely, we ask that finite limits and finite colimits exist and that certain limit type constructions are colimit type constructions and conversely. Similarly, the good notion of morphisms of stable ∞-categories, namely, exact functors, are defined by asking for the property that certain (co)limits are preserved.

This is in contrast to the more classical notion of a triangulated category which addresses the bad categorical properties of derived categories of abelian categories or of homotopy categories of stable model categories or stable ∞-categories by imposing additional *structure*. Given an additive category, the axioms of a triangulated category ask for the existence of some non-canonical additional structure (the suspension functor and the class of distinguished triangles) satisfying certain properties. Similarly, given two triangulated categories $\mathcal{T}$ and $\mathcal{T}'$, a morphism should be an additive functor $F\colon \mathcal{T} \to \mathcal{T}'$ which sends distinguished triangles to distinguished triangles. In order to make this precise, one has to ask for the existence of an exact structure, i.e., a natural isomorphism $F\Sigma \cong \Sigma F$.

Despite their great successes in many areas of pure mathematics, it was obvious from the very beginning on (see for example already the introduction to [54]) that the axioms of a triangulated category suffer certain defects (non-functoriality of the cone construction, no good theory of homotopy limits and homotopy colimits, diagram categories of triangulated categories do not admit canonical triangulations).

There are more traditional attempts to improve the axioms of a triangulated category and the basic idea goes back at least to [11]. The idea is to ask for more structure, leading to *higher triangulations*. This use of the word 'higher' is meant to indicate that one asks for higher octahedron axioms, i.e., that one also encodes iterated (co)fibers associated to longer strings of composable morphisms (see [88] for a precise definition). It can be shown that homotopy categories of stable ∞-categories or stable model categories can be naturally endowed with higher triangulations [50, §13], illustrating the slogan that 'these enhancements encode all the triangulated structure'.

14.6.2 Stabilization and the universality of spectra

In this section we discuss the stabilization process which can be realized by passing to ∞-categories of internal spectrum objects. Similar constructions were also carried out in the language of model categories (for example by Schwede [101] and Hovey [61]) as well as in the framework of derivators by Heller [55].

The basic notion to begin with is the following one [79, Def. 8.1].

Definition 14.6.15. Let $\mathcal{C}$ be a finitely complete, finitely cocomplete, and pointed ∞-category. A *prespectrum object* in $\mathcal{C}$ is a functor

$$X\colon N(\mathbb{Z} \times \mathbb{Z}) \to \mathcal{C}$$

such that for all $i \neq j$ the value $X(i,j)$ is a zero object. The full subcategory of $\mathrm{Fun}(N(\mathbb{Z} \times \mathbb{Z}), \mathcal{C})$ spanned by the prespectrum objects is denoted by $\mathrm{PSp}(\mathcal{C})$.

Here, we consider the poset $\mathbb{Z}$ with the natural ordering as a category. Since only the diagonal entries are possibly non-trivial, we use the shorthand notation $X_m = X(m,m)$. Thus, a part of a prespectrum object $X \in \mathrm{PSp}(\mathcal{C})$ looks like

$$\begin{array}{ccccc} & & 0 & \longrightarrow & X_{m+1} \\ & & \uparrow & & \uparrow \\ 0'' & \longrightarrow & X_m & \longrightarrow & 0' \\ \uparrow & & \uparrow & & \\ X_{m-1} & \longrightarrow & 0'''. & & \end{array}$$

By definition of the suspension and loop functors $(\Sigma, \Omega)\colon \mathcal{C} \rightleftarrows \mathcal{C}$, given $X \in \mathrm{PSp}(\mathcal{C})$ we obtain induced morphisms

$$\alpha_{m-1}\colon \Sigma X_{m-1} \to X_m \qquad \text{and} \qquad \beta_m\colon X_m \to \Omega X_{m+1}. \tag{14.6.4}$$

Definition 14.6.16. Let $\mathcal{C}$ be a finitely complete, finitely cocomplete, and pointed ∞-category and let $X \in \mathrm{PSp}(\mathcal{C})$.

(i) The prespectrum X is a *spectrum below n* if $\beta_m\colon X_m \xrightarrow{\sim} \Omega X_{m+1}$ is an equivalence for all $m < n$. The full subcategory of $\mathrm{PSp}(\mathcal{C})$ spanned by the spectra below n is denoted by $\mathrm{Sp}_n(\mathcal{C})$.

(ii) The prespectrum X is a *spectrum object* if $\beta_m\colon X_m \xrightarrow{\sim} \Omega X_{m+1}$ is an equivalence for all $m \in \mathbb{Z}$. The full subcategory of $\mathrm{PSp}(\mathcal{C})$ spanned by the spectrum objects is denoted by $\mathrm{Sp}(\mathcal{C})$.

Example 14.6.17. An important special case is obtained if we start with the pointed ∞-category $\mathcal{S}_*$ of pointed spaces. In that case, we simplify notation and write $\mathrm{Sp} = \mathrm{Sp}(\mathcal{S}_*)$ for the *∞-category of spectra.*

Theorem 14.6.18. *The ∞-category* Sp *of spectra is stable and presentable.*

We discuss further below the fact that Sp is the *stabilization* of the ∞-category $\mathcal{S}$ of spaces. More generally, given an ∞-category $\mathcal{C}$, we refer to

$$\mathrm{Stab}(\mathcal{C}) = \mathrm{Sp}(\mathcal{C}_*)$$

as the *stabilization* of $\mathcal{C}$.

Under certain assumptions on a pointed ∞-category $\mathcal{C}$ we will now construct a *spectrification functor*, i.e., a left adjoint $L\colon \mathrm{PSp}(\mathcal{C}) \to \mathrm{Sp}(\mathcal{C})$ to the fully faithful inclusion functor $\iota\colon \mathrm{Sp}(\mathcal{C}) \to \mathrm{PSp}(\mathcal{C})$, exhibiting $\mathrm{Sp}(\mathcal{C})$ as a localization of $\mathrm{PSp}(\mathcal{C})$. To begin with there is the following result.

Proposition 14.6.19. *Let $\mathcal{C}$ be a finitely complete, finitely cocomplete, and pointed ∞-category. The fully faithful inclusion $\iota_n\colon \mathrm{Sp}_n(\mathcal{C}) \to \mathrm{PSp}(\mathcal{C})$ admits a left adjoint $L_n\colon \mathrm{PSp}(\mathcal{C}) \to \mathrm{Sp}_n(\mathcal{C})$,*

$$(L_n, \iota_n)\colon \mathrm{PSp}(\mathcal{C}) \rightleftarrows \mathrm{Sp}_n(\mathcal{C}).$$

The idea is of course that spectra below a certain level are somehow determined by the higher levels. And in fact, the left adjoint L_n can be constructed as follows. Given a prespectrum $X \in \mathrm{PSp}(\mathcal{C})$ we restrict it to the full subcategory of $N(\mathbb{Z} \times \mathbb{Z})$ spanned by

$$Q_n = \{(i,j) \in \mathbb{Z} \times \mathbb{Z} \mid i \neq j \text{ or } i = j \geq n\}$$

and then set

$$L_n(X) = \mathrm{RKan}_{Q_n \hookrightarrow N(\mathbb{Z}\times\mathbb{Z})}(X|_{Q_n}).$$

Here RKan stands for an ∞-categorical variant of the usual right Kan extension [82, §4.3]. Under suitable completeness assumptions on the ∞-categories involved, right Kan extensions can again be calculated pointwise, i.e., are given by limits over certain slice categories. In our situation, the corresponding slice categories are cofinally finite and the above right Kan extensions hence exist. The essential image of L_n consists of the spectra below n.

With a bit more care, one can show that there is a sequence of functors

$$\mathrm{id} \to L_0 \to L_1 \to L_2 \to \ldots\colon \mathrm{PSp}(\mathcal{C}) \to \mathrm{PSp}(\mathcal{C}),$$

and it is hence tempting to simply set $L := \mathrm{colim}_n L_n$. This in fact works if one imposes the following conditions on the ∞-category $\mathcal{C}$ ([79, Cor. 8.17]).

Proposition 14.6.20. *Let $\mathcal{C}$ be a finitely complete, countably cocomplete, and pointed ∞-category. If the loop functor $\Omega_{\mathcal{C}}\colon \mathcal{C} \to \mathcal{C}$ commutes with sequential colimits, then*

$$L := \mathrm{colim}_n L_n\colon \mathrm{PSp}(\mathcal{C}) \to \mathrm{PSp}(\mathcal{C})$$

is a localization with essential image $\mathrm{Sp}(\mathcal{C})$. *We refer to L as the* spectrification functor.

An important example of an ∞-category satisfying these assumptions is the ∞-category of pointed spaces. In this case, let $\mathcal{D}_n \subseteq \mathrm{Sp}_n$ be the full subcategory spanned by those X such that $\alpha_m\colon \Sigma X_m \to X_{m+1}$ defined in (14.6.4) is an equivalence for $m \geq n$. Thus, morally, such a prespectrum X is essentially determined by its value X_n. And in fact, the evaluation map $\mathrm{ev}_n\colon \mathcal{D}_n \to \mathcal{S}_*$ is an acyclic Kan fibration. Let us choose a section $s_{\tilde{\Sigma}^{\infty-n}}\colon \mathcal{S}_* \to \mathcal{D}_n$ of ev_n and set

$$\tilde{\Sigma}^{\infty-n}\colon \mathcal{S}_* \xrightarrow{s_{\tilde{\Sigma}^{\infty-n}}} \mathcal{D}_n \to \mathrm{PSp}.$$

(We note again that this is well-defined since the space of sections is a contractible Kan complex; see the discussion around (14.6.2).) Denoting the n-th evaluation functor $\mathrm{PSp} \to \mathcal{S}_*$ by $\tilde{\Omega}^{\infty-n}$ we obtain an adjunction

$$(\tilde{\Sigma}^{\infty-n}, \tilde{\Omega}^{\infty-n})\colon \mathcal{S}_* \rightleftarrows \mathrm{PSp}.$$

Combining this with Proposition 14.6.19 and Proposition 14.6.20 we deduce the following result.

Proposition 14.6.21. *There is the following sequence of adjunctions*

$$(\Sigma_+^{\infty-n}, \Omega_-^{\infty-n})\colon \mathcal{S} \underset{-}{\overset{+}{\rightleftarrows}} \mathcal{S}_* \underset{\tilde{\Omega}^{\infty-n}}{\overset{\tilde{\Sigma}^{\infty-n}}{\rightleftarrows}} \mathrm{PSp} \underset{\iota}{\overset{L}{\rightleftarrows}} \mathrm{Sp}\,.$$

Dropping the first adjunction in the proposition we obtain the adjunction

$$(\Sigma^{\infty-n}, \Omega^{\infty-n})\colon \mathcal{S}_* \rightleftarrows \mathrm{Sp},$$

and it turns out that a similar adjunction exists for arbitrary pointed, presentable ∞-categories. In fact, the evaluation functor $\Omega^{\infty-n}\colon \mathrm{Sp}(\mathcal{C}) \to \mathcal{C}$ clearly makes sense for every finitely complete, finitely cocomplete, and pointed ∞-category $\mathcal{C}$. If $\mathcal{C}$ is moreover presentable, then it can be shown that $\Omega^{\infty-n}$ satisfies the assumptions of the special adjoint functor theorem (Theorem 14.4.13). Thus, there is a left adjoint $\Sigma^{\infty-n}\colon \mathcal{C} \to \mathrm{Sp}(\mathcal{C})$, the *suspension spectrum functor* or the *n-th free spectrum functor*,

$$(\Sigma^{\infty-n}, \Omega^{\infty-n})\colon \mathcal{C} \rightleftarrows \mathrm{Sp}(\mathcal{C}),$$

although in this generality the functor $\Sigma^{\infty-n}$ does not admit such a nice explicit description as in the case of $\mathcal{S}_*$. In the following important result [79, Cor. 15.5] we denote by $\mathrm{Fun}^{\mathrm{L}}(\mathcal{C}, \mathcal{D})$ the ∞-category of colimit-preserving functors between presentable ∞-categories $\mathcal{C}$ and $\mathcal{D}$.

Theorem 14.6.22. *Let $\mathcal{C}, \mathcal{D}$ be pointed, presentable ∞-categories and suppose that $\mathcal{D}$ is stable. Restriction along the suspension spectrum functor $\Sigma^\infty\colon \mathcal{C} \to \mathrm{Sp}(\mathcal{C})$ induces an equivalence of ∞-categories*

$$\mathrm{Fun}^{\mathrm{L}}(\mathrm{Sp}(\mathcal{C}), \mathcal{D}) \xrightarrow{\sim} \mathrm{Fun}^{\mathrm{L}}(\mathcal{C}, \mathcal{D}).$$

As a central special case [79, Cor. 15.6], let us again consider the ∞-category $\mathcal{S}_*$. We refer to the image of the zero sphere $S^0 = \Delta^0_+$ under $\Sigma^\infty\colon \mathcal{S}_* \to \mathrm{Sp}$ as the *sphere spectrum.*

Corollary 14.6.23. *Let $\mathcal{D}$ be a stable, presentable ∞-category. Evaluation at the sphere spectrum induces an equivalence of ∞-categories*

$$\mathrm{Fun}^{\mathrm{L}}(\mathrm{Sp}, \mathcal{D}) \xrightarrow{\sim} \mathcal{D}.$$

In fact, this follows from Theorem 14.6.22 by observing that the evaluation map factors as

$$\mathrm{Fun}^{\mathrm{L}}(\mathrm{Sp}, \mathcal{D}) \xrightarrow{\sim} \mathrm{Fun}^{\mathrm{L}}(\mathcal{S}_*, \mathcal{D}) \xrightarrow{\sim} \mathcal{D}$$

where the second equivalence is given by evaluation on S^0; see Proposition 14.6.3. This corollary makes precise the statement that the ∞-category Sp of spectra is the *free stable ∞-category on one generator*, namely on the sphere spectrum.

14.6.3 Tensor products of presentable ∞-categories

We now turn to the tensor product of presentable ∞-categories which plays a key role in the construction of the smash product on the ∞-category of spectra; see §14.6.4. We begin with the corresponding construction in ordinary category theory.

Let $\mathcal{C}_1$ and $\mathcal{C}_2$ be locally presentable categories. The *tensor product* of $\mathcal{C}_1$ and $\mathcal{C}_2$ is a locally presentable category $\mathcal{C}_1 \otimes \mathcal{C}_2$ together with a universal bilinear map, i.e., a functor

$$\mathcal{C}_1 \times \mathcal{C}_2 \to \mathcal{C}_1 \otimes \mathcal{C}_2 \tag{14.6.5}$$

that preserves colimits in both variables separately and that is universal in the following sense: For any locally presentable category $\mathcal{D}$ restricting along (14.6.5) induces an equivalence of categories

$$\mathrm{Fun}^{\mathrm{L}}(\mathcal{C}_1 \otimes \mathcal{C}_2, \mathcal{D}) \xrightarrow{\sim} \mathrm{Fun}^{\mathrm{L,L}}(\mathcal{C}_1 \times \mathcal{C}_2, \mathcal{D})$$

where $\mathrm{Fun}^{\mathrm{L,L}}(\mathcal{C}_1 \times \mathcal{C}_2, \mathcal{D}) \subseteq \mathrm{Fun}(\mathcal{C}_1 \times \mathcal{C}_2, \mathcal{D})$ is the full subcategory spanned by the functors that preserve colimits in both variables separately.

It can be shown that the tensor product always exists [17, Chapter 5] and that it admits the explicit description

$$\mathcal{C}_1 \otimes \mathcal{C}_2 = \mathrm{Fun}^{\mathrm{R}}(\mathcal{C}_1^{\mathrm{op}}, \mathcal{C}_2), \tag{14.6.6}$$

where $\mathrm{Fun}^{\mathrm{R}}(-,-) \subseteq \mathrm{Fun}(-,-)$ denotes the full subcategory spanned by the *limit-preserving* functors.

In [83] Lurie establishes a variant of this monoidal structure for presentable ∞-categories. Let $\mathcal{P}\mathrm{r}^{\mathrm{L}}$ be the (very large) ∞*-category of presentable* ∞*-categories* and colimit-preserving functors.

Theorem 14.6.24. *The* ∞*-category* $\mathcal{P}\mathrm{r}^{\mathrm{L}}$ *admits a closed symmetric monoidal structure* $\mathcal{P}\mathrm{r}^{\mathrm{L},\otimes} \to N(\mathcal{F}\mathrm{in})$ *such that the following properties are satisfied.*

(i) The tensor product $\mathcal{C}_1 \otimes \mathcal{C}_2$ *corepresents the functor that sends* $\mathcal{D}$ *to the* ∞*-category* $\mathrm{Fun}^{\mathrm{L,L}}(\mathcal{C}_1 \times \mathcal{C}_2, \mathcal{D})$ *of functors that preserve colimits separately in both variables.*

(ii) The ∞*-category* $\mathcal{C}_1 \otimes \mathcal{C}_2$ *is equivalent to* $\mathrm{Fun}^{\mathrm{R}}(\mathcal{C}_1^{\mathrm{op}}, \mathcal{C}_2)$.

(iii) The ∞*-category* $\mathcal{S}$ *of spaces is the monoidal unit.*

(iv) The internal hom is given by $\mathrm{Fun}^{\mathrm{L}}(\mathcal{C}_1, \mathcal{C}_2)$.

We denote by $\mathcal{P}\mathrm{r}^{\mathrm{L,smon}}$ the ∞-category of presentable, symmetric monoidal closed ∞-categories and symmetric monoidal, colimit-preserving functors. The following is a variant of Perspective 14.5.21 and it follows from the results in [83, §4.8] (see [83, Rmk. 4.8.1.8]).

Proposition 14.6.25. *There is an equivalence of* ∞*-categories*

$$\mathrm{Alg}_{\mathbb{E}_\infty}(\mathcal{P}\mathrm{r}^{\mathrm{L},\otimes}) \simeq \mathcal{P}\mathrm{r}^{\mathrm{L,smon}}.$$

Thus, Proposition 14.5.22 applied to the monoidal structure of Theorem 14.6.24 implies that the ∞-category $\mathcal{S}$ of spaces endowed with a certain symmetric monoidal structure is an initial object in $\mathcal{P}\mathrm{r}^{\mathrm{L,smon}}$. As a special case of [83, Cor. 4.8.1.12], it turns out that the monoidal structure is the usual Cartesian monoidal structure and that it can be characterized by the properties that

(i) the pairing $\times \colon \mathcal{S} \times \mathcal{S} \to \mathcal{S}$ preserves colimits separately in both variables and

(ii) the point $\Delta^0 \in \mathcal{S}$ is a monoidal unit.

14.6.4 The smash product

In this subsection we briefly discuss an ∞-categorical version of the smash product monoidal structure on spectra. The stabilization of presentable ∞-categories can be summarized by the following theorem. We denote by $\mathcal{P}r^{\mathrm{L}}_{\mathrm{St}} \subseteq \mathcal{P}r^{\mathrm{L}}_{\mathrm{Pt}} \subseteq \mathcal{P}r^{\mathrm{L}}$ the full subcategories spanned by stable, presentable and pointed and presentable ∞-categories, respectively. The following result is a consequence of the techniques in [83, §4.8] (see, in particular, [83, Prop. 4.8.2.11] and [83, Prop. 4.8.2.18]).

Theorem 14.6.26. *The stabilization* $\mathrm{Stab}\colon \mathcal{P}r^{\mathrm{L}} \to \mathcal{P}r^{\mathrm{L}}_{\mathrm{St}}$ *of presentable ∞-categories factors as a composition of adjunctions*

$$\mathcal{P}r^{\mathrm{L}} \rightleftarrows \mathcal{P}r^{\mathrm{L}}_{\mathrm{Pt}} \rightleftarrows \mathcal{P}r^{\mathrm{L}}_{\mathrm{St}}.$$

It turns out that the symmetric monoidal structure $\mathcal{P}r^{\mathrm{L},\otimes} \to N(\mathcal{F}\mathrm{in})$ has similar variants in the pointed and in the stable context. These results are instances of a more general construction of canonical monoidal structures on smashing localizations associated to idempotent objects (see again [83, §4.8.2]). Let us collect the variant of Theorem 14.6.24 for stable, presentable ∞-categories.

Theorem 14.6.27. *The ∞-category $\mathcal{P}r^{\mathrm{L}}_{\mathrm{St}}$ admits a closed symmetric monoidal structure $\mathcal{P}r^{\mathrm{L},\otimes}_{\mathrm{St}} \to N(\mathcal{F}\mathrm{in})$ such that the following properties are satisfied.*

(i) The tensor product $\mathcal{C}_1 \otimes \mathcal{C}_2$ corepresents the functor that sends $\mathcal{D}$ to the ∞-category $\mathrm{Fun}^{\mathrm{L,L}}(\mathcal{C}_1 \times \mathcal{C}_2, \mathcal{D})$ *of functors that preserve colimits separately in both variables.*

(ii) The ∞-category $\mathcal{C}_1 \otimes \mathcal{C}_2$ is equivalent to $\mathrm{Fun}^{\mathrm{R}}(\mathcal{C}_1^{\mathrm{op}}, \mathcal{C}_2)$.

(iii) The ∞-category Sp *of spectra is the monoidal unit.*

(iv) The internal hom is given by $\mathrm{Fun}^{\mathrm{L}}(\mathcal{C}_1, \mathcal{C}_2)$.

And there is a similar variant for $\mathcal{P}r^{\mathrm{L}}_{\mathrm{Pt}}$ the only difference being that the ∞-category $\mathcal{S}_*$ of pointed spaces is the monoidal unit.

Let $\mathcal{P}r^{\mathrm{L,smon}}_{\mathrm{St}}$ denote the ∞-category of stable, presentable, symmetric monoidal closed ∞-categories and symmetric monoidal, colimit-preserving functors. As a consequence of Perspective 14.5.21 there is the following result.

Proposition 14.6.28. *There is an equivalence of ∞-categories*

$$\mathrm{Alg}_{\mathbb{E}_\infty}(\mathcal{P}r^{\mathrm{L},\otimes}_{\mathrm{St}}) \simeq \mathcal{P}r^{\mathrm{L,smon}}_{\mathrm{St}}.$$

An application of Proposition 14.5.22 to $\mathcal{P}r^{\mathrm{L},\otimes}_{\mathrm{St}}$ implies that the ∞-category Sp of spectra can be endowed with a certain symmetric monoidal structure, the *smash product*, such that the resulting symmetric monoidal ∞-category $\mathrm{Sp}^{\otimes}$ is an initial object in $\mathcal{P}r^{\mathrm{L,smon}}_{\mathrm{St}}$. It turns out that the monoidal structure can be characterized by the properties that

(i) the monoidal structure $\otimes\colon \mathrm{Sp} \times \mathrm{Sp} \to \mathrm{Sp}$ preserves colimits separately in both variables and

(ii) the sphere spectrum is the monoidal unit.

Having the smash product at our disposal we can finally make the following definition.

Definition 14.6.29. (i) The *∞-category of $\mathbb{E}_\infty$-ring spectra* is the ∞-category of commutative algebra objects $\mathrm{Alg}_{\mathbb{E}_\infty}(\mathrm{Sp}^\otimes)$.

(ii) The *∞-category of $\mathbb{A}_\infty$-ring spectra* is the ∞-category of (associative) algebra objects $\mathrm{Alg}_{\mathbb{A}_\infty}(\mathrm{Sp}^\otimes)$.

These are ∞-categorical versions of the more classical model categories of commutative or associative ring spectra. In fact, for a precise statement along these lines using *symmetric spectra* see [81, Example 8.21]. Having these key notions in place one could now begin with an ∞-categorical study of stable homotopy theory [83] which together with the theory of ∞-topoi [82, §§6-7] provides the foundations for Lurie's ∞-categorical approach to derived algebraic geometry. For this we refer the reader to the literature and to David Gepner's chapter in this Handbook.

We conclude this section by a short discussion of monoidal aspects of the stabilization of presentable ∞-categories. References for the remainder of this sections are [83, §8.4.2] and [44, §3].

Theorem 14.6.30. *Let $\mathcal{C}^\otimes$ be a closed symmetric monoidal structure on a presentable ∞-category. The ∞-categories $\mathcal{C}_*$ and $\mathrm{Sp}(\mathcal{C})$ admit closed symmetric monoidal structures, that are uniquely determined by the requirement that the respective free functors from $\mathcal{C}$ are symmetric monoidal. Moreover, also the remaining left adjoint in*

$$\mathcal{C} \to \mathcal{C}_* \to \mathrm{Sp}(\mathcal{C})$$

is uniquely symmetric monoidal.

The monoidal structures in this theorem enjoy the following universal properties. Given closed symmetric monoidal, presentable ∞-categories $\mathcal{C}, \mathcal{D}$ we denote by $\mathrm{Fun}^{\mathrm{L},\otimes}(\mathcal{C},\mathcal{D})$ the ∞-category of symmetric monoidal, colimit-preserving functors from $\mathcal{C}$ to $\mathcal{D}$.

Theorem 14.6.31. *Let $\mathcal{C}, \mathcal{D}$ be closed symmetric monoidal, presentable ∞-categories.*

(i) If $\mathcal{D}$ is pointed, then the symmetric monoidal functor $\mathcal{C} \to \mathcal{C}_*$ induces an equivalence of ∞-categories

$$\mathrm{Fun}^{\mathrm{L},\otimes}(\mathcal{C}_*,\mathcal{D}) \xrightarrow{\sim} \mathrm{Fun}^{\mathrm{L},\otimes}(\mathcal{C},\mathcal{D}).$$

(ii) If $\mathcal{D}$ is stable, then the symmetric monoidal functor $\mathcal{C} \to \mathrm{Sp}(\mathcal{C})$ induces an equivalence of ∞-categories

$$\mathrm{Fun}^{\mathrm{L},\otimes}(\mathrm{Sp}(\mathcal{C}),\mathcal{D}) \xrightarrow{\sim} \mathrm{Fun}^{\mathrm{L},\otimes}(\mathcal{C},\mathcal{D}).$$

One way to summarize some of these results is as follows.

Theorem 14.6.32. *The stabilization $\mathcal{P}\mathrm{r}^{\mathrm{L}} \to \mathcal{P}\mathrm{r}^{\mathrm{L}}_{\mathrm{St}}$ of presentable ∞-categories admits a monoidal refinement which factors as a composition of adjunctions*

$$\mathcal{P}\mathrm{r}^{\mathrm{L},\otimes} \rightleftarrows \mathcal{P}\mathrm{r}^{\mathrm{L},\otimes}_{\mathrm{Pt}} \rightleftarrows \mathcal{P}\mathrm{r}^{\mathrm{L},\otimes}_{\mathrm{St}}.$$

We conclude this course by the following perspective on a refined picture of the stabilization process [44].

Perspective 14.6.33. Let us recall that stable ∞-categories are obtained from pointed ∞-categories by imposing an additional exactness property, asking that pushouts and pullbacks agree. There are two more intermediate steps given by preadditive and additive ∞-categories, respectively, both of which are obtained by adding certain exactness properties to pointed ∞-categories. A *preadditive* ∞-category is a pointed ∞-category with finite biproducts. It follows from these axioms that every object can be canonically turned into an $\mathbb{E}_\infty$-monoid object and this monoid structure is given by the fold map. We say that a preadditive ∞-category is *additive* if these canonical $\mathbb{E}_\infty$-monoid structures actually are $\mathbb{E}_\infty$-group structures.

Focusing again on the context of presentable ∞-categories it can be shown that there are universal examples of such ∞-categories. In fact, the ∞-category $\mathrm{Mon}_{\mathbb{E}_\infty}(\mathcal{S})$ of $\mathbb{E}_\infty$-spaces is a preadditive, presentable ∞-category and the left adjoint $\mathcal{S} \to \mathrm{Mon}_{\mathbb{E}_\infty}(\mathcal{S})$ to the forgetful functor $\mathrm{Mon}_{\mathbb{E}_\infty}(\mathcal{S}) \to \mathcal{S}$ exhibits $\mathrm{Mon}_{\mathbb{E}_\infty}(\mathcal{S})$ as the free preadditive ∞-category on one generator, namely the free $\mathbb{E}_\infty$-space generated by Δ^0. More specifically, for every preadditive, presentable ∞-category $\mathcal{D}$, there are equivalences of ∞-categories

$$\mathrm{Fun}^{\mathrm{L}}(\mathrm{Mon}_{\mathbb{E}_\infty}(\mathcal{S}), \mathcal{D}) \xrightarrow{\sim} \mathrm{Fun}^{\mathrm{L}}(\mathcal{S}, \mathcal{D}) \xrightarrow{\sim} \mathcal{D}.$$

A similar universal property is enjoyed by $\mathcal{C} \to \mathrm{Mon}_{\mathbb{E}_\infty}(\mathcal{C})$ where $\mathcal{C}$ is a general presentable ∞-categories. And if we pass to the context of additive presentable ∞-categories instead, then the universal example is given by the ∞-category $\mathrm{Grp}_{\mathbb{E}_\infty}(\mathcal{S})$ of grouplike $\mathbb{E}_\infty$-spaces.

It turns out that the stabilization of presentable ∞-categories factors through the ∞-category of preadditive, presentable ∞-categories and also through the ∞-category of additive, presentable ∞-categories. More precisely, the obvious forgetful functors admit left adjoints and the stabilization hence factors as

$$\mathcal{P}\mathrm{r}^{\mathrm{L}} \rightleftarrows \mathcal{P}\mathrm{r}^{\mathrm{L}}_{\mathrm{Pt}} \rightleftarrows \mathcal{P}\mathrm{r}^{\mathrm{L}}_{\mathrm{Pre}} \rightleftarrows \mathcal{P}\mathrm{r}^{\mathrm{L}}_{\mathrm{Add}} \rightleftarrows \mathcal{P}\mathrm{r}^{\mathrm{L}}_{\mathrm{St}}.$$

Finally, let us also mention that this factorization admits a monoidal refinement parallel to the one in Theorem 14.6.32. For details and applications we refer the reader to [44].

Bibliography

[1] O. Antolín Camerona. A whirlwind tour of the world of $(\infty, 1)$-categories. arXiv:1303.4669, 2013.

[2] J. Adámek and J. Rosický. *Locally presentable and accessible categories*. Volume 189 of *London Math. Soc. Lecture Note Ser.* Cambridge University Press, 1994.

[3] J. Adámek, J. Rosický, and E.M. Vitale. *Algebraic theories*. Volume 184 of *Cambridge Tracts in Math.* Cambridge University Press, 2011.

[4] J.F. Adams. *Stable homotopy and generalised homology*. Chicago Lectures in Mathematics. University of Chicago Press, Chicago, 1974.

[5] J.F. Adams. *Infinite loop spaces*. Volume 90 of *Ann. of Math. Stud.* Princeton University Press, 1978.

[6] M. Artin, A. Grothendieck, and J. L. Verdier. *Théorie des topos et cohomologie étale des schémas. Tome 1: Théorie des topos*. Lecture Notes in Math., Vol. 269. Springer-Verlag, 1972. Séminaire de

Géométrie Algébrique du Bois-Marie 1963–1964 (SGA 4). Avec la collaboration de N. Bourbaki, P. Deligne et B. Saint-Donat.

[7] M. Artin, A. Grothendieck, and J. L. Verdier. *Théorie des topos et cohomologie étale des schémas. Tome 2.* Lecture Notes in Math., Vol. 270. Springer-Verlag, 1972. Séminaire de Géométrie Algébrique du Bois-Marie 1963–1964 (SGA 4), Avec la collaboration de N. Bourbaki, P. Deligne et B. Saint-Donat.

[8] M. Artin, A. Grothendieck, and J. L. Verdier. *Théorie des topos et cohomologie étale des schémas. Tome 3.* Lecture Notes in Math., Vol. 305. Springer-Verlag, 1973. Séminaire de Géométrie Algébrique du Bois-Marie 1963–1964 (SGA 4). Avec la collaboration de P. Deligne et B. Saint-Donat.

[9] C. Barwick and D. Kan. Relative categories: Another model for the homotopy theory of homotopy theories. Indag. Math. (N.S.) 23(1–2):42–68, (2012).

[10] F. Beckert and M. Groth. Abstract cubical homotopy theory. arXiv:1803.06022, 2018.

[11] A. Beĭlinson, J. Bernstein, and P. Deligne. Faisceaux pervers. In *Analysis and topology on singular spaces, I (Luminy, 1981)*, volume 100 of *Astérisque*, pages 5–171. Soc. Math. France, 1982.

[12] T. Beke. Sheafifiable homotopy model categories. *Math. Proc. Cambridge Philos. Soc.*, 129(3):447–475, 2000.

[13] J. Bénabou. Introduction to bicategories. In *Reports of the Midwest Category Seminar*, pages 1–77. Springer, 1967.

[14] J.E. Bergner. A model category structure on the category of simplicial categories. *Trans. Amer. Math. Soc.*, 359(5):2043–2058, 2007.

[15] J.E. Bergner. A survey of $(\infty, 1)$-categories. In *Towards higher categories*, volume 152 of *IMA Vol. Math. Appl.*, pages 69–83. Springer, 2010.

[16] J.E. Bergner. Workshop on the homotopy theory of homotopy theories. Available at `http://www.math.ucr.edu/~jbergner/HomotopyWorkshop.pdf`, 2011.

[17] G. J. Bird. *Limits in 2-categories of locally-presented categories.* PhD thesis, University of Sydney, 1984.

[18] F. Borceux. *Handbook of categorical algebra. 1: Basic category theory.* Volume 50 of *Encyclopedia Math. Appl.* Cambridge University Press, 1994.

[19] F. Borceux. *Handbook of categorical algebra. 2: Categories and structures.* Volume 51 of *Encyclopedia Math. Appl.* Cambridge University Press, 1994.

[20] F. Borceux. *Handbook of categorical algebra. 3: Categories of sheaves.* Volume 52 of *Encyclopedia Math. Appl.* Cambridge University Press, 1994.

[21] A.K. Bousfield. The localization of spaces with respect to homology. *Topology*, 14:133–150, 1975.

[22] A.K. Bousfield and D. Kan. *Homotopy limits, completions and localizations.* Lecture Notes in Math., Vol. 304. Springer-Verlag,1972.

[23] M. Boardman and R. Vogt. *Homotopy invariant algebraic structures on topological spaces.* Lecture Notes in Math., Vol. 347. Springer-Verlag, 1973.

[24] E.H. Brown, Jr. Cohomology theories. *Ann. of Math. (2)*, 75:467–484, 1962.

[25] J.-M. Cordier. Sur la notion de diagramme homotopiquement cohérent. Third Colloquium on Categories, Part VI (Amiens, 1980). *Cahiers Topologie Géom. Différentielle*, 23(1):93–112, 1982.

[26] J.-M. Cordier and T. Porter. Vogt's theorem on categories of homotopy coherent diagrams. *Math. Proc. Cambridge Philos. Soc.*, 100(1):65–90, 1986.

[27] J.-M. Cordier and T. Porter. Homotopy coherent category theory. *Trans. Amer. Math. Soc.*, 349(1):1–54, 1997.

[28] J. Cranch. Algebraic Theories and $(\infty, 1)$-categories. arXiv:1011.3243, 2010.

[29] B. Day. On closed categories of functors. In *Reports of the Midwest Category Seminar, IV*, Lecture Notes in Mathematics, Vol. 137, pages 1–38. Springer, 1970.

[30] D. Dugger. Combinatorial model categories have presentations. *Adv. Math.*, 164(1):177–201, 2001.

[31] D. Dugger. Universal homotopy theories. *Adv. Math.*, 164(1):144–176, 2001.

[32] D. Dugger. A primer on homotopy colimits. Available at `http://math.uoregon.edu/~ddugger/hocolim.pdf`, 2008.

[33] D. Dugger, S. Hollander, and D. C. Isaksen. Hypercovers and simplicial presheaves. *Math. Proc. Cambridge Philos. Soc.*, 136(1):9–51, 2004.

[34] D. Dugger and D. I. Spivak. Mapping spaces in quasi-categories. *Algebr. Geom. Topol.*, 11(1):263–325, 2011.

[35] D. Dugger and D.I. Spivak. Rigidification of quasi-categories. *Algebr. Geom. Topol.*, 11(1):225–261, 2011.
[36] W. G. Dwyer, P. S. Hirschhorn, D. M. Kan, and J.H. Smith. *Homotopy limit functors on model categories and homotopical categories.* Volume 113 of *Math. Surv. Monogr.* Amer. Math. Soc., 2004.
[37] W. G. Dwyer and D. M. Kan. Function complexes in homotopical algebra. *Topology*, 19(4):427–440, 1980.
[38] C. Faith. *Algebra: rings, modules and categories. I.* Grundlehren Math. Wiss., Vol. 190, Springer-Verlag, 1973.
[39] L. Fajstrup and J. Rosický. A convenient category for directed homotopy. *Theory Appl. Categ.*, 21:No. 1, 7–20, 2008.
[40] P.J. Freyd. Abelian categories. *Repr. Theory Appl. Categ.*, (3):1–190, 2003.
[41] G. Friedman. Survey article: an elementary illustrated introduction to simplicial sets. *Rocky Mountain J. Math.*, 42(2):353–423, 2012.
[42] P. Gabriel and F. Ulmer. *Lokal präsentierbare Kategorien.* Lecture Notes in Math., Vol. 22. Springer-Verlag, 1971.
[43] P. Gabriel and M. Zisman. *Calculus of fractions and homotopy theory.* volume 35 of Ergeb. Math. Grenzgeb. (3), Springer-Verlag, 1967.
[44] David Gepner, Moritz Groth, and Thomas Nikolaus. Universality of multiplicative infinite delooping machines. *Algebr. Geom. Topol.*, 15(6):3107–3153, 2015.
[45] D. Gepner and R. Haugseng. Enriched ∞-categories via non-symmetric ∞-operads. arXiv:1312.3178, 2013.
[46] P.G. Goerss and J.F. Jardine. *Simplicial homotopy theory*, volume 174 of *Progr. Math.* Birkhäuser Verlag, 1999.
[47] J.W. Gray. *Formal category theory: Adjointness for 2-categories.* Lecture Notes in Math., Vol. 391. Springer-Verlag, 1974.
[48] M. Groth and M. Rahn. Higher symmetries in abstract stable homotopy theories. arXiv:1904.00580, 2019.
[49] M. Groth and M. Shulman. Characterizations of abstract stable homotopy theories. arXiv:1704.08084, 2017.
[50] M. Groth and J. Šťovíček. Abstract representation theory of Dynkin quivers of type A. arXiv:1409.5003, 2014.
[51] M. Groth and J. Šťovíček. Tilting theory via stable homotopy theory. J. Reine Angew. Math. 743 (2018), 29–90.
[52] A. Grothendieck. Sur quelques points d'algèbre homologique. *Tôhoku Math. J. (2)*, 9:119–221, 1957.
[53] R. Haugseng. Rectification of enriched ∞-categories. Algebr. Geom. Topol. 15 (2015), no. 4, 1931–1982.
[54] A. Heller. Stable homotopy categories. *Bull. Amer. Math. Soc.*, 74:28–63, 1968.
[55] A. Heller. Stable homotopy theories and stabilization. *J. Pure Appl. Algebra*, 115(2):113–130, 1997.
[56] P.S. Hirschhorn. *Model categories and their localizations.* Volume 99 of *Math. Surv. Monogr.* Amer. Math. Soc., 2003.
[57] A. Hirschowitz and C. Simpson. Descente pour les n-champs, preprint.
[58] S. Hollander. A homotopy theory for stacks. *Israel J. Math.*, 163:93–124, 2008.
[59] T. Holm, P. Jørgensen, and R. Rouquier, editors. *Triangulated categories.* Volume 375 of *London Math. Soc. Lecture Note Ser.* Cambridge University Press, 2010.
[60] M. Hovey. *Model categories.* Volume 63 of *Math. Surv. Monogr.* Amer. Math. Soc., 1999.
[61] M. Hovey. Spectra and symmetric spectra in general model categories. *J. Pure Appl. Algebra*, 165(1):63–127, 2001.
[62] M. Hovey. The Eilenberg–Watts theorem in homotopical algebra. arXiv:0910.3842, 2010.
[63] M. Hovey, B. Shipley, and J. Smith. Symmetric spectra. *J. Amer. Math. Soc.*, 13(1):149–208, 2000.
[64] A. Ilias. A model structure on $\mathsf{Cat}_{\mathsf{top}}$. arXiv:1110.2695, 2011.
[65] J.F. Jardine. Simplicial presheaves. *J. Pure Appl. Algebra*, 47(1):35–87, 1987.
[66] J.F. Jardine. Boolean localization, in practice. *Doc. Math.*, 1(13):245–275, 1996.
[67] A. Joyal. The theory of quasi-categories I,II, in preparation.
[68] A. Joyal. Quasi-categories and Kan complexes. *J. Pure Appl. Algebra*, 175(1–3):207–222, 2002. Special volume celebrating the 70th birthday of Professor Max Kelly.

[69] A. Joyal. The theory of quasi-categories and its applications. Lectures at the CRM (Barcelona). Preprint, 2008.

[70] A. Joyal and M. Tierney. Strong stacks and classifying spaces. In *Category theory (Como, 1990)*, volume 1488 of *Lecture Notes in Math.*, pages 213–236. Springer, 1991.

[71] D.M. Kan. On c. s. s. complexes. *Amer. J. Math.*, 79:449–476, 1957.

[72] M. Kashiwara and P. Schapira. *Categories and sheaves.* Volume 332 of *Grundlehren Math. Wiss.* Springer-Verlag, 2006.

[73] G.M. Kelly. On MacLane's conditions for coherence of natural associativities, commutativities, etc. *J. Algebra*, 1:397–402, 1964.

[74] G.M. Kelly. Doctrinal adjunction. In *Category Seminar (Proc. Sem., Sydney, 1972/1973)*, pages 257–280. Lecture Notes in Math., Vol. 420. Springer, 1974.

[75] G.M. Kelly and S. Lack. $\mathcal{V}$-Cat is locally presentable or locally bounded if $\mathcal{V}$ is so. *Theory Appl. Categ.*, 8:555–575, 2001.

[76] S. Lack. A Quillen model structure for Gray-categories. *J. K-Theory*, 8(2):183–221, 2011.

[77] T. Leinster. *Higher operads, higher categories.* Volume 298 of *London Math. Soc. Lecture Note Ser.* Cambridge University Press, 2004.

[78] J. Lurie. Derived algebraic geometry III: Commutative algebra. arXiv:math/0703204v3, 2007.

[79] J. Lurie. Stable ∞-categories. arXiv:math/0608228, 2007.

[80] J. Lurie. Derived algebraic geometry II: Noncommutative algebra. Available at http://www.math.harvard.edu/~lurie, 2009.

[81] J. Lurie. Derived algebraic geometry III: Commutative algebra. arXiv:math/0703204, 2009.

[82] J. Lurie. *Higher topos theory.* Volume 170 of *Ann. of Math. Stud.* Princeton University Press, 2009.

[83] J. Lurie. Higher algebra. Available at http://www.math.harvard.edu/~lurie, 2011.

[84] S. Mac Lane. Natural associativity and commutativity. *Rice Univ. Studies*, 49(4):28–46, 1963.

[85] S. Mac Lane. *Categories for the working mathematician.* Volume 5 of *Grad. Texts in Math.* Springer-Verlag, second edition, 1998.

[86] S. Mac Lane and I. Moerdijk. *Sheaves in geometry and logic: A first introduction to topos theory.* Universitext. Springer-Verlag, 1994. Corrected reprint of the 1992 edition.

[87] M. Makkai and R. Paré. *Accessible categories: the foundations of categorical model theory.* Volume 104 of *Contemp. Math.* Amer. Math. Soc., 1989.

[88] G. Maltsiniotis. Catégories triangulées supérieures. Available at http://people.math.jussieu.fr/~maltsin/textes.html, 2005.

[89] M. Markl, S. Shnider, and J. Stasheff. *Operads in algebra, topology, and physics.* Volume 96 of *Math. Surv. and Monogr.* Amer. Math. Soc., 2002.

[90] A. Neeman. *Triangulated categories.* Volume 148 of *Ann. Math. Stud.* Princeton University Press, 2001.

[91] D. Puppe. Stabile Homotopietheorie. I. *Math. Ann.*, 169:243–274, 1967.

[92] D.G. Quillen. *Homotopical algebra.* Lecture Notes in Math., Vol. 43. Springer-Verlag, 1967.

[93] C. Rezk. A model category for categories. Available at http://www.math.uiuc.edu/~rezk/papers.html.

[94] C. Rezk. A model for the homotopy theory of homotopy theory. *Trans. Amer. Math. Soc.*, 353(3):973–1007, 2001.

[95] C. Rezk. Toposes and homotopy toposes. Available at http://www.math.uiuc.edu/~rezk/, 2005.

[96] E. Riehl. On the structure of simplicial categories associated to quasi-categories. *Math. Proc. Cambridge Philos. Soc.*, 150(3):489–504, 2011.

[97] Emily Riehl and Dominic Verity. The 2-category theory of quasi-categories. *Adv. Math.*, 280:549–642, 2015.

[98] J. Rosický. On combinatorial model categories. *Appl. Categ. Structures*, 17(3):303–316, 2009.

[99] J. Rosický and W. Tholen. Left-determined model categories and universal homotopy theories. *Trans. Amer. Math. Soc.*, 355(9):3611–3623, 2003.

[100] J.J. Rotman. *An introduction to homological algebra.* Volume 85 of *Pure Applied Math.* Academic Press Inc. [Harcourt Brace Jovanovich Publishers], 1979.

[101] S. Schwede. Spectra in model categories and applications to the algebraic cotangent complex. *J. Pure Appl. Algebra*, 120(1):77–104, 1997.

[102] S. Schwede. Stable homotopical algebra and Γ-spaces. *Math. Proc. Cambridge Philos. Soc.*, 126(2):329–356, 1999.

[103] S. Schwede and B. Shipley. Equivalences of monoidal model categories. *Algebr. Geom. Topol.*, 3:287–334, 2003.

[104] G. Segal. Categories and cohomology theories. *Topology*, 13:293–312, 1974.

[105] M. Shulman. Homotopy limits and colimits and enriched homotopy theory. arXiv:math/0610194, 2006.

[106] C. Simpson. A Giraud-type characterization of the simplicial categories associated to closed model categories as ∞-pretopoi. arXiv:math.AT/9903167, 2007.

[107] C. Simpson. *Homotopy theory of higher categories.* Volume 19 of *New Mathematics Monographs.* Cambridge University Press, 2012.

[108] J.D. Stasheff. Homotopy associativity of H-spaces. I, II. *Trans. Amer. Math. Soc. 108 (1963), 275-292; ibid.*, 108:293–312, 1963.

[109] B. Toën and G. Vezzosi. Homotopical algebraic geometry. I. Topos theory. *Adv. Math.*, 193(2):257–372, 2005.

[110] F. Ulmer. Properties of dense and relative adjoint functors. *J. Algebra*, 8:77–95, 1968.

[111] J.-L. Verdier. Des catégories dérivées des catégories abéliennes. *Astérisque*, (239),(1997), 1996. With a preface by Luc Illusie, Edited and with a note by Georges Maltsiniotis.

[112] A. Vistoli. Grothendieck topologies, fibered categories and descent theory. In *Fundamental algebraic geometry*, volume 123 of *Math. Surveys Monogr.*, pages 1–104. Amer. Math. Soc., 2005.

[113] R. M. Vogt. *Boardman's stable homotopy category.* Lecture Notes Series, No. 21. Matematisk Institut, Aarhus Universitet, Aarhus, 1970.

[114] R.M. Vogt. Homotopy limits and colimits. *Math. Z.*, 134:11–52, 1973.

[115] C.A. Weibel. *An introduction to homological algebra.* Volume 38 of *Cambridge Stud. Adv. Math.* Cambridge University Press, 1994.

[116] H. Wolff. V-cat and V-graph. *J. Pure Appl. Algebra*, 4:123–135, 1974.

Mathematisches Institüt, Universität Bonn, Bonn, Germany
E-mail address: mgroth@math.uni-bonn.de

15

Topological cyclic homology

Lars Hesselholt and Thomas Nikolaus

Topological cyclic homology is a manifestation of Waldhausen's vision that the cyclic theory of Connes and Tsygan should be developed with the initial ring $\mathbb{S}$ of higher algebra as base. In his philosophy, such a theory should be meaningful integrally as opposed to rationally. Bökstedt realized this vision for Hochschild homology [9], and he made the fundamental calculation that

$$\mathrm{THH}_*(\mathbb{F}_p) = \mathrm{HH}_*(\mathbb{F}_p/\mathbb{S}) = \mathbb{F}_p[x]$$

is a polynomial algebra on a generator x in degree two [10]. By comparison,

$$\mathrm{HH}_*(\mathbb{F}_p/\mathbb{Z}) = \mathbb{F}_p\langle x\rangle$$

is the divided power algebra,[1] so Bökstedt's periodicity theorem indeed shows that by replacing the base $\mathbb{Z}$ by the base $\mathbb{S}$, denominators disappear. In fact, the base-change map $\mathrm{HH}_*(\mathbb{F}_p/\mathbb{S}) \to \mathrm{HH}_*(\mathbb{F}_p/\mathbb{Z})$ can be identified with the edge homomorphism of a spectral sequence

$$E^2_{i,j} = \mathrm{HH}_i(\mathbb{F}_p/\pi_*(\mathbb{S}))_j \Rightarrow \mathrm{HH}_{i+j}(\mathbb{F}_p/\mathbb{S}),$$

so apparently the stable homotopy groups of spheres have exactly the right size to eliminate the denominators in the divided power algebra.

The appropriate definition of cyclic homology relative to $\mathbb{S}$ was given by Bökstedt–Hsiang–Madsen [11]. It involves a new ingredient not present in the Connes–Tsygan cyclic theory: a Frobenius map. The nature of this Frobenius map is now much better understood by the work of Nikolaus–Scholze [32]. As in the Connes–Tsygan theory, the circle group $\mathbb{T}$ acts on topological Hochschild homology, and negative topological cyclic homology and periodic topological cyclic homology are defined to be the homotopy fixed points and the Tate construction, respectively, of this action:

$$\mathrm{TC}^-(A) = \mathrm{THH}(A)^{h\mathbb{T}} \qquad \text{and} \qquad \mathrm{TP}(A) = \mathrm{THH}(A)^{t\mathbb{T}}.$$

Mathematics Subject Classification. 14F30, 19D50, 19D55, 55R45.

Key words and phrases. topological cyclic homology, topological Hochschild homology, trace methods, perfectoid rings, Bott periodicity.

[1] The divided power algebra $\mathbb{F}_p\langle x\rangle$ has generators $x^{[i]}$ with $i \in \mathbb{N}$ subject to the relations that $x^{[i]} \cdot x^{[j]} = \binom{i+j}{i} \cdot x^{[i+j]}$ for all $i, j \in \mathbb{N}$ and $x^{[0]} = 1$. So $x^i = (x^{[1]})^i = i! \cdot x^{[i]}$.

There is always a canonical map from the homotopy fixed points to the Tate construction, but, after p-completion, the pth Frobenius gives rise to another such map and topological cyclic homology is the equalizer

$$\mathrm{TC}(A) \longrightarrow \mathrm{TC}^{-}(A) \underset{\mathrm{can}}{\overset{(\varphi_p)}{\rightrightarrows}} \mathrm{TP}(A)^{\wedge} = \textstyle\prod_p \mathrm{TP}(A)_p^{\wedge}.$$

Here "$(-)^{\wedge}$" and "$(-)_p^{\wedge}$" indicates profinite and p-adic completion.

Topological cyclic homology receives a map from algebraic K-theory, called the cyclotomic trace map. Roughly speaking, this map records traces of powers of matrices and may be viewed as a denominator-free version of the Chern character. There are two theorems that concern the behavior of this map applied to cubical diagrams of connective $\mathbb{E}_1$-algebras in spectra. If A is such an n-cube, then the theorems give conditions for the $(n+1)$-cube

$$K(A) \longrightarrow \mathrm{TC}(A)$$

to be cartesian. For $n = 1$, the Dundas–Goodwillie–McCarthy theorem [16] states that this is so provided $\pi_0(A_0) \to \pi_0(A_1)$ is a surjection with nilpotent kernel. And for $n = 2$, the Land–Tamme theorem [26], which strengthens theorems of Cortiñas [15] and Geisser–Hesselholt [21], states that this is so provided A is cartesian and $\pi_0(A_1 \otimes_{A_0} A_2) \to \pi_0(A_{12})$ an isomorphism. For $n \geq 3$, it is an open question to find conditions on A that make $K(A) \to \mathrm{TC}(A)$ cartesian. It is also not clear that the conditions for $n = 1$ and $n = 2$ are optimal. Indeed, a theorem of Clausen–Mathew–Morrow [14] states that for every commutative ring R and ideal $I \subset R$ with (R, I) henselian, the square

$$\begin{array}{ccc} K(R) & \longrightarrow & \mathrm{TC}(R) \\ \downarrow & & \downarrow \\ K(R/I) & \longrightarrow & \mathrm{TC}(R/I) \end{array}$$

becomes cartesian after profinite completion. So in this case, the conclusion of the Dundas–Goodwillie–McCarthy theorem holds under a much weaker assumption on the 1-cube $R \to R/I$. The Clausen–Mathew–Morrow theorem may be seen as a p-adic analogue of Gabber rigidity [20]. Indeed, for prime numbers ℓ that are invertible in R/I, the left-hand terms in the square above vanish after ℓ-adic completion, so the Clausen–Mathew–Morrow recovers and extends the Gabber rigidity theorem.

Absolute comparison theorems between K-theory and topological cyclic homology begin with the calculation that for R a perfect commutative $\mathbb{F}_p$-algebra, the cyclotomic trace map induces an equivalence

$$K(R)_p^{\wedge} \longrightarrow \tau_{\geq 0}\,\mathrm{TC}(R)_p^{\wedge}.$$

The Clausen–Mathew–Morrow theorem then implies that the same is true for every commutative ring R such that (R, pR) is henselian and such that the Frobenius $\varphi\colon R/p \to R/p$ is surjective. Indeed, in this case, $(R/p)^{\mathrm{red}}$ is perfect and $(R/p, \mathrm{nil}(R/p))$ is henselian; see [14, Corollary 6.9]. In particular, this is true for all semiperfectoid rings R.[2]

[2] A commutative $\mathbb{Z}_p$-algebra R is perfectoid (resp. semiperfectoid), for example, if there exists a non-zero-divisor $\pi \in R$ with $p \in \pi^p R$ such that the π-adic topology on R is complete and separated and such that the Frobenius $\varphi\colon R/\pi \to R/\pi^p$ a bijection (resp. surjection).

The starting point for the calculation of topological cyclic homology and its variants is the Bökstedt periodicity theorem, which we mentioned above. Since Bökstedt's paper [10] has never appeared, we take the opportunity to give his proof in Section 15.2 below. The full scope of this theorem was realized only recently by Bhatt–Morrow–Scholze [4], who proved that Bökstedt periodicity holds for every perfectoid ring. More precisely, their result, which we explain in Section 15.3 below states that if R is perfectoid, then[3]

$$\mathrm{THH}_*(R, \mathbb{Z}_p) = R[x]$$

on a polynomial generator of degree 2. Therefore, as is familiar from complex orientable cohomology theories, the Tate spectral sequence

$$E^2_{i,j} = \hat{H}^{-i}(B\mathbb{T}, \mathrm{THH}_j(R, \mathbb{Z}_p)) \Rightarrow \mathrm{TP}_{i+j}(R, \mathbb{Z}_p)$$

collapses, since all non-zero elements are concentrated in even total degree. However, since the $\mathbb{T}$-action on $\mathrm{THH}(R, \mathbb{Z}_p)$ is non-trivial, the ring homomorphism given by the edge homomorphism of the spectral sequence,

$$\mathrm{TP}_0(R, \mathbb{Z}_p) \xrightarrow{\theta} \mathrm{THH}_0(R, \mathbb{Z}_p),$$

does not admit a section, and therefore, we cannot identify the domain with a power series algebra over R. Instead, this ring homomorphism is canonically identified with the universal p-complete pro-infinitesimal thickening

$$A = A_{\mathrm{inf}}(R) \xrightarrow{\theta} R$$

introduced by Fontaine [19], which we recall in Section 15.3 below. In addition, Bhatt–Scholze [6] show that, rather than a formal group over R, there is a canonical p-typical λ-ring structure

$$A \xrightarrow{\lambda} W(A),$$

the associated Adams operation $\varphi\colon A \to A$ of which is the composition of the inverse of the canonical map $\mathrm{can}\colon \mathrm{TC}^-_0(R, \mathbb{Z}_p) \to \mathrm{TP}_0(R, \mathbb{Z}_p)$ and the Frobenius map $\varphi\colon \mathrm{TC}^-_0(R, \mathbb{Z}_p) \to \mathrm{TP}_0(R, \mathbb{Z}_p)$. The kernel $I \subset A$ of the edge homomorphism $\theta\colon A \to R$ is a principal ideal, and Bhatt–Scholze show that the pair $((A, \lambda), I)$ is a prism in the sense that

$$p \in I + \varphi(I)A,$$

and that this prism is perfect in the sense that $\varphi\colon A \to A$ is an isomorphism. We remark that if $\xi \in I$ is a generator, then, equivalently, the prism condition means that the intersection of the divisors "$\xi = 0$" and "$\varphi(\xi) = 0$" is contained in the special fiber "$p = 0$," whence the name. We thank Riccardo Pengo for the Figure 15.0.1, which illustrates $\mathrm{Spec}(A)$.

Bhatt–Morrow–Scholze further show that, given the choice of ξ, one can choose $u \in \mathrm{TC}^-_2(R, \mathbb{Z}_p)$, $v \in \mathrm{TC}^-_{-2}(R, \mathbb{Z}_p)$, and $\sigma \in \mathrm{TP}_2(R, \mathbb{Z}_p)$ in such a way that

$$\begin{aligned} \mathrm{TC}^-_*(R, \mathbb{Z}_p) &= A[u, v]/(uv - \xi), \\ \mathrm{TP}_*(R, \mathbb{Z}_p) &= A[\sigma^{\pm 1}], \end{aligned}$$

[3] We write $\pi_*(X, \mathbb{Z}_p)$ for the homotopy groups of the p-completion of a spectrum X, and we write $\mathrm{THH}_*(R, \mathbb{Z}_p)$ instead of $\pi_*(\mathrm{THH}(R), \mathbb{Z}_p)$. For p-completion, see Bousfield [12].

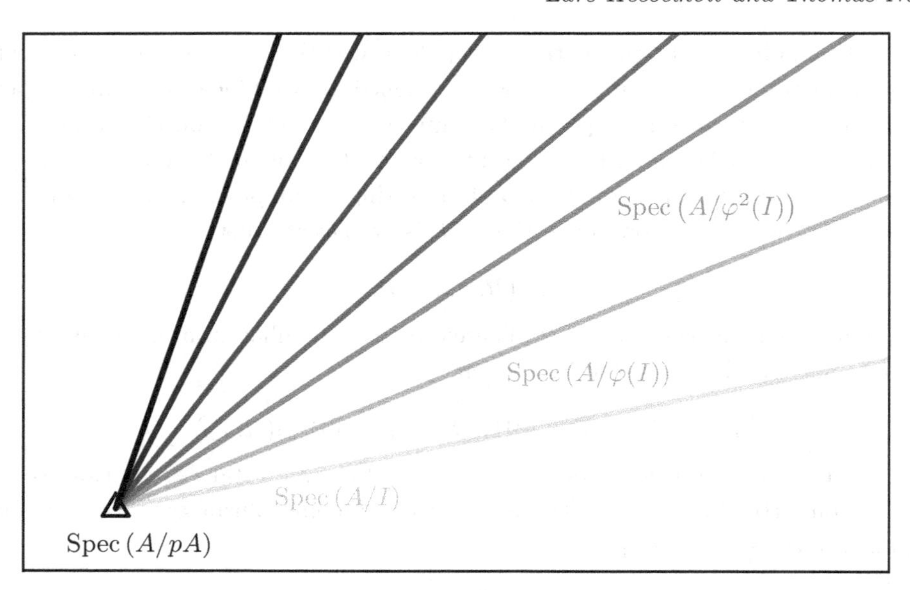

FIGURE 15.0.1. Spec(A) as a prism

and $\varphi(u) = \alpha \cdot \sigma$, $\varphi(v) = \alpha^{-1}\varphi(\xi) \cdot \xi^{-1}$, $\mathrm{can}(u) = \xi \cdot \sigma$, and $\mathrm{can}(v) = \sigma^{-1}$ with $\alpha \in A$ a unit. In these formulas, the unit α can be eliminated, if one is willing to replace the generator $\xi \in I$ by the generator $\varphi^{-1}(\alpha) \cdot \xi \in I$. We use these results for $R = \mathcal{O}_C$, where C is a complete algebraically closed p-adic field, to give a purely p-adic proof of Bott periodicity. In particular, Bökstedt periodicity implies Bott periodicity, but not vice versa.

The Nikolaus–Scholze approach to topological cyclic homology is also very useful for calculations. To wit, Speirs has much simplified the calculation of the topological cyclic homology of truncated polynomial algebras over a perfect $\mathbb{F}_p$-algebra [34], and we have evaluated the topological cyclic homology of planar cuspical curves over a perfect $\mathbb{F}_p$-algebra [23]. Here, we illustrate this approach in Section 15.4, where we identify the cofiber of the assembly map

$$\mathrm{TC}(R) \otimes BC_{p+} \longrightarrow \mathrm{TC}(R[C_p])$$

for R perfectoid in terms of an analogue of the affine deformation to the normal cone along $I \subset A = A_{\mathrm{inf}}(R)$ with p as the parameter.

Finally, we mention that Bhatt–Morrow–Scholze [5] have constructed weight[4] filtrations of topological cyclic homology and its variants such that, on jth graded pieces, the equalizer of p-completed spectra

$$\mathrm{TC}(S, \mathbb{Z}_p) \longrightarrow \mathrm{TC}^-(S, \mathbb{Z}_p) \underset{\mathrm{can}}{\overset{\varphi}{\rightrightarrows}} \mathrm{TP}(S, \mathbb{Z}_p)$$

gives rise to an equalizer

$$\mathbb{Z}_p(j)[2j] \longrightarrow \mathrm{Fil}^j\, \widehat{\Delta}_S\{j\}[2j] \underset{\text{“incl”}}{\overset{\text{“}\frac{\varphi}{\xi^j}\text{”}}{\rightrightarrows}} \widehat{\Delta}_S\{j\}[2j]$$

[4] If S is a smooth $\mathbb{F}_p$-algebra, then, on the jth graded piece of the Bhatt–Morrow–Scholze filtration on $\mathrm{TP}_*(S, \mathbb{Z}_p)[1/p]$, the geometric Frobenius Fr_p acts with pure weight j in the sense that $\mathrm{Fr}_p^* = p^j \varphi_p$, where φ_p is the cyclotomic Frobenius.

Here S is any commutative ring, $\widehat{\Delta}_S = \widehat{\Delta}_S\{0\}$ is an $\mathbb{E}_\infty$-algebra in the derived ∞-category of S-modules, $\widehat{\Delta}_S\{j\}$ is an invertible $\widehat{\Delta}_S$-module, and $\mathrm{Fil}^{\cdot}\,\widehat{\Delta}_S\{j\}$ is the derived complete descending "Nygaard" filtration thereof. The equalizer $\mathbb{Z}_p(j)$ is a version of syntomic cohomology that works correctly for all weights j, as opposed to only for $j < p-1$. The Bhatt–Morrow–Scholze filtration of $\mathrm{TP}(S, \mathbb{Z}_p)$ gives rise to an Atiyah–Hirzebruch type spectral sequence

$$E^2_{i,j} = H^{j-i}(\mathrm{Spec}(S), \widehat{\Delta}_S\{j\}) \Rightarrow \mathrm{TP}_{i+j}(S, \mathbb{Z}_p),$$

and similarly for $\mathrm{TC}(S, \mathbb{Z}_p)$ and $\mathrm{TC}^-(S, \mathbb{Z}_p)$. If R is perfectoid, then

$$H^i(\mathrm{Spec}(R), \widehat{\Delta}_R\{j\}) \simeq \begin{cases} A_{\mathrm{inf}}(R) & \text{if } i = 0 \\ 0 & \text{if } i \neq 0, \end{cases}$$

for all integers j, and methods for evaluating these "prismatic" cohomology groups are currently being developed by Bhatt–Scholze [6].

Acknowledgements

It is a great pleasure to acknowledge the support that we have received while preparing this chapter. Hesselholt was funded in part by the Isaac Newton Institute as a Rothschild Distinguished Visiting Fellow and by the Mathematical Sciences Research Institute as a Simons Visiting Professor, and Nikolaus was funded in part by the Deutsche Forschungsgemeinschaft under Germany's Excellence Strategy EXC 2044 390685587, Mathematics Münster: Dynamics–Geometry–Structure.

15.1 Topological Hochschild homology

We sketch the definition of topological Hochschild homology, topological cyclic homology, and the cyclotomic trace map from algebraic K-theory following Nikolaus–Scholze [32] and Nikolaus [31].

15.1.1 Definition

If R is an $\mathbb{E}_\infty$-algebra in spectra, then we define $\mathrm{THH}(R)$ to be the colimit of the diagram $\mathbb{T} \to \mathrm{Alg}_{\mathbb{E}_\infty}(\mathsf{Sp})$ that is constant with value R, and we write

$$\mathrm{THH}(R) = R^{\otimes \mathbb{T}}$$

to indicate this colimit. Here $\mathbb{T}$ is the circle group. The action of $\mathbb{T}$ on itself by left translation induces a $\mathbb{T}$-action on the $\mathbb{E}_\infty$-algebra $\mathrm{THH}(R)$. In addition, the map of $\mathbb{E}_\infty$-algebras $R \to \mathrm{THH}(R)$ induced by the structure map of the colimit exhibits $\mathrm{THH}(R)$ as the initial $\mathbb{E}_\infty$-algebra with $\mathbb{T}$-action under R.

We let p be a prime number, and let $C_p \subset \mathbb{T}$ be the subgroup of order p. The *Tate diagonal* is a natural map of $\mathbb{E}_\infty$-algebras in spectra

$$R \xrightarrow{\Delta_p} (R^{\otimes C_p})^{tC_p}.$$

Heuristically, this map takes a to the equivalence class of $a \otimes \cdots \otimes a$, but it exists only in higher algebra.[5] Moreover, the map $R \to \mathrm{THH}(R)$ of $\mathbb{E}_\infty$-algebras in spectra extends uniquely to a map $R^{\otimes C_p} \to \mathrm{THH}(R)$ of $\mathbb{E}_\infty$-algebras in spectra with C_p-action, which, in turn, induces a map of Tate spectra

$$(R^{\otimes C_p})^{tC_p} \longrightarrow \mathrm{THH}(R)^{tC_p}.$$

This map also is a map of $\mathbb{E}_\infty$-algebras, and its target carries a residual action of $\mathbb{T}/C_p$, which we identify with $\mathbb{T}$ via the isomorphism given by the pth root. Hence, by the universal property of $R \to \mathrm{THH}(R)$, there is a unique map φ_p of $\mathbb{E}_\infty$-algebras in spectra with $\mathbb{T}$-action which makes the diagram

$$\begin{array}{ccc} R & \xrightarrow{\Delta_p} & (R^{\otimes C_p})^{tC_p} \\ \downarrow & & \downarrow \\ \mathrm{THH}(R) & \xrightarrow{\varphi_p} & \mathrm{THH}(R)^{tC_p} \end{array}$$

in $\mathrm{Alg}_{\mathbb{E}_\infty}(\mathrm{Sp})$ commute (in the ∞-categorical sense). The map φ_p is called the pth cyclotomic Frobenius, and the family of maps $(\varphi_p)_{p\in\mathbb{P}}$ indexed by the set $\mathbb{P}$ of prime numbers makes $\mathrm{THH}(R)$ a *cyclotomic* spectrum in the following sense.

Definition 15.1.1 (Nikolaus–Scholze). A *cyclotomic spectrum* is a pair of a spectrum with $\mathbb{T}$-action X and a family $(\varphi_p)_{p\in\mathbb{P}}$ of $\mathbb{T}$-equivariant maps

$$X \xrightarrow{\varphi_p} X^{tC_p}.$$

The ∞-category of cyclotomic spectra is the pullback of simplicial sets

$$\begin{array}{ccc} \mathrm{CycSp} & \longrightarrow & \prod_{p\in\mathbb{P}} \mathrm{Fun}(\Delta^1, \mathrm{Sp}^{B\mathbb{T}}) \\ \downarrow & & \downarrow {\scriptstyle \prod(\mathrm{ev}_0,\mathrm{ev}_1)} \\ \mathrm{Sp}^{B\mathbb{T}} & \xrightarrow{(\mathrm{id},(-)^{tC_p})_{p\in\mathbb{P}}} & \prod_{p\in\mathbb{P}} \mathrm{Sp}^{B\mathbb{T}} \times \mathrm{Sp}^{B\mathbb{T}}. \end{array}$$

We remark that, in contrast to the earlier notions of cyclotomic spectra in Hesselholt–Madsen [22] and Blumberg–Mandell [8], the Nikolaus–Scholze definition does not require equivariant homotopy theory.

It is shown in [32] that CycSp is a presentable stable ∞-category, and that it canonically extends to a symmetric monoidal ∞-category

$$\mathrm{CycSp}^{\otimes} \longrightarrow \mathrm{Fin}_*$$

with underlying ∞-category CycSp. Now, the construction of THH given above produces a lax symmetric monoidal functor

$$\mathrm{Alg}_{\mathbb{E}_\infty}(\mathrm{Sp}^{\otimes}) \xrightarrow{\mathrm{THH}} \mathrm{Alg}_{\mathbb{E}_\infty}(\mathrm{CycSp}^{\otimes}).$$

Topological Hochschild homology may be defined, more generally, for (small) stable ∞-categories $\mathcal{C}$. If R is an $\mathbb{E}_\infty$-algebra in spectra and $\mathcal{C} = \mathrm{Perf}_R$ is the stable ∞-category of

[5] In fact, if k is a commutative ring, then the space of natural transformations between the corresponding endofunctors on $\mathrm{Alg}_{\mathbb{E}_\infty}(\mathcal{D}(k))$ is empty.

perfect R-modules, there is a canonical equivalence

$$\mathrm{THH}(R) \simeq \mathrm{THH}(\mathrm{Perf}_R)$$

of cyclotomic spectra. The basic idea is to define the underlying spectrum with $\mathbb{T}$-action $\mathrm{THH}(\mathcal{C})$ to be the geometric realization of the cyclic spectrum that, in simplicial degree n, is given by

$$\mathrm{THH}(\mathcal{C})_n = \mathrm{colim}\Big(\bigotimes_{0 \le i \le n} \mathrm{map}_{\mathcal{C}}(x_i, x_{i+1}) \Big),$$

where the colimit ranges over the space of $(n+1)$-tuples in the groupoid core $\mathcal{C}^{\simeq}$ of the ∞-category $\mathcal{C}$, $\mathrm{map}_{\mathcal{C}}$ denotes the mapping spectrum in $\mathcal{C}$, and the index i is taken modulo $n+1$. We indicate the steps necessary to make sense of this definition; see [31] and the forthcoming paper [30] for details.

First, to make sense of the colimit above, one must construct a functor

$$(\mathcal{C}^{\simeq})^{n+1} \longrightarrow \mathrm{Sp}$$

that to $(x_0, \dots, x_n)$ assigns $\bigotimes_{0 \le i \le n} \mathrm{map}_{\mathcal{C}}(x_i, x_{i+1})$. This can be achieved by a combination of the tensor product functor, the mapping spectrum functor $\mathrm{map}_{\mathcal{C}} \colon \mathcal{C}^{\mathrm{op}} \times \mathcal{C} \to \mathrm{Sp}$, and the canonical equivalence $\mathcal{C}^{\simeq} \simeq (\mathcal{C}^{\mathrm{op}})^{\simeq}$. Second, one must lift the assignment $n \mapsto \mathrm{THH}(\mathcal{C})_n$ to a functor $\Lambda^{\mathrm{op}} \to \mathrm{Sp}$ from Connes' cyclic category such that the face and degeneracy maps are given by composing adjacent morphisms and by inserting identities, respectively, while the cyclic operator is given by cyclic permutation of the tensor factors. As explained in [32, Appendix B], for every cyclic spectrum $\Lambda^{\mathrm{op}} \to \mathrm{Sp}$, the geometric realization of the simplicial spectrum $\Delta^{\mathrm{op}} \to \Lambda^{\mathrm{op}} \to \mathrm{Sp}$ carries a natural $\mathbb{T}$-action. Finally, one must construct the cyclotomic Frobenius maps

$$\mathrm{THH}(\mathcal{C}) \xrightarrow{\varphi_p} \mathrm{THH}(\mathcal{C})^{tC_p}.$$

These are defined following [32, Section III.2] as the Tate-diagonal applied levelwise followed by the canonical colimit-Tate interchange map.

15.1.2 Topological cyclic homology and the trace

Taking R to be the sphere spectrum $\mathbb{S}$, we obtain an $\mathbb{E}_\infty$-algebra in cyclotomic spectra $\mathrm{THH}(\mathbb{S})$, which we denote by $\mathbb{S}^{\mathrm{triv}}$. Its underlying spectrum is $\mathbb{S}$, and its cyclotomic Frobenius map φ_p can be identified with a canonical $\mathbb{T}$-equivariant refinement of the composition

$$\mathbb{S} \xrightarrow{\mathrm{triv}} \mathbb{S}^{hC_p} \xrightarrow{\mathrm{can}} \mathbb{S}^{tC_p}$$

of the map triv induced from the projection $BC_p \to \mathrm{pt}$ and the canonical map. Since the ∞-category CycSp is stable, we have associated with every pair of objects $X, Y \in \mathrm{CycSp}$ a mapping spectrum $\mathrm{map}_{\mathrm{CycSp}}(X, Y)$, which depends functorially on X and Y.

Definition 15.1.2. The *topological cyclic homology* of a cyclotomic spectrum X is the mapping spectrum $\mathrm{TC}(X) = \mathrm{map}_{\mathrm{CycSp}}(\mathbb{S}^{\mathrm{triv}}, X)$.

If $X = \mathrm{THH}(R)$ with R an $\mathbb{E}_\infty$-algebra in spectra, then we abbreviate and write $\mathrm{TC}(R)$ instead of $\mathrm{TC}(X)$. Similarly, if $X = \mathrm{THH}(\mathcal{C})$ with $\mathcal{C}$ a stable ∞-category, then we write $\mathrm{TC}(\mathcal{C})$ instead of $\mathrm{TC}(X)$.

This definition of topological cyclic homology as given above is abstractly elegant and useful, but for concrete calculations, a more concrete formula is necessary. Therefore, we unpack Definition 15.1.2. We assume that X is bounded below and write

$$\mathrm{TC}^-(X) \xrightarrow{\mathrm{can}} \mathrm{TP}(X)$$

for the canonical map $X^{h\mathbb{T}} \to X^{t\mathbb{T}}$. We call the domain and target of this map the negative topological cyclic homology and the periodic topological cyclic homology of X, respectively. Since the cyclotomic Frobenius maps

$$X \xrightarrow{\varphi_p} X^{tC_p}$$

are $\mathbb{T}$-equivariant, they give rise to a map of $\mathbb{T}$-homotopy fixed points spectra

$$X^{h\mathbb{T}} \xrightarrow{(\varphi_p^{h\mathbb{T}})} \textstyle\prod_{p\in\mathbb{P}} (X^{tC_p})^{h\mathbb{T}}.$$

Moreover, there is a canonical map

$$\textstyle\prod_{p\in\mathbb{P}} (X^{tC_p})^{h\mathbb{T}} \longleftarrow X^{t\mathbb{T}},$$

which becomes an equivalence after profinite completion, since X is bounded below, by the Tate-orbit lemma [32, Lemma II.4.2]. Hence, we get a map

$$\mathrm{TC}^-(X) \xrightarrow{\varphi} \mathrm{TP}(X)^\wedge,$$

where "$(-)^\wedge$" indicates profinite completion. There is also a canonical map

$$\mathrm{TC}^-(X) \xrightarrow{\mathrm{can}} \mathrm{TP}(X)^\wedge$$

given by the composition of the canonical from the homotopy fixed point spectrum to the Tate construction followed by the completion map. This gives the following description of $\mathrm{TC}(X)$.

Proposition 15.1.3. *For bounded below cyclotomic spectra X, there is natural equalizer diagram*

$$\mathrm{TC}(X) \longrightarrow \mathrm{TC}^-(X) \underset{\mathrm{can}}{\overset{\varphi}{\rightrightarrows}} \mathrm{TP}(X)^\wedge.$$

We now explain the definition of the cyclotomic trace map from K-theory to topological cyclic homology. Let $\mathrm{Cat}_\infty^{\mathrm{stab}}$ be the ∞-category of small, stable ∞-categories and exact functors. The ∞-category of noncommutative motives (or a slight variant thereof) of Blumberg–Gepner–Tabuada [7] is defined to be the initial (large) ∞-category with a functor

$$z : \mathrm{Cat}_\infty^{\mathrm{stab}} \longrightarrow \mathrm{NMot}$$

such that the following hold:

(1) (Stability) The ∞-category NMot is stable.

(2) (Localization) For every Verdier sequence[6] $\mathcal{C} \to \mathcal{D} \to \mathcal{D}/\mathcal{C}$ in $\mathrm{Cat}_\infty^{\mathrm{stab}}$, the image sequence $z(\mathcal{C}) \to z(\mathcal{D}) \to z(\mathcal{D}/\mathcal{C})$ in NMot is a fiber sequence.

(3) (Morita invariance) For every map $\mathcal{C} \to \mathcal{D}$ in $\mathrm{Cat}_\infty^{\mathrm{stab}}$ that becomes an equivalence after idempotent completion, the image map $z(\mathcal{C}) \to z(\mathcal{D})$ in NMot is an equivalence.

The main theorem of op. cit. states[7] that for every (small) stable ∞-category $\mathcal{C}$, there is a canonical equivalence

$$K(\mathcal{C}) \simeq \mathrm{map}_{\mathrm{NMot}}(z(\mathrm{Perf}_{\mathbb{S}}), z(\mathcal{C}))$$

between its nonconnective algebraic K-theory spectrum and the indicated mapping spectrum in NMot. In general, one may view the mapping spectra in NMot as bivariant versions of nonconnective algebraic K-theory. Accordingly, the mapping spectra in CycSp are bivariant versions of TC. As we outlined in the previous section, topological Hochschild homology is a functor

$$\mathrm{Cat}_\infty^{\mathrm{stab}} \xrightarrow{\mathrm{THH}} \mathrm{CycSp},$$

and one can show that it satisfies the properties (1)–(3) above. There is a very elegant proof of (2) and (3) based on work of Keller, Blumberg–Mandell, and Kaledin that uses the trace property of THH, see the forthcoming paper [30] for a summary. Accordingly, the functor THH admits a unique factorization

$$\mathrm{Cat}_\infty^{\mathrm{stab}} \xrightarrow{z} \mathrm{NMot} \xrightarrow{\mathrm{tr}} \mathrm{CycSp}$$

with tr exact. In particular, for every stable ∞-category $\mathcal{C}$, we have an induced map of mapping spectra

$$\mathrm{map}_{\mathrm{NMot}}(z(\mathrm{Perf}_{\mathbb{S}}), z(\mathcal{C})) \xrightarrow{\mathrm{tr}} \mathrm{map}_{\mathrm{CycSp}}(\mathbb{S}^{\mathrm{triv}}, \mathrm{THH}(\mathcal{C})).$$

This map, by definition, is the cyclotomic trace map, which we write

$$K(\mathcal{C}) \xrightarrow{\mathrm{tr}} \mathrm{TC}(\mathcal{C}).$$

More concretely, on connective covers, considered here as $\mathbb{E}_\infty$-groups in spaces, the cyclotomic trace map is given by the composition

$$\begin{aligned}\Omega^\infty K(\mathcal{C}) &\simeq \Omega(\mathrm{colim}_{\Delta^{\mathrm{op}}}(S\mathcal{C}^{\mathrm{idem}})^{\simeq}) \to \Omega(\mathrm{colim}_{\Delta^{\mathrm{op}}}\Omega^\infty\,\mathrm{TC}(S\mathcal{C}^{\mathrm{idem}})) \\ &\to \Omega^\infty\Omega(\mathrm{colim}_{\Delta^{\mathrm{op}}}\mathrm{TC}(S\mathcal{C}^{\mathrm{idem}})) \simeq \Omega^\infty\,\mathrm{TC}(\mathcal{C}),\end{aligned}$$

[6] This means that $\mathcal{C} \to \mathcal{D} \to \mathcal{C}/\mathcal{D}$ is both a fiber sequence and cofiber sequence in $\mathrm{Cat}_\infty^{\mathrm{stab}}$. In particular, $\mathcal{D} \to \mathcal{C}$ is fully faithful and its image in $\mathcal{D}$ is closed under retracts.

[7] In fact, we do not require $z\colon \mathrm{Cat}_\infty^{\mathrm{stab}} \to \mathrm{NMot}$ to preserve filtered colimits, as do [7].

where $S(-)$ and $(-)^{\mathrm{idem}}$ indicate Waldhausen's construction and idempotent completion, respectively, where the second map is induced from the map

$$\mathcal{D}^{\simeq} \simeq \mathrm{Map}_{\mathrm{Cat}_{\infty}^{\mathrm{stab}}}(\mathrm{Perf}_{\mathbb{S}}, \mathcal{D}) \xrightarrow{\mathrm{THH}} \mathrm{Map}_{\mathrm{CycSp}}(\mathbb{S}^{\mathrm{triv}}, \mathrm{THH}(\mathcal{D})) \simeq \Omega^{\infty}\,\mathrm{TC}(\mathcal{D}),$$

and where last equivalence follows from TC satisfying (2) and (3) above.

15.1.3 Connes' operator

The symmetric monoidal ∞-category of spectra with $\mathbb{T}$-action is canonically equivalent to the symmetric monoidal ∞-category of modules over the group algebra $\mathbb{S}[\mathbb{T}]$. The latter is an $\mathbb{E}_{\infty}$-algebra in spectra, and

$$\pi_*(\mathbb{S}[\mathbb{T}]) = (\pi_*\mathbb{S})[d]/(d^2 - \eta d),$$

where d has degree 1 and is obtained from a choice of generator of $\pi_1(\mathbb{T})$ by translating it to the basepoint in the group ring. The relation $d^2 = \eta d$ is a consequence of the fact that, stably, the multiplication map $\mu\colon \mathbb{T}\times\mathbb{T} \to \mathbb{T}$ splits off the Hopf map $\eta \in \pi_1(\mathbb{S})$. From this calculation we conclude that a $\mathbb{T}$-action on an $\mathbb{E}_{\infty}$-algebra in spectra T gives rise to a graded derivation

$$\pi_j(T) \xrightarrow{d} \pi_{j+1}(T),$$

and this is Connes' operator. The operator d is not quite a differential, since we have $d \circ d = d \circ \eta = \eta \circ d$.

The $\mathbb{E}_{\infty}$-algebra structure on T gives rise to power operations in homology. In singular homology with $\mathbb{F}_2$-coefficients, there are power operations

$$\pi_j(\mathbb{F}_2 \otimes T) \xrightarrow{Q^i} \pi_{i+j}(\mathbb{F}_2 \otimes T)$$

for all integers i introduced by Araki–Kudo [25], and in singular homology with $\mathbb{F}_p$-coefficients, where p is odd, there are similar power operations

$$\pi_j(\mathbb{F}_p \otimes T) \xrightarrow{Q^i} \pi_{2i(p-1)+j}(\mathbb{F}_p \otimes T)$$

for all integers i defined by Dyer–Lashof [17]. The power operations are natural with respect to maps of $\mathbb{E}_{\infty}$-rings, but it is not immediately clear that they are compatible with Connes' operator, too. We give a proof that this is the nevertheless the case, following Angeltveit–Rognes [1, Proposition 5.9] and the very nice exposition of Höning [24].

Proposition 15.1.4. *If T is an $\mathbb{E}_{\infty}$-ring with $\mathbb{T}$-action, then*

$$Q^i \circ d = d \circ Q^i$$

for all integers i.

Proof. The adjunct $\tilde{\mu}\colon T \to \mathrm{map}(\mathbb{T}_+, T)$ of the map $\mu\colon T \otimes \mathbb{T}_+ \to T$ induced by the $\mathbb{T}$-action on T is a map of $\mathbb{E}_{\infty}$-rings, as is the canonical equivalence $T \otimes \mathrm{map}(\mathbb{T}_+, \mathbb{S}) \to \mathrm{map}(\mathbb{T}_+, T)$.

Composing the former map with an inverse of the latter map, we obtain a map of $\mathbb{E}_\infty$-rings

$$T \xrightarrow{\tilde{d}} T \otimes \mathrm{map}(\mathbb{T}_+, \mathbb{S}),$$

and hence, the induced map on homology preserves power operations. We identify the homology of the target via the isomorphism

$$\pi_*(\mathbb{F}_p \otimes T) \otimes_{\pi_*(\mathbb{F}_p)} \pi_*(\mathbb{F}_p \otimes \mathrm{map}(\mathbb{T}_+, \mathbb{S})) \longrightarrow \pi_*(\mathbb{F}_p \otimes T \otimes \mathrm{map}(\mathbb{T}_+, \mathbb{S}))$$

and the direct sum decomposition of $\mathrm{map}(\mathbb{T}_+, \mathbb{S})$ induced by the direct sum decomposition above, and under this idenfication, we have

$$\tilde{d}(a) = a \otimes 1 + da \otimes [S^{-1}].$$

Now, the Cartan formula for power operations shows that

$$Q^i(\tilde{d}(a)) = Q^i(a \otimes 1 + da \otimes [S^{-1}]) = Q^i(a) \otimes 1 + Q^i(da) \otimes [S^{-1}],$$

since $Q^0([S^{-1}]) = [S^{-1}]$ and $Q^i([S^{-1}]) = 0$ for $i \neq 0$. But we also have

$$Q^i(\tilde{d}(a)) = \tilde{d}(Q^i(a)) = Q^i(a) \otimes 1 + d(Q^i(a)) \otimes [S^{-1}],$$

so we conclude that $Q^i \circ d = d \circ Q^i$ as stated. □

We finally discuss the HKR-filtration. If k is a commutative ring and A a simplicial commutative k-algebra, then the Hochschild spectrum

$$\mathrm{HH}(A/k) = A^{\otimes_k \mathbb{T}}$$

has a complete and $\mathbb{T}$-equivariant descending filtration

$$\cdots \subseteq \mathrm{Fil}^n \mathrm{HH}(A/k) \subseteq \cdots \subseteq \mathrm{Fil}^1 \mathrm{HH}(A/k) \subseteq \mathrm{Fil}^0 \mathrm{HH}(A/k) \simeq \mathrm{HH}(A/k)$$

defined as follows. If A/k is smooth and discrete, then

$$\mathrm{Fil}^n \mathrm{HH}(A/k) = \tau_{\geq n} \mathrm{HH}(A/k),$$

and in general, the filtration is obtained from this special case by left Kan extension. The filtration quotients are identified as follows. If A/k is discrete, then $\mathrm{HH}(A/k)$ may be represented by a simplicial commutative A-algebra, and hence, its homotopy groups $\mathrm{HH}_*(A/k)$ form a strictly[8] anticommutative graded A-algebra. Moreover, Connes' operator gives rise to a differential on $\mathrm{HH}_*(A/k)$, which raises degrees by one and is a graded k-linear derivation. By definition, the de Rham-complex $\Omega^*_{A/k}$ is the universal example of this algebraic

[8] Here "strictly" indicates that elements of odd degree square to zero. This follows from [13, Théorème 4] by considering the universal case of Eilenberg–Mac Lane spaces.

structure, and therefore, we have a canonical map

$$\Omega^*_{A/k} \longrightarrow \mathrm{HH}_*(A/k),$$

which, by the Hochschild–Kostant–Rosenberg theorem, is an isomorphism, if A/k smooth. By the definition of the cotangent complex, this shows that

$$\mathrm{gr}^j \mathrm{HH}(A/k) \simeq (\Lambda^j_A L_{A/k})[j]$$

with trivial $\mathbb{T}$-action. Here Λ^j_A indicates the non-abelian derived functor of the jth exterior power over A.

15.2 Bökstedt periodicity

Bökstedt periodicity is the fundamental result that $\mathrm{THH}_*(\mathbb{F}_p)$ is a polynomial algebra over $\mathbb{F}_p$ on a generator in degree two. We present a proof, which is close to Bökstedt's original proof in the unpublished manuscript [10]. The skeleton filtration of the standard simplicial model for the circle induces a filtration of the topological Hochschild spectrum. For every homology theory, this gives rise to a spectral sequence, called the Bökstedt spectral sequence, that converges to the homology of the topological Hochschild spectrum. It is a spectral sequence of Hopf algebras in the symmetric monoidal category of quasi-coherent sheaves on the stack defined by the homology theory in question, and to handle this rich algebraic structure, we find it useful to introduce the geometric language of Berthelot and Grothendieck [3, II.1].

15.2.1 The Adams spectral sequence

If $f\colon A \to B$ is a map of anticommutative graded rings, then extension of scalars along f and restriction of scalars along f define adjoint functors

$$\mathrm{Mod}_A \underset{f_*}{\overset{f^*}{\rightleftarrows}} \mathrm{Mod}_B$$

between the respective categories of graded modules. Moreover, the extension of scalars functor f^* is symmetric monoidal, while the restriction of scalars functor f_* is lax symmetric monoidal with respect to the tensor product of graded modules.

We let k be an $\mathbb{E}_\infty$-ring and form the cosimplicial $\mathbb{E}_\infty$-ring

$$\Delta \xrightarrow{k^{\otimes[-]}} \mathrm{Alg}_{\mathbb{E}_\infty}(\mathsf{Sp}).$$

Here, as usual, $[n] \in \Delta$ denotes the finite ordinal $\{0, 1, \dots, n\}$, so $k^{\otimes[n]}$ is an $(n+1)$-fold tensor product. We will assume that the map

$$A = \pi_*(k^{\otimes[0]}) \xrightarrow{d^1} \pi_*(k^{\otimes[1]}) = B$$

is flat, so that $d^0, d^2 \colon k^{\otimes[1]} \to k^{\otimes[2]}$ induce an isomorphism of graded rings

$$B \otimes_A B = \pi_*(k^{\otimes[1]}) \otimes_{\pi_*(k^{\otimes[0]})} \pi_*(k^{\otimes[1]}) \xrightarrow{d^2+d^0} \pi_*(k^{\otimes[2]}).$$

The map $d^1 \colon k^{\otimes[1]} \to k^{\otimes[2]}$ now gives rise to a map of graded rings

$$B \xrightarrow{d^1} B \otimes_A B$$

and the sextuple

$$(A, B,\ B \xrightarrow{s^0} A\ ,\ A \underset{d^1}{\overset{d^0}{\rightrightarrows}} B\ ,\ B \xrightarrow{d^1} B \otimes_A B)$$

forms a cocategory object in the category of graded rings with the cocartesian symmetric monoidal structure. Here the maps $s^0 \colon A \to B$, $d^0, d^1 \colon A \to B$, and $d^1 \colon B \to B \otimes_A B$ are the opposites of the unit map, the source and target maps, and the composition map. Likewise, the septuple, where we also include the map $\chi \colon B \to B$ induced by the unique non-identity automorphism of the set $[1] = \{0, 1\}$, forms a cogroupoid in this symmetric monoidal category. We will abbreviate and simply write (A, B) for this cogroupoid object.

In general, given a cogroupoid object (A, B) in graded rings, we define an (A, B)-module[9] to be a pair (M, ε) of an A-module M and a B-linear map

$$d^{1*}(M) \xrightarrow{\varepsilon} d^{0*}(M)$$

that makes the following diagrams, in which the equality signs indicate the unique isomorphisms, commute.

$$\begin{array}{ccc} s^{0*}d^{1*}(M) & \xrightarrow{s^{0*}(\varepsilon)} & s^{0*}d^{0*}(M) \\ & \diagdown\!\!\diagdown \quad \diagup\!\!\diagup & \\ & M & \end{array}$$

$$\begin{array}{ccccc} & d^{1*}d^{1*}(M) & \xrightarrow{d^{1*}(\varepsilon)} & d^{1*}d^{0*}(M) & \\ \diagup\!\!\diagup & & & & \diagdown\!\!\diagdown \\ (B \otimes d^1)^* d^{1*}(M) & & & & (d^0 \otimes B)^* d^{0*}(M) \\ \searrow^{(B \otimes d^1)^*(\varepsilon)} & & & & \nearrow_{(d^0 \otimes B)^*(\varepsilon)} \\ & (B \otimes d^1)^* d^{0*}(M) & = & (d^0 \otimes B)^* d^{1*}(M) & \end{array}$$

We say that ε is a stratification of the A-module M relative to the cogroupoid (A, B). The map ε, we remark, is necessarily an isomorphism. We define a map of (A, B)-modules $f \colon (M_0, \varepsilon_0) \to (M_1, \varepsilon_1)$ to be a map of A-modules $f \colon M_0 \to M_1$ that makes the diagram of

[9] The cogroupoid (A, B) defines a stack $\mathcal{X}$, and the categories of (A, B)-modules and quasi-coherent $\mathcal{O}_{\mathcal{X}}$-modules are equivalent. For this reason, we prefer to say (A, B)-module instead of (A, B)-comodule, as is more common in the homotopy theory literature.

B-modules

$$\begin{array}{ccc} d^{1*}(M_0) & \xrightarrow{\varepsilon_0} & d^{0*}(M_0) \\ \downarrow{\scriptstyle d^{0*}(f)} & & \downarrow{\scriptstyle d^{1*}(f)} \\ d^{0*}(M_1) & \xrightarrow{\varepsilon_1} & d^{1*}(M_1) \end{array}$$

commute. In this case, we also say that the A-linear map $f\colon M_0 \to M_1$ is horizontal with respect ε. The category $\mathrm{Mod}_{(A,B)}$ of (A,B)-modules admits a symmetric monoidal structure with the monoidal product defined by

$$(M_1, \varepsilon_1) \otimes_{(A,B)} (M_2, \varepsilon_2) = (M_1 \otimes_A M_2, \varepsilon_{12}),$$

where ε_{12} is the unique map that makes the diagram

$$\begin{array}{ccc} d^{1*}(M_1) \otimes_B d^{1*}(M_2) & = & d^{1*}(M_1 \otimes_A M_2) \\ \downarrow{\scriptstyle \varepsilon_1 \otimes \varepsilon_2} & & \downarrow{\scriptstyle \varepsilon_{12}} \\ d^{0*}(M_1) \otimes_B d^{0*}(M_2) & = & d^{0*}(M_1 \otimes_A M_2) \end{array}$$

commute. The unit for the monoidal product is given by the A-module A with its unique structure of (A,B)-module, where $\varepsilon\colon d^{1*}(A) \to d^{0*}(A)$ is the unique B-linear map that makes the diagram

$$\begin{array}{ccc} s^{0*}d^{1*}(A) & \xrightarrow{s^{0*}(\varepsilon)} & s^{0*}d^{0*}(A) \\ & \searrow\!\!\!= \quad =\!\!\!\swarrow & \\ & A & \end{array}$$

commute.

We again let (A,B) be the cogroupoid associated with the $\mathbb{E}_\infty$-ring k. For every spectrum X, we consider the cosimplicial spectrum $k^{\otimes[-]} \otimes X$. The homotopy groups $\pi_*(k^{\otimes[0]} \otimes X)$ and $\pi_*(k^{\otimes[1]} \otimes X)$ form a left A-module and a left B-module, respectively. Moreover, we have A-linear maps

$$\pi_*(k^{\otimes[0]} \otimes X) \xrightarrow{d^i} d^i_*(\pi_*(k^{\otimes[1]} \otimes X))$$

induced by $d^i\colon k^{\otimes[0]} \otimes X \to k^{\otimes[1]} \otimes X$, and their adjunct maps

$$d^{i*}(\pi_*(k^{\otimes[0]} \otimes X)) \xrightarrow{\widetilde{d^i}} \pi_*(k^{\otimes[1]} \otimes X)$$

are B-linear isomorphisms. We now define the (A,B)-module associated with the spectrum X to be the pair (M,ε) with

$$M = \pi_*(k^{\otimes[0]} \otimes X)$$

and with ε the unique map that makes the following diagram commute.

$$\begin{array}{ccc} d^{1*}(M) & \xrightarrow{\varepsilon} & d^{0*}(M) \\ & {\scriptstyle \widetilde{d^1}}\searrow \quad \swarrow{\scriptstyle \widetilde{d^0}} & \\ & \pi_*(k^{\otimes[1]} \otimes X). & \end{array}$$

We often abbreviate and write $\pi_*(k \otimes X)$ for the (A,B)-module (M,ε).

The skeleton filtration of the cosimplicial spectrum $k^{\otimes[-]} \otimes X$ gives rise to the conditionally convergent k-based Adams spectral sequence

$$E^2_{i,j} = \operatorname{Ext}^{-i}_{(\pi_*(k),\pi_*(k\otimes k))}(\pi_*(k), \pi_*(k \otimes X))_j \Rightarrow \pi_{i+j}(\lim_\Delta k^{\otimes[-]} \otimes X),$$

where the Ext-groups are calculated in the abelian category of modules over the cogroupoid $(\pi_*(k), \pi_*(k \otimes k))$. An $\mathbb{E}_\infty$-algebra structure on X gives rise to a commutative monoid structure on $\pi_*(k \otimes X)$ in the symmetric monoidal category of $(\pi_*(k), \pi_*(k \otimes k))$-modules and makes the spectral sequence one of bigraded rings.

If X is a k-module, then the augmented cosimplicial spectrum

$$X \xrightarrow{d^0} k^{\otimes[-]} \otimes X$$

acquires a nullhomotopy. Therefore, the spectral sequence collapses and its edge homomorphism becomes an isomorphism

$$\pi_j(X) \longrightarrow \operatorname{Hom}_{(\pi_*(k),\pi_*(k\otimes k))}(\pi_*(k), \pi_*(k \otimes X))_j.$$

This identifies $\pi_j(X)$ with the subgroup of elements $x \in \pi_j(k\otimes X)$ that are horizontal[10] with respect to the stratification relative to $(\pi_*(k), \pi_*(k \otimes k))$ in the sense that $\varepsilon(1 \otimes_{A,d^1} x) = 1 \otimes_{A,d^0} x$.

15.2.2 The Bökstedt spectral sequence

In general, if R is an $\mathbb{E}_\infty$-ring, then

$$\operatorname{THH}(R) \simeq R^{\otimes \operatorname{colim}_{\Delta^{\mathrm{op}}} \Delta^1[-]/\partial\Delta^1[-]} \simeq \operatorname{colim}_{\Delta^{\mathrm{op}}} R^{\otimes \Delta^1[-]/\partial\Delta^1[-]}.$$

A priori, the right-hand term is the colimit in $\operatorname{Alg}_{\mathbb{E}_\infty}(\mathsf{Sp})$, but since the index category Δ^{op} is sifted [28, Lemma 5.5.8.4], the colimit agrees with the one in Sp. The increasing filtration of $\Delta^1[-]/\partial\Delta^1[-]$ by the skeleta induces an increasing filtration of $\operatorname{THH}(R)$,

$$\operatorname{Fil}_0 \operatorname{THH}(R) \to \operatorname{Fil}_1 \operatorname{THH}(R) \to \cdots \to \operatorname{Fil}_n \operatorname{THH}(R) \to \cdots.$$

We let k be an $\mathbb{E}_\infty$-ring and let (A, B) be the associated cogroupoid in graded rings, where $A = \pi_*(k)$ and $B = \pi_*(k \otimes k)$. We also let C be the commutative monoid $\pi_*(k \otimes R)$ in the symmetric monoidal category of (A, B)-modules. Here we assume that $d^0 \colon A \to B$ and $\eta \colon A \to C$ are flat. The filtration above gives rise to a spectral sequence

$$E^2_{i,j} = \operatorname{HH}_i(C/A)_j \Rightarrow \pi_{i+j}(k \otimes \operatorname{THH}(R)),$$

called the Bökstedt spectral sequence. It is a spectral sequence of C-algebras in the symmetric monoidal category of (A, B)-modules, and Connes' operator on $\pi_*(k\otimes \operatorname{THH}(R))$ induces

[10] In comodule nomenclature, horizontal elements are called primitive elements.

a map

$$E^r_{i,j} \xrightarrow{d} E^r_{i+1,j}$$

of spectral sequences, which, on the E^2-term, is equal to Connes' operator

$$\mathrm{HH}_i(C/A)_j \xrightarrow{d} \mathrm{HH}_{i+1}(C/A)_j.$$

In particular, if $y \in E^2_{i,j}$ and $dy \in E^2_{i+1,j}$ both survive the spectral sequence, and if y represents a homology class $\tilde{y} \in \pi_{i+j}(k \otimes \mathrm{THH}(R))$, then dy represents the homology class $d\tilde{y} \in \pi_{i+j+1}(k \otimes \mathrm{THH}(R))$.

Theorem 15.2.1 (Bökstedt). *The canonical map of graded $\mathbb{F}_p$-algebras*

$$\mathrm{Sym}_{\mathbb{F}_p}(\mathrm{THH}_2(\mathbb{F}_p)) \longrightarrow \mathrm{THH}_*(\mathbb{F}_p)$$

is an isomorphism, and $\mathrm{THH}_2(\mathbb{F}_p)$ is a 1-dimensional $\mathbb{F}_p$-vector space.

Proof. We let $k = R = \mathbb{F}_p$ and continue to write $A = \pi_*(k)$, $B = \pi_*(k \otimes k)$, and $C = \pi_*(k \otimes R)$. We apply the Bökstedt spectral sequence to show that, as a C-algebra in the symmetric monoidal category of (A, B)-modules,

$$\pi_*(k \otimes \mathrm{THH}(R)) = C[x]$$

on a horizontal generator x of degree 2, and use the Adams spectral sequence to conclude that $\mathrm{THH}_*(R) = R[x]$, as desired. Along the way, we will use the fact, observed by Angeltveit–Rognes [1], that the maps

$$S^1 \underset{\phi}{\overset{\psi}{\rightleftarrows}} S^1 \vee S^1 \qquad \{\infty\} \underset{\varepsilon}{\overset{\eta}{\rightleftarrows}} S^1 \qquad S^1 \xrightarrow{\chi} S^1$$

where ψ and ϕ are the pinch and fold maps, η and ε the unique maps, and χ the flip map, give $\pi_*(k \otimes \mathrm{THH}(R))$ the structure of a C-Hopf algebra in the symmetric monoidal category of (A, B)-modules, assuming that the unit map is flat. Moreover, the Bökstedt spectral sequence is a spectral sequence of C-Hopf algebras, provided that the unit map $C \to E^r$ is flat for all $r \geq 2$. We remark that the requirement that the comultiplication on $E = E^r$ be a map of C-modules in the symmetric monoidal category of (A, B)-modules is equivalent to the requirement that the diagram

$$\begin{array}{ccccc} d^{1*}(E) & \xrightarrow{d^{1*}(\psi)} & d^{1*}(E \otimes_C E) & = & d^{1*}(E) \otimes_{d^{1*}(C)} d^{1*}(E) \\ \downarrow{\scriptstyle \varepsilon_E} & & & & \downarrow{\scriptstyle \varepsilon_E \otimes_{\varepsilon_C} \varepsilon_E} \\ d^{0*}(E) & \xrightarrow{d^{0*}(\psi)} & d^{0*}(E \otimes_C E) & = & d^{0*}(E) \otimes_{d^{0*}(C)} d^{0*}(E) \end{array}$$

commutes.

To begin, we recall from Milnor [29] that, as a graded A-algebra,

$$C = \begin{cases} A[\,\bar{\xi}_i \mid i \geq 1\,] & \text{for } p = 2, \\ A[\,\bar{\xi}_i \mid i \geq 1\,] \otimes_A \Lambda_A\{\bar{\tau}_i \mid i \geq 0\} & \text{for } p \text{ odd}, \end{cases}$$

where $\bar{\tau}_i = \chi(\tau_i)$ and $\bar{\xi}_i = \chi(\xi_i)$ are the images by the antipode of Milnor's generators τ_i and ξ_i. Here, for $p = 2$, $\deg(\bar{\xi}_i) = p^i - 1$, while for p odd, $\deg(\bar{\xi}_i) = 2(p^i - 1)$ and $\deg(\bar{\tau}_i) = 2p^i - 1$. The stratification

$$d^{1*}(C) \xrightarrow{\ \varepsilon\ } d^{0*}(C)$$

is given by

$$\varepsilon(1 \otimes_{A,d^1} \bar{\xi}_i) = \sum \bar{\xi}_s \otimes_{A,d^0} \bar{\xi}_t^{p^s}$$
$$\varepsilon(1 \otimes_{A,d^1} \bar{\tau}_i) = 1 \otimes_{A,d^0} \bar{\tau}_i + \sum \bar{\tau}_s \otimes_{A,d^0} \bar{\xi}_t^{p^s}$$

with the sums indexed by $s, t \geq 0$ with $s + t = i$. Moreover, we recall from Steinberger [35] that the power operations on $C = \pi_*(k \otimes R)$ satisfy

$$Q^{p^i}(\bar{\xi}_i) = \bar{\xi}_{i+1},$$
$$Q^{p^i}(\bar{\tau}_i) = \bar{\tau}_{i+1}.$$

A very nice brief account of this calculation is given in [38].

We first consider $p = 2$. The E^2-term of the Bökstedt spectral sequence, as a C-Hopf algebra in (A, B)-modules, takes the form

$$E^2 = \Lambda_C\{d\bar{\xi}_i \mid i \geq 1\}$$

with $\deg(\bar{\xi}_i) = (0, 2^i - 1)$ and $\deg(d\bar{\xi}_i) = (1, 2^i - 1)$, and all differentials in the spectral sequence vanish. Indeed, they are C-linear derivations and, for degree reasons, the algebra generators $d\bar{\xi}_i$ cannot support non-zero differentials. We define x to be the image of $\bar{\xi}_1$ by the composite map

$$\pi_1(k \otimes R) \xrightarrow{\ \eta\ } \pi_1(k \otimes \mathrm{THH}(R)) \xrightarrow{\ d\ } \pi_2(k \otimes \mathrm{THH}(R)),$$

and proceed to show, by induction on $i \geq 0$, that the homology class x^{2^i} is represented by the element $d\bar{\xi}_{i+1}$ in the spectral sequence. The case $i = 0$ follows from what was said above, so we assume the statement has been proved for $i = r - 1$ and prove it for $i = r$. We have

$$x^{2^r} = (x^{2^{r-1}})^2 = Q^{2^r}(x^{2^{r-1}}),$$

which, by induction, is represented by $Q^{2^r}(d\bar{\xi}_r)$. But Proposition 15.1.4 and Steinberger's calculation show that

$$Q^{2^r}(d\bar{\xi}_r) = d(Q^{2^r}(\bar{\xi}_r)) = d\bar{\xi}_{r+1},$$

so we conclude that x^{2^r} is represented by $d\bar{\xi}_{r+1}$. Hence, as a graded C-algebra,

$$\pi_*(k \otimes \mathrm{THH}(R)) = C[x].$$

Finally, we calculate that

$$\begin{aligned}\varepsilon(1\otimes_{A,d^1}x) &= \varepsilon((\mathrm{id}\otimes_{A,d^1}d)(\eta\otimes_{A,d^1}\eta)(1\otimes_{A,d^1}\bar{\xi}_1))\\ &= (\mathrm{id}\otimes_{A,d^0}d)(\eta\otimes_{A,d^0}\eta)(\varepsilon(1\otimes_{A,d^1}\bar{\xi}_1))\\ &= (\mathrm{id}\otimes_{A,d^0}d)(\eta\otimes_{A,d^0}\eta)(\bar{\xi}_1\otimes_{A,d^0}1+1\otimes_{A,d^0}\bar{\xi}_1)\\ &= 1\otimes_{A,d^0}x,\end{aligned}$$

which shows that the element x is horizontal with respect to the stratification of $\pi_*(k\otimes \mathrm{THH}(R))$ relative to (A,B).

We next let p be odd. As a C-Hopf algebra,

$$E^2=\Lambda_C\{d\bar{\xi}_i\mid i\geq 1\}\otimes_C\Gamma_C\{d\bar{\tau}_i\mid i\geq 0\}$$

with $\deg(\bar{\xi}_i)=(0,2p^i-2)$, $\deg(d\bar{\xi}_i)=(1,2p^i-2)$, $\deg(\bar{\tau}_i)=(0,2p^i-1)$, and $\deg(d\bar{\tau}_i)=(1,2p^i-1)$, and with the coproduct given by

$$\begin{aligned}\psi(d\bar{\xi}_i)&=1\otimes d\bar{\xi}_i+d\bar{\xi}_i\otimes 1,\\ \psi((d\bar{\tau}_i)^{[r]})&=\textstyle\sum(d\bar{\tau}_i)^{[s]}\otimes(d\bar{\tau}_i)^{[t]},\end{aligned}$$

where the sum ranges over $s,t\geq 0$ with $s+t=r$. Here $(-)^{[r]}$ indicates the rth divided power. We define x to be the image of $\bar{\tau}_0$ by the composite map

$$\pi_1(k\otimes R)\xrightarrow{\ \eta\ }\pi_1(k\otimes\mathrm{THH}(R))\xrightarrow{\ d\ }\pi_2(k\otimes\mathrm{THH}(R))$$

and see, as in the case $p=2$, that the homology class x^{p^i} is represented by the element $d\bar{\tau}_i$. This is also shows that for $i\geq 1$, the element

$$d\bar{\xi}_i=d(\beta(\bar{\tau}_i))=\beta(d\bar{\tau}_i)$$

represents the homology class $\beta(x^{p^i})=p^ix^{p^i-1}\beta(x)$, which is zero. Hence, this element is annihilated by some differential. We claim that for all $i,s\geq 0$,

$$d^{p-1}((d\bar{\tau}_i)^{[p+s]})=a_i\cdot d\bar{\xi}_{i+1}\cdot(d\bar{\tau}_i)^{[s]}$$

with $a_i\in k^*$ a unit that depends on i but not on s. Granting this, we find as in the case $p=2$ that, as a C-algebra,

$$\pi_*(k\otimes\mathrm{THH}(R))=C[x]$$

with x horizontal of degree 2, which proves the theorem.

To prove the claim, we note that a shortest possible non-zero differential between elements of lowest possible total degree factors as a composition

$$E^r_{i,j}\xrightarrow{\ \pi\ }QE^r_{i,j}\xrightarrow{\ \bar{d}^r\ }PE^r_{i-r,j+r-1}\xrightarrow{\ \iota\ }E^r_{i-r,j+r-1},$$

where π is the quotient by algebra decomposables and ι is the inclusion of the coalgebra primitives. We further observe that $QE^2_{i,j}$ is zero, unless i is a power of p, and that $PE^2_{i,j}$ is zero, unless $i = 1$. Hence, the shortest possible non-zero differential of lowest possible total degree is

$$d^{p-1}((d\bar{\tau}_0)^{[p]}) = a_0 \cdot d\bar{\xi}_1$$

with $a_0 \in k$. In particular, we have $E^{p-1} = E^2$. If $a_0 = 0$, then $d\bar{\xi}_1$ survives the spectral sequence, so $a_0 \in k^*$. This proves the claim for $i = s = 0$.

We proceed by nested induction on $i, s \geq 0$ to prove the claim in general. We first note that if, for a fixed $i \geq 0$, the claim holds for $s = 0$, then it holds for all $s \geq 0$. For let $s \geq 1$ and assume, inductively, that the claim holds for all smaller values of s. One calculates that the difference

$$d^{p-1}((d\bar{\tau}_i)^{[p+s]}) - a_i \cdot d\bar{\xi}_{i+1} \cdot (d\bar{\tau}_i)^{[s]}$$

is a coalgebra primitive element, which shows that it is zero, since all non-zero coalgebra primitives in $E^{p-1} = E^2$ have filtration $i = 1$.

It remains to prove that the claim holds for all $i \geq 0$ and $s = 0$. We have already proved the case $i = 0$, so we let $j \geq 1$ and assume, inductively, that the claim has been proved for all $i < j$ and all $s \geq 0$. The inductive assumption implies that E^p is a subquotient of the C-subalgebra

$$D = \Lambda_C\{d\bar{\xi}_i \mid i \geq j+1\} \otimes_C \Gamma_C\{d\bar{\tau}_i \mid i \geq j\} \otimes_C C[d\bar{\tau}_i \mid i < j]/((d\bar{\tau}_i)^p \mid i < j)$$

of $E^{p-1} = E^2$. Now, since $\pi_*(k \otimes \mathrm{THH}(R))$ is an augmented C-algebra, all elements of filtration 0 survive the spectral sequence. Hence, if $x \in E^r$ with $r \geq p$ supports a non-zero differential, then x has filtration at least $p+1$. But all algebra generators in D of filtration at least $p+1$ have total degree at least $2p^{j+2}$, so either $d^{p-1}((d\bar{\tau}_j)^{[p]})$ is non-zero, or else all elements in D of total degree at most $2p^{j+2} - 2$ survive the spectral sequence. Since we know that the element $d\bar{\xi}_{j+1} \in D$ of total degree $2p^{j+1} - 1$ does not survive the spectral sequence, we conclude that the former is the case. We must show that

$$d^{p-1}((d\bar{\tau}_j)^{[p]}) = a_j \cdot d\bar{\xi}_{j+1}$$

with $a_j \in k^*$, and to this end, we use the fact that E^{p-1} is a C-Hopf algebra in the symmetric monoidal category of (A, B)-modules. We have

$$\begin{aligned}
\varepsilon(1 \otimes_{A,d^1} d\bar{\xi}_i) &= \varepsilon((\mathrm{id} \otimes_{A,d^1} d)(1 \otimes_{A,d^1} \bar{\xi}_i)) = (\mathrm{id} \otimes_{A,d^0} d)(\varepsilon(1 \otimes_{A,d^1} \bar{\xi}_i)) \\
&= (\mathrm{id} \otimes_{A,d^0} d)(\textstyle\sum \bar{\xi}_s \otimes_{A,d^0} \bar{\xi}_t^{p^s}) = 1 \otimes_{A,d^0} d\bar{\xi}_i, \\
\varepsilon(1 \otimes_{A,d^1} d\bar{\tau}_i) &= \varepsilon((\mathrm{id} \otimes_{A,d^1} d)(1 \otimes_{A,d^1} \bar{\tau}_i)) = (\mathrm{id} \otimes_{A,d^0} d)(\varepsilon(1 \otimes_{A,d^1} \bar{\tau}_i)) \\
&= (\mathrm{id} \otimes_{A,d^0} d)(1 \otimes_{A,d^0} \bar{\tau}_i + \textstyle\sum \bar{\tau}_s \otimes_{A,d^0} \bar{\xi}_t^{p^s}) = 1 \otimes_{A,d^0} d\bar{\tau}_i,
\end{aligned}$$

where the sums range over $s, t \geq 0$ with $s+t = i$. Hence, the sub-k-vector space of E^{p-1} that consists of the horizontal elements of bidegree $(1, 2p^{j+1} - 2)$ is spanned by $d\bar{\xi}_{j+1}$. Therefore, it suffices to show that $(d\bar{\tau}_j)^{[p]}$, and hence, $d^{p-1}((d\bar{\tau}_j)^{[p]})$ is horizontal. We have already proved that $d\bar{\tau}_j$ is horizontal, and using the fact that E^{p-1} is a C-algebra in the symmetric monoidal category of (A, B)-modules, we conclude that $(d\bar{\tau}_j)^s$, and therefore, $(d\bar{\tau}_j)^{[s]}$ is horizontal for all $0 \leq s < p$. Finally, we make use of the fact that E^{p-1} is a C-coalgebra in the symmetric monoidal category of (A, B)-modules. Since

$$\psi((d\bar{\tau}_j)^{[p]}) = \sum (d\bar{\tau}_j)^{[s]} \otimes (d\bar{\tau}_j)^{[t]}$$

with the sum indexed by $s, t \geq 0$ with $s + t = p$, and since we have already proved that $(d\bar{\tau}_j)^{[s]}$ with $0 \leq s < p$ are horizontal, we find that $(d\bar{\tau}_j)^{[p]}$ is horizontal. This completes the proof. □

We finally recall the following analogue of the Segal conjecture. This is a key result for understanding topological cyclic homology and its variants.

Addendum 15.2.2. *The Frobenius induces an equivalence*

$$\mathrm{THH}(\mathbb{F}_p) \xrightarrow{\varphi} \tau_{\geq 0}\,\mathrm{THH}(\mathbb{F}_p)^{tC_p}.$$

Proof. See [32, Section IV.4]. □

15.3 Perfectoid rings

Perfectoid rings are to topological Hochschild homology what separably closed fields are to K-theory: they annihilate Kähler differentials. In this section, we present the proof by Bhatt–Morrow–Scholze that Bökstedt periodicity holds for every perfectoid ring R. As a consequence, the Tate spectral sequence

$$E^2_{i,j} = \hat{H}^{-i}(B\mathbb{T}, \mathrm{THH}_j(R, \mathbb{Z}_p)) \Rightarrow \mathrm{TP}_{i+j}(R, \mathbb{Z}_p)$$

collapses and gives the ring $\mathrm{TP}_0(R, \mathbb{Z}_p)$ a complete and separated descending filtration, the graded pieces of which are free R-modules of rank 1. The ring $\mathrm{TP}_0(R, \mathbb{Z}_p)$, however, is not a power series ring over R. Instead, it agrees, up to unique isomorphism over R, with Fontaine's p-adic period ring $A_{\mathrm{inf}}(R)$, the definition of which we recall below. Finally, we use these results to prove that Bökstedt periodicity implies Bott periodicity.

15.3.1 Perfectoid rings

A $\mathbb{Z}_p$-algebra R is perfectoid, for example, if there exists a non-zero-divisor $\pi \in R$ with $p \in \pi^p R$ such that R is complete and separated with respect to the π-adic topology and such that the Frobenius $\varphi\colon R/\pi \to R/\pi^p$ is an isomorphism. We will give the general definition, which does not require $\pi \in R$ to be a non-zero-divisor, below. Typically, perfectoid rings are large and highly non-noetherian. Moreover, the ring R/π is typically not a field, but is also a large non-noetherian ring with many nilpotent elements. An example to keep in mind is the valuation ring $\mathcal{O}_C$ in an algebraically closed field $C/\mathbb{Q}_p$ that is complete with respect to a non-archimedean absolute value extending the p-adic absolute value on $\mathbb{Q}_p$; here we can take π to be a pth root of p.

We recall some facts from [4, Section 3]. If a ring S contains an element $\pi \in S$ such that $p \in \pi S$ and such that π-adic topology on S is complete and separated, then the canonical projections

$$\begin{array}{ccccc}
\lim_{n,F} W(S) & \longrightarrow & \lim_{n,F} W(S/p) & \longrightarrow & \lim_{n,F} W(S/\pi) \\
\downarrow & & \downarrow & & \downarrow \\
\lim_{n,F} W_n(S) & \longrightarrow & \lim_{n,F} W_n(S/p) & \longrightarrow & \lim_{n,F} W_n(S/\pi)
\end{array}$$

all are isomorphisms. Here the limits range over non-negative integers n with the respective Witt vector Frobenius maps as the structure maps. Moreover, since the Witt vector Frobenius for $\mathbb{F}_p$-algebras agrees with the map of rings of Witt vectors induced by the Frobenius, we have a canonical map

$$W(\lim_{n,\varphi} S/p) \longrightarrow \lim_{n,F} W(S/p),$$

and this map, too, is an isomorphism, since the Witt vector functor preserves limits. The perfect $\mathbb{F}_p$-algebra

$$S^\flat = \lim_{n,\varphi} S/p = \lim(S/p \xleftarrow{\varphi} S/p \xleftarrow{\varphi} \cdots)$$

is called the tilt of S, and its ring of Witt vectors

$$A_{\text{inf}}(S) = W(S^\flat)$$

is called Fontaine's ring of p-adic periods. The Frobenius automorphism φ of $S^\flat$ induces the automorphism $W(\varphi)$ of $A_{\text{inf}}(S)$, which, by abuse of notation, we also write φ and call the Frobenius.

We again consider the diagram of isomorphisms at the beginning of the section. By composing the isomorphisms in the diagram with the projection onto $W_n(S)$ in the lower left-hand term of the diagram, we obtain a ring homomorphism $\widetilde{\theta}_n \colon A_{\text{inf}}(S) \to W_n(S)$, and we define

$$A_{\text{inf}}(S) \xrightarrow{\theta_n} W_n(S)$$

to be $\theta_n = \widetilde{\theta}_n \circ \varphi^n$. It is clear from the definition that the diagrams

$$\begin{array}{ccc}
A_{\text{inf}}(S) & \xrightarrow{\theta_n} & W_n(S) \\
{\scriptstyle \text{id}}\downarrow\downarrow{\scriptstyle \varphi} & & {\scriptstyle R}\downarrow\downarrow{\scriptstyle F} \\
A_{\text{inf}}(S) & \xrightarrow{\theta_{n-1}} & W_{n-1}(S)
\end{array}
\qquad
\begin{array}{ccc}
A_{\text{inf}}(S) & \xrightarrow{\widetilde{\theta}_n} & W_n(S) \\
{\scriptstyle \varphi^{-1}}\downarrow\downarrow{\scriptstyle \text{id}} & & {\scriptstyle R}\downarrow\downarrow{\scriptstyle F} \\
A_{\text{inf}}(S) & \xrightarrow{\widetilde{\theta}_{n-1}} & W_{n-1}(S)
\end{array}$$

commute. The map $\theta = \theta_1 \colon A_{\text{inf}}(S) \to S = W_1(S)$ is Fontaine's map from [19], which we now describe more explicitly. There is a well-defined map

$$S^\flat \xrightarrow{(-)^\#} S$$

that to $a = (x_0, x_1, \dots) \in \lim_{n,\varphi} S/p = S^\flat$ assigns $a^\# = \lim_{n\to\infty} y_n^{p^n}$ for any choice of lifts $y_n \in S$ of $x_n \in S/p$. It is multiplicative, but it is not additive unless S is an $\mathbb{F}_p$-algebra.

Using this map, we have

$$\theta(\sum_{i\geq 0}[a_i]\,p^i) = \sum_{i\geq 0} a_i^{\#} p^i,$$

where $[-]\colon S^\flat \to W(S^\flat)$ is the Teichmüller representative. We can now state the general definition of a perfectoid ring that is used in [4], [5], and [6].

Definition 15.3.1. A $\mathbb{Z}_p$-algebra R is *perfectoid* if there exists $\pi \in R$ such that $p \in \pi^p R$, such that the π-adic topology on R is complete and separated, such that the Frobenius $\varphi\colon R/p \to R/p$ is surjective, and such that the kernel of $\theta\colon A_{\mathrm{inf}}(R) \to R$ is a principal ideal.

The ideal $I = \mathrm{Ker}(\theta) \subset A = W(R^\flat)$ is typically not fixed by the Frobenius on A, but it always satisfies the prism property that

$$p \in I + \varphi(I)A.$$

If an ideal $J \subset A$ satisfies the prism property, then the quotient A/J is an untilt of $R^\flat$ in the sense that it is perfectoid and that its tilt is $R^\flat$. In fact, every untilt of $R^\flat$ arises as A/J for some ideal $J \subset A$ that satisfies the prism property. The set of such ideals is typically large, but it has a very interesting p-adic geometry. Indeed, for $R = \mathcal{O}_C$, there is a canonical one-to-one correspondence between orbits under the Frobenius of such ideals and closed points of the Fargues–Fontaine curve FF_C [18]. Among all ideals $J \subset A$ that satisfy the prism property, the ideal $J = pA$ is the only one for which the untilt $A/J = R^\flat$ is of characteristic p; all other untilts A/J are of mixed characteristic $(0,p)$. One can show that every untilt A/J is a reduced ring and that a generator ξ of the ideal $J \subset A$ necessarily is a non-zero-divisor. Hence, such a generator is well-defined, up to a unit in A. An untilt A/J may have p-torsion, but if an element is annihilated by some power of p, then it is in fact annihilated by p. We refer to [6] for proofs of these statements.

Bhatt–Morrow–Scholze prove in [5, Theorem 6.1] that Bökstedt periodicity for $R = \mathbb{F}_p$ implies the analogous result for R any perfectoid ring.

Theorem 15.3.2 (Bhatt–Morrow–Scholze). *If R is a perfectoid ring, then the canonical map is an isomorphism of graded rings*

$$\mathrm{Sym}_R(\mathrm{THH}_2(R,\mathbb{Z}_p)) \longrightarrow \mathrm{THH}_*(R,\mathbb{Z}_p),$$

and $\mathrm{THH}_2(R,\mathbb{Z}_p)$ is a free R-module of rank 1.

Proof. We follow the proof in loc. cit. We first claim that the canonical map

$$\Gamma_R(\mathrm{HH}_2(R/\mathbb{Z},\mathbb{Z}_p)) \longrightarrow \mathrm{HH}_*(R/\mathbb{Z},\mathbb{Z}_p)$$

is an isomorphism and that the R-module $\mathrm{HH}_2(R/\mathbb{Z},\mathbb{Z}_p)$ is free of rank 1. To prove this, we first notice that the base-change map

$$\mathrm{HH}(R/\mathbb{Z}) \longrightarrow \mathrm{HH}(R/A_{\mathrm{inf}}(R))$$

is a p-adic equivalence. Indeed, we always have

$$\mathrm{HH}(R/A_{\mathrm{inf}}(R)) \simeq \mathrm{HH}(R/\mathbb{Z}_p) \otimes_{\mathrm{HH}(A_{\mathrm{inf}}(R)/\mathbb{Z}_p)} A_{\mathrm{inf}}(R),$$

and $\mathrm{HH}(A_{\mathrm{inf}}(R)/\mathbb{Z}) \to A_{\mathrm{inf}}(R)$ is a p-adic equivalence, because

$$\mathrm{HH}(A_{\mathrm{inf}}(R)/\mathbb{Z}) \otimes_{\mathbb{Z}} \mathbb{F}_p \simeq \mathrm{HH}(A_{\mathrm{inf}}(R) \otimes_{\mathbb{Z}} \mathbb{F}_p/\mathbb{F}_p) \simeq \mathrm{HH}(R^\flat/\mathbb{F}_p) \simeq R^\flat.$$

The last equivalence holds since $R^\flat$ is perfect. Now, we write $R \simeq \Lambda_{A_{\mathrm{inf}}(R)}\{y\}$ with $dy = \xi$ to see that $R \otimes_{A_{\mathrm{inf}}(R)} R \simeq \Lambda_R\{y\}$ with $dy = 0$, and similarly, we write $R \simeq \Lambda_R\{y\} \otimes \Gamma_R\{x\}$ with $dx^{[i]} = x^{[i-1]}y$ to see that

$$\mathrm{HH}(R/A_{\mathrm{inf}}(R)) \simeq R \otimes_{R \otimes_{A_{\mathrm{inf}}(R)} R} R \simeq \Gamma_R\{x\},$$

which proves the claim.

It follows, in particular, that the R-module

$$\mathrm{THH}(R, \mathbb{Z}_p) \otimes_{\mathrm{THH}(\mathbb{Z})} \mathbb{Z} \simeq \mathrm{HH}(R/\mathbb{Z}, \mathbb{Z}_p)$$

is pseudocoherent in the sense that it can be represented by a chain complex of finitely generated free R-modules that is bounded below. Since $\mathrm{THH}(\mathbb{Z})$ has finitely generated homotopy groups, we conclude, inductively, that

$$\mathrm{THH}(R, \mathbb{Z}_p) \otimes_{\mathrm{THH}(\mathbb{Z})} \tau_{\leq n} \mathrm{THH}(\mathbb{Z})$$

is a pseudocoherent R-module for all $n \geq 0$. Therefore, also $\mathrm{THH}(R, \mathbb{Z}_p)$ is a pseudocoherent R-module.

Next, we claim that any ring homomorphism $R \to R'$ between perfectoid rings induces an equivalence

$$\mathrm{THH}(R, \mathbb{Z}_p) \otimes_R R' \longrightarrow \mathrm{THH}(R', \mathbb{Z}_p).$$

Indeed, it suffices to prove that the claim holds after extension of scalars along the canonical map $\mathrm{THH}(\mathbb{Z}) \to \mathbb{Z}$. This reduces us to proving that

$$\mathrm{HH}(R, \mathbb{Z}_p) \otimes_R R' \longrightarrow \mathrm{HH}(R', \mathbb{Z}_p)$$

is an equivalence, which follows from the first claim.

We now prove that the map in the statement is an isomorphism. The case of $R = \mathbb{F}_p$ is Theorem 15.2.1, and the case of a perfect $\mathbb{F}_p$-algebra follows from the base-change formula that we just proved. In the general case, we show, inductively, that the map is an isomorphism in degree $i \geq 0$. So we assume that the map is an isomorphism in degrees $< i$ and prove that it is an isomorphism in degree i. By induction, the R-module $\tau_{<i} \mathrm{THH}(R, \mathbb{Z}_p)$ is perfect, and hence, the R-module $\tau_{\geq i} \mathrm{THH}(R, \mathbb{Z}_p)$ is pseudocoherent. It follows that the R-module $\mathrm{THH}_i(R, \mathbb{Z}_p)$ is finitely generated. Since R is perfectoid, the composition

$$R \longrightarrow R/p \longrightarrow \bar{R} = \mathrm{colim}_{n,\varphi} R/p = R/\sqrt{pR}$$

of the canonical projection and the canonical map from the initial term in the colimit is surjective. Since $\bar{R}$ is a perfect $\mathbb{F}_p$-algebra, the base-change formula and the inductive hypothesis show that, in the diagram

$$\begin{array}{ccc} \operatorname{Sym}_R(\operatorname{THH}_2(R,\mathbb{Z}_p)) \otimes_R \bar{R} & \longrightarrow & \operatorname{THH}_*(R,\mathbb{Z}_p) \otimes_R \bar{R} \\ \downarrow & & \downarrow \\ \operatorname{Sym}_{\bar{R}}(\operatorname{THH}_2(\bar{R},\mathbb{Z}_p)) & \longrightarrow & \operatorname{THH}_*(\bar{R},\mathbb{Z}_p), \end{array}$$

the vertical maps are isomorphisms in degrees $\leq i$, and we have already seen that the lower horizontal map is an isomorphism. Hence, the upper horizontal map is an isomorphism in degrees $\leq i$. Since the kernel of the map $R \to \bar{R}$ contains the Jacobson radical, Nakayama's lemma shows that the map in the statement of the theorem is surjective in degrees $\leq i$. To prove that it is also injective, we consider the diagram

$$\begin{array}{ccc} \operatorname{Sym}_R(\operatorname{THH}_2(R,\mathbb{Z}_p)) & \longrightarrow & \operatorname{THH}_*(R,\mathbb{Z}_p) \\ \downarrow & & \downarrow \\ \prod_{\mathfrak{p}} \operatorname{Sym}_R(\operatorname{THH}_2(R,\mathbb{Z}_p)) \otimes_R R_{\mathfrak{p}} & \longrightarrow & \prod_{\mathfrak{p}} \operatorname{THH}_*(R,\mathbb{Z}_p) \otimes_R R_{\mathfrak{p}} \end{array}$$

where the products range over the minimal primes $\mathfrak{p} \subset R$. Since R is reduced, the left-hand vertical map is injective and the local rings $R_{\mathfrak{p}}$ are fields. Hence, it suffices to prove that for every minimal prime $\mathfrak{p} \subset R$, the map

$$\operatorname{Sym}_R(\operatorname{THH}_2(R,\mathbb{Z}_p)) \otimes_R R_{\mathfrak{p}} \longrightarrow \operatorname{THH}_*(R,\mathbb{Z}_p) \otimes_R R_{\mathfrak{p}}$$

is injective in degrees $\leq i$. To this end, we write $\operatorname{Spec}(R)$ as the union of the closed subscheme $\operatorname{Spec}(\bar{R})$ and its open complement $\operatorname{Spec}(R[1/p])$. If $\mathfrak{p}$ belongs to $\operatorname{Spec}(\bar{R})$, then the map in question is an isomorphism in degrees $\leq i$ by what was proved above. Similarly, if $\mathfrak{p}$ belongs to $\operatorname{Spec}(R[1/p])$, then the map in a question is an isomorphism in all degrees, since the map

$$\operatorname{Sym}_R(\operatorname{THH}_2(R,\mathbb{Z}_p)) \otimes_R R[1/p] \longrightarrow \operatorname{THH}_*(R,\mathbb{Z}_p) \otimes_R R[1/p]$$

is so, by the claim at the beginning of the proof. This completes the proof. □

Addendum 15.3.3. *If R is a perfectoid ring, then*

$$\operatorname{THH}(R,\mathbb{Z}_p) \xrightarrow{\varphi} \tau_{\geq 0} \operatorname{THH}(R,\mathbb{Z}_p)^{tC_p}$$

is an equivalence.

Proof. See [5, Proposition 6.2]. □

We show that Fontaine's map $\theta\colon A_{\mathrm{inf}}(R) \to R$ is the universal p-complete pro-infinitesimal thickening following [19, Théorème 1.2.1]. We remark that, in loc. cit., Fontaine defines $A_{\mathrm{inf}}(R)$ to be the $\operatorname{Ker}(\theta)$-adic completion of $W(R^\flat)$. We include a proof here that this is not necessary in that the $\operatorname{Ker}(\theta)$-adic topology on $W(R^\flat)$ is already complete and separated.

Proposition 15.3.4 (Fontaine). *If R is a perfectoid ring, then the map*

$$A_{\mathrm{inf}}(R) \xrightarrow{\theta} R$$

is initial among ring homomorphisms $\theta_D \colon D \to R$ such that D is complete and separated in both the p-adic topology and the $\mathrm{Ker}(\theta_D)$*-adic topology.*

Proof. We first show that $A = A_{\mathrm{inf}}(R)$ is complete and separated in both the p-adic topology and the $\mathrm{Ker}(\theta)$-adic topology. Since $R^\flat$ is a perfect $\mathbb{F}_p$-algebra, we have $p^n A = V^n(A) \subset A$, so the p-adic topology on $A = W(R^\flat)$ is complete and separated. Moreover, since $p \in A$ is a non-zero-divisor, this is equivalent to A being derived p-complete. As we recalled above, a generator $\xi \in \mathrm{Ker}(\theta)$ is necessariy a non-zero-divisor. Therefore, the $\mathrm{Ker}(\theta)$-adic topology on A is complete and separated if and only if A is derived ξ-adically complete, and since A is derived p-adically complete, this, in turn, is equivalent to A/p being derived ξ-adically complete. Now, we have

$$\begin{array}{ccccccc} A/p & \longrightarrow & A/(p,\xi^n) & \xrightarrow{\varphi^{-n}} & A/(p,\xi) & \xrightarrow{\theta} & R/p \\ \| & & \downarrow{\scriptstyle\varphi} & & \downarrow{\scriptstyle\varphi} & & \downarrow{\scriptstyle\varphi} \\ A/p & \longrightarrow & A/(p,\xi^{n-1}) & \xrightarrow{\varphi^{-(n-1)}} & A/(p,\xi) & \xrightarrow{\theta} & R/p \end{array}$$

with the middle- and right-hand maps bijective and with the remaining maps surjective. Taking derived limits, we obtain

$$A/p \longrightarrow (A/p)^\wedge_\xi \longrightarrow (A/\xi)^\flat \longrightarrow R^\flat$$

with the middle- and right-hand maps equivalences. The composite map takes the class of $\sum [a_i]p^i$ to a_0, and therefore, it, too, is an equivalence. This proves that the left-hand map is an equivalence, as desired.

Let $\theta_D \colon D \to R$ be as in the statement. We wish to prove that there is a unique ring homomorphism $f \colon A \to D$ such that $\theta = \theta_D \circ f$. Since A and D are derived p-complete and $L_{(A/p)/\mathbb{F}_p} \simeq 0$, this is equivalent to showing that there is a unique ring homomorphism $\bar{f} \colon A/p \to D/p$ with the property that $\bar{\theta} = \bar{\theta}_D \circ \bar{f}$, where $\bar{\theta} \colon A/p \to R/p$ and $\bar{\theta}_D \colon D/p \to R/p$ are induced by $\theta \colon A \to R$ and $\theta_D \colon D \to R$, respectively. Identifying A/p with $R^\flat$, we wish to show that there is a unique ring homomorphism $\bar{f} \colon R^\flat \to D/p$ such that $a^\# + pR = \bar{\theta}_D(\bar{f}(a))$ for all $a \in R^\flat$. Since the $\mathrm{Ker}(\theta_D)$-adic topology on D is complete and separated, so is the $\mathrm{Ker}(\bar{\theta}_D)$-adic topology on D/p. It follows that for $a = (x_0, x_1, \dots) \in R^\flat$, the limit

$$\bar{f}(a) = \lim_{n \to \infty} \widetilde{x}_n^{p^n} \in D/p,$$

where we choose $\widetilde{x}_n \in D/p$ with $\bar{\theta}_D(\widetilde{x}_n) = x_n \in R/p$, exists and is independent of the choices made. This defines a map $\bar{f} \colon R^\flat \to D/p$, and the uniqueness of the limit implies that it is a ring homomorphism. It satisfies $\bar{\theta} = \bar{\theta}_D \circ \bar{f}$ by construction, and it is unique with this property, since the $\mathrm{Ker}(\bar{\theta}_D)$-adic topology on D/p is separated. □

We identify the diagram of p-adic homotopy groups

$$\mathrm{TC}_*^-(R,\mathbb{Z}_p) \underset{\mathrm{can}}{\overset{\varphi}{\rightrightarrows}} \mathrm{TP}_*(R,\mathbb{Z}_p)$$

for R perfectoid. By Bökstedt periodicity, the spectral sequences

$$E_{i,j}^2 = H^{-i}(B\mathbb{T}, \mathrm{THH}_j(R,\mathbb{Z}_p)) \Rightarrow \mathrm{TC}_{i+j}^-(R,\mathbb{Z}_p)$$
$$E_{i,j}^2 = \hat{H}^{-i}(B\mathbb{T}, \mathrm{THH}_j(R,\mathbb{Z}_p)) \Rightarrow \mathrm{TP}_{i+j}(R,\mathbb{Z}_p)$$

collapse. It follows that the respective edge homomorphisms in total degree 0 satisfy the hypotheses of Proposition 15.3.4, and therefore, there exists a unique ring homomorphism making the diagram

$$\begin{array}{ccccc} A_{\mathrm{inf}}(R) & \longrightarrow & \mathrm{TC}_0^-(R,\mathbb{Z}_p) & \xrightarrow{\mathrm{can}} & \mathrm{TP}_0(R,\mathbb{Z}_p) \\ \downarrow{\scriptstyle\theta} & & \downarrow{\scriptstyle\mathrm{edge}} & & \downarrow{\scriptstyle\mathrm{edge}} \\ R & = & \mathrm{THH}_0(R,\mathbb{Z}_p) & = & \mathrm{THH}_0(R,\mathbb{Z}_p) \end{array}$$

commute. We view $\mathrm{TC}_*^-(R,\mathbb{Z}_p)$ and $\mathrm{TP}_*(R,\mathbb{Z}_p)$ as graded $A_{\mathrm{inf}}(R)$-algebras via the top horizontal maps in the diagram. Bhatt–Morrow–Scholze make the following calculation in [5, Propositions 6.2 and 6.3].

Theorem 15.3.5 (Bhatt–Morrow–Scholze). *Let R be a perfectoid ring, and let ξ be a generator of the kernel of Fontaine's map $\theta\colon A_{\mathrm{inf}}(R) \to R$.*

(1) *There exists $\sigma \in \mathrm{TP}_2(R,\mathbb{Z}_p)$ such that*

$$\mathrm{TP}_*(R,\mathbb{Z}_p) = A_{\mathrm{inf}}(R)[\sigma^{\pm 1}].$$

(2) *There exists $u \in \mathrm{TC}_2^-(R,\mathbb{Z}_p)$ and $v \in \mathrm{TC}_{-2}^-(R,\mathbb{Z}_p)$ such that*

$$\mathrm{TC}_*^-(R,\mathbb{Z}_p) = A_{\mathrm{inf}}(R)[u,v]/(uv-\xi).$$

(3) *The graded ring homomorphisms*

$$\mathrm{TC}_*^-(R,\mathbb{Z}_p) \underset{\mathrm{can}}{\overset{\varphi}{\rightrightarrows}} \mathrm{TP}_*(R,\mathbb{Z}_p)$$

are φ-linear and $A_{\mathrm{inf}}(R)$-linear, respectively, and u, v, and σ can be chosen in such a way that $\varphi(u) = \alpha\cdot\sigma$, $\varphi(v) = \alpha^{-1}\varphi(\xi)\cdot\sigma^{-1}$, $\mathrm{can}(u) = \xi\cdot\sigma$, and $\mathrm{can}(v) = \sigma^{-1}$, where $\alpha \in A_{\mathrm{inf}}(R)$ is a unit.[11]

Proof. The canonical map $A_{\mathrm{inf}}(R) \to \mathrm{TP}_0(R,\mathbb{Z}_p)$ is a map of filtered rings, where the domain and target are given the ξ-adic filtration and the filtration induced by the Tate spectral sequence, respectively. Since both filtrations are complete and separated, the map

[11] If one is willing to replace ξ by $\varphi^{-1}(\alpha)\xi$, then the unit α can be eliminated.

is an isomorphism if and only if the induced maps of filtration quotients are isomorphisms. These, in turn, are R-linear maps between free R-modules of rank 1, and to prove that they are isomorphisms, it suffices to consider the case $R = \mathbb{F}_p$.

The canonical map can: $\mathrm{TC}^-_*(R, \mathbb{Z}_p) \to \mathrm{TP}_*(R, \mathbb{Z}_p)$ induces the map of spectral sequences that, on E^2-terms, is given by the localization map

$$R[t, x] \longrightarrow R[t^{\pm 1}, x],$$

where $t \in E^2_{-2,0}$ and $x \in E^2_{2,0}$ are any R-module generators. It follows that $\mathrm{TP}_*(R, \mathbb{Z}_p)$ is 2-periodic and concentrated in even degrees, so (1) holds for any $\sigma \in \mathrm{TP}_2(R, \mathbb{Z}_p)$ that is an $A_{\mathrm{inf}}(R)$-module generator, or equivalently, is represented in the spectral sequence by an R-module generator of $E^2_{2,0}$. We now fix a choice of $\sigma \in \mathrm{TP}_2(R, \mathbb{Z}_p)$, and let $t \in E^2_{-2,0}$ and $x \in E^2_{0,2}$ be the unique elements that represent $\sigma^{-1} \in \mathrm{TP}_{-2}(R, \mathbb{Z}_p)$ and $\xi\sigma \in \mathrm{TP}_2(R, \mathbb{Z}_p)$, respectively. The latter classes are the images by the canonical map of unique classes $v \in \mathrm{TC}^-_{-2}(R, \mathbb{Z}_p)$ and $u \in \mathrm{TC}^-_2(R, \mathbb{Z}_p)$, and $uv = \xi$. This proves (2) and the part of (3) that concerns the canonical map.

It remains to identify $\varphi\colon \mathrm{TC}^-_*(R, \mathbb{Z}_p) \to \mathrm{TP}_*(R, \mathbb{Z}_p)$. In degree zero, we have fixed identifications of domain and target with $A_{\mathrm{inf}}(R)$, and we first prove that, with respect to these identifications, the map in question is given by the Frobenius $\varphi\colon A_{\mathrm{inf}}(R) \to A_{\mathrm{inf}}(R)$. To this end, we consider the diagram

$$\begin{array}{ccccc}
\mathrm{THH}(R)^{h\mathbb{T}} & \xrightarrow{\varphi^{h\mathbb{T}}} & (\mathrm{THH}(R)^{tC_p})^{h\mathbb{T}} & \longrightarrow & (R^{tC_p})^{h\mathbb{T}} \\
\downarrow & & \downarrow & & \downarrow \\
\mathrm{THH}(R) & \xrightarrow{\varphi} & \mathrm{THH}(R)^{tC_p} & \longrightarrow & R^{tC_p},
\end{array}$$

where, on the right, we view R as an $\mathbb{E}_\infty$-ring with trivial $\mathbb{T}$-action, and where the right-hand horizontal maps both are induced by the unique extension $\mathrm{THH}(R) \to R$ of the identity map of R to a map of $\mathbb{E}_\infty$-rings with $\mathbb{T}$-action. Applying $\pi_0(-, \mathbb{Z}_p)$, we obtain the diagram of rings

$$\begin{array}{ccccc}
A_{\mathrm{inf}}(R) & \xrightarrow{\pi_0(\varphi^{h\mathbb{T}})} & A_{\mathrm{inf}}(R) & \xrightarrow{\theta} & R \\
\downarrow{\scriptstyle \theta} & & \downarrow & & \downarrow{\scriptstyle \mathrm{pr}} \\
R & \xrightarrow{\pi_0(\varphi)} & R/p & = & R/p,
\end{array}$$

where we identify the upper right-hand horizontal map by applying naturality of the edge homomorphism of the Tate spectral sequence to $\mathrm{THH}(R) \to R$. Now, it follows from the proof of Proposition 15.3.4 that the map $\pi_0(\varphi^{h\mathbb{T}})$ is uniquely determined by the map $\pi_0(\varphi)$. Moreover, the latter map is identified in [32, Corollary IV.2.4] to be the map that takes x to the class of x^p. We conclude that $\pi_0(\varphi^{h\mathbb{T}})$ is equal to the Frobenius $\varphi\colon A_{\mathrm{inf}}(R) \to A_{\mathrm{inf}}(R)$, since the latter makes the left-hand square commute. Finally, since

$$\mathrm{TC}^-_2(R, \mathbb{Z}_p) \xrightarrow{\varphi} \mathrm{TP}_2(R, \mathbb{Z}_p)$$

is an isomorphism, we have $\varphi(u) = \alpha \cdot \sigma$ with $\alpha \in A_{\mathrm{inf}}(R)$ a unit, and the relation $uv = \xi$ implies that $\varphi(v) = \alpha^{-1}\varphi(\xi) \cdot \sigma^{-1}$. □

15.3.2 Bott periodicity

We fix a field C that contains $\mathbb{Q}_p$ and that is both algebraically closed and complete with respect to a non-archimedean absolute value that extends the p-adic absolute value on $\mathbb{Q}_p$. The valuation ring

$$\mathcal{O}_C = \{x \in C \mid |x| \leq 1\} \subset C$$

is a perfectoid ring, and we proceed to evaluate its topological cyclic homology. We explain that this calculation, which uses Bökstedt periodicity, gives a purely p-adic proof of Bott periodicity.

The calculation uses a particular choice of generator ξ of the kernel of the map $\theta\colon A_{\mathrm{inf}}(\mathcal{O}_C) \to \mathcal{O}_C$, which we define first. We fix a generator

$$\varepsilon \in \mathbb{Z}_p(1) = T_p(C^*) = \mathrm{Hom}(\mathbb{Q}_p/\mathbb{Z}_p, C^*)$$

of the p-primary Tate module of C^*. It determines and is determined by the sequence $(1, \zeta_p, \zeta_{p^2}, \dots)$ of compatible primitive p-power roots of unity in C, where $\zeta_{p^n} = \varepsilon(1/p^n + \mathbb{Z}_p) \in C$. By abuse of notation, we also write

$$\varepsilon = (1, \zeta_p, \zeta_{p^2}, \dots) \in \mathcal{O}_{C^\flat}.$$

We now define the elements $\mu, \xi \in A_{\mathrm{inf}}(\mathcal{O}_C)$ by

$$\begin{aligned} \mu &= [\varepsilon] - 1 \\ \xi &= \mu/\varphi^{-1}(\mu) = ([\varepsilon] - 1)/([\varepsilon^{1/p}] - 1). \end{aligned}$$

The element ξ lies in the kernel of $\theta\colon A_{\mathrm{inf}}(\mathcal{O}_C) \to \mathcal{O}_C$, since

$$\theta(\xi) = \theta(\textstyle\sum_{0 \leq k < p} [\varepsilon^{1/p}]^k) = \sum_{0 \leq k < p} \zeta_p^k = 0,$$

and it is a generator. More generally, the element

$$\xi_r = \mu/\varphi^{-r}(\mu) = ([\varepsilon] - 1)/([\varepsilon^{1/p^r}] - 1)$$

generates the kernel of $\theta_r\colon A_{\mathrm{inf}}(\mathcal{O}_C) \to W_r(\mathcal{O}_C)$, and the element μ generates the kernel of the induced map[12]

$$A_{\mathrm{inf}}(\mathcal{O}_C) \xrightarrow{(\theta_r)} \lim_{r,R} W_r(\mathcal{O}_C) = W(\mathcal{O}_C).$$

Theorem 15.3.6. *Let C be a field extension of $\mathbb{Q}_p$ that is both algebraically closed and complete with respect to a non-archimedean absolute value that extends the p-adic absolute*

[12] Its cokernel $R^1 \lim_r \mathrm{Ker}(\theta_r)$ is a huge $A_{\mathrm{inf}}(\mathcal{O}_C)$-module that is almost zero.

value on $\mathbb{Q}_p$, and let $\mathcal{O}_C \subset C$ be the valuation ring. The canonical map of graded rings

$$\mathrm{Sym}_{\mathbb{Z}_p}(\mathrm{TC}_2(\mathcal{O}_C, \mathbb{Z}_p)) \longrightarrow \mathrm{TC}_*(\mathcal{O}_C, \mathbb{Z}_p)$$

is an isomorphism, and $\mathrm{TC}_2(\mathcal{O}_C, \mathbb{Z}_p)$ is a free $\mathbb{Z}_p$-module of rank 1. Moreover, the map $\mathrm{TC}_{2m}(\mathcal{O}_C, \mathbb{Z}_p) \to \mathrm{TC}^-_{2m}(\mathcal{O}_C, \mathbb{Z}_p)$ takes a $\mathbb{Z}_p$-module generator of the domain to $\varphi^{-1}(\mu)^m$ times an $A_{\mathrm{inf}}(\mathcal{O}_C)$-module generator of the target.

Proof. We let $\varepsilon \in T_p(C^*)$ and $\xi, \mu \in A_{\mathrm{inf}}(\mathcal{O}_C)$ be as above. According to Theorem 15.3.5, the even (resp. odd) p-adic homology groups of $\mathrm{TC}(\mathcal{O}_C)$ are given by the kernel (resp. cokernel) of the $\mathbb{Z}_p$-linear map

$$A_{\mathrm{inf}}(\mathcal{O}_C)[u, v]/(uv - \xi) \xrightarrow{\varphi - \mathrm{can}} A_{\mathrm{inf}}(\mathcal{O}_C)[\sigma^{\pm 1}]$$

given by

$$\begin{aligned} (\varphi - \mathrm{can})(a \cdot u^m) &= (\alpha^m \varphi(a) - \xi^m a) \cdot \sigma^m \\ (\varphi - \mathrm{can})(a \cdot v^m) &= (\alpha^{-m} \varphi(\xi)^m \varphi(a) - a) \cdot \sigma^{-m}, \end{aligned}$$

where $m \geq 0$ is an integer, and where $\alpha \in A_{\mathrm{inf}}(\mathcal{O}_C)$ is a fixed unit. We need only consider the top formula, since we know, for general reasons, that the p-adic homotopy groups of $\mathrm{TC}(\mathcal{O}_C)$ are concentrated in degrees ≥ -1.

We first prove an element $a \cdot u^m$ in the image of the map

$$\mathrm{TC}_{2m}(\mathcal{O}_C, \mathbb{Z}_p) \longrightarrow \mathrm{TC}^-_{2m}(\mathcal{O}_C, \mathbb{Z}_p)$$

is of the form $a = \varphi^{-1}(\mu)^m b$, for some $b \in A_{\mathrm{inf}}(\mathcal{O}_C)$. The element b is uniquely determined by a, since $\mathcal{O}_C$, and hence, $A_{\mathrm{inf}}(\mathcal{O}_C)$ is an integral domain. This image consists of the elements $a \cdot u^m$, where $a \in A_{\mathrm{inf}}(\mathcal{O}_C)$ satisfies

$$\alpha^m \varphi(a) = \xi^m a.$$

Rewriting this equation in the form

$$a = \varphi^{-1}(\alpha)^{-m} \varphi^{-1}(\xi)^m \varphi^{-1}(a),$$

we find inductively that for all $r \geq 1$,

$$a = \varphi^{-1}(\alpha_r)^{-m} \varphi^{-1}(\xi_r)^m \varphi^{-r}(a),$$

where $\alpha_r = \prod_{0 \leq i < r} \varphi^{-i}(\alpha)$ and $\xi_r = \prod_{0 \leq i < r} \varphi^{-i}(\xi)$. Therefore, we have

$$\varphi(a) \in \bigcap\nolimits_{r \geq 1} \xi_r^m A_{\mathrm{inf}}(\mathcal{O}_C) = \mu^m A_{\mathrm{inf}}(\mathcal{O}_C),$$

as desired. Moreover, the element $a \cdot u^m = \varphi^{-1}(\mu)^m b \cdot u^m$ belongs to the image of the map above if and only if $b \in A_{\mathrm{inf}}(\mathcal{O}_C)$ solves the equation

$$\alpha^m \varphi(b) = b.$$

Indeed, the elements $\mu, \xi \in A_{\mathrm{inf}}(\mathcal{O}_C)$ satisfy $\mu = \xi \varphi^{-1}(\mu)$.

To complete the proof, it suffices to show that the canonical map

$$\mathrm{Sym}_{\mathbb{Z}/p}(\mathrm{TC}_2(\mathcal{O}_C,\mathbb{Z}/p)) \longrightarrow \mathrm{TC}_*(\mathcal{O}_C,\mathbb{Z}/p)$$

is an isomorphism, and that the $\mathbb{Z}/p$-vector space $\mathrm{TC}_2(\mathcal{O}_C,\mathbb{Z}/p)$ has dimension 1. First, to show that $\mathrm{TC}_{2m-1}(\mathcal{O}_C,\mathbb{Z}/p)$ is zero, we must show that for every $c \in \mathcal{O}_{C^\flat}$, there exists $a \in \mathcal{O}_{C^\flat}$ such that

$$\alpha^m a^p - \xi^m a = c.$$

The ring $\mathcal{O}_{C^\flat}$ is complete with respect to the non-archimedean absolute value given by $|x|_{C^\flat} = |x^\#|_C$, and its quotient field $C^\flat$ is algebraically closed. Hence, there exists $a \in C^\flat$ that solves the equation in question, and we must show that $|a|_{C^\flat} \leq 1$. If $|a| = |a|_{C^\flat} > 1$, then

$$|\alpha^m a^p| = |a|^p > |a| \geq |\xi|^m |a| = |\xi^m a|,$$

and since $|\ \ |$ is non-archimedean, we conclude that

$$|\alpha^m a^p - \xi^m a| = |\alpha^m a^p| > 1 \geq |c|.$$

Hence, every solution $a \in C^\flat$ to the equation in question is in $\mathcal{O}_{C^\flat}$. This shows that the group $\mathrm{TC}_{2m-1}(\mathcal{O}_C,\mathbb{Z}/p)$ is zero.

Finally, we determine the $\mathbb{Z}/p$-vector space $\mathrm{TC}_{2m}(\mathcal{O}_C,\mathbb{Z}/p)$, which we have identified with the subspace of $\mathrm{TC}^-_{2m}(\mathcal{O}_C,\mathbb{Z}/p)$ that consists of the elements of the form $b\varphi^{-1}(\mu)^m \cdot u^m$, where $b \in \mathcal{O}_{C^\flat}$ satisfies the equation

$$\alpha^m b^p = b.$$

Since $C^\flat$ is algebraically closed, there are p solutions to this equations, namely, 0 and the $(p-1)$th roots of α^{-m}, all of which are units in $\mathcal{O}_{C^\flat}$. This shows that $\mathrm{TC}_{2m}(\mathcal{O}_C,\mathbb{Z}/p)$ is a $\mathbb{Z}/p$-vector space of dimension 1, for all $m \geq 0$. It remains only to notice that if b_1 and b_2 satisfy $\alpha^{m_1} b_1^p = b_1$ and $\alpha^{m_2} b_2^p = b_2$, respectively, then $b = b_1 b_2$ satisfies $\alpha^{m_1+m_2} b^p = b$, which shows that the map in the statement is an isomorphism. □

We thank Antieau–Mathew–Morrow for sharing the elegant proof of the following statement with us.

Lemma 15.3.7. *With notation as in Theorem 15.3.6, the map*

$$K(\mathcal{O}_C,\mathbb{Z}_p) \xrightarrow{j^*} K(C,\mathbb{Z}_p)$$

is an equivalence.

Proof. For every ring R, the category of coherent R-modules is abelian, and we define $K'(R)$ to be the algebraic K-theory of this abelian category. If R is coherent as an R-module, then the category of coherent R-modules contains the category of finite projective R-modules as a full exact subcategory. So in this situation the canonical inclusion induces a map of

K-theory spectra

$$K(R) \longrightarrow K'(R).$$

If, in addition, the ring R is of finite global dimension, then the resolution theorem [33, Theorem 3] shows that this map is an equivalence. In particular, this is so, if R is a valuation ring. For every finitely generated ideal in R is principal, so R is coherent, and it follows form [2] that R is of finite global dimension.

We now choose any pseudouniformizer $\pi \in \mathcal{O}_C$ and apply the localization theorem [33, Theorem 5] to the abelian category of coherent $\mathcal{O}_C$-modules and the full abelian subcategory of coherent $\mathcal{O}_C/\pi$-modules. The localization sequence then takes the form

$$K'(\mathcal{O}_C/\pi) \xrightarrow{i_*} K(\mathcal{O}_C) \xrightarrow{j^*} K(C),$$

since $\mathcal{O}_C$ and C both are valuation rings. In a similar manner, we obtain, for any pseudouniformizer $\pi^\flat \in \mathcal{O}_{C^\flat}$, the localization sequence

$$K'(\mathcal{O}_{C^\flat}/\pi^\flat) \xrightarrow{i_*} K(\mathcal{O}_{C^\flat}) \xrightarrow{j^*} K(C^\flat).$$

We may choose π and $\pi^\flat$ such that $\mathcal{O}_C/\pi$ and $\mathcal{O}_{C^\flat}/\pi^\flat$ are isomorphic rings, so we conclude that the lemma is equivalent to the statement that

$$K(\mathcal{O}_{C^\flat}, \mathbb{Z}_p) \xrightarrow{j^*} K(C^\flat, \mathbb{Z}_p)$$

is an equivalence. But $\mathcal{O}_{C^\flat}$ and $C^\flat$ are both perfect local $\mathbb{F}_p$-algebras, so the domain and target are both equivalent to $\mathbb{Z}_p$. Finally, the map in question is a map of $\mathbb{E}_\infty$-algebras in spectra, so it is necessarily an equivalence. □

Corollary 15.3.8 (Bott periodicity). *The canonical map of graded rings*

$$\mathrm{Sym}_{\mathbb{Z}}(K_2^{\mathrm{top}}(\mathbb{C})) \longrightarrow K_*^{\mathrm{top}}(\mathbb{C})$$

is an isomorphism, and $K_2^{\mathrm{top}}(\mathbb{C})$ is a free abelian group of rank 1.

Proof. The homotopy groups of $K^{\mathrm{top}}(\mathbb{C})$ are finitely generated[13], so it suffices to prove the analogous statements for the p-adic homotopy groups, for all prime numbers p. We fix p, let C be as in Theorem 15.3.6, and choose a ring homomorphism $f\colon \mathbb{C} \to C$. By Suslin [36, 37], the canonical maps

$$K^{\mathrm{top}}(\mathbb{C}) \longleftarrow K(\mathbb{C}) \xrightarrow{f^*} K(C)$$

become weak equivalences upon p-completion. Moreover, by Lemma 15.3.7, the map $j^*\colon K(\mathcal{O}_C) \to K(C)$ becomes a weak equivalence after p-completion. The ring $\mathcal{O}_C$ is a henselian local ring with algebraically closed residue field k of characteristic p. Therefore,

[13] This follows by a Serre class argument from the fact that the homology groups of the underlying space are finitely generated.

by Clausen–Mathew–Morrow [14], the cyclotomic trace map induces an equivalence

$$K(\mathcal{O}_C, \mathbb{Z}_p) \xrightarrow{\text{tr}} \mathrm{TC}(\mathcal{O}_C, \mathbb{Z}_p),$$

so the statement follows from Theorem 15.3.6. □

15.4 Group rings

Let G be a discrete group. We would like to understand the topological cyclic homology of the group ring $R[G]$, where R is a ring or, more generally, a connective $\mathbb{E}_1$-algebra in spectra. Since the assignment $G \mapsto \mathrm{TC}(R[G])$ is functorial in the 2-category of groups,[14] we get an "assembly" map

$$\mathrm{TC}(R) \otimes BG_+ \longrightarrow \mathrm{TC}(R[G]),$$

and what we will actually do here is to consider the cofiber of this map. By [27, Theorem 1.2], topological cyclic homology for a given group G can be assembled from the cyclic subgroups of G. We will focus on the case $G = C_p$ of a cyclic group of prime order p, but the methods that we present here can be generalized to the case of cyclic p-groups. We will be interested in the p-adic homotopy type of the cofiber. To this end, we remark that the p-completion of $\mathrm{THH}(R)$, which we denote by $\mathrm{THH}(R, \mathbb{Z}_p)$ as before, inherits a cyclotomic structure. For R connective, we have

$$\mathrm{TC}(R, \mathbb{Z}_p) \simeq \mathrm{TC}(\mathrm{THH}(R, \mathbb{Z}_p)).$$

The formula we give involves the non-trivial extension of groups

$$C_p \longrightarrow \mathbb{T}_p \longrightarrow \mathbb{T}.$$

The middle group $\mathbb{T}_p$ is a circle, but the right-hand map is a p-fold cover, and by restriction along this map, a spectrum with $\mathbb{T}$-action X gives rise to a spectrum with $\mathbb{T}_p$-action, which we also write X.

Theorem 15.4.1. *For a connective $\mathbb{E}_1$-algebra R in spectra, there is a natural cofiber sequence of spectra*

$$\mathrm{TC}(R, \mathbb{Z}_p) \otimes BC_{p+} \longrightarrow \mathrm{TC}(R[C_p], \mathbb{Z}_p) \longrightarrow \mathrm{THH}(R, \mathbb{Z}_p)_{h\mathbb{T}_p}[1] \otimes C_p,$$

where C_p is considered as a pointed set with basepoint 1.

We note that the right-hand term in the sequence above is non-canonically equivalent to a $(p-1)$-fold sum of copies of $\mathrm{THH}(R, \mathbb{Z}_p)_{h\mathbb{T}_p}[1]$. We do not determine the boundary map in the sequence. The proof of Theorem 15.4.1 requires some preparation and preliminary

[14] By the latter we mean the full subcategory of the ∞-category of spaces consisting of spaces of the form BG. Concretely objects are groups, morphisms are group homomorphisms and 2-morphisms are conjugations.

results. First, we recall that for a spherical group ring $\mathbb{S}[G]$, there is a natural equivalence

$$\mathrm{THH}(\mathbb{S}[G]) \simeq \mathbb{S} \otimes LBG_+,$$

where $LBG = \mathrm{Map}(\mathbb{T}, BG)$ is the free loop space. Moreover, the equivalence is $\mathbb{T}$-equivariant for the $\mathbb{T}$-action on LBG induced from the action of $\mathbb{T}$ on itself by multiplication. Hence, for general R, we have

$$\mathrm{THH}(R[G]) \simeq \mathrm{THH}(R) \otimes LBG_+,$$

where $\mathbb{T}$ acts diagonally on the right-hand side. Since G is discrete, we further have a $\mathbb{T}$-equivariant decomposition of spaces

$$LBG \simeq \coprod BC_G(x),$$

where x ranges over a set of representatives of the conjugacy classes of elements of G, and where $C_G(x) \subset G$ is the centralizer of $x \in G$. The action by $\mathbb{T} \simeq B\mathbb{Z}$ on $BC_G(x)$ is given by the map $B\mathbb{Z} \times BC_G(x) \to BC_G(x)$ induced by the group homomorphism $\mathbb{Z} \times C_G(x) \to C_G(x)$ that to (n, g) assigns $x^n g$. Specializing to the case $G = C_p$, we get the following description.

Lemma 15.4.2. *There is a natural* $\mathbb{T}$*-equivariant cofiber sequence of spaces*

$$BC_p^{\mathrm{triv}} \to LBC_p \to (BC_p^{\mathrm{res}})_+ \otimes C_p,$$

where BC_p^{triv} *has the trivial* $\mathbb{T}$*-action and* $BC_p^{\mathrm{res}} \simeq (\mathrm{pt})_{hC_p}$ *has the residual action by* $\mathbb{T} = \mathbb{T}_p/C_p$.[15]

As a consequence, we obtain for every $\mathbb{E}_1$-algebra in spectra R, a cofiber sequence of spectra with $\mathbb{T}$-action

$$\mathrm{THH}(R) \otimes BC_{p+}^{\mathrm{triv}} \longrightarrow \mathrm{THH}(R[G]) \longrightarrow (\mathrm{THH}(R) \otimes BC_{p+}^{\mathrm{res}}) \otimes C_p,$$

where the left-hand map is the assembly map. To determine the cyclotomic structure on the terms of this sequence, we prove the following result.

Lemma 15.4.3. *Let* X *be a spectrum with* $\mathbb{T}$*-action that is bounded below, and let* $\mathbb{T}$ *act diagonally on* $X \otimes BC_{p+}^{\mathrm{res}}$*. Then* $(X \otimes BC_{p+}^{\mathrm{res}})^{tC_p} \simeq 0$.

Proof. We write $X \simeq \lim_n \tau_{\leq n} X$ as the limit of its Postnikov tower. The spectra $\tau_{\leq n} X$ inherit a $\mathbb{T}$-action, and the equivalence is $\mathbb{T}$-equivariant. The map induced by the canonical projections,

$$X \otimes BC_{p+}^{\mathrm{res}} \longrightarrow \lim_n(\tau_{\leq n} X \otimes BC_{p+}),$$

is an equivalence, since the connectivity of the fibers tend to infinity with n, and therefore, also the map

$$(X \otimes BC_{p+}^{\mathrm{res}})^{tC_p} \longrightarrow \lim_n((\tau_{\leq n} X \otimes BC_{p+})^{tC_p})$$

[15] This comes from the fact that pt carries a (necessarily trivial) $\mathbb{T}_p$-action. In a point set model, it can be described as $E\mathbb{T}_p/C_p$.

is an equivalence. Indeed, the analogous statements for homotopy fixed points and homotopy orbits is respectively clear and a consequence of the fact that the connectivity of the fibers tend to infinity with n.

Since X is bounded below, we may assume that X is concentrated in a single degree with necessarily trivial $\mathbb{T}$-action. As spectra with $\mathbb{T}$-action,

$$X \otimes BC_{p+}^{\mathrm{res}} \simeq X_{hC_p},$$

where the right-hand side has the residual $\mathbb{T}$-action. But $(X_{hC_p})^{tC_p} \simeq 0$ by the Tate orbit lemma [32, Lemma I.2.1], so the lemma follows. □

Corollary 15.4.4. *Let X be a spectrum with $\mathbb{T}$-action that is bounded below and p-complete. The spectrum $X \otimes BC_{p+}^{\mathrm{res}}$ with the diagonal $\mathbb{T}$-action admits a unique cyclotomic structure, and, with respect to this structure,*

$$\mathrm{TC}(X \otimes BC_{p+}^{\mathrm{res}}) \simeq X_{h\mathbb{T}_p}[1].$$

Proof. A cyclotomic structure on a spectrum with $\mathbb{T}$-action Y consists of a family of $\mathbb{T}$-equivariant maps $\varphi_\ell \colon Y \to Y^{tC_\ell}$, one for every prime number ℓ, including p. In the case of $Y = X \otimes BC_p^{\mathrm{res}}$, the target of this map is contractible for all ℓ. Indeed, for $\ell \neq p$, this follows from Y being p-complete, and for $\ell = p$, it follows from Lemma 15.4.3. Hence, there is a unique such family of maps. In order to evaluate TC, we first note that Lemma 15.4.3 also implies that

$$\mathrm{TP}(X \otimes BC_{p+}^{\mathrm{res}}) \simeq \mathrm{TP}(X \otimes BC_{p+}^{\mathrm{res}})^\wedge \simeq 0.$$

Accordingly,

$$\mathrm{TC}(X \otimes BC_{p+}^{\mathrm{res}}) \simeq \mathrm{TC}^-(X \otimes BC_{p+}^{\mathrm{res}}) = (X \otimes BC_{p+}^{\mathrm{res}})^{h\mathbb{T}},$$

and by the vanishing of $(-)^{t\mathbb{T}}$, this is further equivalent to

$$(X \otimes BC_{p+}^{\mathrm{res}})_{h\mathbb{T}}[1] \simeq (X_{hC_p})_{h\mathbb{T}}[1] \simeq X_{h\mathbb{T}_p}[1],$$

where, in the middle term, $C_p \subset \mathbb{T}_p$ acts trivially on X. □

Proof of Theorem 15.4.1. We have proved that there is a cofiber sequence

$$\mathrm{THH}(R) \otimes BC_{p+}^{\mathrm{triv}} \longrightarrow \mathrm{THH}(R[C_p]) \longrightarrow \mathrm{THH}(R) \otimes BC_{p+}^{\mathrm{res}} \otimes C_p$$

of spectra with $\mathbb{T}$-action in which the left-hand map is the assembly map. We get an induced fiber sequence with p-adic coefficients. Since the forgetful functor from cyclotomic spectra to spectra with $\mathbb{T}$-action creates colimits, this map is also the assembly map in the ∞-category of cyclotomic spectra, and by Corollary 15.4.4, the induced cyclotomic structure on cofiber necessarily is the unique cyclotomic structure, for which

$$\mathrm{TC}(\mathrm{THH}(R, \mathbb{Z}_p) \otimes BC_{p+}^{\mathrm{res}} \otimes C_p) \simeq \mathrm{THH}(R, \mathbb{Z}_p)_{h\mathbb{T}_p}[1] \otimes C_p.$$

Finally, by the universal property of the colimit, there is a canonical map

$$\mathrm{TC}(R, \mathbb{Z}_p) \otimes BC_p \longrightarrow \mathrm{TC}(\mathrm{THH}(R, \mathbb{Z}_p) \otimes BC_p^{\mathrm{triv}}),$$

and by [14, Theorem 2.7], this map is an equivalence. □

Example 15.4.5. For $R = \mathbb{S}$, the sequence in Theorem 15.4.1 becomes

$$\mathrm{TC}(\mathbb{S}, \mathbb{Z}_p) \otimes BC_{p+} \longrightarrow \mathrm{TC}(\mathbb{S}[C_p], \mathbb{Z}_p) \longrightarrow (\mathbb{S}_p \otimes B\mathbb{T}_{p+})[1] \otimes C_p,$$

where $B\mathbb{T}_p \simeq \mathbb{P}^\infty(\mathbb{C})$. One can also give a formula for $\mathrm{TC}(\mathbb{S}[C_p], \mathbb{Z}_p)$ in this case, but this is more complicated than the formula for the cofiber of the assembly map.

Finally, we evaluate the homotopy groups of $\mathrm{THH}(R, \mathbb{Z}_p)_{h\mathbb{T}_p}$ in the case, where R is a p-torsion free perfectoid ring. We consider the diagram

$$\begin{array}{ccc} \pi_0(\mathrm{THH}(R, \mathbb{Z}_p)^{h\mathbb{T}}) & \longrightarrow & \pi_0(\mathrm{THH}(R, \mathbb{Z}_p)^{h\mathbb{T}_p}) \\ \downarrow & & \downarrow \\ \pi_0(\mathrm{THH}(R, \mathbb{Q}_p)^{h\mathbb{T}}) & \longrightarrow & \pi_0(\mathrm{THH}(R, \mathbb{Q}_p)^{h\mathbb{T}_p}), \end{array}$$

where the horizontal maps are given by restriction along $\mathbb{T}_p \to \mathbb{T}$, and where the vertical maps are given by change-of-coefficients. The respective homotopy fixed point spectral sequences endow each of the four rings with a descending filtration, which we refer to as the Nygaard filtration, and they are all complete and separated in the topology. We have identified the top left-hand ring with Fontaine's ring $A = A_{\mathrm{inf}}(R)$. The lower horizontal map is an isomorphism, and the common ring is identified with $A[1/p]^\wedge$, where "$(-)^\wedge$" indicates Nygaard completion. We further have compatible edge homomorphisms

$$\begin{array}{ccc} \pi_0(\mathrm{THH}(R, \mathbb{Z}_p)^{h\mathbb{T}}) & \longrightarrow & \pi_0(\mathrm{THH}(R, \mathbb{Z}_p)^{h\mathbb{T}_p}) \\ \downarrow{\scriptstyle \theta} & & \downarrow{\scriptstyle \theta_p} \\ \mathrm{THH}_0(R, \mathbb{Z}_p) & = & \mathrm{THH}_0(R, \mathbb{Z}_p), \end{array}$$

whose kernels $I \subset A$ and $I_p \subset A_p$ are principal ideals, and we can choose generators ξ and ξ_p such that the top horizontal map takes ξ to $p\xi_p$. This identifies the top right-hand ring A_p with the subring

$$A_p = (\textstyle\sum_{n \geq 0} p^{-n} I^n)^\wedge \subset A[1/p]^\wedge$$

given by the Nygaard completion of the Rees construction.

Proposition 15.4.6. *Let R be a p-torsion free perfectoid ring. The map*

$$\pi_*(\mathrm{THH}(R, \mathbb{Z}_p)^{h\mathbb{T}_p}) \xrightarrow{\mathrm{can}} \pi_*(\mathrm{THH}(R, \mathbb{Z}_p)^{t\mathbb{T}_p})$$

is given by the localization of graded A_p-algebras

$$A_p[u, v_p]/(uv_p - \xi_p) \longrightarrow A_p[u, v_p^{\pm 1}]/(uv_p - \xi_p) = A_p[v_p^{\pm 1}],$$

where u and v_p are homogeneous elements of degree 2 and -2, and where ξ_p is a generator of the kernel I_p of the edge homomorphism $\theta_p \colon A_p \to R$.

Proof. The proof is analogous to the proof of Theorem 15.3.5. □

Corollary 15.4.7. *Let R be a p-torsion free perfectoid ring. For $m \geq 0$,*

$$\pi_{2m}(\mathrm{THH}(R,\mathbb{Z}_p)_{h\mathbb{T}_p}) = A_p/I_p^{m+1} \cdot v_p^{-(m+1)},$$

and the remaining homotopy groups are zero.

Remark 15.4.8. It is interesting to compare the calculation above to the case, where R is a perfect $\mathbb{F}_p$-algebra. In this case, the map

$$\pi_0(\mathrm{THH}(R,\mathbb{Z}_p)^{h\mathbb{T}}) \longrightarrow \pi_0(\mathrm{THH}(R,\mathbb{Z}_p)^{h\mathbb{T}_p})$$

takes $uv - p = 0$ to $(uv_p - 1)p = 0$, and since $1 - uv_p$ is a unit, we conclude that, in the target ring, $p = 0$. Hence, this ring is a power series ring $R[[y]]$ on a generator y that is represented by $t_p x$ in the spectral sequence

$$E^2 = R[t_p, x] \Rightarrow \pi_*(\mathrm{THH}(R,\mathbb{Z}_p)^{h\mathbb{T}_p}).$$

Hence, for $m \geq 0$, we have

$$\pi_{2m}(\mathrm{THH}(R,\mathbb{Z}_p)_{h\mathbb{T}_p}) = R[[y]]/y^{m+1} \cdot v_p^{-(m+1)}$$

and the remaining homotopy groups are zero.

We also consider the case of the semi-direct product $G = \mathrm{Aut}(C_p) \ltimes C_p$. The conjugacy classes of elements in G are represented by the elements $(\mathrm{id}, 0)$, $(\mathrm{id}, 1)$, and $(\alpha, 0)$ with $\alpha \in \mathrm{Aut}(C_p) \smallsetminus \{\mathrm{id}\}$, the centralizers of which are G, C_p, and $\mathrm{Aut}(C_p)$, respectively. Hence, for every $\mathbb{E}_1$-algebra in spectra R, there is a cofiber sequence of spectra with $\mathbb{T}$-action

$$\begin{aligned} &\mathrm{THH}(R) \otimes BG_+^{\mathrm{triv}} \longrightarrow \mathrm{THH}(R[G]) \\ &\longrightarrow \mathrm{THH}(R) \otimes BC_{p+}^{\mathrm{res}} \oplus \mathrm{THH}(R) \otimes B\,\mathrm{Aut}(C_p)_+ \otimes \mathrm{Aut}(C_p), \end{aligned}$$

where the last tensor factor $\mathrm{Aut}(C_p)$ is viewed as a pointed space with id as basepoint. Moreover, as cyclotomic spectra, the right-hand summand splits off. Therefore, after p-completion, we arrive at the following statement.

Theorem 15.4.9. *Let $G = \mathrm{Aut}(C_p) \ltimes C_p$, and let R be a connective $\mathbb{E}_1$-ring. There is a canonical fiber sequence of spectra*

$$\begin{aligned} &\mathrm{TC}(R,\mathbb{Z}_p) \otimes BG_+ \longrightarrow \mathrm{TC}(R[G],\mathbb{Z}_p) \\ &\qquad\longrightarrow \mathrm{THH}(R,\mathbb{Z}_p)_{h\mathbb{T}_p}[1] \oplus \mathrm{TC}(R,\mathbb{Z}_p) \otimes \mathrm{Aut}(C_p), \end{aligned}$$

and moreover, the summand $\mathrm{TC}(R,\mathbb{Z}_p) \otimes \mathrm{Aut}(C_p)$ splits off $\mathrm{TC}(R[G],\mathbb{Z}_p)$.

Bibliography

[1] V. Angeltveit and J. Rognes. Hopf algebra structure on topological Hochschild homology. *Algebr. Geom. Topol.*, 5:1223–1290, 2005.

[2] I. Berstein. On the dimension of modules and algebras. IX. Direct limits. *Nagoya Math. J.*, 13:83–84, 1958.

[3] P. Berthelot. *Cohomologie Cristalline des Schemas de Caracteristique* $p > 0$, volume 407 of *Lecture Notes in Math.* Springer-Verlag,1974.

[4] B. Bhatt, M. Morrow, and P. Scholze. Integral p-adic hodge theory. Publ. Math. Inst. Hautes Études Sci. 128 (2018), 219–397.

[5] B. Bhatt, M. Morrow, and P. Scholze. Topological Hochschild homology and integral p-adic hodge theory. arXiv:1802.03261, 2018.

[6] B. Bhatt and P. Scholze. Prisms and prismatic cohomology. arXiv:1905.08229, 2019.

[7] A. J. Blumberg, D. Gepner, and G. Tabuada. A universal characterization of higher algebraic K-theory. *Geom. Topol.*, 17:733–838, 2013.

[8] A. J. Blumberg and M. A. Mandell. The homotopy theory of cyclotomic spectra. *Geom. Topol.*, 19:3105–3147, 2015.

[9] M. Bökstedt. Topological Hochschild homology. Preprint, Universität Bielefeld, 1985.

[10] M. Bökstedt. Topological Hochschild homology of $\mathbb{Z}$ and $\mathbb{Z}/p$. Preprint, Universität Bielefeld, 1985.

[11] M. Bökstedt, W.-C. Hsiang, and I. Madsen. The cyclotomic trace and algebraic K-theory of spaces. *Invent. Math.*, 111:465–540, 1993.

[12] A. K. Bousfield. The localization of spectra with respect to homology. *Topology*, 18:257–281, 1979.

[13] H. Cartan. Constructions multiplicatives. In *Séminaire Henri Cartan, Vol. 7 (1954–1955)*. Exp. 4, Secrétariat mathématique, 11 rue Pierre Curie, Paris, 1956.

[14] D. Clausen, A. Mathew, and M. Morrow. K-theory and topological cyclic homology of henselian pairs. arXiv:1803.10897, 2018.

[15] G. Cortiñas. The obstruction to excision in K-theory and in cyclic homology. *Invent. Math.*, 164:143–173, 2006.

[16] B. I. Dundas, T. G. Goodwillie, and R. McCarthy. *The local structure of algebraic K-theory*, volume 18 of *Algebr. Appl.* Springer-Verlag, 2013.

[17] E. Dyer and R. K. Lashof. Homology of iterated loop spaces. *Amer. J. Math.*, 84:35–88, 1962.

[18] L. Fargues. La courbe. In *Proc. Int. Cong. of Math.*, volume 1, pages 261–290, Rio de Janeiro (2018).

[19] J.-M. Fontaine. Le corps des périodes p-adiques. With an appendix by P. Colmez. In *Périodes p-adiques* (Bures-sur-Yvette, 1988), volume 223 of *Astérisque*, pages 59–111, 1994.

[20] O. Gabber. K-theory of Henselian local rings and Henselian pairs. In *Algebraic K-theory, commutative algebra, and algebraic geometry (Santa Margherita Ligure, 1989)*, volume 126 of *Contemp. Math.*, pages 59–70, 1992. Amer. Math. Soc.

[21] T. Geisser and L. Hesselholt. Bi-relative algebraic K-theory and topological cyclic homology. *Invent. Math.*, 166:359–395, 2006.

[22] L. Hesselholt and I. Madsen. On the K-theory of finite algebras over Witt vectors of perfect fields. *Topology*, 36:29–102, 1997.

[23] L. Hesselholt and T. Nikolaus. Algebraic K-theory of planar cuspical curves. arXiv:1903.08295, 2019.

[24] E. Höning. Bökstedt's computation of $\mathrm{THH}(\mathbb{F}_p)$. In *Arbeitsgemeinschaft: Topological Cyclic Homology*, volume 15 of *Oberwolfach Rep.*, 2018.

[25] T. Kudo and S. Araki. Topology of H_n-spaces and H-squaring operations. *Mem. Fac. Sci. Kyusyu Univ. Ser. A.*, 10:85–120, 1956.

[26] M. Land and G. Tamme. On the K-theory of pullbacks. arXiv:1808.05559, 2018.

[27] W. Lück, H. Reich, J. Rognes, and M. Varisco. Algebraic K-theory of group rings and the cyclotomic trace map. *Adv. Math.*, 304:930–1020, 2017.

[28] J. Lurie. *Higher topos theory*, *Ann. of Math. Stud.* 170. Princeton University Press, 2009.

[29] J. W. Milnor. The Steenrod algebra and its dual. *Ann. of Math.*, 67:150–171, 1958.

[30] T. Nikolaus. On topological cyclic homology and cyclic K-theory. In preparation.

[31] T. Nikolaus. Topological cyclic homology and cyclic K-theory. In *Arbeitsgemeinschaft: Topological Cyclic Homology*, volume 15 of *Oberwolfach Rep.*, 2018.

[32] T. Nikolaus and P. Scholze. On topological cyclic homology. *Acta Math.*, 221:203–409, 2018.

[33] D. Quillen. Higher algebraic K-theory I. In *Algebraic K-theory I: Higher K-theories, Battelle Memorial Inst., Seattle, Washington, 1972*, volume 341 of *Lecture Notes in Math.*, pages 85–147, Springer-Verlag, 1973.

[34] M. Speirs. On the K-theory of truncated polynomial algebras, revisited. arXiv:1901.10602, 2019.

[35] M. Steinberger. Homology operations for H_∞ and H_n ring spectra. In H_∞*-ring spectra and their applications*, volume 1176 of *Lecture Notes in Math.*, chapter III, pages 56–87. Springer-Verlag, 1986.

[36] A. A. Suslin. On the K-theory of algebraically closed fields. *Invent. Math.*, 73:241–245, 1983.

[37] A. A. Suslin. On the K-theory of local fields. *J. Pure Appl. Alg.*, 34:304–318, 1984.

[38] G. Wang. Dyer-Lashof operations. In *Arbeitsgemeinschaft: Topological Cyclic Homology*, volume 15 of *Oberwolfach Rep.*, 2018.

Nagoya University, Nagoya, Japan, and University of Copenhagen, 1017 Copenhagen K, Denmark

E-mail address: larsh@math.nagoya-u.ac.jp

Universitat Münster, 48149 Münster, Germany

E-mail address: nikolaus@uni-muenster.de

16

Lie algebra models for unstable homotopy theory

Gijs Heuts

16.1 Introduction

Write $\mathrm{Ho}(\mathcal{S}_*)$ for the homotopy category of pointed spaces. The tools of algebraic topology often amount to studying spaces by applying a variety of functors

$$F : \mathrm{Ho}(\mathcal{S}_*) \to \mathrm{Ho}(\mathcal{A}),$$

for $\mathcal{A}$ a homotopy theory that is 'algebraic' in nature. A typical example of such an F is the functor $C^*(- : R)$ taking cochains with values in a commutative ring R. In this case $\mathcal{A}$ can be taken to be (the opposite of) the category of differential graded algebras over R, using the cup product of cochains as multiplication. If one wishes to take the rather subtle commutativity properties of the cup product into account, a more refined choice for $\mathcal{A}$ would be the category of $\mathbf{E}_\infty$-algebras over R.

An optimist might hope to choose $\mathcal{A}$ cleverly enough so that one can construct a functor F that is an equivalence of categories. If one restricts the domain of F to the homotopy category $\mathrm{Ho}(\mathcal{S}_{\mathbb{Q}}^{\geq 2})$ of simply connected rational pointed spaces, then there is the following landmark result [69]:

Theorem 16.1.1 (Quillen [69]). *There are equivalences of categories*

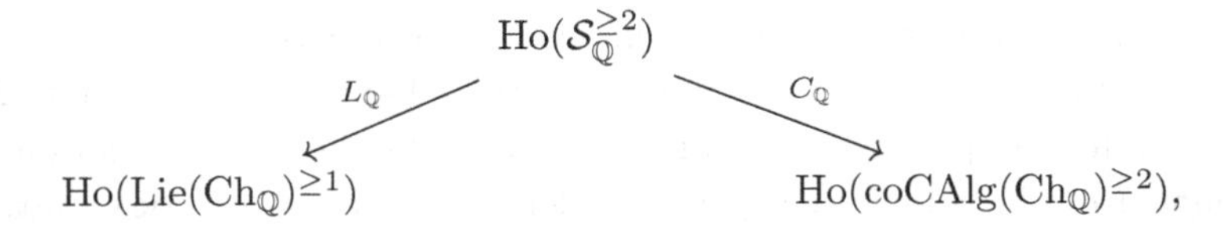

where $\mathrm{Lie}(\mathrm{Ch}_{\mathbb{Q}})^{\geq 1}$ *denotes the category of connected differential graded Lie algebras over* $\mathbb{Q}$ *and* $\mathrm{coCAlg}(\mathrm{Ch}_{\mathbb{Q}})^{\geq 2}$ *the category of simply connected differential graded cocommutative coalgebras over* $\mathbb{Q}$.

The functor $C_{\mathbb{Q}}$ is essentially a version of the rational chains; in particular, the homology of the underlying chain complex of $C_{\mathbb{Q}}(X)$ is the rational homology of the space X and the coproduct is the usual one arising from the diagonal map of X. The functor $L_{\mathbb{Q}}$ is perhaps more surprising. It refines the collection of rational homotopy groups of X, in the sense that

Mathematics Subject Classification. 55Q51, 55Q15, 55P43, 55P60, 55P62.

Key words and phrases. periodic homotopy groups, chromatic homotopy theory, spectral Lie algebras, rational homotopy theory.

there is a natural isomorphism

$$H_*(L_{\mathbb{Q}}(X)) \cong (\pi_{*+1}X) \otimes \mathbb{Q}.$$

The induced Lie bracket on $H_*(L_{\mathbb{Q}}(X))$ corresponds to the classical Whitehead product on the (rational) homotopy groups of X.

Theorem 16.1.1 is a very satisfying result in itself. In conjunction with Sullivan's approach to rational homotopy theory using minimal models [76] it has led to many interesting developments within homotopy theory, as well as striking applications to geometry (see [34, 35] for an overview). Moreover, it raises a number of natural questions. Are there subcategories of $\mathrm{Ho}(\mathcal{S}_*)$ other than that of rational spaces which can be described in terms of Lie algebras and/or coalgebras? Even better, does the entire category $\mathrm{Ho}(\mathcal{S}_*)$ admit such an algebraic model? This survey mostly concerns an answer to the first question, although we include some discussion of the second one at the very end of Section 16.8.

Rationalization is the first step in a hierarchy of localizations of $\mathcal{S}_*$, which we will refer to as the *v_n-periodic localizations.* For each prime p there is a family of these, indexed by the natural numbers $n \geq 0$. The case $n = 0$ is precisely rational homotopy theory. The v_n-periodic localizations are in a precise sense the most elementary (or 'prime') localizations of homotopy theory. Our main objective will be to discuss the following generalization of Quillen's Theorem 16.1.1 (see [42]):

Theorem 16.1.2. *There is an equivalence of homotopy theories*

$$L_{v_n} : \mathcal{S}_{v_n} \longrightarrow \mathrm{Lie}(\mathrm{Sp}_{v_n}),$$

with $\mathcal{S}_{v_n}$ denoting the v_n-periodic localization of $\mathcal{S}_*$ and $\mathrm{Lie}(\mathrm{Sp}_{v_n})$ the v_n-periodic localization of the homotopy theory of spectral Lie algebras*.*

Explaining this result will of course require a discussion of spectral Lie algebras, a concept that has only recently surfaced. It has several promising applications within homotopy theory and interesting connections to derived algebraic geometry.

We now give an overview of the contents of this chapter. In Section 16.2 we review Quillen's rational homotopy theory. In particular, we will highlight the relation between the Lie algebra model $L_{\mathbb{Q}}(X)$ and the coalgebra model $C_{\mathbb{Q}}(X)$, which can be expressed in terms of Koszul duality. Since it plays an important role in this survey, we devote Section 16.4 to a general discussion of this duality. It is preceded by Section 16.3, which gives a short summary of monads and their algebras. In Section 16.5 we discuss spectral Lie algebras and their connection with the Goodwillie tower of the identity functor of $\mathcal{S}_*$. In Section 16.6 we turn to periodic phenomena in homotopy theory. We describe the general philosophy of stable chromatic homotopy theory, in analogy with the theory of localizations in algebra, and its application to unstable homotopy theory through the Bousfield–Kuhn functors. In Section 16.7 we come to Theorem 16.1.2. The reader might note the absence of coalgebras in its statement. There does exist a coalgebra model for $\mathcal{S}_{v_n}$, but unlike the rational case it turns out that the Lie algebra model plays a preferred role for $n > 0$. Loosely speaking, coalgebras can only be used to describe a certain *completion* of $\mathcal{S}_{v_n}$, rather than this homotopy theory itself. We will discuss this point in Section 16.7 as well. In the final Section 16.8 we discuss examples and several open questions which deserve further investigation.

We stress that the style of this survey is rather informal. To counter this we include many pointers to the literature, where most of our statements and their proofs can be found with a more technically precise treatment. Still, we provide sketches of the proofs of most major results (sometimes from a non-standard perspective), hoping to make this chapter self-contained enough to convey to the reader all of the essential ideas involved. The reader might also be interested in the survey of Behrens–Rezk [12], which concerns some of the same topics we describe here.

Conventions and notation. Thus far we have not been specific about the formalism for homotopy theory we use. Most statements in this survey are sufficiently generic for the reader to adapt to their preferred setting. However, for convenience we work in the setting of ∞-categories [47, 61]. This Handbook also includes an expository account by Groth. Thus $\mathcal{S}_*$ and Sp will denote the ∞-categories of pointed spaces and of spectra respectively. We write $\mathcal{S}_*^{\geq n}$ for the full subcategory of $\mathcal{S}_*$ on the $(n-1)$-connected spaces. We will use the standard notation

$$\mathcal{S}_* \underset{\Omega^\infty}{\overset{\Sigma^\infty}{\rightleftarrows}} \mathrm{Sp}$$

for the adjunction between the suspension spectrum Σ^∞ and the zeroth space Ω^∞. All limits and colimits are to be thought of as ∞-categorical ones (unless explicitly stated otherwise); in other words, they are *homotopy* limits and colimits. For ∞-categories $\mathcal{C}$ and $\mathcal{D}$ we denote by $\mathrm{Fun}(\mathcal{C}, \mathcal{D})$ the ∞-category of functors between them. Our reference for algebra in the setting of ∞-categories is [62]. When discussing coalgebras we will often use the adjective *commutative* (rather than cocommutative) when no confusion can arise, i.e., when we are not simultaneously considering an algebra structure.

16.2 Rational homotopy theory

The results of Quillen and Sullivan on rational homotopy theory have been extensively documented [69, 76, 20, 34, 59]; we will therefore feel free to give a somewhat non-traditional exposition with an eye towards the results of later sections.

A map $f\colon X \to Y$ of simply connected pointed spaces is a *rational equivalence* if it induces an isomorphism on rational homotopy groups, i.e., if for every $n \geq 2$ the map

$$\pi_n f \otimes \mathbb{Q}\colon \pi_n X \otimes \mathbb{Q} \to \pi_n Y \otimes \mathbb{Q}$$

is an isomorphism. Equivalently, f induces an isomorphism on rational homology. A simply connected pointed space X is said to be *rational* if for each $n \geq 2$ the abelian group $\pi_n X$ is uniquely divisible, i.e., is a vector space over $\mathbb{Q}$. Equivalently, each homology group $H_n(X; \mathbb{Z})$ is a rational vector space (with $n \geq 2$ again). We write $\mathcal{S}_{\mathbb{Q}}^{\geq 2}$ for the full subcategory of $\mathcal{S}_*^{\geq 2}$ on the rational spaces.

Every simply connected space admits a *rationalization* $\eta_X : X \to X_{\mathbb{Q}}$, which is a map satisfying the following two properties:

(1) The space $X_{\mathbb{Q}}$ is rational.

(2) The map η_X is a rational equivalence.

Moreover, the assignment $X \mapsto X_{\mathbb{Q}}$ can be made functorial and η_X can be made a natural transformation. The following terminology is convenient to describe the situation:

Definition 16.2.1. Suppose $\mathcal{C}$ is an ∞-category and W a class of morphisms in $\mathcal{C}$. A functor $L\colon \mathcal{C} \to \mathcal{D}$ is called a *localization of* $\mathcal{C}$ *at* W if L sends each morphism in W to an equivalence in $\mathcal{D}$ and is universal with this property. More precisely, if $\mathcal{E}$ is any other ∞-category then precomposition by L determines an equivalence of ∞-categories

$$L^* : \mathrm{Fun}(\mathcal{D}, \mathcal{E}) \to \mathrm{Fun}_W(\mathcal{C}, \mathcal{E}),$$

where the codomain denotes the full subcategory of $\mathrm{Fun}(\mathcal{C}, \mathcal{E})$ of functors sending elements of W to equivalences. A localization $L\colon \mathcal{C} \to \mathcal{D}$ is called *reflective* if L admits a fully faithful right adjoint.

Thus, a localization of $\mathcal{C}$ at W is the universal solution to inverting the elements of W. If the localization is reflective, then $\mathcal{D}$ can be thought of as a full subcategory of $\mathcal{C}$. Such reflective localizations are closely related to left Bousfield localizations in the context of model categories.

In the setting of interest to us here, the existence of rationalizations described above implies that the functor

$$\mathcal{S}_*^{\geq 2} \to \mathcal{S}_{\mathbb{Q}}^{\geq 2} : X \mapsto X_{\mathbb{Q}}$$

is a reflective localization of $\mathcal{S}_*^{\geq 2}$ at the class of rational equivalences. An entirely analogous procedure produces the localization

$$\mathrm{Sp} \to \mathrm{Sp}_{\mathbb{Q}} : E \mapsto E_{\mathbb{Q}}$$

of the ∞-category of spectra at the rational equivalences. One can explicitly identify the latter functor as $E_{\mathbb{Q}} \simeq H\mathbb{Q} \wedge E$. The ∞-category $\mathrm{Sp}_{\mathbb{Q}}$ is in fact equivalent to the ∞-category $\mathrm{Ch}_{\mathbb{Q}}$ of rational chain complexes and this equivalence respects the usual symmetric monoidal structures (cf. [75] and Theorem 7.1.2.13 of [62]). Under these identifications, the rational spectrum $(\Sigma^\infty X)_{\mathbb{Q}}$ corresponds (up to natural equivalence) to the reduced rational chains $\widetilde{C}_*(X; \mathbb{Q})$. Applying the universal property of $\mathcal{S}_{\mathbb{Q}}^{\geq 2}$, the functor

$$\mathcal{S}_*^{\geq 2} \to \mathrm{Sp}_{\mathbb{Q}} : X \mapsto (\Sigma^\infty X)_{\mathbb{Q}}$$

factors over a functor for which we write

$$\Sigma^\infty_{\mathbb{Q}} \colon \mathcal{S}_{\mathbb{Q}}^{\geq 2} \to \mathrm{Sp}_{\mathbb{Q}}.$$

We will not drag the equivalence between $\mathrm{Sp}_{\mathbb{Q}}$ and $\mathrm{Ch}_{\mathbb{Q}}$ along and use $\Sigma^\infty_{\mathbb{Q}}$ as our version of the (reduced) rational chains functor.

The ∞-category $\mathcal{S}_{\mathbb{Q}}^{\geq 2}$ carries two evident symmetric monoidal structures, namely the Cartesian product and the rationalization of the smash product, although the second structure does not have a unit (since we insisted that our spaces be simply connected). Any rational space X is a commutative coalgebra with respect to the Cartesian product using the diagonal. The natural map from product to smash product then also makes X a commutative coalgebra with respect to smash product (although without a counit). Since Σ^∞ is a symmetric monoidal functor with respect to smash product, we conclude that $\Sigma^\infty_{\mathbb{Q}}$ may be factored as a composition

$$\mathcal{S}_{\mathbb{Q}}^{\geq 2} \xrightarrow{\widetilde{C}_{\mathbb{Q}}} \mathrm{coCAlg}^{\mathrm{nu}}(\mathrm{Sp}_{\mathbb{Q}}) \to \mathrm{Sp}_{\mathbb{Q}},$$

where the second arrow is the forgetful functor. The middle term denotes the ∞-category of commutative coalgebras without counit in the symmetric monoidal ∞-category $\mathrm{Sp}_{\mathbb{Q}}$. For us a commutative coalgebra without counit in a symmetric monoidal ∞-category $\mathcal{C}$ is by definition a non-unital commutative algebra object of $\mathcal{C}^{\mathrm{op}}$ (cf. Definition 5.3 of [42] or Definition 3.1.1 of [60]).

Remark 16.2.2. There are often good 1-categorical models for ∞-categories of commutative algebra objects; for example, the ∞-category CAlg(Sp) of commutative ring spectra can be built from a model category of commutative monoids (in the usual 1-categorical sense) in any good symmetric monoidal model category of spectra. The situation for coalgebras is much less pleasant; we will ignore the issue and work with the above definition.

We say a coalgebra in $\mathrm{Sp}_{\mathbb{Q}}$ is n-connected if its underlying spectrum is n-connected. One half of Theorem 16.1.1 may then be rephrased as follows:

Theorem 16.2.3 (Quillen [69])**.** *The functor $\widetilde{C}_{\mathbb{Q}}$ gives an equivalence of ∞-categories $\mathcal{S}_{\mathbb{Q}}^{\geq 2} \to \mathrm{coCAlg}^{\mathrm{nu}}(\mathrm{Sp}_{\mathbb{Q}})^{\geq 2}$.*

We will describe two proofs of this theorem; the first is close to the traditional proofs, the second uses Goodwillie calculus and does not appear to be in the literature.

Sketch of first proof of Theorem 16.2.3. To check that $\widetilde{C}_{\mathbb{Q}}$ is fully faithful, first note that it admits a right adjoint $\mathrm{Map}_{\mathrm{coCAlg}}(H\mathbb{Q}, -)$, where $H\mathbb{Q}$ is equipped with the trivial coalgebra structure. We will argue that for any $X \in \mathcal{S}_{\mathbb{Q}}^{\geq 2}$ the unit map of this adjunction

$$X \to \mathrm{Map}_{\mathrm{coCAlg}}(H\mathbb{Q}, \widetilde{C}_{\mathbb{Q}}X)$$

is an equivalence. Here and above, the mapping spaces refer to those in the ∞-category $\mathrm{coCAlg}^{\mathrm{nu}}(\mathrm{Sp}_{\mathbb{Q}})$.

We work by induction on the Postnikov tower of X. Write $\tau_{\leq n}X$ for the nth Postnikov section. Since X is simply connected, its Postnikov tower consists of principal fibrations of the form

$$\tau_{\leq n}X \to \tau_{\leq n-1}X \to K(\pi_n X, n+1).$$

The convergence of the homological Eilenberg–Moore spectral sequence for this fibration implies that $\widetilde{C}_{\mathbb{Q}}$ sends this fiber sequence to a fiber sequence of coalgebras. The reader might be more familiar with the dual statement that the cohomological version of this spectral sequence gives, for a suitable fibration $F \to E \to B$, an equivalence of commutative $H\mathbb{Q}$-algebras

$$H\mathbb{Q} \otimes_{C^*(B;\mathbb{Q})} C^*(E;\mathbb{Q}) \to C^*(F;\mathbb{Q}),$$

but working with cochains would require us to introduce finite type hypotheses. In any case, we may now use induction to reduce to the case where X is an Eilenberg–Mac Lane space $K(V,n)$ for V a rational vector space. Then $\widetilde{C}_{\mathbb{Q}}(K(V,n))$ is the cofree commutative coalgebra (without counit) generated by $\Sigma^n HV$ (see Remark 16.2.4), the latter denoting the rational spectrum whose single nonvanishing homotopy group is V in dimension n. But clearly

$$\mathrm{Map}_{\mathrm{coCAlg}}(H\mathbb{Q}, \mathrm{cofree}(\Sigma^n HV)) \simeq \mathrm{Map}_{\mathrm{Sp}_{\mathbb{Q}}}(H\mathbb{Q}, \Sigma^n HV) \simeq K(V,n),$$

so that $\widetilde{C}_{\mathbb{Q}}$ is indeed fully faithful.

To see it is essentially surjective, one observes that $\mathrm{coCAlg}^{\mathrm{nu}}(\mathrm{Sp}_{\mathbb{Q}})^{\geq 2}$ is generated under colimits by a trivial coalgebra C_2 on a single generator in degree 2; indeed, any simply connected coalgebra can be built from 'cells' (where in this case cell means trivial coalgebra on a generator in degree ≥ 2) in the same way that a space can be built from spheres. Since $\widetilde{C}_{\mathbb{Q}}$ preserves colimits and C_2 is the image of the rational 2-sphere $S^2_{\mathbb{Q}}$, this completes the argument. □

Remark 16.2.4. In general cofree coalgebras are rather hard to understand, in contrast to the case of free algebras (for which there is a simple formula). The difference can be traced back to the fact that the smash product of spectra or tensor product of chain complexes commutes with colimits in each variable separately, but not with limits. However, for degree reasons the cofree commutative coalgebra on a connected rational spectrum E is described by the 'naive' formula

$$\mathrm{cofree}(E) \cong \bigvee_{k\geq 1} (E^{\wedge k})^{h\Sigma_k}.$$

Note that since we work over the rational numbers, it is not important whether we use homotopy fixed points or orbits on the right-hand side. Also note that if $E = \Sigma^n HV$ for a rational vector space V, this formula describes precisely the rational homology of $K(V,n)$. This fact was used in the argument above.

Sketch of second proof of Theorem 16.2.3. We now outline an alternative argument using Goodwillie calculus and some ideas from the theory of descent (or comonadicity). We refer to the chapter by Gregory Arone and Michael Ching in this Handbook for background on Goodwillie calculus. A short reminder on monads and comonads can be found in Section 16.3. The following sets the tone for some of the techniques we will employ in Section 16.7.

The adjoint pair $(\Sigma^\infty_{\mathbb{Q}}, \Omega^\infty)$ between $\mathrm{S}^{\geq 2}_{\mathbb{Q}}$ and $\mathrm{Sp}_{\mathbb{Q}}$ in particular gives a comonad $\Sigma^\infty_{\mathbb{Q}}\Omega^\infty$ on $\mathrm{Sp}_{\mathbb{Q}}$, and $\Sigma^\infty_{\mathbb{Q}}$ gives a comparison functor

$$\mathcal{S}^{\geq 2}_{\mathbb{Q}} \to \mathrm{coAlg}_{\Sigma^\infty_{\mathbb{Q}}\Omega^\infty}(\mathrm{Sp}_{\mathbb{Q}}).$$

The right-hand side denotes the ∞-category of *coalgebras* (sometimes also called *left comodules*) for the comonad $\Sigma^\infty_{\mathbb{Q}}\Omega^\infty$. In fact, this comparison functor is an equivalence; one says that the adjunction is *comonadic*. This follows immediately (by the dual of Lemma 16.3.3 below) from the fact that for any $X \in \mathcal{S}_{\mathbb{Q}}^{\geq 2}$, the limit of the cosimplicial object $(\Omega^\infty\Sigma^\infty_{\mathbb{Q}})^{\bullet+1}X$ (which may be thought of as the Bousfield–Kan $H\mathbb{Q}$-resolution) recovers X, meaning that the natural map

$$X \longrightarrow \mathrm{Tot}(\Omega^\infty\Sigma^\infty_{\mathbb{Q}}X \rightrightarrows (\Omega^\infty\Sigma^\infty_{\mathbb{Q}})^2X \Rrightarrow \cdots)$$

is an equivalence. Indeed, for an Eilenberg-Mac Lane space $X = K(V, n)$, the fact that $X = \Omega^\infty\Sigma^n HV$ implies that this cosimplicial object admits an extra codegeneracy (sometimes called a contracting codegeneracy). For general simply connected X the conclusion follows from induction on its Postnikov tower.

The forgetful-cofree adjoint pair between $\mathrm{coCAlg}^{\mathrm{nu}}(\mathrm{Sp}_{\mathbb{Q}})^{\geq 2}$ and $\mathrm{Sp}_{\mathbb{Q}}$ is comonadic as well, the relevant comonad in this case being the cofree coalgebra functor described in Remark 16.2.4. The universal property of the cofree coalgebra and the fact that the functor $\Sigma^\infty_{\mathbb{Q}}$ takes values in commutative coalgebras imply that the counit $\varepsilon\colon \Sigma^\infty_{\mathbb{Q}}\Omega^\infty \to \mathrm{id}_{\mathrm{Sp}_{\mathbb{Q}}}$ induces a natural transformation

$$\gamma\colon \Sigma^\infty_{\mathbb{Q}}\Omega^\infty \to \mathrm{cofree}.$$

To prove the theorem it suffices to show that γ is an equivalence. This follows easily from the well-known calculation of the homogeneous layers (or derivatives) of $\Sigma^\infty_{\mathbb{Q}}\Omega^\infty$, which states

$$D_n(\Sigma^\infty_{\mathbb{Q}}\Omega^\infty)(E) \simeq (E^{\wedge n})_{h\Sigma_n},$$

together with the fact that the Goodwillie tower of $\Sigma^\infty_{\mathbb{Q}}\Omega^\infty$ converges on connected spectra (see for example Section 3.2 of [51] and Example 2.14 of [5]). Alternatively, the fact that γ is an equivalence is easily deduced from Theorem 2.28 of [52].

□

Remark 16.2.5. In our argument involving homogeneous functors we tacitly used again that $(E^{\wedge n})^{h\Sigma_n} \simeq (E^{\wedge n})_{h\Sigma_n}$ in the ∞-category $\mathrm{Sp}_{\mathbb{Q}}$ of rational spectra. This equivalence between fixed points and orbits will turn out to be a crucial feature when establishing variations on rational homotopy theory later on.

Remark 16.2.6. The argument above works just as well to prove the integral statement

$$\mathcal{S}_*^{\geq 2} \simeq \mathrm{coAlg}_{\Sigma^\infty\Omega^\infty}(\mathrm{Sp})^{\geq 2}.$$

It is only in order to explicitly identify the comonad $\Sigma^\infty_{\mathbb{Q}}\Omega^\infty$ as the cofree commutative coalgebra that we used we are working over the rational numbers.

Remark 16.2.7. So far we have taken a rather abstract perspective. However, much of the power of rational homotopy theory derives from its computability. The linear (or Spanier-Whitehead) dual of the functor $C_{\mathbb{Q}}$ gives a functor from $\mathcal{S}_{\mathbb{Q}}^{\geq 2}$ to the ∞-category of commutative $H\mathbb{Q}$-algebras or, equivalently, the ∞-category of rational commutative differential graded algebras (cdga's). The latter functor can be described explicitly using Sullivan's functor A_{PL} of *polynomial differential forms* [76]. Moreover, every cdga arising in this way is equivalent to a *minimal* cdga A, which is one whose underlying graded commutative algebra is free and for which the differential takes values in decomposable elements. Two such minimal cdga's are equivalent if and only if they are isomorphic, making the theory very rigid. For a rational space X, a minimal replacement of $A_{\mathrm{PL}}(X)$ is called a *minimal model* for X. These minimal models are a very powerful tool; for example, the rational homotopy groups of a space X of finite type can be computed directly as the linear dual of the indecomposables of the corresponding minimal model, with the Whitehead bracket determined by the differential via a simple procedure.

The other half of Theorem 16.1.1 concerns differential graded Lie algebras and arises as follows. If $\mathfrak{g}_*$ is a differential graded Lie algebra, then one can associate to it the (reduced) Chevalley–Eilenberg complex $\mathrm{CE}(\mathfrak{g}_*)$ computing its Lie algebra homology. The underlying graded vector space of $\mathrm{CE}(\mathfrak{g}_*)$ is

$$\mathrm{Sym}^{\geq 1}(\mathfrak{g}_*[1]) := \bigoplus_{n\geq 1} \mathrm{Sym}^n(\mathfrak{g}_*[1]),$$

where $\mathfrak{g}_*[1]$ is the shift defined by $\mathfrak{g}_i[1] := \mathfrak{g}_{i-1}$. Here $\mathrm{Sym}^n(V)$ is the nth symmetric power $(V^{\otimes n})_{\Sigma_n}$ in the graded sense, i.e., it behaves like the exterior power on odd degree elements. The differential d_{CE} is constructed from the differential d of $\mathfrak{g}$ and the Lie bracket (see Appendix B.6 of [69]). In fact, $\mathrm{Sym}^{\geq 1}(\mathfrak{g}_*[1])$ carries an evident coalgebra structure that is compatible with the differential just defined, in which the primitive elements are precisely $\mathfrak{g}_*[1] \subseteq \mathrm{Sym}^{\geq 1}(\mathfrak{g}_*[1])$. (An element x of a coalgebra without counit is primitive if $\Delta x = 0$.) This makes $\mathrm{CE}(\mathfrak{g}_*)$ a differential graded commutative coalgebra without counit. It is not hard to verify that the functor CE preserves quasi-isomorphisms of differential graded Lie algebras. The following then establishes the remaining half of Theorem 16.1.1:

Theorem 16.2.8 (Quillen [69]). *The functor*

$$\mathrm{CE}\colon \mathrm{Lie}(\mathrm{Ch}_{\mathbb{Q}})^{\geq 1} \to \mathrm{coCAlg}^{\mathrm{nu}}(\mathrm{Ch}_{\mathbb{Q}})^{\geq 2}$$

is an equivalence of ∞-categories.

We will place this theorem in a more general context in Section 16.4. The inverse functor takes the derived primitives of a commutative coalgebra (shifted down in degree by 1). We conclude this section by highlighting one observation about Theorem 16.1.1. Let us write

$$\Phi_0\colon \mathcal{S}_{\mathbb{Q}}^{\geq 2} \to \mathrm{Sp}_{\mathbb{Q}}^{\geq 1}$$

for the composition of the functor $L_{\mathbb{Q}}$ with the forgetful functor from differential graded Lie algebras to $\mathrm{Sp}_{\mathbb{Q}}$ (which we have identified with $\mathrm{Ch}_{\mathbb{Q}}$). Also, write Θ_0 for its left adjoint, which is the composition of the free Lie algebra functor with the inverse of the equivalence $L_{\mathbb{Q}}$. The reason for this seemingly strange notation is an analogy with the Bousfield–Kuhn functor, which we discuss in Section 16.6. We now have two different adjunctions between the ∞-categories of rational spaces and of rational spectra, summarized in the following diagram (left adjoints on top):

$$\mathrm{Sp}_{\mathbb{Q}}^{\geq 1} \underset{\Phi_0}{\overset{\Theta_0}{\rightleftarrows}} \mathcal{S}_{\mathbb{Q}}^{\geq 2} \underset{\Omega^\infty}{\overset{\Sigma_{\mathbb{Q}}^\infty}{\rightleftarrows}} \mathrm{Sp}_{\mathbb{Q}}^{\geq 2}.$$

The horizontal composition from right to left computes the derived primitives of the cofree coalgebra, shifted down by one. In fact it is rather easy to see that for a rational vector space V, the primitives of cofree(V) are precisely V. From this one can deduce that $\Phi_0\Omega^\infty$ is precisely the functor Σ^{-1} that shifts down by one. Consequently, the composition of left adjoints $\Sigma^\infty_\mathbb{Q}\Theta_0$ is Σ, the functor shifting up by one. Alternatively, one can verify directly that the Chevalley–Eilenberg homology of a free Lie algebra on a chain complex V is quasi-isomorphic to $V[1]$. Variations on these observations will play an important role in Sections 16.6 and 16.7.

16.3 Monads and their algebras

As shown by the second proof of Theorem 16.2.3, the yoga of monads and comonads can be very useful when attempting to 'model' a given ∞-category $\mathcal{C}$ by some homotopy theory of algebras or coalgebras. We already used it to identify the ∞-category of simply connected pointed spaces with that of coalgebras for the comonad $\Sigma^\infty\Omega^\infty$. Later on, we will relate localizations of the ∞-category of spaces to algebras over the spectral Lie operad using similar techniques. To help the reader we offer this short section, which collects some basic facts about (co)monads and the associated (co)simplicial objects. We will make systematic use of these throughout the rest of this survey. More background can be found in Section 4.7 of [62] and in [73].

Let $\mathcal{C}$ be an ∞-category. Then the ∞-category $\mathrm{Fun}(\mathcal{C},\mathcal{C})$ admits a monoidal structure, with tensor product given by composition of functors. A *monad* (resp. *comonad*) is a monoid (resp. comonoid) in this monoidal ∞-category. If

$$\mathcal{C} \underset{G}{\overset{F}{\rightleftarrows}} \mathcal{D}$$

is an adjoint pair, with F left adjoint, then the composite GF gives a monad on $\mathcal{C}$ (and dually the functor FG gives a comonad on $\mathcal{D}$). Writing $\eta : \mathrm{id}_\mathcal{C} \to GF$ for the unit and $\varepsilon : FG \to \mathrm{id}_\mathcal{D}$ for the counit of the adjunction, the multiplication and unit maps of GF are given by

$$GFGF \xrightarrow{G\varepsilon F} GF \qquad \text{and} \qquad \mathrm{id}_\mathcal{C} \xrightarrow{\eta} GF.$$

In ordinary category theory these maps would then have to satisfy an associativity and a unitality condition. In the setting of ∞-categories, these constraints become extra data rather than properties; a monoid object can be encoded by a certain diagram of shape $\mathbf{\Delta}^{\mathrm{op}}$, cf. Section 4.1 of [62].

A typical example to have in mind is the one where $\mathcal{D}$ is the ∞-category of commutative algebras in $\mathcal{C}$ (and we assume $\mathcal{C}$ to be a sufficiently nice symmetric monoidal ∞-category), with G the forgetful functor and F the free commutative algebra functor. The monad GF then assigns to an object X of $\mathcal{C}$ the underlying object of the free commutative algebra on X. An important example of a comonad, already used in the previous section, is the functor $\Sigma^\infty\Omega^\infty$ on the ∞-category of spectra.

If T is a monad on an ∞-category $\mathcal{C}$, one can speak of T-*algebras* in $\mathcal{C}$. Such an algebra is an object X that is first of all equipped with an 'action'

$$\mu \colon TX \to X.$$

Again, in ordinary category this map would then have to satisfy an associativity condition (with respect to the multiplication of T) and a unitality condition (using the unit map of T). Working with ∞-categories means one has to encode these constraints in a coherent way; we refer to Section 4.2 of [62] for details. One important point for us is that any T-algebra X in particular gives rise to an augmented simplicial object in the ∞-category of T-algebras of the form

$$\cdots \ T^2X \rightrightarrows TX \longrightarrow X.$$

The degeneracy maps use the unit of T, while the face maps describe the monoid multiplication of T and the action map $TX \to X$. This augmented simplicial object is canonically contractible when considered as a diagram in the underlying ∞-category $\mathcal{C}$. Indeed it admits 'extra degeneracies' $X \to TX$, and similarly $T^\bullet X \to T^{\bullet+1}X$, given by the unit of T. It follows that the diagram above expresses X as the colimit of the simplicial object $T^{\bullet+1}X$, thus describing X as a colimit of a diagram of *free* T-algebras. We write $\mathrm{Alg}_T(\mathcal{C})$ for the ∞-category of T-algebras in $\mathcal{C}$. Dually, if Q is a comonad on $\mathcal{C}$, we write $\mathrm{coAlg}_Q(\mathcal{C})$ for the ∞-category of Q-coalgebras.

Remark 16.3.1. The definition of a T-algebra is really a special instance of the definition of a left module M in an ∞-category $\mathcal{C}$ for an associative algebra A in some monoidal ∞-category $\mathcal{D}$ that acts on $\mathcal{C}$. In our case T plays the role of A and $\mathcal{D} = \mathrm{Fun}(\mathcal{C}, \mathcal{C})$. In fact, this interpretation as modules also inspires some notation which will be useful. For an associative algebra A with a right module M and left module N one can form the *two-sided bar construction*, which is the simplicial object

$$\cdots M \otimes A^{\otimes 2} \otimes N \ M \otimes A \otimes N \rightrightarrows M \otimes N$$

for which we write $\mathrm{Bar}(M, A, N)_\bullet := M \otimes A^{\otimes \bullet} \otimes N$. The face maps use the action of A on M and N for the outer faces and the multiplication of A for the inner faces. The colimit of $\mathrm{Bar}(M, A, N)_\bullet$ computes the *relative tensor product* $M \otimes_A N$. With this notation, the simplicial object we used above the remark can be written $\mathrm{Bar}(T, T, X)_\bullet$.

For any adjoint pair

$$\mathcal{C} \underset{G}{\overset{F}{\rightleftarrows}} \mathcal{D},$$

giving a monad $T := GF$ on $\mathcal{C}$, the functor G can canonically be factored as

$$\mathcal{D} \xrightarrow{\varphi} \mathrm{Alg}_T(\mathcal{C}) \xrightarrow{\text{forget}} \mathcal{C}.$$

Indeed, for $X \in \mathcal{D}$ the object $G(X)$ has a natural T-algebra structure with action map given by

$$TG(X) = GFG(X) \xrightarrow{G\varepsilon} G(X).$$

One says that the adjoint pair (F, G) is *monadic* if the comparison functor $\varphi : \mathcal{D} \to \mathrm{Alg}_T(\mathcal{C})$ is an equivalence of ∞-categories. The Barr–Beck theorem is a surprisingly useful recognition criterion for monadic adjunctions, generalized to higher category theory by Lurie [62]. (See [73] for an alternative approach.)

Theorem 16.3.2 (Barr–Beck, Lurie). *An adjoint pair $F : \mathcal{C} \rightleftarrows \mathcal{D} : G$ is monadic if and only if the following two conditions are satisfied:*

(a) The functor G is conservative, *i.e., a morphism f in $\mathcal{D}$ is an equivalence if and only if $G(f)$ is an equivalence.*

(b) If $X_\bullet$ is a simplicial object in $\mathcal{D}$ such that $G(X_\bullet)$ is split (i.e., admits an augmentation with extra degeneracies), then the colimit of $X_\bullet$ exists in $\mathcal{D}$ and is preserved by G.

Condition (b) might seem cryptic at first sight: however, it is automatically satisfied if $\mathcal{D}$ admits all geometric realizations (i.e., colimits of $\mathbf{\Delta}^{\mathrm{op}}$-shaped diagrams) and G preserves these, which will suffice for our applications. We usually denote the colimit of a simplicial object $X_\bullet$ by $|X_\bullet|$. The typical split simplicial object to have in mind is the diagram above Remark 16.3.1 describing $T^{\bullet+1}X$.

Any adjoint pair $F : \mathcal{C} \rightleftarrows \mathcal{D} : G$ gives rise to a simplicial resolution of the identity functor of $\mathcal{D}$. To be precise, it gives a simplicial object $(FG)^{\bullet+1}$ with an augmentation to $\mathrm{id}_{\mathcal{D}}$:

$$\cdots \;\substack{\textstyle\rightarrow\\[-2pt]\textstyle\rightarrow\\[-2pt]\textstyle\rightarrow\\[-2pt]\textstyle\rightarrow}\; (FG)^2 \;\substack{\textstyle\rightarrow\\[-2pt]\textstyle\rightarrow\\[-2pt]\textstyle\rightarrow}\; FG \xrightarrow{\varepsilon} \mathrm{id}_{\mathcal{D}}.$$

In the notation of bar constructions this simplicial object can be written as $\mathrm{Bar}(F, GF, G)_\bullet$. The degeneracy maps use the unit η: $\mathrm{id}_{\mathcal{C}} \to GF$, the face maps use the counit ε. A dual comment applies to give a cosimplicial resolution of the identity functor of $\mathcal{C}$. We used exactly this resolution, associated to the adjunction $(\Sigma^\infty, \Omega^\infty)$, in the second proof of Theorem 16.2.3. The following is a useful criterion for monadicity:

Lemma 16.3.3. *Suppose $F : \mathcal{C} \rightleftarrows \mathcal{D} : G$ is an adjoint pair and that $\mathcal{D}$ admits colimits of G-split simplicial objects. Then this pair is monadic if and only if for every object X of $\mathcal{C}$, the map*

$$|(FG)^{\bullet+1}|(X) \to X$$

arising from the simplicial resolution described above is an equivalence.

16.4 Koszul duality

The duality between Lie algebras and commutative coalgebras expressed by Theorem 16.2.8 is a special case of what is usually called *Koszul duality* or *bar-cobar duality.* We will begin this section with a discussion of this duality for associative algebras and coalgebras. In the setting of differential graded algebras this goes back at least to the work of Adams [1], Priddy [68], and Moore [64]; we will phrase it in a more general context following Lurie (Section 4.3 of [58]). The remainder of the section is a discussion of Koszul duality in the generality of (co)algebras for (co)operads. References for this material are [39, 40, 56] in the differential graded context, [24] in the context of stable homotopy theory, and [36] for a discussion at the more general level of ∞-categories.

For the purposes of this discussion, let $\mathcal{C}$ be a symmetric monoidal ∞-category. We will assume it admits limits and colimits. Suppose A is an associative algebra object of $\mathcal{C}$ equipped with an augmentation $\varepsilon : A \to \mathbf{1}$ to the unit of $\mathcal{C}$. Then the *bar construction* of A, denoted $\mathrm{Bar}(A)$, is the relative tensor product $\mathbf{1} \otimes_A \mathbf{1}$, which by definition is the colimit of the simplicial object

$$\cdots A \otimes A \,\substack{\longrightarrow\\[-1ex]\longrightarrow\\[-1ex]\longrightarrow}\, A \,\substack{\longrightarrow\\[-1ex]\longrightarrow}\, \mathbf{1}.$$

The degeneracy maps of this simplicial object are constructed from the unit of A, whereas the face maps use the multiplication $A \otimes A \to A$ for the 'inner' faces and the augmentation ε for the 'outer' faces. In the notation of Remark 16.3.1 it can be written as $\mathrm{Bar}(\mathbf{1}, A, \mathbf{1})_\bullet$.

The object $\mathrm{Bar}(A)$ admits a comultiplication, using the maps

$$\mathbf{1} \otimes_A \mathbf{1} \simeq \mathbf{1} \otimes_A A \otimes_A \mathbf{1} \xrightarrow{\varepsilon} \mathbf{1} \otimes_A \mathbf{1} \otimes_A \mathbf{1} \to (\mathbf{1} \otimes_A \mathbf{1}) \otimes (\mathbf{1} \otimes_A \mathbf{1}).$$

(The last map arises from the universal property of the colimit defining the term $\mathbf{1} \otimes_A \mathbf{1} \otimes_A \mathbf{1}$; under the additional assumption that the tensor product on $\mathcal{C}$ commutes with geometric realizations in each variable, it is always an equivalence.) However, the situation is much better than just this: the object $\mathrm{Bar}(A)$ can be upgraded to a homotopy-coherent associative coalgebra object of $\mathcal{C}$ equipped with a coaugmentation. Write $\mathrm{Alg}^{\mathrm{aug}}(\mathcal{C})$ for the ∞-category of augmented associative algebras in $\mathcal{C}$ and

$$\mathrm{coAlg}^{\mathrm{aug}}(\mathcal{C}) := \mathrm{Alg}^{\mathrm{aug}}(\mathcal{C}^{\mathrm{op}})^{\mathrm{op}}$$

for the ∞-category of coaugmented associative coalgebras. The generality of the following statement is first articulated by Lurie in [58], but its essential form traces back to Moore [64]:

Theorem 16.4.1. *There is an adjoint pair of functors (in which* Bar *is left adjoint) as follows:*

$$\mathrm{Alg}^{\mathrm{aug}}(\mathcal{C}) \underset{\mathrm{Cobar}}{\overset{\mathrm{Bar}}{\rightleftarrows}} \mathrm{coAlg}^{\mathrm{aug}}(\mathcal{C}).$$

Here Cobar is the bar construction applied to the opposite category, i.e., it sends a coaugmented coalgebra C to the totalization of the cosimplicial object

$$\mathbf{1} \,\substack{\longrightarrow\\[-1ex]\longrightarrow}\, C \,\substack{\longrightarrow\\[-1ex]\longrightarrow\\[-1ex]\longrightarrow}\, C \otimes C \cdots.$$

The proof of Theorem 16.4.1 proceeds by showing that the two mapping spaces $\mathrm{Map}_{\mathrm{Alg}^{\mathrm{aug}}(\mathcal{C})}(A, \mathrm{Cobar}(C))$ and $\mathrm{Map}_{\mathrm{coAlg}^{\mathrm{aug}}(\mathcal{C})}(\mathrm{Bar}(A), C)$ are both equivalent to the space of lifts of the pair (C, A), thought of as an augmented algebra object of the ∞-category $\mathcal{C}^{\mathrm{op}} \times \mathcal{C}$, to an augmented algebra object of the twisted arrow category $\mathrm{TwArr}(\mathcal{C})$. This is closely related to the more classical notion of twisting cochains.

The aim of the rest of this section is to apply this general setup for bar-cobar duality to the case of operads and cooperads. We write $\mathrm{SymSeq}(\mathcal{C})$ for the ∞-category of symmetric sequences in $\mathcal{C}$; its objects are sequences of objects $\mathcal{O} = \{\mathcal{O}(n)\}_{n\geq 1}$ of $\mathcal{C}$ where the nth term $\mathcal{O}(n)$ is equipped with an action of the symmetric group Σ_n. To any such sequence we associate a functor

$$F_{\mathcal{O}}\colon \mathcal{C} \to \mathcal{C}\colon X \mapsto \bigoplus_{n\geq 1}(\mathcal{O}(n) \otimes X^{\otimes n})_{h\Sigma_n}.$$

The ∞-category SymSeq($\mathcal{C}$) carries a monoidal structure, called the *composition product* and usually denoted by $\circ$, which is essentially characterized by the fact that the assignment

$$\mathrm{SymSeq}(\mathcal{C}) \to \mathrm{Fun}(\mathcal{C}, \mathcal{C})\colon \mathcal{O} \mapsto F_{\mathcal{O}}$$

is monoidal [41]. Here the ∞-category on the right is monoidal via the composition of functors. The unit for the composition product is the symmetric sequence $\mathbf{1}$, which has the monoidal unit of $\mathcal{C}$ as its first term and all other terms equal to 0. In these terms, an operad (resp. a cooperad) in $\mathcal{C}$ is a monoid (resp. a comonoid) in SymSeq($\mathcal{C}$) with respect to the composition product.

Since we excluded the case $n = 0$ from our definitions above, all operads and cooperads we consider here are in fact *non-unital.* If $\mathcal{O}$ is an operad, then the associated functor $F_{\mathcal{O}}$ has the structure of a monad (and dually for a cooperad). One can then define an $\mathcal{O}$-algebra in $\mathcal{C}$ to be precisely an $F_{\mathcal{O}}$-algebra.

Remark 16.4.2. Consider a cooperad $\mathcal{O}$. Observe that coalgebras for the comonad $F_{\mathcal{O}}$ are objects X of $\mathcal{C}$ equipped with a 'comultiplication map'

$$X \to \bigoplus_{n\geq 1}(\mathcal{O}(n) \otimes X^{\otimes n})_{h\Sigma_n}$$

and further data recording its coassociativity and counitality. In the special case where $\mathcal{O}$ is the commutative cooperad (which as a symmetric sequence has $\mathcal{O}(n)$ equal to the monoidal unit of $\mathcal{C}$ for each $n \geq 1$), this notion of coalgebra is generally *not* the same as that of a commutative coalgebra. One important difference is that the n-fold comultiplication of a commutative coalgebra C gives a map

$$C \to (C^{\otimes n})^{h\Sigma_n},$$

but this map need not factor through $(C^{\otimes n})_{h\Sigma_n}$. For this reason, the coalgebras for the comonad $F_{\mathcal{O}}$ are sometimes called *conilpotent divided power $\mathcal{O}$-coalgebras* (cf. [36]). The adjective conilpotent refers to the direct sum, the term divided powers refers to the fact that $F_{\mathcal{O}}$ involves coinvariants for the symmetric groups, rather than invariants.

If $f\colon \mathcal{O} \to \mathcal{P}$ is a map of operads in $\mathcal{C}$, one obtains an adjunction between the corresponding ∞-categories of algebras:

$$\mathrm{Alg}_{\mathcal{O}} \underset{f^*}{\overset{f_!}{\rightleftarrows}} \mathrm{Alg}_{\mathcal{P}}.$$

Here the right adjoint f^* is restriction along f. Informally, the left adjoint is the relative tensor product $\mathcal{P} \circ_{\mathcal{O}} -$, using the analogy between operads with their algebras and rings with their modules. More precisely, one first observes that $f_!$ sends free $\mathcal{O}$-algebras to free

$\mathcal{P}$-algebras. As we described in Section 16.3, any $\mathcal{O}$-algebra X admits a natural simplicial resolution in terms of free $\mathcal{O}$-algebras, namely the bar construction $\mathrm{Bar}(F_{\mathcal{O}}, F_{\mathcal{O}}, X)_\bullet$. Then $f_!X$ can be computed as the colimit of the simplicial object

$$(\cdots F_{\mathcal{P}}F_{\mathcal{O}}^2 X \substack{\longrightarrow\\\longrightarrow\\\longrightarrow} F_{\mathcal{P}}F_{\mathcal{O}}X \rightrightarrows F_{\mathcal{P}}X) =: \mathrm{Bar}(F_{\mathcal{P}}, F_{\mathcal{O}}, X).$$

Let us specialize to the case where the ∞-category $\mathcal{C}$ is stable and $\mathcal{O}$ is *reduced*, meaning $\mathcal{O}(1)$ is the monoidal unit of $\mathcal{C}$. The latter assumption gives essentially unique morphisms of operads $i\colon \mathbf{1} \to \mathcal{O}$ and $p\colon \mathcal{O} \to \mathbf{1}$, including the first term and projecting onto it respectively. Note that we may identify $\mathrm{Alg}_{\mathbf{1}}(\mathcal{C})$ with $\mathcal{C}$ itself, since $F_{\mathbf{1}}$ is the identity functor of $\mathcal{C}$. Then i^* is the forgetful functor and $i_!$ the free $\mathcal{O}$-algebra functor. The functor p^* can be identified with the *trivial algebra functor* $\mathrm{triv}_{\mathcal{O}}$, equipping an object X of $\mathcal{C}$ with the $\mathcal{O}$-algebra structure for which all maps

$$(\mathcal{O}(n) \otimes X^{\otimes n})_{h\Sigma_n} \to X$$

are zero when $n > 1$ (recall that $\mathcal{C}$ is stable, so has a zero object). The left adjoint $p_!$ computes the *derived indecomposables* of an $\mathcal{O}$-algebra X. We will denote it by $\mathrm{TAQ}_{\mathcal{O}}$, which stands for *topological André–Quillen homology*. The reason for this terminology is that in the special case $\mathcal{O} = \mathbf{Com}$, this functor is closely related to the homology of commutative rings as defined in terms of the cotangent complex by André and Quillen [70]. The two adjunctions just discussed give a diagram

$$\mathcal{C} \underset{\text{forget}}{\overset{\text{free}_{\mathcal{O}}}{\rightleftarrows}} \mathrm{Alg}_{\mathcal{O}}(\mathcal{C}) \underset{\text{triv}_{\mathcal{O}}}{\overset{\mathrm{TAQ}_{\mathcal{O}}}{\rightleftarrows}} \mathcal{C}$$

in which both horizontal composites are equivalent to the identity (because $pi = \mathrm{id}$).

In fact the functor $\mathrm{TAQ}_{\mathcal{O}}$ can be characterized by a universal property:

Theorem 16.4.3 (Basterra–Mandell [9], Lurie [62, Theorem 7.3.4.7]). *Suppose the symmetric monoidal ∞-category $\mathcal{C}$ is stable, presentable, and the tensor product preserves colimits in each variable separately. Then the adjoint pair $(\mathrm{TAQ}_{\mathcal{O}}, \mathrm{triv}_{\mathcal{O}})$ exhibits $\mathcal{C}$ as the stabilization of $\mathrm{Alg}_{\mathcal{O}}(\mathcal{C})$. In other words, $\mathrm{TAQ}_{\mathcal{O}}$ is the initial colimit-preserving functor from $\mathrm{Alg}_{\mathcal{O}}(\mathcal{C})$ to a presentable stable ∞-category.*

Sketch of proof. Take the diagram above the statement of the theorem, stabilize the ∞-category $\mathrm{Alg}_{\mathcal{O}}(\mathcal{C})$, and replace the functors by their linearizations (or first derivatives) to obtain the following diagram:

$$\mathcal{C} \underset{\partial\text{forget}}{\overset{\partial\text{free}_{\mathcal{O}}}{\rightleftarrows}} \mathrm{Sp}(\mathrm{Alg}_{\mathcal{O}}(\mathcal{C})) \underset{\partial\text{triv}_{\mathcal{O}}}{\overset{\partial\mathrm{TAQ}_{\mathcal{O}}}{\rightleftarrows}} \mathcal{C}.$$

The chain rule $\partial(G \circ F) \cong \partial(G) \circ \partial(F)$ guarantees that the horizontal composites are still equivalent to the identity. Moreover, it is a general fact (and not hard to verify) that the stabilization of a monadic adjunction is monadic, so that the pair of functors on the left exhibits $\mathrm{Sp}(\mathrm{Alg}_{\mathcal{O}}(\mathcal{C}))$ as monadic over $\mathcal{C}$. The relevant monad is the linearization of the functor forget $\circ$ free$_{\mathcal{O}}$; the latter is explicitly given by assigning to $X \in \mathcal{C}$ the object

$$\bigoplus_{n\geq 1}(\mathcal{O}(n) \otimes X^{\otimes n})_{h\Sigma_n}.$$

The linearization of this clearly is the monad

$$X \mapsto \mathcal{O}(1) \otimes X \cong X,$$

from which it follows that $\partial\mathrm{free}_{\mathcal{O}}\colon \mathcal{C} \to \mathrm{Sp}(\mathrm{Alg}_{\mathcal{O}}(\mathcal{C}))$ is an equivalence. This implies the theorem. □

Remark 16.4.4. One can paraphrase Theorem 16.4.3 by saying that $\mathrm{TAQ}_{\mathcal{O}}$ is the universal homology theory for $\mathcal{O}$-algebras. For specific choices of $\mathcal{O}$ it reproduces familiar notions of homology. As an example, take $\mathcal{C}$ to be the ∞-category $\mathrm{Ch}_{\mathbb{Z}}$ of chain complexes of abelian groups. As alluded to above, $\mathrm{TAQ}_{\mathbf{Com}}$ reproduces André–Quillen homology of commutative rings. For associative algebras it gives a degree shift of Hochschild homology (when working over a field) or of Shukla homology (for general rings). For Lie algebras it produces a degree shift of the Chevalley–Eilenberg homology already discussed in Section 16.2. The fact that $\mathrm{TAQ}_{\mathcal{O}}$ preserves colimits, together with the equivalence $\mathrm{TAQ}_{\mathcal{O}} \circ \mathrm{free}_{\mathcal{O}} \cong \mathrm{id}_{\mathcal{C}}$, makes this homology theory a useful tool to understand the 'cell structure' of an algebra, meaning the way that X can be built from free algebras on generators of $\mathcal{C}$ (the cells) by pushouts (the cell attachments). This is similar to how singular homology can be used to investigate the minimal cell structure of a topological space. In case $\mathcal{O}$ is the operad $\mathbf{E}_n$ of little n-discs, this perspective is used to great effect in [38] and its sequels.

The starting point of Koszul duality from the point of view of operads, which originates in [40] and [39], is the following. Again consider an operad $\mathcal{O}$ for which $\mathcal{O}(1)$ is the monoidal unit of $\mathcal{C}$, so that projection to this term gives an augmentation $\varepsilon\colon \mathcal{O} \to \mathbf{1}$. Then $\mathcal{O}$ becomes an augmented associative algebra in $\mathrm{SymSeq}(\mathcal{C})$ and thus it makes sense to speak of its bar construction $\mathrm{Bar}(\mathcal{O})$, which is a cooperad.

Example 16.4.5. One of the most important examples is the duality between the commutative operad and the Lie operad. To be precise, the results of Ginzburg–Kapranov [40] and Getzler–Jones [39] show that the cobar construction of the commutative cooperad in chain complexes of R-modules (for some commutative ring R) give the Lie operad with a degree shift. Dually, the bar construction of the Lie operad produces the commutative cooperad, again with a shift. In Section 16.5 we will return to a variant of this example in the context of stable homotopy theory in order to define spectral Lie algebras. Another example is the bar construction of the associative operad, which is the linear dual of the associative operad with a degree shift. More generally, the bar construction of the operad given by the singular chains of the little discs operad $\mathbf{E}_n$ is the cooperad $C^*\mathbf{E}_n^\vee$ (shifted n times) [37]. This is what is often referred to as the *self-duality* of the $\mathbf{E}_n$-operad. Conjecturally this self-duality already holds for the operad $\Sigma^\infty_+\mathbf{E}_n$ in the ∞-category of spectra. At the time of writing this has not yet been resolved, but at least the underlying symmetric sequence of the cooperad $\mathrm{Bar}(\Sigma^\infty_+\mathbf{E}_n)$ has the expected homotopy type [24].

Remark 16.4.6. In fact, for operads and cooperads in a stable ∞-category $\mathcal{C}$ for which the tensor product is exact in each variable separately, the process of forming bar and cobar constructions is invertible; the unit and counit maps

$$\mathcal{O} \to \mathrm{Cobar}(\mathrm{Bar}(\mathcal{O})) \qquad \mathrm{Bar}(\mathrm{Cobar}(\mathcal{Q})) \to \mathcal{Q}$$

are equivalences (cf. [36] and [24]).

We write $\mathrm{coAlg}^{\mathrm{dp}}_{\mathrm{Bar}(\mathcal{O})}(\mathcal{C})$ for the ∞-category of conilpotent divided power $\mathrm{Bar}(\mathcal{O})$-coalgebras (cf. Remark 16.4.2). More briefly, this is the ∞-category of coalgebras for the associated comonad $F_{\mathrm{Bar}(\mathcal{O})}$.

Theorem 16.4.7 (Francis–Gaitsgory [36]). *The functor* $\mathrm{TAQ}_{\mathcal{O}}$ *factors as a composite*

$$\mathrm{Alg}_{\mathcal{O}}(\mathcal{C}) \xrightarrow{B_{\mathcal{O}}} \mathrm{coAlg}^{\mathrm{dp}}_{\mathrm{Bar}(\mathcal{O})}(\mathcal{C}) \xrightarrow{\text{forget}} \mathcal{C}.$$

In other words, for any $\mathcal{O}$*-algebra* X *the object* $\mathrm{TAQ}_{\mathcal{O}}(X)$ *naturally admits the structure of a conilpotent divided power* $\mathrm{Bar}(\mathcal{O})$*-coalgebra. Moreover, the functor* $B_{\mathcal{O}}$ *admits a right adjoint* $C_{\mathrm{Bar}(\mathcal{O})}$.

Sketch of proof. As explained in Section 16.3, the fact that $\mathrm{TAQ}_{\mathcal{O}}$ is left adjcint formally implies that it factors through the ∞-category of coalgebras for the comonad $\mathrm{TAQ}_{\mathcal{O}} \circ \mathrm{triv}_{\mathcal{O}}$. Thus it suffices to identify this comonad with $F_{\mathrm{Bar}(\mathcal{O})}$. To do this, recall that we may resolve any $\mathcal{O}$-algebra X by free algebras and get an equivalence

$$|\mathrm{Bar}(F_{\mathcal{O}}, F_{\mathcal{O}}, X)| \simeq X.$$

Hence

$$\begin{aligned} \mathrm{TAQ}_{\mathcal{O}}(\mathrm{triv}_{\mathcal{O}} X) &\simeq \mathrm{TAQ}_{\mathcal{O}} |\mathrm{Bar}(F_{\mathcal{O}}, F_{\mathcal{O}}, \mathrm{triv}_{\mathcal{O}} X)| \\ &\simeq |\mathrm{Bar}(\mathrm{id}_{\mathcal{C}}, F_{\mathcal{O}}, \mathrm{id}_{\mathcal{C}})|(X) \\ &\simeq F_{\mathrm{Bar}(\mathcal{O})}(X). \end{aligned}$$

The existence of the right adjoint $C_{\mathrm{Bar}(\mathcal{O})}$ can be deduced from the adjoint functor theorem. Alternatively, it may be constructed explicitly as the *derived primitives* of a $\mathrm{Bar}(\mathcal{O})$-coalgebra, which is a construction formally dual to that of the derived indecomposables functor $\mathrm{TAQ}_{\mathcal{O}}$. □

The real challenge in understanding Koszul duality lies in identifying appropriate subcategories of $\mathrm{Alg}_{\mathcal{O}}(\mathcal{C})$ and $\mathrm{coAlg}^{\mathrm{dp}}_{\mathrm{Bar}(\mathcal{O})}(\mathcal{C})$ on which the functors $B_{\mathcal{O}}$ and $C_{\mathrm{Bar}(\mathcal{O})}$ give an equivalence of ∞-categories. Francis and Gaitsgory [36] describe several cases and generally conjecture that it should suffice to restrict to what are called pro-nilpotent algebras and ind-conilpotent coalgebras. A partial result goes as follows:

Theorem 16.4.8 (Ching–Harper [25]). *Let* R *be a commutative ring spectrum and* $\mathcal{O}$ *an operad in the* ∞*-category* Mod_R *of* R*-module spectra with* $\mathcal{O}(1) = R$*. Assume that* R *and* *the terms $\mathcal{O}(n)$ are connective. Then the restriction*

$$\mathrm{Alg}_{\mathcal{O}}(\mathrm{Mod}_R)^{\geq 1} \underset{C_{\mathrm{Bar}(\mathcal{O})}}{\overset{B_{\mathcal{O}}}{\rightleftarrows}} \mathrm{coAlg}_{\mathrm{Bar}(\mathcal{O})}(\mathrm{Mod}_R)^{\geq 1}$$

of the adjunction of Theorem 16.4.7 to connected objects is an equivalence of ∞-categories.

Remark 16.4.9. In the presence of a useful notion of connectivity (such as a t-structure on $\mathcal{C}$), the strategy of proof of Theorem 16.4.8 goes through much more generally.

In particular, one may apply Theorem 16.4.8 to the case where $R = H\mathbb{Q}$ and $\mathcal{O}$ is the Lie operad to retrieve Quillen's theorem 16.2.8.

16.5 Spectral Lie algebras

In this section we describe an extension of the theory of Lie algebras to the ∞-category Sp of spectra. This extension has only recently begun to be exploited and promises to be very useful.

The non-unital commutative operad **Com** in the ∞-category of spectra is the operad parametrizing non-unital commutative ring spectra: it has $\mathbf{Com}(n) = \mathbb{S}$, the sphere spectrum, in every degree $n \geq 1$. The relevant structure maps

$$\mathbf{Com}(n) \otimes \mathbf{Com}(k_1) \otimes \cdots \otimes \mathbf{Com}(k_n) \to \mathbf{Com}(k_1 + \cdots + k_n)$$

are the canonical equivalences determined by the fact that $\mathbb{S}$ is the unit of the smash product. We can take the termwise Spanier–Whitehead dual $\mathbf{Com}^\vee$ to obtain the non-unital commutative cooperad, which of course still has every term equal to the sphere spectrum $\mathbb{S}$. The following definition is inspired by the duality between the Lie operad and the commutative cooperad in chain complexes described in Example 16.4.5:

Definition 16.5.1. The *spectral Lie operad* **L** is the cobar construction Cobar($\mathbf{Com}^\vee$). We write Lie(Sp) for the ∞-category of **L**-algebras in Sp and refer to its objects as *spectral Lie algebras.*

Remark 16.5.2. The commutative cooperad can be constructed in any symmetric monoidal ∞-category $\mathcal{C}$, since it only uses the unit of the symmetric monoidal structure. Hence one can make the above definition of Lie algebras in essentially any context, although one should usually require $\mathcal{C}$ to be stable for this to be useful.

The operad **L** was first constructed by Salvatore [74] and Ching [23]. The spectra $\mathbf{L}(n)$ can be described very explicitly as the Spanier–Whitehead duals of certain finite simplicial complexes, which we will now explain. These descriptions originate in the work of Johnson on the Goodwillie derivatives of the identity [46] (on which we will have more to say below) and were reformulated in the form we present below by Arone–Mahowald [7]. Write $\mathbf{P}(n)$ for the set of partitions of (meaning equivalence relations on) the set $\{1, \ldots, n\}$. We regard $\mathbf{P}(n)$

as a partially ordered set under refinement of partitions. It has a minimal (resp. maximal) element, namely the trivial partition with only one equivalence class (resp. the discrete partition). We write $\mathbf{P}^+(n)$ (resp. $\mathbf{P}^-(n)$) for the subset of $\mathbf{P}(n)$ obtained by discarding the minimal element (resp. the maximal element). Also, we write $\mathbf{P}^\pm(n)$ for the intersection $\mathbf{P}^+(n) \cap \mathbf{P}^-(n)$. Note that the symmetric group Σ_n naturally acts on $\mathbf{P}(n)$, as well as on the various subsets we have defined.

Definition 16.5.3. The *nth partition complex* Π_n is the Σ_n-space $|\mathbf{P}^\pm(n)|$ obtained as the geometric realization of the nerve of the poset $\mathbf{P}^\pm$. Furthermore, we define

$$K_n := |\mathbf{P}(n)|/(|\mathbf{P}^+(n)| \cup |\mathbf{P}^-(n)|).$$

The space K_n is homotopy equivalent to the double suspension $\Sigma S\Pi_n$ of the partition complex. Here S denotes unreduced suspension, whereas Σ denotes reduced suspension, with $S\Pi_n$ regarded as pointed at one of the two cone points.

Example 16.5.4. The space Π_2 is empty, so that K_2 is S^1 with trivial Σ_2-action. The space Π_3 is discrete and has three points, corresponding to the partition (12)(3) and its permutations. As a Σ_3-space it is isomorphic to Σ_3/Σ_2. Hence K_3 is homotopy equivalent to a wedge of two 2-spheres, although this identification disregards the Σ_3-action. The space Π_4 is a one-dimensional simplicial complex homotopy equivalent to a wedge of six circles, so that K_4 is equivalent to a wedge of six 3-spheres. Again one should be careful that this is an identification of the homotopy type of the underlying space, disregarding the action of Σ_4.

Ching [23] observed that the terms of the cobar construction of the commutative cooperad are precisely the Spanier–Whitehead duals of the K_n:

$$\mathbf{L}(n) \cong (\Sigma^\infty K_n)^\vee.$$

The operad structure of $\mathbf{L}$ is reflected in a cooperad structure on the collection of spaces $\{K_n\}_{\geq 1}$. Roughly speaking it can be described by observing that K_n is homeomorphic to a certain space of weighted rooted trees with n leaves, with comultiplication defined by decomposing trees into smaller subtrees grafted along a common edge. In general the space K_n is homotopy equivalent to a wedge of $(n-1)!$ spheres of dimension $n-1$. Detailed results on the equivariant topology of the partition complexes can be found in the works of Arone–Dwyer [4] and Arone–Brantner [6].

Write **Lie** for the usual Lie operad in abelian groups. Taking the integral homology of the spectra $\mathbf{L}(n)$ gives an operad in graded abelian groups, which is precisely a degree shift of **Lie** thought of as sitting in homological degree zero. Indeed, as graded abelian groups one has

$$H_*\mathbf{L}(n) \cong \mathbf{Lie}(n)[1-n],$$

but it is better to write

$$H_*\mathbf{L}(n) \cong (\mathbf{Lie}(n)[1]) \otimes (\mathbb{Z}[-1])^{\otimes n}$$

to make the action of Σ_n explicit. It acts by permuting the n factors of $\mathbb{Z}[-1]$ on the right-hand side; said differently, the Σ_n-action on $H_*\mathbf{L}(n)$ is the same as the action on $\mathbf{Lie}(n)$ twisted by the sign representation. These identifications are compatible with the operad structures on both sides. One way to prove these facts is to relate the homology of the cobar construction that defines $\mathbf{L}$ to the algebraic cobar construction for the commutative cooperad in graded abelian groups.

Remark 16.5.5. This shifting of degree for an operad is a rather harmless procedure. Indeed, endowing a graded abelian group M with the structure of an algebra for the operad $H_*\mathbf{L}$ is the same thing as giving $M[-1]$ the structure of a graded Lie algebra. The occurrence of these shifts also explains the degree shift occurring in Theorem 16.2.8.

A direct connection between the spectral Lie operad and the homotopy theory of spaces (and indeed, one of the original motivations for studying this operad) is given by the *Goodwillie derivatives of the identity.* We refer to the chapter by Arone and Ching in this Handbook for a survey of Goodwillie calculus, but let us summarize what is relevant for us here. The Goodwillie tower of the identity functor on the ∞-category $\mathcal{S}_*$ gives, for each pointed space X, a tower of spaces interpolating between X and its stable homotopy type:

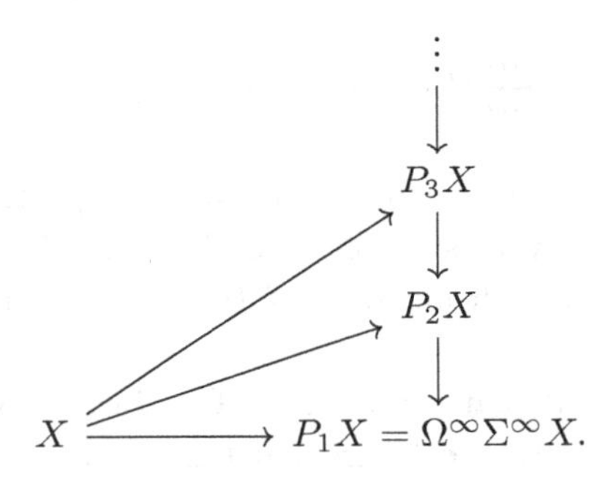

The fiber D_nX of the map

$$P_nX \to P_{n-1}X$$

is called the *nth homogenous layer* of the tower. It turns out to be an infinite loop space $D_nX = \Omega^\infty \mathbf{D}_nX$ associated with a spectrum $\mathbf{D}_nX$ which can be written as

$$\mathbf{D}_nX \simeq (\partial_n\mathrm{id} \otimes \Sigma^\infty X^{\otimes n})_{h\Sigma_n}$$

for some spectrum $\partial_n\mathrm{id}$ with Σ_n-action called the *nth derivative* of the identity functor.

The connection to spectral Lie algebras is that $\partial_n\mathrm{id}$ can be identified with $\mathbf{L}(n)$. The homotopy type of $\partial_n\mathrm{id}$ was first determined by Johnson [46]. We follow a line of reasoning due to Arone–Ching [5], because it relates directly to our earlier discussion of cobar constructions. They prove that the map

$$\mathrm{id} \to \mathrm{Tot}\big(\mathrm{Cobar}(\Omega^\infty, \Sigma^\infty\Omega^\infty, \Sigma^\infty)^\bullet\big)$$

arising from the cosimplicial resolution of the identity functor via $\Omega^\infty\Sigma^\infty$ (see Section 16.3) induces an equivalence of derivatives

$$\partial_*\mathrm{id} \simeq \mathrm{Tot}\big(\mathrm{Cobar}(\partial_*\Omega^\infty, \partial_*(\Sigma^\infty\Omega^\infty), \partial_*\Sigma^\infty)^\bullet\big).$$

To interpret the right-hand side, one uses that the functors Ω^∞ and Σ^∞ are linear (in Goodwillie's sense), that the derivatives of $\Sigma^\infty\Omega^\infty$ can be identified with the commutative cooperad $\mathbf{Com}^\vee$ (with the cooperad structure corresponding to the fact that this functor is a comonad), and that the chain rule for functors from the ∞-category Sp to itself states

$$\partial_*(GF) \simeq \partial_*G \circ \partial_*F.$$

The right-hand side denotes the composition product of symmetric sequences. Putting these ingredients together gives

$$\partial_*\mathrm{id} \simeq \mathrm{Tot}\big(\mathrm{Cobar}(\mathbf{1}, \mathbf{Com}^\vee, \mathbf{1})^\bullet\big) = \mathrm{Cobar}(\mathbf{Com}^\vee)$$

and the right-hand side is precisely our definition of $\mathbf{L}$. Thus, the Goodwillie tower produces a spectral sequence converging to the homotopy groups of X, starting from the homotopy groups of the spectrum

$$\bigoplus_{n\geq 1}(\mathbf{L}(n) \otimes \Sigma^\infty X^{\otimes n})_{h\Sigma_n},$$

which is the *free* spectral Lie algebra on the suspension spectrum of X. This *Goodwillie spectral sequence* has been used very succesfully by Behrens [10], who combines it with the EHP sequence to reproduce a significant part of Toda's calculations of unstable homotopy groups of spheres.

A starting point for most calculations with the Goodwillie tower is the homology of the spectra $\mathbf{D}_nX$, which has been studied in great detail by Arone–Mahowald [7] and Arone–Dwyer [4] in the case where X is a sphere. For example, when this sphere is of odd dimension then the layers $\mathbf{D}_nX$ are contractible whenever n is not a power of a prime. When $n = 2^k$, its cohomology with coefficients in $\mathbb{F}_2$ is free over the subalgebra $\mathcal{A}^*_{k-1}$ of the Steenrod algebra $\mathcal{A}^*$ generated by $\mathrm{Sq}^1, \ldots, \mathrm{Sq}^{2^{k-1}}$ (and a similar statement holds at odd primes). This has very useful consequences for the analysis of the v_n-periodic homotopy groups of the Goodwillie tower. Also, the spaces Π_{p^k} are closely related to Tits buildings for the groups $\mathrm{GL}_k(\mathbb{F}_p)$, and using this Arone–Dwyer relate the calculation of the mod p homology of $\mathbf{D}_{p^k}(X)$ to that of $\mathrm{GL}_k(\mathbb{F}_p)$ with coefficients in the Steinberg module. The reader can find a much more elaborate discussion of these results in the chapter of Arone and Ching in this Handbook. These homology calculations also lead to a theory of power operations for spectral Lie algebras; we refer to Behrens [10] and Antolín-Camarena [3] for the case of $\mathbb{F}_2$-coefficients, Kjaer [48] for $\mathbb{F}_p$-coefficients with $p > 2$, and Brantner [21] for the case of Morava E-theory (in particular including p-complete complex K-theory).

16.6 Periodic unstable homotopy theory

In this section we discuss some fundamental concepts of chromatic homotopy theory and emphasize their role in unstable homotopy theory. The reader can find a much more thorough exposition of the chromatic perspective on stable homotopy theory in the chapter of Barthel and Beaudry in this volume or consult some of the standard references [31, 43, 45, 71, 72].

The rational homotopy groups of a pointed space X are the result of considering the homotopy classes of maps $[S^k, X]_*$ from spheres to X and subsequently inverting the action of the degree p maps $p : S^k \to S^k$ for all primes p. As demonstrated by Serre's calculation of the rational homotopy groups of spheres and the rational homotopy theory of Quillen and Sullivan, these rational homotopy groups are surprisingly tractable invariants.

If one wants to move beyond rational homotopy theory, a natural starting point is the 'mod p homotopy groups' $[S^k/p, X]_* =: \pi_{k-1}(X; \mathbb{Z}/p)$, where S^k/p denotes the cofiber of the degree p map on S^k (i.e. a mod p Moore space). These are groups when $k \geq 2$, which are abelian if $k \geq 3$. The complexity of calculating them is on par with that of the usual homotopy groups. However, it turns out one can again invert the action of a certain map to make the problem more tractable. To be precise, there is a certain map

$$\alpha : \Sigma^d S^k/p \to S^k/p$$

apparently due to Barratt, described by Adams in [2]. Here for p odd one has $d = 2(p-1)$ and $k \geq 3$, and for $p = 2$ one should take $d = 8$ and $k \geq 5$. The crucial feature of this map is that it induces an isomorphism on complex K-theory. In particular, any number of iterates of (suspensions of) α is *not* null-homotopic. In fact, α induces multiplication by the $(p-1)$st power of the Bott class (for p odd) or its fourth power (for $p = 2$). As such one can think of it as a geometric manifestation of Bott periodicity. The map α provides an action of the graded ring $\mathbb{Z}[\alpha]$ (with $|\alpha| = d$) on the graded abelian group $\pi_*(X; \mathbb{Z}/p)$ (with $* \geq 2$) and one defines the *v_1-periodic mod p homotopy groups* of X to be

$$v_1^{-1}\pi_*(X; \mathbb{Z}/p) := \mathbb{Z}[\alpha^{\pm 1}] \otimes_{\mathbb{Z}[\alpha]} \pi_*(X; \mathbb{Z}/p).$$

The rational and v_1-periodic homotopy groups of spaces form the beginning of a hierarchy of *v_n-periodic homotopy groups*, which we will define shortly. This hierarchy is closely related to the sequence of 'prime' localizations $L_{H\mathbb{Q}}$, $L_{K(1)}$, $L_{K(2)}$, ..., of stable homotopy theory at the Morava K-theories, which play the role of prime fields in the ∞-category of spectra. The latter picture is discussed in detail in the chapter of Barthel–Beaudry in this volume. While the focus for them is mostly on these chromatic localizations (and the closely related localizations L_n), for us the fundamental ingredient of chromatic homotopy theory will be the periodicity results of Hopkins and Smith [45] (and the finite localizations L_n^f). We begin with a brief recollection of the thick subcategory theorem.

From now on we fix a prime p and work in the ∞-categories of p-local spectra and spaces. We will often leave the adjective p-local implicit. We say a spectrum X is *of type* $\geq n$ if $K(i)_*X = 0$ for $i < n$, and we say X is *of type* n if it is of type $\geq n$ and additionally

$K(n)_*X \neq 0$. It turns out that if a finite spectrum X has $K(n-1)_*X = 0$ then also $K(i)_*X = 0$ for $i < n-1$, so that the former is a sufficient condition for X to be of type n. Write $\mathrm{Sp}^{\mathrm{fin}}_{(p)}$ for the ∞-category of p-local finite spectra and $\mathrm{Sp}^{\mathrm{fin}}_{\geq n}$ for the subcategory of those X that are of type $\geq n$. This subcategory is *thick*: if two terms in a cofiber sequence are contained in it, then so is the third, and moreover it is closed under retracts. These thick subcategories form a nested sequence

$$\cdots \subseteq \mathrm{Sp}^{\mathrm{fin}}_{\geq n+1} \subseteq \mathrm{Sp}^{\mathrm{fin}}_{\geq n} \subseteq \cdots \subseteq \mathrm{Sp}^{\mathrm{fin}}_{\geq 0} = \mathrm{Sp}^{\mathrm{fin}}_{(p)}.$$

The fact that these are proper inclusions is a highly nontrivial result of Mitchell [63]: for every n there exists a finite spectrum of type n. The following explains the fundamental importance of this notion for stable homotopy theory:

Theorem 16.6.1 (The thick subcategory theorem, Hopkins–Smith [45]). *Every thick subcategory of* $\mathrm{Sp}^{\mathrm{fin}}_{(p)}$ *is of the form* $\mathrm{Sp}^{\mathrm{fin}}_{\geq n}$ *for some* n.

The filtration of the ∞-category of finite p-local spectra above also gives a filtration

$$\cdots \subseteq \mathrm{Sp}_{\geq n+1} \subseteq \mathrm{Sp}_{\geq n} \subseteq \cdots \subseteq \mathrm{Sp}_{\geq 0} = \mathrm{Sp}_{(p)}$$

of the ∞-category of all (not necessarily finite) p-local spectra by localizing subcategories. Here we write $\mathrm{Sp}_{\geq n}$ for the smallest subcategory of Sp that contains $\mathrm{Sp}^{\mathrm{fin}}_{\geq n}$ and is closed under colimits. As with any filtration, the goal is now to understand its 'associated graded' and subsequently investigate how the layers fit together. In this survey we will be almost exclusively concerned with the first aspect.

To define what we mean by associated graded, first observe that the tower of subcategories above gives a corresponding tower of localizations

$$\cdots \to L^f_n\mathrm{Sp} \to L^f_{n-1}\mathrm{Sp} \to \cdots \to L^f_0\mathrm{Sp} = \mathrm{Sp}_{\mathbb{Q}}.$$

Here $L^f_n\mathrm{Sp}$ is the localization (in the sense of Definition 16.2.1) of $\mathrm{Sp}_{(p)}$ at the set of maps $f\colon X \to Y$ whose cofiber is contained in $\mathrm{Sp}_{\geq n+1}$. Said differently, it is the quotient $\mathrm{Sp}_{(p)}/\mathrm{Sp}_{\geq n+1}$ computed in the ∞-category of stable ∞-categories and exact functors. Since $\mathrm{Sp}_{\geq n+1}$ is closed under colimits in $\mathrm{Sp}_{(p)}$, all these localizations are reflective. We write $L^f_n\colon \mathrm{Sp}_{(p)} \to \mathrm{Sp}_{(p)}$ for the composite of the localization functor $\mathrm{Sp}_{(p)} \to L^f_n\mathrm{Sp}$ and its right adjoint. With this notation we have natural transformations $L^f_n \to L^f_{n-1}$. As a consequence of the thick subcategory theorem, one can in fact characterize L^f_n as the localization functor that kills a *single* finite type $n+1$ spectrum V. The layers of our filtration can now be described as follows:

Definition 16.6.2. The ∞-category Sp_{v_n} of *v_n-periodic spectra* is the quotient $\mathrm{Sp}_{\geq n}/\mathrm{Sp}_{\geq n+1}$, meaning it is the universal stable ∞-category that receives an exact functor from $\mathrm{Sp}_{\geq n}$ which is identically zero on the subcategory $\mathrm{Sp}_{\geq n+1}$.

This definition might seem abstract, but we will recast it in more concrete (and perhaps more familiar) terms using another fundamental result of Hopkins and Smith. To state it we need some terminology. A v_n *self-map* of a p-local spectrum X is a map $v\colon \Sigma^d X \to X$, for some integer $d \geq 0$, with the property that $K(n)_*v$ is an isomorphism and $K(m)_*v$ is nilpotent whenever $m \neq n$. If $n = 0$ one always has $d = 0$ and one should require that v acts by multiplication by a rational number on $K(0)_*X = H_*(X;\mathbb{Q})$. For general $n \geq 1$, replacing v by a sufficiently high power if necessary one can always arrange that on $K(n)_*X$ it acts by a power of $v_n \in K(n)_*$ and by zero on $K(m)_*X$ for $m \neq n$. The typical examples to keep in mind are the following: multiplication by p is a v_0 self-map (and exists for any spectrum), whereas the Adams map α is a v_1 self-map of the mod p Moore spectrum, which is of type 1.

Theorem 16.6.3 (The periodicity theorem, Hopkins–Smith [45]). *If X is a finite spectrum of type $\geq n$, then it admits a v_n self-map. Furthermore, v_n self-maps are asymptotically unique in the sense that for $f\colon X \to Y$ a map of finite spectra and v_n-self-maps v, w of X, Y respectively, there exist $M, N \gg 0$ for which the square*

$$\begin{array}{ccc} \Sigma^{Md}X & \xrightarrow{f} & \Sigma^{Ne}Y \\ \downarrow{\scriptstyle v^N} & & \downarrow{\scriptstyle w^M} \\ X & \xrightarrow{f} & Y \end{array}$$

commutes up to homotopy, where d and e are the degrees of v and w respectively. In particular, taking f to be the identity, any two v_n self-maps on X become homotopic after sufficiently many iterations.

Now suppose V is a finite spectrum of type n and $v\colon \Sigma^d V \to V$ is a v_n self-map. For any spectrum X we can define its v-periodic homotopy groups by

$$v^{-1}\pi_*(X;V) := \mathbb{Z}[v^{\pm 1}] \otimes_{\mathbb{Z}[v]} \pi_*\mathrm{Map}(V, X).$$

An equivalent description is as follows. Since V is finite, the mapping spectrum $F(V, X)$ is equivalent to the smash product $V^\vee \otimes X$, with $V^\vee$ the Spanier–Whitehead dual of V. Form the *telescope*

$$v^{-1}V^\vee := \varinjlim(V^\vee \xrightarrow{v^\vee} \Sigma^{-d}V^\vee \xrightarrow{v^\vee} \cdots).$$

Then $v^{-1}\pi_*(X;V) \cong \pi_*(v^{-1}V^\vee \otimes X)$.

Definition 16.6.4. A map of spectra is a v_n-*periodic equivalence* if it induces an isomorphism on v-periodic homotopy groups.

This definition really only depends on n; indeed, the thick subcategory theorem implies that it is independent of the choice of V, whereas the asymptotic uniqueness of v_n self-maps gives independence of the choice of v. It is common to write $T(n) = v^{-1}V^\vee$, so that a v_n-periodic equivalence is by definition a $T(n)_*$-equivalence. The notation is a little ambiguous, because $T(n)$ depends on choices. However, the associated notion of equivalence does not. The basic facts to keep in mind are the following:

(a) As a consequence of the periodicity theorem, any v_n self-map of a finite spectrum of type $\geq n$ is a v_n-periodic equivalence.

(b) If W is a finite spectrum of type $\geq n+1$, then $W \to 0$ is a v_n-periodic equivalence. This is immediate from (a) and the fact that the null map $W \xrightarrow{0} W$ is a v_n self-map.

We now characterize Sp_{v_n} in terms of the v_n-periodic equivalences and describe two ways in which it can be realized as a full subcategory of $\mathrm{Sp}_{(p)}$:

Proposition 16.6.5. *The ∞-category Sp_{v_n} of v_n-periodic spectra is the localization (in the sense of Definition 16.2.1) of $\mathrm{Sp}_{(p)}$ at the v_n-periodic equivalences. It is equivalent to the following two subcategories of $\mathrm{Sp}_{(p)}$:*

(1) The full subcategory $\mathrm{Sp}_{T(n)}$ of $T(n)$-local spectra in the sense of Bousfield.

(2) The full subcategory $M_n^f\mathrm{Sp}$ of spectra of the form $M_n^f X$, where M_n^f denotes the fiber of the natural transformation $L_n^f \to L_{n-1}^f$.

Furthermore, the functors

$$M_n^f\mathrm{Sp} \underset{M_n^f}{\overset{L_{T(n)}}{\rightleftarrows}} \mathrm{Sp}_{T(n)}$$

are mutually inverse equivalences.

Sketch of proof. The ∞-category $\mathrm{Sp}_{T(n)}$ is the localization at the v_n-periodic equivalences (which equal $T(n)_*$-equivalences) by definition. The equivalence between $\mathrm{Sp}_{T(n)}$ and $M_n^f\mathrm{Sp}$ follows from the fact that all of the maps in the diagram

$$\begin{array}{ccccc} & & X & & \\ & & \downarrow & & \\ M_n^f X & \longrightarrow & L_n^f X & \longrightarrow & L_{T(n)}X \end{array}$$

are v_n-periodic equivalences. Finally, Sp_{v_n} is equivalent to $M_n^f\mathrm{Sp}$. Indeed, by construction Sp_{v_n} is the essential image of $\mathrm{Sp}_{\geq n}$ in $L_n^f\mathrm{Sp}$, which we identify with the full subcategory of $\mathrm{Sp}_{(p)}$ consisting of L_n^f-local spectra. This essential image is contained in $M_n^f\mathrm{Sp}$, because for $V \in \mathrm{Sp}_{\geq n}$ the spectrum $L_{n-1}^f V$ is null. It is also all of $M_n^f\mathrm{Sp}$, since the fiber of the map $X \to L_{n-1}^f X$ is a colimit of spectra of type $\geq n$. □

Remark 16.6.6. The spectra $M_n^f X$ are often called the *monochromatic* or *monocular* layers of X (the latter term is used by Bousfield [19], who attributes it to Ravenel). The equivalence between $M_n^f\mathrm{Sp}$ and $\mathrm{Sp}_{T(n)}$ is analogous to the equivalence between p-primary torsion and (derived) p-complete objects of the derived category $D(\mathbb{Z})$.

We now proceed to the unstable case. The theory of localizations of unstable homotopy theory was developed by Bousfield [16, 17, 19] and Dror-Farjoun [33]. More details on the material we discuss below can also be found in [42] and [57].

A pointed space X is of type $\geq n$ precisely if its suspension spectrum is. Similarly, a v_n self-map of such a space is a self-map v such that $\Sigma^\infty v$ is a v_n self-map of the suspension spectrum. Again we would like to formally invert the v_n-periodic equivalences of pointed spaces. One can deduce from results of Bousfield that this is possible:

Theorem 16.6.7 (See [19] and [42]). *The localization of $\mathcal{S}_*$ at the v_n-periodic equivalences (in the sense of Definition 16.2.1) exists; we denote it by $M : \mathcal{S}_* \to \mathcal{S}_{v_n}$. It has the following properties:*

(1) The functor M preserves finite limits and filtered colimits.

(2) The ∞-category $\mathcal{S}_{v_n}$ is compactly generated. If V is any pointed finite type space of type n, then $M(\Sigma V)$ is a compact object and generates $\mathcal{S}_{v_n}$ under colimits.

(3) The stabilization of $\mathcal{S}_{v_n}$ is equivalent to Sp_{v_n}.

The reader should be warned that the localization M is *not* reflective; in particular, it does not arise directly from a left Bousfield localization on the level of model categories. Rather, one should think of it as the composition of a left and then a right localization, as will become apparent. Theorem 16.6.7 is proved by explicitly constructing an ∞-category analogous to $M_n^f\mathrm{Sp}$ in the stable case. A crucial ingredient is Bousfield's classification of the localizations associated to 'nullifying' a finite space. If A is any space, one calls a space X *P_A-local* (or *A-null*) if the evident map

$$X \simeq \mathrm{Map}(*, X) \to \mathrm{Map}(A, X)$$

is a homotopy equivalence. Essentially, the space A is contractible from the point of view of X. A general space X admits a P_A-localization $X \to P_AX$, and the subcategory of $\mathcal{S}_*$ on pointed P_A-local spaces is precisely the localization of $\mathcal{S}_*$ with respect to the map $A \to *$. We write $\langle A \rangle$ for the collection of spaces Y for which $P_AY \simeq *$ and call it the *Bousfield class* of A. These are partially ordered; we write $\langle A \rangle \leq \langle B \rangle$ if every Y for which $P_AY \simeq *$ also satisfies $P_BY \simeq *$. One can regard the following as an unstable analog of the thick subcategory theorem. Its proof relies on the stable version.

Theorem 16.6.8 (Bousfield [16, Theorem 9.15]). *Let W and W' be p-local finite pointed spaces that are also suspensions. Then the following are equivalent:*

(1) $\langle W \rangle \leq \langle W' \rangle$,

(2) $\mathrm{type}(W) \geq \mathrm{type}(W')$ and $\mathrm{conn}(W) \geq \mathrm{conn}(W')$, with $\mathrm{conn}(W)$ denoting the minimal i with $\pi_iW \neq 0$.

Thus, up to keeping track of connectivity, finite suspension spaces are still classified by their type, as was the case with finite spectra. Now pick a finite suspension V_{n+1} of type $n+1$. We will denote $\mathrm{conn}(V_{n+1})$ by d_{n+1} and the localization functor $P_{V_{n+1}}$ by L_n^f. The latter is of course slightly abusive, since L_n^f depends not only on n, but also on the connectivity d_{n+1} (but no more, according to Theorem 16.6.8). More importantly, Theorem 16.6.8 immediately implies that for any pointed space X, the v_i-periodic homotopy groups of L_n^fX vanish for $i > n$. Bousfield also proves that any pointed space X, the map $X \to L_n^fX$ is a v_i-periodic equivalence for $i \leq n$ (see [19]). In the stable case this implication can be reversed. Unstably one has to take a little care, because L_n^f does not affect the homotopy groups of X in dimensions below d_{n+1}. However, the following is true:

Proposition 16.6.9 (Bousfield [19]). *A map $\varphi\colon X \to Y$ of d_{n+1}-connected pointed spaces is a v_i-periodic equivalence for every $0 \leq i \leq n$ if and only if $L_n^f(\varphi)$ is an equivalence of pointed spaces.*

We define M_n^f in analogy with the stable case as the fiber of the natural transformation $L_n^f \to L_{n-1}^f$; the latter exists by Theorem 16.6.8 as long as we have arranged our choices so that $\mathrm{conn}(V_n) \leq \mathrm{conn}(V_{n+1})$, which we will always assume to be the case. For X a pointed space, there are maps

$$X \to L_n^f X \leftarrow M_n^f X$$

that are both v_n-periodic equivalences. Moreover, for $i \neq n$ the v_i-periodic homotopy groups of X vanish. Write $\mathcal{M}_n^f$ for the full subcategory of $\mathcal{S}_*$ on spaces of the form $(M_n^f X)\langle d_{n+1}\rangle$, where the brackets indicate the d_{n+1}-connected cover. The functor

$$X \mapsto (M_n^f X)\langle d_{n+1}\rangle$$

is naturally related to the identity by a zig-zag of v_n-periodic equivalences; moreover, a map between spaces in $\mathcal{M}_n^f$ is an equivalence if and only if it is a v_n-periodic equivalence. From this one deduces the first statement of Theorem 16.6.7. For the remainder of the proof we refer the reader to [42].

We conclude this section with a review of the Bousfield–Kuhn functor and its adjoint. We refer to [19] and [54] for a much more thorough discussion. Suppose V is a finite pointed space of type n with a v_n self-map $v\colon \Sigma^d V \to V$. Then for X a pointed space, one can define a (pre)spectrum $\Phi_v X$ with constituent spaces

$$(\Phi_v X)_0 = \mathrm{Map}_*(V, X),\ (\Phi_v X)_d = \mathrm{Map}_*(V, X),\ \ldots, (\Phi_v X)_{kd} = \mathrm{Map}_*(V, X),\ \ldots$$

and structure maps

$$(\Phi_v X)_{kd} = \mathrm{Map}_*(V, X) \xrightarrow{v^*} \mathrm{Map}_*(\Sigma^d V, X) \cong \Omega^d(\Phi_v X).$$

This defines the *telescopic functor*

$$\Phi_v : \mathcal{S}_* \to \mathrm{Sp}$$

associated to v. By construction it satisfies $\pi_*\Phi_v X \cong v^{-1}\pi_*(X; V)$. In fact this functor takes values in $T(n)$-local spectra, so that one may replace the codomain by $\mathrm{Sp}_{T(n)}$ or, equivalently, Sp_{v_n}. As a consequence of Theorem 16.6.3 the telescopic functor Φ_v does not really depend on the choice of v, but only on V. We will write Φ_V instead of Φ_v. In fact, the dependence on V can be made (contravariantly) functorial using Theorem 16.6.3. This can be used to conveniently package the various telescopic functors into one. The *Bousfield–Kuhn functor* is a functor

$$\Phi_n : \mathcal{S}_* \to \mathrm{Sp}_{v_n}$$

satisfying the following properties (see [54]):

(1) There are equivalences

$$V^\vee \otimes \Phi_n X \simeq \Phi_V X,$$

natural in X and V.

(2) The composition $\Phi_n \Omega^\infty$ is naturally equivalent to the localization functor $L_{T(n)}$. In particular, the $T(n)$-localization of any spectrum E depends only on its zeroth space $\Omega^\infty E$, regardless of its structure as an infinite loop space.

The functor Φ is even characterized up to equivalence by property (1). The construction of Φ_n and the proof of its basic properties rely on the following trick of Kuhn:

Lemma 16.6.10 (Kuhn [50]). *For any $n \geq 0$ there exists a directed system*

$$F(1) \to F(2) \to F(3) \to \cdots$$

of finite spectra of type n equipped with a map

$$\varinjlim_k F(k) \to \mathbb{S}$$

that is a $T(m)_*$-equivalence for every $m \geq n$.

Sketch of proof. In case $n = 1$ this is familiar from algebra: the colimit $\varinjlim_k \mathbb{S}^{-1}/p^k$ is equivalent to the sphere spectrum after p-completion, in the same way that the derived p-completion of $\mathbb{Z}/p^\infty[-1]$ is $\mathbb{Z}_p$ in the ∞-category $D(\mathbb{Z})$. The general proof goes by induction on n: given a finite spectrum V of type $n-1$, one picks a v_{n-1}-self-map v. The cofibers $\Sigma^{-1}V/v^k$ of the maps

$$\Sigma^{-1}V \xrightarrow{v^k} \Sigma^{-1-kd}V$$

form a directed system with a map to V. The map from the colimit $\Sigma^{-1}V/v^\infty$ to V is a $T(m)_*$-equivalence for $m \geq n$ because the telescope $v^{-1}V$ is $T(m)$-acyclic. □

To construct Φ_n one picks a system as in Lemma 16.6.10 and defines $\Phi_n := \varprojlim_k \Phi_{F(k)}$. Property (1) is now fairly easily deduced from the identification

$$V^\vee \otimes \Phi_{F(k)} \simeq F(k)^\vee \otimes \Phi_V.$$

Note also that Lemma 16.6.10 implies that Φ_n is completely determined by the functors $F(k)^\vee \otimes \Phi$, which shows that (1) indeed characterizes Φ_n up to equivalence. This characterization also shows that the choice of $F(k)$ is inessential to the definition of Φ_n. Property (2), striking as it may be, is quite easily proved as well. Again it suffices to check it for Φ_V. We should argue that for any spectrum E there is a natural equivalence $\Phi_V\Omega^\infty(E) \simeq L_{T(n)}V^\vee \otimes E$. Since $\Phi_V\Omega^\infty$ preserves v_n-periodic equivalences, we may without loss of generality assume that E is already $T(n)$-local. But then the maps

$$\mathrm{Map}_*(V, \Omega^\infty E) \xrightarrow{v^*} \Omega^d \mathrm{Map}_*(V, \Omega^\infty E)$$

are equivalences and the formula we wrote down for $\Phi_V\Omega^\infty(E)$ is already an Ω-spectrum. Even better, it is clearly the Ω-spectrum $V^\vee \otimes E$.

Since Φ_n sends v_n-periodic equivalences to equivalences (essentially by construction), it factors through the localization $\mathcal{S}_{v_n}$. We still denote the resulting functor by

$$\Phi_n \colon \mathcal{S}_{v_n} \to \mathrm{Sp}_{v_n}.$$

The following will be crucial:

Proposition 16.6.11 (Bousfield [19]). *The functor Φ_n admits a left adjoint $\Theta_n\colon \mathrm{Sp}_{v_n} \to \mathcal{S}_{v_n}$.*

Sketch of proof. An explicit construction of Θ_n is described in [19] and also in [54]. To argue existence, roughly one can do the following. It suffices to show that for every spectrum E, the functor

$$\mathcal{S}_{v_n} \to \mathcal{S} : X \mapsto \mathrm{Map}(E, \Phi_n X)$$

is corepresentable by some object $\Theta_n(E)$. Moreover, it suffices to consider a collection of objects E that generate Sp_{v_n} under colimits, such as the finite spectra of type n. For such an E, the functor above takes the form $X \mapsto \Omega^\infty \Phi_E X$. Considering the definition of the telescopic functor Φ_E, one sees that for $X \in \mathcal{S}_{v_n}$ the colimit defining this functor becomes eventually constant, showing that indeed our functor is corepresented by (the image in $\mathcal{S}_{v_n}$ of) a finite type n space E' for which $\Sigma^\infty E' \simeq E$ in Sp_{v_n}. □

We now have two adjunctions relating the ∞-category $\mathcal{S}_{v_n}$ of v_n-periodic spaces to its stable counterpart Sp_{v_n}, organized in the following diagram:

$$\mathrm{Sp}_{v_n} \underset{\Phi_n}{\overset{\Theta_n}{\rightleftarrows}} \mathcal{S}_{v_n} \underset{\Omega^\infty_{v_n}}{\overset{\Sigma^\infty_{v_n}}{\rightleftarrows}} \mathrm{Sp}_{v_n}.$$

The adjunction on the right is the stabilization of $\mathcal{S}_{v_n}$, which we already mentioned in Theorem 16.6.7. Property (2) of the Bousfield–Kuhn functor in fact implies that the horizontal composites $\Phi_n\Omega^\infty_{v_n}$ and $\Sigma^\infty_{v_n}\Theta_n$ are both equivalent to the identity functor of Sp_{v_n}. In this way the diagram above is very much analogous to the one at the end of Section 16.2 and the one right above Theorem 16.4.3.

16.7 Lie algebras and v_n-periodic spaces

The aim of this section is to outline a proof of Theorem 16.1.2, relating the ∞-category $\mathcal{S}_{v_n}$ to the ∞-category of spectral Lie algebras. More details can be found in [42]. We also discuss to what extent there is a 'model' for $\mathcal{S}_{v_n}$ in terms of commutative coalgebras, Koszul dual to the Lie algebra model provided by Theorem 16.1.2. This second model is closely related to recent work of Behrens–Rezk [11], as we will explain.

The proof of Theorem 16.1.2 begins with the following, showing that the ∞-category $\mathcal{S}_{v_n}$ can indeed be expressed as some kind of algebras in Sp_{v_n}. Throughout this section we will use the ∞-categories Sp_{v_n} and $\mathrm{Sp}_{T(n)}$ interchangeably.

Theorem 16.7.1 (Eldred–Heuts–Mathew–Meier [32]). *The adjoint pair (Θ_n, Φ_n) is monadic, i.e., the functor*

$$\varphi : \mathcal{S}_{v_n} \to \mathrm{Alg}_{\Phi_n\Theta_n}(\mathrm{Sp}_{v_n})$$

induced by Φ_n is an equivalence of ∞-categories.

Sketch of proof. By Theorem 16.3.2 it suffices to check that Φ_n is conservative (which is essentially immediate from the construction of $\mathcal{S}_{v_n}$) and that it preserves geometric realizations (i.e., colimits of simplicial objects). Since a map of $T(n)$-local spectra is an equivalence if and only if it is an equivalence after smashing with some finite type n spectrum, it suffices to show that Φ_V preserves geometric realizations, for V a finite space of type n. This functor can be expressed as the following colimit:

$$\Phi_V(X) = \varinjlim(\Sigma^\infty \mathrm{Map}_*(V, X) \to \Sigma^{\infty-d}\mathrm{Map}_*(V, X) \to \cdots).$$

Therefore it suffices to show that the functor

$$L_{T(n)}\Sigma^\infty \mathrm{Map}_*(V, -) : \mathcal{S}_{v_n} \to \mathrm{Sp}_{T(n)}$$

preserves geometric realizations. Generally, a functor of the form $\mathrm{Map}_*(V, -)$ only preserves geometric realizations of diagrams of spaces that are at least $\dim(V)$-connected. But this will suffice; we can take the dimension of the space V_{n+1} used to define the localization L_n^f to be at least the dimension of V. □

Remark 16.7.2. Although the proof of Theorem 16.7.1 is quite formal, the conclusion is perhaps surprising; it states that the v_n-periodic part $\mathcal{S}_{v_n}$ of unstable homotopy theory can be completely described in terms of stable homotopy theory, namely the stable ∞-category Sp_{v_n} and the monad $\Phi_n\Theta_n$.

It now remains to argue that $\Phi_n\Theta_n$ is in fact the free spectral Lie algebra monad on Sp_{v_n}. This will use some special features of the ∞-category of functors from the ∞-category of $T(n)$-local spectra to itself. It turns out that in this context the relation between operads and monads is much tighter than in a general symmetric monoidal ∞-category. To explain the situation we introduce some terminology:

Definition 16.7.3. A functor $F : \mathrm{Sp}_{T(n)} \to \mathrm{Sp}_{T(n)}$ is *coanalytic* if it is equivalent to one of the form $F_{\mathcal{O}}$, with $\mathcal{O}$ a symmetric sequence of $T(n)$-local spectra:

$$F(X) \simeq L_{T(n)} \bigoplus_{k\geq 1} (\mathcal{O}(k) \otimes X^{\otimes k})_{h\Sigma_k}.$$

We write $\mathrm{coAn}(\mathrm{Sp}_{T(n)})$ for the full subcategory of $\mathrm{Fun}(\mathrm{Sp}_{T(n)}, \mathrm{Sp}_{T(n)})$ on the coanalytic functors.

We discussed the assignment

$$\mathrm{SymSeq}(\mathcal{C}) \to \mathrm{Fun}(\mathcal{C}, \mathcal{C}) \colon \mathcal{O} \mapsto F_{\mathcal{O}}$$

in Section 16.4. For general $\mathcal{C}$ this is far from fully faithful. However, the $T(n)$-local setting is quite special:

Proposition 16.7.4. *The functor above gives an equivalence of ∞-categories* $\mathrm{SymSeq}(\mathrm{Sp}_{T(n)}) \to \mathrm{coAn}(\mathrm{Sp}_{T(n)})$.

Sketch of proof. The proof consists of two ingredients. First, one needs the fact that any natural transformation

$$(\mathcal{O}(k) \otimes X^{\otimes k})_{h\Sigma_k} \to (\mathcal{P}(l) \otimes X^{\otimes l})_{h\Sigma_l}$$

between such homogeneous functors is null whenever $k \neq l$. This follows from the general theory of Goodwillie calculus when $k > l$. For the case $k < l$ one needs the additional fact that Tate spectra associated with the symmetric groups vanish in the $T(n)$-local category. This is a fundamental result of Kuhn [51]; a short alternative proof is provided by Clausen–Mathew [26]. The second ingredient is that any natural transformation from a k-homogeneous functor

$$X \mapsto (\mathcal{O}(k) \otimes X^{\otimes k})_{h\Sigma_k}$$

to a coanalytic functor factors through a finite sum of layers. This uses a nilpotence argument of Mathew, which ultimately relies on Tate vanishing again (see the appendix of [42]). □

The equivalence of Proposition 16.7.4 sends the composition product of symmetric sequences to the composition of functors. Hence we find the following alternative description of (co)operads in the $T(n)$-local setting:

Corollary 16.7.5. *The ∞-category of operads (resp. of cooperads) in $T(n)$-local spectra is equivalent to the ∞-category of monoids (resp. comonoids) in the ∞-category* $\mathrm{coAn}(\mathrm{Sp}_{T(n)})$. *In other words, a (co)monad on $\mathrm{Sp}_{T(n)}$ whose underlying functor is coanalytic corresponds essentially uniquely to a (co)operad in $\mathrm{Sp}_{T(n)}$.*

The obvious example, which is both an operad and a cooperad, is of course the identity functor of $\mathrm{Sp}_{T(n)}$. The first nontrivial example is the following:

Theorem 16.7.6 (Kuhn [53]). *For $E \in \mathrm{Sp}_{T(n)}$ there is a natural equivalence*

$$\Sigma^\infty_{v_n}\Omega^\infty_{v_n}(E) \simeq L_{T(n)} \bigoplus_{k\geq 1} E^{\otimes k}_{h\Sigma_k}.$$

In particular the comonad $\Sigma^\infty_{v_n}\Omega^\infty_{v_n}$ is coanalytic, hence a cooperad.

This result essentially follows from a theorem of Kuhn [53] on the splitting of the functor $L_{T(n)}\Sigma^\infty\Omega^\infty$ on a suitable class of spectra. Here we outline a different approach:

Sketch of proof. For a commutative ring spectrum R and a spectrum X, there is the well-known adjunction

$$\mathrm{Map}_{\mathrm{CAlg}}(\Sigma^\infty_+\Omega^\infty X, R) \simeq \mathrm{Map}_{\Omega^\infty}(\Omega^\infty X, \mathrm{GL}_1 R)$$

where the right-hand side denotes the space of infinite loop maps from $\Omega^\infty X$ into the space of units $\mathrm{GL}_1 R$. A rather straightforward adaptation of this setup to our context provides an adjunction

$$\mathrm{Map}_{\mathrm{CAlg}^{\mathrm{nu}}}(\Sigma^\infty_{v_n}\Omega^\infty_{v_n} E, R) \simeq \mathrm{Map}_{\Omega^\infty}(\Omega^\infty_{v_n} E, M(\mathrm{GL}_1 R))$$

for a non-unital $T(n)$-local commutative ring spectrum R and $E \in \mathrm{Sp}_{T(n)}$. Here $M : \mathcal{S}_* \to \mathcal{S}_{v_n}$ denotes the localization functor. In fact, $M(\mathrm{GL}_1 R)$ can be identified with $\Omega^\infty_{v_n} R$ as an object of $\mathcal{S}_{v_n}$, but the $\mathbf{E}_\infty$-structure corresponds to the multiplication, rather than addition, on R. Applying Φ_n and using $\Phi_n \Omega^\infty_{v_n} \simeq \mathrm{id}$ then gives a further equivalence

$$\mathrm{Map}_{\Omega^\infty}(\Omega^\infty_{v_n} E, M(\mathrm{GL}_1 R)) \simeq \mathrm{Map}(E, R),$$

so that we have found a natural equivalence

$$\mathrm{Map}_{\mathrm{CAlg}^{\mathrm{nu}}}(\Sigma^\infty_{v_n}\Omega^\infty_{v_n} E, R) \simeq \mathrm{Map}(E, R).$$

On the other hand, the universal property of the free non-unital commutative algebra also provides an equivalence

$$\mathrm{Map}_{\mathrm{CAlg}^{\mathrm{nu}}}(L_{T(n)}\mathrm{Sym}_{\geq 1} E, R) \simeq \mathrm{Map}(E, R)$$

with

$$L_{T(n)}\mathrm{Sym}_{\geq 1} E = L_{T(n)} \bigoplus_{k \geq 1} E^{\otimes k}_{h\Sigma_k}$$

denoting the spectrum of the theorem. The conclusion now follows from the Yoneda lemma. □

Remark 16.7.7. Chasing through the proof above gives an explicit description of the equivalence of the theorem. Applying Φ_n to the unit map $\eta : \Omega^\infty_{v_n} E \to \Omega^\infty_{v_n}\Sigma^\infty_{v_n}\Omega^\infty_{v_n} E$ and using $\Phi_n \Omega^\infty_{v_n} \simeq \mathrm{id}$ gives a natural map

$$\lambda : E \to L_{T(n)}\Sigma^\infty_{v_n}\Omega^\infty_{v_n} E.$$

Since the right-hand side is a non-unital commutative ring spectrum, this map λ naturally extends to a map of non-unital commutative rings

$$L_{T(n)}\mathrm{Sym}_{\geq 1}(E) \to \Sigma^\infty_{v_n}\Omega^\infty_{v_n} E$$

which is an equivalence by the theorem.

In fact, as already suggested by the formula of Theorem 16.7.6, the cooperad $\Sigma^\infty_{v_n}\Omega^\infty_{v_n}$ really plays the role of the commutative cooperad: the terms of the corresponding symmetric sequence are (the $T(n)$-localization of) the sphere spectrum in each degree and one can show (e.g. using Goodwillie calculus) that the cooperad structure maps are as expected. We will use Theorem 16.7.6 in two ways. The first is a very useful characterization of coanalytic functors:

Proposition 16.7.8. *A functor $F \colon \mathrm{Sp}_{T(n)} \to \mathrm{Sp}_{T(n)}$ is coanalytic if and only if it preserves filtered colimits and geometric realizations.*

A proof of this result is given in [42] following an argument of Lurie; essentially, one writes any functor F preserving filtered colimits and geometric realizations as a colimit of functors closely resembling $\Sigma^\infty_{v_n}\Omega^\infty_{v_n}$ and uses the fact that a colimit of coanalytic functors is coanalytic.

Our goal is to analyze the monad $\Phi_n\Theta_n$. First, one observes it is actually corresponds to an operad (cf. Corollary 16.7.5):

Corollary 16.7.9. *The functor $\Phi_n\Theta_n$ is coanalytic.*

Proof. By Proposition 16.7.8 it suffices to check that $\Phi_n\Theta_n$ preserves filtered colimits and geometric realizations. For Θ_n there is nothing to check since it is a left adjoint; for Φ_n, the fact that it preserves geometric realizations was part of the proof of Theorem 16.7.1. For filtered colimits one may reduce to Φ_V as usual, where it is obvious. □

It remains to relate the operad $\Phi_n\Theta_n$ to the spectral Lie operad. In fact, we will indicate how to produce a map of operads

$$\gamma : \Phi_n\Theta_n \to \mathrm{Cobar}(\Sigma^\infty_{v_n}\Omega^\infty_{v_n}).$$

The right-hand side is the $T(n)$-local spectral Lie operad, so we should then prove that γ is an equivalence. To do this we proceed as follows. As explained in Section 16.3, the adjoint pair $(\Sigma^\infty_{v_n}, \Omega^\infty_{v_n})$ gives a 'cosimplicial resolution' of the identity functor of $\mathcal{S}_{v_n}$:

$$\mathrm{id}_{\mathcal{S}_{v_n}} \longrightarrow \Omega^\infty_{v_n}\Sigma^\infty_{v_n} \rightrightarrows \Omega^\infty_{v_n}\Sigma^\infty_{v_n}\Omega^\infty_{v_n}\Sigma^\infty_{v_n} \Rrightarrow \Omega^\infty_{v_n}(\Sigma^\infty_{v_n}\Omega^\infty_{v_n})^2\Sigma^\infty_{v_n} \cdots.$$

Now precompose with Θ_n, postcompose with Φ_n, and apply the equivalence $\Phi_n\Omega^\infty_{v_n} \simeq \mathrm{id}_{\mathrm{Sp}_{T(n)}} \simeq \Sigma^\infty_{v_n}\Theta_n$ to get the coaugmented cosimplicial object

$$\Phi_n\Theta_n \longrightarrow \mathrm{id}_{\mathrm{Sp}_{T(n)}} \rightrightarrows \Sigma^\infty_{v_n}\Omega^\infty_{v_n} \Rrightarrow (\Sigma^\infty_{v_n}\Omega^\infty_{v_n})^2 \cdots.$$

The totalization produces $\mathrm{Cobar}(\Sigma^\infty_{v_n}\Omega^\infty_{v_n})$, the coaugmentation gives the map γ. The following is then the final step in the proof of Theorem 16.1.2:

Theorem 16.7.10. *The map γ is an equivalence, so that $\Phi_n\Theta_n$ is equivalent to the $T(n)$-localization of the spectral Lie operad* $\mathbf{L}$.

Sketch of proof. Checking whether a natural transformation between coanalytic functors is an equivalence can be done at the level of Goodwillie derivatives. On the left-hand side, the fact that Θ_n preserves colimits and Φ_n preserves limits, as well as filtered colimits, implies that

$$D_k(\Phi_n\Theta_n) \simeq \Phi_n \circ D_k\mathrm{id}_{\mathcal{S}_{v_n}} \circ \Theta_n.$$

This reduces the verification to checking that

$$\partial_*\mathrm{id}_{\mathcal{S}_{v_n}} \to \mathrm{Cobar}(\mathbf{1}, \partial_*(\Sigma^\infty_{v_n}\Omega^\infty_{v_n}), \mathbf{1})$$

is an equivalence, which is yet another version of the result of Arone–Ching (Theorem 0.3 of [5]) already mentioned in Section 16.5. □

Remark 16.7.11. Theorem 16.1.2 is proved in the rather abstract setting of $T(n)$-local homotopy theory. However, one can specialize it to obtain a version for $K(n)$-local homotopy theory as well.

Remark 16.7.12. After Theorem 16.1.1 we indicated that the Lie model of a rational space in particular encodes the Whitehead products on rational homotopy groups. Similarly, the spectral Lie algebra model of Theorem 16.1.2 will in particular encode the Whitehead products on v_n-periodic homotopy groups. A concise proof of this can be given by 'differentiating' the Hilton–Milnor theorem.

In the remainder of this section we discuss to what extent there is a model for $\mathcal{S}_{v_n}$ in terms of commutative coalgebras, Koszul dual to the Lie algebra model provided by Theorem 16.1.2. This is closely related to recent work of Behrens–Rezk [11], as we will explain.

As in the rational case (cf. Section 16.2), one can construct a functor

$$C_{v_n} : \mathcal{S}_{v_n} \to \mathrm{coCAlg}^{\mathrm{nu}}(\mathrm{Sp}_{v_n})$$

by observing that each space $X \in \mathcal{S}_{v_n}$ is a non-unital commutative algebra with respect to smash product (using the diagonal as comultiplication) and that the suspension spectrum functor $\Sigma^{\infty}_{v_n}$ preserves smash products. Alternatively, identifying $\mathcal{S}_{v_n}$ with $\mathrm{Lie}(\mathrm{Sp}_{v_n})$ as in Theorem 16.1.2, one can use Theorem 16.4.7 to produce a functor from $\mathcal{S}_{v_n}$ to the ∞-category of commutative coalgebras in Sp_{v_n}. The functor C_{v_n} admits a right adjoint for which we write R_{v_n}. Its existence follows from the adjoint functor theorem; alternatively, it can be constructed much more concretely as the *primitives* of a coalgebra. For this reason we denote the composite functor ΦR_{v_n} by

$$\mathrm{prim}_{v_n} : \mathrm{coCAlg}^{\mathrm{nu}}(\mathrm{Sp}_{v_n}) \to \mathrm{Sp}_{v_n}.$$

We remind the reader that these primitives are computed by a cobar construction, which is formally dual to the derived indecomposables (or TAQ) of a non-unital commutative ring spectrum, as discussed in Section 16.4. The left adjoint of prim_{v_n} is the functor which equips a spectrum $E \in \mathrm{Sp}_{v_n}$ with the trivial non-unital commutative algebra structure.

The unit of the adjunction described above is a map

$$\mathrm{id}_{\mathcal{S}_{v_n}} \to R_{v_n} C_{v_n},$$

that admits the following descriptions:

(1) It is the *Goodwillie completion*, in the sense that the functor $R_{v_n} C_{v_n}$ is the limit of the Goodwillie tower of the identity on the ∞-category $\mathcal{S}_{v_n}$.

(2) After identifying $\mathcal{S}_{v_n}$ with the ∞-category $\mathrm{Lie}(\mathrm{Sp}_{v_n})$, it is the *pronilpotent completion* of a spectral Lie algebra. This is defined as follows. For every $k \geq 1$ one considers the truncation $t_k \colon \mathbf{L} \to \tau_{\leq k}\mathbf{L}$, which is an equivalence in arities up to k and has $\tau_{\leq k}\mathbf{L}(j) \cong 0$ for $j > k$. Any $\mathbf{L}$-algebra X then admits a truncation $t_k^*(t_!)_k X$, by pushing forward and pulling back along t_k. The inverse limit over k is the pronilpotent completion of X. The connection with (1) is that the functor $t_k^*(t_k)_!$ is the k-excisive approximation (in the sense of Goodwillie) of the identity functor of the ∞-category of spectral Lie algebras. (This perspective originates in [67].)

It is not difficult to construct examples of objects in $\mathrm{Lie}(\mathrm{Sp}_{v_n})$ that are *not* complete in the above sense (e.g. most free spectral Lie algebras). Consequently, the functor C_{v_n} cannot be an equivalence of ∞-categories. Nonetheless, the comparison with coalgebras is a very useful way of 'approximating' the ∞-category $\mathcal{S}_{v_n}$ and an effective method of computing the value of the Bousfield–Kuhn functor Φ on many spaces of interest, including spheres. We will conclude this section by making this precise, inspired by the work of Behrens–Rezk [11].

Consider the following diagram of adjoint pairs, with left adjoints on top or on the left:

$$\begin{array}{ccccccc} & & & & \mathcal{S}_* & & \\ & & & & L_n^f \downarrow\uparrow & & \\ \mathrm{Sp}_{T(n)} & \underset{\Phi_n}{\overset{\Theta_n}{\rightleftarrows}} & \mathcal{S}_{v_n} & \underset{M}{\rightleftarrows} & L_n^f\mathcal{S}_* & \underset{R_{L_n^f}}{\overset{C_{L_n^f}}{\rightleftarrows}} & \mathrm{coCAlg}^{\mathrm{nu}}(\mathrm{Sp}_{T(n)}). \end{array}$$

Here L_n^f is the localization away from a finite suspension space of type $n+1$, as in the previous section. The functor $C_{L_n^f}$ assigns to an L_n^f-local space X the $T(n)$-localization of its suspension spectrum $L_{T(n)}\Sigma^\infty X$, with coalgebra structure using the diagonal of X as usual. Recall our slight abuse of notation, writing Φ_n for the functor $\mathcal{S}_{v_n} \to \mathrm{Sp}_{T(n)}$ in the diagram, as well as for the composition

$$\mathcal{S}_* \xrightarrow{L_n^f} L_n^f\mathcal{S}_* \xrightarrow{M} \mathcal{S}_{v_n} \xrightarrow{\Phi_n} \mathrm{Sp}_{T(n)}.$$

Using the unit of the adjoint pair $(C_{L_n^f}, R_{L_n^f})$ we find for every pointed space X the *comparison map*

$$\Phi_n(X) \to \Phi_n R_{L_n^f} C_{L_n^f}(L_n^f X) \cong \mathrm{prim}_{v_n} C_{L_n^f}(L_n^f X).$$

The crucial point is that the right-hand side can be identified with the limit of the Goodwillie tower

$$\cdots \to (P_3\Phi_n)(X) \to (P_2\Phi_n)(X) \to (P_1\Phi_n)(X) \simeq L_{T(n)}\Sigma^\infty X$$

of Φ_n. A variant of this idea was established in the work of Behrens–Rezk [11]. The perspective sketched here is discussed in Section 5 of [42].

Definition 16.7.13. A pointed space X is Φ_n*-good* if the map

$$\Phi_n X \to \varprojlim_k (P_k\Phi_n)(X)$$

is an equivalence. If this map is an equivalence after $K(n)$-localization, then X is $\Phi_{K(n)}$*-good.*

A consequence of the discussion above is therefore:

Theorem 16.7.14. *A pointed space X is Φ-good if and only if the comparison map*

$$\Phi_n(X) \to \mathrm{prim}_{v_n} C_{L_n^f}(L_n^f X)$$

is an equivalence.

The work of Behrens–Rezk [11] is in the $K(n)$-local setting and uses commutative algebras, rather than coalgebras. The dual of the commutative coalgebra $\Sigma^\infty X$ is the commutative non-unital ring spectrum $\mathbb{S}^X$. By taking Spanier–Whitehead duals, the derived primitives of $\Sigma^\infty X$ admit a map

$$\mathrm{prim}(\Sigma^\infty X) \to \mathrm{TAQ}(\mathbb{S}^X)^\vee$$

to the dual of the derived indecomposables of $\mathbb{S}^X$. In this way one gets the comparison map of the following result:

Corollary 16.7.15 (Behrens–Rezk [11]). *If X is a pointed space for which $L_{K(n)}\Sigma^\infty X$ is $K(n)$-locally dualizable, then X is $\Phi_{K(n)}$-good if and only if the comparison map*

$$L_{K(n)}\Phi_n(X) \to L_{K(n)}(\mathrm{TAQ}(\mathbb{S}^X))^\vee$$

is an equivalence.

A discussion of these results from a different perspective is in the chapter ofArone and Ching in this Handbook. The use of Corollary 16.7.15 is that the codomain of the comparison map is amenable to explicit calculation, using the techniques developed by Behrens–Rezk in [11]. Some examples of Φ_n-goodness are the following:

(1) Spheres are Φ_n-good. This is a key result of Arone–Mahowald [7]. They prove something even stronger, namely that the map $\Phi_n(X) \to (P_k\Phi_n)(X)$ is an equivalence if X is a sphere S^l and $k \geq 2p^n$, or even $k \geq p^n$ if l is odd.

(2) Spaces of the form $\Theta_n(\mathbb{S}^l)$ are Φ_n-good.

There are also many non-examples. The following were essentially first observed in [22]:

(1) Suppose V is a type n space with a v_n self-map and write $W = \Sigma^2 V$. Then W is *not* Φ_n-good. Briefly, under these conditions one has $W \simeq \Theta_n(\Sigma^\infty_{v_n} W)$. Translating to $\mathrm{Lie}(\mathrm{Sp}_{v_n})$, W corresponds to a *free* spectral Lie algebra. In particular,

$$\Phi_n(W) \simeq L_{T(n)} \bigoplus_{k\geq 1} (\mathbf{L}(k) \otimes \Sigma^\infty_{v_n} W^{\otimes k})_{h\Sigma_k}.$$

The limit of the Goodwillie tower of Φ_n evaluated on W will give the direct product, rather than the direct sum.

(2) A wedge of spheres is not Φ_n-good. Using the Hilton–Milnor theorem, one can describe $\Phi(S^a \vee S^b)$ as a direct sum of terms of the form $\Phi_n(S^l)$, for l ranging over an infinite set determined by a and b. Again, the limit of the Goodwillie tower will instead be the direct product.

There exist more extreme counterexamples of the following kind. Take X to be the cofiber (in $\mathcal{S}_{v_n}$) of a map that is a $T(n)_*$-equivalence, but not a v_n-periodic equivalence. Such maps were essentially first described for $n = 1$ by Langsetmo–Stanley [55], by modifying the Adams self-map of a Moore space. The Goodwillie tower of Φ evaluated on X will then vanish identically, although X itself is non-trivial.

16.8 Questions

In this final section we list several open questions, loosely grouped by subject.

(A) Localization and completion. The first few questions we list below are closely related to some of those raised by Behrens–Rezk in their survey [12]. In the ∞-category $\mathcal{S}_{v_n}$ there are the following analogues of the usual notions of localization and completion:

(a) Every object $X \in \mathcal{S}_{v_n}$ admits a *$T(n)$-localization*. Note that this is the same as Bousfield localization of objects in $\mathcal{S}_{v_n}$ with respect to the stabilization functor $\Sigma^\infty_{v_n}: \mathcal{S}_{v_n} \to \mathrm{Sp}_{T(n)}$.

(b) The *Q_{v_n}-completion* of an object $X \in \mathcal{S}_{v_n}$ is the limit of the Bousfield–Kan cosimplicial object

$$\Omega^\infty_{v_n}\Sigma^\infty_{v_n}X \rightrightarrows (\Omega^\infty_{v_n}\Sigma^\infty_{v_n})^2X \,\substack{\to\\\to\\\to}\, \cdots.$$

(c) The *Goodwillie completion* of an object $X \in \mathcal{S}_{v_n}$ is the limit of the Goodwillie tower

$$\cdots \to P_3\mathrm{id}_{\mathcal{S}_{v_n}}(X) \to P_2\mathrm{id}_{\mathcal{S}_{v_n}}(X) \to P_1\mathrm{id}_{\mathcal{S}_{v_n}}(X) = \Omega^\infty_{v_n}\Sigma^\infty_{v_n}X.$$

of the identity of $\mathcal{S}_{v_n}$ evaluated on X. Alternatively, identifying $\mathcal{S}_{v_n}$ with $\mathrm{Lie}(\mathrm{Sp}_{v_n})$, the Goodwillie completion is the pronilpotent completion of a spectral Lie algebra.

If X is Q_{v_n}-complete, or Goodwillie complete, then it is also $T(n)$-local. An argument of Bousfield can be adapted to show that every H-space in $\mathcal{S}_{v_n}$ is already $T(n)$-local. An object X is Goodwillie complete if and only if it is Φ_n-good in the sense of Definition 16.7.13.

(A1) It is not hard to argue that the class of Φ_n-good spaces is closed under finite products. What other general closure properties does this class have?

(A2) Do the Q_{v_n}-completion and the Goodwillie completion agree?

(A3) Under what conditions does the $T(n)$-localization of X agree with its Q_{v_n}-completion or its Goodwillie completion?

(A4) What is the relation between the spectral sequence associated with the cosimplicial object of (b) (which is a version of the unstable Adams spectral sequence for the ∞-category $\mathcal{S}_{v_n}$) and the v_n-periodic unstable Adams and Adams–Novikov spectral sequences studied by Bendersky–Curtis–Miller [14], Bendersky [13], and Davis–Mahowald [30]?

We say a spectral Lie algebra X is *nilpotent* if it is in the essential image of the pullback functor

$$t_k^*: \mathrm{Alg}(\tau_{\leq k}\mathbf{L}) \to \mathrm{Alg}(\mathbf{L})$$

for some $k \geq 1$. With this terminology, the $T(n)$-local spectral Lie algebras corresponding to spheres are nilpotent, by the results of Arone–Mahowald discussed in the previous section. We know that these particular examples are Φ_n-good.

(A5) Is any nilpotent $X \in \mathrm{Lie}(\mathrm{Sp}_{T(n)})$ complete in the sense of either (b) or (c)?

This would provide a large class of examples of Φ_n-good spaces, since the class of nilpotent spectral Lie algebras satisfies various closure properties not obviously shared by the class of Φ_n-good spaces.

(B) Exponents. The torsion part of the homotopy groups of S^n has a p-exponent. Better yet, Cohen–Moore–Neisendorfer [28, 27] prove that the p^k-power map of the H-space $\Omega_0^{2k+1}S^{2k+1}$ is nullhomotopic. Consequently, the spectrum $\Phi_n(S^{2k+1})$ has the same p-exponent. As usual, one deduces corresponding results for even-dimensional spheres from this using the EHP sequence. One can speculate about versions of such results for v_i-exponents with $i > 0$. Wang [78] computes the homotopy groups of $L_{K(2)}\Phi_2(S^3)$ at primes $p \geq 5$ and shows that v_1^2 acts trivially on them. However, the situation is more subtle than before, since the element v_1 does *not* act trivially on the Morava E-theory $E_*(\Phi_2(S^3))$.

There are similar results for p-exponents of Moore spaces S^k/p^r, with $k \geq 2$ and $r \geq 1$. If p is odd, then the work of Cohen–Moore–Neisendorfer [29] and Neisendorfer [65] shows that $\Omega^2 S^k/p^r$ has null-homotopic p^{r+1}-power map. For $p = 2$ and $r \geq 2$ there are 2-primary exponent results by Theriault [77]. It seems the remaining case $S^k/2$ is still open. These exponent results for Moore spaces give corresponding exponents for the spectra $\Phi_1 S^k/p^r$. A possible generalization would be the following:

(B1) If V is the suspension of a type n space, does the spectrum $\Phi_n V$ have a v_i-exponent for all $i < n$?

Here we say that a spectrum X has a v_i-exponent if for any finite type i spectrum W with v_i-self-map v, the smash product map $v \otimes X$ is nilpotent. Under the equivalence between $\mathcal{S}_{v_n} \simeq \mathrm{Lie}(\mathrm{Sp}_{v_n})$, the space ΣV corresponds to the free spectral Lie algebra on the spectrum $\Sigma^{\infty+1}V$. This allows for a reformulation of (B1) in terms of the v_i-exponents of the underlying spectrum of that free spectral Lie algebra.

(C) The Bousfield–Kuhn functor.

Although $T(n)$- or $K(n)$-homology equivalences of spaces behave quite differently from v_n-periodic equivalences, there are still many statements for homology that have counterparts for v_n-periodic homotopy groups. One such example is the Whitehead theorem: if a map $f\colon X \to Y$ of simply connected pointed spaces induces an isomorphism in $K(n)$-homology for each $n \geq 0$, then it is a weak homotopy equivalence. This is proved by Bousfield in [15]; an alternative proof is given by Hopkins–Ravenel in [44]. There is a version for v_n-periodic homotopy groups which states the following:

Theorem 16.8.1 (Barthel–Heuts–Meier, [8])**.** *If a map $f\colon X \to Y$ of simply connected finite CW-complexes is a v_n-periodic equivalence for every $n \geq 0$, then f is a p-local homotopy equivalence.*

Note that this result includes a finiteness hypothesis on X and Y. It is well-known that a finite spectrum with $K(n)_*X = 0$ also has $K(n-1)_*X = 0$. (In fact, another result of Bousfield [18] states that a space X with $K(n)_*X = 0$ also has $K(i)_*X = 0$ for all $0 < i \leq n$, without assuming finiteness of X.) This inspires the following question:

(C1) For a finite pointed CW-complex X, does $\Phi_n X \simeq 0$ imply $\Phi_i X \simeq 0$ for $i < n$?

We already discussed the 'two adjunctions' diagram

$$\mathrm{Sp}_{v_n} \underset{\Phi_n}{\overset{\Theta_n}{\rightleftarrows}} \mathcal{S}_{v_n} \underset{\Omega^\infty_{v_n}}{\overset{\Sigma^\infty_{v_n}}{\rightleftarrows}} \mathrm{Sp}_{v_n}$$

at the end of Section 16.6. The adjunction on the right can be characterized by the universal property of stabilization (in the world of presentable ∞-categories); the functor $\Sigma^\infty_{v_n}$ is the initial colimit-preserving functor from $\mathcal{S}_{v_n}$ to a presentable stable ∞-category. A positive answer to the following would give a similar universal property of the Bousfield–Kuhn functor:

(C2) Is the adjoint pair (Θ_n, Φ_n) the costabilization of $\mathcal{S}_{v_n}$? In other words, is Θ_n the terminal colimit-preserving functor from a presentable stable ∞-category to $\mathcal{S}_{v_n}$?

(D) Beyond monochromatic unstable homotopy theory.

The chromatic approach to homotopy theory involves two aspects: (1) understanding monochromatic layers and (2) assembling those layers to reconstruct a space or spectrum. So far we have only discussed (1). We will now give a brief discussion of (2) and pose some questions.

We have seen that for any n there are comparison functors

$$\mathcal{S}_{v_n} \to \mathrm{coCAlg}^{\mathrm{nu}}(\mathrm{Sp}_{T(n)})$$

that, while not fully faithful in general (except for $n = 0$), at least behave reasonably well. Integrally, however, the functor

$$\mathcal{S}_* \to \mathrm{coCAlg}^{\mathrm{nu}}(\mathrm{Sp})$$

is quite far from being fully faithful. To get a much better approximation, one should replace the right-hand side by the ∞-category of coalgebras for the comonad $\Sigma^\infty\Omega^\infty$ (cf. Remark 16.2.6). The price to pay, though, is that it is in general not so clear what has to be done in order to upgrade a commutative coalgebra spectrum to a $\Sigma^\infty\Omega^\infty$-coalgebra. A first step is to consider the pull-back square (see Proposition 1.9 of [51])

$$\begin{array}{ccc} P_2(\Sigma^\infty\Omega^\infty)(X) & \longrightarrow & (X \wedge X)^{h\Sigma_2} \\ \downarrow & & \downarrow \\ X & \xrightarrow{\tau_2} & (X \wedge X)^{t\Sigma_2}. \end{array}$$

If a commutative coalgebra X has a compatible coalgebra structure for $\Sigma^\infty\Omega^\infty$, then the left-hand vertical map has a section. This is equivalent to the existence of a diagonal lift in the square. In other words, the composite of the comultiplication

$$\delta_2 \colon X \to (X \wedge X)^{h\Sigma_2}$$

and the canonical map $(X \wedge X)^{h\Sigma_2} \to (X \wedge X)^{t\Sigma_2}$ is homotopic to the map

$$\tau_2 \colon X \to (X \wedge X)^{t\Sigma_2}.$$

This latter map exists for any spectrum and is called the *Tate diagonal.* One can think of it as the stable (or linear) shadow of the diagonal map of spaces. (See [49, 42, 32] for much more discussion.) More generally, one can define the notion of a *Tate coalgebra* as in [42]. It is first of all a commutative coalgebra, meaning a spectrum X equipped with maps

$$\delta_k \colon X \to (X^{\wedge k})^{h\Sigma_k}$$

for $k \geq 2$ and a coherent system of homotopies relating the δ_k for various k. Secondly, these comultiplications δ_k have to be compatible with certain generalized Tate diagonals

$$\tau_k \colon X \to (X^{\wedge k})^{t\Sigma_k}$$

which are constructed inductively. It is proved in [42] that there is an equivalence between the ∞-category of simply connected pointed spaces $\mathcal{S}_*^{\geq 2}$ and the ∞-category $\mathrm{coAlg}^{\mathrm{Tate}}(\mathrm{Sp})^{\geq 2}$ of simply connected Tate coalgebras in spectra.

In the ∞-category of $T(n)$-local spectra, Tate constructions associated with finite groups are contractible [51]. Hence the theory of Tate coalgebras in $\mathrm{Sp}_{T(n)}$ reduces to that of commutative coalgebras. One can think of the Tate diagonals τ_k above as determining the 'attaching data' between the ∞-categories of commutative coalgebras in $\mathrm{Sp}_{T(n)}$ for varying n, which assembles them together into the ∞-category of Tate coalgebras.

What is less clear is what the Koszul dual side of this picture should be. The ∞-category of spectral Lie algebras cannot be a good model for $\mathcal{S}_*$ without v_n-periodic localization; these ∞-categories have the same stabilization, but the Tate diagonals on Sp one would associate with the ∞-category Lie(Sp) are zero, as opposed to the usual Tate diagonals arising as the stabilization of the product on $\mathcal{S}_*$. Said (very) informally, the k-invariants of the Goodwillie tower of Lie(Sp) are trivial, whereas they are not for $\mathcal{S}_*$. However, one could hope that it is possible to assemble the ∞-categories $\mathrm{Lie}(\mathrm{Sp}_{T(n)})$ for varying n in a more interesting way:

(D1) Does there exist a good theory of 'transchromatic' spectral Lie algebras, related to $\mathcal{S}_*$ (or $L_n^f\mathcal{S}_*$) by an adjoint pair, which after v_n-periodic localization reduces to the theory of $T(n)$-local spectral Lie algebras and the Bousfield–Kuhn functor Φ_n relating it to $\mathcal{S}_{v_n}$?

(D2) If question (D1) admits a reasonable answer, then what is the relation to Mandell's p-adic homotopy theory?

Bibliography

[1] J. F. Adams. On the cobar construction. *Proc. Nat. Acad. Sci. U.S.A.*, 42:409–412, 1956.

[2] J. F. Adams. On the groups $J(X)$. IV. *Topology*, 5:21–71, 1966.

[3] Omar Antolín Camarena. *The mod 2 homology of free spectral Lie algebras.* Ph.D. thesis, Harvard University, 2015.

[4] G. Z. Arone and W. G. Dwyer. Partition complexes, Tits buildings and symmetric products. *Proc. London Math. Soc. (3)*, 82(1):229–256, 2001.

[5] Greg Arone and Michael Ching. Operads and chain rules for the calculus of functors. *Astérisque*, (338), 2011.

[6] Gregory Arone and Lukas Brantner. The action of Young subgroups on the partition complex. arXiv:1801.01491, 2018.

[7] Gregory Arone and Mark Mahowald. The Goodwillie tower of the identity functor and the unstable periodic homotopy of spheres. *Invent. Math.*, 135(3):743–788, 1999.

[8] Tobias Barthel, Gijs Heuts, and Lennart Meier. A whitehead theorem for periodic homotopy groups. arXiv:1811.04030, 2018.

[9] Maria Basterra and Michael A. Mandell. Homology and cohomology of E_∞ ring spectra. *Math. Z.*, 249(4):903–944, 2005.

[10] Mark Behrens. The Goodwillie tower and the EHP sequence. *Mem. Amer. Math. Soc.*, 218(1026), 2012.

[11] Mark Behrens and Charles Rezk. The Bousfield-Kuhn functor and topological André-Quillen cohomology. arXiv:1712.03045, 2017.

[12] Mark Behrens and Charles Rezk. Spectral algebra models of unstable v_n-periodic homotopy theory. arXiv:1703.02186, 2017.

[13] M. Bendersky. The v_1-periodic unstable Novikov spectral sequence. *Topology*, 31(1):47–64, 1992.

[14] M. Bendersky, E. B. Curtis, and H. R. Miller. The unstable Adams spectral sequence for generalized homology. *Topology*, 17(3):229–248, 1978.

[15] A. K. Bousfield. On homology equivalences and homological localizations of spaces. *Amer. J. Math.*, 104(5):1025–1042, 1982.

[16] A. K. Bousfield. Localization and periodicity in unstable homotopy theory. *J. Amer. Math. Soc.*, 7(4):831–873, 1994.

[17] A. K. Bousfield. Homotopical localizations of spaces. *Amer. J. Math.*, 119(6):1321–1354, 1997.

[18] A. K. Bousfield. On $K(n)$-equivalences of spaces. In *Homotopy invariant algebraic structures (Baltimore, MD, 1998)*, volume 239 of *Contemp. Math.*, pages 85–89. Amer. Math. Soc., 1999.

[19] A. K. Bousfield. On the telescopic homotopy theory of spaces. *Trans. Amer. Math. Soc.*, 353(6):2391–2426, 2001.

[20] A. K. Bousfield and V. K. A. M. Gugenheim. On PL de Rham theory and rational homotopy type. *Mem. Amer. Math. Soc.*, 8(179), 1976.

[21] Lukas Brantner. *The Lubin-Tate theory of spectral Lie algebras.* PhD thesis, Harvard University, 2017.

[22] Lukas Brantner and Gijs Heuts. The v_n-periodic Goodwillie tower on wedges and cofibres. arXiv:1612.02694, 2016.

[23] Michael Ching. Bar constructions for topological operads and the Goodwillie derivatives of the identity. *Geom. Topol.*, 9:833–933, 2005.

[24] Michael Ching. Bar-cobar duality for operads in stable homotopy theory. *J. Topol.*, 5(1):39–80, 2012.

[25] Michael Ching and John E. Harper. Derived Koszul duality and TQ-homology completion of structured ring spectra. *Adv. Math.*, 341:118–187, 2019.

[26] Dustin Clausen and Akhil Mathew. A short proof of telescopic Tate vanishing. *Proc. Amer. Math. Soc.*, 145(12):5413–5417, 2017.

[27] F. R. Cohen, J. C. Moore, and J. A. Neisendorfer. The double suspension and exponents of the homotopy groups of spheres. *Ann. of Math. (2)*, 110(3):549–565, 1979.

[28] F. R. Cohen, J. C. Moore, and J. A. Neisendorfer. Torsion in homotopy groups. *Ann. of Math. (2)*, 109(1):121–168, 1979.

[29] F. R. Cohen, J. C. Moore, and J. A. Neisendorfer. Exponents in homotopy theory. In *Algebraic topology and algebraic K-theory (Princeton, N.J., 1983)*, *Ann. of Math. Stud.* 113, pages 3–34. Princeton Univ. Press, 1987.

[30] Donald M. Davis and Mark Mahowald. v_1-periodicity in the unstable Adams spectral sequence. *Math. Z.*, 204(3):319–339, 1990.

[31] Ethan S. Devinatz, Michael J. Hopkins, and Jeffrey H. Smith. Nilpotence and stable homotopy theory. I. *Ann. of Math. (2)*, 128(2):207–241, 1988.

[32] Rosona Eldred, Gijs Heuts, Akhil Mathew, and Lennart Meier. Monadicity of the Bousfield–Kuhn functor. Proc. Amer. Math. Soc. 147 (2019), no. 4, 1789–1796.

[33] Emmanuel Dror Farjoun. *Cellular spaces, null spaces and homotopy localization*, *Lecture Notes in Math.* 1622. Springer-Verlag, 1996.

[34] Yves Félix, Stephen Halperin, and Jean-Claude Thomas. *Rational homotopy theory*, volume 205 of *Graduate Texts in Math.* 205. Springer-Verlag, 2001.

[35] Yves Félix, John Oprea, and Daniel Tanré. *Algebraic models in geometry*, *Oxf. Grad. Texts Math.* 17. Oxford University Press, 2008.

[36] John Francis and Dennis Gaitsgory. Chiral Koszul duality. *Selecta Math. (N.S.)*, 18(1):27–87, 2012.

[37] Benoit Fresse. Koszul duality of E_n-operads. *Selecta Math. (N.S.)*, 17(2):363–434, 2011.

[38] Soren Galatius, Alexander Kupers, and Oscar Randal-Williams. Cellular E_k-algebras. arXiv:1805.07184, 2018.

[39] Ezra Getzler and John D. S. Jones. Operads, homotopy algebra and iterated integrals for double loop spaces. arXiv:hep-th/9403055, 1994.

[40] Victor Ginzburg and Mikhail Kapranov. Koszul duality for operads. *Duke Math. J.*, 76(1):203–272, 1994.

[41] Rune Haugseng. ∞-operads via Day convolution. arXiv:1708.09632, 2017.

[42] Gijs Heuts. Lie algebras and v_n-periodic spaces. arXiv:1803.06325, 2018.

[43] Michael J. Hopkins. Global methods in homotopy theory. In *Homotopy theory (Durham, 1985)*, *London Math. Soc. Lecture Note Ser.* 117, pages 73–96. Cambridge Univ. Press, 1987.

[44] Michael J. Hopkins and Douglas C. Ravenel. Suspension spectra are harmonic. *Bol. Soc. Mat. Mexicana (2)*, 37(1-2):271–279, 1992.

[45] Michael J. Hopkins and Jeffrey H. Smith. Nilpotence and stable homotopy theory. II. *Ann. of Math. (2)*, 148(1):1–49, 1998.

[46] Brenda Johnson. The derivatives of homotopy theory. *Trans. Amer. Math. Soc.*, 347(4):1295–1321, 1995.

[47] A. Joyal. Quasi-categories and Kan complexes. *J. Pure Appl. Algebra*, 175(1-3):207–222, 2002. Special volume celebrating the 70th birthday of Professor Max Kelly.

[48] Jens Jakob Kjaer. On the odd primary homology of free algebras over the spectral Lie operad. *J. Homotopy Relat. Struct.*, 13(3):581–597, 2018.

[49] John R. Klein. Moduli of suspension spectra. *Trans. Amer. Math. Soc.*, 357(2):489–507, 2005.

[50] Nicholas J. Kuhn. Morava K-theories and infinite loop spaces. In *Algebraic topology (Arcata, CA, 1986)*, volume 1370 of *Lecture Notes in Math.*, pages 243–257. Springer, 1989.

[51] Nicholas J. Kuhn. Tate cohomology and periodic localization of polynomial functors. *Invent. Math.*, 157(2):345–370, 2004.

[52] Nicholas J. Kuhn. Localization of André-Quillen-Goodwillie towers, and the periodic homology of infinite loopspaces. *Adv. Math.*, 201(2):318–378, 2006.

[53] Nicholas J. Kuhn. Localization of André-Quillen-Goodwillie towers, and the periodic homology of infinite loopspaces. *Adv. Math.*, 201(2):318–378, 2006.

[54] Nicholas J. Kuhn. A guide to telescopic functors. *Homology Homotopy Appl.*, 10(3):291–319, 2008.

[55] Lisa Langsetmo and Don Stanley. Nondurable K-theory equivalence and Bousfield localization. *K-Theory*, 24(4):397–410, 2001.

[56] Jean-Louis Loday and Bruno Vallette. *Algebraic operads*, *Grundlehren der Math. Wiss.* 346. Springer, 2012.

[57] J. Lurie, M.J. Hopkins, et al. Unstable chromatic homotopy theory. Available at `math.harvard.edu/~lurie/Thursday2017.html`.

[58] Jacob Lurie. Derived Algebraic Geometry X: Formal Moduli Problems. Available at `http://www.math.harvard.edu/~lurie/papers/DAG-X.pdf`.

[59] Jacob Lurie. Derived Algebraic Geometry XIII: Rational and p-adic Homotopy Theory. Available at `http://www.math.harvard.edu/~lurie/papers/DAG-XIII.pdf`.

[60] Jacob Lurie. Elliptic Cohomology II: Orientations. Available at `http://www.math.harvard.edu/~lurie/papers/Elliptic-II.pdf`.

[61] Jacob Lurie. *Higher topos theory*, *Ann. of Math. Stud.* 170 Princeton University Press, 2009.

[62] Jacob Lurie. Higher algebra. Available at `http://www.math.harvard.edu/~lurie/papers/HA.pdf`, 2017.

[63] Stephen A. Mitchell. Finite complexes with $A(n)$-free cohomology. *Topology*, 24(2):227–246, 1985.

[64] J. C. Moore. Differential homological algebra. *Actes de Congrés International des Mathématiciens (Nice, 1970), Tome 1*, pages 335–339, 1971.

[65] Joseph A. Neisendorfer. The exponent of a Moore space. In *Algebraic topology and algebraic K-theory (Princeton, N.J., 1983), Ann. of Math. Stud.* 113, pages 35–71. Princeton Univ. Press, 1987.

[66] Thomas Nikolaus and Peter Scholze. On topological cyclic homology. *Acta Math.*, 221(2):203–409, 2018.

[67] Luis Pereira. *Goodwillie calculus and algebras over a spectral operad.* Ph.D. thesis, MIT, 2013.

[68] Stewart B. Priddy. Koszul resolutions and the Steenrod algebra. *Bull. Amer. Math. Soc.*, 76:834–839, 1970.

[69] Daniel Quillen. Rational homotopy theory. *Ann. of Math. (2)*, 90:205–295, 1969.

[70] Daniel Quillen. On the (co-) homology of commutative rings. In *Applications of Categorical Algebra (Proc. Sympos. Pure Math., Vol. XVII, New York, 1968)*, pages 65–87. Amer. Math. Soc., 1970.

[71] Douglas C. Ravenel. *Complex cobordism and stable homotopy groups of spheres*, volume 121 of *Pure and Applied Mathematics*. Academic Press, Inc., 1986.

[72] Douglas C. Ravenel. *Nilpotence and periodicity in stable homotopy theory*, *Ann. of Math. Stud.* '28 Princeton University Press, 1992.

[73] Emily Riehl and Dominic Verity. Homotopy coherent adjunctions and the formal theory of monads. *Adv. Math.*, 286:802–888, 2016.

[74] Paolo Salvatore. *Configuration operads, minimal models and rational curves.* PhD thesis, Oxford University, 1998.

[75] Stefan Schwede and Brooke Shipley. Stable model categories are categories of modules. *Topology*, 42(1):103–153, 2003.

[76] Dennis Sullivan. Infinitesimal computations in topology. *Inst. Hautes Études Sci. Publ. Math.*, (47):269–331 (1978), 1977.

[77] Stephen D. Theriault. Homotopy exponents of mod 2^r Moore spaces. *Topology*, 47(6):369–398, 2008.

[78] Guozhen Wang. *Unstable chromatic homotopy theory.* Ph.D. thesis, MIT, 2015.

Mathematical Institute, Utrecht University, 3584 CD Utrecht, The Netherlands
E-mail address: g.s.k.s.heuts@uu.nl

17

Equivariant stable homotopy theory

Michael A. Hill

17.1 Introduction

Equivariant stable homotopy theory considers spaces and spectra endowed with the action of a fixed group G. Classically, this group has been taken to be finite or compact Lie, but here we will consider only the case of a finite group acting. Our goal is to produce a broad-strokes overview of the state of equivariant stable homotopy theory, focusing the intuition behind many of the objects and constructions, exploring some of the tools in equivariant algebra, and showing how one can compute with these as easily as one computes classically.

There are many wonderful references for much of the foundational material in equivariant stable homotopy theory. For example [2], [57], [32], [18], [60], and [67, 68] are excellent sources for learning about specific models and their applications. In this, we will focus more on multiplicative and computational aspects, working through various examples along the way.

In all that follows, we work as model independently as possible. We will use the phrase "homotopically meaningful" to signify that a particular functor or construction descends to the underlying ∞-category or lifts to a Quillen functor on appropriate model categories.

Notation and conventions

In this chapter, G will be a finite group. Letters like H and K will most often refer to subgroups of G, and N will refer to normal subgroups. Spaces are always assumed to be compactly generated, weak Hausdorff.

Acknowledgements

This material is based upon work supported by the National Science Foundation under Grant No. 1440140, while the author was in residence at the Mathematical Sciences Research Institute in Berkeley, California, during the Spring semester of 2019. The author was also supported by NSF Grant DMS–1811189. The author also thanks Andrew Blumberg, Tyler Lawson, and Haynes Miller for their careful comments on earlier drafts.

Mathematics Subject Classification. 18B30, 20J05, 20J06, 20J15, 19L47, 55N91, 55N35, 55P91, 55P92, 55Q91, 55S91, 55T25, 55T15, 55T99, 18G15.

Key words and phrases. equivariant, equivariant homotopy, G-space, equivariant stable homotopy, spectral sequence, computations, Künneth spectral sequence, Adams spectral sequence, stabilization, slice spectral sequence, transfer, norm, equivariant commutative ring spectra, N_∞-ring spectrum.

17.2 G-spaces and functors between them

17.2.1 The categories of G-spaces

Definition 17.2.1. A G-space is a topological space X together with a continuous map

$$G \times X \longrightarrow X$$

$$(g, x) \longmapsto g \cdot x$$

such that

1. if $e \in G$ is the identity, then for all $x \in X$, $e \cdot x = x$, and
2. for all $g, h \in G$ and $x \in X$, we have $g \cdot (h \cdot x) = (g \cdot h) \cdot x$.

As is common in mathematics, although a G-space is two pieces of data, we will normally denote them only by the name of the underlying space.

Definition 17.2.2. If X and Y are G-spaces then an equivariant map $f\colon X \to Y$ is a continuous map $f\colon X \to Y$ such that for all $g \in G$ and $x \in X$, we have

$$f(g \cdot x) = g \cdot f(x).$$

It is a useful exercise to check that equivariant maps compose and that the identity is equivariant.

Notation 17.2.3. Let Top^G denote the category of G-spaces and equivariant maps.

We have a forgetful functor from G-spaces to spaces that just forgets the action of G.

Notation 17.2.4. Let $i^*_{\{e\}}\colon \mathsf{Top}^G \to \mathsf{Top}$ be the forgetful functor.

This forgetful functor is faithful, since an equivariant map is just a continuous map with the property that it commutes with the action of G. In particular, we can use the natural topological enrichment of Top to produce a topological enrichment on Top^G. Here the topology on the Hom sets is just the subspace topology given by the faithful inclusions from $i^*_{\{e\}}$.

Example 17.2.5. If V is a finite dimensional orthogonal representation of G, then we have several G-spaces attached to V:

1. Let $D(V) = \big\{\vec{v} \in V \mid ||\vec{v}|| \leq 1\big\}$ be the unit disk in V,
2. let $S(V) = \big\{\vec{v} \in V \mid ||\vec{v}|| = 1\big\}$ be the unit sphere in V, and
3. let $S^V = D(V)/S(V)$ be the one point compactification of V, the "V-sphere".

Example 17.2.6. If V is a representation of G, then let

$$a_V : S^0 \to S^V$$

be the inclusion of the origin and point at infinity. We call this the Euler class of V, since it is the Euler class of the equivariant bundle $V \to *$.

Our forgetful functor down to spaces is just one of a host of forgetful functors wherein we forget the actions of only some of the elements of G.

Definition 17.2.7. Given any subgroup $H \subseteq G$, let

$$i_H^* : \mathsf{Top}^G \to \mathsf{Top}^H$$

be the forgetful functor that forgets the actions of elements of G not in H.

Example 17.2.8. If V is any representation of G, then

$$i_H^* a_V = a_{i_H^* V}.$$

These forgetful functors play an essential role in equivariant unstable and stable homotopy theory. They are categorically very well behaved, commuting with all limits and colimits in the category, and they have both adjoints. The left adjoint is given by a kind of balanced tensor product, while the right adjoint is given by an equivariant function object, just as in ordinary representation theory.

Definition 17.2.9. If X is an H-space, then let

$$G \underset{H}{\times} X = G \times X/\sim,$$

where $\sim$ is the equivalence relation given by $(gh, x) \sim (g, hx)$ for all $g \in G$, $h \in H$, and $x \in X$. This has a G-action given by

$$g \cdot [(g', x)] = [(gg', x)].$$

Definition 17.2.10. If Y is an H-space, then let

$$\mathsf{Top}^H(G, Y)$$

be the space of H-equivariant maps from G (viewed as an H-space with the left action of H on G) to Y. This gets an action of G via the right action of G on itself:

$$(g \cdot f)(g') = f(g'g).$$

Proposition 17.2.11. *The constructions*

$$X \mapsto G \underset{H}{\times} X, \quad Y \mapsto \mathsf{Top}^H(G, Y)$$

extend to functors

$$G \underset{H}{\times} \text{-}, \mathsf{Top}^H(G,\text{-})\colon \mathsf{Top}^H \to \mathsf{Top}^G,$$

called induction and coïnduction respectively.

Induction is left adjoint to the forgetful functor i_H^*, and coïnduction is right adjoint to the forgetful functor i_H^*.

Just as classically, we can test spaces by mapping in points. Here, however, we have to remember at which subgroup our point was "born". Since the point is the terminal object in the category of spaces, there is a unique H-space structure on $\{*\}$: the trivial one. This gives us the G-space

$$G/H \cong G \underset{H}{\times} *.$$

Proposition 17.2.12. *If Y is a G-space, then*

$$\mathsf{Top}^G(G/H, Y) \cong \{y \in Y \mid h \cdot y = y, \forall h \in H\}.$$

Definition 17.2.13. If $H \subseteq G$ and if Y is a G-space, then the H-fixed points of Y are

$$Y^H := \{y \in Y \mid h \cdot y = y, \forall h \in H\}.$$

The key step for Proposition 17.2.12 is that stabilizers of points only grow under an equivariant map. We can use this to describe the left adjoint to the H-fixed points functor.

Notation 17.2.14. Let

$$i_*\colon \mathsf{Top} \to \mathsf{Top}^G$$

be the functor that endows a space X with a trivial G-action: for all $g \in G$ and $x \in X$, $g \cdot x = x$.

The same construction works for other quotient groups.

Notation 17.2.15. Let $Q = G/N$, and let

$$i_*^N\colon \mathsf{Top}^Q \to \mathsf{Top}^G$$

be the functor that views a Q-space as a G-space via the quotient map $G \to Q$.

Underlying this is the observation that every continuous map between spaces with a trivial G-action is equivariant.

Proposition 17.2.16. *The functor i_* is left adjoint to the G-fixed points functor*

$$(\text{-})^G\colon \mathsf{Top}^G \to \mathsf{Top}.$$

More generally, the functor

$$G \underset{H}{\times} i_*\colon \mathsf{Top} \to \mathsf{Top}^G$$

is left adjoint to the H-fixed points functor.

Under the homeomorphisms given by Proposition 17.2.12, maps of orbits correspond to various inclusions and maps between the fixed points for various subgroups of G. This gives us a way to conceptualize a G-space: begin with the fixed points and then begin adding orbits of the form G/H (in families), working our way down the subgroup lattice of G. We shall make this concept increasingly precise.

We also have pointed versions of all of these statements.

Notation 17.2.17. Let Top_*^G be the category of G-spaces equipped with a G-fixed base-point.

Notation 17.2.18. Let $G_+ \underset{H}{\wedge}$- and $\mathsf{Top}_*^H(G_+, \text{-})$ be the (pointed) induction and coïnduction.

17.2.2 Equivariant homotopies and CW-complexes

We define homotopies and CW-structures largely in parallel with the classical ones. We first note that the category of G-spaces has a closed symmetric monoidal structure.

Definition 17.2.19. If X and Y are G-spaces, then let $X \times Y$ be endowed with the action

$$g \cdot (x, y) = (gx, gy).$$

Let $\underline{\mathsf{Top}}(X, Y)$ have the action

$$(gf)(x) = g\big(f(g^{-1}x)\big).$$

Essentially the same definitions can be applied in the pointed cases, giving the smash product and pointed mapping spaces.

Proposition 17.2.20. *The Cartesian product and function spaces with conjugation action give a closed symmetric monoidal structure on* Top^G.

The smash product and pointed function spaces with conjugation action give a closed symmetric monoidal structure on Top_*^G.

Remark 17.2.21. We can auto-enrich Top^G, forming a category $\underline{\mathsf{Top}}$. Our notation is chosen to reflect the fact that the former is the fixed points of the latter.

Definition 17.2.22. If $f_0, f_1 \colon X \to Y$ are equivariant maps, then a homotopy from f_0 to f_1 is an equivariant map

$$F \colon X \times I \to Y$$

such that for all $x \in X$ and $i \in \{0, 1\}$, $F(x, i) = f_i(x)$. We say that f_0 and f_1 are homotopic if there is a homotopy from one to the other.

In the pointed case, we require that the homotopy be relative to the basepoint.

Notation 17.2.23. If X and Y are G-spaces, let $[X, Y]^G$ denote equivariant homotopy classes of maps from X to Y.

If X and Y are pointed G-spaces, let $[X, Y]_*^G$ denote the equivariant homotopy classes of pointed maps from X to Y.

Example 17.2.24. If V is a representation such that $V^G = \{0\}$, then the Euler class a_V is not homotopic to a constant map. Any such homotopy would, applying fixed points, provide a null-homotopy of the identity map on S^0.

Conversely, if $V^G \neq \{0\}$, then the Euler class a_V is null-homotopic, with nullhomotopy given by tracing along a ray in V^G.

The classical arguments that "homotopic is an equivalence relation" go through without change. Here, however, we can already see more rigidity than in the classical case.

Proposition 17.2.25. *If $f\colon X \to Y$ and $f'\colon Y \to X$ are homotopy inverses, then for all $H \subseteq G$, f and f' induce homotopy equivalences*

$$f^H\colon X^H \rightleftarrows Y^H \colon f'^H.$$

Our notion of homotopy and homotopy equivalence then gives the weak one.

Definition 17.2.26. A map $f\colon X \to Y$ is a weak homotopy equivalence if for all $H \subseteq G$,

$$f^H\colon X^H \to Y^H$$

is a weak homotopy equivalence.

These are the weak equivalences in a model structure on Top^G. The fibrations are also defined relative to the fixed points.

Theorem 17.2.27. *There is a model category structure on Top^G in which the weak equivalences are the weak homotopy equivalences, in which the fibrations are those maps $p\colon E \to B$ such that for all $H \subseteq G$,*

$$p^H\colon E^H \to B^H$$

is a [Serre] fibration, and where the cofibrations are what they have to be.

Using Proposition 17.2.12, we can turn our notation of a weak equivalence into a diagramatic statement: a map $f\colon X \to Y$ is a weak homotopy equivalence if for all orbits G/H, the induced map

$$\mathsf{Top}^G(G/H, f)\colon \mathsf{Top}^G(G/H, X) \to \mathsf{Top}^G(G/H, Y)$$

is a weak equivalence of spaces.

Definition 17.2.28. Let Orb^G be the full subcategory of Top^G generated by orbits.

Definition 17.2.29. If X is a G-space, then let

$$\underline{X}\colon (\mathsf{Orb}^G)^{op} \to \mathsf{Top}$$

be the restriction of the Yoneda functor given by

$$\underline{X}(G/H) = \mathsf{Top}^G(G/H, X)$$

to Orb^G.

The Yoneda embedding says that the assignment $X \to \underline{X}$ gives a functor

$$\mathsf{Top}^G \to \mathrm{Fun}\left((\mathsf{Orb}^G)^{op}, \mathsf{Top}\right).$$

Since Top is a cofibrantly generated model category, we have an induced model structure on $\mathrm{Fun}\left((\mathsf{Orb}^G)^{op}, \mathsf{Top}\right)$ in which the weak equivalences and fibrations are determined levelwise.

We also have a functor that turns a diagram of this shape into a G-space.

Notation 17.2.30. Let $J\colon \mathsf{Orb}^G \to \mathsf{Top}^G$ be the inclusion of the orbit category into the category of G-spaces.

Let

$$\text{-} \otimes_{\mathsf{Orb}^G} J\colon \mathrm{Fun}\left((\mathsf{Orb}^G)^{op}, \mathsf{Top}\right) \to \mathsf{Top}^G$$

by the coend with J.

Elmendorf's Theorem is that these two functors are inverse Quillen equivalences.

Theorem 17.2.31. [24] *There is a Quillen equivalence*

$$\underline{(\text{-})}\colon \mathsf{Top}^G \rightleftarrows \mathrm{Fun}\left((\mathsf{Orb}^G)^{op}, \mathsf{Top}\right) \colon \text{-} \otimes_{\mathsf{Orb}^G} J.$$

Remark 17.2.32. Elmendorf's Theorem shows that homotopically, there is a difference between the category of G-spaces and the functor category

$$\mathrm{Fun}(BG, \mathrm{Top}),$$

where BG is the category with one object and morphism set G.

For example, the G-homeomorphism type of X can be read out of $\underline{X}$. The automorphism group of $G/\{e\}$ as a G-space is G^{op}, and $X \cong \underline{X}(G/\{e\})$. This identification need not be homotopically meaningful, however. For example, the map $EG \to *$ induces an equivalence at level $G/\{e\}$, but not of full diagrams.

17.2.2.1 CW-Structures

Our notion of a G-CW complex: we attach cells with various stabilizers inductively.

Definition 17.2.33. A G-CW structure on a G-space X is a a filtration of X

$$\emptyset = X^{[-1]} \subseteq X^{[0]} \subseteq X^{[1]} \subseteq \cdots \subseteq \bigcup_i X^{[i]} = X$$

such that

1. $X^{[0]}$ is a discrete G-set,

2. for each i, there is a discrete G-set T_i and a G-map

$$\theta_i\colon T_i \times S^{i-1} \to X^{[i-1]}$$

such that we have a pushout diagram

$$\begin{array}{ccc} T_i \times S^{i-1} & \xrightarrow{\theta_i} & X^{[i-1]} \\ \downarrow & & \downarrow \\ T_i \times D^i & \longrightarrow & X^{[i]}, \end{array}$$

3. X has the direct limit topology induced by the filtration.

There is a pointed version of a G-CW complex defined analogously; we will use both.

Since every G-set decomposes as a disjoint union of orbits, each of our sets T_i can so be decomposed. This means that attaching data in the second condition could equivalently have been written

$$\theta_i \colon \coprod_{j \in \mathcal{I}_i} G/H_j \times S^{i-1} \to X^{[i-1]}.$$

The choices here then hide some of the naturality: we choose a presentation of T_i as a disjoint union of orbits.

Example 17.2.34. If $\underline{X}$ is such that $\underline{X}(G/H)$ is a CW-complex for all H, and all of the maps are cellular, then the G-space produced by Elmendorf's Theorem is a G-CW complex.

Example 17.2.35. If $X_\bullet$ is a simplicial object in G-sets, then the geometric realization of $X_\bullet$ has the natural structure of a G-CW complex.

Theorem 17.2.36 (Equivariant Whitehead Theorem)**.** *A weak equivalence between G-CW complexes is a homotopy equivalence.*

Showing this uses an obstruction theory that records the stabilizers of individual cells as well as a varying target. To get a sense for what happens, consider attaching a single equivariant cell $G_+ \underset{H}{\wedge} D^i$ to a G-space X along a map $\theta \colon G_+ \underset{H}{\wedge} S^{i-1} \to X$. Consider also a map $f \colon X \to Y$. Then an extension of f over $X \cup (G_+ \underset{H}{\wedge} D^i)$ exists if and only if

$$f \circ \theta \colon G_+ \underset{H}{\wedge} S^{i-1} \to Y$$

is null-homotopic. By Proposition 17.2.11, this is null if and only if the adjoint map

$$\widetilde{f \circ \theta} \colon S^{i-1} \to i_H^* Y$$

is null. Since S^{i-1} has a trivial H-action, any H-equivariant map $S^{i-1} \to Y$ or $S^{i-1} \times I \to Y$ must land in Y^H. Thus the obstruction to extending over a cell of the form $G_+ \wedge_H D^i$ is in $\pi_{i-1}(Y^H)$. As H-varies over the cells, so then does the group in which our extensions live. Coefficient systems and Bredon cohomology record exactly what we see.

17.2.3 Coefficient systems and cohomology

17.2.3.1 The Category of Coefficient Systems

Notation 17.2.37. Let Fin^G denote the category of finite G-sets.

Disjoint union gives Fin^G a co-Cartesian monoidal structure; Cartesian product gives the Cartesian monoidal structure.

Definition 17.2.38. Let $\mathcal{C}$ be a category with finite products. A coefficient system of objects of $\mathcal{C}$ is a product preserving functor

$$\underline{M}\colon (\mathsf{Fin}^G)^{op} \to \mathcal{C}.$$

In this definition, we use the fact that since Fin^G has coproducts given by disjoint union, $(\mathsf{Fin}^G)^{op}$ has products given by disjoint union. The condition is then that

1. for all T_1 and T_2, the inclusions $T_1 \hookrightarrow T_1 \amalg T_2 \hookleftarrow T_2$ induce an isomorphism

$$\underline{M}(T_1 \amalg T_2) \cong \underline{M}(T_1) \times \underline{M}(T_2),$$

2. and $\underline{M}(\emptyset) = *$, the terminal object.

Example 17.2.39. Since the wedge product is the coproduct in pointed spaces, Definition 17.2.29 extends to give a coefficient system $\underline{X}$ of spaces for any G-space X:

$$\underline{X}(T) := \mathsf{Top}_*^G(T_+, X).$$

Remark 17.2.40. We can restate Elmendorf's Theorem as being a Quillen equivalence between G-spaces and coefficient systems of spaces. This is a powerful reinterpretation that is essential in the recent ∞-categorical treatments of equivariant homotopy theory.

Proposition 17.2.41. *Let X be a pointed G-space. Then the assignment*

$$T \mapsto \pi_k\big(\mathsf{Top}_*^G(T_+, X)\big) =: \underline{\pi}_k(X)(T)$$

gives a coefficient system of pointed sets if $k = 0$, of groups if $k = 1$, or of abelian groups if $k \geq 2$.

Definition 17.2.42. Let Coeff be the category whose objects are coefficient systems of abelian groups and whose morphisms are natural transformations.

Coefficient systems are so-named because they are the natural coefficients for equivariant cohomology.

Example 17.2.43. A C_p-coefficient system $\underline{M}$ is the following data:

1. An abelian group $\underline{M}(*)$,
2. a C_p-module $\underline{M}(C_p)$, and
3. a "restriction map"

$$res_e^{C_p}\colon \underline{M}(*) \to \underline{M}(C_p)$$

that factors through the inclusion of the fixed points

$$\underline{M}(C_p)^{C_p} \subseteq \underline{M}(C_p).$$

Proposition 17.2.44. *The category* Coeff *is an abelian category with a finite set of projective generators.*

The projective generators can all be chosen to represent the various evaluation functors

$$\underline{M} \mapsto \underline{M}(T),$$

or even simply to represent evaluation at the orbits G/H, by the product preserving property.

Theorem 17.2.45. [15] *For any coefficient system $\underline{M}$, there is a unique cohomology theory on G-CW pairs for which*

$$H^*(G/H; \underline{M}) \cong \begin{cases} \underline{M}(G/H) & * = 0 \\ 0 & \textit{otherwise.} \end{cases}$$

We build this out of the natural coefficient system of chain complexes we get by composing with Elmemdorf coefficient system with the singular chains functor. Rather than spell this out, we describe one of the main ways we can compute with this: cellular cohomology.

17.2.3.2 Cellular Bredon Homology

The usual suspension axiom shows that the induced spheres $G_+ \wedge_H S^n$ play the role for Bredon cohomology that ordinary spheres play for classical cohomology: they are "Moore spaces" in the sense that the have a single non-vanishing cohomology group. We can now prove the usual results about the cellular cohomology and the corresponding relationship to Bredon cohomology, copying the usual definitions.

Definition 17.2.46. If X is aG-CW complex of finite type, then let

$$C^k_{\text{cell}}(X; \underline{M}) = H^k(X^{[k]}, X^{[k-1]}; \underline{M}).$$

Define a boundary map

$$\delta \colon C^k_{\text{cell}}(X; \underline{M}) \to C^{k+1}_{\text{cell}}(X; \underline{M})$$

via the long exact sequence for the triple $(X^{[k+1]}, X^{[k]}, X^{[k-1]})$.

The standard argument then applies here to show that this complex gives Bredon cohomology.

Proposition 17.2.47. *The cohomology of the cellular cochain complex is the Bredon cohomology of X.*

We further unpack this in the case that X is finite type. In this case, for each k

$$X^{[k]}/X^{[k-1]} \cong T_{k+} \wedge S^k,$$

for some finite G-set T_k. Moreover, the boundary map is the map induced by

$$\partial_k \colon T_{k+} \wedge S^k \cong X^{[k]}/X^{[k-1]} \to \Sigma X^{[k-1]} \to \Sigma X^{[k-1]}/X^{[k-2]} \cong T_{k-1+} \wedge S^k.$$

By definition, this map is an element of

$$\pi_k\big(T_{k-1+} \wedge S^k\big)(T_k).$$

When $k \geq 2$. we can easily describe the bottom homotopy coefficient system of an induced sphere like this.

Definition 17.2.48. For each finite G-set T, let $\underline{\mathbb{Z}}[T]$ be the coefficient system defined by

$$\underline{\mathbb{Z}}[T](T') := \mathbb{Z}\{\mathsf{Top}^G(T',T)\},$$

the free abelian group on the set $\mathsf{Top}^G(T',T)$.

Theorem 17.2.49. *If T is a finite G-set, then for all $n \geq 2$, we have a natural (in T) isomorphism*

$$\underline{\pi}_n(T_+ \wedge S^n) \cong \underline{\mathbb{Z}}[T].$$

Proof. Using the closed monoidal structure on G-spaces, we have a natural isomorphism

$$[T'_+ \wedge S^n, T_+ \wedge S^n]^G \cong [S^n, \mathsf{Top}^G(T',T)_+ \wedge S^n].$$

The right-hand side is non-equivariant homotopy classes of maps, and by the Hurewicz theorem, we have

$$[S^n, \mathsf{Top}^G(T',T)_+ \wedge S^n] \cong H_n\big(\mathsf{Top}^G(T',T)_+ \wedge S^n; \mathbb{Z}\big) \cong \bigoplus_{\mathsf{Top}^G(T',T)} \mathbb{Z}.$$

□

Remark 17.2.50. Since we only test against a finite G-set (and hence compact), Theorem 17.2.49 remains true if T is infinite.

It is helpful to additively enlarge the category Coeff.

Notation 17.2.51. Let $\mathsf{Fin}^G_{\mathbb{Z}}$ be the category with objects finite G-sets and with morphisms the free abelian group on the morphisms in Fin^G:

$$\mathsf{Fin}^G_{\mathbb{Z}}(S,T) := \mathbb{Z}\{\mathsf{Top}^G(S,T)\}.$$

Proposition 17.2.52. *The natural faithful inclusion $\mathsf{Fin}^G \hookrightarrow \mathsf{Fin}^G_{\mathbb{Z}}$ induces an equivalence of categories between coefficient systems of abelian groups and functors $\big(\mathsf{Fin}^G_{\mathbb{Z}}\big)^{op} \to \mathsf{Ab}$ that take disjoint unions to products and which are linear on Hom objects.*

Remark 17.2.53. The coefficient systems $\underline{\mathbb{Z}}[T]$ are the representable functors in this enlarged diagram category, representing the evaluation at T functor. These are projective generators.

This then allows us to determine the effect of the attaching maps on Bredon cohomology (and hence makes computing equivariant cohomology groups as easy as computing the non-equivariant ones). Any map

$$f\colon T'_+ \wedge S^n \to T_+ \wedge S^n,$$

induces a corresponding map of coefficients systems on $\underline{\pi}_n$:

$$\underline{\mathbb{Z}}[T'] \to \underline{\mathbb{Z}}[T].$$

Thus for any coefficient system $\underline{M}$, by the Yoneda Lemma, we have a "restriction along f" map

$$f^*\colon \underline{M}(T) \to \underline{M}(T')$$

given by mapping out of f. This is exactly the Bredon differential induced by the relative attaching map.

Theorem 17.2.54. *The Bredon cellular cochain complex is the complex $C^*_{cell}(X;\underline{M})$ with*

$$C^k_{cell}(X;\underline{M}) := \underline{M}(T_k),$$

and where the coboundary map is

$$C^{k-1}_{cell}(X;\underline{M}) = \underline{M}(T_{k-1}) \xrightarrow{\underline{M}(\partial_k)} \underline{M}(T_k) = C^k_{cell}(X;\underline{M}).$$

Remark 17.2.55. If we work instead with a covariant functor $\mathsf{Fin}^G \to \mathsf{Ab}$ that takes disjoint union to direct sum, then we can mirror the entire argument to build the Bredon homology and Bredon cellular homology.

Example 17.2.56. We close this section with an example of how to compute Bredon homology. Let $G = C_2$, with generator γ, and let $\underline{M}$ be a coefficient system. Let σ be the 1-dimensional sign representation. We compute $H^*(S^{k\sigma};\underline{M})$ for any k as a functor of $\underline{M}$.

A cell structure for $S^{k\sigma}$ is given by

$$S^0 \cup (C_{2+} \wedge e^1) \cup_{1-\gamma} (C_{2+} \wedge e^2) \cup \cdots \cup (C_{2+} \wedge e^k),$$

where the bottom attaching map is the action map

$$C_{2+} \wedge S^0 \to S^0,$$

which induces the restriction map

$$\underline{M}(*) \to \underline{M}(C_2)$$

on Bredon cellular cochains. The attaching map for the ℓ-cell modulo the $(\ell-2)$-skeleton is the map

$$(1 + (-1)^\ell \gamma),$$

and this induces multiplication by this element in $\underline{M}(C_2)$. Our chain complex is therefore

$$\underline{M}(*) \xrightarrow{res_e^{C_2}} \underline{M}(C_2) \xrightarrow{1-\gamma} \underline{M}(C_2) \to \cdots \to \underline{M}(C_2),$$

and the Bredon cohomology is the cohomology of this cochain complex.

17.2.4 Families and isotropy separation

One of the most useful consequences of Elmendorf's theorem is a way to isolate the contribution of cells with a particular stabilizer. This is called "isotropy separation", and stably, it will provide an explanation for several confusing features.

Definition 17.2.57. A family of subgroups is a set $\mathcal{F}$ of subgroups of G such that

1. if $H \in \mathcal{F}$ and if $K \subseteq G$, then $K \in \mathcal{F}$, and
2. if $H \in \mathcal{F}$, then for all $g \in G$, $gHg^{-1} \in \mathcal{F}$.

Remark 17.2.58. We can repackage the two conditions via the orbit category, as together they say that if $H \in \mathcal{F}$, and if we have a map $G/K \to G/H$ in the orbit category, then $K \in \mathcal{F}$. This shows that a family of subgroups is the same data as a sieve on the orbit category.

Definition 17.2.59. If X is a space, then let

$$\Phi_X = \{H \mid X^H \neq \emptyset\}.$$

Then Φ_X is a family, the "geometric isotropy of X". It records which stabilizers can show up in a G-CW decomposition of X.

Associated to a family, there is a universal homotopy type given by Elmendorf's Theorem.

Definition 17.2.60. If $\mathcal{F}$ is a family, then let $\underline{E\mathcal{F}}$ be the coefficient system given by

$$\underline{E\mathcal{F}}(G/H) = \begin{cases} \emptyset & H \notin \mathcal{F} \\ * & H \in \mathcal{F}. \end{cases}$$

Let $E\mathcal{F}$ be the G-space produced by Elmendorf's Theorem from *underlineE$\mathcal{F}$*.

The following proposition is immediate from the coefficient system formulation of the universal space and gives some explanation of the nomenclature.

Proposition 17.2.61. *If X is a G-CW complex, then*

$$[X, E\mathcal{F}]^G = \begin{cases} * & \Phi_X \subseteq \mathcal{F} \\ \emptyset & \Phi_X \not\subseteq \mathcal{F}. \end{cases}$$

Corollary 17.2.62. *The space $E\mathcal{F}$ is determined by the condition that*

$$(E\mathcal{F})^H \simeq \begin{cases} \emptyset & H \notin \mathcal{F} \\ * & H \in \mathcal{F}. \end{cases}$$

Example 17.2.63. Let $\mathcal{F}_e = \{\{e\}\}$. The associated space $E\mathcal{F}_e$ is a G-CW complex such that the fixed points for any non-trivial subgroup are empty and such that the underlying space is contractible. This is exactly the homotopical description of EG.

Example 17.2.64. Let $\mathcal{All} = \{H \mid H \subseteq G\}$. Then a model for $E\mathcal{All}$ is a point.

Example 17.2.65. Let N be a normal subgroup. The collection of subgroups that intersect N trivially forms a family $\mathcal{F}_N$ with associated universal space $E\mathcal{F}_N$.

Example 17.2.66. Let $\mathcal{P} = \{H \mid H \subsetneq G\}$ be the family of proper subgroups. Then a model for the homotopy type of $E\mathcal{P}$ is

$$\operatorname{colim}_n S(n\bar{\rho}_G),$$

where $\bar{\rho}_G$ is the quotient of the regular representation ρ_G by the trivial summand.

Since fixed points commute with products (both being limits), given any family $\mathcal{F}$, we can functorially restrict the isotropies to only be elements of $\mathcal{F}$ by simply crossing with $E\mathcal{F}$.

Proposition 17.2.67. *If X is a G-CW complex, then the geometric isotropy of $X \times E\mathcal{F}$ is given by*

$$\mathcal{F} \cap \Phi_X.$$

Corollary 17.2.68. *If $\mathcal{F}_1$ and $\mathcal{F}_2$ are two families, then*

$$E\mathcal{F}_1 \times E\mathcal{F}_2$$

is the universal space associated to the family $\mathcal{F}_1 \cap \mathcal{F}_2$.

Example 17.2.69. For any G-CW complex X,

$$E\mathcal{F}_e \times X = EG \times X$$

is the Borel space which frees up the action of G.

Definition 17.2.70. If $\mathcal{F}$ is a family, then let

$$E\mathcal{F}_+ \to S^0$$

be the pointed map that sends $E\mathcal{F}$ to the non-basepoint. Let $\tilde{E}\mathcal{F}$ denote the cofiber.

Example 17.2.71. If $\mathcal{F} = \mathcal{All}$, then $\tilde{E}\mathcal{F} \simeq *$.

Example 17.2.72. If $\mathcal{F} = \mathcal{P}$ is the family of proper subgroups, then a model for $\tilde{E}\mathcal{P}$ is

$$\tilde{E}\mathcal{P} = S^{\infty\bar{\rho}_G} = \operatorname{colim}_n S^{n\bar{\rho}_G} = S^0[a_{\bar{\rho}_G}^{-1}],$$

the infinite $\bar{\rho}_G$-sphere.

Definition 17.2.73. For a pointed G-space X, let

$$E\mathcal{F}_+ \wedge X \to X \to \tilde{E}\mathcal{F} \wedge X$$

be the isotropy separation sequence, the result of smashing the defining cofiber sequence for $\tilde{E}\mathcal{F}$ with X.

Considering the fixed points of $E\mathcal{F}$, the following is an immediate, important application of the definitions.

Proposition 17.2.74. *Let X be a pointed G-CW complex and let $\mathcal{F}$ be a family.*

1. *For any $H \in \mathcal{F}$, the map*
$$i_H^*(E\mathcal{F}_+ \wedge X) \to i_H^* X$$
is an H-equivalence and $i_H^*(\tilde{E}\mathcal{F} \wedge X)$ is contractible.

2. *For any $K \notin \mathcal{F}$, the map*
$$X^K \to (\tilde{E}\mathcal{F} \wedge X)^K$$
is an equivalence an $(E\mathcal{F}_+ \wedge X)^K$ is contractible.

Putting this all together, the isotropy separation sequence expresses X as an "extension" of two conceptually simpler spaces:

1. A space $E\mathcal{F}_+ \wedge X$ whose geometric isotropy is contained in $\mathcal{F}$ and
2. A space $\tilde{E}\mathcal{F} \wedge X$ that is contractible when restricted to any of the subgroups in $\mathcal{F}$.

17.3 Stabilization and G-spectra

17.3.1 Conceptual goals for stabilization

There are several different conceptual approaches to stabilization in G-spectra, and somewhat surprisingly, these lead to the same results. There are two dominant themes: one geometric and one algebraic.

Goal (Geometric Stabilization)**.** Have a good theory of Milnor–Spanier–Atiyah duality for G-manifolds.

If we have a manifold on which G acts smoothly, then we can attempt to mirror the Milnor–Spanier–Atiyah explanation of Poincaré duality via an identification of the dual of our manifold with the Thom spectrum of the virtual normal bundle. Almost immediately we run into trouble: if the group action is non-trivial, then we have no equivariant embeddings of our manifold into a Euclidean space with a trivial action. We must instead consider an embedding of our manifold into some representation of G. This gives a kind of S-duality for our manifold, but it requires that we consider suspensions by possibly non-trivial representations.

Goal (Algebraic Stabilization)**.** Universally make pushout diagrams and pullback diagrams agree (and hence finite coproducts should be finite products).

This is one of the usual ∞-categorical formulations of stabilization: we take spaces and universally build a category out of it in which pushout and pullback diagrams agree. In particular, we see that finite coproducts and finite products then necessarily agree. In the equivariant context, we have an added subtlety that is fundamental to the more modern approach to understanding equivariant stable homotopy theory: *the group should be allowed to act on all indexing objects.* For pushouts and pullbacks, this means that the group can act on the indexing diagram, while for finite coproducts being finite products, this includes an identification of induction (the coproduct over G/H) with coïnduction (the product over G/H).

17.3.1.1 The Spanier–Whitehead category

Boardman's stable homotopy category was defined as an extension of the ordinary Spanier–Whitehead category under colimits. The same kind of analysis works equivariantly. We loosely sketch Adams' original treatment of the equivariant Spanier–Whitehead category [2] [62].

Definition 17.3.1. A universe for G is a countably infinite dimensional orthogonal representation U such that

1. the trivial representation $\mathbb{R}\subseteq U$ and
2. if $V\subseteq U$ is a finite dimensional representation, then the infinite orthogonal sum of V with itself also embeds into U.

A complete universe is one that contains all irreducible representations of G.

Equivalently, a complete universe is isomorphic to the countably infinite orthogonal direct sum of copies of the regular representation ρ_G.

Notation 17.3.2. If V is a finite dimensional representation of G, then let

$$\Sigma^V : \mathsf{Top}^G \to \mathsf{Top}^G$$

be the functor

$$X \mapsto S^V \wedge X.$$

Definition 17.3.3. If X and Y are finite, pointed G-CW complexes, then let

$$\{X, Y\}^G = \varinjlim_{V} [\Sigma^V X, \Sigma^V Y]_*^G$$

be the "stable homotopy classes of maps from X to Y", where the direct limit is taken over the poset of finite dimensional representations in a chosen complete G-universe.

Let SW^G denote the category whose objects are finite, pointed G-CW complexes and whose morphisms are these stable homotopy classes of maps.

Proposition 17.3.4. *The smash product of finite, pointed G-CW complexes extends to a symmetric monoidal product on the Spanier–Whitehead category.*

There is a clear extension of Spanier's original notion of "S-duality" to this equivariant Spanier–Whitehead category as described by Spanier–Whitehead [63]. Here, if X is a G-CW complex embedded in the V-sphere, then X is V-dual to the unreduced suspension of its complement. This gives us, for example, that if T is a finite G-set that embeds into V, then T_+ and $S^V \wedge T_+$ are V-dual.

Many of the standard arguments apply without change here.

Proposition 17.3.5. *The equivariant Spanier–Whitehead category is additive: finite wedges and products exist and agree and the morphism sets are naturally abelian group valued. The composition and symmetric monoidal products induce bilinear maps on morphism sets.*

The fact that our universe contains all trivial suspensions also ensures that in the Spanier–Whitehead category, cofiber sequences are also fiber sequences.

Corollary 17.3.6. *For any finite, pointed G-CW complex Y, the functors*

$$X \mapsto \begin{cases} \{X, \Sigma^n Y\}^G & n \geq 0 \\ \{\Sigma^{-n} X, Y\}^G & n \leq 0. \end{cases}$$

give a cohomology theory on finite, G-CW complexes.

The inclusion of G-spaces with a trivial action into G-spaces preserves cofiber sequences and hence by restriction, we have a cohomology theory on CW complexes (viewed as G-CW complexes with a trivial action). We can work a little more generally.

Proposition 17.3.7. *If $Q = G/N$ is a quotient of G, then the pushforward $i_*^N \colon \mathsf{Top}^Q \to \mathsf{Top}^G$ extends to an embedding*

$$i_*^N \colon \mathsf{SW}^Q \to \mathsf{SW}^G$$

that which a G/N-CW complex X to itself, viewed as a G-CW complex, and which on morphisms, is the map induced on colimits by the inclusion of the subsystem of G/N-representations in all G-representations.

When $N = G$, we have just the ordinary Spanier–Whitehead category, and we can then restate our result about cohomology theories.

Corollary 17.3.8. *For any finite, pointed G-CW complex Y, the functors*

$$X \mapsto \begin{cases} \{X, \Sigma^n Y\}^G & n \geq 0 \\ \{\Sigma^{-n} X, Y\}^G & n \leq 0. \end{cases}$$

give a cohomology theory on finite CW complexes.

Remark 17.3.9. These cohomology theories on ordinary CW-pairs are not simply represented by maps in the ordinary, non-equivariant Spanier–Whitehead category from X to Y^G. We will see an explicit example of this in Example 17.3.20 below.

The non-trivial representations in the universe play a different role than the trivial ones: they provide a good theory of duality for manifolds with non-trivial action, as described above, and they produce transfer maps. We first build a slight extension of the Spanier–Whitehead category. The standard cofinality argument gives the following.

Proposition 17.3.10. *For any finite dimensional representation V, the V-fold suspension gives a fully faithful embedding of the equivariant Spanier–Whitehead category into itself.*

Definition 17.3.11. Let the extended Spanier–Whitehead category be the category obtained from the equivariant Spanier–Whitehead category by formally adjoining representing objects $\Sigma^{-V}Y$ for the functors

$$X \mapsto \{\Sigma^V X, Y\}^G.$$

Let $\overline{\mathcal{SW}}^G$ denote the extended Spanier–Whitehead category.

This turns the suspension functors into autoequivalences of the Spanier–Whitehead category, and we can replace V-duality with ordinary, categorical duality.

Corollary 17.3.12. *In the extended Spanier–Whitehead category, all objects are dualizable.*

There is a natural map

$$Y \to \Sigma^{-V}(\Sigma^V Y) \tag{17.3.1}$$

adjoint to the identity map on $\Sigma^V Y$. This map is an equivalence in the extended Spanier–Whitehead category: it represents the suspension map

$$\{X, Y\}^G \to \{\Sigma^V X, \Sigma^V Y\}^G.$$

In particular, we can think of the objects $\Sigma^{-V}Y$ as being the formal desuspension of Y by V. Using these maps, we can extended $\mathbb{Z}$-graded theories.

Definition 17.3.13. Let Y is a pointed G-CW complex and let V and W be representations of G, then we define a functor of finite, pointed G-CW complexes by

$$X \mapsto \{\Sigma^W X, \Sigma^V Y\}^G =: Y^{W-V}(X).$$

The notation hides some of the naturality. We mean here actual pairs of representations, not isomorphism classes. There is also significant naturality in the vector spaces. We first rephrase the full faithfulness of suspension here.

Proposition 17.3.14. *If V, W, and U are finite dimensional representations, then suspension gives a natural isomorphism*

$$Y^{W-V}(X) \xrightarrow{\cong} Y^{(U\oplus W)-(U\oplus V)}(X).$$

These isomorphisms are compatible in the sense that if U' is another finite dimensional representation, then the map

$$Y^{W-V}(X) \xrightarrow{\cong} Y^{((U'\oplus U)\oplus W)-((U'\oplus U)\oplus V)}(X)$$

is the same as the composite

$$Y^{W-V}(X) \xrightarrow{\cong} Y^{(U\oplus W)-(U\oplus V)}(X) \xrightarrow{\cong} Y^{((U'\oplus U)\oplus W)-((U'\oplus U)\oplus V)}(X).$$

Additionally, given any isomorphism of representations $V \to V'$, we have an associated isomorphism of representation spheres $S^V \to S^{V'}$. Smashing with either X or Y will then give us a natural isomorphisms. These new maps clearly depend only on the stable homotopy type of the map $S^V \to S^{V'}$, however.

Definition 17.3.15. Let JO^s be the maximal subgroupoid of the full subcategory of SW^G spanned by the representation spheres S^V.

Proposition 17.3.16. *Given a pointed G-space Y, the assignment*

$$(V, W, X) \mapsto Y^{W-V}(X)$$

extends to a functor

$$\mathsf{JO}^{s,op} \times \mathsf{JO}^s \times \mathsf{SW}^{G,op} \to \mathsf{Ab}.$$

This is the prototype of an "RO(G)-graded cohomology theory". The naming is quite misleading, however, since we are indexing here by pairs of representations (See Adams' extended discussion of this point for a stronger warning [2]). The isomorphism type of the abelian group

$$Y^{W-V}(X, A)$$

naturally depends only on the associated virtual representation $W - V \in RO(G)$. The problem is that the representation spheres can have non-trivial automorphisms. Put another way, in the extended Spanier–Whitehead category, representation spheres are all invertible under the smash product, since this is just a restatement of the map in Equation 17.3.1 applied to $Y = S^0$ being an isomorphism. The description above is a concrete way to describe the grading by the Picard groupoid of this symmetric monoidal category, as in [27]. There is a natural map RO(G) to the Picard group of the symmetric monoidal category, but in forming this, we have thrown away the information given by the isomorphisms.

17.3.1.2 Change of Groups

To compare the Spanier–Whitehead categories for G and its subgroups, we simply note that the set of representations of a subgroup H that are the restriction of a representation of G are cofinal in all representations of H. This gives the following.

Proposition 17.3.17. *If H is a subgroup of G, then the restriction functor i_H^* in pointed spaces extends to a restriction functor*

$$i_H^* \colon \mathsf{SW}^G \to \mathsf{SW}^H.$$

The induction functor $G_+ \wedge_H (\text{-})$ in pointed spaces extends to an induction functor

$$G_+ \underset{H}{\wedge} (\text{-}) \colon \mathsf{SW}^H \to \mathsf{SW}^G$$

which is left-adjoint to the restriction.

Our second goal is realized in the Spanier–Whitehead category via the additional representations: these give us stable maps in the wrong direction to the ordinary action maps $G/H \to *$. A choice of embedding

$$e \colon G/H \hookrightarrow V$$

of G/H into a representation V gives us a Thom collapse map

$$S^V \to G/H_+ \wedge S^V,$$

and hence a stable homotopy class

$$t_H^G \in \{S^0, G/H_+\}^G.$$

Definition 17.3.18. Let t_H^G be the transfer, the stable homotopy class $\{S^0, G/H_+\}^G$ given by the Thom collapse of any choice of embedding $G/H \hookrightarrow V$.

This definition seems to depend on the choices of embeddings, but by consider larger and larger representations (i.e. looking farther down the colimit defining the stable homotopy classes of maps), we see that the connectivity of the space of such embeddings goes to infinity and hence any choices are stably homotopic.

The "wrong-way" transfer map on orbits provides powerful new tools. In particular, it realizes the algebraic goal and makes the Spanier–Whitehead category behave even more like an equivariant algebraic category like the chain complexes of representations.

Theorem 17.3.19 (Wirthmüller isomorphism [71]). *The induction functor $G_+ \wedge_H (\text{-})$ is also right-adjoint to the restriction i_H^*.*

Induction being also the *right* adjoint to the forgetful functor means that we lose a lot of intuition for maps between G-CW complexes. This is what has traditionally made computations in equivariant stable homotopy theory more daunting than their classical counterparts.

Example 17.3.20. For any finite group G, we have an isomorphism

$$\{S^0, G_+\}^G \cong \{S^0, S^0\}^{\{e\}} \cong \mathbb{Z}.$$

In particular, the free G-set G has non-trivial maps from the fixed G-set $*$!

It is possible, however, to give some algebraic information about at least stable maps out of the zero sphere. For this, we exploit some of the additional extra structure.

17.3.2 Mackey functors and Segal–tom Dieck

Combining our adjunctions with ordinary finite sums being finite products also gives us the duality results that we want.

Corollary 17.3.21. *If T is a finite G-set, then T_+ is self-dual in the extended equivariant Spanier–Whitehead category.*

Corollary 17.3.22. *For any finite G-set T, the functors*

$$\{T_+ \wedge (\text{-}), (\text{-})\}^G \colon \mathsf{SW}^{G^{op}} \times \mathsf{SW}^G \to \mathsf{Ab} \text{ and}$$
$$\{(\text{-}), T_+ \wedge (\text{-})\}^G \colon \mathsf{SW}^{G^{op}} \times \mathsf{SW}^G \to \mathsf{Ab}$$

are naturally isomorphic.

This algebraic structure is encoded in a Mackey functor.

Definition 17.3.23. [21] A Mackey functor is a pair of functors: one covariant, $\underline{M}_*$, and one contravariant, $\underline{M}^*$, from the category of finite G-sets to abelian groups such that

1. The functors agree on objects:

$$\underline{M}^*(T) = \underline{M}_*(T) =: \underline{M}(T).$$

2. The functor $\underline{M}^*$ is product preserving.

3. We have a Beck–Chevalley condition: if we have a pullback diagram of finite G-sets

$$\begin{array}{ccc} T' & \xrightarrow{f'} & T \\ {\scriptstyle h'}\downarrow & & \downarrow{\scriptstyle h} \\ S' & \xrightarrow[f]{} & S, \end{array}$$

then we have a commutative diagram

$$\begin{array}{ccc} \underline{M}(T') & \xrightarrow{\underline{M}_*(f')} & \underline{M}(T) \\ {\scriptstyle \underline{M}^*(h')}\uparrow & & \uparrow{\scriptstyle \underline{M}^*(h)} \\ \underline{M}(S') & \xrightarrow[\underline{M}_*(f)]{} & \underline{M}(S). \end{array}$$

The contravariant map $\underline{M}^*(f)$ is called the "restriction" along f, while the covariant map $\underline{M}_*(f)$ is called the "transfer" along f.

Morphisms of Mackey functors are simply collections of homomorphisms of abelian groups that commute with all of the additional structure maps.

Notation 17.3.24. Let $\mathcal{M}ackey^G$ denote the category of G-Mackey functors.

Example 17.3.25. A C_p-Mackey functor is a C_p-coefficient system $\underline{M}$ together with a transfer map

$$tr_e^{C_p} : \underline{M}(C_p/e) \to \underline{M}(C_p/C_p)$$

that factors through the C_p coinvariants of $\underline{M}(C_p/e)$ and such that

$$res_e^{C_p} \circ tr_e^{C_p}(a) = \sum_{g \in C_p} g \cdot a.$$

The push-pull formula, when applied to the case where all of T, S, and S' are orbits, is sometimes called the "double-coset formula". It expresses the various ways we could rewrite

$$\{G/H_+ \wedge X, G/K_+ \wedge Y\}^G$$

as either the set of H-equivariant maps or of K-equivariant maps.

Corollary 17.3.26. *For any finite, pointed G-CW complexes X and Y, the coefficient system*

$$T \mapsto \{T_+ \wedge X, Y\}^G$$

extends to a Mackey functor.

In general, the Beck–Chevalley condition that shows up in the definition of a Mackey functor suggest a reformulation in terms of functors from a correspondence category. This works for Mackey functors too, as shown by Lindner.

Definition 17.3.27. Let $\mathcal{A}_{\mathbb{N}}$ denote the category whose objects are finite G-sets and for which the morphisms from S to T are isomorphisms classes of finite G-sets over $S \times T$:

$$\mathcal{A}_{\mathbb{N}}(S,T) = \{[S \leftarrow U \rightarrow T]\}.$$

Composition is given by pullback.

The identity functor gives an isomorphism $\mathcal{A}_{\mathbb{N}} \cong \mathcal{A}_{\mathbb{N}}^{op}$, and the disjoint union of finite G-sets is both the product and the coproduct in this category. This means that the Hom sets are naturally commutative monoid valued.

Proposition 17.3.28. [52] *A Mackey functor is equivalently described as a product preserving functor* $\mathcal{A}_{\mathbb{N}} \rightarrow \mathsf{Ab}$.

This makes Mackey functors into a kind of diagram category, and morphisms of Mackey functors are just natural transformations.

Proposition 17.3.29. *The category* Mackey^G *of G-Mackey functors is an abelian category with enough projectives.*

In fact, the projective generators are easy to describe.

Definition 17.3.30. Let $\underline{A}_T$ be the functor

$$S \mapsto K\big(\mathcal{A}_{\mathbb{N}}(T,S)\big),$$

where $K(\text{-})$ denotes group completion.

When $T = *$, we call $\underline{A} = \underline{A}_*$ the Burnside Mackey functor.

Remark 17.3.31. The abelian group $\underline{A}(*)$ is the Grothendieck group of the category Fin^G. In particular, objects are finite virtual G-sets. This has a ring structure under Cartesian product, and we call it the Burnside ring.

Proposition 17.3.32. *The Mackey functor* $\underline{A}_T$ *represents the functor*

$$\underline{M} \mapsto \underline{M}(T),$$

and hence is projective.

There is a non-full, product preserving embedding

$$(\mathsf{Fin}^G)^{op} \hookrightarrow \mathcal{A}_{\mathbb{N}}$$

which is also the identity on objects and which sends a map $f\colon S \rightarrow T$ to the correspondence

$$T \xleftarrow{f} S \xrightarrow{=} S.$$

Precomposing with this inclusion gives a forgetful functor

$$U\colon \mathsf{Mackey}^G \to \mathsf{Coeff}^G.$$

This forgetful functor just forgets the transfer maps, recording only the restriction maps. The algebra underlying both Mackey functors and a lot of the equivariant intuition is that this is part of a monadic (and comonadic) adjunction.

Proposition 17.3.33. *The forgetful functor $U\colon \mathsf{Mackey}^G \to \mathsf{Coeff}^G$ has a left adjoint $L\colon \mathsf{Coeff}^G \to \mathsf{Mackey}^G$, the "Mackeyfication" of a coefficient system. The category of Mackey functors is the category of algebras over the associated monad.*

The left adjoint L free adjoins missing transfers, placing only those contraints required by the Mackey functor structure.

Example 17.3.34. Consider a C_p-coefficient system $\underline{M}$. Then $L(\underline{M})(C_p/e) = \underline{M}(C_p/e)$, while

$$L(\underline{M})(C_p/C_p) = \underline{M}(C_p/C_p) \oplus \big(\underline{M}(C_p/e)\big)/C_p,$$

where $\underline{M}(C_p/e)/C_p$ is the coinvariants. The transfer map here is composite of the inclusion with the canonical quotient:

$$\underline{M}(C_p/e) \to \big(\underline{M}(C_p/e)\big)/C_p \hookrightarrow L(\underline{M})(C_p/C_p),$$

while the restriction is given by the restriction in $\underline{M}$ on the summand $\underline{M}(C_p/C_p)$ and the trace

$$\big(\underline{M}(C_p/e)\big)/C_p \xrightarrow{a\mapsto\sum ga} \big(\underline{M}(C_p/e)\big)^{C_p}\subseteq \underline{M}(C_p/e).$$

This connects back to equivariant stable homotopy theory in a very transparent way.

Theorem 17.3.35. *If X is a finite, pointed G-CW complex, then we have a natural isomorphism*

$$\underline{\{S^0, X\}} \cong L\big(\underline{\pi}_0^s(X)\big),$$

where $\underline{\pi}_0^s$ is the coefficient system

$$T \mapsto \pi_0^s\big(\mathsf{Top}_*^G(T_+, X)\big) = \{S^0, \mathsf{Top}_*^G(T_+, X)\}^e$$

of non-equivariant stable maps to the fixed points of X.

Since left adjoints are easy to compute on representable functors, we deduce the stable homotopy groups of the zero sphere.

Corollary 17.3.36. *For any finite G, we have*

$$\underline{\{S^0, S^0\}} \cong \underline{A}.$$

Putting this again into words, equivariant stabilization takes our expected value and freely puts in the transfers. These transfer terms arise as summands, isomorphic to Weyl coinvariants of the value of the coefficient system at other orbits.

An incredible theorem of Segal and tom Dieck vastly generalizes this, allowing us to understand what happens more arbitrarily. In particular, they determined the representing object for the cohomology theory given by Corollary 17.3.8.

Theorem 17.3.37 (Segal–tom Dieck splitting). *Let Y be a fixed finite, pointed G-CW complex. If X is a finite, pointed G-CW complex on which G acts trivially, then the functor*

$$X \mapsto \{X, \Sigma^n Y\}^G$$

is represented by the non-equivariant infinite loop space

$$\prod_{(H)} \Omega^\infty \Sigma^n (EW_G(H)_+ \underset{W_G(H)}{\wedge} \Sigma^\infty Y^H),$$

where the product ranges over all conjugacy classes of subgroups of G and where $W_G(H)$ is the Weyl group of H in G.

17.3.3 Fixed points as a Mackey functor

Even though we have not produced any particular models of G-spectra, the Spanier–Whitehead category provides a list of desirable things. In particular, the abelian group valued cohomology theory given by

$$Y^*(X, A) = \{X/A, \Sigma^n Y\}^G$$

is really just a piece of a Mackey functor valued cohomology theory:

$$\underline{Y}^*(X, A)(T) = \{X/A, \Sigma^n T_+ \wedge Y\}^G.$$

The Segal–tom Dieck splitting describes the representing object for each of these (as T varies):

$$\bigvee_{(H)} \Sigma^\infty EW_G(H)_+ \underset{W_G(H)}{\wedge} (T_+ \wedge Y)^H,$$

and the Yoneda Lemma then gives more structure reflecting the Mackey functor structure on the cohomology groups:

Proposition 17.3.38. *The assignment*

$$T \mapsto \bigvee_{(H)} \Sigma^\infty EW_G(H)_+ \underset{W_G(H)}{\wedge} (T_+ \wedge Y)^H$$

extends to a Mackey functor object in the homotopy category of spectra.

The summand corresponding to G is simply

$$\Sigma^\infty_+ Y^G.$$

This is the one we would expect to have! The others are somewhat more surprising. The Mackey structure gives us a way to interpret these, however: they are the images of the transfers from various subgroups. To see this, note that the assignment

$$Y \mapsto \bigvee_{(H)} \Sigma^\infty EW_G(H)_+ \underset{W_G(H)}{\wedge} Y^H$$

is also functorial in the G-CW complex Y. Moreover, map induced by a map $f\colon Y \to Z$ is the expected one: wedge together the various maps

$$EW_G(H)_+ \underset{W_G(H)}{\wedge} Y^H \xrightarrow{EW_G(H)_+ \underset{W_G(H)}{\wedge} f^H} EW_G(H)_+ \wedge_{W_G(H)} Z^H.$$

Applying this to the maps $G/H_+ \wedge Y \to Y$ gives us an identification of the various other summands. We spell this out for $H = \{e\}$.

Example 17.3.39. Now let $T = G$ as above and consider the unique map $G \to *$. This gives us the transfer from $\{e\}$ to G in the Mackey functor structure, and in the homotopy category, this is witnessed as the map

$$\bigvee_{(H)} \Sigma^\infty EW_G(H)_+ \underset{W_G(H)}{\wedge} (G_+ \wedge Y)^H \to \bigvee_{(H)} \Sigma^\infty EW_G(H)_+ \underset{W_G(H)}{\wedge} Y^H.$$

Since $G_+ \wedge Y$ has a free G-action, the only nontrivial summand is the one corresponding to $\{e\}$. Thus the transfer is represented in the homotopy category by

$$Y \simeq EW_G(\{e\})_+ \underset{W_G(\{e\})}{\wedge} (G_+ \wedge Y) \to \bigvee_{(H)} \Sigma^\infty EW_G(H)_+ \underset{W_G(H)}{\wedge} Y^H.$$

The appearance of the homotopy orbits, rather than just the underlying space, are also explained by the Mackey functor structure. The transfer map factors through the Weyl coïnvariants, since there is a unique map $G \to *$. The representing object then factors through the homotopy coïnvariants.

Remark 17.3.40. The restriction maps in the Mackey functor structure can also be described, but they are a little less intuitive. When restricting all the way to $H = \{e\}$, the map is just the wedge of the ordinary Becker–Gottleib transfer maps

$$\Sigma^\infty EW_G(H)_+ \underset{W_G(H)}{\wedge} Y^H \to \Sigma^\infty Y^H \to \Sigma^\infty Y.$$

Remark 17.3.41. The construction

$$Y \mapsto \bigvee_{(H)} EW_G(H)_+ \underset{W_G(H)}{\wedge} \Sigma^\infty Y^H$$

commutes with filtered colimits, and we can therefore use this to extend to infinite complexes in the usual way.

17.3.4 Categories of G-spectra

There are two main approaches to lifting the homotopical discussions above to a model or ∞-category: extending the geometric, representation theory based spectra or extending the algebra, Mackey functor approach. These all have the same underlying homotopy theory, which is generated by the extended equivariant Spanier–Whitehead category under colimits, just as with the ordinary stable homotopy category.

Continuing our model-independent approach, we list a collection of desired features, which are in some sense defining properties. Any category Sp^G of G-spectra will have these properties.

1. They are closed symmetric monoidal, pointed model / ∞-categories.
2. For every $H \subseteq G$, there is a strong symmetric monoidal restriction functor
$$i_H^*\colon \mathrm{Sp}^G \to \mathrm{Sp}^H$$
which has both adjoints
$$G_+ \underset{H}{\wedge} (\text{-}) \dashv i_H^*(\text{-}) \dashv F_H(G_+, \text{-}).$$
3. The natural map
$$G_+ \underset{H}{\wedge} (\text{-}) \to F_H(G_+, \text{-})$$
descends to an equivalence (the Wirthmüller isomorphism) in the homotopy category.
4. There are functors $\Sigma_+^\infty\colon \mathrm{Top}^G \to \mathrm{Sp}^G$ which commute with the restriction functors.
5. The homotopy functor $\Sigma_+^\infty\colon ho\mathrm{Top}^G \to ho\mathrm{Sp}^G$ factors through the extended Spanier–Whitehead category and induces an equivalence between the extended Spanier–Whitead category and the compact objects in $ho\mathrm{Sp}^G$.
6. The homotopy category is a Brown category [44].

Following the historical designation, we call such a category of G-spectra "genuine". There are many models for a category of genuine G-spectra:

1. Lewis–May–Steinberger consider the clear extension of Adams' original treatment of spectra, viewing G-spectra as sequences of G-spaces, indexed by finite dimensional representations of some fixed G-universe [50].
2. Shimakawa built an equivariant version of Segal's Γ spaces, showing that Γ-G-spaces model connective G-spectra [61].
3. Mandell–May generalized orthogonal spectra to equivariant orthogonal spectra [53, 39].
4. Mandell extended work of Hovey–Shipley–Smith on symmetric spectra to produce equivariant symmetric spectra [55].

5. Blumberg considered continuous functors on G-CW complexes and a version of excision here [10] (See also [20]).

6. Guillou–May took a more algebraic approach, building equivariant spectra as spectral Mackey funtors [35].

7. Barwick produced a genuine ∞-category of spectral Mackey functors in a very general context [8].

Any of these categories will work for our discussion in this section. When discussing multiplicative concerns, there are several subtleties that arise, and only some of the models are known to work well there: equivariant orthogonal spectra [39], equivariant symmetric spectra [38], and the ∞-categorical enrichment for spectral Mackey functors.

17.3.4.1 Equivariant cohomology and Brown representability

Since our category of G-spectra contains fully faithful the extended Spanier–Whitehead category as a symmetric monoidal subcategory, the Picard groupoid of the extended Spanier–Whitehead category is a sub-groupoid of the Picard groupoid of G-spectra. This gives a natural grading for maps in the category and for homology or cohomology theories. Just as classically, G-spectra represent cohomology theories.

Definition 17.3.42 (See [57, XIII]). An RO(G)-graded cohomology theory is a functor

$$E^{(\text{-})}(\text{-})\colon \mathsf{JO}^s \times ho\mathsf{Top}_*^{G,op} \to \mathsf{Ab}$$

together with natural suspension isomorphisms

$$\Sigma^W\colon E^V(X) \to E^{W\oplus V}(\Sigma^W X),$$

for each representation W such that

1. for each fixed V, E^V takes wedges to products and is exact in the middle for cofiber sequences,

2. the suspension isomorphisms are natural in W for maps in JO^s, and

3. for all representations W and U, we have

$$\Sigma^{W\oplus U} = \Sigma^W \circ \Sigma^U.$$

Naturality of the suspension isomorphisms gives us a commutative square for any V and W and any map $f\colon X \to Y$

$$\begin{array}{ccc} E^V(Y) & \xrightarrow{E^V(f)} & E^V(X) \\ {\scriptstyle\Sigma^W}\downarrow & & \downarrow{\scriptstyle\Sigma^W} \\ E^{W\oplus V}(\Sigma^W Y) & \xrightarrow[E^{W\oplus V}(\Sigma^W f)]{} & E^{W\oplus V}(\Sigma^W X). \end{array}$$

In particular, the map $E^V(f)$ is completely determined by $E^{W\oplus V}(\Sigma^W f)$.

Proposition 17.3.43. *If E is an $\mathrm{RO}(G)$-graded cohomology theory, then E descends to a functor*

$$E^{(\text{-})}(\text{-})\colon \mathsf{JO}^s \times \mathsf{SW}^{G,op} \to \mathsf{Ab}.$$

Although we call these $\mathrm{RO}(G)$-graded, as written, these are graded by the actual representations. The suspension isomorphisms allow us to extend to pairs of representations, just as with the Spanier–Whitehead category. Again, the correct approach is to grade on the Picard groupoid. In this more general setting, the $\mathrm{RO}(G)$-graded cohomology theories actually descend to the extended Spanier–Whitehead categories, where we set

$$E^V(\Sigma^{-W}X) := E^{W\oplus V}(X),$$

and the map induced by the map $X \to \Sigma^{-W}\Sigma^W X$ of Equation 17.3.1 is taken to $(\Sigma^W)^{-1}$.

Notation 17.3.44. Following Hu–Kriz [45], we will use $*$ as a wildcard for grading by $\mathbb{Z}$ and $\star$ for $\mathrm{RO}(G)$.

Proposition 17.3.45. *For any G-spectrum E, the assignment*

$$(V, X) \mapsto [\Sigma^\infty X, S^V \wedge E]^G$$

together with the natural suspension isomorphisms

$$[\Sigma^\infty X, S^V \wedge E]^G \xrightarrow{\cong} [\Sigma^\infty \Sigma^W X, S^{W\oplus V} \wedge E]^G$$

give an $\mathrm{RO}(G)$-graded cohomology theory.

The Brown representability theorem allows us to invert this procedure, at least in the homotopy category of spectra. The best reference for this is the treatment in the Alaska notes [57], which uses Lewis–May–Steinberger spectra [50]. However, since the representability takes place in the homotopy category, the result is again model agnostic.

Theorem 17.3.46. [57, XIII.3] *If $E^\star(\text{-})$ is an $\mathrm{RO}(G)$-graded cohomology theory, then there is a G-spectrum E which represents it in the homotopy category.*

Example 17.3.47. In general, Bredon cohomology for a coefficient system $\underline{M}$ only gives a $\mathbb{Z}$-graded cohomology theory. However, if $\underline{M}$ is the underlying coefficient system for a Mackey functor, then Lewis–May–McClure show that Bredon cohomology with coefficients in $\underline{M}$ has a natural extension [49]. The resulting G-spectrum is the Eilenberg–Mac Lane spectrum $H\underline{M}$ and it has the defining property that the homotopy Mackey functors are given by

$$\underline{\pi}_k H\underline{M} = \begin{cases} \underline{M} & k = 0 \\ 0 & \text{otherwise.} \end{cases}$$

To get a sense of why the transfers are needed here, consider the group $G = C_2$. If Bredon cohomology with coefficients in $\underline{M}$ extends to an $\mathrm{RO}(C_2)$-graded theory, then we can take the Bredon cohomology of any virtual representation sphere. Recall that the sign sphere S^σ has a cell structure

$$S^0 \cup C_{2+} \wedge e^1,$$

where the attaching map is the action map

$$C_{2+} \wedge S^0 \to S^0.$$

The dual of the fold map is the transfer map $S^0 \to C_{2+}$, giving a fiber sequence

$$S^{-\sigma} \to S^0 \to C_{2+}.$$

Mapping out of this gives a formula for the $-\sigma$th cohomology of a point with coefficients in $\underline{M}$, and we see that we must have a map

$$\underline{M}(C_2) \to \underline{M}(*)$$

that restricts to the ordinary fold map non-equivariantly. This is equivalent to the data of a C_2-Mackey functor.

As Lewis observed, working with this larger grading allows us to see more structure than we might have just using integral gradings [47].

Example 17.3.48. Let $G = C_2$, and let $\underline{\mathbb{Z}}$ be the constant Mackey functor $\mathbb{Z}$. Then

$$H_*(\mathbb{C}P^1; \underline{\mathbb{Z}}) = \begin{cases} \mathbb{Z} & * = 0 \\ \mathbb{Z}/2 & * = 1 \\ 0 & \textit{otherwise}, \end{cases}$$

while

$$H_\star(\mathbb{C}P^1; \underline{\mathbb{Z}}) = H_\star(pt; \underline{\mathbb{Z}}) \oplus H_{\star-\rho}(pt; \underline{\mathbb{Z}}).$$

Here $\mathbb{C}P^1$ is a C_2-space via complex conjugation, and ρ is the regular representation of C_2.

Example 17.3.49. Consider the equivariant K-theory functor that assigns to a G-space the Grothendieck group of complex, equivariant vector bundles over X. Equivariant Bott periodicity says that for every complex representation V, we have a natural isomorphism

$$\tilde{K}^0_G(X) \cong \tilde{K}^0_G(\Sigma^V X).$$

These isomorphisms allow us to extend to an RO(G)-graded theory. There is an analogous story for real K-theory KO_G or Atiyah's Real K-theory $K_{\mathbb{R}}$.

Building on this example, we can construct homotopical versions of bordism theories via Thom spectra.

Example 17.3.50. Consider the space $BO_G(n)$ that classifies n-dimensional equivariant bundles. There is a universal n-plane bundle over this, and associated Thom space. These assemble to give the Thom spectrum MO_G of homotopy equivariant unoriented bordism.

There is a simple story for equivariant complex bordism; here we use the spaces $BU_G(n)$ that classify n-dimensional complex equivariant bundles. This gives a spectrum MU_G.

There is also a Real version of bordism, where we take the spaces $BU(n)$ as C_2 spaces with the C_2-action by complex conjugation. Here, the associated Thom spectrum is $\mathbf{MU}_{\mathbb{R}}$, the Real bordism spectrum of Fujii–Landweber [26, 46].

These theories are not the geometrically defined bordism theories one would expect from the non-equivariant case. The geometrically defined theories are only $\mathbb{Z}$-graded theories.

17.3.5 Fixed, homotopy fixed, and geometric fixed points

We turn now some of the basic properties and constructions.

17.3.5.1 Fixed points

Proposition 17.3.51. *For any normal subgroup N, there is a strong symmetric monoidal, faithful functor*

$$i_*^N : \mathsf{Sp}^{G/N} \to \mathsf{Sp}^G$$

lifting and extending the functor on the extended Spanier–Whitehead categories of Proposition 17.3.7.

When $N = G$, we will also write this just as i_*.

Remark 17.3.52. The point-set models for the pushforward are all strong symmetric monoidal functors. However, we have no effective control over the homotopical behavior of commutative monoids under it. In Example 17.4.48 below, we show how badly this can go.

Proposition 17.3.53. *The functor i_*^N has a right adjoint*

$$(\text{-})^N : \mathsf{Sp}^G \to \mathsf{Sp}^{G/N},$$

the N-fixed point functor.

In general, this fixed point functor is difficult to understand. Our conditions on the relationship with the Spanier–Whitehead category then describes this on finite G-CW complexes.

Theorem 17.3.54 (Segal–tom Dieck Splitting)**.** *If X is a finite G-CW complex, then*

$$\left(\Sigma_+^\infty X\right)^G \simeq \bigvee_{(H)} EW_G(H)_+ \underset{W_G(H)}{\wedge} \Sigma_+^\infty X^H.$$

The zero sphere S^0 is in the image of the pushforward, so combining the pushforward with induction, we see that the spectrum of maps out of

$$G/H_+ \cong G_+ \underset{H}{\wedge} S^0$$

is the H-fixed points. Mapping out of the maps $G/H_+ \to G/K_+$ for various subgroups H and K then give us restriction and conjugation maps

$$E^K \to E^H.$$

The Wirthmüller isomorphism says that induction is also the right adjoint to the restriction, homotopically, and this gives us transfer maps the other way

$$E^H \to E^K.$$

Proposition 17.3.55. *For any G-spectrum E, the fixed points E^H as H ranges over the subgroups of G assemble into a Mackey functor object in the homotopy category of spectra.*

This proposition can be viewed as a more general one, building on the closed symmetric monoidal structure.

Proposition 17.3.56. *For any G-spectra E and E', the homotopy classes of maps $E \to E'$ assemble into a Mackey functor*

$$T \mapsto [T_+ \wedge E, E']^G$$

which when evaluated at G/H, records the homotopy classes of H-equivariant maps.

This is the key idea in G-spectra: stabilization made our category equivariantly additive. We can not only add maps as usual in a stable setting but also form "twisted" sums, where the group acts on the source. These are the transfer maps. All of the algebraic invariants that we are used to are naturally Mackey functor valued in equivariant stable homotopy.

In particular, our homotopy groups naturally assemble into homotopy Mackey functors. The fully faithful inclusion of the extended Spanier–Whitehead category into the homotopy category allows us to compute this for the sphere spectrum.

Corollary 17.3.57. *There are natural Mackey enrichments of Bredon cohomology with coefficients in $\underline{M}$ and of the equivariant K-theories.*

Just as classically, spheres are appropriately connected.

Proposition 17.3.58. *For all $n < 0$, we have*

$$\underline{\pi}_n(T_+ \wedge \Sigma^\infty S^0) = 0,$$

while

$$\underline{\pi}_0(T_+ \wedge \Sigma^\infty S^0) = \underline{A}_T.$$

The connectivity of the sphere spectrum then allows us to mirror the classical formation of the Postnikov tower: the long exact sequence in stable homotopy shows that we can kill homotopy groups above some fixed dimension k for a G-spectrum by transfinitely coning off all maps from equivariant spectra of the form $T_+ \wedge S^\ell$ with $\ell \geq k$.

Proposition 17.3.59. *There is a t-structure on G-spectra, where $\tau_{\geq 0}^{Post}$ consists of all spectra E with $\underline{\pi}_k(E) = 0$ for $k < 0$ and similarly for $\tau_{\leq -1}^{Post}$. The heart of this t-structure is the category of Mackey functors.*

Corollary 17.3.60. *Every G-spectrum E has a Postnikov tower: there is a functorial tower under E*

$$\dots \to P^n_{Post}(E) \to P^{n-1}_{Post}(E) \to \dots$$

in which the homotopy Mackey functors of $P^n_{Post}(E)$ vanish for $k > n$ and agree with those of E for $k \leq n$. The fibers are Eilenberg–Mac Lane spectra for the homotopy Mackey functors of E.

Remark 17.3.61. The tower under E dual to the Postnikov tower is the Whitehead tower, where here we approximate E by appropriately connective objects.

Proposition 17.3.62. *The t structure is compatible with the symmetric monoidal structure in the sense that if $E \in \tau_{\geq n}^{Post}$ and and $E' \in \tau_{\geq m}^{Post}$, then*

$$E \wedge E' \in \tau^{Post}_{\geq (n+m)}.$$

This is important for knowing the multiplicativity of the Atiyah–Hirzebruch spectral sequence below.

We close this subsection with a comment about describing fixed points for more general spectra. For trivial desuspensions of finite G-CW complexes, we can still understand the fixed points. For more general desuspensions, we have little understanding.

Example 17.3.63. Let $G = C_2$ and let σ denote the sign representation of C_2. The fixed points of $\Sigma^{-\sigma}\Sigma^\infty X$ sit in a fiber sequence

$$\left(\Sigma^{-\sigma}\Sigma^\infty X\right)^{C_2} \to \Sigma^\infty X^{C_2} \vee (\Sigma^\infty EC_{2+} \underset{C_2}{\wedge} X) \to \Sigma^\infty X,$$

where the rightmost map is the restriction in the homotopy Mackey functor for X.

17.3.5.2 Borel and coBorel, free and cofree

The space EG is a universal space on which G acts freely. Using this, we can free up or cofree up the action on any G-space or spectrum. This has the effect of weakening the amount of information we need to remember.

Definition 17.3.64. If E is a G-spectrum, then let

$$E_h := EG_+ \wedge E$$

be the Borel construction on E.

Even though the action on EG is free, we have non-trivial fixed points stably. This is another consequence of the transfer, generalizing what we saw in Example 17.3.20.

Theorem 17.3.65 (Adams' Isomorphism [2])**.** *For any G-spectrum E, the transfer induces an equivalence*

$$EG_+ \underset{G}{\wedge} E \to \left(EG_+ \wedge E\right)^G.$$

More generally, if E is a spectrum on which a normal subgroup N acts freely, then the transfer induces an equivalence of G/N-spectra

$$E_N \to E^N,$$

and the fixed points also become the left adjoint to the pushforward.

Remark 17.3.66. Smashing with the universal space $E\mathcal{F}_{N+}$ of Examples 17.2.65 always frees up the N-action.

Definition 17.3.67. If E is a G-spectrum, then let

$$E^h := F(EG_+, E)$$

be the cofree G-spectrum of pointed maps from EG_+ to E.

Lemma 17.3.68. *The map $EG_+ \to S^0$ gives us canonical maps*

$$E_h \to E \text{ and } E \to E^h,$$

and the map

$$E_h \to E$$

is the left-most map from the isotropy separation sequence smashed with E.

By construction, both the Borel construction and the cofree spectrum care only about underlying equivalences.

Proposition 17.3.69. *If $f\colon E \to E'$ is an equivariant map such that $i_e^* f$ is an equivalence, then*

$$f^h\colon E^h \to E'^h \text{ and } f_h\colon E_h \to E'_h$$

are equivariant equivalences.

Definition 17.3.70. The homotopy fixed points of E are the G-fixed points of E^h:

$$E^{hG} := \left(F(EG_+, E)\right)^G.$$

Since $i_H^* EG = EH$ for any subgroup H,

$$E^{hH} \simeq \left(F(EG_+, E)\right)^H,$$

and the notation is unambiguous.

Example 17.3.71. Let $\underline{M}$ be a Mackey functor. By construction, the homotopy Mackey functors of the spectrum

$$H\underline{M}^h = F(EG_+, H\underline{M})$$

are the Bredon cohomology Mackey functors of EG with coefficients in $\underline{M}$. The standard cellular complex for EG_+ then allows us to determine these via cellular Bredon cohomology. In particular, we see that the homotopy Mackey functors are the ordinary Mackey functors associated to group cohomology, in this case, with coefficients in the G-module $\underline{M}(G)$:

$$\underline{\pi}_{-k}(HM^h)(G/K) \cong H^k(K; \underline{M}(G)).$$

The cofree construction can be viewed in another way, connecting G-spectra to the perhaps more expected "G-objects in spectra". Remark 17.2.32 points out that in spaces, the corresponding categories of spaces are homotopically distinct, and here we see something similiar. The cofree construction gives us a way to compare them.

Definition 17.3.72. A G-object in spectra is a functor $BG \to \mathsf{Sp}$. The category of G-objects in spectra is the functor category $\mathrm{Fun}(BG, \mathsf{Sp})$.

Just as with spaces, a G-object in spectra is an ordinary spectrum E together with an action map of spectra $G_+ \wedge E \to E$ that satisfies the usual axioms for an action. There is an evident forgetful functor which forgets the additional structure.

Proposition 17.3.73. *The restriction map*

$$\mathsf{Sp}^G \to \mathsf{Sp}$$

lifts along the forgetful functor $Fun(BG, \mathsf{Sp}) \to \mathsf{Sp}$ to give a functor

$$\mathsf{Sp}^G \to \mathrm{Fun}(BG, \mathsf{Sp}).$$

The easiest way to see this is actually via the Wirthmüller isomorphisms. The restriction of E can be identified with the fixed points of $G_+ \wedge E$, via induction's equivalence with coinduction. As we have used before, $G^{op} \cong G$ is the automorphism group of G as a G-set, and by naturality of the tensoring operation of G-spaces on G-spectra, this gives an action.

Proposition 17.3.74. *The cofree G-spectrum gives a homotopical functor*

$$F(EG_+, \text{-})\colon \mathrm{Fun}(BG, \mathsf{Sp}) \to \mathsf{Sp}^G$$

from G-objects in spectra to G-spectra.

This allows us to view any G-object in spectra as an actual G-spectrum in a homotopically meaningful way.

Example 17.3.75. The Goerss–Hopkins–Miller theorem says that for any perfect field k and formal group law Γ over k, the Lubin–Tate spectrum $E(k, \Gamma)$ admits the structure of an E_∞ ring spectrum and that the Morava stabilizer group $\mathrm{Aut}(\Gamma)$ acts on this via E_∞ ring maps. Thus for any finite subgroup G of $\mathrm{Aut}(\Gamma)$, $E(k, \Gamma)$ can be viewed as a G-spectrum.

Example 17.3.76. If M is an abelian group, then the cofree spectrum

$$HM^h = F(EG_+, i_*^G HM)$$

represents Borel cohomology:

$$X \mapsto H^*(EG_+ \underset{G}{\wedge} X; M).$$

Remark 17.3.77. The cofree construction is also lax monoidal, so it preserves various kinds of structured products. More surprisingly, it takes E_∞ ring spectra to G-E_∞-ring spectra (See Proposition 17.3.95 below).

Since the map

$$E \to E^h$$

is an underlying equivalence, the map

$$E_h \to (E^h)_h$$

is always an equivalence. If we smash the isotropy separation sequence for the family $\mathcal{F}_e$ with E^h and take fixed points, then we get a cofiber sequence of spectra. The cofiber was described by Greenlees–May.

Definition 17.3.78. [33] If E is a spectrum with G-action, then let

$$E^{tG} = \left(\tilde{E}G \wedge F(EG_+, E)\right)^G$$

be the Tate spectrum of E.

Again, since cofree spectra care only about the underlying spectrum with a G-action, it makes no difference if we consider genuine G or these less strict ones.

Corollary 17.3.79 (Tate sequence). *For any G-spectrum E, we have a cofiber sequence*

$$E_{hG} \to E^{hG} \to E^{tG},$$

where the first map is a lift to spectra of the trace map from group homology to group cohomology.

Remark 17.3.80. The trace map

$$E_{hG} \to E^{hG}$$

is often called the "norm map", especially in trace methods literature. We call it the trace to avoid confusion with the multiplicative norm functors described below.

We can interpret the homotopy fixed points as providing a way to isolate the contribution of the underlying homotopy. This is the key feature of the Segal conjecture, proved by Carlsson: maps out of EG_+ can be viewed as a completion at the ideal given by the restriction to the trivial subgroup [17].

Theorem 17.3.81 (Segal Conjecture [17]). *If X is a finite, pointed G-CW spectrum, then the projection map $X_h \to X$ induces a natural isomorphism*

$$\pi^*(X)^\wedge_I \to \pi^*(EG_+ \wedge X),$$

where $I \subseteq \underline{A}(*)$ is the ideal of the Burnside ring consisting of all elements that restrict to 0 in $\underline{A}(G)$.

It can be helpful here to work more generally.

Proposition 17.3.82. *If $\mathcal{F}$ is a family, then the homotopy type of $F(E\mathcal{F}_+, E)$ depends only on the H-equivariant homotopy type of E for all $H \subseteq G$. Moreover, if we have an inclusion $\mathcal{F}_1 \subseteq \mathcal{F}_2$, then we have a canonical map*

$$F(E\mathcal{F}_{2+}, E) \to F(E\mathcal{F}_{1+}, E).$$

This map is an equivalence when restricted to any $H \in \mathcal{F}_1$.

Applying this to $\mathcal{F}_2 = \mathcal{A}ll$ gives immediately a special case.

Corollary 17.3.83. *For any family $\mathcal{F}$, we have a natural map*

$$E \to F(E\mathcal{F}_+, E).$$

When E is a (commutative, associative, etc.) ring spectrum, this is a map of (commutative, associative, etc.) ring spectra.

Theorem 17.3.84. [3] *If $\mathcal{F}$ is a family of subgroups, and if $I(\mathcal{F}) \subseteq \underline{A}(*)$ is the ideal of all elements of the Burnside ring that restrict to zero in $\underline{A}(G/H)$ for all $H \in \mathcal{F}$, then the projection $E\mathcal{F}_+ \wedge X \to X$ induces an isomorphism*

$$\pi^*(X)^\wedge_{I(\mathcal{F})} \to \pi^*(E\mathcal{F}_+ \wedge X).$$

17.3.5.3 Geometric fixed points

The categorical fixed points are in many ways poorly behaved: they are not strong symmetric monoidal and they do not do what we expect on suspension spectra. In both cases, the underlying problem is that the fixed points for proper subgroups contribute to the fixed points for all of G via the transfer maps. The geometric fixed points fix both of these problems.

Definition 17.3.85. Let

$$\Phi^G(E) = \left(\tilde{E}\mathcal{P} \wedge E\right)^G$$

be the geometric fixed points of E, where again $\mathcal{P}$ is the family of proper subgroups.

Definition 17.3.86. For a G-spectrum E, the isotropy separation sequence for E is the cofiber sequence

$$E\mathcal{P}_+ \wedge E \to E \to \tilde{E}\mathcal{P} \wedge E$$

we get by smashing E with the defining cofiber sequence for $\tilde{E}\mathcal{P}$.

By functoriality of the fixed points functor, the following is immediate.

Proposition 17.3.87. *We have a natural map*

$$E^G \to \Phi^G(E).$$

By definition of $\tilde{E}\mathcal{P}$, if H is a proper subgroup, then

$$\Phi^G(G/H_+) = \left(\tilde{E}\mathcal{P} \wedge G/H_+\right)^G \simeq *,$$

since $i_H^*\tilde{E}\mathcal{P}$ is equivariantly contractible. Now if X is a G-CW complex, then the inclusion of the fixed points $X^G \hookrightarrow X$ is the inclusion of a subcomplex with the property that the quotient X/X^G is built entirely out of cells with a proper stabilizer. This gives use the first desired property.

Proposition 17.3.88. *The geometric fixed points of a suspension spectrum is the suspension spectrum of the fixed points:*

$$\Sigma_+^\infty(X^G) \xrightarrow{\simeq} \Phi^G\Sigma_+^\infty(X),$$

where the map is induced by the inclusion of the fixed points of X.

Remark 17.3.89. We can take the observation about the contractbility of the geometric fixed points for proper subgroups as the *defining feature* of the geometric fixed points. There is a smashing localization that nullifies any G-CW complex built out of cells with proper stabilizers, and the result of applying this to S^0 is just $\tilde{E}\mathcal{P}$. The fixed points on the category of local objects in then an equivalence.

The geometric fixed points is also strong symmetric monoidal.

Proposition 17.3.90. *For G-spectra E and E', there is a natural equivalence*

$$\Phi^G(E \wedge E') \simeq \Phi^G(E) \wedge \Phi^G(E').$$

The intuition behind property is most easily seen via the model of $\tilde{E}\mathcal{P}$ as the infinite reduced regular representation sphere. In this case, we see that we have an equivalence

$$\tilde{E}\mathcal{P} \simeq S^0[a_{\bar{\rho}_G}^{-1}],$$

where the class

$$a_{\bar{\rho}_G} \colon S^0 \to S^{\bar{\rho}_G}$$

is the Euler class of the representation $\bar{\rho}_G$. The right-hand side of this equivalence has the structure of an E_∞ ring spectrum that is solid in the sense that the multiplication map

$$S^0[a_{\bar{\rho}_G^{-1}}] \wedge S^0[a_{\bar{\rho}_G^{-1}}] \to S^0[a_{\bar{\rho}_G^{-1}}]$$

is an equivalence.

These two properties also describe the geometric fixed points for virtual representation spheres.

Example 17.3.91. If V is a virtual representation of G, then

$$\Phi^G S^V \simeq S^{V^G}.$$

Since the geometric fixed points is a strong symmetric monoidal functor, for each conjugacy class of subgroup H, we have a ring map

$$[S^0, S^0]^G \xrightarrow{\Phi^H} [\Phi^H S^0, \Phi^H S^0] \cong [S^0, S^0] \cong \mathbb{Z}.$$

Definition 17.3.92. The mark homomorphism (also called the ghost coordinates) is the map

$$\underline{A}(*) \to \prod_{(H)} \mathbb{Z}$$

given by the product of the various geometric fixed points maps.

This homomorphism also has a purely algebraic description: it assigns to a virtual G-set T the virtual cardinality of T^H.

Theorem 17.3.93. [21] *The mark homomorphism is an injective ring map and a rational isomorphism.*

Example 17.3.94. For $G = C_2$ and the spectrum $\mathbf{MU}_{\mathbb{R}}$, we have

$$\Phi^{C_2}\mathbf{MU}_{\mathbb{R}} \simeq MO.$$

17.3.5.4 Tate squares

We can combine the isotropy separation sequence with the maps on generalized cofree spectra to inductively build G-spectra out of pieces with prescribed isotropy. If $\mathcal{F}_1 \subseteq \mathcal{F}_2$ are families, then we have a natural map of G-spectra

$$F(E\mathcal{F}_{2+}, E) \to F(E\mathcal{F}_{1+}, E).$$

If $\mathcal{F}_0$ is a third family, then we can smash this map with the isotropy separation sequence for $\mathcal{F}_0$, getting a diagram

$$\begin{array}{ccccc} E\mathcal{F}_{0+} \wedge F(E\mathcal{F}_{2+}, E) & \longrightarrow & F(E\mathcal{F}_{2+}, E) & \longrightarrow & \tilde{E}\mathcal{F}_0 \wedge F(E\mathcal{F}_{2+}, E) \\ \downarrow & & \downarrow & & \downarrow \\ E\mathcal{F}_{0+} \wedge F(E\mathcal{F}_{1+}, E) & \longrightarrow & F(E\mathcal{F}_{1+}, E) & \longrightarrow & \tilde{E}\mathcal{F}_0 \wedge F(E\mathcal{F}_{1+}, E). \end{array} \tag{17.3.2}$$

The map

$$F(E\mathcal{F}_{2+}, E) \to F(E\mathcal{F}_{1+}, E).$$

is an H-equivalence for any $H \in \mathcal{F}_1$, so in particular, if $\mathcal{F}_0 \subseteq \mathcal{F}_1$, then the left-most map is an equivalence. This is the generalized Tate diagram.

Proposition 17.3.95. *If $\mathcal{F}_0 \subseteq \mathcal{F}_1 \subseteq \mathcal{F}_2$ are families, then we have a natural pullback diagram*

$$\begin{array}{ccc} F(E\mathcal{F}_{2+}, E) & \longrightarrow & \tilde{E}\mathcal{F}_0 \wedge F(E\mathcal{F}_{2+}, E) \\ \downarrow & & \downarrow \\ F(E\mathcal{F}_{1+}, E) & \longrightarrow & \tilde{E}\mathcal{F}_0 \wedge F(E\mathcal{F}_{1+}, E). \end{array}$$

If E is a ring spectrum, then all maps in the diagram are maps of ring spectra.

Definition 17.3.96. The fixed points of $\tilde{E}\mathcal{F}_0 \wedge F(E\mathcal{F}_{1+}, E)$ are generalized Tate spectra.

Example 17.3.97. Let $G = C_p$. The only interesting case for families is $\mathcal{F}_0 = \mathcal{F}_1 = \mathcal{F}_e$, and $\mathcal{F}_2 = \mathcal{A}ll$. Our diagram becomes

$$\begin{array}{ccc} E & \longrightarrow & \tilde{E}C_p \wedge E \\ \downarrow & & \downarrow \\ F(EC_{p+}, E) & \longrightarrow & \tilde{E}C_p \wedge F(EC_{p+}, E). \end{array}$$

The family $\mathcal{F}_e$ is also the family of proper subgroups, so $\tilde{E}C_p \wedge E$ is the data of an ordinary spectrum: the geometric fixed points. The bottom row depends only on the data of E as a functor $BG \to \mathsf{Sp}$, not as a genuine G-spectrum. So we have reduced the problem of describing a G-spectrum to that of a G-object in spectra, and ordinary spectrum (the geometric fixed points), and a map. This has been generalized to larger finite groups by Abram–Kriz, Glasman, and Ayala–Mazel-Gee–Rozenblyum [1], [28], [7].

17.4 Equivariant commutative ring spectra and norms

The category of spectra has a symmetric monoidal product, the smash product. It makes sense then to ask about monoids and commutative monoids in the category of spectra and to ask about these in a homotopy coherent way. In particular, we can ask for G-spectra R together with maps

$$R^{\wedge n} \to R$$

that satisfy the usual associativity or commutativity diagrams, either strictly or up to coherent homotopy. However, just as additively, we should follow the mandate that in equivariant homotopy, the group should be allowed to act on all indices in a diagram. In particular, we should be able to have the group action on $R^{\wedge n}$ intertwine the action on R and a permutation action.

If R has a strictly commutative multiplication, then the multiplication will extend naturally to any of these twisted products. Homotopically, the question becomes more subtle. This introduces a different flavors of commutative ring spectra, the N_∞-ring spectra, where we always have a (coherently commutative) multiplication and some collections of these twisted products.

17.4.1 Norms

We start by discussing the general construction for the smash products in which the group acts also on the smash factors. Since the group G is finite, any subgroup is of finite index. We can then form the analogue of "tensor induction" in spectra with the smash product. Heuristically, given an H-spectrum E, we smash together G/H-copies of E and have the group act as in induction, permuting the factors and acting.

Theorem 17.4.1. [39, Appendix B] *If $H \subseteq G$, then there are strong symmetric monoidal, homotopically meaningful "norm" functors*

$$N_H^G \colon \mathsf{Sp}^H \to \mathsf{Sp}^G$$

such that

1. *the norm commutes with sifted colimits,*
2. *if X is an H-space, then*

$$N_H^G(\Sigma_+^\infty X) \simeq \Sigma_+^\infty(\mathsf{Top}^H(G, X)),$$

3. *if V is a virtual representation of H, then*

$$N_H^G S^V \simeq S^{Ind_H^G V}.$$

The properties listed for the norm allow us to compute it for any H-spectrum. Since suspensions of finite G-CW complexes generate the category under colimits, we can resolve any H-spectrum as a colimit (in fact, directed) of shifts of suspension spectra. The norm then commutes with the colimit and has the listed formulae for the other pieces.

Remark 17.4.2. The norm is not an additive functor. However, considering the case of X is a discrete H-space above, we can easily describe a formula for the norm of wedges as wedges of norms.

Example 17.4.3. Let $G = C_2$. Then we have

$$N_e^{C_2}(S^0 \vee S^0) \simeq N_e^{C_2}(\Sigma_+^\infty\{a,b\}) \simeq \Sigma_+^\infty(\mathsf{Top}(C_2, \{a,b\})) \simeq S^0 \vee S^0 \vee C_{2+}.$$

The two copies of S^0 correspond to the norms of the two summands. The C_{2+} here is collectively the first sphere smashed with the second and the second sphere smashed with the first (i.e. the two different orders in which to do this).

There is an internal version of the twisted smash product.

Definition 17.4.4. If H is a subgroup of G, then let

$$N^{G/H}(E) = N_H^G i_H^* E.$$

If T is a finite G-set, then let

$$N^T(E) = \bigwedge_T E = \bigwedge_{G/H \subseteq T} N^{G/H}(E)$$

be the smash product over the orbits in T of the norm for that orbit.

Proposition 17.4.5. *The functors $N^T \colon \mathrm{Sp}^G \to \mathrm{Sp}^G$ are strong symmetric monoidal, homotopical functors.*

The geometric fixed points of norms are surprisingly easy to describe. This is essentially the "Tate diagonal" used in trace methods. For $G = C_p$, this is a key part of the Segal conjecture and of the topological Singer construction [58].

Proposition 17.4.6. *The diagonal map induces an equivalence*

$$\Phi^H E \to \Phi^G N_H^G(E).$$

In general, fixed and homotopy fixed points of norms are very difficult to understand. We have some connectivity estimates, however.

Proposition 17.4.7. *If E is in $\tau_{\geq k}^{Post}$, then $N_H^G E$ is in $\tau_{\geq k}^{Post}$ as well.*

In general, this is the best we can do for a connectivity estimate. Since we are smashing together several $(k-1)$-connected things, one might expect to see the connectivity scale, just as with the smash product. This need not be the case.

Example 17.4.8. Consider the sphere S^k, which is in $\tau_{\geq k}^{Post}$ for $H = \{e\}$. The norm to G is the regular representation sphere $S^{k\rho_G}$, where $\rho_G = \mathbb{R}[G]$ is the regular representation. The map

$$a_{\bar{\rho}_G} \colon S^k \to S^{k\rho_G}$$

is essential, since it induces an isomorphism on geometric fixed points. This shows that $\underline{\pi}_k(S^{k\rho_G}) \neq 0$.

17.4.2 N_∞-ring spectra

17.4.2.1 N_∞ operads

One of the key benefits for the point-set models of spectra is having good, homotopically meaningful categories commutative ring spectra and their modules. In EKMM style S-modules, a commutative ring spectrum is the same data as an E_∞ ring spectrum, and the same is true for symmetric or orthogonal spectra [25, 54].

In all cases, a key computation is that for a nice spectrum E, the action of Σ_n on the nth smash powers of E is actually free, and hence the natural map

$$E\Sigma_{n+} \underset{\Sigma_n}{\wedge} E^{\wedge n} \to E^{\wedge n}/\Sigma_n$$

is a weak equivalence. The homotopy symmetric powers assemble to give the free E_∞ algebra on E, while the actual ones give the free commutative algebra, and this provides a comparison between the two concepts.

Equivariantly, we have choices for exactly what the homotopy symmetric powers mean, since we can also include an action of G on the Σ_n-free spaces. We begin with two important examples.

Definition 17.4.9. Let $E\Sigma_n$ be $i_*^G E\Sigma_n$. This is $E\Sigma_n$ with a $G\times\Sigma_n$-action via the projection $G\times\Sigma_n\to\Sigma_n$. This is the universal space for the family

$$\mathcal{F}_{\Sigma_n}^{tr}=\{H\times\{e\}\mid H\subseteq G\}$$

of subgroups of G.

Definition 17.4.10. Let $E_G\Sigma_n=E\mathcal{F}_{\Sigma_n}$ be the universal space the family of subgroups Γ such that

$$\Gamma\cap(\{e\}\times\Sigma_n)=\{(e,e)\}.$$

Example 17.4.11. If $G=C_2$, then the space $E_G\Sigma_2$ can be modeled as an infinite sphere:

$$E_G\Sigma_2\simeq S(\infty\rho\otimes\tau),$$

where τ is the sign representation of Σ_2 and ρ is the regular representation of C_2.

The space $E\Sigma_2$ can be modeled as a different infinite sphere:

$$E\Sigma_2\simeq S(\infty\tau).$$

The C_2-fixed points of $E_G\Sigma_2/\Sigma_2$ are $\mathbb{R}P^\infty\amalg\mathbb{R}P^\infty$, while those of $B\Sigma_2$ are just $\mathbb{R}P^\infty$.

We generalize the classical notion of an E_∞ operad by allowing any such space, building a general class of N_∞ operads [11].

Definition 17.4.12. An N_∞ operad is a symmetric operad $\mathcal{O}$ in G-spaces such that for each n, the $G\times\Sigma_n$-space $\mathcal{O}_n$ is equivalent to $E\mathcal{F}_n(\mathcal{O})$ for some family $\mathcal{F}_n(\mathcal{O})$ with

$$\mathcal{F}_{\Sigma_n}^{tr}\subseteq\mathcal{F}_n(\mathcal{O})\subseteq\mathcal{F}_{\Sigma_n}.$$

An N_∞ operad $\mathcal{O}$ is a G-E_∞ operad if for each n, $\mathcal{O}_n\simeq E\mathcal{F}_{\Sigma_n}$.

The family $\mathcal{F}_{\Sigma_n}$ is the largest family such that Σ_n acts freely on the universal space. The conditions $\mathcal{F}_n(\mathcal{O})\subseteq\mathcal{F}_{\Sigma_n}$ then means Σ_n acts freely. The conditions that $\mathcal{F}_{\Sigma_n}^{tr}\subseteq\mathcal{F}_n(\mathcal{O})$ will guarantee enough G-fixed points (which will be necessary for us to have a coherent product in the operadic algebras).

Many of the classical constructions of operads immediately give N_∞ operads.

Example 17.4.13. If U is a universe, then the linear isometries operad defined by

$$\mathcal{L}_n=\mathrm{Isom}(U^{\oplus n},U)$$

is an N_∞ operad.

Example 17.4.14. If U is a universe, then the little disks operad for U is an N_∞ operad.

Example 17.4.15. If $\mathcal{O}$ is an ordinary E_∞ operad, then endowing it with the trivial action, we get an N_∞-operad we can $\mathcal{O}^{tr}$. This corresponds to the family $\mathcal{F}^{tr}_{\Sigma_n}$.

Just as non-equivariantly, there are model categories of N_∞-ring spectra and of commutative ring spectra.

Definition 17.4.16. Let $\mathcal{C}omm^G$ be the category of commutative ring spectra in Sp^G.

If $\mathcal{O}$ is an N_∞-operad, let $\mathcal{O}$-Alg be the category of $\mathcal{O}$-algebras in Sp^G.

Proposition 17.4.17. *The categories $\mathcal{C}omm^G$ and $\mathcal{O}$-Alg are enriched in spaces, are complete and co-complete, and are tensored and cotensored over Top^G. They can also be enriched in Top^G.*

Theorem 17.4.18. *The categories $\mathcal{C}omm^G$ and $\mathcal{O}$-Alg admit model structures such that the forgetful functor*

$$\mathcal{C}omm^G \to \mathsf{Sp}^G \text{ and } \mathcal{O}\text{-}\mathsf{Alg} \to \mathsf{Sp}^G$$

are homotopical right adjoint and compatible with the enrichments.

Definition 17.4.19. If $\mathcal{O}$ is an N_∞ operad, then let

$$\mathbb{P}_{\mathcal{O}} \colon \mathsf{Sp}^G \to \mathcal{O}\text{-}\mathsf{Alg}$$

be the free $\mathcal{O}$-algebra functor.

Moreover, the categories of commutative ring spectra and G-E_∞ ring spectra are closely connected. For any G-E_∞ operad $\mathcal{O}$, there is a canonical map of operads $\mathcal{O} \to \mathcal{C}omm$. The operadic base-change gives a Quillen equivalence in nice cases.

Theorem 17.4.20. *The categories $\mathcal{C}omm^G$ and of G-E_∞ ring spectra in Sp^G are Quillen equivalent.*

Finally, we have a somewhat surprising additional feature of the cofree construction.

Proposition 17.4.21. *If R is an E_∞ ring spectrum on which G acts by E_∞ ring maps, then $F(EG_+, R)$ naturally has the structure of a G-E_∞-ring spectrum.*

17.4.3 Multiplicative norms

17.4.3.1 Norms in operadic algebras

The subgroups that arise in an N_∞ operad parameterize various kinds of twisted products. To see this, we unpack the conditions a little.

Let $\Gamma \subseteq G \times \Sigma_n$ be in $\mathcal{F}_n$. Since the intersection with Σ_n is trivial, the projection onto G, when restricted to Γ is an injection. The image is some subgroup H, and we observe that Γ is then the graph of a homomorphism $f \colon H \to \Sigma_n$. Any homomorphism $H \to \Sigma_n$ defines an H-set structure on the set $\{1, \ldots, n\}$, and we see that the subgroups correspond to presentations of H-set structures of cardinality n. This gives a name to these subgroups: graph subgroups.

Definition 17.4.22. Let $\mathcal{O}$ be an N_∞ operad. A finite H-set T of cardinality n is admissible if the graph of a homomorphism $H \to \Sigma_n$ defining it is in $\mathcal{F}_n(\mathcal{O})$.

Example 17.4.23. For any N_∞-operad $\mathcal{O}$ and for any cardinality n, the set $\{1, \dots, n\}$ with a trivial G action is admissible. This is the G-set corresponding to the subgroup $G \times \{e\}$.

Since for all n, $\mathcal{F}_n(\mathcal{O})$ is a universal space, we have an equivalent formulation.

Proposition 17.4.24. *Let T be a finite H set of cardinality n and let Γ be the corresponding graph subgroup. Then T is admissible if and only if*

$$\mathsf{Top}^{G\times\Sigma_n}(G \times \Sigma_n/\Gamma, \mathcal{O}_n) \neq \emptyset.$$

Now let R be an $\mathcal{O}$-algebra. We can then consider the effect of composing any such map with the operadic structure map, getting a contractible space of maps

$$(G \times \Sigma_n/\Gamma)_+ \underset{\Sigma_n}{\wedge} R^{\wedge n} \to \mathcal{O}_{n+} \underset{\Sigma_n}{\wedge} R^{\wedge n} \to R.$$

In the source, the presence of Γ intertwines the Σ_n action on the indexing set for the smash powers with the H action in G: every $h \in H$ acts both on the individual factors of R and on the indexing set via $H \to \Sigma_n$. This is therefore an example of the norm.

Proposition 17.4.25. *Let T be a finite H-set of cardinality n, and let $\Gamma \subseteq G \times \Sigma_n$ be the graph of a homomorphism $H \to \Sigma_n$ presenting T. Then for any G-spectrum E, we have a G-equivariant equivalence*

$$(G \times \Sigma_n/\Gamma)_+ \underset{\Sigma_n}{\wedge} E^{\wedge n} \simeq G_+ \underset{H}{\wedge} N^T E.$$

Putting this all together, we see exactly what kinds of structure we see.

Theorem 17.4.26. *Let $\mathcal{O}$ be an N_∞ operad and let R be an $\mathcal{O}$-algebra. For each admissible H-set T, the operad gives a contractible space of maps*

$$N^T i_H^* R \to i_H^* R.$$

When $T = * \amalg *$, this is the contractible space of multiplications, describing an underlying E_∞ ring structure.

Since they come from the operadic structure, these are natural for maps of operadic algebras.

Theorem 17.4.26 used only the operadic action on R and said nothing about the compatibility or constraints imposed by the operadic composition. In fact, this puts huge constraints on the collection of admissible sets, allowing for a complete classification of the homotopy category of N_∞ operads.

Since for each n, $\mathcal{F}_n(\mathcal{O})$ is a family, we deduce several consequences.

Proposition 17.4.27. *Let $\mathcal{O}$ be an N_∞ operad.*

1. *If T is an admissible H-set and if $q\colon G/K \to G/H$ is a map of G-sets, then q^*T is an admissible K-set.*

2. *If T is an admissible H-set and T' is isomorphic to T, then T' is an admissible H-set.*

3. *The trivial H-set of cardinality n is admissible for all n.*

The operadic composition connects the admissible sets for various cardinalities.

Proposition 17.4.28. *Let $\mathcal{O}$ be an N_∞ operad.*

1. *If $T = T_1 \amalg T_2$, then T_1, T_2 are admissible H-sets if and only if T is.*

2. *If K/H is an admissible K-set and T is an admissible H-set, then $K \underset{H}{\times} T$ is an admissible K-set.*

Definition 17.4.29. An indexing system is a collection of full subcategories $\mathcal{O}(H) \subseteq \mathsf{Fin}^H$ for each $H \subseteq G$, such that for each H, the objects of $\mathcal{O}(H)$ satisfy the conditions of Propositions 17.4.27 and 17.4.28.

There is a poset of indexing systems, ordered by inclusion for all H, and the map that sends an N_∞ operad to its admissibles gives a functor to this poset. Blumberg–Hill showed this is a fully-faithful inclusion on the homotopy category, and Gutiérrez–White showed it is essentially surjective.

Theorem 17.4.30. [11, 36] *There is an equivalence of categories between the homotopy category of N_∞-operads and the poset of indexing systems.*

There are several additional approaches to building N_∞ operads out of the combinatorial data. Rubin considers explicit categorical models [59], while Bonventre–Pereira describe a genuine equivariant extension of operads [13].

Remark 17.4.31. Instead of working multiplicatively, we could have asked for a classification of the possible extension of coefficient systems that include only some transfers. These are again parameterized by indexing systems [12]. This us ways to talk about variants of G-spectra interpolating between the $\mathbb{Z}$-graded theory built out of coefficient systems and the genuine one considered so far here.

17.4.3.2 Norms in commutative rings

For commutative ring spectra, we can make even more intuitive statements. The smash product is the coproduct in $\mathcal{C}omm^G$, and the norm N_H^G is induction built out of the smash product. In particular, it is a strong symmetric monoidal functor, and hence lifts to a functor on $\mathcal{C}omm^H$.

Theorem 17.4.32. *The norm*

$$N_H^G \colon \mathcal{C}omm^H \to \mathcal{C}omm^G$$

is left adjoint to the forgetful functor $i_H^* \colon \mathcal{C}omm^G \to \mathcal{C}omm^H$.

Corollary 17.4.33. *If R is a commutative ring spectrum, then for each H, we have a natural map of equivariant commutative ring spectra*

$$N_H^G i_H^* R \to R.$$

The identification of the adjunction also demonstrates some unexpected behavior for commutative ring spectra, showing that these are more than just naive E_∞ ring spectra. As a consequence of the corollary, we have two features:

1. For each $H \subseteq G$, the G-geometric fixed points of the left-adjoint to the forgetful functor are
$$\Phi^G N_H^G i_H^* R \simeq \Phi^H(R),$$
via the diagonal, and
2. we have a map of (ordinary) E_∞-rings
$$\Phi^H(R) \to \Phi^G(R).$$

The first of these can fail quite spectacularly for naive E_∞-rings.

Example 17.4.34. The left-adjoint L_H^G to the forgetful functor $i_H^* \colon \mathcal{O}\text{-}\mathsf{Alg}^G \to \mathcal{O}\text{-}\mathsf{Alg}^H$ can be computed easily on frees by the universal property:

$$L_H^G \mathbb{P}_{\mathcal{O}^{tr}}(E) \simeq \mathbb{P}_{\mathcal{O}^{tr}}\Big(G_+ \underset{H}{\wedge} E\Big).$$

Since G acts trivially on $\mathcal{O}_n^{tr}$ for all n, we have that the unit map gives an equivalence

$$S^0 \to \Phi^G \mathbb{P}_{\mathcal{O}^{tr}}\Big(G_+ \underset{H}{\wedge} E\Big).$$

This in turn shows that the geometric fixed points of the left adjoint is just the constant functor S^0.

Using the fact that the smash product is the coproduct in $\mathcal{C}omm^G$, we have

Corollary 17.4.35. *If R is an equivariant commutative ring spectrum, then the assignment*

$$T \mapsto N^T R$$

extends to a functor from Fin^G to $\mathcal{C}omm^G$.

17.4.4 Tambara functors

17.4.4.1 Norms in equivariant algebra

These norm maps endow the homotopy Mackey functors of an $\mathcal{O}$-algebra with extra structure. Since these are also the groups associated to a homology theory, stable homotopy, and since equivariant homology theories are RO(G)-graded, we have an RO(G)-graded extension.

Notation 17.4.36. If V and W are representations, let

$$\pi_{V-W}(E) = \pi_{V-W}^G(E) = [S^V, \Sigma^W E]^G.$$

Let

$$\underline{\pi}_{V-W}(E) = \big(T \mapsto [T_+ \wedge S^V, \Sigma^W E]^G\big)$$

be the obvious Mackey enrichment.

Smashing together representative maps gives the following.

Proposition 17.4.37. *If R is an equivariant commutative ring spectrum, then $\pi_\star(R)$ is an equivariant graded commutative ring.*

Here the adjective "equivariant" for the graded commutativity refers to the fact that $\pi_0 S^0 = A(G)$ can have more units than just ± 1, and these units can show up in the formulae for moving representation spheres past each other. In general, the swap map

$$S^V \wedge S^W \to S^W \wedge S^V$$

is a unit in $\pi_0 S^0$; this is the element that controls commutativity.

Example 17.4.38. For $G = C_2$, the sign sphere S^σ swaps with itself by the element $1 - [C_2] \in A(C_2)$. We can see this by checking against all geometric fixed points: the underlying map is the interchange of the 1-sphere with itself (so -1) while the C_2-geometric fixed points is the interchange of the 0-sphere with itself (so 1).

Remark 17.4.39. This technique works quite generally: the ghost coordinates of Definition 17.3.92 give us an injective map

$$[S^{V\oplus W}, S^{W\oplus V}] \to \prod_H [S^{V^H \oplus W^H}, S^{W^H \oplus V^H}] \cong \prod_H \mathbb{Z}.$$

For each coordinate, we have a non-equivariant computation of the degree of moving the W^H-sphere past the V^H-sphere, and this is either ± 1. The corresponding unit in the Burnside ring is the one corresponding to this sequence of units in $\mathbb{Z}$.

Here, we have only used the multiplication in the homotopy category. If R is an $\mathcal{O}$-algebra, then we have more structure.

Proposition 17.4.40. *Let $\mathcal{O}$ be an N_∞ operad. If R is an equivariant commutative ring spectrum, and if G/H is an admissible G-set for $\mathcal{O}$, then we have norm maps*

$$N_H^G \colon \pi_V^H(R) \to \pi^G_{Ind_H^G V}(R)$$

defined by composing the norm functor with the norm in R as an $\mathcal{O}$-algebra:

$$(f \colon S^V \to i_H^* R) \mapsto \left(S^{Ind_H^G V} \xrightarrow{N_H^G f} N_H^G i_H^* R \to R\right).$$

Remark 17.4.41. Note that in the formation of the norms, we did not need V to be a representation of G: the norm fixed this, producing a representation of G for its index. This suggests an even further extension of the RO(G)-grading: indexing on pairs consisting of subgroups together with a virtual representation for that subgroup. This was used in [39] and was developed more fully in [4].

17.4.4.2 Green and Tambara functors

When $V = \{0\}$, then the grading issues become much simpler: inducing the zero representation gives the zero representation and adding the zero representation to itself gives the zero representation. In particular, our homotopy Mackey functors will have extra structure.

Proposition 17.4.42. *The category of Mackey functors admits a closed symmetric monoidal structure with the symmetric monoidal product the box product* $-\Box-$, *and internal Hom* $\underline{\mathrm{Hom}}$. *The symmetric monoidal unit is the Burnside Mackey functor* $\underline{A}$.

The description of Mackey functors as a diagram category on the Burnside category shows how to build the box product: we form the Day convolution product of the tensor product on abelian groups with the Cartesian product in the Burnside category.

Definition 17.4.43. A [commutative] Green functor is a commutative monoid for the box product.

Since the zero sphere commutes with itself on-the-nose, we have a ordinary commutativity for the multiplication on $\underline{\pi}_0$.

Proposition 17.4.44. *If* R *in an equivariant spectrum with a homotopy commutative and associative multiplication, then* $\underline{\pi}_0(R)$ *is a commutative Green functor.*

The norm maps on an $\mathcal{O}$-algebra given multiplicative maps on $\underline{\pi}_0$.

Proposition 17.4.45. *If* R *is an* $\mathcal{O}$ *algebra and if* H/K *is an admissible* H*-set, then we have a multiplicative norm map*

$$\underline{\pi}_0(R)(G/K) \to \underline{\pi}_0(R)(G/H).$$

When $H = G$, these are the expected maps

$$\underline{\pi}_0(R)(G/H) \to \underline{\pi}_0(R)(*).$$

More generally, for admissible H-sets, we identified the corresponding groups in the G-Mackey functor $\underline{\pi}_0(R)$ with the corresponding groups in the H-Mackey functors associated to the restriction $i_H^*\underline{R}$.

These norm maps are not linear, but satisfy instead a kind of twisted distributive law, identical to that studied by Tambara in his original formulation of TNR functors [66] (See also [64]). These formulae for the norms of sums or of transfers are built out of decomposing a coinduced G-set into constituent pieces, and this gives an externalized version of the axioms Tambara considered [40].

Example 17.4.46. For $G = C_2$, we have

$$N_e^{C_2}(a+b) = N_e^{C_2}(a) + N_e^{C_2}(b) + tr_e^{C_2}(a \cdot \gamma(b)),$$

where γ is the non-trivial element of C_2 acting on $\underline{R}(C_2)$.

We can generalize Tambara's construction, considering only certain norms, namely those arising from an indexing system. This gives a category of incomplete Tambara functors.

Theorem 17.4.47. [16, 11] *If R is an $\mathcal{O}$-algebra, then Mackey functor $\underline{\pi}_0(R)$ naturally has the structure of an incomplete Tambara functor with norms for the indexing system of admissibles for $\mathcal{O}$.*

Because of this, Angeltveit–Bohmann call the resulting structure an RO-graded Tambara functor.

Example 17.4.48. With the Tambara structure, we can show that the pushforward from trivial spectra to C_2-spectra cannot be homotopically compatible with the commutative ring structure. Consider the Eilenberg–Mac Lane spectrum $H\mathbb{F}_2$. Then the pushforward has

$$\underline{\pi}_0(i_* H\mathbb{F}_2) \cong \underline{A}/2$$

is the reduction of the Burnside ring modulo 2. There is no way to put a norm map on this making it into a Tambara functor.

17.5 Computational techniques

We close with several methods for computing in equivariant stable homotopy theory.

17.5.1 Classical spectral sequences

17.5.1.1 Atiyah–Hirezebruch

Just as classically, there is an Atiyah–Hirzebruch spectral sequences computing the homotopy classes of maps between two G-spectra.

Theorem 17.5.1. *There is a multiplicative spectral sequence with E_2 term*

$$E_2^{p,q} = H^p\big(X; \underline{\pi}_q(E)\big)$$

and converging to $[\Sigma^{q-p}X, E]^G$. The d_r differential changes degree by $(r, 1-r)$.

We get this filtration by either filtering the source by the skeletal filtration or the target by the Postnikov filtration. The latter also shows that all of the differentials are secondary cohomology operations, just as classically.

More generally, since all of our algebra invariants are naturally Mackey functors, we have a natural Mackey enrichment.

Theorem 17.5.2. *There is a spectral sequence of Green functors with E_2-term*

$$\underline{E}_2^{p,q} = \underline{H}^p\big(X; \underline{\pi}_q(E)\big)$$

and converging to the homotopy Mackey functors $\underline{E}^{p-q}(X)$.

Remark 17.5.3. In general, even if E is a commutative ring spectrum, this will only be a spectral sequence of graded Green functors. This means that the differentials satisfy the usual Leibnitz rule for products, but we have no control over the norms of classes. The multiplicativity of the spectral sequence is the essential heart of Proposition 17.3.62, while the failure of this to be compatible with norms is Proposition 17.4.7.

As a particular example, we have the homotopy fixed points spectral sequence computing the homotopy groups of the homotopy fixed points E^{hG}. We classically describe this via the skeletal filtration of the source EG_+, but we can equivalently filter the spectrum E by its Postnikov tower.

Theorem 17.5.4. *There is a multiplicative spectral sequence*

$$E_2^{s,t} \cong H^s\big(G; \pi_t(i_e^* E)\big) \Rightarrow \pi_{t-s} E^{hG}.$$

These homotopy fixed points spectral sequences are one of the primary tools used in the trace methods approach to algebra K-theory, together with the generalized Tate diagrams of Proposition 17.3.95. Here the right-hand side will be known by induction (using the cyclotomic structure) while the bottom row will be built out of this homotopy fixed points spectral sequence.

More generally, this can also be done in Mackey functors.

Theorem 17.5.5. *There is a multiplicative spectral sequence of Mackey functors*

$$\underline{E}_2^{s,t}(G/K) \cong H^s\big(K; \pi_t(i_e^* E)\big) \Rightarrow \underline{\pi}_{t-s} E^h.$$

17.5.1.2 Künneth and Universal Coefficients

Lewis and Mandell proved several versions of a Künneth and universal coefficients spectral sequence [48]. To describe the spectral sequences, we have to do more homological algebra in Mackey functors.

Definition 17.5.6. If $\underline{R}$ is a commutative Green functor, then an $\underline{R}$-module is a Mackey functor $\underline{M}$ together with an action map

$$\underline{R} \Box \underline{M} \to \underline{M}$$

making the usual module associativity and unitality diagrams commute. Let $\underline{R}$-Mod be the category of $\underline{R}$-modules.

The standard arguments about commutative monoids in a symmetric monoidal category show the following.

Proposition 17.5.7. *If $\underline{R}$ is a commutative Green functor, then the closed symmetric monoidal structure on Mackey functors induces a closed symmetric monoidal structure on $\underline{R}$-modules.*

Since the category of Mackey functors has enough projectives (Proposition 17.3.29), the category of $\underline{R}$-modules has enough projectives for any $\underline{R}$. We can therefore consider derived functors.

Definition 17.5.8. Let $\underline{\mathrm{Tor}}^{R}_{-s}(\underline{M}, \underline{N})$ be the $-s$th derived functor of $\underline{M} \Box_{\underline{R}} \underline{N}$, and let $\underline{\mathrm{Ext}}^{s}_{\underline{R}}(\underline{M}, \underline{N})$ be the sth derived functor of $\underline{\mathrm{Hom}}_{\underline{R}}(\underline{M}, \underline{N})$.

Theorem 17.5.9. [48] *Let R be an E_∞ ring spectrum and let M and N be R-modules.*
There is a spectral sequence

$$E_2^{s,t} = \underline{\mathrm{Tor}}^{-s,t}_{\underline{\pi}_* R}(\underline{\pi}_* M, \underline{\pi}_* N) \Rightarrow \underline{\pi}_{t-s}(M \underset{R}{\wedge} N).$$

There is a spectral sequence

$$E_2^{s,t} = \underline{\mathrm{Ext}}^{s,t}_{\underline{\pi}_* R}(\underline{\pi}_* M, \underline{\pi}_* N) \Rightarrow \underline{\pi}_{t-s}\big(\mathrm{Hom}_R(M, N)\big).$$

These are both Adams graded.

Lewis–Mandell also work with RO(G)-graded versions.

Theorem 17.5.10. [48] *Let R be an E_∞ ring spectrum and let M and N be R-modules.*
There is a spectral sequence

$$\underline{E}_2^{s,V} = \underline{\mathrm{Tor}}^{-s,V}_{\underline{\pi}_\star R}(\underline{\pi}_\star M, \underline{\pi}_\star N) \Rightarrow \underline{\pi}_{V-s}(M \underset{R}{\wedge} N).$$

There is a spectral sequence

$$\underline{E}_2^{s,V} = \underline{\mathrm{Ext}}^{s,V}_{\underline{\pi}_\star R}(\underline{\pi}_\star M, \underline{\pi}_\star N) \Rightarrow \underline{\pi}_{V-s}\big(\mathrm{Hom}_R(M, N)\big).$$

These are both Adams graded.

Even for ordinary Bredon homology or cohomology, so $R = H\underline{A}$, this is much trickier than the classical, non-equivariant cases of homology over a PID. Greenlees has shown that the category of Mackey functors has projective dimension 0, 1, or ∞ [30], so the E_2-terms for the ordinary Künneth and universal coefficients spectral sequence in general have infinitely many lines.

Example 17.5.11. Let $G = C_p$, and consider the constant Mackey functor $\underline{\mathbb{Z}}$. Then we have an exact sequence

$$0 \to \underline{\mathbb{Z}} \to \underline{A}_{C_p} \xrightarrow{1-\gamma} \underline{A}_{C_p} \xrightarrow{1} \underline{A} \xrightarrow{p-[C_p]} \underline{A} \to \underline{\mathbb{Z}} \to 0.$$

In particular, we have a periodic resolution of $\underline{\mathbb{Z}}$ with period 4, and we see that having infinitely many Tor or Ext groups will be generic here.

Many of the recent computations in algebraic topology have focused on computations with the constant Mackey functor $\underline{\mathbb{Z}}$ for cyclic groups. Here, the category of modules has projective dimension 3 [6] (See also [14]), and computations are simpler. Zeng has used this to determine the RO(G)-graded homology of a point for cyclic 2-groups [72].

17.5.2 Adams and Adams–Novikov spectral sequences

When we study free G-spectra, then we can equivalently consider Borel theories.

Definition 17.5.12. Let

$$HM_c = EG_+ \wedge HM^h$$

be the freed up Borel cohomology spectrum. This represents "co-Borel cohomology".

Proposition 17.3.69 shows that Borel and co-Borel cohomologies agree on free G-spectra. Greenlees has used these theories to build an Adams spectral sequence which is computable out of essentially non-equivariant information.

Theorem 17.5.13. [29] *If G is a finite p-group and Y is a free G-spectrum with which p-complete, bounded below, and locally finite, then there is a convergent Adams spectral sequence for any X:*

$$E_2^{s,t} = \mathrm{Ext}^{s,t}_{(H\mathbb{F}_p^h)^*(H\mathbb{F}_p^h)}\left((H\mathbb{F}_p^h)^*(Y), (H\mathbb{F}_{p_c})^*(X)\right) \Rightarrow [X,Y]^G_{t-s}.$$

Moreover, we have an isomorphism

$$(H\mathbb{F}_p^h)^*(H\mathbb{F}_p^h) \cong H^*(BG;\mathbb{F}_p)\tilde{\otimes}\mathcal{A}_p,$$

where $\mathcal{A}_p$ is the mod-p Steenrod algebra, and where $\tilde{\otimes}$ refers to the fact that we twist the multiplication in the tensor product by the natural action of $\mathcal{A}_p$ on $H^*(BG;\mathbb{F}_p)$.

Even though this looks like it gives us information only for free spectra, this actually gives much more control over even non-free spectra. The Segal conjecture links these. Moreover, for G a finite p-group, the natural map

$$\pi^*(X)^\wedge_p \to \left(\pi^*(X)^\wedge_I\right)^\wedge_p$$

is an isomorphism. So the Greenlees spectral sequence can be used also to just compute stable cohomotopy, and more generally, to compute maps into any finite, p-complete G-spectrum Y (free or not). This was used by Szymik to compute equivariant stable stems in low degrees [65].

17.5.2.1 Real versions

For $G = C_2$, there is another form of the Adams spectral sequence due to Hu–Kriz [45].

Definition 17.5.14. Let $\underline{\mathbb{F}}_2$ be the constant Mackey functor with value $\mathbb{F}_2$. Let $\mathbb{M}_2$ be $\pi_\star H(\underline{\mathbb{F}}_2)$.

The ring $\mathbb{M}_2$ was originally computed by Stong (unpublished), and a complete description can be found in [19]. A very nice way to determine this ring is given by Greenlees [31]. This is a non-Noetherian ring, but can be easily described. C. May has recently shown that this ring is injective as a module over itself, simplifying many structural questions [56].

Proposition 17.5.15. *We have*

$$H_1(S^\sigma;\underline{\mathbb{F}}_2) \cong \mathbb{F}_2,$$

generated by a class u_σ.

As a module over $\mathbb{F}_2[a_\sigma, u_\sigma]$*, we have*

$$\mathbb{M}_2 \cong \mathbb{F}_2[a_\sigma, u_\sigma] \oplus \theta \cdot \mathbb{F}_2[a_\sigma^{\pm 1}, u_\sigma^{\pm 1}]/\mathbb{F}_2[a_\sigma, u_\sigma].$$

The class θ *is in degree* $2\sigma - 2$ *and squares to zero.*

Remark 17.5.16. The ring $\mathbb{M}_2$ is isomorphic to the cohomology ring of an $RO(G)$-weighted $\mathbb{P}^1$ with coefficients in the powers of the canonical bundle. The generators are a_σ and u_σ, and the class θ is the Serre dual class. Work of Greenlees–Meier generalizes this [34].

Hu–Kriz computed the $RO(C_2)$-graded Hopf algebroid of stable co-operations for Bredon homology with coefficients in $\underline{\mathbb{F}}_2$, showing that in fact, this algebra is a free module over the $RO(C_2)$-graded homology of a point. This is a striking example of the power of the $RO(C_2)$-grading, since the structure of the algebra with just $\mathbb{Z}$-grading is overly complicated.

Theorem 17.5.17. [45] *We have an isomorphism*

$$\mathcal{A}_\star := H(\underline{\mathbb{F}}_2)_\star H(\underline{\mathbb{F}}_2) \cong \mathbb{M}_2[\bar{\xi}_1, \dots][\bar{\tau}_0, \dots]/\big(\bar{\tau}_i^2 = a_\sigma \bar{\tau}_{i+1} + (u_\sigma + a_\sigma \bar{\tau}_0)\bar{\xi}_{i+1}\big).$$

The degrees of the elements are

$$|\bar{\xi}_i| = (2^i - 1)\rho_2 \quad |\bar{\tau}_i| = (2^i - 1)\rho_2 + 1,$$

and the coproducts of the generators are the classical coproducts. The left and right units on a_σ *are the obvious inclusions. The right unit on* u_σ *is* $u_\sigma + a_\sigma \bar{\tau}_0$*.*

Theorem 17.5.18. [45] *There is an Adams spectral sequence*

$$E_2^{s,V} = \operatorname{Ext}_{\mathcal{A}_\star}^{s,V}\left(H_\star(X), H_\star(Y)\right) \Rightarrow [X, Y_I^\wedge],$$

where I *is the augmentation ideal of the Burnside ring.*

Moreover, Araki showed that the Fujii–Landweber spectrum of Real bordism $\mathbf{MU}_\mathbb{R}$ is flat in the sense that

$$\mathbf{MU}_{\mathbb{R}\star}\mathbf{MU}_\mathbb{R} = \pi_\star \mathbf{MU}_\mathbb{R}[\bar{b}_1, \dots],$$

where $|\bar{b}_i| = i\rho_2$ [5]. Hu–Kriz analyzed this as a Hopf algebroid, producing their Real Adams–Novikov spectral sequence.

Theorem 17.5.19. [45] *There is a spectral sequence*

$$E_2^{s,V} = \operatorname{Ext}^{s,V}_{(\mathbf{MU}_{\mathbb{R}\star}, \mathbf{MU}_{\mathbb{R}\star}\mathbf{MU}_\mathbb{R})}\left(\mathbf{MU}_{\mathbb{R}\star}(X), \mathbf{MU}_{\mathbb{R}\star}(Y)\right) \Rightarrow [X, Y_I^\wedge].$$

For other cyclic p-groups, determining the structure of the dual Steenrod algebra and even of the algebraic category in which we should work is ongoing work.

17.5.3 Slice spectral sequence

Building on motivic intuition and on work of Dugger, Hill–Hopkins–Ravenel produced an equivariant refinement of the Postnikov tower that uses various induced representation spheres in place of ordinary trivial ones [23], [39]. We describe here the version using regular representations first described by Ullman [69].

Definition 17.5.20. [22] A localizing subcategory is a full subcategory closed under homotopy colimits and such that for any cofiber sequence

$$X \to Y \to Z,$$

if X and either Y or Z is in the subcategory, then the third is.

Definition 17.5.21. Let $\tau_{\geq n}$ be the smallest localizing subcategory containing all

$$G_+ \underset{H}{\wedge} S^{k\rho_H},$$

where ρ_H is the regular representation of H and where $k \cdot |H| \geq n$.

By construction, these are nested: if $m \geq n$, then

$$\tau_{\geq m} \subseteq \tau_{\geq n}.$$

Definition 17.5.22. Let $P^n\colon \mathcal{S}p^G \to \mathcal{S}p^G$ be the endofunctor of $\mathcal{S}p^G$ that nullifies the localizing category $\tau_{\geq n}$. The slice tower for E is the tower of spectra under E:

$$\dots \to P^{n+1}(E) \to P^n(E) \to \dots.$$

The fiber $P_n^n(E)$ of $P^n(E) \to P^{n-1}(E)$ is the n-slice of E.

Recent work of Hill–Yarnall and of Wilson has given an equivalent description of the filtration in terms of the geometric fixed points for various subgroups [43] [70]. This gives an interpretation of this tower as running the Postnikov towers for the geometric fixed points of various subgroups at speeds proportionate to the index of the subgroup.

The slice tower of E gives a spectral sequence computing the homotopy Mackey functors of E, or more generally the Mackey functor of maps from any spectrum X into E.

Proposition 17.5.23. *If E is a G-spectrum and X is a finite G-CW complex, then there is a strongly convergent, Adams graded spectral sequence*

$$\underline{E}_2^{s,V} = \underline{\pi}_{V-s} F(X, P_{\dim V}^{\dim V}(E)) \Rightarrow \underline{\pi}_{V-s} F(X, E) = \underline{E}^{s-V}(X).$$

Restricted to (-1)-connected G-spectra, the slice tower has another, essentially defining feature. The sequence of categories of slice $\geq n$ spectra is the smallest sequence of localizing subcategories $\mathcal{C}_n$ such that

1. All finite G-sets are in the zeroth category,
2. S^1 is in the first category,
3. if $X \in \mathcal{C}_n$ and $Y \in \mathcal{C}_m$, then $X \wedge Y \in \mathcal{C}_{n+m}$, and

4. if $X \in \mathcal{C}_n$, then for all finite H-sets T

$$(G_+ \underset{H}{\wedge} N^T i_H^* X) \wedge Y \in \mathcal{C}_{|T|n+m}.$$

The first three properties are shared by the Whitehead filtration for the Postnikov tower, which can be reinterpreted as the smallest sequence of localizing subcategories satisfying just these. These conditions plus the last guarantee that the slice spectral sequence has unexpectedly strong multiplicative properties.

Proposition 17.5.24. *If R is a commutative ring spectrum, then the slice tower computing the* $\mathrm{RO}(G)$*-graded homotopy of R is a spectral sequence of graded Tambara functors.*

In general, it can be difficult to describe the slice associated graded for a spectrum. For the group C_2 (or more generally, for C_p), we have complete control of the slices, and they depend only on the homotopy Mackey functors in the dimensions of regular representation spheres or regular representations spheres plus 1.

Theorem 17.5.25. [39, 41] *If E is a C_2-spectrum, then the slices are given by*

$$P_{2k}^{2k}(E) = \Sigma^{k\rho} H\underline{\pi}_{k\rho}(E) \text{ and } P_{2k+1}^{2k+1}(E) = \Sigma^{k\rho+1} H\mathcal{P}^0\underline{\pi}_{k\rho+1}(E),$$

where $\mathcal{P}^0$ is the endofunctor on C_2 Mackey functors that kills the kernel of the restriction.

More generally, for cyclic 2-groups we understand the slices for the norms of $\mathbf{MU}_{\mathbb{R}}$. This was the key computational step in the solution to the Kervaire invariant one problem [39].

Theorem 17.5.26. [39] *If $E = N_{C_2}^{C_{2^n}} \mathbf{MU}_{\mathbb{R}}$, then the odd slices of E are contractible, while the even slices are wedges of* $\mathrm{RO}(C_{2^n})$*-graded suspensions of $H\underline{\mathbb{Z}}$, where $\underline{\mathbb{Z}}$ is the constant Mackey functor with value $\mathbb{Z}$.*

More recently, the slice spectral sequence has been applied to questions about the Hopkins–Miller higher real K-theory spectra $EO_n(G)$ for G a cyclic 2-group. Recall from Example 17.3.75 that if G is a finite subgroup of $\mathrm{Aut}(\Gamma)$ for some formal group Γ of height n over a perfect field, then we can view E_n as a G-spectrum by the Hopkins–Miller theorem. Restricting attention to the prime 2, there is a canonical order 2 automorphism of any formal group (in fact over any ring): inversion. Thus every Lubin–Tate spectrum is canonically a C_2-spectrum.

Theorem 17.5.27. [37] *Let E_n be the Lubin–Tate spectrum for a height n formal group over a perfect field of characteristic* 2. *Then there is a Real orientation*

$$\mathbf{MU}_{\mathbb{R}} \to E_n$$

lifting an underlying orientation.

Now if G is a finite subgroup of $\mathrm{Aut}(\Gamma)$ that contains C_2, then E_n is also a G-spectrum. Since it is the cofreed up spectrum for an E_∞-ring spectrum, it is canonically a G-E_∞-ring spectrum, and hence has norms.

Corollary 17.5.28. *If G is a finite subgroup of* $\mathrm{Aut}(\Gamma)$ *that contains C_2, then any Real orientation of E_n gives a G-equivariant map*

$$N_{C_2}^{G}\mathbf{MU}_{\mathbb{R}} \to E_n.$$

The slice machinery can then be used to understand the Lubin–Tate spectra computationally as G-spectra, showing how to unpack the Hurewicz image [51], how to compute the homotopy groups [42], and how to describe the Picard group [9].

Bibliography

[1] William C. Abram and Igor Kriz. The equivariant complex cobordism ring of a finite abelian group. *Math. Res. Lett.*, 22(6):1573–1588, 2015.

[2] J. F. Adams. Prerequisites (on equivariant stable homotopy) for Carlsson's lecture. In *Algebraic topology, Aarhus 1982 (Aarhus, 1982)*, volume 1051 of *Lecture Notes in Math.*, pages 483–532. Springer, 1984.

[3] J. F. Adams, J.-P. Haeberly, S. Jackowski, and J. P. May. A generalization of the Segal conjecture. *Topology*, 27(1):7–21, 1988.

[4] Vigleik Angeltveit and Anna Marie Bohmann. Graded Tambara functors. *J. Pure Appl. Algebra*, 222(12):4126–4150, 2018.

[5] Shôrô Araki. Orientations in τ-cohomology theories. *Japan. J. Math. (N.S.)*, 5(2):403–430, 1979.

[6] James E. Arnold, Jr. Homological algebra based on permutation modules. *J. Algebra*, 70(1):250–260, 1981.

[7] David Ayala, Aaron Mazel-Gee, and Nick Rozenblyum. A naive approach to genuine G-spectra and cyclotomic spectra. arXiv.org:1710.06416, 2017.

[8] Clark Barwick. Spectral Mackey functors and equivariant algebraic K-theory (I). *Adv. Math.*, 304:646–727, 2017.

[9] Agnes Beaudry, Irina Bobkova, Michael Hill, and Vesna Stojanoska. Invertible $K(2)$-local E-modules in C_4-spectra. arXiv:1901.02109, 2019.

[10] Andrew J. Blumberg. Continuous functors as a model for the equivariant stable homotopy category. *Algebr. Geom. Topol.*, 6:2257–2295, 2006.

[11] Andrew J. Blumberg and Michael A. Hill. Operadic multiplications in equivariant spectra, norms, and transfers. *Adv. Math.*, 285:658–708, 2015.

[12] Andrew J. Blumberg and Michael A. Hill. Incomplete Tambara functors. *Algebr. Geom. Topol.*, 18(2):723–766, 2018.

[13] Peter Bonventre and Luis A. Pereira. Genuine equivariant operads. arXiv:1707.02226, 2017.

[14] Serge Bouc, Radu Stancu, and Peter Webb. On the projective dimensions of Mackey functors. *Algebr. Represent. Theory*, 20(6):1467–1481, 2017.

[15] Glen E. Bredon. *Equivariant cohomology theories.* Volume 34 of *Lecture Notes in Math.* Springer-Verlag, 1967.

[16] M. Brun. Witt vectors and equivariant ring spectra applied to cobordism. *Proc. Lond. Math. Soc. (3)*, 94(2):351–385, 2007.

[17] Gunnar Carlsson. Equivariant stable homotopy and Segal's Burnside ring conjecture. *Ann. of Math. (2)*, 120(2):189–224, 1984.

[18] Gunnar Carlsson. A survey of equivariant stable homotopy theory. *Topology*, 31(1):1–27, 1992.

[19] Jeffrey L. Caruso. Operations in equivariant $\mathbf{Z}/p$-cohomology. *Math. Proc. Cambridge Philos. Soc.*, 126(3):521–541, 1999.

[20] Emanuele Dotto. Higher equivariant excision. *Adv. Math.*, 309:1–96, 2017.

[21] Andreas W. M. Dress. *Notes on the theory of representations of finite groups. Part I: The Burnside ring of a finite group and some AGN-applications.* Universität Bielefeld, Fakultät für Mathematik, Bielefeld, 1971. With the aid of lecture notes, taken by Manfred Küchler.

[22] E. Dror Farjoun. Cellular inequalities. In *The Čech centennial (Boston, MA, 1993)*, volume 181 of *Contemp. Math.*, pages 159–181. Amer. Math. Soc., 1995.

[23] Daniel Dugger. An Atiyah-Hirzebruch spectral sequence for KR-theory. *K-Theory*, 35(3-4):213–256 (2006), 2005.

[24] A. D. Elmendorf. Systems of fixed point sets. *Trans. Amer. Math. Soc.*, 277(1):275–284, 1983.

[25] A. D. Elmendorf, I. Kriz, M. A. Mandell, and J. P. May. *Rings, modules, and algebras in stable homotopy theory*, volume 47 of *Math. Surveys Monogr.* Amer. Math. Soc., 1997. With an appendix by M. Cole.

[26] Michikazu Fujii. Cobordism theory with reality. *Math. J. Okayama Univ.*, 18(2):171–188, 1975/76.

[27] David Gepner and Tyler Lawson. Brauer groups and Galois cohomology of commutative ring spectra. arXiv:1607.01118, 2016.

[28] Saul Glasman. Stratified categories, geometric fixed points and a generalized Arone–Ching theorem. arXiv:1507.01976, 2017.

[29] J. P. C. Greenlees. Stable maps into free G-spaces. *Trans. Amer. Math. Soc.*, 310(1):199–215, 1988.

[30] J. P. C. Greenlees. Some remarks on projective Mackey functors. *J. Pure Appl. Algebra*, 81(1):17–38, 1992.

[31] J. P. C. Greenlees. Four approaches to cohomology theories with reality. In *An alpine bouquet of algebraic topology*, volume 708 of *Contemp. Math.*, pages 139–156. Amer. Math. Soc., 2018.

[32] J. P. C. Greenlees and J. P. May. Equivariant stable homotopy theory. In *Handbook of algebraic topology*, pages 277–323. North-Holland, 1995.

[33] J. P. C. Greenlees and J. P. May. Generalized Tate cohomology. *Mem. Amer. Math. Soc.*, 113(543):viii+178, 1995.

[34] J. P. C. Greenlees and Lennart Meier. Gorenstein duality for real spectra. *Algebr. Geom. Topol.*, 17(6):3547–3619, 2017.

[35] Bertrand J. Guillou and J. Peter May. Equivariant iterated loop space theory and permutative G-categories. *Algebr. Geom. Topol.*, 17(6):3259–3339, 2017.

[36] Javier J. Gutiérrez and David White. Encoding equivariant commutativity via operads. *Algebr. Geom. Topol.*, 18(5):2919–2962, 2018.

[37] Jeremy Hahn and XiaoLin Danny Shi. Real orientation of Lubin–Tate spectra. arXiv:1707.03413, 2017.

[38] Markus Hausmann. G-symmetric spectra, semistability and the multiplicative norm. *J. Pure Appl. Algebra*, 221(10):2582–2632, 2017.

[39] M. A. Hill, M. J. Hopkins, and D. C. Ravenel. On the nonexistence of elements of Kervaire invariant one. *Ann. of Math. (2)*, 184(1):1–262, 2016.

[40] Michael A. Hill and Kristen Mazur. An equivariant tensor product on Mackey functors. *J. Pure Appl. Algebra*, 2019.

[41] Michael A. Hill and Lennart Meier. The C_2-spectrum $\mathrm{Tmf}_1(3)$ and its invertible modules. *Algebr. Geom. Topol.*, 17(4):1953–2011, 2017.

[42] Michael A. Hill, XiaoLin Danny Shi, Guozhen Wang, and Zhouli Xu. The slice spectral sequence of a C_4-equivariant height-4 Lubin–Tate theory. arXiv:1811.07960, 2018.

[43] Michael A. Hill and Carolyn Yarnall. A new formulation of the equivariant slice filtration with applications to C_p-slices. *Proc. Amer. Math. Soc.*, 146(8):3605–3614, 2018.

[44] Mark Hovey, John H. Palmieri, and Neil P. Strickland. Axiomatic stable homotopy theory. *Mem. Amer. Math. Soc.*, 128(610),1997.

[45] Po Hu and Igor Kriz. Real-oriented homotopy theory and an analogue of the Adams-Novikov spectral sequence. *Topology*, 40(2):317–399, 2001.

[46] Peter S. Landweber. Conjugations on complex manifolds and equivariant homotopy of MU. *Bull. Amer. Math. Soc.*, 74:271–274, 1968.

[47] L. Gaunce Lewis, Jr. The $RO(G)$-graded equivariant ordinary cohomology of complex projective spaces with linear $\mathbf{Z}/p$ actions. In *Algebraic topology and transformation groups (Göttingen, 1987)*, volume 1361 of *Lecture Notes in Math.*, pages 53–122. Springer, 1988.

[48] L. Gaunce Lewis, Jr. and Michael A. Mandell. Equivariant universal coefficient and Künneth spectral sequences. *Proc. London Math. Soc. (3)*, 92(2):505–544, 2006.

[49] G. Lewis, J. P. May, and J. McClure. Ordinary $RO(G)$-graded cohomology. *Bull. Amer. Math. Soc. (N.S.)*, 4(2):208–212, 1981.

[50] L. G. Lewis, Jr., J. P. May, M. Steinberger, and J. E. McClure. *Equivariant stable homotopy theory*, volume 1213 of *Lecture Notes in Math.*. Springer-Verlag, 1986. With contributions by J. E. McClure.

[51] Guchuan Li, XiaoLin Danny Shi, Guozhen Wang, and Zhouli Xu. Hurewicz images of real bordism theory and real Johnson-Wilson theories. *Adv. Math.*, 342:67–115, 2019.

[52] Harald Lindner. A remark on Mackey-functors. *Manuscripta Math.*, 18(3):273–278, 1976.

[53] M. A. Mandell and J. P. May. Equivariant orthogonal spectra and S-modules. *Mem. Amer. Math. Soc.*, 159(755),2002.

[54] M. A. Mandell, J. P. May, S. Schwede, and B. Shipley. Model categories of diagram spectra. *Proc. London Math. Soc. (3)*, 82(2):441–512, 2001.

[55] Michael A. Mandell. Equivariant symmetric spectra. In *Homotopy theory: relations with algebraic geometry, group cohomology, and algebraic K-theory*, volume 346 of *Contemp. Math.*, pages 399–452. Amer. Math. Soc., 2004.

[56] Clover May. A structure theorem for $RO(C_2)$-graded Bredon cohomology. arXiv:1804.03691, 2018.

[57] J. P. May. *Equivariant homotopy and cohomology theory*, volume 91 of *CBMS Reg. Conf. Ser. Math.* Amer. Math. Soc., 1996. With contributions by M. Cole, G. Comezaña, S. Costenoble, A. D. Elmendorf, J. P. C. Greenlees, L. G. Lewis, Jr., R. J. Piacenza, G. Triantafillou, and S. Waner.

[58] Sverre Lunø e Nielsen and John Rognes. The topological Singer construction. *Doc. Math.*, 17:861–909, 2012.

[59] Jonathan Rubin. Combinatorial N_∞ operads. arXiv:1705.03585, 2017.

[60] Stafan Schwede. Lectures on equivariant stable homotopy theory. Available at `http://www.math.uni-bonn.de/people/schwede/equivariant.pdf`.

[61] Kazuhisa Shimakawa. Infinite loop G-spaces associated to monoidal G-graded categories. *Publ. Res. Inst. Math. Sci.*, 25(2):239–262, 1989.

[62] E. H. Spanier and J. H. C. Whitehead. A first approximation to homotopy theory. *Proc. Nat. Acad. Sci. U. S. A.*, 39:655–660, 1953.

[63] E. H. Spanier and J. H. C. Whitehead. Duality in homotopy theory. *Mathematika*, 2:56–80, 1955.

[64] Neil Strickland. Tambara functors. arXiv:1205.2516, 2012.

[65] Markus Szymik. Equivariant stable stems for prime order groups. *J. Homotopy Relat. Struct.*, 2(1):141–162, 2007.

[66] D. Tambara. On multiplicative transfer. *Comm. Algebra*, 21(4):1393–1420, 1993.

[67] Tammo tom Dieck. *Transformation groups and representation theory*, volume 766 of *Lecture Notes in Math.*. Springer, 1979.

[68] Tammo tom Dieck. *Transformation groups*, volume 8 of *De Gruyter Stud. Math.*. Walter de Gruyter & Co.,1987.

[69] John Ullman. On the slice spectral sequence. *Algebr. Geom. Topol.*, 13(3):1743–1755, 2013.

[70] Dylan Wilson. On categories of slices. arXiv:1711.03472, 2017.

[71] Klaus Wirthmüller. Equivariant homology and duality. *Manuscripta Math.*, 11:373–390, 1974.

[72] Mingcong Zeng. Equivariant Eilenberg–Mac Lane spectra in cyclic p-groups. arXiv:1710.01769, 2017.

Department of Mathematics, University of California Los Angeles, Los Angeles, CA 90025, U.S.A.

E-mail address: `mikehill@math.ucla.edu`

18

Motivic stable homotopy groups

Daniel C. Isaksen and Paul Arne Østvær

18.1 Introduction

Motivic homotopy theory was developed by Morel and Voevodsky [64] [69] in the 1990s. The original motivation for the theory was to import homotopical techniques into algebraic geometry. For example, it allowed for the powerful theory of Steenrod operations in algebro-geometric cohomology theories. Motivic homotopy theory was essential to the solution of several long-standing problems in algebraic K-theory, such as the the Milnor conjecture [101] and the Bloch-Kato conjecture [103].

The past twenty years have witnessed a great expansion of motivic homotopy theory. Just as in classical homotopy theory, the motivic version of stable homotopy groups is among the most fundamental invariants. The goal of this article is to describe what is known about motivic stable homotopy groups and to suggest directions for further study. This topic is just one of several active directions within motivic homotopy theory.

There are at least two motivations for studying motivic stable homotopy groups closely. First, these groups ought to carry interesting arithmetic information about the structure of the ground field; this is essentially the same as the original purpose of motivic homotopy theory.

But another motivation has arisen in recent years. It turns out that the richer structure of motivic stable homotopy theory sheds new light on the structure of the classical stable homotopy groups. In other words, motivic stable homotopy theory is of interest in classical homotopy theory, independently of the applications to algebraic geometry.

The motivic stable homotopy category $\mathbf{SH}(k)$ is constructed roughly as follows. Start with a ground field k, and consider the category of smooth schemes over k. Expand this category to a larger category with better formal properties. Finally, impose homotopical relations, especially that the affine line $\mathbb{A}^1$ is contractible. The result of this process is the unstable motivic homotopy category over k.

Next, identify a bigraded family $S^{p,q}$ of unstable spheres for $p \geq q \geq 0$. These spheres are smash products of the topological circle $S^{1,0}$ and the algebraic circle $S^{1,1} = \mathbb{A}^1 - 0$. Then stabilize with respect to this bigraded family to obtain motivic stable homotopy theory.

A key property of the classical stable homotopy category is that every spectrum is (up to homotopy) built from spheres by homotopy colimits. The motivic analogue of this property does not hold. In other words, there exist motivic spectra that are not cellular. Non-trivial

Mathematics Subject Classification. 14F42, 55Q45, 55T15.

Key words and phrases. motivic stable homotopy group, Adams spectral sequence, effective slice spectral sequence.

field extensions of the base field are examples of non-cellular motivic spectra. Elliptic curves are another source of such examples. In both cases, interesting arithmetic properties of the algebraic object interfere.

One important consequence is that the motivic stable homotopy groups $\pi_{p,q}X = [S^{p,q}, X]$ do not detect equivalences. One solution to this problem is to consider *motivic stable homotopy sheaves.* The idea is to keep track of maps not only from spheres, but maps out of all smooth schemes. By construction of the motivic stable homotopy category $\mathbf{SH}(k)$, these smooth schemes serve as generators for the category. Therefore, motivic stable homotopy sheaves do detect motivic equivalences.

We will not pursue the motivic stable homotopy sheaf perspective further in this article.[1] Instead, we will focus just on the groups $\pi_{p,q} = [S^{p,q}, S^{0,0}]$. The precise relationship is that the motivic stable homotopy groups are the global sections of the motivic stable homotopy sheaves. For cellular motivic spectra, the motivic stable homotopy groups do detect equivalences, and the most commonly studied motivic spectra are typically cellular. Moreover, the stable homotopy sheaves are "unramified", which means that they are actually determined by their sections over fields. So a thorough understanding of motivic stable homotopy groups over arbitrary fields leads back to complete information about the sheaves as well.

18.1.1 Completions

Classically, Serre's finiteness theorem says that in positive dimensions, every stable homotopy group is finite. Therefore, it is enough to compute the p-completions of the stable homotopy groups for all primes p. Tools such as the Adams spectral sequence and the Adams-Novikov spectral sequence allow for the calculation of these p-completions.

As one would expect, the motivic situation is more involved. The motivic stable homotopy groups are not finite in general. For example, the group $\pi_{-1,-1}$ contains a copy of the multiplicative group $k^\times$ of the base field. When $k = \mathbb{C}$, this is an uncountable group that is infinitely divisible. Therefore, all of the p-completions of $\pi_{-1,-1}$ vanish, even though $\pi_{-1,-1}$ itself is non-zero. Nevertheless, we will compute p-completions, even though there is a certain loss of information.

A full accounting of the machinery of completion in the motivic context is beyond the scope of this article. We refer to [43] and [24]. The completions arise naturally when considering convergence for the Adams and Adams-Novikov spectral sequences.

The situation with completions is in fact even more complicated. The first motivic Hopf map η is the projection $\mathbb{A}^2 - 0 \to \mathbb{P}^1$, i.e., a map $S^{3,2} \to S^{2,1}$ representing an element in the motivic stable homotopy group $\pi_{1,1}$. We need to complete with respect to η in order to obtain convergence for the relevant spectral sequences. The arithmetic square

$$\begin{array}{ccc} \mathbf{1} & \longrightarrow & \mathbf{1}[\eta^{-1}] \\ \downarrow & & \downarrow \\ \mathbf{1}^\wedge_\eta & \longrightarrow & \mathbf{1}^\wedge_\eta[\eta^{-1}] \end{array}$$

[1]See the chapter in this Handbook by Kirsten Wickelgren and Ben Williams for a survey of unstable motivic homotopy theory.

is a homotopy pullback for the motivic sphere spectrum $\mathbf{1}$, as discussed in [83, Lemma 3.9]. Thus, information about the η-completion and the η-localization (and how they fit together) leads to information about the uncompleted motivic stable homotopy groups.

Because of our computational perspective, we will mostly be interested in torsion in the motivic stable homotopy groups. The articles [67] and [3] study the rationalized motivic stable homotopy groups. After inverting 2, the motivic sphere spectrum splits into two summands, and the rationalizations of these summands can be described in terms of motivic cohomology with rational coefficients. We will not pursue this perspective further in this article.

18.1.2 Organization

In Section 18.2, we introduce the motivic Adams spectral sequence, which is one of the key tools for computing stable motivic homotopy groups. The discussion includes background on Milnor K-theory of fields, the motivic cohomology of a point, the structure of the motivic Steenrod algebra, and Ext groups over the motivic Steenrod algebra.

In Section 18.3, we introduce the motivic effective slice spectral sequence, which is the other key tool for computing stable motivic homotopy groups. We explain how the effective slice filtration arises in the stable motivic homotopy category $\mathbf{SH}(k)$, how to compute the layers of this filtration for the motivic sphere spectrum, and how to compute the layers of a few other related motivic spectra.

In Sections 18.4 and 18.5, we specialize in the base fields $\mathbb{C}$ and $\mathbb{R}$. We describe results derived from the motivic Adams spectral sequence. This includes calculations of the η-periodic motivic stable homotopy groups, significant implications for the calculation of classical stable homotopy groups, and exotic nilpotence and periodicity properties.

In Section 18.6, we discuss calculations over other fields. This includes computations over finite fields, a general discussion of graded commutativity, Milnor-Witt K-theory and its relationship to the groups $\pi_{n,n}$, and some results on the groups $\pi_{n+1,n}$.

Section 18.7 covers calculations of stable homotopy groups of other motivic spectra, including the K-theory spectra **KGL**, **kgl**, **KQ**, and **kq**; motivic (truncated) Brown-Peterson spectra $\mathbf{MBP}\langle n\rangle$; and the Witt theory spectrum **KW**. The key tool here is the motivic effective slice spectral sequence. We also describe how computations involving **KW** are related to Milnor's conjecture on quadratic forms.

Finally, Section 18.8 provides some open-ended projects that may motivate further work in the subject of stable motivic homotopy groups.

18.1.3 Notation

Our convention for motivic grading is to use the notation (i, j), where i is the topological dimension and j is the motivic weight. Some authors use the notation $(i - j) + j\alpha$, which illuminates the analogy to equivariant homotopy theory.

The following summary of notations is provided for the reader's convenience. Detailed descriptions are provided throughout the manuscript.

- $K_*^M(k)$ is the Milnor K-theory of the ground field k. (Section 18.2.1)
- $\mathbb{M}_p = H^{*,*}(k;\mathbb{F}_p)$ is the motivic cohomology of the ground field k with $\mathbb{F}_p$ coefficients. (Theorem 18.2.7)
- $A_* = A_*^k$ is the dual Steenrod algebra of motivic cohomology operations. The prime is implicit in this notation. (Section 18.2.2)
- $\mathrm{Ext}_k = \mathrm{Ext}_A(\mathbb{M}_p, \mathbb{M}_p)$ is the cohomology of the motivic Steenrod algebra, which serves as the E_2-page of the motivic Adams spectral sequence that converges to the completed motivic stable homotopy groups. (Section 18.2.3)
- $\pi_{i,j} = \pi_{i,j}^k$ is the motivic stable homotopy group over k in degree (i,j), completed at some prime that is implicit and also at η.
- $\Pi_m^k = \bigoplus_{n\in\mathbb{Z}} \pi_{m+n,n}^k$ is the mth Milnor-Witt stem. Sometimes it is more convenient to study all of the groups in Π_m^k at once, rather than one at a time. (Sections 18.5.1 and 18.6.3)
- $K_*^{MW}(k)$ is the Milnor-Witt K-theory of a field k. (Section 18.6.2)
- $\mathsf{f}_q(\mathbf{E})$ is the qth effective slice cover of a motivic spectrum $\mathbf{E}$. (Section 18.3.1)
- $\mathsf{s}_q(\mathbf{E})$ is the qth layer of a motivic spectrum $\mathbf{E}$ with respect to the effective slice filtration, i.e., the qth slice of $\mathbf{E}$. (Section 18.3.1)
- $\tilde{\mathsf{f}}_q(\mathbf{E})$ is the qth very effective slice cover of a motivic spectrum $\mathbf{E}$. (Section 18.3.1)
- $\mathbf{SH}(k)$ is the motivic stable homotopy category over the base field k. (Section 18.3)
- $\mathbf{1} = S^{0,0}$ is the motivic sphere spectrum that serves as the unit object of the stable motivic homotopy category $\mathbf{SH}(k)$.
- $S^{i,j}$ is the motivic sphere spectrum of dimension (i,j), and $\Sigma^{i,j}$ represents the suspension functor given by smashing with $S^{i,j}$.
- $\mathbf{M}R$ is the motivic Eilenberg-Mac Lane spectrum that represents motivic cohomology with coefficients in R. Usually R is $\mathbb{Z}$ or $\mathbb{Z}/n$. (Sections 18.3.2 and 18.7.1)
- $\mathbf{KGL}$ is the $(2,1)$-periodic algebraic K-theory motivic spectrum. (Sections 18.3.3 and 18.7.2)
- $\mathbf{kgl} = \mathsf{f}_0\mathbf{KGL} = \tilde{\mathsf{f}}_0\mathbf{KGL}$ is the 0th effective slice cover of $\mathbf{KGL}$, or equivalently the 0th very effective slice cover of $\mathbf{KGL}$. (Sections 18.3.1 and 18.7.4)
- $\mathbf{KQ}$ is the $(8,4)$-periodic Hermitian K-theory motivic spectrum. (Sections 18.3.3 and 18.7.4)
- $\mathbf{kq} = \tilde{\mathsf{f}}_0(\mathbf{KQ})$ is the 0th very effective slice cover of $\mathbf{KQ}$. (Sections 18.3.1 and 18.7.4)
- $\mathbf{KW}$ is the Witt theory spectrum obtained from $\mathbf{KQ}$ by inverting the first motivic Hopf map η. (Sections 18.3.3 and 18.7.5)

- **MGL** is the motivic algebraic cobordism spectrum, and **MU** is the analogous classical complex cobordism spectrum. (Section 18.3.2)
- **MBP** is the motivic version of the Brown-Peterson spectrum, and $\mathbf{MBP}\langle n\rangle$ is the truncated version, while **BP** and $\mathbf{BP}\langle n\rangle$ are the classical analogues. These constructions depend on the choice of some prime, which is implicit in the notation. (Sections 18.3.2 and 18.7.3)

18.1.4 Acknolwedgements

The authors appreciate constructive feedback from J. Hornbostel, O. Röndigs, M. Spitzweck, and G. Wilson.

18.2 The motivic Adams spectral sequence

Our techniques for computing motivic stable homotopy groups are fundamentally cohomological in nature. So our discussion begins with the computational properties of motivic cohomology.

18.2.1 Milnor K-theory and the cohomology of a point

Let k be a field. The *Milnor K-theory* $K_*^M(k)$ of k is a graded ring constructed as follows [62]. Let $K_0^M(k)$ be $\mathbb{Z}$, and let $K_1^M(k)$ be the abelian group $k^\times$. This presents a notational confusion, because the group structure on $k^\times$ is multiplicative, but we would like to think of $k^\times$ as an additive group. We use the symbol $[a]$ to represent the element of $K_1^M(k)$ corresponding to the element a of $k^\times$. Then we have identities such as

$$[ab] = [a] + [b].$$

Now $K_*^M(k)$ is defined to be the graded commutative ring generated by the elements of $K_1^M(k)$ subject to the *Steinberg relations*

$$[a] \cdot [1-a] = 0$$

in $K_2^M(k)$ for all a in $k - \{0,1\}$.

The Milnor K-theory of familiar fields can be wildly complicated. For example, $K_1^M(\mathbb{C})$ is an uncountable infinitely divisible abelian group. Fortunately, we will usually consider the much simpler Milnor K-theory modulo p, i.e., $K_*^M(k)/p$, for a fixed prime p.

We give several standard examples of Milnor K-theory computations. More details can be found in [62], [105, Section III.7], or [56, Chapter 14E].

Example 18.2.1. $K_*^M(\mathbb{C})/p = \mathbb{F}_p$, concentrated in degree 0, since every element of $\mathbb{C}^\times$ is a pth power of some other element.

Example 18.2.2. $K_*^M(\mathbb{R})/2 = \mathbb{F}_2[\rho]$, where ρ is another name for the element $[-1]$ in $K_1^M(\mathbb{R})$. This calculation follows from the observation that every non-negative real number is a square. On the other hand, $K_*^M(\mathbb{R})/p = \mathbb{F}_p$ if p is odd, since every element of $\mathbb{R}^\times$ is a pth power of some other element.

In general, let ρ be the element $[-1]$ in $K_1^M(k)/2$ for any field k. Note that ρ is zero if and only if -1 is a square in k. We will see later in Section 18.2.2 that ρ plays a central role in the structure of the motivic Steenrod algebra.

Example 18.2.3. If $p \equiv 1 \bmod 4$, then $K_*^M(\mathbb{Q}_p)/2 = \mathbb{F}_2[\pi, u]/(\pi^2, u^2)$, where $\pi = [p]$ and $u = [a]$ for an element a in $\mathbb{Q}_p^\times$ that maps to a non-square in $\mathbb{F}_p^\times$.

Example 18.2.4. If $p \equiv 3 \bmod 4$, then $K_*^M(\mathbb{Q}_p)/2 = \mathbb{F}_2[\pi, \rho]/(\rho^2, \rho\pi + \pi^2)$, where $\pi = [p]$ and $\rho = [-1]$.

Example 18.2.5. $K_*^M(\mathbb{Q}_2)/2 = \mathbb{F}_2[\pi, \rho, u]/(\rho^3, u^2, \pi^2, \rho u, \rho\pi, \rho^2 + u\pi)$, where $\pi = [2]$, $\rho = [-1]$, and $u = [5]$.

Example 18.2.6. $K_*^M(\mathbb{F}_q)$ equals $\mathbb{Z}[u]/u^2$, where $u = [a]$ for any generator a of $\mathbb{F}_q^\times$.

For us, the point of these Milnor K-theory calculations is that we can use them describe the motivic cohomology of a point. Motivic cohomology is bigraded, where the first grading corresponds to the classical topological degree, and the second corresponds to the motivic weight. If k contains a primitive pth root of unity ζ_p, we let τ denote the corresponding generator of $H^{0,1}(k; \mathbb{F}_p) \cong \mathbb{Z}/p$.

Theorem 18.2.7. [101, 103] *Suppose that p and $\mathrm{char}(k)$ are coprime, and that k contains a primitive pth root of unity. Then the motivic cohomology $\mathbb{M}_p = H^{*,*}(k; \mathbb{F}_p)$ with coefficients in $\mathbb{F}_p$ is isomorphic to*

$$\frac{K_*^M(k)}{p}[\tau],$$

where $K_n^M(k)/p$ has degree (n, n), and τ has degree $(0, 1)$.

Figure 18.2.1 graphically depicts the calculation in Theorem 18.2.7.

18.2.2 The motivic Steenrod algebra

Our next task is to record the structure of the motivic Steenrod algebra [40], [102]. We continue to assume that the base field k contains a primitive pth root of unity and that $\mathrm{char}(k)$ and p are coprime, so that we rely on the calculation of Theorem 18.2.7.

In classical topology, the dual Steenrod algebra is easier to describe than the Steenrod algebra itself. This asymmetry arises from the fact that the coproduct structure (i.e., the Cartan formula) is simpler than the product structure (i.e., the Adem relations).

In the motivic context, the advantages of the dual Steenrod algebra are even more pronounced. The motivic Adem relations possess some non-trivial complications that make them difficult to even write down correctly. Beware that more than one published version of the motivic Adem relations has mistakes. See [40, Theorem 5.1] for the correctly formulated relations.

Let p be an odd prime. The dual motivic Steenrod algebra A_* is a commutative $\mathbb{M}_p$-algebra, where $\mathbb{M}_p = H^{*,*}(k; \mathbb{F}_p)$ is the cohomology of a point (see Theorem 18.2.7). It is generated by elements τ_i for $i \geq 0$ and ξ_j for $j \geq 1$, subject to the relations $\tau_i^2 = 0$. Here τ_i

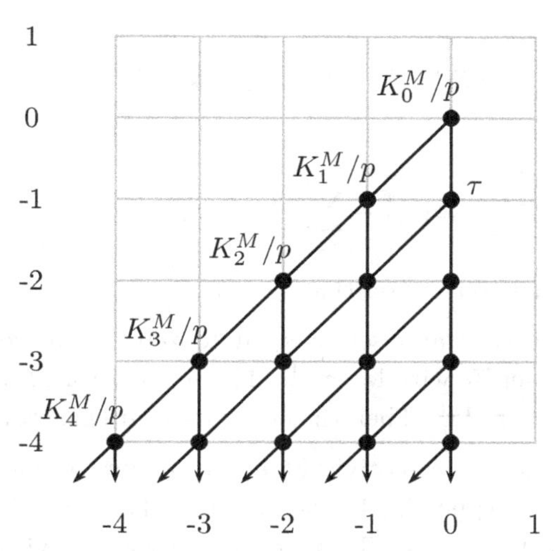

FIGURE 18.2.1. The cohomology of a point. The grading is homological to align with later homotopy group calculations. The topological degree is on the x-axis, and the motivic weight is on the y-axis. Each dot in column $-n$ represents a copy of $K_n^M(k)/p$. Vertical lines indicate multiplication by τ. Diagonal lines connect the source and target of multiplications by elements of $K_1^M(k)/p$.

has degree $(2p^i - 1, p^i - 1)$, and ξ_j has degree $(2p^j - 2, p^j - 1)$. The coproduct Δ on A_* is described by

$$\Delta(\tau_i) = \tau_i \otimes 1 + \sum_{k=0}^{i} \xi_{i-k}^{p^k} \otimes \tau_k$$

and

$$\Delta(\xi_j) = \sum_{k=0}^{j} \xi_{j-k}^{p^k} \otimes \xi_k.$$

In order to properly interpret these formulas, we adopt the usual convention that $\xi_0 = 1$.

The structure of the motivic Steenrod algebra at odd primes is essentially the same as the structure of the classical Steenrod algebra. In fact, in the odd primary case, the motivic Steenrod algebra is the classical Steenrod algebra tensored over $\mathbb{F}_p$ with a larger coefficient ring. Consequently, motivic Ext calculations over A_* are essentially identical to classical Ext calculations.

The situation at $p = 2$ is more interesting. In this case, the dual motivic Steenrod algebra A_* is the commutative $\mathbb{M}_2$-algebra generated by elements τ_i for $i \geq 0$ and ξ_j for $j \geq 1$, subject to the relations

$$\tau_i^2 = \tau\xi_{i+1} + \rho\tau_{i+1} + \rho\tau_0\xi_{i+1}.$$

Here τ_i has degree $(2^{i+1}-1, 2^i-1)$, and ξ_j has degree $(2^{j+1}-2, 2^j-1)$. The coproduct Δ on A_* is described by

$$\Delta(\tau_i) = \tau_i \otimes 1 + \sum_{k=0}^{i} \xi_{i-k}^{2^k} \otimes \tau_k$$

and

$$\Delta(\xi_j) = \sum_{k=0}^{j} \xi_{j-k}^{2^k} \otimes \xi_k.$$

As before, we adopt the usual convention that $\xi_0 = 1$.

Remark 18.2.8. If we invert τ and set ρ equal to zero, then we obtain the classical Steenrod algebra tensored over $\mathbb{F}_2$ with $\mathbb{M}_2[\tau^{-1}]$. The point is that ξ_{i+1} is no longer needed as a generator because $\xi_{i+1} = \tau^{-1}\tau_i^2$. Then the elements τ_i are generators, with no relations.

Unlike in the classical case, the motivic Steenrod algebra is more accurately a Hopf algebroid (in the sense of [81, Appendix 1], for example). For sake of tradition, we will not use the term "algebroid" in this context. This means that the motivic Steenrod algebra acts non-trivially on the motivic cohomology of a point in general. In other words, the right unit $\eta_R : \mathbb{M}_p \to A_*$ is not the same as the left unit $\eta_L : \mathbb{M}_p \to A_*$. More precisely, η_L is the obvious inclusion, while $\eta_R(\tau) = \tau + \rho\tau_0$ and $\eta_R(\rho) = \rho$. This complication occurs only if $\rho = [-1]$ is non-zero in $K_*^M(k)/p$, i.e., only if -1 is not a pth power in the ground field k.

For definiteness, we restate the action of the Steenrod squares on $\mathbb{M}_p$ in more concrete terms. At $p = 2$, we have $\mathrm{Sq}^1(\tau) = \rho$. There are also non-zero squaring operations on other elements, but they can all be derived from this basic formula.

The motivic Cartan formula takes the form

$$\mathrm{Sq}^{2k}(xy) = \sum_{a+b=2k} \tau^{\varepsilon}\mathrm{Sq}^a(x)\mathrm{Sq}^b(y),$$

where ε is 0 if a and b are even, while ε is 1 if a and b are odd [102, Proposition 9.7]. (There is also a motivic Cartan formula for odd squares, but it has some additional complications involving ρ.) This implies that

$$\mathrm{Sq}^2(\tau^2) = \tau\mathrm{Sq}^1(\tau)\mathrm{Sq}^1(\tau) = \rho^2\tau.$$

At odd primes, we have that $\beta(\tau) = \rho$, where β is the Bockstein operation.

18.2.3 The cohomology of the Steenrod algebra

The next step is to compute the cohomology of the motivic Steenrod algebra, i.e.,

$$\mathrm{Ext}_k = \mathrm{Ext}_A(\mathbb{M}_p, \mathbb{M}_p).$$

This object is a trigraded commutative ring with higher structure in the form of Massey products. Two of the gradings correspond to the degrees familiar in the classical situation. The additional grading corresponds to the motivic weight, which has no classical analogue.

The cobar resolution is the standard way to compute Ext groups. This works just as well motivically as it does classically. The cobar resolution is useful for (very) low-dimensional explicit calculations and for general properties. In a larger range, one must attack the calculation with more sophisticated tools such as the May spectral sequence [59] [24].

In the modern era, the best way to determine Ext in a range is by computer. These entirely algebraic invariants are effectively computable in a large range. Computer algorithms typically rely on minimal resolutions rather than the cobar resolution because it grows more slowly. In practice, computer calculations of Ext far outpace our ability to interpret the data with the Adams spectral sequence [17] [18] [19] [71] [47].

Of course, explicit calculations over a field k depend on specific knowledge of $K^M_*(k)$.

Remark 18.2.9. Over $\mathbb{C}$ and at $p = 2$, the cohomology of the motivic Steenrod algebra was computed with the motivic May spectral sequence through the 70-stem [44]. The most recent computer calculations over $\mathbb{C}$ extend far beyond the 100-stem, with ongoing progress into even higher stems [47].

Remark 18.2.10. Over $\mathbb{F}_q$ and at $p = 2$, the cohomology of the motivic Steenrod algebra is studied in detail through the 20-stem [107]. The Milnor K-theory $K^M_*(\mathbb{F}_q)$ is relatively simple, as described in Example 18.2.6; this makes explicit computations practical.

If $\operatorname{char} \mathbb{F}_q \equiv 1 \bmod 4$, then -1 is a square and ρ is zero in $\mathbb{M}_2$. Therefore, the situation is essentially identical to the calculation over $\mathbb{C}$.

However, if $\operatorname{char} \mathbb{F}_q \equiv 3 \bmod 4$, then ρ is non-zero in $\mathbb{M}_2$. This case is more complicated (and more interesting).

18.2.4 The ρ-Bockstein spectral sequence

Over $\mathbb{R}$, the cohomology $\mathbb{M}_2$ of a point is $\mathbb{F}_2[\tau, \rho]$. As discussed in Section 18.2.2, the motivic Steenrod algebra acts non-trivially on a point. This complication significantly increases the difficulty of Ext calculations. In other terms, the relation $\tau_i^2 = \tau\xi_{i+1} + \rho\tau_{i+1} + \rho\tau_0\xi_{i+1}$ is the source of these difficulties.

In this type of situation, the standard approach is to impose a filtration that hides these complications (or rather, pushes them into higher filtration). Then one obtains a spectral sequence that computes Ext.

In this case, we filter by powers of ρ and obtain a ρ-Bockstein spectral sequence that converges to $\mathrm{Ext}_{\mathbb{R}}$ [35] [26]. The associated graded object of $A^{\mathbb{R}}_*$ is easily identified with $\mathbb{A}^{\mathbb{C}}_*[\rho]$. From this observation, one can deduce that the spectral sequence takes the form

$$\mathrm{Ext}_{\mathbb{C}}[\rho] \Rightarrow \mathrm{Ext}_{\mathbb{R}}\,.$$

In order for the ρ-Bockstein spectral sequence to be practical, one needs a method for computing differentials. The first step in this program is to consider the effect of inverting ρ [26, Theorem 4.1]. If one inverts ρ in A_*, then the relation

$$\tau_i^2 = \tau\xi_{i+1} + \rho\tau_{i+1} + \rho\tau_0\xi_{i+1}$$

can be rewritten in the form

$$\tau_{i+1} = \rho^{-1}\tau_i^2 + \rho^{-1}\tau\xi_{i+1} + \tau_0\xi_{i+1}.$$

This shows that $A_*[\rho^{-1}]$ can be simply described as $\mathbb{M}_2[\rho^{-1}][\tau_0, \xi_1, \xi_2, \ldots]$. The cohomology of $A_*[\rho^{-1}]$ is then straightforward to compute. One obtains the isomorphism

$$\mathrm{Ext}_{\mathbb{R}}[\rho^{-1}] \cong \mathrm{Ext}_{\mathrm{cl}}[\rho^{\pm 1}].$$

In other words, the ρ-periodic part of $\mathrm{Ext}_{\mathbb{R}}$ is identified with classical Ext groups. Beware that this isomorphism changes degrees. For example, the motivic element h_{i+1} in the $(2^{i+1}-1)$-stem maps to the classical element h_i in the $(2^i - 1)$-stem.

Remark 18.2.11. In fact, this computational observation extends to an equivalence of homotopy categories. Namely, the ρ-periodic $\mathbb{R}$-motivic stable homotopy category $\mathbf{SH}(\mathbb{R})[\rho^{-1}]$ is equivalent to the classical stable homotopy category [10].

Inverting ρ on an $\mathbb{R}$-motivic spectrum is analogous to taking the geometric fixed points of a C_2-equivariant spectrum.

It remains to compute the ρ^r-torsion in $\mathrm{Ext}_{\mathbb{R}}$ for all r, i.e., to compute the ρ-Bockstein d_r differentials for all r. It turns out that these differentials are forced by combinatorial considerations in a large range. Every element x of the ρ-Bockstein E_1-page falls into one of three categories:

1. x survives, and $\rho^k x$ is not hit by any differential for all k. These elements are completely known because of the calculation of $\mathrm{Ext}_{\mathbb{R}}[\rho^{-1}]$.

2. x survives and $\rho^k x$ is hit by some differential for some $k \geq 1$.

3. x does not survive.

Somewhat surprisingly, these considerations determine many ρ-Bockstein differentials. In other words, in a large range, there is only one pattern of differentials that is consistent with sorting all elements into these three classes.

See [26, Section 5] for a description of this process in low dimensions. Preliminary results indicate that the same naive method works into much higher dimensions.

Remark 18.2.12. The ρ-Bockstein spectral sequence is useful for fields other than $\mathbb{R}$, in which -1 is not a square and therefore ρ is non-zero. Some examples are finite fields $\mathbb{F}_q$ for which $\operatorname{char} \mathbb{F}_q \equiv 3 \bmod 4$ [107].

18.2.5 The motivic Adams spectral sequence

The point of computing Ext_k is that it serves as the input to the motivic Adams spectral sequence

$$E_2 = \mathrm{Ext}_k \Rightarrow \pi^k_{*,*}.$$

We will not give a detailed construction of the spectral sequence. Just as in the classical case, the idea is to construct an Adams resolution of the motivic sphere spectrum, using copies of the motivic Eilenberg-Mac Lane $\mathbf{M}\mathbb{F}_2$ and a few of its basic properties [63] [24] [43].

A certain amount of care must be taken with convergence of the motivic Adams spectral sequence. In particular, one must complete not only with respect to a prime, but also with respect to the first Hopf map η. Convergence results are worked out in [43] [42] [58] [53], and we will not discuss them further here.

There are three major phases to carrying out an Adams spectral sequence computation:

1. Compute the E_2-page. This is an algebraic problem. See Section 18.2.3.
2. Compute Adams differentials. This is the hardest of the three steps.
3. The E_∞-page gives the filtration quotients in a filtration on $\pi_{*,*}$, but these quotients can hide some of the structure of $\pi_{*,*}$. We may have two elements α and β of $\pi_{*,*}$ detected by elements a and b of the E_∞-page with filtrations i and j respectively. Sometimes, the product $\alpha\beta$ is non-zero in $\pi_{*,*}$, but is detected in filtration greater than $i + j$. In this case, the product ab equals zero in the E_∞-page, even though $\alpha\beta$ is non-zero. Some work is required to reconstruct $\pi_{*,*}$ itself from its filtration quotients. See [44, Section 4.1.1] for a careful discussion of these issues.

18.3 The motivic effective slice spectral sequence

In this section we review the effective slice filtration of the stable motivic homotopy category $\mathbf{SH}(k)$. This filtration was originally proposed by Voevodsky [99] under the name "slice filtration". The goal is to understand the associated spectral sequence for the sphere. The latter approximates the stable motivic homotopy groups in a way we will make precise.

The effective slice spectral sequence (18.3.5) and its very effective version introduced in [93] are algebro-geometric analogues of the topological Atiyah-Hirzebruch spectral sequence [9]. Among the applications of the effectice filtration we note Voevodsky's construction of the effective slice spectral sequence relating motivic cohomology to algebraic K-theory [100], Levine's proof of full faithfulness of the constant functor from the stable topological homotopy category to the stable motivic homotopy category over algebraically closed fields [55], a new proof of Milnor's conjecture on quadratic forms in [86], and the identification of Morel's plus part of the rational sphere spectrum with rational motivic cohomology along with finiteness for the motivic stable homotopy groups over finite fields in [83].

The equivariant slice filtration has become an important tool classical homotopy theory. For example, it is central to major progress on elements of Kervaire invariant one [36]. The reader should beware that the equivariant slice filtration does not directly correspond to the motivic effective slice filtration. However, the filtrations do happen to align in many cases related to K-theory and cobordism.

18.3.1 The effective slice filtration

Let $\mathbf{SH}^{\mathsf{eff}}(k)$ be the localizing subcategory of the stable motivic homotopy category $\mathbf{SH}(k)$ generated by the motivic suspension spectra of all smooth schemes over k, i.e., the smallest full triangulated subcategory of $\mathbf{SH}(k)$ that contains suspension spectra of smooth schemes and is closed under coproducts [99]. These subcategories form the effective slice filtration

$$\cdots \subset \Sigma^{2q+2,q+1}\mathbf{SH}^{\mathsf{eff}}(k) \subset \Sigma^{2q,q}\mathbf{SH}^{\mathsf{eff}}(k) \subset \Sigma^{2q-2,q-1}\mathbf{SH}^{\mathsf{eff}}(k) \subset \cdots. \tag{18.3.1}$$

Since $\mathbf{SH}^{\mathsf{eff}}(k)$ is closed under simplicial suspsension and desuspension, the subcategory $\Sigma^{2q,q}\mathbf{SH}^{\mathsf{eff}}(k)$ is equal to the subcategory $\Sigma^{0,q}\mathbf{SH}^{\mathsf{eff}}(k)$. The effective slice filtration is exhaustive in the sense that $\mathbf{SH}(k)$ is the smallest triangulated subcategory that is closed under coproducts and contains $\Sigma^{2q,q}\mathbf{SH}^{\mathsf{eff}}(k)$ for all q.

The inclusion $\Sigma^{2q,q}\mathbf{SH}^{\mathsf{eff}}(k) \subset \mathbf{SH}(k)$ is left adjoint to a functor that takes a motivic spectrum to its "qth effective slice cover". (This is analogous to the classical inclusion of q-connected spectra into all spectra, which is left adjoint to the q-connected cover.) We write f_q for the functor that takes a motivic spectrum to its qth effective slice cover.

Any motivic spectrum $\mathbf{E}$ has an effective slice filtration

$$\cdots \to \mathsf{f}_{q+1}\mathbf{E} \to \mathsf{f}_q\mathbf{E} \to \mathsf{f}_{q-1}\mathbf{E} \to \cdots. \tag{18.3.2}$$

The associated graded object of this filtration is $\mathsf{s}_*\mathbf{E}$, where the qth component $\mathsf{s}_q\mathbf{E}$ is the cofiber of $\mathsf{f}_{q+1}\mathbf{E} \to \mathsf{f}_q\mathbf{E}$.

A related construction employs the very effective subcategory $\mathbf{SH}^{\mathsf{veff}}(k)$ introduced in [93, §5]. This is the smallest full subcategory of $\mathbf{SH}(k)$ that contains all suspension spectra of smooth schemes of finite type, is closed under homotopy colimits, and is closed under extensions in the sense that if $X \to Y \to Z$ is a cofiber sequence such that X and Z belong to $\mathbf{SH}^{\mathsf{veff}}(k)$, then so does Y.

We note that $\mathbf{SH}^{\mathsf{veff}}(k)$ is contained in $\mathbf{SH}^{\mathsf{eff}}(k)$ but it is not a triangulated subcategory of $\mathbf{SH}(k)$ since it is not closed under simplicial desuspension. The very effective slice filtration takes the form

$$\cdots \subset \Sigma^{2q+2,q+1}\mathbf{SH}^{\mathsf{eff}}(k) \subset \Sigma^{2q,q}\mathbf{SH}^{\mathsf{eff}}(k) \subset \Sigma^{2q-2,q-1}\mathbf{SH}^{\mathsf{eff}}(k) \subset \cdots, \tag{18.3.3}$$

and we obtain notions of very effective slice covers $\tilde{\mathsf{f}}_q$ and very effective slices $\tilde{\mathsf{s}}_q$ for all $n \in \mathbb{Z}$. To connect this to motivic homotopy groups we mention the fact that if $\mathbf{E} \in \mathbf{SH}^{\mathsf{veff}}(k)$ then $\pi_{m,n}\mathbf{E} = 0$ for $m < n$ [93, Lemma 5.10].Some examples of very effective motivic spectra are algebraic cobordism $\mathbf{MGL}$, the effective slice cover $\mathbf{kgl} := \mathsf{f}_0\mathbf{KGL}$ of algebraic K-theory [93], and the very effective slice cover $\mathbf{kq} := \tilde{\mathsf{f}}_0\mathbf{KQ}$ of hermitian K-theory [4]. One advantage of the very effective slice filtration is that it maps to the topological Postnikov tower under Betti realization [34, §3.3].

Higher structural properties of motivic spectra such as A_∞- and E_∞-structures interact well with these filtrations. In [34] it is shown that if $\mathbf{E}$ is an A_∞- or E_∞-algebra in motivic symmetric spectra [49] then $\mathsf{f}_0\mathbf{E}$ and $\tilde{\mathsf{f}}_0\mathbf{E}$ are naturally equipped with the structure of an A_∞- resp. E_∞-algebra. Moreover, the canonical maps $\mathsf{f}_0\mathbf{E} \to \mathbf{E}$ and $\tilde{\mathsf{f}}_0\mathbf{E} \to \mathbf{E}$ can accordingly be modelled as maps of A_∞- and E_∞-algebras. For every $q \in \mathbb{Z}$ the functors f_q, s_q, $\tilde{\mathsf{f}}_q$, and $\tilde{\mathsf{s}}_q$ respect module structures over E_∞-algebras.

Applying motivic homotopy groups to the filtration (18.3.2) yields by standard techniques an exact couple and the effective slice spectral sequence

$$E^1_{m,q,n}(\mathbf{E}) = \pi_{m,n}\mathsf{s}_q\mathbf{E} \Longrightarrow \pi_{m,n}\mathbf{E}. \tag{18.3.4}$$

For calculations it is important to note that the first effective slice differential is induced by the composition $\mathsf{s}_q\mathbf{E} \to \Sigma^{1,0}\mathsf{f}_{q+1}\mathbf{E} \to \Sigma^{1,0}\mathsf{s}_{q+1}\mathbf{E}$.

18.3.2 The slices of 1

By [83], the effective slice spectral sequence for the sphere $\mathbf{1}$ over any field has E_1-page given by

$$E^1_{m,q,n} = \pi_{m,n}\mathsf{s}_q(\mathbf{1}). \tag{18.3.5}$$

This spectral sequence converges conditionally, in the sense of Boardman [14] to the η-completed stable motivic homotopy groups. Technically, there is no need to complete at a prime, i.e., the effective slice spectral sequence is defined integrally. However, for practical purposes it is best to further complete one prime at a time. Here $\mathsf{s}_q(\mathbf{1})$ is the qth slice of $\mathbf{1}$, i.e., the qth layer of the effective slice filtration on $\mathbf{1}$.

The first step in understanding the effective slice spectral sequence (18.3.5) is to understand the slices $\mathsf{s}_q(\mathbf{1})$. Voevodsky [99, Conjectures 9 and 10] predicted correctly that these slices are governed by the structure of the Adams-Novikov E_2-page (see (18.3.7) below). It turns out that $\mathsf{s}_0(\mathbf{1})$ identifies with the integral motivic Eilenberg-Mac Lane spectrum $\mathbf{M}\mathbb{Z}$, while $\mathsf{s}_q(\mathbf{1})$ for $q \geq 1$ are $\mathsf{s}_0(\mathbf{1})$-modules, i.e., $\mathbf{M}\mathbb{Z}$-modules.

We will now sketch how the slices $\mathsf{s}_q(\mathbf{1})$ can be computed. First, a geometric argument shows that the cone of the unit map $\mathbf{1} \to \mathbf{MGL}$ lies in $\Sigma^{2,1}\mathbf{SH}^{\mathsf{eff}}(k)$. It follows that $\mathsf{s}_0(\mathbf{1}) \to \mathsf{s}_0(\mathbf{MGL})$ is an isomorphism since s_0 is trivial on $\Sigma^{2,1}\mathbf{SH}^{\mathsf{eff}}(k)$.

The slices of $\mathbf{MGL}$ are described in Theorem 18.3.2. It turns out to be more convenient to describe all the slices of $\mathbf{MGL}$ simultaneously, rather than one at a time. So we will describe the entire graded motivic spectrum $\mathsf{s}_*\mathbf{MGL}$. We use the term "effective slice degree" to refer to this external grading.

Remark 18.3.1. For the following discussion, if the base field k is of positive characteristic, one must invert its characteristic in the motivic spectra $\mathbf{MGL}$ and $\mathbf{M}\mathbb{Z}$. See [83, §2.1-2.2] for precise statements. For legibility we will not make any notational changes.

Theorem 18.3.2. ([90, Corollary 4.7],[39, §8.3], [92, Theorem 3.1]) *Over a base field of characteristic zero the graded motivic spectrum* $\mathsf{s}_*\mathbf{MGL}$ *is* $\mathbf{M}\mathbb{Z}[x_1, x_2, \ldots]$, *where* x_j *has motivic degree* $(2j, j)$ *and effective slice degree* j.

The expression $\mathbf{M}\mathbb{Z}[x_1, x_2, \ldots]$ in Theorem 18.3.2 bears some explanation. This object is a direct sum of copies of suspensions $\mathbf{M}\mathbb{Z}$, indexed by monomials in the x_j. Each monomial $x_1^{i_1}x_2^{i_2}\ldots x_r^{i_r}$ contributes a copy of $\Sigma^{2q,q}\mathbf{M}\mathbb{Z}$ to $\mathsf{s}_q\mathbf{MGL}$, where $q = \sum_{j=1}^{r} ji_j$. Note that q is the degree of the monomial of the same name in $\pi_*\mathbf{MU} = \mathbb{Z}[x_1, x_2, \ldots]$, where $\mathbf{MU}$ is the classical complex cobordism spectrum and x_j has degree $2j$ [61].

The positive slices of $\mathbf{1}$ are determined in several steps, starting with the standard cosimplicial $\mathbf{MGL}$-resolution

$$\mathbf{1} \longrightarrow \mathbf{MGL} \rightrightarrows \mathbf{MGL}^{\wedge 2} \substack{\longrightarrow\\\longrightarrow\\\longrightarrow} \mathbf{MGL}^{\wedge 3} \cdots,$$

which induces a natural isomorphism

$$\mathsf{s}_q(\mathbf{1}) \xrightarrow{\cong} \operatorname*{holim}_{\Delta} \mathsf{s}_q(\mathbf{MGL}^{\wedge(\bullet+1)}). \tag{18.3.6}$$

In order to describe the slices of $\mathbf{MGL}^{\wedge n}$, first recall the smash product decompositions

$$\mathbf{MGL} \wedge \mathbf{MGL} = \mathbf{MGL}[b_1, b_2, b_3, \ldots]$$

and

$$\mathbf{MU} \wedge \mathbf{MU} = \mathbf{MU}[b_1, b_2, b_3, \ldots]$$

from [72, Lemma 6.4(i)] and [2, p. 87] respectively. Here b_j is the standard choice of generator in the dual Landweber-Novikov algebra [81, Theorem 4.1.11], and the motivic degree of b_j is $(2j, j)$, while the classical degree of b_j is $2j$. These decompositions extend to higher smash powers in an obvious way.

From Theorem 18.3.2 and these smash product decompositions, it follows that the graded motivic spectrum $\mathsf{s}_*(\mathbf{MGL}^{\wedge n})$ is $\mathbf{M}\mathbb{Z} \otimes \pi_*(\mathbf{MU}^{\wedge n})$ [99, Conjecture 6]. More precisely, $\pi_*(\mathbf{MU}^{\wedge n})$ is a free abelian group on a set of generators, concentrated in even degrees. Each generator of $\pi_{2q}(\mathbf{MU}^{\wedge n})$ contributes a summand $\Sigma^{2q,q}\mathbf{M}\mathbb{Z}$ to $\mathsf{s}_q(\mathbf{MGL}^{\wedge n})$.

With additional work detailed in [83, §2.2], the above produces an isomorphism

$$\mathsf{s}_*(\mathbf{1}) \simeq \mathbf{M}\mathbb{Z} \otimes \mathrm{Ext}^{*,*}_{\mathbf{MU}_*\mathbf{MU}}(\mathbf{MU}_*, \mathbf{MU}_*). \tag{18.3.7}$$

More precisely, each copy of $\mathbb{Z}/n$ in $\mathrm{Ext}^{p,2q}_{\mathbf{MU}_*\mathbf{MU}}(\mathbf{MU}_*, \mathbf{MU}_*)$ contributes a summand $\Sigma^{2q-p,q}\mathbf{M}\mathbb{Z}/n$ to $\mathsf{s}_q(\mathbf{1})$. Here p is the Adams-Novikov filtration, i.e., the homological degree of the Ext group, and $2q$ is the total degree, i.e., the stem plus the Adams-Novikov filtration.

We note the slices are $\mathbf{M}\mathbb{Z}$-modules and the effective slice d_1-differentials are maps between motivic Eilenberg-Mac Lane spectra. These facts make the effective slice spectral sequence amenable to calculations over a base scheme affording an explicit description of its motivic cohomology ring along with the action of the motivic Steenrod algebra.

The Ext groups in (18.3.7) form the E_2-page of the Adams-Novikov spectral sequence, which has been extensively studied by topologists [81]. From an algebraic viewpoint, this is the cohomology of the "Hopf algebroid" corepresenting the functor carrying a commutative ring to the groupoid of formal group laws over it and their strict isomorphisms [2]. If $(s,t) \neq (0,0)$, then $\mathrm{Ext}^{s,t}_{\mathbf{MU}_*\mathbf{MU}}(\mathbf{MU}_*, \mathbf{MU}_*)$ is finite.

In practice, $\mathrm{Ext}_{\mathbf{MU}_*\mathbf{MU}}(\mathbf{MU}_*, \mathbf{MU}_*)$ is most easily studied one prime at a time. For the Brown-Peterson spectrum $\mathbf{BP}$ at a prime p, we have

$$\mathrm{Ext}^{s,t}_{\mathbf{MU}_*\mathbf{MU}}(\mathbf{MU}_*, \mathbf{MU}_*) = \bigoplus_p \mathrm{Ext}^{s,t}_{\mathbf{BP}_*\mathbf{BP}}(\mathbf{BP}_*, \mathbf{BP}_*)$$

when $(s,t) \neq (0,0)$, so one should really study $\mathrm{Ext}^{s,t}_{\mathbf{BP}_*\mathbf{BP}}(\mathbf{BP}_*, \mathbf{BP}_*)$ separately for each prime.

This input provides a systematic way of keeping track of the direct summands of $\mathsf{s}_q(\mathbf{1})$. For example, at $p = 2$, the well-known elements α_1^q in $\mathrm{Ext}^{q,2q}_{\mathbf{BP}_*\mathbf{BP}}(\mathbf{BP}_*, \mathbf{BP}_*)$ contribute $\Sigma^{q,q}\mathbf{M}\mathbb{Z}/2$ summands to $\mathsf{s}_q(\mathbf{1})$, and the element $\alpha_{2/2}$ in $\mathrm{Ext}^{1,4}_{\mathbf{BP}_*\mathbf{BP}}(\mathbf{BP}_*, \mathbf{BP}_*)$ contributes a $\Sigma^{3,2}\mathbf{M}\mathbb{Z}/4$ summand to $\mathsf{s}_2(\mathbf{1})$. At $p = 3$, the element α_1 in $\mathrm{Ext}^{1,4}_{\mathbf{BP}_*\mathbf{BP}}(\mathbf{BP}_*, \mathbf{BP}_*)$ contributes $\Sigma^{3,2}\mathbf{M}\mathbb{Z}/3$ to $\mathsf{s}_2(\mathbf{1})$. These copies of $\Sigma^{3,2}\mathbf{M}\mathbb{Z}/4$ and $\Sigma^{3,2}\mathbf{M}\mathbb{Z}/3$ assemble into a copy of $\Sigma^{3,2}\mathbf{M}\mathbb{Z}/12$ in $\mathsf{s}_2(\mathbf{1})$.

Remark 18.3.3. Our notation for $\mathsf{s}_*(\mathbf{1})$ suggests that we are describing $\mathsf{s}_*(\mathbf{1})$ as a ring object, but the notation is somewhat deceptive. For example, consider the map $\mathsf{s}_1(\mathbf{1}) \wedge \mathsf{s}_1(\mathbf{1}) \to \mathsf{s}_2(\mathbf{1})$ at $p = 2$, where $\mathsf{s}_1(\mathbf{1})$ is equivalent to $\Sigma^{1,1}\mathbf{M}\mathbb{Z}/2$ corresponding to α_1, and $\mathsf{s}_2(\mathbf{1})$ is equivalent to $\Sigma^{2,2}\mathbf{M}\mathbb{Z}/2 \vee \Sigma^{3,2}\mathbf{M}\mathbb{Z}/4$ corresponding to α_1^2 and $\alpha_{2/2}$ respectively.

The source $\Sigma^{1,1}\mathbf{M}\mathbb{Z}/2 \wedge \Sigma^{1,1}\mathbf{M}\mathbb{Z}/2$ splits as a wedge of suspensions of copies of $\mathbf{M}\mathbb{Z}/2$, where the wedge is indexed by a basis for the motivic Steenrod algebra. Thus we want to calculate the map

$$\Sigma^{2,2} \bigvee_{b \in B} \Sigma^{d_b}\mathbf{M}\mathbb{Z}/2 \to \Sigma^{2,2}\mathbf{M}\mathbb{Z}/2 \vee \Sigma^{3,2}\mathbf{M}\mathbb{Z}/4, \tag{18.3.8}$$

where B is a basis for the motivic Steenrod algebra, and d_b is the bidegree of b.

If we restrict to the summand of the source of (18.3.8) corresponding to the element 1 of the motivic Steenrod algebra, then we have a map

$$\Sigma^{2,2}\mathbf{M}\mathbb{Z}/2 \to \Sigma^{2,2}\mathbf{M}\mathbb{Z}/2 \vee \Sigma^{3,2}\mathbf{M}\mathbb{Z}/4.$$

This map is the identity on the first summand of the target and zero on the second summand of the target, corresponding to the multiplicative relation $\alpha_1 \cdot \alpha_1 = \alpha_1^2$.

However, if we restrict to the summand of the source of (18.3.8) corresponding to the element Sq^1 of the motivic Steenrod algebra, then we have a map

$$\Sigma^{3,2}\mathbf{M}\mathbb{Z}/2 \to \Sigma^{2,2}\mathbf{M}\mathbb{Z}/2 \vee \Sigma^{3,2}\mathbf{M}\mathbb{Z}/4.$$

This map turns out to be non-trivial when restricted to the second summand of the target, and this has explicit computational consequences for the effective slice spectral sequence.

This warning about multiplicative structures applies to other slice calculations in this manuscript.

It is natural to ask how η acts on the slices of $\mathbf{1}$. A complete description can be extracted from the work of Andrews-Miller [6] that calculates the α_1-periodic Ext groups over $\mathbf{BP}_*\mathbf{BP}$ at the prime 2. This shows that $\mathsf{s}_*(\mathbf{1})[\eta^{-1}]$ is

$$\mathbf{M}\mathbb{Z}/2[\alpha_1^{\pm 1}, \alpha_3, \alpha_4]/\alpha_3^2, \tag{18.3.9}$$

where α_1 has motivic degree $(1,1)$ and effective slice degree 1; α_3 has motivic degree $(5,3)$ and effective slice degree 3; and α_4 has motivic degree $(7,4)$ and effective slice degree 4.

18.3.3 The slices of other motivic spectra

Related to (18.3.7), the slices of algebraic K-theory $\mathbf{KGL}$, hermitian K-theory $\mathbf{KQ}$, and higher Witt-theory $\mathbf{KW}$ can be identified.

Theorem 18.3.4. [54, 86]

1. *The graded motivic spectrum $\mathsf{s}_*\mathbf{KGL}$ is equivalent to $\mathbf{M}\mathbb{Z}[\beta^{\pm 1}]$, where β has motivic degree $(2,1)$ and effective slice degree 1.*

2. *When* $\mathrm{char}(k) \neq 2$*, the graded motivic spectrum* $\mathsf{s}_*\mathbf{KQ}$ *is equivalent to*

$$\mathbf{M}\mathbb{Z}[\alpha_1, v_1^{\pm 2}]/2\alpha_1,$$

where α_1 *has motivic degree* $(1,1)$ *and effective slice degree* 1*, and* v_1^2 *has motivic degree* $(4,2)$ *and effective slice degree* 2*.*

3. *When* $\mathrm{char}(k) \neq 2$*, the graded motivic spectrum* $\mathsf{s}_*\mathbf{KW}$ *is equivalent to*

$$\mathbf{M}\mathbb{Z}/2[\alpha_1^{\pm 1}, v_1^{\pm 2}],$$

where α_1 *has motivic degree* $(1,1)$ *and effective slice degree* 1*, and* v_1^2 *has motivic degree* $(4,2)$ *and effective slice degree* 2*.*

Part (1) of Theorem 18.3.4 says that $\mathsf{s}_q\mathbf{KGL}$ is $\Sigma^{2q,q}\mathbf{M}\mathbb{Z}$.

In Part (2) of Theorem 18.3.4, the monomials v_1^{2q} contribute copies of $\Sigma^{4q,2q}\mathbf{M}\mathbb{Z}$ to $\mathsf{s}_{2q}\mathbf{KQ}$. Also, for $p \geq 1$, monomials $\alpha_1^p v_1^{2q}$ contribute copies of $\Sigma^{4q+p,2q+p}\mathbf{M}\mathbb{Z}/2$ to $\mathsf{s}_{2q+p}\mathbf{KQ}$. Beware that the notation for $\mathsf{s}_*\mathbf{KQ}$ only partly describes the multiplicative structure, as in Remark 18.3.3.

We obtain Part (3) of Theorem 18.3.4 from Part (2) since $\mathbf{KW}$ is obtained from $\mathbf{KQ}$ by inverting η in the same way that $\mathbf{1}[\eta^{-1}]$ is obtained from $\mathbf{1}$.

Remark 18.3.5. Spitzweck's work on motivic cohomology in [91] shows the isomorphisms in (18.3.7), (18.3.9), and all three parts of Theorem 18.3.4 hold not only over fields, but also over Dedekind domains of mixed characteristic with no residue fields of characteristic 2 (see [83, §2.3]).

Thanks to the slice calculations reviewed in this section we know precisely how the E^1-pages of the effective slice spectral sequences for $\mathbf{1}$, $\mathbf{1}[\eta^{-1}]$, $\mathbf{KGL}$, $\mathbf{KQ}$, and $\mathbf{KW}$ are given in terms of motivic cohomology groups of the base scheme. Now all the fun can begin with determining the corresponding differentials and resolving the information hidden by the associated graded structure of the effective slice E_∞-pages.

18.4 $\mathbb{C}$-motivic stable homotopy groups

In this section, we fix the base field $k = \mathbb{C}$. This special case is made easier by the fact that $\mathbb{C}$ has trivial arithmetic properties, i.e., $K_*^M(\mathbb{C})/p$ is trivial. In fact, the calculations for any algebraically closed field of characteristic zero work out identically.

To begin, the stable homotopy groups $\pi_{*,0}^{\mathbb{C}}$ are isomorphic to the classical stable homotopy groups π_* [55]. This isomorphism occurs even without completion. However, complicated exotica phenomena occur in other weights.

In this section, we also fix the prime $p = 2$. There are interesting phenomena to study at odd primes, but those cases have not yet been studied as extensively. See [94] for some $\mathbb{C}$-motivic results at odd primes.

As mentioned in Remark 18.2.9, the cohomology of the $\mathbb{C}$-motivic Steenrod algebra is completely known in a large range. Current computer calculations extend beyond the 100-

stem, with partial information out to the 140-stem. We recommend that the reader view $\mathbb{C}$-motivic Adams charts throughout this discussion [45].

18.4.1 $\mathbb{C}$-motivic and classical stable homotopy groups

As observed in Remark 18.2.8, inverting τ takes the $\mathbb{C}$-motivic Steenrod algebra to the classical Steenrod algebra with $\tau^{\pm 1}$ adjoined. In fact, this principle extends to Ext groups, so that there is an isomorphism

$$\mathrm{Ext}_{\mathbb{C}}(\mathbb{M}_2, \mathbb{M}_2)[\tau^{-1}] \cong \mathrm{Ext}_{\mathrm{cl}}(\mathbb{F}_2, \mathbb{F}_2)[\tau^{\pm 1}],$$

where $\mathrm{Ext}_{\mathrm{cl}}(\mathbb{F}_2, \mathbb{F}_2)$ represents Ext groups over the classical Steenrod algebra. Even further, there is an isomorphism between the τ-periodic $\mathbb{C}$-motivic Adams spectral sequence and the classical Adams spectral sequence. Consequently, the τ-periodic $\mathbb{C}$-motivic stable homotopy groups are isomorphic to the classical stable homotopy groups with $\tau^{\pm 1}$ adjoined.

This comparison between the $\mathbb{C}$-motivic and classical situations is induced by the Betti realization functor that takes a complex variety to its underlying topological space of $\mathbb{C}$-valued points.

18.4.2 η-periodic-phenomena

Classically, the element h_1^4 of $\mathrm{Ext}_{\mathrm{cl}}$ is zero. From Section 18.4.1, it follows that $\mathbb{C}$-motivically, $\tau^k h_1^4$ equals zero for some k. Classically, there is a relation $h_0^2 h_2 = h_1^3$. In the $\mathbb{C}$-motivic setting, the weight of $h_0^2 h_2$ is 2, while the weight of h_1^3 is 3. Consequently, the correct motivic relation is $h_0^2 h_2 = \tau h_1^3$. Therefore,

$$\tau h_1^4 = h_0^2 h_1 h_2 = 0,$$

but it turns out that h_1^4 is not zero. Moreover, every element h_1^k is non-zero for all $k \geq 0$.

There is a naive explanation for this phenomenon. Recall that the element h_1 is detected by $[\xi_1]$ in the $\mathbb{C}$-motivic cobar resolution. Unlike in the classical case, the element ξ_1 is indecomposable in the $\mathbb{C}$-motivic dual Steenrod algebra. It is this property of the $\mathbb{C}$-motivic dual Steenrod algebra that ultimately allows h_1 to be not nilpotent.

This exotic behavior of h_1 leads us inevitably to ask about the effect of inverting h_1 on the cohomology of the $\mathbb{C}$-motivic Steenrod algebra, and about the effect of inverting the first Hopf map η in the $\mathbb{C}$-motivic stable homotopy groups. Inspection of a motivic Adams chart [45] reveals many classes, such as c_0, d_0, and e_0 that are non-zero after inverting h_1.

We draw particular attention to the element B_1 (also known as Mh_1) in the 46-stem. From the perspective of the May spectral sequence, it is a surprise that this element survives h_1-inversion because it is unrelated to any of the many elements in lower stems that survive h_1-inversion. This observation eventually led to a clean calculation.

Theorem 18.4.1. [32] *The h_1-periodic cohomology of the $\mathbb{C}$-motivic Steenrod algebra is*

$$\mathrm{Ext}_{\mathbb{C}}[h_1^{-1}] \cong \mathbb{F}_2[h_1^{\pm 1}][v_1^4, v_2, v_3, \cdots],$$

where v_1^4 has degree $(8, 4, 4)$ and v_n has degree $(2^n - 2, 1, 2^{n-1} - 1)$.

The article [32] conjectured that there are Adams differentials $d_2(v_n) = h_1 v_{n-1}^2$ for all $n \geq 3$. This pattern of Adams differentials would determine the η-periodic $\mathbb{C}$-motivic stable homotopy groups. The conjecture was proved by Andrews and Miller [6], who computed the α_1-periodic E_2-page of the Adams-Novikov spectral sequence. This Adams-Novikov computation immediately determines the η-periodic $\mathbb{C}$-motivic stable homotopy groups, using the circle of ideas discussed below in Section 18.4.3.

Theorem 18.4.2. [6] *The η-periodic $\mathbb{C}$-motivic stable homotopy groups are*

$$\mathbb{F}_2[\eta^{\pm 1}][\mu, \sigma]/\sigma^2,$$

where μ has degree $(9,5)$ and σ has degree $(7,4)$.

Theorem 18.4.2 holds even without completions.

18.4.3 Adams differentials

The hardest part of a classical or motivic Adams spectral sequence computation is the determination of Adams differentials. Some techniques include:

1. Relations obtained by shuffling Toda brackets, which then imply that differentials must occur.

2. Moss's theorem [70], which can show that certain elements in the Adams spectral sequence must survive to detect certain Toda brackets.

3. Bruner's theorem on the interaction between algebraic squaring operations and Adams differentials [16].

4. Comparison to other information about stable homotopy groups, such as the image of J [1], the homotopy groups of *tmf* [21], and the Adams-Novikov spectral sequence.

Many examples of these techniques appear in [11] [16] [44] [48] [57] [104]. The manuscript [44] contains an exhaustive discussion of Adams differentials up to the 59-stem. With much effort, this bound was pushed slightly further to the 61-stem [104]. However, all of these techniques become impractical in higher stems.

Recent work on the comparison between the $\mathbb{C}$-motivic and classical stable homotopy categories has provided a new tool for computing Adams differentials that has allowed us to extend computations into a much larger range [47]. Current calculations extend beyond the 90-stem, with ongoing progress into even higher stems.

Two new ingredients allow for this program to succeed. The first is computer data for the Adams-Novikov E_2-page. The second ingredient is the following theorem.

Theorem 18.4.3. [30, 31, 80] *In $\mathbb{C}$-motivic stable homotopy theory,*

1. *The cofiber $\mathbf{1}/\tau$ of τ is an E_∞-ring spectrum, in an essentially unique way.*

2. *With suitable finiteness conditions, the homotopy category of $\mathbf{1}/\tau$-modules is equivalent to the derived category of $\mathbf{BP}_*\mathbf{BP}$-comodules.*

3. *The Adams spectral sequence for $\mathbf{1}/\tau$ is isomorphic to the algebraic Novikov spectral sequence [75] [60] that converges to the Adams-Novikov E_2-page.*

It is remarkable and unexpected that a homotopical category such as $\mathbf{1}/\tau$-modules would have a purely algebraic description. The proof of part (3) of Theorem 18.4.3 is that the filtrations associated to the two spectral sequences correspond under the equivalence of part (2).

A new, more powerful approach to computing Adams differentials is summarized in the following steps:

1. Compute the cohomology of the $\mathbb{C}$-motivic Steenrod algebra by machine.
2. Compute by machine the algebraic Novikov spectral sequence, including all differentials and multiplicative structure.
3. Use Theorem 18.4.3 to completely describe the motivic Adams spectral sequence for $\mathbf{1}/\tau$.
4. Use the cofiber sequence

$$\Sigma^{0,-1}\mathbf{1} \xrightarrow{\tau} \mathbf{1} \to \mathbf{1}/\tau \to \Sigma^{1,-1}\mathbf{1}$$

and naturality of Adams spectral sequences to pull back and push forward Adams differentials for $\mathbf{1}/\tau$ to Adams differentials for the motivic sphere.

5. Apply a variety of ad hoc arguments to deduce additional Adams differentials for the motivic sphere, as described at the beginning of this section.
6. Use a long exact sequence in homotopy groups to deduce hidden τ extensions in the motivic Adams spectral sequence for the sphere.
7. Invert τ to obtain the classical Adams spectral sequence and the classical stable homotopy groups.

The weak link in this algorithm is step (5), where ad hoc arguments come into play. The method will continue to calculate new stable homotopy groups until the ad hoc arguments become too complicated to resolve. It is not yet clear when this will occur.

18.4.4 Motivic nilpotence and periodicity

Section 18.4.2 discussed the non-nilpotence of the element η of $\pi_{1,1}^{\mathbb{C}}$. This non-nilpotent behavior led Haynes Miller to propose that there might be an infinite family of periodicity operators w_n such that w_0 is η, in analogy to the v_n-periodicity operators that begin with the non-nilpotent element $v_0 = 2$.

In fact, Miller's guess turned out to be correct [5] [29]. In more detail, there exist $\mathbb{C}$-motivic ring spectra $\mathbf{K}(w_n)$ whose motivic stable homotopy groups are of the form $\mathbb{F}_2[w_n^{\pm 1}]$. Also, for each n, there exist finite $\mathbb{C}$-motivic complexes that possess w_n-self-maps. More precisely, there exist complexes $\mathbf{X}_n$ equipped with maps of the form

$$\Sigma^{d|w_n|}\mathbf{X}_n \xrightarrow{f_n} \mathbf{X}_n$$

that induce isomorphisms in $\mathbf{K}(w_n)$-homology. One can then study w_n-periodic families of elements in $\pi^{\mathbb{C}}_{*,*}$ by considering compositions of the form

$$\Sigma^{p,q}\mathbf{1} \longrightarrow \Sigma^{dm|w_n|}\mathbf{X}_n \xrightarrow{f_n} \cdots \xrightarrow{f_n} \mathbf{X}_n \longrightarrow \mathbf{1}.$$

Andrews [5] found the first explicit examples of w_1-periodic families.

It turns out that there are even more exotic $\mathbb{C}$-motivic periodicities [52]. Roughly speaking, these periodicities correspond to the elements h_{ij} of the May spectral sequence for $i \geq j$. The v_n-periodicities correspond to the elements $h_{n+1,0}$, while the w_n-periodicities correspond to $h_{n+1,1}$. While there do exist $\mathbb{C}$-motivic spectra of the form $\mathbf{K}(h_{ij})$, these objects do not all possess ring structures.

18.5 $\mathbb{R}$-motivic stable homotopy groups

In this section, we fix the base field $k = \mathbb{R}$. This special case introduces new phenomena not seen in the $\mathbb{C}$-motivic situation. The first complication is that $K^M_*(\mathbb{R})/2$ is now a polynomial algebra on one class ρ. Studying the $\mathbb{R}$-motivic case allows us to grapple with the difficulties presented by the non-zero element ρ. However, it avoids even further complications that would be created by relations involving ρ.

In this section, we also fix the prime $p = 2$. The $\mathbb{R}$-motivic case at odd primes is not so interesting, since $K^M_*(\mathbb{R})/p$ is zero.

As mentioned in Section 18.2.4, the ρ-Bockstein spectral sequence is an effective tool for computing the cohomology of the $\mathbb{R}$-motivic Steenrod algebra. Preliminary work shows that the spectral sequence can be completely analyzed in a large range. See [26] for a sense of the structure of the calculation, although the range studied there is just a tiny portion of what is possible.

18.5.1 η-periodic phenomena

Just as in the $\mathbb{C}$-motivic case, one can study the effect of inverting h_1 on the cohomology of the $\mathbb{R}$-motivic Steenrod algebra, and about the effect of inverting the first Hopf map η in the $\mathbb{R}$-motivic stable homotopy groups. This study was carried out to completion in [33].

The first step is to consider the h_1-periodic version of the ρ-Bockstein spectral sequence discussed in Section 18.2.4, which takes the form

$$\mathrm{Ext}_{\mathbb{C}}[\rho][h_1^{-1}] \Rightarrow \mathrm{Ext}_{\mathbb{R}}[h_1^{-1}].$$

This spectral sequence can be completely analyzed.

The next step is to consider the h_1-periodic version of the $\mathbb{R}$-motivic Adams spectral sequence, which takes the form

$$\mathrm{Ext}_{\mathbb{R}}[h_1^{-1}] \Rightarrow \pi^{\mathbb{R}}_{*,*}[\eta^{-1}].$$

Once again, this spectral sequence can be completely analyzed, but it is much more interesting than the $\mathbb{C}$-motivic case discussed in Section 18.4.2, which collapses at the E_3-page.

In fact, in this case there are non-trivial differentials on every page of the spectral sequence. These differentials are deduced by analyzing higher homotopical structure, i.e., Toda brackets.

Before we can state the conclusion of the calculation, we need some additional notation. This complication arises because we want to invert an element η that has non-zero degree $(1,1)$. We write $\Pi_m^{\mathbb{R}}$ for the direct sum

$$\bigoplus_{n\in\mathbb{Z}} \pi^{\mathbb{R}}_{m+n,n},$$

which we call the mth Milnor-Witt stem. Then η acts on each $\Pi_m^{\mathbb{R}}$, and we can consider $\Pi_m^{\mathbb{R}}[\eta^{-1}]$.

Theorem 18.5.1. [33] *The η-periodic $\mathbb{R}$-motivic stable homotopy groups $\pi_{*,*}[\eta^{-1}]$ are given by:*

1. $\Pi_0^{\mathbb{R}}[\eta^{-1}] = \mathbb{Z}_2[\eta^{\pm1}]$.
2. $\Pi_{4m-1}^{\mathbb{R}}[\eta^{-1}] = \mathbb{Z}/2^{u+1}[\eta^{\pm1}]$ *for $m > 1$, where u is the 2-adic valuation of $4m$.*
3. $\Pi_m^{\mathbb{R}}[\eta^{-1}] = 0$ *otherwise.*

The groups that appear in Theorem 18.5.1 are reminiscent of the groups that appear in the classical image of J. One might speculate on a more direct connection between these η-periodic $\mathbb{R}$-motivic stable homotopy groups and the image of J. However, the Toda bracket structures are different.

18.6 Motivic stable homotopy groups over general fields

Sections 18.4 and 18.5 discussed what is known about $\mathbb{C}$-motivic and $\mathbb{R}$-motivic stable homotopy groups. We now consider motivic stable homotopy groups over larger classes of fields. Naturally, specific information is harder to obtain when the base field is allowed to vary widely.

In addition to the phenomena described in the following sections, we also mention the article [25], which discusses the existence of motivic Hopf maps and some relations satisfied by these maps.

18.6.1 Motivic graded commutativity

In the classical case, the stable homotopy groups π_* are graded commutative, in the sense that

$$\alpha\beta = (-1)^{|\alpha||\beta|}\beta\alpha$$

for all α and β. In other words, α and β anti-commute if they are both odd-dimensional classes, and they strictly commute otherwise. Ultimately, the graded commutativity arises from the fact that the twist map

$$S^1 \wedge S^1 \to S^1 \wedge S^1$$

has degree -1.

Remark 18.6.1. The graded commutativity of the classical stable homotopy groups has non-trivial consequences for the structure of the Adams spectral sequence. For example, consider the third Hopf map σ in π_7, which is detected by h_3 in the Adams spectral sequence. Graded commutativity implies that $2\sigma^2$ must equal zero in π_{14}. Now $h_0h_3^2$ detects $2\sigma^2$, and it is non-zero in the Adams E_2-page. Therefore, there must be an Adams differential $d_2(h_4) = h_0h_3^2$. This is the first differential in the Adams spectral sequence.

In the motivic situation, graded commutativity takes a slightly more complicated form. The twist map

$$S^{1,0} \wedge S^{1,0} \to S^{1,0} \wedge S^{1,0}$$

represents -1 in $\pi^k_{0,0}$, but the twist map

$$S^{1,1} \wedge S^{1,1} \to S^{1,1} \wedge S^{1,1}$$

represents a different element, usually called ε, in $\pi^k_{0,0}$. This description of the twist maps leads to the following theorem on motivic graded commutativity.

Theorem 18.6.2. ([23, Proposition 1.18], [25, Proposition 2.5], [65, Corollary 6.1.2]) *Let k be any base field. For α in $\pi^k_{a,b}$ and β in $\pi^k_{c,d}$,*

$$\alpha\beta = (-1)^{(a-b)(c-d)}\varepsilon^{bd}\beta\alpha.$$

Remark 18.6.3. Classically, the element 2 of π_0 plays the role of the "zeroth Hopf map". Motivically, it is $1-\varepsilon$ that plays this role. The cofiber of $1-\varepsilon$ is a 2-cell complex. In the cohomology of this 2-cell complex, there is a Sq^1 operation that connects the two cells. In the cofiber of 2, there is a Sq^1 operation, but there is also a non-trivial Sq^2 operation when ρ is non-zero.

18.6.2 Milnor-Witt K-theory

In this section, we will recall the work of Morel on the motivic stable homotopy groups $\pi^k_{n,n}$ over arbitrary base fields k. For more details, see [65] [66] [68]. For related work, see also [22] [38, Appendix] [74] [96].

Definition 18.6.4. The Milnor-Witt K-theory $K^{MW}_*(k)$ of a field k is the graded associative ring generated by elements $[u]$ for all u in $k^\times$, and the element η, subject to the relations

1. $[u][1-u] = 0$ for all u in k except for 0 and 1.
2. $[uv] = [u] + [v] + \eta[u][v]$ for all u and v in $k^\times$.
3. $[u]\eta = \eta[u]$ for all u in $k^\times$.
4. $\eta(\eta[-1] + 2) = 0$.

The degree of $[u]$ is 1, and the degree of η is -1.

Setting $\eta = 0$ in Milnor-Witt K-theory recovers ordinary Milnor K-theory (see Section 18.2.1), so we can view Milnor-Witt K-theory as a kind of deformation of Milnor K-theory. The first relation in Definition 18.6.4 is precisely the Steinberg relation, while the second relation is analogous to the additivity relation in Milnor K-theory, with an error term involving η.

Theorem 18.6.5. ([65, Theorem 6.4.1],[66, Theorem 6.2.1], [68, Corollary 1.25]) *For any field k, the motivic stable homotopy group $\pi^k_{n,n}$ is isomorphic to the nth Milnor-Witt K-theory group $K^{MW}_n(k)$.*

Note, in particular, that $\pi^k_{0,0}$ is isomorphic to the Grothendieck-Witt ring $GW(k)$ of quadratic forms over k. Unlike most of the other results in this article, there is no need for completions in the statement of Theorem 18.6.5.

The element ε that governs graded commutativity corresponds here to $-1 - \eta[-1]$, so the last relation of Definition 18.6.4 is equivalent to the relation $\eta(1 - \varepsilon) = 0$. The element ρ in $\pi_{-1,-1}$ corresponds to $[-1]$, as in Section 18.2.1.

18.6.3 The first Milnor-Witt stem

For simplicity let k be a field of characteristic zero. By the work reviewed in Section 18.6.2 a next logical step is to compute the first Milnor-Witt stem $\Pi^k_1 = \bigoplus_{n\in\mathbb{Z}} \pi^k_{n+1,n}$. One of the major inspirations for this calculation is Morel's π_1-conjecture in weight zero, which states that there is a short exact sequence

$$0 \to K^M_2(k)/24 \to \pi^k_{1,0} \to k^\times/2 \oplus \mathbb{Z}/2 \to 0. \tag{18.6.1}$$

This conjecture is proved in [83].

The kernel $K^M_2(k)/24$ in (18.6.1) is generated by the second motivic Hopf map ν in $\pi^k_{3,2}$, in the sense that its elements are all of the form $\alpha\nu$, where α is an element of $\pi^k_{-2,-2}$. Such elements α correspond to elements of $K^M_2(k)$, as in Theorem 18.6.5.

On the other hand, the image $k^\times/2 \oplus \mathbb{Z}/2$ has two generators. The second factor is generated by η_{top}, i.e., the image of the classical first Hopf map in $\pi^k_{1,0}$. The first factor is generated by $\eta\eta_{\text{top}}$, in the sense that its elements are all of the form $\alpha\eta\eta_{\text{top}}$, where α is an element of $\pi^k_{-1,-1}$. These generators are subject to the relations $24\nu = 0$ and $12\nu = \eta^2\eta_{\text{top}}$, which are related to the corresponding classical relations $24\nu = 0$ and $12\nu = \eta^3$ in π_3.

It turns out the surjection in (18.6.1) arises from the unit map for the hermitian K-theory spectrum $\mathbf{KQ}$ of quadratic forms. One may speculate that its kernel is the image of a motivic J-homomorphism $K^M_2(k) = \pi_{1,0}GL \to \pi_{1,0}\mathbf{1}$ for the general linear group GL. The Hopf construction should witness that ν is in the image of a motivic J-homomorphism, so the relation $24\nu = 0$ may be a shadow of some motivic version of the Adams conjecture.

More generally, for every $n \in \mathbb{Z}$, there is an exact sequence [83]

$$0 \to K^M_{2-n}/24(k) \to \pi^k_{n+1,n} \to \pi_{n+1,n}\mathsf{f}_0\mathbf{KQ}. \tag{18.6.2}$$

Here $\mathsf{f}_0\mathbf{KQ}$ is the effective slice cover of hermitian K-theory (see Section 18.3). Note that the homotopy groups of $\mathbf{KQ}$ and $\mathsf{f}_0\mathbf{KQ}$ agree in nonnegative weight. The rightmost map in (18.6.2) is surjective for $n \geq -4$. In fact, the rightmost map is surjective for all n if $\mathsf{f}_0\mathbf{KQ}$ is replaced by the very effective slice cover $\mathbf{kq} = \tilde{\mathsf{f}}_0\mathbf{KQ}$ [84].

The proof of (18.6.2) is achieved by performing calculations with the effective slice spectral sequence for the sphere $\mathbf{1}$, converging conditionally to the homotopy of the η-completion $\mathbf{1}^{\wedge}_{\eta}$. For all integers n, there is a canonically induced isomorphism

$$\pi_{n+1,n}\mathbf{1} \xrightarrow{\cong} \pi_{n+1,n}\mathbf{1}^{\wedge}_{\eta}$$

noted in [83, Corollary 5.2].

The exact sequence (18.6.2) vastly generalizes computations in [79] for fields of cohomological dimension at most two, i.e., if Milnor K-theory is concentrated in degrees 0, 1, and 2. Examples include algebraically closed fields, finite fields, p-adic fields, and totally imaginary number fields (but not $\mathbb{Q}$ or $\mathbb{R}$).

It is interesting to compare the above with the computations of unstable motivic homotopy groups of punctured affine spaces in [7] and [8]. If $d > 3$, the extension for $\pi_d(\mathbf{A}^d \smallsetminus \{0\})$ conjectured by Asok-Fasel [28, Conjecture 7, p. 1894] coincides with (18.6.2). As noted in [7], the sheaf version of the exact sequence (18.6.2) and a conjectural Freudenthal $\mathbb{P}^1$-suspension theorem imply Murthy's conjecture on splittings of vector bundles [7, Conjecture 1].

18.6.4 Finite fields

This section describes work of Wilson and Østvær [107] on stable motivic homotopy groups over finite fields of order $q = p^n$, where p is an odd prime.

As always, the starting point is the motivic cohomology of a point. Using Example 18.2.6 and Theorem 18.2.7, we obtain that $H^{*,*}(\mathbb{F}_q; \mathbb{F}_2) = \mathbb{F}_2[u]/u^2$.

At this point, the discussion splits into two cases. If $p \equiv 1 \bmod 4$, then -1 is a square in $\mathbb{F}_q$, and ρ equals 0. Consequently, Ext calculations (i.e., motivic Adams E_2-pages) are essentially the same as in the $\mathbb{C}$-motivic case discussed in Section 18.2.3.

On the other hand, if $p \equiv 3 \bmod 4$, then u can be taken to be ρ. The Ext groups in this case can be computed with the ρ-Bockstein spectral sequence as in Section 18.2.4.

However, there are significant differences to the $\mathbb{R}$-motivic case that arise from the relation $\rho^2 = 0$.

In either case, there are interesting motivic Adams differentials for finite fields that have no $\mathbb{C}$-motivic nor $\mathbb{R}$-motivic analogue [107, Corollary 7.12]. More specifically, there are differentials of the form

$$d_r(\tau^k) = u\tau^{k-1}h_0^r$$

for some values of r and k that depend on the order q of the finite field $\mathbb{F}_q$. These differentials are remarkable because they occur already at the very beginning of the spectral sequence in the 0-stem! The proofs of these differentials depend on a priori knowledge of the motivic cohomology of $\mathbb{F}_q$ via its étale cohomology [89], as discussed in Section 18.7.1.

An analysis of the Adams spectral sequence leads to an isomorphism

$$\pi^{\mathbb{F}_q}_{n,0} \cong \pi^s_n \oplus \pi^s_{n+1}.$$

for $0 \leq n \leq 18$. In particular, the groups $\pi^{\mathbb{F}_q}_{4,0}$ and $\pi^{\mathbb{F}_q}_{12,0}$ are trivial.

18.6.5 η-periodic phenomena

Relatively little is known about η-periodic phenomena over fields other than $\mathbb{C}$ and $\mathbb{R}$. Sections 18.4.2 and 18.5.1 describe the η-periodic groups $\pi^{\mathbb{C}}_{*,*}[\eta^{-1}]$ and $\pi^{\mathbb{R}}_{*,*}[\eta^{-1}]$. Recent work of Wilson [106] calculates $\pi^{k}_{*,*}[\eta^{-1}]$ for $\mathbb{Q}$ and for fields of cohomological dimension at most 2, i.e., for fields whose Milnor K-theory vanishes above degree 2. All finite fields satisfy this condition. The η-periodic groups are described in terms of the Witt group $W(k)$ of quadratic forms over k.

For more general fields, the following theorem summarizes what we do know. Recall the notation

$$\Pi^k_m = \bigoplus_{n \in \mathbb{Z}} \pi^k_{m+n,n}$$

from Section 18.5.1.

Theorem 18.6.6. [3] *Let k be a field such that* $\operatorname{char} k \neq 2$. *Then*

$$\Pi^k_m[\eta^{-1}] \otimes \mathbb{Q} = 0$$

for all $m > 0$.

Theorem 18.6.6 leaves open the question of torsion in $\Pi^k_m[\eta^{-1}]$. This torsion is likely to be quite interesting.

Additionally, we have some low-dimensional information.

Theorem 18.6.7. [82] *Let k be a field such that* $\operatorname{char} k \neq 2$. *Then Π^k_1 and Π^k_2 are both zero.*

18.7 Other motivic spectra

One way to obtain information about stable motivic homotopy groups is to consider other motivic spectra that are equipped with unit maps from the motivic sphere spectrum. The homotopy groups of these other motivic spectra can give information about the motivic stable homotopy groups by passing along the unit map. We will discuss a few examples in this section.

Each motivic spectrum discussed below is of fundamental interest in its own right. Their associated cohomology theories detect interesting phenomena in algebraic geometry, but that is not the focus of this discussion. We will not discuss their construction and geometric origins because these topics go beyond the scope of this article.

The unit map for algebraic cobordism factors through $\mathbf{1}/\eta$, so every module over an oriented motivic ring spectrum is η-complete [85, Lemma 2.1].Hence the motivic Eilenberg-Mac Lane spectrum $\mathbf{M}\mathbb{Z}$, the K-theory spectrum **KGL** and its covers, and the truncated Brown-Peterson spectra $\mathbf{MBP}\langle n\rangle$ are all η-complete. On the other hand, hermitian K-theory **KQ** and higher Witt-theory **KW** do not coincide with their respective η-completions.

18.7.1 Motivic Eilenberg-Mac Lane spectra

Suppose ℓ is prime to the characteristic of the base field k. Our first example is the motivic Eilenberg-Mac Lane spectrum $\mathbf{M}\mathbb{Z}/\ell^\nu$ that represents motivic cohomology with $\mathbb{Z}/\ell^\nu$ coefficients [98] [87]. The stable homotopy groups of $\mathbf{M}\mathbb{Z}/\ell^\nu$ are the same as the cohomology of a point with coefficients in $\mathbb{Z}/\ell^\nu$, as described in Section 18.2.1. The mod-ℓ motivic Steenrod algebra is the ring of operations on $\mathbf{M}\mathbb{Z}/\ell$, as described in Section 18.2.3. This circle of ideas leads eventually to the motivic Adams spectral sequence of Section 18.2.5.

The Beilinson-Lichtenbaum conjecture, which is a consequence of the Milnor and Bloch-Kato conjectures, offers the following comparison isomorphism due to Voevodsky.

Theorem 18.7.1. [103] *For $p \leq q$ and X a smooth k-scheme, the étale sheafification functor induces an isomorphism*

$$H^{p,q}(X;\mathbb{Z}/\ell^\nu) \xrightarrow{\cong} H^p_{\acute{e}t}(X;\mu_{\ell^\nu}^{\otimes q}) \tag{18.7.1}$$

between motivic and étale cohomology groups, where μ_{ℓ^ν} is the sheaf of ℓ^νth roots of unity.

This important isomorphism identifies the mod-ℓ^ν motivic cohomology of k with the classical cohomology groups of the absolute Galois group of k [88].

Let e be the exponent of the multiplicative group $(\mathbb{Z}/\ell^\nu)^\times$ of units. The sheaf $\mu_{\ell^\nu}^{\otimes e}$ is constant, so $H^{0,e}(k;\mathbb{Z}/\ell^\nu)$ is isomorphic to $H^{0,0}(k;\mathbb{Z}/\ell^\nu) = \mathbb{Z}/\ell^\nu$. Let us choose a generator τ_{ℓ^ν} of $H^{0,e}(k;\mathbb{Z}/\ell^\nu)$. Using Theorem 18.7.1, we deduce an isomorphism

$$H^{p,q}(X;\mathbb{Z}/\ell^\nu)[\tau_{\ell^\nu}^{-1}] \cong H^p_{\acute{e}t}(X;\mu_{\ell^\nu}^{\otimes q}) \tag{18.7.2}$$

for all integers $p,q \in \mathbb{Z}$ and all smooth k-schemes X. The periodicity discussed here is related to Thomason's seminar paper [95], as well as to a recent generalization to motivic spectra such as algebraic cobordism [27].

In general, the motivic cohomology groups $H^{p,q}(k;\mathbb{Z}/\ell^\nu)$ do not have a simple description in terms of Milnor K-theory. The case $\nu = 1$ is discussed in Section 18.2.1. If k contains all ℓ^νth roots of unity, then the sheaf μ_{ℓ^ν} is trivial, and Theorem 18.7.1 gives a practial way of computing $H^{p,q}(k;\mathbb{Z}/\ell^\nu)$. One way to view these difficulties is that the ℓ-Bockstein spectral sequence that starts with $H^{*,*}(k;\mathbb{Z}/\ell)$ and converges to $H^{*,*}(k;\mathbb{Z}/\ell^\nu)$ is non-trivial.

In addition to $\mathbf{M}\mathbb{Z}/\ell$, we may also consider the motivic Eilenberg-Mac Lane spectrum $\mathbf{M}\mathbb{Z}$ that represents integral motivic cohomology. The cofiber sequence

$$\mathbf{M}\mathbb{Z} \xrightarrow{\ell^\nu} \mathbf{M}\mathbb{Z} \longrightarrow \mathbf{M}\mathbb{Z}/\ell^\nu$$

provides a tool for understanding $\mathbf{M}\mathbb{Z}$ in terms of $\mathbf{M}\mathbb{Z}/\ell^\nu$ for each rational prime ℓ. Its rational part $\mathbf{M}\mathbb{Q}$ identifies with its étale counterpart as in [101, Lemma 6.8].

18.7.2 Algebraic K-theory

Our next family of examples arises from K-theory. In this section, we will discuss the motivic analogues of the classical periodic K-theory spectrum **KU** and its connective cover **ku**.

The motivic spectrum **KGL** represents algebraic K-theory [98] in the following sense. The homotopy groups $\pi_{n,0}\mathbf{KGL}$ are isomorphic to the algebraic K-theory $K_n(k)$ of the base field k. In addition, there is a Bott element β in $\pi_{2,1}\mathbf{KGL}$, and **KGL** is periodic in β. This determines all of the stable homotopy groups of **KGL**, in terms of classical algebraic K-theory. The motivic spectrum **KGL** is analogous to **KU** in the classical case.

We next wish to consider the effective slice cover $\mathbf{kgl} = \mathsf{f}_0\mathbf{KGL}$, which is analogous to **ku** in the classical case. In the case of **KGL**, the very effective slice cover $\tilde{\mathsf{f}}_0\mathbf{KGL}$ coincides with **kgl** because **kgl** is already very effective [93, Corollary 5.13].In the specific $\mathbb{C}$-motivic case, there is another way to construct the same motivic spectrum **kgl** that is more in the spirit of the classical Postnikov tower [46].

The motivic spectrum **kgl** is best understood via its connection to the motivic truncated Brown-Peterson spectrum $\mathbf{MBP}\langle 1\rangle$, discussed below in Section 18.7.3.

18.7.3 Motivic Brown-Peterson spectra and truncations

Let **MBP** denote the motivic Brown-Peterson spectrum at the prime 2 over a characteristic 0 base field k constructed in [41] and [97]. This is the universal 2-typical oriented motivic ring spectrum. It turns out that the 2-localized effective slice cover $\mathbf{kgl}_{(2)}$ is the truncated Brown-Peterson spectrum $\mathbf{MBP}\langle 1\rangle$ sitting in a tower

$$\mathbf{MBP} = \mathbf{MBP}\langle\infty\rangle \to \cdots \to \mathbf{MBP}\langle n\rangle \to \mathbf{MBP}\langle n-1\rangle \to \cdots \to \mathbf{MBP}\langle 0\rangle$$

of **MBP**-modules. The maps in this tower come from the fact that $\mathbf{MBP}\langle n-1\rangle$ is the cofiber of the map

$$v_n : \mathbf{MBP}\langle n\rangle \to \mathbf{MBP}\langle n\rangle.$$

This picture is entirely analogous to the classical situation [51], in which $\mathbf{BP}\langle n-1\rangle$ is the cofiber of

$$\mathbf{BP}\langle n\rangle \to \mathbf{BP}\langle n\rangle.$$

The motivic spectrum $\mathbf{MBP}\langle 0\rangle = \mathbf{M}\mathbb{Z}_{(2)}$ is the 2-local motivic Eilenberg-Mac Lane spectrum by the theorem of Hopkins, Morel, and Hoyois [39]. (At odd primes, $\mathbf{MBP}\langle 0\rangle$ is an Adams summand of localized connective algebraic K-theory [73, §4].) For $n > 1$ we can view the groups $\pi_*\mathbf{MBP}\langle n\rangle$ as higher height generalizations of the algebraic K-theory groups of the base field.

In order to understand the homology of $\mathbf{MBP}\langle n\rangle$ as a comodule over A_* we employ the auxiliary Hopf algebroids

$$\begin{aligned}\mathcal{E}(n)_* &= (\mathbb{M}_2, A_*/(\xi_1, \xi_2, \ldots) + (\tau_{n+1}, \tau_{n+2}, \ldots)) \\ &= (\mathbb{M}_2, \mathbb{M}_2[\tau_0, \ldots, \tau_n]/(\tau_i^2 - \rho\tau_{i+1}, \tau_n^2)).\end{aligned}$$

We permit $n = \infty$, in which case

$$\begin{aligned}\mathcal{E}(\infty)_* &= (\mathbb{M}_2, A_*/(\xi_1, \xi_2, \ldots)) \\ &= (\mathbb{M}_2, \mathbb{M}_2[\tau_0, \tau_1, \ldots]/(\tau_i^2 - \rho\tau_{i+1})).\end{aligned}$$

By [77] there is an isomorphism of Hopf algebroids

$$\mathbf{M}\mathbb{Z}/2_*\mathbf{MBP}\langle n\rangle \cong A_* \,\square_{\mathcal{E}(n)_*}\, \mathbb{M}_2.$$

By a standard change-of-rings isomorphism, the Adams E_2-page for $\mathbf{MBP}\langle n\rangle$ identifies with

$$\mathrm{Ext}_{\mathcal{E}(n)}(\mathbb{M}_2, \mathbb{M}_2).$$

For explicit calculations of $\pi_*\mathbf{MBP}\langle n\rangle$ over $\mathbb{C}$, $\mathbb{R}$, $\mathbb{Q}_p$, and $\mathbb{Q}$ we refer to [78].

18.7.4 Hermitian K-theory

In this section, we consider the motivic versions of the classical real K-theory spectrum **KO** and its connective cover **ko**.

The motivic spectrum **KQ** represents Karoubi's hermitian K-theory [37]. Its effective slice cover $\mathsf{f}_0\mathbf{KQ}$ does not coincide with its very effective slice cover $\mathbf{kq} := \tilde{\mathsf{f}}_0\mathbf{KQ}$ studied in [4]. We note that **kq** and **kgl** are connected via the motivic Hopf map $\eta\colon \mathbb{A}^2 \smallsetminus \{0\} \to \mathbb{P}^1$ in the cofiber sequence

$$\Sigma^{1,1}\mathbf{kq} \xrightarrow{\eta} \mathbf{kq} \to \mathbf{kgl}. \tag{18.7.3}$$

An analogous cofiber sequence for the effective slice covers of **KQ** and **KGL** does not exist by the proof of [85, Corollary 5.1].By (18.7.3) the Betti realization of **kq** identifies with **ko** and one can calculate the mod-2 motivic homology $\mathbf{M}\mathbb{Z}/2_*\mathbf{kq}$ as

$$A_* \,\square_{A(1)_*}\, \mathbb{M}_2,$$

where $A(1)_*$ is the Hopf algebroid

$$\begin{aligned}A(1)_* &= (\mathbb{M}_2, A_*/(\xi_1^2, \xi_2, \xi_3 \ldots) + (\tau_2, \tau_3, \ldots)) \\ &= (\mathbb{M}_2, \mathbb{M}_2[\tau_0, \tau_1, \xi_1]/(\tau_0^2 + \tau\xi_1 + \rho\tau_1 + \rho\tau_0\xi_1, \xi_1^2, \tau_1^2).\end{aligned}$$

By change-of-rings, the Adams E_2-page for **kq** takes the form

$$\mathrm{Ext}^{*,*}_{A(1)}(\mathbf{M}_2, \mathbf{M}_2). \tag{18.7.4}$$

As usual, this Adams spectral sequence computes the homotopy groups of **kq** completed at 2 and η. For explicit calculations with (18.7.4) over $\mathbb{C}$ and $\mathbb{R}$, we refer to [46] and [35].

18.7.5 Higher Witt theory

In this section we review the proof of Milnor's conjecture on quadratic forms based on the effective slice spectral sequence for higher Witt-theory **KW** [86]. Recall that **KW** is defined by inverting η on **KQ**.

Suppose k is a field of characteristic unequal to 2. Recall the Milnor K-theory of k described in Section 18.2.1. In degrees zero, one, and two, these groups agree with Quillen's K-groups, but for higher degrees they differ in general. Milnor [62] proposed two conjectures relating $K^M_*(k)/2$ to the mod-2 Galois cohomology ring $H^*(F;\mu_2)$ and the graded Witt ring $GrW^I_*(k) = \oplus_{q\geq 0} I(k)^q/I(k)^{q+1}$ for the fundamental ideal $I(k)$ of even dimensional forms in the Witt ring $W(k)$, in the form of two graded ring homomorphisms:

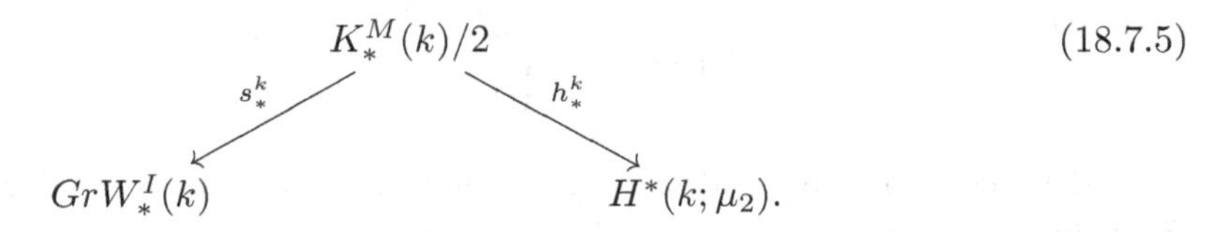

(18.7.5)

The Milnor conjecture on Galois cohomology states that h^k_* is an isomorphism, while the Milnor conjecture on quadratic forms states that s^k_* is an isomorphism. The proofs of both conjectures [101] [76] are two striking applications of motivic homotopy theory.

The slices of **KW** were described in Part (3) of Theorem 18.3.4. We record the first differentials in the effective slice spectral sequence for **KW** as maps between motivic spectra. The differential

$$d_1^{\mathbf{KW}}(q)\colon \mathsf{s}_q\mathbf{KW} \to \Sigma^{1,0}\mathsf{s}_{q+1}\mathbf{KW}$$

is a map of the form

$$\bigvee_{i\in\mathbb{Z}} \Sigma^{2i+q,q}\mathbf{M}\mathbb{Z}/2 \to \Sigma^{2,1}\bigvee_{j\in\mathbb{Z}} \Sigma^{2j+q,q}\mathbf{M}\mathbb{Z}/2.$$

Let $d_1^{\mathbf{KW}}(q,i)$ denote the restriction of $d_1^{\mathbf{KW}}(q)$ to the ith summand $\Sigma^{2i+q,q}\mathbf{M}\mathbb{Z}/2$ of $\mathsf{s}_q\mathbf{KW}$. By comparing with motivic cohomology operations of weight one, it suffices to consider $d_1^{\mathbf{KW}}(q,i)$ as a map from $\Sigma^{2i+q,q}\mathbf{M}\mathbb{Z}/2$ to

$$\Sigma^{2i+q+4,q+1}\mathbf{M}\mathbb{Z}/2 \vee \Sigma^{2i+q+2,q+1}\mathbf{M}\mathbb{Z}/2 \vee \Sigma^{2i+q,q+1}\mathbf{M}\mathbb{Z}/2.$$

The latter map affords the closed formula

$$d_1^{\mathbf{KW}}(q,i) = \begin{cases} (\mathrm{Sq}^3\mathrm{Sq}^1, \mathrm{Sq}^2, 0) & i-2q \equiv 0 \bmod 4 \\ (\mathrm{Sq}^3\mathrm{Sq}^1, \mathrm{Sq}^2+\rho\mathrm{Sq}^1, \tau) & i-2q \equiv 2 \bmod 4. \end{cases} \tag{18.7.6}$$

This sets the stage for the proof by Röndigs and the second author [86] of Milnor's conjecture on quadratic forms formulated in [62, Question 4.3].For fields of characteristic zero this conjecture was first shown by Orlov, Vishik and Voevodsky in [76], and by Morel [63] using different approaches.

According to Part (3) of Theorem 18.3.4, the effective slice spectral sequence for **KW** fills out the entire upper half-plane. A strenuous computation using (18.7.6), Adem relations, and the action of the Steenrod squares on the mod-2 motivic cohomology ring of k shows that it collapses. A key point from Theorem 18.2.7 is that if $0 \leq p \leq q$ and a belongs to $H^{p,q}(k;\mathbb{F}_2)$, then a equals $\tau^{q-p}c$ for some c in $H^{p,p}(k;\mathbb{F}_2)$. The action of the Steenrod operations in weight at most 1 is then completely described by the formulas

$$\mathrm{Sq}^1(\tau^n c) = \begin{cases} \rho\tau^{n-1}c & n \equiv 1 \bmod 2 \\ 0 & n \equiv 0 \bmod 2 \end{cases}$$

$$\mathrm{Sq}^2(\tau^n c) = \begin{cases} \rho^2\tau^{n-1}c & n \equiv 2,3 \bmod 4 \\ 0 & n \equiv 0,1 \bmod 4. \end{cases}$$

With the above in hand, a long calculation shows that the effective slice E_2-page for **KW** in degree (p,q) equals $H^{p,p}(k;\mathbb{F}_2)$ if p mod 4, and equals zero otherwise.]

To connect this computation with the theory of quadratic forms, one shows the spectral sequence converges to the filtration of the Witt ring $W(k)$ by the powers of the fundamental ideal $I(k)$ of even dimensional forms. By identifying motivic cohomology with Galois cohomology for fields we arrive at the following result.

Theorem 18.7.2. [86, Theorem 1.1] *When* $\mathrm{char}(k) \neq 2$, *the effective slice spectral sequence for* **KW** *converges and furnishes a complete set of invariants*

$$I(k)^q/I(k)^{q+1} \xrightarrow{\cong} H^q(k;\mu_2)$$

for quadratic forms over k *with values in the mod-2 Galois cohomology ring.*

18.8 Future directions

The purpose of this section is to encourage further research into motivic stable homotopy groups by describing some specific projects.

Problem 18.8.1. Classical periodicity consists of existence results, such as the existence of Morava K-theories and the existence of v_n-self-maps; and uniqueness results, such as the fact that a finite complex possesses a unique v_n-self-map (at least up to taking powers). In the $\mathbb{C}$-motivic situation, many of the analogous existence results have been established.

On the other hand, much work remains to be done on the uniqueness results. In particular, it is not known whether there are any periodicities in addition to those discovered by Andrews [5], Gheorghe [29], and Krause [52]. Also, it turns out that a finite complex can possess more than one type of periodic self-map. It is not yet understood how periodicities of different types can co-exist, nor what that means for the structure of $\mathbb{C}$-motivic stable homotopy groups.

To date, there is little work on exotic nilpotence and periodicity over other base fields. At least some of the same phenomena must occur. See [50] and [38] for some first results.

Problem 18.8.2. Currently available techniques allow for many more $\mathbb{R}$-motivic computations than have so far been carried out. Preliminary data suggests many interesting connections to other aspects of stable homotopy theory. In particular, it appears that the $\mathbb{R}$-motivic image of J has order greater than the classical image of J [13]. A full accounting of this phenomenon is needed. Both the motivic Adams spectral sequence and the slice spectral sequence ought to be important tools.

The $\mathbb{R}$-motivic stable homotopy category $\mathbf{SH}(\mathbb{R})$ appears to be closely related to the C_2-equivariant stable homotopy category. Data suggests that the C_2-equivariant stable homotopy category is the same as "τ-periodic $\mathbb{R}$-motivic stable homotopy theory".

Recent on-going work of Behrens and Shah addresses this issue.

Bruner and Greenlees [20] showed how to reformulate the classical Mahowald invariant [12] in terms of C_2-equivariant stable homotopy groups. The $\mathbb{R}$-motivic stable homotopy groups are somewhat easier to study than the C_2-equivariant stable homotopy groups, and they also ought to be useful for Mahowald invariants.

Problem 18.8.3. In classical homotopy, the topological modular forms spectrum *tmf* is an essential tool for studying stable homotopy groups. The cohomology of *tmf* is $A//A(2)$, where A is the classical Steenrod algebra and $A(2)$ is the subalgebra of A generated by Sq^1, Sq^2, and Sq^4. This means that the homotopy groups of *tmf* can be computed by an Adams spectral sequence whose $E(2)$-page is the cohomology of $A(2)$. The Adams spectral sequence for *tmf* can be completely analyzed, and this analysis provides much information about the classical stable homotopy groups.

It is likely that a similar story plays out in the motivic situation. We would like to know that a "motivic modular forms" spectrum *mmf* exists over an arbitrary field k. The motivic stable homotopy groups of *mmf* ought to be computable by an Adams spectral sequence whose E_2-page is the cohomology of the subalgebra $A(2)$ of the k-motivic Steenrod algebra A.

Problem 18.8.4. Formula (18.3.7) describes the E_1-page of the effective slice spectral sequence, in terms of the classical Adams-Novikov E_2-page and in terms of motivic cohomology with coefficients in $\mathbb{Z}$ and $\mathbb{Z}/2^n$. Preliminary calculations indicate that this E_1-page has interesting "exotic" products that are not simply seen by the classical Adams-Novikov E_2-page. This product structure deserves careful study in low dimensions. See Remark 18.3.3 for more discussion.

Problem 18.8.5. The E_1-page of the effective slice spectral sequence can be described over a general base field k, at least in terms of the Milnor K-theory of k. Some differentials have been understood in general in very low dimensions [83]. These results about effective slice differentials ought to be accessible in a larger range of dimensions, at least relative to arithmetic input from k.

Problem 18.8.6. Section 18.7.5 discusses how the effective slice filtration for $\mathbf{KW}$ informs us about quadratic forms over a field k. It is possible that these ideas can be extended to study quadratic forms over smooth k-schemes.

Bibliography

[1] J. F. Adams. On the groups $J(X)$. IV. *Topology*, 5:21–71, 1966.

[2] J. F. Adams. *Stable homotopy and generalised homology.* Chicago Lectures in Mathematics. University of Chicago Press, 1974.

[3] A. Ananyevskiy, M. Levine, and I. Panin. Witt sheaves and the η-inverted sphere spectrum. *J. Topol.*, 10(2):370–385, 2017.

[4] A. Ananyevskiy, O. Röndigs, and P. A. Østvær On very effective hermitian K-theory. *Math. Z.* https://doi.org/10.1007/s00209-019-02302-z, 2019.

[5] M. J. Andrews. New families in the homotopy of the motivic sphere spectrum. *Proc. Amer. Math. Soc.*, 146(6):2711–2722, 2018.

[6] M. J. Andrews and H. Miller. Inverting the Hopf map. *J. Topol.*, 10(4):1145–1168, 2017.

[7] A. Asok and J. Fasel Splitting vector bundles outside the stable range and A^1-homotopy sheaves of punctured affine spaces. *J. Amer. Math. Soc.*, 28(4):1031–1062, 2015.

[8] A. Asok, K. Wickelgren, and B. Williams The simplicial suspension sequence in $\mathbb{A}^1$-homotopy. *Geom. Topol.*, 21(4):2093–2160, 2017.

[9] M. F. Atiyah and F. Hirzebruch Vector bundles and homogeneous spaces. In Proc. Sympos. Pure Math., Vol. III, pages 7–38. Amer. Math. Soc., 1961.

[10] T. Bachmann Motivic and real étale stable homotopy theory. *Compos. Math.*, 154(5):883–917, 2018.

[11] M. G. Barratt, M. E. Mahowald, and M. C. Tangora Some differentials in the Adams spectral sequence. ii. *Topology*, 9:309–316, 1970.

[12] M. Behrens Some root invariants at the prime 2. In Proceedings of the Nishida Fest (Kinosaki 2003). Geom. Topol. Monogr. 10, 2007, pages 1–40.

[13] E. Belmont and D. C. Isaksen The $\mathbb{R}$-motivic Adams spectral sequence. In preparation.

[14] J. M. Boardman Conditionally convergent spectral sequences. Homotopy invariant algebraic structures (Baltimore, MD, 1998) Contemp. Math., 239, Amer. Math. Soc., 1999. pages 49–84.

[15] S. Borghesi Algebraic Morava K-theories. *Invent. Math.*, 151(2):381–413, 2003.

[16] R. R. Bruner A new differential in the Adams spectral sequence. *Topology*, 23(3):271–276, 1984.

[17] R. R. Bruner Calculation of large Ext modules. Computers in geometry and topology (Chicago, IL, 1986), Lecture Notes in Pure and Appl. Math. 114, Dekker, 1989. pages 79–104.

[18] R. R. Bruner Ext in the nineties. Algebraic topology (Oaxtepec, 1991), Contemp. Math., 146, Amer. Math. Soc., 1993. pages 71–90.

[19] R. R. Bruner The cohomology of the mod 2 Steenrod algebra: A computer calculation. Available at http://www.math.wayne.edu/~rrb/papers/cohom.pdf, 1997.

[20] R. R. Bruner and J. Greenlees The Bredon-Löffler conjecture. *Experiment. Math.*, 4(4):289–297, 1995.

[21] C. L. Douglas, J. Francis, A. G. Henriques, and M. A. Hill, editors. *Topological modular forms*, Math. Surv. Monogr., 201. Amer. Math. Soc., 2014.

[22] A. Druzhinin and J. I. Kylling. Framed correspondences and the zeroth stable motivic homotopy group in odd characteristic. Preprint, 2018.

[23] D. Dugger Coherence for invertible objects and multigraded homotopy rings. *Algebr. Geom. Topol.*, 14(2):1055–1106, 2014.

[24] D. Dugger and D. C. Isaksen. The motivic Adams spectral sequence. *Geom. Topol.*, 14(2):967–1014, 2010.

[25] D. Dugger and D. C. Isaksen. Motivic Hopf elements and relations. *New York J. Math.*, 19:823–871, 2013.

[26] D. Dugger and D. C. Isaksen. Low-dimensional Milnor-Witt stems over $\mathbb{R}$. *Ann. K-Theory*, 2(2):175–210, 2017.

[27] E. Elmanto, M. Levine, M. Spitzweck, and P. A. Østvær. Motivic Landweber exact theories and étale cohomology. Preprint, 2017.

[28] T. Geisser, A. Huber-Klawitter, U. Jannsen, and M. Levine, organizers. Algebraic K-theory and motivic cohomology. *Oberwolfach Rep.*, 10(2):1861–1913, 2013. Abstracts from the workshop held June 23–29, 2013.

[29] B. Gheorghe. Exotic motivic periodicities. arXiv:1709.00915, 2017.

[30] B. Gheorghe. The motivic cofiber of τ. *Doc. Math.* 23 (2018), 1077–1127.

[31] B. Gheorghe, G. Wang, and Z. Xu. The special fiber of the motivic deformation of the stable homotopy category is algebraic. arXiv:1809.09290, 2018.

[32] B. J. Guillou and D. C. Isaksen. The η-local motivic sphere. *J. Pure Appl. Algebra*, 219(10):4728–4756, 2015.

[33] B. J. Guillou and D. C. Isaksen. The η-inverted $\mathbb{R}$-motivic sphere. *Algebr. Geom. Topol.*, 16(5):3005–3027, 2016.

[34] J. J. Gutiérrez, O. Röndigs, M. Spitzweck, and P. A. Østvær, Motivic slices and coloured operads. *J. Topol.*, 5(3):727–755, 2012.

[35] M. A. Hill Ext and the motivic Steenrod algebra over $\mathbb{R}$. *J. Pure Appl. Algebra*, 215(5):715–727, 2011.

[36] M. A. Hill, M. J. Hopkins, and D. C. Ravenel. On the nonexistence of elements of Kervaire invariant one. *Ann. of Math. (2)*, 184(1):1–262, 2016.

[37] J. Hornbostel. $\mathbb{A}^1$-representability of Hermitian K-theory and Witt groups. *Topology*, 44(3):661–687, 2005.

[38] J. Hornbostel. Some comments on motivic nilpotence. *Trans. Amer. Math. Soc.*, 370(4):3001–3015, 2018. With an appendix by Marcus Zibrowius.

[39] M. Hoyois. From algebraic cobordism to motivic cohomology. *J. Reine Angew. Math.*, 702:173–226, 2015.

[40] M. Hoyois, S. Kelly, and P. A. Østvær. The motivic Steenrod algebra in positive characteristic. *J. Eur. Math. Soc. (JEMS)*, 19(12):3813–3849, 2017.

[41] P. Hu and I. Kriz. Some remarks on real and algebraic cobordism. *K-Theory*, 22(4):335–366, 2001.

[42] P. Hu, I. Kriz, and K. Ormsby. Convergence of the motivic Adams spectral sequence. *J. K-Theory*, 7(3):573–596, 2011.

[43] P. Hu, I. Kriz, and K. Ormsby. Remarks on motivic homotopy theory over algebraically closed fields. *J. K-Theory*, 7(1):55–89, 2011.

[44] D. C. Isaksen. Stable stems. arXiv:1407.8418, 2014.

[45] D. C. Isaksen. Classical and motivic Adams charts. arXiv:1408.0248, 2014.

[46] D. C. Isaksen and A. Shkembi. Motivic connective K-theories and the cohomology of $A(1)$. *J. K-Theory*, 7(3):619–661, 2011.

[47] D. C. Isaksen, G. Wang, and Z. Xu. More stable stems. In preparation.

[48] D. C. Isaksen and Z. Xu. Motivic stable homotopy and the stable 51 and 52 stems. *Topology Appl.*, 190:31–34, 2015.

[49] J. F. Jardine. Motivic symmetric spectra. *Doc. Math.*, 5:445–552, 2000.

[50] R. Joachimi. Thick ideals in equivariant and motivic stable homotopy categories. arXiv:1503.08456, 2015.

[51] D. C. Johnson and W. S. Wilson. Projective dimension and Brown-Peterson homology. *Topology*, 12:327–353, 1973.

[52] A. Krause. Periodicity in motivic homotopy theory and over BP_*BP. Preprint, 2018.

[53] J. I. Kylling and G. M. Wilson. Strong convergence in the motivic Adams spectral sequence. arXiv:1901.03399, 2018.

[54] M. Levine. The homotopy coniveau tower. *J. Topol.*, 1(1):217–267, 2008.

[55] M. Levine. A comparison of motivic and classical stable homotopy theories. *J. Topol.*, 7(2):327–362, 2014.

[56] B. A. Magurn. *An algebraic introduction to K-theory*, Encyclopedia Math. Appl. 87. Cambridge University Press, 2002.

[57] M. Mahowald and M. Tangora. Some differentials in the Adams spectral sequence. *Topology*, 6:349–369, 1967.

[58] L. Mantovani. Localizations and completions in motivic homotopy theory. Preprint, 2018.

[59] J. P. May. The cohomology of restricted Lie algebras and of Hopf algebras. *Bull. Amer. Math. Soc.*, 71:372–377, 1965.

[60] H. R. Miller. *Some algebraic aspects of the Adams-Novikov spectral sequence*. Ph.D. Thesis, Princeton University. ProQuest LLC, Ann Arbor, MI, 1975.

[61] J. Milnor. On the cobordism ring Ω^* and a complex analogue. I. *Amer. J. Math.*, 82:505–521, 1960.

[62] J. Milnor. Algebraic K-theory and quadratic forms. *Invent. Math.*, 9:318–344, 1969/1970.

[63] F. Morel. Suite spectrale d'Adams et invariants cohomologiques des formes quadratiques. *C. R. Acad. Sci. Paris Sér. I Math.*, 328(11):963–968, 1999.

[64] F. Morel. Théorie homotopique des schémas. *Astérisque* 256, 1999.

[65] F. Morel. An introduction to $\mathbb{A}^1$-homotopy theory. Contemporary developments in algebraic K-theory, ICTP Lect. Notes, XV, Abdus Salam Int. Cent. Theoret. Phys., Trieste. Pages 357–441. 2004.

[66] F. Morel. On the motivic π_0 of the sphere spectrum. Axiomatic, enriched and motivic homotopy theory, NATO Sci. Ser. II Math. Phys. Chem., 131, Kluwer Acad. Publ., Dordrecht, 2004. Pages 219–260. 2004.

[67] F. Morel. Rationalized motivic sphere spectrum and rational motivic cohomology. Preprint, 2006.

[68] F. Morel. *$\mathbb{A}^1$-algebraic topology over a field, Lecture Notes in Math.* 2052. Springer, 2012.

[69] F. Morel and V. Voevodsky. $\mathbf{A}^1$-homotopy theory of schemes. *Inst. Hautes Études Sci. Publ. Math.*, 90:45–143 (2001), 1999.

[70] R. M. F. Moss. Secondary compositions and the Adams spectral sequence. *Math. Z.*, 115:283–310, 1970.

[71] C. Nassau. `http://www.nullhomotopie.de/charts/index.html`.

[72] N. Naumann, M. Spitzweck, and P. A. Østvær. Motivic Landweber exactness. *Doc. Math.*, 14:551–593, 2009.

[73] N. Naumann, M. Spitzweck, and P. A. Østvær. Existence and uniqueness of E_∞ structures on motivic K-theory spectra. *J. Homotopy Relat. Struct.*, 10(3):333–346, 2015.

[74] A. Neshitov. Framed correspondences and the Milnor-Witt K-theory. *J. Inst. Math. Jussieu*, 17(4):823–852, 2018.

[75] S. P. Novikov. Methods of algebraic topology from the point of view of cobordism theory. *Izv. Akad. Nauk SSSR Ser. Mat.*, 31:855–951, 1967.

[76] D. Orlov, A. Vishik, and V. Voevodsky. An exact sequence for $K^m_*/2$ with applications to quadratic forms. *Ann. of Math. (2)*, 165(1):1–13, 2007.

[77] K. M. Ormsby. Motivic invariants of p-adic fields. *J. K-Theory*, 7(3):597–618, 2011.

[78] K. M. Ormsby and P. A. Østvær. Motivic Brown-Peterson invariants of the rationals. *Geom. Topol.*, 17(3):1671–1706, 2013.

[79] K. M. Ormsby and P. A. Østvær. Stable motivic π_1 of low-dimensional fields. *Adv. Math.*, 265:97–131, 2014.

[80] P. Pstragowski. Synthetic spectra and the cellular motivic category. arXiv:1803.01804, 2018.

[81] D. C. Ravenel. *Complex cobordism and stable homotopy groups of spheres, Pure and Applied Mathematics* 121. Academic Press, Inc., 1986.

[82] O. Röndigs On the η-inverted sphere. *Tata Institute of Fundamental Research Publications* 19. arXiv:1602.08798.

[83] O. Röndigs, M. Spitzweck, and P. A. Østvær. The first stable homotopy groups of motivic spheres. *Ann. of Math. (2)* 189 (2019), no. 1, 1–74.

[84] O. Röndigs, M. Spitzweck, and P. A. Østvær. The second stable homotopy groups of motivic spheres. In prepration.

[85] O. Röndigs, M. Spitzweck, and P. A. Østvær. The motivic Hopf map solves the homotopy limit problem for K-theory. *Doc. Math.*, 23:1405–1424, 2018.

[86] O. Röndigs and P. A. Østvær. Slices of Hermitian K-theory and Milnor's conjecture on quadratic forms. *Geom. Topol.*, 20(2):1157–1212, 2016.

[87] O. Röndigs and P. A. Østvær. Modules over motivic cohomology. *Adv. Math.*, 219(2):689–727, 2008.

[88] J.-P. Serre, *Galois cohomology.* Springer Monogr. Math. Springer-Verlag, corrected reprint of the 1997 English edition, 2002. Translated from the French by Patrick Ion and revised by the author.

[89] C. Soulé. K-théorie des anneaux d'entiers de corps de nombres et cohomologie étale. *Invent. Math.*, 55(3):251–295, 1979.

[90] M. Spitzweck. Relations between slices and quotients of the algebraic cobordism spectrum. *Homology Homotopy Appl.*, 12(2):335–351, 2010.

[91] M. Spitzweck. A commutative $\mathbb{P}^1$-spectrum representing motivic cohomology over Dedekind domains. Mém. Soc. Math. Fr. (N.S.) No. 157 (2018).

[92] M. Spitzweck. Algebraic cobordism in mixed characteristic. arXiv:1404.2542, 2014.

[93] M. Spitzweck and P. A. Østvær. Motivic twisted K-theory. *Algebr. Geom. Topol.*, 12(1):565–599, 2012.

[94] S.-T. Stahn. The motivic Adams-Novikov spectral sequence at odd primes over $\mathbb{C}$ and $\mathbb{R}$. arXiv:1606.06085, 2016.

[95] R. W. Thomason. Algebraic K-theory and étale cohomology. *Ann. Sci. École Norm. Sup. (4)*, 18(3):437–552, 1985.

[96] R. Thornton. The homogeneous spectrum of Milnor-Witt K-theory. *J. Algebra*, 459:376–388, 2016.

[97] G. Vezzosi. Brown-Peterson spectra in stable $\mathbb{A}^1$-homotopy theory. *Rend. Sem. Mat. Univ. Padova*, 106:47–64, 2001.

[98] V. Voevodsky. $\mathbb{A}^1$-homotopy theory. In *Proceedings of the International Congress of Mathematicians, Vol. I (Berlin, 1998)*, volume Extra Vol. I, 1998. Pages 579–604.

[99] V. Voevodsky. Open problems in the motivic stable homotopy theory. I. Motives, polylogarithms and Hodge theory, Part I (Irvine, CA, 1998), Int. Press Lect. Ser., 3, I, Int. Press, 2002. Pages 3–34.

[100] V. Voevodsky. A possible new approach to the motivic spectral sequence for algebraic K-theory. Recent progress in homotopy theory (Baltimore, MD, 2000), Contemp. Math., 293, Amer. Math. Soc., 2002. Pages 371–379.

[101] V. Voevodsky. Motivic cohomology with $\mathbf{Z}/2$-coefficients. *Publ. Math. Inst. Hautes Études Sci.*, 98:59–104, 2003.

[102] V. Voevodsky. Reduced power operations in motivic cohomology. *Publ. Math. Inst. Hautes Études Sci.*, 98:1–57, 2003.

[103] V. Voevodsky. On motivic cohomology with $\mathbf{Z}/l$-coefficients. *Ann. of Math. (2)*, 174(1):401–438, 2011.

[104] G. Wang and Z. Xu. On the uniqueness of the smooth structure of the 61-sphere. *Ann. of Math. (2)*, 186(2):501-580, 2017.

[105] C. A. Weibel. *The K-book*, *Grad. Stud. Math.* 145. Amer. Math. Soc., 2013.

[106] G. M. Wilson. The eta-inverted sphere over the rationals. *Algebr. Geom. Topol.*, 18(3):1857–1881, 2018.

[107] G. M. Wilson and P. A. Østvær. Two-complete stable motivic stems over finite fields. *Algebr. Geom. Topol.*, 17(2):1059–1104, 2017.

Department of Mathematics, Wayne State University, Detroit, MI 48202, U.S.A.

E-mail address: isaksen@wayne.edu

Department of Mathematics, University of Oslo, 0316 Oslo, Norway

E-mail address: paularne@math.uio.no

19

E_n-spectra and Dyer-Lashof operations

Tyler Lawson

19.1 Introduction

Cohomology operations are absolutely essential in making cohomology an effective tool for studying spaces. In particular, the mod-p cohomology groups of a space X are enhanced with a binary cup product, a Bockstein derivation, and Steenrod's reduced power operations; these satisfy relations such as graded-commutativity, the Cartan formula, the Adem relations, and the instability relations [90]. The combined structure of these cohomology operations is very effective in homotopy theory because of three critical properties.

These operations are natural. We can exclude the possibility of certain maps between spaces because they would not respect these operations.

These operations are constrained. We can exclude the existence of certain spaces because the cup product and power operations would be incompatible with the relations that must hold.

These operations are complete. Because cohomology is *representable*, we can determine all possible natural operations which take an n-tuple of cohomology elements and produce a new one. All operations are built, via composition, from these basic operations. All relations between these operations are similarly built from these basic relations.

In particular, this last property makes the theory reversible: there are mechanisms which take cohomology as input and converge to essentially complete information about homotopy theory in many useful cases, with the principal examples being the stable and unstable Adams spectral sequences. The stable Adams spectral sequence begins with the Ext-groups $\mathrm{Ext}(H^*(Y), H^*(X))$ in the category of modules with Steenrod operations and converges to the stable classes of maps from X to a p-completion of Y [1]. The unstable Adams spectral sequence is similar, but it begins with nonabelian Ext-groups that are calculated in the category of graded-commutative rings with Steenrod operations [19, 18].

Our goal is to discuss multiplicative homotopy theory: spaces, categories, or spectra with extra multiplicative structure. In this situation, we will see that the *Dyer–Lashof operations* play the role that the Steenrod operations did in ordinary homotopy theory.

In ordinary algebra, commutativity is an extremely useful *property* possessed by certain monoids and algebras. This is no longer the case in multiplicative homotopy theory

Mathematics Subject Classification. primary: 55P43; secondary 55S12, 55S20.

Key words and phrases. ring spectrum, Dyer-Lashof operations or category theory. In category theory, commutativity becomes *structure*: to give symmetry to a monoidal category C we must make a choice of a natural twist isomorphism $\tau\colon A \otimes B \to B \otimes A$. Moreover, there are more degrees of symmetry possible than in algebra because we can ask for weaker or stronger identities on τ. By asking for basic identities to hold we obtain the notion of a braided monoidal category, and by asking for very strong identities to hold we obtain the notion of a symmetric monoidal category. In homotopy theory and higher category theory we rarely have the luxury of imposing identities, and these become replaced by extra structure. One consequence is that there are many degrees of commutativity, parametrized by operads.

The most classical such structures arose geometrically in the study of iterated loop spaces. For a pointed space X, the n-fold loop space $\Omega^n X$ has algebraic operations parametrized by certain configuration spaces $\mathcal{E}_n(k)$, which assemble into an *E_n-operad*; moreover, there is a converse theorem due to Boardman–Vogtand May that provides a recognition principle for what structure on Y is needed to express it as an iterated loop space. As n grows, these spaces possess more and more commutativity, reflected algebraically in extra Dyer–Lashof operations on the homology H_*Y that are analogous to the Steenrod operations.

In recent years there is an expanding library of examples of ring spectra that only admit, or only naturally admit, these intermediate levels of structure between associativity and commutativity. Our goal in this chapter is to give an outline of the modern theory of highly structured ring spectra, particularly E_n ring spectra, and to give a toolkit for their study. One of the things that we would like to emphasize is how to usefully work in this setting, and so we will discuss useful tools that are imparted by E_n ring structures, such as operations on them that unify the study of Steenrod and Dyer–Lashof operations. We will also introduce the next stage of structure in the form of secondary operations. Throughout, we will make use of these operations to show that structured ring spectra are heavily constrained, and that many examples do not admit this structure; we will in particular discuss our proof in [47] that the 2-primary Brown–Peterson spectrum does not admit the structure of an E_∞ ring spectrum, answering an old question of May [60]. At the close we will discuss some ongoing directions of study.

19.2 Acknowledgements

The author would like to thank Andrew Baker, Tobias Barthel, Clark Barwick, Robert Bruner, David Gepner, Saul Glasman, Gijs Heuts, Nick Kuhn, Michael Mandell, Akhil Mathew, Lennart Meier, and Steffen Sagave for discussions related to this material, and Haynes Miller for a careful reading of an earlier draft of this paper. The author also owes a significant long-term debt to Charles Rezk for what understanding he possesses.

The author was partially supported by NSF grant 1610408 and a grant from the Simons Foundation. The author would like to thank the Isaac Newton Institute for Mathematical Sciences for support and hospitality during the programme HHH when work on this paper was undertaken. This work was supported by: EPSRC grant numbers EP/K032208/1 and EP/R014604/1.

19.3 Operads and algebras

Throughout this section, we will let C be a fixed symmetric monoidal topological category. For us, this means that C is enriched in the category S of spaces, that there is a functor $\otimes\colon \mathsf{C}\times\mathsf{C}\to\mathsf{C}$ of enriched categories, and that the underlying functor of ordinary categories is extended to a symmetric monoidal structure. We will write $\mathrm{Map}_{\mathsf{C}}(X,Y)$ for the mapping space between two objects, and $\mathrm{Hom}_{\mathsf{C}}(X,Y)$ for the underlying set. Associated to C there is the (ordinary) homotopy category $h\mathsf{C}$, with morphisms $[X,Y]=\pi_0\mathrm{Map}_{\mathsf{C}}(X,Y)$.

19.3.1 Operads

Associated to any object $X\in\mathsf{C}$ there is an *endomorphism operad* $\mathrm{End}_{\mathsf{C}}(X)$. The k'th term is

$$\mathrm{Map}_{\mathsf{C}}(X^{\otimes k},X),$$

with an operad structure given by composition of functors. For any operad $\mathcal{O}$, this allows us to discuss $\mathcal{O}$-algebra structures on the objects of C, maps of $\mathcal{O}$-algebras, and further structure.

If $\mathcal{O}$ is the associative operad $\mathcal{A}ssoc$, then $\mathcal{O}$-algebras are monoid objects in the symmetric monoidal structure on C. If $\mathcal{O}$ is the commutative operad $\mathcal{C}omm$, then $\mathcal{O}$-algebras are strictly commutative monoids in C. However, these operads are highly rigid and do not take any space-level structure into account. Mapping spaces allow us to encode many different levels of structure.

Example 19.3.1. There is a sequence of operads $\mathcal{A}_1\to\mathcal{A}_2\to\mathcal{A}_3\to\cdots$ built out of the Stasheff associahedra [89]. An $\mathcal{A}_2$-algebra has a unital binary multiplication; an $\mathcal{A}_3$-algebra has a chosen homotopy expressing associativity, and has Massey products; an $\mathcal{A}_4$-algebra has a homotopy expressing a juggling formula for Massey products; and so on. Moreover, each operad is simply built from the previous: extension from an $\mathcal{A}_{n-1}$-structure to an $\mathcal{A}_n$-structure roughly asks to extend a certain map $S^{n-3}\times X^n\to X$ to a map $D^{n-2}\times X^n\to X$ expressing an n-fold coherence law for the multiplication [3]. This gives $\mathcal{A}_n$ a *perturbative* property: if $X\to Y$ is a homotopy equivalence, then $\mathcal{A}_n$-algebra structures on one space can be transferred to the other.

Example 19.3.2. The colimit of the $\mathcal{A}_n$-operads is called $\mathcal{A}_\infty$, and it is equivalent to the associative operad. It satisfies a *rectification* property: In a well-behaved category like the category S of spaces or the category Sp of spectra, any $\mathcal{A}_\infty$-algebra is equivalent in the homotopy category of $\mathcal{A}_\infty$-algebras to an associative object.

Example 19.3.3. There is a sequence of operads $\mathcal{E}_1\to\mathcal{E}_2\to\mathcal{E}_3\to\cdots$, where the space $\mathcal{E}_n(k)$ is homotopy equivalent to the configuration space of unordered k-tuples of points in $\mathbb{R}^n$. These have various models, such as the *little cubes* or *little discs* operads. The $\mathcal{E}_1$-operad is equivalent to the associative operad, and the $\mathcal{E}_\infty$-operad is equivalent to the commutative operad. We refer to an algebra over any operad equivalent to $\mathcal{E}_n$ as an *E_n-algebra.* These play an important role in the *recognition principle* [62, 16]: given an E_n-algebra X we can construct an n-fold classifying space B^nX; and if the binary multiplication makes $\pi_0(X)$ into a group then $X \simeq \Omega^n B^n X$. The relation between E_n-algebra structures and iteration of the functor Ω is closely related to an additivity result of Dunn [25], who showed that E_{n+1}-algebras are equivalent to E_1-algebras in the category of E_n-algebras.

Example 19.3.4. Associated to a topological monoid M, there is an operad $\mathcal{O}_M$ whose only nonempty space is $\mathcal{O}_M(1) = M$. An algebra over this operad is precisely an object with M-action. This operad is usually not perturbative. However, M can be resolved by a *cellular* topological monoid $\widetilde{M} \to M$ such that $\mathcal{O}_{\widetilde{M}}$-algebras are perturbative and can be rectified to $\mathcal{O}_M$-algebras. This construction is a recasting of Cooke's obstruction theory for lifting homotopy actions of a group G to honest actions [24]; stronger versions of this were developed by Dwyer–Kan and Badzioch [26, 4].

Example 19.3.5. There is a free-forgetful adjunction between operads and symmetric sequences. Given any sequence of spaces Z_n with Σ_n-actions, we can construct an operad $\text{Free}(Z)$ such that a $\text{Free}(Z)$-algebra structure is the same as a collection of Σ_n-equivariant maps $Z_n \to \text{Map}_{\mathsf{C}}(A^{\otimes n}, A)$.

If, further, Z_1 is equipped with a chosen point e, we can construct an operad $\text{Free}(Z, e)$ such that a $\text{Free}(Z, e)$-algebra structure is the same as a $\text{Free}(Z)$-algebra structure such that e acts as the identity: $\text{Free}(Z, e)$ is a pushout of a diagram $\text{Free}(Z) \leftarrow \text{Free}(\{e\}) \to \text{Free}(\emptyset)$ of operads.

Example 19.3.6. In the previous example, let Z_2 be S^1 with the antipodal action of Σ_2 and let all other Z_n be empty, freely generating an operad $\mathcal{Q}_1$ that we call the *cup-1 operad*. A $\mathcal{Q}_1$-algebra is an object A with a Σ_2-equivariant map $S^1 \to \text{Map}_{\mathsf{C}}(A^{\otimes 2}, A)$. The Σ_2-equivariant cell decomposition of S^1 allows us to describe $\mathcal{Q}_1$-algebras as objects with a binary multiplication m and a chosen homotopy from the multiplication m to the multiplication in the opposite order $m \circ \sigma$. In particular, any homotopy-commutative multiplication lifts to a $\mathcal{Q}_1$-algebra structure.

In the category Sp of spectra, one of the main applications of E_n-algebras is that they have well-behaved categories of modules, whose homotopy categories are triangulated categories.

Theorem 19.3.7 (Mandell [57]). *An E_1-algebra R in Sp has a category of left modules LMod_R. An E_2-algebra structure on R makes the homotopy categories of left modules and right modules equivalent, and gives the homotopy category of left modules a monoidal structure $\otimes_R$. An E_3-algebra structure on R extends this monoidal structure to a braided monoidal structure. An E_4-algebra structure on R makes this braided monoidal structure into a symmetric monoidal structure.*

Theorem 19.3.8. *An E_1-algebra R in Sp has a monoidal category of bimodules. An E_∞-algebra R in Sp has a symmetric monoidal category of left modules.*

19.3.2 Monads

If C is not just enriched, but is *tensored* over spaces, an $\mathcal{O}$-algebra structure on X is expressible in terms internal to C. An $\mathcal{O}$-algebra structure is equivalent to having *action* maps

$$\gamma_k\colon \mathcal{O}(k) \otimes X^{\otimes k} \to X$$

that are invariant under the action of Σ_k and respect composition in the operad $\mathcal{O}$. If C has colimits, we can define *extended power constructions*

$$\mathrm{Sym}^k_{\mathcal{O}}(X) = \left(\mathcal{O}(k) \otimes_{\Sigma_k} X^{\otimes k}\right),$$

and an associated free $\mathcal{O}$-algebra functor

$$\mathrm{Free}_{\mathcal{O}}(X) = \coprod_{k\geq 0} \mathrm{Sym}^k_{\mathcal{O}}(X).$$

An $\mathcal{O}$-algebra structure on X is then determined by a single map $\mathrm{Free}_{\mathcal{O}}(X) \to X$. To say more, we need C to be compatible with enriched colimits in the sense of [42, §3].

Definition 19.3.9. A symmetric monoidal category C is *compatible with enriched colimits* if the monoidal structure on C preserves enriched colimits in each variable separately.

Compatibility with enriched colimits is necessary to give composite action maps

$$\mathcal{O}(k) \otimes \left(\bigotimes_{i=1}^{k} \mathcal{O}(n_i) \otimes X^{\otimes n_i}\right) \to X^{\otimes \Sigma n_i}$$

and make them assemble into a monad structure $\mathrm{Free}_{\mathcal{O}} \circ \mathrm{Free}_{\mathcal{O}} \to \mathrm{Free}_{\mathcal{O}}$ on this free functor. In this case, $\mathcal{O}$-algebras are equivalent to $\mathrm{Free}_{\mathcal{O}}$-algebras, and $\mathrm{Sym}^k_{\mathcal{O}}$ and $\mathrm{Free}_{\mathcal{O}}$ are enriched functors.

When these functors are enriched functors, they also give rise to a monad on the homotopy category hC. We refer to algebras over it as *homotopy $\mathcal{O}$-algebras.* This is strictly stronger than being an $\mathcal{O}$-algebra in the homotopy category; the latter asks for compatible maps $\pi_0\mathcal{O}(n) \to [A^{\otimes n}, A]$, whereas the former asks for compatible elements in $[\mathcal{O}(n) \otimes_{\Sigma_n} A^{\otimes n}, A]$ that use $\mathcal{O}$ before passing to homotopy. In the case of the E_n-operads, such a structure in the homotopy category is what is classically known as an *$\mathcal{H}_n$-algebra* [20].

This type of structure can be slightly rigidified using pushouts of free algebras. For any operad $\mathcal{O}$ with identity $e \in \mathcal{O}(1)$, we can construct a homotopy coequalizer diagram

$$\mathrm{Free}(\mathrm{Free}(\mathcal{O}, e), e) \rightrightarrows \mathrm{Free}(\mathcal{O}, e) \to \mathcal{O}^h$$

in the category of operads. An object A has an $\mathcal{O}^h$-algebra structure if and only if there are Σ_k-equivariant maps $\mathcal{O}(k) \to \mathrm{Map}_{\mathrm{C}}(A^{\otimes k}, A)$ so that the associativity diagram homotopy commutes and so that e acts by the identity. In particular, A has an $\mathcal{O}^h$-algebra structure if and only if it has a homotopy $\mathcal{O}$-algebra structure; the $\mathcal{O}^h$-structure has a *chosen* homotopy for the associativity of composition. For example, there is an operad parametrizing objects with a unital binary multiplication, a chosen associativity homotopy, and a chosen commutativity homotopy.

19.3.3 Connective algebras

In the category of spectra, the Eilenberg–Mac Lane spectra HA are characterized by a useful mapping property. We refer to a spectrum as *connective* if it is (-1)-connected. For any connective spectrum X, the natural map

$$\mathrm{Map}_{\mathsf{Sp}}(X, HA) \to \mathrm{Hom}_{\mathsf{Ab}}(\pi_0 X, A)$$

is a weak equivalence.

This has a number of strong consequences. For example, we get an equivalence of endomorphism operads $\mathrm{End}_{\mathsf{Sp}}(HA) \to \mathrm{End}_{\mathsf{Ab}}(A)$, obtained by taking π_0:

$$\mathrm{End}_{\mathsf{Sp}}(HA)_k = \mathrm{Map}(HA^{\otimes k}, HA) \simeq \mathrm{End}_{\mathsf{Ab}}(A)_k = \mathrm{Hom}(A^{\otimes k}, A).$$

Thus, an action of an operad $\mathcal{O}$ on HA is equivalent to an action of $\pi_0\mathcal{O}$ on A, and this equivalence is natural. This technique also generalizes, using the equivalences

$$\mathrm{Hom}(H\pi_0 R^{\otimes n}, H\pi_0 R) \xrightarrow{\sim} \mathrm{Map}(R^{\otimes n}, H\pi_0 R).$$

Proposition 19.3.10. *Suppose R is a connective spectrum and $\mathcal{O}$ is an operad acting on R. Then the map $R \to H\pi_0(R)$ can be given, in a functorial way, the structure of a map of $\mathcal{O}$-algebras.*

Example 19.3.11. If A is given the structure of a commutative ring, HA inherits an essentially unique structure of an E_∞-algebra. If R is a connective and homotopy commutative ring spectrum, then it can be equipped with an action of the cup-1 operad $\mathcal{Q}_1$ from 19.3.6. Any ring homomorphism $\pi_0 R \to A$ lifts to a map of $\mathcal{Q}_1$-algebras $R \to HA$.

19.3.4 Example algebras

Example 19.3.12. There exist models for the category of spectra so that the function spectrum

$$F(\Sigma^\infty_+ X, A) = A^X$$

is a lax monoidal functor $\mathcal{S}^{op} \times \mathsf{Sp} \to \mathsf{Sp}$, with the homotopy groups of A^X being the unreduced A-cohomology groups of X. The diagonal Δ makes any space X into a commutative monoid in $\mathcal{S}^{op}$. If A is an $\mathcal{O}$-algebra in Sp, then A^X then becomes an $\mathcal{O}$-algebra.

Example 19.3.13. For any spectrum E, composition of functions naturally gives the *endomorphism algebra* spectrum $\mathrm{End}(E) = F(E, E)$ the structure of an $\mathcal{A}_\infty$-algebra, and E is a left module over $\mathrm{End}(E)$. The homotopy groups of $\mathrm{End}(E)$ are sometimes called the *E-Steenrod algebra* and they parametrize operations on E-cohomology.

Example 19.3.14. The suspension spectrum functor

$$X \mapsto \Sigma^\infty_+ X = \mathbb{S}[X]$$

is strong symmetric monoidal. As a result, it takes $\mathcal{O}$-algebras to $\mathcal{O}$-algebras. For example, any topological group G has an associated *spherical group algebra* $\mathbb{S}[G]$.

Example 19.3.15. For any pointed space X, the n-fold loop space $\Omega^n X$ is an E_n-algebra in spaces, and $\mathbb{S}[\Omega^n X]$ is an E_n-algebra. For any spectrum Y the space $\Omega^\infty Y$ is an E_∞-algebra in spaces, and $\mathbb{S}[\Omega^\infty Y]$ is an E_∞-algebra.

Example 19.3.16. The Thom spectra MO and MU have E_∞ ring structures [63]. At any prime p, MU decomposes into a sum of shifts of the Brown–Peterson spectrum BP, which has the structure of an E_4-ring spectrum [10].

Example 19.3.17. The smash product being symmetric monoidal implies that it is also a strong symmetric monoidal functor $\mathrm{Sp} \times \mathrm{Sp} \to \mathrm{Sp}$. If A and B are $\mathcal{O}$-algebras then so is $A \otimes B$.

Example 19.3.18. For a map $Q \to R$ of E_∞ ring spectra, there is an adjunction

$$\mathrm{Mod}_Q \rightleftarrows \mathrm{Mod}_R$$

between the extension of scalars functor $M \mapsto R \otimes_Q M$ and the forgetful functor. The left adjoint is strong symmetric monoidal and the right adjoint is lax symmetric monoidal, and hence both functors preserve $\mathcal{O}$-algebras.

This allows us to narrow our focus. For example, if E has an E_∞-algebra structure and we are interested in understanding operations on the E-homology of $\mathcal{O}$-algebras, we can restrict our attention to those operations on the homotopy groups of $\mathcal{O}$-algebras in Mod_E rather than considering all possible operations on the E-homology.

19.3.5 Multicategories

A *multicategory* (or colored operad) encodes the structure of a category where functions have multiple input objects. They serve as a useful way to encode many multilinear structures in stable homotopy theory: multiplications, module structures, graded rings, and coherent structures on categories. In this section we will give a quick introduction to them, and will return in §19.7.

Definition 19.3.19. A *multicategory* $\mathcal{M}$ consists of the following data:

1. a collection $Ob(\mathcal{M})$ of objects;
2. a set $\mathrm{Mul}_{\mathcal{M}}(\mathbf{x}_1, \ldots, \mathbf{x}_d; \mathbf{y})$ of *multimorphisms* for any objects $\mathbf{x}_i$ and $\mathbf{y}$ of $\mathcal{M}$, or more generally a set $\mathrm{Mul}_{\mathcal{M}}(\{\mathbf{x}_s\}_{s\in S}; \mathbf{y})$ for any finite set S and objects $\mathbf{x}_s$, $\mathbf{y}$;
3. composition operations
$$\circ\colon\ \mathrm{Mul}_{\mathcal{M}}(\{\mathbf{y}_t\}_{t\in T}; \mathbf{z}) \times \prod_{t\in T} \mathrm{Mul}_{\mathcal{M}}(\{\mathbf{x}_s\}_{s\in f^{-1}(t)}; \mathbf{y}_t) \to \mathrm{Mul}_{\mathcal{M}}(\{\mathbf{x}_s\}_{s\in S}; \mathbf{z})$$
for any map $f\colon S \to T$ of finite sets and objects $\mathbf{x}_s$, $\mathbf{y}_t$, and $\mathbf{z}$ of $\mathcal{M}$; and
4. identity morphisms $\mathrm{id}_X \in \mathrm{Mul}_{\mathcal{M}}(\mathbf{x}; \mathbf{x})$ for any object $\mathbf{x}$.

These are required to satisfy two conditions:

1. unitality: $\mathrm{id}_{\mathbf{y}} \circ g = g \circ (\mathrm{id}_{\mathbf{x}_s}) = g$ for any $g \in \mathrm{Mul}_{\mathcal{M}}(\{\mathbf{x}_s\}_{s\in S}; \mathbf{Y})$; and
2. associativity: $h \circ (g_u \circ (f_t)) = (h \circ (g_u)) \circ f_t$ for any $S \to T \to U$ of finite sets.

The *underlying ordinary category* of $\mathcal{M}$ is the category with the same objects as $\mathcal{M}$ and $\mathrm{Hom}_{\mathcal{M}}(\mathbf{x}, \mathbf{y}) = \mathrm{Mul}_{\mathcal{M}}(\mathbf{x}; \mathbf{y})$.

If the sets of multimorphisms are given topologies so that composition is continuous, we refer to $\mathcal{M}$ as a *topological multicategory.*

A (topological) *multifunctor* $F\colon \mathcal{M} \to \mathcal{N}$ is a map $F\colon Ob(\mathcal{M}) \to Ob(\mathcal{N})$ on the level of objects, together with (continuous) maps

$$\mathrm{Mul}_{\mathcal{M}}(\mathbf{x}_1, \ldots, \mathbf{x}_d; \mathbf{y}) \to \mathrm{Mul}_{\mathcal{M}}(F\mathbf{x}_1, \ldots, F\mathbf{x}_d; F\mathbf{y})$$

that preserve identity morphisms and composition.

Example 19.3.20. An operad is equivalent to a single-object multicategory. For any object $\mathbf{x}$ in a multicategory $\mathcal{M}$, the full sub-multicategory spanned by $\mathbf{x}$ is an operad called the endomorphism operad of $\mathbf{x}$.

Example 19.3.21. A symmetric monoidal topological category $\mathcal{M}$ can be regarded as a multicategory by defining

$$\mathrm{Mul}_{\mathcal{M}}(X_1, \ldots, X_d; Y) = \mathrm{Map}_{\mathcal{M}}(X_1 \otimes \cdots \otimes X_d, Y).$$

This recovers the definition of the endomorphism operad of an object X.

The notion of an algebra over a multicategory will extend the notion of an algebra over an operad.

Definition 19.3.22. For (topological) multicategories $\mathcal{M}$ and C, the category $\mathsf{Alg}_{\mathcal{M}}(\mathsf{C})$ of *$\mathcal{M}$-algebras* in C is the category of (topological) multifunctors $\mathcal{M} \to \mathsf{C}$ and natural transformations.

For any object $\mathbf{x} \in \mathcal{M}$, the *evaluation* functor $\mathrm{ev}_{\mathbf{x}}\colon \mathsf{Alg}_{\mathcal{M}}(\mathsf{C}) \to \mathsf{C}$ sends an algebra A to the value $A(\mathbf{x})$.

Example 19.3.23. The multicategory Mod parametrizing "ring-module pairs" has two objects, $\mathbf{a}$ and $\mathbf{m}$, and

$$\mathrm{Mul}_{\mathrm{Mod}}(\mathbf{x}_1, \ldots, \mathbf{x}_d; \mathbf{y}) = \begin{cases} * & \text{if } \mathbf{y} = \mathbf{a} \text{ and all } \mathbf{x}_i \text{ are } \mathbf{a}, \\ * & \text{if } \mathbf{y} = \mathbf{m} \text{ and exactly one } \mathbf{x}_i \text{ is } \mathbf{m}, \\ \emptyset & \text{otherwise.} \end{cases}$$

A multifunctor $\mathrm{Mod} \to \mathsf{C}$ is equivalent to a pair (A, M) of a commutative monoid A of C and an object M with an action of A.

Example 19.3.24. A commutative monoid Γ can be regarded as a symmetric monoidal category with no non-identity morphisms, and in the associated multicategory we have

$$\mathrm{Mul}_{\Gamma}(g_1, \ldots, g_d; g) = \begin{cases} * & \text{if } \sum g_s = g, \\ \emptyset & \text{otherwise.} \end{cases}$$

A multifunctor $\Gamma \to \mathsf{C}$ determines objects X_g of C, a map from the unit to X_0, and multiplication maps $X_{g_1} \otimes \cdots \otimes X_{g_d} \to X_{g_1+\cdots+g_d}$: these multiplications are collectively unital, symmetric, and associative. We refer to such an object as a Γ-graded commutative monoid.

Example 19.3.25. The addition of natural numbers makes the partially ordered set $(\mathbb{N}, \geq)$ into a symmetric monoidal category. In the associated multicategory we have

$$\mathrm{Mul}_{\mathbb{N}}(n_1, \ldots, n_d; m) = \begin{cases} * & \text{if } \sum n_i \geq m, \\ \emptyset & \text{otherwise.} \end{cases}$$

A multifunctor $\Gamma \to \mathsf{C}$ determines a sequence of objects

$$\cdots \to X_2 \to X_1 \to X_0$$

of C and multiplication maps $X_{n_1} \otimes \cdots \otimes X_{n_d} \to X_{n_1+\cdots+n_d}$: these multiplications are collectively unital, symmetric, and associative, as well as being compatible with the inverse system. We refer to such an object *strongly filtered* commutative monoid in C.

Remark 19.3.26. If $\mathcal{M}_1$ and $\mathcal{M}_2$ are multicategories, there is a product multicategory $\mathcal{M}_1 \times \mathcal{M}_2$, obtained by taking products of objects and products of multimorphism spaces. Products allow us to extend the above constructions. For example, taking the product of an operad $\mathcal{O}$ with the multicategories of the previous examples, we construct multicategories that parametrize: pairs (A, M) of an $\mathcal{O}$-algebra and an $\mathcal{O}$-module; Γ-graded $\mathcal{O}$-algebras; and strongly filtered $\mathcal{O}$-algebras.

Example 19.3.27. Let $\mathcal{M}$ be the multicategory whose objects are integers, and define $\mathrm{Mul}_{\mathcal{M}}(m_1, \ldots, m_d; n)$ to be the set of natural transformations

$$\theta\colon H^{m_1}(X) \times \cdots \times H^{m_d}(X) \to H^n(X)$$

of contravariant functors on the category S of spaces; composition is composition of natural transformations. The category $\mathcal{M}$ is a category of multivariate cohomology operations. Any fixed space X determines an evaluation multifunctor $\mathrm{ev}_X\colon \mathcal{M} \to \mathsf{Sets}$, sending n to $H^n(X)$; any homotopy class of map $X \to Y$ of spaces determines a natural transformation of multifunctors in the opposite direction. Stated concisely, this is a functor

$$hS^{op} \to \mathsf{Alg}_{\mathcal{M}}(\mathsf{Sets})$$

that takes a space to an encoding of its cohomology groups and cohomology operations.

More generally, a category D with a chosen set of functors $\mathsf{D} \to \mathsf{Sets}$ determines a multicategory $\mathcal{M}$ spanned by them: we can define $\mathrm{Mul}(F_1, \ldots, F_d; G)$ to be the set of natural transformations $\prod F_i \to G$, so long as there is always a set (rather than a proper class) of natural transformations. If we view a functor F as assigning an invariant to each object of D, a multimorphism $\prod F_i \to G$ is a natural operation of several variables on such invariants. Evaluation on objects of D takes the form of a functor

$$\mathsf{D} \to \mathsf{Alg}_{\mathcal{M}}(\mathsf{Sets}),$$

encoding both the invariants assigned by these functors and the natural operations on them. These are examples of *multi-sorted algebraic theory* in the sense of Bergner [13], closely related to the work of [17, 88]. We will return to the discussion of this structure in §19.4.4.

Just as with ordinary operads, there are often free-forgetful adjunctions between objects of C and algebras over a multicategory.

Proposition 19.3.28. *Suppose that $\mathcal{M}$ is a small topological multicategory and that C is a symmetric monoidal topological category with compatible colimits in the sense of Definition 19.3.9.*

1. *For objects $\mathbf{x}$ and $\mathbf{y}$ of $\mathcal{M}$, there are extended power functors*

$$\mathrm{Sym}^k_{\mathcal{M},\mathbf{x}\to\mathbf{y}}\colon \mathsf{C}_{\mathbf{x}} \to \mathsf{C}_{\mathbf{y}},$$

given by

$$\mathrm{Sym}^k_{\mathcal{M},\mathbf{x}\to\mathbf{y}}(X) = \mathrm{Mul}_{\mathcal{M}}(\underbrace{\mathbf{x},\mathbf{x},\ldots,\mathbf{x}}_{k};\mathbf{y}) \otimes_{\Sigma_k} X^{\otimes k}.$$

2. *The evaluation functor $\mathrm{ev}_{\mathbf{x}}\colon \mathsf{Alg}_{\mathcal{M}}(\mathsf{C}) \to \mathsf{C}$ has a left adjoint*

$$\mathrm{Free}_{\mathcal{M},\mathbf{x}}\colon \mathsf{C} \to \mathsf{Alg}_{\mathcal{M}}(\mathsf{C}).$$

The value of $\mathrm{Free}_{\mathcal{M},\mathbf{x}}(X)$ on any object $\mathbf{y}$ of $\mathcal{M}$ is

$$\mathrm{ev}_{\mathbf{y}}(\mathrm{Free}_{\mathcal{M},\mathbf{x}}(X)) = \coprod_{k\geq 0} \mathrm{Sym}^k_{\mathcal{M},\mathbf{x}\to\mathbf{y}}(X).$$

Remark 19.3.29. These generalize the constructions of extended powers and free algebras from §19.3.2. If $\mathcal{M}$ has a single object $\mathbf{x}$, encoding an operad $\mathcal{O}$, then $\mathrm{Sym}^k_{\mathcal{M},\mathbf{x}\to\mathbf{x}} = \mathrm{Sym}^k_{\mathcal{O}}$ and $\mathrm{Free}_{\mathcal{M},\mathbf{x}}$ encodes $\mathrm{Free}_{\mathcal{O}}$.

Example 19.3.30. The free $\mathbb{Z}$-graded commutative monoid on an object X in degree $n \neq 0$ is equal to the symmetric product $\mathrm{Sym}^k(X)$ in degree kn for $k \geq 0$. All other gradings are the initial object.

Example 19.3.31. The free strongly filtered commutative monoid on an object X_1 in degree 1 is a filtered object of the form

$$\cdots \to \coprod_{k\geq 2} \mathrm{Sym}^k X_1 \to \coprod_{k\geq 1} \mathrm{Sym}^k X_1 \to \coprod_{k\geq 0} \mathrm{Sym}^k X_1.$$

If we have a strongly filtered commutative algebra $\cdots \to X_2 \to X_1 \to X_0$, then this gives action maps $\mathrm{Sym}^k X_1 \to X_k$. More generally, there are action maps $\mathrm{Sym}^k X_n \to X_{kn}$ that are compatible in n.

19.4 Operations

In this section we will fix a spectrum E, viewed as a coefficient object.

19.4.1 E-homology and E-modules

We can study $\mathcal{O}$-algebras through their E-homology.

Definition 19.4.1. Given a spectrum E, an *E-homology operation* for $\mathcal{O}$-algebras is a natural transformation of functors $\theta\colon E_m(-) \to E_{m+d}(-)$ of functors on the homotopy category of $\mathcal{O}$-algebras.

Such operations can be difficult to classify in general. However, if E has a commutative ring structure then we can do more. In this case, any $\mathcal{O}$-algebra A has an *E-homology object* $E \otimes A$ which is an $\mathcal{O}$-algebra in Mod_E, and any space X has an *E-cohomology object* E^X which is an E_∞-algebra object in Mod_E. By definition, we have

$$E_m(A) = [S^m, E \otimes A]_{\mathsf{Sp}}$$

and

$$E^m(X) = [S^{-m}, E^X]_{\mathsf{Sp}}.$$

Therefore, we can construct natural operations on the E-homology of $\mathcal{O}$-algebras or the E-cohomology of spaces by finding natural operations on the homotopy groups of $\mathcal{O}$-algebras in Mod_E.

Example 19.4.2. If X is an $\mathcal{O}$-algebra in spaces, then $E[X] = E \otimes \Sigma^\infty_+ X$ is an $\mathcal{O}$-algebra in Mod_E.

19.4.2 Multiplicative operations

In this section we will construct our first operations on the homotopy groups of $\mathcal{O}$-algebras over a fixed commutative ring spectrum E.

The functor π_* from the homotopy category of spectra to graded abelian groups is lax symmetric monoidal under the Koszul sign rule. The induced functor π_* from $\mathsf{Alg}_{\mathcal{O}}(\mathsf{Sp})$ or $\mathsf{Alg}_{\mathcal{O}}(\mathrm{Mod}_E)$ to graded abelian groups naturally takes values in the category of graded abelian groups, or graded E_*-modules, with an action of the operad $\pi_0\mathcal{O}$ in sets.

Example 19.4.3. In the case of an E_n-operad, $\pi_0\mathcal{O}$ is isomorphic to the associative operad when $n = 1$ and the commutative operad when $n \geq 2$. The E-homology groups of an E_n-algebra in Sp form a graded E_*-algebra. If $n \geq 2$, this algebra is graded-commutative.

By applying E_* to the action maps in the operad, we stronger information.

Proposition 19.4.4. *The homology groups $E_*\mathcal{O}(k)$ form an operad $E_*\mathcal{O}$ in graded E_*-modules, and the functor π_* from $\mathsf{Alg}_{\mathcal{O}}(\mathrm{Mod}_E)$ to graded abelian groups has a natural lift to the category of graded $E_*\mathcal{O}$-modules.*

Example 19.4.5. The homotopy groups of E_n-algebras have a natural bilinear *Browder bracket*

$$[-,-]\colon \pi_q(A) \otimes \pi_r(A) \to \pi_{q+(n-1)+r}(A).$$

This satisfies the following formulas.

Antisymmetry: $[\alpha, \beta] = -(-1)^{(|\alpha|+n-1)(|\beta|+n-1)}[\beta, \alpha]$.

Leibniz rule: $[\alpha, \beta\gamma] = [\alpha, \beta]\gamma + (-1)^{|\beta|(|\alpha|+n-1)}\alpha[\beta, \gamma]$.

Graded Jacobi identity:

$$\begin{aligned} 0 = \quad & (-1)^{(|\alpha|+n-1)(|\gamma|+n-1)}[\alpha, [\beta, \gamma]] \\ & + (-1)^{(|\beta|+n-1)(|\alpha|+n-1)}[\beta, [\gamma, \alpha]] \\ & + (-1)^{(|\gamma|+n-1)(|\beta|+n-1)}[\gamma, [\alpha, \beta]]. \end{aligned}$$

In the case of E_1-algebras, this reduces to the ordinary bracket

$$[\alpha, \beta] = \alpha\beta - (-1)^{|\alpha||\beta|}\beta\alpha$$

in the graded ring $\pi_*(A)$.

The Browder bracket is defined, just as it was defined in homology [23], using the image of the generating class $\lambda \in \pi_{n-1}\mathcal{E}_n(2) \cong \pi_{n-1}S^{n-1}$ coming from the little cubes operad. The antisymmetry and Jacobi identities are obtained by verifying identities in the graded operad $\pi_*(\Sigma^\infty_+\mathcal{E}_n)$. For example, if σ is the 2-cycle in Σ_2 we have

$$\lambda \circ \sigma = (-1)^n\lambda,$$

and if τ is a 3-cycle in Σ_3 we have

$$\lambda \circ (1 \otimes \lambda) \circ (1 + \tau + \tau^2) = 0.$$

However, the signs indicate that there is some care to be taken. In particular, the Browder bracket of elements $\alpha \in \pi_q(A)$ and $\beta \in \pi_r(A)$ is defined to be the following composite:

$$\begin{aligned} S^q \otimes S^{n-1} \otimes S^r & \to A \otimes \Sigma^\infty_+\mathcal{E}_n(2) \otimes A \\ & \to \Sigma^\infty_+\mathcal{E}_n(2) \otimes A \otimes A \\ & \to A \end{aligned}$$

This order is chosen because it is more consistent with writing the Browder bracket as an inline binary operation $[x, y]$ than with writing it as an operator $\lambda(x, y)$ on the left. The subscript on the range $\pi_{q+(n-1)+r}(A)$ reflects this choice (cf. [73]). This gives us the definition

$$[\alpha, \beta] = (-1)^{(n-1)|\alpha|}\gamma(\lambda \otimes \alpha \otimes \beta),$$

where γ is the action map of the operad $\pi_*(\Sigma^\infty_+\mathcal{E}_n)$ on π_*A. Both the verification of the identities on λ in the stable homotopy groups of configuration spaces, and the verification of the consequent antisymmetry, Leibniz, and Jacobi identities, are reasonable but error-prone exercises from this point; compare [22].

19.4.3 Representability

We will ultimately be interested in natural operations on homotopy and homology groups. However, it is handy to use a more general definition that replaces S^m by a general object. This accounts for the possibility of operations of several variables, and can also help reduce difficulties involving naturality in the input S^m.

Definition 19.4.6. For spectra M and X, we define the *M-indexed homotopy* of X to be

$$\pi_M(X) = [M, X]_{\mathsf{Sp}} \cong [E \otimes M, X]_{\mathrm{Mod}_E}.$$

For spectra M, X, and E we define the M-indexed E-homology of X to be

$$E_M(X) = \pi_M(E \otimes X).$$

If M is S^m, we instead use the more standard notation $\pi_m(-)$ for $\pi_{S^m}(-)$ or $E_m(-)$ for $E_{S^m}(-)$.

Definition 19.4.7. Let E be a commutative ring spectrum. A *homotopy operation* for $\mathcal{O}$-algebras over E is a natural transformation

$$\theta \colon \pi_M \to \pi_N$$

of functors on the homotopy category of $\mathsf{Alg}_{\mathcal{O}}(\mathrm{Mod}_E)$. When $\mathcal{O}$ and E are understood, we just refer to such natural transformations as homotopy operations.

We refer to the resulting operation $E_M(-) \to E_N(-)$ on the E-homology groups of $\mathcal{O}$-algebras as the induced E-homology operation.

As in Example 19.3.27, we can assemble operations with varying numbers of inputs into an algebraic structure.

Definition 19.4.8. Fix an operad $\mathcal{O}$ and a commutative ring spectrum E. The multicategory $\mathrm{Op}^E_{\mathcal{O}}$ of *operations* for $\mathcal{O}$-algebras in Mod_E has, as objects, spectra N. For any $M_1, \ldots, M_d$ and N, the group of multimorphisms

$$\mathrm{Op}^E_{\mathcal{O}}(M_1, \ldots, M_d; N)$$

is the group of natural transformations $\prod \pi_{M_i} \to \pi_N$ of functors $h\mathsf{Alg}_{\mathcal{O}}(\mathrm{Mod}_E) \to \mathsf{Sets}$. If E or $\mathcal{O}$ are understood, we drop them from the notation.

In the unary case, we write $\mathrm{Op}^E_{\mathcal{O}}(M; N)$ for the set of homotopy operations $\pi_M \to \pi_N$ for $\mathcal{O}$-algebras in Mod_E.

The free-forgetful adjunction between spectra and $\mathcal{O}$-algebras in Mod_E allows us to exhibit the functor π_M as representable.

Proposition 19.4.9. *Suppose that E is a commutative ring spectrum, $\mathcal{O}$ is an operad with associated free algebra monad $\mathrm{Free}_{\mathcal{O}}$. Then there is a natural isomorphism*

$$\pi_M(A) \cong [E \otimes \mathrm{Free}_{\mathcal{O}}(M), A]_{\mathsf{Alg}_{\mathcal{O}}(\mathrm{Mod}_E)}$$

for A in the homotopy category of $\mathsf{Alg}_{\mathcal{O}}(\mathrm{Mod}_E)$. In particular, the object $E \otimes \mathrm{Free}_{\mathcal{O}}(M)$ is a representing object for the functor π_M.

Proof. The forgetful functor $\mathsf{Alg}_{\mathcal{O}}(\mathrm{Mod}_E) \to \mathsf{Sp}$ can be expressed as a composite $\mathsf{Alg}_{\mathcal{O}}(\mathrm{Mod}_E) \to \mathsf{Alg}_{\mathcal{O}}(\mathsf{Sp}) \to \mathsf{Sp}$, and as such has a composite left adjoint $M \mapsto \mathrm{Free}_{\mathcal{O}}(M) \mapsto E \otimes \mathrm{Free}_{\mathcal{O}}(M)$; this adjunction passes to the homotopy category. Therefore, applying this adjunction we find

$$\begin{aligned}\pi_M(A) &\cong [\mathrm{Free}_{\mathcal{O}}(M), A]_{\mathsf{Alg}_{\mathcal{O}}(\mathsf{Sp})} \\ &\cong [E \otimes \mathrm{Free}_{\mathcal{O}}(M), A]_{\mathsf{Alg}_{\mathcal{O}}(\mathrm{Mod}_E)}\end{aligned}$$

as desired. □

Remark 19.4.10. It is possible to index more generally. Given an E-module L, we also have functors $\pi_L^E(-) = [L, -]_{\mathrm{Mod}_E}$; the free $\mathcal{O}$-algebra $\mathrm{Free}_{\mathcal{O}}(L)$ in the category of E-modules is then a representing object for π_L^E in $\mathsf{Alg}_{\mathcal{O}}(\mathrm{Mod}_E)$. We recover the above case by setting $L = E \otimes M$.

The Yoneda lemma now gives the following.

Corollary 19.4.11. *Let F be a functor from $h\mathsf{Alg}_{\mathcal{O}}(\mathrm{Mod}_E)$ to the category of sets. Natural transformations of functors $\pi_M \to F$ are in bijective correspondence with $F(E \otimes \mathrm{Free}_{\mathcal{O}}(M))$.*

In particular, there is an isomorphism

$$\mathrm{Op}_{\mathcal{O}}^E(M_1, \ldots, M_d; N) \cong E_N(\mathrm{Free}_{\mathcal{O}}(\oplus M_i))$$

from the group of natural transformations $\prod \pi_{M_i} \to \pi_N$ to the E-homology group of the free algebra.

The canonical decomposition of §19.3.2 for the monad $\mathrm{Free}_{\mathcal{O}}$ into extended powers gives us a canonical decomposition of operations.

Definition 19.4.12. For $k \geq 0$, the group of *operations of weight k* is the subgroup

$$\mathrm{Op}_{\mathcal{O}}^E(M_1, \ldots, M_d; N)^{\langle k \rangle} = E_N(\mathrm{Sym}_{\mathcal{O}}^k(\oplus M_i))$$

of $\mathrm{Op}_{\mathcal{O}}^E(M_1, \ldots, M_d; N) \cong E_N(\mathrm{Free}_{\mathcal{O}}(\oplus M_i))$.

A *power operation of weight k* is a unary operation of weight k: an element of the subgroup

$$\mathrm{Op}_{\mathcal{O}}^E(M, N)^{\langle k \rangle} \cong E_N(\mathrm{Sym}_{\mathcal{O}}^k(M))$$

of $\mathrm{Op}_{\mathcal{O}}^E(M, N)$.

Remark 19.4.13. Composition multiplies weight. Furthermore, if the object N is dualizable, the group of all operations is a direct sum: every operation decomposes canonically as a sum of operations of varying weights.

19.4.4 Structure on operations

Even when restricted to ordinary homotopy groups, these operations between the homotopy groups of $\mathcal{O}$-algebras in Mod_E form a rather rich algebraic structure [13], whose characteristics should be discussed; we learned most of this from Rezk [77, 76]. Recall

$$\mathrm{Op}(m_1, \ldots, m_d; n) = \mathrm{Op}^E_{\mathcal{O}}(m_1, \ldots, m_d; n) \cong \pi_n(E \otimes \mathrm{Free}_{\mathcal{O}}(\oplus S^{m_i})).$$

Here are some characteristics of this algebraic theory.

1. We think of the elements in these groups as operators, in the sense that they can *act*. Given $\alpha \in \mathrm{Op}(m_1, \ldots, m_d; n)$, an $\mathcal{O}$-algebra R in Mod_E and $x_i \in \pi_{m_i} R$, we can apply α to get a natural element
$$\alpha \propto (x_1, \ldots, x_d) \in \pi_n R.$$
This action is associative with respect to composition, but only distributes over addition on the left.

2. For each $1 \leq k \leq d$, there is a *fundamental generator* $\iota_k \in \mathrm{Op}(m_1, \ldots, m_d; m_k)$ that acts by projecting:
$$\iota_k \propto (x_1, \ldots, x_d) = x_k.$$

3. These operators can *compose*: given $\alpha \in \mathrm{Op}(m_1, \ldots, m_d; n)$ and $\beta_i \in \mathrm{Op}(\ell_1, \ldots, \ell_c; m_i)$, there is a composite operator
$$\alpha \propto (\beta_1, \ldots, \beta_d) \in \mathrm{Op}(\ell_1, \ldots, \ell_c; n).$$
Composition is unital. It is also associative, both with itself and with acting on elements. Again, it only distributes over addition on the left.

4. Composition respects weight: if α is in weight a and β_i are in weights b_i, then $\alpha \propto (\beta_i)$ is in weight $a \cdot (\sum b_i)$.

Example 19.4.14. Take $E = HR$ for a commutative ring R and let $\mathcal{O}$ to be the associative operad. Then the graded group

$$Op(m_1, \ldots, m_d; *) = \oplus_n \mathrm{Op}(m_1, \ldots, m_d; n) \cong H_*(\mathrm{Free}_{\mathcal{O}}(\oplus S^{m_i}; R))$$

is the free associative graded R-algebra on the fundamental generators $\iota_1 \ldots \iota_d$ with ι_i in degree m_i, and the composition operations are *substitution*. For example, the element $\iota_1 + \iota_2 \in \mathrm{Op}(n, n; n)$ acts by the binary addition operation in degree n; the elements $\iota_1\iota_2$ and $\iota_2\iota_1$ in $\mathrm{Op}(n_1, n_2; n_1 + n_2)$ represent binary multiplication in either order; the element $(\iota_1)^2 \in \mathrm{Op}(n; 2n)$ represents the squaring operation; for $r \in R$ the element $r\iota_1 \in \mathrm{Op}(n; n)$ represents scalar multiplication by r; combinations of these operations are represented by identities such as

$$\iota_1^2 \propto (\iota_1 + \iota_2) = \iota_1^2 + \iota_1\iota_2 + \iota_2\iota_1 + \iota_2^2.$$

In this structure, each monomial has constant weight equal to its degree.

Example 19.4.15. Take $\mathcal{O}$ to be an E_n-operad. Then, for any p and q, the Browder bracket is a natural transformation $\pi_p \times \pi_q \to \pi_{p+(n-1)+q}$, and it is realized by an element $[\iota_1, \iota_2]$ in $\mathrm{Op}(p, q; p + (n-1) + q)$ of weight two. Relations between the product and the Browder bracket are expressed universally by relations between compositions: for example, antisymmetry is expressed by an identity

$$[\iota_1, \iota_2] = -(-1)^{(p+n-1)(q+n-1)}[\iota_2, \iota_1].$$

Remark 19.4.16. Inside the collection of all unary operations, there is a subgroup of *additive* operations: those operations f that satisfy

$$f \propto (\iota_1 + \iota_2) = f \propto \iota_1 + f \propto \iota_2.$$

Composition of such operations is bilinear, and so the collection of objects and additive operations form a category enriched in abelian groups. In some cases, the additive operations can be used to determine the general structure [77].

19.4.5 Power operations

We will begin to narrow our study of power operations and focus on unary operations, of fixed weight, between integer gradings.

Definition 19.4.17. Fix an operad $\mathcal{O}$ and a commutative ring spectrum E. The group of *power operations of weight k on degree m* for $\mathcal{O}$-algebras in Mod_E is the graded abelian group

$$\mathrm{Pow}_{\mathcal{O}}^E(m, k) = \pi_*(F(S^m, E \otimes \mathrm{Sym}_{\mathcal{O}}^k(S^m))) \cong \bigoplus_{r \in \mathbb{Z}} \mathrm{Op}_{\mathcal{O}}^E(m, m + r)^{\langle k \rangle}.$$

If $\mathcal{O}$ or E are understood, we drop them from the notation.

An element of $\mathrm{Pow}(m, k)$ in grading r represents a weight-k natural transformation $\pi_m \to \pi_{m+r}$ on the homotopy category of $\mathcal{O}$-algebras in Mod_E, and induces a natural transformation $E_m \to E_{m+r}$ on the homotopy category of $\mathcal{O}$-algebras. (While we index these group by integers, they depend on a choice of representing object and in particular on an orientation of S^m; making implicit identifications will result in sign issues.)

Remark 19.4.18. These operations, and the relations between them, are still possessed by homotopy $\mathcal{O}$-algebras in the sense of §19.3.2.

Remark 19.4.19. Suppose that Σ_k acts freely and properly discontinuously on $\mathcal{O}(k)$. Let $V \subset \mathbb{R}^k$ be the subspace of elements that sum to 0, with associated vector bundle $\overline{\rho} \to B\Sigma_k$ of dimension $k-1$. For any m there is an associated virtual bundle $\mathbb{R}^m \otimes \overline{\rho}$. If we define

$$P(k) = \mathcal{O}(k)/\Sigma_k,$$

then there is a virtual bundle $m\overline{\rho}$ on $P(k)$. The Thom spectrum $P(k)^{m\overline{\rho}}$ of this virtual bundle is canonically equivalent to the spectrum $\Sigma^{-m}\Sigma^\infty_+\mathcal{O}(k)\otimes_{\Sigma_k}(S^m)^{\otimes k}$ that appears in the definition of $\mathrm{Pow}(m,k)$.

This allows us to give a more concise expression

$$\mathrm{Pow}(m,k) = E_*(P(k)^{m\overline{\rho}}),$$

which is particularly useful in cases where we can apply a Thom isomorphism for E-homology.

Example 19.4.20. Consider the case of operations of weight 2 for E_n-algebras. The space $P(2) = \mathcal{C}_n(2)/\Sigma_2$ is homotopy equivalent to the real projective space $\mathbb{RP}^{n-1}$, the line bundle $\overline{\rho} = \sigma$ is associated to the sign representation of Σ_2, and the Thom spectrum $(\mathbb{RP}^{n-1})^{m\sigma}$ is commonly known as the *stunted projective space* $\mathbb{RP}^{m+n-1}_m$ which has a cell decomposition with one cell in each dimension between m and $m+n-1$. (When $m \geq 0$ this is literally the suspension spectrum of $\mathbb{RP}^{m+n-1}/\mathbb{RP}^{m-1}$.) Therefore, the operations of weight 2 on degree m are parametrized by the E-homology group

$$\mathrm{Op}^E_m(2) = E_*(\mathbb{RP}^{m+n-1}_m).$$

Example 19.4.21. When $E = H\mathbb{F}_2$, we find $H_*(\mathbb{RP}^{m+n-1}_m)$ is $\mathbb{F}_2$ in degrees m through $(m+n-1)$, and so we obtain unique *Dyer–Lashof operations* Q^r for $m \leq r \leq m+n-1$ that send elements in π_m to elements in π_{m+r}.

Example 19.4.22. Consider the cup-1 operad $\mathcal{Q}_1$ defined in Example 19.3.6. Then the weight-2 operations on the E-homology of $\mathcal{Q}_1$-algebras are parametrized by $E_*(\mathbb{RP}^{m+1}_m)$. This stunted projective space is the Thom spectrum of m times the Möbius line bundle over S^1.

For example, we can take E to be the sphere spectrum. If $m = 2k$ there is a splitting

$$\mathbb{RP}^{2k+1}_{2k} \simeq S^{2k} \oplus S^{2k+1}.$$

Chosen generators in $\pi_{2k}(S^{2k}\oplus S^{2k+1})$ and $\pi_{2k+1}(S^{2k}\oplus S^{2k+1})$ give operations that increase degree by $2k$ and $2k+1$, respectively. A choice of splitting $S^{2k+1} \to \mathbb{RP}^{2k+1}_{2k}$ determines an operation $\mathrm{Sq}_1\colon \pi_{2k}(-) \to \pi_{4k+1}(-)$ called the *cup-1 square*. It satisfies $2\,\mathrm{Sq}_1(a) = [a,a]$.

In the case that we have an E_∞ ring spectrum, this has been studied in [20, §V] and [12], and can be chosen in such a way that it satisfies the following addition and multiplication identities on even-degree homotopy elements:

$$2\,\mathrm{Sq}_1(a) = 0$$

$$\mathrm{Sq}_1(a+b) = \mathrm{Sq}_1(a) + \mathrm{Sq}_1(b) + (\tfrac{|a|}{2}+1)ab\eta$$

$$\mathrm{Sq}_1(ab) = a^2\,\mathrm{Sq}_1(b) + \mathrm{Sq}_1(a)b^2 + \tfrac{|ab|}{4}a^2b^2\eta.$$

For example, $\mathrm{Sq}_1(n) = \binom{n}{2}\eta$ for $n \in \mathbb{Z}$. In the absence of higher commutativity, these identities should have correction terms involving the Browder bracket.

19.4.6 Stability

In this section we will consider compatibility relations between operations on different homotopy degrees.

Recall from §19.3.2 that the monad $\mathrm{Free}_{\mathcal{O}}$ decomposed into the homogeneous functors defined by

$$\mathrm{Sym}^k_{\mathcal{O}}(X) = \Sigma^\infty_+ \mathcal{O}(k) \otimes_{\Sigma_k} X^{\otimes k}.$$

In particular, these functors are *continuous*: they induce functions

$$\mathrm{Map}(X, Y) \to \mathrm{Map}(\mathrm{Sym}^k_{\mathcal{O}}(X), \mathrm{Sym}^k_{\mathcal{O}}(Y))$$

between mapping spaces, and for $k > 0$ they have the property that they are *pointed*: $\mathrm{Sym}^k_{\mathcal{O}}(*) = *$ and hence the functor $\mathrm{Sym}^k_{\mathcal{O}}$ induces continuous maps of *pointed* mapping spaces.

Definition 19.4.23. For any spectrum M, any pointed space Z, and any $k > 0$, the *assembly map*

$$\mathrm{Sym}^k_{\mathcal{O}}(M) \otimes \Sigma^\infty Z \to \mathrm{Sym}^k_{\mathcal{O}}(M \otimes \Sigma^\infty Z)$$

is adjoint to the composite map of pointed spaces

$$\begin{aligned} Z &\to \mathrm{Map}_{\mathrm{Sp}}(S^0, \Sigma^\infty Z) \\ &\to \mathrm{Map}_{\mathrm{Sp}}(M, M \otimes \Sigma^\infty Z) \\ &\to \mathrm{Map}_{\mathrm{Sp}}(\mathrm{Sym}^k_{\mathcal{O}}(M), \mathrm{Sym}^k_{\mathcal{O}}(M \otimes \Sigma^\infty Z)). \end{aligned}$$

The *suspension map*

$$\sigma_n \colon \ \mathrm{Pow}(m, k) \to \mathrm{Pow}(m+n, k)$$

is induced by the composite map of function spectra

$$\begin{aligned} F(S^m, E \otimes \mathrm{Sym}^k_{\mathcal{O}}(S^m)) &\to F(S^m \otimes S^n, E \otimes \mathrm{Sym}^k_{\mathcal{O}}(S^m) \otimes S^n) \\ &\to F(S^m \otimes S^n, E \otimes \mathrm{Sym}^k_{\mathcal{O}}(S^m \otimes S^n)). \end{aligned}$$

Remark 19.4.24. The operation $\sigma = \sigma_1$ has a concrete meaning: it is designed for *compatibility with the Mayer–Vietoris sequence.* To illustrate this, first recall that for a homotopy commutative diagram

$$\begin{array}{ccc} A & \longrightarrow & B \\ \downarrow & & \downarrow \\ C & \longrightarrow & D \end{array}$$

of spectra, we have natural maps $A \to P \leftarrow \Sigma^{-1} D$ where P is the homotopy pullback.

Now suppose that we are given a diagram of $\mathcal{O}$-algebras as above that is a homotopy pullback, inducing a boundary map $\partial \colon \Sigma^{-1} D \to P \simeq A$. Given maps $\theta \colon N \to E \otimes \mathrm{Sym}^k_{\mathcal{O}}(M)$ and $\alpha \colon \Sigma M \to D$, we can map in a trivial homotopy pullback diagram to the above, then apply action maps and naturality of the connecting homomorphisms. We get a commuting diagram:

$$\begin{array}{ccccc} N & \xrightarrow{\theta} & E \otimes \mathrm{Sym}^k_{\mathcal{O}} M & \xrightarrow{\partial\alpha} & A \\ \downarrow \wr & & \downarrow & & \downarrow \wr \\ \Sigma^{-1}\Sigma N & \longrightarrow & P' & \longrightarrow & P \\ \uparrow \wr & & \uparrow & & \uparrow \partial \\ \Sigma^{-1}\Sigma N & \xrightarrow[\sigma\theta]{} & \Sigma^{-1} E \otimes \mathrm{Sym}^k_{\mathcal{O}}(\Sigma M) & \xrightarrow[\Sigma^{-1}\alpha]{} & \Sigma^{-1} D. \end{array}$$

Therefore, for an operation $\theta\colon [M,-] \to [N,-]$ for $\mathcal{O}$-algebras in Mod_E, we find that

$$\partial \circ \sigma\theta \sim \theta \circ \partial.$$

This description makes implicit choices about the orientation of the circle that appears in the operation Ω when taking homotopy pullbacks, and this can result in sign headaches.

Proposition 19.4.25. *For $k, r > 0$, the suspension $\sigma_r\colon \mathrm{Pow}(m,k) \to \mathrm{Pow}(m+r,k)$ is the map*

$$E_*(P(k)^{m\overline{\rho}}) \to E_*(P(k)^{(m+r)\overline{\rho}})$$

on E-homology induced by the inclusion of virtual bundles $m\overline{\rho} \to m\overline{\rho} \oplus r\overline{\rho}$.

Proof. The assembly map $\mathrm{Sym}^k_{\mathcal{O}}(S^m) \otimes S^n \to \mathrm{Sym}^k_{\mathcal{O}}(S^{m+n})$ is the map

$$(\Sigma^\infty_+ \mathcal{O}(k) \otimes_{\Sigma_k} S^{m\rho}) \otimes S^n \to (\Sigma^\infty_+ \mathcal{O}(k) \otimes_{\Sigma_k} S^{(m+n)\rho}),$$

which is the map

$$P(k)^{m\rho} \otimes S^r \to P(k)^{(m+r)\rho}$$

induced by the direct sum inclusion $m\rho \oplus r \to (m\rho \oplus r) \oplus r\overline{\rho}$ of virtual bundles. The map σ_r is obtained by desuspending both sides $(m+r)$ times, which gives the map induced by the direct sum inclusion $m\overline{\rho} \to m\overline{\rho} \oplus r\overline{\rho}$ of virtual bundles. □

Example 19.4.26. The Dyer–Lashof operations for E_n-algebras are explicitly *unstable*. For example, in weight two the n-fold suspension maps $\mathbb{RP}^{m+n-1}_m \to \mathbb{RP}^{(m+n)+n-1}_{m+n}$ are trivial, and so the map $\mathrm{Op}^E_m(2) \to \mathrm{Op}^E_{m+n}(2)$ is trivial. This recovers the well-known fact that all Dyer–Lashof operations for E_n-algebras map to zero under n-fold suspension.

By contrast, the Dyer–Lashof operations for E_∞-algebras are *stable*: the maps $H_*\mathbb{RP}^\infty_m \to H_*\mathbb{RP}^\infty_{m+1}$ are surjections, and so the quadratic operations all lift to elements in the homotopy of

$$\lim_m (H \otimes \mathbb{RP}^\infty_m).$$

By [32, 16.1], this is the desuspended Tate spectrum $(\Sigma^{-1}H)^{t\Sigma_2}$.

Remark 19.4.27. More generally, the fully stable operations of prime weight p on the homotopy of E_∞ E-algebras are detected by the p-localized Tate spectrum

$$(\Sigma^{-1}E_{(p)})^{t\Sigma_p}.$$

See [20, II.5.3] and [30].

19.4.7 Pro-representability

Suppose that $E = \operatorname{colim} E_\alpha$ is an expression of E as a filtered colimit of finite spectra. Then there is an identification

$$E_m A = \operatorname*{colim}_\alpha [S^m, E_\alpha \otimes A] = \operatorname*{colim}_\alpha [S^m \otimes DE_\alpha, A],$$

where D is the Spanier–Whitehead dual. We cannot move the colimit inside, but we can view $\{S^m \otimes DE_\alpha\}$ as a *pro-object* in the category of spectra. This makes the functor E_m representable by embedding the category of spectra into the category of pro-spectra.

For algebras over an operad $\mathcal{O}$, we can go even further and find that

$$E_m(A) = [\{\mathrm{Free}_{\mathcal{O}}(S^m \otimes DE_\alpha)\}, A]_{\text{pro-}\mathcal{O}}$$

is now a representable functor in the homotopy category of pro-$\mathcal{O}$-algebras, and in this category we can determine all the natural operations $E_m \to E_n$:

$$\begin{aligned}
Nat_{\text{pro-}\mathcal{O}}(E_m(-), E_n(-)) &= [\{\mathrm{Free}_{\mathcal{O}}(S^n \otimes DE_\alpha)\}, \{\mathrm{Free}_{\mathcal{O}}(S^m \otimes DE_\beta)\}]_{\text{pro-}\mathcal{O}} \\
&= \pi_0 \lim_\beta \operatorname*{colim}_\alpha \mathrm{Map}_{\mathcal{O}}(\mathrm{Free}_{\mathcal{O}}(S^n \otimes DE_\alpha), \mathrm{Free}_{\mathcal{O}}(S^m \otimes DE_\beta)) \\
&= \pi_0 \lim_\beta \operatorname*{colim}_\alpha \mathrm{Map}_{\mathrm{Sp}}(S^n \otimes DE_\alpha, \mathrm{Free}_{\mathcal{O}}(S^m \otimes DE_\beta)) \\
&= \pi_0 \lim_\beta \mathrm{Map}_{\mathrm{Sp}}(S^n, E \otimes \mathrm{Free}_{\mathcal{O}}(S^m \otimes DE_\beta)) \\
&= \pi_n \lim_\beta E \otimes (\mathrm{Free}_{\mathcal{O}}(S^m \otimes DE_\beta)).
\end{aligned}$$

The algebra of natural transformations has natural maps in from the group

$$[S^m \otimes E, S^n \otimes E]$$

of cohomology operations (and these maps are isomorphisms if $\mathcal{O}$ is trivial), and it has a natural map to the limit

$$\lim_\beta E_n(\mathrm{Free}_{\mathcal{O}}(S^n \otimes DE_\beta)).$$

This map to the limit is an isomorphism if no higher derived functors intrude. We can think of this as the algebra of *continuous* operations on E-homology.

19.5 Classical operations

19.5.1 E_n Dyer–Lashof operations at $p = 2$

We will now specialize to the case of ordinary mod-2 homology. When we do so, we have Thom isomorphisms for many bundles and we have explicit computations of the homology of configuration spaces due to Cohen [23]. Similar results with more complicated identities hold at odd primes.

Proposition 19.5.1. *Let $H = H\mathbb{F}_2$ be the mod-2 Eilenberg–Mac Lane spectrum. Then the group $\mathrm{Op}^H_m(2)$ of weight-2 operations for E_n-algebras has exactly one nonzero operation in each degree between m and $m+n-1$, and no others.*

Proof. By Remark 19.4.19, this is a calculation $H_*(\mathbb{RP}^{n+m-1}_m)$ of the mod-2 homology of stunted projective spaces. □

Theorem 19.5.2. [20, III.3.1, III.3.2, III.3.3] *Let $H = H\mathbb{F}_2$ be the mod-2 Eilenberg–Mac Lane spectrum. Then E_n-algebras in* Mod_H *have* Dyer–Lashof operations

$$Q_i \colon \pi_m \to \pi_{2m+i}$$

for $0 \le i \le n-1$. These satisfy the following formulas.

Additivity: $Q_r(x+y) = Q_r(x) + Q_r(y)$ *for* $r < n-1$.

Squaring: $Q_0 x = x^2$.

Unit: $Q_j 1 = 0$ *for* $j > 0$.

Cartan formula: $Q_r(xy) = \sum_{p+q=r} Q_p(x)Q_q(y)$ *for* $r < n-1$.

Adem relations: $Q_rQ_s(x) = \sum \binom{j-s-1}{2j-r-s} Q_{r+2s-2j}Q_j(x)$ *for* $r > s$.

Stability: $\sigma Q_0 = 0$, *and* $\sigma Q_r = Q_{r-1}$ *for* $r > 0$.

Extension: *If an E_n-algebra structure extends to an E_{n+1}-algebra structure, the operations Q_r for E_{n+1}-algebras coincide with the operations Q_r for E_n-algebras.*

There is also a bilinear Browder bracket

$$[-,-] \colon \pi_r \otimes \pi_s \to \pi_{r+(n-1)+s}$$

satisfying the following formulas.

Antisymmetry: $[x, y] = [y, x]$ *and* $[x, x] = 0$.

Unit: $[x, 1] = 0$.

Leibniz rule: $[x, yz] = [x, y]z + y[x, z]$.

Jacobi identity: $[x, [y, z]] + [y, [z, x]] + [z, [x, y]] = 0$.

Dyer–Lashof vanishing: $[x, Q_r y] = 0$ *for* $r < n-1$.

Top additivity: $Q_{n-1}(x+y) = Q_{n-1}x + Q_{n-1}y + [x,y]$.

Top Cartan formula: $Q_{n-1}(xy) = \sum_{p+q=n-1} Q_p(x)Q_q(y) + x[x,y]y$.

Adjoint identity: $[x, Q_{n-1}y] = [y,[y,x]]$.

Extension: *If an E_n-algebra structure extends to an E_{n+1}-algebra structure, the bracket is identically zero.*

E_1-bracket: $[x,y] = xy + yx$ *if* $n = 1$.

Remark 19.5.3. There are two common indexing conventions for the Dyer–Lashof operations. This lower-indexing convention is designed to emphasize the range where the operations are defined, and is especially useful for E_n-algebras. The upper-indexing convention defines $Q^s x = Q_{s-|x|}x$ so that Q^s is always a natural transformation $\pi_m \to \pi_{s+m}$, with the understanding that $Q^s x = 0$ for $s < |x|$.

Example 19.5.4. Suppose that X is an n-fold loop space, so that $H[X]$ is an E_n-algebra in left H-modules. Then we recover the classical Dyer–Lashof operations

$$Q_r\colon H_n(X) \to H_{2n+r}(X)$$

in the homology of iterated loop spaces.

Theorem 19.5.5. [20, IX.2.1], [23, III.3.1] *For any spectrum X and any $1 \le n \le \infty$, $H_*(\mathrm{Free}_{E_n}(X))$ is the free object $\mathbb{Q}_{E_n}(H_*X)$ in the category of graded $\mathbb{F}_2$-algebras with Dyer–Lashof operations and Browder bracket satisfying the identities of Theorem 19.5.2.*

Remark 19.5.6. This theorem is the analogue of the calculation of the cohomology of Eilenberg–Mac Lane spaces as free algebras in a category of algebras with Steenrod operations. As such, it means that we have a *complete* theory of homotopy operations for E_n-algebras over H.

Example 19.5.7. In the case $n < \infty$ we can give a straightforward description of $\mathbb{Q}_{E_n}V$ if V has a basis with a single generator e. In this case, the antisymmetry, unit, and Dyer–Lashof vanishing axioms can be used to show that the free algebra has trivial Browder bracket, and so the free algebra $\mathbb{Q}_{E_n}(V)$ is a graded polynomial algebra

$$\mathbb{F}_2[Q_J e]$$

as we range over generators $Q_J e = (Q_1)^{j_1}(Q_2)^{j_2}\ldots(Q_{n-1})^{j_{n-1}}e$.

19.5.2 E_∞ Dyer–Lashof operations at $p = 2$

When $n = \infty$, the results of the previous section become significantly simpler, and it is worth expressing using the upper indexing for Dyer–Lashof operations.

Theorem 19.5.8. [20, III.1.1] *Let $H = H\mathbb{F}_2$ be the mod-2 Eilenberg–Mac Lane spectrum. Then E_∞-algebras in* Mod_H *have* Dyer–Lashof operations

$$Q^r\colon \pi_m \to \pi_{m+r}$$

for $r \in \mathbb{Z}$. These satisfy the following formulas.

Additivity: $Q^r(x+y) = Q^r(x) + Q^r(y)$.

Instability: $Q^r x = 0$ *if* $r < |x|$.

Squaring: $Q^r x = x^2$ *if* $r = |x|$.

Unit: $Q^r 1 = 0$ *for* $r \neq 0$.

Cartan formula: $Q^r(xy) = \sum_{p+q=r} Q^p(x)Q^q(y)$.

Adem relations: $Q^r Q^s = \sum \binom{i-s-1}{2i-r} Q^{s+r-i}Q^i$ *for* $r > 2s$.

Stability: $\sigma Q^r = Q^r$.

Example 19.5.9. For any space X, H^X is an E_∞-algebra in the category of left H-modules, and hence it has Dyer–Lashof operations

$$Q^i \colon H^n(X) \to H^{n-i}(X).$$

It turns out that these are precisely the *Steenrod operations*:

$$Sq^i = Q^{-i}.$$

From this point of view, the identity $Q^0 x = x$ is not obvious. In fact, Mandell has shown that this identity is characteristic of algebras that come from spaces: the functor $X \mapsto (H\overline{\mathbb{F}}_p)^X$ from spaces to E_∞-algebras over the Eilenberg–Mac Lane spectrum $H\overline{\mathbb{F}}_p$ is fully faithful, and the essential image is detected in terms of the coefficient ring being generated by classes that are annihilated by the analogue at arbitrary primes of the identity $(Q^0 - 1)$ [56].

Example 19.5.10. In the case $n = \infty$ there is always a straightforward basis for the free algebra. If $\{e_i\}$ is a basis of a graded vector space V over $\mathbb{F}_2$, then the free algebra $\mathbb{Q}_{E_\infty}(V)$ is a graded polynomial algebra

$$\mathbb{F}_2[Q^J e_i]$$

as we range over generators $Q^J e_i = Q^{j_1} \cdots Q^{j_p} e_i$ such that $j_i \leq 2j_{i+1}$ and $j_1 - j_2 - \cdots - j_p > |e_i|$.

19.5.3 Iterated loop spaces

The following is an unpointed group-completion theorem for E_n-spaces.

Theorem 19.5.11. [23, III.3.3] *For any space X and any $1 \leq n \leq \infty$, the map $X \to \Omega^n\Sigma^n X_+$ induces a map* $\mathrm{Free}_{E_n}(X) \to \Omega^n\Sigma^n X_+$*, and the resulting ring map*

$$\mathbb{Q}_{E_n}(H_*X) = H_*(\mathrm{Free}_{E_n}(X)) \to H_*(\Omega^n\Sigma^n X_+)$$

is a localization that inverts the images of $\pi_0(X)$.

Remark 19.5.12. A pointed version of the group-completion theorem, involving $\Omega^n\Sigma^n X$, is much more standard and implies this one. This theorem holds for $\Omega^n\Sigma^n$ if we replace Free_{E_n} with a version that takes the basepoint to a unit and we replace $\mathbb{Q}_{E_n}(H_*X)$ with either $\mathbb{Q}_{E_n}(\widetilde{H}_* X)$ a reduced version $\widetilde{\mathbb{Q}}_{E_n}$ that sends a chosen element to the unit. However, we wanted to give a version that de-emphasizes implicit basepoints for comparison with §19.8.2.

Proposition 19.5.13. *Suppose Y is a pointed space. Then the suspension map*

$$\sigma\colon \widetilde{H}_*(\Omega^n Y) \to \widetilde{H}_{*+1}(\Omega^{n-1} Y),$$

induced by the map $\Sigma\Omega^n Y \to \Omega^{n-1} Y$, is compatible with the Dyer–Lashof operations and the Browder bracket:

$$\sigma(Q^r x) = Q^r(\sigma x)$$
$$\sigma[x, y] = [\sigma x, \sigma y].$$

In particular, in the bar spectral sequence

$$\mathrm{Tor}_{**}^{H_*\Omega^n Y}(\mathbb{F}_2, \mathbb{F}_2) \Rightarrow H_*\Omega^{n-1} Y,$$

the operations on the image $\widetilde{H}_*\Omega^n Y \twoheadrightarrow \mathrm{Tor}_1^{H_*\Omega^n Y}(\mathbb{F}_2, \mathbb{F}_2)$ are representatives for the operations on $H_*\Omega^{n-1} Y$.

This provides some degree of conceptual interpretation for the bracket and the Dyer–Lashof operations. Since $H_*\Omega^2 Y$ is commutative, the Tor-algebra is also commutative even though it is converging to the possibly noncommutative ring $H_*\Omega Y$, and so the noncommutativity is tracked by multiplicative extensions in the spectral sequence [73]. The Browder bracket in $H_*\Omega^n Y$ exists to remember that, after $n-1$ deloopings, there are commutators $xy \pm yx$ in $H_*\Omega Y$.

Similarly, elements in positive filtration in the Tor-algebra of a commutative ring always satisfy $x^2 = 0$, even though this may not be the case in $H_*\Omega^{n-1} Y$. The element $Q_0 x$ is x^2; the elements $Q_1 x, Q_2 x, \ldots, Q_{n-1} x$ determine the line of succession for the property of being x^2 as the delooping process is iterated.

Remark 19.5.14. The group-completion theorem allows us to relate the homology of a delooping to certain nonabelian derived functors [68]. Similar spectral sequences computing E_n-homology of chain complexes have been studied by Richter and Ziegenhagen [78].

Associated to the n-fold loop space $\Omega^n Y$ of an $(n-1)$-connected space, which is an E_n-algebra (or an infinite loop space $\Omega^\infty Y$ associated to a connective spectrum), we can construct three augmented simplicial objects:

$$\cdots \mathrm{Free}_{E_n}\mathrm{Free}_{E_n}\mathrm{Free}_{E_n}\Omega^n Y \Rrightarrow \mathrm{Free}_{E_n}\mathrm{Free}_{E_n}\Omega^n Y \rightrightarrows \mathrm{Free}_{E_n}\Omega^n Y \to \Omega^n Y$$
$$\cdots \Omega^n\Sigma^n_+ \mathrm{Free}_{E_n}\mathrm{Free}_{E_n}\Omega^n Y \Rrightarrow \Omega^n\Sigma^n_+ \mathrm{Free}_{E_n}\Omega^n Y \rightrightarrows \Omega^n\Sigma^n_+\Omega^n Y \to \Omega^n Y$$
$$\cdots \Sigma^n_+ \mathrm{Free}_{E_n}\mathrm{Free}_{E_n}\Omega^n Y \Rrightarrow \Sigma^n_+ \mathrm{Free}_{E_n}\Omega^n Y \rightrightarrows \Sigma^n_+\Omega^n Y \to Y$$

These are, respectively, two-sided bar constructions: $B(\mathrm{Free}_{E_n}, \mathrm{Free}_{E_n}, \Omega^n Y)$, $B(\Omega^n\Sigma^n_+, \mathrm{Free}_{E_n}, \Omega^n Y)$, and $B(\Sigma^n_+, \mathrm{Free}_{E_n}, \Omega^n Y)$.

The first augmented bar construction $B(\mathrm{Free}_{E_n}, \mathrm{Free}_{E_n}, \Omega^n Y)$ has an extra degeneracy, and so its geometric realization is homotopy equivalent to $\Omega^n Y$ as E_n-spaces. Therefore, it is a group-complete E_n-space.

There is a natural map

$$B(\mathrm{Free}_{E_n}, \mathrm{Free}_{E_n}, \Omega^n Y) \to B(\Omega^n \Sigma^n_+, \mathrm{Free}_{E_n}, \Omega^n Y)$$

which is, levelwise, a group-completion map [29, Appendix Q], [66], and induces a group-completion map on geometric realization. However, the source is already group-complete, and so this map is an equivalence on geometric realizations. Thus, the augmentation $|B(\Omega^n \Sigma^n_+, \mathrm{Free}_{E_n}, \Omega^n Y)| \to \Omega^n Y$ is an equivalence.

The bar construction $B(\Sigma^n_+, \mathrm{Free}_{E_n}, \Omega^n Y)$ is a simplicial diagram of $(n-1)$-connected pointed spaces, and so by a theorem of May [62] we can commute Ω^n across geometric realization. The natural augmentation

$$\Omega^n |B(\Sigma^n_+, \mathrm{Free}_{E_n}, \Omega^n Y)| \to |B(\Omega^n \Sigma^n_+, \mathrm{Free}_{E_n}, \Omega^n Y)| \to \Omega^n Y$$

is an equivalence. By assumption, Y is $(n-1)$-connected and so $|B(\Sigma^n_+, \mathrm{Free}_{E_n}, \Omega^n Y)| \to Y$ is also an equivalence. Therefore, the simplicial object $B(\Sigma^n_+, \mathrm{Free}_{E_n}, \Omega^n Y)$ can be used to compute $H_* Y$.

Let $A = H_*(\Omega^n Y)$. The reduced homology of $B(\Sigma^n_+, \mathrm{Free}_{E_n}, \Omega^n Y)$ is

$$\cdots \Sigma^n \mathbb{Q}_{E_n} \mathbb{Q}_{E_n} A \Rrightarrow \Sigma^n \mathbb{Q}_{E_n} A \Rightarrow \Sigma^n A,$$

which is a bar complex $\Sigma^n B(Q, \mathbb{Q}_{E_n}, A)$ computing nonabelian derived functors. These are specifically the derived functors of an *indecomposables* functor Q, which takes an augmented $\mathbb{Q}_{E_n}$-algebra $A \to \mathbb{F}_2$ and returns the quotient of the augmentation ideal by all products, brackets, and Dyer–Lashof operations. The result is a Miller spectral sequence that begins with nonabelian derived functors of Q on $H_*(\Omega^n Y)$ and converges to $\widetilde{H}_* Y$.

19.5.4 Classical groups

The Dyer–Lashof operations on the homology of the spaces BO and BU, and hence on the homology of the Thom spectra MO and MU, was determined by work of Kochman [43]; here we will state a form due to Priddy [74].

Theorem 19.5.15. *The ring $H_* MO \cong H_* BO$ is a polynomial algebra on classes a_i in degree i. The Dyer–Lashof operations are determined by the identities of formal series*

$$\sum_j Q^j a_k = \left(\sum_{n=k}^{\infty} \sum_{u=0}^{k} \binom{n-k+u-1}{u} a_{n+u} a_{k-u} \right) \left(\sum_{n=0}^{\infty} a_n \right)^{-1},$$

where $a_0 = 1$ by convention. In particular, $Q^n a_k \equiv \binom{n-1}{k} a_{n+k}$ mod decomposable elements.

The ring $H_*MU \cong H_*BU$ is a polynomial algebra on classes b_i in degree $2i$. The Dyer–Lashof operations are determined by the identities of formal series

$$\sum_j Q^j b_k = \left(\sum_{n=k}^{\infty}\sum_{u=0}^{k}\binom{n-k+u-1}{u} b_{n+u}b_{k-u}\right)\left(\sum_{n=0}^{\infty} b_n\right)^{-1},$$

where $b_0 = 1$ by convention. In particular, $Q^{2n}b_k \equiv \binom{n-1}{k}b_{n+k}$ mod decomposable elements, and $Q^{2n+1}b_k = 0$.

Remark 19.5.16. Implicit in this calculation is the fact that the Thom isomorphisms $H_*MO \cong H_*BO$ and $H_*MU \cong H_*BU$ preserve Dyer–Lashof operations. Lewis showed that, for an E_n-map $f\colon X \to BGL_1(\mathbb{S})$, the Thom isomorphism $H_*X \cong H_*Mf$ lifts to an equivalence of E_n ring spectra

$$H[X] \to H \otimes Mf$$

called the Thom diagonal [49, 7.4]. As a result, the Thom isomorphism is automatically compatible with Dyer–Lashof operations for H-algebras.

Example 19.5.17. We have explicit calculations of the first few Dyer–Lashof operations in H_*MO:

$$\begin{aligned}
Q^2a_1 &= a_1^2\\
Q^4a_1 &= a_3 + a_1a_2 + a_1^3\\
Q^6a_1 &= a_1^4\\
Q^8a_1 &= a_5 + a_1a_4 + a_2a_3 + a_1^2a_3 + a_1a_2^2 + a_1^3a_2 + a_1^5\\
Q^6a_2 &= a_5 + a_1a_4 + a_2a_3 + a_1a_2^2
\end{aligned}$$

These same formulas hold for the b_i in H_*MU.

19.5.5 The Nishida relations and the dual Steenrod algebra

Recall that, if R is an E_n-algebra in Sp, $H\otimes R$ is an E_n-algebra in Mod_H whose homotopy groups are the homology groups of R. As a result, there are two types of operations on H_*R:

- The E_n-algebra structure gives $H_*(R)$ Dyer–Lashof operations $Q_0, \ldots, Q_{n-1}$ and a Browder bracket.
- The property of being homology gives $H_*(R)$ Steenrod operations $P_d\colon H_mR \to H_{m-d}R$. To make these dual to the Steenrod operations Sq^d in cohomology, $P_d(x)$ is defined as a composite

  $$S^{m-d} \xrightarrow{\Sigma^{-d}x} (\Sigma^{-d}H) \otimes R \xrightarrow{\chi Sq^d} H \otimes R.$$

  This implicitly reverses multiplication order: for example, the Adem relation $Sq^3 = Sq^1Sq^2$ becomes $P_3 = P_2P_1$.

The Nishida relations express how these structures interact.

Theorem 19.5.18. [20, III.1.1, III.3.2] *Suppose that R is an E_n-algebra in* Sp. *Then the Steenrod operations in homology satisfy relations as follows.*

Cartan formula: $P_r(xy) = \sum_{p+q=r} P_p(x)P_q(y)$.

Nishida relations: $P_rQ^s = \sum \binom{s-r}{r-2i} Q^{s-r+i}P_i$.

Browder Cartan formula: $P_r[x,y] = \sum_{p+q=r}[P_px, P_qy]$.

Remark 19.5.19. By contrast with the Adem relations, the Nishida relations behave very differently if we use lowerindexing. We find

$$P_rQ_s(x) = \sum \binom{|x|+s-r}{r-2i} Q_{s-r+i}P_i(x).$$

In particular, the lower-indexed Nishida relations depend on the degree of x [21].

Remark 19.5.20. If we use the pro-representability of homology as in §19.4.7, we can obtain a combined algebraic object that encodes both the Q^r and the P_d together with the Nishida relations.

19.5.6 Eilenberg–Mac Lane objects

If the homology H_*R is easily described a module over the Steenrod algebra, the Nishida relations can completely determine the Dyer–Lashof operations. This was applied by Steinberger to compute the Dyer–Lashof operations in the dual Steenrod algebra explicitly. (Conversely, Baker showed that the Nishida relations themselves are completely determined by the Dyer–Lashof operation structure of the dual Steenrod algebra [5].)

Theorem 19.5.21. [20, III.2.2, III.2.4] *Let A_* be the dual Steenrod algebra*

$$\mathbb{F}_2[\xi_1, \xi_2, \ldots]$$

where $|\xi_i| = 2^i - 1$, with conjugate generators $\overline{\xi}_i$ (here ξ_i is denoted by ζ_i in [70]). Then the Dyer–Lashof operations on the generators are determined by the following formulas.

1. *There is an identity of formal series*

$$(1 + \xi_1 + Q^1\xi_1 + Q^2\xi_1 + Q^3\xi_1 + \ldots) = (1 + \xi_1 + \xi_2 + \xi_3 + \ldots)^{-1}.$$

2. *For any i, we have*

$$Q^s\overline{\xi}_i = \begin{cases} Q^{s+2^i-2}\xi_1 & \text{if } s \equiv 0, -1 \mod 2^i, \\ 0 & \text{otherwise.} \end{cases}$$

3. *In particular, $Q^{2^i-2}\xi_1 = \overline{\xi}_i$, and $Q_1\overline{\xi}_i = \overline{\xi}_{i+1}$.*

Remark 19.5.22. This allows us to say that the dual Steenrod algebra can be re-expressed as follows:

$$A_* \cong \mathbb{F}_2[x, Q_1x, (Q_1)^2x, \ldots]$$

This is the same as the homology of $\Omega^2 S^3$: both are identified with the homology of the free E_2-algebra on a generator $x = \xi_1$ in degree 1. Mahowald showed that it was possible to realize this isomorphism of graded algebras: he constructed a Thom spectrum over $\Omega^2 S^3$ such that the Thom isomorphism realizes the isomorphism $A_* \cong H_*\Omega^2 S^3$ [55]. This has a rather remarkable interpretation: there exists a construction of the Eilenberg–Mac Lane spectrum H as the free E_2-algebra R such that the unit map $\mathbb{S} \to R$ has a chosen nullhomotopy of the image of 2. This result has been extended to odd primes by Blumberg–Cohen–Schlichtkrull [14].

Proposition 19.5.23. *Let Hk be the Eilenberg–Mac Lane spectrum for an algebra k over $\mathbb{F}_2$. Then there is an isomorphism*

$$H_* H\mathbb{F} \cong A_* \otimes k$$

of graded rings, and under this identification the Dyer–Lashof operation Q^r on H_*Hk is given by $Q^r \otimes \varphi$, where φ is the Frobenius on k.

Proof. For any H-module N, the action $H \otimes N \to N$ induces an isomorphism $H_*H \otimes \pi_* N \to H_*N$. We already know $Q^0(1 \otimes \alpha) = 1 \otimes \alpha^2$, and so by the Cartan formula it suffices to show that $Q^s(1 \otimes \alpha) = 0$ for $s > 0$.

We now proceed inductively by applying the Nishida relations. If we know $Q^t(1 \otimes \alpha) = 0$ for $0 < t < s$, we find that for all $r > 0$ we have

$$\begin{aligned} P_r Q^s(1 \otimes \alpha) &= \sum \binom{s-r}{r-2i} Q^{s-r+i} P_i (1 \otimes \alpha) \\ &= \binom{s-r}{r} Q^{s-r}(1 \otimes \alpha). \end{aligned}$$

By the inductive hypothesis, this vanishes unless $s = r$, but in the case $s = r$ the binomial coefficient vanishes. However, the only elements in H_*Hk that are acted on trivially by all the Steenrod operations are the elements in the image of π_*Hk, and those are concentrated in degree zero. Thus $Q^s(1 \otimes \alpha) = 0$. □

Remark 19.5.24. The same proof can be used to show that the Browder bracket is trivial on H_*Hk.

Example 19.5.25. The composite map $MU \to MO \to H$, on homology, is given in terms of the generators of Theorem 19.5.15 by $b_1 \mapsto a_1^2 \mapsto \xi_1^2$ and $b_2 \mapsto 0$. The image of H_*MU in A_* is $\mathbb{F}_2[\xi_1^2, \xi_2^2, \ldots]$, the homology of the Brown–Peterson spectrum BP.

In H_*MU, Example 19.5.17 implies we have the identities

$$Q^6 b_2 = b_5 + b_1 b_4 + b_2 b_3 + b_1 b_2^2 = Q^8 b_1 + b_1^2 Q^4(b_1).$$

By contrast, in the dual Steenrod algebra we have the identity $0 = Q^8(\xi_1^2) + \xi_1^4 Q^4(\xi_1^2)$. Even though the map $H_*MU \to H_*BP$ splits as a map of algebras, and the target is closed under the Dyer–Lashof operations, we have

$$Q^8 b_1 + b_1^2 Q^4(b_1) = Q_6(b_1) + b_1^2 Q_2(b_1) \neq 0$$

but its image is zero. This implies that the map $H_*MU \to H_*BP$ does not have a splitting that respects the Dyer–Lashof operations for E_7-algebras. As a result, there exists no map $BP \to MU_{(2)}$ of E_7-algebras. This result, and its analogue at odd primes, is due to Hu–Kriz–May [39].

19.5.7 Nonexistence results

The tremendous amount of structure present in the homology of a ring spectrum allows us to produce a rather large number of nonexistence results. The following is a generalization of the classical result that the mod-2 Moore spectrum does not admit a multiplication due to the existence of a nontrivial Steenrod operation Sq^2 in its cohomology; we learned this line of argument from Charles Rezk.

Proposition 19.5.26. *Suppose that R is a homotopy associative ring spectrum containing an element u in nonzero degree such that $P_k(u)$ vanishes either in the range $k > |u|$ or in the range $0 < k < |u|$. Then either $P_{|u|}(u)$ is nilpotent or H_*R is nonzero in infinitely many degrees.*

Proof. We find, by the Cartan formula, that

$$P_{d|u|}(u^d) = (P_{|u|}u)^d.$$

Therefore, either the elements u^d are nonzero for all d or the element $P_{|u|}u$ is nilpotent. □

Corollary 19.5.27. *Suppose that R is a connective homotopy associative ring spectrum such that $H_0(R) = \pi_0(R)/2$ has no nilpotent elements. If any nonzero element in $H_0(R)$ is in the image of the Steenrod operations, then H_*R must be nonzero in infinitely many degrees.*

Corollary 19.5.28. *Suppose that R is a homotopy associative ring spectrum and that some Hopf invariant element 2, η, ν, or σ maps to zero under the unit map $\mathbb{S} \to R$. Then either $H_*R = 0$ or H_*R is infinite-dimensional.*

Proof. Writing h for Hopf invariant element in degree $2^k - 1$ with trivial image, the unit $\mathbb{S} \to R$ extends to a map $f\colon C(h) \to R$ from the mapping cone. The homology of $C(h)$ has a basis of elements 1 and v with one nontrivial Steenrod operation acting via $P_{2^k}v = 1$, and $u = f_*(v)$ has the desired properties. □

Recall from §19.3.3 that, for R connective, a map $\pi_0 R \to A$ of commutative rings automatically extends to a map $R \to HA$ compatible with the multiplicative structure that exists on R; e.g., if R is homotopy commutative then the map $R \to HA$ has the structure of a map of $\mathcal{Q}_1$-algebras. This has the following consequence.

Proposition 19.5.29. *Suppose that R is a connective ring spectrum with a ring homomorphism $\pi_0 R \to k$ where k is an $\mathbb{F}_2$-algebra (equivalently, a map $H_0 R \to k$). Then there is a map $R \to Hk$ that induces a homology map $H_*R \to A_* \otimes k$ with the following properties.*

1. *The map $H_*R \to A_* \otimes k$ is a map of rings that is surjective in degree zero.*
2. *If R is homotopy commutative, then there is an operation Q_1 on H_*R that is compatible with the operation Q_1 on $A_* \otimes k$.*
3. *If R has an E_n-algebra structure, the map $R \to Hk$ is a map of E_n-algebras and so $H_*R \to A_* \otimes k$ is compatible with the Dyer–Lashof operations $Q_0, \ldots, Q_{n-1}$.*

In particular, the image of H_*R in $A_* \otimes k$ is a subalgebra $B_* \subset A_*$ closed under multiplication and some number of Dyer–Lashof operations.

Example 19.5.30. For $n > 0$ there are connective Morava K-theories $k(n)$, with coefficient ring $\mathbb{F}_2[v_n]$, that have homology

$$\mathbb{F}_2[\bar{\xi}_1, \ldots, \bar{\xi}_n, \bar{\xi}_{n+1}^2, \bar{\xi}_{n+2}, \ldots]$$

as a subalgebra of the dual Steenrod algebra. This subring is not closed under the Dyer–Lashof operation Q_1 unless $n = 0$, and so the connective Morava K-theories are not homotopy–commutative. (By convention we often define the connective Morava K-theory $k(0)$ to be $H\mathbb{Z}_2$, which is commutative.)

Similarly, for $n > 0$ the integral connective Morava K-theories $k_{\mathbb{Z}}(n)$, with coefficient ring $\mathbb{Z}_2[v_n]$, have homology

$$\mathbb{F}_2[\bar{\xi}_1^2, \bar{\xi}_2, \ldots, \bar{\xi}_n, \bar{\xi}_{n+1}^2, \bar{\xi}_{n+2}, \ldots]$$

as a subalgebra of the dual Steenrod algebra. This subring is not closed under the Dyer–Lashof operation Q_1 unless $n = 1$, and so the only possible homotopy-commutative integral Morava K-theory is $k_{\mathbb{Z}}(1)$—the connective complex K-theory spectrum.

There are obstruction-theoretic proofs which show that all of these have A_∞ structures [3, 48].

Example 19.5.31. The Dyer–Lashof operations satisfy $Q_2(\bar{\xi}_i^2) = \bar{\xi}_{i+1}^2$, and so the smallest possible subring of A_* that contains $\xi_1^2 = \bar{\xi}_1^2$ and is closed under Q_2 is an infinite polynomial algebra $\mathbb{F}_2[\bar{\xi}_i^2] = \mathbb{F}_2[\xi_i^2]$. If R is a connective ring spectrum with a quotient map $\pi_0 R \to \mathbb{F}$ such that the Hopf element $\eta \in \pi_1(\mathbb{S})$ maps to zero in $\pi_* R$, then there is a commutative diagram

$$\begin{array}{ccc} C(\eta) & \longrightarrow & R \\ \downarrow & & \downarrow \\ H\mathbb{Z}/2 & \longrightarrow & H\mathbb{F}. \end{array}$$

We conclude that ξ_1^2 is in the image of the map $H_*R \to H_*H\mathbb{F}$.

The spectra $X(n)$ appearing in the nilpotence and periodicity theorems of Devinatz–Hopkins–Smith fit into a sequence

$$X(1) \to X(2) \to X(3) \to \ldots$$

of Thom spectra on the spaces $\Omega SU(n)$. They have E_2-ring structures, and each ring $H_*X(n)$ is a polynomial algebra $\mathbb{F}_2[x_1, \ldots, x_{n-1}]$ on finitely many generators. For $n = 2$ the map

$H_*X(2) \to A_*$ is the map $\mathbb{F}_2[\xi_1^2] \to A_*$, and this implies that each $X(n)$ has ξ_1^2 in the image of its homology. As $H_*X(n)$ is finitely generated as an algebra, its image in the dual Steenrod algebra is too small to be closed under the operation Q_2. This excludes the possibility that $X(n)$ has an E_3-structure.

19.5.8 Ring spaces

Associated to an E_∞ ring spectrum E, there is a sequence of infinite loop spaces $\{E_n\}_{n\in\mathbb{Z}}$ in an Ω-spectrum representing E. These spaces are extremely strongly structured: they inherit both *additive* structure from the spectrum structure on E, and *multiplicative* structure from the E_∞ ring structure. In the case of the sphere spectrum, these operations were investigated in-depth in relationship to surgery theory [67, 54, 61]. Ravenel and Wilson discussed the structure coming from a ring spectrum E extensively in [75], encoding it in the structure of a *Hopf ring*, and the interaction between additive and multiplicative operations is developed in-depth in [23, §II]. These structures are very tightly wound.

1. Because the E_n are spaces, the diagonals $E_n \to E_n \times E_n$ gives rise to a *coproduct*

   $$\Delta\colon H_*(E_n) \to H_*(E_n) \otimes H_*(E_n),$$

   For an element x we write $\sum x' \otimes x''$ for its coproduct. The path components $E^n = \pi_0 E_n$ also give rise to elements $[\alpha] \in H_0 E_n$.

2. The homology groups $H_* E_n$ have Steenrod operations P_r.

3. The suspension maps $\Sigma E_n \to E_{n+1}$ in the spectrum structure give stabilization maps

   $$H_m(E_n) \to H_{m+1}(E_{n+1}),$$

4. The infinite loop space structure on E_n gives $H_* E_n$ an *additive* Pontryagin product

   $$\#\colon H_*(E_n) \otimes H_*(E_n) \to H_*(E_n)$$

   making it into a Hopf algebra, and it has *additive* Dyer–Lashof operations

   $$Q^r\colon H_m(E_n) \to H_{m+r}(E_n).$$

5. If E has a ring spectrum structure, the multiplication $E \otimes E \to E$ gives *multiplicative* Pontryagin products

   $$\circ\colon H_*(E_n) \otimes H_*(E_m) \to H_*(E_{n+m}).$$

   These are appropriately unital, associative, or graded-commutative if E has these properties.

6. If E has an E_∞ ring spectrum structure, there are *multiplicative* Dyer–Lashof operations

   $$\widetilde{Q}^r\colon H_m(E_0) \to H_{m+r}(E_0)$$

on the homology of the 0'th space. In general, we cannot say more. If E has further structure—an H_∞^d-structure—there are also multiplicative Dyer–Lashof operations outside degree zero [47, §4.1].

These are subject to a large number of identities discussed in [75, 1.12, 1.14], [23, II.1.5, II.1.6, II.2.5], and [45, 1.5]. Here are the most fundamental identities:

Distributive rule: $(x \# y) \circ z = \sum (x \circ z') \# (y \circ z'')$

Projection formula: $x \circ Q^s y = \sum Q^{s+k}(P_k x \circ y)$

Mixed Cartan formula:

$$\widetilde{Q}^n(x \# y) = \sum_{p+q+r=n} \widetilde{Q}^p(x') \# Q^q(x'' \circ y') \# \widetilde{Q}^r(y'')$$

Mixed Adem relations:

$$\widetilde{Q}^r Q^s x = \sum_{i+j+k+l=r+s} \binom{r-i-2l-1}{j+s-i-l} Q^i \widetilde{Q}^j x' \# Q^k \widetilde{Q}^l x''$$

Example 19.5.32. There is an identity

$$Q^1[a] \# [-2a] = \eta \cdot a$$

which allows us to determine information about the multiplication-by-η map $\pi_0 R \to \pi_1 R \to H_1 \Omega^\infty R$ from the additive Dyer–Lashof structure. Similarly $\widetilde{Q}^1$ determines information about its multiplicative version $\eta_m \colon \pi_0(R) \to \pi_1(R)$. For example, the mixed Cartan formula implies that

$$\eta_m(x + y) = \eta_m(x) + \eta_m(y) + \eta \cdot xy$$

in $H_1(R)$. In particular, $\widetilde{Q}^1[n] = \binom{n}{2}\eta \# [n^2]$ for $n \in \mathbb{Z}$ (cf. Example 19.3.11).

19.6 Higher-order structure

19.6.1 Secondary composites

Secondary operations, at their core, arise when there are relations between relations. Suppose that we are a sequence $X_0 \xrightarrow{f} X_1 \xrightarrow{g} X_2 \xrightarrow{h} X_3$ of maps such that the double composites are nullhomotopic. Then hgf is nullhomotopic for two reasons. Choosing nullhomotopies of gf and hg, we can glue the nullhomotopies together to determine a loop in the space of maps $X \to W$: a *value* of the associated secondary operation. Because we must make choices of nullhomotopy, there is some natural indeterminacy in this construction, and so it typically takes a set of values $\langle h, g, f\rangle$. To construct secondary operations, we minimally need to work in a category C with mapping spaces; we also need canonical basepoints of the spaces $\mathrm{Map}_{\mathsf{C}}(X_i, X_j)$ for $j \geq i + 2$ that are preserved under composition [47, §2].

Example 19.6.1. Suppose that A is a subspace of X and $\alpha \in H^n(X, A)$ is a cohomology element that restricts to zero in $H^n(X)$. Then the long exact sequence in cohomology implies that we can lift α to an element in $H^{n-1}(A)$, but there are multiple choices of lift. This can be represented by a sequence of maps

$$A \to X \to X/A \to K(\mathbb{Z}, n)$$

where the double composites are nullhomotopic; the secondary operation is then a map $A \to \Omega K(\mathbb{Z}, n) = K(\mathbb{Z}, n-1)$.

Secondary operations enrich the homotopy category $h\mathsf{C}$ with extra structure.

1. Every test object $T \in \mathsf{C}$ represents a functor $[T, -] = \pi_0 \mathrm{Map}_{\mathsf{C}}(T, -)$ on $h\mathsf{C}$, and if T has an augmentation $T \to 0$ to an initial object then this functor has a canonical null element. If the values of $[T, -]$ differ on X and Y, X and Y cannot be equivalent in $h\mathsf{C}$.
2. Every map of test objects $\Theta\colon S \to T$ determines an operation: a natural transformation of functors $\theta\colon [T, -] \to [S, -]$ on $h\mathsf{C}$. If S and T are augmented and the map Θ is compatible with the augmentations, then θ preserves the null element. If θ has different behaviour for X and Y, X and Y cannot be equivalent.
3. Given an augmented map $\Phi\colon R \to S$ and a map $\Theta\colon S \to T$ such that the double composite $\Theta\Phi\colon R \to T$ is trivial, we get an identity $\varphi\theta = 0$ of associated operations. There is an associated secondary operation $\langle -, \Theta, \Phi \rangle$. It is only defined on those elements $\alpha \in [T, X]$ with $\theta(\alpha) = 0$; it takes values in $\pi_1 \mathrm{Map}_{\mathsf{C}}(R, X)$; it is only well-defined up to indeterminacy.
4. We can also associate information to maps in the same way. Suppose we have an augmented map $\Theta\colon S \to T$ of test objects representing an operation θ. Given any map $f\colon X \to Y$, there is an associated functional operation $\langle f, -, \Theta \rangle$. It is only defined on those elements $\alpha \in [T, X]$ such that $f(\alpha) = 0$ and $\theta(\alpha) = 0$; it takes values in $\pi_1 \mathrm{Map}_{\mathsf{C}}(S, Y)$; it is only well-defined up to indeterminacy.

Applying this to the test objects S^n in the category of pointed spaces, we get Toda's bracket construction that enriches the homotopy groups of spaces with secondary composites. Applying this to the test objects $K(A, n)$ in the opposite of the category of spaces, we get Adams' secondary operations that enrich the cohomology groups of spaces with secondary cohomology operations.

19.6.2 Secondary operations for algebras

We recall from §19.4.3 that, for a spectrum M and an $\mathcal{O}$-algebra in Mod_E, we have

$$\pi_M(A) = [M, A]_{\mathsf{Sp}} \cong [E \otimes \mathrm{Free}_{\mathcal{O}}(M), A]_{\mathsf{Alg}_{\mathcal{O}}(\mathrm{Mod}_E)}.$$

Using free algebras as our test objects, we already used this representability of homotopy groups to classify the natural operations on the homotopy groups of $\mathcal{O}$-algebras in Mod_E. The space of maps now means that we can construct secondary operations.

Proposition 19.6.2. *Suppose that we have zero-preserving operations $\theta\colon \pi_M \to \pi_N$ and $\varphi\colon \pi_N \to \pi_P$ on the homotopy category of $\mathcal{O}$-algebras in* Mod_E, *and that there is a relation $\varphi \circ \theta = 0$. Then there exists a secondary operation*

$$\langle -, \Theta, \Phi \rangle \colon \pi_M(A) \supset \mathrm{Ker}\theta \to \pi_{P+1}(A)/Im(\sigma\varphi),$$

where $\sigma(\varphi)$ is a suspended operation (see §19.4.6).

Such a secondary operation is constructed from a sequence

$$E \otimes \mathrm{Free}_{\mathcal{O}}(P) \xrightarrow{\Phi} E \otimes \mathrm{Free}_{\mathcal{O}}(N) \xrightarrow{\Theta} E \otimes \mathrm{Free}_{\mathcal{O}}(M) \to A$$

where the double composites are null; the nullhomotopy of $\Theta \circ \Phi$ is chosen once and for all, while the second nullhomotopy is allowed to vary. This produces elements in

$$\pi_1 \mathrm{Map}_{\mathrm{Alg}_{\mathcal{O}}(\mathrm{Mod}_E)}(E \otimes \mathrm{Free}_{\mathcal{O}}(P), A) \cong \pi_1 \mathrm{Map}_{\mathrm{Sp}}(P, A) \cong [\Sigma P, A].$$

Example 19.6.3. Every Adem relation between Dyer–Lashof operations produces a secondary Dyer–Lashof operation. For example, the relation $Q^{2n+2}Q^n + Q^{2n+1}Q^{n+1} = 0$ produces a natural transformation

$$\pi_m(A) \supset \mathrm{Ker}(Q^n, Q^{n+1}) \to \pi_{m+2n+3}(A)/Im(Q^{2n+2}, Q^{2n+1})$$

on the homotopy of H-algebras.

Example 19.6.4. Relations involving operations other than composition and addition can also produce secondary operations, and the canonical examples of these are *Massey products*. An $\mathcal{A}_2$-algebra R has a binary multiplication operation $R \otimes R \to R$, and if R is an $\mathcal{A}_3$-algebra it has a chosen associativity homotopy. As a result, if we have elements x, y, and z in $\pi_* R$ such that $xy = yz = 0$, then we can glue together two nullhomotopies of xyz to obtain a *bracket* $\langle x, y, z \rangle$ that specializes to definitions of Massey products or Toda brackets.

In trying to express nonlinear relations as secondary operations, however, we rapidly find that we want to move into a *relative* situation. A Massey product is defined on the kernel of the map $\pi_p \times \pi_q \times \pi_r \to \pi_{p+q} \times \pi_{q+r}$ sending (x, y, z) to (xy, yz). However, the relation $x(yz) = (xy)z$ is not expressible solely as some operation on xy and yz: we need to remember x and z as well, but we *do not* want to enforce that they are zero.

We find that the needed expression is homotopy commutativity of the following diagram:

$$\begin{array}{ccc}
\mathrm{Free}_{\mathcal{A}_3}(S^p \oplus S^{p+q+r} \oplus S^r) & \xrightarrow{\Phi} & \mathrm{Free}_{\mathcal{A}_3}(S^p \oplus S^{p+q} \oplus S^{q+r} \oplus S^r) \\
\downarrow & & \downarrow \Theta \\
\mathrm{Free}_{\mathcal{A}_3}(S^p \oplus S^r) & \longrightarrow & \mathrm{Free}_{\mathcal{A}_3}(S^p \oplus S^q \oplus S^r)
\end{array}$$

The right-hand map classifies the operation $\theta(x, y, z) = (x, xy, yz, z)$, and the top map classifies the operation $\varphi(x, u, v, z) = (x, xv - uz, z)$. The bottom-left object is not the initial object in the category of $\mathcal{A}_3$-algebras, so we *enforce* this by switching to the category C of $\mathcal{A}_3$-algebras under $\mathrm{Free}_{\mathcal{A}_3}(S^p \oplus S^r)$. In this category, we genuinely have augmented objects with a nullhomotopic double composite

$$\mathrm{Free}_{\mathsf{C}}(S^{p+q+r}) \to \mathrm{Free}_{\mathsf{C}}(S^{p+q} \oplus S^{q+r}) \to \mathrm{Free}_{\mathsf{C}}(S^q)$$

that defines a Massey product.

19.6.3 Juggling

Secondary operations are part of the homotopy theory of C, and there is typically no method to determine secondary operations purely in terms of the homotopy category. However, there are many composition-theoretic tools that use one secondary operation to determine information about another: typically, one starts with a 4-fold composite

$$X \xrightarrow{f} Y \xrightarrow{g} Z \xrightarrow{h} U \xrightarrow{k} V,$$

with some assortment of double-composites being nullhomotopic, and relates various associated secondary operations. This process is called *juggling*, and learning to juggle secondary operations is one of the main steps in applying them. For instance, one of the main juggling formulas—the Peterson–Stein formula—asserts that the sets $\langle k, h, g\rangle f$ and $k\langle h, g, f\rangle$ are inverse in π_1 when both sides make sense.

Example 19.6.5. The Adem relations $Q^{2n+1}Q^n$ and $Q^{4n+3}Q^{2n+1}$ give rise to a secondary operation $\langle \mathbf{Q}^n, \mathbf{Q}^{2n+1}, \mathbf{Q}^{4n+3}\rangle$, an element of $\pi_{7n+5+m}(H \otimes \mathrm{Free}_{E_\infty}(S^m))$ representing an operation that increases degree by $7n+5$. The juggling formula says that, for any element $\alpha \in \pi_m(A)$ with $Q^n(A) = 0$, we have

$$Q^{4n+3}\langle \alpha, \mathbf{Q}^n, \mathbf{Q}^{2n+1}\rangle = \langle \mathbf{Q}^n, \mathbf{Q}^{2n+1}, \mathbf{Q}^{4n+3}\rangle(\alpha).$$

In other words, this secondary composite of operations gives a universal formula for how to apply Q^{4n+3} to this secondary operation.

19.6.4 Application to the Brown–Peterson spectrum

In this section we will give a brief account of the main result of [47], which uses secondary operatons to show that the 2-primary Brown–Peterson spectrum BP does not admit the structure of an E_{12} ring spectrum. These results have been generalized by Senger to show that, at the prime p, BP does not have an E_{2p^2+4} ring structure [86].

As in §19.3.3, if the Brown–Peterson spectrum has an E_n-algebra structure then the map

$$BP \to H\mathbb{Z}_{(2)} \to H$$

can be given the structure of a map of E_n-algebras. On homology, this would then induce a monomorphism

$$\mathbb{F}_2[\xi_1^2, \xi_2^2, \dots] \to \mathbb{F}_2[\xi_1, \xi_2, \dots]$$

of algebras equipped with E_n Dyer–Lashof operations and secondary Dyer–Lashof operations. The dual Steenrod algebra, on the right, has operations that are completely forced.

Therefore, if we can calculate enough to show that the subalgebra H_*BP is not closed under secondary operations for E_n-algebras, we arrive at an obstruction to giving BP the structure of E_n ring spectrum.

The calculation of secondary operations in H_*H is accomplished with judicious use of juggling formulas, ultimately reducing questions about secondary operations to questions about primary ones.

- There is a pushout diagram of E_∞ ring spectra

$$\begin{array}{ccc} H \otimes MU & \xrightarrow{i} & H \otimes H \\ \downarrow & & \downarrow j \\ H & \longrightarrow & H \otimes_{MU} H. \end{array}$$

  This makes $H \otimes MU$ into an augmented H-algebra, and gives a nullhomotopy of the composite $H \otimes MU \to H \otimes H \to H \otimes_{MU} H$. The elements α in H_*MU that map to zero in H_*H are then candidates for secondary operations: we can construct $\langle j, i, \alpha\rangle$ in the MU-dual Steenrod algebra $\pi_*(H \otimes_{MU} H)$.

- These elements are concretely detected: they have explicit representatives on the 1-line of a two-sided bar spectral sequence

$$\mathrm{Tor}^{H_*MU}(H_*H, \mathbb{F}_2) \Rightarrow \pi_*(H \otimes_{MU} H).$$

- If we can determine *primary* operations $\theta(\langle i, j, \alpha\rangle)$ in the MU-dual Steenrod algebra, the juggling formulas of §19.6.3 tell us about functional operations $\langle j, \alpha, \Theta\rangle$ in the ordinary dual Steenrod algebra.

- Steinberger's calculations of primary operations φ in the dual Steenrod algebra then allow us to determine the values of $\varphi\langle j, \alpha, \Theta\rangle$, and juggling formulas again allow us to determine information about secondary operations $\langle \alpha, \Theta, \Phi\rangle$ in the dual Steenrod algebra.

This method, then, reduces us to carrying out some key computations.

We must determine primary operations in the MU-dual Steenrod algebra. Some of these, by work of Tilson [92], are determined by Kochman's calculations from Theorem 19.5.15: the Künneth spectral sequence

$$\mathrm{Tor}^{\pi_*MU}(\mathbb{F}_2, \mathbb{F}_2) \Rightarrow \pi_*(H \otimes_{MU} H)$$

calculating the MU-dual Steenrod algebra is compatible with Dyer–Lashof operations. However, there are remaining extension problems in the Tor, and these turn out to be precisely what we are interested in when juggling.

The MU-dual Steenrod algebra is an exterior algebra, whose generators correspond to the indecomposables in π_*MU. The extension problems in the Tor spectral sequence arise because some generators in π_*MU have nontrivial image in H_*MU and are detected by Tor_1, while others have trivial image in H_*MU and are detected by Tor_0. The solution is to find an algebra R mapping to MU that does not have this problem. If we can find one so that the map $\pi_*R \to \pi_*MU$ is surjective, the map from the R-dual Steenrod algebra to the MU-dual Steenrod algebra is surjective. If the generators of π_*R have nontrivial image in H_*R, then the spectral sequence

$$\mathrm{Tor}^{H_*R}(H_*H, \mathbb{F}_2) \Rightarrow \pi_*(H \otimes_R H)$$

detects all needed classes with Tor_1 and hence eliminates the extension problem.

For this purpose, we used the spherical group algebra $\mathbb{S}[SL_1(MU)]$. The Dyer–Lashof operations in $H_*SL_1(MU)$ are derived from the multiplicative Dyer–Lashof operations $\widetilde{Q}^n$ in $\Omega^\infty MU$. This is a lengthy calculation of power operations in the Hopf ring, and it is ultimately determined by calculations of Johnson–Noel of power operations in the formal group theory of MU [40].

Finally, we must determine a candidate secondary operation in H_*H to which we can apply this procedure—there are many candidate operations and many dead ends. The secondary operation is rather large: it was found using a calculation in Goerss–Hopkins obstruction theory that is detailed at length in [46].

19.7 Coherent structures

In §19.3 we discussed algebras over an operad in a general topological category, or more generally algebras over a multicategory $\mathcal{M}$, including extended power and free algebra functors. The definitions we used made heavy use of a strict symmetric monoidal structure on the category of spectra.

In this section we will discuss the coherent viewpoint on these constructions that makes use of the machinery of Lurie [53] and with the goal of connecting different strata in the literature. To begin, we should point some of the problems that this discussion is meant to solve.

We would like to demonstrate that our constructions are model-independent. There are several different symmetric monoidal categories of spectra [27, 38, 82] with several different model structures, and there is a nontrivial amount of work involved in showing that an equivalence between two different categories of spectra gives an equivalence between categories of algebras [84]. These issues are compounded when we attempt to relate notions of commutative algebras in different categories, even if they have equivalent homotopy theory [93].

We would also like to allow weaker structure than a symmetric monoidal structure. For example, given a fixed E_n-algebra R we will use this to discuss the classification of power operations on E_n-algebras under R. Our natural home for this discussion will be the category of E_n R-modules (as in Example 19.6.4).

19.7.1 Structured categories

As discussed in §19.1, classical symmetric monoidal categories are analogues of commutative monoids with the difference that they require natural isomorphisms to express associativity, commutativity, and the like. We can express this structure using simplicial operads. For any categories C and D, there is a groupoid $\mathrm{Fun}(\mathsf{C}, \mathsf{D})^{\simeq}$ of functors and natural isomorphisms. Taking the nerve, we get a simplicially enriched category $\mathcal{C}at$, and it makes sense to ask whether C has the structure of an algebra over a simplicial operad $\mathcal{O}$.

Example 19.7.1. A symmetric monoidal category can be expressed as an algebra over the Barratt–Eccles operad [6].

Example 19.7.2. In classical category theory, a *braided monoidal category* in the sense of [41] can be encoded by a sequence of maps

$$NP_n \to \mathrm{Fun}(\mathsf{C}^n, \mathsf{C})$$

from the nerves of the pure braid groups to the categories of functors $\mathsf{C}^n \to \mathsf{C}$. The required compatibilities between these maps can be concisely expressed by noting that these nerves assemble into an E_2-operad, and that a braided monoidal category is an algebra over this operad.

We would like to discuss E_n-analogues of these structures in the context of categories with morphism spaces. We will give some definitions in this section, on the point-set level, with the purpose of interpolating the older and newer definitions. We would like to say that an $\mathcal{O}$-monoidal category is an algebra over the operad $\mathcal{O}$ in $\mathcal{C}at$, but this requires us to be clever enough to have a well-behaved definition of a *space* of functors between two enriched categories; the failure of enriched categories to have a well-behaved enriched functor category is a principal motivation for the use of quasicategories.

Until further notice, all categories and multicategories are assumed to be enriched in spaces and all functors are functors of enriched categories.

Definition 19.7.3. Suppose that $p\colon \mathsf{C} \to \mathcal{M}$ is a multifunctor, and write $\mathsf{C}_{\mathbf{x}}$ for the category $p^{-1}(\mathbf{x})$. Given objects $X_i \in \mathsf{C}_{\mathbf{x}_i}$ and a map

$$\alpha\colon A \to \mathrm{Mul}_{\mathcal{M}}(\mathbf{x}_1, \ldots, \mathbf{x}_d; \mathbf{y})$$

of spaces, an *α-twisted product* is an object $Y \in \mathsf{C}_{\mathbf{y}}$ and a map $A \to \mathrm{Mul}_{\mathsf{C}}(X_1, \ldots, X_d; Y)$ such that, for any $Z \in \mathsf{C}$ with $p(Z) = \mathbf{z}$, the diagram

$$\begin{array}{ccc} \mathrm{Map}_{\mathsf{C}}(Y, Z) & \longrightarrow & \mathrm{Map}(A, \mathrm{Mul}_{\mathsf{C}}(X_1, \ldots, X_d; Z)) \\ \downarrow & & \downarrow \\ \mathrm{Map}_{\mathcal{M}}(\mathbf{y}, \mathbf{z}) & \longrightarrow & \mathrm{Map}(A, \mathrm{Mul}_{\mathcal{M}}(\mathbf{x}_1, \ldots, \mathbf{x}_d; \mathbf{z})) \end{array}$$

is a pullback. If it exists, we denote it by $A \ltimes_{\alpha} (X_1, \ldots, X_d)$.

Definition 19.7.4. A *weakly $\mathcal{M}$-monoidal category* is a multifunctor $p\colon \mathsf{C} \to \mathcal{M}$ that has α-twisted product for any inclusion

$$\alpha\colon \{f\} \subset \mathrm{Mul}_{\mathcal{M}}(\mathbf{x}_1, \dots, \mathbf{x}_d; \mathbf{y}).$$

A *strongly $\mathcal{M}$-monoidal category* is a category that has α-twisted products for all α.

Remark 19.7.5. In particular, for any point $f \in \mathrm{Mul}_{\mathcal{M}}(\mathbf{x}_1, \dots, \mathbf{x}_d; \mathbf{y})$, this universal property can be used to produce a functor

$$\{f\} \ltimes (-)\colon \mathsf{C}_{\mathbf{x}_1} \times \cdots \times \mathsf{C}_{\mathbf{x}_d} \to \mathsf{C}_{\mathbf{y}},$$

and these are compatible with composition (up to natural isomorphism). A weakly $\mathcal{M}$-monoidal category determines, up to natural equivalence, a multifunctor $\mathcal{M} \to \mathcal{C}at$.

Example 19.7.6. Every multicategory C has a multifunctor to the one-object multicategory $\mathcal{C}omm$ associated to the commutative operad. The multicategory is $\mathcal{C}omm$-monoidal if and only if multimorphisms $(X_1, \dots, X_d) \to Y$ are always representable by an object $X_1 \otimes \cdots \otimes X_d$, which is precisely when C comes from a symmetric monoidal category. It is strongly $\mathcal{C}omm$-monoidal only if it is also tensored over spaces in a way compatible with the monoidal structure as in Definition 19.3.9.

Example 19.7.7. Associated to a monoidal category C we can build a multicategory: multimorphisms $(X_1, \dots, X_d) \to Y$ are pairs of a permutation $\sigma \in \Sigma_d$ and a map $f\colon X_{\sigma(1)} \otimes \cdots \otimes X_{\sigma(d)} \to Y$. There is a multifunctor from this category to the multicategory $\mathcal{A}ssoc$ corresponding to the associative operad: it sends all objects to the unique object, and sends each multimorphism (σ, f) as above to the permutation σ. Conversely, an $\mathcal{A}ssoc$-monoidal category comes from a monoidal category.

Example 19.7.8. Suppose that A is a commutative ring and B is an A-algebra. Then there is a multicategory C as follows.

1. An object of C is either an A-module or a right B-module.
2. The set $\mathrm{Mul}_{\mathsf{C}}(M_1, \dots, M_d; N)$ of multimorphisms is

$$\begin{cases} \mathrm{Hom}_A(M_1 \otimes_A \cdots \otimes_A M_d, N) & \text{if } N \text{ and all } M_i \text{ are } A\text{-modules,} \\ \mathrm{Hom}_B(M_1 \otimes_A \cdots \otimes_A M_d, N) & \text{if } N \text{ and exactly one } M_i \text{ are } B\text{-modules,} \\ \emptyset & \text{otherwise.} \end{cases}$$

This comes equipped with a functor from C to the multicategory Mod from Example 19.3.23 that parametrizes ring-module pairs: any A-module is sent to $\mathbf{a}$ and any B-module is sent to $\mathbf{m}$. This makes C into a Mod-monoidal category, expressing the fact that Mod_A has a tensor product and that objects of RMod_B can be tensored with objects of Mod_A. This makes RMod_B *left-tensored* over Mod_A.

Example 19.7.9. Fiberwise homotopy theory studies the category $\mathsf{S}_{/B}$ of spaces over B. Let $\mathcal{O}$ be an operad and B be a space with the structure of an $\mathcal{O}$-algebra. Then $\mathsf{S}_{/B}$ has the structure of a strongly $\mathcal{O}$-monoidal category in the following way. For spaces $X_1, \ldots, X_d$ and Y over B, the space of multimorphisms is the pullback

$$\begin{array}{ccc} \mathrm{Mul}_{/B}(X_1, \ldots, X_d; Y) & \longrightarrow & \mathrm{Map}(X_1 \times \cdots \times X_d, Y) \\ \downarrow & & \downarrow \\ \mathcal{O}(d) \longrightarrow \mathrm{Map}(B^d, B) & \longrightarrow & \mathrm{Map}(X_1 \times \cdots \times X_d, B). \end{array}$$

That is, a multimorphism consists of a point $f \in \mathcal{O}(d)$ and a commutative diagram

$$\begin{array}{ccc} X_1 \times \cdots \times X_d & \longrightarrow & Y \\ \downarrow & & \downarrow \\ B^d & \xrightarrow[f]{} & B. \end{array}$$

With this definition, it is straightforward to verify that for $\alpha\colon A \to \mathcal{O}(d)$, the α-twisted product $A \ltimes_\alpha (X_1, \ldots, X_d)$ is the following space over B:

$$A \times X_1 \times \cdots \times X_d \to \mathcal{O}(d) \times B^d \to B.$$

In general, this should not be expected to be part of a symmetric monoidal structure on the category of spaces over B, even up to equivalence.

Example 19.7.10. Let $\mathcal{L}$ be the category of *universes*: an object is a countably infinite dimensional inner product space U. These objects have an associated multicategory: the space $\mathrm{Mul}_{\mathcal{L}}(U_1, \ldots, U_d; V)$ of multimorphisms is the (contractible) space of linear isometric embeddings $U_1 \oplus \cdots \oplus U_d \hookrightarrow V$. Over $\mathcal{L}$, there is a category $\mathrm{Sp}_{\mathcal{L}}$ of *indexed spectra*. An object is a pair (U, X) of a universe U and a spectrum X (in the Lewis–May–Steinberger sense [49]) indexed on U; a multimorphism $((U_1, X_1), \ldots, (U_d, X_d)) \to (V, Y)$ is a pair of a linear isometric embedding $i\colon U_1 \oplus \cdots \oplus U_d \to V$ and a map $i_*(X_1 \wedge \cdots \wedge X_d) \to Y$ of spectra indexed on V.

This does not describe the topology on the multimorphisms in this category. Given a map $A \to \mathcal{L}(U_1, \ldots, U_d; V)$ and spectra X_i indexed on U_i, there is a *twisted half-smash product* $A \ltimes (X_1, \ldots, X_d)$ indexed on V [49, §VI], equivalent to the smash product $A_+ \wedge X_1 \wedge \cdots \wedge X_d$. There exists a topology on the multimorphisms so that a continuous map in from A is equivalent to a map $A \to \mathcal{L}(U_1, \ldots, U_d; V)$ and a map $A \ltimes (X_1, \ldots, X_d) \to Y$. By design, then, the projection $\mathrm{Sp}_{\mathcal{L}} \to \mathcal{L}$ makes the category of indexed spectra strongly $\mathcal{L}$-monoidal.

Example 19.7.11. Fix an E_n-algebra A in Sp, and consider the category of E_n-algebras R with a factorization $A \to R \to A$ of the identity map. This has an associated *stable category*, serving as the natural target for Goodwillie's calculus of functors: the category of E_n A-modules [28]. This category should also not be expected to have a symmetric monoidal structure, but the tensor product over R does give it the structure of an E_n-monoidal category. For example, for an associative algebra A in Sp, the tensor product over A gives the category of A-bimodules a monoidal structure.

19.7.2 Multi-object algebras

Just as we cannot make sense of a commutative monoid in a nonsymmetric monoidal category, we need relationships between an operad $\mathcal{O}$ and any multiplicative structure on a category C before $\mathcal{O}$ can act on objects.

Definition 19.7.12. Suppose $p\colon \mathsf{C} \to \mathcal{N}$ and $\mathcal{M} \to \mathcal{N}$ are multifunctors. An *$\mathcal{M}$-algebra in* C is a lift in the diagram

$$\begin{array}{ccc} & & \mathsf{C} \\ & \nearrow & \downarrow p \\ \mathcal{M} & \longrightarrow & \mathcal{N} \end{array}$$

of multifunctors. We write $\mathsf{Alg}_{\mathcal{M}/\mathcal{N}}(\mathsf{C})$ for this category of $\mathcal{M}$-algebras.

Example 19.7.13. If C and $\mathcal{M}$ are arbitrary multicategories, then using the unique maps from C and $\mathcal{M}$ to the terminal multicategory $\mathcal{C}omm$ we recover the definition of $\mathsf{Alg}_{\mathcal{M}}(\mathsf{C})$, the category of $\mathcal{M}$-algebras in C from Definition 19.3.22.

Example 19.7.14. Let the space B be an algebra over an operad $\mathcal{O}$ and consider the fiberwise category $\mathsf{S}_{/B}$ of spaces over B with the strongly $\mathcal{O}$-monoidal structure from Example 19.7.9. An $\mathcal{O}$-algebra in $\mathsf{S}_{/B}$ is an $\mathcal{O}$-algebra X with a map of $\mathcal{O}$-algebras $X \to B$.

Example 19.7.15. Consider the category of indexed spectra $\mathsf{Sp}_{\mathcal{L}}$ from Example 19.7.10. The fact that the external smash product $(X_1 \wedge \cdots \wedge X_n)$ is naturally indexed on the direct sum of the associated universes obstructed making the category of spectra indexed on any individual universe Sp strictly symmetric monoidal, and so we cannot ask about commutative monoids in $\mathsf{Sp}_{\mathcal{L}}$—but the structure available is still enough to do multiplicative homotopy theory. An $\mathcal{L}$-algebra in $\mathsf{Sp}_{\mathcal{L}}$ recovers the classical definition of an E_∞ ring spectrum from [63]. Similarly we can define $\mathcal{O}$-algebras for any operad $\mathcal{O}$ with an augmentation to $\mathcal{L}$ [49, VII.2.1].

Proposition 19.7.16. *Suppose that C is strongly $\mathcal{N}$-monoidal and that $\mathcal{M} \to \mathcal{N}$ is a map of multicategories. In addition, suppose that C has enriched colimits and that formation of α-twisted products preserves enriched colimits in each variable.*

1. *For objects $\mathbf{x}$ and $\mathbf{y}$ of $\mathcal{M}$, there are extended power functors*

$$\mathrm{Sym}^k_{\mathcal{M},\mathbf{x}\to\mathbf{y}}\colon \mathsf{C}_{\mathbf{x}} \to \mathsf{C}_{\mathbf{y}},$$

given by

$$\mathrm{Sym}^k_{\mathcal{M},\mathbf{x}\to\mathbf{y}}(X) = \mathrm{Mul}_{\mathcal{M}}(\underbrace{\mathbf{x},\mathbf{x},\ldots,\mathbf{x}}_{k};\mathbf{y}) \ltimes (X, X, \ldots, X)/\Sigma_k.$$

2. *The evaluation functor* $\mathrm{ev}_{\mathbf{x}}\colon \mathsf{Alg}_{\mathcal{M}}(\mathsf{C}) \to \mathsf{C}_{\mathbf{x}}$ *has a left adjoint*

$$\mathrm{Free}_{\mathcal{M},\mathbf{x}}\colon \mathsf{C}_{\mathbf{x}} \to \mathsf{Alg}_{\mathcal{M}}(\mathsf{C}).$$

The value of $\mathrm{Free}_{\mathcal{M},\mathbf{x}}(X)$ *on any object* $\mathbf{y}$ *of* $\mathcal{M}$ *is*

$$\mathrm{ev}_{\mathbf{y}}(\mathrm{Free}_{\mathcal{M},\mathbf{x}}(X)) = \coprod_{k\geq 0} \mathrm{Sym}^k_{\mathcal{M},\mathbf{x}\to\mathbf{y}}(X).$$

Example 19.7.17. Let B be a space with an action of an operad $\mathcal{O}$, and let X a space over B. Then the extended powers are

$$\mathrm{Sym}^k_{\mathcal{O}}(X) = \left(\mathcal{O}(k) \times_{\Sigma_k} X^k \to \mathcal{O}(k) \times_{\Sigma_k} B^k \to B\right).$$

Example 19.7.18. Suppose that Γ is a commutative monoid and that X is a Γ-graded E_∞ ring spectrum, as in Example 19.3.24. Then there are action maps $\mathrm{Sym}^k X_g \to X_{kg}$. These give rise to Dyer–Lashof operations $Q^i\colon H_*X_g \to H_{*+i}(X_{2g})$.

Example 19.7.19. Suppose that $\cdots \to X_2 \to X_1 \to X_0$ is a strongly filtered E_∞ ring spectrum, as in Example 19.3.25. Then there are action maps $\mathrm{Sym}^k X_n \to X_{kn}$ that are compatible. These give rise to power operations $Q^i\colon H_*X_n \to H_{*+i}X_{2n}$ that are compatible as n varies, and there are induced power operations on the associated spectral sequence.

Example 19.7.20. Given a spectrum X indexed on a universe U as in Example 19.7.10, the extended powers are modeled by twisted half-smash products:

$$\mathrm{Sym}^k_{U \to U}(X) \simeq E\Sigma_k \ltimes_{\Sigma_k} (X^{\wedge k}).$$

This recovers the machinery that was put to effective use in the 1970s and 1980s for studying E_∞ ring spectra and H_∞-ring spectra, before the development of strictly monoidal categories of spectra.

19.7.3 ∞-operads

The point-set discussion of the previous sections provides a library of examples. As the basis for a theory it relies on the existence of rigid models and preservation of colimits.

Example 19.7.21. Consider the fiberwise category of spaces over a fixed base space B. This category has a symmetric monoidal fiber product $X \times_B Y$. The fiber product typically needs fibrant input to represent the homotopy fiber product; the fiber product typically does not produce cofibrant output. This makes it difficult to use the standard machinery to study algebras and modules in this category. These problems have received significant attention in the setting of parametrized stable homotopy theory [64, 51, 52].

Example 19.7.22. The category of nonnegatively graded chain complexes over a commutative ring R is equivalent to the category of simplicial R-modules via the Dold–Kan correspondence. This correspondence is lax symmetric monoidal in one direction, but only lax monoidal in the other. Moreover, while both sides have morphism spaces, the Dold–Kan correspondence only preserves these up to weak equivalence, even for fibrant-cofibrant objects.

Example 19.7.23. In the standard models of equivariant stable homotopy theory the notion of strict G-commutativity is equivalent to one encoded by equivariant operads rather than ordinary ones [58, 35, 15]. This means that an E_∞-algebra A (in the sense of an ordinary E_∞ operad) may not have a strictly commutative model [65, 34], and this makes it more difficult to construct a symmetric monoidal model for the category of A-modules.

The framework of ∞-operads [53] (or, alternatively, that of dendroidal sets [71]) is one method to express coherent multiplicative structures. Here are some of the salient points.

- This generalization takes place in the theory of ∞-categories (specifically quasicategories), equivalent to the study of categories enriched in spaces. Every category enriched in spaces gives rise to an ∞-category; every ∞-category has morphism spaces between its objects.

- In this framework, for ∞-categories C and D there is a space $\mathrm{Fun}(\mathsf{C}, \mathsf{D})$ encoding the structure of functors and natural equivalences.

- In an ∞-category, homotopy limits and colimits are intrinsic notions rather than arising from a particular construction. Many common constructions produce presentable ∞-categories, which have all homotopy limits and colimits.

- Multicategories generalize to so-called *∞-operads*. These have an underlying ∞-category, and there are spaces of multimorphisms to an object from a tuple of objects. Every topological multicategory gives rise to an ∞-operad; every ∞-operad can be realized by a topological multicategory. The precise definitions are similar in spirit to Segal's encoding of E_∞-spaces [85].

- An ∞-operad $\mathcal{O}$ has an associated notion of an $\mathcal{O}$-monoidal ∞-category. An $\mathcal{O}$-monoidal ∞-category is expressed in terms of maps $\mathsf{C} \to \mathcal{O}$ of ∞-operads with properties analogous to that from Definition 19.7.4, with the main difference that spaces of morphisms are respected. An $\mathcal{O}$-monoidal ∞-category is also equivalent to a functor from $\mathcal{O}$ to a category of categories: each object $\mathbf{x}$ of $\mathcal{O}$ has an associated category $\mathsf{C}_{\mathbf{x}}$, and one can associate a map
$$\mathrm{Mul}_{\mathcal{O}}(\mathbf{x}_1, \ldots, \mathbf{x}_d; \mathbf{y}) \to \mathrm{Fun}(\mathsf{C}_{\mathbf{x}_1}, \ldots, \mathsf{C}_{\mathbf{x}_d}; \mathsf{C}_{\mathbf{y}})$$
of spaces.

- We can discuss algebras and modules in terms of sections, just as in Definition 19.7.12.

All of this structure is systematically invariant under equivalence. Equivalent ∞-operads give rise to equivalent notions of an $\mathcal{O}$-algebra structure on C; ∞-categories equivalent to C have equivalent notions of $\mathcal{O}$-algebra structures to those on C; equivalent $\mathcal{O}$-monoidal ∞-categories have equivalent categories of $\mathcal{M}$-algebras for any map $\mathcal{M} \to \mathcal{O}$ of ∞-operads.

Example 19.7.24. An E_n-operad has an associated ∞-operad $\mathcal{O}$, and as a result we can define an E_n-monoidal ∞-category C to be an $\mathcal{O}$-monoidal ∞-category. When $n = 1$, 2, or ∞ we can recover monoidal, braided monoidal, and symmetric monoidal structures.

19.7.4 Modules

Mandell's theorem (19.3.7), which is about structure on the homotopy category of left modules over an E_n-algebra, is a reflection of higher structure on the category of left modules itself.

Theorem 19.7.25. [53, 5.1.2.6, 5.1.2.8] *Suppose that* C *is an* E_k*-monoidal* ∞*-category that has geometric realization of simplicial objects, and such that the tensor product preserves such geometric realizations in each variable separately. Then the category of left modules over an* E_k*-algebra* A *is* E_{k-1}*-monoidal, and has all colimits that exist in* C.

As previously discussed, the category of left modules over an associative algebra R is not made monoidal under the tensor product over R, but the category of bimodules is. The generalization of this result to E_n-algebras is the following.

Theorem 19.7.26. [53, 3.4.4.2] *Suppose* C *is an* E_n*-monoidal presentable* ∞*-category such that the monoidal structure preserves homotopy colimits in each variable separately. Then for any* E_n*-algebra* R *in* C*, there is a category* $\mathrm{Mod}_R^{E_n}(\mathsf{C})$ *of* E_n R*-modules. This is a presentable* E_n*-monoidal* ∞*-category whose underlying monoidal operation is the tensor product over* R.

In particular, if C *is a presentable* ∞*-category with a symmetric monoidal structure that preserves colimits in each variable, and* R *is an* E_n*-algebra in* C*, the category of* E_n R*-modules in* C *has an* E_n*-monoidal structure that preserves colimits in each variable.*

Roughly, an E_n R-module M has multiplication operations $R^{\otimes k} \otimes M \to M$ parametrized by $(k+1)$-tuples of points of configuration space, where one point is marked by M and the rest by R. This has the more precise description of E_n-modules as left modules.

Theorem 19.7.27. ([53, 5.5.4.16], [28]) *Suppose that* C *is a symmetric monoidal* ∞*-category and that the monoidal product preserves colimits in each variable separately. For an* E_n*-algebra* R *in* C*, the factorization homology* $\int_{D^n \setminus 0} R$ *has the structure of an* E_1*-algebra, and the category of* E_n R*-modules is equivalent to the category of left modules over* $\int_{D^n \setminus 0} R$.

Remark 19.7.28. In the category of spectra, this could be regarded as a consequence of the Schwede–Shipley theorem [83] or its generalizations. There is a free-forgetful adjunction between E_n R-modules and Sp, and the image $\mathrm{Free}_{E_n\text{-}R}(\mathbb{S})$ of the sphere spectrum under the left adjoint is a compact generator for the category of E_n R-modules. Therefore, E_n R-modules are equivalent to the category of modules over the endomorphism ring

$$F_{E_n\text{-}R}(\mathrm{Free}_{E_n\text{-}R}(\mathbb{S}), \mathrm{Free}_{E_n\text{-}R}(\mathbb{S})) \simeq \mathrm{Free}_{E_n\text{-}R}(\mathbb{S}).$$

This theorem, then, is an identification of the free E_n R-module.

Example 19.7.29. When $n = 1$, the category of E_1 R-modules is the category of left modules over $R \otimes R^{op}$. When $n = 2$, the category of E_2-R-modules is the category of left modules over the topological Hochschild homology $\mathrm{THH}(R)$.

19.7.5 Coherent powers

In the classical case, we described an $\mathcal{O}$-algebra structure on A in terms of action maps

$$\mathrm{Sym}^k_{\mathcal{O}}(A) = \mathcal{O}(k) \otimes_{\Sigma_k} A^{\otimes k} \to A$$

from extended power constructions to A, and gave a formula

$$\mathrm{Free}_{\mathcal{O}}(X) = \coprod_{k \geq 0} \mathrm{Sym}^k_{\mathcal{O}}(A)$$

for the free $\mathcal{O}$-algebra on an object in the case where the monoidal structure is compatible with enriched colimits; we also discussed the multi-object analogue in §19.7.2. The analogous constructions for ∞-operads are carried out in [53, §3.1.3], and we will sketch these results here.

Fix an ∞-operad $\mathcal{O}$. For any objects $\mathbf{x}_1, \ldots, \mathbf{x}_d, \mathbf{y}$ of $\mathcal{O}$, we can construct a space

$$\mathrm{Mul}_{\mathcal{O}}(\mathbf{x}_1, \ldots, \mathbf{x}_d; \mathbf{y})$$

of multimorphisms in $\mathcal{O}$; if the $\mathbf{x}_i$ are equal, this further can be given a natural action of the symmetric group.

Let C be an $\mathcal{O}$-monoidal ∞-category C. In particular, C encodes categories $\mathsf{C}_{\mathbf{x}}$ parametrized by the objects $\mathbf{x}$ of $\mathcal{O}$, and functors $f\colon \mathsf{C}_{\mathbf{x}_1} \times \cdots \times \mathsf{C}_{\mathbf{x}_d} \to \mathsf{C}_{\mathbf{y}}$ parametrized by the multimorphisms $f\colon (\mathbf{x}_1, \ldots, \mathbf{x}_d) \to \mathbf{y}$ of $\mathcal{O}$. Suppose that the categories $\mathsf{C}_{\mathbf{x}}$ have homotopy colimits and the functors preserve homotopy colimits in each variable. Then there exist *extended power functors*

$$\mathrm{Sym}^k_{\mathcal{O},\mathbf{x}\to\mathbf{y}}\colon \mathsf{C}_{\mathbf{x}} \to \mathsf{C}_{\mathbf{y}},$$

whose value on $X \in \mathsf{C}_{\mathbf{u}}$ is a homotopy colimit

$$\left(\operatorname*{hocolim}_{\alpha \in \mathrm{Mul}_{\mathcal{O}}(\mathbf{x},\ldots,\mathbf{x};\mathbf{y})} \alpha(X \oplus \cdots \oplus X)\right)_{h\Sigma_k}.$$

These extended powers have the property that an $\mathcal{O}$-algebra A has natural maps $\mathrm{Sym}^k_{\mathcal{O},\mathbf{x}\to\mathbf{y}}(A(\mathbf{x})) \to A(\mathbf{y})$. Moreover, there is a free-forgetful adjunction between $\mathcal{O}$-algebras and $\mathsf{C}_{\mathbf{x}}$, and the free object $\mathrm{Free}_{\mathcal{O},\mathbf{x}}(X)$ on $X \in \mathsf{C}_U$ has the property that its value on $\mathbf{y}$ is exhibited as the coproduct

$$\mathrm{ev}_{\mathbf{y}}(\mathrm{Free}_{\mathcal{O},\mathbf{x}}(X)) \simeq \coprod_{k \geq 0} \mathrm{Sym}^k_{\mathcal{O},\mathbf{x}\to\mathbf{y}}(X).$$

Remark 19.7.30. Composing with the diagonal $\mathsf{C}_{\mathbf{x}} \to \prod \mathsf{C}_{\mathbf{x}}$ gives a Σ_k-equivariant map

$$\mathrm{Mul}_{\mathcal{O}}(\underbrace{\mathbf{x}, \ldots, \mathbf{x}}_{k}; \mathbf{y}) \to \mathrm{Fun}(\mathsf{C}_{\mathbf{x}} \times \cdots \times \mathsf{C}_{\mathbf{x}}, \mathsf{C}_{\mathbf{y}}) \to \mathrm{Fun}(\mathsf{C}_{\mathbf{x}}, \mathsf{C}_{\mathbf{y}})$$

that factors through the homotopy orbit space

$$P(k) = \mathrm{Mul}_{\mathcal{O}}(\mathbf{x}, \ldots, \mathbf{x}; \mathbf{y})_{h\Sigma_k}.$$

This space $P(k)$ then serves as a parameter space for tensor-power functors $\mathsf{C}_{\mathbf{x}} \to \mathsf{C}_{\mathbf{y}}$.

In the case of an ordinary single-object ∞-operad $\mathcal{O}$ such as an E_n-operad, we can rephrase in terms of $P(k)$. Such an ∞-operad $\mathcal{O}$ is equivalent to an ordinary operad in spaces and an $\mathcal{O}$-monoidal ∞-category is equivalent to an ∞-category C with a map $\mathcal{O} \to \mathrm{End}(\mathsf{C})$.

We recover a formula

$$\mathrm{Free}_{\mathcal{O}}(X) \simeq \coprod_{k \geq 0} \operatorname*{hocolim}_{\alpha \in P(k)} \alpha(X, \ldots, X)$$

for the free algebra on X. When $X = S^m$, this is the Thom spectrum

$$\coprod_{k \geq 0} P(k)^{m\rho},$$

closely related to Remark 19.4.19.

When $\mathcal{O}$ is an E_n-operad, the space $P(k)$ is equivalent to the space $\mathcal{C}_n(k)/\Sigma_k$, a model for the space of unordered configurations of k points in $\mathbb{R}^n$. When $n = \infty$ the space $P(k)$ is a model for $B\Sigma_k$, and we find that the we recover the ordinary homotopy symmetric power:

$$\mathrm{Sym}^k_{E_\infty}(X) \simeq (X^{\otimes k})_{h\Sigma_k}.$$

Example 19.7.31. Fix a space B and consider the fiberwise category $\mathcal{S}_{/B}$. The homotopy fiber product $X \times^h_B Y$ gives this the structure of a symmetric monoidal ∞-category, breaking up independently over the components of B. If B is path-connected, then the extended power and free functors on $(X \to B)$ are those obtained by applying the extended power and free functors to the fiber.

Example 19.7.32. Given an E_n R-module M, the free E_n R-algebra on an E_n R-module M is

$$\coprod_{k \geq 0} \operatorname*{hocolim}_{\alpha \in C_n(k)/\Sigma_k} M^{\otimes_\alpha k},$$

where each point α of configuration space determines a functor $M^{\otimes_\alpha k} \simeq M \otimes_R \cdots \otimes_R M$.

More can be said under the identification between E_n-modules and modules over factorization homology. If M is the free E_n R-module on S^m, then we obtain an identification of the free E_n-algebra under R on S^m:

$$R \amalg^{E_n} \mathrm{Free}_{E_n}(S^m) \simeq \coprod_{k \geq 0} \left(\int_{\mathbb{R}^k \setminus \{p_1, \ldots, p_k\}} R \right) \otimes_{\Sigma_k} S^{m\rho_k}.$$

Remark 19.7.33. The interaction between connective objects and their Postnikov truncations from §19.3.3 generalizes to the case where we have an $\mathcal{O}$-monoidal ∞-category $\mathcal{C}$ with a *compatible t-structure* in the sense of [53, 2.2.1.3]. This means that the categories $\mathcal{C}_{\mathbf{x}}$ indexed by the objects $\mathbf{x}$ of $\mathcal{O}$ all have t-structures, and the functors induced by the morphisms in $\mathcal{O}$ are all additive with respect to connectivity. Then [53, 2.2.1.8] implies that connective $\mathcal{O}$-algebras have Postnikov towers: the collection of truncation functors $\tau_{\leq n}$ is compatible with the $\mathcal{O}$-monoidal structure on $\mathcal{C}_{\geq 0}$.

19.8 Further invariants

19.8.1 Units and Picard spaces

Definition 19.8.1. For an E_n-monoidal ∞-category C with unit $\mathbb{I}$, the *Picard space* $\mathrm{Pic}(\mathsf{C})$ is the full subgroupoid of C spanned by the *invertible objects*: objects X for which there exists an object Y such that $Y \otimes X \simeq X \otimes Y \simeq \mathbb{I}$.

Remark 19.8.2. The classical Picard group of the homotopy category $h\mathsf{C}$ is the set $\pi_0 \mathrm{Pic}(\mathsf{C})$ of path components.

In particular, $\mathrm{Pic}(\mathsf{C})$ is closed under the E_n-monoidal structure on C, giving it a canonical E_n-space structure. Moreover, by construction $\pi_0 \mathrm{Pic}(\mathsf{C}) = (\pi_0 \mathsf{C})^\times$ is a group, and so $\mathrm{Pic}(\mathsf{C})$ is an n-fold loop space. The loop space $\Omega \mathrm{Pic}(\mathsf{C})$ is the space of homotopy self-equivalences of the unit $\mathbb{I}$; in the case of the category LMod_R of left modules, it is homotopy equivalent to the unit group $GL_1(R)$ of R.

Proposition 19.8.3. [2, §7] *If R is an E_n ring spectrum, then the space $GL_1(R)$ of homotopy self-equivalences of the left module R has an n-fold delooping. If $n \geq 2$, the space* $\mathrm{Pic}(R) = \mathrm{Pic}(\mathsf{LMod}_R)$ *has an $(n-1)$-fold delooping.*

19.8.2 Topological André–Quillen cohomology

Topological André-Quillen homology and cohomology are invariants of ring spectra developed by Kriz and Basterra [44, 7]. For a fixed map of E_∞ ring spectra $A \to B$, we can define a topological André–Quillen homology object $\mathrm{TAQ}(A \to R \to B)$ for any object R in the category of E_∞ rings between A and B. This is characterized by the following properties [8]:

1. It naturally takes values in the category of B-modules.
2. It takes homotopy colimits of E_∞ ring spectra between A and B to homotopy colimits of B-modules.
3. There is a natural map $B \otimes_A (R/A) \to \mathrm{TAQ}(A \to R \to B)$.
4. For a left A-module X with a map $X \to B$, the composite natural map
$$B \otimes_A X \to B \otimes_A \mathrm{Free}^A_{E_\infty}(X) \to \mathrm{TAQ}(A \to \mathrm{Free}^A_{E_\infty}(X) \to B)$$
of B-modules is an equivalence.
5. Under the above equivalence, the natural map
$$\mathrm{TAQ}(A \to \mathrm{Free}^A_{E_\infty} \mathrm{Free}^A_{E_\infty}(X) \to B) \to \mathrm{TAQ}(A \to \mathrm{Free}^A_{E_\infty}(X) \to B)$$
is equivalent to the map
$$B \otimes_A \mathrm{Free}^A_{E_\infty}(X) \to B \otimes_A X$$
that collapses $B \otimes_A (\amalg \mathrm{Sym}^k(X))$ to the factor with $k = 1$.

Topological André-Quillen homology measures how difficult it is to build R as an A-algebra: any description of R as an iterated pushout along maps of free of E_∞-algebras, starting from A, determines a description of the topological André–Quillen cohomology of R as an iterated pushout of B-modules. Basterra showed that TAQ-cohomology groups

$$\mathrm{TAQ}^n(R; M) = [\mathrm{TAQ}(\mathbb{S} \to R \to R), \Sigma^n M]_{\mathrm{Mod}_R}$$

plays the role for Postnikov towers of E_∞ ring spectra that ordinary cohomology does for spectra.

From this point of view, TAQ also has natural generalizations to $\mathrm{TAQ}^{\mathcal{O}}$ for algebras over an arbitrary operad [8, 33], although there may be a choice of target category that takes more work to describe. In particular, for E_n-algebras these are related to an iterated bar construction [9].

Topological André–Quillen homology also enjoys the following properties, proved in [7, 8].

Base-change: For a map $B \to C$, we have a natural equivalence

$$C \otimes_B \mathrm{TAQ}(A \to R \to B) \simeq \mathrm{TAQ}(A \to R \to C).$$

In particular, if we define $\Omega_{R/A} = \mathrm{TAQ}(A \to R \to R)$, then

$$\mathrm{TAQ}(A \to R \to B) = B \otimes_R \Omega_{R/A}.$$

Transitivity: For a composite $A \to R \to S \to B$, there is a natural cofiber sequence

$$\mathrm{TAQ}(A \to R \to B) \to \mathrm{TAQ}(A \to S \to B) \to \mathrm{TAQ}(R \to S \to B).$$

In particular, for $A \to R \to S$ we have cofiber sequences

$$S \otimes_R \Omega_{R/A} \to \Omega_{S/A} \to \Omega_{S/R}.$$

Representability: Suppose that there is a functor h^* from the category of pairs $(R \to S)$ of E_∞ ring spectra between A and B to the category of graded abelian groups. Suppose that this is a cohomology theory on the category of E_∞ ring spectra between A and B: it satisfies homotopy invariance, has a long exact sequence, satisfies excision for homotopy pushouts of pairs, and takes coproducts to products. Then there is a B-module M with a natural isomorphism

$$\begin{aligned} h^n(S, R) &\cong \mathrm{TAQ}^n(S, R; M) \\ &= [\mathrm{TAQ}(R \to S \to B), \Sigma^n M]_{\mathrm{Mod}_B} \end{aligned}$$

of abelian groups.

For any E_∞ ring spectrum B, algebras mapping to B have TAQ-homology $\mathrm{TAQ}(\mathbb{S} \to R \to B)$, valued in the category of B-modules. The square-zero algebras

$$B \oplus M$$

are representing objects for TAQ-cohomology $\mathrm{TAQ}^*(R; M)$.

Representability allows us to construct and classify operations in TAQ-cohomology by B-algebra maps between such square-zero extensions.

Proposition 19.8.4. *Any element in $[\Sigma \operatorname{Sym}^2 M, N]_{\mathrm{Mod}_B}$ has a naturally associated map $B \oplus M \to B \oplus N$ of augmented commutative B-algebras and hence gives rise to a natural* TAQ*-cohomology operation* $\mathrm{TAQ}(-; M) \to \mathrm{TAQ}(-; N)$ *for commutative algebras mapping to B.*

Proof. By viewing B as concentrated in grading 0 and M as concentrated in grading 1, we can give a $\mathbb{Z}$-graded construction (as in Example 19.3.24) of $B \oplus M$ as an iterated sequence of pushouts along maps of free algebras. The first such pushout is

$$\mathrm{Free}^B_{E_\infty}(M) \leftarrow \mathrm{Free}^B_{E_\infty}(\operatorname{Sym}^2 M) \to B.$$

Further pushouts only alter gradings 3 and higher.

We now view $B \oplus N$ as graded by putting N in grading 2. We find that homotopy classes of maps of graded algebras $B \oplus M \to B \oplus N$ are equivalent to maps $\Sigma \operatorname{Sym}^2 M \to N$. □

Example 19.8.5. Letting $M = B \otimes S^m$, we have

$$\Sigma \operatorname{Sym}^2(M) \simeq B \otimes \Sigma^{m+1} \mathbb{RP}^\infty_m.$$

Therefore, we get a map from the B-cohomology $B^n(\Sigma^{m+1}\mathbb{RP}^\infty_m)$ of stunted projective spaces to the group of natural cohomology operations $TAQ^m(-; B) \to TAQ^n(-; B)$.

Remark 19.8.6. The fact that elements in the B-homology of stunted projective spaces produce homotopy operations while elements in their B-cohomology produce TAQ-cohomology operations with a shift is a reflection of Koszul duality.

Example 19.8.7. Letting $M = (B \otimes S^q) \oplus (B \otimes S^r)$, and using the projection

$$\begin{aligned}\Sigma \operatorname{Sym}^2(B \otimes (S^q \oplus S^r)) &\simeq \Sigma \operatorname{Sym}^2(B \otimes S^q) \oplus \Sigma \operatorname{Sym}^2(B \otimes S^r) \oplus \Sigma(B \otimes S^q \otimes S^r)\\ &\to B \otimes S^{q+1+r},\end{aligned}$$

we get a binary operation

$$[-,-]\colon\ \mathrm{TAQ}^q(-; B) \times \mathrm{TAQ}^r(-; B) \to \mathrm{TAQ}^{q+1+r}(-; B)$$

that (up to a normalization factor) we call the TAQ-*bracket.*

Example 19.8.8. If $B = H\mathbb{F}_2$, then there are TAQ-cohomology operations

$$R^a\colon\ \mathrm{TAQ}^m(-; H\mathbb{F}_2) \to \mathrm{TAQ}^{m+a}(-; H\mathbb{F}_2)$$

for $a \geq m+1$, and a bracket

$$\mathrm{TAQ}^q(-;H\mathbb{F}_2) \times \mathrm{TAQ}^r(-;H\mathbb{F}_2) \to \mathrm{TAQ}^{q+1+r}(-;H\mathbb{F}_2).$$

In this form, the operation R^{a+1} is Koszul dual to Q^a, in the sense that nontrivial values of R^{a+1} in TAQ-cohomology detect relations on the operator Q^a in homology. Similarly, the bracket in TAQ is Koszul dual to the multiplication.

The operations were constructed by Basterra–Mandell [10]. In further unpublished work, they showed that these operations (and their odd-primary analogues) generate all the natural operations on TAQ-cohomology with values in $H\mathbb{F}_p$ and determined the relations between them. In particular, the operations R^a above satisfy the same Adem relations that the Steenrod operations Sq^a do; the TAQ-bracket has the structure of a shifted restricted Lie bracket, whose restriction is the bottommost defined operation R^a.

Basterra–Mandell's proof uses a variant of the Miller spectral sequence from [68]. We will close out this section with a sketch of how such spectral sequences are constructed, parallel to the delooping spectral sequence from Remark 19.5.14.

Proposition 19.8.9. *Suppose that R is an E_∞ ring spectrum with a chosen map $R \to H\mathbb{F}_p$. Then there is a Miller spectral sequence*

$$AQ^{DL}_*(\pi_*(H\mathbb{F}_p \otimes R)) \Rightarrow \mathrm{TAQ}_*(\mathbb{S} \to R \to H\mathbb{F}_p),$$

where the left-hand side are the nonabelian derived functors of an indecomposables functor Q that sends an augmented graded-commutative $\mathbb{F}_p$-algebra with Dyer–Lashof operations to the quotient of the augmentation ideal by all products and Dyer–Lashof operations.

Proof. We construct an augmented simplicial object:

$$\cdots \mathrm{Free}_{E_\infty} \mathrm{Free}_{E_\infty} \mathrm{Free}_{E_\infty} R \Rrightarrow \mathrm{Free}_{E_\infty} \mathrm{Free}_{E_\infty} R \rightrightarrows \mathrm{Free}_{E_\infty} R \to R.$$

If U is the forgetful functor, from commutative ring spectra mapping to $H\mathbb{F}_p$ to spectra mapping to $H\mathbb{F}_p$, this is the bar construction $B(\mathrm{Free}_{E_\infty}, U\,\mathrm{Free}_{E_\infty}, UR)$. The underlying simplicial spectrum $B(U\,\mathrm{Free}_{E_\infty}, U\,\mathrm{Free}_{E_\infty}, UR)$ has an extra degeneracy, so its geometric realization is equivalent to R. Moreover, the forgetful functor from E_∞ rings to spectra preserves sifted homotopy colimits, and hence geometric realization because the simplicial indexing category is sifted. Therefore, applying the homotopy colimit preserving functor $\mathrm{TAQ} = \mathrm{TAQ}(\mathbb{S} \to (-) \to H\mathbb{F}_p)$ and the natural equivalence $\mathrm{TAQ} \circ \mathrm{Free}_{E_\infty}(R) \simeq H\mathbb{F}_p \otimes R$, we get an equivalence

$$|B(H\mathbb{F}_p \otimes (-), U\,\mathrm{Free}_{E_\infty}, UR)| \simeq \mathrm{TAQ}(R).$$

However, this bar construction is a simplicial object of the form

$$\cdots H\mathbb{F}_p \otimes \mathrm{Free}_{E_\infty} \mathrm{Free}_{E_\infty} R \Rrightarrow H\mathbb{F}_p \otimes \mathrm{Free}_{E_\infty} R \rightrightarrows H\mathbb{F}_p \otimes R.$$

Taking homotopy groups, we get a simplicial object

$$\mathbb{Q}_{E_\infty}\mathbb{Q}_{E_\infty}H_*R \Rrightarrow \mathbb{Q}_{E_\infty}H_*R \Rightarrow H_*R.$$

Moreover, the structure maps make this the bar construction

$$B(Q, \mathbb{Q}_{E_\infty}, H_*R)$$

that computes derived functors of Q on graded-commutative algebras with Dyer–Lashof operations. Therefore, the spectral sequence associated to the geometric realization computes $\mathrm{TAQ}_*(\mathbb{S} \to R \to H\mathbb{F}_p)$ and has the desired E_2-term. □

Remark 19.8.10. We can also apply cohomology rather than homology and get a spectral sequence computing topological André–Quillen cohomology.

This leaves open a hard algebraic part of Basterra–Mandell's work: actually calculating these derived functors, and in particular finding relations amongst the operations R^a and the bracket $[-,-]$ that give a complete description of TAQ-cohomology operations.

19.9 Further questions

We will close this paper with some problems that we think are useful directions for future investigation.

Problem 19.9.1. Develop useful obstruction theories which can determine the existence of or maps between E_n-algebras in a wide variety of contexts.

The obstruction theory due to Goerss–Hopkins [31] is the prototype for these results. In unpublished work [87], Senger has given a development of this theory for E_∞-algebras where the obstructions occur in nonabelian Ext-groups calculated in the category of graded-commutative rings with Dyer–Lashof operations and Steenrod operations satisfying the Nishida relations, and provided tools for calculating with them. This played a critical role in [47, 46].

In closely related situations, the tools available remain rudimentary. For example, there is essentially no workable obstruction theory for the construction of commutative rings of any type in equivariant stable theory. Tools arising from the Steenrod algebra have been essential in most of the deep results in homotopy theory, such as the Segal conjecture [50] and the Sullivan conjecture [69]. Without the analogues, there is a limit to how much structure can be revealed.

Problem 19.9.2. Give a modern redevelopment of homology operations for E_∞ ring spaces and E_n ring spaces.

The observant reader may have noticed that, despite the rich structure present, the principal material that we have referenced for E_∞ ring spaces is several decades old. Several major advances have happened in multiplicative stable homotopy theory since then, and the author feels that there is still a great deal to be mined. Having this material accessible to modern toolkits would be extremely useful.

For one example, the theory of E_∞ ring spaces from the point of view of symmetric spectra has been studied in detail by Sagave and Schlichtkrull [81, 79, 80]. For another, the previous emphasis on E_∞ ring spaces should be tempered by the variety of examples that we now know only admit A_∞ or E_n ring structures.

Problem 19.9.3. Give a unified theory of graded Hopf algebras and Hopf rings, capable of encoding some combination of non-integer gradings, power operations, group-completion theorems, and the interaction with the unit.

Ravenel–Wilson's theory of Hopf rings is integer-graded. We now know many examples—motivic homotopy theory, equivariant homotopy theory, $K(n)$-local theory, modules over E_n ring spectra—that may have natural gradings of a much wider variety than this, such as a Picard group. Moreover, multiplicative theory should involve much more structure: we should have a sequence of spaces graded not just by a Picard group, but by the Picard space that also encodes structure nontrivial higher interaction between gradings and the unit group.

Problem 19.9.4. Give a precise general description of the Koszul duality relationship between homotopy operations and TAQ-cohomology operations. Give a complete construction of the algebra of operations on TAQ-cohomology for E_n-algebras with coefficients in Hk, for k a commutative ring. Give complete descriptions of the TAQ-cohomology for a large library of Eilenberg–Mac Lane spectra Hk and Morava's forms of K-theory.

Because TAQ-cohomology governs the construction of ring spectra via their Postnikov tower, essentially any information that we can provide about these objects is extremely useful.

Problem 19.9.5. Determine an algebro-geometric expression for power operations and their relationship to the Steenrod operations.[1] Do the same for the operations that appear in the Hopf ring associated to an E_∞ ring space.

At the prime 2, it has been known for some time that the action of the Steenrod algebra can be concisely packaged as a coaction of the dual Steenrod algebra, a Hopf algebra corresponding to the group scheme of automorphisms of the additive formal group over $\mathbb{F}_2$. The Dyer–Lashof operations on infinite loop spaces generate an algebra analogous to the Steenrod algebra, and its dual was described by Madsen [54]; the result is closely related to Dickson invariants. However, the full action of the Dyer–Lashof operations or the interaction between the Dyer–Lashof algebra and the Steenrod algebra do not yet have a geometric packaging.

Conjecture 19.9.6. *For Lubin–Tate cohomology theories E and F of height n, there is a natural algebraic structure parametrizing operations from continuous E-homology to continuous F-homology for certain E_∞ ring spectra, expressed in terms of the algebraic geometry of* isogenies *of formal groups.*

---

[1] See the chapter by Nathaniel Stapleton in this Handbook for related material.

This is complete: there is an obstruction theory for the construction of and mapping between $K(n)$-local E_∞ ring spectra whose algebraic input is completed E-homology equipped with these operations.

In this paper we have not really touched on the extensive study of power operations in chromatic homotopy theory, but see the chapters in this Handbook by Nathaniel Stapleton and by Tobias Barthel and Agnès Beaudry. Given Lubin–Tate cohomology theories E and F associated to formal groups of height n at the prime p, we have both cohomology operations and power operations. In [37] the algebra of cohomology operations is expressed in terms of isomorphisms of formal groups. Extensive work of Ando, Strickland, and Rezk has shown that power operations are expressed in terms of quotient operations for subgroups of the formal groups. It has been known for multiple decades [91, §28] that the natural home combining these two types of operations is the theory of *isogenies* of formal groups. However, there are important details about formal topologies which have never been resolved.[2]

Problem 19.9.7. Determine the natural instability relations for operations in *unstable* elliptic cohomology and in unstable Lubin–Tate cohomology.

Strickland states that isogenies are a natural interpretation for unstable cohomology operations in E-theory. However, isogenies encode the analogue of the cohomological Steenrod operations, the multiplicative Dyer–Lashof operations, and the Nishida relations between them. They do *not* encode any analogue of the instability relation $Sq^n = Q^{-n}$ that we see in the cohomology of spaces.

In chromatic theory, our only accessible example so far is K-theory. For p-completed K-theory, the cohomology operations are generated by the Adams operations ψ^k for $k \in \mathbb{Z}_p^\times$. For torsion-free algebras, the power operations are controlled by the operation ψ^p and its congruences [36, 77]. The *unstable* operations in the K-theory of spaces, by contrast, arise from the algebra of symmetric polynomials and are essentially governed by the ψ^n for $n \in \mathbb{N}$; the fact that the other ψ^k are determined by these enforces some form of continuity. This is also closely tied to the question of whether there are geometric interpretations of some type for elliptic cohomology theories or Lubin–Tate cohomology theories.[3]

Problem 19.9.8. Determine a useful way to encode secondary operation structures on E_∞ or E_n rings.

In the case of secondary Steenrod operations, there is a useful formulation due to Baues of an extension of the Steenrod algebra that can be used to encode all of the secondary operation structure [11, 72]. No such systematic descriptions are known for secondary Dyer–Lashof operations, especially since the Dyer–Lashof operations are expressed in a more complicated way than the action of an algebra on a module.

Problem 19.9.9. Determine useful relationships between the homotopy types of an E_n ring spectrum, the unit group GL_1 and the Picard space $\mathrm{Pic}(R)$, and the spaces $BGL_n(R)$.

---

[2]The reader should be advised that, even at height 1, there are difficult issues with E-theory here involving left-derived functors of completion.

[3]One possible viewpoint is that we could interpret $\mathbb{N}$ as the monoid of endomorphisms of the *multiplicative monoid* M_1, which contains the unit group GL_1.

This is closely tied to orientation theory, algebraic K-theory, and the study of spaces involved in surgery theory.

Investigations in these directions due to Mathew–Stojanoska revealed that there is a nontrivial relationship between the k-invariants for R and the unit spectrum $gl_1(R)$ at the edge of the stable range at the prime 2 [59], and forthcoming work of Hess has shown that this relation can be recovered from the mixed Cartan formula. The odd-primary analogues of this are not yet known.

Problem 19.9.10. Find an odd-primary formula for the mixed Adem relations similar to the Kuhn–Tsuchiya formula.

There is a description of the mixed Adem relations [23], valid at any prime, but it is difficult to apply in concrete examples. The 2-primary formula described in §19.5.8 is much more direct; it was originally stated by Tsuchiya and proven by Kuhn [45]. There is no known odd-primary analogue of this formula.

Bibliography

[1] J. F. Adams. *Stable homotopy and generalised homology.* Chicago Lectures in Mathematics. University of Chicago Press, 1974.

[2] Matthew Ando, Andrew J. Blumberg, and David Gepner. Parametrized spectra, multiplicative Thom spectra and the twisted Umkehr map. *Geom. Topol.*, 22(7):3761–3825, 2018.

[3] Vigleik Angeltveit. Topological Hochschild homology and cohomology of A_∞ ring spectra. *Geom. Topol.*, 12(2):987–1032, 2008.

[4] Bernard Badzioch. Algebraic theories in homotopy theory. *Ann. of Math. (2)*, 155(3):895–913, 2002.

[5] Andrew Baker. Power operations and coactions in highly commutative homology theories. *Publ. Res. Inst. Math. Sci.*, 51(2):237–272, 2015.

[6] M. G. Barratt and Peter J. Eccles. Γ^+-structures. I. A free group functor for stable homotopy theory. *Topology*, 13:25–45, 1974.

[7] M. Basterra. André-Quillen cohomology of commutative S-algebras. *J. Pure Appl. Algebra*, 144(2):111–143, 1999.

[8] Maria Basterra and Michael A. Mandell. Homology and cohomology of E_∞ ring spectra. *Math. Z.*, 249(4):903–944, 2005.

[9] Maria Basterra and Michael A. Mandell. Homology of E_n ring spectra and iterated THH. *Algebr. Geom. Topol.*, 11(2):939–981, 2011.

[10] Maria Basterra and Michael A. Mandell. The multiplication on BP. *J. Topol.*, 6(2):285–310, 2013.

[11] Hans-Joachim Baues. *The algebra of secondary cohomology operations*, volume 247 of *Progr. Math.* Birkhäuser Verlag, 2006.

[12] Hans-Joachim Baues and Fernando Muro. Toda brackets and cup-one squares for ring spectra. *Comm. Algebra*, 37(1):56–82, 2009.

[13] Julia E. Bergner. Rigidification of algebras over multi-sorted theories. *Algebr. Geom. Topol.*, 6:1925–1955, 2006.

[14] Andrew Blumberg, Ralph Cohen, and Christian Schlichtkrull. THH of Thom spectra and the free loop space. *Geom. Topol.* 14(2):1165–1242, 2010.

[15] Andrew J. Blumberg and Michael A. Hill. Operadic multiplications in equivariant spectra, norms, and transfers. *Adv. Math.*, 285:658–708, 2015.

[16] J. M. Boardman and R. M. Vogt. *Homotopy invariant algebraic structures on topological spaces.* Lecture Notes in Math. 347. Springer-Verlag, 1973.

[17] James Borger and Ben Wieland. Plethystic algebra. *Adv. Math.*, 194(2):246–283, 2005.

[18] A. K. Bousfield, E. B. Curtis, D. M. Kan, D. G. Quillen, D. L. Rector, and J. W. Schlesinger. The mod-p lower central series and the Adams spectral sequence. *Topology*, 5:331–342, 1966.

[19] A. K. Bousfield and D. M. Kan. The homotopy spectral sequence of a space with coefficients in a ring. *Topology*, 11:79–106, 1972.

[20] R. R. Bruner, J. P. May, J. E. McClure, and M. Steinberger. *H_∞ ring spectra and their applications*, Lecture Notes in Math. 1176. Springer-Verlag, 1986.

[21] H. E. A. Campbell, F. P. Peterson, and P. S. Selick. Self-maps of loop spaces. I. *Trans. Amer. Math. Soc.*, 293(1):1–39, 1986.

[22] F. R. Cohen. On configuration spaces, their homology, and Lie algebras. *J. Pure Appl. Algebra*, 100(1-3):19–42, 1995.

[23] Frederick R. Cohen, Thomas J. Lada, and J. Peter May. *The homology of iterated loop spaces*. Lecture Notes in Math. 533. Springer-Verlag, 1976.

[24] George Cooke. Replacing homotopy actions by topological actions. *Trans. Amer. Math. Soc.*, 237:391–406, 1978.

[25] Gerald Dunn. Tensor product of operads and iterated loop spaces. *J. Pure Appl. Algebra*, 50(3):237–258, 1988.

[26] W. G. Dwyer and D. M. Kan. An obstruction theory for diagrams of simplicial sets. *Nederl. Akad. Wetensch. Indag. Math.*, 46(2):139–146, 1984.

[27] A. D. Elmendorf, I. Kriz, M. A. Mandell, and J. P. May. *Rings, modules, and algebras in stable homotopy theory*. Math. Surveys Monogr. 47. Amer. Math. Soc. 1997. With an appendix by M. Cole.

[28] John Francis. The tangent complex and Hochschild cohomology of E_n-rings. *Compos. Math.*, 149(3):430–480, 2013.

[29] Eric M. Friedlander and Barry Mazur. Filtrations on the homology of algebraic varieties. *Mem. Amer. Math. Soc.*, 110(529), 1994. With an appendix by Daniel Quillen.

[30] Saul Glasman and Tyler Lawson. Stable power operations. Available at `http://www-users.math.umn.edu/~tlawson/papers/power.pdf`.

[31] Paul G. Goerss and Michael J. Hopkins. Moduli spaces of commutative ring spectra. In *Structured ring spectra, London Math. Soc. Lecture Note Ser.* 315, pages 151–200. Cambridge Univ. Press, 2004.

[32] J. P. C. Greenlees and J. P. May. Generalized Tate cohomology. *Mem. Amer. Math. Soc.*, 113(543), 1995.

[33] J. E. Harper. Bar constructions and Quillen homology of modules over operads. Algebr. Geom. Topol. 10 (2010), no. 1, 87–136.

[34] M. A. Hill and M. J. Hopkins. Equivariant multiplicative closure. In *Algebraic topology: applications and new directions*, volume 620 of *Contemp. Math.*, pages 183–199. Amer. Math. Soc., 2014.

[35] M. A. Hill, M. J. Hopkins, and D. C. Ravenel. On the nonexistence of elements of Kervaire invariant one. *Ann. of Math. (2)*, 184(1):1–262, 2016.

[36] Michael J. Hopkins. $K(1)$-local E_∞-ring spectra. In *Topological modular forms*, Math. Surveys Monogr. 201, pages 287–302. Amer. Math. Soc., 2014.

[37] Mark Hovey. Operations and co-operations in Morava E-theory. *Homology Homotopy Appl.*, 6(1):201–236, 2004.

[38] Mark Hovey, Brooke Shipley, and Jeff Smith. Symmetric spectra. *J. Amer. Math. Soc.*, 13(1):149–208, 2000.

[39] P. Hu, I. Kriz, and J. P. May. Cores of spaces, spectra, and E_∞ ring spectra. In *Equivariant stable homotopy theory and related areas* (Stanford, CA, 2000). *Homology Homotopy Appl.*, 3(2):341–354, 2001.

[40] Niles Johnson and Justin Noel. For complex orientations preserving power operations, p-typicality is atypical. *Topology Appl.*, 157(14):2271–2288, 2010.

[41] André Joyal and Ross Street. Braided tensor categories. *Adv. Math.*, 102(1):20–78, 1993.

[42] G. M. Kelly. Basic concepts of enriched category theory. *Repr. Theory Appl. Categ.*, (10), 2005. Reprint of the 1982 original [Cambridge Univ. Press].

[43] Stanley O. Kochman. Homology of the classical groups over the Dyer-Lashof algebra. *Trans. Amer. Math. Soc.*, 185:83–136, 1973.

[44] I. Kriz. Towers of E_∞-ring spectra with an application to BP. Unpublished, 1995.

[45] Nicholas J. Kuhn. The homology of the James-Hopf maps. *Illinois J. Math.*, 27(2):315–333, 1983.

[46] T. Lawson. Calculating obstruction groups for E-infinity ring spectra. Homotopy Theory: Tools and Applications, *Contemp. Math.* 729, pages 179–203. Amer. Math. Soc., 2019.

[47] Tyler Lawson. Secondary power operations and the Brown–Peterson spectrum at the prime 2. *Ann. of Math. (2)*, 188(2):513–576, 2018.

[48] A. Lazarev. Homotopy theory of A_∞ ring spectra and applications to MU-modules. *K-Theory*, 24(3):243–281, 2001.

[49] L. G. Lewis, Jr., J. P. May, M. Steinberger, and J. E. McClure. *Equivariant stable homotopy theory*, volume 1213 of Lecture Notes in Math. 1213. Springer-Verlag, 1986. With contributions by J. E. McClure.

[50] W. H. Lin, D. M. Davis, M. E. Mahowald, and J. F. Adams. Calculation of Lin's Ext groups. *Math. Proc. Cambridge Philos. Soc.*, 87(3):459–469, 1980.

[51] John A. Lind. Bundles of spectra and algebraic K-theory. *Pacific J. Math.*, 285(2):427–452, 2016.

[52] John A. Lind and Cary Malkiewich. The Morita equivalence between parametrized spectra and module spectra. In *New directions in homotopy theory*, volume 707 of *Contemp. Math.*, pages 45–66. Amer. Math. Soc., 2018.

[53] Jacob Lurie. Higher Algebra. Available at `http://www.math.harvard.edu/~lurie/papers/higheralgebra.pdf`, 2017.

[54] Ib Madsen. On the action of the Dyer-Lashof algebra in $H_*(G)$. *Pacific J. Math.*, 60(1):235–275, 1975.

[55] Mark Mahowald. Ring spectra which are Thom complexes. *Duke Math. J.*, 46(3):549–559, 1979.

[56] Michael A. Mandell. E_∞ algebras and p-adic homotopy theory. *Topology*, 40(1):43–94, 2001.

[57] Michael A. Mandell. The smash product for derived categories in stable homotopy theory. *Adv. Math.*, 230(4-6):1531–1556, 2012.

[58] M. A. Mandell and J. P. May. Equivariant orthogonal spectra and S-modules. *Mem. Amer. Math. Soc.*, 159(755), 2002.

[59] Akhil Mathew and Vesna Stojanoska. The Picard group of topological modular forms via descent theory. *Geom. Topol.*, 20(6):3133–3217, 2016.

[60] J. P. May. Problems in infinite loop space theory. In *Conference on homotopy theory (Evanston, Ill., 1974)*, volume 1 of *Notas Mat. Simpos.*, pages 111–125. Soc. Mat. Mexicana, México, 1975.

[61] J. Peter May. Homology operations on infinite loop spaces. Algebraic Topology (Proc. Sympos. Pure Math., Vol. XXII, Univ. Wisconsin, Madison, Wis., 1970), pages 171–185. Amer. Math. Soc., 1971.

[62] J. Peter May. *The geometry of iterated loop spaces.* Lecture Notes in Math. 271. Springer-Verlag, 1972.

[63] J. Peter May. *E_∞ ring spaces and E_∞ ring spectra.* Lecture Notes in Math. 577. Springer-Verlag, 1977. With contributions by Frank Quinn, Nigel Ray, and Jørgen Tornehave.

[64] J. P. May and J. Sigurdsson. *Parametrized homotopy theory.* Math. Surveys and Monogr. 132. Amer. Math. Soc., 2006.

[65] J. E. McClure. E_∞-ring structures for Tate spectra. *Proc. Amer. Math. Soc.*, 124(6):1917–1922, 1996.

[66] D. McDuff and G. Segal. Homology fibrations and the "group-completion" theorem. *Invent. Math.*, 31(3):279–284, 1975/76.

[67] R. James Milgram. The mod 2 spherical characteristic classes. *Ann. of Math. (2)*, 92:238–261, 1970.

[68] Haynes Miller. A spectral sequence for the homology of an infinite delooping. *Pacific J. Math.*, 79(1):139–155, 1978.

[69] Haynes Miller. The Sullivan conjecture on maps from classifying spaces. *Ann. of Math. (2)*, 120(1):39–87, 1984.

[70] John Milnor. The Steenrod algebra and its dual. *Ann. of Math. (2)*, 67:150–171, 1958.

[71] Ieke Moerdijk and Ittay Weiss. Dendroidal sets. *Algebr. Geom. Topol.*, 7:1441–1470, 2007.

[72] Christian Nassau. On the secondary Steenrod algebra. *New York J. Math.*, 18:679–705, 2012.

[73] Xianglong Ni. The bracket in the bar spectral sequence for a finite-fold loop space. ArXiv:1908.09233, 2019.

[74] Stewart Priddy. Dyer-Lashof operations for the classifying spaces of certain matrix groups. *Quart. J. Math. Oxford Ser. (2)*, 26(102):179–193, 1975.

[75] Douglas C. Ravenel and W. Stephen Wilson. The Hopf ring for complex cobordism. *Bull. Amer. Math. Soc.*, 80:1185–1189, 1974.

[76] Charles Rezk. Lectures on power operations. Available at `http://www.math.uiuc.edu/~rezk/papers.html`.

[77] Charles Rezk. The congruence criterion for power operations in Morava E-theory. *Homology, Homotopy Appl.*, 11(2):327–379, 2009.
[78] Birgit Richter and Stephanie Ziegenhagen. A spectral sequence for the homology of a finite algebraic delooping. *J. K-Theory*, 13(3):563–599, 2014.
[79] Steffen Sagave and Christian Schlichtkrull. Diagram spaces and symmetric spectra. *Adv. Math.*, 231(3–4):2116–2193, 2012.
[80] Steffen Sagave and Christian Schlichtkrull. Group completion and units in I-spaces. *Algebr. Geom. Topol.*, 13(2):625–686, 2013.
[81] Christian Schlichtkrull. Units of ring spectra and their traces in algebraic K-theory. *Geom. Topol.*, 8:645–673, 2004.
[82] Stefan Schwede. Stable homotopical algebra and Γ-spaces. *Math. Proc. Cambridge Philos. Soc.*, 126(2):329–356, 1999.
[83] Stefan Schwede and Brooke Shipley. Stable model categories are categories of modules. *Topology*, 42(1):103–153, 2003.
[84] Stefan Schwede and Brooke E. Shipley. Algebras and modules in monoidal model categories. *Proc. London Math. Soc. (3)*, 80(2):491–511, 2000.
[85] Graeme Segal. Categories and cohomology theories. *Topology*, 13:293–312, 1974.
[86] A. Senger. The Brown-Peterson spectrum is not $E_{2(p^2+2)}$ at odd primes. arXiv:1710.09822, 2017.
[87] Andrew Senger. On the realization of truncated Brown–Peterson spectra as E_∞ ring spectra. unpublished.
[88] Andrew Stacey and Sarah Whitehouse. The hunting of the Hopf ring. *Homology Homotopy Appl.*, 11(2):75–132, 2009.
[89] James Dillon Stasheff. Homotopy associativity of H-spaces. I, II. *Trans. Amer. Math. Soc.* 108:275–292, 1963; ibid., 108:293–312, 1963.
[90] N. E. Steenrod. *Cohomology operations.* Lectures by N. E. Steenrod written and revised by D. B. A. Epstein. Ann. of Math. Stud. 50. Princeton University Press, 1962.
[91] Neil Strickland. Functorial philosophy for formal phenomena. Available at `http://hopf.math.purdue.edu/Strickland/fpfp.pdf`.
[92] Sean Tilsòn. Power operations in the Kunneth spectral sequence and commutative HF_p-algebras. arXiv:1602.06736, 2016.
[93] David White. Model structures on commutative monoids in general model categories. *J. Pure Appl. Algebra* 221(12):3124–3168, 2017.

School of Mathematics, University of Minnesota, Minneapolis, MN 55455, U.S.A.

E-mail address: `tlawson@math.umn.edu`

20

Assembly maps

Wolfgang Lück

20.1 Introduction

20.1.1 The homotopy theoretic description of assembly maps

The quickest and probably for a homotopy theorist most convenient approach to assembly maps is via homotopy colimits as explained in Subsection 20.7.3. Let $\mathcal{F}$ be a family of subgroups of G, i.e., a collection of subgroups closed under conjugation and passing to subgroups. Let $\mathsf{Or}(G)$ be the orbit category and $\mathsf{Or}_{\mathcal{F}}(G)$ be the full subcategory consisting of objects G/H satisfying $H \in \mathcal{F}$. Consider a covariant functor $\mathbf{E}^G \colon \mathsf{Or}(G) \to \mathsf{Spectra}$ to the category of spectra. We get from the inclusion $\mathsf{Or}_{\mathcal{F}}(G) \to \mathsf{Or}(G)$ and the fact that G/G is a terminal object in $\mathsf{Or}(G)$ a map

$$\operatorname{hocolim}_{\mathsf{Or}_{\mathcal{F}}(G)} \mathbf{E}^G|_{\mathsf{Or}_{\mathcal{F}}(G)} \to \operatorname{hocolim}_{\mathsf{Or}(G)} \mathbf{E}^G = \mathbf{E}^G(G/G). \tag{20.1.1}$$

It is called *assembly map* since we are trying to assemble the values of $\mathbf{E}^G$ on homogeneous spaces G/H for $H \in \mathcal{F}$ to get $\mathbf{E}(G/G)$.

On homotopy groups this assembly map can also be described as the map

$$H_n^G(\mathrm{pr}; \mathbf{E}^G) \colon H_n^G(E_{\mathcal{F}}(G); \mathbf{E}^G) \to H_n^G(G/G; \mathbf{E}^G) = \pi_n(\mathbf{E}^G(G/G)) \tag{20.1.2}$$

induced by the projection $\mathrm{pr} \colon E_{\mathcal{F}}(G) \to G/G$ of the classifying G-space $E_{\mathcal{F}}(G)$ for the family $\mathcal{F}$, see Section 20.6, to G/G, where $H_n^G(-; \mathbf{E}^G)$ is the G-homology theory in the sense of Definition 20.3.1 associated to $\mathbf{E}^G$, see Lemma 20.3.4.

In all interesting situations one can take a global point of view. Namely, one starts with a covariant functor respecting equivalences $\mathbf{E} \colon \mathsf{Groupoids} \to \mathsf{Spectra}$ and defines for a group G the functor $\mathbf{E}^G$ to be the composite of $\mathbf{E}$ with the functor $\mathsf{Or}(G) \to \mathsf{Groupoids}$ given by the transport groupoid of a G-set, see Subsection 20.7.4.

20.1.2 Isomorphism Conjectures

The Meta Isomorphism Conjecture for G, $\mathcal{F}$ and $\mathbf{E}^G$, see Section 20.7, says that the assembly map of (20.1.1) is a weak homotopy equivalence, or, equivalently, that the map (20.1.2) is bijective for all $n \in \mathbb{Z}$.

2010 *Mathematics Subject Classification.* primary: 18F25, secondary: 46L80, 55P91, 57N99.

Key words and phrases. assembly maps, equivariant homology and homotopy theory, Farrell-Jones Conjecture, Baum-Connes Conjecture.

If we take for $\mathbf{E}$ an appropriate functor modelling the algebraic K-theory or the algebraic L-theory with decoration $\langle -\infty \rangle$ of the group ring RG and for $\mathcal{F}$ the family of virtually cyclic subgroups, we obtain the Farrell-Jones Conjecture 20.8.1. It assembles $K_n(RG)$ and $L_n^{\langle -\infty \rangle}(RG)$ in terms of $K_n(RH)$ and $L_n^{\langle -\infty \rangle}(RH)$, where H runs through the virtually cyclic subgroups of G.

If we take for $\mathbf{E}$ an appropriate functor modelling the topological K-theory of the reduced group C^*-algebra $C_r^*(G)$ and for $\mathcal{F}$ the family of finite subgroups, we obtain the Baum-Connes Conjecture 20.9.1. It assembles $K_n^{\mathrm{top}}(C_r^*(G))$ in terms of $K_n^{\mathrm{top}}(C_r^*(H))$, where H runs through the finite subgroups of G.

The Farrell-Jones Conjecture 20.8.1 and the Baum-Connes Conjecture 20.9.1 are very powerful conjectures and are the main motivation for the study of assembly maps. A survey of a lot of striking applications such as the ones to the conjectures of Bass, Borel, Gromov-Lawson-Rosenberg, Kadison, Kaplansky, and Novikov is given in Subsections 20.8.2 and 20.9.4. The Farrell-Jones Conjecture 20.8.1 and the Baum-Connes Conjecture 20.9.1 are known to be true for a surprisingly large class of groups, as explained in Subsections 20.8.7 and 20.9.5. All of this is an impressive example how homotopy theoretic methods can be used for problems in other fields such as algebra, geometry, manifold theory and operator algebras.

20.1.3 Other interpretations of assembly maps

The homotopy theoretic approach is the best for structural purposes. The applications and the proofs of the Farrell-Jones Conjecture 20.8.1 and the Baum-Connes Conjecture 20.9.1 require sophisticated analytic, topological and geometric interpretations of the homotopy theoretic assembly maps, for instance in terms of surgery theory, see Subsection 20.8.3, as forget control maps, see Subsection 20.8.4, and in terms of index theory, see Subsection 20.9.3. This presents an intriguing interaction between homotopy theory, geometry and operator theory.

20.1.4 The universal property of the assembly map

In Section 20.5 we characterize the assembly map in the sense that it is the universal approximation from the left by an excisive functor of a given homotopy invariant functor $G\text{-}\mathsf{CW}^2 \to \mathsf{Spectra}$. This is the key ingredient in the difficult identification of the various assembly maps mentioned in Subsection 20.1.3 above. It reflects the fact that in all of the Isomorphism Conjectures the hard and interesting object is the target and the source is given by the G-homology of the classifying G-spaces for a specific family of subgroups. The source is more accessible than the target since one can apply standard methods from algebraic topology such as spectral sequences and equivariant Chern characters.

20.1.5 Relative assembly maps

Relative assembly maps are studied in Section 20.12. They address the problem to make the families appearing in the various Isomorphism Conjectures as small as possible.

20.1.6 Further aspects of assembly maps

The homotopy theoretic approach to assembly allows to relate assembly maps for various theories, such as the algebraic K-theory and A-theory via linearization, see Subsection 20.8.6, algebraic K-theory of groups rings and topological K-theory of reduced group C^*-algebras, see Subsection 20.9.6, and algebraic K-theory of groups rings and the topological cyclic homology of the spherical group ring via cyclotomic traces, see Subsection 20.10.3. How assembly maps can be used for computations is illustrated in Section 20.13 which is based on the global point of view described in Section 20.11. Finally we formulate the challenge of extending equivariant homotopy theory for finite groups to infinite groups in Section 20.14.

The idea of the geometric assembly map is due to Quinn [101, 102] and its algebraic counterpart was introduced by Ranicki [104]. See also Loday [68].

20.1.7 Conventions

Throughout this paper G denotes a (discrete) group. Ring means associative ring with unit. All spectra are non-connective.

Acknowledgements

This paper is dedicated to Andrew Ranicki. It is financially supported by the ERC Advanced Grant “KL2MG-interactions” (no. 662400) of the author granted by the European Research Council, and by the Cluster of Excellence “Hausdorff Center for Mathematics” at Bonn.

20.2 Some basic categories

In this section we recall some well-known basic categories.

20.2.1 G-CW-complexes

Definition 20.2.1 (G-CW-complex)**.** A *G-CW-complex* X is a G-space together with a G-invariant filtration

$$\emptyset = X_{-1} \subseteq X_0 \subset X_1 \subseteq \cdots \subseteq X_n \subseteq \cdots \subseteq \bigcup_{n\geq 0} X_n = X$$

such that X carries the colimit topology with respect to this filtration (i.e., a set $C \subseteq X$ is closed if and only if $C \cap X_n$ is closed in X_n for all $n \geq 0$) and X_n is obtained from X_{n-1}

for each $n \geq 0$ by attaching equivariant n-dimensional cells, i.e., there exists a G-pushout

$$\begin{array}{ccc} \coprod_{i \in I_n} G/H_i \times S^{n-1} & \xrightarrow{\coprod_{i \in I_n} q_i^n} & X_{n-1} \\ \downarrow & & \downarrow \\ \coprod_{i \in I_n} G/H_i \times D^n & \xrightarrow{\coprod_{i \in I_n} Q_i^n} & X_n. \end{array}$$

A map $f\colon X \to Y$ between G-CW-complexes is called *cellular* if $f(X_n) \subseteq Y_n$ holds for all $n \geq 0$. We denote by G-CW the category of G-CW-complexes with cellular G-maps as morphisms and by G-CW^2 the corresponding category of G-CW-pairs. For basic information about G-CW-complexes we refer for instance to [69, Chapter 1 and 2].

20.2.2 The orbit category

The orbit category $\mathsf{Or}(G)$ has as objects homogeneous spaces G/H and as morphisms G-maps. It can be viewed as the category of 0-dimensional G-CW-complexes, whose G-quotient space is connected. In particular we can think of $\mathsf{Or}(G)$ as a full subcategory of G-CW.

20.2.3 Spectra

In this paper we can work with the most elementary category $\mathsf{Spectra}$ of spectra. A *spectrum* $\mathbf{E} = \{(E(n), \sigma(n)) \mid n \in \mathbb{Z}\}$ is a sequence of pointed spaces $\{E(n) \mid n \in \mathbb{Z}\}$ together with pointed maps called *structure maps* $\sigma(n)\colon E(n) \wedge S^1 \longrightarrow E(n+1)$. A *map of spectra* $\mathbf{f}\colon \mathbf{E} \to \mathbf{E}'$ is a sequence of maps $f(n)\colon E(n) \to E'(n)$ that are compatible with the structure maps $\sigma(n)$, i.e., we have $f(n+1) \circ \sigma(n) = \sigma'(n) \circ (f(n) \wedge \mathrm{id}_{S^1})$ for all $n \in \mathbb{Z}$. Maps of spectra are sometimes called functions in the literature, they should not be confused with the notion of a map of spectra in the stable category, see [1, III.2.].

The homotopy groups of a spectrum are defined by

$$\pi_i(\mathbf{E}) \quad := \quad \operatorname{colim}_{k \to \infty} \pi_{i+k}(E(k)), \tag{20.2.1}$$

where the ith structure map of the system $\pi_{i+k}(E(k))$ is given by the composite

$$\pi_{i+k}(E(k)) \xrightarrow{S} \pi_{i+k+1}(E(k) \wedge S^1) \xrightarrow{\sigma(k)_*} \pi_{i+k+1}(E(k+1))$$

of the suspension homomorphism S and the homomorphism induced by the structure map. A *weak equivalence* of spectra is a map $\mathbf{f}\colon \mathbf{E} \to \mathbf{F}$ of spectra inducing an isomorphism on all homotopy groups.

20.3 G-homology theories and $\mathsf{Or}(G)$-spectra

Let Λ be a commutative ring. Next we recall the obvious generalization of the notion of a (generalized) homology theory to a G-homology theory.

Definition 20.3.1 (*G*-homology theory). A *G-homology theory $\mathcal{H}_*^G$ with values in Λ-modules* is a collection of covariant functors $\mathcal{H}_n^G$ from the category G-CW^2 of G-CW-pairs to the category of Λ-modules indexed by $n \in \mathbb{Z}$ together with natural transformations

$$\partial_n^G(X,A)\colon \mathcal{H}_n^G(X,A) \to \mathcal{H}_{n-1}^G(A) := \mathcal{H}_{n-1}^G(A,\emptyset)$$

for $n \in \mathbb{Z}$ such that the following axioms are satisfied:

- *G-homotopy invariance*

  If f_0 and f_1 are G-homotopic G-maps of G-CW-pairs $(X,A) \to (Y,B)$, then $\mathcal{H}_n^G(f_0) = \mathcal{H}_n^G(f_1)$ for $n \in \mathbb{Z}$;

- *Long exact sequence of a pair*

  Given a pair (X,A) of G-CW-complexes, there is a long exact sequence

$$\cdots \xrightarrow{\mathcal{H}_{n+1}^G(j)} \mathcal{H}_{n+1}^G(X,A) \xrightarrow{\partial_{n+1}^G} \mathcal{H}_n^G(A) \xrightarrow{\mathcal{H}_n^G(i)} \mathcal{H}_n^G(X) \xrightarrow{\mathcal{H}_n^G(j)} \mathcal{H}_n^G(X,A) \xrightarrow{\partial_n^G} \cdots,$$

  where $i\colon A \to X$ and $j\colon X \to (X,A)$ are the inclusions;

- *Excision*

  Let (X,A) be a G-CW-pair and let $f\colon A \to B$ be a cellular G-map of G-CW-complexes. Equip $(X \cup_f B, B)$ with the induced structure of a G-CW-pair. Then the canonical map $(F,f)\colon (X,A) \to (X \cup_f B, B)$ induces an isomorphism

$$\mathcal{H}_n^G(F,f)\colon \mathcal{H}_n^G(X,A) \xrightarrow{\cong} \mathcal{H}_n^G(X \cup_f B, B);$$

- *Disjoint union axiom*

  Let $\{X_i \mid i \in I\}$ be a family of G-CW-complexes. Denote by $j_i\colon X_i \to \coprod_{i\in I} X_i$ the canonical inclusion. Then the map

$$\bigoplus_{i\in I} \mathcal{H}_n^G(j_i)\colon \bigoplus_{i\in I} \mathcal{H}_n^G(X_i) \xrightarrow{\cong} \mathcal{H}_n^G\left(\coprod_{i\in I} X_i\right)$$

  is bijective.

If $\mathbf{E}$ is a spectrum, then one gets a (non-equivariant) homology theory $H_*(-;\mathbf{E})$ by defining

$$H_n(X,A;\mathbf{E}) = \pi_n\left((X_+ \cup_{A_+} \operatorname{cone}(A_+)) \wedge \mathbf{E}\right)$$

for a CW-pair (X,A) and $n \in \mathbb{Z}$, where X_+ is obtained from X by adding a disjoint base point and cone denotes the (reduced) mapping cone. Its main property is $H_n(\{\bullet\};\mathbf{E}) = \pi_n(\mathbf{E})$. This extends to G-homology theories as follows. Since the building blocks of G-spaces are homogeneous spaces, we will have to consider a covariant $\mathsf{Or}(G)$-spectrum, i.e., a covariant functor $\mathbf{E}^G\colon \mathsf{Or}(G) \to \mathsf{Spectra}$, instead of a spectrum.

Definition 20.3.2 (Excisive). We call a covariant functor

$$\mathbf{E}\colon G\text{-}\mathsf{CW}^2 \to \mathsf{Spectra}$$

homotopy invariant if it sends G-homotopy equivalences to weak homotopy equivalences of spectra.

The functor $\mathbf{E}$ is *excisive* if it has the following four properties:

- It is homotopy invariant;
- The spectrum $\mathbf{E}(\emptyset)$ is weakly contractible;
- It respects homotopy pushouts up to weak homotopy equivalence, i.e., if the G-CW-complex X is the union of G-CW-subcomplexes X_1 and X_2 with intersection X_0, then the canonical map from the homotopy pushout of $\mathbf{E}(X_2) \longleftarrow \mathbf{E}(X_0) \longrightarrow \mathbf{E}(X_2)$ to $\mathbf{E}(X)$ is a weak homotopy equivalence of spectra;
- It respects disjoint unions up to weak homotopy, i.e., the natural map $\bigvee_{i\in I} \mathbf{E}(X_i) \to \mathbf{E}(\coprod_{i\in I} X_i)$ is a weak homotopy equivalence for all index sets I.

One easily checks

Lemma 20.3.3. *Suppose that the covariant functor $\mathbf{E}\colon G\text{-}\mathsf{CW}^2 \to \mathsf{Spectra}$ is excisive. Then we obtain a G-homology theory with values in $\mathbb{Z}$-modules by assigning to G-CW-pair (X,A) and $n \in \mathbb{Z}$ the abelian group $\pi_n(\mathbf{E}(X,A))$.*

A G-space X defines a contravariant $\mathsf{Or}(G)$-space $O^G(X)$ by sending G/H to $\operatorname{map}_G(G/H, X) = X^H$. Given a contravariant pointed $\mathsf{Or}(G)$-space Y and a covariant pointed $\mathsf{Or}(G)$-space Z, there is the pointed space $Y \wedge_{\mathsf{Or}(G)} Z$. Its construction is explained for instance in [26, Section 1]. This construction is natural in Y and Z. Its main property is that one obtains for every pointed space X an adjunction homeomorphism

$$\operatorname{map}(Y \wedge_{\mathsf{Or}(G)} Z, X) \xrightarrow{\cong} \operatorname{mor}_{\mathsf{Or}(G)}(Y, \operatorname{map}(Z,X))$$

where the source is the pointed mapping space and the target is the topological space of natural transformations from Y to the contravariant pointed $\mathsf{Or}(G)$-space $\operatorname{map}(Z,X)$ sending G/H to the pointed mapping space $\operatorname{map}(Z(G/H), X)$. If $\mathbf{E}^G$ is a covariant $\mathsf{Or}(G)$-spectrum, then one obtains a spectrum $Y \wedge_{\mathsf{Or}(G)} \mathbf{E}^G$. Hence we can extend a covariant functor $\mathbf{E}^G\colon \mathsf{Or}(G) \to \mathsf{Spectra}$ to a covariant functor

$$(\mathbf{E}^G)_\%\colon G\text{-}\mathsf{CW}^2 \to \mathsf{Spectra}, \quad (X,A) \mapsto O^G(X_+ \cup_{A_+} \operatorname{cone}(A_+)) \wedge_{\mathsf{Or}(G)} \mathbf{E}^G. \qquad (20.3.1)$$

The easy proofs of the following two results are left to the reader.

Lemma 20.3.4. *If $\mathbf{E}^G$ is a covariant $\mathsf{Or}(G)$-spectrum, then $(\mathbf{E}^G)_\%$ is excisive and we obtain a G-homology theory $H_*^G(-;\mathbf{E}^G)$ by*

$$H_n^G(X,A;\mathbf{E}^G) = \pi_n((\mathbf{E}^G)_\%(X,A)) = \pi_n\big(O^G(X_+ \cup_{A_+} \operatorname{cone}(A_+)) \wedge_{\mathsf{Or}(G)} \mathbf{E}^G\big)$$

satisfying $H_n^G(G/H;\mathbf{E}^G) = \pi_n(\mathbf{E}^G(G/H))$ for $n \in \mathbb{Z}$ and $H \subseteq G$.

Lemma 20.3.5. *Let $\mathbf{t}\colon \mathbf{E} \to \mathbf{F}$ be a natural transformation of covariant functors $G\text{-}\mathsf{CW}^2 \to \mathsf{Spectra}$. Suppose that $\mathbf{E}$ and $\mathbf{F}$ are excisive and $\mathbf{t}(G/H)$ is a weak homotopy equivalence for any homogeneous G-space G/H.*

Then $\mathbf{t}(X,A)\colon \mathbf{E}(X,A) \to \mathbf{F}(X,A)$ *is a weak homotopy equivalence for every G-CW-pair* (X,A).

20.4 Approximation by an excisive functor

The following result follows from [26, Theorem 6.3]. Its non-equivariant version is due to Weiss-Williams [127].

Theorem 20.4.1 (Approximation by an excisive functor). *Let* $\mathbf{E}\colon G\text{-}\mathsf{CW}^2 \to \mathsf{Spectra}$ *be a covariant functor that is homotopy invariant. Let* $\mathbf{E}|\colon \mathsf{Or}(G) \to \mathsf{Spectra}$ *be its composite with the obvious inclusion* $\mathsf{Or}(G) \to G\text{-}\mathsf{CW}^2$.

Then there exists a covariant functor

$$\mathbf{E}^{\%}\colon G\text{-}\mathsf{CW}^2 \to \mathsf{Spectra}$$

and natural transformations

$$\begin{aligned} \mathbf{A_E}\colon \mathbf{E}^{\%} &\to \mathbf{E}; \\ \mathbf{B_E}\colon \mathbf{E}^{\%} &\to \mathbf{E}|_{\%}, \end{aligned}$$

satisfying:

1. *The functor* $\mathbf{E}^{\%}$ *is excisive;*
2. *The map* $\mathbf{A_E}(G/H)\colon \mathbf{E}^{\%}(G/H) \to \mathbf{E}(G/H)$ *is a weak homotopy equivalence for every homogeneous space* G/H;
3. *The map* $\mathbf{B_E}(X,A)\colon \mathbf{E}^{\%}(X,A) \to \mathbf{E}|_{\%}(X,A)$ *is a weak homotopy equivalence for every G-CW-pair* (X,A);
4. *The functor* $\mathbf{E}$ *is excisive if and only if* $\mathbf{A_E}(X,A)$ *is a weak homotopy equivalence for every G-CW-pair* (X,A);
5. *The transformations* $\mathbf{A_E}$ *and* $\mathbf{B_E}$ *are functorial in* $\mathbf{E}$.

Although one does not need to understand the explicit construction of $\mathbf{E}^{\%}$, $\mathbf{A_E}$ and $\mathbf{B_E}$ and the proof of Theorem 20.4.1 for the applications of Theorem 20.4.1 and for the reminder of this paper, we make some comments about it for the interested reader.

As an illustration we firstly present a naive suggestion in the non-equivariant case, which turns out to require too restrictive assumptions on $\mathbf{E}$, and therefore will not be the final solution, but conveys a first idea. Namely, we can define a map of pointed sets $X_+ \wedge \mathbf{E}(\{\bullet\}) \to \mathbf{E}(X)$ for a CW-complex X by sending an element in the target represented by (x,e) for $x \in X$ and $e \in \mathbf{E}(\{\bullet\})$ to $\mathbf{E}(c_x\colon \{\bullet\} \to X)(e)$, where $c_x\colon \{\bullet\} \to X$ is the constant map with value x. The problem is that the only reasonable way of ensuring the continuity of this map is to require that $\mathbf{E}$ itself is continuous, i.e., the map $\mathrm{map}(X,Y) \to \mathrm{map}(E(X)_n, E(Y)_n)$ sending f to $E(f)_n$ has to be continuous for all $n \in \mathbb{Z}$. But this assumption is not satisfied for the functors $\mathbf{E}$ which are of interest for us and will be considered below.

The solution is to take homotopy invariance into account and to work simplicially. Let us consider the special case, where G is trivial and X is a simplicial complex. For any simplex σ of X we have the inclusion $i[\sigma]\colon \sigma \to X$ and can therefore define maps

$$A_{\mathbf{E}}[\sigma]_n\colon \sigma_+ \wedge E(\sigma)_n \xrightarrow{\mathrm{pr}_+ \wedge \mathrm{id}_{E(\sigma)_n}} \{\bullet\}_+ \wedge E(\sigma)_n = E(\sigma)_n \xrightarrow{E(i[\sigma])_n} E(X)_n;$$
$$B_{\mathbf{E}}[\sigma]_n\colon \sigma_+ \wedge E(\sigma)_n \xrightarrow{\mathrm{id}_{\sigma_+} \wedge E(\mathrm{pr})_n} \sigma_+ \wedge E(\{\bullet\})_n.$$

where pr denotes the projection onto $\{\bullet\}$. Now define a space $E^{\%}(X)_n$ by gluing the spaces $\sigma_+ \wedge E(\sigma)_n$ for σ running over the simplices of X together according to the simplicial structure, more precisely, for an inclusion $j\colon \tau \to \sigma$ of simplices we identify a point in $\tau_+ \wedge E(\tau)_n$ with its image in $\sigma_+ \wedge E(\sigma)_n$ under the obvious map $j_+ \wedge E(j)_n$. One easily checks that the various maps $A_{\mathbf{E}}[\sigma]_n$ and $B_{\mathbf{E}}[\sigma]_n$ fit together to maps of pointed spaces

$$\begin{aligned} A_{\mathbf{E}}(X)_n\colon E^{\%}(X)_n &\to E(X)_n; \\ B_{\mathbf{E}}(X)_n\colon E^{\%}(X)_n &\to E|_{\%}(X)_n := X_+ \wedge E(\{\bullet\})_n, \end{aligned}$$

and thus to maps of spectra

$$\begin{aligned} \mathbf{A}_{\mathbf{E}}(X)\colon \mathbf{E}^{\%}(X) &\to \mathbf{E}(X)_n; \\ \mathbf{B}_{\mathbf{E}}(X)\colon \mathbf{E}^{\%}(X) &\to \mathbf{E}|_{\%}(X) := X_+ \wedge \mathbf{E}(\{\bullet\}). \end{aligned}$$

Notice that each map $\mathbf{E}(\mathrm{pr})\colon \mathbf{E}(\sigma) \to \mathbf{E}(\{\bullet\})$ is by assumption a weak homotopy equivalence. This implies that $\mathbf{B}_{\mathbf{E}}(X)\colon \mathbf{E}^{\%}(X) \to \mathbf{E}|_{\%}(X)$ is a weak homotopy equivalence. Since the functor $\mathbf{E}_{\%}$ is excisive, the functor $\mathbf{E}^{\%}$ is excisive. If $X = \{\bullet\}$, the map $\mathbf{A}_{\mathbf{E}}(\{\bullet\})\colon \mathbf{E}^{\%}(\{\bullet\}) \to \mathbf{E}(\{\bullet\})$ is an isomorphism and in particular a weak homotopy equivalence.

Now we see, where the name assembly map comes from. In the case of a simplicial complex X we want to assemble $\mathbf{E}(X)$ by its values $\mathbf{E}(\sigma)$ for the various simplices of X, which leads to the definition of $\mathbf{E}^{\%}(X)$. Intuitively it is clear that $\mathbf{E}(X)$ carries the same information as $\mathbf{E}|^{\%}(X)$ if and only if $\mathbf{E}$ is excisive since the condition excisive allows to compute the values of $\mathbf{E}$ on X by its values on the simplices taking into account how the simplices are glued together to yield X.

Finally one wants a definition that is independent of the simplicial structure and actually applies to more general spaces X than simplicial complexes. Therefore one uses simplicial sets and in particular the singular simplicial set $S.X$. Recall that $S.X$ is the functor from the category of finite ordered sets $\mathbf{\Delta}$ to the category of sets Sets sending the finite ordered set $[p]$ to the set $\mathrm{map}(\Delta_p, Y)$ for Δ_p the standard p-simplex. For the equivariant version one has to bring the orbit category into play. So one considers for a G-space X the functor

$$\mathsf{Or}(G) \times \mathbf{\Delta} \to \mathsf{Sets}, \quad (G/H, [p]) \mapsto \mathrm{map}_G(G/H \times \Delta_p, X).$$

In some sense one uses free resolution of contravariant functors $\mathsf{Or}(G) \times \mathbf{\Delta} \to \mathsf{Spaces}$ to get the right construction of $\mathbf{E}^{\%}$ and of the desired transformations $\mathbf{A}_{\mathbf{E}}$ and $\mathbf{B}_{\mathbf{E}}$ so that the claims appearing in Theorem 20.4.1 can be proved. Details can be found in [26].

20.5 The universal property

Next we explain why Theorem 20.4.1 characterizes the assembly map in the sense that $\mathbf{A_E}\colon \mathbf{E}^{\%} \longrightarrow \mathbf{E}$ is the universal approximation from the left by an excisive functor of a homotopy invariant functor $\mathbf{E}\colon G\text{-}\mathsf{CW}^2 \to \mathsf{Spectra}$. Namely, let $\mathbf{T}\colon \mathbf{F} \to \mathbf{E}$ be a transformation of covariant functors $G\text{-}\mathsf{CW}^2 \to \mathsf{Spectra}$ such that $\mathbf{F}$ is excisive. Then for any G-$\mathcal{F}$-CW-pair (X, A) the following diagram commutes

$$\begin{array}{ccc} \mathbf{F}^{\%}(X) & \xrightarrow[\simeq]{\mathbf{A_F}(X)} & \mathbf{F}(X) \\ {\scriptstyle \mathbf{T}^{\%}(X)}\downarrow & & \downarrow{\scriptstyle \mathbf{T}(X)} \\ \mathbf{E}^{\%}(X) & \xrightarrow[\mathbf{A_E}(X)]{} & \mathbf{E}(X) \end{array}$$

and $\mathbf{A_F}(X)$ is a weak homotopy equivalence by Theorem 20.4.1 (4). Hence $\mathbf{T}(X)$ factorizes over $\mathbf{A_E}(\mathbf{X})$ up to natural weak homotopy equivalence.

Suppose additionally that $\mathbf{T}(G/H)$ is a weak homotopy equivalence for every subgroup $H \subseteq G$. Then both $\mathbf{T}^{\%}(X)$ and $\mathbf{A_F}(X)$ are weak homotopy equivalences by Lemma 20.3.5 and Theorem 20.4.1 (4), and hence $\mathbf{T}(X)$ can be identified with $\mathbf{A_E}(X)$ up to natural weak homotopy equivalence.

Recall that there is a natural weak equivalence $\mathbf{B_E}(X)\colon \mathbf{E}^{\%}(X) \xrightarrow{\simeq} \mathbf{E}|_{\%}(X)$, so that one may replace in the considerations above $\mathbf{E}^{\%}(X)$ by $\mathbf{E}|_{\%}(X)$, which depends on the values of $\mathbf{E}$ on homogeneous spaces only. This universal property will be the key ingredient for the identification of various versions of assembly maps.

20.6 Classifying spaces for families of subgroups

We recall the notion classifying space for a family which was introduced by tom Dieck [115].

Definition 20.6.1 (Family of subgroups)**.** A *family $\mathcal{F}$ of subgroups of a group G* is a set of subgroups of G that is closed under conjugation with elements of G and under passing to subgroups.

Our main examples of families are the trivial family $\mathcal{TR}$ consisting of the trivial subgroup, the family $\mathcal{ALL}$ of all subgroups, and the families $\mathcal{FCY}$, $\mathcal{CY}$, $\mathcal{FIN}$, and $\mathcal{VCY}$ of finite cyclic subgroups, of cyclic subgroups, of finite subgroups, and of virtually cyclic subgroups.

Definition 20.6.2 (Classifying G-space for a family of subgroups)**.** Let $\mathcal{F}$ be a family of subgroups of G. A model $E_{\mathcal{F}}(G)$ for the *classifying spaces for the family $\mathcal{F}$ of subgroups of G* is a G-CW-complex $E_{\mathcal{F}}(G)$ that has the following properties:

1. All isotropy groups of $E_{\mathcal{F}}(G)$ belong to $\mathcal{F}$;

2. For any G-CW-complex Y, whose isotropy groups belong to $\mathcal{F}$, there is up to G-homotopy precisely one G-map $Y \to X$.

We abbreviate $\underline{E}G := E_{\mathcal{FIN}}(G)$ and call it the *universal G-space for proper G-actions.* We also write $\underline{\underline{E}}G := E_{\mathcal{VCY}}(G)$.

Equivalently, $E_{\mathcal{F}}(G)$ is a terminal object in the G-homotopy category of G-CW-complexes, whose isotropy groups belong to $\mathcal{F}$. In particular two models for $E_{\mathcal{F}}(G)$ are G-homotopy equivalent and for two families $\mathcal{F}_0 \subseteq \mathcal{F}_1$ there is up to G-homotopy precisely one G-map $E_{\mathcal{F}_0}(G) \to E_{\mathcal{F}_1}(G)$. There are functorial constructions for $E_{\mathcal{F}}(G)$ generalizing the bar construction, see [26, Section 3 and Section 7].

Theorem 20.6.3 (Homotopy characterization of $E_{\mathcal{F}}(G)$). *A G-CW-complex X is a model for $E_{\mathcal{F}}(G)$ if and only if for every subgroup $H \subseteq G$ its H-fixed point set X^H is weakly contractible if $H \in \mathcal{F}$, and is empty if $H \notin \mathcal{F}$.*

A model for $E_{\mathcal{ALL}}(G)$ is G/G. A model for $E_{\mathcal{TR}}(G)$ is the same as a model for EG, i.e., the universal covering of BG, or, equivalently, the total space of the universal G-principal bundle. There are many interesting geometric models for classifying spaces $\underline{E}G = E_{\mathcal{FIN}}(G)$, e.g., the Rips complex for a hyperbolic group, the Teichmüller space for a mapping class group, and so on. The question whether there are finite-dimensional models, models of finite type or finite models has been studied intensively during the last decades. For more information about classifying spaces for families we refer for instance to [77].

20.7 The Meta-Isomorphism Conjecture

In this section we formulate the Meta-Isomorphism Conjecture, from which all other Isomorphism Conjectures such as the one due to Farrell-Jones and Baum-Connes are obtained by specifying the parameters $\mathbf{E}$ and $\mathcal{F}$.

20.7.1 The Meta-Isomorphism Conjecture for G-homology theories

Let $\mathcal{H}_*^G$ be a G-homology theory with values in Λ-modules for some commutative ring Λ. The projection $\mathrm{pr}\colon E_{\mathcal{F}}(G) \to G/G$ induces for all integers $n \in \mathbb{Z}$ a homomorphism of Λ-modules

$$\mathcal{H}_n^G(\mathrm{pr})\colon \mathcal{H}_n^G(E_{\mathcal{F}}(G)) \to \mathcal{H}_n^G(G/G) \tag{20.7.1}$$

which is called the *assembly map.*

Conjecture 20.7.1 (Meta-Isomorphism Conjecture for G-homology theories). *The group G satisfies the* Meta-Isomorphism Conjecture *with respect to the G-homology theory $\mathcal{H}_*^G$ and the family $\mathcal{F}$ of subgroups of G, if the assembly map*

$$\mathcal{H}_n^G(\mathrm{pr})\colon \mathcal{H}_n^G(E_{\mathcal{F}}(G)) \to \mathcal{H}_n^G(G/G)$$

of (20.7.1) *is bijective for all $n \in \mathbb{Z}$.*

If we choose $\mathcal{F}$ to be the family $\mathcal{ALL}$ of all subgroups, then G/G is a model for $E_{\mathcal{ALL}}(G)$ and the Meta-Isomorphism Conjecture 20.7.1 is obviously true. The point is to find an as small as possible family $\mathcal{F}$. The idea of the Meta-Isomorphism Conjecture 20.7.1 is that one wants to compute $\mathcal{H}_n^G(G/G)$, which is the unknown and the interesting object, by assembling it from the values $\mathcal{H}_n^G(G/H)$ for $H \in \mathcal{F}$.

20.7.2 The Meta-Isomorphism Conjecture on the level of spectra

Often the construction of the assembly map is done already on the level of spectra or can be lifted to this level. Consider a covariant functor

$$\mathbf{E}^G \colon \mathsf{Or}(G) \to \mathsf{Spectra}.$$

Conjecture 20.7.2 (Meta-Isomorphism Conjecture for spectra). *The group G satisfies the* Meta-Isomorphism Conjecture *with respect to the covariant functor* $\mathbf{E}^G \colon \mathsf{Or}(G) \to \mathsf{Spectra}$ *and the family $\mathcal{F}$ of subgroups of G, if the projection* $\mathrm{pr}\colon E_{\mathcal{F}}(G) \to G/G$ *induces a weak homotopy equivalence*

$$(\mathbf{E}^G)_{\%}(\mathrm{pr})\colon (\mathbf{E}^G)_{\%}(E_{\mathcal{F}}(G)) \to (\mathbf{E}^G)_{\%}(G/G) = \mathbf{E}^G(G/G).$$

Notice that $(\mathbf{E}^G)_{\%}(\mathrm{pr})\colon (\mathbf{E}^G)_{\%}(E_{\mathcal{F}}(G)) \to (\mathbf{E}^G)_{\%}(G/G) = \mathbf{E}^G(G/G)$ is a weak homotopy equivalence if and only if for every $n \in \mathbb{Z}$ the map

$$H_n^G(\mathrm{pr}; \mathbf{E}^G)\colon H_n^G(E_{\mathcal{F}}(G); \mathbf{E}^G) \to H_n^G(G/G; \mathbf{E}^G)$$

is a bijection, where $H_*^G(-; \mathbf{E}^G)$ is the G-homology theory associated to $\mathbf{E}^G$, see Lemma 20.3.4. In other words, Conjecture 20.7.2 is equivalent to Conjecture 20.7.1 if we take for $\mathcal{H}_*^G$ the G-homology theory associated to $\mathbf{E}^G$.

20.7.3 The assembly map in terms of homotopy colimits

The assembly map appearing in Conjecture 20.7.2 can be interpreted in terms of homotopy colimits as follows. The $\mathcal{F}$-restricted orbit category $\mathsf{Or}_{\mathcal{F}}(G)$ be the full subcategory of $\mathsf{Or}(G)$ consisting of those objects G/H for which H belongs to $\mathcal{F}$. Let $\mathbf{E}^G|_{\mathsf{Or}_{\mathcal{F}}(G)}$ be the restriction of $\mathbf{E}^G$ to $\mathsf{Or}_{\mathcal{F}}(G)$. Then we get from the inclusion $\mathsf{Or}_{\mathcal{F}}(G) \to \mathsf{Or}(G)$ and the fact that G/G is a terminal object in $\mathsf{Or}(G)$ a map

$$\mathrm{hocolim}_{\mathsf{Or}_{\mathcal{F}}(G)} \mathbf{E}^G|_{\mathsf{Or}_{\mathcal{F}}(G)} \to \mathrm{hocolim}_{\mathsf{Or}(G)} \mathbf{E}^G = \mathbf{E}^G(G/G).$$

This map can be identified with $(\mathbf{E}^G)_{\%}(\mathrm{pr})\colon (\mathbf{E}^G)_{\%}(E_{\mathcal{F}}(G)) \to (\mathbf{E}^G)_{\%}(G/G)$, see [26, Section 5.2]. Again this explains the name assembly map: we try to put the values of $\mathbf{E}^G$ on homogeneous spaces G/H for $H \in \mathcal{F}$ together to get its value at G/G.

20.7.4 Spectra over $\mathsf{Groupoids}$

In all interesting cases we will obtain $\mathbf{E}^G$ as follows. Let $\mathsf{Groupoids}$ be the category of small groupoids. Consider a covariant functor

$$\mathbf{E}\colon \mathsf{Groupoids} \to \mathsf{Spectra}$$

that *respects equivalences*, i.e., it sends equivalences of groupoids to weak equivalences of spectra. Given a G-set S, its *transport groupoid* $\mathcal{T}^G(S)$ has S as set of objects and the set of morphism from s_0 to s_1 is $\{g \in G \mid gs_0 = s_1\}$. Composition comes from the multiplication in G. We get for every group G a functor

$$\mathbf{E}^G \colon \mathsf{Or}(G) \to \mathsf{Spectra}$$

by composing $\mathbf{E}$ with the functor $\mathsf{Or}(G) \to \mathsf{Groupoids}$, $G/H \mapsto \mathcal{T}^G(G/H)$.

Notice that a group G can be viewed as a groupoid with one object and G as the set of automorphisms of this object, and hence we can consider $\mathbf{E}(G)$. We have the obvious identifications $\mathbf{E}(G) = \mathbf{E}^G(G/G) = (\mathbf{E}^G)_{\%}(G/G)$. Moreover, for every subgroup $H \subseteq G$ there is an equivalence of groupoids $H \to \mathcal{T}^G(G/H)$ sending the unique object of H to the object eH, which induces a weak homotopy equivalence $\mathbf{E}(H) \to \mathbf{E}^G(G/H)$.

The various prominent Isomorphism Conjectures such as the one due to Farrell-Jones and Baum-Connes are now obtained by specifying $\mathbf{E} \colon \mathsf{Groupoids} \to \mathsf{Spectra}$, the group G and the family $\mathcal{F}$.

20.8 The Farrell-Jones Conjectures

20.8.1 The Farrell-Jones Conjecture for K-and L-theory

Let R be a ring (with involution). There exist covariant functors respecting equivalences

$$\mathbf{K}_R \colon \mathsf{Groupoids} \to \mathsf{Spectra}; \qquad (20.8.1)$$
$$\mathbf{L}_R^{\langle -\infty \rangle} \colon \mathsf{Groupoids} \to \mathsf{Spectra}, \qquad (20.8.2)$$

such that for every group G and all $n \in \mathbb{Z}$ we have

$$\pi_n(\mathbf{K}_R(G)) \cong K_n(RG);$$
$$\pi_n(\mathbf{L}_R^{\langle -\infty \rangle}(G)) \cong L_n^{\langle -\infty \rangle}(RG).$$

Here $K_n(RG)$ is the n-th algebraic K-group of the group ring RG and $L_n^{\langle -\infty \rangle}(RG)$ is the nth quadratic L-group with decoration $\langle -\infty \rangle$ of the group ring RG equipped with the involution sending $\sum_{g \in G} r_g g$ to $\sum_{g \in G} \overline{r_g} g^{-1}$.

The details of this construction can be found in [26, Section 2]. If we now take these functors and the family $\mathcal{VCY}$ of virtually cyclic subgroups, we obtain

Conjecture 20.8.1 (Farrell-Jones Conjecture). *A group G satisfies the K-*theoretic or L-theoretic Farrell-Jones Conjecture *if for every ring (with involution) R the assembly maps induced by the projection* $\mathrm{pr} \colon \underline{\underline{E}}G \to G/G$

$$H_n^G(\mathrm{pr}; \mathbf{K}_R) \colon H_n^G(\underline{\underline{E}}G; \mathbf{K}_R) \to H_n^G(G/G; \mathbf{K}_R) = K_n(RG);$$
$$H_n^G(\mathrm{pr}; \mathbf{L}_R^{\langle -\infty \rangle}) \colon H_n^G(\underline{\underline{E}}G; \mathbf{L}_R^{\langle -\infty \rangle}) \to H_n^G(G/G; \mathbf{L}_R^{\langle -\infty \rangle}) = L_n^{\langle -\infty \rangle}(RG),$$

are bijective for all $n \in \mathbb{Z}$.

It is crucial that we use non-connective K-spectra and that the decoration for the L-theory is $\langle -\infty \rangle$, see [39].

The original version of the Farrell-Jones Conjecture appeared in [38, 1.6 on page 257]. A detailed exposition on the Farrell-Jones Conjecture will be given in [79], see also [82].

20.8.2 Applications of the Farrell-Jones Conjecture

Here are some consequences of the Farrell-Jones Conjecture. For more information about these and other applications we refer for instance to [9, 79, 82].

20.8.2.1 Computations

One can carry out *explicit computations* of K and L-groups of group rings by applying methods from algebraic topology to the left side given by a G-homology theory and by finding small models for the classifying spaces of families using the topology and geometry of groups, see Section 20.13.

20.8.2.2 Vanishing of lower and middle K-groups

If G is a torsionfree group satisfying the K-theoretic Farrell-Jones Conjecture 20.8.1, then $K_n(\mathbb{Z}G)$ for $n \leq -1$, the reduced projective class group $\widetilde{K}_0(\mathbb{Z}G)$, and the Whitehead group $\mathrm{Wh}(G)$ vanish.

This has the following consequences. Every homotopy equivalence $f\colon X \to Y$ of connected CW-complexes with $\pi_1(Y) \cong G$ is simple. Every h-cobordism over a closed manifold M of dimension ≥ 5 and $G \cong \pi_1(M)$ is trivial. Every finitely generated projective $\mathbb{Z}G$-module is stably free. Every finitely dominated connected CW-complex X with $\pi_1(X) \cong G$ is homotopy equivalent to a finite CW-complex.

20.8.2.3 Kaplansky's Idempotent Conjecture

If the torsionfree group G satisfies the K-theoretic Farrell-Jones Conjecture 20.8.1, then G satisfies the *Idempotent Conjecture* that for a commutative integral domain R the only idempotents of RG are 0 and 1.

20.8.2.4 Novikov Conjecture

If G satisfies the L-theoretic Farrell-Jones Conjecture 20.8.1, then G satisfies the *Novikov Conjecture* about the homotopy invariance of higher signatures. For more information about the Novikov Conjecture we refer for instance to [42, 43, 61].

20.8.2.5 Borel Conjecture

If G is a torsionfree group satisfying the K-theoretic and the L-theoretic Farrell-Jones Conjecture 20.8.1, then G satisfies the *Borel Conjecture* in dimensions ≥ 5, i.e., if M and N are closed aspherical manifolds of dimension ≥ 5 with $\pi_1(M) \cong \pi_1(N) \cong G$, then M and N are homeomorphic and every homotopy equivalence from M to N is homotopic to a homeomorphism.

20.8.2.6 Bass Conjecture

If G satisfies the K-theoretic Farrell-Jones Conjecture 20.8.1, then G satisfies the *Bass Conjecture*, see [9, 13].

20.8.2.7 Automorphism groups

The Farrell-Jones Conjecture 20.8.1 yields rational computations of the homotopy groups and homology groups of the automorphisms groups of an aspherical closed manifold in the topological, PL and smooth category, see for instance instance [36], [37, Section 2] and [35, Lecture 5].

For instance, if M is an aspherical orientable closed (smooth) manifold of dimension > 10 with fundamental group G such that G satisfies the Farrell-Jones Conjecture 20.8.1, then we get for $1 \leq i \leq (\dim M - 7)/3$

$$\pi_i(\mathrm{Top}(M)) \otimes_{\mathbb{Z}} \mathbb{Q} = \begin{cases} \mathrm{center}(G) \otimes_{\mathbb{Z}} \mathbb{Q} & \text{if } i = 1; \\ 0 & \text{if } i > 1, \end{cases}$$

and

$$\pi_i(\mathrm{Diff}(M)) \otimes_{\mathbb{Z}} \mathbb{Q} = \begin{cases} \mathrm{center}(G) \otimes_{\mathbb{Z}} \mathbb{Q} & \text{if } i = 1; \\ \bigoplus_{j=1}^{\infty} H_{(i+1)-4j}(M; \mathbb{Q}) & \text{if } i > 1 \text{ and } \dim M \text{ odd}; \\ 0 & \text{if } i > 1 \text{ and } \dim M \text{ even}. \end{cases}$$

For a survey on automorphisms of manifolds we refer to [128].

20.8.2.8 Boundary of hyperbolic groups

In [11] a proof of a conjecture of Gromov is given in dimensions $n \geq 6$ using the Farrell-Jones Conjecture 20.8.1 that a torsionfree hyperbolic group with S^n as boundary is the fundamental group of an aspherical closed topological manifold. This manifold is unique to homeomorphism. The stable Cannon Conjecture is treated in [41].

20.8.2.9 Poincaré duality groups

If G is a Poincaré duality group of dimension $n \geq 6$ and satisfies the Farrell-Jones Conjecture 20.8.1, then it is the fundamental group of an aspherical closed homology ANR-manifold, see [11]. It is unique up to s-cobordism. Whether it can be chosen to be an aspherical closed topological manifold, depends on its Quinn obstruction.

20.8.2.10 Tautological classes and aspherical manifolds

The vanishing of tautological classes for many bundles with fiber an aspherical manifold is proved in [45].

20.8.2.11 Fibering manifolds

The problem when a map from some closed connected manifold to an aspherical closed manifold approximately fibers, i.e., is homotopic to Manifold Approximate Fibration, is analyzed in [40].

20.8.3 The interpretation of the Farrell-Jones assembly map for L-theory in terms of surgery theory

So far we have given a homotopy theoretic approach to the assembly map. This is the easiest approach and well-suited for structural questions such as comparing the assembly maps of various different theories, as explained below. For concrete applications it is important to give geometric or analytic interpretations. For instance, one key ingredient in the proof that the Borel Conjecture follows from the Farrell-Jones Conjecture is a geometric interpretation of the assembly for the trivial family in terms of surgery theory, notably the surgery exact sequence, which we briefly sketch next.

Definition 20.8.2 (The structure set)**.** Let N be a closed topological manifold of dimension n. We call two simple homotopy equivalences $f_i\colon M_i \to N$ from closed topological manifolds M_i of dimension n to N for $i = 0, 1$ equivalent if there exists a homeomorphism $g\colon M_0 \to M_1$ such that $f_1 \circ g$ is homotopic to f_0.

The *structure set* $\mathcal{S}(N)$ of N is the set of equivalence classes of simple homotopy equivalences $M \to X$ from closed topological manifolds of dimension n to N. This set has a preferred base point, namely, the class of the identity $\mathrm{id}\colon N \to N$.

One easily checks that the Borel Conjecture holds for $G = \pi_1(N)$ for a closed aspherical manifold N if and only if $\mathcal{S}(N)$ consists of precisely one element, namely, the class of $\mathrm{id}_N\colon N \to N$. The surgery exact sequence, which we will explain next, gives a way of calculating the structure set.

Definition 20.8.3 (Normal map of degree one)**.** A *normal map of degree one* with target the connected closed manifold N of dimension n consists of:

- A connected closed n-dimensional manifold M;
- A map of degree one $f\colon M \to N$;
- A $(k+n)$-dimensional vector bundle ξ over N;
- A bundle map $\overline{f}\colon TM \oplus \underline{\mathbb{R}^k} \to \xi$ covering f.

There is an obvious normal bordism relation, and we denote by $\mathcal{N}(N)$ the set of bordism classes of normal maps with target N. One can assign to a normal map $f\colon M \to N$ its surgery obstruction $\sigma(f) \in L_n^s(\mathbb{Z}G)$ taking values in the nth quadratic L-group with decoration s, where $G = \pi_1(N)$ and $n = \dim(N)$. If $n \geq 5$, the surgery obstruction vanishes if and only if one can find (by doing surgery) a representative in the normal bordisms class, whose underlying map f is a simple homotopy equivalence. It yields a map $\sigma\colon \mathcal{N}(N) \to L_n^s(\mathbb{Z}G)$. There is a map $\eta\colon \mathcal{S}_n^{\mathrm{top}}(N) \to \mathcal{N}(N)$ that assigns to the class of a simple homotopy equivalence $f\colon M \to N$ with a closed manifold M as source the normal map given by f itself and the bundle data coming from TM and $\xi = (f^{-1})^*TM$ for some homotopy inverse f^{-1} of f. We denote by $\mathcal{N}(N \times [0,1], N \times \partial[0,1])$ the normal bordism classes of normal maps relative boundary. Essentially these are normal maps $(M, \partial M) \to (N \times [0,1], N \times \partial[0,1])$ of degree one which are simple homotopy equivalences on the boundary. There is a surgery obstruction relative boundary which yields a map $\sigma\colon \mathcal{N}(N \times [0,1], N \times \partial[0,1]) \to L^s_{n+1}(\mathbb{Z}G)$. There is a also a map $\partial\colon L^s_{n+1}(\mathbb{Z}G) \to \mathcal{S}^{\mathrm{top}}_n(N)$ that sends an element $x \in L^s_{n+1}(\mathbb{Z}G)$ to the class of a simple homotopy equivalence $f\colon M \to N$ for which there exists a normal map relative boundary of triads $(F, f_0, \mathrm{id}_N)\colon (W; M, N) \to (N \times [0,1]; N \times \{0\}, N \times \{1\})$ whose relative surgery obstruction is x. If $n \geq 5$, then one obtains a long exact sequence of abelian groups, the *surgery exact sequence* due to Browder, Novikov, Sullivan, and Wall

$$\mathcal{N}(N \times [0,1], N \times \partial[0,1]) \xrightarrow{\sigma_{n+1}} L^s_{n+1}(\mathbb{Z}G) \xrightarrow{\partial} \mathcal{S}(N) \xrightarrow{\eta} \mathcal{N}(N) \xrightarrow{\sigma_n} L^s_n(\mathbb{Z}G). \qquad (20.8.3)$$

If we can show that σ_{n+1} is surjective and σ_n is injective, then the Borel Conjecture holds for $G = \pi_1(N)$, if N is an aspherical closed manifold of dimension $n \geq 5$.

Let $\mathbf{L}$ be the L-theory spectrum. It has the property $\pi_n(\mathbf{L}) = L^{\langle -\infty \rangle}_n(\mathbb{Z})$. Denote by $\mathbf{L}\langle 1 \rangle$ its 1-*connective cover*. It comes with a natural map of spectra $\mathbf{L}\langle 1 \rangle \to \mathbf{L}$, which induces on π_i an isomorphism for $i \geq 1$, and we have $\pi_i(\mathbf{L}\langle 1 \rangle) = 0$ for $i \leq 0$. There are natural identifications coming among other things from the Pontrjagin-Thom construction

$$\begin{aligned} u_n\colon \mathcal{N}(N) &\xrightarrow{\cong} H_n(N; \mathbf{L}\langle 1 \rangle) = \pi_n(N_+ \wedge \mathbf{L}\langle 1 \rangle); \\ u_{n+1}\colon \mathcal{N}(N \times [0,1], N \times \{0,1\}) &\xrightarrow{\cong} H_{n+1}(N; \mathbf{L}\langle 1 \rangle) = \pi_{n+1}\big(N_+ \wedge \mathbf{L}\langle 1 \rangle\big). \end{aligned}$$

An easy spectral sequence argument shows that the canonical map

$$v_n\colon H_n(N; \mathbf{L}\langle 1 \rangle) \to H_n(N; \mathbf{L})$$

is injective and the canonical map

$$v_{n+1}\colon H_{n+1}(N; \mathbf{L}\langle 1 \rangle) \to H_{n+1}(N; \mathbf{L})$$

is bijective for $n = \dim(N)$.

For the remainder of this subsection we assume additionally that N is aspherical. There is a natural identification for $m = n, n+1$, see Definition 20.11.1,

$$w_m\colon H^G_m(EG; \mathbf{L}^{\langle -\infty \rangle}_{\mathbb{Z}}) \xrightarrow{\cong} H_m(BG; \mathbf{L}) = H_m(N; \mathbf{L}).$$

The K-theoretic Farrell-Jones Conjecture applied to the torsionfree group G implies that $K_n(\mathbb{Z}G)$ for $n \leq -1$, the reduced projective class group $\widetilde{K}_0(\mathbb{Z}G)$, and the Whitehead group $\mathrm{Wh}(G)$ vanish. One concludes from the so called *Rothenberg sequences*, see [105, Theorem 17.2 on page 146], that for $m = n, n+1$ the canonical map

$$r_m\colon L^s_m(\mathbb{Z}G) \xrightarrow{\cong} L^{\langle -\infty \rangle}_m(\mathbb{Z}G)$$

is bijective. The up to G-homotopy unique G-map $i\colon EG = E_{\mathcal{TR}}(G) \to \underline{\underline{E}}G$ induces for all

$m \in \mathbb{Z}$ an isomorphism, see Theorem 20.12.2 (5),

$$H_m^G(i; \mathbf{L}_{\mathbb{Z}}^{\langle -\infty \rangle}) \colon H_m^G(EG; \mathbf{L}_{\mathbb{Z}}^{\langle -\infty \rangle}) \xrightarrow{\cong} H_m^G(\underline{\underline{E}}G; \mathbf{L}_{\mathbb{Z}}^{\langle -\infty \rangle}).$$

The L-theoretic Farrell-Jones Conjecture predicts the bijectivity of the assembly map

$$H_n^G(\mathrm{pr}, \mathbf{L}_{\mathbb{Z}}^{\langle -\infty \rangle}) \colon H_n^G(\underline{\underline{E}}G; \mathbf{L}_{\mathbb{Z}}^{\langle -\infty \rangle}) \xrightarrow{\cong} H_n^G(G/G; \mathbf{L}_{\mathbb{Z}}^{\langle -\infty \rangle}) = L_n^{\langle -\infty \rangle}(\mathbb{Z}G).$$

The following diagram commutes

$$\begin{array}{ccccc}
\mathcal{N}(N) & & \xrightarrow{\quad\quad\quad\sigma_n\quad\quad\quad} & & L_n^s(\mathbb{Z}G) \\
u_n \big\downarrow \cong & & & & \\
H_n(N; \mathbf{L}\langle 1 \rangle) & & & & \\
v_n \big\downarrow & & & & \cong \big\uparrow r_n \\
H_n(N; \mathbf{L}) & & & & \\
\cong \big\uparrow w_n & & & & \\
H_n^G(EG; \mathbf{L}_{\mathbb{Z}}^{\langle -\infty \rangle}) & \xrightarrow[H_n^G(i, \mathbf{L}_{\mathbb{Z}}^{\langle -\infty \rangle})]{\cong} & H_n^G(\underline{\underline{E}}G; \mathbf{L}_{\mathbb{Z}}^{\langle -\infty \rangle}) & \xrightarrow[H_n^G(\mathrm{pr}, \mathbf{L}_{\mathbb{Z}}^{\langle -\infty \rangle})]{\cong} & L_n^{\langle -\infty \rangle}(\mathbb{Z}G).
\end{array}$$

If we replace everywhere n by $n+1$ and in the upper left corner $\mathcal{N}(N)$ by $\mathcal{N}(N \times [0,1], N \times \partial[0,1])$, we get the analogous commutative diagram. The proof of the commutativity of these diagrams is rather involved and we refer for a proof for instance to [62]. We conclude from these two diagrams that σ_n is injective and σ_{n+1} is bijective since v_n is injective and v_{n+1} is bijective. Recall that this implies the vanishing of the structure set $\mathcal{S}(N)$. Hence the Farrell-Jones Conjecture 20.8.1 implies the Borel Conjecture in dimensions ≥ 5.

For more information about L-groups and surgery theory and the arguments and facts above we refer for instance to [18, 19, 25, 70, 104, 124].

20.8.4 The interpretation of the Farrell-Jones assembly map in terms of controlled topology

We have defined the assembly map appearing in the Farrell-Jones Conjecture as a map induced by the projection $\underline{\underline{E}}G \to G/G$ for a G-homology theory $H_*^G(X; \mathbf{E}^G)$ or for the functor $(\mathbf{E}^G)_\% \colon G\text{-}\mathsf{CW} \to \mathsf{Spectra}$. We have also given a homotopy theoretic interpretation in terms of homotopy colimits and described its universal property to be the best approximation from the left by an excisive functor. This interpretation is good for structural and computational aspects, but it turns out that it is not helpful for the proof that the assembly maps are weak homotopy equivalences. There is no direct homotopy theoretic construction of an inverse up to weak homotopy equivalence known to the author.

For the actual proofs that the assembly maps are weak homotopy equivalences, the interpretation of the assembly map as a *forget control map* is crucial. This fundamental idea is due to Quinn.

Roughly speaking, one attaches to a metric space certain categories, to these categories spectra and then takes their homotopy groups, where everything depends on a choice of certain control conditions which in some sense measure sizes of cycles. If one requires certain control conditions, one obtains the source of the assembly map. If one requires no control conditions, one obtains the target of the assembly map. The assembly map itself is forgetting the control condition.

One of the basic features of a homology theory is excision. It often comes from the fact that a representing cycle can be arranged to have arbitrarily good control. An example is the technique of subdivision which allows us to make the representing cycles for singular homology arbitrarily controlled, i.e., the diameter of the image of a singular simplex appearing in a singular chain with non-zero coefficient can be arranged to be arbitrarily small. This is the key ingredient in the proof that singular homology satisfies excision. In general one may say that requiring control conditions amounts to implementing homological properties.

With this interpretation it is clear what the main task in the proof of surjectivity of the assembly map is: *achieve control*, i.e., manipulate cycles without changing their homology class so that they become sufficiently controlled. There is a general principle that a proof of surjectivity also gives injectivity. Namely, proving injectivity means that one must construct a cycle whose boundary is a given cycle, i.e., one has to solve a surjectivity problem in a relative situation. The actual implementation of this idea is rather technical. The proof that this forget control version of the assembly map agrees up to weak homotopy equivalence with the homotopy theoretic one appearing in the Farrell-Jones Conjecture 20.8.1 is a direct application of Section 20.5. The same is true also for the version of the assembly map appearing in [38, 1.6 on page 257], as explained in [26, page 239].

To achieve control one can now use geometric methods. The key ingredients are contracting maps and open coverings, transfers, flow spaces, and the geometry of the group G.

For more information about the general strategy of proofs we refer for instance to [2, 78, 79].

20.8.5 The Farrell-Jones Conjecture for Waldhausen's A-theory

Waldhausen has defined for a CW-complex X its algebraic K-theory space $A(X)$ in [120, Chapter 2]. As in the case of algebraic K-theory of rings it will be necessary to consider a non-connective version. Vogell [117] has defined a delooping of $A(X)$ yielding a non-connective spectrum $\mathbf{A}(X)$ for a CW-complex X. This construction actually yields a covariant functor from the category of topological spaces to the category of spectra. We can assign to a groupoid $\mathcal{G}$ its classifying space $B\mathcal{G}$. Thus we obtain a covariant functor

$$\mathbf{A}\colon \mathsf{Groupoids} \to \mathsf{Spectra}, \quad \mathcal{G} \mapsto \mathbf{A}(B\mathcal{G}), \tag{20.8.4}$$

denoted by $\mathbf{A}$ again. It respects equivalences, see [120, Proposition 2.1.7] and [117]. If we now take this functor and the family $\mathcal{VCY}$ of virtually cyclic subgroups, we obtain the A-theoretic Farrell-Jones Conjecture

Conjecture 20.8.4 (A-theoretic Farrell-Jones Conjecture)**.** *A group G satisfies the A-*theoretic Farrell-Jones Conjecture *if the assembly maps induced by the projection* $\mathrm{pr}\colon \underline{\underline{E}}G \to$

G/G

$$H_n^G(\mathrm{pr};\mathbf{A})\colon H_n^G(\underline{\underline{E}}G;\mathbf{A}) \to H_n^G(G/G;\mathbf{A}) = \pi_n(\mathbf{A}(BG))$$

is bijective for all $n \in \mathbb{Z}$.

The A-theoretic Farrell-Jones Conjecture 20.8.4 is an important ingredient in the computation of the group of self-homeomorphisms of an aspherical closed manifold in the stable range using the machinery of Weiss-Williams [128]. Moreover, it is related to Whitehead spaces, pseudo-isotopy spaces and spaces of h-cobordisms, see for instance [32, 119, 120, 121, 122, 123, 128].

20.8.6 Relating the assembly maps for K-theory and for A-theory

Let X be a connected CW-complex with fundamental group $\pi = \pi_1(X)$. Essentially by passing to the cellular $\mathbb{Z}\pi$-chain complex of the universal covering one obtains a natural map of (non-connective) spectra, natural in X, called *linearization map*

$$\mathbf{L}(X)\colon \mathbf{A}(X) \quad\to\quad \mathbf{K}(\mathbb{Z}\pi_1(X)). \tag{20.8.5}$$

The next result follows by combining [118, Section 4] and [119, Proposition 2.2 and Proposition 2.3].

Theorem 20.8.5 (Connectivity of the linearization map)**.** *Let X be a connected CW-complex. Then:*

1. *The linearization map $\mathbf{L}(X)$ of (20.8.5) is 2-connected, i.e., the map*

   $$L_n := \pi_n(\mathbf{L}(X))\colon A_n(X) \to K_n(\mathbb{Z}\pi_1(X))$$

   is bijective for $n \leq 1$ and surjective for $n = 2$;

2. *Rationally the map L_n is bijective for all $n \in \mathbb{Z}$, provided that X is aspherical.*

Thus one obtain a transformation $\mathbf{L}\colon \mathbf{A} \to \mathbf{K}_{\mathbb{Z}}$ of covariant functors $\mathsf{Groupoids} \to \mathsf{Spectra}$, where $\mathbf{K}_{\mathbb{Z}}$ and $\mathbf{A}$ have been defined in (20.8.1) and (20.8.4). It induces a commutative diagram

$$\begin{array}{ccc} H_n^G(\underline{\underline{E}}G;\mathbf{A}) & \xrightarrow{H_n^G(\mathrm{pr};\mathbf{A})} & H_n^G(G/G;\mathbf{A}) = \pi_n(\mathbf{A}(BG)) \\ \downarrow{\scriptstyle H_n^G(\mathrm{id}_{\underline{\underline{E}}G};\mathbf{L})} & & \downarrow{\scriptstyle \pi_n(\mathbf{L}(BG))} \\ H_n^G(\underline{\underline{E}}G;\mathbf{K}_{\mathbb{Z}}) & \xrightarrow[H_n^G(\mathrm{pr};\mathbf{K}_R)]{} & K_n(\mathbb{Z}G) \end{array} \tag{20.8.6}$$

where the upper horizontal arrow is the assembly map appearing in A-theoretic Farrell-Jones Conjecture 20.8.4, the lower horizontal arrow is the assembly map appearing in the K-theoretic Farrell-Jones Conjecture 20.8.1, and both vertical arrows are bijective for $n \leq 1$, surjective for $n = 2$ and rationally bijective for all $n \in \mathbb{Z}$. In particular the K-theoretic Farrell-Jones Conjecture 20.8.1 for $R = \mathbb{Z}$ and the A-theoretic Farrell-Jones Conjecture 20.8.4 are rationally equivalent.

20.8.7 The status of the Farrell-Jones Conjecture

There is a more general version of the Farrell-Jones Conjecture, the so called *Full Farrell-Jones Conjecture*, where one allows coefficients in additive categories and the passage to finite wreath products, It implies the Farrell-Jones Conjectures 20.8.1. For A-theory there is a so called *fibered* version which implies Conjecture 20.8.4. Let $\mathcal{FJ}$ be the class of groups for which the Full Farrell-Jones Conjecture and the fibered A-theoretic Farrell-Jones Conjecture holds. Notice that then any group in $\mathcal{FJ}$ satisfies in particular Conjectures 20.8.1 and 20.8.4.

Theorem 20.8.6 (The class $\mathcal{FJ}$).

1. *The following classes of groups belong to $\mathcal{FJ}$:*

   (a) *Hyperbolic groups;*

   (b) *Finite dimensional* CAT(0)*-groups;*

   (c) *Virtually solvable groups;*

   (d) *(Not necessarily cocompact) lattices in second countable locally compact Hausdorff groups with finitely many path components;*

   (e) *Fundamental groups of (not necessarily compact) connected manifolds (possibly with boundary) of dimension ≤ 3;*

   (f) *The groups $GL_n(\mathbb{Q})$ and $GL_n(F(t))$ for $F(t)$ the function field over a finite field F;*

   (g) *S-arithmetic groups;*

   (h) *mapping class groups;*

2. *The class $\mathcal{FJ}$ has the following inheritance properties:*

   (a) Passing to subgroups
   Let $H \subseteq G$ be an inclusion of groups. If G belongs to $\mathcal{FJ}$, then H belongs to $\mathcal{FJ}$;

   (b) Passing to finite direct products
   If the groups G_0 and G_1 belong to $\mathcal{FJ}$, then also $G_0 \times G_1$ belongs to $\mathcal{FJ}$;

   (c) Group extensions
   Let $1 \to K \to G \to Q \to 1$ be an extension of groups. Suppose that for any cyclic subgroup $C \subseteq Q$ the group $p^{-1}(C)$ belongs to $\mathcal{FJ}$ and that the group Q belongs to $\mathcal{FJ}$. Then G belongs to $\mathcal{FJ}$;

   (d) Directed colimits
   Let $\{G_i \mid i \in I\}$ be a direct system of groups indexed by the directed set I (with arbitrary structure maps). Suppose that for each $i \in I$ the group G_i belongs to $\mathcal{FJ}$. Then the colimit $\operatorname{colim}_{i\in I} G_i$ belongs to $\mathcal{FJ}$;

   (e) Passing to finite free products
   If the groups G_0 and G_1 belong to $\mathcal{FJ}$, then $G_0 * G_1$ belongs to $\mathcal{FJ}$;

   (f) Passing to overgroups of finite index
   Let G be an overgroup of H with finite index $[G : H]$. If H belongs to $\mathcal{FJ}$, then G belongs to $\mathcal{FJ}$;

Proof. See [3, 4, 5, 7, 8, 10, 34, 54, 58, 109, 125, 126]. □

It is not known whether all amenable groups belong to $\mathcal{FJ}$.

20.9 The Baum-Connes Conjecture

20.9.1 The Baum-Connes Conjecture

Recall that the *reduced group C^*-algebra* $C_r^*(G)$ is a certain completion of the complex group ring $\mathbb{C}G$. Namely, there is a canonical embedding of $\mathbb{C}G$ into the space $B(l^2(G))$ of bounded operators $L^2(G) \to L^2(G)$ equipped with the supremums norm given by the right regular representation, and $C_r^*(G)$ is the norm closure of $\mathbb{C}G$ in $B(L^2(G))$. There is a covariant functor respecting equivalences

$$\mathbf{K}^{\text{top}}\colon \mathsf{Groupoids} \to \mathsf{Spectra}, \tag{20.9.1}$$

such that for every group G and all $n \in \mathbb{Z}$ we have

$$\pi_n(\mathbf{K}^{\text{top}}(G)) \cong K_n^{\text{top}}(C_r^*(G)),$$

where $K_n^{\text{top}}(C_r^*(G))$ is the topological K-theory of the reduced group C^*-algebra $C_r^*(G)$, see [52]. If we now take this functors and the family $\mathcal{FIN}$ of finite subgroups, we obtain

Conjecture 20.9.1 (Baum-Connes Conjecture). *A group G satisfies the* Baum-Connes Conjecture *if the assembly maps induced by the projection* $\text{pr}\colon \underline{E}G \to G/G$

$$H_n^G(\text{pr}; \mathbf{K}^{\text{top}})\colon H_n^G(\underline{E}G; \mathbf{K}^{\text{top}}) \to H_n^G(G/G; \mathbf{K}^{\text{top}}) = K_n^{\text{top}}(C_r^*(G)).$$

is bijective for all $n \in \mathbb{Z}$.

The original version of the Baum-Connes Conjecture is stated in [14, Conjecture 3.15 on page 254]. There is also a version, where the ground field $\mathbb{C}$ is replaced by $\mathbb{R}$. The complex version of the Baum-Connes Conjecture 20.9.1 implies automatically the real version, see [16, 112].

20.9.2 The Baum-Connes Conjecture with coefficients

There is also a more general version of the Baum-Connes Conjecture 20.9.1, where one allows twisted coefficients.However, there are counterexamples to this more general version, see [48, Section 7]. There is a new formulation of the Baum-Connes Conjecture with coefficients in [15], where these counterexamples do not occur anymore. At the time of writing no counterexample to the Baum-Connes Conjecture 20.9.1 or to the version of [15] is known to the author.

20.9.3 The interpretation of the Baum Connes assembly map in terms of index theory

For applications of the Baum-Connes Conjecture 20.9.1 it is essential that the Baum-Connes assembly maps can be interpreted in terms of indices of equivariant operators with values in C^*-algebras. Namely, one assigns to a Kasparov cycle representing an element in the equivariant KK-group $KK_n^G(C_0(X), \mathbb{C})$ in the sense of Kasparov [55, 56, 57] its C^*-valued index in $K_n(C_r^*(G))$ in the sense of Mishchenko-Fomenko [92], thus defining a map

$$KK_n^G(C_0(X), \mathbb{C}) \to K_n^{\text{top}}(C_r^*(G)),$$

provided that X is proper and cocompact and $C_0(X)$ is the C^*-algebra (possibly without unit) of continuous function $X \to \mathbb{C}$ vanishing at infinity. This is the approach appearing in [14].

The other equivalent approach is based on the Kasparov product. Given a proper cocompact G-CW-complex X, one can assign to it an element $[p_X] \in KK_0^G(\mathbb{C}, C_0(X) \rtimes_r G)$, where $C_0(X) \rtimes_r G$ denotes the reduced crossed product C^*-algebra associated to the G-C^*-algebra $C_0(X)$. Now define the Baum-Connes assembly map by the composition of a descent map and a map coming from the Kasparov product

$$\begin{aligned} KK_n^G(C_0(X), \mathbb{C}) &\xrightarrow{j_r^G} KK_n(C_0(X) \rtimes_r G, C_r^*(G)) \\ &\xrightarrow{[p_X]\widehat{\otimes}_{C_0(X)\rtimes_r G} -} KK_n(\mathbb{C}, C_r^*(G)) = K_n^{\text{top}}(C_r^*(G)). \end{aligned}$$

This extends to arbitrary proper G-CW-complexes X by defining the source by

$$K_n^G(C_0(X), \mathbb{C}) := \operatorname{colim}_{C \subseteq X} K_n^G(C_0(C), \mathbb{C}),$$

where C runs through the finite G-CW-subcomplexes of Y directed by inclusion. Hence we can take $X = \underline{E}G$ above without assuming any finiteness conditions on $\underline{E}G$. For some information about these two approaches and their identification, at least for torsionfree G, we refer to [64].

One can identify the original assembly map of [14] with the assembly map appearing in Conjecture 20.9.1 using Section 20.9.1 and the fact that

$$\operatorname{colim}_{C \subseteq X} H_n^G(C; \mathbf{K}^{\text{top}}) \xrightarrow{\cong} H_n^G(X; \mathbf{K}^{\text{top}})$$

is an isomorphism. This is explained in [26, page 247-248], Unfortunately, the proof is based on an unfinished preprint by Carlsson-Pedersen-Roe [20], where the assembly map appearing in [14, Conjecture 3.15 on page 254] is implemented on the spectrum level. Another proof of the identification is given in [44, Corollary 8.4] and [94, Theorem 1.3].

20.9.4 Applications of the Baum-Connes Conjecture

20.9.4.1 Computations

One can carry out *explicit computations* of topological K-groups of group C^*-algebras and related C^*-algebras by applying methods from algebraic topology to the left side given by a G-homology theory and by finding small models for the classifying spaces of families using the topology and geometry of groups. This leads to classification results about certain C^*-algebras, see for instance [28, 33, 66, 67].

20.9.4.2 (Modified) Trace Conjecture

The Baum-Connes Conjecture 20.9.1 implies the *Trace Conjecture for torsionfree groups* that for a torsionfree group G the image of

$$\operatorname{tr}_{C_r^*(G)} \colon K_0(C_r^*(G)) \to \mathbb{R}$$

consists of the integers. If one drops the condition torsionfree, there is the so called *Modified Trace Conjecture*,which is implied by Baum-Connes Conjecture 20.9.1, see [73].

20.9.4.3 Kadison Conjecture

The Baum-Connes Conjecture 20.9.1 implies the *Kadison Conjecture* that for a torsionfree group G the only idempotent elements in $C_r^*(G)$ are 0 and 1.

20.9.4.4 Novikov Conjecture

The Baum-Connes Conjecture 20.9.1 implies the *Novikov Conjecture.*

20.9.4.5 The Zero-in-the-spectrum Conjecture

The *Zero-in-the-spectrum Conjecture* says, if $\widetilde{M}$ is the universal covering of an aspherical closed Riemannian manifold M, then zero is in the spectrum of the minimal closure of the pth Laplacian on $\widetilde{M}$ for some $p \in \{0, 1, \dots, \dim M\}$. It is a consequence of the Strong Novikov Conjecture and hence of the Baum-Connes Conjecture 20.9.1, see [72, Chapter 12].

20.9.4.6 The Stable Gromov-Lawson-Rosenberg Conjecture

Let $\Omega_n^{\mathrm{Spin}}(BG)$ be the bordism group of closed Spin-manifolds M of dimension n with a reference map to BG. Given an element $[u \colon M \to BG] \in \Omega_n^{\mathrm{Spin}}(BG)$, we can take the $C_r^*(G;\mathbb{R})$-valued index of the equivariant Dirac operator associated to the G-covering $\overline{M} \to M$ determined by u. Thus we get a homomorphism

$$\operatorname{ind}_{C_r^*(G;\mathbb{R})} \colon \Omega_n^{\mathrm{Spin}}(BG) \quad \to \quad K_n^{\mathrm{top}}(C_r^*(G;\mathbb{R})). \tag{20.9.2}$$

A *Bott manifold* is any simply connected closed Spin-manifold B of dimension 8 whose $\widehat{A}$-genus $\widehat{A}(B)$ is 1. We fix such a choice; the particular choice does not matter for the sequel. Notice that $\operatorname{ind}_{C_r^*(\{1\};\mathbb{R})}(B) \in K_8^{\mathrm{top}}(\mathbb{R}) \cong \mathbb{Z}$ is a generator and the product with this element induces the Bott periodicity isomorphisms $K_n^{\mathrm{top}}(C_r^*(G;\mathbb{R})) \xrightarrow{\cong} K_{n+8}^{\mathrm{top}}(C_r^*(G;\mathbb{R}))$. In particular

$$\mathrm{ind}_{C_r^*(G;\mathbb{R})}(M) \quad = \quad \mathrm{ind}_{C_r^*(G;\mathbb{R})}(M \times B), \tag{20.9.3}$$

if we identify $K_n^{\mathrm{top}}(C_r^*(G;\mathbb{R})) = K_{n+8}^{\mathrm{top}}(C_r^*(G;\mathbb{R}))$ via Bott periodicity.

Conjecture 20.9.2 (Stable Gromov-Lawson-Rosenberg Conjecture). *Let M be a closed connected* Spin*-manifold of dimension $n \geq 5$. Let $u_M \colon M \to B\pi_1(M)$ be the classifying map of its universal covering. Then $M \times B^k$ carries for some integer $k \geq 0$ a Riemannian metric with positive scalar curvature if and only if*

$$\mathrm{ind}_{C_r^*(\pi_1(M);\mathbb{R})}([M, u_M]) = 0 \quad \in K_n^{\mathrm{top}}(C_r^*(\pi_1(M);\mathbb{R})).$$

If M carries a Riemannian metric with positive scalar curvature, then the index of the Dirac operator must vanish by the Bochner-Lichnerowicz formula [106]. The converse statement that the vanishing of the index implies the existence of a Riemannian metric with positive scalar curvature is the hard part of the conjecture. The unstable version of Conjecture 20.9.2, where one does not stabilize with B^k, is not true in general; see [110].

A sketch of the proof of the following result can be found in Stolz [114, Section 3].

Theorem 20.9.3 (The Baum-Connes Conjecture implies the Stable Gromov-Lawson-Rosenberg Conjecture). *If the assembly map for the real version of the Baum-Connes Conjecture 20.9.1 is injective for the group G, then the Stable Gromov-Lawson-Rosenberg Conjecture 20.9.2 is true for all closed* Spin*-manifolds of dimension ≥ 5 with $\pi_1(M) \cong G$.*

The space of positive scalar curvature metrics is analyzed for instance in [100, 129, 130].

20.9.4.7 Homotopy invariance of L^2-Rho-invariants and L^2-signature theorems

Results about the homotopy invariance of L^2-Rho-invariants and L^2-signature theorems based on the Baum-Connes Conjecture can be found in [22, 59, 60, 87, 88].

20.9.4.8 Knot theory

Cochran-Orr-Teichner give in [24] new obstructions for a knot to be a slice which are sharper than the Casson-Gordon invariants. They use L^2-signatures and the Baum-Connes Conjecture 20.9.1. We also refer to the survey article [23] about non-commutative geometry and knot theory.

20.9.5 The status of the Baum-Connes Conjecture

Let $\mathcal{BC}$ be the class of groups for which the Baum-Connes Conjecture with coefficients, which implies the Baum-Connes Conjecture 20.9.1, is true.

Theorem 20.9.4 (Status of the Baum-Connes Conjecture 20.9.1).

1. *The following classes of groups belong to $\mathcal{BC}$.*

    (a) A-T-menable groups;

    (b) Hyperbolic groups;

    (c) One-relator groups;

    (d) Fundamental groups of compact 3-manifolds (possibly with boundary);

2. *The class $\mathcal{BC}$ has the following inheritance properties:*

    (a) Passing to subgroups
    Let $H \subseteq G$ be an inclusion of groups. If G belongs to $\mathcal{BC}$, then H belongs to $\mathcal{BC}$;

    (b) Passing to finite direct products
    If the groups G_0 and G_1 belong to $\mathcal{BC}$, the also $G_0 \times G_1$ belongs to $\mathcal{BC}$;

    (c) Group extensions
    Let $1 \to K \to G \to Q \to 1$ be an extension of groups. Suppose that for any finite subgroup $F \subseteq Q$ the group $p^{-1}(F)$ belongs to $\mathcal{BC}$ and that the group Q belongs to $\mathcal{BC}$. Then G belongs to $\mathcal{BC}$;

    (d) Directed unions
    Let $\{G_i \mid i \in I\}$ be a direct system of subgroups of G indexed by the directed set I such that $G = \bigcup_{i \in I} G_i$. Suppose that G_i belongs to $\mathcal{BC}$ for every $i \in I$. Then G belongs to $\mathcal{BC}$;

    (e) Actions on trees
    Let G be a countable discrete group acting without inversion on a tree T. Then G belongs to $\mathcal{BC}$ if and only if the stabilizers of each of the vertices of T belong to $\mathcal{BC}$. In particular $\mathcal{BC}$ is closed under amalgamated products and HNN-extensions.

Proof. See [4, 21, 47, 63, 91, 96, 97]. □

It is not known whether finite-dimensional CAT(0)-groups and $SL_n(\mathbb{Z})$ for $n \geq 3$ belong to $\mathcal{BC}$.

For more information about the Baum-Connes Conjecture and its applications we refer for instance to [14, 46, 49, 50, 51, 79, 82, 93, 99, 108, 111, 116].

20.9.6 Relating the assembly maps of Farrell-Jones to the one of Baum-Connes

One can construct the following commutative diagram

$$\begin{array}{ccc}
H_n^G(\underline{E}G;\mathbf{L}_{\mathbb{Z}}^{\langle -\infty\rangle})[1/2] & \longrightarrow & L_n^{\langle -\infty\rangle}(\mathbb{Z}G)[1/2] \\
{\scriptstyle l}\downarrow\cong & & {\scriptstyle \mathrm{id}}\downarrow\cong \\
H_n^G(\underline{E}G;\mathbf{L}_{\mathbb{Z}}^{\langle -\infty\rangle}[1/2]) & \longrightarrow & L_n^{\langle -\infty\rangle}(\mathbb{Z}G)[1/2] \\
{\scriptstyle i_0}\uparrow\cong & & \cong\uparrow{\scriptstyle j_0} \\
H_n^G(\underline{E}G;\mathbf{L}_{\mathbb{Z}}^{p}[1/2]) & \longrightarrow & L_n^{p}(\mathbb{Z}G)[1/2] \\
{\scriptstyle i_1}\downarrow\cong & & \cong\downarrow{\scriptstyle j_1} \\
H_n^G(\underline{E}G;\mathbf{L}_{\mathbb{Q}}^{p}[1/2]) & \longrightarrow & L_n^{p}(\mathbb{Q}G)[1/2] \\
{\scriptstyle i_2}\downarrow\cong & & \downarrow{\scriptstyle j_2} \\
H_n^G(\underline{E}G;\mathbf{L}_{\mathbb{R}}^{p}[1/2]) & \longrightarrow & L_n^{p}(\mathbb{R}G)[1/2] \\
{\scriptstyle i_3}\downarrow\cong & & \downarrow{\scriptstyle j_3} \\
H_n^G(\underline{E}G;\mathbf{L}_{C_r^*(?;\mathbb{R})}^{p}[1/2]) & \longrightarrow & L_n^{p}(C_r^*(G;\mathbb{R}))[1/2] \\
{\scriptstyle i_4}\uparrow\cong & & \cong\uparrow{\scriptstyle j_4} \\
H_n^G(\underline{E}G;\mathbf{K}_{\mathbb{R}}^{\mathrm{top}}[1/2]) & \longrightarrow & K_n^{\mathrm{top}}(C_r^*(G;\mathbb{R}))[1/2] \\
\cong\uparrow{\scriptstyle l} & & {\scriptstyle \mathrm{id}}\uparrow\cong \\
H_n^G(\underline{E}G;\mathbf{K}_{\mathbb{R}}^{\mathrm{top}})[1/2] & & K_n^{\mathrm{top}}(C_r^*(G;\mathbb{R}))[1/2] \\
{\scriptstyle i_5}\downarrow & & \downarrow{\scriptstyle j_5} \\
H_n^G(\underline{E}G;\mathbf{K}_{\mathbb{C}}^{\mathrm{top}})[1/2] & \longrightarrow & K_n(C_r^*(G))[1/2]
\end{array} \tag{20.9.4}$$

where all horizontal maps are assembly maps and the vertical arrows are induced by transformations of functors Groupoids $\to$ Spectra. These transformations are induced by change of rings maps except the one from $\mathbf{K}_{\mathbb{R}}^{\mathrm{top}}[1/2]$ to $\mathbf{L}_{C_r^*(?;\mathbb{R})}^{p}[1/2]$ which is much more complicated and carried out in [65]. This sophisticated and key ingredient was missing in [61, Lemma 22.13 on page 196], where the existence of such a dagram was claimed. The same remark applies also to [107, Theorem 2.7], see [65, Subsection 1.1]. Actually, it does not exist without inverting two on the spectrum level. Since it is a weak equivalence, the maps i_4 and j_4 are bijections.

For any finite group H each of the following maps is known to be a bijection because of [104, Proposition 22.34 on page 252] and $\mathbb{R}H = C_r^*(H;\mathbb{R})$

$$L_n^p(\mathbb{Z}H)[1/2] \xrightarrow{\cong} L_n^p(\mathbb{Q}H)[1/2] \xrightarrow{\cong} L_n^p(\mathbb{R}H)[1/2] \xrightarrow{\cong} L_n^p(C_r^*(H;\mathbb{R})).$$

The natural map $L_n^p(RG)[1/2] \to L_n^{\langle -\infty\rangle}(RG)[1/2]$ is an isomorphism for any $n \in \mathbb{Z}$, group G and ring with involution R by the Rothenberg sequence, see [105, Theorem 17.2 on page 146]. Hence we conclude from the equivariant Atiyah Hirzebruch spectral sequence that the vertical arrows i_1, i_2, and i_3 are isomorphisms. The arrow j_1 is bijective by [103, page 376]. The maps l are isomorphisms for general results about localizations.

The lowermost vertical arrows i_5 and j_5 are known to be split injective because the inclusion $C_r^*(G;\mathbb{R}) \to C_r^*(G;\mathbb{C})$ induces an isomorphism $C_r^*(G;\mathbb{R}) \to C_r^*(G;\mathbb{C})^{\mathbb{Z}/2}$ for the $\mathbb{Z}/2$-operation coming from complex conjugation $\mathbb{C} \to \mathbb{C}$. The following conjecture is already raised as a question in [61, Remark 23.14 on page 197], see also [65, Completion Conjecture in Subsection 5.2].

Conjecture 20.9.5 (Passage for L-theory from $\mathbb{Q}G$ to $\mathbb{R}G$ to $C_r^*(G;\mathbb{R})$)**.** *The maps j_2 and j_3 appearing in diagram* (20.9.4) *are bijective.*

One easily checks

Lemma 20.9.6. *Let G be a group.*

1. *Suppose that G satisfies the L-theoretic Farrell-Jones Conjecture 20.8.1 with coefficients in the ring R for $R = \mathbb{Q}$ and $R = \mathbb{R}$ and the Baum-Connes Conjecture 20.9.1. Then G satisfies Conjecture 20.9.5;*

2. *Suppose that G satisfies Conjecture 20.9.5. Then G satisfies the L-theoretic Farrell-Jones Conjecture 20.8.1 for the ring $\mathbb{Z}$ after inverting* 2*, if and only if G satisfies the real version of the Baum-Connes Conjecture 20.9.1 after inverting* 2*;*

3. *Suppose that the assembly map appearing in the Baum-Connes Conjecture 20.9.1 is (split) injective after inverting* 2*. Then the assembly map appearing in L-theoretic Farrell-Jones Conjecture 20.8.1 with coefficients in the ring for $R = \mathbb{Z}$ is (split) injective after inverting* 2.

20.10 Topological cyclic homology

Let $\mathbf{R}$ be a (well-pointed connective) symmetric ring spectrum and p be a prime. There are covariant functors respecting equivalences

$$\begin{aligned} \mathbf{THH}_{\mathbf{R}} \colon \mathsf{Groupoids} &\to \mathsf{Spectra}; \\ \mathbf{TC}_{\mathbf{R},p} \colon \mathsf{Groupoids} &\to \mathsf{Spectra}, \end{aligned}$$

such that for every group G and all $n \in \mathbb{Z}$ we have

$$\begin{aligned} \pi_n(\mathbf{THH}_{\mathbf{R}}(G)) &\cong \pi_n(\mathbf{THH}(\mathbf{R}[G])); \\ \pi_n(\mathbf{TC}_{\mathbf{R};p}(G)) &\cong \pi_n(\mathbf{TC}(\mathbf{R}[G];p)), \end{aligned}$$

where $\mathbf{THH}(\mathbf{R}[G])$ is the topological Hochschild homology and $\mathbf{THH}(\mathbf{R}[G];p)$ is the topological cyclic homology of the group ring spectrum $\mathbf{R}[G]$. [1]

[1]See the chapter in this Handbook by Lars Hesselholt and Thomas Nikolaus for more information about topological Hochschild and cyclic homology.

20.10.1 Topological Hochschild homology

If we now take the functor $\mathbf{THH_R}$ and the family $\mathcal{CY}$ of cyclic subgroups, we obtain from [84, Theorem 1.19] that the Farrell-Jones Conjecture for topological Hochschild homology is true for all groups.

Theorem 20.10.1 (Topological Hochschild homology). *The assembly maps induced by the projection* $\mathrm{pr}\colon E_{\mathcal{CY}}(G) \to G/G$

$$H_n^G(\mathrm{pr}; \mathbf{THH_R})\colon H_n^G(E_{\mathcal{CY}}(G); \mathbf{THH_R}) \to H_n^G(G/G; \mathbf{THH_R}) = \pi_n(\mathbf{THH}(\mathbf{R}[G]))$$

is bijective for all $n \in \mathbb{Z}$.

20.10.2 Topological cyclic homology

If we take the functor $\mathbf{TC_{R;p}}$ and the family $\mathcal{FIN}$ of cyclic subgroups, we obtain from [83, Theorem 1.5] that the injectivity part of the Farrell-Jones Conjecture for topological cyclic homology is true under certain finiteness assumptions

Theorem 20.10.2 (Split injectivity for topological cyclic homology). *Assume that one for the following conditions hold for the family* $\mathcal{F}$*:*

1. *We have* $\mathcal{F} = \mathcal{FIN}$ *and there is a model for* $\underline{E}G$ *of finite type;*

2. *We have* $\mathcal{F} = \mathcal{VCY}$ *and* G *is hyperbolic or virtually abelian.*

   Then the assembly maps induced by the projection $\mathrm{pr}\colon \underline{E}G \to G/G$

$$H_n^G(\mathrm{pr}; \mathbf{TC_{R;p}})\colon H_n^G(\underline{E}G; \mathbf{TC_{R;p}}) \to H_n^G(G/G; \mathbf{THH_{R;p}}) = \pi_n(\mathbf{THH}(\mathbf{R}[G]; p))$$

is split injective for all $n \in \mathbb{Z}$.

Moreover, we also have, see [83, Theorem 1.2].

Theorem 20.10.3 (Topological cyclic homology and finite groups). *If* G *is a finite group, then the assembly map for the family* $\mathcal{CY}$ *of cyclic subgroups*

$$H_n^G(\mathrm{pr}; \mathbf{TC_{R;p}})\colon H_n^G(E_{\mathcal{CY}}(G); \mathbf{TC_{R;p}}) \to H_n^G(G/G; \mathbf{TC_{R;p}}) = \pi_n(\mathbf{TC}(\mathbf{R}[G]; p))$$

is bijective for all $n \in \mathbb{Z}$.

Remark 20.10.4 (The Farrell Jones Conjecture for topological cyclic homology is not true in general). There are examples, where the assembly map

$$H_n^G(\mathrm{pr}; \mathbf{TC_{R;p}})\colon H_n^G(\underline{\underline{E}}G; \mathbf{TC_{R;p}}) \to \pi_n(\mathbf{THH}(\mathbf{R}[G]; p))$$

not surjective, see [83, Theorem 1.6]. At least there is a pro-isomorphism for $\mathbf{TC_{R;p}}$ with respect to the family $\mathcal{CY}$, see [83, Theorem 1.4]. The complications occurring with topological cyclic homology are due to the fact that smash products and homotopy inverse limit do not commute in general, see [85].

20.10.3 Relating the assembly maps of Farrell-Jones to the one for topological cyclic homology via the cyclotomic trace

There is an important transformation from algebraic K-theory to topological cyclic homology, the so called *cyclotomic trace*. It relates the assembly maps for the algebraic K-theory of $\mathbb{Z}G$ to the cyclic topological homology of the spherical group ring of G and is a key ingredient in proving the rational injectivity of K-theoretic assembly maps. The construction of the cyclotomic trace and the proof of the K-theoretic Novikov conjecture is carried out in the celebrated paper by Boekstedt-Hsiang-Madsen [17]. The passage from $\mathcal{TR}$ to $\mathcal{FIN}$, thus detecting a much larger portion of the algebraic K-theory of $\mathbb{Z}G$ and proving new results about the Whitehead group $\mathrm{Wh}(G)$, is presented in [84].

For more information about topological cyclic homology we refer for instance to [17, 31, 95].

20.11 The global point of view

At various occasions it has turned out that one should take a global point of view, i.e., one should not consider each group separately, but take into account that in general there is a theory which can be applied to every group and the values for the various groups are linked. This appears for instance in the following definition taken from [71, Section 1].

Let $\alpha\colon H \to G$ be a group homomorphism. Given an H-space X, define the *induction of X with α* to be the G-space $\mathrm{ind}_\alpha X := G \times_\alpha X$, which is the quotient of $G \times X$ by the right H-action $(g,x)\cdot h := (g\alpha(h), h^{-1}x)$ for $h \in H$ and $(g,x) \in G \times X$.

Definition 20.11.1 (Equivariant homology theory). An *equivariant homology theory with values in Λ-modules* $\mathcal{H}^?_n$ assigns to each group G a G-homology theory $\mathcal{H}^G_*$ with values in Λ-modules (in the sense of Definition 20.3.1) together with the following so called *induction structure*:

Given a group homomorphism $\alpha\colon H \to G$ and a H-CW-pair (X,A), there are for every $n \in \mathbb{Z}$ natural homomorphisms

$$\mathrm{ind}_\alpha\colon \mathcal{H}^H_n(X,A) \to \mathcal{H}^G_n(\mathrm{ind}_\alpha(X,A)) \tag{20.11.1}$$

satisfying:

- *Compatibility with the boundary homomorphisms*
  $\partial^G_n \circ \mathrm{ind}_\alpha = \mathrm{ind}_\alpha \circ \partial^H_n$;
- *Functoriality*
  Let $\beta\colon G \to K$ be another group homomorphism. Then we have for $n \in \mathbb{Z}$
  $$\mathrm{ind}_{\beta\circ\alpha} = \mathcal{H}^K_n(f_1) \circ \mathrm{ind}_\beta \circ \mathrm{ind}_\alpha\colon \mathcal{H}^H_n(X,A) \to \mathcal{H}^K_n(\mathrm{ind}_{\beta\circ\alpha}(X,A)),$$
  where $f_1\colon \mathrm{ind}_\beta \mathrm{ind}_\alpha(X,A) \xrightarrow{\cong} \mathrm{ind}_{\beta\circ\alpha}(X,A)$, $\quad (k,g,x) \mapsto (k\beta(g),x)$ is the natural K-homeomorphism;

- *Compatibility with conjugation*

  For $n \in \mathbb{Z}$, $g \in G$ and a (proper) G-CW-pair (X, A) the homomorphism $\operatorname{ind}_{c(g)\colon G \to G}\colon \mathcal{H}_n^G(X, A) \to \mathcal{H}_n^G(\operatorname{ind}_{c(g)\colon G \to G}(X, A))$ agrees with $\mathcal{H}_n^G(f_2)$ for the G-homeomorphism $f_2\colon (X, A) \to \operatorname{ind}_{c(g)\colon G \to G}(X, A)$ that sends x to $(1, g^{-1}x)$ in $G \times_{c(g)} (X, A)$;

- *Bijectivity*

  If $\operatorname{Ker}(\alpha)$ acts freely on $X \setminus A$, then $\operatorname{ind}_\alpha\colon \mathcal{H}_n^H(X, A) \to \mathcal{H}_n^G(\operatorname{ind}_\alpha(X, A))$ is bijective for all $n \in \mathbb{Z}$.

Because of the following theorem it will pay off that in Subsection 20.7.4 we considered functors defined on $\mathsf{Groupoids}$ and not only on $\mathsf{Or}(G)$.

Theorem 20.11.2 (Constructing equivariant homology theories using spectra). *Consider a covariant functor* $\mathbf{E}\colon \mathsf{Groupoids} \to \mathsf{Spectra}$ *respecting equivalences.*

Then there is an equivariant homology theory $H_*^?(-; \mathbf{E})$ *satisfying*

$$H_n^G(G/H; \mathbf{E}) \cong H_n^H(\{\bullet\}; \mathbf{E}) \cong \pi_n(\mathbf{E}(H))$$

for every subgroup $H \subseteq G$ *of every group* G *and every* $n \in \mathbb{Z}$.

Proof. See [82, Proposition 5.6 on page 793]. □

The global point of view has been taken up and pursued by Stefan Schwede on the level of spectra in his book [113], where global equivariant homotopy theory for compact Lie groups is developed. To deal with spectra is much more advanced and sophisticated than with equivariant homology.[2]

20.12 Relative assembly maps

In the formulations of the Isomorphism Conjectures above such as the one due to Farrell-Jones and Baum-Connes it is important to make the family $\mathcal{F}$ as small as possible. The largest family we encounter is $\mathcal{VCY}$, but there are special cases, where one can get smaller families. In particular it is desirable to get away with $\mathcal{FIN}$, since there are often finite models for $\underline{E}G = E_{\mathcal{FIN}}(G)$, whereas conjecturally there is a finite model for $\underline{\underline{E}}G = E_{\mathcal{VCY}}(G)$ only if G itself is virtually cyclic, see [53, Conjecture 1].

The general problem is to study and hopefully to prove bijectivity of relative assembly map associated to two families $\mathcal{F} \subseteq \mathcal{F}'$, i.e., of the map induced by the up to G-homotopy unique G-map $E_{\mathcal{F}}(G) \to E_{\mathcal{F}'}(G)$

$$\operatorname{asmb}_{\mathcal{F} \subseteq \mathcal{F}'}\colon H_n^G(E_{\mathcal{F}}(G)) \to H_n^G(E_{\mathcal{G}}(G))$$

for a G-homology theory H_*^G with values in Λ-modules. In studying this the global point of view becomes useful.

[2]See the chapter by Mike Hill in this Handbook for more information about equivariant stable homotopy theory.

The main technical result is the so called Transitivity Principle, which we explain next. For a family $\mathcal{F}$ of subgroups of G and a subgroup $H \subset G$ we define a family of subgroups of H

$$\mathcal{F}|_H = \{K \cap H \mid K \in \mathcal{F}\}.$$

Theorem 20.12.1 (Transitivity Principle). *Let $\mathcal{H}^?_*(-)$ be an equivariant homology theory with values in Λ-modules. Suppose $\mathcal{F} \subset \mathcal{F}'$ are two families of subgroups of G. If for every $H \in \mathcal{F}'$ and every $n \in \mathbb{Z}$ the assembly map*

$$\mathrm{asmb}_{\mathcal{F}|_H \subseteq \mathcal{ALL}} \colon \mathcal{H}^H_n(E_{\mathcal{F}|_H}(H)) \to \mathcal{H}^H_n(\{\bullet\})$$

is an isomorphism, then for every $n \in \mathbb{Z}$ the relative assembly map

$$\mathrm{asmb}_{\mathcal{F} \subseteq \mathcal{F}'} \colon \mathcal{H}^G_n(E_{\mathcal{F}}(G)) \to \mathcal{H}^G_n(E_{\mathcal{F}'}(G))$$

is an isomorphism.

Proof. See [82, Theorem 65 on page 742]. □

One has the following results about diminishing the family of subgroups. Denote by $\mathcal{VCY}_I$ the family of subgroups of G that are either finite or admit an epimorphism onto $\mathbb{Z}$ with finite kernel. Obviously $\mathcal{FIN} \subseteq \mathcal{VCY}_I \subseteq \mathcal{VCY}$.

Theorem 20.12.2 (Relative assembly maps).

1. *The relative assembly map for K-theory*

$$\mathrm{asmb}_{\mathcal{TR} \subseteq \mathcal{VCY}} \colon H^G_n(E_{\mathcal{TR}}(G); \mathbf{K}_R) \to H^G_n(E_{\mathcal{VCY}}(G); \mathbf{K}_R)$$

   is bijective for all $n \in \mathbb{Z}$, provided that G is torsionfree and R is regular;

2. *The relative assembly map for K-theory*

$$\mathrm{asmb}_{\mathcal{FIN} \subseteq \mathcal{VCY}} \colon H^G_n(E_{\mathcal{FIN}}(G); \mathbf{K}_R) \to H^G_n(E_{\mathcal{VCY}}(G); \mathbf{K}_R)$$

   is bijective for all $n \in \mathbb{Z}$, provided that R is a regular ring containing $\mathbb{Q}$;

3. *The relative assembly map for K-theory*

$$\mathrm{asmb}_{\mathcal{VCY}_I \subseteq \mathcal{VCY}} \colon H^G_n(E_{\mathcal{VCY}_I}(G); \mathbf{K}_R) \xrightarrow{\cong} H_n(E_{\mathcal{VCY}}(G); \mathbf{K}_R)$$

   is bijective for all $n \in \mathbb{Z}$;

4. *The relative assembly map for K-theory*

$$\mathrm{asmb}_{\mathcal{FIN} \subseteq \mathcal{VCY}} \otimes_{\mathbb{Z}} \mathrm{id}_{\mathbb{Q}} \colon H^G_n(E_{\mathcal{FIN}}(G); \mathbf{K}_R) \otimes_{\mathbb{Z}} \mathbb{Q} \to H^G_n(E_{\mathcal{VCY}}(G); \mathbf{K}_R) \otimes_{\mathbb{Z}} \mathbb{Q}$$

   is bijective for all $n \in \mathbb{Z}$, provided that R is regular;

5. *The relative assembly map for L-theory*

$$\mathrm{asmb}_{\mathcal{TR} \subseteq \mathcal{VCY}} \colon H^G_n(E_{\mathcal{TR}}(G); \mathbf{L}^{\langle -\infty \rangle}_R) \to H^G_n(E_{\mathcal{VCY}}(G); \mathbf{L}^{\langle -\infty \rangle}_R)$$

is an isomorphism for all $n \in \mathbb{Z}$, provided that G is torsionfree;

6. *There relative assembly map for L-theory*

$$\operatorname{asmb}_{\mathcal{FIN}\subseteq\mathcal{VCY}_I}\colon H_n^G(E_{\mathcal{FIN}}(G);\mathbf{L}_R^{\langle-\infty\rangle}) \to H_n^G(E_{\mathcal{VCY}_I}(G);\mathbf{L}_R^{\langle-\infty\rangle})$$

is bijective for all $n \in \mathbb{Z}$;

7. *The relative assembly map for L-theory*

$$\operatorname{asmb}_{\mathcal{FIN}\subseteq\mathcal{VCY}}[1/2]\colon H_n^G(E_{\mathcal{FIN}}(G);\mathbf{L}_R^{\langle-\infty\rangle})[1/2] \to H_n^G(E_{\mathcal{VCY}}(G);\mathbf{L}_R^{\langle-\infty\rangle})[1/2]$$

is bijective for all $n \in \mathbb{Z}$;

8. *The relative assembly map for topological K-theory*

$$\operatorname{asmb}_{\mathcal{FCY}\subseteq\mathcal{FIN}}\colon H_n^G(E_{\mathcal{FCY}}(G);\mathbf{K}^{\mathrm{top}}) \to H_n^G(E_{\mathcal{FIN}}(G);\mathbf{K}^{\mathrm{top}})$$

is bijective for all $n \in \mathbb{Z}$. This is also true for the real version;

9. *The relative assembly maps for K-theory and L-theory*

$$\begin{array}{rcl}\operatorname{asmb}_{\mathcal{FIN}\subseteq\mathcal{VCY}}\colon H_n^G(E_{\mathcal{FIN}}(G);\mathbf{K}_R) & \to & H_n^G(E_{\mathcal{VCY}}(G);\mathbf{K}_R);\\ \operatorname{asmb}_{\mathcal{FIN}\subseteq\mathcal{VCY}}\colon H_n^G(E_{\mathcal{FIN}}(G);\mathbf{L}_R^{\langle-\infty\rangle}) & \to & H_n^G(E_{\mathcal{VCY}}(G);\mathbf{L}_R^{\langle-\infty\rangle}),\end{array}$$

are split injective for all $n \in \mathbb{Z}$.

Proof. See [12, Theorem 1.3], [6, Theorem 0.5], [29], [76, Lemma 4.2], [82, Section 2.5], [89, Theorem 0.1 and Theorem 0.3], and [90]. □

Remark 20.12.3 (Torsionfree groups). A typical application is that for a torsionfree group G and a regular ring the K-theoretic Farrell-Jones Conjecture 20.8.1 implies together with Theorem 20.12.2 (1) that the assembly map

$$H_n(BG;\mathbf{K}(R)) = \pi_n(BG_+ \wedge \mathbf{K}(R) \to K_n(RG)$$

is bijective for all $n \in \mathbb{Z}$. Analogously, for a torsionfree group G the L-theoretic Farrell-Jones Conjecture 20.8.1 implies together with Theorem 20.12.2 (5) that the assembly map

$$H_n(BG;\mathbf{L}^{\langle-\infty\rangle}(R)) = \pi_n(BG_+ \wedge \mathbf{L}^{\langle-\infty\rangle}) \to L_n^{\langle-\infty\rangle}(RG)$$

is bijective for all $n \in \mathbb{Z}$.

20.13 Tools for computations

Most computations of K- and L-groups of group rings are done using the Farrell-Jones Conjecture 20.8.1 and the Baum-Connes Conjecture 20.9.1. The situation in the Farrell-Jones Conjecture 20.8.1 is more complicated than in the Baum-Connes setting, since the family $\mathcal{VCY}$ is much harder to handle than the family $\mathcal{FIN}$. One can consider $H_n^G(\underline{E}G;\mathbf{K}_R)$ and $H_n^G(\underline{\underline{E}}G,\underline{E}G;\mathbf{K}_R)$ separately because of Theorem 20.12.2 (9), where one considers $\underline{E}G$ as a G-CW-subcomplex of $\underline{\underline{E}}G$. The term $H_n^G(\underline{\underline{E}}G,\underline{E}G;\mathbf{K}_R)$ involves Nil-terms and UNil-terms, which are hard to determine. For $H_n^G(\underline{E}G;\mathbf{K}_R)$, $H_n^G(\underline{E}G;\mathbf{L}_R^{\langle -\infty\rangle})$ and $H_n^G(\underline{E}G;\mathbf{K}^{\text{top}})$ one can use the equivariant Atyiah-Hirzebruch spectral sequence or the p-chain spectral sequence, see Davis-Lueck [27]. Rationally these groups can often be computed explicitly using equivariant Chern characters, see [71, Section 1]. Notice that these can only be constructed since we take on the global point of view as explained in Section 20.11. Often an important input is that one obtains from the geometry of the underlying group nice models for $\underline{E}G$ and can construct $\underline{\underline{E}}G$ from $\underline{E}G$ by attaching a tractable family of equivariant cells.

Here are two example, where these ideas lead to a explicit computation, whose outcome is as simple as one can hope.

Theorem 20.13.1 (Farrell-Jones Conjecture for torsionfree hyperbolic groups for K-theory)**.** *Let G be a torsionfree hyperbolic group.*

1. *We obtain for all $n \in \mathbb{Z}$ an isomorphism*

$$H_n(BG;\mathbf{K}(R)) \oplus \bigoplus_C (NK_n(R) \oplus NK_n(R)) \xrightarrow{\cong} K_n(RG),$$

where C runs through a complete system of representatives of the conjugacy classes of maximal infinite cyclic subgroups. If R is regular, we have $NK_n(R) = 0$ for all $n \in \mathbb{Z}$;

2. *The abelian groups $K_n(\mathbb{Z}G)$ for $n \leq -1$, $\widetilde{K}_0(\mathbb{Z}G)$, and $\mathrm{Wh}(G)$ vanish;*

3. *We get for every ring R with involution and $n \in \mathbb{Z}$ an isomorphism*

$$H_n(BG;\mathbf{L}^{\langle -\infty\rangle}(R)) \xrightarrow{\cong} L_n^{\langle -\infty\rangle}(RG).$$

For every $j \in \mathbb{Z}, j \leq 2$, and $n \in \mathbb{Z}$, the natural map

$$L_n^{\langle j\rangle}(\mathbb{Z}G) \xrightarrow{\cong} L_n^{\langle -\infty\rangle}(\mathbb{Z}G)$$

is bijective;

4. *We get for every $n \in \mathbb{Z}$ an isomorphism*

$$K_n^{\text{top}}(BG) \xrightarrow{\cong} K_n^{\text{top}}(C_r^*(G)).$$

Proof. See [86, Theorem 1.2]. □

Theorem 20.13.2. *Suppose that G satisfies the Baum-Connes Conjecture 20.9.1 and K-theoretic Farrell-Jones Conjecture 20.8.1 with coefficients in the ring $\mathbb{C}$. Let $\operatorname{con}(G)_f$ be the set of conjugacy classes (g) of elements $g \in G$ of finite order. We denote for $g \in G$ by $C_G\langle g\rangle$ the centralizer in G of the cyclic subgroup generated by g.*

Then we get the following commutative square, whose horizontal maps are isomorphisms and complexifications of assembly maps, and whose vertical maps are induced by the obvious change of theory homomorphisms

$$\begin{array}{ccc}
\bigoplus_{p+q=n} \bigoplus_{(g)\in \mathrm{con}(G)_f} H_p(C_G\langle g\rangle;\mathbb{C}) \otimes_{\mathbb{Z}} K_q(\mathbb{C}) & \xrightarrow{\cong} & K_n(\mathbb{C}G) \otimes_{\mathbb{Z}} \mathbb{C} \\
\downarrow & & \downarrow \\
\bigoplus_{p+q=n} \bigoplus_{(g)\in \mathrm{con}(G)_f} H_p(C_G\langle g\rangle;\mathbb{C}) \otimes_{\mathbb{Z}} K_q^{\mathrm{top}}(\mathbb{C}) & \xrightarrow{\cong} & K_n^{\mathrm{top}}(C_r^*(G)) \otimes_{\mathbb{Z}} \mathbb{C}.
\end{array}$$

Proof. See [71, Theorem 0.5]). □

20.14 The challenge of extending equivariant homotopy theory to infinite groups

We have seen that it is important to study equivariant homology and homotopy also for groups that are not necessarily finite. In particular equivariant KK-theory has developed into a whole industry, which, however, does not really take the point of view of spectra instead of K-groups and cycles into account. So we encounter

Problem 20.14.1. Extend equivariant homotopy theory for finite groups to infinite groups, at least in the case of proper G-actions.

A few first steps are already in the literature. We have already explained the notion of an equivariant homology and the existence of equivariant Chern characters, see [71], where the global point of view enters. There is also a cohomological version, see [75]. Topological K-theory has systematically been studied in [80, 81], and an attempt of defining Burnside rings and equivariant cohomotopy for proper G-spaces is presented in [74]. These do include multiplicative structures.

An important ingredient in equivariant homotopy theory for finite groups is to stabilize with unit spheres in finite-dimensional orthogonal representations. However, there are infinite groups such that any finite-dimensional representation is trivial and therefore one has to stabilize with equivariant vector bundles, see [74, Remark 6.17]. Or one may have to pass even to Hilbert bundles and equivariant Fredholm operator between these, see [98] and also [81].

There are various interesting pairings on the group level in the literature, such as Kasparov products, the action of Swan groups on algebraic K-theory and so on. They all should be implemented on the spectrum level. So a systematical study of higher structures for equivariant spectra over infinite groups has to be carried out and one has to find the right equivariant homotopy category. This applies also to multiplicative structures and smash products. First steps will be presented in [30] using orthogonal spectra. This seems to work well for topological K-theory, but is probably not adequate for algebraic K-theory. This remark also holds for global equivariant homotopy theory.

A general description of Mackey structures and induction theorems in the sense of Dress is described in [6]. There are more sophisticated Mackey structure and transfers in the equivariant homotopy of finite groups, but it is not at all clear whether and how they extend to infinite groups.

Topological K-theory and the Baum-Connes Conjecture make sense and are studied also for topological groups, e.g., reductive p-adic groups and Lie groups. It is conceivable that also the Farrell-Jones Conjecture has an analogue for Hecke algebras of totally disconnected groups, see [82, Conjecture 119 on page 773]. So one can ask Problem 20.14.1 also for (not necessarily compact) topological groups instead of infinite (discrete) groups.

Bibliography

[1] J. F. Adams. *Stable homotopy and generalised homology.* Chicago Lectures in Mathematics. University of Chicago Press, 1974.

[2] A. Bartels. On proofs of the Farrell-Jones conjecture. In *Topology and geometric group theory*, volume 184 of *Springer Proc. Math. Stat.*, pages 1–31. Springer, 2016.

[3] A. Bartels and M. Bestvina. The Farrell-Jones Conjecture for mapping class groups. *Invent. Math.* 215(2):651–712, 2019.

[4] A. Bartels, S. Echterhoff, and W. Lück. Inheritance of isomorphism conjectures under colimits. In Cortinaz, Cuntz, Karoubi, Nest, and Weibel, editors, *K-Theory and noncommutative geometry*, EMS Ser. Congr. Rep., pages 41–70. Eur. Math. Soc., 2008.

[5] A. Bartels, F. T. Farrell, and W. Lück. The Farrell-Jones Conjecture for cocompact lattices in virtually connected Lie groups. *J. Amer. Math. Soc.*, 27(2):339–388, 2014.

[6] A. Bartels and W. Lück. Induction theorems and isomorphism conjectures for K- and L-theory. *Forum Math.*, 19:379–406, 2007.

[7] A. Bartels and W. Lück. The Borel conjecture for hyperbolic and CAT(0)-groups. *Ann. of Math. (2)*, 175:631–689, 2012.

[8] A. Bartels, W. Lück, and H. Reich. The K-theoretic Farrell-Jones conjecture for hyperbolic groups. *Invent. Math.*, 172(1):29–70, 2008.

[9] A. Bartels, W. Lück, and H. Reich. On the Farrell-Jones Conjecture and its applications. *J. Topol.*, 1:57–86, 2008.

[10] A. Bartels, W. Lück, H. Reich, and H. Rüping. K- and L-theory of group rings over $\mathrm{GL}_n(\mathbf{Z})$. *Publ. Math., Inst. Hautes Étud. Sci.*, 119:97–125, 2014.

[11] A. Bartels, W. Lück, and S. Weinberger. On hyperbolic groups with spheres as boundary. *J. Differential Geom.*, 86(1):1–16, 2010.

[12] A. C. Bartels. On the domain of the assembly map in algebraic K-theory. *Algebr. Geom. Topol.*, 3:1037–1050 (electronic), 2003.

[13] H. Bass. Euler characteristics and characters of discrete groups. *Invent. Math.*, 35:155–196, 1976.

[14] P. Baum, A. Connes, and N. Higson. Classifying space for proper actions and K-theory of group C^*-algebras. In *C^*-algebras: 1943–1993 (San Antonio, TX, 1993), Contemp. Math.*, 167, pages 240–291. Amer. Math. Soc., 1994.

[15] P. Baum, E. Guentner, and R. Willett. Expanders, exact crossed products, and the Baum-Connes conjecture. *Ann. K-Theory*, 1(2):155–208, 2016.

[16] P. Baum and M. Karoubi. On the Baum-Connes conjecture in the real case. *Q. J. Math.*, 55(3):231–235, 2004.

[17] M. Bökstedt, W. C. Hsiang, and I. Madsen. The cyclotomic trace and algebraic K-theory of spaces. *Invent. Math.*, 111(3):465–539, 1993.

[18] S. Cappell, A. Ranicki, and J. Rosenberg, editors. *Surveys on surgery theory. Vol. 1. Ann. of Math. Stud.*, 145, Princeton University Press, 2000. Papers dedicated to C. T. C. Wall.

[19] S. Cappell, A. Ranicki, and J. Rosenberg, editors. *Surveys on surgery theory. Vol. 2. Ann. of Math. Stud.*, 149, Princeton University Press, 2001. Papers dedicated to C. T. C. Wall.

[20] G. Carlsson and E. Pedersen. Controlled algebra and the Baum-Connes conjecture. in preparation, 1996.

[21] J. Chabert and S. Echterhoff. Permanence properties of the Baum-Connes conjecture. *Doc. Math.*, 6:127–183, 2001.

[22] S. Chang and S. Weinberger. On invariants of Hirzebruch and Cheeger-Gromov. *Geom. Topol.*, 7:311–319, 2003.

[23] T. D. Cochran. Noncommutative knot theory. *Algebr. Geom. Topol.*, 4:347–398, 2004.

[24] T. D. Cochran, K. E. Orr, and P. Teichner. Knot concordance, Whitney towers and L^2-signatures. *Ann. of Math. (2)*, 157(2):433–519, 2003.

[25] D. Crowley, W. Lück, and T. Macko. Surgery Theory: Foundations. book, in preparation, 2019.

[26] J. F. Davis and W. Lück. Spaces over a category and assembly maps in isomorphism conjectures in K- and L-theory. *K-Theory*, 15(3):201–252, 1998.

[27] J. F. Davis and W. Lück. The p-chain spectral sequence. *K-Theory*, 30(1):71–104, 2003. Special issue in honor of Hyman Bass on his seventieth birthday. Part I.

[28] J. F. Davis and W. Lück. The topological K-theory of certain crystallographic groups. *Journal of Non-Commutative Geometry*, 7:373–431, 2013.

[29] J. F. Davis, F. Quinn, and H. Reich. Algebraic K-theory over the infinite dihedral group: a controlled topology approach. *J. Topol.*, 4(3):505–528, 2011.

[30] D. Degrijse, M. Hausmann, W. Lück, I. Patchkoria, and S. Schwede. Proper equivariant stable homotopy theory. in preparation, 2019.

[31] B. I. Dundas, T. G. Goodwillie, and R. McCarthy. *The local structure of algebraic K-theory*, volume 18 of *Algebra and Applications*. Springer-Verlag, 2013.

[32] W. Dwyer, M. Weiss, and B. Williams. A parametrized index theorem for the algebraic K-theory Euler class. *Acta Math.*, 190(1):1–104, 2003.

[33] S. Echterhoff, W. Lück, N. C. Phillips, and S. Walters. The structure of crossed products of irrational rotation algebras by finite subgroups of $\mathrm{SL}_2(\mathbb{Z})$. *J. Reine Angew. Math.*, 639:173–221, 2010.

[34] N.-E. Enkelmann, W. Lück, M. Pieper, M. Ullmann, and C. Winges. On the Farrell–Jones conjecture for Waldhausen's A–theory. *Geom. Topol.*, 22(6):3321–3394, 2018.

[35] F. T. Farrell. The Borel conjecture. In F. T. Farrell, L. Göttsche, and W. Lück, editors, *High dimensional manifold theory*, number 9 in ICTP Lecture Notes, pages 225–298. Abdus Salam International Centre for Theoretical Physics, Trieste, 2002. Proceedings of the summer school "High dimensional manifold theory" in Trieste May/June 2001, Number 1. http://www.ictp.trieste.it/~pub_off/lectures/vol9.html.

[36] F. T. Farrell and W. C. Hsiang. On the rational homotopy groups of the diffeomorphism groups of discs, spheres and aspherical manifolds. In *Algebraic and geometric topology (Proc. Sympos. Pure Math., Stanford Univ., Stanford, Calif., 1976), Part 1*, Proc. Sympos. Pure Math., XXXII, pages 325–337. Amer. Math. Soc., 1978.

[37] F. T. Farrell and L. E. Jones. Rigidity in geometry and topology. In *Proceedings of the International Congress of Mathematicians, Vol. I, II (Kyoto, 1990)*, pages 653–663, 1991. Math. Soc. Japan.

[38] F. T. Farrell and L. E. Jones. Isomorphism conjectures in algebraic K-theory. *J. Amer. Math. Soc.*, 6(2):249–297, 1993.

[39] F. T. Farrell, L. E. Jones, and W. Lück. A caveat on the isomorphism conjecture in L-theory. *Forum Math.*, 14(3):413–418, 2002.

[40] T. Farrell, W. Lück, and W. Steimle. Approximately fibering a manifold over an aspherical one. *Math. Ann.*, 370(1-2):669–726, 2018.

[41] S. Ferry, W. Lück, and S. Weinberger. On the stable Cannon conjecture. arXiv:1804.00738, 2018.

[42] S. C. Ferry, A. A. Ranicki, and J. Rosenberg, editors. *Novikov conjectures, index theorems and rigidity. Vol. 1.* Cambridge University Press, 1995. Including papers from the conference held at the Mathematisches Forschungsinstitut Oberwolfach, Oberwolfach, September 6–10, 1993.

[43] S. C. Ferry, A. A. Ranicki, and J. Rosenberg, editors. *Novikov conjectures, index theorems and rigidity. Vol. 2.* Cambridge University Press, 1995. Including papers from the conference held at the Mathematisches Forschungsinstitut Oberwolfach, Oberwolfach, September 6–10, 1993.

[44] I. Hambleton and E. K. Pedersen. Identifying assembly maps in K- and L-theory. *Math. Ann.*, 328(1-2):27–57, 2004.

[45] F. Hebestreit, W. Lück, M. Land, and O. Randal-Williams. A vanishing theorem for tautological classes of aspherical manifolds. arXiv:1705.06232, 2017.

[46] N. Higson. The Baum-Connes conjecture. In *Proceedings of the International Congress of Mathematicians, Vol. II (Berlin, 1998)*, pages 637–646, 1998.

[47] N. Higson and G. Kasparov. E-theory and KK-theory for groups which act properly and isometrically on Hilbert space. *Invent. Math.*, 144(1):23–74, 2001.

[48] N. Higson, V. Lafforgue, and G. Skandalis. Counterexamples to the Baum-Connes conjecture. *Geom. Funct. Anal.*, 12(2):330–354, 2002.

[49] N. Higson and J. Roe. Mapping surgery to analysis. I. Analytic signatures. *K-Theory*, 33(4):277–299, 2005.

[50] N. Higson and J. Roe. Mapping surgery to analysis. II. Geometric signatures. *K-Theory*, 33(4):301–324, 2005.

[51] N. Higson and J. Roe. Mapping surgery to analysis. III. Exact sequences. *K-Theory*, 33(4):325–346, 2005.

[52] M. Joachim. K-homology of C^*-categories and symmetric spectra representing K-homology. *Math. Ann.*, 327(4):641–670, 2003.

[53] D. Juan-Pineda and I. J. Leary. On classifying spaces for the family of virtually cyclic subgroups. In *Recent developments in algebraic topology*, volume 407 of *Contemp. Math.*, pages 135–145. Amer. Math. Soc., 2006.

[54] H. Kammeyer, W. Lück, and H. Rüping. The Farrell–Jones conjecture for arbitrary lattices in virtually connected Lie groups. *Geom. Topol.*, 20(3):1275–1287, 2016.

[55] G. G. Kasparov. Operator K-theory and its applications: elliptic operators, group representations, higher signatures, C^*-extensions. In *Proceedings of the International Congress of Mathematicians, Vol. 1, 2 (Warsaw, 1983)*, pages 987–1000, Warsaw, 1984.

[56] G. G. Kasparov. Equivariant KK-theory and the Novikov conjecture. *Invent. Math.*, 91(1):147–201, 1988.

[57] G. G. Kasparov. K-theory, group C^*-algebras, and higher signatures (conspectus). In *Novikov conjectures, index theorems and rigidity, Vol. 1 (Oberwolfach, 1993)*, volume 226 of *London Math. Soc. Lecture Note Ser.*, pages 101–146. Cambridge Univ. Press, 1995.

[58] D. Kasprowski, M. Ullmann, C. Wegner, and C. Winges. The A-theoretic Farrell-Jones conjecture for virtually solvable groups. *Bull. Lond. Math. Soc.*, 50(2):219–228, 2018.

[59] N. Keswani. Homotopy invariance of relative eta-invariants and C^*-algebra K-theory. *Electron. Res. Announc. Amer. Math. Soc.*, 4:18–26, 1998.

[60] N. Keswani. Von Neumann eta-invariants and C^*-algebra K-theory. *J. London Math. Soc. (2)*, 62(3):771–783, 2000.

[61] M. Kreck and W. Lück. *The Novikov conjecture: Geometry and algebra*, volume 33 of *Oberwolfach Seminars*. Birkhäuser Verlag, 2005.

[62] P. Kühl, T. Macko, and A. Mole. The total surgery obstruction revisited. *Münster J. Math.*, 6:181–269, 2013.

[63] V. Lafforgue. The Baum-Connes conjecture with coefficients for hyperbolic groups. (La conjecture de baum-connes à coefficients pour les groupes hyperboliques.). *J. Noncommut. Geom.*, 6(1):1–197, 2012.

[64] M. Land. The analytical assembly map and index theory. *J. Noncommut. Geom.*, 9(2):603–619, 2015.

[65] M. Land and T. Nikolaus. On the relation between K- and L-theory of C^*-algebras. *Math. Ann.*, 371(1-2):517–563, 2018.

[66] M. Langer and W. Lück. Topological K-theory of the group C^*-algebra of a semi-direct product $\mathbb{Z}^n \rtimes \mathbb{Z}/m$ for a free conjugation action. *J. Topol. Anal.*, 4(2):121–172, 2012.

[67] X. Li and W. Lück. K-theory for ring C^*-algebras – the case of number fields with higher roots of unity. *J. Topol. Anal.* 4(4):449–479, 2012.

[68] J.-L. Loday. K-théorie algébrique et représentations de groupes. *Ann. Sci. École Norm. Sup. (4)*, 9(3):309–377, 1976.

[69] W. Lück. *Transformation groups and algebraic K-theory*, volume 1408 of *Lecture Notes in Math.*. Springer-Verlag, 1989.

[70] W. Lück. A basic introduction to surgery theory. In F. T. Farrell, L. Göttsche, and W. Lück, editors, *High dimensional manifold theory*, number 9 in ICTP Lecture Notes, pages 1–224. Abdus Salam International Centre for Theoretical Physics, Trieste, 2002. Proceedings of the summer school "High dimensional manifold theory" in Trieste May/June 2001, Number 1. `http://www.ictp.trieste.it/~pub_off/lectures/vol9.html`.

[71] W. Lück. Chern characters for proper equivariant homology theories and applications to K- and L-theory. *J. Reine Angew. Math.*, 543:193–234, 2002.

[72] W. Lück. *L^2-Invariants: Theory and applications to geometry and K-theory*, volume 44 of *Ergeb. Math. Grenzgeb.* (3), Springer-Verlag, 2002.

[73] W. Lück. The relation between the Baum-Connes conjecture and the trace conjecture. *Invent. Math.*, 149(1):123–152, 2002.

[74] W. Lück. The Burnside ring and equivariant stable cohomotopy for infinite groups. *Pure Appl. Math. Q.*, 1(3):479–541, 2005.

[75] W. Lück. Equivariant cohomological Chern characters. *Internat. J. Algebra Comput.*, 15(5-6):1025–1052, 2005.

[76] W. Lück. K- and L-theory of the semi-direct product of the discrete 3-dimensional Heisenberg group by $\mathbb{Z}/4$. *Geom. Topol.*, 9:1639–1676, 2005.

[77] W. Lück. Survey on classifying spaces for families of subgroups. In *Infinite groups: geometric, combinatorial and dynamical aspects*, volume 248 of *Progr. Math.*, pages 269–322. Birkhäuser, 2005.

[78] W. Lück. Survey on aspherical manifolds. In A. Ran, H. te Riele, and J. Wiegerinck, editors, *Proceedings of the 5-th European Congress of Mathematics Amsterdam 14 -18 July 2008*, pages 53–82. Eur. Math. Soc., 2010.

[79] W. Lück. Isomorphism conjectures in K- and L-theory. in preparation, 2020.

[80] W. Lück and B. Oliver. Chern characters for the equivariant K-theory of proper G-CW-complexes. In *Cohomological methods in homotopy theory (Bellaterra, 1998)*, volume 196 of *Progr. Math.*, pages 217–247. Birkhäuser, 2001.

[81] W. Lück and B. Oliver. The completion theorem in K-theory for proper actions of a discrete group. *Topology*, 40(3):585–616, 2001.

[82] W. Lück and H. Reich. The Baum-Connes and the Farrell-Jones conjectures in K- and L-theory. In *Handbook of K-theory. Vol. 1, 2*, pages 703–842. Springer, 2005.

[83] W. Lück, H. Reich, J. Rognes, and M. Varisco. Assembly maps for topological cyclic homology of group algebras. arXiv:1607.03557, 2016.

[84] W. Lück, H. Reich, J. Rognes, and M. Varisco. Algebraic K-theory of group rings and the cyclotomic trace map. *Adv. Math.*, 304:930–1020, 2017.

[85] W. Lück, H. Reich, and M. Varisco. Commuting homotopy limits and smash products. *K-Theory*, 30(2):137–165, 2003. Special issue in honor of Hyman Bass on his seventieth birthday. Part II.

[86] W. Lück and D. Rosenthal. On the K- and L-theory of hyperbolic and virtually finitely generated abelian groups. *Forum Math.*, 26(5):1565–1609, 2014.

[87] W. Lück and T. Schick. Various L^2-signatures and a topological L^2-signature theorem. In *High-dimensional manifold topology*, pages 362–399. World Sci. Publ., 2003.

[88] W. Lück and T. Schick. Approximating L^2-signatures by their compact analogues. *Forum Math.*, 17(1):31–65, 2005.

[89] W. Lück and W. Steimle. Splitting the relative assembly map, Nil-terms and involutions. *Ann. K-Theory*, 1(4):339–377, 2016.

[90] M. Matthey and G. Mislin. Equivariant K-homology and restriction to finite cyclic subgroups. Preprint, 2003.

[91] I. Mineyev and G. Yu. The Baum-Connes conjecture for hyperbolic groups. *Invent. Math.*, 149(1):97–122, 2002.

[92] A. S. Miščenko and A. T. Fomenko. The index of elliptic operators over C^*-algebras. *Mathematics of the USSR-Izvestiya*, 15(1):87–112, 1980.

[93] G. Mislin and A. Valette. *Proper group actions and the Baum-Connes conjecture.* Advanced Courses in Mathematics. CRM Barcelona. Birkhäuser Verlag, 2003.

[94] P. D. Mitchener. C^*-categories, groupoid actions, equivariant KK-theory, and the Baum-Connes conjecture. *J. Funct. Anal.*, 214(1):1–39, 2004.

[95] T. Nikolaus and P. Scholze. On topological cyclic homology. *Acta Math.* 221(2):203–409, 2018, and *Acta Math.* 222(1):215–218, 2019).

[96] H. Oyono-Oyono. Baum-Connes Conjecture and extensions. *J. Reine Angew. Math.*, 532:133–149, 2001.

[97] H. Oyono-Oyono. Baum-Connes conjecture and group actions on trees. *K-Theory*, 24(2):115–134, 2001.

[98] N. C. Phillips. *Equivariant K-theory for proper actions.* Longman Scientific & Technical, 1989.

[99] P. Piazza and T. Schick. Bordism, rho-invariants and the Baum-Connes conjecture. *J. Noncommut. Geom.*, 1(1):27–111, 2007.

[100] P. Piazza and T. Schick. Rho-classes, index theory and Stolz' positive scalar curvature sequence. *J. Topol.*, 7(4):965–1004, 2014.

[101] F. Quinn. $B_{(\mathrm{TOP}_n)}\tilde{}$ and the surgery obstruction. *Bull. Amer. Math. Soc.*, 77:596–600, 1971.

[102] F. Quinn. Assembly maps in bordism-type theories. In *Novikov conjectures, index theorems and rigidity, Vol. 1 (Oberwolfach, 1993)*, volume 226 of *London Math. Soc. Lecture Note Ser.*, pages 201–271. Cambridge Univ. Press, 1995.

[103] A. A. Ranicki. *Exact sequences in the algebraic theory of surgery*, volume 26 of *Math. Notes.* Princeton University Press, 1981.

[104] A. A. Ranicki. *Algebraic L-theory and topological manifolds*, volume 102 of *Cambridge Tracts in Math.* Cambridge University Press, 1992.

[105] A. A. Ranicki. *Lower K- and L-theory*, volume 178 of *London Math. Soc. Lecture Note Ser.* Cambridge University Press, 1992.

[106] J. Rosenberg. C^*-algebras, positive scalar curvature and the Novikov conjecture. III. *Topology*, 25:319–336, 1986.

[107] J. Rosenberg. Analytic Novikov for topologists. In *Novikov conjectures, index theorems and rigidity, Vol. 1 (Oberwolfach, 1993)*, volume 226 of *London Math. Soc. Lecture Note Ser.*, pages 338–372. Cambridge Univ. Press, 1995.

[108] J. Rosenberg. Structure and applications of real C^*-algebras. In *Operator algebras and their applications*, volume 671 of *Contemp. Math.*, pages 235–258. Amer. Math. Soc., 2016.

[109] H. Rüping. The Farrell–Jones conjecture for S-arithmetic groups. *J. Topol.*, 9(1):51–90, 2016.

[110] T. Schick. A counterexample to the (unstable) Gromov-Lawson-Rosenberg conjecture. *Topology*, 37(6):1165–1168, 1998.

[111] T. Schick. Index theory and the Baum-Connes conjecture. In *Geometry Seminars. 2001-2004 (Italian)*, pages 231–280. Univ. Stud. Bologna, 2004.

[112] T. Schick. Real versus complex K-theory using Kasparov's bivariant KK-theory. *Algebr. Geom. Topol.*, 4:333–346, 2004.

[113] S. Schwede. *Global homotopy theory*, volume 34 of *New Mathematical Monographs.* Cambridge University Press, 2018.

[114] S. Stolz. Manifolds of positive scalar curvature. In T. Farrell, L. Göttsche, and W. Lück, editors, *High dimensional manifold theory*, number 9 in ICTP Lecture Notes, pages 661–708. Abdus Salam International Centre for Theoretical Physics, Trieste, 2002. Proceedings of the summer school "High dimensional manifold theory" in Trieste May/June 2001, Number 2. `http://www.ictp.trieste.it/~pub_off/lectures/vol9.html`.

[115] T. tom Dieck. Orbittypen und äquivariante Homologie. I. *Arch. Math. (Basel)*, 23:307–317, 1972.

[116] A. Valette. *Introduction to the Baum-Connes conjecture.* Birkhäuser Verlag, 2002. From notes taken by Indira Chatterji, With an appendix by Guido Mislin.

[117] W. Vogell. Algebraic K-theory of spaces, with bounded control. *Acta Math.*, 165(3-4):161–187, 1990.

[118] W. Vogell. Boundedly controlled algebraic K-theory of spaces and its linear counterparts. *J. Pure Appl. Algebra*, 76(2):193–224, 1991.

[119] F. Waldhausen. Algebraic K-theory of topological spaces. I. In *Algebraic and Geometric Topology, Stanford Univ., Stanford, Calif., 1976), Part 1*, Proc. Sympos. Pure Math., XXXII, pages 35–60. Amer. Math. Soc., 1978.

[120] F. Waldhausen. Algebraic K-theory of spaces. In *Algebraic and geometric topology (New Brunswick, N.J., 1983)*, volume 1126 of *Lecture Notes in Math.*, pages 318–419. Springer-Verlag, 1985.

[121] F. Waldhausen. Algebraic K-theory of spaces, concordance, and stable homotopy theory. In *Algebraic Topology and Algebraic K-theory (Princeton, N.J., 1983)*, Ann. of Math. Stud., 113, pages 392–417. Princeton Univ. Press, 1987.

[122] F. Waldhausen. An outline of how manifolds relate to algebraic K-theory. In *Homotopy theory (Durham, 1985)*, volume 117 of *London Math. Soc. Lecture Note Ser.*, pages 239–247. Cambridge Univ. Press, 1987.

[123] F. Waldhausen, B. Jahren, and J. Rognes. *Spaces of PL manifolds and categories of simple maps*, volume 186 of *Ann. of Math. Stud.* Princeton University Press, 2013.

[124] C. T. C. Wall. *Surgery on compact manifolds*, volume 69 of *Math. Surveys Monogr.* Amer. Math. Soc., second edition, 1999. Edited and with a foreword by A. A. Ranicki.

[125] C. Wegner. The K-theoretic Farrell-Jones conjecture for CAT(0)-groups. *Proc. Amer. Math. Soc.*, 140(3):779–793, 2012.

[126] C. Wegner. The Farrell-Jones conjecture for virtually solvable groups. *J. Topol.*, 8(4):975–1016, 2015.

[127] M. Weiss and B. Williams. Assembly. In *Novikov conjectures, index theorems and rigidity, Vol. 2 (Oberwolfach, 1993)*, pages 332–352. Cambridge Univ. Press, 1995.

[128] M. Weiss and B. Williams. Automorphisms of manifolds. In *Surveys on surgery theory, Vol. 2*, volume 149 of *Ann. of Math. Stud.*, pages 165–220. Princeton Univ. Press, 2001.

[129] Z. Xie and G. Yu. A relative higher index theorem, diffeomorphisms and positive scalar curvature. *Adv. Math.*, 250:35–73, 2014.

[130] Z. Xie and G. Yu. Higher rho invariants and the moduli space of positive scalar curvature metrics. *Adv. Math.*, 307:1046–1069, 2017.

MATHEMATISCHES INSTITUT DER UNIVERSITÄT BONN, 53115 BONN, GERMANY
E-mail address: wolfgang.lueck@him.uni-bonn.de

21

Lubin-Tate theory, character theory, and power operations

Nathaniel Stapleton

21.1 Introduction

Chromatic homotopy theory decomposes the category of spectra at a prime p into a collection of categories according to certain periodicities. There is one of these categories for each natural number n and it is called the $K(n)$-local category. Unfortunately, when $n > 1$, even the $K(n)$-local categories are often quite difficult to understand computationally.[1] When $n = 2$, significant progress has been made but above $n = 2$ many basic computational questions are open. Even the most well-behaved ring spectra in the $K(n)$-local category, the height n Morava E-theories, still hold plenty of mysteries. In order to understand E-cohomology, Hopkins, Kuhn, and Ravenel constructed a character map for each E-theory landing in a form of rational cohomology. They proved that the codomain of their character map serves as an approximation to E-cohomology in a precise sense. It turns out that many of the deep formal properties of the $K(n)$-local category can be expressed in terms of simple formulas or relations satisfied by simple formulas on the codomain of the character map. This is intriguing for several reasons: The codomain of the character map is not $K(n)$-local, it is a rational approximation to E-cohomology so it removes all of the torsion from E-cohomology. The codomain of the character map for the E-cohomology of a finite group is a simple generalization of the ring of class functions on the group. Thus, in this case, these deep properties of the $K(n)$-local category are reflected in combinatorial and group theoretic properties of certain types of conjugacy classes in the group. Finally, a certain $\mathbb{Q}$-algebra, known as the Drinfeld ring of infinite level structures, arises from topological considerations as the coefficients of the codomain of the character map. This $\mathbb{Q}$-algebra plays an important role in the local Langlands program and the group actions that feed into the local Langlands correspondence are closely related to fundamental properties of the Morava E-theories.

One of the main goals of this article is to explain the relationship between the power operations

$$P_m\colon E^0(BG) \to E^0(BG \times B\Sigma_m)$$

Mathematics Subject Classification. 55N20, 55N22, 55S25.

Key words and phrases. Morava E-theory, power operations, character theory, Lubin-Tate theory, level structures, formal groups.

[1]See the chapter in this Handbook by Tobias Barthel and Agnès Beaudry for an extensive survey of this subject.

on the Morava E-cohomology of finite groups and Hopkins-Kuhn-Ravenel character theory. Morava E-theory is built out of the Lubin-Tate ring associated to a finite height formal group over a perfect field of characteristic p by using the Landweber exact functor theorem. Lubin-Tate theory plays an important role in local arithmetic geometry and so it is not too surprising that other important objects from arithmetic geometry, such as the Drinfeld ring, that are closely related to the Lubin-Tate ring arise in the construction of the character map. The power operations on Morava E-theory arise as a consequence of the Goerss-Hopkins-Miller theorem that implies that each Morava E-theory spectrum has an essentially unique E_∞-ring structure. However, this multiplicative structure plays no role in the construction of the character map, so it may be surprising that the two structures interact as well as they do. The Drinfeld ring at infinite level picks up extra symmetry that does not exist at finite level. This extra symmetry turns out to be the key to understanding the relationship between character theory and power operations.

To give a careful statement of the results, we need some setup. Fix a prime p and a height n formal group $\mathbb{F}$ over κ, a perfect field of characteristic p. Associated to this data is an E_∞-ring spectrum $E = E(\mathbb{F}, \kappa)$ known as Morava E-theory. The coefficient ring E^0 is the Lubin-Tate ring. Morava E-theory has the astounding property that the automorphism group of E as an E_∞-ring spectrum is the discrete extended Morava stabilizer group $\mathrm{Aut}(\mathbb{F}/\kappa) \rtimes \mathrm{Gal}(\kappa/\mathbb{F}_p)$. The homotopy fixed points of E for this action by the extended stabilizer group are precisely the $K(n)$-local sphere - the unit in the category of $K(n)$-local spectra. Therefore, a calculation of the E-cohomology of a space is the starting point for the calculation of the $K(n)$-local cohomotopy of the space.

Hopkins, Kuhn, and Ravenel made significant progress towards understanding the E-cohomology of finite groups by constructing an analogue of the character map in representation theory for these rings. Recall that the classical character map is a map of commutative rings from the representation ring of G into class functions on G taking values in the complex numbers. Hopkins, Kuhn, and Ravenel constructed a generalization of the ring of class functions on G, called $\mathrm{Cl}_n(G, C_0)$. It is defined to be the ring of C_0-valued functions on the set of conjugacy classes of n-tuples of pairwise commuting p-power order elements of G. The ring C_0 is an important object in arithmetic geometry that appears here because of it carries the universal isomorphism of p-divisible groups between $(\mathbb{Q}_p/\mathbb{Z}_p)^n$ and the p-divisible group associated to the universal deformation of $\mathbb{F}/\kappa$. The character map is a map of commutative rings

$$\chi\colon E^0(BG) \to \mathrm{Cl}_n(G, C_0).$$

The ring $\mathrm{Cl}_n(G, C_0)$ comes equipped with a natural action of $\mathrm{GL}_n(\mathbb{Z}_p)$. Hopkins, Kuhn, and Ravenel prove that the character map induces an isomorphism

$$\mathbb{Q} \otimes E^0(BG) \cong \mathrm{Cl}_n(G, C_0)^{\mathrm{GL}_n(\mathbb{Z}_p)}.$$

Thus $\mathbb{Q} \otimes E^0(BG)$ admits a purely algebraic description.

The E_∞-ring structure on E gives rise to power operations. For each $m \geq 0$, these are multiplicative non-additive natural transformations

$$P_m\colon E^0(X) \to E^0(X \times B\Sigma_m).$$

When $X = BG$, one might hope for an operation on generalized class functions $\mathrm{Cl}_n(G, C_0) \to \mathrm{Cl}_n(G \times \Sigma_m, C_0)$ compatible with P_m through the character map.

One of the main goals of this article is to give a formula for such a map. To describe the formula, we need to introduce some more notation. Let $\mathbb{L} = \mathbb{Z}_p^n$ and let $\mathbb{T} = \mathbb{L}^*$, the Pontryagin dual of $\mathbb{L}$, so that $\mathbb{T}$ is isomorphic to $(\mathbb{Q}_p/\mathbb{Z}_p)^n$. Let $\mathrm{Aut}(\mathbb{T})$ be the group of automorphisms of $\mathbb{T}$ and let $\mathrm{Isog}(\mathbb{T})$ be the monoid of endoisogenies (endomorphisms with finite kernel) of $\mathbb{T}$. The group $\mathrm{Aut}(\mathbb{T})$ is isomorphic to $\mathrm{GL}_n(\mathbb{Z}_p)$. There is a right action of $\mathrm{Isog}(\mathbb{T})$ by ring maps on C_0. Given a finite group G, let $\mathrm{Hom}(\mathbb{L}, G)$ be the set of continuous homomorphisms from $\mathbb{L}$ to G. This set is naturally in bijective correspondence with n-tuples of pairwise commuting p-power order elements in G. The group G acts $\mathrm{Hom}(\mathbb{L}, G)$ by conjugation and we will write $\mathrm{Hom}(\mathbb{L}, G)/\sim$ for the set of conjugacy classes. The ring of generalized class functions $\mathrm{Cl}_n(G, C_0)$ is just the ring of C_0 valued functions on $\mathrm{Hom}(\mathbb{L}, G)/\sim$. There is a canonical bijection

$$\mathrm{Hom}(\mathbb{L}, \Sigma_m)/\sim \;\cong \mathrm{Sum}_m(\mathbb{T}),$$

where

$$\mathrm{Sum}_m(\mathbb{T}) = \{\oplus_i H_i | H_i \subset \mathbb{T} \text{ and } \sum_i |H_i| = m\},$$

the set of formal sums of subgroups of $\mathbb{T}$ such that the sum of the orders is precisely m. Finally let $\mathrm{Sub}(\mathbb{T})$ be the set of finite subgroups of $\mathbb{T}$ and let $\pi\colon \mathrm{Isog}(\mathbb{T}) \to \mathrm{Sub}(\mathbb{T})$ be the map sending an isogeny to its kernel. Choose a section ϕ of π so that, for $H \subset \mathbb{T}$, ϕ_H is an isogeny of $\mathbb{T}$ with kernel H.

Now we are prepared to describe the formula for the "power operation" on class functions

$$P_m^\phi\colon \mathrm{Cl}_n(G, C_0) \to \mathrm{Cl}_n(G \times \Sigma_m, C_0).$$

This formula depends on our choice of section ϕ. Because of the canonical bijection described above, an element in $\mathrm{Hom}(\mathbb{L}, G \times \Sigma_m)/\sim$ may be described as a pair $([\alpha], \oplus_i H_i)$, where $\alpha\colon \mathbb{L} \to G$. Given $f \in \mathrm{Cl}_n(G, C_0)$, we define

$$P_m^\phi(f)([\alpha], \oplus_i H_i) = \prod_i (f([\alpha\phi_{H_i}^*])\phi_{H_i}) \in C_0,$$

where $\phi_{H_i}^*\colon \mathbb{L} \to \mathbb{L}$ is the Pontryagin dual of ϕ_{H_i} and we are making use of the right action of $\mathrm{Isog}(\mathbb{T})$ on C_0. Using this construction, we prove the following theorem:

Theorem 21.1.1. *For any section ϕ of $\pi\colon \mathrm{Isog}(\mathbb{T}) \to \mathrm{Sub}(\mathbb{T})$, there is a commutative diagram*

$$\begin{array}{ccc} E^0(BG) & \xrightarrow{P_m} & E^0(BG \times B\Sigma_m) \\ \chi \downarrow & & \downarrow \chi \\ \mathrm{Cl}_n(G, C_0) & \xrightarrow{P_m^\phi} & \mathrm{Cl}_n(G \times \Sigma_m, C_0), \end{array}$$

where the vertical arrows are the Hopkins-Kuhn-Ravenel character map.

Several ingredients go into the proof of this theorem. The ring C_0 is the rationalization of the Drinfeld ring at infinite level. It is closely related to a certain moduli problem over Lubin-Tate space. The first ingredient is an understanding of the symmetries of this moduli problem. These symmetries play an important role in the local Langlands correspondence, do not go into the proof of the Goerss-Hopkins-Miller theorem, and show up here because of the close relationship between power operations and isogenies of formal groups. The second ingredient is a result of Ando, Hopkins, and Strickland, which gives an algebro-geometric description of a simplification of P_m applied to abelian groups. The third ingredient is the fact that, rationally, the E-cohomology of finite groups can be detected by the E-cohomology of the abelian subgroups. This is a consequence of Hopkins, Kuhn, Ravenel character theory.

Recall that $\mathrm{GL}_n(\mathbb{Z}_p)$ acts on $\mathrm{Cl}_n(G, C_0)$. We prove that the choice of section ϕ disappears after restricting to the $\mathrm{GL}_n(\mathbb{Z}_p)$-invariant class functions.

Theorem 21.1.2. *For each section ϕ of $\pi\colon \mathrm{Isog}(\mathbb{T}) \to \mathrm{Sub}(\mathbb{T})$, the power operation P_m^ϕ sends $\mathrm{GL}_n(\mathbb{Z}_p)$-invariant class functions to $\mathrm{GL}_n(\mathbb{Z}_p)$-invariant class functions and the resulting operation*

$$\mathrm{Cl}_n(G, C_0)^{\mathrm{GL}_n(\mathbb{Z}_p)} \to \mathrm{Cl}_n(G \times \Sigma_m, C_0)^{\mathrm{GL}_n(\mathbb{Z}_p)}$$

is independent of the choice of ϕ.

Both of these results are simplifications of the main results of [5]. By proving simpler statements, we are able to bypass work understanding conjugacy classes of n-tuples of commuting elements in wreath products and also the application of the generalization of Strickland's theorem in [18]. This simplifies the argument and allows us to make use of more classical results in stable homotopy theory thereby easing the background required of the reader.

A further goal of this article is to describe an action of the stabilizer group on generalized class functions that is compatible with the action of the stabilizer group on $E^0(BG)$ through the character map. There is a natural action of $\mathrm{Aut}(\mathbb{F}/\kappa)$ on C_0 and we show that the diagonal action of $\mathrm{Aut}(\mathbb{F}/\kappa)$ on $\mathrm{Cl}_n(G, C_0)$ makes the character map into an $\mathrm{Aut}(\mathbb{F}/\kappa)$-equivariant map. Further, we show that the diagonal action of the stabilizer group on $\mathrm{Cl}_n(G, C_0)$ commutes with the power operations P_m^ϕ exhibiting an algebraic analogue of the fact that $\mathrm{Aut}(\mathbb{F}/\kappa)$ acts on E by E_∞-ring maps.

Acknowledgements

Most of this article is based on [5], joint work with Tobias Barthel. My understanding of this material is a result of our work together and it is a pleasure to thank him for being a great collaborator. I learned much of the material in the first several sections from Charles Rezk. His work has continued to provide inspiration and guidance. I also offer my sincere thanks to Peter Bonventre, Haynes Miller, and Tomer Schlank for their helpful comments and careful reading of the text.

21.2 Notation and conventions

We will use Hom to denote the set of morphisms between two objects in a category. Sometimes the set of morphisms may have extra structure and, in that case, we will feel free to use it. We will capitalize the first letter, such as in Hom, Level, and Sub, to denote functors. Given a (possibly formal) scheme X, we will write $\mathcal{O}_X$ to denote the ring of functions on X. Given a complete local ring $(R, \mathfrak{m})$, we will write $\mathrm{Spf}(R)$ to denote the functor $\mathrm{Hom}(R, -)$ from the category of complete local rings to sets corepresented by R. Thus $\mathcal{O}_{\mathrm{Spf}(R)} \cong R$.

21.3 Lubin-Tate theory

In this section we recall the Lubin-Tate moduli problem and describe several well-behaved moduli problems that live over the Lubin-Tate moduli problem.

We begin with a brief overview of the theory of (1-dimensional, commutative) formal groups over complete local rings. Let CompRings be the category of complete local rings. Objects in this category are local commutative rings $(R, \mathfrak{m})$ such that $R \cong \lim_k R/\mathfrak{m}^k$ and maps are local homomorphisms of rings. We will denote the quotient map $R \to R/\mathfrak{m}$ by π. Note that if κ is a field, then $(\kappa, (0))$ is a complete local ring. Given a complete local ring $(R, \mathfrak{m})$, let $\mathrm{CompRings}_{R/}$ be the category of complete local R-algebras. The formal affine line over R,

$$\hat{\mathbb{A}}^1_R \colon \mathrm{CompRings}_{R/} \to \mathrm{Set}_*,$$

is the functor to pointed sets corepresented by the complete local R-algebra $(R[\![x]\!], \mathfrak{m} + (x))$. Thus, for a complete local R-algebra $(S, \mathfrak{n})$,

$$\hat{\mathbb{A}}^1_R(S, \mathfrak{n}) = \mathrm{Hom}_{R\text{-alg}}(R[\![x]\!], S) \cong \mathfrak{n},$$

where $\mathrm{Hom}_{R\text{-alg}}(-,-)$ denotes the set of morphisms in $\mathrm{CompRings}_{R/}$ and the last isomorphism is the map that takes a morphism to the image of x in S.

A formal group $\mathbb{G}$ over R is a functor

$$\mathbb{G} \colon \mathrm{CompRings}_{R/} \to \text{Abelian Groups}$$

that is abstractly isomorphic to $\hat{\mathbb{A}}^1_R$ when viewed as a functor to pointed sets. A homomorphism of formal groups is a natural transformation of functors.

A coordinate on a formal group $\mathbb{G}$ over R is a choice of isomorphism $\mathbb{G} \xrightarrow{\cong} \hat{\mathbb{A}}^1_R$. Given a coordinate, the multiplication map

$$\mu \colon \mathbb{G} \times \mathbb{G} \to \mathbb{G}$$

gives rise to a map

$$\mu^*\colon R[\![x]\!] \to R[\![x_1, x_2]\!]$$

and $\mu^*(x) = x_1 +_{\mathbb{G}} x_2$ is the formal group law associated to the coordinate. In other words, given a coordinate, we have an identification of the abelian group $\mathbb{G}(S, \mathfrak{n})$ with $(\mathfrak{n}, +_{\mathbb{G}})$, the maximal ideal of S with addition given by the formal group law associated to the coordinate.

Example 21.3.1. The additive formal group $\mathbb{G}_a$ over R associates to a complete local R-algebra $(S, \mathfrak{n})$ the abelian group $\mathfrak{n}$ with its usual additive structure. An isomorphism $\mathbb{G}_a \cong \hat{\mathbb{A}}^1_R$ is given by taking the underlying pointed set and the associated formal group law is

$$x_1 +_{\mathbb{G}_a} x_2 = x_1 + x_2.$$

Example 21.3.2. The multiplicative formal group $\mathbb{G}_m$ over R associates to a complete local R-algebra $(S, \mathfrak{n})$ the set of units of S of the form $1 + n$ for $n \in \mathfrak{n}$ with its usual multiplicative structure. An isomorphism $\mathbb{G}_m \cong \hat{\mathbb{A}}^1_R$ is given by sending $1 + n$ to n and the associated formal group law is

$$x_1 +_{\mathbb{G}_m} x_2 = x_1 + x_2 + x_1 x_2.$$

For $i \in \mathbb{N}$, we define the i-series of a formal group law $x_1 +_{\mathbb{G}} x_2$ to be the power series

$$[i]_{\mathbb{G}}(x) = \overbrace{x +_{\mathbb{G}} \ldots +_{\mathbb{G}} x}^{i \text{ terms}}.$$

Given a formal group over a complete local ring R and a map of complete local rings $j\colon R \to S$, we will write $j^*\mathbb{G}$ for the restriction of the functor $\mathbb{G}$ to the category of complete local S-algebras.

Let κ be a perfect field of characteristic p. "Perfect" means that the Frobenius $\sigma\colon \kappa \to \kappa$ is an isomorphism of fields. A formal group $\mathbb{F}$ over κ is said to be of height n if, after choosing a coordinate, the associated formal group law $x +_{\mathbb{F}} y$ has the property that the p-series $[p]_{\mathbb{F}}(x)$ has first term ux^{p^n}, where $u \in \kappa^\times$. The additive formal group over κ has height ∞ and the multiplicative formal group over κ has height 1.

Fix a formal group $\mathbb{F}$ over κ of height n. All of the constructions in this section will depend on this choice. Associated to $\mathbb{F}/\kappa$ is a moduli problem (i.e. a functor to groupoids)

$$\mathrm{LT}\colon \mathrm{CompRings} \to \mathrm{Groupoids}$$

studied by Lubin and Tate in [13]. We will take a coordinate-free approach to this moduli problem that is closely related to the description in [20, Section 6]. The functor LT takes values in groupoids. Next we will describe the objects and morphisms of these groupoids.

A deformation of $\mathbb{F}/\kappa$ to a complete local ring $(R, \mathfrak{m})$ is a triple of data

$$(\mathbb{G}/R, i\colon \kappa \to R/\mathfrak{m}, \tau : \pi^*\mathbb{G} \xrightarrow{\cong} i^*\mathbb{F}),$$

where $\mathbb{G}$ is a formal group over R, $i\colon \kappa \to R/\mathfrak{m}$ is a map of fields, and τ is an isomorphism of formal groups over $R/\mathfrak{m}$. We will often use shorthand and refer to a deformation as a triple $(\mathbb{G}, i, \tau)$. The deformations of $\mathbb{F}/\kappa$ to $(R, \mathfrak{m})$ form the objects of $\mathrm{LT}(R, \mathfrak{m})$. The morphisms of $\mathrm{LT}(R,\mathfrak{m})$ are known as $\star$-isomorphisms. A $\star$-isomorphism δ between two deformations $(\mathbb{G},i,\tau)$ and $(\mathbb{G}',i',\tau')$ only exists when $i=i'$ and then it is an isomorphism of formal groups $\delta\colon \mathbb{G}\xrightarrow{\cong}\mathbb{G}'$ compatible with the isomorphisms τ and τ' in the sense that the square

$$\begin{array}{ccc} \pi^*\mathbb{G} & \xrightarrow{\tau} & i^*\mathbb{F} \\ \downarrow{\scriptstyle \pi^*\delta} & & \downarrow{\scriptstyle \mathrm{id}} \\ \pi^*\mathbb{G}' & \xrightarrow{\tau'} & i^*\mathbb{F} \end{array}$$

commutes.

Given a map of complete local rings $j\colon (R,\mathfrak{m})\to(S,\mathfrak{n})$, there is an induced map of groupoids $\mathrm{LT}(R,\mathfrak{m})\to\mathrm{LT}(S,\mathfrak{n})$ defined by sending

$$(\mathbb{G},i,\tau)\mapsto(j^*\mathbb{G},\kappa\xrightarrow{i}R/\mathfrak{m}\xrightarrow{j/\mathfrak{m}}S/\mathfrak{n},(j/\mathfrak{m})^*\tau).$$

This makes LT into a functor to the category of groupoids. Let $W(\kappa)$ be the ring of p-typical Witt vectors on κ. This is a torsion-free complete local ring of characteristic 0 with the property that $W(\kappa)/p\cong\kappa$. When $\kappa=\mathbb{F}_p$, $W(\kappa)=\mathbb{Z}_p$ (the p-adic integers). Now we are prepared to state the Lubin-Tate theorem.

Theorem 21.3.3. [13] *The functor* LT *takes values in fine groupoids (i.e. groupoids whose objects have no nontrivial automorphisms) and is corepresented by a complete local ring* $\mathcal{O}_{\mathrm{LT}}$ *non-canonically isomorphic to* $W(\kappa)[\![u_1,\ldots,u_{n-1}]\!]$ *called the Lubin-Tate ring.*

This theorem was proved in [13]. The homotopy theorist may be interested in reading [17, Chapters 4 and 5] or [14, Lecture 21].

Part of the content of the Lubin-Tate theorem is that, for any complete local ring (R,Mod), the groupoid $\mathrm{LT}(R,\mathsf{Mod})$ is fine and thus equivalent to the functor to sets obtained by taking the set of isomorphism classes of objects in each groupoid. It follows that the identity map on $\mathcal{O}_{\mathrm{LT}}$ classifies the universal deformation $(\mathrm{G}_u,\mathrm{id}_\kappa,\mathrm{id}_\mathbb{F})$. This means that, given any deformation $(\mathbb{G},i,\tau)$ over a complete local ring R there exists a map $j\colon\mathcal{O}_{\mathrm{LT}}\to R$ such that $(j^*\mathrm{G}_u,j/\mathfrak{m},\mathrm{id}_{(j/\mathfrak{m})^*\mathbb{F}})$ is $\star$-isomorphic to $(\mathbb{G},i,\tau)$.

It is possible to add various bells and whistles to the Lubin-Tate functor. We will now explain three ways of doing this, in each case we add some additional data to each deformation. First, we will add the choice of a homomorphism from a fixed abelian group to $\mathbb{G}$ to each deformation. Second, we will add the choice of a "level structure" from a fixed abelian group to $\mathbb{G}$. Third, we will add the choice of a subgroup scheme of $\mathbb{G}$ of a fixed order.

We construct a modification of the Lubin-Tate functor depending on a finite abelian group A. If $\mathbb{G}$ is a formal group over a complete local ring $(R,\mathfrak{m})$, then a homomorphism $A\to\mathbb{G}$ over R is a map of abelian groups $A\to\mathbb{G}(R)$. We will write $\mathrm{Hom}(A,\mathbb{G}(R))$ for this abelian group. Define

$$\mathrm{Hom}(A,\mathrm{G}_u)\colon \mathrm{CompRings}\to\mathrm{Groupoids}$$

to be the functor with $\mathrm{Hom}(A,\mathrm{G}_u)(R,\mathfrak{m})$ equal to the groupoid with objects triples $(f\colon A\to\mathbb{G},i,\tau)$, where $(\mathbb{G},i,\tau)$ is a deformation of $\mathbb{F}/\kappa$ to $(R,\mathfrak{m})$ and f is a homomorphism from A to $\mathbb{G}$ over R, and morphisms $\star$-isomorphisms that commute with this structure. In other words, a $\star$-isomorphism from $(f\colon A \to \mathbb{G}, i, \tau)$ to $(f'\colon A \to \mathbb{G}', i, \tau')$ is a $\star$-isomorphism δ from $(\mathbb{G}, i, \tau)$ to $(\mathbb{G}', i, \tau')$ such that the diagram

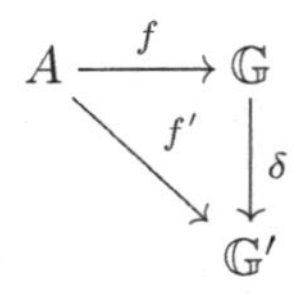

commutes.

The functor $\operatorname{Hom}(A, \mathrm{G}_u)$ lives over LT. There is a forgetful natural transformation $\operatorname{Hom}(A, \mathrm{G}_u) \to \mathrm{LT}$ given by sending an object $(f\colon A \to \mathbb{G}, i, \tau)$ to $(\mathbb{G}, i, \tau)$. It follows from the definition that the forgetful functor $\operatorname{Hom}(A, \mathrm{G}_u)(R, \mathfrak{m}) \to \mathrm{LT}(R, \mathfrak{m})$ is faithful. Thus $\operatorname{Hom}(A, \mathrm{G}_u)(R, \mathfrak{m})$ is a fine groupoid.

Corollary 21.3.4. *The functor* $\operatorname{Hom}(A, \mathrm{G}_u)$ *is corepresentable by a complete local ring* $\mathcal{O}_{\operatorname{Hom}(A,\mathrm{G}_u)}$ *that is finitely generated and free as a module over* $\mathcal{O}_{\mathrm{LT}}$.

Proof. The idea of the proof is that we will first construct an $\mathcal{O}_{\mathrm{LT}}$-algebra with the correct algebraic properties and then use Theorem 21.3.3 to show that the underlying complete local ring corepresents the moduli problem $\operatorname{Hom}(A, \mathrm{G}_u)$.

Recall that $\mathrm{CompRings}_{\mathcal{O}_{\mathrm{LT}}/}$ is the category of complete local $\mathcal{O}_{\mathrm{LT}}$-algebras. Consider the functor

$$\mathrm{G}_u^A\colon \mathrm{CompRings}_{\mathcal{O}_{\mathrm{LT}}/} \to \text{Abelian Groups}$$

that sends a complete local $\mathcal{O}_{\mathrm{LT}}$-algebra $j\colon \mathcal{O}_{\mathrm{LT}} \to R$ to the abelian group $\operatorname{Hom}(A, j^* \mathrm{G}_u(R))$, which is isomorphic to $\operatorname{Hom}(A, \mathrm{G}_u(R))$. When $A = C_{p^k}$, this functor is corepresented by $\mathcal{O}_{\mathrm{LT}}[\![x]\!]/[p^k]_{\mathrm{G}_u}(x)$, where $[p^k]_{\mathrm{G}_u}(x)$ is the p^k-series for the formal group law associated to a choice of coordinate on G_u. This is a complete local ring with maximal ideal $\mathfrak{m} + (x)$, where $\mathfrak{m}$ is the maximal ideal of $\mathcal{O}_{\mathrm{LT}}$, and it follows from the Weierstrass preparation theorem that it is a free module of rank p^{kn}. If $A \cong \prod_i C_{p^{k_i}}$, then G_u^A is corepresented by a tensor product of these sorts of $\mathcal{O}_{\mathrm{LT}}$-algebras and thus is finitely generated and free as an $\mathcal{O}_{\mathrm{LT}}$-module. Thus $\mathcal{O}_{\mathrm{G}_u^A}$ is a complete local $\mathcal{O}_{\mathrm{LT}}$-algebra that is finitely generated and free as an $\mathcal{O}_{\mathrm{LT}}$-module.

We will now show that the functor corepresented by the underlying complete local ring

$$\operatorname{Hom}(\mathcal{O}_{\mathrm{G}_u^A}, -)\colon \mathrm{CompRings} \to \mathrm{Set}$$

is naturally isomorphic to $\operatorname{Hom}(A, \mathrm{G}_u)$.

Let $(R, \mathfrak{m})$ be a complete local ring. A map of complete local rings $\mathcal{O}_{\mathrm{G}_u^A} \to R$ is equivalent to the choice of an $\mathcal{O}_{\mathrm{LT}}$-algebra structure on R, $j\colon \mathcal{O}_{\mathrm{LT}} \to R$, and a homomorphism $A \to j^* \mathrm{G}_u$. Using this data, we may form the deformation

$$(A \to j^* \mathrm{G}_u, j/\mathfrak{m}, \mathrm{id}_{(j/\mathfrak{m})^* \mathbb{F}}).$$

Let $(f\colon A \to \mathbb{G}, i, \tau)$ be an object in the groupoid $\mathrm{Hom}(A, \mathrm{G}_u)(R, \mathfrak{m})$. By Theorem 21.3.3, there is a map $j\colon \mathcal{O}_{\mathrm{LT}} \to R$ and a unique $\star$-isomorphism

$$(\mathbb{G}, i, \tau) \cong (j^* \mathrm{G}_u, j/\mathfrak{m}, \mathrm{id}_{(j/\mathfrak{m})^*\mathbb{F}})$$

which includes the data of an isomorphism of formal groups $t\colon \mathbb{G} \cong j^* \mathrm{G}_u$. Thus the deformation (f, i, τ) is (necessarily uniquely) $\star$-isomorphic to the deformation

$$(A \xrightarrow{f} \mathbb{G} \xrightarrow{t} j^* \mathrm{G}_u, j/\mathfrak{m}, \mathrm{id}_{(j/\mathfrak{m})^*\mathbb{F}}).$$

Now the data of j and $A \to j^* \mathrm{G}_u$ is equivalent to the data of a map $\mathcal{O}_{\mathrm{G}_u^A} \to R$. These two constructions are clearly inverse to each other. □

Let A be a finite abelian p-group. Level structures on formal groups were introduced by Drinfeld in [8, Section 4]. A level structure $f\colon A \to \mathbb{G}$ is a homomorphism from A to $\mathbb{G}$ over R with a certain property. There is a divisor on $\mathbb{G}$ associated to every homomorphism from A to $\mathbb{G}$ and the idea is that a level structure is a homomorphism over R for which the associated divisor is a subgroup scheme of $\mathbb{G}$. More explicitly, fix a coordinate on $\mathbb{G}$, which provides an isomorphism $\mathcal{O}_{\mathbb{G}} \cong R[\![x]\!]$. For $f\colon A \to \mathbb{G}(R)$, let $g_f(x) = \prod_{a\in A}(x - f(a))$. The divisor associated to f is $\mathrm{Spf}(R[\![x]\!]/(g_f(x)))$ and f is a level structure if this is a subgroup scheme of $\mathbb{G}$ and the rank of A is less than or equal to n (the height). It is useful to think about level structures as injective maps, even though they are not necessarily injective in any sense.

We define

$$\mathrm{Level}(A, \mathrm{G}_u)\colon \mathrm{CompRings} \to \mathrm{Groupoids}$$

to be the functor that assigns to a complete local ring $(R, \mathfrak{m})$ the groupoid $\mathrm{Level}(A, \mathrm{G}_u)(R, \mathfrak{m})$ with objects triples $(l\colon A \to \mathbb{G}, i, \tau)$, where l is a level structure, and morphisms $\star$-isomorphisms that commute with the level structures in the same way as in the definition of $\mathrm{Hom}(A, \mathrm{G}_u)(R, \mathfrak{m})$. Again, there is a faithful forgetful natural transformation $\mathrm{Level}(A, \mathrm{G}_u) \to \mathrm{LT}$. The proof of the next corollary is along the same lines as the proof of Corollary 21.3.4 with some extra work needed to show that the complete local ring is a domain.

Corollary 21.3.5. [8, Proposition 4.3] *The functor* $\mathrm{Level}(A, \mathrm{G}_u)$ *is corepresentable by a complete local domain* $\mathcal{O}_{\mathrm{Level}(A,\mathrm{G}_u)}$ *that is finitely generated and free as a module over* $\mathcal{O}_{\mathrm{LT}}$.

A particular example of this functor will play an important role. Let $\mathbb{L}$ be a free $\mathbb{Z}_p$-module of rank n and let $\mathbb{T} = \mathbb{L}^*$, the Pontryagin dual of $\mathbb{L}$. Thus $\mathbb{L}$ is abstractly isomorphic to $\mathbb{Z}_p^{\times n}$ and $\mathbb{T}$ is abstractly isomorphic to $(\mathbb{Q}_p/\mathbb{Z}_p)^{\times n}$. We use the notation $\mathbb{L}$ to indicate that we have fixed a (p-adic) lattice and $\mathbb{T}$ for the Pontryagin dual indicate that it is (p-adic) torus. Let $\mathbb{T}[p^k]$ be the p^k-torsion in the torus and let

$$\mathbb{L}_{p^k} = p^k\,\mathbb{L} \subset \mathbb{L}.$$

By construction, there is a canonical isomorphism $\mathbb{T}[p^k] \cong (\mathbb{L}/\mathbb{L}_{p^k})^*$. Both $\mathbb{T}[p^k]$ and $\mathbb{L}/\mathbb{L}_{p^k}$ are abstractly isomorphic to $(\mathbb{Z}/p^k)^{\times n}$. The functors $\mathrm{Level}(\mathbb{T}[p^k], \mathrm{G}_u)$ will be particularly important.

Finally we introduce one last moduli problem over LT. Define

$$\mathrm{Sub}_{p^k}(\mathbb{G}_u)\colon \mathrm{CompRings} \to \mathrm{Groupoids}$$

to be the functor that assigns to a complete local ring $(R, \mathfrak{m})$ the groupoid $\mathrm{Sub}_{p^k}(\mathbb{G}_u)(R, \mathfrak{m})$ with objects triples $(H \subset \mathbb{G}, i, \tau)$, where H is a subgroup scheme of $\mathbb{G}$ of order p^k. Morphisms are $\star$-isomorphisms that send the chosen subgroup to the chosen subgroup. Just as with the other moduli problems, there is a faithful forgetful natural transformation $\mathrm{Sub}_{p^k}(\mathbb{G}_u) \to \mathrm{LT}$. Once again, the proof of the next corollary is along the same lines as the proof of Corollary 21.3.4 with some extra work needed to show that the complete local ring is finitely generated and free as an $\mathcal{O}_{\mathrm{LT}}$-module.

Corollary 21.3.6. [20, Theorem 42] *The moduli problem* $\mathrm{Sub}_{p^k}(\mathbb{G}_u)$ *is corepresented by a complete local ring* $\mathcal{O}_{\mathrm{Sub}_{p^k}(\mathbb{G}_u)}$ *that is finitely generated and free as a module over* $\mathcal{O}_{\mathrm{LT}}$.

21.4 Symmetries of certain moduli problems over Lubin-Tate space

Besides being corepresentable, the moduli problems described in the last section are highly symmetric. We begin by describing the action of the stabilizer group on each of the moduli problems and then we will explain how to take the quotient of a deformation by a finite subgroup and how a certain limit of these moduli problems picks up extra symmetry. A reference that includes a concise exposition of much of the material in this section is [7, Section 1].

The group $\mathrm{Aut}(\mathbb{F}/\kappa)$ of automorphisms of $\mathbb{F}$ over κ is known as the (small) Morava stabilizer group. This group acts on the left on all of the moduli problems that we have described in a very simple way. The action of $s \in \mathrm{Aut}(\mathbb{F}/\kappa)$ is given by

$$s \cdot (\mathbb{G}, i, \tau) = (\mathbb{G}, i, (i^*s)\tau), \tag{21.4.1}$$

where $i^*s\colon i^*\mathbb{F} \to i^*\mathbb{F}$ is the induced isomorphism. This is well-defined: if δ is a $\star$-isomorphism from $(\mathbb{G}, i, \tau)$ to $(\mathbb{G}', i, \tau')$ then f is still a $\star$-isomorphism between $s \cdot (\mathbb{G}, i, \tau)$ and $s \cdot (\mathbb{G}', i, \tau')$ since

$$\begin{array}{ccccc} \pi^*\mathbb{G} & \xrightarrow{\tau} & i^*\mathbb{F} & \xrightarrow{i^*s} & i^*\mathbb{F} \\ \downarrow{\scriptstyle \pi^*\delta} & & \downarrow{\scriptstyle =} & & \downarrow{\scriptstyle =} \\ \pi^*\mathbb{G}' & \xrightarrow{\tau'} & i^*\mathbb{F} & \xrightarrow{i^*s} & i^*\mathbb{F} \end{array}$$

commutes. The actions of $\mathrm{Aut}(\mathbb{F}/\kappa)$ on $\mathrm{Hom}(A, \mathbb{G}_u)$, $\mathrm{Level}(A, \mathbb{G}_u)$, and $\mathrm{Sub}_{p^k}(\mathbb{G}_u)$ are defined similarly.

There is a right action of $\mathrm{Aut}(A)$ on both $\mathrm{Hom}(A, \mathbb{G}_u)$ and $\mathrm{Level}(A, \mathbb{G}_u)$ given by precomposition and a left action of $\mathrm{Aut}(A)$ on both $\mathrm{Hom}(A, \mathbb{G}_u)$ and $\mathrm{Level}(A, \mathbb{G}_u)$ given by precomposition with the inverse. Since neither of these actions affects $\mathbb{G}$, i, or τ, the corresponding actions of $\mathrm{Aut}(A)$ on $\mathcal{O}_{\mathrm{Hom}(A,\mathbb{G}_u)}$ and $\mathcal{O}_{\mathrm{Level}(A,\mathbb{G}_u)}$ are by $\mathcal{O}_{\mathrm{LT}}$-algebra maps. We mention the left action by precomposition with the inverse because that is the action that we will generalize and that will play the most important role in our constructions. Specializing to $A = \mathbb{T}[p^k]$, we see that there is a left action of $\mathrm{Aut}(\mathbb{T}[p^k]) \cong \mathrm{GL}_n(\mathbb{Z}/p^k)$ on $\mathrm{Level}(\mathbb{T}[p^k], \mathrm{G}_u)$ given by precomposition with the inverse.

There is a forgetful natural transformation $\mathrm{Level}(\mathbb{T}[p^k], \mathrm{G}_u) \to \mathrm{Level}(\mathbb{T}[p^{k-1}], \mathrm{G}_u)$ induced by the map sending a level structure $\mathbb{T}[p^k] \to \mathbb{G}$ to the level structure given by the composite $\mathbb{T}[p^{k-1}] \subset \mathbb{T}[p^k] \to \mathbb{G}$. Let

$$\mathrm{Level}(\mathbb{T}, \mathrm{G}_u) = \lim_k \mathrm{Level}(\mathbb{T}[p^k], \mathrm{G}_u).$$

We will refer to the ring of functions $\mathcal{O}_{\mathrm{Level}(\mathbb{T},\mathrm{G}_u)} = \mathrm{colim}_k\, \mathcal{O}_{\mathrm{Level}(\mathbb{T}[p^k],\mathrm{G}_u)}$ as the Drinfeld ring. The groupoid $\mathrm{Level}(\mathbb{T}, \mathrm{G}_u)(R, \mathfrak{m})$ can be thought of as follows: It has objects consisting of tuples $(l\colon \mathbb{T} \to \mathbb{G}, i, \tau)$ and morphisms given by $\star$-isomorphisms, where l is a map from $\mathbb{T}$ to $\mathbb{G}$ over R such that, for all $k \geq 0$, the induced map $\mathbb{T}[p^k] \to \mathbb{G}$ is a level structure. If $H \subset \mathbb{T}$ is a subgroup of order p^k, then H determines a map $\mathrm{Level}(\mathbb{T}, \mathrm{G}_u) \to \mathrm{Sub}_{p^k}(\mathrm{G}_u)$ defined by

$$(l, i, \tau) \mapsto (l(H) \subset \mathbb{G}, i, \tau).$$

There is a left action of $\mathrm{Aut}(\mathbb{T})$ on $\mathrm{Level}(\mathbb{T}, \mathrm{G}_u)$ given by precomposition with the inverse. It turns out that the action of $\mathrm{Aut}(\mathbb{T})$ on $\mathrm{Level}(\mathbb{T}, \mathrm{G}_u)$ extends to an action of $\mathrm{Isog}(\mathbb{T})$, the monoid of endoisogenies of $\mathbb{T}$, a much larger object! An isogeny $\mathbb{T} \to \mathbb{T}$ is a homomorphism with finite kernel. To describe the action of $\mathrm{Isog}(\mathbb{T})$, we first need to describe the quotient of a deformation by a finite subgroup.

Let $\sigma\colon \mathrm{Spec}(\kappa) \to \mathrm{Spec}(\kappa)$ be the Frobenius endomorphism on κ and let $\sigma_{\mathbb{F}}\colon \mathbb{F} \to \mathbb{F}$ be the Frobenius endomorphism on the formal group $\mathbb{F}$. Let $\sigma^*\mathbb{F}$ be the pullback of $\mathbb{F}$ along σ. The relative Frobenius is an isogeny of formal groups of degree p

$$\mathrm{Frob}\colon \mathbb{F} \to \sigma^*\mathbb{F}.$$

It is constructed using the universal property of the pullback:

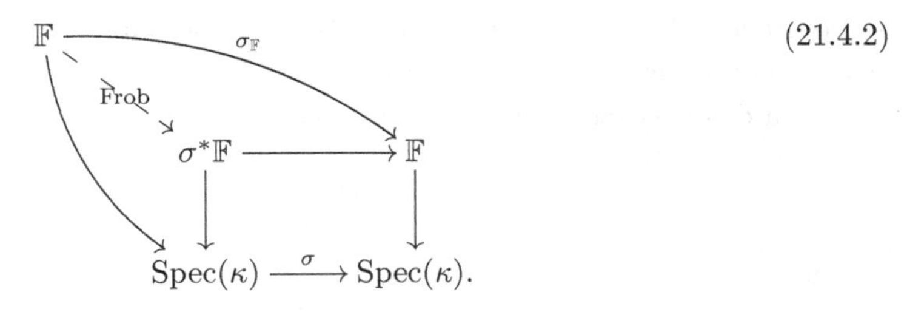

(21.4.2)

The formal group $\mathbb{F}$ lives over κ, so there is a canonical map $\mathbb{F} \to \mathrm{Spec}(\kappa)$. This fits together with the $\sigma_{\mathbb{F}}$ to produce the map $\mathrm{Frob}\colon \mathbb{F} \to \sigma^*\mathbb{F}$. Using the kth power of the Frobenius σ^k and $\sigma_{\mathbb{F}}^k$ in the diagram above instead of σ and $\sigma_{\mathbb{F}}$, we construct the kth relative Frobenius $\mathrm{Frob}^k\colon \mathbb{F} \to (\sigma^k)^*\mathbb{F}$. The kernel of the kth relative Frobenius is a subgroup scheme of $\mathbb{F}$ of order p^k. A formal group over a field of characteristic p has a unique subgroup scheme of order p^k for each $k \geq 0$. Thus $\mathrm{Ker}(\mathrm{Frob}^k) \subset \mathbb{F}$ is the unique subgroup scheme of $\mathbb{F}$ of order p^k.

Let $(\mathbb{G}, i, \tau)$ be a deformation of $\mathbb{F}/\kappa$ to a complete local ring R and let $H \subset \mathbb{G}$ be a subgroup scheme of order p^k. The quotient of $\mathbb{G}$ by H, $\mathbb{G}/H$, was first defined in [12, Section 1]. In [20, Section 5], Strickland shows that the quotient is the formal group

$$\mathbb{G}/H = \operatorname{Coeq}(\mathbb{G} \times H \underset{\text{proj.}}{\overset{\text{act}}{\rightrightarrows}} \mathbb{G})$$

constructed as the coequalizer of the action map and the projection map.

There is a canonical way to extend the quotient formal group $\mathbb{G}/H$ to a deformation of $\mathbb{F}$ to R. Let $q\colon \mathbb{G} \to \mathbb{G}/H$ be the quotient map. Consider the following diagram of formal groups over R/m

$$\begin{array}{ccc} \pi^*\mathbb{G} & \xrightarrow{\tau} & i^*\mathbb{F} \\ \downarrow{\scriptstyle \pi^* q} & & \downarrow{\scriptstyle i^* \mathrm{Frob}^k} \\ \pi^*(\mathbb{G}/H) & \overset{\tau/H}{\dashrightarrow} & i^*(\sigma^k)^*\mathbb{F}. \end{array}$$

Since both $\operatorname{Ker}(\pi^* q) \subset \pi^*\mathbb{G}$ and $\operatorname{Ker}(i^* \mathrm{Frob}^k \circ \tau) \subset \pi^*\mathbb{G}$ have the same order, we must have $\operatorname{Ker}(\pi^* q) = \operatorname{Ker}(i^* \mathrm{Frob}^k \circ \tau)$. Thus, by the first isomorphism theorem for formal groups [12, Theorem 1.5], there is a unique isomorphism (the dashed arrow) making the diagram commutes. We will call this isomorphism τ/H. Thus, given the deformation $(\mathbb{G}, i, \tau)$ over R and $H \subset \mathbb{G}$, we may form the deformation

$$(\mathbb{G}/H, i \circ \sigma^k, \tau/H).$$

Now we will produce the promised left action of the monoid $\operatorname{Isog}(\mathbb{T})$ on $\operatorname{Level}(\mathbb{T}, \mathrm{G}_u)$. First note that $\operatorname{Isog}(\mathbb{T})$ really is much larger than $\operatorname{Aut}(\mathbb{T})$. Pontryagin duality produces an anti-isomorphism between $\operatorname{Aut}(\mathbb{T})$ and $\mathrm{GL}_n(\mathbb{Z}_p)$ and an anti-isomorphism between $\operatorname{Isog}(\mathbb{T})$ and the monoid of $n \times n$ matrices with coefficients in $\mathbb{Z}_p$ and non-zero determinant. Let $H \subset \mathbb{T}$ be a finite subgroup of order p^k, let $\phi_H \in \operatorname{Isog}(\mathbb{T})$ be an endoisogeny of $\mathbb{T}$ with kernel H, and let $q_H\colon \mathbb{T} \to \mathbb{T}/H$ be the quotient map. Recall that if $l\colon \mathbb{T}[p^k] \to \mathbb{G}$ is a level structure and $H \subset \mathbb{T}[p^k]$, then $l(H) \subset \mathbb{G}$ is a subgroup scheme of order $|H|$. We will abuse notation and refer to $l(H) \subset \mathbb{G}$ as H.

For a deformation equipped with level structure

$$(l\colon \mathbb{T} \to \mathbb{G}, i, \tau),$$

we set

$$\phi_H \cdot (l, i, \tau) = (l/H \circ \psi_H^{-1}, i \circ \sigma^k, \tau/H), \tag{21.4.3}$$

where $l/H\colon \mathbb{T}/H \to \mathbb{G}/H$ is the induced level structure and ψ_H is the unique isomorphism that makes the square

$$\begin{array}{ccc} \mathbb{T} & & \\ \downarrow{\scriptstyle q_H} & \searrow{\scriptstyle \phi_H} & \\ \mathbb{T}/H & \underset{\cong}{\xrightarrow{\psi_H}} & \mathbb{T} \end{array} \tag{21.4.4}$$

commute. We leave it to the reader to check that ϕ_H preserves $\star$-isomorphisms and to check that the resulting formula does in fact give a left action of $\operatorname{Isog}(\mathbb{T})$.

Proposition 21.4.1. *The formula of* (21.4.3) *defines a left action of* $\mathrm{Isog}(\mathbb{T})$ *on* $\mathrm{Level}(\mathbb{T}, \mathbb{G}_u)$.

One subtlety worth mentioning is that this action is weak: for $\phi, \phi' \in \mathrm{Isog}(\mathbb{T})$, there is a (necessarily unique) $\star$-isomorphism between $\phi \cdot \phi'(l, i, \tau)$ and $(\phi \circ \phi') \cdot (l, i, \tau)$. This induces a strict action on $\star$-isomorphism classes.

Restricting the above action to the case that $H = e$, the trivial subgroup, we see that q_e is the identity map and that $\psi_e = \phi_e$ so that

$$\phi_e \cdot (l, i, \tau) = (l \circ \phi_e^{-1}, i, \tau).$$

This shows that the action of $\mathrm{Isog}(\mathbb{T})$ on $\mathrm{Level}(\mathbb{T}, \mathbb{G}_u)$ extends the action of $\mathrm{Aut}(\mathbb{T})$ given by precomposition with the inverse. Since we have a left action of $\mathrm{Isog}(\mathbb{T})$ on $\mathrm{Level}(\mathbb{T}, \mathbb{G}_u)$, we have a right action of $\mathrm{Isog}(\mathbb{T})$ on $\mathcal{O}_{\mathrm{Level}(\mathbb{T}, \mathbb{G}_u)}$.

21.5 Morava E-theory

In this section we construct Morava E-theory using the Landweber exact functor theorem, we calculate the E-cohomology of finite abelian groups, we describe the Goerss-Hopkins-Miller theorem, and we describe the resulting power operations on E-cohomology.

Morava E-theory is a cohomology theory that is built using the Landweber exact functor theorem applied to the universal formal group over the Lubin-Tate ring associated to the height n formal group $\mathbb{F}/\kappa$. Theorem 21.3.3 states that there is a non-canonical isomorphism

$$\mathcal{O}_{\mathrm{LT}} \cong W(\kappa)[\![u_1, \ldots, u_{n-1}]\!].$$

It follows from [20, Section 6] that a coordinate can be chosen on the universal deformation formal group $\mathbb{G}_u$ such that the associated formal group law has the property that

$$[p]_{\mathbb{G}_u}(x) = u_i x^{p^i} \mod (p, u_1, \ldots u_{i-1}, x^{p^i+1}).$$

Since $(p, u_1, \ldots, u_{n-1})$ is a regular sequence in $W(\kappa)[\![u_1, \ldots, u_{n-1}]\!]$, this property allows the Landweber exact functor theorem to be applied producing a homology theory $E_*(-) = E(\mathbb{F}, \kappa)_*(-)$, called Morava E-theory, by the formula

$$E_*(X) = \mathcal{O}_{\mathrm{LT}} \otimes_{MUP_0} MUP_*(X),$$

where MUP is the 2-periodic complex cobordism spectrum $\bigvee_{k \in \mathbb{Z}} \Sigma^{2k} MU$. By construction, this spectrum has the property that $\pi_0 MUP \cong \oplus_{k \in \mathbb{Z}} \pi_{2k} MU$. Good introductions to the Landweber exact functor theorem can be found in [10] and [14, Lecture 16]. For a somewhat different perspective on Morava E-theory and a broader explanation of its role in chromatic homotopy theory, see Section 3 of the chapter by Tobias Barthel and Agnès Beaudry in this Handbook.

Example 21.5.1. When $(\mathbb{F}, \kappa) = (\mathbb{G}_m, \mathbb{F}_p)$, the formal multiplicative group over the field with p elements, then $\mathbb{G}_u = \mathbb{G}_m$ over $W(\kappa) = \mathbb{Z}_p$. The resulting cohomology theory is precisely p-adic K-theory.

There is a cohomology theory associated to this homology theory. On spaces equivalent to a finite CW-complex, it satisfies

$$E^*(X) = \mathcal{O}_{\mathrm{LT}} \otimes_{MUP_0} MUP^*(X).$$

From now on, we will always write E^0 rather than $\mathcal{O}_{\mathrm{LT}}$ for the Lubin-Tate ring. We will primarily be interested in spaces of the form BG for G a finite group and these are almost always not equivalent to a finite CW complex. To get a grip on the E-cohomology of these large spaces, we boot strap from the calculation of $E^*(BS^1)$ using the Atiyah-Hirzebruch spectral sequence, the Milnor sequence, and the Mittag-Leffler condition to control $\lim^1$ as described very clearly in both [15, Chapter 2.2] and [2, Section 2 of Part II].

Proposition 21.5.2. [2, Lemma 2.5 of Part II] *There is an isomorphism of E^*-algebras*

$$E^*(BS^1) \cong E^*[\![x]\!].$$

Hopkins, Kuhn, and Ravenel observe in [11, Lemma 5.7] that the Gysin sequence associated to the circle bundle $S^1 \to BC_{p^k} \to BS^1$ can be used to calculate $E^*(BC_{p^k})$ from Proposition 21.5.2.

Proposition 21.5.3. *The isomorphism of Proposition 21.5.2 induces an isomorphism of E^*-algebras*

$$E^*(BC_{p^k}) \cong E^*[\![x]\!]/([p^k]_{\mathbb{G}_u}(x)).$$

Proof. The Gysin sequence associated to the circle bundle $S^1 \to BC_{p^k} \to BS^1$ is the exact sequence

$$E^*[\![x]\!] \xrightarrow{\times [p^k]_{\mathbb{G}_u}(x)} E^*[\![x]\!] \longrightarrow E^*(BC_{p^k}),$$

where $[p^k]_{\mathbb{G}_u}(x)$ is in degree -2. Since $E^*[\![x]\!]$ is a domain, multiplication by $[p^k]_{\mathbb{G}_u}(x)$ is injective and the result follows. $\square$

As we noted in the proof of Corollary 21.3.4, the Weierstrass preparation theorem implies that $E^*[\![x]\!]/([p^k]_{\mathbb{G}_u}(x))$ is a finitely generated free module over E^*. Since $E^*(BC_{p^k})$ is a finitely generated free module, we can use the Künneth isomorphism to boot strap our calculation to finite abelian groups.

Proposition 21.5.4. *Let $A \cong \prod_{i=1}^{m} C_{p^{k_i}}$, then there is an isomorphism of E^*-algebras*

$$E^*(BA) \cong E^*[\![x_1, \ldots, x_m]\!]/([p^{k_1}]_{\mathbb{G}_u}(x), \ldots, [p^{k_m}]_{\mathbb{G}_u}(x)).$$

In this article, we will restrict our attention to the 0th cohomology ring $E^0(X)$. Since E is even periodic, this commutative ring contains plenty of information. The isomorphism of Proposition 21.5.4 induces an isomorphism of E^0-algebras

$$E^0(BA) \cong E^0[\![x_1, \ldots, x_m]\!]/([p^{k_1}]_{\mathbb{G}_u}(x), \ldots, [p^{k_m}]_{\mathbb{G}_u}(x)).$$

Given the proof of Corollary 21.3.4, the careful reader may already recognize a close relationship between $E^0(BA)$ and $\mathrm{Hom}(A, \mathrm{G}_u)$.

Now that we have calculated the E-cohomology of finite abelian groups we will turn our attention to the multiplicative structure enjoyed by Morava E-theory. The most important theorem regarding Morava E-theory is due to Goerss, Hopkins, and Miller [9] and we will endeavor to describe it now. Given a pair $(\mathbb{F}, \kappa)$, we have produced a homology theory $E(\mathbb{F}, \kappa)_*(-)$. Let FG denote the category with objects pairs $(\mathbb{F}, \kappa)$ of a perfect field of characteristic p and a finite height formal group over κ. A morphism from $(\mathbb{F}, \kappa)$ to $(\mathbb{F}', \kappa')$ is given by a map $j\colon \kappa \to \kappa'$ and an isomorphism of formal group laws $j^*\mathbb{F} \to \mathbb{F}'$. This category is equivalent to the category called FGL in [9]. Goerss, Hopkins, and Miller produce a fully faithful functor

$$\mathrm{FG} \xrightarrow{E(-,-)} E_\infty\text{-ring spectra}$$

such that the underlying homology theory of the E_∞-ring spectrum $E(\mathbb{F}, \kappa)$ is $E(\mathbb{F}, \kappa)_*(-)$. Moreover, they show that the space of E_∞-ring spectra R such that $E(\mathbb{F}, \kappa)_*R$ an algebra object in graded $E(\mathbb{F}, \kappa)_*E(\mathbb{F}, \kappa)$-comodules that is isomorphic to $E(\mathbb{F}, \kappa)_*E(\mathbb{F}, \kappa)$ is equivalent to $B\,\mathrm{Aut}_{\mathrm{FG}}(\mathbb{F}, \kappa)$. Thus the group of automorphisms of $E(\mathbb{F}, \kappa)$ as an E_∞-ring spectrum is the extended Morava stabilizer group $\mathrm{Aut}_{\mathrm{FG}}(\mathbb{F}, \kappa) \cong \mathrm{Aut}(\mathbb{F}/\kappa) \rtimes \mathrm{Gal}(\kappa, \mathbb{F}_p)$. It follows that $\mathrm{Aut}(\mathbb{F}/\kappa)$ acts on $E(\mathbb{F}, \kappa)^*(X)$ by ring maps for any space X.

In the homotopy category, the E_∞-ring structure on Morava E-theory gives rise to power operations. For an E_∞-ring spectrum E, these are constructed as follows: Let X be a space and let $\Sigma^\infty_+ X \to E$ be a an element of $E^0(X)$. Applying the extended powers functor $(E\Sigma_m)_+ \wedge_{\Sigma_m} (-)^{\wedge m}$ to both sides produces a map of spectra

$$(E\Sigma_m)_+ \wedge_{\Sigma_m} (\Sigma^\infty_+ X)^{\wedge m} \simeq \Sigma^\infty_+(E\Sigma_m \times_{\Sigma_m} X^{\times m}) \to (E\Sigma_m)_+ \wedge_{\Sigma_m} E^{\wedge m}.$$

Now the E_∞-ring structure on E gives us a map

$$(E\Sigma_m)_+ \wedge_{\Sigma_m} E^{\wedge m} \to E.$$

Composing these two maps produces the mth total power operation

$$\mathbb{P}_m\colon E^0(X) \to E^0(E\Sigma_m \times_{\Sigma_m} X^{\times m}),$$

a multiplicative, non-additive map. We will solely be concerned with the case $X = BG$, when G is a finite group. There is an equivalence

$$E\Sigma_m \times_{\Sigma_m} BG^{\times m} \simeq BG \wr \Sigma_m,$$

where $G \wr \Sigma_m = G^{\times m} \rtimes \Sigma_m$ is the wreath product. In this case, the total power operation takes the form

$$\mathbb{P}_m\colon E^0(BG) \to E^0(BG \wr \Sigma_m). \tag{21.5.1}$$

We will make use of variants of the total power operation, which we will introduce after a brief interlude regarding transfers.

As $E^*(-)$ is a cohomology theory, it has a theory of transfer maps for finite covers. For $H \subset G$, the map $BH \to BG$ is a finite cover and thus there is a transfer map in E-cohomology of the form

$$\mathrm{Tr}_E \colon E^0(BH) \to E^0(BG).$$

This is a map of $E^0(BG)$-modules for the $E^0(BG)$-module structure on the domain coming from the restriction map. We will see that these transfers play an important role in understanding the additive properties of power operations and that they interact nicely with the character maps of Hopkins, Kuhn, and Ravenel. In particular, these transfers can be used to define several important ideals. Assume $i, j > 0$ and that $i + j = m$, then there is a transfer map

$$\mathrm{Tr}_E \colon E^0(BG \wr (\Sigma_i \times \Sigma_j)) \to E^0(BG \wr \Sigma_m)$$

induced by the inclusion $G \wr (\Sigma_i \times \Sigma_j) \subset G \wr \Sigma_m$. Summing over these maps as i and j vary gives a map of $E^0(BG \wr \Sigma_m)$-modules

$$\bigoplus_{i,j} \mathrm{Tr}_E \colon \bigoplus_{i,j} E^0(BG \wr (\Sigma_i \times \Sigma_j)) \to E^0(BG \wr \Sigma_m).$$

Define the ideal $I_{\mathrm{tr}} \subset E^0(BG \wr \Sigma_m)$ by

$$I_{\mathrm{tr}} = \mathrm{im}\left(\bigoplus_{i,j} \mathrm{Tr}_E\right).$$

When G is trivial, we will still refer to this ideal as $I_{\mathrm{tr}} \subset E^0(B\Sigma_m)$.

Since Σ_m acts trivially on the image of the diagonal $\Delta \colon G \to G^m$, we have a map of spaces $BG \times B\Sigma_m \to BG \wr \Sigma_m$. Restriction along this map gives us the power operation

$$P_m \colon E^0(BG) \xrightarrow{\mathbb{P}_m} E^0(BG \wr \Sigma_m) \to E^0(BG \times B\Sigma_m).$$

Since $E^0(B\Sigma_m)$ is a free E^0-module by [19, Theorem 3.2], we have a Künneth isomorphism

$$E^0(BG \times B\Sigma_m) \cong E^0(BG) \otimes_{E^0} E^0(B\Sigma_m).$$

We will also denote the resulting map by

$$P_m \colon E^0(BG) \to E^0(BG) \otimes_{E^0} E^0(B\Sigma_m)$$

and refer to it as the power operation as well. Both the total power operation $\mathbb{P}_m$ and the power operation P_m are multiplicative non-additive maps. Let $\iota \colon * \to B\Sigma_m$ be the inclusion of base point. The maps $\mathbb{P}_m$ and P_m are called power operations because the composite

$$E^0(BG) \xrightarrow{P_m} E^0(BG) \otimes_{E^0} E^0(B\Sigma_m) \xrightarrow{\mathrm{id} \otimes \iota^*} E^0(BG)$$

is the mth power map $(-)^m \colon E^0(BG) \to E^0(BG)$.

The failure of the total power operation to be additive is controlled by the ideal I_{tr} ([6, VIII Proposition 1.1.(vi)]). In other words, $I_{\mathrm{tr}} \subset E^0(BG \wr \Sigma_m)$ is the smallest ideal such that the quotient

$$\mathbb{P}_m/I_{\mathrm{tr}} \colon E^0(BG) \to E^0(BG \wr \Sigma_m)/I_{\mathrm{tr}} \tag{21.5.2}$$

is a map of commutative rings. The same relationship holds for the ideal I_{tr} and the power operation P_m, the quotient map

$$P_m/I_{\mathrm{tr}} \colon E^0(BG) \to E^0(BG) \otimes_{E^0} E^0(B\Sigma_m)/I_{tr}$$

is a map of commutative rings. Because of this, we will refer to P_m/I_{tr} as the additive mth power operation.

The total power operation and its variants satisfy several identities that hold for any E_∞-ring spectrum. These are described in great detail in Chapters I and VIII of [6]. For the purposes of this paper, we will need the following identity, which follows from [6, VIII Proposition 1.1.(i)], that will come in handy for the induction arguments in Section 21.9: Let $i, j > 0$ such that $i+j = m$ and let $\Delta_{i,j} \colon \Sigma_i \times \Sigma_j \to \Sigma_m$ be the canonical map and also recall that $\Delta \colon G \to G \times G$ is the diagonal. The following diagram commutes

$$\begin{array}{ccc} E^0(BG) & \xrightarrow{\quad P_m \quad} & E^0(BG \times B\Sigma_m) \\ \downarrow{\scriptstyle P_i \times P_j} & & \downarrow{\scriptstyle \Delta^*_{i,j}} \\ E^0(BG \times B\Sigma_i \times BG \times B\Sigma_j) & \xrightarrow{\quad \Delta^*\tau^* \quad} & E^0(BG \times B\Sigma_i \times B\Sigma_j), \end{array} \tag{21.5.3}$$

where $P_i \times P_j$ is the external product

$$E^0(BG) \to E^0(BG \times B\Sigma_i) \times E^0(BG \times B\Sigma_j) \to E^0(BG \times B\Sigma_i \times BG \times B\Sigma_j)$$

and $\Delta^*\tau^*$ is induced by the map of spaces

$$BG \times B\Sigma_i \times B\Sigma_j \to BG \times BG \times B\Sigma_i \times B\Sigma_j \to BG \times B\Sigma_i \times BG \times B\Sigma_j.$$

21.6 Character theory

Since Morava E-theory is a well-behaved generalization of p-adic K-theory, one might wonder if some of the more geometric properties of K-theory can be extended to E-theory. The character map in classical representation theory is an injective map of commutative rings from the representation ring of G to the ring of class functions on G taking values in $\mathbb{C}$

$$\chi \colon R(G) \to \mathrm{Cl}(G, \mathbb{C}).$$

It is the map taking a representation $\rho \colon G \to \mathrm{GL}_n(\mathbb{C})$ to the class function sending g to $\mathrm{Tr}(\rho(g))$, the trace of $\rho(g)$. It has the important property that

$$\mathbb{C} \otimes \chi \colon \mathbb{C} \otimes_{\mathbb{Z}} R(G) \to \mathrm{Cl}(G, \mathbb{C})$$

is an isomorphism.

These ideas can be transported to the K-cohomology of BG. In [1], Adams showed that the p-adic K-theory of BG also has a well-behaved character theory. His result makes use of Atiyah's completion theorem [4], which states that $K^0(BG)$ is isomorphic to the completion of the representation ring $R(G)$ at the augmentation ideal.

In [11], Hopkins, Kuhn, and Ravenel tackled the question of generalizing the character map to Morava E-theory. They did not have geometric cocycles (such as G-representations) available to them, but they did have a generalization of the representation ring of G: $E^0(BG)$.

Following the construction of the classical character map, they produce an E^0-algebra C_0 that plays the role of $\mathbb{C}$ above, a ring of "generalized" class functions $\mathrm{Cl}_n(G, C_0)$, and a character map

$$\chi\colon E^0(BG) \to \mathrm{Cl}_n(G, C_0)$$

with the property that the induced map

$$C_0 \otimes_{E^0} E^0(BG) \to \mathrm{Cl}_n(G, C_0)$$

is an isomorphism. In this section we will give a short exposition of their work on character theory and explain the relationship between their character map and the action of the stablizer group on $E^0(BG)$. A longer exposition on character theory can be found in [16, Appendix A.1], written by the author, and of course, the best source is the original [11].

We begin by describing Proposition 5.12 in [11], which draws a connection between the group cohomology calculations of Section 21.5 and the moduli problem $\mathrm{Hom}(A, \mathbb{G}_u)$ of Section 21.3. Recall that $A^* = \mathrm{Hom}(A, S^1)$ is the Pontryagin dual of A.

Proposition 21.6.1. [11, Proposition 5.12] *Let A be a finite abelian group. There is a canonical isomorphism of E^0-algebras*

$$E^0(BA) \cong \mathcal{O}_{\mathrm{Hom}(A^*, \mathbb{G}_u)}$$

compatible with the action of the stabilizer group.

The ring C_0 is defined to be the rationalization of the Drinfeld ring $\mathcal{O}_{\mathrm{Level}(\mathbb{T}, \mathbb{G}_u)}$. It has the property that for all $k \geq 0$, there is a canonical isomorphism of group schemes

$$\mathrm{Spec}(C_0) \times_{\mathrm{Spec}(E^0)} \mathbb{G}_u[p^k] \cong \mathbb{T}[p^k].$$

In fact, this can be taken as the defining property of C_0, but we will not need this fact in this document. Since C_0 is the rationalization of the Drinfeld ring, it comes equipped with a right action of $\mathrm{Isog}(\mathbb{T})$ by ring maps and an action of $\mathrm{Aut}(\mathbb{T})$ by $\mathbb{Q} \otimes E^0$-algebra maps. In fact, C_0 is an $\mathrm{Aut}(\mathbb{T})$-Galois extension of $\mathbb{Q} \otimes E^0$ [11, Propositions 6.5 and 6.6] so there is an isomorphism

$$C_0^{\mathrm{Aut}(\mathbb{T})} \cong \mathbb{Q} \otimes E^0.$$

Recall that $\mathbb{L} \cong \mathbb{Z}_p^{\times n}$ and that G is a finite group. Let $\mathrm{Hom}(\mathbb{L}, G)$ be the set of continuous homomorphisms from $\mathbb{L}$ to G. This set is in bijective correspondence with the set of n-tuples of pairwise commuting p-power order elements of G. There is an action of G on this set by conjugation. We define

$$\begin{aligned}\mathrm{Cl}_n(G, C_0) &= \{C_0\text{-valued functions on the set } \mathrm{Hom}(\mathbb{L}, G)/\!\sim\}\\ &\cong \prod_{\mathrm{Hom}(\mathbb{L},G)/\sim} C_0,\end{aligned}$$

where $\mathrm{Hom}(\mathbb{L}, G)/\!\sim$ is the set of conjugacy classes of maps from $\mathbb{L}$ to G. Thus $\mathrm{Cl}_n(G, C_0)$ is the ring of conjugation invariant functions on $\mathrm{Hom}(\mathbb{L}, G)$ taking values in C_0.

The character map is defined as follows: Given a conjugacy class $\alpha\colon \mathbb{L} \to G$, there exists a $k \in \mathbb{N}$ such that α factors through $\mathbb{L}/\mathbb{L}_{p^k}$. Applying the classifying space functor to this map gives a map of spaces

$$B\,\mathbb{L}/\mathbb{L}_{p^k} \to BG.$$

Applying E-cohomology to this map gives a map of E^0-algebras

$$\alpha^*\colon E^0(BG) \to E^0(B\,\mathbb{L}/\mathbb{L}_{p^k}).$$

By Proposition 21.6.1, the codomain is canonically isomorphic to $\mathcal{O}_{\mathrm{Hom}(\mathbb{T}[p^k],\mathbb{G}_u)}$, which is the ring corepresenting $\mathrm{Hom}(\mathbb{T}[p^k], \mathbb{G}_u)$. Recall that there are canonical forgetful natural transformations $\mathrm{Level}(\mathbb{T}, \mathbb{G}_u) \to \mathrm{Level}(\mathbb{T}[p^k], \mathbb{G}_u) \to \mathrm{Hom}(\mathbb{T}[p^k], \mathbb{G}_u)$ over LT that induce maps of E_0-algebras

$$\mathcal{O}_{\mathrm{Hom}(\mathbb{T}[p^k],\mathbb{G}_u)} \to \mathcal{O}_{\mathrm{Level}(\mathbb{T}[p^k],\mathbb{G}_u)} \to \mathcal{O}_{\mathrm{Level}(\mathbb{T},\mathbb{G}_u)} \to C_0.$$

Composing these gives an E^0-algebra map

$$\begin{aligned}\chi_{[\alpha]}\colon E^0(BG) \xrightarrow{\alpha^*} E^0(B\,\mathbb{L}/\mathbb{L}_{p^k}) &\cong \mathcal{O}_{\mathrm{Hom}(\mathbb{T}[p^k],\mathbb{G}_u)}\\ &\to \mathcal{O}_{\mathrm{Level}(\mathbb{T}[p^k],\mathbb{G}_u)} \to \mathcal{O}_{\mathrm{Level}(\mathbb{T},\mathbb{G}_u)} \to C_0.\end{aligned} \tag{21.6.1}$$

Putting these together for all of the conjugacy classes in $\mathrm{Hom}(\mathbb{L}, G)$ gives the character map

$$\chi\colon E^0(BG) \to \mathrm{Cl}_n(G, C_0).$$

Example 21.6.2. Let A be a finite abelian group and let $\alpha\colon \mathbb{L} \to A$ be a map. In this case, $\chi_{[\alpha]}$ admits an interpretation completely in terms of the moduli problems in Section 21.3. The map in (21.6.1) implies that we may view $\chi_{[\alpha]}$ as landing in $\mathcal{O}_{\mathrm{Level}(\mathbb{T},\mathbb{G}_u)}$. Proposition 21.6.1 implies that the domain is canonically isomorphic to $\mathcal{O}_{\mathrm{Hom}(A^*,\mathbb{G}_u)}$. Putting these observations together, $\chi_{[\alpha]}$ gives us a map

$$\mathcal{O}_{\mathrm{Hom}(A^*,\mathbb{G}_u)} \to \mathcal{O}_{\mathrm{Level}(\mathbb{T},\mathbb{G}_u)}$$

or on the level of moduli problems, a map

$$\mathrm{Level}(\mathbb{T}, \mathbb{G}_u) \to \mathrm{Hom}(A^*, \mathbb{G}_u).$$

Unwrapping the definition of $\chi_{[\alpha]}$, when this map is applied to a complete local ring (R, m), it sends a deformation with level structure

$$(l\colon \mathbb{T} \to \mathbb{G}, i, \tau)$$

to the deformation equipped with homomorphism

$$(A^* \xrightarrow{\alpha^*} \mathbb{T} \xrightarrow{l} \mathbb{G}, i, \tau).$$

There is a left action of $\mathrm{Aut}(\mathbb{T})$ on $\mathrm{Hom}(\mathbb{L}, G)/\sim$ given by precomposition with the Pontryagin dual. Combining this with the right action of $\mathrm{Aut}(\mathbb{T})$ on C_0, there is a right action on the set of C_0-valued functions on $\mathrm{Hom}(\mathbb{L}, G)/\sim$ (ie. $\mathrm{Cl}_n(G, C_0)$). Explicitly this action is defined as follows: Let $\phi \in \mathrm{Aut}(\mathbb{T})$, let $f \in \mathrm{Cl}_n(G, C_0)$, and let $[\alpha] \in \mathrm{Hom}(\mathbb{L}, G)/\sim$ then

$$(f\phi)([\alpha]) = (f([\alpha\phi^*]))\phi.$$

To see that this is an action note that, for $\phi, \tau \in \mathrm{Aut}(\mathbb{T})$,

$$((f\phi)\tau)([\alpha]) = (f\phi)([\alpha\tau^*])\tau = f([\alpha\tau^*\phi^*])\phi\tau = f([\alpha(\phi\tau)^*])\phi\tau.$$

It turns out that the base change of the character map χ to C_0

$$C_0 \otimes \chi\colon C_0 \otimes_{E^0} E^0(BG) \to \mathrm{Cl}_n(G, C_0)$$

is equivariant with respect to the right $\mathrm{Aut}(\mathbb{T})$-action on the source given by the action of $\mathrm{Aut}(\mathbb{T})$ on the left tensor factor and the right $\mathrm{Aut}(\mathbb{T})$-action on the target given above.

Theorem 21.6.3. [11, Theorem C] *The character map induces an isomorphism*

$$C_0 \otimes_{E^0} E^0(BG) \xrightarrow{\cong} \mathrm{Cl}_n(G, C_0)$$

and taking $\mathrm{Aut}(\mathbb{T})$*-fixed points gives an isomorphism*

$$\mathbb{Q} \otimes E^0(BG) \xrightarrow{\cong} \mathrm{Cl}_n(G, C_0)^{\mathrm{Aut}(\mathbb{T})}.$$

The primary goal of this document is to give an exposition of the relationship between the power operations of Section 21.5 and the character map above. However, we are already in the position to explain the relationship between the stabilizer group action on $E^0(BG)$ and the character map, so we will do that now.

Since the stabilizer group $\mathrm{Aut}(\mathbb{F}/\kappa)$ acts on E by E_∞-ring maps, it acts on the function spectrum E^X by E_∞-ring maps for any space X and given a map of spaces $X \to Y$, the induced map of E_∞-ring spectra

$$E^Y \to E^X$$

is $\mathrm{Aut}(\mathbb{F}/\kappa)$-equivariant. Recall from Section 21.3 that $\mathrm{Aut}(\mathbb{F}/\kappa)$ also acts on $\mathcal{O}_{\mathrm{Hom}(\mathbb{T}[p^k], \mathbb{G}_u)}$, $\mathcal{O}_{\mathrm{Level}(\mathbb{T}[p^k], \mathbb{G}_u)}$ and $\mathcal{O}_{\mathrm{Level}(\mathbb{T}, \mathbb{G}_u)}$ by commutative ring maps and thus there is also an action of $\mathrm{Aut}(\mathbb{F}/\kappa)$ on C_0. Proposition 21.6.1 implies that the canonical isomorphism

$$E^0(B\,\mathbb{L}/\mathbb{L}_{p^k}) \cong \mathcal{O}_{\mathrm{Hom}(\mathbb{T}[p^k], \mathbb{G}_u)}$$

is $\mathrm{Aut}(\mathbb{F}/\kappa)$-equivariant. Putting all of this together, we see that the composite of (21.6.1)

$$E^0(BG) \to E^0(B\,\mathbb{L}/\mathbb{L}_{p^k}) \cong \mathcal{O}_{\mathrm{Hom}(\mathbb{T}[p^k],\mathbb{G}_u)} \\ \to \mathcal{O}_{\mathrm{Level}(\mathbb{T}[p^k],\mathbb{G}_u)} \to \mathcal{O}_{\mathrm{Level}(\mathbb{T},\mathbb{G}_u)} \to C_0$$

is $\mathrm{Aut}(\mathbb{F}/\kappa)$-equivariant. Since the character map is a product of maps of this form, we may conclude the following proposition.

Proposition 21.6.4. *The Hopkins-Kuhn-Ravenel character map*

$$\chi\colon E^0(BG) \to \mathrm{Cl}_n(G, C_0)$$

is $\mathrm{Aut}(\mathbb{F}/\kappa)$*-equivariant for the canonical action of* $\mathrm{Aut}(\mathbb{F}/\kappa)$ *on* $E^0(BG)$ *and the diagonal action of* $\mathrm{Aut}(\mathbb{F}/\kappa)$ *on* $\mathrm{Cl}_n(G, C_0)$.

21.7 Transfers and conjugacy classes of tuples in symmetric groups

In keeping with the theme of this article, in this section we will describe the relationship between transfer maps for Morava E-theory and the character map. In Theorem D of [11], Hopkins, Kuhn, and Ravenel show that a surprisingly simple function between generalized class functions, inspired by the formula for induction in representation theory, is compatible with the transfer map for Morava E-theory.

To describe Theorem D of [11], we need a bit of group-theoretic set-up. Let $H \subset G$ be a subgroup and let $[\alpha] \in \mathrm{Hom}(\mathbb{L}, G)/\sim$. The set of cosets G/H has a left action of G and we can take the fixed points for the action of the image of α on G/H to get the set $(G/H)^{\mathrm{im}\,\alpha}$. We leave the following useful group-theoretic lemma to the reader.

Lemma 21.7.1. *Let* $H \subset G$ *be a subgroup and let* $\alpha\colon \mathbb{L} \to G$*. An element* $g \in G$ *has the property that* $\mathrm{im}(g^{-1}\alpha g) \subset H$ *if and only if* $gH \in (G/H)^{\mathrm{im}\,\alpha}$.

Let $[\alpha] \in \mathrm{Hom}(\mathbb{L}, G)/\sim$ and let $f \in \mathrm{Cl}_n(H, C_0)$ be a generalized class function. Mimicking the formula for the transfer in representation theory, Hopkins, Kuhn, and Ravenel define a transfer map on generalized class functions,

$$\mathrm{Tr}_{C_0}\colon \mathrm{Cl}_n(H, C_0) \to \mathrm{Cl}_n(G, C_0),$$

for $H \subset G$ by the formula

$$\mathrm{Tr}_{C_0}(f)([\alpha]) = \sum_{gH \in (G/H)^{\mathrm{im}\,\alpha}} f([g^{-1}\alpha g]). \tag{21.7.1}$$

Note that this formula is independent of the implicit choice of system of representatives of G/H.

Theorem 21.7.2. [11, Theorem D] *For any finite group G and subgroup $H \subset G$, there is a commutative diagram*

$$\begin{array}{ccc} E^0(BH) & \xrightarrow{\mathrm{Tr}_E} & E^0(BG) \\ \chi \downarrow & & \downarrow \chi \\ \mathrm{Cl}_n(H, C_0) & \xrightarrow{\mathrm{Tr}_{C_0}} & \mathrm{Cl}_n(G, C_0). \end{array}$$

Since the additive power operations are ring maps, they are easier to understand than the power operations. To study the additive power operations, we will understand the analogue of the ideal $I_{\mathrm{tr}} \subset E^0(B\Sigma_{p^k})$ in the ring of generalized class functions $\mathrm{Cl}_n(\Sigma_{p^k}, C_0)$. We will make use of the following calculation.

Given $\alpha\colon \mathbb{L} \to G$, let $1_{[\alpha]} \in \mathrm{Cl}_n(G, C_0)$ be the class function with value 1 on $[\alpha]$ and 0 elsewhere. Applying (21.7.1), we may conclude from Lemma 21.7.1 that if $\alpha\colon \mathbb{L} \to G$ has the property that $\mathrm{im}\, g^{-1}\alpha g \subset H$, then

$$\mathrm{Tr}_{C_0}(1_{[g^{-1}\alpha g]})([\alpha])$$

is a positive natural number, where we are viewing $1_{[g^{-1}\alpha g]}$ as a class function in $\mathrm{Cl}_n(H, C_0)$. Also, if $[\beta] \neq [\alpha] \in \mathrm{Hom}(\mathbb{L}, G)/\sim$, then

$$\mathrm{Tr}_{C_0}(1_{[g^{-1}\alpha g]})([\beta]) = 0.$$

Otherwise, β would be conjugate to $g^{-1}\alpha g$, which is conjugate to α.

Definition 21.7.3. Let

$$\mathrm{Sum}_m(\mathbb{T}) = \{\oplus_i H_i | H_i \subset \mathbb{T}, \sum_i |H_i| = m\}$$

be the set of formal sums of subgroups $H_i \subset \mathbb{T}$ such that the sum of the orders is m. These sums are unordered and repetitions are allowed.

Remark 21.7.4. Another way to describe the set $\mathrm{Sum}_m(\mathbb{T})$ is as the set of multisets of finite subgroups of $\mathbb{T}$ whose orders sum to m. We prefer the description in the definition above because it suggests that $\mathrm{Sum}_m(\mathbb{T})$ consists of certain "effective Weil divisors" on the set of finite subgroups of $\mathbb{T}$.

Proposition 21.7.5. *There is a canonical bijection of sets*

$$\mathrm{Hom}(\mathbb{L}, \Sigma_m)/\sim\, \cong \mathrm{Sum}_m(\mathbb{T}).$$

Proof. Since $\Sigma_m = \mathrm{Aut}(\underline{m})$, where $\underline{m}$ is a fixed m-element set, a map $\alpha\colon \mathbb{L} \to \Sigma_m$ gives $\underline{m}$ the structure of an $\mathbb{L}$-set. Maps $\alpha\colon \mathbb{L} \to \Sigma_m$ and $\beta\colon \mathbb{L} \to \Sigma_m$ are conjugate if and only if the corresponding $\mathbb{L}$-sets are isomorphic. Thus $\mathrm{Hom}(\mathbb{L}, \Sigma_m)/\sim$ is in bijective correspondence with the set of isomorphism classes of $\mathbb{L}$-sets of size m.

Given α, decompose $\underline{m} = \coprod_i \underline{m}_i$, where $\underline{m}_i$ is a transitive $\mathbb{L}$-set and let $\mathbb{L}_i$ be the stabilizer of any point in $\underline{m}_i$. Since $\mathbb{L}$ is abelian, the stabilizer does not depend on the choice of point. We define $H_i = (\mathbb{L} / \mathbb{L}_i)^* \subset \mathbb{L}^* = \mathbb{T}$. This construction only depends on the conjugacy class of α, so send $[\alpha]$ to $\oplus_i H_i$. We leave it to the reader to check that this map is a bijection. □

Let $\mathrm{Hom}(\mathbb{L}, \Sigma_m)^{\mathrm{trans}}$ be the set of transitive homomorphisms $\mathbb{L} \to \Sigma_m$. Note that $\mathbb{L} \to \Sigma_m$ is transitive if and only if the image is a transitive abelian subgroup of Σ_m. Since $\mathbb{L} = \mathbb{Z}_p^n$, this can only occur when m is a power of p.

Proposition 21.7.6. *There is a canonical bijection*

$$\mathrm{Hom}(\mathbb{L}, \Sigma_{p^k})^{\mathrm{trans}}/\!\sim\, \cong \mathrm{Sub}_{p^k}(\mathbb{T}),$$

where $\mathrm{Sub}_{p^k}(\mathbb{T})$ *is the set of subgroups* $H \subset \mathbb{T}$ *of order* p^k.

Proof. Follows immediately from the construction of the bijection of Proposition 21.7.5. □

The next lemma will be useful when we try to understand the relation between additive power operations and character theory. Recall the inclusion $\Delta_{i,j} \colon \Sigma_i \times \Sigma_j \hookrightarrow \Sigma_m$, where $i+j=m$ and $i,j>0$. Proposition 21.7.5 gives an isomorphism

$$\mathrm{Hom}(\mathbb{L}, \Sigma_i \times \Sigma_j)/\!\sim\, \cong \mathrm{Sum}_i(\mathbb{T}) \times \mathrm{Sum}_j(\mathbb{T}).$$

We leave the proof of the next lemma to the reader.

Lemma 21.7.7. *Assume* $i,j>0$ *and* $i+j=m$, *then the map*

$$\mathrm{Sum}_i(\mathbb{T}, G) \times \mathrm{Sum}_j(\mathbb{T}, G) \to \mathrm{Sum}_m(\mathbb{T}, G)$$

induced by $\Delta_{i,j}$ *sends a pair of sums of subgroups*

$$(\oplus_l H_l, \oplus_l K_l)$$

to the sum

$$(\oplus_l H_l) \oplus (\oplus_l K_l).$$

Corollary 21.7.8. *Let* $\sum_j a_j p^j$ *be the* p*-adic expansion of* m. *The inclusion* $\prod_j \Sigma_{p^j}^{\times a_j} \subseteq \Sigma_m$ *induces a surjection*

$$\prod_j \mathrm{Sum}_{p^j}(\mathbb{T})^{\times a_j} \twoheadrightarrow \mathrm{Sum}_m(\mathbb{T}).$$

Proof. Follows from Lemma 21.7.7. □

Finally, we use these ideas to give a description $\mathrm{Cl}_n(\Sigma_m, C_0)/I_{\mathrm{tr}}$, where $I_{\mathrm{tr}} \subset \mathrm{Cl}_n(\Sigma_m, C_0)$ is the image of the sum of the transfer maps

$$\bigoplus_{i,j} \mathrm{Tr}_{C_0} \colon \mathrm{Cl}_n(\Sigma_i \times \Sigma_j, C_0) \to \mathrm{Cl}_n(\Sigma_m, C_0)$$

along the inclusions $\Delta_{i,j}$ for all $i,j>1$ with $i+j=m$.

Proposition 21.7.9. *There is a canonical isomorphism*

$$\mathrm{Cl}_n(\Sigma_m, C_0)/I_{\mathrm{tr}} \cong \prod_{\mathrm{Sub}_m(\mathbb{T})} C_0.$$

Idea of the proof. Proposition 21.7.5 implies that we have a canonical isomorphism

$$\mathrm{Cl}_n(\Sigma_m, C_0) \cong \prod_{\mathrm{Sum}_m(\mathbb{T})} C_0.$$

We will show that $I_{\mathrm{tr}} \subset \mathrm{Cl}_n(\Sigma_m, C_0)$ consists of the functions supported on the set of sums of subgroups with more than one summand.

First we will show that functions with support not contained in the set of sums of subgroups with more than one summand cannot be hit by the transfer map. The transfer map Tr_{C_0} is a sum over elements that conjugate a conjugacy class $[\mathbb{L} \to G]$ into the subgroup $H \subset G$. In our case $G = \Sigma_m$ and $H = \Sigma_i \times \Sigma_j$, where $i, j > 0$ and $i + j = m$. If $\mathbb{L} \to \Sigma_m$ is transitive then every conjugate of the map is also transitive and so no conjugate can land inside $\Sigma_i \times \Sigma_j$.

Now let $\alpha\colon \mathbb{L} \to \Sigma_m$ be a non-transitive map. Thus $[\alpha]$ corresponds to a sum of subgroups with more than one summand. Therefore there exists $i, j > 0$ with $i + j = m$ such that α is conjugate to a map $\beta\colon \mathbb{L} \to \Sigma_m$ that lands in $\Sigma_i \times \Sigma_j \subset \Sigma_m$. The discussion after Theorem 21.7.2 implies that $\mathrm{Tr}_{C_0}(1_{[\beta]})$ is concentrated on $[\alpha]$ and there it is a non-zero natural number (which is invertible in C_0). Since the transfer map is a map of C_0-modules, I_{tr} contains the factor of $\mathrm{Cl}_n(\Sigma_m, C_0)$ corresponding to $[\alpha]$. □

By Theorem 21.7.2, there is a commutative diagram

$$\begin{array}{ccc} \bigoplus_{i,j} E^0(B\Sigma_i \times B\Sigma_j) & \xrightarrow{\oplus \mathrm{Tr}_E} & E^0(B\Sigma_m) \\ \downarrow & & \downarrow \\ \bigoplus_{i,j} \mathrm{Cl}_n(\Sigma_i \times \Sigma_j, C_0) & \xrightarrow{\oplus \mathrm{Tr}_{C_0}} & \mathrm{Cl}_n(\Sigma_m, C_0), \end{array}$$

where the sum is over all $i, j > 0$ such that $i + j = m$. Therefore we have an induced map of E^0-algebras

$$\chi\colon E^0(B\Sigma_m)/I_{\mathrm{tr}} \to \mathrm{Cl}_n(\Sigma_m, C_0)/I_{\mathrm{tr}} \cong \prod_{\mathrm{Sub}_{p^k}(\mathbb{T})} C_0$$

that we will continue to call χ. Theorem 21.6.3 above implies that this map gives rise to an isomorphism

$$C_0 \otimes_{E^0} E^0(B\Sigma_m)/I_{\mathrm{tr}} \xrightarrow{\cong} \mathrm{Cl}_n(\Sigma_m, C_0)/I_{\mathrm{tr}}.$$

Finally, we will conclude this section by introducing convenient notation. Assume that $H \in \mathrm{Sub}_{p^k}(\mathbb{T})$ corresponds to the transitive conjugacy class $[\beta\colon \mathbb{L} \to \Sigma_{p^k}]$. We define χ_H to be the composite

$$\chi_H\colon E^0(B\Sigma_m)/I_{\mathrm{tr}} \to \mathrm{Cl}_n(\Sigma_m, C_0)/I_{\mathrm{tr}} \xrightarrow{\pi_H} C_0. \tag{21.7.2}$$

Since the quotient map $\mathrm{Cl}_n(\Sigma_m, C_0) \to \mathrm{Cl}_n(\Sigma_m, C_0)/I_{\mathrm{tr}}$ is just a projection, the map χ_H and the map $\chi_{[\beta]}$ of (21.6.1) are related by the following commutative diagram:

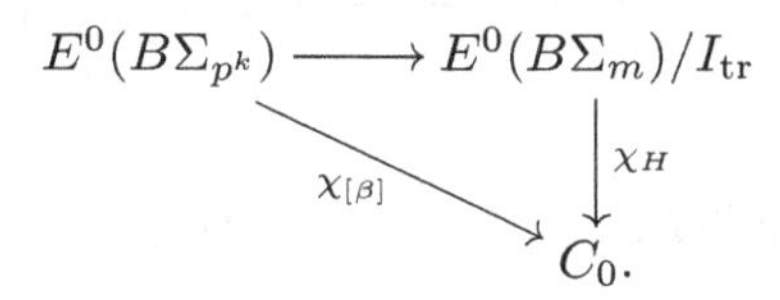

21.8 A theorem of Ando-Hopkins-Strickland

An important ingredient in understanding the relationship between power operations and character theory is a result of Ando, Hopkins, and Strickland that gives an algebro-geometric interpretation of a special case of the power operation in terms of Lubin-Tate theory. Their result indicates that there is a connection between power operations and the $\mathrm{Isog}(\mathbb{T})$-action on C_0.

One starting point for making the connection between algebraic geometry and power operations is Strickland's theorem [19] which gives an algebro-geometric interpretation of $E^0(B\Sigma_{p^k})/I_{\mathrm{tr}}$.

Theorem 21.8.1. [19, Theorem 9.2] *There is a canonical isomorphism of E^0-algebras*

$$E^0(B\Sigma_{p^k})/I_{\mathrm{tr}} \cong \mathcal{O}_{\mathrm{Sub}_{p^k}(\mathrm{G}_u)}.$$

Idea of the proof. It is worth understanding the origin of this isomorphism. Let

$$\mathrm{Div}_{p^k}(\mathrm{G}_u)\colon \mathrm{CompRings}_{E^0/} \to \mathrm{Set}$$

be the functor that assigns to $j\colon E^0 \to R$ the set of effective divisors on $j^*\,\mathrm{G}_u$. These divisors are just subschemes of $j^*\,\mathrm{G}_u$ of the form

$$\mathrm{Spf}(R[\![x]\!]/(a_0 + a_1 x + \cdots + a_{p^k-1}x^{p^k-1} + x^{p^k})),$$

where a_i is an element of the maximal ideal of R. Since a subgroup scheme has an underlying divisor, $\mathrm{Sub}_{p^k}(\mathrm{G}_u)$ is a closed subscheme of $\mathrm{Div}_{p^k}(\mathrm{G}_u)$. Proposition 8.31 in [21] gives a canonical isomorphism of formal schemes

$$\mathrm{Div}_{p^k}(\mathrm{G}_u) \cong \mathrm{Spf}\, E^0(BU(p^k)),$$

where $U(p^k)$ is the unitary group.

It is not hard to check that the standard representation $s_{p^k}\colon \Sigma_{p^k} \to U(p^k)$ fits into a commutative diagram

$$\begin{array}{ccc} E^0(BU(p^k)) & \xrightarrow{\cong} & \mathcal{O}_{\mathrm{Div}_{p^k}(\mathbb{G}_u)} \\ \downarrow{\scriptstyle s^*_{p^k}} & & \downarrow \\ E^0(B\Sigma_{p^k})/I_{\mathrm{tr}} & & \mathcal{O}_{\mathrm{Sub}_{p^k}(\mathbb{G}_u)} \\ \downarrow & & \downarrow \\ \mathrm{Cl}_n(\Sigma_{p^k}, C_0)/I_{\mathrm{tr}} & \xrightarrow{\cong} & \prod_{\mathrm{Sub}_{p^k}(\mathbb{T})} C_0. \end{array}$$

The map

$$\mathcal{O}_{\mathrm{Sub}_{p^k}(\mathbb{G}_u)} \to \prod_{\mathrm{Sub}_{p^k}(\mathbb{T})} C_0$$

is just the map to the base change

$$\mathcal{O}_{\mathrm{Sub}_{p^k}(\mathbb{G}_u)} \to C_0 \otimes_{E^0} \mathcal{O}_{\mathrm{Sub}_{p^k}(\mathbb{G}_u)} \cong \mathcal{O}_{\mathrm{Sub}_{p^k}(\mathbb{T})} \cong \prod_{\mathrm{Sub}_{p^k}(\mathbb{T})} C_0$$

and it is an injection since $\mathcal{O}_{\mathrm{Sub}_{p^k}(\mathbb{G}_u)}$ is a finitely generated free E^0-module (Corollary 21.3.6).

The map $E^0(B\Sigma_{p^k})/I_{\mathrm{tr}} \to \mathrm{Cl}_n(\Sigma_{p^k}, C_0)/I_{\mathrm{tr}}$ is also induced by base change and it is an injection by Theorem 8.6 of [19], which states that $E^0(B\Sigma_{p^k})/I_{\mathrm{tr}}$ is a free E^0-module of finite rank.

A diagram chase in the commutative diagram above gives us an injective map

$$\mathcal{O}_{\mathrm{Sub}_{p^k}(\mathbb{G}_u)} \hookrightarrow E^0(B\Sigma_{p^k})/I_{\mathrm{tr}}.$$

Proving that this map is an isomorphism requires some work. We refer the reader to [20] or [18] to see two different ways that this can be accomplished. □

The isomorphism in Theorem 21.8.1 is built in such a way that the following diagram commutes

$$\begin{array}{ccc} E^0(B\Sigma_{p^k})/I_{\mathrm{tr}} & \xrightarrow{\cong} & \mathcal{O}_{\mathrm{Sub}_{p^k}(\mathbb{G}_u)} \\ \downarrow & & \downarrow \\ \mathrm{Cl}_n(\Sigma_{p^k}, C_0)/I_{\mathrm{tr}} & \xrightarrow{\cong} & \prod_{\mathrm{Sub}_{p^k}(\mathbb{T})} C_0 \\ \downarrow{\scriptstyle \pi_H} & & \downarrow{\scriptstyle \pi_H} \\ C_0 & \xrightarrow{=} & C_0. \end{array} \tag{21.8.1}$$

The composite of the left vertical arrows is χ_H from (21.7.2). The composite of the right vertical arrows factors through $\mathcal{O}_{\mathrm{Level}(\mathbb{T},\mathbb{G}_u)}$. The resulting map $\mathcal{O}_{\mathrm{Sub}_{p^k}(\mathbb{G}_u)} \to \mathcal{O}_{\mathrm{Level}(\mathbb{T},\mathbb{G}_u)}$, which depends on $H \subset \mathbb{T}$, is the map of moduli problems that sends

$$(l, i, \tau) \mapsto (l(H) \subset \mathbb{G}, i, \tau).$$

Recall that the additive power operation applied to BA for A a finite abelian group is the ring map

$$P_{p^k}/I_{\mathrm{tr}} \colon E^0(BA) \to E^0(BA) \otimes_{E^0} E^0(B\Sigma_{p^k})/I_{\mathrm{tr}}.$$

The domain and codomain both admit interpretations in terms of moduli problems over Lubin-Tate space. Applying $\mathrm{Spf}(-)$ gives a map

$$\mathrm{Hom}(A^*, \mathrm{G}_u) \times_{\mathrm{LT}} \mathrm{Sub}_{p^k}(\mathrm{G}_u) \to \mathrm{Hom}(A^*, \mathrm{G}_u).$$

There is an obvious guess for what this map might do when applied to a complete local ring R. An object in the groupoid

$$\mathrm{Hom}(A^*, \mathrm{G}_u)(R) \times_{\mathrm{LT}(R)} \mathrm{Sub}_{p^k}(\mathrm{G}_u)(R)$$

is a deformation of $\mathbb{F}$ to R, $(\mathbb{G}, i, \tau)$, equipped with a homomorphism $A^* \to \mathbb{G}$ and a subgroup scheme of order p^k, $H \subset \mathbb{G}$. We need to produce a deformation of $\mathbb{F}$ equipped with a map from A^*. This can be accomplished by taking the composite

$$A^* \to \mathbb{G} \to \mathbb{G}/H$$

and recalling that $\mathbb{G}/H$ is a deformation in a canonical way. This is the content of Proposition 3.21 of [3]:

Theorem 21.8.2. [3, Proposition 3.21] *The power operation P_{p^k}/I_{tr} is the ring of functions on the map of moduli problems*

$$\mathrm{Sub}_{p^k}(\mathrm{G}_u) \times_{\mathrm{LT}} \mathrm{Hom}(A^*, \mathrm{G}_u) \to \mathrm{Hom}(A^*, \mathrm{G}_u)$$

that, when applied to a complete local ring (R, m), sends a deformation equipped with subgroup of order p^k and homomorphism $A^* \to \mathbb{G}$

$$(H \subset \mathbb{G}, A^* \to \mathbb{G}, i, \tau)$$

to the $\star$-isomorphism class of the tuple

$$(A^* \to \mathbb{G} \to \mathbb{G}/H, i \circ \sigma^k, \tau/H).$$

This theorem can be used to understand the relationship between P_{p^k}/I_{tr} applied to finite abelian groups and the character maps. Recall the construction of $\chi_{[\alpha]} \colon E^0(BG) \to C_0$ in (21.6.1), the construction of $\chi_H \colon E^0(B\Sigma_{p^k})/I_{\mathrm{tr}} \to C_0$ in (21.7.2), and that $\mathrm{Isog}(\mathbb{T})$ acts on C_0 through ring maps.

Proposition 21.8.3. *Let $H \subset \mathbb{T}$ be a subgroup of order p^k, let $\alpha \colon \mathbb{L} \to A$ be a group homomorphism, and let $\phi_H \colon \mathbb{T} \to \mathbb{T}$ be an isogeny such that* $\mathrm{Ker}(\phi_H) = H$. *There is a* *commutative diagram*

$$\begin{array}{ccc} E^0(BA) & \xrightarrow{P_{p^k}/I_{\mathrm{tr}}} & E^0(BA)\otimes_{E^0} E^0(B\Sigma_{p^k})/I_{\mathrm{tr}} \\ \Big\downarrow{\scriptstyle \chi_{[\alpha\phi_H^*]}} & & \Big\downarrow{\scriptstyle \chi_{[\alpha]}\otimes\chi_H} \\ C_0 & \xrightarrow{\phi_H} & C_0. \end{array}$$

Proof. Since the image of the character map lands in $\mathcal{O}_{\mathrm{Level}(\mathbb{T},\mathbb{G}_u)}$, it suffices to prove that

$$\begin{array}{ccc} E^0(BA) & \xrightarrow{P_{p^k}/I_{\mathrm{tr}}} & E^0(BA)\otimes_{E^0} E^0(B\Sigma_{p^k})/I_{\mathrm{tr}} \\ \Big\downarrow{\scriptstyle \chi_{[\alpha\phi_H^*]}} & & \Big\downarrow{\scriptstyle \chi_{[\alpha]}\otimes\chi_H} \\ \mathcal{O}_{\mathrm{Level}(\mathbb{T},\mathbb{G}_u)} & \xrightarrow{\phi_H} & \mathcal{O}_{\mathrm{Level}(\mathbb{T},\mathbb{G}_u)} \end{array}$$

commutes. Recall the definition of ψ_H from Diagram (21.4.4). Now applying Example 21.6.2 to the maps $\chi_{[\alpha]}$ and $\chi_{[\alpha\phi_H^*]}$, applying the discussion around Diagram (21.8.1) to χ_H, and applying Theorem 21.8.2 to the top arrow, we just need to check that a certain diagram of moduli problems commutes. Going around the bottom direction gives

$$(l,i,\tau)\mapsto((l/H)\psi_H^{-1},i\sigma^k,\tau/H)\mapsto((l/H)\psi_H^{-1}(\alpha\phi_H^*)^*,i\sigma^k,\tau/H)$$

and going around the top direction gives

$$(l,i,\tau)\mapsto(l\alpha^*,i,\tau)\times(H\subset\mathbb{G},i,\tau)\mapsto(A^*\xrightarrow{l\alpha^*}\mathbb{G}\to\mathbb{G}/H,i\sigma^k,\tau/H).$$

We want to show that the two resulting deformations are equal. Taking the quotient with respect to $H\subset\mathbb{T}$ and its image in $\mathbb{G}$ the level structure l gives the commutative diagram

$$\begin{array}{ccc} \mathbb{T} & \xrightarrow{l} & \mathbb{G} \\ \Big\downarrow{\scriptstyle q_H} & & \Big\downarrow \\ \mathbb{T}/H & \xrightarrow{l/H} & \mathbb{G}/H. \end{array}$$

Thus the composite $A^*\xrightarrow{l\alpha^*}\mathbb{G}\to\mathbb{G}/H$ is equal to $(l/H)q_H\alpha^*$. Now

$$(l/H)\psi_H^{-1}(\alpha\phi_H^*)^*=(l/H)\psi_H^{-1}\phi_H^*\alpha^*=(l/H)q_H\alpha^*$$

by Diagram (21.4.4), so we are finished. □

21.9 The character of the power operation

With all of these tools in hand, we can finally get down to business and prove Theorem 21.1.1. We'd like to construct a "power operation" on generalized class functions that is compatible with the power operation on E through the character map of Hopkins, Kuhn, and Ravenel. We can say this diagramatically: we'd like to construct an operation

$$\mathrm{Cl}_n(G, C_0) \to \mathrm{Cl}_n(G \times \Sigma_m, C_0)$$

making the diagram

$$\begin{array}{ccc} E^0(BG) & \xrightarrow{P_m} & E^0(BG \times B\Sigma_m) \\ \downarrow{\scriptstyle \chi} & & \downarrow{\scriptstyle \chi} \\ \mathrm{Cl}_n(G, C_0) & \longrightarrow & \mathrm{Cl}_n(G \times \Sigma_m, C_0) \end{array}$$

commute. Proposition 21.8.3 takes us close to this goal. The work that remains is to remove the quotient by the transfer ideal from Proposition 21.8.3 and extend our power operation on class functions from finite abelian group to all finite groups.

Why would we like to construct a "power operation" on class functions? Since the ring of generalized class functions is just the C_0-valued functions on a set, such a formula should have a particularly simple form. Ideally it would only include information from group theory and information about endomorphisms of the ring C_0. This turns out to be the case. Further, Theorem 21.6.3 above implies that class functions knows quite a bit about $E^0(BG)$. For instance, the rationalization of $E^0(BG)$ sits inside class functions as the $\mathrm{Aut}(\mathbb{T})$-invariants. We will begin by describing the formula for the power operation on class functions as well as some of its consequences and then we will give a description of the proof of the fact that it is compatible with the power operations for E-theory.

We will produce more than one power operation on generalized class functions. Let $\mathrm{Sub}(\mathbb{T})$ be the set of finite subgroups of $\mathbb{T}$. Let $\pi\colon \mathrm{Isog}(\mathbb{T}) \to \mathrm{Sub}(\mathbb{T})$ be the surjective map sending an endoisogeny of $\mathbb{T}$ to its kernel. The map π makes $\mathrm{Isog}(\mathbb{T})$ into an $\mathrm{Aut}(\mathbb{T})$-principal bundle over $\mathrm{Sub}(\mathbb{T})$ and we will write $\Gamma(\mathrm{Sub}(\mathbb{T}), \mathrm{Isog}(\mathbb{T}))$ for the set of sections. For each $\phi \in \Gamma(\mathrm{Sub}(\mathbb{T}), \mathrm{Isog}(\mathbb{T}))$,

$$\begin{array}{c} \mathrm{Isog}(\mathbb{T}) \\ {\scriptstyle \pi}\downarrow \uparrow{\scriptstyle \phi} \\ \mathrm{Sub}(\mathbb{T}), \end{array}$$

we will produce an operation

$$P_m^{\phi}\colon \mathrm{Cl}_n(G, C_0) \to \mathrm{Cl}_n(G \times \Sigma_m, C_0)$$

compatible with the power operation on E through the character map.

Given a subgroup $H \subset \mathbb{T}$, let $\phi_H\colon \mathbb{T} \to \mathbb{T}$ be the corresponding isogeny of $\mathbb{T}$, so $\phi_H \in \mathrm{Isog}(\mathbb{T})$ with $\mathrm{Ker}(\phi_H) = H$. Let $f \in \mathrm{Cl}_n(G, C_0)$ be a generalized class function. To define

P^ϕ_m we only need to give a value for $P^\phi_m(f)$ on a conjugacy class $[\mathbb{L} \to G \times \Sigma_m]$. Proposition 21.7.5 implies that this conjugacy class corresponds to a pair $([\alpha\colon \mathbb{L} \to G], \oplus_i H_i)$. We set

$$P^\phi_m(f)([\alpha], \oplus_i H_i) = \prod_i \big(f([\alpha \circ \phi^*_{H_i}])\phi_{H_i}\big).$$

Example 21.9.1. As we have seen, when $(\mathbb{F}, \kappa) = (\mathbb{G}_m, \mathbb{F}_p)$, $E = K_p$. In this case, $n = 1$ and $\mathbb{T} \cong \mathbb{Q}_p/\mathbb{Z}_p$. Hence there is a canonical element in $\Gamma(\mathrm{Sub}(\mathbb{T}), \mathrm{Isog}(\mathbb{T}))$, the section sending $\mathbb{T}[p^k]$ to the multiplication by p^k map on $\mathbb{T}$. It turns out that, in this case, the action of these isogenies on C_0 is trivial. Thus the formula for P^ϕ_m simplifies for this choice of section. We get

$$P^\phi_m(f)([\alpha], \oplus_i H_i) = \prod_i f([\alpha \circ \phi^*_{H_i}]).$$

Since subgroups of $\mathbb{T}$ are all cyclic when $n = 1$, we can be even more explicit. A conjugacy class $[\alpha]$ corresponds to a conjugacy class $[g]$ in G such that $g^{p^l} = e$ for some $l \in \mathbb{N}$. If $H_i = \mathbb{T}[p^{k_i}]$, then $\phi^*_{H_i}\colon \mathbb{L} \to \mathbb{L}$ is just multiplication by p^{k_i}. Thus $[\alpha \circ \phi^*_{H_i}]$ corresponds to the conjugacy class $[g^{p^{k_i}}]$ and we get the formula

$$P^\phi_m(f)([\alpha], \oplus_i H_i) = \prod_i f([g^{p^{k_i}}]).$$

The formula for P^ϕ_m may look like it has come out of the blue. Let us verify that it makes sense in the case that we understand best:

Proposition 21.9.2. *Let A be a finite abelian group and let $\phi \in \Gamma(\mathrm{Sub}(\mathbb{T}), \mathrm{Isog}(\mathbb{T}))$, then there is a commutative diagram*

$$\begin{array}{ccc}
E^0(BA) & \xrightarrow{P_{p^k}/I_{\mathrm{tr}}} & E^0(BA) \otimes_{E^0} E^0(B\Sigma_{p^k})/I_{\mathrm{tr}} \\
\downarrow{\scriptstyle \chi} & & \downarrow{\scriptstyle \chi\otimes\chi} \\
\mathrm{Cl}_n(A, C_0) & \xrightarrow{P^\phi_{p^k}/I_{\mathrm{tr}}} & \mathrm{Cl}_n(A, C_0) \otimes_{C_0} \mathrm{Cl}_n(\Sigma_{p^k}, C_0)/I_{\mathrm{tr}}.
\end{array}$$

Proof. Let $([\alpha\colon \mathbb{L} \to A], H)$ correspond to a conjugacy class $[\mathbb{L} \to A \times \Sigma_{p^k}]$ which is transitive on the factor Σ_{p^k}. Let $f \in \mathrm{Cl}_n(A, C_0)$, then the definition of $P^\phi_{p^k}/I_{\mathrm{tr}}$ applied to f is

$$P^\phi_{p^k}/I_{\mathrm{tr}}(f)([\alpha], H) = f([\alpha\phi^*_H])\phi_H.$$

Thus the value of $P^\phi_{p^k}/I_{\mathrm{tr}}(f)$ on the conjugacy class $([\alpha], H)$ is determined by the value of f on $[\alpha\phi^*_H]$. Hence, it suffices to prove that the diagram

$$\begin{array}{ccc}
E^0(BA) & \xrightarrow{P_{p^k}/I_{\mathrm{tr}}} & E^0(BA) \otimes_{E^0} E^0(B\Sigma_{p^k})/I_{\mathrm{tr}} \\
\downarrow{\scriptstyle \chi_{[\alpha\phi^*_H]}} & & \downarrow{\scriptstyle \chi_{[\alpha]}\otimes\chi_H} \\
C_0 & \xrightarrow{\phi_H} & C_0
\end{array}$$

commutes. This is Proposition 21.8.3. □

Now that we have checked our sanity, we will prove that these operations satisfy basic properties that we've come to expect from operations called power operations. We give a complete proof of the next lemma in order to help the reader get used to manipulating P_m^ϕ.

Lemma 21.9.3. *For all* $\phi \in \Gamma(\mathrm{Sub}(\mathbb{T}), \mathrm{Isog}(\mathbb{T}))$*, the power operation* P_m^ϕ *is natural in the group variable.*

Proof. Let $\gamma\colon G \to K$ be a group homomorphism so that

$$\gamma^*\colon \mathrm{Cl}_n(K, C_0) \to \mathrm{Cl}_n(G, C_0)$$

is defined by $(\gamma^* f)([\alpha]) = f([\gamma\alpha])$ for $[\alpha\colon \mathbb{L} \to G]$. We wish to show that

$$P_m^\phi(\gamma^* f) = (\gamma \times \mathrm{id}_{\Sigma_m})^* P_m^\phi(f).$$

This follows because

$$\begin{aligned} P_m^\phi(\gamma^* f)([\alpha], \oplus_i H_i) &= \prod_i (\gamma^* f)([\alpha\phi_{H_i}^*])\phi_{H_i} \\ &= \prod_i f([\gamma\alpha\phi_{H_i}^*])\phi_{H_i} \\ &= P_m^\phi(f)([\gamma\alpha], \oplus_i H_i) \\ &= (\gamma \times \mathrm{id}_{\Sigma_m})^* P_m^\phi(f). \end{aligned}$$

□

Given two power operations on class function P_i^ϕ and P_j^ϕ, their external product is the map

$$P_i^\phi \times P_j^\phi\colon \mathrm{Cl}_n(G, C_0) \to \mathrm{Cl}_n(G \times \Sigma_i \times G \times \Sigma_j, C_0)$$

given by the formula

$$(P_i^\phi \times P_j^\phi)(f)(([\alpha_1], \oplus_i H_i), ([\alpha_2], \oplus_i K_i)) = P_i^\phi(f)([\alpha_1], \oplus_i H_i) P_j^\phi(f)([\alpha_2], \oplus_i K_i).$$

Just as at the end of Section 21.5, let $\tau\Delta$ be the map of groups

$$G \times \Sigma_i \times \Sigma_j \to G \times G \times \Sigma_i \times \Sigma_j \to G \times \Sigma_i \times G \times \Sigma_j,$$

where the first map is induced by the diagonal and the second by the twist isomorphism. Restricting $P_i^\phi \times P_j^\phi$ along the $\tau\Delta$ gives us a map

$$\Delta^*\tau^*(P_i^\phi \times P_j^\phi)\colon \mathrm{Cl}_n(G, C_0) \to \mathrm{Cl}_n(G \times G \times \Sigma_i \times \Sigma_j, C_0).$$

The following lemma is an analogue of the identity involving power operations in Diagram (21.5.3). We leave the proof to the reader.

Lemma 21.9.4. *Let $\phi \in \Gamma(\mathrm{Sub}(\mathbb{T}), \mathrm{Isog}(\mathbb{T}))$ and let $i, j > 0$ with $i+j = m$, then restriction along the inclusion $\Delta_{i,j}\colon \Sigma_i \times \Sigma_j \subset \Sigma_m$ induces the commutative diagram*

$$\begin{array}{ccc} \mathrm{Cl}_n(G, C_0) & \xrightarrow{P_m^\phi} & \mathrm{Cl}_n(G \times \Sigma_m, C_0) \\ & \searrow^{\Delta^*\tau^*(P_i^\phi \times P_j^\phi)} & \downarrow \Delta_{i,j}^* \\ & & \mathrm{Cl}_n(G \times \Sigma_i \times \Sigma_j, C_0). \end{array}$$

Now that we have developed some basic properties of the operations P_m^ϕ, we describe three injections that will each play a role in the proof of Theorem 21.9.8.

Lemma 21.9.5. *Let $\sum_j a_j p^j$ be the p-adic expansion of m. The inclusion $\prod_j \Sigma_{p^j}^{\times a_j} \subseteq \Sigma_m$ induces an injection*

$$\mathrm{Cl}_n(\Sigma_m, C_0) \hookrightarrow \bigotimes_j \mathrm{Cl}_n(\Sigma_{p^j}, C_0)^{\otimes a_j}.$$

Proof. This follows immediately from Corollary 21.7.8. □

We leave the proof of the next lemma to the reader.

Lemma 21.9.6. *Let G be a finite group. The product of the restriction maps to abelian subgroups of G is an injection*

$$\mathrm{Cl}_n(G, C_0) \hookrightarrow \prod_{A \subset G} \mathrm{Cl}_n(A, C_0),$$

where the product ranges over all abelian subgroups of G.

Lemma 21.9.7. *For all $k > 0$, there is an injection*

$$\mathrm{Cl}_n(\Sigma_{p^k}, C_0) \hookrightarrow \mathrm{Cl}_n(\Sigma_{p^{k-1}}^{\times p}, C_0) \times \mathrm{Cl}_n(\Sigma_{p^k}, C_0)/I_{\mathrm{tr}},$$

where the map to the left factor is the restriction along $\Sigma_{p^{k-1}}^{\times p} \subset \Sigma_{p^k}$ and the map to the right factor is the quotient map.

Proof. Applying Proposition 21.7.5 and Corollary 21.7.6 as well as Lemma 21.7.7, we see that this map is the C_0-valued functions on the map of sets

$$\mathrm{Sub}_{p^k}(\mathbb{T}) \coprod \mathrm{Sum}_{p^{k-1}}(\mathbb{T})^{\times p} \to \mathrm{Sum}_{p^k}(\mathbb{T})$$

sending a subgroup in the first component to itself and the p-tuple of sums of subgroups in the second component to their sum. This is a surjective map of sets. □

Finally we can state the result that we have been working towards:

Theorem 21.9.8. *Let $\phi \in \Gamma(\mathrm{Sub}(\mathbb{T}), \mathrm{Isog}(\mathbb{T}))$ and let G be a finite group. There is a commutative diagram*

$$\begin{array}{ccc} E^0(BG) & \xrightarrow{P_m} & E^0(BG \times B\Sigma_m) \\ \chi \downarrow & & \downarrow \chi \\ \mathrm{Cl}_n(G, C_0) & \xrightarrow{P_m^\phi} & \mathrm{Cl}_n(G \times \Sigma_m, C_0). \end{array}$$

It may seem unsatisfying that any choice of section gives us a power operation on generalized class functions that is compatible with the power operation on Morava E-theory. This is essentially a consequence of the fact that C_0 is an $\mathrm{Aut}(\mathbb{T})$-Galois extension of $\mathbb{Q}\otimes E^0$. It turns out that the choice disappears after taking the $\mathrm{Aut}(\mathbb{T})$-invariants of the ring of generalized class functions!

The rest of this section constitutes a proof of Theorem 21.9.8. The proof of Theorem 21.9.8 has three steps. We will first reduce to the case that $m = p^k$, then we will reduce to the case that G is abelian and finally we will perform an induction on k.

We begin with the reduction to the case $m = p^k$. Consider the commutative diagram

$$\begin{array}{ccccc} E^0(BG) & \xrightarrow{P_m} & E^0(BG \times B\Sigma_m) & \longrightarrow & E^0(BG \times \prod_j B\Sigma_{p^j}^{a_j}) \\ \downarrow & & \downarrow & & \downarrow \\ \mathrm{Cl}_n(G, C_0) & \xrightarrow{P_m^{\phi}} & \mathrm{Cl}_n(G \times \Sigma_m, C_0) & \hookrightarrow & \mathrm{Cl}_n(G \times \prod_j \Sigma_{p^j}^{a_j}, C_0). \end{array}$$

The composite along the top is the external product of power operations $\Delta^*\tau^*(\prod_j P_{p_j}^{\times a_j})$. The bottom composite is the external product $\Delta^*\tau^*(\prod_j (P_{p_j}^{\phi})^{\times a_j})$. Since the bottom right arrow is an injection by Lemma 21.9.5, to prove that the left hand square commutes, it suffices to prove the large square commutes. The large square is built only using power operations of the form P_{p^k} and $P_{p^k}^{\phi}$.

Now we describe the reduction to finite abelian groups. Assume that the diagram

$$\begin{array}{ccc} E^0(BA) & \xrightarrow{P_{p^k}} & E^0(BA \times B\Sigma_{p^k}) \\ \chi \downarrow & & \downarrow \chi \\ \mathrm{Cl}_n(A, C_0) & \xrightarrow{P_{p^k}^{\phi}} & \mathrm{Cl}_n(A \times \Sigma_{p^k}, C_0) \end{array}$$

commutes for any choice of ϕ and any finite abelian group. Consider the cube

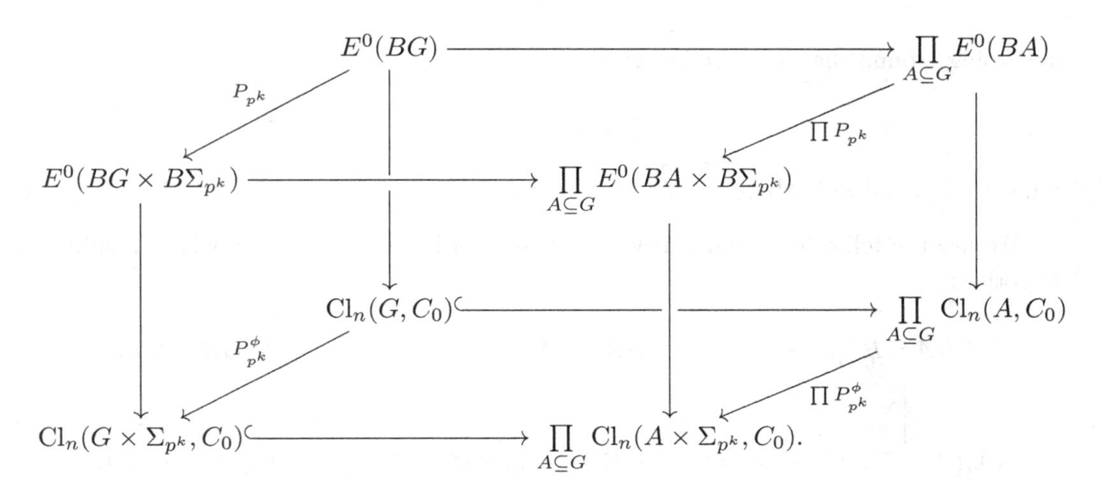

We do not claim that this cube commutes yet. Since Lemma 21.9.6 implies that the horizontal arrows of the bottom face are injections, a diagram chase shows that the left face commutes if the right face commutes. This reduces us to the case of finite abelian groups.

Now we come to the inductive part of the proof. The base case of our induction, $k = 0$, is the following proposition:

Proposition 21.9.9. *For each section $\phi \in \Gamma(\mathrm{Sub}(\mathbb{T}), \mathrm{Isog}(\mathbb{T}))$, there is a commutative diagram*

$$\begin{array}{ccc} E^0(BA) & \xrightarrow{P_1} & E^0(BA) \\ \downarrow \chi & & \downarrow \chi \\ \mathrm{Cl}_n(A, C_0) & \xrightarrow{P_1^{\phi}} & \mathrm{Cl}_n(A, C_0). \end{array}$$

Proof. Recall that P_1 is the identity map and let $e \subset \mathbb{T}$ be the trivial subgroup so ϕ_e is an automorphism of $\mathbb{T}$. The power operation P_1^{ϕ} is defined by

$$P_1^{\phi}(f)([\alpha]) = f([\alpha\phi_e])\phi_e.$$

Thus, working one factor at a time, it suffices to show that the diagram

$$\begin{array}{ccc} E^0(BA) & \xrightarrow{\mathrm{id}} & E^0(BA) \\ \downarrow \chi_{[\alpha\phi_e^*]} & & \downarrow \chi_{[\alpha]} \\ \mathcal{O}_{\mathrm{Level}(\mathbb{T},\mathbb{G}_u)} & \xrightarrow{\phi_e} & \mathcal{O}_{\mathrm{Level}(\mathbb{T},\mathbb{G}_u)} \end{array}$$

commutes. Now the proof follows the same lines as the proof of Proposition 21.8.3. Since $H = e$, the map ψ_H of Diagram (21.4.4) is just ϕ_e. Applying Example 21.6.2 to each of the vertical arrows and (21.4.3) to the bottom arrow, we see that, on the level of moduli problems, going around the bottom direction gives

$$(l, i, \tau) \mapsto (l\phi_e^{-1}, i, \tau) \mapsto (l\phi_e^{-1}(\alpha\phi_e^*)^*, i, \tau)$$

and going around the top direction gives

$$(l, i, \tau) \mapsto (l\alpha^*, i, \tau).$$

Since $l\phi_e^{-1}(\alpha\phi_e^*)^* = l\phi_e^{-1}\phi_e\alpha^* = l\alpha^*$, we are done. □

We use the following commutative diagram in order to be able to apply our induction hypothesis

$$\begin{array}{ccc} E^0(BA \times B\Sigma_{p^k}) & \longrightarrow & E^0(BA \times B\Sigma_{p^{k-1}}^{\times p}) \times (E^0(BA) \otimes_{E^0} E^0(B\Sigma_{p^k})/I_{\mathrm{tr}}) \\ \downarrow \chi & & \downarrow \\ \mathrm{Cl}_n(A \times \Sigma_{p^k}, C_0) & \hookrightarrow & \mathrm{Cl}_n(A \times \Sigma_{p^{k-1}}^{\times p}, C_0) \times (\mathrm{Cl}_n(A, C_0) \otimes_{C_0} \mathrm{Cl}_n(\Sigma_{p^k}, C_0)/I_{\mathrm{tr}}), \end{array}$$

where both of the horizontal arrows are the product of the restriction and quotient maps and the bottom horizontal arrow is an injection by Lemma 21.9.7.

Now Diagram (21.5.3) and Lemma 21.9.4 imply that it is suffices to prove that the two diagrams

$$\begin{array}{ccc} E^0(BA) & \xrightarrow{\Delta^*\tau^*(P^{\times p}_{p^{k-1}})} & E^0(BA)\otimes_{E^0} E^0(B\Sigma^{\times p}_{p^{k-1}}) \\ \downarrow & & \downarrow \\ \mathrm{Cl}_n(A,C_0) & \xrightarrow{\Delta^*\tau^*((P^{\phi}_{p^{k-1}})^{\times p})} & \mathrm{Cl}_n(A,C_0)\otimes_{C_0}\mathrm{Cl}_n(\Sigma^{\times p}_{p^{k-1}},C_0) \end{array}$$

and

$$\begin{array}{ccc} E^0(BA) & \xrightarrow{P_{p^k}/I_{\mathrm{tr}}} & E^0(BA)\otimes_{E^0} E^0(B\Sigma_{p^k})/I_{\mathrm{tr}} \\ \downarrow & & \downarrow \\ \mathrm{Cl}_n(A,C_0) & \xrightarrow{P^{\phi}_{p^k}/I_{\mathrm{tr}}} & \mathrm{Cl}_n(A,C_0)\otimes_{C_0}\mathrm{Cl}_n(\Sigma_{p^k},C_0)/I_{\mathrm{tr}} \end{array}$$

commute. The first commutes by the induction hypothesis and the second commutes by Proposition 21.9.2.

21.10 Aut($\mathbb{T}$)-fixed points and the total power operation

In this final section we will describe the restriction of P^{ϕ}_m to Aut($\mathbb{T}$)-invariant class functions thereby proving Theorem 21.1.2, we will discuss the relationship between the power operations P^{ϕ}_m and the stabilizer group action, and we will describe some of the changes necessary to produce a total power operation on class functions.

In order to show that Aut($\mathbb{T}$)-invariant class functions are sent to Aut($\mathbb{T}$)-invariant class functions by P^{ϕ}_m, we need the following lemma, which describes the Aut($\mathbb{T}$)-action induced on $\mathrm{Sum}_m(\mathbb{T})$ through the isomorphism of Proposition 21.7.5.

In order to show that Aut($\mathbb{T}$)-invariant class functions are sent to Aut($\mathbb{T}$)-invariant class functions by P^{ϕ}_m, we need to understand the Aut($\mathbb{T}$)-action induced on $\mathrm{Sum}_m(\mathbb{T})$ through the isomorphism of Proposition 21.7.5. The automorphism $\gamma \in \mathrm{Aut}(\mathbb{T})$ acts on $\mathrm{Hom}(\mathbb{L},\Sigma_m)/{\sim}$ by $\gamma\cdot[\alpha] = [\gamma^*\alpha]$. The corresponding action of Aut($\mathbb{T}$) on $\mathrm{Sum}_m(\mathbb{T})$ through the isomorphism of Proposition 21.7.5 is given by

$$\gamma \cdot \oplus_i H_i = \oplus_i \gamma H_i.$$

To see this, recall from the proof of Proposition 21.7.5 that $H_i = (\mathbb{L}\,/\,\mathbb{L}_i)^* \subset \mathbb{T}$, where $\mathbb{L}_i$ is the stabilizer of the transitive $\mathbb{L}$-set $\underline{m}_i$. The stabilizer of the $\mathbb{L}$-set structure on $\underline{m}_i$ given by acting through γ^* is $(\gamma^*)^{-1}\,\mathbb{L}_i$ and there is a canonical isomorphism

$$(\mathbb{L}\,/(\gamma^*)^{-1}\,\mathbb{L}_i)^* \cong \gamma((\mathbb{L}\,/\,\mathbb{L}_i)^*) = \gamma H_i.$$

We leave the following calculation to the reader.

Proposition 21.10.1. *For any section $\phi \in \Gamma(\mathrm{Sub}(\mathbb{T}), \mathrm{Isog}(\mathbb{T}))$, the power operation*

$$P_m^{\phi}\colon \mathrm{Cl}_n(G, C_0) \to \mathrm{Cl}_n(G \times \Sigma_m, C_0)$$

sends $\mathrm{Aut}(\mathbb{T})$*-invariant class functions to* $\mathrm{Aut}(\mathbb{T})$*-invariant class functions and the resulting map*

$$\mathrm{Cl}_n(G, C_0)^{\mathrm{Aut}(\mathbb{T})} \to \mathrm{Cl}_n(G \times \Sigma_m, C_0)^{\mathrm{Aut}(\mathbb{T})}$$

is independent of the choice of ϕ.

Theorem 21.6.3 states that there is an isomorphism

$$\mathbb{Q} \otimes E^0(BG) \cong \mathrm{Cl}_n(G, C_0)^{\mathrm{Aut}(\mathbb{T})}.$$

Thus we have produced a "rational power operation"

$$P_m^{\mathbb{Q}}\colon \mathbb{Q} \otimes E^0(BG) \to \mathbb{Q} \otimes E^0(BG \times B\Sigma_m)$$

and, at the same time, given a formula for it.

Proposition 21.10.2. *For any $\phi \in \Gamma(\mathrm{Sub}(\mathbb{T}), \mathrm{Isog}(\mathbb{T}))$, the diagonal action of the stabilizer group* $\mathrm{Aut}(\mathbb{F}/\kappa)$ *on generalized class functions commutes with the power operation P_m^{ϕ}. That is, for $s \in \mathrm{Aut}(\mathbb{F}/\kappa)$, there is a commutative diagram*

$$\begin{array}{ccc} \mathrm{Cl}_n(G, C_0) & \xrightarrow{P_m^{\phi}} & \mathrm{Cl}_n(G \times \Sigma_m, C_0) \\ \downarrow{\scriptstyle s} & & \downarrow{\scriptstyle s} \\ \mathrm{Cl}_n(G, C_0) & \xrightarrow{P_m^{\phi}} & \mathrm{Cl}_n(G \times \Sigma_m, C_0). \end{array}$$

Proof. Recall that $\mathrm{Aut}(\mathbb{F}/\kappa)$ acts on C_0 on the right through ring maps and that $\mathrm{Isog}(\mathbb{T})$ acts on C_0 on the right through ring maps. Tracing through the diagram in the proposition, we want to show that, for $f \in \mathrm{Cl}_n(G, C_0)$ and $s \in \mathrm{Aut}(\mathbb{F}/\kappa)$,

$$\prod_i f([\alpha\phi_{H_i}^*])s\phi_{H_i} = \prod_i f([\alpha\phi_{H_i}^*])\phi_{H_i}s.$$

Thus it suffices to show that the actions of $\mathrm{Aut}(\mathbb{F}/\kappa)$ and $\mathrm{Isog}(\mathbb{T})$ on C_0 commute. Since $C_0 = \mathbb{Q} \otimes \mathcal{O}_{\mathrm{Level}(\mathbb{T}, \mathbb{G}_u)}$, it suffices to show that the actions commute on $\mathcal{O}_{\mathrm{Level}(\mathbb{T}, \mathbb{G}_u)}$ and this is a question about moduli problems over Lubin-Tate space.

Recall the formulas of (21.4.1) and (21.4.3). Since $\mathrm{Aut}(\mathbb{F}, \kappa)$ does not affect the data of $\mathbb{G}$ or i in a deformation, it suffices to show that

$$((i\sigma^k)^*s)(\tau/H) = (i^*s\tau)/H$$

and this is a statement about isogenies of formal groups over κ. Unwrapping this equality, we want to show that the diagram

$$\begin{array}{ccccc} \pi^*\mathbb{G} & \xrightarrow{\tau} & i^*\mathbb{F} & \xrightarrow{i^*s} & i^*\mathbb{F} \\ \downarrow & & \downarrow {\scriptstyle i^*\mathrm{Frob}^k} & & \downarrow {\scriptstyle i^*\mathrm{Frob}^k} \\ (\pi^*\mathbb{G})/H & \xrightarrow[\tau/H]{} & (i\sigma^k)^*\mathbb{F} & \xrightarrow[(i\sigma^k)^*s]{} & (i\sigma^k)^*\mathbb{F} \end{array}$$

commutes. The left hand square commutes by construction and the right hand square is i^* applied to the square

$$\begin{array}{ccc} \mathbb{F} & \xrightarrow{s} & \mathbb{F} \\ {\scriptstyle \mathrm{Frob}^k}\downarrow & & \downarrow{\scriptstyle \mathrm{Frob}^k} \\ (\sigma^k)^*\mathbb{F} & \xrightarrow{(\sigma^k)^*s} & (\sigma^k)^*\mathbb{F}. \end{array}$$

It is not hard to see that this commutes. Choose a coordinate so that we have formal group laws $x +_{\mathbb{F}} y$ and $x +_{(\sigma^k)^*\mathbb{F}} y$ and automorphisms of formal group laws $f_s(x)$ and $f_{(\sigma^k)^*s}(x)$ such that

$$(x +_{\mathbb{F}} y)^{p^k} = x^{p^k} +_{(\sigma^k)^*\mathbb{F}} y^{p^k}$$

and

$$(f_s(x))^{p^k} = f_{(\sigma^k)^*s}(x^{p^k}).$$

Then the induced diagram of κ-algebras

$$\begin{array}{ccc} \kappa[\![x]\!] & \longleftarrow & \kappa[\![x]\!] \\ \uparrow & & \uparrow \\ \kappa[\![x]\!] & \longleftarrow & \kappa[\![x]\!] \end{array}$$

sends

$$\begin{array}{ccc} f_{(\sigma^k)^*s}(x^{p^k}) = (f_s(x))^{p^k} & \longleftarrow\!\shortmid & x^{p^k} \\ \uparrow & & \uparrow \\ f_{(\sigma^k)^*s}(x) & \longleftarrow\!\shortmid & x \end{array}$$

and thus commutes. □

Finally, we describe a few of the changes necessary to construct a total power operation, rather than just a power operation, on generalized class functions and prove that it is compatible with the total power operation on Morava E-theory of (21.5.1). Just as in the case of P^{ϕ}_m, the construction of the total power operation on generalized class functions

$$\mathbb{P}^{\phi}_m\colon \mathrm{Cl}_n(G, C_0) \to \mathrm{Cl}_n(G \wr \Sigma_m, C_0)$$

depends on the choice of a section $\phi \in \Gamma(\mathrm{Sub}(\mathbb{T}), \mathrm{Isog}(\mathbb{T}))$.

Diagram (21.4.4) plays a more important role in the definition of $\mathbb{P}^{\phi}_{m}$. For $H \subset \mathbb{T}$, let $\mathbb{L}_H = (\mathbb{T}/H)^*$ so that we have an inclusion $q^*_H : \mathbb{L}_H \hookrightarrow \mathbb{L}$ and an isomorphism

$$\psi^*_H : \mathbb{L} \xrightarrow{\cong} \mathbb{L}_H .$$

Consider the set

$$\mathrm{Sum}_m(\mathbb{T}, G) = \{\oplus_i(H_i, [\alpha_i]) | H_i \subset \mathbb{T}, \sum_i |H_i| = m, \text{ and } [\alpha_i : \mathbb{L}_{H_i} \to G]\}.$$

When $G = e$, this is $\mathrm{Sum}_m(\mathbb{T})$. Proposition 21.7.5 may be generalized to give the following proposition.

Proposition 21.10.3. *There is a canonical bijection*

$$\mathrm{Hom}(\mathbb{L}, G \wr \Sigma_m)/\sim \; \cong \mathrm{Sum}_m(\mathbb{T}, G).$$

Using this proposition, we define $\mathbb{P}^{\phi}_{m}$ as follows:

$$\mathbb{P}^{\phi}_{m}(f)(\oplus_i(H_i, [\alpha_i])) = \prod_i f([\alpha_i \psi^*_{H_i}])\phi_{H_i}$$

for $f \in \mathrm{Cl}_n(G, C_0)$.

The main result of [5] is the following:

Theorem 21.10.4. [5, Theorem 9.1] *Let* $\phi \in \Gamma(\mathrm{Sub}(\mathbb{T}), \mathrm{Isog}(\mathbb{T}))$ *and let* G *be a finite group. There is a commutative diagram*

$$\begin{array}{ccc} E^0(BG) & \xrightarrow{\mathbb{P}_m} & E^0(BG \wr \Sigma_m) \\ \downarrow{\scriptstyle \chi} & & \downarrow{\scriptstyle \chi} \\ \mathrm{Cl}_n(G, C_0) & \xrightarrow{\mathbb{P}^{\phi}_{m}} & \mathrm{Cl}_n(G \wr \Sigma_m, C_0). \end{array}$$

To prove Theorem 21.10.4, we need generalizations of the ingredients that were required for the proof of Theorem 21.9.8. We require a generalization of Theorem 21.8.2 to the additive total power operation (21.5.2) applied to abelian groups:

$$\mathbb{P}_m : E^0(BA) \to E^0(BA \wr \Sigma_m)/I_{\mathrm{tr}}.$$

This is provided by [5, Theorem 8.4]. It makes use of a generalization of Theorem 21.8.1 to rings of the form $E^0(BA \wr \Sigma_{p^k})/I_{\mathrm{tr}}$, which is the main result of [18].

Bibliography

[1] J. F. Adams. Maps between classifying spaces. II. *Invent. Math.*, 49(1):1–65, 1978.

[2] J. F. Adams. *Stable homotopy and generalised homology.* Chicago Lectures in Mathematics. University of Chicago Press, 1995. Reprint of the 1974 original.

[3] Matthew Ando, Michael J. Hopkins, and Neil P. Strickland. The sigma orientation is an H_∞ map. *Amer. J. Math.*, 126(2):247–334, 2004.

[4] Michael F. Atiyah. Characters and cohomology of finite groups. *Publ. Math., Inst. Hautes Etud. Sci.*, 9:247–288, 1961.

[5] Tobias Barthel and Nathaniel Stapleton. The character of the total power operation. *Geom. Topol.*, 21(1):385–440, 2017.

[6] R. R. Bruner, J. P. May, J. E. McClure, and M. Steinberger. H_∞ *ring spectra and their applications*, volume 1176 of *Lecture Notes in Math.*. Springer-Verlag, 1986.

[7] H. Carayol. Nonabelian Lubin-Tate theory. In *Automorphic forms, Shimura varieties, and L-functions, Vol. II (Ann Arbor, MI, 1988)*, volume 11 of *Perspect. Math.*, pages 15–39. Academic Press, 1990.

[8] V. G. Drinfel′d. Elliptic modules. *Mat. Sb. (N.S.)*, 94(136):594–627, 656, 1974.

[9] P. G. Goerss and M. J. Hopkins. Moduli spaces of commutative ring spectra. In *Structured ring spectra*, volume 315 of *London Math. Soc. Lecture Note Ser.*, pages 151–200. Cambridge Univ. Press, 2004.

[10] Michael Hopkins. Complex oriented cohomology theories and the language of stacks. Available at `http://www.math.rochester.edu/people/faculty/doug/otherpapers/coctalos.pdf`.

[11] Michael J. Hopkins, Nicholas J. Kuhn, and Douglas C. Ravenel. Generalized group characters and complex oriented cohomology theories. *J. Am. Math. Soc.*, 13(3):553–594, 2000.

[12] Jonathan Lubin. Finite subgroups and isogenies of one-parameter formal Lie groups. *Ann. of Math. (2)*, 85:296–302, 1967.

[13] Jonathan Lubin and John Tate. Formal moduli for one-parameter formal Lie groups. *Bull. Soc. Math. France*, 94:49–59, 1966.

[14] Jacob Lurie. Chromatic homotopy theory (252x). Available at `http://www.math.harvard.edu/~lurie/252x.html`.

[15] Haynes Miller. Notes on cobordism. Available at `http://www-math.mit.edu/~hrm/papers/cobordism.pdf`.

[16] Eric Peterson. *Formal Geometry and Bordism Operations*. Volume 177 of *Cambridge Stud. Adv. Math.*, Cambridge University Press, 2018.

[17] Charles Rezk. Notes on the Hopkins-Miller theorem. In *Homotopy theory via algebraic geometry and group representations (Evanston, IL, 1997)*, volume 220 of *Contemp. Math.*, pages 313–366. Amer. Math. Soc., 1998.

[18] Tomer M. Schlank and Nathaniel Stapleton. A transchromatic proof of Strickland's theorem. *Adv. Math.*, 285:1415–1447, 2015.

[19] N. P. Strickland. Morava E-theory of symmetric groups. *Topology*, 37(4):757–779, 1998.

[20] N. P. Strickland. Finite subgroups of formal groups. *J. Pure Appl. Algebra*, 121(2):161–208, 1997.

[21] N. P. Strickland. Formal schemes and formal groups. In *Homotopy invariant algebraic structures (Baltimore, MD, 1998)*, volume 239 of *Contemp. Math.*, pages 263–352. Amer. Math. Soc., 1999.

Department of Mathematics, University of Kentucky, Lexington, KY 40506, U.S.A.

E-mail address: `nat.j.stapleton@gmail.com`

22

Unstable motivic homotopy theory

Kirsten Wickelgren and Ben Williams

22.1 Introduction

Morel–Voevodsky's $\mathbb{A}^1$-homotopy theory transports tools from algebraic topology into arithmetic and algebraic geometry, allowing us to draw arithmetic conclusions from topological arguments. Comparison results between classical and $\mathbb{A}^1$-homotopy theories can also be used in the reverse direction, allowing us to infer topological results from algebraic calculations. For example, see the article by Isaksen and Østvær on motivic stable homotopy groups in this volume. The present article will introduce unstable $\mathbb{A}^1$-homotopy theory and give several applications.

Underlying all $\mathbb{A}^1$-homotopy theories is some category of schemes. A special case of a scheme is that of an *affine scheme*, $\operatorname{Spec} R$, which has an associated topological space, the points of which are the prime ideals of R, and on which there is a sheaf of rings essentially provided by R itself. For example, when R is a finitely generated k-algebra, R can be written as $k[x_1, \ldots, x_n]/\langle f_1, \ldots, f_m\rangle$, and $\operatorname{Spec} R$ can be thought of as the common zero locus of the polynomials $f_1, f_2, \ldots, f_m$, that is to say, $\{(x_1, \ldots, x_n) : f_i(x_1, \ldots, x_n) = 0 \text{ for } i = 1, \ldots, m\}$. Indeed, for a commutative k-algebra A, the set

$$(\operatorname{Spec} R)(A) := \{(x_1, \ldots, x_n) \in A^n : f_i(x_1, \ldots, x_n) = 0 \text{ for } i = 1, \ldots, m\}$$

is the set of A-points of $\operatorname{Spec} R$, where an A-point is a map $\operatorname{Spec} A \to \operatorname{Spec} R$. We remind the reader that a scheme X is a locally ringed space that is locally isomorphic to affine schemes. The topology on a scheme X is called the *Zariski topology*, with basis given by subsets U of affines $\operatorname{Spec} R$ of the form $U = \{\mathfrak{p} \in \operatorname{Spec} R : g \notin \mathfrak{p}\}$ for some g in R. It is too coarse to be of use for classical homotopy theory; for instance, the topological space of an irreducible scheme is contractible, having the generic point as a deformation retract. Standard references for the theory of schemes include [50], [99], and for the geometric view that motivates the theory [37].

Both to avoid certain pathologies and to make use of technical theorems that can be proved under certain assumptions, $\mathbb{A}^1$-homotopy theory restricts itself to considering subcategories of the category of all schemes. In the seminal [80] the restriction is already made to $\mathbf{Sm}_S$, the full subcategory of smooth schemes of finite type over a finite dimensional

Mathematics Subject Classification. 14F42, 19E15, 11E81, 14J60, 55R25, 55P91, 14N10.

Key words and phrases. unstable Motivic homotopy theory, unstable A^1-homotopy theory, algebraic K-theory, Milnor–Witt K-theory, Milnor K-Theory, Euler class, enumerative geometry, classification of vector bundles, Murthy's conjecture.

noetherian base scheme S. The smoothness condition, while technical, is geometrically intuitive. For example, when $S = \operatorname{Spec}\mathbb{C}$, the smooth S-schemes are precisely those whose $\mathbb{C}$-points form a smooth manifold. The "finite-type" condition is most easily understood if $S = \operatorname{Spec} R$ is affine, in which case the finite-type S-schemes are those covered by finitely many affine schemes, each of which is determined by the vanishing of finitely many polynomials in finitely many variables over the ring R. The most important case, and the best understood, is when $S = \operatorname{Spec} k$ is the spectrum of a field. In this case $\mathbf{Sm}_k$ is the category of smooth k-varieties.

In this article, we will use the notation S to indicate a base scheme, always assumed noetherian and finite dimensional. For the more sophisticated results later in this article, it will be necessary to assume $S = \operatorname{Spec} k$ for some field k. We have not made efforts to be precise about the maximal generality of base scheme S for which a particular result is known to hold.

To see the applicability of $\mathbb{A}^1$-homotopy theory, consider the following example. The topological Brouwer degree map

$$\deg : [S^n, S^n] \to \mathbb{Z}$$

from pointed homotopy classes of maps of the n-sphere to itself can be evaluated on a smooth map

$$f : S^n \to S^n$$

in the following manner. Choose a regular value p in S^n and consider its finitely many preimages $f^{-1}(p) = \{q_1, \dots, q_m\}$. At each point q_i, choose local coordinates compatible with a fixed orientation on S^n. The induced map on tangent spaces can then be viewed as an $\mathbb{R}$-linear isomorphism $T_{q_i} f : \mathbb{R}^n \to \mathbb{R}^n$, with an associated Jacobian determinant $\operatorname{Jac} f(q_i) = \det(\frac{\partial (T_{q_i} f)_j}{x_k})_{j,k}$. The assumption that p is a regular value implies that $\operatorname{Jac} f(q_i) \neq 0$, and therefore there is a local degree $\deg_{q_i} f$ of f at q_i such that

$$\deg_{q_i} f = \begin{cases} +1 \text{ if } \operatorname{Jac} f(q_i) > 0, \\ -1 \text{ if } \operatorname{Jac} f(q_i) < 0. \end{cases}$$

The degree $\deg f$ of f is then given

$$\deg f = \sum_{q \in f^{-1}(p)} \deg_q f,$$

as the appropriate sum of $+1$'s and -1's.

Lannes and Morel suggested a modification of this formula to give a degree for an algebraic function $f : \mathbb{P}^1 \to \mathbb{P}^1$ valued in nondegenerate symmetric bilinear forms. Let k be a field and let $\mathrm{GW}(k)$ denote the Grothendieck–Witt group, whose elements are formal differences of k-valued, non-degenerate, symmetric, bilinear forms on finite dimensional k-vector spaces. We will say more about this group in Section 22.4. For a in $k^*/(k^*)^2$, denote by $\langle a \rangle$ the element of $\mathrm{GW}(k)$ determined by the isomorphism class of the bilinear form $(x, y) \mapsto axy$. For simplicity, assume that p is a k-point of $\mathbb{P}^1$, i.e., p is an element of k or ∞, and that the points q of $f^{-1}(p)$ are also k-points such that $\operatorname{Jac} f(q) \neq 0$. Then the $\mathbb{A}^1$-degree of f in $\mathrm{GW}(k)$ is given by

$$\deg f = \sum_{q \in f^{-1}(p)} \langle \operatorname{Jac} f(q) \rangle.$$

In other words, the $\mathbb{A}^1$-degree counts the points of the inverse image weighted by their Jacobians, instead of weighting only by the signs of their Jacobians.

Morel shows that this definition extends to an $\mathbb{A}^1$-degree homomorphism

$$\deg : [\mathbb{P}^n_k/\mathbb{P}^{n-1}_k, \mathbb{P}^n_k/\mathbb{P}^{n-1}_k] \to \mathrm{GW}(k) \tag{22.1.1}$$

from $\mathbb{A}^1$-homotopy classes of endomorphisms of the quotient $\mathbb{P}^n_k/\mathbb{P}^{n-1}_k$ to the Grothendieck–Witt group. We will use this degree to enrich results in classical enumerative geometry over $\mathbb{C}$ to equalities in $\mathrm{GW}(k)$ in Section 22.4.6. For (22.1.1) to make sense, we must define the quotient $\mathbb{P}^n_k/\mathbb{P}^{n-1}_k$. It is not a scheme; it is a space in the sense of Morel–Voevodsky. In Section 22.2, we sketch the construction of the homotopy theory of spaces, including Thom spaces and the Purity Theorem. In Section 22.3, we discuss realization functors to topological spaces, which allow us to see how $\mathbb{A}^1$-homotopy theory combines phenomena associated to the real and complex points of a variety. An example of this is the degree, discussed in the following section, along with Euler classes. The Milnor–Witt K-theory groups are also introduced in Section 22.4. These are the global sections of certain unstable homotopy sheaves of spheres. The following section, Section 22.5, discusses homotopy sheaves of spaces, characterizing important properties. It also states the unstable connectivity theorem. The last section describes some beautiful applications to the study of algebraic vector bundles.

Some things we do not do in this article include Voevodsky's groundbreaking work on the Bloch-Kato and Milnor conjectures. A superb overview of this is given [76]. We also do not deal with any stable results, which have their own chapter in this Handbook, by Daniel Isaksen and Paul Arne Østvær.

In our presentation, we concentrate on those aspects and applications of the theory that relate to the calculation of the unstable homotopy sheaves of spheres. The most notable of the applications, at present, is the formation of an $\mathbb{A}^1$-obstruction theory of $B\,\mathrm{GL}_n$, and from there, the proof of strong results about the existence and classification of vector bundles on smooth affine varieties. These can be found in Section 22.6.

Acknowledgements

We wish to thank Aravind Asok, Tom Bachmann, Jean Fasel, Raman Parimala, Joseph Rabinoff and Krishanu Sankar, as well as the organizers of Homotopy Theory Summer Berlin 2018, the Newton Institute and the organizers of the Homotopy Harnessing Higher Structures programme. The first-named author was partially supported by National Science Foundation Award DMS-1552730.

22.2 Overview of the construction of unstable $\mathbb{A}^1$-homotopy theory

22.2.1 Homotopy theory of Spaces

Morel and Voevodsky [80] constructed a "homotopy theory of schemes", in a category sufficiently general to include both schemes and simplicial sets. By convention, the objects in such a category may be called a "motivic space" or an "$\mathbb{A}^1$-space" or, most commonly, simply a "space". The underlying category of spaces is a category of simplicial presheaves on a category of schemes, where schemes themselves are embedded by means of a Yoneda functor. Two localizations are then performed: a Nisnevich localization and an $\mathbb{A}^1$-localization. Before describing these, we clarify the notion of "a homotopy theory" that we use. Two standard choices are that a homotopy theory is a simplicial model category or a homotopy theory is an ∞-category, a.k.a., a quasi-category. Background on simplicial model categories can be found in [52] and [88]. Lurie's *Higher Topos Theory* [70] contains many tools used for doing $\mathbb{A}^1$-homotopy theory with ∞-categories. A short introduction to ∞-categories is contained in Moritz Groth's chapter in this volume. In the present case, either the model-categorical or the ∞-categorical approach can be chosen, and once this choice has been made, presheaves of simplicial sets and the resulting localizations take on two different, although compatible, meanings, either of which produces a homotopy theory of schemes as described below.

To fix ideas, we use simplicial model categories. Let S be a noetherian scheme of finite dimension, and let $\mathbf{Sm}_S$ denote the category of smooth schemes of finite type over S, defined as a full subcategory of schemes over S. Let $\mathbf{sPre}(\mathbf{Sm}_S)$ denote the category of functors from the opposite category of $\mathbf{Sm}_S$ to simplicial sets, i.e., $\mathbf{sPre}(\mathbf{Sm}_S) = \mathrm{Fun}(\mathbf{Sm}_S^{\mathrm{op}}, \mathbf{sSet})$. Objects of $\mathbf{sPre}(\mathbf{Sm}_S)$ are called *simplicial presheaves*. The properties of $\mathbf{sPre}(\mathbf{Sm}_S)$ that make it desirable for homotopy theoretic purposes are the following:

- There is a *Yoneda embedding* $\eta : \mathbf{Sm}_S \to \mathbf{sPre}(\mathbf{Sm}_S)$, sending a scheme X to the simplicial presheaf $U \mapsto \mathbf{Sm}_S(U, X)$, where the set of maps is understood as a 0-dimensional simplicial set.

- The category $\mathbf{sPre}(\mathbf{Sm}_S)$ has all small limits and colimits. In this it is different from the category $\mathbf{Sm}_S$ itself, which is far from containing all the colimits one might wish for in doing homotopy theory. The Yoneda embedding preserves limits but not colimits.

- The category $\mathbf{sPre}(\mathbf{Sm}_S)$ is tensored over $\mathbf{sSet}$.

The Yoneda embedding will be tacitly used throughout to identify a smooth scheme X with the presheaf it represents, a 0-dimensional object of $\mathbf{sPre}(\mathbf{Sm}_S)$. We remark in passing that S itself represents a terminal object of $\mathbf{sPre}(\mathbf{Sm}_S)$.

Objectwise model structures

One can put several, Quillen equivalent, model structures on $\mathbf{sPre}(\mathbf{Sm}_S)$ such that weak equivalences are detected objectwise, meaning that a map $X \to Y$ of simplicial presheaves is a weak equivalence if and only if the map $X(U) \to Y(U)$ of simplicial sets is a weak equivalence for all objects U of $\mathbf{Sm}_S$. These model structures are sometimes called *objectwise* structures. Some standard choices for such a model structure are the injective [58], projective [26] and flasque [57] [34].

To get to $\mathbb{A}^1$-homotopy theory from this point, we carry out two left Bousfield localizations, discussed in 22.2.1 and 22.2.1 below. One can learn about Bousfield localization in [52].

The first localization is analogous to the passage from presheaves to sheaves and depends on a choice of Grothendieck topology on $\mathbf{Sm}_S$. We remind the reader that a Grothendieck topology is not a topology in the point–set sense, rather it is sufficient data to allow one to speak meaningfully about locality and sheaves. We refer to [71] for generalities on sheaves. A category equipped with a choice of Grothendieck topology is called a *site*, but we do not use this term again in this article.

The Nisnevich topology

The standard choice of Grothendieck topology for $\mathbb{A}^1$-homotopy theory is the *Nisnevich* topology [82], although the étale topology is also used, producing a different homotopy theory of spaces. If one wishes to study possibly nonsmooth schemes, then different topologies, for instance the cdh topology, can be used. The theory in the nonsmooth case is less well developed, but [105] establishes that the theory for all schemes and the cdh topology is a localization of the theory for smooth schemes and the Nisnevich topology, so that there is a localization functor L from one homotopy category to the other. Strikingly, the same paper proves that that for a map $f : X \to Y$ in the smooth, Nisnevich theory, Lf is a weak equivalence if and only if Σf, the suspension, is a weak equivalence.

In order to describe the Nisnevich topology, it is first necessary to describe étale maps: A map $Y \to X$ in $\mathbf{Sm}_S$ is *étale* if for every point y of Y, the induced map on tangent spaces is an isomorphism [27, 2.2 Proposition 8 and Corollary 10]. Many other characterizations of étale maps can be found in [47, §17] . A finite collection of maps $\{U_i \to X\}$ is a *Nisnevich cover* if the maps $U_i \to X$ are étale and for every point $x \in X$, there is a point u in some U_i mapping to x such that the induced map on residue fields $\kappa(x) \to \kappa(u)$ is an isomorphism. The topology generated by this pretopology is the Nisnevich topology*Nisnevich topology* on $\mathbf{Sm}_S$. We remark that the additional constraint imposed by the Nisnevich condition is not vacuous even when $S = \operatorname{Spec} \bar{k}$ is the spectrum of an algebraically closed field. For instance, when $|n| > 1$, the n-th power map $\mathbb{C}^* \to \mathbb{C}^*$—or more correctly $\mathbb{G}_{m,\mathbb{C}} \to \mathbb{G}_{m,\mathbb{C}}$—is an étale cover but not a Nisnevich cover because the map on the residue fields of the generic points is the homomorphism $\mathbb{C}(x) \to \mathbb{C}(x)$ sending $x \mapsto x^n$, which is not an isomorphism. One sees that given a Nisnevich cover $f : \coprod U_i \to X$, there must be a dense open subset V_0 of X on which f admits a section, and a dense open subset V_1 of the complement of V_0 on which f admits a (possibly different) section and so on, so that X has a stratification such that f admits sections on the open complements of the strata; this definition is used in [32, Section 3.1].

Local model structures

A *hypercover* of X is a simplicial object $V_\bullet \to X$ generalization of the Čech nerve

$$\cdots \mathrel{\substack{\to\\\to\\\to}} \coprod_i (U_i \cap U_j) \rightrightarrows \coprod_i U_i \to X$$

associated to the cover $\{U_i \to X\}$. See, for example [45, Chapter 3] or [34, §4]. One may therefore define hypercovers for the Nisnevich topology.

One carries out a Bousfield localization, forcing the maps $\operatorname{hocolim}_n V_n \to X$ to be weak equivalences for every hypercover $V_\bullet \to X$. This localization is called *Nisnevich localization* when the corresponding topology is the Nisnevich topology. The resulting simplicial model structures are referred to as *local model structures*, and they are all Quillen equivalent via identity functors, and the weak equivalences are called *local weak equivalences*. The local injective model structure on presheaves was originally constructed by Jardine in [58, Theorem 2.3], in a slight generalization of a construction due to Joyal [60] who considered only sheaves rather than presheaves.

Distinguished squares

The theory constructed by Joyal and Jardine is very general, applying to all sites. We refer to [59] for more about local homotopy theory *per se*. In the cases called for by $\mathbb{A}^1$-homotopy, there is a more elementary approach to the theory. Building on ideas of Brown and Gersten [29], one defines a *distinguished square* of schemes to be a diagram

$$\begin{array}{ccc} U \times_X V & \longrightarrow & V \\ \downarrow & & \downarrow p \\ U & \xrightarrow{\ i\ } & X \end{array}$$

of schemes where i is an open immersion, p is an étale morphism and p restricts to an isomorphism on the closed complements $p^{-1}(X - U) \to X - U$ (given the reduced induced subscheme structures). Then a functor $\mathcal{F} : \mathbf{Sm}_S^{\mathrm{op}} \to \mathbf{Set}$ is a Nisnevich sheaf if $\mathcal{F}$ takes distinguished squares to cartesian squares of sets, and a functor $\mathcal{X} : \mathbf{Sm}_S^{\mathrm{op}} \to \mathbf{sSet}$ is, loosely speaking, suitable for $\mathbb{A}^1$-homotopy theory if it takes distinguished squares to homotopy cartesian diagrams of simplicial sets. The following proposition, [26, Lemma 4.1] is the simplest in a family of many such results.

Proposition 22.2.1. *A simplicial presheaf* $\mathcal{X} : \mathbf{Sm}_S^{\mathrm{op}} \to \mathbf{sSet}$ *is fibrant in the projective local model structure if* $\mathcal{X}(\cdot)$ *takes values in Kan complexes and* $\mathcal{X}$ *takes distinguished squares to homotopy cartesian squares.*

Variations on this idea are considered in [57, Section 4.2], [14, Section 3.2] and [104], where we have listed the references in order from least to most general.

$\mathbb{A}^1$ model structures

After localizing to turn the objectwise model structures into local model structures, one localizes a second time, in order to make the projection maps $X \times \mathbb{A}^1 \to X$ into weak equivalences for all smooth schemes X in $\mathbf{Sm}_S$. Any resulting simplicial model structure is an $\mathbb{A}^1$-*model structure*—depending on which objectwise structure one began with, one arrives at the injective, projective, flasque, etc., $\mathbb{A}^1$-model structure. A pleasant overview of the relationships between the various model categories appearing in this story can be found in [57].

Definition 22.2.2. We will say that an object $\mathcal{X}$ is $\mathbb{A}^1$-*local* if there exists a local weak equivalence $\mathcal{X} \to \mathcal{Y}$ where $\mathcal{Y}$ is fibrant in an $\mathbb{A}^1$-model structure.

The condition on $\mathcal{X}$ above is equivalent to that of [80, Definition 2.1, p. 106] or of [52, Definition 3.1.4(a)] applied to the class of all projections $X \times \mathbb{A}^1 \to X$.

Associated to the $\mathbb{A}^1$-homotopy theory over the base S, however it is constructed, there is an $\mathbb{A}^1$-homotopy category. This is a category in which the objects are the objects of $\mathbf{sPre}(\mathbf{Sm}_k)$ and in which $\mathbb{A}^1$-weak equivalences are isomorphisms. This is denoted $\mathcal{H}(S)$.

Remark 22.2.3. The term "local" appears in the theory in at least two ways (to say nothing of localization of rings, which also appears). Namely, "local homotopy theory" is a homotopy theory of sheaves reflecting the local character of sheaf theory, and $\mathbb{A}^1$-homotopy theory is a further localization, in the sense inherited from Bousfield and Kan [28], of the local theory. In certain sources, notably [80] and [79], the local homotopy theory is called "simplicial homotopy" and what we have called "local weak equivalences" are called "simplicial (weak) equivalences" and similarly for other terms. This reflects the fact that the local homotopy theory is much closer to classical homotopy theory than the $\mathbb{A}^1$-homotopy theory is—for instance, the $\mathbb{A}^1$-homotopy theory elevates certain varieties to the status of "sphere" whereas in the local theory, only simplicial spheres $\Delta^n/\partial\Delta^n$ are granted this title. The $\mathbb{A}^1$-homotopy theory is also called "motivic homotopy theory", but the connection to motives is loose.

These simplicial model categories of spaces allows us to carry out homotopy theory on schemes. We can form limits and colimits, in particular smash products of schemes or of schemes and simplicial sets, and give meaning to the spaces in (22.1.1). Many results from the classical homotopy theory of simplicial sets carry over. For example, we have excision: suppose that $Z \to X$ is a closed immersion in $\mathbf{Sm}_k$ and $Y \to X$ is an open subset of X. Then there is a pushout square of schemes

$$\begin{array}{ccc} Y - (Z \cap Y) & \longrightarrow & X - Z \\ \downarrow & & \downarrow \\ Y & \longrightarrow & X \end{array} \tag{22.2.1}$$

corresponding to a Nisnevich cover of X. The Nisnevich localization procedure guarantees that Nisnevich covers give rise to pushouts in spaces, causing (22.2.1) to be a homotopy pushout, giving the excision weak-equivalence

$$X/(X - Z) \simeq Y/(Y - (Z \cap Y)).$$

We also can take any vector bundle $p : E \to X$, where X is a smooth scheme, and decompose X by an open cover of subschemes U_i such that the induced maps $p^{-1}(U_i) \to U_i$ is isomorphic to the projection $\mathbb{A}^n \times U_i \to U_i$, which is a weak equivalence. A colimiting argument then shows that $p : E \to X$ is itself an $\mathbb{A}^1$-weak equivalence. This works for any sort of Nisnevich-locally-trivial fibration with $\mathbb{A}^1$-contractible fibers. In particular, for quasiprojective varieties over a field k, the Jouanolou trick [106, Proposition 4.3] produces a map $p : \operatorname{Spec} R \to X$ which is an $\mathbb{A}^1$-equivalence having affine source. Any quasiprojective smooth k-variety is therefore $\mathbb{A}^1$-equivalent to an affine variety.

22.2.2 Homotopy Sheaves

An inconvenience in $\mathbb{A}^1$-homotopy theory is varying availability of basepoints. By a "basepoint" we might mean a map $p : S \to \mathcal{X}$ over S, viz., a morphism from the terminal object of the category $\mathbf{sPre}(\mathbf{Sm}_S)$. Unfortunately, since $\mathcal{X}$ amounts to a family of simplicial sets parametrized by $\mathbf{Sm}_S^{\mathrm{op}}$, for certain objects $U \to S$ of $\mathbf{Sm}_S^{\mathrm{op}}$ there may be path components of $\mathcal{X}(U)$ that are not in the image of the map $\pi_0\mathcal{X}(S) \to \pi_0\mathcal{X}(U)$. For instance, if V is the closed subscheme of $\mathbb{A}^2$ over $\mathbf{R}$ determined by the vanishing of $x^2 + y^2 + 1$, then $V(\mathbf{R})$ is empty[1], while $V(\mathbb{C})$ is a discrete set of uncountably many points.

In order to handle this technicality, one must allow "basepoints" $x_0 \in \mathcal{X}(U)$ that are not in the image of $\mathcal{X}(S) \to \mathcal{X}(U)$. That is, for any object U of $\mathbf{Sm}_S$, and any $x_0 \in (\mathcal{X}(U))_0$ one defines a sequence of presheaves on the slice category $\mathbf{Sm}_S/U$ by

$$(V \xrightarrow{f} U) \mapsto \pi_n(|\mathcal{X}(V)|, f^*(x_0)). \tag{22.2.2}$$

Here $|\mathcal{X}(V)|$ denotes the geometric realization of the simplicial set.

Definition 22.2.4. The Nisnevich sheaves associated to the functors of (22.2.2) are the *homotopy sheaves* of $\mathcal{X}$ for the basepoint x_0, and will be denoted $\boldsymbol{\pi}_n(\mathcal{X}, x_0)$. If $n \geq 1$, then the sheaves are sheaves of groups, and they are abelian if $n \geq 2$. One may define an unpointed $\boldsymbol{\pi}_0(\mathcal{X})$ similarly.

We define $\boldsymbol{\pi}_n^{\mathbb{A}^1}$, the $\mathbb{A}^1$*-homotopy sheaves* , by first replacing $\mathcal{X}$ by an $\mathbb{A}^1$ local object—for instance, $R_{\mathbb{A}^1}\mathcal{X}$—and then calculating $\boldsymbol{\pi}_n$ of the resulting object.

Both the homotopy sheaves and the $\mathbb{A}^1$-homotopy sheaves satisfy a Whitehead theorem. (In fact, this is taken as the definition of local weak equivalence in [58].) The following proposition follows by combining [59, §4.1]] with [34].

Proposition 22.2.5. *If $f : \mathcal{X} \to \mathcal{Y}$ is a morphism in $\mathbf{sPre}(\mathbf{Sm}_S)$, then f is a local weak equivalence if and only if the morphisms*

$$f_* : \boldsymbol{\pi}_0(\mathcal{X}) \to \boldsymbol{\pi}_0(\mathcal{Y})$$

and

$$f_* : \boldsymbol{\pi}_n(\mathcal{X}, x_0) \to \boldsymbol{\pi}_n(\mathcal{Y}, f(x_0))$$

are isomorphisms for all choices of basepoint $x_0 \in (\mathcal{X}(U))_0$ as U ranges over the objects of $\mathbf{Sm}_S$.

[1]It is a standard abuse of notation to write $V(R)$ for $V(\operatorname{Spec} R)$.

For the $\mathbb{A}^1$-case, one can apply the above to the map $R_{\mathbb{A}^1}\mathcal{X} \to R_{\mathbb{A}^1}\mathcal{Y}$. Most simply, when $\mathcal{X}$ is $\mathbb{A}^1$-connected, i.e., when $\boldsymbol{\pi}_0^{\mathbb{A}^1}(\mathcal{X})$ is a singleton, only one basepoint need be considered. This version appears as [80, Proposition 2.14 , §3, p. 110].

Proposition 22.2.6. *Let $f : \mathcal{X} \to \mathcal{Y}$ be a morphism of $\mathbb{A}^1$-connected objects in* $\mathbf{sPre}(\mathbf{Sm}_S)_\bullet$. *Then f is an $\mathbb{A}^1$-weak equivalence if and only if*

$$f_* : \boldsymbol{\pi}_n^{\mathbb{A}^1}(\mathcal{X}, x_0) \to \boldsymbol{\pi}_n^{\mathbb{A}^1}(\mathcal{Y}, y_0)$$

is an isomorphism for all $i \geq 1$.

One can define homotopy groups of $\mathcal{X}$ by taking sections of the homotopy sheaves. We abuse notation and write $\mathcal{X}(k)$ for what should strictly be written $\mathcal{X}(\operatorname{Spec} k)$.

Proposition 22.2.7. [109, Corollary 2.8] *Let k be a field and let $\mathcal{X}$ be an object of* $\mathbf{sPre}(\mathbf{Sm}_k)$. *Choose a basepoint $x_0 \in (\mathcal{X}(k))_0$. Then*

$$\boldsymbol{\pi}_n^{\mathbb{A}^1}(\mathcal{X}, x_0)(k) = [S^n \wedge (\operatorname{Spec} k)_+, \mathcal{X}]_{\mathbb{A}^1}$$

22.2.3 The $\mathrm{Sing}^{\mathbb{A}^1}$-functor and naive $\mathbb{A}^1$-homotopy

If $\mathcal{F}$ is a functor defined on rings, or some subcategory of rings, then it is possible to define *elementary homotopy* between two elements in $\mathcal{F}(R)$. One considers the two evaluation maps $R[t] \to R$ given by evaluation at $t = 0$ and $t = 1$. An elementary homotopy between a and b in $\mathcal{F}(R)$ is an element $H \in \mathcal{F}(R[t])$ such that under the two evaluation maps $H \mapsto a$ and $H \mapsto b$. There is no reason in general to this should be an equivalence relation, so one takes the equivalence relation generated by it in order to define the *naive homotopy class* of $a \in \mathcal{F}(R)$.

If $\operatorname{Spec} R$ is an object of $\mathbf{Sm}_S$ and $\mathcal{F}$ is a functor on $\mathbf{Sm}_S$, then the elementary homotopy H between a and b above can be restated using the Yoneda Lemma as a map $H : \operatorname{Spec} R \times \mathbb{A}^1 \to \mathcal{F}$. We extend this define a *elementary homotopy* between two morphisms $f, g : \mathcal{X} \to \mathcal{Y}$ in $\mathbf{sPre}(\mathbf{Sm}_S)$: namely, a map $H : \mathcal{X} \times \mathbb{A}^1 \to \mathcal{Y}$ such that the two composite maps $\mathcal{X} \to \mathcal{Y}$ induced by $\{0\} \to \mathbb{A}^1$ and $\{1\} \to \mathbb{A}^1$ are f and g.

One may expand on this construction. Let us define the *algebraic n-simplex* over $\mathbb{Z}$:

$$\Delta^n_{\mathbb{Z}} = \frac{\mathbb{Z}[t_0, \ldots, t_n]}{\left(\sum_{i=0}^n t_i - 1\right)}$$

For a general base S, we may define $\Delta^n_S = S \times_{\mathbb{Z}} \Delta^n_{\mathbb{Z}}$, which is to say, using the same variables and relations, but over S. The objects Δ^n_S assemble to form a cosimplicial object in the category of $\mathbf{Sm}_S$, where the coface and codegeneracy maps mimic the coface and codegeneracy maps of the standard topological cosimplicial object $\Delta^\bullet_{\mathbf{Top}}$ for which

$$\Delta^n_{\mathbf{Top}} = \left\{ (t_0, \ldots, t_n) \in \mathbf{R}^{n+1} \mid \sum_{i=0}^n t_i = 1,\ 0 \leq t_i \leq 1 \forall i \right\}$$

—in the algebraic case, it is not possible to impose inequality conditions. We remark that, although the scheme Δ^n_S is constructed by analogy with the simplicial set Δ^n, the two are different objects of $\mathbf{sPre}(\mathbf{Sm}_k)$.

For any presheaf $\mathcal{F} : \mathbf{Sm}_S^{\mathrm{op}} \to \mathbf{Set}$, one can produce the simplicial presheaf $\mathrm{Sing}^{\mathbb{A}^1} \mathcal{F} : \mathbf{Sm}_S^{\mathrm{op}} \to \mathbf{sSet}$ by

$$\mathrm{Sing}^{\mathbb{A}^1}(\mathcal{F})(U)_n = \mathcal{F}(U \times \Delta^n_S).$$

This is sometimes called the Suslin–Voevodsky construction. If $\mathcal{F}$ is already a simplicial sheaf, then $\mathrm{Sing}^{\mathbb{A}^1} \mathcal{F}$ is constructed by forming the bisimplicial presheaf $\mathrm{Sing}^{\mathbb{A}^1}(\mathcal{F}_\bullet)$ and then taking a diagonal construction. There is a natural monomorphism $\mathrm{id} \to \mathrm{Sing}^{\mathbb{A}^1}$.

The following notable result, [80, Prop. 3.4, p.89] may explain some of the importance of the $\mathrm{Sing}^{\mathbb{A}^1}$-functor in the theory:

Proposition 22.2.8. *Let $f, g : \mathcal{X} \to \mathcal{Y}$ be two morphisms in $\mathbf{sPre}(\mathbf{Sm}_k)$ and let $H : \mathcal{X} \times \mathbb{A}^1 \to \mathcal{Y}$ be an elementary homotopy between f and g. Then there is a homotopy $\mathrm{Sing}^{\mathbb{A}^1} X \times \Delta^1 \to \mathrm{Sing}^{\mathbb{A}^1} Y$ from $\mathrm{Sing}^{\mathbb{A}^1}(f)$ to $\mathrm{Sing}^{\mathbb{A}^1}(g)$.*

It is worth noting that the naive $\mathbb{A}^1$-homotopy classes of maps $U \to X$ are exactly the path components of the $\mathbb{A}^1$ singular construction $\mathrm{Sing}^{\mathbb{A}^1} X$ evaluated at U, i.e., $\pi_0(\mathrm{Sing}^{\mathbb{A}^1} X(U))$. In general, $\mathrm{Sing}^{\mathbb{A}^1} X$ is not an $\mathbb{A}^1$-local object, so that naive $\mathbb{A}^1$-homotopy is not the same as genuine $\mathbb{A}^1$-homotopy. This is perhaps the great challenge in unstable $\mathbb{A}^1$-homotopy—the naive, geometrically appealing construction does not coincide with the genuine construction possessed of good theoretical properties. We refer to Subsection 22.6.6 for examples where naive and genuine $\mathbb{A}^1$-homotopy classes agree. One construction of $R_{\mathbb{A}^1}$, the $\mathbb{A}^1$-fibrant replacement functor, [80, Lemma 3.20, p. 93], is by means of an infinite composite of alternately applying $\mathrm{Sing}^{\mathbb{A}^1}$ and fibrant replacement in local homotopy,

For an example where the naive and genuine $\mathbb{A}^1$-homotopy classes of maps differ, we refer to [30], in which the naive $\mathbb{A}^1$-homotopy classes of maps $\mathbb{P}^1 \to \mathbb{P}^1$ are calculated when $S = \operatorname{Spec} k$, the spectrum of a field. In order to put this in context, we should mention that in [79, Theorem 7.36 and Remark 7.37], Morel calculated

$$[\mathbb{P}^1, \mathbb{P}^1]_{\mathbb{A}^1} = \mathrm{GW}(k) \times_{k^*/k^{*2}} k^*. \tag{22.2.3}$$

It is to be remarked that although $\mathbb{P}^1$ is a motivic sphere, this calculation lies outside the main body of calculations of self-maps of spheres appearing in (22.5.3) below, since $\mathbb{P}^1 \simeq S^1 \wedge \Gamma$ is not $\mathbb{A}^1$-simply-connected. The main theorem of [30] is that the naive homotopy classes of maps of spheres make up the submonoid of $\mathrm{GW}(k) \times_{k^*/k^{*2}} k^*$ corresponding to elements in $\mathrm{GW}(k)$ of positive degree, so that the map from the set of naive $\mathbb{A}^1$-homotopy classes of maps $\mathbb{P}^1 \to \mathbb{P}^1$ to the set of all $\mathbb{A}^1$-homotopy classes is, in fact, a group completion.

22.2.4 Spheres

Let Γ denote the punctured affine line $\mathbb{G}_m = \mathbb{A}^1 - \{0\}$, given the basepoint 1, and let S^1 denote the pointed simplicial circle. The role of the spheres S^n in classical algebraic topology is now played by smash products of the spaces S^1 and Γ. We use the notation

$$S^{p+q\alpha} = S^{p+q,q} := (S^1)^{\wedge p} \wedge \mathbb{G}_m^{\wedge q},$$

and such spaces will be called *motivic spheres*. The presence of two different indexing conventions for spheres in $\mathbb{A}^1$-homotopy theory can be confusing, but both appear in the literature, so we give both as well. The notation ΣX (or $\Sigma_{S^1} X$ if there is possible confusion) denotes $X \wedge S^1$. The question of which schemes have the $\mathbb{A}^1$-homotopy type of spheres has been studied in [5], and the most common examples are the following.

Example 22.2.9. The scheme $\mathbb{P}^1$ is $\mathbb{A}^1$-homotopy equivalent to $S^{1+\alpha} = S^{2,1}$. Specifically, the pushout diagram

$$\begin{array}{ccc} \mathbb{G}_m & \longrightarrow & \mathbb{A}^1 \\ \downarrow & & \downarrow \\ \mathbb{A}^1 & \longrightarrow & \mathbb{P}^1 \end{array}$$

and the weak equivalence $\mathbb{A}^1 \simeq *$, induce a weak equivalence

$$\mathbb{P}^1 \simeq \Sigma \mathbb{G}_m.$$

Example 22.2.10. The scheme $\mathbb{A}^n - \{0\}$ is $\mathbb{A}^1$-homotopy equivalent to $S^{n-1+n\alpha} = S^{2n-1,n}$. The case $n = 1$ is immediate, and then one may proceed by induction on n: the pushout diagram

$$\begin{array}{ccc} \mathbb{G}_m \times (\mathbb{A}^{n-1} - \{0\}) & \longrightarrow & \mathbb{A}^1 \times (\mathbb{A}^{n-1} - \{0\}) \\ \downarrow & & \downarrow \\ \mathbb{G}_m \times \mathbb{A}^n & \longrightarrow & \mathbb{A}^n - \{0\} \end{array}$$

and the weak equivalences $\mathbb{G}_m \times \mathbb{A}^n \simeq \mathbb{G}_m$ and $\mathbb{A}^1 \times (\mathbb{A}^{n-1} - \{0\}) \simeq \mathbb{A}^{n-1} - \{0\}$, induce a weak equivalence between $\mathbb{A}^n - \{0\}$ and the homotopy pushout of the diagram $Y \leftarrow X \times Y \rightarrow X$ of projection maps for $X = \mathbb{G}_m$ and $Y = \mathbb{A}^{n-1} - \{0\}$. Since this latter homotopy pushout is identified with $\Sigma(X \wedge Y)$ by pushing out the rows of the diagram

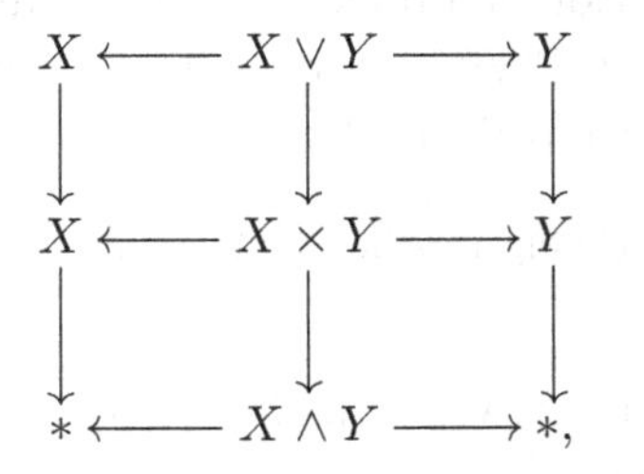

the result follows.

An important related example is the identification of the $\mathbb{A}^1$-homotopy type of $\mathbb{P}^n/\mathbb{P}^{n-1}$ with $S^{n+n\alpha} = S^{2n,n}$.

Example 22.2.11. There is a standard closed immersion[2] $\mathbb{P}^{n-1} \hookrightarrow \mathbb{P}^n$ sending homogeneous coordinates $[x_0 : \ldots : x_n]$ to $[0 : x_0 : \ldots : x_n]$. The homotopy type of the quotient

[2]We use the algebro-geometric term "closed immersion" for a map isomorphic to the inclusion of a closed subscheme. The usual term in topology is "closed embedding", which is also used in [80], but is not widespread in the literature on $\mathbb{A}^1$-homotopy theory.

space $\mathbb{P}^n/\mathbb{P}^{n-1}$ is identified by the chain of weak equivalences

$$\mathbb{P}^n/\mathbb{P}^{n-1} \simeq \mathbb{P}^n/(\mathbb{P}^n - \{0\}) \simeq \mathbb{A}^n/(\mathbb{A}^n - \{0\}) \simeq \Sigma(\mathbb{A}^n - \{0\})$$

with $\Sigma(\mathbb{A}^n - \{0\})$. Therefore, by example 22.2.10, we have $\mathbb{P}^n/\mathbb{P}^{n-1} \simeq S^{n+n\alpha}$.

22.2.5 Thom spaces and purity

Let $V \to X$ be a vector bundle and let $X \hookrightarrow V$ be the zero section.

Definition 22.2.12. The *Thom space*, denoted $\mathrm{Th}(V)$ or X^V, of V is defined

$$\mathrm{Th}(V) := V/(V - X).$$

A parametrized version of Example 22.2.11 shows that the $\mathbb{A}^1$-homotopy type of the Thom space over a scheme can be alternatively described as the cofiber of a closed immersion. Namely, let 1 denote the trivial bundle on X. For any vector bundle V on X, let $\mathbb{P}(V)$ denote the projective space bundle given by the fiberwise projectivization. As in Example 22.2.11, there is a standard closed immersion $\mathbb{P}(V) \hookrightarrow \mathbb{P}(V \oplus 1)$. The complementary open subscheme is isomorphic to V.

Proposition 22.2.13. *There is an $\mathbb{A}^1$-weak equivalence* $\mathrm{Th}(V) \simeq \mathbb{P}(V \oplus 1)/\mathbb{P}(V)$.

This presentation of the Thom space appears in the case of complex bundles on compact spaces in [19, p. 79].

Proof. By excision, there is an $\mathbb{A}^1$-weak equivalence $\mathrm{Th}(V) \simeq \mathbb{P}(V \oplus 1)/(\mathbb{P}(V \oplus 1) - X)$. The scheme $\mathbb{P}(V \oplus 1) - X$ is the total space of the vector bundle $\mathcal{O}(1)$ over $\mathbb{P}(V)$, giving the claimed weak equivalence. □

Example 22.2.14. A trivialization of a rank n vector bundle V on a scheme X determines an $\mathbb{A}^1$-weak equivalence $\mathrm{Th}(V) \simeq (\mathbb{P}^n/\mathbb{P}^{n-1}) \wedge X_+$. In particular, for a point z on a smooth scheme X, there is an $\mathbb{A}^1$-weak equivalence $\mathrm{Th}(T_z X) \simeq \mathbb{P}^n_{k(z)}/\mathbb{P}^{n-1}_{k(z)}$.

The analogue of the tubular neighborhood theorem from classical topology is Morel–Voevodsky's purity theorem.

Purity Theorem 22.2.15 (Morel, Voevodsky). *Let $Z \hookrightarrow X$ be a closed immersion in* $\mathbf{Sm}_S$. *Then there is a natural $\mathbb{A}^1$-weak equivalence*

$$\mathrm{Th}(N_Z X) \simeq X/(X - Z)$$

where $N_Z X \to Z$ denotes the normal bundle.

The original proof is [80, Section 3 Theorem 2.23 p.115], and a particularly accessible exposition, done over an algebraically closed field, may be found in [1, Section 7], using unpublished notes of A. Asok and [55].

Here, we will give an outline of a proof and draw some pictures in the case where $Z \hookrightarrow X$ is the inclusion of the point into the affine line $\{0\} \hookrightarrow \mathbb{A}^1_k$.

First, we explain the concept of "blow-up" in algebraic geometry briefly. For a detailed treatment, we refer to [50]. Given a closed immersion $Z \hookrightarrow X$, the *blow-up* $\pi : \mathrm{Bl}_Z X \to X$ of X along Z is a map with the properties that:

- The restriction of π to $\mathrm{Bl}_Z X - \pi^{-1}(Z) \to X - Z$ is an isomorphism,
- $\pi^{-1}(Z) \to Z$ is isomorphic to the projectivization of the normal bundle $\mathbb{P}N_Z X \to Z$.

The subvariety $\pi^{-1}(Z)$ is called the *exceptional divisor* of the blow-up.

In other words, we cut out Z and glue in a projectivization of the normal bundle. In the topology of manifolds, this can be accomplished by first removing an open tubular neighborhood of Z from X, so that one has introduced a boundary component S homeomorphic to a sphere bundle in the normal bundle $N_Z X$, and then taking a fiberwise quotient of S to replace it by a projectivization of $N_Z X$. Algebraically, one takes the sheaf of ideals $\mathcal{I}$ determining the closed immersion $Z \to X$, forms the sheaf of graded algebras $\oplus_{i=0}^{\infty} \mathcal{I}^i$ and takes the relative $\underline{\mathrm{Proj}}$ construction, i.e., $\mathrm{Bl}_Z X \to X$ is the canonical map $\underline{\mathrm{Proj}}_X \oplus_{i=0}^{\infty} \mathcal{I}^i \to X$. The morphism $\pi : \mathrm{Bl}_Z X \to X$ satisfies the property that the sheaf of ideals $(\pi^{-1}\mathcal{I})\mathcal{O}_{\mathrm{Bl}_Z X}$ associated to $\pi^{-1}Z$ is the sheaf of sections of a line bundle, and any other morphism $Y \to X$ with this property factors through π, [50, II 7.14].

Example 22.2.16. The blow-up $\mathrm{Bl}_0 \mathbb{A}^n$ of affine n-space $\mathbb{A}^n$ at the origin is the closed subscheme of $\mathbb{A}^n \times \mathbb{P}^{n-1}$ determined by the ideal generated by $\{x_i y_j - x_j y_i : i, j = 1, \ldots, n\}$ where $(x_1, \ldots, x_n)$ are the coordinates on $\mathbb{A}^n$ and $[y_1, \ldots, y_n]$ are homogeneous coordinates on $\mathbb{P}^{n-1}$. The restriction to $\mathrm{Bl}_0 \mathbb{A}^n$ of the projection $\mathbb{A}^n \times \mathbb{P}^{n-1} \to \mathbb{P}^{n-1}$ is the projection of the tautological bundle on $\mathbb{P}^{n-1}$, denoted $\mathcal{O}(-1)$, to $\mathbb{P}^{n-1}$. Note that in the case where $n = 2$, the blow-up $\mathrm{Bl}_0 \mathbb{A}^2$ can be embedded in three-space (meaning $\mathbb{P}^3$), which makes for good visualizations: See [50, Figure 3 p. 29] or the map π in Figure 22.2.2. In the latter, the solid black vertical line is the exceptional divisor $\mathbb{P}^1$, and the dashed and dot-dashed lines are both fibers of the line bundle $\mathcal{O}(-1)$. Blow-ups of $\mathbb{A}^2$ are sometimes used as jungle gyms: See Figure 22.2.1.

An essential component of the proof of the purity theorem 22.2.15 is the "deformation to the normal bundle", a pre-existing idea in intersection theory [46, Chapter 5]. The input to this construction is a closed immersion $Z \hookrightarrow X$ in $\mathbf{Sm}_k$ and the output is a family of closed immersions over $\mathbb{A}^1$:

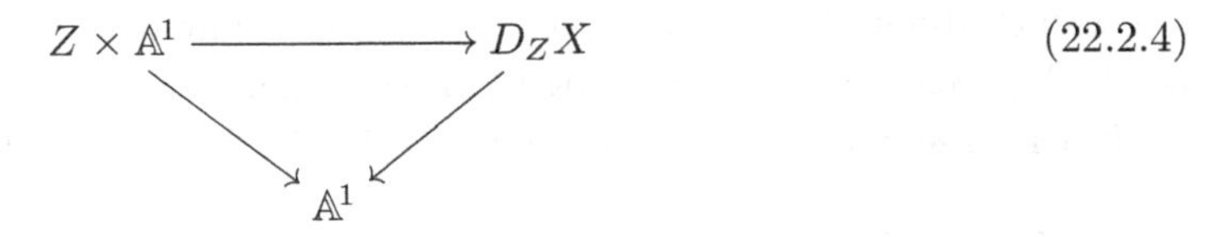

(22.2.4)

such that over $t = 1$, the fiber of the family is the original closed immersion $Z \hookrightarrow X$, and over $t = 0$, the fiber of the family is the inclusion of the zero section into $N_Z X$.

To form this family, first consider $\mathrm{Bl}_{Z \times 0}(X \times \mathbb{A}^1) \xrightarrow{\pi} X \times \mathbb{A}^1 \to \mathbb{A}^1$. The inverse image of $t = 0$ under the composition is $\pi^{-1}(X \times \{0\})$. It is the union of the exceptional divisor $\mathbb{P}(N_{Z \times 0}(X \times \mathbb{A}^1))$ and a copy of $\mathrm{Bl}_{Z \times 0}(X \times 0)$. These are glued by identifying the copy

Figure 22.2.1. $\mathrm{Bl}_0\,\mathbb{A}^2$ Jungle Gym

of $\mathbb{P}N_Z X$ embedded in $\mathrm{Bl}_{Z\times 0}(X \times 0)$ as the exceptional divisor, and the following copy of $\mathbb{P}N_Z X$ in $\mathbb{P}(N_{Z\times 0}(X\times\mathbb{A}^1))$: the bundle $N_{Z\times 0}(X\times\mathbb{A}^1)$ is canonically isomorphic to $N_Z X\oplus 1$. Thus $\mathbb{P}(N_{Z\times 0}(X\times\mathbb{A}^1))$ is the union of $N_Z X$ and $\mathbb{P}N_Z X$, and the latter $\mathbb{P}N_Z X$ is the copy we seek. Then let

$$D_Z X := \mathrm{Bl}_{Z\times 0}(X\times\mathbb{A}^1) - \mathrm{Bl}_{Z\times 0}(X\times 0),$$

Since the copy of $Z\times\mathbb{A}^1$ in $\mathrm{Bl}_{Z\times 0}(X\times\mathbb{A}^1)$ intersects the exceptional divisor as the zero section of $N_Z X$ in $\mathbb{P}(N_{Z\times 0}(X\times\mathbb{A}^1))$, the scheme $D_Z X$ provides the claimed family. $D_0\mathbb{A}^1$ is shown in Figure 22.2.2.

To prove purity, one shows that the fibers of (22.2.4) above $t=0$ and $t=1$ induce $\mathbb{A}^1$-weak equivalences

$$i_0 : N_Z X/(N_Z X - Z) \to D_Z X/(D_Z X - Z\times\mathbb{A}^1_k)$$

and

$$i_1 : X/(X-Z) \to D_Z X/(D_Z X - Z\times\mathbb{A}^1),$$

respectively. Like the étale topology, the Nisnevich topology is fine enough to reduce many arguments about closed immersions in $\mathbf{Sm}_S$ to arguments for the standard inclusion $\mathbb{A}^n\hookrightarrow\mathbb{A}^{n+c}$. A precise theorem enabling such reductions may be found in [48, II 4.10].

In the case where $Z\hookrightarrow X$ is $0\hookrightarrow\mathbb{A}^n$, the deformation to the normal bundle

$$D_Z X = \mathrm{Bl}_0(\mathbb{A}^n\times\mathbb{A}^1) - \mathrm{Bl}_0(\mathbb{A}^n\times 0) \cong \mathcal{O}_{\mathbb{P}^n_k}(-1) - \mathcal{O}_{\mathbb{P}^{n-1}_k}(-1) \cong \mathcal{O}_{\mathbb{P}^n_k}(-1)|_{\mathbb{A}^n_k}$$

is the restriction of the total space of the tautological bundle on $\mathbb{P}^n_k$ restricted to the standard copy of $\mathbb{A}^n$. The fiber above 0 is the zero-section of this bundle. The fiber above 1 also defines a section. We can see directly that i_0 and i_1 are $\mathbb{A}^1$-weak equivalences.

Example 22.2.17. Let z be a point of a smooth scheme X. A trivialization of $T_z X$ gives an $\mathbb{A}^1$-weak equivalence $X/(X-z)\simeq(\mathbb{P}^n/\mathbb{P}^{n-1})\wedge\operatorname{Spec}k(z)_+$, cf. Example 22.2.14.

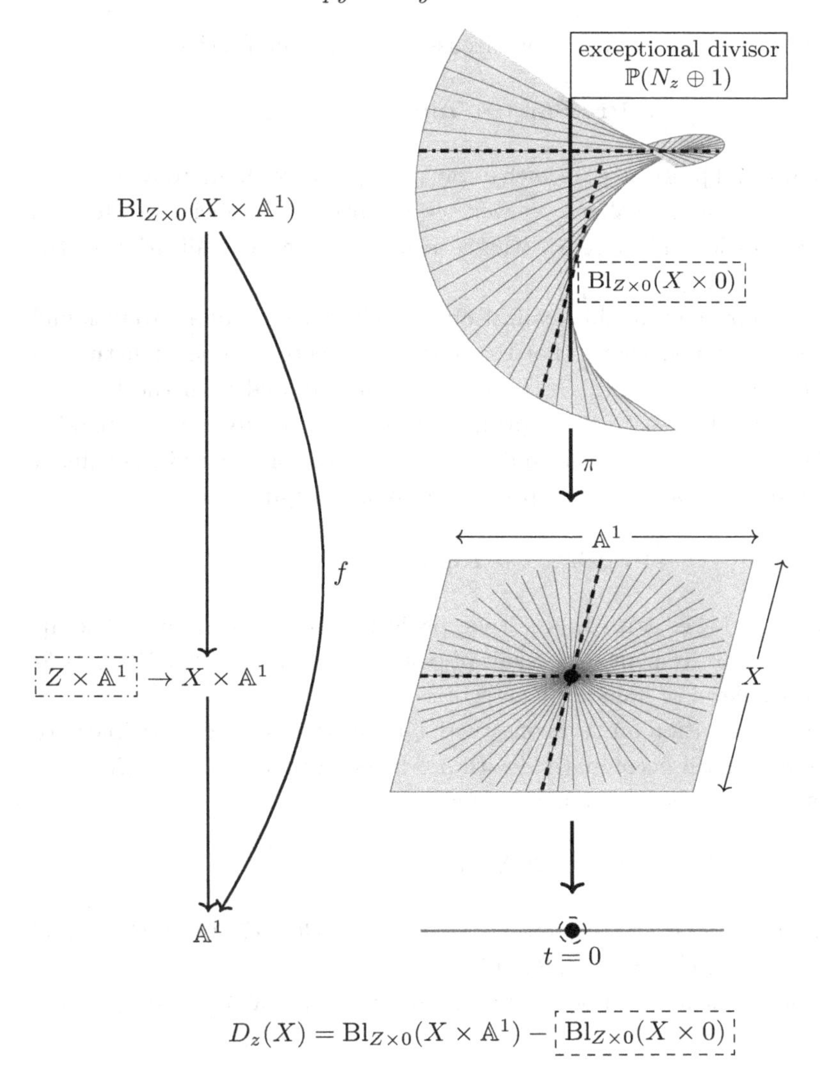

FIGURE 22.2.2. Deformation to the normal bundle

22.3 Realizations

22.3.1 Complex Realization

If $k \subset \mathbb{C}$ is a subfield of the complex numbers, then one may view the category $\mathbf{Sm}_k$ of smooth k-schemes as a category of complex varieties. Let us abuse notation and write $X(\mathbb{C})$ for the complex manifold determined by a smooth k-scheme X in $\mathbf{Sm}_k$. The object $X(\mathbb{C})$, being a manifold, is the sort of thing to which classical homotopy theory is well adapted. One therefore might hope for a functor $|\cdot| : \mathbf{sPre}(\mathbf{Sm}_k) \to \mathbf{Top}$ with good homotopical properties. Such a functor can be constructed in the most obvious way: for a smooth k-scheme X, one defines $|X| = X(\mathbb{C})$, as discussed above. The functor

$$|\cdot| : \mathbf{sPre}(\mathbf{Sm}_k) \to \mathbf{Top}$$

is then extended to all simplicial presheaves in such a way as to preserve (homotopy) colimits. For formal reasons, if K is a simplicial set, viewed as a constant simplicial presheaf in $\mathbf{sPre}(\mathbf{Sm}_k)$, then this realization $|K|$ agrees with the usual topological realization of the simplicial set K.

The technical details in showing that this realization procedure works out are to be found in [33], [34] and [35]. The first paper shows that the motivic projective model structure on $\mathbf{sPre}(\mathbf{Sm}_k)$ is an $\mathbb{A}^1$-localization of a "universal" model structure, and then the last two show that the relations imposed in the motivic projective model structure, namely making hypercovers and trivial line bundles into weak equivalences, are relations that yield ordinary homotopy equivalences after realization. The upshot is a Quillen adjunction

$$|\cdot| : \mathbf{sPre}(\mathbf{Sm}_k) \leftrightarrows \mathbf{Top} : S,$$

when the left hand side is endowed with the motivic projective model structure. The approach outlined above is an improvement on a result of [80] where a functor $t^{\mathbb{C}} : \mathcal{H}(\mathbb{C}) \to \mathcal{H}$ is produced, but only on the level of homotopy categories.

Being a left adjoint, the realization functor constructed in this way is not compatible with fiber sequences, but one sees that it is compatible with products, in that $|X \times Y| \approx |X| \times |Y|$, [84, Section A.4] and as a consequence, one may deduce

$$|B(X, G, Y)| \approx B(|X|, |G|, |Y|)$$

for two-sided bar constructions, so that a sequence of spaces $G \to B(G, G, X) \to B(*, G, X)$ has the realization $|G| \to B(|G|, |G|, |X|) \to B(*, |G|, |X|)$.

Since realization is homotopically well behaved, one obtains maps $[X, Y]_{\mathbb{A}^1_{\mathbb{C}}} \to [|X|, |Y|]$.

22.3.2 Real Realization

Suppose now that $k = \mathbf{R}$. Take $X(\mathbb{C}) = \mathrm{Mor}_{\mathbf{R}}(\operatorname{Spec} \mathbb{C}, X)$ and recall that the Galois action $C_2 = \mathrm{Gal}(\mathbb{C}/\mathbf{R})$ endows this manifold with a C_2 action. Extending the functor $X \mapsto X(\mathbb{C})$ in such a way as to preserve homotopy colimits, one produces an adjunction:

$$|\cdot|_{\mathrm{equi}} : \mathbf{sPre}(\mathbf{Sm}_k) \leftrightarrows C_2 - \mathbf{Top} : S_{\mathbf{R}}.$$

A map $f : X \to Y$ of C_2-spaces is a weak equivalence if and only if it induces weak equivalences after taking H fixed points $f^H : X^H \to Y^H$ for both $H = C_2$ and $H = \{e\}$. With this definition, the functor $|\cdot|_{\mathrm{equi}}$ is compatible with the homotopy—specifically [35, Section 5.3] establishes that $|\cdot|_{\mathrm{equi}}$ is a left Quillen functor when the source is endowed with the motivic projective model structure, the arguments being largely the same as in the complex case.

Composing $|\cdot|_{\mathrm{equi}}$ with the functor taking C_2 fixed points, which is a right adjoint, one arrives at a functor $X \mapsto |X|_{\mathrm{equi}}^{C_2} := |X|_{\mathbf{R}}$, and if X is a scheme, then $|X|_{\mathbf{R}}$ is the analytic space $X(\mathbf{R})$ endowed with the analytic topology. Since the functor $|\cdot|_{\mathbf{R}}$ involves the taking of fixed points it does not preserve all colimits, and therefore it cannot be a left adjoint. Nonetheless, it preserves smash products.

22.3.3 Étale realization

In the case of an arbitrary Noetherian base scheme, S, there is an *étale realization* functor, constructed in [56], operating along similar lines to the above two functors. In this case, one begins with a functor $\mathbf{Sm}_S \to \text{pro-}\mathbf{sPre}(\mathbf{Sm}_k)$, taking a smooth S-scheme X to the étale topological type of X, as defined in [45] following [3]. One then extends this functor to all of $\mathbf{sPre}(\mathbf{Sm}_S)$ by requiring it to commute with homotopy colimits. The result is again a left Quillen functor

$$|\cdot|_{\text{ét}} : \mathbf{sPre}(\mathbf{Sm}_S) \leftrightarrows \text{pro-}\mathbf{Top} : S$$

with a certain extra complication since the target category is a category of pro-objects.

The étale realization functor can transport theorems in $\mathbb{A}^1$-homotopy theory to the older and very successful theory of étale homotopy or cohomology. In étale cohomology over a field, the Galois representations given by certain cohomology groups are the subject of beautiful theorems and conjectures, e.g., [44] [43]. Thus, it might be hoped that $\text{Gal}(\bar{k}/k)$ should act on $|X_{\bar{k}}|_{\text{ét}}$ and the target category might be enriched to $\text{Gal}(\bar{k}/k)$-equivariant-pro-spaces. There is active work in this direction [21] [22] [49] [63] [85] [86].

22.4 Degree

A major accomplishment in $\mathbb{A}^1$-homotopy theory is Morel's $\mathbb{A}^1$-degree homomorphism

$$\deg : [S^n \wedge \Gamma^{\wedge n}, S^n \wedge \Gamma^{\wedge n}] \to \text{GW}(k),$$

This generalizes to a computation of certain stable and unstable $\mathbb{A}^1$-homotopy groups of spheres as Milnor–Witt $\mathbf{K}$-theory groups. In this section, we introduce these groups, as well as some computational tools to understand them, and discuss their relationship to $\mathbb{A}^1$-homotopy theory.

22.4.1 The Grothendieck–Witt group

The Grothendieck–Witt group $\text{GW}(k)$ is both complicated enough to support interesting invariants and simple enough to allow explicit computations. It is defined as follows. The isomorphism classes of non-degenerate, symmetric, bilinear forms $\beta : V \times V \to k$ over k, where V is a finite dimensional k-vector space, admit the operations of perpendicular direct sum $\oplus$ and tensor product $\otimes$. These operations give the set of such isomorphism classes the structure of a semi-ring. Taking the group completion, i.e., introducing formal differences of isomorphism classes, defines the ring $\text{GW}(k)$.

As in the introduction, let $\langle a\rangle$ denote the element of $\mathrm{GW}(k)$ corresponding to the bilinear form $k \times k \to k$ taking (x, y) to axy.

The ring $\mathrm{GW}(k)$ has a presentation [65, II Theorem 4.1 pg 39] [39, Theorem 4.7 p. 23] with the generators given by $\langle a\rangle$ for a in k^* and the relations

1. $\langle a\rangle = \langle ab^2\rangle$ for all $a, b \in k^*$.
2. $\langle a\rangle\langle b\rangle = \langle ab\rangle$ for all $a, b \in k^*$.
3. $\langle a\rangle + \langle b\rangle = \langle a+b\rangle + \langle ab(a+b)\rangle$ for all $a, b \in k^*$ such that $a + b \neq 0$.

The relations (2) and (3) imply that $\langle a\rangle + \langle -a\rangle$ is independent of a. This form is called the hyperbolic form, denoted

$$h := \langle 1\rangle + \langle -1\rangle.$$

The fact that $\{\langle a\rangle : a \in k^*/(k^*)^2\}$ is a set of generators is equivalent to the statement from bilinear algebra that a symmetric bilinear form can be diagonalized stably. In characteristic different from 2, any bilinear form can be diagonalized. In characteristic 2, it may be necessary to add a diagonal form before diagonalization is possible. For example, the form given by the matrix $\begin{bmatrix} 0 & 1 \\ 1 & 0 \end{bmatrix}$ is not isomorphic to the diagonal form $\begin{bmatrix} 1 & 0 \\ 0 & -1 \end{bmatrix}$, but they are stably isomorphic, and thus equal in the Grothendieck–Witt group.

The *rank* of the bilinear form $\beta : V \times V \to k$ is the dimension of V as a k-vector space. So for example, the generators $\langle a\rangle$ for a in $k^*/(k^*)^2$ are the isomorphism classes of all non-degenerate rank one forms. The rank defines a ring homomorphism

$$\mathrm{rank} : \mathrm{GW}(k) \to \mathbb{Z}.$$

The kernel of the rank homomorphism

$$I := \mathrm{Ker}(\mathrm{rank})$$

is called the fundamental ideal . It gives rise to a filtration of GW

$$\mathrm{GW} \supseteq I \supseteq I^2 \supseteq \cdots$$

related to étale cohomology and Milnor K-theory by the Milnor Conjecture [75], which is now a theorem of Voevodsky proven using $\mathbb{A}^1$-homotopy theory [102] [103]. Specifically, assume that the characteristic of k is not 2 and let $\mathrm{H}^n(k, \mathbb{Z}/2)$ denote the étale cohomology of $\mathrm{Spec}\, k$ with $\mathbb{Z}/2$ coefficients, or equivalently, the (continuous) cohomology of the Galois group of the separable closure of k with $\mathbb{Z}/2$ coefficients. Let $\mathrm{K}_n^{\mathrm{M}}(k)$ denote the degree-n summand of the Milnor K-theory of k, [75]. For present purposes, we only state that $\mathrm{K}_n^{\mathrm{M}}(k)$ is the quotient of the n-fold tensor product $k^* \otimes k^* \otimes \ldots \otimes k^*$ by the group generated by tensors $a_1 \otimes a_2 \otimes \ldots \otimes a_n$ where there is some i such that $a_i + a_{i+1} = 1$. In degrees 0, 1 one has $\mathrm{K}_0^{\mathrm{M}}(k) = \mathbb{Z}$ and $\mathrm{K}_1^{\mathrm{M}}(k) = k^*$. The Kummer map $k^* \to H^1(k, \mathbb{Z}/2)$ is the first boundary map obtained by taking the Galois cohomology of the exact sequence

$$1 \to \{\pm 1\} \to k_s^* \xrightarrow{z \mapsto z^2} k_s^* \to 1,$$

where k_s denotes the separable closure of k. It extends to a ring homomorphism $\mathrm{K}^{\mathrm{M}}_*(k) \otimes \mathbb{Z}/2 \to \mathrm{H}^*(k, \mathbb{Z}/2)$. The Milnor conjecture says there are isomorphisms

$$I^n/I^{n+1} \cong \mathrm{K}^{\mathrm{M}}_n(k) \otimes \mathbb{Z}/2 \cong \mathrm{H}^n(k, \mathbb{Z}/2),$$

where the second isomorphism is induced by the map just discussed coming from the Kummer map and the first sends

$$a_1 \otimes a_2 \otimes \cdots \otimes a_n$$

in $\mathrm{K}^{\mathrm{M}}_n(k)$ to

$$(\langle a_1 \rangle - \langle 1 \rangle)(\langle a_2 \rangle - \langle 1 \rangle) \cdots (\langle a_n \rangle - \langle 1 \rangle)$$

in I^n/I^{n+1}.

The maps $I^n \to I^n/I^{n+1} \to \mathrm{H}^n(k, \mathbb{Z}/2)$ define invariants on bilinear forms in I^n. The 0th of these is by definition the rank. The first is the *discriminant*

$$\mathrm{Disc} : \mathrm{GW}(k) \to \mathrm{H}^1(k, \mathbb{Z}/2) \cong k^*/(k^*)^2,$$

which sends the bilinear form $\beta : V \times V \to k$ to the determinant of a Gram matrix for β

$$\mathrm{Disc}(\beta) = (\beta(v_i, v_j))_{i,j}$$

where $\{v_1, \ldots v_n\}$ is a basis of V. The second is the Hasse–Witt invariant

$$\gamma : \mathrm{GW}(k) \to \mathrm{H}^2(k, \mathbb{Z}/2) \cong \mathrm{K}^{\mathrm{M}}_2(k) \otimes \mathbb{Z}/2,$$

which takes $\langle a_1 \rangle + \langle a_2 \rangle + \cdots + \langle a_m \rangle$ to the image of

$$\sum_{i<j} a_i \otimes a_j$$

in $\mathrm{K}^{\mathrm{M}}_2(k) \otimes \mathbb{Z}/2$. If we identify $\mathrm{H}^2(k, \mathbb{Z}/2)$ with the 2-torsion subgroup of the Brauer group of equivalence classes of central simple k-algebras [51, Chapter 4], the class $\gamma(\langle a_1 \rangle + \langle a_2 \rangle + \cdots + \langle a_m \rangle)$ is that of the tensor product of the quaternion algebras (a_i, a_j) as (i, j) runs over pairs of integers from 1 to m with $i < j$. The third is the Arason invariant [2]. The higher invariants are not explicitly named, but do not vanish in general. For example, in the Grothendieck–Witt group of the field $\mathbb{C}((x_1, x_2, \ldots, x_n))$, the nth quotient I^n/I^{n+1} is not 0.

For many fields k, there are explicit computations of $\mathrm{GW}(k)$ and algorithms for deciding when two sums of the given generators represent the same element. See [65].

Example 22.4.1. When $k = \mathbb{C}$ or any algebraically closed field, $k^*/(k^*)^2$ is the one element group, so by the previously given presentation of $\mathrm{GW}(k)$, we see that $\mathrm{rank} : \mathrm{GW}(k) \to \mathbb{Z}$ is an isomorphism.

Example 22.4.2. Let $k = \mathbb{R}$. For a bilinear form $\beta : V \times V \to \mathbb{R}$ over the real numbers, Sylvester's law of inertia states that there is a basis of V so that β is diagonal with only 1's, -1's and 0's on the diagonal. The *signature* of β is the number of 1's minus the number of -1's. The signature determines a homomorphism $\text{sign} : \mathrm{GW}(\mathbb{R}) \to \mathbb{Z}$. Non-degenerate bilinear forms over $\mathbb{R}$ are classified by their rank and signature, and thus the homomorphism

$$\text{rank} \times \text{sign} : \mathrm{GW}(\mathbb{R}) \to \mathbb{Z} \times \mathbb{Z}$$

determines an isomorphism from $\mathrm{GW}(\mathbb{R})$ onto its image. The rank and the signature must have the same parity, but this condition determines the image: We have an isomorphism

$$\mathrm{GW}(\mathbb{R}) \cong \mathbb{Z} \times \mathbb{Z}.$$

Example 22.4.3. For a finite field $\mathbb{F}_q$, the rank and the discriminant define a group isomorphism $\mathrm{GW}(\mathbb{F}_q) \cong \mathbb{Z} \times \mathbb{F}_q^*/(\mathbb{F}_q^*)^2$. When q is odd, note that this means that the Grothendieck–Witt group is isomorphic to $\mathbb{Z} \times \mathbb{Z}/2$.

Springer's theorem [65, VI Theorem 1.4] computes the Grothendieck–Witt group of $\mathbb{Q}_p$ and $\mathbb{C}((x_1, x_2, \ldots, x_n))$. A theorem of Milnor gives the analogous computation for function fields [75, Theorem 5.3]. For $\mathrm{GW}(\mathbb{Q})$, see [65, VI 4].

In summary, the Grothendieck–Witt group is interesting and computable.

22.4.2 Milnor–Witt groups and Degree

The ring $\mathrm{GW}(k)$ is the degree 0 subring of a graded ring $\mathbf{K}_*^{\mathrm{MW}}(k)$ developed by Morel and M. J. Hopkins [79] to describe part of the motivic homotopy groups of spheres. For a field k, the ring $\mathbf{K}_*^{\mathrm{MW}}(k)$ is the associative, graded ring with the following presentation. The generators consist of one element η of degree -1 and elements $[a]$ of degree 1 for each a in k^*.

These generators are subject to the relations

- (Steinberg relation) $[a][1-a] = 0$ for each a in $k - \{0,1\}$.
- $[ab] = [a] + [b] + \eta[a][b]$ for every a, b in k^*.
- $[a]\eta = \eta[a]$ for every a in k^*.
- $\eta h = 0$,

where h is the hyperbolic element $h = \eta[-1] + 2$. One can identify $\mathrm{GW}(k)$ with $\mathbf{K}_0^{\mathrm{MW}}(k)$ by identifying $\langle a \rangle$ with $1 + \eta[a]$.

Moreover, $\mathbf{K}_*^{\mathrm{MW}}(k)$ can be recovered from $\mathrm{GW}(k)$ and $\mathrm{K}_*^{\mathrm{M}}(k)$ as the fiber product

$$\begin{array}{ccc} \mathbf{K}_*^{\mathrm{MW}}(k) & \longrightarrow & I(k)^n \\ \downarrow & & \downarrow \\ \mathrm{K}_n^{\mathrm{M}}(k) & \longrightarrow & I(k)^n/I(k)^{n+1}, \end{array} \tag{22.4.1}$$

where the map $\mathrm{K}_n^{\mathrm{M}}(k) \to I(k)^n/I(k)^{n+1}$ comes from the Milnor Conjecture.

Milnor–Witt $\mathbf{K}$-theory appears as $\mathbb{A}^1$-homotopy groups of spheres. As in classical algebraic topology, it is useful to pass to a stable homotopy category, as there are stronger tools and sometimes cleaner statements, even if the end goal is an unstable result. As both S^1 and $\mathbb{G}_{m,k}$ are spheres, one can arrange for different notions of stability. For Poincaré duality, a useful choice is to force $(-)\wedge\mathbb{P}^1$ to be invertible, resulting in $\mathbb{P}^1$-stable $\mathbb{A}^1$-homotopy theory, denoted $\mathcal{SH}(k)$ [100, Definition 5.7].

Morel constructs an isomorphism of graded rings

$$\mathbf{K}_*^{\mathrm{MW}}(k) \to [S^0, \mathbb{G}_{m,k}^{\wedge *}]_{\mathcal{SH}(k)}, \tag{22.4.2}$$

where the ring structure on the target is induced by smash product. The map (22.4.2) is defined by sending η to the Hopf map $\mathbb{A}_k^2 - \{0\} \to \mathbb{P}_k^1$ (given by $(x, y) \mapsto [x, y]$), and sending $[a]$ to the map $S^0 \to \mathbb{G}_{m,k}$ that preserves base points and that sends the non-basepoint of $S^0 = \operatorname{Spec} k \coprod \operatorname{Spec} k$ to the rational point a in k^*. It follows that $\langle a\rangle$ is sent to the map $\mathbb{P}_k^1 \to \mathbb{P}_k^1$ given by $[x, y] \mapsto [x, ay]$ [79, Lemma 3.43].

To prove that (22.4.2) is an isomorphism requires substantial work and uses the theory of strictly $\mathbb{A}^1$-invariant sheaves. The latter is discussed in Section 22.5 and a proof of the theorem is in [79, Corollary 6.43]. See also [77, Section 6.3] for a sketch.

Given that (22.4.2) is an isomorphism, we obtain the degree map using its inverse.

There is an unstable result analogous to the isomorphism (22.4.2): for $n \geq 2$ and $j \geq 1$, we have

$$[S^n \wedge \mathbb{G}_m^{\wedge r}, S^n \wedge \mathbb{G}_m^{\wedge j}] \cong \mathbf{K}_{j-r}^{\mathrm{MW}}(k), \tag{22.4.3}$$

which, moreover, extends to an isomorphism of sheaves discussed in Section 22.5. For $n = 1$, we are led to consider the unstable $\mathbb{A}^1$-homotopy classes of maps $\mathbb{P}_k^1 \to \mathbb{P}_k^1$, discussed in Section 22.2.3; see Equation (22.2.3).

Realization functors produce maps from the homotopy groups of (22.4.2) and (22.4.3), which are understood for $k = \mathbb{C}$ and $\mathbb{R}$ in an appropriate range.

First, let $k = \mathbb{C}$. Recall from Section 22.3.1 that complex realization induces a map $[X, Y]_{\mathbb{A}^1_{\mathbb{C}}} \to [|X|, |Y|]$ for spaces X and Y. Let $X = S^{n+r\alpha} = S^n \wedge \mathbb{G}_m^{\wedge r}$ and $Y = S^{n+j\alpha} = S^n \wedge \mathbb{G}_m^{\wedge j}$. Assume $n \geq 2$ and $j \geq 1$. Then one has maps

$$\mathbf{K}_{j-r}^{\mathrm{MW}}(\mathbb{C}) \to \pi_{n+r}(S^{n+j}). \tag{22.4.4}$$

For dimensional reasons, this map vanishes when $j > r$. When $j = r$, the identity map generates both source and target, and one has an isomorphism $\mathbb{Z} = \mathrm{GW}(\mathbb{C}) \to \mathbb{Z} = \pi_{n+j}(S^{n+j})$. In the cases where $j < r$, one knows from [79] that $\mathbf{K}_{j-r}^{\mathrm{MW}}(\mathbb{C}) = W(\mathbb{C})$ contains a single nontrivial class, that of η^{j-r}, the iterated Hopf map, and so the image of the realization map (22.4.4) is the group generated by $\eta_{\mathrm{top}}^{j-r}$, which vanishes if $j-r$ is sufficiently large.

The analogous maps induced by real realization are quite different in character from those of the complex case (22.4.4). Namely, $|S^{n+r\alpha}|_{\mathbf{R}} \simeq S^n \wedge |\Gamma^{\wedge r}|_{\mathbf{R}} \simeq S^n$, and so there are induced maps

$$\pi_{n+r\alpha}(S^{n+j\alpha})(\mathbf{R}) = \mathbf{K}_{j-n}^{\mathrm{MW}}(\mathbf{R}) \to \pi_n(S^n) = \mathbb{Z}.$$

Under $\mathbb{R}$-realization, $[a] \mapsto 0$ if $a > 0$, and $[a] \mapsto 1$ if $a < 0$, and $\eta \mapsto -2$, for a proof, see [13, Proposition 3.1.3].

It follows that Morel's degree has the agreeable feature that for subfields k of $\mathbb{R}$, it records the ordinary topological degrees of both the complex and real realizations. Specifically, the following diagram commutes.

$$\begin{array}{ccccc} [S^{2n}, S^{2n}] & \xleftarrow{|\cdot|_{\mathbb{R}}} & [\mathbb{P}^n_k/\mathbb{P}^{n-1}_k, \mathbb{P}^n_k/\mathbb{P}^{n-1}_k] & \xrightarrow{|\cdot|} & [S^n, S^n] \,. \\ \downarrow \deg & & \downarrow \deg & & \downarrow \deg \\ \mathbb{Z} & \xleftarrow{\text{rank}} & \mathrm{GW}(k) & \xrightarrow{\text{sign}} & \mathbb{Z} \end{array}$$

22.4.3 Local Degree

Since the degree homomorphism given above factors through the inverse of a map $[S^0, S^0]_{\mathcal{SH}(k)} \to \mathrm{GW}(k)$, it does not give a method to compute the degree of a map given by algebraic equations. For this, a notion of local degree is useful, and we discuss this now.

Let U and W be Zariski open subsets of $\mathbb{A}^n_k$, and let $f : U \to W$ be a map. Suppose x is a closed point of U such that $y = f(x)$ is a k-point (meaning that the coordinates of y are elements of k), and such that x is isolated in $f^{-1}(y)$. By possibly shrinking U, we may assume that f maps $U - \{x\}$ to $W - \{y\}$, and therefore f induces a map

$$f : U/(U - \{x\}) \to W/(W - \{y\}).$$

The Purity Theorem identifies the quotients $W/(W - \{y\})$ and $U/(U - \{x\})$ with the sphere $(\mathbb{P}^1)^{\wedge n}$ and the smash product $(\mathbb{P}^1)^{\wedge n} \wedge \operatorname{Spec} k(x)_+$, respectively. The *local degree* $\deg_x f$ of f at x is the degree of the composite

$$\mathbb{P}^n/\mathbb{P}^{n-1} \to \mathbb{P}^n/(\mathbb{P}^n - \{x\}) \cong U/(U - \{x\}) \xrightarrow{f} W/(W - \{y\}) \cong (\mathbb{P}^1)^{\wedge n} \cong \mathbb{P}^n/\mathbb{P}^{n-1}$$

in $\mathrm{GW}(k)$. If x is also a k-point, it is more natural to take the degree of the composite

$$(\mathbb{P}^1)^{\wedge n} \to U/(U - \{x\}) \xrightarrow{f} W/(W - \{y\}) \to (\mathbb{P}^1)^{\wedge n}.$$

These definitions are equivalent [61, Proposition 11]. In the presence of Nisnevich local coordinates (see [62, Definition 17]) and appropriate orientations, the definition of local degree can be generalized.

There is an explicit algorithm to compute the local degree [38] [36] [61], which can be implemented with a computer algebra package, or even by hand in good circumstances. Namely, consider a map $f : U \to W$, and a point x in U as above. Since W is a subset of $\mathbb{A}^n_k$, we can express f as an n-tuple of functions $f = (f_1, \dots, f_n)$ defined on U. Since U is an open subset of $\mathbb{A}^n_k$, the point x corresponds to a prime ideal p in $k[x_1, \dots, x_n]$. The f_i can be viewed as elements of the localization $k[x_1, \dots, x_n]_p$. The assumption that x is isolated in $f^{-1}(y)$ implies that the quotient

$$Q = \frac{k[x_1, \dots, x_n]_p}{\langle f_1, \dots, f_n \rangle}$$

is finite dimensional as a k-vector space. The $\mathbb{A}^1$-local degree of f at p is represented by the following bilinear form on Q. The Jacobian

$$J = \det\left(\frac{\partial f_i}{\partial x_j}\right)_{i,j}$$

determines an element of Q. For simplicity, we assume that p is a k-point, and that $\operatorname{rank}_k Q$ is not divisible by the characteristic of k. Choose any k-linear function

$$\eta : Q \to k$$

such that

$$\eta(J) = \operatorname{rank}_k Q.$$

Then define the bilinear form $\omega^{\mathrm{EKL}} : Q \times Q \to k$ by

$$\omega^{\mathrm{EKL}}(g,h) = \eta(gh).$$

It can be shown that the isomorphism class of ω^{EKL} does not depend on the choice of η [38, Proposition 3.5] and that the class of ω^{EKL} in $\mathrm{GW}(k)$ is the $\mathbb{A}^1$-local degree [61]:

$$\deg_x f = \omega^{\mathrm{EKL}}.$$

For the analogous algorithm without the two assumptions, see the work of Scheja and Storch [93, Section 3].

As in classical algebraic topology, the degree of a map can be expressed as a sum of local degrees. For example, let $f : \mathbb{P}^n_k \to \mathbb{P}^n_k$ be a finite map. Let $\mathbb{A}^n_k$ denote the Zariski open subset of $\mathbb{P}^n_k$ which is the complement of the standard closed immersion $\mathbb{P}^{n-1}_k \hookrightarrow \mathbb{P}^n_k$ described in Example 22.2.11. Suppose that $f^{-1}(\mathbb{A}^n_k) = \mathbb{A}^n_k$, so there is an induced map

$$F : \mathbb{P}^n_k/\mathbb{P}^{n-1}_k \to \mathbb{P}^n_k/\mathbb{P}^{n-1}_k.$$

Then, for any k-point y of $\mathbb{A}^n_k$,

$$\deg F = \sum_{x \in f^{-1}(y)} \deg_x f.$$

This is well-known and the proof is very similar to a classical argument. A reference is [61, Proposition 13].

22.4.4 Residue and Transfer maps

The fact that $\mathrm{GW}(k)$ and $\mathbf{K}^{\mathrm{MW}}_*(k)$ are global sections of stable $\mathbb{A}^1$-homotopy sheaves $\pi^{\mathbb{A}^1,s}_*$ implies the existence of extra structure, for example residue and transfer maps. Residue maps are closely related to the unramified sheaves discussed in more detail in the next section. Transfer maps exist for 2-fold $\mathbb{G}_m$-loop sheaves of $\mathbb{A}^1$-homotopy sheaves, i.e., $\pi^{\mathbb{A}^1}_*$ for $* \geq 2$ [79, Chapter 4]. In particular, transfer maps exist for stable $\mathbb{A}^1$-homotopy sheaves and GW. We discuss these maps in more detail now.

We have a residue map in the following situation. Let $v : K \to \mathbb{Z} \cup \infty$ be a valuation on a field K containing k, and let $\mathcal{O}_v$ denote the corresponding valuation ring of elements of non-negative valuation. The residue field $k(v)$ is the quotient of $\mathcal{O}_v$ by its maximal ideal consisting of the elements of K of positive valuation. Let π be a uniformizer (i.e., an element such that $v(\pi) = 1$). Then there is a unique homomorphism ∂_v^π called a *residue map*

$$\partial_v^\pi : \mathbf{K}_{n+1}^{\mathrm{MW}}(K) \to \mathbf{K}_n^{\mathrm{MW}}(k(v))$$

commuting with η and such that

$$\partial_v^\pi([\pi][a_1][a_2]\dots[a_n]) = [\overline{a_1}][\overline{a_2}]\dots[\overline{a_n}],$$

$$\partial_v^\pi([a_0][a_1][a_2]\dots[a_n]) = 0,$$

where $a_i \in \mathcal{O}_v^*$ and $\overline{a_i}$ denotes the image of a_i in $k(v)^*$. The kernel of the residue maps define the sections of the sheaf $\mathbf{K}_*^{\mathrm{MW}}$ over $\operatorname{Spec} \mathcal{O}_v$, see the discussion following [79, Lemma 3.19].

We have a transfer map in the following situation. Let $K \subseteq L$ be a field extension of finite rank, where K is finite type over k (i.e., where K can be generated as a ring over k by finitely many elements).

We first note that the inclusion $K \subseteq L$ corresponds to a map of schemes $\operatorname{Spec} L \to \operatorname{Spec} K$ in the opposite direction. The sheaf property gives restriction maps $\mathbf{K}_*^{\mathrm{MW}}(K) \to \mathbf{K}_*^{\mathrm{MW}}(L)$. Not surprisingly, these maps are given by $[a] \mapsto [a]$, $\eta \mapsto \eta$, and correspond to pullback of bilinear forms when restricted to GW.

Transfer maps for such field extensions can be constructed by producing a stable map

$$\operatorname{Spec} K \to \operatorname{Spec} L$$

in the direction opposite to the map of spaces, closely analogous to the umkehr construction or Becker–Gottlieb transfer[79, 4.2]: When L is generated by a single element over K, we can choose a closed point z of $\mathbb{P}_K^1$ with residue field L, or equivalently, the data of a closed immersion

$$z : \operatorname{Spec} L \hookrightarrow \mathbb{P}_K^1.$$

Using Purity (c.f. Examples 22.2.17 and 22.2.14), we then have the cofiber sequence

$$(\mathbb{P}_K^1 - \{z\})/\infty \to \mathbb{P}_K^1/\infty \to \mathbb{P}_K^1/(\mathbb{P}_K^1 - \{z\}) \simeq \mathrm{Th}(N_z\mathbb{P}_K^1) \simeq \mathbb{P}_k^1 \wedge (\operatorname{Spec} L)_+.$$

The quotient map, or Thom collapse map, in this sequence

$$\mathbb{P}_k^1 \wedge (\operatorname{Spec} K)_+ \cong \mathbb{P}_K^1/\infty \to \mathbb{P}_k^1 \wedge (\operatorname{Spec} L)_+$$

is the desired stable map $\operatorname{Spec} K \to \operatorname{Spec} L$, and induces a transfer map

$$\tau_{L/K}^z : \mathbf{K}_*^{\mathrm{MW}}(L) \to \mathbf{K}_*^{\mathrm{MW}}(K),$$

called the *geometric transfer*. This transfer depends on the chosen generator of L over K. Although L may not be generated over K by a single element, we can always choose a finite list of generators. Given an ordered such list, we define a transfer $\mathbf{K}_*^{\mathrm{MW}}(L) \to \mathbf{K}_*^{\mathrm{MW}}(K)$ by composing the transfers just constructed.

For simplicity, assume that the characteristic of k is not 2. The dependency of the transfer map on the chosen generators can be eliminated by modifying the definition in the following manner: Suppose z is a generator of L over K. The monic minimal polynomial of z can be expressed canonically as $P(x^{p^m})$ where P is a separable polynomial. The derivative $P'(z^{p^m})$ of P evaluated at z^{p^m} is an element of L^*. The *cohomological transfer*

$$\mathrm{Tr}_{L/K} : \mathbf{K}_*^{\mathrm{MW}}(L) \to \mathbf{K}_*^{\mathrm{MW}}(K) \tag{22.4.5}$$

is then defined

$$\mathrm{Tr}_{L/K}(\beta) = \tau^z_{L/K}(\langle P'(z^{p^m})\rangle \beta),$$

and this map is independent of the chosen generator [79, Theorem 4.27]. More generally, for any finite extension $K \subseteq L$, one has a cohomological transfer (22.4.5) by composing the the transfers just defined for a sequence of generators, and the resulting map is again independent of the chosen sequence. This independence of the choice of generators is more naturally understood terms of twisted Milnor–Witt K-theory, where the twist is by the canonical sheaf ([41, 1.5], [79, Chapter 5]).

From the algebraic perspective, there are many possible transfers $\mathrm{GW}(L) \to \mathrm{GW}(K)$ for $[L : K] \leq \infty$ defined as follows: for any nonzero K-linear map $f : L \to K$, and non-degenerate bilinear form $\beta : V \times V \to L$, the composite

$$f \circ \beta : V \times V \xrightarrow{\beta} L \xrightarrow{f} K$$

is a non-degenerate bilinear form. Thus we may define a map $\mathrm{Tr}_f : \mathrm{GW}(L) \to \mathrm{GW}(K)$ that takes the isomorphism class of β to the isomorphism class of $f \circ \beta$, where in the former case, V is viewed as an L-vector space, and in the latter case, V is viewed as a K-vector space.

This abundance of transfers for GW implies the same for $\mathbf{K}_*^{\mathrm{MW}}$ by the fiber product square (22.4.1) and the canonical transfer on Milnor K-theory.

When $K \subset L$ is a finite separable extension, there is a canonical choice of such an f, namely, the trace map $L \to K$ from Galois theory, given by summing the Galois conjugates of an element of L. The resulting transfer map is the cohomological transfer (22.4.5) defined above. This transfer arises naturally when studying the local degree at non-k-rational points. For example, let $f : U \to W$ be a map between open subsets of $\mathbb{A}^n_k$. Suppose that x in U maps to a k-point y of W and that x is an isolated in $f^{-1}(y)$, so the local degree $\deg_x^{\mathbb{A}^1} f$ exists. If the extension $k \subseteq k(x)$ is separable and f is étale at x, then

$$\deg_x^{\mathbb{A}^1} f = \mathrm{Tr}_{k(x)/k}\langle J(x)\rangle,$$

where J denotes the Jacobian of f. A proof of this is in [61, Proposition 14], and this proof relies on Hoyois's work in [54].

Example 22.4.4. $\mathrm{Tr}_{\mathbb{C}/\mathbb{R}}\langle 1\rangle = \begin{bmatrix} 2 & 0 \\ 0 & -2 \end{bmatrix} = \langle 1\rangle + \langle -1\rangle = h$, where the central expression is the Gram matrix with respect to the $\mathbb{R}$-basis $\{1, i\}$ of $\mathbb{C}$.

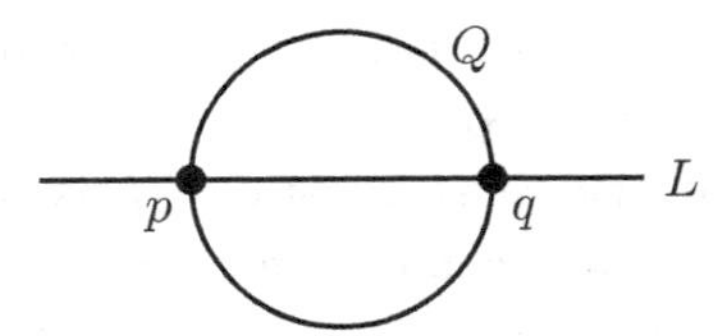

FIGURE 22.4.1. $T_pX \cap X \subset T_pX$

22.4.5 Euler class

Given an oriented vector bundle of rank r on an oriented $\mathbb{R}$-manifold of dimension r, one can define an Euler number. It is an element of $\mathbb{Z}$ that counts the number of zeros of a section weighted by a local index. This local index is defined as follows: Let σ denote a section and suppose that x is an isolated zero. Choose local coordinates around x and a local trivialization of the vector bundle such that both are compatible with the appropriate orientations. With these choices, the section can be viewed as a function $\mathbb{R}^r \to \mathbb{R}^r$. The index $\operatorname{ind}_x \sigma$ is then the local degree of this function at x, where x is viewed as a point of $\mathbb{R}^r$, by a slight abuse of notation. The Euler number e is then

$$e = \sum_{x:\sigma(x)=0} \operatorname{ind}_x \sigma, \tag{22.4.6}$$

when σ has only isolated zeros.

With Morel's $\mathrm{GW}(k)$-valued degree, this Euler number can be enriched to an element of $\mathrm{GW}(k)$. Namely, one can again define a local index using the local degree and then use (22.4.6) to define the Euler number, now an element of $\mathrm{GW}(k)$ [62]. There are subtleties involved in choosing coordinates and identifying σ with a function, due to difficulties in defining algebraic functions. One must then show that the Euler number is well-defined. There are other approaches avoiding these difficulties, using oriented Chow or Chow–Witt groups of Barge and Morel [20]. Barge and Morel construct an Euler class (loc. cit.) in oriented Chow and it can be pushed forward under certain conditions to land in $\mathrm{GW}(k)$. See the work of Jean Fasel [40] and Marc Levine [66]. Morel gives an alternate construction of an Euler class using obstruction theory [79, Section 8.2]. This Euler class lies in the same oriented Chow group, and will be discussed further in Section 22.6.4. Some comparison results are available between these two Euler classes [10] [66, Proposition 11.6]. Further approaches to defining an Euler class using $\mathbb{A}^1$-homotopy theory are found in [31] [69].

22.4.6 Applications to enumerative geometry

Questions in enumerative geometry ask to count a set of algebro-geometric objects satisfying certain conditions, or more generally, to describe this set. A classical example is the question "How many lines intersect four general lines in three dimensional space?" The questions are typically posed so there is a fixed answer, such as "2," as opposed to "sometimes 2 and sometimes 0," and to get such "invariance of number," one needs to work over an algebraically closed field; the number of solutions to a polynomial equation of degree n is always n if one works over $\mathbb{C}$ (at least if you count with multiplicity, or restrict your attention to a general polynomial), but not over $\mathbb{R}$, and the same phenomenon appears in the classical question quoted above about the number of lines. However, a feature of $\mathbb{A}^1$-homotopy theory is its applicability to general fields k. Moreover, there are classical theorems in enumerative geometry where a count of geometric objects is identified with an Euler number, which is then computed using characteristic class techniques. Equipped with an enriched Euler class in $\mathrm{GW}(k)$, as in Section 22.4.5, we may take these theorems and hope to perform the analogous counts over other fields.

What sorts of results appear? Here is an example taken from [62]. A cubic surface X is the zero locus in $\mathbb{P}^3_k$ of a degree 3 homogenous polynomial f in $k[x_0, x_1, x_2, x_3]$. It is a lovely 19th century result of Salmon and Cayley that when $k = \mathbb{C}$ and X is smooth, the number of lines in X is exactly 27. Over the real numbers, the number of real lines is either 3, 7, 15, or 27, and in particular, the number depends on the surface. A classification was obtained by Segre [90], but it is a recent observation of Benedetti–Silhol [23], Finashin–Kharlamov [42], Horev–Solomon [53] and Okonek–Teleman [83] that a certain signed count of lines is always 3. Specifically, Segre distinguished between two types of real lines on $X = \{[x_0, x_1, x_2, x_3] \in \mathbb{RP}^3 : f(x_0, x_1, x_2, x_3) = 0\}$, called *hyperbolic* and *elliptic*. The distinction is as follows. Let L be a real line on X. For every point p of L, consider the intersection $T_pX \cap X$ of the tangent plane to the cubic surface at p with the cubic surface itself. Since T_pX is a plane, the intersection is a curve in the plane, which by Bézout's theorem has degree $3 \cdot 1 = 3$. The line L must be contained in this plane curve. Algebraic curves can be decomposed into irreducible components, and it follows that we can express $T_pX \cap X$ as a union $T_pX \cap X = L \cup Q$, where Q is a plane curve of degree $3 - 1 = 2$. Applying Bézout's theorem again shows that the intersection $L \cap Q$ of the plane curves L and Q is degree 2, or in other words, generically consists of two points, which we call $\{p, q\}$. See Figure 22.4.1. We may thus define an involution $I : L \to L$ by sending p to q.

The points of the intersection $L \cap Q = \{p, q\}$ are precisely the points x on L such that $T_xX = T_pX$. To see this, note that T_qX contains the span of a vector along L and a vector along Q. At least generically, it follows that T_qX contains a 2-dimensional subspace of T_pX. Since T_qX is 2-dimensional (X is smooth), it follows that $T_qX = T_pX$. Similar reasoning applies in the reverse direction as well. So we can characterize the involution I as the unique map exchanging points on L with the same tangent space to X.

An automorphism of $L \cong \mathbb{RP}^1$ is a conjugacy class of element of $\mathrm{PGL}_2\,\mathbb{R}$, and the elements of $\mathrm{PGL}_2\,\mathbb{R}$ are classified as elliptic, hyperbolic, or parabolic by the behavior of the fixed points. If I has a complex conjugate pair of fixed points, I is elliptic, if I has two real fixed points, it is hyperbolic, and if I has one fixed point, it is parabolic. Involutions are never parabolic. Segre classified the line L as elliptic (respectively hyperbolic) if I is.

Another description of this distinction involves (S)pin structures. The tangent plane T_pX rotates around L as p travels along L, describing a loop in the frame bundle of $\mathbb{P}^3$. The frames in this loop consist of the line, a normal to L in TX, and a normal to X in $\mathbb{P}^3$, obtained by monodromy of an initial such choice. L is hyperbolic if such a loop lifts to the double cover, and hyperbolic if it does not. In other words, place your index finger along the line, hold your palm on the tangent plane, and have your hand travel around $L \simeq S^1$. If your hand returns twisted up, the line is elliptic.

The signed count referred to above is that

$$\#\{\text{hyperbolic lines}\} - \#\{\text{elliptic lines}\} = 3. \tag{22.4.7}$$

It is a general principle that analogous results over $\mathbb{C}$ and $\mathbb{R}$ may be realizations of a more general result in $\mathbb{A}^1$-homotopy theory. In the case of the count of lines on cubic surface, this is indeed the case: Consider a cubic surface X over a field k. Suppose L is a line in $\mathbb{P}^3_k$ which lies in X. The coefficients of L determine a field extension $k(L)$. Moreover, the previously given definition of the involution I carries over in this generalized situation, determining an involution I of the line, and thus a conjugacy class in $\mathrm{PGL}_2\, k(L)$. Such a conjugacy class has a well-defined determinant $\det I$ in $k(L)^*/(k(L)^*)^2$. The *Type* of L is $\mathrm{Type}(L) = \langle \det I \rangle$ in $\mathrm{GW}(k(L))$. Alternatively, the type of L may be described as $\mathrm{Type}(L) = \langle D \rangle$, where D is the unique element of $k(L)^*/(k(L)^*)^2$ so that the fixed points of the involution I are a conjugate pair of points defined over the field $k(L)[\sqrt{D}]$. There is a third description of the type as $\mathrm{Type}(L) = \langle -1 \rangle \deg I$ the multiplication of the $\mathbb{A}^1$-degree of I and $\langle -1 \rangle$ in $\mathrm{GW}(k(L))$. The theorem of Salmon and Cayley and (22.4.7) then are realizations of the following theorem [62, Theorem 1].

Theorem 22.4.5. *Let X be a smooth cubic surface over a field k of characteristic not* 2. *Then*

$$\sum_{\textit{lines } L \textit{ in } X} \mathrm{Tr}_{k(L)/k}\, \mathrm{Type}(L) = 15\langle 1 \rangle + 12\langle -1 \rangle.$$

This is proven by identifying the left hand side with the Euler class of the third symmetric power of the dual tautological bundle on the Grassmannian of lines in $\mathbb{P}^3$.

Other results along these lines include [54], [66], [67], [68], [108], [96], [24], and this is an active area of research.

22.5 Homotopy sheaves and the connectivity theorem

In the monograph [79], Morel establishes a number of extremely strong results describing the $\mathbb{A}^1$-homotopy sheaves of objects of $\mathbf{sPre}(\mathbf{Sm}_k)$ when k is an infinite perfect field. The homotopy sheaves, (22.2.4), are sheaves $\mathbf{Sm}_k$ for the Nisnevich topology. The assumptions on the field are probably both unnecessary, but the literature does not currently contain proofs of certain necessary statements for finite or imperfect fields.

Let $\mathcal{X}$ denote an object of $\mathbf{sPre}(\mathbf{Sm}_k)_\bullet$, i.e., a based object of $\mathbf{sPre}(\mathbf{Sm}_k)$. The $\mathbb{A}^1$-homotopy sheaves $\boldsymbol{\pi}_i^{\mathbb{A}^1}(\mathcal{X})$—when $i \geq 1$—satisfy a strong $\mathbb{A}^1$-invariance property.

Definition 22.5.1. Let $\mathcal{G}$ denote a sheaf of groups for the Nisnevich topology on $\mathbf{Sm}_k$. The sheaf $\mathcal{G}$ is *strongly* $\mathbb{A}^1$*-invariant* if the map

$$\mathrm{H}^n(U, \mathcal{G}) \to \mathrm{H}^n(U \times \mathbb{A}^1{}_k, \mathcal{G}) \tag{22.5.1}$$

is an isomorphism for $n \in \{0, 1\}$ and for all smooth k-schemes U.

Definition 22.5.2. Let $\mathcal{G}$ denote a sheaf of abelian groups for the Nisnevich topology on $\mathbf{Sm}_k$. The sheaf $\mathcal{G}$ is *strictly* $\mathbb{A}^1$*-invariant* if the map of (22.5.1) is an isomorphism for $n \in \{0, 1, \dots\}$ and for all smooth k-schemes U.

Theorem 22.5.3 (Morel, [79, Corollary 6.2])**.** *Let k be an infinite perfect field. Let $\mathcal{X}$ be an object of $\mathbf{sPre}(\mathbf{Sm}_k)_\bullet$. Then $\boldsymbol{\pi}_1^{\mathbb{A}^1}(\mathcal{X})$ is strongly $\mathbb{A}^1$-invariant and for any $i \geq 1$, the sheaf of abelian groups $\boldsymbol{\pi}_i^{\mathbb{A}^1}(\mathcal{X})$ is strictly $\mathbb{A}^1$-invariant.*

Over an infinite perfect field, a strongly $\mathbb{A}^1$-invariant sheaf of abelian groups is strictly $\mathbb{A}^1$-invariant, [79, Theorem 5.46], so for us the distinction is chiefly useful for defining a category of "strictly $\mathbb{A}^1$-invariant sheaves" , which necessarily consist of abelian groups.

We remark in passing that

$$\mathrm{H}^0(U, \boldsymbol{\pi}_i^{\mathbb{A}^1}(\mathcal{X})) = \boldsymbol{\pi}_i^{\mathbb{A}^1}(\mathcal{X})(U),$$

so that either condition implies $\mathbb{A}^1$-invariance: $\boldsymbol{\pi}_i^{\mathbb{A}^1}(\mathcal{X})(U \times \mathbb{A}^1) = \boldsymbol{\pi}_i^{\mathbb{A}^1}(\mathcal{X})(U)$.

There exists an abelian group that is $\mathbb{A}^1$-invariant but not strongly $\mathbb{A}^1$-invariant—for instance, if one defines $\tilde{\mathbb{Z}}[\Gamma]$ to be the abelian group generated by the sheaf of sets Γ, modulo the relation that $1 \in \Gamma$ is identified with the 0 element, then $\tilde{\mathbb{Z}}[\Gamma]$ is $\mathbb{A}^1$-invariant—since Γ itself is $\mathbb{A}^1$-invariant—but the free strongly $\mathbb{A}^1$-invariant sheaf of abelian groups generated by $\tilde{\mathbb{Z}}[\Gamma]$ is $\mathbf{K}_1^{\mathrm{MW}}$ by [79, Theorem 3.37].

A sheaf of groups, $\mathcal{G}$, is strongly $\mathbb{A}^1$-invariant if and only if it appears as an $\mathbb{A}^1$-homotopy sheaf $\boldsymbol{\pi}_1^{\mathbb{A}^1}$, since it appears as the $\mathbb{A}^1$-fundamental sheaf of $B\mathcal{G}$. The strongly $\mathbb{A}^1$-invariant sheaves $\mathcal{G}$ are *unramified* sheaves, which is to say, briefly, that $\mathcal{G}(U)$ is the product of $\mathcal{G}(U_\alpha)$ as U_α range over the irreducible components, and that for any dense open $V \subset U$, $\mathcal{G}(U) \to \mathcal{G}(V)$ is an injection which is even an isomorphism if $V - U$ is everywhere of codimension at least 2 in U, see [79, Section 2.1]. A feature of unramified sheaves $\mathcal{G}$ in this sense is that $\mathcal{G}$ can be recovered from the values $\mathcal{G}(F)$ where F ranges over fields of finite transcendence degree over k, along with subsets $\mathcal{G}(\mathcal{O}_v) \subset \mathcal{G}(F)$ associated to any discrete valuation v on F, and specialization maps $\mathcal{G}(\mathcal{O}_v) \to \mathcal{G}(\kappa(v))$ mapping to the residue fields, all satisfying certain compatibility axioms, [79, Section 2.1]. Most strikingly, if $\mathcal{G}(F)$ vanishes for all field extensions of k, then $\mathcal{G} = 0$.

The category of strictly $\mathbb{A}^1$-invariant sheaves forms a full abelian subcategory of the category of sheaves of abelian groups on $\mathbf{Sm}_k$, being the heart of a t-structure—this appears as [79, Corollary 6.24].

One can rephrase the statement that the sheaves $\boldsymbol{\pi}_i^{\mathbb{A}^1}(\mathcal{X})$ are strongly $\mathbb{A}^1$-invariant for $i \geq 1$ as follows: if $\mathcal{X}$ is $\mathbb{A}^1$-local, then $\mathcal{X} \to R_{\mathbb{A}^1}\mathcal{X}$ is a local equivalence and so the local homotopy sheaves $\boldsymbol{\pi}_i(\mathcal{X})$ for $i \geq 1$ are strongly $\mathbb{A}^1$-invariant. There is a partial converse:

Theorem 22.5.4. *Suppose $\mathcal{X}$ is a pointed connected object of* $\mathbf{sPre}(\mathbf{Sm}_k)$. *Then $\mathcal{X}$ is $\mathbb{A}^1$-local if and only if the local homotopy sheaf $\boldsymbol{\pi}_i(\mathcal{X})$ is strongly $\mathbb{A}^1$-invariant for all $i \geq 1$.*

This holds because $\mathcal{X}$ is $\mathbb{A}^1$-local if the functor $\mathrm{Map}(\cdot, \mathcal{X})$ does not distinguish between U and $U \times \mathbb{A}^1$ up to homotopy. For a connected $\mathcal{X}$, the calculation of maps $\mathrm{Map}(\cdot, \mathcal{X})$ can be reduced to the calculation of $\mathrm{Map}(\cdot, K(\boldsymbol{\pi}_i(\mathcal{X}), j))$ for varying i, j by means of the Postnikov tower, and strong $\mathbb{A}^1$-invariance of $\boldsymbol{\pi}_i(\mathcal{X})$ amounts to the same thing as $\mathbb{A}^1$-locality of $K(\boldsymbol{\pi}_i(\mathcal{X}), j)$. The result is stated as [79, Corollary 6.3].

The assumption of connectivity, while it may seem mild at first, is a great inconvenience. In contrast to classical homotopy theory, where one can generally work component-by-component, it is a feature of the homotopy of sheaves that π_0, a sheaf of sets, can be extremely complicated.

In fact, the difficulties at π_0 are substantial. If $\mathcal{G}$ is a discrete sheaf of groups, viewed as a space in dimension 0, then $\mathcal{G}$ is $\mathbb{A}^1$-local if and only if $\mathcal{G}$ is $\mathbb{A}^1$-invariant. If $\mathcal{G}$ is $\mathbb{A}^1$-invariant without being strongly $\mathbb{A}^1$-invariant, then $R_{\mathbb{A}^1} B\mathcal{G}$ cannot be (locally) equivalent to $B\mathcal{G}$, and so $\pi_0 \Omega R_{\mathbb{A}^1} B\mathcal{G} \not\cong \pi_0 \mathcal{G}$. In particular, $B\mathcal{G}$ does not have the $\mathbb{A}^1$-homotopy type of a delooping of $\mathcal{G}$. It follows from this that $\mathbb{A}^1$-homotopy theory does not correspond to an ∞–topos in the sense of [70, Section 6.1]. This observation, due to J. Lurie, appears as [95, Remark 3.5].

As a heuristic, aside from problems at π_0, results that hold in ∞-topoi (or in the "model topoi" of C. Rezk [97]) can generally be expected to hold in $\mathbb{A}^1$-homotopy theory. For instance:

Theorem 22.5.5. *Suppose*

$$\mathcal{F} \to \mathcal{E} \to \mathcal{B} \tag{22.5.2}$$

is a fiber sequence in a local model structure on $\mathbf{sPre}(\mathbf{Sm}_k)_\bullet$ *where* $\pi_0(B)$ *is* $\mathbb{A}^1$*-invariant. Then* (22.5.2) *is an* $\mathbb{A}^1$*-homotopy fiber sequence.*

This is a corollary of [16, Theorem 2.1.5] (see also Remark 2.1.6), a development of [79, Theorem 6.53].

22.5.1 Contractions

The following construction appears originally in [101], but applied there only to "presheaves with transfers". In the current context, it is due to [79, Section 2.2, pp. 33–36]. Given a presheaf of groups $\mathcal{G}$, define

$$\mathcal{G}_{-1}(U) = \mathrm{Ker}(\mathcal{G}(\Gamma \times U) \xrightarrow{ev_1} \mathcal{G}(U)).$$

Here the map is induced inclusion of $U \times \{1\}$ in $U \times \Gamma$; when the presheaf is applied, this appears as a kind of "evaluation at 1", hence the notation. This construction is functorial in $\mathcal{G}$.

The functor $\mathcal{G} \mapsto \mathcal{G}_{-1}$ is sometimes called *contraction.* It may be iterated, in which case one writes $\mathcal{G}_{-n}$. The functor $(\cdot)_{-1}$ has a number of excellent properties. It restricts to give a functor of (Nisnevich) sheaves, for instance, and it preserves the property of being abelian. It is also left exact in general. When applied to $\mathbb{A}^1$-homotopy sheaves, it has the following striking description, which appears as [79, Theorem 6.13]

Theorem 22.5.6 (Morel). *If X is a pointed, connected $\mathbb{A}^1$-local space, then the (derived) mapping space* $\Omega_\Gamma :=$ Map(Γ, X) *is also pointed, connected and $\mathbb{A}^1$-local, and there is a canonical isomorphism*

$$\pi_n^{\mathbb{A}^1}(\Omega_\Gamma \mathcal{X}) \to \pi_n^{\mathbb{A}^1}(\mathcal{X})_{-1}.$$

This theorem, of course, is known only for homotopy sheaves over $\mathbf{Sm}_k$ where k is a field satisfying the running assumptions of [79].

Any short exact sequence of strictly $\mathbb{A}^1$-invariant sheaves may be realized as the homotopy sheaves of an $\mathbb{A}^1$-homotopy fiber sequence, for instance

$$0 \to \pi_2^{\mathbb{A}^1}(K(\mathcal{A}, 2)) \to \pi_2^{\mathbb{A}^1}(K(\mathcal{B}, 2)) \to \pi_2^{\mathbb{A}^1}(K(\mathcal{C}, 2)) \to 0$$

and since the (derived) mapping space $\mathrm{Map}(\Gamma, \cdot)$ preserves homotopy fiber sequences, one deduces that the functor $\mathcal{A} \mapsto \mathcal{A}_{-1}$ is exact on the category of strictly $\mathbb{A}^1$-invariant sheaves.

It is immediate that if $\mathcal{C}$ is a constant presheaf of abelian groups, then $\mathcal{C}_{-1} = 0$. One might fantasize that if $\mathcal{G}$ is strictly $\mathbb{A}^1$-invariant, then $\mathcal{G}_{-1} = 0$ should imply that $\mathcal{G}$ is constant, but this is far from the case. In fact, from the analysis furnished by [79, Chapter 2] of $(\cdot)_{-1}$, one can deduce that $\mathcal{G}_{-1} = 0$ if and only if $\mathcal{G}$ is *birational* in the sense of converting dense open inclusions of schemes $V \to U$ into isomorphisms $\mathcal{G}(U) \to \mathcal{G}(V)$. In [17, Section 6], a study is made of such sheaves, and the category of all such sheaves over $\mathbf{Sm}_k$ is seen to be equivalent to a very large category of functors on the category of field extensions of k.

22.5.2 Unramified K-theories

Certain groups that were previously known to be functors of fields and field extensions are known to extend to give strictly $\mathbb{A}^1$-invariant Nisnevich sheaves on $\mathbf{Sm}_k$. For instance, there is a strictly $\mathbb{A}^1$-invariant sheaf $\mathbf{K}^Q_n$ such that $\mathbf{K}^Q_n(F) = K^Q_n(F)$Quillen's algebraic K-theory—here the field extension F/k is supposed to be a separable extension of an extension of finite transcendence degree. The sheaf $\mathbf{K}^Q_n$ arises as $\pi_n^{\mathbb{A}^1}(B\,\mathrm{GL}_N)$ for $N \geq n + 2$, this being a consequence of the representability of K-theory for smooth schemes as proved in [80, Theorem 3.13, p. 140]. Analogous constructions for hermitian K-groups are made in [94].

Another notable sheaf is $\mathbf{K}^M_n$, which can be recovered as a quotient of $\mathbf{K}^{\mathrm{MW}}_n = \pi_{n-1}^{\mathbb{A}^1}(\mathbf{A}^n - \{0\})$, the *unramified Milnor–Witt K-theory sheaf* as constructed by Morel in [79, Section 3.2]. In this case, the group of sections for a field is $K^M_n(F)$, as defined in Section 22.4.

In the cases of $\mathbf{K}^Q_n$, $\mathbf{K}^{MW}_n$ and $\mathbf{K}^M_n$, the phenomenon of $\mathbb{P}^1$-stability for the associated theories implies $(\mathbf{K}^Q_n)_{-1} = \mathbf{K}^Q_{n-1}$ and similarly for the other two theories; a proof is outlined in [7, Lemma 2.7, Proposition 2.9].

One may define $\mathbf{K}^{\mathrm{ind}}_3$, the cokernel of a natural map $\phi : \mathbf{K}^M_3 \to \mathbf{K}^Q_3$. By virtue of Matsumoto's theorem, one knows that

$$(\phi)_{-1} : \mathbf{K}^{\mathrm{M}}_2 \to \mathbf{K}^Q_2$$

is an isomorphism, so that $(\mathbf{K}^{\mathrm{ind}}_3)_{-1} \cong 0$. It is known, for instance by [74], that $\mathbf{K}^{\mathrm{ind}}_3$ is not constant, so this furnishes a specific example of a nonconstant strictly $\mathbb{A}^1$-invariant sheaf, the contraction of which is 0.

22.5.3 $\mathbb{A}^1$-homology and the connectivity theorem

In [79, Definition 6.29], Morel defines an "$\mathbb{A}^1$-homology theory" $\mathrm{H}_n^{\mathbb{A}^1}(\mathcal{X})$, by means of the following: for any simplicial (pre)sheaf $\mathcal{X}$ one may define $\mathbb{Z}(\mathcal{X})$, the free abelian group on $\mathcal{X}$, which one converts to a chain complex $C_*(\mathcal{X})$ via the Dold–Kan correspondence. One can take the category of (pre)sheaves of chain complexes on $\mathbf{Sm}_k$, viewed as a setting for homotopy theory in its own right, and then localize with respect to $\mathbb{A}^1$, that is, with respect to the maps $C_* \otimes C_*(\mathbb{A}^1) \to C_*$. One may now replace chain complexes $C_*(\mathcal{X})$ by "abelian $\mathbb{A}^1$-local replacements"—for Morel, these are the fibrant objects in the localized model category—denoted $C_*^{\mathbb{A}^1}(\mathcal{X})$. The homology of such an object is $\mathrm{H}_*^{\mathbb{A}^1}(\mathcal{X})$, and these homology sheaves are all strictly $\mathbb{A}^1$-invariant.

This homology theory is quite distinct from the "motivic homology theory" defined by Voevodsky[101]; it is much closer to the unstable $\mathbb{A}^1$-homotopy theory, and little is known about it. It does enjoy the following three properties:

1. It is S^1 stable: $\tilde{\mathrm{H}}_{n+1}^{\mathbb{A}^1}(\mathcal{X} \wedge S^1) \cong \tilde{\mathrm{H}}_n^{\mathbb{A}^1}(\mathcal{X})$.

2. If $\mathcal{F}$ is concentrated in simplicial degree 0, i.e., $\mathcal{F}_n = \mathcal{F}_0$ for all $n \geq 0$, then $\mathrm{H}_0^{\mathbb{A}^1}(\mathcal{F})$ is the strictly $\mathbb{A}^1$-invariant sheaf freely generated by the sheaf $\mathcal{F}$, in the sense that this construction is left adjoint to an obvious forgetful functor.

3. For any $\mathbb{A}^1$-simply-connected pointed object $\mathcal{X}$, a Hurewicz isomorphism holds. If $\boldsymbol{\pi}_i^{\mathbb{A}^1}(\mathcal{X}) = 0$ for $i \leq n-1$, and $n \geq 2$, then a natural map $\boldsymbol{\pi}_n^{\mathbb{A}^1}(\mathcal{X}) \to \mathrm{H}_n^{\mathbb{A}^1}(\mathcal{X})$ is an isomorphism. A modification of this holds for $\boldsymbol{\pi}_1$, involving abelianization and "$\mathbb{A}^1$-strictification". See [79, Theorems 6.35, 6.37].

It is this theory, and the properties above, that allows Morel to compute the unstable $\mathbb{A}^1$-homotopy sheaves of the spheres. Specifically, provided $n \geq 2$, so that $\boldsymbol{\pi}_n^{\mathbb{A}^1}$ is known to be abelian and strictly $\mathbb{A}^1$ invariant, then $\boldsymbol{\pi}_n^{\mathbb{A}^1}(S^n \wedge \Gamma^{\wedge m})$ is the free strictly $\mathbb{A}^1$-invariant sheaf generated by the set $\Gamma^{\wedge m}$. That is

$$\boldsymbol{\pi}_n(S^n \wedge \Gamma^{\wedge m}) = \mathbf{K}_m^{\mathrm{MW}} \tag{22.5.3}$$

provided $n \geq 2$ and $m \geq 1$.

Combined with further results of Morel's on the contractions of $\mathbf{K}_\bullet^{\mathrm{MW}}$, one deduces

$$\boldsymbol{\pi}_{n+i\alpha}^{\mathbb{A}^1}(S^n \wedge \Gamma^{\wedge m}) = \mathbf{K}_{m-i}^{\mathrm{MW}} \tag{22.5.4}$$

provided $n \geq 2$ and $m \geq 1$. These calculations appear as [79, Remark 6.42].

The following result of Morel is known as the "unstable connectivity theorem".

Theorem 22.5.7. [79, Theorem 6.38] *Let $n > 0$ be an integer and let $\mathcal{X}$ be a pointed $(n-1)$-connected object in $\mathbf{sPre}(\mathbf{Sm}_k)$. Then its $\mathbb{A}^1$-localization is simplicially $(n-1)$-connected.*

22.6 Application to G-bundles

22.6.1 $\mathbb{A}^1$-classifying spaces

A feature of $\mathbb{A}^1$-homotopy theory is that there are two different notions of classifying space of an algebraic group scheme G. In both cases, one wishes to begin with a contractible object EG on which G acts, and then to form the quotient $(EG)/G$. The difference arises because there are two distinct notions of quotient.

The *simplicial* or *Nisnevich* classifying space BG is constructed by taking a contractible object EG on which G acts freely, and then forming the quotient $BG = (EG)/G$ in the category of simplicial Nisnevich sheaves. Standard homotopical devices for the construction of BG, such as found in [73] for instance, will invariably produce this classifying space, up to homotopy.

One may produce a new space, $B_{\text{ét}}G$, by taking a (derived) pushforward of the pullback of BG along the morphism of sites

$$(\mathbf{Sm}_k)_{\text{ét}} \to (\mathbf{Sm}_k)_{\text{Nis}},$$

see [80, Section 4.1].

The *geometric* classifying space, $B_{gm}G$, is constructed in [80, Section 4.2], using what they term an *admissible gadget*, which is a generalization of the construction of [98]. The construction of [80] applies to étale sheaves of groups over a base S, but for the sake of the exposition here, we restrict to a reductive algebraic group G over a field k. While EG cannot be constructed as a variety, one may construct a G-equivariant sequence $U_i \hookrightarrow U_{i+1}$ of increasingly highly-$\mathbb{A}^1$-connected varieties on which G acts freely. The construction is accomplished by considering a suitable family of larger and larger representations of G on affine spaces over k, and discarding a locus including all points where the action of G is not free. The quotients U_i/G may then be taken in the category of algebraic spaces, or if one is particularly careful with the discarded loci, in schemes. The object $B_{gm}G$ is then defined to be the colimit of the ind-algebraic-space $\{U_i/G\}$. The quotient that is constructed here is "geometric", in that it agrees with classically-existing notions of quotient of a variety by a group scheme action, as in, e.g., [81]; it is also the quotient in the big site of étale sheaves. It is then the case, [80, Proposition 4.2.6], that $B_{\text{ét}}G$ is $\mathbb{A}^1$-equivalent to $B_{gm}G$.

If G is a sheaf of étale group schemes, e.g., if G is representable, and if $\mathrm{H}^1_{\text{Nis}}(\cdot, G) \cong \mathrm{H}^1_{\text{ét}}(\cdot, G)$, then the constructions of $B_{\text{ét}}G$ and BG above are naturally simplicially equivalent, [80, Lemma 4.1.18]. In particular, in the case of the *special* algebraic groups of [92, Section 4.1], including GL_n, SL_n and Sp_n, all notions of classifying space considered above are $\mathbb{A}^1$-weakly equivalent.

In contrast, in the case of a nontrivial finite group, both BG and $B_{\text{ét}}G$ are $\mathbb{A}^1$-local objects already, [80, Section 4.3], and one may easily construct a G-torsor $\pi : U \to U/G$ where both U and U/G are smooth k schemes but where π is not Nisnevich-locally trivial. In this case, it follows from [80, Lemma 4.1.8] again that for a nontrivial finite group, the map $BG \to B_{\text{ét}}G$ is not an $\mathbb{A}^1$-equivalence.

In the study of vector bundles, or equivalently of GL_n-bundles, the spaces $B_\gamma \mathrm{GL}_n$ appearing above are approximated by the ordinary Grassmannians, $\mathrm{Gr}_n(\mathbb{A}^r)$. The notation Gr_n is adopted for the colimit as $r \to \infty$, and the $\mathbb{A}^1$-weak equivalences

$$\mathrm{Gr}_n \simeq_{\mathbb{A}^1} B\,\mathrm{GL}_n$$

can be interpreted as a relation between a geometric construction on the left and a homotopical construction on the right.

22.6.2 Classification of Vector Bundles

The following metatheorem lies at the heart of the applications of $\mathbb{A}^1$-homotopy to the classification of vector bundles:

Theorem 22.6.1. *Let k be a sufficiently pleasant base ring, and let $n \in \mathbb{N}$. The functor assigning to a smooth affine k-scheme X its set of rank-n vector bundles, denoted $X \mapsto V_n(X)$, is represented in the $\mathbb{A}^1$-homotopy category by the infinite Grassmannian of n-planes, Gr_n. That is, there is a natural bijection of sets*

$$V_n(X) \leftrightarrow [X_+, \mathrm{Gr}_n]_{\mathbb{A}^1}. \tag{22.6.1}$$

between $V_n(X)$ and morphisms between X and Gr_n in the $\mathbb{A}^1$-homotopy category.

This metatheorem was proved by Morel as [79, Theorem 8.1] in the case where $n \neq 2$ and subject to the running assumptions of [79], namely, that k be an infinite perfect field.[3] Subsequently this was vastly generalized, to the case where k is a regular noetherian ring over a Dedekind domain with perfect residue fields—this includes, in particular, all fields—and allowing the case $n = 2$ in [14, Theorem 5.2.3].

We remark that the restriction to smooth affine k-schemes cannot be weakened. Once the "affine" hypothesis is dropped, the set of isomorphism classes of vector bundles is no longer $\mathbb{A}^1$-invariant, and so cannot be represented in the $\mathbb{A}^1$-homotopy category. One can consult [4] for a proof that $V_n(\mathbb{P}^1) \to V_n(\mathbb{P}^1 \times \mathbb{A}^1)$ is not an isomorphism or that there exists a smooth quasiaffine $X \simeq_{\mathbb{A}^1} *$ such that $V_n(X) \neq V_n(*)$.

One can use the metatheorem to deduce facts about $V_n(X)$. The main computational tool is the existence of a Postnikov tower in $\mathbb{A}^1$-homotopy theory ; which reduces the calculation of maps $X \to \mathrm{Gr}_n$, where X is a smooth affine variety of dimension d, to a succession of lifting problems. These lifting problems are all trivial above dimension d since this is also the Nisnevich cohomological dimension of X.

It should be noted that $\boldsymbol{\pi}_1^{\mathbb{A}^1}(\mathrm{Gr}_n) \cong \Gamma$, and this $\boldsymbol{\pi}_1$ acts nontrivially on the higher homotopy sheaves of Gr_r, so that the lifting problems one encounters for GL_n are Γ-twisted lifting problems. The essentials of the twisted sheaf-theoretic obstruction are set out in [8, Section 6] and [79, Appendix B].

Once the obstruction theory has been set up, two related problems remain: the calculation of $\pi_i^{\mathbb{A}^1}(\mathrm{Gr}_n)$ for various values of i—including the Γ-action—and the interpretation of the resulting lifting problems.

[3] Although the case $n = 2$ can presumably be proved using similar methods, the work on the subject has not been published.

22.6.3 Interpretation of lifting problems

We indicate how the lifting problems may be interpreted in practice. The problem amounts to studying $\mathbb{A}^1$-homotopy classes of maps $X \to K(\mathcal{F}, n)$ where $\mathcal{F}$ is some $\mathbb{A}^1$-homotopy sheaf of GL_n, and often $\mathcal{F}$ is expressed in terms of K-theory.

It is a basic fact, [59, Theorem 8.25] that when X is a smooth scheme, $[X, K(\mathcal{F}, n)] = \mathrm{H}^n_{\mathrm{Nis}}(X, \mathcal{F})$, and for K-theories, these cohomology groups often admit geometric description. For example

$$\mathrm{H}^n_{\mathrm{Nis}}(X, \mathbf{K}^{\mathrm{M}}_n) \cong \mathrm{H}^n_{\mathrm{Zar}}(X, \mathbf{K}^{\mathrm{M}}_n) \cong CH^n(X), \tag{22.6.2}$$

the first identity being [79] and the second being [89, Corollary 6.5]. The group $CH^n(X)$ is the Chow group of codimension-n cycles modulo rational equivalence, see [46]. The calculation by Rost is by means of the Gersten complexes, and is a variation on the "formula of Bloch" $\mathrm{H}^n_{\mathrm{Zar}}(X, \mathbf{K}^{\mathrm{Q}}_n) \cong CH^n(X)$ [87, Theorem 5.19]. These ideas are extended to general strictly $\mathbb{A}^1$-invariant sheaves in place of the various K-theories in [79, Section 5.3]. Over a quadratically closed field, the identity

$$\mathrm{H}^n_{\mathrm{Nis}}(X, \mathbf{K}^{\mathrm{MW}}_n) \cong CH^n(X)$$

is calculated in [8].

22.6.4 The splitting problem

The remainder of this section is devoted to the "splitting problem" for vector bundles, which shows the power of $\mathbb{A}^1$-homotopy. Let $X = \operatorname{Spec} A$ denote a smooth affine k-scheme of dimension d and let V denote a rank n vector bundle on X, i.e., a projective module. The problem is to determine necessary and sufficient conditions for there to be an isomorphism $V \cong V' \oplus A$.

There is an $\mathbb{A}^1$-homotopy fiber sequence,

$$\mathbb{A}^n - \{0\} \longrightarrow \mathrm{Gr}_{n-1} \longrightarrow \mathrm{Gr}_n$$

by [79, Remark 8.15]. This means that the splitting problem can be recast as a relative lifting problem where the fiber is $\mathbb{A}^n - \{0\}$, i.e., a sphere.

When $n < d$

The fiber $\mathbb{A}^n - \{0\}$ is $n-2$-connected, by results of [79, Chapter 6]. It follows that the relative lifting problem that if the dimension d of X is less than n, then one can always split $V \cong V' \oplus A$. This gives a "geometric" argument for a result of Serre, [91].

When $n = d$

More interesting is what happens when one encounters the first obstruction. The problem is slightly easier if all vector bundles appearing are assumed to be equipped with a trivialization of their determinant bundles, so that $B\,\mathrm{SL}_n$ may be substituted for $\mathrm{Gr}_n = B\,\mathrm{GL}_n$. The first obstruction to the splitting problem is an *Euler class*class $\tilde{c}_n(V) \in \mathrm{H}^n_{\mathrm{Nis}}(X; \mathbf{K}^{\mathrm{MW}}_n)$, as calculated in [79, Theorem 8.14], arising directly from the obstruction theory in $\mathbb{A}^1$-homotopy theory, and if the dimension of the base, d, is equal to the rank n of the bundle, then this is the only obstruction.

Related to the above, Nori gave a construction of an "Euler class group" $E(A)$ of a noetherian ring A in [72, Section 1], and defined the Euler class $e(P)$ of a projective module. Bhatwadekar and Sridharan [25] established that the Euler class vanishes precisely when P admits a 1-dimensional trivial summand. There exists a surjective map $\eta : E(A) \to \mathrm{H}^n_{\mathrm{Nis}}(\operatorname{Spec} A; \mathbf{K}^{\mathrm{MW}}_n)$, where A is a smooth k-algebra of dimension n, [40, Chapitre 17], taking one Euler class to the other. According to recent work of Asok and Fasel [11], the map η is an isomorphism for smooth affine varieties over an infinite perfect field of characteristic different from 2.

When $n > d$

One can, with effort, go past the first nonvanishing homotopy sheaf of $\mathbb{A}^n - \{0\}$, and obtain more subtle splitting results.

The main result of [8] is the following:

Theorem 22.6.2. *Suppose X is a smooth affine 3-fold over an algebraically closed field k having characteristic unequal to 2. The map assigning to a rank-2 vector bundle E on X the pair $(c_1(E), c_2(E))$ of Chern classes gives a pointed bijection*

$$V_2(X) \xrightarrow{\cong} \operatorname{Pic}(X) \times CH^2(X).$$

The surjectivity of the map above had previously been established in [64]. The method of proof is a calculation—at least in part—of the first three nontrivial $\mathbb{A}^1$-homotopy groups of GL_n. Of these, two are calculated by Morel in [79]. Specifically $\boldsymbol{\pi}_0^{\mathbb{A}^1}(\mathrm{GL}_n) \cong \Gamma$, for all $n \geq 0$: the reason being that SL_n is $\mathbb{A}^1$-connected [79, 6.52], so that

$$1 \to \mathrm{SL}_n \to \mathrm{GL}_n \to \Gamma \to 1$$

is an $\mathbb{A}^1$-homotopy fiber sequence. It is moreover split, and Γ is strongly $\mathbb{A}^1$-invariant, so it follows that for $n \geq 1$

$$\boldsymbol{\pi}_i^{\mathbb{A}^1}(\mathrm{GL}_n) = \begin{cases} \Gamma & \text{if } i = 0, \\ \boldsymbol{\pi}_i^{\mathbb{A}^1}(\mathrm{SL}_n) & \text{otherwise.} \end{cases}$$

It is then also possible to calculate $\boldsymbol{\pi}_1^{\mathbb{A}^1}(\mathrm{GL}_2) = \boldsymbol{\pi}_1^{\mathbb{A}^1}(\mathrm{SL}_2) = \boldsymbol{\pi}_1^{\mathbb{A}^1}(\mathbb{A}^2 - \{0\}) \cong \mathbf{K}^{\mathrm{MW}}_2$—this belonging in the family of calculations of $\boldsymbol{\pi}_n^{\mathbb{A}^1}(S^{n+m\alpha})$ of [79].

The action of Γ does have to be taken into account, so the calculation is more involved than simply the observations

$$[X, B\Gamma]_{\mathbb{A}^1} = \operatorname{Pic}(X), \quad [X, K(\mathbf{K}^{\mathrm{MW}}_2(\mathcal{L}), 2)]_{\mathbb{A}^1} = CH^2(X),$$

for a smooth X—here $\mathbf{K}^{\mathrm{MW}}_2(\mathcal{L})$ is a twisted form of $\mathbf{K}^{\mathrm{MW}}_2$—, but over a quadratically closed field the ultimate result is the same. For a 3-fold, there is no further obstruction to lifting to a map $[X, B\,\mathrm{GL}_2]_{\mathbb{A}^1}$, but what there might be *a priori* is a choice of different liftings.

The set of such choices is $\mathrm{H}^3(X, \pi_3^{\mathbb{A}^1}(B\,\mathrm{GL}_2)(\mathcal{L}))$, and [8] shows that this is trivial over an algebraically closed field.

The methods outlined can establish a number of similar results to the above. It is a corollary of Theorem 22.6.2 that a rank-2 vector bundle on a smooth affine 3-fold over an algebraically closed field has a trivial line-bundle summand if and only if c_2 vanishes. In [9], the generalization to dimension n is named "Murthy's Splitting Conjecture":

"On an affine variety of dimension n over an algebraically closed field, a rank $n-1$ vector bundle admits a trivial summand if and only if c_{n-1} vanishes."

The conjecture is proved for $n = 4$, again by analysis of a group in the 1-stem: $\pi_3^{\mathbb{A}^1}(\mathbb{A}^3 - \{0\})$. In spite of considerable attention, [6], [12], [18], the groups $\pi_n^{\mathbb{A}^1}(\mathbb{A}^n - \{0\})$ have not been calculated for $n \geq 4$, and the problem remains open.

22.6.5 Other groups

Other groups than GL_n and SL_n have also been studied. For instance, Wendt shows in [107] that if G is a smooth split reductive group over an infinite field k, and if $E \to B$ is a G-torsor satisfying certain local triviality conditions—local triviality in the Zariski topology and triviality over the basepoint of B—then the fiber sequence

$$G \to E \to B$$

is an $\mathbb{A}^1$-homotopy fiber sequence . In [16, 4.1.3], the authors establish, again over an infinite field, that for an isotropic reductive k group G, the functor $\mathrm{H}^1_{\mathrm{Nis}}(\cdot, G)$ of isomorphism classes of Nisnevich-trivial G-torsors is represented in the $\mathbb{A}^1$-homotopy category by BG. These results generalize to a wide class of groups the results of Morel's that are foundational to the study of vector bundles. Unfortunately for the applicability of the theory, however, for many algebraic groups G of interest, say for PGL_n or O_n or the finite groups, the set of bundles of interest is $\mathrm{H}^1_{\text{ét}}(\cdot, G)$ rather than $\mathrm{H}^1_{\mathrm{Zar}}(\cdot, G)$ or $\mathrm{H}^1_{\mathrm{Nis}}(\cdot, G)$; therefore the theory is most interesting in the case of special algebraic groups in the sense of [92]—these groups include SL_n, GL_n and Sp_n. In [15], the same authors apply their techniques to G_2, where $\mathrm{H}^1_{\mathrm{Nis}}(\cdot, G_2)$ classifies "split octonion algebras".

22.6.6 Applications to naive calculations

As a consequence of the representability results of [16], the authors obtain striking statements saying that certain naive $\mathbb{A}^1$-homotopy calculations, as discussed in Subsection 22.2.3, coincide with the calculations produced by the heavy machinery. Over a quasicompact and quasiseparated base S, [16, Theorem 2.3.2] establishes the following.

Theorem 22.6.3. *Suppose G is a finitely presented smooth S-group scheme and that H is a finitely presented closed subgroup scheme such that G/H is an S-scheme, and such that $G \to G/H$ is Nisnevich locally split, and suppose that $\mathrm{H}^1_{\mathrm{Nis}}(\cdot, G)$ and $\mathrm{H}^1_{\mathrm{Nis}}(\cdot, H)$ are $\mathbb{A}^1$-invariant on affine schemes. Then for an affine scheme* $\operatorname{Spec} R$*, the natural map*

$$\pi_0(\mathrm{Sing}^{\mathbb{A}^1}(G/H)(R)) \to [\operatorname{Spec} R, G/H]_{\mathbb{A}^1}$$

is a bijection.

That is, for G/H, provided the source is affine, the naive and genuine $\mathbb{A}^1$-homotopy classes of maps agree. The restrictions on G, H are not perhaps as inconvenient as might be feared, since, in contrast to the case of the étale topology, the Nisnevich cohomology $\mathrm{H}^1_{\mathrm{Nis}}(\cdot, G)$ is liable to be $\mathbb{A}^1$-invariant. In fact, in [16, Theorem 3.3.6], the authors establish $\mathbb{A}^1$ invariance for all isotropic reductive group schemes over infinite fields. The more severe restriction in the theorem is that $G \to G/H$ should be Nisnevich locally split, but nonetheless, the theorem provides a wide range of notable spaces for which naive and genuine $\mathbb{A}^1$-homotopy are the same.

Example 22.6.4. For instance, the theorem applies when $G = \mathrm{GL}_n$ and $H = \mathrm{GL}_{n-r}$ for some $r < n$, embedded via

$$A \mapsto \begin{bmatrix} I_r & \\ & A \end{bmatrix}.$$

In this case, G/H is the variety $V_{r,n}$ of pairs of $n \times r$-matrices (A, B) satisfying $B^T A = I_r$. In the very special case of $r = 1$, the variety $V_{1,n}$ is the $2n-1$-dimensional affine quadric

$$Q_{2n-1} = \operatorname{Spec} k[x_1, \ldots, x_n, y_1, \ldots, y_n]/(\sum_{i=1}^{n} x_i y_i - 1)$$

which is $\mathbb{A}^1$-homotopy equivalent to the sphere $\mathbf{A}^n - \{0\}$. The theorem asserts that if $\operatorname{Spec} R$ is a smooth affine k-variety, then the set $[\operatorname{Spec} R, V_{1,n}]_{\mathbb{A}^1}$ consists of naive $\mathbb{A}^1$-homotopy classes of maps $\operatorname{Spec} R \to V_{1,n}$.

Bibliography

[1] Benjamin Antieau and Elden Elmanto. A primer for unstable motivic homotopy theory. In *Surveys on recent developments in algebraic geometry, Proc. Sympos. Pure Math.*, 95, pages 305–370. Amer. Math. Soc., 2017.

[2] Jón Kr. Arason. Cohomologische invarianten quadratischer Formen. *J. Algebra*, 36(3):448–491, 1975.

[3] M. Artin and B. Mazur. *Étale Homotopy.* Lecture Notes in Math., Vol. 100. Springer-Verlag, 1969.

[4] Aravind Asok and Brent Doran. Vector bundles on contractible smooth schemes. *Duke Math. J.*, 143(3):513–530, 2008.

[5] Aravind Asok, Brent Doran, and Jean Fasel. Smooth models of motivic spheres and the clutching construction. *Int. Math. Res. Not. IMRN*, 2017(6):1890–1925, 2017.

[6] Aravind Asok and Jean Fasel. Toward a meta-stable range in $\mathbb{A}^1$-homotopy theory of punctured affine spaces. In *Oberwolfach Conference*, Oberwolfach, 2013. Mathematisches Forschungsinstitut Oberwolfach.

[7] A. Asok and J. Fasel. Algebraic vector bundles on spheres. *J. Topol.*, 7(3):894–926, 2014.

[8] Aravind Asok and Jean Fasel. A cohomological classification of vector bundles on smooth affine threefolds. *Duke Math. J.*, 163(14):2561–2601, 2014.

[9] Aravind Asok and Jean Fasel. Splitting vector bundles outside the stable range and $\mathbb{A}^1$-homotopy sheaves of punctured affine spaces. *J. Amer. Math. Soc.*, 28(4):1031–1062, 2015.

[10] A. Asok and J. Fasel. Comparing Euler classes. *The Q. J. Math.*, 67(4):603–635, 2016.

[11] Aravind Asok and Jean Fasel. Euler class groups and motivic stable cohomotopy. arXiv:1601.05723, 2016.

[12] Aravind Asok and Jean Fasel. An explicit KO-degree map and applications. *J. Topol.*, 10(1):268–300, 2017.

[13] Aravind Asok, Jean Fasel, and Ben Williams. Motivic spheres and the image of Suslin's Hurewicz map. arXiv:1804.05030, 2018.

[14] Aravind Asok, Marc Hoyois, and Matthias Wendt. Affine representability results in $\mathbb{A}^1$-homotopy theory, I: Vector bundles. *Duke Math. J.*, 166(10):1923–1953, 2017.

[15] Aravind Asok, Marc Hoyois, and Matthias Wendt. Generically split octonion algebras and $\mathbb{A}^1$-homotopy theory. arXiv:1704.03657, 2017.

[16] Aravind Asok, Marc Hoyois, and Matthias Wendt. Affine representability results in $\mathbb{A}^1$-homotopy theory, II: Principal bundles and homogeneous spaces. *Geom. Topol.*, 22(2):1181–1225, 2018.

[17] Aravind Asok and Fabien Morel. Smooth varieties up to $\mathbb{A}^1$-homotopy and algebraic h-cobordisms. *Adv. Math.*, 227(5):1990–2058, 2011.

[18] A. Asok, K. Wickelgren, and B. Williams. The simplicial suspension sequence in $\mathbb{A}^1$-homotopy. *Geom. Topol.*, 21(4):2093–2160, 2017.

[19] M. F. Atiyah *K-Theory*. Lecture notes by D. W. Anderson. W. A. Benjamin, 1967.

[20] Jean Barge and Fabien Morel. Groupe de Chow des cycles orientés et classe d'Euler des fibrés vectoriels. *C. R. Math. Acad. Sci. Paris*, 330(4):287–290, 2000.

[21] Ilan Barnea, Yonatan Harpaz, and Geoffroy Horel. Pro-categories in homotopy theory. *Algebr. Geom. Topol.*, 17(1):567–643, 2017.

[22] Ilan Barnea and Tomer M. Schlank. A projective model structure on pro-simplicial sheaves, and the relative étale homotopy type. *Adv. Math.*, 291:784–858, 2016.

[23] R. Benedetti and R. Silhol. Spin and Pin^- structures, immersed and embedded surfaces and a result of Segre on real cubic surfaces. *Topology*, 34(3):651–678, 1995.

[24] Candace Bethea, Jesse Kass, and Kirsten Wickelgren. An example of wild ramification in an enriched Riemann–Hurwitz formula. arXiv:1812.03386, 2018.

[25] S. M. Bhatwadekar and Raja Sridharan. The Euler class group of a Noetherian ring. *Compos. Math.*, 122(2):183–222, 2000.

[26] Benjamin A. Blander. Local projective model structures on simplicial presheaves. *K-Theory*. 24(3):283–301, 2001.

[27] Siegfried Bosch, Werner Lütkebohmert, and Michel Raynaud. *Néron models*, *Ergeb. Math. Grenzgeb. (3)*. 21. Springer-Verlag, 1990.

[28] A. K. Bousfield and D. M. Kan. *Homotopy limits, completions and localizations.* Lecture Notes in Math., Vol. 304. Springer-Verlag, 1972.

[29] Kenneth S. Brown and Stephen M. Gersten. Algebraic K-theory as generalized sheaf cohomology. In *Algebraic K-Theory, I: Higher K-Theories (Proc. Conf., Battelle Memorial Inst., Seattle, Wash., 1972)*, pages 266–292. Lecture Notes in Math., Vol. 341. Springer, 1973.

[30] Christophe Cazanave. Homotopy classes of rational functions. *C. R. Math. Acad. Sci. Paris*, 346(3–4):129, 2008.

[31] Frédéric Déglise, Fangzhou Jin, and Adheel Khan. Fundamental classes in motivic homotopy theory. arXiv:1805.05920, 2018.

[32] Pierre Deligne. Lectures on motivic cohomology 2000/2001. Available at `www.math.uiuc.edu/K-theory/0527`, 2001.

[33] Daniel Dugger. Universal homotopy theories. *Adv. Math.*, 164(1):144–176, 2001.

[34] Daniel Dugger, Sharon Hollander, and Daniel C. Isaksen. Hypercovers and simplicial presheaves. *Math. Proc. Camb. Phil. Soc.*, 136(1):9–51, 2004.

[35] Daniel Dugger and Daniel C. Isaksen. Topological hypercovers and $\mathbb{A}^1$-realizations. *Math. Zeit.*, 246(4):667–689, 2004.

[36] David Eisenbud. An algebraic approach to the topological degree of a smooth map. *Bull. Amer. Math. Soc.*, 84(5):751–764, 1978.

[37] David Eisenbud and Joe Harris. *The Geometry of schemes.* Grad. Texts in Math. 197, Springer,2000.

[38] David Eisenbud and Harold I. Levine. An algebraic formula for the degree of a C^∞ map germ. *Ann. Math. (2)*, 106(1):19–44, 1977. With an appendix by Bernard Teissier, "Sur une inégalité à la Minkowski pour les multiplicités".

[39] Richard Elman, Nikita Karpenko, and Alexander Merkurjev. *The algebraic and geometric theory of quadratic forms*, volume 56 of *Amer. Math. Soc. Colloq. Publ.*, 2008.

[40] Jean Fasel. Groupes de Chow-Witt. *Mém. Soc. Math. Fr.*, 113, 2008.

[41] Jean Fasel. Lectures on Chow–Witt groups. *Preprint*, following lectures at Motivic Homotopy Theory and Refined Enumerative Geometry, University of Duisburg-Essen, May 14-18, 2018, 2018.

[42] Sergey Finashin and Viatcheslav Kharlamov. Abundance of real lines on real projective hypersurfaces. *Int. Math. Res. Not. IMRN*, 2013(16):3639–3646, 2013.

[43] Jean-Marc Fontaine and Barry Mazur. Geometric Galois representations. In *Elliptic curves, modular forms, & Fermat's last theorem (Hong Kong, 1993)*, Ser. Number Theory, I, pages 41–78. Int. Press, Cambridge, MA, 1995.

[44] Eberhard Freitag and Reinhardt Kiehl. *Étale cohomology and the Weil conjecture, Ergeb. Math. Grenzgeb. (3)*, 13. Springer-Verlag, 1988.

[45] E. M. Friedlander. *Etale homotopy of simplicial schemes.* Ann. of Math. Stud., 104, Princeton University Press, 1982.

[46] William Fulton. *Intersection theory, Ergeb. Math. Grenzgeb. (3).* 2, Springer Verlag, 1984.

[47] Alexander Grothendieck. Éléments de géométrie algébrique: IV. Étude locale des schémas et des morphismes de schémas, Quatrième partie. *Publ. Math. Inst. Hautes Études Sci.*, 32:5–361, 1967.

[48] A. Grothendieck and M. Raynaud. *Revêtements étales et groupe fondamental.* Lect. Notes in Math., Vol. 224, Springer-Verlag, 1971.

[49] Yonatan Harpaz and Tomer M. Schlank. Homotopy obstructions to rational points. In *Torsors, étale homotopy and applications to rational points, London Math. Soc. Lecture Note Ser.*, 405:280–413. Cambridge University Press, 2013.

[50] Robin Hartshorne. *Algebraic geometry, Grad. Texts in Math.* 52, Springer-Verlag, 1977.

[51] I. N. Herstein. *Noncommutative rings, Carus Math. Monogr.* 15 Math. Assoc. of America, 1994.

[52] Philip S. Hirschhorn. *Model categories and their localizations. Math. Surveys Monogr.* 99 Amer. Math. Soc., 2003.

[53] Asaf Horev and Jake P. Solomon. The open Gromov–Witten–Welschinger theory of blowups of the projective plane. arXiv:1210.4034, 2012.

[54] Marc Hoyois. A quadratic refinement of the Grothendieck-Lefschetz-Verdier trace formula. *Algebr. Geom. Topol.*, 14(6):3603–3658, 2014.

[55] Marc Hoyois. The six operations in equivariant motivic homotopy theory. *Adv. Math.*, 305:197–279, 2017.

[56] Daniel C. Isaksen. Etale realization on the $\mathbb{A}^1$-homotopy theory of schemes. *Adv. Math.*, 184(1):37–63, 2004.

[57] Daniel C. Isaksen. Flasque model structures for simplicial presheaves. *K-Theory*, 36(3-4):371–395, 2005.

[58] J. F. Jardine. Simplicial presheaves. *J. Pure Appl. Alg.*, 47(1):35–87, 1987.

[59] John F. Jardine, *Local homotopy theory.* Springer Monogr. Math., 2015.

[60] André Joyal. Letter to Grothendieck, 1983, Available at `https://webusers.imj-prg.fr/~georges.maltsiniotis/ps/lettreJoyal.pdf`.

[61] Jesse Kass and Kirsten Wickelgren. The class of Eisenbud–Khimshiashvili–Levine is the local $\mathbb{A}^1$-brouwer degree. Duke Math. J. 168(3):429–469, 2016.

[62] Jesse Kass and Kirsten Wickelgren. An arithmetic count of the lines on a smooth cubic surface. arXiv:1708.01175, 2017.

[63] Jesse Leo Kass and Kirsten Wickelgren. An étale realization which does NOT exist. In *New Directions in Homotopy Theory, Contemp. Math.* 707, pages 11–30. Amer. Math. Soc., 2018

[64] N. Mohan Kuma and M. Pavaman Murthy. Algebraic cycles and vector bundles over affine three-folds. *Ann. of Math.*, 116(3):579–591, 1982.

[65] T. Y. Lam. *Introduction to quadratic forms over fields, Grad. Stud. Math.* 67, Amer. Math. Soc., 2005.

[66] Marc Levine. Toward a new enumerative geometry. arXiv:1703.03049, 2017.

[67] Marc Levine. Motivic Euler characteristics and Witt-valued characteristic classes. arXiv:1806.10108, 2018.

[68] Marc Levine. Toward an algebraic theory of Welschinger invariants. arXiv:1808.02238, 2018.

[69] Marc Levine and Arpon Raksit. Motivic Gauss–Bonnet formulas. arxiv:1808.08385, 2018.

[70] Jacob Lurie. *Higher topos theory, Ann of Math. Stud.*, 170. Princeton University Press, 2009.

[71] Saunders Mac Lane and Ieke Moerdijk. *Sheaves in Geometry and Logic: A First Introduction to Topos Theory.* Universitext. Springer-Verlag, 1992.

[72] Satya Mandal and Raja Sridharan. Euler classes and complete intersections. *J. Math. Kyoto Univ.*, 36(3):453–470, 1996.

[73] J. Peter May. Classifying spaces and fibrations. *Mem. Amer. Math. Soc.*, 1(1,155), 1975.

[74] A. S. Merkur'ev and A. A. Suslin. The group K_3 for a field. (Russian) Izv. Akad. Nauk SSSR Ser. Mat. 54(3)L522-545, 1990; translation in Math. USSR-Izv. 36(3):541–565, 1991.

[75] John Milnor. Algebraic K-theory and quadratic forms. *Invent. Math.*, 9(4):318–344, 1970.

[76] Fabien Morel. Voevodsky's proof of Milnor's conjecture. *Bull Amer. Math. Soc. (N.S.)*, 35(2):123, 1998.

[77] Fabien Morel. On the motivic π_0 of the sphere spectrum. In *Axiomatic, enriched and motivic homotopy theory, NATO Sci. Ser. II Math. Phys. Chem.*, 131, pages 219–260. Kluwer Acad. Publ., 2004.

[78] Fabien Morel. The stable $\mathbb{A}^1$-connectivity theorems. K-theory, 35:1–68, 2005.

[79] Fabien Morel. *$\mathbb{A}^1$-Algebraic Topology over a Field*, Lecture Notes in Math., Vol. 2052, Springer, 2012.

[80] Fabien Morel and Vladimir Voevodsky. $\mathbb{A}^1$-homotopy theory of schemes. *Publ. Math. Inst. Hautes Études Sci.*, 90(1):45–143, 1999.

[81] D. Mumford, J. Fogarty, and F. Kirwan. *Geometric Invariant Theory, Ergeb. Math. Grenzgeb. (2)*, 34, Springer-Verlag, third edition, 1994.

[82] Ye A. Nisnevich. The completely decomposed topology on schemes and associated descent spectral sequences in algebraic K-theory. In J. F. Jardine and V. P. Snaith, editors, *Algebraic K-Theory: Connections with Geometry and Topology, Lake Louise, AB, 1987* NATO Adv. Sci. Inst. Ser. C Math. Phys. Sci., 279, pages 241–342, Kluwer Acad. Publ., 1989.

[83] Christian Okonek and Andrei Teleman. Intrinsic signs and lower bounds in real algebraic geometry. *J. Reine Angew. Math.*, 688:219–241, 2014.

[84] Ivan Panin, Konstantin Pimenov, and Oliver Röndigs. On Voevodsky's algebraic K-theory spectrum. In Nils Baas, Eric M. Friedlander, Björn Jahren, and Paul Arne Østvær, editors, *Algebraic Topology: The Abel Symposium 2007*, Abel Symposia, pages 279–330. Springer, 2009.

[85] Gereon Quick. Continuous group actions on profinite spaces. *J. Pure Appl. Algebra*, 215(5):1024–1039, 2011.

[86] Gereon Quick. Existence of rational points as a homotopy limit problem. *J. Pure Appl. Algebra*, 219(8):3466–3481, 2015.

[87] Daniel Quillen. Higher algebraic K-theory: I. In H. Bass, editor, *Higher K-Theories*, Lect. Notes in Math., Vol. 341, pages 85–147. Springer, 1973.

[88] Emily Riehl. *Categorical homotopy theory, New Math. Monogr.*, 24, Cambridge University Press, 2014.

[89] Markus Rost. Chow groups with coefficients. *Doc.Math.*, 1(16):319–393, 1996.

[90] B. Segre. *The non-singular cubic surfaces.* Oxford University Press, 1942.

[91] Jean-Pierre Serre. Modules projectifs et espaces fibrés à fibre vectorielle. In *Séminaire P. Dubreil, M.-L. Dubreil-Jacotin et C. Pisot, 1957/58, Fasc. 2, Exposé 23*, page 18. Secrétariat mathématique, 1958.

[92] Jean-Pierre Serre. Espaces fibrés algébriques (d'après André Weil). In *Séminaire Bourbaki, Vol. 2*, pages Exp. No. 82, 305–311. Soc. Math. France, 1995.

[93] Günter Scheja and Uwe Storch. Über Spurfunktionen bei vollständigen Durchschnitten. *J. Reine Angew. Math.*, 278/279:174–190, 1975.

[94] Marco Schlichting and Girja S. Tripathi. Geometric models for higher Grothendieck–Witt groups in $\mathbb{A}^1$-homotopy theory. *Math. Ann.*, 362(3-4):1143–1167, 2015.

[95] Markus Spitzweck and Paul Arne Østvær. Motivic twisted K-theory. *Algebr. Geom. Topol.*, 12(1):565–599, 2012.

[96] Padmavathi Srinivasan and Kirsten Wickelgren. An arithmetic count of the lines meeting four lines in $\mathbb{P}^3$. arXiv:1810.03503, 2018.

[97] Bertrand Toën and Gabriele Vezzosi. Homotopical algebraic geometry. I. Topos theory. *Adv. Math.*, 193(2):257–372, 2005.

[98] Burt Totaro. The Chow ring of a classifying space. In *Algebraic K-theory (Seattle, WA, 1997)*, volume 67 of *Proc. Sympos. Pure Math.*, pages 249–281. Amer. Math. Soc., 1999.

[99] Ravi Vakil. The Rising Sea: Foundations of algebraic geometry notes, Available at `http://math.stanford.edu/~vakil/216blog/FOAGnov1817public.pdf`.

[100] Vladimir Voevodsky. $\mathbf{A}^1$-homotopy theory. Proceedings of the International Congress of Mathematicians, Vol. I 1998). Doc. Math. 1998, Extra Vol. I, 579–604.

[101] Vladimir Voevodsky. Cohomological theory of presheaves with transfers. In *Cycles, Transfers and Motivic Homology Theories*, Ann. of Math. Stud., 143, Princeton University Press, 2000.

[102] Vladimir Voevodsky. Motivic cohomology with $\mathbf{Z}/2$-coefficients. *Publ. Math. Inst. Hautes Études Sci.*, 98(1):59–104, 2003.

[103] Vladimir Voevodsky. Reduced power operations in motivic cohomology. *Publ. Math. Inst. Hautes Études Sci.*, 98(1):1–57, 2003.

[104] Vladimir Voevodsky. Homotopy theory of simplicial sheaves in completely decomposable topologies. *J. Pure Appl. Algebra*, 214(8):1384–1398, 2010.

[105] Vladimir Voevodsky. Unstable motivic homotopy categories in Nisnevich and cdh-topologies. *J. Pure Appl. Algebra*, 214(8):1399–1406, 2010.

[106] Charles A. Weibel. Homotopy algebraic K-theory. In *Algebraic K-Theory and Algebraic Number Theory (Honolulu, HI, 1987), Contemp. Math.* 83, pages 461–488. Amer. Math. Soc., 1989.

[107] Matthias Wendt. Rationally trivial torsors in $\mathbb{A}^1$-homotopy theory. *Journal of K-Theory*, 7(3):541–572, 2011.

[108] Matthias Wendt. Oriented Schubert calculus in Chow–Witt rings of Grassmannians. arXiv:1808.07296, 2018.

[109] Kristen Wickelgren and Ben Williams. The simplicial EHP sequence in $\mathbb{A}^1$ algebraic topology. arXiv:1411.5306, 2014.

Department of Mathematics, Duke University, Durham, NC 27708, USA
E-mail address: kirsten.wickelgren@duke.edu

Department of Mathematics, University of British Columbia, Vancouver BC, Canada V6T 1Z4
E-mail address: tbjw@math.ubc.ca

Index

$\mathcal{A}_n$-algebra, 795
A_∞-structure, 768
Alg, 800, 833
$\mathcal{A}ll$, 712, 860
$\mathbb{A}^1$-enumerative geometry, 956
$\mathbb{A}^1$-homotopy
  elementary, 939, 940
  naive, 939, 940, 967
$\mathbb{A}^1$-local detection, 959
$\mathcal{A}ssoc$, 795, 831
$\mathfrak{C}$, 490
CMon(CAT_∞), 497
$\mathrm{Cat}_\infty, \mathrm{CAT}_\infty$, 492
$\mathrm{Cat}_{\mathrm{dg}}$, 491
$\mathcal{C}^\kappa$, 499
$\mathcal{C}omm$, 795, 831, 833
Cut, 498
$\mathcal{D}(\mathcal{A})$, 506
$\mathcal{D}^b(\mathcal{A})$, 505
$\mathrm{Der}(A, M)$, 539
$\mathrm{Der}(\mathcal{C})$, 538
$\mathcal{D}^-(\mathcal{A})$, 506
E-theory, Morava, 176, 178, 246, 256, 676, 903
$E(n)$, 246
EG, 711
E_n, 23
E_n-algebra, 58, 795, 796, 799, 814, 835, 843
E_n-module, 829, 832, 838
E_n-operad, 797
E_∞-structure, 768
Emb, 32
Esc_n, 4
Exc_*, 502
F-isomorphism, 335
$\mathcal{F}_N$, 712
Fin^G, 706
Fun, 4, 490, 495, 835
GL_1, 839, 845
Gpd_∞, 492
H^d_∞-structure, 824
$\mathcal{H}_n$-algebra, 797
Hom, 795
Ind_κ, 498
J, cokernel of, 256, 477
J, image of, 774, 777, 779, 787
J-construction, 116
JO^s, 717
K-theory
  p-completed, 845
  algebraic, 846, 961
    of ring, 26–28
    of ring spectrum, 26
    of space, 7, 24–26
    of stable ∞-category, 627
  complex, 188, 189
  equivariant topological, 727
  form of, 844
  Hermitian, 961
  higher real, 732
  indecomposable, 961
  Milnor, 761, 779, 948, 955
  Milnor-Witt, 778, 950, 961
    as fiber product, 950
    as homotopy sheaf of spheres, 951, 962
  Morava, 12, 150, 170, 246, 677, 822
  real, 188
$K(n)$, 12, 246
$K3$ cohomology, 257
KE_n, 24
LFun, 500
LPr, 499
$\mathrm{LPr}_{\mathrm{st}}$, 503
MO, 799, 817, 820
MU, 242, 799, 817, 820, 829
MUP, 242, 903
Map, 795
Mod, 538, 748, 800, 831
Mon(CAT_∞), 498
Mul, 799
N, 490
$\mathcal{O}$-Alg, 741
$\mathcal{O}_C$, 646
  K-theory of, 648
  topological cyclic homology of, 646
Ω-spectrum, 508, 823
Op, 805
Orb^G, 704
$\mathcal{O}^{top}$, 228, 252, 254
$\mathcal{P}$, 493
PSp, 510
Φ, 540
Pic, 839
Pow, 808
RFun, 500
RPr, 499
$\mathcal{S}, \mathcal{S}_*$, 492
Σ^∞_+, 504, 724
Σ^∞, 511
$\mathrm{Sing}^{\mathbb{A}^1}$, 940
S^{-1}, 518
$\mathrm{Sp}(\mathcal{C})$, 502
$\mathrm{Sym}^{\leq 1}$, 538
Sym^n_R, 527
TAF, 254
TAQ, 670, 689, 839–843
THH, 26, 836
TMF, 221, 231
T_nF, 31
$\mathrm{Ten}_R(M)$, 527
Θ-structure, 444
  spherical, 450
Θ_2, 285
Θ_2-Segal [0]-category, 290
Θ_2-Segal category, 290
Θ_2-Segal precategory, 290
Θ_2-diagram, underlying simplicial space of, 288
Θ_2-quasi-category, 291
Θ_2-set, 291
Θ_2-space, 288
Tmf, 229
a_V, 701
$^\heartsuit$, 506
∞-category, 44, 553, 835
  $\otimes$-presentable symmetric monoidal, 57

additive, 613
closed symmetric monoidal, 530
cocomplete, 577
complete, 577
derived, 506
exit-path, 85
left tensored, 516
monoidal, 498, 591
of $\mathbb{A}_\infty$-ring spectra, 612
of $\mathbb{E}_\infty$-ring spectra, 612
of ∞-categories, 571, 610
of cyclotomic spectra, 624
of pointed spaces, 602
of presheaves, 493, 582
of spaces, 565
of spectra, 607
pointed, 501, 601
preadditive, 613
presentable, 51, 499, 583
presentable symmetric monoidal, 499
right tensored, 517
sifted, 68
Spanier-Whitehead, 508
stable, 502, 604
symmetric monoidal, 39, 497, 598
unbounded derived, 505
underlying simplicial model category of, 565
∞-group, 515
∞-groupoid, 267, 559
fundamental, 491, 551
∞-monoid, 515
∞-operad, 834–835
∞-topos, 960
$(\infty, 0)$-category, 267
$(\infty, 1)$-category, 269
$(\infty, 2)$-category, 277
κ-compact object, 499
κ_n, 210
mmf, 787
μ, 646
p-compact group, 355
p-completion, Bousfield-Kan, 338
p-divisible group, 251
p-local compact group, 357
p-local finite group, 348
p-local subgroup, 336
p-series, 239
p-toral group, discrete, 356
q-expansion, 228
$\bar{\rho}_G$, 712
t-structure, 505, 838
bounded, 506
heart of, 506
left complete, 506
Postnikov, 505
truncation associated to, 506
τ, cofiber of, 774
$\tau_{\leq n}$, 500
$\otimes^n$, 503
θ-structure, 456
tmf, 231, 774, 787
tt-category, 146
tt-ideal, 147
prime, 147
tt-spectrum, of stable homotopy category, 150
v_n-periodic equivalence, 679
v_n-periodic localization, 658
v_n-periodicity, 678
v_n-self-map, 678
ζ_n, 182

action map, 796, 803, 836
active map, 496
Adams conjecture, 779
Adem relation, 793, 813, 815, 818, 819, 826, 827, 842
mixed, 824, 846
adjoint functor theorem, 499
adjoint identity, 814
adjunction
comonadic, 21, 663
free-forgetful, 721, 799, 836, 837
monadic, 667, 684
admissible gadget, 963
affine spectral scheme, 520
algebra
n-Poisson, 406, 427
n-disk, 39
differential graded Lie, 657, 664
divided power, 619
dual Steenrod, 818–821, 827, 844
for MU, 828
endomorphism, 798
factorization, 60
filtered, 801, 802
for a monad, 665
free, 526, 797, 802, 806, 815, 825, 829, 833, 837, 841
free associative R-, 527
free commutative R-, 527
graded, 801, 802, 834, 841
homotopy, 797
Lie, 80
over multicategory, 800, 833
over operad, 795
restricted Lie, 842
spectral Lie, 20, 658, 673, 684
spherical group, 798, 829
square-zero, 841
Steenrod, 798
cohomology of motivic, 764
dual motivic, 762
symmetric, 527
tensor, 527
universal enveloping, 81
algebraic n-simplex, 939
algebraic group, special, 963
Alperin's fusion theorem, 337, 357
antisymmetry, 804, 813
approximation
by excisive functor, 857
excisive, 5, 510
Arason invariant, 949
Arnold presentation, 426, 427
arrow, cartesian, 495
Artin group, right-angled, 127
Artin representability theorem, 547
assembly map, 650, 810
and cyclotomic trace, 879
and linearization map, 869
Farrell-Jones and Baum-Connes, 875
for G-homology theories, 860
relative, 880
universal property, 859
via forgetting control, 867
via homotopy colimits, 861
via index theory, 872
via surgery theory, 865
associahedron, 795
asymptotic algebraicity, 213

bar construction, 666, 668, 816, 842
barcode, 303

Barr-Beck-Lurie theorem, 521, 667
base-change, 840
étale, 546
basepoint, in $\mathbb{A}^1$-homotopy theory, 938
Baskakov's Theorem, 119, 129
Bass conjecture, 864
Baum-Connes conjecture, 871
status, 874
Beilinson-Lichtenbaum conjecture, 782
Betti numbers, bigraded, 115
bilinear form, rank of, 948
bimodule, 796, 832, 836
biset, 351
$\mathcal{F}$-stable, 352
characteristic, 352
minimal characteristic, 353
bisimplicial space, 279
underlying simplicial space of, 281
Blakers-Massey theorem, 7
Bloch-Kato conjecture, 782
blow-up, 943
bordism, real, 736
Borel conjecture, 863
Borel construction, 730
Bott element, 189
bottleneck distance, 309
bracket
Browder, 804–805, 808, 809, 813, 816, 818, 820
in TAQ, 841, 842
braided monoidal structure, 412–414
Browder Cartan formula, 819
Brown representability theorem, 509
Burnside ring, 351

Cartan formula, 793, 813–815, 819
mixed, 824, 846
motivic, 764
Cartesian product, 703
category
$\mathcal{F}$-restricted orbit, 861
accessible, 580
autoenrichment of G-spaces, 703
braided monoidal, 794, 796, 830, 835
Burnside, 720
differential graded, 491
elegant Reedy, 267
equivariant Spanier-Whitehead, 714
flow, 379–385, 388, 389, 392, 393, 395
fundamental, 555
Grothendieck abelian, 580
left-tensored, 831
locally presentable, 580
module, 796, 829
monoidal, over operad, 830
of G-spaces, 700
of G-spectra, 724
of coefficient systems, 707
of Mackey functors, 720
of modules over E_n-algebra, 835–836
of pointed G-spaces, 703
orbit, 338, 347, 357, 704, 854
Reedy, 267
Segal, 273
simplicial, 269, 560
singularity, 153, 154
stable, 832
stable module, 153
strongly $\mathcal{M}$-monoidal, 831
symmetric monoidal topological, 795, 800
tensor-triangulated, 146
topological, 269
transporter, 334
weakly $\mathcal{M}$-monoidal, 831, 835
chain rule, 19
character theory, HKR, 910
characteristic element, 353
characteristic function, 108
characteristic number, 449
Chern class, 966
Chow group, 965
oriented, 956
chromatic convergence theorem, 172
chromatic fracture square, 174, 194, 247
chromatic splitting conjecture, 174, 194, 205
chromatic tower, 171
classifying space
n-fold, 796
for family of subgroups, 711, 859
geometric, 963
Nisnevich, 963
of saturated fusion system, 348
simplicial, 963
Clausen-Mathew-Morrow theorem, 620, 650
closed immersion, 941
clustering, 299
coalgebra
cofree, 662, 663
commutative, 661
differential graded, 657
Tate, 695
cobar construction, 668, 673
cocone, on simplicial set, 573
coefficient system, 707
coevaluation map, 530
cofiber of τ, 774
cofiber sequence, relative, 537
cofree construction, 731
cogroupoid, 631
associated with $\mathbb{E}_\infty$-ring, 632
coherent multiplication, 829–838
cohomological dimension, virtual, 185
cohomology, 509
RO(G)-graded, 717, 725
Borel, 733
Bredon, 708
continuous, 185
elliptic, 226, 845
equivariant, 708
equivariant cellular, 708
equivariant elliptic, 257
étale, 780, 782, 947, 948
factorization, 87
Galois, 785
group, 333
Lubin-Tate, 845
motivic of a point, 762
prismatic, 623
syntomic, 623
theory for ring spectra, 840
cohomotopy, stable, 715
coinduction, 701, 724
cokernel of J, 256, 477
colimit
enriched, 837
in ∞-category, 577

collection, of subgroups, 336
  ample, 339
commutativity, motivic graded, 778
comparison map, in tt-geometry, 159
complete intersection, 466
completion, η-, 759
complex
  G-CW, 706, 853
  alpha, 306
  Čech, 305
  chain, 834
  finite, of type n, 12
   strongly, 13
  flag, 127
  hairy graph, 434–436
  Kan, 553
  Kontsevich graph, 437–438
  moment-angle, 104
   cohomology of, 114
   unstable splitting of, 120
  partition, 674
  perfect, 151
  Vietoris-Rips, 299, 306
  witness, 307, 319
composition product, 669
concordance, 444
cone, on simplicial set, 573
connective cover, 513
connectivity theorem, unstable, 962
Connes' operator, 628
contraction, 960
cooperad, 424, 668
  n-Poisson, 427
  graph, 430
  Hopf cochain dg, 425
copresheaf, 63
cosimplicial object, standard algebraic, 939
costabilization, 694
cotangent complex, 536
Coxeter group, right-angled, 127
cubical diagram, 3
cup-1 square, 809
cyclotomic Frobenius, 624
cyclotomic identity, 124
cyclotomic trace, 25, 620, 626
  on connective covers, 627

Day convolution product, 500
deformation
  of commutative algebra spectrum, 542
  to the normal bundle, 943
degree, 932, 951
  $\mathbb{A}^1$-, 933
  as sum of local degrees, 953
  local, 952
dendrogram, 299
dendroidal set, 835
Denham-Suciu fibration, 117
derivation, 539
  universal, 543
derivative, of reduced 1-excisive functor, 6
determinant, 182
Dickson invariant, 844
diffeomorphism group, 447, 454
dimension axiom, 509
discriminant, 230, 949
dissimilarity measure, 305
distinguished square, 936
distributive rule, 824
divided congruences, ring of, 247
Dold-Kan correspondence, 491, 834
Drinfeld associator, 421
Drinfeld ring, 901
duality
  S-, 715
  Benson-Carlson, 362
  Brown-Comenetz, 172
  Gross-Hopkins, 212
  Koszul, 20, 24, 87, 658, 667, 671, 689, 841, 842, 844
  local, 173
  Milnor-Spanier-Atiyah, 713
  Poincaré/Koszul, 88
  Spanier-Whitehead, 212, 452, 530, 812
Dundas-Goodwillie-McCarthy theorem, 620
Dunn additivity, 796
Dwyer-Kan equivalence, 270, 563
  of Θ_2-spaces, 289
  of bisimplicial spaces, 281
  of enriched categories, 277
  of Segal precategories, 274
  of Segal spaces, 276

Eilenberg-Steenrod axioms, 77, 509
elliptic curve
  ordinary, 239
  supersingular, 249
embedding tower, 32
embedding, pointed framed, 24
endomorphism ring, 836
enriched colimit, 797
equivalence
  in ∞-category, 558
  local, 266
  of ∞-categories, 563
étale map, 545, 935
Euler class, 701, 956, 966
evaluation map, 530
exceptional divisor, 943
excision
  equivariant, 35
  in $\mathbb{A}^1$-homotopy theory, 937
extended power construction, 797, 802, 829, 833, 837
extension
  n-small, 540
  of scalars, 630
  split square-zero, 538
  square-zero, 540
extension problem, 828
external smash product, 833

face ring, 114
family, of subgroups, 711, 859
Fargues-Fontaine curve, 640
Farrell-Jones conjecture
  A-theoretic, 868
  K- and L-theoretic, 862
  fibered, 870
  for topological cyclic homology, 878
  for topological Hochschild homology, 878
  full, 870
  status of, 870
fat diagonal, 9, 22, 32
fiber product, 834, 838
fiber sequence, $\mathbb{A}^1$-homotopy, 960, 961, 967
fibration
  cartesian, 495
  cocartesian, 495, 590
  inner, 564
  presentable, 499
  right, 51

fibre bundle, 444
fibre integration, 448
field, random, 326
filtration
  Bhatt-Morrow-Scholze, 622
  cardinality, 84
  chromatic, 171
  effective slice, 768
  fat wedge, 136
  geometric, 172
  Goodwillie, 87
  Hochschild-Kostant-Rosenberg, 629
  Nygaard, 653
  stable rank, 34
  very effective slice, 768, 780
fixed points, of G-spectrum, 728
  geometric, 734
  homotopy, 731
Fontaine's ring of p-adic periods, 639, 642
formal group, 844, 895
  height of, 239
  isogeny of, 844
  moduli stack, 168
formal group law, 223
  homomorphism of, 225
  Honda, 176, 179
formula of Bloch, 965
functor
  accessible, 580
  analytic, 7
  Bousfield-Kuhn, 658, 664, 682, 693
  coanalytic, 685, 687
  cofiber, 604
  cohomological, 509
  derivative of, 8
  derivative of the identity, 9, 11
  evaluation, 800, 801, 833
  exact, 502
  excisive, 4, 31, 502, 725, 856
  fiber, 604
  final, 51
  finitary, 4
  forgetful, 700
  Green, 746
  homogeneous, 8
  homological, 506
  indecomposables, 817, 828, 842
  localization, 518
  loop, 603
  Mackey, 719
    box product of, 746
    Burnside ring, 720
    fixed points for suspension spectrum, 722
    homotopy, of spheres, 729
    stable homotopy classes of maps, 720, 729
  monadic, 521
  monoidal, 498, 595
  nonabelian derived, 793, 816, 842
  polynomial, 4, 24
  reduced, 4, 6, 502
  restriction, 700
  spectral Mackey, 725
  spectrification, 608
  stabilization, 724
  stably excisive, 7
  suspension, 603
  symmetric monoidal, 497, 599
  t-exact, 505
  Tambara, 746
  truncation, 500
  unreduced excisive, 535
fundamental generator, 807
fundamental ideal, 948
fusion system, 335, 341
  $\mathcal{H}$-generated, 337, 343
  $\mathcal{H}$-saturated, 343
  centralizer, 344
  constrained, 344
  morphism of, 345, 350
  normalizer, 344
  realizable, 342
  saturated, 341, 356
    classifying space of, 348
    classifying spectrum of, 353
    exotic, 342
    unitary representation of, 350
fusion, control of, 336, 337

Galois representation, 947
general linear group, 528
generalized class function, 909
genus, 449
geometric isotropy, 711
Golod ring, 135
Goodwillie calculus, 87, 832
Goodwillie tower, of the identity, 658, 675, 689
Gram matrix, 949
graph product, 127
Grassmannian, 964
Gromov-Hausdorff metric, 309
Gromov-Lawson-Rosenberg conjecture, stable, 874
Gromov-Prokhorov metric, 325
Gross-Hopkins period map, 184
Grothendieck construction, 495, 593
Grothendieck opfibration, 587
Grothendieck topology, 935
Grothendieck-Teichmüller group, 415
  graded, 421, 438
  profinite, 415, 416, 418
  rational, 415–418
Grothendieck-Teichmüller Lie algebra, 438
Grothendieck-Witt group, 932, 947, 949
group completion, 515, 815, 817, 844
group ring, 528
  twisted, 187
group scheme, trivializable, 225

Hasse square, 166
Hasse-Witt invariant, 949
Hochschild cochain, 98
Hochster's theorem, 115
homology
  $\mathbb{A}^1$-, 962
  M-indexed, 805
  continuous, 185
  equivariant, 855, 879
  factorization, 58, 836, 838
  Floer, 369–372, 385, 388, 394–396, 398
  Hochschild, 67, 629
  Khovanov, 389, 392
  motivic, 962
  periodic topological cyclic, 619
  persistent, 301
  topological André-Quillen, 670, 689, 839
  topological cyclic, 620, 625–628
    of $\mathcal{O}_C$, 646
    of group ring, 650–654

of perfectoid ring, 644
topological Hochschild, 26, 623–625, 836
of $\mathbb{E}_\infty$-algebra in spectra, 623
of stable ∞-category, 624
trace property of, 627
homology decomposition, 339
homotopy category, 795, 825
$\mathbb{A}^1$-, 937
equivariant stable, 155, 787
motivic, 937
of ∞-category, 557
of Θ_2-Segal space, 288
of simplicial category, 562
homotopy coherent diagram, 569
homotopy colimit, 835
homotopy equivalence, of G-spaces, 704
homotopy group, 506
M-indexed, 805
of spectrum, 854
p-adic, 621
representability of, 805
homotopy hypothesis, 268, 492
homotopy limit, 835
homotopy theory
$K(n)$-local, 844
chromatic, 174, 658, 677, 845
equivariant stable, 834, 843
étale, 947
fiberwise, 831, 833, 834, 838
local, 937
motivic, 844
of G-spaces, 704
of modules over E_n-algebra, 844
parametrized, 834
rational, 658, 659
homotopy type
Floer, 372, 376, 377, 384, 385, 388, 394, 397–401
Khovanov, 371, 388, 389, 392, 394
homotopy, equivariant, 703
Hopf map, 105, 821
Hopf ring, 823, 829, 844
horizontal element, with respect to stratification, 633
horn, 269
horn inclusion, 268
hyperbolic form, 948
hyperbolicity, weak, 112
hypercover, 936
hypersurface, 466

idempotent conjecture, 863
image of J, 774, 777, 779, 787
indeterminacy, 824
induction, 701, 724
inert map
in Δ or Δ^{op}, 498
in Fin_*, 496
instability relation, 845
isotropy separation sequence, 712, 734

Jacobi identity, 804, 813
join
iterated, 116
of simplicial sets, 572
Jouanolou trick, 938
juggling formula, 827–828

Kadison conjecture, 873
Kan complex, 268
inner, 270
Kan extension, 50, 59
Kasparov KK-theory, 157
Kontsevich formality, 81
Kummer map, 948

Lagrangian, 369, 372, 398–402
Land-Tamme theorem, 620
Landweber exact functor theorem, 169, 176, 226
law of inertia, Sylvester's, 949
Leibniz rule, 804, 813
level structure, 232, 899
limit, in ∞-category, 577
linearization hypothesis, 213
linking system, 346, 358
centric, 339, 346
localization, 55, 519, 660
v_n-periodic, 658
Bousfield, 171, 936
left Ore, 518
Nisnevich, 936
reflective, 581, 660
Zariski, 519
loop space, 126, 139, 355, 357
Lubin-Tate ring, 897, 903
Lubin-Tate theorem, 844, 897

Mackeyfication, 721
Madsen-Weiss theorem, 461
Mahowald invariant, 787
Mahowald uncertainty principle, 168
Malcev completion, 414, 423
manifold, 444
Bott, 873
framed, 383, 384, 392, 393
moment-angle, 110
pointed framed, 24
string, 255
symplectic, 369, 370, 385, 398
toric, 108
zero-pointed, 24
manifold calculus, 31
mapper, 308
mapping class group, 476
Massey product, 795, 826
Matsumoto's theorem, 961
Mayer-Vietoris sequence, 810
meta-isomorphism conjecture
for G-homology theories, 860
for spectra, 861
metric measure space, 325
metric space, tree-like, 311
Miller-Morita-Mumford class, 448
Milnor conjecture, 767, 782, 785, 949
model category, 265
cartesian, 266
localization of, 267
simplicial, 266
model structure
$\mathbb{A}^1$, 937
Bergner, 270, 563
for G-spaces, 704
for Θ_2-quasi-categories, 291
for Θ_2-Segal [0]-categories, 290
for Θ_2-Segal categories, 290
for Θ_2-Segal spaces, 287
for Θ_2-spaces, 288
for Θ_n-spaces, 292
for 2-fold complete Segal spaces, 281
for categories enriched in complete Segal spaces, 277
for categories enriched in quasi-categories, 278
for categories enriched in Segal categories, 278

for complete Segal spaces, 276
for double Segal spaces, 279
for quasi-categories, 270
for Segal 2-categories, 283
for Segal categories, 274
for Segal category objects, 282
for Segal spaces, 273
for simplicial categories, 270
injective, 267
Joyal, 564
local, 936
objectwise, 934
projective, 267
Reedy, 267
modular form, 227
p-adic, 248
motivic, 787
module
cogroupoid, 631
differential graded, 408, 423
divided power, 22
endotrivial, 360
finitely generated, 523
flat, 524
free, 523, 526
Morava, 188
perfect, 523
projective, 523
pseudocoherent, 641
right $\partial_* I$-, 21
moduli space
of Θ-manifolds, 445, 447
of null bordisms, 457
monad, 520, 665, 796–797, 806
monadicity, 521, 667
monochromatic layer, 171
monoid ring, 528
monoid, topological, 796
Moore-Postnikov stage, 453
Morava stabilizer group, 177, 179, 246, 905
action of, 183
Morita theory, 522
morphism spaces, in ∞-category, 557
morphism, strongly saturated class of, 584
multicategory, 799–802, 829–835
product of, 801
topological, 800, 835
underlying category of, 800
multifunctor, 800
multimorphism, 799
Mumford Conjecture, 461
Murthy's splitting conjecture, 967

nerve
coherent, 271, 490–492, 561
of category, 552
nilpotence theorem, 170
nilpotence, motivic, 775, 786
Nishida relation, 818–820, 845
Nisnevich cover, 935
noncommutative motive, 626
norm, 738
as adjoint, 743
for commutative rings, 742
for operadic algebras, 741
map, 745
normal bundle, 942, 943
normal map of degree one, 865
Novikov conjecture, 863

object
$\mathbb{A}^1$-local, 937
κ-compact, 494
n-truncated, 500
algebra, 498
cofibrant, 265
cohomology, 803
commutative algebra, 497
completely compact, 45
dualizable, 530
fibrant, 265
field, 170
final, in ∞-category, 576
function, 530
homology, 803
invertible, 839
left module, 517
local, 266, 518
mapping, 277, 280, 703
persistence, 301
projective, 494, 516
right module, 517
spectrum, 502
split augmented simplicial, 521
zero, 501
obstruction class, deformation, 542
obstruction theory, 796, 829, 843, 845
operad, 53, 668, 686, 794–796, 800, 830, 835
E_2, 405, 409, 417
E_∞, trivial, 741
E_n, 23, 405
N_∞, 740
associative, 412, 795, 803, 807
Barratt-Eccles, 830
chord diagram, 420
colored, 799
colored braid, 409–411, 414
commutative, 795, 803
cup-1, 796, 798, 809
Drinfeld-Kohno Lie algebra, 418–420, 428
endomorphism, 795, 798, 800
equivariant, 834
formality of little discs, 406, 431, 433
free, 796, 797
linear isometries, 740, 832
little cubes, 795, 804
little disks, 23, 48, 741, 795, 804
parenthesized braid, 411–414
permutation, 412
Poisson, 81, 406, 427
spectral Lie, 19, 23, 673, 688
operation, 807, 825
Adams, 189, 248, 621, 845
additive, 808
Araki-Kudo, 628
cohomology, 793, 801, 845
representability of, 793
continuous, 812
Dyer-Lashof, 256, 628, 793–846
construction of, 809
in dual Steenrod algebra, 819
indexing of, 814, 819
multiplicative, 823, 829
on loop space, 794
properties of, 813, 814
functional, 825, 828
homology, 803–812, 843
homotopy, 805, 814, 841
in TAQ cohomology, 841, 844
power, 628, 793, 806, 808, 834, 844, 845, 891, 905
additive, 912

total, 905
secondary, 824–829, 845
Steenrod, 814, 815, 818, 823, 842, 844, 845
suspension of, 810, 826
unstable, 845
weight of, 806
Ore condition, 518
orientation theory, 846
orthogonal calculus, 33
orthogonal sequence, 33

Panov's conjecture, 113
perfectoid ring, 620, 638–646
periodicity, 188
v_n-, 678
Bökstedt, 619, 621, 633–638
for perfectoid ring, 640
Bott, 646–650
motivic, 775, 786
persistence diagram, 303
persistence images, 315
persistence landscapes, 314
persistence, multidimensional, 319
perturbative property, 796
of $\mathcal{A}_n$, 795
Peterson-Stein formula, 827
phylogenetic hypothesis, 312
Picard group, 210, 839, 844
exotic, 210
of affine scheme, 966
Poincaré duality group, virtual, 185
polyhedral product, 104
aspherical, 127
Poincaré series of, 122
smash, 104
spectral sequence of, 129
stable splitting of, 120
star product, 129
unstable splitting of, 120
wedge of spheres, 136
Pontryagin product, 823
Pontryagin-Thom construction, 448, 452
Postnikov tower, 501, 730, 838, 840, 844
in $\mathbb{A}^1$-homotopy theory, 964
prespectrum, 510
prism, 621, 640
pro-Galois extension, of ring spectra, 177
pro-infinitesimal thickening, p-complete, 621, 642
pro-object, 812
pro-space, 947
projection formula, 824
pullback, in ∞-category, 577
purity theorem, 942, 943
pushforward, 702, 728
pushout, in ∞-category, 577

quadrics, intersection of, 112
quasi-category, 41, 270, 835
Quillen equivalence, 265
Quillen pair, 265

rank, 949
rationalization
of little discs operad, 432, 436
of operads, 426
of spaces, 424
realization
complex, 946
and homotopy sheaves of spheres, 951
equivariant, 946
étale, 947
geometric, 817, 836, 842
real, 947
and homotopy sheaves of spheres, 951
recognition principle, 794, 795
recollement, 534
rectification, 795
Rees construction, p-adic, 653
representability
by pro-object, 819
of TAQ-cohomology, 840
residue map, 953
resolution
Adams-Novikov, 243
algebraic, 196
centralizer, 203
cobar, 764
cosimplicial, 769
duality, 202
finite, 189, 195
Gersten, 965
hybrid, 204
of $\underline{\mathbb{Z}}$, 749
topological, 199
restriction, 701, 724
restriction of scalars, 630
ring space, 823–824, 843
ring spectrum, 794, 798, 799, 803, 821, 823, 827, 833, 834, 839, 845
associative, 514
commutative, 514, 741
complex orientable, 223
even periodic, 224
Gorenstein map of, 363
Landweber, 243
Landweber exact, 242
local commutative, 520
locally Landweber, 244
locally of finite presentation, 516
strongly filtered E_∞, 834
unit group of, 839, 845
robust statistics, 325
root invariant, 787

scheme, affine, 931
Schwede-Shipley theorem, 836
Segal n-category, 292
Segal Θ_2-space, 287
complete, 288
Segal Θ_n-space, complete, 292
Segal 2-category, 283
Segal 2-precategory, 283
Segal category, 273
Segal category object, in complete Segal spaces, 282
Segal condition, 273
Segal conjecture, 843
Segal map, 272
Segal object
in complete Segal spaces, 280
in Segal spaces, 279
Segal precategory, 271
Segal space, 272
n-fold complete, 292
2-fold complete, 281
complete, 275
double, 278
Segal-tom Dieck splitting, 722
algebraic, 721
semi-orthogonal decomposition, 504
semiperfectoid ring, 620
sensor nets, 322
sheaf, 44
$\mathbb{A}^1$-homotopy, 938
of spheres, 962
birational, 961

homotopy, 31, 228, 938
motivic homotopy, 938
$\mathbb{A}^1$-invariant, 959
strictly $\mathbb{A}^1$-invariant, 959, 961
strongly $\mathbb{A}^1$-invariant, 958
unramified, 959
sifted colimit, 57
sign rule, 803, 804, 808, 811
signature, 949
simplicial complex
shifted, 120
totally fillable, 139
simplicial nerve, 271
simplicial presheaf, 934
simplicial set, 41, 266
κ-filtered, 494
κ-small, 494
marked, 571
sifted, 494
site, 935
slice tower, 752
smash product, 832
of spaces, 492
of spectra, 611
smash product theorem, 172
space
$\mathbb{A}^1$-, 934
G-, 700
Φ-good, 690
n-truncated, 500
configuration, 51, 794, 795, 813, 836, 838
Davis-Januszkiewicz, 126
iterated loop, 794, 799, 814, 815, 839
motivic, 934
motivic Thom, 942
Picard, 839, 844, 845
Ran, 85
real projective, 809
simplicial, 267
stratified, 61
stunted projective, 809, 813, 841
spectral algebraic geometry, 256
spectral sequence, 525
C_2-Adams, 751
$K(n)$-local E_n-based Adams-Novikov, 178, 186
k-based Adams, 633
ρ-Bockstein, 765, 776, 780
Adams, 14, 168, 244, 256, 630–633, 758, 773, 793
Adams-Greenlees, 750
Adams-Novikov, 168, 233, 244, 758, 769, 770, 774
algebraic Novikov, 775
algebraic resolution, 198, 200
Atiyah-Hirzebruch, 528, 747
Bökstedt, 633
bar, 816, 828
Bockstein, 782
descent, 177, 186, 229
effective slice, 768
Goodwillie, 7
homotopy fixed point, 177, 186, 200, 748
Künneth, 749, 828
Mackey Atiyah-Hirzebruch, 747
Mackey homotopy fixed point, 748
Mackey Künneth, 749
Miller, 817, 842
motivic Adams, 761–767, 774
motivic May, 765
of polyhedral product, 129
Real Adams-Novikov, 751
slice, 752
topological resolution, 200
universal coefficient, 749
unstable Adams, 692
spectrum
G-, 724
trivial, 728
algebraic K-theory, 768, 772, 783
algebraic cobordism, 768, 769
Barwick equivariant, 725
Brown-Peterson, 168, 770, 794, 799, 820, 827–829
motivic, 783
complex cobordism, 769
connective, 513, 798
cyclotomic, 624
Eilenberg-Mac Lane, 508, 726, 798, 813, 815, 844
ring structure on, 798
as E_2-algebra, 820
cofree, 731
motivic, 769, 782
function, 798
hermitian K-theory, 768, 772, 779, 784
higher Witt-theory, 772, 785
in tt-geometry, 34, 147
indexed, 832, 833
Johnson-Wilson, 171, 246
left module, 517
Lewis-May-Steinberger, 724
Lubin-Tate, 176, 732
Moore, 821
of topological automorphic forms, 254
of topological modular forms, 221
orthogonal, 724
smash product of, 503
suspension, 798
symmetric, 724
Tate, 811
Thom, 529, 727, 799, 809, 817, 820, 822, 838
sphere
$K(1)$-local, 189
generalized homology, 111
homotopy, 460, 477
motivic, 941
homotopy of, 967
homotopy sheaf of, 962
polytopal, 111
representation, 700
splitting problem, for vector bundles, 965
stability, 810–812
stabilization, 823
algebraic, 713
geometric, 713
of ∞-category, 607
stable element, 334, 347, 354
stable envelope, 535
stable splitting, equivariance of, 123
stack, 226
associated to ring spectrum, 243
moduli of elliptic curves, 227
Shimura, 253
Stanley-Reisner ring, 114
exterior, 125
generalized, 131
Steinberg relation, 761, 779, 950
straightening, 495

stratification, 336, 349, 358, 362
  relative to cogroupoid, 631
Strickland's theorem, 915
structure set, 865
subgroup
  $\mathcal{F}$-centric, 343
  $\mathcal{F}$-quasicentric, 345
  $\mathcal{F}$-radical, 343
  p-centric, 337
  fully centralized, 341
  fully normalized, 341
  subcentric, 345, 348
subspace arrangement, 119
Sullivan conjecture, 843
Sullivan model
  for operad, 425
  for space, 423
support
  in tt-geometry, 147
support variety, of finite group, 153
surgery exact sequence, 866
surgery theory, 846
Suslin-Voevodsky construction, 940
suspension isomorphism, 509
symmetric monoidal product, of modules over cogroupoid, 632
symmetric monoidal structure, cartesian, 497
symmetric polynomial, 845
symmetric sequence, 796

tame function, 310
tangent bundle, 535
tangent correspondence, 537
Tate diagonal, 29, 623
Tate diagram, 737
Tate motive, 157
Tate orbit lemma, 626, 652
Taylor tower
  convergence of, 7
  layer of, 8
  of ∞-category, 28
  of functor, 6
telescope conjecture, 172
tensor product
  of locally presentable categories, 610
  of presentable ∞-categories, 500, 610
  relative, 522
tensor-triangulated ideal, 145
theory, multi-sorted, 802, 807
thick subcategory, 678
thick subcategory theorem, 170
Thom isomorphism, 809, 813, 818
tilt, 639
Toda bracket, 774, 825
topological modular forms, 229, 231, 774, 787
  with level structure, 232
topology
  cdh, 935
  étale, 935
  Nisnevich, 935, 944
  Zariski, 931
tor-amplitude, 525
trace conjecture, 873
transfer, 333, 354, 358, 718, 911, 954
transitivity principle, 881
transitivity, for TAQ, 840
triangle, 501
  in pointed ∞-category, 602
triangulation, 372, 394, 396–398
tubular neighborhood, 942, 943
twisted half-smash product, 832, 834
twisted product, 830

ultrafilter, 214
ultraproduct, 214
unitary representation, 359
universal space, for a family of subgroups, 711
universe, 832–834
unstraightening, 93, 495
untilt, 640

vanishing conjecture, 186
vanishing line, horizontal, 186, 200
vector bundle, on an affine variety, 964
Verdier sequence, 504
vertical tangent bundle, 444
virtually polycyclic group, 459, 478
Voronoi cell, 306

Wasserstein distance, 309
weak equivalence
  local, 936
  simplicial, 937
wedge construction, 116
wedge decomposable pair, 133
Weierstrass curve, 244
Whitehead product, 105, 658, 689
  higher, 138
Whitehead theorem, 706
  in $\mathbb{A}^1$-homotopy theory, 938
Wirthmuller isomorphism, 724
Witt vector, 176
Witten genus, 255

Yoneda embedding, 493, 554, 934
  ∞-categorical, 582
Yoneda extension, 555
Yoneda lemma, 806

zero-in-the-spectrum conjecture, 873
zig-zag persistence, 317